Fatigue Data Book: Light Structural Alloys

Scott D. Henry, Manager of Reference Development
Grace M. Davidson, Manager Reference Book Production
Steven R. Lampman, Technical Editor
Faith Reidenbach, Chief Copy Editor
Randall L. Boring, Production Coordinator
William W. Scott, Jr., Director of Technical Publications

Editorial Assistance
Kathleen S. Dragolich
Nikki D. DiMatteo

First printing, November 1995

Library of Congress Cataloging-in-Publication Data

ASM International

Fatigue data book: light structural alloys.

Includes bibliographical references.

1. Light metal alloys—Fatigue.
I. ASM International.
TA484.F37 1995 620.1'66—dc20
p. cm. SAN 204-7586 95-39481

ISBN 0-87170-507-9

ASM International®
Materials Park, OH 44073-0002

Printed in the United States of America

Acknowledgments and Preface

ASM International would like to thank Robert Bucci (ALCOA), Glenn Nordmark (ALCOA, retired), Ralph Stephens (University of Iowa, Mechanical Engineering), and Harold Margolin (Polytechnic University, retired) for their assistance and advice in collecting information for this publication. This book also would not have been possible without the continued commitment by production at ASM International.

This ASM International publication should be a useful supplement to *Atlas of Fatigue Curves* (ASM International, 1986) by providing more coverage of fatigue data for light structural alloys. Due to length restrictions, coverage of aluminum alloy fatigue was limited to stress-controlled *(S-N)* data.

S. Lampman

Table of Contents

Aluminum Alloy Fatigue Data

Aluminum Alloy S-N Fatigue

High-cycle fatigue characteristics commonly are examined on the basis of cyclic *S-N* plots of rotating-beam, axial, or flexure-type sheet tests. Many thousands of tests have been performed, and early work on rotating-beam tests is summarized in Fig. 1. There seems to be greater spread in fatigue strengths for unnotched specimens than for notched specimens. This appears to be evidence that the presence of a notch minimizes differences, thus suggesting similar crack propagation after crack initiation with a sharp notch. In this context, the spread in smooth fatigue life is partly associated with variations in crack initiation sources (at surface imperfections or strain localizations). In general, however, the *S-N* approach does not provide clear distinctions in characterizing the crack initiation and crack propagation stages of fatigue.

When comparing rotating-beam fatigue strength of unnotched aluminum alloy specimens, the *S-N* response curves tend to level out as the number of applied cycles approaches 500 million. This allows some rating of fatigue endurance, and estimated fatigue limits from rotating-beam tests have been tabulated for many commercial aluminum alloys (Table 1). Fatigue limits should not be expected in aggressive environments, as *S-N* response curves don't tend to level out when corrosion fatigue occurs. Rotating-beam strengths determined in the transverse direction are not significantly different from test results in the longitudinal direction. The scatter band limits in Fig. 2 show relatively small effects attributable to working direction, particularly for the notched fatigue data.

Rotating-beam data have also been analyzed to determine whether fatigue strength can be correlated with static strength. From a plot of average endurance limits (at 5×10^8 cycles) plotted against various tensile properties (Fig. 3), there does not appear to be any well-defined quantitative relation between fatigue limit and static strength. This well-known result is common among most nonferrous alloys. It should be noted that proportionate increases in fatigue strength from tensile strengths do appear lower for age-hardened aluminum alloys than for annealed alloys (Fig. 4). A similar trend appears evident for fatigue strength at 5×10^7 cycles (Fig. 5).

Effect of Environment

Another key source of variability in *S-N* data is environment (Ref 1-3). Even atmospheric moisture is recognized to have a corrosive effect on fatigue performance of aluminum alloys. Much high-cycle *S-N* testing has been carried out in uncontrolled ambient lab air environments, thereby contributing to varied amounts of scatter in existing data. This factor should be recognized when comparing results of different investigations.

Most aluminum alloys experience some reduction of fatigue strength in corrosive environments such as seawater, especially in low-stress, long-life tests (e.g., Fig. 6). Unlike sustained-load SCC, fatigue degradation by environment may be true even when the direction of principal loading with respect to grain flow is other than short-transverse. Fatigue response to environment varies with alloy, and therefore final alloy selection for design should address this important interaction. When accumulating data for this purpose, it is recommended that any testing be conducted in a controlled environment, and preferably the environment of the intended application. Often an environment known to be more severe than that encountered in service is used to conservatively establish baseline data and design guidelines. Because environmental interaction with fatigue is a rate-controlled process, interaction of time-dependent fatigue parameters such as frequency, waveform, and load history should be factored into the fatigue analysis (Ref 1-3).

Typically, the fatigue strength of the more corrosion-resistant 5XXX and 6XXX aluminum alloys and tempers are less affected by corrosive environments than are higher-strength 2XXX and 7XXX alloys, as indicated by Fig. 7. Corrosion fatigue performance of 7XXX alloys may, in general, be upgraded by overaging to more corrosion-resistant T7 tempers (Ref 4-9), as indicated by results shown in Fig. 8 and 9. With 2XXX alloys, more corrosion-resistant, precipitation-hardened T8-type tempers provide a better combination of strength and fatigue resistance at high endurances than naturally aged T3 and T4 tempers. However, artificial aging of 2XXX alloys is accompanied by loss in toughness with resultant decrease in fatigue crack growth resistance at intermediate and high stress intensities (Ref 7, 8).

Interaction of a clad protective system with fatigue strength of alloys 2024-T3 and 7075-T6 in air and seawater environments is shown in Fig.

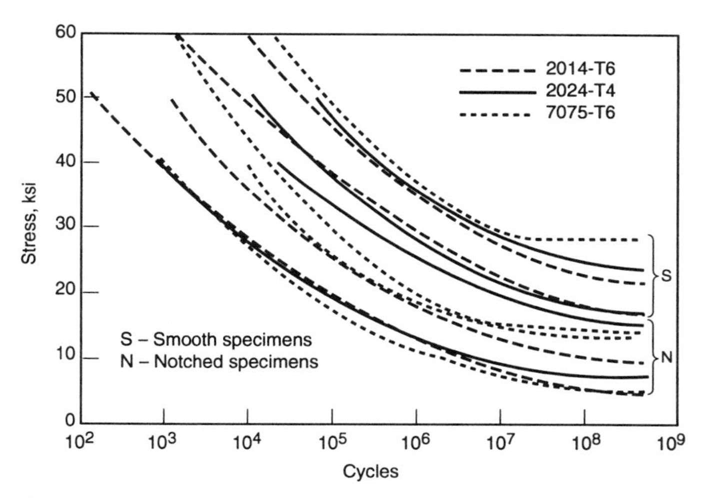

Fig. 1 Comparison of fatigue strength bands for 2014-T6, 2024-T4, and 7075-T6 aluminum alloys for rotating-beam tests. Source: R. Templin, F. Howell, and E. Hartmann, "Effect of Grain-Direction on Fatigue Properties of Aluminum Alloys," Alcoa, 1950

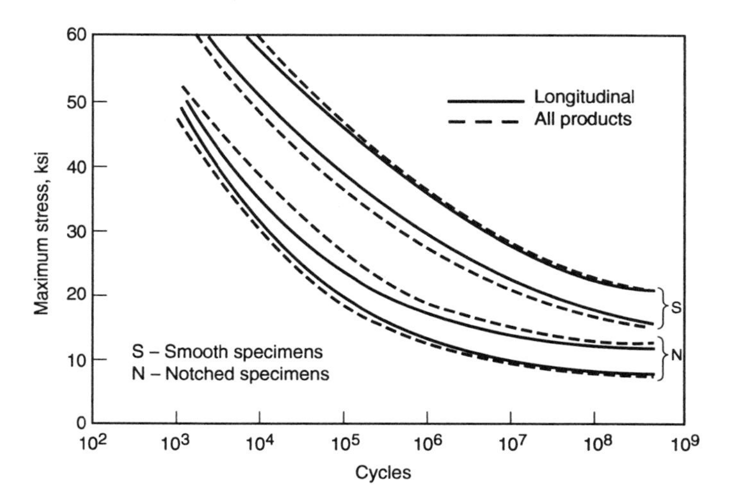

Fig. 2 Comparison of fatigue strength bands for 2014-T6 aluminum alloy products, showing effects of direction. Source: *ASTM Proceedings*, Vol 64, p 581-593

Table 1 Typical tensile properties and fatigue limit of aluminum alloys

Alloy and temper	Ultimate tensile strength		Tensile yield strength		Elongation in 50 mm (2 in.), %		Fatigue endurance limit(a)	
					1.6 mm (1/16 in.) thick specimen	1.3 mm (1/2 in.) diam specimen		
	MPa	ksi	MPa	ksi			MPa	ksi
1060-0	70	10	30	4	43	...	20	3
1060-H12	85	12	75	11	16	...	30	4
1060-H14	95	14	90	13	12	...	35	5
1060-H16	110	16	105	15	8	...	45	6.5
1060-H18	130	19	125	18	6	...	45	6.5
1100-0	90	13	35	5	35	45	35	5
1100-H12	110	16	105	15	12	25	40	6
1100-H14	125	18	115	17	9	20	50	7
1100-H16	145	21	140	20	6	17	60	9
1100-H18	165	24	150	22	5	15	60	9
1350-0	85	12	30	4	...	(d)	...	...
1350-H12	95	14	85	12	...	...	...	...
1350-H14	110	16	95	14	...	...	...	...
1350-H16	125	18	110	16	...	...	...	...
1350-H19	185	27	165	24	...	(e)	50	7
2011-T3	380	55	295	43	...	15	125	18
2011-T8	405	59	310	45	...	12	125	18
2014-0	185	27	95	14	...	18	90	13
2014-T4, T451	425	62	290	42	...	20	140	20
2014-T6, T651	485	70	415	60	...	13	125	18
Alclad 2014-0	175	25	70	10	21	...	...	...
Alclad 2014-T3	435	63	275	40	20	...	...	...
Alclad 2014-T4, T451	420	61	255	37	22	...	...	...
Alclad 2014-T6, T651	470	68	415	60	10	...	...	...
2017-0	180	26	70	10	...	22	90	13
2017-T4, T451	425	62	275	40	...	22	125	18
2018-T61	420	61	315	46	...	12	115	17
2024-0	185	27	75	11	20	22	90	13
2024-T3	485	70	345	50	18	...	140	20
2024-T4, T351	470	68	325	47	20	19	140	20
2024-T361(b)	495	72	395	57	13	...	125	18
Alclad 2024-0	180	26	75	11	20	...	...	...
Alclad 2024-T3	450	65	310	45	18	...	...	...
Alclad 2024-T4, T351	440	64	290	42	19	...	...	...
Alclad 2024-T361(b)	460	67	365	53	11	...	...	...
Alclad 2024-T81, T851	450	65	415	60	6	...	...	...
Alclad 2024-T861(b)	485	70	455	66	6	...	...	...
2025-T6	400	58	255	37	...	19	125	18
2036-T4	340	49	195	28	24	...	125(c)	18(c)
2117-T4	295	43	165	24	...	27	95	14
2125	...	...	...	...	...	...	90	13(d)
2124-T851	485	70	440	64	...	8	...	...
2214	...	...	...	...	...	...	103	15(d)
2218-T72	330	48	255	37	...	11	...	...
2219-0	175	25	75	11	18	...	...	...
2219-T42	360	52	185	27	20	...	...	...
2219-T31, T351	360	52	250	36	17	...	...	...
2219-T37	395	57	315	46	11	...	...	...
2219-T62	415	60	290	42	10	...	105	15
2219-T81, T851	455	66	350	51	10	...	105	15
2219-T87	475	69	395	57	10	...	105	15
2618-T61	440	64	370	54	...	10	125	18
3003-0	110	16	40	6	30	40	50	7
3003-H12	130	19	125	18	10	20	55	8
3003-H14	150	22	145	21	8	16	60	9
3003-H16	180	26	170	25	5	14	70	10
3003-H18	200	29	185	27	4	10	70	10
Alclad 3003-0	110	16	40	6	30	40	...	...
Alclad 3003-H12	130	19	125	18	10	20	...	...
Alclad 3003-H14	150	22	145	21	8	16	...	...
Alclad 3003-H16	180	26	170	25	5	14	...	...
Alclad 3003-H18	200	29	185	27	4	10	...	...
3004-0	180	26	70	10	20	25	95	14

a) Based on 500,000,000 cycles of completely reversed stress using the R.R. Moore type of machine and specimen. (b) Tempers T361 and T861 were formerly designated T36 and T86, respectively. (c) Based on 10 cycles using flexural type testing of sheet specimens. (d) Unpublished Alcoa data. (e) Data from CDNSWRC-TR619409, 1994, cited below. (f) T7451, although not previously registered, has appeared in literature and some specifications as T73651. (g) Sheet flexural. Sources: *Aluminum Standards and Data*, Aluminum Association, and E. Czyryca and M. Vassilaros, *A Compilation of Fatigue Information for Aluminum Alloys*, Naval Ship Research and Development Center, CDNSWC-TR619409, 1994

(*continued*)

Table 1 Typical tensile properties and fatigue limit of aluminum alloys ***(continued)***

Alloy and temper	Ultimate tensile strength		Tensile yield strength		Elongation in 50 mm (2 in.), %		Fatigue endurance limit(a)	
	MPa	ksi	MPa	ksi	1.6 mm (1/16 in.) thick specimen	1.3 mm (1/2 in.) diam specimen	MPa	ksi
3004-H32	215	31	170	25	10	17	105	15
3004-H34	240	35	200	29	9	12	105	15
3004-H36	260	38	230	33	5	9	110	16
3004-H38	285	41	250	36	5	6	110	16
Alclad 3004-0	180	26	70	10	20	25	...	...
Alclad 3004-H32	215	31	170	25	10	17	...	...
Alclad 3004-34	240	35	200	29	9	12	...	...
Alclad 3004-H36	260	38	230	33	5	9	...	...
Alclad 3004-H38	285	41	250	36	5	6	...	...
3105-0	115	17	55	8	24	...	...	...
3105-H12	150	22	130	19	7	...	...	...
3105-H14	170	25	150	22	5	...	...	...
3105-H16	195	28	170	25	4	...	...	...
3105-H18	215	31	195	28	3	...	...	...
3105-H25	180	26	160	23	8	...	...	...
4032-T6	380	55	315	46	...	9	110	16
4043-0	...	...	...	...	...	...	40	6(d)
4043-H38	...	...	...	...	...	...	55	8(d)
5005-0	125	18	40	6	25	...	...	...
5005-H12	140	20	130	19	10	...	...	...
5005-H14	160	23	150	22	6	...	...	...
5005-H16	180	26	170	25	5	...	...	...
5005-H18	200	29	195	28	4	...	...	...
5005-H32	140	20	115	17	11	...	...	...
5005-H34	160	23	140	20	8	...	...	...
5005-H36	180	26	165	24	6	...	...	...
5005-H38	200	29	185	27	5	...	...	...
5005-0	145	21	55	8	24	...	85	12
5050-H32	170	25	145	21	9	...	90	13
5050-H34	195	28	165	24	8	...	90	13
5050-H36	205	30	180	26	7	...	95	14
5050-H38	220	32	200	29	6	...	95	14
5052-0	195	28	90	13	25	30	110	16
5052-H32	230	33	195	28	12	18	115	17
5052-H34	260	38	215	31	10	14	125	18
5052-H36	275	40	240	35	8	10	130	19
5052-H38	290	42	255	37	7	8	140	20
5056-0	290	42	150	22	...	35	140	20
5056-H18	435	63	405	59	...	10	150	22
5056-H38	415	60	345	50	...	15	150	22
5083-0	290	42	145	21	...	22	160	23
5083-H11	303	44	193	28	...	16	150	22(e)
5083-H112	295	43	160	23	...	20	150	22(e)
5083-H113	317	46	227	33	...	16	160	23(e)
5083-H32	317	46	227	33	...	16	150	22(e)
5083-H34	358	52	283	41	...	8	...	...
5083-H321, H116	315	46	230	33	...	16	160	23
5086-0	260	38	115	17	22	...	145	21(e)
5086-H32, H116	290	42	205	30	12	...	50	22(e)
5086-H34	325	47	255	37	10	...	...	...
5086-H112	270	39	130	19	14	...	...	...
5086-H111	270	39	170	25	17	...	145	21(e)
5086-H343	325	47	255	37	10-14	...	160	23(e)
5154-0	240	35	115	17	27	...	115	17
5154-H32	270	39	205	30	15	...	125	18
5154-H34	290	42	230	33	13	...	130	19
5154-H36	310	45	250	36	12	...	140	20
5154-H38	330	48	270	39	10	...	145	21
5154-H112	240	35	115	17	25	...	115	17
5252-H25	235	34	170	25	11	...	...	...
5252-H38, H28	285	41	240	35	5	...	...	...
5254-0	240	35	115	17	27	...	115	17
5254-H32	270	39	205	30	15	...	125	18
5254-H34	290	42	230	33	13	...	130	19
5254-H36	310	45	250	36	12	...	140	20

a) Based on 500,000,000 cycles of completely reversed stress using the R.R. Moore type of machine and specimen. (b) Tempers T361 and T861 were formerly designated T36 and T86, respectively. (c) Based on 10 cycles using flexural type testing of sheet specimens. (d) Unpublished Alcoa data. (e) Data from CDNSWRC-TR619409, 1994, cited below. (f) T7451, although not previously registered, has appeared in literature and some specifications as T73651. (g) Sheet flexural. Sources: *Aluminum Standards and Data*, Aluminum Association, and E. Czyryca and M. Vassilaros, *A Compilation of Fatigue Information for Aluminum Alloys*, Naval Ship Research and Development Center, CDNSWRC-TR619409, 1994

(continued)

Table 1 Typical tensile properties and fatigue limit of aluminum alloys *(continued)*

Alloy and temper	Ultimate tensile strength, MPa	Ultimate tensile strength, ksi	Tensile yield strength, MPa	Tensile yield strength, ksi	Elongation in 50 mm (2 in.), %: 1.6 mm (1/16 in.) thick specimen	Elongation in 50 mm (2 in.), %: 1.3 mm (1/2 in.) diam specimen	Fatigue endurance limit(a), MPa	Fatigue endurance limit(a), ksi
5254-H38	330	48	270	39	10	...	145	21
5254-H112	240	35	115	17	25	...	115	17
5454-0	250	36	115	17	22	...	140	20(e)
5454-H32	275	40	205	30	10	...	140	20(e)
5454-H34	305	44	240	35	10	...	...	...
5454-H111	260	38	180	26	14	...	...	...
5454-H112	250	36	125	18	18	...	...	...
5456-0	310	45	160	23	...	24	150	22(e)
5456-H112	310	45	165	24	...	22	...	...
5456-H321, H116, H32	350	51	255	37	...	16	160	23(e)
5457-0	130	19	50	7	22	...	...	...
5457-H25	180	26	160	23	12	...	...	...
5457-H38, H28	205	30	185	27	6	...	...	...
5652-0	195	28	90	13	25	30	110	16
5652-H32	230	33	195	28	12	18	115	17
5652-H34	260	38	215	31	10	14	125	18
5652-H36	275	40	240	35	8	10	130	19
5652-H38	290	42	255	37	7	8	140	20
5657-H25	160	23	140	20	12	...	...	...
5657-H38, H28	195	28	165	24	7	...	...	...
6061-0	125	18	55	8	25	30	60	9
6061-T4, T451	240	35	145	21	22	25	95	14
6061-T6, T651	310	45	275	40	12	17	95	14
Alclad 6061-0	115	17	50	7	25	...	...	...
Alclad 6061-T4, T451	230	33	130	19	22	...	...	...
Alclad 6061-T6, T651	290	42	255	37	12	...	...	...
6063-0	90	13	50	7	...	...	55	8
6063-T1	150	22	90	13	20	...	60	9
6063-T4	170	25	90	13	22	...	...	...
6063-T5	185	27	145	21	12	...	70	10
6063-T6	240	35	215	31	12	...	70	10
6063-T83	255	37	240	35	9	...	...	...
6063-T831	205	30	185	27	10	...	...	...
6063-T832	290	42	270	39	12	...	...	...
6066-0	150	22	85	12	...	18	...	...
6066-T4, T451	360	52	205	30	...	18	...	...
6066-T6, T651	395	57	360	52	...	12	110	16
6070-T6	380	55	350	51	10	...	95	14
6101-H111	95	14	75	11	...	...	...	...
6101-T6	220	32	195	28	15	...	...	...
6151-T6	...	...	...	...	...	...	83	12
6201-T81	...	...	...	...	...	...	105	15
6262-T9	...	...	...	...	...	...	95	14
6351-T4	250	36	150	22	20	...	...	...
6351-T6	310	45	285	41	14	...	90	13
6463-T1	150	22	90	13	20	...	70	10
6463-T5	185	27	145	21	12	...	70	10
6463-T6	240	35	215	31	12	...	70	10
7002-T6	440	64	365	53	9-12	...	...	...
7039-T6	415	60	345	50	14	...	...	...
7049-T73	515	75	450	65	...	12	...	...
7049-T7352	515	75	435	63	...	11	...	...
7050-T73510, T73511	495	72	435	63	...	12	...	...
7050-T7451(f)	525	76	470	68	...	11	...	...
7050-T7651	550	80	490	71	...	11	...	...
7075-0	230	33	105	15	17	16	117	17(e)
7075-T6, T651	570	83	505	73	11	11	160	23
7072-H14	...	...	...	...	...	...	35	5(g)
7075-T73	503	73	435	63	13	...	150	22(e)
7076-T6	...	...	...	...	...	...	138	20(d)
Alclad 7075-0	220	32	95	14	17	...	...	...
Alclad 7075-T6, T651	525	76	460	67	11	...	...	...
7079-T6	490	71	428	62	10	...	160	23(e)

(a) Based on 500,000,000 cycles of completely reversed stress using the R.R. Moore type of machine and specimen. (b) Tempers T361 and T861 were formerly designated T36 and T86, respectively. (c) Based on 10 cycles using flexural type testing of sheet specimens. (d) Unpublished Alcoa data. (e) Data from CDNSWRC-TR619409, 1994, cited below. (f) T7451, although not previously registered, has appeared in literature and some specifications as T73651. (g) Sheet flexural. Sources: *Aluminum Standards and Data*, Aluminum Association, and E. Czyryca and M. Vassilaros, *A Compilation of Fatigue Information for Aluminum Alloys*, Naval Ship Research and Development Center, CDNSWC-TR619409, 1994

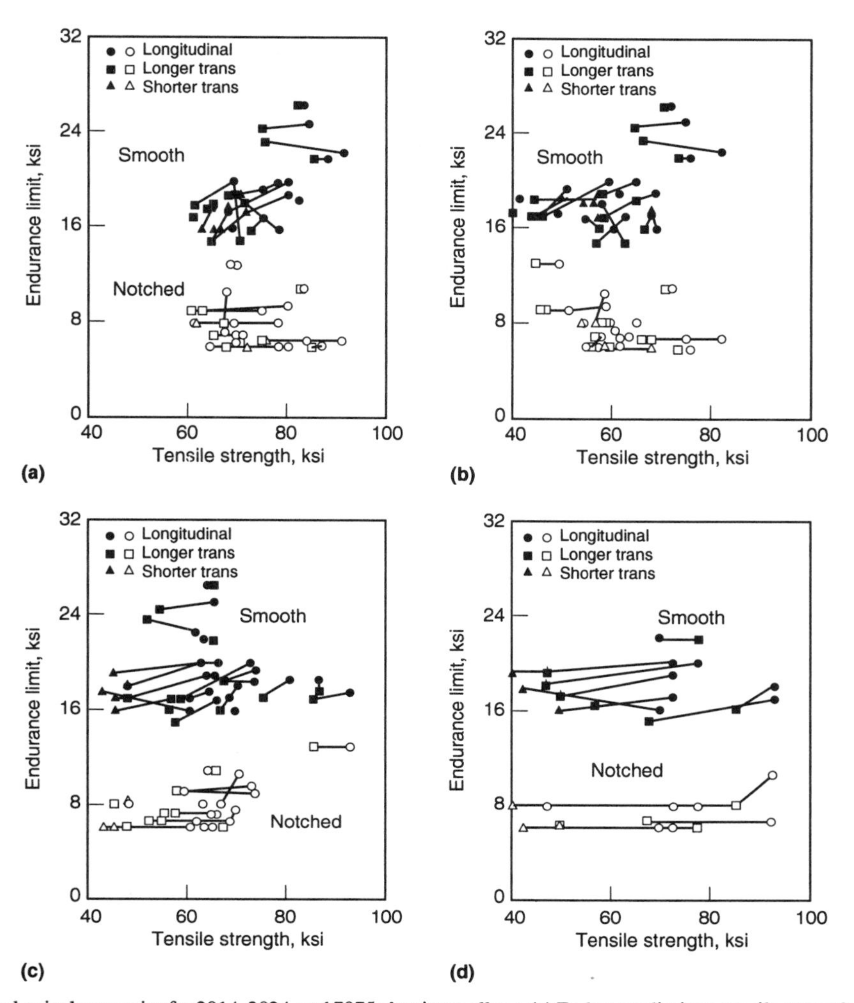

Fig. 3 Plots of fatigue with static mechanical properties for 2014, 2024, and 7075 aluminum alloys. (a) Endurance limit vs. tensile strength. (b) Endurance limit vs. yield strength. (c) Endurance limit vs. elongation. (d) Endurance limit vs. reduction of area. Sharp notches ($K_t > 12$). Source: R. Templin, F. Howell, and E. Hartmann, "Effect of Grain-Direction on Fatigue Properties of Aluminum Alloys," Alcoa, 1950

6. In seawater, benefits of the cladding are readily apparent. In air, the cladding appreciably lowers fatigue resistance.

*Effect of Microporosity**

The size of microporosity in commercial products is affected by the forming processes used in their production. A recent program was undertaken to determine whether the fatigue strength could be improved by the control of microporosity. Because the stress concentrations in products often nullify the effects of metallurgical differences, the program used both unnotched specimens and specimens containing an open hole.

The goal of the program was to effect change in metallic aircraft life assessment methodology through quantitative understanding of how material microstructure affects fatigue durability performance. Various studies have shown that most metal cracking problems encountered in service involve fatigue. Further studies have shown that metallurgical discontinuities and/or manufacturing imperfections often tend to exacerbate such problems by causing cracks to occur sooner than expected. This program concentrated on the initiation and early growth stages of fatigue cracks, where the majority of structural life is spent. The program had two general objectives: (1) quantifying the effect of aluminum alloy

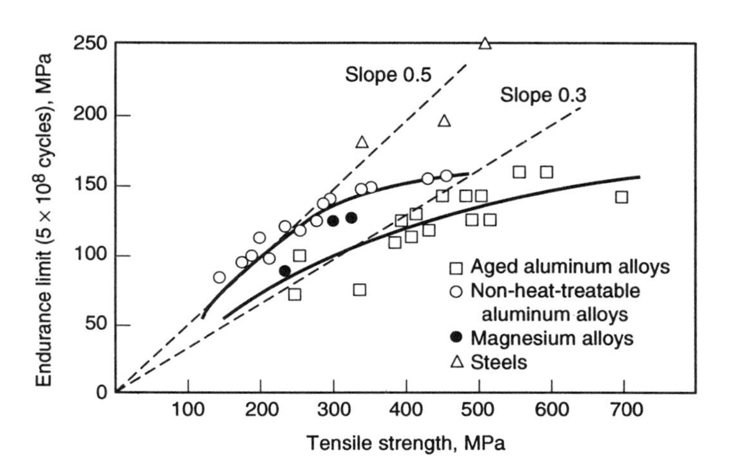

Fig. 4 Fatigue ratios (endurance limit/tensile strength) for aluminum alloys and other materials. Source: P.C. Varley, *The Technology of Aluminum and Its Alloys*, Newnes-Butterworths, London, 1970

*"Effect of Microporosity" is adapted from J.R. Brockenbrough, R.J. Bucci, A.J. Hinkle, J. Liu, P.E. Magnusen, and S.M. Mixasato, "Role of Microstructure on Fatigue Durability of Aluminum Aircraft Alloys," Progress Report, ONR Contract N00014-91-C-0128, 15 April 1993.

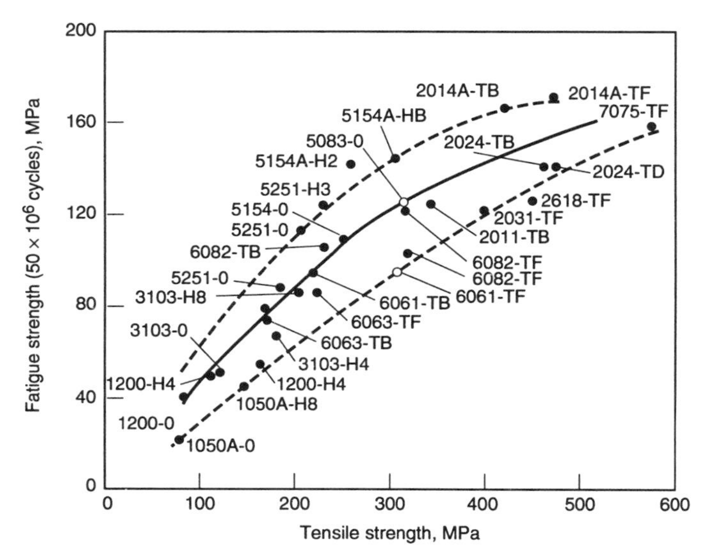

Fig. 5 Relationships between the fatigue strength and tensile strength of some wrought aluminum alloys

microstructure on early-stage fatigue damage evolution and growth; and (2) establishing an analytical framework to quantify structural component life benefits attainable through modification of intrinsic material microstructure. The modeling approach taken coupled quantitative characterizations of representative material microstructures with concepts of probabilistic fracture mechanics.

Reduced Microporosity Materials. Five variants of 7050 plate were selected to provide a range of microstructures to quantify the effects of intrinsic microstructural features on fatigue durability (Table 2). The first material, designated "old-quality" material, was produced using production practices typical of those used in 1984. The material is characterized by extensive amounts of centerline microporosity. Despite the centerline microporosity, this material still meets all existing mechanical property specifications for thick 7050 plate. Current quality production material, designated "new-quality" material, was also used, characterized by reduced levels of centerline microporosity compared to the old-quality material. The new-quality material represents the current benchmark for commercially available material. The processing methods used in the production of the new-quality material are a result of a statistical quality control effort to improve 7050 alloy thick plate (Ref 10). Material taken from two plant-scale production lots of each quality level provided the material for this program. Both materials are 5.7 in. thick 7050-T7451 plate. Static mechanical property characterization of the two 7050 plate pedigrees showed no significant differences in properties other than an increase in short transverse elongation for the new-quality material (Ref 38), and both materials meet the AMS material specification minimums. The fact that both materials meet the property requirements of the AMS specification underscores the limitation of existing specifications in that they do not differentiate intrinsic metal quality.

The next material variant of alloy 7050 plate was specially processed to further reduce the amount of centerline microporosity. The process used to produce the material is currently not used for commercial production and involves process steps that add to current production costs. The production methods are considered Alcoa proprietary. This material is 150 mm (6 in.) thick and is denoted "low-porosity" plate.

The fourth material variant was selected to have minimal microporosity and a significantly smaller constituent particle distribution than either of the previous materials, yet maintain a thick product grain structure. For this purpose, material was selected from the quarter-thickness plane (T/4) of 150 mm (6 in.) thick plate that was produced from ingot that had controlled composition, which limited the formation of coarse constituent particles. Constituent particles in 7050 plate are typically of the types Al_7Cu_2Fe and Mg_2Si. These form during ingot solidification and are insoluble, so they remain in the material through processing to the final product. The mechanical work of processing may break up the particles, which produces stringers of smaller particles. This material is denoted "low-particle" plate.

The final microstructural variant of 7050 plate used in this program was selected to have no microporosity and have a refined constituent particle size distribution and refined grain size compared to standard production thick plate. For these characteristics, thinner plate 1 in. thick was selected. The increased amount of deformation required to produce the thinner plate acts to heal microporosity and break up the coarse constituent particles that are present in the cast ingot. Fatigue testing was conducted on the five 7050 alloy microstructural variants used in this study. The fatigue tests included both smooth-round axial stress specimens and flat bar specimens containing open holes, as described below.

Effect of Microporosity on Fatigue. Smooth axial stress fatigue tests were performed for both the old-quality and the new-quality plate materials. The tests were done on round bars with a gage diameter of 12.7 mm (0.5 in.). Gage sections were sanded longitudinally to remove circumferential machining marks. Testing was done at a maximum stress of 240 MPa (35 ksi), a stress ratio $R = 0.1$, and cyclic frequency of 10 Hz in laboratory air. The specimen orientation was long-transverse (L-T) relative to the parent plate. The specimens were removed from the midthick-

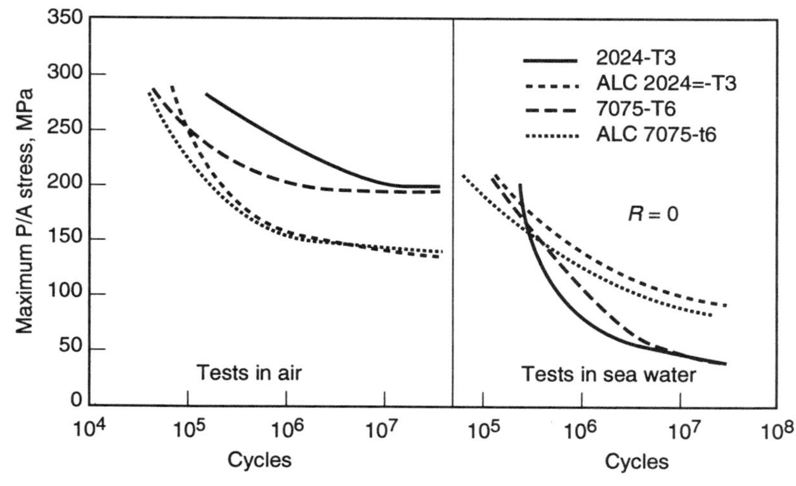

Fig. 6 Axial stress fatigue strength of 0.8 mm 2024, 7075, and clad sheet in air and seawater, $R = 0$. Source: Ref 22

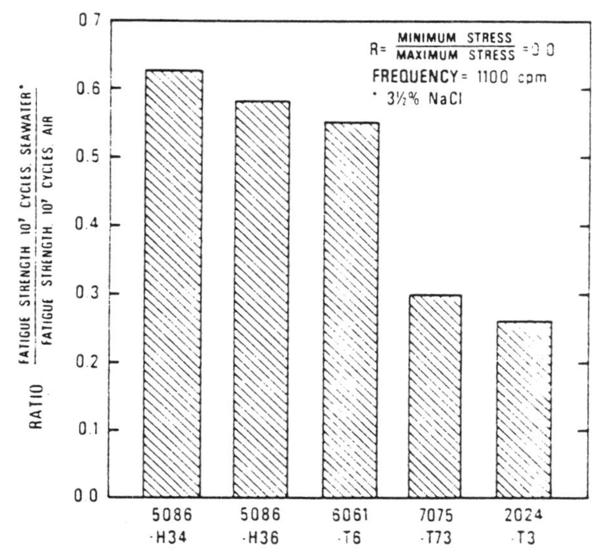

Fig. 7 Comparison of axial-stress fatigue strengths of 0.032 in. aluminum alloy sheet in seawater and air. Source: Ref 22

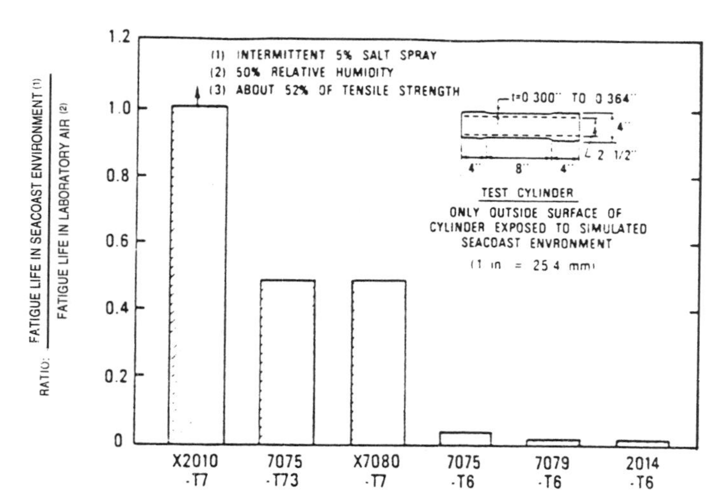

Fig. 8 Comparisons of fatigue lives of pressurized hydraulic cylinders in laboratory air and simulated seacoast environments at 80% design stress. Sources: Ref 3, 31

ness (T/2) plane of the plate where microporosity concentration is the greatest (Ref 11). The lifetimes of the specimens are plotted in Fig. 10 as a cumulative failure plot, where the data are sorted in order of ascending lifetime and ordinate is the percentile ranking of the specimens relative to the total number of tests. Thus the lifetime corresponding to the 50% point on the ordinate represents the median lifetime, where half of the specimens failed prior to that lifetime and half failed at longer lifetimes. The data show that the cumulative distribution of fatigue lifetimes for the new-quality material is longer than for the old-quality material.

Fatigue tests were also performed for the old- and new-quality materials using flat specimens containing open holes. Tests were performed at four stress levels for each material pedigree at a stress ratio of $R = 0.1$ and cyclic frequency of 25 Hz in laboratory air. As with the round specimens, the specimen orientation was L-T and the specimens were removed from the T/2 plane of the plate. The holes were deburred by polishing with diamond compound prior to testing. The polishing was done only on the corners and not in the bore of the hole, and it resulted in slight rounding of the corners. The fatigue lifetime data are plotted in Fig. 11 as an *S-N* plot. Also plotted for both materials are the 95% confidence limits for the *S-N* curves. The confidence limits were obtained from a Box-Cox analysis of the data, which enables statistical determination of the mean

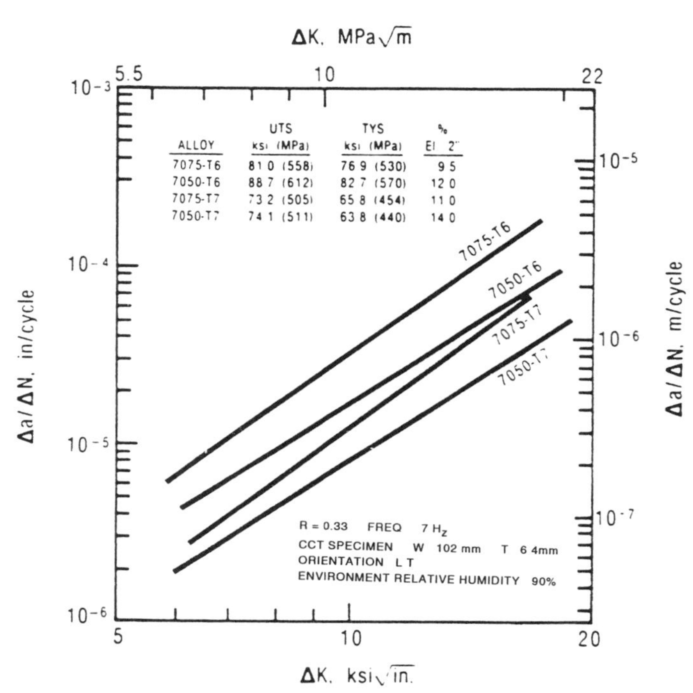

Fig. 9 Cyclic stress intensity range, ΔK, vs. cyclic fatigue crack growth rate, $\Delta a/\Delta N$, of laboratory-fabricated high-strength 7XXX aluminum alloys

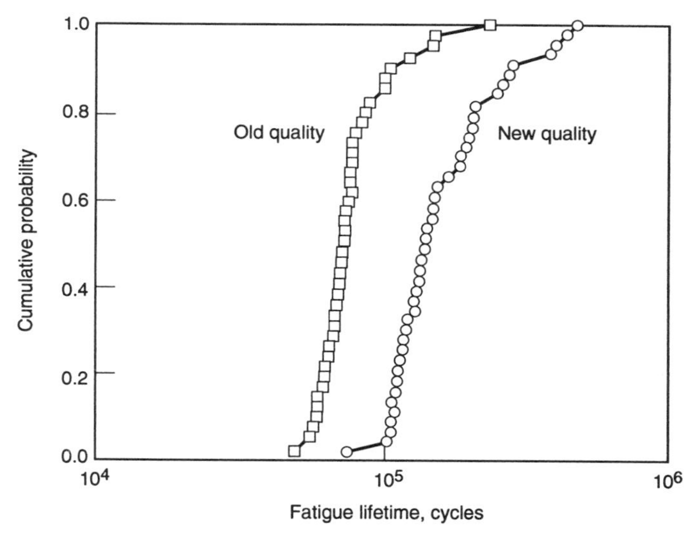

Fig. 10 Cumulative smooth fatigue lifetime distributors for old-quality and new-quality plate (see text for definitions). Tests conducted at 240 MPa (35 ksi) max stress, $R = 0.1$

Table 2 Summary of the 7050 plate materials used in the study of the effect of microporosity on fatigue

Material	Product thickness, in.	Key microstructural features
Old-quality plate	5.7	Large porosity
New-quality plate	5.7	Porosity
Low-porosity plate	6.0	Small porosity, constituent particles
Low-particle plate	6.0 (T/4)	Small constituents, thick plate grain structure
Thin plate	1.0	Refined grain size and constituent particles

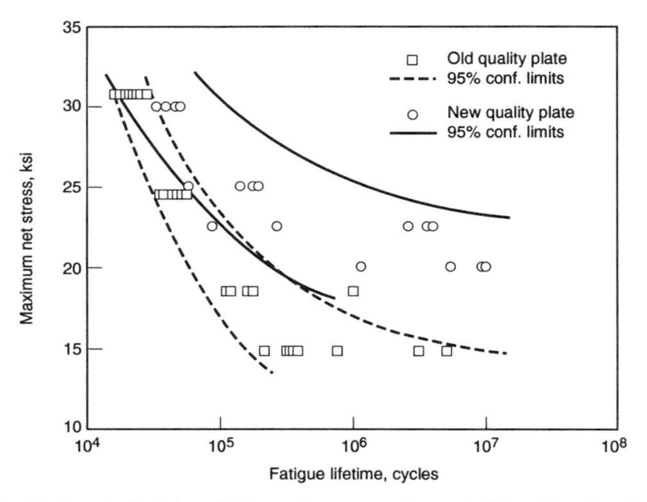

Fig. 11 Open-hole fatigue lifetimes for new-quality and old-quality plate (see text for definitions). Tests conducted at $R = 0.1$

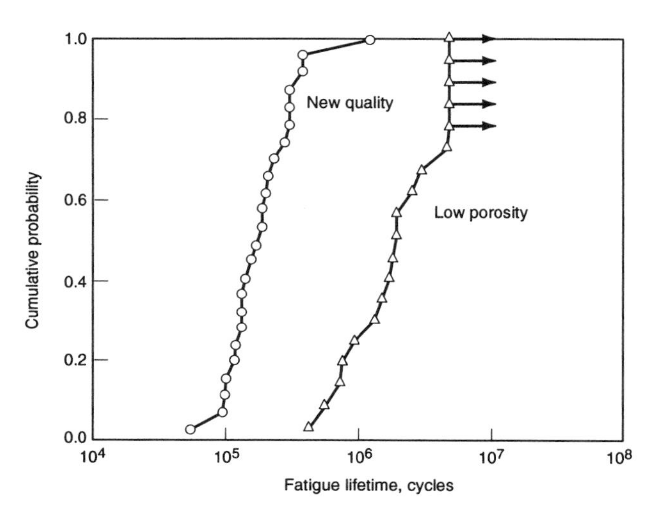

Fig. 12 Cumulative smooth fatigue life distributions for new-quality and low-porosity thick plate (see text for definitions). Tests conducted at 275 MPa (40 ksi) max stress, $R = 0.1$

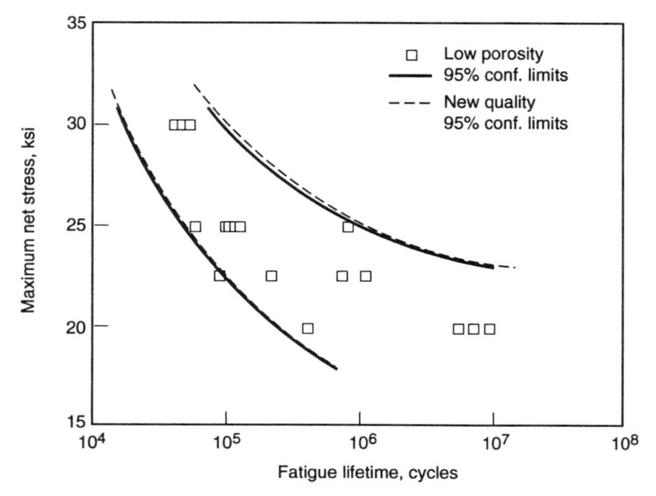

Fig. 13 Open-hole fatigue lifetimes for low-porosity plate and 95% confidence limits for new-quality plate (see text for definitions). Tests conducted at $R = 0.1$

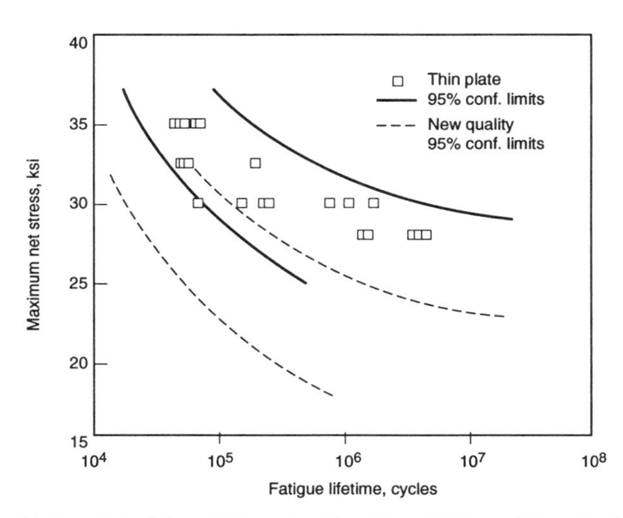

Fig. 14 Open-hole fatigue lifetimes for thin plate and 95% confidence limits for new-quality plate (see text for definitions). Tests conducted at $R = 0.1$

Table 3 Comparison of the calculated material fatigue strengths and improvements in open-hole fatigue for 7050 plate materials

Material	Fatigue stress at 10^5 cycles, MPa	Improvement, %	Failure mechanism
Old-quality plate	110	...	Microporosity
New-quality plate	125	14%	Microporosity
Low-porosity plate	125	14%	Consituent particles
Low-particle plate	150	36%	Micropores/constituent particles

Table 4 Hierarchy of fatigue-initiating features in 7050-T7451 plate

Material	Dominant microstructural feature: Smooth fatigue (round bars)	Open-hole fatigue
Old-quality plate	Coarse microporosity	Coarse microporosity
New-quality plate	Microporosity	Microporosity
Low-porosity plate	Fine microporosity	Constituent particles
Low-particle plate	...	Grain structure
Thin plate	Grain structure/constituent particles	Grain structure/constituent particles

S-N response and the 95% confidence limits (Ref 12). The data clearly show that at equivalent stresses the new-quality material exhibited longer lifetimes than the old-quality material.

Cumulative smooth fatigue lifetime distributions were obtained for the new-quality plate and the low-porosity plate tested at a maximum cyclic stress of 275 MPa (40 ksi), as opposed to the 240 MPa (35 ksi) maximum stress in Fig. 10. The stress was increased over the previous tests because the lifetimes of the low-porosity plate would be too long to practically test at the lower stress level. All other testing conditions were the same as in the previous tests. The fatigue lifetime data are shown in Fig. 12. It can be seen that the low-porosity plate shows considerably longer smooth fatigue lifetimes than the new-quality plate.

Open-hole specimen *S-N* fatigue results were also obtained for the low-porosity plate for comparison to the new-quality and old-quality plate open-hole S-N curves. The fatigue data are plotted in Fig. 13 along with the bounds for the old-quality and new-quality plate determined from the Box-Cox analysis. The open-hole data for the low-porosity plate lies within the bounds for the new-quality plate. This occurs despite the significant improvement in the smooth specimen fatigue lifetime for the low-porosity plate. An investigation into the mechanisms for the observed behavior are given in the following section, "Fractography."

A limited amount of fatigue testing was completed on the low-particle thick plate. Five open-hole specimen fatigue tests were performed at a maximum net stress of 200 MPa (30 ksi). A comparison of the calculated fatigue strengths for the thick plate specimens is given in Table 3.

The final microstructural variant of 7050, the thin plate, was tested using the same open-hole *S-N* fatigue test conditions as for the other materials. The data are shown in Fig. 14 with the confidence limits for the new-quality plate for comparison. The open-hole data for the thin plate are considerably longer than for the new-quality plate.

Fractography of the failed fatigue specimens was performed to identify the microstructural features that affect the fatigue damage process for each of the material microstructural conditions. The fractography was performed using a scanning electron microscope.

As summarized in Table 4, the failures for both the old-quality and new-quality materials in both the smooth and open-hole tests were controlled by porosity. The sizes of pores in the new-quality material are smaller than in the old-quality material, and hence the fatigue lifetimes are longer. The smooth fatigue failures for the low-porosity material are dominated by pores, which are most often associated with particles. These may have a combined effect on the propensity to initiate fatigue damage. The failure of the open-hole specimens for low-porosity plate are controlled by the Al_7Cu_2Fe particles, but one failure initiated at a Mg_2Si particle. At low stress, two failures are seen to initiate as Stage 1 fatigue cracks. Despite the change in initiation mechanism from the new-quality plate where pores dominated, the open-hole lifetimes of the low-porosity plate and those of the new-quality plate are similar. The limited smooth data for the thin plate show Stage 1 fatigue as the dominant mechanism with one occurrence of initiation from a Mg_2Si particle. The initiation mechanisms observed for thin plate in the open-hole test appear to be stress-dependent. Stage 1 initiations occur at the low stresses, with a mix of initiation from Al_7Cu_2Fe and Mg_2Si particles at the higher stresses.

The fatigue initiation site characterizations provide an understanding of the mechanisms of initiation and how the material microstructure affects the performance of a material. These characterizations have led to the development of an understanding of the hierarchy of mechanisms that control the fatigue. This information is used as input into probabilistic models that enable prediction of material performance based on microstructure. In addition, this understanding can aid material and process designers in the optimization of alloys for structural longevity.

REFERENCES

1. C.M. Hudson and S.K. Seward, A Literature Review and Inventory of the Effects of Environment on the Fatigue Behavior of Metals, *Eng. Fracture Mech.*, Vol 8 (No. 2), 1976, p 315-329
2. "Corrosion Fatigue of Aircraft Materials," AGARD Report 659, North Atlantic Treaty Organization, 1977
3. C.Q. Bowles, "The Role of Environment, Frequency, and Wave Shape during Fatigue Crack Growth in Aluminum Alloys," Report LR-270, Delft University of Technology, The Netherlands, 1978
4. G.E. Nordmark, B.W. Lifka, M.S. Hunter, and J.G. Kaufman, "Stress Corrosion and Corrosion Fatigue Susceptibility of High Strength Alloys," Technical Report AFML-TR-70-259, Wright-Patterson Air Force Base, 1970
5. T.H. Sanders, R.R. Sawtell, J.T. Staley, R.J. Bucci, and A.B. Thakker, "Effect of Microstructure on Fatigue Crack Growth of 7XXX Aluminum Alloys under Constant Amplitude and Spectrum Loading," Final Report, Contract N00019-76-C-0482, Naval Air Systems Command, 1978
6. J.T. Staley, "How Microstructure Affects Fatigue and Fracture of Aluminum Alloys," paper presented at the International Symposium on Fracture Mechanics (Washington, DC), 1978
7. W.G. Truckner, J.T. Staley, R.J. Bucci, and A.B. Thakker, "Effects of Microstructure on Fatigue Crack Growth of High Strength Aluminum Alloys," Report AFML-TR-76-169, U.S. Air Force Materials Laboratory, 1976
8. J.T. Staley, W.G. Truckner, R.J. Bucci, and A.B. Thakker, Improving Fatigue Resistance of Aluminum Aircraft Alloys, *Aluminum*, Vol 53, 1977, p 667-669
9. M.V. Hyatt, "Program to Improve the Fracture Toughness and Fatigue Resistance of Aluminum Sheet and Plate for Airframe Applications," Technical Report AFML-TR-73-224, Wright-Patterson Air Force Base, 1973
10. C.R. Owen, R.J. Bucci, and R.J. Kegarise, Aluminum Quality Breakthrough for Aircraft Structural Reliability, *Journal of Aircraft*, Vol 26 (No. 2), Feb 1989, p 178-184
11. P.E. Magnusen, A.J. Hinkle, W.T. Kaiser, R.J. Bucci, and R.L. Rolf, Durability Assessment Based on Initial Material Quality, *Journal of Testing and Evaluation*, Vol 18 (No. 6), Nov 1990, p 439-445
12. A.J. Hinkle and M.R. Emptage, Analysis of Fatigue Life Data Using the Box-Cox Transformation, *Fatigue and Fracture of Engineering Materials and Structures*, Vol 14 (No. 5), 1991, p 591-600

Aluminum Alloy S-N Data

Aluminum and 2xxx Alloys

Unalloyed Aluminum

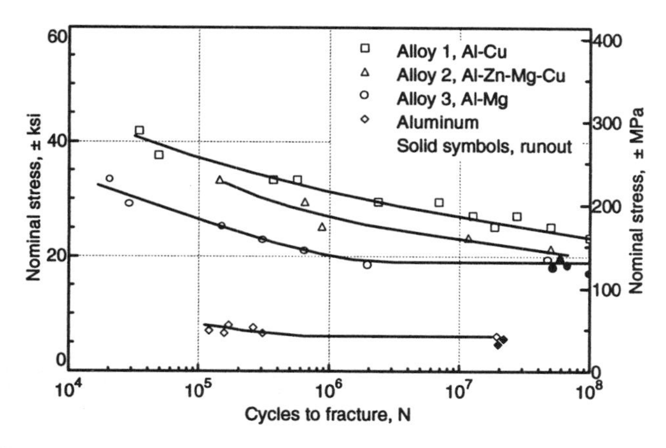

Fig. 1 Fatigue of 99.25% aluminum and various aluminum alloys. Alloy 1, fully heat treated Al-4.5Cu; Alloy 2, fully heat treated Al-6Zn-2.2%Mg-1.33Cu; Alloy 3, Al-5%Mg. Solid symbols are runout (no failure). Source: *Aluminum Labs Reseach Bulletin*, No. 1, 1952

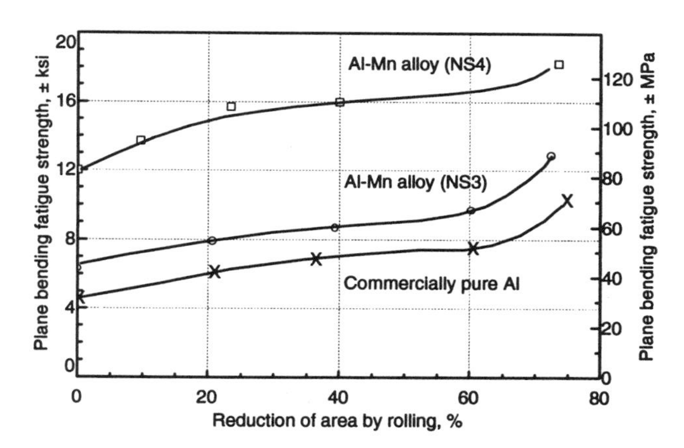

Fig. 2 Effect of cold work on aluminum fatigue. Source: G. Forrest, *Metal Fatigue*, 1959

2008

2008-T4 and -T62: Room-temperature fatigue strength

Temper	Form and thickness	Specimen	Test mode	Stress ratio	Atmosphere	Fatigue strength, MPa (ksi) 10^4	10^5	10^6	10^7	10^8	5×10^8
T4	Sheet	Unnotched sheet	Axial	0.1	Air	...	(36)	(23)	(21.5)	...	...
	Sheet	Sheet notch, $K_t=4.5$	Axial	0.1	Air	...	(16)	(10)	...	...	...
	Sheet	Sheet notch, $K_t>12$	Axial	0.1	Air	...	(14.5)	(9)	38 (5.5)	...	...
	Sheet	Unnotched sheet	Axial	0.1	3.5% NaCl	...	(27)	(20)	...	...	...
	Sheet	Sheet notch; $K_t=4.5$	Axial	0.1	3.5% NaCl	...	(13)	(6.5)	(4)	...	...
	Sheet	Sheet notch, $K_t>12$	Axial	0.1	3.5% NaCl	...	...	(5)	(3.5)	...	...
T62	Sheet	Unnotched sheet	Axial	0.1	Air	...	(39)	(31)	(29)	...	...
	Sheet	Sheet notch, $K_t=4.5$	Axial	0.1	Air	...	(16)	(9)	(8)	...	...
	Sheet	Sheet notch, $K_t>12$	Axial	0.1	Air	...	(14.5)	(7)	(6.5)	...	...
	Sheet	Unnotched sheet	Axial	0.1	3.5% NaCl	...	(34)	(16.5)	(9)	...	...
	Sheet	Sheet notch, $K_t=4.5$	Axial	0.1	3.5% NaCl	...	(14)	(7)	(5)	...	...
	Sheet	Sheet notch, $K_t>12$	Axial	0.1	3.5% NaCl	...	(13)	(5)	(3)	...	...
T4	Sheet	Unnotched	Axial	0.1	Air	...	(36)	(23)	(21)	...	...
	Sheet	Unnotched	Axial	0.1	3.5% NaCl	(35)	(27)	(20)	...	...	...
	Sheet	Notched, $K_t=4.5$	Axial	0.1	Air	...	(16)	(10)	...	...	...
	Sheet	Notched, $K_t=4.5$	Axial	0.1	3.5% NaCl	...	(13)	(6)	(4)	...	...
	Sheet	Notched, $K_t>12$	Axial	0.1	Air	(32)	(14)	(9)	(5)	...	...

(continued)

2008-T4 and -T62: Room-temperature fatigue strength (*continued*)

Temper	Form and thickness	Specimen	Test mode	Stress ratio	Atmosphere	Fatigue strength, MPa (ksi)					
						10^4	10^5	10^6	10^7	10^8	5×10^8
	Sheet	Notched, $K_t >$ 12	Axial	0.1	3.5% NaCl	...	...	(5)	(3)	...	...
T62	Sheet	Unnotched	Axial	0.1	Air	(52)	(39)	(31)	(29)	...	...
	Sheet	Unnotched	Axial	0.1	3.5% NaCl	...	(34)	(16)	(9)	...	...
	Sheet	Notched, $K_t =$ 4.5	Axial	0.1	Air	...	...	...	...	...	...
	Sheet	Notched, $K_t =$ 4.5	Axial	0.1	3.5% NaCl	(34)	(14)	(9)	8...	...	...
	Sheet	Notched, $K_t =$ 4.5	Axial	0.1	3.5% NaCl	(23)	...	(7)	(5)	...	...
	Sheet	Notched, $K_t >$ 12	Axial	0.1	Air	(33)	(13)	(7)	(6)	...	...
	Sheet	Notched, $K_t >$ 12	Axial	0.1	3.5% NaCl	(21)	(9)	(5)	(3)	...	...

Source: Unpublished Alcoa

2011

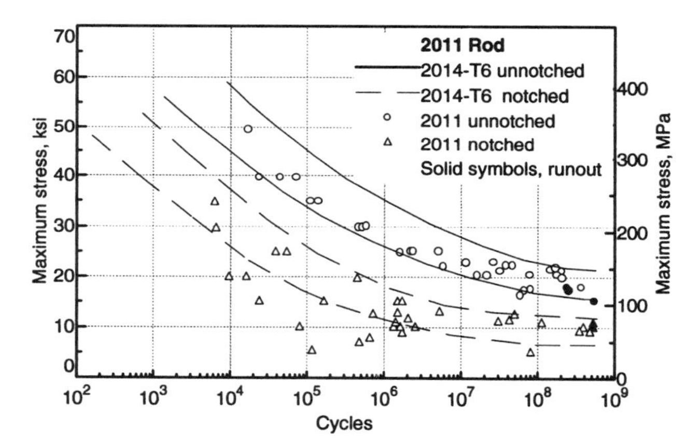

Fig. 3 2011 rotating beam fatigue (tempers T3, T6, and T8 combined) compared with upper and lower bands for 2014-T6 (notched and unnotched). Specimens were taken from rolled and drawn rod. R.R. Moore specimens with 9-7/8 in. surface radius and 0.300 in. minimum diameter for unnotched specimens. Notched specimens had a 0.330 in. diameter at the notch and a 0.480 in. diameter outside the 60° notch. Notched specimens had a sharp V-notch with a radius r < 0.001 in. at notch root. Solid symbols indicate runout (no failure). Source: Alcoa, 1962

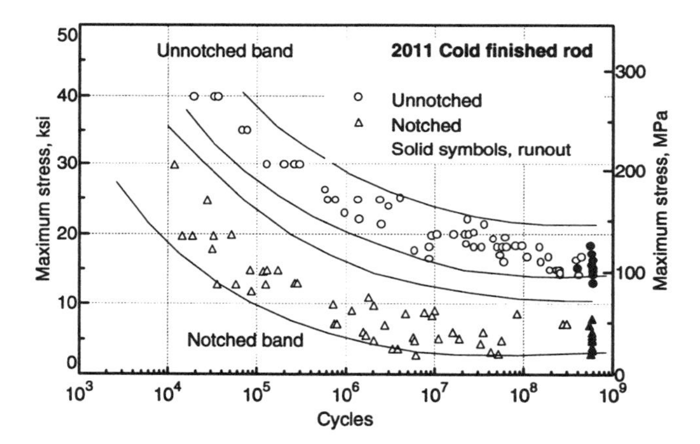

Fig. 4 2011 rotating beam fatigue. Specimens were taken from cold finished rod (rolled-and-drawn, or extruded-rolled-and drawn). R.R. Moore specimens with 9-7/8 in. surface radius and 0.300 in. minimum diameter for unnotched specimens. Notched specimens had a 0.330 in. diameter at the notch and a 0.480 in. diameter outside the 60° notch. Notched specimens had a sharp V-notch with a radius r < 0.001 in. at notch root. Solid symbols indicate runout (no failure). Source: Alcoa, 1967

2011-T4, -T6, and -T8: Room-temperature fatigue strength in air

Temper	Form and thickness	Specimen	Test mode	Stress ratio	Fatigue strength, MPa (ksi), at cycles of:					
					10^4	10^5	10^6	10^7	10^8	5×10^8
T4	Rod	RR Moore	Rotating beam	−1	...	...	(24)	(19)	(15.5)	(14)
	Rod	RR Moore notch	Rotating beam	−1	(28)	(18)	(11)	(9)	(8.5)	(8.5)
T6	Rod	RR Moore	Rotating beam	−1	...	(28)	(22)	(16)	(13.5)	(13)
	Rod	RR Moore notch	Rotating beam	−1	(26)	(18)	(12)	(9)	(8.5)	(8.5)
T8	Rod	RR Moore	Rotating beam	−1	...	(37)	(27)	(23)	(20)	(19)
	Rod	RR Moore notch	Rotating beam	−1	...	(17)	(11)	(9.5)	(9)	(8.5)

Source: Unpublished Alcoa data

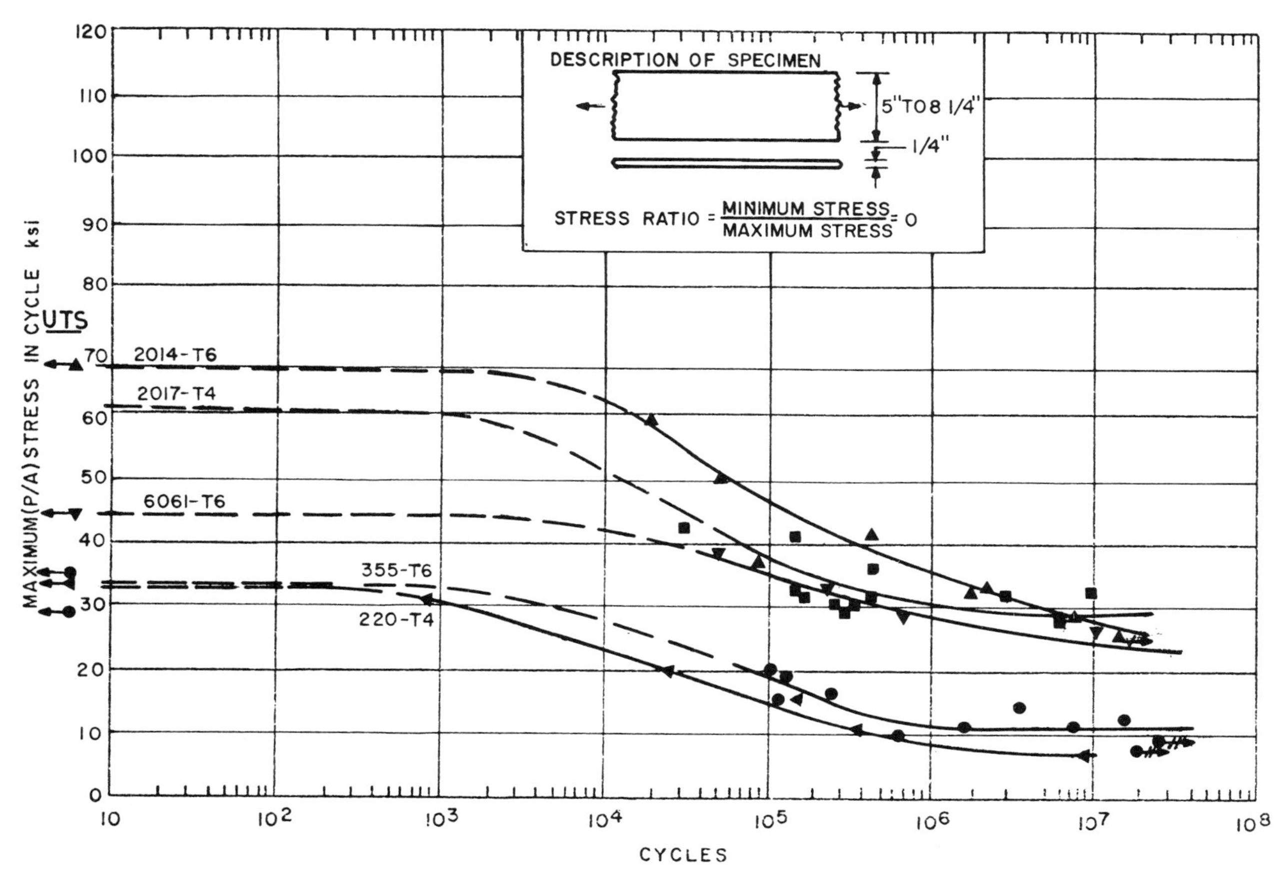

Fig. 5 Axial fatigue of 2014-T6 plate and other Al-alloy plate. Source: Alcoa, 1958, reported in Naval Research Lab Report CDNSWC-TR619409, 1994

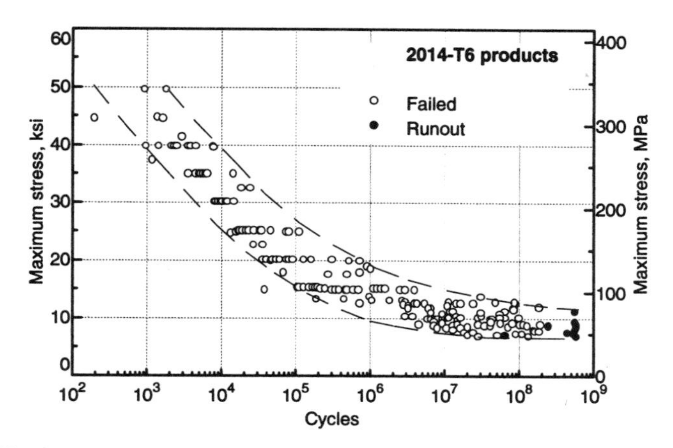

Fig. 6 2014-T6 notched rotating beam fatigue of (radius at notch root <0.001 in.). Notched specimens had a 0.330 in. diameter at the notch and a 0.480 in. diameter outside the 60° notch. Source: Alcoa, 1957

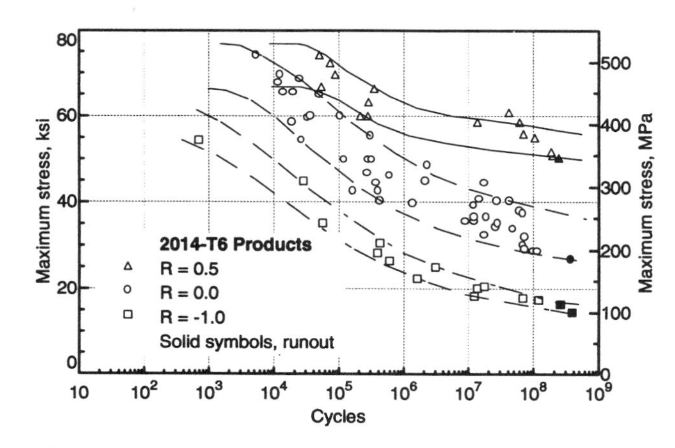

Fig. 7 2014-T6 unnotched axial fatigue. Solid symbols indicate runout (no failure). Unnotched axial specimens with 9-7/8 in. surface radius and a minimum diameter of 0.160 or 0.200 in. Source: Alcoa, 1955

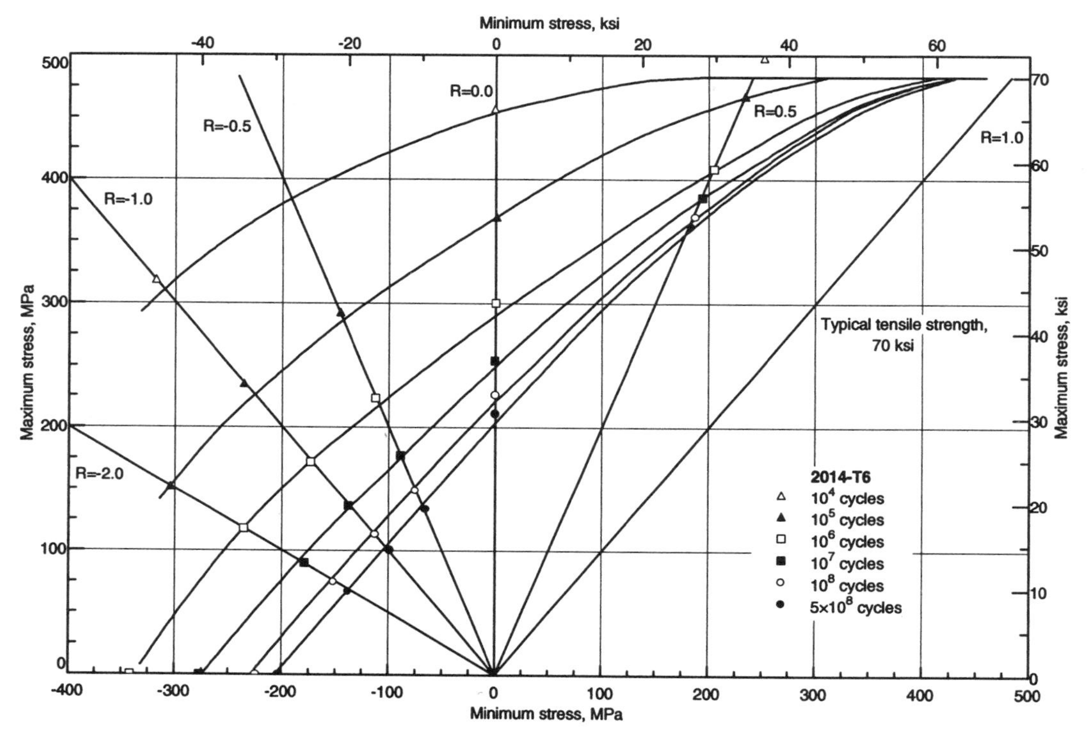

Fig. 8 2014-T6 modified Goodman diagram . Source: Alcoa, 1955

2014-T6: Room-temperature fatigue strength in air

Form and thickness	Specimen	Test mode	Stress ratio	Fatigue strength (ksi) at cycles of:					
				10^4	10^5	10^6	10^7	10^8	5×10^8
Plate, 0.25 in.	Polished bar, 0.2 in. diam	Axial	0	(66)	(50)	(41)	(36)	(33)	...
Plate, 0.25 in.	Flat plate, 1 in. wide , as rolled	Axial	0	(53)	(38)	(28)	(26)	(25)	...
Plate, 0.25 in.	Flat plate, 5 in. wide, machined	Axial	0	(62)	(46)	(35)	(27)	(25)	...
Alclad plate, 3/8 in.	Flat plate	Axial	0	...	(38)	(21)	(18)	(16)	...
Plate, 0.25 in.	Flat plate, 7.5 in. central hole	Axial	0	(37)	(19)	(14)	(12)	(11)	...
Alclad plate, 0.25 in.	Flat plate, 5 in wide transverse butt weld, bead off	Axial	0	(26)	(17)	(11)	(7)	...	...
Sheet, 0.16 in.	Transverse single V bouu weld, bead on	Axial	0	(32)	(18.5)	(14.5)	(14.5)	...	...

Source: M.G. Vassilaros and E.J. Czyryca, Naval Research Lab Report CDNSWC-TR619409, A Compilation of Fatigue Information for Aluminum Alloys, 1994

2017

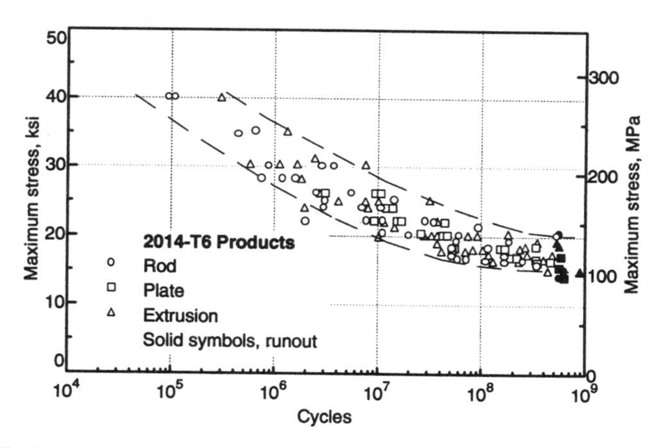

Fig. 9 2017-T4 unnotched rotating beam fatigue (plate and rod). Unnotched axial specimens with 9-7/8 in. surface radius and 0.300 in. minimum diameter. Solid symbols indicate runout (no failure). Source: Alcoa, 1959

2017-O and -T4: Room-temperature fatigue strength in air

Temper	Form and thickness	Specimen	Test mode	Stress ratio	Fatigue strength (ksi) at cycles of: 10^4	10^5	10^6	10^7	10^8	5×10^8
O	Rod	RR Moore	Rotating beam	−1	...	(21)	(19)	(16)	(14)	(13)
	Rod	RR Moore notch	Rotating beam	−1	(15)	(10)	(8)	(7)	(6.5)	(6.5)
T4	1/4 in. plate	Polished, 0.2 in. D bar	Axial	0	...	...	(38)	(30)	(22)	...
	1/4 in. plate	Flat plate, 5 in. wide	Axial	0	(50)	(38)	(30)	(27)	...	...
	1/4 in. plate	Flat plate, 71/2 in. wide	Axial	0	(30)	(18)	(11)	(9)	(8)	...
	W	RR Moore	Rotating beam	−1	...	(42)	(33)	(24)	(18)	(18)
	R	Round	Axial	0	...	(49)	(41)	(35)	(30)	(26)
	R	Round	Axial	−1	...	(31)	(25)	(20)	(17)	(16)

Sources: Unpublished Alcoa data and M.G. Vassilaros and E.J. Czyryca, Naval Research Lab Report CDNSWC-TR619409, A Compilation of Fatigue Information for Aluminum Alloys, 1994

2024-T3

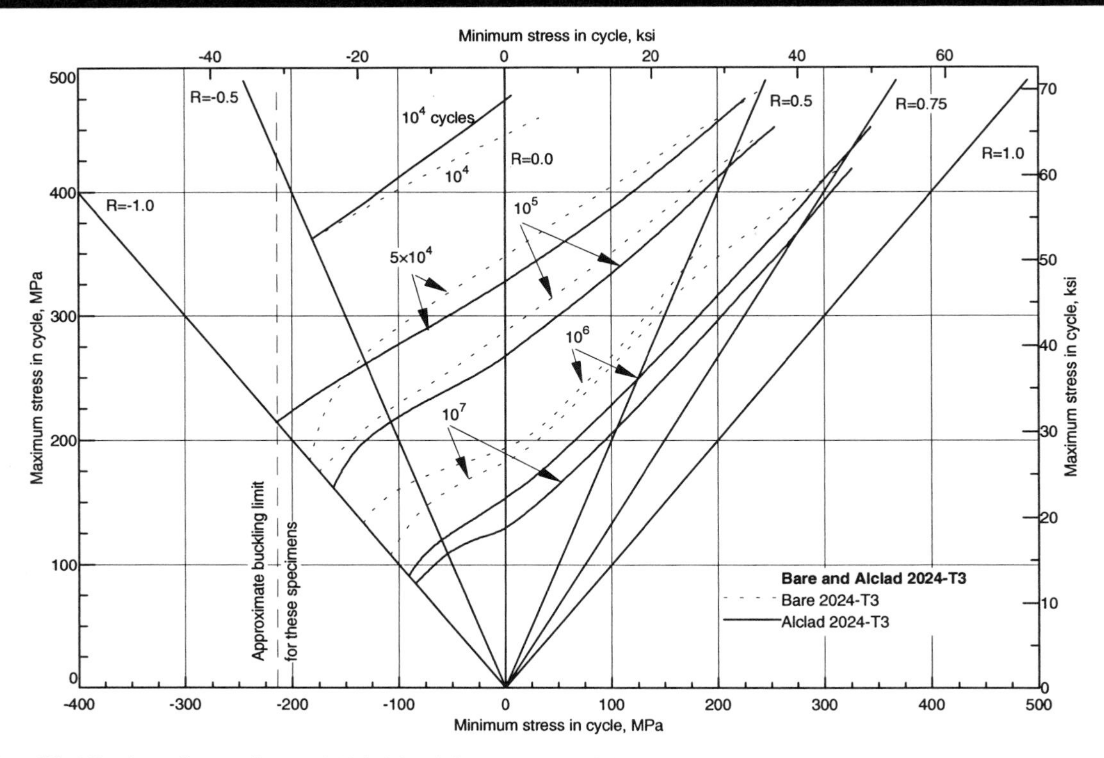

Fig. 10 2024-T3 modified Goodman diagram (bare and Alclad sheet). Source: Alcoa, 1957

2024-T3 sheet: Room-temperature fatigue strength in air

Specimen	Test mode	Stress ratio	Fatigue strength (ksi) at cycles of: 10^4	10^5	10^6	10^7	10^8	5×10^8
Unnotched	Flexure	–1	...	(34)	(27)	(21)	(18)	(17)
Unnotched	Axial	0.75	...	...	(72)	(62)	(62)	...
Unnotched	Axial	0.50	...	(62)	(50)	(49)	(49)	(49)
Unnotched	Axial	0.0	(65)	(42)	(27)	(26)	(26)	...
Unnotched	Axial	–0.5	(53)	(32)	(24)	(22)	(22)	...
Unnotched	Axial	–1.0	...	(25)	(19)	(17)	(17)	...
Unnotched	Axial	0.5	...	...	(54)	...	...	...
Unnotched	Axial	0.4	...	(68)	(48)	...	...	...
Unnotched	Axial	0.02	...	(51)	(36)	...	...	...
Unnotched	Axial	–0.3	(66)	(43)	...	...	...	...
Unnotched	Axial	–0.6	(58)	(38)	(27)	...	...	...
Unnotched	Axial	–0.8	...	(35)	(25)	...	...	...
Unnotched	Axial	–1.0	(52)	(33)	(23)	(18)	...	...
Notched, $K_t = 1.5$	Axial	–1.0	(42)	(24)	(18)	(16)	...	...
Notched, $K_t = 1.5$	Axial	10(a)	(49)	(31)	(25)	(24)	...	...
Notched, $K_t = 1.5$	Axial	20(a)	(56)	(39)	(33)	(31)	...	...
Notched, $K_t = 1.5$	Axial	30(a)	(63)	(47)	(42)	(40)	...	...
Notched, $K_t = 2.0$	Axial	–1.0	(30)	(19)	(13)	(11)	...	...
Notched, $K_t = 2.0$	Axial	10(a)	(37)	(26)	(21)	(18)	...	...
Notched, $K_t = 2.0$	Axial	20(a)	(45)	(34)	(29)	(28)	...	...
Notched, $K_t = 2.0$	Axial	30(a)	(53)	(43)	(39)	...	...	...
Notched, $K_t = 4.0$	Axial	–1.0	(18)	(12)	(9)	(7)	...	...
Notched, $K_t = 4.0$	Axial	10(a)	(25)	(19)	(16)	...	...	...
Notched, $K_t = 4.0$	Axial	20(a)	(33)	(27)	(25)	(24)	...	...
Notched, $K_t = 4.0$	Axial	30(a)	(42)	(36)	(34)	(33)	...	...
Notched, $K_t = 5.0$	Axial	0	(15)	(10)	(7)	(5)	...	...
Notched, $K_t = 5.0$	Axial	10(a)	(21)	(17)	...	(13)	...	...
Notched, $K_t = 5.0$	Axial	20(a)	(29)	(25)	...	(23)	...	...
Notched, $K_t = 5.0$	Axial	30(a)	(37)	(34)	...	(33)	...	...
Unnotched	Flexure	–1.0	(32)	(17)	(15)	(14)	(14)	...
Unnotched	Axial	0.75	...	...	(63)	(55)	(54)	...
Unnotched	Axial	0.50	...	(61)	(37)	(33)	(32)	...
Unnotched	Axial	0.0	(70)	(39)	(24)	(18)	(18)	...
Unnotched	Axial	–0.5	(52)	(31)	(17)	(16)	(16)	...
Unnotched	Axial	–1.0	...	(24)	(13)	(12)	(12)	...

(a) Mean stress in ksi. Sources: Unpublished Alcoa data; M.G. Vassilaros and E.J. Czyryca, Naval Research Lab Report CDNSWC-TR619409, A Compilation of Fatigue Information for Aluminum Alloys, 1994

2024-T4

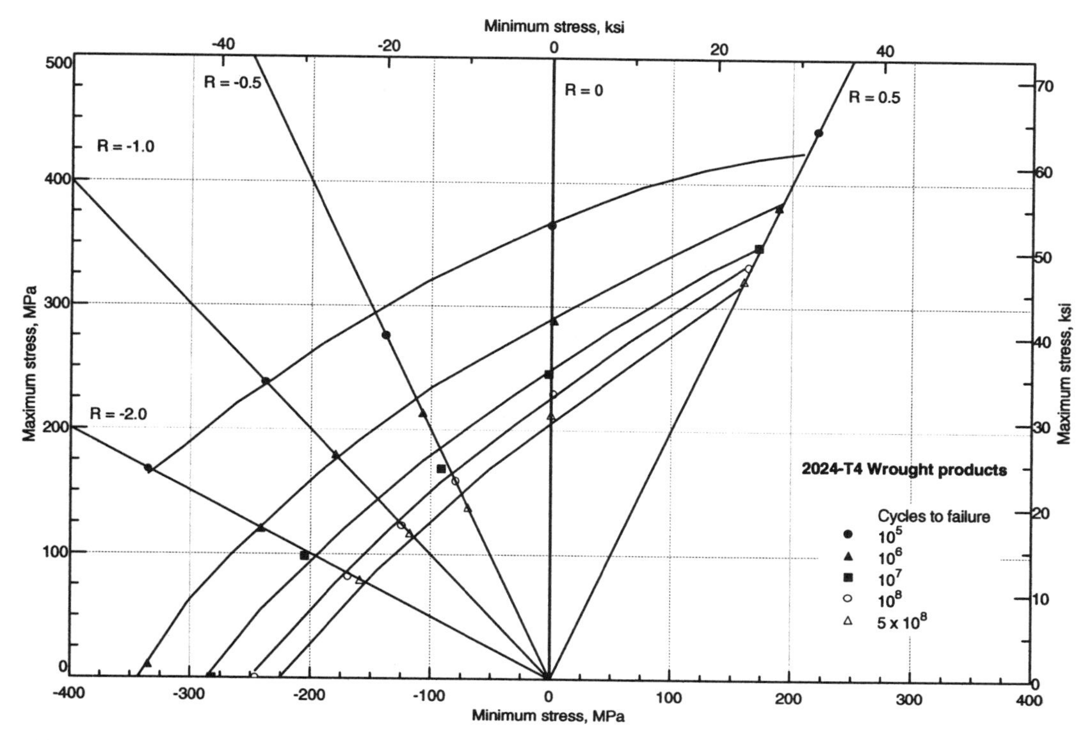

Fig. 11 2024-T4 modified Goodman diagram. Source: *Proceedings ASTM*, Vol 55, 1955, p 955

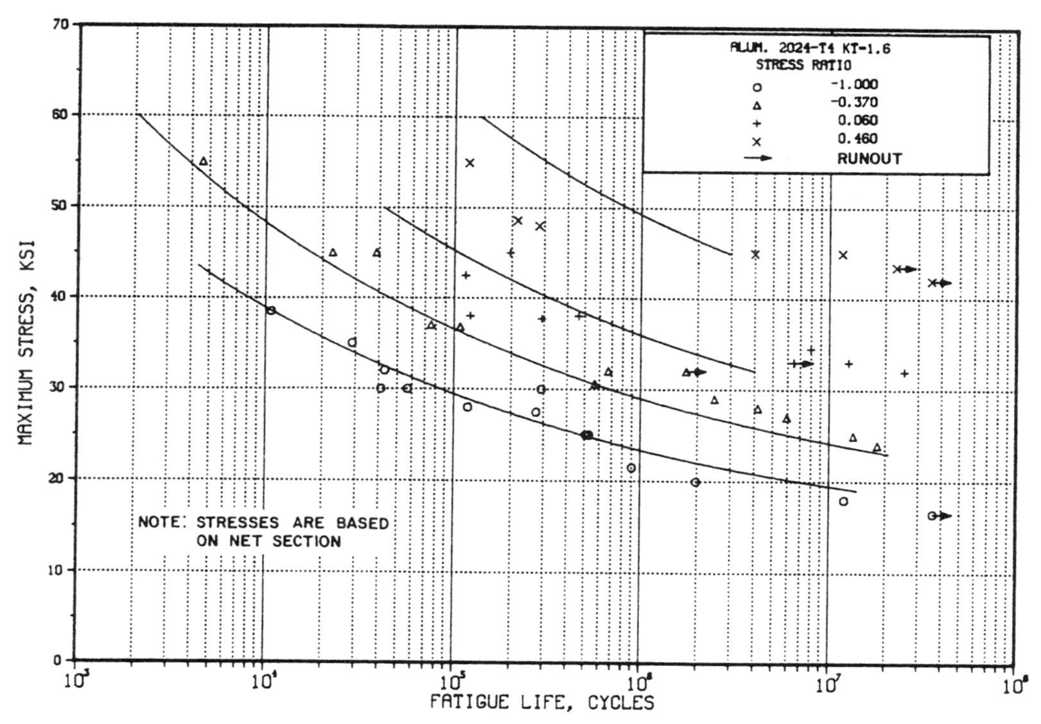

Fig. 12 2024-T4 notched axial fatigue ($K_t = 1.6$) from bar in longitudinal direction. Source: MIL-HDBK-5

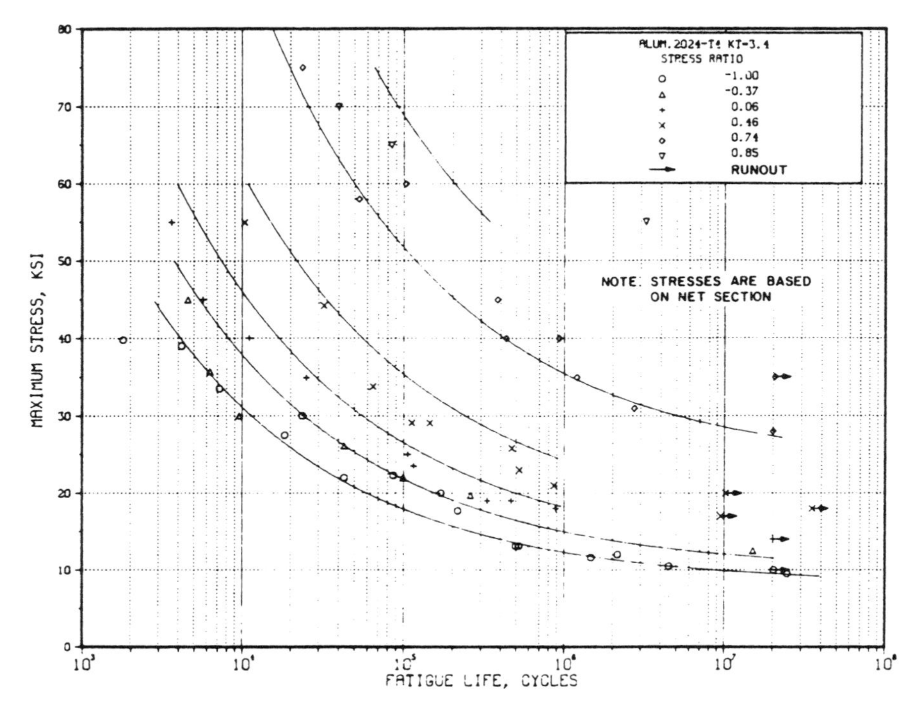

Fig. 13 2024-T4 notched axial fatigue ($K_t = 3.4$) from bar in longitudinal direction. Source: MIL-HDBK-5

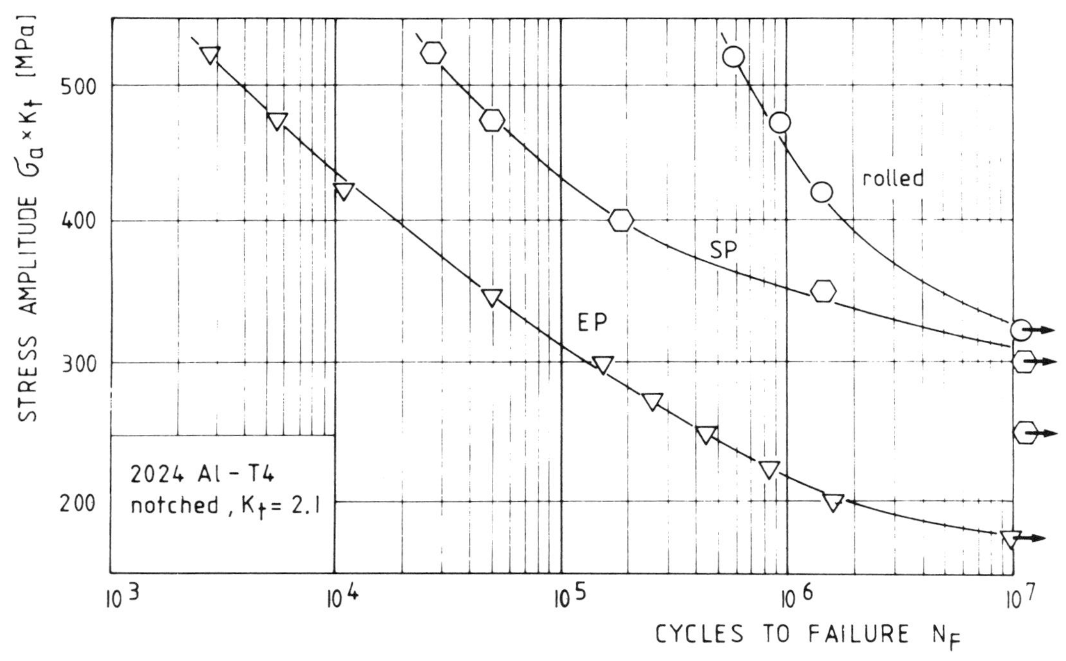

Fig. 14 2024-T4 notched rotating beam fatigue ($K_t = 2.1$) for rolled, shot peened (SP), and electrolytically polished (EP) specimens cut from plate in longitudinal direction. Notched specimens were surface rolled using a three-roll device with an optimum rolling force of 0.8 kN reported by L. Wagner *et al.*, Influence of Surface Rolling on Notched Fatigue Strength of Al 2024 in Two Age Hardening Conditions, *Fatigue 93*, MCPE. Figure source: *Surface Engineering*, DGM, 1993

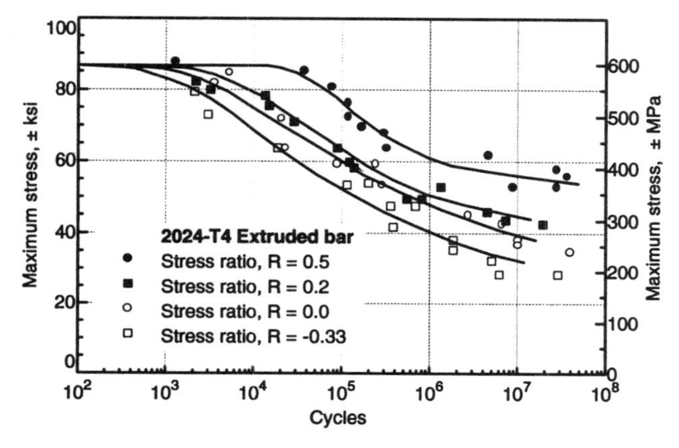

Fig. 15 2024-T4 unnotched axial fatigue (extruded bar). Unnotched axial specimens with 9-7/8 in. surface radius and 0.158 in. minimum diameter. Source: Alcoa, 1954

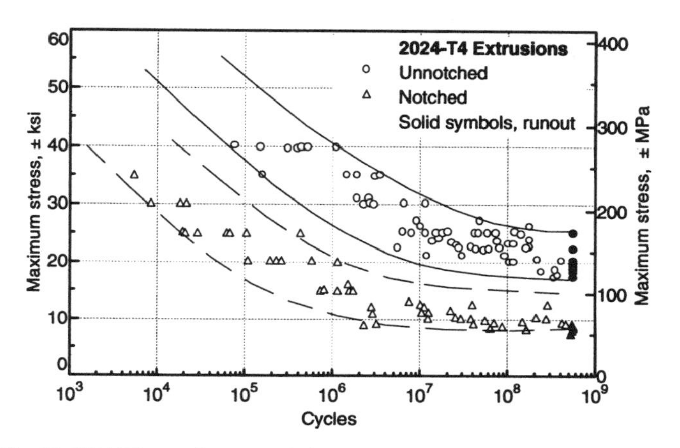

Fig. 16 2024-T4 rotating beam fatigue for unnotched and notched specimens from extrusions (radius at notch root <0.001 in.). R.R. Moore specimens with 9-7/8 in. surface radius and 0.300 in. minimum diameter for unnotched specimens. Notched specimens had a 0.330 in. diameter at the notch and a 0.480 in. diameter outside the 60° notch. Solid symbols indicate runout (no failure). Source: Alcoa, 1955

2024-T4: Room-temperature fatigue strength in air

Form and thickness	Specimen	Test mode	Stress ratio	Fatigue strength (ksi) at cycles of: 10^4	10^5	10^6	10^7	10^8	5×10^8
Wire	Round, unnotched	Axial	−1.00	(49)	(38)	(30)	(22)	(18)	...
	Round, unnotched	Axial	−0.50	...	(44)	(34)	(27)	(21)	...
	Round, unnotched	Axial	0.00	(70)	(54)	(42)	(33)	(27)	...
	Round, unnotched	Axial	0.06	(72)	(56)	(43)	(34)	...	...
	Round, unnotched	Axial	0.50	...	(78)	(60)	(50)	...	...
	Round, notched, $K_t = 1.6$	Axial	−1.0	(39)	(20)	(23)	(19)	...	...
	Round, notched, $K_t = 1.6$	Axial	0.06	...	(45)	(36)	(32)	...	...
	Round, notched, $K_t = 1.6$	Axial	0.46	...	(62)	(49)	...	...	...
	Round, notched, $K_t = 2.4$	Axial	−1.0	(31)	(23)	(17)	(12)	...	...
	Round, notched, $K_t = 2.4$	Axial	0.06	(45)	(32)	(24)	...	...	...
	Round, notched, $K_t = 2.4$	Axial	0.46	(59)	(42)	(31)	...	...	...
	Round, notched, $K_t = 3.4$	Axial	−1.0	(31)	(18)	(12)	(10)	...	...
	Round, notched, $K_t = 3.4$	Axial	0.06	(46)	(27)	(18)	...	...	...
	Round, notched, $K_t = 3.4$	Axial	0.46	(62)	(36)	(24)	...	...	...
Plate	Round, notched	Axial	0	(62)	(47)	(36)	(33)	(32)	...
	Flat, notched, $K_t = 2.76$	Axial	0	(36)	(21)	(16)	(13)	(12)	...
Extrusion	Round, unnotched	Rotating beam	−1	...	(45)	(36)	(26)	(23)	...
	Round, notched	Rotating beam	−1	(36)	(24)	(16)	(11)	(10)	...

Sources: Unpublished Alcoa data; M.G. Vassilaros and E.J. Czyryca, Naval Research Lab Report CDNSWC-TR619409, A Compilation of Fatigue Information for Aluminum Alloys, 1994

2024-T6

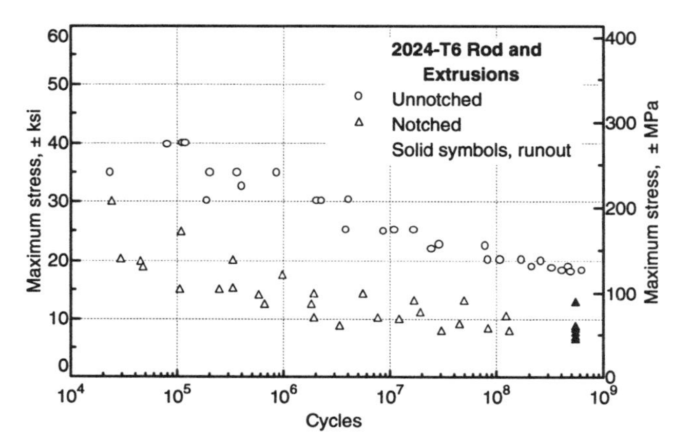

Fig. 17 2024-T6 rotating beam fatigue for unnotched and notched specimens from extrusion and rolled-and-drawn rod (radius at notch root <0.001 in.). R.R. Moore specimens with 9-7/8 in. surface radius and 0.300 in. minimum diameter for unnotched specimens. Notched specimens had a 0.330 in. diameter at the notch and a 0.480 in. diameter outside the 60° notch. Solid symbols indicate runout (no failure). Source: Alcoa, 1954

2024-T6 extruded rod: Rotating beam fatigue strength in air at room temperature

Specimen	Stress ratio	Fatigue strength (ksi) at cycles of: 10^4	10^5	10^6	10^7	10^8	5×10^8
R.R.Moore, unnotched	–1	...	(37)	(30)	(25)	(20)	(18)
R.R. Moore, notched	–1	...	(18)	(13)	(11)	(10)	(9)

Source: Unpublished Alcoa data

2024-T36, -T351 and -T361

2024-T36, -T351 and -T361: Room-temperature fatigue strength in air

Temper	Form and thickness	Specimen	Test mode	Stress ratio	Fatigue strength (ksi) at cycles of: 10^4	10^5	10^6	10^7	10^8	5×10^8
T36	Sheet, 0.165 in.	Flat plate	Bending	–1	...	(36)	(19)	(18)	(16)	...
	Alclad sheet	Flat sheet with 3/16 in. rivet	Axial	–1	...	(12)	(7)	(3)	(2)	...
T351	Plate	Round, Unnotched	Rotating beam	–1	...	(43)	(32)	(24)	(17)	(16)
		Round, Notched	Rotating beam	–1	(30)	(21)	(14)	(10)	(8)	(7)
T361	Sheet	Flat	Bending	–1	...	(36)	(19)	(18)	(16)	...
	Plate	R.R. Moore, Unnotched	Rotating beam	–1	(55)	(39)	(27)	(21)	(17)	(16)
	Plate	R.R. Moore, Notched	Rotating beam	–1	...	(21)	(12)	(9)	(8)	(8)

Sources: Unpublished Alcoa data

2024-T86, -T851, -T852, -T861

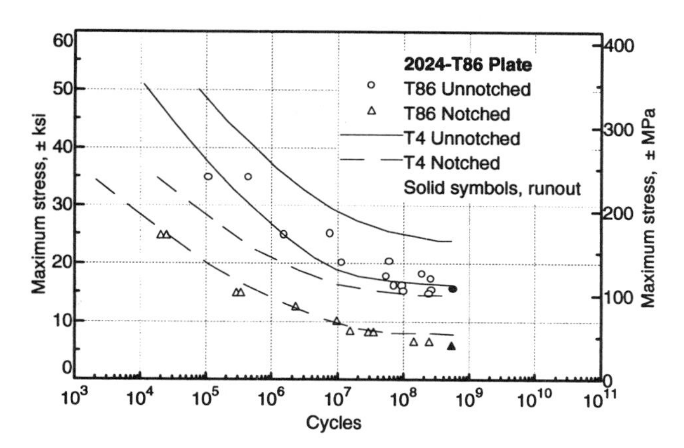

Fig. 18 2024-T86 and 2024-T4 rotating beam fatigue for unnotched and notched specimens from 5/8 in. plate (radius at notch root <0.001 in.). R.R. Moore specimens with 9-7/8 in. surface radius and 0.300 in. minimum diameter for unnotched specimens. Notched specimens had a 0.330 in. diameter at the notch and a 0.480 in. diameter outside the 60° notch. Solid symbols indicate runout (no failure). Source: Alcoa, 1962

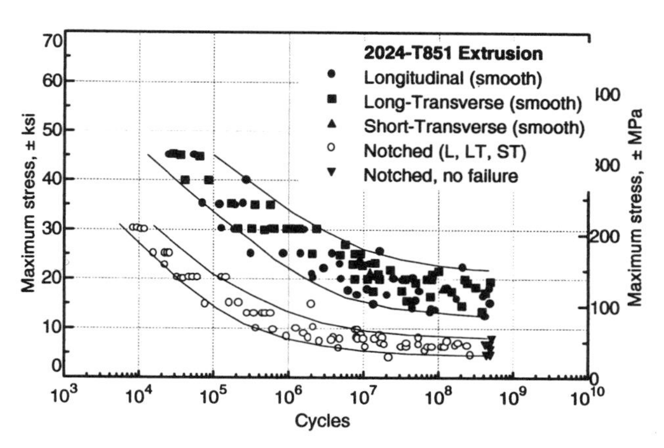

Fig. 19 2024 rotating beam fatigue for unnotched and notched (K_t > 12) specimens from temper conditions T851(rod, plate, forgings), T852 (forgings), and T8511 (extrusion). Thicknesses up to 3.5 in. R.R. Moore specimens with 9-7/8 in. surface radius and 0.300 in. minimum diameter for unnotched specimens. Notched specimens had a 0.330 in. diameter at the notch and a 0.480 in. diameter outside the 60° notch. Source: Alcoa

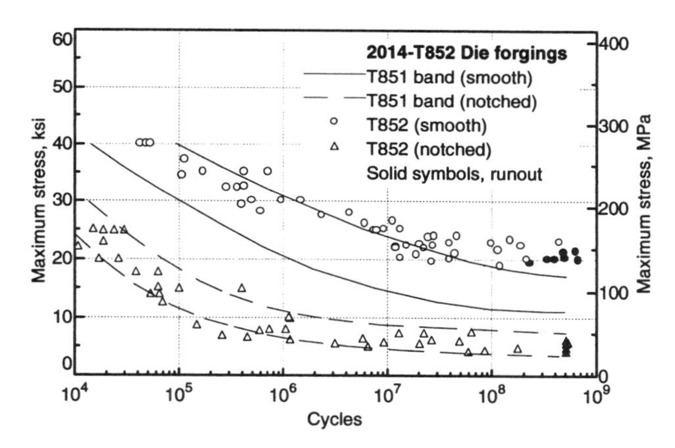

Fig. 20 2024-T852 rotating beam fatigue for unnotched and notched specimens from web and flange sections of die forgings (radius at notch root <0.001 in.). R.R. Moore specimens with 9-7/8 in. surface radius and 0.300 in. minimum diameter for unnotched specimens. Notched specimens had a 0.330 in. diameter at the notch and a 0.480 in. diameter outside the 60° notch. Solid symbols indicate runout (no failure). Band lines are for 2024-T851 preforged plate, 1.5 to 5 in. thick. Source: Alcoa

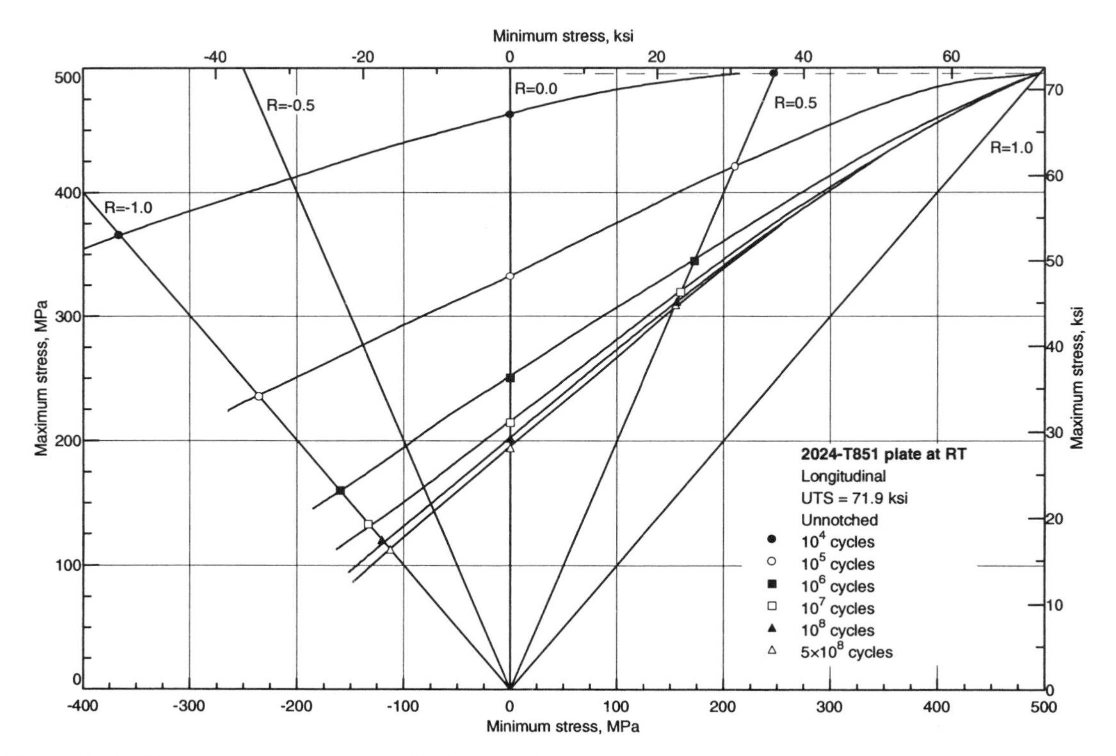

Fig. 21 2024-T851 unnotched axial fatigue at room temperature (7/8 in. plate). Source: Alcoa

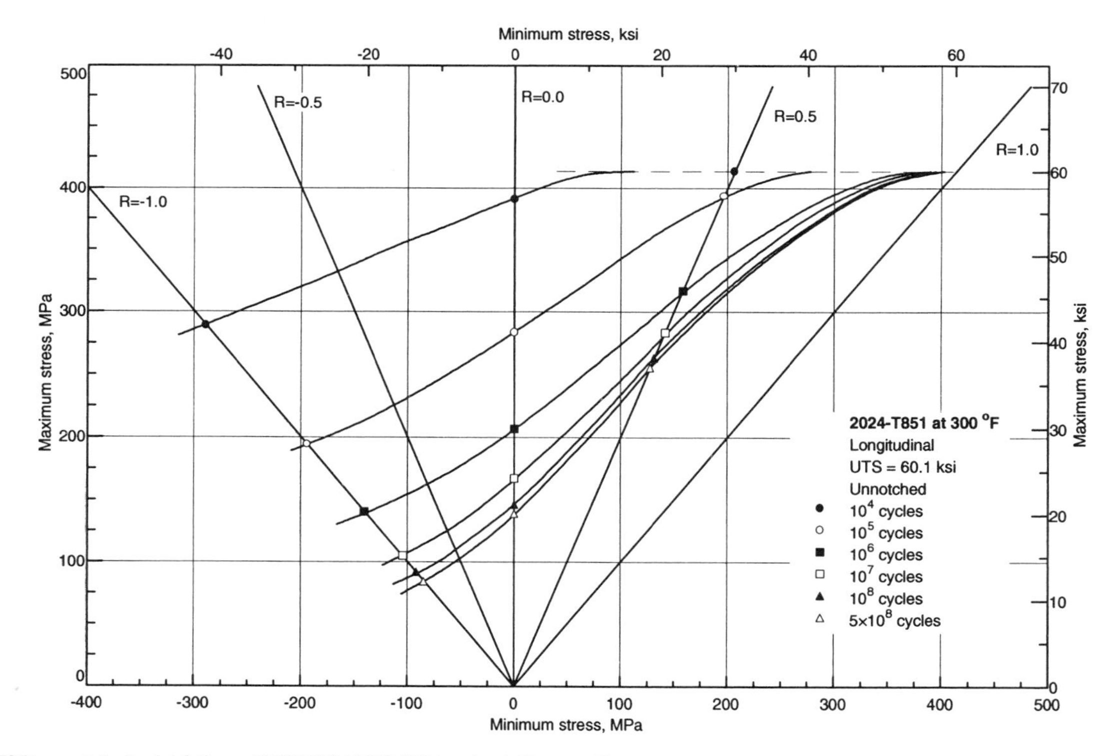

Fig. 22 2024-T851 unnotched axial fatigue at 150 °C(300 °F) (7/8 in. plate). Source: Alcoa

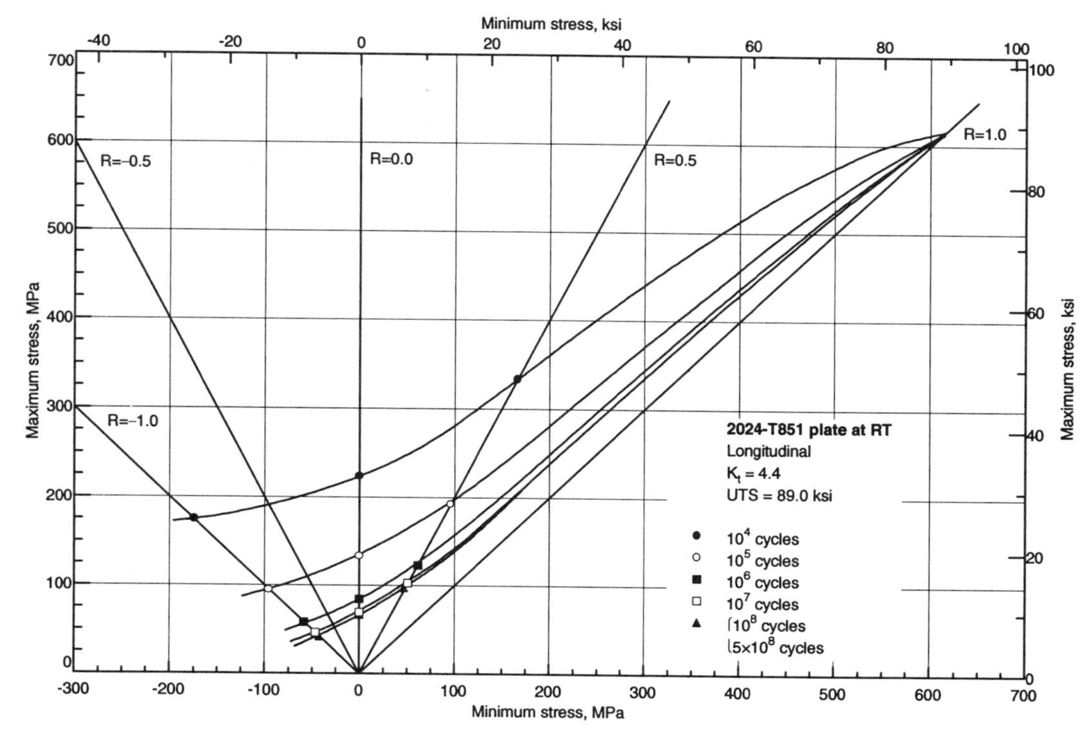

Fig. 23 2024-T851 notched axial fatigue ($K_t = 4.4$, $r = 0.005$ in.) at room temperature (7/8 in. plate)

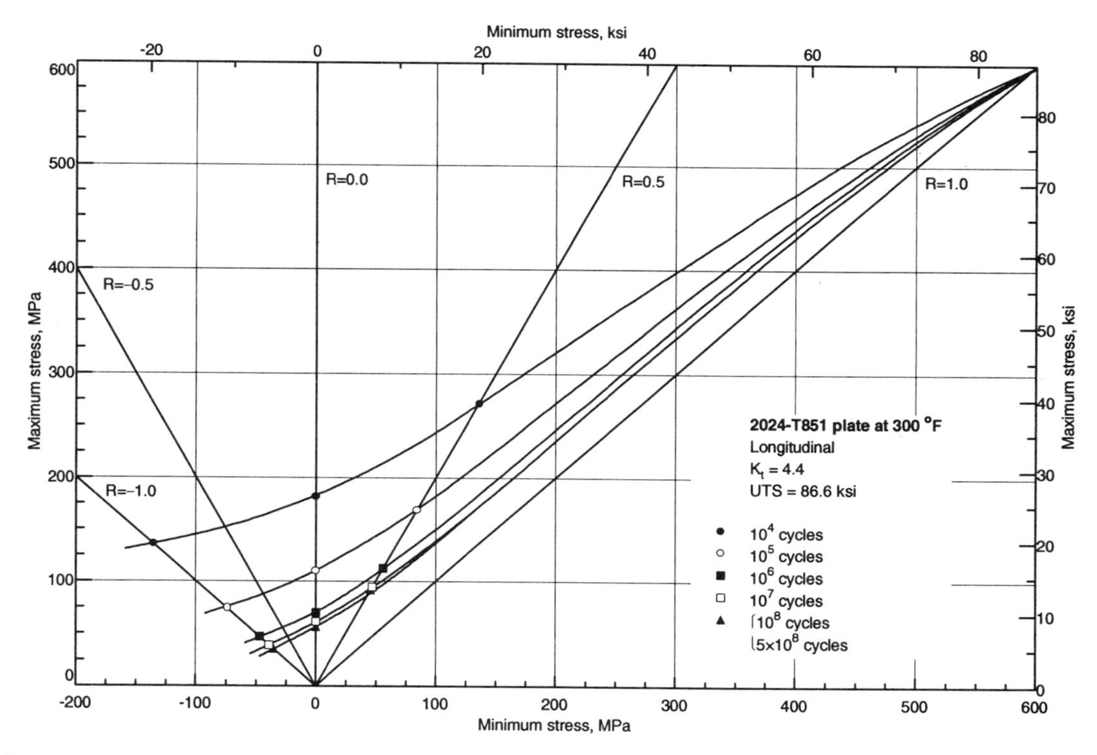

Fig. 24 2024-T851 notched axial fatigue ($K_t = 4.4$, $r = 0.005$ in.) at 150 °C (300 °F). Source: Alcoa, 1965

2024-T86, -T851, -T852, and -T861: Room-temperature fatigue strength in air

Temper	Form and thickness	Specimen	Test mode	Stress ratio	Fatigue strength (ksi) at cycles of:					
					10^4	10^5	10^6	10^7	10^8	5×10^8
T86	Sheet, 0.11 in.	Flat plate	Bending	−1	...	(30)	(23)	(20)	(17)	...
	Plate, 0.25 in.	Flat plate	Bending	−1	...	(28)	(19)	(16)	(15)	...
	Alclad sheet	Flat sheet with 3/16 in. rivet	Axial	−1	...	(11)	(5)	(2.5)	(2)	...
T851	Plate	Round, unnotched	Axial	0.5	...	(62)	(50)	(46)	(45)	(45)
			Axial	0.0	(67)	(44)	(36)	(31)	(29)	(28)
			Axial	−1.0	(53)	(34)	(23)	(19)	(17)	(16)
T851	Plate	Round, notched	Axial	0.5	(40)	(22)	(15)	(10)	(9)	...
			Axial	0.0	(25)	(14)	(9)	(7)	(6)	...
			Axial	−1.0	(19)	(10)	(6)	(4)	...	...
T852	Forging	Round	Axial	0.0	(63)	(43)	(38)	(36)	(35)	...
		R.R.Moore, unnotched	Rotating beam	−1.0	...	(37)	(30)	(24)	(21)	(20)
		R.R.Moore, notched	Rotating beam	−1.0	(28)	(14)	(8)	(6)	(6)	(5)
T861	Plate	Flat plate	Bending	−1	...	(30)	(23)	(20)	(17)	...
		Flat plate	Bending	−1	...	(28)	(19)	(16)	(15)	...
		Round, notched, $K_t = 4.4$	Axial	0.5	...	(16)	(16)	(12)	(12)	...
		Round, notched, $K_t = 4.4$	Axial	0.0	(29)	(11)	(11)	(9)	(9)	...
		Round, notched, $K_t = 4.4$	Axial	−1.0	(22)	(8)	(8)	(6)	(5.5)	...
		Round, notched, $K_t = 12$	Axial	0.5	...	(12)	(12)	(10)	(9.5)	...
		Round, notched, $K_t = 12$	Axial	0.0	...	(10)	(10)	(9)	(8)	...
		Round, notched, $K_t = 12$	Axial	−1.0	(25)	(8)	(8)	(7)	(6)	...

Sources: Unpublished Alcoa data; M.G. Vassilaros and E.J. Czyryca, Naval Research Lab Report CDNSWC-TR619409, A Compilation of Fatigue Information for Aluminum Alloys, 1994

2025

2025-T6 forging: Room-temperature fatigue strength in air

Specimen	Test mode	Stress ratio	Fatigue strength (ksi) at cycles of:					
			10^4	10^5	10^6	10^7	10^8	5×10^8
RR Moore	Rotating beam	−1	...	(37)	(27)	(21)	(17)	(16)
RR Moore notch	Rotating beam	−1	(30)	(20)	(12)	(9)	(8)	(8)
Round	Axial	−1	...	(31)	(25)	(21)	(18)	...
Round notch	Axial	−1	(19)	(11)	(6)	(5)	(5)	...

Source: Unpublished Alcoa data

2124

2124-T351 and -T851 plate: Room-temperature fatigue strength in air

Temper	Specimen	Test mode	Stress ratio	Fatigue strength (ksi) at cycles of:					
				10^4	10^5	10^6	10^7	10^8	5×10^8
T351	Round	Axial	0	(64)	(45)	(37)	(32)	...	...
T851	RR Moore	Rotating beam	−1	...	(37)	(26)	(19)	...	...
	RR Moore notch	Rotating beam	−1	(28)	(16)	(9)	(7)	(6)	(6)
	Round	Axial	0	(64)	(44)	(26)	(12)	...	...
	Roundnotch, $K_t = 3$	Axial	0	(31)	(15)	(11)	(11)	...	...

Source: Unpublished Alcoa data

2219-T6 and -T8

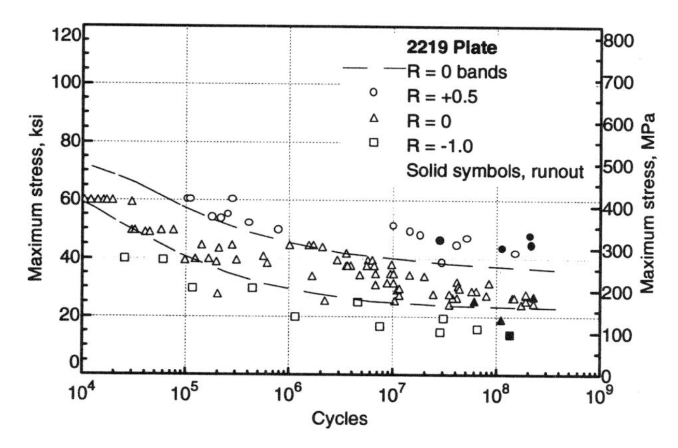

Fig. 25 2219-T8 unnotched axial fatigue at room temperature (plate, L, LT, ST). Unnotched axial specimens with 9-7/8 in. surface radius and 0.300 in. minimum diameter. Solid symbols indicate runout (no failure). Source: Alcoa, 1965

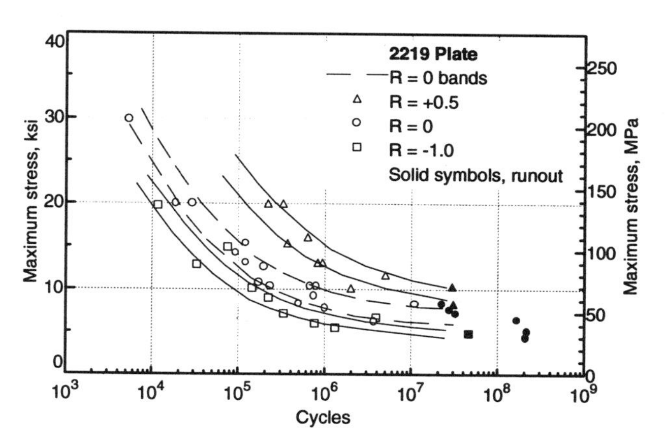

Fig. 26 2219-T8 notched axial fatigue ($K_t > 12$) at room temperature (plate). Notched axial specimens with 0.300 in. diameter at the notch and a 0.360 in. diam outside the 60° notch. Solid symbols indicate runout (no failure). Source: Alcoa, 1965

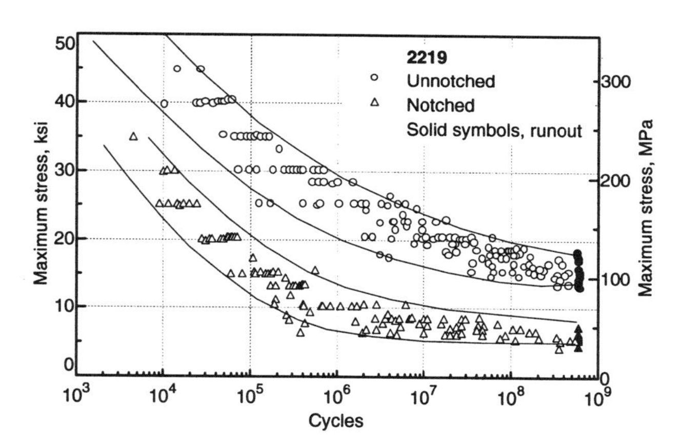

Fig. 27 2219-T8 rotating beam fatigue for unnotched and notched specimens at room temperature from plate, forgings, and extrusions (radius at notch root <0.001 in.). R.R. Moore specimens with 9-7/8 in. surface radius and 0.300 in. minimum diameter for unnotched specimens. Notched specimens had a 0.330 in. diameter at the notch and a 0.480 in. diameter outside the 60° notch. Solid symbols indicate runout (no failure). Source: Alcoa, 1965

2219: High-temperature fatigue strength in air

Temper	Form and thickness	Specimen	Test mode	Stress ratio	Temperature °C (°F)	Fatigue strength (ksi) at cycles of: 10^4	10^5	10^6	10^7	10^8	5×10^8
T6	Forging	R.R. Moore	Rotating beam	−1	(RT)	(40)	(30)	(25)	(21)	(19)	...
		Cantilever, Notched, K_t >12	Bending	−1	150 °C (300 °F)	...	(27)	(21)	(18)	(15)	(13)
		Cantilever, Unnotched	Bending	−1	200 °C (400 °F)	...	(25)	(20)	(15)	(12)	(11)
		Cantilever, Notched, K_t >12	Bending	−1	260 °C (500 °F)	(27)	(22)	(17)	(12)	(9)	(8)
		Cantilever, Unnotched	Bending	−1	315 °C (600 °F)	(24)	(18)	(13)	(9)	(8)	...
		Cantilever, Notched, K_t >12	Bending	−1	315 °C (600 °F)	(22)	(14)	(8)	(5)	(4)	(4)
T31	Plate	RR Moore	Rotating beam	−1	(RT)	...	(36)	(27)	(20)	(15)	(14)
	Rod	RR Moore	Rotating beam	−1	(RT)	(28)	(18)	(12)	(9)	(7)	(6)

Source: *Metals Progress*, Sept 1961

2219-T62

2219-T62: Room-temperature fatigue strength in air

Form and thickness	Specimen	Test mode	Stress ratio	Fatigue strength (ksi) at cycles of: 10^4	10^5	10^6	10^7	10^8	5×10^8
Plate, extrusion	RR Moore	Rotating beam	−1	...	(34)	(26)	(21)	(17)	(15)
Plate, extrusion	RR Moore notch	Rotating beam	−1	(25)	(15)	(9)	(6)	(6)	(5)
Plate	Round	Axial	0	(56)	(40)	(38)	(37)	(35)	...
0.06 in. Sheet	Trans. single-V butt weld, 2319 filler, bead-on	Axial	0	(32)	(18.5)	(14.5)	(14.5)	...	...
As above	As above, bead-off	Axial	0	(37)	(28)	(22.5)	(20)	...	...
0.06 in. Sheet	Trans. single-V butt weld, 2319 filler, bead-on	Axial	0	(43)	(26)	(22)	(20)	...	...
0.06 in. Sheet	As above	Axial	0	(32)	(18.5)	(14.5)	(14.5)	...	...
As above	As above, bead-off	Axial	0	...	(28)	(25)	(25)	...	...

Sources: Unpublished Alcoa data and M.G. Vassilaros and E.J. Czyryca, Naval Research Lab Report CDNSWC-TR619409, A Compilation of Fatigue Information for Aluminum Alloys, 1994

2219-T87

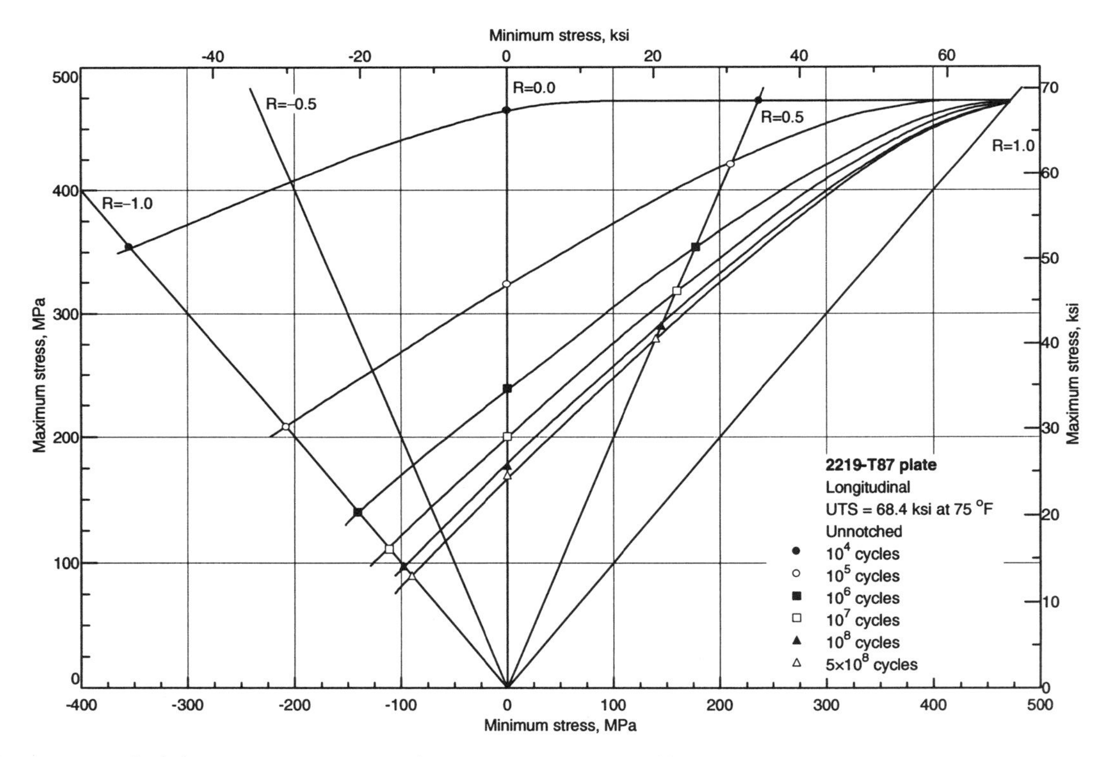

Fig. 28 2219-T87 unnotched axial fatigue at room temperature (1 in. plate). Source: Alcoa, 1966

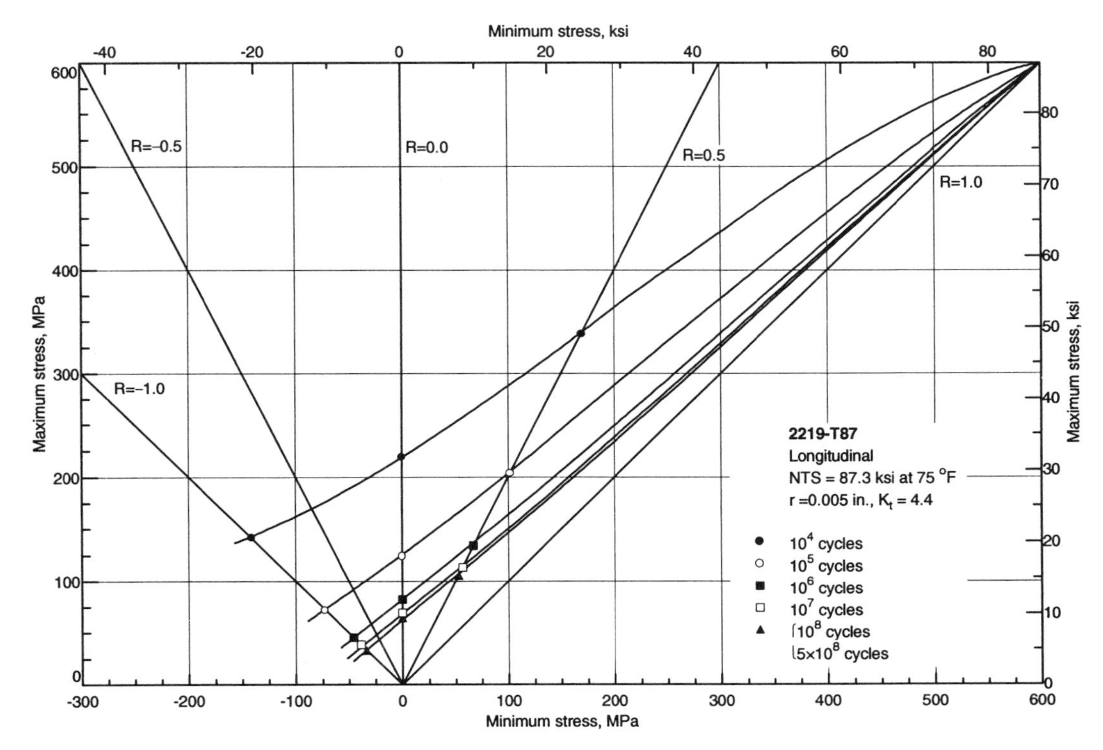

Fig. 29 2219-T87 notched axial fatigue (K_t = 4.4, r = 140.005 in.) at room temperature. Source: Alcoa, 1966

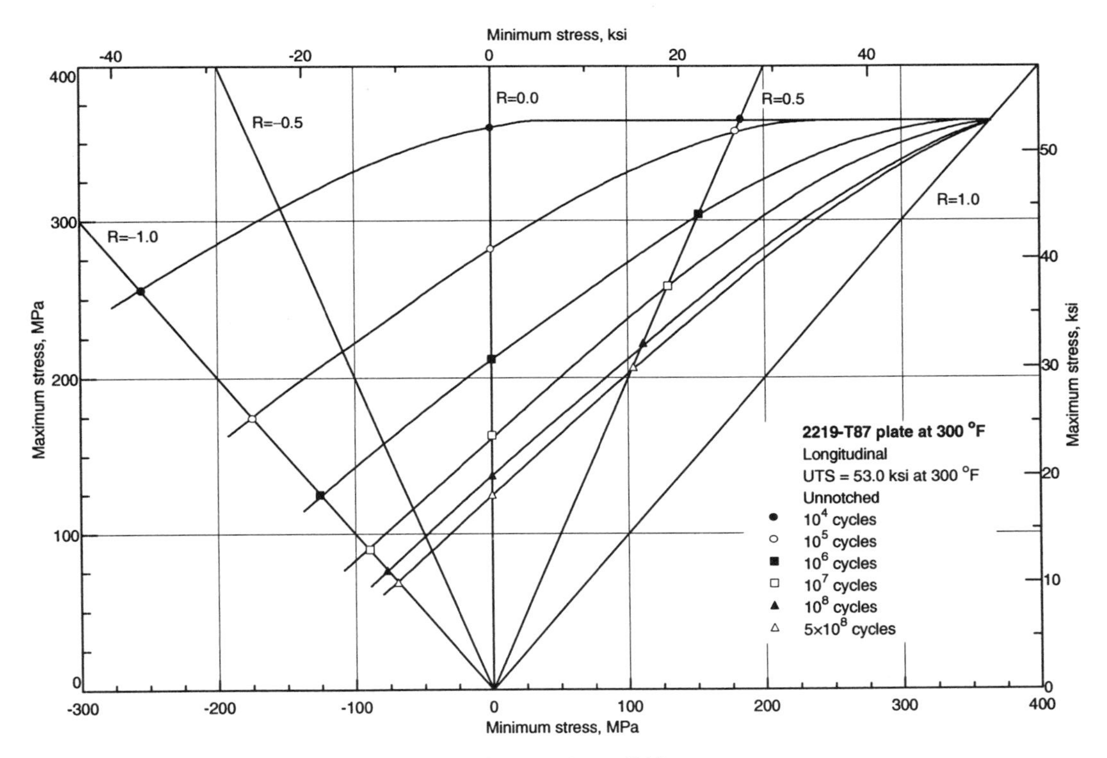

Fig. 30 2219-T87 unnotched axial fatigue at 150 °C (300 °F) (1 in. plate). Source: Alcoa, ,1966

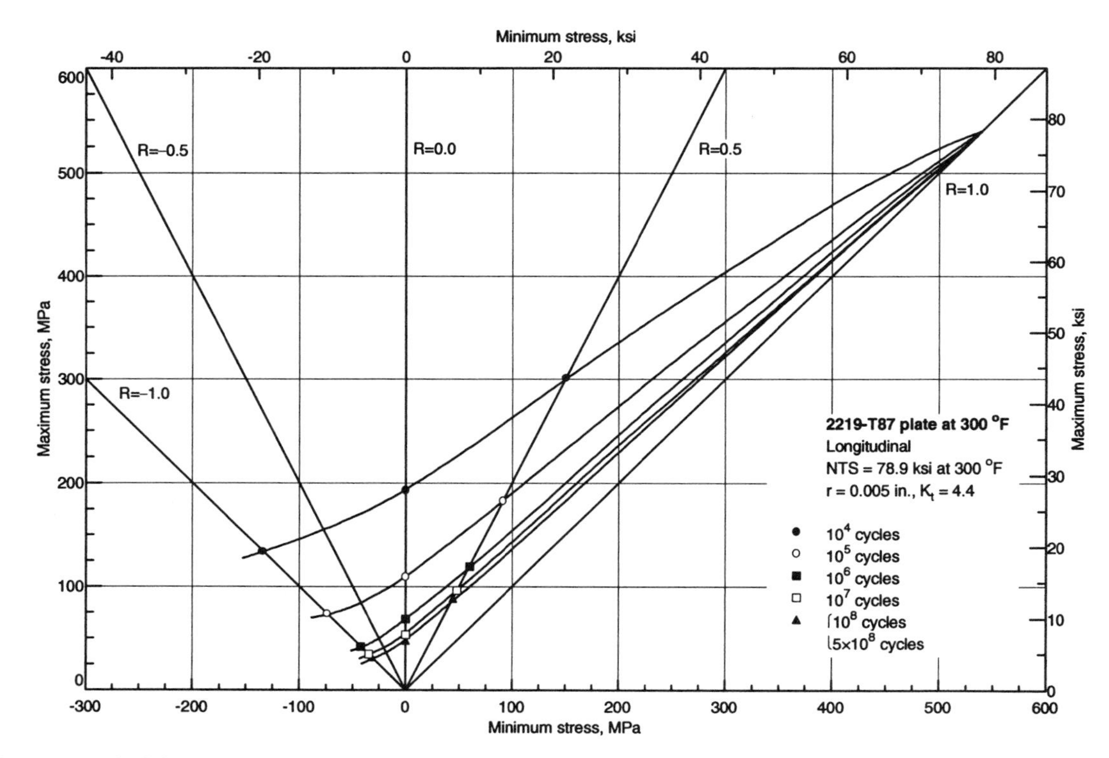

Fig. 31 2219-T87 notched axial fatigue ($K_t = 4.4$, $r = 0.005$ in.) at 150 °C (300 °F). Source: Alcoa, 1966

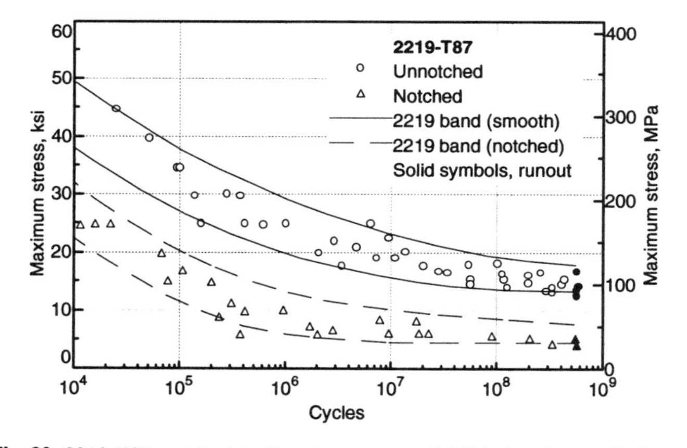

Fig. 32 2219-T87 notched (radius at notch root <0.001 in.) and unnotched rotating beam fatigue (plate). R.R. Moore specimens with 9-7/8 in. surface radius and 0.300 in. minimum diameter for unnotched specimens. Notched specimens had a 0.330 in. diameter at the notch and a 0.480 in. diameter outside the 60° notch. Solid symbols indicate runout (no failure). Band lines are for 2219 products in various tempers except annealed.

2219-T87 plate: Room-temperature fatigue strength in air

Specimen	Test mode	Stress ratio	Fatigue strength (ksi) at cycles of:					
			10^4	10^5	10^6	10^7	10^8	5×10^8
RR Moore	Rotating beam	−1	...	(37)	(27)	(21)	(17)	(16)
RR Moore notch	Rotating beam	−1	(30)	(20)	(12)	(9)	(8)	(8)
Round	Axial	−1	...	(31)	(25)	(21)	(18)	...
Round notch	Axial	−1	(19)	(11)	(6)	(5)	(5)	...

Source: Unpublished Alcoa data

2219-T851

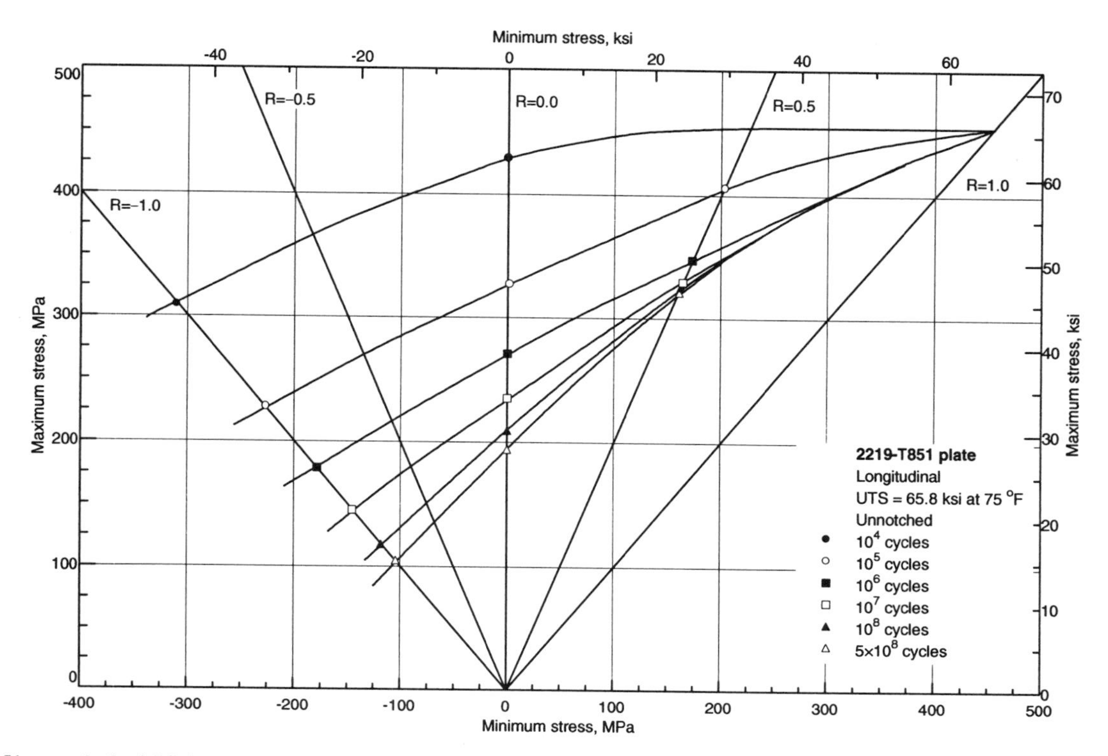

Fig. 33 2219-T851 unnotched axial fatigue at room temperature (1.25 in. plate). Source: Alcoa, 1966

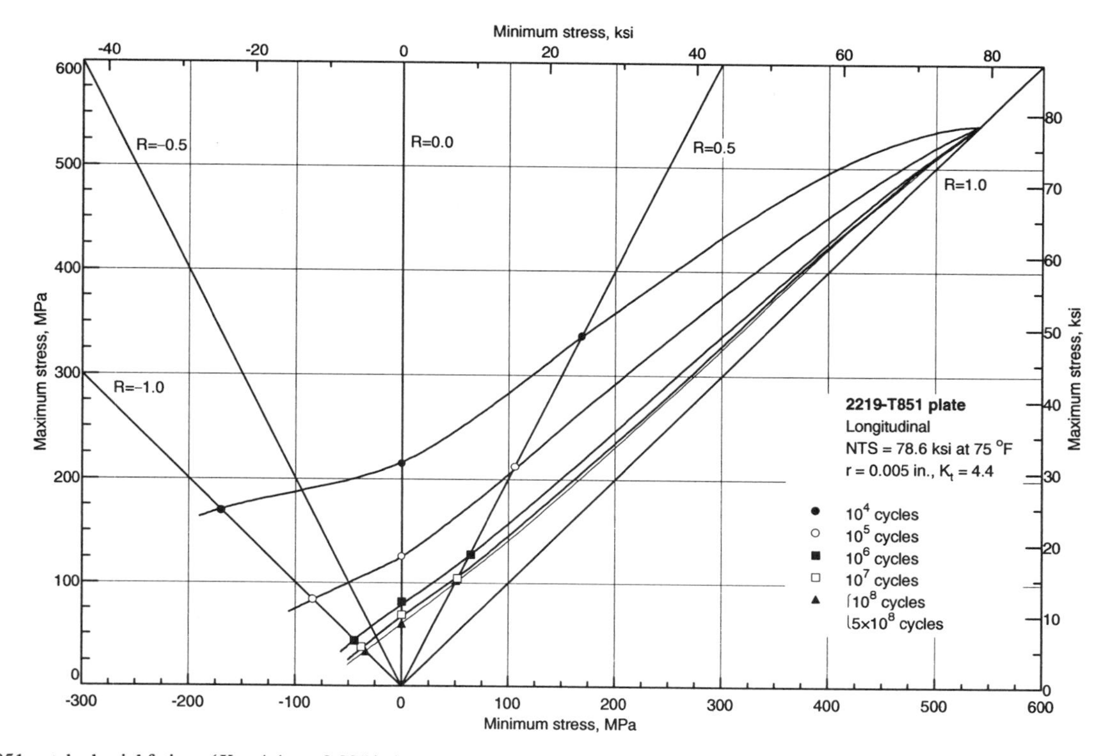

Fig. 34 2219-T851 notched axial fatigue ($K_t = 4.4$, $r = 0.005$ in.) at room temperature. Source: Alcoa, 1966

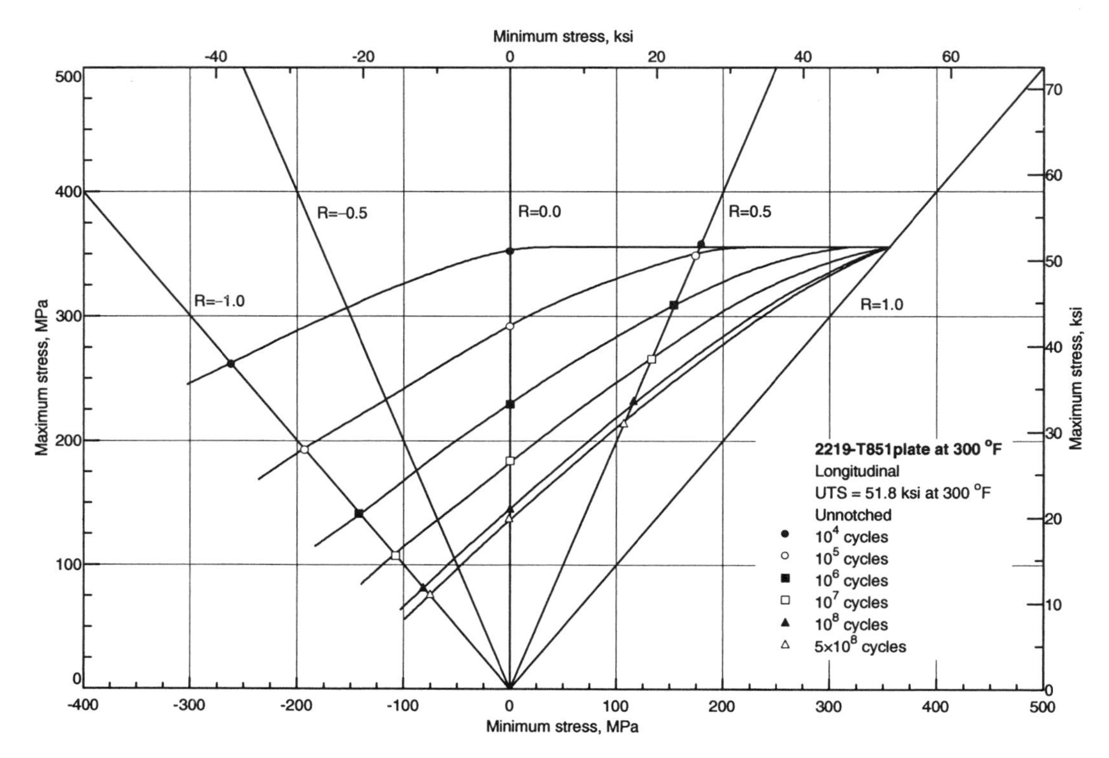

Fig. 35(a) 2219-T851 unnotched axial fatigue at 150 °C (300 °F) (1.25 in. plate). Source: Alcoa, 1966

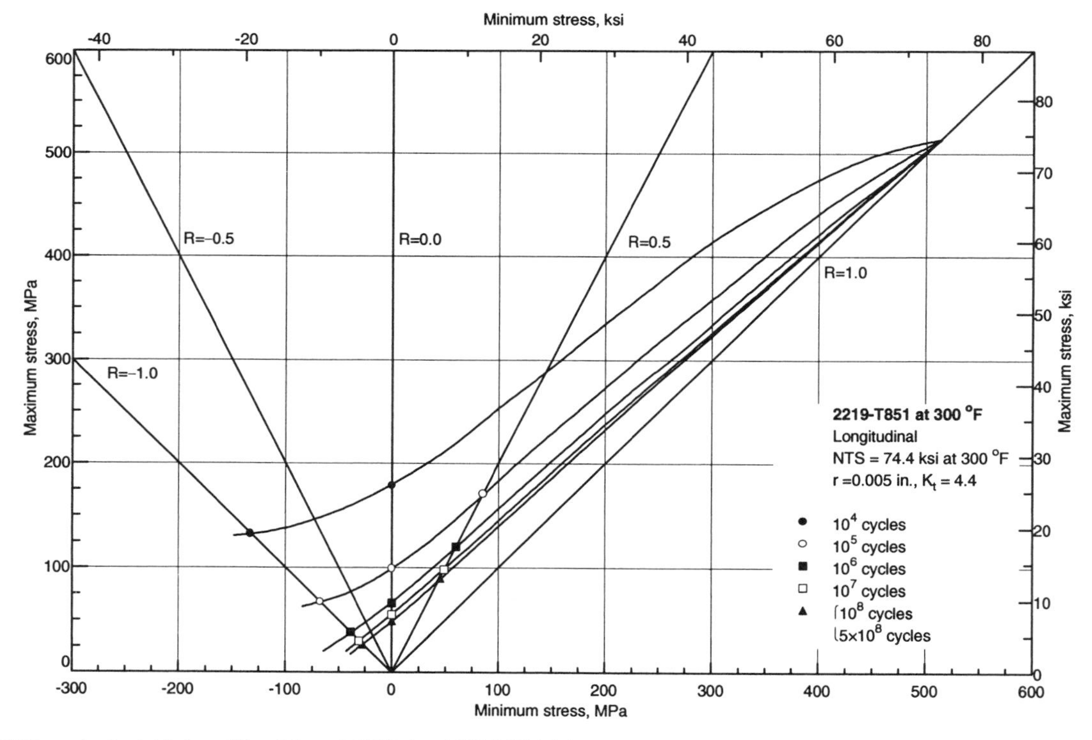

Fig. 35(b) 2219-T851 notched axial fatigue ($K_t = 4.4$, $r = 0.005$ in.) at 150 °C (300 °F). Source: Alcoa, 1966

2618-T6 and -T651

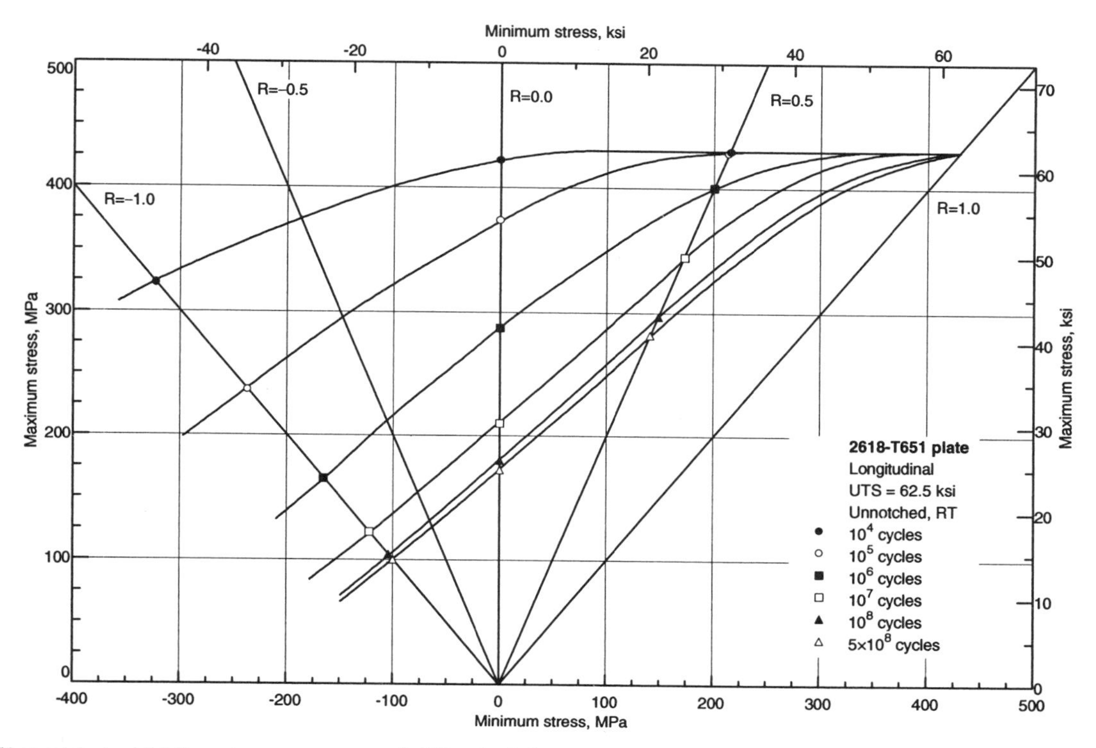

Fig. 36 2618-T651 unnotched axial fatigue at room temperature (1.35 in. plate). Source: Alcoa, 1968

2618-T6 and -T651: Room-temperature fatigue strength in air

Form and thickness	Specimen	Test mode	Stress ratio	Fatigue strength (ksi) at cycles of:					
				10^4	10^5	10^6	10^7	10^8	5×10^8
Extrusion, forging	R.R. Moore, Unnotched	Rotating beam	−1	(50)	(36)	(27)	(22)	(26)	(18)
	R.R. Moore, Notched	Rotating beam	−1	(30)	(19)	(12)	(10)	(9)	(8)
Plate	Round	Axial	0.5	...	(60)	(58)	(50)	(43)	(41)
			0.0	(60)	(54)	(42)	(30)	(26)	(25)
			−1.0	(47)	(34)	(24)	(18)	(9)	...

Source: Unpublished Alcoa data

3xxx and 4xxx Alloys

3003

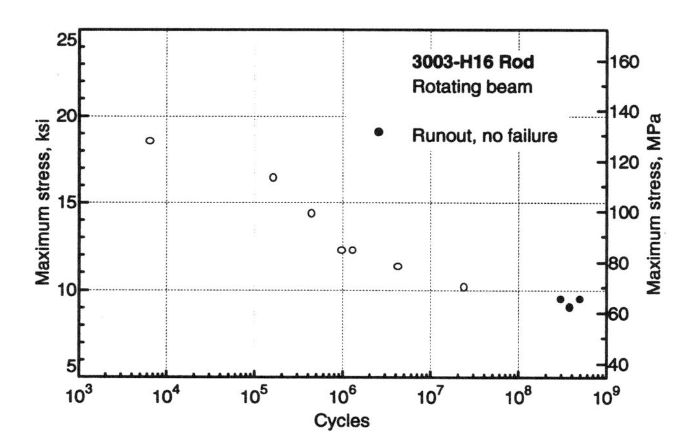

Fig. 37(a) 3003-H16 unnotched rotating beam fatigue at room temperature (1 and 1/8 in. rod). Unnotched axial specimens with 9-7/8 in. surface radius and 0.300 in. minimum diameter.

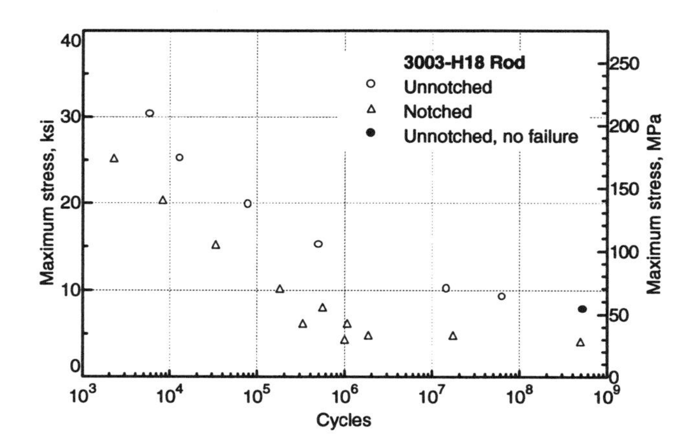

Fig. 37(b) 3003-H18 notched and unnotched rotating beam fatigue at room temperature (0.75 in. rolled and drawn rod). Radius at notch root was <0.001 in. for notched specimens. R.R. Moore specimens with 9-7/8 in. surface radius and 0.300 in. minimum diameter for unnotched specimens. Notched specimens had a 0.330 in. diameter at the notch and a 0.480 in. diameter outside the 60° notch. Source: Alcoa

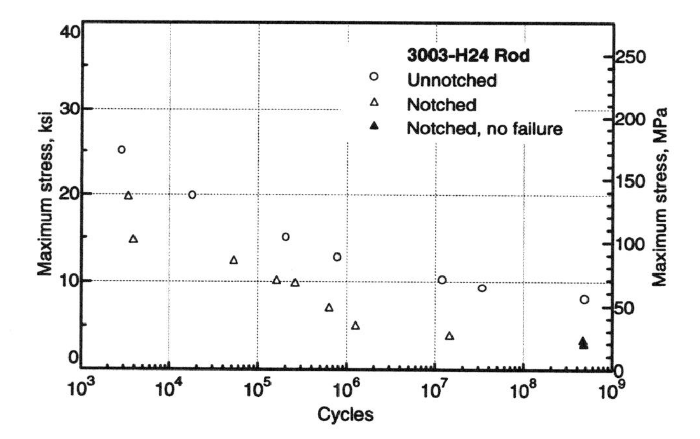

Fig. 37(c) 3003-H24 notched and unnotched rotating beam fatigue at room temperature (0.75 in. rolled and drawn rod). Radius at notch root was <0.001 in. for notched specimens. R.R. Moore specimens with 9-7/8 in. surface radius and 0.300 in. minimum diameter for unnotched specimens. Notched specimens had a 0.330 in. diameter at the notch and a 0.480 in. diameter outside the 60° notch. Source: Alcoa

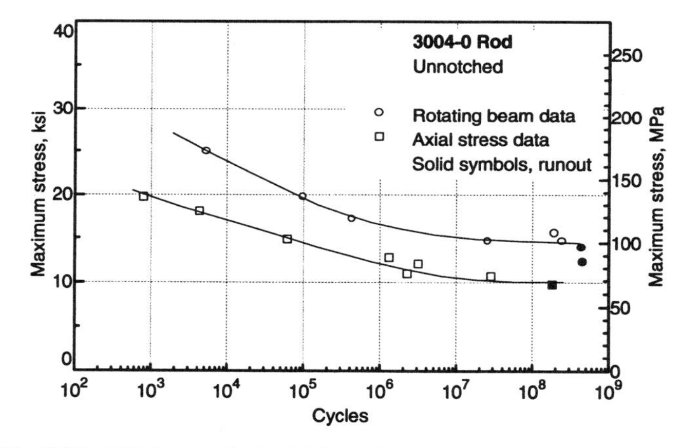

Fig. 37(d) 3003-0 unnotched axial (R = –1) and rotating beam (R = –1) fatigue (0.75 in. diam rolled and drawn rod). Unnotched rotating beam specimens with 9-7/8 in. surface radius and 0.300 in. minimum diameter. Unnotched axial specimens with 9-7/8 in. surface radius and 0.200 in. minimum diameter. Source: Alcoa

3003: Room-temperature fatigue strength in air

Temper	Form and thickness	Specimen	Test mode	Stress ratio	Fatigue strength (ksi) at cycles of:					
					10^4	10^5	10^6	10^7	10^8	5×10^8
F	¼ in. Plate	Flat plate, transverse butt-weld, bead-off	Axial	0	(17)	(12)	(9)	(8)	...	...
	Plate	Butt-weld, bead-on	Axial	0	(16)	(11)	(7)	(6)	(6)	...
H16	Rod	RR Moore	Rotating beam	–1	...	(17)	(12)	(11)	(10)	(9)
H18	Rod	RR Moore	Rotating beam	–1	(27)	(19)	(13)	(10)	(9)	(9)
		RR Moore notched	Rotating beam	–1	(19)	(12)	(6)	(5)	(4)	(4)
H24	Rod	RR Moore	Rotating beam	–1	(22)	(16)	(12)	(10)	(9)	(8)
		RR Moore notched	Rotating beam	–1	(17)	(11)	(6)	(4)	(4)	(4)

Sources: Unpublished Alcoa data and M.G. Vassilaros and E.J. Czyryca, Naval Research Lab Report CDNSWC-TR619409, A Compilation of Fatigue Information for Aluminum Alloys, 1994

3004

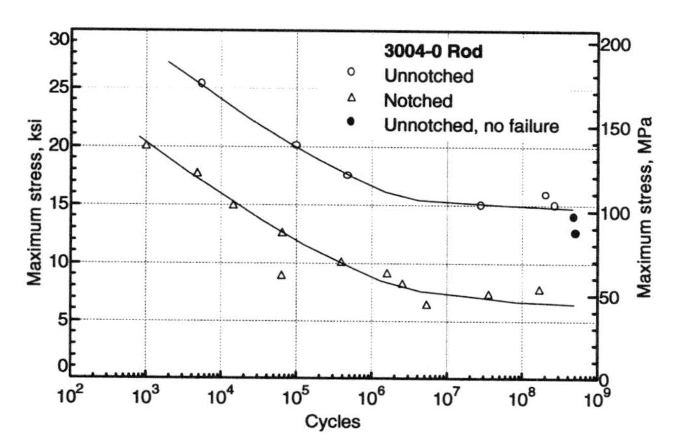

Fig. 37(e) 3004-0 notched and unnotched rotating beam fatigue at room temperature (0.75 in. diam rod). Radius at notch root was <0.001 in. for notched specimens. R.R. Moore specimens with 9-7/8 in. surface radius and 0.300 in. minimum diameter for unnotched specimens. Notched specimens had a 0.330 in. diameter at the notch and a 0.480 in. diameter outside the 60° notch. Source: Alcoa

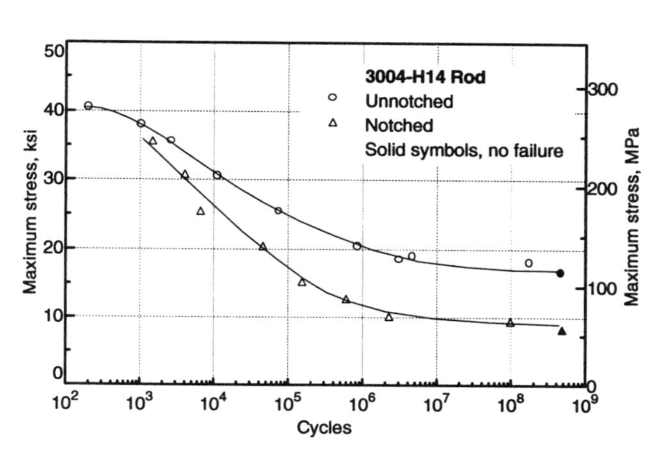

Fig. 37(f) 3004-H14 notched and unnotched rotating beam fatigue at room temperature (0.75 in. rolled and drawn rod). Radius at notch root was <0.001 in. for notched specimens. R.R. Moore specimens with 9-7/8 in. surface radius and 0.300 in. minimum diameter for unnotched specimens. Notched specimens had a 0.330 in. diameter at the notch and a 0.480 in. diameter outside the 60° notch. Source: Alcoa

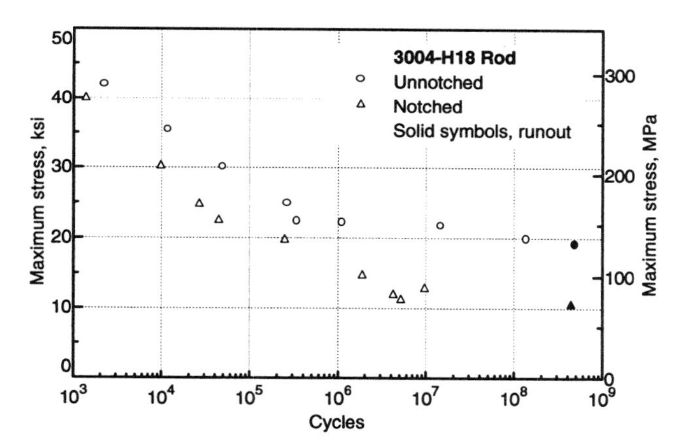

Fig. 37(g) 3004-H18 notched and unnotched rotating beam fatigue at room temperature (0.75 in. diam rod). Radius at notch root was <0.001 in. for notched specimens. R.R. Moore specimens with 9-7/8 in. surface radius and 0.300 in. minimum diameter for unnotched specimens. Notched specimens had a 0.330 in. diameter at the notch and a 0.480 in. diameter outside the 60° notch. Solid symbols indicate runout (no failure). Source: Alcoa

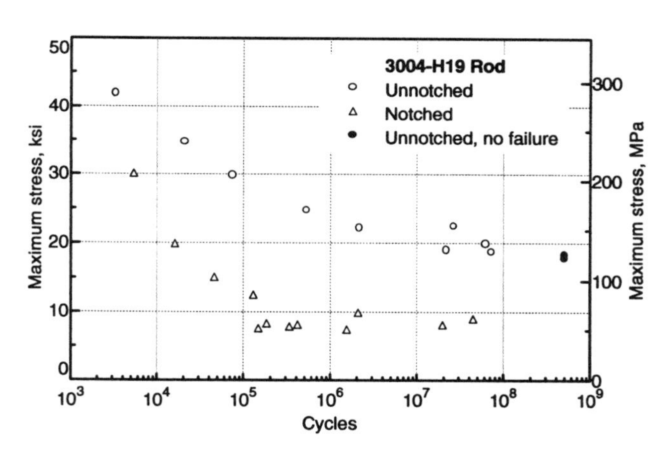

Fig. 37(h) 3004-H19 notched and unnotched rotating beam fatigue at room temperature (0.75 in. diam rod). Radius at notch root was <0.001 in. for notched specimens. R.R. Moore specimens with 9-7/8 in. surface radius and 0.300 in. minimum diameter for unnotched specimens. Notched specimens had a 0.330 in. diameter at the notch and a 0.480 in. diameter outside the 60° notch. Source: Al-

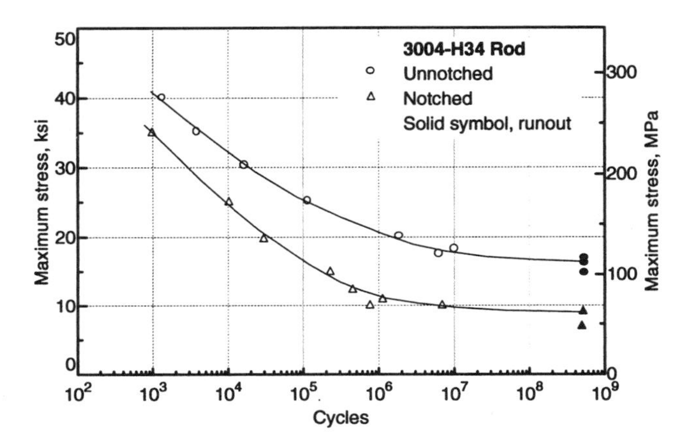

Fig. 37(i) 3004-H34 notched and unnotched rotating beam fatigue at room temperature (0.75 in. diam rod). Radius at notch root was <0.001 in. for notched specimens. R.R. Moore specimens with 9-7/8 in. surface radius and 0.300 in. minimum diameter for unnotched specimens. Notched specimens had a 0.330 in. diameter at the notch and a 0.480 in. diameter outside the 60° notch. Solid symbols indicate runout (no failure). Source: Alcoa

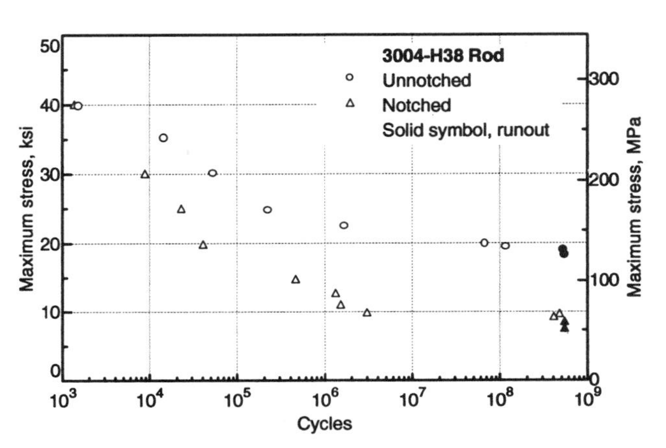

Fig. 37(j) 3004-H38 notched and unnotched rotating beam fatigue at room temperature (0.75 in. diam rolled and drawn rod). Radius at notch root was <0.001 in. for notched specimens. R.R. Moore specimens with 9-7/8 in. surface radius and 0.300 in. minimum diameter for unnotched specimens. Notched specimens had a 0.330 in. diameter at the notch and a 0.480 in. diameter outside the 60° notch. Solid symbols indicate runout (no failure). Source: Alcoa

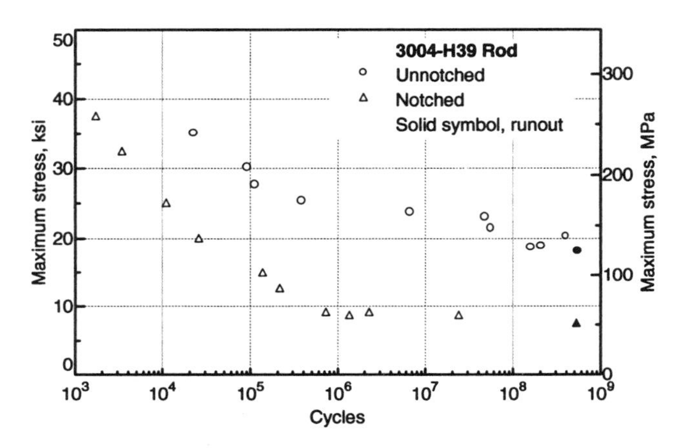

Fig. 37(k) 3004-H39 notched and unnotched rotating beam fatigue at room temperature (0.75 in. diam rod). Radius at notch root was <0.001 in. for notched specimens. R.R. Moore specimens with 9-7/8 in. surface radius and 0.300 in. minimum diameter for unnotched specimens. Notched specimens had a 0.330 in. diameter at the notch and a 0.480 in. diameter outside the 60° notch. Solid symbols indicate runout (no failure). Source: Alcoa

3004 rod: Room-temperature fatigue strength in air

Temper	Form and thickness	Specimen	Test mode	Stress ratio	Fatigue strength (ksi) at cycles of: 10^4	10^5	10^6	10^7	10^8	5×10^8
0	Rod	R.R. Moore, unnotched	Rotating beam	–1	(24)	(20)	(16)	(15)	(15)	(14)
		R.R. Moore, notched	Rotating beam	–1	(26)	(12)	(9)	(7)	(7)	(6)
		Round	Axial	–1	(17)	(15)	(12)	(11)	(10)	(16)
H14	Rod	R.R. Moore, unnotched	Rotating beam	–1	(31)	(24)	(20)	(18)	(17)	(17)
		R.R. Moore, notched	Rotating beam	–1	(26)	(17)	(11)	(10)	(9)	(9)
H18	Rod	R.R. Moore, unnotched	Rotating beam	–1	(35)	(27)	(22)	(21)	(20)	(20)
		R.R. Moore, notched	Rotating beam	–1	(30)	(21)	(16)	(13)	(12)	(11)
H19	Rod	R.R. Moore, unnotched	Rotating beam	–1	(38)	(29)	(23)	(20)	(19)	(19)
		R.R. Moore, notched	Rotating beam	–1	(22)	(13)	(10)	(9)	(8)	(8)
H34	Rod	R.R. Moore, unnotched	Rotating beam	–1	(32)	(25)	(20)	(18)	(17)	(16)
		R.R. Moore, notched	Rotating beam	–1	(25)	(16)	(11)	(10)	(9)	(9)
H38	Rod	R.R. Moore, unnotched	Rotating beam	–1	(34)	(28)	(23)	(21)	(19)	(19)
		R.R. Moore, notched	Rotating beam	–1	(29)	(19)	(12)	(10)	(9)	(9)
H39	Rod	R.R. Moore, unnotched	Rotating beam	–1	(39)	(29)	(23)	(21)	(19)	(19)
		R.R. Moore, notched	Rotating beam	–1	(26)	(15)	(10)	(8)	(8)	(8)

Source: Unpublished Alcoa data

4xxx and 5xxx Alloys

4032-T6

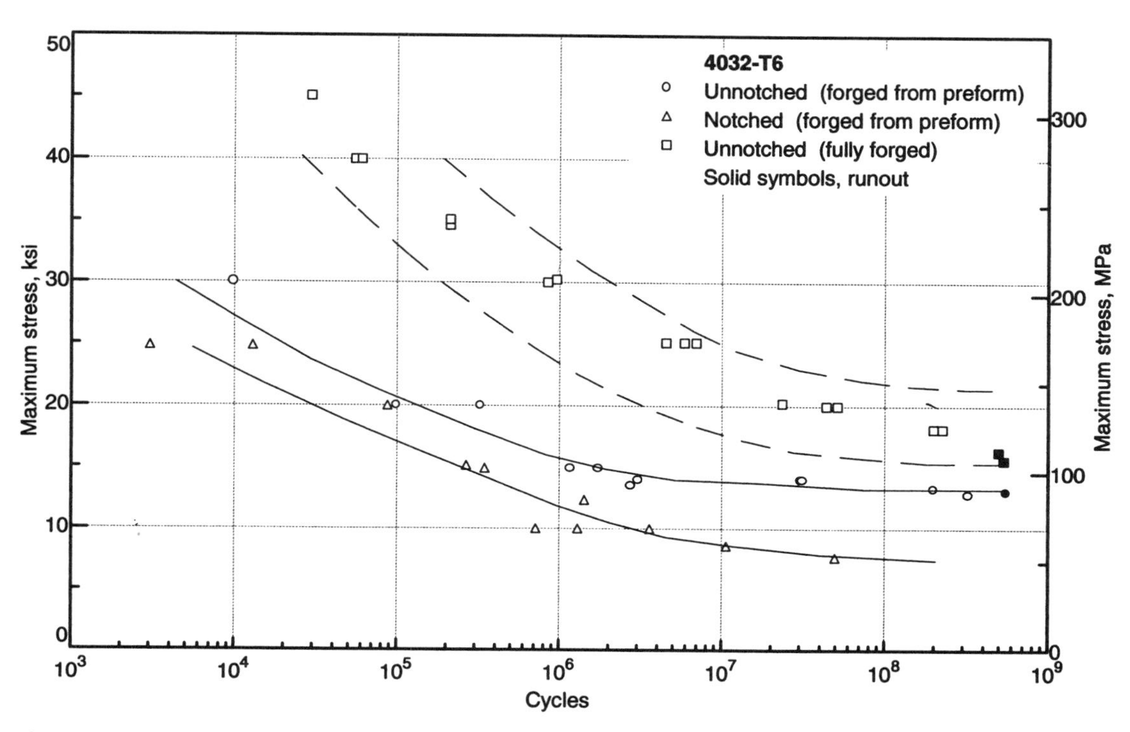

Fig. 37(l) 4032-T6 notched ($K_t > 12$) and unnotched rotating beam fatigue for specimens from die forged pistons (fully forged and forged from cast preforms as indicated). Solid symbols indicate runout (no failure). Source: Alcoa

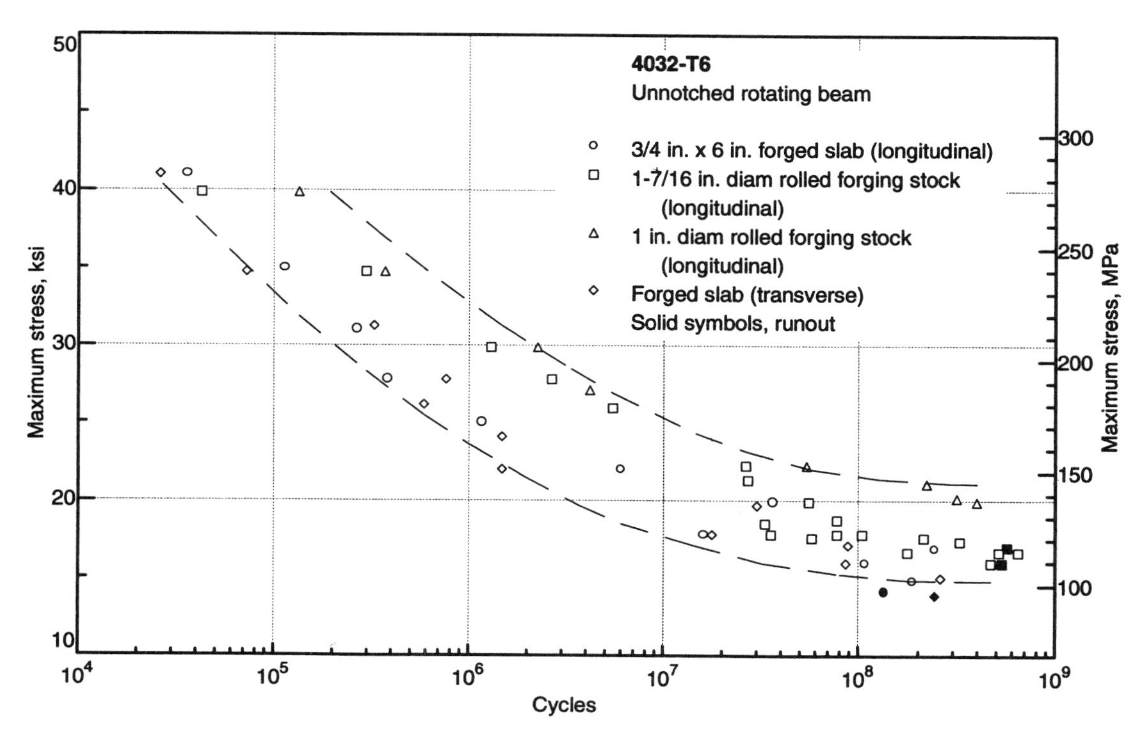

Fig. 37(m) 4032-T6 unnotched rotating beam fatigue. R.R. Moore specimens with 9-7/8 in. surface radius and 0.300 in. minimum diameter for unnotched specimens. Source: Alcoa

4043

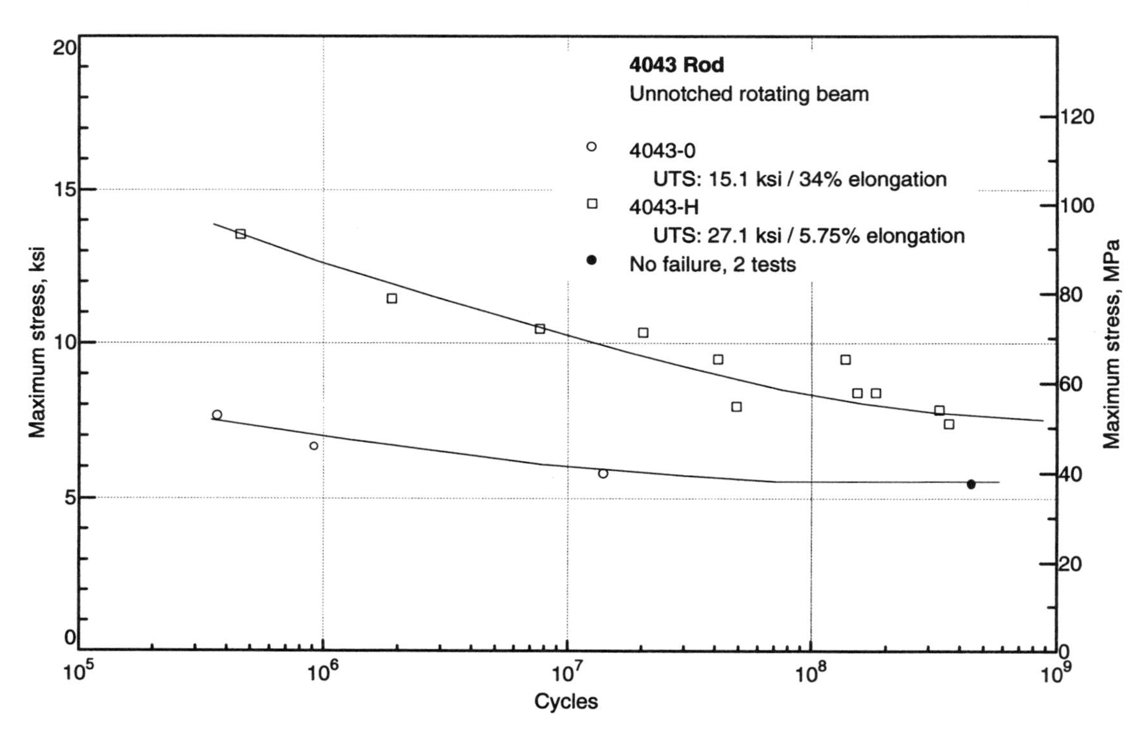

Fig. 37(n) 4043 unnotched rotating beam fatigue (0.75 in. diam rod). Source Alcoa

5005

5005-0 rod: Room-temperature fatigue strength in air

Specimen	Test mode	Stress ratio	Fatigue strength (ksi) at cycles of: 10^4	10^5	10^6	10^7	10^8	5×10^8
RR Moore	Rotating beam	−1	(24)	(20)	(16)	(15)	(15)	(14)
RR Moore notch	Rotating beam	−1	(16)	(12)	(9)	(7)	(7)	(6)
Round	Axial	−1	(17)	(15)	(12)	(11)	(10)	(16)
RR Moore	Rotating beam	−1	(31)	(24)	(20)	(18)	(17)	(17)
RR Moore notch	Rotating beam	−1	(26)	(17)	(11)	(10)	(9)	(9)
RR Moore	Rotating beam	−1	(35)	(27)	(22)	(21)	(20)	(20)
RR Moore notch	Rotating beam	−1	(30)	(21)	(16)	(13)	(12)	(11)
RR Moore	Rotating beam	−1	(38)	(29)	(23)	(20)	(19)	(19)
RR Moore notch	Rotating beam	−1	(22)	(13)	(10)	(9)	(8)	...
RR Moore	Rotating beam	−1	(32)	(25)	(20)	(18)	(17)	(16)
RR Moore notch	Rotating beam	−1	(25)	(16)	(11)	(10)	(9)	(9)
RR Moore	Rotating beam	−1	(34)	(28)	(23)	(21)	(19)	(19)
RR Moore notch	Rotating beam	−1	(29)	(19)	(12)	(10)	(9)	(9)
RR Moore	Rotating beam	−1	(39)	(29)	(23)	(21)	(19)	(19)
RR Moore notch	Rotating beam	−1	(26)	(15)	(10)	(8)	(8)	(8)

Sources: Unpublished Alcoa data

5005-H14, -H18, -H19: Fatigue strength in air

Temper	Form and thickness	Specimen	Test mode	Stress ratio	Temperature °C (°F)	Fatigue strength (ksi) at cycles of: 10^4	10^5	10^6	10^7	10^8	5×10^8
H14	Rod	RR Moore	Rotating beam	−1	(RT)	(25)	(20)	(15)	(13)	(12)	(12)
		RR Moorenotched	Rotating beam	−1	(RT)	(17)	(12)	(8)	(5)	(4)	(4)
H18	Plate	Cantilever beam	Bending	−1	24 °C (75 °F)	(29)	(22)	(17.5)	(15.5)	(15)	(15)
	Plate	Cantilever beam	Bending	−1	100 °C (212 °F)	(28)	(21.5)	(17)	(15.5)	(15)	(15)
	Plate	Cantilever beam	Bending	−1	150 °C (300 °F)	(27)	(21)	(16.5)	(14.5)	(13)	(12)
	Rod	RR Moore	Rotating beam	−1	(RT)	...	(22)	(17)	(15)	(14)	(14)
H19	Rod	RR Moore	Rotating beam	−1	(RT)	(30)	(23)	(18)	(16)	(15)	(15)
		RR Moorenotched	Rotating beam	−1	(RT)	...	(14)	(6)	(3)	(3)	(3)

Sources: Unpublished Alcoa data and M.G. Vassilaros and E.J. Czyryca, Naval Research Lab Report CDNSWC-TR619409, A Compilation of Fatigue Information for Aluminum Alloys, 1994

5050

5050: Fatigue strength in air

Temper	Form and thickness	Specimen	Test mode	Stress ratio	Temperature °C (°F)	Fatigue strength (ksi) at cycles of: 10^4	10^5	10^6	10^7	10^8	5×10^8
O	Plate	Cantilever beam	Bending	−1	24 °C (75 °F)	(24)	(17.5)	(14.5)	(13.5)	(13)	(12.5)
	Plate	Cantilever beam	Bending	−1	150 °C (300 °F)	(17)	(12.5)	(9.5)	(8.5)	(8.0)	(8.0)
	Plate	Cantilever beam	Bending	−1	200 °C (400 °F)	(15)	(10.5)	(8.0)	(6.5)	(6.0)	(6.0)
	Plate	Cantilever beam	Bending	−1	260 °C (500 °F)	(13)	(9.5)	(6.5)	(4.5)	(4.0)	(4.0)
	Rod	RR Moore	Rotating beam	−1	(RT)	(23)	(18)	(15)	(14)	(13)	(13)
	Rod	RR Moore, notched $K_t = 12$	Rotating beam	−1	(RT)	(15)	(10)	(6)	(55)	(5)	(5)
H34	Plate	Cantilever beam	Bending	−1	24 °C (75 °F)	(30)	(22.5)	(19)	(17.5)	(16.5)	(16)
	Plate	Cantilever beam	Bending	−1	150 °C (300 °F)	(25)	(21)	(16)	(12)	(9.0)	(8)
	Plate	Cantilever beam	Bending	−1	200 °C (400 °F)	(21)	(15.5)	(11.5)	(8.0)	(6.0)	(6.0)
	Plate	Cantilever beam	Bending	−1	260 °C (500 °F)	(16)	(11.5)	(7.0)	(5.0)	(4.0)	(4.0)
	Rod	RR Moore	Rotating beam	−1	(RT)	(30)	(23)	(19)	(18)	(16)	(16)
	Rod	RR Moore, notched $K_t = 12$	Rotating beam	−1	(RT)	(25)	(16)	(11)	(8)	(7)	(7)
H38	Plate	Cantilever beam	Bending	−1	24 °C (75 °F)	(38)	(25)	(20.5)	(19)	(18.5)	(16)
H38	Plate	Cantilever beam	Bending	−1	150 °C (300 °F)	(27)	(22)	(17.5)	(13.5)	(10.5)	(8)
	Plate	Cantilever beam	Bending	−1	200 °C (400 °F)	(21)	(15.5)	(11.5)	(8.0)	(6.0)	(6)
	Plate	Cantilever beam	Bending	−1	260 °C (500 °F)	(16)	(11.5)	(7.0)	(5.0)	(4.0)	(4)
	Rod	RR Moore	Rotating beam	−1	(RT)	(33)	(25)	(21)	(19)	(18)	(18)
	Rod	RR Moore, notched $K_t = 12$	Rotating beam	−1	(RT)	(26)	(15)	(10)	(8)	(7)	(7)

Sources: Unpublished Alcoa data and M.G. Vassilaros and E.J. Czyryca, Naval Research Lab Report CDNSWC-TR619409, A Compilation of Fatigue Information for Aluminum Alloys, 1994

5052

5052-O, -H14, -H16, -H18: Fatigue strength in air

Temper	Form and thickness	Specimen	Test mode	Stress ratio	Temperature °C (°F)	Fatigue strength (ksi) at cycles of: 10^4	10^5	10^6	10^7	10^8	5×10^8
O	Plate	Cantilever beam	Bending	−1	24 °C (75 °F)	(31)	(23.5)	(19.5)	(17.5)	(16.5)	(16)
	Plate	Cantilever beam	Bending	−1	150 °C (300 °F)	(21)	(17)	(14)	(11.5)	(10)	(10)
	Plate	Cantilever beam	Bending	−1	200 °C (400 °F)	(19.5)	(14.5)	(11)	(9.5)	(8.5)	(8)
	Plate	Cantilever beam	Bending	−1	260 °C (500 °F)	(17)	(12)	(9)	(7.5)	(7)	(7)
H14	Plate	Cantilever beam	Bending	−1	24 °C (75 °F)	(38)	(27)	(21)	(19)	(18)	(18)
	Plate	Cantilever beam	Bending	−1	150 °C (300 °F)	(29)	(24)	(19)	(16)	(14.5)	(14)
	Plate	Cantilever beam	Bending	−1	200 °C (400 °F)	(26)	(20)	(15)	(13)	(10.5)	(10)
	Plate	Cantilever beam	Bending	−1	260 °C (500 °F)	(23)	(15)	(10)	(7)	(6.5)	(6)
H16	Rod	RR Moore	Rotating beam	−1	(RT)	(37)	(28)	(21)	(19)	(19)	(18.5)
	Rod	RR Moore, notched $K_t = 12$	Rotating beam	−1	(RT)	(21)	(14)	(8)	(6)	(6)	(6)
H18	Plate	Cantilever beam	Bending	−1	24 °C (75 °F)	(41)	(29.5)	(24)	(22.5)	(21)	(20)
	Plate	Cantilever beam	Bending	−1	150 °C (300 °F)	(34)	(24)	(19)	(16)	(14.5)	(14)
	Plate	Cantilever beam	Bending	−1	200 °C (400 °F)	(29)	(20)	(15)	(13)	(10.5)	(10)
H18	Plate	Cantilever beam	Bending	−1	260 °C (500 °F)	(23)	(15)	(10)	(7)	(6.5)	(6)
	Rod	RR Moore	Rotating beam	−1	(RT)	(40)	(30)	(24)	(21)	(20)	(20)
	Rod	RR Moore, notched $K_t = 12$	Rotating beam	−1	(RT)	(30)	(19)	(11)	(9)	(8)	(8)

Sources: Unpublished Alcoa data and M.G. Vassilaros and E.J. Czyryca, Naval Research Lab Report CDNSWC-TR619409, A Compilation of Fatigue Information for Aluminum Alloys, 1994

5052-H32 and -H34: Fatigue strength in various environments

Temper	Form and thickness	Specimen	Test mode	Stress ratio	Temperature °C (°F)	Atmosphere	Fatigue strength (ksi) at cycles of: 10^4	10^5	10^6	10^7	10^8	5×10^8
H32	Plate		Bending	–1	(RT)	Air	(33)	(28)	(22)	(16)	0	( )
	Plate		Bending	–1	(RT)	Salt water	(27)	(21)	0	0	0	0
		Flat plate	Axial	0	(RT)	Air	(33)	(33)	(27)	(26)	0	0
H34	Plate	Cantilever	Bending	–1	24 °C (75 °F)	Air	(26)	(26)	(20)	(19)	(18)	(18)
		Cantilever	Bending	–1	150 °C (300 °F)	Air	(24)	(24)	(19)	(16)	(14.5)	(14)
		Cantilever	Bending	–1	200 °C (400 °F)	Air	(20)	(20)	(15)	(13)	(10.5)	(10)

Sources: Unpublished Alcoa data and M.G. Vassilaros and E.J. Czyryca, Naval Research Lab Report CDNSWC-TR619409, A Compilation of Fatigue Information for Aluminum Alloys, 1994

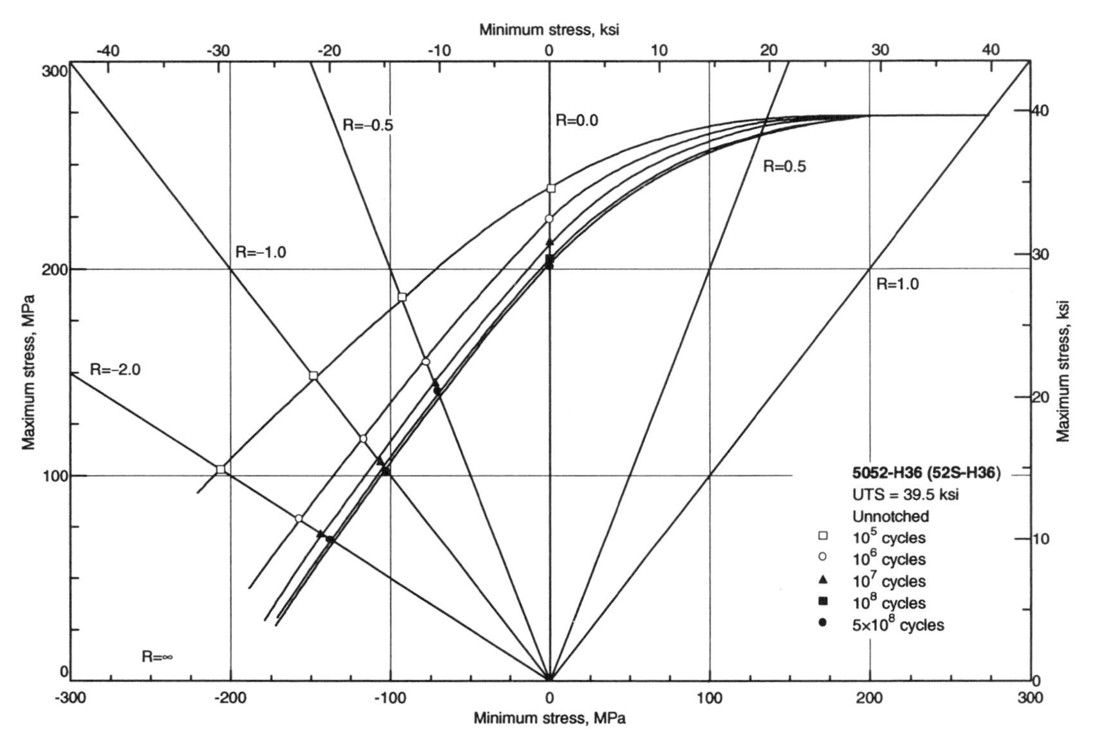

Fig. 38 5052-H36 (52S-H36) unnotched axial fatigue at room temperature. Source: Alcoa, 1952

5052-H36: Room-temperature fatigue strength in air

Form and thickness	Specimen	Test mode	Stress ratio	Fatigue strength (ksi) at cycles of: 10^4	10^5	10^6	10^7	10^8	5×10^8
Plate	Flat plate	Axial	+0.5	(39)	(38.5)	(38.5)	(38)	(38)	(37.5)
Plate	Flat plate	Axial	0	...	(34)	(32)	(31)	(30)	(30)
Plate	Flat plate	Axial	–0.5	...	(27)	(22.5)	(21)	(20.5)	(20)
Plate	Flat plate	Axial	–1	...	(21.5)	(17)	(15.5)	(15)	(15)
Plate	Flat plate	Axial	–2	...	(15)	(11.5)	(10.5)	(10)	(10)
¾ in. diam Rod	0.330 in. diam	Alternating torsion	+0.5	...	...	...	(30)	(29)	(28)
¾ in. diam Rod	0.330 in. diam.	Alternating torsion	0	...	(26)	(24)	(22)	(20)	(19)
¾ in. diam Rod	0.330 in. diam	Alternating torsion	–1	(21)	(18)	(15)	(13)	(11)	(10)
Rod	RR Moore	Rotating beam	–1	...	...	(23)	(21)	(20)	(20)
Rod	RR Moore, notched $K_t = 12$	Rotating beam	–1	...	...	(6)	(4)	(3.5)	(3.5)
Rod	Round	Axial	0		(34)	(33)	(31)	(30)	(30)
			–0.5	...	(27)	(23)	(22)	(21)	
			–1.0		(22)	(16)	(18)	(15)	(15)
			–2.0		(15)	(11)	(10.5)	(8)	...

Sources: Unpublished Alcoa data and M.G. Vassilaros and E.J. Czyryca, Naval Research Lab Report CDNSWC-TR619409, A Compilation of Fatigue Information for Aluminum Alloys, 1994

5052-H38: Fatigue strength in air at various temperatures

Form and thickness	Specimen	Test mode	Stress ratio	Temperature °C (°F)	Fatigue strength (ksi) at cycles of: 10^4	10^5	10^6	10^7	10^8	5×10^8
Plate	Cantilever beam	Bending	−1	24 °C (75 °F)	(41)	(29.5)	(24)	(22.5)	(21)	(20)
Plate	Cantilever beam	Bending	−1	150 °C (300 °F)	(34)	(24)	(19)	(16)	(14.5)	(14)
Plate	Cantilever beam	Bending	−1	200 °C (400 °F)	(29)	(20)	(15)	(13)	(10.5)	(10)
Plate	Cantilever beam	Bending	−1	260 °C (500 °F)	(23)	(15)	(10)	(7)	(6.5)	(6)
Rod	RR Moore	Rotating beam	−1	(RT)	(41)	(31)	(24)	(22)	(21)	(21)
	RR Moore, notched, $K_t = 12$	Rotating beam	−1	(RT)	(29)	(18)	(13)	(9)	(8)	(8)

Sources: Unpublished Alcoa data and M.G. Vassilaros and E.J. Czyryca, Naval Research Lab Report CDNSWC-TR619409, A Compilation of Fatigue Information for Aluminum Alloys, 1994

5053

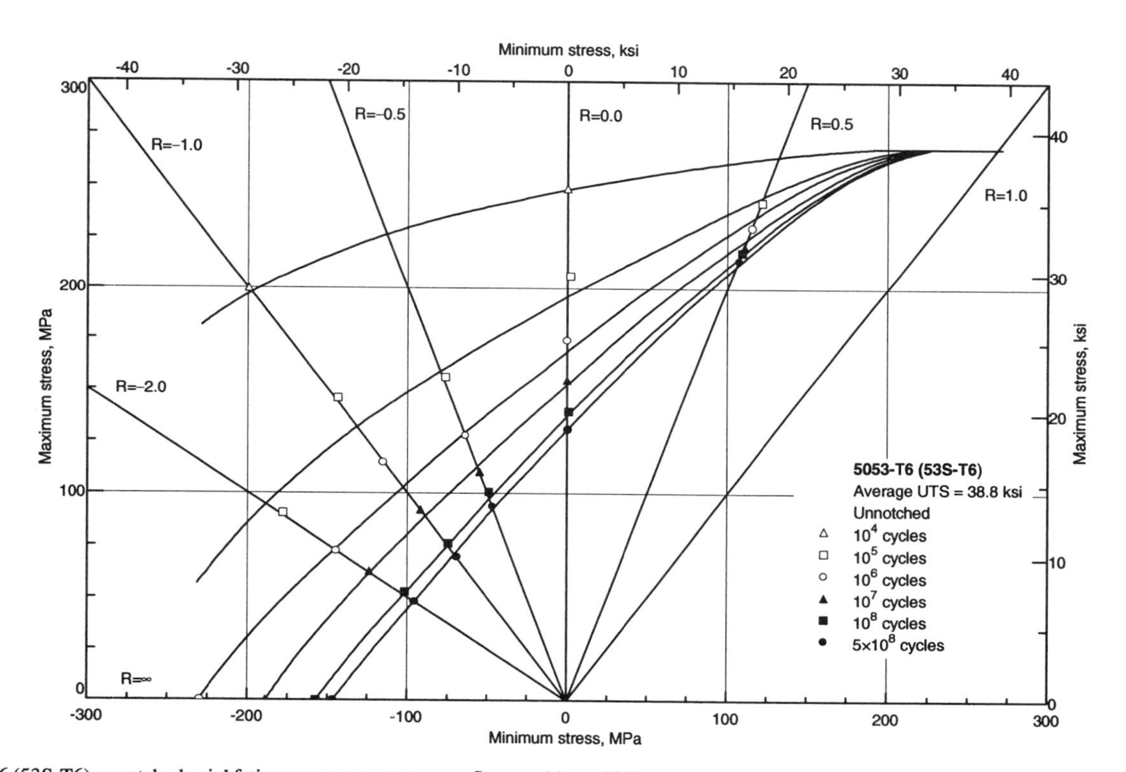

Fig. 39 5053-T6 (53S-T6) unnotched axial fatigue at room temperature. Source: Alcoa, 1952

5056

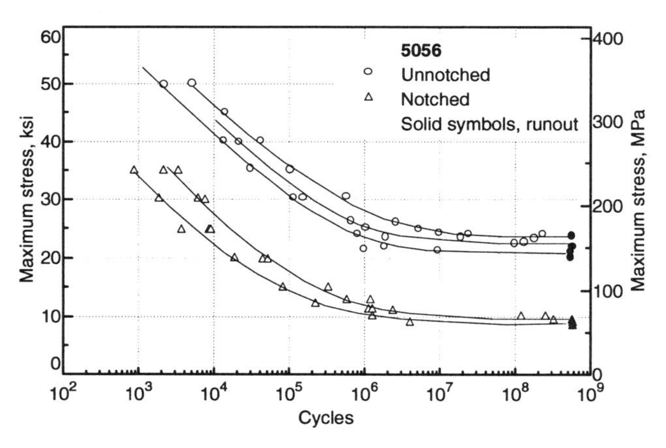

Fig. 40 5056-H32 and -H34 notched (radius at notch root <0.001 in.) and unnotched rotating beam fatigue (3/4 in. rod, rolled and drawn). R.R. Moore specimens with 9-7/8 in. surface radius and 0.300 in. minimum diameter for unnotched specimens. Notched specimens had a 0.330 in. diameter at the notch and a 0.480 in. diameter outside the 60° notch. Source: Alcoa, 1961

5056: Fatigue strength in air at various temperatures

Temper	Form and thickness	Specimen	Test mode	Stress ratio	Temperature °C (°F)	Fatigue strength (ksi) at cycles of:					
						10^4	10^5	10^6	10^7	10^8	5×10^8
0	Plate	Cantilever	Bending	−1	24 °C (75 °F)	(42)	(30)	(23)	(21)	(20)	(20)
					150 °C (300 °F)	(32)	(22.5)	(17)	(15.5)	(15)	(15)
					200 °C (400 °F)	(25)	(17)	(12)	(10)	(9)	(8.5)
					260 °C (500 °F)	(22)	(15)	(10.5)	(8.5)	(7.5)	(7)
H32	Plate	Cantilever	Bending	−1	24 °C (75 °F)	(44)	(32)	(25)	(23)	(22)	(22)
					150 °C (300 °F)	(34)	(24)	(19.5)	(17)	(16.5)	(16)
					260 °C (500 °F)	(23)	(16.5)	(11.5)	(8)	(6.5)	(6)
H34	Plate	Cantilever	Bending	−1	24 °C (75 °F)	(46)	(35)	(27.5)	(24.5)	(23.5)	(23.5)
					150 °C (300 °F)	(36)	(27.5)	(20.5)	(17)	(15.5)	(15)
					260 °C (500 °F)	(26)	(16.5)	(11)	(8)	(6.5)	(6)
0	**Rod**	R.R. Moore, unnotched	Rotating beam	−1	(RT)	(41)	(30)	(23)	(21)	(20)	(20)
		R.R. Moore, notched, $K_t = 12$	Rotating beam	−1	(RT)	(22)	(14)	(10)	(9)	(8.5)	(8.5)
H32	Rod	R.R. Moore, unnotched	Rotating beam	−1	(RT)	(43)	(36)	(28)	(25)	(24)	(23.5)
		R.R. Moore, notched, $K_t = 12$	Rotating beam	−1	(RT)	(22)	(17)	(12)	(10)	(9)	(9)
H34	Rod	R.R. Moore, unnotched	Rotating beam	−1	(RT)	(46)	(32)	(25)	(23)	(22)	(22)
		R.R. Moore, notched, $K_t = 12$	Rotating beam	−1	(RT)	(7)	(17)	(12)	(10)	(9)	(9)

Sources: Unpublished Alcoa data; M.G. Vassilaros and E.J. Czyryca, Naval Research Lab Report CDNSWC-TR619409, A Compilation of Fatigue Information for Aluminum Alloys, 1994

5082

5082-H11: Room-temperature fatigue strength in air

Form and thickness	Specimen	Test mode	Stress ratio	Fatigue strength (ksi) at cycles of:					
				10^4	10^5	10^6	10^7	10^8	5×10^8
½ in. plate	RR Moore	Rotating beam	–1	...	(31)	(28)	(26)	(24)	(23)
½ in. plate	Flat plate	Reversed plane bending	–1	...	...	(18)	(15)	...	...
½ in. plate	Flat plate, ¼ in. machined from one surface	Reversed plane bending	–1	...	(26)	(19)	(17)	...	...
½ in. plate	Flat plate, 1/8 in. machined from both surfaces	Reversed plane bending	–1	...	(27.5)	(23)	(18.5)	...	...
¼ in. plate	Longitudinal butt weld, 5056 filler, bead-on	Reversed plane bending	–1	...	(22.5)	(12.5)	(8)	...	...
¼ in. plate	Longitudinal. butt weld, 5056 filler bead-off	Reversed plane bending	–1	...	(22.5)	(15)	(10)	...	...
¼ in. plate	Transverse butt weld, 5056 filler, bead-off	Reversed plane bending	–1	...	(22)	(14)	(11)	...	...
¼ in. plate	As above but bead-on	Reversed plane bending	–1	...	(18)	(13)	(10.5)	...	...
¾ in. plate	RR Moore	Rotating beam	–1	...	(33)	(29)	(27)	(26)	(25)
¾ in. plate	Transverse butt weld, 5056 filler	Rotating beam	–1	...	(25)	(18)	(14)	(14)	(14)
¾ in. plate	As above but heat affected zone stressed	Rotating beam	–1	...	(30)	(22)	(18)	(17.5)	(17.5)
Plate	Double fillet of 5056 filler, welds parallel to stress	Reversed plane bending	–1	...	(18)	(13)	(10)	...	...
½ in. plate	Longitudinally fillet welded beams, 5083 filler	Plane bending 3- point load	.5	...	...	(28.5)	(25)	...	...
½ in. plate	Longitudinally fillet welded beams, 5083 filler	Plane bending 3- point load	0	...	...	(16)	(13)	...	...
½ in. plate	Longitudinally fillet welded beams, 5083 filler	Plane bending 3- point load	–1	...	...	(8)	(6)	...	...
½ in. plate	As above but with 4043 filler	Plane bending 3- point load	–1	...	...	(8)	(6)	...	...

Source: M.G. Vassilaros and E.J. Czyryca, Naval Research Lab Report CDNSWC-TR619409, A Compilation of Fatigue Information for Aluminum Alloys, 1994

5083-0 and -H11

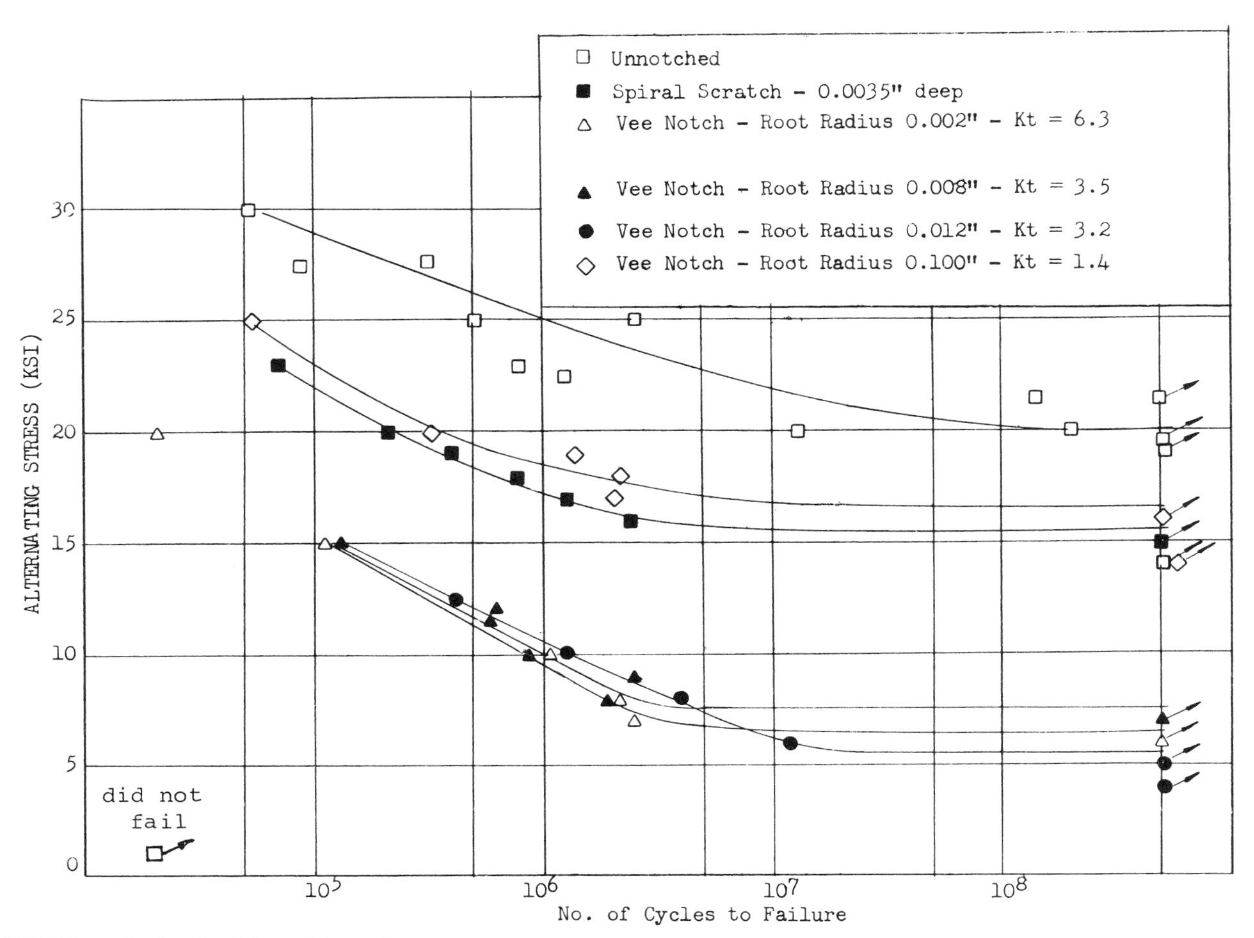

Fig. 41 5083-H11 rotating bending fatigue with machined notches and spiral scratches. Source: Naval Research Lab Report CDNSWC-TR619409, 1994

5083-0: Fatigue strength in air

Form and thickness	Specimen	Test mode	Stress ratio	Temperature °C (°F)	Fatigue strength (ksi) at cycles of: 10^4	10^5	10^6	10^7	10^8	5×10^8
¾ in. plate	RR Moore	Rotating beam	−1	(RT)	(47)	(32)	(23)	(22)	(21)	(20.5)
00.375 in. plate	Flat plate	Axial	0	(RT)	...	...	(24)	(22)	...	...
¼ in. plate	Flat plate ¾ in. wide	Reverse bending	−1	(RT)	...	(21)	(15)	(12.5)	(12)	...
¼ in. plate	Flat plate ¾ in. wide, transverse butt weld bead-off, annealed	Reverse bending	−1	(RT)	(30)	(20)	(12)	(10.5)	(10)	...
¼ in. plate	Flat plate, ¾ in. wide, transverse butt weld, bead-off	Reverse bending	−1	(RT)	...	(20)	(14)	(13)	(12.5)	...
¼ in. plate	Flat plate, ¾ in. wide, transverse butt weld, bead-on quenched	Reverse bending	−1	(RT)	(23)	(18)	(13.5)	(12.5)	(12)	...
¼ in. plate	Flat plate, ¾ in. wide, transverse butt weld, bead-on	Reverse bending	−1	(RT)	(25)	(14)	(9.5)	(9)	(8.5)	...
¾ in. plate	RR Moore, transverse butt weld, bead-off	Rotating beam	−1	(RT)	...	(26)	(17)	(13)	(12)	...
00.375 in. plate	Transverse butt weld, bead-on	Axial	0	(RT)	...	(19)	(15)	(12)	...	...
00.375 in. plate	Longitudinal butt weld, bead-on	Axial	0	(RT)	...	...	(16.5)	(12)	...	...
00.375 in. plate	Transverse butt weld, bead-off	Axial	0	(RT)	...	...	(19)	(17)	(16)	...
Plate	Cantilever beam	Bending	−1	24 °C (75 °F)	(42)	(31)	(24)	(22.5)	(22)	(22)
Plate	Cantilever beam	Bending	−1	150 °C (300 °F)	(32)	(24)	(18)	(16)	(15)	(14.5)
Plate	Cantilever beam	Bending	−1	200 °C (400 °F)	(28)	(21)	(16.5)	(13)	(11.5)	(10.5)
Plate	Cantilever beam	Bending	−1	260 °C (500 °F)	(24)	(17.5)	(13)	(9)	(7)	(6.5)
Plate	Cantilever beam	Bending	−1	315 °C (600 °F)	(20)	(15)	(10.5)	(7)	(5.5)	(5.5)
¼ in. plate	Transverse double-V butt weld, bead-on	Axial	0	(RT)	(18)	(15)	(13)	(11)	...	...
¼ in. plate	As above, bead-off	Axial	0	(RT)	...	(16.5)	(13)	(11)	...	...

(continued)

5083-0: Fatigue strength in air *(continued)*

Form and thickness	Specimen	Test mode	Stress ratio	Temperature °C (°F)	Fatigue strength (ksi) at cycles of: 10^4	10^5	10^6	10^7	10^8	5×10^8
0.157 in. sheet	Longitudinal single-V butt weld, bead-off	Axial	0	(RT)	...	(19)	(15.5)	...	...	...
Plate	RRMoore	Rotating beam	−1	(RT)	(43)	(31)	(24)	(22)	(21)	(21)
Plate	RR Moore, notched, $K_t = 12$	Rotating beam	−1	(RT)	(25)	(17)	(11)	(9)	(8)	(8)
Plate	Round, notched, $K_t = 3$	Axial	0.5	(RT)	...	(28)	(19)	(18)		
			0.0		(24)	(15)	(12)	...	...	...

Sources: Unpuplished Alcoa data and M.G. Vassilaros and E.J. Czyryca, Naval Research Lab Report CDNSWC-TR619409, A Compilation of Fatigue Information for Aluminum Alloys, 1994

5083-H11: Room-temperature fatigue strength

Form and thickness	Specimen	Test mode	Stress ratio	Atmosphere	Fatigue strength (ksi) at cycles of: 10^4	10^5	10^6	10^7	10^8	5×10^8
¾ in. plate	RR Moore	Rotating beam	−1	Air	...	(28)	(25)	(22)	(20)	(20)
¾ in. plate	RR Moore	Rotating beam	−1	Distilled water	...	...	(20)	(13)	(7)	(6)
¾ in. plate	RR Moore	Rotating beam	−1	3% salt water	...	...	(17)	(12)	(8)	(5)
5⁄16 in. plate	Transverse butt weld, 1 pass 5056 filler	Reverse plane bending	−1	Air	...	(18)	(14)	(12)	...	...
5⁄16 in. plate	Transverse butt weld, 2 passes 5056 filler	Reverse plane bending	−1	Air	...	(17)	(14)	(11)	...	...
½ in. plate	RR Moore	Rotating beam	−1	Air	...	(32)	(28)	(27)	(26)	(26)
1 in. plate	RR Moore	Rotating beam	−1	Air	...	(33)	(28)	(25)	(24)	(24)
½ in. plate	RR Moore	Rotating beam	−1	Air	...	(32)	(29)	(28)	(25)	(24)
½ in. plate	RR Moore	Rotating beam	−1	Air	...	...	(28)	(25)	(24)	(24)
¾ in. plate	RR Moore	Rotating beam	−1	Air	...	(28)	(25)	(22)	(20.5)	(20)
¾ in. plate	As above, with V-notch, $K_t = 1.4$	Rotating beam	−1	Air	...	(23)	(18)	(17)	(17)	(17)
¾ in. plate	As above, with spiral scratch, 0.0035 in. deep	Rotating beam	−1	Air	...	(22)	(17)	(16)	(16)	(16)
¾ in. plate	As above, with V-notch, $K_t = 3.5$	Rotating beam	−1	Air	...	...	(10)	(8)	(8)	(8)
¾ in. plate	As above, with V-notch, $K_t = 6.3$	Rotating beam	−1	Air	...	...	(9)	(7)	(7)	(7)
1.66 in. plate	RR Moore	Rotating beam	−1	Air	...	(33)	(24.5)	(21.5)	(21)	(21)

Sources: Unpublished Alcoa data and M.G. Vassilaros and E.J. Czyryca, Naval Research Lab Report CDNSWC-TR619409, A Compilation of Fatigue Information for Aluminum Alloys, 1994

5083-H112

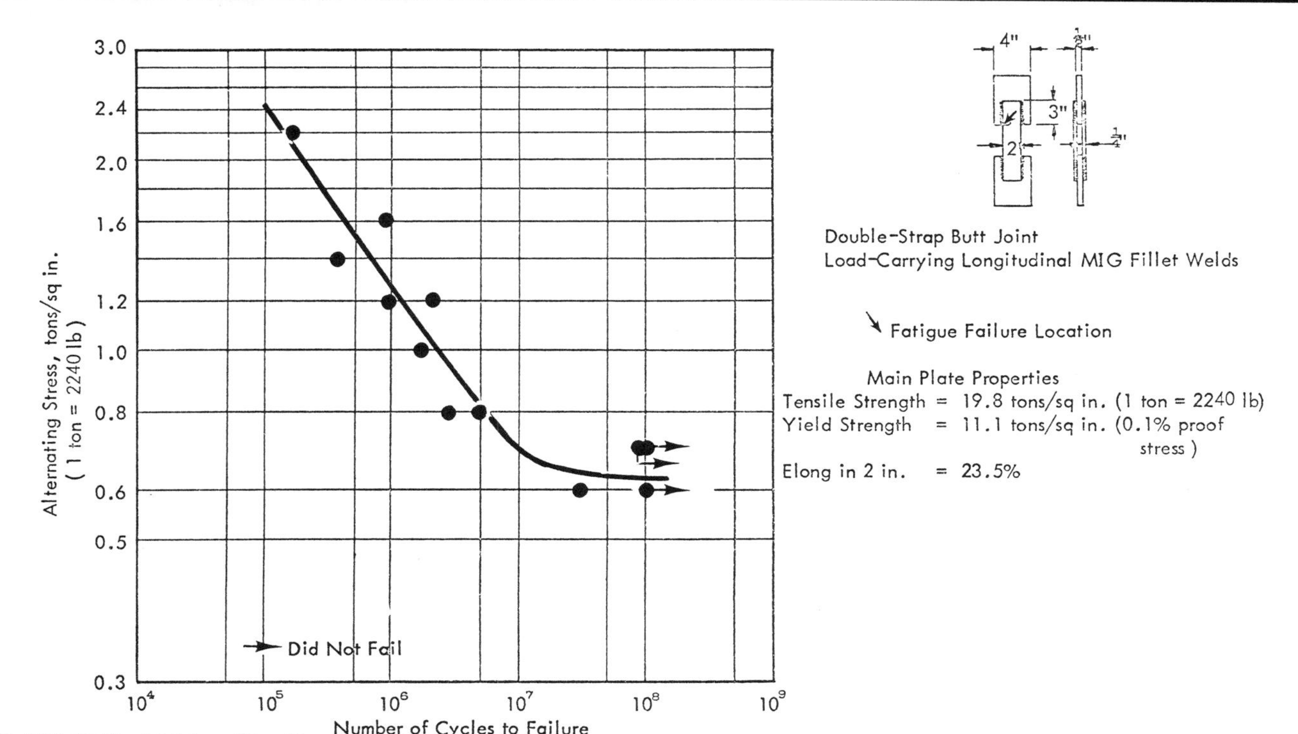

Fig. 42 5083-H112 axial fatigue ($R = -1$) results for for double-strap butt welded channels with longitudinal fillet welds, 5183 filler. Source: Kaiser Aluminum, 1966, reported in Naval Research Lab Report CDNSWC-TR619409, 1994

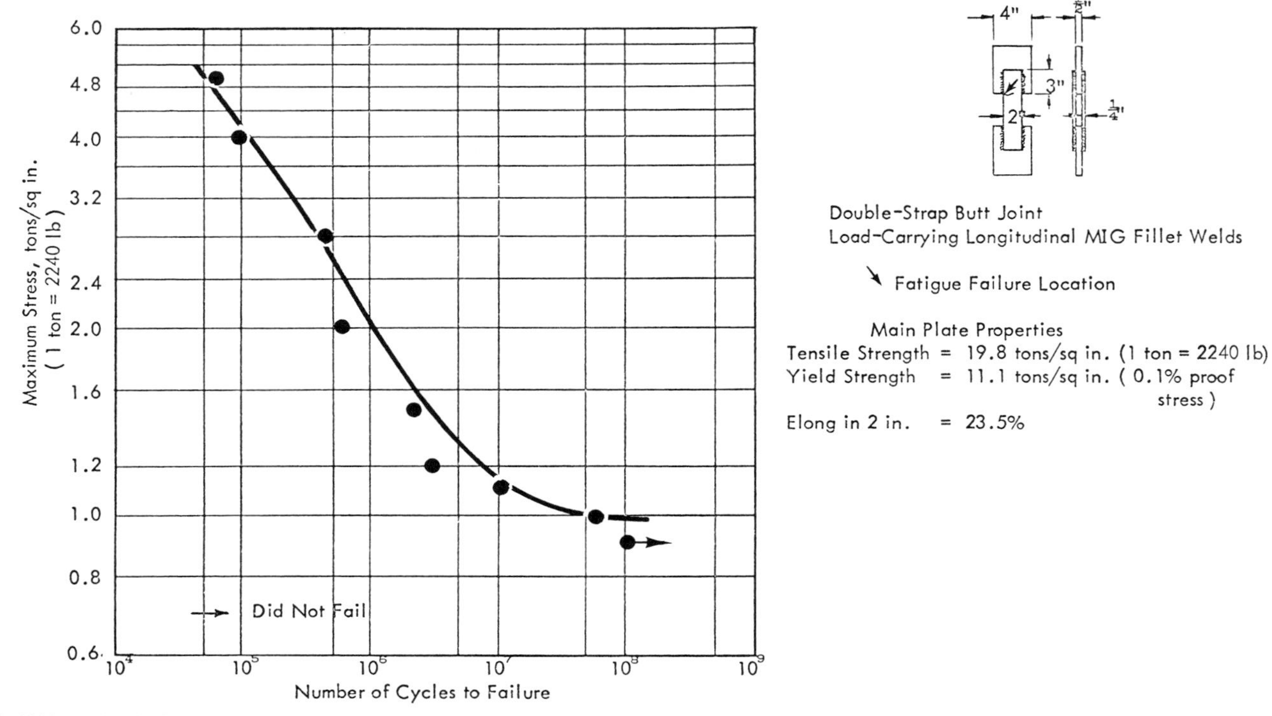

Fig. 43 5083-H112 axial fatigue ($R = 0$) for double-strap butt welded plate with longitudinal fillet welds, 5183 filler. Source: Kaiser Aluminum, 1966, reported in Naval Research Lab Report CDNSWC-TR619409, 1994

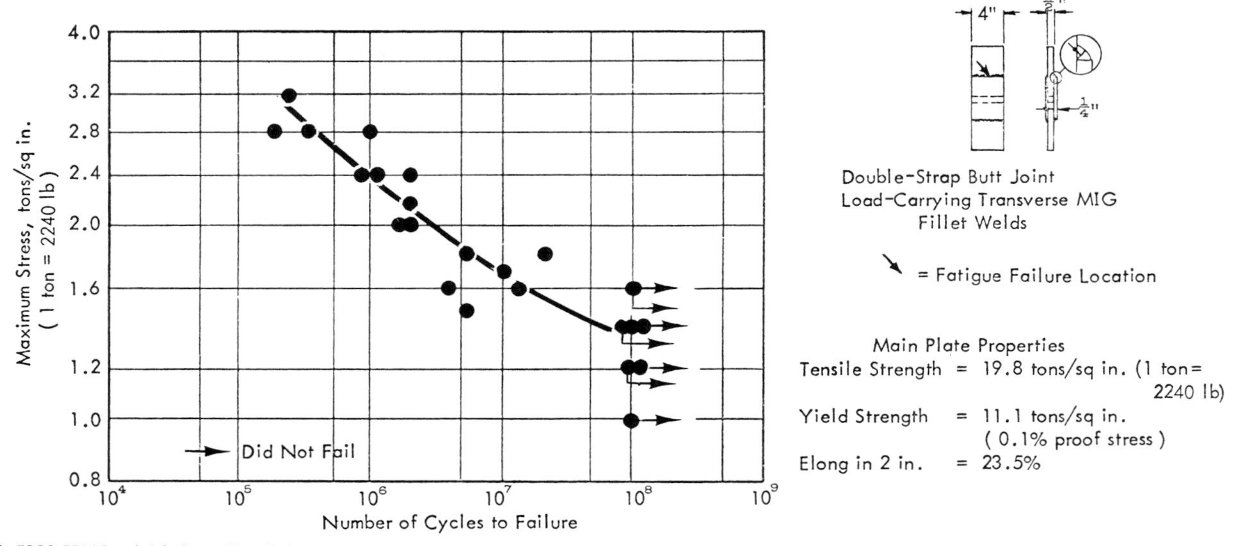

Fig. 44 5083-H112 axial fatigue ($R = 0$) for double-strap butt weld with transverse fillet welds, 5183 filler. Source: Kaiser Aluminum, 1966, reported in Naval Research Lab Report CDNSWC-TR619409, 1994

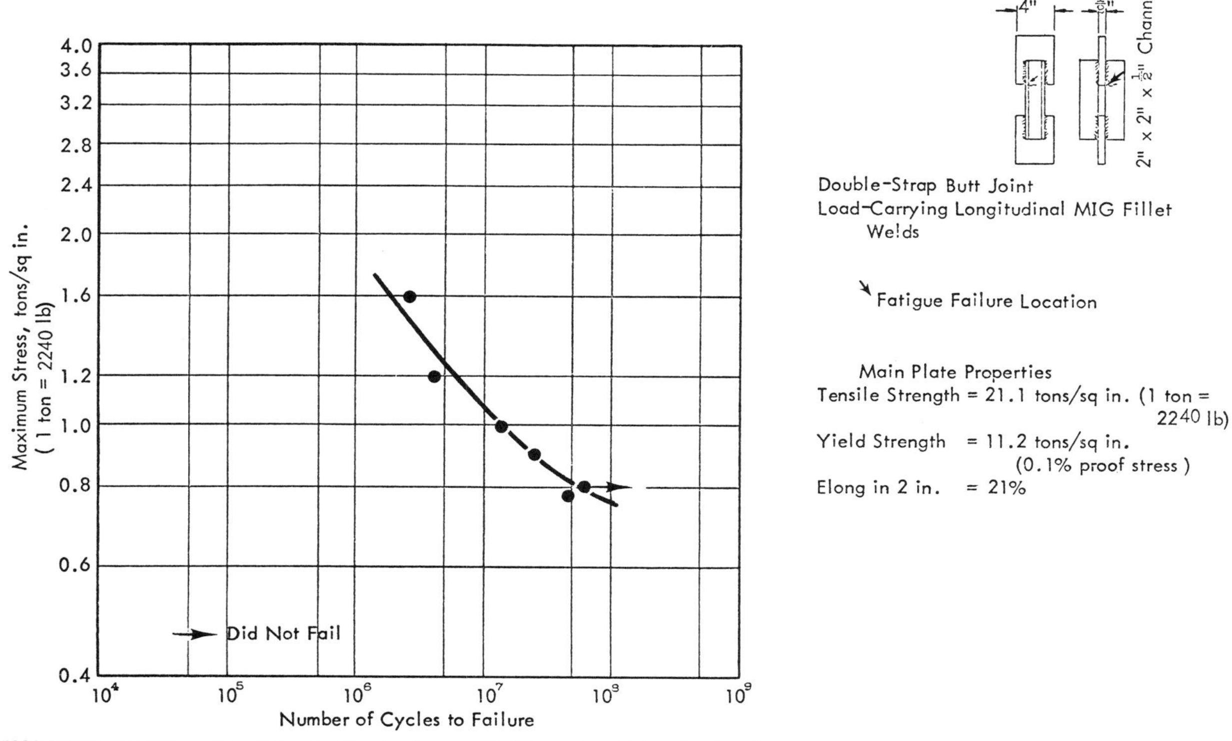

Fig. 45 5083-H112 axial fatigue ($R = 0$) for double-strap butt welded channels with longitudinal fillet welds, 5183 filler. Source: Kaiser Aluminum, 1966, reported in Naval Research Lab Report CDNSWC-TR619409, 1994

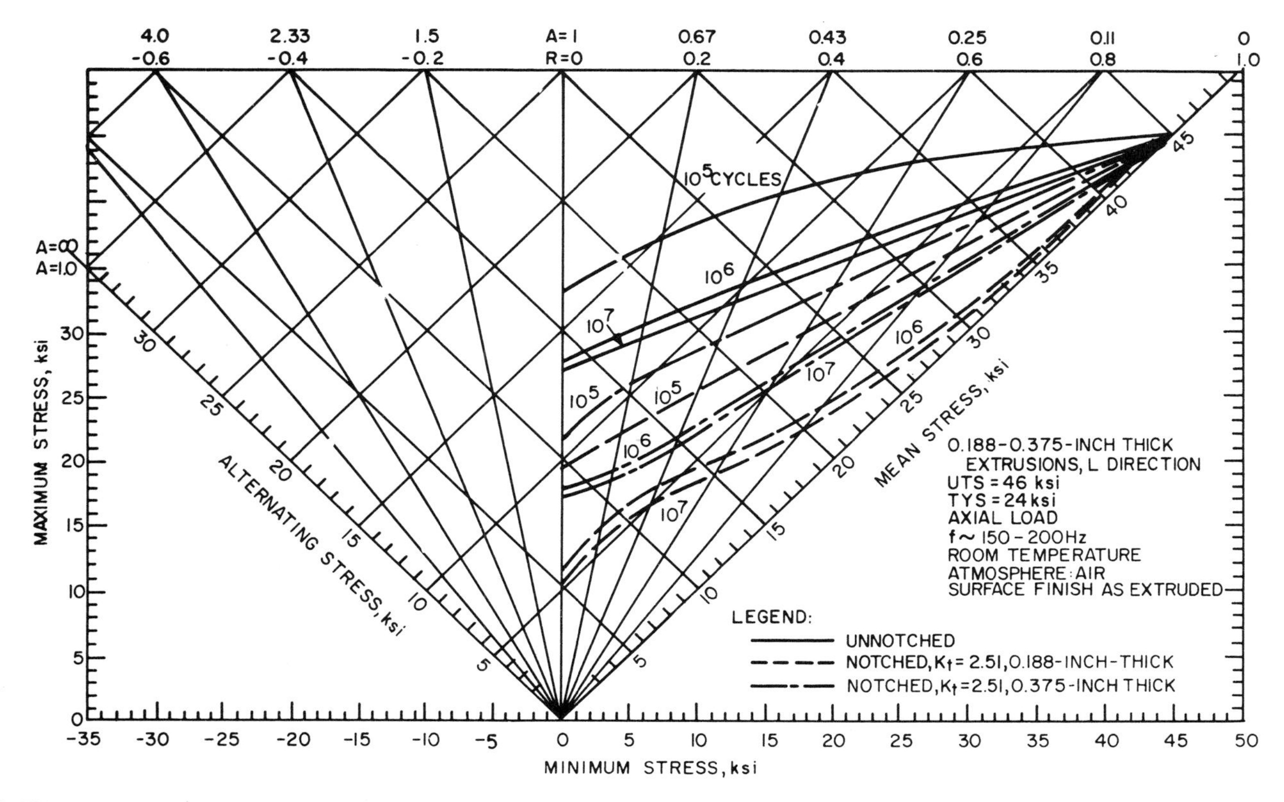

Fig. 46 5083-H112 constant life diagram (extrusions). Source: Naval Research Lab Report CDNSWC-TR619409, 1994

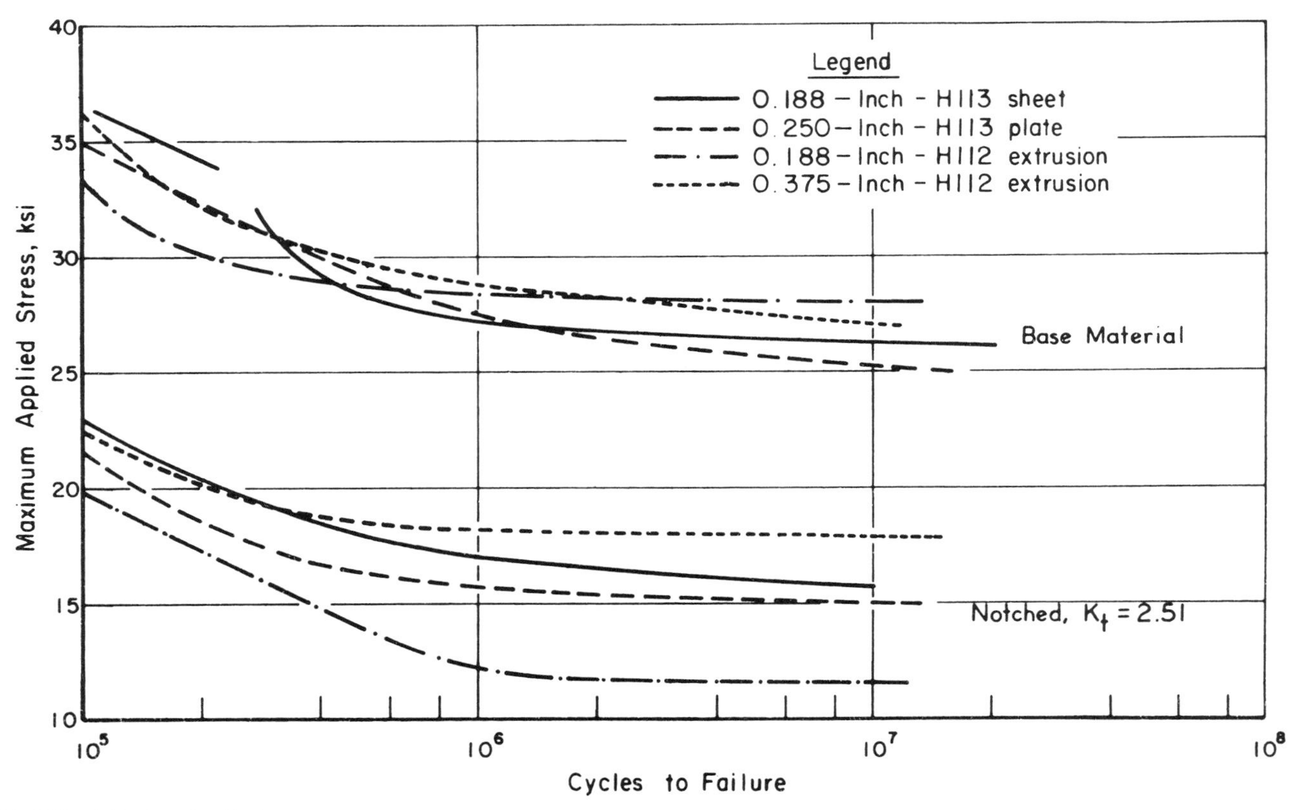

Fig. 47 5083 axial fatigue ($R = 0$) of sheet, plate, and extrusions. Source: Naval Research Lab Report CDNSWC-TR619409, 1994

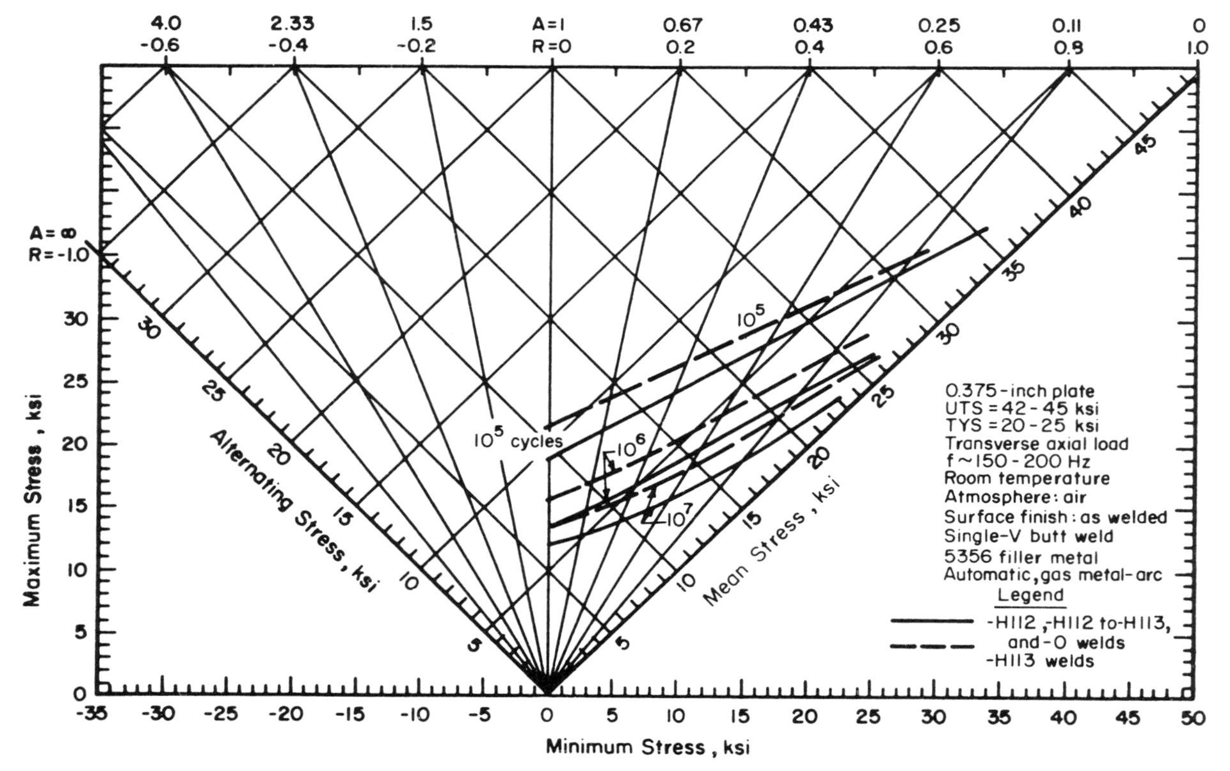

Fig. 48 5083 constant life diagram for butt welds (various tempers). Source: Naval Research Lab Report CDNSWC-TR619409, 1994

5083-H112: Room-temperature fatigue strength in air

Form and thickness	Specimen	Test mode	Stress ratio	Fatigue strength (ksi) at cycles of: 10^4	10^5	10^6	10^7	10^8	5×10^8
½ in. plate	RR Moore	Rotating beam	–1	...	(32)	(26)	(24)	(24)	(24)
Extrusion	Flat plate	Reverse bending	–1	(30)	(21)	(16)	(14)	(14)	...
⅜ in. plate	Flat plate	Axial	0	...	(35)	(29)	(25)	...	...
Sheet	Smooth	Axial	0	...	...	(27)	(26)	...	...
Sheet	Notched	Axial	0	...	...	(17)	(25)	...	...
⅜ in. plate	Smooth	Axial	0	...	(36)	(27)	(15)	...	...
¼ in. plate	Notched	Axial	0	...	(22)	(16)	...	...	...
⅜ in. plate	Longitudinal single-V butt weld, 5356 filler, bead-on	Axial	0	...	...	(16)	(13)	...	...
⅜ in. plate	Transverse single-V butt weld, 5356 filler, bead-on	Axial	0	...	(20)	(14)	(12)	...	...
⅜ in. plate	As above, bead-off	Axial	0	...	(27)	(20)	(19)	...	...
¼ in. plate	Transverse butt weld, 5356 filler, bead-on	Axial	–1	...	...	(9)	(8)	(7.5)	...
⅜ in. plate	Attachments longitudinal fillet-welded, 5183 filler, as welded	Axial	0	...	(8.5)	(4.5)	...	...	...
⅜ in. plate	Attachments longitudinal fillet-welded, 5183 filler, bead-on, annealed	Axial	0	...	(11.2)	(6.0)	...	...	...
¼ in. plate	Attachments longitudinal fillet-welded, 5183 filler, as welded	Axial	–1	...	(9)	(3.5)	(2)	...	...
¼ in. plate	Attachments butt welded to main plate edges, 5183 filler	Axial	–1	...	...	(5.0)	(3.0)	...	...
¼ in. plate	Single attachment longitudinal fillet-welded flat against main plate, 5183 filler	Axial	–1	...	38 (5.5)	(2.0)	(1.5)	(1.3)	...
¼ in. plate	Attachments longitudinal fillet-welded flat against main plate, 5183 filler	Axial	–1	...	...	(3.6)	(2.9)	(2.5)	...
¼ in. plate	Attachments transverse fillet-welded to main plate, 5183 filler	Axial with 16 ksi compressive mean stress	...	...	(5.6)	(3.3)	(2.9)	(2.7)	...
¼ in. plate	As above	Axial	–1	...		(11.2)	(9.0)	(9.0)	...
½ in. plate	Double strap butt joint, longitudinal fillet-weld, 5183 filler as welded	Axial	0	...	(3.8)	(4.0)	(2.5)	(1.6)	...
⅜ in. plate	As above	Axial	0	...	(6)	(3.2)	(1.4)	...	...
½ in. plate	Similar to above	Axial with 16 ksi mean stress	...	...	(3.8)	(2)	(1.1)	(0.90)	...
½ in. plate	Similar to above	Axial with 6.7 ksi mean stress	...	...	(4)	(2)	(1.1)	(0.9)	...
½ in. plate	Similar to above	Axial	–1	...	(5.4)	(2.9)	(1.5)	(1.4)	
½ in. plate	Similar to above	Axial	0	...	(9.1)	(4.7)	(2.6)	(2.2)	...
½ in. plate	Similar to above	Axial with 2.2 ksi mean stress	...	...	(4.9)	(2.2)	(1.2)	(1.1)	...
½ in. plate	Double strap butt joint transverse fillet-weld, 5183 filler	Axial	0	...	...	(4)	(2.6)	...	...
½ in. plate	Similar to above	Axial	0	...		(5.4)	(3.8)	(3.2)	...
3/8 in. plate	Transverse butt-weld, machined round	Rotating beam	–1	...	(25)	(16)	(14)	(13)	...
¾ in. plate	Butt-weld 3 in. dia, bead-off	Rotating beam	–1	...	(17)	(11)	(9)	...	...

Sources: Unpublished Alcoa data and M.G. Vassilaros and E.J. Czyryca, Naval Research Lab Report CDNSWC-TR619409, A Compilation of Fatigue Information for Aluminum Alloys, 1994

5083-H31, -H32, and -H34

5083-H31, -H32, and -H34: Room-temperature fatigue strength in air

Temper	Form and thickness	Specimen	Test mode	Stress ratio	Fatigue strength (ksi) at cycles of: 10^4	10^5	10^6	10^7	10^8	5×10^8
H31	¼ in. plate	Butt-weld, 5056 filler, bead-on	Reverse bending	–1	...	...	(17)	(14)	...	...
	½ in. plate	Butt-weld, bead-on	Axial	0.5	...	...	(12)	(9)	...	...
	½ in. plate	Butt-weld, bead-on	Axial	0	...	(13.5)	(8.5)	(7)	...	...
	½ in. plate	Butt-weld, bead-on	Axial	–1	...	...	(5)	(4.5)	...	...
	½ in. plate	Lap joints with transverse fillet welds	Axial	0	...	...	(3.8)	(2.6)	(1.5)	...
	½ in. plate	Longitudinally fillet-welded beams	3-point bending	–1	...	...	(8)	(6)	...	...
H32	⅛ in. sheet	Flat sheet	Reverse bending	–1	(35)	(25)	(17.5)	(16.5)	(16)	...
	¼ in. plate	Smooth	Axial	0	...	...	(29)	(28)	...	...
	¼ in. plate	Transverse butt-weld, double-V, 5056 filler, bead-on	Axial	0	...	...	(13)	(10)	...	...
	¼ in. plate	Transverse butt-weld, single-V with backing strip, 5056 filler, bead-on	Axial	0	...	...	(8.5)	(6.5)	...	...
H34	0.08 in. sheet	Flat sheet	Reverse bending	–1	...	(26.5)	(20)	(19)	(18)	...
	¼ in. plate	Transverse square butt-weld, 5056 filler, bead-on	Axial	0	...	...	(13)	(10)	(8.5)	...

Source: Unpublished Alcoa data and M.G. Vassilaros and E.J. Czyryca, Naval Research Lab Report CDNSWC-TR619409, A Compilation of Fatigue Information for Aluminum Alloys, 1994

5083-H113

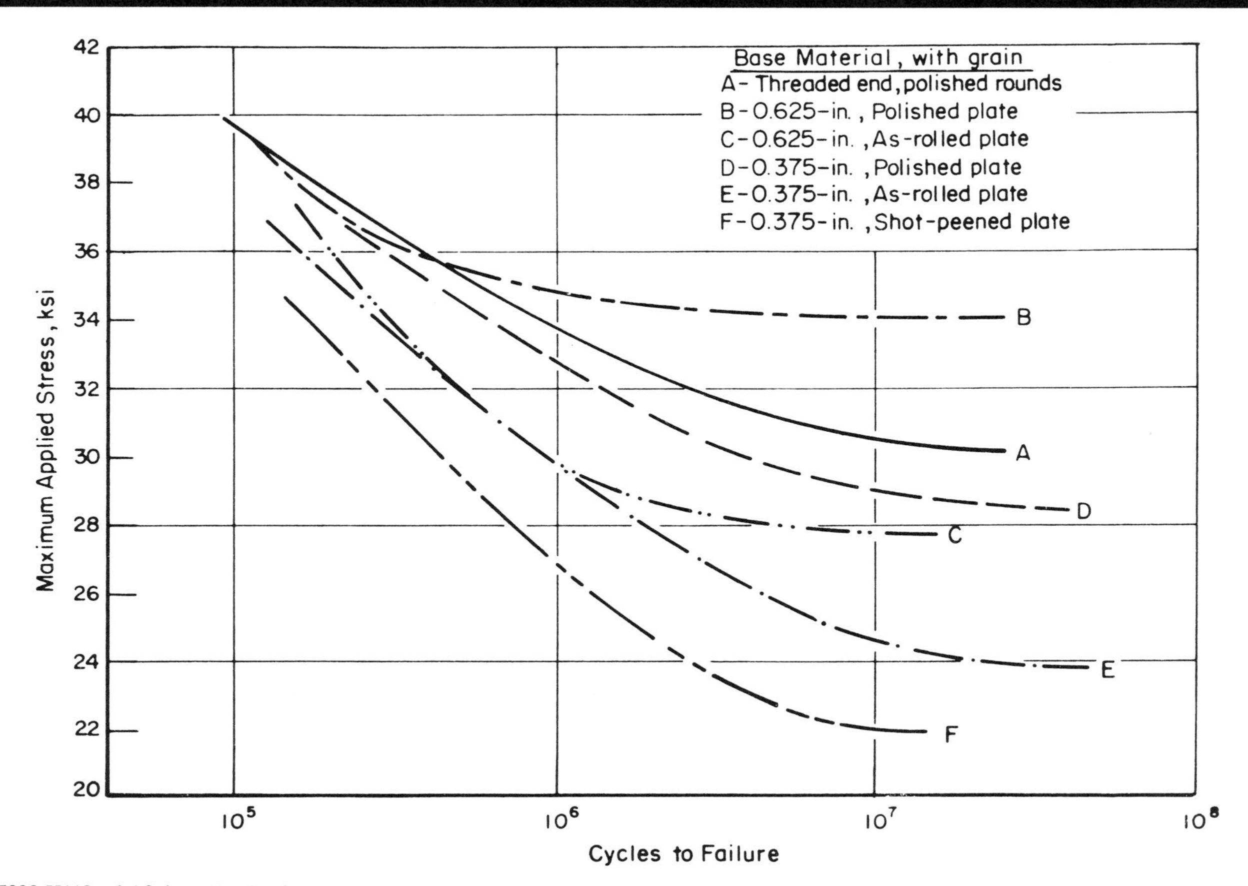

Fig. 49 5083-H113 axial fatigue ($R = 0$) of plate with various surface conditions. Source: Naval Research Lab Report CDNSWC-TR619409, 1994

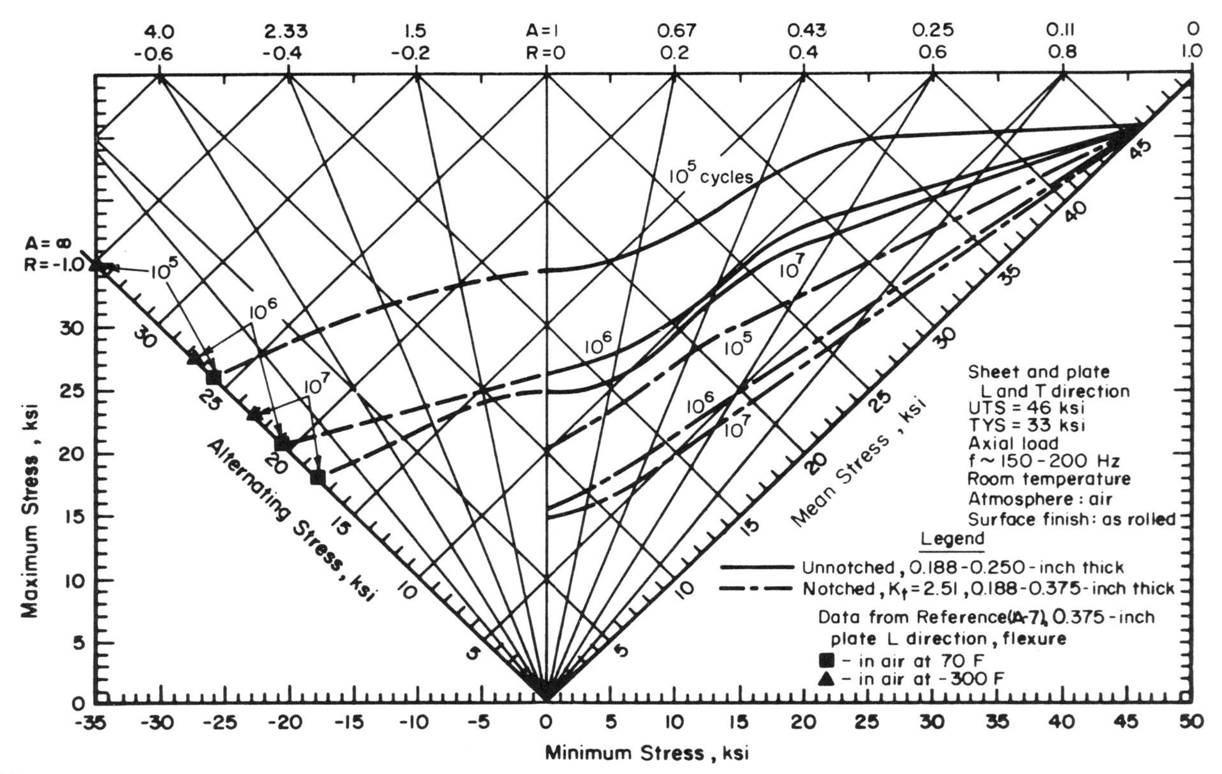

Fig. 50 5083-H113 typical constant life diagram (sheet, plate). Source: Naval Research Lab Report CDNSWC-TR619409, 1994

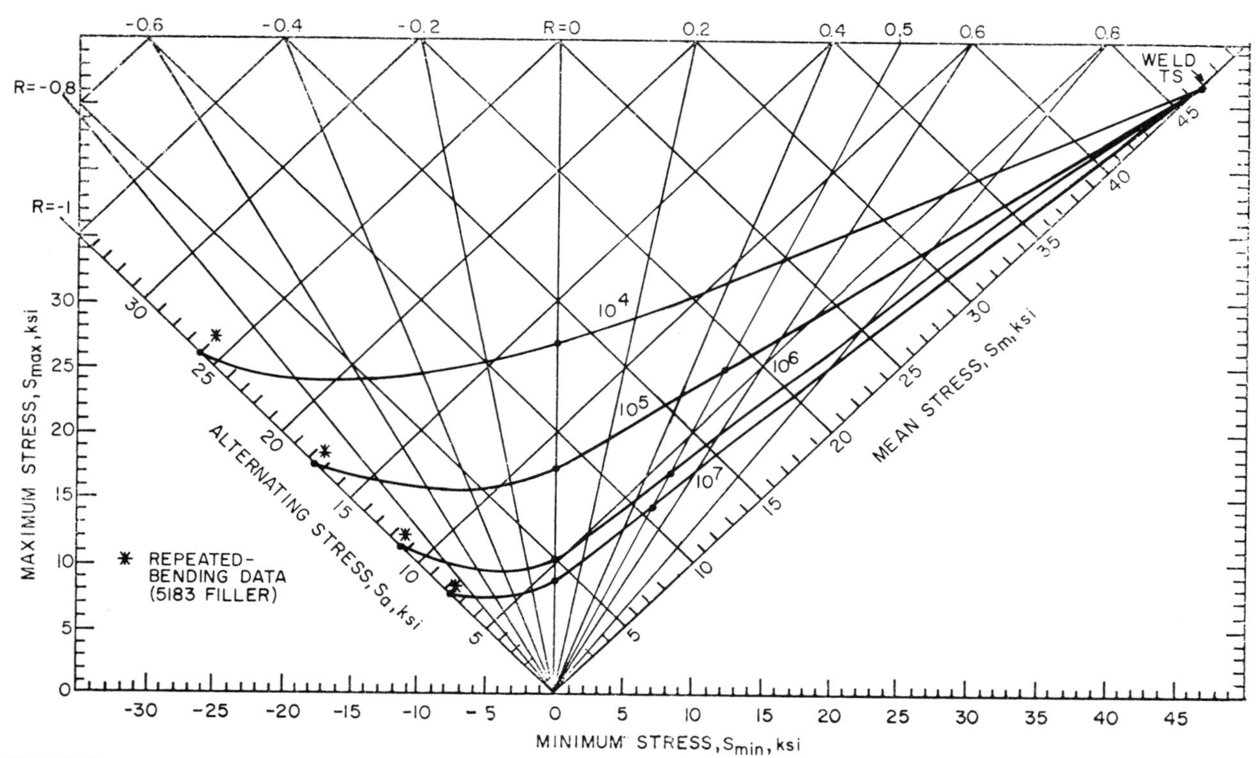

Fig. 51 5083-H113 constant life diagram for butt welds (3/8 in., bead on, 5356 filler). Source: Naval Research Lab Report CDNSWC-TR619409, 1994

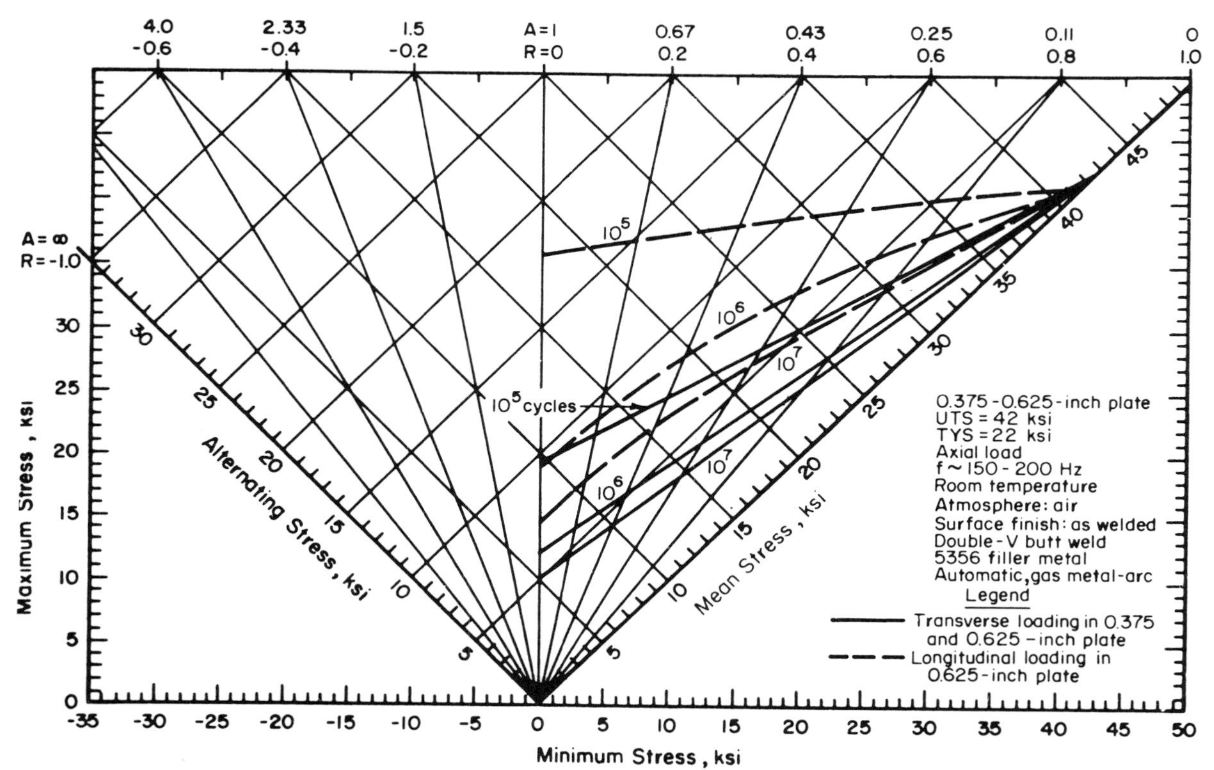

Fig. 52 5083-H113 constant life diagram for butt welded plate Source: Naval Research Lab Report CDNSWC-TR619409, 1994

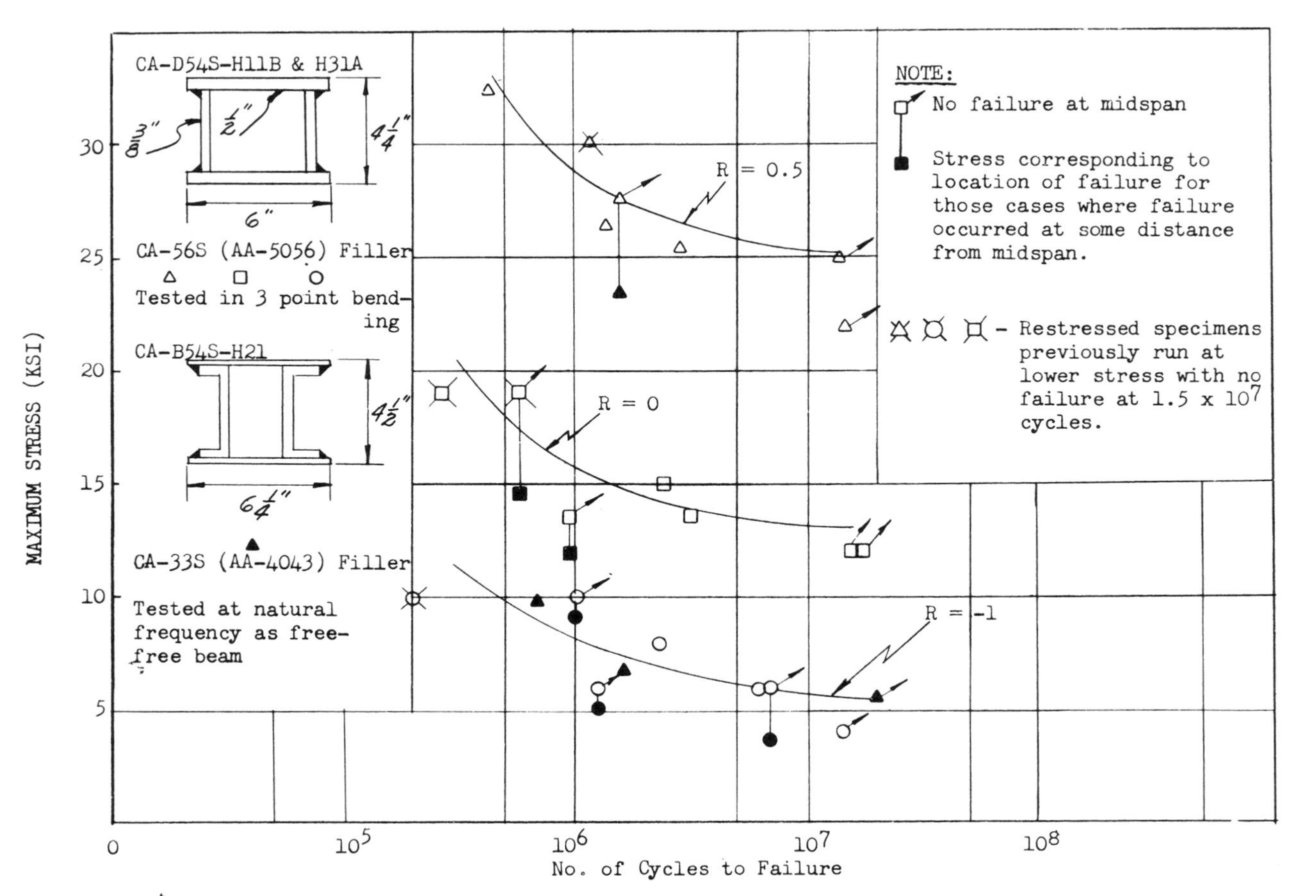

Fig. 53 5083 and 5082 plane bending fatigue of longitudinal fillet welded beams. Source: Naval Research Lab Report CDNSWC-TR619409, 1994

5083-H113: Fatigue strength in air

Form and thickness	Specimen	Test mode	Stress ratio	Temperature °C (°F)	Fatigue strength (ksi) at cycles of:					
					10^4	10^5	10^6	10^7	10^8	5×10^8
½ in. plate	RR Moore, longitudinal	Rotating beam	–1	(RT)	...	(32)	(26)	(25)	(24.5)	(24)
½ in. plate	RR Moore, transverse	Rotating beam	–1	(RT)	...	(35)	(27)	(24.5)	(23.5)	(23)
⅜ in. plate	Flat plate, transverse	Reverse bending	–1	(RT)	(40)	(26)	(21)	(18)	(16)	...
⅜ in. plate	Flat plate, transverse	Reverse bending	–1	(–300°F)	...	(35)	(27.5)	(23)	...	...
⅛ in. sheet	Flat plate, longitudinal	Reverse bending	–1	(RT)	(35)	(21)	(17)	(16)	(15)	...
⅜ in. plate	Flat plate, longitudinal	Axial	–1	(RT)	...	(43)	(35)	(32)	...	...
¾ in. plate	Round , 0.3 in. diam	Axial	0	(RT)	(49)	(40)	(36)	(35)	...	...
¾ in. plate	Round, 0.3 in. diam	Axial	0	(–320°F)	(64)	(44)	(40)	(39)	...	...
¼ in. plate	Flat plate	Axial	0	(RT)	...	(35)	(27.5)	(25)	...	...
0.188 in. sheet	Flat plate	Axial	0	(RT)	...	(36)	(27)	(26)	...	...
¼ in. plate	Notched plate, $K_t = 2.5$	Axial	0	(RT)	...	(22)	(16)	(15)	...	...
0.188 in. sheet	Notched plate, $K_t = 2.5$	Axial	0	(RT)	...	(23)	(17)	(16)	...	...
Plate	Flat plate	Axial	0	(RT)	(50)	(39)	(33)	(32)	(32)	(32)
0.625 in. plate	Polished round	Axial	0	(RT)	...	(40)	(34)	(31)	(30)	...
0.625 in. plate	Polished plate	Axial	0	(RT)	...	(40)	(35)	(34)	(34)	...
0.625 in. plate	As rolled plate	Axial	0	(RT)	...	(39)	(30)	(28)	...	...
00.375 in. plate	Polished plate	Axial	0	(RT)	...	(39)	(31)	(29)	(28)	...
00.375 in. plate	As rolled plate	Axial	0	(RT)	...	(38)	(30)	(24.5)	(23.5)	...
00.375 in. plate	Shot-peened plate	Axial	0	(RT)	...	(36)	(27)	(22)	...	...
Plate	Cantilever beam	Bending	–1	24 °C (75 °F)	(42)	(31)	(25)	(23.5)	(23)	(23)
Plate	Cantilever beam	Bending	–1	150 °C(300 °F)	(32)	(26)	(19)	(16)	(14)	(13.5)
Plate	Cantilever beam	Bending	–1	200 °C (400 °F)	(28)	(22.5)	(17.5)	(13.5)	(11)	(10.5)
Plate	Cantilever beam	Bending	–1	260 °C (500 °F)	(24)	(17)	(12)	(9)	(7)	(6.5)
Plate	Cantilever beam	Bending	–1	315 °C (600 °F)	(18.5)	(12.5)	(7.5)	38 (5.5)	(4.5)	(4.5)
⅜ in. plate	Transverse butt-weld, 5183 filler, bead-on	Reverse bending	–1	(RT)	(26)	(17.5)	(11)	(7.5)	(7)	...
⅜ in. plate	Transverse butt-weld, 5183 filler, bead-on	Reverse bending	–1	(–300 °F)	(29)	(21)	(14)	(10)	(9)	...
⅜ in. plate	Transverse butt-weld, 5183 filler, bead-off	Reverse bending	–1	(RT)	(30)	(23.5)	(15)	(10)	(10)	...
⅜ in. plate	Transverse butt-weld, 5183 filler, bead-off	Reverse bending	–1	(–300 °F)	(33)	(28)	(23.5)	(18)	...	...

(continued)

5083-H113: Fatigue strength in air *(continued)*

Form and thickness	Specimen	Test mode	Stress ratio	Temperature °C (°F)	Fatigue strength (ksi) at cycles of:					
					10^4	10^5	10^6	10^7	10^8	5×10^8
⅛ in. plate	Transverse butt-weld, 5183 filler, bead-on	Reverse bending	–1	(RT)	(26)	(17)	(9)	(8)	(7)	...
⅜ in. plate	Transverse butt-weld, 5356 filler, bead-on	Axial	0	(RT)	(27)	(17)	(10)	(9)	...	...
⅜ in. plate	Transverse butt-weld, 5356 filler, bead-on	Axial	0.5	(RT)	...	...	(17)	(15)	...	...
⅜ in. plate	As above, bead-off	Axial	0	(RT)		(27)	(18)	(14)	...	...
⅜ in. plate	Transverse butt-weld, 5183 filler, bead-on	Axial	0	(RT)	(27)	(17)	(13)	(11)	...	...
⅜ in. plate	As above, bead-off	Axial	0	(RT)	...	(28)	(19)	(15)	...	...
¾ in. plate	0.3 diam round, transverse butt-weld, 5556 filler, bead-off	Axial	0	(RT)	(41)	(28)	(22)	(22)	...	...
¼ in. plate	Transverse single-V butt-weld, 5356 filler, bead-on	Axial	0	(RT)	...	(27)	(20)	(18)	(18)	...
⅜ in. plate	Transverse single-V butt-weld, 5356 filler, bead-off	Axial	0	(RT)	...	(23)	(17)	(16)	...	...
⅜ in. plate	Transverse double-V butt-weld, 5356 filler, bead-off	Axial	0	(RT)	...	(25)	(19)	(17)	(16)	...
0.188 in. plate	Transverse single-V butt-weld, 5356 filler, bead-on	Axial	0	(RT)	...	(17.5)	(12.5)	(10)	...	...
⅜ in. plate	Transverse butt-weld to 6016-T6 plate, 5356 filler, bead-off	Axial	0.5	(RT)	...	...	(26)	(21)	...	...
⅜ in. plate	Single fillet lap joint, 5356 filler	Axial	0	(RT)	(8)	(4.5)	(3)	(2)	...	...
⅜ in. plate	As above, double	Axial	0	(RT)	(9)	(4.5)	(2.5)	(2)	...	...
⅜ in. plate	Lap joint to 6016-T6 with 5356 filler, single fillet	Axial	0	(RT)	(11)	(5)	(3.5)	(2)	...	...
⅜ in. plate	As above, double fillet	Axial	0	(RT)	(16)	(7.5)	(4.5)	(2.5)	...	...
⅜ in. plate	Tee-joint, single fillet, 5356 filler	Axial	0	(RT)	(19.5)	(3.5)	(2)	(1)	...	...
⅜ in. plate	As above, double fillet	Axial	0	(RT)	...	(13)	(7.5)	(4.5)	...	...
⅞ in. plate	Transverse single-V butt-weld, 5556 filler, machined round	Axial	0	(–320°F)	...	(30)	(22)	...	...	...
⅞ in. plate	Transverse single-V butt-weld, 5556 filler, machined round	Axial	0	(RT)	...	(35)	(28)	...	...	...
⅜ in. plate	Transverse butt-weld, 5356 filler, bead-on	Axial	0	(RT)	(27)	(16)	(9)	(8)	...	...
0.188 in. sheet	Transverse butt-weld, 5356 filler, bead-on	Axial	0	(RT)	...	(21)	(14)	(11)	(11)	...
0.188 in. sheet	Transverse butt-weld, 5356 filler, bead-on	Axial	0.25	(RT)	...	(23)	(16)	(13)	...	...
0.188 in. sheet	Transverse butt-weld, 5356 filler, bead-on	Axial	0.5	(RT)	...	...	(20)	(16)	(15)	...
¼ in. plate	Transverse single-V butt-weld, 5356 filler, bead-on	Axial	0	(RT)	...	(21)	(16)	(13)	(12)	...
¼ in. plate	Transverse single-V butt-weld, 5356 filler, bead-on	Axial	0.25	(RT)	...	(25)	(19)	(15)	(14)	...
¼ in. plate	Transverse single-V butt-weld, 5356 filler, bead-on	Axial	0.5	(RT)	...	...	(22)	(19)	(18)	...
⅜ in. plate	Transverse single-V butt-weld, 5356 filler, bead-on	Axial	0	(RT)	...	(21)	(15)	(13)	(12)	...
⅜ in. plate	Transverse single-V butt-weld, 5356 filler, bead-on	Axial	0.25	(RT)	...	(24)	(18)	(15)	(14)	...
⅜ in. plate	Transverse single-V butt-weld, 5356 filler, bead-on	Axial	0.5	(RT)	...	...	(20)	(17)	(16)	...
⅜ in. plate	Transverse double-V butt-weld, 5356 filler, bead-on	Axial	0	(RT)	...	(18)	(11)	(9)	(8)	...
⅜ in. plate	Transverse double-V butt-weld, 5356 filler, bead-on	Axial	0.25	(RT)	...	(21)	(14)	(11)	(10)	...
⅜ in. plate	Transverse double-V butt-weld, 5356 filler, bead-on	Axial	0.5	(RT)	...	...	(18)	(14)	(12)	...
0.188 in. sheet	Longitudinal butt-weld, 5356 filler, bead-on	Axial	0	(RT)	...	(24)	(14)	(9)	...	...
0.188 in. sheet	Longitudinal butt-weld, 5356 filler, bead-on	Axial	0.25	(RT)	...	(28)	(18)	(12)	...	...
0.188 in. sheet	Longitudinal butt-weld, 5356 filler, bead-on	Axial	0.5	(RT)	...	...	(27)	(19)	(18)	...
¼ in. plate	Longitudinal butt-weld, 5356 filler, bead-on	Axial	0	(RT)	...	(24)	(15)	(11)	(10)	...
¼ in. plate	Longitudinal butt-weld, 5356 filler, bead-on	Axial	0.25	(RT)	...	...	(22)	(16)	(14)	...
¼ in. plate	Longitudinal butt-weld, 5356 filler, bead-on	Axial	0.5	(RT)	...	...	(28)	(20)	(20)	...
⅜ in. plate	Longitudinal double-V butt-weld, 5356 filler, bead-on	Axial	0	(RT)	...	...	(19)	(15)	(14)	...
⅜ in. plate	Longitudinal double-V butt-weld, 5356 filler, bead-on	Axial	0.25	(RT)	...	...	(24)	(19)	(18)	...
⅜ in. plate	Longitudinal double-V butt-weld, 5356 filler, bead-on	Axial	0.5	(RT)	...	...	(31)	(24)	...	...

(continued)

5083-H113: Fatigue strength in air (*continued*)

Form and thickness	Specimen	Test mode	Stress ratio	Temperature °C (°F)	Fatigue strength (ksi) at cycles of: 10^4	10^5	10^6	10^7	10^8	5×10^8
0.188 in. sheet	Transverse butt-weld, single-V, 5356 filler, bead-off	Axial	0	(RT)	...	(29)	(21)	(20)	...	...
0.625 in. plate	Transverse butt-weld, double-V, 5356 filler, bead-on	Axial	0	(RT)	...	(21)	(14)	(12)	...	...
0.625 in. plate	As above, bead-off	Axial	0	(RT)	...	(30)	(18)	(15)	...	...
⅜ in. plate	Transverse butt-weld, square edge, 5356 filler, bead-on	Axial	0	(RT)	...	(16)	(10)	(8)	...	...
¼ in. plate	As above	Axial	0	(RT)	...	(17)	(11.5)	(10)	...	...
⅜ in. plate	Transverse butt-weld, single-V, 5356 filler, shot peened	Axial	0	(RT)	...	(23)	(18)	(15)	...	...
0.188 in. sheet	Transverse butt-weld, single-V, 4043 filler, bead-on	Axial	0	(RT)	...	(22)	(14)	(12)	...	...
0.188 in. sheet	As above, bead-off	Axial	0	(RT)	...	...	(23)	(22)	...	...
0.625 in. plate	Transverse butt-weld, double-V, 5556 filler, bead-on	Axial	0	(RT)	...	(17)	(15)	(13)	...	...
0.625 in. plate	As above, bead-off	Axial	0	(RT)	...	(20)	(18)	(17)	...	...
0.625 in. plate	Transverse butt-weld, double-V, 4043 filler, bead-on	Axial	0	(RT)	...	...	(16.5)	(13)	...	...
0.625 in. plate	As above, bead-off	Axial	0	(RT)	...	...	(22)	(18)	...	...
⅜ in. plate	Flat plate, 1 in. wide, transverse GMA butt-weld to 6061-T6, 5356 filler, bead-on	Axial	0	(RT)	(26)	(15)	(9)	(8)	(7)	...
⅜ in. plate	Flat plate, 1 in. wide, transverse GMA butt-weld to 6061-T6, 5356 filler, bead-on	Axial	0.5	(RT)	...	(28)	(18.5)	(15)	...	...
⅜ in. plate	As above, bead-off	Axial	0	(RT)	...	(27)	(20)	(15)	...	...

Sources: Unpublished Alcoa data and M.G. Vassilaros and E.J. Czyryca, Naval Research Lab Report CDNSWC-TR619409, A Compilation of Fatigue Information for Aluminum Alloys, 1994

5083-H321, -H323

5083-H321 and -H323: Room-temperature fatigue strength in air

Temper	Form and thickness	Specimen	Test mode	Stress ratio	Fatigue strength (ksi) at cycles of: 10^4	10^5	10^6	10^7	10^8	5×10^8
H321	Plate, 0.25 in.	R.R. Moore	Rotating bending	−1	()	(32.5)	(26.5)	(24.5)	(23)	(22)
H323	Sheet, 1/8 in.	Flat sheet	Rotating bending	−1	()	(27)	(19)	(18.5)	()	()
H323	Plate, 3/8 in.	Tee joint, single fillet weld	Axial	0	(9)	(5)	(3)	(2)	()	()
		Tee joint, double fillet	Axial	0	(26)	(17)	(10)	(9)	()	()
		Tee joint to 6061-T6, single fillet	Axial	0	(11)	(7)	(5)	(4)	()	()
		Tee joint to 6061-T6, double fillet	Axial	0	(26)	(19)	(8)	(5)	()	()

Sources: M.G. Vassilaros and E.J. Czyryca, Naval Research Lab Report CDNSWC-TR619409, A Compilation of Fatigue Information for Aluminum Alloys, 1994

5086

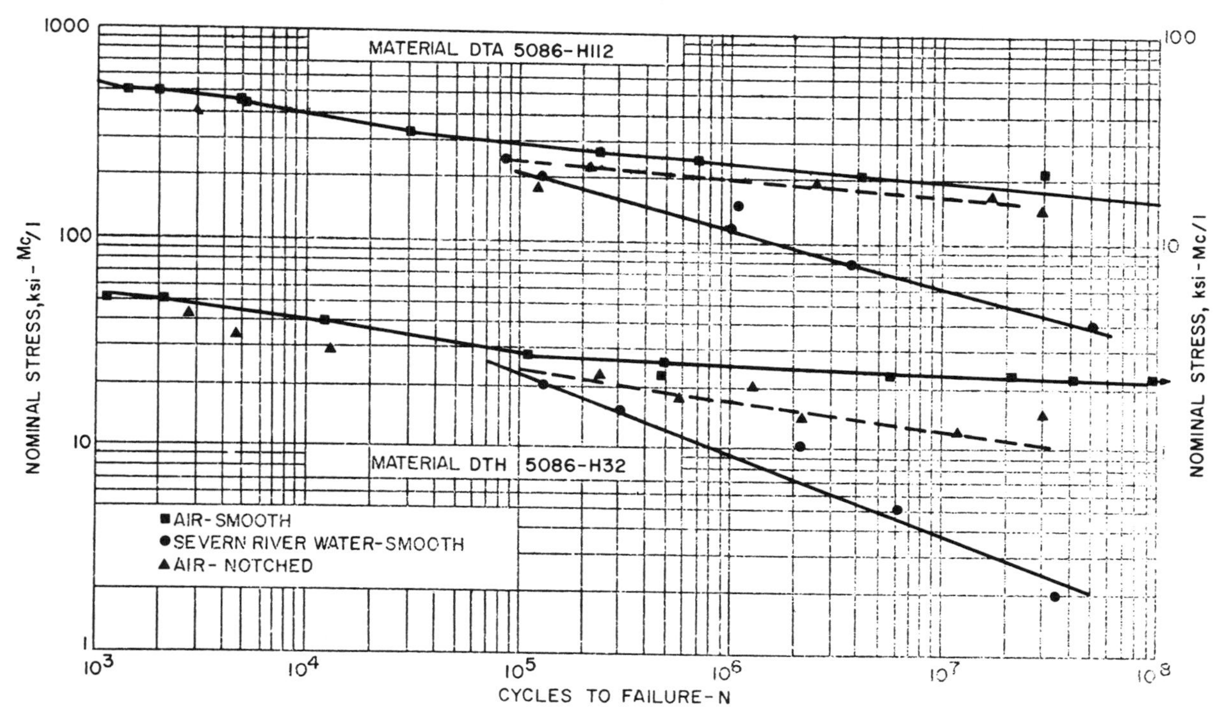

Fig. 54 5086-H112 and H32 rotating beam fatigue

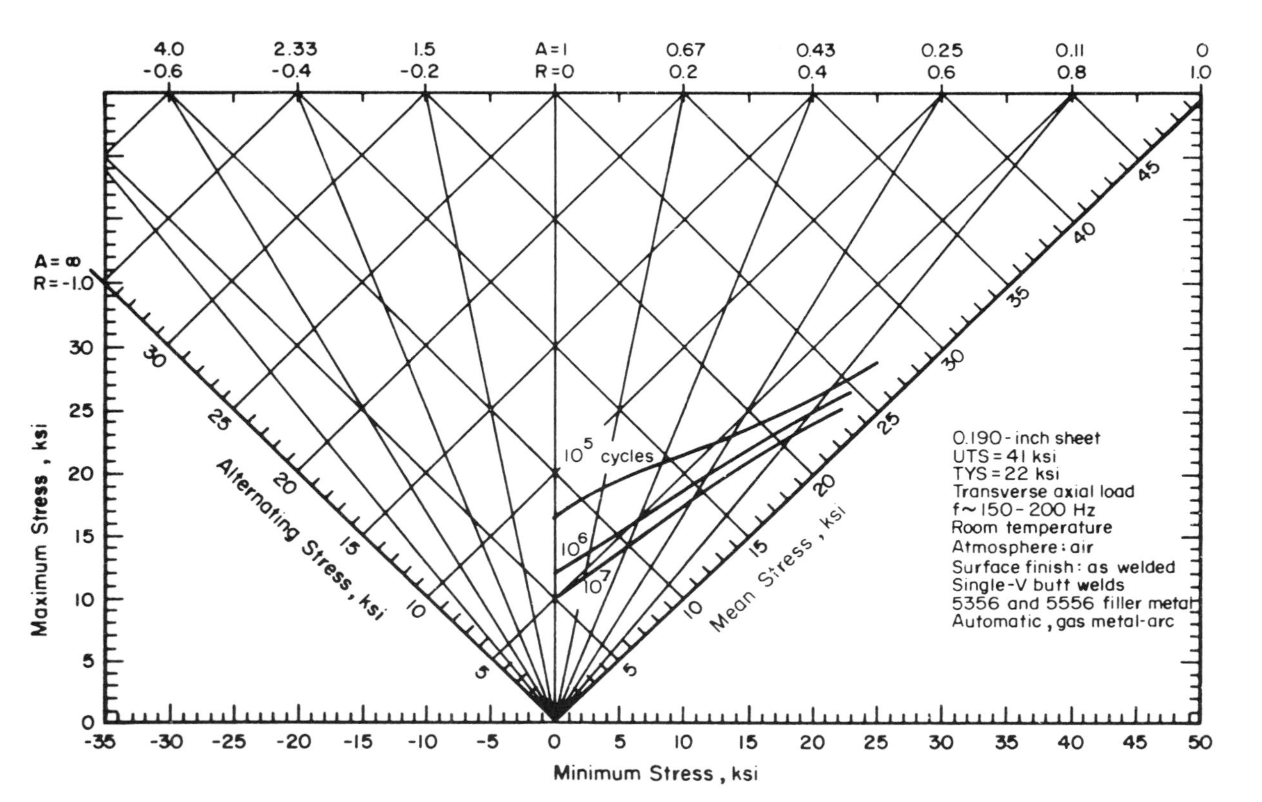

Fig. 55 5086-H32 typical constant life diagram of butt welded sheet .Source: Reynolds, 1968, in Naval Research Lab Report CDNSWC-TR619409, 1994

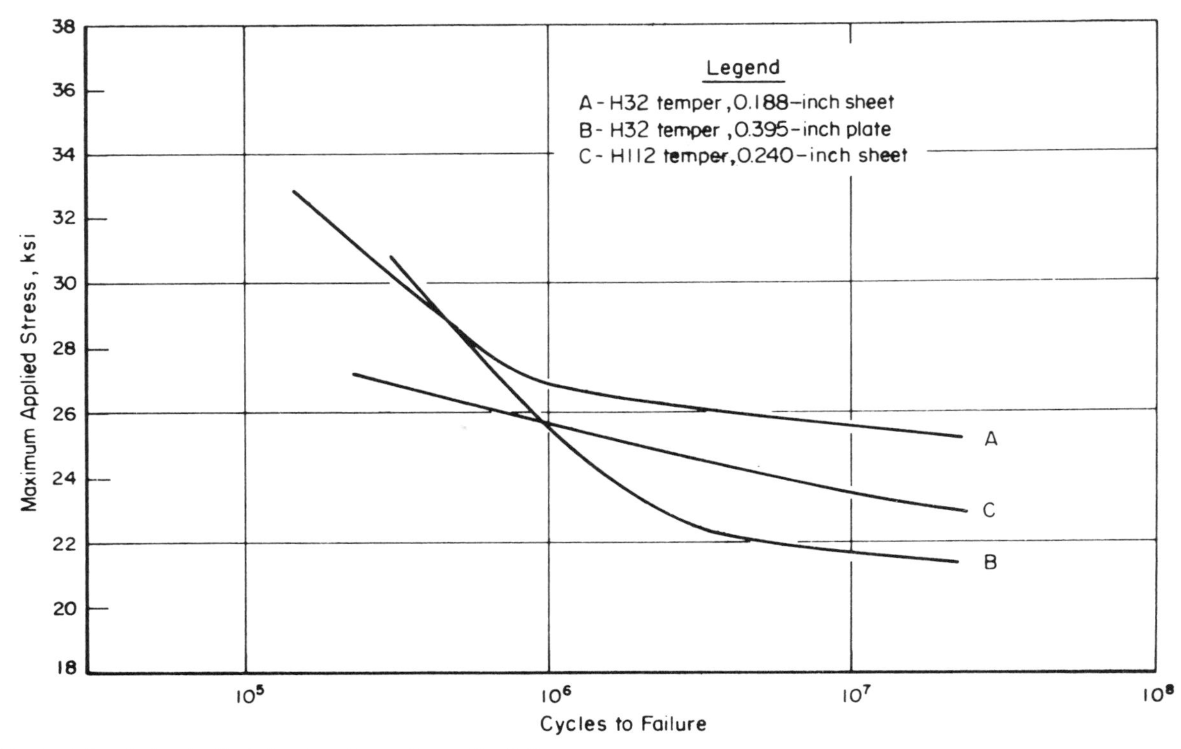

Fig. 56 5086 axial fatigue in longitudinal direction. Source: Reynolds, 1968

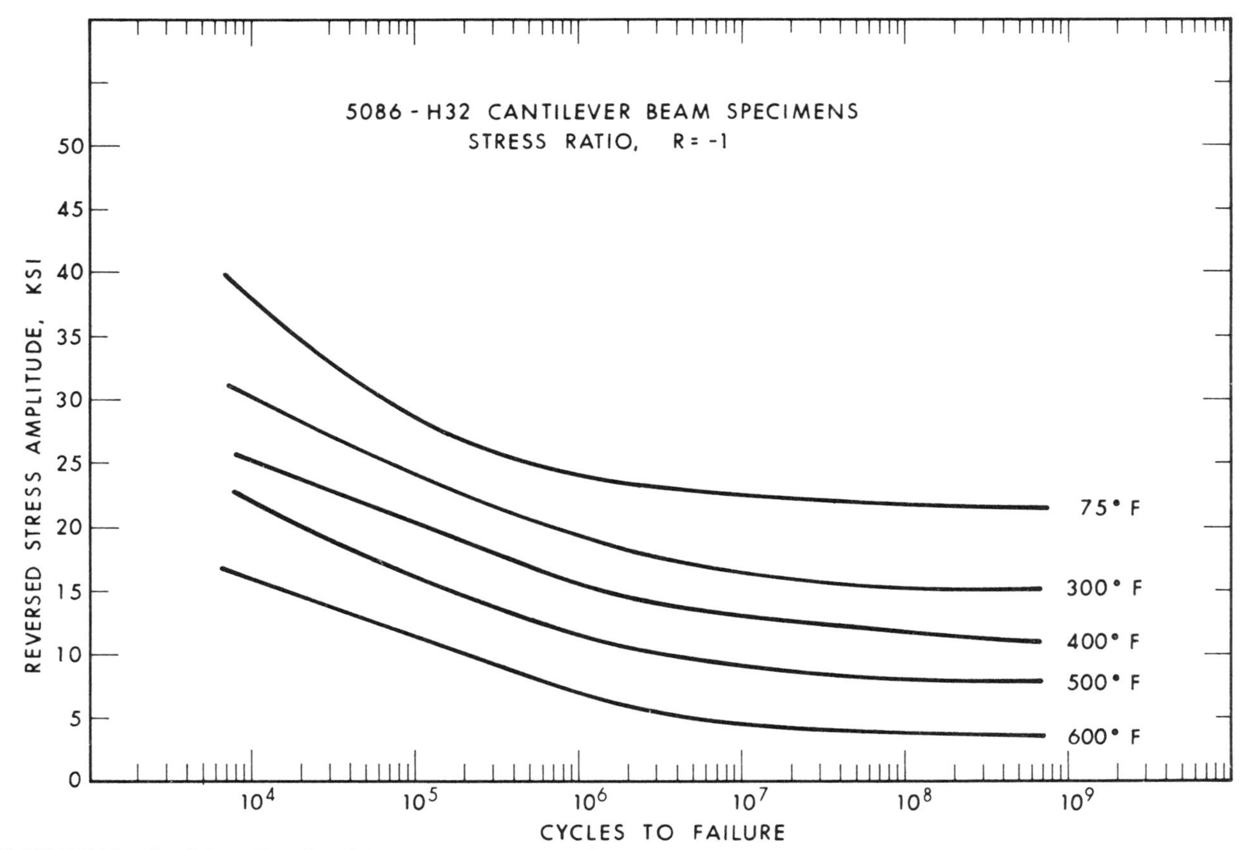

Fig. 57 5086-H32 bending fatigue ($R = -1$) at high temperatures, Alcoa, 1964

5086-0: Fatigue strength in air

Form and thickness	Specimen	Test mode	Stress ratio	Temperature °C (°F)	Fatigue strength (ksi) at cycles of: 10^4	10^5	10^6	10^7	10^8	5×10^8
Plate	Smooth	Axial	0.5	(RT)	(41.5)	(41.5)	(39)	(37)	(35.5)	(35.5)
Plate	Smooth	Axial	0.0	(RT)	(41.5)	(38)	(26)	(25)	(25)	(25)
Plate	Smooth	Axial	–0.5	(RT)	(35.5)	(28.5)	(21.5)	(20)	(19.5)	(19.5)
Plate	Smooth	Axial	–1	(RT)	(29.5)	(24)	(19.5)	(17.5)	(17)	(16.5)
Plate	Cantilever beam	Reverse bending	–1	24 °C (75 °F)	(38)	(28.5)	(24)	(22.5)	(22)	(21.5)
Plate	Cantilever beam	Reverse bending	–1	150 °C (300 °F)	(30)	(24)	(19)	(16)	(15.5)	(15)
Plate	Cantilever beam	Reverse bending	–1	200 °C (400 °F)	(25)	(20.5)	(15.5)	(13)	(12)	(11)
Plate	Cantilever beam	Reverse bending	–1	260 °C (500 °F)	(22)	(16)	(11.5)	(9)	(8)	(8)
Plate	Cantilever beam	Reverse bending	–1	315 °C (600 °F)	(16)	(11.5)	(7)	(4.5)	(4)	(3.5)
¾ in. plate	0.2 in. diam round	Axial	0	(RT)	...	(39)	(26)	(25)	(24)	(24)
0.197 in. sheet	Tee-joint fillet welded with 5086 filler	Repeated bending with 5.7 ksi mean stress	...	(RT)	...	...	...	38 (5.5)	(5)	...
0.197 in. sheet	Lap joint with single fillet weld of 5086 filler	As above	...	(RT)	...	...	(6)	(4.5)	(4)	...
0.197 in. sheet	Similar to above but double fillet	As above	...	(RT)	...	...	(6.5)	(5.8)	(5.6)	...
Plate	RR Moore	Rotating beam	–1	(RT)	(37)	(28)	(24)	(23)	(22)	(22)
Plate	RR Moore, Notch, K_t = 12	Rotating beam	–1	(RT)	(21)	(12)	(9)	(8)	(7)	(7)

Sources: Unpublished Alcoa data and M.G. Vassilaros and E.J. Czyryca, Naval Research Lab Report CDNSWC-TR619409, A Compilation of Fatigue Information for Aluminum Alloys, 1994

5086-H111: Room-temperature fatigue strength in air and salt water

Form and thickness	Specimen	Test mode	Stress ratio	Atmosphere	Fatigue strength (ksi) at cycles of: 10^4	10^5	10^6	10^7	10^8	5×10^8
Extrusion	Smooth, longitudinal	Axial	0.10	Air	...	(29)	(18)	(12.5)	...	...
Extrusion	Smooth, longitudinal	Axial	0.10	Salt water	...	(25)	(15)	(10.5)	...	...
Extrusion	Smooth, transverse	Axial	0.10	Air	(42)	(30)	(20)	...	...	...
Extrusion	Smooth, transverse	Axial	0.10	Salt water	(39)	(28)	(18)	...	...	...
Extrusion	Longitudinal weldment	Axial	0.10	Air	(32)	(22)	(14.5)	(11.5)	...	...
Extrusion	Longitudinal weldment	Axial	0.10		(30)	(20)	(13)	(10)	...	...
Extrusion	Transverse weldment	Axial	0.10	Air	...	(27)	(17)	(11.5)	...	...
Extrusion	Transverse weldment	Axial	0.10	Salt water	...	(25)	(15)	(10)	...	...

Source: M.G. Vassilaros and E.J. Czyryca, Naval Research Lab Report CDNSWC-TR619409, A Compilation of Fatigue Information for Aluminum Alloys, 1994

5086-H112: Room-temperature fatigue strength in air and water

Form and thickness	Specimen	Test mode	Stress ratio	Atmosphere	Fatigue strength (ksi) at cycles of: 10^4	10^5	10^6	10^7	10^8	5×10^8
¾ in. plate	Longitudinal 0.30 in. diam round	Reverse bending	–1	Air	...	(30)	(22.5)	(22)	(22)	(22)
¾ in. plate	As above, transverse butt weld, 5083 filler, bead-off		–1	Air	...	(25)	(16)	(13.5)	(12.5)	...
.240 in. sheet	Smooth, longitudinal	Axial	0	Air	...	...	(25.5)	(24)	...	...
6 in. bar	Smooth, .5 in. diam round	Reverse bending	–1	Air	(38)	(28)	(23)	(20)	(17)	...
6 in. bar	Same as above	Reverse bending	–1	Severn River Water	...	(21)	(12)	(6)	...	...
6 in. bar	Same as above, but notched, K_t = 3	Reverse bending	–1	Air	...	(24)	(20)	(17)	...	...
¾ in. plate	Transverse single-V butt weld, 5056 filler, machined round	Reverse bending	–1	Air	...	(25)	(17.5)	(15)	...	...
Extrusion	RR Moore	Rotating beam	–1	Air	(40)	(28)	(23)	(12)	(12)	(12)
Extrusion	RR Moore, notched, K_t = 12	Rotating beam	–1	Air	(23)	(16)	(11)	(9)	(9)	(8.5)

Sources: Unpublished Alcoa data and M.G. Vassilaros and E.J. Czyryca, Naval Research Lab Report CDNSWC-TR619409, A Compilation of Fatigue Information for Aluminum Alloys, 1994

5086-H32: Fatigue strength at various temperatures in air and water

Form and thickness	Specimen	Test mode	Stress ratio	Temperature °C (°F)	Atmosphere	Fatigue strength (ksi) at cycles of: 10^4	10^5	10^6	10^7	10^8	5×10^8
Plate	3 in. diam round smooth	Axial	0	24 °C (75 °F)	Air	...	(40)	(35)	(35)	...	...
Plate	3 in. diam round smooth	Axial	0	(−320 °F)	Air	...	(45)	(40)	...	...	...
Plate	As above with butt weld, bead off 5356 filler	Axial	0	24 °C (75 °F)	Air	...	(30)	(20)	(17)	...	...
Plate	As above with butt weld, bead off 5356 filler	Axial	0	(−320 °F)	Air	...	(35)	(25)	...	...	...
Plate	Cantilever beam	Bending	−1	24 °C (75 °F)	Air	(38)	(28.5)	(24)	(22.5)	(22)	(21.5)
Plate	Cantilever beam	Bending	−1	150 °C (300 °F)	Air	(30)	(24)	(19)	(16)	(15.5)	(15)
Plate	Cantilever beam	Bending	−1	200 °C (400 °F)	Air	(25)	(20.5)	(15.5)	(13)	(12)	(11)
Plate	Cantilever beam	Bending	−1	260 °C (500 °F)	Air	(22)	(16)	(11.5)	(9)	(8)	(8)
Plate	Cantilever beam	Bending	−1	315 °C (600 °F)	Air	(16)	(11.5)	(7)	(4.5)	(4)	(3.5)
0.375 in. plate	Smooth transverse	Axial	0.10	(RT)	Air	...	(29)	(16)	...	...	...
0.375 in. plate	Smooth transverse	Axial	0.10	(RT)	Salt water	...	(26)	(14)	(10)	...	...
0.375 in. plate	Smooth longitudinal	Axial	0.10	(RT)	Air	...	(31)	(19)	(13)	...	...
0.375 in. plate	Smooth longitudinal	Axial	0.10	(RT)	Salt water	(43)	(29)	(16)	(10)	...	...
0.375 in. plate	Smooth transverse weld	Axial	0.10	(RT)	Air	...	(27)	(18)	(13)	...	...
0.375 in. plate	Smooth transverse weld	Axial	0.10	(RT)	Salt water	...	(23)	(14)	(11)	...	...
0.375 in. plate	Smooth longitudinal weld	Axial	0.10	(RT)	Air	...	(27)	(16)	...	...	...
0.375 in. plate	Smooth longitudinal weld	Axial	0.10	(RT)	Salt water	...	(23)	(13)	...	...	...
0.188 in. sheet	Smooth	Axial	0	(RT)	Air	...	...	(27)	(26)	...	...
0.188 in. sheet	Notched, $K_t = 2.5$	Axial	0	(RT)	Air	...	(22)	(15.5)	(14.5)	...	...
0.750 in. plate	RR Moore	RB	−1	(RT)	Air	...	(31)	(25.5)	(24.5)	(24)	(24)
3/8 in. plate	GMA butt weld, bead-on, 5356 filler	Axial	0	(RT)	Air	...	(22)	(12)	(14)	...	...
3/8 in. plate	As above, bead-off	Axial	0	(RT)	Air	...	(27)	(19)	(10)	...	...
7/8 in. plate	Smooth .300 in. diam	Axial	0	(−320 °F)	Air	...	(47)	(40)	...	...	...
7/8 in. plate	Smooth .300 in. diam	Axial	0	(75 °F)	Air	...	(41)	(34)	(34)	...	...
7/8 in. plate	Welded, bead-off .300 in. diam	Axial	0	(−320 °F)	Air	...	(34)	(25)	...	...	...
7/8 in. plate	Welded, bead-off .300 in. diam	Axial	0	(75 °F)	Air	...	(28)	(19)	(19)	...	...
0.188 in. sheet	Welded bead-on	Axial	0	(RT)	Air	...	...	(12)	(11)	...	...
0.188 in. sheet	Welded bead-off	Axial	0	(RT)	Air	...	...	(17)	(16)	...	...
0.19 in. sheet	Welded, bead-off, transverse, loaded 5556 filler	Axial	0	(RT)	Air	...	...	(18)	(16)	...	...
0.19 in. sheet	As above but 5356 filler	Axial	0	(RT)	Air	...	...	(16)	(15)	...	...
0.19 in. sheet	As welded, longitudinal, loaded 5356 filler	Axial	0	(RT)	Air	...	...	(16)	(12)	...	...
0.19 in. sheet	As above but 5556 filler	Axial	0	(RT)	Air	...	...	(15)	(10)	...	...
0.19 in. sheet	As welded, transverse, loaded 5356 filler	Axial	0	(RT)	Air	...	...	(11)	(11)	...	...
0.19 in. sheet	As above but 5556 filler	Axial	0	(RT)	Air	...	...	(12)	(9)	...	...
0.375 in. plate	Welded, single V, 5356 filler	Axial	0	(RT)	Air	...	...	(17)	(11)	...	...
0.375 in. plate	Transverse double-V butt-weld, 5356 filler	Axial	0	(RT)	Air	...	...	(13)	(8)	...	...
0.188 in. plate	Flat sheet	Axial	0	(RT)	Air	...	...	(27)	(25.5)	...	...
0.395 in. plate	Flat sheet	Axial	0	(RT)	Air	...	...	(25.5)	(21.5)	...	...
2 in. plate	½ in. diam cantilever beam	Reverse bending	−1	(RT)	Air	(40)	(30)	(24)	(21)	(20)	...
2 in. plate	½ in. diam cantilever beam	Reverse bending	−1	(RT)	Severn River Water	...	(22)	(9)	(4)	...	...
2 in. plate	As above, but notched, $K_t = 3$	Reverse bending	−1	(RT)	Air	...	(22)	(16)	(11)	...	...
Plate	RR Moore	Rotating beam	−1	(RT)	Air	(39)	(29)	(23)	(22)	(21)	(21)
Plate	RR Moore	Rotating beam	−1	(RT)	Air	(24)	(16)	(12)	(9)	(7)	(7)

Sources: Unpublished Alcoa data and M.G. Vassilaros and E.J. Czyryca, Naval Research Lab Report CDNSWC-TR619409, A Compilation of Fatigue Information for Aluminum Alloys, 1994

5154

5154-0: Fatigue strength in air

Form and thickness	Specimen	Test mode	Stress ratio	Temperature °C (°F)	Fatigue strength (ksi) at cycles of:					
					10^4	10^5	10^6	10^7	10^8	5×10^8
¾ in. plate	RR Moore	RB	−1	(RT)	...	(26)	(22)	(21)	(21)	(21)
⅜ in. plate	Flat plate	Reverse bending	−1	(RT)	...	...	(14)	(12)	...	...
⅜ in. plate	Plate type	Axial	0	(RT)	...	(30)	(19)	(18)	...	...
Plate	Cantilever beam	Bending	−1	24 °C (75 °F)	(36)	(27)	(20.5)	(18)	(17.5)	(17)
Plate	Cantilever beam	Bending	−1	150 °C (300 °F)	(26)	(19.5)	(14.5)	(12)	(11.5)	(11)
Plate	Cantilever beam	Bending	−1	200 °C (400 °F)	(22)	(17.5)	(13)	(10.5)	(10)	(10)
Plate	Cantilever beam	Bending	−1	260 °C (500 °F)	(18.5)	(14)	(10)	(8)	(7)	(6.5)
⅜ in. plate	Smooth	Axial	.5	(RT)	(34.5)	(34.5)	(31.5)	(29.0)	(28.5)	(28.5)
⅜ in. plate	Smooth	Axial	0.0	(RT)	(34.5)	(33.0)	(22.0)	(19.5)	(18.5)	(18.5)
⅜ in. plate	Smooth	Axial	−.5	(RT)	(34.5)	(25.0)	(18.0)	(15.5)	(14.5)	(14.5)
⅜ in. plate	Smooth	Axial	−1.0	(RT)	(34.0)	(21.0)	(16.5)	(14.5)	(14.0)	(14.0)
¼ in. plate	Single-V transverse butt weld, argon arc, bead-on	Axial	0	(RT)	...	(22)	(15)	(11.5)	...	...
¼ in. plate	As above, self adjusting arc	Axial	0	(RT)	...	(15)	(10)	(9)	...	...
¼ in. plate	Single-V transverse butt-weld, bead-off	Reverse plate bending	−1	(RT)	(26)	(15)	(11)	(10)	...	...
⅜ in. plate	Longitudinal fillet weld double strap joint	Axial	0	(RT)	...	(6)	(3)	...	...	...
⅜ in. plate	As above, spot-heated	Axial	0	(RT)	...	...	(9)	...	...	...
⅜ in. plate	Longitudinal fillet welded attachments, both sides	Axial	0	(RT)	...	(8)	(4.5)	(3)	...	...
⅜ in. plate	As above, annealed	Axial	0	(RT)	...	(11)	(6)	(4)	...	...
Rod	RR Moore	Rotating beam	−1	(RT)	(37)	(27)	(20)	(18)	(17)	(17)
Rod	RR Moore, notched, $K_t = 12$	Rotating beam	−1	(RT)	(20)	(13)	(9)	(8)	(8)	(7.5)

Sources: Unpublished Alcoa data and M.G. Vassilaros and E.J. Czyryca, Naval Research Lab Report CDNSWC-TR619409, A Compilation of Fatigue Information for Aluminum Alloys, 1994

5154-H34:Fatigue strength in air at various temperatures

Form and thickness	Specimen	Test mode	Stress ratio	Temperature °C (°F)	Fatigue strength (ksi) at cycles of:					
					10^4	10^5	10^6	10^7	10^8	5×10^8
⅜ in. plate	Plate type	Axial	0	(RT)	...	(31)	(28)	(28)	...	...
⅜ in. plate	Butt-welded, bead-on	Axial	0	(RT)	(25)	(16)	(10.5)	(10)	...	...
Plate	Butt-welded, bead-on	Axial	0.75	(RT)	(38)	(36)	(26)	(21)	...	...
Plate	Butt-welded, bead-on	Axial	.5	(RT)	(37)	(27)	(17)	(16)	...	...
Plate	Butt-welded, bead-on	Axial	−1	(RT)	(19)	(9)	(6)	(6)	...	...
Plate	Cantilever beam	Bending	−1	24 °C (75 °F)	(39)	(29)	(23)	(21)	(20)	(20)
Plate	Cantilever beam	Bending	−1	150 °C (300 °F)	(33)	(24)	(18)	(16)	(14.5)	(14)
Plate	Cantilever beam	Bending	−1	200 °C (400 °F)	(28)	(20)	(14.5)	(11.5)	(10.5)	(10)
Plate	Cantilever beam	Bending	−1	260 °C (500 °F)	(23)	(15)	(10)	(8)	(7)	(6.5)
½ in. plate	Fillet-welded I-beam, 5154 filler	4-point bending	0	(RT)	...	...	(17)	(12)	...	...
½ in. plate	Fillet-welded I-beam, 5154 filler	4-point bending	−1	(RT)	...	(18.5)	(11)	(6)	...	...
⅜ in. plate	Transverse single-V butt weld with transverse fillet welded backing strap 5154 filler	Axial	0	(RT)	(24)	(16)	(11)	(10)	...	...
⅜ in. plate	As above, 5056 filler	Axial	0	(RT)	...	(15)	(10)	(9)	...	...
⅜ in. plate	As above, 5154 filler and 6061 -T6 strap	Axial	0	(RT)	(21)	(13)	(10)	(9)	...	...
⅜ in. plate	Transverse single-V butt weld, 5154 filler	Axial	0	(RT)	(25)	(14)	(9)	(8.5)	...	...
⅜ in. plate	As above, 5056 filler, bead-on	Axial	0	(RT)	(29)	(17)	(11)	(10)	...	...
⅜ in. plate	As above, 5056 filler, bead off	Axial	0	(RT)	...	(29)	(17)	(11)	...	...
Rod, plate	RR Moore	Rotating beam	−1	(RT)	(37)	(28)	(23)	(21)	(20)	(70)
Rod, plate	RR Moore, notched, $K_t = 12$	Rotating beam	−1	(RT)	(24)	(14)	(9)	(8)	(8)	(8)
H38	RR Moore	Rotating beam	−1	(RT)	(43)	(32)	(26)	(24)	(24)	(23.5)
Rod	RR Moore, notched, $K_t = 12$	Rotating beam	−1	(RT)	(23)	(12)	(9)	(8)	(7.5)	(7.5)

Sources: Unpublished Alcoa data and M.G. Vassilaros and E.J. Czyryca, Naval Research Lab Report CDNSWC-TR619409, A Compilation of Fatigue Information for Aluminum Alloys, 1994

5182-0 Sheet

5182-0 sheet: Room-temperature fatigue strength in air and salt solution

Specimen	Test mode	Stress ratio	Atmosphere	Fatigue strength (ksi) at cycles of: 10^4	10^5	10^6	10^7	10^8	5×10^8
Sheet	Axial	0.1	Air	...	(32)	(24)	(20)	...	...
Sheet	Axial	0.1	3.5% NaCl	...	(26)	(19)	(15)	...	...
Sheet, notched, $K_t \geq 12$	Axial	0.1	Air	(19)	(14)	(9)	(7)	...	...
Sheet, notched, $K_t \geq 12$	Axial	0.1	3.5% NaCl	(16)	(10)	(6)	(5)	...	...

Source: Unpublished Alcoa data

5356

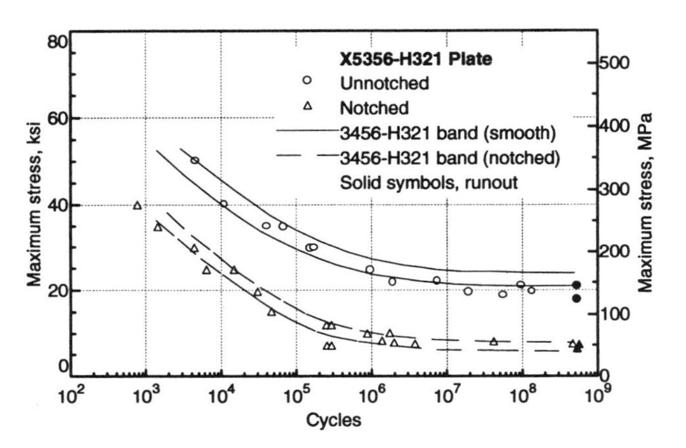

Fig. 58 5356-H321 notched (radius at notch root <0.001 in.) and unnotched rotating beam fatigue at room temperature (3/4 in. plate). R.R. Moore specimens with 9-7/8 in. surface radius and 0.300 in. minimum diameter for unnotched specimens. Notched specimens had a 0.330 in. diameter at the notch and a 0.480 in. diameter outside the 60° notch. Solid symbols indicate runout (no failure). Band lines are for 5456-H321 plate. Source: Alcoa, 1959

5356:Room-temperature fatigue strength in air

Temper	Form and thickness	Specimen	Test mode	Stress ratio	Fatigue strength (ksi) at cycles of: 10^4	10^5	10^6	10^7	10^8	5×10^8
H321	3/8 in. plate	Plate Type	Axial	0	...	(34)	(26)	(25)	(25)	...
	3/8 in. plate	Plate Type	Axial	0	...	(30)	(19)	(18)	(18)	...
	3/8 in. plate	Transverse single-V butt weld, 5356 filler, bead-on	Axial		(31)	(19)	(10)	(10)	(10)	(10)
	3/8 in. plate	As above, bead-off	Axial	Axial	(29)	(23)	(17.5)	(11.5)	(11.5)	...
0	Plate	RR Moore	Rotating beam	−1	(40)	(30)	(22)	(19)	(19)	(17)
		RR Moore, notched, $K_t = 12$	Rotating beam		(23)	(15)	(10)	(8)	(8)	(8)
	Plate	RR Moore	Rotating beam	−1	(43)	(31)	(24)	(22)	(22)	(21)
		RR Moore, notched, $K_t = 12$	$K_t = 12$	Rotating beam	(24)	(14)	(10)	(9)	(9)	(8)

Sources: Unpublished Alcoa data and M.G. Vassilaros and E.J. Czyryca, Naval Research Lab Report CDNSWC-TR619409, A Compilation of Fatigue Information for Aluminum Alloys, 1994

5454

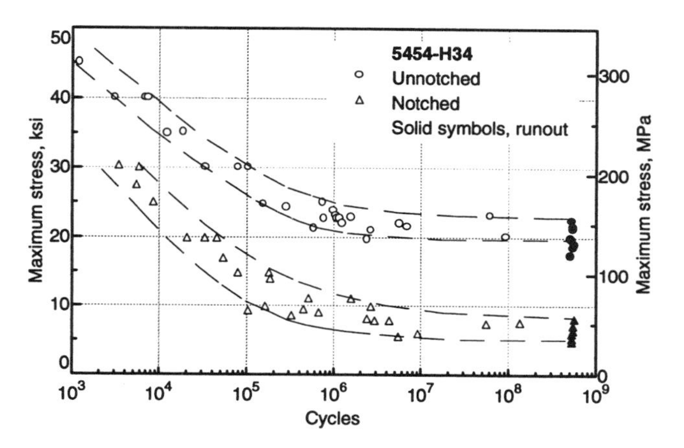

Fig. 59 5454-H34 notched (radius at notch root <0.001 in.) and unnotched rotating beam fatigue at room temperature (plate and rolled-and-drawn rod). R.R. Moore specimens with 9-7/8 in. surface radius and 0.300 in. minimum diameter for unnotched specimens. Notched specimens had a 0.330 in. diameter at the notch and a 0.480 in. diameter outside the 60° notch. Solid symbols indicate runout (no failure). Source: Alcoa, 1959

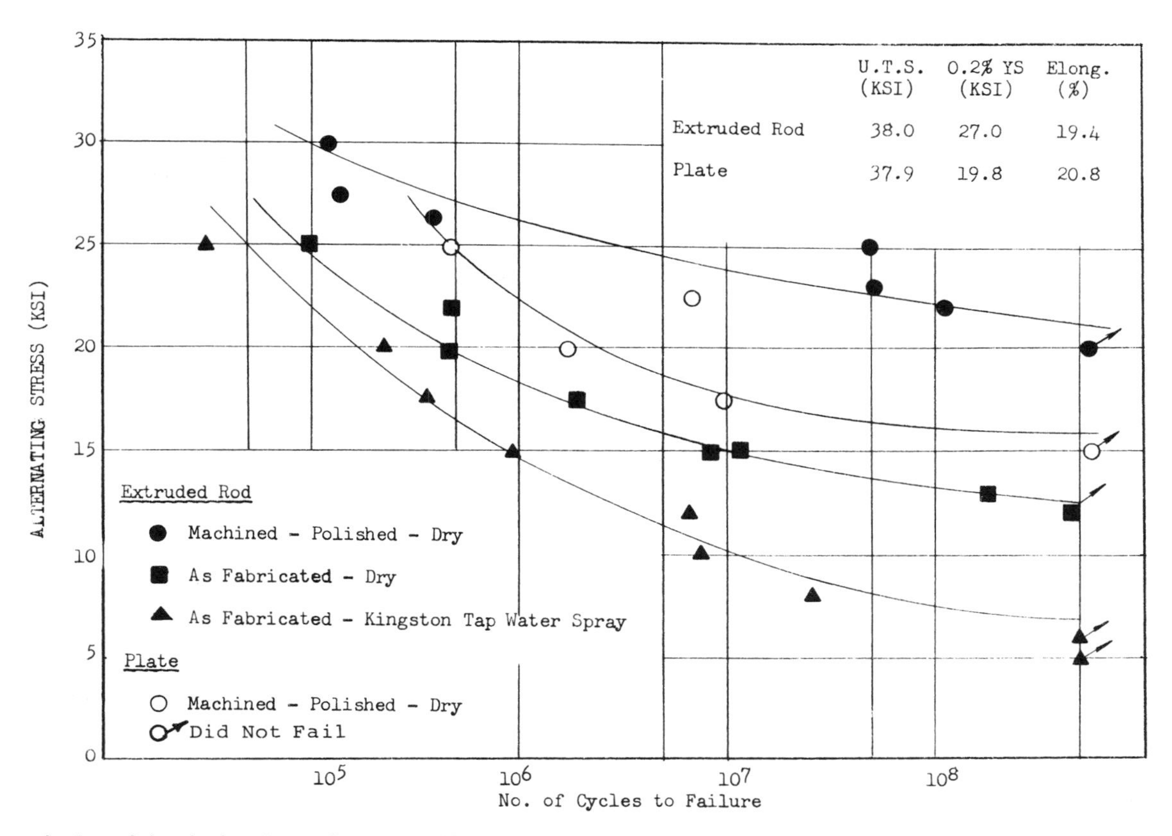

Fig. 60 5454 rotating beam fatigue in air and water. Source: Naval Research Lab Report CDNSWC-TR619409, 1994

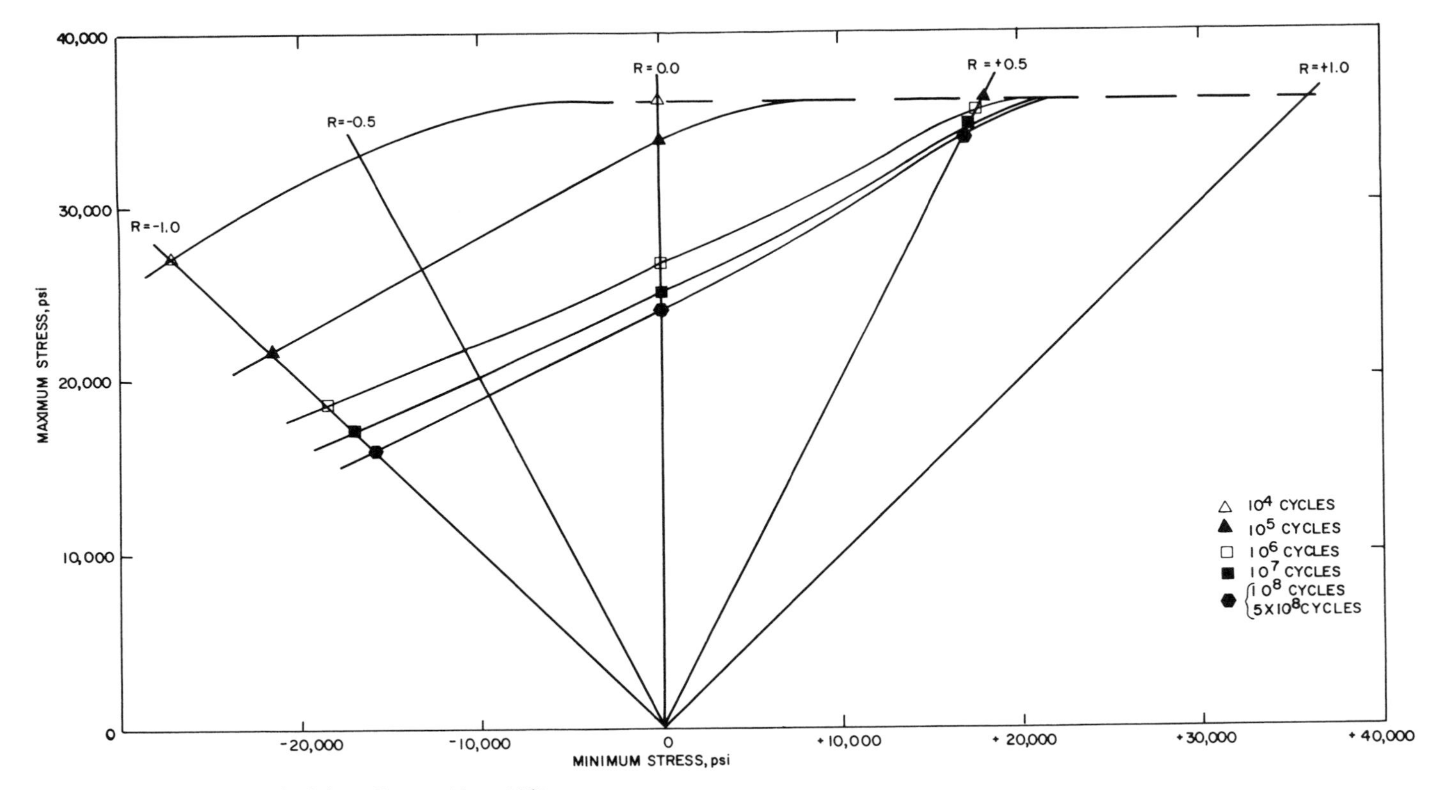

Fig. 61 5454-O unnotched axial fatigue. Source: Alcoa, 1964

5454: Room-temperature fatigue strength

Temper	Form and thickness	Specimen	Test mode	Stress ratio	Atmosphere	Fatigue strength (ksi) at cycles of: 10^4	10^5	10^6	10^7	10^8	5×10^8
0	¾ in. plate	0.2 in. diam round	Axial	0	Air	...	(34)	(26)	(25)	(24)	(23)
	Plate	Flat plate	Axial	+0.5	Air	(36)	(36)	(35)	(34.5)	(34)	(34)
	Plate	Flat plate	Axial	0	Air	(36)	(34)	(27)	(25)	(24)	(24)
	Plate	Flat plate	Axial	–0.5	Air	(33)	(26.5)	(21.5)	(20)	(19)	(19)
	Plate	Flat plate	Axial	–1	Air	(27)	(21.5)	(18.5)	(17)	(16)	(16)
	0.188 in. sheet	Transverse square butt weld, 5456 filler, bead-on	Reverse plate bending	–1	Air	...	(17.5)	(12)	(9)	...	...
	0.188 in. sheet	As above, 5154 filler	Reverse plate bending	–1	Air	...	(17)	(12)	(10)	(8)	...
	Weld	RR Moore	RB	–1	Air	(33)	(25)	(22)	(21)	(20)	(20)
		RR Moore, notched, K_t = 12	Rotating beam	–1	Air	(22)	(15)	(10)	(8)	(7)	(7)
–0	Sheet	Sheet	Axial	0.10	Air	...	(32)	(23)	(20)	...	...
		Sheet	Axial	0.10	3.5% NaCl	...	...	(17)	...	...	...
		Sheet, notched, $K_t \geq 12$	Axial	0.10	Air	(18)	(13)	(9)	(7)	...	...
		Sheet, notched, $K_t \geq 12$	Axial	0.10	3.5% NaCl	(17)	(10)	(7)	(6)	...	...
H32	Sheet	Sheet	Axial	0.10	Air	...	(39)	(28)	(26)	...	...
		Sheet	Axial	0.10	3.5% NaCl	...	...	(21)	...	...	...
		Sheet, notched, $K_t \geq 12$	Axial	0.10	Air	(25)	(15)	(9)	(6)	...	...
		Sheet, notched, $K_t \geq 12$	Axial	0.10	3.5% NaCl	(20)	(12)	(7)	(5)	...	...

Sources: Unpublished Alcoa data and M.G. Vassilaros and E.J. Czyryca, Naval Research Lab Report CDNSWC-TR619409, A Compilation of Fatigue Information for Aluminum Alloys, 1994

5454-H32: Fatigue strength in air

Form and thickness	Specimen	Test mode	Stress ratio	Temperature °C (°F)	Fatigue strength (ksi) at cycles of: 10^4	10^5	10^6	10^7	10^8	5×10^8
Plate	Flat plate	Axial	0	(RT)	(40)	(39)	(31)	(30)	(30)	(30)
0.188 in. sheet	Transverse square butt weld, 5154 filler, bead-on	Axial	0	(RT)	...	(19)	(13.5)	(11)	(10)	...
0.188 in. sheet	As above, 5456 filler	Axial	0	(RT)	...	(17)	(12)	(10)	(9)	...
⅞ in. plate	Transverse single-V butt weld, 5554 filler, machined round	Axial	0	(RT)	...	(31)	(21)	...	...	...
⅞ in. plate	Transverse single-V butt weld, 5554 filler, machined round	Axial	0	(–320 °F)	...	(38)	(29)	...	...	...
Plate, rod	RR Moore	Rotating beam	–1	(RT)	(37)	(28)	(23)	(22)	(21)	(21)
Plate, rod	RR Moore, notched, $K_t = 12$	Rotating beam	–1	(RT)	(24)	(14)	(9)	(8)	(7)	(7)

Sources: Unpublished Alcoa data and M.G. Vassilaros and E.J. Czyryca, Naval Research Lab Report CDNSWC-TR619409, A Compilation of Fatigue Information for Aluminum Alloys, 1994

5454-F, -H11, and -H111: Room-temperature fatigue strength in air

Temper	Form and thickness	Specimen	Test mode	Stress ratio	Fatigue strength (ksi) at cycles of: 10^4	10^5	10^6	10^7	10^8	5×10^8
H11	0.188 in. sheet	Transverse square butt weld, 5154 filler, bead-on	Rotating beam	–1	...	...	(13.5)	(11)	(10)	...
	0.188 in. sheet	As above, 5456 filler	Rotating beam	–1	(20)	(15)	(11)	(10)	(9.5)	...
H111	⅜ in. plate	Flat plate	Rotating beam	–1	...	(22)	(17)	(15.5)	(15)	...
	¾ in. plate	As above, polished	Rotating beam	–1	...	(30)	(22)	(17.5)	(16)	(15)
F	⅜ in. diam extrusion rod	RR Moore, polished	Rotating beam	–1	...	(30)	(26)	(24)	(22)	(21)
	⅜ in. diam extrusion rod	As above, as fabricated	Rotating beam	–1	...	(25)	(18)	(15)	(13)	(12)
	⅜ in. diam extrusion rod	As above, but tested in tap water	Rotating beam	–1	...	(22)	(15)	(10)	(7)	(5)

Sources: Unpublished Alcoa data and M.G. Vassilaros and E.J. Czyryca, Naval Research Lab Report CDNSWC-TR619409, A Compilation of Fatigue Information for Aluminum Alloys, 1994

5456

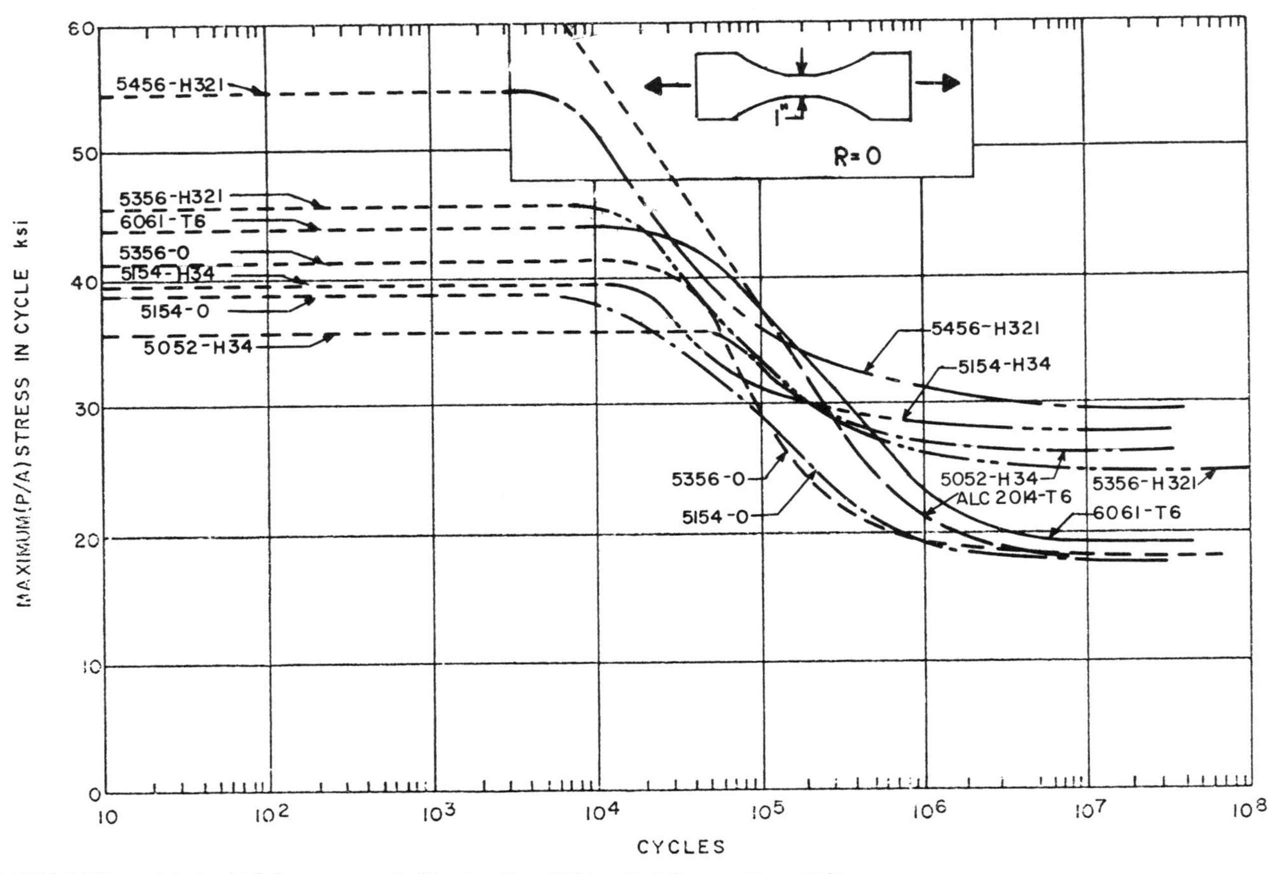

Fig. 62 5456-H321 unnotched axial fatigue compared with other alloys (3/8 in. plate). Source: Alcoa, 1961

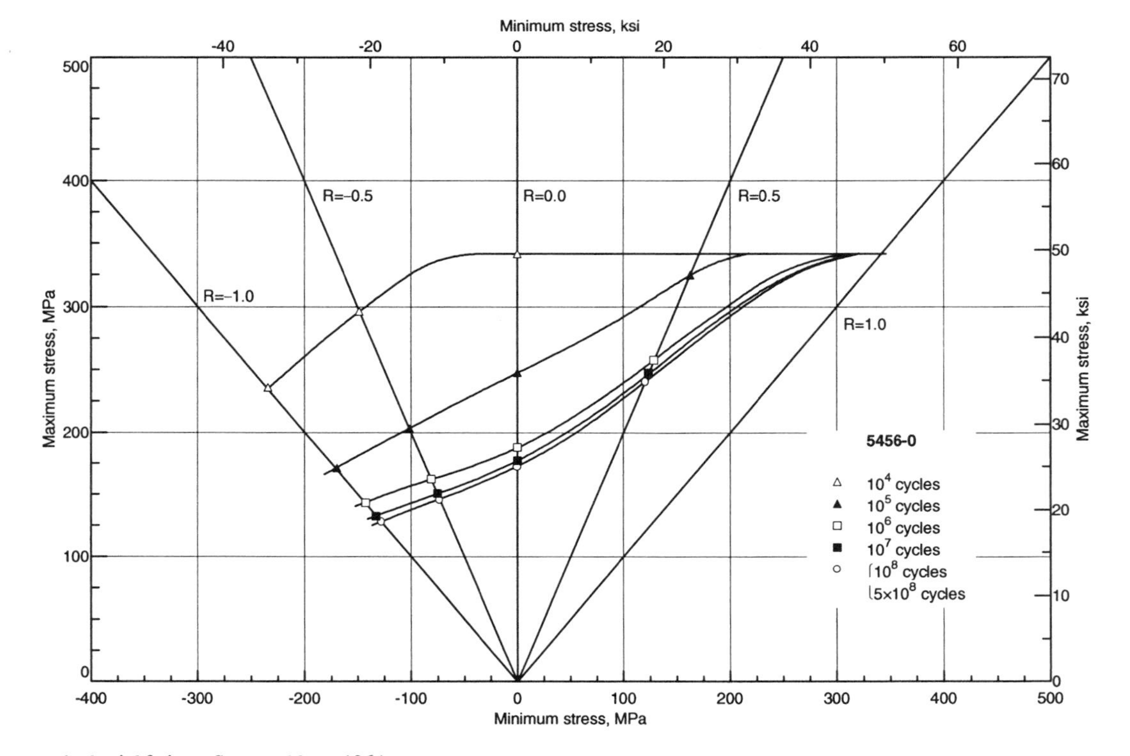

Fig. 63 5456-O unnotched axial fatigue. Source: Alcoa, 1964

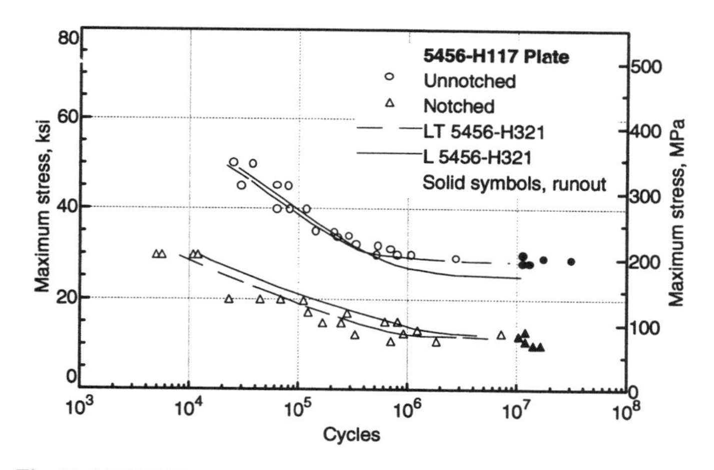

Fig. 64 5456-H117 notched ($K_t = 3$, $r = 0.013$ in.) and unnotched axial fatigue ($R = 0$). Solid symbols indicate runout (no failure). Data points are for 5456-H117 plate (0.75 and 1 in. thick). Curve lines are for 5456-H321 plate (1 in. thick). Unnotched axial specimens with 9-7/8 in. surface radius and 0.300 in. minimum diameter. Notched axial specimens with 0.253 in. diameter at the notch and a 0.303 in. diam outside the 60° notch.

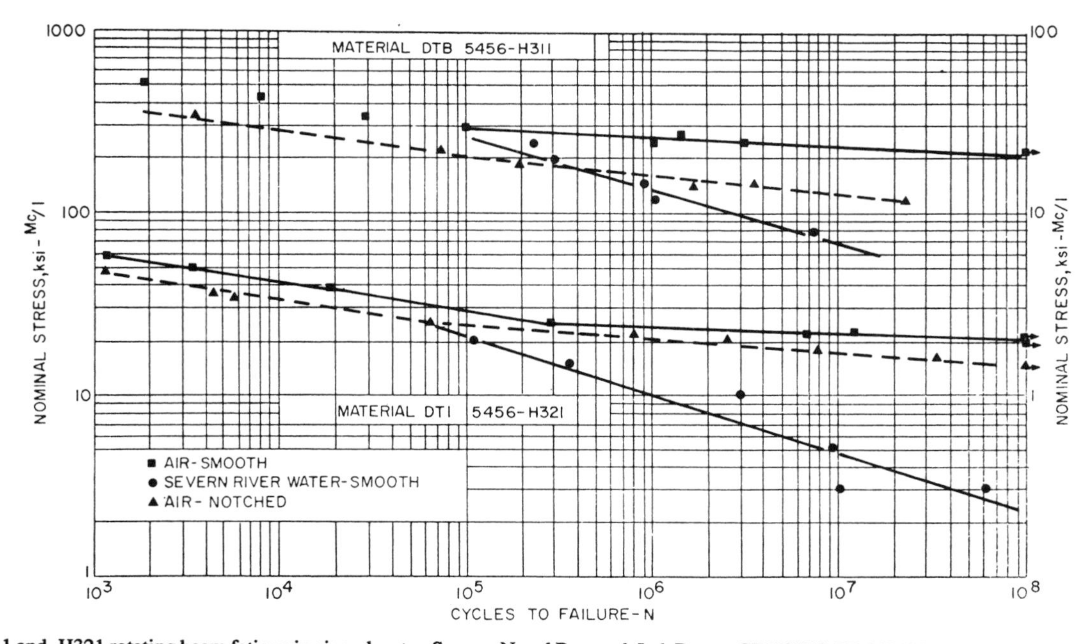

Fig. 65 5456-H311 and -H321 rotating beam fatigue in air and water. Source: Naval Research Lab Report CDNSWC-TR619409, 1994

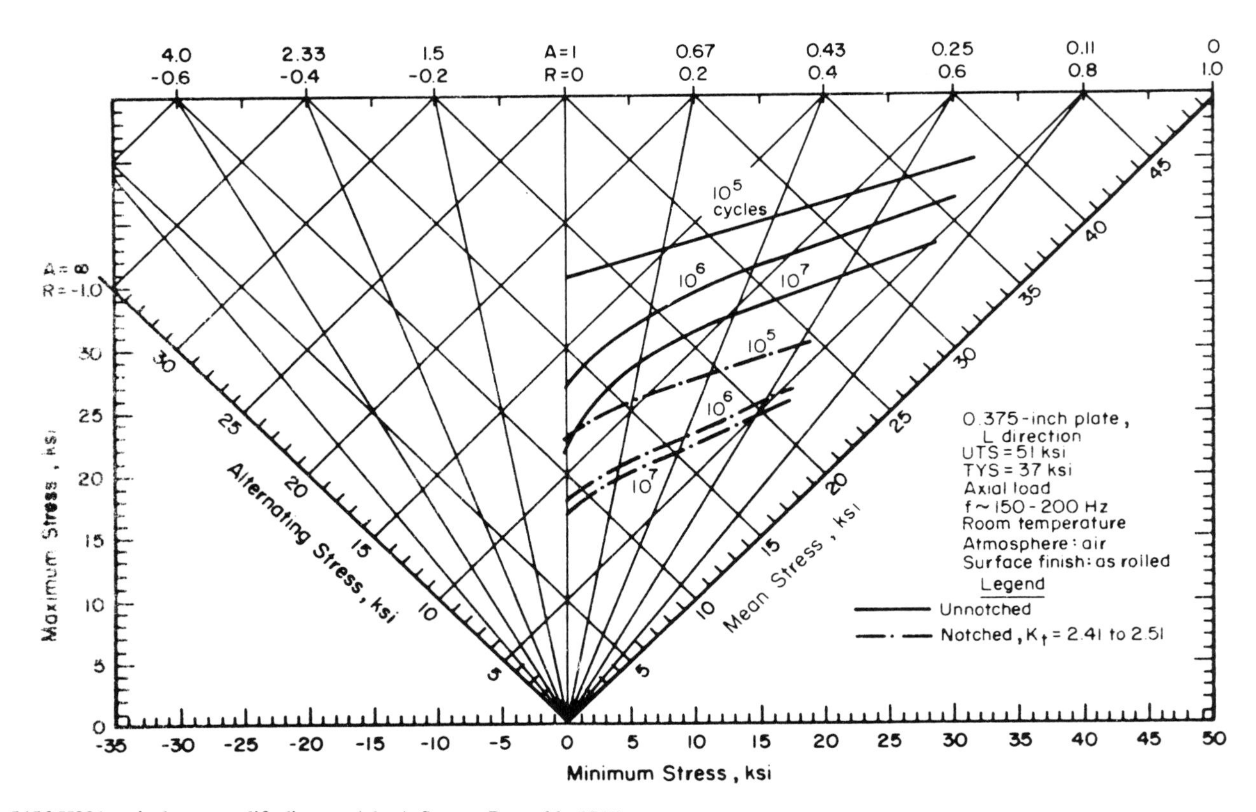

Fig. 66 5456-H321 typical constant life diagram (plate). Source: Reynolds, 1968

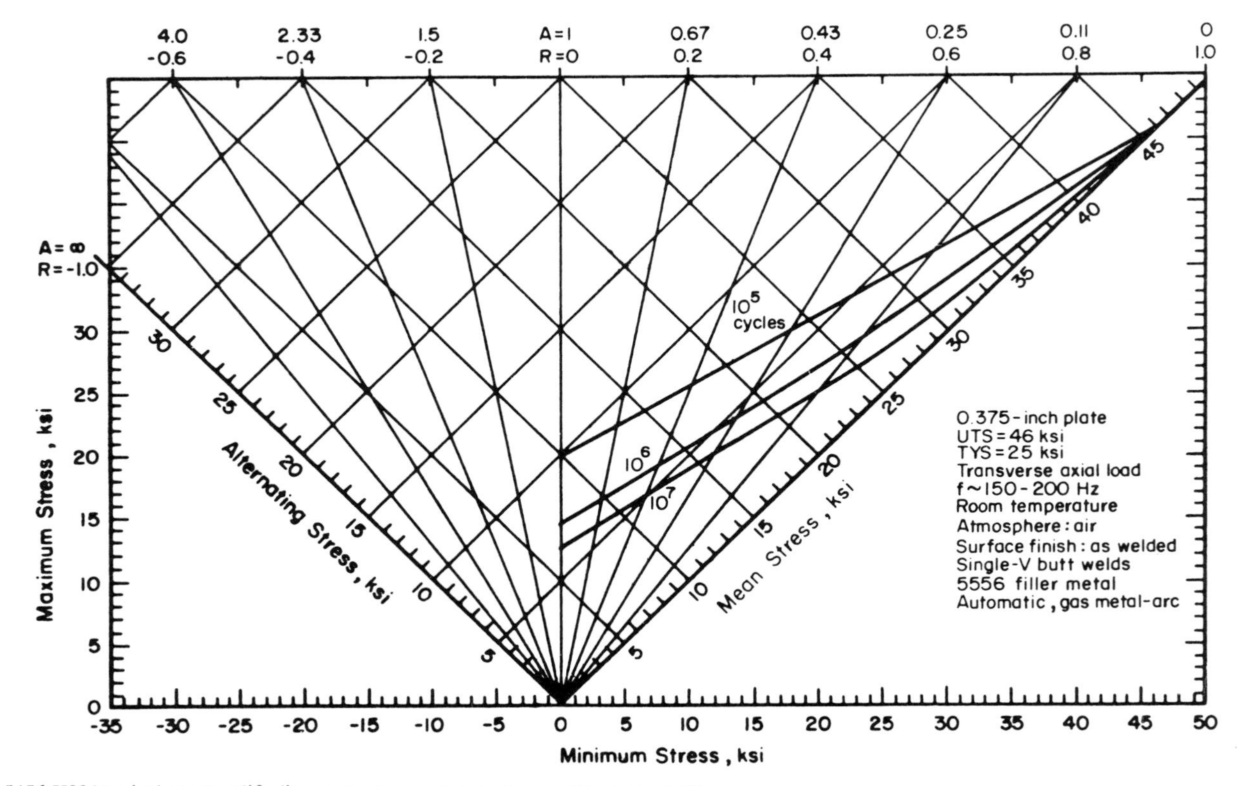

Fig. 67 5456-H321 typical constant life diagram butt welded plate. Source: Reynolds, 1968

5456-H112 and -H311: Room-temperature fatigue strength

Temper	Form and thickness	Specimen	Test mode	Stress ratio	Atmosphere	Fatigue strength (ksi) at cycles of:					
						10^4	10^5	10^6	10^7	10^8	5×10^8
H112	Plate	Flat plate	Axial	0	Air	(52)	(8)	(9)	(27.5)	(27)	(27)
	Plate	Round	Axial	0	Air	(55)	(39)	(29)	(29)	...	...
	Plate	Round, notched, $K_t = 3$	Axial	0	Air	(29)	(20)	(13)	(12)	...	...
H311	6 in.diam bar	.5 in round cantilever	Rotating beam	–1	Air	(40)	(30)	(26)	(23)	(21)	...
	6 in. diam bar	.5 in round cantilever	Rotating beam	–1	Severn River Water	(40)	(29)	(12)	(7)	...	
	6 in. diam bar	As above, but notched, $K_t = 3$	Rotating beam	–1	Air	(29)	(21)	(17)	(12)	...	...
	Plate	Flat plate	Axial	0	Air	(52)	(38)	(29)	(27.5)	(27)	(27)

Sources: Unpublished Alcoa data and M.G. Vassilaros and E.J. Czyryca, Naval Research Lab Report CDNSWC-TR619409, A Compilation of Fatigue Information for Aluminum Alloys, 1994

Alloy Name: 5456

5456-H32 and -H321: Fatigue strength

Form	Specimen	Test mode	Stress ratio	Temperature °C (°F)	Atmosphere	Fatigue strength (ksi) at cycles of:					
						10^4	10^5	10^6	10^7	10^8	5×10^8
H321 plate	.30 in. diam round	Axial	0	(75 °F)	Air	...	(42)	(36)	(35)	...	...
H321 plate	.30 in. diam round	Axial	0	(–320 °F)	Air	...	(47)	(40)	...	...	...
H321 plate	As above but welded bead-off 5556 filler	Axial	0	(75 °F)	Air	(44)	(30)	(25)	...	...	...
H321 plate	As above but welded bead-off 5556 filler	Axial	0	(–320 °F)	Air	(50)	(37)	(29)	...	...	...
H32 plate ⅞ in.	0.330 in. diam round	Axial	0	(75 °F)	Air	...	(40)	(31)	(31)	(30)	...
H32 plate ⅞ in.	0.330 in. diam round	Axial	0	(–320 °F)	Air	...	(47)	(39)	...	...	...
H32 plate ⅞ in.	As above but welded, bead-off	Axial	0	(75 °F)	Air	(33)	(30)	(21)	(20)	...	...
H32 plate ⅞ in.	As above but welded, bead-off	Axial	0	(–320 °F)	Air	(40)	(37)	(29)	...	...	...
⅞ in. plate	0.330 in. diam round	Torsion	0	(RT)	Air	...	(32)	(28)	(26)	(25)	...
⅞ in. plate	0.330 in. diam round	Torsion	–1	(RT)	Air	...	(22)	(19)	(17)	(16)	...
0.375 in. plate	Butt-weld, bead-off 5556 filler	Axial	0	(RT)	Air	...	...	(18)	(17)	...	...
0.375 in. plate	Butt-weld, bead-off 5556 filler	Axial	0	(RT)	Air	...	(20)	(14)	(12)	...	...
0.500 in. plate	Butt weld, bead-off 5556 filler, double-V	Axial	0	(RT)	Air	...	...	(20)	(16)	...	...
0.500 in. plate	Butt-weld, bead-on, 5556 filler, double-V	Axial	0	(RT)	Air	...	(17)	(11)	(9)	...	...
0.375 in. plate	Butt-weld, single-V longitudinal (5556 filler), bead-on	Axial	0	(RT)	Air	...	(20.5)	(12)	(11)	...	...
0.375 in. plate	Butt-weld, single-V longitudinal (5556 filler), bead-on	Axial	0	(RT)	Air	...	(20)	(14)	(13)	...	...
0.375 in. plate	Butt-weld, double-V longitudinal (5556 filler), bead-on	Axial	0	(RT)	Air	...	...	(14)	(12)	...	...
0.375 in. plate	Butt-weld, double-V transverse (5556 filler), bead-on	Axial	0	(RT)	Air	...	(18)	(14)	(12)	...	...
⅜ in. plate	Flat beam, transverse	Repeated bending	–1	(75 °F)	...	...	(31)	(19)	(18)	(17)	...
⅜ in. plate	As above but butt weld bead-off, 5556 filler	Repeated bending	–1	(75 °F)	...	(33.5)	(24)	(15)	(13)	(12)	...
⅜ in. plate	As above	Repeated bending	–1	(–300 °F)	...	...	(18)	(21)	...	...	...
⅜ in. plate	As above but bead-on	Repeated bending	–1	(75 °F)	...	(27)	(15)	(9)	(7)	(7)	...
⅜ in. plate	As above	Repeated bending	–1	(–300 °F)	...	(29)	(21)	(13)	(8)	...	...
⅜ in. plate	Butt-weld, bead-on single-V 5356 filler	Axial	0	(RT)	Air	(34)	(21)	(11)	(11)	...	...
⅜ in. plate	Smooth plate	Axial	0	(RT)	Air	(50)	(36)	(31)	(29)	(29)	...
Plate	Welded non-load carrying appendage	Axial	0	(RT)	Air	...	(15)	(6)	...	...	...
Plate	Welded, bead-on 5556 filler	Axial	0	(RT)	Air	(26)	(16)	(10)	(10)	...	...
Plate	Welded to cast 355-T6 (4043 filler)	Axial	0	(RT)	Air	(26)	(17)	(9)	(8.5)	...	...

(continued)

5456-H32 and -H321: Fatigue strength *(continued)*

Form	Specimen	Test mode	Stress ratio	Temperature °C (°F)	Atmosphere	Fatigue strength (ksi) at cycles of: 10^4	10^5	10^6	10^7	10^8	5×10^8
3 in. plate	0.50 in. diam round cantilever	Reverse bending	–1	(RT)	Air	(40)	(30)	(22)	(21)	(20)	...
3 in. plate	As above	Reverse bending	–1	(RT)	Severn River Water	...	(20)	(10)	(5)	(2)	...
3 in. plate	As above but notched K_t = 3.0	Reverse bending	–1	(RT)	Air	(32)	(23)	(20)	(17)	(16)	...
Plate	Smooth	Axial	0	(RT)	Air	(52.5)	(42)	(36)	(35)	(34)	(34)
0.4 in. plate	Longitudinal weld bead-on, as welded (5556 filler)	Axial	.5	(RT)	Air	...	...	(22)	(15)	...	...
3/8 in. plate	Longitudinal weld bead-on (5556 filler) but stress relieved	Axial	.5	(RT)	Air	...	(30)	(23)	(17)	...	...
3/8 in. plate	As above but as welded	Axial	0.0	(RT)	Air	...	...	(12)	(8)	...	...
3/8 in. plate	As above but stress relieved	Axial	0.0	(RT)	Air	...	...	(15)	(11)	...	...
3/8 in. plate	As above but as welded	Axial	–.5	(RT)	Air	...	(15)	(8)	(6)	...	...
3/8 in. plate	As above but stress relieved	Axial	–.5	(RT)	Air	...	...	(12)	(9)	...	...
3/8 in. plate	As above but shot-peened	Axial	0	(RT)	Air	...	(23)	(18.5)	(16)	...	...
3/8 in. plate	As above but hammer-peened	Axial	0.5	(RT)	Air	...	...	(23)	(15.5)	...	...
3/8 in. plate	Double strap lap joint with transverse fillet welds, 5556 filler	Axial	0	(RT)	Air	...	(10)	(7)	...	...	...
3/8 in. plate	As above, peened	Axial	0	(RT)	Air	...	...	(11)	...	...	...
3/8 in. plate	Double strap lap joint with longitudinal fillet welds, 5556 filler	Axial	0	(RT)	Air	...	(7)	(4)	...	...	...
3/8 in. plate	As above, peened	Axial	0	(RT)	Air	...	(9)	(4)	...	...	...
3/8 in. plate	As above, thermal stress relieved	Axial	0	(RT)	Air	...	(6.5)	(4)	...	...	...
3/8 in. plate	Transverse double-V butt-weld, 5556 filler, bead-on	Axial	0.25	(RT)	Air	...	(20.5)	(14.5)	(13)	...	...
3/8 in. plate	Transverse single-V butt-weld, 5556 filler, bead-on	Axial	0.25	(RT)	Air	...	(23)	(17)	...	...	...
3/8 in. plate	Transverse single-V butt-weld, 5556 filler, bead-on	Axial	0.5	(RT)	Air	...	(30)	(21)	(18)	...	...
3/8 in. plate	Transverse single-V butt-weld, 5556 filler, bead-on	Axial	0	(RT)	Air	(32)	(21)	(13)	(11)	...	...
3/8 in. plate	As above, bead-off	Axial	0	(RT)	Air	(38)	(29)	(20)	(15)	...	...
1/4 in. plate	Transverse single-V butt-weld, 5556 filler, bead-on	Axial	0	(RT)	Air	(26)	(16)	(10)	(10)	...	...
1/4 in. plate	As above, bead-off	Axial	0	(RT)	Air	...	(17)	...	...	...	...
1/2 in. plate	Transverse double-V butt-weld, 5556 filler, bead-on	Axial	0.25	(RT)	Air	...	(22)	(12)	(10)	...	...
Plate	RR Moore	Rotating beam	–1	(RT)	Air	(43)	(32)	(26)	(23)	(23)	(22)
Plate	RR Moore, notched, K_t = 12	Rotating beam	–1	(RT)	Air	(25)	(14)	(9)	(7)	(6)	(6)
Plate	Round	Axial	0	(RT)	Air	(54)	(40)	(35)	(34)	(33)	(33)

Sources: Unpublished Alcoa data and M.G. Vassilaros and E.J. Czyryca, Naval Research Lab Report CDNSWC-TR619409, A Compilation of Fatigue Information for Aluminum Alloys, 1994

5456-H343

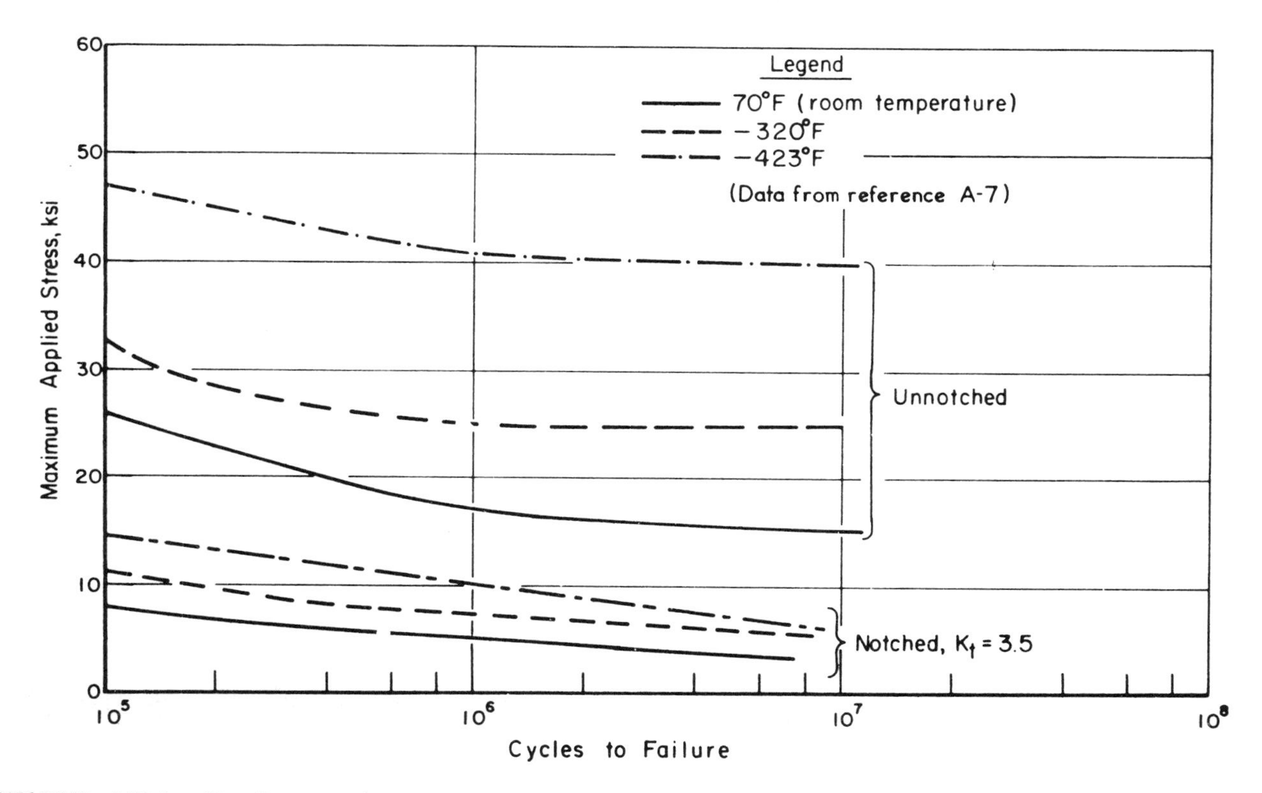

Fig. 68 5456-H343 axial fatigue ($R = -1$) at cryogenic temperatures. Source: Reynolds, 1968

5456-H343: Fatigue strength in air at low temperatures

Form and thickness	Specimen	Test mode	Stress ratio	Temperature °C (°F)	Fatigue strength (ksi) at cycles of: 10^4	10^5	10^6	10^7	10^8	5×10^8
0.125 in. sheet	Flat sheet	Axial	0.1	(RT)	...	(44)	(30)	(25)	...	...
0.100 in. sheet	Flat sheet	Axial	−1	(RT)	...	(26)	(17)	(15)	...	...
0.100 in. sheet	Flat sheet	Axial	−1	(−320 °F)	...	(32)	(26)	(25)	...	...
0.100 in. sheet	Flat sheet	Axial	−1	(−423 °F)	...	(47)	(41)	(40)	...	...
0.100 in. sheet	Flat sheet notched	Axial	−1	(RT)	...	(8)	(5)	(3)	...	...
0.100 in. sheet	Flat sheet notched	Axial	−1	(−320 °F)	...	(11)	(8)	(5)	...	...
0.100 in. sheet	Flat sheet notched	Axial	−1	(−423 °F)	...	(14)	(10)	(6)	...	...
0.100 in. sheet	Transverse butt weld, 5556 filler, as-welded	Axial	−1	(RT)	(24)	(16)	(11)	(10)	...	...
0.100 in. sheet	Transverse butt weld, 5556 filler, as-welded	Axial	−1	(−320 °F)	...	(21)	(13)	(12)	...	...
0.100 in. sheet	Transverse butt weld, 5556 filler, as-welded	Axial	−1	(−423 °F)	...	(28)	(23)	(18)	...	...

Source: M.G. Vassilaros and E.J. Czyryca, Naval Research Lab Report CDNSWC-TR619409, A Compilation of Fatigue Information for Aluminum Alloys, 1994

6xxx Alloys

6009

6009 sheet: Room-temperature fatigue strength in various environments

Temper	Specimen	Test mode	Stress ratio	Atmosphere	Fatigue strength (ksi) at cycles of:					
					10^4	10^5	10^6	10^7	10^8	5×10^8
T4	Unnotched sheet	Axial	0.1	Air	...	(36)	(26)	(23)	...	...
	Unnotched sheet	Axial	0.1	3.5% NaCl	...	(30)	(23)	(19)	...	...
	Notched, $K_t > 12$	Axial	0.1	Air	(26)	(13)	(7)	(5)	...	...
	Notched, $K_t > 12$	Axial	0.1	3.5% NaC	(22)	(11)	(5)	(4)	...	...
T62	Unnotched sheet	Axial	0.5	Air	...	...	(40)	(37)	...	...
	Unnotched sheet	Axial	0.5	3.5% NaCl	...	(45)	(35)	(30)	...	...
	Unnotched sheet	Axial	0.25	Air	...	(41)	(33)	(30)	...	...
	Unnotched sheet	Axial	0.25	3.5% NaCl	...	(41)	(25)	...	...	...
	Unnotched sheet	Axial	0.1	Air	...	(36)	(26)	(24)	...	...
	Unnotched sheet	Axial	0.1	3.5% NaCl	...	(25)	(13)	(9)	...	...
T62	Notched, $K_t = 3$	Axial	0.1	Air	...	(19)	(12)	(10)	...	...
	Notched, $K_t = 3$	Axial	0.1	3.5% NaCl	...	(12)	...	...	...	...
	Notched, $K_t = 4.5$	Axial	0.1	Air	(35)	(13)	(10)	(8)	...	...
	Notched, $K_t > 12$	Axial	0.1	Air	(33)	(12)	(8)	(7)	...	...
	Notched, $K_t > 12$	Axial	0.1	3.5% NaCl	(22)	(10)	(5)	(4)	...	...
	Unnotched sheet	Axial	–0.5	Air	(42)	(30)	(20)	(18)	...	...
	Unnotched sheet	Axial	–0.5	3.5% NaCl	...	(24)	(11)	...	...	...

Sources: Unpublished Alcoa data

6010 and 6013 sheet

6010-T62 sheet: Room-temperature fatigue strength in various environments

Specimen	Test mode	Stress ratio	Atmosphere	Fatigue strength (ksi) at cycles of:					
				10^4	10^5	10^6	10^7	10^8	5×10^8
Sheet, unnotched	Axial	0.1	Air	...	(37)	(23)	(21)	...	...
Sheet, unnotched	Axial	0.1	3.5% NaCl	...	(25)	(14)	...	...	...
Sheet, notched, $K_t = 3$	Axial	0.1	Air	(30)	(21)	(14)	(12)	...	...
Sheet, notched, $K_t = 3$	Axial	0.1	3.5% NaCl	...	...	(11)	(10)	...	...
Sheet, notched, $K_t \geq 12$	Axial	0.1	Air	(24)	(14)	(10)	(8)	...	...
Sheet, notched, $K_t \geq 12$	Axial	0.1	3.5% NaCl	(21)	(10)	(7)	...	...	...

Sources: Unpublished Alcoa data

6013-T62 sheet: Room-temperature fatigue strength in various environments

Specimen	Test mode	Stress ratio	Atmosphere	Fatigue strength (ksi) at cycles of:					
				10^4	10^5	10^6	10^7	10^8	5×10^8
Unnotched sheet	Axial	0.1	Air	...	(39)	(30)	(26)	...	...
Unnotched sheet	Axial	0.1	3.5% NaCl	...	(31)	(14)	...	...	...
Notched, $K_t = 3$	Axial	0.1	Air	...	(21)	(15)	(14)	...	...
Notched, $K_t > 12$	Axial	0.1	Air	(25)	(13)	(9)	(8)	...	...
Notched, $K_t > 12$	Axial	0.1	3.5% NaCl	...	(10)	(7)	(5)	...	...

Sources: Unpublished Alcoa data

6053- T6

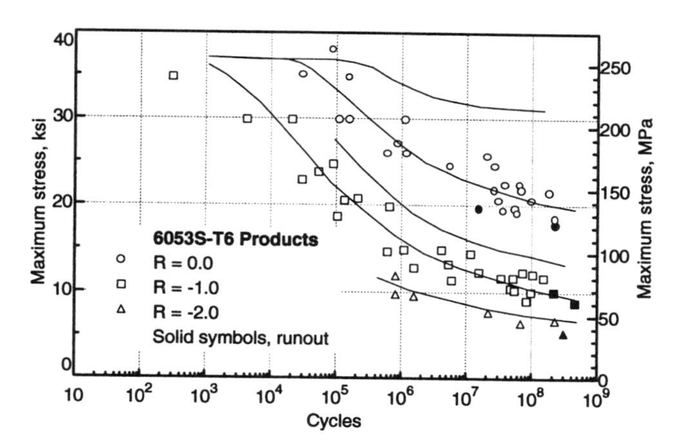

Fig. 69 6053-T6 unnotched axial fatigue at room temperature. Unnotched axial specimens with 9-7/8 in. surface radius and 0.200 in. minimum diameter. Source: Alcoa

6053-T6 wire: Unnotched axial fatigue in room-temperature air

Specimen	Stress ratio	Fatigue strength (ksi) at cycles of: 10^4	10^5	10^6	10^7	10^8	5×10^8
Unnotched round	0.5	...	...	(34)	(32)	(31)	(31)
	0.0	(37)	(33)	(27)	(23)	(21)	(20)
	–0.5	...	(27)	(20)	(17)	(15)	(14)
	–1.0	(30)	(22)	(16)	(13)	(11)	(10)
	–2.0	...	...	(11)	(9)	(8)	(7)

Sources: Unpublished Alcoa data

6061

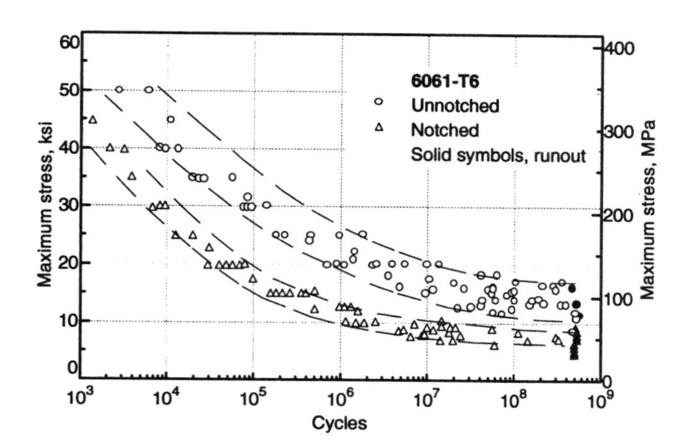

Fig. 70 6061-T6 notched (radius at notch root <0.001 in.) and unnotched rotating beam fatigue at room temperature. Solid symbols indicate runout (no failure). Longitudinal and transverse specimens from extruded bar (5/8 x3.5 in.), rolled-and-drawn rod (0.75 in.), and rolled plate (1.25 in. thick). R.R. Moore specimens with 9-7/8 in. surface radius and 0.300 in. minimum diameter for unnotched specimens. Notched specimens had a 0.330 in. diameter at the notch and a 0.480 in. diameter outside the 60° notch. Source: Alcoa, 1960

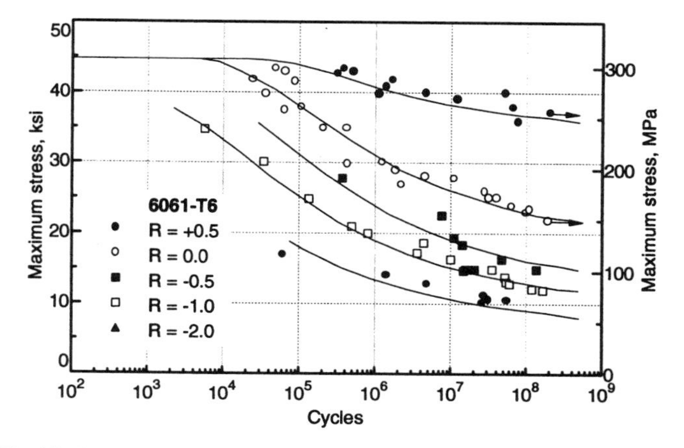

Fig. 71 6061-T6 unnotched axial fatigue at room temperature. Typical tensile strength, 310 MPa (45 ksi). Unnotched axial specimens with 9-7/8 in. surface radius and 0.200 in. minimum diameter. Source: Alcoa

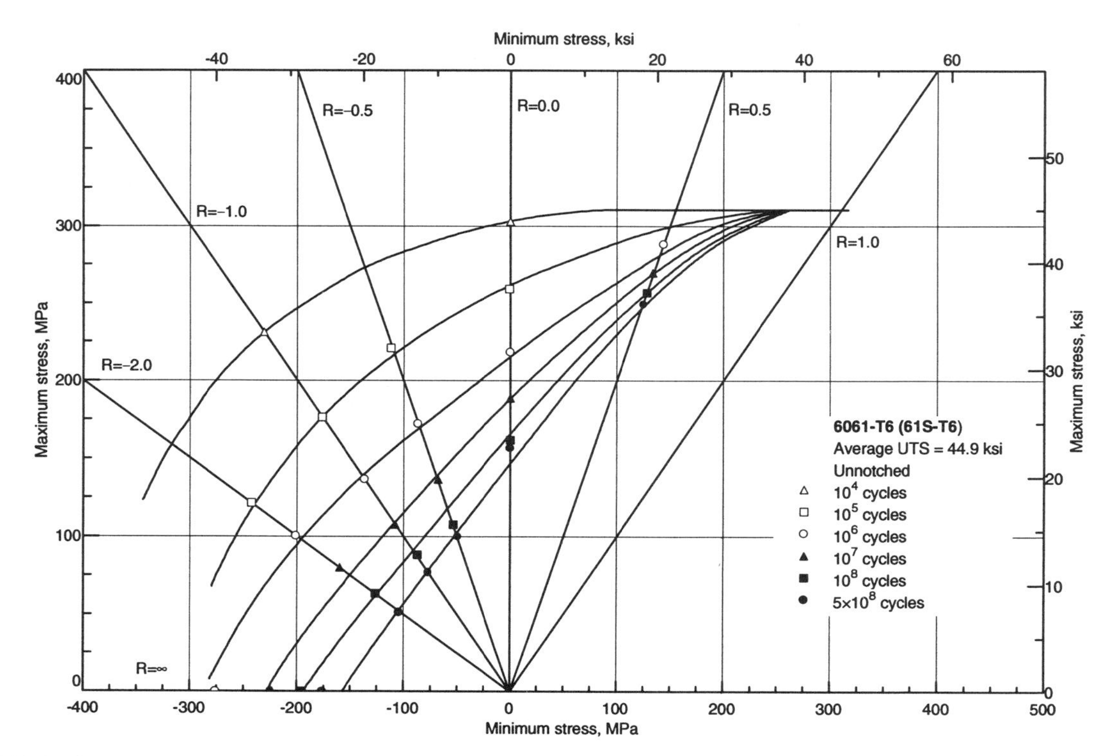

Fig. 72 6061-T6 (61S-T6) unnotched axial fatigue. Source: Alcoa

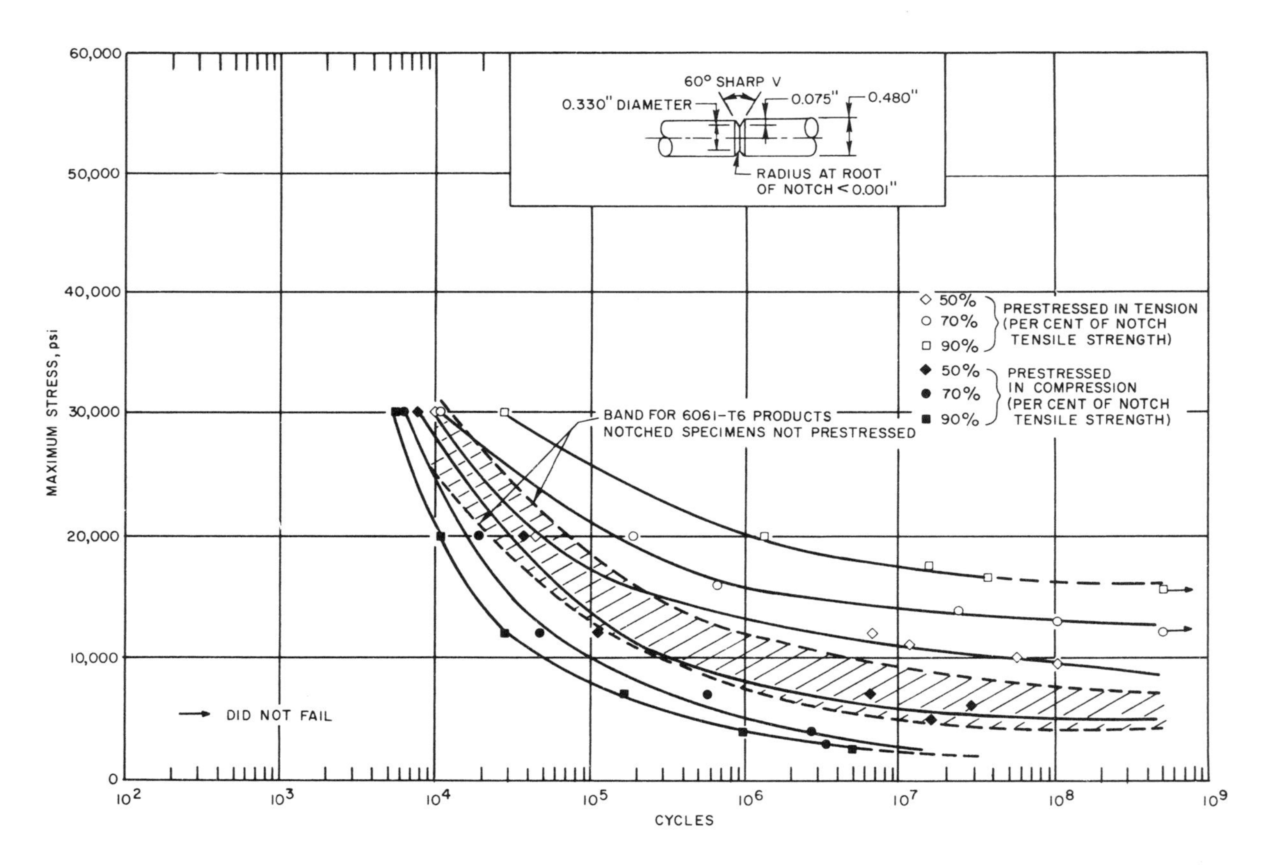

Fig. 73 6061-T6 rotating beam fatigue with residual stress. Source: Alcoa, 1968, reported in Naval Research Lab Report CDNSWC-TR619409, 1994

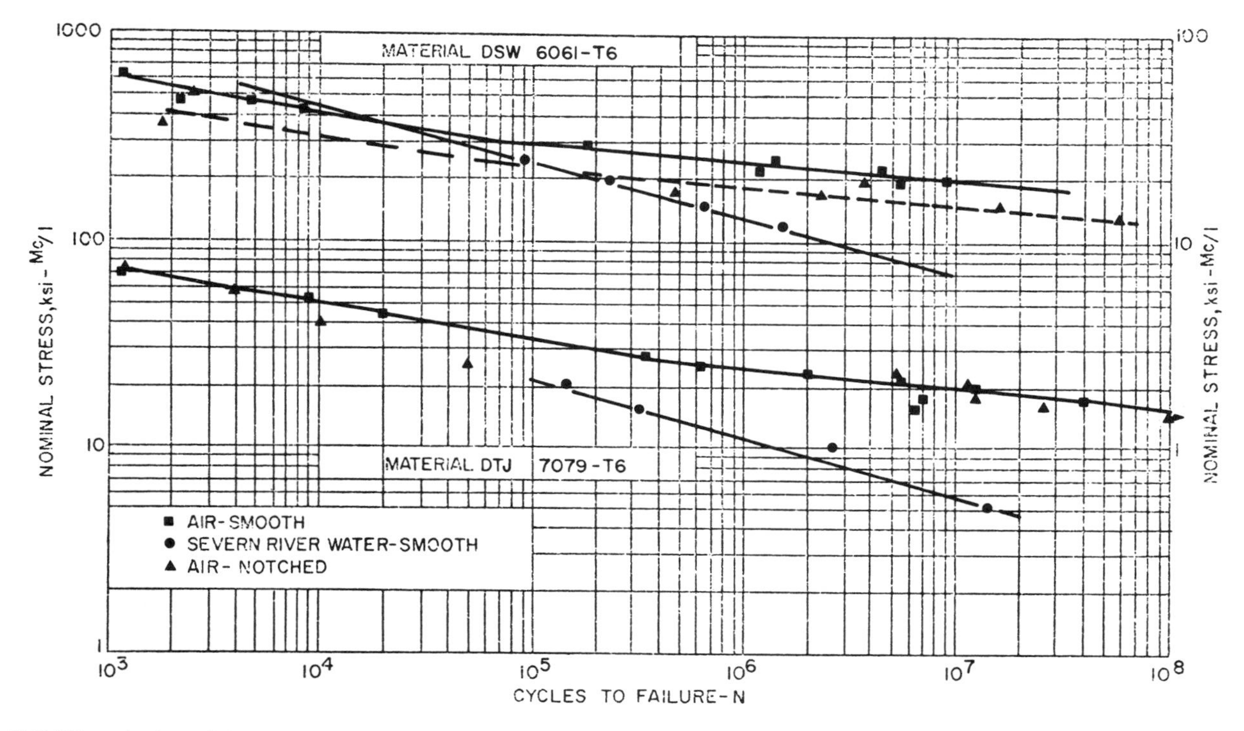

Fig. 74 6061-T6 rotating beam fatigue in air and water. Source: Naval Research Lab Report CDNSWC-TR619409, 1994

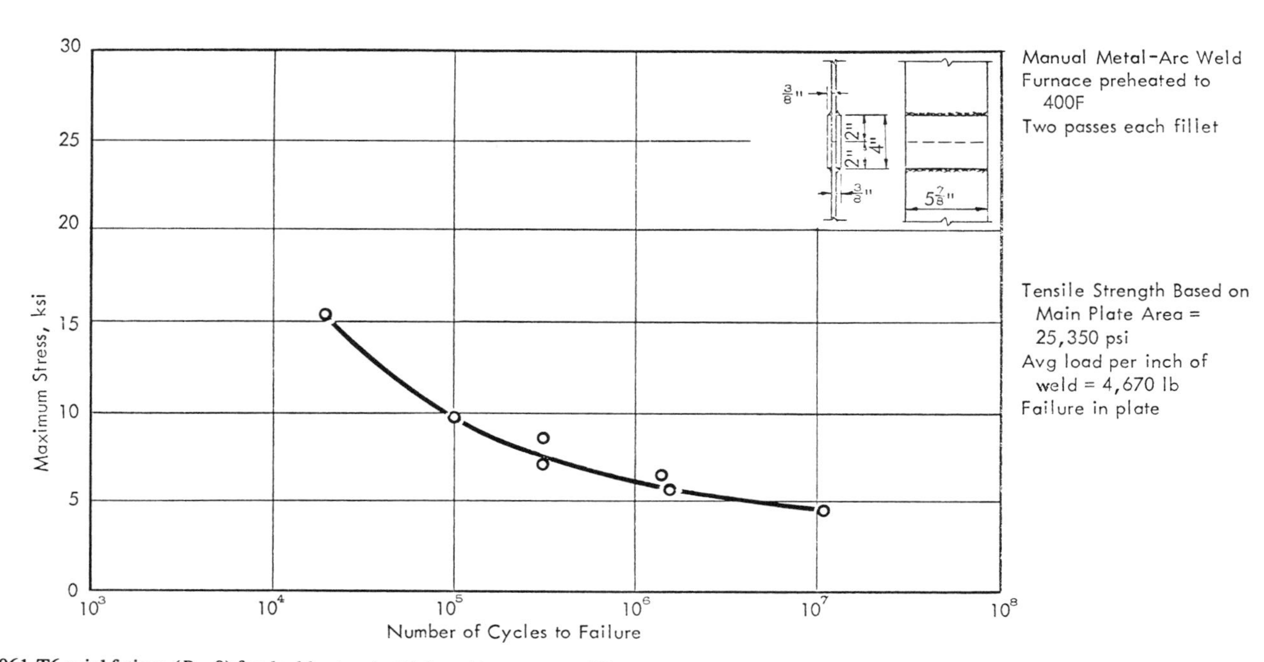

Fig. 75 6061-T6 axial fatigue ($R = 0$) for double strap butt joint with transverse fillet welds (4043 filler). Source: Kaiser, 1966

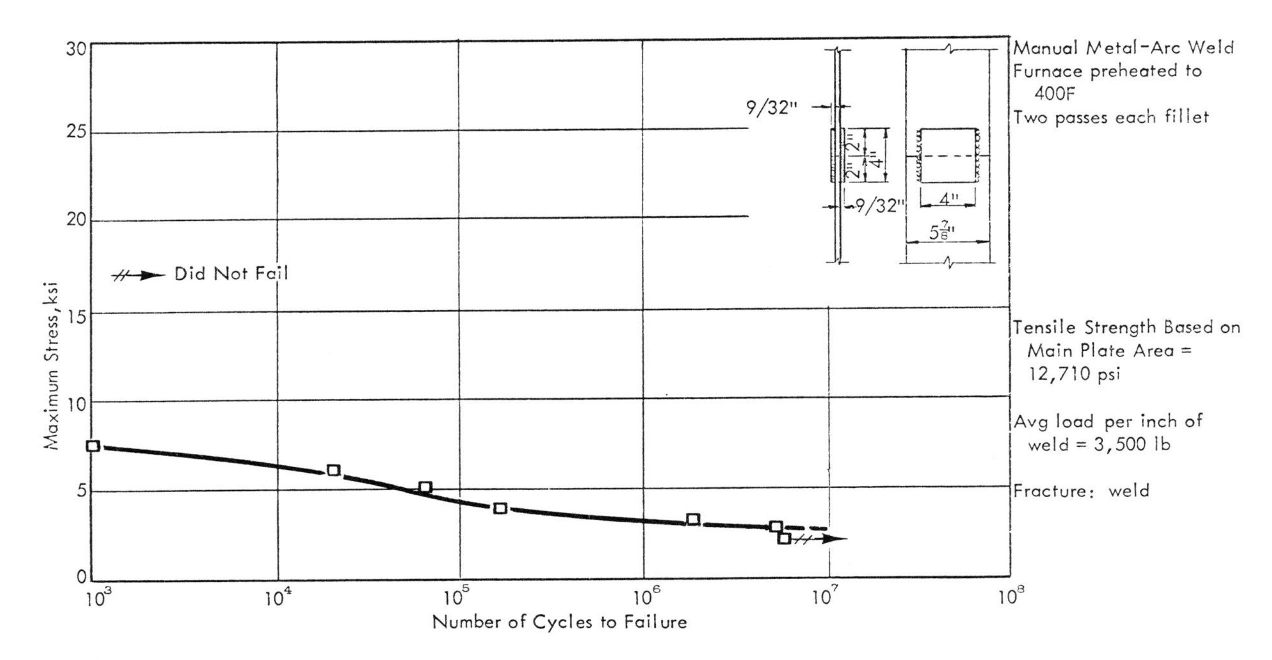

Fig. 76 6061-T6 axial fatigue ($R = 0$) for double strap butt joint with longitudinal fillet welds (4043 filler). Source: Kaiser, 1966

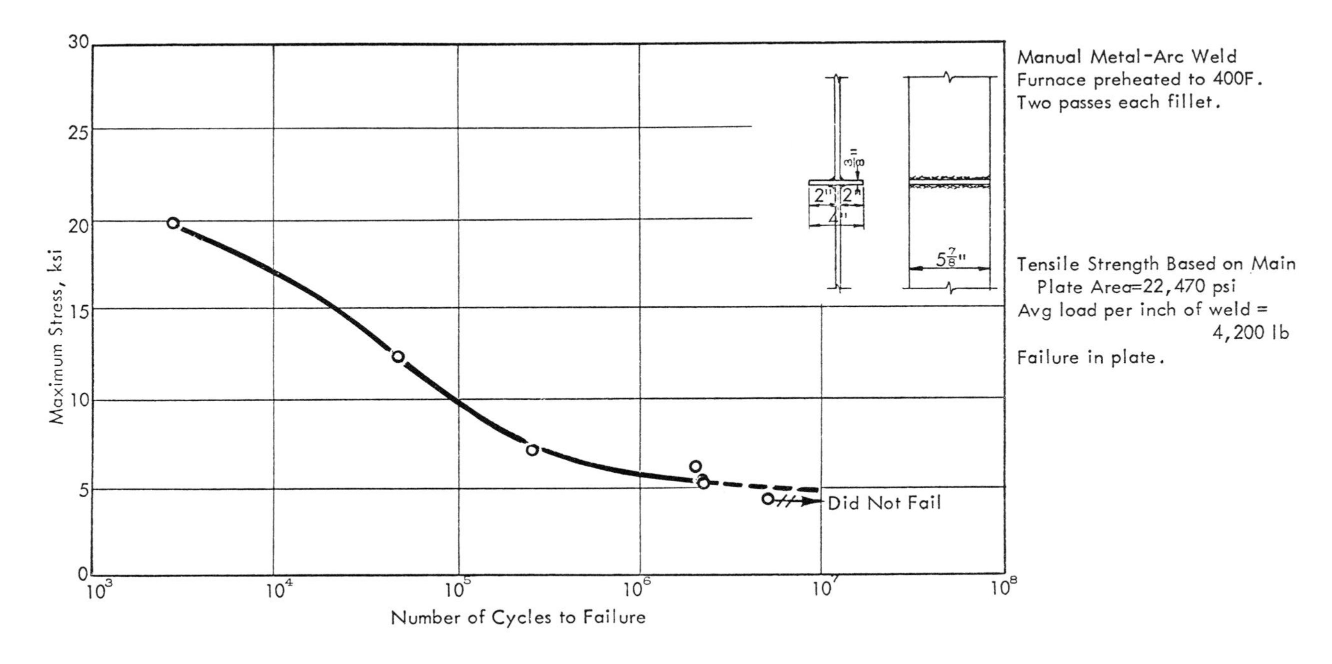

Fig. 77 6061-T6 axial fatigue ($R = 0$) for cruciform weld (4043 filler). Source: Kaiser, 1966

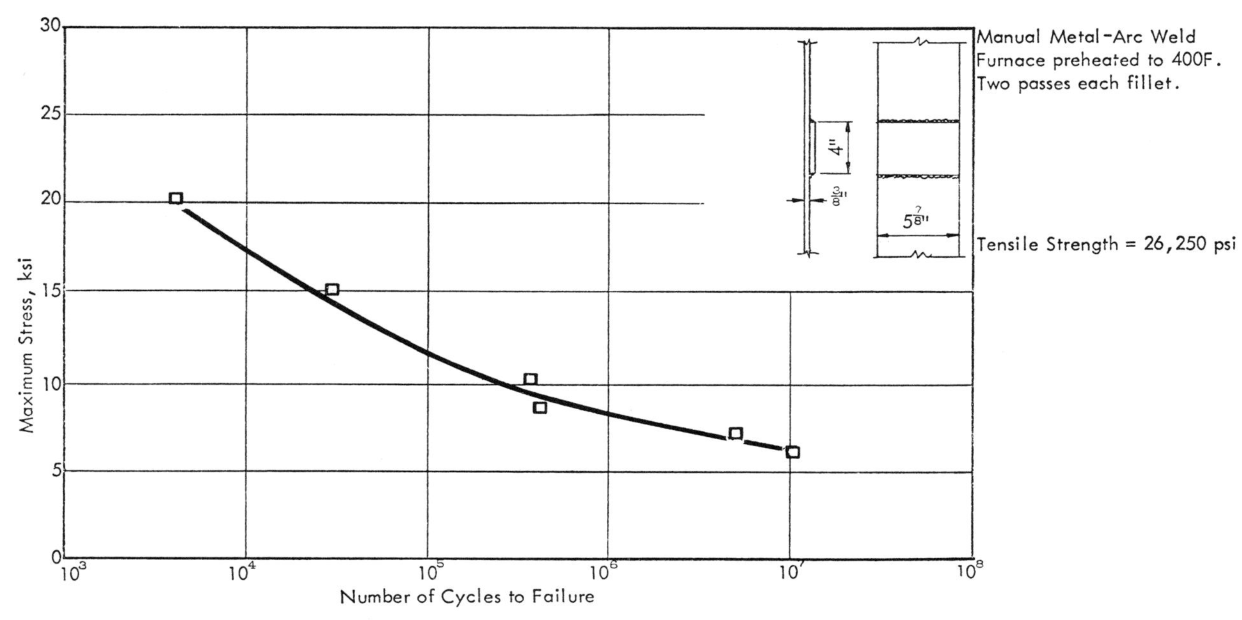

Fig. 78 6061-T6 axial fatigue ($R = 0$) with welded plate attachment (parallel attachment, transverse fillet weld with 4043 filler). Source: Kaiser, 1966

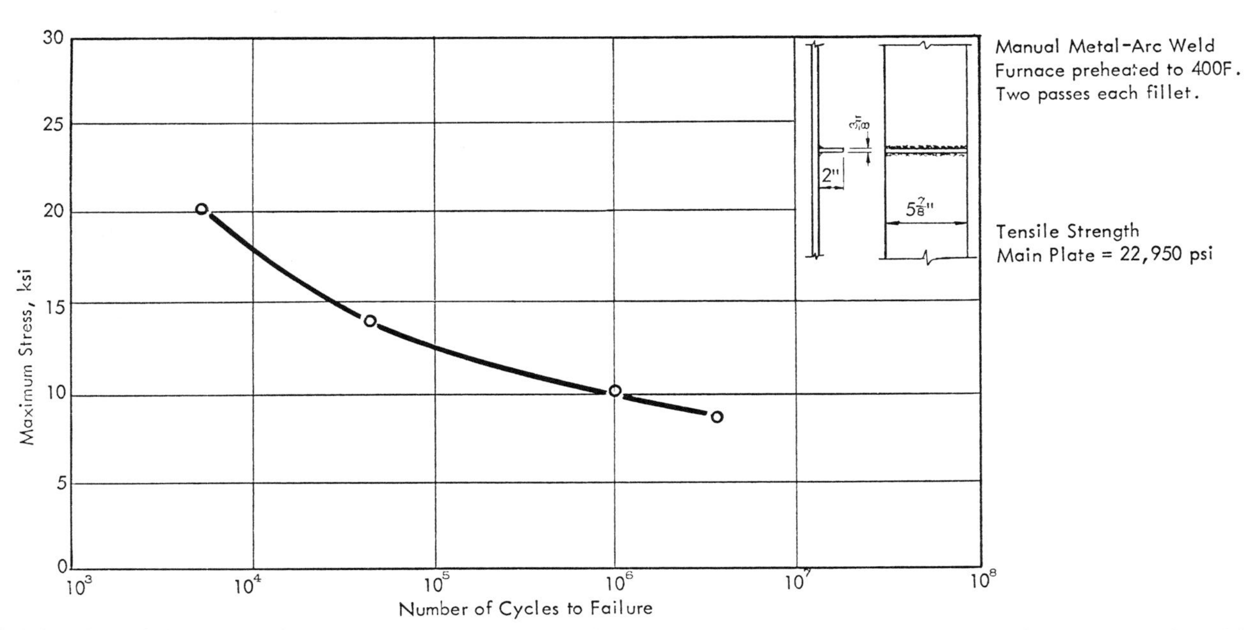

Fig. 79 6061-T6 axial fatigue ($R = 0$) with welded plate attachment (perpendicular attachment, transverse fillet weld with 4043 filler). Source: Kaiser, 1966

6061: Room-temperature fatigue strength in various environments

Temper	Form and thickness	Specimen	Test mode	Stress ratio	Atmosphere	Fatigue strength (ksi) at cycles of: 10^4	10^5	10^6	10^7	10^8	5×10^8
T4	Sheet	Sheet, unnotched	Axial	0.1	Air	...	(37)	(30)	(28)	...	...
	Sheet	Sheet, unnotched	Axial	0.1	3.5% NaCl	...	(31)	(16)	(9)	...	...
	Sheet	Sheet, notched, $K_t = 3$	Axial	0.1	Air	(30)	(20)	(13)	(13)	...	...
	Sheet	Sheet, notched, $K_t = 3$	Axial	0.1	3.5% NaCl	...	(16)	(13)	...	...	...
	Sheet	Sheet, notched, $K_t \geq 12$	Axial	0.1	Air	(23)	(14)	(9)	(8)	...	...
	Sheet	Sheet, notched, $K_t \geq 12$	Axial	0.1	3.5% NaCl	(22)	(1)	(6)	(5)	...	...
T6	Sheet	Sheet, unnotched	Axial	0.1	Air	...	(46)	(31)	(29)	...	...
	Sheet	Sheet, unnotched	Axial	0.1	3.5% NaCl	...	...	(23)	...	...	...
	Sheet	Sheet, notched, $K_t = 3$	Axial	0.1	3.5% NaCl	...	(14)	...	...	...	...
	Sheet	Sheet, notched, $K_t \geq 12$	Axial	0.1	Air	(26)	(16)	(10)	(8)	...	...
	Sheet	Sheet, notched, $K_t \geq 12$	Axial	0.1	3.5% NaCl	(21)	(10)	(7)	(6)	...	...
T4	3/4 in. plate	Butt welded with 4043 filler, machined round	Reverse bending	−1	Air	...	(21)	(16.5)	(14)	(13)	...
	3/4 in. plate	As above but 5056 filler	Reverse bending	−1	Air	...	(24)	(17)	(13)	(12.5)	...
	Rod	RR Moore	Rotating beam	−1	Air	...	(29)	(23)	(18)	(14)	(13)
	Rod	RR Moore, notched, $K_t = 12$	Rotating beam	−1	Air	(26)	(19)	(13)	(10)	(9)	(9)

Sources: Unpublished Alcoa data and M.G. Vassilaros and E.J. Czyryca, Naval Research Lab Report CDNSWC-TR619409, A Compilation of Fatigue Information for Aluminum Alloys, 1994

Alloy Name: 6061

6061-T6: Room-temperature fatigue strength in various environments

Form and thickness	Specimen	Test mode	Stress ratio	Atmosphere	Fatigue strength (ksi) at cycles of: 10^4	10^5	10^6	10^7	10^8	5×10^8
Plate	0.2 in. diam round polished	Axial	0	Air	(44)	(37)	(28)	(26)	(24)	...
¼ in. plate	Flat plate, 5 in. wide	Axial	0	Air	(40)	(35)	(28)	(24)	(22)	...
⅜ in. plate	Flat plate	Axial	0	Air	...	(30)	(17)	(14)	...	...
Plate	0.3 in. diam round	Reverse bending	−1	Air	...	(31)	(22.5)	(17.5)	(15)	(13)
⅜ in. plate	Flat plate	Reverse bending	−1	Air	...	(26.5)	(17.5)	(14)	(13)	...
¾ in. plate	RR Moore	Rotating beam	−1	Air	...	(32)	(24)	(17.5)	(13.5)	(12)
6 in. diam bar	½ in. diam cantilever beam	Reverse bending	−1	Air	(42)	(30)	(24)	(20)	(17)	...
6 in. diam bar	½ in. diam cantilever beam	Reverse bending	−1	Severn River Water	...	(23)	(13)	(7)	...	...
6 in. diam bar	As above, notched, $K_t = 3$	Reverse bending	−1	Air	(32)	(23)	(18)	(15)	(12)	...
0.125 in. sheet	Flat sheet	Reverse bending	−1	Air	...	(23)	(18)	(15)	(13)	...
¼ in. plate	Flat plate, 7 ½ in. wide, central hole	Axial	0	Air	(30)	(18)	(14)	(10)	(8)	...
¼ in. plate	Flat plate, 7 in. wide, 2 rows (4) 5/8 in. rivets	Axial	0	Air	(32)	(25)	(17)	(9.5)	...	...
⅜ in. plate	Flat plate	Axial	0	Air	(44)	(38)	(23)	(19)	...	...
¼ in. plate	Flat plate, 5 in. wide transverse butt weld, bead-on 4043 filler	Axial	0	Air	(23)	(16)	(11)	(8.5)	(8)	...
¼ in. plate	Flat plate, 5 in. wide, transverse butt weld, bead-on 4043 filler	Axial	−0.5	Air	...	(10)	(8.5)	(8)	(6)	...
¼ in. plate	Flat plate, 5 in. wide. transverse butt weld, bead-on 4043 filler	Axial	+0.5	Air	(26)	(25)	(16)	(13)	(12)	...
¼ in. plate	Flat plate, 5 in. wide. transverse butt weld, bead-on 4043 filler	Axial	+0.75	Air	(26)	(26)	(25)	(18)	(17)	...
3 in. plate	½ in. diam cantilever beam, transverse butt weld, bead-off	Reverse bending	−1	Air	(30)	(14)	(8)	(5)	...	...
3 in. plate	½ in. diam cantilever beam, transverse butt weld, bead-off	Reverse bending	−1	Severn River Water	...	(11)	(7)	(3)	...	...
⅜ in. plate	Flat plate, 6 in. wide transverse butt weld, 4043 filler bead-on	Axial	0	Air	(12)	(8)	(6)	(5.5)	...	...
⅜ in. plate	As above, bead-off	Axial	0	Air	(23)	(15)	(9)	(7)	...	...
¾ in. plate	0.03 in. diam round, transverse butt weld, bead-off	Reverse bending	−1	Air	(25)	(19)	(14.5)	(13)	(12)	...
⅜ in. plate	Flat plate, 6 in. wide, scalloped bead-on butt welds, 4043 filler	Axial	0	Air	(21)	(13)	(7)	(6)	...	...
⅜ in. plate	Flat plate, 5 ⅞ in. wide single strap butt joint, single transverse fillet weld, 4043 filler	Axial	0	Air	(4)	(2.5)	(2)	(1.5)	...	...
⅜ in. plate	Flat plate, 5 ⅞ in. wide single strap butt joint, single transverse fillet weld, 4043 filler	Axial	0	Air	(6)	(2.5)	(2)	(1.5)	...	...

(continued)

6061-T6: Room-temperature fatigue strength in various environments *(continued)*

Form and thickness	Specimen	Test mode	Stress ratio	Atmosphere	Fatigue strength (ksi) at cycles of: 10^4	10^5	10^6	10^7	10^8	5×10^8
3/8 in. plate	Flat plate, 5 7/8 in. wide single strap butt joint, single transverse fillet weld, 4043 filler	Axial	0	Air	(4)	(2.5)	(2)	(2)	...	...
3/8 in. plate	Flat plate, 6 in. wide single strap butt joint longitudinal fillet welds 4043 filler	Axial	0	Air	(3)	(2.5)	(1)	(0.5)	...	...
3/8 in. plate	Flat plate, 5 7/8 in. wide double strap butt joint, longitudinal fillet welds, 4043 filler	Axial	0	Air	...	(10)	(6)	(4.5)	...	...
3/8 in. plate	As above, but non-symmetrical	Axial	0	Air	(12)	(6)	(4)	(3)	(2)	...
3/8 in. plate	As above, but staggered	Axial	0	Air	(12)	(6)	(4)	(3)	(2)	...
3/8 in. plate	Flat plate, 5 7/8 in. wide double strap butt joint, longitudinal fillet welds, 4043 filler	Axial	0	Air	(6)	(4)	(3)	(2)	...	...
3/8 in. plate	Flat plate, 5 7/8 in. wide, cruciform joint, 4043 filler	Axial	0	Air	(17)	(10)	(6)	(5)	...	...
3/8 in. plate	Flat plate, 5 7/8 in. wide, parallel plate weld, transverse fillet welds, 4043 filler	Axial	0	Air	(17)	(12)	(8)	(6)	...	...
3/8 in. plate	Flat plate, 5 7/8 in. wide, perpendicular plate weld, transverse fillet weld, 4043 filler	Axial	0	Air	(18)	(12.5)	(10)	...	...	...
1/4 in. plate	Transverse square butt weld, 5056 filler, bead-on	Reverse plate bending	–1	Air	...	(20.5)	(13)	(8)	...	...
1/4 in. plate	As above, 4043 filler	Reverse plate bending	–1	Air	...	(21)	(15)	(10)	...	...
3/8 in. plate	Transverse single-V butt welded to 5083-H113, 5356 filler, bead-on	Axial	0	Air	(26)	(16)	(9.5)	(7.5)	...	...
3/8 in. plate	As above, bead-off	Axial	0	Air	...	(27)	(20)	(15)	...	...
3/8 in. plate	Longitudinal single-V butt weld, 4043 filler, bead-on	Axial	0	Air	...	(21)	(13)	(11)	...	...
3/4 in. plate	Square butt weld 4043 filler, machined round	Reverse bending	–1	Air	...	(21)	(16.5)	(14)	(12)	...
3/4 in. plate	As above, 5056 filler	Reverse bending	–1	Air	...	(22.5)	(16)	(14)	(13)	...
1/4 in. plate	Longitudinal square butt weld 4043 filler, bead-on	Reverse plate bending	–1	Air	...	(23)	(14)	(9.5)	...	...
1/4 in. plate	As above, bead-off	Reverse plate bending	–1	Air	...	(20)	(14)	(11)	...	...
1/4 in. plate	Longitudinal square butt weld, 5056 filler, bead-on	Reverse plate bending	–1	Air	...	(23)	(15)	(9)	...	...
1/4 in. plate	As above, bead-off	Reverse plate bending	–1	Air	...	(24.5)	(16)	(11)	...	...
1/4 in. plate	Transverse fillet welded lap joint to 5052-H32, 5356 filler	Axial	0	Air	(9)	(4.5)	(3.5)	(2)	...	...
1/4 in. plate	As above but welded to 5083-H113	Axial	0	Air	(11)	(7.5)	(4)	(2)	...	...
1/2 in. plate	Fillet welded I-beam, 4043 filler	4-point	0.2	Air	...	(27.5)	(17.5)	(7.5)	...	...
1/4 in. plate	Fillet-welded tee-joint to 5083-H321 base plate, 5356 filler	Axial	0	Air	(26)	(17.5)	(10)	(9)	...	...
Weld	RR Moore	Rotating beam	–1	Air	(45)	(32)	(22)	(17)	(14)	(13)
Weld	RR Moore, notched, $K_t = 12$	Rotating beam	–1	Air	(29)	(17)	(11)	(9)	(8)	(7)
Weld	Round	Axial	0.50	Air	...	...	(48)	(38)	...	...
			0.00		...	(39)	(31)	(25)	(22)	...
			–0.50		...	(31)	(24)	(19)	(15)	...
			–1.00		(33)	(25)	(20)	(16)	(12)	...

Sources: Unpublished Alcoa data and M.G. Vassilaros and E.J. Czyryca, Naval Research Lab Report CDNSWC-TR619409, A Compilation of Fatigue Information for Aluminum Alloys, 1994

6063

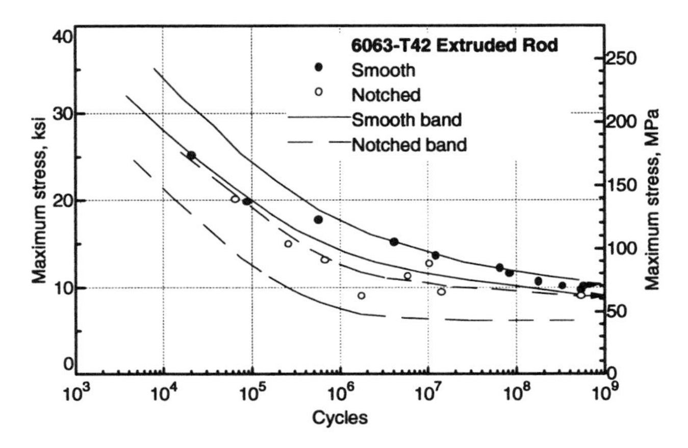

Fig. 80 6063-T42 notched (radius at notch root <0.001 in.) and unnotched rotating beam fatigue at room temperature (extruded rod, 3/4 in. diam). R.R. Moore specimens with 9-7/8 in. surface radius and 0.300 in. minimum diameter for unnotched specimens. Notched specimens had a 0.330 in. diameter at the notch and a 0.480 in. diameter outside the 60° notch. Source: Alcoa, 1953

6063: Room-temperature fatigue strength in air

Temper	Form and thickness	Specimen	Test mode	Stress ratio	Fatigue strength (ksi) at cycles of: 10^4	10^5	10^6	10^7	10^8	5×10^8
T63	3/8 in. plate	Transverse butt weld 4043 filler, bead-on	Reverse plate bending	−1	...	...	(14.5)	(8)	...	...
	3/8 in. plate	As above, 5056 filler	Reverse plate bending	−1	...	(17)	(11.5)	(8)	(6.5)	...
	0.188 in. sheet	Transverse butt weld 4043 filler, bead-on	Reverse plate bending	−1	...	...	(13)	(9.5)	(8)	...
	0.188 in. sheet	As above, 5056 filler	Reverse plate bending	−1	...	...	(13)	(9.5)	(8)	...
T42	Extrusion	RR Moore	Rotating beam	−1	...	(22)	(16)	(13)	(11)	(10)
		RR Moore, notched, $K_t = 12$	Rotating beam	−1	(25)	(16)	(10)	(8)	(8)	...
T6	Extrusion	RR Moore	Rotating beam	−1	(34)	(24)	(17)	(14)	(11)	(10)
		RR Moore, notched, $K_t = 12$	Rotating beam	−1	(21)	(17)	(11)	(9)	(8)	(7)

Sources: Unpublished Alcoa data and M.G. Vassilaros and E.J. Czyryca, Naval Research Lab Report CDNSWC-TR619409, A Compilation of Fatigue Information for Aluminum Alloys, 1994

7xxx Alloys

7002

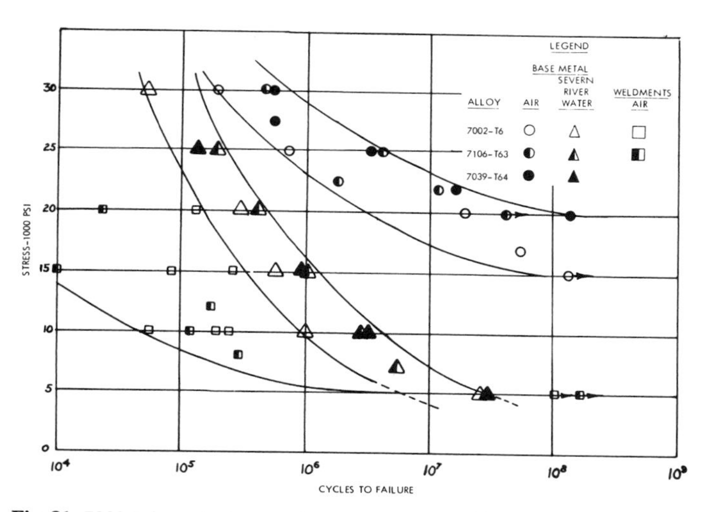

Fig. 81 7002-T6 rotating beam fatigue compared with 7036-T64 and 7106-T63. Source: Naval Research Lab Report CDNSWC-TR619409, 1994

7002-T6 plate: Room-temperature fatigue strength in air and Severn River water

Form and thickness	Specimen	Test mode	Stress ratio	Atmosphere	Fatigue strength (ksi) at cycles of:					
					10^4	10^5	10^6	10^7	10^8	5×10^8
2 ½ in. plate	0.5 in. diam cantilever beam	Reverse bending	−1	Air	...	...	(23)	(17)	(10)	...
2 ½ in. plate	As above	Reverse bending	−1	Severn River Water	...	(23)	(10)	...	...	...
2 ½ in. plate	As above, but butt welded single-V	Reverse bending	−1	Air	...	(7.5)	...	...	...	...

Source: M.G. Vassilaros and E.J. Czyryca, Naval Research Lab Report CDNSWC-TR619409, A Compilation of Fatigue Information for Aluminum Alloys, 1994

7005

7005: Room-temperature fatigue strength in air

Temper	Form and thickness	Specimen	Test mode	Stress ratio	Fatigue strength (ksi) at cycles of:					
					10^4	10^5	10^6	10^7	10^8	5×10^8
T53	Extrusion	RR Moore	Rotating beam	−1	...	(35)	(27)	(23)	(21)	(19)
	Extrusion	RR Moore, notched, $K_t = 12$	Rotating beam	−1	(27)	(17)	(10)	(6)	(5)	(5)
T6, T63	Extrusion	RR Moore	Rotating beam	−1	...	(36)	(28)	(23)	(21)	(19)
	Extrusion	RR Moore, notched, $K_t = 12$	Rotating beam	−1	(25)	(18)	(10)	(5)	(5)	(4)
T6351	Plate	RR Moore	Rotating beam	−1	...	(34)	(26)	(21)	(18)	(17)
	Plate	RR Moore, notched, $K_t = 12$	Rotating beam	−1	...	(14)	(8)	(6)	(5)	(5)
T6, T6351	Sheet, plate	Sheet	Sheet flexure	−1	...	(29)	(20)	(16)	(14)	(13)

Sources: Unpublished Alcoa data

7039

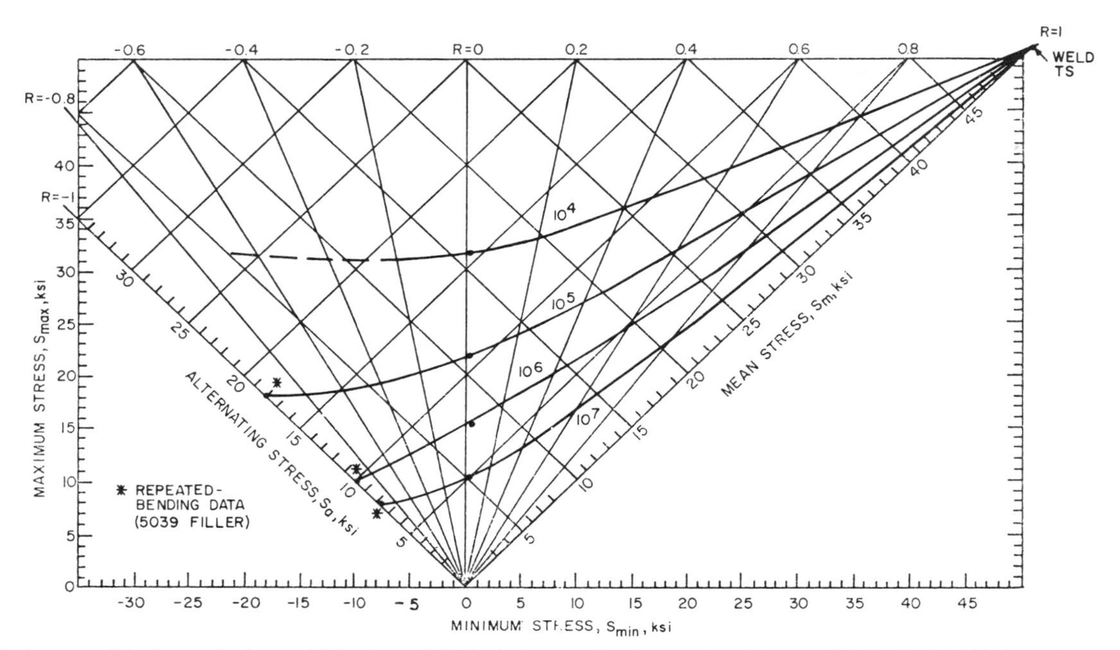

Fig. 82 7039-T61 constant life diagram for butt weld (bead on, 5183 filler). Source: Naval Research Lab Report CDNSWC-TR619409, 1994

7039: Room-temperature fatigue strength in air

Temper	Form and thickness	Specimen	Test mode	Stress ratio	Fatigue strength (ksi) at cycles of: 10^4	10^5	10^6	10^7	10^8	5×10^8
T64	⅜ in. plate	Smooth	Axial	0.1	(65)	(37)	(27)	(25)	...	...
T61	⅜ in. plate	Smooth	Axial	0.0	(59)	(34)	(29)	(26)	...	...
	⅜ in. plate	Smooth	Axial	0.1	(59)	(40)	(29)	(26)	...	...
T6	⅛ in. plate	Smooth	Axial	0.1	(58)	(44)	(33)	...	...	...
T64	0.375 in. plate	Flat plate	Reverse bending	–1	...	(26)	(20)	(18)	(17)	...
T61	0.750 in. plate	RR Moore	Rotating beam	–1	...	(31)	(26)	(22)	(20)	...
T64	0.750 in. plate	RR Moore	Rotating beam	–1	...	(35)	(27.5)	(25)	(24)	(24)
T61	0.375 in. plate	Flat plate	Reverse bending	–1	...	(28)	(15)	(14)	(13)	...
	T61 ⅜ in. plate	GMA butt weld bead-on (5183) filler	Axial	0	(32)	(22)	(15)	(10)	...	...
	T61 ⅜ in. plate	GMA butt weld bead-off (5183 filler)	Axial	0	...	(30)	(20)	(15)	...	...
T64	3 in. plate	0.500 in. round cantilever	Reverse bending	–1	...	...	(28)	(23)	(20)	...
T64	3 in. plate	0.500 round cantilever	Reverse bending	–1	...	...	(16)	(7)	...	...
T61	⅜ in. plate	Transverse butt weld, 5039 filler	Axial	0	(33)	(22)	(17)	(14)	...	...
T61	⅜ in. plate	As above, bead-off	Axial	0	...	(30)	(21)	(17)	...	...
	⅜ in. plate	Lap joint, double fillet of 5039 filler	Axial	0	(9)	(5)	(2)	(1)	...	...
	⅜ in. plate	Tee-joint, double fillet of 5039 filler	Axial	0	(25)	(14)	(6)	(4)	...	...
	⅜ in. plate	As above, single fillet	Axial	0	(7)	(4)	(2)	(2)	...	...
	¼ in. plate	Transverse electron beam weld, as welded	Axial	0	...	(25)	(19.5)	(11.5)	...	...
T61	¼ in. plate	Transverse electron beam weld, as welded	Reverse plate bending	–1	...	(23)	(16)	(12)	(10)	...
T64	⅛ in. sheet	Transverse butt-weld, 5183 filler, bead-on	Reverse plate bending	–1	...	(17.5)	(8.5)	(7.5)	(7)	...
	⅜ in. plate	Transverse single-V butt-weld, X5039 filler, bead-on	Reverse plate bending	–1	...	(18)	(10)	(8)	(7)	...
T6, T63, T651	Plate, extrusion	RR Moore	Rotating beam	–1	...	(39)	(28)	(23)	(22)	(22)
T6, T63, T651	Plate, extrusion	RR Moore, notched, $K_t = 12$	Rotating beam	–1	(24)	(14)	(9)	(6)	(6)	(5)
T6, T63, T651	Sheet	Sheet	Sheet flexure	–1	...	(32)	(22)	(19)	(17)	(16)

Sources: Unpublished Alcoa data and M.G. Vassilaros and E.J. Czyryca, Naval Research Lab Report CDNSWC-TR619409, A Compilation of Fatigue Information for Aluminum Alloys, 1994

7049

7049-T73 forgings: Room-temperature fatigue strength in air

Specimen	Test mode	Stress ratio	Fatigue strength (ksi) at cycles of: 10^4	10^5	10^6	10^7	10^8	5×10^8
Round	Axial	0.10	(68)	(53)	(44)	(39)	...	...
		0.00	(65)	(51)	(42)	(37)	...	...
Round, notched, $K_t = 2.4$	Axial	0.10	(42)	(24)	(14)	(8)	...	...
		−0.40	(29)	(17)	(16)	...	...	...
Round, notched, $K_t = 3.0$	Axial	0.10	(37)	(19)	(10)	(6)	...	...

Source: Unpublished Alcoa data

7050

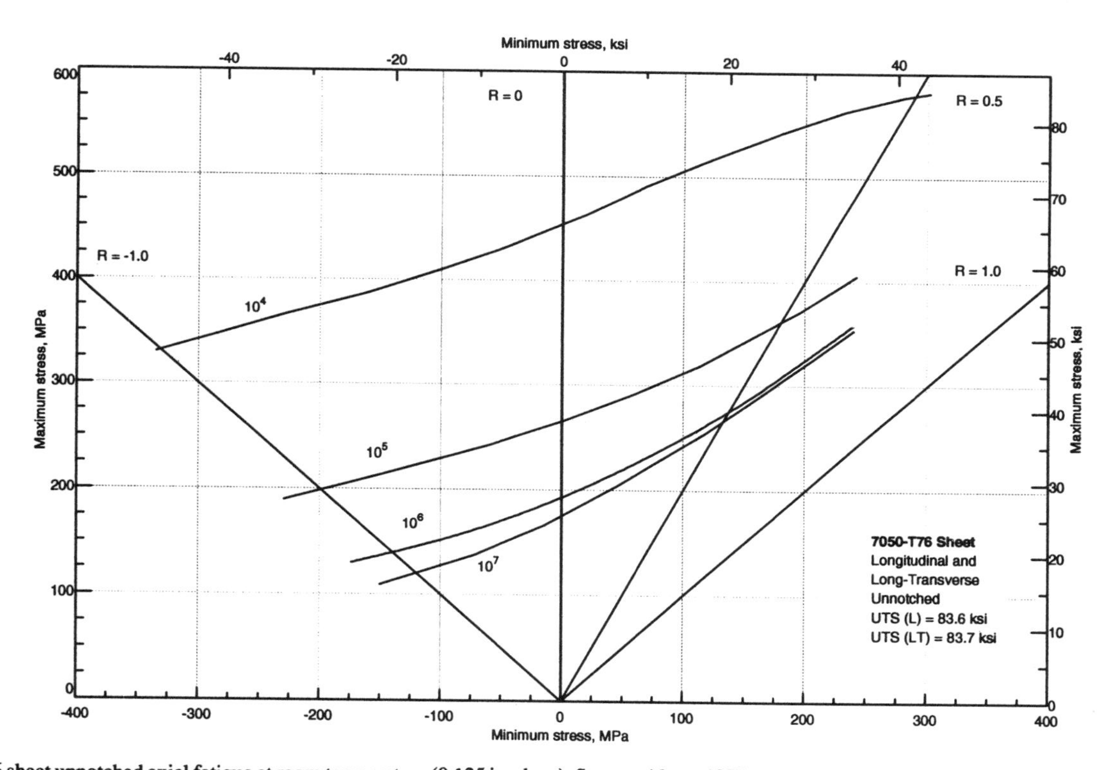

Fig. 83 7050-T6 sheet unnotched axial fatigue at room temperature (0.125 in. sheet). Source: Alcoa, 1974

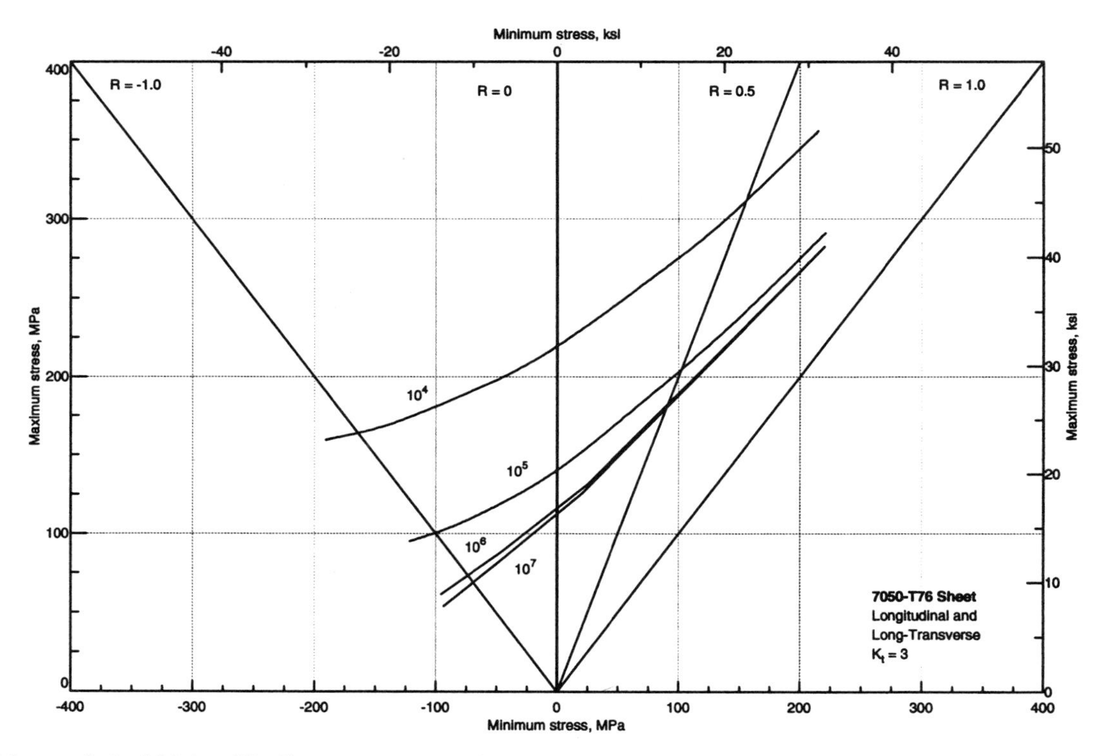

Fig. 84 7050-T6 sheet notched axial fatigue ($K_t = 3$) at room temperature.Source: Alcoa, 1974

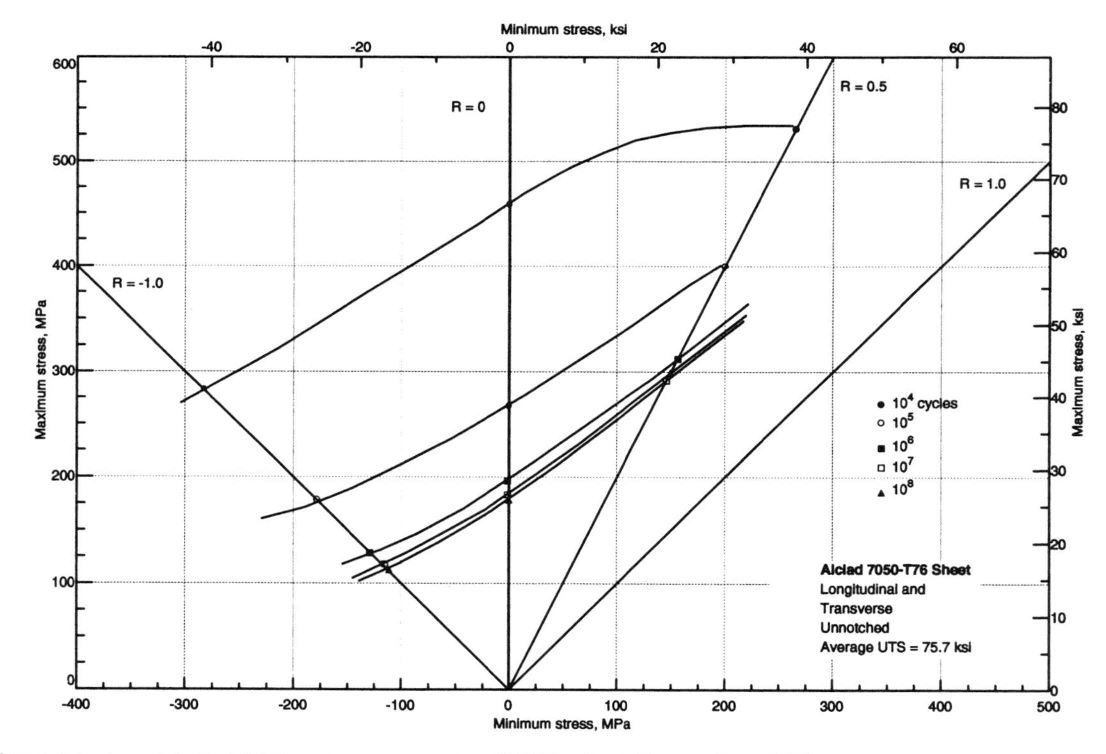

Fig. 85 7050-T6 Alclad sheet unnotched axial fatigue at room temperature (0.125 in. sheet). Source: Alcoa, 1975

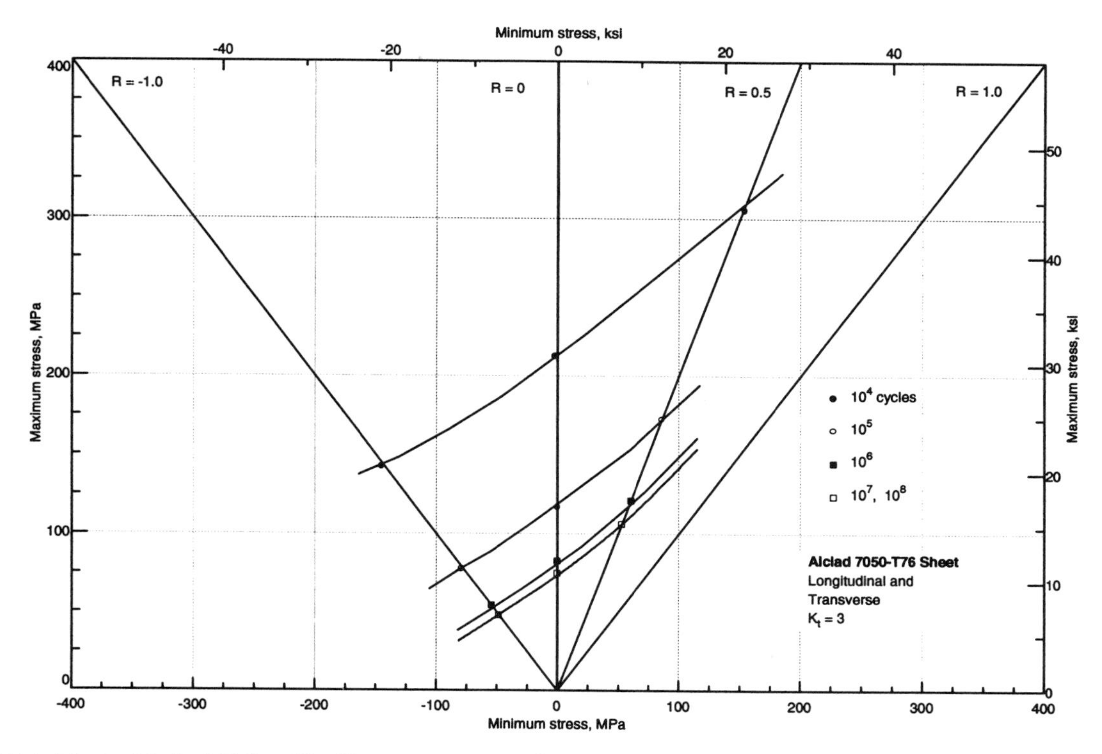

Fig. 86 7050-T6 Alclad sheet notched axial fatigue ($K_t = 3$) at room temperature. Source: Alcoa, 1975

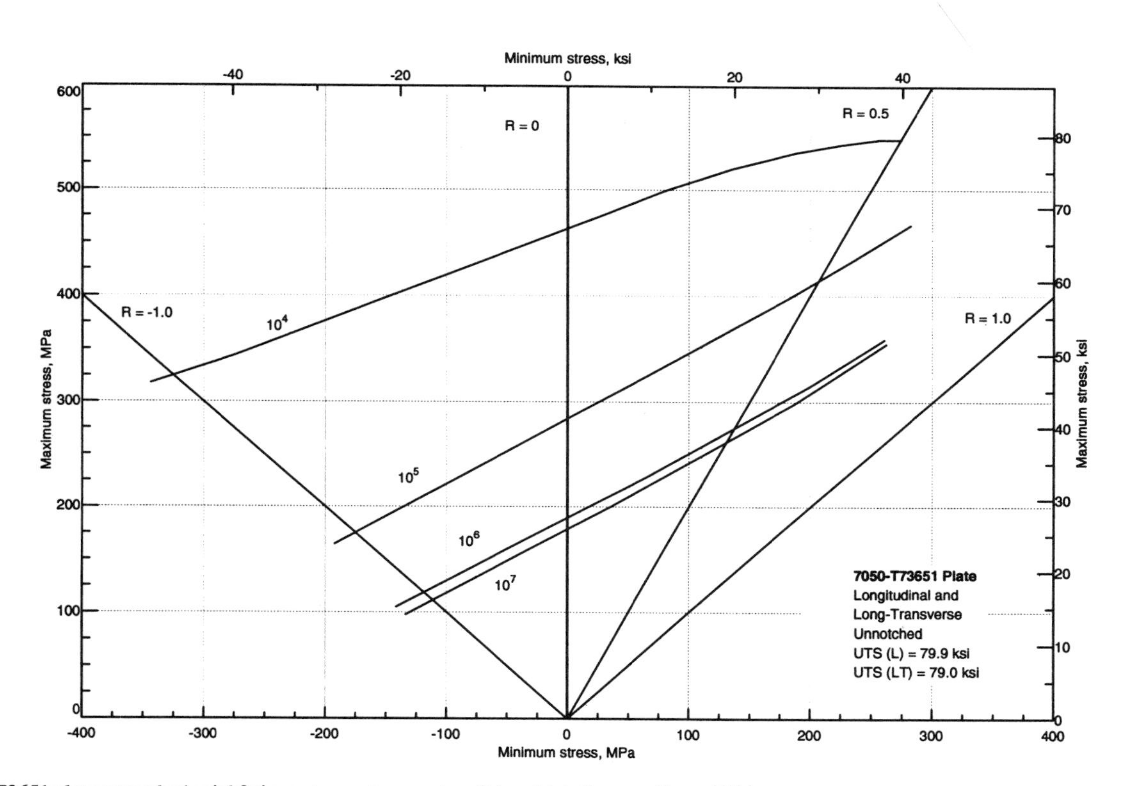

Fig. 87 7050-T73651 plate unnotched axial fatigue at room temperature (1 in. plate). Source: Alcoa, 1974

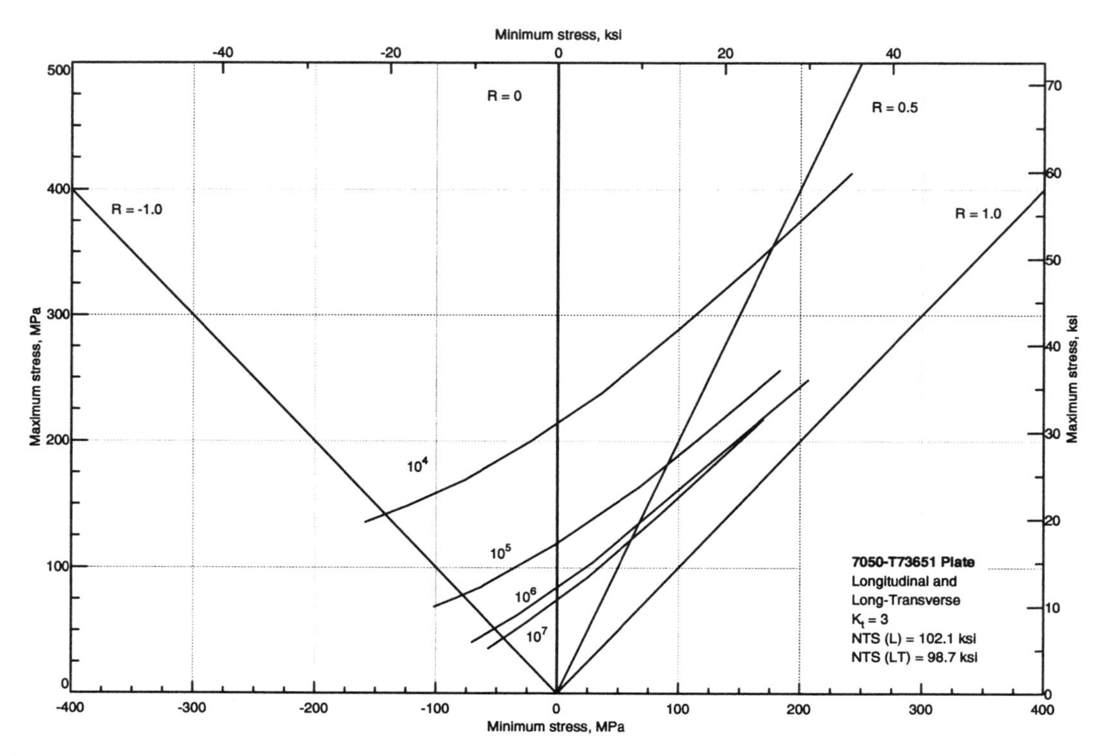

Fig. 88 7050-T73651 plate notched axial fatigue ($K_t = 3$) at room temperature. Source: Alcoa, 1974

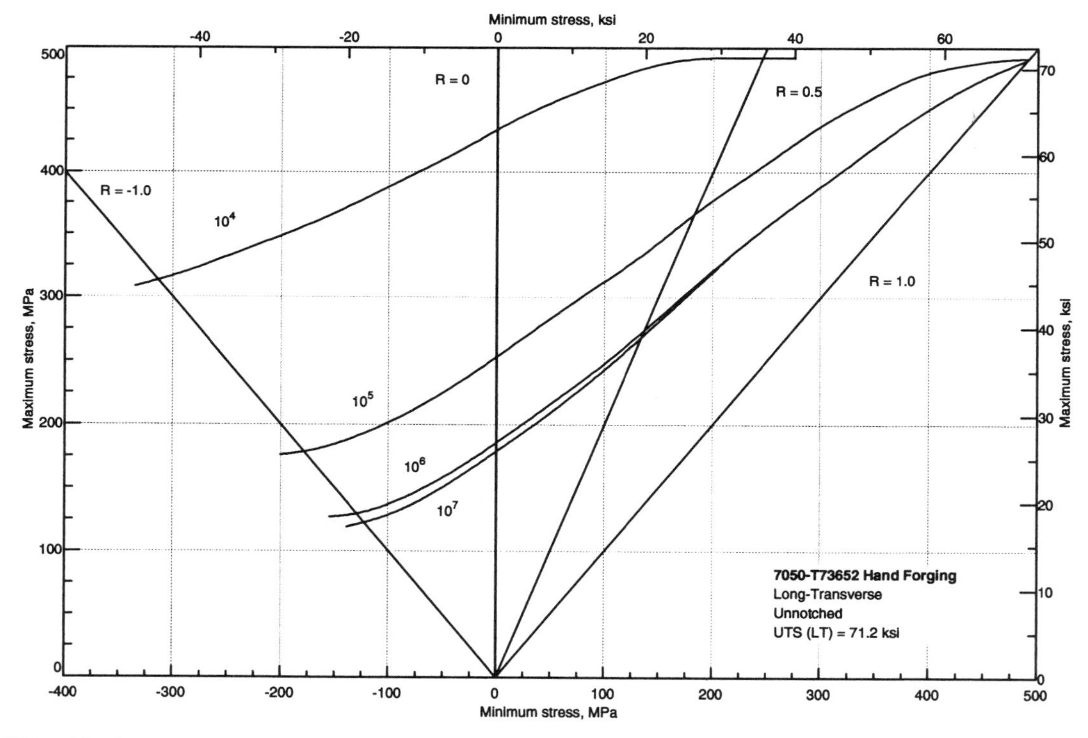

Fig. 89 7050-T73652 hand forging unnotched axial fatigue (long transverse, room temperature). Source: Alcoa, 1974

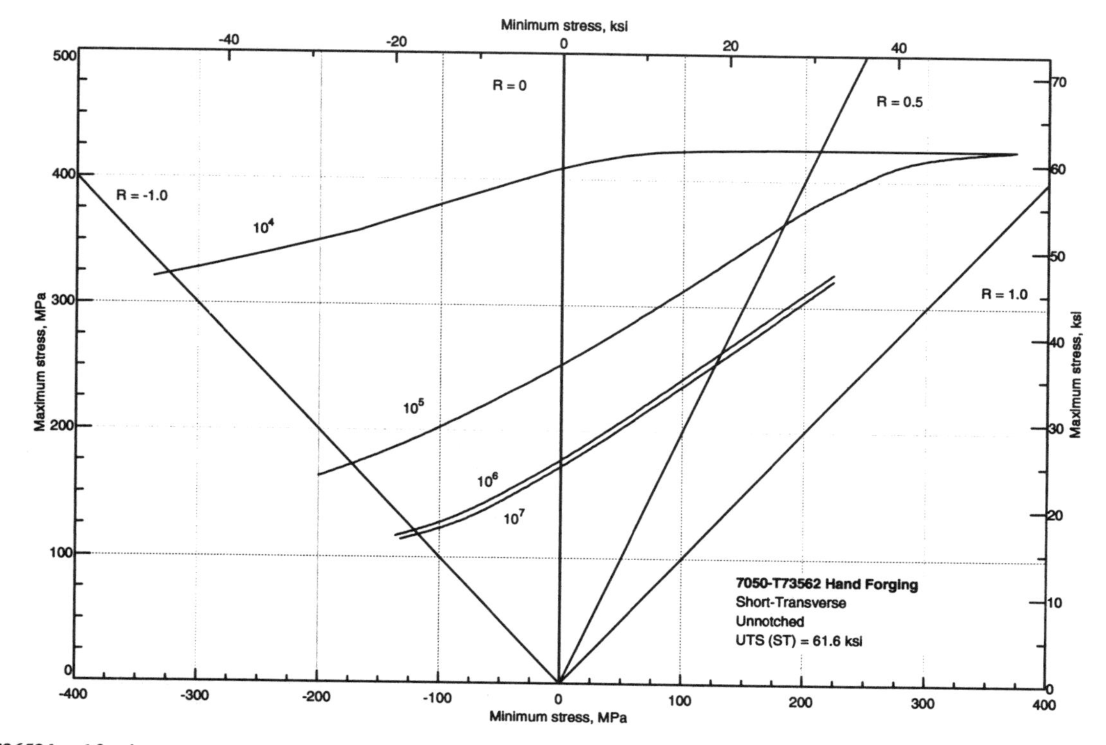

Fig. 90 7050-T73652 hand forging unnotched axial fatigue (short transverse, room temperature). Source: Alcoa, 1974

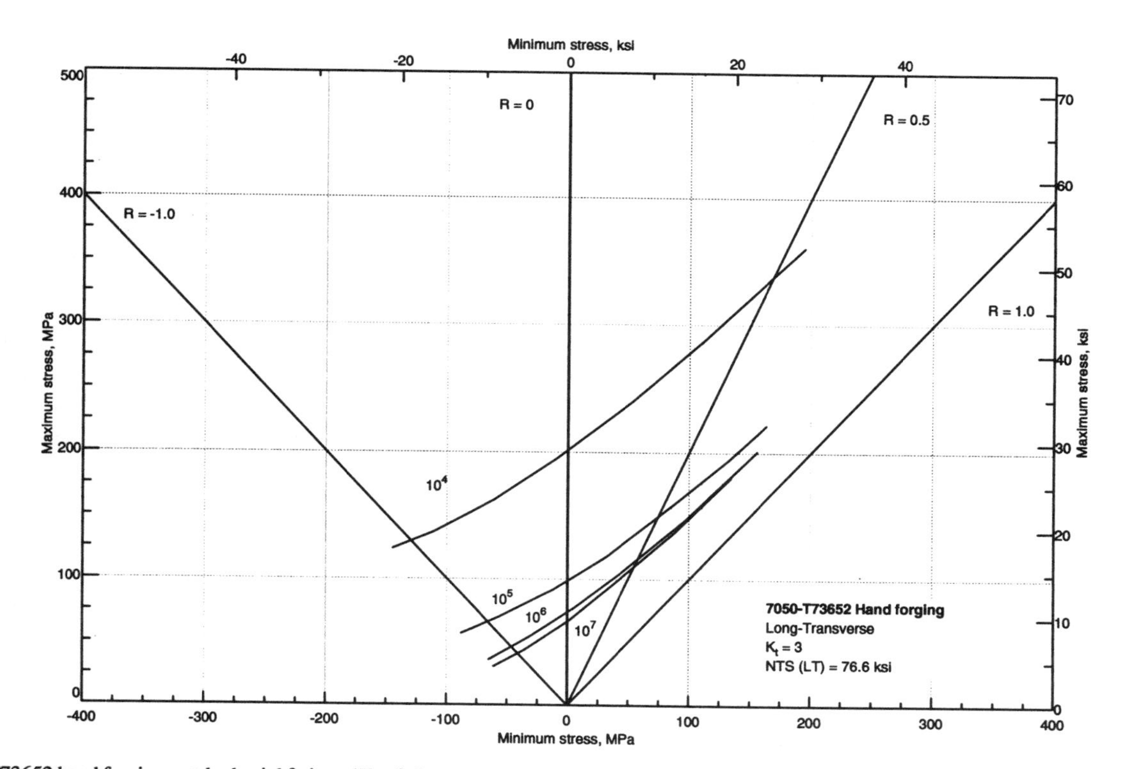

Fig. 91 7050-T73652 hand forging notched axial fatigue ($K_t = 3$, long transverse, room temperature). Source: Alcoa, 1974

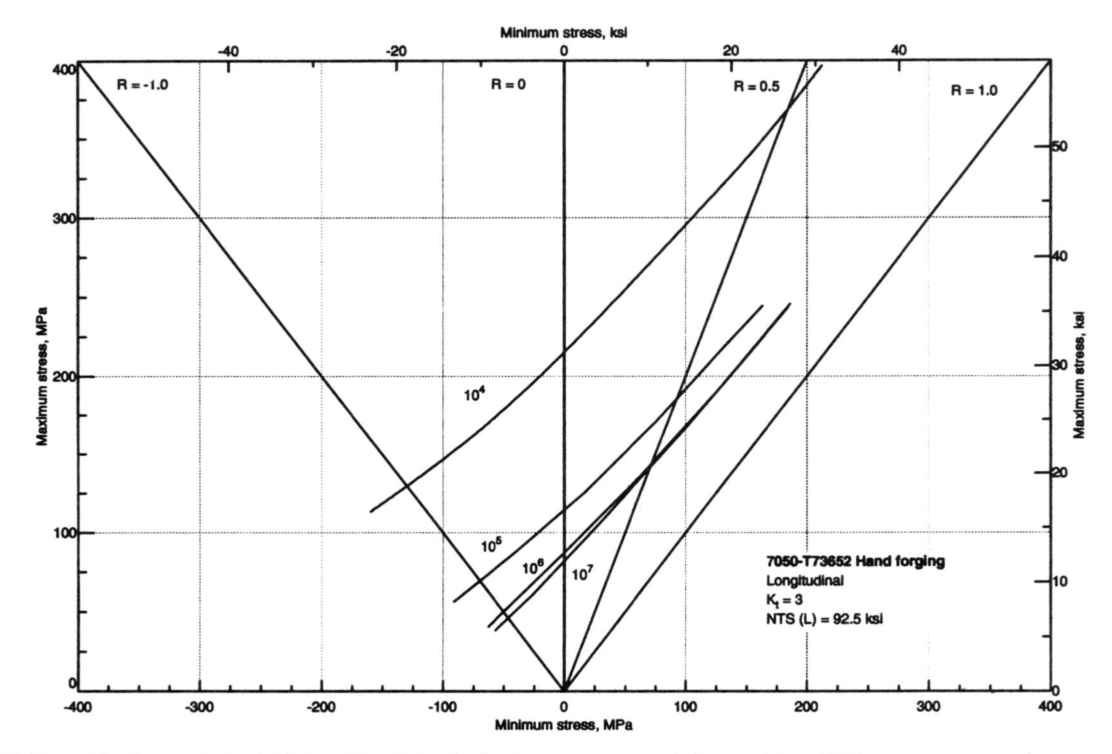

Fig. 92 7050-T73652 hand forging notched axial fatigue ($K_t = 3$, longitudinal, room temperature). Source: Alcoa, 1974

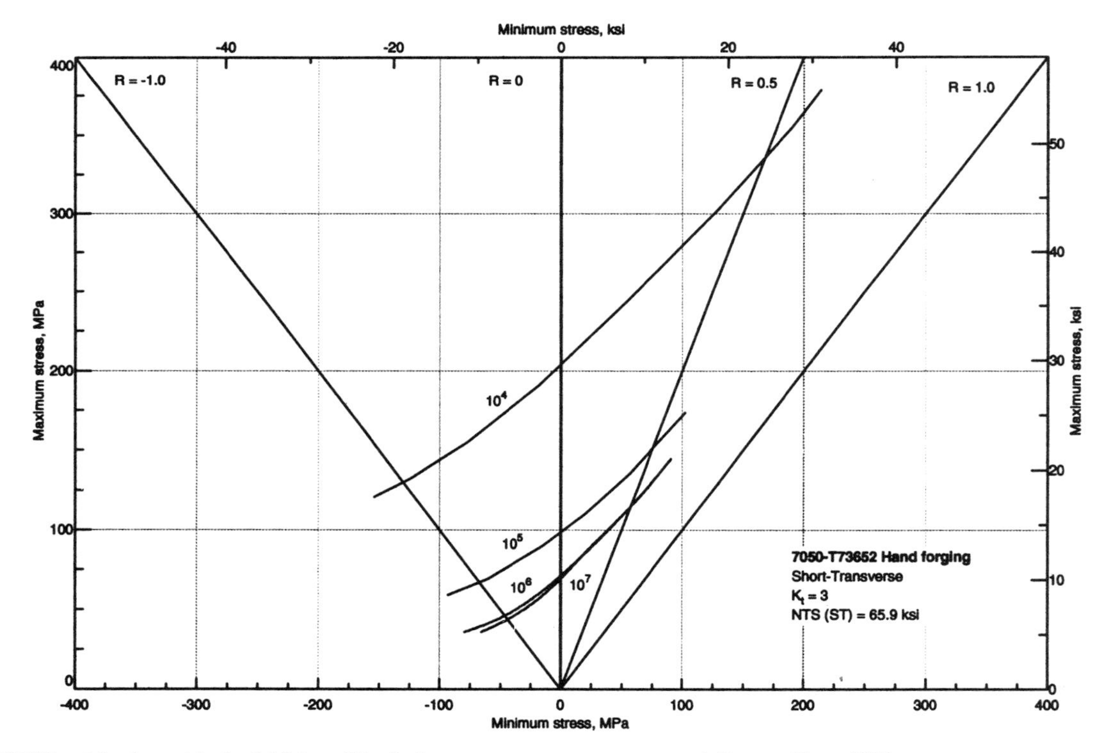

Fig. 93 7050-T73652 hand forging notched axial fatigue ($K_t = 3$, short transverse, room temperature). Source: Alcoa, 1974

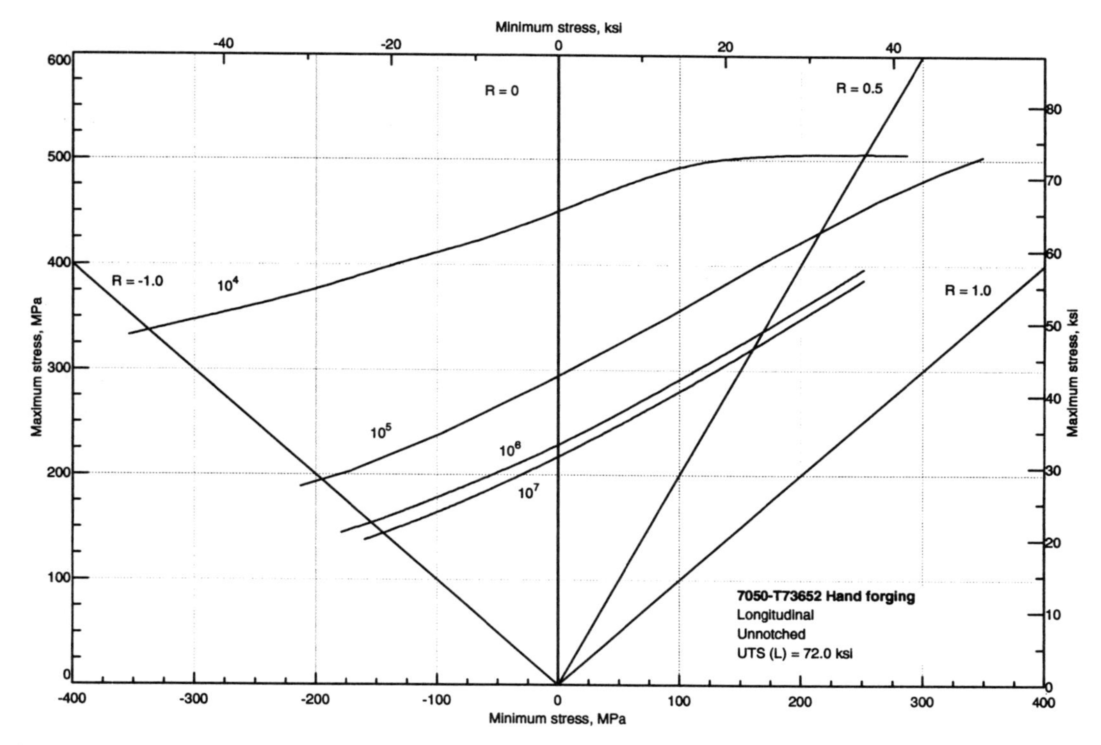

Fig. 94 7050-T73652 hand forging unnotched axial fatigue (longitudinal, room temperature). Source: Alcoa, 1974

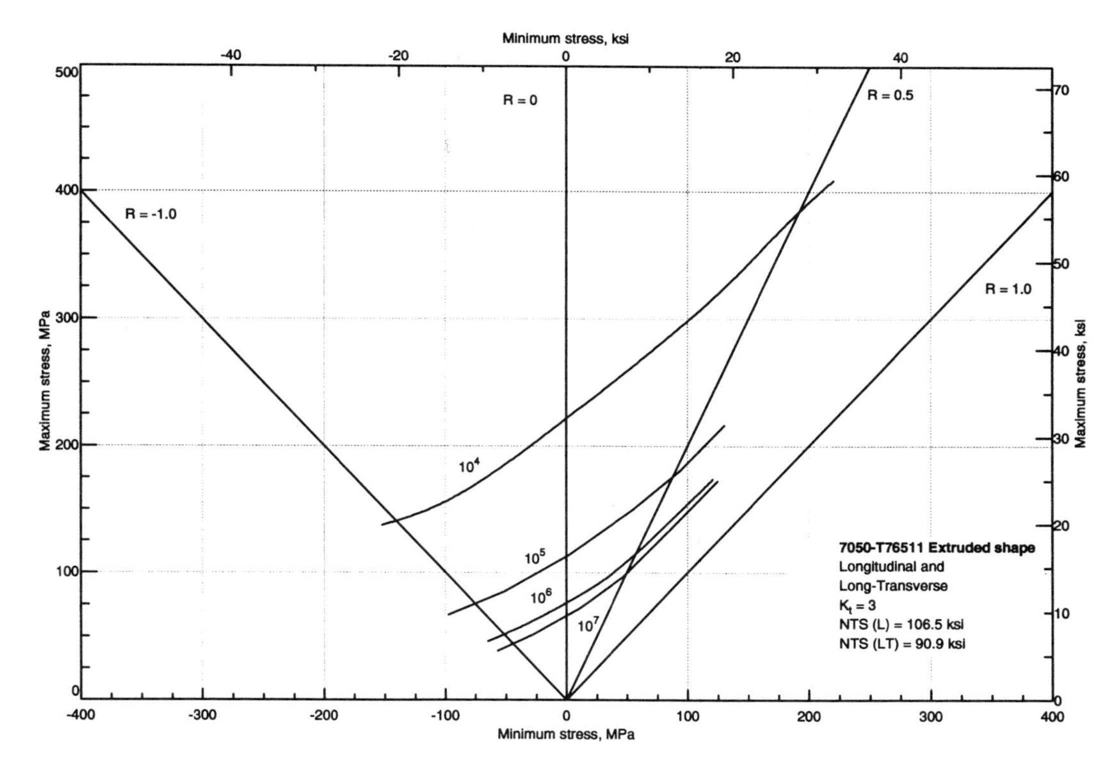

Fig. 95 7050-T76511 extruded shape notched axial fatigue ($K_t = 3$) at room temperature (longitudinal and long transverse). Source: Alcoa, 1974

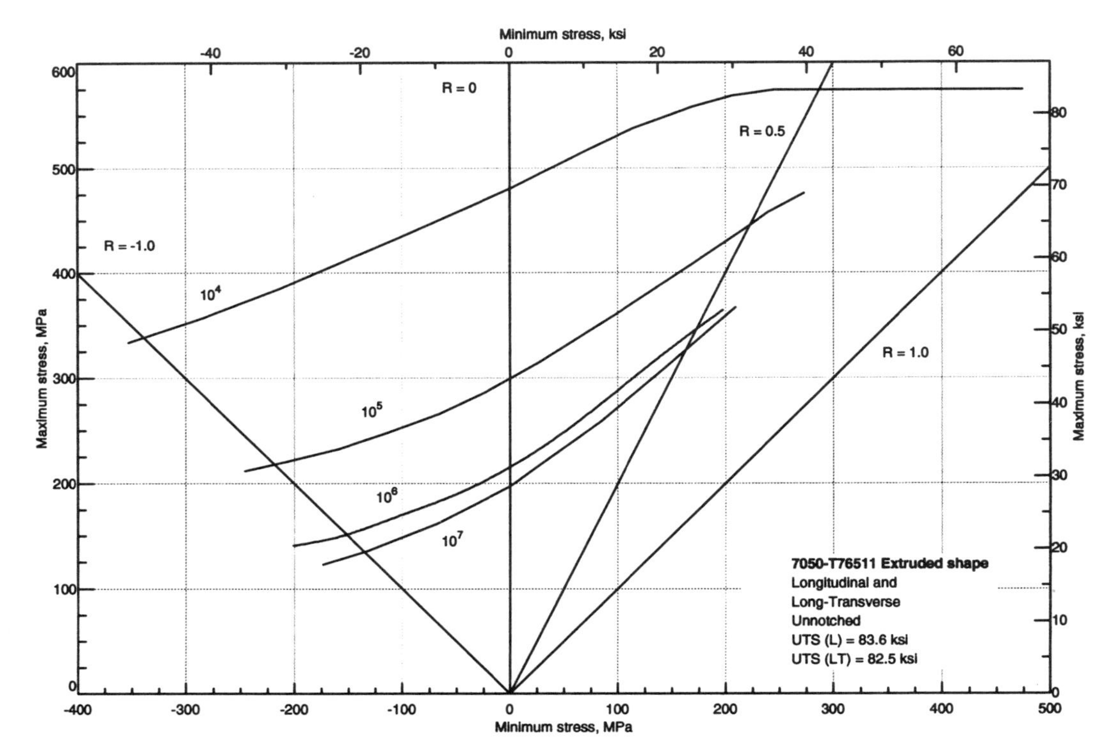

Fig. 96 7050-T76511 extruded shape unnotched axial fatigue (longitudinal and long transverse). Source: Alcoa, 1974

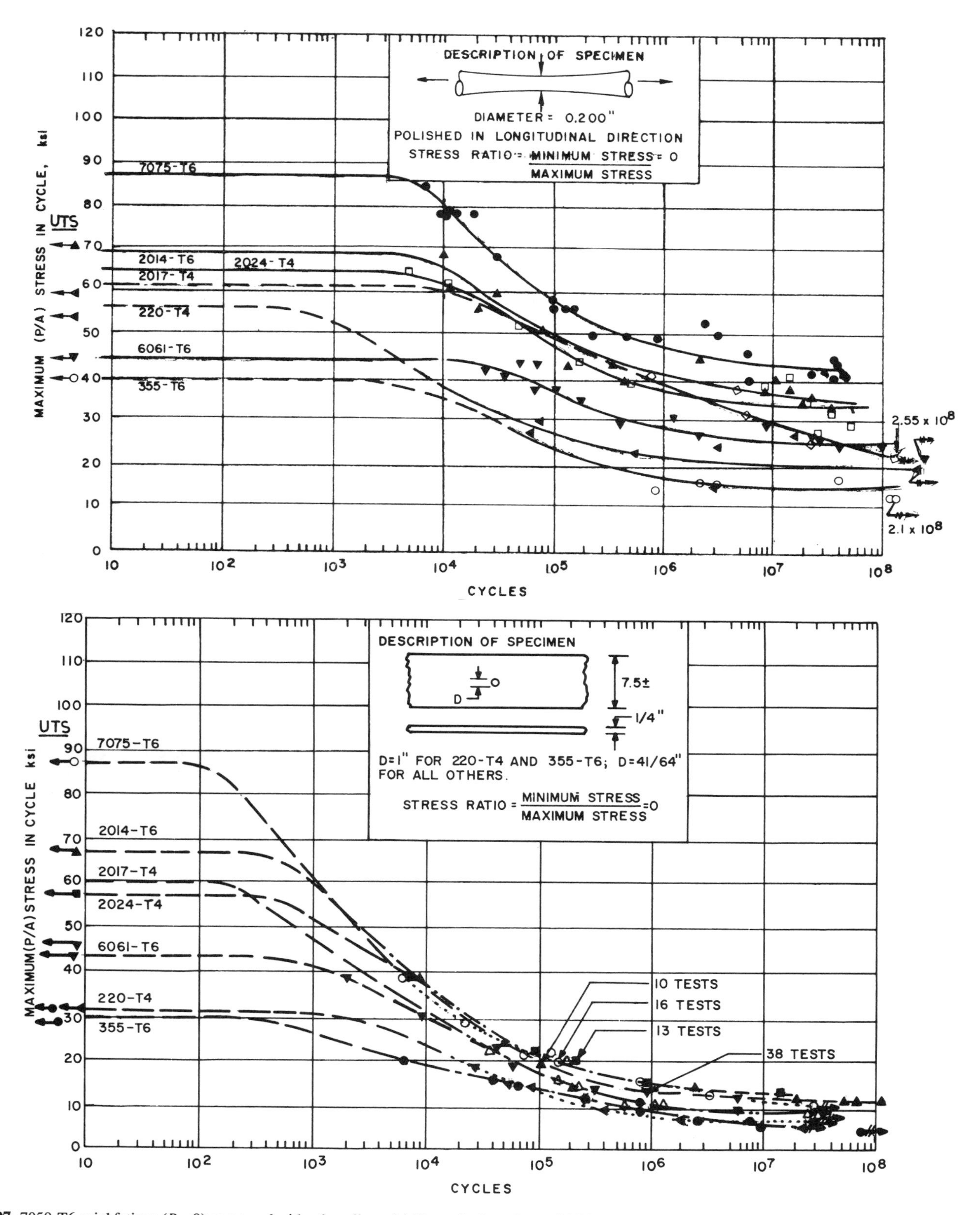

Fig. 97 7050-T6 axial fatigue ($R = 0$) compared with other alloys. (a) Unnotched specimen. (b) Plate containing a hole. Source: Alcoa, 1958

7050: Room-temperature fatigue strength in air

Temper	Form and thickness	Specimen	Test mode	Stress ratio	Fatigue strength (ksi) at cycles of:					
					10^4	10^5	10^6	10^7	10^8	5×10^8
T7	Forging	RR Moore	Rotating beam	–1	(52)	(38)	(28)	(25)	(23)	(22)
	Forging	RR Moore, notched, $K_t = 12$	Rotating beam	–1	(30)	(18)	(12)	(9)	(8)	(8)
	Extrusion	Round	Axial	0.10	(72)	(47)	(34)	(27)	...	...
				–1.00	(46)	(30)	(22)	(17)	(15)	...
	Extrusion	Round, notched, $K_t = 3.0$	Axial	0.10	(40)	(18)	(10)	...	......	...
				0.00	(36)	(17)	(9)	(8)	(7)	...
				–1.00	(22)	(11)	(7)	(5)	...	...
T7451	Plate	Round	Axial	0.50	...	(68)	(46)	(35)	...	...
				0.00	(74)	(45)	(30)	(23)	...	...
				–1.00	(48)	(28)	(19)	(15)	...	...
	Plate	Round, notched, $K_t = 3.0$	Axial	0.50	(50)	(28)	(16)	(9)	...	...
				0.10	(35)	(20)	(11)	(7)	...	...
				0.00	(33)	(18)	(10)	(6)	...	...
				–1.00	(21)	(12)	(7)	(4)	...	...
T4751X	Extrusion	Flat, hole $K_t = 2.6$	Axial	0.20	(45)	(26)	(17)	(14)	...	...
				–0.40	(30)	(22)	(15)	...	...	...
T7452	Forging	Round	Axial	0.50	...	(64)	(50)	(47)	...	...
				0.00	(70)	(42)	(34)	(31)	...	...
				–1.00	(47)	(38)	(22)	(20)	...	...
	Forging	Round, notched, $K_t = 3.0$	Axial	0.50	...	(28)	(17)	(12)	...	...
				0.00	(31)	(15)	(11)	(8)	...	...
				–1.00	(20)	(11)	(7)	(5)	...	...
T74	Forging	Round	Axial	0.00	(68)	(49)	(35)	(25)	...	...
	Forging	Round, notched	Axial	0.33	(49)	(28)	(16)	(9)	...	...
				0.00	(38)	(22)	(12)	(7)	...	...
				–1.00	(24)	(14)	(8)	...	...	...
	Extrusion	Round	Axial	0.50	...	(73)	(51)	(38)	...	...
				0.10	...	(51)	(36)	(27)	...	..
				0.00	(73)	(48)	(34)	(25)	...	...
				–1.00	(48)	(32)	(22)	(17)	...	...
	Extrusion	Round, notched, $K_t = 3.0$	Axial	0.50	(46)	(28)	(16)	(9)	...	...
				0.10	(34)	(20)	(12)	(7)	...	...
				0.00	(32)	(19)	(11)	(6)	...	...
				–1.00	(22)	(13)	(7)	(21)	...	...

Source: Unpublished Alcoa data

7075-0 and -T6

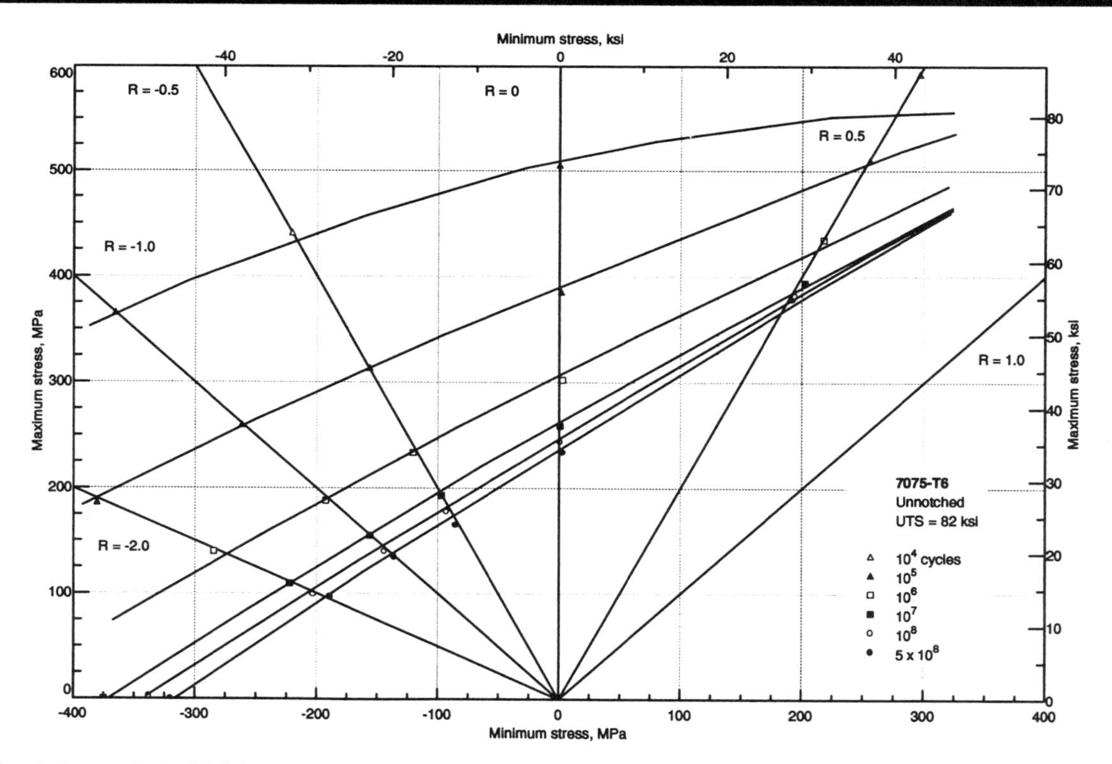

Fig. 98 7075-T6 typical unnotched axial fatigue at room temperature. Unnotched axial specimens with 9-7/8 in. surface radius and 0.300 in. minimum diameter. Source: Alcoa, 1955

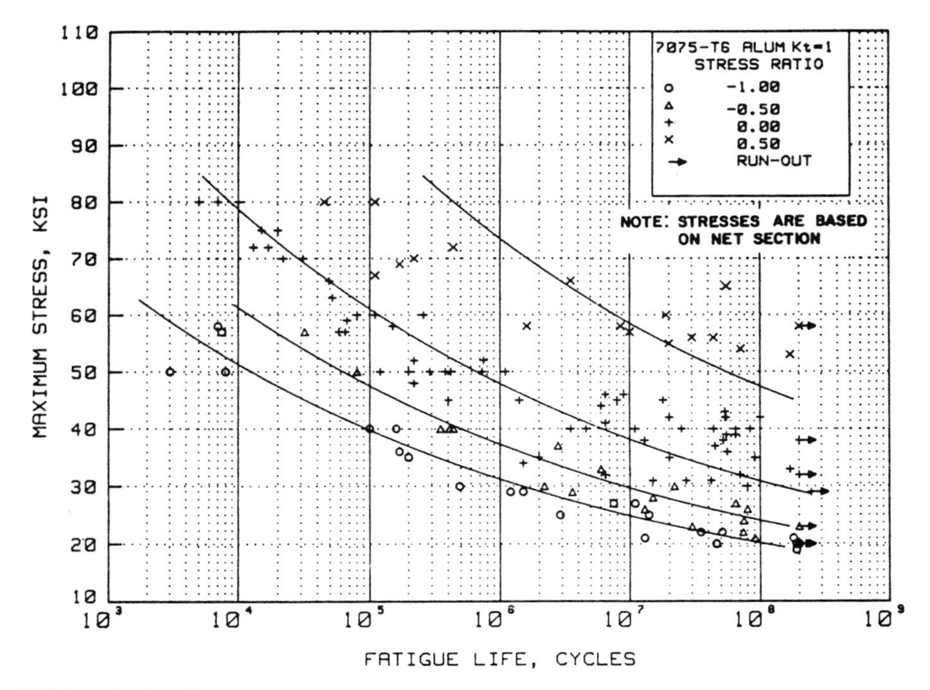

Fig. 99 7075-T6 unnotched axial fatigue (various forms, longitudinal direction), Source: MIL-HDBK-5

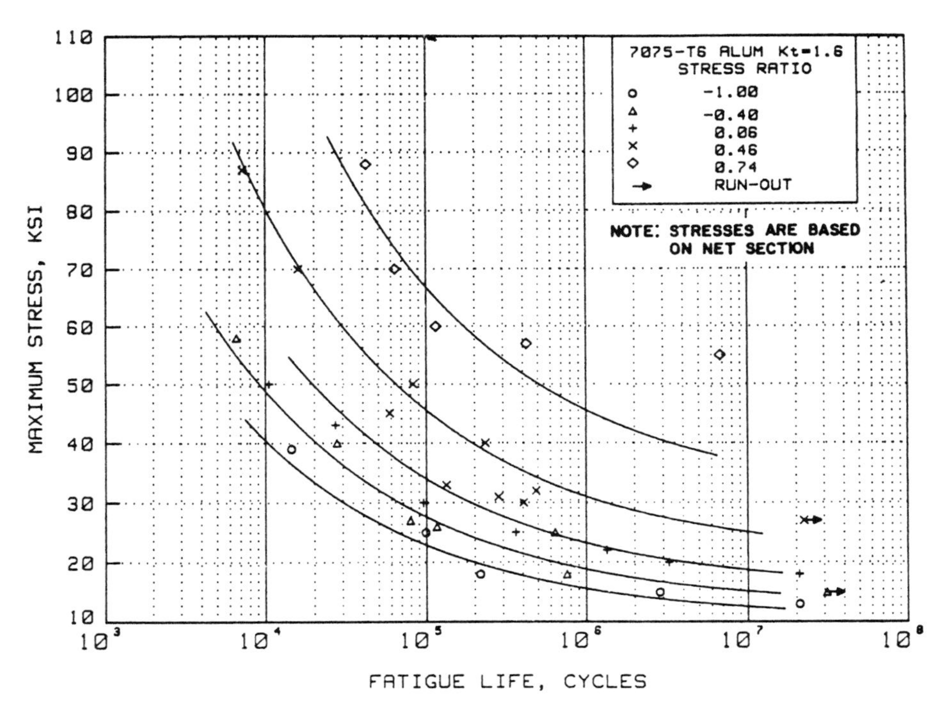

Fig. 100 7075-T6 notched axial fatigue ($K_t = 1.6$) longitudinal direction (rolled bar). Source: MIL-HDBK-5

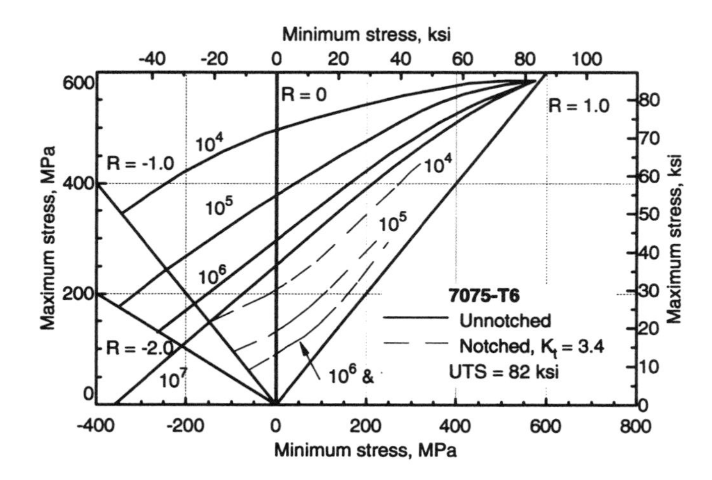

Fig. 101 7075-T6 constant life diagram for notched (K_t = 3.4) and unnotched specimens. Typical tensile strength, 565 MPa (82 ksi). Source: H. Grover, *Fatigue of Aircraft Structures*, NAVAIR 01-1A-13, 1966

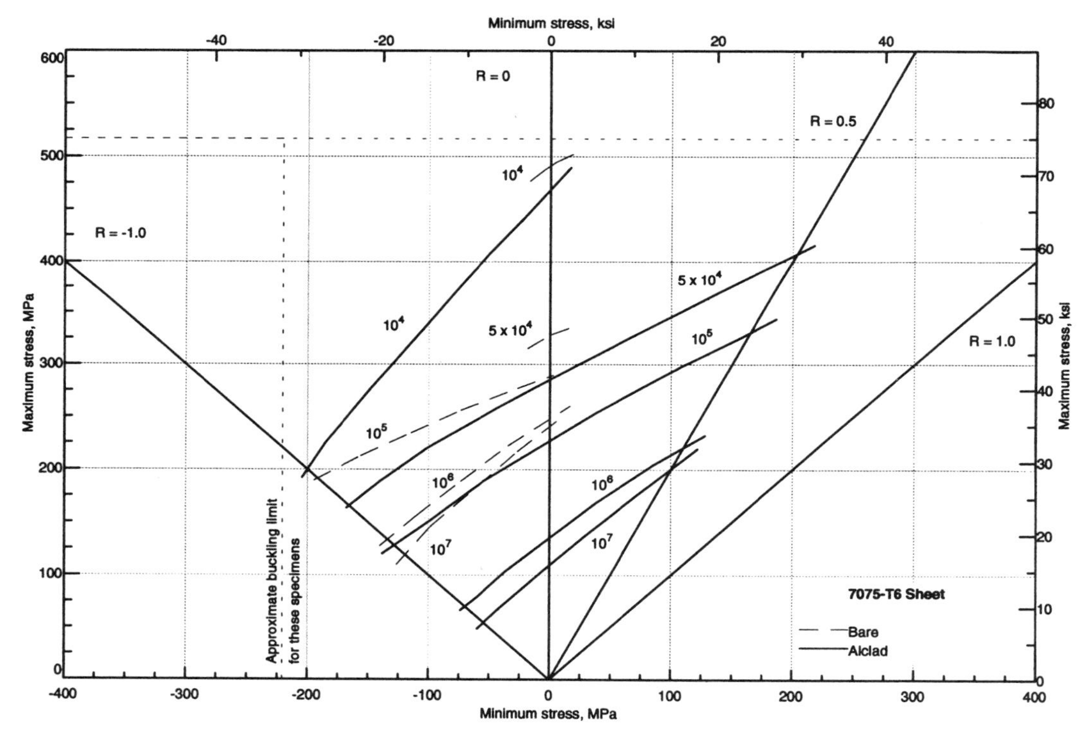

Fig. 102 7075-T6 unnotched axial fatigue at room temperature (sheet). Source: Alcoa, 1957

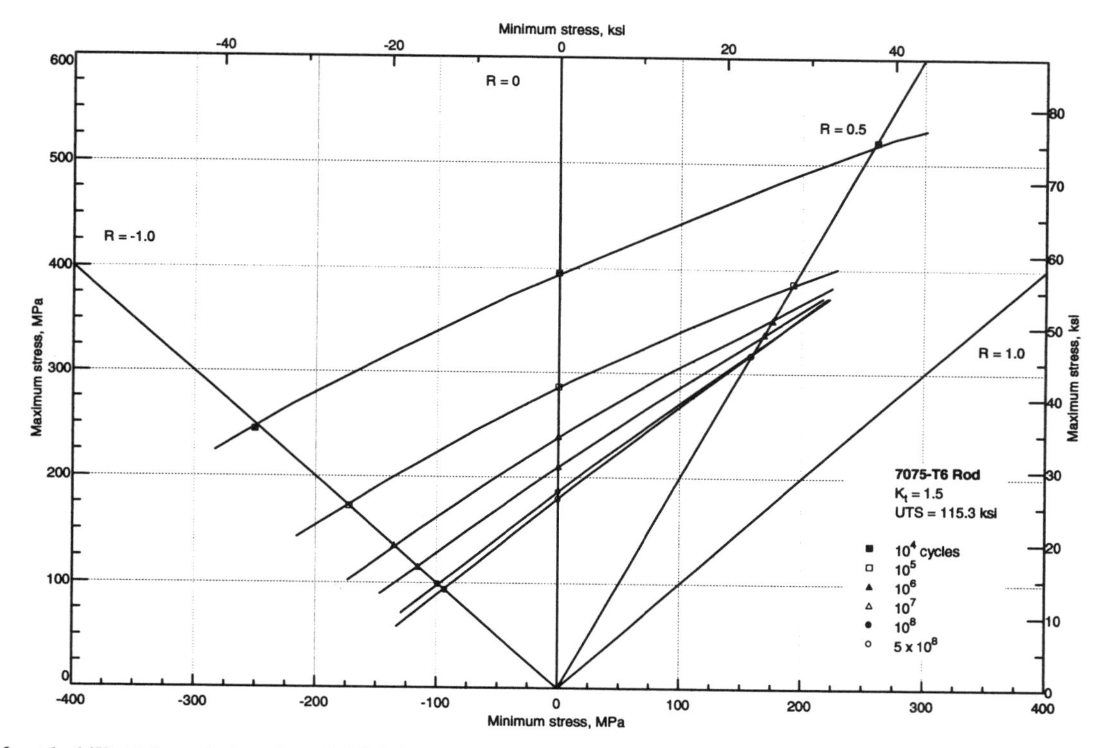

Fig. 103 7075-T6 notched (K_t = 1.5, notch-tip radius of 0.062 in.) axial fatigue at room temperature (rolled and drawn rod)

7075-0: Room-temperature fatigue strength in air

Form and thickness	Specimen	Test mode	Stress ratio	Fatigue strength (ksi) at cycles of: 10^4	10^5	10^6	10^7	10^8	5×10^8
0.157 in. sheet	Transverse single-V butt weld, bead-on	Reverse plate bending	−1	...	(14)	...	...	...	...
0.157 in. sheet	Transverse single-V butt weld, bead-on	Reverse plate bending	0	...	(17)	(10.5)	...	...	...
0.157 in. sheet	As above, shot-peened	Reverse plate bending	−1	...	(20)	(9.5)	...	...	...
0.157 in. sheet	As above, shot-peened	Reverse plate bending	0	...	(24.5)	(15)	...	...	...
0.157 in. sheet	Transverse double-V butt weld, bead-on	Axial	0	...	(13)	...	...	...	...
0.157 in. sheet	As above, shot-peened	Axial	0	...	(21)	(16.5)	...	...	...

Source: M.G. Vassilaros and E.J. Czyryca, Naval Research Lab Report CDNSWC-TR619409, A Compilation of Fatigue Information for Aluminum Alloys, 1994

7075-0 butt weld: Room-temperature fatigue strength in air

Form and thickness	Specimen	Test mode	Stress ratio	Fatigue strength (ksi) at cycles of: 10^4	10^5	10^6	10^7	10^8	5×10^8
Sheet, 0.157 in.	Transverse single butt weld, bead on	Reverse bending	−1	...	(14)	...	...	...	...
			0	...	(17)	(10.5)	...	...	...
Sheet, 0.157 in.	Same as above and shot peened	Reverse bending	−1	...	(20)	(9.5)	...	...	...
			0	...	(24.5)	(15)	...	...	...
Sheet, 0.157 in.	Tranverse double butt weld, bead on	Axial	0	...	(13)	...	...	...	...
	Same as above and shot peened		0	...	(21)	(16.5)	...	...	...

Sources: Unpublished Alcoa data

7075-T6: Room-temperature fatigue strength in air

Form and thickness	Specimen	Test mode	Stress ratio	Fatigue strength (ksi) at cycles of:					
				10^4	10^5	10^6	10^7	10^8	5×10^8
0.064 in. Alclad sheet	Flat sheet, 1 in. wide, (1) 3/16 in. rivet	Axial	−1	...	(11.6)	(5.7)	(3.1)	(2.3)	...
1 in. rolled bar	Flat plate with central hole	Axial	0	(28)	(18)	(13)	(11)	(10)	...
Plate	0.2 in. diam round	Axial	0	(42)	(37)	(28)	(26)	(24)	...
1/4 in. plate	Flat plate, 7 1/2 in. wide with central hole	Axial	0	(34)	(24)	(16)	(12)	(10)	...
0.090 in. sheet	Flat sheet	Reverse bending	−1	...	(26)	(18)	(16)	(15)	...
1/4 in. plate	Flat sheet	Reverse bending	−1	...	(31)	(19)	(18)	(17)	...
Bar, extrusion	Round	Axial	0.50	...	...	(73)	(57)	(47)	...
			0.00	(79)	(61)	(47)	(28)	(31)	...
			−0.50	(61)	(47)	(37)	(20)	(24)	...
			−1.00	(51)	(40)	(31)	(15)	(20)	...
Bar	Round, notched, $K_t = 1.6$	Axial	0.74	...	(67)	(45)	...	...	...
			0.46	(80)	(45)	(31)	(25)	...	...
			0.06	...	(34)	(23)	(19)	...	...
			−0.40	(49)	(27)	(19)	(15)	...	...
			−1.00	(40)	(23)	(15)	(12)	...	...
Bar	Round, notched	Axial	0.46	(41)	(25)	(22)	(11)	...	...
			0.06	(35)	(20)	(16)	(9)	...	...
			−0.40	(28)	(17)	(13)	(8)	...	...
			−1.00	(25)	(15)	(10)	(7)	...	...
Sheet	Sheet	Axial	0.40	...	(64)	...	...	...	...
			0.00	(75)	(50)	...	...	...	...
			−0.50	(61)	(41)	(28)	...	...	...
			−0.60	(60)	(40)	...	...	...	...
			−0.80	(56)	(30)	...	...	...	...
			−1.00	(53)	(36)	(24)	(16)	...	...
Sheet	Sheet, notched, $K_t = 1.5$	Axial	30.0(a)	(58)	(46)	(40)	...	...	...
			20.0(a)	(51)	(39)	(33)	...	...	...
			10.0(a)	(45)	(33)	(26)	...	...	...
			−1.0	(40)	(27)	(20)	(17)	...	...
Sheet	Sheet, notched, $K_t = 2.0$	Axial	30.0(a)	(50)	(41)	(37)	...	...	...
			20.0(a)	(43)	(33)	(29)	...	...	...
			10.0(a)	(37)	(26)	(22)	...	...	...
			−1.0	(31)	(20)	(15)	(13)	...	...
Sheet	Sheet, notched, $K_t = 4.0$	Axial	30.0(a)	(39)	(35)	...	...	...	...
			20.0(a)	(31)	(26)	(24)	(22)	...	...
			10.0(a)	(25)	(14)	(16)	...	...	...
			−1.0	(19)	(13)	(9)	(7)	...	...
Sheet	Sheet, notched, $K_t = 5.0$	Axial	30.0(a)	(38)	(34)	...	...	...	...
			20.0(a)	(29)	(24)	(23)	(22)	...	...
			10.0(a)	(22)	(16)	(14)	(13)	...	...
			−1.0	(15)	(9)	(7)	(5)	...	...
Sheet	Sheet	Axial	0.75	...	...	(66)	(65)	...	...
			0.50	...	(51)	(32)	(30)	...	...
			0.0	(68)	(33)	(20)	(16)	...	...
			−0.5	(51)	(28)	(19)	(14)	...	...
			−1.0	...	(19)	(16)	(8)	...	...

(a) Mean stress. Sources: Unpuplished Alcoa data and M.G. Vassilaros and E.J. Czyryca, Naval Research Lab Report CDNSWC-TR619409, A Compilation of Fatigue Information for Aluminum Alloys, 1994

7075-T73

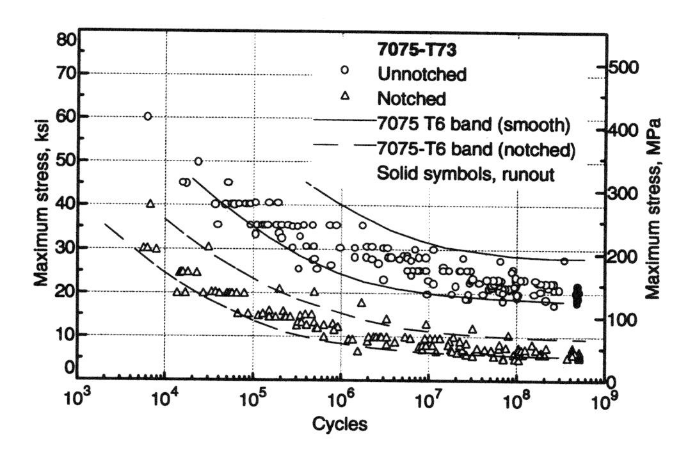

Fig. 104 7075-T73 notched (radius at notch root <0.001 in.) and unnotched rotating beam fatigue at room temperature. Solid symbols indicate runout (no failure). Data points are for 7075-T73 specimens from plate, rod, and forgings in L, LT, and ST directions. Band lines are for stress relieved 7075-T6 type products (L and LT directions). R.R. Moore specimens with 9-7/8 in. surface radius and 0.300 in. minimum diameter for unnotched specimens. Notched specimens had a 0.330 in. diameter at the notch and a 0.480 in. diameter outside the 60° notch. Source: Alcoa, 1971

7075-T73: Room-temperature fatigue strength in air

Form and thickness	Specimen	Test mode	Stress ratio	Fatigue strength (ksi) at cycles of:					
				10^4	10^5	10^6	10^7	10^8	5×10^8
Plate	0.3 in. diam polished round	Reverse bending	–1	(47)	(36)	(27)	(22)	(19)	(18)
Plate	As above, but with 60° V-notch	Reverse bending	–1	(27)	(16)	(11)	(8)	(6)	(6)
Plate	As above	Axial	0	(62)	(43)	(28)	(25)	(23)	...
0.09 in. sheet	Transverse electron beam weld, as welded	Axial	0	...	(24.5)	(20.5)	(18)	...	...
0.09 in. sheet	Transverse electron beam weld, as welded	Reverse bending	–1	...	(23)	(13)	(11)	...	...
Wire	RR Moore	Rotating beam	–1	(52)	(39)	(30)	(26)	(24)	(21)
Wire	RR Moore, notched, $K_t = 12$	Rotating beam	–1	(30)	(17)	(12)	(9)	(7)	(6)
Forging	Round	Axial	0	(60)	(41)	(39)	(38)	(37)	...
Plate, Extrusion	Round	Axial	0	...	(42)	(39)	(37)	(36)	...

Sources: Unpublished Alcoa data and M.G. Vassilaros and E.J. Czyryca, Naval Research Lab Report CDNSWC-TR619409, A Compilation of Fatigue Information for Aluminum Alloys, 1994.

7075-T6510, -T7351, and -T73510

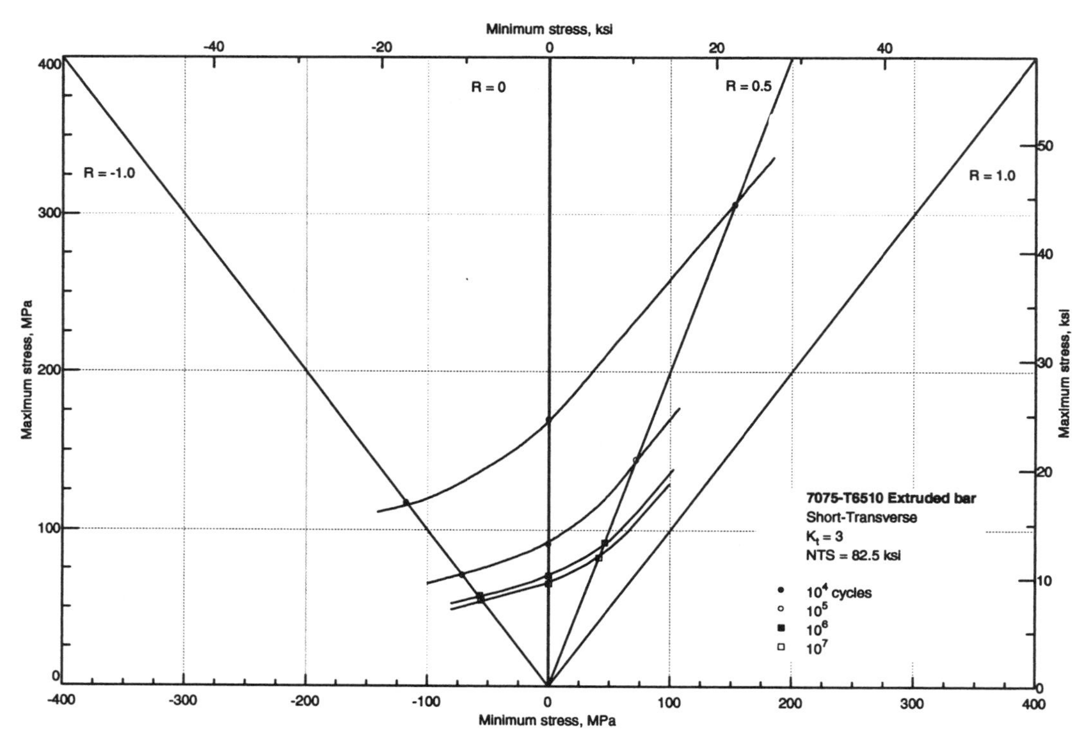

Fig. 105 7075-T6510 short-transverse notched axial fatigue ($K_t = 3$) at room temperature (extruded bar, 3.5 x7.5 in.). Source: Alcoa, 1969

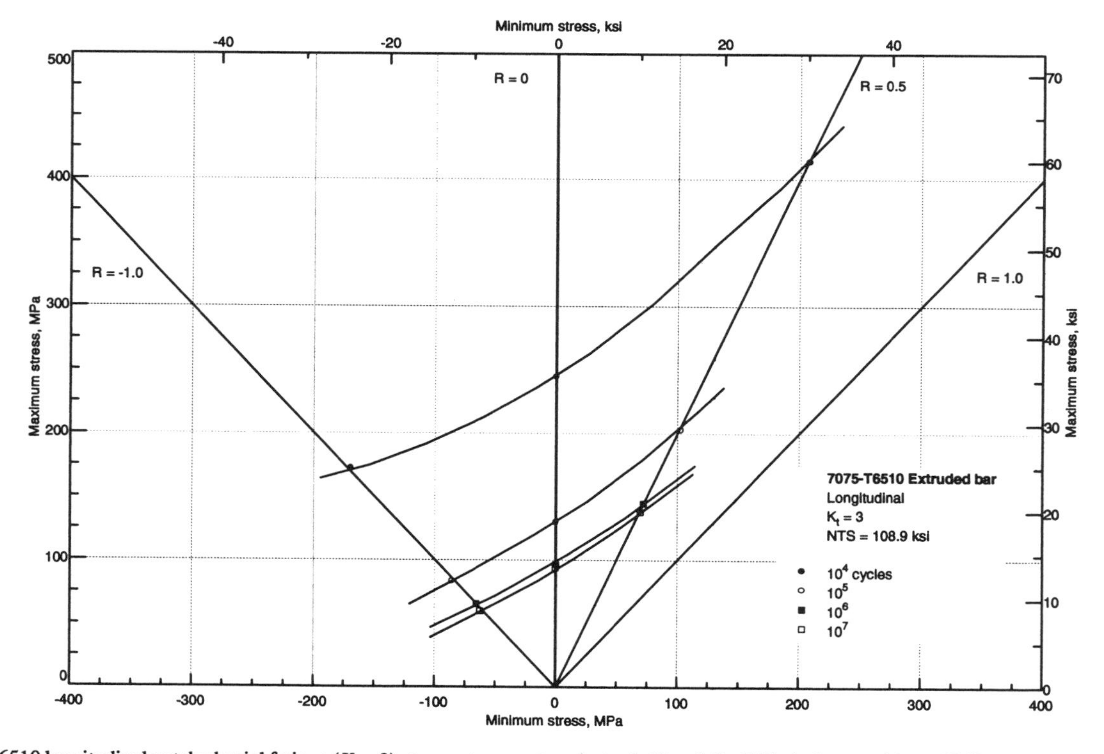

Fig. 106 7075-T6510 longitudinal notched axial fatigue ($K_t = 3$) at room temperature (extruded bar, 3.5 x 7.5 in.). Source: Alcoa, 1969

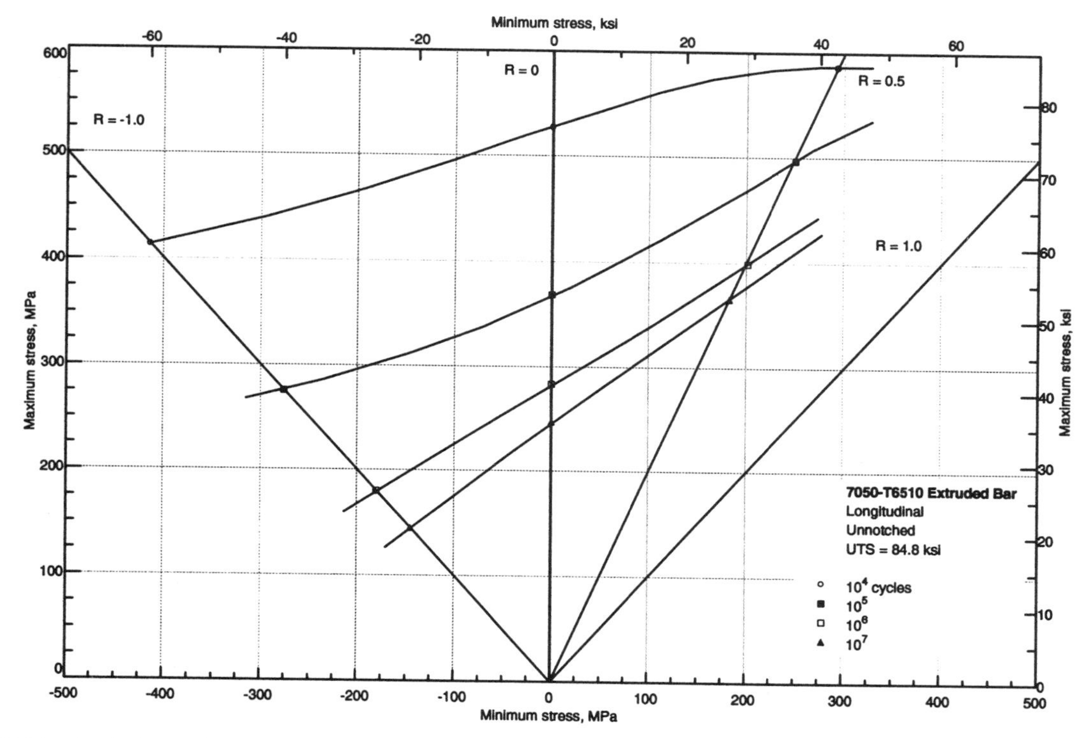

Fig. 107 7075-T6510 longitudinal unnotched axial fatigue (extruded bar)

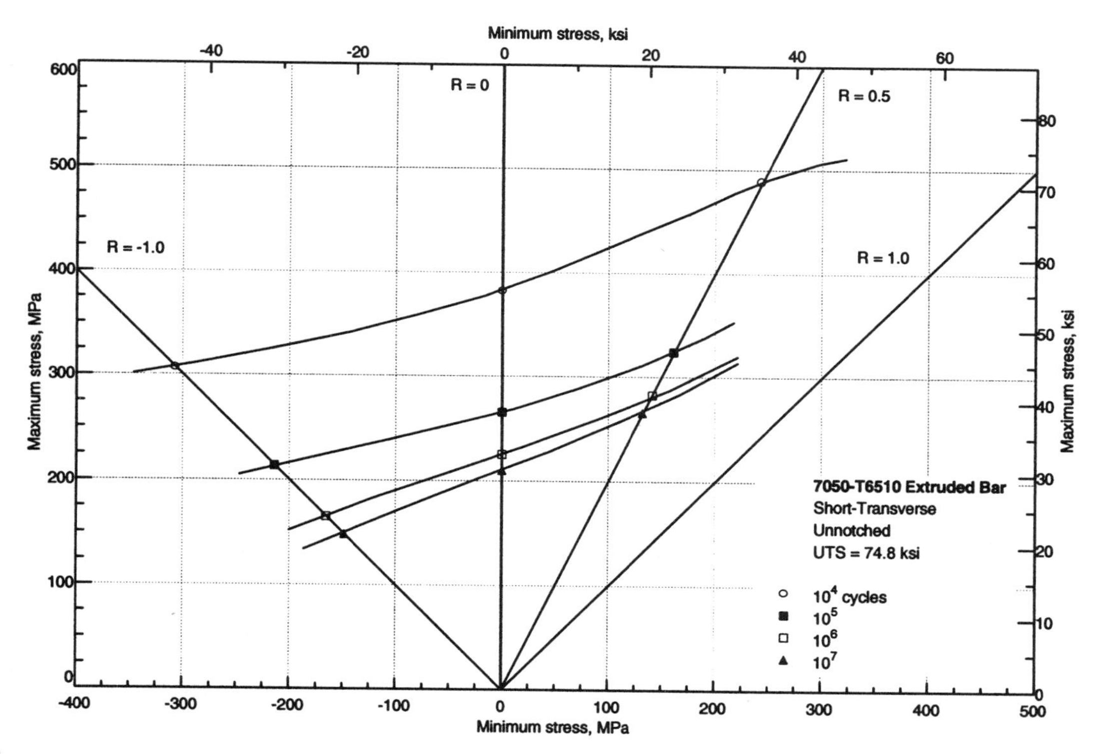

Fig. 108 7075-T6510 short-transverse unnotched axial fatigue (extruded bar)

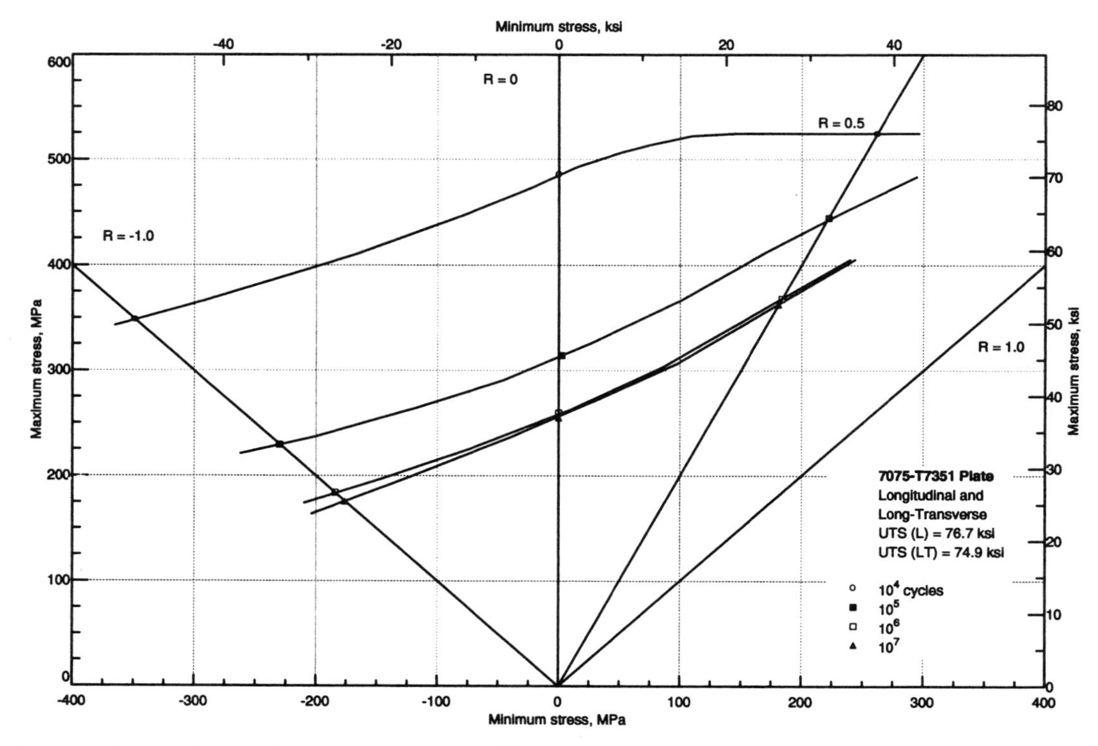

Fig. 109 7075-T7351 unnotched axial fatigue (1 in. plate). Source: Alcoa, 1970

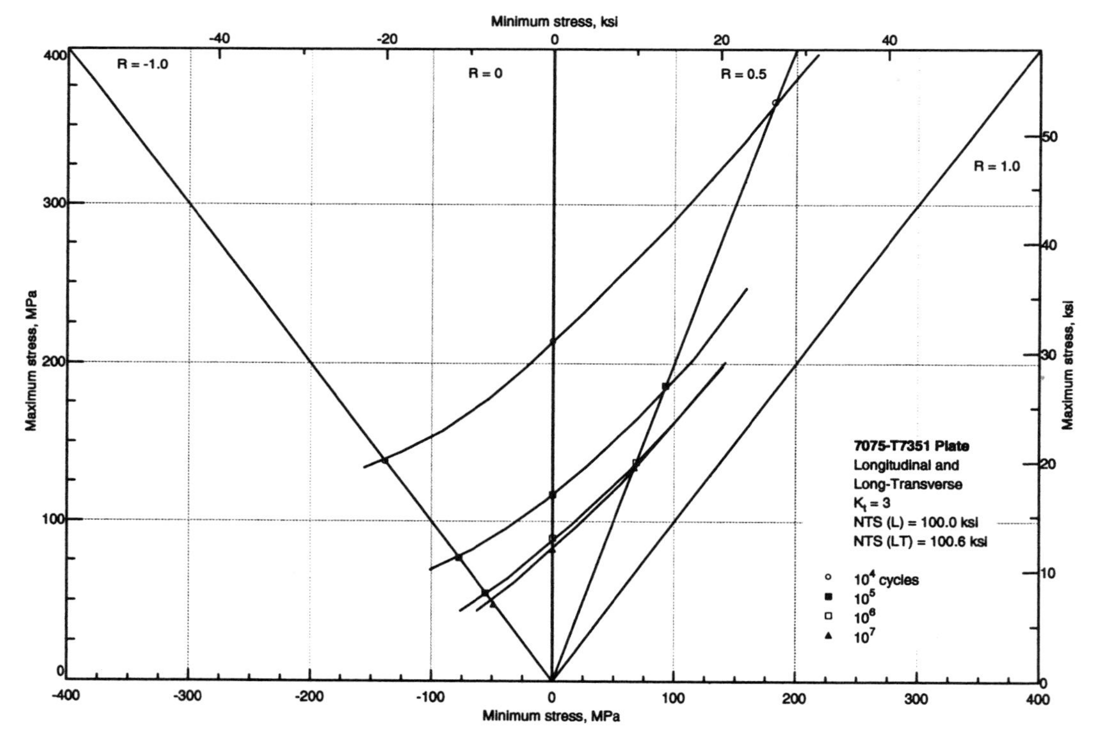

Fig. 110 7075-T7351 notched axial fatigue ($K_t = 3$). Source: Alcoa, 1970

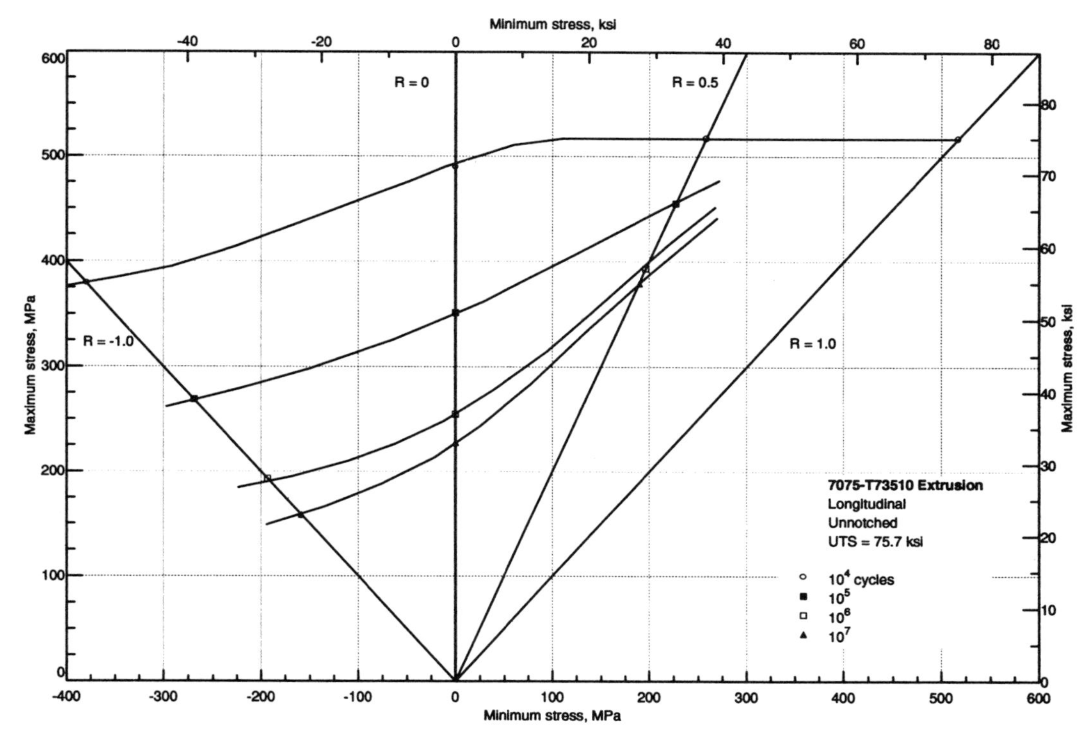

Fig. 111 7075-T73510 longitudinal unnotched axial fatigue (extrusion)

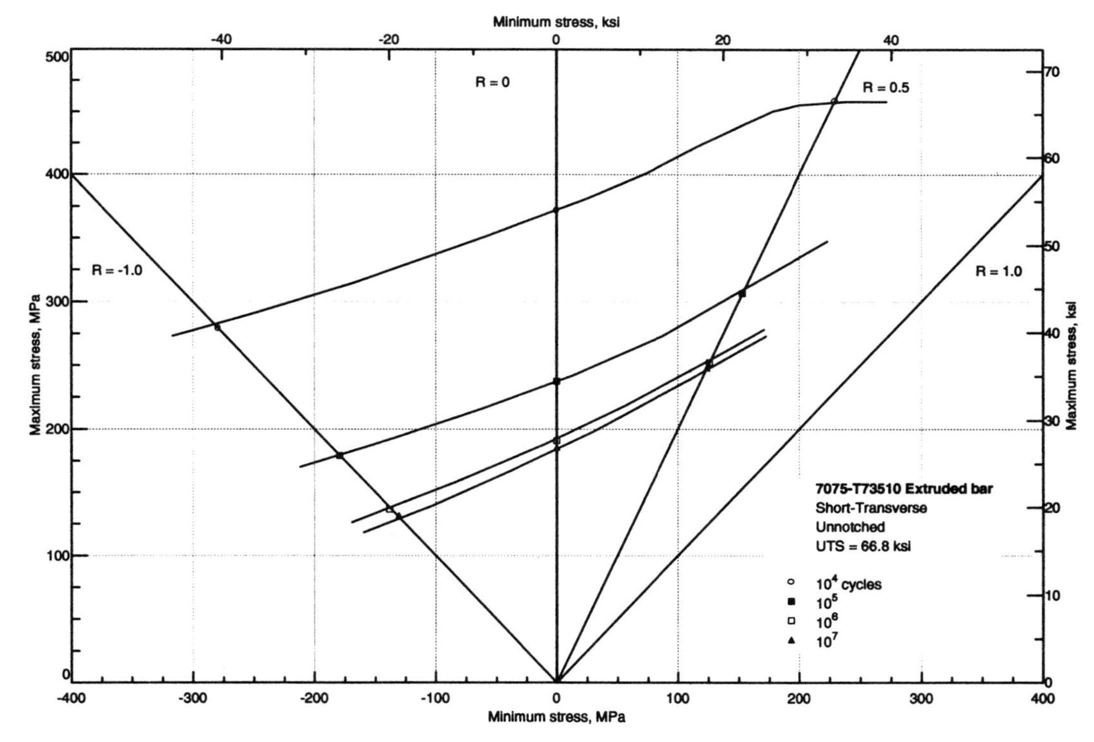

Fig. 112 7075-T73510 short-transverse unnotched axial fatigue (extrusion)

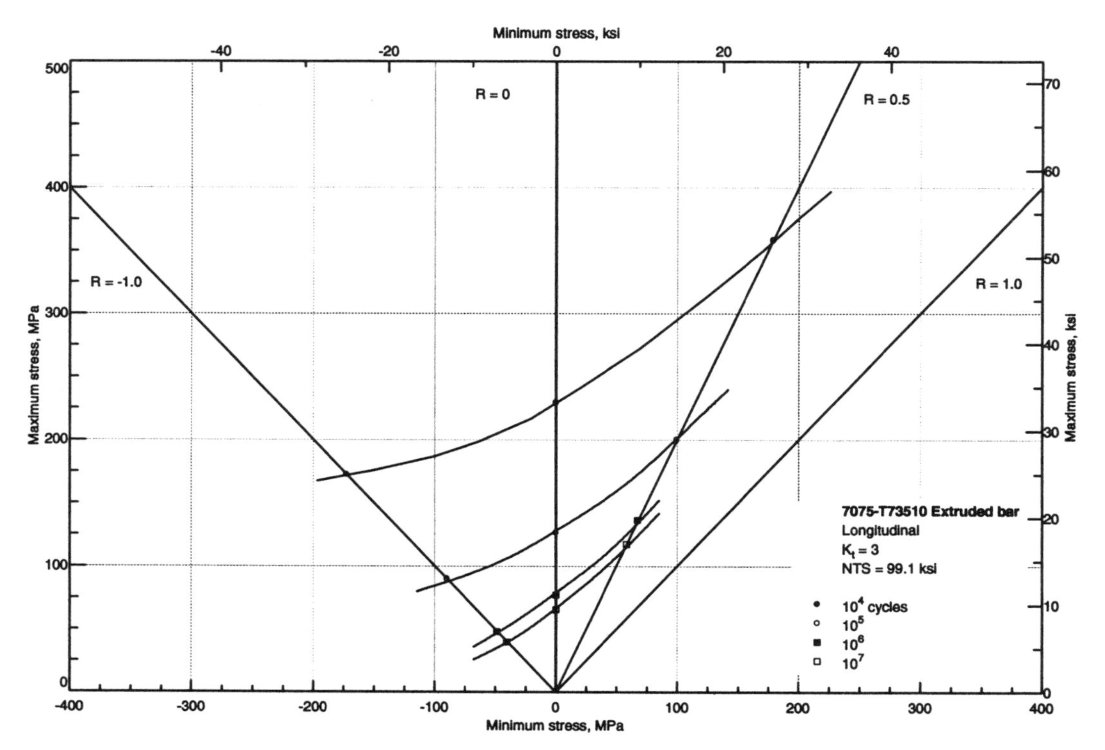

Fig. 113 7075-T3510 longitudinal notched fatigue ($K_t = 3$) specimens (extruded bar, 3.5 x7.5 in.)

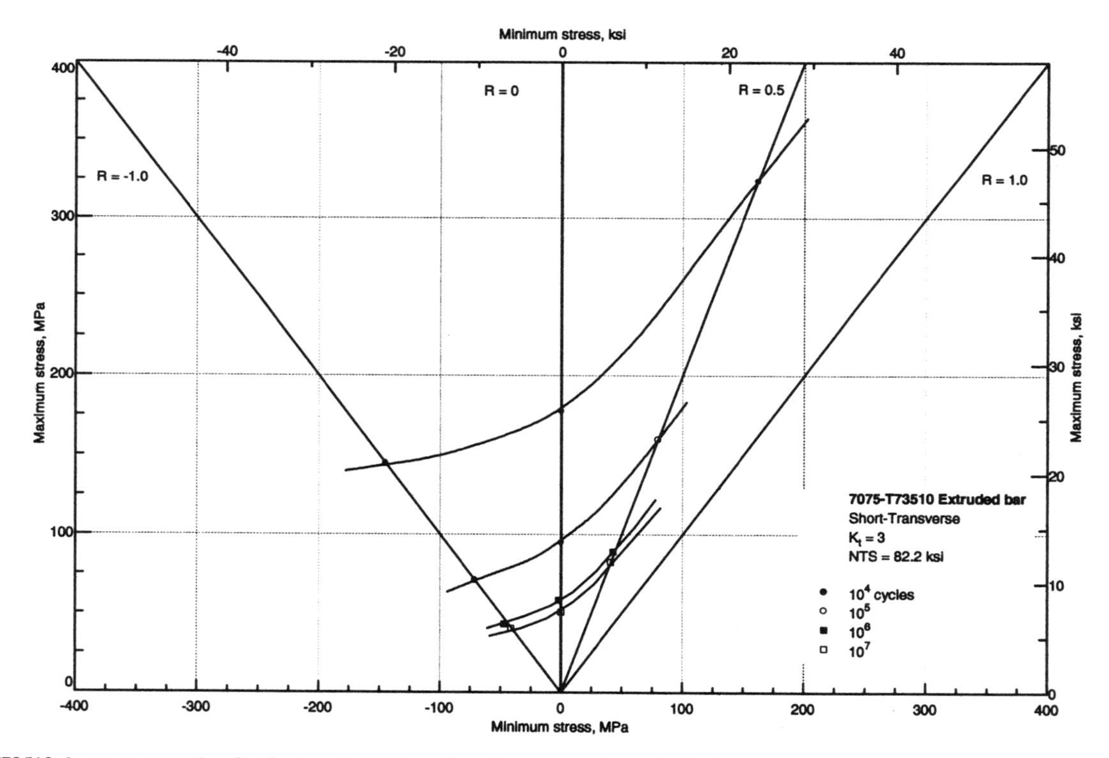

Fig. 114 7075-T73510 short-transverse longitudinal notched fatigue ($K_t = 3$) specimens (extruded bar)

7075 Corrosion Fatigue

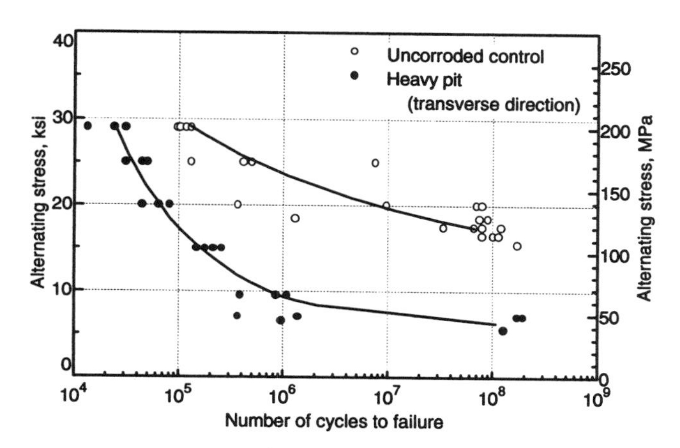

Fig. 115 7075-T6 intergranular corrosion effect on fatigue (bending fatigue, $R = -1$, 0.125 in. sheet). Source: *Mat. Perform.*, Vol 14, 1975, p 22

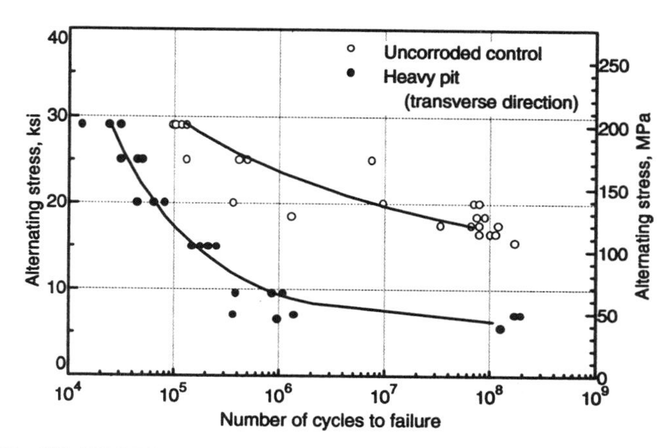

Fig. 116 7075-T6 pitting corrosion effect on fatigue (bending fatigue, $R = -1$, 0.125 in. sheet). Source: *Mat. Perform.*, Vol 14, 1975, p 22

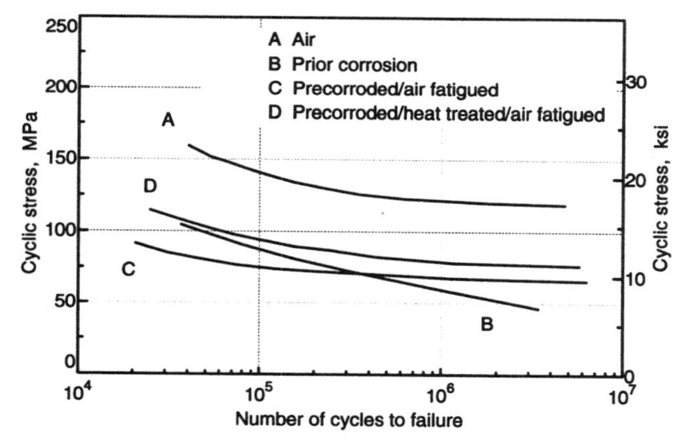

Fig. 117 7075-T6 fatigue in air after precorrosion in 0.5 M NaCl at room temperature (mean stress 40 ksi, 275 MPa)

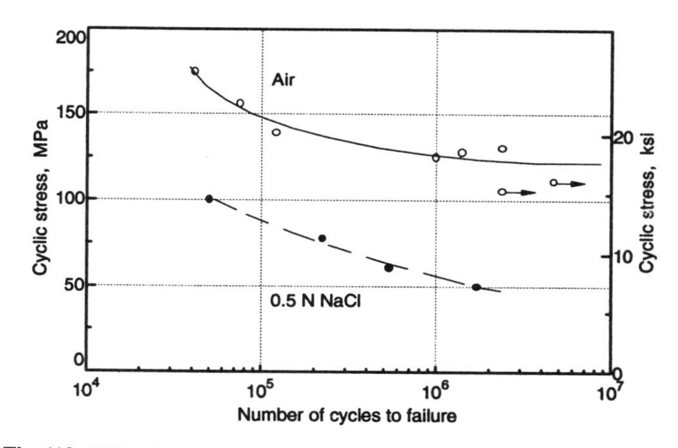

Fig. 118 7075-T6 fatigue lives in 0.5 N NaCl solution. Mean stress 40 ksi (275 MPa). Source: RPI Technical report to ONR, AD763455, April 1973

7076, 7079, and 7106

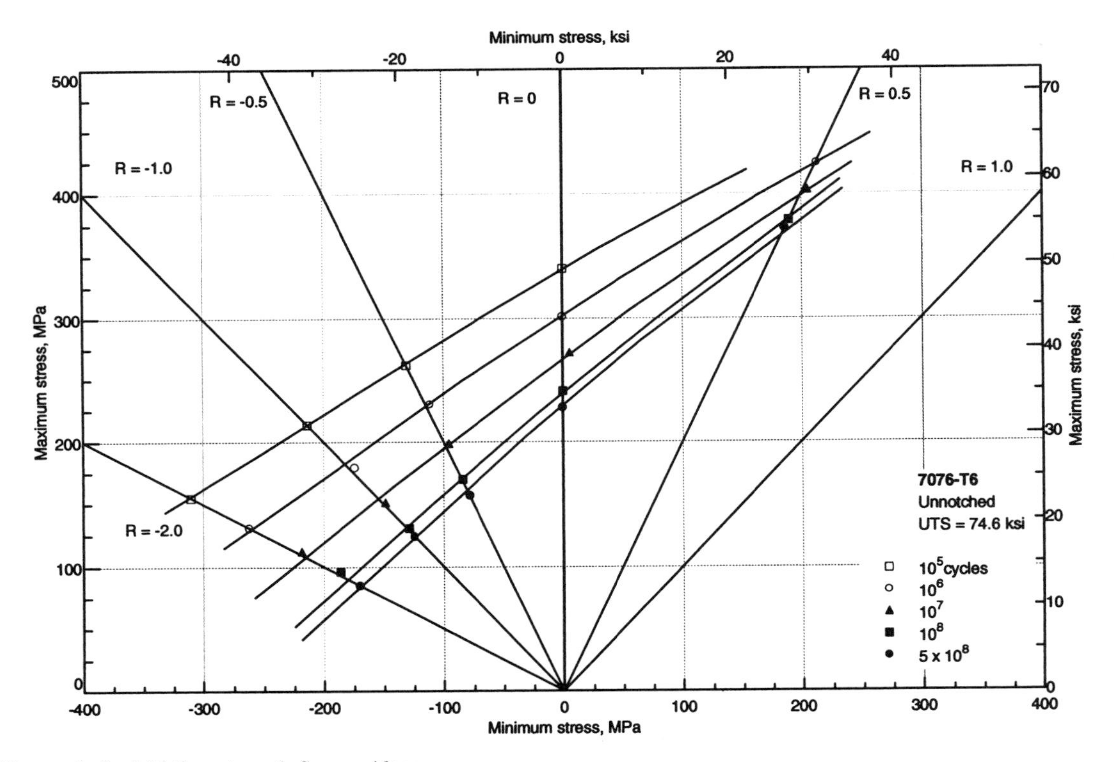

Fig. 119 7076-T6 unnotched axial fatigue strength. Source: Alcoa

7079-T6: Room-temperature fatigue strength in air and Severn River water

Form and thickness	Specimen	Test mode	Stress ratio	Atmosphere	Fatigue strength (ksi) at cycles of: 10^4	10^5	10^6	10^7	10^8	5×10^8
3 in. plate	0.500 in. round cantilever	Reverse bending	−1	Air	(49)	(35)	(25)	(18)	(15)	...
3 in. plate	0.500 in. round cantilever	Reverse bending	−1	Severn River Water	...	(25)	(11)	(5)	(2)	...
3 in. plate	As above but notched, $K_t = 3.0$	Reverse bending	−1	Air	...	(20)	(11)	(8)	(6)	...
3 in. plate	As above	Reverse bending	−1	Severn River Water	...	(18)	(10)	(6)	(4)	...
0.1 in. sheet	Transverse electron beam weld, as welded	Axial	0	Air	...	(21.5)	(17)	...	...	...
0.1 in. sheet	Transverse electron beam weld, as welded	Reverse plate bending	−1	Air	...	(23)	(15)	(13)	(12.5)	...
Forging	RR Moore	Rotating beam	−1	Air	(58)	(37)	(28)	(26)	(25)	(24)
Forging	RR Moore, notched, $K_t = 12$	Rotating beam	−1	Air	(29)	(18)	(12)	(9)	(8)	(8)

Sources: Unpublished Alcoa data and M.G. Vassilaros and E.J. Czyryca, Naval Research Lab Report CDNSWC-TR619409, A Compilation of Fatigue Information for Aluminum Alloys, 1994

7106-T63: Room-temperature fatigue strength in air and Severn River water

Form and thickness	Specimen	Test mode	Stress ratio	Atmosphere	Fatigue strength (ksi) at cycles of: 10^4	10^5	10^6	10^7	10^8	5×10^8
3 in. plate	½ in. Diam cantilever beam	Reverse bending	−1	Air	...	...	(28)	(23)	(20)	...
3 in. plate	½ in. Diam cantilever beam	Reverse bending	−1	Severn River Water	...	(28)	(16)	(6)	...	...
3 in. plate	As above, transverse butt weld, bead-off 5180 filler	Reverse bending	−1	Air	...	(12)	(7)	(6)	(5)	...
⅜ in. plate	Transverse single-V butt weld, 5180 filler, bead-on	Axial	0	Air	(28.5)	(20)	(13)	(12)	...	...
⅜ in. plate	As above, bead-off	Axial	0	Air	...	(30)	(21)	(19)	...	...

Source: M.G. Vassilaros and E.J. Czyryca, Naval Research Lab Report CDNSWC-TR619409, A Compilation of Fatigue Information for Aluminum Alloys, 1994

7079 and X7080

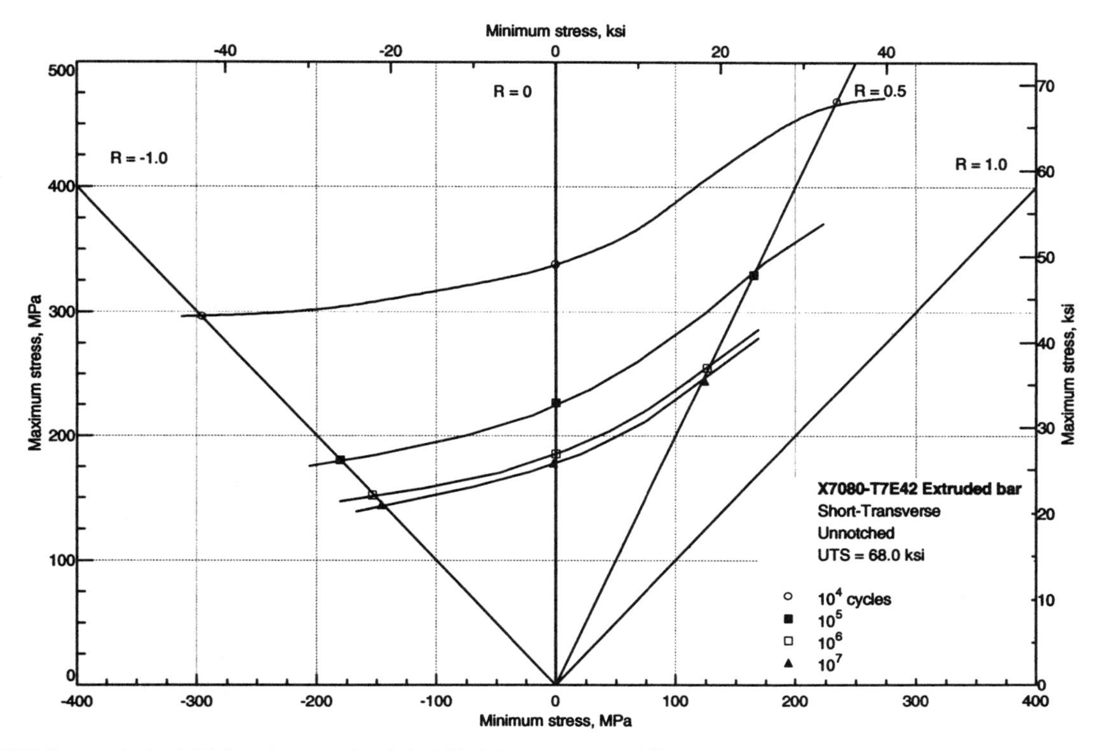

Fig. 120 X7080-T7E42 unnotched axial fatigue (extruded bar, 3.5 x7.5 in.). Source: Alcoa, 1968

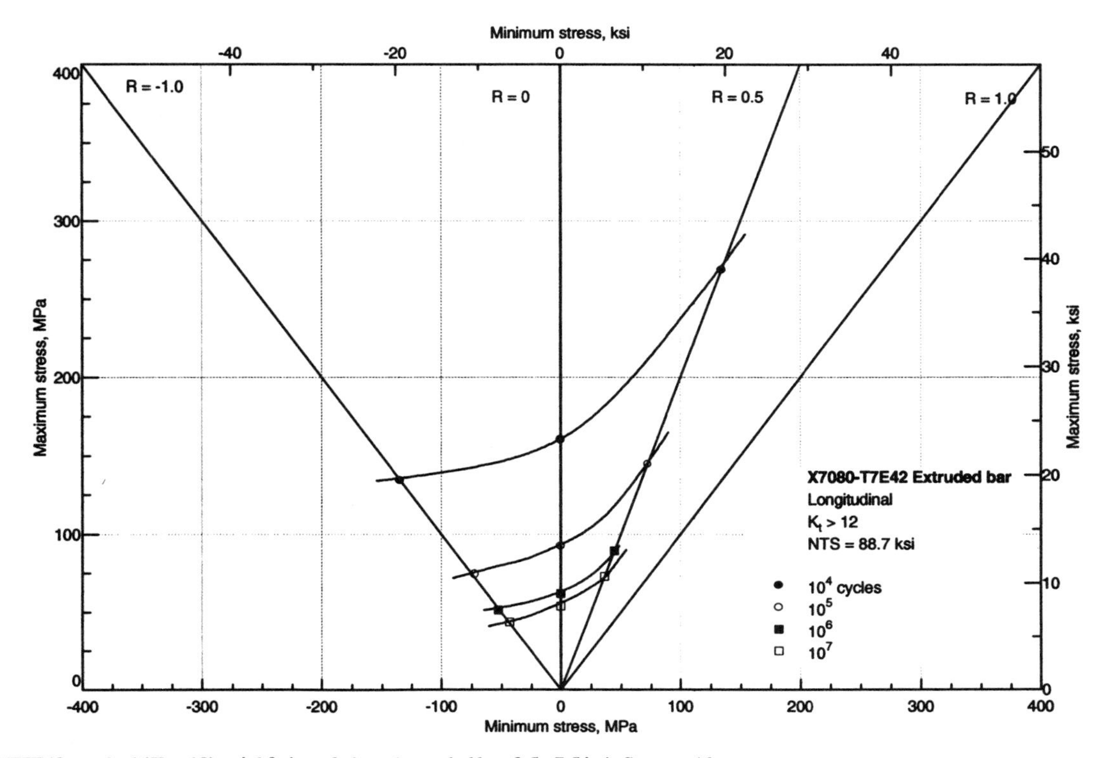

Fig. 121 X7080-T7E42 notched ($K_t > 12$) axial fatigue fatigue (extruded bar, 3.5 x7.5 in.). Source: Alcoa

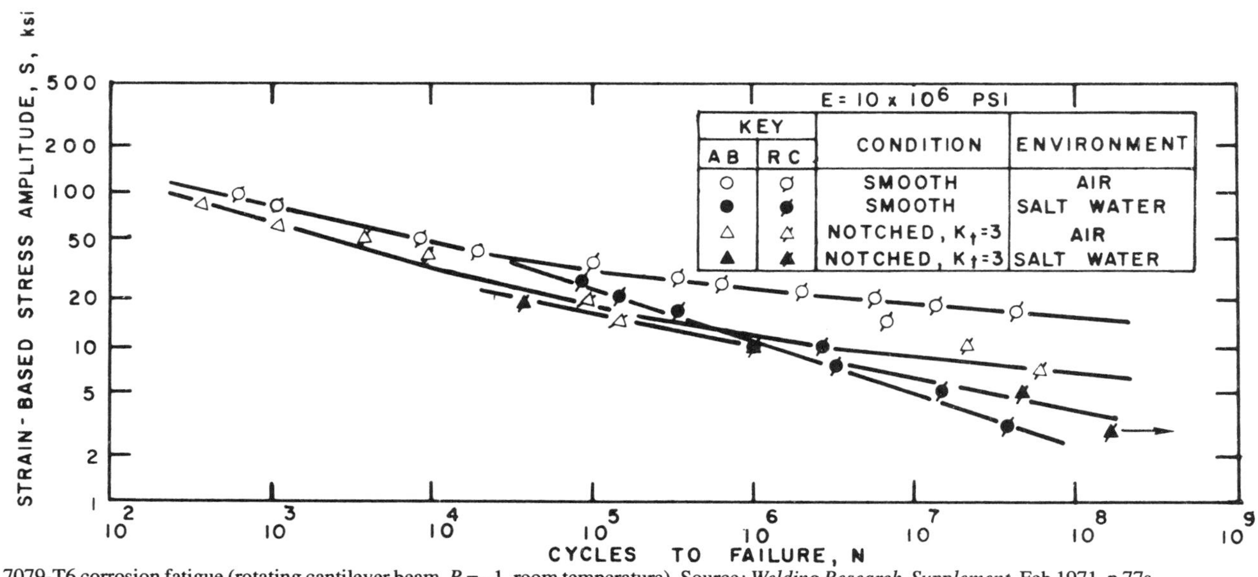

Fig. 122 7079-T6 corrosion fatigue (rotating cantilever beam, $R = -1$, room temperature). Source: *Welding Research Supplement*, Feb 1971, p 77s

7149-T73

7149-T73: Room-temperature fatigue strength in air

Form and thickness	Specimen	Test mode	Stress ratio	Fatigue strength (ksi) at cycles of: 10^4	10^5	10^6	10^7	10^8	5×10^8
Forging	Unnotched round	Axial	0.5	...	(68)	(50)	(42)	...	...
		Axial	0.1	...	(54)	(40)	(34)	...	...
		Axial	–0.5	(65)	(44)	(33)	(28)	...	...
Forging	Notched, $K_t = 3$	Axial	0.5	(48)	(30)	(20)	(15)	...	...
		Axial	0.1	(37)	(24)	(16)	(12)	...	...
		Axial	–0.5	(30)	(19)	(13)	(10)	...	...

Source: MIL-HDBK 5

7175

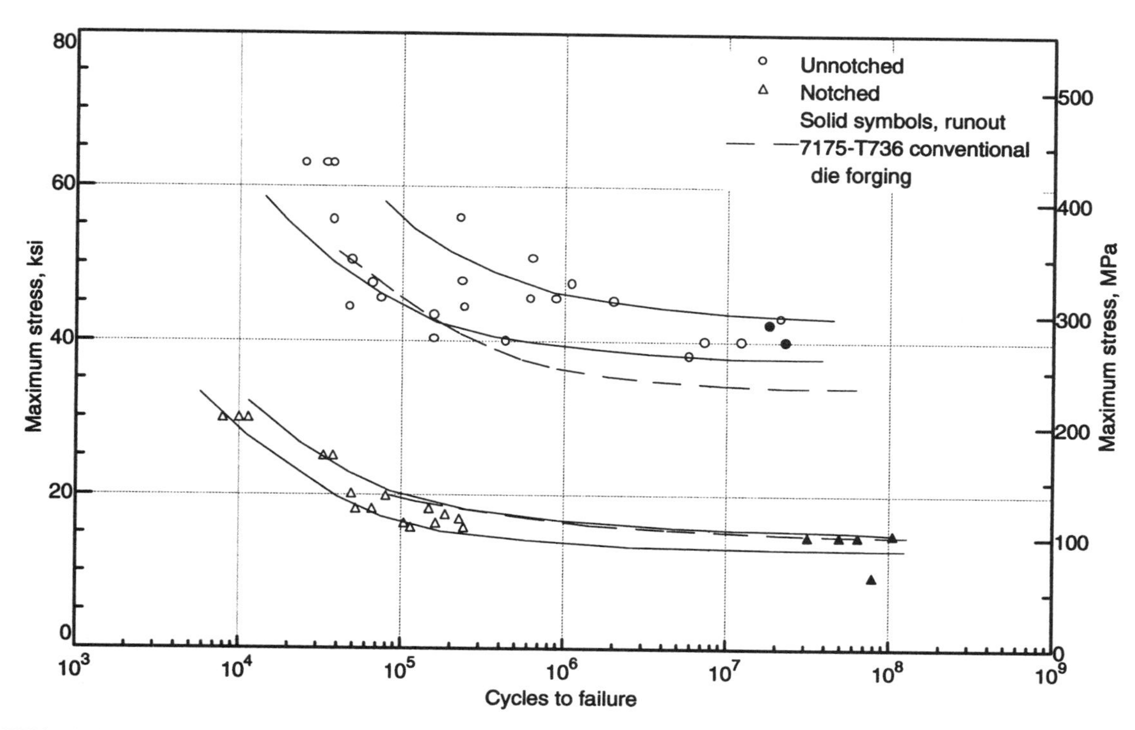

Fig. 123 7175-T736 axial fatigue ($R = 0$) of precision and conventional forgings. Solid symbols indicate runout (no failure). $K_t = 3$ for notched data.

7175: Room-temperature fatigue strength in air

Temper	Form and thickness	Specimen	Test mode	Stress ratio	Fatigue strength (ksi) at cycles of: 10^4	10^5	10^6	10^7	10^8	5×10^8
T66		Round, notched, $K_t = 12$	Axial	0	(21)	(16)	(15)	(13)	...	...
T7651X	Extrusion	Round	Axial	0	(76)	(46)	(36)	(35)	...	...
T736	Forging	RR Moore	Rotating beam	–1	(68)	(50)	(44)	(42)	...	...
		RR Moore, notched, $K_t = 12$	Rotating beam	–1	(28)	(17)	(14)	(13)	(13)	...
T6	Forging	RR Moore	Rotating beam	–1	...	(44)	(35)	(27)	(26)	(26)
		RR Moore, notched, $K_t = 12$	Rotating beam	–1	(26)	(14)	(11)	(8)	(7)	(7)

Source: Unpublished Alcoa data

7178

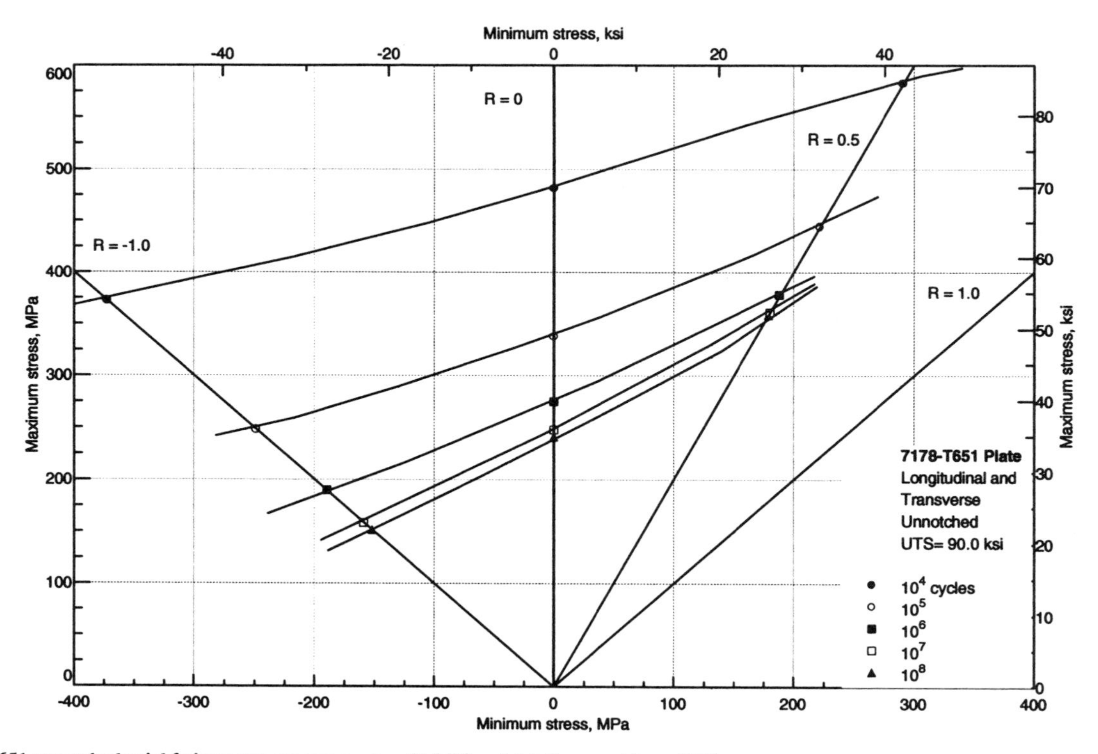

Fig. 124 7178-T651 unnotched axial fatigue at room temperature (1-3/8 in. plate). Source: Alcoa, 1968

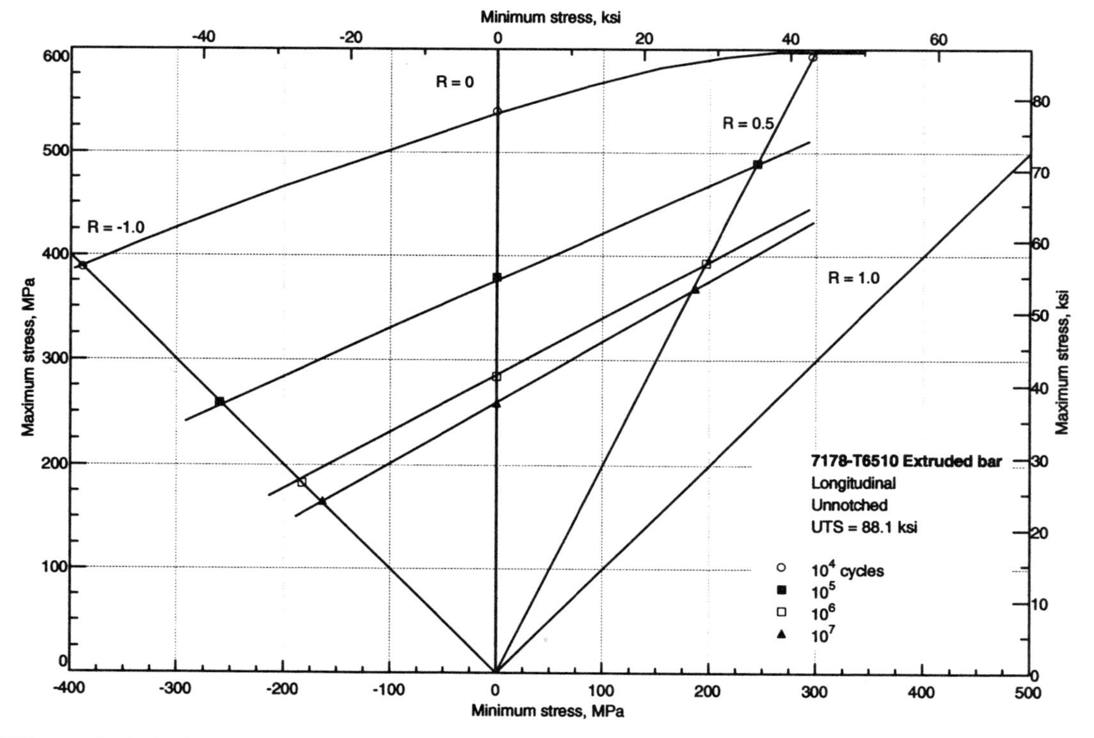

Fig. 125 7178-T6510 unnotched axial fatigue at room temperature (3.5 x7.5 in. extruded bar). Source: Alcoa

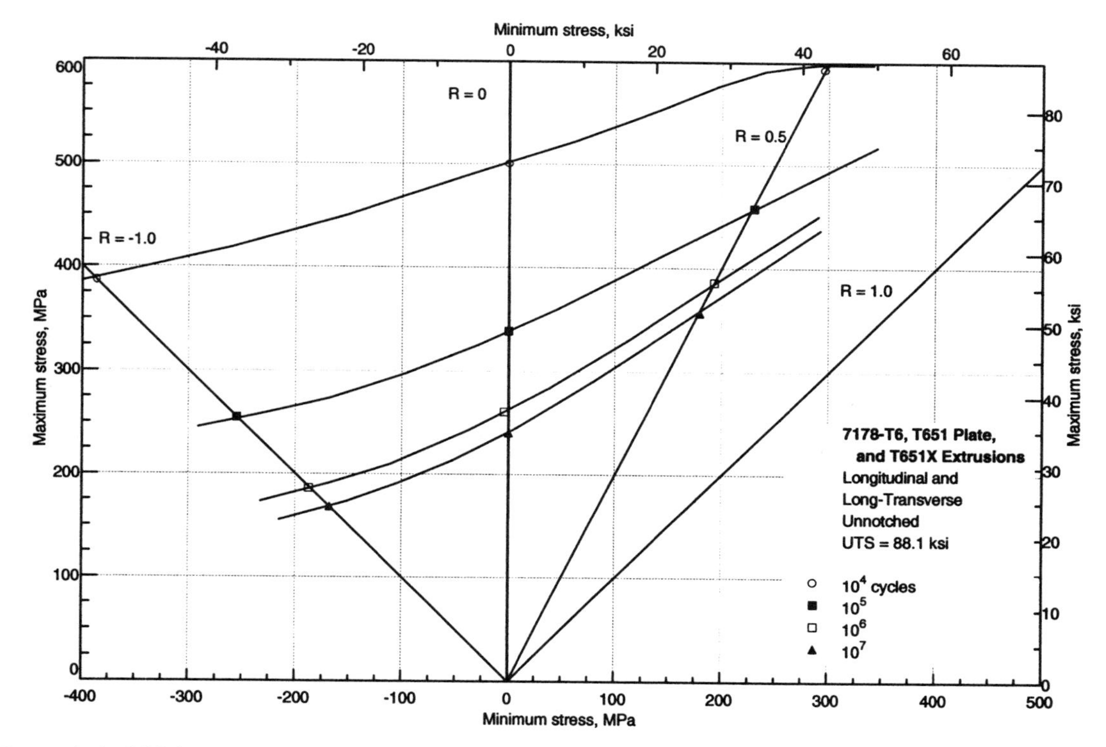

Fig. 126 7178-T6 unnotched axial fatigue at room temperature. Source: Alcoa

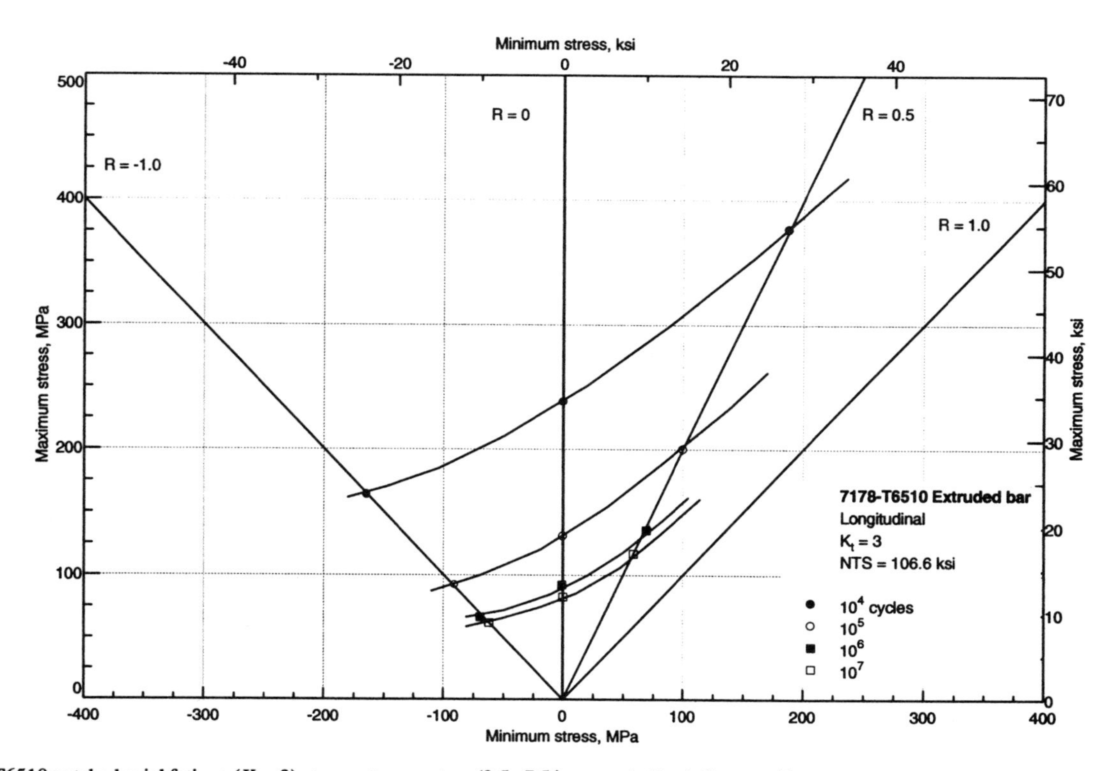

Fig. 127 7178-T6510 notched axial fatigue ($K_t = 3$) at room temperature (3.5 x 7.5 in. extruded bar). Source: Alcoa

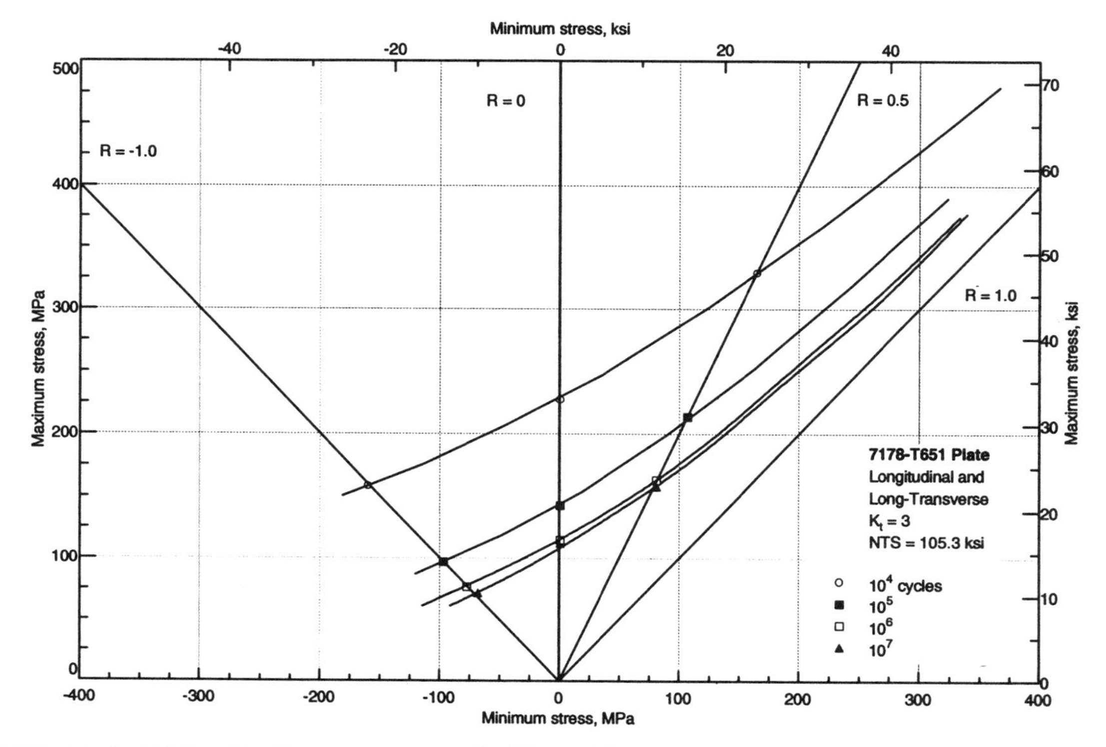

Fig. 128 7178-T6510 notched axial fatigue ($K_t = 3$) at room temperature (1-3/8 in. plate) Source: Alcoa

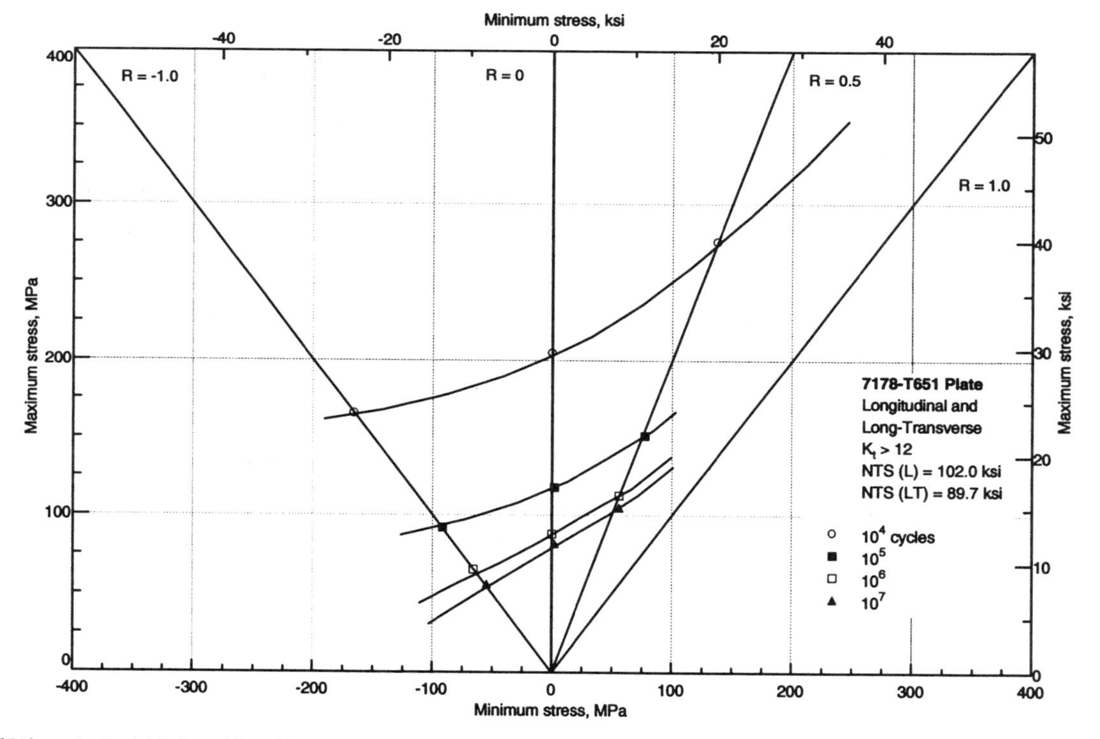

Fig. 129 7178-T6510 notched axial fatigue ($K_t > 12$) at room temperature (1-3/8 in. plate) Source: Alcoa

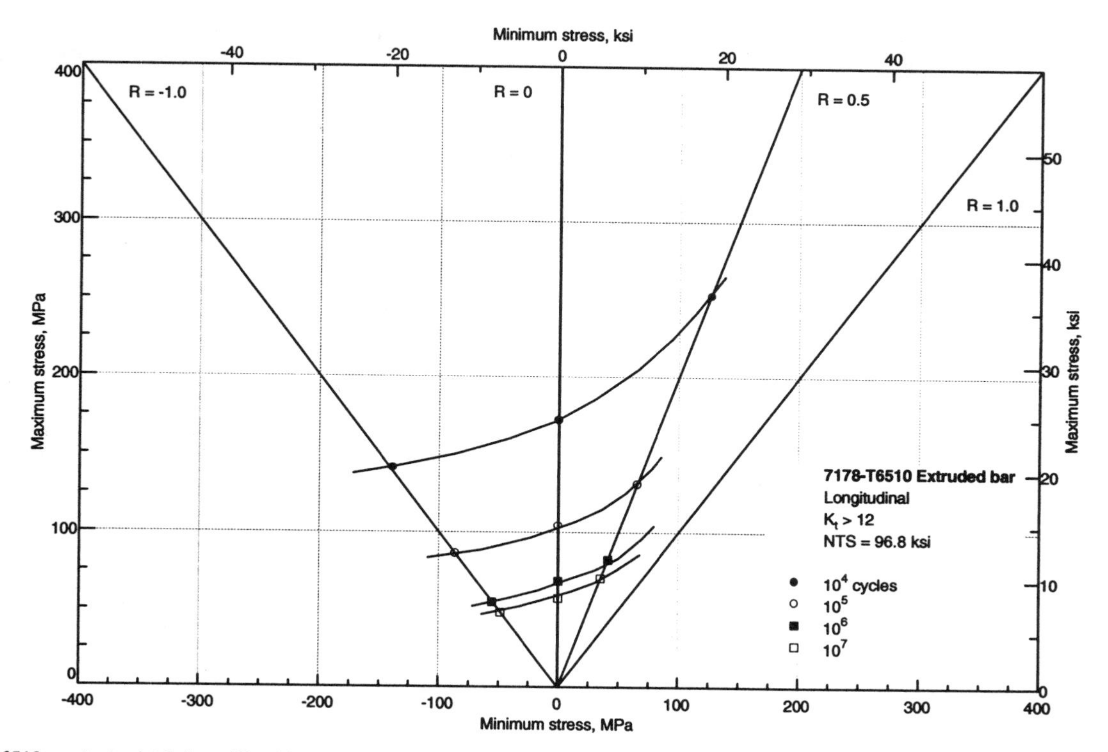

Fig. 130 7178-T6510 notched axial fatigue ($K_t > 12$) at room temperature (3.5 x7.5 in. extruded bar). Source: Alcoa

7178: Room-temperature fatigue strength in air

Temper	Form and thickness	Specimen	Test mode	Stress ratio	Fatigue strength (ksi) at cycles of: 10^4	10^5	10^6	10^7	10^8	5×10^8
T6	Wire	RR Moore	Rotating beam	–1	...	(48)	(36)	(29)	(25)	(24)
		RR Moore, notched, $K_t = 12$	Rotating beam	–1	(28)	(17)	(12)	(9)	(8)	(7)
T76	Plate, Extrusion	Round	Axial	0	(70)	(48)	(39)	(38)	(37)	...
	Forging	RR Moore	Rotating beam	–1	...	(40)	(32)	(30)	(28)	(28)
		RR Moore, notched, $K_t = 12$	Rotating beam	–1	(32)	(19)	(14)	(9)	(7)	(6)
T7651, T7651X	Plate, Extrusion	RR Moore	Rotating beam	–1	...	(36)	(29)	(26)	(24)	(23)
		RR Moore, notched, $K_t = 12$	Rotating beam	–1	(29)	(17)	(10)	(8)	(7)	(7)

Sources: Unpublished Alcoa data

7475

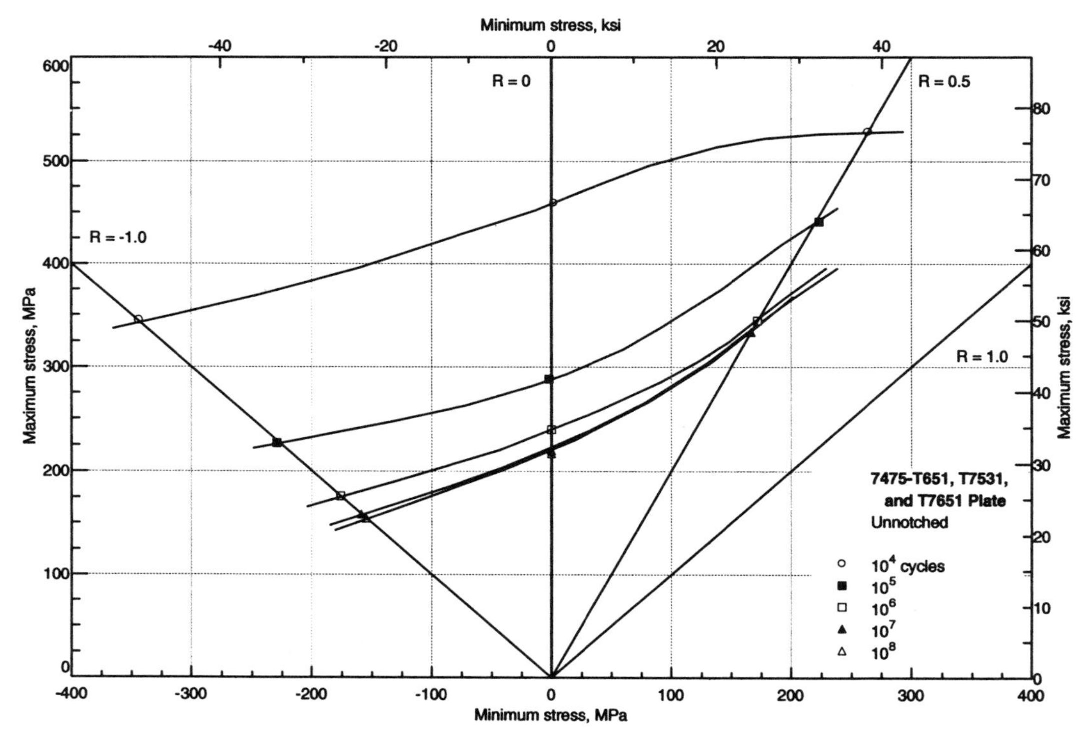

Fig. 131 7475-T651, -T7351, and -T7651 unnotched axial fatigue (plate). Source: Alcoa

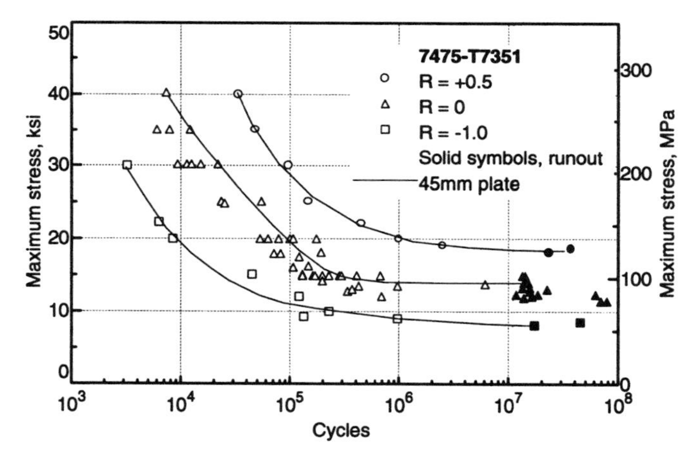

Fig. 132 7475-T7351 notched axial fatigue (K_t = 3, notch radius of 0.013 in.). Solid symbols indicate runout (no failure). Data are for longitudinal and long-transverse specimens of plate varying in thicknesses from 19 to 90 mm (0.75 to 3.50 in.). Plotted lines are for 45-mm (1.75-in.) plate data.

7475 sheet and forging: Room-temperature fatigue strength in air

Temper	Form and thickness	Specimen	Test mode	Stress ratio	Fatigue strength (ksi) at cycles of:					
					10^4	10^5	10^6	10^7	10^8	5×10^8
T61, T761	Sheet	Sheet	Axial	0.00	(68)	(49)	(35)	(25)	...	...
	Sheet	Sheet	Axial	0.00	(73)	(58)	(46)	(36)	(30)	...
	Sheet	Sheet, notched, $K_t = 3.0$	Axial	0.00	(31)	(22)	(15)	(11)	...	...
T736	Forging	RR Moore	Rotating beam	−1	...	(45)	(35)	(31)	(30)	...
		RR Moore, notched, $K_t = 12$	Rotating beam	−1	(31)	(14)	(13)	(11)	...	...

Sources: Unpublished Alcoa data

7475 plate: Room-temperature fatigue strength in air

Temper	Form and thickness	Specimen	Test mode	Stress ratio	Fatigue strength (ksi) at cycles of:					
					10^4	10^5	10^6	10^7	10^8	5×10^8
T7351	Plate	Round	Axial	0.50	...	(58)	(42)	(31)	...	...
	Plate	Round	Axial	0.00	(60)	(44)	(32)	(24)	...	...
	Plate	Round	Axial	−1.00	(46)	(33)	(24)	(18)	...	...
T7351 and T7651	Plate	Round, notched, $K_t = 3.0$	Axial	0.50	...	(32)	(22)	(18)	...	...
	Plate	Round, notched, $K_t = 3.0$	Axial	0.20	(37)	(23)	(16)	(12)	...	...
	Plate	Round, notched, $K_t = 3.0$	Axial	0.10	(35)	(21)	(14)	(11)	...	...
	Plate	Round, notched, $K_t = 3.0$	Axial	0.00	(32)	(20)	(13)	(10)	...	...
	Plate	Round, notched, $K_t = 3.0$	Axial	−1.00	(20)	(12)	(8)	(7)	...	...

Source: Unpublished Alcoa data

7475-T7351 plate: Room-temperature fatigue strength in air from MIL-HDBK 5

Specimen	Test mode	Stress ratio	Fatigue strength (ksi) at cycles of:					
			10^4	10^5	10^6	10^7	10^8	5×10^8
Unnotched round	Axial	0.5	...	(58)	(42)	(31)	...	...
		0.0	(60)	(44)	(32)	(24)	...	...
		−1.0	(46)	(33)	(24)	(18)	...	...
Notched, $K_t = 3$	Axial	0.5	...	(32)	(22)	(18)	...	...
		0.2	(37)	(23)	(16)	(12)	...	...
		0.1	(35)	(21)	(14)	(11)	...	...
		0.0	(32)	(20)	(13)	(10)	...	...
		−1.0	(20)	(12)	(8)	(7)	...	...

Source: MIL-HDBK 5

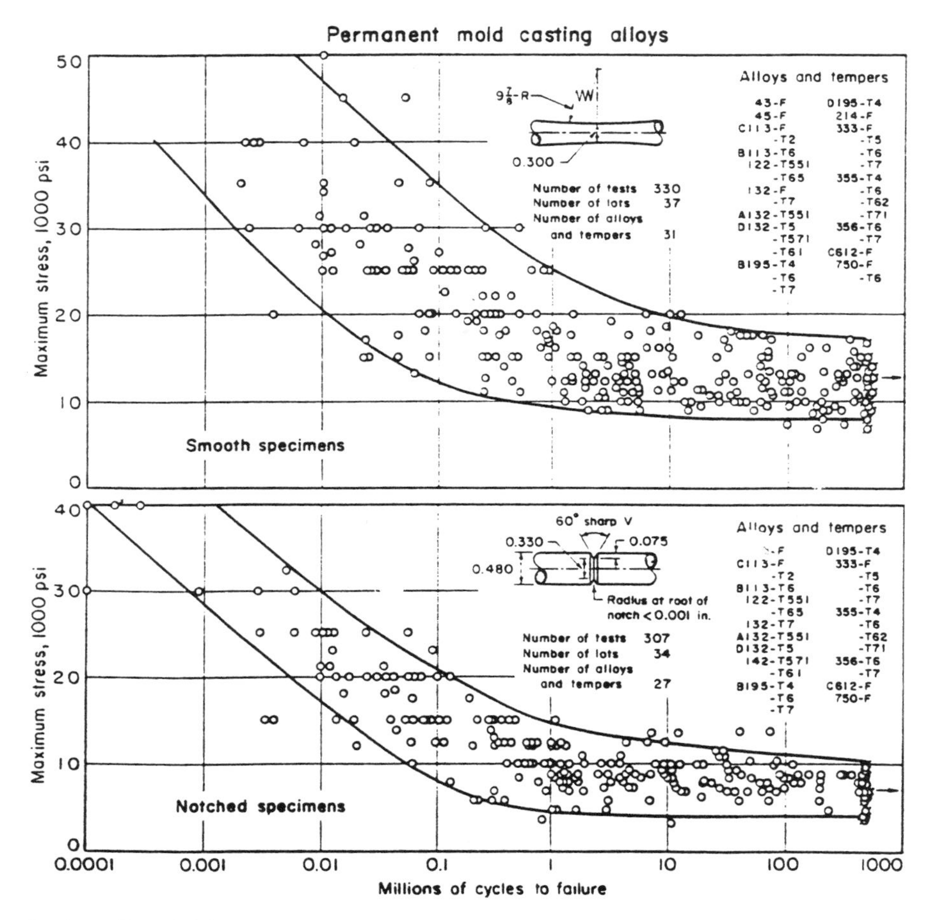

Fig. 133 Scatterbands for rotating beam ($R = -1$) fatigue strength of permanent mold aluminum casting alloys

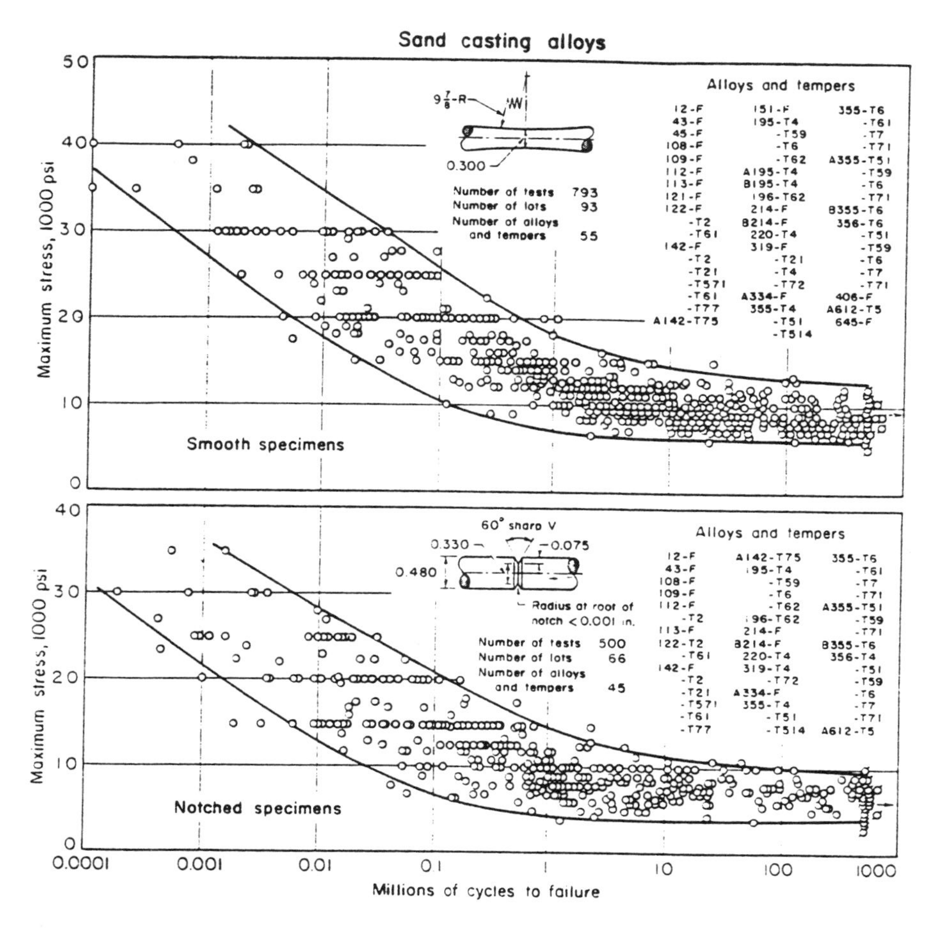

Fig. 134 Scatterbands for rotating beam ($R = -1$) fatigue strength of sand cast aluminum alloys

Cast aluminum alloys: Miscellaneous room-temperature fatigue strength in air

Alloy and temper	Form and thickness	Specimen	Test mode	Stress ratio	Fatigue strength (ksi) at cycles of:					
					10^4	10^5	10^6	10^7	10^8	5×10^8
354	1 in. cast plate	Polished cantilever beam, 1/2 in. diam	Reverse bending	–1	(38)	(26)	(17)	(15)	(13)	...
355-T6	1/4 in. plate	Polished, 0.2 in. diam bar	Axial	0	(34)	(24)	(16)	(14)	(14)	...
	1/4 in. plate	Flat plate, 5 in. wide	Axial	0	...	(18)	(10)	(10)	(9)	...
	1/4 in. plate	Flat plate, 7 1/2 in. wide, 1 in. diam central hole	Axial	0	(20)	(12)	(9)	(6)	(5)	...
	1/4 in. plate	Flat plate, 7 1/2 in. wide, 2 rows, (4) 5/8 in. rivets	Axial	0	(24)	(19)	(15)	(11)	(9)	...
	1/4 in. plate	Flat plate, 5 in. wide, transverse butt-weld, bead-off	Axial	0	(22)	(17)	(9)	(7)	(6)	...
A356-T61	1 in. cast plate	Polished cantilever beam, 1/2 in. diam	Reverse bending	–1	(33)	(22)	(13)	(11)	(10)	(31)
A356-T61		Round	Axial	0.5	...	...	(34)	(32)	(31)	(20)
		Round	Axial	0.0	(37)	(33)	(27)	(23)	(21)	(14)
		Round	Axial	–0.5	...	(27)	(20)	(17)	(15)	(10)
		Round	Axial	–1.0	(30)	(22)	(16)	(13)	(11)	...
		Round	Axial	–2.0	...	...	(11)	(9)	(8)	(7)
A356-T61		Round	Axial	0.50	...	(68)	(50)	(42)	...	...
		Round	Axial	0.10	...	(54)	(40)	(34)	...	...
		Round	Axial	–0.50	(65)	(44)	(33)	(28)	...	...
		Round	Axial	0.50	(48)	(30)	(20)	(15)	...	...
		Round	Axial	0.10	(37)	(24)	(16)	(12)	...	...
		Round	Axial	–0.50	(30)	(19)	(13)	(10)	...	...
359-T62	1 in. cast plate	Polished cantilever beam, 1/2 in. diam	Reverse bending	–1	(34)	(23)	(14)	(10)	(8)	

Sources: Unpublished Alcoa data and M.G. Vassilaros and E.J. Czyryca, Naval Research Lab Report CDNSWC-TR619409, A Compilation of Fatigue Information for Aluminum Alloys, 1994

Magnesium Alloy Fatigue Data

Magnesium Alloys Fatigue and Fracture*

Magnesium possesses the lowest density of all structural metals, having about 25% the density of iron and approximately 33% that of aluminum. Because of this low density, both cast and wrought magnesium alloys have been developed for a wide variety of structural applications in which low weight is important, if not a requirement.

Magnesium and magnesium alloys are used in structural applications for automotive, industrial, materials-handling, commercial, and aerospace equipment. In the aerospace market, air frame applications in new designs have virtually disappeared and the most significant use of magnesium alloys has been confined largely to cast engine and transmission housings, notably for helicopters. Historically, the Volkswagon Beetle motor car has represented the largest single application of magnesium alloys, which were used for crank case and transmission housing castings that weighed a total of 17 kg. This was said to represent a saving of 50 kg when compared with using cast iron and was critical for the stability of this rear-engined vehicle.

More recently, weight reduction by material substitution is an application area for increased automotive fuel economy and reduced exhaust emissions. This is the purpose, for example, of the U.S. Corporate Average Fuel Economy (CAFE) legislation, which requires manufacturers to achieve progressive improvements in the fuel economies of vehicles they produce, and which is providing much of the stimulus to programs of weight reduction through materials substitution. Table 1 presents a selection of magnesium alloy components in various new production vehicles. The other major areas in which the use of magnesium is expanding are appliances and sporting goods. The trend here has again been for an increase in the use of magnesium die castings, for example in computer housing and mobile telephone cases where lightness, capability for thin-wall casting, and provision of electromagnetic shielding are special advantages.

This introductory article briefly reviews the physical metallurgy, alloy types, and fatigue and fracture toughness of magnesium alloys. The sections on fatigue and fracture toughness in this article provide a general overview, while subsequent articles are compilations of alloy-specific fatigue data.

Table 1 Selection of magnesium alloy components in new production motor cars and trucks

Company	Part	Model	Alloy
Ford	Clutch housing, oil pan, steering column	Ranger	AZ91HP AZ91B
	Four-wheel drive transfer case housing	Aerostar 1994	AZ91D
	Manual transmission case housing	Bronco	AZ91D
General Motors	Valve cover, air cleaner, clutch housing (manual)	Corvette	AZ91HP
	Induction cover	North Star V-8 1992	AZ91D
	Clutch pedal, brake pedal, steering column brackets	"W" Oldsmobile, Pontiac, Buick	AZ91D
Chrysler	Drive brackets, oil pan	Jeep 1993	...
	Steering column brackets	LH Midsize 1993	...
	Drive brackets, oil pan	Viper	...
Daimler-Benz	Seat frames	500 SL	AM20/50
Alfa-Romeo	Miscellaneous components (45 kg)	GTV	AZ91B
Porsche AG	Miscellaneous components (53 kg)	911	...
	Wheels (7.44 kg each)	944 Turbo	AZ91D
Honda	Cylinder head cover	City Turbo	AZ91D
	Wheels (5.9 kg each)	Prelude	AM60B
Toyota	Steering wheel	Lexus	AM60B
Feuling Engineering	Cylinder block, oil sump, camshaft cover, front cover assembly	5HQ Quad 4 Aerotech	AZ91E ZE41A ZC63

Source: *Materials Australasia*, Vol 20 (No. 9), 1992, p 12.

The discussions on individual alloy systems of practical importance focus mainly on casting alloys. However, some new developments involving novel magnesium-base alloys prepared by rapid solidification processing, and metal matrix composites, are also considered. In all cases, special emphasis is placed on relationships between microstructure and properties. A more general account of the metallurgy of magnesium alloys is given in the classic book by Emley (Ref 1) and developments in alloy theory are discussed in Ref 2 to 4. Microstructural features are considered in more detail in the recent book by Polmear (Ref 5). Special attention to properties and the design of magnesium products is given by Busk (Ref 6). The proceedings of a recent comprehensive conference on magnesium alloys are also available (Ref 7).

Alloy Designations and Tempers. No international code for designating magnesium alloys exists, although there has been a trend toward adopting the method used by the American Society for Testing and Materials. In this system, the first two letters indicate the principal alloying elements according to the following code:

- A—aluminum
- B—bismuth
- C—copper
- D—cadmium
- E—rare earths
- F—iron
- H—thorium
- K—zirconium
- L—lithium
- M—manganese
- N—nickel
- P—lead
- Q—silver
- R—chromium
- S—silicon
- T—tin
- W—yttrium
- Y—antimony
- Z—zinc

The letter corresponding to the element present in greater quantity in the alloy is used first; if the elements are equal in quantity the letters are listed alphabetically. The two (or one) letters are followed by numbers that represent the nominal compositions of these principal alloying elements in weight percent, rounded off to the nearest whole number (e.g., AZ91 indicates the alloy Mg-9Al-1Zn, the actual composition ranges being 8.3 to 9.7 wt% Al and 0.4 to 1.0 wt%

*Adapted from the following articles: I.J. Polmear, Magnesium Alloys and Applications, submitted for publication in the *Materials Properties Handbook* series, ASM International; G.L. Makar and J. Kruger, Corrosion of Magnesium, *International Materials Review*, Vol 38, 1993, p 138-153; and W.K. Miller, Stress Corrosion Cracking of Magnesium Alloys, *Stress-Corrosion Cracking: Materials Performance and Evaluation*, ASM International, 1992.

Zn. A limitation is that information concerning other intentionally added elements is not given, and the system may need to be modified on this account. Suffix letters A, B, C, and so on refer to variations in composition and purity within the specified range, and X indicates that the alloy is experimental.

For heat-treated or work-hardened conditions, the designations are specified by the same system used for aluminum alloys. Commonly used tempers are T5 (alloys that were artificially aged after casting), T6 (alloys that were solution treated, quenched, and artificially aged), and T7 (alloys that were solution treated and stabilized).

Physical Metallurgy

In contrast to aluminum, magnesium has a close-packed hexagonal crystal structure, with parameters of $a = 3.202$ Å, $c = 5.199$ Å, and $c/a = 1.624$ Å (which is very close to the ratio of 1.633 obtained by piling spheres in the same arrangement). This structure is basic to much of the physical metallurgy of magnesium and magnesium alloys. At room temperature, slip occurs mainly on (0001) ($\langle 11\bar{2}0 \rangle$), with a small amount sometimes seen on pyramidal planes such as $(10\bar{1}1)\langle 11\bar{2}0 \rangle$. As the temperature is raised, pyramidal slip becomes easier and more prevalent. However, note that the slip directions, whether associated with basal or the pyramidal planes, are coplanar with (0001), a general observation for all observed slip in magnesium and magnesium alloys. Therefore, it is impossible for a polycrystalline piece of magnesium to deform without cracking unless deformation mechanisms other than slip are available. These mechanisms are twinning, banding, and grain-boundary deformation.

Another key feature that dominates the physical metallurgy of magnesium alloys is the fact that the atomic diameter (0.320 nm) of magnesium enjoys favorable size factors with a diverse range of solute elements. In this regard, atoms with sizes that place them within the favorable ±15% range with respect to magnesium are shown in the shaded band in Fig. 1 (Ref 8, 9). Further restrictions in solid solubility are imposed through differences in valency (relative valency effect) and because of the chemical affinity of the highly electropositive magnesium with elements such as silicon and tin, which leads to the formation of a number of stable compounds.

Aluminum, zinc, cerium, yttrium, silver, thorium, and zirconium are examples of widely differing metals that may be present in commercial alloys. Apart from magnesium and cadmium, which form a continuous series of solid solutions, the magnesium-rich sections of binary phase diagrams (Ref 10, 11) show peritectic (or, more commonly, eutectic) systems. Of the wide range of intermetallic compounds that may form (Ref 3, 4), the three most frequent types of structures are:

- AB: Simple cubic CsCl structure. Examples are MgTl, MgAg, CeMg, and SnMg. It will be seen that magnesium can be either the electropositive or electronegative component.
- AB_2: Laves phases with ratio $R_A/R_B = 1.23$ preferred. Three types exist: $MgCu_2$ (face-centered cubic, stacking sequence abcabc), $MgZn_2$ (hexagonal, stacking sequence ababab), and $MgNi_2$ (hexagonal, stacking sequence abacaba).
- CaF_2: Face-centered cubic. This group contains Group IV elements; examples are Mg_2Si and Mg_2Sn.

Another characteristic of alloy systems in which solubility is strongly influenced by atomic size factors is that solid solubility generally decreases with decreasing temperature (Ref 10). Such a feature is a necessary requirement for precipitation hardening, and most magnesium alloys are amenable to this phenomenon, although the responses are significantly less than is observed in some aluminum alloys. Precipitation processes are usually complex (Ref 5), and a feature common to

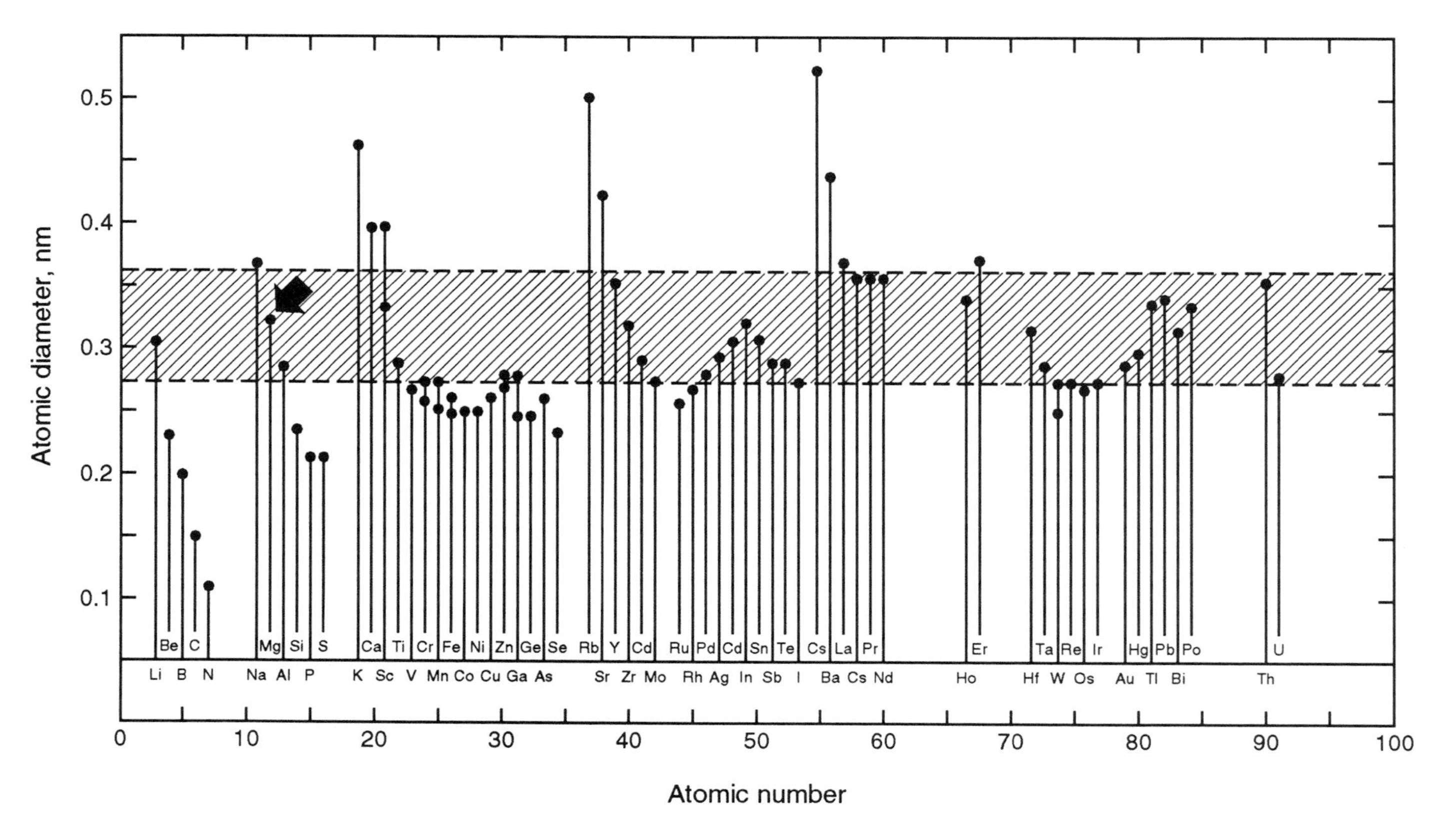

Fig. 1 Atomic diameters of the elements and the favorable size factor (shaded area) with respect to magnesium

Alloy system	Precipitation process				
Mg–Al	SSSS	→ $Mg_{17}Al_{12}$ equilibrium precipitate nucleated on $(0001)_{Mg}$ (incoherent)			
Mg–Zn(–Cu)	SSSS	→ GP zones discs $\parallel\{0001\}_{Mg}$ (coherent)	→ $MgZn_2$ rods $\perp\{0001\}_{Mg}$ hcp $a = 0{\cdot}52$ nm $c = 0{\cdot}85$ nm (coherent)	→ $MgZn_2$ discs $\parallel\{0001\}_{Mg}$ $(11\bar{2}0)_{MgZn_2}\parallel\{10\bar{1}0\}_{Mg}$ hcp $a = 0{\cdot}52$ nm $c = 0{\cdot}848$ nm (semicoherent)	→ Mg_2Zn_3 trigonal $a = 1{\cdot}724$ nm $b = 1{\cdot}445$ nm $c = 0{\cdot}52$ nm $\gamma = 138^\circ$ (incoherent)
Mg–RE(Nd)	SSSS	→ GP zones (Mg–Nd) plates $\parallel\{10\bar{1}0\}_{Mg}$ (coherent)	→ β'' Mg_3Nd? hcp DO_{19} superlattice plates $(0001)_{\beta''}\parallel(0001)_{Mg}$ $\{1010\}_{\beta''}\parallel\{10\bar{1}0\}_{Mg}$	→ β' Mg_3Nd fcc plates $a = 0{\cdot}736$ nm $(011)_{\beta'}\parallel(0001)_{Mg}$ $(\bar{1}1\bar{1})_{\beta'}\parallel\{\bar{2}110\}_{Mg}$ (semicoherent)	→ β $Mg_{12}Nd$ bct $a = 1{\cdot}03$ nm $c = 0{\cdot}593$ nm (incoherent)
Mg–Y–Nd	SSSS	→ ?	→ β'' DO_{19} superlattice hexagonal plates $(0001)_{\beta''}\parallel(0001)_{Mg}$ $[01\bar{1}0]_{\beta''}\parallel[01\bar{1}0]_{Mg}$	→ β' $Mg_{12}NdY$? bc orthorhombic plates $(0001)_{\beta'}\parallel(0001)_{Mg}$ $[100]_{\beta'}\parallel[\bar{2}110]_{Mg}$ $[010]_{\beta'}\parallel[01\bar{1}0]_{Mg}$	→ β $Mg_{11}NdY_2$? bcc $(011)_{\beta}\parallel(0001)_{Mg}$ $[1\bar{1}1]_{\beta}\parallel[1\bar{2}10]_{Mg}$ (incoherent)
Mg–Th	SSSS	→ β'' Mg_3Th? hcp DO_{19} superlattice discs $\parallel\{10\bar{1}0\}_{Mg}$ (coherent)		→ Mg_2Th (i) β_1' hexagonal (ii) β_2' fcc (both semicoherent)	→ β $Mg_{23}Th_6$ fcc $a = 1{\cdot}43$ nm (incoherent)
Mg–Ag–RE(Nd)	SSSS	→ Rodlike GP zones $\perp(0001)_{Mg}$ (coherent)		→ γ hexagonal rods $a = 0{\cdot}963$ nm $c = 1{\cdot}024$ nm $\parallel$ to $[0001]_{Mg}$ (coherent)	
	SSSS	→ Ellipsoidal GP zones $\parallel(0001)_{Mg}$ (coherent)		→ β hexagonal equiaxed $a = 0{\cdot}556$ nm $c = 0{\cdot}521$ nm $(0001)_{\beta}\parallel(0001)_{Mg}$ $(11\bar{2}0)_{\beta}\parallel(10\bar{1}0)_{Mg}$ (semicoherent)	→ $Mg_{12}Nd_2Ag$ complex hexagonal laths (incoherent)

SSSS supersaturated solid solution.

Fig. 2 Probable precipitation processes in magnesium alloys. Source: Ref 5

most is that one stage involves formation of an ordered, hexagonal precipitate with a DO_{19} (Mg_3Cd) structure that is coherent with the magnesium lattice (Fig. 2). This structure appears to be analogous to the well-known θ″ phase that may form in aged Al-Cu alloys, and it appears to be present when alloys show a maximum response to age hardening. The DO_{19} cell has an *a*-axis twice the length of the *a*-axis of the magnesium matrix, whereas the *c*-axes are the same. The precipitate forms as plates or discs parallel to the $<0001>_{Mg}$ directions, which lie along the $(10\bar{1}0)_{Mg}$ and $\{11\bar{2}0\}_{Mg}$ planes. In this regard it is significant to note that alternate $(10\bar{1}0)$ and $(11\bar{2}0)$ planes in a structure of composition Mg_3X consist entirely of magnesium atoms. Thus, the formation of a low-energy interface along these planes is to be expected, because only second-nearest neighbor bonds need to be altered. This structural feature could account for the fact that the phase is relatively stable over a wide temperature range, and it seems likely to be a key factor in promoting creep resistance in those magnesium alloys in which it occurs.

Castings

Cast magnesium alloys have always predominated over wrought alloys, particularly in Europe, where traditionally they have comprised 85 to 90% of all products. The earliest commercially used alloying elements were aluminum, zinc, and manganese; the Mg-Al-Zn system is still the one most widely used for castings. The first wrought alloy was Mg-1.5Mn, which was used for sheet, extrusions, and forgings, but this material has largely been superseded.

Early Mg-Al-Zn castings suffered severe corrosion in wet or moist conditions until the discovery in 1925 that small additions (0.2 wt%) of manganese increase corrosion resistance (Ref 12). The role of manganese was to remove iron and certain other heavy metal impurities as relatively harmless intermetallic compounds, some of which separate out during melting. In this regard, classic work by Hanawalt et al. (Ref 13) showed that the corrosion rate increased abruptly once so-called "tolerance limits" were exceeded, which were 5, 170, and 1300 ppm for nickel, iron, and copper, respectively.

Another problem with earlier magnesium alloy castings was that grain size tended to be large and variable, often resulting in poor mechanical properties, microporosity and, in the case of wrought products, excessive directionality of properties (Ref 1, 12). Values of yield strength also tended to be particularly low relative to ultimate tensile strength (UTS). In 1937, Sauerwald in Germany discovered that zirconium had an intense grain-refining effect on magnesium, although several years elapsed before a reliable method was developed to alloy this metal. Here it is interesting to note that the lattice parameters of hexagonal α-zirconium ($a = 0.323$ nm, $c = 0.514$ nm) are very close to those of magnesium (a = 0.320 nm, c = 0.520 nm). This suggests the possibility that zirconium particles may provide sites for the heterogeneous nucleation of magnesium grains during solidification. Paradoxically, zirconium could not be used in most existing alloys because it was removed from solid solution because of the formation of stable compounds with aluminum and manganese. This led to the evolution of a completely new series of cast and wrought zirconium-containing alloys that were found to have much improved mechanical properties at both room and elevated temperatures. Such alloys are now widely used in the aerospace industries. Compositions of the major commercial magnesium casting alloys are shown in Table 2. The zirconium-free and zirconium-containing series of alloys are considered separately below.

Zirconium-Free Casting Alloys

Alloys Based on the Mg-Al System. The binary Mg-Al system was the basis for early magnesium casting alloys. The maximum solid solubility of aluminum is 12.7 wt% at 437 °C, decreasing to around 2% at room temperature (Ref 10, 11). In the as-cast condition, the β-phase $Mg_{17}Al_{12}$ forms around grain boundaries, being most prevalent in more slowly cooled sand or permanent mold castings. Annealing or solution treating at temperatures around 430 °C will cause all or part of the β-phase to dissolve, and it might be expected that subsequent quenching and aging would induce significant precipitation hardening. However, aging results in transformation of the supersaturated solid solution directly to a coarsely dispersed, equilibrium precipitate β-phase without the appearance of Guinier-Preston (GP) zones or intermediate precipitates (Ref 14) (Fig. 2). Moreover, the β-phase may form by discontinuous precipitation in which even coarser cells spread out from grain boundaries. Because the response to aging is relatively poor, alloys based on the Mg-Al system are generally used in the as-cast condition.

The most widely used alloy is AZ91C (Mg-9Al-0.7Zn-0.2Mn) in the form of die castings. As mentioned above, the corrosion resistance of this alloy is adversely affected by the presence of cathodic impurities such as iron and nickel, and for some purposes strict limits have now been placed on these elements. Higher-purity versions such as AZ91D (0.004% max Fe, 0.001% max Ni, 0.015% max Cu, 0.17% min Mn) have corrosion rates in salt fog tests that are as much as 100 times lower than those for AZ91C, so that they become comparable with those for aluminum casting alloys (Ref 15). Other commonly used alloys of the same type are AZ81 and AZ63.

Requirements for specific property improvements have stimulated the development of alternative die casting alloys. For applications where greater ductility and fracture toughness are required, a series of high-purity alloys with reduced aluminum contents is available. Examples are AM60, AM50, and AM20 (Table 2). The improved properties arise because of a reduction in the amount of $Mg_{17}Al_{12}$ around grain boundaries (Ref 16, 17). Such alloys are used for automotive parts such as wheels, seat frames, and steering wheels.

The mechanical properties of the AZ and AM series of alloys decrease rapidly at temperatures above 120 to 130 °C (250 to 265 °F) (Ref 18 and 19). This behavior is attributed to the fact that magnesium alloys undergo creep mainly by grain-boundary sliding. The phase $Mg_{17}Al_{12}$, which has a melting point of approximately 460 °C (860 °F) and is comparatively soft at lower temperatures, does not serve to pin boundaries. Accordingly, commercial requirements have led to the investigation of other alloys based on the Mg-Al system.

The addition of 1% Ca improves creep strength of Mg-Al alloys but makes them prone to hot cracking (Ref 17). Creep properties are also improved by lowering the aluminum content and introducing silicon (Ref 18, 19). This has the effect of reducing the amount of $Mg_{17}Al_{12}$ and, for die castings that cool relatively quickly, silicon combines with magnesium to form fine and relatively hard particles of the compound Mg_2Si in grain boundaries (Ref 17, 19). Two examples are the alloys AS41 (Mg-4.5Al-1Si-0.3Mn) and AS21 (Mg-2.2Al-1Si-0.3Mn), both of which have creep properties superior to those of AZ91 at temperatures above 130 °C (265 °F). Alloy AS21, with the lower content of aluminum, performs better than AS41 but is more difficult to cast because of reduced fluidity. These alloys were exploited on a large scale in the various generations of the famous Volkswagon Beetle engine. The creep properties of Mg-Al-Si alloys still fall well below those of competing die cast aluminum alloys such as A380 (Fig. 3), and attention has recently been directed at Mg-Al alloys containing rare earth (RE) elements added as naturally occurring cerium mischmetal (commonly 55Ce-20La-15Nd-5Pr) (Ref 17, 20). Again, the alloys are suitable only for die castings because slower cooling results in the formation of coarse particles of Al_2RE compounds. One composition, AE42 (Mg-4Al-2RE-0.3Mn), has a good combination of properties, including creep strength superior to that of Mg-Al-Si alloys (Fig. 3). The mechanism by which creep properties are influenced by RE additions is unclear, although finely dispersed precipitates have been detected in aged binary Mg-1.3RE alloy (Ref 21). In addition, nucleation of a stable $Mg_{12}Ce$ in grain boundaries has been observed during creep and is considered to reduce deformation by grain-boundary sliding (Ref 1). It should be noted, however, that use of mischmetal does raise the cost of the alloy because this addition is several times more expensive than an equal weight of silicon.

Alloys Based on the Mg-Zn System. Binary Mg-Zn alloys also respond to age hardening (Ref 22) and, unlike Mg-Al alloys, form coherent GP zones and semicoherent intermediate precipitates (Fig. 2). However, these alloys are difficult to grain refine and are susceptible to microporosity, so they are not used for commercial castings.

Recent work (Ref 23) has shown that a ternary addition of copper results in a marked increase in both ductility and response to age hardening (Ref 24). Moreover, mechanical properties are similar to those of AZ91 at room temperature, but these properties are more reproducible and elevated-temperature stability is increased. One typical sand casting alloy is ZC63 (Mg-6Zn-3Cu-0.5Mn). The progressive addition of copper to Mg-Zn alloys has been found to raise the eutectic temperature, which is important because it permits the use of higher solution treatment temperatures, thereby maximizing solution of zinc and copper. The structure of the eutectic is also changed from being completely divorced in binary Mg-Zn alloys, with the Mg-Zn compound distributed around grain boundaries and between dendrite arms, to truly lamellar in the ternary copper-containing alloy (Fig. 4). A typical heat treatment cycle involves solution treatment at 440 °C, hot water quench, and age 16 to 24 h at 180 to 200 °C. Hardening is associated with two main precipitates, β_2' (rods) and β_2' (plates or discs), which appear to be similar to the phases observed in aged Mg-Zn alloys (Fig. 2). However, the concentration of at least one of these precipitates is greater when copper is present. Typical room-temperature properties are 0.2% yield strength of 150 MPa, UTS of 235 MPa, and elongation of 5%. Although the presence of copper in Mg-Al-Zn alloys has a very detrimental effect on corrosion resistance, this does not seem to be the case with Mg-Zn-Cu alloys, presumably because much of the copper is incorporated in the eutectic phase $Mg(Cu, Zn)_2$ (Ref 24). Castings made from these alloys are being promoted for use in motor car engines.

Zirconium-Containing Casting Alloys

The maximum solubility of zirconium in molten magnesium is 0.6%, and the addition of other elements has been necessary because binary Mg-Zr alloys are not sufficiently strong for commercial applications. These elements have been selected on the basis of compatibility with zirconium, founding characteristics, and desired properties. In this latter regard, the two principal objectives have been improved tensile properties (including higher ratios of yield strength to tensile strength) and increased creep resistance. These requirements have been dictated by the aerospace industries. Nominal compositions of commercial alloys are included in Table 2.

Mg-Zn-Zr Alloys. The ability to grain refine Mg-Zn alloys with zirconium led to the introduction of ternary alloys such as ZK51 (Mg-4.5Zn-0.7Zr). However, because these alloys are also susceptible to microporosity and are not weldable, they have found little practical application.

Table 2 Nominal composition, typical tensile properties, and characteristics of selected magnesium casting alloys

ASTM designation	British designation	Nominal composition: Al	Zn	Mn	Si	Cu	Zr	RE (MM)	RE (Nd)	Th	Y	Ag	Condition	Tensile properties: 0.2% yield strength, MPa	Ultimate tensile strength, MPa	Elongation, %	Characteristics
AZ63	...	6	3	0.3	...	...	...	...	...	...	...	...	As-sand cast T6	75	180	4	Good room-temperature strength and ductility
														110	230	3	
AZ81	A8	8	0.5	0.3	...	...	...	...	...	...	...	...	As-sand cast T4	80	140	3	Tough, leak-tight castings.
														80	220	5	With 0.0015 Be, used for pressure die casting
AZ91	AZ91	9.5	0.5	0.3	...	...	...	...	...	...	...	...	As-sand cast	95	135	2	General-purpose alloys used for sand and die castings
													T4	80	230	4	
													T6	120	200	3	
													As-chill cast	100	170	2	
													T4	80	215	5	
													T6	120	215	2	
AM50	...	5	...	0.3	...	...	...	...	...	...	...	...	As-die cast	125	200	7*	High-pressure die castings
AM20	...	2	...	0.5	...	...	...	...	...	...	...	...	As-die cast	105	135	10*	Good ductility and impact strength
AS41	...	4	...	0.3	1	...	...	...	...	...	...	...	As-die cast	135	225	4.5*	Good creep properties to 150 °C
AS21	...	2	...	0.4	1	...	...	...	...	...	...	...	As-die cast	110	170	4*	Good creep properties to 150 °C
ZK51	Z5Z	...	4.5	...	...	...	0.7	...	...	...	...	...	T5	140	235	5	Sand castings, good room-temperature strength and ductility
ZK61	...	...	6	...	...	...	0.7	...	...	...	...	...	T5	175	275	5	As for ZK51
ZE41	RZ5	...	4.2	...	...	...	0.7	1.3	...	...	...	...	T5	135	180	2	Sand castings, good room-temperature strength, improved castability
ZC63	ZC63	...	6	0.5	...	3	...	...	...	...	...	...	T6	145	240	5	Pressure-tight castings, good elevated-temperature strength, weldable
EZ33	ZRE1	...	2.7	...	...	...	0.7	3.2	...	...	...	...	Sand cast T5	95	140	3	Good castability, pressure tight, weldable, creep resistant to 250 °C
													Chill cast T5	100	155	3	
HK31	MTZ	...	...	...	...	...	0.7	...	...	3.2	...	...	Sand cast T6	90	185	4	Sand castings, good castability, weldable, creep resistant to 350 °C
HZ32	ZT1	...	2.2	...	...	...	0.7	...	...	3.2	...	...	Sand or chill cast T5	90	185	4	As for HK31
QE22	MSR	...	...	...	...	...	0.7	...	2.5	...	...	2.5	Sand or chill cast T6	185	240	2	Pressure tight and weldable, high yield strength to 250 °C
QH21	QH21	...	...	...	...	...	0.7	...	1	1	...	2.5	As-sand cast T6	185	240	2	Pressure tight, weldable, good creep resistance and yield strength to 300 °C
WE54	WE54	...	...	...	...	...	0.5	...	3.25	...	5.1	...	T6	200	285	4	High strength at room and elevated temperatures. Good corrosion resistance, weldable
WE43	WE43	...	...	...	...	...	0.5	...	3.25	...	4	...	T6	190	250	7	

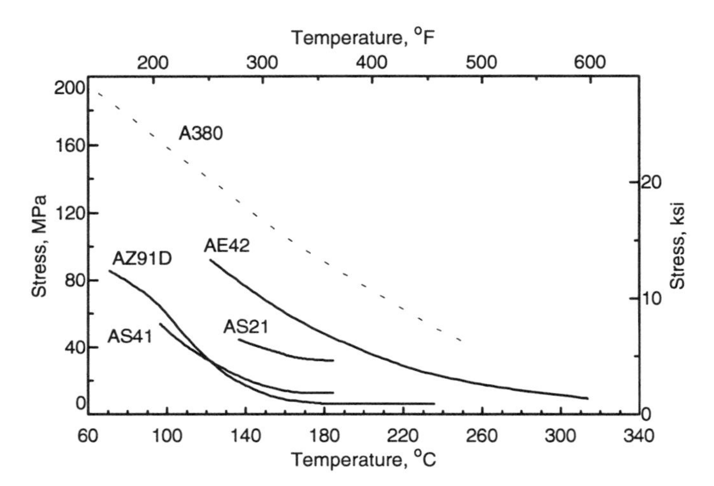

Fig. 3 Stress for 0.1% creep strain in 100 h for cast alloys based on the Mg-Al system and for the aluminum casting alloy A380. Source: Ref 17

Mg-RE Alloys. Magnesium forms solid solutions with a number of RE elements, and the magnesium-rich regions of the respective binary systems all show simple eutectics. The same applies for alloys with additions of the relatively cheaper mischmetals based on cerium or neodymium (e.g., 80Nd-16Pr-2Gd) (Ref 1). As a consequence, the alloys have good casting characteristics, because the presence of the relatively low-melting-point eutectics as networks in grain boundaries tends to suppress microporosity. In the as-cast condition, the alloys generally have cored α-grains surrounded by grain-boundary networks. Aging causes precipitation to occur within the grains (Ref 21), and as mentioned above, the generally good creep resistance they display is attributed to both the strengthening effect of this precipitate and the presence of the grain-boundary phases that reduce grain-boundary sliding (Ref 1).

The properties of Mg-RE alloys are enhanced by adding zirconium to refine grain size, and further increases in strength occur if zinc is added as well. The most widely used of these alloys is ZE41 (Mg-4.2Zn-1.3Ce-0.6Zr), which has moderate strength when given a T5 aging treatment that is maintained up to 150 °C (300 °F). One popular application has been helicopter transmission housings (Ref 25). Higher tensile properties, combined with good creep strength at temperatures up to 250 °C (480 °F), have been achieved with the alloy EZ33 (Mg-3RE-2.5Zn-0.6Zr) (Fig. 5). Further increases in strength might be expected with even higher zinc contents, but a massive grain-boundary phase containing zinc and RE elements is formed that both causes embrittlement and lowers the solidus temperature, thereby reducing the opportunity to solution treat the alloys prior to aging. Fisher (Ref 26) has shown that this latter phase can be dissociated by a specialized treatment involving prolonged heating in a hydrogen atmosphere, and this treatment has been successfully applied to thin-wall castings made from the alloy ZE63 (Mg-5.3Zn-2.5RE-0.7Zr).

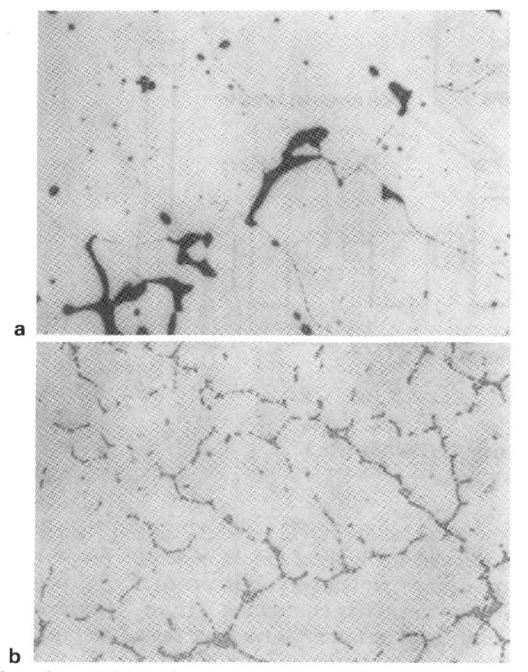

Fig. 4 Effect of the addition of copper on the morphology of the eutectic for alloy Mg-6Zn. (a) Binary alloy solution treated 8 h at 330 °C. 100×. (b) Ternary alloy Mg-6Zn-1.5Cu, solution treated 8 h at 430 °C. 100×

One recent development has sought to take advantage of the particularly high solid solubility of yttrium in magnesium (maximum of 12.5 wt%) and the capacity of Mg-Y alloys to age harden. A series of Mg-Y-Nd-Zr alloys has been produced combining high strength at ambient temperatures with good creep resistance at temperatures up to 300 °C (570 °F) (Ref 15, 23, 27). At the same time, the heat-treated alloys have a resistance to corrosion that is superior to that of other high-temperature magnesium alloys and comparable to that of many aluminum-base casting alloys (Ref 25, 28, 29). From a practical viewpoint, pure yttrium is expensive; it is also difficult to alloy with magnesium because of its high melting point (1500 °C) and its strong affinity for oxygen. It has been found that a cheaper yttrium-containing mischmetal containing around 75% of this element, together with heavy RE metals such as gadolinium and erbium, can be substituted for pure yttrium (Ref 15). Melting practices have also been changed so that the alloys can be processed in an inert atmosphere of argon and SF_6.

Precipitation in Mg-Y-Nd alloys is also complex (Fig. 2). Extremely fine β″-plates having the DO_{19} structure are formed upon aging below 200 °C (390 °F). However, the T6 treatment normally involves aging at 250 °C (480 °F), which is above the solvus for β″ and leads to precipitation of fine plates of the body-centered orthorhombic phase β′, which is thought to have the composition $Mg_{12}NdY$ (Ref 24). Maximum strengthening combined with an adequate level of ductility have been found to occur in an alloy containing approximately 6% Y and 2% Nd. The first commercially available alloy was WE54 (Mg-5.25Y-3.5RE(1.5-2Nd)-0.45Zr) (Ref 23). In the T6 condition, typical tensile properties at room temperature are 0.2% yield strength of 200 MPa, UTS of 275 MPa, and elongation of 4%, and it showed elevated-temperature properties superior to those of existing magnesium alloys (Fig. 5). It was revealed, however, that prolonged exposure to temperatures around 150 °C led to a gradual reduction in ductility to levels that were unacceptable (Ref 15), and this change was found to arise from the slow secondary precipitation of the β″-phase throughout the grains (Ref 24). Subsequently, King et al. (Ref 25, 28) showed that adequate ductility can be

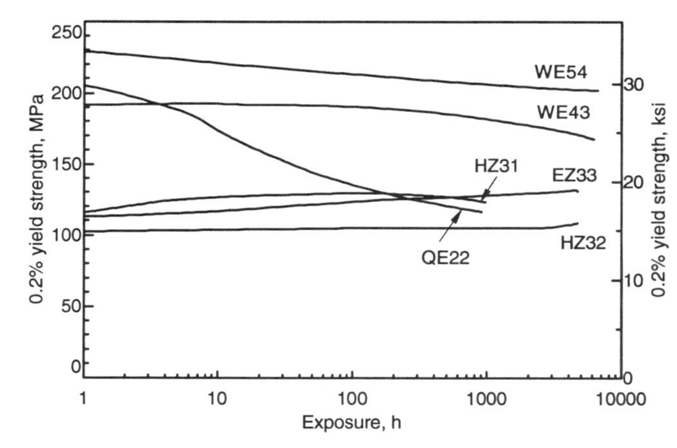

Fig. 5 Effect of exposure at 250 °C on 0.2% yield strength at room temperature for several cast magnesium alloys containing rare earth elements

retained with only a slight reduction in overall strength if the yttrium content is reduced and the neodymium content is increased. On the basis of this work, an alternative composition WE43 (Mg-4Y-2.25Nd-1 Heavy RE-0.4 min Zr) was developed (Fig. 5). (The principal heavy RE elements are ytterbium, erbium, dysprosium, and gadolinium.)

Alloys Based on the Mg-Th System. The addition of thorium also increases creep resistance in magnesium alloys, and cast and wrought alloys have been used in service at temperatures up to 350 °C (660 °F). As with the RE elements, thorium improves casting properties and the alloys are weldable (Ref 1).

Ternary compositions such as HK31 (Mg-3Th-0.7Zr) have been available for some time and have microstructures similar to those of the Mg-RE-Zr alloys. Precipitation hardening again leads to the formation of an ordered DO19 phase, which is probably Mg3Th or Mg23Th6 (Fig 2). The good creep resistance is attributed to the presence of fine dispersions of a phase such as this within the grains, together with another thorium-containing phase that forms discontinuously in grain boundaries. Again in parallel with other alloys based on the Mg-RE system, thorium-containing alloys have been developed to which zinc has been added, such as HZ32 (Mg-3Th-2.2Zn-0.7Zr), and ZH62 (Mg-5.7Zn-1.8Th-0.7Zr). The presence of zinc further increases creep strength (Fig. 5), and this is attributed, at least in part, to the introduction of an acicular phase that forms along grain boundaries. However, little appears to be known about the precise influence of zinc on precipitation in the Mg-Th system.

Although thorium-containing alloys have found applications in missiles and spacecraft (Ref 28), they are now losing favor because of environmental considerations and are generally considered to be obsolete. In Britain, for example, alloys containing as little as 2% of this element are classified as radioactive materials that require special handling, thereby increasing the cost and complexity of manufacture.

Alloys Based on the Mg-Ag System. The potential importance of the Mg-Ag system was recognized by Payne and Bailey (Ref 30), who discovered that the relatively low tensile properties of Mg-RE-Zr alloys could be much increased by the addition of silver. Substituting neodymium mischmetal for cerium mischmetal gave a further increase in strength, and several compositions have been developed for service at elevated temperatures.

The most widely used alloy has been QE 22 (Mg-2.5Ag-2RE[Nd]-0.7Zr), for which the optimal heat treatment is solution treatment for 4 to 8 h at 525 °C (980 °F), cold water quench, and aging 8 to 16 h at 200 °C (390 °F). This alloy has been used for a number of aerospace applications, including landing wheels, gearbox housings, and rotor heads for helicopters (Ref 15). If the silver content is below 2%, the precipitation process appears to be similar to that occurring in Mg-RE alloys and involves formation of Mg-Nd precipitates (Ref 24). However, for higher amounts of silver, two independent precipitation processes have been reported, both of which lead ultimately to the formation of an equilibrium phase of probable composition $Mg_{12}Nd_2Ag$ (Fig. 2). The presence of a phase with the DO_{19} structure has not been confirmed, although one precipitate, designated γ, has characteristics that suggest it may be such a phase. Precipitate size is also refined by the addition of silver, and maximum age hardening and creep resistance appear to be associated with the presence of the $\gamma + \beta$ precipitates.

Elevated-temperature properties may be further enhanced by the partial substitution of the RE (Nd) component by thorium. One alloy, QH21 (Mg-2.5Ag-1Th-1RE[Nd]-0.7Zr), showed the highest values of tensile properties and creep resistance at temperatures up to 250 °C (480 °F), prior to the development of the alloys containing yttrium. However, it will be recalled that these latter alloys have the advantage of high corrosion resistance, which is not shared by alloys such as QE22 or QH21 due to the presence of the noble metal silver. Moreover, QH21 is also becoming obsolete because of the presence of the radioactive element thorium.

Production of Castings

Most magnesium alloy components are produced by high-pressure die casting. Cold-chamber machines are used for the largest castings, and molten shot weights of 10 kg or more can now be injected in less than 100 ms at pressures that may be as high as 1500 bars (Ref 31, 32). Hot-chamber machines are used for most applications and are more competitive for smaller sizes because shorter cycle times are obtainable. Magnesium alloys offer particular advantages for both these processes (Ref 33, 34):

- Most alloys show high fluidity, which allows casting of intricate and thin-wall parts (e.g., 2 mm).
- Magnesium has a low specific heat per unit volume when compared with other metals. For example, comparative ratios for magnesium, aluminum, and zinc are 1 to 1.36 to 1.53. This means that magnesium castings cool more quickly, allowing faster cycle times and reducing die wear.
- High gate pressures can be achieved at moderate pressures because of the low density of magnesium.
- Iron from the dies has very low solubility in magnesium alloys, which is beneficial because it reduces any tendency to sticking.

Magnesium alloy components can be successfully prepared by sand casting and by gravity casting into permanent molds. However, conventional pouring practices can cause problems, as turbulent metal flow may introduce oxides and dross because of the reactive nature of magnesium. One solution has been to introduce the molten metal into the bottom of the mold cavity, thereby allowing unidirectional filling of the mold, and to apply controlled pressure to improve metal flow (Ref 25, 33). Such a process has been adapted for the production of automotive wheels (Ref 33, 35). Squeeze casting has also been used to prepare higher-quality castings from existing alloys such as AZ91 (Ref 17) and to produce castings in alloys that could not be successfully cast by conventional processes. One example (Ref 36) is the alloy Mg-12Zn-1Cu-1Si, which exhibits good room-temperature properties (e.g., 0.2% yield strength of 200 MPa) that are sustained at relatively high levels through the temperature range 100 to 200 °C (212 to 390 °F).

Magnesium alloys are also amenable to thixotropic casting, which offers the opportunity to produce good-quality fine-grain products more cheaply than by high-pressure die casting (Ref 17). One promising technique involves heating alloy granules to approximately 20 °C (36 °F) below the liquidus temperature and injection molding the resulting slurry into a die by means of a high-torque screw drive. Tests with the alloy AZ91 have shown that although tensile properties are little changed, impact toughness is significantly improved (Ref 37).

Mechanical Properties

Typical mechanical properties of various cast and wrought magnesium alloys are listed in Table 3 (Ref 38). More thorough compilations of typical mechanical properties are in Ref 6 and in Volume 2 of the *ASM Handbook, Properties and Selection: Nonferrous Alloys and Special-Purpose Materials.*

For castings, these values may be obtained by testing separately cast specimens. Tensile strengths of investment mold and shell mold castings compare favorably with those of sand and permanent mold castings. Yield strength, tensile strength, and percentage elongation may vary with cooling rate and generally are lower than those of separately cast sand mold test bars. Likewise, tensile properties will vary from specimens machined from different sections of castings of varying thickness. Some specifications permit a 25% reduction in tensile strength and a 75% reduction in elongation for specimens machined from castings, as compared with requirements for separately cast bars. Minimum tensile properties are shown in Tables 4 and 5.

Fatigue strength of magnesium casting alloys, as determined using laboratory test samples, covers a relatively wide scatter band, which is characteristic of other metals as well. The stress-life (*S-N*) curves have a gradual change in slope and become essentially parallel to the horizontal axis at 10 to 100 million cycles (Fig. 6) (Ref 39). Most of the fatigue data

Table 3 Representative mechanical properties of magnesium alloys

Alloy	Temper	%E	Tensile strength at 25 °C, MPa			Creep strength, MPa Stress for 0.2 % TE in 100 h			Rotating beam fatigue, MPa Stress for failure at indicated cycles:			
			YS	CYS	UTS	95 °C	200 °C	250 °C	10^5	10^6	10^7	5×10^7
Gravity castings												
AZ91C	F	2	95	95	165	...	...	...	110	105	95	90
	T4	14	85	85	275	...	...	...	135	115	105	100
	T6	5	130	275	...	...	...	...	125	105	90	85
AZ92A	F	2	95	95	165	...	...	...	115	105	95	92
	T4	9	95	95	275	...	...	...	125	110	105	100
	T5	2	110	110	180	...	...	...	125	110	105	100
	T6	2	145	145	275	55	10	...	130	115	100	97
EQ21A	T6	4	195	195	261	...	95(a)	36	...	100	98	96
QE22A	T6	4	205	205	275	...	85(a)	30	150	125	115	105
WE54A	T6	4	200	200	275	...	...	56	...	102	99	97
ZE41A	T5	5	140	140	205	...	...	...	124	97	96	94
ZK51A	T5	8	165	165	275	65	30	...	105	75	70	75
Die castings												
AM60A	F	8	130	130	220	...	...	...	...	...	...	...
AS41A	F	6	140	140	210	...	40(a, c)	...	110	90	90	...
AZ91A or D	F	3	160	160	230	...	25(a, c)	...	...	...	...	...
Extrusions												
AZ31B	F	15	200	95	260	50	...	...	180	160	145	140
AZ80A	T5	7	275	240	380	...	...	...	190	175	160	155
HM31A	T5	10	270	160	305	...	75	70	125	105	90	90
ZK30A	F	18	239	213	309	...	28(a)	...	152	131	124	124
ZK60A	T5	12	295	215	360	20	5	...	210	170	150	145
ZM21A	F	11	162	...	255	...	...	...	...	...	...	...
Sheet												
AZ31B	0	21	150	110	255	55	...	...	...	...	...	...
	H24	15	220	180	290	30	...	...	155(b)	120(b)	103(b)	103(b)
HM21A	T8	11	160	130	250	...	80	50	...	...	...	...
ZK30A	...	8	185	154	270	...	...	...	...	...	...	...
ZM21A	0	13	131	...	232	...	...	...	...	...	...	...

Note: %E, percent elongation; YS, yield strength; CYS, compressive yield strength; UTS, ultimate tensile strength; TE, total elongation. (a) Creep extension only. (b) Cantilever bending fatigue with $R = -1$. (c) 150 °C. Source: Ref 38

Table 4 Minimum tensile properties from designated areas of sand castings

Alloy	Temper	Class 1			Class 2			Class 3		
		%E	YS, MPa (ksi)	UTS, MPa (ksi)	%E	YS, MPa (ksi)	UTS, MPa (ksi)	%E	YS, MPa (ksi)	UTS, MPa (ksi)
AM100A	T6	3	140 (20)	260 (38)	1.5	125 (18)	240 (35)	1	110 (16)	205 (30)
AZ91C	T6	4	125 (18)	240 (35)	3	110 (16)	200 (29)	2	95 (14)	185 (27)
AZ92A	T6	3	170 (25)	275 (40)	1	140 (20)	235 (34)	0.75	125 (18)	205 (30)
HK31A	T6	6	110 (16)	230 (33)	3	95 (14)	200 (29)	1	85 (12)	170 (25)
QE22A	T6	4	195 (28)	275 (40)	2	180 (26)	255 (37)	4	160 (23)	230 (33)
ZE63A	T6	6	195 (28)	290 (42)	5	180 (26)	275 (40)	2	165 (24)	255 (37)
ZH62A	T5	5	160 (23)	260 (38)	3	145 (21)	235 (34)	2	130 (19)	215 (31)
ZK51A	T5	6	145 (21)	250 (36)	4	130 (19)	220 (32)	3	115 (17)	200 (29)
ZK61A	T6	6	200 (29)	290 (42)	4	180 (26)	255 (37)	2	160 (23)	235 (34)

Note: %E, percent elongation; YS, yield strength; UTS, ultimate tensile strength. Source: MIL-M-46062B, "Magnesium Alloy Castings—High Strength"

for magnesium alloys are *S-N* curves dating from the 1930s to the 1960s. A substantial portion of the early *S-N* data was summarized by H.J. Grover et al. (see Table 6) (Ref 40). The effect of different surface conditions is shown in Fig. 7.

The effect of temperature on fatigue limits is shown in Fig. 8 for alloys developed for high-temperature service, such as EZ33, HZ32, and REP14. The fatigue limit of QE22A is also shown. The reduction in fatigue appears to be much less for QE22A at high temperatures.

Wrought Alloys

The hexagonal crystal structure of magnesium places limitations on the amount of deformation that can be tolerated, particularly at low temperatures. At room temperature, deformation occurs mainly by slip on the basal planes in the close-packed <$11\bar{2}0$> directions and by twinning on the pyramidal {$10\bar{1}2$} planes. With stresses parallel to the basal planes, twinning of this type is only possible in compression, whereas with stresses perpendicular to the basal planes, it is only possible in tension. Above about 250 °C (480 °F), additional pyramidal {$10\bar{1}1$} slip

Table 5 Mechanical properties of permanent mold castings

Alloy	Temper	Temperature, °C	Minima separate bars					Average from castings				
			%E	YS		UTS		%E	YS		UTS	
				MPa	ksi	MPa	ksi		MPa	ksi	MPa	ksi
AM100A	F	25	...	70	10	140	20	...	...	...	...	...
	T4	25	6	70	10	235	34	1.5	70	10	175	25
	T6	25	2	105	15	235	34	...	95	14	175	25
	T61	25	...	115	17	235	34	...	95	14	175	25
AZ63A	F	25	4	75	11	180	26	...	...	...	...	...
	T4	25	7	75	11	235	34	2	70	10	175	25
	T5	25	2	85	12	180	26	...	...	...	...	...
	T6	25	3	110	16	235	34	0.8	95	14	175	25
AZ81A	T4	25	7	75	11	235	34	1.8	70	10	175	25
AZ91C	F	25	...	75	11	160	23	...	...	...	...	...
	T4	25	7	75	11	235	34	1.8	70	10	175	25
	T5	25	2	85	12	160	23	...	...	...	...	...
	T6	25	3	110	16	235	34	0.8	100	14	175	25
AZ92A	F	25	...	75	11	160	23	...	...	...	...	...
	T4	25	6	75	11	235	34	2.5	70	10	175	25
	T5	25	...	85	12	160	34	...	...	...	...	...
	T6	25	...	125	18	235	34	...	110	16	175	25
EZ33A	T5	25	2	95	14	140	20	0.5	85	12	105	15
		260	...	55	8	90	13	...	40	6	70	10
HK31A	T6	25	4	90	13	185	27	1.0	80	12	160	23
		260	...	90	13	145	21	...	70	10	95	14
HZ32A	T5	25	4	90	13	185	27	1.0	80	12	160	23
		260	...	55	8	90	13	...	40	6	70	10
QE22A	T6	25	2	170	25	240	35	0.5	160	23	220	32
		315	...	70	10	90	13	...	55	8	70	10

Note: %E, percent elongation; YS, yield strength; UTS, ultimate tensile strength. Source: ASTM B199 and federal specification QQ-M-55C

planes become operative so that deformation becomes much easier and twinning is less important. Production of wrought magnesium alloy products, therefore, is normally carried out by hot working.

Wrought materials are produced mainly by extrusion, rolling, and press forging at temperatures in the range 300 to 500 °C (570 to 930 °F), and detailed accounts of deformation and fracture behavior are available (Ref 1, 3, 4). Two general comments can be made concerning directionality effects in wrought products:

- Because the elastic modulus does not show much variation in different directions of the hexagonal magnesium crystal, preferred orientation has relatively little effect on the modulus of wrought products.
- Because twinning readily occurs when compressive stresses are parallel to the basal plane, wrought magnesium alloys tend to show lower values of longitudinal yield strength in compression than in tension. The ratio may lie between 0.5 and 0.7 and is an important characteristic of magnesium alloys because the design of lightweight structures involves buckling properties, which in turn are strongly dependent on compressive strength. The value varies with different alloys and is increased by promoting fine grain size because the contribution of grain boundaries to overall strength becomes proportionately greater.

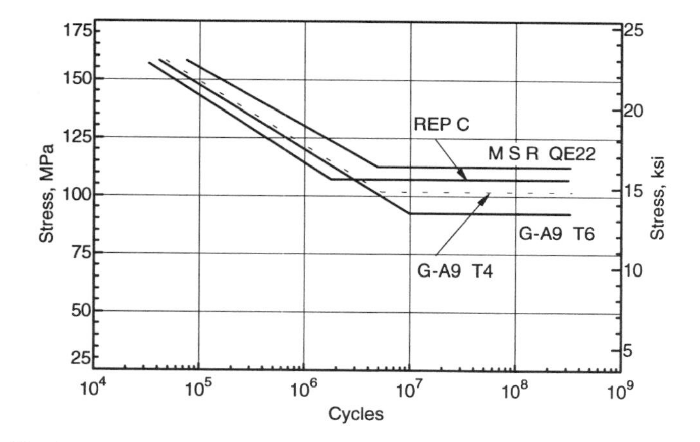

Fig. 6 Fatigue strength of magnesium alloys at room temperature. Source: Ref 39

As with cast alloys, the wrought alloys may be divided into two groups according to whether or not they contain zirconium (Table 7). Specific alloys have been developed that are suitable for wrought products, most of which fall into the same categories as the casting alloys already discussed (Ref 2, 6). Examples of sheet and plate alloys include AZ31 (Mg-3Al-1Zn-0.3Mn), which is the most widely used because it offers a good combination of strength, ductility, and corrosion resistance, and thorium-containing alloys such as HM21 (Mg-2Th-0.6Mn), which show good creep resistance to temperatures up to 350 °C. Magnesium alloys can be easily extruded into either solid or hollow sections at speeds that depend on alloy content. Higher-strength alloys such as AZ81 (Mg-8Al-1Zn-0.7Mn), ZK61 (Mg-6Zn-0.7Zr), and the more recent composition ZCM711 (Mg-6.5Zn-1.25Cu-0.75Mn) all have strength-to-weight ratios comparable to those of the strongest wrought aluminum alloys. ZM21 (Mg-2Zn-1Mn) can be extruded at high speeds and is the lowest-cost magnesium extrusion alloy available. Again, thorium-containing alloys such as HM31 (Mg-3Th-1Mn) show the best elevated-temperature properties. Magnesium forgings are less common and are often pre-extruded to refine microstructure, and alloy compositions are similar to those used for other wrought products.

One alloy system confined to specialty wrought components is Mg-Li, which has been exploited to produce lightweight materials (e.g., specific gravity 1.35) that have particularly high values of specific modulus. Lithium has a high solid solubility in magnesium (Ref 10, 11), and binary alloys containing more than 11 wt% of this element have the more desirable body-centered cubic structures, thereby offering the prospect of extensive cold formability (Ref 41). Such alloys are also amenable to age hardening, although they overage and soften at relatively low temperatures (e.g., 50 to 70 °C) as a consequence of the abnormally high mobility of lithium atoms and vacancies. Somewhat greater stability has been achieved by adding other elements, such as aluminum (e.g.,

Table 6 Cast magnesium alloy fatigue strength

Material	Designation	Condition	UTS, ksi	YS, ksi	%E	Fatigue loading		Fatigue specimen		K_t	Fatigue strength (max stress, ksi) at indicated cycles					Comments
						Type	Condition	Type	Size, in.		10^4	10^5	10^6	10^7	10^8	
Sand castings and permanent mold casting alloys																
AZ63-AC	AZ63A-F	...	29.0	14.0	6.0	RB	−1 *R*	Unnotched	0.30*d*	...	...	18.0	...	10.5	10.5	...
AZ63-AC	AZ63A-F	...	29.0	14.0	6.0	RB	−1 *R*	Circular notch	0.0295*r*, 0.295*d*, 0.354*D*	2	...	13.5	...	7.0	7.0	...
AZ63-AC	AZ63A-F	...	29.0	14.0	6.0	RB	−1 *R*	V-notch	60° notch, 0.002*r*, 0.30*d*, 0.48*D*	5	...	8.5	...	5.0	4.0	...
AZ63-ACS	AZ63A-T2	...	29.0	14.0	5.0	RB	−1 *R*	Unnotched	0.30*d*	...	...	23.0	...	15.0	12.0	...
AZ63-ACS	AZ63A-T2	...	29.0	14.0	5.0	RB	−1 *R*	Circular notch	0.0295*r*, 0.295*d*, 0.354*D*	2	...	12.5	...	9.5	8.5	...
AZ63-ACS	AZ63A-T2	...	29.0	17.7	5.0	RB	−1 *R*	Unnotched	0.30*d*	...	...	23.0	...	15.0	12.0	...
AZ63-ACS	AZ63A-T2	...	29.5	16.7	6.0	RB	−1 *R*	Unnotched	0.30*d*	...	...	...	18.5	14.6	12.0	All data at 150 °F
AZ63-ACS	AZ63A-T2	...	28.0	13.7	9.0	RB	−1 *R*	Unnotched	0.30*d*		...	...	10.9	9.5	8.4	All data at 275 °F
AZ63-HT	AZ63A-T4	...	40.0	14.0	12.0	RB	−1 *R*	Unnotched	0.30*d*	...	...	21.5	...	16.5	16.5	...
AZ63-HT	AZ63A-T4	...	40.0	14.0	12.0	RB	−1 *R*	Circular notch	0.0295*r*, 0.295*d*, 0.354*D*	2	...	17.5	...	9.0	8.5	...
AZ63-HT	AZ63A-T4	...	40.0	14.0	12.0	RB	−1 *R*	V-notch	60° notch, 0.002*r*, 0.30*d*, 0.48*D*	5	...	8.5	...	7.0	6.5	...
AZ63-HT	AZ63A-T4	...	39.0	15.0	11.0	RB	−1 *R*	Unnotched	0.30*d*	...	...	...	19.1	18.3	17.5	Fatigue data at 150 °F
AZ63-HT	AZ63A-T4	...	29.0	12.0	15.0	RB	−1 *R*	Unnotched	0.30*d*	...	...	...	11.9	10.7	9.5	Fatigue data at 275 °F
AZ63-HTA	AZ63A-T6	...	40.0	19.0	5.0	RB	−1 *R*	Unnotched	0.30*d*	...	...	18.0	...	16.0	15.0	...
AZ63-HTA	AZ63A-T6	...	40.0	19.0	5.0	RB	−1 *R*	Circular notch	0.0295*r*, 0.295*d*, 0.354*D*	2	...	11.5	...	9.5	9.0	...
AZ63-HTA	AZ63A-T6	...	40.0	19.0	5.0	RB	−1 *R*	V-notch	60° notch, 0.002*r*, 0.30*d*, 0.48*D*	5	...	8.5	...	5.0	3.5	...
AZ63-HTA	AZ63A-T6	...	42.0	19.0	7.7	RB	−1 *R*	Unnotched	0.30*d*	...	...	...	16.8	16.0	15.0	Fatigue data at 150 °F
AZ63-HTA	AZ63A-T6	...	33.5	16.5	13.0	RB	−1 *R*	Unnotched	0.30*d*	...	...	17.2	...	12.5	11.0	Fatigue data at 250 °F
AZ63-HTA	AZ63A-T6	...	33.5	16.5	13.0	RB	−1 *R*	Circular notch	0.0295*r*, 0.295*d*, 0.354*D*	2	...	11.5	...	7.0	7.0	All data at 250 °F
AZ63-HTA	AZ63A-T6	...	25.0	15.0	15.0	RB	−1 *R*	Unnotched	0.30*d*	...	...	...	11.6	9.2	7.4	Fatigue data at 275 °F
AZ63-HTA	AZ63A-T6	...	24.5	15.0	15.0	RB	−1 *R*	Unnotched	0.30*d*	...	...	...	...	9.0	7.5	All data at 300 °F
AZ63-HTA	AZ63A-T6	...	24.5	15.0	15.0	RB	−1 *R*	Circular notch	0.0295*r*, 0.295*d*, 0.354*D*	2	...	9.5	...	6.0	5.0	All data at 300 °F
AZ63-HTS	AZ63A-T7	...	39.0	18.0	8.0	RB	−1 *R*	Unnotched	0.30*d*	...	...	21.0	...	16.0	15.0	...
AZ63-HTS	AZ63A-T7	...	39.0	18.0	8.0	RB	−1 *R*	Circular notch	0.0295*r*, 0.295*d*, 0.354*D*	2	...	14.0	...	10.5	9.5	...
AZ63-HTS	AZ63A-T7	...	39.0	18.0	8.0	RB	−1 *R*	V-notch	60° notch, 0.002*r*, 0.30*d*, 0.48*D*	5	...	7.5	...	6.0	5.5	...
AZ63-HTS	AZ63A-T7	...	38.3	17.0	11.0	RB	−1 *R*	Unnotched	0.30*d*	...	...	...	18.0	16.0	15.0	Fatigue data at 150 °F
AZ63-HTS	AZ63A-T7	...	38.0	16.5	15.0	RB	−1 *R*	Unnotched	0.30*d*	...	...	17.5	14.0	12.5	12.0	All data at 200 °F
AZ63-HTS	AZ63A-T7	...	...	...	...	RB	−1 *R*	Unnotched	0.30*d*	...	...	17.4	...	12.5	12.3	Fatigue data at 200 °F
AZ63-HTS	AZ63A-T7	...	...	...	...	RB	−1 *R*	Circular notch	0.0295*r*, 0.295*d*, 0.354*D*	2	...	12.2	...	7.8	7.0	Fatigue data at 200 °F

Note: UTS, ultimate tensile strength; YS, yield strength; %E, percent elongation; B, bending; RB, rotating bending; AX, axial; *d*, inside diameter; *D*, outside diameter; *t*, thickness; *r*, radius. Source: H.J. Grover, S.A. Gordon, and L.R. Jackson, *Fatigue of Metals and Structures*, Department of the Navy, 1960, p 354-359

(continued)

Table 6 Cast magnesium alloy fatigue strength (continued)

Material	Designation	Condition	UTS, ksi	YS, ksi	%E	Fatigue loading		Fatigue specimen		K_t	Fatigue strength (max stress, ksi) at indicated cycles					Comments
						Type	Condition	Type	Size, in.		10^4	10^5	10^6	10^7	10^8	
Sand castings and permanent mold casting alloys (continued)																
AZ63-HTS	AZ63A-T7	...	30.3	15.2	18.0	RB	−1 *R*	Unnotched	0.30*d*	...	...	...	11.5	10.7	9.6	All data at 275 °F
AZ63-HTS	AZ63A-T7	...	27.0	14.3	26.0	RB	−1 *R*	Unnotched	0.30*d*	...	...	13.5	11.6	9.5	9.5	All data at 300 °F
AZ63-HTS	AZ63A-T7	...	27.0	14.3	26.0	RB	−1 *R*	Circular notch	0.0295*r*, 0.295*d*, 0.354*D*	2	...	10.0	...	5.5	5.0	All data at 300 °F
AZ63-HTS	AZ63A-T7	0.20 in. thick	39.0	18.0	8.0	B	−1 *R*	Unnotched	0.30*t*	...	...	15.0	12.8	11.0	...	...
AZ63-HTS	AZ63A-T7	0.20 in. thick	38.0	16.5	15.0	B	−1 *R*	Unnotched	0.22*t*	...	...	13.4	10.6	8.6	...	All data at 200 °F
AZ63-HTS	AZ63A-T7	0.20 in. thick	27.0	14.3	26.0	B	−1 *R*	Unnotched	0.20*t*	...	...	11.5	8.6	6.2	...	All data at 300 °F
AZ63-HTS	AZ63A-T7	0.20 in. thick	18.0	11.5	40.0	B	−1 *R*	Unnotched	0.20*t*	...	...	11.0	6.3	5.0	...	All data at 400 °F
AZ63-HTS	AZ63A-T7	0.20 in. thick	7.5	4.0	77.0	B	−1 *R*	Unnotched	0.20*t*	...	...	7.3	5.3	3.0	...	All data at 600 °F
AZ92-AC	AZ92A-F	...	24.0	14.0	2.0	RB	−1 *R*	Unnotched	0.30*d*	...	...	18.5	...	14.5	13.0	...
AZ92-AC	AZ92A-F	...	24.0	14.0	2.0	RB	−1 *R*	Circular notch	0.0295*r*, 0.295*d*, 0.3540*D*	2	...	11.0	...	8.5	7.5	...
AZ92-AC	AZ92A-F	...	24.0	14.0	2.0	RB	−1 *R*	V-notch	60° notch, 0.002*r*, 0.30*d*, 0.48*D*	5	...	8.5	...	5.0	4.0	...
AZ92-ACS	AZ92A-T2	...	25.0	19.0	8.0	RB	−1 *R*	Unnotched	0.30*d*	...	...	...	15.8	14.5	13.5	Fatigue data at 150 °F
AZ92-ACS	AZ92A-T2	...	23.8	14.5	3.0	RB	−1 *R*	Unnotched	0.30*d*	...	...	...	8.5	8.2	8.0	Fatigue data at 275 °F
AZ92-HT	AZ92A-T4	...	40.0	16.0	10.0	RB	−1 *R*	Unnotched	0.30*d*	...	...	20.0	...	16.5	16.0	...
AZ92-HT	AZ92A-T4	...	40.0	16.0	10.0	RB	−1 *R*	Circular notch	0.0295*r*, 0.295*d*, 0.354*D*	2	...	15.5	...	13.5	12.5	...
AZ92-HT	AZ92A-T4	...	40.0	16.0	10.0	RB	−1 *R*	V-notch	60 ° notch, 0.002*r*, 0.03*d*, 0.48*D*	5	...	10.0	...	6.0	5.5	...
AZ92-HT	AZ92A-T4	...	38.5	19.0	7.7	RB	−1 *R*	Unnotched	0.30*d*	...	...	...	18.2	17.0	16.0	Fatigue data at 150 °F
AZ92-HT	AZ92A-T4	...	31.0	17.0	30.0	RB	−1 *R*	Unnotched	0.30*d*	...	...	...	12.0	10.8	9.0	All data at 275 °F
AZ92-HTA	AZ92A-T6	...	40.0	23.0	2.0	RB	−1 *R*	Unnotched	0.30*d*	...	...	19.5	...	16.0	15.0	...
AZ92-HTA	AZ92A-T6	...	40.0	23.0	2.0	RB	−1 *R*	Circular notch	0.0295*r*, 0.295*d*, 0.354*D*	2	...	14.0	...	9.5	9.5	...
AZ92-HTA	AZ92A-T6	...	40.0	23.0	2.0	RB	−1 *R*	V-notch	60° notch, 0.002*r*, 0.30*d*, 0.48*D*	5	...	7.5	...	5.0	4.5	...
AZ92-HTA	AZ92A-T6	...	42.0	26.0	2.0	RB	−1 *R*	Unnotched	0.30*d*	...	...	...	17.7	16.5	15.0	Fatigue data at 150 °F
AZ92-HTA	AZ92A-T6	...	35.0	20.5	31.0	RB	−1 *R*	Unnotched	0.30*d*	...	...	14.5	...	11.5	10.5	All data at 250 °F
AZ92-HTA	AZ92A-T6	...	35.0	20.5	31.0	RB	−1 *R*	Circular notch	0.0295*r*, 0.295*d*, 0.354*D*	2	...	13.5	...	8.0	6.0	All data at 250 °F
AZ92-HTA	AZ92A-T6	...	28.0	18.0	38.0	RB	−1 *R*	Unnotched	0.30*d*	...	...	...	13.4	12.0	10.0	All data at 275 °F
AZ92-HTA	AZ92A-T6	...	28.0	18.0	35.0	RB	−1 *R*	Unnotched	0.30*d*	...	...	16.0	...	11.0	10.0	All data at 300 °F
AZ92-HTA	AZ92A-T6	...	28.0	18.0	35.0	RB	−1 *R*	Circular notch	0.0295*r*, 0.295*d*, 0.354*D*	2	...	11.5	...	6.5	6.0	All data at 300 °F
AZ92-HTS	AZ92A-T7	...	40.7	22.0	4.0	RB	−1 *R*	Unnotched	0.30*d*	...	...	20.5	...	15.0	13.0	...
AZ92-HTS	AZ92A-T7	...	40.7	22.0	4.0	RB	−1 *R*	Circular notch	0.0295*r*, 0.295*d*, 0.354*D*	2	...	14.5	...	10.0	9.0	...
AZ92-HTS	AZ92A-T7	...	40.0	20.0	3.0	RB	−1 *R*	V-notch	60° notch, 0.002*r*, 0.30*d*, 0.48*D*	5	...	8.0	...	6.5	6.0	...
AZ92-HTS	AZ92A-T7	...	27.5	15.5	23.0	RB	−1 *R*	Unnotched	0.30*d*	...	...	15.5	...	10.0	9.0	All data at 300 °F

Note: UTS, ultimate tensile strength; YS, yield strength; %E, percent elongation; B, bending; RB, rotating bending; AX, axial; *d*, inside diameter; *D*, outside diameter; *t*, thickness; *r*, radius. Source: H.J. Grover, S.A. Gordon, and L.R. Jackson, *Fatigue of Metals and Structures*, Department of the Navy, 1960, p 354-359

(continued)

Table 6 Cast magnesium alloy fatigue strength (continued)

Material	Designation	Condition	UTS, ksi	YS, ksi	%E	Fatigue loading		Fatigue specimen		K_t	Fatigue strength (max stress, ksi) at indicated cycles					Comments
						Type	Condition	Type	Size, in.		10^4	10^5	10^6	10^7	10^8	
Sand castings and permanent mold casting alloys (continued)																
AZ92-HTS	AZ92A-T7	...	27.5	15.5	23.0	RB	−1 *R*	Circular notch	0.0295*r*, 0.295*d*, 0.354*D*	2	...	11.0	...	5.5	4.5	All data at 300 °F
AZ92-HTS	AZ92A-T7	...	40.0	21.0	1.0	RB	−1 *R*	Unnotched	0.30*d*	...	...	...	17.3	15.0	13.0	Fatigue data at 150 °F
AZ92-HTS	AZ92A-T7	...	28.0	17.0	38.0	RB	−1 *R*	Unnotched	0.30*d*	...	...	...	12.5	10.0	9.0	All data at 275 °F
EM61-ACS	EM61XA-T2	...	18.0	14.0	0.5	RB	−1 *R*	Unnotched	0.30*d*	...	...	...	8.3	...	7.0	Fatigue data at 150 °F
EM61-ACS	EM61XA-T2	...	16.0	12.0	1.2	RB	−1 *R*	Unnotched	0.30*d*	...	...	...	6.8	5.3	4.1	All data at 275 °F
EM61-HTA	EM61XA-T6	...	19.3	17.7	1.2	RB	−1 *R*	Unnotched	0.30*d*	...	...	11.0	9.8	7.8	6.5	...
EM61-HTA	EM61XA-T6	...	19.7	15.6	1.3	RB	−1 *R*	Unnotched	0.20*t*	...	...	12.5	9.3	8.0	7.5	All data at 300 °F
EM61-HTA	EM61XA-T6	0.20 in. thick	19.3	17.7	1.2	B	−1 *R*	Unnotched	0.20*t*	...	...	12.5	9.5	7.3	...	...
EM61-HTA	EM61XA-T6	0.20 in. thick	19.7	15.6	1.3	B	−1 *R*	Unnotched	0.20*t*	...	14.0	10.0	7.0	7.0	...	All data at 300 °F
EM61-HTA	EM61XA-T6	0.20 in. thick	...	...	...	B	−1 *R*	Unnotched	0.20*t*	...	...	7.7	5.3	3.0	...	Fatigue data at 600 °F
EM61-HTA	EM61XA-T6	...	...	...	...	RB	−1 *R*	Unnotched	0.30*d*	...	...	11.2	...	7.9	6.5	...
EM61-HTA	EM61XA-T6	...	...	...	...	RB	−1 *R*	Circular notch	0.0295*r*, 0.295*d*, 0.354*D*	2	...	9.6	...	6.5	6.5	...
E10-HTA	E10XA-T6	...	19.0	...	...	RB	−1 *R*	Unnotched	0.30*d*	...	...	11.0	...	...	9.0	...
E10-HTA	E10XA-T6	...	19.0	...	...	RB	−1 *R*	Circular notch	0.0295*r*, 0.295*d*, 0.354*D*	2	...	9.0	...	...	5.5	...
E10-HTA	E10XA-T6	...	19.7	16.5	1.0	RB	−1 *R*	Unnotched	0.30*d*	...	...	12.0	...	...	7.5	Fatigue data at 275 °F
E10-HTA	E10XA-T6	...	19.7	16.5	1.0	RB	−1 *R*	Circular notch	0.0295*r*, 0.295*d*, 0.354*D*	2	...	9.0	...	...	6.5	All data at 275 °F
E10-HTA	E10XA-T6	0.20 in. thick	19.0	...	...	B	−1 *R*	Unnotched	0.20*t*	...	...	8.0	...	6.5	...	...
E10-HTA	E10XA-T6	0.20 in. thick	16.5	9.5	7.0	B	−1 *R*	Unnotched	0.20*t*	...	...	13.0	...	4.0	...	All data at 600 °F
EM101-HTA	EM101A-T6	...	20.0	18.5	0.8	RB	−1 *R*	Unnotched	0.30*d*	...	...	10.5	...	...	7.5	...
EM101-HTA	EM101A-T6	...	20.0	18.5	0.8	RB	−1 *R*	Circular notch	0.0295*r*, 0.295*d*, 0.354*D*	2	...	10.0	...	...	5.5	...
EM101-HTA	EM101A-T6	...	20.0	15.2	0.5	RB	−1 *R*	Unnotched	0.30*d*	...	...	12.0	...	...	8.0	All data at 275 °F
EM101-HTA	EM101A-T6	...	20.0	15.2	0.5	RB	−1 *R*	Circular notch	0.0029*r*, 0.295*d*, 0.354*D*	2	...	10.0	...	...	5.0	All data at 275 °F
AZ63X-AC	AZ63A-F	...	29.0	14.0	6.0	Ax	0.0 *R*	Unnotched	0.30*d*	...	...	22.0	20.0	18.0	...	...
AZ63X-AC	AZ63A-F	...	29.0	14.0	6.0	Ax	0.25 *R*	Unnotched	0.30*d*	...	...	24.0	21.5	19.0	...	...
AZ63X-ACS	AZ63A-T2	...	29.0	14.0	5.0	Ax	0.0 *R*	Unnotched	0.30*d*	...	...	23.0	21.0	19.0	...	...
AZ63X-ACS	AZ63A-T2	...	29.0	14.0	5.0	Ax	0.25 *R* *	Unnotched	0.30*d*	...	...	26.0	23.0	21.0	...	...
AZ63-HT	AZ63A-T4	...	40.0	14.0	12.0	Ax	0.0 *R*	Unnotched	0.30*d*	...	...	23.0	19.0	15.0	...	...
AZ63-HT	AZ63A-T4	...	40.0	14.0	12.0	Ax	0.25 *R*	Unnotched	0.30*d*	...	...	25.0	22.5	21.0	...	...
AZ63-HTS	AZ63A-T7	...	39.0	18.0	8.0	Ax	0.25 *R*	Unnotched	...	...	...	31.0	25.7	24.0	...	...
AZ63-HTS	AZ63A-T7	...	40.0	17.0	7.0	Ax	0.0 *R*	Unnotched	0.30*d*	...	...	24.0	21.0	21.0	...	...
AZ63-HTS	AZ63A-T7	...	40.0	17.0	7.0	Ax	0.25 *R*	Unnotched	0.30*d*	...	...	29.0	26.0	24.0	...	...
AZ63-HTS	AZ63A-T7	...	38.0	16.5	15.0	Ax	0.25 *R*	Unnotched	...	...	...	26.2	23.8	21.5	...	All data at 200 °F
AZ63-HTS	AZ63A-T7	...	30.3	15.2	18.0	Ax	0.0 *R*	Unnotched	0.30*d*	...	...	21.5	20.0	18.5	...	All data at 275 °F
AZ63-HTS	AZ63A-T7	...	30.3	15.2	18.0	Ax	0.25 *R*	Unnotched	0.30*d*	...	...	24.0	22.0	20.0	...	All data at 275 °F
AZ63-HTS	AZ63A-T7	...	27.0	14.2	40.3	Ax	0.25 *R*	Unnotched	...	...	...	24.4	22.0	20.0	...	All data at 300 °F
AZ63-HTA	AZ63A-T6	...	40.0	19.0	5.0	Ax	0.0 *R*	Unnotched	0.30*d*	...	...	28.0	21.0	20.0	...	...
AZ63-HTA	AZ63A-T6	...	40.0	19.0	5.0	Ax	0.25 *R*	Unnotched	0.30*d*	...	...	25.0	23.0	21.0	...	...

Note: UTS, ultimate tensile strength; YS, yield strength; %E, percent elongation; B, bending; RB, rotating bending; AX, axial; *d*, inside diameter; *D*, outside diameter; *t*, thickness; *r*, radius. Source: H.J. Grover, S.A. Gordon, and L.R. Jackson, *Fatigue of Metals and Structures*, Department of the Navy, 1960, p 354-359

(continued)

Table 6 Cast magnesium alloy fatigue strength (continued)

Material	Designation	Condition	UTS, ksi	YS, ksi	%E	Fatigue loading		Fatigue specimen		K_t	Fatigue strength (max stress, ksi) at indicated cycles					Comments
						Type	Condition	Type	Size, in.		10^4	10^5	10^6	10^7	10^8	
Sand castings and permanent mold casting alloys (continued)																
AZ63-HTA	AZ63A-T6	...	28.5	15.5	15.0	Ax	0.0 *R*	Unnotched	0.30*d*	...	...	22.0	20.0	18.0	...	All data at 275 °F
AZ63-HTA	AZ63A-T6	...	28.5	15.5	15.0	Ax	0.25 *R*	Unnotched	0.30*d*	...	...	25.0	24.0	23.0	...	All data at 275 °F
AZ92-HTA	AZ92A-T6	...	40.0	23.0	2.0	RB	−1 *R*	Unnotched	0.30*d*	...	...	19.5	...	16.0	15.0	...
AZ92-HTA	AZ92A-T6	...	40.0	23.0	2.0	RB	−1 *R*	Circular notch	0.0295*r*, 0.295*d*, 0.354*D*	2.0	...	14.0	...	9.5	9.5	...
AZ92-HTS	AZ92A-T7	...	40.0	23.0	3.2	RB	−1 *R*	Unnotched	0.30*d*	...	...	...	15.0	12.5	11.0	All data at 150 °F
AZ92-HTS	AZ92A-T7	...	29.0	17.0	33.0	RB	−1 *R*	Unnotched	0.30*d*	...	...	...	8.7	7.3	6.0	All data at 275 °F
Casting alloys (tested with cast skin present)																
AZ63-HTA	AZ63A-T6	I-beam, cantilever specimen	44.0	...	6.0	B	−1 *R*	Unnotched	...	...	26.0	9.7	7.0	6.6	...	...
AZ63-HTA	AZ63A-T6	I-beam, cantilever specimen	26.0	14.0	31.0	B	−1 *R*	Unnotched	...	...	15.0	9.4	6.8	...	...	All data at 300 °F
AZ63-HTAS	AZ63A-T7	I-beam, cantilever specimen	17.5	8.5	37.0	B	−1 *R*	Unnotched	...	...	9.8	6.7	4.6	4.6	...	All data at 400 °F
AZ63-HTAS	AZ63A-T7	I-beam, cantilever specimen	11.0	6.3	39.0	B	−1 *R*	Unnotched	...	...	8.5	5.8	...	...	...	All data at 500 °F
AZ92-HT	AZ92A-T4	Longitudinal, 0.200 in. thick	40.0	16.0	10.0	B	−1 *R*	Unnotched	...	...	...	15.0	...	10.0	10.0	...
CM62	EM62A-F	I-beam, cantilever specimen	15.5	...	1.0	B	−1 *R*	Unnotched	...	...	14.0	9.0	5.7	...	...	...
CM62	EM62A-F	I-beam, cantilever specimen	15.3	7.5	7.0	B	−1 *R*	Unnotched	...	...	10.5	5.6	4.0	...	...	All data at 500 °F
CM62	EM62A-F	I-beam, cantilever specimen	12.0	6.2	34.0	B	−1 *R*	Unnotched	...	...	8.1	5.8	3.8	...	...	All data at 600 °F
Die casting alloys																
AZ91-ACS	AZ91A-T2	...	26.0	18.0	1.0	RB	−1 *R*	Unnotched	0.03*d*	...	...	...	16.7	14.0	12.5	All data at 150 °F
AZ91-ACS	AZ91A-T2	...	25.0	13.0	7.0	RB	−1 *R*	Unnotched	0.03*d*	...	...	...	11.5	10.0	9.0	All data at 150 °F
AZ91-HT	AZ91A-T4	...	...	...	...	RB	−1 *R*	Unnotched	0.03*d*	...	...	...	18.2	16.5	15.0	Fatigue data at 150 °F
AZ91-HT	AZ91A-T4	...	...	...	...	RB	−1 *R*	Unnotched	0.03*d*	...	...	...	13.8	12.0	10.0	Fatigue data at 275 °F
AZ91-HTA	AZ91A-T6	...	44.0	22.0	4.5	RB	−1 *R*	Unnotched	0.03*d*	...	...	...	15.8	13.0	9.6	All data at 150 °F
AZ91-HTA	AZ91A-T6	...	27.0	14.0	40.0	RB	−1 *R*	Unnotched	0.03*d*	...	...	...	10.8	10.0	7.0	All data at 275 °F
AZ91-HTS	AZ91A-T7	...	39.0	20.0	3.5	RB	−1 *R*	Unnotched	0.03*d*	...	...	...	14.6	12.6	11.8	All data at 150 °F
AZ91-HTS	AZ91A-T7	...	27.0	15.0	30.0	RB	−1 *R*	Unnotched	0.03*d*	...	...	...	11.8	11.0	10.1	All data at 275 °F
Die casting alloy (tested with as-cast skin present)																
AZ91-AC	AZ91A-F	...	33.0	22.0	3.0	RB	−1 *R*	Unnotched	0.25*d*	...	...	17.5	...	15.0	15.0	...

Note: UTS, ultimate tensile strength; YS, yield strength; %E, percent elongation; B, bending; RB, rotating bending; AX, axial; *d*, inside diameter; *D*, outside diameter; *t*, thickness; *r*, radius. Source: H.J. Grover, S.A. Gordon, and L.R. Jackson, *Fatigue of Metals and Structures*, Department of the Navy, 1960, p 354-359

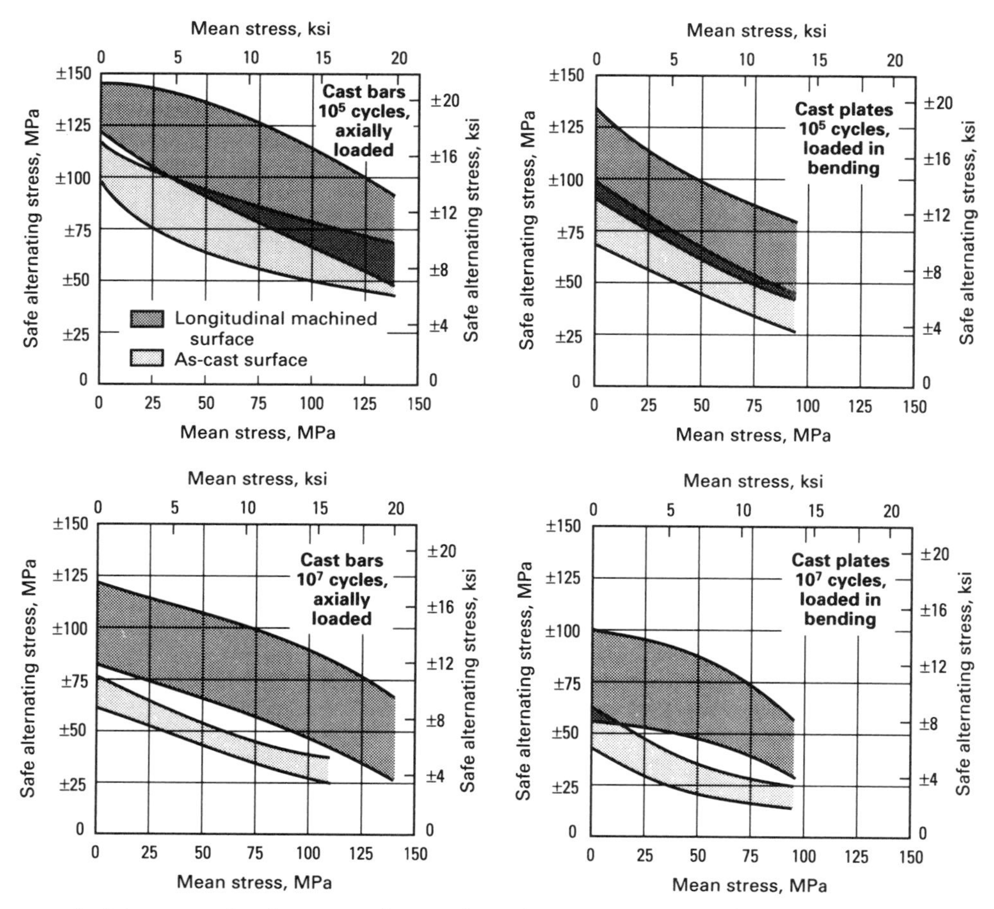

Fig. 7 Effect of surface type on the fatigue properties of cast magnesium-aluminum-zinc alloys. Source: *Metals Handbook*, 9th ed., Vol 2, ASM International, 1979, p 461

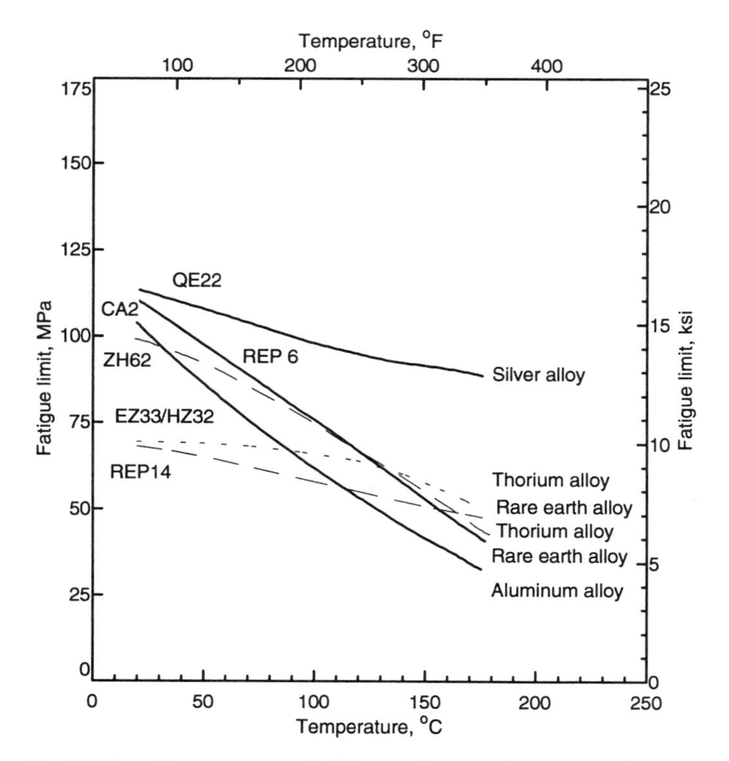

Fig. 8 Effect of temperature on the fatigue endurance of some magnesium casting alloys. Source: Ref 39

LA141, Mg-14Li-1Al), but uses for these alloys have been limited to special applications such as components for spacecraft and body armor.

Wrought Forms

Extruded bars and shapes are made of several types of magnesium alloys. For normal strength requirements, one of the magnesium-aluminum-zinc (AZ) alloys is usually selected. The strength of these alloys increases as aluminum content increases. Alloy AZ31B is a widely used moderate-strength grade with good formability; it is used extensively for cathodic protection. Alloy AZ31C is a lower-purity commercial variation of AZ31B for lightweight structural applications that do not require maximum corrosion resistance. The M1A and ZM21A alloys can be extruded at higher speeds than AZ31B, but they have limited use because of their lower strength. Alloy AZ10A has a low aluminum content and thus is of lower strength than AZ31B, but it can be welded without subsequent stress relief. The AZ61A and AZ80A alloys can be artificially aged for additional strength (with a sacrifice in ductility); AZ80A is not available in hollow shapes. Alloy AZ21X1 is designed specially for use in battery applications.

Alloy ZK60A is used where high strength and good toughness are required. This alloy is heat treatable and is normally used in the artificially aged (T5) condition. ZK21A and ZK40A alloys are of lower strength and are more readily extrudable than ZK60A; they have had limited use in hollow tubular strength requirements.

Alloy ZC71 is a member of a new family of magnesium alloys containing neither aluminum nor zirconium. The alloy can be extruded at high rates and exhibits good strength properties. The corrosion resis-

Table 7 Nominal composition, typical tensile properties, and characteristics of selected wrought magnesium alloys

ASTM designation	British designation	Nominal composition: Al	Zn	Mn	Zr	Th	Cu	Li	Condition	Tensile properties: 0.2% yield strength, MPa	Ultimate tensile strength, MPa	Elongation, %	Characteristics
M1	AM503	...	...	...	1.5	...	...	...	Sheet, plate F	70	200	4	Low- to medium-strength alloy, weldable, corrosion resistant
									Extrusions F	130	230	4	
									Forgings F	105	200	4	
AZ31	AZ31	3	1	0.3 (0.20 min)	...	...	...	...	Sheet, plate O	120	240	11	Medium-strength alloy, weldable, good formability
									H24	160	250	6	
									Extrusions F	130	230	4	
									Forgings F	105	200	4	
AZ61	AZM	6.5	1	0.3 (0.15 min)	...	...	...	...	Extrusions F	180	260	7	High-strength alloy, weldable
									Forgings F	160	275	7	
AZ80	AZ80	8.5	0.5	0.2 (0.12 min)	...	...	...	...	Forgings T6	200	290	6	High-strength alloy
ZM21	ZM21	...	2	1	...	...	...	...	Sheet, plate O	120	240	11	Medium-strength alloy, good formability, good damping capacity
									H24	165	250	6	
									Extrusions	155	235	8	
									Forgings	125	200	9	
ZMC711	...	...	6.5	0.75	...	...	1.25	...	Extrusions T6	300	325	3	High-strength alloy
LA141	...	1.2	...	0.15 min	...	...	...	14	Sheet, plate T7	95	115	10	Ultralight weight (specific gravity 1.35)
ZK31	ZW3	...	3	...	0.6	...	...	...	Extrusions T5	210	295	8	High-strength alloy, some weldability
									Forgings T5	205	290	7	
ZK61	...	...	6	...	0.8	...	...	...	Extrusions F	210	285	6	High-strength alloy
									T5	240	305	4	
									Forgings T5	?160	275	7	
HK31	...	...	...	...	0.7	3.2	...	...	Sheet, plate H24	170	230	4	High creep resistance to 350 °C, weldable
									Extrusions T5	180	255	4	
HM21	...	...	...	0.8	...	2	...	...	Sheet, plate T8	135	215	6	High creep resistance to 350 °C, short time exposure to 425 °C, weldable
									T81	180	255	4	
									Forgings T5	175	225	3	

tance of ZC71 is similar to that of AZ91C, but it falls far short of that of AZ91E.

Alloy HM31A is of moderate strength. It is suitable for use in applications requiring good strength and creep resistance at temperatures in the range of 150 to 425 °C (300 to 800 °F).

Forgings are made of AZ31B, AZ61A, AZ80A, M1A, and ZK60A. Alloy HM21A is also a good forging alloy. Alloys M1A and AZ31B may be used for hammer forgings (whereas the other alloys are almost always press forged); however, there has been a gradual decline in the use of the magnesium-manganese alloy M1A. The AZ80A alloy has greater strength than AZ61A and requires the slowest rate of deformation of the magnesium-aluminum-zinc alloys. Alloy ZK60A has essentially the same strength as AZ80A but with greater ductility. To develop maximum properties, both AZ80A and ZK60A are heat treated to the T5 condition; AZ80A may be given the T6 solution heat treatment, followed by artificial aging to provide maximum creep stability. Alloy HM21A is given the T5 temper. It is useful at elevated temperatures up to 370 to 425 °C (700 to 800 °F) for applications in which good creep resistance is needed.

Hydraulic and mechanical processes are both used for the forging of magnesium. A slow and controlled rate of deformation is desirable because it facilitates control of the plastic flow of metal; therefore, hydraulic press forging is the most commonly used process. Magnesium, which has a hexagonal crystal structure, is more easily worked at elevated temperature. Consequently, forging stock (ingot or billet) is heated to a temperature between 350 and 500 °C (650 and 950 °F) prior to forging.

Sheet and plate are rolled from magnesium-aluminum-zinc (AZ and photoengraving grade, or PE) and magnesium-thorium (HK and HM) alloys.

Alloy AZ31B is the most widely used alloy for sheet and plate and is available in several grades and tempers. It can be used at temperatures up to 100 °C (200 °F). The HK31A and HM21A alloys are suitable for use at temperatures up to 315 and 345 °C (600 and 650 °F), respectively. However, HM21A has superior strength and creep resistance. Alloy PE is a special-quality sheet with excellent flatness, corrosion resistance, and etchability. It is used in photoengraving.

Wrought Mechanical Properties

Typical mechanical properties data for various wrought alloys are shown in Tables 3 and 7. More detailed information on property minimums is covered in Ref 6 and in Volume 2 of the *ASM Handbook, Properties and Selection: Nonferrous Alloys and Special-Purpose Materials*. In general, the direction, temperature, and speed at which an alloy is fabricated have a significant effect on the mechanical properties of wrought parts. For example, extrusions produced at higher temperatures and speeds have lower strength than those produced under normal operating conditions.

Table 8 Wrought magnesium alloy fatigue strength

Material	Designation	UTS, ksi	YS, ksi	%E	Fatigue loading		Fatigue specimen		K_t	Fatigue strength (max stress), ksi, at indicated cycles:					Comments
					Type	Condition	Type	Size, in.		10^4	10^5	10^6	10^7	10^8	
AZ31B-F (extruded)															
AZ31X	AZ31B-F	40.1	29.9	16.0	B	$-1R$	Unnotched	...	...	30.0	24.0	22.0	19.0	...	
AZ31X	AZ31B-F	45.6	39.6	15.0	B	$-1R$	Unnotched	...	...	42.0	30.0	26.0	23.0	...	At −108 °F
AZ31X	AZ31B-F	62.8	48.4	6.5	B	$-1R$	Unnotched	...	...	45.0	32.0	...	...	...	At −320 °F
AZ31X	AZ31B-F	40.0	30.0	15.0	RB	$-1R$	Unnotched	0.30*d*	...	...	27.0	...	21.0	19.0	...
AZ31X	AZ31B-F	39.5	29.0	16.0	RB	$-1R$	Unnotched	...	...	...	23.5	21.0	18.0	16.0	...
AZ31X	AZ31B-F	37.0	24.0	19.0	RB	$-1R$	Unnotched	...	...	...	...	21.5	20.1	...	At 150 °F
AZ31X	AZ31B-F	40.0	30.0	15.0	B	$-1R$	Unnotched	0.117*t*, 1.0*w*	...	...	17.0	...	10.0	10.0	...
AZ31X	AZ31B-F	40.0	30.0	15.0	RB	$-1R$	Circular notch	0.0295*r*, 0.295*d*, 0.354*D*	2	...	16.0	...	10.5	9.0	...
AZ61A-F (extruded)															
AZ61X	AZ61A-F	45.0	32.0	15.0	RB	$-1R$	Unnotched	0.30*d*	...	...	24.5	...	20.5	20.0	...
AZ61X	AZ61A-F	44.5	29.0	18.0	RB	$-1R$	Unnotched	0.30*d*	...	...	...	27.0	24.0	21.0	At 150 °F
AZ61X	AZ61A-F	45.0	32.0	15.0	RB	$-1R$	Circular notch	0.0295*r*, 0.295*d*, 0.354*D*	2	...	16.0	...	11.5	10.0	...
AZ61X	AZ61A-F	45.0	32.0	15.0	RB	$-1r$	V-notch	60° notch, 0.002*r*, 0.30*d*, 0.480*D*	5	...	11.5	...	6.5	6.0	...
AZ61X	AZ61A-F	45.0	32.0	15.0	B	$-1R$	Unnotched	0.114*t*, 1.0*w*	...	...	19.0	...	13.5	13.5	...
AZ80A-F (extruded)															
AZ80X	AZ80A-F	46.5	33.5	12.0	RB	$-1R$	Unnotched	0.30*d*	...	...	29.0	26.0	22.5	19.0	...
AZ80X	AZ80A-F	48.5	32.3	15.0	RB	$-1R$	Unnotched	0.30*d*	...	...	...	28.6	24.0	22.4	At 150 °F
AZ80X	AZ80A-F	35.0	26.0	39.0	RB	$-1R$	Unnotched	0.30*d*	...	...	...	15.6	13.5	12.0	At 275 °F
AZ80X	AZ80A-F	49.0	33.0	11.0	RB	$-1R$	Unnotched	...	...	...	16.5	...	12.5	11.0	...
AZ80X	AZ80A-F	49.0	33.0	11.0	RB	$-1R$	Circular notch	0.0295*r*, 0.295*d*, 0.354*D*	2	...	10.5	...	7.0	6.0	...
AZ80A-T51 (extruded)															
AZ80X-A	AZ80A-T51	43.0	29.0	23.0	RB	$-1R$	V-notch	60° notch, 0.002*r*, 0.30*d*, 0.480*D*	5	...	...	24.0	21.5	20.0	At 150 °F
AZ80X-A	AZ80A-T51	32.0	21.0	41.0	RB	$-1R$	Unnotched	0.30*d*	...	...	...	16.3	14.0	10.1	At 275 °F
AZ80A-T51 (longitudinal, extruded)															
AZ80X-A	AZ80A-T51	50.0	34.0	6.0	B	$-1R$	Unnotched	0.116*t*, 1.0*w*	...	...	19.5	...	13.0	13.0	...
AZ80A-T51 (extruded)															
AZ80X-HTA	AZ80A-T51	50.0	34.0	7.0	RB	$-1R$	Unnotched	0.30*d*	...	...	27.0	...	22.5	20.5	...
AZ80X-HTA	AZ80A-T51	54.0	37.5	13.0	RB	$-1R$	Unnotched	0.30*d*	...	...	...	25.4	23.0	20.4	At 150 °F
AZ80X-HTA	AZ80A-T51	48.5	32.0	20.0	RB	$-1R$	Unnotched	0.30*d*	...	...	23.0	17.9	13.0	13.0	At 200 °F
AZ80X-HTA	AZ80A-T51	37.0	23.0	31.0	RB	$-1R$	Unnotched	0.30*d*	...	...	...	16.3	14.0	11.3	At 275 °F
AZ80X-HTA	AZ80A-T51	33.5	21.5	35.0	RB	$-1R$	Unnotched	0.30*d*	...	...	20.6	15.2	11.0	10.5	At 300 °F
AZ80A-F (longitudinal, extruded)															
AZ80X-HTA	AZ80A-F	50.0	34.0	7.0	B	$-1R$	Unnotched	0.117*t*, 1.0*w*	...	...	20.0	...	12.5	12.5	...
AZ80X-HTA	AZ80A-T51	50.0	34.0	5.0	Ax	$0R$	Unnotched	0.30*d*	...	...	39.0	35.5	32.0	...	...
AZ80A-T51 (wrought)															
AZ80X-HTA	AZ80A-T51	50.0	34.0	5.0	Ax	$0.25R$	Unnotched	0.30*d*	...	...	39.0	36.5	34.0	...	...

Note: UTS, ultimate tensile strength; YS, yield strength; %E, percent elongation; B, bending; RB, rotating bending; Ax, axial; *d*, inside diameter; *D*, outside diameter; *t*, thickness; *r*, radius; *w*, width; *W*, notch width. Source: H.J. Grover, S.A. Gordon, and L.R. Jackson, *Fatigue of Metals and Structures*, Department of the Navy, 1960, p 346-353

(continued)

Table 8 Wrought magnesium alloy fatigue strength (continued)

Material	Designation	UTS, ksi	YS, ksi	%E	Fatigue loading Type	Fatigue loading Condition	Fatigue specimen Type	Fatigue specimen Size, in.	K_t	Fatigue strength (max stress), ksi, at indicated cycles: 10^4	10^5	10^6	10^7	10^8	Comments
AZ80A-T51 (wrought) (continued)															
AZ80X-HTA	AZ80A-T51	49.0	30.0	7.0	Ax	0.25*R*	Unnotched	...	...	...	36.8	34.3	33.0	...	...
AZ80X-HTA	AZ80A-T51	30.0	11.0	35.0	Ax	0.25*R*	Unnotched	...	...	...	26.5	24.8	23.0	...	At 300 °F
AZ80X-HTA	AZ80A-T51	...	...	...	Ax	0.25*R*	Unnotched	...	...	...	24.5	16.5	11.0	...	Fatigue data at 400 °F
AZ80A-T51 (extruded)															
AZ80X-HTA	AZ80A-T51	57.5	40.5	4.0	RB	−1*R*	Circular notch	0.0295*r*, 0.295*d*, 0.354*D*	2	...	17.0	...	13.0	11.0	...
AZ80X-HTA	AZ80A-T51	57.5	40.5	4.0	RB	−1*R*	V-notch	60° notch, 0.002*r*, 0.30*d*, 0.48*D*	5	...	11.0	...	8.0	7.0	...
MI-F (extruded)															
M-1	MI-F	38.0	26.0	10.0	RB	−1*R*	Unnotched	0.30*d*	...	...	15.5	...	10.9	10.5	...
M-1	MI-F	38.0	26.0	10.0	B	−1*R*	Unnotched	0.125*t*, 1.0*w*, 0.0295*r*, 0.295*d*	...	...	14.0	...	10.0	10.0	...
M-1	MI-F	38.0	26.0	10.0	RB	−1*R*	Circular notch	0.354*D*	2	...	11.0	...	7.0	5.5	...
MI-O (sheet)															
M-1a	MI-O	34.0	19.0	12.0	Ax	0.25*R*	Unnotched	...	...	25.8	23.3	21.0	21.0	...	...
M-1a	MI-O	23.0	14.0	25.0	Ax	0.25*R*	Unnotched	...	...	...	22.0	18.6	16.6	...	At 200 °F
M-1a	MI-O	20.0	12.0	30.0	Ax	0.25*R*	Unnotched	...	...	26.2	21.0	16.5	13.0	...	At 300 °F
EM51XA-T6 (wrought)															
EM51-HTA	EM61XA-T6	37.0	28.0	5.0	Ax	0.25*R*	Unnotched	...	...	35.0	31.0	28.0	25.0	...	...
EM51-HTA	EM51XA-T6	25.0	15.9	18.0	Ax	0.25*R*	Unnotched	...	...	35.0	27.5	23.0	22.0	...	All data at 300 °F
EM51-HTA	EM51XA-T6	24.0	15.3	18.0	Ax	0.25*R*	Unnotched	...	...	29.0	23.4	18.0	15.6	...	All data at 400 °F
EM61XA-F (extruded)															
EM61	EM61XA-F	...	...	...	RB	−1*R*	Unnotched	...	...	...	...	18.5	17.5	15.0	Fatigue data at 150 °F
EM61XA-T2 (extruded)															
EM61-S	EM61XA-T2	...	...	...	RB	−1*R*	Unnotched	...	...	...	...	19.0	18.2	16.0	Fatigue data at 150 °F
EM61-S	EM61XA-T2	30.0	26.0	14.0	RB	−1*R*	Unnotched	...	...	...	...	17.6	15.0	12.6	All data at 275 °F
AM100A-F (forged)															
A-10	AM100A-F	...	...	...	RB	−1*R*	Unnotched	0.30*d*	...	...	...	17.6	15.0	12.0	Fatigue data at 150 °F
A-10	AM100A-F	...	...	...	RB	−1*R*	Unnotched	0.30*D*	...	...	...	13.0	9.9	7.7	Fatigue data at 275 °F
AZ61A-F (forged)															
AZ61X	AZ61A-F	43.0	26.0	12.0	RB	−1*R*	Unnotched	0.30*d*	...	...	27.5	...	21.5	19.5	...
AZ61X	AZ61A-F	43.0	26.0	12.0	RB	−1*R*	Circular notch	0.0295*r*, 0.295*d*, 0.354*D*	2	...	14.5	...	11.0	9.5	...
AZ80A-F (forged)															
AZ80X	AZ80A-F	46.0	31.0	8.0	RB	−1*R*	Unnotched	0.30*d*	...	...	28.0	...	22.0	19.5	...
AZ80X	AZ80A-F	46.0	31.0	8.0	RB	−1*R*	Circular notch	0.0295*r*, 0.295*d*, 0.354*D*	2	...	16.5	...	11.5	10.5	...
AZ80X	AZ80A-F	51.0	31.0	15.0	RB	−1*R*	Unnotched	0.30*d*	...	...	...	24.6	22.0	19.5	Fatigue data at 150 °F
AZ80X	AZ80A-F	29.0	19.0	56.0	RB	−1*R*	Unnotched	0.30*d*	...	...	...	15.8	12.7	9.3	Fatigue data at 275 °F

Note: UTS, ultimate tensile strength; YS, yield strength; %E, percent elongation; B, bending; RB, rotating bending; Ax, axial; *d*, inside diameter; *D*, outside diameter; *t*, thickness; *r*, radius; *w*, width; *W*, notch width. Source: H.J. Grover, S.A. Gordon, and L.R. Jackson, *Fatigue of Metals and Structures*, Department of the Navy, 1960, p 346-353

(continued)

Table 8 Wrought magnesium alloy fatigue strength (continued)

Material	Designation	UTS, ksi	YS, ksi	%E	Fatigue loading Type	Fatigue loading Condition	Fatigue specimen Type	Fatigue specimen Size, in.	K_t	Fatigue strength (max stress), ksi, at indicated cycles: 10^4	10^5	10^6	10^7	10^8	Comments
AZ80A-T5 (forged)															
AZ80X-A	AZ80A-T5	50.0	34.0	6.0	RB	−1*R*	Unnotched	0.30*d*	...	...	25.0	...	20.0	18.0	...
AZ80X-A	AZ80A-T5	50.0	34.0	6.0	RB	−1*R*	Circular notch	0.0295*r*, 0.295*d*, 0.354*D*	2	...	16.5	...	9.5	9.5	...
AZ80X-A	AZ80A-T5	50.0	34.0	8.5	RB	−1*R*	Unnotched	0.30*d*	...	...	...	22.3	20.0	17.7	Fatigue data at 150 °F
AZ80X-A	AZ80A-T5	...	...	...	RB	−1*R*	Unnotched	0.30*d*	...	...	...	14.2	11.8	10.0	Fatigue data at 275 °F
AZ80X-A	AZ80A-T5	26.0	14.9	52.0	RB	−1*R*	Unnotched	0.30*d*	...	...	20.5	...	11.0	10.5	All data at 300 °F
AZ80X-A	AZ80A-T5	26.0	14.9	52.0	RB	−1*R*	Circular notch	0.0295*r*, 0.295*d*, 0.354*D*	2	...	10.5	...	6.5	5.0	All data at 300 °F
AZ80A-T6 (forged)															
AZ80X-HTA	AZ80A-T6	49.0	30.0	7.0	RB	−1*R*	Unnotched	0.30*d*	...	...	26.0	...	16.0	16.0	...
AZ80X-HTA	AZ80A-T6	49.0	30.0	7.0	RB	−1*R*	Circular notch	0.0295*r*, 0.295*d*, 0.354*D*	2	...	14.5	...	9.5	9.5	...
AZ80A-T6 (forged)															
AZ80X-HTA	AZ80A-T6	45.7	17.0	9.0	RB	−1*R*	Unnotched	0.30*d*	...	...	...	20.0	18.0	16.0	All data at 150 °F
AZ80X-HTA	AZ80A-T6	43.0	14.3	14.0	RB	−1*R*	Unnotched	0.30*d*	...	...	23.0	17.5	13.0	13.0	All data at 200 °F
AZ80X-HTA	AZ80A-T6	43.0	14.3	14.0	RB	−1*R*	Circular notch	0.0295*r*, 0.295*d*, 0.354*D*	2	...	12.7	...	7.5	7.0	All data at 200 °F
AZ80X-HTA	AZ80A-T6	36.0	12.5	30.0	RB	−1*R*	Unnotched	0.30*d*	...	...	...	15.1	12.5	10.5	All data at 275 °F
AZ80X-HTA	AZ80A-T6	49.0	30.0	7.0	B	−1*R*	Unnotched	0.20*t*	...	...	20.5	16.8	16.0	...	...
AZ80X-HTA	AZ80A-T6	30.0	11.0	35.0	RB	−1*R*	Unnotched	...	...	...	20.6	15.2	11.0	10.5	All data at 300 °FAZ80X-HTA
AZ80X-HTA	AZ80A-T6	30.0	11.0	35.0	B	−1*R*	Unnotched	0.20*t*	...	20.0	12.0	7.5	6.0	...	All data at 300 °F
AZ80X-HTA	AZ80A-T6	...	...	...	B	−1*R*	Unnotched	0.20*t*	...	20.0	12.0	7.5	5.0	...	Fatigue data at 400 °F
AZ80X-HTA	AZ80A-T6	30.0	11.0	35.0	RB	−1*R*	Unnotched	0.30*d*	...	...	20.6	15.2	11.0	10.5	All data at 300 °F
AZ80X-HTA	AZ80A-T6	30.0	11.0	35.0	RB	−1*R*	Circular notch	0.0295*r*, 0.295*d*, 0.354*D*	2	...	10.8	...	6.4	5.0	All data at 300 °F
EM21XA-F (forged)															
EM21	EM21XA-F	...	...	...	RB	−1*R*	Unnotched	0.30*d*	...	...	...	16.0	15.0	13.0	Fatigue data at 150 °F
EM21	EM21XA-F	...	...	...	RB	−1*R*	Unnotched	0.30*d*	...	...	...	13.6	12.6	12.0	Fatigue data at 275 °F
EM41XA-F (forged)															
EM41	EM41XA-F	37.0	27.0	8.0	RB	−1*R*	Unnotched	0.30*d*	...	...	...	17.7	16.2	14.0	Fatigue data at 150 °F
EM41	EM41XA-F	27.0	22.0	14.0	RB	−1*R*	Unnotched	0.30*d*	...	...	...	17.8	16.6	16.0	Fatigue data at 275 °F
EM51XA-T6 (forged)															
EM51-HTA	EM51XA-T6	37.0	28.0	5.0	RB	−1*R*	Unnotched	0.30*d*	...	...	16.5	15.0	13.3	12.0	...
EM51-HTA	EM51XA-T6	...	...	...	RB	−1*R*	Circular notch	0.0295*r*, 0.295*d*, 0.354*D*	2	...	12.0	...	8.3	7.0	...
EM51-HTA	EM51XA-T6	37.0	28.0	5.0	B	−1*R*	Unnotched	...	...	23.0	18.3	14.5	12.5	...	...
EM51-HTA	EM51XA-T6	25.0	15.9	18.0	B	−1*R*	Unnotched	0.20*t*	...	18.0	15.0	12.2	9.4	...	All data at 300 °F
EM51-HTA	EM51XA-T6	24.0	15.3	18.0	B	−1*R*	Unnotched	0.20*t*	...	15.5	12.7	9.8	8.0	...	All data at 400 °F
EM51-HTA	EM51XA-T6	13.8	7.4	18.0	B	−1*R*	Unnotched	...	...	13.0	8.8	6.3	3.7	...	All data at 600 °F

Note: UTS, ultimate tensile strength; YS, yield strength; %E, percent elongation; B, bending; RB, rotating bending; Ax, axial; *d*, inside diameter; *D*, outside diameter; *t*, thickness; *r*, radius; *w*, width; *W*, notch width. Source: H.J. Grover, S.A. Gordon, and L.R. Jackson, *Fatigue of Metals and Structures*, Department of the Navy, 1960, p 346-353

(continued)

Table 8 Wrought magnesium alloy fatigue strength (continued)

Material	Designation	UTS, ksi	YS, ksi	%E	Fatigue loading		Fatigue specimen		K_t	Fatigue strength (max stress), ksi, at indicated cycles:					Comments
					Type	Condition	Type	Size, in.		10^4	10^5	10^6	10^7	10^8	
M1A-F (forged)															
M1-a	M1A-F	34.0	19.0	12.0	B	−1R	Unnotched	...	...	19.0	13.0	9.0	7.0	...	...
M1-a	M1A-F	23.0	13.8	26.0	B	−1R	Unnotched	0.20t	...	...	11.7	8.0	5.8	...	All data at 200 °F
M1-a	M1A-F	20.0	12.0	31.0	B	−1R	Unnotched	0.20t	...	...	18.0	7.4	6.0	...	All data at 300 °F
M1-a	M1A-F	17.2	9.1	34.0	B	−1R	Unnotched	0.20t	...	13.9	9.5	6.5	5.0	...	All data at 400 °F
EM51XA-F (forged)															
EM51	EM51XA-F	...	...	...	B	−1R	Unnotched	0.20t	...	18.0	14.2	11.0	8.9	...	All data at 300 °F
EM51	EM51XA-F	...	...	...	B	−1R	Unnotched	0.20t	...	16.7	12.2	9.2	7.5	...	All data at 600 °F
EM51	EM51XA-F	...	...	...	B	−1R	Unnotched	0.20t	...	10.4	8.3	6.5	5.0	...	All data at 300 °F
EM51	EM51XA-F	...	...	...	Ax	0.25R	Unnotched	0.20t	...	32.0	30.0	27.5	25.5	...	...
EM51	EM51XA-F	...	...	...	Ax	0.25R	Unnotched	0.20t	...	30.0	24.2	19.7	16.0	...	...
AZ31A-O (longitudinal, sheet)															
AZ31X-a	AZ31A-O	37.0	22.0	21.0	B	−1R	Unnotched	0.02t, 0.50w	...	...	19.5	...	15.5	15.5	...
AZ31X-a	AZ31X-O	37.0	22.0	21.0	B	−1R	Edge notch	0.062r, 0.020t, 0.525w, 0.650W	1.6	...	17.0	...	13.0	13.0	...
AZ31A-H24 (longitudinal, sheet)															
AZ31X-h	AZ31A-H24	43.0	33.0	11.0	B	−1R	Unnotched	0.020t, 0.50w	...	...	20.0	...	19.5	17.5	...
AZ31X-h	AZ31A-H24	43.0	33.0	11.0	B	−1R	Edge notch	0.062r, 0.020t, 0.525w, 0.65W	1.6	...	19.5	...	13.0	11.0	...
AZ31A-O (longitudinal, sheet)															
AZ31X-a	AZ31A-O	37.0	22.0	21.0	B	−1R	Unnotched	0.064t, 0.65w	...	...	15.0	...	14.0	13.5	...
AZ31X-a	AZ31A-O	37.0	22.0	21.0	B	−1R	Surface notch	60° notch, 0.001r, 0.65W, 0.003d, 0.064t	2	...	18.0	...	13.0	...	...
AZ31A-H24 (longitudinal, sheet)															
AZ31X-h	AZ31A-H24	43.0	33.0	11.0	B	−1R	Unnotched	0.064t, 0.65w	...	...	17.0	...	15.5	15.0	...
AZ31X-h	AZ31A-H24	43.0	33.0	11.0	B	0.25R	Surface notch	60° notch, 0.001r, 0.65W, 0.003d, 0.064t	...	...	21.0	...	15.0	...	...
AZ31A-O (sheet)															
AZ31X-a	AZ31A-O	37.0	22.0	21.0	Ax	0.25R	Unnotched	0.30d	...	...	21.5	21.0	21.0	...	...
AZ31X-a	AZ31A-O	37.0	22.0	21.0	Ax	0.50R	Unnotched	0.30d	...	...	28.0	27.0	26.0	...	...
AZ31X-a	AZ31A-O	37.0	20.0	21.0	Ax	0.25R	Unnotched	0.064t, 1.0w	...	32.0	26.0	21.0	20.0	...	...
AZ31A-H24 (sheet)															
AZ31X-h	AZ31A-H24	43.0	37.0	11.0	Ax	0.25R	Unnotched	0.064t, 1.0w	...	...	36.0	35.0	34.0	...	...
AZ31X-h	AZ31A-H24	43.0	37.0	11.0	Ax	0.50R	Unnotched	0.064t, 1.0w	...	...	39.0	36.0	35.0	...	...
AZ31X-h	AZ31A-H24	46.0	38.0	5.0	Ax	0.25R	Unnotched	0.064t, 1.0w	...	40.0	28.0	25.0	24.0	...	...
AZ31X-h	AZ31A-H24	46.0	38.0	5.0	Ax	0.50R	Unnotched	0.064t, 1.0w	...	...	37.0	30.0	29.0	...	...
AZ31X-h	AZ31A-H24	43.0	33.0	11.0	Ax	0.25R	Unnotched	0.064t, 1.0w	...	...	24.0	23.0	23.0	...	...
AZ31X-h	AZ31A-H24	43.0	33.0	11.0	Ax	0.50R	Unnotched	0.064t, 1.0w	...	...	28.0	26.0	26.0	...	...

Note: UTS, ultimate tensile strength; YS, yield strength; %E, percent elongation; B, bending; RB, rotating bending; Ax, axial; *d*, inside diameter; *D*, outside diameter; *t*, thickness; *r*, radius; *w*, width; *W*, notch width. Source: H.J. Grover, S.A. Gordon, and L.R. Jackson, *Fatigue of Metals and Structures*, Department of the Navy, 1960, p 346-353

(continued)

Table 8 Wrought magnesium alloy fatigue strength (continued)

Material	Designation	UTS, ksi	YS, ksi	%E	Fatigue loading		Fatigue specimen		K_t	Fatigue strength (max stress), ksi, at indicated cycles:					Comments
					Type	Condition	Type	Size, in.		10^4	10^5	10^6	10^7	10^8	
AZ51XA-O (longitudinal, sheet)															
AZ51X-a	AZ51XA-O	41.0	22.0	18.0	B	−1*R*	Unnotched	0.064*t*, 1.0*w*	...	...	19.5	...	14.5	14.0	...
AZ51XA-H24 (longitudinal, sheet)															
AZ51X-h	AZ51XA-H24	44.0	33.0	9.0	B	−1*R*	Unnotched	0.064*t*, 1.0*w*	...	...	24.0	...	16.0	16.0	...
AZ51X-h	AZ51XA-H24	44.0	33.0	9.0	B	−1*R*	Surface notch	60° notch, 0.003*d*, 0.001*r*, 0.064*t*, 0.65*W*	2.0	...	22.5	...	15.0	...	...
AZ51X-h	AZ51XA-H24	44.0	33.0	9.0	Ax	0.25*R*	Unnotched	0.064*t*, 1.0*w*	...	...	34.0	29.0	28.0	...	...
AZ61A-O (longitudinal, sheet)															
AZ61X-a	AZ61A-O	43.0	26.0	16.0	B	−1*R*	Unnotched	0.020*t*, 0.50*w*	...	...	23.5	...	15.0	14.0	...
AZ61X-a	AZ61A-O	43.0	26.0	16.0	B	−1*R*	Edge notch	0.062*r*, 0.020*t*, 0.525*w*, 0.65*W*	1.6	...	15.0	...	11.5	11.5	...
AZ61A-H24 (longitudinal, sheet)															
AZ61X-h	AZ61A-H24	47.0	34.0	9.0	B	−1*R*	Unnotched	0.020*t*, 0.50*w*	...	...	30.0	...	21.0	18.0	...
AZ61X-h	AZ61A-H24	47.0	34.0	9.0	B	−1*R*	Edge notch	0.062*r*, 0.020*t*, 0.525*w*, 0.65*W*	1.6	...	15.0	...	10.5	10.5	...
AZ61A-O (longitudinal, sheet)															
AZ61X-a	AZ61A-O	43.0	26.0	16.0	B	−1*R*	Unnotched	0.064*t*, 1.0*w*	...	...	18.0	...	14.5	14.0	...
AZ61X-a	AZ61A-O	43.0	26.0	16.0	B	−1*R*	Edge notch	0.062*r*, 0.020*t*, 0.525*w*, 0.65*W*	1.6	...	24.0	...	25.5*	...	...
AZ61A-H24 (longitudinal, sheet)															
AZ61X-h	AZ61A-H24	47.0	34.0	9.0	B	−1*R*	Unnotched	0.064*t*, 1.0*w*	...	...	19.0	...	16.0	15.0	...
AZ61X-h	AZ61A-H24	47.0	34.0	9.0	B	−1*R*	Edge notch	0.062*r*, 0.020*t*, .525*w*, 0.65*W*	1.6	...	21.0	...	16.5	...	...
AZ61A-O (sheet)															
AZ61X-a	AZ61A-O	43.0	26.0	16.0	Ax	0.25*R*	Unnotched	0.064*t*, 1.0*w*	...	...	24.0	20.0	18.0	...	...
AZ61X-a	AZ61A-O	43.0	26.0	16.0	Ax	0.50*R*	Unnotched	0.064*t*, 1.0*w*	...	...	31.0	24.0	29.0	...	...
AZ61X-a	AZ61A-O	43.0	26.0	16.0	Ax	0.50*R*	Unnotched	0.064*t*, 0.50*w*	...	...	31.0	24.0	23.0	...	...
AZ61A-H24 (sheet)															
AZ61X-h	AZ61A-H24	47.0	34.0	9.0	Ax	0.25*R*	Unnotched	0.064*t*, 1.0*w*	...	...	24.0	21.0	21.0	...	...
AZ61X-h	AZ61A-H24	47.0	34.0	9.0	Ax	0.50*R*	Unnotched	0.064*t*, 1.0*w*	...	...	34.0	25.0	23.0	...	...
AZ61X-h	AZ61A-H24	46.0	35.0	12.0	Ax	0.25*R*	Unnotched	0.064*t*, 0.50*w*	...	40.0	25.0	24.0	23.0	...	...
AZ61X-h	AZ61A-H24	46.0	35.0	12.0	Ax	0.50*R*	Unnotched	0.064*t*, 0.50*w*	...	41.0	33.0	26.0	25.0	...	...
AZ61X-h	AZ61A-H24	46.0	35.0	12.0	Ax	0.75*R*	Unnotched	0.064*t*, 0.50*w*	...	44.0	37.0	31.0	30.0	...	...
M1A-O (sheet)															
M1-a	M1A-0	33.0	15.0	17.0	Ax	0.25*R*	Unnotched	0.064*t*, 1.0*w*	...	...	17.0	14.5	14.5	...	...
M1-a	M1A-O	33.0	15.0	17.0	Ax	0.50*R*	Unnotched	0.064*t*, 1.0*w*	...	...	27.0	22.0	22.0	...	...

Note: UTS, ultimate tensile strength; YS, yield strength; %E, percent elongation; B, bending; RB, rotating bending; Ax, axial; *d*, inside diameter; *D*, outside diameter; *t*, thickness; *r*, radius; *w*, width; *W*, notch width. Source: H.J. Grover, S.A. Gordon, and L.R. Jackson, *Fatigue of Metals and Structures*, Department of the Navy, 1960, p 346-353

(continued)

Table 8 Wrought magnesium alloy fatigue strength (continued)

Material	Designation	UTS, ksi	YS, ksi	%E	Fatigue loading		Fatigue specimen		K_t	Fatigue strength (max stress), ksi, at indicated cycles:					Comments
					Type	Condition	Type	Size, in.		10^4	10^5	10^6	10^7	10^8	
M1A-H24 (sheet)															
M1-h	M1A-H24	37.0	29.0	8.0	Ax	0.25*R*	Unnotched	0.064*t*, 1.0*w*	...	...	27.0	23.5	23.0	...	...
M1-h	M1A-H24	37.0	29.0	8.0	Ax	0.50*R*	Unnotched	0.064*t*, 1.0*w*	...	...	32.0	27.0	26.0	...	...
M1A-O (longitudinal, sheet)															
M1-a	M1A-O	33.0	15.0	17.0	B	−1*R*	Unnotched	0.020*t*, 1.0*w*	...	...	16.0	...	10.0	10.0	...
M1-a	M1A-O	33.0	15.0	17.0	B	−1*R*	Edge notch	0.062*r*, 0.020*t*, 0.525*w*, 0.65*W*	1.6	...	10.0	...	5.0	5.0	...
M1-a	M1A-O	33.0	15.0	17.0	B	−1*R*	Unnotched	0.064*t*, 1.0*w*	...	...	13.5	...	9.0	7.5	...
M1-a	M1A-O	33.0	15.0	17.0	B	−1*R*	Surface notch	60° notch, 0.001*r*, 0.064*t*, 0.003*d*, 0.65*W*	2.0	...	15.5	...	9.5	...	...
M1A-H24 (longitudinal, sheet)															
M1-h	M1A-H24	37.0	29.0	8.0	B	−1*R*	Unnotched	0.020*t*, 0.50*w*	...	...	16.0	...	10.5	10.5	...
M1-h	M1A-H24	37.0	29.0	8.0	B	−1*R*	Edge notch	0.062*r*, 0.020*t*, 0.525*w*, 0.65*W*	1.6	...	10.5	...	5.0	4.5	...
M1-h	M1A-H24	3.0	29.0	8.0	B	−1*R*	Unnotched	0.064*t*, 1.0*w*	...	...	16.0	...	10.5	10.5	...
M1-h	M1A-H24	37.0	29.0	8.0	B	−1*R*	Surface notch	60° notch, 0.001*r*, 0.064*t*, 0.003*d*, 0.65*W*	2.0	...	15.5	...	10.0	...	...
ZK60A-T5 (longitudinal, extruded)															
ZK60A-T5	...	48.0	40.9	...	Ax	−1*R*	Unnotched	0.30*d*	...	36.0	24.5	20.0	18.5	18.0	...
ZK60A-T5	...	48.0	40.9	...	B	−1*R*	Notched	60° notch, 0.010*r*, 0.300*d*, 0.350*D*	2.8	20.0	15.0	11.5	8.5	7.0	...
ZK60A-T5	...	48.0	40.9	...	Ax	0*R*	Unnotched	0.350*d*	...	50.0	40.0	33.0	32.0	32.0	...
ZK60A-T5	...	48.0	40.9	...	Ax	0*R*	Notched	60° notch, 0.010*r*, 0.350*d*, 0.400*D*	2.9	25.0	13.5	12.5	12.0	12.0	...
ZK60A-T5	...	48.0	40.9	...	Ax	−1*R*	Unnotched	0.350*d*	...	30.5	23.5	20.5	19.9	19.0	...
ZK60A-T5	...	48.0	40.9	...	Ax	−1*R*	Notched	60° notch, 0.010*r*, 0.350*d*, 0.400*D*	2.9	15.0	11.0	9.0	8.0	8.0	...

Note: UTS, ultimate tensile strength; YS, yield strength; %E, percent elongation; B, bending; RB, rotating bending; Ax, axial; *d*, inside diameter; *D*, outside diameter; *t*, thickness; *r*, radius; *w*, width; *W*, notch width. Source: H.J. Grover, S.A. Gordon, and L.R. Jackson, *Fatigue of Metals and Structures*, Department of the Navy, 1960, p 346-353

Mechanical properties of forgings depend on the orientation of the tested specimen in relation to the flow patterns developed during forging.

Fatigue Strength. Most fatigue properties information for wrought magnesium alloys consists of *S-N* data generated in the 1940s to 1960s (Ref 42), as is the case for cast magnesium alloys. Table 8 lists the early data compiled by Grover et al. (Ref 40). More recent data are included in the fatigue datasheets following this article. In general, fatigue strengths of smooth specimens from wrought alloys tend to be higher than those of smooth specimens from castings. However, notched fatigue strengths for castings and wrought forms are more comparable.

Novel Magnesium Alloys

Metal Matrix Composites. Although most studies of metal matrix composites (MMCs) have been concerned with aluminum alloy matrices, various combinations of magnesium alloys reinforced with ceramic particulates such as SiC, Al2O3, and graphite are being investigated with the aim of attaining properties not available in conventional alloys. The favored route for producing magnesium alloy MMCs is to stir the particulates into the melt, followed by either high-pressure die casting or the application of squeeze casting methods (Ref 43, 44). The requirement that the molten metal wet the ceramic reinforcement is facilitated with magnesium alloys because of the ability of this metal to react with, and absorb, oxygen and nitrogen that may be present on the surfaces of the particulates. In this regard, magnesium does offer an advantage over aluminum. The wettability of SiC particulates, in particular, is further assisted by reaction with magnesium to form layers of the compound Mg2Si.

Squeeze castings have been made for the alloy AZ91 reinforced with different fibers, including SiC, glass, and a proprietary form of Al_2O_3 known as Saffil. As an example, the creep life of squeeze-cast AZ91 containing 16 vol% Saffil fibers at 180 °C is an order of magnitude longer than the unreinforced alloy, and the fatigue endurance limit at this temperature may be doubled (Ref 43). Elastic modulus at room temperature increases linearly with volume percent of fibers, according to the standard rule of mixtures, and is twice that of the alloy AZ91 if the level of Saffil fibers is raised to 30 vol%. Values for ductility and fracture toughness are, however, very low once the fiber content increases beyond 10 to 15 vol%. Improvements in mechanical properties can also be obtained in wrought magnesium alloys containing particulates. For example, the elastic modulus of the medium-strength extruded alloy AZ31 (Mg-3Al-1Zr) can be more than doubled by the presence of 20% SiC whiskers, but the elongation is reduced from 15% to 1% (Ref 23).

Experiments have also been conducted with the ultralight Mg-Li alloy matrices containing fibers of SiC, Al_2O_3, or graphite (Ref 45). However, significant fiber degradation occurs during heating and fabrication because of the reaction of lithium with all but SiC whiskers. Moreover, mechanical properties have been found to be unstable at quite low temperatures as a consequence of the abnormally high mobility of lithium and vacancies in the alloy matrices, as mentioned earlier. In this regard, the desirable localized stress gradients that normally develop close to the ends of fibers have been found to relax continually, even at relatively high strain rates (e.g., 10^{-2} s^{-1}).

Rapidly Solidified Alloys. Rapid solidification processing of alloys has attracted much interest during the past decade because the extreme rates of cooling involved (e.g., 10^5 to 10^6 °C/s) can lead to the production of fine, homogeneous microstructures, extended solid solubility, and the introduction of new phases. Special attention has been paid to light alloys, and it has been shown that mechanical properties may be enhanced, particularly at elevated temperatures. Corrosion resistance may also be improved, because the more homogeneous microstructures tend to disperse elements and particulates that normally act as cathodic centers, and because the extended solubility of various elements may shift the electrode potentials of light alloys to more noble values (Ref 46).

A number of magnesium alloys have been produced by RSP in the form of melt-spun ribbons that are then usually mechanically ground to powders, sealed in cans, and extruded to produce bars (Ref 46, 47). One recent European study involved more than 80 compositions based on the Mg-Al system with additions of zinc, mischmetal, silicon, strontium, and calcium. Another joint program involving Magnesium Elektron in Britain and Allied Signal in the United States led to an alloy EA55RS (Mg-5Al-5Zn-5Nd) that is available commercially (Ref 15). Microstructures of the bulk products tend to be similar and comprise fine grains 0.3 to 5 mm in size and dispersoids of compounds such as Mg17Al12, Al2Ca, Mg3Nd, and Mg12Ce. Tensile strengths may exceed 500 MPa, which compares with maximum values of 250 to 300 MPa for conventionally cast magnesium alloys. Some alloys show improved creep resistance at moderately elevated temperatures, but others undergo accelerated creep deformation, presumably due to enhanced grain-boundary sliding in the fine-grain microstructures (Ref 46). Because of this, some alloys are amenable to superplastic forming at temperatures as low as 150 °C.

Typical mechanical properties of EA55RS are shown in Table 9 and Fig. 9. Annealing at 350 °C (660 °F) improves fracture toughness with some loss of strength (Table 9). It appears that this annealing treatment dissolves some thermally unstable intermetallic phases, thus improving ductility and toughness (Ref 48). The effect of heat treatment on fracture mechanic properties of rapidly solidified magnesium alloys is shown in Fig. 10. The fatigue resistance of EA55RS in rotating bend tests (Fig. 9) is excellent. The runout stress at 5×10^7 cycles for EA55RS is 195 MPa, which is significantly higher than that of currently available magnesium alloys (e.g., 100 to 105 MPa for QH21A-T6, QE22A-T6, and EQ21A-T6). The higher fatigue strength achieved in EA55RS is due to the small subgrain size obtained by rapid solidification, which retards fatigue-induced crack initiation.

Amorphous Alloys. It is well known that the mechanical properties of alloys produced with an amorphous (or glassy) atomic structure by liquid quenching techniques (such as melt spinning) may be notably greater than those obtained for the normal crystalline state. Following a report that enhanced strength and ductility could be achieved in certain amorphous aluminum alloys (Ref 49), similar results have been obtained for a number of magnesium alloys (Ref 50, 51). Most promising have been ternary compositions of the general formula Mg-M-Ln, where M is nickel or copper and Ln is a lanthanide element such a syttrium (Ref 50). The presence of the lanthanide element is a key feature, and it may be noted that Mg-Y, for example, has a high negative enthalpy of mixing. Moreover, the lanthanides have a larger atomic size than magnesium, whereas copper and nickel atoms are smaller. This suggests that localized strain energy will be reduced if the three atoms cluster together. These two factors are presumed to reduce overall atomic diffusivity during cooling from the molten state, suppressing nucleation of crystalline phases.

Tests on ribbons of some amorphous Mg-M-Ln alloys have shown values of tensile strength, elastic modulus, and hardness in the ranges 610 to 850 MPa, 40 to 61 GPa, and 193 to 237 DPH, respectively, which greatly exceed the maximum values (around 300 MPa, 45 GPa, and 85 DPH) for the strongest conventionally cast magnesium alloys. Most compositions show good bending ductility, although tensile fracture strains (including elastic strains) lie in the range of 0.014 to 0.018, indicating little or no capacity for plastic deformation. In a later development, even higher values of tensile strength and significant amounts of

Table 9 Typical room-temperature mechanical properties of EA55RS extrusions

Temper	Extrusion size, mm	YS, MPa	UTS, MPa	%E	K_{Ic} (MPa $\sqrt{m}$) Across grain	K_{Ic} (MPa $\sqrt{m}$) Along grain
F	102 × 32	385	461	14	6	5
T4	...	371	434	16	15	9

Note: YS, yield strength; UTS, ultimate tensile strength; %E, percent elongation; T4, annealed at 350 °C for 90 min, water quenched. Source: Ref 48

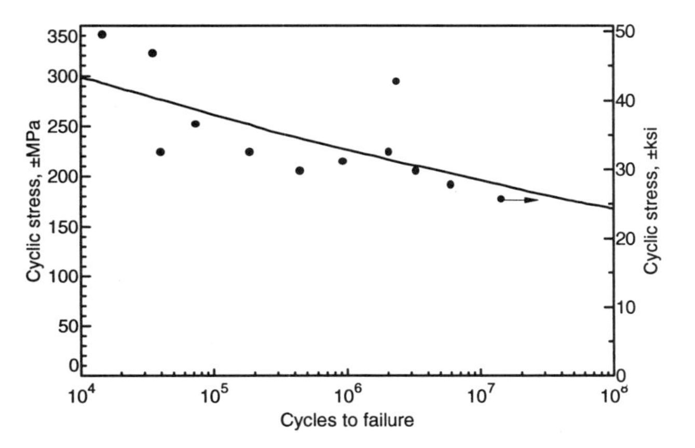

Fig. 9 Rotating bending fatigue properties of EA55RS at 2800 Hz. Source: Ref 48

plastic deformation prior to fracture have been obtained by rapidly solidified ribbons of the alloy $Mg_{85}Zn_{12}Ce_3$, which has a mixed amorphous/crystalline microstructure (Ref 52). The crystalline phase comprises ultrafine particles of hexagonal close-packed magnesium, which appears to be supersaturated with zinc and cerium and is uniformly dispersed throughout the amorphous matrix. In the quenched condition, the tensile strength of the ribbon is 655 MPa, and this may be increased to over 930 MPa by annealing for 20 s at 110 °C, which increases the average particle size from 3 to 20 nm. The presence of the particles appears to facilitate homogeneous plastic deformation rather than the usual localized shear, and tensile elongations of 7% and 3%, respectively, have been recorded (Ref 52).

Amorphous alloys are in a metastable condition, and heating will induce crystallization at a certain critical temperature Tx. It has been found that Tx tends to increase as solute content is raised, and ratios of Tx to melting temperature (in K) as high as 0.64 have been recorded, which is equivalent to 326 °C (Ref 51). Thus, these amorphous magnesium alloys display relatively high thermal stability, suggesting the exciting prospect of obtaining such structures in bulk castings as well as in ribbons. In this regard Inoue and colleagues (Ref 51) have succeeded in obtaining amorphous structures in chill cast cylinders prepared by injecting molten alloys into a copper mold. Amorphous structures have been confirmed in 2 mm diameter bars for the alloy MgCu10Y10, and in bars in excess of 5 mm for MgCu25Y10. Mechanical properties were found to be similar to those obtained for rapidly cooled melt-spun ribbons, even though cooling rates for casting in the copper mold were as low as 100 °C/s (180 °F/s).

Fatigue Strength

Most of the fatigue data for magnesium alloys are *S-N* curves dating from the 1930s to 1960s. A substantial portion of early *S-N* data has been summarized by H.J. Grover et al. (see Tables 6 and 8). Additional data may be found in Ref 6 and the fatigue data compilation following this article. Strain-life (ε-*N*) curves for magnesium alloys are very rare, and most fatigue crack growth behavior data have originated from work conducted in the former Soviet Union.

Fatigue strength of magnesium alloys, as determined using laboratory test samples, covers a relatively wide scatter band, which is characteristic of other metals as well. The *S-N* curves exhibit a gradual change in slope and become essentially parallel to the horizontal axis at 10 to 100 million cycles.

Fatigue strengths are higher for wrought products than for cast test bars. The ranges of fatigue strength (S_f) in rotating bending ($R = -1$) for 10^8 cycles are given as follows in the *Aerospace Structural Metals Handbook*, Vol 3, Belfour Stulen, 1987:

- S_f = 60 to 100 MPa (8 to 14 ksi) for magnesium alloys

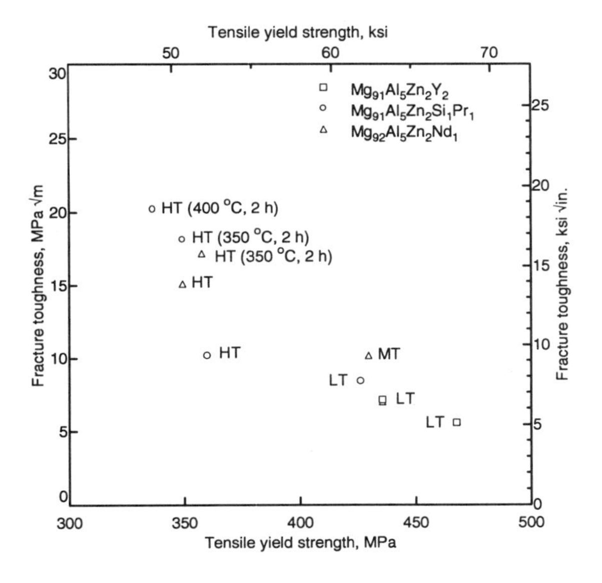

(a)

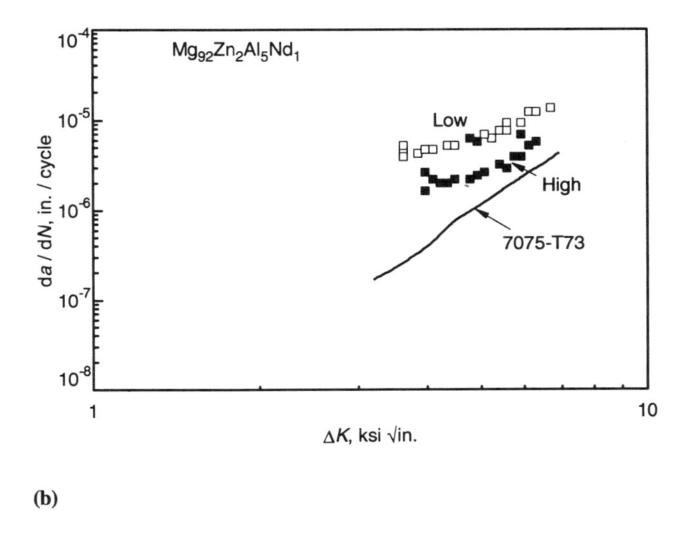

(b)

Fig. 10 Effect of heat treatment on fracture mechanic properties of rapidly solidified magnesium alloys. (a) Fracture toughness vs. yield strength of as-extruded and heat-treated rapid solidification/powder metallurgy magnesium alloys extruded at various temperatures (LT—low temperature ≈150 °C; MT—medium temperature ≈200 °C; HT—high temperature ≈250 °C). Data in parentheses refer to annealing temperature and time. (b) Fatigue crack propagation rate for rapid-solidification-processed Mg-5Al-1Nd (at.%) alloy extruded at low and higher temperatures compared with the propagation rate for 7075-T73 aluminum alloy. Sources: S.K. Das, Rapidly Solidified P/M Aluminum and Magnesium Alloys—Recent Developments; and F. Hehmann and H. Jones, Developments in Magnesium Alloys by Rapid Solidification Processing: An Update, *Advanced Aluminum and Magnesium Alloys*, ASM, 1990

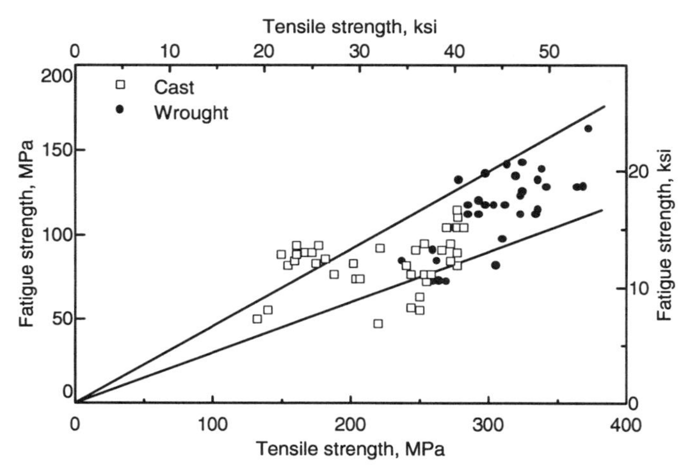

Fig. 11 Rotating bending fatigue strength vs. ultimate tensile strength of magnesium alloys (small smooth specimens). Source: R.B. Heywood, *Designing Against Fatigue of Metals*, Reinhold, 1962

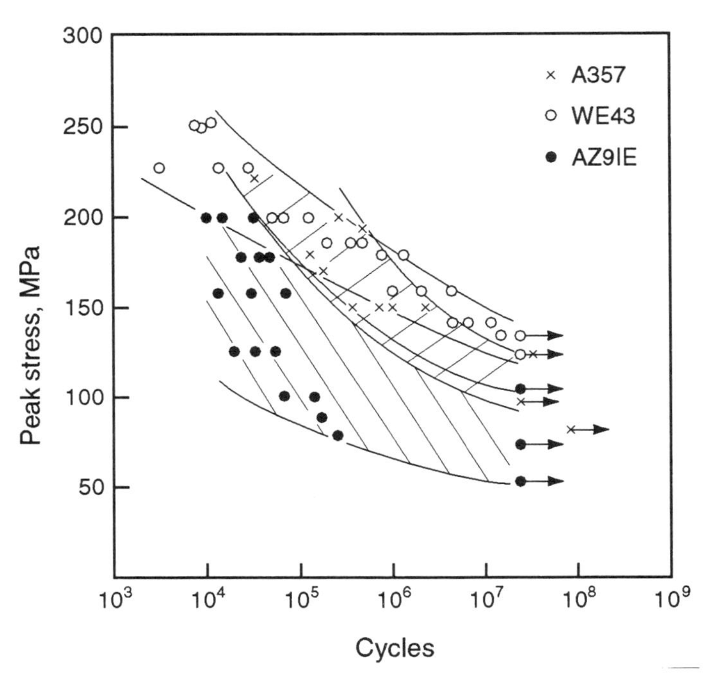

Fig. 12 Fatigue properties of A357, AZ91E, and WE43. R = 0.1. Source: B. Geary, Corrosion Resistant Magnesium Casting Alloys, *Advanced Aluminum and Magnesium Alloys*, ASM, 1990

- S_f = 110 to 170 MPa (16 to 24 ksi) for wrought magnesium alloys

These general ranges agree with a plot of fatigue strength versus ultimate tensile strength (Fig. 11).

The ratio of figure strength to tensile strength is not as well defined for magnesium alloys as for steels. This is due, in part, to the effect of strengthening mechanisms on fatigue strength. For example, solid-solution strengthening increases the fatigue strength of magnesium alloys, whereas cold working and precipitation strengthening produce little improvement in fatigue strength at longer lives (Ref 42).

Axial fatigue *S-N* curves for AZ91E and WE43 are shown in Fig. 12, along with comparative data for A357 aluminum. The flat curve typical of magnesium alloys contrasts with that of aluminum where there is a marked change in slope between low- and high-cycle regimes. These different shapes of curve indicate that, although A357 performs well at low cycles, the situation changes so that WE43 has the better properties at high cycles. AZ91E has significantly lower properties in the low-cycle regime as a result of lower strength and porosity, but at high cycles the difference is not so marked.

Fatigue Mechanisms. The initiation of fatigue cracks in magnesium alloys is related to slip in preferably oriented grains and is often related to the existence of micropores. For pure magnesium, crack orientation is more strongly influenced by grain boundaries than the slip (Ref 42).

The initial stage of fatigue crack growth usually occurs from quasicleavage, which is common in hexagonal close-packed structures such as magnesium. Further crack growth micromechanisms can be brittle or ductile and trans- or intergranular, depending on metallurgical structure and environmental influence. Some magnesium alloys can have either a hexagonal or body-centered cubic structure, depending on their chemical composition.

Effect of Surface Condition. High-cycle fatigue strength is influenced primarily by surface condition. Sharp notches, small radii, fretting, and corrosion are more likely to reduce fatigue life than variations in chemical compositions or heat treatment. For example, removing the relatively rough as-cast surfaces of castings by machining improves fatigue properties of the castings (see Fig. 7).

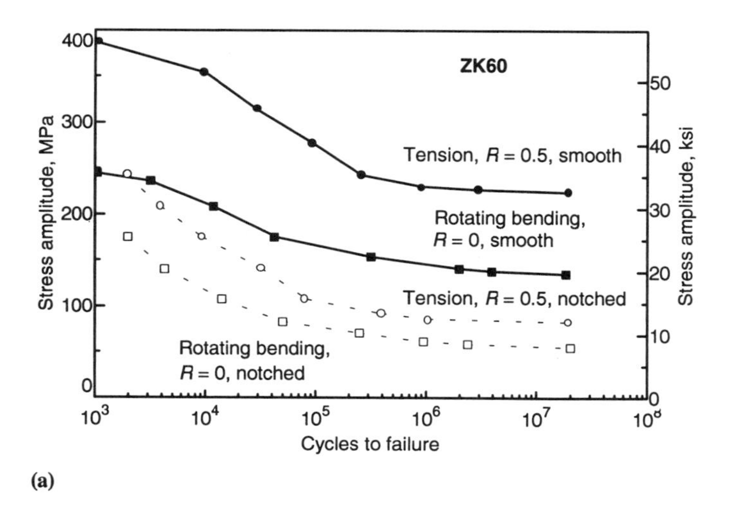

(a)

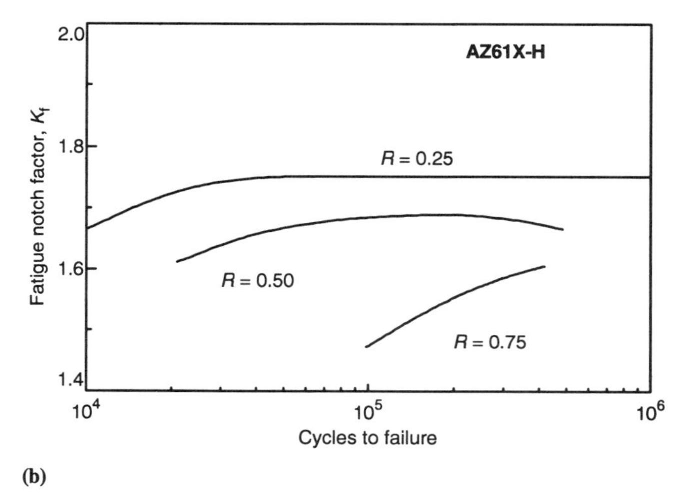

(b)

Fig. 13 Effect of stress ratio and notches on fatigue of two magnesium alloys. (a) Rotating bending and tension-compression *S-N* curves of ZK60. (b) Fatigue life of AZ61X-H with different notch factors. Sources: *Prod. Eng.*, Vol 22, 1951, p 159-163, and *Proceed. ASTM*, Vol 46, 1946, p 783-798

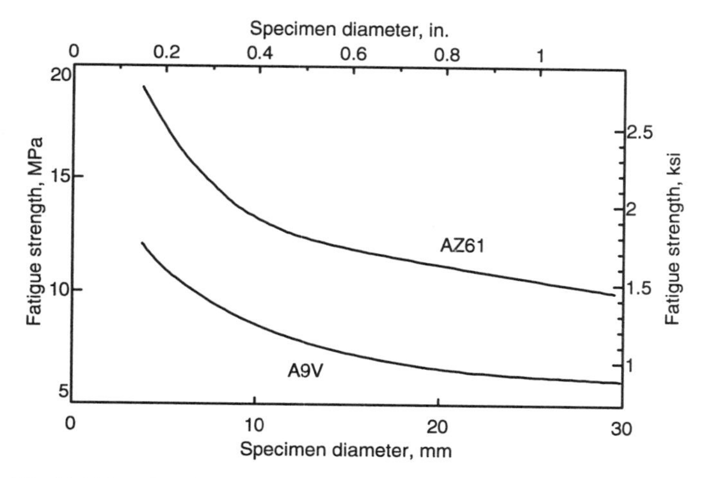

Fig. 14 Effect of specimen size on fatigue strength of magnesium alloys (smooth, rotating bending specimens). Source: *Prevention of the Failure of Metals Under Repeated Stress*, John Wiley & Sons, 1941

When fatigue is the controlling factor in design, every effort should be made to decrease the severity of stress raisers. Use of generous fillets in re-entrant corners and gradual changes of section greatly increase fatigue life. Conditions in which the effects of one stress raiser overlap those of another should be eliminated. Further improvement in fatigue strength can be obtained by inducing stress patterns conducive to long life. Cold working the surfaces of critical regions by rolling or peening to achieve appreciable plastic deformation produces residual compressive surface stress and increases fatigue life.

Surface rolling of radii is especially beneficial to fatigue resistance because radii generally are the locations of higher-than-normal stresses. In surface rolling, the size and shape of the roller, as well as the feed and pressure, are controlled to obtain definite plastic deformation of the surface layers for an appreciable depth (0.25 to 0.38 mm, or 0.010 to 0.015 in.). In all surface working processes, caution must be exercised to avoid surface cracking, which decreases fatigue life. For example, if shot peening is used, the shot must be smooth and round. The use of broken shot or grit can result in surface cracks.

Test Effects. As with other alloys, several test variables affect the fatigue strength of magnesium alloys. As expected, notched specimens and increasing R ratios decrease fatigue strength (Fig. 13a and 13b). The size of parts also reduces bending fatigue strength.

Generally, thicker portions of castings have greater microporosity and thus reduced fatigue strength, and thick extended bars (>75 mm diameter) and large forgings can experience reduced fatigue strength and increased notch sensitivity. Fatigue strength is also affected by specimen size (Fig. 14), because larger specimens provide greater surface area for crack initiation.

Fracture Toughness and Crack Growth

Fracture Toughness. Typical values of magnesium alloy toughness are summarized in Table 10. The critical stress intensity factor, K_{Ic}, a material constant, is the largest stress intensity the material will support, under conditions of plane strain, without failing catastrophically. If K_{Ic} is known for the material, and the geometry and stress are known for the part, the largest crack that can be tolerated can be calculated. The larger the critical stress intensity factor, the larger the flaw size that can be tolerated.

One of the most difficult problems in fracture mechanics is the prediction of failure when section stresses approach or exceed yield values. Under these conditions, the critical stress intensity (K_c) lies outside the domain of linear elastic fracture mechanics and is not a material constant. In such cases, the *apparent* K_{Ic} depends on specimen geometry and flaw size. An example of variations in apparent K_{Ic} values outside the domain of linear conditions is shown in Fig. 15 for magnesium alloy HM21A-T8 (Ref 53).

The J-integral method has been used as a fracture criterion for nonlinear fracture mechanics. From tests on various alloys including magnesium alloy AZ31B, the J-integral is a valid fracture criterion for monotonic loading of thin section metals by Mode I stress systems (Ref 54).

Table 10 Typical toughness of magnesium alloys

Alloy	Temper	Temperature, °C	Tensile strength, ksi: Unnotched	Tensile strength, ksi: Notched	Tensile strength, ksi: Ratio	Charpy V-notch, J	K_{Ic}, ksi $\sqrt{(in.)}$
Sand castings							
AZ81A	T4	25	...	...	...	6.1	...
AZ91C	F	25	...	...	...	0.79	...
	T4	25	...	...	0.90	4.1	...
	T6	25	...	...	0.86(a)	1.4	10.4
AZ92A	F	25	...	...	...	0.7	...
	T4	25	...	...	...	4.1	...
	T6	25	...	...	...	1.4	...
EQ21A	T6	20	...	...	...	...	14.9
EZ32A	T5	20	...	...	...	1.5(b)	...
HZ32A	T5	20	...	...	...	2.2(b)	...
QE22A	T6	25	...	...	1.06(a)	2.0	12.0
WE54A	T6	20	...	...	...	...	10.4
QH21A	T6	25	...	...	...	...	17.0
ZE41A	T5	25	...	...	...	1.4	14.1
ZE63A	T6	25	...	...	...	0.07	19.1
ZH62A	T5	25	...	...	...	3.4(b)	...
ZK51A	T5	20	...	...	...	3.5(b)	...
Extruded alloys							
AZ31B	F	25	...	...	...	3.4	25.5
AZ61A	F	25	...	...	...	4.4	27.3
AZ80A	F	25	46	34(c)	0.75	1.3	26.4
		−195	61	25(c)	0.40	...	...

(continued)

Table 10 Typical toughness of magnesium alloys *(continued)*

Alloy	Temper	Temperature, °C	Tensile strength, ksi			Charpy V-notch, J	K_{Ic}, ksi $\sqrt{in.}$
			Unnotched	Notched	Ratio		
	T5	25	50	22(c)	0.45	1.4	14.75
		−195	65	14(c)	0.22	...	...
	T6	25	...	...	...	1.4	...
HM31A	T5	25	44	38(c)	0.87	...	...
		−78	51	38(c)	0.75	...	...
		−195	59	39(c)	0.66	...	...
ZK30	F	25	...	...	...	4.0	41.8(d)
ZK60A	T5	25	51	49(c)	0.96	3.4	31.4
		0	...	...	...	2.2	...
		−78	...	...	...	2.2	...
		−195	74	45(c)	0.61	...	...
Sheet							
AZ31B-O	...	24	38	31(e)	0.83	5.9	...
	...	−196	58	33(e)	0.53	...	...
AZ31B-H24	...	24	41	33(e)	0.81	...	26
	...	−196	60	24(e)	0.40	...	...
HK31A-O	...	24	31	27(e)	0.86	4.0	30
	...	−196	52	30(e)	0.57	...	...
HK31A-H24	...	24	39	33(e)	0.85	3.0	23
	...	−196	58	36(e)	0.63	...	...
HM21A-T8	...	24	36	33(e)	0.94	...	23
	...	−78	46	30(e)	0.67	...	...
	...	−196	54	32(e)	0.59	...	...
ZE10-O	...	24	33	29(e)	0.87	6.6	21
	...	−78	44	30(e)	0.69	...	...
	...	−196	53	31(e)	0.59	...	...
ZE10A-H24	...	24	38	38(e)	1.00	...	28
	...	−78	46	38(e)	0.83	...	...
	...	−196	54	30(e)	0.56	...	...
ZH11A-H24	...	20	...	...	...	4.4(b)	...

(a) Notched/unnotched tensile strength ratio with a notch radius of 0.008 mm. (b) Izod specimen. (c) The notched specimen has a reduced section of 0.06 in. × 1 in., a 60° V-notch, a 0.700 in. notched width, and a notch root radius of 0.0003 in. (d) Value is for J_{Ic} since specimen was too small for an accurate K_{Ic} value; true value for K_{Ic} is lower. (e) Specimen dimensions: Total width, 1 in.; notched width, 0.700 in.; thickness, 0.60 in.; 60° V-notch with 0.0003 in. radius. Source: Adapted from Ref 6

The results indicate that for a wide range of material behavior and specimen size J_c is not a function of crack length or specimen geometry. Additional details of the results are given in Tables 11 and 12. Statistical data for CT (compact tension) specimens are presented in Table 11 and compared with the mean values for data fromCC and DEC (double edge cracked) specimens in Table 12. Standard deviations for J_c from CT specimens are seen to be on the order of ±11% of the mean for most of the alloys. These data can be put in the perspective of linear elastic fracture mechanics by the conversion:

$$K_c = \sqrt{(EJ_c)}$$

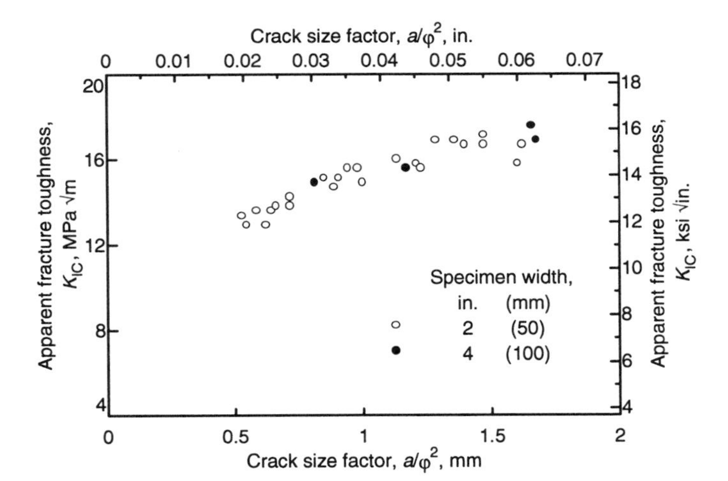

Fig. 15 Variation of apparent fracture toughness (K_{Ic}) with crack size. Source: Ref 53

Fatigue Crack Growth. As previously mentioned, availability of fatigue crack growth data on magnesium alloys is rather limited because most fatigue crack growth data were generated in the former Soviet Union. However, in examining fatigue crack growth rate curves for many materials exhibiting very large differences in microstructure, the striking

Table 11 Fracture toughness of various alloys

Alloy	Mean value, J_c, J/mm²	Standard deviation, J_c, J/mm²	Mean value, K_c, MPa$\sqrt{m}$	Standard deviation, K_c, MPa$\sqrt{m}$
6061-0	0.125	0.009	93.0	3.4
7075-0	0.075	0.008	71.7	3.5
70/30	0.282	0.029	176.0	9.1
AZ31B	0.052	0.003	48.4	1.4
1018	0.342	0.039	266.0	15.0
4130	0.218	0.021	212.0	10.0
HP9-4-20	0.245	0.023	218.0	11.0

Compact tension (CT) specimens

Table 12 Comparison of mean values of J_c for various specimen geometries and alloys

Alloy	Mean values of J_c, J/mm²		
	Compact tension (CT) specimens	Center cracked (CC) specimens	Double edge cracked (DEC) specimens
6061-0	0.125	0.115	...
7075-0	0.075	0.078	0.065
70/30	0.282	0.285	...
AZ31B	0.052	0.054	0.048
1018	0.342	0.368	...
4130	0.218	0.248	0.216

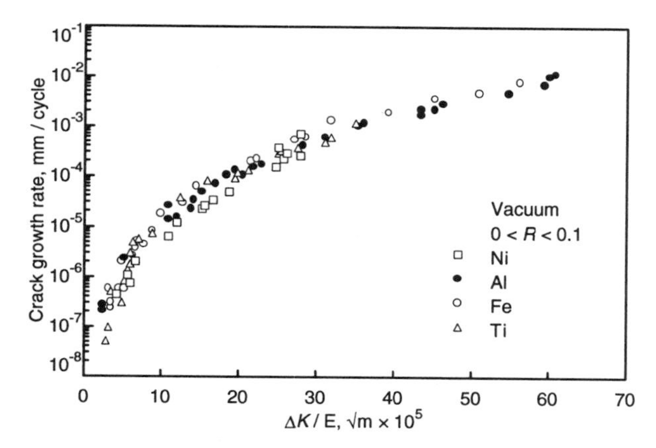

Fig. 16 Crack growth rate curves for several metals compared on the basis of driving force normalized by modulus. Original work includes a much larger range of materials, including polymers. Source: Ref 55

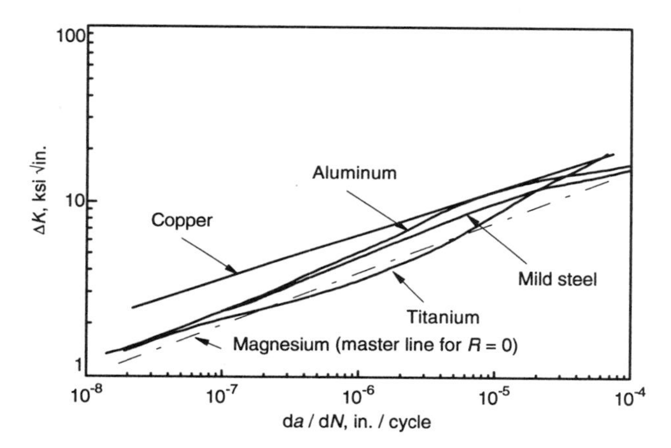

Fig. 17 Comparison of crack propagation curves. ΔK reduced to modulus of aluminum

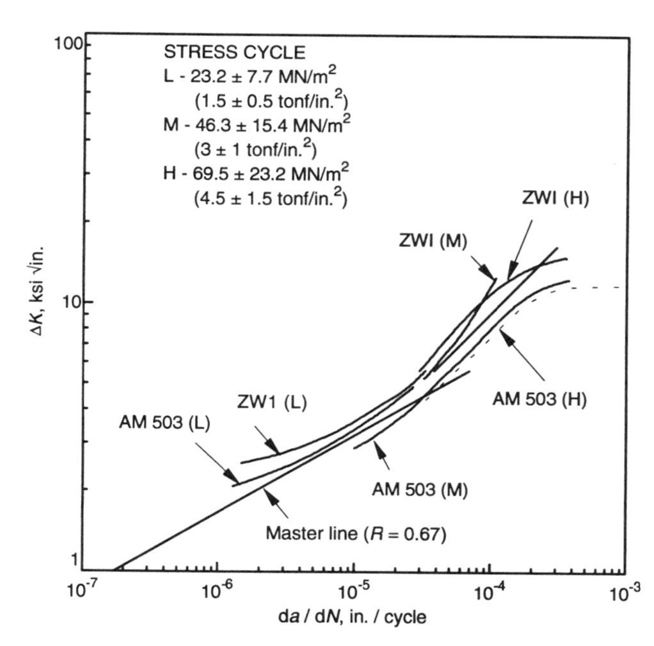

Fig. 18 ΔK vs. da/dN at $R = 0.5$ for two magnesium alloys

feature is the similarities between these curves, not the differences (although from an engineering viewpoint, the differences in crack growth rate between alloys can be important when integrating along da/dN curves to obtain the lifetime of a structure). However, from the mechanistic point of view, these differences are small. For example, a large range of metals can be represented by a single curve if the driving force (DK) is normalized by modulus, as illustrated in Fig. 16. A similar result is observed for crack growth rates of two magnesium alloys (Fig. 17) when DK is normalized to the modulus of elasticity (in this case, the modulus of aluminum). Crack growth test results and tensile properties for the two magnesium alloys are shown in Fig. 18 and Table 13 (Ref 56).

The similarities illustrated in Fig. 16 and 17 lead to the conclusion that the mechanisms of fatigue crack growth through a wide variety of monolithic materials are likely to be the same. Therefore, if those mechanisms can be determined for a limited number of materials, they should be representative of most materials, at least metals. Research on the mechanisms and models of fatigue crack growth is reviewed in Ref 57.

However, it must be mentioned that the data in Fig. 16 and 17 exclude the effect of environment, which is important because one of the major factors affecting fatigue crack growth rates is the environment, particularly water vapor (see Fig. 19 for magnesium alloy ZK60A). Also note that Fig. 16 and 17 include only simple metals and no complex alloys. Thus, it may not be possible to represent data along a single line for cracks grown in complex alloys in humid air or other aggressive environments.

More information on fatigue crack growth of magnesium alloys is available from the articles listed below, which contain crack growth data for the indicated materials. The list was derived from the extensive bibliographic listing of fracture mechanic sources given in Ref 58 to 61. Ad-

Table 13 Composition and mechanical properties of two magnesium alloys

Material	Chemical analysis, %		Static mechanical properties(a)						
			Tensile strength		0.2% yield strength		Elongation on 50.8 mm (2 in.), %	Modulus of elasticity, E	
			MN/m²	tonf/in.²	MN/m²	tonf/in.²		MN/m² × 10³	lbf/in.² × 10⁶
High-strength magnesium alloy ZW1	Zn 1.45	Si 0.005 max	250	16.2	165	10.6	7	43	6.2
	Al 0.01 max	Fe 0.005 max							
	Mn 0.07 max	Ni 0.003 max							
	Zr 0.60	Mg bal							
	Cu 0.015 max								
Medium-strength magnesium alloy AM 503	Zn 0.01 max	Si 0.006	200	13.0	107	6.9	6	43	6.2
	Al 0.01 max	Fe 0.007							
	Mn 1.62	Ni 0.003 max							
	Cu 0.002	Mg bal							

(a) Determined from specimens cut longitudinally from the sheet material

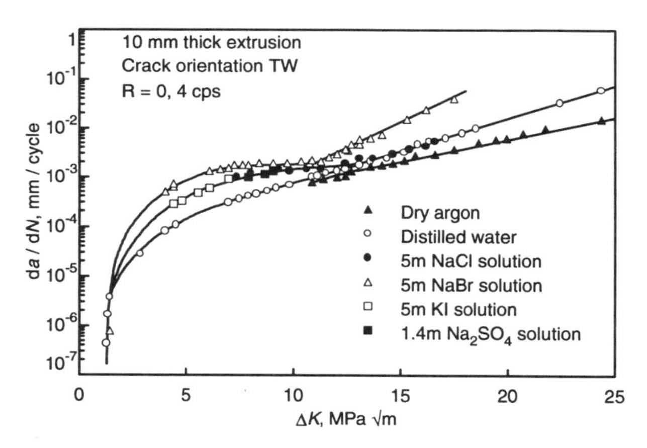

Fig. 19 Corrosion-fatigue crack growth curves for ZK60A-T5 in different environments. Source: M.O. Speidel, M.J. Blackburn, T.R. Beck, and J.A. Feeney, *Proc. Corrosion Fatigue: Chem. Mech. Microstructure*, Storrs, CT, 1971, p 324-325

ditional data can be found in the appendix in this Volume entitled "Magnesium Alloy Fatigue Data Compilation."

- EZ33A-T5: T.M. Morton et al., *Experimental Mechanics*, Vol 14, 1974, p 208-213
- ZK60A-T5: C.J. McMahon, *Journal of Engineering Materials Technology*, Vol 95, 1973, p 142-149
- Pure Mg: H.H. Johnson and P.C. Paris, *Engineering Fracture Mechanics*, Vol 1, 1968, p 3-45
- ZW1: L.P. Pook and A.F. Greenan, *Engineering Fracture Mechanics*, Vol 5, 1973, p 935-946
- AM503: P. Pook, *Developments in Fracture Mechanics—1*, Applied Science Publishers, 1979, p 183-220
- MA12: N.M. Grinberg, *International Journal of Fatigue*, Vol 4, 1982, p 83-95
- MA12: N.M. Grinberg and V.A. Serdyuk, *International Journal of Fatigue*, Vol 3, p 143-148
- IMV6: N.M. Grinberg and V.A. Serdyuk, *International Journal of Fatigue*, Vol 3, p 143-148
- ZW1: L.P. Pook, *Developments in Fracture Mechanics—1*, Applied Science Publishers, 1979, p 183-220

Stress Corrosion and Corrosion Fatigue*

Magnesium has a normal electrode potential at 25 °C of –2.30 V, with respect to the hydrogen electrode potential taken as zero, which places it high in the electrochemical series. However, its solution potential is lower (e.g., –1.7 V in dilute chloride solution) with respect to a normal calomel electrode because of polarization of the surface with a film of $Mg(OH)_2$. The oxide film on magnesium offers considerable surface protection in rural and some industrial environments, and the corrosion rate of magnesium lies between that of aluminum and low-carbon steels. Essentially, a high sensitivity to impurities and lack of passive film stability below pH 10.5 accounts for most of magnesium corrosion problems (Ref 62). Tarnishing occurs readily, and some general surface roughening may take place after long periods, but unlike some aluminum alloys, magnesium and its alloys are virtually immune from intercrystalline attack (Ref 63).

Magnesium is readily attacked by all mineral acids except chromic and hydrofluoric acids, the latter actually producing a protective film of MgF_2 that prevents attack by most other acids. In contrast, magnesium is very resistant to corrosion by alkalis if the pH exceeds 10.5, which corresponds to that of a saturated $Mg(OH)_2$ solution. Chloride ions promote rapid attack of magnesium in aqueous solutions, as do sulphate and nitrate ions, whereas soluble fluorides are chemically inert. With organic solutions, methyl alcohol and glycol attack magnesium, whereas ethyl alcohol, oils, and degreasing agents are inert.

The corrosion behavior of alloys varies with composition. Where alloying elements form grain-boundary phases, as is generally the case in casting alloys, corrosion rates are likely to be greater than those occurring with pure magnesium. Magnesium alloys can suffer rapid attack in moist conditions, mainly because of the presence of more noble metal impurities, notably iron, nickel, and copper. Each of these elements, or the compounds they form, acts as a minute cathode in the presence of a corroding medium, creating microgalvanic cells with the relatively anodic magnesium matrix. Nickel and copper are not usually a problem in current alloys because they are present only in very low levels in primary magnesium. Iron tends to be more troublesome because there is always a risk of pickup from crucibles that are made from mild steel. However, the potential detrimental effect of iron in zirconium-free alloys is reduced by adding manganese (as $MnCl_2$) to the melt. This element combines with the iron and settles to the bottom of the melt or forms intermetallic compounds that, depending on the iron-manganese ratio, reduce the cathodic effect of the iron.

Mg-Al-Zn and Mg-Al alloys are particularly susceptible to the presence of impurities, and the widely used alloy AZ91C has largely been superseded by higher-purity versions (AZ91D for pressure die casting and AZ91E for gravity die casting), which have strict limits for the nickel, iron, and copper contents. Further improvements are possible by applying a T6 aging treatment (Ref 64). Alloys AZ91D and AZ91E exhibit corrosion rates that are similar to those of comparable cast aluminum alloys (Ref 25).

The yttrium-containing alloys WE43 and WE54 have also shown corrosion resistance comparable to that of many cast aluminum alloys and superior to that of other magnesium alloys in the group that contain zirconium (Ref 25, 28, 29). Further significant improvements in corrosion resistance have been observed with several magnesium alloys prepared by rapid solidification processing (Ref 15, 24, 46). As mentioned earlier, this result has been attributed to the more homogeneous microstructures obtained in these materials and to the fact that the solid solubility of the various alloying elements is extended, which may shift corrosion potentials to more positive values. There is also evidence that the alloys containing aluminum have more protective surface films (Ref 65).

Stress-Corrosion Cracking

Wrought and cast magnesium alloys (particularly those containing aluminum) are susceptible to stress-corrosion cracking (SCC) when statically loaded below their yield strengths in some environments. Early studies found susceptibility for wrought forms but little or no susceptibility for cast forms. However, the presence of aluminum as an alloying element is the main factor affecting SCC susceptibility, rather than cast or wrought form. For aluminum-containing magnesium alloys, SCC susceptibility is comparable to that of wrought and cast forms of similar composition (Fig. 20) (Ref 66).

While all magnesium alloys will stress corrode to some extent, the most susceptible are those containing aluminum as an alloying element. Magnesium alloys containing aluminum should not be designed for prolonged exposure to stresses near the yield point. Residual stresses from

*Adapted from "Corrosion of Magnesium" (G.L. Makar and J. Kruger, *International Materials Review*, Vol 38, No. 3, 1993, p 138-153) and "Stress-Corrosion Cracking of Magnesium Alloys" in *Stress-Corrosion Cracking: Materials Performance and Evaluation*, ASM International, 1992.

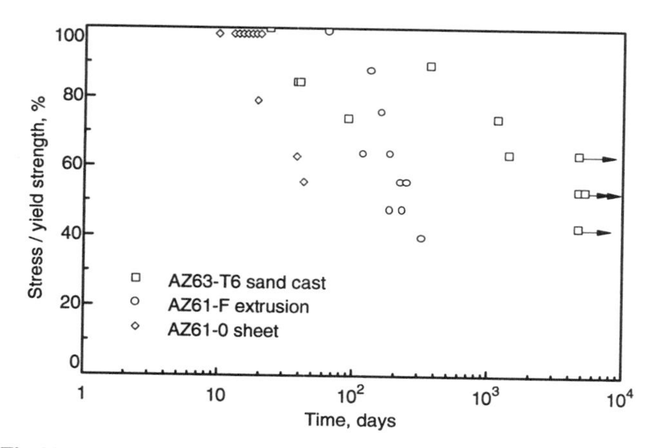

Fig. 20 Data comparing similar cast and wrought magnesium alloys during long-term stress-corrosion cracking (SCC). Long-term rural-atmosphere SCC data compare similar-composition AZ61 sheet, extruded AZ61, and sand-cast AZ63. Although there is a great deal of scatter in these data, all three materials exhibited similar behavior at the higher stress levels. At lower stress levels, the cast alloys became more resistant to SCC.

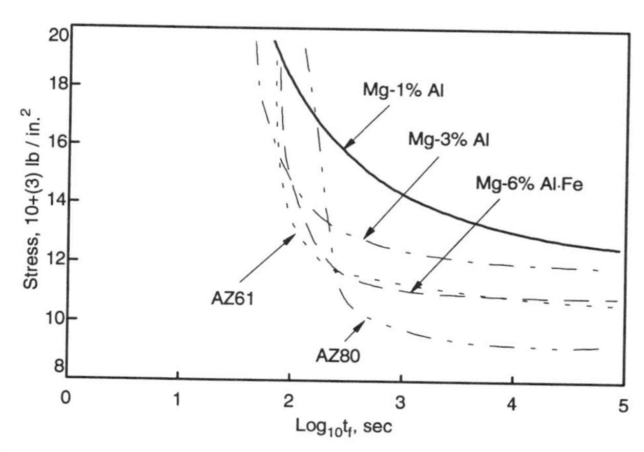

Fig. 21 Stress vs. time-to-failure (t_f) for magnesium-aluminum alloys in aqueous 40 g/L NaCl + 40 g/L Na_2CrO_4. Source: J.A. Beavers, G.H. Koch, and W.E. Berry, "Corrosion of Metals in Marine Environments," Metals and Ceramics Information Center, Battelle Columbus Laboratories, July 1986

operations such as welding or machining must also be relieved by heat treatment (Ref 6). Service failures normally result from excessive residual stress produced during fabrication (Ref 67-73). Speidel (Ref 74), in a comprehensive review of more than 3000 unclassified failure reports from aerospace companies, government agencies, and research laboratories in the United States and five Western European countries, estimated that approximately 10 to 60 magnesium aerospace component SCC service failures occurred each year from 1960 to 1970. Of this total, more than 70% involved either cast alloy AZ91-T6 or wrought alloy AZ80-F, both of which contain aluminum. In contrast, magnesium alloys without aluminum do not, in practice, have a stress-corrosion problem (Ref 6). The broad class of alloys containing zirconium are also sufficiently insensitive that SCC is not a problem in practice.

Stress-corrosion cracking of magnesium alloys can occur in many environments. In general, the only solutions that do not induce SCC are either those that are nonactive to magnesium, such as dilute alkalies, concentrated hydrofluoric acid, and chromic acid, or those that are highly active, in which general corrosion predominates. More information on the effect of air, water, and aqueous solutions is contained in the article "SCC of Magnesium Alloys" in *Stress-Corrosion Cracking* (ASM International, 1992).

Effects of Alloy Composition. Pure magnesium is not susceptible to SCC when loaded up to its yield strength in atmospheric and most aqueous environments (Ref 68, 75-78). The only reports of SCC of pure magnesium have emanated from laboratory tests in which specimens were immersed in very severe SCC solutions (Ref 79-81).

As previously mentioned, aluminum-containing magnesium alloys have the highest SCC susceptibility, with the sensitivity increasing with increasing aluminum content, as illustrated in Fig. 21. An aluminum content above a threshold level of 0.15 to 2.5% is reportedly required to induce SCC behavior (Ref 69, 82, 83), with the effect peaking at approximately 6% Al (Ref 1). The aluminum- and zinc-bearing AZ alloys, which are the most commonly used magnesium alloys, have the greatest susceptibility to SCC. Alloys with higher aluminum content, such as AZ61, AZ80, and AZ91, can be very susceptible to SCC in atmospheric (Fig. 22) and more severe environments (Fig. 23), while lower-aluminum AZ31 is generally more resistant (Fig. 24 and 25). However, it too can suffer SCC under certain conditions.

Magnesium-zinc alloys that are alloyed with either zirconium or RE elements, but not with aluminum, such as ZK60 and ZE10, have intermediate SCC resistance (Fig. 26), and in some cases SCC has not been a serious problem (Ref 1). However, SCC can still occur in atmospheric environments at stresses as low as 50% of the yield strength, although life is significantly longer than for Mg-Al-Zn alloys.

Magnesium alloys that contain neither aluminum nor zinc are the most SCC resistant. Magnesium-manganese alloys, such as M1, are among the alloys with the highest resistance to SCC, and they are generally considered to be immune when loaded up to the yield strength in normal environments. In fact, SCC of Mg-Mn alloys has been reported only in tests involving stresses higher than the yield strength (Ref 66, 84) and/or exposure to very severe laboratory environments (Ref 85). Alloys QE22, HK31, and HM21 are also resistant to SCC, exhibiting SCC thresholds at approximately 70 to 80% of the yield strength in rural-atmosphere tests (Ref 66).

Magnesium-lithium alloys are of commercial interest because of their higher stiffness and lower density compared with other magnesium alloys. Tests in humid air have resulted in SCC failures of Mg-Li-Al alloys, but SCC did not occur during testing of Mg-Li alloys strengthened with zinc, silicon, and/or silver instead of aluminum (Ref 86).

Recent Investigations. The earliest investigations of SCC of magnesium and magnesium alloys focused on the influence of alloy chemistry and microstructure, which were just gaining recognition as controlling factors in magnesium corrosion behavior (Ref 87, 88). The overwhelming majority of these studies used a chromate-chloride electrolyte (typically 40 g–1 each) because of its relevance to service conditions, in which chloride ions present in the environment attempt to penetrate a chromate-inhibited magnesium surface. These solutions cause especially severe cracking, but magnesium is also affected similarly by neutral solutions containing only chlorides or even distilled water (Ref 89).

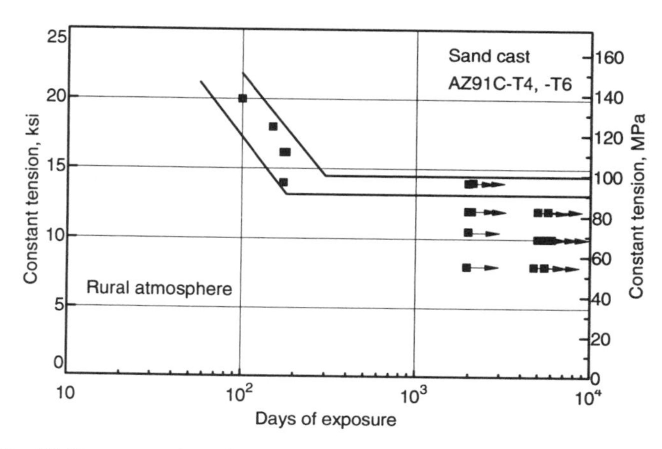

Fig. 22 Stress corrosion of sand-cast AZ91C (T4 and T6) in rural atmosphere. Source: Ref 6

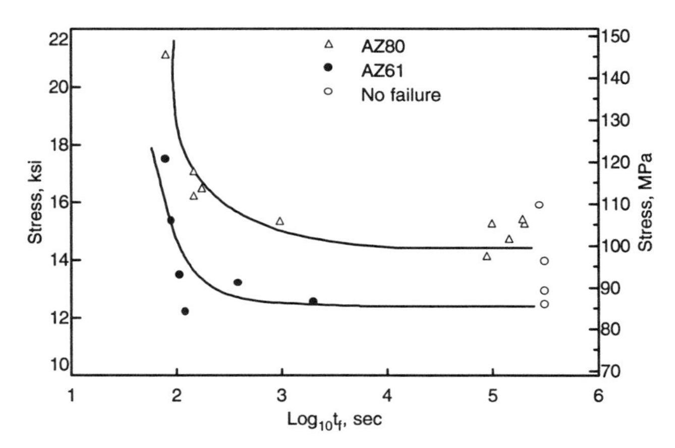

Fig. 23 Stress vs. time-to-failure (t_f) for the two-phase alloys AZ80 (Mg-8.5Al-0.5Zn) and AZ61 (Mg-6Al-1Zn) in aqueous 40 g/L NaCl + 40 g/L Na_2CrO_4. Source: J.A. Beavers, G.H. Koch, and W.E. Berry, "Corrosion of Metals in Marine Environments," Metals and Ceramics Information Center, Battelle Columbus Laboratories, July 1986

Studies of Mg-Al systems have shown that SCC is normally a transgranular phenomenon (Ref 90). Cracking occurs along twin interfaces or various crystallographic planes, and there is general agreement that hydrogen embrittlement is the dominant mechanism.

Recent investigations of pure magnesium (Ref 91) studied the crystallography of transgranular SCC in chromate-chloride electrolyte, identifying cleavage facets propagating on $\{2\bar{2}03\}$ planes joined by steps on $\{2\bar{2}03\}$ planes. No role of hydrogen was investigated; rather, the authors proposed a fracture process consistent with both the hydride and lattice decohesion models for hydrogen embrittlement. Stampella and co-workers (Ref 92) claimed decohesion is responsible for transgranular cracking in pH 10 sodium sulphate, and that pitting is a necessary precursor to embrittlement because hydrogen cannot penetrate the $Mg(OH)_2$ film. They ruled out stable hydride formation because no such precipitates appeared on fracture surfaces and the effect was reversed by room -temperature desiccation of the samples.

A recent study (Ref 93) sought to compare the stress corrosion behavior of rapidly solidified alloys to that of cast Mg-Al alloys, paying special attention to the role played by hydrogen and repassivation kinetics. This investigation, which was the first for rapidly solidified Mg-Al alloys, showed that all the alloys, as well as pure magnesium, failed by transgranular SCC in a chromate-chloride electrolyte at displacement rates between 5×10^{-5} and 9×10^{-3} mm s^{-1}. This failure mode was manifested in quasicleavage on the fracture surfaces and in lower maxima in stress intensity and displacement, and it was concluded that transgranular SCC probably aoccurs in these materials s a result of hydrogen embrittlement. Results from constant displacement rate testing were explained by a hydride formation model using realistic estimates for the diffusivity of hydrogen in magnesium. Based on repassivation results, dissolution appears incapable of achieving the observed crack growth rates. Potential pulse and scratching electrode experiments demonstrated superior repassivation behavior for rapidly solidified Mg-Al alloys compared with their as-cast counterparts, indicating that homogeneity retards pit nucleation and thereby retards the development of local environments that impair repassivation. Increasing the aluminum content from 1 to 9% improved the repassivation rate of rapidly solidified alloys. This study also showed that repassivation participates in this SCC mechanism, probably by localizing the corrosion reactions and controlling the amount of hydrogen that enters the unprotected alloy surface when film rupture occurs.

The studies described above establish several characteristics of the SCC of magnesium and magnesium-aluminum alloys. Cracking is usually transgranular, though certain heat treatments can activate the intergranular mode by affecting grain size and grain-boundary precipitation. Transgranular cracking involves quasicleavage on microstructural features identified as twin boundary interfaces or various crystallographic planes. Secondary phases appear to act by precipitating at these microstructural features and initiating localized galvanic attack through interaction with the magnesium matrix. There is general agreement that hydrogen embrittlement is the dominant mechanism for transgranular SCC in these materials, but its influence may be limited to conditions under which a bare surface exists to provide entry, thereby making repassivation rate an important parameter in the cracking process. Competition between the disruption of the surface film and repassivation constitutes an important factor in the SCC of Mg-Al in a mixture of chloride and chromate (Ref 94). Investigations by Frankenthal indicate a relationship between pitting and SCC (Ref 95).

Corrosion Fatigue

Substantial reductions in fatigue strength are shown in laboratory tests using NaCl spray or drops. Such tests are useful for comparing alloys, heat treatments, and protective coatings. Effective coatings, by excluding the corrosive environment, provide the primary defense against corrosion fatigue. The effect of coatings in an air environment is shown in Fig. 27. The effect of air on corrosion fatigue of commercially pure magnesium (9980A) is shown in Fig. 28.

Fatigue Crack Growth and Corrosion. An example of a fundamental study of the corrosion fatigue of magnesium alloys is that of Speidel et al. (Ref 96). They chose the extensively used commercial high-strength magnesium alloy ZK60A (Mg-5.2Zn-0.45Zr-0.22Mn)

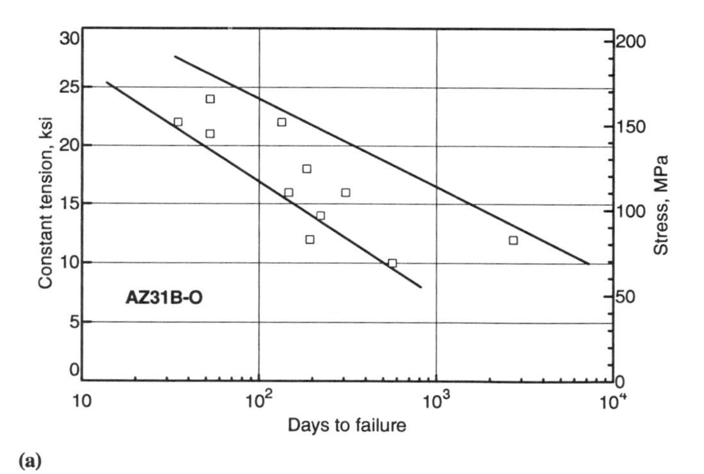

(a)

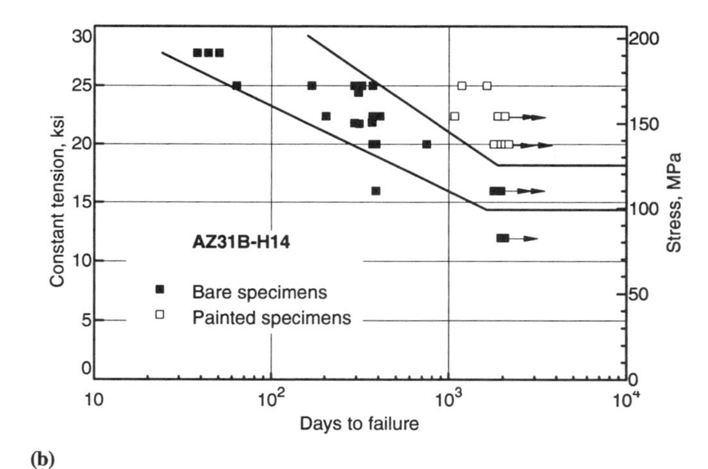

(b)

Fig. 24 Stress-corrosion resistance for AZ31B sheet in rural atmosphere. (a) AZ31B-O sheet. (b) AZ31B-H14 sheet. Source: Ref 6

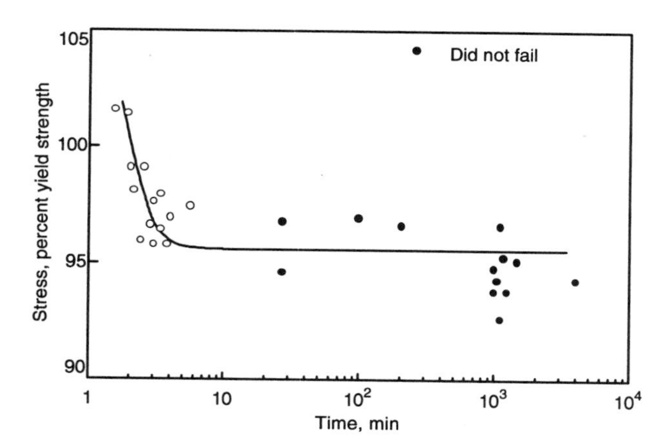

Fig. 25 Time-to-failure for AZ31 (Mg-3Al-1Zn) magnesium alloy exposed in a 3.5% NaCl + 2%K_2CrO_4 aqueous solution at 30 °C. Source: H.L. Logan, What We Don't Know About Stress Corrosion Cracking, *The Coupling of Basic and Applied Corrosion Research—A Dialogue*, National Association of Corrosion Engineers, 1969, p 57-62

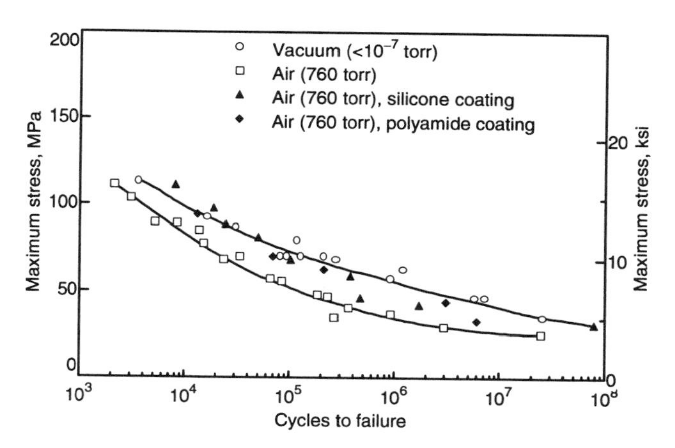

Fig. 27 Fatigue behavior comparison of coated and uncoated magnesium alloy specimens at room temperature. Source: H.T. Sumsion, *J. Spacecr. Rockets*, Vol 5 (No. 6), 1968, p 70

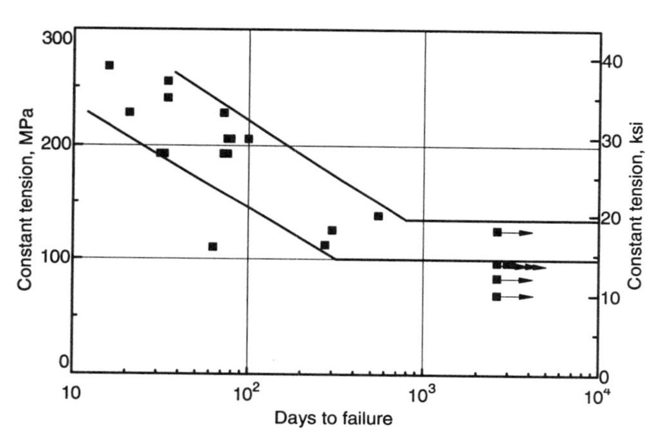

Fig. 26 Stress corrosion of ZK60A-T5 extrusion in rural atmosphere

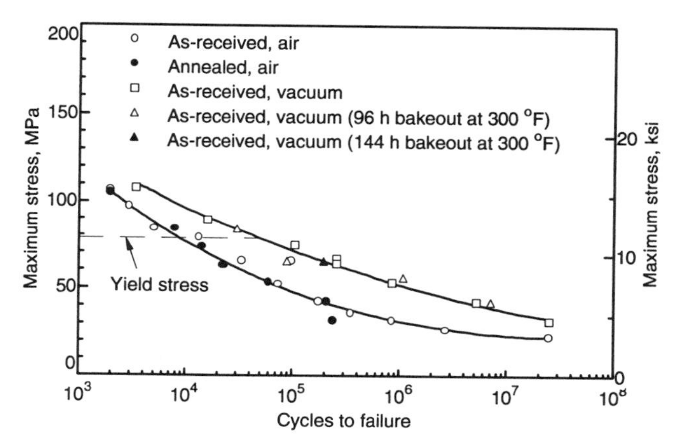

Fig. 28 Fatigue of commercial pure 9980A magnesium (UNS M19980) in air and in vacuum. Conditions: cantilever bending, $R = -1$, 30 Hz, room temperature. Source: *J. Spacecraft Rockets*, Vol 5, 1968, p 700-704

for their fracture mechanics corrosion fatigue study, because in the artificially aged (T5) condition, this alloy has a plateau (region II) in its *V-K* (crack growth velocity vs. stress intensity) curve that is lower than in other magnesium alloys. All magnesium alloys behave similarly with respect to environmentally enhanced subcritical crack growth, according to Speidel et al. They found that both stress-corrosion and corrosion-fatigue cracks propagate in a mixed transgranular-intergranular mode.

Speidel et al. (Ref 96) measured the corrosion-fatigue crack growth rate (da/dN) as a function of the cyclic stress intensity (ΔK) for ZK60A-T5 for all of the aqueous environments shown in Fig. 19 and compared the corrosion fatigue with stress-corrosion behavior. They found that:

- Corrosion-fatigue crack growth rate is accelerated by the same environments as those that accelerate stress-corrosion crack growth (i.e., sulphate and halide ions).
- The boundary between regions II and III in NaBr solutions of the da/dN versus ΔK curve is higher than the stress-corrosion threshold (K_{Iscc}), which occurs at a much lower stress intensity.
- There is a distinct boundary between regions II and III for all the media given in Fig. 19 (except dry argon). This boundary occurs at about the same stress intensity (~14 MPa $\sqrt{m}$) as K_{Issc} in distilled water.

It is common practice to protect the surface of magnesium and its alloys, and such protection is essential where contact occurs with other structural metals because this may lead to severe galvanic corrosion. Methods available for magnesium (Ref 1, 5, 6, 7, 25, 90) are summarized below. Additional information is contained in Volume 5 of the *ASM Handbook*.

- *Fluoride anodizing* involves alternating current anodizing at up to 120 V in a bath of 25% ammonium bifluoride, which removes surface impurities and produces a thin, pearly white film of MgF_2. This film is normally stripped in boiling chromic acid before further treatment because it gives poor adhesion to organic treatments.
- *Chemical treatments* involve pickling and conversion of the oxide coating. Components are dipped in chromate solutions, which clean and passivate the surface to some extent through formation of a film of $Mg(OH)_2$ and a chromium compound. Such films have only slight protective value, but they form a good base for subsequent organic coatings.
- *Electrolytic anodizing* includes proprietary treatments that deposit a hard ceramic-like coating, which offers some abrasion resistance in addition to corrosion protection (e.g., Dow 17, HEA, and MGZ treatments). Such films are very porous and provide little protection in the unsealed state,

but they may be sealed by immersion in a solution of hot dilute sodium dichromate and ammonium bifluoride, followed by draining and drying. A better method is to impregnate with a high-temperature curing epoxy resin (see below). Resin-sealed anodic films offer very high resistance to both corrosion and abrasion, and in some instances they can even be honed to provide a bearing surface. Impregnation is also used to achieve pressure tightness in casting that are susceptible to microporosity.

- *Sealing with epoxy resins:* The component is heated to 200 to 220 °C to remove moisture, cooled to approximately 60 °C, and dipped in the resin solution. After removal from this solution, draining, and air drying to evaporate solvents, the component is baked at 200 to 220 °C to polymerize the resin. Heat treatment may be repeated once or twice to build up the desired coating thickness, which is commonly 0.025 mm.
- *Standard paint finishes:* The surface of the component should be prepared as in the methods described above, after which it is preferable to apply a chromate-inhibited primer followed by a good-quality top coat.
- *Vitreous enameling* can be applied to alloys that do not possess too low a solidus temperature. Surface preparation involves dipping the work in a chromate solution before applying the frit.
- *Electroplating:* Several stages of surface cleaning and the application of pretreatments, such as a zinc conversion coating, are required before depositing chromium, nickel, or some other metal.

Magnesium alloy components for aerospace applications require maximum protection. Schemes involving chemical cleaning by fluoride anodizing, pretreatment by chromating or anodizing, and sealing with epoxy resin, a chromate primer, and a top coat are sometimes mandatory.

REFERENCES

1. E.F. Emley, *Principles of Magnesium Technology*, Pergamon Press, 1966
2. W. Hume-Rothery and G.V. Raynor, *The Structure of Metals & Alloys*, Institute of Metals, 1956
3. G.V. Raynor, *The Physical Metallurgy of Magnesium & Its Alloys*, Pergamon Press, 1959
4. C.S. Roberts, *Magnesium & Its Alloys*, John Wiley & Sons, 1960
5. I.J. Polmear, *Light Alloys: Metallurgy of the Light Metals*, 2nd ed., Edward Arnold, 1989
6. R.S. Busk, *Magnesium Products Design*, Marcel Dekker, 1987
7. B.L. Merdike and F. Hehmann, Ed., *Magnesium Alloys and Their Applications*, Proceedings of an International Conference (Garmisch-Partenkirchen, Germany), Vol 3, DGM Informationsgesellschaft, 1992
8. W. Hume-Rothery, *The Structure of Metals & Alloys*, Institute of Metals, 1944
9. L.A. Carapella, *Metal. Prog.*, Vol 48, 1945, p 297
10. T.R. Massalski, *Binary Phase Diagrams*, 2nd ed., Vol 1-4, ASM International, 1990
11. A.A. Nayeb-Hashemi and J.B. Clark, *Phase Diagrams of Binary Magnesium Alloys*, ASM International, 1988
12. A. Beck, *Magnesium und seine Legierungen*, Springer Verlag, 1939; *Technology of Magnesium Alloys*, F.A. Hughes & Co. Ltd., 1940
13. J.D. Hanawalt, C.E. Nelson, and J.A. Peloubet, *Trans. AIME*, Vol 147, 1942, p 273
14. J.B. Clark, *Acta Metall.*, Vol 16, 1968, p 141
15. J.F. King, in *Advanced Materials Technology International*, G.B. Brook, Ed., Sterling Publications, 1990, p 12
16. T.Kr. Aune, D.L. Albright, and H. Westengen, Technical Paper 900792, SAE Int. Congress and Exposition (Detroit, MI), 1990
17. H. Westengen, in *Proc. Int. Conf. on Recent Advances in Science & Engineering of Light Metals* (Sendai, Japan), Japan Institute of Metals, 1991, p 77
18. F. Hollrigl-Rosta, E. Just, J. Kohler, and H.J. Melzer, *Proc. 37th Annual World Magnesium Conf.*, Int. Magnesium Assoc., 1980, p 38
19. J.S. Waltrip, *Proc. 47th Annual World Magnesium Conf.*, Int. Magnesium Assoc., 1990, p 124
20. H. Mercer, SAE Technical Report 900788, March 1990
21. L.Y. Wei and G.L. Dunlop, in *Magnesium Alloys and Their Applications*, B.L. Merdike and F. Hehmann, Ed., DGM Informationsgesellschaft, 1992, p 335
22. J.B. Clark, *Acta Metall.*, Vol 13, 1965, p 1281
23. W. Unsworth and J.F. King, in *Proc. Int. Conf. on Magnesium Technology*, Institute of Metals, 1986, p 25
24. G.W. Lorimer, in *Proc. Int. Conf. on Magnesium Technology*, Institute of Metals, 1986, p 47
25. P. Lyon, J.F. King, and G.A. Fowler, *Int. Gas Turbine & Aerospace Congress & Exposition* (Orlando, FL), ASME, 1991
26. P.A. Fisher, *Foundry*, Vol 95 (No. 8), 1967, p 68
27. B.L. Mordike and I. Stulikova, *The Metallurgy of Light Alloys*, Institute of Metals, 1983, p 145
28. J.F. King, G.A. Fowler, and P. Lyon, *Proc. Conf. Light Weight Alloys for Aerospace Applications II*, E.W. Lee and N.J. Kim, Ed., Minerals, Metals & Materials Society, 1991, p 423
29. W. Durako and L. Joesten, *Proc. 49th Annual World Magnesium Conf.*, Int. Magnesium Society, 1992, p 87
30. R.S.M. Payne and N. Bailey, *J. Inst. Metals*, Vol 88, 1959-1960, p 417
31. Klein, in *Magnesium Alloys and Their Applications: Proceedings of an International Conference* (Garmisch-Partenkirchen, Germany), B.L. Merdike and F. Hehmann, Ed., DGM Informationsgesellschaft, 1992, p 53
32. T.Kr. Aune and H. Westengen, in *Magnesium Alloys and Their Applications: Proceedings of an International Conference* (Garmisch-Partenkirchen, Germany), B.L. Merdike and F. Hehmann, Ed., DGM Informationsgesellschaft, 1992, p 221
33. H. Westengen and J. Bolstad, in *Proc. Conf. DVM Tag 91*, DVM, 1991, p 225
34. R. Busk, *Aluminum Alloys—Contemporary Research and Applications*, A.K. Vasudevan and R.D. Doherty, Ed., Vol 31 of *Treatise on Materials Science and Technology*, Chapter 35, Academic Press, 1989
35. Ebbesen, in *Magnesium Alloys and Their Applications: Proceedings of an International Conference* (Garmisch-Partenkirchen, Germany), B.L. Merdike and F. Hahmann, Ed., DGM Informationsgesellschaft, 1992, p 267
36. Chadwick and Bloyce, in *Magnesium Alloys and Their Applications: Proceedings of an International Conference* (Garmisch-Partenkirchen, Germany), B.L. Merdike and F. Hehmann, Ed., DGM Informationsgesellschaft, 1992, p 93
37. Carahan et al., in *Magnesium Alloys and Their Applications: Proceedings of an International Conference* (Garmisch-Partenkirchen, Germany), B.L. Merdike and F. Hehmann, Ed., DGM Informationsgesellschaft, 19e2, p 69
38. R. Busk, Magnesium Alloys, *Aluminum Alloys—Contemporary Research and Applications*, A.K. Vasudevan and R.D. Doherty, Ed., Vol 31 of *Treatise on Materials Science and Technology*, Academic Press, 1989, p 663-680
39. *Modern Casting*, Dec 1967, p 84-90
40. H.J. Grover, S.A. Gordon, and L.R. Jackson, *Fatigue of Metals and Structures*, Department of the Navy, 1960

41. W.A. Freeth and G.V. Raynor, *J. Inst. Metals*, Vol 82, 1953-1954, p 575
42. R.I. Stechens and V.V. Ogarevic, *Ann. Rev. Mater. Sci.*, Vol 20, 1990, p 141-177
43. G.A. Chadwick, *Acta Metall.*, Vol 16, 1968, p 75
44. S. Gullberg et al., *Acta Metall.*, Vol 16, 1968, p 355
45. J.H. Mason, C.W. Warwick, P.J. Smith, J.A. Charles, and T.W. Clyne, *Mater. Sci.*, Vol 24, 1989, p 3934
46. S.K. Das and G.F. Chang, in *Rapidly Solidified Alloys*, S.K. Das, B.H. Kear, and C.M. Adam, Ed., AIME, 1985, p 137
47. R.E. Lewis, A. Joshi, and H. Jonas, in *Processing of Structural Metals by Rapid Solidification*, F.H. Froes and S.J. Savage, Ed., ASM International, 1986, p 367
48. S.K. Das et al., in Ref 7: *Magnesium Alloys and Their Applications: Proceedings of an International Conference* (Garmisch-Partenkirchen, Germany), B.L. Merdike and F. Hehmann, Ed., DGM Informationsgesellschaft, 1992, p 487-494
49. A. Inoue, M. Yamamoto, H.M. Kimura, and T. Matsumoto, *J. Mater. Sci. Letters*, Vol 6, 1987, p 194
50. S.G. Kim, A. Inoue, and T. Matsumoto, *Mat. Trans. Japan Inst. Met.*, Vol 31, 1990, p 929
51. A. Inoue, A. Kato, T. Zhong, S.G. Kim, and T. Matsumoto, *Mat. Trans. Japan Inst. Met.*, Vol 32, 1991, p 609
52. A. Inoue and T. Matsumoto, *Mat. Trans. Japan Inst. Metals*, Vol 32, 1991, p 875
53. T.W. Orange, *Engr. Fracture Mechanics*, Vol 3, 1971, p 53-69
54. J.P. Hickerson, Jr., *Engr. Fracture Mechanics*, Vol 9, 1977, p 75-85
55. M.O. Speidel, in *High-Temperature Materials in Gas Turbines*, P.R. Sahm and M.O. Speidel, Ed., Elsevier, 1974, p 207-251
56. L.P. Pook and A.F. Greenan, *Engr. Fracture Mechanics*, Vol 5, 1973, p 935-946
57. D.L. Davidson and J. Lankford, *International Materials Reviews*, Vol 37 (No. 2), 1992, p 45-76
58. C.M. Hudson and S.K. Seward, *Int. Journal of Fracture*, Vol 14, 1978, p 151-184
59. C.M. Hudson and S.K. Seward, *Int. Journal of Fracture*, Vol 20, 1982, p 57-117
60. C.M. Hudson and S.K. Seward, *Int. Journal of Fracture*, Vol 39, 1989, p 43-63
61. C.M. Hudson and J.J. Ferrainlo, *Int. Journal of Fracture*, Vol 48, 1991, p 19-43
62. G.L. Maker and J. Kruger, *International Metals Review*, Vol 38 (No. 3), 1993, p 138
63. H.P. Godard et al., *The Corrosion of Metals*, John Wiley, 1967, p 259
64. H. Hoy-Petersen, *Proc. 47th Annual World Magnesium Conference* (Cannes, 1990), Int. Magnesium Association
65. C.B. Baliga, P. Tsakiropolous, and J.F. Watts, *Int. Journal of Rapid Solidification*, Vol 4, 1989, p 231
66. "Exterior Stress Corrosion Resistance of Commercial Magnesium Alloys," Report Mt 19622, Dow Chemical USA, 8 March 1966
67. W.S. Loose and H.A. Barbian, Stress-Corrosion Testing of Magnesium Alloys, *Symp. Stress-Corrosion Cracking of Metals*, ASTM, 1945, p 273-292
68. W.S. Loose, Magnesium and Magnesium Alloys, *The Corrosion Handbook*, H.H. Uhlig, Ed., John Wiley and Sons, 1948, p 232-250
69. H.L. Logan, Magnesium Alloys, *The Stress Corrosion of Metals*, John Wiley and Sons, 1966, p 217-237
70. *Magnesium: Designing Around Corrosion*, Dow Chemical Co., 1982, p 16
71. J.D. Hanawalt, Joint Discussion on Aluminum and Magnesium, *Symp. Stress-Corrosion Cracking of Metals*, ASTM, 1945
72. M. Vialatte, Study of the SCC Behavior of the Alloy Mg-8% Al, *Symp. Engineering Practice to Avoid Stress Corrosion Cracking*, NATO, 1970, p 5-1 to 5-10
73. J.J. Lourens, Failure Analysis as a Basis for Design Modification of Military Aircraft, *Fracture and Fracture Mechanics Case Studies*, R.B. Tait and G.G. Garrett, Ed., Pergamon Press, 1985, p 47-56
74. M.O. Speidel, Stress Corrosion Cracking of Aluminum Alloys, *Metall. Trans. A*, Vol 6, 1975, p 631-651
75. G. Siebel, The Influence of Stress on the Corrosion of Electron Metals, *Jahrbuch der deutschen Luftfahrtforschung*, Part 1, 1937, p 528-531
76. A. Beck, *The Technology of Magnesium and Its Alloys*, 2nd ed., F.A. Hughes and Co., 1940, p 294-297
77. N.D. Tomashov, *Theory of Corrosion and Protection of Metals* (transl.), B.H. Tytell, I. Geld, and H.S. Preiser, Ed., MacMillan, 1966, p 626
78. M.J. Blackburn and M.O. Speidel, The Influence of Microstructure on the Stress Corrosion Cracking of Light Alloys, *Electron Microscopy and Structure of Materials*, G. Thomas, Ed., University of California Press, 1972, p 905-919
79. E.I. Meletis and R.F. Hochman, Crystallography of Stress Corrosion Cracking in Pure Magnesium, *Corrosion*, Vol 40 (No. 1), 1984, p 39-45
80. R.S. Stampella, R.P.M. Procter, and V. Ashworth, Environmentally-Induced Cracking of Magnesium, *Corros. Sci.*, Vol 24 (No. 4), 1984, p 325-341
81. S.P. Lynch and P. Trevena, Stress Corrosion Cracking and Liquid Metal Embrittlement in Pure Magnesium, *Corrosion*, Vol 44 (No. 2), 1988, p 133-124
82. *Metals Handbook*, 9th ed., Vol 13, *Corrosion*, ASM International, 1987, p 745
83. R.D. Heidenreich, C.H. Gerould, and R.E. McNulty, Electron Metallographic Methods and Some Results for Magnesium Alloys, *Trans AIME*, Vol 166, 1946, p 15
84. H.L. Logan and H. Hessing, Stress Corrosion of Wrought Magnesium Base Alloys, *J. Res. Natl. Bur. Stand.*, Vol 44, 1950, p 233-243
85. E.C.W. Perryman, Stress-Corrosion of Magnesium Alloys, *J. Inst. Met.*, Vol 78, 1951, p 621-642
86. J.C. Kiszka, Stress Corrosion Tests of Some Wrought Magnesium-Lithium Base Alloys, *Mater. Protect.*, Vol 4 (No. 2), 1965, p 28-29
87. D.K. Priest, F.H. Beck, and M.G. Fontana, in *Trans. ASM*, Vol 48, 1955, p 473-492
88. R.D. Heidenreich, C.H. Gerould, and R.E. McNulty, in *Trans. AIME*, Vol 166, 1946, p 15-36
89. W.S. Loose, in *Magnesium*, Proceedings of lecture series presented at National Metal Cong. and Exposition, Cleveland, OH, 1946, American Society for Metals, 244
90. G.L. Makar and J. Kruger, *Int. Met. Rev.*, Vol 38 (No. 3), 1993, p 138
91. E.I. Meletis and R.F. Hochman, *Corrosion*, Vol 40, 1984, p 39-45
92. R.S. Stampella, R.P.M. Procter, and V. Ashworth, *Corros. Sci.*, Vol 24, 1984, p 325-341
93. G.L. Makar, J. Kruger, and K. Sieradzki, *Corros. Sci.*, 1993
94. E.N. Pugh et al., *Proceedings of the Second International Conference on Fracture*, Chapman Hall, 1969, p 387
95. R.P. Frankenthal, *Corros. Sci.*, Vol 8, 1988, p 811
96. M.O. Speidel, M.J. Blackburn, T.R. Beck, and J.A. Feeney, in *Corrosion Fatigue: Chemistry, Mechanics and Microstructure*, O. Devereux et al., Ed., National Association of Corrosion Engineers, 1986, p 331

Magnesium Alloy Fatigue Data

Collected by R.I. Stephens and C.D. Schrader, Mechanical Engineering Department, The University of Iowa

Mg-Al Casting Alloys

AM100A

Composition: 9.9Al-0.10Mn-bal Mg
Product form: Casting, permanent mold cast
Heat treatment: F, T4, T6
Modulus of elasticity (avg at RT): 45 GPa (6.5×10^6 psi)
RT tensile strength/elongation: F: 150 MPa (21 ksi)/2%.
T4: 275 MPa (40 ksi)/10%.
T6: 275 MPa (40 ksi)/4%
RT yield strength: F: 85 MPa (12 ksi).
T4: 90 MPa (13 ksi).
T6: 110 MPa (16 ksi)
Test temperature: Room temperature
Test environment: Air
Failure criterion: Fracture
Loading condition: Rotating bending ($R = -1$)
Specimen geometry: Same notched, $K_t = 1$ and 2
Surface: Smooth
Source: J.D. Hanawalt, C.E. Nelson, and R.S. Busk, Properties and Characteristics of Common Magnesium Casting Alloys, *AFS*, Vol 53, 1945, p 77-86

Table 1 AM100A: Rotating bending, $R = -1$ fatigue strength for permanent mold castings

	Fatigue strength at:			
	10^6 cycles		10^8 cycles	
Condition	MPa	ksi	MPa	ksi
Smooth, $K_t = 1$				
F	110	16	82	12
T4	124	18	96	14
T6	110	16	76	11
Notched, $K_t = 2$				
F	69	10	48	7
T4	82	12	55	8
T6	76	11	62	9

AZ63A (UNS M11630)

Composition: 6.0Al-0.15Mn-3.0Zn-bal Mg
Product form: Sand cast
Heat treatment: F, T4, T6
Modulus of elasticity (avg at RT): 45 GPa (6.5×10^6 psi)
RT tensile strength/elongation: F: 200 MPa (29 ksi)/6%.
T4: 275 MPa (40 ksi)/12%.
T6: 275 MPa (40 ksi)/5%
RT yield strength: F: 95 MPa (14 ksi).
T4: 90 MPa (13 ksi)
T6: 130 MPa (19 ksi)
Test temperature: Room temperature
Test environment: Air
Failure criterion: Fracture
Loading condition: Rotating beam ($R = -1$), axial ($R = 0.25$)
Specimen geometry: Unnotched
Surface: Smooth
Source: "Dow Data Sheet on Fatigue Properties," Dow Chemical Co., 20 May 1958; "Aerospace Structural Metals Handbook," Battelle Columbus Laboratories, 1991, p 2

Table 2 AZ63A: Fatigue strength of sand cast test bars

	Fatigue strength at:							
	10^5 cycles		10^6 cycles		10^7 cycles		10^8 cycles	
Condition	MPa	ksi	MPa	ksi	MPa	ksi	MPa	ksi
Rotating beam, $R = -1$								
F	103-124	15-18	82-103	12-15	69-89	10-13	65.5-82	9.5-12
T4	124-158	18-23	110-144	16-21	96-131	14-19	89-117	13-17
T6	117-138	17-20	103-124	15-18	82-110	12-16	76-103	11-15
Direct stress, $R = 0.25$								
F	151-172	22-25	138-158	20-23	124-144	18-21	...	...
T4	158-186	23-27	144-165	21-24	124-144	18-21	...	...
T6	172-207	25-30	158-172	23-25	151-172	22-25	...	...

AZ63A, Notched Fatigue

Composition: 6.8Al-0.33Mn-3.1Zn-0.01C-0.009Cu-0.009Fe-0.001Ni-0.005Pb-0.04Si-0.01Sn
Heat treatment: T4
Modulus of elasticity (avg at RT): 45 GPa (6.5×10^6 psi)
RT tensile strength/elongation: 260 MPa (37.8 ksi)/6.9%
RT yield strength: 168 MPa (24.4 ksi)
Test temperature: Room temperature
Test environment: Air
Failure criterion: Fracture
Loading condition: Rotating beam, $R = -1$
Specimen geometry: Notched, $K_t = 1.69$
Surface: Smooth
Frequency: 10,000 rpm
Source: R.B. Clapper and J.A. Watz, Determination of Fatigue Crack Initiation and Propagation in a Magnesium Alloy, *ASTM STP 196*, 1956, p 111-119

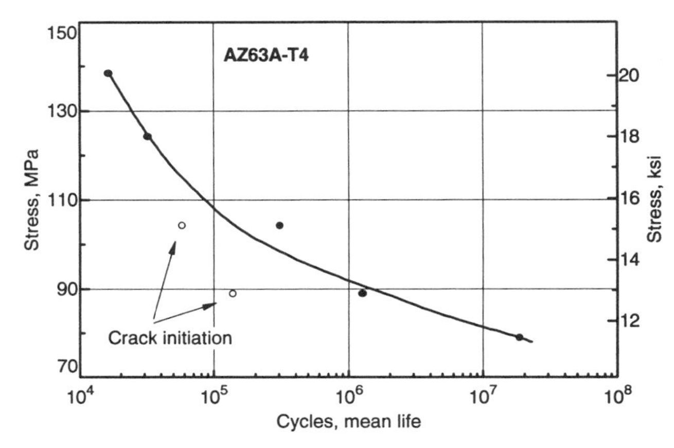

Fig. 1 AZ63A-T4: *S-N* curve for cast and notched specimens

AZ91B (UNS M11912) Axial Fatigue

Composition: 9.0Al-0.13Mn-0.68Zn-bal Mg
Product form: High-pressure die cast
Heat treatment: F
Modulus of elasticity (avg at RT): 45 GPa (6.5×10^6 psi)
RT tensile strength/elongation: 230 MPa (33 ksi)/3%
RT yield strength: 160 MPa (23 ksi)
Test temperature: Room temperature
Test environment: Air
Failure criterion: Fracture
Loading condition: Axial
Specimen geometry: Separately cast test bars
Surface: As cast
Source: "AZ91A-F Die Castings, R.R. Moore Fatigue Curves," Dow Chemical Co., TS&D Letter Enclosure, 3 March 1957; R.S. Busk, *Magnesium Products Design*, Marcel Dekker, 1987, p 280

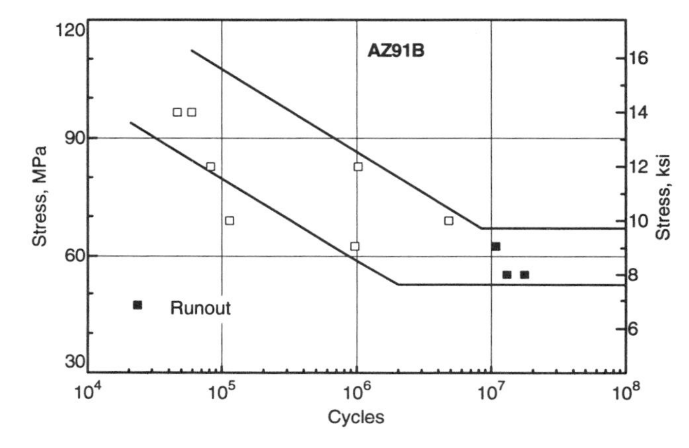

Fig. 2 AZ91B: Axial fatigue of die cast bar

AZ91B, Rotating Beam

Composition: 9.0Al-0.13Mn-0.68Zn-bal Mg
Product form: High-pressure die cast
Heat treatment: F
Modulus of elasticity (avg at RT): 45 GPa (6.5×10^6 psi)
RT tensile strength/elongation: 230 MPa (33 ksi)/3%
RT yield strength: 160 MPa (23 ksi)
Test temperature: Room temperature
Test environment: Air
Failure criterion: Fracture
Loading condition: Rotating beam
Specimen geometry: Separately cast test bars
Surface: As cast
Source: "AZ91A-F Die Castings, R.R. Moore Fatigue Curves," Dow Chemical Co., TS&D Letter Enclosure, 3 March 1957; R.S. Busk, *Magnesium Products Design*, Marcel Dekker, 1987, p 279

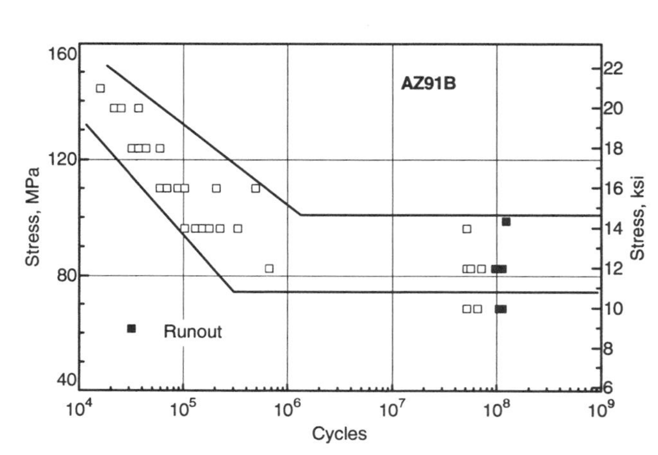

Fig. 3 AZ91B: Rotating beam fatigue of die cast bar

AZ91B, Plate Bending

Composition: 9.0Al-0.13Mn-0.68Zn-bal Mg
Product form: High-pressure die cast
Heat treatment: F
Modulus of elasticity (avg at RT): 45 GPa (6.5×10^6 psi)
RT tensile strength/elongation: 230 MPa (33 ksi)/3%
RT yield strength: 160 MPa (23 ksi)
Test temperature: Room temperature
Test environment: Air
Failure criterion: Fracture
Loading condition: Plate bending
Specimen geometry: Cast panels
Surface: As cast, edges machined
Source: "AZ91A-F Die Castings, R.R. Moore Fatigue Curves," Dow Chemical Co., TS&D Letter Enclosure, 3 March 1957; R.S. Busk, *Magnesium Products Design*, Marcel Dekker, 1987, p 281

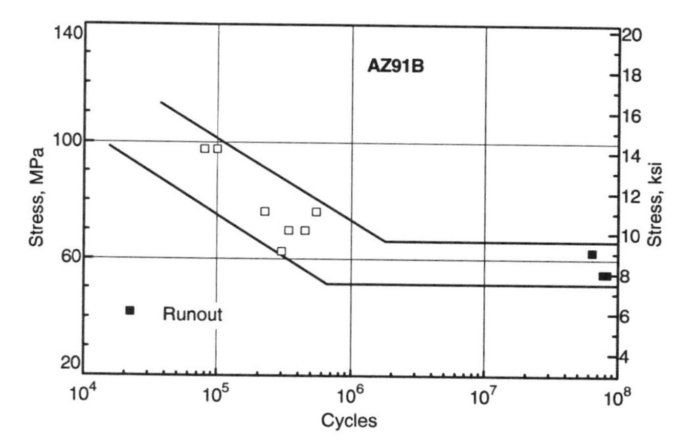

Fig. 4 AZ91B: Plate bending fatigue of die cast bar

AZ91C (UNS M1194), Unnotched Fatigue

Composition: 9.0Al-0.13Mn-0.68Zn-bal Mg
Product form: Sand cast
Heat treatment: F, T4, T6
Modulus of elasticity (avg at RT): 45 GPa (6.5×10^6 psi)
RT tensile strength/elongation: F: 165 MPa (24 ksi)/2%.
T4: 275 MPa (40 ksi)/14%.
T6: 275 MPa (40 ksi)/5%
RT yield strength: F: 95 MPa (13 ksi).
T4: 85 MPa (12 ksi).
T6: 130 MPa (19 ksi)
Test temperature: 25 °C (77 °F)
Test environment: Air
Failure criterion: Fracture
Loading condition: Krouse plate bending, $R = -1$
Surface: Machined, polished
Source: "Tables of Fatigue Strength of Sand Cast Magnesium Alloys," Dow Chemical Co., TS&D Letter Enclosure, 19 Nov 1965

Table 3 AZ91C: Bending fatigue of cast specimens

	Stress at:					
	10^5 cycles		10^6 cycles		10^7 cycles	
Temper	MPa	ksi	MPa	ksi	MPa	ksi
F	85	12	75	11	60	8
T4	105	15	85	12	75	11
T6	105	15	85	12	65	9

AZ91C, Notched Fatigue

Composition: 9.0Al-0.13Mn-0.68Zn-bal Mg
Product form: Sand cast
Heat treatment: F, T4, T6
Modulus of elasticity (avg at RT): 45 GPa (6.5×10^6 psi)
RT tensile strength/elongation: F: 165 MPa (24 ksi)/2%.
T4: 275 MPa (40 ksi)/14%.
T6: 275 MPa (40 ksi)/5%
RT yield strength: F: 95 MPa (13 ksi).
T4: 85 MPa (12 ksi).
T6: 130 MPa (19 ksi)
Test temperature: 25 °C (77 °F)
Test environment: Air
Loading condition: Rotating beam, $R = -1$
Specimen geometry: Some notched, $K_t = 2$
Surface: Smooth, machined, and polished
Source: "Tables of Fatigue Strength of Sand Cast Magnesium Alloys," Dow Chemical Co., TS&D Letter Enclosure, 19 Nov 1965

Table 4 AZ91C: Rotating beam fatigue strength

	Fatigue strength at:											
	Smooth								Notched			
	10^5 cycles		10^6 cycles		10^7 cycles		10^8 cycles		10^6 cycles		10^8 cycles	
Temper	MPa	ksi	MPa	ksi	MPa	ksi	MPa	ksi	MPa	ksi	MPa	ksi
F	110	16	105	15	95	13.5	85	12	70	10	55	8
T4	135	19	115	16	105	15	95	13.5	75	11	60	9
T6	125	18	105	15	90	13	80	11.5	75	11	55	8

AZ91C, Strain-Life Fatigue

Composition: 8.7Al-0.13Mn-0.70Zn-bal Mg
Product form: Sand cast
Heat treatment: T6
Hardness: 61 HB
Modulus of elasticity (avg at RT): 45 GPa (6.5×10^6 psi)
RT tensile strength/elongation: 140 MPa (20 ksi)/3%
RT yield strength: 110 MPa (16 ksi)
Test temperature: Room temperature
Test environment: Air
Loading condition: Strain control, $R = -1$
Source: R. Chernenkoff, private communication

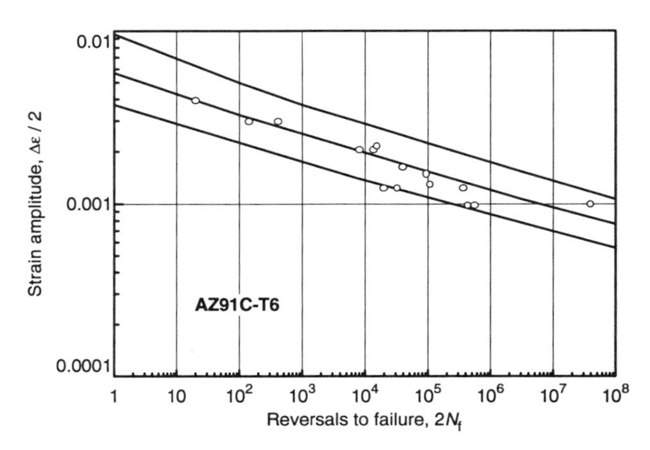

Fig. 5 AZ91C-T6: Strain-life diagram for cast specimens

AZ91D-HP (UNS M11916), Strain-Life Fatigue

Composition: 9.0Al-0.13Mn-0.68Zn-bal high-purity Mg
Product form: Sand cast
Hardness: 58 HB
Modulus of elasticity (avg at RT): 45 GPa (6.5×10^6 psi)
RT tensile strength/elongation: 235 MPa (34 ksi)/5%
RT yield strength: 160 MPa (23 ksi)
Test temperature: Room temperature
Test environment: Air
Loading condition: Strain control, $R = -1$
Specimen geometry: 3.175 mm (0.125 in.) thick × 6.35 mm (0.25 in.) wide
Surface: Skin on
Gauge length: 7.62 mm (0.3 in.)
Source: R. Chernenkoff, private communication

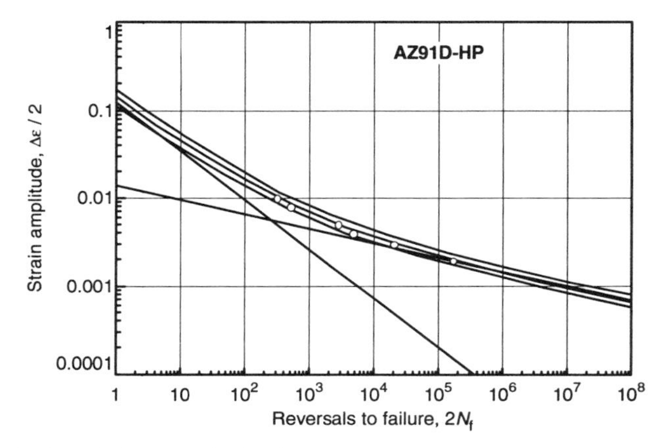

Fig. 6 AZ91D-HP: Strain-life diagram

AZ91E (UNS M11919), Strain-Life Fatigue

Composition: 8.97Al-0.54Zn-0.12Mn-0.010Cu-0.01Si-0.003Fe-0.0010Ni
Product form: Sand cast blocks
Heat treatment: T6
Modulus of elasticity (avg at RT): 45 GPa (6.5×10^6 psi)
RT tensile strength/elongation: 318 MPa (46 ksi)/12%
RT yield strength: 142 MPa (20 ksi)
Test temperature: Room temperature
Test environment: Laboratory air
Failure criterion: Fracture or 20% load drop
Loading condition: Axial strain control, $R = -1$, 0, and -2
Specimen geometry: Uniform gauge section, 6.35 mm (0.25 in.) diameter
Surface: Machined, longitudinal, and polished
Gauge length: 13 mm (0.5 in.)
Frequency: 0.5 to 30 Hz
Strain rate: 0.015 to 0.33 s^{-1}
Waveform: Triangular
Source: D. Goodenberger, "Fatigue and Fracture Behavior of AZ91E-T6 Sand Cast Magnesium Alloy," Master's thesis, The University of Iowa, Dec 1990

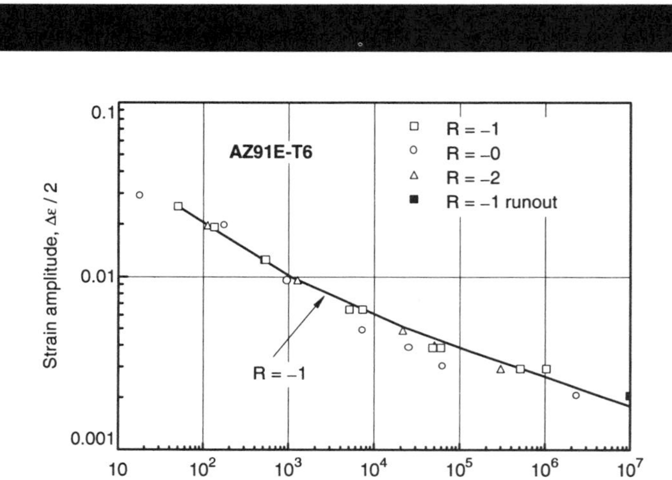

Fig. 7 AZ91E-T6: Low-cycle fatigue

AZ91E Fatigue Crack Growth

Composition: 8.97Al-0.54Zn-0.17Mn-0.010Cu-0.01Si-0.003Fe-0.0010Ni
Product form: Sand cast
Heat treatment: T6
Modulus of elasticity (avg at RT): 45 GPa (6.5×10^6 psi)
RT tensile strength/elongation: 318 MPa (46 ksi)/12%
RT yield strength: 142 MPa (20 ksi)
Test temperature: Room temperature
Test environment: Laboratory air
Loading condition: Load control, $R = 0.5$ and 0.05
Specimen geometry: Compact type
Surface: Polished
Frequency: 20 to 40 Hz
Waveform: Haversine
Source: D. Goodenberger, "Fatigue and Fracture Behavior of AZ91E-T6 Sand Cast Magnesium Alloy," Master's thesis, The University of Iowa, Dec 1990

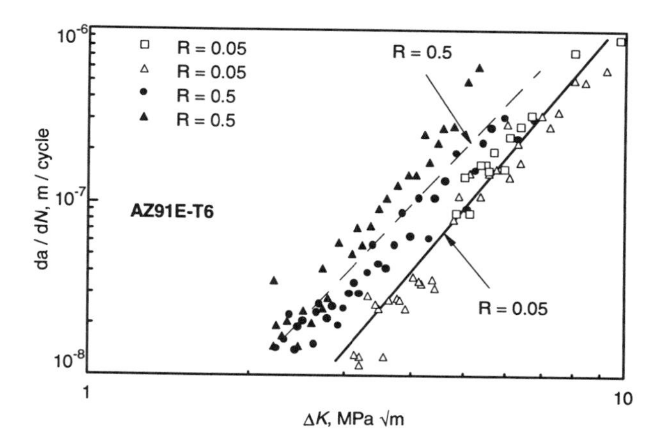

Fig. 8 AZ91E-T6: Fatigue crack growth behavior

AZ91E, Corrosion Fatigue

Composition: 8.97Al-0.54Zn-0.17Mn-0.010Cu-0.01Si-0.003Fe-0.0010Ni
Product form: Sand cast
Heat treatment: T6
Modulus of elasticity (avg at RT): 45 GPa (6.5×10^6 psi)
RT tensile strength/elongation: 318 MPa (46 ksi)/12%
RT yield strength: 142 MPa (20 ksi)
Test temperature: Room temperature
Test environment: 3.5% salt water solution
Failure criterion: 20% load drop or fracture
Loading condition: Strain control, axial $R = -1$, 0, and -2
Specimen geometry: Uniform gage section, 6.35 mm (0.025 in.)
Surface: Machined, longitudinal polish
Gauge length: 13 mm (½ in.)
Frequency: 0.5, 1, and 2 Hz
Strain rate: 0.004 to 0.026 s^{-1}
Test specifications: 12-h presoak
Source: C.D. Schrader, "Corrosion Fatigue and Stress Corrosion Cracking of AZ291E-T6 Sand Cast Magnesium Alloy," Master's thesis, The University of Iowa, Dec 1992

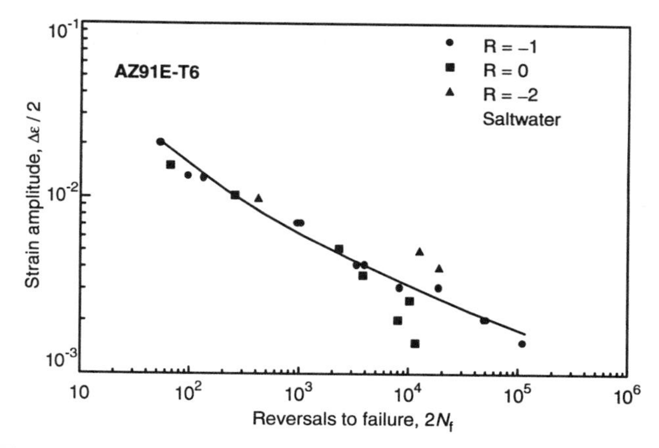

Fig. 9 AZ91E-T6: Low-cycle fatigue in salt-water solution

AZ91E, Crack Growth with Corrosion

Composition: 8.97Al-0.54Zn-0.17Mn-0.010Cu-0.01Si-0.003Fe-0.0010Ni
Product form: Sand cast
Heat treatment: T6
Modulus of elasticity (avg at RT): 45 GPa (6.5×10^6 psi)
RT tensile strength/elongation: 318 MPa (46 ksi)/12%
RT yield strength: 142 MPa (20 ksi)
Test temperature: 23 °C (74 °F)
Test environment: 3.5% salt water solution
Loading condition: Load control, $R = 0.5$ and 0.05
Specimen geometry: Compact type
Surface: Polished
Frequency: 3 Hz, Haversine
Source: C.D. Schrader, "Corrosion Fatigue and Stress Corrosion Cracking of AZ291E-T6 Sand Cast Magnesium Alloy," Master's thesis, The University of Iowa, Dec 1992

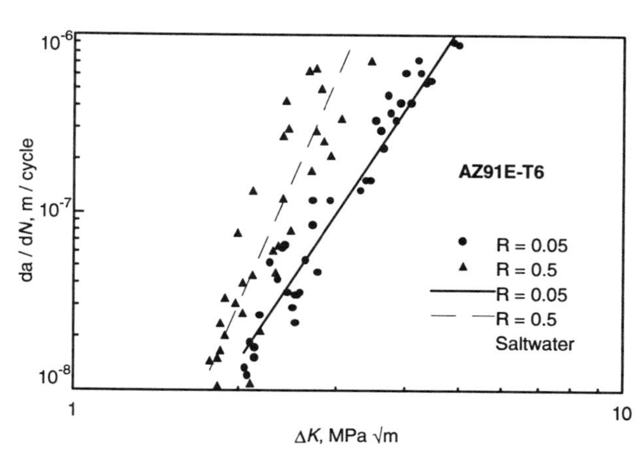

Fig. 10 AZ91E: Fatigue crack growth behavior in salt water solution

AZ92A (UNS M11920), Unnotched Fatigue

Composition: 9.0Al-0.10Mn-2.0Zn-bal Mg
Product form: Sand cast
Heat treatment: F, T4, T6
Modulus of elasticity (avg at RT): 45 GPa (6.5×10^6 psi)
RT tensile strength/elongation: F: 165 MPa (24 ksi)/2%
T4: 275 MPa (40 ksi)/9%
T6: 275 MPa (40 ksi)/2%
RT yield strength: F: 95 MPa (13 ksi)
T4: 95 MPa (13 ksi)
T6: 145 MPa (21 ksi)
Test temperature: 25 °C (77 °F)
Test environment: Air
Failure criterion: Fracture
Loading condition: Krouse plate bending, axial
Surface: Machined and polished
Source: "Tables of Fatigue Strength of Sand Cast Magnesium Alloys," Dow Chemical Co., TS&D Letter Enclosure, 19 Nov 1965

Table 5 AZ92A: Fatigue of sand castings

	Stress at:											
	Krouse, $R = -1$						Axial, $R = 0.25$					
	10^5 cycles		10^6 cycles		10^7 cycles		10^5 cycles		10^6 cycles		10^7 cycles	
Temper	MPa	ksi	MPa	ksi	MPa	ksi	MPa	ksi	MPa	ksi	MPa	ksi
F	90	13	75	11	65	9	165	24	160	23	150	21
T4	105	15	85	12	60	8	235	34	215	31	205	29
T6	125	18	100	14	90	13	210	30	195	28	190	27

AZ92A, Notched Fatigue

Composition: 9.0Al-0.2Mn-2.0Zn-bal Mg
Product form: Sand cast
Heat treatment: F, T4, T6
Modulus of elasticity (avg at RT): 45 GPa (6.5×10^6 psi)
RT tensile strength/elongation: F: 165 MPa (24 ksi)/2%
T4: 275 MPa (40 ksi)/9%
T6: 275 MPa (40 ksi)/2%
RT yield strength: F: 95 MPa (13 ksi)
T4: 95 MPa (13 ksi)
T6: 145 MPa (21 ksi)
Test temperature: Room temperature
Test environment: Air
Failure criterion: Fracture
Loading condition: Rotating beam, $R = -1$
Specimen geometry: Some specimens notched
Surface: Smooth, machined, and polished; also notched
Test specifications: $R = -1$
Source: "Tables of Fatigue Strength of Sand Cast Magnesium Alloys," Dow Chemical Co., TS&D Letter Enclosure, 19 Nov 1965

Table 6 AZ92A: Fatigue of notched sand castings

	Fatigue strength at:																	
	Smooth										Notched							
											$K_t = 2$				$K_t = 5$			
	10^5		10^6		10^7		10^8		5×10^8		10^6		10^8		10^6		10^8	
Temper	MPa	ksi	MPa	ksi	MPa	ksi	MPa	ksi	MPa	ksi	MPa	ksi	MPa	ksi	MPa	ksi	MPa	ksi
F	115	16.5	105	15	95	13.5	90	13	85	12	80	11.5	55	8	...	...	...	...
T4	125	18	110	16	105	15	95	13.5	90	13	100	14	75	11	55	8	40	6
T6	130	19	115	16.5	100	14	95	13.5	85	12	80	11.5	60	8.5	40	6	30	4

Mg-Al Wrought Alloys

AZ31B (UNS M11311)

Composition: 3.0Al-0.20Mn-1.0Zn-bal Mg
Product form: Extrusion
Heat treatment: F
Modulus of elasticity (avg at RT): 44.8 GPa (6.5×10^6 psi)
RT tensile strength/elongation: 260 MPa (37.7 ksi)/15%
RT yield strength: 200 MPa (29 ksi)
Test temperature: Room temperature
Test environment: Air
Failure criterion: Fracture
Loading condition: Rotating beam ($R = -1$), plate bending ($R = -1$), axial load ($R = 0.25$)
Surface: Polished, etched
Source: "Magnesium in Design," Bulletin 141-213, Dow Chemical Company, 1967; R.S. Busk, *Magnesium Products Design*, Marcel Dekker, New York, 1987, p 352

Table 1 Fatigue strengths of AZ31B-F extrusions

Alloy	Type of test	R value	Surface or stress concentration	Fatigue limit (MPa) at cycles: 10^5	10^6	10^7	10^8
AZ31B-F	Rotating beam	−1	Polished	180	160	145	130
	Plate bending	−1	Etched	105	95	90	85
	Axial load	0.25	Polished	175	160	150	

AZ31B, Plate Bending Fatigue

Composition: 3.0Al-0.20Mn-1.0Zn-bal Mg
Product form: Plate
Heat treatment: H24
Modulus of elasticity (avg at RT): 44.8 GPa (6.5×10^6 psi)
RT tensile strength/elongation: 270 MPa (39.1 ksi)/19%
RT yield strength: 185 MPa (26.8 ksi)
Test temperature: Room temperature
Test environment: Air
Failure criterion: Fracture
Loading condition: Cantilever bending, $R = -1, 0$
Source: R.S. Busk, *Magnesium Products Design*, Marcel Dekker, New York, 1987, p 452

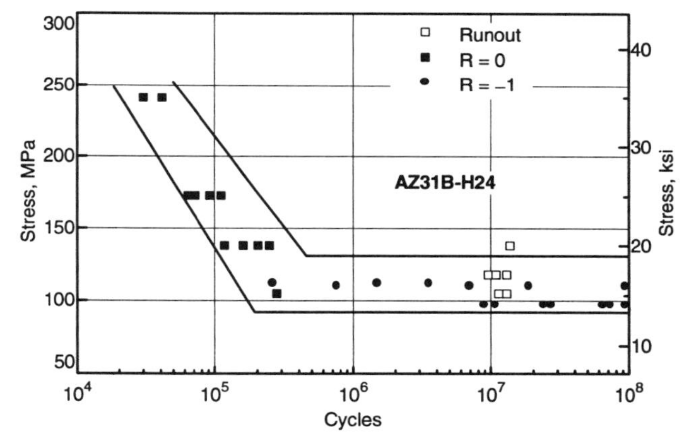

Fig. 1 Cantilever bending fatigue of AZ31B-H24 plate

AZ31B, Sheet, Bending Fatigue

Composition: 3.0Al-0.20Mn-1.0Zn-bal Mg
Product form: Sheet
Heat treatment: H24
Modulus of elasticity (avg at RT): 44.8 GPa (6.5×10^6 psi)
RT tensile strength/elongation: 290 MPa (42 ksi)/15%
RT yield strength: 220 MPa (31.9 ksi)
Test temperature: Room temperature
Test environment: Air
Failure criterion: Fracture
Loading condition: Cantilever bending, $R = -1$
Surface: Anodic coated
Source: S.J. Ketcham, "Investigation of Anodic Coatings for Magnesium Alloys," Report No. NAMC-AML-1347, Aeronautical Materials Laboratory, Naval Materials Center, Philadelphia, PA, 10 Jan 1962; R.S. Busk, *Magnesium Products Design*, Marcel Dekker, New York, 1987, p 449

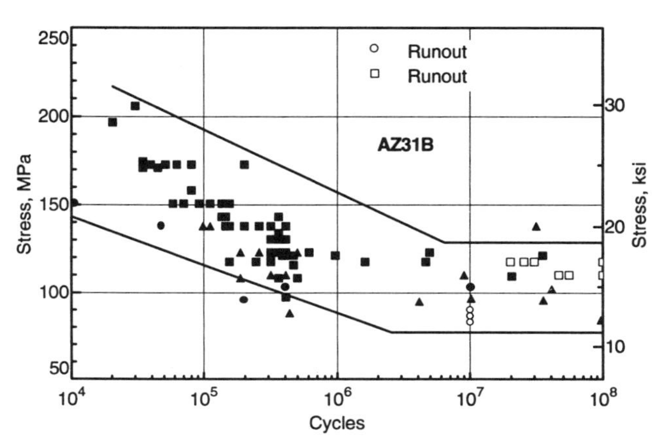

Fig. 2 Bending fatigue of AZ31B-H24 sheet

AZ31B, Strain Life Fatigue

Composition: 3.0Al-0.20Mn-1.0Zn-bal Mg
Product form: Forged
Heat treatment: F
Modulus of elasticity (avg at RT): 44.8 GPa (6.5×10^6 psi)
RT tensile strength/elongation: 260 MPa (37.7 ksi)/9%
RT yield strength: 195 MPa (28.2 ksi)
Test temperature: Room temperature
Test environment: Air
Failure criterion: Fracture
Loading condition: Strain control ($R = -1$)
Surface: Smooth
Source: S.S. Manson, Fatigue: A Complex Subject–Some Simple Approximations, *Exp. Mech.*, July 1965

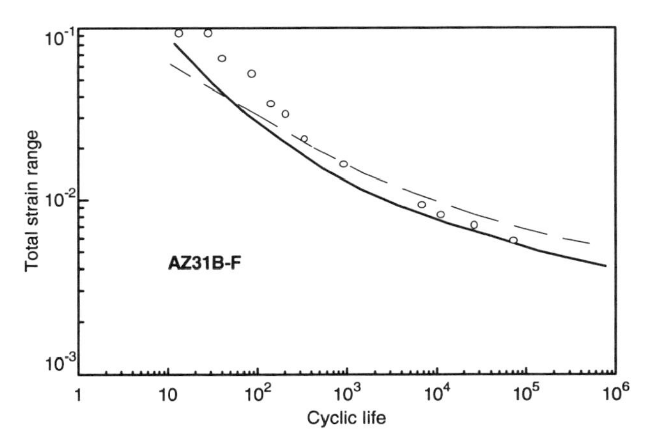

Fig. 3 Strain-life diagram for AZ31B

AZ31B, Corrosion Fatigue

Composition: 3.0Al-0.20Mn-1.0Zn-bal Mg
Product form: Extrusion
Heat treatment: F
Modulus of elasticity (avg at RT): 44.8 GPa (6.5×10^6 psi)
RT tensile strength/elongation: 260 MPa (37.7 ksi)/15%
RT yield strength: 200 MPa (29 ksi)
Test temperature: Room temperature
Test environment: Desiccator, water immersion, condensed water in air and various substances
Failure criterion: Fracture
Loading condition: Axial load, $R = 0.25$
Surface: Smooth
Source: Dow Chemical Company, unpublished; R.S. Busk, *Magnesium Products Design*, Marcel Dekker, New York, 1987, p 356

Table 2 Effect of corrosion on fatigue properties of AZ31B (Axial load; $R = 0.25$)

		Fatigue limit at cycles:					
		MPa			ksi		
Alloy	Atmosphere of test	10^5	10^6	10^7	10^5	10^6	10^7
Fatigue limits							
AZ31B-F	Desiccator	195	170	165	28	25	24
	Water immersion		145	145	...	21	21
	Condensed water in air	130	90	85	19	13	12

Life of AZ31B-F at a constant stress of 140 MPa (20 ksi)

Atmosphere of test	Cycles to failure
Desiccator	$>10^8$
Condensed water in:	
Air	6×10^4
Oxygen	3.5×10^5
Nitrogen	6×10^5
Argon	2×10^6
Argon + CO_2	10^5
Argon + SO_2	10^5
Argon + ammonia	5×10^5
Air + ammonia	5×10^5
Air + ammonia + SO_2	10^5

AZ31B, Fatigue Crack Growth

Composition: 3.0Al-0.20Mn-1.0Zn-bal Mg
Product form: 0.5 in. plate
Heat treatment: H24
Modulus of elasticity (avg at RT): 44.8 GPa (6.5 × 10^6 psi)
RT tensile strength/elongation: 250 MPa (36.2 ksi)/21%
RT yield strength: 150 MPa (21.7 ksi)
Test temperature: Room temperature
Test environment: Laboratory air
Loading condition: Load control, $R = 0.1, 0.4, 0.7$
Specimen orientation: T-L
Specimen geometry: C (T) 0.5 in. thick
Frequency: 5-50 Hz
Source: R.G. Forman, unpublished data

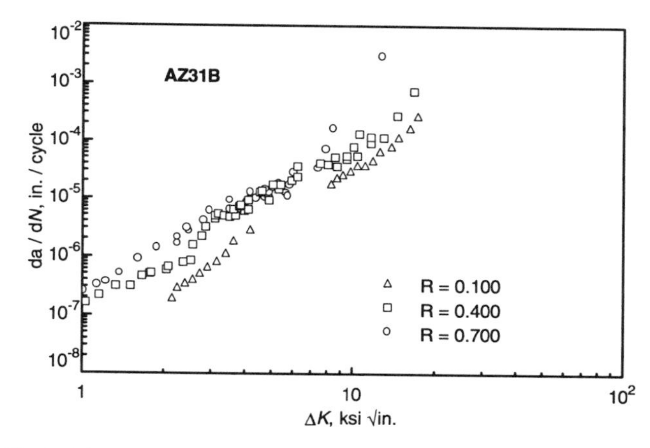

Fig. 4 *da*/d*N* data for AZ31B magnesium (H24)

AZ261A (UNS M11610)

Composition: 6.5Al-0.15Mn-1.0Zn-bal Mg
Product form: Pancake forging
Heat treatment: F
Modulus of elasticity (avg at RT): 44.8 GPa (6.5 × 10^6 psi)
RT tensile strength/elongation: 195 MPa (28.2 ksi)/%
RT yield strength: 180 MPa (26.1 ksi)
Test temperature: Room temperature
Test environment: Air
Failure criterion: Fracture
Loading condition: Rotating beam ($R = -1$), Flexure ($R = -1$)
Specimen orientation: Longitudinal
Surface: Polished, as forged
Source: "Magnesium in Design," Bulletin 141-213, Dow Chemical Company, 1967; R.S. Busk, *Magnesium Products Design*, Marcel Dekker, New York, 1987, p 489

Table 3 Fatigue strength of AZ61A forgings

Alloy	Temper	Forging	Orientation	Fatigue test	Surface	Stress (MPa) at cycles: 10^5	10^6	10^7	10^8
AZ61A	F	Pancake	Longitudinal	Rotating beam	Polished	180	150	145	140
				Flexure	As-forged	110	85	75	...

AZ61A, Extrusion

Composition: 6.5Al-0.15Mn-1.0Zn-bal Mg
Product form: Extrusion
Heat treatment: F
Modulus of elasticity (avg at RT): 44.8 GPa (6.5 × 10^6 psi)
RT tensile strength/elongation: 310 MPa (44.9 ksi)/16%
RT yield strength: 230 MPa (33.3 ksi)
Test temperature: Room temperature
Test environment: Air
Failure criterion: Fracture
Loading condition: Rotating beam ($R = -1$), plate bending ($R = -1$), Axial ($R = 0.25$)
Specimen geometry: Some notched ($K_t = 2.0$)
Surface: Polished, extruded
Source: "Magnesium in Design," Bulletin 141-213, Dow Chemical Company, 1967; R.S. Busk, *Magnesium Products Design*, Marcel Dekker, New York, 1987, p 352

Table 4 Fatigue strength of AZ61A-F extrusions

Type of test	*R* value	Surface or stress concentration	Fatigue limit (MPa) at cycles: 10^5	10^6	10^7	5×10^7	10^8
Rotating beam	−1	Polished	185	170	155	...	140
		$K_t = 2$	...	...	...	130	...
Plate bending	1	Extruded	125	95	85	...	...
Axial load	0.25	Polished	170	140	130	...	...

AZ61A Bar

Composition: 6.5Al-0.15Mn-1.0Zn-bal Mg
Product form: Bar
Heat treatment: F
Modulus of elasticity (avg at RT): 44.8 GPa (6.5 × 10^6 psi)
RT tensile strength/elongation: 310 MPa (44.9 ksi)/16%
RT yield strength: 230 MPa (33.3 ksi)
Test temperature: Room temperature
Test environment: Air
Failure criterion: Fracture
Loading condition: Rotating bend, $R = -1.0$
Surface: Smooth
Source: *Metallic Materials and Elements for Flight Vehicle Structures*, MIL-HDBK-5, Department of Defense, Aug 1962; *Aerospace Structural Metals Handbook*, Battelle Metals and Ceramics Information Center, Columbus, OH, 1991, code 3603, p 7

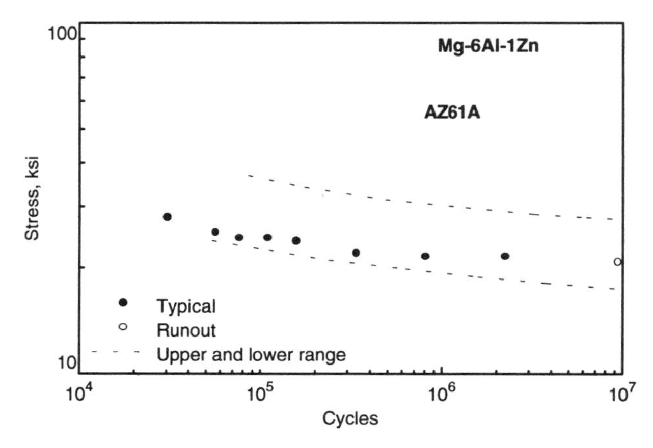

Fig. 5 S-N curves for AZ61A bar

AZ61A, Plate

Composition: 6.5Al-0.15Mn-1.0Zn-bal Mg
Product form: Plate
Heat treatment: F
Modulus of elasticity (avg at RT): 44.8 GPa (6.5 × 10^6 psi)
RT tensile strength/elongation: 310 MPa (44.9 ksi)/16%
RT yield strength: 230 MPa (33.3 ksi)
Test temperature: Room temperature
Test environment: Air
Failure criterion: Fracture
Loading condition: Rotating bend, $R = -1.0$
Surface: Smooth
Source: *Metallic Materials and Elements for Flight Vehicle Structures*, MIL-HDBK-5, Department of Defense, Aug 1962; *Aerospace Structural Metals Handbook*, Battelle Metals and Ceramics Information Center, Columbus, OH, 1991, code 3603, p 7

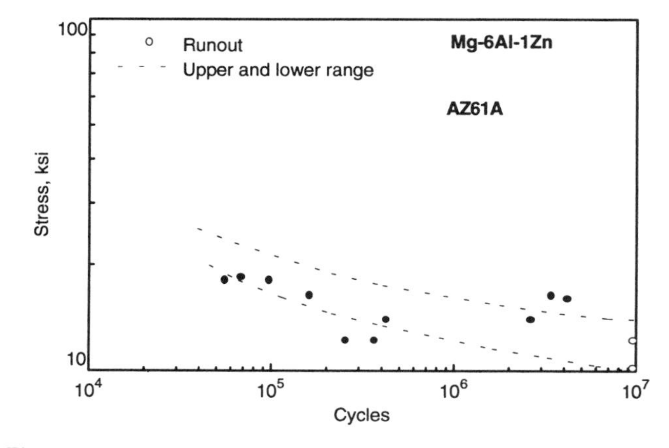

Fig. 6 S-N curves for AZ61A plate

AZ80A (UNS M11800)

Composition: 8.5Al-0.12Mn-0.5Zn-bal Mg
Product form: Extrusion
Heat treatment: F
Modulus of elasticity (avg at RT): 44.8 GPa (6.5 × 10^6 psi)
RT tensile strength/elongation: 340 MPa (49.3 ksi)/11%
RT yield strength: 250 MPa (36.2 ksi)
Test temperature: Room temperature
Test environment: Air
Failure criterion: Fracture
Loading condition: Rotating beam ($R = -1$) and plate bending ($R = 0.25$)
Surface: Polished, extruded
Source: "Magnesium in Design," Bulletin 141-213, Dow Chemical Company, 1967; R.S. Busk, *Magnesium Products Design*, Marcel Dekker, New York, 1987, p 352

Table 5 Fatigue strength of AZ80A-F extrusions

Alloy	Type of test	R value	Surface or stress concentration	Fatigue limit (MPa) at cycles:			
				10^5	10^6	10^7	10^8
AZ80A-F	Rotating beam	–1	Polished	190	175	160	150
	Plate bending	0.25	Extruded	140	100	90	...

AZ80A (F, T4, T5, and T6 Tempers)

Composition: 8.5Al-0.12Mn-0.5Zn-bal Mg
Product form: Extrusions and forgings
Heat treatment: F, T4, T5, T6
Modulus of elasticity (avg at RT): 44.8 GPa (6.5 × 10^6 psi)
RT tensile strength/elongation: Extrusion F: 340 MPa (49.3 ksi)/11%; Forging: F: 315 MPa (45.6 ksi)/8%; T5: 345 MPa (50 ksi)/6%; T6: 345 MPa (50 ksi)/5%
RT yield strength: Extrusion F: 250 MPa (36.2 ksi); Forging: F: 215 MPa (31.1 ksi); T5: 235 MPa (34 ksi); T6: 250 MPa (36.2 ksi)
Test temperature: Room temperature
Test environment: Air
Failure criterion: Fracture
Loading condition: Rotating beam ($R = -1$), reversed bending ($R = -1$)
Surface: Smooth, as extruded and as forged
Source: "Strength of Metal Aircraft Elements," ANC-5, March 1955; *Aerospace Structural Metals Handbook,* Battelle Metals and Ceramics Information Center, Columbus, OH, 1991, code 3501, p 2

Table 6 Fatigue strength of AZ80A

Condition	Temperature	Stress concentration	Fatigue strength, ksi, at:			
			10^5 cycles	10^6 cycles	10^7 cycles	10^8 cycles
Rotating beam, $R = -1$						
F(a)	Room temperature	Smooth	25-30	23-28	21-26	20-24
F(b)	Room temperature	Smooth	28-30	24-26	20-22	18-20
T4(b)	Room temperature	Smooth	...	21-24	18-21	16-18
T5(b)	Room temperature	Smooth	26-30	22-25	19-21	16-19
T6(b)	Room temperature	Smooth	23-27	19-22	16-19	14-16
Reversed bending, $R = -1$						
F(a)	Room temperature	As-extruded	19-21	13-16	12-14	...
T6(b)	Room temperature	As-forged	16-20	13-16	12-15	...

(a) Extrusion. (b) Forging

AZ80A-T5, Notched

Composition: 8.5Al-0.12Mn-0.5Zn-bal Mg
Product form: Extrusions and forgings
Heat treatment: T5
Modulus of elasticity (avg at RT): 44.8 GPa (6.5 × 10^6 psi)
RT tensile strength/elongation: Extrusion: 340 MPa (49.3 ksi)/11%; Forging: 345 MPa (50 ksi)/6%
RT yield strength: Extrusion: 250 MPa (36.2 ksi); Forging: 235 MPa (34 ksi)
Test temperature: Room temperature and150 °C (300 °F)
Test environment: Air
Failure criterion: Fracture
Loading condition: Reversed bending (R = −1), rotating beam ($R = -1$)
Specimen geometry: Some notched, $K_t = 2$
Surface: Smooth
Source: *Ordnance Materials Handbook, Magnesium and Magnesium Alloys,* ORDP 20-303, Sept 1956; Battelle Metals and Ceramics Information Center, Columbus, OH, 1991, code 3501, p 2

Table 7 Fatigue strength of notched AZ80A

Condition	Temperature, °F	Stress concentration	Fatigue strength, ksi, at:		
			10^5 cycles	10^7 cycles	10^8 cycles
Reversed bending, $R = -1$					
T5(a)	Room temperature	Smooth, $K_t = 1$	20	12.5	...
Rotating beam, $R = -1$					
T5(b)	300	Smooth	25	20	18
		Notched, $K_t = 2$	16.5	9.5	9.5
		Smooth	20.5	11	10.5
		Notched, $K_t = 2$	10.5	6.5	5

(a) Extrusion. (b) Forging

AZ80A, Bending Fatigue

Composition: 8.5Al-0.12Mn-0.5Zn-bal Mg
Product form: Extrusion (longitudinal)
Modulus of elasticity (avg at RT): 44.8 GPa (6.5×10^6 psi)
RT tensile strength/elongation: 372 MPa (53.9 ksi)
RT yield strength: 284 MPa (41.2 ksi)
Test temperature: Room temperature
Test environment: Air
Failure criterion: Fracture
Loading condition: Reversed bending ($R = -1$)
Specimen orientation: Longitudinal
Notch geometry: 60°, 0.025 in. deep, 0.010 in. radius
Surface: Polished
Frequency: 90, 2000, 3450 cpm
Source: T.T. Oberg and W.J. Trapp, High Stress Fatigue of Aluminum and Magnesium Alloys, *Prod. Eng.*, Vol 22 (No. 2), Feb 1951, p 163

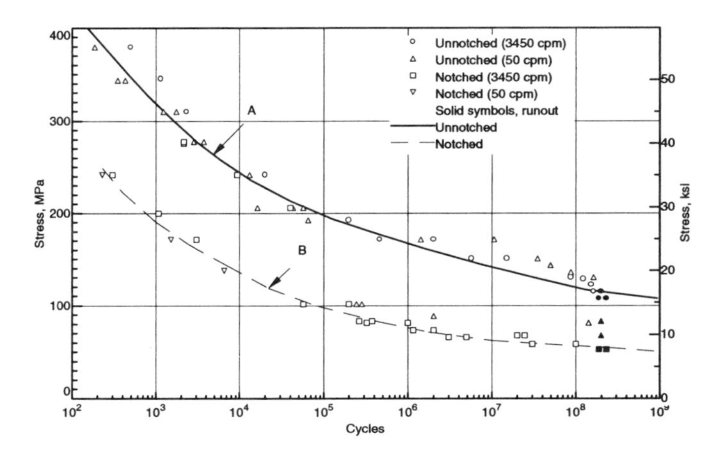

Fig. 7 High stress fatigue of AZ80 magnesium alloy

AZ80, Axial Fatigue

Composition: 8.5Al-0.12Mn-0.5Zn-bal Mg
Product form: Extrusion (longitudinal)
Modulus of elasticity (avg at RT): 44.8 GPa (6.5×10^6 psi)
RT tensile strength/elongation: 372 MPa (53.9 ksi)
RT yield strength: 284 MPa (41.2 ksi)
Test temperature: Room temperature
Test environment: Air
Failure criterion: Fracture
Loading condition: Axial ($R = 0$ and -1)
Specimen orientation: Longitudinal
Specimen geometry: 60°, 0.025 in. deep, 0.010 in. radius
Surface: Polished
Frequency: 90, 2000, 3450 cpm
Source: T.T. Oberg and W.J. Trapp, High Stress Fatigue of Aluminum and Magnesium Alloys, *Prod. Eng.*, Vol 22 (No. 2), Feb 1951, p 163

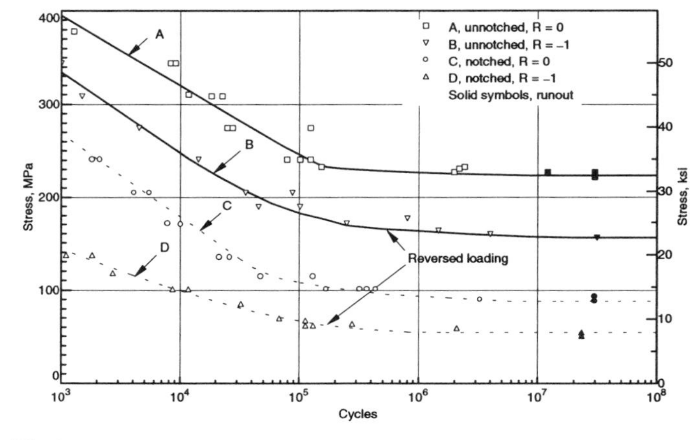

Fig. 8 High stress fatigue of AZ80 magnesium alloy

AZ81A (UNS M11810)

Composition: 7.5Al-0.13Mn-0.7Zn-bal Mg
Product form: Pancake forging
Heat treatment: F, T5, T6
Modulus of elasticity (avg at RT): 44.8 GPa (6.5×10^6 psi)
Test temperature: Room temperature
Test environment: Air
Failure criterion: Fracture
Loading condition: Rotating beam ($R = -1$) and flexure ($R = -1$)
Specimen orientation: Longitudinal
Specimen geometry: Pancake
Surface: Polished, as-forged
Remarks: Typically classified as a casting alloy
Source: "Magnesium in Design," Bulletin 141-213, Dow Chemical Company, 1967

Table 8 Fatigue of AZ81A forgings

Alloy	Temper	Forging	Orientation	Fatigue test	Surface	Stress (MPa) at cycles: 10^5	10^6	10^7	10^8
AZ81A	F	Pancake	Longitudinal	Rotating beam	Polished	200	170	145	130
	T5					195	160	140	125
						170	140	125	105
	T6			Flexure	As-forged	125	95	95	...

Mg-Zn Alloys

ZE41A Casting Alloy (UNS M16410)

Composition: 3.5-5.0 Zn, 0.75-1.75 R.E., 0.4-1.0 Zr, 0.10 Cu (max), 0.01 Ni (max), 0.30 other (max), bal Mg
Product form: Sand casting
Heat treatment: T5
Modulus of elasticity (avg at RT): 44.8 GPa (6.5×10^6 psi)
RT tensile strength/elongation: 200 MPa (29 ksi)/2.5%
RT yield strength: 134 MPa (19.5 ksi)
Test temperature: Room temperature
Test environment: Air
Failure criterion: Fracture
Loading condition: Rotating beam, $R = -1$
Specimen geometry: Some notched, U-notch, $K_t = 2.0$
Surface: Smooth, machined, and polished
Frequency: 2,960 cpm
Source: M.J. Miles, "ZE41: A Magnesium Alloy with Improved Mechanical Properties for Use in Helicopter Transmission Castings," American Helicopter Society, May 1977

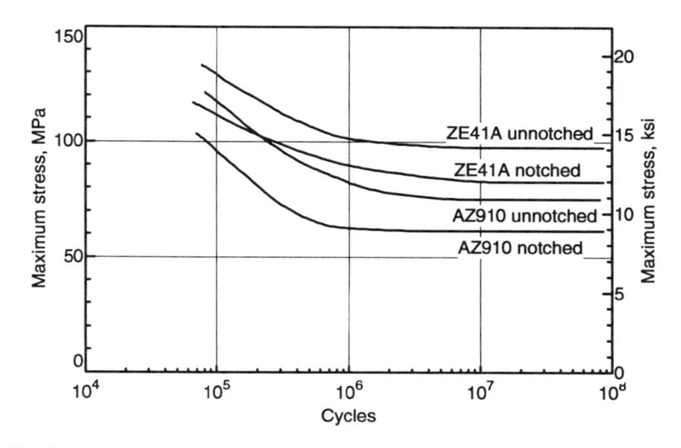

Fig. 1 Fatigue properties of ZE41A, notched and unnotched

ZK61A Casting Alloy (UNS M1660)

Composition: 6.0Zn-0.8Zr-bal Mg
Product form: Sand cast
Heat treatment: F, T6
Modulus of elasticity (avg at RT): 44.8 GPa (6.5×10^6 psi)
RT tensile strength/elongation: T6: 275 MPa (39.8 ksi)/5%
RT yield strength: 180 MPa (26.1 ksi)
Test temperature: Room temperature
Test environment: Air
Failure criterion: Fracture
Loading condition: Rotating beam ($R = -1$)
Specimen geometry: Some notched, $K_t = 2$
Surface: Smooth
Source: J.W. Meier, Characteristics of High-Strength Magnesium Casting Alloy ZK61, *Trans. AFS*, Vol 61, 1953, p 719-728

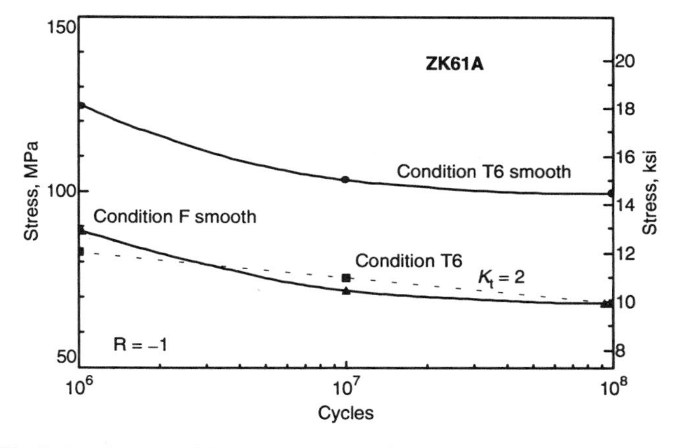

Fig. 2 Rotating beam fatigue strength of ZK61A

ZH 62A Casting Alloy (UNS M16620)

Composition: 1.8Th-5.7Zn-0.7Zr-bal Mg
Product form: Sand cast
Heat treatment: T5
Modulus of elasticity (avg at RT): 44.8 GPa (6.5 × 10^6 psi)
RT tensile strength/elongation: 275 MPa (39.8 ksi)/6%
RT yield strength: 170 MPa (24.6 ksi)
Test temperature: Room temperature
Test environment: Air
Failure criterion: Fracture
Loading condition: Krouse plate bending ($R = -1$)
Surface: Machined and polished
Source: "Tables of Fatigue Strength of Sand Cast Magnesium Alloys," Dow Chemical Co., TS&D Letter Enclosure, 19 Nov 1965; "Mechanical Properties and Chemical Compositions of Cast Magnesium Alloys," Bulletin 440, Magnesium Elektron Ltd., March 1981

Table 1 ZH62A and ZK61A fatigue strength compilation

Alloy and temper	Form and thickness	Specimen	Test mode	Stress ratio	Temperature	Atmosphere	Fatigue strength, MPa (ksi), at cycles:				
							10^4	10^5	10^6	10^7	5×10^7
ZK61A-T6	Sand cast	Machined and polished, unnotched	Krouse plate bending	−1	Room temperature	Air	...	585 (85)	415 (60)	345 (50)	...
ZH62A-T5	Sand cast	Machined and polished, unnotched	Krouse plate bending	−1	Room temperature	Air	...	860 (125)	585 (85)	517 (75)	...
ZH62A-T5	Sand cast	Machined and polished, unnotched	Rotating beam	−1	Room temperature	Air	...	825 (120)	585 (85)	570 (83)	565 (82)
ZH62A-T5	Sand cast	Machined and polished, notched $K_t = 1.8$	Rotating beam	−1	Room temperature	Air	...	...	550 (80)	...	525 (76)

Sources: TSD Letter, Nov 1965, and Magnesium Elektron Bulletin 440, 1981

ZE10A Sheet (UNS M16100)

Composition: 0.17 Rare Earths-1.2Zn-bal Mg
Product form: Sheet
Heat treatment: F
Modulus of elasticity (avg at RT): 44.8 GPa (6.5 × 10^6 psi)
RT tensile strength/elongation: 260 MPa (37.7 ksi)/10%
RT yield strength: 170 MPa (24.6 ksi)
Test temperature: Room temperature
Test environment: Air
Failure criterion: Fracture
Loading condition: Axial load, $R = 0, 0.5$
Surface: Smooth
Source: "ZE10A Sheet & Plate Magnesium Alloy," TS&D Letter Enclosure," Dow Chemical Co., 1 June 1984

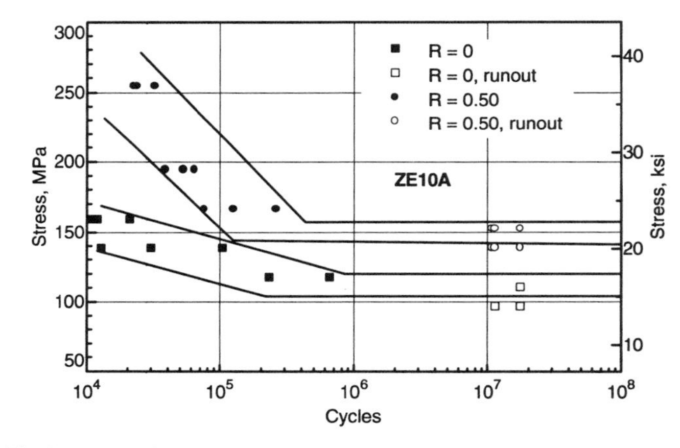

Fig. 3 Fatigue of ZE10A sheet

ZK60A Extrusions

Composition: 5.5Zn-0.5Zr-bal Mg
Product form: Extrusion
Heat treatment: F, T5
Modulus of elasticity (avg at RT): 44.8 GPa (6.5×10^6 psi)
RT tensile strength/elongation: F: 340 MPa (49.3 ksi)/14%; T5: 365 MPa (52.9 ksi)/11%
RT yield strength: F: 260 MPa (37.7 ksi); T5: 305 MPa (44.2 ksi)
Test temperature: Room temperature
Test environment: Air
Failure criterion: Fracture
Loading condition: Rotating beam ($R = -1$), axial load ($R = 0.25$)
Specimen orientation: Longitudinal
Surface: Machined and polished
Source: "Magnesium in Design," Form No. 141-213-67, Dow Chemical Co., Metal Products Dept., 1967; *Aerospace Structural Materials Handbook*, Battelle Metals and Ceramics Information Center, Columbus, OH, 1991, code 3506, p 11

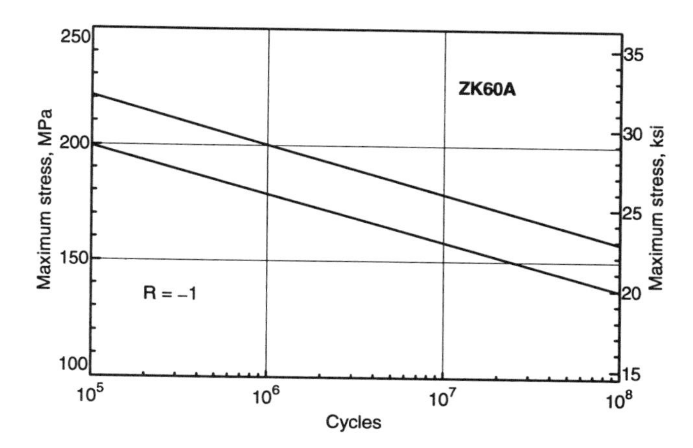

Fig. 4 ZK60A (F temper): rotating beam fatigue ($R = -1$)

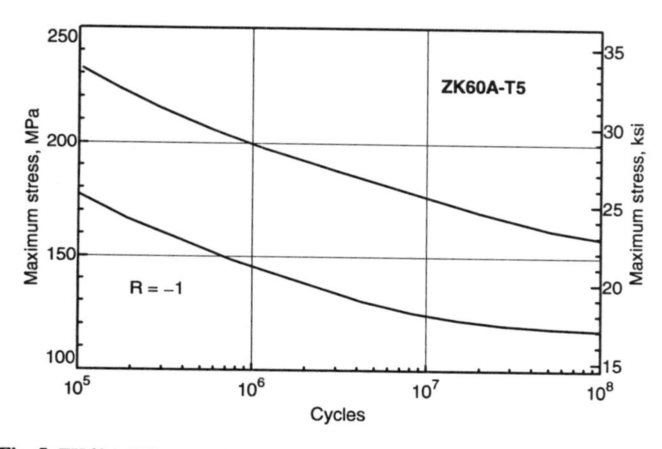

Fig. 5 ZK60A (T5 temper): rotating beam fatigue ($R = -1$)

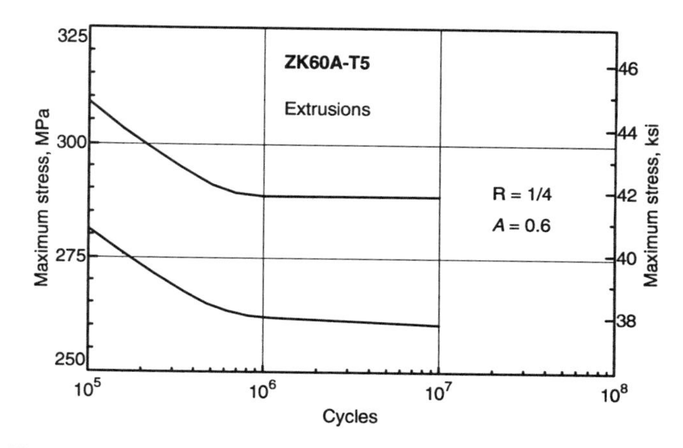

Fig. 6 ZK60A (F temper) extrusions: axial fatigue ($R = 0.25$)

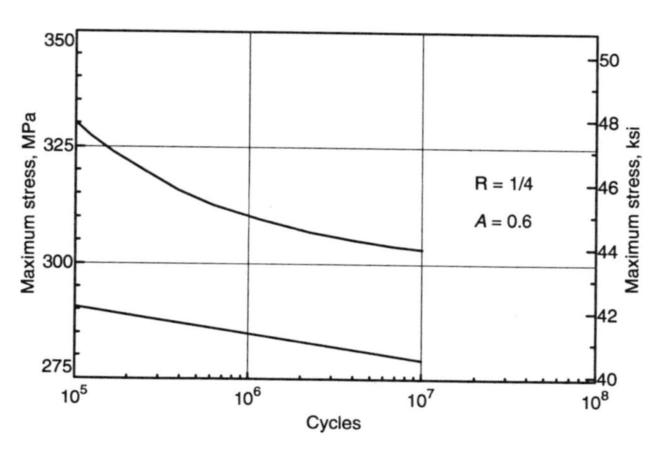

Fig. 7 ZK60A (T5 temper) extrusions: axial fatigue ($R = 0.25$)

ZK60A (UNS M16600) Fatigue Strength

Composition: 5.5Zn-0.5Zr-bal Mg
Product form: Forging
Heat treatment: T5
Modulus of elasticity (avg at RT): 44.8 GPa (6.5 × 10^6 psi)
RT tensile strength/elongation: 305 MPa (44.2 ksi)/16%
RT yield strength: 205 MPa (29.7 ksi)
Test temperature: Room temperature
Test environment: Air
Failure criterion: Fracture
Loading condition: Rotating beam, flexure, $R = -1$
Specimen geometry: Some notched, $K_t = 2$
Surface: Machined and polished
Source: H.C. Buckelew, Magnesium Alloy Cuts Aircraft Wheel Cost, Weight, *SAE J.*, Vol 72, April 1964, p 90-94; *Aerospace Structural Materials Handbook*, Battelle Metals and Ceramics Information Center, Columbus, OH, 1991, code 3506, p 11

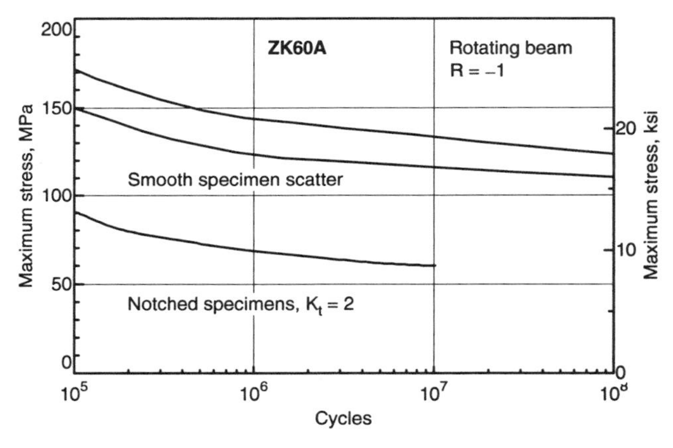

Fig. 8a Rotating beam fatigue strength of ZK60A-T5 forgings

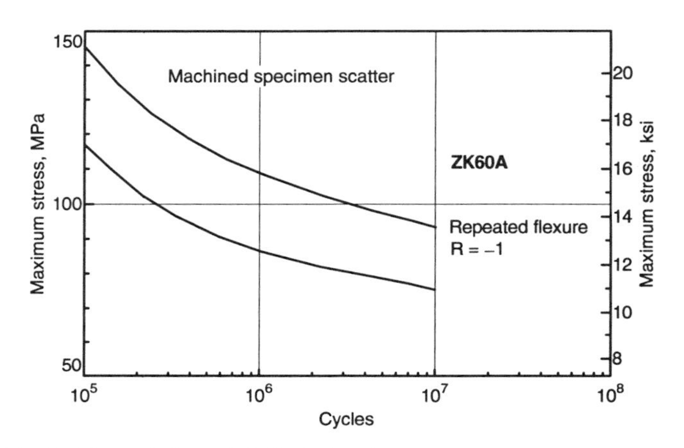

Fig. 8b Flexure fatigue of ZK60A-T5 forgings

ZK60A Fatigue Strength

Composition: 5.5Zn-0.4Zr-bal Mg
Product form: Forging—wheel rim
Heat treatment: T5, T6
Modulus of elasticity (avg at RT): 44.8 GPa (6.5 × 10^6 psi)
RT tensile strength/elongation: 305 MPa (44.2 ksi)/16%
RT yield strength: 205 MPa (29.7 ksi)
Test temperature: Room temperature
Test environment: Air
Failure criterion: Fracture
Loading condition: Rotating beam and flexure
Specimen orientation: Tangential and axial
Specimen geometry: Wheel rim
Surface: Polished
Source: "Magnesium ZK60A-T6 Forging Alloy," TS&D Letter Enclosure," Dow Chemical Co.; R.S. Busk, *Magnesium Products Design*, Marcel Dekker, New York, 1987, p 490

Table 2 Fatigue strength of forged ZK60A wheel rims

Alloy	Temper	Forging	Orientation	Fatigue test	Surface	Stress, MPa, at cycles: 10^5	10^6	10^7	10^8
ZK60A	T5	Wheel rim	Tangential	Rotating beam	Polished	160	140	125	125
			Axial		Polished	160	140	125	125
					Notched(b)	95	85	70	60
			Axial	Flexure(a)	Polished	130	95	90	85
	T6	Wheel rim	Tangential	Rotating beam	Polished	185	150	130	115
			Axial		Polished	185	150	130	115
					Notched(b)	115	90	70	60
			Axial	Flexure(a)	Polished	140	125	110	105

(a) $R = -1$. (b) $K_t = 2$

ZK60A (UNS M16600) Plate, Fatigue Crack Growth

Composition: 5.5Zn-0.4Zr-bal Mg
Product form: 0.5 in. (12.7 mm) plate
Heat treatment: T5
Modulus of elasticity (avg at RT): 44.8 GPa (6.5×10^6 psi)
RT tensile strength/elongation: 305 MPa (44.2 ksi)/16%
RT yield strength: 205 MPa (29.7 ksi)
Test temperature: Room temperature
Test environment: Laboratory air
Loading condition: Load control, $R = 0.1, 0.4, 0.7$
Specimen orientation: T-L
Specimen geometry: C(T), 0.5 in. (12.7 mm) thick
Frequency: 5-50 Hz
Source: R.G. Forman, unpublished data

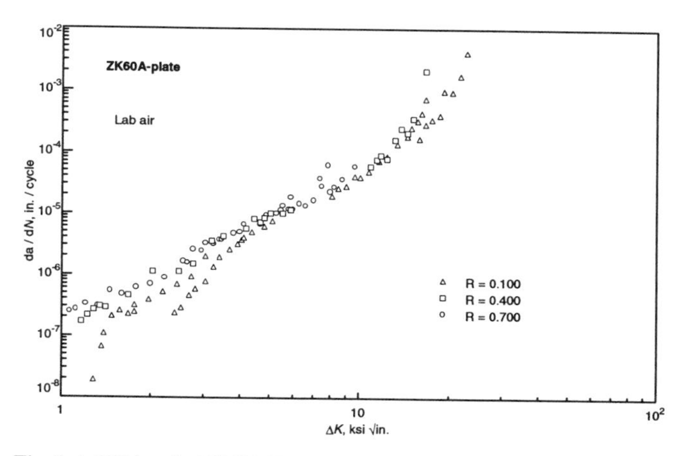

Fig. 9 da/dN data for ZK60A-T5

Table 3 $R = 0.1$ data

Sample number	da/dN, in./cycle	ΔK, ksi√in.
1	0.1974E-04	8.070
2	0.2578E-04	8.530
3	0.2844E-04	9.030
4	0.4119E-04	9.570
5	0.4008E-04	10.190
6	0.4874E-04	10.860
7	0.7328E-04	11.600
8	0.8680E-04	12.430
9	0.1398E-03	13.360
10	0.1900E-03	14.420
11	0.2642E-03	15.030
12	0.3274E-03	15.650
13	0.4365E-03	16.330
14	0.7291E-03	16.800
15	0.1747E-03	15.970
16	0.2882E-03	16.660
17	0.3487E-03	17.540
18	0.4102E-03	18.530
19	0.9857E-03	19.460
20	0.9443E-03	20.450
21	0.1736E-02	21.610
22	0.4208E-02	22.790
23	0.4729E-01	40.430
24	0.2260E-06	2.390
25	0.2870E-06	2.500
26	0.4490E-06	2.680
27	0.5740E-06	2.800
28	0.7830E-06	3.010
29	0.1309E-05	3.220
30	0.1884E-05	3.410
31	0.2549E-05	3.660
32	0.3097E-05	3.890
33	0.3982E-05	4.090
34	0.4840E-05	4.420
35	0.6060E-05	4.780
36	0.7601E-05	5.110
37	0.1900E-07	1.260
38	0.6200E-07	1.330
39	0.1030E-06	1.370
40	0.2010E-06	1.460
41	0.2560E-06	1.550
42	0.2150E-06	1.660
43	0.2390E-06	1.760
44	0.3140E-06	1.760
45	0.3720E-06	1.950
46	0.4970E-06	2.180
47	0.6940E-06	2.410
48	0.9250E-06	2.740
49	0.1992E-05	3.000
50	0.3752E-05	4.080

Table 4 $R = 0.4$ data

Sample number	da/dN, in./cycle	ΔK, ksi√in.
1	0.1620E-06	1.140
2	0.2080E-06	1.200
3	0.2630E-06	1.250
4	0.3010E-06	1.320
5	0.2850E-06	1.380
6	0.4500E-06	1.650
7	0.1097E-05	1.980
8	0.1104E-05	2.460
9	0.1485E-05	2.780
10	0.3476E-05	3.120
11	0.4065E-05	3.510
12	0.5707E-05	4.160
13	0.6056E-04	10.890
14	0.7624E-04	11.420
15	0.8830E-04	11.930
16	0.8009E-04	12.520
17	0.1662E-03	13.150
18	0.2510E-03	13.840
19	0.2089E-03	14.600
20	0.3535E-03	15.440
21	0.2109E-02	16.810
22	0.8108E-05	4.400
23	0.7317E-05	4.620
24	0.8634E-05	4.750
25	0.9651E-05	4.960
26	0.9932E-05	5.560
27	0.1150E-04	5.860

Table 5 $R = 0.7$ **data**

Sample number	da/dN, in./cycle	ΔK, ksi√in.
1	0.2070E-06	0.930
2	0.2300E-06	0.990
3	0.2350E-06	1.040
4	0.2700E-06	1.090
5	0.3230E-06	1.190
6	0.3010E-06	1.300
7	0.5200E-06	1.420
8	0.4620E-06	1.580
9	0.6080E-06	1.760
10	0.6700E-06	1.970
11	0.8520E-06	2.220
12	0.1583E-05	2.540
13	0.2455E-05	2.920
14	0.3613E-05	3.390
15	0.1127E-04	5.460
16	0.1198E-04	5.740
17	0.1175E-04	6.000
18	0.1412E-04	6.310
19	0.1346E-04	6.620
20	0.1609E-04	6.980
21	0.2816E-04	7.370
22	0.2816E-04	7.370
23	0.2257E-04	7.870
24	0.2700E-04	8.250
25	0.3560E-04	8.740
26	0.6002E-04	9.640
27	0.1566E-05	2.640
28	0.2569E-05	2.770
29	0.2413E-05	2.910
30	0.3213E-05	3.010
31	0.3043E-05	3.220
32	0.3726E-05	3.450
33	0.3975E-05	3.520
34	0.4725E-05	3.770
35	0.4909E-05	3.950
36	0.6441E-05	4.120
37	0.6513E-05	4.600
38	0.9299E-05	4.830
39	0.9709E-05	5.330
40	0.1357E-04	5.560
41	0.1862E-04	5.880
42	0.3760E-04	7.290
43	0.6330E-04	7.800
44	0.2770E-01	10.250

ZK60A Extrusion, Fatigue Crack Growth in Water

Composition: 5.5Zn-0.4Zr-bal Mg
Product form: Extrusion
Heat treatment: T5
Modulus of elasticity (avg at RT): 44.8 GPa (6.5×10^6 psi)
RT tensile strength/elongation: 365 MPa (52.9 ksi)/16%
RT yield strength: 305 MPa (44.2 ksi)
Test temperature: Room temperature
Test environment: Distilled water and dry argon
Loading condition: Load control, $R = 0$
Specimen orientation: S-T
Specimen geometry: MC(T)
Frequency: 4 Hz
Source: E.U. Lee, "Corrosion Fatigue and Stress Corrosion Cracking of 7475-T7351 Aluminum Alloy," *Corrosion Cracking, Int. Conf. Expos. Fatigue, Corrosion Cracking, Fracture Mechanics, and Failure Analysis*, V.S. Goel, Ed., 1985, p 123-128

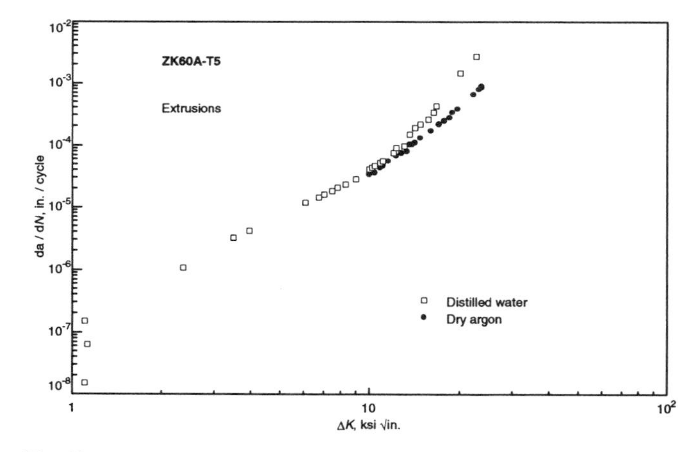

Fig. 10 da/dN data for ZK60A magnesium

Table 6 Dry argon data, $R = 0$

Sample number	da/dN, in./cycle	ΔK, ksi$\sqrt{in.}$
1	0.3365E-04	9.806
2	0.3826E-04	10.340
3	0.4240E-04	10.766
4	0.4694E-04	11.032
5	0.5336E-04	11.511
6	0.6886E-04	12.256
7	0.7445E-04	12.737
8	0.8251E-04	13.163
9	0.9861E-04	13.641
10	0.1122E-03	14.228
11	0.1278E-03	15.083
12	0.1649E-03	15.828
13	0.2133E-03	16.948
14	0.2490E-03	17.695
15	0.2763E-03	18.283
16	0.3390E-03	18.975
17	0.3962E-03	19.883
18	0.6639E-03	22.392
19	0.7555E-03	23.032
20	0.8582E-03	23.404
21	0.9046E-03	23.832

Table 7 Distilled water data, $R = 0$

Sample number	da/dN in./cycle	ΔK, ksi$\sqrt{in.}$
1	0.1579E-07	1.114
2	0.6086E-07	1.129
3	0.1423E-06	1.105
4	0.1058E-05	2.335
5	0.3121E-05	3.485
6	0.4017E-05	3.908
7	0.1163E-04	6.025
8	0.1390E-04	6.664
9	0.1579E-04	7.036
10	0.1748E-04	7.355
11	0.2037E-04	7.834
12	0.2314E-04	8.206
13	0.2767E-04	8.845
14	0.3947E-04	9.694
15	0.4264E-04	10.068
16	0.4605E-04	10.388
17	0.5101E-04	10.760
18	0.5509E-04	11.080
19	0.7287E-04	11.878
20	0.8700E-04	12.248
21	0.1015E-03	12.996
22	0.1411E-03	13.630
23	0.1865E-03	14.320
24	0.2228E-03	14.798
25	0.2667E-03	15.599
26	0.3268E-03	16.130
27	0.4320E-03	16.820
28	0.1465E-02	20.328
29	0.2767E-02	22.135

Mg-Th Alloys

Table 1 Magnesium-thorium alloys: Miscellaneous fatigue strength data

Alloy and temper	Form and thickness	Specimen	Test mode	Stress ratio	Temperature °C (°F)	Temperature Atmosphere	Fatigue strength, MPa (ksi), at cycles of: 10^4	10^5	10^6	10^7	$5x10^7$	10^8
HK31A-T6	Sand cast	Machined and polished	Krouse plate bending	−1	25 (75)	Air	...	125 (18)	75 (10.8)	60 (8.7)	...	...
	Sand cast	Machined and polished	Krouse plate bending	−1	200 (390)	Air	...	100 (14.5)	55 (8)	50 (7.25)	...	...
	Sand cast	Machined and polished	Krouse plate bending	−1	315 (600)	Air	...	60 (8.7)	45 (6.5)	35 (5)	...	...
HM31A-T5	Extrusion	Polished	Rotating beam	−1	RT	Air	...	125 (18)	105 (15)	90 (13)	...	95 (13.7)
	Extrusion	Notched, K_t=2.0	Rotating beam	−1	RT	Air		...	90 (13)	55 (8)	45 (6.5)	...
HZ32A-T5	Sand cast	Unnotched, machined and polished	Rotating beam	−1	RT	Air	...	105 (15.2)	90 (13)	75 (10.8)	70 (10)	70 (10)
	Sand cast	Notched, K_t=2	Rotating beam	−1	RT	Air	...	...	60 (8.7)	...	60 (8.7)	...
	Sand cast	Notched, K_t=5	Rotating beam	−1	RT	Air	...	...	40 (5.8)	...	...	30 (4.35)

Sources: Dow Chemical and Magnesium Elektron

HK31A (UNS M13310)

Composition 3.2Th-0.7Zr-bal Mg
Product form Sheet
Heat treatment H24
Modulus of elasticity (avg at RT) 44.8 GPa (6.5×10^6 psi)
RT tensile strength/elongation 255 MPa (36.9 ksi)/10%
RT yield strength 205 MPa (29.7 ksi)
Test temperature Room temperature, 149 °C (300 °F), 260 °C (500 °F)
Test environment Air
Failure criterion Fracture
Loading condition Axial loading (R = 0.33, and 0.01)
Specimen geometry Notched, K_t = 3., 5
Surface Smooth
Source A.A. Blatcherwich and A.E. Cers, "Fatigue, Creep, and Stress Rupture of Several Superalloys," AFML Technical Report 69-12, Jan 1969

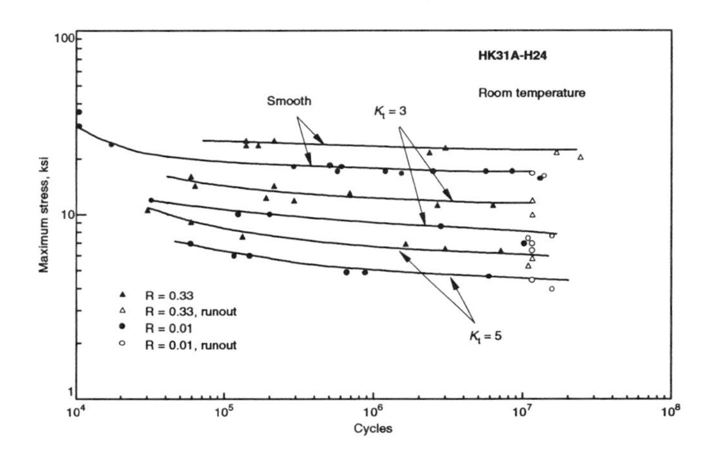

Fig. 1 Room temperature and axial fatigue of HK31A

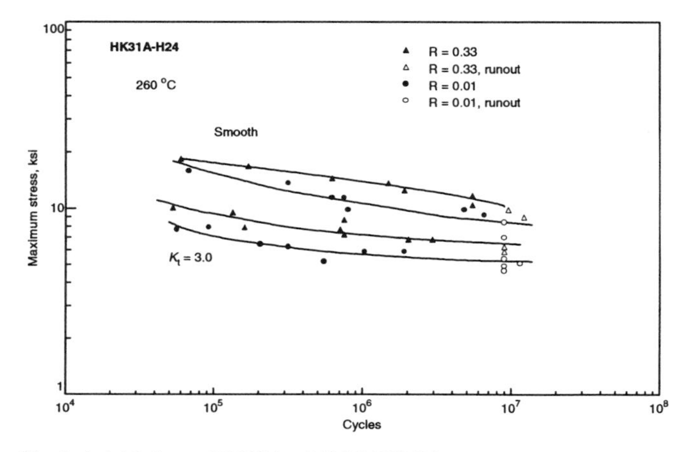

Fig. 2 Axial fatigue of HK31A at 260 °C (500 °F)

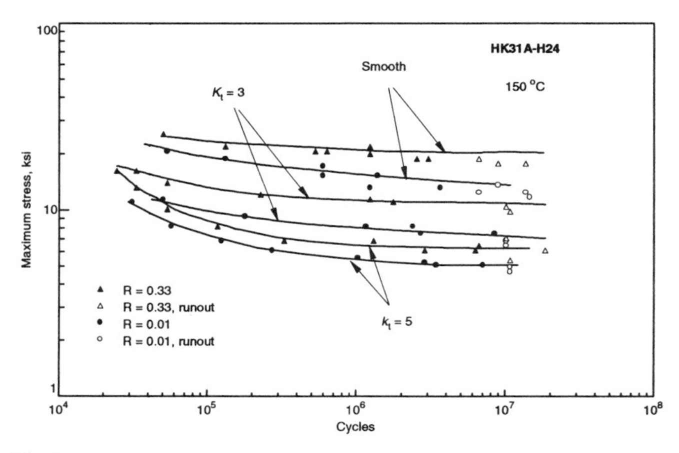

Fig. 3 Axial fatigue of HK31A at 150 °C (300 °F)

HZ32A (UNS M13320)

Composition 3.2Th-2.1Zn-0.7Zr-bal Mg
Product form Sand-cast test bars
Heat treatment T5
Modulus of elasticity (avg at RT) 44.8 GPa (6.5×10^6 psi)
RT tensile strength/elongation 205 MPa (29.7 ksi)/7%
RT yield strength 95 MPa (13.7 ksi)
Test temperature 150 to 370 °C (300 to 700 °F)
Test environment Air
Failure criterion Fracture
Loading condition Reverse bending ($R = -1$)
Surface Smooth
Source C.J.P. Ball, et al., Further Progress in the Development of Magnesium-Zirconium Alloys to Give Good Creep and Fatigue Properties between 500 and 650 °F, *Trans. AIME*, Vol 197, 1953, p 924

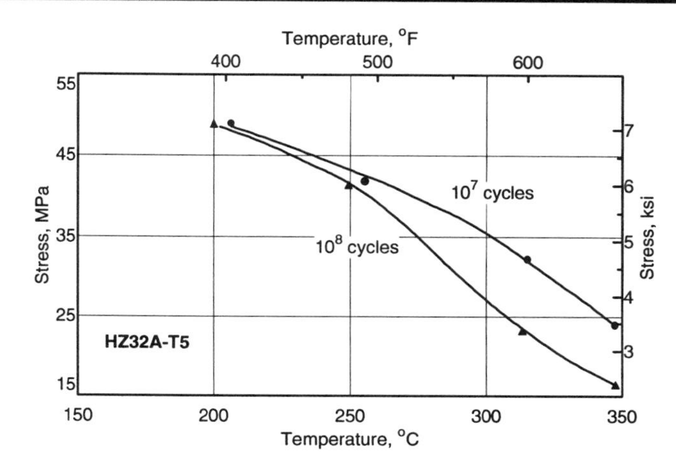

Fig. 4 The effect of temperature on the fatigue strength of HZ32A

HM21A (UNS M13210)

Composition 0.8Mn-2.0Th-bal Mg
Product form Forging—wheel rim
Heat treatment T5
Modulus of elasticity (avg at RT) 44.8 GPa (6.5×10^6 psi)
RT tensile strength/elongation 235 MPa (34 ksi)/9%
RT yield strength 150 MPa (21.7 ksi)
Test temperature Room temperature, 200 °C (390 °F), 315 °C (600 °F)
Test environment Air
Failure criterion Fracture
Loading condition Rotating beam, axial ($R = 0.25$)
Specimen orientation Tangential
Specimen geometry Wheel rim, some notched $K_t = 2.0$
Surface Polished
Source "Magnesium Forging Alloys for Elevated Temperature Service," TS&D Letter Enclosure, Dow Chemical Co., 24 April 1982

Table 2 Fatigue properties of HM21A-T5 forgings

Forging	Orientation	Fatigue test	Surface	Stress, MPa, at cycles: 10^5	10^6	10^7	10^8
Wheel rim	Tangential	Rotating beam	Polished	110	85	60	60
			Notched	70	55	40	35
		Axial	Polished	145	115	95	...
		(tested at 200 °C)		115	110	105	...
		(tested at 315 °C)		95	85	70	...

HM21A, Air and Vacuum Fatigue

Composition 2.0% Th-0.57% Mg- bal Mg
Heat treatment As-received
Modulus of elasticity (avg at RT) 44.8 GPa (6.5×10^6 psi)
RT tensile strength/elongation 214 MPa (31 ksi)
Test temperature Room temperature
Test environment Air and vacuum
Failure criterion Fracture
Loading condition Fully reversed ($R = -1$) plane bending
Specimen geometry Flat cantilever, ¼ in. thick
Surface Polished
Frequency 30 Hz
Test specifications Some specimens underwent outgassing
Source H.T. Sumsion, Vacuum Effects on Fatigue Properties of Magnesium and Two Magnesium Alloys, *J. Spacecraft*, Vol 5 (No. 6), 1988, p 700-704

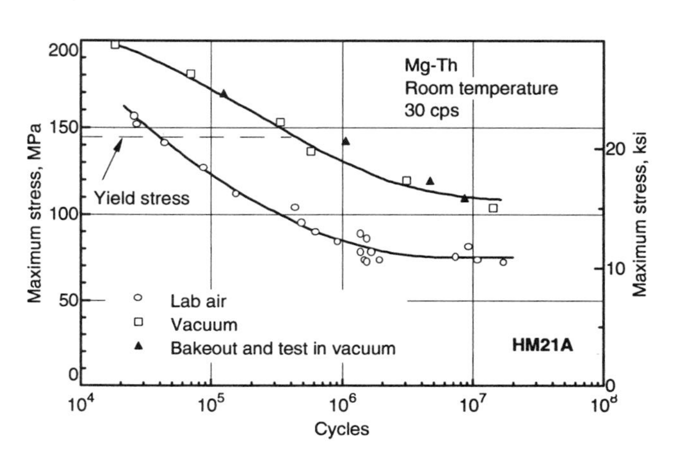

Fig. 5 Fatigue of HM21A in air and vacuum

HM21A Sheet

Composition 0.8Mn-2.0Th-bal Mg
Product form Sheet
Heat treatment T8
Modulus of elasticity (avg at RT) 44.8 GPa (6.5 × 10^6 psi)
RT tensile strength/elongation 255 MPa (36.9 ksi)/11%
RT yield strength 186 MPa (26.8 ksi)
Test temperature Room, 204 °C (400 °F), 315 °C (600 °F)
Test environment Air
Failure criterion Fracture
Loading condition Axial load ($R = 0.25$)
Specimen geometry 1.63 mm sheet
Source "HM21A-T8 Magnesium Alloy Sheet and Plate," Dow Chemical Co., Revised: 24 April 1964

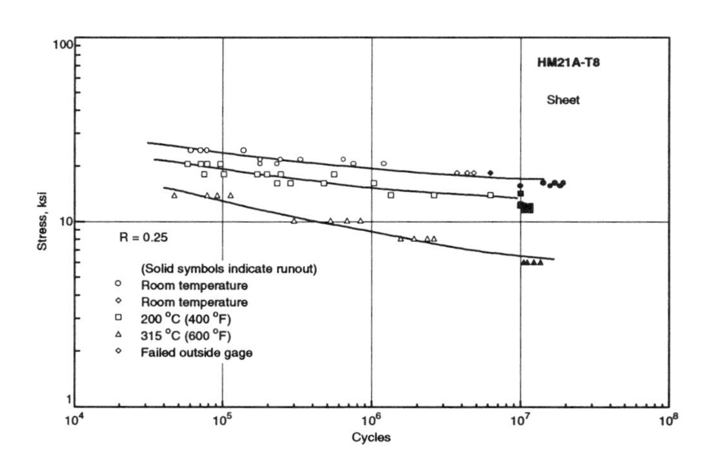

Fig. 6 Fatigue of HM21A sheet

HM31A (UNS M13312)

Composition 0.8Mn-3.0Th-bal Mg
Product form Extrusion
Heat treatment F
Modulus of elasticity (avg at RT) 44.8 GPa (6.5 × 10^6 psi)
Test temperature Room
Test environment Air
Failure criterion Fracture
Loading condition Rotating beam ($R = -1$)
Specimen orientation L, T, LT
Specimen geometry Some notched $K_t = 2$
Surface Polished
Source "Magnesium in Design," Form No. 141-213-67, Dow Chemical Co., 1967

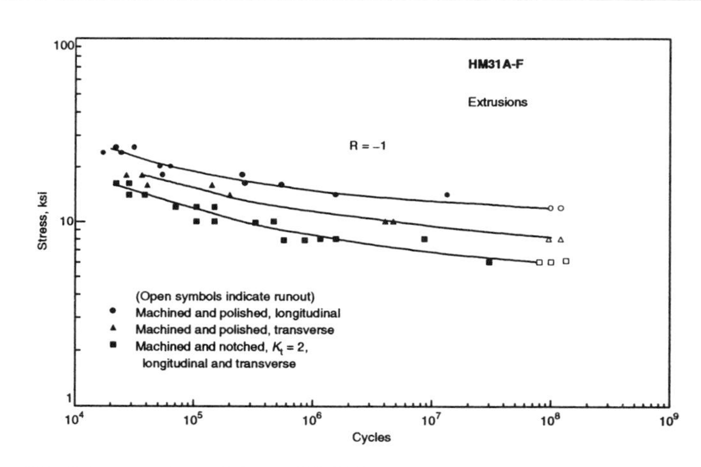

Fig. 7 Rotating beam fatigue of HM31A

Miscellaneous Mg Alloys

Magnesium-Silver Alloys: Fatigue Strength at Room Temperature

Table 1 Magnesium-silver alloys: fatigue strength at room temperature

Alloy and temper	Form and thickness	Specimen	Test mode	Stress ratio	Typical UTS MPa (ksi)	Atmosphere	Fatigue strength, MPa (ksi), at cycles of:					
							10^4	10^5	10^6	10^7	5×10^7	10^8
EQ21A-T6	Sand casting	Unnotched, machined and polished	Rotating beam	−1	260 (37.8)	Air	...	113 (16)	100 (14.5)	98 (14)	96 (13.9)	...
	Sand casting	Notched, $K_t = 2.0$	Rotating beam	−1	...	Air	...	...	58 (8.4)	...	...	54 (7.8)
QH21A-T6	Sand cast	Unnotched, machined and polished	Rotating beam	−1	240 (34.8)	Air	...	135 (19.5)	110 (16)	105 (15.2)	105 (15.2)	...
	Sand cast	Notched, $K_t = 2.0$	Rotating beam	−1	...	Air	...	...	65 (9.4)	...	65 (9.4)	...
QE22A-T6	Sand cast	Unnotched, machined and polished	Rotating beam	−1	275 (40)	Air	...	150 (21.7)	125 (18)	115 (16.6)	105 (15.2)	105 (15.2)
	Sand cast	Notched, $K_t = 2.0$	Rotating beam	−1	...	Air	...	...	85 (12.3)	...	65 (9.4)	60 (8.7)

Sources: "Electron EQ21A: Another Casting Alloy Developed by Magnesium Elektron Ltd.," Bulletin 464, Magnesium Elektron, Ltd., July 1984; "Mechanical Properties and Chemical Compositions of Cast Magnesium Alloys," Bulletin 440, Magnesium Elektron, Ltd., March 1981; "Tables of Fatigue Strength of Sand Cast Magnesium Alloys," TS&D Letter Enclosure, Dow Chemical Co., 19 Nov 1965

QH21 (UNS 18210)

Composition 2.5%Ag-1.0%Th-1.0%Nd-0.6%Zr
Product form Cast
Heat treatment T6
Modulus of elasticity (avg at RT) 44 GPa (6.3×10^6 psi)
RT tensile strength/elongation 276 MPa (40.0 ksi)/4%
RT yield strength 210 MPa (30.5 ksi)
Test temperature Room temperature
Test environment Air
Failure criterion Fracture
Loading condition Rotating bending, $R = -1$
Specimen geometry Unnotched and U-notched, $K_t = 2.0$
Frequency 2960 cpm
Source W. Unsworth, J.F. King, and S.L. Brashaw, "QH21—A High Performance Magnesium Casting Alloy for Aerospace Applications"

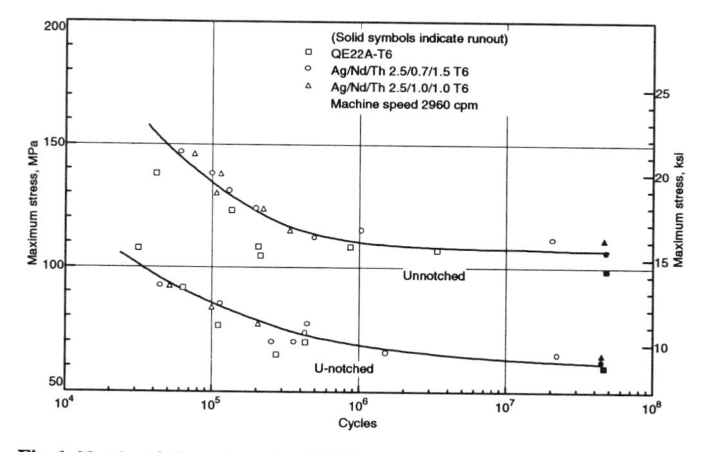

Fig. 1 Notched fatigue strength of QH21

QE22A, High-Temperature S-N Data

Composition 2.2R.E.-2.5Ag-0.7Zr
Product form Sand cast
Heat treatment T6
Modulus of elasticity (avg at RT) 44.8 GPa (6.5×10^6 psi)
RT tensile strength/elongation 275 MPa (39.8 ksi)/4%
RT yield strength 205 MPa (29.7 ksi)
Test temperature Room temperature, 20 °C (68 °F), 200 °C (392 °F), 250 °C (482 °F)
Test environment Air
Failure criterion Fracture
Loading condition Rotating beam, $R = -1$
Source M. Marrien, Magnesium Casting Alloys for Aircraft Structures, *Mod. Cast.*, Vol 51, March 1967, p 60-62; *Aerospace Structural Metals Handbook*, Battelle Metals and Ceramics Information Center, Columbus, OH, 1991, code 3406, p 6

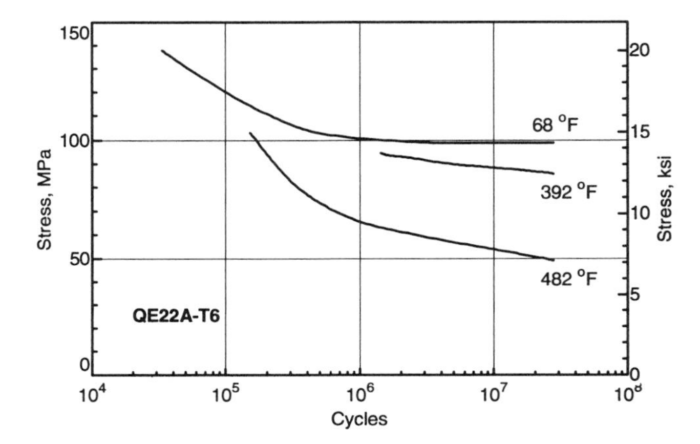

Fig. 2 Fatigue strength of QE22A at high temperature

QE22A Fatigue Crack Growth

Composition 2.2R.E.-2.5Ag-0.7Zr
Product form Plate
Heat treatment T6
Modulus of elasticity (avg at RT) 44.8 GPa (6.5×10^6 psi)
RT tensile strength/elongation 275 MPa (39.8 ksi)/4%
RT yield strength 205 MPa (29.7 ksi)
Test temperature Room temperature
Test environment Air
Loading condition Load control, $R = 0$

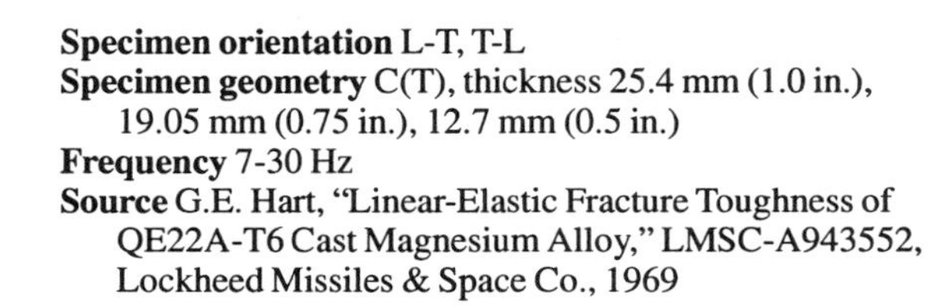

Specimen orientation L-T, T-L
Specimen geometry C(T), thickness 25.4 mm (1.0 in.), 19.05 mm (0.75 in.), 12.7 mm (0.5 in.)
Frequency 7-30 Hz
Source G.E. Hart, "Linear-Elastic Fracture Toughness of QE22A-T6 Cast Magnesium Alloy," LMSC-A943552, Lockheed Missiles & Space Co., 1969

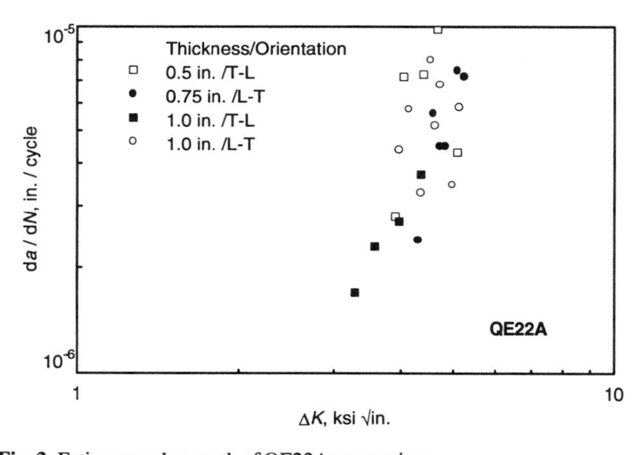

Fig. 3 Fatigue crack growth of QE22A magnesium

Table 2 QE22A crack growth data ($R = 0$)

Sample number	da/dN, in./cycle	ΔK, ksi$\sqrt{in.}$
Thickness: 0.5 in.; orientation: *T-L*		
1	0.4330E-05	5.100
2	0.2800E-05	3.880
3	0.7200E-05	4.060
4	0.7300E-05	4.400
5	0.1000E-04	4.650
Thickness: 0.75 in.; orientation: *L-T*		
1	0.7500E-05	5.070
2	0.7250E-05	5.250
3	0.4500E-05	4.760
4	0.4500E-05	4.880
5	0.2400E-05	4.360
6	0.5600E-05	4.600
Thickness: 1.0 in.; orientation: *T-L*		
1	0.3700E-05	4.370
2	0.2700E-05	3.940
3	0.2300E-05	3.600
4	0.1710E-05	3.300
Thickness: 0.5 in.; orientation: *L-T*		
1	0.5250E-05	4.600
2	0.5800E-05	4.150
3	0.4380E-05	3.940
4	0.8170E-05	4.500
5	0.6900E-05	4.760
6	0.5900E-05	5.120
7	0.3500E-05	5.000
8	0.3300E-05	4.380

Magnesium-Rare Earth Casting Alloy EZ33A (UNS M12330)

Composition 2.8R.E.-2.6Zn-0.7Zr-bal Mg
Product form Sand cast
Heat treatment T5
Modulus of elasticity (avg at RT) 44.8 GPa (6.5×10^6 psi)
RT tensile strength/elongation 160 MPa (23.2 ksi)/3%
RT yield strength 105 MPa (15.2 ksi)
Test temperature Room temperature
Test environment Air
Failure criterion Fracture
Loading condition Rotating beam, $R = -1$
Specimen geometry Some notched, $K_t = 2$ and 5
Surface Smooth, machined, and polished
Source "Tables of Fatigue Strength of Sand Cast Magnesium Alloys," TS&D Letter Enclosure, Dow Chemical Co., 19 Nov 1965

Table 3 Fatigue strength of EZ33A

Specimen	Fatigue strength, MPa (ksi), at: 10^5	10^6	10^7	5×10^7	10^8
Unnotched	100 (14.5)	85 (12.3)	75 (10.8)	70 (10)	70 (10)
Notched, $K_t = 2$	...	50 (7.25)	...	50 (7.25)	45 (6.5)
Notched, $K_t = 5$	...	40 (5.8)	...	...	30 (4.35)

EZ33A, High-Temperature Fatigue

Composition 2.25R.E.-2.25Zn-0.6Zr-bal Mg
Product form Cast
Modulus of elasticity (avg at RT) 44.8 GPa (6.5×10^6 psi)
RT tensile strength/elongation 152 MPa (22 ksi)/3%
RT yield strength 103 MPa (15 ksi)
Test temperature Room temperature, 204 °C (400 °F), and 260 °C (500 °F)
Test environment Air
Failure criterion Fracture
Loading condition Rotating bending, $R = -1$
Test specifications Temperature comparisons
Source W. Unsworth, Meeting the High Temperature Aerospace Challenge, *Light Metal Age*, Vol 44, 1986

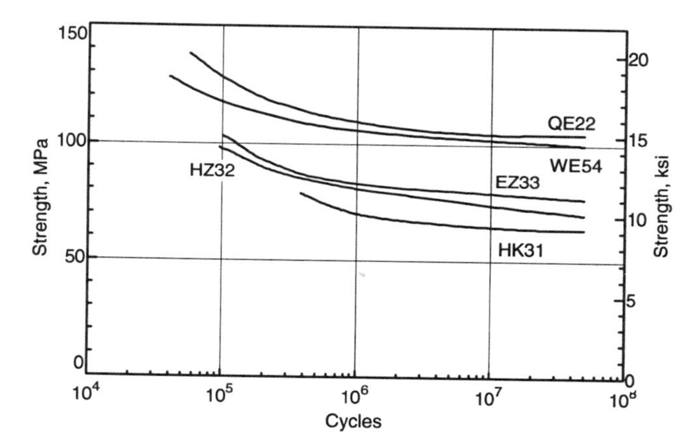

Fig. 4 Fatigue strength of EZ33A at room temperature

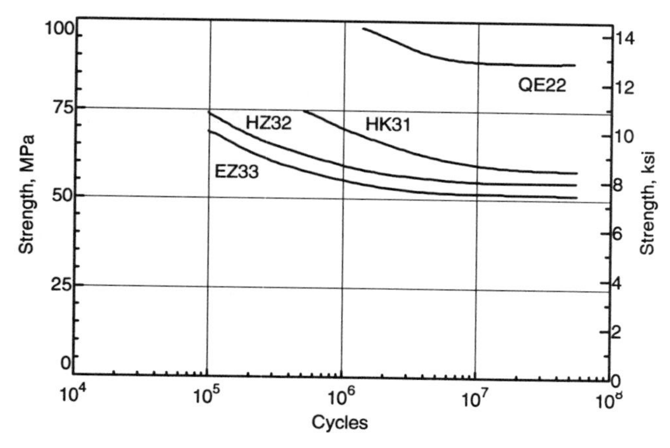

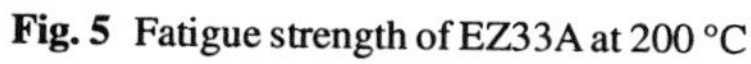
Fig. 5 Fatigue strength of EZ33A at 200 °C

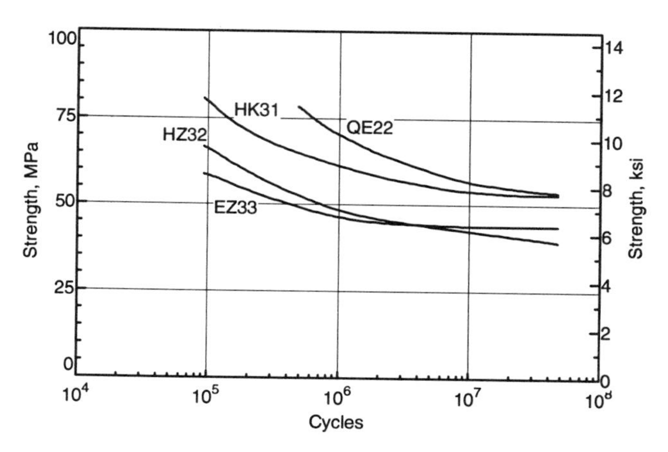

Fig. 6 Fatigue strength of EZ33A at 260 °C

EZ33A Fatigue Crack Growth

Composition 2.7Zn-0.5Zr-0.1Cr-0.01Ni-0.3R.E.-bal Mg
Product form Cast
Heat treatment T5, aged 5 h at 216 °C
Modulus of elasticity (avg at RT) 44.8 GPa (6.5×10^6 psi)
RT yield strength 110 MPa (15.9 ksi)
Test temperature Room temperature
Test environment Laboratory air, humidity, 38%
Loading condition Load control, $R = 0.1$, 0.8, and −1.0
Specimen geometry Side notched, 152 mm × 50.8 mm × 6.4 mm
Frequency and waveform 40-50 Hz, sinusoidal, secant method
Source P.K. Liaw, T.L. Ho, and J.K. Donald, Near-Threshold Fatigue Crack Growth Behavior in a Magnesium Alloy, *Scr. Metall.*, Vol 18, 1984, p 821-824

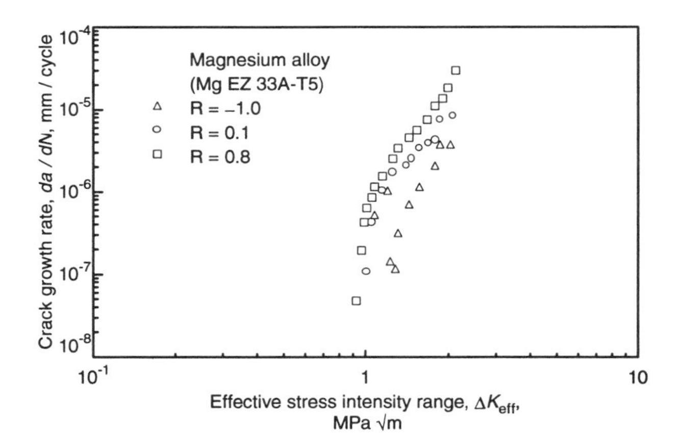

Fig. 7 Fatigue crack growth of EZ33A-T5 magnesium

Magnesium-Lithium Wrought Alloy LA 141A (UNS M14141)

Composition 14% Li-1.0% Al-bal Mg
Heat treatment As received
Modulus of elasticity (avg at RT) 44.8 GPa (6.5×10^6 psi)
RT tensile strength/elongation 131 MPa (19.0 ksi)
Test temperature Room temperature
Test environment Air and vacuum
Failure criterion Fracture
Loading condition Fully reversed ($R = -1$) plane bending
Specimen geometry Flat cantilever, 6 mm (0.25 in.) thick
Surface Polished
Frequency 30 cps
Test specifications Some specimens underwent outgassing
Source H.T. Sumsion, Vacuum Effects on Fatigue Properties of Magnesium and Two Magnesium Alloys, *J. Spacecraft*, Vol 5 (No. 6), 1968, p 700-704

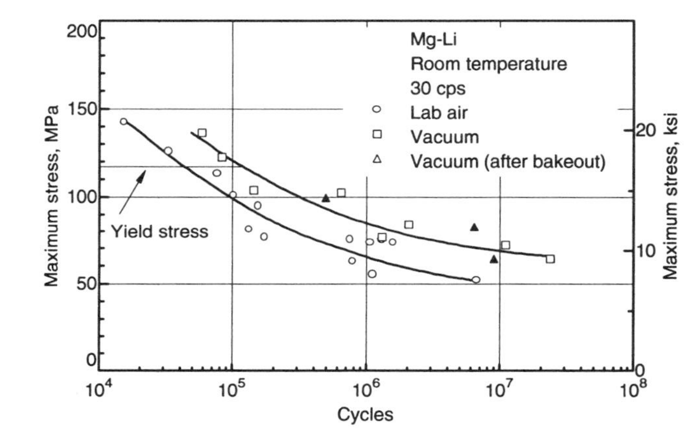

Fig. 8 Fatigue of LA141A in air and vacuum

LA141A Sheet

Composition 14.0Li-1.0Al-bal Mg
Product form Sheet
Modulus of elasticity (avg at RT) 44.8 GPa (6.5×10^6 psi)
RT tensile strength/elongation 130 MPa (19 ksi)/5%
RT yield strength 96.5 MPa (14 ksi)
Test temperature Room temperature
Test environment Air
Failure criterion Fracture
Loading condition Reversed bending ($R = -1$)
Specimen geometry 1.58 mm thick
Surface Degreased, epoxy coated, and urethane coated
Frequency 2000 cpm
Source "Magnesium-Lithium Products," Brooks and Perkins, Inc.

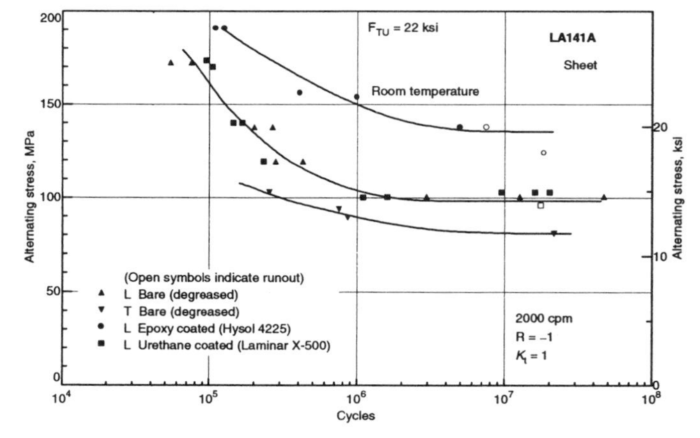

Fig. 9 Fatigue of LA141A with various coatings

M1A (UNS M15100)

Composition 1.2Mn-bal Mg
Product form Extrusion
Heat treatment F
Modulus of elasticity (avg at RT) 44.8 GPa (6.5×10^6 psi)
Test temperature Room temperature
Test environment Air
Failure criterion Fracture
Loading condition Rotating beam ($R = -1$)
Specimen geometry Some notched, $K_t = 1.8$
Surface Polished, S.C.F. = 1.8
Test specifications $R = -1$
Source "Magnesium in Design," Bulletin 141-213, Dow Chemical Co., 1967

Table 4 Rotating beam ($R = -1$) fatigue strength of M1A-F extrusions

Alloy	Type of test	*R* value	Surface or stress concentration	Fatigue limit, MPa, at cycles:			
				10^5	10^6	10^7	5×10^7
M1A-F	Rotating beam	−1	Polished	107	88	85	83
			1.8	76	54	50	48

M1A Fatigue Crack Growth

Composition Zn-0.01 max, Al-0.01 max, Mn-1.62, Cu-0.02, Si-0.006, Fe-0.007, Ni-0.003 max, Mg rem
Product form Sheet
Modulus of elasticity (avg at RT) 43 GPa (6.2×10^6 psi)
RT tensile strength/elongation 200 MPa (29 ksi)/6%
RT yield strength 107 MPa (15.5 ksi)
Test temperature Room temperature
Test environment Laboratory air
Loading condition Load control, $R = 0$, 0.5, 0.67, and 0.78
Specimen orientation L-T
Specimen geometry M(T) 0.1 in. (2.54 mm) thick, 10 in. (254 mm) wide
Frequency 3.5-10 Hz
Source G. Pook, Fatigue Crack Growth Characteristics of Two Magnesium Alloys, *Eng. Fract. Mech.*, Vol 5 (No. 4), 1973, p 935-936

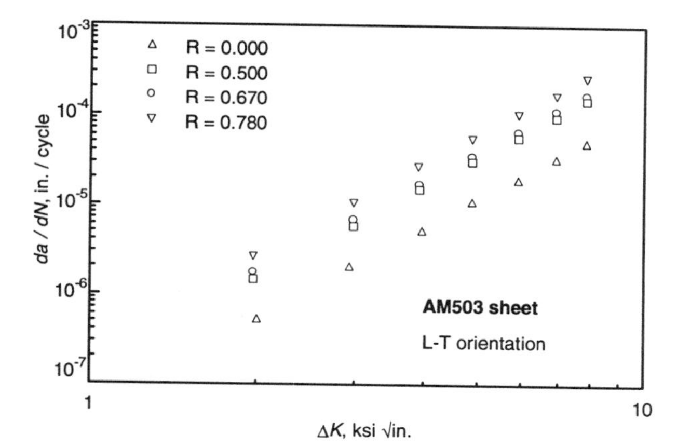

Fig. 10 *da*/*d*N data for AM503 magnesium

Table 5 M1A (AM503) crack growth data

Sample number	*da*/*d*N, in./cycle	ΔK, ksi$\sqrt{in.}$
***R* = 0**		
1	0.4940E-06	2.000
2	0.1920E-05	3.000
3	0.5030E-05	4.000
4	0.1060E-04	5.000
5	0.1960E-04	6.000
6	0.3280E-04	7.000
7	0.5130E-04	8.000
***R* = 0.5**		
1	0.1430E-05	2.000
2	0.5550E-05	3.000
3	0.1460E-04	4.000
4	0.3070E-04	5.000
5	0.5660E-04	6.000
6	0.9490E-04	7.000
7	0.1480E-03	8.000
***R* = 0.67**		
1	0.1660E-05	2.000
2	0.6460E-05	3.000
3	0.1690E-04	4.000
4	0.3580E-04	5.000
5	0.6590E-04	6.000
6	0.1100E-03	7.000
7	0.1730E-03	8.000
***R* = 0.78**		
1	0.2560E-05	2.000
2	0.9950E-05	3.000
3	0.2610E-04	4.000
4	0.5510E-04	5.000
5	0.1020E-03	6.000
6	0.1700E-03	7.000
7	0.2660E-03	8.000

GA3Z1 Strain Life

Composition 3Al, 1Zn, bal Mg
Heat treatment Annealed for various grain sizes (see table)
Condition Hexagonal crystal structure with some deformation by twinning
Hardness 50 HB
Modulus of elasticity (avg at RT) 43 GPa (6.2×10^6 psi)
RT tensile strength/elongation 240 to 290 MPa (35 to 42 ksi)
RT yield strength 180 to 220 MPa (26 to 32 ksi)

Test environment Room-temperature air
Failure criterion Fracture
Loading condition Axial and torsional strain control
Specimen geometry Hollow specimens
Gauge length and thickness 25 mm long, 1.5 mm thick
Frequency 0.03 Hz, triangular
Source Cyclic Plastic Deformation by Twinning of a Magnesium Alloy GA3Z1, *Advanced Al and Mg Alloys*, ASM International, p 837-846

Table 6 GA3Z1: Plastic and elastic fatigue parameters

Heat treatment	Grain size	Elastic parameters		Plastic parameters		Regime of plastic parameters
		σ_f/E	b	ε_f	c	
Annealed at 320 °C for 2 h	3 μm	0.0314	−0.262	76.99	−1.56	Before breaking
				0.094	−0.58	After breaking
Annealed at 540 °C for 2 h	30 μm	0.016	−0.182	1.084	−0.81	Before breaking
				0.169	−0.61	After breaking
Recrystallization annealed at 540 °C for 24 h	80 μm	0.0238	−0.245	27.93	−1.451	No breaking

E = 43 GPa. Composition (wt%): 2.5-3.5 Al, 0.5-1.5 Zn, 0.2 Mn (min), 0.1 Si, 0.05 Cu, 0.005 Fe, and 0.005 Ni (max)

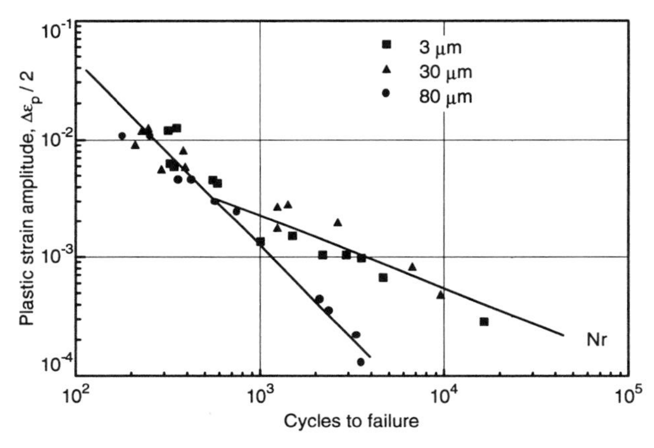

Fig. 11 Plastic strain life of GA3Z1. For testing under tension-compression, the presence of twinning was noticed for any deformation rate for the grain sizes equal to 30 and 80 μm. Twinning is the cause of a modification of the slope of the plastic line. The plastic deformation for the larger grain sizes presents a common slope for deformation zones where twinning is noticeable. The second line corresponds to the 3 μm specimen, for which twinning practically does not exist.

Russian Alloys: Fatigue Crack Growth

Table 7 Chemical composition of some wrought Russian magnesium alloys, wt%

Alloy designation	Al	Zn	Zr	Y	Cd	Mn	Ca	Ce	La	Nd	Li
VMD-10	<0.04	0.75-0.81	0.49-0.50	7.1-7.9	0.59-0.63						
ML8	...	5.5-6.6	0.7-1.1		0.2-0.8	...	...	...	...	...	...
IMV6	0.12	...	...	7.8	...	0.55	0.49	0.11	...		...
MA2-1	4.17	0.85	...	...	...	0.5	...	...	...	...	...
MA15	...	3.15	...	...	...	...	1.88	...	0.83	...	...
MA12	...	0.44	...	...	...	...	...	...	...	2.9	...
MA21	5.4	1.0	...	...	...	...	4.7	...	...		8.6
MA18	...	...	...	...	...	...	...	...	...	...	...

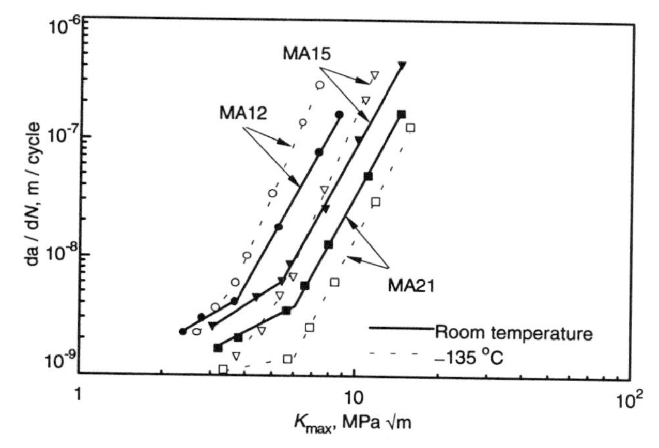

Fig. 12 Fatigue crack growth rate curves for magnesium alloys at room temperature and –135 °C. Source: V.A. Serdyuk, *Probl. Prochn.*, Vol 11, 1980, p 18-23

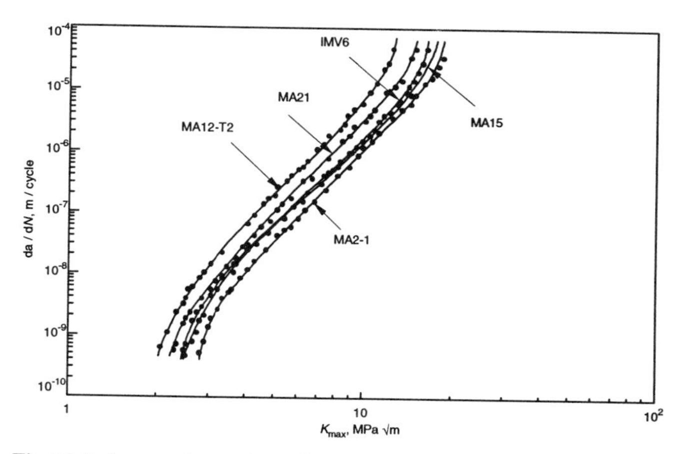

Fig. 13 Fatigue crack growth rate for magnesium alloys. Source: N.M. Grinberg, V.A. Serdyuk, T.I. Malinkina, and A.S. Kamishkov, *Probl. Prochn.*, Vol 1, 1982, p 61–67

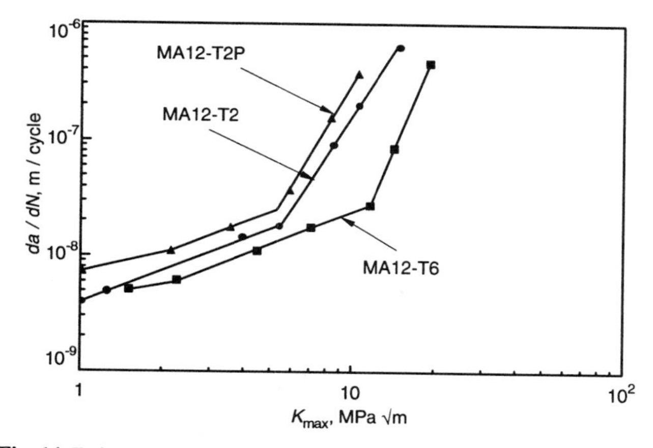

Fig. 14 Fatigue crack growth rate data for MA12 in different structural states. Source: N.M. Grinberg and V.A. Serdyuk, *Probl. Prochn.*, Vol 10, 1978, p 16–52

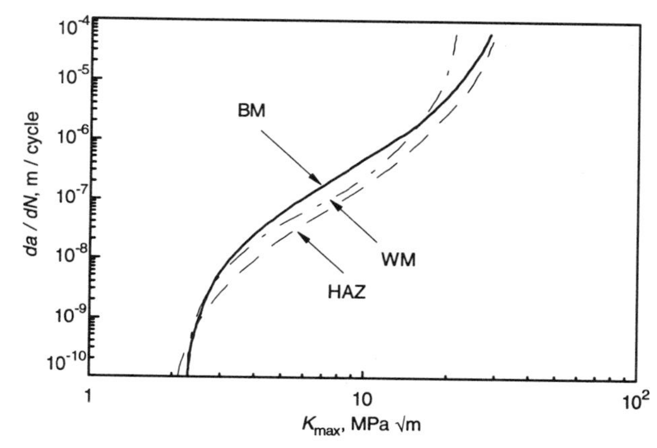

Fig. 15 A comparison of fatigue crack growth rate curves for different parts of an MA15 weld joint. BM, base metal; WM, weld metal; HAZ, heat-affected zone. Source: S.Y. Yarema, O.D. Zinyuk, and T.I. Malinkina, *Fiz.-Khim. Mekh. Mater.*, Vol 17 (No. 5), 1981, p 72–76

Titanium Alloy Fatigue Data

Titanium Alloys Fatigue and Fracture

Adapted from the article "Fracture Properties of Titanium Alloys" in *Application of Fracture Mechanics for Selection of Structural Materials* (ASM, 1982) with revision by H. Margolin, Polytechnic University

Titanium is used for two primary reasons: (a) structural efficiency, which derives from its combination of high strength and low density; and (b) resistance to corrosion by chlorides and oxidizing media, which derives from its strong passivation tendencies. Titanium, like most structural materials, is supplied in all mill product forms, and several titanium alloys are available to meet specific needs. Most important among titanium alloys is Ti-6Al-4V. Offering a strength-to-density ratio of 25×10^6 mm (1×10^6 in.), Ti-6Al-4V has found application for a wide variety of aerospace hardware. Jet aircraft manufacturers are the principal consumers of titanium in this market. The Ti-6Al-4V alloy is often specified for critical parts, the failure of which could result in loss of an entire system. In these situations, the higher acquisition cost of titanium can be more than offset by its reduced costs of ownership. Because of its great popularity, Ti-6Al-4V has become the best understood of all titanium alloys, and much of the property data obtained on this alloy have been stored in computer data banks and is available for statistical analysis.

Several metallurgical and environmental variables have been identified that influence the fracture behavior of titanium alloys in general and of alloy Ti-6Al-4V in particular; the effects of these variables will be discussed in detail in this chapter. It is beyond the scope of this chapter, however, to provide design-type data to the user requiring the ultimate in performance. That type of information must either be generated by the user or be obtained from other sources.

The references and the Appendix at the end of this chapter contain additional mechanical property data. The purpose of this chapter is to provide the reader with general guidelines that indicate what variables have what effects on toughness and fatigue crack propagation. Among the metallurgical variables of importance are composition, microstructure (as it depends on processing and heat treatment), and crystallographic texture. Environmental factors are discussed also. Due to its importance, the metallurgy of alloy Ti-6Al-4V is treated first, followed by a brief discussion indicating how certain other titanium alloys differ metallurgically from Ti-6Al-4V. The remainder of this chapter reviews, in highlight fashion, what is known about fracture toughness and fatigue crack propagation, including the effects of testing environments, for the most common titanium alloys.

Summary. Titanium and titanium alloys have a well-earned reputation for reliability in service. In no small measure, this is a result of the double and triple vacuum arc melting procedures employed throughout the industry in producing the alloys. That reputation is protected also by the excellent corrosion resistance exhibited by titanium. Titanium does not corrode in salt water. Crack initiation in titanium is almost always mechanically induced; only under very special circumstances will cracks initiate due to a combination of an environment and static stress.

Because of the many possible effects of chemistry, microstructure, texture, environment, and loading, it is not possible to quantify the crack growth behavior of titanium alloys unless these factors are closely controlled. Alloys within a given class, such as alpha-beta alloys, show parallel trends in their fracture toughness and crack propagation behaviors. To the extent that they have been studied, the trends for interstitial effects are similar for all alloys, the higher levels of interstitials leading to faster fatigue crack propagation (FCP) and lower K_{Ic}. A similar trend is observed for variations in microstructure. Those microstructures (Widmanstätten or recrystallization annealed) that give the highest K_{Ic} values generally yield the lowest crack growth rates whether under fatigue or sustained loads. Moreover, the environmental media studied tend to exhibit similar rank orders of severity among K_{Ic}, FCP, and sustained load crack propagation. Salt water appears to be the most severe of the media studied. Readers interested in additional quantitative comparisons may consult the original references from which this chapter was drawn or may perform their own tests. Finally, those readers who are interested in the available design information or the underlying metallurgy may consult the general references list at the end of this article. The references that are cited refer primarily to alloys for which standard specifications exist.

Metallurgy of Titanium Alloys

Titanium exists in two crystalline states. In pure titanium, the low-temperature α phase is stable at temperatures below about 883 °C (1621 °F) and crystallizes in the hexagonal close packed structure with a c/a ratio of 1.58, which is slightly less than the ideal ratio for packing of rigid spheres. The high-temperature β form is a body centered cubic phase that is stable from about 883 °C (1621 °F) to the melting point.

The transformation temperature and phase compositions of titanium can be altered by alloying additions. Elements that increase the transformation temperature are known as alpha stabilizers, and those that decrease it are called beta stabilizers. Other, sparingly soluble elements, when present in excess of their solubility limits, may form compounds or second phases of essentially pure solute. Of the elements commonly present in Ti-6Al-4V, carbon is a compound former; vanadium, iron and hydrogen are beta stabilizers; and aluminum, oxygen, and nitrogen are alpha stabilizers.

Metallurgy of Ti-6Al-4V

Standard grade Ti-6Al-4V becomes 100% beta phase at temperatures above about 1000 °C (1832 °F). Below this temperature, alpha and beta phases coexist. Ti-6Al-4V is thus a two-phase alloy with the beta phase present even at cryogenic temperatures.* This comes about because 4 wt% vanadium exceeds the alpha solubility limit. When additional phases occur, it is usually because the alloy has been contaminated with an impurity (such as boron) or because an element such as yttrium has been added for grain refining purposes. To avoid problems caused by impurities, maximum impurity levels are limited by specifications that cover composition limits. At room temperature, Ti-6Al-4V is about 90 vol% alpha phase. Thus, the alpha phase dominates the physical, chemical, and mechanical properties of this alloy.

Alloy Ti-6Al-4V may be obtained in two basic ranges of composition: the standard grade and the "extra low interstitial" (ELI) grade. In the ELI grade, oxygen is held to less than 0.13 wt%, whereas the maximum oxygen content of the standard grade is commonly 0.20 wt%. The two grades have the following typical composition ranges:

*See later discussion of beta alloys.

Element	Composition, wt %	
	Standard	ELI
Aluminum	5.75-6.75	5.50-6.50
Vanadium	3.5-4.5	3.5-4.5
Iron	0.25 max	0.25 max
Oxygen	0.20 max	0.13 max
Nitrogen	0.05 max	0.05 max
Hydrogen	0.015 max	0.015 max
Carbon	0.08 max	0.08 max

Oxygen, nitrogen, hydrogen, and carbon are the interstitial elements. Except for carbon, they are all readily soluble in titanium. In general, they increase strength and decrease ductility, and in this sense have effects quite similar to those of metallic alloying additions. Carbon has limited solubility and is a strong compound former, but carbon levels are so low in commercial products that carbides are virtually nonexistent. Hydrogen, aside from being a beta stabilizer, has other unique features. It is soluble as well as highly mobile in titanium. Hydrogen can, therefore, be picked up during processing operations such as forging, heat treating, and pickling. By the same token, hydrogen can be removed from titanium by vacuum annealing operations at temperatures on the order of 700 to 900 °C (1292 to 1652 °F). In vacuum annealing operations, both the metal and the furnace surfaces must be clean to ensure effective outgassing. Depending in part on the amount of beta phase present, hydrogen at sufficiently high levels is manifested by hydride precipitation and embrittlement. Embrittlement may occur as a delayed reaction. Residual stress gradients lead to hydrogen gradients which may localize the hydrides (Ref 1 and 2). Aluminum tends to increase the apparent solubility of hydrogen in alpha titanium (Ref 3).

However, measurements of hydrogen activity (Ref 4 and 5) have shown that Al activity reduces the solubility of hydrogen in alpha (see discussion in section on beta alloys). The findings of Ref 3 have been explained on the basis that Al additions increase the strength of alpha, making it more difficult for hydrides to form because of the increased work required to accommodate the expansion associated with hydride formations. There is recent evidence that hydrides contain Al (Ref 6 and 7).

Hydrogen also is highly soluble in beta titanium and in the beta phase of alpha + beta alloys. For these reasons, welding of unalloyed titanium to alloys such as Ti-6Al-4V is not recommended lest the hydrogen normally present in Ti-6Al-4V migrates to the unalloyed titanium and causes embrittlement (Ref 1). Hydrogen in titanium is readily controlled, and specifications limit the maximum allowable content.

Microstructure and Transformation Behavior. Control of microstructure is the primary key to successful application of alloy Ti-6Al-4V. It depends on both processing history and heat treatment. The microstructure that combines highest strength and ductility is not the microstructure that provides optimum fracture toughness or resistance to crack growth. The over-all effects of processing history and heat treatment on microstructure are very complex. However, the present discussion will illustrate those features most likely to be found in the alloy by the user.

Figure 1 illustrates the effect of solution temperature on the microstructure obtained at a cooling rate equivalent to that observed in parts of moderate thickness. In Fig. 1(a), the Widmanstätten-like transformed

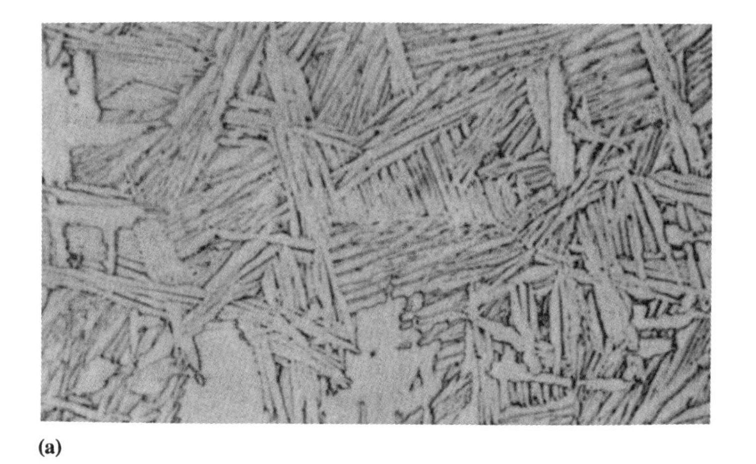

(a)

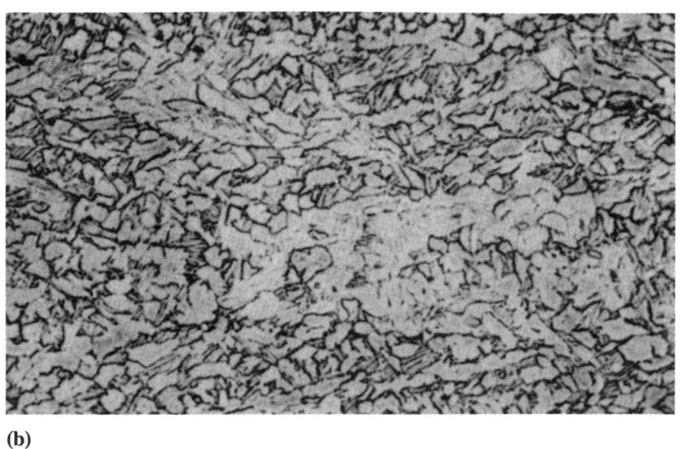

(b)

(c)

Fig. 1 Typical microstructures of alloy Ti-6Al-4V, showing effect of solution temperature. (a) 1010 °C (1850 °F), 1 h, encapsulated cool; 500×. (b) 982 °C (1800 °F), 1 h, encapsulated cool; 500×. (c) 927 °C (1700 °F), 1 h, encapsulated cool; 435×

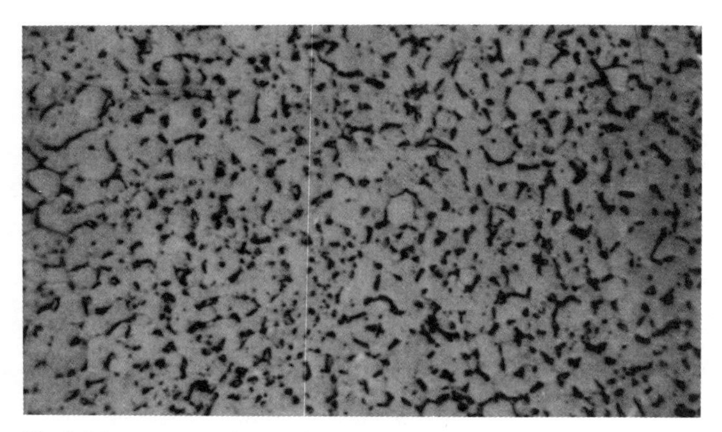

Fig. 2 Microstructure of alloy Ti-6Al-4V after recrystallization annealing. 927 °C (1700 °F), 1 h, very slow cool; 400×

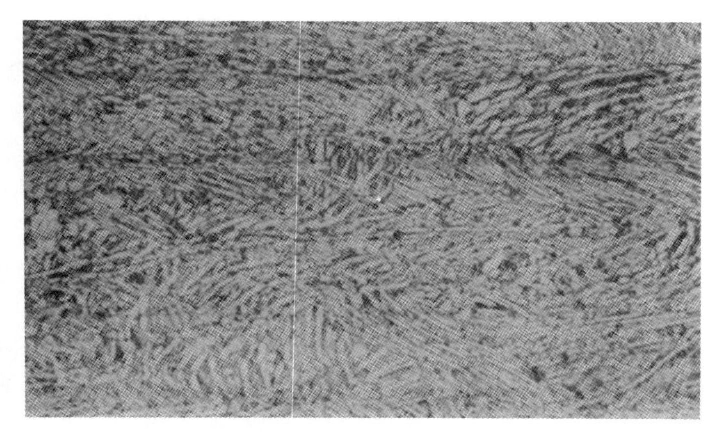

Fig. 3 Distorted Widmanstätten alpha remaining as a result of limited working in the α+β field. Rolled at 955 °C (1750 °F); 75×

microstructure arises from $\beta \rightarrow \alpha + \beta$ transformation during cooling from the beta-phase field. The light, platelike, and patchy areas are composed of alpha phase that nucleated and grew during cooling, whereas the retained beta phase is restricted primarily to the dark-etching outlines between the plates. Platelet colonies (that is, groups of platelets) in various orientations are apparent in the microstructure. The prior beta grain size exceeded the entire field of view in Fig. 1(a). The alloy in Fig. 1(b) was cooled to room temperature from about 20 °C (36 °F) below the beta transus. In this photomicrograph, the more or less equiaxed light areas are primary alpha phase, which is the alpha phase that was present at the solution temperature. The background is a transformed structure similar to that in Fig. 1(a) except that the platelet colonies are smaller. In Fig. 1(c), after cooling to room temperature from about 70 °C (126 °F) below the transformation temperature, the predominant feature is equiaxed primary alpha. The transformed background is a minor constituent.

Another microstructure of special interest in obtaining maximum fracture toughness is one produced by so-called recrystallization annealing. In this process, the alloy is heated to a temperature about 70 °C (126 °F) below the transformation temperature, held for a time and then very slowly cooled. The resulting microstructure is shown in Fig. 2. Most of this structure is continuous primary alpha phase and regrowth alpha (that is, alpha which has formed on the pre-existing primary alpha phase). The primary alpha existed at the upper temperature and nucleated the regrowth alpha during cooling. Stable beta phase decorates the alpha grain boundary triple points. This particular microstructure provides an excellent combination of fracture toughness and ductility. Another significant feature of the recrystallization annealed microstructure in Fig. 2 is that

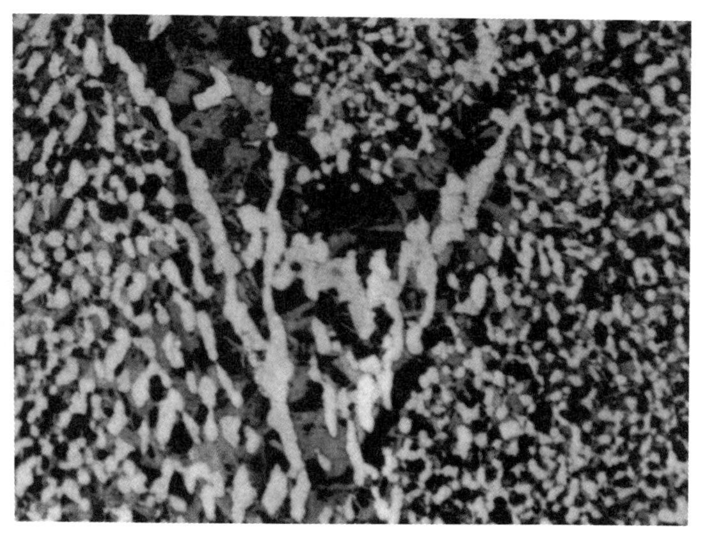

Fig. 4 Grain boundary alpha remnants which were not broken up after forging because of improper cooling from the β field. Rolled at 940 °C (1725 °F); 182.5×

the alpha phase is fully recrystallized and has a very low dislocation density.

The effects of prior processing on microstructure are quite varied. Extensive hot working in the alpha + beta field is required to produce the microstructure typified by Fig. 1(c). If hot working of the alpha + beta phases has been limited, microstructures such as that shown in Fig. 3 will occur. As may be observed, the Widmanstätten platelets are quite distorted but are not yet broken up. This condition is not particularly detrimental to fracture toughness, although it may affect fatigue crack propagation. This worked alpha recrystallizes and forms grains which grow entirely across the alpha platelets. Surface tension requirements cause the beta to wet, i.e., penetrate the alpha/alpha boundaries (Ref 8). This behavior breaks up the platelets, and recrystallization of beta causes further rearrangement of alpha to produce the equiaxed structure.

When alloy Ti-6Al-4V is processed improperly after heating into the beta field, alpha phase can form preferentially along the prior beta grains. Extensive hot work is required to break up such structures. An example of grain boundary alpha not completely broken up is shown in Fig. 4. Because cracks tend to propagate in or near interfaces (Ref 9, 10, 11), this type of structure can provide loci for crack initiation and propagation and thereby lead to premature failure.

Microstructural control is effected by using proper combinations of hot work and heat treatment. Heat treatment alone does not suffice to convert the Widmanstätten structure to an equiaxed form; therefore, heat treatment alone is not used unless a transformed structure is desired. Grain refinement cannot be obtained by heat treatment, and, after one $\beta \rightarrow \alpha + \beta$ sequence has been accomplished, additional $\alpha + \beta \rightarrow \beta \rightarrow \alpha + \beta$ cycles have no effect on the basic crystallographic texture although the texture may be coarsened as a consequence of beta grain growth.

Alpha arising from the $\beta \rightarrow \alpha + \beta$ transformation has a crystallographic (Burgers) relationship with the parent beta. Generally, two or more variants are present (Ref 12).

At the highest cooling rates, Ti-6Al-4V can form hexagonal close packed martensite which contains vanadium in supersaturation. Crystallographically, the martensite constituent and the alpha phase are nearly identical and are difficult to distinguish by optical means. Electron or x-ray metallography is required to distinguish quantitatively between martensite and fine Widmanstätten alpha. Figure 5 illustrates the effect that cooling rate has on the microstructure of Ti-6Al-4V when quenching is effected from the beta and high alpha + beta fields.

Electron microscopy has often revealed a phase at alpha/beta interfaces, a phase which has been referred to as interface phase. It has been

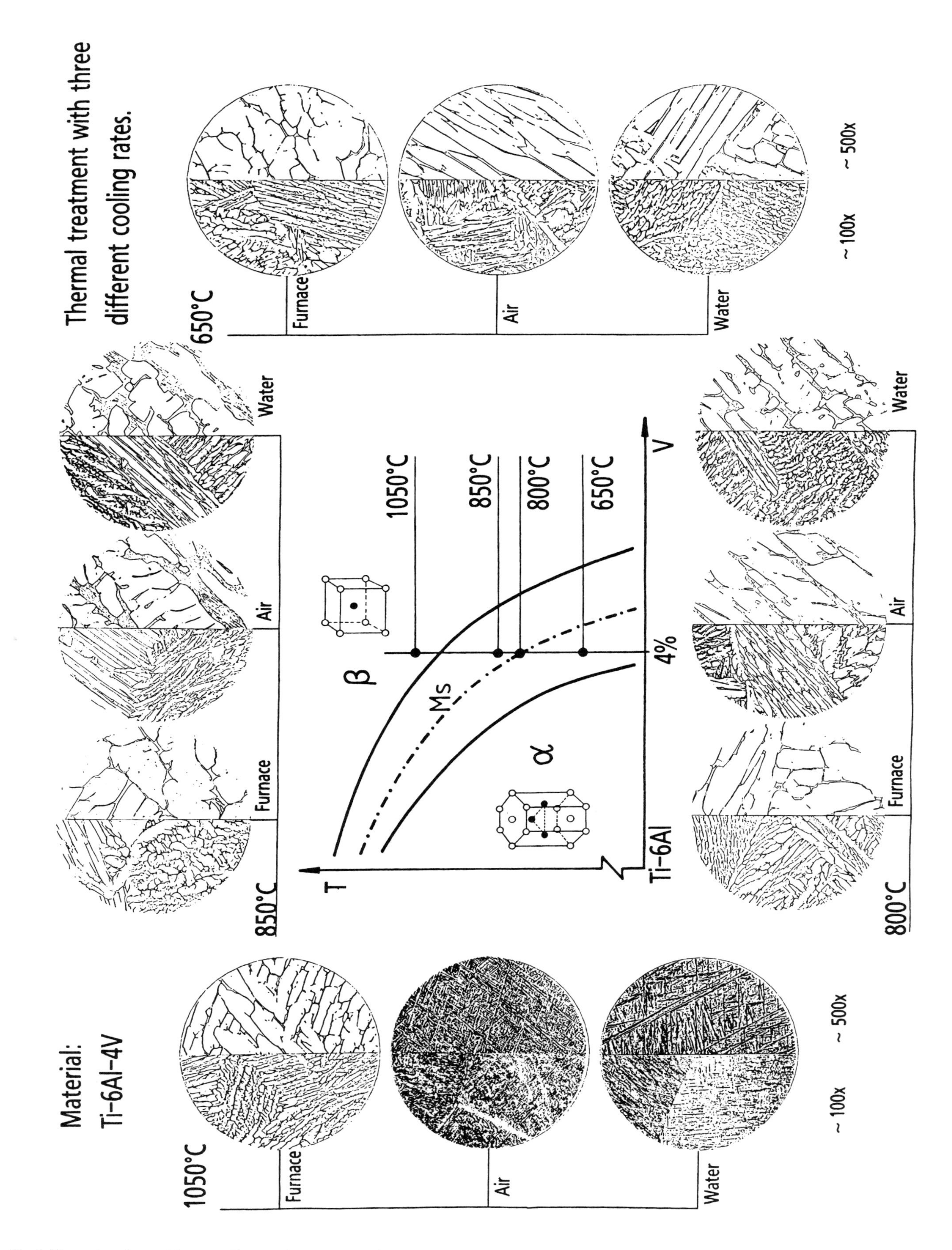

Fig. 5 Illustration of quenching rate effect on microstructures of alloy Ti-6Al-4V

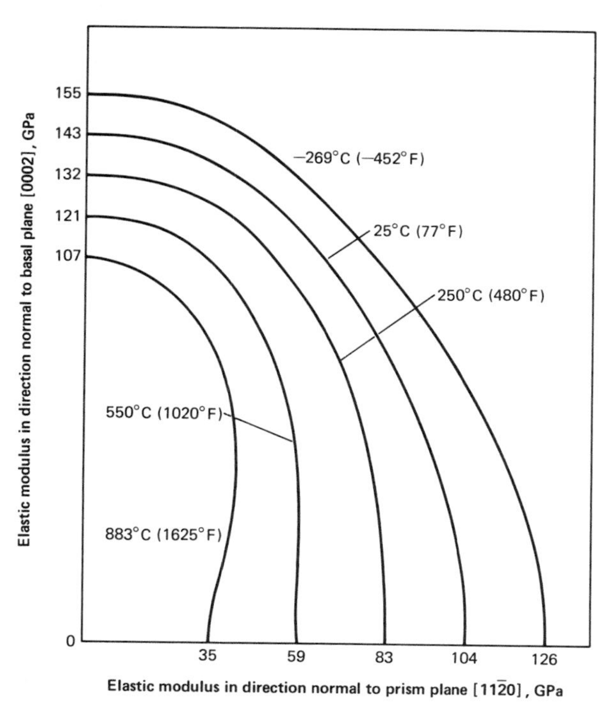

Fig. 6 Plot of elastic modulus vs direction in single crystal of titanium for various temperatures. Distance from origin to isotherm in direction of interest is equal to modulus. Source: Ref 14

shown that this phase is an artifact produced during the preparation of foils for electron microscopy (Ref 13).

Deformation Modes. Crucial to the toughness question is the size of the plastic zone that can form ahead of a propagating crack. That size, simplistically, depends on the yield strength, and this in turn depends on the factors discussed in the upcoming discussion on sources of strength. Having a high yield strength and a relatively low elastic modulus, Ti-6Al-4V can store more energy elastically than most metals before plastic deformation begins. The elastic modulus depends on the direction in which it is measured in a single crystal of titanium. This feature is illustrated in Fig. 6 (Ref 14).

The flow stress also depends on the direction of measurement. In directions normal to the hexagonal axis of an alpha titanium crystal, $\bar{a}$ slip, on the prism, pyramidal, and basal planes, is the primary deformation mode. In directions parallel, to the hexagonal axis, twinning and c + a slip act to accommodate the plastic strain. Each slip and twinning mode has its own unique flow stress and amount of strain that can be accommodated.

The plastic zone ahead of an advancing crack is, therefore, not uniform in cross section. It varies in a macroscopic sense in response to the microstructure (phase, shape) discussed in the previous section. It also varies from grain to grain in accordance with crystal type, whether alpha or beta, and with crystal orientation. To complicate matters further, Poisson's ratio and its plastic counterparts necessarily depend on direction on both the macro and micro scales. Furthermore, the true strain at ultimate load is not a linear function of alloy content even in the Ti-Al binary system (Ref 15). Events occurring in and around a crack tip and its associated plastic zone in Ti-6Al-4V are, for these reasons, complex indeed. Toughness in the Ti-6Al-4V alloy is not yet quantifiable from first principles. Nevertheless, there is a great deal of empirical information available for Ti-6Al-4V from which some general rules can be developed.

Summary. The mechanical properties of titanium alloys thus depend on alloy chemistry, microstructure, and metallographic texture through its influence on elastic and plastic anisotropy. The influence of these factors on strength, toughness, and resistance to environmental effects on crack propagation is discussed further in the following sections.

The Ti-6Al-4V alloy derives its annealed strength from several sources, the principal source being substitutional and interstitial alloying of elements in solid solution in both alpha and beta phases. Aluminum is the most important substitutional solid solution strengthener. Its effect on strength is linear (Ref 16). Other, less important sources of strengthening are interstitial solid solution strengthening, grain size effects, second phase (beta) effects, ordering in alpha, age hardening, and effects of crystallographic texture (Ref 17). Aluminum in Ti-6Al-4V, as suspected by Williams and Blackburn (Ref 18), gives rise to some tendency toward ordering in the alpha phase, the ordered product being Ti_3Al (Ref 19). Ordering in the alpha phase contributes perhaps 15 to 35 MPa (2 to 5 ksi) to the strength of standard Ti-6Al-4V, and contributes less than this to the strength of the ELI grade. Ordering also appears to degrade toughness. The effect of crystallographic texture is to introduce directionality into the strength equation. Relative to the hexagonal axis in alpha, strength (and modulus) is high in the parallel direction and low normal to that direction. Because metalworking operations tend to produce preferred crystallographic orientations in alpha grains, strength becomes an anisotropic quantity in most product forms. This feature can be minimized by proper processing and is rarely of direct concern. In some instances, it can be an advantage.

Because the beta phase present in alloy Ti-6Al-4V can be manipulated in amount and composition by heat treatment, the alloy is responsive to heat treatment. The $\beta \rightarrow \alpha + \beta$ reaction at low temperature leads to increased strength. The key is to quench from high in the $\alpha + \beta$ field and then age at a lower temperature. A typical strengthening heat treatment consists of heating for 1 h at 955 °C (1750 °F) and water quenching, followed by heating for 4 h at 540 °C (1000 °F) and air cooling. Response is limited in a practical sense, however, by two factors: (a) the small amount of beta in Ti-6Al-4V and (b) section size. The first factor puts an intrinsic ceiling on the increased strengthening response available—about 280 MPa (40 ksi) in thin-gage material. The second factor relates to depth of hardening, because Ti-6Al-4V is not effectively hardenable in sections greater than 25 mm (1 in.) in thickness. The Ti-6Al-4V alloy is, therefore, most commonly used in the annealed condition.

Other Alpha-Beta and Alpha Alloys

Metallurgy of High-Strength Alpha-Beta Alloys. Two alloys that fall in the high-strength alpha-beta class are Ti-6Al-6V-2Sn, which is used in airframes, and Ti-6Al-2Sn-4Zr-6Mo, which is used in jet engines. Alloy Ti-6Al-2Sn-4Zr-6Mo is also often classified as a super alpha alloy. Both of these alloys are stronger and more readily heat treated than Ti-6Al-4V. These features arise from the increased solid solution strengthening afforded by tin and zirconium, which have relatively small effects on the transformation temperature, and from the increased amounts of beta phase that result from the larger vanadium and molybdenum additions. Both vanadium and molybdenum are beta stabilizers. The Ti-6Al-6V-2Sn alloy contains the beta stabilizers copper and iron in combined amounts up to 1.4 wt% for enhanced strength and response to aging. Alloy Ti-6Al-2Sn-4Zr-6Mo is also useful at the moderately elevated temperatures from 425 to 480 °C (800 to 900 °F). This alloy combines high tensile strength with good creep resistance. The alpha phase tends to order more readily in these alloys than in alloy Ti-6Al-4V. Moreover, the transformed alpha platelets in Ti-6Al-2Sn-4Zr-6Mo tend to be narrower than those in Ti-6Al-4V, and formation of packets of parallel platelets is less likely. For both Ti-6Al-6V-2Sn and Ti-6Al-2Sn-4Zr-6Mo, the nose of the C curve defining the $\beta \rightarrow \alpha + \beta$ transformation as it depends on time and temperature is shifted to lower temperatures and longer times than for Ti-6Al-4V. Martensite does not form in ordinary situations.

Alpha is the dominant phase in these alloys but to a lesser extent than in Ti-6Al-4V. The physical metallurgy of these alloys is otherwise very similar to that of Ti-6Al-4V.

Metallurgy of "Super Alpha" Alloys. Alloys Ti-6Al-2Sn-4Zr-2Mo and Ti-8Al-1Mo-1V are in the "super alpha" class. They are used primarily in jet engine applications and are useful at temperatures above the normal range for Ti-6Al-4V. Alloy Ti-6Al-2Sn-4Zr-2Mo may be modified with silicon additions of up to 0.1%, and, when beta annealed (i.e., annealed by heating above the transformation temperature), the modified alloy provides the highest creep strength and temperature capability of all commercial titanium alloys currently (1980) produced in the United States. The Ti-8Al-1Mo-1V alloy has the highest modulus and lowest density of any commercial titanium alloy. Each of these alloys tends to order in the alpha phase more readily than does Ti-6Al-4V. Also, the nose of the C curve defining the $\beta \rightarrow \alpha + \beta$ transformation is shifted upward and to the left, or to higher temperatures and shorter times, in comparison with Ti-6Al-4V. Martensite forms more readily in either of these alloys than in Ti-6Al-4V.

Ti-8Al-1Mo-1V has limited usefulness and both Ti-6242 and Ti-6242S have wider applicability. A further modification of Ti-6242 is TiMetal® 1100 (Ti-6Al-2.7Sn-4Zr-0.4Mo-0.45Si-0.03Fe-0.07O_2), which offers about a 55% creep advantage over Ti-6242 (Ref 20). TiMetal® 1100 is not in wide service, but it offers the advantage of relatively easy processing. Creep resistance is enhanced by producing a transformed, i.e., Widmanstätten alpha structure in the super alpha alloys. For TiMetal® 1100, final forging is carried out in the beta field, and this is followed by a stabilization heat treatment at 600 °C (1110 °F) for 8 h.

Another relatively new super alpha alloy is IMI 834 (Ti-5.7Al-3.9Sn-3.5Zr-0.49Mo-0.84Nb-0.33Si-0.14Fe-0.126O_2-0.058C). TiMetal® 1100 has better creep resistance but lower fatigue resistance than IMI 834. This latter alloy is processed high in the alpha-beta field for a good combination of creep resistance and fatigue strength. A comparison of the two alloys is reported in Timet HTL Report "High-Temperature Alloy Comparison Ti-1100 and IMI 834" (P. Bania, July 1990).

Generally speaking, these alloys contain less beta phase than Ti-6Al-4V. Age hardening treatments are thus not very effective and are, moreover, deleterious to creep resistance. These alloys therefore are usually employed as solution annealed and stabilized. Solution annealing may be done at a temperature some 35 °C (63 °F) below the transformation temperature, and stabilization is commonly produced by heating for 8 h at about 590 °C (1100 °F).

At high temperatures, dynamic strain aging arising from aluminum, silicon and tin, and perhaps oxygen and zirconium, is thought to contribute to the creep resistance of these materials.

The alpha phase dominates the properties of these alloys to a greater extent than it does in Ti-6Al-4V. The metallurgy of the super alpha alloys is otherwise similar to that of Ti-6Al-4V.

Metallurgy of Beta Alloys

An alloy is considered to be a beta alloy if it contains sufficient beta stabilizer alloying element to retain the beta phase without transformation to martensite on quenching to room temperature. A number of Ti alloys contain more than this minimum amount of beta stabilizer alloy addition. The current status of beta alloys is thoroughly reviewed in Ref 21.

Alloying. Beta isomorphous elements such as Mo and V require more alloy addition on a weight percent basis to retain beta than do alloying elements such as Cr and Fe, which are eutectoid formers. Although less alloying element is required to retain the beta phase when eutectoid formers are used, they cannot be used alone to retain beta, because on long-time holding at elevated temperatures, these alloys decompose to form alpha plus compound. These decomposed structures have much poorer ductility than the alpha-beta alloys have prior to decomposition. One such early alloy was the Ti-3Al-5Cr alloy.

As a consequence of this decomposition, beta alloys are usually combinations of beta isomorphous and eutectoid former elements. The presence of beta isomorphous alloying elements reduces the tendency for compound formation by increasing their solubility in the beta phase (Ref 22). In addition, because segregation during melting is less of a problem for beta isomorphous elements than for eutectoid formers, beta alloys tend to have larger amounts of beta isomorphous than eutectoid former elements.

Table 1 Beta stabilizing elements

Beta stabilizer	Type	β_c, wt % (a)	β_t(b) suppression, °C (°F)
Mo	Isomorphous	10	9.4 (17)
V	Isomorphous	15	12.2 (22)
W	Isomorphous	22.5	3.9 (7)
Nb	Isomorphous	36.0	7.2 (13)
Ta	Isomorphous	45.0	2.2 (4)
Fe	Eutectoid	3.5	17.8 (32)
Cr	Eutectoid	6.5	15.0 (27)
Cu	Eutectoid	13.0	12.2 (22)
Ni	Eutectoid	9.0	22.2 (40)
Co	Eutectoid	7.0	21.1 (38)
Mn	Eutectoid	6.5	22.2 (40)
Si	Eutectoid	...	38.8 (70)

(a) Approximate wt% needed to retain 100% beta upon quenching. (b) Approximate amount of beta transus reduction per wt% addition. Note: Bania has ascribed to Al a negative value for retaining beta and has put this value equivalent to the value of Mo in retaining beta. Source: Ref 23

Table 2 Beta alloys of current interest

Composition, wt%	Common name	Principal uses
Ti-3Al-8V-6Cr-4Zr-4Mo	Beta C or 38-6-44	Springs
Ti-10V-2Fe-3Al	Ti-10-2-3	Air frames
Ti-15V-3Cr-3Sn-3Al	Ti-15-3	Strip producible, cold formable, age hardenable, weldable
Ti-15Mo-2.7Nb-3Al-0.2Si	Beta 21S	Oxidation resistant and candidate for composite matrix

Source: Ref 23

Since the ability to retain the beta phase depends on the rate of working through the alpha + beta phase field, the retention of undecomposed beta is also a function of section size. Thus, for larger section sizes, greater beta alloy content is required.

Bania (Ref 23) has compiled a list of beta stabilizing elements and their ability to retain beta on quenching (Table 1). Table 1 indicates that, on the whole, eutectoid former elements, in comparison to isomorphous elements have a larger tendency to lower the beta transus per wt% addition and require a smaller wt% addition to retain the beta phase.

It is possible to sum the beta retaining power of a group of alloying elements by adding the fractional equivalents of the beta retention wt% required for each element. For example, 5 wt% of Mo represents half the Mo required to retain beta and 1.75 wt% of Fe represents half the amount to retain beta (Table 1). Thus, an alloy containing 5 wt% Mo and 1.75 wt% Fe should be able to retain beta on quenching.

There is no truly stable beta alloy because even the most highly alloyed beta will, on holding at elevated temperatures, begin to precipitate omega, alpha, Ti_3Al, or silicides, depending on temperature, time, and alloy composition (Ref 21, p 173-185). All beta alloys contain a small amount of Al, an alpha stabilizer, in order to strengthen alpha which may be present after heat treating. The composition of the precipitating alpha is not constant and will depend on the temperature of heat treatment. The higher the temperature in the alpha + beta phase field, the higher will be the Al content of alpha.

Beta Alloy Compositions of Present Interest. There is no single beta alloy that has the same broad applicability as Ti-6Al-4V. Consequently, specific alloys are used because their properties suit a particular application. In general, retained beta alloys are used for workability, corrosion resistance, and the ability to heat treat larger section sizes in which beta has been retained. Beta alloys also tend to have higher density and lower elastic modulus values than alpha alloys. Beta alloys also have a tendency to alloy segregation (Ref 23). Table 2 lists some beta alloys of current interest and their principal uses. At present, beta alloys constitute a small fraction of titanium usage.

There is current interest also in using the eutectoid forming alloys to reduce the cost of beta alloys. A Ti-4.5Fe-6.8Mo-1.5Al alloy is currently being studied because of the cost advantages of using a ferro-molybdenum master alloy. An early alloy with high iron contents was the Ti-1Al-8V-5Fe alloy developed for use as fasteners which require high strengths. Still another alloy being developed for fasteners also has high iron contents, Ti-6V-6.2Mo-5.7Fe-3Al (Ref 23).

Beta alloys are being employed in the McDonnell Douglas C-17 and the Boeing 777 (Ref 23). Ti-10-2-3 forgings have been used extensively on the Boeing 777, but particularly for the landing gear (Ref 21, p 335-345). The Beta C and the Ti-15-3 alloys are being used and it is anticipated that Beta-21S will find service in the nacelle area. Beta C, which is hardenable to section sizes of 115 mm (4.5 in.) has potential for use in a water brake for aircraft carriers (Ref 21, p 361-374).

Mechanical Properties: Alpha and Alpha-Beta Alloys

Table 3 lists typical minimum property guarantees for titanium alloy mill products. In Table 4, the effects of temperature on strength are shown for the same alloys, and data for unalloyed titanium are included to illustrate that the alloys not only have higher room-temperature strengths but also retain much larger fractions of that strength at elevated temperatures.

Table 5 lists typical specifications for the alloys discussed here. These alloys are covered also by numerous commercial specifications, and design information is readily available. More extensive listings of specifications are given in the reference book *Materials Properties Handbook: Titanium Alloys* (ASM, 1994).

In terms of the principal heat treatments used for titanium, beta annealing decreases strength by 35 to 100 MPa (5 to 15 ksi) depending on prior grain size, average crystallographic texture, and testing direction. Solution treating and aging can be used to enhance strength at the expense of fracture toughness in alloys containing sufficient beta stabilizer (that is, 4 wt% V or more).

Fracture Toughness

Fracture toughness can be varied within a nominal titanium alloy by as much as a multiple of two or three by manipulating alloy chemistry, microstructure, and texture. Some trade-offs of other desired properties may be necessary to achieve high fracture toughness. Plane-strain fracture toughness, K_{Ic}, is of special interest because the critical crack size at which unstable growth can occur is proportional to $(K_{Ic})^2$ and strength is often achieved in titanium alloys at the expense of K_{Ic}.

The main purpose of this section is to indicate the scope of possibilities, as well as some of the property trade-offs required, for obtaining high levels of fracture toughness in titanium alloys. A further purpose is to review some of the specific variables that are known to affect fracture toughness.

Effects of Alloy Chemistry. There are significant differences among titanium alloys (Ref 24), but there is also appreciable overlap in their properties. Table 6 gives examples of typical plane-strain fracture toughness ranges for alpha-beta titanium alloys. From these data it is apparent that the basic alloy chemistry affects the relationship between strength and toughness. From Table 6 it is also evident that transformed microstructures may greatly enhance toughness while only slightly reducing strength.

Within the permissible range of chemistry for a specific titanium alloy and grade, oxygen is the most important variable insofar as its effect on toughness is concerned. This is readily shown by the data for Ferguson and Berryman (Ref 25), who reported strength and K_{Ic} values for specimens of alpha-beta processed and recrystallization annealed Ti-6Al-4V. Regression analysis of their data shows that for each 0.01% increase in oxygen, toughness is reduced by about 3.7 MPa$\sqrt{m}$ (3.4 ksi$\sqrt{in.}$). Whether this is a direct effect or an indirect effect, in the sense that oxygen increases strength and the strength increase reduces K_{Ic}, remains to be determined. Multiple regression analysis of the Ferguson and Berryman data, where both oxygen content and tensile strength are assumed to be independent variables, shows that tensile strength is the dominant variable (the residual effect of oxygen does not reach statisti-

Table 3 Typical mill-guaranteed room-temperature tensile properties for selected titanium alloys

Alloy	Ultimate strength		Yield strength		Ductility	
	MPa	ksi	MPa	ksi	Elongation, %	Reduction in area, %
Ti-6Al-4V	895	130	825	120	10	20
Ti-6Al-6V-2Sn	1065	155	995	145	10	20
Ti-6Al-2Sn-4Zr-6Mo	1030	150	965	140	10	20
Ti-6Al-2Sn-4Zr-2Mo	895	130	825	120	10	25
Ti-8Al-1Mo-1V	895	130	825	120	10	20

Table 4 Fraction of room-temperature strength retained at elevated temperature for several titanium alloys(a)

Temperature		Unalloyed Ti		Ti-6Al-4V		Ti-6Al-6V-2Sn		Ti-6Al-2Sn-4Zr-6Mo		Ti-6Al-2Sn-4Zr-2Mo		Ti-1100(a)		IMI-834	
°C	°F	TS	YS	TS	YS	TS	YS	TS	YS	TS	YS	TS	YS	TS	YS
93	200	0.80	0.75	0.90	0.87	0.91	0.89	0.90	0.89	0.93	0.90	0.93	0.92	...	...
204	400	0.57	0.45	0.78	0.70	0.81	0.74	0.80	0.80	0.83	0.76	0.81	0.85	0.85	0.78
316	600	0.45	0.31	0.71	0.62	0.76	0.69	0.74	0.75	0.77	0.70	0.76	0.79	...	...
427	800	0.36	0.25	0.66	0.58	0.70	0.63	0.69	0.71	0.72	0.65	0.75	0.76	...	...
482	900	0.33	0.22	0.60	0.53	...	...	0.66	0.69	0.69	0.62	0.72	0.74	...	...
538	1000	0.30	0.20	0.51	0.44	...	...	0.61	0.66	0.66	0.60	0.69	0.69	...	...
593	1100	...	...	...	...	...	...	...	...	...	...	0.66	0.63	0.63	0.61

(a) Short time tensile test with less than 1 h at temperature prior to test

Table 5 Typical specifications for titanium and titanium alloys

Material	AMS	Military MIL-I
Unalloyed Ti	4900	9046, 9047
Ti-6Al-4V	4911	9046, 9047
Ti-6Al-4V ELI	4907	9046, 9047
Ti-6Al-6V-2Sn	4918	9046, 9047
Ti-6Al-2Sn-4Zr-6Mo	4981	9047
Ti-6Al-2Sn-4Zr-2Mo	4929	9046, 9047
Ti-8Al-1Mo-1V	4915, 4916	9046, 9047

Table 6 Typical fracture toughness of high-strength titanium alloys

Alloy	Alpha morphology	Yield strength MPa	Yield strength ksi	Fracture toughness (K_{Ic}) MPa$\sqrt{m}$	Fracture toughness (K_{Ic}) ksi$\sqrt{in.}$
Ti-6Al-4V	Equiaxed	910	130	44-66	40-60
	Transformed	875	125	88-110	80-100
Ti-6Al-6V-2Sn	Equiaxed	1085	155	33-55	30-50
	Transformed	980	140	55-77	50-70
Ti-6Al-2Sn-4Zr-6Mo	Equiaxed	1155	165	22-23	20-30
	Transformed	1120	160	33-55	30-50

Source: Ref 24

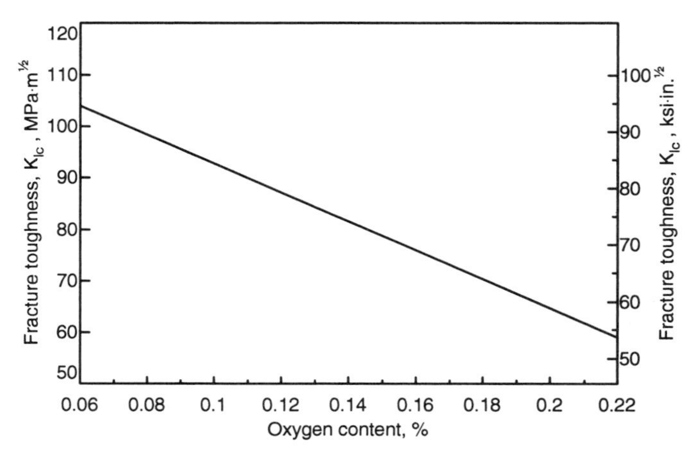

Fig. 7 Influence of oxygen content on fracture toughness of recrystallization annealed alloy Ti-6Al-4V. Source: Ref 25

Table 7 Effect of hydrogen content on room-temperature K_{Ic} in alloy Ti-6Al-4V after furnace cooling from 927 °C (1700 °F)

Hydrogen content, ppm	K_{Ic} at room temperature(a) MPa$\sqrt{in.}$	K_{Ic} at room temperature(a) ksi$\sqrt{in.}$
At 0.16 wt% oxygen		
8	145	132
36	118	107
53	104	95
122	100	91
At 0.05 wt% oxygen		
9	133	121
36	125	114
50	96	87
125	101	92

(a) Specimens were tested in accord with ASTM E399, but were loaded rapidly (total testing time = 10 s). Source: Ref 30

Table 8 Relationship between K_{Ic} and fraction of transformed structure in alloy Ti-6Al-4V

Heat treating temperature(a) °C	Heat treating temperature(a) °F	Fraction of transformed structure, %	K_{Ic} MPa$\sqrt{m}$ (b)	K_{Ic} ksi$\sqrt{in.}$
1050	1922	100	69.0 (69.9)	64
950	1742	70	61.5 (60.4)	55
850	1562	20	46.5 (44.6)	40
750	1382	10	39.5 (41.5)	38

(a) Heated for 1 h at indicated temperature and then air cooled. (b) Values in parentheses calculated from linear least-squares expression relating % transformation to K_{Ic}. Source: Ref 33

cal significance). This implies that, if oxygen affects K_{Ic}, it does so through its strengthening effect. The solid solution strengthening effect of oxygen is further complicated by the fact that oxygen tends to promote the formation of Ti_3Al (Ref 19). Finally, the precision and accuracy to which oxygen can be analyzed in titanium do not compare with the relative precision and accuracy common to strength measurements.

The work of Rosenberg and Parris (Ref 26) on the Ti-xAl-2Mo system (x varies from 4 to 6%) gives some evidence that both aluminum and oxygen can exert influences on toughness independent of strength. These authors, however, did not employ recrystallization annealing, which should minimize formation of Ti_3Al in the continuous alpha which dominates the toughness properties. For the recrystallization annealed condition, the continuous alpha is regrowth alpha, which is low in both aluminum and oxygen; the solute-rich alpha in which Ti_3Al can form is thus isolated at the core of each primary grain.

Whatever that situation, the data of Ferguson and Berryman (Ref 25) on the oxygen effect are given in Fig. 7. In essence, if high fracture toughness is required, oxygen must be kept low, other things being equal. Reducing aluminum, as in Ti-6Al-4V ELI (extra-low interstitial), is also indicated, but the effect is not as strong as it is for oxygen (Ref 26).

The effect of chemistry on K_{Ic} for Ti-6Al-4V was shown in another way by Cooper (Ref 27), who summed the expected changes in the β transus temperature, ΔT, with alloy additions, and correlated that with K_{Ic}. He found a negative effect of ΔT on K_{Ic}. Since oxygen and aluminum additions have the effect of increasing ΔT, the data are consistent.

The effect of oxygen on K_{Ic} is not limited to alpha-beta alloys at ambient temperatures. Van Stone and his coworkers (Ref 28 and 29) reported a much higher K_{Ic} value for Ti-5Al-2.5Sn ELI (which has a low oxygen content) than for the standard grade in the temperature range from −253 to +22 °C (−423 to +72 °F). Slow cooling of Ti-5Al-2.5Sn ELI from the solution temperature was found to decrease K_{Ic}, whereas this effect of cooling rate was absent for the standard grade. Van Stone attributed the change in toughness to a change in deformation mode morphology. The combination of ordering and normal interstitial content did not significantly change the slip character, whereas ordering in ELI plate resulted in coarser slip bands.

As might be expected, hydrogen also has an effect on toughness. The work of Meyn (Ref 30) shows that very low hydrogen contents (less than about 40 ppm) enhance toughness. This effect is particularly dramatic with hydrogen contents below 10 ppm. Table 7 illustrates the essential results for Ti-6Al-4V at two different oxygen levels. Meyn used a high loading rate. However, the effect of hydrogen content on K_{Ic} has been confirmed by Chen (Ref 31). The work of Chesnutt et al. (Ref 32) on hydrogen effects with nonvalid K_{Ic} tests shows the same trend; in this study, however, the hydrogen effect may have depended on microstructure.

Effects of Microstructure. Improvements in K_{Ic} can be obtained by providing either of two basic types of microstructures: (a) transformed structures, or structures transformed as much as possible, because such structures provide tortuous crack paths; and (b) equiaxed structures composed mainly of regrowth alpha that have both low dislo-

Table 9 Effect of primary alpha dispersion on K_{Ic} for alloy Ti-4Al-2Sn-4Mo-0.5Si (IMI550) plate(a)

d(b), μm	K_{Ic} MPa√m (c)	K_{Ic} ksi√in.
7.58	59.3 (59.2)	54
5.83	62.9 (62.3)	57
3.78	63.1 (65.9)	60
2.50	67.4 (68.1)	62
3.71	67.5 (66.0)	60
2.26	67.8 (68.6)	62
1.96	71.1 (69.1)	63

(a) 30-mm plate heated for 1 h at 900 °C (1652 °F) and air cooled, then heated for 24 h at 500 °C (932 °F) and air cooled. (b) d is mean phase boundary intercept distance. (c) Values in parentheses were calculated from the linear least-squares expression relating K_{Ic} to d. Source: Ref 9

Table 10 Effect of forging procedure on fracture toughness of alloy Ti-6Al-2Sn-4Zr-6Mo

Forging temperature	K_{Ic} MPa√m (c)	K_{Ic} ksi√in.
55 °C (100 °F) below beta transus	41, 40	37, 36
40 °C (70 °F) above beta transus	71, 72	65, 66

Note: Heated for 1 h at 885 °C (1625 °F) and air cooled, then heated for 8 h at 595 °C (1100 °F) and air cooled. Source: Ref 32

cation densities and low concentrations of aluminum and oxygen (the so-called "recrystallization annealed" structures). It is not yet known (in 1994) whether or not combinations of these two types of structures would further enhance K_{Ic} values.

Transformed structures appear to be tough primarily because fractures in such structures must proceed along tortuous, many-faceted crack paths. According to the work of Hall and Hammond (Ref 33), K_{Ic} is proportional to the fraction of transformed microstructure in alloy Ti-6Al-4V (see Table 8). These authors, however, propose that it is strain induced transformation of the retained laths of beta phase that leads to enhanced fracture toughness. Evidently, their idea is that this "TRIP" mechanism enhances "ductility" in front of each crack tip. However, in comparing beta alloys deformed by either slip or "TRIP" mechanisms, Wardlaw et al. (Ref 34) could find no advantage in ductility for the "TRIP" alloys. Curtis and Spurr (Ref 35) suggested that it is primarily the alpha platelet size and efficient dispersion of the beta phase that enhance toughness. In any event, the work of Chesnutt and Spurling (Ref 11) provides direct evidence that crack tortuosity is an important factor in determining the form of fracture topography—microstructure correlations for the same sample. Hall and Pierce (Ref 36) made similar observations concerning microstructures for alloy Ti-6Al-6V-2Sn. Hall et al. (Ref 37) recommended that, for the best combination of fracture toughness and tensile ductility in Ti-6Al-2Sn-4Zr-6Mo, a microstructure containing 10% primary alpha be employed. There is strong evidence that crack tortuosity is an important variable affecting K_{Ic}.

Rogers (Ref 9) looked at the toughness relationship in another way. The alloy he studied was IMI550 (Ti-4Al-2Sn-4Mo-0.5Si) that had been heated for 1 h at 900 °C (1625 °F) and air cooled, then heated for 24 h at 500 °C (932 °F) and air cooled. His data show that an increase in the mean phase boundary intercept distance between primary alpha and transformed microstructure diminishes fracture toughness. Rogers' data are presented in Table 9. Whether this relationship holds true for all alpha-beta titanium alloys is not known. Rogers' rationale is that, because crack loci and void formation tend to occur at the interfaces between alpha and transformed beta, then a decrease in the distance between phase boundaries causes an increase in the spatial frequency of microvoid formation so as to make blunting and arresting of cracks more likely.

In a similar vein, Gerberich and Baker (Ref 38) proposed that a proper balance between platelet thickness and spacing in the transformed microstructure is required to achieve highest toughness. Platelets need to be thick enough to turn a crack while being spaced such that turns are frequent.

Greenfield and Margolin (Ref 39) studied a complex experimental alpha-beta alloy for which strength was held constant in both equiaxed alpha and transformed microstructural conditions. For the equiaxed alpha data, toughness increased with beta grain boundary area per unit volume. For the transformed condition, the data showed that toughness increased with grain boundary alpha thickness up to about 5.5 microns, after which fracture toughness revealed no further increase up to an alpha thickness of 10 microns. It must be recognized that the observed fracture surface occurred during catastrophic crack propagation. This implies that the features which gave rise to increased tortuosity operated at the onset of catastrophic failure. It may well be that the processes involve crack blunting either by entering the grain boundary alpha along which crack propagation took place or by increasing the amount of plastic deformation required before the crack would propagate. Since alpha is usually the softer phase the effect of closer alpha spacing (Ref 9) may have been to permit greater deformation before the crack could propagate. In order to understand the particular processes that affect fracture toughness in a given alloy, it is necessary to observe the interaction between the crack and the microstructure at various stages preceding catastrophic crack propagation.

In any case, beta forging can be substituted for beta heat treating. See Chesnutt et al. (Ref 32) for data on alloy Ti-6Al-2Sn-4Zr-6Mo, Petrak (Ref 40) for data on alloy Ti-6Al-4V, Ulitchny et al. (Ref 41) for data on alloy Ti-6Al-6V-2Sn, and Chesnutt et al. (Ref 42) for data on all three of these alloys. The results of Chesnutt et al. (Ref 32) are presented in Table 10. Curtis and Spurr (Ref 35) reported a similar effect of rolling temperature on Ti-6Al-4V; beta rolling enhances K_{Ic}. Chesnutt et al. (Ref 43) and Berryman et al. (Ref 44) presented similar results for the experimental alpha-beta alloy Corona 5 (Ti-4.5Al-5Mo-1.5Cr). Bohanek (Ref 45) demonstrated the same effect of transformed structure enhancement of toughness in Ti-6Al-4V billet. He also showed that this effect does not necessarily carry through to a forged part.

Because welds in alloy Ti-6Al-4V will contain transformed products, one would expect such welds to be relatively high in toughness. This is, in fact, the case, as the data of Ferguson and Berryman (Ref 25) show. Their data are summarized in Table 11.

Grain size apparently is not always a definite variable. Margolin et al. (Ref 46) showed that toughness first decreased, and then increased, as

Table 11 Fracture toughness of alloy Ti-6Al-4V (0.11 wt% O_2) in welds and heat-affected zones

Postweld stress relief	K_{Ic} Weld MPa√m	Weld ksi√in.	Heat-affected zone MPa√m	Heat-affected zone ksi√in.	Base metal(a) MPa√m	Base metal(a) ksi√in.
2 h at 590 °C (1100 °F), AC	87(b)	79(b)	81(c)	74(c)	92	84
1 h at 650 °C (1200 °F), AC	85(d)	77(d)	77(d)(e)	70(d)(e)	92	84
1 h at 760 °C (1400 °F), AC	...	...	76(d)	69(d)	92	84

(a) Recrystallization anneal from Fig. 7.7 at 0.11 wt% O_2. (b) Based on data from 2 samples. (c) Based on data from 20 samples. (d) Based on data from 1 sample. (e) Annealed for 2 h at 650 °C (1200 °F), AC. Source: Ref 25

Table 12 Effect of test direction on mechanical properties of textured Ti-6Al-2Sn-4Zr-6Mo plate

Test direction(a)	Tensile strength, MPa	Yield strength, MPa	Elongation, %	Reduction in area, %	Elastic modulus, GPa	K_{Ic}, MPa$\sqrt{m}$	K_{Ic}, ksi$\sqrt{in.}$	K_{Ic} specimen orientation
L	1027	952	11.5	18.0	107	75	68	L-T
T	1358	1200	11.3	13.5	134	91	83	L-T
S	938	924	6.5	26.0	104	49	45	S-T

(a) High basal pole intensities reported in the transverse direction, 90° from normal, and also intensity nodes in positions 45° from the longitudinal (rolling) direction and about 40° from the plate normal. Source: Ref 48

beta grain size decreased in equiaxed Ti-5.25Al-5.5V-0.9Fe-0.5Cu at a yield strength of 1240 MPa (180 ksi), whereas at 1140 MPa (165 ksi) the toughness of the same equiaxed alloy decreased continuously with decreasing grain size (Ref 39). Mahajan and Margolin (Ref 47) also showed that cracks tend to develop at interfaces between primary alpha and transformed beta and along slip bands in alpha on surfaces of Ti-6Al-2Sn-4Zr-6Mo.

Recrystallization annealing is a very efficient method of obtaining high fracture toughness. Also, the relatively precise temperature control necessary to limit primary alpha to a low volume fraction is not needed. As discussed above, a recrystallization anneal is effected simply by heating to a temperature about 70 °C (126 °F) below the transformation temperature, holding for a time sufficient to reach equilibrium, and then cooling slowly to about 700 °C (1300 °F). A transformed structure is thereby completely avoided; the resulting microstructure is nearly free of dislocations, and it is this characteristic that justifies the term "recrystallization anneal." This procedure is well established for the Ti-6Al-4V ELI alloy, and typical data for the recrystallization annealed alloy are shown in Fig. 7. The slow cooling should be terminated at about 700 °C (1300 °F) to avoid ordering and Ti_3Al formation. This is less important for the ELI grade than for the standard grade.

A real virtue of the recrystallization annealed microstructure is that substantial toughness is achieved while maintaining ductility and fatigue resistance at high strength levels. Such a microstructure also tends to reduce scatter in the data, thus permitting higher design criteria.

Effects of Texture. Effects of texture arise from preferred crystallographic orientation in the material of interest. With all c axes of the alpha grains tending to lie along one (or a few) direction(s) in the product, the physical and mechanical property values necessarily will depend on the direction in which they are measured. Toughness is no exception. The effects that crystallographic texture can have on properties of Ti-6Al-2Sn-4Zr-6Mo (Ref 48) are shown in Table 12. Table 12 is not typical of mill or forged products; the data are given only for illustration. The Ti-6Al-4V alloy shows similar behavior (Ref 49 to 53). Harmsworth (Ref 54) gives similar data for Ti-6Al-6V-2Sn.

Tchorzewski and Hutchinson (Ref 53) showed that effects of alpha phase crystallographic texture can override effects of microstructure and grain size in Ti-6Al-4V. They found that texture affects both the onset of crack extension and the maximum load sustainable in the presence of a fatigue crack. Moreover, texture was found to influence the shape of the plastic zone ahead of the crack. Finally, these authors concluded that the conditions for plane strain fracture may be more stringent in materials having certain textures than in isotropic materials; this remains to be verified.

Toughness directionality is not restricted to alpha or to alpha-beta alloys. Williams et al. (Ref 55) demonstrated that toughness in near-beta alloys is a directional property. In this context, near-beta alloys are those alloys in which primary alpha is the minor phase. Such alloys commonly contain beta-stabilizing additions of 8 wt% or more.

Because the basic crystallographic texture arises during hot working operations and cannot be entirely eliminated by heat treatment, there are no known methods of producing wrought titanium parts having completely random texture and thus zero directionality. However, by balancing the hot working operations as much as possible along all three reference axes, it is possible to reduce directionality to acceptable levels. But every part is different, and it has become usual practice in the aerospace industry for critical parts to be qualified by the fabricator. On the other hand, by tailoring the alpha phase crystallographic texture to a specific need—say, modulus along a fiber axis—texturing offers a significant potential for enhancing properties. In practice to date (1980), attempts to develop specific textures have been somewhat inconsistent. Paton et al. (Ref 56) suspect that textured material may have increased susceptibility to environmental effects in specific directions.

Effects of Environment. Effects of temperature on toughness are usually less abrupt for titanium than for common low-alloy steel. For example, Tobler (Ref 57) reported a gradual K_{Ic} transition temperature between −196 and −143 °C (−320 and −215 °F) for recrystallization annealed Ti-6Al-4V ELI. For temperatures at and above −143 °C (−215 °F), his K_{Ic} values were typically about 90 MPa · $\sqrt{m}$ (82 ksi · $\sqrt{in.}$). At −196 °C (−320 °F), his values were typically 60 to 65 MPa$\sqrt{m}$ (55 to 60 ksi$\sqrt{in.}$). The loss is about 30%. The early conclusion by Christian and Hurlich (Ref 58) that Ti-6Al-4V ELI may be used to cryogenic temperatures thus has some justification. The same may not be true of standard grade Ti-6Al-4V. Tobler's product contained 62 ppm hydrogen, whereas that of Christian and Hurlich contained 70 to 100 ppm. Other authors who have provided temperature-dependent toughness results for alloy Ti-6Al-4V are Cervay (Ref 59) and Hall et al. (Ref 60).

Van Stone et al. (Ref 28) found no significant K_{Ic} transition temperature for Ti-5Al-2.5Sn ELI at temperatures down to −263 °C (−442 °F). Values for K_{Ic} did decrease with temperature, however.

Chait and Lum (Ref 61) determined the toughness trend (Charpy energy per unit area) with temperature for Ti-6Al-6V-2Sn. Because of the rich beta content, their low-temperature toughness values were quite low. There was no sharp transition temperature. The toughness vs temperature curve had a simple "S" shape over the temperature range from −196 to +377 °C (−320 to +700 °F). Harmsworth (Ref 54) reported K_{Ic} data for alloy Ti-6Al-6V-2Sn over the temperature range from −54 to +93 °C (−65 to +200 °F) the trends of which were in agreement with those of Chait and Lum.

In aerospace applications, there is a natural concern that chemical environmental factors such as water, salt water, or jet fuel will alter toughness in critical components. Data obtained at McDonnell Aircraft (Ref 62) on annealed standard grade Ti-6Al-4V indicate the following environmental effects on the apparent value of fracture toughness:

- Laboratory air, 56 MPa$\sqrt{m}$ (51 ksi$\sqrt{in.}$)
- JP4 fuel, 47 MPa$\sqrt{m}$ (43 ksi$\sqrt{in.}$)
- 3.5% salt solution, 34 MPa$\sqrt{m}$ (31 ksi$\sqrt{in.}$)

The results obtained by Ferguson and Berryman (Ref 25) and by Hall et al. (Ref 60) are similar in that salt water degrades K_{Ic}. The early results of Hatch et al. (Ref 63) also showed that salt water can degrade crack growth resistance even in thin sheet where plane strain conditions do not exist. These authors also found significant alloy effects on crack growth in salt water.

Curtis and Spurr (Ref 35) charted the effects of quenching temperature on K_{Ic} and K_{Iscc} (3.5% salt solution) for alloys Ti-6Al-4V and Ti-4Al-3Mo-1V. The K_{Ic} and K_{Iscc} curves parallel each other for each alloy.

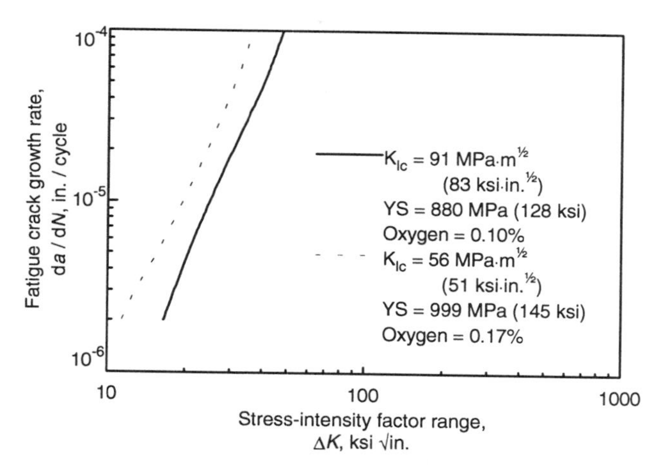

Fig. 8 Variation of fatigue crack propagation rate with yield strength, K_{Ic} and other variables for annealed Ti-6Al-4V forgings. $R = 0.02$; 10Hz; air, argon and JP4 environments pooled; average of six tests per trend line. Source: Ref 62

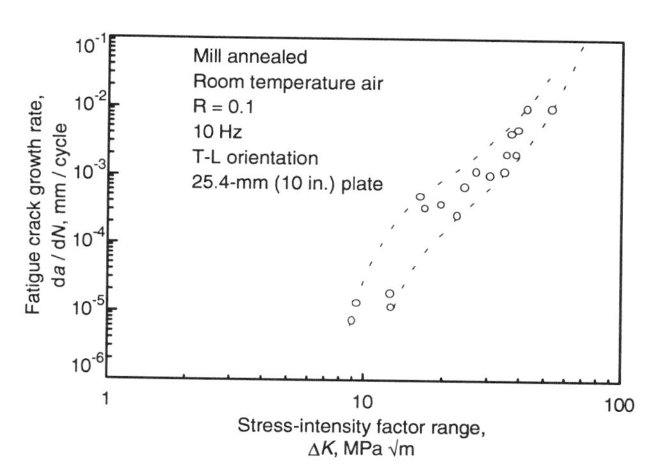

Fig. 9 Illustration of the scatter that can occur in fatigue crack propagation measurements. The data are for six heats of mill annealed alloy Ti-6Al-4V. Source: Ref 49

The K_{Ic} and K_{Iscc} curves are also parallel for Ti-6Al-4V where rolling temperature is the independent variable and for Ti-4Al-3Mo-1V where either aging temperature or aging time is the independent variable. These authors also showed beta annealing to benefit both K_{Ic} and K_{Iscc} in analogous ways. For both alloys, however, $K_{Iscc} < K_{Ic}$. Although these experiences are limited, one might expect that most factors which enhance K_{Ic} also enhance K_{Iscc}.

Fatigue Crack Propagation

Just as K_{Ic} is important in calculating loads that a structural member can carry in the presence of a flaw, so also it is important in many cases to know what the remaining fatigue life is in the presence of a fatigue crack or other sharp crack. In a very general way, fatigue crack propagation (FCP) behavior in titanium parallels fracture toughness—that is, for a given alloy those conditions giving highest toughness tend also to give lowest cyclic growth rates under fatigue loading. Amateau et al. (Ref 64) have collected some crack propagation models, a few of which recognize an inverse K_{Ic}-FCP relationship. The correlation, however, is certainly not so good that one variable can be calculated from the other. Furthermore, there may be an additional variable, such as strength, that affects (or even controls) both K_{Ic} and FCP in similar ways. This possibility is illustrated in Fig. 8. It is also quite possible that fatigue crack propagation and toughness do not arise from the same mechanism, especially when the stress-intensity factor range ΔK is much less than K_{Ic}.

In FCP measurements, there can be a significant amount of scatter in the data. The example shown in Fig. 9 is one of the more extreme cases encountered and arises for the mill annealed condition where uncertainties of microstructure, texture, and strength may exist. Part of the scatter is due to test reproducibility. There may also be a point-to-point material variability. The latter is due to minor processing and material inhomogeneities. Variations in chemistry, microstructure, and texture effects within a given lot may, in some cases, be additive even under controlled conditions. There are, of course, lot-to-lot differences. For design purposes, users are well advised to use statistical data derived from information in digital form from several lots.

Much less scatter than that illustrated in Fig. 9 was observed for most other conditions reported by Bjeletich (Ref 49). See also Tobler (Ref 57) and Chesnutt et al. (Ref 32) for additional data. Like Bjeletich, these authors present data in scatter-plot form. One purpose of showing Fig. 9 is to give the reader a feel for what is required before an effect can be confirmed. For example, two separate curves within the scatter band do not necessarily indicate a trend, and more data would be required.

The basic Paris equation, $da/dN = C(\Delta K)^n$, is only approximately obeyed for titanium. Deviations from the Paris equation, while often fairly reproducible in very specific material forms under very specific conditions, are not systematic. Generally, log-log plots of da/dN and ΔK yield three regions. A Stage I threshold region usually lies somewhere below about $\Delta K = 12$ MPa$\sqrt{m}$ (11 ksi$\sqrt{in.}$) and is characterized by steep slopes where da/dN increases very rapidly with ΔK. The Stage II Paris law region of more moderate slope extends upward from ΔK values of about 12 MPa$\sqrt{m}$. At still higher ΔK values, a Stage III portion of the curve slopes more and more steeply as ΔK approaches K_{Ic}.

Studies of fracture surfaces have shown that there is a transition from cyclic cleavage to striations as ΔK increases beyond the threshold stress. For the microstructures studied by Yuen et al. (Ref 65), the transition occurs at about 13 MPa$\sqrt{m}$ (12 ksi$\sqrt{in.}$). Paton et al. (Ref 12), however, found the transition to occur over a range of ΔK. Yuen et al. (Ref 65) proposed that the transition occurs when the cyclic plastic zone size is on the order of the alpha grain size and corresponds to a transition from single to multiple slip. During a study of Widmanstätten structures, Yoder et al. (Ref 66) found that the transition occurs when the mean packet size equals the cyclic plastic zone size.

Table 13 Selected data on effect of alloy type on fatigue crack propagation resistance in room-temperature air at 0.6 Hz

Fatigue crack propagation, da/dN			Stress-intensity factor range, ΔK					
			Ti-8Al-1Mo-1V		Ti-6Al-2Sn-4Zr-2Mo		Ti-6Al-2Sn-4Zr-6Mo	
m/cycle	in./cycle	R(a)	MPa$\sqrt{m}$	ksi$\sqrt{in.}$	MPa$\sqrt{m}$	ksi$\sqrt{in.}$	MPa$\sqrt{m}$	ksi$\sqrt{in.}$
2.54×10^{-7}	1×10^{-5}	0.1	22	20	23	21	17	15
		0.5	17	15	18	16	15	14
2.54×10^{-8}	1×10^{-6}	0.1	13	12	11(b)	10(b)	...	...
		0.5	10	9	10	9	7	6

(a) Load ratio. (b) Extrapolated. Source: Ref 68

Table 14 Effect of oxygen on fatigue crack propagation in room-temperature air for Ti-6Al-4V, RA or FA/DB(a), plate at $R = 0.3$ and frequencies of 1 to 6 Hz(b)

Fatigue crack propagation, *da/dN*		Stress-intensity factor range, ΔK			
		0.20% O_2		0.08-0.10% O_2	
M/cycle	in./cycle	MPa$\sqrt{m}$	ksi$\sqrt{in.}$	MPa$\sqrt{m}$	ksi$\sqrt{in.}$
2.54×10^{-5}	10^{-3}	56, 53, 46	51, 48, 42	60, 64	55, 58
2.54×10^{-6}	10^{-4}	39, 33, 33, 33	35, 30, 30, 30	43, 43	39, 39
2.54×10^{-7}	10^{-5}	19, 19, 18	17, 17, 16	21, 21	19, 19
2.54×10^{-8}	10^{-6}	11, 11, 12	10, 10, 11	12, 15	11, 14

(a) RA = recrystallization annealed; DB = diffusion bonded or DB thermal cycle. (b) No frequency effect over range used. Source: Ref 25

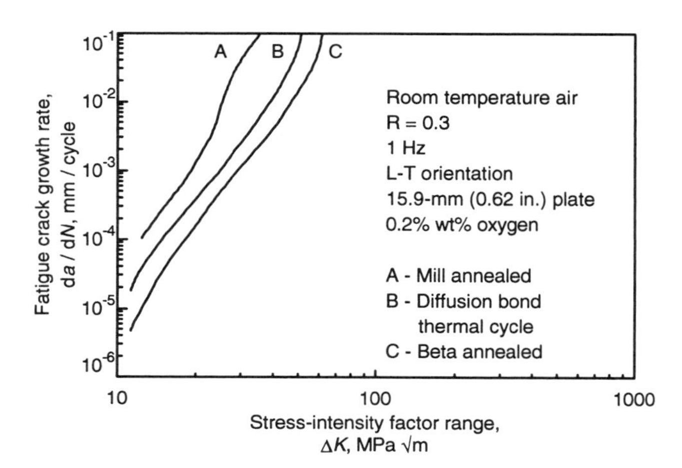

Fig. 10 Effect of heat treatment on fatigue crack growth rate of alloy Ti-6Al-4V. Source: Ref 25

Wells (Ref 67) reported that there is a material thickness effect on FCP, but, because his data lie within a scatter band similar to that illustrated in Fig. 9, his results require confirmation.

The following section addresses the variables that influence fatigue crack propagation, FCP, in titanium. As in the previous section, the effects of chemistry, microstructure, texture, and environment are discussed in order.

Effects of Alloy Chemistry. Titanium alloys have different FCP characteristics just as they have different K_{Ic} characteristics. Selected data from Beyer et al. (Ref 68), summarized in Table 13, indicate that fatigue cracks propagate more rapidly in Ti-6Al-2Sn-4Zr-6Mo than in Ti-8Al-1Mo-1V or Ti-6Al-2Sn-4Zr-2Mo under the same test conditions. This may be a simple effect of strength. However, the relative amounts of beta phase may lead to intrinsically different fatigue crack propagation characteristics. The Ti-6Al-2Sn-4Zr-6Mo alloy is also more easily textured. In addition, different phases, such as orthorhombic martensite, may exist and could effect FCP problems after aging.

Just as for K_{Ic}, it appears that oxygen has an effect on FCP. Results of one investigation are presented in Fig. 8. In addition, the results obtained by Ferguson and Berryman (Ref 25), which are summarized in Table 14, show that a lower oxygen content improves FCP resistance. The effect of oxygen in these data is significant only in a statistical sense for ΔK values above about 20 MPa$\sqrt{m}$ (18 ksi$\sqrt{in.}$). In that region, increases in oxygen content cause increases in crack propagation rate. Such a trend might be anticipated on the premise that K_{Ic} approximates a limiting values for ΔK in FCP and that K_{Ic} is influenced negatively by oxygen.

Yoder et al. (Ref 69) proposed that an apparent effect of oxygen content on FCP in beta annealed Ti-6Al-4V can be ascribed to an influence of oxygen on the Widmanstätten α packet or colony size and that it is the latter that controls FCP. Lower *da/dN* values were apparently associated with higher oxygen contents and larger Widmanstätten α colonies. This idea needs confirmation. However, Yoder and coworkers (Ref 66, 70, and 71) have adequately established the packet size effect.

Chesnutt et al. (Ref 32) reported similar trends and, moreover, showed that the effect of oxygen in Ti-6Al-4V extends to at least 317 °C (600 °F) for the beta annealed condition. Variations in oxygen content between 0.082 and 0.185% had essentially no effect on FCP for the recrystallization annealed condition.

Chesnutt et al. (Ref 32) also reported effects of hydrogen content on FCP. Hydrogen in the 5 to 325 ppm range appears to accelerate FCP in beta annealed Ti-6Al-4V but has no discernible effect on FCP in recrystallization annealed Ti-6Al-4V. At 317 °C (600 °F), hydrogen has no effect on FCP in either microstructure. Additional FCP studies by Mahoney and Paton (Ref 72) on a Si-containing alloy at ambient temperature and with a hold at K_{max} during fatigue cycling have shown that significant accelerations can occur when hydrogen content is in the 150 to 350 ppm range. Further studies by Chesnutt and Paton (Ref 73) have shown that a combination of subambient temperature with a hold at K_{max} during fatigue cycling can cause significant accelerations in FCP in Ti – 6 wt% Al and Ti-6Al-4V alloys in the recrystallization annealed condition. Kennedy et al. (Ref 74) showed the same effect in Ti-6Al-4V and also demonstrated that hydrogen migrates into the region of the crack tip under the influence of the stress gradient.

Effects of Microstructure. Here also, there are general parallels between K_{Ic} and FCP. Like K_{Ic}, FCP is favorably influenced by transformation microstructure and also by application of recrystallization-anneal-type thermal cycles. Microstructure in a given lot of Ti-6Al-4V can affect FCP by a factor of more than ten, and can affect ΔK by 5 to 30 MPa$\sqrt{m}$ (4 to 27 ksi$\sqrt{in.}$), depending on where these parameters are measured on the *da/dN* curve; see, for example, Paton et al. (Ref 12), who reported the work of Williams and Rauscher. Generally speaking, beta annealed microstructures have the lowest fatigue crack growth rates, whereas mill annealed microstructures yield the highest growth rates. A typical example of such behavior is shown in Fig. 10, after Ferguson and Berryman (Ref 25). Wells (Ref 67) also reported lower FCP rates in Ti-6Al-4V after beta annealing. Bjeletich (Ref 49) and Tobler (Ref 57) reported results for weldments in Ti-6Al-4V. The welds were equivalent in FCP to the base metal. Yoder et al. (Ref 75) reported a 50-fold difference in Region II FCP, but both microstructure and oxygen level varied in the specimens they studied.

Amateau et al. (Ref 64) showed that the effects of Widmanstätten α microstructure on FCP in Ti-6Al-6V-2Sn is entirely analogous to that in Ti-6Al-4V. However, Yoder et al. (Ref 76) found crack branching in Ti-6Al-4V but not in Ti-6Al-6V-2Sn. Chesnutt et al. (Ref 32) found that transformed microstructures provide generally lower rates of FCP in Ti-6Al-2Sn-4Zr-6Mo.

Effects of Crystallographic Texture. Ferguson and Berryman (Ref 25) found effects of FCP test direction to be minimal in Ti-6Al-4V plate and forgings for a range of oxygen levels and heat treatment conditions. Because section sizes were generally above 14 mm (1/2 in.) in this work, it can be surmised that crystallographic texture was not strongly developed. Hall et al. (Ref 60) reported a similar result for beta annealed Ti-6Al-4V ELI involving a single comparison at 60 Hz.

Table 15 Tensile properties of equiaxed α–β Ti-Mn alloys

Alloy No.	0.2 Pct TYS (MPa) LT(a)	0.2 Pct TYS (MPa) TL(b)	UTS (MPa) LT	UTS (MPa) TL	Elongation, % LT	Elongation, % TL	Reduction in area, % LT	Reduction in area, % TL	UTS/TYS LT	UTS/TYS TL
1	296	372	423	429	32	38	65	66	1.43	1.15
2	378	438	525	534	26	29	56	56	1.39	1.22
3	430	532	652	715	28	25	48	46	1.52	1.34
4	600	704	823	864	26	22	44	47	1.37	1.23
5	841	907	943	960	20	20	33	42	1.12	1.06
6	898	929	898	935	20	20	44	28	1.00	1.10

(a) LT: load axis parallel to the longitudinal rolling direction. (b) TL: load axis parallel to the long transverse direction. Source: *Met. Trans.*, Vol 15A, 1984, p 157

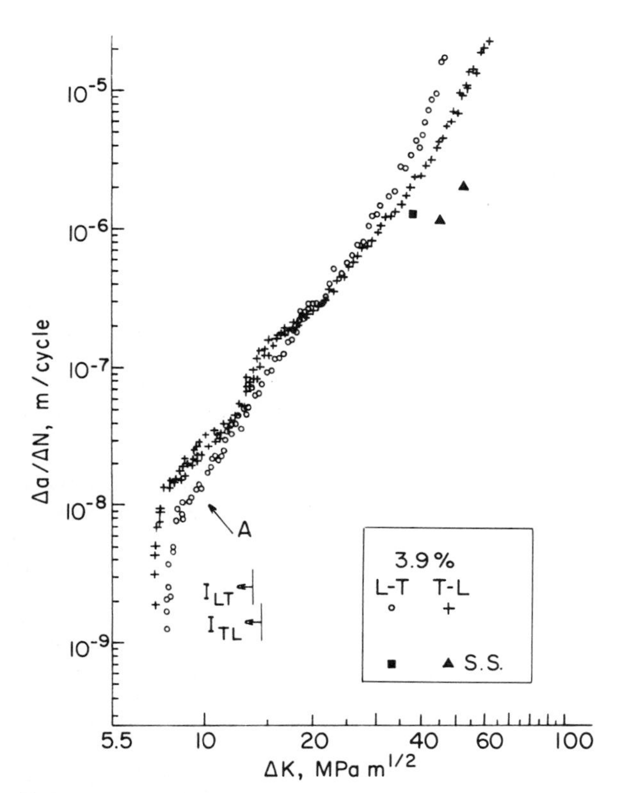

Fig. 11 Crack propagation data for the 3.9 Mn alloy in the LT and TL directions. Arrow indicates end of Stage I. Letter indicates ΔK value where SEM photomicrograph was taken. S.S. refers to crack propagation rates determined from striation spacings. Source: *Met. Trans.*, Vol 15A, Jan 1984, p 160

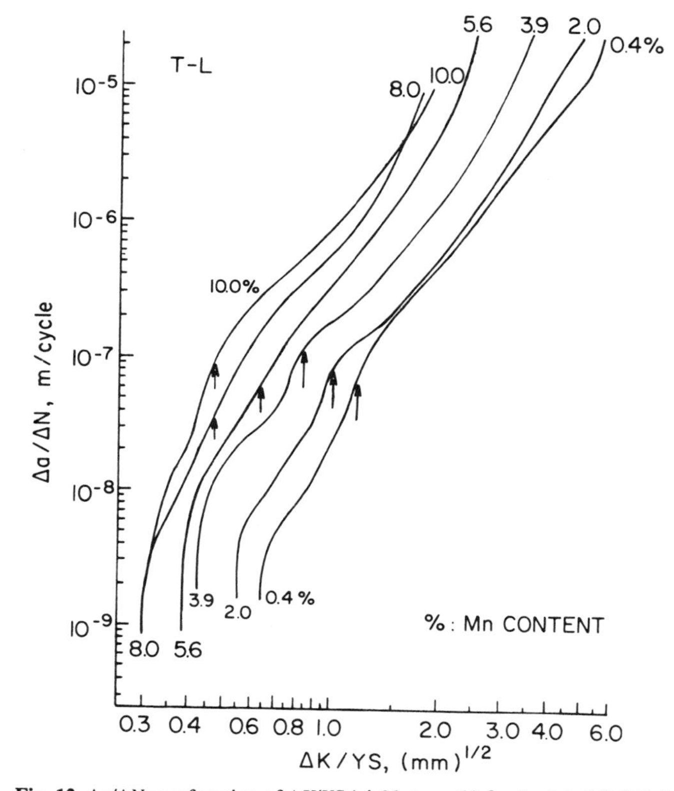

Fig. 12 $\Delta a/\Delta N$ as a function of $\Delta K/YS$ (yield strength) for the 0.4, 2.0, 3.9, 5.6, 8.0, and 10.0 Mn alloys for the TL direction. Arrows indicate end of Stage I. Source: *Met. Trans.*, Vol 15A, Jan 1984, p 161

Stubbington (Ref 10) described his own work and the results of Bowen (Ref 77) which showed a significant effect of test direction in a strongly basal textured Ti-6Al-4V bar at room temperature. He concluded that, whereas a five-minute dwell at maximum load had no effect when the maximum stress was parallel to the basal plane, this was not the case when the maximum stress was normal to the basal plane. In the latter instance, a five-minute hold caused a seven-fold increase in FCP rate. Moreover, he found that this increase depended on the volume fraction of primary alpha phase.

Stubbington (Ref 10) also found the factor n in the Paris equation to vary with test direction. For the L-S, S-L, and T-S orientations, n was about 2.5 and the fatigue striations were normal to the direction of growth. For the L-T, S-T and T-L orientations, n values ranged from 3.1 to 4.1. For these orientations, the fracture topography was complex and no single crack growth model could be proposed.

In summary, the effects of texture exist, but, at least for certain mill product forms, they may not be discernible unless the crystallographic texture is fairly well developed.

The effects of texture, volume fraction of phases, and yield strength were investigated for a series of Ti-Mn alloys containing 0.4 to 10% Mn (Ref 77), and finished rolled in the alpha + beta phase field to produce equiaxed alpha structures. The dimensions of the test specimens met ASTM specifications E399-1974 for compact tension testing and fatigue crack propagation. An R ratio of 0.1 was used. For these alloys the volume fraction of beta varied from 0.19 for the 0.4% Mn alloy to 0.759 for the 10% Mn alloy with the compositions of the alpha and beta each essentially constant. The tensile properties are shown in Table 15. For both the L-T and T-L directions the beta phase has approximately three times the yield strength of alpha, as was found in another study (Ref 78).

In general texturing increased with increasing Mn content. The alpha texture can be described as one in which the basal planes are largely normal to the rolling plane and lie within a region of ± 25° around the transverse direction. Within this scatter the texture can be expressed as $(1\bar{2}10)\ [\bar{1}100]$. The beta texture could be described as $(001)\ [\bar{1}10]$ with a scatter around the transverse direction.

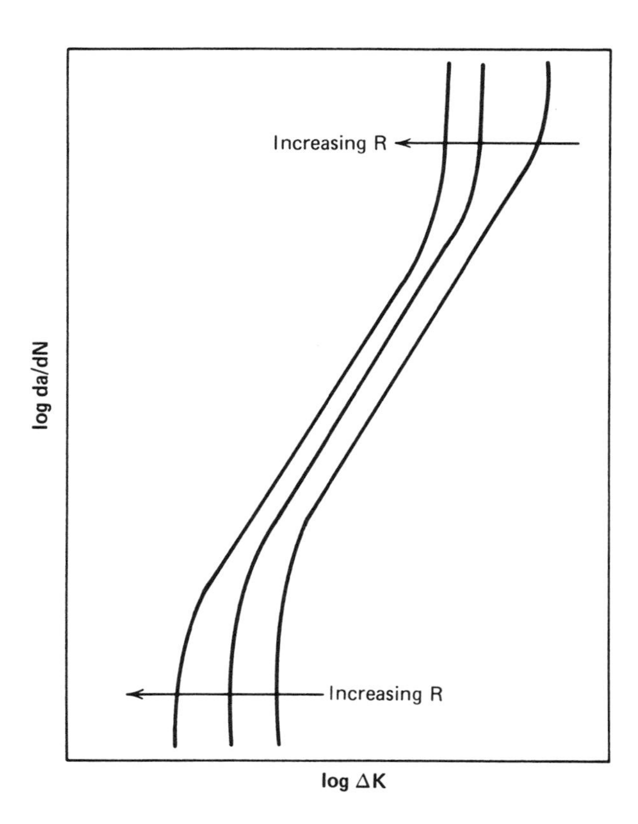

Fig. 13 Schematic illustration of *da/dN*-Δ*K* curve behavior as a function of increasing *R* in titanium alloys. Note that the effect of *R* on *da/dN* is more pronounced in Stage I and Stage III. Source: Ref 32

A typical $\Delta a/\Delta N$ vs ΔK curve is shown in Fig. 11. There is generally little scatter in the data and relatively little difference in the L-T and T-L propagation rates. When beta became the matrix phase at 5.6 Mn the tendency was for crack propagation to be slightly greater in the T-L direction.

When the ΔK values are normalized for yield strength the $\Delta a/\Delta N$ curves take the form shown in Fig. 12. From this figure, it can be seen that crack propagation rates increase with Mn content and therefore with volume fraction of beta phase. The shift of the T-L test direction to the more rapid crack propagation rate was considered to occur because the beta texture permitted greater crack extension per cycle than did the alpha texture. It may well be that the more rapid crack propagation rate associated with the beta phase is related to the presence of a pre-omega structure in the beta, a structure which would raise the yield strength of beta. The omega phase tends to dissolve when slip passes through it (Ref 79). The same can be expected of the pre-omega structure. Thus, the local slipped region would be softer, permitting further slip and making crack propagation easier.

It is also of interest to note that the crack propagation rate of Stage II could be related to the square root of the plastic zone site, which could be measured on a cross section normal to the crack propagation plane.

***R*-Ratio and Frequency Effects.** The most often studied mechanical variables are load ratio *R* (minimum load/maximum load) and load application frequency. Numerous authors (Ref 10, 25, 60, 68, and 72) are in agreement that increasing *R* at the same ΔK level generally leads to increased FCP rates. A schematic of this effect appears in Fig. 13. Load ratio effects are usually strongest in the Stage I and Stage III portions of the FCP curves.

Yuen et al. (Ref 65) reported that this effect extends into the region of negative *R*, at least to $R = -5.0$ for Ti-6Al-4V. The results of Beyer et al. (Ref 68) on Ti-8Al-1Mo-1V, Ti-6Al-2Sn-4Zr-2Mo and Ti-6Al-2Sn-4Zr-6Mo are not in agreement, however; here, increasing negative *R* ratios gave lower FCP rates.

The data of Ferguson and Berryman (Ref 25) on Ti-6Al-4V support the results of Chesnutt et al. (Ref 32). In these cases, increasing positive *R* ratios increased the FCP rate.

Hall et al. (Ref 60) agree with both of the above references and show that the extent to which Stage III diverges as *R* changes depends on test conditions and environment.

Frequency effects are complex. Lower frequencies lead to higher FCP rates only in a general, over-all sense. Many comparisons do not show frequency effects, whereas others do. There appear to be some reasons for this.

The work of Beyer et al. (Ref 68) addresses one important aspect. In their work on Ti-8Al-1Mo-1V at 482 °C (900 °F), frequency had little or no effect at $R = 0.1$ but had rather large effects at $R = 0.5$ and 0.7. Similar but not identical trends were noted for Ti-6Al-2Sn-4Zr-2Mo and Ti-6Al-2Sn-4Zr-6Mo alloys at high temperatures by the same authors.

Bjeletich (Ref 49) studied another aspect. In the heat-affected zones of electron beam weldments in Ti-6Al-4V, he found no frequency effect over the range from 0.1 to 10 Hz at $R = 0.1$ in laboratory air at 79 °C (175 °F). For these test conditions, however, the introduction of a 3.5% salt solution produced a significantly higher FCP rate for the lower frequency.

Chesnutt et al. (Ref 32) reported a frequency effect (1 vs 20 Hz) on FCP in Ti-6Al-4V in several microstructures in a 3.5% salt solution at $R = 0.1$ at room temperature, whereas Ferguson and Berryman (Ref 25) reported little or no effect of frequency over the range from 1 to 30 Hz for testing in air. The latter authors also reported that neither frequency (1 to 6 Hz range) nor *R* ratio (0.08 to 0.5) affects the FCP rate in mill annealed Ti-6Al-4V sheet.

Chesnutt et al. (Ref 32) also reported results of tests at 20 Hz on Ti-6Al-4V (in six microstructural conditions) in dry air that were interrupted periodically for five-minute holds at maximum load. For three of the six microstructures, hold time had no effect. Two other microstructures were well within the scatter band. Only for the recrystallization annealed condition were the results conceivably indicative of an effect, with hold times leading to higher FCP rates. Even in that case, however, four of seven data points were within the scatter band.

Frequency effects, however, may depend on factors such as chemistry and environment. See the works of Chesnutt and Paton (Ref 73) and Mahoney and Paton (Ref 72) for hydrogen and temperature interactions in this regard. Bania and Eylon (Ref 81) reported a possible beneficial effect of a five-minute dwell at maximum load for Ti-6Al-4V that, if real, depends on heat treatment. This effect is suppressed for the solution treated and aged condition, and is present for three different annealing treatments.

Thermal and Chemical Environment Effects. The data of Bjeletich (Ref 49) show insignificant differences in FCP rates over the temperature range from −53 to +77 °C (−64 to +170 °F) for electron beam weldments in Ti-6Al-4V plate. The results of Ferguson and Berryman (Ref 25) are similar except that, in the threshold region, FCP is faster at room temperature than at −53 °C (−64 °F).

Chesnutt et al. (Ref 32), however, studied a wider range of temperatures (−83 to +317 °C; −118 to +600 °F) and found that increasing temperature accelerates FCP at 20 Hz and $R = 0.7$ in Ti-6Al-4V pancake forgings given recrystallization anneals or alpha-beta solution treatments followed by aging. For a beta-finish forging given a beta solution treatment and an over-aging treatment, however, these authors found little or no temperature effect. For $R = 0.3$, the converse of the foregoing microstructure effects appears to apply.

Tobler (Ref 82) examined the FCP behavior of Ti-6Al-4V ELI over the temperature range from −269 to +22 °C (−452 to +72 °F) and found that all data fell within the scatter band.

In summary, the effect of test temperature is not always present, but when it is it seems to depend on both load ratio and microstructure. At least in the literature cited, the effect of increasing temperature, when it exists, is to increase the FCP rate. In theory, however, the opposite effect

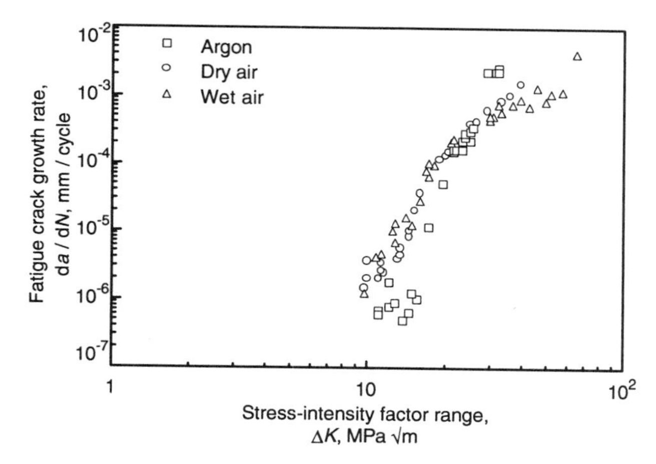

Fig. 14 Dependence of FCP in α-β forged Ti-6Al-4V pancake forging on gaseous environment. Testing at 20 Hz and $R = 0.1$. Pancake alpha-beta forged and then heated at 973 °C (1780 °F) for 4 h, cooled at 50 °C (90 °F) per hour to 760 °C (1400 °F), and air cooled. Source: Ref 32

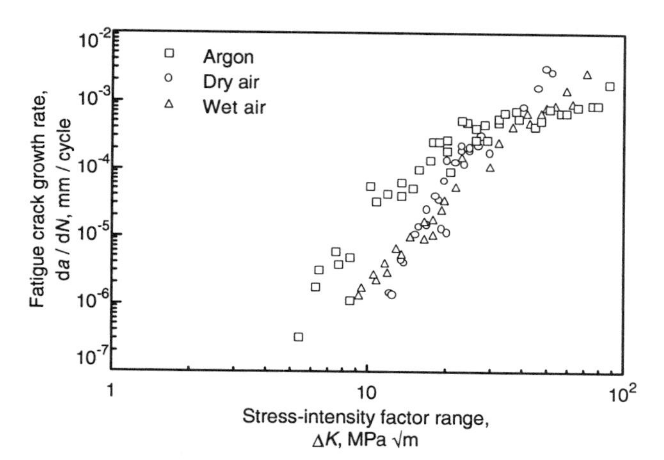

Fig. 15 Dependence of FCP in β forged Ti-6Al-4V pancake forging on gaseous environment. Tested at 20 Hz and $R = 0.1$. Pancake beta forged and then heated at 1037 °C (1900 °F) for 1/2 h, water quenched, reheated to 704 °C (1300 °F), and air cooled. Source: Ref 32

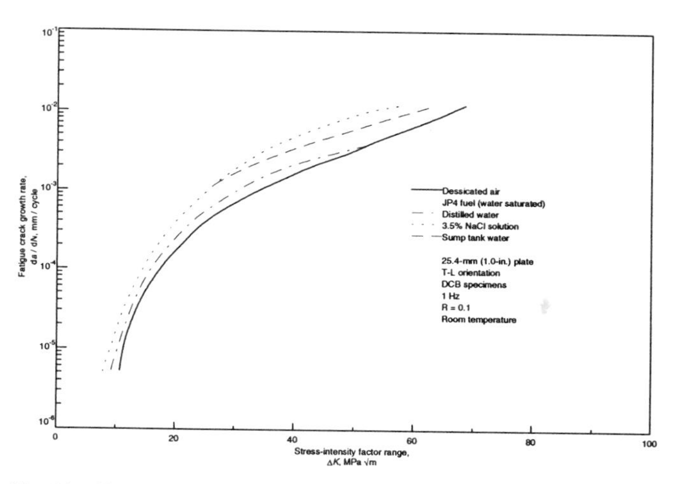

Fig. 16 Effect of environment on fatigue crack growth rates at $R = 0.1$ in recrystallization annealed alloy Ti-6Al-4V (1 Hz). Source: Ref 60

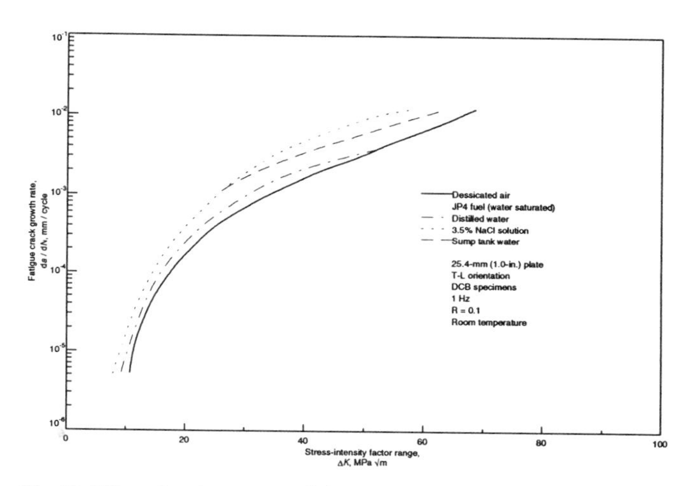

Fig. 17 Effect of environment on fatigue crack growth rates at $R = 0.50$ in recrystallization annealed alloy Ti-6Al-4V (0.1 Hz). Source: Ref 60

could be observed if hydrogen were present and its effect diminished with increasing temperature.

Being a reactive metal, titanium will react with all nonmetallic elements excluding the inert gases. For titanium oxides, nitrides, and carbides, the free energies of compound formation are higher than for most other metals. Titanium will also readily absorb any nascent hydrogen that happens to be reduced on a fresh surface, such as at the root of an advancing fatigue crack. Such surfaces also absorb oxygen, and perhaps nitrogen, if environmental oxygen becomes depleted.

Because of the reactive nature of titanium, therefore, it should come as no surprise that environment affects FCP rates in titanium just as it affects fracture toughness. The only surprise in the available data is that the chemical environment does not have a larger effect than it does. In general, only the more severe environments (such as a 3.5% NaCl solution) affect FCP rates by an order of magnitude or more, and these effects may in part depend on mechanical factors as discussed in the previous section. Considerably larger effects of environment on crack growth rates may be found near the fatigue threshold.

Chesnutt et al. (Ref 32) tested specimens of Ti-6Al-4V in several environments. The effects of argon, dry air, and wet air environments seem to depend at least on microstructure, as shown in Fig. 14 and 15. The effect of a mildly reactive environment is not always negative.

Ferguson and Berryman (Ref 25) studied FCP rates in recrystallization annealed Ti-6Al-4V plate as affected by sump tank water and JP4 fuel environments using low-humidity air as a basis for comparison. Although the effect of sump tank water was a slight increase in FCP rate in most cases, this effect was less than that of an increase in R from 0.3 to 0.5. Where material chemistry was reported, the effect of sump tank water was independent of alloy oxygen content beyond the 0.08 to 0.15% range. The work of Ferguson and Berryman on FCP rates in JP4 fuel is less complete, but the available data indicate that JP4 fuel is a mild environment.

Hall et al. (Ref 60) found FCP rates for beta annealed Ti-6Al-4V ELI plate in JP4 fuel, water, and sump tank water environments to lie between those in air and those in salt water. Salt water is a relatively harsh environment, and its effect is to enhance FCP rates. Increasing R or decreasing frequency enhances the effect of salt water on FCP. Figures 16 and 17 illustrate the effect of frequency rather dramatically. These authors also found that dye penetrant type ZL-2A has no influence on FCP rate relative to the rate in dry air for both beta annealed Ti-6Al-4V ELI.

Data obtained at McDonnell Aircraft (Ref 62) also show JP4 fuel to affect FCP in Ti-6Al-4V to a lesser degree than does salt water (3.5% NaCl).

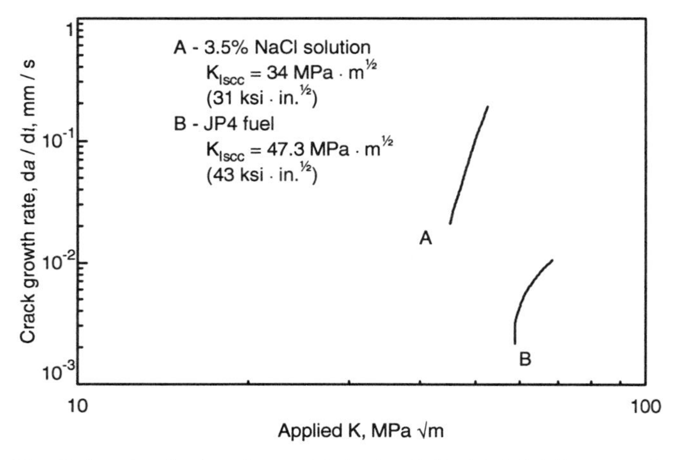

Fig. 18 Sustained load crack growth behavior of alloy Ti-6Al-4V in two environments. $K_{Ic} = 56\ MPa\sqrt{m}$ (51 ksi$\sqrt{in.}$). Source: Ref 62

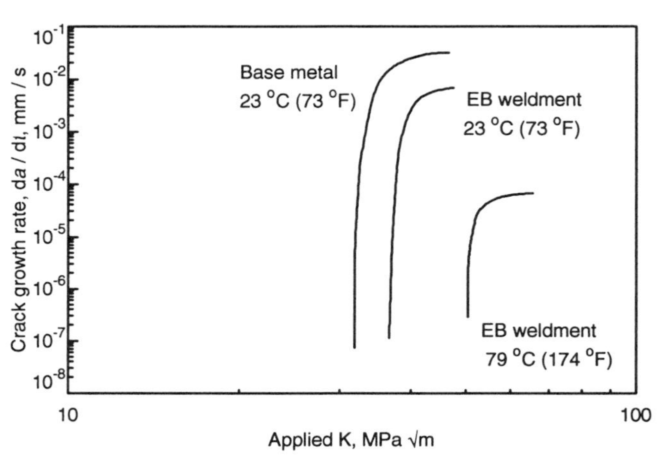

Fig. 19 Effect of microstructure and temperature on sustained load crack propagation in Ti-6Al-4V in 3.5% NaCl solution. Source: Ref 49

Harmsworth (Ref 54) reported data showing that water and JP4 fuel are relatively mild environments for Ti-6Al-6V-2Sn. As for Ti-6Al-4V, the testing frequency and a salt water environment have a combined effect on FCP for Ti-6Al-6V-2Sn.

Sustained-Load Crack Propagation

Although rising load and dynamic load spectra are important considerations in vehicle design, the ability of a material to sustain a load without failure in the presence of a fatigue crack or other sharp crack in a specific environment also may be important.

Only a few studies have been reported on sustained-load crack propagation for titanium alloys, and these studies have been concerned primarily with static load crack growth in the presence of JP4 fuel or a 3.5% NaCl solution. Results obtained at McDonnell Aircraft (Ref 62) are presented in Fig. 18. The rank order of severity of effect of JP4 fuel and 3.5% NaCl on sustained-load crack propagation rate in Ti-6Al-4V is the same as for fatigue crack propagation rate: 3.5% NaCl has the more severe effect.

Bjeletich (Ref 49) reported similar results for Ti-6Al-4V weldments compared with mill annealed base metal. He also found a microstructure and temperature effect, which is shown in Fig. 19. Just as is the case for K_{Ic} and FCP, transformed microstructures (in the welds) lead to lower propagation rates. The negative effect of temperature is probably due to crack blunting via enhanced corrosion at the crack tip.

For the Ti-4Al-3Mo-1V alloy, Williams (Ref 83) found no difference between vacuum and moist air environments in time to failure under the static load. In limited work, he also found little effect of salt water. This lack of effect is probably related to the low aluminum content of this alpha-beta alloy.

In early work, Hatch et al. (Ref 63) pointed out that sustained-load crack propagation in 3.5% NaCl solution is not limited to thick materials. These authors showed that a number of alloys are susceptible in sheet form and moreover that gage effects can exist within a given alloy.

It is also evident that some care should be taken during sustained-load testing. Petrak (Ref 84) found that exposure to room air between the termination of fatigue precracking and static-load testing in salt water may affect K_{Iscc} results in Ti-5Al-2.5Sn. Overnight (or longer) exposures in room air appeared to have the effect of blunting the crack and increasing the K_{Iscc} value. The data are too sparse to indicate the magnitude of the effect.

Mechanical Properties: Beta Alloys

The β- and β-rich α/β-alloys (hereinafter referred to as beta alloys), in that they can be heat treated to a wide range of strength levels, offer the opportunity to tailor the strength-toughness properties combinations to a specific application. That is, moderate strength with high toughness or high strength with moderate toughness can be achieved. This is generally not possible for other types of titanium alloys as they cannot be heat treated over a very wide range. At moderate strength levels, say 965 MPa (140 ksi) and above, the fracture toughness of the beta alloys can be processed to achieve higher values than for the other types (α- and α/β-alloys).

To accomplish these higher toughnesses, however, the processing window is tighter than that normally used for the other alloy types. For the less highly β-stabilized alloys, such as Ti-10V-2Fe-3Al, Ti-17, and Beta-CEZ for example, the thermomechanical process is critical to the properties combinations achieved as this has a strong influence on the final microstructure and the resultant tensile strength and fracture toughnesses which may be achieved.

This is somewhat less important in the more highly β-stabilized alloys such as Ti-3Al-8V-6Cr-4Mo-4Zr (Beta C) and Ti-15V-3Cr-3Al-3Sn. In these the final microstructure, precipitated α, is so fine that microstructural manipulation through thermomechanical processing is not as effective. In these cases the aging heat treatments—sequence and temperature—are more critical. The key is to obtain a uniform precipitation. This may be obtained by a low-high aging sequence, or, with residual cold or warm work, possibly a high-low aging sequence. When highly alloyed beta alloys, such as Beta C, are cold worked prior to aging, high strength can be obtained with good ductility because cold work induces finer and more uniform precipitation.

The thermomechanical processing must, however, be controlled to provide a uniform microstructure throughout the cross-section of the material, and, in conjuction with the heat treatment, avoid the occurrence of extensive grain boundary α or a precipitate free zone near the grain boundaries.

Some characteristic mechanical properties of β alloys are given in Table 16. Comparison of Table 16 with Table 6 indicates that the β alloys have a higher fracture toughness at higher yield strength than the α/β alloys listed in Table 6. More complete property data are provided in Ref 21 and the reference book *Materials Properties Handbook: Titanium Al-*

Table 16 Typical mechanical properties of selected beta alloys

Alloy	Tensile yield strength, MPa	Tensile yield strength, ksi	Ultimate tensile strength, MPa	Ultimate tensile strength, ksi	Elongation, %	Reduction in area, %	K_{Ic}, MPa$\sqrt{m}$	K_{Ic}, ksi$\sqrt{in.}$
Beta C(a)	1090	158	1143	165.8	5.5	8.6	72.2	65.6
Beta 21S	1150	164	1057	151	8	...	101(b)	92(b)
Ti-15-3(c)	...	...	1520	220	...	16	59	53.6
Beta CEZ(d)	1200	174	1315	191	10	26	75	69
Ti-10-2-3 (high strength)(e)	1185	172	1250	181	8	18	52	47
Ti-10-2-3 (medium strength)(e)	1080	157	1160	168	16	44	73	66
Ti-10-2-3 (low strength)(e)	940	136	1020	148	22	61	102	93

(a) 457 mm (18 in.) diam × 238 mm (9 ⅜ in.) I.D. extrusion, air cooled from 815 °C (1500 °F) (above the transus), aged 24 h at 565 °C (1050 °F). (b) K_c for 1.3 mm (0.050 in.) strip, aged 8 h at 595 °C (1100 °F). (c) Solution treated (above the beta transus) at 850 °C (1560 °F), aged for 11 ks at 600 °C (1112 °F), re-aged at 500 °C (932 °F) for 43 ks. (d) Nominal composition, Ti-5Al-2Sn-4Zr-4Mo-2Cr-1Fe; 300 mm diam pancake, processed through the beta transus to 600 °C (1112 °F), reheated to 830 °C (1472 °F), water quench, plus 570 °C (1054 °F) 8 h, air cool. (e) For forgings, Ti-10-2-3 is heat treated 16 to 33 °C (60-100 °F) below the beta transus (744-765 °C, or 1370-1410 °F) and water quenched. Aging temperatures range from 480 to 620 °C (900-1100 °F) depending on desired strength.

loys (ASM, 1994). Some representative data for selected β alloys are also given in the Appendix following this article.

Fatigue of Beta Alloys. The high cycle fatigue strength of beta titanium alloys is superior to alpha + beta or near-alpha titanium alloys. This is the result of the higher strength level (150-230 ksi) in comparison to alpha + beta alloys (130-160 ksi), and also the result of the microstructural refinement typical of solution treated and aged (STA) material. While the grain size range of near-alpha or alpha + beta alloys is 10 to 50 microns, the alpha grain size of STA beta alloys could be under 1 micron. The very high tensile strength, coupled with microstructural refinement, result in some cases a 10^6 fatigue strength at 965 MPa (140 ksi), which is almost twice as high as that of mill annealed Ti-6Al-4V. This good combination of strength and fine structure also leads to fatigue ratios *s* (*s* = fatigue strength/tensile strength) of up to 0.8.

For this reason, in recent years, many fatigue critical applications for titanium, such as helicopter rotors, landing gear components, and airframe structures, are considering the use of high strength beta alloys such as the Ti-10V-2Fe-3Al. However, there is a down side to the fatigue behavior of beta titanium alloys. Because these alloys are typically used in the high strength STA condition, they have poor tolerance to material defects such as casting porosity or occasional large grain boundary alpha grains. Such large grains can be found along prior boundaries of beta grains when the alloy is slowly cooled during beta solution treatment. These grains are known as grain boundary alpha (GBα) and when their thickness exceeds few microns, they may cause a severe degradation of the fatigue life. Grain boundary alpha grains are typically very long and planar and may cause fatigue crack initiation along their α/β interfaces. On the fatigue fracture surface, they are noticeable as large, flat, and shiny initiation sites. When such fatigue failure behavior is observed, any effort should be made to modify the process in such a way that will eliminate the GBα through faster cooling. If such elimination is not possible, such as in large forgings or thick section castings, strength level in excess of 1170 MPa (170 ksi) should be avoided.

An optimized beta alloy for fatigue application should be processed in such a way to result in defect-free material with small prior beta grains, no visible grain boundary alpha phase, and a fine and uniform dispersion of the precipitated alpha phase. Because of the high strength condition of these alloys, a special consideration should also be given to the machined surface, as the alloy sensitivity to defects is high.

Another issue is the sensitivity to interstitial elements, such as hydrogen or oxygen. The beta phase has much higher diffusivity and solubility for such elements and surface contamination is possible during casting, heat treating, and high-temperature application which may lead to a loss of fatigue strength. Of course, most beta titanium alloys are not used at temperatures high enough to have a concern over this point. In addition, beta alloy parts are, typically, chem milled after casting or heat treatment to remove surface contamination.

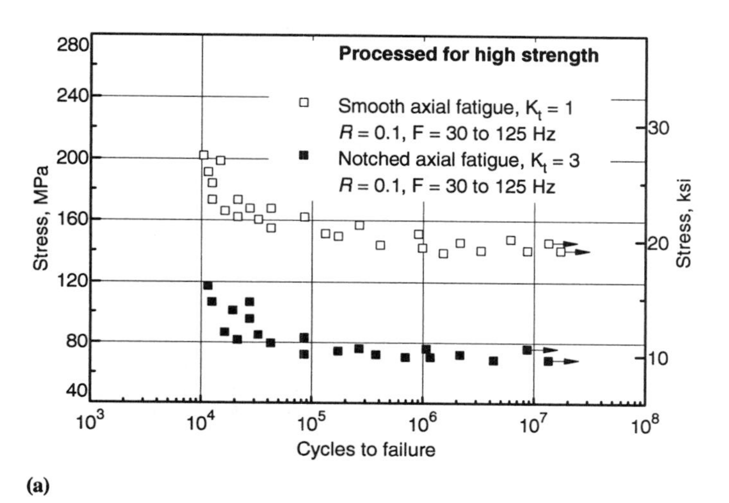

(a)

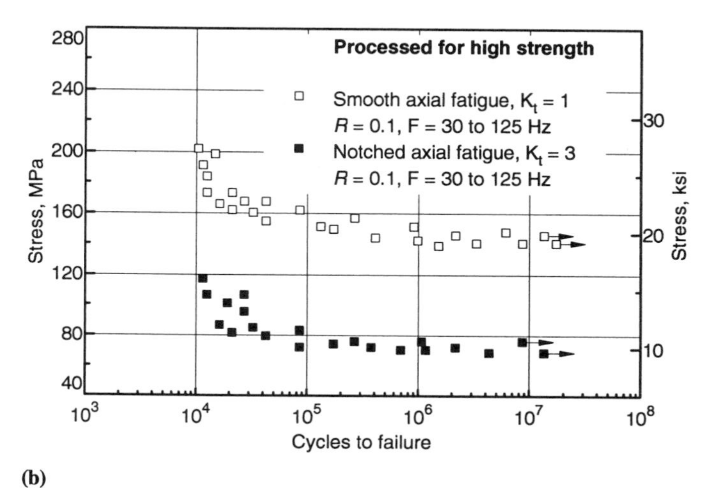

(b)

Fig. 20 Smooth and notched axial fatigue data Ti-10V-2Fe-3Al precision and conventional forgings. (a) Processed for high strength level, 173/180 ksi (1193/1240 MPa) UTS, AMS 4983A/4984. (b) Processed for low strength 140 ksi (965 MPa) UTS per AMS 4987. Source: Ref 21, p 508, 509

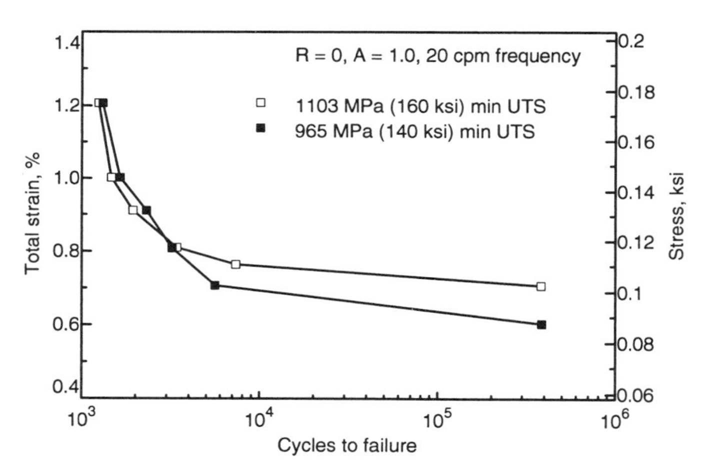

Fig. 21 Strain controlled, low-cycle fatigue data Ti-10V-2Fe-3Al forgings, processed to intermediate strength levels, 160 ksi (1103 MPa) and 140 ksi (965 MPa) UTS. Source: Ref 21, p 511

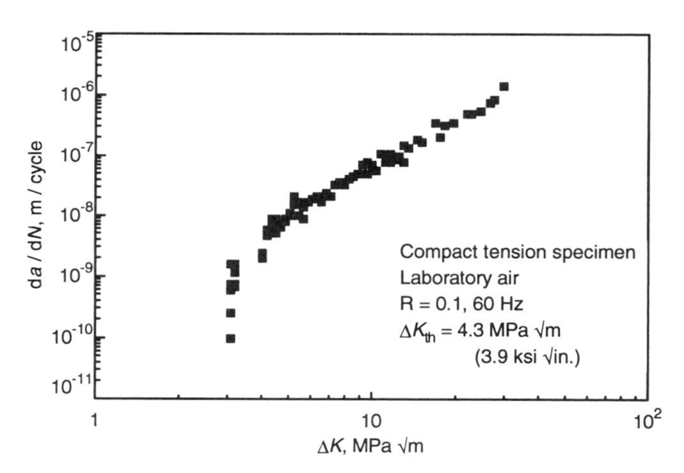

Fig. 22 Fatigue crack growth rate data. Ti-10V-2Fe-3Al forgings. Processed to high strength level. 180 ksi (1240 MPa) UTS. Source: Ref 21, p 511

Typical fatigue life curves for two strength levels of Ti-10V-2Fe-3Al are shown in Fig. 20 and 21. Fatigue crack propagation results for approximately the same strength level of 1240 MPa (180 ksi) are given in Fig. 22. The data for Fig. 22, referring to tests in air, show a higher resistance to crack propagation for Ti-10Al-2Fe-3Al at a higher strength level than would be reached for Ti-6Al-4V tested in air and given the indicated heat treatment in Fig. 14. Figure 15 presents data for Ti-6Al-4V pancaked forged in the β field, quenched and subsequently reheated in the α + β phase field. The resultant Widmanstätten α + β structure is known to have greater crack resistance than equiaxed structures. Nonetheless, for similar test conditions, beta alloy Ti-10V-2Fe-3Al reveals slower crack propagation rates than the lower strength Ti-6Al-4V alloy.

Smooth bar fatigue life of alpha/beta Ti alloys has shown that for a given strength level fatigue crack initiation is easier in Widmanstätten structures than in equiaxed structures and takes place more readily for coarser microstructures (Ref 85, 86). In this connection it is interesting to note that laser treatment plus aging of the surface of 10 mm thick plate of Ti-15V-3Cr-3Al-3Sn (Ti-15-3) has produced a finer α + β structure at the surface than at the interior (Ref 87). This finer structure would tend to make the alloy more resistant to crack nucleation, other factors remaining the same, while the coarser inner structure would make the alloy more resistant to crack propagation. Thermomechanical modification of surface microstructures is also used to enhance fatigue resistance of alpha and alpha/beta alloys (Ref 88).

Hydrogen Effects. Beta alloys can absorb large quantities of hydrogen, on the order of 4000-5000 ppm (R. Schutz in Ref 21, p 75-91). However, this is a metastable solubility, since the addition of alloying elements to the beta phase increases the vapor pressure of hydrogen at a given hydrogen content (Ref 89). Hydrides do not form in beta despite the high degree of supersaturation. However, when alpha is present, hydrides can form at alpha-beta interfaces at relatively low hydrogen contents at room temperature (Ref 90). For example, a Ti-2Fe-2Cr-2Mo alpha-beta alloy (not in use today) with 40 ppm H showed a 55% reduction of area at room temperature. In the presence of 290 ppm H, the reduction of area was reduced to 10%. Direct evidence of hydride precipitation at relatively low hydrogen contents in alpha-beta alloys has been presented in Ref 90 and 91.

It is also of interest to note that the evidence of hydrogen trapping increased in beta Ti alloys after aging (Ref 92). Since aging tends to produce alpha, and since hydrides tend to precipitate at alpha + beta interfaces, these results are not surprising. This interpretation of hydrogen trapping after aging is supported by the observations of Young and Scully (Ref 93) for alloy Ti-15-3 and Beta 21S. For Ti-15-3 hydrogen in solution had essentially no influence on effective plastic strain to fracture, which remained close to 30% for hydrogen contents up to 3000 ppm. However, for the Beta-21S alloy, increasing hydrogen contents from 850 ppm to 2800 ppm reduced the effective plastic strain for fracture from 14% to 3%. After peak aging, effective plastic strain for unhydrogenated Ti-15-3 dropped to 6% and to 3% for Beta 21S. In the aged condition, during which alpha formed, 3800 ppm reduced the effective plastic strain essentially to zero for Ti-15-3 while 900 ppm was sufficient to reduce it to zero for Beta 21S.

The difference in behavior of beta alloys, Beta 21S and Ti-15-3, containing hydrogen raises the interesting question as to why these differences occur, particularly for the cases when both alpha and beta phases are present together. Recent evidence (Ref 94) indicates that hydrogen in an α-Ti/β-Ti-13Mn couple shifted from alpha to beta during cooling. Such a shift would occur only if, during cooling, the hydrogen activity in alpha increased relative to the activity of hydrogen in beta. It has also been shown (Ref 91) that a Ti-4Mn-4Al alloy has greater resistance to room temperature slow strain hydrogen embrittlement than a Ti-8Mn alloy. For the Ti-8Mn alloy a sharp drop in ductility occurs at 200 ppm H, while for the Ti-4Mn-4Al alloy, it requires 1000 ppm H before the sharp drop in ductility is seen.

The alpha of the Ti-8Mn alloy probably contains on the order of 0.1-0.2% Mn, while the alpha of the Ti-4Mn-4Al contains at least 4Al and possibly more. The Al in alpha of the Ti-4Mn-4Al alloy would increase the activity of hydrogen relative to the alpha of the Ti-8Mn alloy (Ref 95, 96). If, during room-temperature straining hydrogen moved from beta to the alpha/beta interface to precipitate, where the hydride has been seen (Ref 91, 92), the increased activity of the Al-alloyed alpha would require a larger content of hydrogen in beta before hydrogen would move to the interface. This would account for the greater tolerance for hydrogen in the Ti-4Mn-4Al alloy.

Most likely the same process operates in the beta alloys when alpha has precipitated. Thus, the composition of alpha must be significant in the movement of hydrogen from alpha to beta and from beta to alpha. The difference in tolerance of the solution treated and annealed beta alloys to hydrogen under constrained conditions of tensile testing, i.e., in the presence of notches, may be related to differences in microstructure resulting from the solution anneal. There is insufficient data available to comment on these differences. Considerable work remains to be done to understand the behavior of hydrogen in both alpha/beta and beta alloys.

REFERENCES

1. J.L. Waisman, R. Toosky, and G. Sines, Uphill Diffusion and Progressive Embrittlement—Hydrogen in Titanium, *Metallurgical Transactions A*, Vol 8 (No. 8), 1977, p 1249-1256
2. P.N. Adler and R.L. Schulte, Stress-Induced Hydrogen Migration in Beta-Phase Titanium, *Scripta Metallurgica*, Vol 12 (No. 7), 1978, p 669-672
3. N.E. Paton et al., *Met. Trans.*, Vol 2, Oct 1971, p 2791-2796
4. J.R. Kennedy, P.N. Adler, and H. Margolin, *Met. Trans. A*, Vol 24A, 1993, p 2763-2771
5. J.L. Waisman, R. Toosky, and G. Sines, *Met. Trans. A*, Vol 8A, 1977, p 1249-1256
6. S. Nourbakhsh et al., *Scripta Met et Mater.*, Vol 30, 1994, p 209-212
7. D.S. Schwartz et al., *Acta Met*, Vol 39, 1991, p 2799-2803
8. H. Margolin and P. Cohen, "Evolution of the Equiaxed Morphology of Phases in Ti-6Al-4V," in *Titanium Science and Technology*, Kimura and Izum, Eds, TMS, 1980, p 1555-1561
9. D.H. Rogers, The Effect of Microstructure and Composition on the Fracture Toughness of Titanium Alloys, in *Titanium Science and Technology*, Vol 3, Plenum Press, 1973, p 1719-1730
10. C.A. Stubbington, Metallurgical Aspects of Fatigue and Fracture in Titanium Alloys, in *Alloy Design for Fatigue and Fracture Resistance*, AGARD-CP-185, NATO, Brussels, Belgium, 1976, p 3-1
11. J.C. Chesnutt and R.A. Spurling, Fracture Topography—Microstructure Correlations in SEM, *Metallurgical Transactions A*, Vol 8 (No. 1), 1977, p 216-218
12. N.E. Paton, J.C. Williams, J.C. Chesnutt and A.W. Thompson, The Effects of Microstructure on the Fatigue and Fracture of Commercial Titanium Alloys, *Alloy Design for Fatigue and Fracture Resistance*, AGARD-CP-185, NATO, Brussels, Belgium, 1976, p 4-1
13. D. Banerjee, et al., *Acta Met*, Vol 36, 1988, p 125-141
14. H.W. Rosenberg, Deformation Mechanics of Alpha Titanium Alloys, Ph.D. thesis, Stanford University, May 1971
15. H.W. Rosenberg, Room Temperature Flow Stress and Strain, *Scientific and Technological Aspects of Titanium and Titanium Alloys*, Vol 1, edited by J.C. Williams and G.F. Belov, Plenum Press, New York, 1982, p 739-745
16. H.W. Rosenberg and W.D. Nix, Solid Solution Strengthening in Ti-Al Alloys, *Metallurgical Transactions*, Vol 4 (No. 5), May 1973, p 1333-1338
17. H.W. Rosenberg, Heat Treating of Titanium Alloys, Paper EM77-16, Society for Manufacturing Engineers, 1976
18. J.C. Williams and M.J. Blackburn, A Comparison of Phase Transformations in 3 Commercial Titanium Alloys, *ASM Transactions Quarterly*, Vol 60 (No. 3), Sept 1967, p 373-383
19. G. Welch et al., Deformation Characteristics of Age Hardened Ti-6Al-4V, *Metallurgical Transactions A*, Vol 8 (No. 1), 1977, p 169-177
20. P.J. Bania, Next Generation Titanium Alloys for Elevated-Temperature Service, *ISIJ International*, Vol 3, 1994, p 840-847
21. D. Eylon, R. Boyer, D. Koss, Eds, *Beta Titanium in the 1990s*, The Metallurgical Society, 1993
22. H. Margolin and J.P. Nielsen, *Titanium Metallurgy*, Modern Materials, Vol 2, Academic Press, 1960, p 225-325
23. P. Bania, "Beta Titanium Alloys and Their Role in the Titanium Industry," *Beta Titanium Alloys in the 1990s*, The Metallurgical Society, 1993, p 3-14
24. M.A. Greenfield, C.M. Pierce, and J.A. Hall, The Effect of Microstructure on the Control of Mechanical Properties in Alpha-Beta Titanium Alloys, *Titanium Science and Technology*, Vol 3, Plenum Press, 1973, p 1731
25. R.R. Ferguson and R.G. Berryman, Fracture Mechanics Evaluation of B-1 Materials, Report AFML-TR-76-137, Vol 1, Rockwell International, Los Angeles, 1976
26. H.W. Rosenberg and W.M. Parris, Alloy, Texture and Microstructural Effects on Yield Stress and Mixed Mode Fracture Toughness of Titanium, *Fatigue and Fracture Toughness—Cryogenic Behavior*, STP 556, American Society for Testing and Materials, Philadelphia, 1974, p 26-43
27. D.E. Cooper, Correlation Study of Fracture Toughness of Airframe Forgings, TIMET Internal Report, Toronto Quality Control Dept., Jan 31, 1974
28. R.H. Van Stone et al., Toughness of Ti-5Al-2.5Sn Plate at Cryogenic Temperatures, *Toughness and Fracture Behavior of Titanium*, STP 651, American Society for Testing and Materials, Philadelphia, 1978, p 154-179
29. R.H. Van Stone, J.R. Low, and J.L. Shannon, Investigation of Fracture Mechanism of Ti-5Al-2.5Sn at Cryogenic Temperatures, *Metallurgical Transactions A*, Vol 9 (No. 4), 1978, p 539-552
30. D.A. Meyn, Effect of Hydrogen on Fracture and Inert-Environment Sustained Load Cracking Resistance of Alpha-Beta Titanium Alloys, *Metallurgical Transactions A*, Vol 5 (No. 11), Nov 1974, p 2405-2414
31. C. Chen (private communication), Wyman Gordon Co. data
32. J.C. Chesnutt, A.W. Thompson, and J.C. Williams, Influence of Metallurgical Factors on the Fatigue Crack Growth Rate in Alpha-Beta Titanium Alloys, Report AFML-TR-78-68, Rockwell International, Thousand Oaks, CA, 1978
33. T.W. Hall and C. Hammond, The Relation Between Crack Propagation Characteristics and Fracture Toughness in Alpha+Beta Titanium Alloys, *Titanium Science and Technology*, Vol 2, Plenum Press, 1973, p 1365-1376
34. T.L. Wardlaw, H.W. Rosenberg, and W.M. Parris, Development of Economical Sheet Titanium Alloy, Report AFML-TR-73-296
35. R.E. Curtis and W.F. Spurr, Effect of Microstructure on Fracture Properties of Titanium Alloys in Air and Salt Solution, *ASM Transactions Quarterly*, Vol 61 (No. 1), March 1968, p 115-127
36. J.A. Hall and C.M. Pierce, Improved Properties of Ti-6Al-6V-2Sn Through Microstructure Modification, Report AFML-TR-70-312, Air Force Flight Dynamics Laboratory, Wright-Patterson AFB, OH, Feb 1971
37. J.A. Hall et al., Property-Microstructure Relationships in Titanium Alloy Ti-6Al-2Sn-4Zr-6Mo, Report AFML-TR-71-206, Air Force Materials Laboratory, Wright-Patterson AFB, OH, Nov 1971
38. W.W. Gerberich and G.S. Baker, On the Toughness of Two-Phase 6Al-4V Titanium Microstructures, *Applications Related Phenomena in Titanium Alloys*, STP 432, American Society for Testing and Materials, Philadelphia, 1967, p 80-99
39. M.A. Greenfield and H. Margolin, Interrelationship of Fracture Toughness and Microstructure in a Ti-5.25Al-5.5V-0.9Fe-0.5Cu Alloy, *Metallurgical Transactions*, Vol 2 (No. 3), March 1971, p 841-847
40. G.J. Petrak, Mechanical Property Evaluation of Beta Forged Ti-6Al-4V, Report AFML-TR-70-291, Dayton University Research Institute, OH, Jan 1971
41. M.G. Ulitchny, H.J. Rack, and D.B. Dawson, Mechanical and Microstructural Properties Characterization of Heat-Treated, Beta-Extruded Ti-6Al-6V-2Sn, *Toughness and Fracture Behavior of Titanium*, STP 651, American Society for Testing and Materials, Philadelphia, 1978, p 17-42
42. J.C. Chesnutt, C.G. Rhodes, and J.C. Williams, The Relationship Between Mechanical Properties, Microstructure

and Fracture Topography in Alpha-Beta Titanium Alloys, *Fractography-Microscopic Cracking Processes*, STP 600, American Society for Testing and Materials, Philadelphia, 1976, p 99-138

43. J.C. Chesnutt et al., The Effect of Microstructure on Fracture of a New High Toughness Titanium Alloy, *Fracture 1977*, Vol 2, University of Waterloo Press, Ontario, Canada, 1977, p 195-202
44. R.G. Berryman et al., A New Titanium Alloy With Improved Fracture Toughness: Ti-4.5Al-5Mo-1.5Cr, *Journal of Aircraft*, Vol 14 (No. 12), 1977, p 1182-1185
45. E. Bohanek, Relationship of Fracture Toughness to Ti-6Al-4V Billet Microstructure and Processing, TIMET Technical Report No. 49, Project 99-5, TIMET, Pittsburgh, Aug 2, 1971
46. H. Margolin et al., Fracture Toughness, Aging Behavior, Grain Growth, and Hardness of Alpha-Beta Titanium Alloys, Report AFML-TR-73-172, Sept 1973
47. Y. Mahajan and H. Margolin, Surface Cracking in Alpha-Beta Titanium Alloys Under Unidirectional Loading, *Metallurgical Transactions A*, Vol 9 (No. 3), 1978, p 427-431
48. M.J. Harrigan et al., The Effect of Rolling Texture on the Fracture Mechanics of Ti-6Al-2Sn-4Zr-6Mo Alloy, *Titanium Science and Technology*, Vol 2, Plenum Press, 1973, p 1297-1317
49. J.G. Bjeletich, Development of Engineering Data on Thick-Section Electron Beam Welded Titanium, Report AFML-TR-73-197, Aug 1973
50. P.L. Hendricks, Effects of Heat Treatment and Microstructure on Tensile and Fracture Toughness Properties of Titanium 6Al-4V Alloy Plate, AIAA Paper 74-374, Las Vegas, NV, April 1974
51. G.S. Hall, S.R. Seagle, and H.R. Bomberger, Effect of Specimen Width on Fracture Toughness of Ti-6Al-4V Plate, *Toughness and Fracture Behavior of Titanium*, STP 651, American Society for Testing and Materials, Philadelphia, 1978, p 227-245
52. A.W. Bowen, Influence of Crystallographic Orientation on Fracture Toughness of Strongly Textured Ti-6Al-4V, *Acta Metallurgica*, Vol 26 (No. 9), 1978, p 1423-1433
53. R.M. Tchorzewski and W.B. Hutchinson, Anisotropy of Fracture Toughness in Textured Titanium-6Al-4V Alloy, *Metallurgical Transactions A*, Vol 9 (No. 8), Aug 1978, p 1113-1123
54. C.L. Harmsworth, Fracture Toughness and Fatigue Properties of Ti-6Al-6V-2Sn Annealed for Possible F-15 Applications, Report No. LA71-1, Wright-Patterson AFB, OH, Feb 2, 1971
55. J.C. Williams et al., Development of High Fracture Toughness Titanium Alloys, *Toughness and Fracture Behavior of Titanium*, STP 651, American Society for Testing and Materials, Philadelphia, 1978, p 64-114
56. N.E. Paton, J.C. Williams, J.C. Chesnutt, and A.W. Thompson, Paper SC-PP-75-60, AGARD Conference on Alloy Design for Fatigue and Fracture Resistance, Belgium, April 1975
57. R.L. Tobler, Low-Temperature Fracture Behavior of a Ti-6Al-4V Alloy and the Electron-Beam Welds, *Toughness and Fracture Behavior of Titanium*, STP 651, American Society for Testing and Materials, Philadelphia, 1978, p 267-294
58. J.L. Christian and A. Hurlich, Physical and Mechanical Properties of Pressure Vessel Materials for Applications in a Cryogenic Environment, Report ASD-TDR-628, Part 2, April 1963
59. R.R. Cervay, Mechanical Properties of Ti-6Al-4V Annealed Forgings, Report AFML-TR-74-4, March 1974
60. L.R. Hall, R.W. Finger, and W.F. Spurr, Corrosion Fatigue Crack Growth in Aircraft Structural Materials, Report AFML-TR-73-204, Sept 1973
61. R. Chait and P.T. Lum, Influence of Test Temperatures on Fracture Toughness and Tensile Properties of Ti-8Mo-8V-2Fe-3Al and Ti-6Al-6V-2Sn Alloys Heat-Treated to High-Strength Levels, *Toughness and Fracture Behavior of Titanium*, STP 651, American Society for Testing and Materials, Philadelphia, 1978, p 180-199
62. Report MDC-A0913, Phase B Test Program, McDonnell Aircraft Co., McDonnell Douglas Corp., St. Louis, May 18, 1971
63. A.J. Hatch, H.W. Rosenberg, and E.F. Erbin, Effects of Environment on Cracking in Titanium Alloys, *Stress Corrosion Cracking of Titanium*, STP 397, American Society for Testing and Materials, Philadelphia, 1966, p 122-136
64. M.F. Amateau, W.D. Hanna, and E.G. Kendall, The Effect of Microstructure on Fatigue Crack Propagation in Ti-6Al-6V-2Sn Alloy, International Conference on Mechanical Behavior of Materials, Kyoto, Japan, Aug 17, 1971, p 77-89
65. A. Yuen et al., Correlations Between Fracture Surface Appearance and Fracture Mechanics Parameters for Stage-II Fatigue Crack-Propagation in Ti-6Al-4V, *Metallurgical Transactions*, Vol 5 (No. 8), 1974, p 1833-1842
66. G.R. Yoder, L.A. Cooley, and T.W. Crooker, Quantitative Analysis of Microstructural Effects of Fatigue Crack Growth in Widmanstätten Ti-6Al-4V and Ti-8Al-1Mo-1V, *Engineering Fracture Mechanics*, Vol 11 (No. 4), 1979, p 805-816
67. R.R. Wells, New Alloys for Advanced Metallics Fighter-Wing Structures, AIAA Paper 74-372, Las Vegas, NV, April 1974
68. J.R. Beyer, D.L. Sims, and R.M. Wallace, Titanium Damage Tolerant Design Data for Propulsion Systems, Report AFML-TR-77-101, Air Force Materials Laboratory, Wright-Patterson AFB, OH, June 1977
69. G.R. Yoder, L.A. Cooley, and T.W. Crooker, Fatigue Crack Propagation Resistance of Beta-Annealed Ti-6Al-4V Alloys of Differing Interstitial Oxygen Contents, *Metallurgical Transactions A*, Vol 9 (No. 10), 1978, p 1413-1420
70. G.R. Yoder, L.A. Cooley, and T.W. Crooker, Observations on Microstructurally Sensitive Fatigue Crack Growth in a Widmanstätten Ti-6Al-4V Alloy, *Metallurgical Transactions A*, Vol 8A (No. 11), Nov 1977, p 1737-1743
71. G.R. Yoder and D. Eylon, On the Effect of Colony Size on Fatigue Crack Growth in Widmanstätten Structure $\alpha+\beta$ Titanium Alloys, *Metallurgical Transactions A*, Vol 10A (No. 11), Nov 1979, p 1808-1810
72. M.W. Mahoney and N.E. Paton, Fatigue and Fracture Characteristics of Silicon-Bearing Titanium Alloys, *Metallurgical Transactions A*, Vol 9 (No. 10), 1978, p 1497-1501
73. J.C. Chesnutt and N.E. Paton, Fatigue Crack Propagation in Ti-6Al and Ti-6Al-4V Containing 100-300 ppm Hydrogen at Ambient and Sub-ambient Temperatures, (in preparation)
74. J.R. Kennedy, P.N. Adler, and R.L. Schulte, Hydrogen Related Fatigue Fracture of Ti-6Al-4V, *Scripta Metallurgica*, Vol 14 (No. 3), March 1980, p 299-301
75. G.R. Yoder, L.A. Cooley, and T.W. Crooker, 50-Fold Difference in Region-II Fatigue Crack Propagation Resistance in Titanium Alloys, A Grain Size Effect, *Journal of Engineering Materials and Technology, Transactions of ASME*, Vol 101 (No. 1), Jan 1979, p 86-91
76. G.R. Yoder, L.A. Cooley, and T.W. Crooker, Enhancement of Fatigue Crack Growth and Fracture Resistance in Ti-6Al-4V and Ti-6Al-6V-2Sn Through Microstructural Modification, *Journal of Engineering Materials and Technology, Transactions of ASME*, Series H, Vol 199 (No. 4), Oct 1977, p 313-318
77. J.S. Park and H. Margolin, *Met. Trans.*, Vol 15A, 1984, p 155-171

78. S. Ankem and H. Margolin, *Met. Trans.*, Vol 17A, 1986, p 2209-2226
79. E. Levine, S. Hayden, and H. Margolin, *Acta Met*, Vol 22, 1974, p 1443-1448
80. A.W. Bowen, Some Relationships Between Crystallography and Stage II Fatigue Crack Growth in a Ti-6Al-4V Alloy, Cambridge U.K. Conference on Microstructure and Design of Alloys, Aug 1973, Institute of Metals, 1973, p 446-450
81. P.J. Bania and D. Eylon, Fatigue Crack-Propagation of Titanium-Alloys Under Dwell-Time Conditions, *Metallurgical Transactions A*, Vol 9 (No. 6), 1978, p 847-855
82. R.L. Tobler, Fatigue Crack Growth and J-Integral Fracture Parameters of Ti-6Al-4V at Ambient and Cryogenic Temperatures, *Cracks and Fracture*, STP 601, American Society for Testing and Materials, Philadelphia, 1976, p 346-370
83. D.N. Williams, Subcritical Crack Growth Under Sustained Load, *Metallurgical Transactions*, Vol 5 (No. 11), Nov 1974, p 2351-2358
84. G.J. Petrak, Crack Arrest and Crack Initiation in a Titanium Alloy, *Engineering Fracture Mechanics*, Vol 4, 1972, p 347-355
85. Y. Suleh and H. Margolin, *Met. Trans. A*, Vol 13A, 1982, p 1275-1281
86. Y. Mahajan and H. Margolin, *Met. Trans. A*, Vol 13A, 1982, p 269-274
87. H. Fujii and H.G. Suzuk, Effect of Solution Treatment Conditions on Aging Response, in Ref 21, p 249-259
88. Thermomechanical Surface Treatment in *Materials Properties Handbook: Titanium Alloys*, ASM International, 1994, p 1156-1158
89. A. Margolin, P. Adler, and R. Schulte, "Effects of Alloying Element Gradient on Hydrogen Segregation in Beta Phase Ti," *Scripta Met*, 1985, Vol 19, p 699-702
90. D.N. Williams, "Hydrogen in Titanium and Titanium Alloys," Titanium Metallurgical Report #100, Battelle Memorial Institute, May 16, 1958
91. G.A. Lenning and R.I. Jaffee, "Effect of Hydrogen on the Properties of Titanium and Titanium Alloys," Titanium Metallurgy Laboratory, Report #27, Dec 27, 1955
92. B.G. Pound, "The Effect of Aging on Hydrogen Trapping in β-Titanium Alloys," *Acta Met et Materialia*, Vol 42, 1994, p 1551-1559
93. G.A. Young, Jr. and J.R. Scully, "The Effects of Hydrogen on the Room Temperature Mechanical Properties of Ti-15V-3Cr-3Al-3Sn and Ti-15Mo-3Nb-3Al," *Beta Titanium Alloys in the 1990s*, D. Eylon, R.R. Buyer and D.A. Koss, Eds., TMS, Warrendale, PA, 1993, p 147-158
94. P.N. Adler, R.L. Schulte, and H. Margolin, "Hydrogen Distribution Within the Alpha/Beta Interface Region," *Proceedings of the Harold Margolin Symposium on Microstructure/Property Relations in Alpha/Beta Titanium Alloys*, TMS Annual Meeting, San Francisco, Feb 27-Mar 3, 1994
95. J.R. Kennedy et al., *Met. Trans. A*, Vol 24A, 1993, p 2763-2771
96. J.L. Waisman et al., *Met. Trans. A*, Vol 8A, 1977, p 1249-1256
97. J.S. Grauman, "Effects of Aircraft Fluid on TiMetal 21S," Ibid, p 127-135

SELECTED REFERENCES: Properties and Fabrication

- *Materials Properties Handbook: Titanium Alloys*, ASM International, 1994
- *Damage Tolerant Design Handbook*, MCIC-HB-01, Metals and Ceramics Information Center, Battelle-Columbus, Ohio. Part I deals with fracture toughness; Part 2 deals with fatigue crack growth
- *Aerospace Structural Metals Handbook*, AFML-TR-68-115, Mechanical Properties Data Center, Belfour Stulen, Traverse City, Michigan. Vol 3 deals with titanium
- *Metals Handbook*, ASM International
- *Military Standardization Handbook—Metallic Materials and Elements for Aerospace Vehicle Structures,* MIL-HDBK-5, Department of Defense, Washington, DC 20025, Obtainable from Naval Publications and Forms Center, 5801 Taber Avenue, Philadelphia, PA
- AMS (Aerospace Material Specification), Society of Automotive Engineers, 400 Commonwealth Drive, Warrendale, PA. Specifications cover specific alloy products
- *Annual Book of ASTM Standards,* American Society for Testing and Materials, 1916 Race Street, Philadelphia, PA. Current Edition. Parts 8, 9, 10, 11 and 41 are useful in various ways
- ASME Boiler and Pressure Vessel Code, American Society of Mechanical Engineers, United Engineering Center, 345 East Forty-Seventh Street, New York, NY 10017. Section VIII provides rules for construction of pressure vessels
- *Titanium Alloys Handbook*, MCICHB-62, Battelle

SELECTED REFERENCES: Metallurgy of Titanium

- E.W. Collings, *Physical Metallurgy of Titanium Alloys*, ASM International
- A.D. McQuillan and M.K. McQuillan, *Titanium*, Academic Press, New York, NY, 1956, Now out of print but still a useful source of information
- R.I. Jaffee and N.E. Promisel, Ed., *The Science, Technology, and Application of Titanium*, Pergamon Press, New York, NY, 1970. Reports the First International Conference on Titanium
- R.I. Jaffee and H.M. Burte, Ed., *Titanium Science and Technology*, Plenum Press, New York, NY, 1973. Reports the Second International Conference on Titanium
- Sazhin et al., Ed., *Titanium Alloys for Modern Technology*, NASA TT F-596, Clearinghouse for Federal Scientific and Technical Information, Springfield, Virginia. Translated from Russian
- Kornilov et al., Ed., *Physical Metallurgy of Titanium*, NASA TT F-338, Clearinghouse for Federal Scientific and Technical Information, Springfield, Virginia. Translated from Russian
- *Applications Related Phenomena in Titanium Alloys*, STP 432, American Society for Testing and Materials, Philadelphia, 1968. Environmental effects dealt with include salt water
- *Stress Corrosion Cracking of Titanium*, STP 397, American Society for Testing and Materials, Philadelphia, 1966. Deals mainly with hot salt stress corrosion
- *Stress Corrosion Cracking: Materials Performance and Evaluation*, ASM International, 1992

Commercially Pure and Modified Titanium

Commercially pure titanium has been available as mill products since 1950 and is used for applications that require moderate strength combined with good formability and corrosion resistance. Production was developed largely because of aerospace demands for a material lighter than steel and more heat resistant than aluminum alloys. However, commercially pure titanium is very useful when high corrosion resistance and good weldability are desired.

Commercially pure titanium is available in several grades, which have varying amounts of impurities such as carbon, hydrogen, iron, nitrogen, and oxygen. Some modified grades also contain small palladium additions (Ti-0.2 Pd) and nickel-molybdenum additions (Ti-0.3Mo-0.8Ni). These alloy additions allow improvements in corrosion resistance and/or strength.

Commercial purity titanium generally has more than 1000 ppm oxygen and iron, nitrogen, carbon, and silicon as principal impurities. Because small amounts of interstitial impurities greatly affect the mechanical properties of pure titanium, it is not convenient to distinguish between the various grades of unalloyed titanium on the basis of chemical analysis. Titanium mill products are more readily distinguished by mechanical properties. For example, the four ASTM grades of unalloyed titanium are grouped as follows:

ASTM grade	Minimum tensile strength		0.2% yield strength	
	MPa	ksi	MPa	ksi
Grade 1	240	35	170-310	25-45
Grade 2	345	50	275-450	40-65
Grade 3	440	64	380-550	55-80
Grade 4	550	80	480-655	70-95

Density. 4.51 g/cm^3 (0.16 3 lb/in.^{3})

Unalloyed Ti Grade I, R50250

Unalloyed titanium is available as four different ASTM grades, which are classified by their levels of impurities (primarily oxygen) and the resultant effect on strength and ductility. ASTM Grade 1 has the highest purity, lowest strength, and best room-temperature ductility and formability of the four ASTM unalloyed titanium grades.

ASTM titanium Grade 1 should be used where maximum formability is required and where low iron and interstitial contents might enhance corrosion resistance. It exhibits excellent corrosion resistance in highly oxidizing to mildly reducing environments, including chlorides. Grade 1 can be used in continuous service up to 425 °C (800 °F) and in intermittent service up to 540 °C (1000 °F). In addition, Grade 1 has good impact properties at low temperatures.

Chemistry

ASTM Grade 1 titanium has impurity limits of 0.18 O, 0.20 Fe, 0.03 N, and 0.10 C wt.% max. Equivalent compositions from other specifications are best determined by mechanical properties, because small variations in interstitial contents may raise yield strengths above maximum permitted values or lower ductility below minimum specifications.

Hydrogen content as low as 30 to 40 ppm can induce severe hydrogen embrittlement in commercially pure titanium (see the section "Hydrogen Damage" in this datasheet).

Product Forms and Condition

Unalloyed titanium Grade 1 is available in all wrought forms and has the best formability of the four ASTM grades. Like the other unalloyed titanium grades, Grade 1 can be satisfactorily welded, machined, cold worked, hot worked, and cast.

Unalloyed titanium typically has an annealed alpha structure in wrought, cast, and P/M forms. The yield strength of Grade 1 is comparable to that of fully annealed 304 stainless steel.

Applications

Typical uses for Grade 1 titanium include chemical, marine, and similar applications, heat exchangers, components for chemical processing and desalination equipment, condenser tubing, pickling baskets and anodes of various types. In the chemical and engineering industries, Grade 1 is an ideal material for a wide variety of chemical reactor vessels because of its resistance to attack by seawater, moist chlorine, moist metallic chlorides, chlorite and hypochlorite solutions, nitric and chromic acids. It lacks resistance to biofouling.

Unalloyed titanium grade 1 and equivalents: Specifications and compositions

Specification	Designation	Description	C	Fe	H	N	O	Si	OE	OT	Other
UNS	R50100		0.03 max	0.1 max	0.005 max	0.012 max	0.1 max				bal Ti
UNS	R50120		0.05	0.2	0.008	0.02	0.1				bal Ti
UNS	R50125		0.05	0.2	0.008	0.02	0.1-0.15				bal Ti
UNS	R50250		0.1	0.2	0.015	0.03	0.18				bal Ti
China											
GB 3620	TA-1		0.05	0.15	0.015	0.03	0.15	0.1			bal Ti
Europe											
AECMA prEN2525	P01	Sh Strp	0.08	0.2	0.0125	0.05	0.2		0.1	0.6	bal Ti
AECMA prEN3441	P01	Sh Strp Ann HR	0.08	0.2	0.0125	0.05	0.2		0.1	0.6	bal Ti
AECMA prEN3487	P01	Sh Strp Ann CR	0.08	0.2	0.0125	0.05	0.2		0.1	0.6	bal Ti
France											
AIR 9182	T-35	Sh Strp	0.08	0.12	0.01	0.05			0.04		bal Ti

(continued)

Unalloyed titanium grade 1 and equivalents: Specifications and compositions (continued)

Specification	Designation	Description	C	Fe	H	N	O	Si	OE	OT	Other
Germany											
DIN 17850	3.7025	Plt Sh Strp Rod Wir Frg Ann	0.08	0.2	0.013	0.05	0.1				bal Ti
DIN 17850	Ti I	Sh Strp Plt Rod Wir Frg Ann	0.08	0.2	0.013	0.05	0.1				bal Ti
DIN 17860	3.7025	Sh Strp	0.08	0.2	0.013	0.05	0.1				bal Ti
DIN 17862	3.7025	Rod	0.08	0.2	0.013	0.05	0.1				bal Ti
DIN 17863	3.7025	Wire	0.08	0.2	0.013	0.05	0.1				bal Ti
DIN 17864	3.7025	Frg	0.08	0.2	0.013	0.05	0.1				bal Ti
Japan											
JIS Class 1	Ti Class 1			0.2	0.015	0.05	0.15				bal Ti
JIS H4600	TP28H/C Class 1	HR CR Sh		0.2	0.013	0.05	0.15				bal Ti
JIS H4600	TR28H/C Class 1	HR CR Strp		0.2	0.013	0.05	0.15				bal Ti
JIS H4630	TTP28D/E Class 1	Smls pipe		0.2	0.015	0.05	0.15				bal Ti
JIS H4630	TTP28W/WD Class 1	As-weld/weld & drawn pipe		0.2	0.015	0.05	0.15				bal Ti
JIS H4631	TTH28D Class 1	Smls tube for heat exch		0.2	0.015	0.05	0.15				bal Ti
JIS H4631	TTH28W/WD Class 1	Weld tube for heat exch		0.2	0.015	0.05	0.15				bal Ti
JIS H4650	TB28C/H Class 1	HW CD Bar		0.2	0.015	0.05	0.15				bal Ti
JIS H4670	TW28 Class 1	Wire		0.2	0.015	0.05	0.15				bal Ti
Russia											
	Weld el		0.12	0.006	0.04	0.12	0.1				bal Ti
	Weld el		0.03	0.008	0.04	0.15					bal Ti
			0.1	0.3	0.015	0.04	0.15	0.15			bal Ti
	All forms		0.03	0.01	0.04	0.15	0.15				bal Ti
			0.05	0.3	0.01	0.04	0.15	0.15			bal Ti
			0.05	0.3	0.01	0.04	0.15	0.15			bal Ti
OST 1.90013-71	VT1-00	Sh Plt Strp Foil Rod Frg Ann	0.05	0.2	0.008	0.04	0.1	0.08		0.1	
UK											
BS 2TA.1	2TA.1	Sh Strp HT		0.2	0.01						Ti 99.78 min;
DTD 5013		Bar Bil		0.2	0.013						bal Ti
USA											
AMS 4951E	AMS 4951	Fill met gas-met W arc weld	0.08	0.2	0.005	0.05	0.18			0.6	bal Ti
ASME SB-265	Ti Grade 1	Sh Strp Plt Ann	0.1	0.2	0.015	0.03	0.18		0.1	0.4	bal Ti
ASME SB-381	F-1	Frg Ann	0.1	0.2	0.015	0.03	0.18		0.1	0.4	bal Ti
ASTM B265-79	Ti Grade 1	Sh Strp Plt Ann	0.1	0.2	0.015	0.03	0.18			0.4	bal Ti
ASTM B337-87	Ti Grade 1	Weld smls pipe Ann	0.1	0.2	0.015	0.03	0.18			0.4	bal Ti
ASTM B338-87	Ti Grade 1	Smls weld tube Exch Conds Ann	0.1	0.2	0.015	0.03	0.18			0.4	bal Ti
ASTM B348-87	Ti Grade 1	Bar Bil Ann	0.1	0.2	0.01-0.0125	0.03	0.18			0.4	bal Ti
ASTM B381-87	F-1	Frg Ann	0.1	0.2	0.015	0.03	0.18			0.4	bal Ti
ASTM F467-84a	Ti Grade 1	Nut	0.1	0.2	0.0125	0.05	0.18				bal Ti
ASTM F467M-84b	Ti Grade 1	Metric Nut	0.1	0.2	0.0125	0.05	0.18				bal Ti
ASTM F468-84a	Ti Grade 1	Bolt Screw Stud	0.1	0.2	0.0125	0.05	0.18				bal Ti
ASTM F468M-84b	Ti Grade 1	Metric Bolt Screw Stud	0.1	0.2	0.0125	0.05	0.18				bal Ti
ASTM F67-88	Ti Grade 1	Surg imp HW CW Frg Ann	0.1	0.2	0.0125-0.015	0.03	0.18				bal Ti
AWS A5.16-70	ERTi-1	Weld fill met	0.03	0.1	0.005	0.012	0.1				bal Ti
AWS A5.16-70	ERTi-2	Weld fill met	0.05	0.2	0.008	0.02	0.1				bal Ti
AWS A5.16-70	ERTi-3	Weld fill met	0.05	0.2	0.008	0.02	0.1-0.15				bal Ti
MIL T-81556A	CP-4	Ext Bar Shap Ann	0.08	0.2	0.015	0.05	0.15			0.3	bal Ti
MIL T-81915A		Invest Cast	0.08	0.2	0.015	0.05	0.2			0.6	bal Ti
MIL T-9046J	CP-4	Sh Strp Plt Ann	0.08	0.2	0.015	0.05	0.15			0.3	bal Ti

Unalloyed titanium grade 1 compositions: Producer specifications

Specification	Designation	Description	C	Fe	H	N	O	Si	OE	OT	Other
Germany											
Deutsche T	Contimet 30	Sh Strp Plt Bar Wir Frg Pip	0.06	0.15	0.13	0.05	0.12				bal Ti
Fuchs	T2	Frg									
Japan											
Daido	DT1	Rod Bar Sh Strp Frg Ann	0.1	0.2	0.0125	0.03	0.18				bal Ti
Kobe	KS40	Sh Strp Tu Plt Wir Bar Pip Ann		0.1	0.01	0.03	0.1				bal Ti
Kobe	KS40LF	Low Fe grade		0.05	0.01	0.03	0.1				bal Ti
Kobe	KS40S	Ann		0.1	0.01	0.03	0.08				bal Ti
Kobe	KS50	Ann		0.15	0.01	0.03	0.15				bal Ti
Kobe	KS50LF	Low Fe grade		0.05	0.01	0.03	0.15				bal Ti
Sumitomo	ST-40										
UK											
Imp. Metal	IMI 110	Rod									Ti 99.8
Imp. Metal	IMI 115	All forms	0.1	0.2	0.013	0.03	0.15				bal Ti

(continued)

Unalloyed titanium grade 1 compositions: Producer specifications (continued)

Specification	Designation	Description	C	Fe	H	N	O	Si	OE	OT	Other
USA											
Chase Ext.	CDX GR-1										
OREMET	Ti-1										
RMI	RMI 25	Chemical/marine/airframe apps	0.08	0.2	0.015	0.03	0.18				bal Ti
Tel.Rodney	A35										
TIMET	TIMETAL 35A	Ann	0.1 max	0.02 max	0.15 max	0.03 max	0.18 max				bal Ti
TMCA	Ti-1										

Unalloyed Ti Grade 2, R50400

Grade 2 titanium is the "workhorse" for industrial applications, having a guaranteed minimum yield strength of 275 MPa (40 ksi) and good ductility and formability. The yield strength of Grade 2 is comparable to those of annealed austenitic stainless steels, and it is used where excellent formability is required and where low interstitial contents might enhance corrosion resistance.

Grade 2 also has good impact properties at low temperatures and excellent resistance to erosion and to corrosion by seawater and marine atmospheres. Grade 2 can be used in continuous service up to 425 °C (800 °F) and in intermittent service up to 540 °C (1000 °F).

Chemistry

ASTM Grade 2 titanium has the same nitrogen content limits as ASTM Grade 1 (0.03% max), the same iron content limits as ASTM Grade 3 (0.30% max), and a maximum oxygen concentration of 0.25% that is approximately midway between the 0.18 to 0.40% range in the other three ASTM unalloyed titanium grades.

Effect of Impurities. The increased iron and oxygen concentrations of ASTM Grade 2 compared to ASTM Grade 1 impart additional tensile strength (345 vs 240 MPa, or 50 vs 35 ksi) and yield strength (275 vs 170 MPa, or 40 vs 25 ksi) to Grade 2 but at the expense of ductility (20% elongation for Grade 2 vs 24% elongation for Grade 1). Higher iron and interstitial contents also may degrade corrosion resistance relative to Grade 1.

Hydrogen content as low as 30 to 40 ppm can induce hydrogen embrittlement in CP titanium (see the section "Hydrogen Damage" in this datasheet).

Product Forms and Condition

Titanium Grade 2 is available in all wrought product forms. In cast form, ASTM Grade 2 constitutes about 5% of cast titanium products. Like other unalloyed titanium grades, Grade 2 can be welded, machined, cast, and cold worked.

Titanium Grade 2 typically has an annealed alpha structure in wrought, cast, and P/M forms. It is not heat treatable.

Applications

Typical uses for titanium Grade 2 include chemical, marine, and similar applications, airframe skin and nonstructural components, heat exchangers, cryogenic vessels, components for chemical processing and desalination equipment, condenser tubing, pickling baskets, anodes, shafting, pumps, vessels, and piping systems. Grade 2 offers high ductility for fabrication and moderate strength in service.

Aircraft applications include exhaust-pipe shrouds, fireproof bulkheads, gas-turbine bypass ducts, hot-air ducts, engine cowlings, formed brackets and skins for hot areas. Other aircraft applications include galley equipment, chemical toilets and floor supports under these areas.

Reaction vessels and heat exchangers are a major application of Grade 2 titanium because of its resistance to attack by seawater, moist chlorine, moist metallic chlorides, chlorite and hypochlorite solutions, nitric and chromic acids, organic acid, sulfides, and many industrial gaseous environments. Grade 2 titanium also has excellent resistance to deposit, impingement, and crevice attack even in highly polluted waters, and is therefore used extensively in tubular and plate-type heat exchangers for condensers, evaporators, and other components of marine vessels, power stations, oil refineries, offshore platforms, and water-purification plants.

Electrochemical Processing Equipment. The insulating property of the anodic film on titanium makes it an ideal and cost-efficient material for anodizing jigs and plating baskets. Other applications include high-efficiency heat-exchanger systems for electrolytes. A very thin coating of a precious metal such as platinum enables Grade 2 titanium anode to operate at high current density in many electrolytes. Consequently, non-consumable noble-metal coated Grade 2 titanium anodes are in demand for chlorine-production cells, electrodialysis plants, electroplating equipment, and cathodic protection of condensers, seagoing rigs, and jetties.

Most electrodeposits do not adhere well to commercial purity Grade 2 titanium. This characteristic has led to the widespread use of Grade 2 titanium for cathodes or starter-sheet blanks in many electrochemical metal-refining operations.

Unalloyed titanium grade 2 and equivalents: Specifications and compositions

Specification	Designation	Description	C	Fe	H	N	O	Si	OE	OT	Other
UNS	R50130		0.05	0.3	0.008	0.02	0.15-0.25				bal Ti
UNS	R50400		0.1	0.3	0.015	0.03	0.25				bal Ti
China											
GB 3620	TA-2		0.1 max	0.3 max	0.015 max	0.05 max	0.2 max	0.15 max			bal Ti
Europe											
AECMA prEN2518	Ti-PO2	Sh Strp Bar	0.08	0.2	0.01	0.06	0.25			0.6	bal Ti
AECMA prEN2526	Ti-PO2	Sh Strp	0.08 max	0.25 max	0.0125 max	0.05 max	0.25 max		0.1 max	0.6 max	bal Ti

(continued)

Unalloyed titanium grade 2 and equivalents: Specifications and compositions (continued)

Specification	Designation	Description	C	Fe	H	N	O	Si	OE	OT	Other
Europe (continued)											
AECMA prEN3378	Ti PO2	Wir	0.08 max	0.25 max	0.0125 max	0.05 max	0.25 max		0.1 max	0.6 max	bal Ti
AECMA prEN3442	Ti-PO2	Sh Strp Ann HR	0.08 max	0.25 max	0.0125 max	0.05 max	0.25 max		0.1 max	0.6 max	bal Ti
AECMA prEN3451	Ti-PO2	Frg NHT	0.08 max	0.25 max	0.0125 max	0.05 max	0.25 max		0.1 max	0.6 max	bal Ti
AECMA prEN3452	Ti-PO2	Frg Ann	0.08 max	0.25 max	0.0125 max	0.05 max	0.25 max		0.1 max	0.6 max	bal Ti
AECMA prEN3460	Ti-PO2	Bar Ann	0.08 max	0.25 max	0.0125 max	0.05 max	0.25 max		0.1 max	0.6 max	bal Ti
AECMA prEN3498	Ti-PO2	Sh Strp Ann CR	0.08 max	0.25 max	0.0125 max	0.05 max	0.25 max		0.1 max	0.6 max	bal Ti
France											
AIR 9182	T-35	Sh CR	0.08	0.12	0.015	0.05		0.04			Ti 99.69 min
AIR 9182	T-40	Sh	0.08	0.12	0.015	0.05		0.04			Ti 99.69 min
Germany											
DIN 17850	Ti II	Sh Strp Plt Rod Wir Frg Ann	0.08	0.25	0.013	0.06	0.2				bal Ti
DIN 17850	Ti III	Sh Strp Plt Rod Wir Frg Ann	0.1	0.3	0.013	0.06	0.25				bal Ti
DIN 17850	WL 3.7035	Plt Sh Strp Rod Wir Frg Ann	0.08	0.25	0.013	0.06	0.2				bal Ti
DIN 17850	WL 3.7055	Sh Plt Strp Rod Wir Frg Ann	0.1	0.3	0.013	0.06	0.25				bal Ti
DIN 17860	3.7035	Sh Strp	0.08 max	0.25 max	0.013 max	0.06 max	0.2 max				bal Ti
DIN 17862	3.7035	Rod	0.08 max	0.25 max	0.013 max	0.06 max	0.2 max				bal Ti
DIN 17863	3.7035	Wir	0.08 max	0.25 max	0.013 max	0.06 max	0.2 max				bal Ti
DIN 17864	3.7035	Frg	0.08 max	0.25 max	0.013 max	0.06 max	0.2 max				bal Ti
WL 3.7024		Sh Wir Ann	0.08	0.2	0.0125	0.05	0.2			0.6	bal Ti
WL 3.7034		Sh Bar Frg Wir Ann	0.08	0.25	0.0125	0.06	0.25			0.6	bal Ti
Japan											
	Class 2			0.25	0.015	0.05	0.2				bal Ti
JIS H4361	TTH 35D Class 2	Smls Tub		0.25	0.015	0.05	0.2				bal Ti
JIS H4600	TP 35 H/C Class 2	Sh HR CR		0.25	0.013	0.05	0.2				bal Ti
JIS H4600	TR 35 H/C Class 2	Strp HR CR		0.25	0.013	0.05	0.2				bal Ti
JIS H4630	TTP 35 D/E Class 2	Smls Pip		0.25	0.015	0.05	0.2				bal Ti
JIS H4630	TTP 35 W/WD Class 2	Weld Pip		0.25	0.015	0.05	0.2				bal Ti
JIS H4631	TTH 35 W/WD Class 2	Weld Tub		0.25	0.015	0.05	0.2				bal Ti
JIS H4650	TB 35 C/H Class 2	Bar Rod HW CD		0.25	0.015	0.05	0.2				bal Ti
JIS H4670	TW 35 Class 2	Wir		0.25	0.015	0.05	0.2				bal Ti
Russia											
OST 1.90000-76	VT1-O	Mult Forms Ann	0.07	0.3	0.01	0.04	0.2	0.1		0.3	bal Ti
OST 1.90060-72	VT1L	Cast	0.15	0.3	0.015	0.05	0.2	0.15		0.3	W 0.2; bal Ti
Spain											
UNE 38-711	L-7001	Sh Plt Strp Bar Wir Ext Ann	0.08	0.2	0.0125	0.05	0.2				bal Ti
UNE 38-712	L-7002	Sh Plt Strp Bar Wir Ext Ann	0.08	0.25	0.0125	0.05	0.25				bal Ti
UK											
BS 2TA.2		Sh Strp HT		0.2	0.01						Ti 99.78 min
BS 2TA.3	2TA.3	Bar HT		0.2	0.01						Ti 99.78 min
BS 2TA.4	2TA.4	Frg HT		0.2	0.01						Ti 99.79 min
BS 2TA.5	2TA.5	Frg HT		0.2	0.01						Ti 99.78 min
DTD 5073		Tub	0.01 max	0.2 max	0.015 max						bal Ti
USA											
AMS 4902E		Sh Strp Plt Ann	0.08	0.3	0.015	0.05	0.2			0.3	bal Ti
AMS 4941C		Weld Tub Ann	0.1	0.2	0.015	0.05	0.25			0.15	bal Ti
ASM 4942C		Smls Tube Ann	0.1	0.3	0.015	0.03	0.25			0.3	bal Ti
ASME SB-265	Ti Grade 2	Sh Strp Plt Ann	0.1 max	0.3 max	0.015 max	0.03 max	0.25 max		0.1 max	0.4 max	bal Ti
ASME SB-381	F-2	Frg Ann	0.1 max	0.3 max	0.015 max	0.03 max	0.25 max		0.1 max	0.4 max	bal Ti
ASTM B 265	Ti Grade 2	Sh Strp Plt Ann	0.1	0.3	0.015	0.03	0.25			0.4	bal Ti
ASTM B 337	Ti Grade 2	Pip Ann	0.1	0.3	0.015	0.03	0.25			0.4	bal Ti
ASTM B 338	Ti Grade 2	Tube for heat exch/cond									bal Ti
ASTM B 348	Ti Grade 2	Bar Bil Ann	0.1	0.3	0.0125-0.01	0.03	0.25			0.4	bal Ti
ASTM B 367-87	Ti Grade 2	Cast	0.1 max	0.2 max	0.015 max	0.05 max	0.4 max		0.1 max	0.4 max	bal Ti
ASTM B 381	Ti Grade F-2	Frg Ann	0.1	0.3	0.015	0.03	0.25				bal Ti
ASTM F467-84	Ti Grade 2	Nut	0.1 max	0.3 max	0.0125 max	0.05 max	0.25 max				bal Ti
ASTM F467M-84a	Ti Grade 2	Nut Met	0.1 max	0.3 max	0.0125 max	0.05 max	0.25 max				bal Ti
ASTM F468-84	Ti Grade 2	Blt Scr Std	0.1 max	0.3 max	0.0125 max	0.05 max	0.25 max				bal Ti
ASTM F468M-84b	Ti Grade 2	Blt Scr Std Met	0.1 max	0.3 max	0.0125 max	0.05 max	0.25 max				bal Ti
ASTM F67	Ti Grade 2	Surg imp HW CW Frg Ann	0.1	0.3	0.015-0.0125	0.03	0.25				bal Ti
AWS A5.16-70	ERTi-4	Weld Fill Met	0.05	0.3	0.008	0.02	0.15-0.25				bal Ti
MIL T-81556A	Code CP-3	Ext Bar Shp Ann	0.08	0.3	0.015	0.05	0.2			0.3	bal Ti
MIL T-81915	Type I Comp A	Air/chem/marine apps Cast Ann	0.08	0.2	0.015	0.05	0.2			0.6	bal Ti
MIL T-9046J	Code CP-3	Sh Strp Plt Ann	0.08	0.3	0.015	0.05	0.2			0.3	bal Ti

Unalloyed titanium grade 2 compositions: Producer specifications

Specification	Designation	Description	C	Fe	H	N	O	Si	OE	OT	Other
France											
Ugine	UT35	Sh Plt Bar Frg Ann	0.08		0.0125	0.05	0.2				bal Ti
Ugine	UT40	Sh Plt Bar Frg Ann	0.08	0.25	0.0125	0.06	0.25				bal Ti
Germany											
Otto Fuchs	T3	Frg									Ti 99.5
Thyssen	Contimet 35	Sh Strp Plt Bar Wir Pip Ann	0.06	0.2	0.013	0.05	0.18				bal Ti
Thyssen	Contimet 35 D	Mult forms Ann	0.06	0.25	0.013	0.05	0.25				bal Ti
Japan											
Daido	DT 2	Rod Bar Sh Strp Frg Ann	0.1	0.3	0.0125	0.03	0.2				bal Ti
Kobe	KS60	Sh Strp Tub Plt Wir Bar Pi Ann		0.3	0.01	0.03	0.2				bal Ti
Kobe	KS60LF	Low Fe Ann		0.05	0.01	0.03	0.2				bal Ti
Nippon	T1X			0.5		0.1	0.2				bal Ti
Sumitomo	ST-50										
Sumitomo	ST6										
Toho	TIB		0.03	0.1	0.005	0.015	0.15				Ti 99.7 min
Toho	TIBLF	Low Fe	0.02	0.05	0.005	0.01	0.15				Ti 99.7 min
Toho	TIC		0.03	0.15	0.005	0.02	0.25				Ti 99.6 min
Toho	TICLF	Low Fe	0.02	0.05	0.005	0.01	0.25				Ti 99.6 min
UK											
Imp. Metal	IMI 125	Mult forms	0.1	0.2	0.013	0.03	0.2				bal Ti
Imp. Metal	IMI 130	Sh Bar	0.1	0.2	0.013	0.03	0.25				bal Ti
USA											
Chase Ext.	CDX GR-2										
OREMET	Ti-2										
RMI	RMI 40	Mult forms Ann	0.08	0.25	0.015	0.03	0.2				bal Ti
Tel.Rodney	A40										
TIMET	TIMETAL 50A	Ann	0.08 max	0.2 max	0.0125 max	0.05 max					bal Ti
TMCA	Ti 2										

Unalloyed Ti Grade 3, R50550

Grade 3 titanium is a general-purpose grade of commercially pure titanium that has excellent corrosion resistance in highly oxidizing to mildly reducing environments, including chlorides, and an excellent strength-to-weight ratio. Thus, like other titanium metals and alloys, Grade 3 bridges the design gap between aluminum and steel and provides many of the desirable properties of each. Grade 3 also has good impact toughness at low temperatures.

Chemistry

ASTM Grade 3 titanium has lower iron limits than ASTM Grade 4 (0.3 wt% vs 0.5 wt% max) and the second highest oxygen contents (0.35 wt%) of the four ASTM grades for unalloyed titanium. Only Grade 4 has higher strength levels than Grade 3.

Effect of Impurities. Excessive impurity levels may raise yield strength above maximum permitted values and decrease elongation or reduction in area below minimum values. Higher iron and interstitial contents may affect corrosion resistance.

Hydrogen content as low as 30 to 40 ppm can induce hydrogen embrittlement in commercially pure titanium (see the section "Hydrogen Damage" in this datasheet).

Product Forms and Condition

Like other unalloyed titanium grades, Grade 3 is available in all wrought product forms and can be satisfactorily welded, machined, and cast. Most forming operations can be carried out at room temperature but warm forming reduces springback and power requirements.

Titanium Grade 3 typically has an annealed alpha structure for wrought, cast, and P/M forms.

Applications

Grade 3 is used for nonstructural aircraft parts and for all types of applications requiring corrosion resistance. Typical uses for CP titanium include chemical and marine applications, airframe skin and nonstructural components, heat exchangers, cryogenic vessels, components for chemical processing and desalination equipment, condenser tubing, and pickling baskets.

Unalloyed titanium grade 3 and equivalents: Specifications and compositions

Specification	Designation	Description	C	Fe	H	N	O	Si	OE	OT	Other
UNS	R50550		0.1	0.3	0.015	0.05	0.35			0.4	bal Ti
France											
AIR 9182	T-50	Sh Ann	0.08	0.25	0.015	0.07		0.04			Ti 99.54 min

(continued)

Unalloyed titanium grade 3 and equivalents: Specifications and compositions (continued)

Specification	Designation	Description	C	Fe	H	N	O	Si	OE	OT	Other
Germany											
DIN 17850	Ti IV	Sh Strp Plt Rod Wir Frg Ann	0.1	0.35	0.013	0.07	0.3				bal Ti
DIN 17850	WL 3.7065	Plt Sh Strp Rod Wir Frg Ann	0.1	0.35	0.013	0.07	0.3				bal Ti
DIN 17860	3.7055	Sh Strp	0.1 max	0.3 max	0.013 max	0.06 max	0.25 max				bal Ti
DIN 17862	3.7055	Rod	0.1 max	0.3 max	0.013 max	0.06 max	0.25 max				bal Ti
DIN 17863	3.7055	Wir	0.1 max	0.3 max	0.013 max	0.06 max	0.25 max				bal Ti
DIN 17864	3.7055	Frg	0.1 max	0.3 max	0.013 max	0.06 max	0.25 max				bal Ti
Japan											
JIS	Class 3			0.3	0.015	0.07	0.3				bal Ti
JIS H4600	TP 49 H/C Class 3	Sh HR CR		0.3	0.013	0.07	0.3				bal Ti
JIS H4600	TR 49 H/C Class 3	Strp HR CR		0.3	0.013	0.07	0.3				bal Ti
JIS H4630	TTP 49 D/E Class 3	Smls Pip Hot Ext CD		0.3	0.015	0.07	0.3				bal Ti
JIS H4630	TTP 49 W/WD Class 3	Weld Pip		0.3	0.015	0.07	0.3				bal Ti
JIS H4631	TTH 49 D Class 3	Smls Tub CD		0.3	0.015	0.07	0.3				bal Ti
JIS H4631	TTH 49 W/WD Class 3	Weld Tub		0.3	0.015	0.07	0.3				bal Ti
JIS H4650	TB 49 C/H Class 3	Bar HW CD		0.3	0.015	0.07	0.3				bal Ti
JIS H4670	TW 49 Class 3	Wir		0.3	0.015	0.07	0.3				bal Ti
UK											
BS 2TA.6		Sh Strp HT		0.2	0.01						Ti 99.78 min
BS 2TA.7		Bar HT		0.2	0.01						Ti 99.78 min
BS 2TA.8		Frg		0.2	0.01						Ti 99.79 min
BS 2TA.9		Frg HT		0.2	0.015						Ti 99.78 min
DTD 5023		Sh Strp		0.2 max	0.0125 max						bal Ti
DTD 5273		Bar		0.2 max	0.0125 max						bal Ti
DTD 5283		Frg		0.2 max	0.0125 max						bal Ti
USA											
AMS 4900J		Sh Strp Plt Ann	0.08	0.3	0.015	0.05	0.3			0.3	bal Ti
AMS 4951E		Weld Wir	0.08 max	0.2 max	0.005 max	0.05 max	0.18 max		0.1 max	0.6 max	bal Ti
ASME SB-265	Grade 3	Sh Strp Plt Ann	0.1 max	0.3 max	0.015 max	0.05 max	0.35 max		0.1 max	0.4 max	bal Ti
ASME SB-381	F-3	Frg An	0.1 max	0.3 max	0.015 max	0.05 max	0.35 max		0.1 max	0.4 max	bal Ti
ASTM B 265	Grade 3	Sh Strp Plt Ann	0.1	0.3	0.015	0.05	0.35			0.4	bal Ti
ASTM B 337	Grade 3	Weld Smls Pip Ann	0.1	0.3	0.015	0.05	0.35			0.4	bal Ti
ASTM B 338	Grade 3	Smls Weld Tub Ann	0.1	0.3	0.015	0.05	0.35			0.4	bal Ti
ASTM B 348	Grade 3	Bar Bil Ann	0.1	0.3	0.0125	0.05	0.35			0.4	bal Ti
ASTM B 381	Grade F-3	Frg Ann	0.1	0.3	0.015	0.05	0.35			0.4	bal Ti
ASTM B 367-87	C-3	Cast	0.1 max	0.25 max	0.015 max	0.05 max	0.4 max		0.1 max	0.4 max	bal Ti
ASTM F 67	Grade 3	Surg Imp	0.1	0.3	0.015-0.0125	0.05	0.35				bal Ti
MIL T-81556A	Code CP-2	Ext Bar Shp Ann	0.08	0.3	0.015	0.05	0.3			0.3	bal Ti
MIL T-9046J	Code CP-2	Sh Strp Plt Ann	0.08	0.3	0.015	0.05	0.3			0.3	bal Ti

Unalloyed titanium grade 3 compositions: Producer specifications

Specification	Designation	Description	C	Fe	H	N	O	Si	OE	OT	Other
France											
Ugine	UT50	Sh Bar Frg Ann	0.08	0.25	0.0125	0.07	0.35				bal Ti
Germany											
Thyssen	Contimet 55	Mult Forms Ann	0.06	0.3	0.013	0.05	0.35				bal Ti
Titan	RT 20		0.1	0.35	0.013	0.07	0.3				bal Ti
Japan											
Daido	DT 3	Mult Forms Ann	0.1	0.3	0.0125	0.05	0.35				bal Ti
Kobe	KS70	Ann		0.3	0.01	0.05	0.3				bal Ti
Kobe	KS70LF	Low Fe Mult Forms Ann		0.05	0.01	0.05	0.3				bal Ti
Sumitomo	ST-70										
Toho	TID		0.05	0.2	0.01	0.04	0.3				Ti 99.4 min
UK											
Imp. Metal	IMI 130										
USA											
Chase Ext.	CDX GR-32										
OREMET	Ti-3										
RMI	RMI 55	Mult Forms Ann	0.08	0.25	0.015	0.05	0.3				bal Ti
Tel.Rodney	A55										
TIMET	TIMETAL 65A	Ann	0.1 max	0.2 max	0.015 max	0.05 max	0.35 max				bal Ti
TMCA	Ti 3										

Unalloyed Ti Grade 4, R50700

Grade 4 has the highest strength of the four ASTM unalloyed titanium grades in addition to good ductility and moderate formability. The benefits of strength and lightness of Grade 4 are retained at moderate temperatures. Its strength-to-weight ratio is higher than that of AISI type 301 stainless steel at temperatures up to 315 °C (600 °F). Grade 4 also has outstanding resistance to corrosion fatigue in salt water. The stress required to cause failure in several million cycles is 50% higher for this material than for K-Monel or AISI type 431 stainless steel.

Chemistry

ASTM Grade 4 has the highest oxygen (0.40 wt%) and iron (0.50 wt%) content of the four unalloyed titanium ASTM grades. The higher content of iron and interstitials may reduce corrosion resistance.

Hydrogen content as low as 30 to 40 ppm can induce hydrogen embrittlement in commercially pure titanium (see the section "Hydrogen Damage" in this datasheet).

Product Forms and Condition

Commercially pure Grade 4 is available in all wrought product forms and can be satisfactorily machined, cast, welded, and cold worked. Most forming operations are performed at room temperature but warm forming (150 to 425 °C, 300 to 800 °F) is often done to reduce springback and power requirements. Complex forms must be produced by warm forming.

Grade 4 typically has an annealed alpha structure in wrought, cast, and P/M forms.

Applications

Because Grade 4 has excellent resistance to corrosion and erosion applications, it is suitable for a wide range of chemical and marine applications, where it often can be used interchangeably with Grade 3. It can be used in continuous service at temperatures up to 425 °C (800 °F), and intermittent service to 540 °C (1000 °F).

Unalloyed titanium grade 4 and equivalents: Specifications and compositions

Specification	Designation	Description	C	Fe	H	N	O	Si	OE	OT	Other
UNS	R50700		0.1	0.5	0.015	0.05	0.4			0.4	bal Ti
China											
GB 3620	TA-3		0.1 max	0.4 max	0.015 max	0.05 max	0.3 max	0.15 max			bal Ti
Europe											
AECMA prEN2519	Ti-PO4	Bar Frg Sh Strp	0.08	0.35	0.01-0.0125	0.07	0.4			0.6	bal Ti
AECMA prEN2520	Ti-PO4	Frg	0.08 max	0.2 max	0.0125 max	0.07 max	0.4 max		0.1 max	0.6 max	bal Ti
AECMA prEN2527	Ti-PO4	Sh Strp	0.08 max	0.2 max	0.0125 max	0.07 max	0.4 max		0.1 max	0.6 max	bal Ti
AECMA prEN3443	Ti-PO4	Strp Sh Ann CR	0.08 max	0.2 max	0.0125 max	0.07 max	0.4 max		0.1 max	0.6 max	bal Ti
AECMA prEN3453	Ti-PO4	Frg NHT	0.08 max	0.2 max	0.0125 max	0.07 max	0.4 max		0.1 max	0.6 max	bal Ti
AECMA prEN3461	Ti-PO4	Bar Ann	0.08 max	0.2 max	0.0125 max	0.07 max	0.4 max		0.1 max	0.6 max	bal Ti
AECMA prEN3496	Ti-PO4	Frg Ann	0.08 max	0.2 max	0.0125 max	0.07 max	0.4 max		0.1 max	0.6 max	bal Ti
AECMA prEN3499	Ti-PO4	Sh Strp Ann CR	0.08 max	0.2 max	0.0125 max	0.07 max	0.4 max		0.1 max	0.6 max	bal Ti
France											
AIR 9182	T-60	Sh Ann	0.08	0.3	0.015	0.08		0.04			Ti 99.56 min
Germany											
DIN	3.7064	Sh Rod Bar Frg Ann	0.08	0.35	0.0125	0.07	0.4				bal Ti
DIN 17860	3.7065	Sh Strp	0.1 max	0.35 max	0.013 max	0.07 max	0.3 max				bal Ti
DIN 17862	3.7065	Rod	0.1 max	0.35 max	0.013 max	0.07 max	0.3 max				bal Ti
DIN 17863	3.7065	Wir	0.1 max	0.35 max	0.013 max	0.07 max	0.3 max				bal Ti
DIN 17864	3.7065	Frg	0.1 max	0.35 max	0.013 max	0.07 max	0.3 max				bal Ti
Spain											
UNE 38-714	L-7004	Mult Forms Ann	0.1	0.4	0.0125	0.07	0.4				bal Ti
UK											
BS 2TA6		Sh Strp	0.08 max	0.2 max	0.0125 max						bal Ti
BS 2TA7		Bar	0.08 max	0.2 max	0.0125 max						bal Ti
BS 2TA8		Frg	0.08 max	0.2 max	0.01 max						bal Ti
BS 2TA9		Frg		0.2 max	0.015 max						bal Ti
USA											
AMS 4901L		Sh Strp Plt Ann	0.08	0.5	0.015	0.05	0.4			0.3	bal Ti
AMS 4921F		Bar Wir Frg Bil Rng Ann	0.08	0.5	0.0125	0.05	0.4			0.3	bal Ti
ASTM B 265	Grade 4	Sh Plt Strp Ann	0.1	0.5	0.015	0.05	0.4			0.4	bal Ti
ASTM B 348	Grade 4	Bar Bil Ann	0.1	0.5	0.0125-0.01	0.05	0.4			0.4	bal Ti
ASTM B 367	Grade C-2	Cast	0.1	0.2	0.015	0.05	0.4			0.4	bal Ti
ASTM B 367	Grade C-3	Cast	0.1	0.25	0.015	0.05	0.4			0.4	bal Ti
ASTM B 381	Grade F-4	Frg Ann	0.1	0.5	0.015	0.05	0.4			0.4	bal Ti
ASTM F467-84	Grade 4	Nut	0.1 max	0.5 max	0.0125 max	0.07 max	0.4 max				bal Ti
ASTM F468-84	Grade 4	Blt Scrw Std	0.1 max	0.5 max	0.0125 max	0.07 max	0.4 max				bal Ti
ASTM F67	Grade 4	Sh Strp Bar HR CR Ann Frg	0.1	0.5	0.015-0.0125	0.05	0.4				bal Ti
MIL F-83142	Comp 1	Frg Ann	0.08	0.5	0.0125	0.05	0.4			0.3	bal Ti
MIL T-81556A	Code CP-1	Ext Bar Shp Ann	0.08	0.5	0.015	0.05	0.4			0.3	bal Ti
MIL T-9046J	Code CP-1	Sh Strp Plt Ann	0.08	0.5	0.015	0.05	0.4			0.3	bal Ti
MIL T-9047-G	SP-70	Bar	0.08 max	0.5 max	0.015 max	0.05 max	0.4 max			0.3 max	bal Ti
MIL T-9047G	Ti-CP-70	Bar Bil Ann	0.08	0.5	0.0125	0.05	0.4 max			0.3	Y 0.005; bal Ti

Unalloyed titanium grade 4 commercial equivalents: Compositions

Specification	Designation	Description	C	Fe	H	N	O	Si	OE	OT	Other
France											
Ugine	UT60	Bar Frg Sh Plt Ann	0.1	0.35	0.0125	0.07	0.4				bal Ti
Germany											
Otto Fuchs	T6	Frg									Ti 99
Japan											
Daido	DT 4	Bar Rod Shp Frg Ann	0.1	0.5	0.0125	0.05	0.5				bal Ti
Kobe	KS85	Sh Strp Plt Wir Bar Ann		0.4	0.01	0.05	0.4				bal Ti
Sumitomo	ST-80										
UK											
Imp. Metal	IMI 155	Sh	0.1	0.2	0.013	0.03	0.38				bal Ti
Imp. Metal	IMI 160	Rod Bar Bil Wir	0.1	0.2	0.017	0.05	0.4				bal Ti
USA											
Chase Ext.	CDX GR-4										
Crucible	A-70	Ann	0.05-0.15			0.07 max					bal Ti
OREMET	Ti-4										
RMI	RMI 70	Mult Forms Ann	0.08	0.5	0.015	0.05	0.4				bal Ti
Tel.Rodney	A40										
TIMET	Ti-75A	Ann	0.1 max	0.3 max	0.015 max		0.4 max				bal Ti
TIMET	TIMETAL 100A	Ann	0.01 max	0.3 max	0.01 max	0.05 max	0.4 max				bal Ti
TMCA	Ti 4										

Ti-0.2Pd, R52400 (Grade 7), R52250 (Grade 11)

The two Ti-0.2Pd ASTM grades (7 and 11) have better resistance to crevice corrosion at low pH and elevated temperatures than that of ASTM Grades 1, 2, and 12, and they are recommended for chemical-industry applications involving environments that are moderately reducing or that fluctuate between oxidizing and reducing. The palladium-containing alloys extend the range of titanium applications in hydrochloric, phosphoric, and sulfuric acid solutions. Their good fabricability, weldability, and strength are similar to those of corresponding grades of unalloyed titanium. Ti-0.2 Pd Grade 7 is comparable to Grade 2 in strength, while Grade 11 is comparable to unalloyed Grade 1 in strength.

Chemistry

A relatively small addition of palladium (0.15 to 0.20 wt%) to unalloyed titanium permits its use in stronger reducing media such as mild sulfuric and hydrochloric acids.

The higher oxygen content (0.25 wt%) and higher iron content (0.30 wt%) of the Grade 7 alloy results in lower ductility and cold formability but higher strength than Grade 11 which has a maximum oxygen content of 0.18 wt% and a maximum iron content of 0.20 wt%.

Hydrogen content as low as 30 to 40 ppm can induce hydrogen embrittlement in commercially pure titanium (see the section "Hydrogen Damage" in this datasheet).

Product Forms and Condition

Both Grade 7 and Grade 11 alloys are flat rolled products, extrusions, wires, tubing, and pipe. Ti-0.2Pd grades can be satisfactorily cast, welded, machined, and cold worked. Most forming operations are performed at room temperature, but warm forming (150 to 425 °C, or 300 to 800 °F) is sometimes employed.

Ti-0.2Pd products typically have an annealed alpha structure.

Applications

Ti-0.2Pd, Grade 7 and Grade 11 are used for chemical-industry equipment and for special corrosion applications. These alloys have excellent corrosion resistance for chemical processing applications. They are also used for storage applications involving media that are mildly reducing or that fluctuate between oxidizing and reducing. The palladium-containing alloys are also used where high cold formability in component fabrication is required, such as cold pressed plates for plate/frame heat exchangers and chlor-alkali anodes. ASTM Grades 7 and 11 can be used in continuous service up to 425 °C (800 °F) and in intermittent service up to 540 °C (1000 °F).

Ti-0.2Pd grades 7 and 11 and equivalents: Specifications and compositions

Specification	Designation	Description	C	Fe	H	N	O	Pd	Si	OT	Other
UNS	R52250	Grade 11	0.1	0.2	0.015	0.03	0.18	0.12-0.25			bal Ti
UNS	R52400	Grade 7	0.1	0.3	0.015	0.03	0.25	0.12-0.25			bal Ti
UNS	R52401	Filler	0.05	0.25	0.008	0.02	0.15	0.15-0.25			bal Ti
Germany											
DIN 17851	3.7225		0.06 max	0.15 max	0.0013 max	0.05 max	0.12 max	0.12-0.25		0.4 max	bal Ti
DIN 17851	3.7235		0.06 max	0.2 max	0.0013 max	0.05 max	0.18 max	0.12-0.25		0.4 max	bal Ti
DIN 17851	3.7255		0.06 max	0.25 max	0.0013 max	0.05 max	0.25 max	0.12-0.25		0.4 max	bal Ti

(continued)

Ti-0.2Pd grades 7 and 11 and equivalents: Specifications and compositions (continued)

Specification	Designation	Description	C	Fe	H	N	O	Pd	Si	OT	Other
Japan											
JIS H 4635 type 11	TTP28PdD	Smls Pip CD		0.2 max	0.015 max	0.05 max	0.15 max	0.12-0.25			bal Ti
JIS H 4635 type 11	TTP28PdE	Smls Pip HE		0.2 max	0.015 max	0.05 max	0.15 max	0.12-0.25			bal Ti
JIS H 4635 type 11	TTP28PdW	Weld Pip		0.2 max	0.015 max	0.05 max	0.15 max	0.12-0.25			bal Ti
JIS H 4635 type 11	TTP28PdWD	Weld Pip CD		0.2 max	0.015 max	0.05 max	0.15 max	0.12-0.25			bal Ti
JIS H 4635 type 12	TTP35PdD	Smls Pip CD		0.25 max	0.015 max	0.05 max	0.2 max	0.12-0.25			bal Ti
JIS H 4635 type 12	TTP35PdE	Smls Pip HE		0.25 max	0.015 max	0.05 max	0.2 max	0.12-0.25			bal Ti
JIS H 4635 type 12	TTP35PdW	Weld Pip		0.25 max	0.015 max	0.05 max	0.2 max	0.12-0.25			bal Ti
JIS H 4635 type 12	TTP35PdWD	Weld Pip CD		0.25 max	0.015 max	0.05 max	0.2 max	0.12-0.25			bal Ti
JIS H 4635 type 13	TTP49PdD	Smls Pip CD		0.3 max	0.015 max	0.07 max	0.3 max	0.12-0.25			bal Ti
JIS H 4635 type 13	TTP49PdE	Smls Pip HE		0.3 max	0.015 max	0.07 max	0.3 max	0.12-0.25			bal Ti
JIS H 4635 type 13	TTP49PdW	Weld Pip		0.3 max	0.015 max	0.07 max	0.3 max	0.12-0.25			bal Ti
JIS H 4635 type 13	TTP49PdWD	Weld Pip CD		0.3 max	0.015 max	0.07 max	0.3 max	0.12-0.25			bal Ti
JIS H 4636 type 11	TTH28PdD	Smls Pip CD		0.2 max	0.015 max	0.05 max	0.15 max	0.12-0.25			bal Ti
JIS H 4636 type 11	TTH28PdW	Weld Pip		0.2 max	0.015 max	0.05 max	0.15 max	0.12-0.25			bal Ti
JIS H 4636 type 11	TTH28PdWD	Weld Pip CD		0.2 max	0.015 max	0.05 max	0.15 max	0.12-0.25			bal Ti
JIS H 4636 type 12	TTH35PdD	Smls Pip CD		0.25 max	0.015 max	0.05 max	0.2 max	0.12-0.25			bal Ti
JIS H 4636 type 12	TTH35PdW	Weld Pip		0.25 max	0.015 max	0.05 max	0.2 max	0.12-0.25			bal Ti
JIS H 4636 type 12	TTH35PdWD	Weld Pip CD		0.25 max	0.015 max	0.05 max	0.2 max	0.12-0.25			bal Ti
JIS H 4636 type 13	TTH49PdD	Smls Pip CD		0.3 max	0.015 max	0.07 max	0.3 max	0.12-0.25			bal Ti
JIS H 4636 type 13	TTH49PdW	Weld Pip		0.3 max	0.015 max	0.07 max	0.3 max	0.12-0.25			bal Ti
JIS H 4636 type 13	TTH49PdWD	Weld Pip CD		0.3 max	0.015 max	0.07 max	0.3 max	0.12-0.25			bal Ti
JIS H 4655 type 11	TB28PdC	Rod Bar CD		0.2 max	0.015 max	0.05 max	0.15 max	0.12-0.25			bal Ti
JIS H 4655 type 11	TB28PdH	Rod Bar HW		0.2 max	0.015 max	0.05 max	0.15 max	0.12-0.25			bal Ti
JIS H 4655 type 12	TB35PdC	Bar Rod CD		0.25 max	0.015 max	0.05 max	0.2 max	0.12-0.25			bal Ti
JIS H 4655 type 12	TB35PdH	Bar Rod HW		0.25 max	0.015 max	0.05 max	0.2 max	0.12-0.25			bal Ti
JIS H 4655 type 13	TB49PdC	Bar Rod CD		0.3 max	0.015 max	0.07 max	0.3 max	0.12-0.25			bal Ti
JIS H 4655 type 13	TB49PdH	Bar Rod HW		0.3 max	0.015 max	0.07 max	0.3 max	0.12-0.25			bal Ti
JIS H 4675 type 11	TW28Pd	Wir		0.2 max	0.015 max	0.05 max	0.15 max	0.12-0.25			bal Ti
JIS H 4675 type 12	TW35Pd	Wir		0.25 max	0.015 max	0.05 max	0.2 max	0.12-0.25			bal Ti
JIS H 4675 type 13	TW49Pd	Wir		0.3 max	0.015 max	0.07 max	0.25 max	0.12-0.25			bal Ti
Russia											
	4200		0.07	0.18	0.01	0.04	0.12	0.15-0.3	0.1	0.3	bal Ti
Spain											
UNE 38-715	L-7021	Sh Plt Strp Bar Wir Ext Ann	0.08	0.25	0.0125	0.05	0.25	0.12-0.25			bal Ti
USA											
ASTM B 265	Grade 11	Sh Plt Strp Ann	0.1	0.2	0.015	0.03	0.18	0.12-0.25		0.4	bal Ti
ASTM B 265	Grade 7	Sh Strp Plt Ann	0.1	0.3	0.015	0.03	0.25	0.12-0.25		0.4	bal Ti
ASTM B 337	Grade 11	Smls Weld Pip	0.1	0.2	0.015	0.03	0.18	0.12-0.25		0.4	bal Ti
ASTM B 337	Grade 7	Wld Smls Pip Ann	0.1	0.3	0.015	0.03	0.25	0.12-0.25		0.4	bal Ti
ASTM B 338	Grade 11	Smls Weld Tub Ann	0.1	0.2	0.015	0.03	0.18	0.12-0.25		0.4	bal Ti
ASTM B 338	Grade 7	Smls Weld Tub Ann	0.1	0.3	0.015	0.03	0.25	0.12-0.25		0.4	bal Ti
ASTM B 348	Grade 11	Bar Bil Ann	0.1	0.2	0.0125-0.01	0.03	0.18	0.12-0.25		0.4	bal Ti
ASTM B 348	Grade 7	Bar Bil Ann	0.1	0.3	0.0125	0.03	0.25	0.12-0.25		0.4	bal Ti
ASTM B 367	Grade Ti-Pd 7B	Cast	0.1	0.2	0.015	0.05	0.4	0.12		0.4	bal Ti
ASTM B 381	Grade F-11	Frg Ann	0.1	0.2	0.015	0.03	0.18	0.12-0.25		0.4	bal Ti
ASTM B 381	Grade F-7	Frg Ann	0.1	0.3	0.015	0.03	0.25	0.12-0.25		0.4	bal Ti
ASTM F467-84	Grade 7	Nut	0.1 max	0.3 max	0.0125 max	0.05 max	0.25 max	0.12-0.25			bal Ti
ASTM F467M-84a	Grade 7	Met Nut	0.1 max	0.3 max	0.0125 max	0.05 max	0.25 max	0.12-0.25			bal Ti
ASTM F468-84	Grade 7	Blt Scrw Std	0.1 max	0.3 max	0.0125 max	0.05 max	0.25 max	0.12-0.25			bal Ti
ASTM F468M-84b	Grade 7	Met Blt Scrw Std	0.1 max	0.3 max	0.0125 max	0.05 max	0.25 max	0.12-0.25			bal Ti
AWS A5.16-70	ERTi-0.2Pd	Weld Fill Met	0.05	0.25	0.008	0.02	0.15	0.15-0.25			bal Ti

Ti-0.2Pd grades 7 and 11 compositions: Producer specifications

Specification	Designation	Description	C	Fe	H	N	O	Pd	Si	OT	Other
France											
Ugine	UT35-02	Sh Plt Bar Frg Ann	0.08	0.2	0.015	0.05	0.2	0.2			bal Ti
Germany											
Deutsche T	Contimet Pd 02/30	Mult Forms Ann	0.06	0.15	0.013	0.05	0.12	0.15-0.25			bal Ti
Deutsche T	Contimet Pd 02/35	Mult Forms Ann	0.06	0.2	0.013	0.05	0.18	0.15-0.25			bal Ti
Deutsche T	Contimet Pd 02/35 D	Mult Forms Ann	0.06	0.25	0.013	0.05	0.25	0.15-0.25			bal Ti
Deutsche T	RT 12(Pd)	Sh Strp Bar Frg	0.08	0.2	0.013	0.05	0.1	0.15-0.25			bal Ti
Deutsche T	RT 15(Pd)		0.08	0.25	0.013	0.06	0.2	0.15-0.25			bal Ti
Deutsche T	RT 18(Pd)	Frg	0.1	0.3	0.013	0.06	0.25	0.15-0.25			bal Ti

(continued)

Ti-0.2Pd grades 7 and 11 compositions: Producer specifications (continued)

Specification	Designation	Description	C	Fe	H	N	O	Pd	Si	OT	Other
Japan											
Kobe	KS40PdA	Mult Forms Ann		0.05	0.01	0.03	0.1	0.12-0.2			bal Ti
Kobe	KS40PdB	Mult Forms Ann		0.05	0.01	0.03	0.1	0.17-0.25			bal Ti
Kobe	KS50PdA	Mult Forms Ann		0.05	0.01	0.03	0.15	0.12-0.2			bal Ti
Kobe	KS50PdB	Mult Forms Ann		0.05	0.01	0.03	0.15	0.17-0.25			bal Ti
Kobe	KS70PdA	Mult Forms Ann		0.05	0.01	0.05	0.3	0.12-0.2			bal Ti
Kobe	KS70PdB	Mult Forms Ann		0.05	0.01	0.05	0.3	0.17-0.25			bal Ti
Sumitomo	ST-40P										
Sumitomo	ST-50P										
Sumitomo	ST-60P										
Toho	15PAT		0.02	0.05	0.005	0.01	0.1	0.15 min			bal Ti
Toho	15PBT		0.03	0.08	0.005	0.015	0.15	0.15 min			bal Ti
Toho	20PAT		0.02	0.05	0.005	0.01	0.1	0.2 min			bal Ti
Toho	20PBT		0.03	0.08	0.005	0.015	0.15	0.2 min			bal Ti
UK											
Imp. Metal	IMI 260	Sh						0.15			bal Ti
Imp. Metal	IMI 262	Mult Forms						0.15			bal Ti
USA											
Crucible	A-40 Pd										
OREMET	Ti-11										
OREMET	Ti-17										
RMI	RMI 0.2%Pd	Mult Forms Ann	0.08	0.3	0.015	0.03	0.2	0.2			bal Ti
TIMET	Ti-0.2Pd										
TIMET	TIMETAL 35A Pd										
TIMET	TIMETAL 50A Pd										
TMCA	Ti-7										
TMCA	Ti-11										

Ti-0.3Mo-0.8Ni, R53400

Ti-0.3Mo-0.8Ni (ASTM grade 12), introduced in 1974 for corrosion-resistant applications, is considerably superior to unalloyed titanium in several respects. It exhibits better resistance to crevice corrosion in hot brines (similar to that of Ti-Pd but at much lower cost) and is more resistant than unalloyed Ti (but not Ti-0.2Pd) to corrosion by acids. It also offers significantly greater strength than unalloyed grades for use in high temperature, high pressure applications. This often permits the use of thinner wall sections in pressure vessels and piping, that often translates into cost advantages. Ti-0.3Mo-0.8Ni is less expensive than Ti-0.2Pd grades but does not offer the same crevice corrosion resistance at low pH (<3 pH). In near-neutral brines, crevice corrosion is similar to Ti-0.2Pd.

Chemistry

CP Grade 12 has allowable nitrogen, carbon, hydrogen, iron, and oxygen levels comparable to Grade 2 and Grade 7 except for a lower carbon content (0.08 wt% vs 0.10 wt% max). The titanium content in Grade 12 is lowered through the addition of two beta stabilizers, molybdenum and nickel.

Hydrogen content as low as 30 to 40 ppm can induce hydrogen embrittlement in commercially pure titanium (see the section "Hydrogen Damage" in this datasheet).

Product Forms and Condition

Grade 12 can be readily forged and can be cold worked on equipment used for stainless steels. It is available in all wrought forms and can be cast, welded, and machined.

Ti-0.3Mo-0.8Ni products typically have an annealed alpha structure. The tensile and yield strengths of Ti-0.3Mo-0.8Ni exceed those of either the Grade 2 alloy or the Grade 7 alloy. Compared to palladium-containing grades (ASTM Grade 11), Grade 12 has double the tensile and yield strengths of Grade 11.

Applications

Grade 12 is used in applications requiring moderate strength and enhanced corrosion resistance, such as equipment for chemical, marine, and other industries. Recommended environments for ASTM Grade 12 include seawater, brines, moist chlorine above 120 °C (250 °F), hot process streams containing chlorides where crevices may be pres-ent, oxidizing acids, dilute reducing acids, organic acids, and combinations of these with hot, brackish, or saline cooling waters. This material is used for equipment such as heat exchangers, pressure vessels, chlorine cells, salt evaporators, piping, pollution-control equipment, and other fabrications.

Fatigue Properties

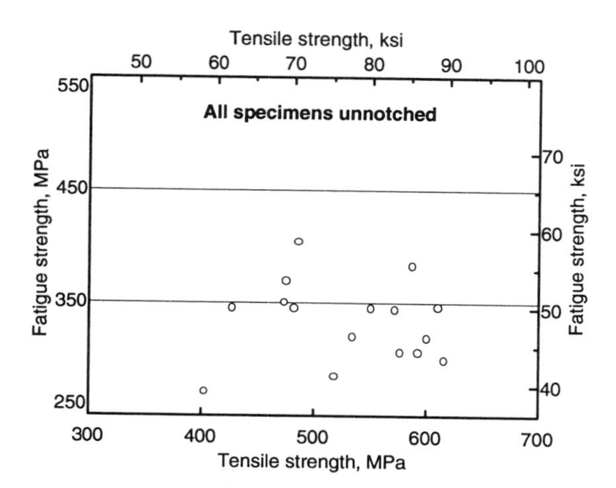

CP Ti: Fatigue strength at 10^7 cycles. Source: *Metals Handbook*, Vol 1, *Properties and Selection*, 8th ed., American Society for Metals, 1961

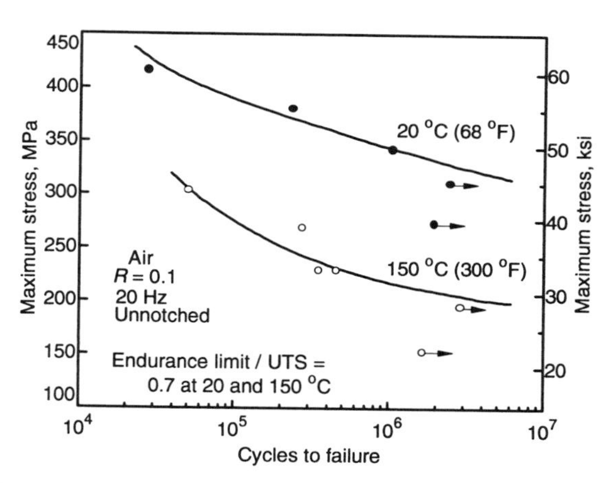

Low-iron grade 2 Ti: Fatigue at 150 °C. Commercially pure low-iron grade 2 titanium plates containing 0.03% iron or less were double vacuum melted by the consumable electrode arc melting process. The ingots were forged to slabs and rolled to 44.4 or 50.8 mm (1.7 or 2 in.) thick plate product. The plate product was annealed at 730 °C (1345 °F) for 30 min and air cooled. Source: P.A. Russo and J.D. Schöbel, "Mechanical Properties of Commercially Pure Titanium Containing Low Iron," presentation at ACHEMA 82, 1982

ASTM grade 4: RT rotating and axial fatigue

Product form	Test method	Stress ratio (R)	Stress concentration	Fatigue strength at cycles 10^5 MPa	10^5 ksi	10^6 MPa	10^6 ksi	10^7 MPa	10^7 ksi
Bar	Rotating beam	-1	Smooth, $K = 1$	517	75	469	68	427	62
			Notched, $K = 2.7$	289	42	262	38	248	36
Sheet	Direct stress	0.6	Smooth, $K = 1$	...	...	...	...	538	78

Annealed Ti-70 sheet and bar. Source: *Aerospace Structural Metals Handbook*, Vol 4, Code 3701, Battelle, 1963

ASTM grade 3: Reverse bending fatigue

Temperature °C	Temperature °F	Stress concentration	Fatigue strength at cycles 10^6 MPa	10^6 ksi	10^7 MPa	10^7 ksi
−191	−312	Smooth, $K_t = 1$	...	...	689	100
		Notched, $K_t = 2.7$	...	...	317	46
RT	RT	Smooth, $K_t = 1$	289	42	282	41
		Notched, $K_t = 2.7$	151	22	241	35
315	600	Smooth, $K_t = 1$	...	...	144	21

Annealed Ti-55 bar; $R = -1$. Source: *Aerospace Structural Metals Handbook*, Vol 4, Code 3701, Battelle, 1963

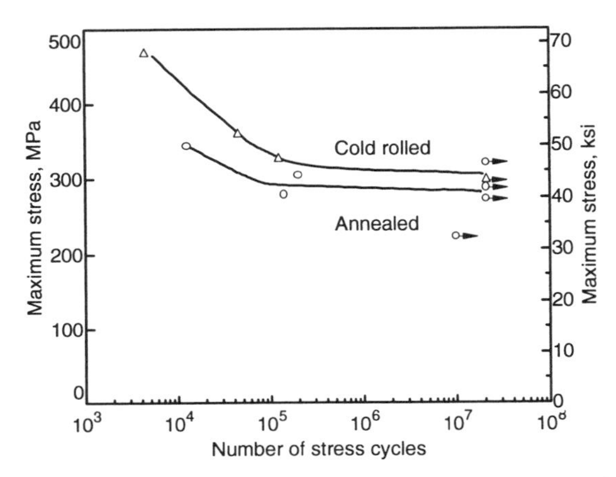

ASTM grade 3 Ti: RT rotating beam fatigue strength. Source: *Metals Handbook*, Vol 3, *Properties and Selection of Stainless Steels, Tool Materials and Special Purpose Metals*, 9th ed., American Society for Metals, 1980, p 376

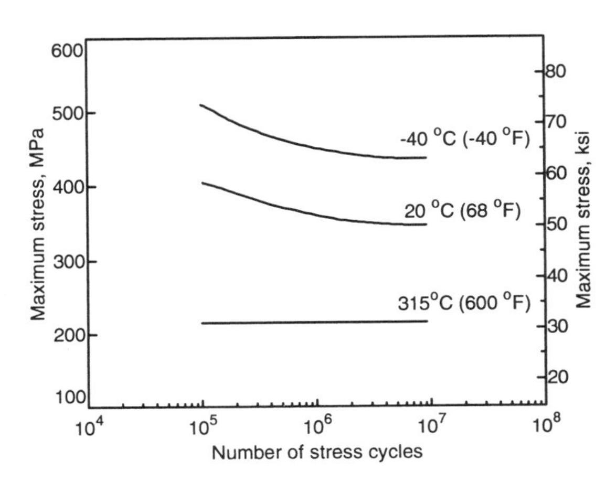

ASTM grade 4 Ti: Rotating-beam fatigue strength. Data are for unnotched, polished specimens machined from annealed bar stock. Source: *Metals Handbook*, Vol 3, 9th ed., American Society for Metals, 1980, p 378

Fracture Properties

Commercially pure titanium, like lowcarbon steel, is considered ductile and tough. Because of the low strength and high toughness of commercially pure titanium, planestrain fracture toughness tests are not meaningful unless specimen thickness is greater than about 150 mm (6 in.). This type of test is not practical for routine testing of commercially pure product. Unlike lowcarbon steel, titanium does not exhibit a good correlation between absorbed impact energy and other more technically rigorous toughness tests such as dynamic tear energy and fracture toughness. Also titanium, unlike steel, does not exhibit a ductile/brittle transition temperature. However, the notched impact test does allow comparison within a given material type and can be used as a quality control tool.

CP Ti: Fracture toughness in air and 3.5% NaCl solution at 25 °C

Alloy	Thickness		Heat treatment(a)	Tensile yield strength		K_Q		K_{ISCC} or K_{SCC}	
	mm	in.		MPa	ksi	MPa$\sqrt{m}$	ksi$\sqrt{in.}$	MPa$\sqrt{m}$	ksi$\sqrt{in.}$
Ti, grade 2	...	...	αA, AC	...	...	66	60	66	60
Ti, grade 2 (0.06% O_2)	19	0.75	αA, AC	...	...	58	53	57	52
Ti, grade 3 (0.020% O_2)	19	0.75	αA, AC	...	...	79	72	74	68
Ti, grade 4 (0.40% O_2)	19	0.75	αA, AC	...	...	99	90	58	53
Ti, grade 4	13	0.50	αA, AC	572	83	135	123	36	33
			αA, WQ	586	85	140	128	37	34
			STA	565	82	124	113	43	39
			βA,WQ	524	76	115	105	77	70
			βSTA	530	77	105	96	52	48

(a) αA,alpha anneal; βA,beta anneal; βSTA, beta solution treated and annealed. Source: R. Schutz, *Stress-Corrosion Cracking of Titanium Alloys*, in Stress-Corrosion Cracking: Materials Performance and Evaluation, ASM International,1992

CP Ti: Charpy V-notch impact toughness

Alloy	Absorbed energy at 20 °C (68 °F)(a)	
	J	ft · lb
Ti-3Al-2.5V	48	35
	48	35
	48	35
	54	40
Grade 1	302	223
	309	228
	305	225
	312	230
Grade 2	114	84
	148	109
	167	123
	171	126
Grade 3	30	22
	39	29
	43	32
	66	49

(a) Longitudinal test direction. Source: *Industrial Applications of Titanium and Zirconium: Third Conference*,R.T. Webster and C.S. Young, Ed., STP 830, ASTM,1984

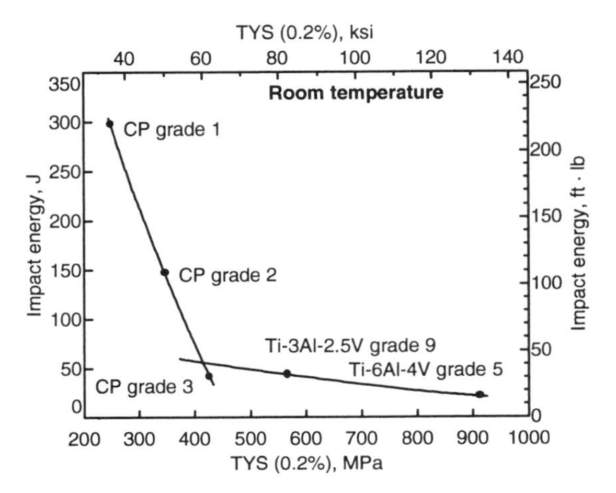

CP Ti: Charpy V-notch impact toughness vs yield strength. Source: *Industrial Applications of Titanium and Zirconium: Third Conference*, R.T. Webster and C.S. Young, Ed., STP 830, ASTM, 1984

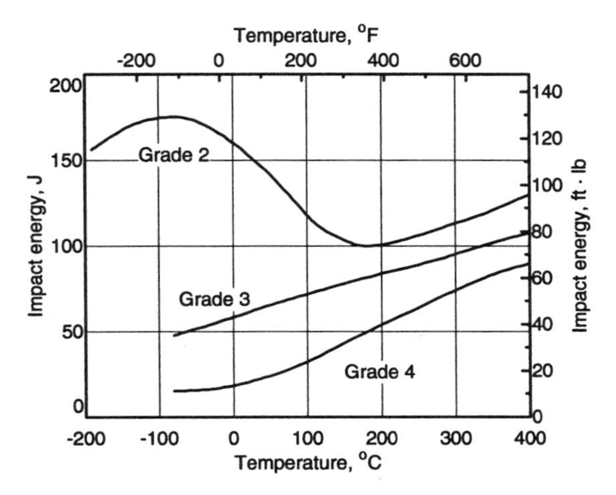

CP Ti: Charpy V-notch impact toughness vs temperature. Source: *Metals Handbook*, Vol 3, 9th ed., American Society for Metals, 1980, p 375

Fracture Mechanism Maps

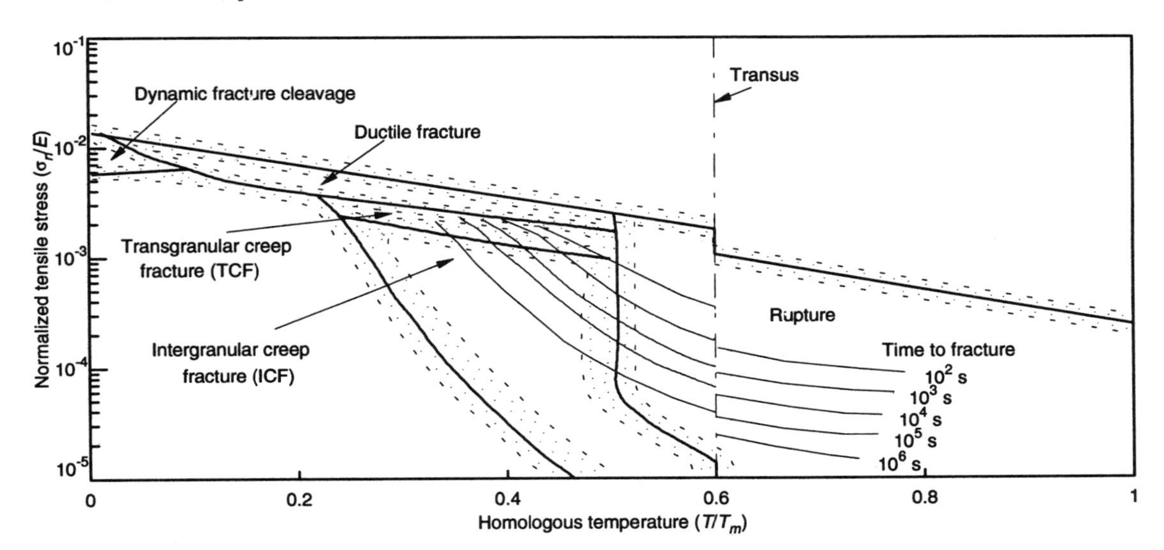

CP Ti: Fracture mechanisms plotted by stress and temperature. Although the TCF field in this figure appears very narrow compared to the intergranular field, this does not reflect the fact that TCF and rupture, and not ICF, dominate the high-temperature fracture behavior of CP titanium. The time-temperature diagram of fracture (see other figure) does not give rise to such an ambiguous impression.

Ti-3Al-2.5V

Common Name: Tubing Alloy, ASTM Grade 9
UNS Number: R56320

Ti3Al2.5V, which is intermediate in strength between unalloyed titanium and Ti6Al4V, has excellent cold formability required for production of seamless tubing, strip and foil. Like Ti6Al4V, 3Ti3Al2.5V has a high strengthtoweight ratio and is lighter than stainless steel. Ti3Al2.5V has 20 to 50% higher strength than unalloyed titanium at both room and elevated temperatures. It has comparable weldability, and is much more amenable to cold working than Ti6Al4V (which does not have good cold forming properties).

Chemistry and Density

With 3 wt% aluminum as an alpha stabilizer 4and 2.5 wt% vanadium as a beta stabilizer, Ti3Al2.5V is sometimes referred to as "half 64." High impurity levels may raise yield strength above maximum permitted values or decrease elongation or reduction in area below minimum values.

Density. 4.48 g/cm^3 (0.162 lb/in.^{3})

Product Forms

2Ti3Al2.5V is available as foil, seamless tubing, pipe, forgings, and rolled products. Ti3Al2.5V was developed for tubing and foil applications. Seamless tubing made of Ti3Al2.5V is readily cold formed on the same type of conventional tubebending equipment used for forming stainless steel. Cold worked and stress relieved tubing generally is not bent to radii less than 3 times the outer diameter in production shops, although radially textured tubing can be bent to 1.5. Relatively thinwall tubing should be bent using tubing fillers or other inside diameter constraints. Ti3Al2.5V tubing is readily welded by standard gas tungsten arc welding with inertgas shielding and by use of automatic welding tools with built in inert gas purge chambers.

Product Condition/Microstructure

Ti3Al2.5V is a nearalpha alphabeta alloy that is generally used in the coldworked and stressrelieved condition. Ti3Al2.5V can be heat treated to high strength, but it has very limited hardenability.

Applications

Ti3Al2.5V seamless tubing was originally developed for aircraft hydraulic and fuel systems and has a proven performance record in hightechnology military aircraft, spacecraft, and commercial aircraft. The Lockheed C5A was the first military production program in which Ti3Al2.5V tubing was employed. This tubing was also selected for the hydraulic system of the Concorde Supersonic Transport. Its first application in subsonic commercial aircraft was the Boeing 767. Since then, Ti3Al2.5V tubing has been chosen for most of the other commercial transports, commuter aircraft, and spacecraft. This alloy also can be readily rolled in strip and foil, the latter of which is used as the honeycomb layer between face sheets of Ti6Al4V sheet in sandwich structures.

Ti3Al2.5V is also employed, mostly in tubular form, in various nonaerospace applications such as sports equipment (golfclub shafts,

Ti-3Al-2.5V: Specifications and compositions

Specification	Designation	Description	Composition, wt%								
			Al	C	Fe	H	N	O	V	OT	Other
UNS	R56320		2.5-3.5	0.05	0.25	0.013	0.02	0.12	2-3		bal Ti
UNS	R56321	Weld Fill Wir	2.5-3.5	0.04	0.25	0.005	0.012	0.1	2-3		bal Ti
China											
	Ti-3Al-2.5V		2.5-3.5	0.08 max	0.3 max	0.015 max	0.05 max	0.12 max	2-3		Si 0.15 max; bal Ti
Europe											
AECMA Ti-P69	prEN3120	Tub CW SR	2.5-3.5	0.05 max	0.3 max	0.015 max	0.02 max	0.12 max	2.5-3.5	0.4 max	Y 0.005 max; OE 0.1 max; bal Ti
Russia											
GOST	AK2		3					0.25-0.35	2.5		bal Ti
GOST	IMP-7	Powd	3		0.3	0.01	0.03	0.16	2		Si 0.6; bal Ti
USA											
AMS 4943D		Tub Ann	2.5-3.5	0.05	0.3	0.015	0.02	0.12	2-3	0.4	Y 0.005; bal Ti
AMS 4944D		Smls Tub CW SR	2.5-3.5	0.05 max	0.3 max	0.015 max	0.02 max	0.12 max	2-3	0.4 max	Y 0.005 max; OE 0.1 max; bal Ti
AMS 4944D		Tub CW SR	2.5-3.5	0.05	0.3	0.015	0.02	0.12	2-3	0.4	Y 0.005; bal Ti
AMS 4945		Smls Tub	2.5-3.5	0.05 max	0.3 max	0.015 max	0.02 max	0.12 max	2-3	0.4 max	Y 0.005 max; OE 0.1 max; bal Ti
ASTM B 337	Grade 9	Smls Weld Pip Ann	2.5-3.5	0.05	0.25	0.013	0.02	0.12	2-3	0.4	bal Ti
ASTM B 338	Grade 9	Smls Weld Tub CW SR	2.5-3.5	0.1	0.25	0.013	0.02	0.12	2-3	0.4	bal Ti
ASTM B 348	Grade 9	Bar Bil Ann	2.5-3.5	0.05	0.25	0.0125	0.02	0.12	2-3	0.4	bal Ti
ASTM B 381	Grade F-9	Frg Ann	2.5-3.5	0.05	0.25	0.015	0.02	0.12	2-3	0.4	bal Ti
ASTM B265-79		Sh Strp Plt	2.5-3.5	0.1 max	0.25 max	0.015 max	0.02 max	0.15 max	2-3		bal Ti
AWS A5.16-70	ERTi-3Al-2.5V-1	Weld Fill Met	2.5-3.5	0.04	0.25	0.005	0.012	0.1	2-3		bal Ti
AWS A5.16-70	ERTi-3Al-2.5V	Weld Fill Met	2.5-3.5	0.05	0.25	0.008	0.02	0.12	2-3		bal Ti
MIL T-9046J	Code AB-5	Sh Strp Plt Ann	2.5-3.5	0.05	0.3	0.015	0.02	0.12	2-3	0.4	bal Ti
MIL T-9047G	Ti-3Al-2.5V	Bar Bil Ann	2.5-3.5	0.05	0.3	0.015	0.02	0.12	2-3	0.4	Y 0.005; bal Ti

tennis racquets, and bicycle frames), medical and dental implants, and expensive ballpointpen casings. In addition to its high strength-to-weight ratio, Ti3Al2.5V is being used in such applications because of its excellent torsion resistance (golfclub shafts and tennis racquets) and corrosion resistance (medical and dental products). Golfclub shafts of Ti3Al2.5V have been heat treated to tensile strengths of approximately 1140 MPa (165 ksi). Other sports products for which Ti3Al2.5V tubing is being investigated include ski poles, fishing poles, and tent stakes.

Use Limitations. The rotary flexure fatigue life of pressurized Ti3Al2.5V tubing is influenced by its crystallographic texture by residual stresses produced in straightening operations, surface roughness, and ovality. Flattening during bending operations reduces the impulse fatigue life of tubing as a result of the superposition of three additive stresses: residual stresses due to flattening, membrane stresses following pressurization, and bending stresses in the flattened tube wall. Overpressurization of tubing (autofrettage) can decrease flattening, thus increasing the impulse fatigue life. Use of improper support assemblies may cause end fitting displacement with attendant installation stresses on the final system, outweighing the beneficial effect of overpressurization.

The reliability of tubing is adversely affected by cracking in service resulting from internal and surface irregularities. Production defects may be inclusions, separations in the tubing wall, or fissures at the inner and outer surfaces. Surface damage usually takes the form of chafing or denting.

Fatigue Properties

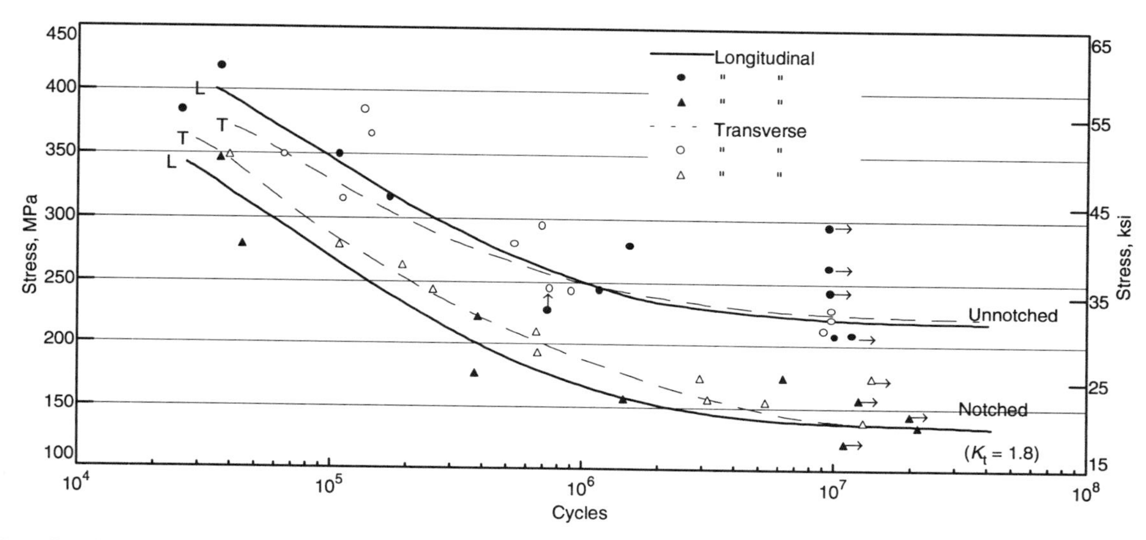

Ti-3Al-2.5V: Smooth and notched bending fatigue. Specimens were annealed 1.0 mm (0.040 in.) sheet.

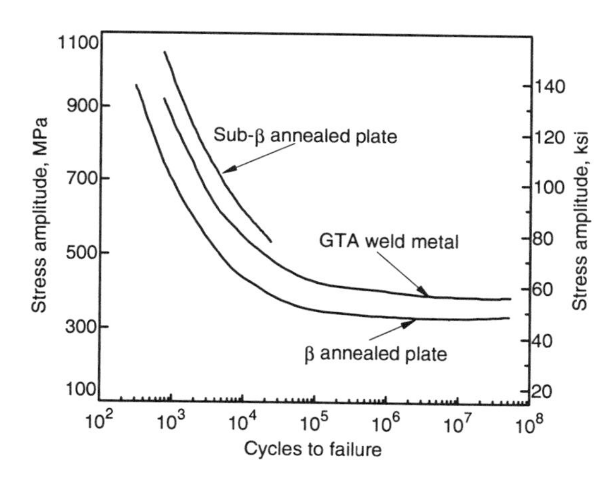

Ti-3Al-2.5V: Fatigue of plate and GTA weld metal. Strain-controlled low-cycle and load-controlled high-cycle fatigue tests were performed on β annealed, sub-β annealed and welded Ti-3Al-2.5V extruded plate. The axial low-cycle fatigue hourglass specimens had a minimum diameter of 6.35 mm (0.25 in.) and were tested at 2 cycles/min according to procedures outlined in ASTM E606. Rotating cantilever beam high-cycle fatigue specimens had a minimum diameter of 4.75 mm (0.187 in.) and were tested at a frequency of 6000 cycles/min.

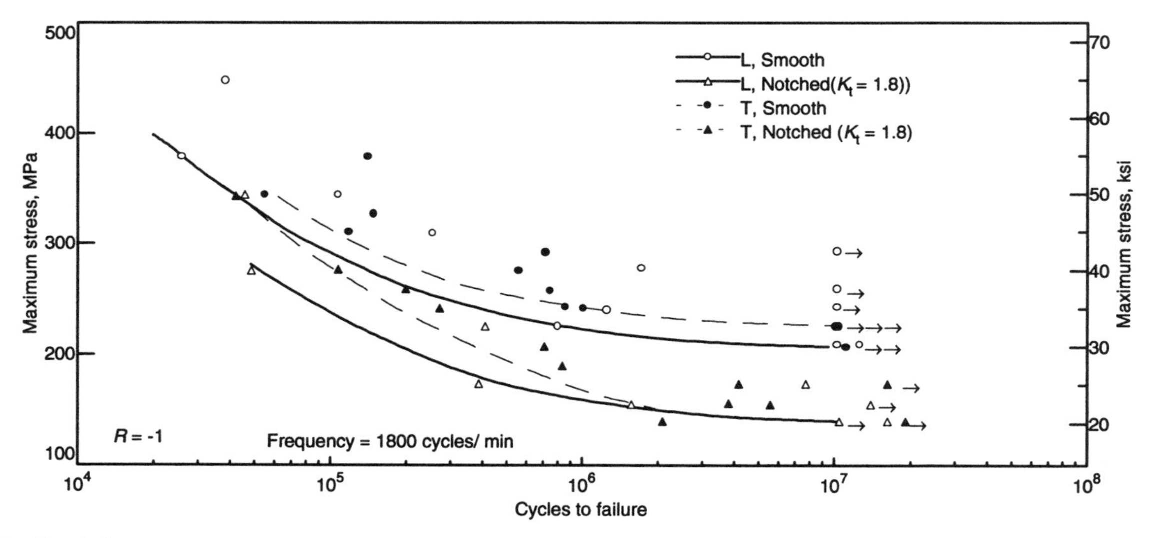

Ti-3Al-2.5V: Bending fatigue strength of annealed sheet. Test material was 1.0 mm (0.040 in.) sheet, annealed at 785 °C (1450 °F), 2 h, vacuum cooled; ultimate tensile strength (L and T) 538 MPa (78 ksi). Source: Bridgeport Brass Co. Report 1000R436, M.O. 83025, Dec 21, 1962; reported in *Aerospace Structural Metals Handbook*, Code 3725, Battelle Columbus Laboratories, 1980, p 28

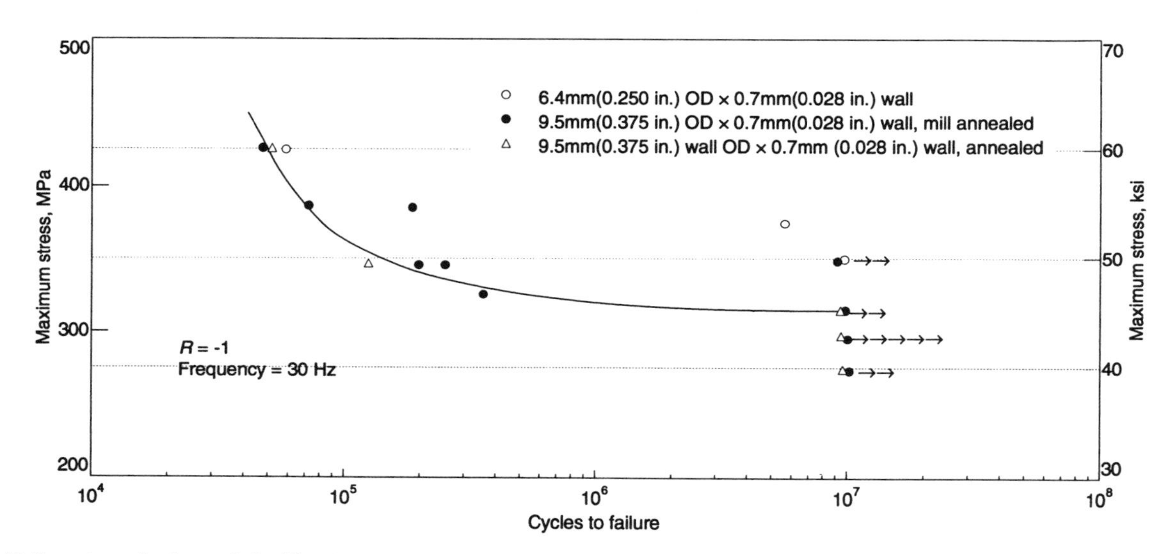

Ti-3Al-2.5V: Fatigue strength of annealed tubing. Source: Pratt & Whitney Aircraft Group Internal Report; June 1973; reported in *Aerospace Structural Metals Handbook*, Code 3725, Battelle Columbus Laboratories, 1980, p 28

Fracture Properties

Impact Toughness

Ti-3Al-2.5V: Charpy V-notch impact strength of extruded plate and welds

As-extruded α - β and α heat treated materials exhibit excellent impact toughness, about twice that of the β annealed plate.

As extruded	101 J (74 ft · lbf)
β annealed	44 J (32 ft · lbf)
α + β annealed(I)	82 J (60 ft · lbf)
α + β annealed(II)	87 J (64 ft · lbf)
α annealed	86 J (63 ft · lbf)
Weld metal	82 J (60 ft · lbf)

Note: TL orientation; test temperature, 0 °C (32 °F). Source: I. Caplan, "Ti-3Al-2.5V for Seawater Piping Applications," in *Industrial Applications of Titanium and Zirconium: Fourth Volume*, ASTM STP 917, C. Young and J. Durham, Ed., ASTM, Philadelphia, 1986, p 43

Ti-3Al-2.5V: Charpy V-notch impact strength of 25 mm (1 in.) extruded plate

Test temperature		As extruded		β anneal 955 °C (1755 °F), 30 min, AC		High-temperature α + β anneal 915 °C (1680 °F), 30 min, AC		High-temperature α + β anneal 915 °C (1680 °F) 30 min, WQ		Low-temperature α + β anneal 805 °C (1475 °F) 30 min, AC	
°C	°F	J	ft · lbf	J	ft · lbf	J	ft · lbf	J	ft · lbf	J	ft · lbf
93	200	...	...	48	36	118	87	123	91	116	86
RT		107	79	46	34	86	64	92	68	101	75
0	32	100	74	43	32	81	60	81	60	86	64
−62	−80	...	...	38	28	69	51	61	45	69	51

Source: *Aerospace Structural Metals Handbook*, Code 3725, Battelle Columbus Laboratories,1980, p 20

Fracture Toughness

Seawater Stress Corrosion. Notched, dead-weight loaded, cantilever beam specimens measuring 25 mm (1 in.) by 50 mm (2 in.) by 330 mm (13 in.) were used to evaluate the seawater stress-corrosion cracking performance of heat treated and welded plates. The specimens were step-loaded in seawater to a given stress intensity and held until failure occurred, or for a maximum of 1000 h. None of the materials displayed any stress-corrosion cracking susceptibility based on fractographic examination of failed specimens. However, the β and sub-β annealed material did exhibit time-dependent sustained load failures. The sustained load cracking threshold stress-intensity value in seawater (K_{ISLC}) was defined as the average of the minimum time-dependent failure and the maximum runout for a given material condition. The weld metal did not exhibit any time-dependent failure up to a maximum stress intensity of 123 MPa$\sqrt{m}$ (112 ksi$\sqrt{in.}$).

Ti-3Al-2.5V: Sustained load cracking of heat treated plate in seawater

Heat treatment/ condition	Threshold (K_{ISLC}) MPa$\sqrt{m}$	ksi$\sqrt{in.}$
β annealed(a)	75	68
Sub-β annealed	88	80

Note: Test duration 1000 h. (a) K_{max} in air = 81 MPa$\sqrt{m}$ (74 ksi$\sqrt{in.}$). Source: I. Caplan, "Ti-3Al-2.5V for Seawater Piping Applications," in *Industrial Applications of Titanium and Zirconium: Fourth Volume*, C. Young and J. Durham, Ed., ASTM STP 917, ASTM, Philadelphia, 1986, p 43

Ti-3Al-2.5V: Fracture toughness of extrusions in several heat treated conditions compared to weld metal

Condition	Fracture toughness (J_{IC}) kJ/m²	in. · lb/in.²	Tear modulus(a)
As-extruded	40	230	7
β annealed	70	400	10
α + β annealed (near β transus)	93	530	24
α + β annealed (near α transus)	123	700	26
α annealed	100	570	31
Weld metal	151	860	27

Note: Chemical composition of extrusions: 2.71% Al, 0.011% C, 0.005% Cu, 0.191% Fe, 0.0014% H, 0.005% Mn, 0.013% N, 0.099% O, 0.015% Si, and 2.56% V. Weld metal composition: 0.033% H, 0.009% N, and 0.096% O. Fracture toughness was determined according to ASTM E813 using computer-interactive unloading compliance procedures. (a) Nondimensional. Source: I. Caplan, "Ti-3Al-2.5V for Seawater Piping Applications," in *Industrial Applications of Titanium and Zirconium: Fourth Volume*, C. Young and J. Durham, Ed., ASTM STP 917, ASTM, Philadelphia, 1986, p 43

Seamless Tubing

Ti-3Al-2.5V is used primarily as seamless tubing, which can exhibit variations in crystallographic orientation ranging from a radial texture to a circumferential texture (see figure). Texture variations of Ti-3Al-2.5V tubing provide a useful means of tailoring properties, and a radial texture has the characteristic of increasing both tensile yield strength and elongation (see figures).

Typical Properties

Extruded tube intermediates can be cold worked to a moderately high-strength ductile product (see table). For higher strengths and potential weight savings on aircraft hydraulic tubing, a seamless Ti-6Al-4V tubing product has been developed. Strength comparisons of Ti-6Al-4V and Ti-3Al-2.5V seamless tubing are provided below.

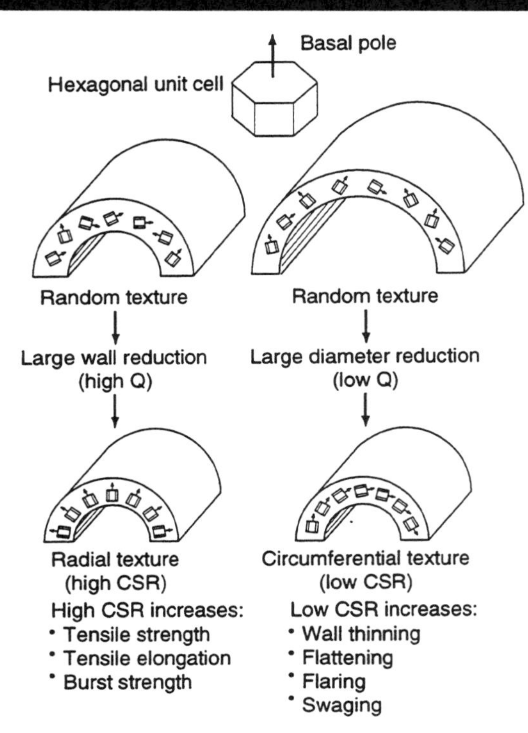

Ti-3Al-2.5V: Effect of tube reductions on texture and properties. The contractile strain ratio (CSR) is the ratio of diametral to radial strain from a given stress. AMS 4945 specified a 1.3 minimum CSR for most tube sizes.. Low and high CSR both reduce fatigue strength, whereas a midrange CSR increases fatigue strength.

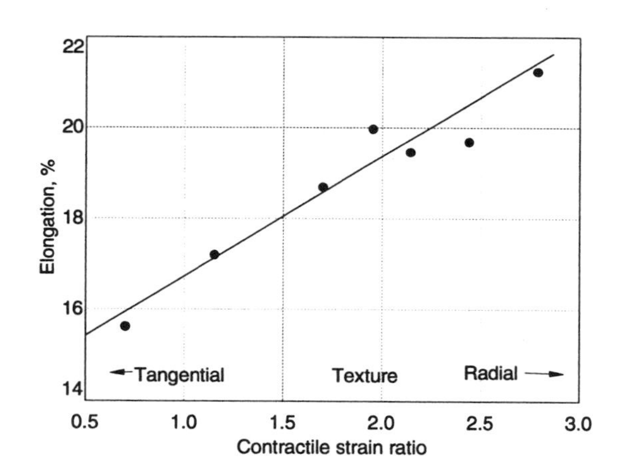

(a)

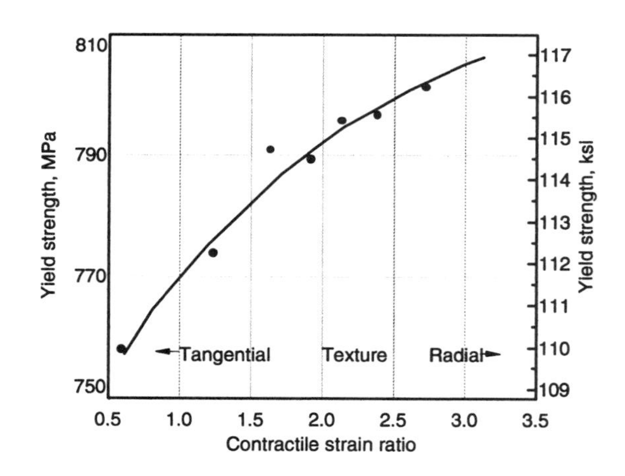

(b)

Ti-3Al-2.5V: Effect of texture on tensile properties. Data are an average of 71 samples.

Ti-3Al-2.5V: Tensile properties of tubing

Condition	Basis	Ultimate tensile strength MPa	Ultimate tensile strength ksi	Tensile yield strength MPa	Tensile yield strength ksi	Elongation in 50 mm (2 in.), %
Cold worked	Typical	1034	150	896	130	7-11
	Full hard(a)		137		118	9
	Half hard(a)		126		98	12
Cold worked plus stress relieved	Typical(b)	909	132	792	115	19
	Minimum	861	125	723	105	16
Annealed	Typical(b)	648	94	579	84	29
	Minimum	620	90	517	75	15(c)

(a) Values reported in *Titanium Alloys Handbook*, MCIC-HB-02, 1972. (b) Typical values are an average from C.E. Forney, Jr.,and J.H. Schemel, *Ti-3Al-2.5V Seamless Tubing Engineering Guide*, 2nd ed., Sandvik Special Metals, 1987. (c) 14% minimum for 6.35 and 9.5 mm (0.25 and 0.375 in.) OD sizes

Seamless tubing comparison

Alloy	System pressure MPa	System pressure ksi	Wall mm	Wall in.	Theoretical burst(a) psi	Actual burst	lb/ft	Weight savings, %
6.35 mm (0.25 in.) OD								
Ti-3-2.5	34	5	0.55	0.022	23, 900	...	0.0306	...
Ti-6-4	34	5	0.38	0.015	19, 071	...	0.0213	30.04
	34	5	0.40	0.016	20, 417	...	0.0226	26.14
Ti-3-2.5(b)	20	3	0.40	0.016	17, 014	...	0.0229	...
Ti-6-4(b)	20	3	0.40	0.016	20, 417	...	0.0226	1.31
13 mm (0.5 in.) OD								
Ti-3-2.5	34	5	1.09	0.043	23, 317	...	0.1200	...
Ti-6-4	34	5	0.66	0.026	16, 406	20, 800(d)	0.0743	38.08
Ti-6-4	34	5	0.84	0.033	21, 094	...	0.0930	22.50
Ti-3-2.5(b)	20	3	0.66	0.026	13, 672	...	0.0753	...
Ti-6-4(c)	20	3	0.55	0.022	13, 778	...	0.0634	15.8
19 mm (0.75 in.) OD								
Ti-3-2.5	34	5	1.65	0.065	23, 511	...	0.2719	...
Ti-6-4	34	5	0.99	0.039	16, 406	18, 200(e)	0.1673	38.47
Ti-3-2.5	20	3	0.99	0.039	13, 672	...	0.1693	...
Ti-6-4(c)	20	3	0.84	0.033	13, 778	...	0.1427	15.71
25 mm (1 in.) OD								
Ti-3-2.5	34	5	2.2	0.088	23, 900	...	0.4901	...
Ti-6-4	34	5	1.29	0.051	16, 076	...	0.2919	40.44
Ti-6-4	34	5	1.67	0.066	21, 094	...	0.3718	24.14
Ti-3-2.5(b)	20	3	1.29	0.051	13, 397	...	0.2956	...
Ti-6-4(c)	20	3	1.09	0.043	13, 452	...	0.2482	16.04

(a) Pressures calculated as: $P = \text{UTS} \times (\text{OD}^2 - \text{ID}^2)/(\text{OD}^2 + \text{ID}^2)$. (b) Minimum UTS of 860 MPa (125 ksi). (c) Minimum UTS of 1035 MPa (150 ksi). (d) Contractile strain ratio (CSR) 2.2. (e) Annealed to lower mechanical properties: 999 MPa (145 ksi) ultimate tensile strength, 862 MPa (125 ksi) yield strength CSR 1.08. Source: C. Forney, Sandvik Special Metals

Ti-5Al-2.5Sn

Common Name: Ti-5-2½ and Ti-5-2½ ELI
UNS Numbers: R54520/R54521

Developed by Battelle for RemCru (later called Crucible Steel) as an intermediate-strength, weldable alloy, Ti-5Al-2.5Sn was first manufactured in 1950. Its primary use was in applications requiring moderate strength and excellent weldability. It was one of the first alloys to be developed commercially and is one of the few original alloys still in commercial use. Although it is still available from all producers, it is being replaced by Ti-6Al-4V in many applications.

Chemistry and Density

As interstitial element content increases, both yield and tensile strengths increase and fracture toughness decreases. The extra low interstitial (ELI) grade of Ti-5Al-2.5Sn (UNS R54521) is especially well suited for service at cryogenic temperatures and exhibits an excellent combination of strength and toughness at –250 °C (–420 °F).

Density. 4.48 g/cm^3 (0.162 lb/in.^{3})

Product Forms

Ti-5Al-2.5Sn is available as bar, plate, sheet, strip, wire, forgings, and extrusions. The ELI grade is quite difficult to hot work into some product forms, particularly when converting from ignot to billet because of shear cracking, often referred to as strain-induced porosity. Ti-5Al-2.5Sn can be cast, machined, and welded.

Product Condition/Microstructure

Ti-5Al-2.5Sn is a medium-strength, all-alpha titanium alloy. It has very high fracture toughness at both room temperature and elevated temperatures and is used only in the annealed condition.

Applications

Ti-5Al-2.5Sn is used for gas turbine engine castings and rings, rocket motor casings, aircraft forgings and extrusions, aerospace structural members in hot spots (near engines and leading edges of wings), ordnance equipment, chemical-processing equipment requiring elevated-temperature strength superior to that of unalloyed titanium and excellent weldability, and other applications demanding good weld fabricability, oxidation resistance, and intermediate strength at service temperatures up to 480 °C (900 °F).

Ti-5Al-2.5Sn ELI is employed for liquid hydrogen tankage and high-pressure vessels at temperatures below –195 °C (–320 °F), structural members for aircraft, and gas turbine parts. It is used in applications requiring ductility and toughness greater than those of the standard grade, although at some sacrifice in strength, particularly in hardware for service at cryogenic temperatures.

Precautions in Use. The elevated temperature stress-corrosion resistance of this alloy in the presence of solid salt is lower than those of other commonly used titanium alloys. Use of Ti-5Al-2.5Sn (like all titanium alloys) in contact with liquid oxygen, or in contact with gaseous oxygen at pressures above approximately 345 kPa (50 psi), constitutes severe fire and explosion hazard.

Ti-5Al-2.5Sn: Specifications and compositions

			Composition, wt% (a)								
Specification	Designation	Description	Al	C	Fe	H	N	O	Sn	OT	Other
UNS	R54520		4-6	0.1	0.5	0.02	0.05	0.2	2-3		bal Ti
UNS	R54521	ELI	5						2.5		bal Ti
UNS	R54522	Weld Fill Met	4.7-5.6	0.05	0.4	0.008	0.03	0.12	2-3		bal Ti
UNS	R54523	ELI Weld Fill Met	4.7-5.6	0.04	0.25	0.005	0.012	0.1	2-3		bal Ti
China											
GB 3620	TA-7		4-6	0.1 max	0.3 max	0.015 max	0.05 max	0.2 max	2-3		Si 0.15 max; bal Ti
Germany											
	WL 3.7114		4.5-5.5	0.08	0.5	0.015-0.02	0.05	0.2	2-3	0.4	bal Ti
DIN 17851	Ti-5Al-2.5Sn	Sh Strp Plt Rod Wir	4-6	0.08	0.5	0.02	0.05	0.2	2-3		bal Ti
DIN 17851	WL 3.7115	Plt Sh Strp Ann	4-6	0.08	0.5	0.02	0.05	0.2	2-3		bal Ti
Russia											
GOST	VT5-1KT		4-5.5	0.05	0.2	0.008	0.04	0.12	2-3		Zr 0.2; Mn 0.1; Si 0.1; bal Ti
GOST 19807-74	VT5-1	Sh Plt Strp Rod Frg Ann	4-6	0.1	0.3	0.015	0.05	0.15	2-3	0.3	Zr 0.3; Si 0.15; bal Ti
Spain											
UNE 38-716	L-7101	Sh Strp Plt Bar Frg Ext	4.5-5.5	0.15	0.5	0.02	0.07	0.2	2-3		bal Ti
UK											
BS TA14		Sh	4-6	0.08 max	0.5 max	0.0125 max			2-3		bal Ti
BS TA15		Bar	4-6	0.08 max	0.5 max	0.0125 max			2-3		bal Ti
BS TA16		Frg	4-6	0.08 max	0.5 max	0.0125 max			2-3		bal Ti
BS TA17		Frg	4-6		0.5 max	0.015 max			2-3		bal Ti
USA											
AMS 4909D		ELI Sh Strp Plt Ann	4.5-5.75	0.05	0.25	0.0125	0.035	0.12	2-3	0.3	Y 0.005; O + Fe = 0.32; bal Ti
AMS 4910J		Sh Strp Plt Ann	4.5-5.75	0.08	0.5	0.02	0.05	0.2	2-3	0.4	Y 0.005; bal Ti
AMS 4924D		ELI Bar Frg Rng Ann	4.7-5.6	0.05	0.25	0.0125	0.035	0.12	2-3	0.4	Y 0.005; O + Fe = 0.32; bal Ti
AMS 4926H		Bar Wir Bil Rng Ann	4-6	0.08	0.5	0.02	0.05	0.2	2-3	0.4	Y 0.005; bal Ti
AMS 4953D		Weld Fill Wir	4.5-5.75	0.08	0.5	0.015	0.05	0.175	2-3	0.4	Y 0.005; bal Ti
AMS 4966J		Frg Ann	4-6	0.08	0.5	0.02	0.05	0.2	2-3	0.4	Y 0.005; bal Ti
ASTM B 265	Grade 6	Sh Strp Plt Ann	4-6	0.1	0.5	0.02	0.05	0.2	2-3	0.4	bal Ti
ASTM B 348	Grade 6	Bar Bil Ann	4-6	0.1	0.5	0.0125	0.05	0.2	2-3	0.4	bal Ti
ASTM B 367	Grade C-6	Cast	4-6	0.1	0.5	0.015	0.05	0.2	2-3	0.4	bal Ti
ASTM B 381	Grade F-6	Frg Ann	4-6	0.1	0.5	0.02	0.05	0.3	2-3	0.4	bal Ti
AWS A5.16-70	ERTi-5Al-2.5Sn-1	ELI Weld Fill Met	4.7-5.6	0.04	0.25	0.005	0.012	0.1	2-3		bal Ti
AWS A5.16-70	ERTi-5Al-2.5Sn	Weld Fill Met	4.7-5.6	0.05	0.4	0.008	0.03	0.12	2-3		bal Ti
MIL F-83142A	Comp 2	Frg Ann	4.5-5.75	0.08	0.5	0.02	0.05	0.2	2-3	0.4	bal Ti
MIL F-83142A	Comp 2	Frg HT	4.5-5.75	0.08	0.5	0.02	0.05	0.2	2-3	0.4	bal Ti
MIL F-83142A	Comp 3	ELI Frg Ann	4.5-5.75	0.05	0.25	0.0125	0.035	0.12	2-3	0.3	bal Ti
MIL T-81556A	Code A-1	Ext Bar Shp Ann	4.5-5.75	0.08	0.5	0.02	0.05	0.2	2-3	0.4	bal Ti
MIL T-81556A	Code A-2	ELI Ext Bar Shp Ann	4.5-5.75	0.05	0.25	0.0125	0.035	0.12	2-3	0.3	bal Ti
MIL T-81915	Type II Comp A	Cast Ann	4.5-5.75	0.08	0.5	0.02	0.05	0.2	2-3	0.4	bal Ti
MIL T-9046J	Code A-1	Sh Strp Plt Ann	4.5-5.75	0.08	0.5	0.02	0.05	0.2	2-3	0.4	bal Ti
MIL T-9046J	Code A-2	ELI Sh Strp Plt Ann	4.5-5.75	0.05	0.25	0.0125	0.035	0.12	2-3	0.3	bal Ti
MIL T-9047G	Ti-5Al-2.5Sn	Bar Bil Ann	4.5-5.75	0.08	0.5	0.02	0.05	0.2	2-3	0.4	Y 0.005; bal Ti
MIL T-9047G	Ti-5Al-2.5Sn ELI	ELI Bar Bil Ann	4.5-5.75	0.05	0.25	0.0125	0.035	0.12	2-3	0.3	Y 0.005; bal Ti

(a) Maximum unless a range is specified

Ti-5Al-2.5Sn: Compositions

Specification	Designation	Description	Composition, wt% (a) Al	C	Fe	H	N	O	Sn	OT	Other
France											
Ugine	UTA5E	Sh Bar Ann	4.5-5.5	0.15	0.5	0.02	0.07	0.2	2-3		bal Ti
Ugine	UTA5EL	ELI Bar Ann	4.5-5.75	0.05	0.25	0.0125	0.035	0.12	2-3		bal Ti
Germany											
Deutsche T	Contimet AlSn 52	Sh Strp Plt Bar Frg Pip Ann	4.5-5.5	0.08	0.5	0.02	0.05	0.2	2-3		bal Ti
Deutsche T	Contimet AlSn 52 ELI	ELI Plt Bar Frg Pip Ann	4.7-5.6	0.06	0.15	0.013	0.05	0.12	2-3		bal Ti
Fuchs	TL52	Frg	5						2.5		bal Ti
Japan											
Kobe	KS5-2.5	Ann	4-6		0.5	0.02	0.05	0.2	2-3		bal Ti
Kobe	KS5-2.5ELI	ELI Ann	4.7-5.6		0.25	0.0125	0.035	0.12	2-3		bal Ti
MMA	5137										
Sumitomo	SAT-525										
Toho	525AT		4-6	0.03	0.5	0.02	0.05	0.3	2-3		bal Ti
USA											
OREMET	Ti 5-2.5										
RMI	RMI 5Al-2.5Sn	Mult Forms Ann	4-6	0.08	0.5	0.0175-0.02	0.05	0.2	2-3		
RMI	RMI 5Al-2.5Sn ELI	ELI Mult Forms Ann	4.7-5.75	0.08	0.25	0.0125-0.015	0.03	0.13	2-3		bal Ti
TIMET	TIMETAL 5-2.5	Ann	4-6	0.1 max	0.5 max	0.02 max	0.05 max	0.2 max	2-3		bal Ti
TIMET	TIMETAL 5-2.5 ELI	Ann	4.5-5.75	0.05 max	0.25 max	0.0125 max	0.035 max	0.12 max	2-3		bal Ti

(a) Maximum unless a range is specified

Fatigue Life

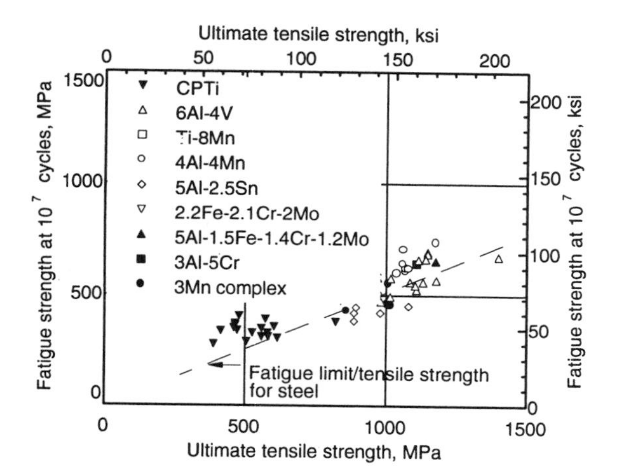

Ti-5Al-2.5Sn: Fatigue endurance ratio comparison. Source: TML Report No. 77, Battelle, 1957

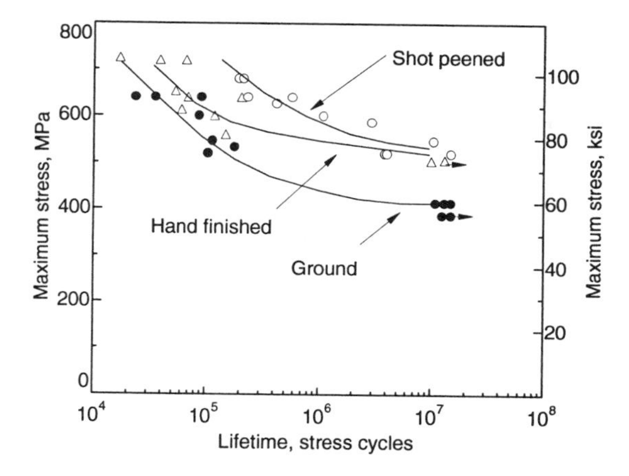

Ti-5Al-2.5Sn: Rotating-beam fatigue strength. Effect of surface finish. Source: *Metals Handbook, Properties and Selection: Stainless Steels, Tool Materials, and Special-Purpose Materials,* Vol 3, 9th ed., ASM, 1980

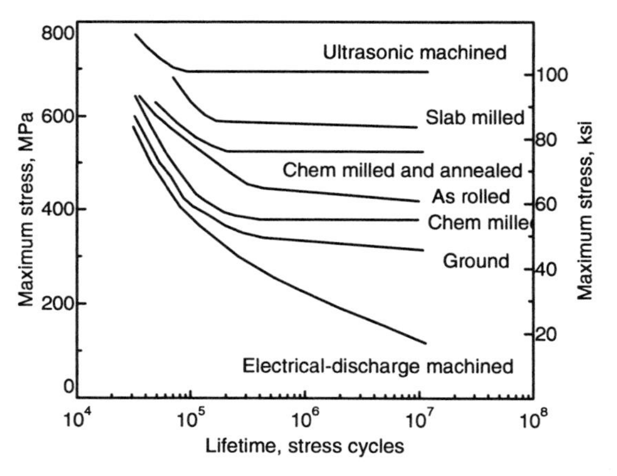

Ti-5Al-2.5Sn: Rotating-beam fatigue strength. Effect of surface finish. Source: *Metals Handbook, Properties and Selection: Stainless Steels, Tool Materials, and Special-Purpose Materials,* Vol 3, 9th ed., ASM, 1980

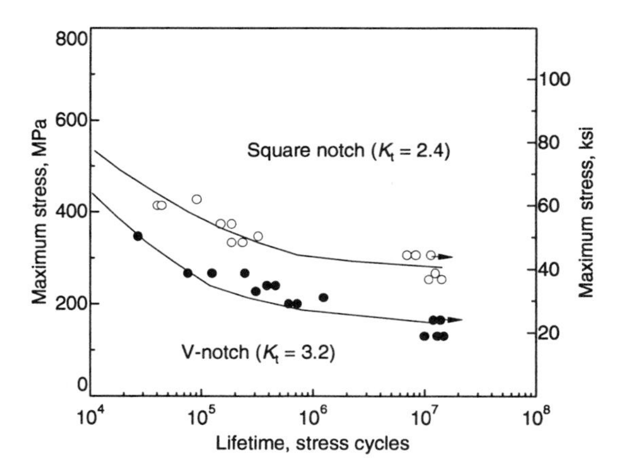

Ti-5Al-2.5Sn: Rotating-beam fatigue strength. Notch fatigue strength for two notches. Source: *Metals Handbook, Properties and Selection: Stainless Steels, Tool Materials, and Special-Purpose Materials,* Vol 3, 9th ed., ASM, 1980

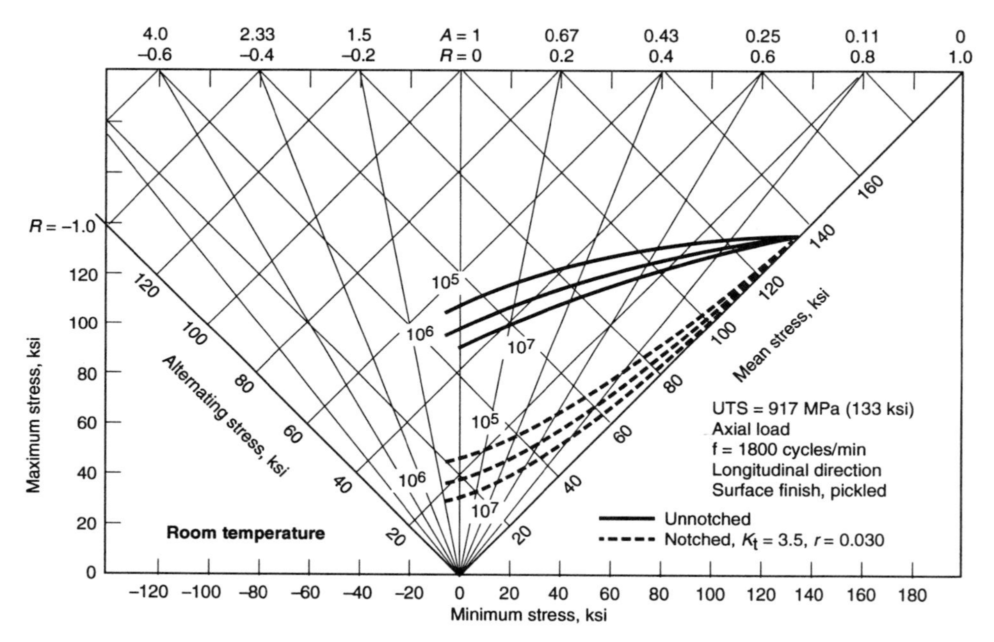

Ti-5Al-2.5Sn: Constant life diagram of mill annealed sheet. Source: MIL-HDBK-5, 1972

Low-Temperature Fatigue Data

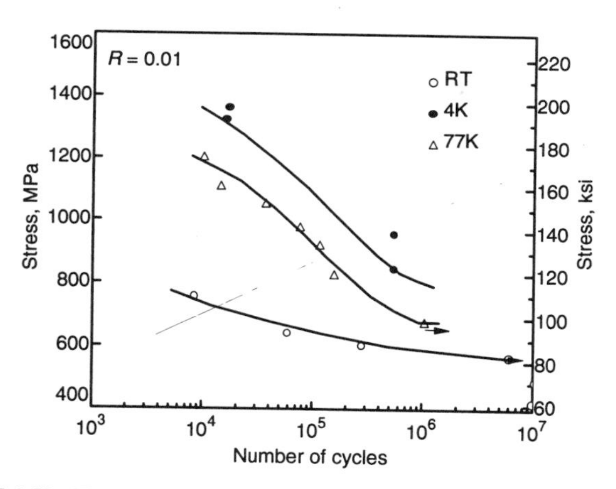

Ti-5Al-2.5Sn ELI: Fatigue strength vs temperature. Source: Titanium Product Literature, Kobe Steel

Ti-5Al-2.5Sn: Low-temperature fatigue life of welded sheet

Alloy and condition	Stressing mode	Stress ratio, R	K_t	Fatigue strengths at 10^6 cycles: 24 °C (75 °F) MPa	24 °C (75 °F) ksi	−196 °C (−320 °F) MPa	−196 °C (−320 °F) ksi	−253 °C (−423 °F) MPa	−253 °C (−423 °F) ksi
Ti-5Al-2.5Sn (ELI) sheet, annealed	Axial	0.01	1	495	72	815	118	760	110
			3.5	220	32	205	30	160	23
Ti-5Al-2.5Sn (ELI) sheet(a)	Axial	0.01	1	485	70	565	82	425	62
Ti-5Al-2.5Sn (ELI) bar, annealed(b)	Axial	0	1	760	110	985	143	925	134
Ti-6Al-4V (ELI) sheet(c)	Axial	0.01	1	505	73	675	98	895	130
			3.5	285	41	295	43	275	40
Ti-6Al-4V (ELI) sheet(a)	Axial	0.01	1	600	87	595	86	560	81
Ti-6Al-4V sheet, annealed	Flex	−1.0	1	345	50	550	80	530	77
			3.1	170	25	185	27	255	37

(a) Gas tungsten arc welded, base metal filler. (b) Cyclic frequency, 28 Hz. (c) STA: 900 °C (1650 °F) 5 min, WQ; 540 °C (1000 °F) 4 h, AC. Source: *Metals Handbook*, Properties and Selection: Stainless Steels, Tool Materials, and Special-Purpose Materials, Vol 3, 9th ed., American Society for Metals, 1980

Fatigue Crack Growth

Ti-5Al-2.5Sn: Fatigue crack growth of annealed sheet at room temperature

Test environment	Fatigue crack growth, µm/cycle (µin./cycle), at: ΔK = 5.5 MPa $\sqrt{m}$ (5 ksi $\sqrt{in.}$), R = 0.67, and 55-58 Hz	ΔK = 11 MPa $\sqrt{m}$ (10 ksi $\sqrt{in.}$), R = 0.67, and 55-58 Hz	ΔK = 22 MPa $\sqrt{m}$ (20 ksi $\sqrt{in.}$), R = 0.1, and 30-50 Hz(a)	ΔK = 55 MPa $\sqrt{m}$ (50 ksi$\sqrt{in.}$), R = 0.1, 30 Hz
L-T specimen orientation				
Dry argon	0.00075 (0.03)	0.00685 (0.27)	0.12-0.14 (4.77-5.56)	2.40 (94.7)
Lab air	0.0038 (0.15)	(2.13)	(11.6-11.7)	(124)
Distilled water	0.00635 (0.25)	(3.49)	(11.8)(b)	(124)
3.5% NaCl	0.0074 (0.29)	(7.97)	(23.5-30.2)	(157)
T-L specimen orientation				
Dry argon	...	(0.49)	(5.35-5.38)	(114)
Lab air	0.0038 (0.15)	(3.08)	(11.8-11.9)	(141)
Distilled water	0.009 (0.36)	(3.72)	(12.0-12.5)	(130)
3.5% NaCl	0.025 (0.98)	(14.6)	(24.5)(b)	(176)

(a) The higher measured values correspond to tests at 30 Hz. (b) 50 Hz. Source: J. Gallagher, *Damage Tolerant Design Handbook*, MCIC-HB-O1R, Battelle, 1983

Ti-5Al-2.5Sn: Fatigue crack growth rate compared to Ti-6Al-4V

Crack growth parameters per the relation $da/dN = C(\Delta K)^n$

Alloy and condition(a)	Orientation	Test temperature, °C	Test temperature, °F	C: da/dN:mm/cycle, ΔK:MPa $\sqrt{m}$	C: da/dN:in./cycle, ΔK:ksi$\sqrt{in.}$	n	Estimated range for ΔK, MPa$\sqrt{m}$	Estimated range for ΔK, ksi$\sqrt{in.}$
Ti-5Al-2.5Sn (NI), annealed bar	T-S	24, −196, −269	75, −320, −452	5.1×10^{-11}	3.2×10^{-12}	4.8	14-30	13-27
Ti-5Al-2.5Sn (LI), annealed bar	T-L	24, −196, −269	75, −320, −452	4.9×10^{-10}	2.8×10^{-11}	4.0	10-60	9-54
Ti-6Al-4V (NI), annealed bar	T-L	24, −196, −269	75, −320, −452	3.1×10^{-12}	2.2×10^{-13}	6.0	14-30	13-27
Ti-6Al-4V (ELI), recrystallization annealed bar	T-L	24, −196, −269	75, −320, −452	1.9×10^{-13}	1.4×10^{-14}	7.0	10-20	9-18
		24, −196	75 −320	3.0×10^{-8}	1.6×10^{-9}	3.0	20-40	18-36

Note: Stress ratio: $R = 0.1$, at 20 to 28 Hz; compact specimens. (a) NI = normal interstitial, LI = low interstitial, ELI = extra low interstitial. Source: R. L. Tobler and R.P. Reed, "Fatigue Crack Growth Resistance of Structural Alloys at Cryogenic Temperatures," in *Advances in Cryogenic Engineering*, K.D. Timmerhaus *et al.*, Ed., Vol 24, Plenum Press, 1978, p 82-90

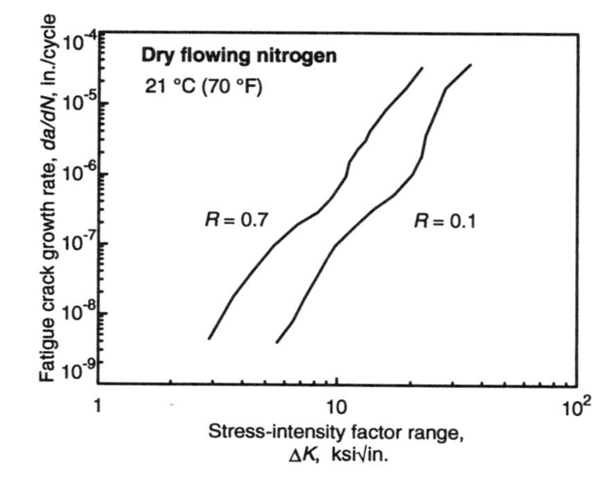

Ti-5Al-2.5Sn ELI: Crack growth at room temperature. Source: D.E. Matejczyk *et al.*, Fatigue Crack Retardation Following Overloads in Inconel 718, Ti-5Al-2.5Sn, and Haynes 188, *Advanced Earth-to-Orbit Propulsion Technology 1986*, Vol 2, NASA Conference Publication 2437, 1986, p 205-219

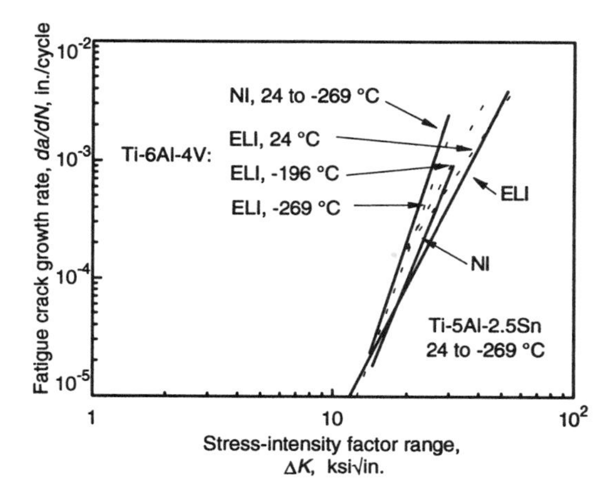

Ti-5Al-2.5Sn: Fatigue crack growth rates. NI = normal interstitial content; ELI = extra-low interstitial content. Source: R.L. Robler and R.P. Reed, in *Advances in Cryogenic Engineering*, Vol 24, K.D. Timmerhaus *et al.*, Ed., Plenum Press, 1978, p 82-90

Fracture Properties

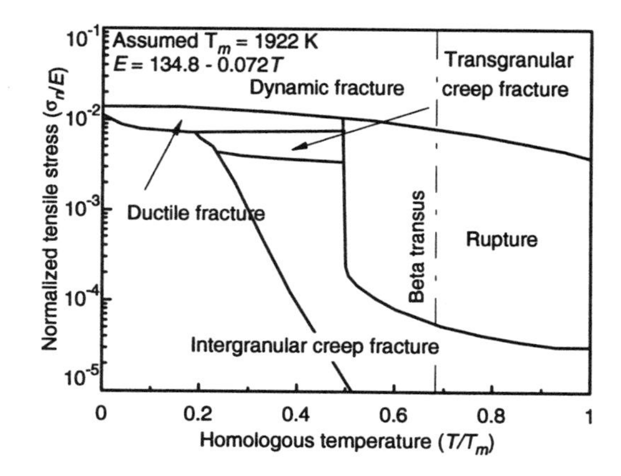

Ti-5Al-2.5Sn: Fracture mechanism map. Source: Krishnamohanrao *et al.*, Fracture Mechanism Maps for Titanium and Its Alloys, *Acta Metall.*, Vol 34, 1986, p 1783-1806

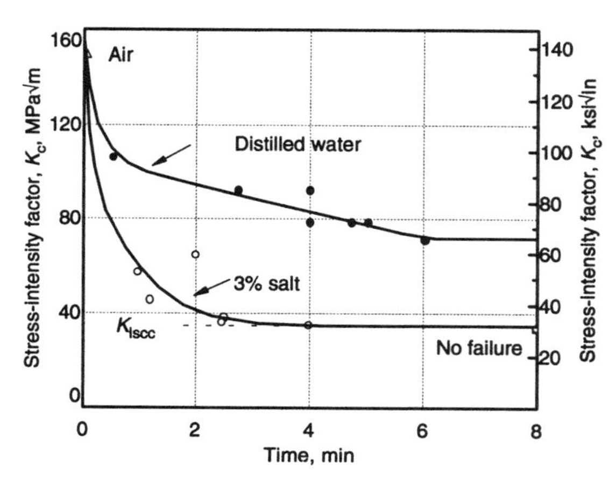

Ti-5Al-2.5Sn: Time-to-fracture. Effect of initial stress intensity on time-to-fracture at ambient temperature. Source: R. Wood and R. Favor, *Titanium Alloys Handbook*, MCIC-HB-02, Battelle Columbus Laboratories, 1972

Low Temperature Toughness (Standard and ELI)

Ti-5Al-2.5Sn: Fracture toughness

Heat treatment variable(a)	Test temperature		Stress intensity, K_{IC}		Specimen, orientation(b) and type(c)	Yield strength(d)	
	K	°F	MPa $\sqrt{m}$	ksi $\sqrt{in.}$		MPa	ksi
Standard grade							
Air cooled	295	72	71.4	65	LT-CT	876	127
	77	−320	53.8	49	LT-CT	1338	194
	20	−423	51.6	47	LT-B	1482	215
	20	−423	50.5	46	LS-B	...	...
Furnace cooled	295	72	65.9	60	LT-CT	882	128
	77	−320	57.1	52	LT-CT	1379	200
	20	−423	47.2	43	LT-B	1517	220
	20	−423	52.7	48	LS-B	...	...
ELI grade							
Air cooled	295	72	118.7	108(e)	LT-CT	703	102
	77	−320	111.0	101	LT-CT	1179	171
	20	−423	91.2	83	LT-B	1303	189
	20	−423	106.6	97	LS-B	...	...
Furnace cooled	295	72	115.4	105(e)	LT-CT	682	99
	77	−320	82.4	75	LT-CT	1179	171
	20	−423	68.1	62	LT-B	1303	189
	20	−423	80.2	73	LS-B	...	...

(a) Air cooled or furnace cooled from annealing temperature. (b) Orientation notation per ASTM E399-74. (c) CT, compact tension specimen; B, bend specimen. (d) 0.2% offset. (e) Invalid toughness values (not 100% plane-strain conditions). Source: *Metals Handbook*, Vol 3, 9th ed., American Society for Metals, 1980, p 384

Ti-5Al-2.5Sn: Comparison of fracture toughness of two titanium alloys

Alloy and condition(a)	Form	Room-temperature yield strength		Specimen design	Orientation	Fracture toughness, K_{IC}							
						24 °C (75 °F)		−196 °C (−320 °F)		−253 °C (−423 °F)		−269 °C (−452 °F)	
		MPa	ksi			MPa$\sqrt{m}$	ksi$\sqrt{in.}$	MPa$\sqrt{m}$	ksi$\sqrt{in.}$	MPa$\sqrt{m}$	ksi$\sqrt{in.}$	MPa$\sqrt{m}$	ksi$\sqrt{in.}$
Ti-5Al-2.5Sn(NI), annealed	Plate	876	127	CT	L-T	71.8	65.4	53.4	48.6	...	...	...	...
		876	127	Bend	L-T	...	...	...	...	51.4	46.8	...	...
		876	127	Bend	L-S	...	...	...	...	50.2	45.7	...	...
	Bar	871	126	CT	T-S	77.2	70.3	42.1	38.3	...	...	42.0	38.2
Ti-5Al-2.5Sn(ELI) annealed	Plate	703	102	CT	L-T	...	...	111	101	...	...	...	...
		703	102	Bend	L-T	...	...	...	...	89.6	81.5	...	...
Ti-5Al-2.5Sn(ELI) as forged	Forging	760	110	CT	R-L	...	...	...	...	79.4	72.3	...	...
					R-C	...	...	...	...	58.5	53.2	...	...
Ti-5Al-2.5Sn(ELI)	Forging(b)	779	113	CT		...	...	...	...	54.4-75.3	49.5-68.5	...	...
Ti-6Al-4V (NI), annealed	Bar	942	136	CT	T-L	47.4	43.2	38.8	35.3	...	...	38.5	35.1
Ti-6Al-4V (ELI), as forged	Forging	830	120	CT	T-L	...	...	61.0	55.5	...	...	54.1	49.2
Ti-6Al-4V (ELI), RA	Forging	830	120	CT	M-L(c)	...	...	62.8	57.2	...	...	...	...
					M-R(c)	...	...	62.0	56.4	...	...	...	...
Ti-6Al-4V (ELI), RA, electron beam welded, SR	Forging	830	120	CT	M-R(c)	...	...	61.1(d)	55.6(d)	...	...	...	...
	Weldment	...	...	...	M-L(c)	...	...	56.9(d)	51.8(d)	...	...	...	...
					M-R(c)	...	...	57.1(e)	52.0(e)	...	...	...	...
					M-R(c)	...	...	51.0(f)	46.4(f)	...	...	...	...

(a) SR = stress relieved: 540 °C (1000 °F) 50 h, AC. FC = furnace cool. AC = air cool. NI = normal interstitial content. ELI = extra low interstitial content. RA = recrystallization annealed: 930 °C (1700 °F) 4 h, FC to 810 °C (1400 °F) 3 h, cooled to 480 °C (900 °F) in ¾ h, AC. (b) Range for 18 tests. (c) M-L and M-R are specific orientations in a spherical forging. (d) Fusion zone. (e) Heat affected zone. (f) Heat affected zone boundary. Source: Metals Handbook,Properties and Selection: Stainless Steels, Tool Materials, *and Special-Purpose Materials*, Vol 3, 9th ed., American Society for Metals, 1980

Ti-5Al-2.5Sn (ELI): Fracture toughness of 13 mm (0.50 in.) thick plate

Direction	Tensile yield strength MPa	ksi	K_{IC} MPa√m	ksi√in.
At –195 °C (–320 °F)				
LS	1206	175	73.0	67.0
	1206	175	76.0	69.8
	1206	175	67.0	60.8
TS	1199	174	58.0	52.8
	1199	174	51.0	46.4
	1199	174	56.5	51.5
LD	1206	175	60.6	55.2
	1206	175	63.0	58.0
	1206	175	60.0	55.0
TD	1199	174	60.6	55.2
	1199	174	64.0	58.4
	1199	174	59.5	54.2
At –252 °C (–423 °F)				
LS	1413	205	56.4	51.4
	1413	205	65.0	59.6
	1413	205	59.7	54.4
TS	1441	209	53.0	48.8
	1441	209	59.3	54.0
	1441	209	59.5	54.2
LD	1413	205	58.9	53.6
	1413	205	59.9	54.6
	1413	205	51.8	47.2
TD	1441	209	65.0	60.0
	1441	209	69.0	63.6
	1441	209	65.0	59.2

Note: Bend bar specimens 6.4 by 13 mm (¼ by ½ in.). Chemical composition: 5.0 wt% Al, 0.023 wt% C, 0.16 wt% Fe, 0.001 wt% H, 0.006 wt% Mn, 0.010 wt% N, 0.086 wt% O, and 2.6 wt% Sn. Plate was annealed by furnace cooling from 815 °C (1500 °F). The TS specimen orientation had a crack direction parallel to the rolling direction. Source: C. Carman and J. Katlin, "Plane Strain Fracture Toughness and Mechanical Properties of 5Al-2.5Sn ELI and Commercial Titanium Alloys at Room and Cryogenic Temperatures," in *Applications Related Phenomena in Titanium Alloys*, ASTM STP 432, ASTM, 1968, p 124

Ti-5Al-2.5Sn: Fracture toughness of 13 mm (0.5 in) thick plate

Direction	Tensile yield strength MPa	ksi	K_{IC} MPa√m	ksi√in.
At –195 °C (–320 °F)				
LS	1399	203	33.2	30.2
	1399	203	28.5	26.0
	1399	203	24.9	22.7
TS	1406	204	64.1	58.4
	1406	204	50.4	45.9
LD	1399	203	44.9	40.9
	1399	203	43.6	39.7
	1399	203	26.9	24.5
TD	1406	204	27.8	25.3
	1406	204	30.1	27.4
	1406	204	24.6	22.4
At –252 °C (–423 °F)				
LS	1606	233	21.5	19.6
	1606	233	29.4	26.8
	1606	233	30.1	27.4
TS	1634	237	58.3	53.1
	1634	237	30.2	27.5
LD	1606	233	52.7	48.0
	1606	233	34.8	31.7
	1606	233	21.3	19.4
TD	1634	237	24.0	21.9
	1634	237	25.6	23.3
	1634	237	23.3	21.2

Note: Bend bar specimens 6.4 by 13 mm (¼ by ½ in.). Chemical composition: 5.1 wt% Al, 0.023 wt% C, 0.34 wt% Fe, 0.017 wt% H, 0.006 wt% Mn, 0.015 wt% N, and 2.3 wt% Sn. Plate was annealed by furnace cooling from 815 °C (1500 °F). Source: C. Carman and J. Katlin, "Plane Strain Fracture Toughness and Mechanical Properties of 5Al-2.5Sn ELI and Commercial Titanium Alloys at Room and Cryogenic Temperatures," in Applications Related Phenomena in Titanium Alloys, ASTM STP 432, ASTM, 1968, p 124

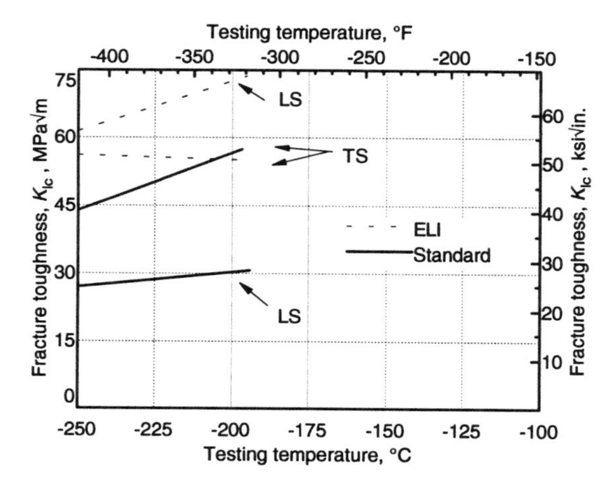

Ti-5Al-2.5Sn: Fracture toughness plate. LS and TS orientations. Plate (13 mm, or 0.5 in.) thick. The TS specimen orientation has crack growth parallel to the rolling direction. All other crack orientations (LS, LD, TD,) are perpendicular to the rolling direction. Source: C. Carman, "Influence of Purity on the Fracture Properties of High-Strength Aluminum, Titanium, and Steel," in *Fracture Toughness of High-Strength Materials: Theory and Practice*, ISI Publication 120, The Iron and Steel Institute, p 116

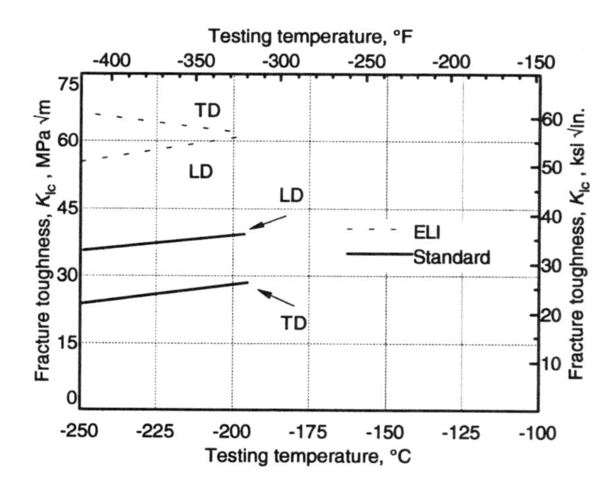

Ti-5Al-2.5Sn: Fracture toughness plate. LD and TD orientations. Plate (13 mm, or 0.5 in.) thick. The TS specimen orientation has crack growth parallel to the rolling direction. All other crack orientations (LS, LD, TD) are perpendicular to the rolling direction. Source: C. Carman, "Influence of Purity on the Fracture Properties of High-Strength Aluminum, Titanium, and Steel," in *Fracture Toughness of High-Strength Materials: Theory and Practice*, ISI Publication 120, The Iron and Steel Institute, p116

ELI Fracture Toughness

Ti-5Al-2.5Sn (ELI): Fracture toughness of 6.4 mm (0.25 in.) thick plate

Direction	Tensile yield strength MPa	Tensile yield strength ksi	K_{IC} MPa√m	K_{IC} ksi√in.
At −195 °C (−320 °F)				
LS	1172	170.5	68.1	62.0
	1172	170.5	67.4	61.4
	1172	170.5	69.4	63.2
TS	1199	174.5	54.1	49.3
	1199	174.5	62.8	57.2
	1199	174.5	63.7	58.0
LS	1172	170.5	67.0	61.0
	1172	170.5	70.9	64.6
	1172	170.5	70.8	64.5
	1172	170.5	74.6	67.9
TS	1199	174.5	66.9	60.9
	1199	174.5	72.5	66.0
	1199	174.5	64.4	58.6
	1199	174.5	70.3	64.0
At −252 °C (−423 °F)				
LS	1344	195	61.9	56.4
	1344	195	59.3	54.0
	1344	195	56.7	51.6
TS	1248	181	53.6	48.8
	1248	181	54.7	49.8
	1248	181	45.9	41.8

Note: Chemical composition: 5.0 wt% Al, 0.023 wt% C, 0.16 wt% Fe, 0.009 wt% H, 0.006 wt% Mn, 0.010 wt% N, 0.080 wt% O, and 2.6 wt% Sn. Plate was annealed by furnace cooling from 815 °C (1500 °F). Source: C. Carman and J. Katlin, "Plane Strain Fracture Toughness and Mechanical Properties of 5Al-2.5Sn ELI and Commercial Titanium Alloys at Room and Cryogenic Temperatures," in *Applications Related Phenomena in Titanium Alloys*, ASTM STP 432, ASTM, 1968, p 124

Ti-5Al-2.5Sn (ELI): Fracture toughness of 25 mm (1 in.) thick plate

Direction	Tensile yield strength MPa	Tensile yield strength ksi	K_{IC} MPa√m	K_{IC} ksi√in.
At −195 °C (−320 °F)				
LS	1213	176	62.5	56.9
	1213	176	57.2	52.1
	1213	176	50.9	46.4
TS	1213	176	50.5	46.0
	1213	176	55.7	50.7
LD	1213	176	65.4	59.5
	1213	176	67.3	61.3
	1213	176	58.2	53.0
TD	1213	176	70.5	64.2
	1213	176	81.9	74.6
	1213	176	66.3	60.4
At −252 °C (−423 °F)				
LS	1399	203	49.4	45.0
	1399	203	69.9	63.7
	1399	203	62.6	57.0
TS	1399	203	51.3	46.7
	1399	203	60.9	55.5
	1399	203	60.6	55.2
	1399	203	60.7	55.3
LD	1399	203	50.5	46.0
	1399	203	61.1	55.6
TD	1399	203	67.0	61.0
	1399	203	83.8	76.3
	1399	203	70.5	64.2

Note: Bend bar specimens. Chemical composition: 5.1 wt% Al, 0.026 wt% C, 0.14 wt% Fe, 0.003 wt% H, 0.004 wt% Mn, 0.101 wt% O, and 2.4 wt% Sn. Plate was annealed by furnace cooling from 815 °C (1500 °F). Source: C. Carman and J. Katlin, "Plane Strain Fracture Toughness and Mechanical Properties of 5Al-2.5Sn ELI and Commercial Titanium Alloys at Room and Cryogenic Temperatures," in *Applications Related Phenomena in Titanium Alloys*, ASTM STP 432, ASTM, 1968, p 124

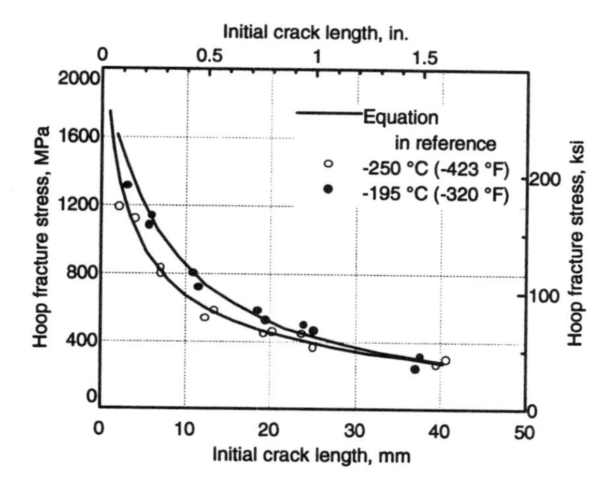

Ti-5Al-2.5Sn ELI: Fracture strength of cracked cylinders. Source: T. Sullivan, "Texture Strengthening and Fracture Toughness of Titanium Alloy Sheet at Room and Cryogenic Temperatures," ASTM STP 432, ASTM, 1968

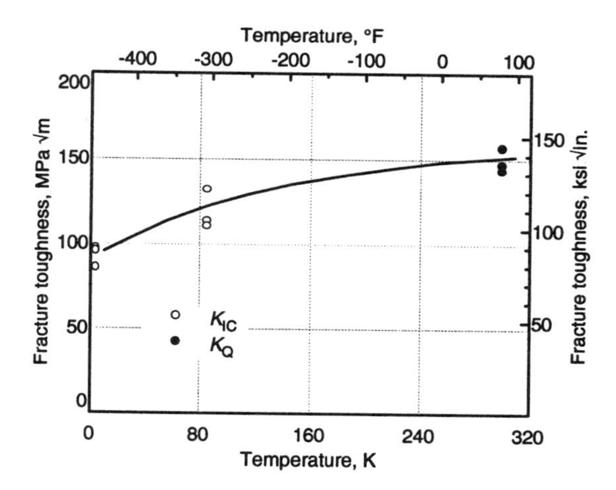

Ti-5Al-2.5Sn ELI: Fracture toughness at several temperatures. Alloy tested was extra-low interstitial grade with typical composition of 4.70 to 5.60 wt% Al, 0.25 wt% Fe max, 0.0125 wt% H max, 0.035 wt% N max, 0.12 wt% O max, and 2.0 to 3.0 wt% Sn. Source: Titanium Product Literature, Kobe Steel

Ti-6Al-2Sn-4Zr-2Mo-0.08Si

Common Name: Ti-6242S Ti-6242Si
UNS Number: R54620

Ti-6Al-2Sn-4Zr-2Mo-0.08Si (Ti-6242S or Ti-6242Si), developed in the late 1960s as an elevated-temperature alloy, has an outstanding combination of tensile strength, creep strength, toughness, and high-temperature stability for long-term applications at temperatures up to 425 °C (800 °F). Ti-6242S is one of the most creep-resistant titanium alloys and is recommended for use up to 565 °C (1050 °F). Proper heat treatment is important in allowing the alloy to develop its maximum creep resistance.

Chemistry and Density

The 6 percent aluminum addition in the Ti-6Al-2Sn-4Zr-2Mo composition is a potent alpha-phase stabilizer, while the 2 percent molybdenum addition represents only a moderate quantity of this potent beta-phase stabilizer. The tin and zirconium additions are solid-solution strengthening elements that are neutral with respect to phase stabilization. The net effect of this combination of alloying elements is the generation of a weakly beta-stabilized, alpha-beta alloy. Since it is weakly beta stabilized, the alloy is also properly described as a near-alpha, alpha-beta alloy.

The original composition of this alloy contained no silicon, but RMI introduced a nominal 0.08% silicon content which allowed the alloy to meet the creep requirements for its intended jet-engine applications. Before any major commercial applications were developed, all producers had added silicon to the original Ti-6242 composition.

Density. 4.54 g/cm^3 (0.164 lb/in.^{3})

Product Forms

Available mill forms include billet, bar, plate, sheet, strip, and extrusions. Cast Ti-6242S products constitute about 7% of cast titanium products. Some forming operations can be carried out at room temperature, and warm forming (425 to 705 °C, or 800 to 1300 °F) is employed when necessary. Ti-6242S has fair weldability. The molten weld metal and adjacent heated zones must be shielded from active gases (nitrogen, oxygen, and hydrogen).

Product Condition/Microstructure

Ti-6242S is sometimes described as a near-alpha or superalpha alloy, but in its normal heat treated condition this alloy has a structure better described as alpha-beta. Proper treatment is needed to develop good creep resistance. Limited hardening of Ti-6242S can be done by solution treating and aging.

Applications

Ti-6242S is used primarily for gas turbine components such as compressor blades, disks, and impellers, and also in sheet-metal form for engine afterburner structures and for various "hot" airframe skin applications, where high strength and toughness, excellent creep resistance, and stress stability at temperatures up to 565 °C (1050 °F) are required.

Ti-6Al-2Sn-4Zr-2Mo-0.08Si: Specifications and compositions

Specification	Designation	Description	Composition, wt%								
			Al	Fe	H	Mo	N	O	Sn	Zr	Other
UNS	R54620		6			2			2	4	bal Ti
UNS	R54621	Weld Fill Met	5.5-6.5	0.05	0.015	1.8-2.2	0.15	0.3	1.8-2.2	3.6-4.4	C 0.04; Cr 0.25; bal Ti
Germany											
	WL 3.7144		5.5-6.5	0.25	0.015	1.8-2.2	0.05	0.15	1.8-2.2	3.6-4.4	C 0.05; bal Ti
Spain											
UNE 38-718	L-7103	Sh Strp Plt Ann	5.5-6.5	0.25	0.015	1.8-2.2	0.05	0.12	1.8-2.2	3.6-4.4	C 0.05; OT 0.4; bal Ti
UNE 38-718	L-7103	Sh Strp Plt HT	5.5-6.5	0.25	0.015	1.8-2.2	0.05	0.12	1.8-2.2	3.6-4.4	C 0.05; OT 0.4; bal Ti
USA											
AMS 4919C		Sh Strp Plt	5.5-6.5	0.25 max	0.015 max	1.8-2.2	0.05 max	0.12 max	1.8-2.2	3.6-4.4	C 0.05 max; Si 0.06-0.1; Y 0.005 max; OE 0.1 max; OT 0.3 max; bal Ti
AMS 4919G		Sh Strp Plt DA	5.5-6.5	0.25	0.015	1.8-2.2	0.05	0.12	1.8-2.2	3.6-4.4	C 0.05; Si 0.1; Y 0.005; OT 0.3; bal Ti
AMS 4975E		Bar Wir Rng Bil STA	5.5-6.5	0.25	0.0125	1.8-2.2	0.05	0.15	1.8-2.2	3.6-4.4	C 0.05; Y 0.005; OT 0.3; Si 0.1; bal Ti
AMS 4975F		Bar Rng HT	5.5-6.5	0.1 max	0.0125 max	1.8-2.2	0.05 max	0.15 max	1.8-2.2	3.6-4.4	C 0.05 max; Si 0.06-0.1; Y 0.005 max; OE 0.1 max; OT 0.3 max; bal Ti
AMS 4976C		Frg STA	5.5-6.5	0.25	0.0125	1.8-2.2	0.05	0.15	1.8-2.2	3.6-4.4	C 0.05; Y 0.005; OT 0.3; Si 0.1; bal Ti
USA (continued)											
AMS 4976D		Frg HT	5.5-6.5	0.1 max	0.0125 max	1.8-2.2	0.05 max	0.15 max	1.8-2.2	3.6-4.4	C 0.05 max; Si 0.06-0.1; Y 0.005 max; OE 0.1 max; OT 0.3 max; bal Ti
MIL T-81556A	Code AB-4	Ext Bar Shp Ann	5.5-6.5	0.25	0.015	1.8-2.2	0.04	0.15	1.8-2.2	3.6-4.4	C 0.05; Y 0.005; Si 0.06-0.1; OT 0.3; bal Ti
MIL T-81556A	Code AB-4	Ext Bar Shp STA	5.5-6.5	0.25	0.015	1.8-2.2	0.04	0.15	1.8-2.2	3.6-4.4	C 0.05; Si 0.06-0.1; Y 0.005; OT 0.3; bal Ti
MIL T-81915	Type III Comp B	Cast Ann	5.5-6.5	0.35	0.015	1.5-2.5	0.05	0.12	1.5-2.5	3.6-4.4	C 0.08; OT 0.4; bal Ti
MIL T-9046J	Code AB-4	Sh Strp Plt DA	5.5-6.5	0.25	0.015	1.8-2.2	0.04	0.15	1.8-2.2	3.6-4.4	C 0.05; OT 0.3; bal Ti
MIL T-9046J	Code AB-4	Sh Strp Plt TA	5.5-6.5	0.25	0.015	1.8-2.2	0.04	0.15	1.8-2.2	3.6-4.4	C 0.05; OT 0.3; bal Ti
MIL T-9047G	Ti-6Al-2Sn-4Zr-2Mo	Bar Bil DA	5.5-6.5	0.25	0.015	1.8-2.2	0.04	0.15	1.8-2.2	3.6-4.4	C 0.05; OT 0.3; Y 0.005; bal Ti
MIL T-9047G	Ti-6Al-2Sn-4Zr-2Mo	Bar Bil STA	5.5-6.5	0.25	0.015	1.8-2.2	0.04	0.15	1.8-2.2	3.6-4.4	C 0.05; Y 0.005; OT 0.3; bal Ti

Ti-6Al-2Sn-4Zr-2Mo-0.08Si : Compositions

Specification	Designation	Description	Al	Fe	H	Mo	N	O	Sn	Zr	Other
			Composition, wt%								
France											
Ugine	UT6242	Bar Frg Ann	5.5-6.5			1.8-2.2			1.8-2.2	3.6-4.4	
Germany											
Deutsche T	Contimet AlSnZrMo 6-2-4-2	Plt Bar Frg Ann	5.5-6.5	0.25	0.015	1.8-2.2	0.05	0.15	1.8-2.2	3.6-4.4	C 0.05; Si 0.06-0.12; bal Ti
Deutsche T	Contimet AlSnZrMo 6-2-4-2	Plt Bar Frg STA	5.5-6.5	0.25	0.015	1.8-2.2	0.05	0.15	1.8-2.2	3.6-4.4	C 0.05; Si 0.06-0.12; bal Ti
Deutsche T	LT 24	Aged	5.5-6.5	0.25	0.015	1.8-2.2	0.05	0.12	1.8-2.2	3.6-4.4	C 0.05; bal Ti
Fuchs	TL62	Frg	6			2			2	4	bal Ti
Japan											
Kobe	KS6-2-4-2	Bar Frg STA	5.5-6.5	0.25	0.015	1.8-2.2	0.05	0.15	1.8-2.2	3.6-4.4	bal Ti
USA											
OREMET	Ti-6242										
RMI	RMI 6Al-2Sn-4Zr-2Mo-0.10Si	Bar Bil Plt Sh STA	5.5-6.5	0.25	0.01-0.0125	1.75-2.25	0.05	0.12	1.75-2.25	3.5-4.5	C 0.08; Si 0.1; bal Ti
TIMET	TIMETAL 6-2-4-2	Bar Bil Plt Sh STA	5.5-6.5	0.25	0.01-0.0125	1.75-2.25	0.05	0.12	1.75-2.25	3.5-4.5	C 0.08; Si 0.1; bal Ti

Phases and Structures

The structures of Ti-6Al-2Sn-4Zr-2Mo alloy are typically equiaxed α in a transformed β matrix, or a fully transformed structure that maximizes creep resistance. The equiaxed α grains found in sheet products tend to be smaller than those found in forgings, as with other alloys, and are present in greater proportion than in forgings. Primary α is typically about 80 to 90% of the structure in sheet products and can be somewhat lower than this in forged products, because the final forging temperature is normally higher than the final rolling temperature used for sheet. As in other near-α alloys, small amounts of residual β phase can be observed metallographically within the transformed β portion of the structure, typically between the acicular α grains of the transformed phase. Breakup of lamellar α into equiaxed α occurs during working (see figure).

Beta Transus. 995 ± 15 °C (1825 ± 25 °F)

Fatigue Properties

Duplex Annealed Sheet

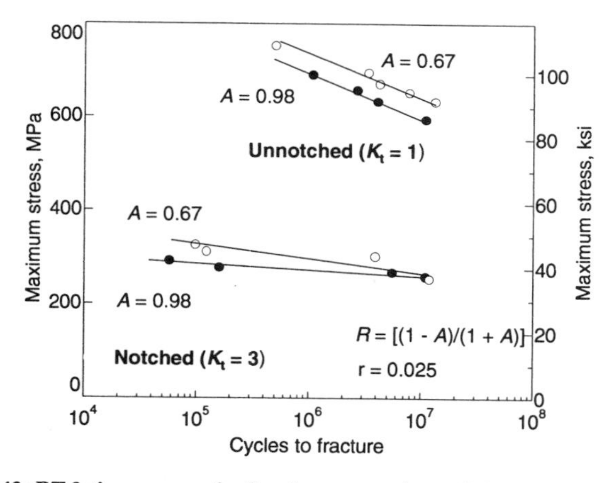

Ti-6242: RT fatigue properties Specimens were 1 mm (0.040 in.) sheet duplex annealed at 900 °C (1650 °F), 30 min, AC + 785 °C (1450 °F), 15 min, AC. Axial fatigue, tension (sinusoidal). Surface: mill finish. Frequency: 2500 cycles/min. Test temperature, 21 °C (70 °F). Source: AFML-TR-67-41; Apr 1967; reported in *Aerospace Structural Metals Handbook*, Vol 4, Code 3718, Battelle Columbus Laboratories, 1978, p 82

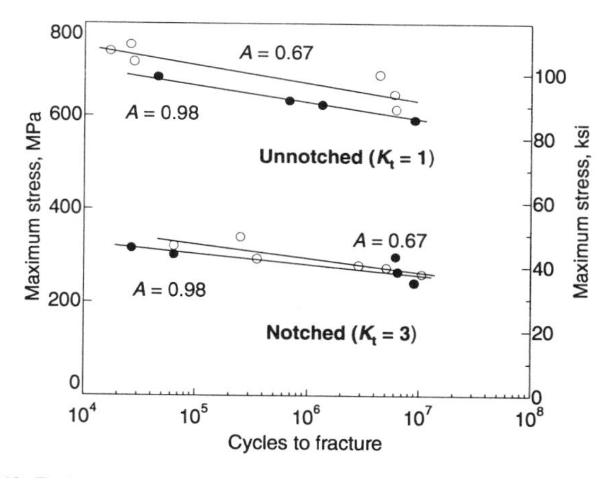

Ti-6242: Fatigue properties at 205 °C Specimens were 1 mm (0.040 in.) sheet duplex annealed at 900 °C (1650 °F), 30 min, AC + 785 °C (1450 °F), 15 min, AC. Axial fatigue, tension (sinusoidal). Surface: mill finish. Frequency: 2500 cycles/min. Test temperature, 205 °C (400 °F). Source: AFML-TR-67-41, Apr 1967, reported in *Aerospace Structural Metals Handbook*, Vol 4, Code 3718, Battelle Columbus Laboratories. 1978, p 82

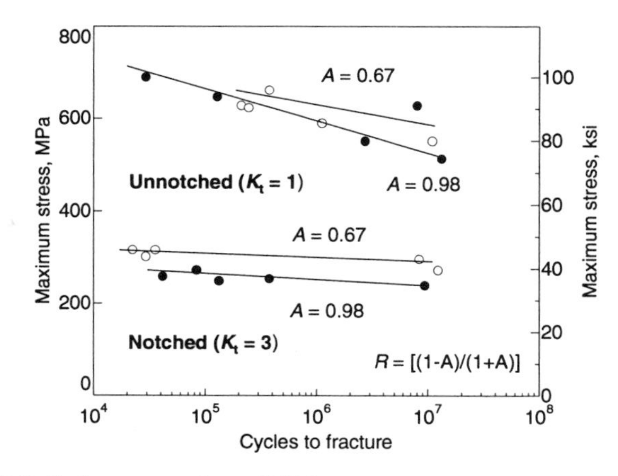

Ti-6242: Fatigue properties at 425 °C Specimens were 1 mm (0.040 in.) sheet duplex annealed at 900 °C (1650 °F), 30 min, AC + 785 °C (1450 °F), 15 min, AC. Axial fatigue: tension (sinusoidal). Surface: mill finish. Frequency: 2500 cycles/min. Test temperature, 425 °C (800 °F). Source: AFML-TR-67-41; Apr 1967; reported in *Aerospace Structural Metals Handbook*, Vol 4, Code 3718, Battelle Columbus Laboratories, 1978, p 83

Duplex Annealed Bar

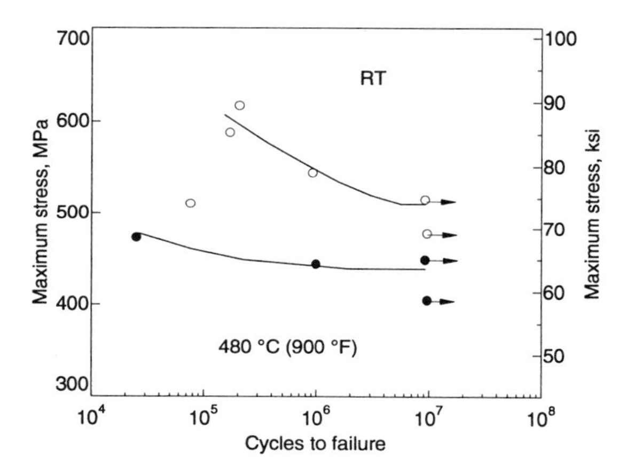

Ti-6242: RT and 480 °C fatigue properties Specimens were 28.5 mm (1.125 in.) diameter bar duplex annealed at 970 °C (1775 °F), 1 h, AC + 595 °C (1100 °F), 8 h, AC. Ultimate tensile strength, 1006 MPa (146 ksi); tensile yield strength, 958 MPa (139 ksi); smooth, rotating beam tests. Source: DMIC Data Sheet; Jan 1967, reported in *Aerospace Structural Metals Handbook*, Vol 4, Code 3718, Battelle Columbus Laboratories, 1978, p 83

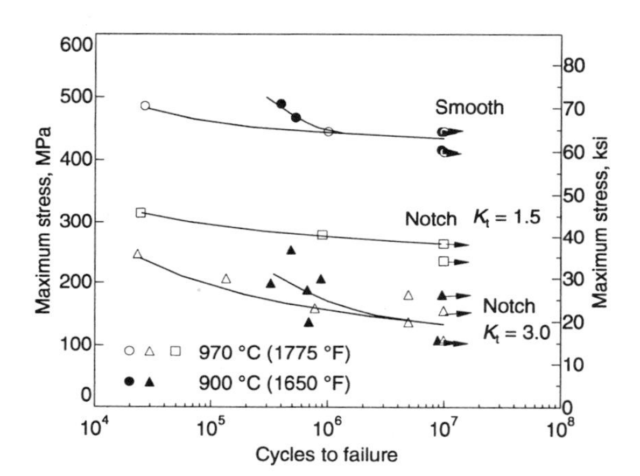

Ti-6242: Fatigue properties at 480 °C Specimens were 28.5 mm (1.125 in.) bar, duplex annealed as indicated. Rotating beam test results reported at 480 °C (900 °F). Source: DMIC Data Sheet; Jan 1967; reported in *Aerospace Structural Metals Handbook*, Vol 4, Code 3718, Battelle Columbus Laboratories, 1978, p 84

Duplex Annealed Forgings

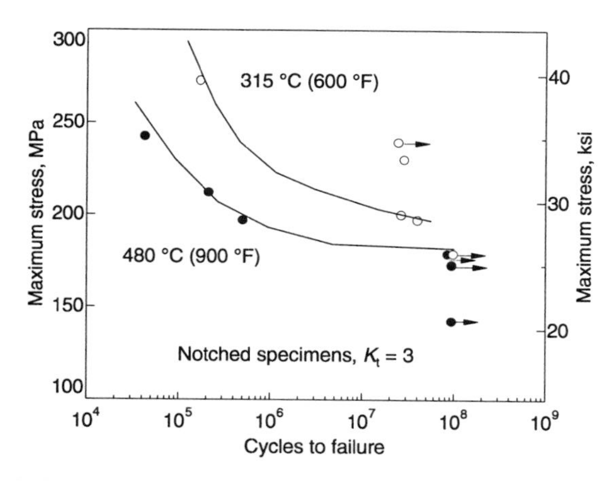

Ti-6242: Fatigue strength at 315 and 480 °C Specimens were compressor disk forgings duplex annealed at 955 °C (1750 °F), 1 h, AC + 595 °C (1100 °F), 8 h, AC. Specimens were stress relieved at 540 °C (1000 °F), 2 h in vacuum before testing. Reverse bending (cantilever); $R = -1$; frequency, 120 cycles/s. Source: Pratt & Whitney Data; June 1972; reported in *Aerospace Structural Metals Handbook*, Vol 4, Code 3718, Battelle Columbus Laboratories, 1978, p 84

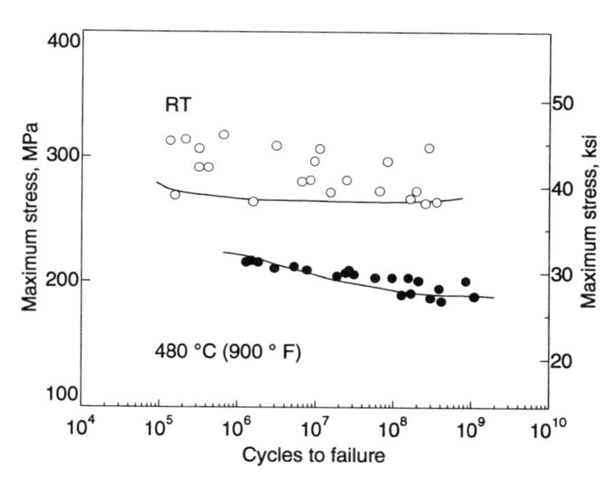

Ti-6242: High-frequency fatigue properties Specimens were 19 mm (0.75 in.) bar duplex annealed at 955 °C (1750 °F), 1 h, AC + 595 (1100 °F), 8 h, AC. Axial fatigue, $R = -1$. Frequency: 13.0 kHz at RT, 13.4 kHz at 480 °C (900 °F). Failure criterion: Crack grown to nearly half of specimen cross section. Source: NASA TR-72618; July 1969; reported in *Aerospace Structural Metals Handbook*, Vol 4, Code 3718, Battelle Columbus Laboratories, 1978, p 84

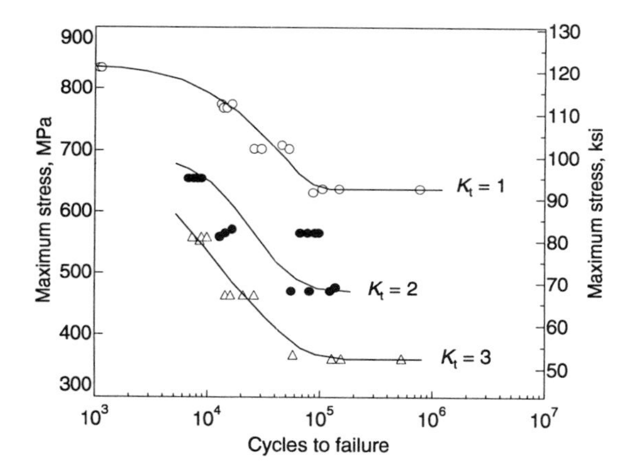

Ti-6242: Fatigue strength at 200 °C Specimens were compressor disk forgings (three forgings from three different heats) duplex annealed at 955 °C (1750 °F), 1 h, AC + 595 °C (1100 °F), 8 h, AC. Axial tension, $R = 0$; frequency, 1800 cycles/min. Source: Pratt & Whitney Data; June 1972; reported in *Aerospace Structural Metals Handbook*, Vol 4, Code 3718, Battelle Columbus Laboratories, 1978, p 84

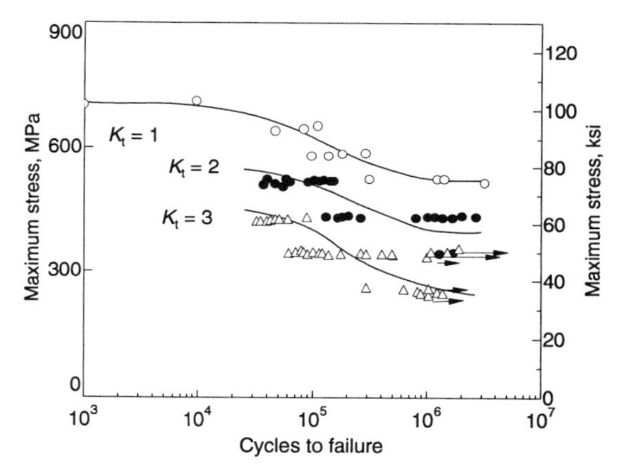

Ti-6242: Fatigue strength at 455 °C Compressor disk forgings from three different heats, treated at 955 °C (1750 °F) for 1 h, AC, 595 °C (1100 °F) for 8 h, AC. Axial tension, $R = 0$; frequency, 1800 cycles/min. Source: Pratt & Whitney Data; June 1972; reported in *Aerospace Structural Metals Handbook*, Vol 4, Code 3718, Battelle Columbus Laboratories, 1978, p 84

Fracture Properties

Impact Toughness

Ti-6242: Effect of heat treatment on RT impact toughness of cast specimens

Heat treatment	Specimen type(a)	Charpy V-notch impact toughness J	ft · lbf
595 °C (1100 °F), 8 h, AC	(b)	30	22
		28	21
		28	21
955 °C (1750 °F), 1 h, AC + 595 °C (1100 °F), 8 h, AC	(b)	38	28
		34	25
		35	26
1035 °C (1900 °F), 1 h, WQ + 595 °C (1100 °F), 8 h, AC	(b)	8	6
		8	6
		8	6
1035 °C (1900 °F), 1 h, HeC + 595 °C (1100 °F), 4 h, HeC	(c) (d)	28	21
	(c) (e)	23-25	17-26

(a) Consumable electrode skull melted in water-cooled copper crucibles and cast into tungsten-lined ceramic molds unless otherwise indicated. (b) Standard Charpy-V bars cast 0.25 to 0.38 mm (0.010-0.015 in.) oversize and machined to final dimensions following heat treatment. (c) Standard Charpy-V bars machined from three thickest sections of heat-treated, variable thickness step castings. (d) Average of nine values from three castings. (e) Range of nine values from three castings. Source: *Aerospace Structural Metals Handbook*, Vol 4, Code 3718, Battelle Columbus Laboratories, June 1978

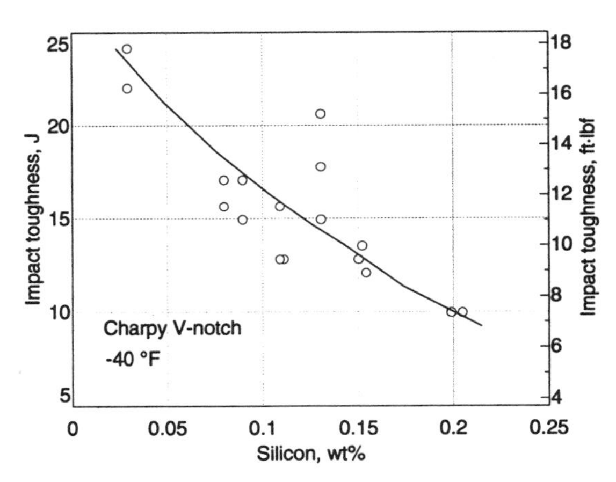

Ti-6242: Impact toughness vs silicon content at –40 °C Effect of silicon content on –40 °C (–40 °F) Charpy-V-notch impact energy of duplex annealed 15.8 mm (5/8 in.) bar treated 15 °C (25 °F) below transus for 1 h, AC, 595 °C (1100 °F), for 8 h, AC. Source: *Aerospace Structural Metals Handbook*, Vol 4, Code 3718, Battelle Columbus Laboratories, June 1978

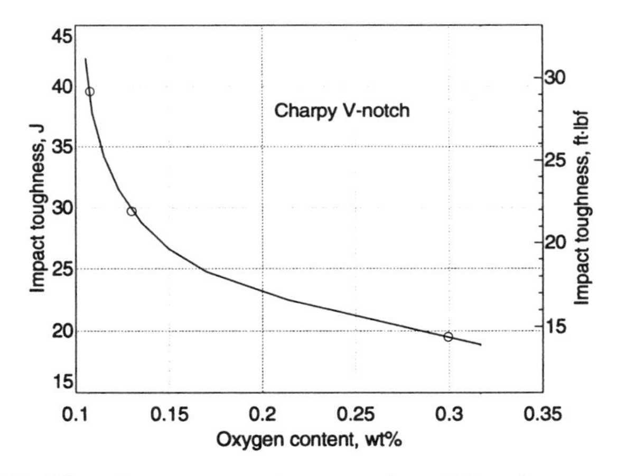

Ti-6242: Effect of oxygen on cast impact toughness Effect of oxygen content on room-temperature Charpy V-notch impact energy for consumable electrode melted castings. Specimens were consumable electrode skulls melted in water-cooled copper crucibles and cast into tungsten-lined ceramic molds. Standard Charpy V-notch bars cast 0.25 to 0.38 mm (0.010 to 0.015 in.) oversize and machined to final dimensions following heat treatment of 595 °C (1100 °F), 8 h, AC or HeC. Source: *Aerospace Structural Metals Handbook*, Vol 4, Code 3718, Battelle Columbus Laboratories, June 1978

Fracture Toughness

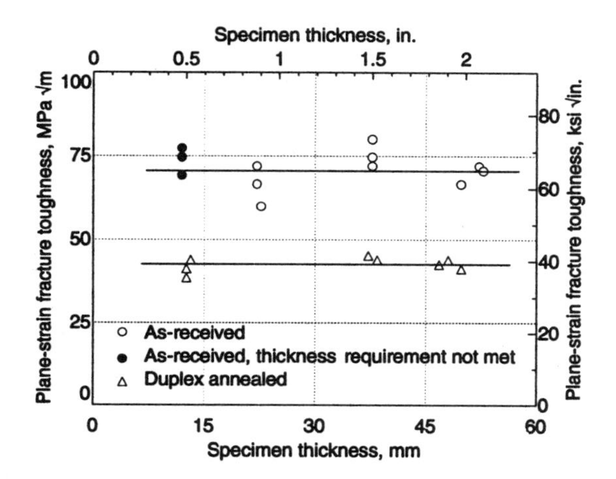

Ti-6242: Fracture toughness of plate As-received 50 mm (2 in.) plate had a tensile yield strength of 855 MPa (124 ksi). Specimens were duplex annealed at 955 °C (1750 °F) for 30 min, in vacuum, water quenched, then at 455 °C (850 °F) for 17 h, and air cooled. Heat treated specimens had a tensile yield strength of 1041 MPa (151 ksi). Fracture toughness was determined by the three-point bend method, according to ASTM E399-74, except for 13 mm (0.5 in) thick specimens where the thickness requirement was not met. Specimen crack orientation was LT; specimens were symmetrical about the central plane of the plate. Source: C. Carman, Frankford Arsenal data, 1970; reported in *Aerospace Structural Metals Handbook*, Vol 4, Code 3718, Battelle Columbus Laboratories, 1978, p 67

Ti-6242: Fracture toughness of forgings

Condition	Tensile yield strength		Elongation,	K_{Ic}	
	MPa	ksi	%	MPa√m	ksi√in.
α + β forged, βST + aged	903	131	12.5	81	73
β forged, α + βST + aged	896	130	11.0	84	76

Source: J.C. Williams and E.A. Starke, Jr., The Role of Thermomechanical Processing in Tailoring the Properties of Aluminum and Titanium Alloys, Deformation, Processing, and Structure, American Society for Metals, 1984

Ti-6242S: Fracture toughness of forgings from various thermomechanical processing (TMP) routes

Property	TMP route 1(a)	TMP route 2(b)	TMP route 3(c)
Tensile yield strength, MPa (ksi)	937 (136)	903 (131)	917 (133)
Ultimate tensile strength, MPa (ksi)	979 (142)	986 (143)	1006 (146)
Elongation (in 4D), %	16	12	12
Reduction of area, %	34	24	26
Fracture toughness, MPa√m (ksi√in.)	56 (51)	78 (71)	76 (69)
Creep at 510 °C, 241 MPa (950 °F, 35 ksi), h to 0.1%	117	447	430
Fatigue crack growth rate ($da/dn \times 10^{-6}$ in./cycle at ΔK 10 ksi√in.)	1.2	0.7	0.8

(a) TMP route 1 (baseline): α-β hot die forged plus sub-β transus duplex heat treatment. (b) TMP route 2: β hot die forged plus duplex heat treatment. (c) TMP route 3: β hot die forged plus direct age. Source: G. Kuhlman, T. Yu, A. Chakrabarti, and R. Pishko, "Mechanical Property Tailoring Titanium Alloys for Jet Engine Applications," in *Titanium 1986: Products and Applications*, Titanium Development Association, 1987, p 122

Ti-6242: Fracture toughness of duplex annealed bar

Alloy Condition	Tensile yield strength		Crack plane orientation	Three-point bend specimen: Thickness		Width		Crack length		K_Q	
	MPa	ksi		mm	in.	mm	in.	mm	in.	MPa√m	ksi√in.
900 °C (1650 °F), 1 h, AC + 595 °C (1100 °F), 8 h, AC	855	124	LT	12.77	0.503	26.47	1.003	13.31	0.524	68(a,b)	62(a,b)
				12.80	0.504	25.42	1.001	12.54	0.494	69(a,b)	63(a,b)
				12.85	0.506	25.50	1.004	13.38	0.527	70(b)	64(b)
	848	123	TL	12.80	0.504	25.50	1.004	12.87	0.507	56(a)	51(a)
				12.75	0.502	25.42	1.001	13.28	0.523	74(b)	68(b)
				12.82	0.505	25.42	1.001	12.98	0.511	61(b)	56(b)
	848	123	TS	12.72	0.501	24.66	0.971	12.42	0.489	60(a)	55(a)
				12.72	0.501	25.65	1.010	12.98	0.511	65(a,b)	59(a,b)
				12.75	0.502	25.65	1.010	13.23	0.521	57(c)	52(c)
975 °C (1790 °F), 1 h, AC + 595 °C (1100 °F), 8 h, AC	820	119	LT	12.80	0.504	25.42	1.001	12.65	0.498	78(a,b)	71(a,b)
				12.80	0.504	25.42	1.001	14.45	0.569	89(b)	81(b)
				12.80	0.504	25.42	1.001	14.91	0.587	70(b)	64(b)
	848	123	TL	12.85	0.506	25.45	1.002	13.53	0.533	82(a,b)	75(a,b)
				12.85	0.506	25.42	1.001	14.09	0.555	77(b)	70(b)
				12.85	0.505	25.42	1.001	13.28	0.523	73(a,b)	67(a,b)
	848	123	TS	12.75	0.502	24.94	0.982	12.72	0.501	67(a,b)	61(a,b)
				12.75	0.502	24.66	0.971	13.66	0.538	71(a,b)	65(a,b)

Note: Specimens were duplex annealed bar 50 by 114 mm (2 by 4.5 in.). Fracture toughness was determined by the three-point bend method. (a) Excessive crack front curvature according to ASTM E-399-74. (b) Insufficient specimen thickness according to ASTM E-399-74. (c) Valid K_{Ic}. Source: C. Hickey, Jr., and T. DeSisto, "Mechanical Properties and Fracture Toughness of Ti-6Al-2Sn-4Zr-2Mo," reported in *Aerospace Structural Metals Handbook*, Vol 4, Code 3718, Battelle Columbus Laboratories, June 1987, p 67

Ti-8Al-1Mo-1V

Common Name: Ti-811
UNS Number: R54810

Ti-8Al-1Mo-1V (Ti-811) was developed around 1954 for high-temperature gas turbine engine applications—specifically, compressor blades and wheels. It is now available from most titanium alloy producers. Ti-811 has the highest tensile modulus of all the commercial titanium alloys and exhibits good creep resistance at temperatures up to 455 °C (850 °F). Ti-811 has a room-temperature tensile strength similar to that of Ti-6Al-4V, but its elevated-temperature tensile strength and creep resistance are superior to those of other commonly available alpha and alpha+beta titanium alloys.

Chemistry and Density

The Ti-8Al-1Mo-1V alloy contains a relatively large amount of the alpha stabilizer, aluminum, and fairly small amounts of the beta stabilizers, molybdenum and vanadium (plus iron as an impurity). Although this alloy is metallurgically an alpha-beta alloy, the small amount of beta stabilizer in this grade (1Mo + 1V) permits only small amounts of the beta phase to become stabilized.

Density. 4.37 g/cm^3 (0.158 lb/in.^{3})

Product Forms

Ti-811 was developed for engine use, principally as forgings. Available forms include billet, bar, plate, sheet, and extrusions. Forming of sheet at room temperature is more difficult than for Ti-6Al-4V, and severe forming operations must be done hot. Ti-811 has good weldability like other alpha or near-alpha alloys. Weldments have similar strength but lower ductility in comparison with the base metal.

Product Condition/Microstructure

Ti-811 is characterized as a near-alpha alloy with several alpha-alloy characteristics such as good creep strength and weldability. However, the alloy does have alpha-beta characteristics such as a mild degree of hardenability. Ti-811 is generally used in the annealed condition, where lamellar alpha morphology from transformed beta is produced by duplex and triplex annealing for enhanced creep resistance.

Applications

Ti-811 is used for airframe and turbine engine applications demanding short-term strength, long-term creep resistance, thermal stability, and stiffness. Ti-811 is predominantly an engine alloy and is available in three grades, including a "premium grade" (triple melted) and a "rotating grade," for use in rotating engine components.

Use Limitations. Like the alpha-beta alloys, Ti-811 is susceptible to hydrogen embrittlement in hydrogenating solutions at room temperature, in air or reducing atmospheres at elevated temperatures, and even in pressurized hydrogen at cryogenic temperatures. Oxygen and nitrogen contamination can occur in air at elevated temperatures and such contamination becomes more severe as exposure time and temperature increase. Ti-811 is susceptible to stress-corrosion cracking in hot salts (especially chlorides) and to accelerated crack propagation in aqueous solutions at ambient temperatures. The environment in which this alloy is to be used should be selected carefully to prevent material degradation.

Ti-8Al-1Mo-1V: Specifications and compositions

Specification	Designation	Description	Composition, wt%								
			Al	C	Fe	H	Mo	N	O	V	Other
UNS	R54810			8			1			1	bal Ti
China											
	Ti-8Al-1Mo-1V		7.5-8.5	0.1 max	0.3 max	0.015 max	0.75-1.25	0.04 max	0.15 max	0.75-1.25	Si 0.15 max; bal Ti
Spain											
UNE 38-717	L-7102	Sh Strp Plt Bar Ext Ann	7.35-8.35	0.08	0.3	0.015	0.75-1.25	0.05	0.12	0.75-1.25	OT 0.4; bal Ti
USA											
AMS 4915C		Sh Strp Plt Ann	7.35-8.35	0.08 max	0.3 max	0.015 max	0.75-1.25	0.05 max	0.12 max	0.75-1.25	OT 0.4 max; Y 0.005 max; OE 0.1 max; bal Ti
AMS 4915F		Sh Strp Plt Ann	7.35-8.35	0.08	0.3	0.015	0.75-1.25	0.05	0.12	0.75-1.25	OT 0.4; Y 0.005; bal Ti
AMS 4916E		Sh Strp Plt Dup Ann	7.35-8.35	0.08	0.3	0.015	0.75-1.25	0.05	0.12	0.75-1.25	OT 0.4; Y 0.005; bal Ti
AMS 4933A		Ext Rng SHT/Stab	7.35-8.35	0.08	0.3	0.015	0.75-1.25	0.05	0.12	0.75-1.25	OT 0.4; Y 0.005; bal Ti
AMS 4955B		Weld Fill Wir	7.35-8.35	0.08	0.3	0.01	0.75-1.25	0.05	0.12	0.75-1.25	OT 0.4; Y 0.005; bal Ti
AMS 4972C		Bar Wir Rng Bil SHT/Stab	7.35-8.35	0.08	0.3	0.015	0.75-1.25	0.05	0.12	0.75-1.25	OT 0.4; Y 0.005; bal Ti
USA (continued)											
AMS 4973C		Frg Bil SHT/Stab	7.35-8.35	0.08	0.3	0.015	0.75-1.25	0.05	0.12	0.75-1.25	OT 0.4; Y 0.005; bal Ti
AWS A5.16-70	ERTi-8Al-1Mo-1V	Weld Fill Met	7.35-8.35	0.05	0.25	0.008	0.75-1.25	0.03	0.12	0.75-1.25	bal Ti
MIL F-83142A	Comp 5	Frg Ann	7.35-8.35	0.08	0.3	0.015	0.75-1.25	0.05	0.15	0.75-1.25	OT 0.4; bal Ti
MIL T-81556A	Code A-4		7.35-8.35	0.08	0.3	0.015	0.75-1.25	0.05	0.15	0.75-1.25	OT 0.4; bal Ti
MIL T-9046J	Code A-4	Sh Strp Plt Ann	7.35-8.35	0.08	0.3	0.015	0.75-1.25	0.05	0.15	0.75-1.25	OT 0.4; bal Ti
MIL T-9047G	Ti-8Al-1Mo-1V	Bar Bil Dup Ann	7.35-8.35	0.08	0.3	0.015	0.75-1.25	0.05	0.15	0.75-1.25	OT 0.4; Y 0.005; bal Ti
SAE J467	Ti-8-1-1		8	0.04 max	0.15 max		1	0.02 max		1	Si 0.07 max; Ni 0.008 max; bal Ti

OT, others total

Ti-8Al-1Mo-1V: Commercial compositions

Specification	Designation	Description	Al	C	Fe	H	Mo	N	O	V	Other
France											
Ugine	UTA8DV	Bar Frg DA	7.3-8.5	0.08	0.3	0.006-0.015	0.75-1.25	0.05	0.12	0.75-1.25	bal Ti
Germany											
Deutsche T	Contimet AlMoV 8-1-1	Plt Bar Frg Ann	7.5-8.5	0.08	0.3	0.015	0.75-1.25	0.05	0.12	0.75-1.25	bal Ti
Japan											
Kobe	KS8-1-1	Bar Frg STA	7.35-8.35		0.3	0.015	0.75-1.25	0.05	0.12	0.75-1.25	bal Ti
USA											
Chase Ext.	8Al-1Mo-1V										
OREMET	Ti-8-1-1										
RMI	RMI 8Al-1Mo-1V	Mult Forms DA	7.5-8.5	0.08		0.015	0.75-1.25	0.05	0.12	0.75-1.25	bal Ti
Timet	TIMETAL 8-1-1	Ann	7.35-8.35	0.08 max	0.3 max	0.015 max	0.75-1.25	0.05 max	0.12 max	0.75-1.25	bal Ti

Fatigue Properties

Ti-8Al-1Mo-1V: Typical rotating beam fatigue of rolled barstock

Condition	Stress MPa	Stress ksi	Cycles to failure
Simplex anneal(a) with a	724	105	45,000 and 55,000
937 MPa (136 ksi) UTS	689	100	50,000
and 9903 MPa (131 ksi)	655	95	200,000
TYS	620	90	140,000
Duplex anneal(b) with a	724	105	85,000
1013 MPa (147 ksi) UTS	689	100	140,000
and 951 MPa (138 ksi)	655	95	200,000 and 300,000
TYS	620	90	1,100,000 and 3,000,000

(a) 760 °C (1400 °F), 24 h, AC. (b) 980 °C (1800 °F), 4 h, AC + 540 °C (1000 °F), 24 h, AC. Source: *Alloy Digest*, Jan 1962

Unnotched Fatigue Life

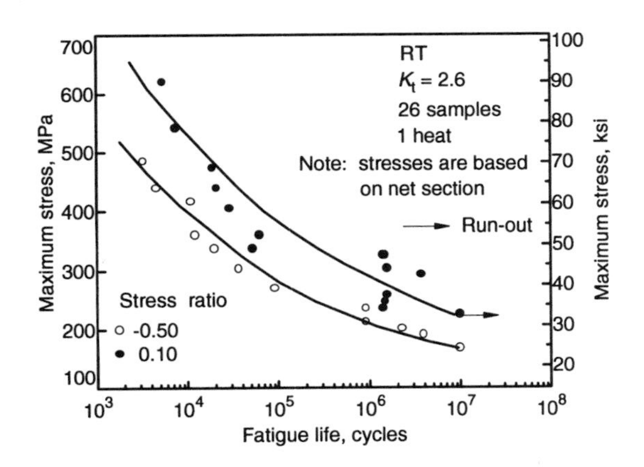

Ti-8Al-1Mo-1V: Best-fit S/N curves for unnotched sheet at RT See table. *Caution*: The equivalent stress model may provide unrealistic life predictions for stress ratios beyond those represented.

Test conditions for best-fit S/N curves

Test Material Common Name:	Ti-8Al-1Mo-1V
Specification Designation:	N/A
Composition:	N/A
Product form:	1.3 mm (0.050 in.) sheet
Heat treatment:	Duplex annealed
Condition/microstructure:	N/A
Hardness:	N/A
Modulus of elasticity (avg at RT):	N/A
RT tensile strength/elongation:	1014.9 MPa (147.2 ksi)
RT yield strength:	934.9 MPa (135.6 ksi)
Test Parameters	
Test temperature:	RT, 200 °C (400 °F) and 345 °C (675 °F)
Test environment:	Air
Failure criterion:	Fracture
Loading condition:	Axial (see figures for *R* ratio)
Specimen orientation:	Long-transverse direction
Specimen geometry:	Unnotched, 19 mm (0.75 in.) net width
Surface:	HNO_3/HF pickled
Gauge length:	N/A
Frequency:	1800 cycles/min
Strain rate:	N/A
Waveform:	N/A
Test specifications:	N/A
Remarks/Reference:	MIL-HDBK-5, Dec 1991

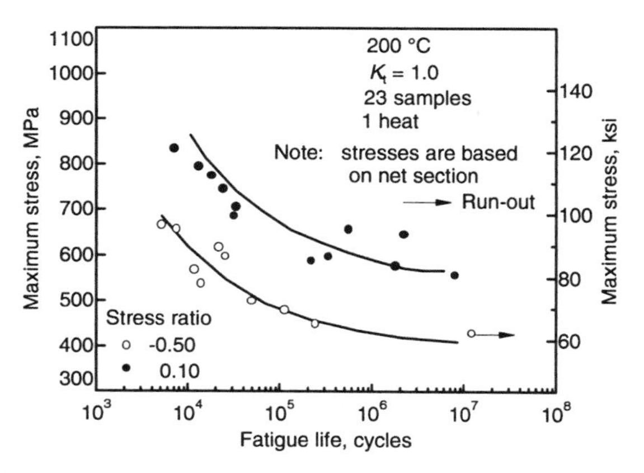

Ti-8Al-1Mo-1V: Best-fit S/N curves at 200 °C See table on previous page for test conditions. UTS at 200 °C (400 °F) was 825 MPa (119.5 ksi). *Caution*: The equivalent stress model may provide unrealistic life predictions for stress ratios beyond those represented.

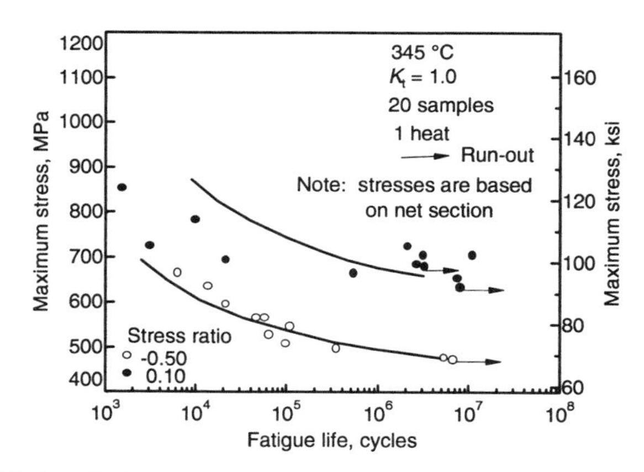

Ti-8Al-1Mo-1V: Best-fit S/N curves at 345 °C See table on previous page for test conditions. UTS at 345 °C (650 °F) was 760 MPa (110.2 ksi). *Caution*: The equivalent stress model may provide unrealistic life predictions for stress ratios beyond those represented.

Notched Fatigue Life

Ti-8Al-1Mo-1V: Fatigue crack growth data

Duplex annealed 1.27 mm (0.05 in.) specimens

Design mean stress			Crack initiation(a)		Crack growth(b)	
MPa	ksi	Fatigue life, flights	Period, flights	Percentage of total life, flights	Period, flights	Percentage of total life, flights
Accelerated tests at room temperature(c)						
172	25	137 158	127 000	93	10 160	7
207	30	18 540	13 300	72	5 240	28
241	35	7 472	5 300	71	2 170	29
172	25	105 988	78 000	74	27 990	26
207	30	57 290	40 300	70	16 990	30
241	35	22 290	12 000	54	10 290	46
Accelerated tests at 560 K(c)						
172	25	36 243	25 500	70	10 740	30
207	30	12 498				
241	35	4 580	2 850	62	1 730	38
Real-time tests(d)						
172	25	19 014	12 000	63	7 010	37
207	30	10 420	8 100	78	2 320	22
241	35	5 093				

(a) For cracks extending 1 mm (0.04 in.) from the notch. (b) For cracks from 1 mm (0.04 in.) long until failure. (c) One test at each design stress. (d) Median value from test. Source: L.A. Imig, "Crack Growth in Ti-8Al-1Mo-1V with Real-Time and Accelerated Flight-by-Flight Loading," *Fatigue Crack Growth under Spectrum Loads*, ASTM STP 595, ASTM, 1976, p 251-264

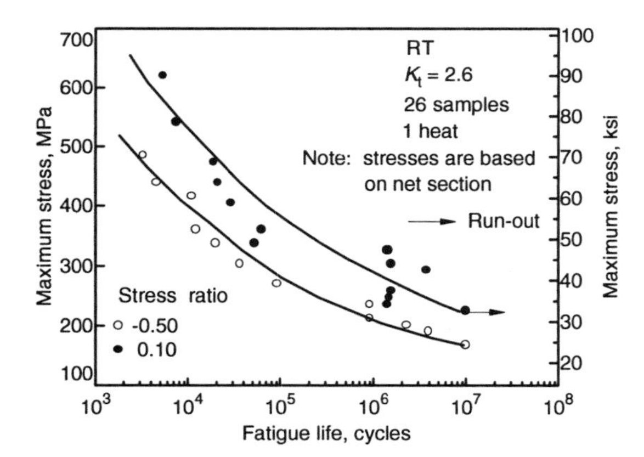

Ti-8Al-1Mo-1V: Best-fit S/N curves for notched sheet at RT See table for test conditions. *Caution*: The equivalent stress model may provide unrealistic life predictions for stress ratios beyond those represented. Source: MIL-HDBK-5, Dec 1991

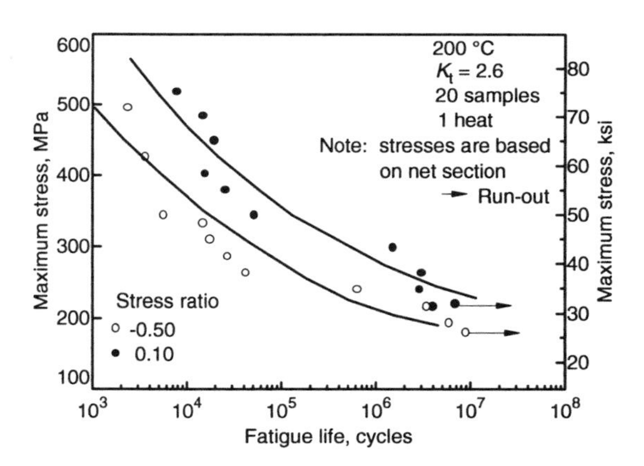

Ti-8Al-1Mo-1V: Best-fit S/N curves at 200 °C See table for test conditions. *Caution*: The equivalent stress model may provide unrealistic life predictions for stress ratios beyond those represented. Source: MIL-HDBK-5, Dec 1991

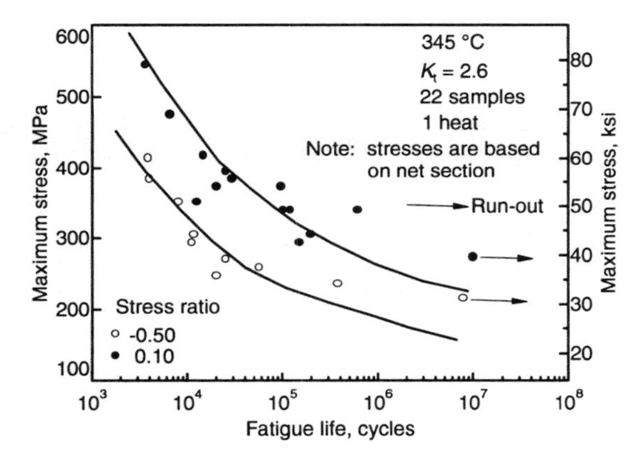

Ti-8Al-1Mo-1V: Best-fit S/N curves at 345 °C See table. *Caution*: The equivalent stress model may provide unrealistic life predictions for stress ratios beyond those represented. Source: MIL-HDBK-5, Dec 1991

Test conditions for best-fit SN curves

Test Material Common Name:	Ti-8Al-1Mo-1V
Specification Designation:	N/A
Composition:	N/A
Product form:	1.33 mm (0.050 in.) sheet
Heat treatment:	Duplex annealed
RT tensile strength/elongation:	1014.9 MPa (147.2 ksi)
RT yield strength:	934.9 MPa (135.6 ksi)
Test Parameters	
Test temperature:	RT, 200 °C (400 °F), and 345 °C (675 °F)
Test environment:	Air
Failure criterion:	Fracture
Loading condition:	Axial (see figures for *R* ratios)
Specimen orientation:	Long transverse grain direction
Specimen geometry:	Notched, hole type, K_t = 2.6; 38 mm (1.500 in.) gross width; 31.7 mm (1.250 in.) net width; 6.4 mm (0.250 in.) diameter hole
Surface:	HNO_3/HF pickled
Gauge length:	N/A
Frequency:	1800 cycles/min
Remarks/Reference:	MIL-HDBK-5, Dec 1991

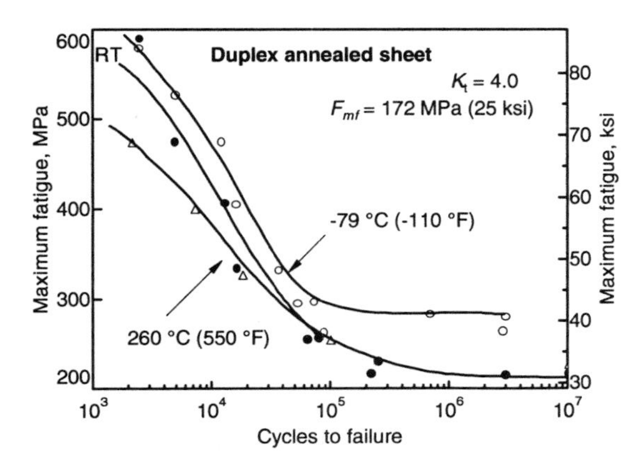

Ti-8Al-1Mo-1V: Notched fatigue at low temperatures 1 mm (0.04 in.) sheet duplex annealed 1010 °C (1850 °F), 5 min, AC + 745 °C (1375 °F), 15 min, AC. Source: *Aerospace Structural Metals Handbook*, Vol 4, Code 3709, Battelle Columbus Laboratories, 1966

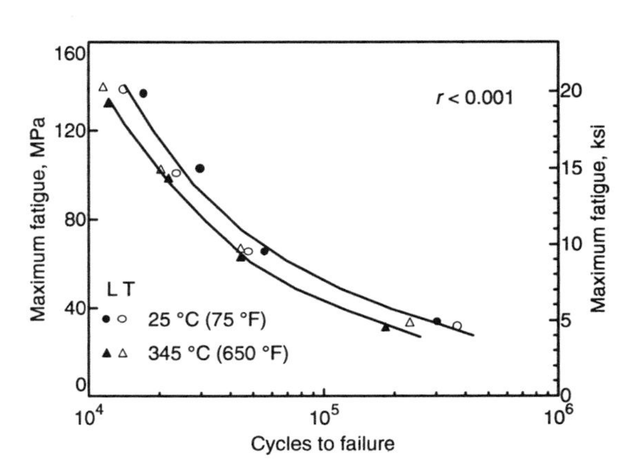

Ti-8Al-1Mo-1V: Axial load sharp notch fatigue Duplex annealed 0.635 mm (0.025 in.) sheet, 1010 °C (1850 °F), 5 min, AC + 745 °C (1375 °F), 15 min, AC. Notched on both sides after annealing with 60° notch. Source: *Aerospace Structural Metals Handbook*, Vol 4, Code 3709, Battelle Columbus Laboratories, 1966

Implant Material Fatigue

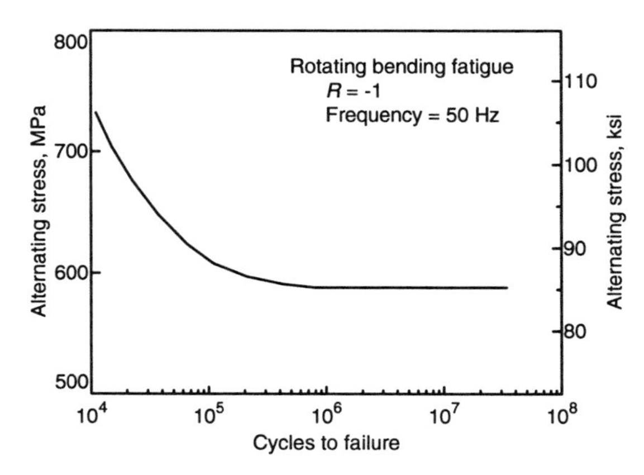

Ti-8Al-1Mo-1V:Fatigue in salt solution Source: M. Levy, *et al.*, The Corrosion Behavior of Titanium Alloys in Chloride Solutions: Materials for Surgical Implants, in *Titanium, Science and Technology*, R.I. Jaffee and H.M. Burte, Ed., 1973, p 2459-2474

Fatigue-Crack Growth

Forged Fan Blades

Fatigue-crack growth rate of Ti-811 at ambient temperatures in both room air and 3.5% NaCl solution for compact-tension specimens made from forged first-stage-turbine fan blades are shown in the accompanying figures. Environment/lower frequency had little effect at low and high ΔK levels, but caused a significant increase in crack-growth rate at intermediate ΔK levels. Comparison of the data with results for conventionally rolled Ti-811 material revealed that the fatigue-crack growth rates for forged specimens were nearly the same as those for the conventionally rolled specimens.

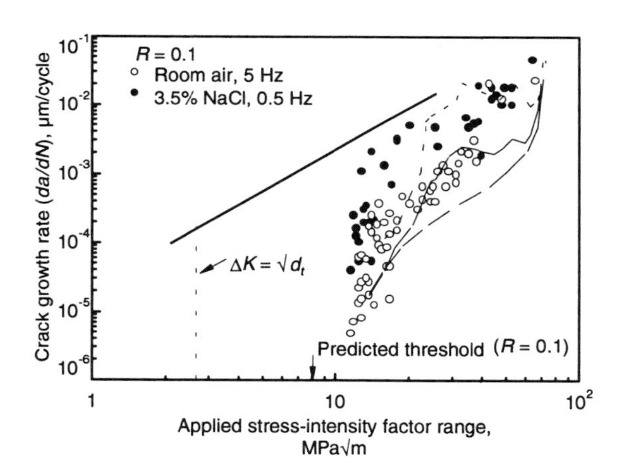

Ti-8Al-1Mo-1V: Crack growth in fan blade specimens Specimens were cut from blades to simulate suspected in-service fracture path. Source: Reported in MCIC-81-42 from W.H. Cullen and F.R. Stonesifer, "Fatigue-Crack-Growth Analysis of Titanium Gas-Turbine Fan Blades," NAVAIR, NRL-MR-3378, Oct 1976, ADA031836

Environmental Effects

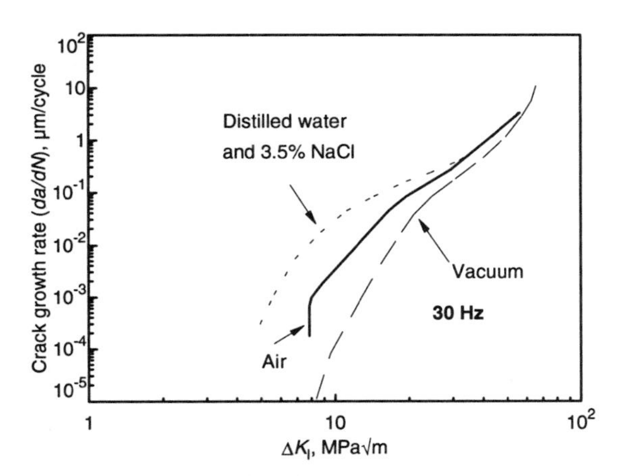

Ti-8Al-1Mo-1V: Crack growth vs environment Source: H. Doker and D. Munz, Influence of Environment on the Fatigue Crack Propagation of Two Titanium Alloys, in *The Influence of Environment on Fatigue*, Institution of Mechanical Engineers, London, 1977, p 123-130

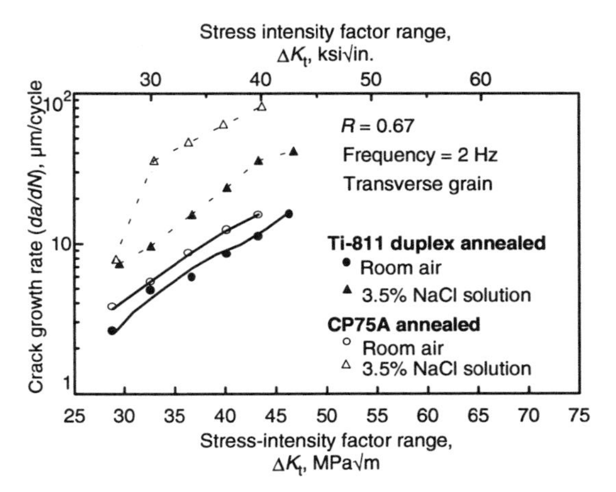

Ti-8Al-1Mo-1V: Crack growth vs CP titanium Source: M.J. Blackburn *et al.*, Boeing Report D1-82-1054, June 1970

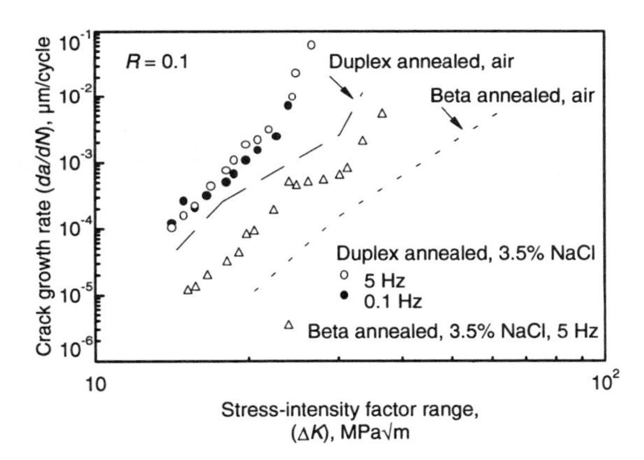

Ti-8Al-1Mo-1V: Microstructure and corrosion fatigue Corrosion-fatigue crack growth rates in 3.5% NaCl solution for duplex annealed and beta annealed microstructures show the expected improvement from transformed structures of beta treatment. Source: G.R. Yoder, L.A. Cooley, and T.W. Crooker, "Improvement of Environmental Crack Propagation Resistance in Ti-8Al-1Mo-1V Through Microstructural Modification," Final Report No. NRL-MR-3955, Mar 1979, ADA069084

Effect of Frequency

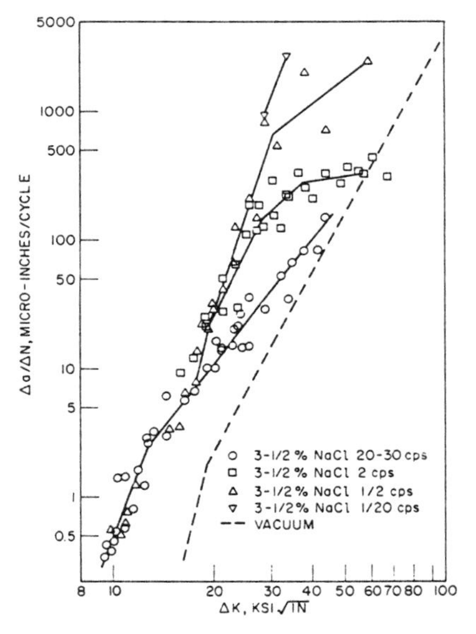

Ti-8Al-1Mo-1V: Effects of frequency in a stress-corrosion-inducing environment Source: D.A. Meyn, *Metall. Trans.*, Vol 2, Mar 1971, p 853-865

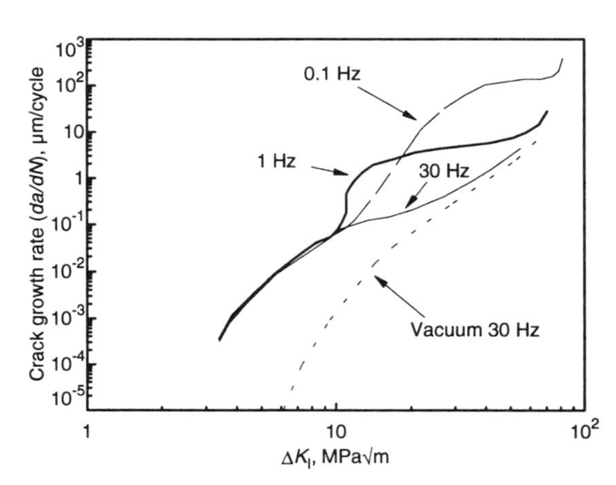

Ti-8Al-1Mo-1V: Effect of frequency on *da/dN* vs ΔK Effect of frequency on *da/dN* versus ΔK curves in 3.5% sodium chloride solution. All fatigue tests were run at $R = 0.1$ with sinusoidal waveform. Source: H. Doker and D. Munz, Influence of Environment on the Fatigue Crack Propagation of Two Titanium Alloys, in *The Influence of Environment on Fatigue*, Institution of Mechanical Engineers, London, 1977, p 123-130

Fracture Properties

Impact Toughness

Ti-8Al-1Mo-1V: Effects of rolling temperature on Charpy impact toughness

Hot working temperature		Test temperature		Impact toughness	
°C	°F	°C	°F	J	ft · lbf
1010	1850		−40	11.5	8.5
		RT		12.9	9.5
		93	200	16.9	12.5
		205	400	25.1	18.5
		315	600	39.3	29.0
1035	1950		−40	16.9	12.5
		RT		23.0	17.0
		93	200	37.3	27.5
		205	400	46.7	34.5
		315	600	59.6	44.0

Source: *Alloy Digest*, Jan 1962

Fracture Properties

An acicular α from β fabrication or heat treatment improves fracture toughness in air but has less of an effect on fracture toughness in salt water. Toughness in salt water is directionally sensitive (see figure).

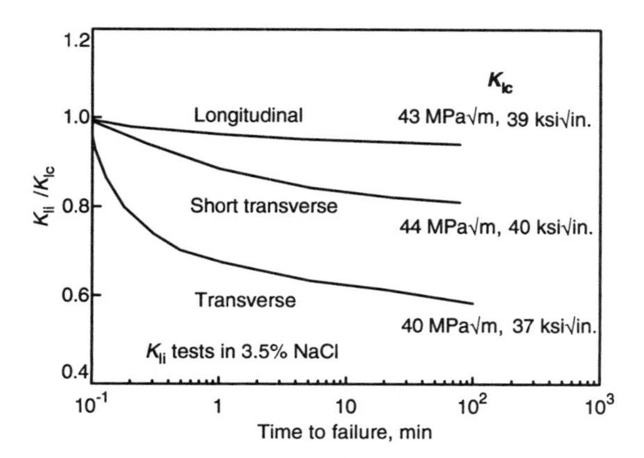

Ti-8Al-1Mo-1V: Stress-corrosion susceptibility Stress-corrosion susceptibility as a function of specimen orientation in 13 mm (0.5 in.) annealed plate. Three-point-loaded notched bend specimens. Source: R. Wood and R. Favor, *Titanium Alloys Handbook*, MCIC-HB-02, Battelle Columbus Laboratories, 1972

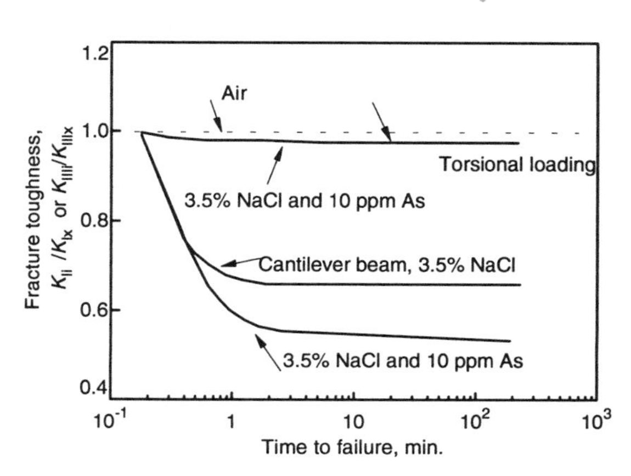

Ti-8Al-1Mo-1V: SCC resistance for bending or torsion Arsenic additions were used to facilitate hydrogen entry and were carried out at –500 mV (SCE). The lower resistance in Mode I to stress-corrosion cracking was taken as an indication that hydrogen played an important role in the cracking process because in Mode I, a state of hydrostatic tensile stress exists ahead of the crack tip, which was thought to promote the concentration of hydrogen. Source: M. Fontana and R. Staehle, Ed., *Advances in Corrosion Science and Technology*, Vol 7, Plenum Press, 1980

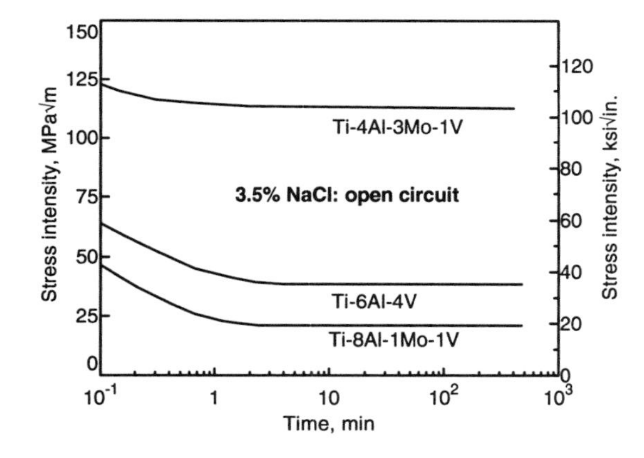

Ti-8Al-1Mo-1V: Time to failure compared to Ti-6Al-4V at 24 °C Mill annealed specimens. Source: T.L. MacKay, C.B. Gilpin, and N.A. Tiner, "Stress-Corrosion Cracking of Titanium Alloys at Ambient Temperatures in Aqueous Solutions," Contract NAS 7-488, Report SM-49105-F1, McDonnell Douglas Corp., July 1967; P. Finden, "Comparative Data—Titanium Alloy Screening Tests," D6-24541-TN, Boeing Company

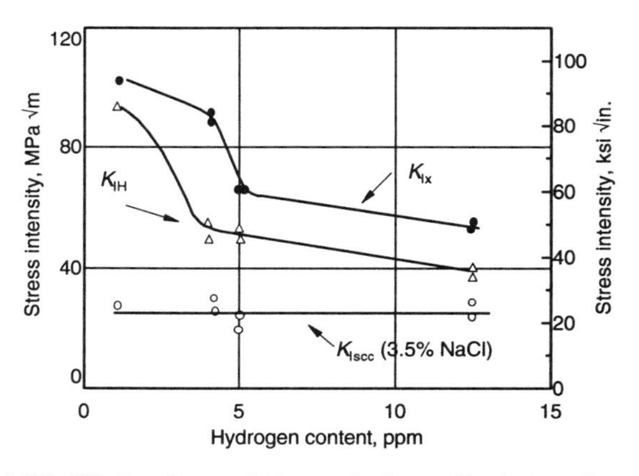

Ti-8Al-1Mo-1V: K_{IX}, K_{IH}, and K_{Iscc} vs hydrogen Notched cantilever beam specimens at room temperature. Source: M.J. Blackburn *et al.*, Boeing Report D1-82-1054, June 1970

Ti-8Al-1Mo-1V: Typical toughness at room temperature

Thickness		Heat	Yield strength		K_{Ic} or K_c		K_{Iscc} or K_{scc}	
mm	in.	treatment	MPa	ksi	MPa√m	ksi√in.	MPa√m	ksi√in.
1.3	0.05	Mill annealed	999	145	82	75	33	30
		Duplex annealed	930	135	176	160	55	50
13	0.50	Mill annealed	999	145	52	48	22	20
		Duplex annealed	930	135	110	100	35	32
		Mill annealed, WQ	841	122	>110	>100	46	42
		βST, WQ	868	126	>110	>100	>110	>100

Source: R.W. Schutz, Stress Corrosion of Titanium Alloys, in *Stress Corrosion*, ASM International, 1992

Ti-8Al-1Mo-1V: Plane-stress toughness-(K_c)

Condition	Ultimate tensile strength(a)		Tensile yield strength(a)		Elongation(a),	Fracture toughness, (K_c)	
	MPa	ksi	MPa	ksi	%	MPa$\sqrt{m}$	ksi$\sqrt{in.}$
Mill annealed	999	145	930	135	8-10	151	138
Duplex annealed	930	135	862	125	8-10	274	250

(a) Guaranteed minimums. Source: R. Wood and R. Favor, *Titanium Alloys Handbook*, MCIC-HB-02, Battelle Columbus Laboratories, 1972

Ti-8Al-1Mo-1V: Effect of heat treatment on impact toughness

Material/condition	Tensile yield strength		Impact toughness			
			Air		Saltwater	
	MPa	ksi	MPa$\sqrt{m}$	ksi$\sqrt{in.}$	MPa $\sqrt{m}$	ksi$\sqrt{in.}$
Annealed	937	136	70	64	26	24
β fabricated	958	139	107	98	25	23
β heat treated	924	134	104	95	32	29
Annealed plus exposed 48 h at 550-600 °C (1020-1110 °F)	965	140	36	33	16	15
Mill annealed	972	141	50	46	19	18

Source: R. Wood and R. Favor, *Titanium Alloys Handbook*, MCIC-HB-02, Battelle Columbus Laboratories, 1972

TIMETAL® 1100

Ti-6Al-2.75Sn-4Zr-0.4Mo-0.45Si Ti-1100
UNS No.: Unassigned

Tom O'Connell, TIMET

Ti-1100 is a near-alpha alloy developed for elevated-temperature use up to 600 °C (1100 °F). It was developed to be used primarily in the beta-processed (beta-worked or beta-annealed) condition. Ti-1100 offers the highest combination of strength, creep resistance, fracture toughness, and stability of any commercially available titanium alloy. It is also recommended for castings.

Effects of Alloying and Impurities. The effects of tin, iron, oxygen, silicon, zirconium, molybdenum, and aluminum on creep, strength, and stability of Ti-1100 have been determined. The alloy development program began with the screening of over 250 compositions of button (250-g) heats. These studies identified compositions that were scaled to 45-kg (100-lb) heats to provide forged product for evaluation. The most promising of these alloys were then scaled to several 815-kg (1800-lb) heats for melting and conversion studies as well as thermomechanical processing (TMP) studies. This successful progression culminated with the production and evaluation of a production-sized 3630-kg (8000-lb) ingot. The outcome of this alloy development study was a composition consisting of Ti-6Al-2.75Sn-4Zr-0.4Mo-0.45Si-0.07O_2-0.02Fe(max).

This alloy is clearly a modification of the Ti-6242-Si alloy that is so widely used today. Although the chemistry differences would appear to be subtle, they are quite dramatic in their effect on creep response, as indicated below:

- **Silicon:** Creep resistance is significantly enhanced up to 0.5% silicon, but beyond that point post-creep ductility (stability) is compromised with no further creep enhancement.
- **Tin:** A similar relationship exists for tin, with stability sacrificed above the 3% level.
- **Iron:** Iron demonstrates a strong effect on time to 0.2% creep strain at the 510 °C (950 °F), 410 MPa (59 ksi) test condition, necessitating iron levels well below those typically encountered in the Ti-6242-Si alloy.
- **Aluminum:** The aluminum level in the new alloy was kept at 6% due to stability problems at higher levels and strength problems at lower levels.
- **Zirconium:** Zirconium was kept high to promote a uniform distribution of silicides in light of the high silicon level. Thus, the chemistry of this alloy was optimized not only for creep strength, but also for stability, strength, and uniformity.

Sponge and Melting Practice. Ti-1100, due to its extremely low iron limit, requires a select grade of titanium sponge. However, sponge containing roughly 100 ppm iron (0.01%) has been produced on a commercial basis, and no problems exist concerning raw material supply. In terms of melting, the high silicon content of this alloy calls for special controls during vacuum arc remelting, especially on the third and final melt. However, Ti-550 (Ti-4Al-2Sn-4Mo-0.5Si) has a comparable silicon content, and this alloy has been successfully melted for several years.

Product Forms. Ti-1100 has been processed successfully to billet, bar, sheet, and weld wire. Forgings have been produced using isothermal and warm die methods, and foil has been produced for use in metal matrix composites.

Investment castings have been produced. The lack of a quench requirement from the solution treatment temperature may enhance the producibility of castings. No data are available on P/M products.

Product Condition. The two standard conditions recommended for the alloy are (1) beta processed ($T >$ 1065 °C, 1950 °F) and annealed (T = 595 °C, 1100 °F) and (2) alpha-beta processed; beta annealed ($T >$ 1065 °C, 1950 °F) plus anneal (T = 595 °C, 1100 °F).

The alloy has only a slight response to cooling rate or section size from the solution treatment (or processing) temperature. Very rapid quenches increase strength and decrease elevated-temperature creep resistance. Ti-1100 generally is used in the beta-processed or beta heat-treated condition, but it is provided in an equiaxed alpha-beta condition for the product forms to enhance processibility.

Applications. Ti-1100 is designed for applications requiring excellent creep strength or fracture properties at temperatures up to approximately 600 °C (1100 °F). High-pressure compressor disks, low-pressure turbine blades, and automotive valves are typical examples.

Ti-1100: Typical composition range

	Composition, wt%								
	Al	Sn	Zr	Fe	Mo	Si	O_2	N_2	C
Minimum	5.7	2.4	3.5	...	0.35	0.35	...	...	...
Maximum	6.3	3.0	4.5	0.02	0.50	0.50	0.09	0.03	0.04
Nominal	6.0	2.7	4.0	...	0.40	0.45	0.07	...	...

Physical Properties

Phases and Structures. Typical microstructures for Ti-1100 include equiaxed α-β for billet and sheet stock. It also transforms to a Widmanstätten or colony α + β structure depending on cooling rate. The effects of cooling rate on the transformed β structure are as follows: alpha-beta processing with a normal cooling rate results in equiaxed primary α plus transformed β with a colony structure plus silicides; beta processing with rapid cooling results in a Widmanstätten structure, whereas slower cooling after β processing results in a colony structure.

In addition to α and β phases, various silicides exist for both α + β or β processed conditions. The silicide solvus has been measured at between 1030 and 1065 °C (1885 and 1950 °F). The α-2 solvus is approximately 740 °C (1365 °F). The β transus is nominally 1015 °C (1860 °F).

Ti-1100: Summary of typical physical properties

Ti_3Al (α_2) solvus	~740 °C (1365 °F)
Beta transus (nominal)	1015 °C (1860 °F)
Silicide solvus	1030 to 1065 °C (1885 to 1950 °F)
Calculated liquidus point	1637 °C (2978 °F)
Density(a)	4.5 g/cm^3 (0.163 lb/in.^{3})
Modulus of elasticity(a)	107 to 117 GPa (15.5 to 17 × 10^6 psi)
Electrical resistivity(a)	1.8 μΩ · m
Magnetic permeability	Nonmagnetic
Specific heat capacity(a)	545 J/kg · K (0.13 Btu/lb · °F)
Thermal conductivity(a)	7 W/m · K (4 Btu/ft · h · °F)
Coefficient of linear expansion	8.5 × 10^{-6}/ °C (4.7 × 10^{-6}/ °F)
Calculated solidus	1615 °C (2939 °F)

(a) Typical values at room temperature of about 20 to 25 °C (68 to 78 °F)

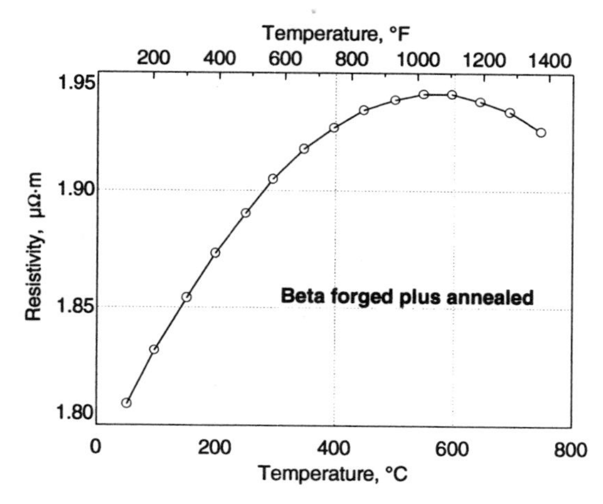

Ti-1100: Electrical resistivity vs temperature Resistivity (R) for beta forged plus annealed material between 25 and 750 °C (77 and 1380 °F) has been determined to fit the expression: $R(10^{-8}\ \Omega \cdot m) = 178 + 0.057\,T + 5 \times 10^{-5}\,T^2$

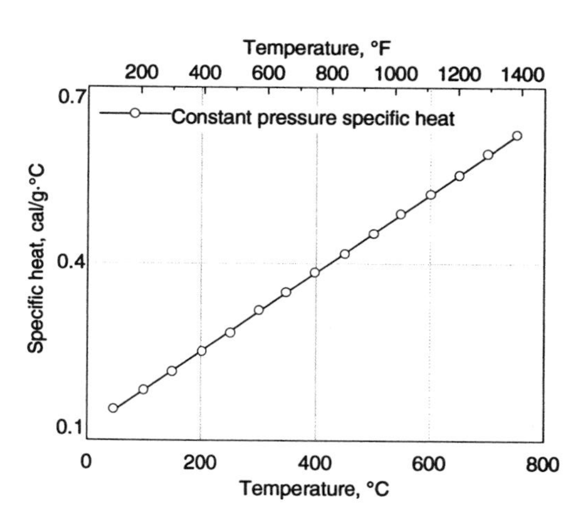

Ti-1100: Specific heat vs temperature The specific heat (C_p) for beta-processed material between 25 and 750 °C (77 and 1380 °F) can be expressed as: C_p (cal/g · °C) = 0.117 + 6.7 × 10^{-4} (T)

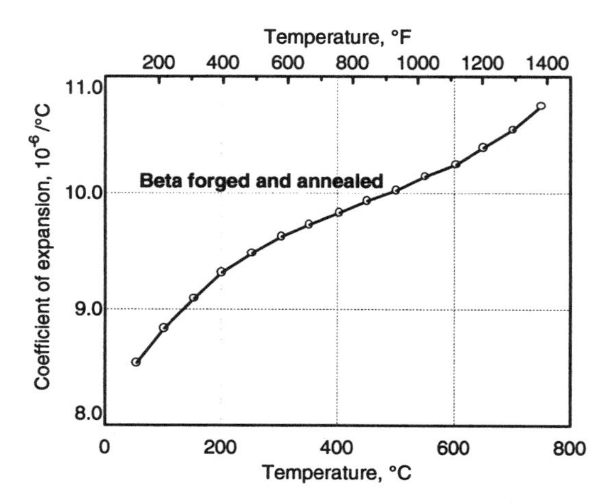

Ti-1100: Thermal coefficient of linear expansion The thermal coefficient of linear expansion (α) between 25 and 750 °C (77 and 1380 °F) for beta-processed material is given by: $\alpha\ (\text{ppm/°C}) = 8.12 + 8.17 \times 10^{-3} T - 1.37 \times 10^{-5} T^2 + 10^{-8} T^3$

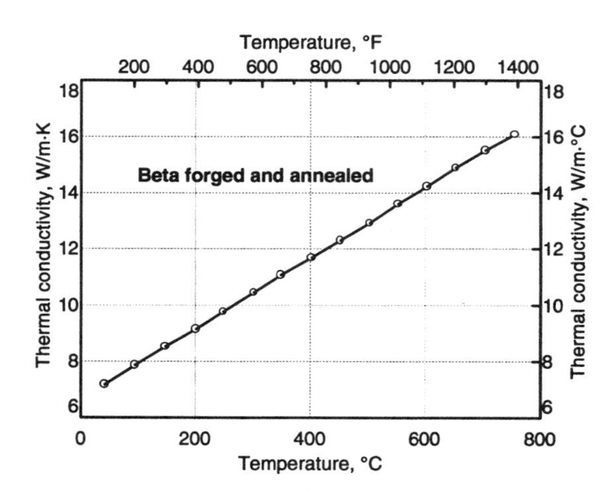

Ti-1100: Thermal conductivity Thermal conductivity (Q) between 25 and 750 °C (77 and 1380 °F) for beta-processed material follows the equation: $Q\ (\text{W/m} \cdot \text{°C}) = 6.62 + 1.27 \times 10^{-2}\ T$

Mechanical Properties

Creep Properties. It has been determined that beta processing greatly improves the creep resistance of Ti-1100 and that the quench rate from the beta forging or annealing temperature will subtly affect the creep resistance. Faster cooling (i.e., oil quench versus air cool) will improve the high-stress, low-temperature portion of the Larson-Miller plot at the expense of the low-stress, high-temperature portion of the plot. Alpha-beta processed material has higher strength and ductility at low temperatures, but has decreased strength at high temperatures (600 °C, or 1110 °F).

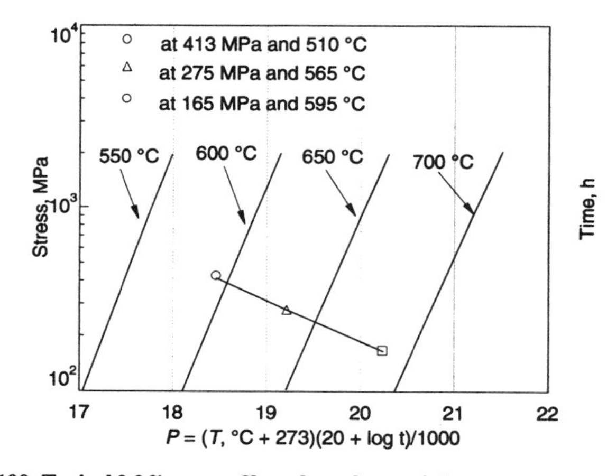

Ti-1100: Typical 0.2 % creep of beta forged material

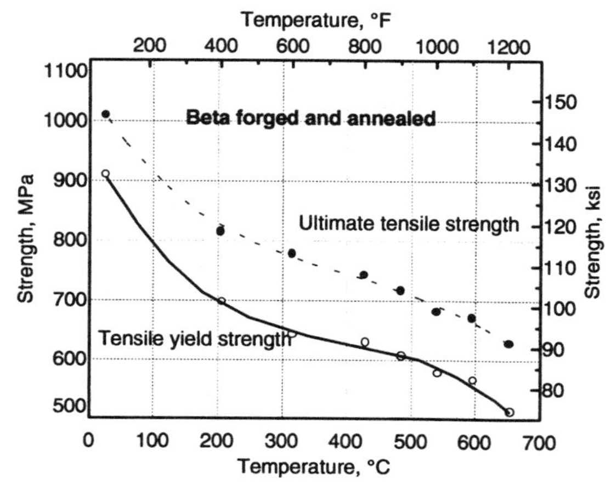

Ti-1100: Yield and tensile strength vs temperature

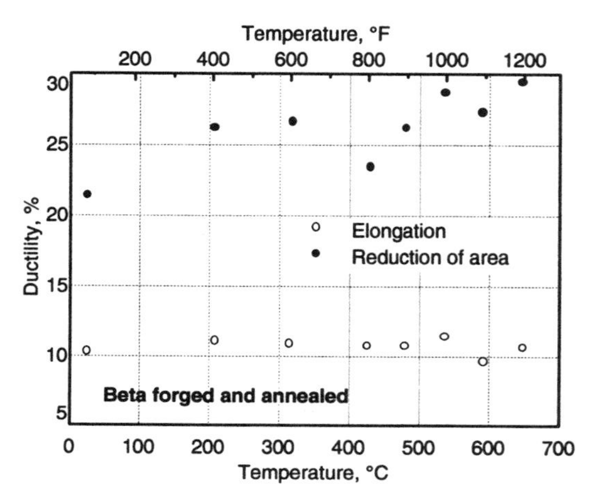

Ti-1100: Tensile ductility vs temperature

Fatigue Properties

Fatigue Crack Growth. Room temperature data are shown. High-temperature crack growth is reported in *Met. Trans.*, Vol 24 A, p 1321.

Ti-1100: Fatigue strength at 10^7 cycles

Temperature			Fatigue strength(a)	
°C	°F	K_t	MPa	ksi
22	71	1.0	655	95
		3.0	250	36
480	895	1.0	500	72
		3.0	235	34

(a) Beta forged and annealed; tested at 30 Hz; $R = 0.1$

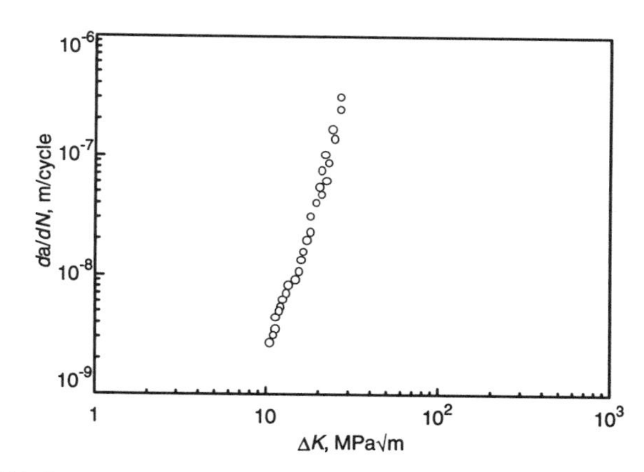

Ti-1100: Room-temperature fatigue crack growth Ti-1100 demonstrates excellent crack growth resistance compared to other conventional alloys from RT to 600 °C (1110 °F). Isothermally forged plus annealed. Tested at 23 °C (75 °F); 20 Hz; $R = 0.1$

Fracture Properties

Ti-1100: Fracture toughness of beta forged and annealed material

Heat treatment	Fracture toughness (K_{Ic}) As processed		Exposed at 650 °C (1200 °F) for 300 h	
	MPa√m	ksi√in.	MPa√m	ksi√in.
1095 °C (2000 °F), FAC + 595 °C (1100 °F), 8 h	62.9	57.2	43.5	39.6
1095 °C (2000 °F), OQ + 595 °C (1100 °F), 8 h	63.7	57.9	40.2	36.6
1150 °C (2100 °F), FAC + 705 °C (1300 °F), 8 h	53.5	48.7	45.7	41.6
1095 °C (2000 °F), FAC + 995 °C (1825 °F), 1 h + 595 °C (1100 °F), 8 h	64.1	58.3	53.2	48.4
1095 °C (2000 °F), FAC + 1095 °C (2000 °F), 0.5 h + 595 °C (1100 °F), 8 h	71.0	64.6	48.3	43.9
995 °C (1825 °F), FAC + 995 °C (1825 °F), 1 h + 595 °C (1100 °F), 8 h	39.4	35.8	30.3	27.5
995 °C (1825 °F), FAC + 1095 °C (2000 °F), 0.5 h + 595 °C (1100 °F), 8 h	75.9	69.0	44.4	40.4

Note: As expected, the alpha-beta heat treated material has the lowest fracture toughness.

Processing

Casting. In a casting study, the strength of Ti-1100 was found to be equivalent to cast Ti-6Al-2Sn-4Zr-2Mo at 540 °C (1000 °F), but was stronger at 595 °C (1100 °F). Ti-1100 is somewhat weaker in thick cross sections than thin ones and exhibits a significant creep advantage relative to Ti-6Al-2Sn-4Zr-2Mo. It also has higher low-cycle fatigue strength at 550 °C (1022 °F) than Ti-6Al-2Sn-4Zr-2Mo.

Forming. Ti-1100 possesses limited cold form-ability and behaves similarly to Ti-6Al-2Sn-4Zr-2Mo in cold and hot forming. Although the $\alpha + \beta$ window is relatively small, the alloy has demonstrated superplasticity in a simulated manufacturing environment.

Machining. Ti-1100 machines essentially the same as Ti-6Al-2Sn-4Zr-2Mo.

Joining. Welding and brazing of Ti-1100 is similar to Ti-6Al-2Sn-4Zr-2Mo.

Rolling characteristics and texture formation are similar to Ti-6Al-2Sn-4Zr-2Mo.

Surface Treatments. Although material-specific surface treatments have not been fully explored, Ti-1100 should behave essentially the same as Ti-6Al-2Sn-4Zr-2Mo.

Forging

Ti-1100 may be hammer forged or press forged using isothermal, warm die, or conventional die methods. The resulting properties will vary depending on the effective cooling rate and strain rate of the deformation process. Finer structures will result in higher tensile strength at the expense of creep strength at high temperatures. Forging below the β transus followed by a beta anneal to obtain the appropriate microstructure generally is not recommended. However, early forging operations (e.g., preform block) may be conducted in the subtransus field with the finish forging conducted in the β field.

As with other difficult to fabricate near-α alloys, precoats or other surface coating techniques are essential on billet stock and intermediate forging shapes during furnacing for forging operations. Ti-1100 may be sensitive to the excessive formation of a case during reheating processes, which may lead to undue surface cracking in forging deformation. As with other α alloys, care must be exercised in the use of dry abrasive grinding techniques used for crack repair.

Recommended Forging Temperatures. The recommended beta forging range is 1090 to 1120 °C (1990 to 2050 °F). Conventional forging is not recommended for this alloy. Die temperatures are listed in "Technical Note 4: Forging."

To achieve desired elevated-temperature and creep performance characteristics, Ti-1100 is designed to be beta processed, creating a transformed, Widmanstätten α-type microstructure, with minimum grain boundary α. To date, thermomechanical processing work with the alloy suggests that β forging, followed by an appropriate post-forging cooling process based on section size, and final stabilization thermal treatment provides optimum properties.

Subtransus forging and beta heat treatment is not currently recommended because Ti-1100, as a near-alpha alloy, is characterized by high unit pressures. However, hot working above the β transus is not cumulative; thus, if multiple forging steps are required (e.g., preform, block, and finish), early forging operations may be conducted subtransus, with the finish forging being conducted above the transus with a sufficiently high level of work.

Supra-transus, beta forging of Ti-1100 significantly reduces unit pressure requirements and crack sensitivity and is conducted from a temperature above the silicide solvus—1040 °C (1905 °F)—to avoid excessive silicide formation. Typically, beta forging reductions of 50 to 75% are recommended for Ti-1100. Low levels of deformation above the β transus should be avoided.

Post-Forging Treatment. The post-forging cooling rate is not highly critical, and generally an air cool is sufficient. However, for thicker section forgings, fan cooling or oil quenching may be required to achieve final part mechanical properties. Final stabilization age thermal treatments may be adjusted to modify final strength properties. Stabilization treatments are generally in the range of 500 to 650 °C (930 to 1200 °F).

IMI 834 Ti-5.8Al-4Sn-3.5Zr-0.7Nb-0.5Mo-0.35Si

IMI 834 is a near-alpha titanium alloy of medium strength (typically 1050 MPa, or 152 ksi) and temperature capability up to about 600 °C (1110 °F) combined with good fatigue resistance. The alloy derives its properties from solid-solution strengthening, and heat treatment high in the alpha + beta phase field. The addition of carbon facilitates treatment by widening the heat treatment window (see figure). IMI 834 has a low beta stabilizer content and therefore has limited hardenability. It retains a good level of properties in sections up to around 75 mm (3 in.) diameter, with small reductions in strength in larger sections.

Product Forms and Condition. IMI 834 is available in the form of bar, billet, plate, sheet, wire, and castings. IMI 834 is weldable using all of the established titanium welding techniques. It is normally alpha + beta solution treated (15% α) and aged. Microstructural characterization of IMI 834 is discussed in *Met. Trans.*, Vol 24A, June 1993, p 1273-1280.

Applications. The major use for IMI 834 is compressor discs and blades in the aerospace industry. General purpose use is intended for IMI 417.

IMI 834: Typical composition range (wt%) and density

	Al	Sn	Zr	Nb	Mo	Si	C	Fe	O_2	N_2	H_2
Minimum	5.5	3.0	3.0	0.5	0.25	0.20	0.04	...	0.075	...	...
Maximum	6.1	5.0	5.0	1.0	0.75	0.60	0.08	0.05	0.150	0.03	0.006
Nominal	5.8	4.0	3.5	0.7	0.5	0.35	0.06	...	0.10	...	...

Density of IMI 834 is 4.55 g/cm^3 (0.164 lb/in.^{3}).

Physical Properties

IMI 834: Summary of typical physical properties

Beta transus	1045 ± 10 °C (1915 ± 20 °F)
Melting (liquidus point)	Not Available
Density(a)	4.55 g/cm^3 (0.164 lbf/in.^{3})
Electrical resistivity(a)	Not Available
Magnetic permeability	Nonmagnetic
Specific heat capacity(a)	Not Available
Thermal conductivity(a)	Not Available
Thermal coefficient of linear expansion(b)	10.6 × 10^{-6}/ °C (5.9 × 10^{-6}/ °F)

(a) Typical values at room temperature of about 20 to 25 °C (68 to 78 °F). (b) Mean coefficient from room temperature to 200 °C (390 °F)

IMI 834: Thermal coefficient of linear expansion

Temperature range		Mean coefficient of thermal expansion	
°C	°F	10^{-6}/°C	10^{-6}/°F
20-200	68-392	10.6	5.9
20-400	68-752	10.9	6.1
20-600	68-1112	11.0	6.1
20-800	68-1472	11.2	6.2
20-1000	68-1832	11.3	6.3

The thermal expansion coefficient of IMI 834 is typical of other titanium alloys. Heat treated bar

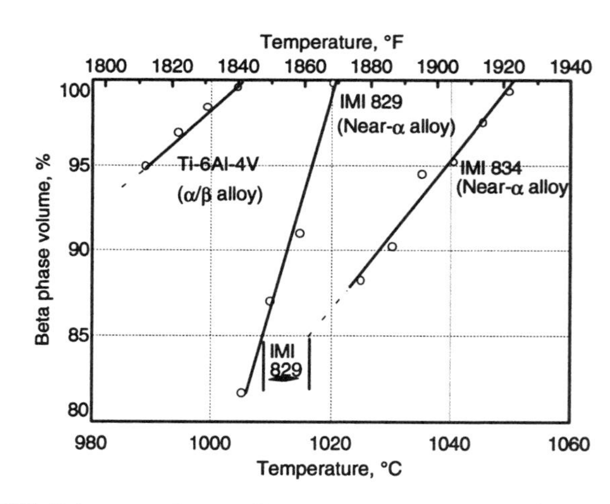

IMI 834: Beta approach curve Beta transus approach curves of IMI 834, IMI 829, and Ti-6Al-4V

Mechanical Properties

Hardness of heat treated IMI 834 is typically 350 HV (20 kg load) or about 35 HRC.

Notch tensile ratio is typically 1.45 ($K_t = 3$).

Impact Strength. Typical Charpy (U-notch) impact strength is 15 J (11 ft · lbf) at room temperature.

Fracture toughness of IMI 834 is typically 45 MPa$\sqrt{m}$ (40 ksi$\sqrt{in.}$) in heat treated discs.

IMI 834: Minimum tensile properties
Typical UTS is 1050 MPa (152 ksi).

Property	Room-temperature minimum
0.2% PS MPa (ksi)	910 (132)
U.T.S. MPa (ksi)	1030 (149)
Elongation (in 5D), %	6
Reduction in area, %	15

Heat treated discs

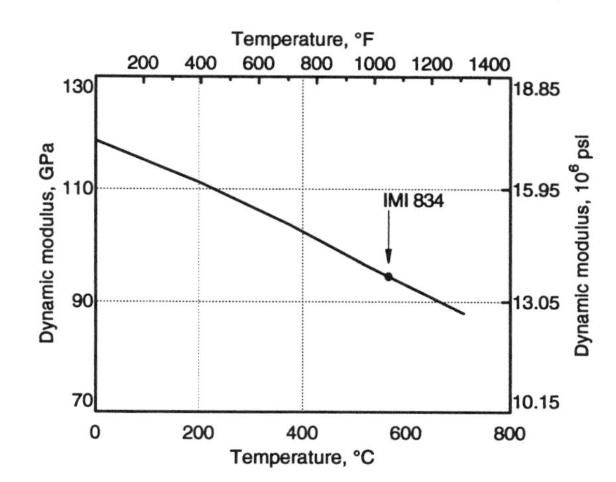

IMI 834: Young's modulus (dynamic) The dynamic modulus of IMI 834 is typical of other near-alpha titanium alloys. Heat treated bar. Source: IMI Titanium "High-Temperature Alloys" brochure

High-Temperature Strength

IMI 834 has useful strength up to 600 °C (1110 °F). IMI 834 is regarded as having long term creep performance up to around 600 °C (1110 °F) and good short term performance up to significantly higher temperatures. Typically, the alloy gives less than 0.1% total plastic strain in 100 hours at 600 °C (1110 °F) under a stress of 150 MPa (21.8 ksi).

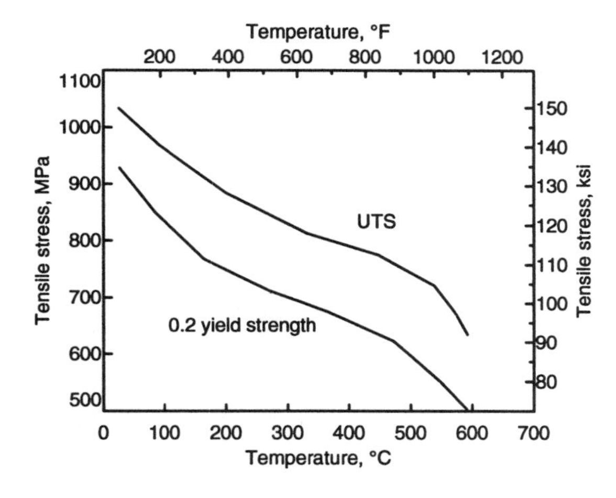

(a)

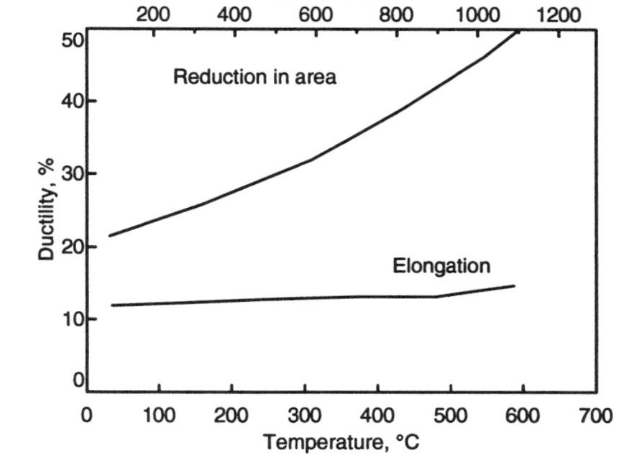

(b)

Heat treated discs. IMI 834: Typical tensile properties

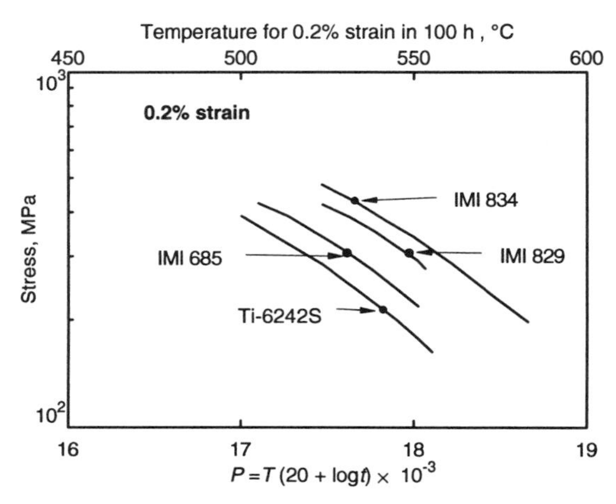

IMI 834: 0.2 % creep strain conditions Heat treated discs or bars.

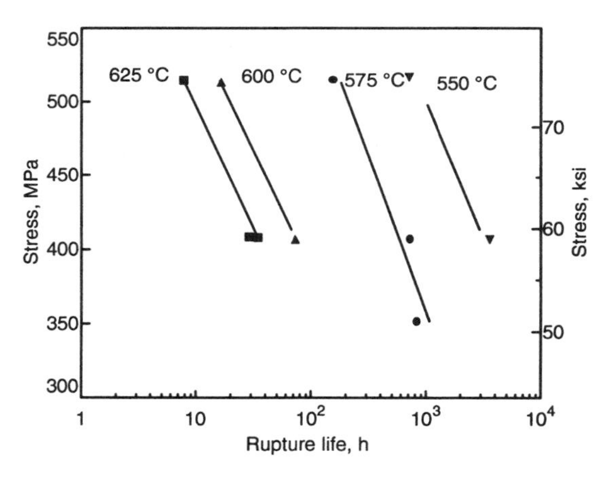

IMI 834: Stress rupture properties Heat treated bar

Fatigue Properties

Low-Cycle Fatigue

Cast IMI 834: Fatigue strength at 10^5 cycles

Condition	Fatigue strength at 10^5 cycles MPa (ksi)
Cast, alpha+beta HIP, plus ½ h at 1070 °C, OQ plus 2 h at 700 °C	700 ± 50 (101.6 ± 7.26)
Cast, beta HIP, plus 2 h at 700 °C	500 ± 50 (72.6 ± 7.26)
Wrought 50 mm (2 in.) diam bar	800 (116.1)

Direct stress, zero minimum ($R = 0$)

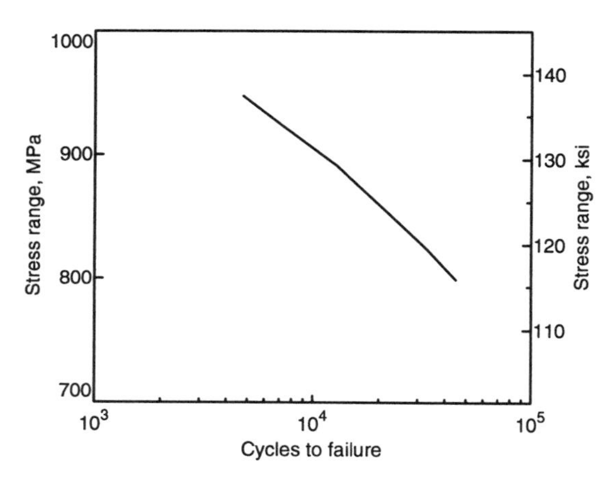

IMI 834: Low-cycle fatigue (R = 0) Unnotched specimens from heat treated bar, direct (axial) stress, room-temperature tests. Source: IMI Titanium "High-Temperature Alloys" brochure

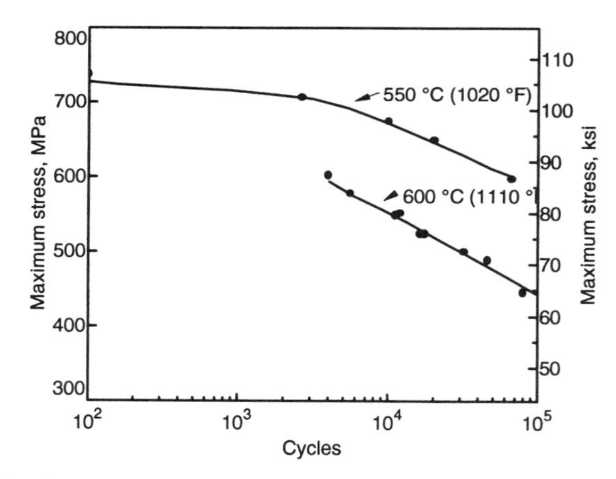

IMI 834: Elevated-temperature low-cycle fatigue Unnotched specimens from heat treated bars, direct (axial) stress, zero minimum stress (R = 0)

High-Cycle Fatigue

Cast IMI 834: Fatigue strength at 10^7 cycles

Condition	Fatigue strength at 10^7 cycles MPa (ksi)
Cast, alpha+beta HIP, plus ½ h at 1070 °C, AC plus 2 h at 700 °C	500 ± 25 (72.6 ± 3.6)
Cast, beta HIP, plus 2 h at 700 °C	400 ± 25 (58.1 ± 3.6)
Wrought, 50 mm (2 in.) diam bar	500 (79.8)

Unnotched specimens, direct stress, zero minimum (R = 0)

Cast IMI 834: Notched fatigue strength

Condition	K_t	Fatigue strength at 10^7 cycles MPa (ksi)
Cast, alpha+beta HIP, plus ½ at 1070 °C, AC, plus 2 h at 700 °C	3.0	250 ± 25 (36.3 ± 3.6)
Wrought, 50 mm (2 in.) Ø bar	2.0	340 (49.3)

Direct stress, zero minimum (R = 0)

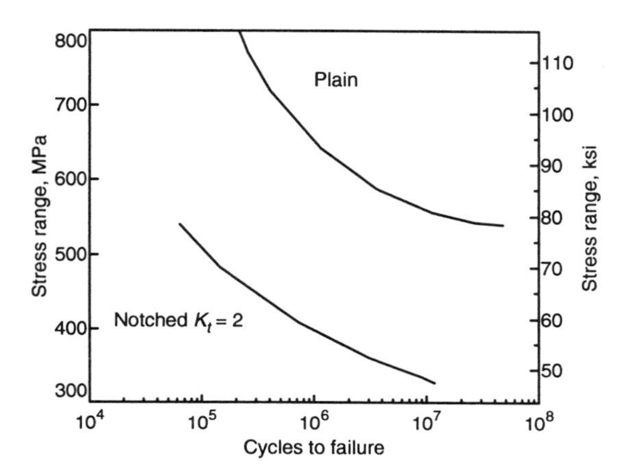

IMI 834: High-cycle fatigue properties (R = 0) Heat treated bar, direct (axial) stress at room temperature.

Crack Propagation

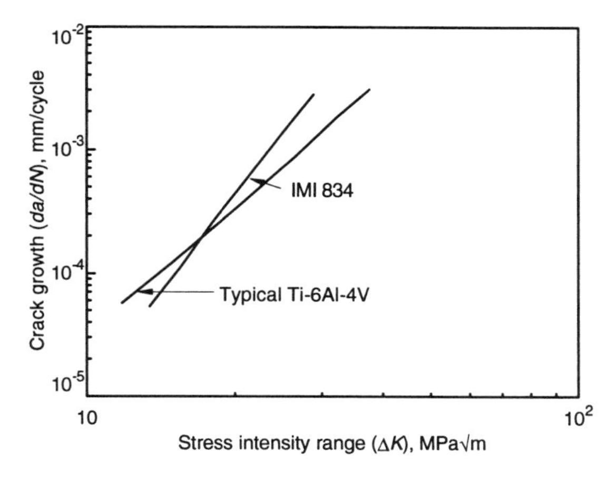

IMI 834: Crack propagation (R = 0) Heat treated bar, longitudinal crack direction room-temperature tests.

Processing

Casting

IMI 834 can be cast using the normal techniques developed for titanium alloys. Typical tensile properties of cast IMI 834 at room temperature and at 600 °C (1110 °F) are given in tables. Cast IMI 834 gives lower tensile ductility than the alpha-beta wrought product but gives better creep performance.

Cast IMI 834: Room-temperature tensile properties

Bar condition	Yield strength (0.2%)		Ultimate tensile strength		Elongation	Reduction in area
	MPa	ksi	MPa	ksi	%	%
Cast + (α+β) HIP + 1070 °C AC + 2 h 700 °C	944	137.0	1071	155.4	5	7
	966	140.2	1072	155.6	5	9
Cast + β HIP + 2 h 700 °C	898	130.3	1040	147.2	6	10
	901	130.8	1025	148.8	4	9
Wrought (15% alpha) OQ + 2 h 700 °C 50 mm bar (2 in.)	950	137.9	1070	155.2	13	23

Cast IMI 834: Tensile properties at 600 °C

Bar condition	Yield strength (0.2%)		Ultimate tensile strength		Elongation, %	Reduction in area, %
	MPa	ksi	MPa	ksi		
Cast + (α+β) HIP + 1070 °C AC + 2 h 700 °C	526	76.3	663	96.2	6	16
	515	74.7	669	97.1	10	29
Cast + β HIP + 2 h 700 °C	467	67.8	566	82.1	6	16
	472	68.5	575	83.5	7	16
Wrought (15% alpha) OQ + 2 h 700 °C 50 mm (2 in.) bar	518	75.2	682	99.0	23	52

Forging

IMI 834 is readily forgeable using conventional hammer, press, or isothermal techniques. Typical forging temperature is around 1010 °C (1850 °F). IMI 834 is stiffer than most other titanium alloys, but it has good forgeability at its recommended forging temperature.

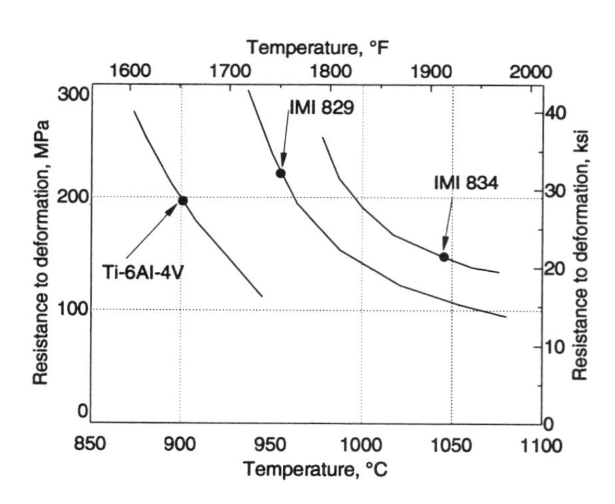

IMI 834: Flow stress As-rolled bar tested with a plastometer up to 1100 °C (2010 °F) at a strain rate of 15/s.

Forming

IMI 834 has very limited cold formability, but good hot formability. It can be produced in sheet and plate form. Typical sheet properties are shown (see table). Superplastic forming is also possible at about 990 °C (1814 °F).

IMI 834: Properties of 2 mm sheet

Material condition(a)	Orientation	Yield strength (0.2%)		Ultimate tensile strength		Elongation (50 mm), %	Creep strain(b), %
		MPa	ksi	MPa	ksi		
Room-temperature properties							
Rolled + *Annealed (800 °C)	L	996	144.6	1114	161.7	11.5	...
	T	1014	147.2	1120	162.5	12	...
*1025 °C (α/β) AC + 2 h 700 °C	L	998	144.8	1145	166.2	11.5	...
	T	1009	146.4	1111	161.2	11	...
*1060 °C (β) AC + 2 h 700 °C	L	947	137.4	1098	159.4	6	...
	T	963	139.8	1103	160.1	6	...
High-temperature (600 °C) properties							
Rolled + *Annealed (800 °C)	L	473	68.7	671	97.4	18	...
	T	510	74.0	720	104.5	14	...
*1025 °C (α/β) AC + 2 h 700 °C	L	518	75.2	702	100.6	16	0.213
	T	546	79.2	728	105.7	18	0.247
*1060 °C (β) AC + 2 h 700 °C	L	554	80.4	716	103.9	12	0.055
	T	532	77.2	729	105.8	12	0.064

(a) An asterisk * indicates a heating duration of 30 minutes. (b) Total plastic strain after exposure of 150 MPa (21.8 ksi) at 600 °C (1110 °F) for 100 hours

Heat Treatment

IMI 834: Recommended heat treatments

Treatment	Temperature		Duration	Cooling method
	°C	°F		
Solution treatment (15% alpha)	1015 ± 5	1860 ± 9	2 hours	Oil quench(a)
Aging	700	1290	2 hours	Air cool

(a) For sections less than about 15 mm (0.6 in.), air cooling is recommended.

Ti-1100: Fatigue strength at 10^7 cycles

Temperature		K_t	Fatigue strength(a)	
°C	°F		MPa	ksi
22	71	1.0	655	95
		3.0	250	36
480	895	1.0	500	72
		3.0	235	34

(a) Beta forged and annealed; tested at 30 Hz; $R = 0.1$

Ti-5Al-2Sn-2Zr-4Mo-4Cr

Common Name: Ti-17
UNS Number: R58650

Ti-5Al-2Sn-2Zr-4Cr-4Mo (Ti-17) is a high-strength, deep hardenable, forging alloy that was developed primarily for gas turbine engine components, such as disks for fan and compressor stages. Ti-17 has strength properties superior to those of Ti-6Al-4V, and also exhibits higher creep resistance at intermediate temperatures.

Product Conditions/ Microstructure

Ti-17 can be heat treated to yield strengths of 1030 to 1170 MPa (150 to 170 ksi). It is more ductile than Ti-6Al-6V-2Sn, and it is superior to Ti-6Al-4V in creep behavior. With hardenability characteristics comparable to those of some beta type alloys, Ti-17 is lower in density and higher in modulus and creep strength than the beta alloys.

Chemistry and Density

Ti-17 may be classified as a "beta-rich" alpha-beta alloy, because it has a beta-stabilizer (Mo + Cr) content of 8%.

Density. 4.65 g/cm^3 (0.168 lb/in.^{3})

Product Forms

Ingot, billet, forgings

Product Conditions/ Microstructure

Ti-17 can be processed in either the beta or alpha plus beta region, and subsequent heat treatment depends on processing history. Special ingot melting conditions are required, particularly during the final melt, to minimize segregation of beta stabilizers (primarily chromium) during solidification. Excessive segregation of beta stabilizers can cause "beta flecks" during forging or upon heat treatment, which constitute microregions of subnormal fracture toughness and ductility. Both forging and heat treating practices must be controlled carefully to minimize the effects of microsegregation (beta flecks).

Applications

Ti-17 is used for heavy-section forgings up to 150 mm (6 in. thick) for gas turbine engine components and other elevated-temperature applications demanding high tensile strength and good fracture toughness. It is used only by General Electric.

Ti-5Al-2Sn-2Zr-4Mo-4Cr: Specifications and Compositions

Specification	Designation	Description	Al	Cr	Fe	H	Mo	N	Sn	Zr	Other
UNS	R58650		4.5-5.5	3.5-4.5	0.3 max	0.0125 max	3.5-4.5	0.04 max	1.5-2.5	1.5-2.5	Mn 0.1 max; Cu 0.1 max; O 0.08-0.13; C 0.05 max; OT 0.3 max; OE 0.1 max; Y 0.005 max; bal Ti
USA											
AMS 4995		Bil STA	4.5-5.5	3.5-4.5	0.3	0.0125	3.5-4.5	0.04	1.5-2.5	1.5-2.5	Mn 0.1 max; Cu 0.1 max; O 0.08-0.13; C 0.05; OT 0.3; Y 0.005; bal Ti

Ti-5Al-2Sn-2Zr-4Mo-4Cr: Commercial Compositions

Specification	Designation	Description	Al	Cr	Fe	H	Mo	N	Sn	Zr	Other
Japan											
Kobe	KS5-2-2-4-4	Bar Frg STA	4.5-5.5	3.5-4.5	0.3	0.0125	3.5-4.5	0.04	1.5-2.5	1.5-2.5	O 0.08-0.13; bal Ti
USA											
OROMET	Ti-17										
TIMET	TIMETAL 17										

Fatigue Properties

Ti-17: Typical STA high-cycle fatigue (unnotched)

Load control, $A = 0.95$

Temperature		Maximum stress		Alternating stress		Cycles to failure
°C	°F	MPa	ksi	MPa	ksi	
24	75	965	140	470	68.2	21,000
		827	120	403	58.5	75,000
		758	110	370	53.6	7,000,000
		724	105	353	51.2	6,000,000
315	600	758	110	370	53.6	37,000
		690	100	336	48.7	50,000
		676	98	329	47.7	88,000
		655	95	319	46.3	15,000,000
		621	90	43.8	302	12,000,000

Source: *Beta Titanium Alloys in the 1980's*, R.R. Boyer and H.W. Rosenberg, Ed., TMS/AIME, 1984, p 438.

Ti-17: Typical STA low-cycle fatigue (unnotched)

Strain control, $A = 1.0$

Temperature		Strain, %			Cycles to failure
°C	°F	Plastic	Elastic	Total	
24	75	0.38	1.49	1.87	3,180
		0.10	1.365	1.465	5,040
		0.03	1.23	1.26	9,650
		0.02	1.135	1.155	15,400
		0.01	1.04	1.05	25,700
		0.025	0.97	0.995	60,600
315	600	0.23	1.30	1.54	3,600
		0.133	1.20	1.34	5,500
		0.044	0.982	1.03	>12,700
		0.055	0.92	0.98	>56,300
		0.017	0.9075	0.93	>86,000
		0.045	0.88	0.93	>16,000

Source: Beta Titanium Alloys in the 1980's, R.R. Boyer and H.W. Rosenberg, Ed., TMS/AIME, 1984, *p 437*

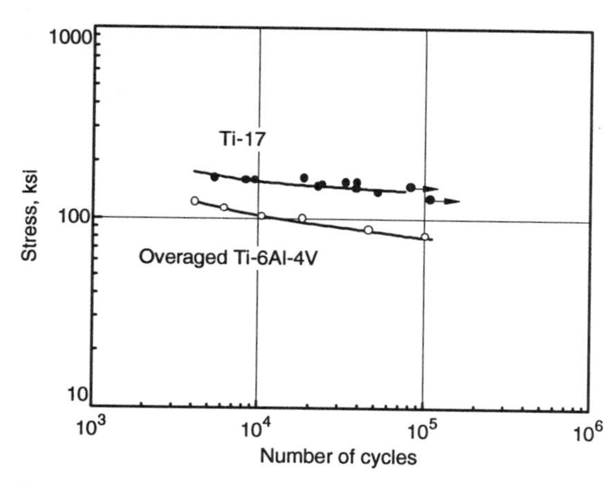

Ti-17: Axial fatigue of STA disk forgings Ti-17 heat treatment: 860 °C (1575 °F), 4 h, AC, 800 °C (1475 °F), 4 h, FAC, 620 °C (1150 °F), 8 h, AC. Axial loaded, $R = 0$, $K_t = 1$; frequency, 20 cycles/min. Source: *Aerospace Structural Metals Handbook*, Vol 4, Code 3724, Battelle Columbus Laboratories, 1976

Ti-17: Typical STA low-cycle fatigue (unnotched)
Load control, ***A*** = 1.0

Temperature		Stress range		Cycles to failure
°C	°F	MPa	ksi	
24	75	1103	160	5,000
		1069	155	10,000
		1000	145	25,000
		931	135	35,000
		896	130	170,000
		827	120	290,000
315	600	896	130	5,000
		862	125	6,000
		841	122	7,000
		834	121	13,000
		827	120	64,000

Source: *Beta Titanium Alloys in the 1980's*, R.R. Boyer and H.W. Rosenberg, Ed., TMS/AIME, 1984, p 437

Fatigue Crack Growth

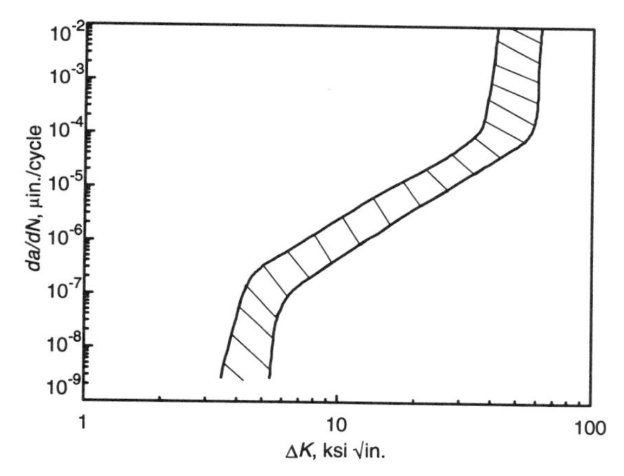

Ti-17: Fatigue crack growth at room temperature Alpha-beta processed spool forgings were heat treated at 860 °C (1575 °F), 4 h, AC + 800 °C (1475 °F), 4 h, FAC + 620 °C (1150 °F), 8 h, AC. Tensile yield strength, 1075 MPa (156 ksi); $B = 1$; $W/B = 2$; L-R orientation. Source: *Aerospace Structural Metals Handbook*, Vol 4, Code 3724, Battelle Columbus Laboratories, 1976

Fracture Properties

Ti-17: Plane-strain fracture toughness at room temperature STA

Yield strength		K_{IC}	
MPa	ksi	MPa√m	ksi√in.
Alpha-beta processed			
1172	170	33	30
1103	160	40	36
1034	150	50	45
Beta-processed			
1172	170	53	48
1103	160	65	59
1034	150	88	80

Source: *Beta Titanium Alloys in the 1980's*, R.R. Boyer and H.W. Rosenberg, Ed., TMS/AIME, 1984, p 438

Ti-17: Effect of reduction ratio on fracture toughness of disk forgings

Reduction ratio	Tensile yield strength(a)		Fracture toughness(b) (K_{IC})	
	MPa	ksi	MPa√m	ksi√in.
Alpha-beta forged + STA(c)				
2:1	1150	167	41.5	37.8
3:1	1145	166	36	32.9
4:1	1165	169	37.2	33.9
Beta forged + STA(d)				
2:1	1117	162	68.8	62.2
3:1	1103	160	61	55.5
4:1	1110	161	55.2	50.2

(a) Average of two tests. (b) 25 mm (1 in.) thick compact tension specimen. (c) 845 °C (1550 °F) for 4 h, FAC, 800 °C (1475 °F) for 4 h, FAC; and 620 °C (1150 °F) for 8 h, AC. (d) 800 °C (1475 °F) 4 h, FAC; 620 °C (1150 °F) 8 h, AC. Source: *Aerospace Structural Metals Handbook*, Vol 4, Code 3724, Battelle Columbus Laboratories, 1976

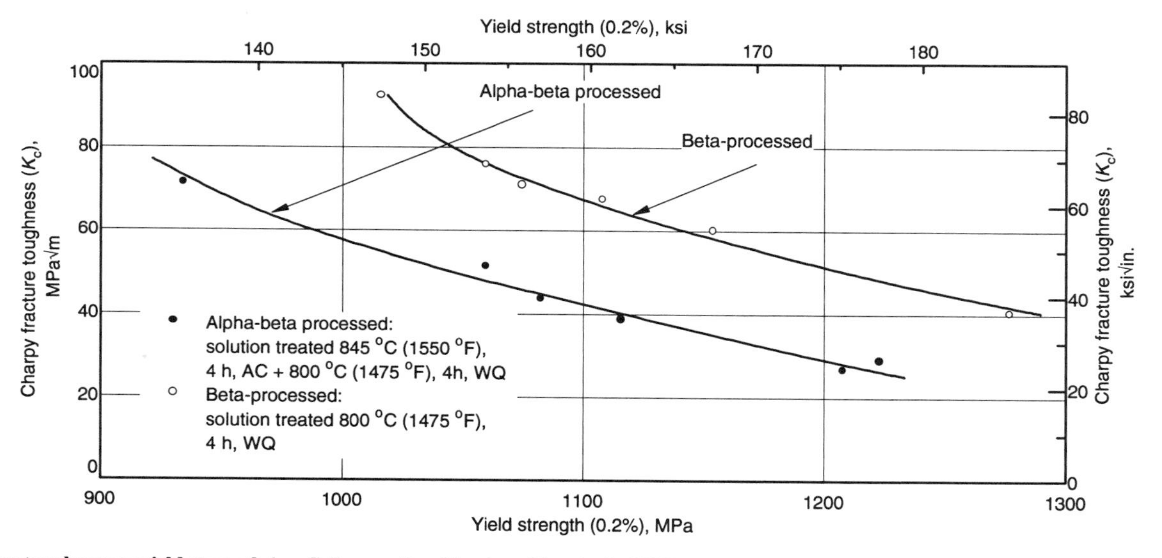

Ti-17: Fracture toughness vs yield strength (aged) Source: *Beta Titanium Alloys in the 1980's*, R.R. Boyer and H.W. Rosenberg, Ed., TMS/AIME, 1984, p 245. Aged to strength (8 h with temperatures from 900 to 1300 °F)

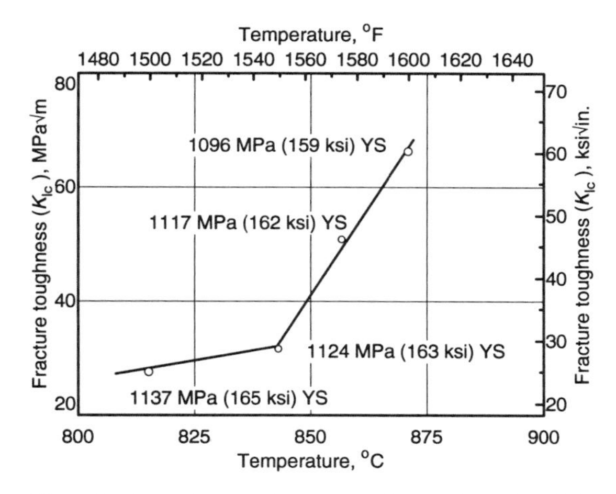

Ti-17: Effect of solution temperature on toughness 25 mm (1 in.) thick compact tension specimen from 457 mm (18 in.) diam × 50 mm (2 in.) thick disk forging. Indicated solution temperature plus 785 °C (1450 °F), 4 h, WQ + 620 °C (1150 °F), 8 h. Source: *Beta Titanium Alloys in the 1980's*, R.R. Boyer and H.W. Rosenberg, Ed., TMS/AIME, 1984, p 246

Forging

G.W. Kuhlman, ALCOA, Forging Division

Ti-17 is a high-strength, highly beta-stabilized, α-β (near-beta) alloy whose primary commercial application is turbine engine rotating components. It can be fabricated into all forging product types, although closed die forgings and rings predominate. Ti-17 is commercially fabricated on all types of forging equipment. Turbine engine disks are frequently produced using hot die or isothermal forging techniques, resulting in near-net closed die forgings with reduced final machining.

Ti-17 is a highly forgeable alloy with lower unit pressures (flow stresses), improved forgeability, and less crack sensitivity than the α-β alloy Ti-6Al-4V. The final microstructure of Ti-17 forgings is developed by thermomechanical processing in forging manufacture tailored to achieve specific microstructural and mechanical-property objectives. Thermomechanical processes use combinations of subtransus and/or supra-transus forging followed by subtransus thermal treatments to fulfill critical mechanical-property criteria.

Final thermal treatments for Ti-17 forgings include two- or three-step practices of single or two-step solution treatments followed by quenching and aging. Solution treatment is subtransus, at 800 °C (1475 °F), followed by water quench or fan air cool for thin sections. For forgings fabricated conventionally, a solution anneal at 855 °C (1575 °F), followed by an air cool, may be used to improve toughness and creep properties. Aging treatment is conducted at 620 °C (1150 °F). Subtransus thermomechanical processes (forging and thermal treatment) for Ti-17 forgings achieve equiaxed (20 to 30%) α in transformed β matrix microstructures that enhance strength, ductility, and particularly low-cycle fatigue properties. Supra-transus thermomechanical processes (beta forging followed by subtransus thermal treatments) achieve transformed, Widmanstätten α microstructures that enhance creep and/or fracture-related properties (T.K. Redden, Ref 1).

Conventional Forging. The objectives in forging Ti-17 are to obtain the final forging shape and desired final microstructure at least cost. Conventional subtransus (α + β) forging thermomechanical processes are most widely used in commercial engine disk forging manufacture. To achieve conventional equiaxed α structures, subtransus reduction of 50 to 75%, accumulated through one or more forging steps, are required. Supra-transus (β) forging for Ti-17 may be used in early forging operations, including upsetting and open die preforming, to reduce unit pressures and ease forging fabrication. However, higher temperature initial forging operations must be followed by sufficient subtransus reduction to achieve the desired predominately equiaxed α structure. Conventionally forged Ti-17 is then subtransus solution treated, quenched, and aged as noted above.

Supra-transus thermomechanical processes for Ti-17 are used for selected disk applications to achieve transformed, Widmanstätten α structures for improved creep and fracture-related properties. Successful β thermomechanical processes for Ti-17 forgings include controlled β forging processes followed by subtransus solution treatment and aging. The β forging thermomechanical processes are particularly well suited to isothermal or hot die forging technology. Beta forging requires subtransus reduction (e.g., 20 to 50%) in early forging (blocker die) stages followed by a controlled, single β forging step, that achieves 30 to 50% reductions. Beta forging Ti-17 requires careful control of forging process conditions, particularly preheat times at temperature, to avoid excessive prior β grain growth. Beta forged Ti-17 is then subtransus heat treated as noted above. Because of inherent variations in forging conditions, β forged Ti-17 may exhibit more final forging product variation than conventionally subtransus forged and heat treated Ti-17 forged product.

Hot die and/or isothermal forging techniques are important commercial methods for fabrication of Ti-17 rotating turbine engine disks to reduce final component cost (from less machining) and/or improve final component microstructural and property uniformity through improved control of forging process conditions. The axisymmetric shapes and designs of such engine components are very well suited to these forging methods. Isothermal forging of Ti-17 disks is frequently accomplished in a single forging step from bar or billet stock, under carefully controlled supra- or subtransus metal and die temperatures, levels of strain, and strain-rate profiles. Hot die forging, where die temperature approaches but is not equivalent to metal temperature, is also used to reduce unit pressures, enhance forgeability, and produce more sophisticated final shapes in fewer forging operations. With either subtransus or supra-transus forging via both of these "hot die" processes and controlled post-forging cooling rates, desired tensile strength, fracture toughness, and creep properties can be achieved in Ti-17 using direct aging, thus eliminating the solution treatment processes (G.W. Kuhlman, Ref 2).

References

1. T.K. Redden, Processing and Properties of Ti-17 Alloy for Aircraft and Turbine Applications, *Beta Titanium in the 1980's*, R.R. Boyer and H.W. Rosenberg, Ed., TMS/AIME, 1984, p 239-254
2. G.W. Kuhlman, *et al.*, "Mechanical Property Tailoring Titanium Alloys for Jet Engine Applications," *Proc. 1986 Int. Conf. Titanium Products and Applications*, Titanium Development Association, 1987, p 122-153

Ti-6Al-2Sn-4Zr-6Mo

Common Name: Ti-6246
UNS Number: R56260

Ti-6Al-2Sn-4Zr-6Mo (Ti-6246) is a heat-treatable alpha-beta alloy designed to combine the long-term, elevated-temperature strength properties of Ti-6Al-2Sn-4Zr-2Mo-0.08Si (Ti-6242S) with much-improved short-term strength properties of a fully hardened alpha-beta alloy. It is used for forgings in intermediate-temperature sections of gas turbine engines, particularly in compressor disks and fan blades. This alloy is used at lower temperatures than Ti-6242S, but should be considered for long-term load-carrying applications at temperatures up to 400 °C (750 °F) and short-term load-carrying applications at temperatures up to 540 °C (1000 °F).

Chemistry and Density

Ti-6246 is a solid-solution-strengthened alloy that responds to heat treatment as a result of the beta-stabilizing effect of its 6% molybdenum content. Silicon additions (0.08 wt%) improve creep resistance. As for all alpha-beta alloys, excessive amounts of aluminum, oxygen, and nitrogen can decrease ductility and fracture toughness.

Density. 4.65 g/cm^3 (0.168 lb/in.^{3})

Product Forms

Ti-6246 is produced by all U.S. melters as billets and bars for forging stock. It has also been produced and evaluated in sheet and plate form.

Product Condition/ Microstructure

Special ingot melting practices must be employed, particularly during final melting, to minimize microsegregation of the beta-stabilizing element, molybdenum, which could result in "beta flecks" (see Technical Note 1). Forging and heat treating practices require special controls to minimize beta flecks, which could result in microregions of high strength and low fracture toughness. Beta flecks are less of a problem for Ti-6246 than for Ti-17.

Applications

Ti-6246 is used for forgings in intermediate-temperature sections of gas turbine engines, particularly for compressor disks and fan blades and also for seals and airframe components. Ti-6246 is also under evaluation for deep, sour-well applications.

Ti-6Al-2Sn-4Zr-6Mo: Specifications and compositions

Specification	Designation	Description	Al	Fe	H	Mo	N	O	Sn	Zr	Other
UNS	R56260		6			6			2	4	bal Ti
USA											
AMS 4981B		Bar Wir Frg Bil STA	5.5-6.5	0.15	0.0125	5.5-6.5	0.04	0.15	1.75-2.25	3.5-4.5	C 0.04; OT 0.4; Y 0.005; bal Ti
MIL F-83142A	Comp 11	Frg Ann	5.5-6.5	0.15	0.0125	5.5-6.5	0.04	0.15	1.75-2.25	3.6-4.4	C 0.04; bal Ti
MIL F-83142A	Comp 11	Frg HT	5.5-6.5	0.15	0.0125	5.5-6.5	0.04	0.15	1.75-2.25	3.6-4.4	C 0.04; bal Ti

Ti-6Al-2Sn-4Zr-6Mo: Commercial compositions

Specification	Designation	Description	Al	Fe	H	Mo	N	O	Sn	Zr	Other
Japan											
Kobe	KS6-2-4-6	Bar Frg STA	5.5-6.5	0.15	0.0125	5.5-6.5	0.04	0.15	1.75-2.25	3.5-4.5	bal Ti
					USA						
Astro	Ti-6Al-2Sn-4Zr-6Mo	Bar	5.5-6.5	0.15 max	0.0125	5.5-6.5	0.04 max	0.15 max	1.8-2.2	3.6-4.4	C 0.1 max; bal Ti
Howmet											
Martin Mar											
Oremet	Ti-6246										
RMI	6Al-2Sn-4Zr-6Mo	Bar Bil STA	5.5-6.5	0.15	0.0125	5.5-6.5	0.04	0.15	1.75-2.25	3.5-4.5	C 0.04; bal Ti
Tel.AllVac											
Timet	TIMETAL 6-2-4-6	DA	5.5-6.5	0.15 max	0.0125 max	5.5-6.5	0.04 max	0.15 max	1.75-2.25	3.5-4.5	C 0.04 max; bal Ti

Phases and Structures

The microstructure of Ti-6246 is typically equiaxed primary α in a transformed β matrix; this can vary, depending on processing and heat treatment history. A microstructure with an optimum combination of strength, ductility, and toughness contains about 10% equiaxed α (primary α) plus a transformed β matrix with relatively coarse secondary α and aged β.

Beta Transus: 935 °C (1715 °F). The 1020 °C transus in figure is suspect.

Transformation Products

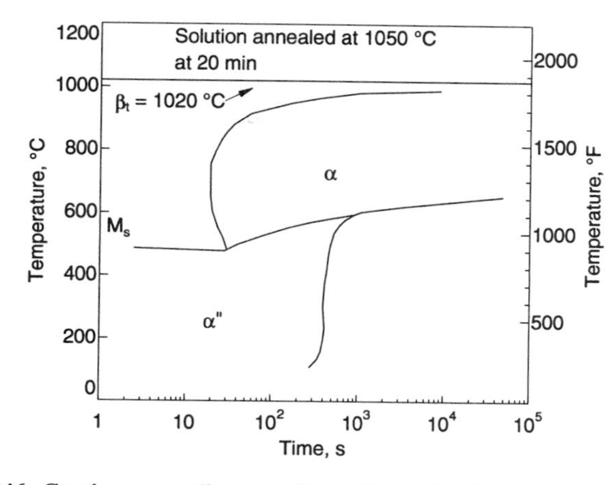

Ti-6246: Continuous cooling transformation and aging diagram Source: W.W. Cias, "Phase Transformation Kinetics, Microstructures, and Hardenability of the Ti-6Al-2Sn-4Zr-6Mo Titanium Alloy," Rp-27-71-02, Climax Molybdenum, 2 March 1972

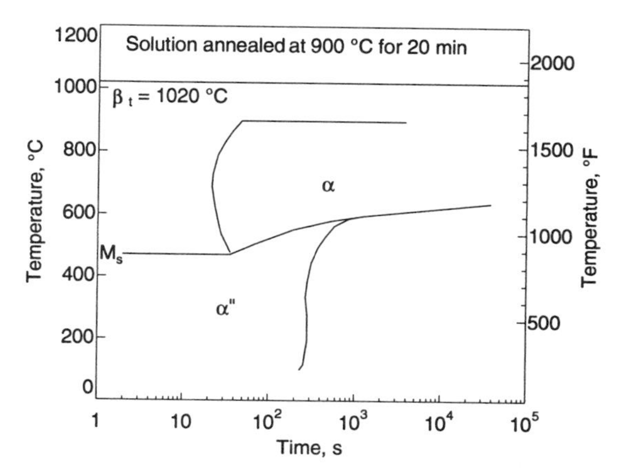

Ti-6246: Continuous cooling transformation diagram Source: W.W. Cias, "Phase Transformation Kinetics, Microstructures, and Hardenability of the Ti-6Al-2Sn-4Zr-6Mo Titanium Alloy," Rp-27-71-02, Climax Molybdenum, 2 March 1972

Fatigue Properties

High-Cycle Fatigue

Ti-6246: Room-temperature axial fatigue strength at 10^7 cycles

Heat treatment	Axial strength at K_t = 1		Fatigue strength 10^7 cycles for: K_t = 3.8	
	MPa	ksi	MPa	ksi
870 °C (1600 °F), 1 h, AC + 595 °C (1100 °F), 8 h, AC	793	115	380	55
910 °C (1675 °F), 1 h, AC + 595 °C (1100 °F), 8 h, AC	825	120	345	50

Note: 25 mm (1 in.) round duplex annealed forgings. Source: *Aerospace Structural Metals Handbook*, Code 3714, Vol 4, Battelle Columbus Laboratories, 1972

Ti-6246: Fatigue and tensile data for various microstructural conditions

Condition	Tensile yield strength		Ultimate tensile strength		Elongation, %	Reduction of area, %	Stress at 10^7 cycles: Smooth		Notched	
	MPa	ksi	MPa	ksi			MPa	ksi	MPa	ksi
10% equiaxed primary α + annealed(a)	1020	148	1109	161	15	37	620	90	289	42
10% equiaxed primary α + STA(b)	1116	162	1213	176	13	37	620	90	248	36
50% equiaxed primary α + annealed	1061	154	1130	164	13	34	620	90	282	41
50% equiaxed primary α + STA	1151	167	1240	180	14	42	675	98	262	40
50% equiaxed primary α + STOA(c)	1068	155	1144	166	14	41	620	90	262	38
50% elongated primary α + STA	1096	159	1206	175	10	23	751	109	276	40
20% elongated primary α + STA	1109	161	1206	175	11	26	620	90	282	41
β forged + STA	1047	152	1199	174	7	13	675	98	262	38

(a) Annealed = 705 °C (1300 °F), 1 h, AC. (b) STA = 885 °C (1630 °F), 1 h, AC + 595 °C (1100 °F), 8 h, AC. (c) STOA = 885 °C (1630 °F), 1 h, AC + 705 °C (1300 °F), 1 h, AC. Source: J.C. Williams and E.A. Starke, in Deformation, Processing, *and Structure*, G. Krauss, Ed., American Society for Metals, 1984, p 332

Low-Cycle Fatigue

Low-cycle fatigue (LCF) behavior of Ti-6246 has been studied to determine the effect of microstructure on cyclic deformation and LCF initiation (Mahajan and Margolin, *Metall. Trans. A*, Vol 13, 1982, p 257-268). Widmanstätten + grain boundary α and equiaxed structures of different α particle sizes were produced in smooth bar specimens of a Ti-6246 alloy, heat treated to produce a 0.2% yield stress of about 1100 MPa (159 ksi). Specimens were cycled at room temperature under total strain control.

At low strains for both Widmanstätten + grain boundary and equiaxed α structures, crack initiation took place at α-β interfaces and in the aged β matrix. In Widmanstätten + grain boundary α structures, profuse extrusion formation was noted as well. At higher strains, cracking was more predominant at slip bands within α.

In Widmanstätten + grain boundary α structures, Widmanstätten α and grain boundary α particles provided sites at which ready crack formation and link-up could take place, thus leading to much longer surface cracks in the Widmanstätten + grain boundary α than in equiaxed α structures for given cycling conditions. Beta grain size played an indirect role in development of fatigue cracks. Larger β grains permitted longer Widmanstätten and grain boundary α particles to form. These longer particles provided longer paths where crack growth could take place preferentially and longer surface cracks could develop.

At larger plastic strains, Widmanstätten α colonies at large angles to the crack propagation direction served to produce multiple cracking along the α-β interfaces and to slow or change the direction of crack propagation at both surface and interior locations. Coarse α particles, which have a small aspect ratio, are favorable for multiple slip and associated multiple cracking at the crack tip.

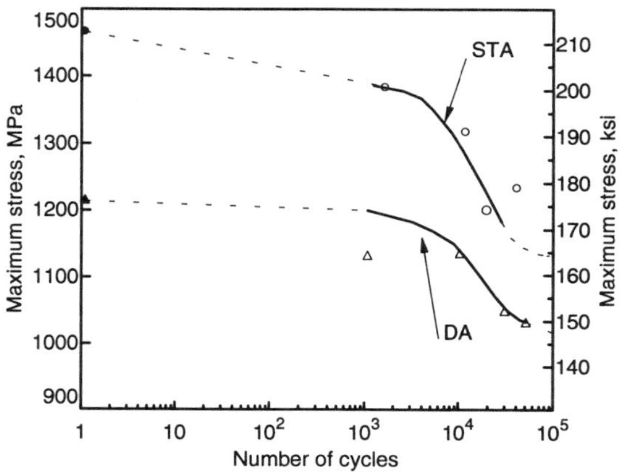

Ti-6246: Low-cycle fatigue STA condition: 1 h at 870 °C (1600 °F), water quench, age 8 h at 595 °C (1100 °F) and air cool. DA (duplex annealed) condition: 15 min at 870 °C, air cool, then 8 h at 540 °C (1000 °F) and air cool. All fatigue tests conducted at a stress ratio of $R = 0.1$. Open symbols indicate fatigue tests; solid symbols, tension tests. Source: *Aerospace Structural Metals Handbook*, Code 3714, Vol 4, Battelle Columbus Laboratories, 1972

Fatigue Crack Growth

A Ti-6246 alloy containing 68 ppm H with basal texture was tested to determine the influence of dwell time at maximum tensile stress on the fatigue crack growth rates (see table). All of the fatigue tests were conducted using displacement-controlled constant stress intensity (K), in air (relative humidity not specified), at room temperature. The Ti-6246 alloy exhibited a nominal two- to threefold increase in the total fatigue crack growth rate as a result of the 10-min dwell at $\Delta K = 38.5$ MPa$\sqrt{\text{m}}$ (35 ksi$\sqrt{\text{in.}}$), but there was little effect on the fatigue crack growth at the lower values of ΔK. The small changes in the fatigue crack growth rate were due to the crack advance by cleavage during the dwell periods. The cleavage fracture was the result of hydrogen embrittlement.

Ti-6246: Fatigue crack growth vs dwell time

Basal transverse textured titanium alloys tested at 21 °C ± 1 °C; R = 0.01; TL orientation; 0.3 Hz

Fatigue crack growth (ΔK)		Dwell time, min	da/dN prior to dwell, μm/cycle	da/dt during dwell, μm/min	da/dN after dwell, μm/cycle	Total da/dN(a), μm/cycle
MPa$\sqrt{\text{m}}$	ksi$\sqrt{\text{in.}}$					
38.46	35.0	10	6.25	44.5	4.62	20.1
27.80	25.3	45	1.38	7.29	1.28	2.87
23.52	21.4	45	0.9	1.0	0.9	1.01

(a) Includes crack advance during the dwell time. Source: *Metall. Trans. A*, Vol 14, 1983, p 2179

Fracture Properties

Ti-6246: Impact toughness

Forging temperature		Solution temperature(a)		Cooling from solution	Aging temperature(b)		Charpy V-notch impact toughness		Ultimate tensile strength	
°C	°F	°C	°F		°C	°F	J	ft · lbf	MPa	ksi
885	1625	830	1525	Air cool	540	1000	12.2	9	1241	180
					595	1100	13.5	10	1255	182
				Oil quench	540	1000	10.1	7.5	1296	188
					595	1100	14.9	11	1268	184
		870	1600	Air cool	540	1000	11.5	8.5	1310	190
					595	1100	9.5	7	1248	181
				Oil quench	540	1000	8.1	6	1489	216
					595	1100	8.1	6	1324	192
915	1675	830	1525	Air cool	540	1000	10.8	8	1193	173
					595	1100	10.8	8	1165	169
				Oil quench	540	1000	10.8	8	1337	194
					595	1100	12.9	9.5	1241	180
		870	1600	Air cool	540	1000	12.2	9	1255	182
					595	1100	12.9	9.5	1275	185
				Oil quench	540	1000	9.5	7	1461	212
					595	1100	8.8	6.5	1350	197

Note: 45 mm (1.75 in.) thick upset forgings. (a) 1 h at temperature. (b) 8 h at temperature. Source: *Aerospace Structural Metals Handbook*, Code 3714, Vol 4, Battelle Columbus Laboratories, 1972

Fracture Toughness

Ti-6246: Fracture toughness of forgings

Condition	Tensile yield strength		Ultimate tensile strength		Elongation,	K_{IC} (K_Q)	
	MPa	ksi	MPa	ksi	%	MPa$\sqrt{m}$	ksi$\sqrt{in.}$
$\alpha + \beta$ forged + STA(a) (10% primary α)	1116	162	1213	176	13	34	31
$\alpha + \beta$ forged + STA(a) (50% primary α)	1150	166	1240	180	14	26	23
$\alpha + \beta$ forged + annealed(b) (50% primary α)	1061	154	1130	164	13	26	23
β forged + STA(a)	1047	152	1199	174	7	57	52

(a) 885 °C (1630 °F), 1 h, AC + 595 °C (1100 °F), AC. (b) 705 °C (1300 °F), 1 h, AC. Source: J.C. Williams and E.A. Starke, in Deformation, *Processing, and Structure,* American Society for Metals, 1984

Ti-6246: Fracture toughness of forgings of several forging and heat treatment conditions and section thicknesses

Forging conditions	Heat treatment conditions	Section thickness		Ultimate tensile strength		Fracture toughness K_{Ic}	
		mm	in.	MPa	ksi	MPa$\sqrt{m}$	ksi$\sqrt{in.}$
885 °C (1625 °F), AC	870 °C (1600 °F), 2 h, AC + 595 °C (1100 °F), 8 h, AC	50	2	1144	166	36.7	33.4
				...	...	33.5	30.5
900 °C (1650 °F), AC	900 °C (1650 °F), 1 h, WQ + 650 °C(1200 °F), 8 h, AC	75	3	1303	189	20.9	19.1
	900 °C (1650 °F), 1 h, AC + 650 °C (1200 °F), 8 h, AC	75	3	1172	170	26.6	24.2
885 °C (1625 °F), AC	915 °C (1675 °F), 1 h, AC + 595 °C (1100 °F), 8 h, AC	50	2	1158	168	50.4	45.9
980 °C (1800 °F), WQ		50	2	1158	168	65.8	59.9
885 °C (1625 °F), AC	915 °C (1675 °F), AC + 525 °C (975 °F), 8 h, AC	25	1	1220	177	32.6	29.7
980 °C (1800 °F), WQ	95 °C (1675 °F), 1 h, AC + 525 °C (975 °F), 8 h , AC	25	1	1220	177	46.0	41.9
980 °C (1800 °F), WQ	915 °C (1675 °F), 1 h, AC + 595 °C (1100 °F), 8 h, AC	25	1	1186	172	47.5	43.3
900 °C (1650 °F), AC	915 °C (1675 °F), 1 h, AC + 845 °C (1550 °F), 8 h, OQ + 595 °C (1100 °F), 8 h, AC	75	3	1220	177	39.0	35.5
		54	2.125	1255	182	36.3	33.1
980 °C (1800 °F), AC	845 °C (1550 °F), 1 h, OQ + 595 °C (1100 °F), 8 h, AC	75	3	1186	172	37.2	33.9
		54	2.125	1268	184	35.4	32.2
		38	1.5	1296	188	27.7	25.2
885 °C (1625 °F), AC	915 °C (1675 °F), 1 h, AC + 595 °C (1100 °F), 4 h, AC	75	3	1186	172	33.7	30.7
		50	2	1165	169	33.3	30.3
		25	1	1234	179	29.9	27.3

Source: W. Heil; reported in *Aerospace Structural Metals Handbook*, Code 3714, Vol 4, Battelle Columbus Laboratories, 1972

Ti-6246: Fracture toughness of STA forgings of two forging conditions and specimen locations

	Fracture toughness (K_{Ic}) for material:			
	α + β forged at 880 °C (1620 °F)		β forged at 1010 °C (1850 °F)	
Specimen location	MPa√m	ksi√in.	MPa√m	ksi√in.
Center tangential	28.28	25.74	...	...
Outside tangential	24.47	22.27	21.81	19.85
	26.49	24.11	...	...
Center diametral	...	...	21.45	19.52

Note: K_{Ic} values determined with precracked three-point notched bend specimens. Heat treatment was at 870 °C (1600 °F) for 1 h, water quench, then at 595 °C (1100 °F) for 8 h, air cool. Source: M. Greenlee and W. Heil, Wyman-Gordon Co. data, 1968; reported in *Aerospace Structural Metals Handbook*, Code 3714, Vol 4, Battelle Columbus Laboratories, 1972

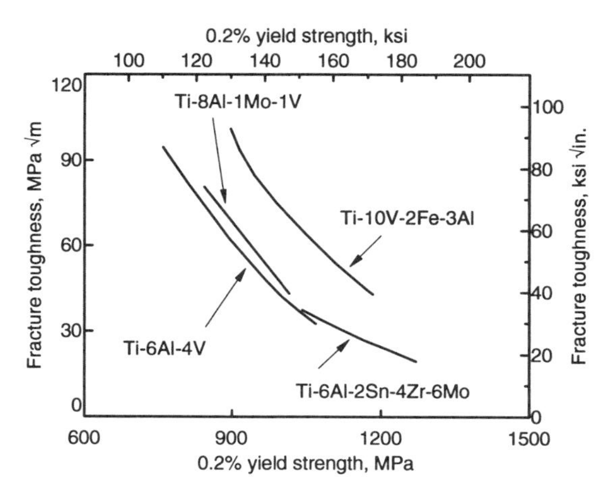

Ti-6246: Fracture toughness vs yield strength Fracture toughness is in part dependent on microstructure and is higher in the presence of an acicular structure. Source: "Titanium," Kobe Steel

Ti-6Al-4V

Ti64, 6Al-4V, 6-4
UNS Number: R56400 (normal interstitial grade); R56401 (extra-low interstitial grade); R56402 (filler metal)

Introduction

Ti-6Al-4V presently is the most widely used titanium alloy, accounting for more than 50% of all titanium tonnage in the world. To date, no other titanium alloy threatens its dominant position. The aerospace industry accounts for more than 80% of this usage. The next largest application of Ti-6Al-4V is medical prostheses, which accounts for 3% of the market. The automotive, marine, and chemical industries also use small amounts of Ti-6Al-4V (see the section "Applications" in this introduction).

Chemistry

Effects of Impurities and Alloying. Ti-6Al-4V is produced in a number of formulations. Depending on the application, the oxygen content may vary from 0.08 to more than 0.2% (by weight), the nitrogen content may be adjusted up to 0.05%, the aluminum content may reach 6.75%, and the vanadium content may reach 4.5%. The higher the content of these elements, particularly oxygen and nitrogen, the higher the strength. Conversely, lower additions of oxygen, nitrogen, and aluminum will improve the ductility, fracture toughness, stress-corrosion resistance, and resistance against crack growth.

ELI Grade. Ti-6Al-4V is available in ELI (extra-low interstitial) grades with high damage-tolerance properties, especially at cryogenic temperatures. The principal compositional characteristics are low oxygen and iron contents.

Ti-6Al-4V-Pd is a grade that has palladium additions (about 0.2 wt% Pd) for enhanced corrosion resistance. Sumitomo Titanium has produced this grade.

Product Forms

Ti-6Al-4V is available in wrought, cast, and powder metallurgy (P/M) forms, with wrought products accounting for more than 95% of the market. The properties of these various product forms will vary depending on their interstitial contents and thermal-mechanical processing. Processing methods and characteristics of Ti-6Al-4V are discussed in a separate section entitled "Processing."

Wrought Product Forms. Ti-6Al-4V is available in a wide range of wrought product forms (see Table).

The aircraft industry uses all wrought product forms. Forgings are used to fabricate various attachment fittings, and sheet and plate are used to fabricate numerous clips, brackets, skins, bulkheads, etc. Extrusions are not used extensively, but are used for parts such as wing chords and other parts with long, constant cross-sections. Wire is used to produce the numerous fasteners found on wings. Ti-6Al-4V tubing has been used for components such as torque tubes. In missile and space applications, Ti-6Al-4V has been used for rocket engine and motor cases, pressure vessels, wings, and generally in applications where weight is critical.

Castings. Ti-6Al-4V of the same chemistry as for wrought materials has excellent casting characteristics. However, the high reactivity of titanium in the molten state requires suitable casting technology and has limited the number of titanium foundries. In general terms, the mechanical and fatigue properties of castings will be slightly lower than for the wrought product, but fracture toughness, stress-corrosion resistance, and crack growth resistance will be comparable to that of annealed wrought Ti-6Al-4V.

Ti-6Al-4V castings are about two to three times the cost of superalloy castings. The cost effectiveness depends on the size, complexity, and the number of castings. Major application is in aerospace and marine use. Other industrial applications include well-logging hardware for the petroleum industry, special automotive parts, boat deck hardware, and medical implants.

P/M Products. The major reason for using the P/M products is to produce near-net shapes. Most of the titanium P/M effort has been with

Ti-6Al-4V: Wrought products

Product	Size and weight ranges	Price comparison(a)
Ingot	3200 to 13,600 kg (7000 to 30,000 lb)	...
Billet	Normally 100 mm (4 in.) diam to about 355 mm (14 in.) diam or square. Billets up to 5000 lb have been sold, but this is not necessarily the upper limit.	...
Bar	Cross-sections up to 0.4 × 0.4 m (16 × 16 in.)	...
Die forging	From <0.5 kg to >1300 kg (<1 lb to >3000 lb)	Ti, $30/lb; Al, $10/lb; stainless steel, $8/lb
Plate	Typical dimensions: Thickness: 5 to 75 mm (0.1875 to 3 in.); Width: 915 and 1220 mm (36 and 48 in.); Length: 1.8, 2.4, and 3 m (72, 96, and 120 in.)	...
Sheet	Typical dimensions: Thickness: 0.4 to 4.75 mm (0.016 to 0.187 in.); Width: 915 and 1220 mm (36 and 48 in.); Length: 1.8, 2.4, and 3 m (72, 96, and 120 in.)	Ti, $16/lb; stainless steel, $3/lb; Al, $2-4/lb; Inco 718, $10/lb
Tube	Specialty item	...
Forged block	Available in a wide range of sizes, with maximum size related to ingot size and the amount of work that can be imparted to the forged block	Ti, $8/lb; stainless steel and Al, $2.50-3/lb
Extrusion	From circle sizes of about 25 to 760 mm (1 to 30 in.) diam. Minimum thickness of about 3 mm (1/8 in.) for small circle sizes, and about 13 mm (1/2 in.) for large circle sizes	Ti, $13-15 lb; 300 series stainless steel, $3-4/lb; 15-5PH, $4-5/lb; 13-8PH, $9-12/lb; Al, $2-4/lb
Wire	Typically manufactured in sizes ranging from 0.28 to 12.2 mm (0.011 to 0.480 in.) diam	1/4 in. wire: Ti, $26/lb; A283, $6/lb; stainless steel, $7.50/lb; 8740, $1/lb; Al 7075, $2.30/lb

(a) Due to its lower density, 1 lb of titanium is approximately 1.7 to 1.8 more material by volume than 1 lb of steel or nickel-base alloy.

Ti-6Al-4V because it is the most widely used alloy having a large data base for comparison.

The two general approaches to titanium P/M are the blended elements (BE) method and the prealloyed (PA) approach. Blended elemental powders cost $6 to $30/lb, depending on the chloride impurity content. Chloride content ranges from 10 to 2000 ppm; powders with low amounts of chloride are more expensive. High-chloride powders cannot be used if good fatigue strength is needed.

Blended elemental P/M parts of Ti-6Al-4V are currently in production for aerospace and nonaerospace applications where full wrought properties are not required and where there is economic advantage to this approach. (See the section "Applications" for examples). The PA approach, however, has been less successful in establishing a commercial market. Prealloyed powders are not cold compactable, and their cost is high ($60 to $100/lb).

Product Condition/ Microstructure

Wrought Ti-6Al-4V is most commonly used in the mill-annealed condition, where it has a good combination of strength, toughness, ductility, and fatigue. Its minimum yield strength may vary from 760 to 895 MPa (110 to 130 ksi), depending on processing, heat treatment, section size, and chemistry (primarily oxygen).

Almost all titanium castings are hot isostatically pressed (HIP'ed) to heal internal porosity not linked to the surface. This minimizes the amount of weld repair, improves the consistency of mechanical properties, and enhances the fatigue performance. Ti-6Al-4V castings are generally used in the $(\alpha + \beta)$-annealed condition, although some special heat treatments can be used to enhance the performance of the castings in comparison to the β anneal.

Annealed Condition. Although Ti-6Al-4V is commonly used in the mill-annealed condition, other annealing treatments are also utilized. For example, annealing just above the beta transus, or annealing high in the $\alpha + \beta$ phase field, creates a Widmanstätten or lamellar $\alpha + \beta$ microstructure with good fracture toughness, stress-corrosion resistance, and crack growth resistance, and creep resistance. Recrystallization annealing of wrought alloy improves tensile ductility and fatigue performance.

Solution Treated, Quenched and Aged Ti-6Al-4V Alloy. Solution-treated and quenched alloys may either have an acicular α'-martensite structure (quenched from above β-transus) or mixed $\alpha' + \alpha$ microstructure (quenched from 900-1000 °C) or mixed $\alpha'' + \alpha$ microstructure (quenched from 800-900 °C), of which the latter is exceptionally soft and ductile. They serve as starting conditions for subsequent aging treatments. Quenched components contain high residual stresses which may not be fully relieved upon aging at low temperatures. Such components may distort during machining. Ti-6Al-4V has excellent hardenability in sections up to about 25 mm (1 in.) thick; strengths as high as 1140 MPa (165 ksi) may be achieved at aging temperatures between 300 and 600 °C.

Applications

Designed primarily for high strength at low to moderate temperatures, Ti-6Al-4V has a high specific strength (strength/density), stability at temperatures up to 400 °C (750 °F), and good corrosion resistance. Cost continues to be an inhibitive factor for its use in industries where weight and corrosion are not critical considerations.

Aerospace Applications. Ti-6Al-4V was developed in the 1950s and initially used for compressor blades in gas turbine engines. Today, wrought Ti-6Al-4V is used extensively for turbine engine and airframe applications. Engine components include blades, discs, and wheels. Wrought forms are used for airframe components. In addition, the superplastic characteristics of fine-grained, equiaxed Ti-6Al-4V is being used increasingly for aerospace applications. It also has good diffusion-bonding characteristics, which, when combined with superplastic forming, enables the fabrication of very complex structures. Significant amounts of superplastically formed and diffusion-bonded structures are used today, particularly for military aircraft.

Aerospace casting applications include the range from major structural components weighing more than 135 kg (300 lb) each to small switch guards weighing less than 30 g (1 oz).

Ti-6Al-4V castings are used extensively for large, complex housings in the turbine engine industry. They are used in a variety of airframe applications, including cargo-handling equipment, flow diverters, torque tubes for brakes, and helicopter rotor hubs. In missile and space applications, they are used for wings, missile bodies, optical sensor housings, and ordnance. Also, Ti-6Al-4V castings are used to attach the main external fuel tanks to the Space Shuttle and the boosters to the external tanks.

Surgical Implants. Wrought Ti-6Al-4V is a useful material for surgical implants because of its low modulus, good tensile and fatigue strength, and biological compatibility. It is used for bone screws and for partial and total hip, knee, elbow, jaw, finger, and shoulder replacement joints. Where fatigue properties are not an issue, the cast alloy also has had minor use as an implant product.

Automotive Applications. In the automotive industry, wrought Ti-6Al-4V is used in special applications in high-performance and racing cars where weight is critical, usually in reciprocating and rotating parts, such as valves, valve springs, connecting rods, and rocker arms. It also has been used for drive shafts and suspension springs. Cast Ti-6Al-4V also has had minor use in automotive applications.

Marine applications of wrought Ti-6Al-4V include armaments, sonar equipment, deep-submergence applications, hydrofoils, and capsules for telephone-cable repeater stations. Casting applications include water-jet inducers for hydrofoil propulsion and seawater ball valves for nuclear submarines.

P/M Applications. The BE method produces a product with less than full density that can be as strong as wrought material, but that generally has lower ductility, toughness, and fatigue strength. Process modifications can improve these latter properties, even making them comparable to wrought, but they increase costs.

The BE approach has found a niche for the production of near-net-shape components or of low-cost preforms for subsequent processing, such as forging. Applications include sidewinder missile housing, missile fins, connecting rods, turbine blade preforms, hex stock preforms for fittings, nuts, mirror hubs, and lens housings.

High cost has thus far limited potential applications of PA technology to, for the most part, the manufacture of critical aerospace components. A number of demonstration parts are now flying in the F-15 and the F-18 airplanes, but none is made on a production basis. The increased demand for titanium aluminides in higher-temperature applications is creating interest in PA technology of P/M titanium.

Ti-6Al-4V and equivalents: specifications and compositions

Specification	Designation	Description	Al	C	Fe	H	N	O	V	OT	Other
UNS	R56400	Weld Wir	5.5-6.75	0.1	0.4	0.015	0.05	0.2	3.5-4.5		bal Ti
UNS	R56401		6						4		bal Ti
UNS	R56402	Fill Met	5.5-6.75	0.04	0.15	0.005	0.012	0.1	3.5-4.5		bal Ti
Europe											
AECMA prEN2517	Ti-P63	Sh Strp Plt Bar Ann	5.5-6.75	0.08	0.3	0.01	0.05	0.2	3.5-4.5	0.4	bal Ti
AECMA prEN2530		Bar Ann	5.5-6.75	0.08 max	0.3 max	0.0125 max	0.05 max	0.2 max	3.5-4.5	0.4 max	OE 0.1 max; bal Ti
AECMA prEN2531		Frg Ann	5.5-6.75	0.08 max	0.3 max	0.0125 max	0.05 max	0.2 max	3.5-4.5	0.4 max	OE 0.1 max; bal Ti
AECMA prEN3310		Frg NHT	5.5-6.75	0.08 max	0.3 max	0.0125 max	0.05 max	0.2 max	3.5-4.5	0.4 max	OE 0.1 max; bal Ti
AECMA prEN3311		Bar Ann	5.5-6.75	0.08 max	0.3 max	0.0125 max	0.05 max	0.2 max	3.5-4.5	0.4 max	OE 0.1 max; bal Ti
AECMA prEN3312		Frg Ann	5.5-6.75	0.08 max	0.3 max	0.0125 max	0.05 max	0.2 max	3.5-4.5	0.4 max	OE 0.1 max; bal Ti
AECMA prEN3313		Frg NHT	5.5-6.75	0.08 max	0.3 max	0.0125 max	0.05 max	0.2 max	3.5-4.5	0.4 max	OE 0.1 max; bal Ti
AECMA prEN3314		Bar STA	5.5-6.75	0.08 max	0.3 max	0.0125 max	0.05 max	0.2 max	3.5-4.5	0.4 max	OE 0.1 max; bal Ti
AECMA prEN3315		Frg STA	5.5-6.75	0.08 max	0.3 max	0.0125 max	0.05 max	0.2 max	3.5-4.5	0.4 max	OE 0.1 max; bal Ti
AECMA prEN3352		Inv Cast Ann HIP	5.5-6.75	0.1 max	0.3 max	0.015 max	0.05 max	0.22 max	3.5-4.5	0.4 max	OE 0.1 max; bal Ti
AECMA prEN3353		Bar Wir STA	5.5-6.75	0.08 max	0.3 max	0.0125 max	0.05 max	0.2 max	3.5-4.5	0.4 max	OE 0.1 max; bal Ti
AECMA prEN3354		Sh Ann	5.5-6.75	0.08 max	0.3 max	0.0125 max	0.05 max	0.2 max	3.5-4.5	0.4 max	OE 0.1 max; bal Ti
AECMA prEN3355		Ext Ann	5.5-6.75	0.08 max	0.3 max	0.0125 max	0.05 max	0.2 max	3.5-4.5	0.4 max	OE 0.1 max; bal Ti
AECMA prEN3456		Sh Strp Ann	5.5-6.75	0.08 max	0.3 max	0.0125 max	0.05 max	0.2 max	3.5-4.5	0.4 max	OE 0.1 max; bal Ti
AECMA prEN3457		Frg NHT	5.5-6.75	0.08 max	0.3 max	0.0125 max	0.05 max	0.2 max	3.5-4.5	0.4 max	OE 0.1 max; bal Ti
AECMA prEN3458		Bar Wir Ann	5.5-6.75	0.08 max	0.3 max	0.0125 max	0.05 max	0.2 max	3.5-4.5	0.4 max	OE 0.1 max; bal Ti
AECMA prEN3464		Plt Ann	5.5-6.75	0.08 max	0.3 max	0.0125 max	0.05 max	0.2 max	3.5-4.5	0.4 max	OE 0.1 max; bal Ti
AECMA prEN3467		Remelt NHT	5.5-6.75	0.08 max	0.3 max	0.0125 max	0.05 max	0.2 max	3.5-4.5	0.4 max	OE 0.1 max; bal Ti
France											
AIR 9183	T-A6V	Bar Rod Frg	5.5-7	0.08	0.25	0.012	0.07	0.2	3.5-4.5		bal Ti
AIR 9184	T-A6V	Blt	5.5-7	0.08 max	0.25 max	0.12 max	0.07 max	0.2 max	3.5-4.5		bal Ti
Germany											
DIN	3.7164	Sh Strp Plt Bar Frg Ann	5.5-6.75	0.08	0.3	0.0125-0.015	0.05	0.2	3.5-4.5	0.4	bal Ti
DIN	3.7264	Cast Ann	5.5-6.75	0.1	0.3	0.015	0.05	0.2	3.5-4.5	0.4	bal Ti
DIN 17850	3.7165	Plt Sh Strp Rod Wir Ann	5.5-6.75	0.08	0.3	0.015	0.05	0.2	3.5-4.5		bal Ti
DIN 17851	3.7165	Sh Plt Strp Rod Wir Ann	5.5-6.75	0.08	0.3	0.015	0.05	0.2	3.5-4.5		bal Ti
DIN 17860	3.7615	Sh Strp	5.5-6.75	0.2 max	0.3 max	0.015 max	0.05 max		3.5-4.5		bal Ti
DIN 17862	3.7615	Rod	5.5-6.75	0.08 max	0.3 max	0.015 max	0.05 max	0.2 max	3.5-4.5		bal Ti
DIN 17864	3.7615	Frg	5.5-6.75	0.08 max	0.3 max	0.015 max	0.05 max	0.2 max	3.5-4.5		bal Ti
Russia											
GOST 19807-74	VT6S	Sh Plt Strp Foil Rod Ann	5.3-6.8	0.08	0.25	0.007	0.05	0.015	3.5-4.5	0.3	Zr 0.3; Si 0.15; bal Ti
OST 1.90000-70	VT6	Sh Plt Strp Foil Rod Frg Ann	5.5-7	0.1	0.3	0.015	0.05	0.2	4.2-6	0.3	Si 0.15; bal Ti
OST 1.90060-72	VT6L	Cast	5-6.5	0.1	0.3	0.015	0.05	0.15	3.5-4.5	0.3	Zr 0.3; Si 0.15; W 0.2; bal Ti
Spain											
UNE 38-723	L-7301	Sh Plt Strp Bar Ex Ann	5.5-6.75	0.1	0.3	0.125	0.05	0.2	3.5-4.5	0.4	bal Ti
UNE 38-723	L-7301	Sh Plt Strp Bar Ex HT	5.5-6.75	0.1	0.3	0.125	0.05	0.2	3.5-4.5	0.4	bal Ti
UK											
BS 2TA.10		Sh Strp HT	5.5-6.75		0.3	0.01					V. 3.5-4.5; Ti 88.18 max; O+N=0.25
BS 2TA.11		Bar	5.5-6.75		0.3	0.01	0.05	0.2			V. 3.5-4.5; Ti 88.18 max;
BS 2TA.12		Frg	5.5-6.75		0.3	0.01	0.05	0.2	3.5-4.5		Ti 88.19 max;
BS 2TA.13		Frg HT	5.5-6.75		0.3	0.01		0.2	3.5-4.5		Ti 88.18 max;
BS 2TA.28		Wir Frg HT Quen	5.5-6.75		0.3	0.01	0.05	0.2	3-5		Ti 88.19 max;
BS 3531 Part 2		Srg Imp	5.5-6.75	0.08 max	0.3 max	0.015 max		0.2 max	3.5-4.5		bal Ti
BS TA.56		Plt to 100 mm HT	5.5-6.75		0.3				3.5-4.5		Ti 88.2 max; O+N=0.25

(continued)

Ti-6Al-4V and equivalents: specifications and compositions (continued)

Specification	Designation	Description	Al	C	Fe	H	N	O	V	OT	Other
UK (continued)											
BS TA.59		Sh Strp	5.5-6.75	0.08 max	0.3 max	0.0125 max			3.5-4.5		N+O=0.25; bal Ti
DTD 5303		Bar Ann	5.5-6.75	0.2 max	0.3 max	0.0125 max	0.05 max		3.5-4.5		bal Ti
DTD 5313		Frg Ann	5.5-6.75		0.3 max	0.01 max	0.05 max	0.2 max	3.5-4.5		bal Ti
DTD 5323		Frg Ann	5.5-6.75		0.3 max	0.015 max	0.05 max	0.2 max	3.5-4.5		bal Ti
DTD 5363		Cast	5.5-6.75		0.3 max	0.15 max	0.05 max	0.25 max	3.5-4.5		N+O=0.27; bal Ti
USA											
AMS 4905A		ELI Plt	5.6-6.3	0.05 max	0.25 max	0.0125 max	0.03 max	0.12 max	3.6-4.4	0.4 max	Y 0.005 max; OE 0.1 max; bal Fe
AMS 4905A		Plt Beta Ann	5.6-6.3	0.05	0.25	0.0125	0.03	0.12	3.6-4.4	0.4	Y 0.005; bal Ti
AMS 4906		Sh Strp	5.5-6.75	0.08 max	0.3 max	0.0125 max	0.05 max	0.2 max	3.5-4.5	0.4 max	Y 0.005 max; bal Ti
AMS 4907D		ELI Sh Strp Plt Ann	5.5-6.5	0.08	0.25	0.0125	0.05	0.13	3.5-4.5	0.3	Y 0.005; bal Ti
AMS 4911F		Sh Strp Plt Ann	5.5-6.75	0.08	0.3	0.015	0.05	0.2	3.5-4.5	0.4	bal Ti
AMS 4920		Frg Ann	5.5-6.75	0.1	0.3	0.0125	0.05	0.2	3.5-4.5	0.4	Y 0.005; bal Ti
AMS 4928K		Bar Wir Frg Bil Rng Ann	5.5-6.75	0.1	0.3	0.0125	0.05	0.2	3.5-4.5	0.4	bal Ti
AMS 4930C		ELI Bar Wir Frg Bil Rng Ann	5.5-6.5	0.08	0.25	0.0125	0.05	0.13	3.5-4.5	0.4	Y 0.005; bal Ti
AMS 4931		ELI Bar Frg Bil Rng	5.5-6.5	0.08	0.25	0.0125	0.03	0.13	3.5-4.5	0.4	Y 0.005; bal Ti
AMS 4934A		Ex Rng STA	5.5-6.75	0.1	0.3	0.0125	0.05	0.2	3.5-4.5	0.4	Y 0.005; bal Ti
AMS 4935E		Ex Rng Ann	5.5-6.75	0.1	0.3	0.0125	0.05	0.2	3.5-4.5	0.4	Y 0.005; bal Ti
AMS 4954D		Fill met gas-met/W-arc weld	5.5-6.75	0.05	0.3	0.015	0.03	0.18	3.5-4.5	0.4	Y 0.005; bal Ti
AMS 4956B		ELI Fill Met Wir	5.5-6.75	0.03	0.15	0.005	0.012	0.08	3.5-4.5	0.1	Y 0.005; bal Ti
AMS 4965E		Bar Frg Rng STA/Mach Press ves	5.5-6.75	0.08	0.3	0.0125	0.05	0.2	3.5-4.5	0.4	Y 0.005; bal Ti
AMS 4967F		Bar Frg Rng Mach/STA Press ves	5.5-6.75	0.08	0.3	0.0125	0.05	0.2	3.5-4.5	0.4	Y 0.005; bal Ti
AMS 4985A		Cast Ann	5.5-6.75	0.1	0.3	0.015	0.05	0.2	3.5-4.5	0.4	Y 0.005; bal Ti
AMS 4991A		Cast Ann	5.5-6.75	0.1	0.3	0.015	0.05	0.2	3.5-4.5	0.4	Y 0.005; bal Ti
AMS 4993A		Powd Sint Nuts	5.5-6.75	0.1	0.3	0.01	0.05	0.3	3.5-4.5	0.4	Si 0.05; Na 0.15; Cl 0.15; bal Ti
AMS 4996		Bill Powd Ann	5.5-6.75	0.1	0.3	0.0125	0.04	0.13-0.19	3.5-4.5	0.2	Mo 0.1 max; Sn 0.1 max; Zr 0.1 max; Mn 0.1 max; Cu 0.1 max; Y 0.001; bal Ti
AMS 4996		ELI Bil	5.5-6.75	0.1 max	0.3 max	0.0125 max	0.04 max	0.13-0.19	3.5-4.5	0.2 max	Y 0.001 max; OE 0.1 max; bal Ti
AMS 4998		ELI Powd	5.5-6.75	0.1 max	0.3 max	0.0125 max	0.04 max	0.13-0.19		0.2 max	Y 0.001 max; OE 0.1 max; bal Ti
AMS 4998		Powd	5.5-6.75	0.1	0.3	0.012	0.04	0.13-0.18	3.5-4.5	0.2	Mo 0.1 max; Sn 0.1 max; Zr 0.1 max; Mn 0.1 max; Cu 0.1 max; Y 0.001; bal Ti
ASTM B 265	Grade 5	Sh Strp Plt Ann	5.5-6.75	0.1	0.4	0.015	0.05	0.2	3.5-4.5	0.4	bal Ti
ASTM B 348	Grade 5	Bar Bil Ann	5.5-6.75	0.1	0.4	0.0125	0.05	0.2	3.5-4.5	0.4	bal Ti
ASTM B 367	Grade C-5	Cast	5.5-6.75	0.1	0.4	0.015	0.05	0.25	3.5-4.5	0.4	bal Ti
ASTM B 381	Grade F-5	Frg Ann	5.5-6.75	0.1	0.4	0.0125	0.05	0.2	3.5-4.5	0.4	bal Ti
ASTM F136		ELI Wrought Ann for Surg Imp	5.5-6.5	0.08	0.25	0.012	0.05	0.13	3.5-4.5		bal Ti
ASTM F467-84	Grade 5	Blt Scr Std	5.5-6.75	0.1 max	0.4 max	0.0125 max	0.05 max	0.2 max	3.5-4.5		bal Ti
ASTM F468-84		Blt Scr Std	5.5-6.75	0.1 max	0.4 max	0.0125 max	0.05 max	0.2 max	3.5-4.5		bal Ti
AWS A5.16-70	ERTi-6Al-4V	Weld fill met	5.5-6.75	0.05	0.25	0.008	0.02	0.15	3.5-4.5		bal Ti
AWS A5.16-70	ERTi-6Al-4V-1	ELI Fill Met Wir Rod	5.5-6.75	0.04	0.15	0.005	0.012	0.1	3.5-4.5		bal Ti
MIL A-46077D		Weld armor plt Ann	5.5-6.5	0.04	0.25	0.0125	0.02	0.14	3.5-4.5	0.4	bal Ti
MIL F-83142A	Comp 6	Frg Ann	5.5-6.75	0.08	0.3	0.015	0.05	0.2	3.5-4.5	0.4	bal Ti
MIL F-83142A	Comp 6	Frg HT	5.5-6.75	0.08	0.3	0.015	0.05	0.2	3.5-4.5	0.4	bal Ti
MIL F-83142A	Comp 7	ELI Frg Ann	5.5-6.5	0.08	0.2-0.25	0.0125	0.05	0.13	3.5-4.5	0.3	bal Ti
MIL F-83142A	Comp 7	ELI Frg HT	5.5-6.5	0.08	0.25	0.0125	0.05	0.13	3.5-4.5	0.3	bal Ti
MIL T-81556A	Code AB-1	Ex Bar Shp Ann	5.5-6.75	0.08	0.3	0.0125	0.05	0.2	3.5-4.5	0.4	bal Ti
MIL T-81556A	Code AB-1	EX Bar Shp STA	5.5-6.75	0.08	0.3	0.0125	0.05	0.2	3.5-4.5	0.4	bal Ti
MIL T-81556A	Code AB-2	ELI Ext Bar Ann	5.5-6.5	0.08	0.25	0.0125	0.05	0.13	3.5-4.5	0.3	bal Ti
MIL T-81915	Type III Comp A	Cast Ann	5.5-6.75	0.08	0.3	0.015	0.05	0.2	3.5-4.5	0.4	bal Ti
MIL T-9046J	Code AB-1	Sh Strp Plt Ann	5.5-6.75	0.08	0.3	0.0125	0.05	0.2	3.5-4.5	0.4	bal Ti
MIL T-9046J	Code AB-1	Sh Strp Plt STA	5.5-6.75	0.08	0.3	0.0125	0.05	0.2	3.5-4.5	0.4	bal Ti
MIL T-9046J	Code AB-2	ELI Sh Strp Plt Ann	5.5-6.5	0.08	0.25	0.0125	0.05	0.13	3.5-4.5		bal Ti
MIL T-9047G		Bar Bil STA	5.5-6.75	0.08	0.3	0.015	0.05	0.2	3.5-4.5	0.4	Y 0.005; bal Ti
MIL T-9047G		ELI Bar Bil Ann	5.5-6.5	0.08	0.25	0.0125	0.05	0.13	3.5-4.5	0.3	Y 0.005; bal Ti
MIL T-9047G	MIL-T-9047G	Bar Bil Ann	5.5-6.75	0.08	0.3	0.015	0.05	0.2	3.5-4.5	0.4	Y 0.005; bal Ti
SAE J467		ELI	6.18	0.023	0.22	0.008	0.026	0.097			bal Ti

Ti-6Al-4V commercial equivalents: compositions

Specification	Designation	Description	Al	C	Fe	H	N	O	V	OT	Other
France											
Ugine	UTA6V	Sh Strp Plt Bar Frg Ann	5.5-6.75	0.08	0.3	0.015	0.07	0.2	3.5-4.5		bal Ti
Ugine	UTA6V	Sh Strp Plt Bar Frg STA	5.5-6.75	0.08	0.3	0.015	0.07	0.2	3.5-4.5		bal Ti
Germany											
DeutscheT.	LT 31	Ann	5.5-6.5	0.08	0.25	0.013	0.07	0.2	3.5-4.5		bal Ti
Fuchs	TL64	Frg	6						4		bal Ti
Fuchs	TL64 ELI	ELI Frg	6						4		bal Ti
Thyssen	Contimet AlV 64	Plt Bar Frg Ann	5.5-6.75	0.1	0.3	0.015	0.05	0.2	3.5-4.5		bal Ti
Thyssen	Contimet AlV 64	Plt Bar Frg STA	5.5-6.75	0.1	0.3	0.015	0.05	0.2	3.5-4.5		bal Ti
Thyssen	Contimet AlV 64 ELI	ELI Plt Bar Frg Pip Ann	5.5-6.75	0.06	0.15	0.013	0.05	0.13	3.5-4.5		bal Ti
Japan											
Daido	DAT 5	Rod Bar Rng Frg Ann	5.5-6.75	0.1	0.3	0.015		0.05	3.5-4.5		bal Ti
Daido	DAT 5	Rod Bar Rng Frg STA	5.5-6.75	0.1	0.3	0.015	0.05	0.2	3.5-4.5		bal Ti
Daido	DT 5	Rod Bar Frg Rng STA	5.5-6.75	0.1	0.3	0.015	0.05	0.2	3.5-4.5		bal Ti
Kobe	KS6-4	Plt Sh Wir Bar Ann	5.5-6.75		0.3	0.0125	0.05	0.2	3.5-4.5		bal Ti
Kobe	KS6-4ELI	ELI Plt Sh Ann	5.5-6.5		0.25	0.0125	0.05	0.13	3.5-4.5		bal Ti
Toho	64AT		5.5-6.75	0.03	0.4	0.0125	0.05	0.2	3-5		bal Ti
Toho	64AT	STA	5.5-6.75	0.03	0.4	0.0125	0.05	0.2	3-5		bal Ti
UK											
IMI	IMI 318	Sh Rod Bar Bil Wir Plt Ex	6						4		bal Ti
USA											
Oremet	Ti-6Al-4V										
RMI	6Al-4V-ELI	ELI Bar Bil Ex Plt Sh Strp Ann	5.5-6.5	0.08	0.25	0.01-0.015		0.13	3.5-4.5		bal Ti
RMI	6Al-4V	Bar Bil Ex Plt Sh Strp Wir Ann	5.6-6.75	0.08	0.25	0.01-0.015	0.05	0.2	3.5-4.5		bal Ti
RMI	6Al-4V	Bar Bil Ex Plt Sh Strp Wir STA	5.6-6.75	0.08	0.25	0.01-0.015	0.05	0.2	3.5-4.5		bal Ti
Tel.Allvac	Allvac 6-4		6	0.18					4		bal Ti
TIMET	TIMETAL 6-4	Ann	5.5-6.75	0.1 max	0.4 max	0.015 max	0.05 max	0.2 max	3.5-4.5		bal Ti
TIMET	TIMETAL 6-4 ELI	ELI Ann	5.5-6.5	0.08 max	0.25 max	0.0125 max	0.05 max	0.13 max	3.5-4.5		bal Ti
TIMET	TIMETAL 6-4 STA	Ing Bil Bar Plt Sh Str STA	5.5-6.75	0.1	0.4	0.015	0.05	0.2	3.5-4.5		bal Ti

Phases and Structures

As an alpha beta alloy, Ti-6Al-4V may have different volume fractions of alpha and beta phases, depending on heat treatment and interstitial (primarily oxygen) content. Beta is stable at room temperature only if it is enriched with more than 15 wt.% vanadium. Such enrichment is obtained when the alloy is slow cooled or annealed below about 750 °C (1400 °F). Slow cooled Ti-6Al-4V contains up to about 90 vol% of the alpha phase.

In addition, Ti-6Al-4V can acquire a large variety of microstructures with different geometrical arrangements of the alpha and beta phases, depending on the particular thermomechanical treatment. These different alpha "morphologies" and microstructures can be roughly classified into three different categories: lamellar, equiaxed, or a mixture of both (bimodal).

Lamellar structures can be readily controlled by heat treatment. Slow cooling into the two-phase region from above the β transus leads to nucleation and growth of the α-phase in plate form starting from β-grain boundaries. The resulting lamellar structure is fairly coarse and is often referred to as plate-like alpha. Air cooling results in a fine needle-like alpha phase referred to as acicular alpha. Certain intermediate cooling rates develop Widmanstätten structures. Water-quenching from the β-phase field followed by annealing in the (α + β)-phase region leads to a much finer lamellar structure. Quenching from temperatures greater than 900 °C (1650 °F) results in a needle-like hcp martensite (α′), while quenching from the 750 to 900 °C (1380 to 1650 °F) temperature range produces an orthorhombic martensite (α″).

Equiaxed microstructures are obtained by extensive (>75% reduction) mechanical working the material in the (α + β)-phase field, where the breakup of lamellar alpha into equiaxed alpha depends on the exact deformation procedure (*e.g.*, see figure). Subsequent annealing at about 700 °C (1300 °F) produces the so-called "mill-annealed" microstructure, which gives microstructure that is very dependent upon previous working. A more reproducible equiaxed structure is obtained by a recrystallization anneal of 4 h at 925 °C (1700 °F) followed by slow cooling. The resulting structure is fairly coarse with an α-grain size of about 15-20 μm.

Bimodal type microstructures consist of isolated primary α-grains in a transformed beta matrix. These microstructures are best obtained by a 1 h anneal at 955 °C (1750 °F) followed by water quenching (or more commonly an air cool) and aging at 600 °C (1100 °F). The resulting primary α-grain size is usually about 15-20 μm in such "solution treated and aged" microstructures. Aging below 650 °C (1200 °F) can also produce precipitates of alpha in previously quenched beta.

Interface Phase. An FCC crystal structure, often called the "interface phase," is frequently observed in thin foils for electron microscopy at lamellar boundaries between alpha phase and beta phase. The interpretation of the interface phase is still controversial. It has been reported to be a phenomenon of thin foil preparation while others claim that it occurs in bulk material as well.

Typical lattice parameters of alpha phase in slow cooled or aged Ti-6Al-4V alloy are $a = 0.2925 \pm 0.0002$ nm, $c = 0.4670 \pm 0.0005$ nm. The lattice parameters vary only slightly as a function of heat treatment because the composition of alpha is relatively constant. The room-temperature lattice parameter of the beta phase in furnace cooled Ti-6Al-4V has been measured as $a = 0.319$ nm $\pm$ 0.001 nm (G. Welsch *et al.*, *Met. Trans. A*, Vol 8A, 1977, p 169-177). Increasing vanadium concentrations decrease the lattice parameter of beta, while interstitial elements increase the lattice parameters of alpha by occupying a fraction of the octahedral interstitial sites. For oxygen concentrations less than 6 at.%, oxygen increases the lattice parameters of alpha as follows (S. Anderson *et al.*, *Acta Chem. Scand.*, Vol 11, 1957, p 1641):

$$a = a_o + 7 \times 10^{-5} \text{ nm/at.\% O}$$

$$c = c_o + 36 \times 10^{-5} \text{ nm/at.\% O}$$

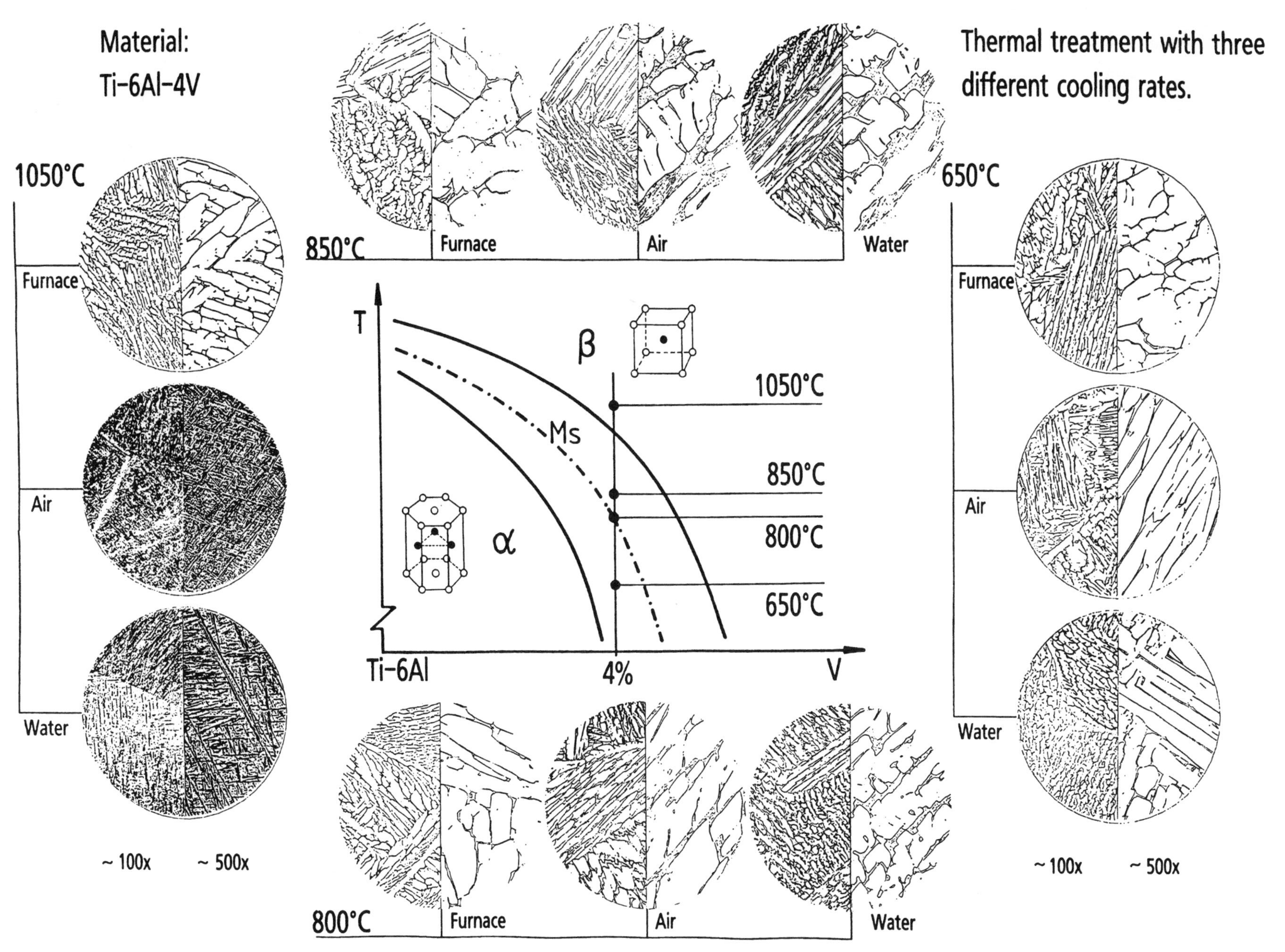

Illustration of quenching rate effect on microstructures of alloy Ti-6Al-4V

Ti-6Al-4V: Lattice parameters after quenching from various temperatures

Quench temperature (°C)	Phase α a nm	Phase α c nm	Phase α c/a	Phase β a nm
950	0.29313	0.46813	0.15969	...
900	0.29320	0.46798	0.15962	...
850	0.29288	0.46750	0.15963	...
800	0.29281	0.46729	0.15960	...
750	0.29259	0.46711	0.15966	...
725	0.29261	0.46706	0.15962	0.32530
700	0.29255	0.46709	0.15966	0.32510
650	0.29243	0.46706	0.15972	0.32295
600	0.29254	0.46711	0.15969	0.22250
550	0.29254	0.46716	0.15970	0.32160
500	0.29245	0.46718	0.15974	0.32145
450	0.29246	0.46705	0.15970	0.32150
400	0.29240	0.46718	0.15977	0.32120

Source: R. Castro and L. Seraphin, *Memoires Scientifique Rev. Metallurg.*, Vol LXIII, No.12, 1966, p 1036

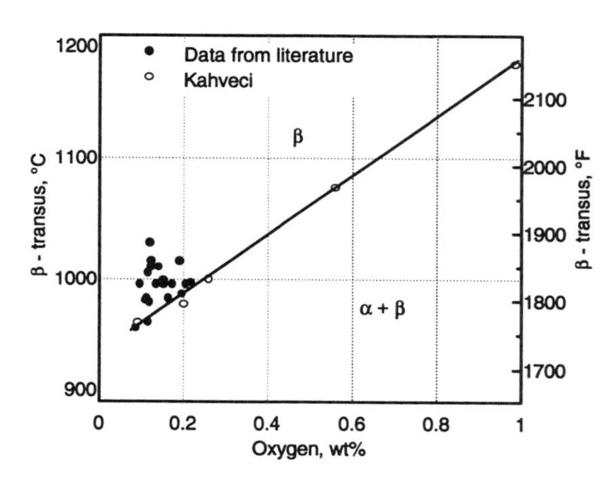

Ti-6Al-4V: Beta transus vs oxygen content Source: A. Kahveci, Thesis, Case Western Reserve University

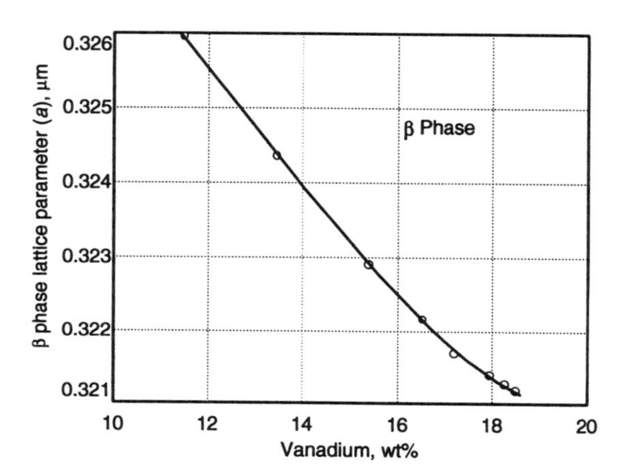

Ti-6Al-4V: Lattice parameter of β phase Variation in lattice parameter beta phase in Ti-6Al-4V alloy after quenching from various heat treatment temperatures. Source: R. Castro and L. Seraphin, *Memoires Scientifique Rev. Metall.*, Vol 63, No. 12, 1966, p 1036

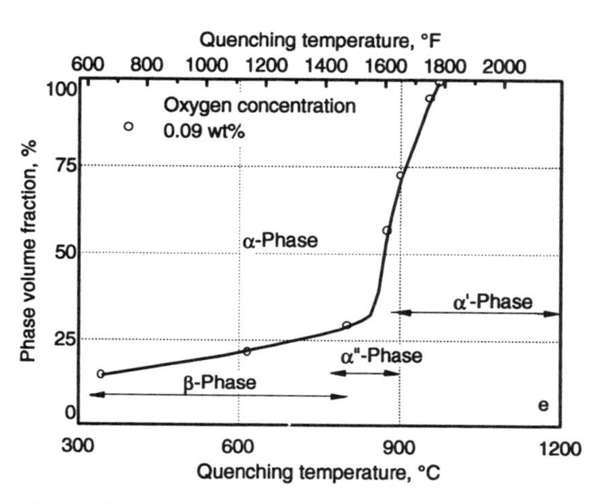

Ti-6Al-4V: Fraction of phase constituents after quenching

Transformation Structures

Hexagonal close packed martensite (α′) is obtained by quenching from above 900 °C (1650 °F) and has an acicular or sometimes fine-lamellar microstructure. It is related crystallographically to the alpha phase and has similar lattice parameters as the alpha phase.

Orthorhombic martensite (α″) is a rather soft martensite that forms during quenching of beta phase with 10 ± 2 wt% vanadium. This occurs when Ti-6Al-4V is quenched from temperatures between 750 and 900 °C (1380 and 1650 °F). The α″ martensite can also form as a stress-induced product by straining metastable beta.

Omega (ω) Precipitation. Oxygen suppresses omega formation, and it does not occur in Ti-6Al-4V alloy of commercial purity. If the β phase is highly enriched with vanadium (over 15 wt%), ω-precipitates might occur during low-temperature aging (200 to 350 °C) or during cooling through the same temperature range. However, no such precipitation has been reported in Ti-6Al-4V.

Ti_3Al Precipitation

The formation of Ti_3Al (α_2) has been experimentally verified in Ti-6Al-4V containing less than 0.2 wt% oxygen (*Met. Trans.*, Vol 8A, 1977,

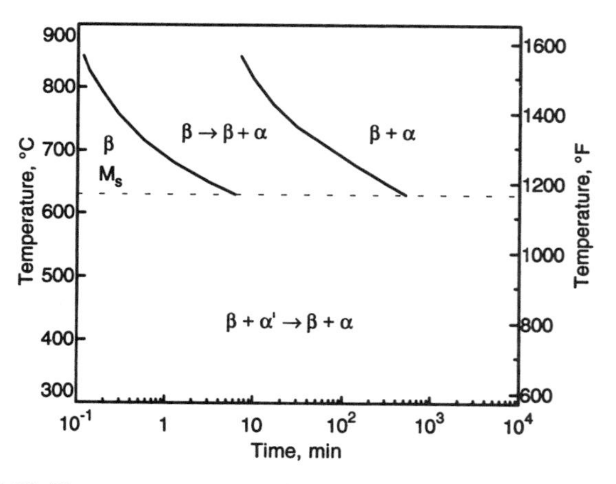

Ti-6Al-4V: Time-temperature-transformation diagram Solution annealed at 1020 °C (1868 °F), and quenched directly to reaction temperatures. Source: L.E. Tanner, "Time-Temperature-Transformation Diagrams of the Titanium Sheet-Rolling-Program Alloys," DMIC Report 46G, Battelle Memorial Institute, Columbus, OH, October 1959

p 169-177), and it occurs in Ti-6Al-4V at aging temperatures from 500 to 600 °C (930 to 1100 °F) when oxygen concentrations (still within specification limits for Ti-6Al-4V) are increased. Oxygen is known to restrict the solubility limit of aluminum in the alpha phase of titanium, thus enhancing the likelihood of Ti_3Al formation. Vanadium also restricts aluminum solubility in alpha titanium. However, no quantitative dependency on oxygen concentration has been established for the $\alpha/(\alpha + \alpha_2)$ solvus line.

The Ti_3Al precipitates are known to promote coarse planar glide on $\{10\bar{1}0\}$ prismatic planes, and one should be able to produce a predominance of either planar pyramidal glide or coarse planar prismatic glide in the α-phase of a Ti-6Al-4V alloy with sufficient oxygen by the choice of aging treatment. At high aging temperatures, ordering should not be expected because oxygen would have a high jump frequency. (G. Welsch and W. Bunk, *Met. Trans. A,* Vol 13A, 1982, p 889-899). On the other hand, an alloy with a very low oxygen concentration (ELI grade) should exhibit only a predominance of prismatic slip, regardless of aging treatment.

General Fatigue Behavior

When evaluating fatigue behavior, comparisons should account for differences in yield strength, grain size, and microstructure. There is evidence regarding Ti-6Al-4V which indicates that superior high-cycle (10^7 cycles) smooth-bar fatigue is obtained when the slip length is small (Ref 1-5). Small slip lengths accompany a fine-grain equiaxed material or by quenching from the β-phase field to produce fine, acicular α′. There is general agreement that the Widmanstätten or colony α + β microstructure has decidedly poorer fatigue strength. In the coarser, equiaxed microstructure the fatigue strength is significantly lower, but it is still better than in the colony microstructure (Ref 6). In general, all microstructural parameters that increase yield strength and/or reduce slip length should improve HCF strength. However, variations in texture, test method (axial versus bending), test conditions (load ratio, frequency), and surface-preparation methods may make comparisons difficult.

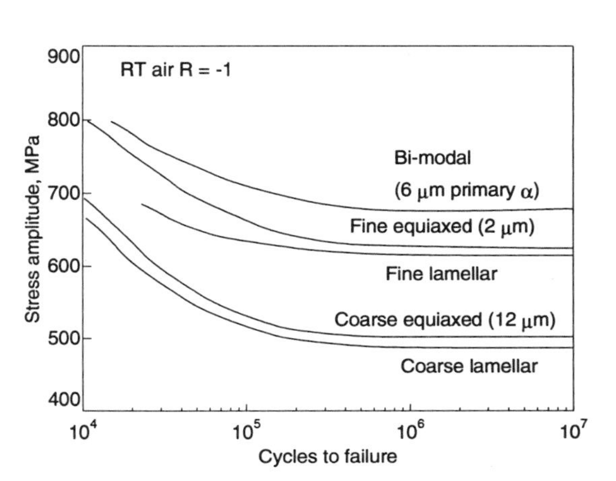

Aged Ti-6Al-4V: HCF strength in air for structures with a basal texture. The bimodal microstructure exhibited the highest HCF strength in air, because the basal planes of the strongly textured α grains are separated from each other. The aggressive effect of laboratory air (about 50% relative humidity) is thought to be due to hydrogen, which is most damaging along the basal plane. Material/Test Parameters: Solution annealed at 800 °C (1470 °F) for 1 h, water quenched, and aged at 500 °C (930 °F) for 24 h. Source: G. Lütjering and A. Gysler, *Titanium Science and Technology*, Vol 4, Deutsche Gesellschaft für Metallkunde e.V., 1985, p 2068

References

1. M. Peters, A. Gysler, and G. Lütjering, in *Titanium '80, Science and Technology*, edited by H. Kimura and O. Izumi, Vol 1, TMS-AIME, 1981, p 1777.
2. J.J. Lucas, in *Titanium Science and Technology*, edited by R.I. Jaffee and H.M. Burte, Plenum Press, 1973, p 2081.

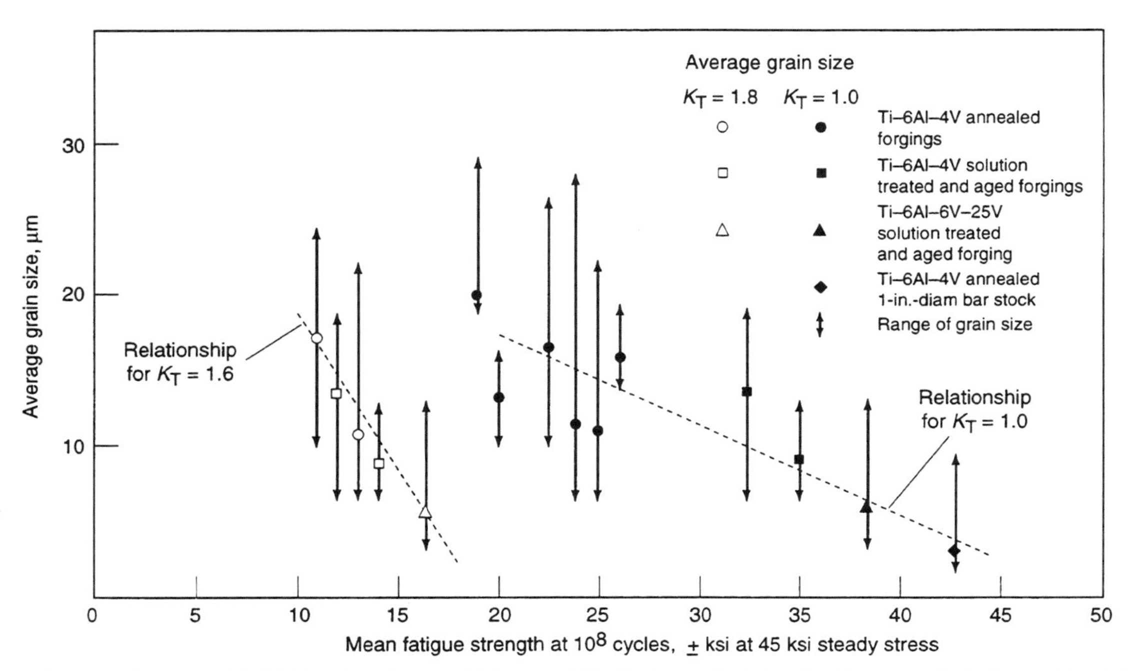

Ti-6Al-4V: Effect of α grain size on crack-initiation stress. Source: J.J. Lucas and P.P. Konieczny, Relationships Between α Grain Size and Crack-Initiation Fatigue Strength of Ti-6Al-4V, *Metall. Trans.*, Vol 2, 1971, p 911-912

3. C.A. Stubbington, 1976 AGARD Conferences Proceedings (No. 185), Reprinted in *Titanium and Titanium Alloys Source Book*, American Society for Metals, 1982, p 140-158.
4. C.A. Stubbington and A.W. Bowen, *J. Mater. Sci.*, Vol 9, 1974, p 941.
5. J.C. Williams and G. Lütjering, in *Titanium '80, Science and Technology*, edited by H. Kimura and O. Izumi, Vol 1, TMS-AIME, p 671.
6. J.C. Chesnutt, A.W. Thompson and J.C. Williams, AFML-TR-78-68, 1978.

Low-Cycle Fatigue

Low-cycle fatigue (LCF) is the regime characterized by a high maximum stress in a cyclic loading situation. It is also characterized by the existence of significant plastic deformation during the fatigue cycle, at least on a localized scale. Low cycle fatigue failure often occurs with a cyclic life of less than about 10^4 to 10^6 cycles.

The LCF life of Ti-6Al-4V is quite sensitive to heat treatment and microstructural details. The tables illustrate the wide range of fatigue properties obtainable with this alloy. As expected, the variations in fatigue properties are accompanied by variations in other properties, making compromises and optimization for specific uses feasible with alloys like Ti-6Al-4V. Crack initiation, known to consume a large portion of total LCF life, is reported (Gilmore and Imam, see table) to be very sensitive to microstructure, with an $\alpha + \beta$ anneal providing the most resistance. However, an STOA treatment of an $\alpha + \beta$ forging of a beta quenched billet offers promise but with hints of some short life tests (Chakrabarti *et al.*, see table).

Ti-6Al-4V: Fatigue crack initiation vs heat treatment

Heat treatment	Strain range(a)	Cycles to crack initiation(b)		
		Mean	Minimum	Standard deviation
α-β anneal: 800 °C (1472 °F) for 3 h, furnace cool (FC) to 600 °C (1112 °F), followed by air cool (AC) to room temperature	±0.006 torsional strain	84,370	73,690	9,875
	±0.012 torsional strain	9,697	9,110	386
Recrystallization anneal: 928 °C (1702 °F) for 4 h, FC to 760 °C (1400 °F) at 180 °C/h, FC to 482 °C (900 °F) at 372 °C (702 °F)/h, AC	±0.006 torsional strain	52,840	40,920	10,637
	±0.012 torsional strain	6,232	4,560	1,025
β anneal	±0.006 torsional strain	42,720	35,240	5,684
	±0.012 torsional strain	963	705	184

(a) Sinusoidal strain amplitude with a frequency of 28.6 Hz for ±0.006 and 0.2 Hz for ±0.012 strain. (b) Based on four specimens for each heat treatment with fatigue life taken as the cycles to initiation of a circumferential crack as indicated by an increase in axial elongation. Source: C.M. Gilmore and M.A. Imam, *Titanium and Titanium Alloys*, J.C. Williams and A.F. Belov, Ed., Plenum Press, 1982, p 637

LCF and fracture toughness of Ti-6Al-4V pancake forgings

Prior stock treatment	Forging condition (soaking condition)	Heat treatments	Number of cycles to failure(a)	Yield strength		Ultimate tensile strength		Fracture toughness	
				MPa	ksi	MPa	ksi	MPa√m	ksi√in.
As received	T_β − 100 °C for 0.5 h/Press/OQ	965 °C for 0.5 h/OQ + 705 °C for 2 h/AC	22,325(b)	1038	150	1071	155	40.5	37
			18,808(b)	1038	150	1071	155	40.5	37
As received	T_β − 100 °C for 0.5 h/Press/OQ	801 °C for 1 h/OQ + 500 °C for 24 h/AC	13,934(b)	1113	161	1126	163.3	33.3	30.3
			16,769(b)	1113	161	1126	163.3	33.3	30.3
Beta quenched	T_β − 165 °C for 0.5 h/Press/OQ	975 °C for 0.5 h/FAC + 801 °C for 1 h/OQ + 500 °C for 24 h/AC	32,581(c)	1087	157	1124	163	52.8	48.1
			17,960(d)	1087	157	1124	163	52.8	48.1

(a) Tested with a stress ratio of R = 0.05, a frequency of 20 cycles/min, and a maximum stress of 880 MPa (127.7 ksi). (b) Failed in gauge. (c) Run-out. (d) Failed at interface of radius and uniform section. *Material parameters:* Bar stock of 7.6 cm diameter had the composition in weight % of Al:6.1, C:0.04, Fe:0.23, H:0.0061, N:0.036, O:0.187, V:4.1, Y:<0.0050. *Reference:* "Microstructure and Mechanical Property Optimization Through Thermomechanical Processing in Ti-6-4 and Ti-6-2-4-6 Alloys," A. Chakrabarti, M. Burn, D. Fournier, and G. Kuhlman, in *Sixth World Conference on Titanium*, 1989, p 1339

Strain Life

Ti-6Al-4V extruded rod: LCF from shear strain
With high shear strains, unstable microstructures, such as those provided by quenching from the solution treatment, provide greater LCF life when compared to annealed or stable microstructures. However, such unstable microstructures are not recommended for cyclic loading applications.

Thermal treatment	Mean life(a)	Minimum life	Standard deviation
α-β Annealed	944	429	443
843 °C (1550 °F) + WQ	2497	1837	717
900 °C (1650 °F) + WQ	9616	8917	758
927 °C (1700 °F) + WQ	2223	2142	605
β solution – 1065 °C (1950 °F) + WQ	2396	1633	487

(a) Results based on four specimens (6.4 mm diam extruded rod) for a shear strain of ±0.02 at 0.2 Hz. The α-β anneal involved holding at 800 °C (1470 °F) for 3 h followed by furnace cooling to 600 °C (1110 °F). Source: M.A. Imam and C.M. Gilmore, Fatigue and Microstructural Properties of Quenched Ti-6Al-4V, *Metall. Trans. A*, Vol 14A, 1983, p 233-240

Stress-Controlled LCF

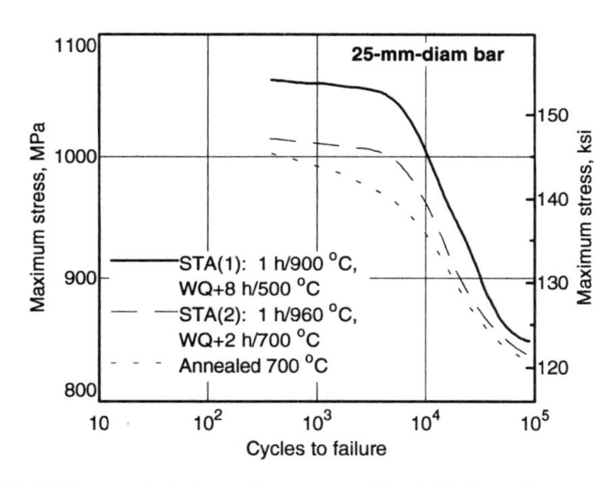

Ti-6Al-4V: Low-cycle fatigue of heat-treated bar. This figure illustrates the effect of tensile strength on stress controlled LCF of Ti-6Al-4V where the high strength STA condition gives the greatest life. Although strain controlled test results are not presented, the conclusions may be different with strong correlation expected with true ductility. Material/Test Parameters: See table for material condition. Testing was performed at constant maximum load, zero minimum, frequency = 10 cycles/min. Source: Data from J.R.B. Gilbert, IMI Titanium Ltd., 1988

Ti-6Al-4V: Bar tensile properties for LCF figure

Heat treatment	0.2% proof stress ksi	Tensile strength ksi	Elongation on 5D, %
900 °C WQ + 500 °C (STA 1)	146	163	15
960 °C WQ + 700 °C (STA 2)	141	156	14
Annealed 700 °C	141	149	14

Cast and P/M

Castings and powder metallurgy (P/M) products are being applied with increasing frequency, primarily for their net shape manufacturing capabilities. There is an LCF penalty for castings and P/M products (see figures), however, these debits can be reduced significantly by hot isostatic pressing (HIP) and heat treatment processing as well as improved PM processing. A compilation from 260 °C LCF tests (see figure) gives results for polished, notched, cast test bars of Ti-6Al-4V with no HIP consolidation.

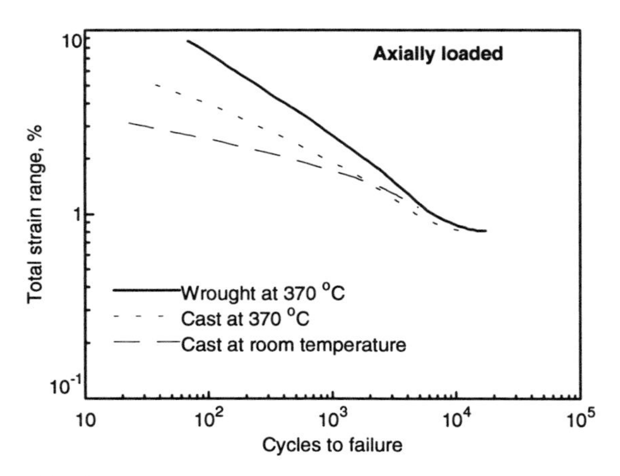

Ti-6Al-4V: LCF properties of cast and wrought STA alloy. Specimens tested at room temperature were subjected to 20 cycles/min, while the elevated-temperature tests used a testing frequency of 5 cycles/min. Material/Test Parameters: Cast specimens taken from thick sections. Source: Data from U. Hellmann and T. Tsareff, General Motors Corp., Detroit Diesel Allison Division, 1971, reported in *Titanium Alloys Handbook*, R. Wood and R. Favor, MCIC-HB-02, Battelle, reprinted 1985, p 2:72-35

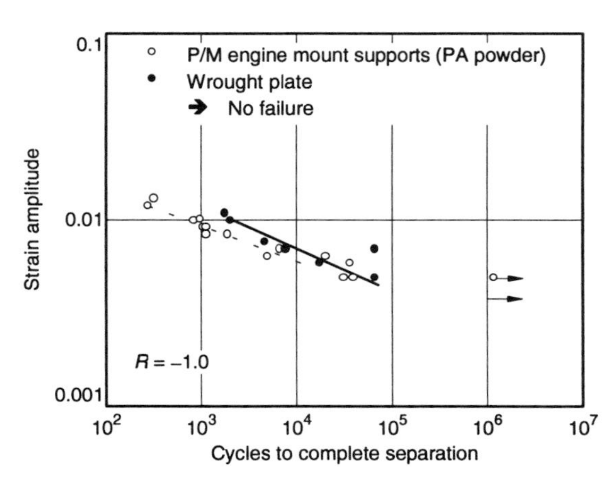

Ti-6Al-4V: LCF of PREP-HIP P/M components. Source: A.S. Sheinker *et al.*, Evaluation and Application of Prealloyed Titanium P/M Parts for Airframe Structures, *Int. J. Powder*, Vol 23 (No. 3), 1987, p 171-176

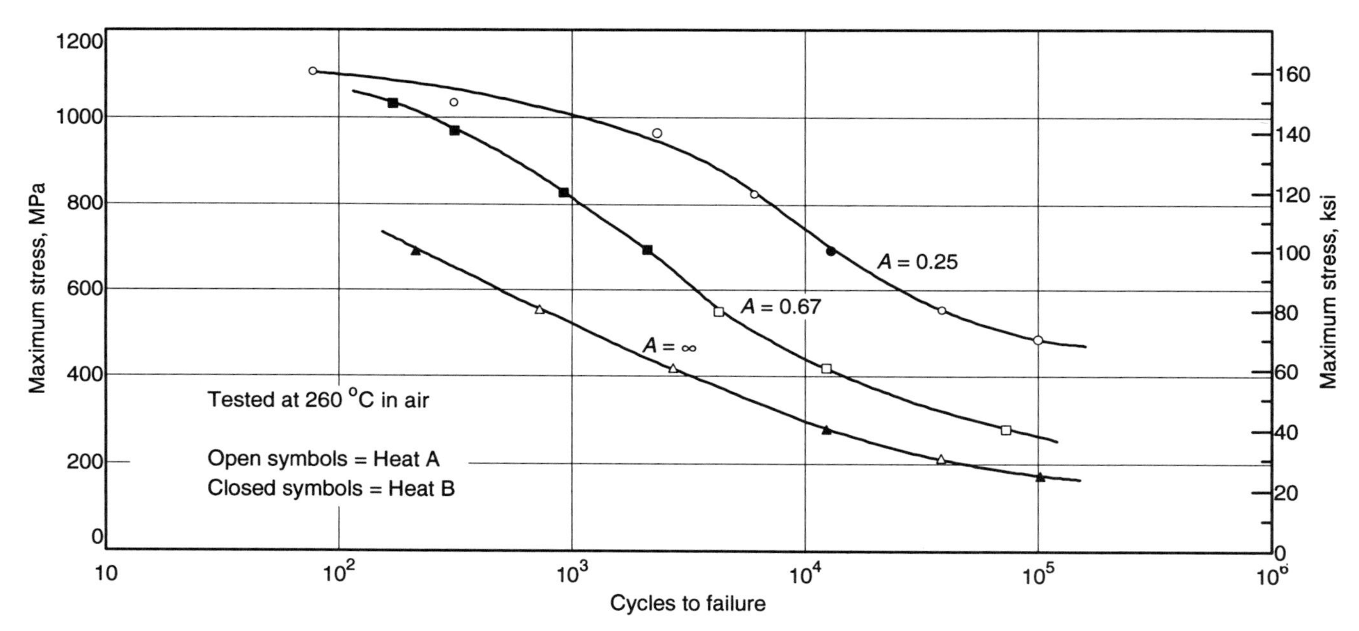

Ti-6Al-4V: Low-cycle axial fatigue for notched (K_t = 3.5) annealed castings (without HIP). Material/Test Parameters: Cast-to-size specimens were annealed at 700 °C (1300 °F) for 2 h, air cooled, then polished with wire and diamond paste, and finish machined. Test frequency was 300 cycles/min. Source: Data from R. Dalal, AVCO Corp., reported in *Aerospace Structural Metals Handbook*, Battelle, Code 3801, p 18

Fatigue Limits and Endurance Ratios

Fatigue limits (or endurance limits) represent the value of stress below which a material can presumably endure an infinite number of cycles. For many variable-amplitude loading conditions, fatigue limits may be observed in the regime of 10^7 cycles or more. Fatigue limits generally are influenced by surface conditions because 80 to 90% of fatigue life in the high-cycle regime (about 10^4 cycles or more) involves nucleation of fatigue cracks at the surface. Fatigue limits are also influenced by microstructure (see figures for fatigue strength in air and vacuum). In some cases, stress-controlled fatigue limits are not observed. These cases are generally attributed to periodic overstrains and the absence of interstitial hardening as with very low oxygen levels (see figure on effect of yield strength). Also, the absence of a fatigue limit in this alloy is often associated with subsurface initiations, especially at cryogenic temperatures (see figure).

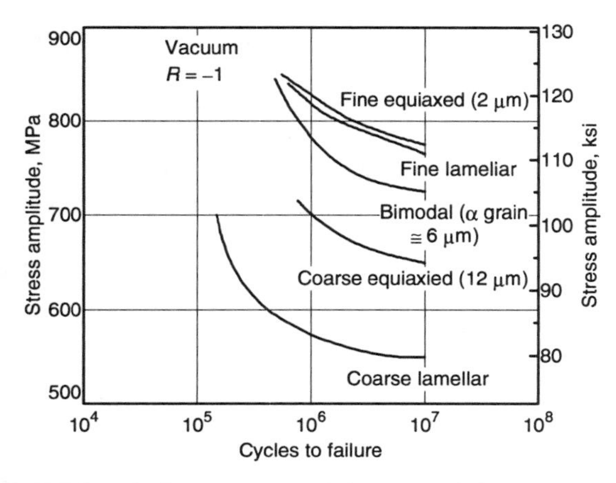

Ti-6Al-4V: Effect of microstructure on fatigue strength in vacuum. Grain size of primary α shown. The prior β grain size, which limits the length of individual lamellae of bimodal and lamellar microstructures, was 6 to 10 μm for the bimodal microstructure and 300-600 μm for the fine lamellar structure. Material/Test Parameters: Annealed at 800 °C (1470 °F) for 1 h, water quenched, and aged at 500 °C (930 °F) for 24 h. Source: G. Lütjering and A. Gysler, *Titanium Science and Technology*, Vol 4, Deutsche Gesellschaft für Metallkunde e.V., 1985, p 2068

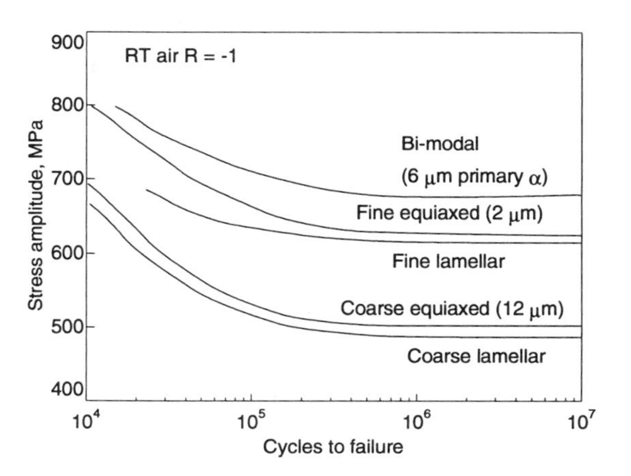

Ti-6Al-4V: Effect of microstructure on fatigue strength in air. Grain size of primary alpha shown. Annealed at 800 °C (1470 °F) for 1 h, WQ, aged at 500 °C for 24 h. Source: G. Lütjering and A. Gysler, *Titanium Science and Technology*, Vol 4, DGM, 1985, p 2068

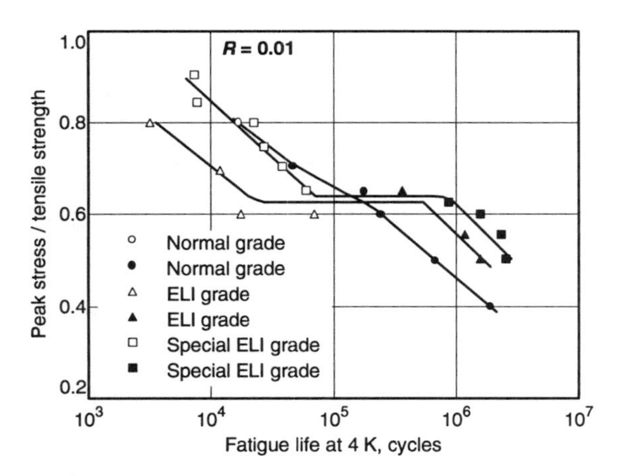

Ti-6Al-4V ELI: Fatigue strength at cryogenic temperatures. Open symbols indicate fracture from surface; closed symbols indicate internal fatigue initiation near surface. Material/Test Parameters: Testing was performed at 4.2 K in a liquid-helium-cooled servohydraulic testing apparatus with a sinusoidal cyclic load. Specimens were taken from α-β forged bars (70-mm, 2.75-in., square) that were annealed at 700 °C (1290 °F), for 2 h. Source: Y. Ito *et al.*, Cryogenic Properties of Extra-Low Oxygen Ti-6Al-4V Alloy, in *6th World Conf. Titanium*, 1989, p 87

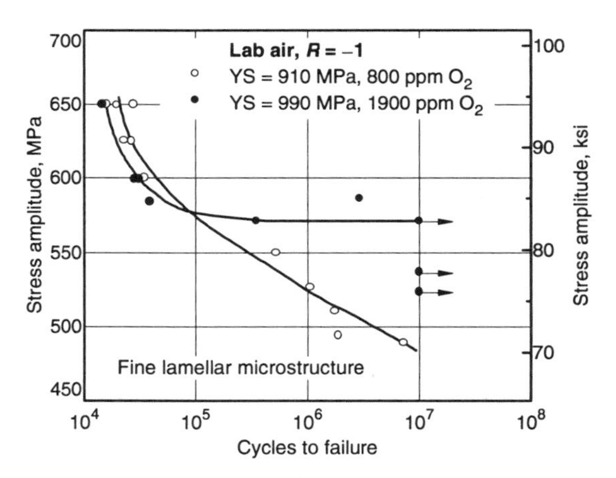

Ti-6Al-4V: Effect of yield strength (YS) on fatigue strength. Decreasing the oxygen from the typical value of 0.19 wt% to 0.08 wt% lowers the yield strength and thus the fatigue strength. Oxygen levels influence the mechanisms of precipitation and hardening, which improves resistance to dislocation motion (and thus increases yield and fatigue strength). Material/Test Parameters: Water quenched from 800 °C (1470 °F) and aged at 500 °C (930 °F) for 4 h. Source: E.A. Starke, Jr. and G. Lütjering, Cyclic Plastic Deformation and Microstructure, in *Fatigue and Microstructure*, American Society for Metals, 1979, p 237

Endurance Ratio

Fatigue Strength and Fatigue Limits. Fatigue limits may be related to tensile strength, although the fatigue limit-to-tensile strength ratio of titanium alloys may reveal more scatter than quenched and tempered low-alloy steels (see figure, *Metals Handbook*). For alloy Ti-6Al-4V, extensive tensile and smooth-bar fatigue data are presented for different alloy conditions by Sparks and Long and summarized by Williams and Starke (see table). Using their data, regression analysis has been performed to see if a correlation exists between 10^7-cycle fatigue strength and yield or tensile strength. In both cases the coefficient of correlation was smaller than 0.1, indicating that essentially no correlation exists. This tends to point out an important difference between Ti alloys and steels—namely, that the effects of microstructure and strength can be offsetting factors so that no change in fatigue performance might be observed even when strength is increased. Thus, fatigue strength may not correlate with tensile strength alone.

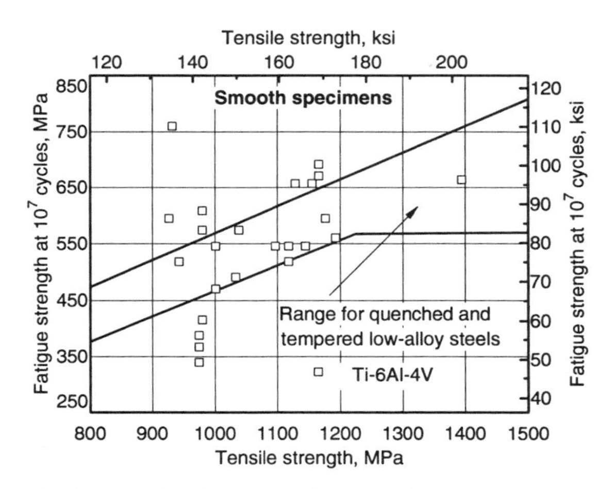

Ti-6Al-4V: Scatter of fatigue strength vs. tensile strength. Source: *Metals Handbook*, Vol 1, 8th ed., *Properties and Selection of Metals*, American Society for Metals, 1961, p 529

Ti-6Al-4V: Fatigue and tensile data for various microstructural conditions

Condition	Yield strength		Tensile strength		Elongation,	Reduction in	Stress (smooth) at 10^7 cycles		Stress (notched) at 10^7 cycles	
	MPa	ksi	MPa	ksi	%	area, %	MPa	ksi	MPa	ksi
10% equiaxed primary α + ann(a)	971	141	1068	155	14	35	537	78	214	31
40% equiaxed primary α + ann	930	135	1013	147	15	41	579	84	255	37
10% equiaxed primary α + STOA(b)	978	142	1061	154	15	41	489	71	220	32
10% equiaxed primary α + ann(c)	958	138	1040	151	14	37	606	88	262	38
50% elongated primary α + ann(c)	923	134	1020	148	13	32	620	90	227	33
β forge + ann	882	128	992	144	11	20	565	82	220	32
β forge/water quench + ann	951	138	1054	153	10	21	606	88	186	27
β forge + STOA	978	142	1075	156	10	20	586	85	220	32
10% equiaxed primary α + ann(d)	882	128	985	143	13	33	620	90	214	31

(a) ann = 705 °C/2 h/AC. (b) STOA = 955 °C/1 h/WQ + 705 °C/2 h/AC. (c) Water quenched off forging press. (d) Low-oxygen material. Source: R.B. Sparks and J.R. Long, AFML-TR-73-301, February 1974. Data summary reported by J.C. Williams and E.A. Starke, in Deformation, Processing, *and Structure*, American Society for Metals, 1984, p 326

Variation of Endurance Ratio

The spread in the endurance ratio (fatigue limit/ultimate tensile strength) is also documented in ASTM STP 459 (see figure). Endurance ratios varied from 0.42 to 0.62 for the unnotched condition. Several of the data points which make up the low side of the band represent slow cooling rates for the beta phase field. The data point for the 1350 °F treatment on the low side of the band was from material containing coarse plate-like alpha, which is further evidence that the coarse plate-like alpha structures lower the endurance ratio of the Ti-6Al-4V alloy. Therefore, heavy sections of Ti-6Al-4V, which contain coarse plate-like or even coarse equiaxed alpha, might be expected to have endurance ratios of 0.4 to 0.45. Fine grained alpha-beta structures or structures produced by water quenching or quenching and aging can be expected to have higher endurance ratios (e.g., between 0.55 and 0.62). However, this trend does not appear to hold true for the notched condition as evidenced by the scatter present in the lower band. The notched endurance ratios varied between 0.17 and 0.3.

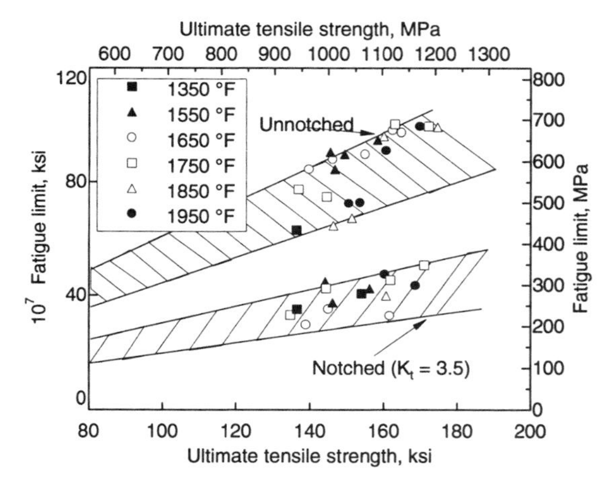

Ti-6Al-4V: Variation in RT endurance ratio. 6.35 mm (0.25 in.) specimens cut from as-rolled bar, solution treated at indicated temperatures, and cooled at various rates (furnace, air, water quench). Rotating-beam fatigue at 8000 rpm. Fatigue limits at 10^7 cycles determined by highest stress amplitude at which three specimens ran 10^7 cycles without failure. Source: L.J. Bartlo, Effect of Microstructure on Fatigue of Ti-6Al-4V Bar, ASTM STP 459, 1969

Surface and Texture Effects on Fatigue

Effect of Residual Stress

Surface residual stress is a predominant factor influencing fatigue, and the residual stress effect is most pronounced in the infinite-life stress range of the endurance-limit regime (10^7 cycles or more). Residual stress is even a more potent indicator than surface roughness in this regime, although residual stress and surface roughness are closely related in many instances. This accounts for the traditional correlation between fatigue strength and surface roughness within reasonable ranges of 2.5 to 5 μm (100 to 200 μin., arithmetic average). However, the correlation between fatigue limits and surface roughness may not be as strong as residual stress.

Surface residual stress is important, but the effect is not simple because of the combined influences of residual stress, cold worked structure and the surface roughness on HCF strength. Surface effects on components of HCF are as follows:

Surface effect	Crack nucleation	Crack propagation
Surface roughness	Accelerates	No effect
Cold work	Retards	Accelerates
Residual compressive stress	Minor or no effect	Retards

Surface roughness and cold work are often associated with the process of inducing residual stress. Because HCF is often largely dominated by fatigue crack nucleation and small crack growth to an observable size, applications of general rules of thumb regarding residual stresses must be exercised with care.

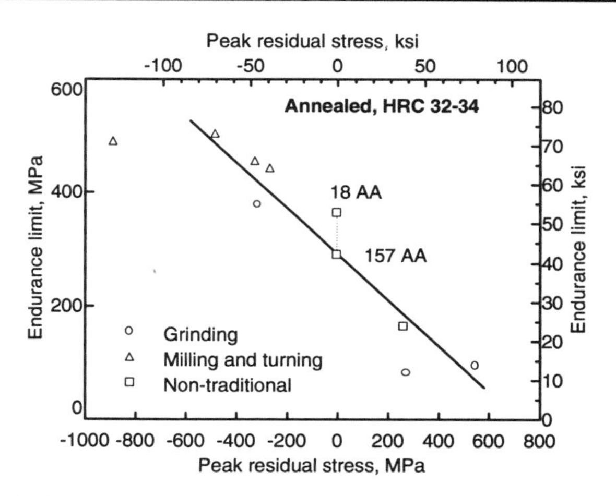

Ti-6Al-4V: Correlation between endurance limit and peak residual stress. Source: W.P. Koster, *Effect of Residual Stress on Fatigue of Structural Alloys, Practical Applications of Residual Stress Technology* (Clayton Rudd, Ed.), ASM International, 1991

Effect of Texture

There are two major texture considerations. There is a microstructural texture in which directly observable microstructural features are aligned on a scale large compared to the size of the individual features. Elongated beta grains in the rolling or forging direction is a typical example of microstructure texture.

There is also a crystallographic texture in which most of the alpha phase grains are aligned such that a unique direction in the hexagonal unit cell of alpha titanium is oriented close to the same direction relative to some physical attribute of the titanium sample, such as the rolling direction for sheet or plate. An example of the effect of crystallographic texture is shown (see figure from Zarkades and Larson), where fatigue life in the various test directions in a plate product are large.

Crystallographic texture is seldom reduced by sub-beta transus heat treatments, and the texture of the α phase in equiaxed and bi-modal microstructures can be a factor on fatigue limits. In a vacuum environment (see figure), the application of stress in the transverse direction resulted in higher HCF strength values as compared to tests in rolling direction. This can be directly correlated to the higher yield stress value observed when the stress is applied perpendicular to the basal planes. The opposite ranking was observed when the tests were performed in laboratory air (see figure), which is explained by hydrogen damage on basal planes. If no shear or normal stresses are acting on the basal planes (tests in rolling direction), then the highest fatigue strength values are observed.

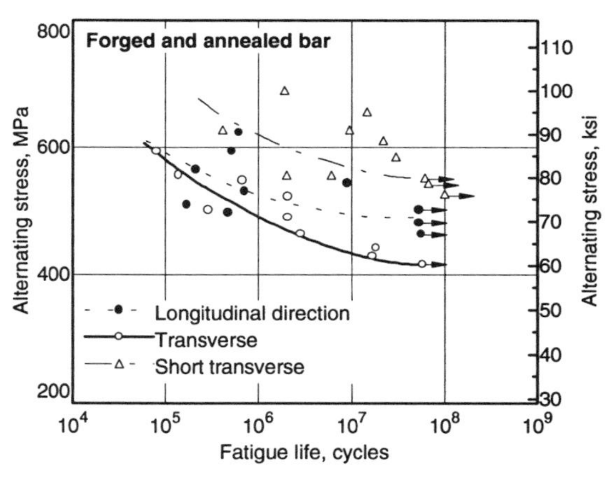

Ti-6Al-4V: Effect of test orientation on fatigue. Although much scatter is evident, a large difference in endurance limit is shown for different test orientations of forged bar (57-mm, or 2.25-in., thick) having a mild and varying texture. Source: A. Bower, The Effect of Testing Direction on the Fatigue and Tensile Properties of a Ti-6Al-4V Bar, *2nd Int. Conf. Titanium*, reported in *Properties of Textured Titanium Alloys*, F. Larson and A. Zarkades, Ed., MCIC-74-20, Battelle Columbus Laboratories, 1974, p 67

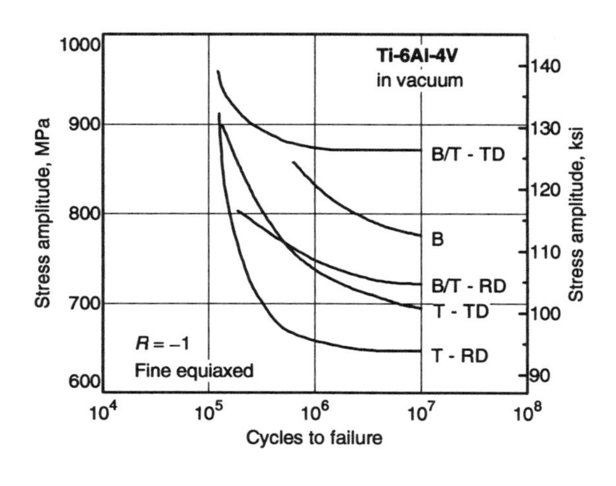

(a)

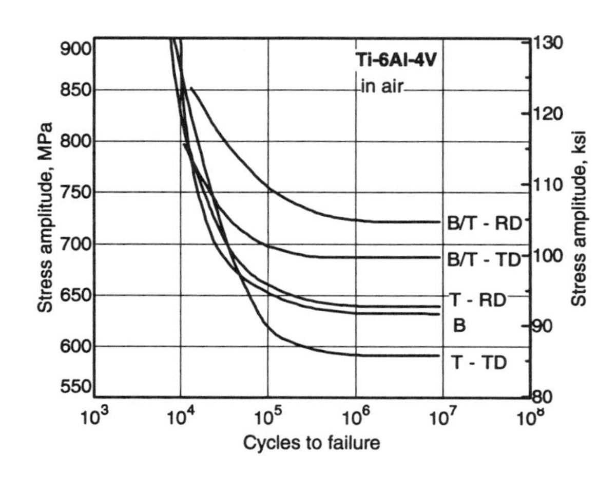

(b)

Ti-6Al-4V: Influence of texture and test direction. B, basal; T, transverse texture; B/T, basal-transverse texture; RD, test in rolling direction; TD, transverse test direction. Material/Test Parameters: Fine equiaxed Ti-6Al-4V, R = –1. Source: A.W. Bowen, *Titanium 80, Science and Technology*, AIME, 1980, p 947

Effect of Surface Treatment

Shot peening and surface finish interact in a complex way to influence HCF. Starting with an undisturbed electropolished surface of a fine lamellar microstructure, shot peening will enhance room-temperature fatigue strength whereas additional electropolishing adds to this enhancement. But, stress relief of the shot peened specimens reduces fatigue strength to levels below the baseline electropolished samples. Adding an electropolish to the stress relieved samples restores most of the benefit. Tests at 500 °C (see two-part figure, Gray *et al.*) indicate that the effect of shot peening is negative without restoration of the original electropolished surface. Similar effects are seen with fine, equiaxed, microstructures (see figure, Wagner *et al*).

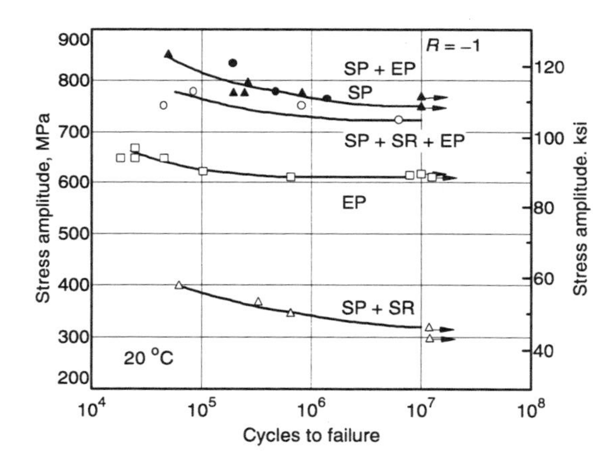

(a)

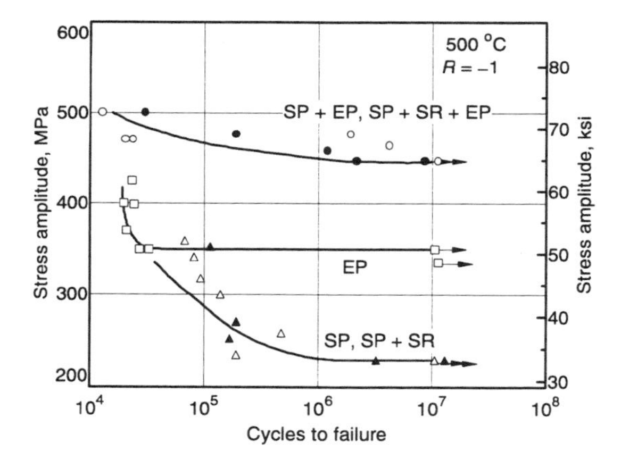

(b)

Ti-6Al-4V: Effect of shot peening and electrolytic polishing. At (a) 20 °C (68 °F) and (b) 500 °C (930 °F). EP, electrolytically polished; SP, shot peened; SR, stress relieved. Material/Test Parameters: Alloy with a fine lamellar (β-quenched) microstructure was machined into blanks 7 × 7 mm (0.28 × 0.28 in.), annealed at 800 °C (1470 °F) for 1 h and quenched, then heat treated at 600 °C (1110 °F) for 24 h. Specimens were electrolytically polished to remove a layer about 100 μm thick. Shot peening was performed using an Almen intensity of I = 0.28 A (mm). Fatigue tests were performed on smooth hourglass-shaped specimens in rotating-beam loading (*R*

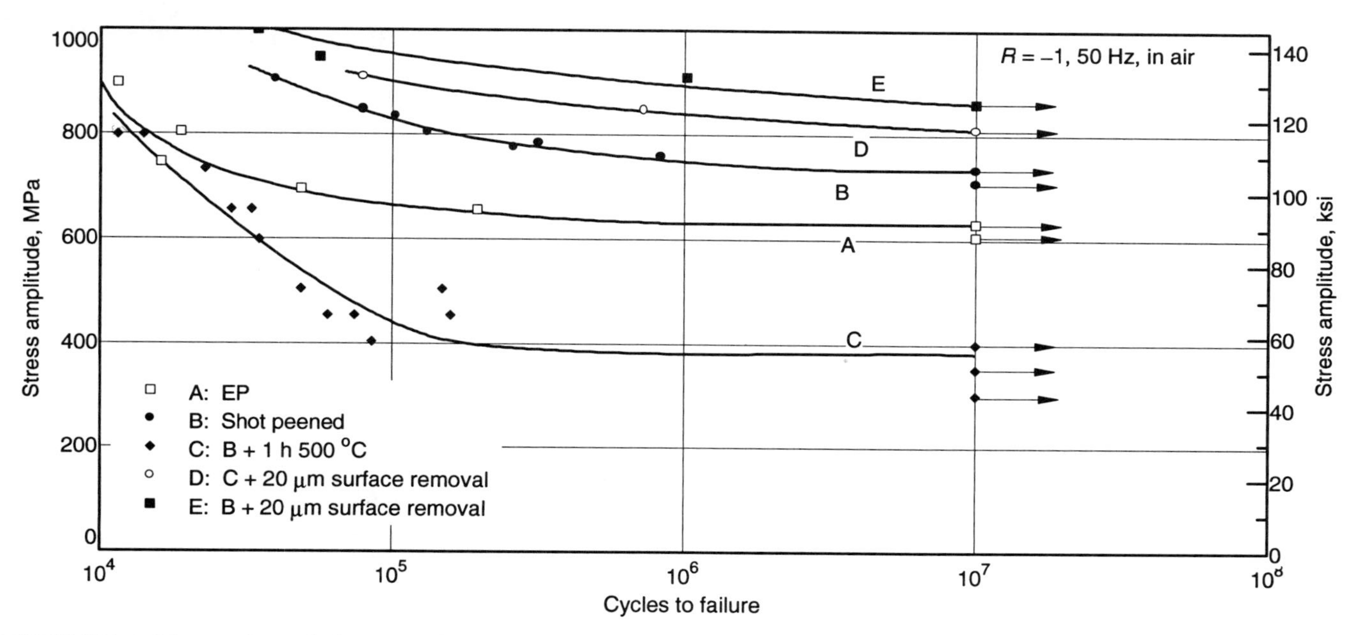

Ti-6Al-4V: Effect of shot peening on fatigue strength. EP, electrolytically polished. Material/Test Parameters: Test material had a fine equiaxed microstructure and a yield strength of 1100 MPa (160 ksi). Shot peening was performed with S230 steel balls (0.6-mm diam) and an Almen intensity of 15A mm/100. Fatigue testing was done on hourglass-shaped specimens with 3.8-mm (0.15-in.) gage diameter in air on a rotating-beam apparatus ($R = -1$, f = 50 Hz). Source: L. Wagner *et al.*, Influence of Surface Treatment on Fatigue Strength of Ti-6Al-4V," in *Titanium Science and Technology*, Deutsche Gesellschaft für Metallkunde e.V., 1985, p 2147

Fretting Fatigue

Titanium has notoriously poor wear resistance when there is sliding contact with itself and other materials. The resultant fretting has a strong effect on fatigue strengths and limits.

Ti-6Al-4V: Fretting fatigue in shot-peened and coated conditions

Fatigue specimen surface treatment	Normal pressure MPa	Fretting fatigue strength 3 × 10⁶ cycles, 20 °C, MPa	20 °C, ksi	400 °C, MPa	400 °C, ksi
None (as received)	35	220	32	220	32
Shot peened	35	270	39	240	35
Shot peened + Cu-Ni-In	35	320	46	245	36
None	140	215	31	190	28

Cyclic tensile load on fretting pad. Source: "Fretting Fatigue in High Temperature Oxidizing Gases," D. Taylor, in *Fretting Fatigue*, R. Waterhouse, Ed., Applied Science Publishers, Ltd., London, 1981, p 177

Ti-6Al-4V: Fretting fatigue at room temperature and 350 °C (660 °F) for alloy in polished and shot-peened conditions

Specimen treatment	Temperature °C	Temperature °F	Fatigue strength Unfretted MPa	Unfretted ksi	After 10^7 cycles Fretted MPa	Fretted ksi	Percent reduction of basic room-temperature properties Unfretted	Fretted
Plain polished	20	68	650	95	140	20	0	78
Plain polished	350	660	580	84	140	20	11	78
Shot-peened Almen A7	20	68	600	87	340	49	7	48
Shot-peened Almen A7	350	660	540	78	310	45	17	52

Cyclic tensile load on fretting pad. 93 MPa fretting stress. Source: "Fretting Fatigue in High Temperature Oxidizing Gases," D. Taylor, in *Fretting Fatigue*, R. Waterhouse, Ed., Applied Science Publishers, Ltd., London, 1981, p 177

Influence of Mean Stress

There are indications that the Ti-6Al-4V alloy exhibits an anomalous mean stress dependence of HCF strength if the material was forged in the (α + β) phase field in contrast to a normal mean stress dependence if the material was β forged or β heat treated (Ref 1-3). The results of an investigation on a fine lamellar structure and a bi-modal structure with a pronounced mixed B/T type of texture tested in RD and TD are shown (see figure, Influence of mean stress). It can be seen that the fine lamellar and the bi-modal structure tested in RD exhibited a normal mean stress dependence of HCF strength whereas the bi-modal structure tested in TD showed the anomalous mean stress dependence, i.e., much lower fatigue strength values with increasing mean stress. No reasonable explanation for this effect is given.

References

1. R.K. Steele and A.J. McEvily, *Eng. Fracture Mech., 8* (1976), p 31.
2. J. Broichhausen and H. van Kann, "Titanium Science and Technology," Plenum Press (1973), 1785.
3. A. Atrens, M. Müller, H. Meyer, G. Faber, and M.O. Speidel, "Corrosion Fatigue of Steam Turbine Blade Material," Pergamon Press (1983), p 4-50.

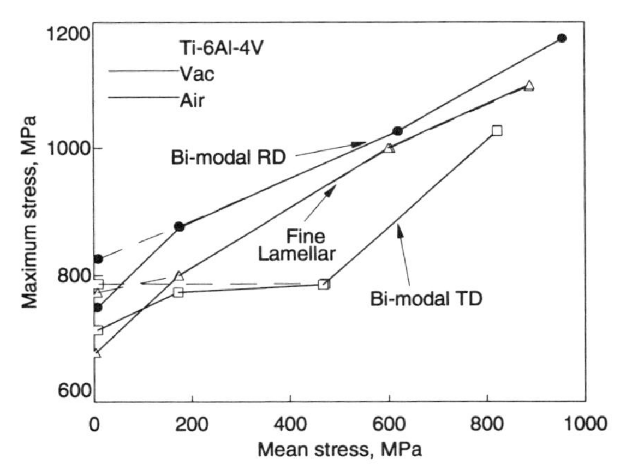

Ti-6Al-4V: Influence of mean stress on HCF strength (10^7 cycles). Source: G. Lütjering and A. Gysler, *Titanium Science and Technology*, Vol 4, Deutsche Gesellschaft für Metallkunde e.V., 1985, p 2072

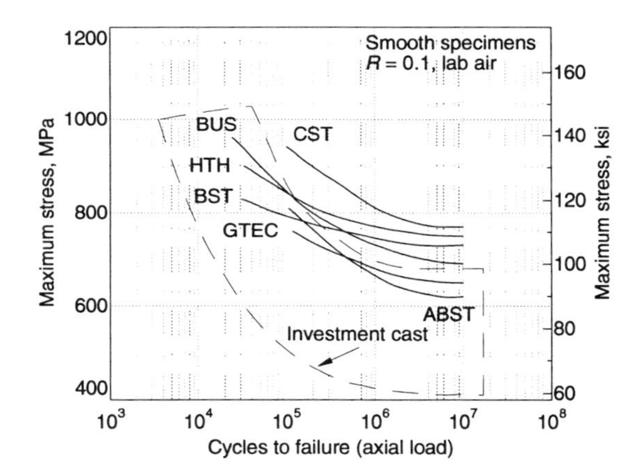

Ti-6Al-4V: Fatigue of investment castings after treatments. ABST, α-β solution treatment; BST, β solution treatment; BUS, broken-up structure; CST, constitutional solution treatment; GTEC, Garrett treatment (long-time, low-temperature anneal); HTH, high-temperature hydrogenation. Material/Test Parameters: 5-Hz triangular wave form. Source: D. Eylon and R. Boyer, "Titanium Alloy Net-Shape Technologies," in *Proc. Int. Conf. Titanium and Aluminum*, Paris, Feb 1990

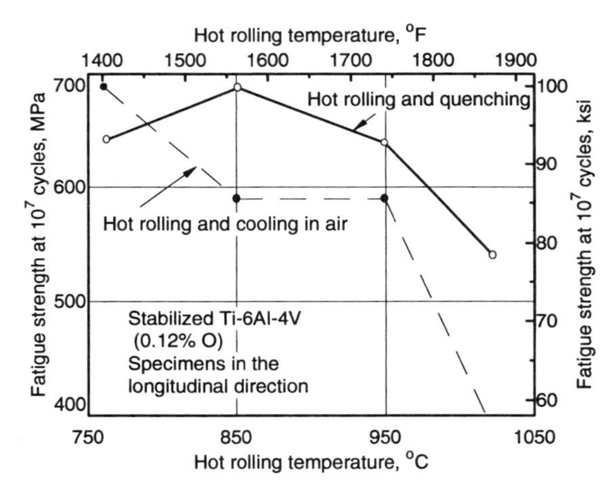

Ti-6Al-4V: Effect of rolling temperature on HCF strength. Material/Test Parameters: Material was 22 mm (0.865 in.) in diameter and was produced by 65% hot rolling at indicated temperatures, followed by annealing at 700 °C (1300 °F) for 1 h, furnace cooled to 500 °C (930 °F), held for 12 h, and air cooled.

Effect of Processing

Effect of Thermomechanical Processing

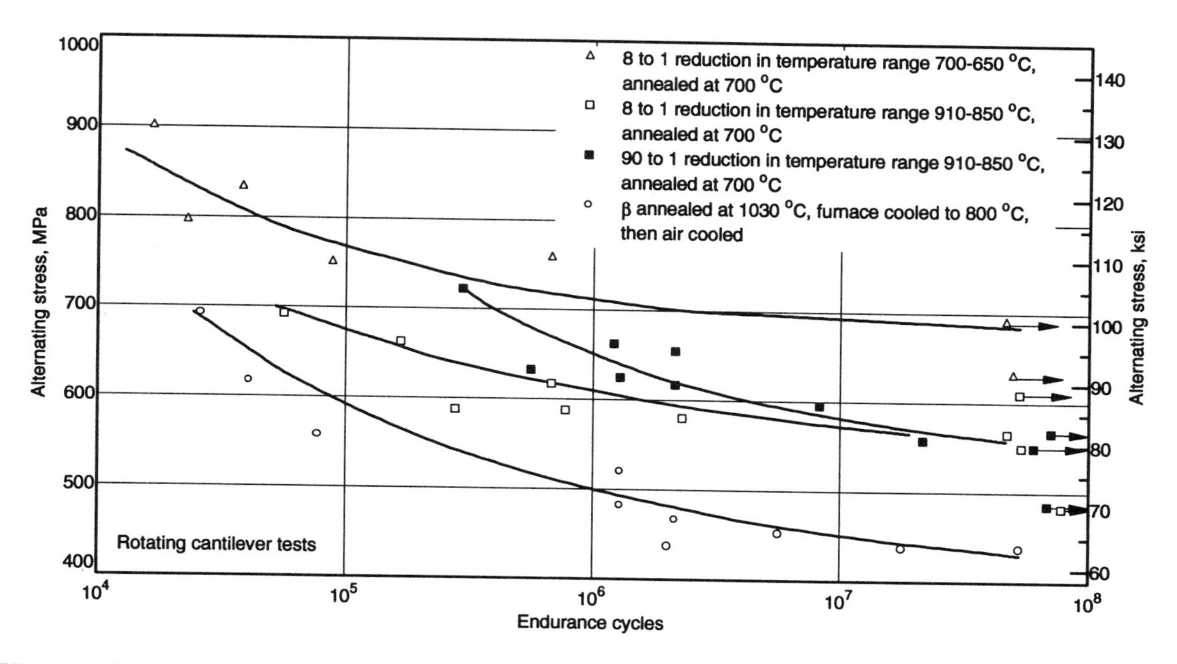

Ti-6Al-4V: Effect of working temperatures on HCF strength. Material/Test Parameters: Ti-6Al-4V bar worked at a high and low temperature in the α-β field are compared with β-annealed material. Source: C.A. Stubbington and A.W. Bowen, Improvements in the Fatigue Life of Ti-6Al-4V Through Microstructural Control, *J.*

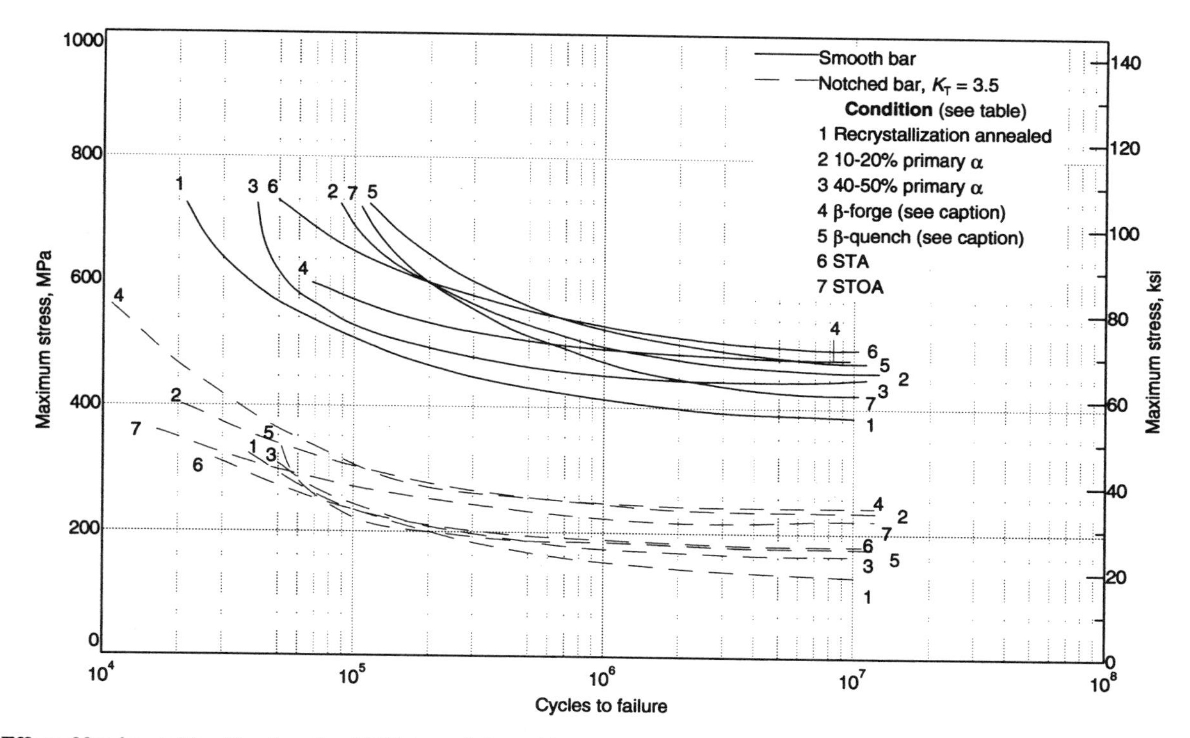

Ti-6Al-4V: Effect of forging and heat treatment on HCF strength (see table on next page for treatments)

Ti-6Al-4V: Caption table for bottom figure on previous page

Condition	Fabrication	Heat treatment
1	α-β forge (β_t – 75 °F)/AC; α-β finish (β_t – 100 °F)/AC	925 °C (1700 °F)/4 h/Cool at 50 °C/h (90 °F/h) to 760 °C (1400 °F)/AC
2	α-β forge (β_t – 50 °F)/AC; α-β finish (β_t – 25 °F)	955 °C (1750 °F)/1 h/AC + 700 °C (1300 °F)/2 h/AC
3	α-β forge (β_t – 75 °F)/AC; α-β finish (β_t – 100 °F)/AC	870 °C (1600 °F)/2 h/AC + 700 °C (1300 °F)/2 h/AC
4 (β-forge)	β forge (β_t + 75 °F)/AC; β finish (β_t + 75 °F)/WQ	700 °C (1300 °F)/2 h/AC
5 (β-quench)	β forge (β_t + 75 °F)/AC; β finish (β_t + 75 °F)/AC	1040 °C (1900 °F)/30 min/WQ + 700 °C (1300 °F)/2 h/AC
6	α-β forge (β_t – 75 °F)/AC; α-β finish (β_t – 75 °F)/AC	955 °C (1750 °F)/1 h/WQ + 595 °C (1100 °F)/4 h/AC
7	α-β forge (β_t – 75 °F)/AC; α-β finish (β_t – 75 °F)/AC	955 °C (1750 °F)/1 h/WQ + 595 °C (1100 °F)/24 h/AC

Source: J.C. Chesnutt *et al.*, "Influence of Metallurgical Factors on Fatigue Crack Growth Rate in Alpha-Beta Titanium Alloys," AFML-TR-78-68, 1978

Effect of Heat Treatment on Fatigue

Annealing

High-cycle fatigue strength is lowered during annealing due to the coarsening of grain sizes. During α-β annealing, for example, a longer annealing time decreases HCF strength because of the increase in the grain size of equiaxed α. Annealing above the β transus reduces fatigue strength still further.

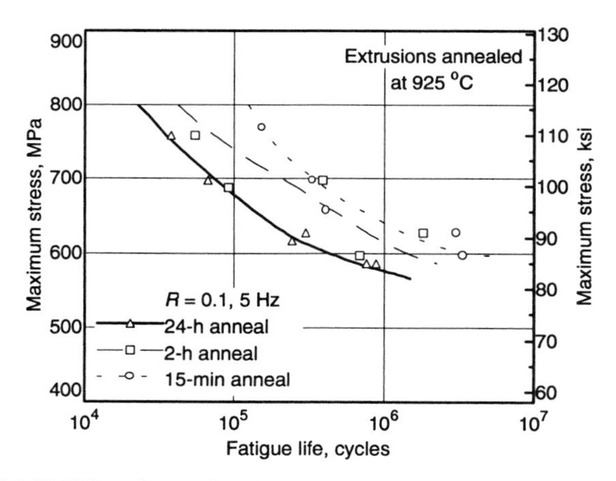

Ti-6Al-4V: Effect of annealing time. Material/Test Parameters: Extrusions had a composition (wt%) of 6.7 Al, 0.01 C, 0.18 Fe, 0.0044 H, 0.013 N, 0.164 O, and 4.1 N. Mill-annealed extrusions as 76-mm (3-in.) diam cylinders were annealed at 925 °C (1700 °F) for times indicated, followed by slow cooling at 50 °C/h (90 °F/h). Source: I. Weiss *et al.*, Recovery, Recrystallization, and Mechanical Properties of Ti-6Al-4V Alloy, in *Proc. 8th Int. Conf. Strength of Metals and Alloys*, H. McQueen *et al.*, Ed., Pergamon Press, 1985, p 1073

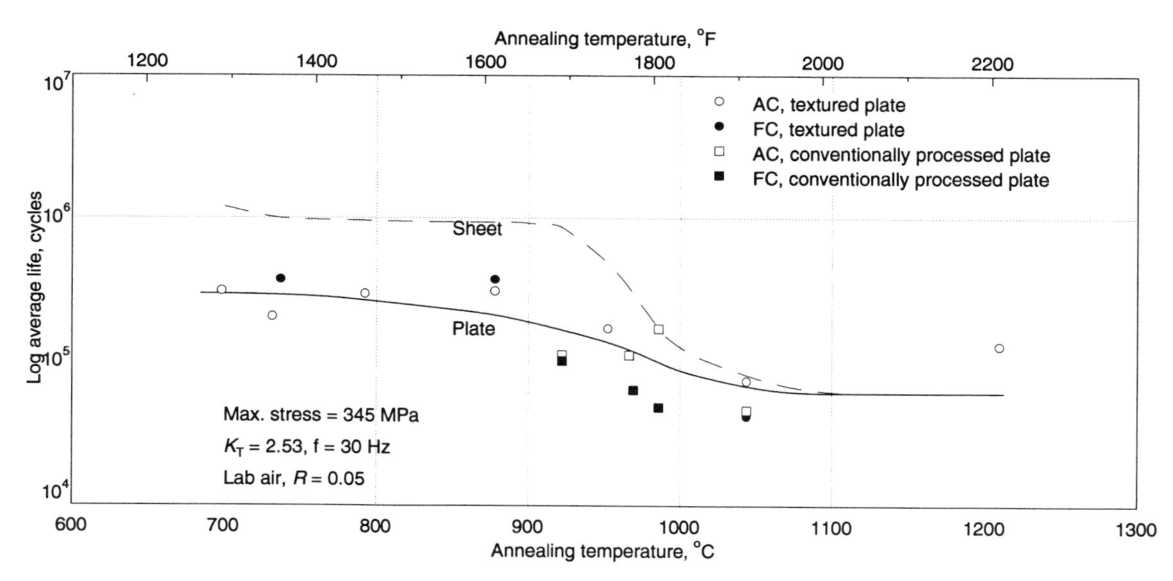

Ti-6Al-4V: Effect of annealing temperature on fatigue strength. Comparison between sheet and plate shows the difference due to the finer grain sizes of sheet (however, the comparison is only general because gage, composition, and rolling practices are different for sheet). Material/Test Parameters: Time and temperature for all anneals below 1040 °C (1900 °F) was 2 h. Above 1040 °C (1900 °F), annealing time was 20 min. All specimens having annealing temperatures above 870 °C (1600 °F) were heated to 730 °C (1350 °F) for 2 h and air cooled. Source: R. Boyer *et al.*, The Effects of Thermal Processing Variations on the Properties of Ti-6Al-4V, in *Microstructure, Fracture Toughness, and Fatigue Crack Growth Rate in Titanium Alloys*, A. Chakrabarti and J.C. Chesnutt, Ed., The Metallurgical Society, 1987, p 149

Effect of Cooling

Fatigue strength is improved by rapid cooling from either the α-β region or from above the β transus. Fast cooling leads to the martensitic formation of α′, which improves fatigue strength. Water quenching without further aging may result in a low HCF strength if the retained β phase is unstable against stress-induced martensitic transformation.

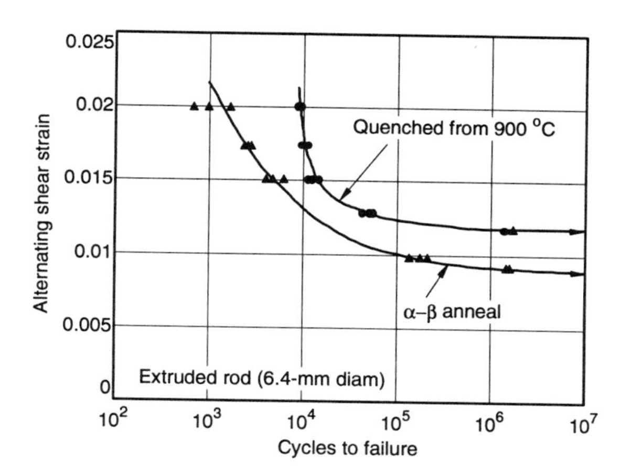

Ti-6Al-4V: Effect of cooling from α-β region on HCF strength. Material/Test Parameters: The α-β anneal alloy was annealed at 800 °C (1470 °F) for 3 h, furnace cooled to 600 °C (1110 °F), and vacuum cooled to room temperature. Source: M.A. Imam and C.M. Gilmore, Fatigue and Microstructural Properties of Quenched Ti-6Al-4V, *Metall. Trans. A*, Vol 14A, 1983, p 233-240

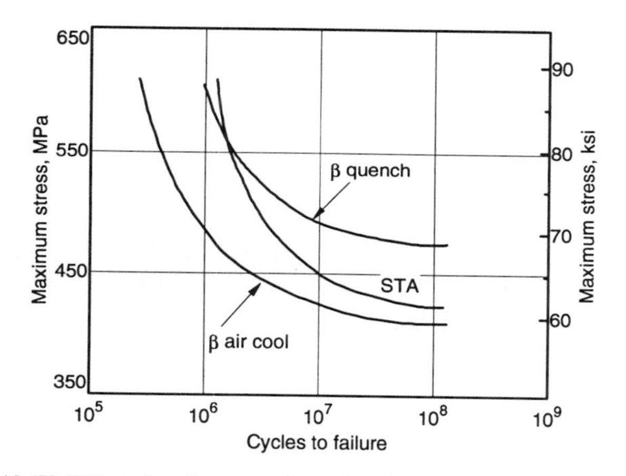

Ti-6Al-4V: Effect of cooling rates from β region. HCF comparison for three conditions for general comparison only.

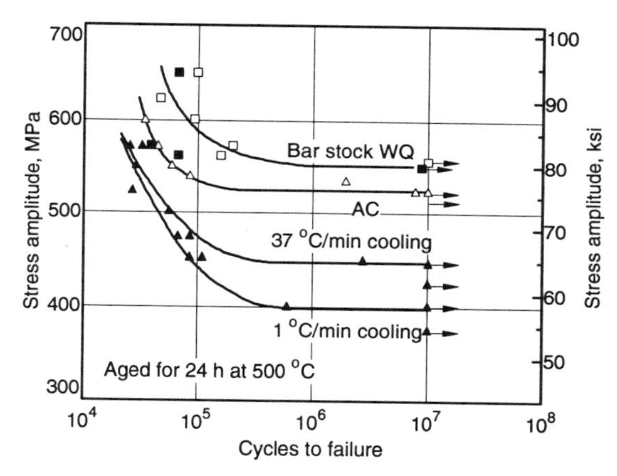

Ti-6Al-4V: Effect of cooling rate from solution anneal on aged HCF strength. Material/Test Parameters: Rotating-beam fatigue tests were performed on hourglass-shaped specimens with gage diameter of 3.8 mm (0.15 in.), electrolytically polished surface, f = 50 Hz, at room temperature in the longitudinal direction. See table for alloy condition. Source: R. Jaffee *et al.*, The Effect of Cooling Rate From the Solution Anneal on the Structure and Properties of Ti-6Al-4V, in *6th World Conf. Titanium*, 1989, p 1501

Ti-6Al-4V: Material condition for HCF strength (see above figure)

Treatment	Primary α, vol%	Primary α, 0.2% μm	Yield strength, MPa	Yield strength, ksi	Reduction in area, %
965 °C/WQ/aged	35	8	1035	150	50
965 °C/AC/800 °C for 1 h/AC/aged	35	8	985	143	47
965 °C/37 °C per min/800 °C for 1 h/AC/aged	45	10	975	141	34
965 °C/1 °C per min/800 °C for 1 h/AC/aged	75	12	955	139	38

STA Condition

Solution-treated and aged (STA) material has good fatigue strength but not as good as that of the fine equiaxed or β quenched materials. Age hardening results in the strengthening of the β phase by the precipitation of small α grains and/or the strengthening of the α phase by Ti_3Al precipitates. The degree of age hardening depends on the solution-anneal temperature and cooling rates.

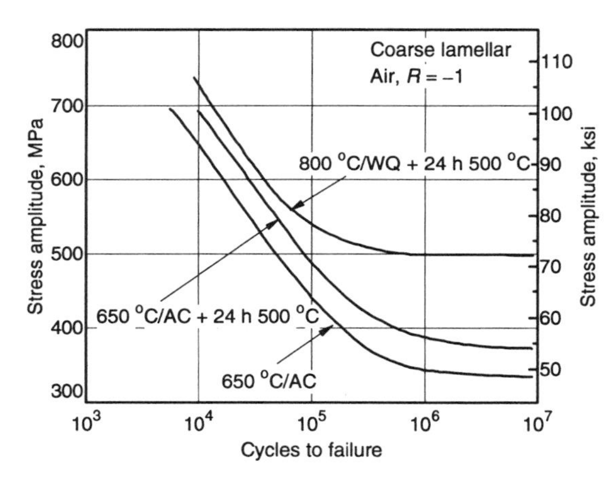

Ti-6Al-4V: Influence of age hardening on HCF strength. Water quenching from 800 °C (1470 °F) without further aging resulted in a low HCF strength value of 290 MPa (42 ksi) because (in addition to the absence of age hardening) the β phase is unstable against stress-induced martensitic transformation, resulting in crack nucleation at low applied stresses. Annealing at 650 °C (1200 °F) instead of 800 °C (1470 °F) leads to stabilization of the β phase due to the higher vanadium content, but the effect of subsequent age hardening at 500 °C (930 °F) is smaller for the β phase and also for the α phase. This is due to the lower vacancy concentration after the 650 °C (1200 °F) anneal as compared to the 800 °C (1470 °F) treatment and its effect on Ti_3Al precipitation. Source: G. Lütjering and A. Gysler, Critical Review of Fatigue, in *Titanium Science and Technology*, Deutsche Gesellschaft für Metallkunde e.V., 1985, p 2069

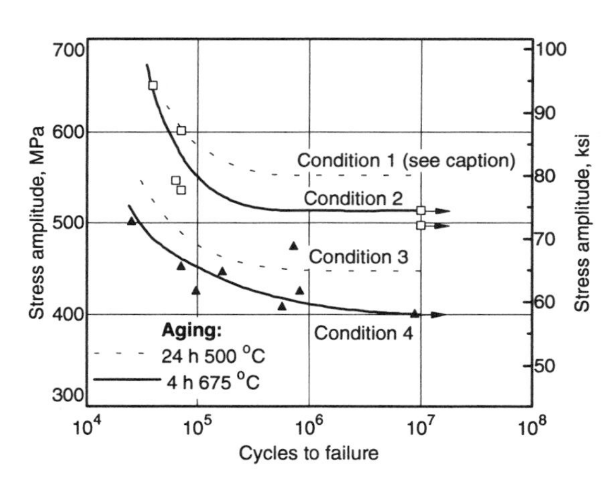

Ti-6Al-4V: HCF strength of age-hardened bar. Material/Test Parameters: Rotating-beam fatigue tests were performed on hourglass-shaped specimens with 3.8-mm (0.15-in.) gage diameter, electrolytically polished surface, f = 50 Hz, at room temperature in the longitudinal direction. See table below for conditions. Source: R. Jaffee *et al.*, The Effect of Cooling Rate From the Solution Anneal on the Structure and Properties of Ti-6Al-4V, in *6th World Conf. Titanium*, 1989, p 1501

Ti-6Al-4V: HCF strength of age-hardened bar (see above figure)

Condition	Solution treatment
1	1 h 955 °C/WQ + 1 h 800 °C/WQ
2	1 h 955 °C/WQ + 1 h 800 °C/AC + 4 h 675 °C/AC
3	1 h 965 °C/37 °C/min to 800 °C + 1 h 800 °C/AC
4	1 h 965 °C/37 °C/min to 800 °C + 1 h 800 °C/AC + 2 h 650 °C/AC

Constant-Life Fatigue Diagrams

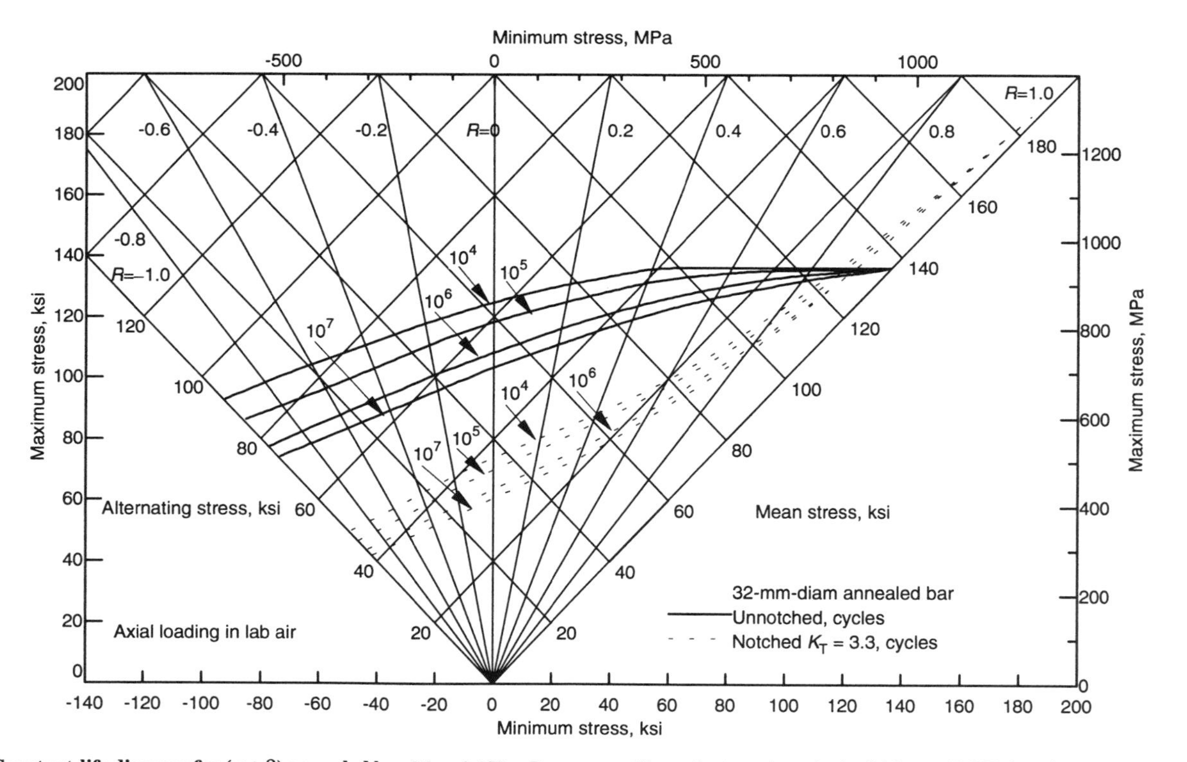

Ti-6Al-4V: Constant-life diagram for (α + β) annealed bar. Material/Test Parameters: Unnotched specimen had a 5.15-mm (0.203-in.) diameter, a tensile strength of 940 MPa (136.5 ksi), and was polished longitudinally with 240, 400, and 600 emery belts. Notched specimen had a gross diameter of 8.4 mm (0.331 in.), a net diameter of 6.4 mm (0.252 in.), and was machined into a V-groove followed by polishing notch root with 600-grit slurry and rotating copper wire. Test frequency: 1750 cycles/min. Source: R. Wood and R. Favor, *Titanium Alloys Handbook*, MCIC-HB-02, Battelle Columbus Laboratories, p 5-4:72-23

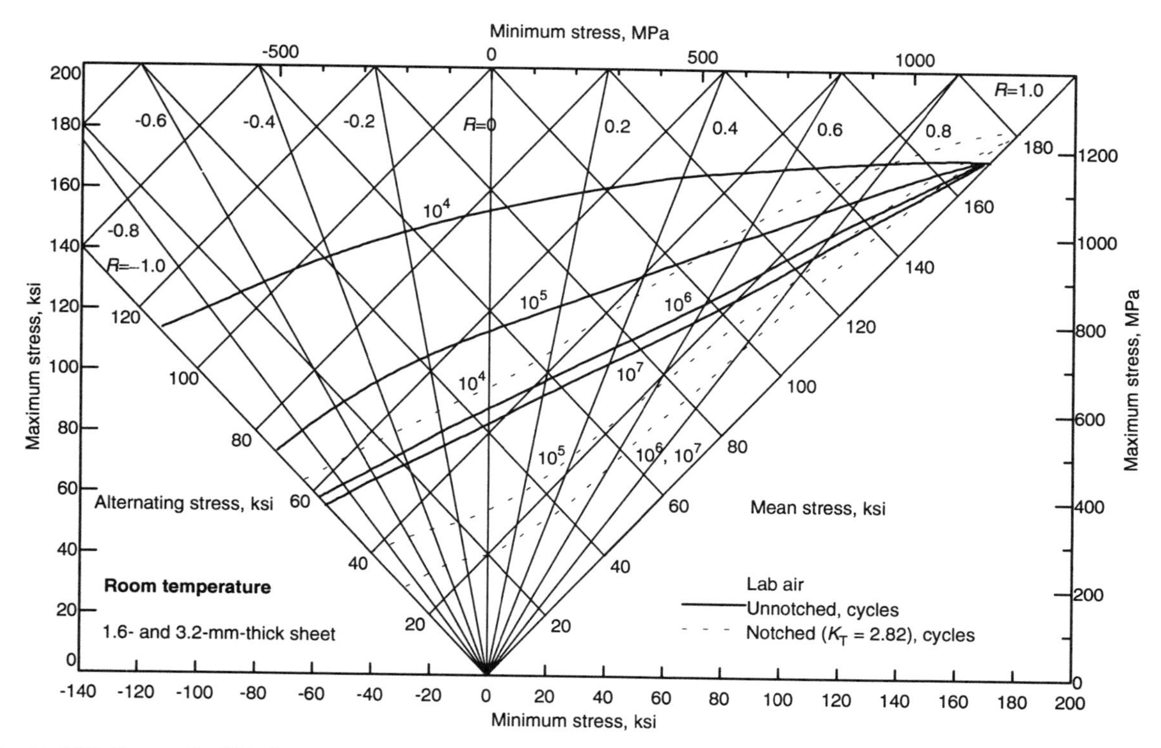

Ti-6Al-4V: Constant-life diagram for STA sheet. Material/Test Parameters: *Unnotched:* As-rolled surface, but edges machined and hand polished with No. 1 and 00 grit emery paper, cleaned with methyl ethyl ketone. *Notched:* Surface and edges prepared as above; 1.587-mm (0.0625-in.) diam hole drilled and reamed. Test frequency, 1500-2200 cycles/min. Source: R. Wood and R. Favor, *Titanium Alloys Handbook*, MCIC-HB-02, Battelle Columbus Laboratories, 1972, reprinted 1985, p 5-4:72-27, 28, 29

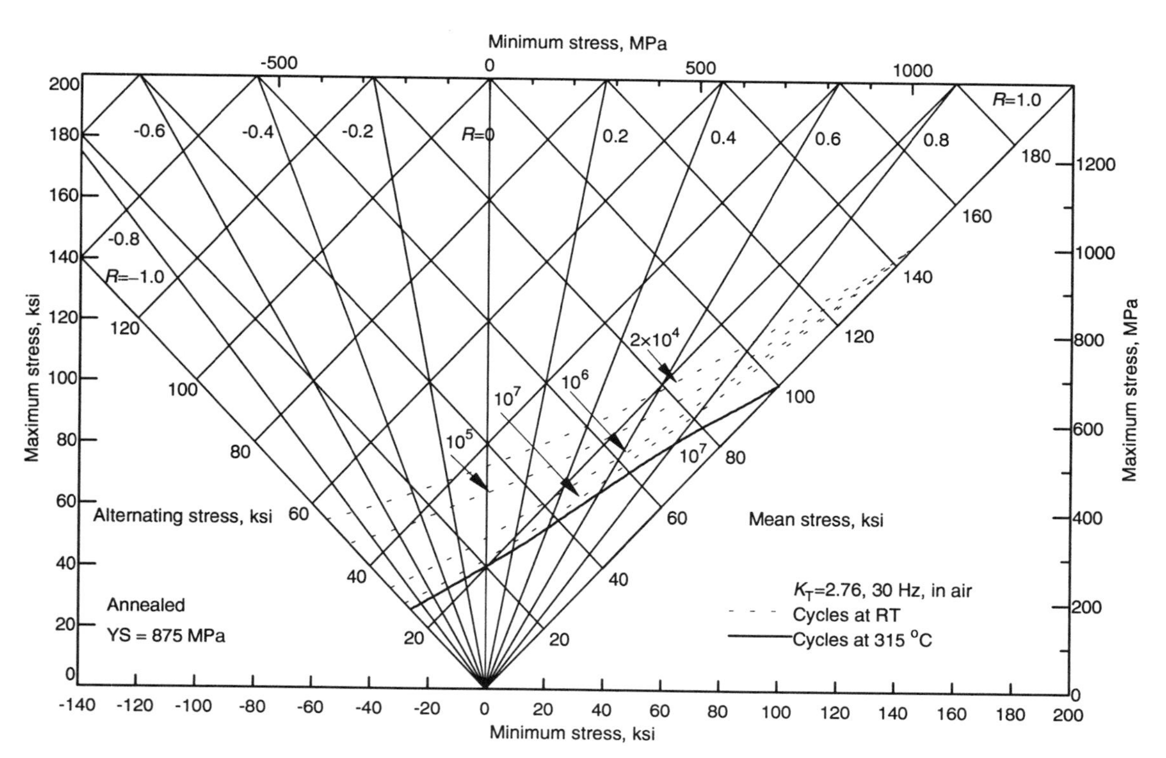

Ti-6Al-4V: Constant-life diagram of extrusions at room temperature (RT). Material/Test Parameters: Specimens were from 7.6- and 14-mm (0.300- and 0.560-in.) thick extrusions with a tensile strength of 985 MPa (143 ksi) and an RMS surface roughness of 1.6 μm (63 μin.). Grain direction was longitudinal. Source: R. Wood and R. Favor, *Titanium Alloys Handbook*, MCIC-HB-02, Battelle Columbus Laboratories, p 5-4:72-24

Duplex Annealed Sheet

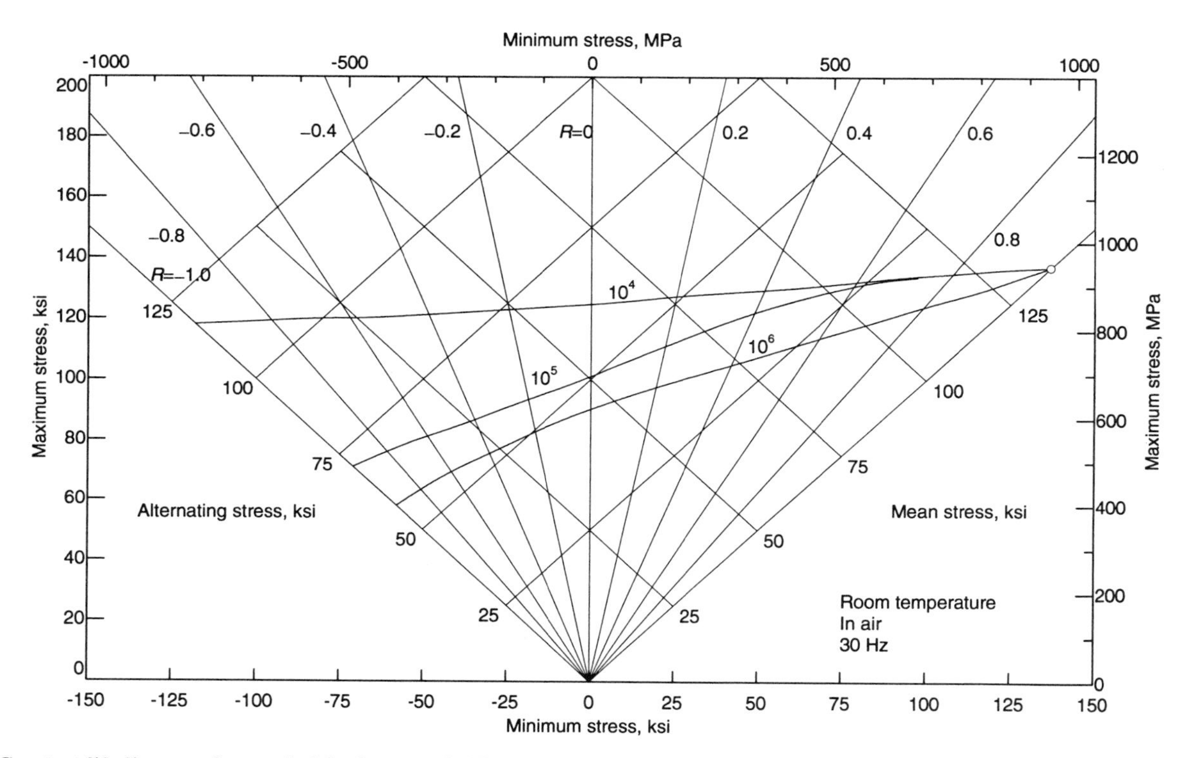

Ti-6Al-4V: Constant-life diagram of unnotched duplex-annealed ELI sheet. Material/Test Parameters: Duplex anneal (DA) sheet = 910 °C (1675 °F) for 10 min/AC + 730 °C (1350 °F) for 4 h/AC. Sheet composition (wt%): 0.11 O, 6.0 Al, 4.0 V, 0.19 Fe, 0.01 N, 0.02 C, and 0.0034 H. Source: R.R. Boyer and R. Bajoraitis, "Standardization of Ti-6Al-4V Processing Conditions," AFML-TR-78-131, 1978

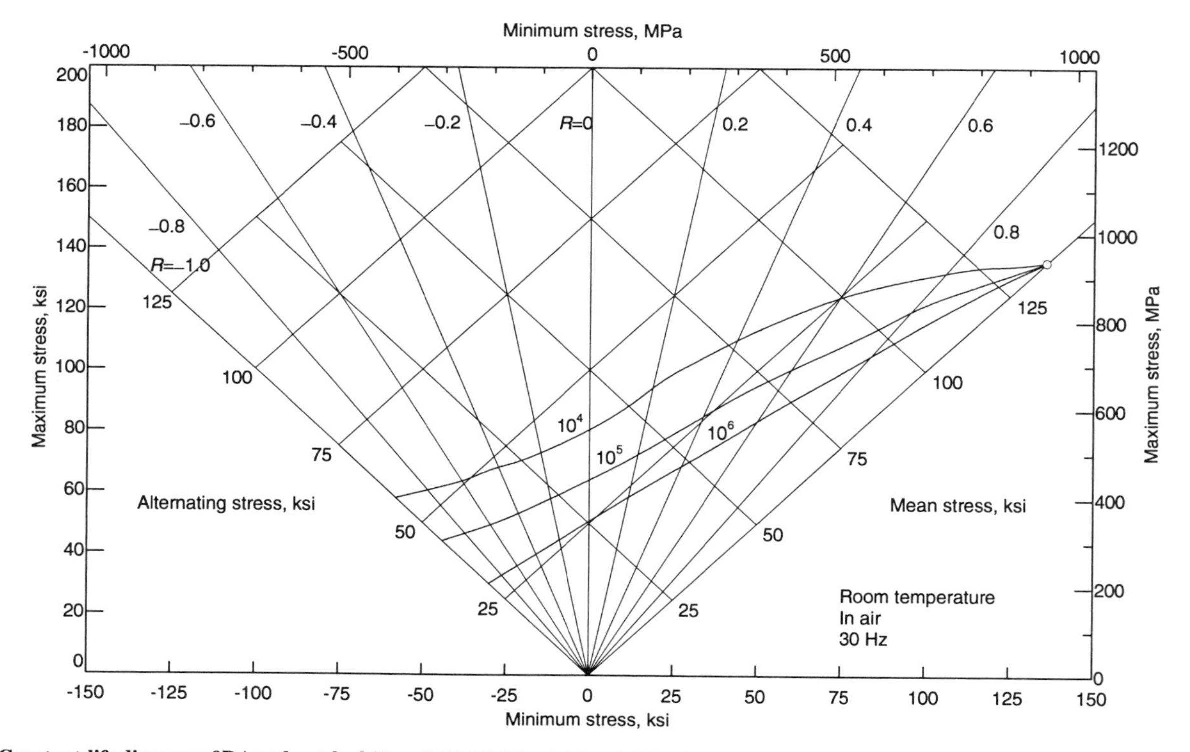

Ti-6Al-4V: Constant-life diagram of DA and notched (K_T = 2.53) ELI sheet. Material/Test Parameters: Duplex anneal (DA) sheet = 910 °C (1675 °F) for 10 min/AC + 730 °C (1350 °F) for 4 h/AC. Sheet composition (wt%): 0.11 O, 6.0 Al, 4.0 V, 0.19 Fe, 0.01 N, 0.02 C, and 0.0034 H. Source: R.R. Boyer and R. Bajoraitis, "Standardization of Ti-6Al-4V Processing Conditions," AFML-TR-78-131, 1978

Beta Annealed Plate

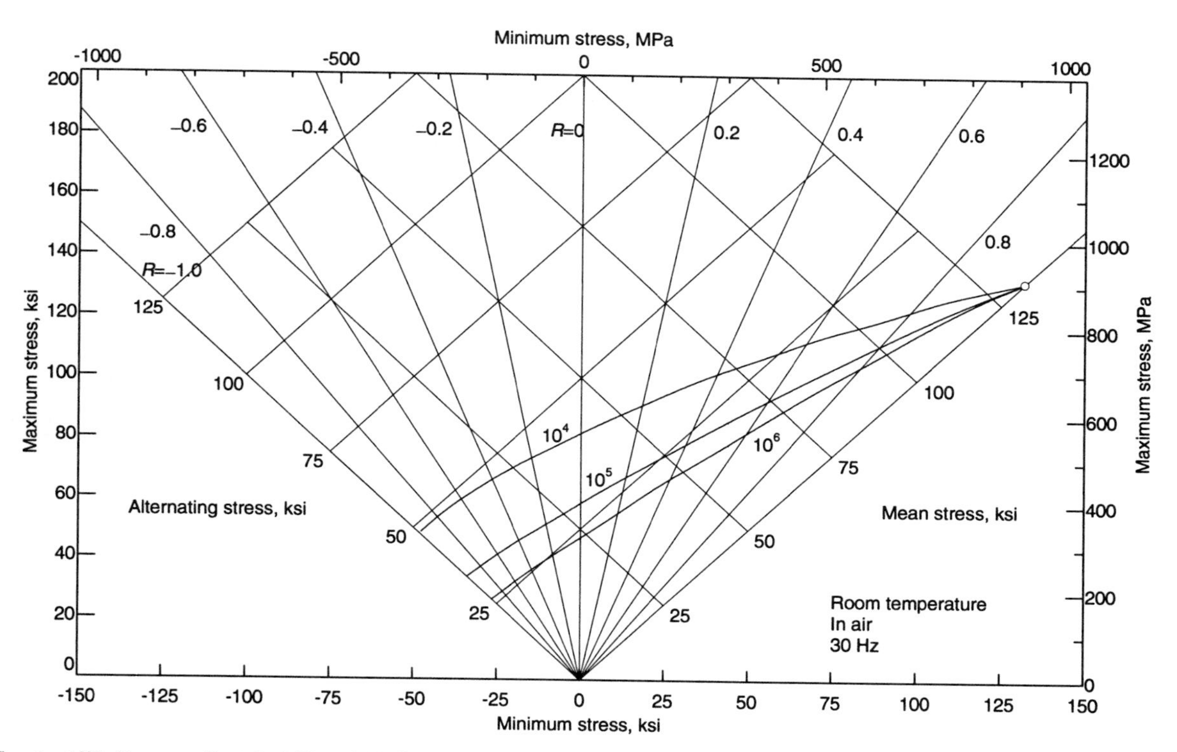

Ti-6Al-4V: Constant-life diagram of notched (K_T = 2.53) β-annealed ELI plate. Material/Test Parameters: Beta anneal (BA) = 1000 °C (1845 °F) for 20 min/AC + 730 °C (1350 °F) for 2 h/AC. Material composition (wt%): 0.10 O, 6.1 Al, 4.0 V, 0.15 Fe, 0.01 N, 0.02 C, and 0.048 H. Source: R.R. Boyer and R. Bajoraitis, "Standardization of Ti-6Al-4V Processing Conditions," AFML-TR-78-131, 1978

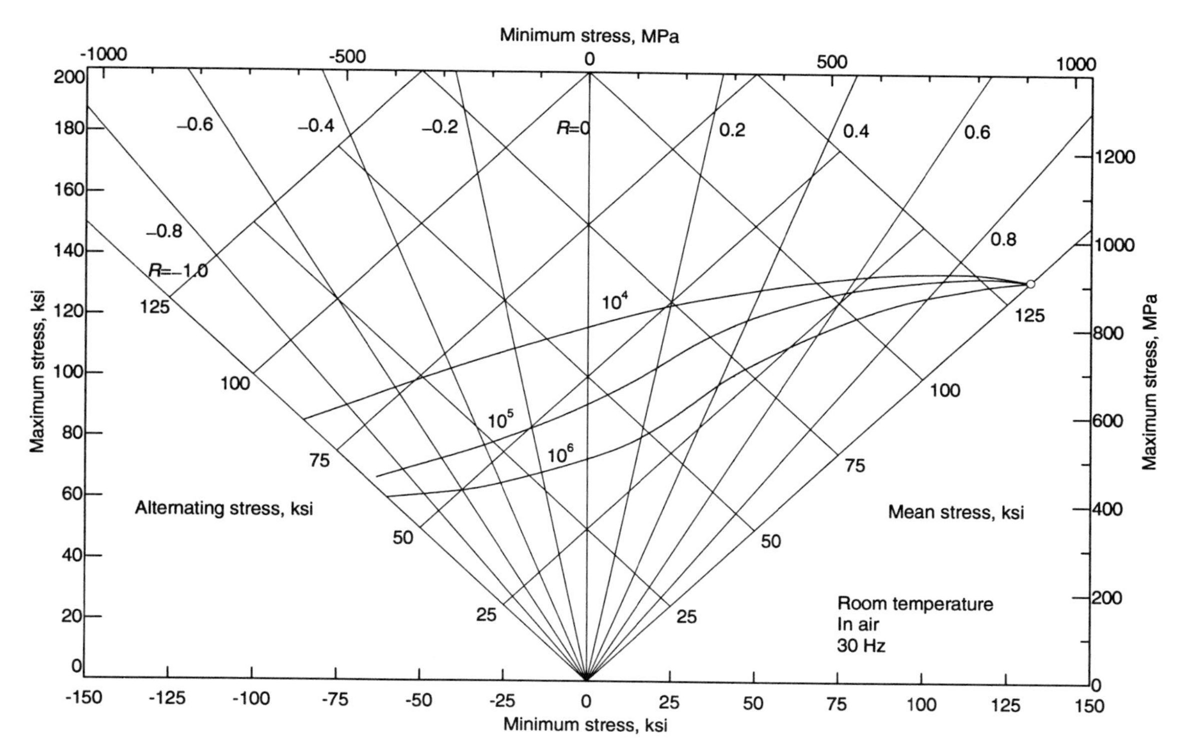

Ti-6Al-4V: Constant-life diagram for unnotched β-annealed ELI plate. Material/Test Parameters: Beta anneal (BA) = 1000 °C (1845 °F) for 20 min/AC + 730 °C (1350 °F) for 2 h/AC. Material composition (wt%): 0.10 O, 6.4 Al, 4.0 V, 0.15 Fe, 0.01 N, 0.02 C, and 0.048 H. Source: R.R. Boyer and R. Bajoraitis, "Standardization of Ti-6Al-4V Processing Conditions," AFML-TR-78-131, 1978

At 315 °C

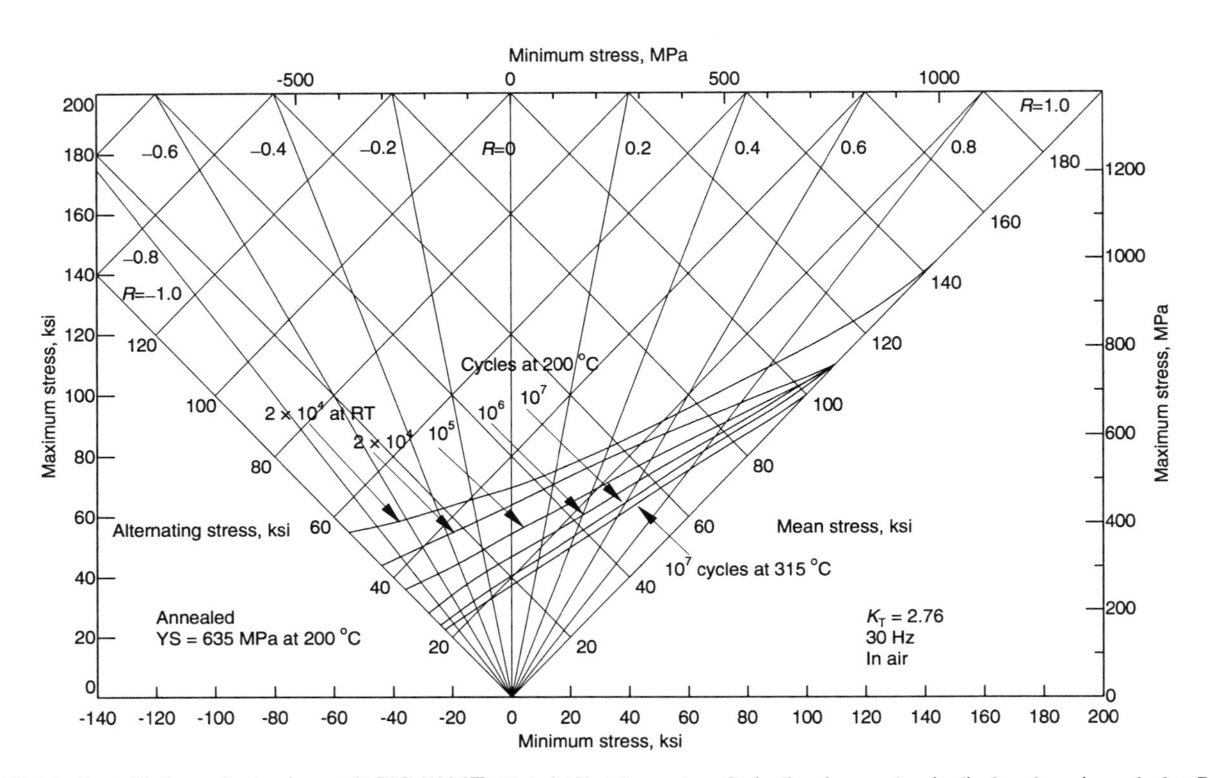

Ti-6Al-4V: Notched axial fatigue of extrusions at 315 °C (400 °F). Material/Test Parameters: Grain direction was longitudinal, and specimens had an RMS surface roughness of 1.6 μm (63 μin.). Tensile strength at 200 °C (400 °F) was 770 MPa (112 ksi). Source: R. Wood and R. Favor, *Titanium Alloys Handbook*, MCIC-HB-02, Battelle Columbus Laboratories, p 5-4:72-25

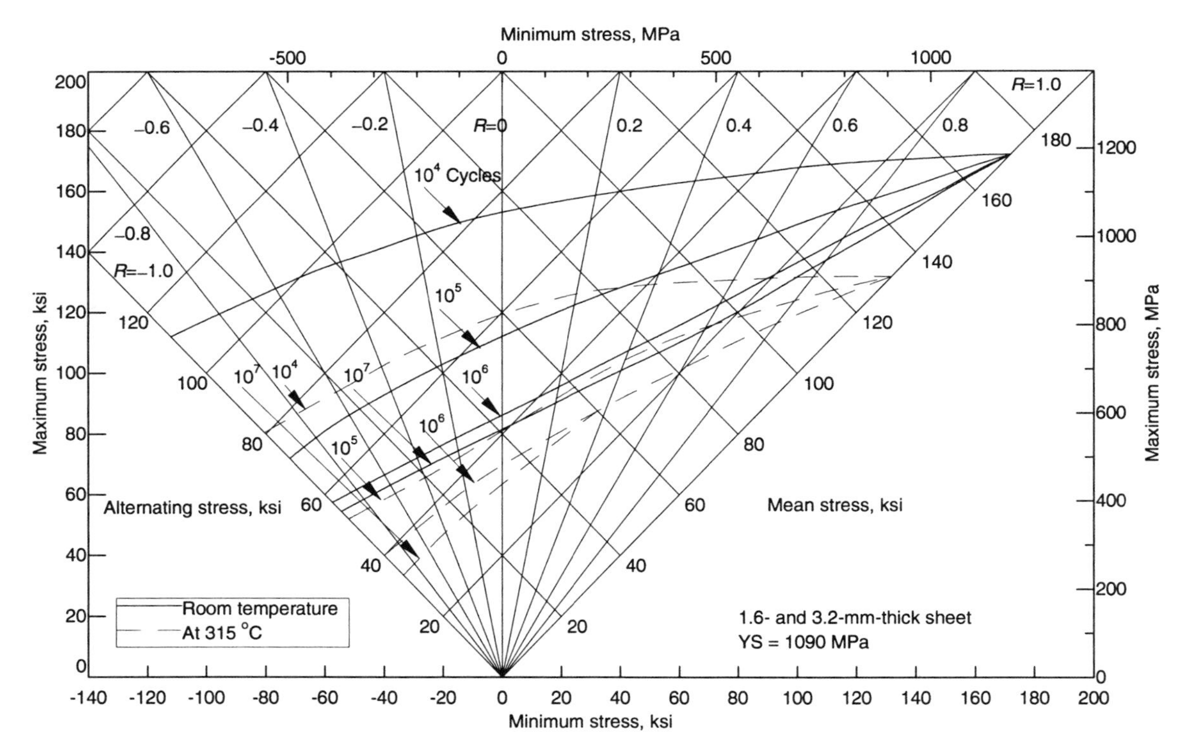

Ti-6Al-4V: Typical constant-life diagram for unnotched STA sheet at 315 °C (600 °F). Material/Test Parameters: As-rolled surfaces, but edges machined and hand polished with No. 1 and 00 grit emery paper, cleaned with methyl ethyl ketone. Axial loading in air; test frequency, 1500-2200 cycles/min. Source: R. Wood and R. Favor, *Titanium Alloys Handbook*, MCIC-HB-02, Battelle Columbus Laboratories, 1972, reprinted 1985, p 5-4:72-27, 28, 29

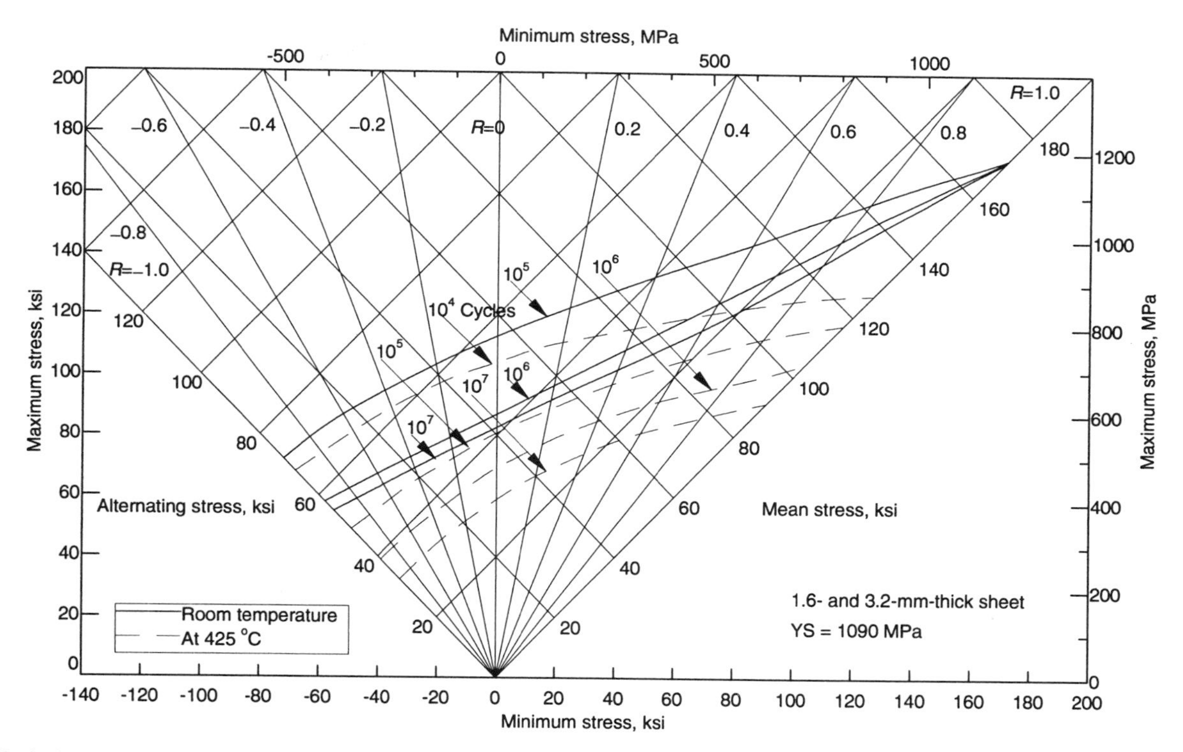

Ti-6Al-4V: Typical constant-life diagram for unnotched STA sheet at 425 °C (800 °F). Material/Test Parameters: As-rolled surfaces, but edges machined and hand polished with No. 1 and 00 grit emery paper, cleaned with methyl ethyl ketone. Axial loading in air; test frequency, 1500-2200 cycles/min. Source: R. Wood and R. Favor, *Titanium Alloys Handbook*, MCIC-HB-02, Battelle Columbus Laboratories, 1972, reprinted 1985, p 5-4:72-27, 28, 29

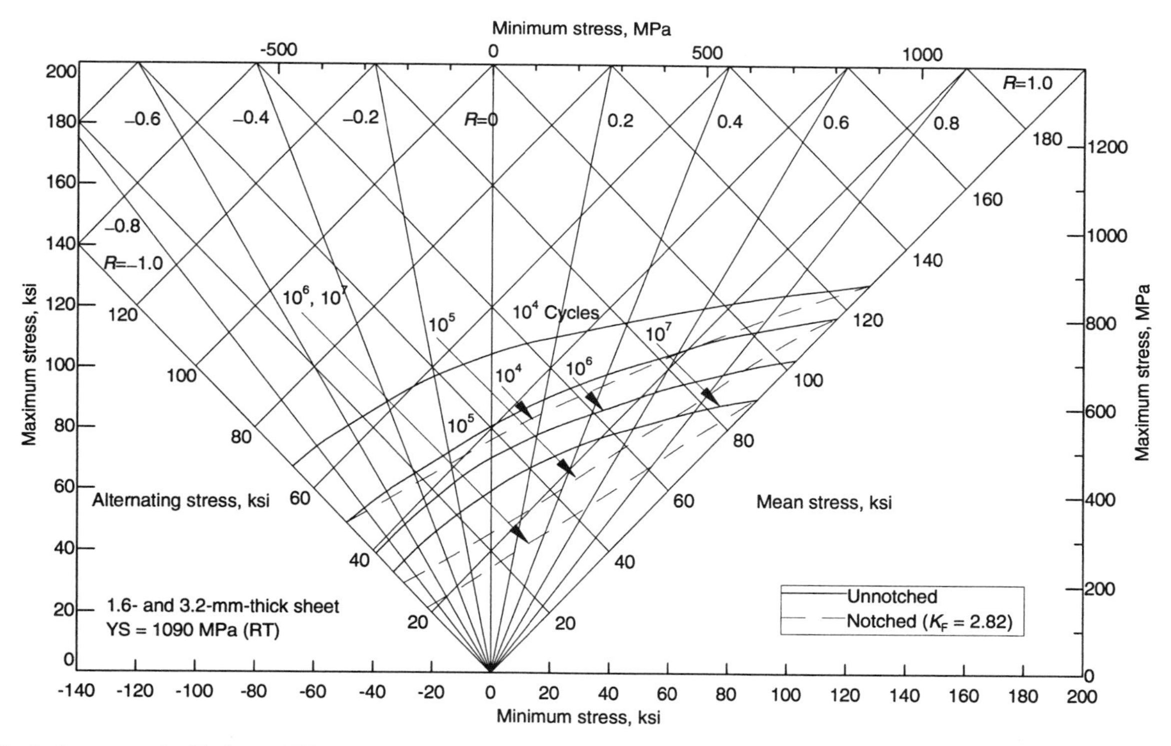

Ti-6Al-4V: Notched vs. unnotched fatigue of STA sheet at 425 °C (800 °F). Material/Test Parameters: *Unnotched:* As-rolled surface, but edges machined and hand polished with No. 1 and 00 grit emery paper, cleaned with methyl ethyl ketone. *Notched:* Surface and edges prepared as above; 1.587-mm (0.0625-in.) diam hole drilled and reamed. Test frequency, 1500-2200 cycles/min. Source: R. Wood and R. Favor, *Titanium Alloys Handbook*, MCIC-HB-02, Battelle Columbus Laboratories, 1972, reprinted 1985, p 5-4:72-27, 28, 29

Unnotched Fatigue Strength

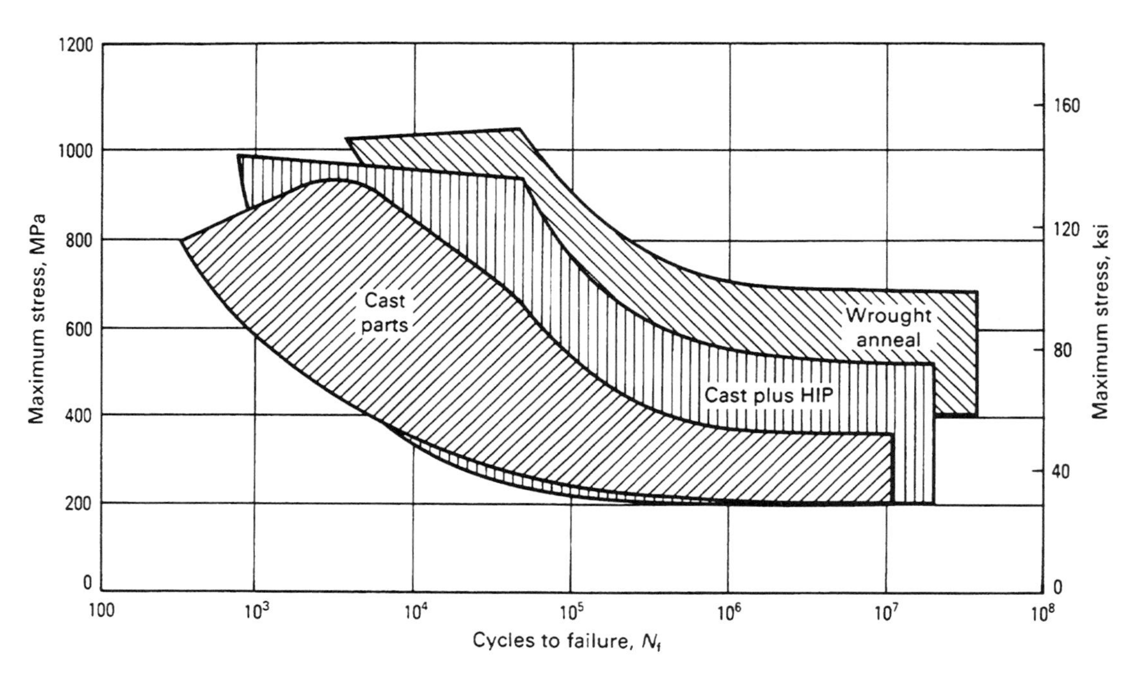

Ti-6Al-4V: Axial fatigue of cast and wrought forms (R = 0.1, unnotched). Source: D. Eylon and R. Boyer, "Titanium Alloy Net-Shape Technologies," in *Proc. Int. Conf. Titanium and Aluminum*, Paris, Feb 1990

Plate

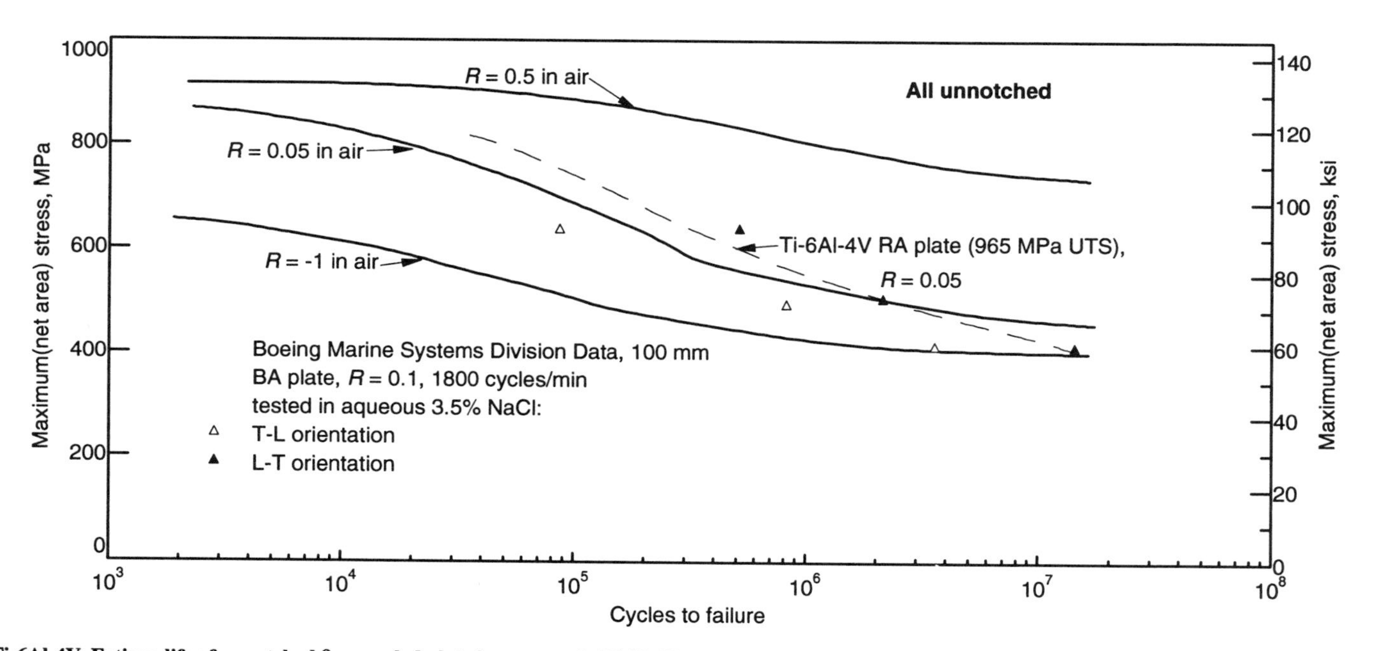

Ti-6Al-4V: Fatigue life of unnotched β-annealed plate in aqueous 3.5% NaCl. Material/Test Parameters: Beta anneal (BA) = 1000 °C (1845 °F) for 20 min/AC + 730 °C (1350 °F) for 2 h/AC. Material composition (wt%): 0.10 O, 6.1 Al, 4.0 V, 0.15 Fe, 0.01 N, 0.02 C, and 0.048 H. Source: Reported in R.R. Boyer and R. Bajoraitis, "Standardization of Ti-6Al-4V Processing Conditions," AFML-TR-78-131, 1978; recrystallization annealed (RA) data from "B-1 Airframe Fatigue Design Properties Manual," Technical Report NA-72-1088, Rockwell International, 1975

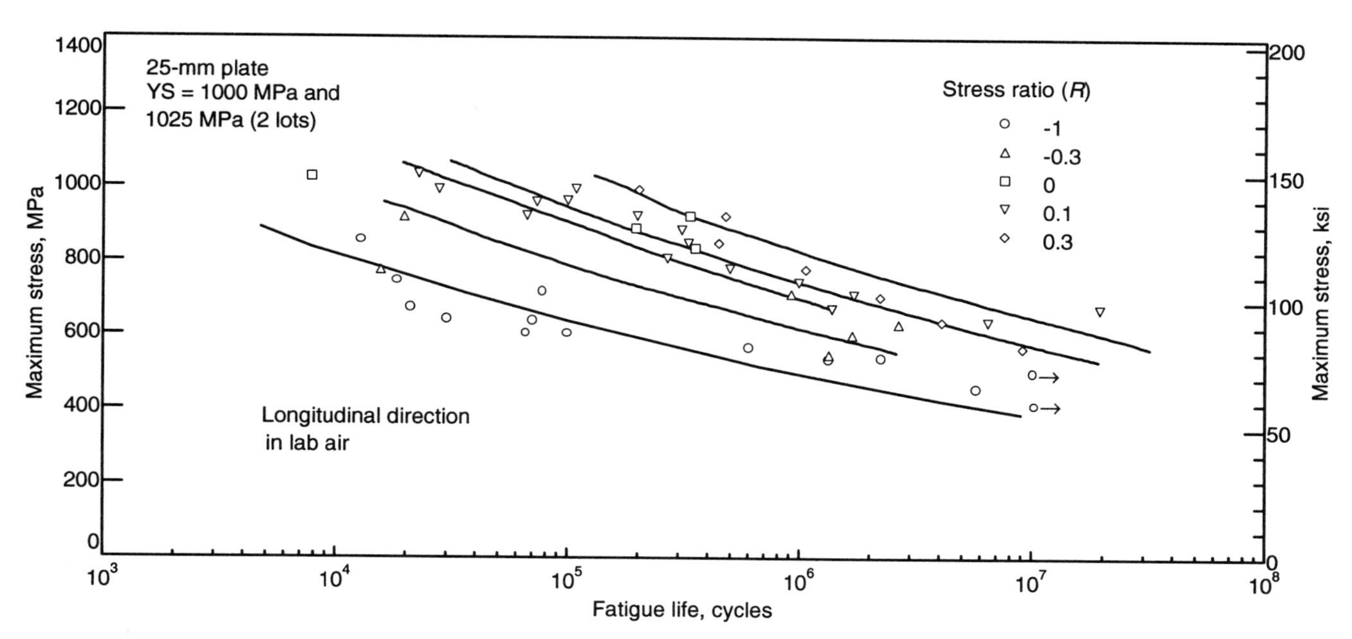

Ti-6Al-4V: Axial fatigue of unnotched STA plate. Material/Test Parameters: Specimens were longitudinally polished with No. 000 emery paper, removing all circumferential marks. Test frequency, 30 to 300 Hz. Source: Data from A. Sommer and G. Martin, North American Rockwell, TR-69-161, 1969; and A. Marrocco, Grumann Aircraft, "Fatigue Characteristics of Ti-6Al-4V and Ti-6Al-6V-2Sn Sheet and Plate," 1968, reported in *MIL-HDBK-5E*, U.S. Dept. of Defense, 1987, p 5-88b

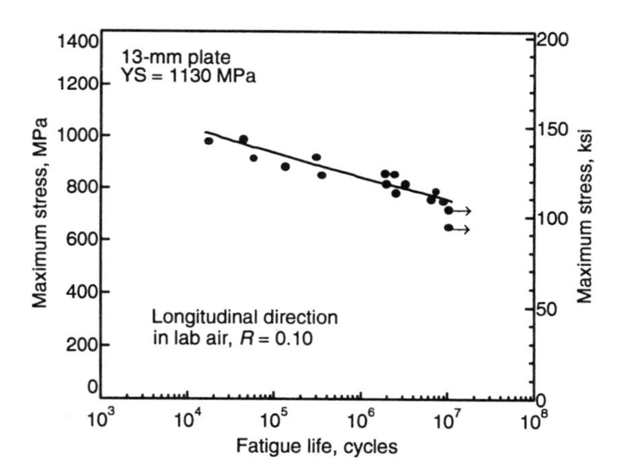

Ti-6Al-4V: Axial fatigue of unnotched STA plate. Machined RMS surface roughness of 1.6 μm (63 μin.). Data from A. Marrocco, Grumann Aircraft, "Fatigue Characteristics of Ti-6Al-4V and Ti-6Al-6V-2Sn Sheet and Plate," 1968, reported in *MIL-HDBK-5E*, U.S. Dept. of Defense, 1987, p 5-88c

Sheet

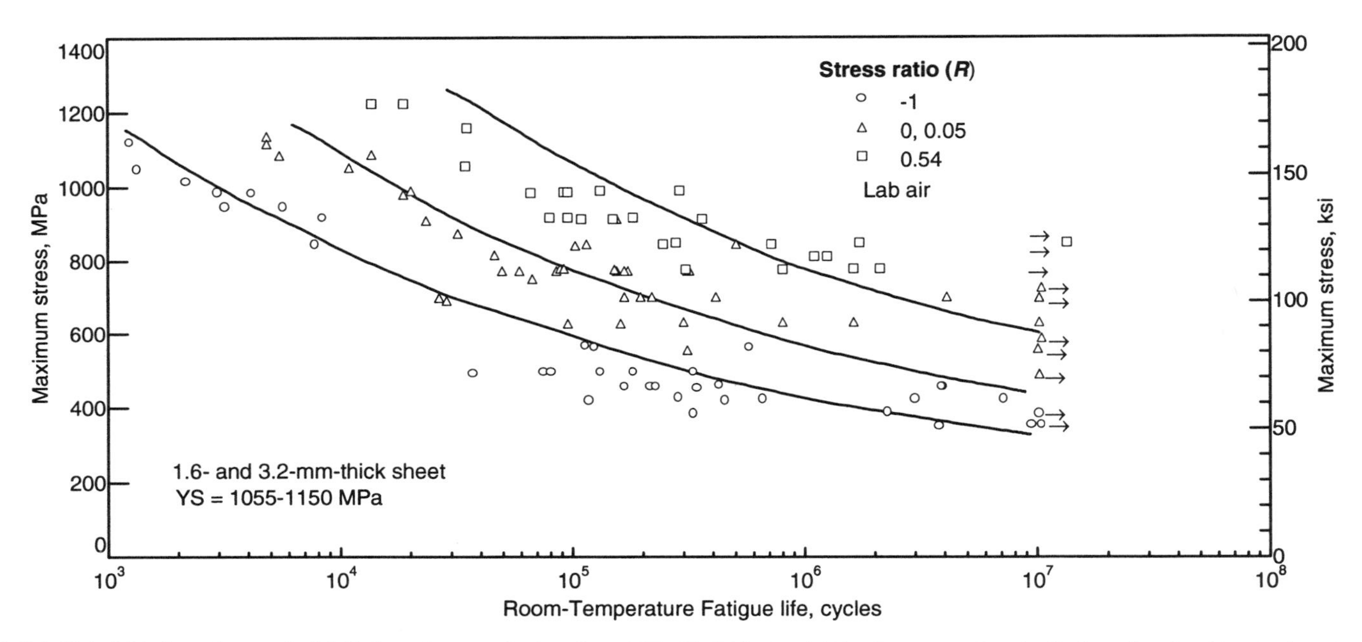

Ti-6Al-4V: Axial fatigue of unnotched STA sheet. Conservative best fit per MIL-HDBK 5. Material/Test Parameters: Longitudinal test direction. Machined specimens were cleaned with methyl ethyl ketone. Edges polished with No. 1 and 00 grit emery paper, recleaned with methyl ethyl ketone. Test frequency not specified. Source: Data from C. Hickey, Jr., Watertown Arsenal, 1962, and from North American Aviation Report, 1960, reported in *MIL-HDBK-5E*, U.S. Dept. of Defense, 1987, p 5-84

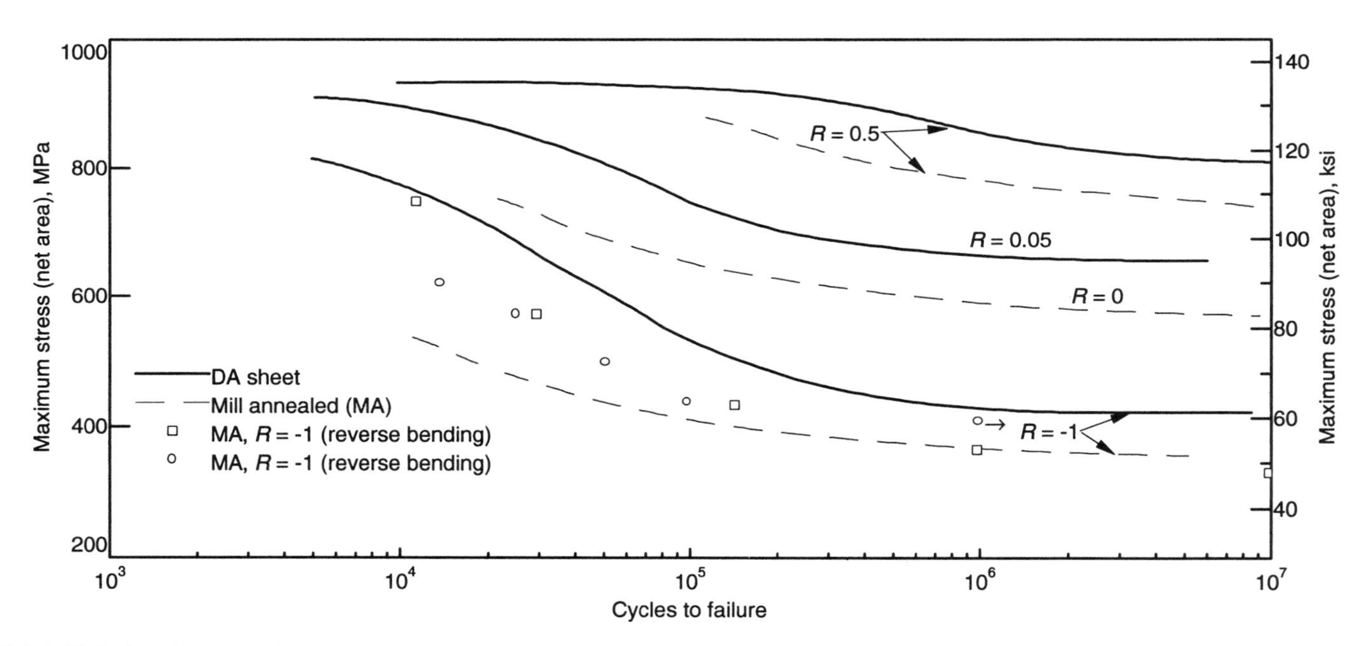

Ti-6Al-4V: Fatigue life of annealed unnotched sheet. Material/Test Parameters: Axial test except as indicated. Duplex anneal (DA) sheet = 910 °C (1675 °F) for 10 min/AC + 730 °C (1350 °F) for 4 h/AC. Sheet composition (wt%): 0.11 O, 6.0 Al, 4.0 V, 0.19 Fe, 0.01 N, 0.02 C, and 0.0034 H. Source: R.R. Boyer and R. Bajoraitis, "Standardization of Ti-6Al-4V Processing Conditions," AFML-TR-78-131, 1978

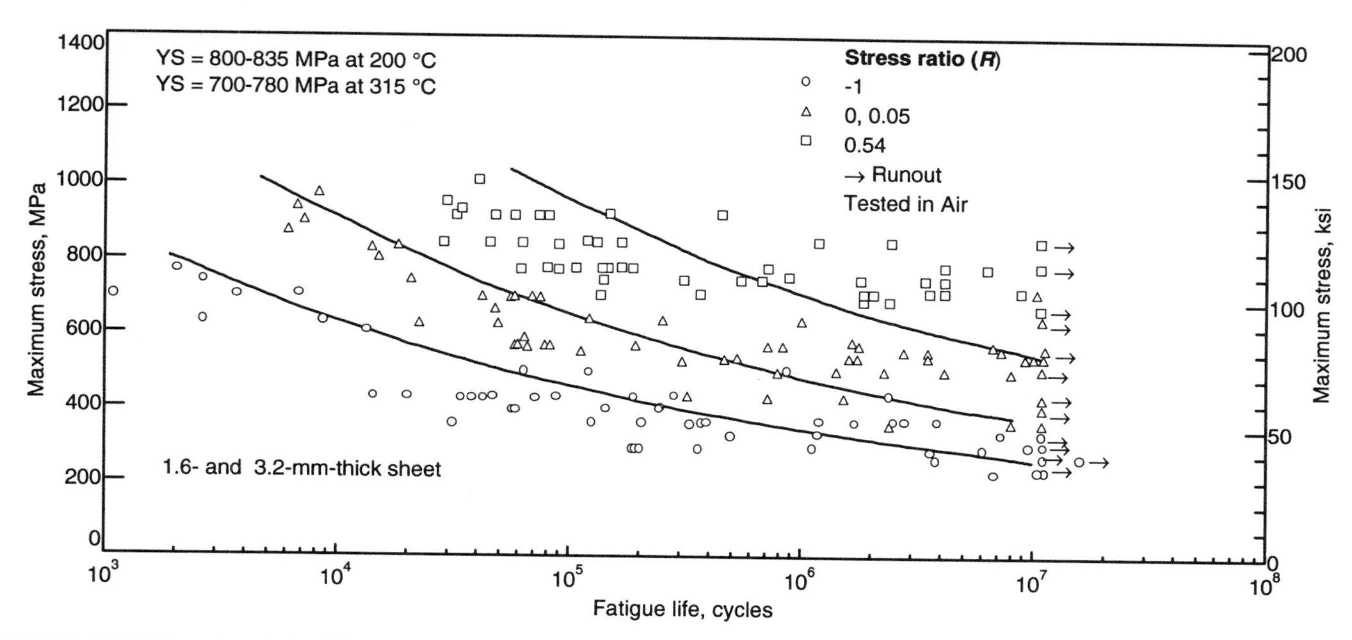

Ti-6Al-4V: Axial fatigue of unnotched STA sheet at 200 and 315 °C (400 and 600 °F). Conservative best-fit per MIL-HDBK 5. Material/Test Parameters: Longitudinal test direction. Machined specimens were cleaned with methyl ethyl ketone. Edges polished with No. 1 and 00 grit emery paper, recleaned with methyl ethyl ketone. Test frequency not specified. Source: Data from C. Hickey, Jr., Watertown Arsenal, 1962, and from North American Aviation Report, 1960, reported in *MIL-HDBK-5E*, U.S. Dept. of Defense, 1987, p 5-86

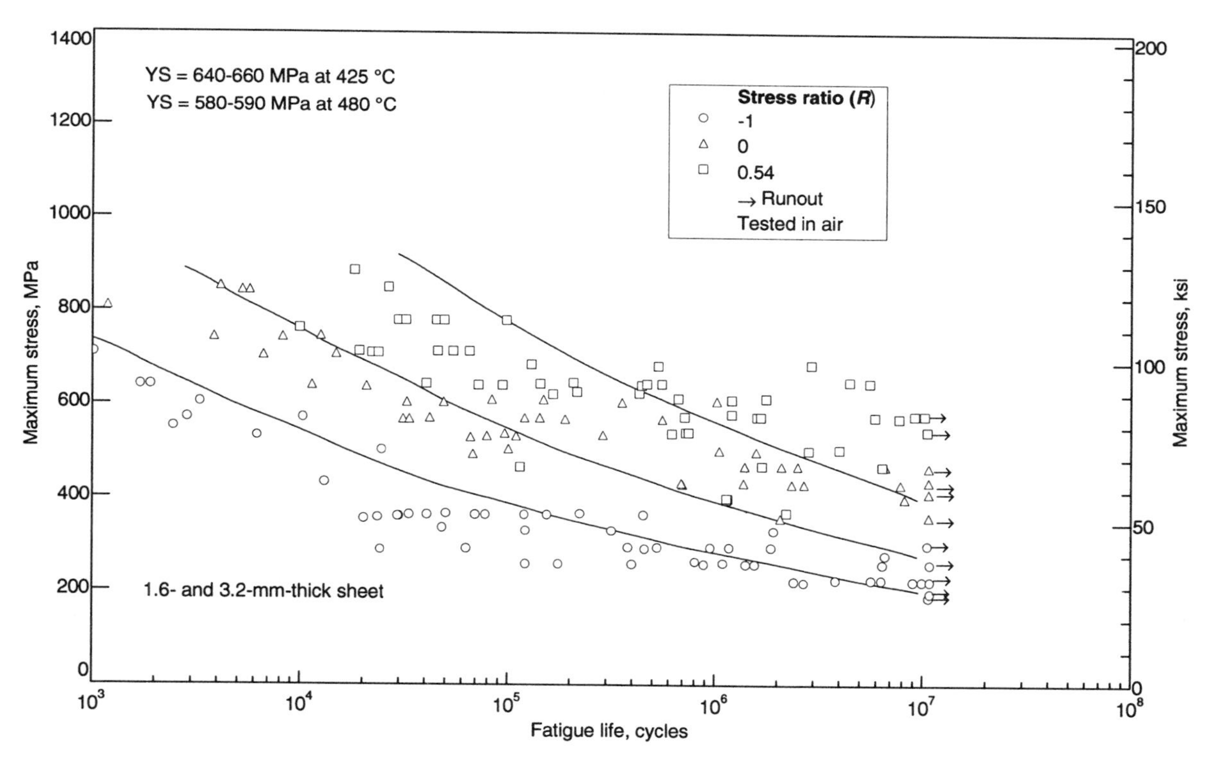

Ti-6Al-4V: Axial fatigue of unnotched sheet at 425 and 480 °C (800 and 900 °F). Conservative best-fit per MIL-HDBK 5. Material/Test Parameters: Longitudinal test direction. Machined specimens were cleaned with methyl ethyl ketone. Edges polished with No. 1 and 00 grit emery paper, recleaned with methyl ethyl ketone. Test frequency not specified. Source: Data from Lockheed-Georgia Co., "Determination of Design Data for Heat Treated Titanium Sheet," 1962, reported in *MIL-HDBK-5E*, U.S. Dept. of Defense, 1987, p 5-88

Strain Life

Initial cyclic strain hardening has been found to occur in Ti-6Al-4V material forged and heat-treated below the beta-transus (*Trans ASM*, 1972, p 263-270). This initial cyclic strain hardening, caused by an increase in dislocation density, was followed by subsequent strain softening.

In the beta-annealed material, initial strain hardening has not been observed (see figure). Softening, which is often associated with localized strains, can be enhanced for material with large slip length. In the beta-annealed (BA) titanium microstructure condition, the alpha platelets in a single colony tend to act as a single slip unit especially in the absence of a significant amount of beta phase between the alpha platelets as shown by Hack and Leverant (*Met Trans*, 13A, 1982, p 1729-1738).

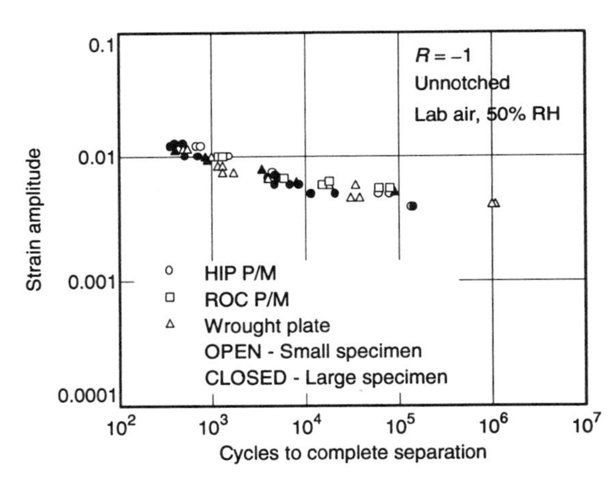

Ti-6Al-4V: Wrought and P/M strain life fatigue. P/M bars consolidated from plasma rotating electrode process (PREP) powder with consolidation by hot isostatic pressing (HIP) and rapid omnidirectional compaction (ROC). Oxygen for P/M bar and wrought plate was near average for standard Ti-6Al-4V. The fatigue lives of the large specimens from the HIP P/M bars were equal to those of the I/M plate. The fatigue lives of the large HIP P/M specimens were all shorter than those of the small P/M specimens (both HIP and ROC), except for one specimen tested at the lowest strain amplitude, 0.004, where they were nearly equal, suggesting a size effect in the HIP P/M materials. Source: SAMPE Quarterly, Oct 1988, p 15-19

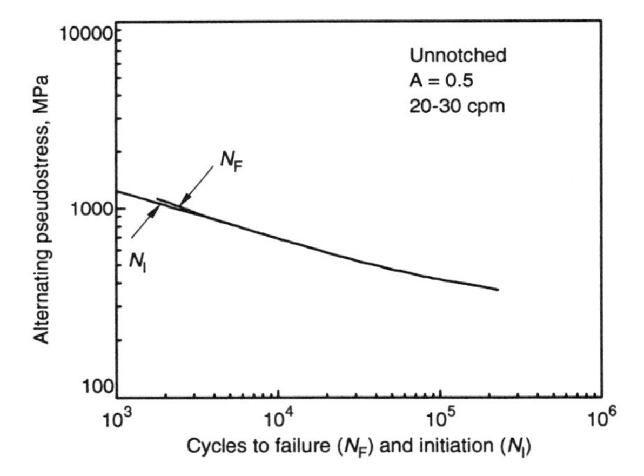

Ti-6Al-4V: LCF at room temperature. Axial-axial longitudinal strain control for forgings up to 150 mm (6 in.) thick; smaller size forgings may actually demonstrate higher pseudostress levels. Tangential orientation, test elastic modulus of 118.6 GPa (17.2×10^6 psi). Source: C. Shamblen, GE Aircraft

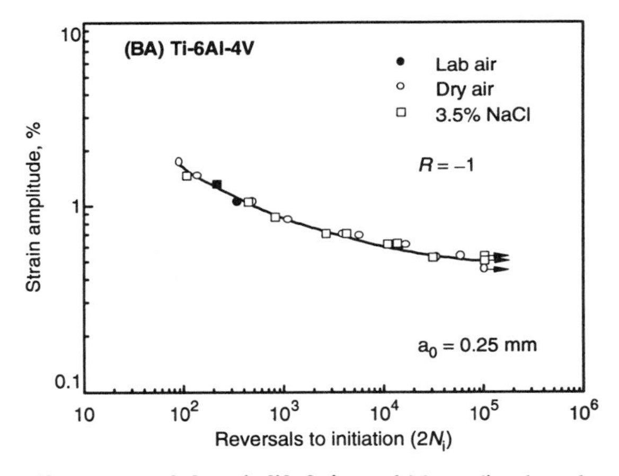

Ti-6Al-4V: Beta annealed strain life fatigue. 6.35 mm diam hourglass specimens from 22.2 mm plate. Cyclic softening occurred in the beta-annealed Ti-6Al-4V alloy at higher strain amplitudes. Both the tensile stress and compressive stress amplitudes decreased as a function of cycling. The percentage decrease in compressive stress amplitude was used to measure cyclic softening occurring and thus allowing the effects of "load-shedding" and cyclic softening to be separated in the measurements of maximum tensile stress. The onset of a 0.250 mm deep fatigue crack initiation was defined as the number of cycles when the tensile stress amplitude showed a 2% greater decrease than the compressive stress amplitude. Source: *Corrosion Cracking*, American Society for Metals, 1986, p 157-165

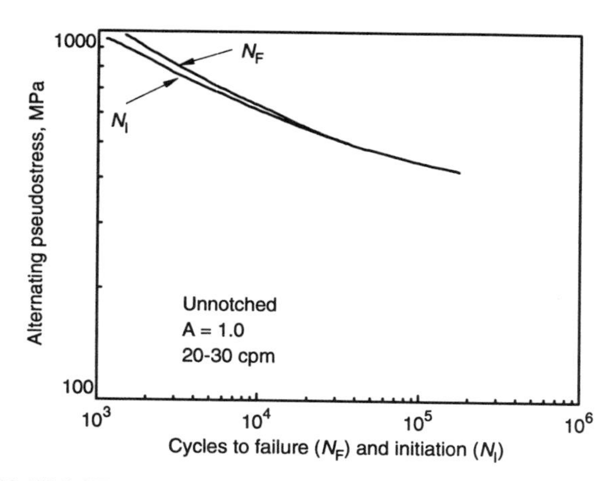

Ti-6Al-4V: LCF at 315 °C (600 °F). Axial-axial longitudinal strain control for forgings up to 150 mm (6 in.) section thickness; smaller size forgings would demonstrate higher pseudostress levels. Tangential orientation, elastic modulus 103 GPa (15×10^6 psi) at test temperature. Source: C. Shamblen, GE Aircraft Engines

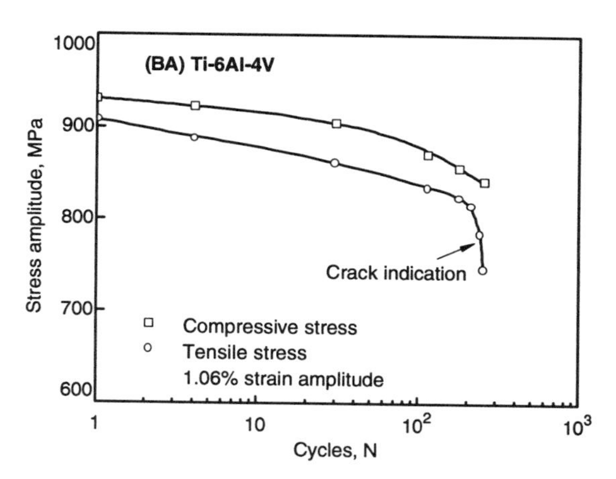

Ti-6Al-4V: Cyclic softening in beta annealed condition. Typical cyclic stress response curves in beta annealed (BA) Ti-6Al-4V. The significant softening is attributed to the alpha colony size, which was quite large (about 100 μm). Source: *Corrosion Cracking,* American Society for Metals, 1986, p 157-165

Notched Fatigue Strength

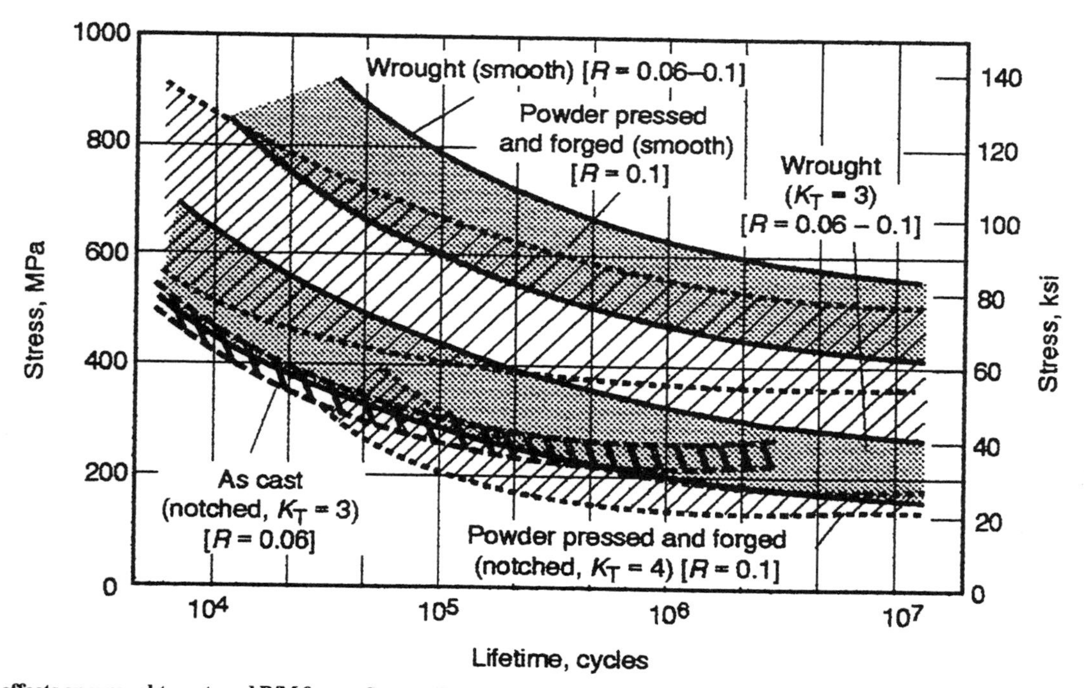

Ti-6Al-4V: Notch effects on wrought, cast, and P/M forms. Source: *Titanium and Titanium Alloys*, MIL-HDBK-697A, U.S. Dept. of Defense, 1974, p 46

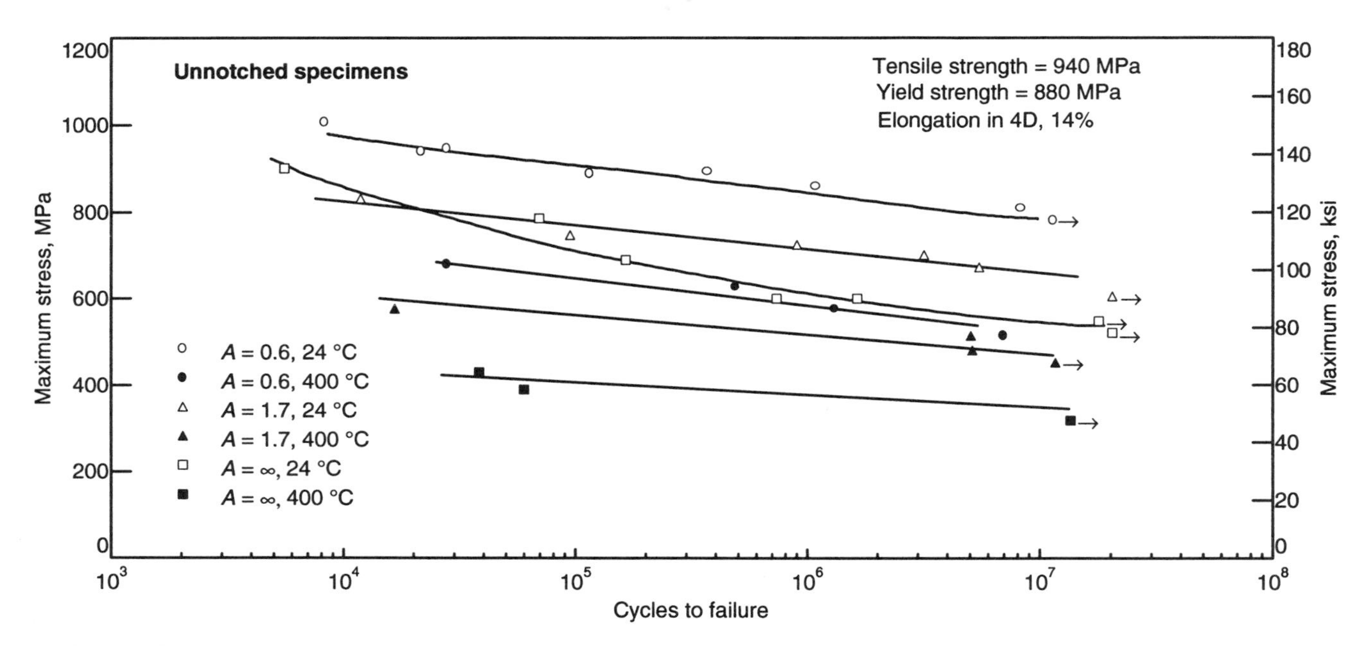

Annealed Ti-6Al-4V: Smooth axial fatigue. See figure below for condition. Source: *Metals Handbook*, Vol 1, 8th ed., *Properties and Selection of Metals*, American Society for Metals, 1961, p 530

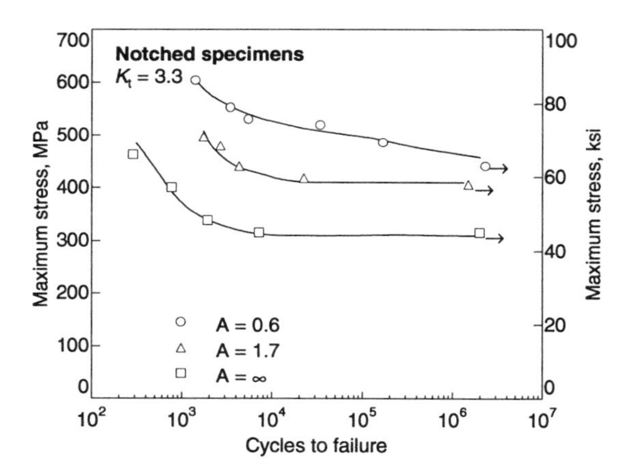

Annealed Ti-6Al-4V: Notched axial fatigue. Annealed sheet, UTS 136 ks TYS, 128 ksi; Elongation (in 4D), 14%. Source: *Metals Handbook*, Vol 1, 8th ed., *Properties and Selection of Metals*, American Society for Metals, 1961, p 530

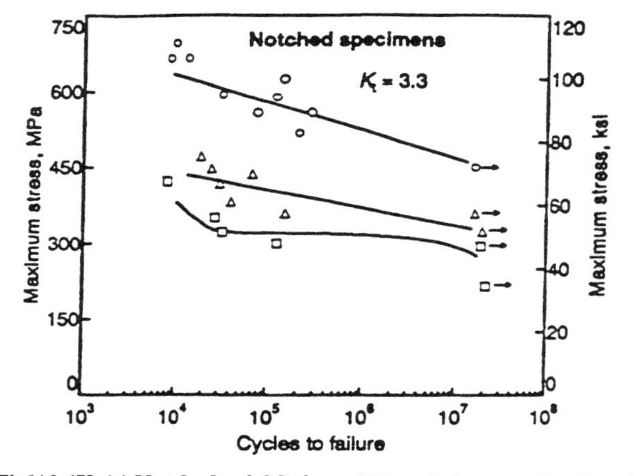

STA Ti-6Al-4V: (a) Notched axial fatigue. STA, solution treated and aged. See part (b) for conditions. Source: *Metals Handbook*, Vol 1, 8th ed., *Properties and Selection of Metals*, American Society for Metals, 1961, p 530

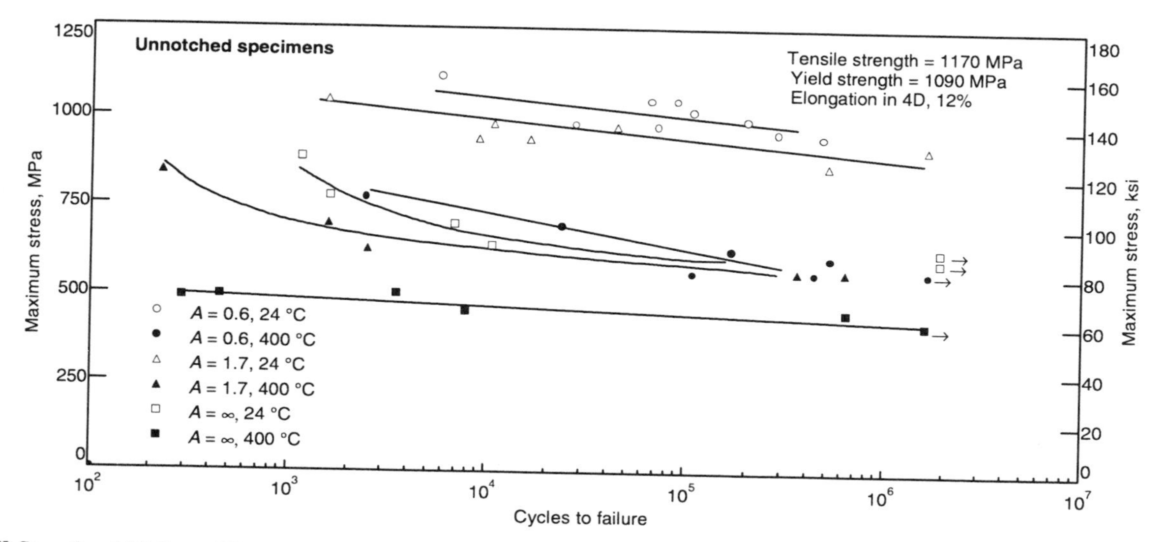

STA Ti-6Al-4V: Smooth axial fatigue. STA, solution treated and aged sheet

Plate

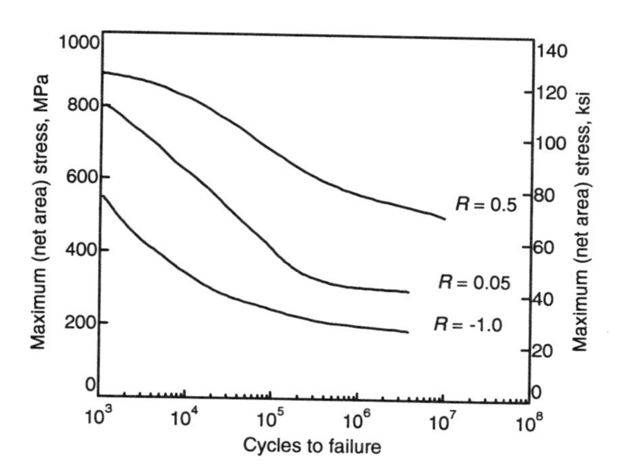

Ti-6Al-4V: Fatigue of notched (K_t = 2.53) β-annealed plate. Stresses based on net section. Material/Test Parameters: Beta anneal (BA) = 1000 °C (1845 °F) for 20 min/AC + 730 °C (1350 °F) for 2 h/AC. Material composition (wt%): 0.10 O, 6.1 Al, 4.0 V, 0.15 Fe, 0.01 N, 0.02 C, and 0.048 H. UTS of 895 MPa (130 ksi). Source: R.R. Boyer and R. Bajoraitis, "Standardization of Ti-6Al-4V Processing Conditions," AFML-TR-78-131, 1978

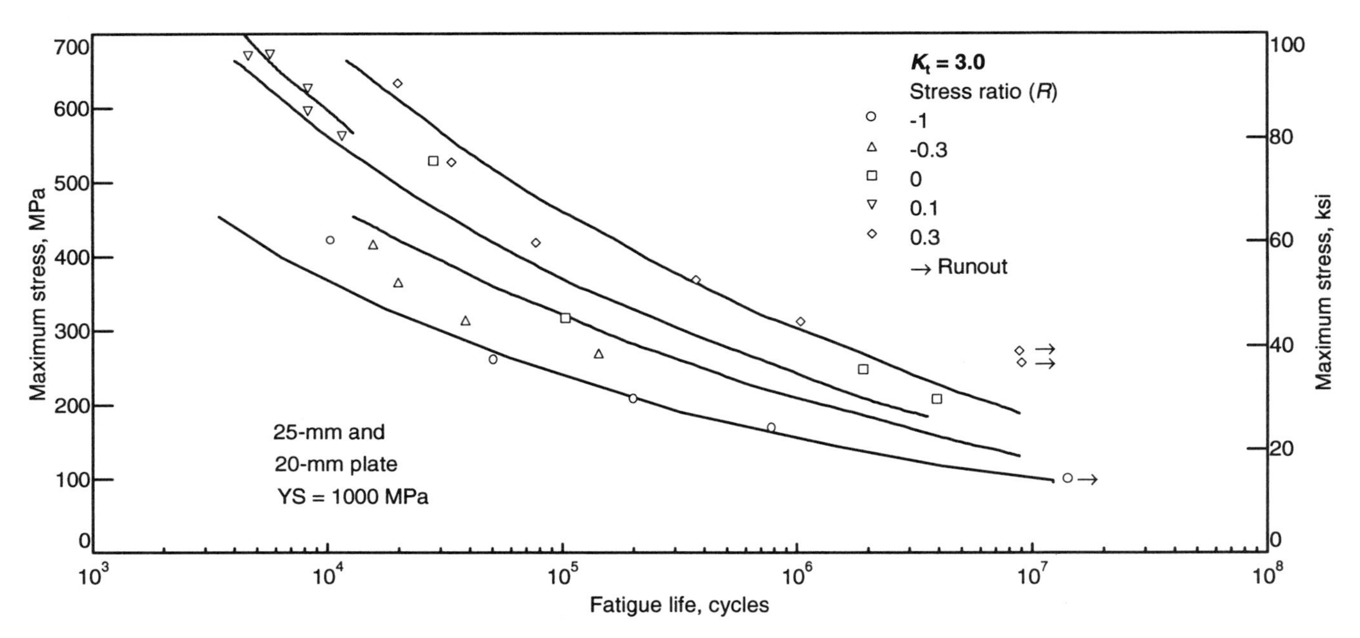

Ti-6Al-4V: Axial fatigue of notched STA plate. Note: Stresses are based on net section. Conservative best-fit curve from MIL-HDBK. Material/Test Parameters: Longitudinal test direction in laboratory air with a test frequency of 30-300 Hz. Source: Data from A. Sommer and G. Martin, North American Rockwell, TR-69-161, 1969; and M. Sargent, General Dynamics, "Fatigue Characteristics of Ti-6Al-4V Plate and Forgings," 1965, reported in *MIL-HDBK-5E*, U.S. Dept. of Defense, 1987, p 5-88d

Bar and Extrusions

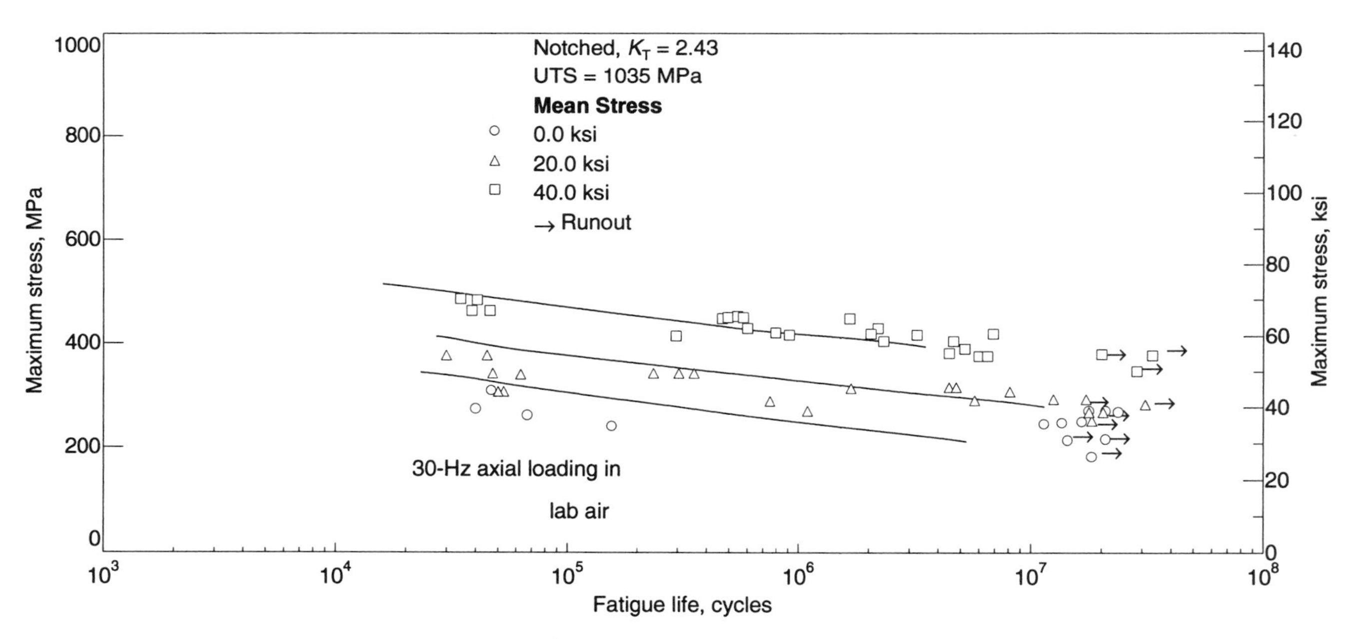

Ti-6Al-4V: Axial notched fatigue in longitudinal direction of bar. Note: Stresses are based on net section. Material/Test Parameters: Specimens were taken from 25-mm (1-in.) diam annealed bar and had a 2.5-μm (100-μin.) machined surface roughness (R_{RMS}). Source: Data from Sikorsky Aircraft, "Fatigue Evaluation of Ti-6Al-4V Bar Stock," 1970, reported in *MIL-HDBK-5E*, U.S. Dept. of Defense, 1987, p 5-71

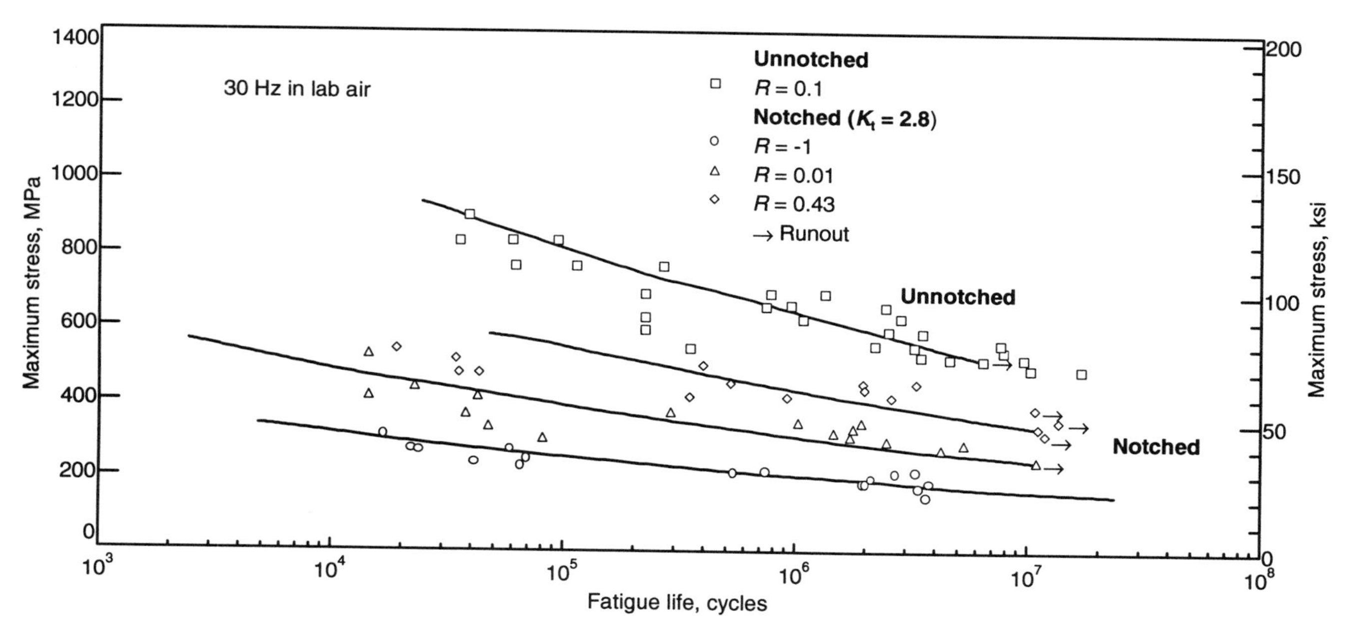

Ti-6Al-4V: Axial fatigue of annealed extrusions (YS = 875 MPa), smooth and notched. Material/Test Parameters: Unnotched and notched specimens were from 7.5- and 14-mm (0.300- and 0.560-in.) thick extrusions with a tensile strength of 985 MPa (143 ksi) and an RMS surface roughness of 1.6 μm (63 μin.). Testing was performed in longitudinal direction at room temperature. Source: Data from R. Brockett and J. Gottbrath, Lockheed-California Co., AFML-TR-67-189, 1967, reported in *MIL-HDBK-5E*, U.S. Dept. of Defense, 1987, p 5-72

Sheet

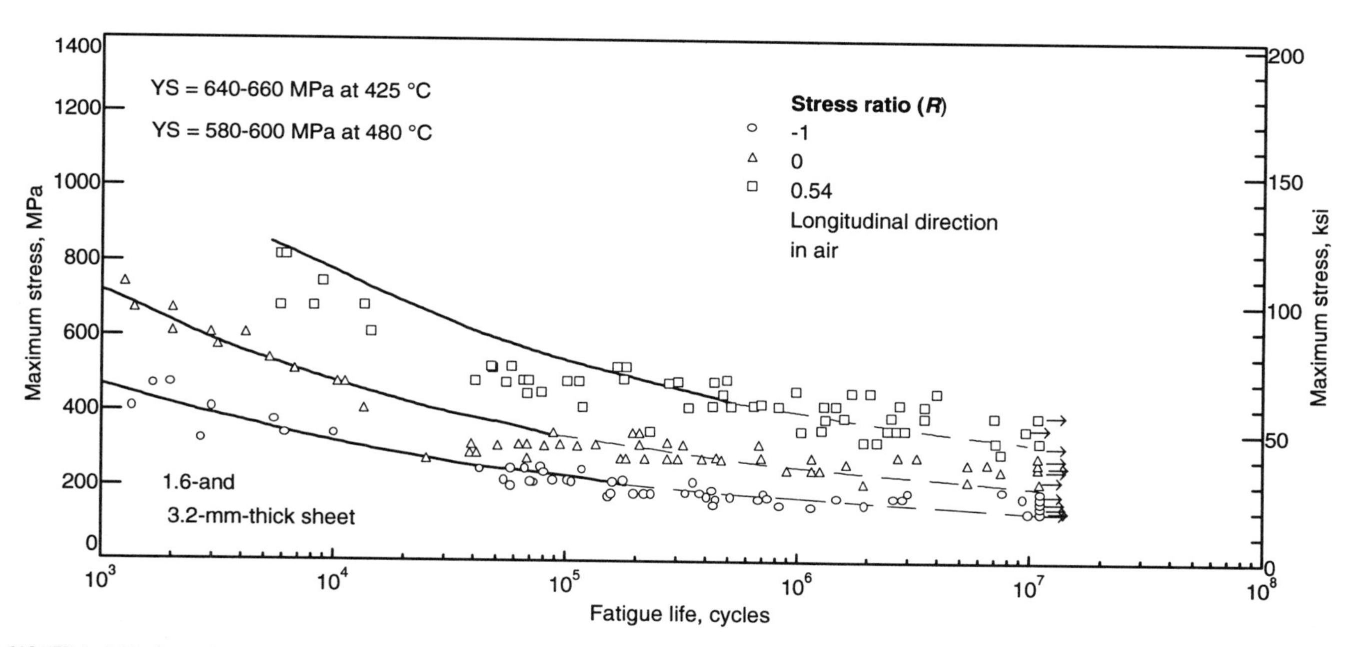

Ti-6Al-4V: Axial fatigue of notched (K_t =2.8) STA sheet at 425 and 480 °C (800 and 900 °F). Note: Stresses are based on net section. Material/Test Parameters: Machined specimens were cleaned with methyl ethyl ketone. Edges polished with No. 1 and 00 grit emery paper, recleaned with methyl ethyl ketone. Test frequency, 1500-2200 cycles/min. Source: Data from Lockheed-Georgia Co., "Determination of Design Data for Heat Treated Titanium Sheet," 1962, reported in *MIL-HDBK-5E*, U.S. Dept. of Defense, 1987, p 5-88a

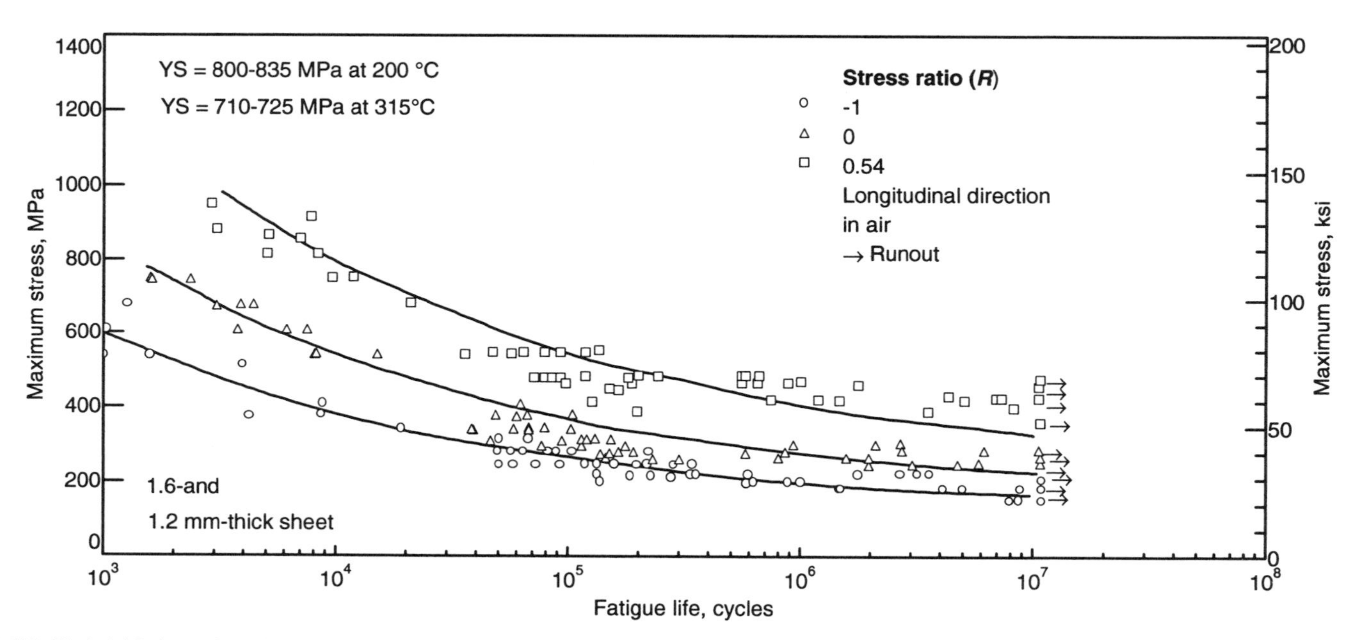

Ti-6Al-4V: Axial fatigue of notched (K_t =2.8) STA sheet at 200 and 315 °C (400 and 600 °F). Note: Stresses are based on net section. Material/Test Parameters: Machined specimens were cleaned with methyl ethyl ketone. Edges polished with No. 1 and 00 grit emery paper, recleaned with methyl ethyl ketone. Test frequency, 1500-2200 cycles/min. Source: Data from Lockheed-Georgia Co., "Determination of Design Data for Heat Treated Titanium Sheet," 1962, reported in *MIL-HDBK-5E*, U.S. Dept. of Defense, 1987, p 5-87

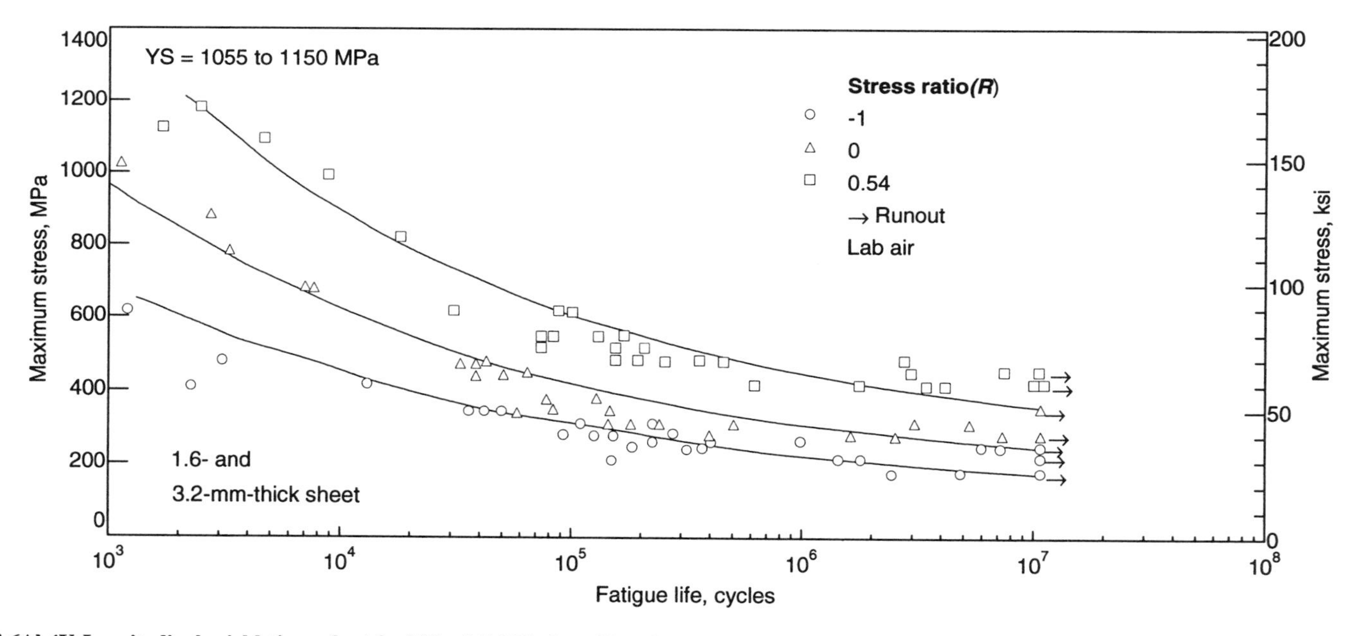

Ti-6Al-4V: Longitudinal axial fatigue of notched (K_t = 2.8) STA sheet. Note: Stresses are based on net section. UTS 1145 to 1220 MPa (166 to 177 ksi). Material/Test Parameters: Machined specimens were cleaned with methyl ethyl ketone. Edges polished with No. 1 and 00 grit emery paper, recleaned with methyl ethyl ketone. Test frequency, 1500-2200 cycles/min. Source: Data from Lockheed-Georgia Co., "Determination of Design Data for Heat Treated Titanium Sheet," 1962, reported in *MIL-HDBK-5E*, U.S. Dept. of Defense, 1987, p 5-85

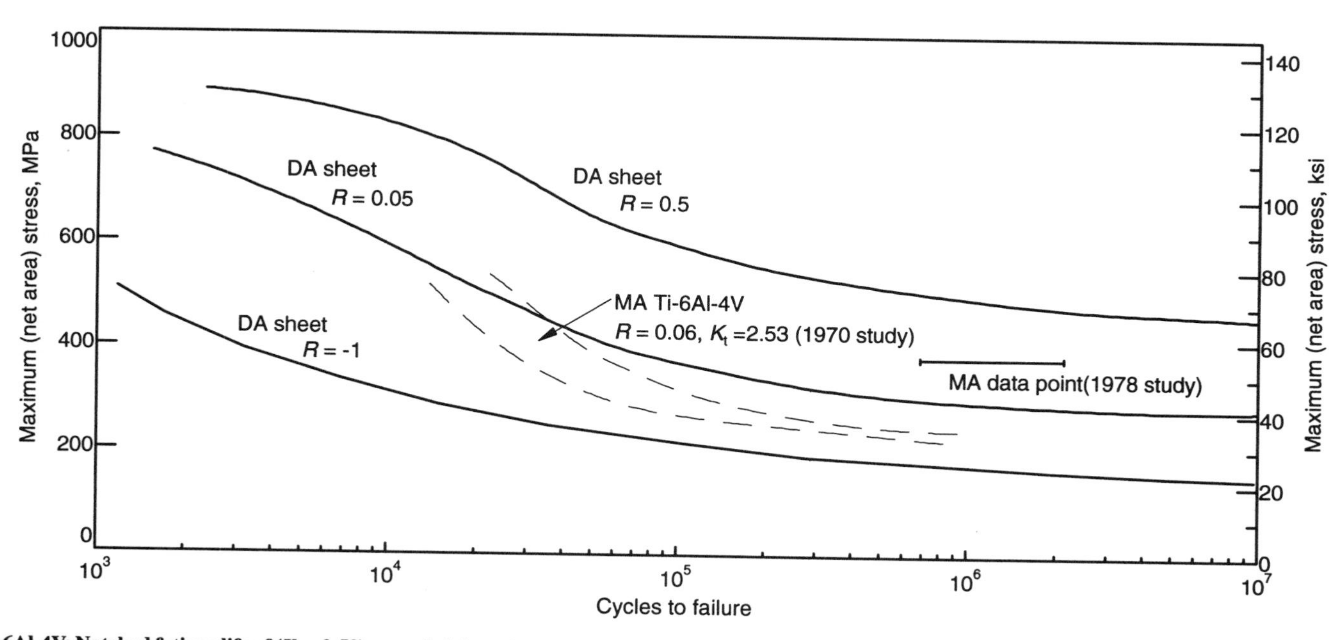

Ti-6Al-4V: Notched fatigue life of (K_t = 2.53) annealed sheet. Material/Test Parameters: Duplex anneal (DA) sheet = 910 °C (1675 °F) for 10 min/AC + 730 °C (1350 °F) for 4 h/AC. Sheet composition (wt%): 0.11 O, 6.0 Al, 4.0 V, 0.19 Fe, 0.01 N, 0.02 C, and 0.0034 H. Source: R.R. Boyer and R. Bajoraitis, "Standardization of Ti-6Al-4V Processing Conditions," AFML-TR-78-131, 1978

Cast and P/M Fatigue

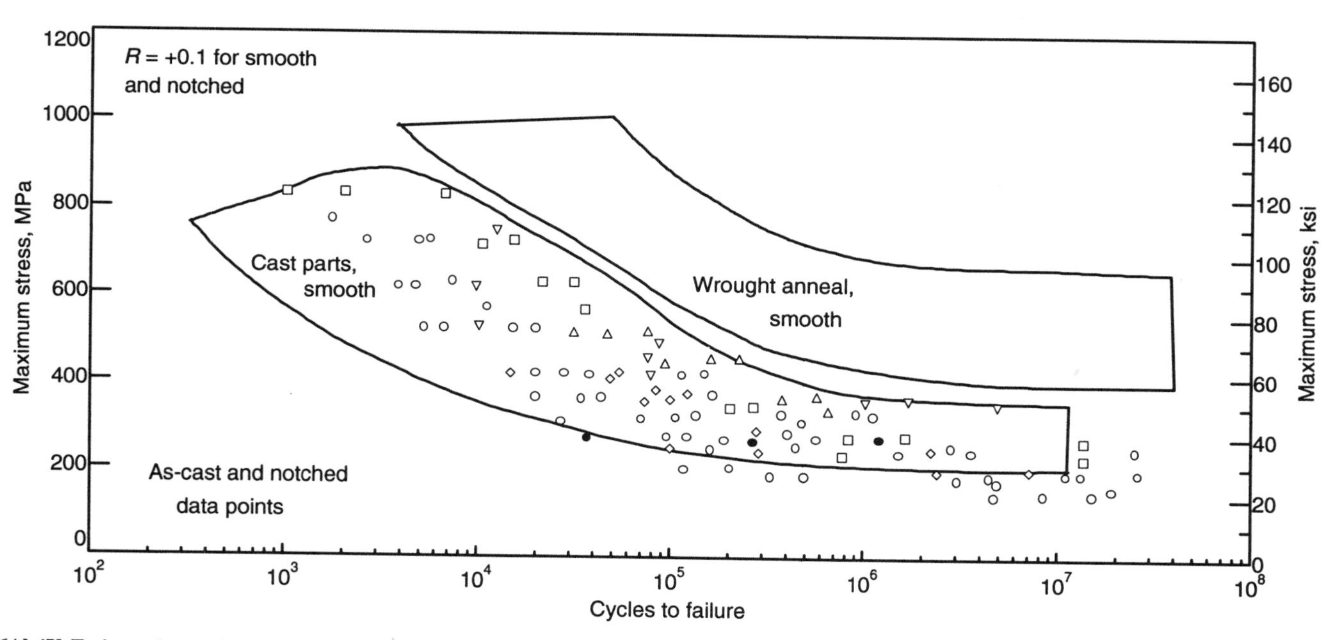

Ti-6Al-4V: Fatigue of smooth and notched castings compared. Data points for notched castings are replotted here. Source: *Metals Handbook*, Vol 3, 9th ed., *Properties and Selection: Stainless Steels, Tool Materials and Special-Purpose Metals*, American Society for Metals, 1980, p 411. Most data points are based on K_t = 3.0, but some are for smooth specimens (K_t = 1). Each symbol represents data from a different source.

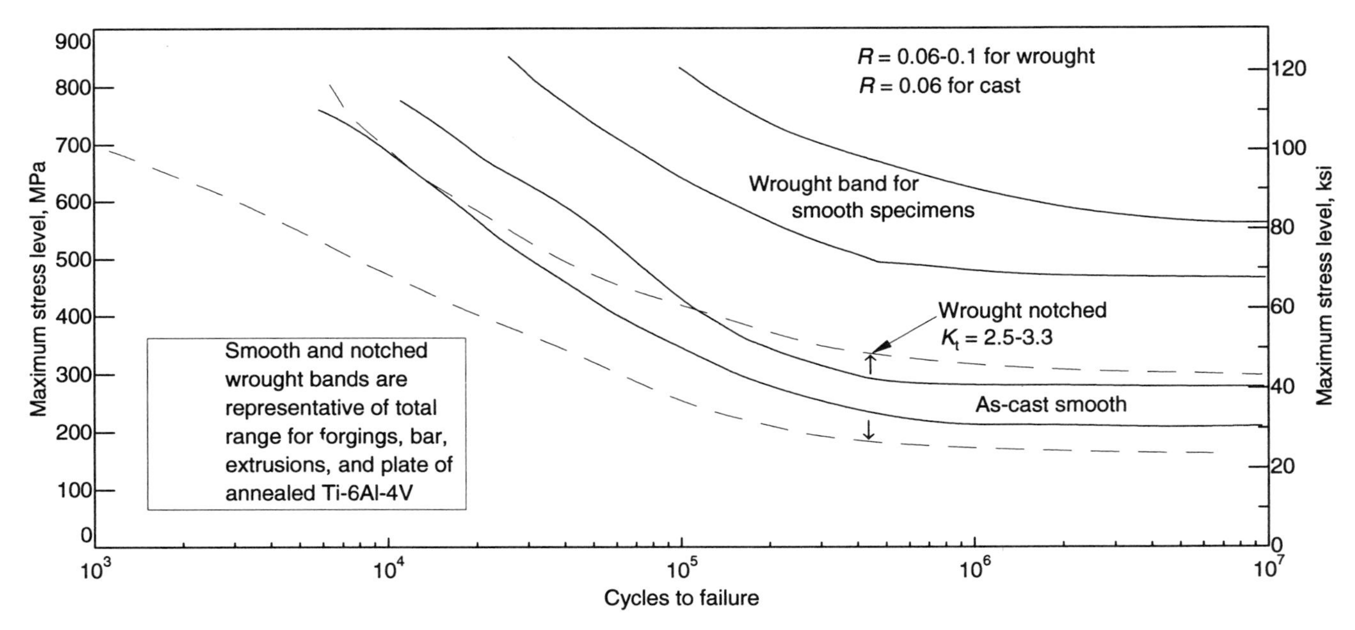

Ti-6Al-4V: Smooth fatigue of castings compared to wrought Ti-6Al-4V. Source: Data from Titech International, Inc., Research Report 114, 1979, reported in *Titanium Alloys Handbook*, MCIC-HB-02, Battelle Columbus Laboratories, 1972, reprinted 1985, p 2-72:37

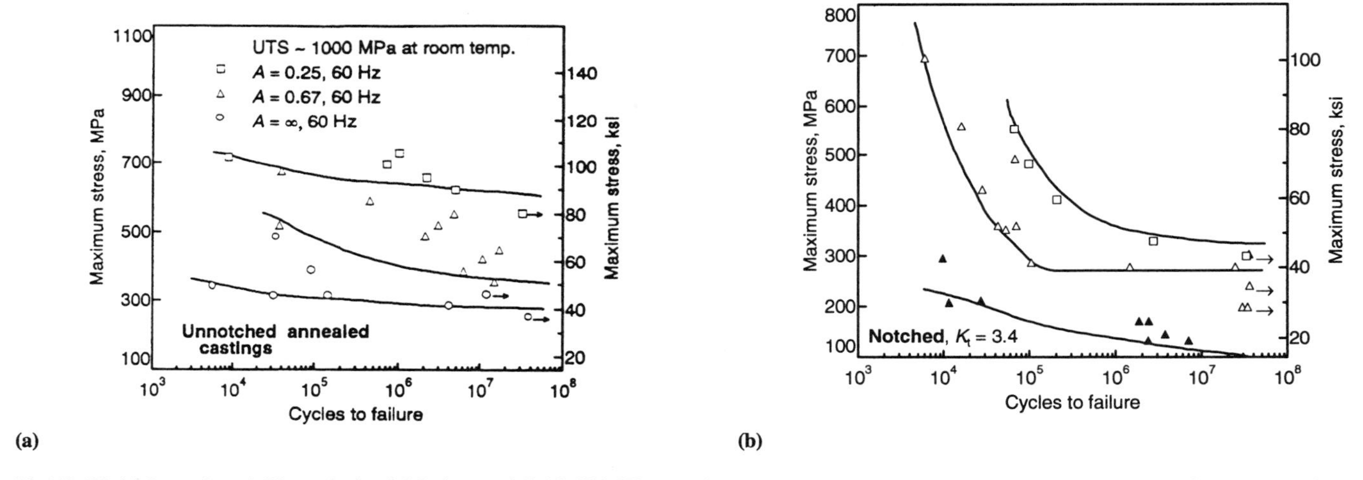

Ti-6Al-4V: (a) Smooth and (b) notched axial fatigue at 260 °C (500 °F). Material/Test Parameters: Data were obtained from two heats of annealed castings having room-temperature ultimate tensile strengths of 997 and 1000 MPa (144.6 and 145.1 ksi). Cast-to-size specimens were annealed at 700 °C (1300 °F) for 2 h and air cooled. Longitudinal polish of smooth specimen had an RMS roughness of less than 0.1 μm (4 μin.). Source: Data from R. Dalal, AVCO Corp., reported in *Aerospace Structural Metals Handbook*, Battelle Columbus Laboratories, Code 3801, p 19

P/M

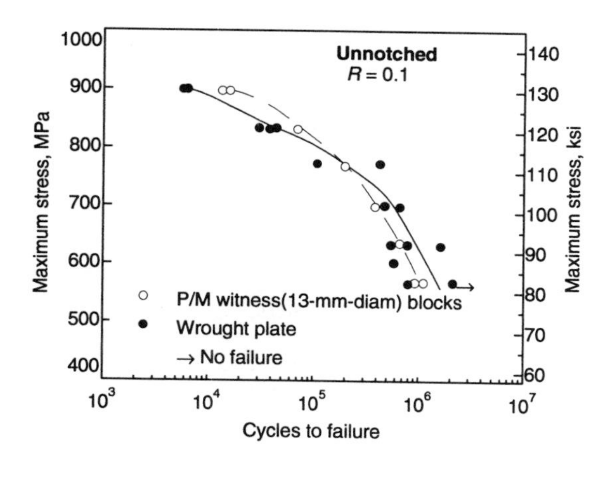

(a)

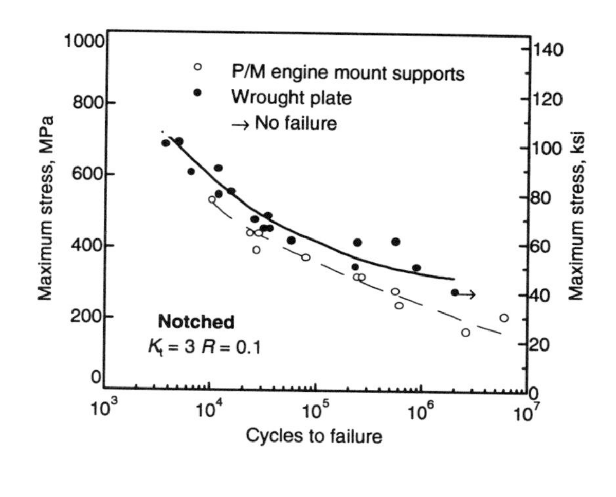

(b)

Prealloyed Ti-6Al-4V P/M parts: (a) Smooth and (b) notched fatigue. Source: A.S. Sheinker *et al.*, Evaluation and Application of Prealloyed Titanium P/M Parts for Airframe Structures, *Int. J. Powder*, Vol 23 (No. 3), 1987, p 171-176

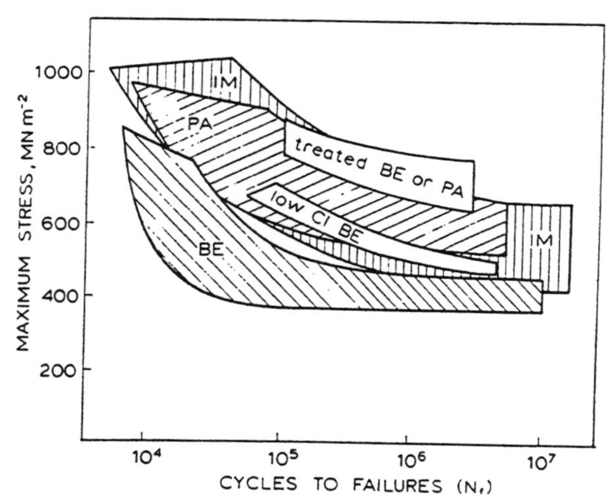

Ti-6Al-4V: Fatigue bands of I/M and P/M products. Note: I/M, ingot metallurgy; PA, prealloyed P/M products; BE, blended elemental P/M products. Source: F.H. Froes and D. Eylon, Powder Metallurgy of Ti Alloy, *Int. Mat. Rev.*, Vol 35 (No. 3), 1990

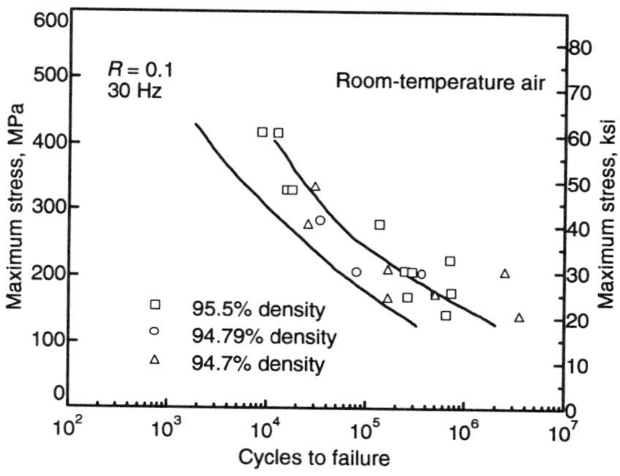

Ti-6Al-4V: Axial fatigue of notched ($K_t = 2.16$) BE powder specimens. Note: Combined middle-density cycles to failure results. Material/Test Parameters: Pressed and sintered without post-sinter compaction. Source: J.A. Miller and G. Brodi, "Consolidation of Blended Elemental Ti-6Al-4V Powder to Near Net Shapes," AFML-TR-79-4028, 1979

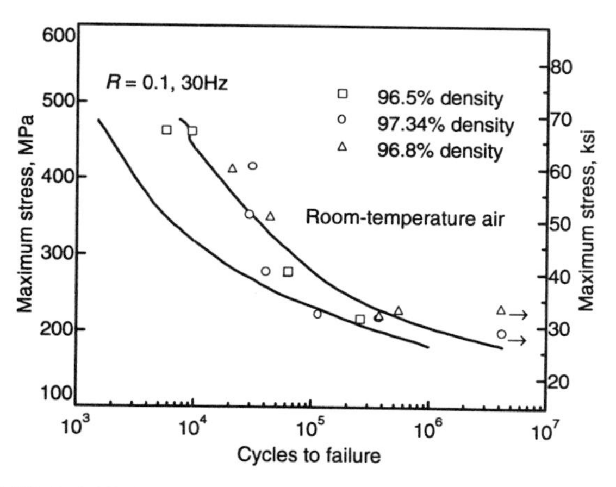

Ti-6Al-4V: Axial fatigue of notched ($K_t = 2.16$) BE powder specimens. Note: Combined high-density cycles to failure results. Material/Test Parameters: Pressed and sintered with no post-sinter compaction. Source: J.A. Miller and G. Brodi, "Consolidation of Blended Elemental Ti-6Al-4V Powder to Near Net Shapes," AFML-TR-79-4028, 1979

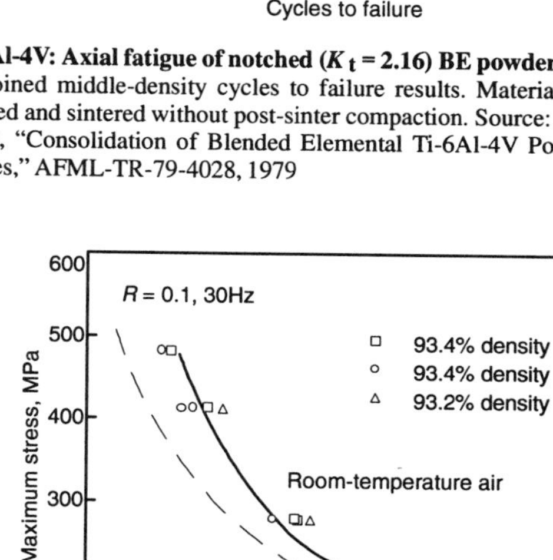

Ti-6Al-4V: Axial fatigue of notched ($K_t = 2.16$) BE powder specimens with porosity. Note: Combined cycles-to-failure results for low-density P/M. Material/Test Parameters: Pressed and sintered without post-sinter compaction. Source: J.A. Miller and G. Brodi, "Consolidation of Blended Elemental Ti-6Al-4V Powder to Near Net Shapes," AFML-TR-79-4028, 1979

Corrosion Fatigue

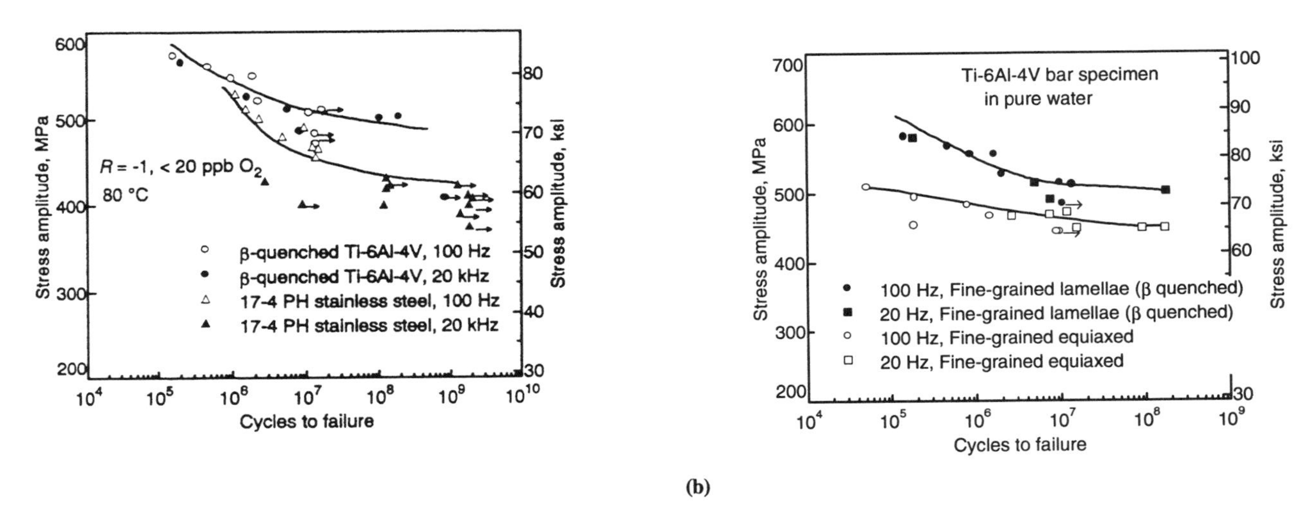

Ti-6Al-4V: Fatigue strength in pure water. (a) β-quenched material compared to a stainless steel in water with dissolved O_2. (b) Fine-grained equiaxed microstructure compared to a fine-grained transformed structure. Source: Adapted from L.D. Roth and L.E. Willertz, in *Environment Sensitive Fracture: Evaluation and Comparison of Test Methods*, STP 821, E.N. Pugh and G.M. Ugiansky, Ed., American Society for Testing and Materials, 1984, p 497; and L.E. Willertz *et al.*, High and Low Frequency Corrosion Fatigue of Some Steam Turbine Blade Alloys, in *Ultrasonic Fatigue*, J.M. Wells *et al.*, Ed., AIME, 1982, p 333-348

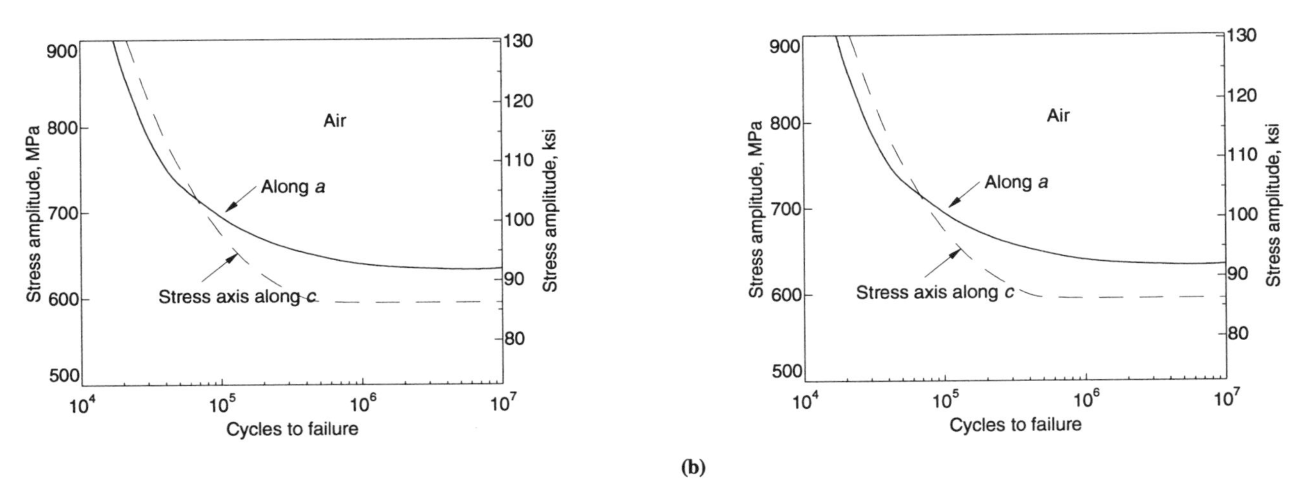

Ti-6Al-4V: Effect of texture and environment on fatigue. The *S-N* curves for Ti-6Al-4V processed in the α-β field show the effects of texture and environment on fatigue strength in (a) air and (b) 3.5% NaCl. These data show that testing in an aqueous 3.5% NaCl solution reduces fatigue strength when the stress axis is along [0001]. Source: J.C. Williams and E.A. Starke, Jr., The Role of Thermomechanical Processing in Tailoring the Properties of Aluminum and Titanium Alloys, in *Deformation, Processing, and Structure*, G. Krauss, Ed., American Society for Metals, 1984, p 335

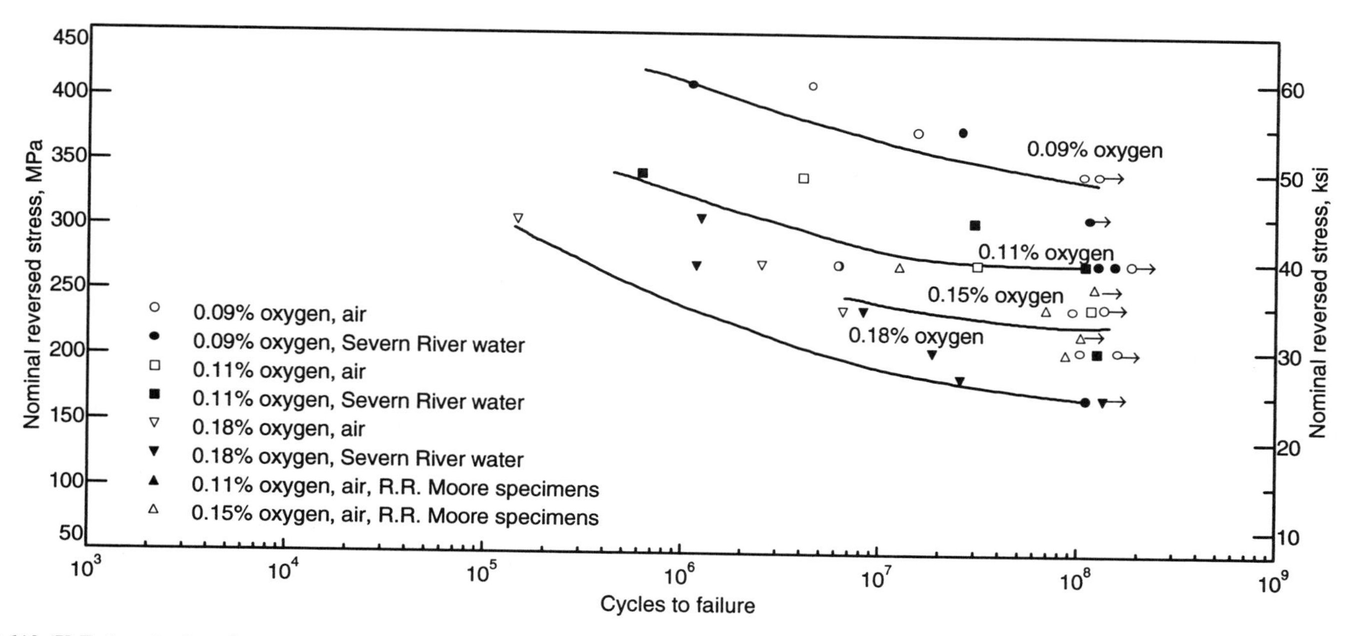

Ti-6Al-4V: Fatigue in air and water of castings with varying oxygen content. Material/Test Parameters: Rotating cantilever beam notched test specimens were machined to 10.7 mm (0.420 in.) square by 12.5 mm (5 in.) long, containing a 0.005-mm (0.0002-in.) root radius notch. Source: A. Morton and I. Lane, Jr., Titanium Castings for Marine Propellers, in *Titanium Science and Technology*, R. Jaffee and H. Burte, Ed., Plenum Press, 1973, p 119

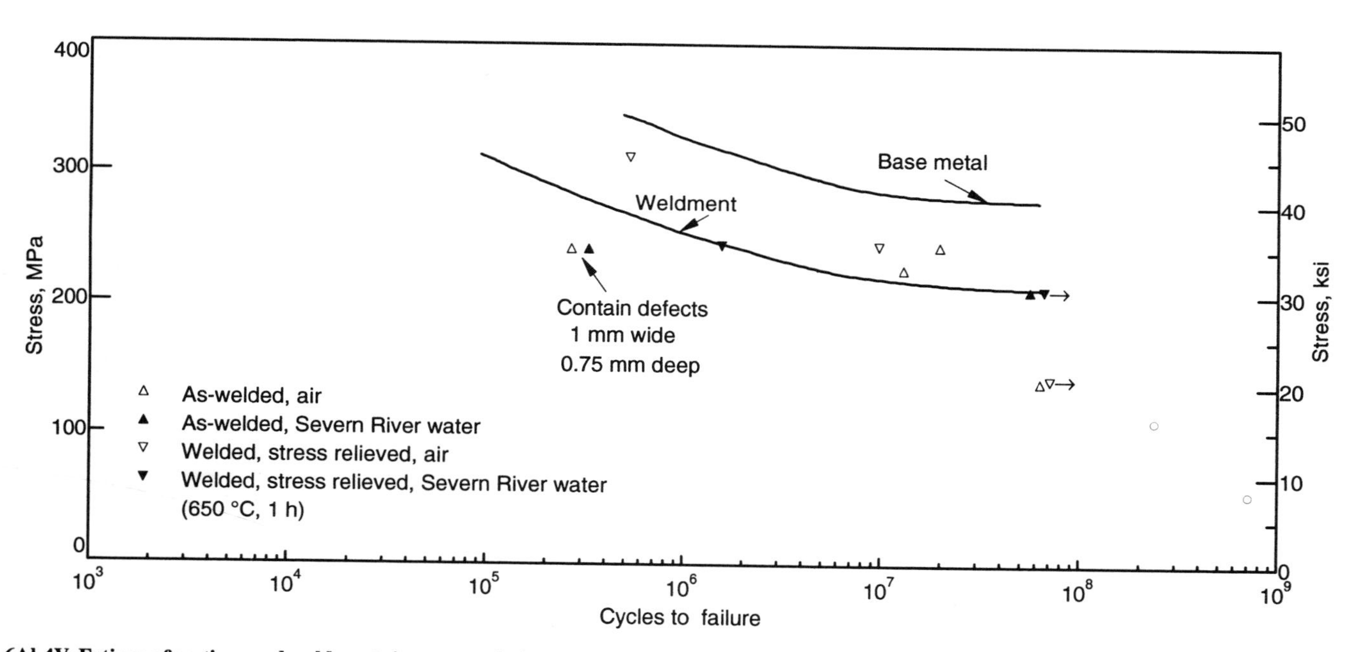

Ti-6Al-4V: Fatigue of castings and weldments in water and air. Material/Test Parameters: Composition (wt%) of castings was 6.02 Al, 0.029 C, 0.11 Fe, 0.0035 H, 0.015 N, 0.11 O, and 4.12 V. Weldments were made by gas metal arc welding with Ti-6Al-4V filler metal containing 0.08% O. Rotating cantilever beam test specimens were machined. Source: A. Morton and I. Lane, Jr., Titanium Castings for Marine Propellers, in *Titanium Science and Technology*, R. Jaffee and H. Burte, Ed., Plenum Press, 1973, p 119

Compared to Stainless Steel

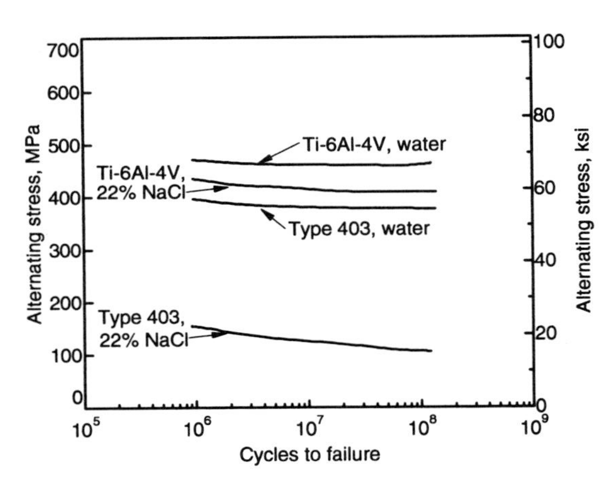

Ti-6Al-4V corrosion fatigue: Compared with type 403 stainless steel. Material/Test Parameters: Corrosion fatigue strength comparison of Ti-6Al-4V and type 403 stainless steel in 22% NaCl solution at 80 °C (176 °F), with <20 ppb O_2, pH 4

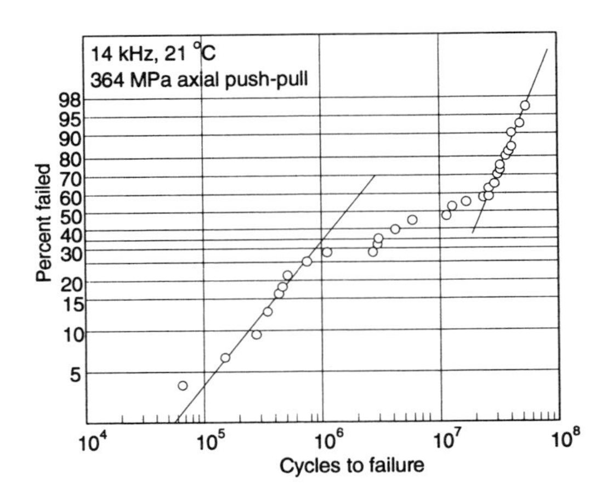

Ti-6Al-4V: Weibull plot for 30 specimens in water. Annealed Ti-6Al-4V in water cooling bath. Source: A. Thiruvengadam, Corrosion Fatigue at High Frequencies and High Hydrostatic Pressures, in *Stress Corrosion Cracking of Metals—A State of the Art*, STP 518, ASTM, 1972, p 139-154

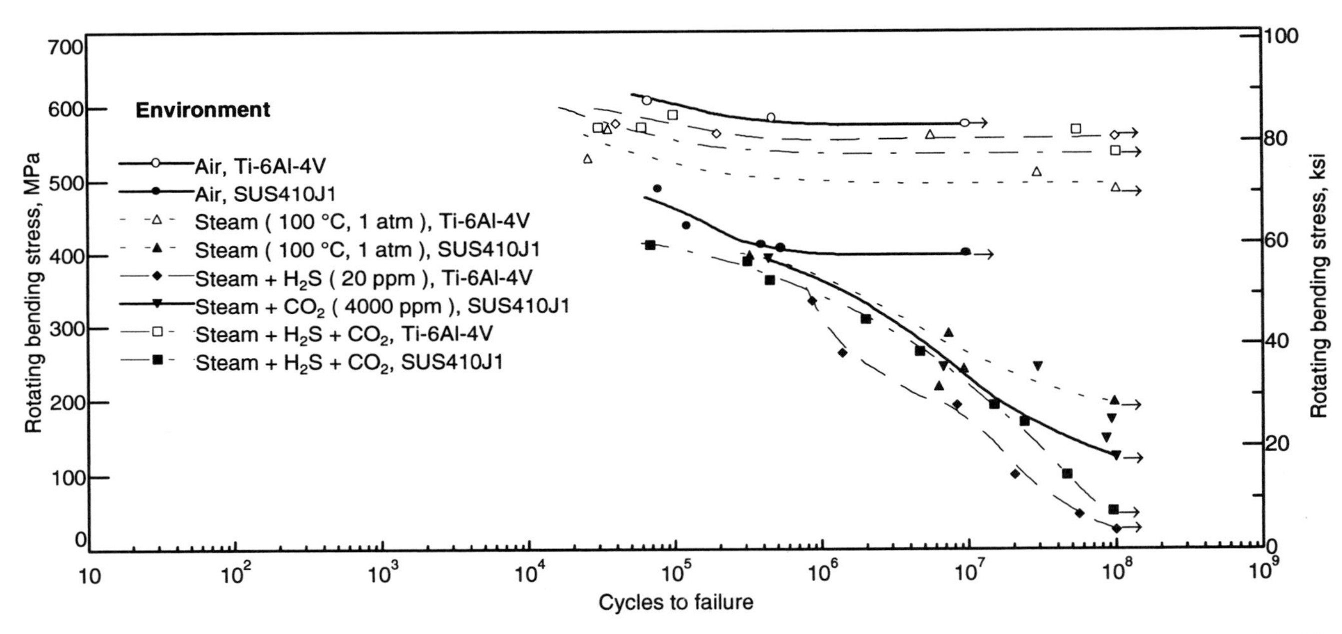

Ti-6Al-4V: Fatigue compared with a stainless steel. Comparison of the effect of industrial gas in steam on the fatigue strength of Ti-6Al-4V and an 11.5-14% Cr stainless steel (SUS410J1)

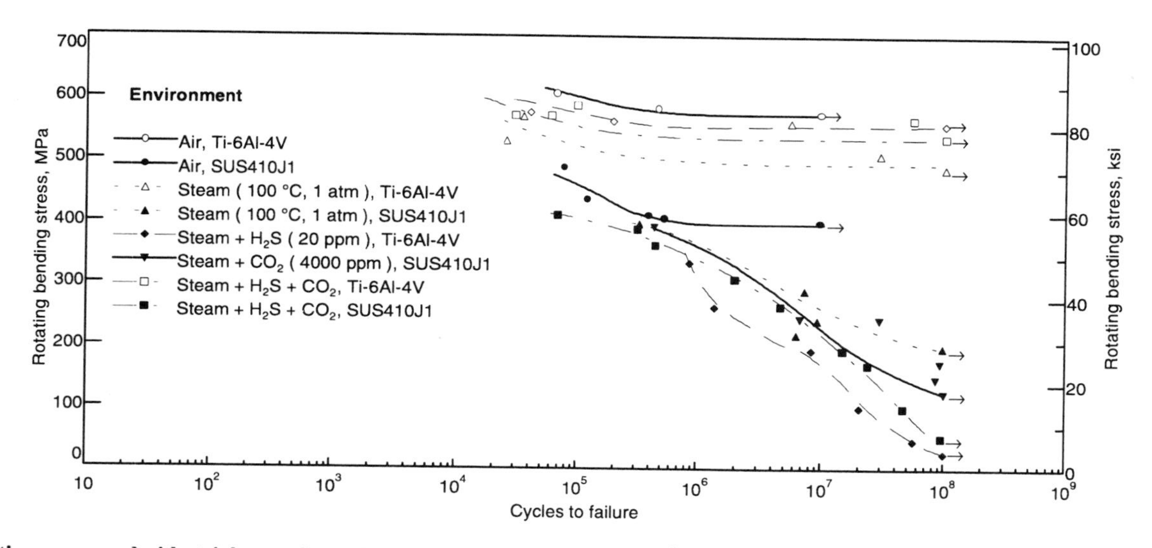

Ti-6Al-4V: Fatigue compared with stainless steel. Asterisk (*) denotes extrapolation from 10^8 cycles. Fatigue strength of Ti-6Al-4V in the last environment (denoted by *) is meaningless because the material dissolves. H, high. L, low. Source: L.E. Willertz *et al.*, in *Ultrasonic Fatigue*, J.M. Wells *et al.*, Ed., TMS-AIME, 1982, p 333

Fatigue Crack Growth in Air

Because fatigue crack propagation (FCP) resistance generally is improved with microstructures containing increased amounts of transformed β morphology, the slowest crack growth rates are frequently observed in products such as castings and β annealed or β processed parts. Age hardening reduces FCP resistance due to lower intrinsic ductility associated with increased strength of the STA condition.

Microstructure morphology can have significant effects on FCP resistance in inert environments, and it must be cautioned that laboratory air is not inert, even for room-temperature FCP testing.

Ti-6Al-4V has slower crack growth rates than aluminum and somewhat faster rates than steel. In aggressive environments such as seawater,

Ti-6Al-4V: FCP data compared with aluminum and steel

Alloy and condition	da/dN(a) µm/cycle	da/dN(a) µin./cycle	Test frequency cycles/min
Ti-6Al-4V, mill annealed	0.5	20	60
Ti-6Al-4V, β annealed	0.2	8	60
7075-T7351	2.5	100	360
HP-9-4-30(b)	0.1	4	360
300M steel	0.13	5	360

(a) All data are for a stress-intensity range (ΔK) of 22 MPa$\sqrt{m}$ (20 ksi$\sqrt{in.}$) and a stress ratio of $R = 0.3$ tested in low-humidity air. (b) High fracture toughness steel having a quenched and tempered yield strength of 1515 to 1655 MPa (220 to 240 ksi). Source: Fracture Mechanics Evaluation of B-1 Materials AFML-TR-76-137, Oct 1976

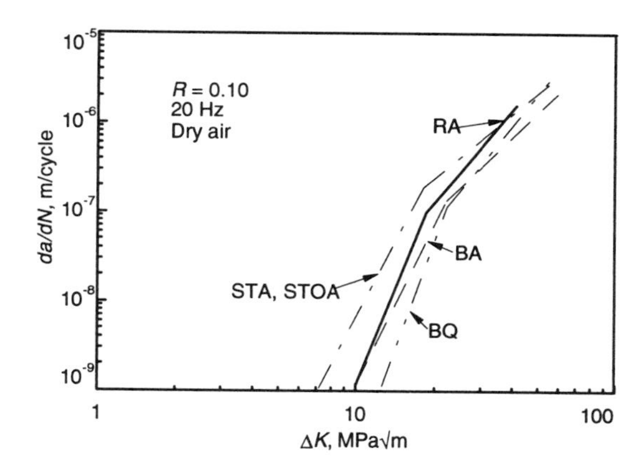

Ti-6Al-4V: Beta-quenched FCP rates vs. other heat treatments. Material/Test Parameters: The five microstructures were produced from pancake forgings to obtain material that was uniformly and very weakly textured. Microstructures selected were either β forged and annealed (BA) or water quenched (BQ), or were forged in the α + β region and heat treated to produce microstructures that were recrystallization annealed (RA), solution treated and aged (STA) at 590 °C (1095 °F), or solution treated and overaged (STOA). These microstructures were characterized in some detail with transmission microscopy. Cited: A.W. Thompson *et al.*, The Effect of Microstructure on Fatigue Crack Propagation Rate in Ti-6Al-4V, in *Titanium and Titanium Alloys, Scientific and Technological Aspects;* Vol 1, J.C. Williams and A.F. Belov, Ed.

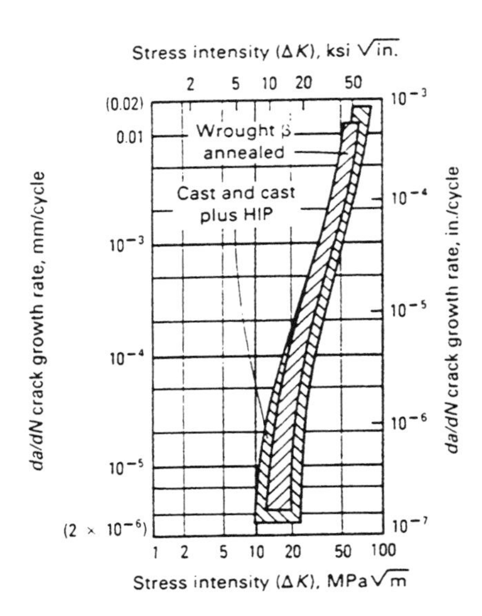

Ti-6Al-4V: Scatterbands for FCP rates of cast and β-annealed wrought products. Source: *Metals Handbook*, Vol 2, 10th ed., *Properties and Selection: Nonferrous Alloys and Special-Purpose Materials*, ASM International, 1990, p 641

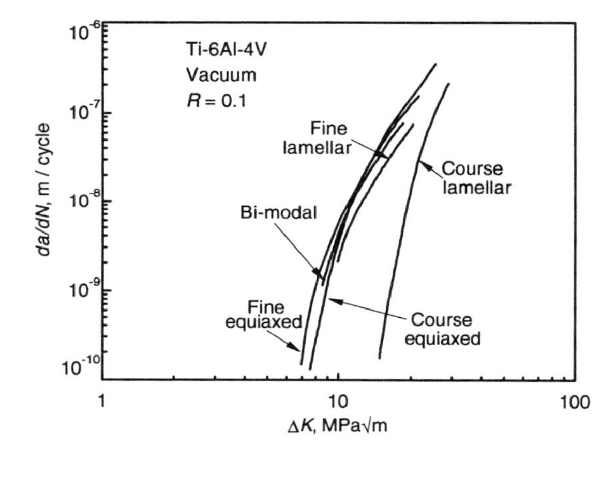

(a)

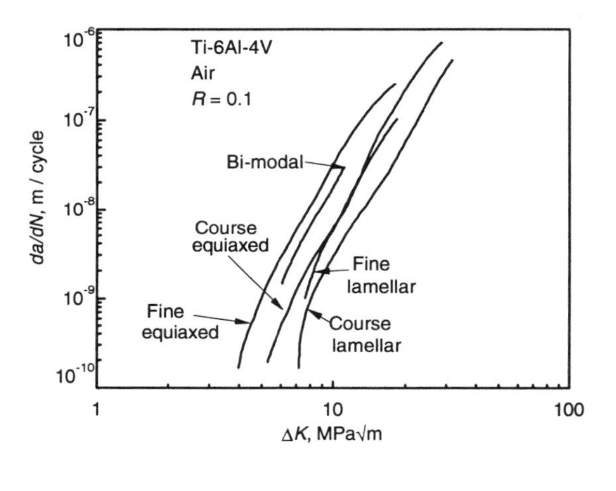

(b)

Ti-6Al-4V: Influence of α morphology on FCP resistance at room temperature. (a) In vacuum. (b) In air. Material/Test Parameters: The equiaxed and bi-modal microstructures had a mixed basal/transverse texture after rolling. The fine equiaxed structure had an average α grain size of 2 μm, and the coarse equiaxed structure had an average α grain size of 12 μm. Source: G Lütjering and A. Gysler, *Titanium Science and Technology*, Vol 4, Deutsche Gesellschaft für Metallkunde e.V., 1985, p 2077

the comparative advantage of Ti-6Al-4V improves because seawater has more of an effect on the FCP rates of steel.

Effect of Texture. Texture and specimen orientation are considered to be more important for crack nucleation than for crack propagation. However, crystallographic orientation can have significant effects when aggressive environments are involved (see the next section "Crack Growth with Corrosion"). Smaller effects of texture and test orientation are observed in laboratory air.

Effect of α-β Processing

Thermal processing or working in the α + β field generally produces an equiaxed microstructure which has less FCP resistance than transformed β structures. Significant amounts of scatter can occur because of variations in microstructure (grain size, morphology), texture, and strength levels. Besides lot-to-lot variations, mill annealed products may also have significant heat-to-heat variations (see figure for mill annealed plate). Improved FCP resistance at somewhat reduced levels of strength can be obtained by recrystallization annealing. Compared to the mill annealed condition, recrystallization annealing softens the alpha phase, thereby increasing its capacity for strain energy absorption at the tip of a propagating crack. As a result, the material has an increased resistance to crack propagation with monotonic loading (intrinsic fracture toughness) and with fatigue loading (FCP resistance). A secondary, beneficial effect of recrystallization anneal is a measurable microstructural coarsening which serves to retard crack growth under centain conditions.

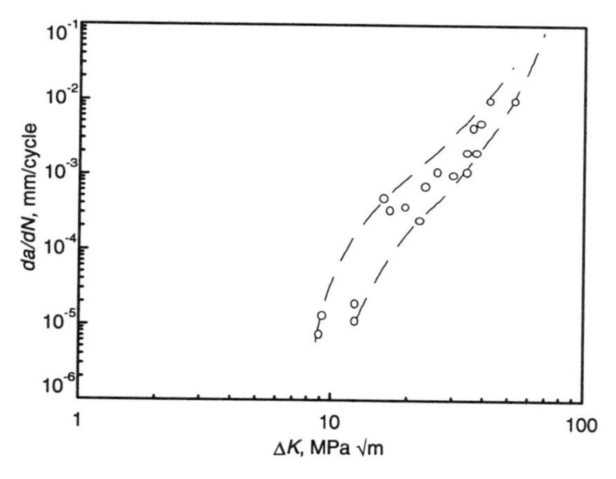

Ti-6Al-4V: Scatter of FCP data for mill-annealed plate. This is one of the more extreme cases of FCP data scatter, which can arise for the mill-annealed condition due to inconsistencies of microstructure, texture, and strength. Data are for six heats of mill-annealed Ti-6Al-4V. Material/Test Parameters: Room-temperature air, $R = 0.1$, 10-Hz frequency, and T-L test orientation of 25-mm (1-in.) plate. Source: *Titanium: A Technical Guide*, ASM International, 1988, p 180

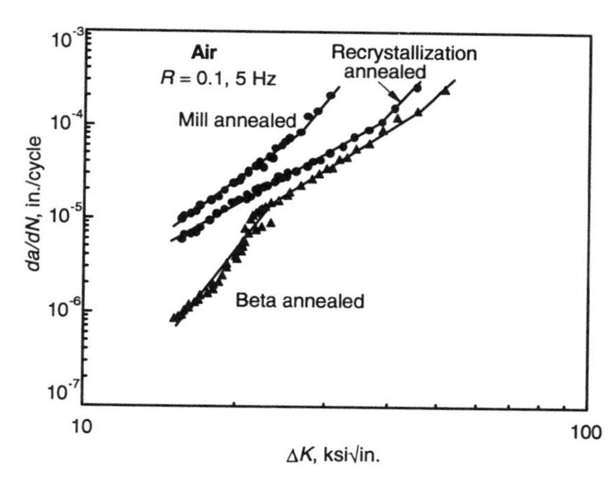

Ti-6Al-4V: Effect of recrystallization annealing on FCP in plate. Material/Test Parameters: Room-temperature air, $R = 0.1$, 5-Hz frequency, and haversine wave form. Source: G.R. Yoder *et al.*, "Effects of Microstructure and Frequency on Corrosion-Fatigue Crack Growth in Ti-7Al-1Mo-1V and Ti-6Al-4V, in *Corrosion Fatigue: Mechanics, Metallurgy, Electrochemistry, and Engineering*, STP 801, T.W. Crooker and B.N. Leis, Ed., American Society for Testing and Materials, 1983, p 159-174

FCP Resistance of Transformed β

Like fracture toughness, FCP resistance improves when a microstructure of transformed β is obtained during β annealing or β solution treatment. In β annealed products, FCP rates below the transition point (ΔK_t) are related to the average Widmanstätten packet size (see figure). Interstitial oxygen (acting as an α stabilizer) and prior β grain size have an indirect influence on FCP rates by affecting the size of the average Widmanstätten packets that form upon cooling from above the β transus. The slowest FCP rates are obtained by slow (furnace) cooling from above the β transus.

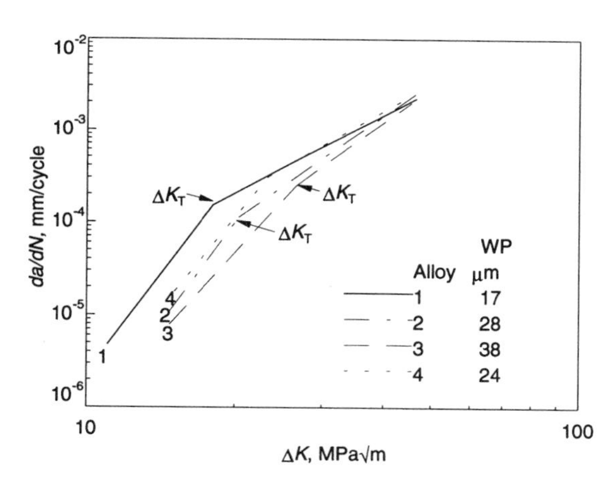

Beta-annealed Ti-6Al-4V: Effect of Widmanstätten packet size on FCP rates. Each plot shows a significant change in slope at a transition point (ΔK_T) at which the reversed plastic zone appears to attain the average Widmanstätten packet (WP) size. Material/Test Parameters: Room-temperature tests in laboratory air at $R = 0.1$ and a frequency of 5 Hz with haversine wave form. Source: *Metall Trans A*, Vol 9A (1978), p 1413-1420

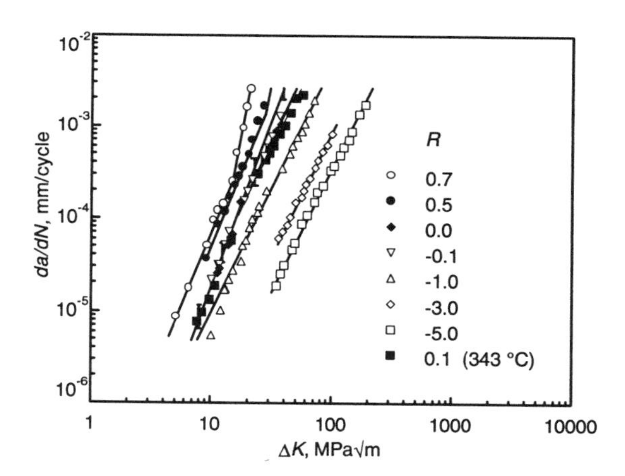

Ti-6Al-4V forgings: FCP rates for various R ratios. The data for $R = 0.1$ is an average of 10 tests. STOA. Material/Test Parameters: Pancake forgings had a composition (wt%) of 6.3-6.43 Al, 0.02-0.033 C, 0.10-0.18 Fe, 0.0050-0.0062 H, 0.013-0.015 N, 0.172-0.183 O, and 4.28-4.3 V. Forged at 970 °C (1775 °F) in the α-β range, solution treated at 955 °C (1750 °F) for 1 h, water quenched, then aged at 700 °C (1300 °F) for 2 h. Specimen dimensions were 15.2 × 5.1 × 0.2 cm (6 × 2 × 0.08 in.) with a 0.76-cm (0.3-in.) slot (electric discharge machined) in center, normal to loading direction. Source: A. Yuen *et al.*, Correlations Between Fracture Surface Appearance and Fracture Mechanics Parameters for Stage II Fatigue Crack Propagation in Ti-6Al-4V, *Metall. Trans.*, Vol 5, 1974, p 1833

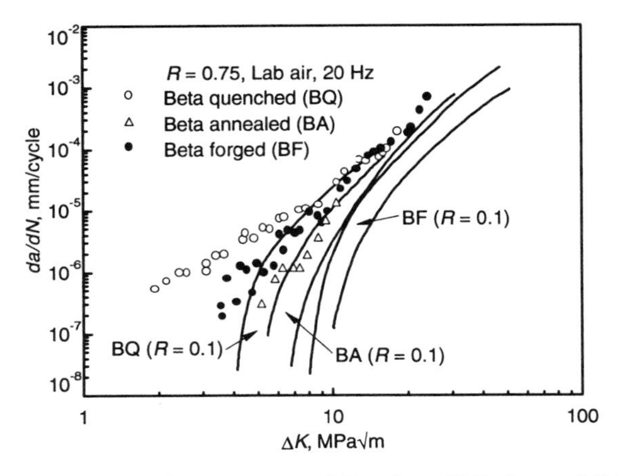

Ti-6Al-4V: Effect of microstructure and *R* ratio on FCP. Source: J.C. Chesnutt, Fatigue Crack Propagation in Titanium Alloys, *Titanium Science and Technology,* Vol 4, Deutsche Gesellschaft für Metallkunde e.V., 1985, p 2227-2233

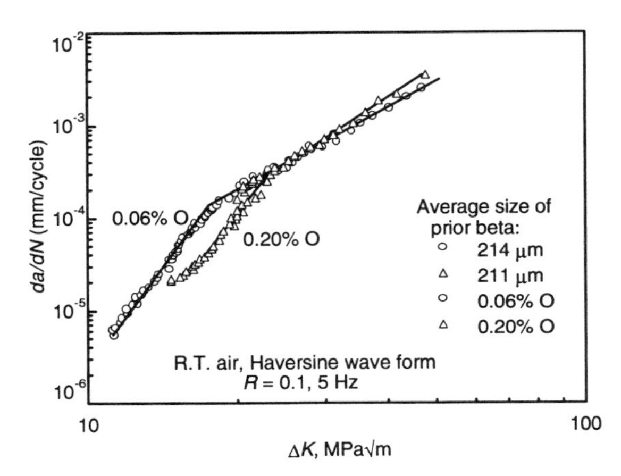

Beta-annealed Ti-6Al-4V: FCP rates with different oxygen contents. The slower crack growth rates for 0.20% O are associated with a larger Widmanstätten packet (WP) size (24 μm) vs a WP size of 17 μm for the specimens with 0.06% O. Source: G.R. Yoder *et al.*, Fatigue Crack Propagation Resistance of Beta Annealed Ti-6Al-4V Alloys of Different Interstitial Oxygen Contents, *Metall Trans A*, Vol 9A, 1978, p 1413-1420

Crack Growth and Corrosion

The test environment can have a very strong influence on crack growth rate depending on the environment, alloy microstructure, alloy texture, and testing parameters (such as cycling frequency, dwell-time, stress-intensities, and electrodynamical potentials). Air is a more aggressive environment than a vacuum, and high-oxygen steam is slightly more detrimental than low-oxygen steam. Aqueous halide solutions are known contributors to stress-corrosion cracking (SCC), which in turn affects FCP rates of titanium alloys.

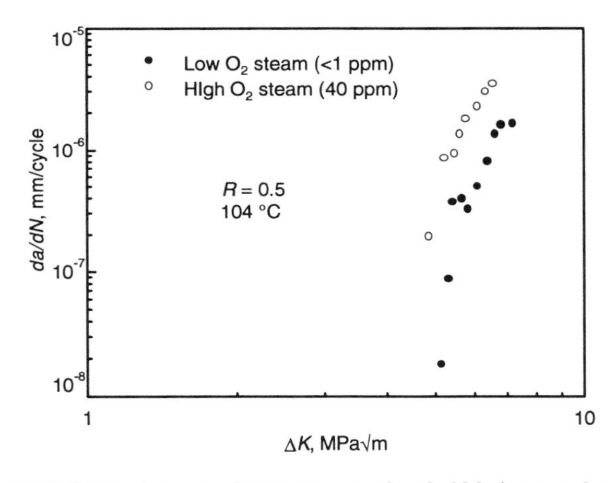

Ti-6Al-4V: Effect of oxygen of steam on near-threshold fatigue crack growth rates in Ti-6Al-4V. Source: P.K. Liaw, J. Anello, and J.K. Donald, Effects of Corrosive Environments on Near-Threshold Fatigue Crack Growth Behavior of Ti-6Al-4V, *Eng. Fract. Mech.*, Vol 19 (No. 6), 1984, p 1047-1056

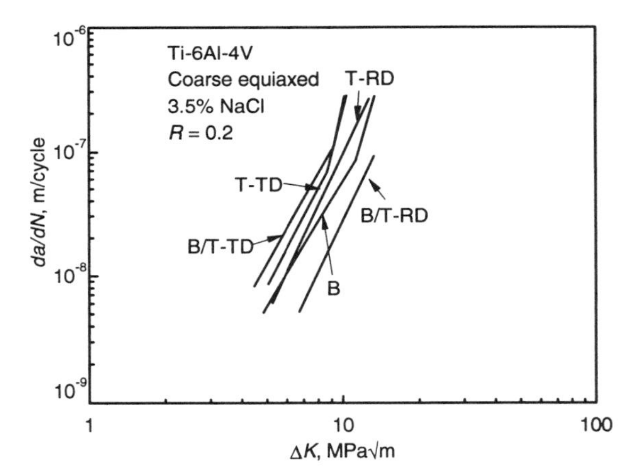

Ti-6Al-4V: Effect of texture on near-threshold FCP in NaCl solution. B, basal texture; B/T, basal/transverse texture; T, transverse texture; RD, rolling direction of test; TD, transverse direction of test. Source: G Lütjering and A. Gysler, *Titanium Science and Technology*, Vol 4, Deutsche Gesellschaft für Metallkunde e.V., 1985, p 2078

Ti-6Al-4V: Crack-growth rates for continuously rolled textured sheet in air and salt water

Stress-intensity range		Environment	Grain direction	Crack growth ($\Delta 2a/\Delta N$)			
				μm/cycle		10^{-6} in./cycle	
MPa√m	ksi√in.			Range	Avg	Range	Avg
55	50	Air	Longitudinal	0.43-0.71	0.5	17-28	20
			Transverse	0.43-0.74	0.53	17-29	21
55	50	3.5% NaCl	Longitudinal	0.53-1.4	0.8	21-56	32
			Transverse	1.6-6.2	3.7	64-245	146
55	50	Distilled H_2O	Longitudinal	0.5-1.3	0.79	20-53	31
			Transverse	1.3-4.0	2.7	53-158	105
66	60	Air	Longitudinal	0.7-1.1	0.84	27-44	33
			Transverse	0.7-1.1	0.84	27-44	33
66	60	3.5% NaCl	Longitudinal	0.94-2.1	1.5	37-84	60
			Transverse	3.6-7.1	5.8	142-280	228
66	60	Distilled H_2O	Longitudinal	0.84-1.9	1.3	33-74	53
			Transverse	2.1-4.6	3.2	83-182	125

Material/test parameters: Preponderance of basal poles parallel to the transverse direction. Tested at $R = 0.67$ and 2 Hz. The specimen was continuously rolled sheet having a thickness of 1.3 mm (0.050 in.) and an area of 305 × 910 mm (12 × 36 in.). The maximum gross-area stress was 170 MPa (25 ksi). Source: Data from F. Parkinson, Boeing Co., 1972, reported in *Properties of Textured Titanium Alloys*, F. Larson and A. Zarkades, Ed., MCIC-74-20, Battelle Columbus Laboratories, 1974, p 70

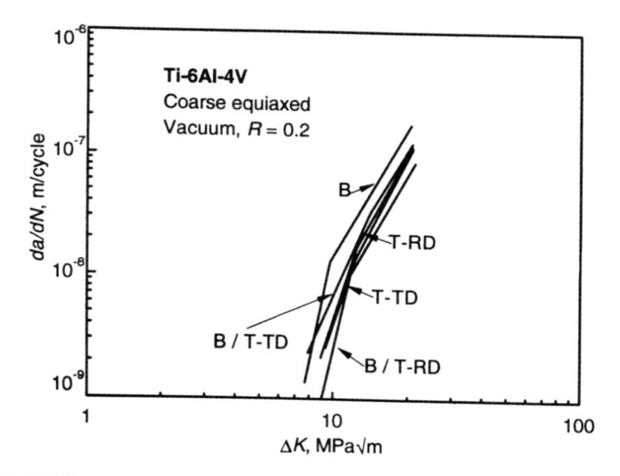

Ti-6Al-4V: Effect of test direction and texture on FCP rates. Texture has little effect in a totally inert environment. B, basal texture; B/T, basal/transverse texture; T, transverse texture; RD, rolling direction of test; TD, transverse direction of test. Source: G Lütjering and A. Gysler, *Titanium Science and Technology*, Vol 4, DGM, e.V., 1985, p 2078

Aqueous Halide Solutions

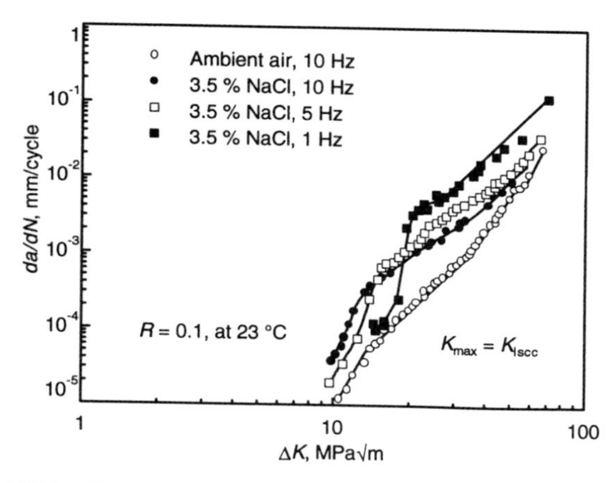

Ti-6Al-4V: Typical FCP rates in salt water. Mill annealed sheet. Source: D.B. Dawson and R.M. Pelloux, Corrosion Fatigue Crack Growth of Titanium Alloys in Aqueous Environments, *Metall. Trans. A*, Vol 5A, 1974, p 723-731

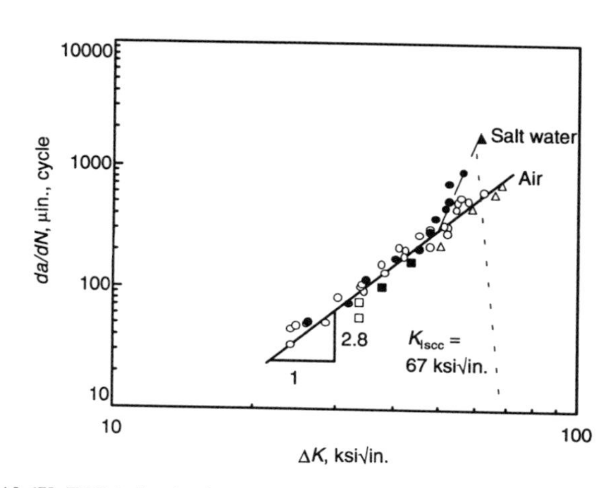

Ti-6Al-4V: FCP behavior in salt water. Salt water has little effect below the K_{ISCC} threshold. Material/Test Parameters: Stress ratio, test frequency, and microstructure unspecified. Source: N.G. Feige and R.L. Kane, Service Experience With Titanium Structures, in *Proceedings of the 26th Conference*, National Association of Corrosion Engineers, 1970, p 194-199

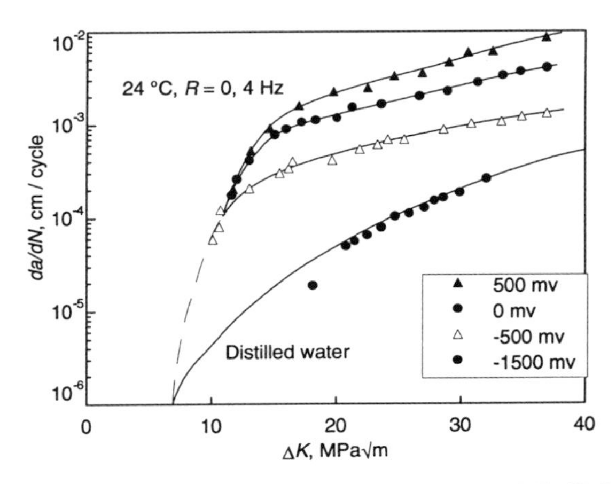

Mill-annealed Ti-6Al-4V: FCP rates in iodide solutions and distilled water. Electrode potentials must be closely monitored during testing. At a potential of –1500 mV, FCP rates in iodide solution are equal to those in distilled water. Source: *Corrosion Fatigue: Chemistry, Mechanics and Microstructure*, Vol 2, National Association of Corrosion Engineers, 1972

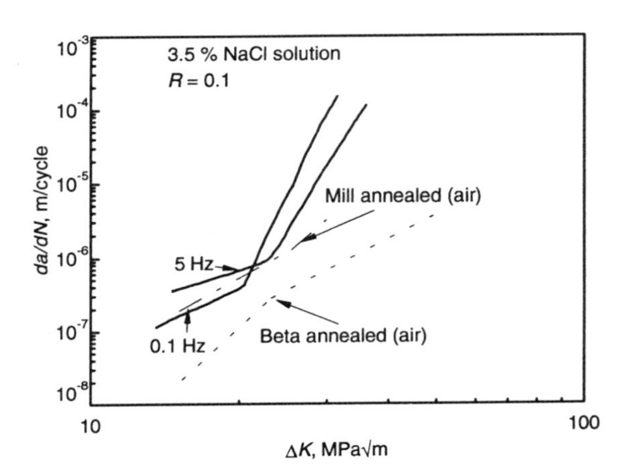

Mill-annealed Ti-6Al-4V: Effect of testing frequency and ΔK on FCP rates in salt water. Source: G.R. Yoder *et al.*, "Effects of Microstructure and Frequency on Corrosion-Fatigue Crack Growth in Ti-8Al-1Mo-1V and Ti-6Al-4V, in *Corrosion Fatigue: Mechanics, Metallurgy, Electrochemistry, and Engineering*, STP 801, T.W. Crooker and B.N. Leis, Ed., American Society for Testing and Materials, 1983, p 159-174

Effect of Test Frequency

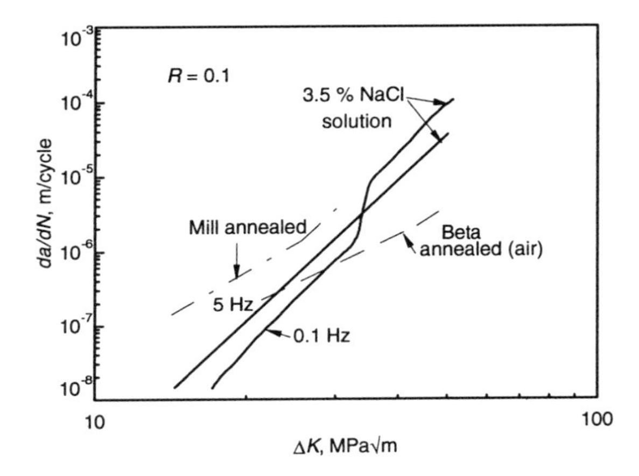

Beta-annealed Ti-6Al-4V: Effect of testing frequency and ΔK on FCP rates in salt water. Source: G.R. Yoder *et al.*, "Effects of Microstructure and Frequency on Corrosion-Fatigue Crack Growth in Ti-8Al-1Mo-1V and Ti-6Al-4V, in *Corrosion Fatigue: Mechanics, Metallurgy, Electrochemistry, and Engineering*, STP 801, T.W. Crooker and B.N. Leis, Ed., American Society for Testing and Materials, 1983, p 159-174

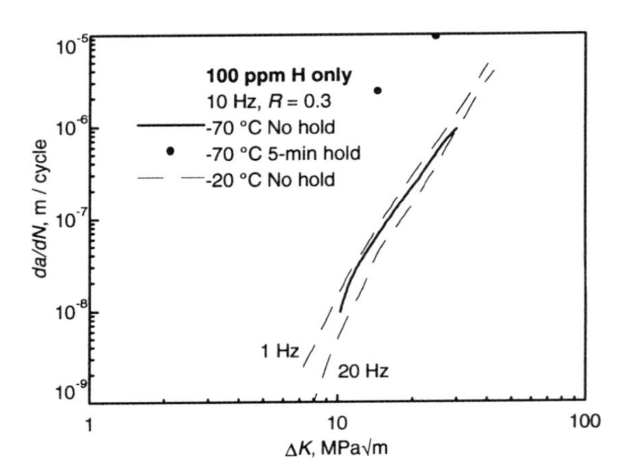

Recrystallization-annealed Ti-6Al-4V: Effect of dwell loading and hydrogen on FCP rates. The RA microstructure, at 100 ppm H_2, shows a significant acceleration with a 5 minute hold time at maximum load when tested at –70 °C. There is even a slight increase in growth rate comparing 1 Hz and 20 Hz test frequencies at –20 °C. The acceleration, in both cases, is attributed to migration of hydrogen to the crack tip during the hold and precipitation of hydrides. This results in quasicleavage fracture of the α phase. Source: J.C. Chesnutt and N.E. Paton, Hold Time Effects on Fatigue Crack Propagation in Ti-6Al and Ti-6Al-4V, in *Microstructure, Fracture Toughness, and Fatigue Crack Growth Rate in Titanium Alloys*, A.K. Chakrabarti and J.C. Chesnutt, Ed., The Metallurgical Society, 1987

Holding the specimen at the maximum load for some time period (or dwell) during the cycling can have a very significant effect on the crack growth rate. The effect is related to migration of hydrogen to the crack and the formation of hydrides; hence factors such as test temperature, microstructure and texture are important. As the fracture mechanism involves quasi-cleavage of the α-phase, which occurs on a plane near the basal plane, this phenomenon is strongly influenced by crystallographic texture. When testing a moderately textured plate where the direction of crack propagation was normal to the principal orientation of the basal planes, no acceleration was observed under any conditions, whereas a significant hold time effect was observed when the crack was propagating along the basal planes (J.C. Chesnutt and N.E. Paton, in figure reference above).

There are three conditions which all must be satisfied in order to observe acceleration of growth rate due to dwell effects: (1) hydrogen contents >100 ppm, (2) a temperature below room temperature (which is dependent on hydrogen level), and (3) a significant hold time in the tensile portion of the loading cycle. Stubbington and Pearson, however, demonstrated that an acceleration could be observed at temperatures up to 60 °C with 200 ppm hydrogen using 45 minute hold times (*Engr. Fracture Mechanics*, Vol 10, 1978, p 723-756).

Cyclic load frequency is an important variable that influences corrosion fatigue for most material, environment, and stress intensity conditions. The effect of frequency (or dwell-time effects) is directly related to the time dependence of the mass transport and chemical reaction steps required for environmental cracking.

Generally, the rate of environmental cracking above that produced in a vacuum increases with decreasing frequency. For titanium alloys and many other materials, however, the effects of frequency also depend on stress-intensity levels (see figures). At low ΔK's the lower frequency grows at a slower rate as repassivation of the surface can reduce the effect of the environment, and at the slower frequency, there is more time for repassivation. The crossover occurs at a ΔK level associated with the onset of cyclic stress corrosion cracking (SCC), and has been referred to as ΔK_{SCC} (see Dawson, D.B. and Pelloux, R.M., *Met. Trans.*, Vol 5, No. 3, March 1974, p 723-731). Above ΔK_{SCC} the effects of frequency are consistent with a cyclic SCC mechanism. With the lower frequency the crack has a longer residence time at the higher ΔK's than at the lower frequencies, resulting in more stress corrosion cracking and faster growth rates. This has been interpreted by Stubbington and Pearson (*Engr. Fracture Mechanics*, 1978, ibid) that above a hydrogen embrittlement mechanism dominates, and da/dN is related to hydrogen mobility to the plastically deformed region near the crack tip. The lower frequency then exhibits higher crack growth rates as there is more time for hydrogenation embrittlement to occur during each cycle.

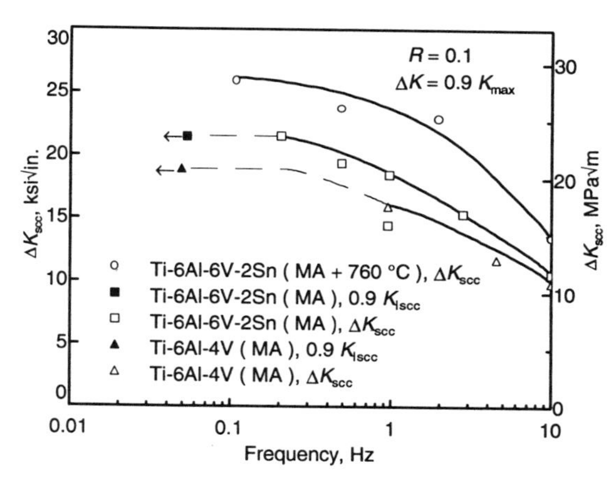

Effect of test frequency on ΔK_{SCC}, the transition stress-intensity factor range for cyclic stress-corrosion cracking. The ΔK level for which $K_{max} = K_{ISCC}$ is shown for reference. ΔK_{SCC} is lower than K_{ISCC} (the stress-corrosion cracking threshold under sustained loading), and it decreases with increasing frequency due to repeated rupture of the passive film at the crack tip. Source: D.B. Dawson and R.M. Pelloux, Corrosion Fatigue Cracking of Titanium Alloys in Aqueous and Methanolic Environments, in *1972 Tri-Service Conference on Corrosion*, MCIC 73-19, M.M. Jacobson and A. Gallaccio, Ed., Battelle Columbus Laboratories, 1972, p 77-94

Impact Toughness

Impact properties are important in that they can be related to the residual strength or fracture toughness of a material. Also, impact tests are less expensive to conduct than compact-tension fracture toughness tests. Frequently, they are run as a screening test to evaluate alloys, heat treatment procedures, and other variables to select tougher materials prior to fracture toughness testing. The correlation between impact energy and fracture toughness, however, is not precisely proportional. Typical room-temperature impact strength of Ti-6Al-4V is about 20 J (15 ft · lbf) for Izod specimens and about 24 J (18 ft · lbf) for standard grade Charpy specimens.

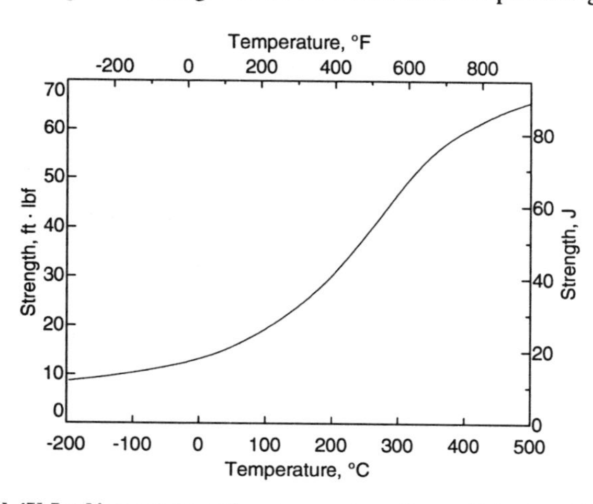

Ti-6Al-4V: Izod impact strength vs temperature. Source: "Titanium Alloy IMI 318," IMI Titanium, Ltd.

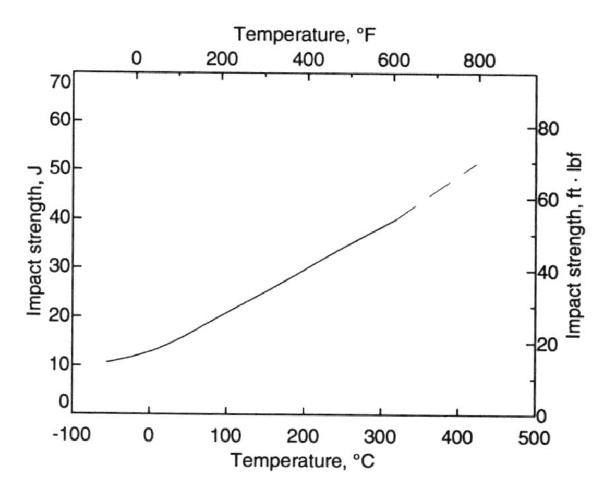

Ti-6Al-4V: Charpy impact strength of annealed bar vs temperature. At –40 °C (–40 °F), annealed bars exhibit average Charpy V-notch impact energy of 24 J (18 ft · lbf). Alloy annealing conditions: holding at 730 °C (1350 °F) for 0.5 to 2 h, depending on section size, air cool. For maximum ductility, furnace cool at a rate of less than 2 °C (50 °F)/min to 540 °C (1000 °F), then air cool. Source: Crucible Specialty Metals, Colt Industries

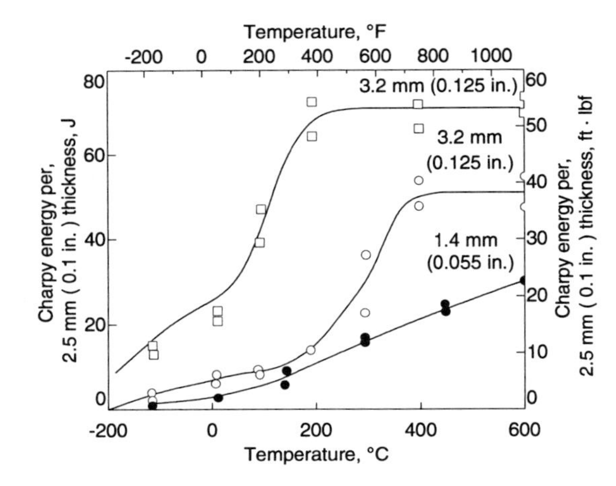

Ti-6Al-4V: Charpy impact strength of sheet. Test results from various heats of annealed sheet. Transverse specimen orientation.

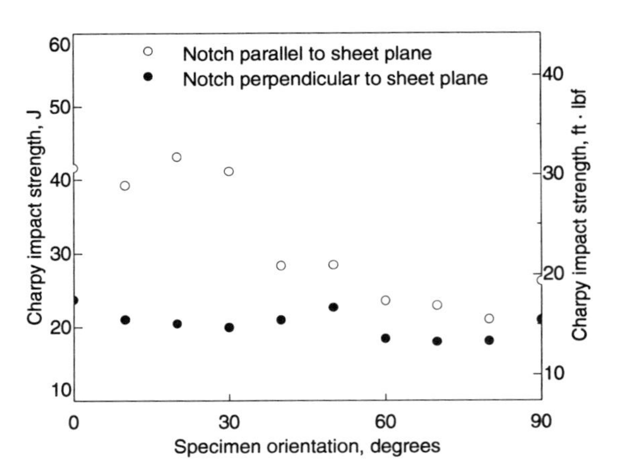

Ti-6Al-4V: Impact strength of textured plate. The highest impact strengths occur in a prism pole specimen (notch length parallel to *c*-axis) with the crack propagating perpendicular to the *c* crystal axis. Specimen orientation refers to the angle between the specimen axis and the rolling direction. Plate thickness was 13 mm (0.525 in.) with 0.2% yield strengths of 840 and 925 MPa (122 and 135 ksi) in the L and T directions, respectively. Source: F. Larson and A. Zarkades, Properties of Textured Titanium Alloys, MCIC 74-20, Battelle, 1974, p 55

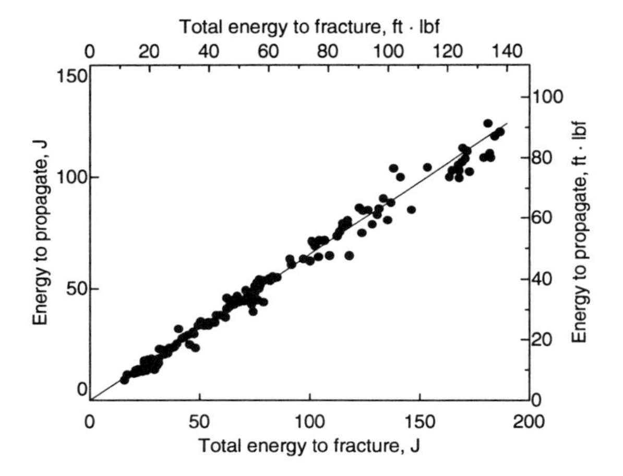

Ti-6Al-4V: Energy to propagate and fracture. Relationship between total energy-to-fracture and energy-to-propagate in Kahn-type tensile-tear test.

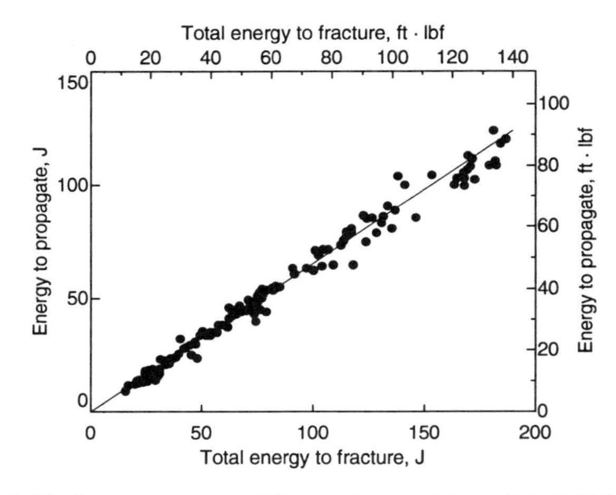

Ti-6Al-4V: Charpy energy per 2.5 mm. The material was Ti-6Al-4V 1.6 mm (0.063 in.). Note that the notch toughness index increases more rapidly with temperature than Charpy energy due to the decrease in yield strength with increasing temperature.

Ti-6Al-4V: Charpy impact strength and ultimate tensile strength compared to titanium alloys

Alloy	Ultimate tensile strength		Charpy impact toughness	
	MPa	ksi	J	ft · lbf
Unalloyed titanium	345	50	54	40
Unalloyed titanium	550	80	27	20
Ti-5Al-2.5Sn	825	120	25	19
Ti-8Al-1Mo-1V	895	130	27	20
Ti-6Al-6V-2Sn	1035	150	16	12
Ti-3Al-8V-6Cr-4Mo-4Zr	1240	180	9.5	7

Source: "The Titanium Industry in the mid-1970's," MCIC-75-26, Battelle, 1975, p 59

Ti-6Al-4V: Charpy impact strength vs temperature for 15.9 mm (5/8 in.) rounds in various heat treated conditions

Temperature		Charpy V-notch impact toughness, for indicated heat treatment: J (ft · lbf)		
°C	°F	705 °C (1300 °F), 2 h, mill annealed	845 °C (1550 °F), 1 h, water quenched, 480 °C (900 °F), 8 h, air cooled	940 °C (1725 °F), 1 h, water quenched, 480 °C (900 °F), 8 h, air cooled
−73	−100	18.9 (14.0), 16.2 (12.0)	20.3 (15.0), 18.9 (14.0)	18.3 (13.5), 15.6 (11.5)
17	0	24.4 (18.0), 25.1 (18.5)	21.7 (16.0), 24.4 (18.0)	18.9 (14.0), 23.0 (17.0)
RT	RT	24.4 (18.0), 23.0 (17.0)	20.3 (15.0), 24.4 (18.0)	20.3 (15.0), 21.7 (16.0)
93	200	28.4 (21.0), 31.8 (23.5)	28.4 (21.0), 28.4 (21.0)	25.7 (19.0), 23.0 (17.0)
150	300	41.3 (30.5), 47.5 (35.0)	36.6 (27.0), 28.4 (21.0)	37.9 (28.0), 28.4 (21.0)
205	400	69.1 (51.0), 45.4 (33.5)	42.0 (31.0), 50.8 (37.5)	43.3 (32.0), 51.5 (38.0)
260	500	66.4 (49.0), 84.1 (62.0)	47.4 (35.0), 46.1 (34.0)	56.9 (42.0), 60.3 (44.5)
315	600	70.5 (52.0), 90.1 (66.5)	64.4 (47.5), 52.8 (39.0)	71.2 (52.5), 65.0 (48.0)
370	700	103.7 (76.5), 93.5 (69.0)	69.8 (51.5), 94.9 (70.0)	56.2 (41.5), 62.3 (46.0)

Source: M. Mote, R. Hooper, and P. Frost, *Engineering Properties of Commercial Titanium Alloys*, TML Report 92, Battelle, 1958, p G-6

Charpy impact strength for castings in several heat treated conditions

Heat treatment	Charpy V-notch impact toughness	
	J	ft · lbf
Conventional anneal at 845 °C (1550 °F), 2 h in vacuum/argon fan cool	34.7 33.7 33.6	25.6 24.9 24.8
Below beta solution treatment and age, consisting of a solution treatment at 985 °C (1810 °F), 1 h in vacuum, an inert gas fan cool at a rate equivalent to air cooling, and an age at 540 °C (1010 °F), 8 h in vacuum/GFC	30.6 28.7 28.2	22.6 21.2 20.8
Above beta solution treatment and age, consisting of 1015 °C (1860 °F), 1 h in vacuum/GFC + 540 °C (1000 °F), 8 h in vacuum/GFC	27.1 26.0 28.7	20.0 19.2 21.2
Below beta cyclic exposure involving six cycles of 980 °C (1800 °F), 10 min in vacuum/GFC to 540 °C (1000 °F) + 540 °C (1000 °F), 10 min in vacuum, heat to 980 °C (1800 °F), followed by a gas fan cool to 21 °C (70 °F)	40.5 29.9 35.5	29.9 22.1 26.2
Nontraditional, proprietary thermal treatment process known as CST-I	24.9 23.8 23.8	18.4 17.6 17.6

Note: Materials from three heats were centrifugally cast into preheated MonoShell investment molds. Composition range was 5.9 to 6.1 wt.% Al, 0.001 to 0.003 wt.% H, 0.01 to 0.02 wt.% N, 0.16 to 0.19 wt.% O, and 3.8 to 4.1 wt.% V. Source: R.J. Smickley, and L.E. Dardi, "The Thermal Processing Response of HIP'ed Investment Cast Ti-6Al-4V Alloy," *Titanium Net Shape Technologies*, F.H. Froes and D. Eylon, Ed., The Metallurgical Society of AIME, 1984, p 201-209

Fracture Toughness

The fracture toughness (K_{Ic}) of Ti-6Al-4V is higher than that of aluminum alloys, but lower than steels. In very general terms, fracture toughness increases as the amount of transformed β (lamellar α/β) structure increases, with β annealing providing the highest fracture toughness. The exception to this is the recrystallization annealed structure, which contains no transformed β, but has a fracture toughness almost as good as β annealed material. Also, coarser microstructures generally provide higher toughness values. The ELI grade alloy is used for fracture-critical applications, due to its superior toughness.

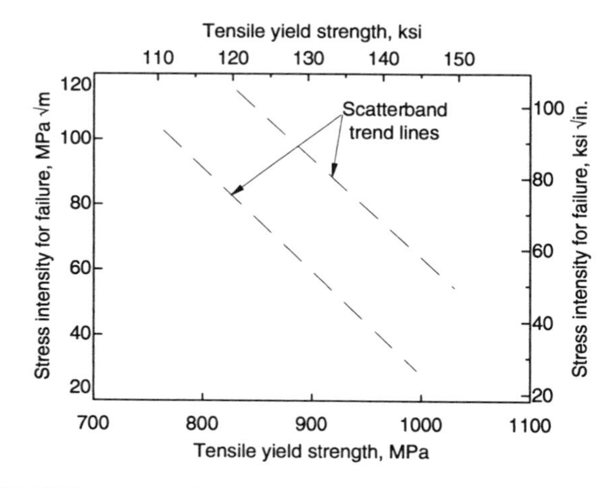

Ti-6Al-4V: Fracture toughness scatter bands. Annealed specimens (bar, plate, and forgings) tested were all within specification limitations. A large number of specimens from numerous heats were tested by compact tension or four-point bending tests.. Source: *Titanium and Titanium Alloys*, MIL-HDBK-697A, 1974, p 43

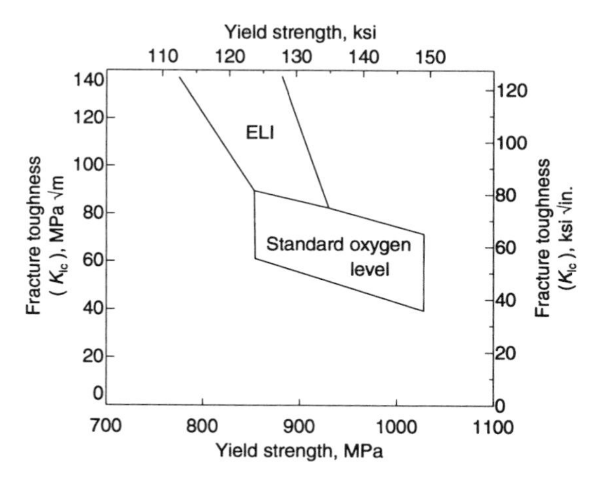

Ti-6Al-4V: Range of yield strength and fracture toughness

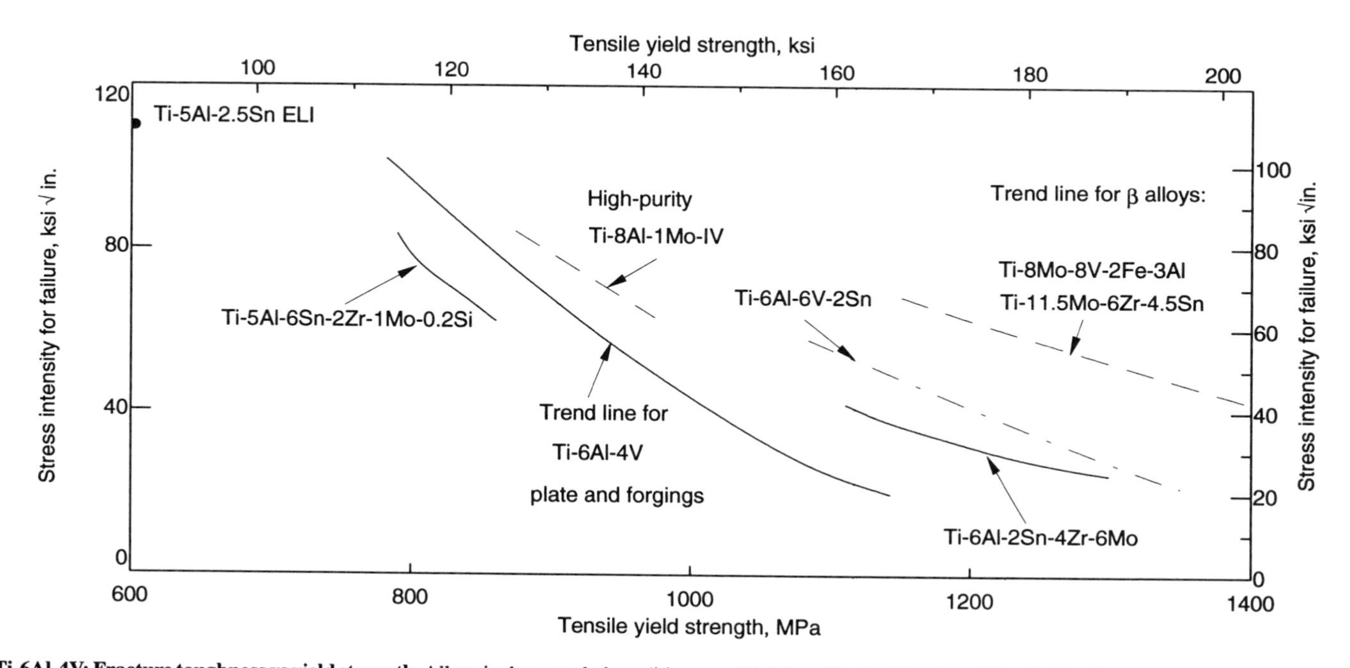

Ti-6Al-4V: Fracture toughness vs yield strength. Alloys in the annealed condition were Ti-5Al-2.5Sn, Ti-8Al-1Mo-1V, and alloys in the solution heat treated and aged conditions were Ti-6Al-2Sn-4Zr-6Mo, Ti-6Al-6V-2Sn, and β alloys Ti-11.5Mo-6Zr-4.5Sn and Ti-8Mo-8V-2Fe-3Al. Ti-6Al-4V trend line represents both annealed and STA conditions.

Effects of Processing

The fracture toughness of Ti-6Al-4V ranges from about 33 to more than 110 MPa$\sqrt{m}$ (30 to 100 ksi$\sqrt{in.}$), depending on oxygen content and heat treatment. At similar strengths, β processed material has a fracture toughness on the order of 50% greater than α-β processed material.

Effect of Welding. Generally, the fracture toughness (and crack-growth resistance) of Ti-6Al-4V welds depends on both fusion zone microstructure and interstitial content. The martensitic microstructure characteristic of electron beam and laser welds with high depth-to-width ratios is characterized by toughnesses below that of the mill annealed base metal. Welds produced at lower cooling rates exhibit increasing toughness levels that are comparable or superior to that of the mill annealed base metal.

Obata *et al.*, studied the fracture toughness of electron beam welds as a function of filler metal thickness and stress relief temperature. The thickness of the filler metal was shown to have a strong influence on the toughness of the weld metal, but no effect in the heat-affected zone (HAZ). Furthermore, the welding sequence did not have a significant effect on the toughness of the weld metal, but heat treatment after welding, rather than prior to welding, was beneficial in the HAZ.

Typical fracture toughness of several alloys

Alloy	Fracture toughness, MPa$\sqrt{m}$	Fracture toughness, ksi$\sqrt{in.}$
Ti-6Al-4V (annealed)	65	60
Ti-6Al-4V (β annealed)	90	80
7075 aluminum (T73)	35	32
4140 steel(a)	115	105
4340 steel(a)	115	105
15-5 PH (a)	80	75

(a) Hardened to 1240 to 1380 MPa (180 to 200 ksi)

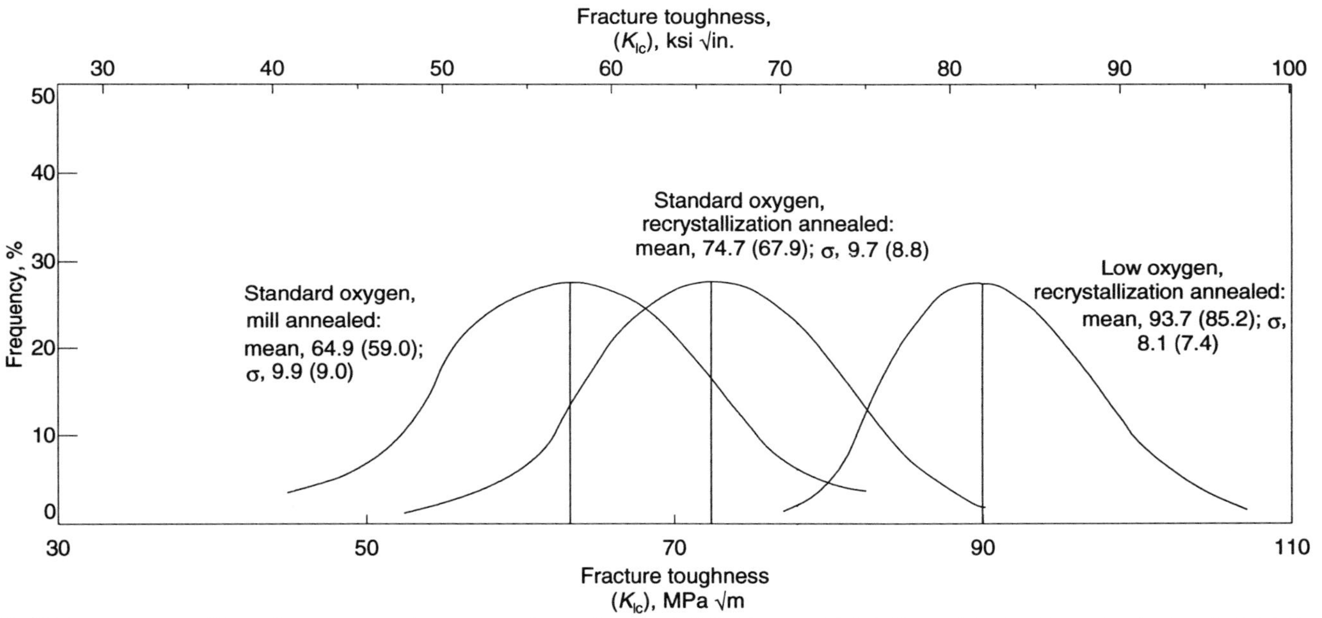

Ti-6Al-4V: Oxygen content/thermal treatment vs fracture toughness. Source: G.W. Kuhlman and F.R. Billman, Selecting Processing Options for High-Fracture Toughness Titanium Airframe Forgings, *Met. Prog.*, March 1987

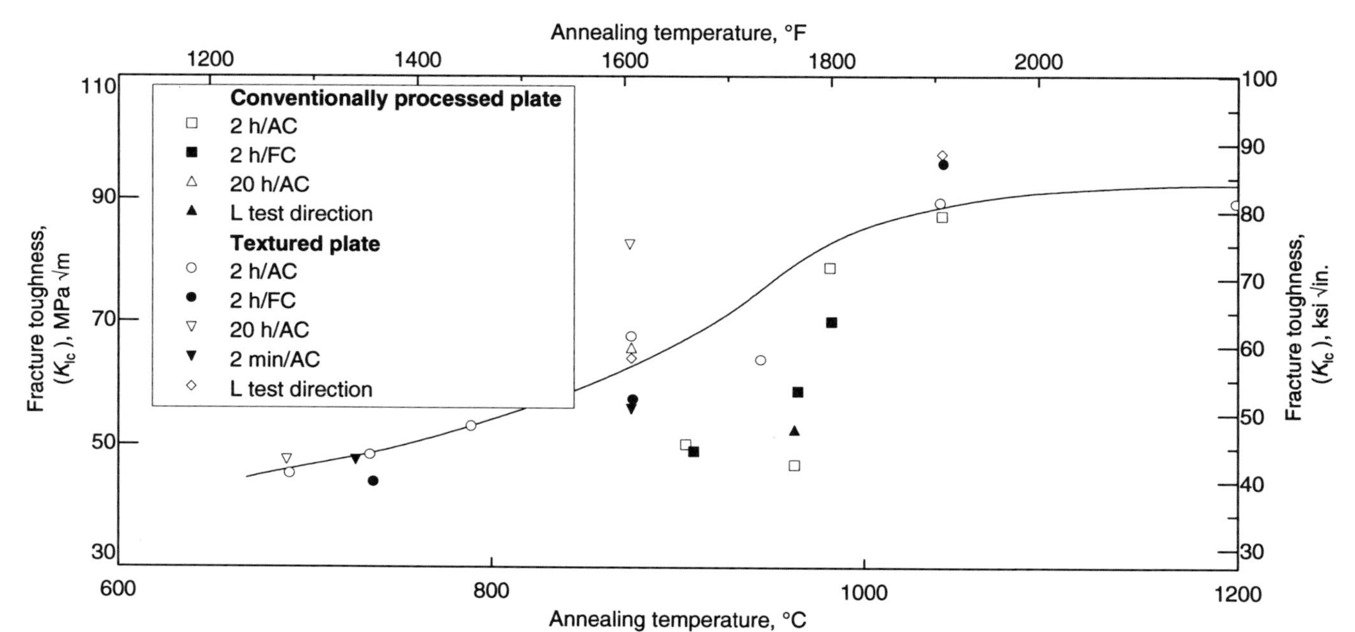

Material description	Oxygen content, wt%	Heat treatment	Grain direction	Tensile strength MPa	Tensile strength ksi	Yield strength MPa	Yield strength ksi	Elongation, %	Reduction of area, %
Conventional plate	0.13	MA	L	986	143	910	132	13	22
		MA	T	1034	150	972	141	15	33
		BA	T	951	138	876	127	11	26
Textured plate	0.126	As rolled	L	945	137	869	126	15	27
		As rolled	T	1034	150	979	142	15	27
		BA	T	931	135	855	124	9	17

Note: Composition 6.3% Al, 0.026% C, 0.07% Fe, 0.005% H, 0.015% N, 0.13% O, 4.10% V. Conventional plate (Air Force plate, 30.5 mm or 1.2 in. thick) was conventionally a-b cross rolled and mill annealed; textured plate (31.75 mm or 1.25 in. thick) was unidirectionally rolled to produce a moderately transverse basal plane (TD) texture. Specimens were annealed 2 min to 20 h at 690 to 1040 °C (1275 to 1900 °F). R.R. Boyer, R. Bajoraitis, and W.F. Spurr, The Effects of Thermal Processing Variations on the *Properties of Ti-6Al-4V, in Microstructure, Fracture Toughness, and Fatigue Crack Growth Rate in Titanium Alloys,* Proceedings of the 1987 TMS-AIME annual symposia on *Effect of Microstructure on Fracture Toughness and Fatigue Crack Growth Rate in Titanium Alloys,* Denver, A.K. Chakrabarti and J.C. Chesnutt, Ed., The Metallurgical Society, 1987, p 149-170

Ti-6Al-4V: (ELI): Annealing temperature vs fracture toughness of plate

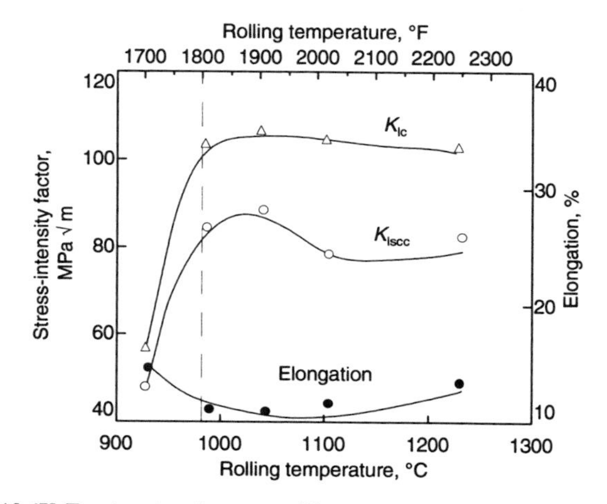

Ti-6Al-4V: Fracture toughness vs rolling temperature tensile strength. Solution treated and aged, 940 °C (1725 °F), and water quenched plus 8 h at 675 °C (1250 °F) and air cooled, after 30% reduction by rolling. Source: R.E. Curtis and W.F. Spurr, Effect of Microstructure on the Fracture Properties of Titanium Alloys in Air and Salt Solution, *Trans. ASM,* Vol 61, p 115-127

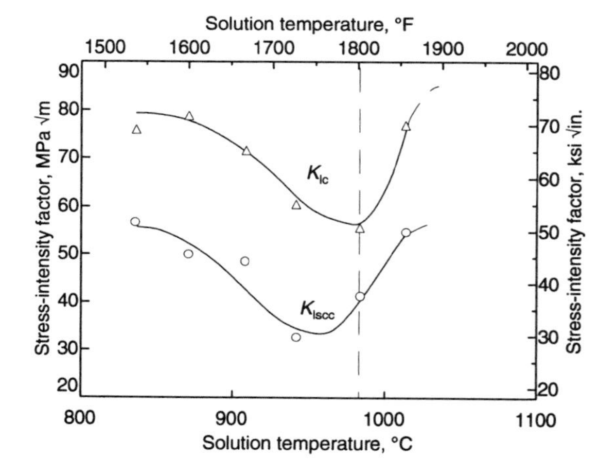

Ti-6Al-4V: Fracture toughness vs solution temperature on tensile/fracture strength. Specimens for tensile and fracture tests were aged 4 h at 675 °C (1250 °F) and air cooled after solution treatment. Source: R.E. Curtis and W.F. Spurr, Effect of Microstructure on the Fracture Properties of Titanium Alloys in Air and Salt Solution, *Trans. ASM,* Vol 61, p 115-127

Ti-6Al-4V: Fracture toughness of powder compacts compared to corresponding wrought alloys

	Powder compact				Wrought alloy			
	Fracture toughness (K_{Ic})		Yield strength		Fracture toughness (K_{Ic})		Yield strength	
Alloy	MPa√m	ksi√in.	MPa	ksi	MPa √m	ksi√in.	MPa	ksi
Ti-10V-2Fe-3Al	51	46	854	124	60	55	1100	160
Ti-6Al-6V-2Sn	44	40	910	132	65	59	965	140
Ti-6Al-2Sn-4Zr-6Mo	31	28	1068	155	33	30	1100	160
Ti-6Al-4V*	35	32	900	130	66	60	860	125

Note: Compositions of cold isopressed and vacuum sintered billets were as follows. Ti-10-2-3: 2.98 wt.% Al, 0.010 wt.% C, 2.03 wt.% Fe, 0.0008 wt.% H, 0.011 wt.% N, 0.13 wt.% O, and 10.25 wt.% V. Ti-6-6-2: 6.00 wt.% Al, 0.019 wt.% C, 0.55 wt.% Cu, 0.48 wt.% Fe, 0.0016 wt.% H, 0.0045 wt.% N, 0.27 wt.% O, 1.65 wt.% Sn, and 5.31 wt.% V. Ti-6-2-4-6: 6.03 wt.% Al, 0.024 wt.% C, 0.060 wt.% Fe, 0.0009 wt.% H, 5.60 wt.% Mo, 0.0055 wt.% N, 0.032 wt.% O, 2.20 wt.% Sn, and 3.60 wt.% Zr. Data for Ti-6-4 powder compact were taken from P. Andersen *et al.*, *Modern Developments in Powder Metallurgy*, Vol 13, 1981. Data for wrought alloys taken from AMS specifications, except Ti-10-2-3 (TIMET data). Source: J. Smugeresky and N. Moody, Properties of High Strength Blended Elemental Powder Metallurgy Titanium Alloys, in *Titanium Net Shape Technologies*, F. Froes and D. Eylon, Ed., 1984, p 131

Ti-6Al-4V (ELI): Fracture toughness of recrystallization annealed forgings

	Yield strength		Fracture toughness	
Direction	MPa	ksi	MPa√m	ksi√in.
LT	841	122	91	83
TL	890	129	92	84

Note: Specimens were recrystallization annealed at 925 °C (1700 °F), 4 h FC to 760 °C (1400 °F), AC. Room-temperature data from section 38.5 mm (1.5 in.) thick. Oxygen in 0.10 to 0.13% range. Source: R. Judy Jr., and B. Goode, "Stress Corrosion Cracking Characteristics of Alloys of Titanium in Salt Water," Interim Report 6564, Naval Research Laboratory, Washington DC, July 1967, as reported in *Damage Tolerant Design Handbook*, Vol 1, MCIC-HB-01, Metals and Ceramics Information Center, Battelle, Columbus, 1972

Ti-6Al-4V: Effect of heat treatment on fracture toughness, strength, and elongation

Heat treatment	Temperature		Fracture toughness (K_{Ic})		Tensile yield strength (0.2%)		Ultimate tensile strength		Elongation,
	°C	°F	MPa√m	ksi√in.	MPa	ksi	MPa	ksi	%
Air cooled	750	1380	67.5	61.4	950	138	1020	148	13.5
	850	1560	69.0	62.8	920	133	1010	146	14.0
	950	1740	73.0	66.4	935	136	1020	148	13.5
	1050	1920	78.0	71.0	940	136	1050	152	10.5
Air cooled and aged	750	1380	58.5	53.2	985	143	1030	149	13.0
	850	1560	55.5	50.5	960	139	1030	149	13.0
	950	1740	85.0	77.4	965	140	1030	149	13.5
	1050	1920	105.0	95.6	945	137	1020	148	2.5
Quenched	750	1380	49.0	44.6	970	141	1050	152	12.5
	850	1560	69.5	63.2	830	120	990	144	16.5
	950	1740	59.0	53.7	1050	152	1150	167	12.0
	1050	1920	59.0	53.7	1070	155	1170	170	11.5
Quenched and aged	750	1380	69.0	62.8	975	141	1030	149	12.0
	850	1560	78.0	71.0	935	136	1010	146	13.0
	950	1740	59.5	54.1	1060	154	1125	163	12.0
	1050	1920	68.0	61.9	1085	157	1140	165	3.5

Note: Composition: 6.35% Al, 0.05% C, 0.07% Fe, 45 ppm H, 0.01% N, 0.18% O, and 3.78% V. $\beta/\alpha + \beta$ transus, 995 ± 15 °C (1825 ± 27 °F). Source: I.W. Hall and C. Hammond, Fracture Toughness, Strength and Microstructure in Alpha + Beta Titanium Alloys, *Titanium and Titanium Alloys*, Vol 1, J.C. Williams and A.F. Belov, Ed., Plenum Press, 1976, p 601-613

Ti-6Al-4V: Effects of overaging on fracture toughness and tensile strength

Type of processing	Heat treatment	Fracture toughness (K_{Ic})		Tensile strength	
		MPa√m	ksi√in.	MPa	ksi
Alpha-beta	940 °C (1725 °F), WQ + aged 4 h at 540 °C (1000 °F), AC	54	49	1195	173
Alpha-beta	940 °C (1725 °F), WQ + overaged 4 h at 675 °C (1250 °F), AC	66	60	1080	157
Beta	940 °C (1725 °F), WQ + aged 4 h at 540 °C (1000 °F), AC	91	83	1160	168
Beta	940 °C (1725 °F), WQ + overaged 4 h at 675 °C (1250 °F), AC	101	92	1055	153

Source: R.A. Wood and R.J. Favor, *Titanium Alloys Handbook*, MCIC, Battelle, Columbus, 1972

Ti-6Al-4V: Fracture toughness

Condition/ heat treatment	Product form	Thicknesses, mm	Thicknesses, in.	Fracture toughness (K_{Ic}), MPa$\sqrt{m}$ (ksi$\sqrt{in.}$), L-T, Mean	L-T, Standard deviation	T-L, Mean	T-L, Standard deviation
Alpha-beta forged, mill annealed	Forging	57	2.25	...	...	38.9 (35.4)	2.9 (2.7)
Annealed	Forging	75	3.00	92.7 (84.4)	1.9 (1.8)	91.6 (83.4)	10.9 (9.9)
	Extrusion	100	4.00	...	...	102.5 (93.3)	2.5 (2.3)
	Billet	100	6.0	87.4 (79.6)	10.5 (9.6)	...	...
Annealed 540 °C (1000 °F) 2 h, AC	Billet	58	2.30	55.9 (50.9)	0.65 (0.6)	...	...
Annealed 705 °C (1300 °F) 4 h, AC	Forging	58	2.30	63.8 (58.1)	1.3 (1.2)	68.3 (62.2)	3.3 (3.0)
Annealed 745 °C (1375 °F) 3 h, AC	Plate	70	2.75	66.3 (60.4)	6.0 (5.5)	...	...
As received	Forged bar	25-89	1.00-3.50	62.7 (57.1)	11.4 (10.4)	60.3 (54.9)	11.8 (10.8)
Forged, beta forged reheated to 1065 °C (1950 °F) drawn to size, annealed 705 °C (1300 °F)	Forged bar	57	2.25	...	...	46.8 (42.6)	4.7 (4.3)
Beta forged, mill annealed 705 °C (1300 °F) 2 h, AC	Forging	50	2.00	77.5 (70.6)	5.4 (4.9)	78 (71.0)	0.43 (0.4)
β processed-mill annealed	Plate	75	3.00	104.2 (94.9)	5.2 (4.8)	...	...
Diffusion bond annealed	Billet	25-89	1.00-3.50	74.9 (68.2)	10.6 (9.7)	70.5 (64.2)	12.9 (11.8)
Mill annealed	Plate	25-38	1.00-1.50	61.1 (55.6)	1.4 (1.3)	...	...
Mill annealed	Plate	31-50	1.25-2.00	...	...	110.5 (100.6)	7.4 (6.8)
	Extrusion	45-100	1.80-4.00	91.7 (83.5)	3.4 (3.1)	96.1 (87.5)	4.5 (4.1)
Mill annealed 705 °C (1300 °F) 2 h, AC	Forging	50	2.00	52.4 (47.7)	3.2 (2.9)	54.4 (49.5)	4.3 (3.9)
	Billet	58	2.30	92.3 (84.0)	3.7 (3.4)	...	...
Recrystallize anneal	Plate	25-63	1.00-2.50	90.9 (82.8)	8.5 (7.8)	88.8 (80.8)	11.8 (10.8)
	Forging	30-170	1.20-6.70	91.8 (83.6)	6.0 (5.5)	92.2 (83.9)	7.6 (6.9)
STA	Plate	15	0.62	...	...	46.6 (42.6)	2.2 (2.0)
925 °C (1700 °F) 6 h, AC, 760 °C (1400 °F) 6 h, AC	Forging	35	1.40	83.4 (75.9)	4.6 (4.2)	89.2 (81.2)	6.3 (5.8)
955 °C (1750 °F) 1 h, WQ, 540 °C (1000 °F) 4 h	Forging	75	3.00	...	...	87.1 (79.3)	5.4 (4.9)
955 °C (1750 °F) 1 h, FC to 595 °C (1100 °F), AC	Plate	38	1.50	...	...	100.5 (91.5)	2.3 (2.1)
955 °C (1750 °F) 1 h, FC to RT	Plate	38	1.50	78.6 (71.8)	3.5 (3.2)	100.6 (91.6)	1.4 (1.3)
955 °C (1750 °F) 2 h, WQ, 540 °C (1000 °F) 2 h, AC, 705 °C (1300 °F) 2 h, AC, STA	Plate	15	0.62	45.5 (41.4)	2.5 (2.3)	...	...

Source: *Damage Tolerant Design Handbook*, J. Gallagher, Battelle, 1983. FC, fan cooled

Weldments

Ti-6Al-4V: Effects of specimen type, orientation, and test temperature on fracture toughness of electron beam welds

Specimen type and orientation(a)	Specimen thickness, mm	Specimen thickness, in.	Test temperature, °C	Test temperature, °F	Fracture toughness(b), K_Q, MPa$\sqrt{m}$	K_Q, ksi$\sqrt{in.}$	K_{Ic}, MPa$\sqrt{m}$	K_{Ic}, ksi$\sqrt{in.}$
BM/T-L	50.8	2.0	RT		115.9	105.5	115.9	105.5
BM/L-T	50.8	2.0	RT		123.1	112.0	119.3	108.6
BM/S-L	50.8	2.0	RT		96.7	88.0	...	...
HAZ/T-L	50.8	2.0	RT		131.6	119.8	...	...
HAZ/L-T	50.8	2.0	RT		129.5	117.9	...	...
HAZ/S-L	50.8	2.0	RT		97.9	89.1	...	...
WM/T-L	50.8	2.0	RT		90.8	82.6	...	...
BM-T-L	50.8	2.0	−54	−65	96.6	87.9	96.6	87.9
HAZ/T-L	50.8	2.0	−54	−65	118.7	108.0	...	...
WM/T-L	50.8	2.0	−54	−65	102.2	93.0	...	...
BM/T-L	50.8	2.0	79	175	128.6	117.0	...	...
HAZ/T-L	50.8	2.0	79	175	149.4	136.0	...	...
WM/T-L	50.8	2.0	79	175	122.2	111.2	...	...
WM/T-L	25.4	1.0	RT		63.1	57.4	...	...

Note: Compact tension specimens were tested to ASTM E399; values shown are averages of three specimens per condition. All specimens were stress relieved for 5 h at 705 °C (1300 °F) prior to testing. (a) BM, base metal; HAZ, heat-affected zone; WM, weld metal; T-L and S-L indicate base-metal orientations. (b) Most WM and HAZ tests were invalid because of insufficient crack length or specimen thickness. Occasional high values in HAZ resulted from unstable crack propagation at large angles from desired path. Source: J.G. Bjeletich, "Development of Engineering Data on Thick Section Electron Beam Welded Titanium," AFML-TR-73-197, U.S. Air Force Materials Laboratory, 1973

Ti-6Al-4V: Fracture toughness of electron beam welds as a function of processing order and stress-relief temperature

Location of notch	Filler-metal thickness, mm	Filler-metal thickness, in.	Welding position(a)	No. of pass	Fracture toughness (K_Q)(b)(c), STA → EBW → SR for SR temperature of: 545 °C (1015 °F), MPa√m	545 °C (1015 °F), ksi√in.	645 °C (1195 °F), MPa√m	645 °C (1195 °F), ksi√in.	EBW → STA, MPa√m	EBW → STA, ksi√in.
Weld metal	None		H	1	...	...	54.9	50.0	71.0*	64.6*
				2	69.5	63.2	71.0	64.6	71.0*	64.6*
			F	1	65.7	59.8	60.2	54.8	71.0*	64.6*
				2	59.2	53.9	65.7	59.8	71.0*	64.6*
	0.5	0.020	H	1	76.0*	69.2*	90.2*	82.1*	...	...
				2	74.4	67.7	76.3*	69.4*	...	...
			F	1	86.5*	78.7*	74.7*	68.0*	...	...
				2	89.0*	81.0*	87.7*	79.8*	...	...
	1.0	0.039	H	1	84.7*	77.1*	...	...	90.2*	82.1*
				2	87.4*	79.5*	89.6*	81.5*	90.2*	82.1*
			F	1	83.4*	75.9*	88.1*	80.2*	90.2*	82.1*
				2	90.2*	82.1*	87.1*	79.3*	90.2*	82.1*
Heat-affected zone	None		H	1	58.0	52.8	49.0	44.6	69.8	63.5
				2	60.8	55.3	55.2	50.2	69.8	63.5
			F	1	61.4	55.9	58.0	52.8	69.8	63.5
				2	60.8	55.3	54.6	49.7	69.8	63.5
	0.5	0.020	H	1	54.6	49.7	59.2	53.9	...	...
				2	67.0	61.0	58.0	52.8	...	...
			F	1	...	...	58.6	53.3	...	...
				2	61.1	55.6	51.2	46.6	...	...
	1.0	0.039	H	1	57.7	52.5	58.9	53.6	76.9*	70.0*
				2	65.7*	59.8*	52.4	47.7	76.9*	70.0*
			F	1	67.0*	61.0*	52.7	48.0	76.9*	70.0*
				2	65.4*	59.5*	51.5	46.9	76.9*	70.0*

Note: Composition, 6.17% Al, 0.017% C, 0.117% Fe, 0.0058% H, 0.0024% N, 0.149% O, and 4.04% V. Focal length, 350 mm (13.8 in.) for horizontal specimens, 400 mm (15.7 in.) for flat specimens. Work distance, 300 mm (11.8 in.). Accelerating voltage, 50 kV. Beam current, 150 mA. Welding speed, 300 mm/min (11.8 in./min). (a) H, horizontal; F, flat. (b) STA, solution treated and aged; EBW, electron beam welded; SR, stress relief. (c) Asterisks indicate invalid K_{Ic}. Source: Y. Obata *et al.*, Fracture Behavior in Electron Beam Welds of Ti-6Al-4V Alloy, *Titanium Science and Technology*, G. Lütjering, U. Zwicker, and W. Bunk, Ed., Deutsche Gesellschaft für Metallkunde eV, 1985, p 807-813

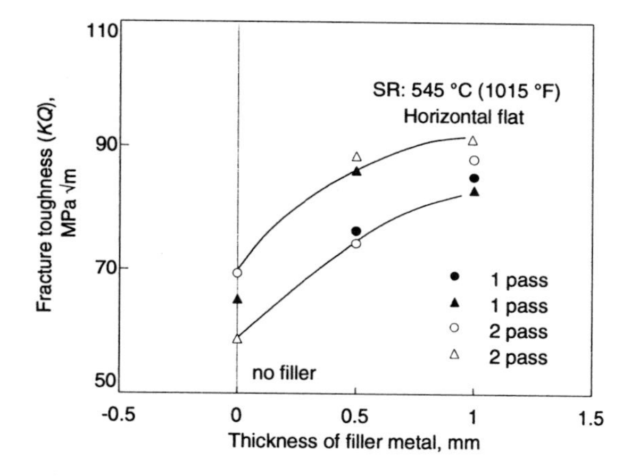

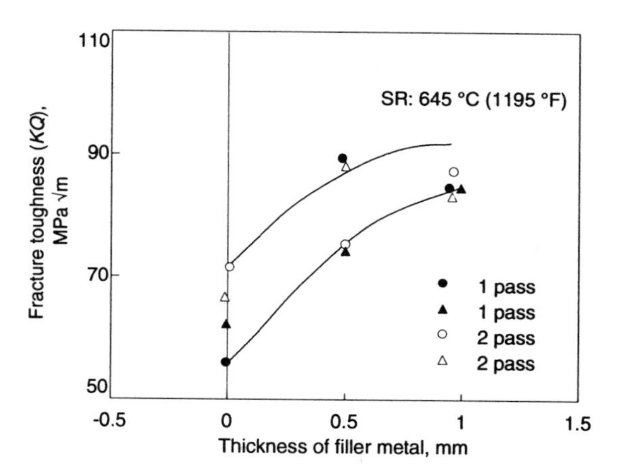

Ti-6Al-4V: Fracture toughness of electron beam welds. Effect of filler metal thickness and stress relief temperature on fracture toughness. Composition: 6.17% Al, 0.017% C, 0.117 % Fe, 0.0058% H, 0.0024%, N, 0.149% O, and 4.04% V. Focal length, 350 mm (13.8 in.) for horizontal specimens, 400 mm (15.7 in.) for flat specimens. Work distance, 300 mm (11.8 in.). Accelerating voltage, 50 kV. Beam current, 150 mA. Welding speed, 300 mm/min (11.8 in./min). Source: Y. Obata *et al.*, Fracture Behavior in Electron Beam Welds of Ti-6Al-4V Alloy, *Titanium Science and Technology*, G. Lütjering, U. Zwicker, and W. Bunk, Ed., Deutsche Gesellschaft für Metallkunde eV, Germany, 1985, p 807-813

Effect of Temperature

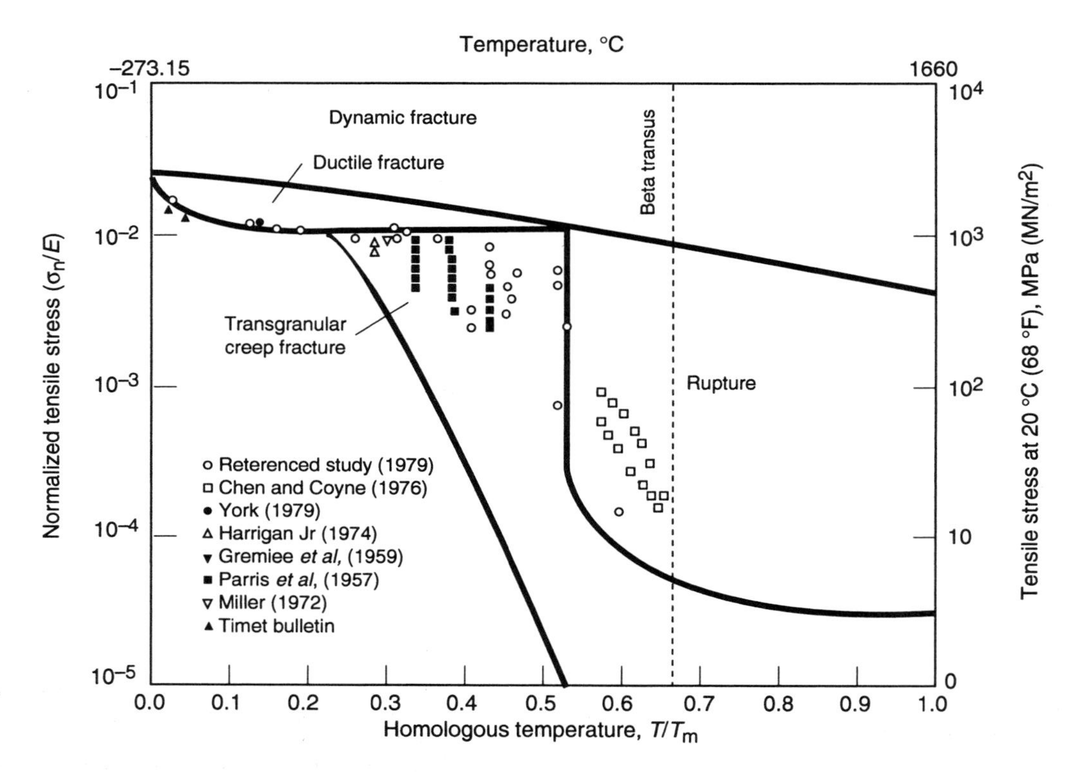

Ti-6Al-4V: Fracture mechanism map. Normalizing parameters: T_m = 1933 K and E(GPa) = 122.7 – 0.056T(in K). Source: Krishnamohanrao *et al.*, Fracture Mechanism Maps for Titanium and Its Alloys, *Acta Metall.*, Vol 34, No. 9, 1986, p 1783-1806

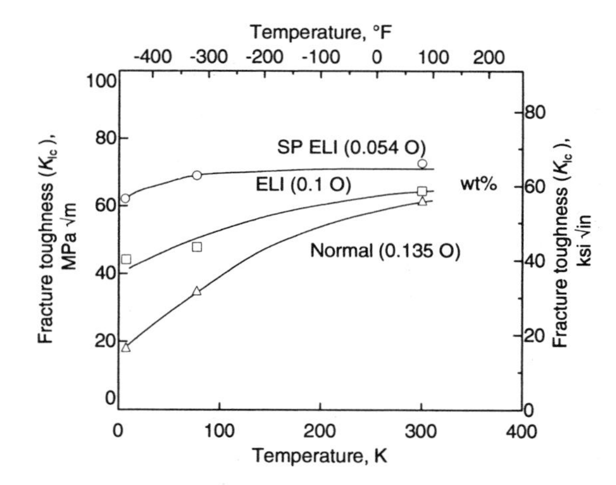

Ti-6Al-4V: Fracture toughness vs oxygen content/temperature.

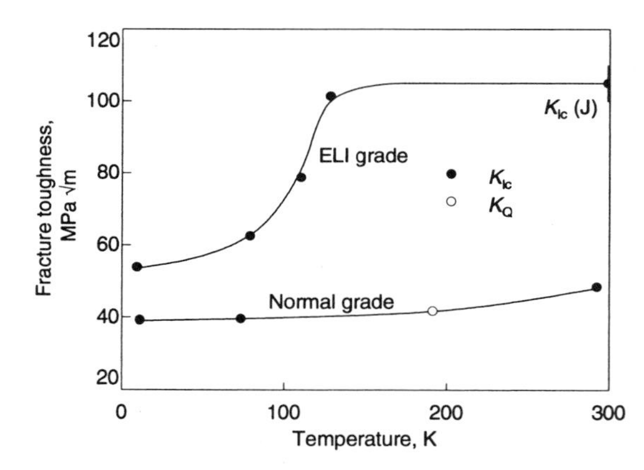

Ti-6Al-4V/Ti-6Al-4V(ELI): Low-temperature fracture toughness. Source: C. Fowlkes and R. Tobler, "Fracture Testing and Results for Ti-6Al-4V Alloy at Liquid Helium Temperature," *Eng. Fract. Mech.*, Vol 8, 1976, p 487, reported in *Materials at Low Temperatures*, R. Reed and A. Clark, Ed., American Society for Metals, 1983, p 407

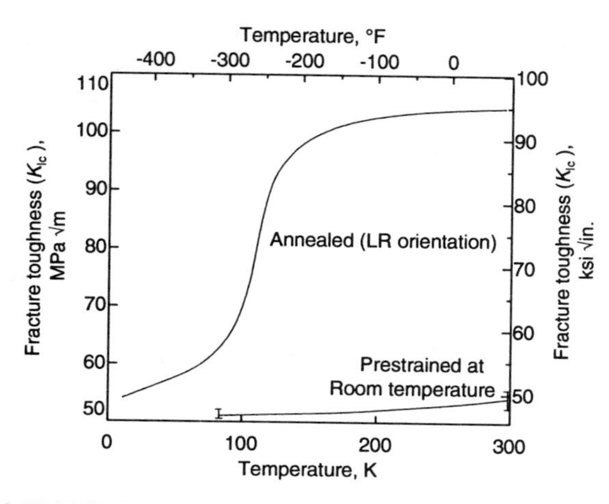

Ti-6Al-4V (ELI): Fracture toughness. Effect of test temperature and prestraining in biaxial tension on recrystallization annealed alloy. Composition, 5.91% Al, 0.018% C, 0.103% Fe, 52 ppm H, 0.014% N, 0.110% O, and 3.95% V. R.L. Tobler, Low Temperature Fracture Behavior of a Ti-6Al-4V Alloy and Its Electron Beam Welds, in *Toughness and Fracture Behavior of Titanium*, STP 651, American Society for Testing and Materials, 1978, p 267-294

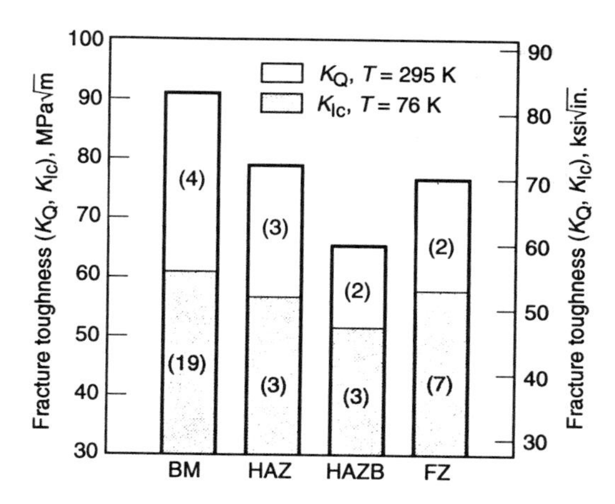

Ti-6Al-4V: Fracture toughness of electron beam welds. Effect of test temperature and test location. Number of tests performed are indicated in parentheses. Source: R.L. Tobler, Low Temperature Fracture Behavior of a Ti-6Al-4V Alloy and Its Electron Beam Welds, in *Toughness and Fracture Behavior of Titanium*, STP 651, American Society for Testing and Materials, 1978, p 267-294

Hydrogen Embrittlement

There are two basic types of hydrogen embrittlement: gaseous hydrogen embrittlement, which basically assumes an infinite source of external hydrogen, and embrittlement involving internal hydrogen contained within the material. The mechanisms involved are similar, but the resulting fracture topography can be different.

Gaseous Hydrogen Embrittlement. In a report on the effects of hydrogen pressure, stress intensity, and temperature on crack-growth rate under gaseous hydrogen embrittlement conditions in a lamellar microstructure, Nelson (*Metall. Trans. A*, Vol 7, 1976, p 621-627) proposes that the rate of growth depends on migration of the hydrogen ahead of the crack tip and on saturation of the β phase with hydrogen to form hydrides. Propagation of the crack occurs when the hydride grows to a thickness that is not capable of withstanding the applied load. Thus, a reduction of hydrogen reduces the growth rate by prolonging the time required for the film to grow to a given thickness; an increase in the stress intensity increases the growth rate by reducing the film thickness that must be achieved to cause rupture of the hydride. A reduction in temperature, however, reduces the growth rate by slowing the diffusion rate of the hydrogen. Gaseous hydrogen embrittlement results in a brittle, terraced fracture.

Microstructure is also an important factor in this type of embrittlement. The rate of hydrogen diffusion is higher by several orders of magnitude in the β phase than in the α phase. Therefore, microstructures with the more continuous β phase, such as β annealed material, are more severely embrittled than those with discontinuous β, such as the fine equiaxed α in the α matrix (Nelson *et al.*, *Metall. Trans.*, Vol 3, 1972, p 469-475). Transport of the hydrogen to the critical location is much more rapid with continuous β; consequently, the effect is more severe.

Embrittlement from Internal Hydrogen. Hydrogen effects involving the presence of internal hydrogen are similar, but it may take longer for these effects to be observed because of the longer transport distances involved. The hydrogen must diffuse to the stress raiser from within the matrix, because the infinite source from the atmosphere is not available. Meyn (*Metall. Trans.*, Vol 5, 1974, p 2405-2414) shows a decrease in fracture toughness due to hydrogen in concentrations of up to 50 ppm. The rate of sustained load cracking increases with increasing hydrogen concentrations above this level. Additionally, hydrogen contents up to 50 ppm affect the sustained load cracking, whereas higher hydrogen levels do not, as reported by Meyn.

Another study, Boyer and Spurr, (*Metall. Trans. A*, Vol 9, 1978, p 23) demonstrates the effect of hydrogen and temperature on sustained load cracking (see figure). The higher the hydrogen content, the higher the temperature for the onset of sustained load cracking. The only specimens that did not exhibit any significant crack growth at temperatures down to –68 °C (–90 °F) were those with hydrogen contents of 122 ppm or lower. Crack growth was occurring at stress intensities as low as about 12 MPa$\sqrt{m}$ (10.9 ksi$\sqrt{in.}$) at –68 °C, whereas material with hydrogen contents of about 50 to 60 ppm did not exhibit crack growth at stress intensities of about 50 MPa$\sqrt{m}$ (45.5 ksi$\sqrt{in.}$).

Crack-growth rate is also a function of hydrogen content and temperature. The crack-growth rate increases with temperature until it reaches a peak, and then the rate decreases. Prior to reaching the peak (from the high-temperature side), hydride nucleation becomes easier,

Ti-6Al-4V: Variation in crack-growth rate and step width as a function of hydrogen pressure

Hydrogen pressure		Crack-growth rate (da/dt)		Step width	
kPa	psi	m/s	ft/s	m	ft
90	13.1	1×10^{-4}	3.3×10^{-4}	3.25	10.7
90.6	13.1	9×10^{-5}	3.0×10^{-4}	3.5	11.5
75	10.9	8×10^{-5}	2.6×10^{-4}	3.5	11.5
60.8	8.8	3.6×10^{-5}	1.2×10^{-4}	4.25	13.9
48.8	7.1	4.75×10^{-5}	1.56×10^{-4}	4.25	13.9
6.1	0.9	7.2×10^{-6}	2.4×10^{-5}	5.25	17.2

Note: DT specimens. Constant stress intensity of 76 MPa$\sqrt{m}$ (69.2 ksi$\sqrt{in.}$) in Stage II. Test temperature, 24 °C (75 °F). Source: H.G. Nelson, A Film-Rupture Model of Hydrogen-Induced, Slow Crack-Growth in Acicular Alpha-Beta Titanium, *Metall. Trans. A*, Vol 7 (No. 5), May 1976, p 621-627

but the diffusion rate decreases. As long as nucleation is the rate-controlling process, crack-growth rate increases with decreasing temperature. When the temperature gets low enough for hydrogen diffusion to become the rate-controlling process, the growth rate decreases with further reductions in temperature.

Boyer and Spurr (*Metall. Trans. A*, Vol 9, 1978, p 23) also report a significant effect of texture on sustained load cracking. When an attempt was made to test specimens in which cracks would have to grow in a direction normal to the basal planes, the cracks would rotate 90° and would not grow in the intended orientation.

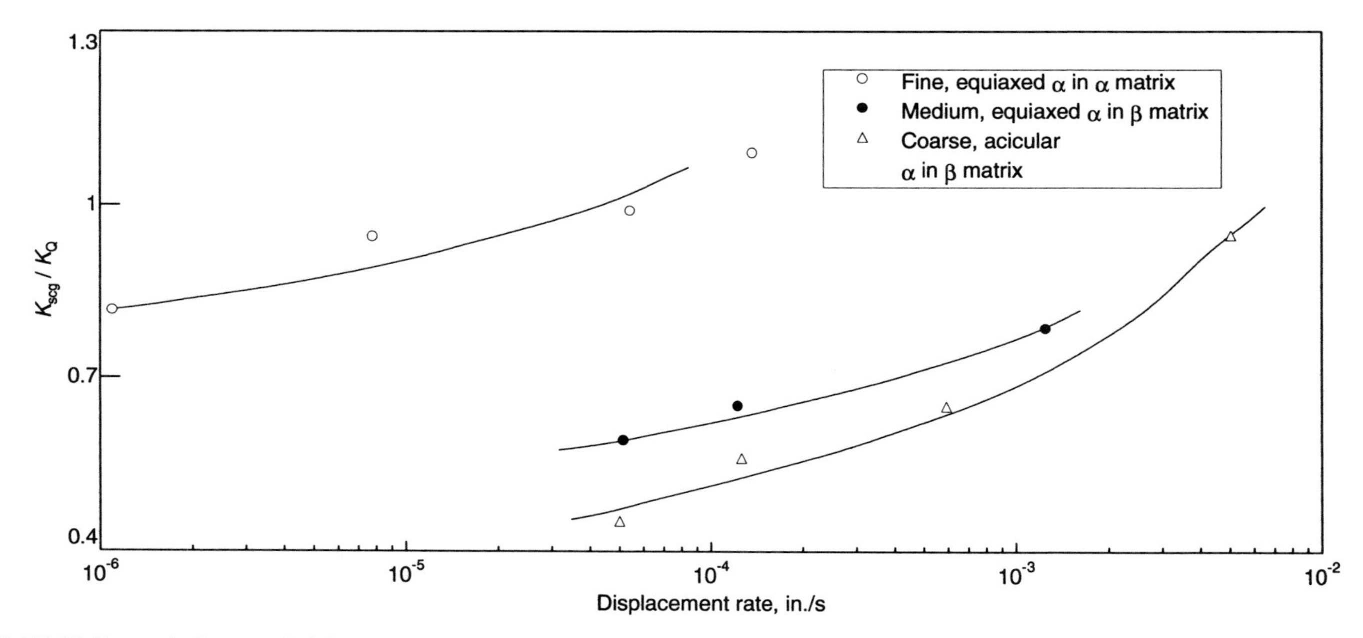

Ti-6Al-4V: Gaseous hydrogen embrittlement. Effect of test displacement rate on hydrogen embrittlement for three microstructures. Hydrogen pressure, 90.7 kPa (680 torr). Test temperature, 24 °C (75 °F) K_{scg}, stress intensity for subcritical crack growth. Source: H.G. Nelson, D.P. Williams, and J.E. Stein. Environmental Hydrogen Embrittlement of an Alpha-Beta Titanium Alloy: Effect of Microstructure, *Metall. Trans.*, Vol 3 (No.2), Feb 1972, p 469-475

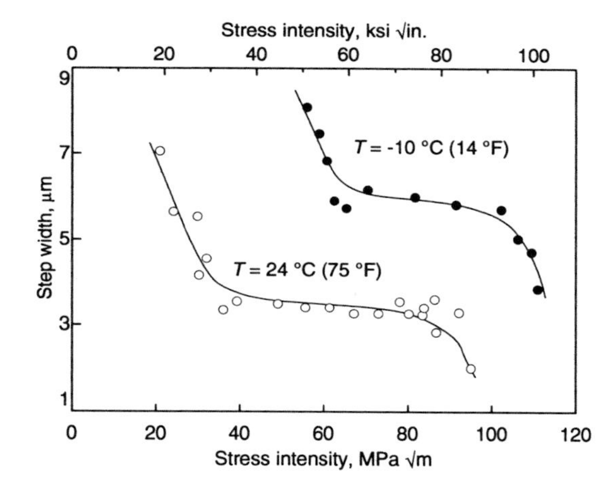

Ti-6Al-4V: Step width vs applied stress intensity. Double cantilever beam specimens. Hydrogen pressure, 90.6 kPa (13.1 psi). Source: H.G. Nelson, A Film-Rupture Model of Hydrogen-Induced, Slow Crack-Growth in Acicular Alpha-Beta Titanium, *Metall. Trans. A*, Vol 7 (No. 5), May 1976, p 621-627

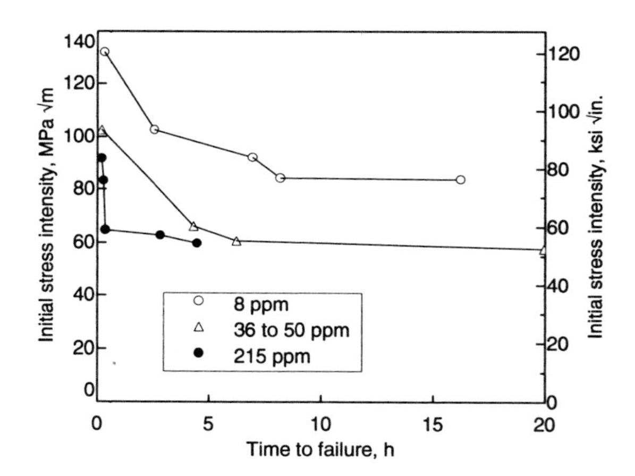

Ti-6Al-4V: Sustained load cracking behavior. Effect of hydrogen content on sustained load cracking of precracked cantilever beam, dead weight loaded specimens. Source: D.A. Meyn, Effect of Hydrogen on Fracture and Inert-Environment Sustained Load Cracking Resistance of Alpha-Beta Titanium Alloys, *Metall. Trans.*, Vol 5 (No. 11), Nov 1974, p 2405-2414

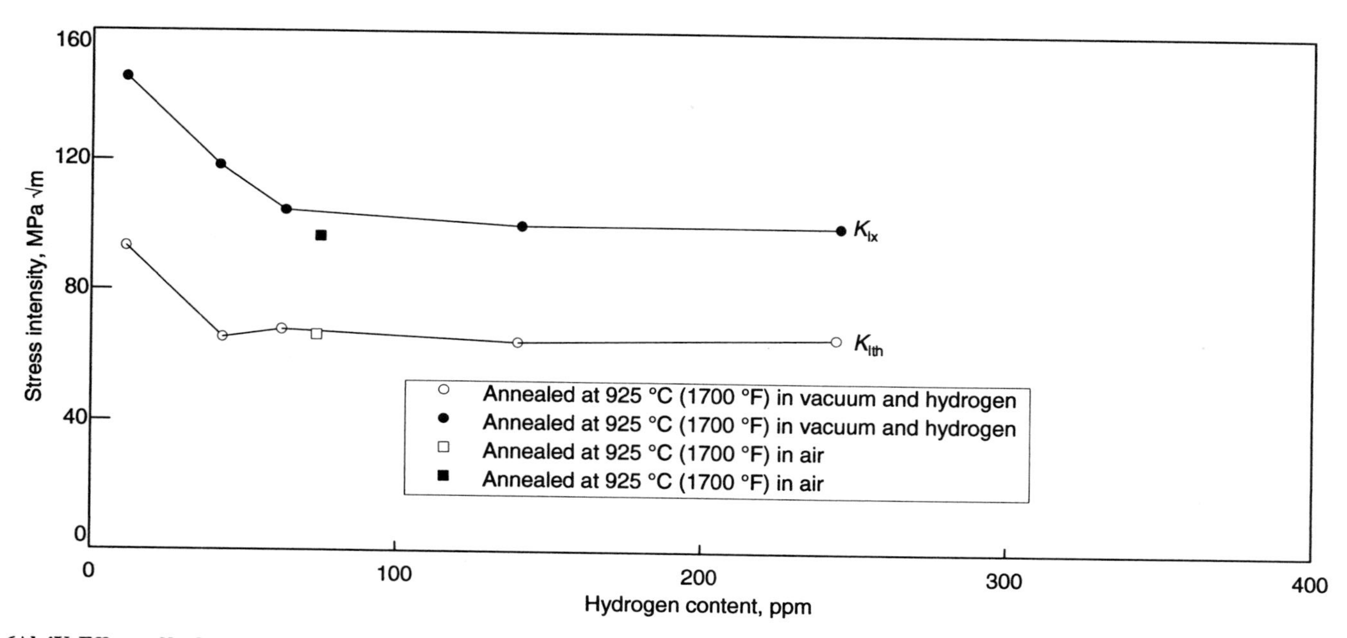

Ti-6Al-4V: Effects of hydrogen content. Fracture toughness (K_{Ix}) vs hydrogen content sustained load crack propagation threshold (K_{Ith}). Source: D.A. Meyn, Effect of Hydrogen on Fracture and Inert-Environment Sustained Load Cracking Resistance of Alpha-Beta Titanium Alloys, *Metall. Trans.*, Vol 5 (No. 11), Nov 1974, p 2405-2414

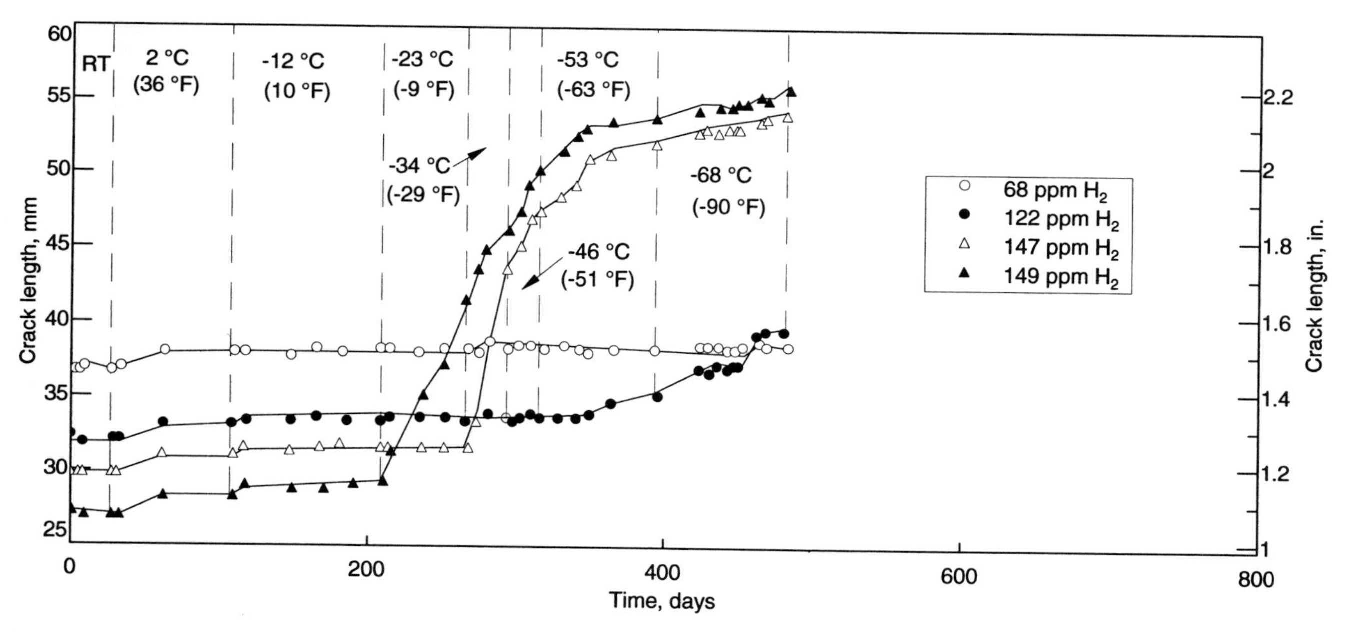

Ti-6Al-4V: Crack length as a function of time. At various test temperatures for hydrogen contents from 68 to 149 ppm. Source: R.R. Boyer and W.F. Spurr, Characterization of Sustained Load Cracking and Hydrogen Effects in Ti-6Al-4V, *Metall. Trans. A*, Vol 9, Jan 1978

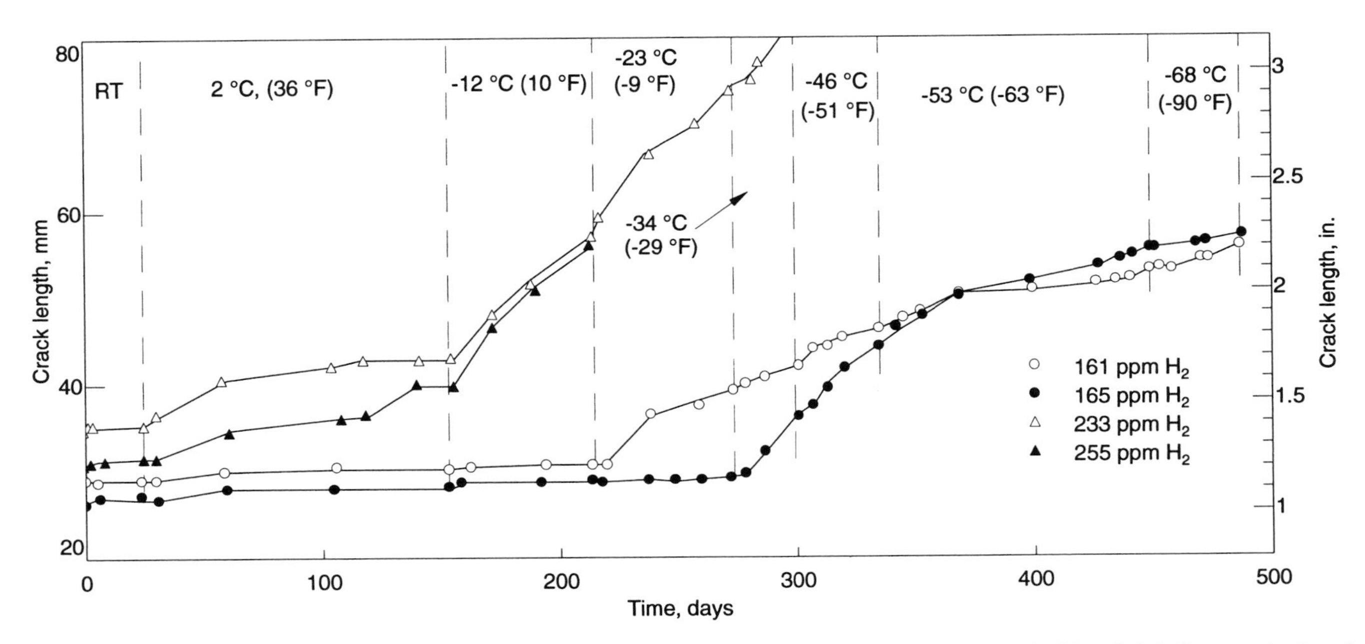

Ti-6Al-4V: Crack length as a function of time. At various test temperatures for hydrogen contents from 161 to 255 ppm. Solid symbols indicate termination of test. Source: R.R. Boyer and W.F. Spurr, Characterization of Sustained Load Cracking and Hydrogen Effects in Ti-6Al-4V, *Metall. Trans. A*, Vol 9, Jan 1978

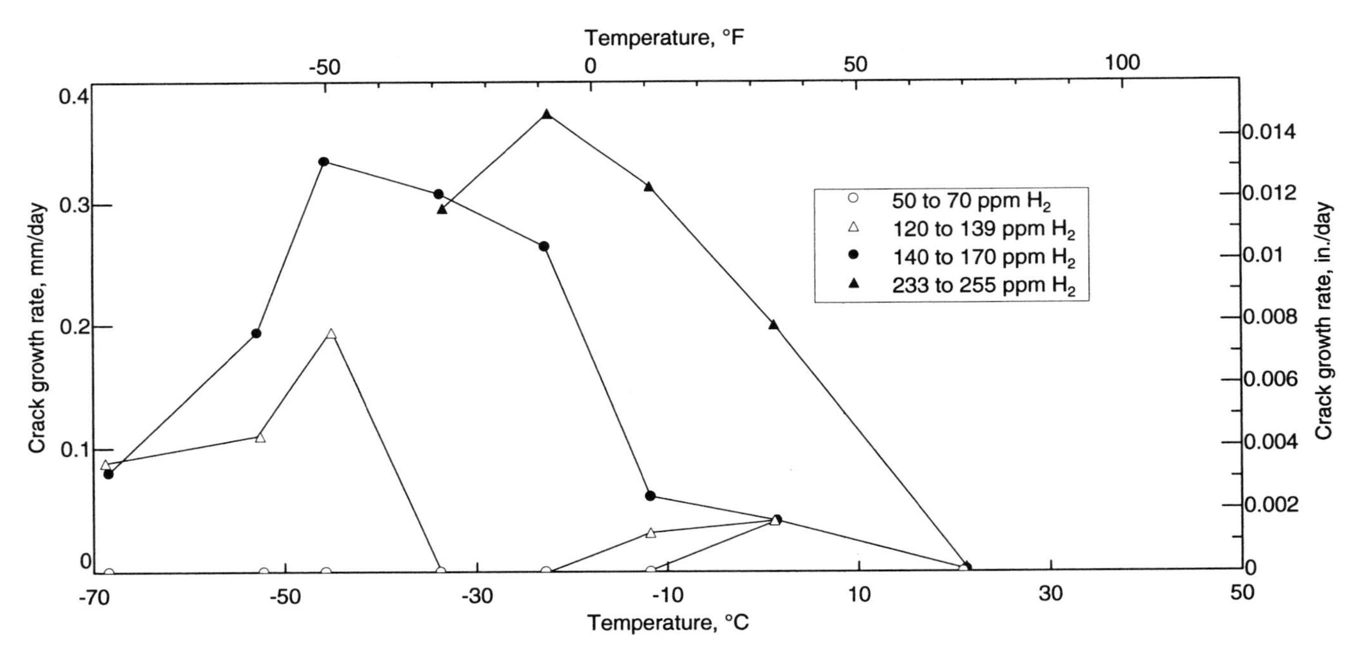

Ti-6Al-4V: Average maximum crack-growth rate vs temperature. These results ignore stress-intensity effects, but they demonstrate the competition between diffusion of hydrogen and nucleation of hydrides. General shape of curve confirmed by W.J. Pardee and N.E. Paton in *Metall. Trans. A*, Vol 11, 1980, p 1391-1400. Source: R.R. Boyer and W.F. Spurr, Characterization of Sustained Load Cracking and Hydrogen Effects in Ti-6Al-4V, *Metall. Trans. A*, Vol 9, Jan 1978

Ti-6Al-6V-2Sn

Common Name: Ti-662
UNS Number: R56620

Ti-6Al-6V-2Sn was developed at New York University on a U.S. Army contract as a higher-strength version of Ti-6Al-4V. It is a corrosion-resistant, high-strength alloy which offers an ultimate tensile strength of 1200 MPa (175 ksi) in the heat treated condition in sizes up to 25 mm (1 in.) diameter. This grade is used in applications requiring high strength-to-weight ratios at temperatures up to 315 °C (600 °F).

Chemistry and Density

Ti-662 contains a total of about 1% (Cu + Fe) in approximately equal proportions, which give it much-improved heat treatability. Its nominal 6% aluminum content stabilizes the alpha phase and increases the hot workability range by raising the beta transus temperature to approximately 945 °C (1735 °F). Cooling from above this temperature with little concurrent or subsequent deformation generally results in inferior ductility. As a neutral stabilizer, the 2% tin strengthens both the alpha and beta phases, and in combination with the aluminum, provides better room- and elevated-temperature strength properties than those of Ti-6Al-4V and other lower-alloy alpha-beta compositions. Beta stabilization is accomplished by nominal additions of 6% vanadium, 0.5% copper, and 0.5% iron. Acting together, these elements permit heat treatment of the alloy to high strength levels by solution treatment and aging.

Density. 4.54 g/cm^3 (0.164 lb/in.^{3})

Alloy Segregation. Ingot composition must be controlled within specified limits, and special melting practices, particularly for the final melt, are required to minimize segregation during solidification. Excessive macrosegregation results in "beta flecks," which are harder, less-ductile areas after heat treatment. Detrimental effects of beta flecks have not been demonstrated for this alloy.

Exceeding Composition Limits. As for all alpha-beta alloys, excessive amounts of aluminum, oxygen, and nitrogen can decrease ductility and fracture toughness. Excessive amounts of beta stabilizers (molybdenum and vanadium) affect the stability of the alloy and increase its heat treatability, therefore making control of properties more difficult. Excessive impurity levels may raise yield strength above maximum permitted values or decrease elongation or reduction in area below minimum values.

Product Forms

Ti-662 is produced by all U.S. titanium melters as bar and billet for forging stock. Plate, sheet, wire, and extrusions are also available.

Product Conditions/Microstructure

In forged sections and plate up to 25 mm (1 in.) thick, solution-treated-and-aged material has a guaranteed minimum ultimate tensile strength of 1170 MPa (170 ksi). For forged sections between 75 and 100 mm (3 and 4 in.) thick, the corresponding ultimate tensile strength is 1035 MPa (150 ksi). The response to heat treatment may vary from heat to heat and the correct aging temperature is best determined by tests on the heat in question. Cooling from above the beta transus with concurrent or subsequent deformation generally results in inferior ductility.

Annealing treatments at temperatures of 640 to 790 °C (1200 to 1450 °F) are applied to produce maximum stability at temperatures up to 450 °C (600 °F). The strengthening response to the precipitation-hardening reaction is dependent on the ability to retain the beta phase during quenching from the solution temperature, and this alloy is sufficiently beta stabilized to attain heat treated properties through section thicknesses up to 100 mm (4 in.).

Applications

Ti-662 is used in applications requiring high strength at temperatures up to 315 °C (600 °F) in the forms of sheet, light-gage plate, extrusions, and small forgings. This alloy is used for airframe structures where strength higher than that of Ti-6Al-4V is required. Usage is generally limited to secondary structures because the attractiveness of higher strength efficiency is minimized by lower fracture toughness and fatigue properties. Ti-662 is used for aircraft structural members, centrifuge parts, and rocket-engine parts.

Limitations in Use. As is characteristic of other titanium alloys, exposure to stress at elevated temperature produces changes in the retained mechanical properties. The stress and temperature limits below which these changes will not occur have not been established for this alloy. Structural applications should be based on a knowledge of the low toughness characterizing the higher-strength conditions of this alloy and the limited toughness of welds. Particular attention should be given to the influence of aggressive environments in the presence of cracks. Such environments include aqueous solutions of chlorides and possibly certain organic solvents such as methanol.

Ti-662: Equivalent specifications

Specification UNS	Designation R56620	Description	Al 5.5	Cu	Fe	H	N	O	Sn 2	V 5.5	Other bal Ti(a)
China											
GB 3620	TC-10		5.5-6.5	0.5	0.5	0.015 max	0.04 max	0.2 max	1.5-2.5	5.5-6.5	C 0.1 max; Si 0.15 max; bal Ti
Europe											
AECMA Ti-P64	prEN3316	Sh Strp Ann	5-6	0.35-1	0.35-1	0.0125 max	0.04 max	0.2 max	1.5-2.5	5-6	C 0.05 max; OT 0.4 max; OE 0.1 max; bal Ti
AECMA Ti-P64	prEN3317	Plt Ann	5-6	0.35-1	0.35-1	0.0125 max	0.04 max	0.2 max	1.5-2.5	5-6	C 0.05 max; OT 0.4 max; OE 0.1 max; bal Ti
AECMA Ti-P64	prEN3318	Frg NHT	5-6	0.35-1	0.35-1	0.0125 max	0.04 max	0.2 max	1.5-2.5	5-6	C 0.05 max; OT 0.4 max; OE 0.1 max; bal Ti
AECMA Ti-P64	prEN3319	Bar Ann	5-6	0.35-1	0.35-1	0.0125 max	0.04 max	0.2 max	1.5-2.5	5-6	C 0.05 max; OT 0.4 max; OE 0.1 max; bal Ti

(continued)

Ti-662: Equivalent specifications

Specification	Designation	Description	Al	Cu	Fe	H	N	O	Sn	V	Other
UNS	R56620		5.5						2	5.5	bal Ti(a)
AECMA Ti-P64	prEN3320	Frg Ann	5-6	0.35-1	0.35-1	0.0125 max	0.04 max	0.2 max	1.5-2.5	5-6	C 0.05 max; OT 0.4 max; OE 0.1 max; bal Ti
Germany											
WL 3.7174		Sh Strp Plt Bar Frg Ann	5-6	0.35-1	0.35-1	0.0125-0.015	0.04	0.2	1.5-2.5	5-6	C 0.05; OT 0.4; bal Ti
WL 3.7174		Sh Strp Plt Bar Frg STA	5-6	0.35-1	0.35-1	0.0125-0.015	0.04	0.2	1.5-2.5	5-6	C 0.05; OT 0.4; bal Ti
Spain											
UNE 38-725	L-7303	Sh Strp Plt Bar Ext Ann	5-6	0.35-1	0.35-1	0.0125	0.04	0.2	1.5-2.5	5-6	C 0.05; OT 0.4; bal Ti
UNE 38-725	L-7303	Sh Strp Plt Bar Ext HT	5-6	0.35-1	0.35-1	0.0125	0.04	0.2	1.5-2.5	5-6	C 0.05; OT 0.4; bal Ti
USA											
AMS 4918F		Sh Strp Plt Ann	5-6	0.35-1	0.35-1	0.015	0.04	0.2	1.5-2.5	5-6	C 0.05; OT 0.4; Y 0.005; bal Ti
AMS 4936B		Ext Rng Ann	5-6	0.35-1	0.35-1	0.015	0.04	0.2	1.5-2.5	5-6	C 0.05; OT 0.4; Y 0.005; bal Ti
AMS 4936B		Ext Rng STA	5-6	0.35-1	0.35-1	0.015	0.04	0.2	1.5-2.5	5-6	C 0.05; OT 0.4; Y 0.005; bal Ti
AMS 4936C		Beta Ext Ann Rng Flsh Wld	5-6	0.35-1	0.35-1	0.015 max	0.04 max	0.2 max	1.5-2.5	5-6	C 0.05 max; OT 0.4 max; Y 0.005 max; OE 0.1 max; bal Ti
AMS 4971C		Bar Frg Wir Rng Bil Ann	5-6	0.35-1	0.35-1	0.015	0.04	0.2	1.5-2.5	5-6	C 0.05; OT 0.4; Y 0.005; bal Ti
AMS 4978B		Bar Wir Frg Bil Rng Ann	5-6	0.35-1	0.35-1	0.015	0.04	0.2	1.5-2.5	5-6	C 0.05; OT 0.4; Y 0.005; bal Ti
AMS 4978C		Bar Frg Rng Ann	5-6	0.35-1	0.35-1	0.015 max	0.04 max	0.2 max	1.5-2.5	5-6	C 0.05 max; OT 0.4 max; Y 0.005 max; OE 0.1 max; bal Ti
AMS 4979B			5-6	0.35-1	0.35-1	0.015	0.04	0.2	1.5-2.5	5-6	C 0.05; OT 0.4; Y 0.005; bal Ti
MIL F-83142A	Comp 8	Frg Ann	5-6	0.35-1	0.35-1	0.015	0.04	0.2	1.5-2.5	5-6	C 0.05; OT 0.3; bal Ti
MIL F-83142A	Comp 8	Frg HT	5-6	0.35-1	0.35-1	0.015	0.04	0.2	1.5-2.5	5-6	C 0.05; OT 0.3; bal Ti
MIL T-81556A	Code AB-3	Ext Bar Shp Ann	5-6	0.35-1	0.35-1	0.015	0.04	0.2	1.5-2.5	5-6	C 0.05; OT 0.3; bal Ti
MIL T-81556A	Code AB-3	Ext Bar Shp STA	5-6	0.35-1	0.35-1	0.015	0.04	0.2	1.5-2.5	5-6	C 0.05; OT 0.3; bal Ti
MIL T-9046J	Code AB-3	Sh Strp Plt Ann	5-6		0.35-1	0.015	0.04	0.2	1.5-2.5	5-6	C 0.05; OT 0.3; bal Ti
MIL T-9046J	Code AB-3	Sh Strp Plt ST	5-6		0.35-1	0.015	0.04	0.2	1.5-2.5	5-6	C 0.05; OT 0.3; bal Ti
MIL T-9046J	Code AB-3	Sh Strp Plt STA	5-6		0.35-1	0.015	0.04	0.2	1.5-2.5	5-6	C 0.05; OT 0.3; bal Ti
MIL T-9047G	Ti-6Al-6V-2Sn	Bar Bil Ann	5-6	0.35-1	0.35-1	0.015	0.04	0.2	1.5-2.5	5-6	C 0.05; OT 0.3; Y 0.005; bal Ti
MIL T-9047G	Ti-6Al-6V-2Sn	Bar Bil STA	5-6	0.35-1	0.35-1	0.015	0.04	0.2	1.5-2.5	5-6	C 0.05; OT 0.3; Y 0.005; bal Ti
USA (continued)											
SAE J467	Ti662		5.5 (nom)	0.7 (nom)	0.7 (nom)	...	0.02 max		2 (nom)	5.5 (nom)	C 0.02 max; Ni 0.006 max; Si 0.1 max; bal Ti

Ti-662: Commercial compositions

Specification	Designation	Description	Al	Cu	Fe	H	N	O	Sn	V	Other
France											
Ugine	UT662	Sh Plt Frg Ann	5-6	0.35-1	0.35-1	0.015	0.04	0.2	1.5-2.5	5-6	bal Ti
Ugine	UT662	Sh Plt Frg QA	5-6	0.35-1	0.35-1	0.015	0.04	0.2	1.5-2.5	5-6	bal Ti
Germany											
Deutsche T	Contimet AlVSn 6-6-2	Plt Bar Frg Pip Ann	5-6	0.35-1	0.35-1	0.015	0.04	0.2	1.5-2.5	5-6	C 0.05; bal Ti
Deutsche T	Contimet AlVSn 6-6-2	Plt Bar Frg Pip STA	5-6	0.35-1	0.35-1	0.015	0.04	0.2	1.5-2.5	5-6	C 0.05; bal Ti
Deutsche T	LT 33	Frg Aged	5-6	0.35-1	0.35-1	0.015	0.04	0.2	1.5-2.5	5-6	C 0.05; bal Ti
Deutsche T	LT 33	Frg Ann	5-6	0.35-1	0.35-1	0.015	0.04	0.2	1.5-2.5	5-6	C 0.05; bal Ti
Japan											
Kobe	KS6-6-2	Plt Sh Ann	5-6	0.35-1	0.35-1	0.0125	0.04	0.2		5-6	bal Ti
Kobe	KS6-6-2	Plt Sh STA	5-6	0.35-1	0.35-1	0.0125	0.04	0.2		5-6	bal Ti
Sumitomo	Ti-6Al-6V-2Sn										
Toho	662AT	STA	5-6		0.35-1	0.015	0.04	0.12-0.2	1.5-2.5	5-6	C 0.05; bal Ti
USA											
OREMET	Ti 6-6-2										
RMI	RMI 6Al-6V-2Sn	Mult Forms Ann	5-6	0.35-1	0.35-1	0.0125-0.015	0.04	0.2	1.5-2.5	5-6	C 0.08; bal Ti
RMI	RMI 6Al-6V-2Sn	Mult Forms STA	5-6	0.35-1	0.35-1	0.0125-0.015	0.04	0.2	1.5-2.5	5-6	C 0.08; bal Ti
Timet	TIMETAL 6-6-2	Ann	5-6		0.35-1	0.015	0.05 max	0.2 max	1.5-2.5	5-6	C 0.05 max; bal Ti
Timet	TIMETAL 6-6-2 STA	Bil Bar Plt Sh Str STA	5-6	0.35-1	0.35-1	0.015	0.04	0.2	1.5-2.5	5-6	C 0.05; bal Ti

Single values are maximums.

Phases and Structures

Alloy Ti-662 is normally processed in the $\alpha + \beta$ two-phase field, resulting in primary equiaxed α and some β. For example, annealing treatments (~760 °C or 1400 °F) moderately low in the two-phase $\alpha + \beta$ field after normal $\alpha + \beta$ processing result in microstructures with a high volume percentage of primary α with stabilized β at the equiaxed α grain boundaries. If the processing involves less exposure time or less working in the $\alpha + \beta$ region and is subsequently annealed at approximately 760 °C (1400 °F), the primary α grains appear more elongated, and the volume percentage is high. Both structures develop acceptable mechanical properties.

Totally transformed β structures are often considered unacceptable, although acicular products do have advantages. Annealing temperatures and cooling rates determine the presence and the coarseness of secondary α (transformed β). For solution treatments up to 825 °C (1515 °F), β is sufficiently enriched with vanadium to prevent decomposition into martensitic α. At temperatures above 900 °C (1650 °F), β decomposes completely to martensitic α. Between these two temperatures, partial transformation of β occurs (see the isothermal TTT diagram after quenching from 850 °C or 1560 °F). From above the β transus, the M_s temperature is about 420 °C (790 °F).

Crystal Structure

Beta Transus: 945 ± 10 °C (1733 ± 20 °F) to 955 ± 5 °C (1750 ± 10 °F)

Transformation Products

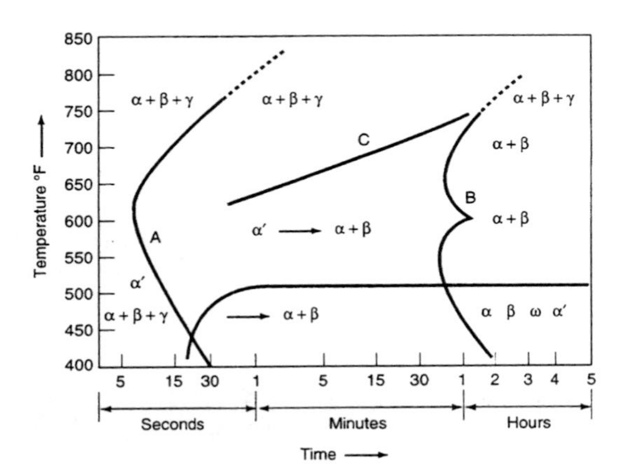

Ti-662: Time-temperature transformations from 850 °C (1560 °F). Dilatometric tests indicated M_S temperature of 640 °C (1185 °F), and X-ray measurements indicated that α′ martensite formed when isothermal holds were stopped by quenching before line A. Beyond line B, β is sufficiently enriched with vanadium to prevent martensitic transformation. Measurements indicated the disappearance of $Ti_3Al(\gamma)$ beyond line C. 25 mm (1 in.) diam specimens solution treated at 850 °C (1560 °F) for 1 h. Composition (wt%): 5.5 V, 5.65 Al, 2.35 Sn, 0.5 Cu, 0.62 Fe. Source: B. Hocheid *et al.*, Isothermal Transformation of Ti-6Al-6V-2Sn Alloy After Preheating in the α-β Range, *Titanium Science and Technology*, R.I. Jaffee and H.M. Burte, Ed., TMS-AIME, 1973, p 1609-1619

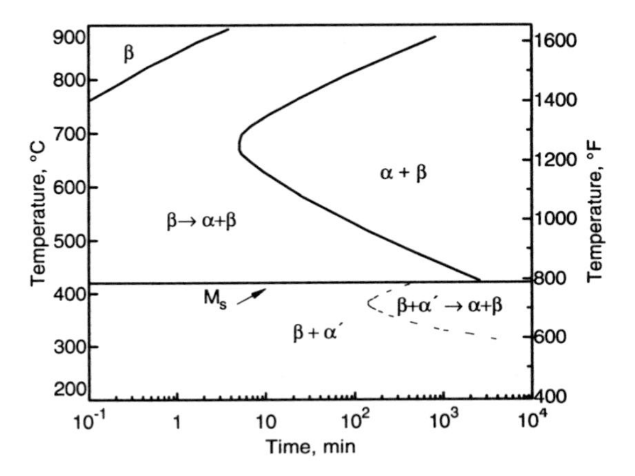

Ti-662: Isothermal transformation diagram. Quenched from β field to temperature indicated. Source: *Titanium Alloy Handbook*, MCIC-HB-02, Battelle Columbus Laboratories, 1972

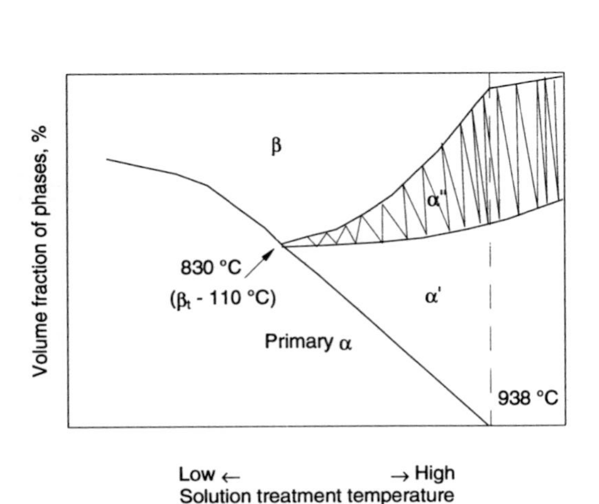

Ti-662: Phase transformation diagram. Source: Y. Murakami *et al.*, Phase Transformation and Heat Treatment in Ti Alloys, *Titanium Science and Technology*, G. Lütjering, U. Zwicker, and W. Bunk, Ed., Deutsche Gesellschaft für Metallkunde, Germany, 1985, p 1405

Low-Cycle Fatigue

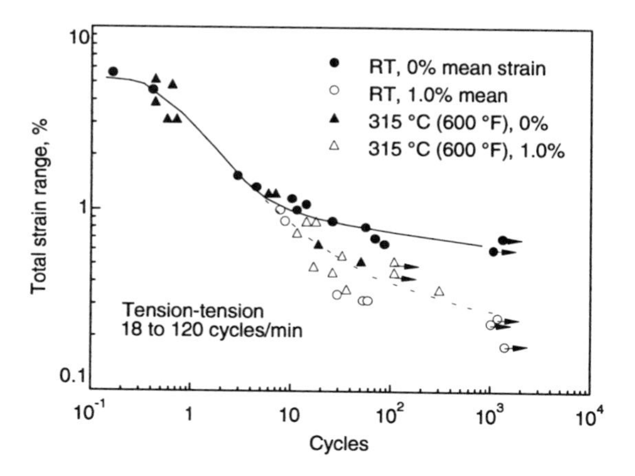

Ti-662: Strain cycling for annealed bar. Specimens were 25 mm (1 in.) diam bar vacuum annealed at 705 °C (1300 °F), 2 h, FC. Source: *Aerospace Structural Metals Handbook*, Vol 4, Code 3715, Battelle Columbus Laboratories, 1975

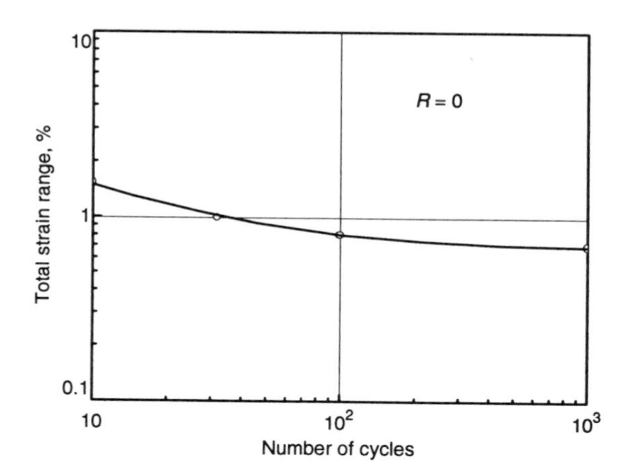

Ti-662: Low-cycle axial fatigue. Stock annealed 2 h at 700 °C (1300 °F) and furnace cooled. Source: *Metals Handbook*, Vol 3, 9th ed., American Society for Metals, 1978

High-Cycle Fatigue

Ti-662: Axial fatigue strength of notched specimens ($R = -1$)

Tensile strengths, MPa (ksi): UTS	TYS	Test conditions(a), Notch factor (K_t)	Fatigue strength, MPa (ksi), at: 10^5 cycles	10^6 cycles	10^7 cycles
RT test temperature					
1100 (160)	1055 (153)	3.4	186 (27)	145 (21)	138 (20)
		5.7	131 (19)	82 (12)	76 (11)
		10.0	138 (20)	69 (10)	...
315 °C (600 °F) test temperature					
875 (127)	717 (104)	3.4	144 (21)	138 (20)	131 (19)
		5.7	110 (16)	76 (11)	69 (10)
		10.0	...	69 (10)	62 (9)

(a) Specimen from 25 mm (1 in.) bar; specimen vacuum annealed at 700 °C (1300 °F) for 2 h, FC. 60° notch with a radius of 0.025, 0.13, and 0.38 mm (0.001, 0.005, and 0.015 in.). Source: *Aerospace Structural Metals Handbook*, Vol 4, Code 3715, Battelle Columbus Laboratories, Dec 1975

Ti-662: Axial fatigue strength of STA forging ($R = 0.1$)

Material condition	Test condition	Fatigue strength, MPa (ksi), at: 10^4 cycles	10^5 cycles	10^6 cycles	10^7 cycles
0.16% O_2, 0.66% Fe	Smooth, RT	1068 (155)	896 (130)	744 (108)	...
	Smooth, 285 °C (550 °F)	827 (120)	724 (105)	641 (93)	551 (80)
	$K_t = 3.9$, RT	537 (78)	289 (42)	207 (30)	172 (25)
0.10% O_2, 1.0% Fe	Smooth, RT	1068 (155)	896 (130)	724 (105)	620 (90)
	$K_t = 3.9$, RT	537 (78)	310 (45)	241 (35)	172 (25)

Note: Tests were conducted on 125 × 150 mm (5 × 6 in.) forged section with specimens heat treated as follows: 870 °C (1600 °F) for 1 h, WQ; plus 595 °C (1100 °F) for 4 h. Source: AFML-TR-65-206

Ti-662: Axial fatigue strength of extrusions ($R = 0.1$)

Material condition	Test condition	Fatigue strength, MPa (ksi), at: 10^4 cycles	10^5 cycles	10^6 cycles	10^7 cycles
Mill annealed, yield strength 945 to 993 MPa (137 to 144 ksi)	Smooth, RT	965 (140)	827 (120)	731 (106)	689 (100)
	$K_t = 4$, RT	482 (70)	289 (42)	220 (32)	220 (32)
T-extrusion, STA(a), L direction	Smooth, RT	...	724 (105)	655 (95)	620 (90)
	$K_t = 3.3$, RT	...	310 (45)	207 (30)	200 (29)

(a) 845 °C (1550 °F), WQ; plus 565 °C (1050 °F) for 4 h, AC. Source: *Aerospace Structural Metals Handbook*, Vol 4, Code 3715, Battelle Columbus Laboratories, Dec 1975

Ti-662: RT axial fatigue strength of annealed plate ($R = 0.1$)

Product form/condition	Test conditions	Fatigue strength, MPa (ksi), at: 10^4 cycles	10^5 cycles	10^6 cycles	10^7 cycles
25 mm (1 in.), mill annealed, 1055 MPa (153 ksi) yield strength	Smooth	1034 (150)(a)	848 (123)	689 (100)	606 (88)
0.18% O_2, 3.2 mm (1.25 in.), annealed(b)	Smooth, L and T direction	...	...	758 (110)	593 (86)
	$K_t = 3.5$, L and T direction	...	241 (35)	193 (28)	179 (26)
0.11% O_2, 25 mm (1 in.), annealed(b)	Smooth, L and T direction	...	...	793 (115)	620 (90)
	$K_t = 3.5$, L and T direction	...	275 (40)	310 (45)	207 (30)

(a) Extrapolated value. (b) 730 °C (1350 °F) annealed for 8 h, AC. Source: *Aerospace Structural Metals Handbook*, Vol 4, Code 3715, Battelle Columbus Laboratories, Dec 1975

Ti-662: Axial RT fatigue strength of STA plate

Product form/condition	Test conditions	Fatigue strength, MPa (ksi), at: 10^4 cycles	10^5 cycles	10^6 cycles	10^7 cycles
25 mm (1 in.) STA plate(a), 1180 MPa (171 ksi) yield strength	$R = 0.1$, Smooth	1068 (155) (b)	951 (138)	827 (120)	744 (108)
	$R = -1$, Smooth	882 (128)	620 (90)	551 (80)	386 (56)
0.18% O_2, 3.2 mm (1.25 in.), STA condition(c)	$R = 0.1$, Smooth, L direction	...	...	917 (133)	724 (105)
	Smooth, T direction	...	...	827 (120)	655 (95)
	$K_t = 3.5$, L direction	...	275 (40)	220 (32)	193 (28)
	$K_t = 3.5$, T direction	...	207 (30)	193 (28)	172 (25)
0.11% O_2, 25 mm (1 in.), STA condition(d)	Smooth, L and T direction	...	...	...	793 (115)
	$K_t = 3.5$, L and T direction	...	344 (50)	296 (43)	275 (40)
0.18% O_2, 50 mm (2 in.), STA condition(e)	Smooth, L and T direction	...	...	...	655 (95)
	$K_t = 3.5$, L and T direction	...	207 (30)	193 (28)	172 (25)

(a) 870 °C (1600 °F), WQ, plus 565 to 595 °C (1050 to 1100 °F) for 4 h, AC. (b) Extrapolated value. (c) 730 °C (1350 °F), 8 h, AC. (d) 845 °C (1550 °F), 1 h, WQ + 650 °C (1200 °F), 4 h. (e) 885 °C (1625 °F), WQ + 565 °C (1050 °F), 4 h. Source: *Aerospace Structural Metals Handbook*, Vol 4, Code 3715, Battelle Columbus Laboratories, Dec 1975

Ti-662: Axial fatigue strength of sheet

RT tensile strengths, MPa (ksi): UTS	TYS	Test conditions(a)	Fatigue strength, MPa (ksi), at: 10^4 cycles	10^5 cycles	10^6 cycles	10^7 cycles
Mill annealed sheet(b)						
...	1090 (158)	$R = 0.1$, smooth, RT	1070(c) (155)	860 (125)	758 (110)	710 (103)
		$R = 0.1$, $K_t = 4$, RT	358 (52)	220 (32)	193 (28)	193 (28)
		$R = 0.25$, smooth, RT	...	862 (125)	793 (115)	758 (110)
		$R = 0.1$, smooth, 450 °F	882 (128)	786 (114)	710 (103)	668 (97)
		$R = -0.1$, $K_t = 4$, 450 °F	372 (54)	193 (28)	165 (24)	165 (24)
2.5 mm (0.1 in.) sheet, **annealed**						
1130 (164)	1070 (155)	$R = 0.1$, smooth, RT	1103 (160)	862 (125)	758 (110)	703 (102)
		$R = 0.1$, smooth, 450 °F	895 (130)	772 (112)	689 (100)	655 (95)
		$R = 0.1$, $K_t = 4.2$, RT	344 (50)	207 (30)	193 (28)	193 (28)
		$R = 0.1$, $K_t = 4.2$, 450 °F	365 (53)	193 (28)	165 (24)	165 (24)
3.2 mm (0.125 in.) STA sheet						
1248 (181)	1220 (177)	$R = 0.1$, smooth, RT	...	827 (120)	786 (114)	786 (114)
		$R = 0.1$, $K_t = 4.2$, RT	310 (45)	213 (31)	207 (30)	207 (30)

(a) Conditions include load type/stress ratio/notch factor (K_t)/ test temperature/etc. (b) Sheet 0.5, 2.5, and 3.2 mm (0.02, 0.10, and 0.125 in.) thick, annealed at 700 to 760 °C (1300 to 1400 °F). (c) Extrapolated. Source: *Aerospace Structural Metals Handbook*, Vol 4, Code 3715, Battelle Columbus Laboratories, Dec 1975

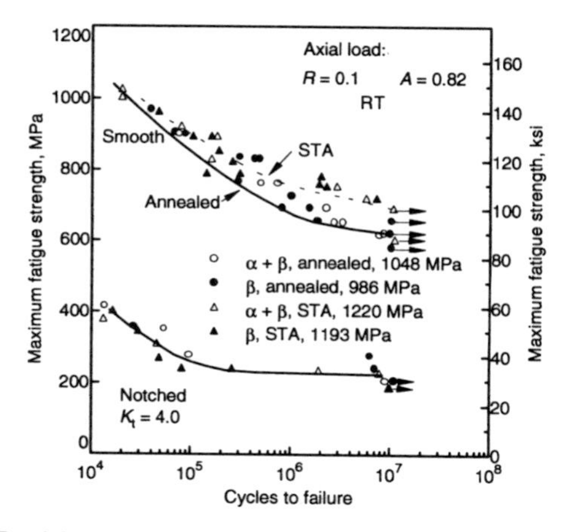

Ti-662: RT axial fatigue strength of forgings. Smooth and notched fatigue strength at room temperature for α + β and for β processed forging. Beta forging involved beta block forging followed by α- β finish forging. Heat treatments were as follows: annealed at 705 to 760 °C (1300 to 1400 °F), 2 h, AC; solution treated and aged at 855 °C (1575 °F), 1 h, WQ + 565 °C (1050 °F), 4 h, AC. Source: *Aerospace Structural Metals Handbook*, Vol 4, Code 3715, Battelle Columbus Laboratories, 1975

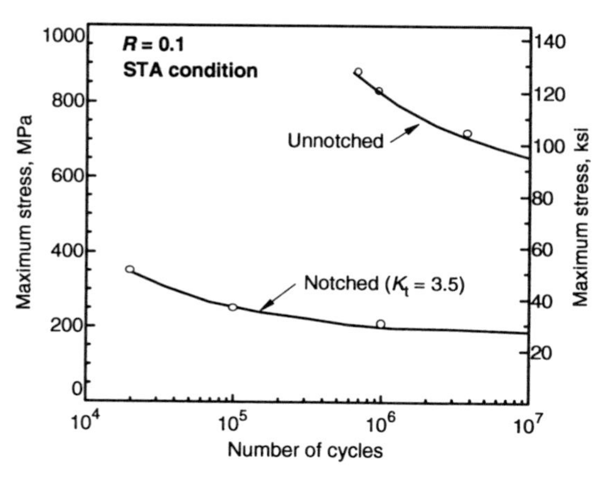

Ti-662: Typical axial fatigue strength. Source: *Metals Handbook*, Vol 3, 9th ed., American Society for Metals, 1978

Constant Lifetime Diagrams

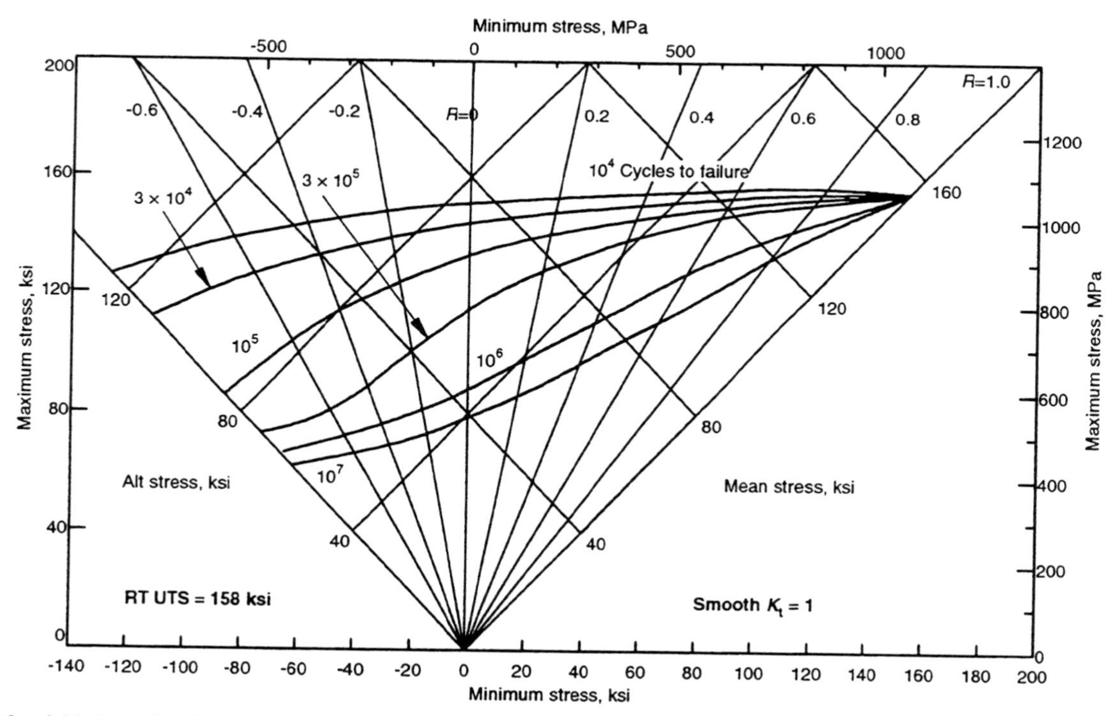

Ti-662: RT smooth axial fatigue of mill annealed plate. Source: *Aerospace Structural Metals Handbook*, Vol 4, Code 3715, Battelle Columbus Laboratories, 1975

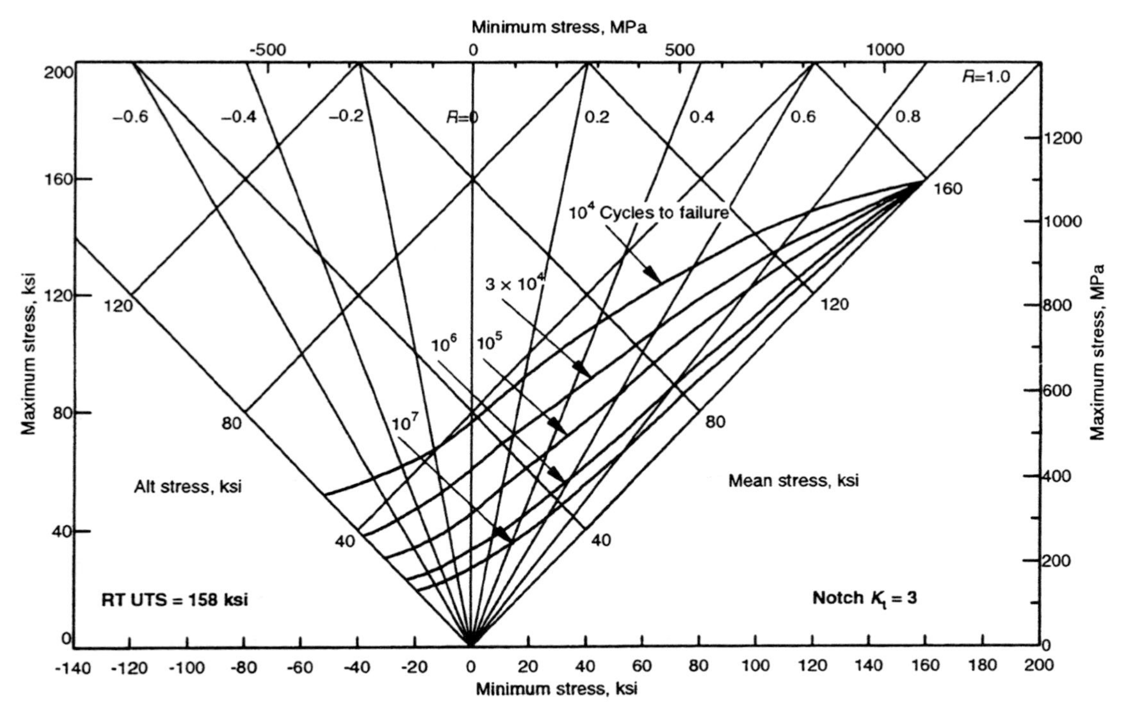

Ti-662: RT notched axial fatigue of mill annealed plate. Source: *Aerospace Structural Metals Handbook*, Vol 4, Code 3715, Battelle Columbus Laboratories, 1975

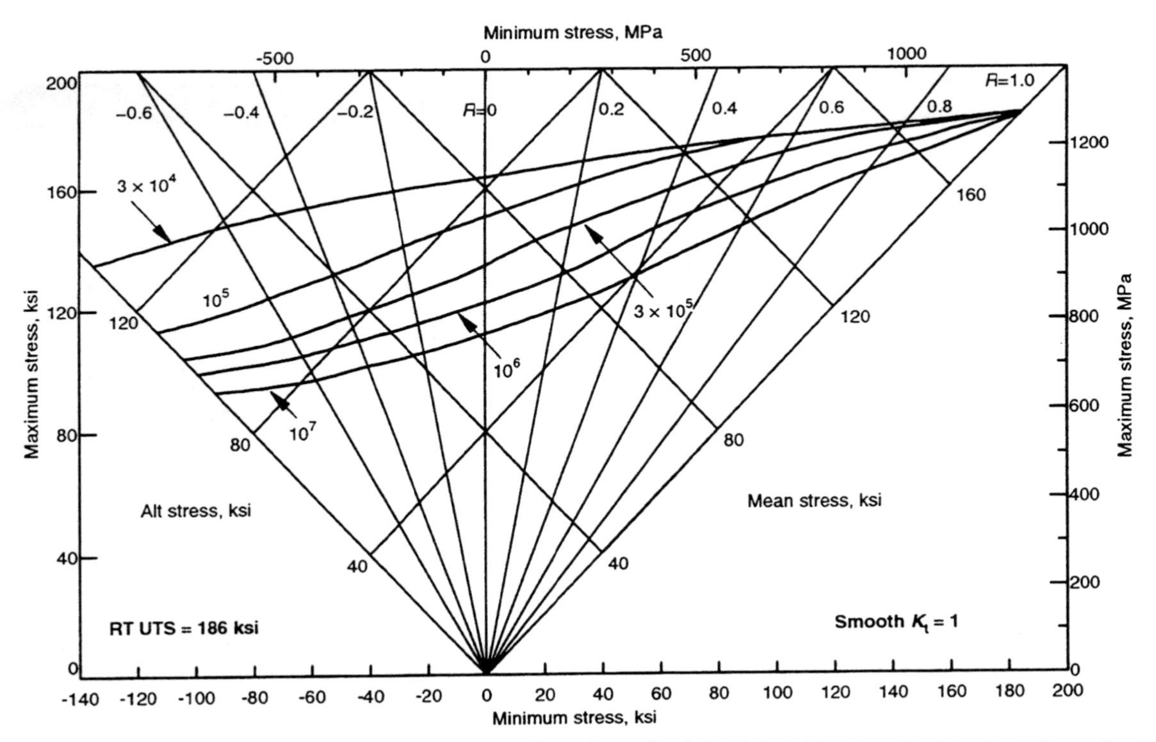

Ti-662: RT smooth axial fatigue of STA plate. Source: *Aerospace Structural Metals Handbook*, Vol 4, Code 3715, Battelle Columbus Laboratories, 1975

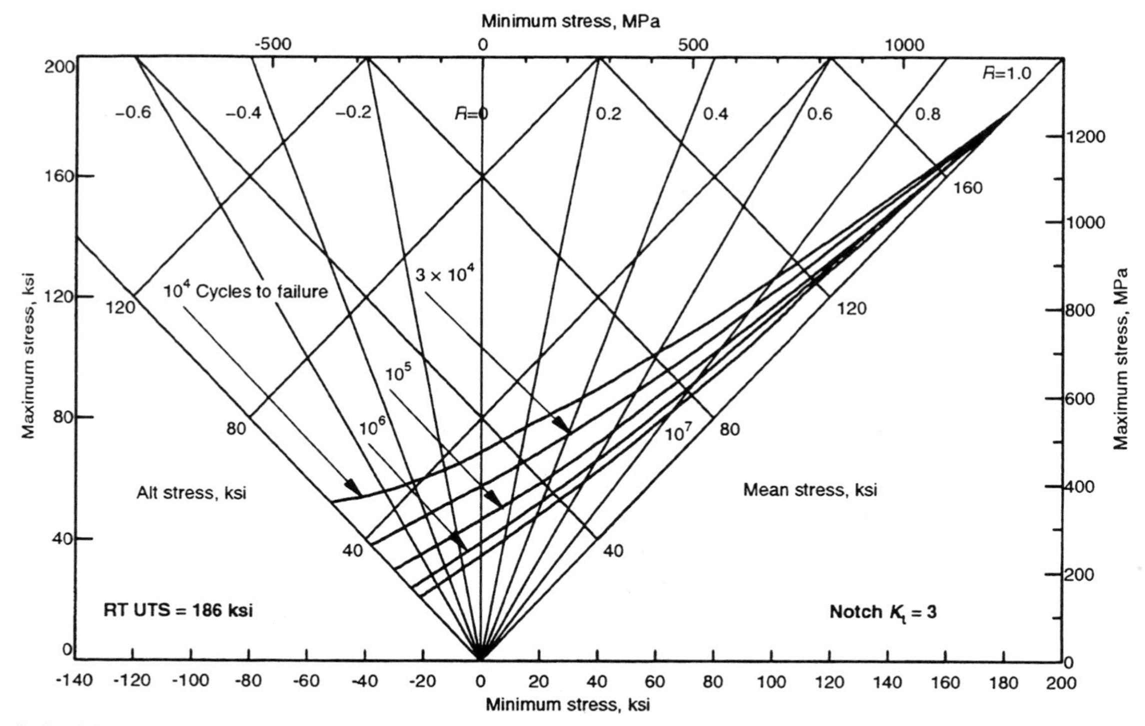

Ti-662: RT notched axial fatigue of STA plate. Source: *Aerospace Structural Metals Handbook*, Vol 4, Code 3715, Battelle Columbus Laboratories, 1975

Fatigue Crack Propagation

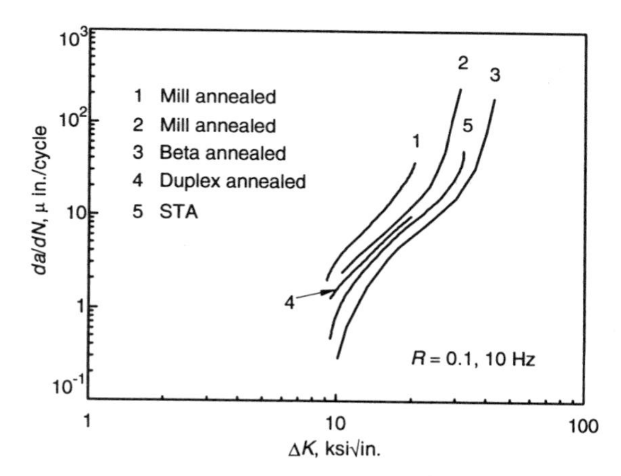

Ti-662: Crack growth rates for annealed plate. 13 mm (0.5 in.) mill annealed plate was tested at room temperature in air at 50 to 70% relative humidity. Source: *Aerospace Structural Metals Handbook*, Vol 4, Code 3715, Battelle Columbus Laboratories, 1975

Line No.	Treatment	Yield strength MPa	ksi
1	Mill anneal	1095	159
2	Mill anneal	1124	163
3	1010 °C (1850 °F) in vacuum	965	140
4	925 °C (1700 °F) + 760 °C (1400 °F)	1041	151
5	915 °C (1675 °F), WQ, 595 °C (1100 °F)	1193	173

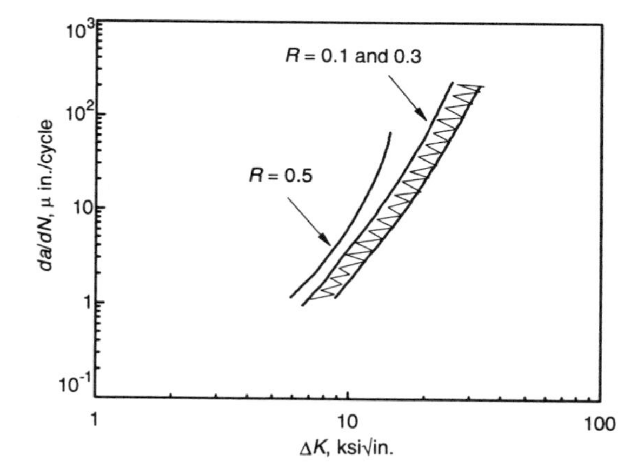

Ti-662: Average crack growth rates. Fatigue crack growth rates at room temperature, tested in laboratory air at 50 to 70% relative humidity. See table for treatments and yield strengths. Source: *Aerospace Structural Metals Handbook*, Vol 4, Code 3715, Battelle Columbus Laboratories, 1975

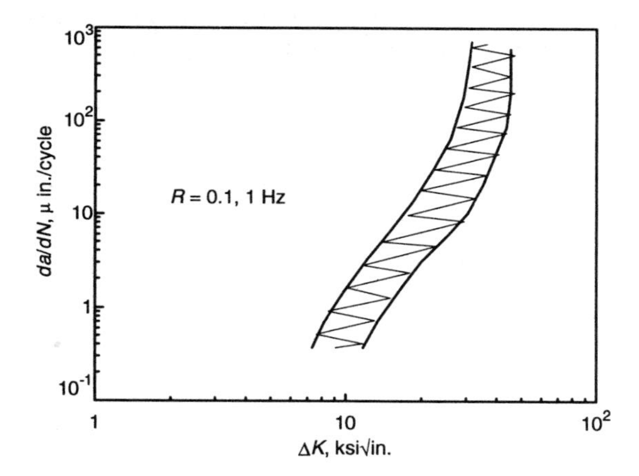

Ti-662: Crack growth of β annealed plate. 13 mm (0.5 in.) β annealed plate was tested at room temperature in air and 3.5% NaCl. Source: *Aerospace Structural Metals Handbook*, Vol 4, Code 3715, Battelle Columbus Laboratories, 1975

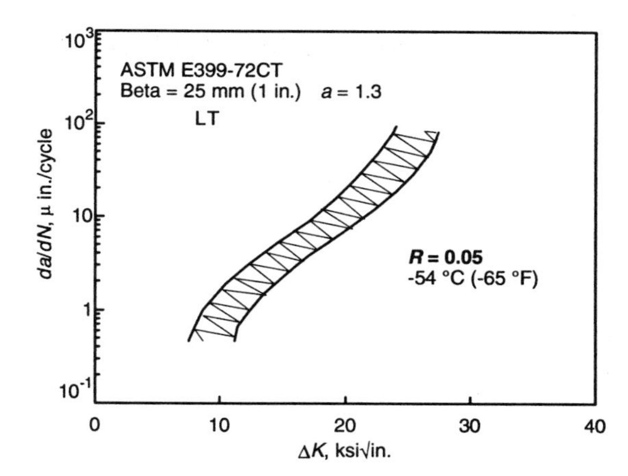

Ti-662: Crack growth rates at –54 °C for STA specimens. 96 mm (3.8 in.) square forged bar heat treated at 870 °C (1600 °F), 30 min, WQ + 540 °C (1000 °F), 6 h. Source: *Aerospace Structural Metals Handbook*, Vol 4, Code 3715, Battelle Columbus Laboratories, 1975

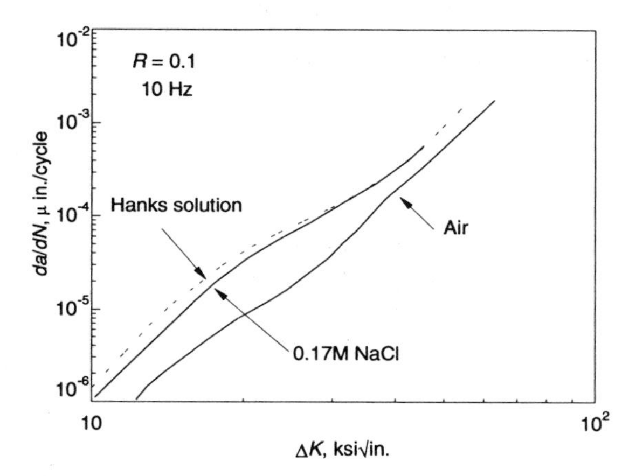

Ti-662: Crack growth in simulated body environments. Annealed sheet at room temperature. Tensile yield strength, 986 MPa (143 ksi). Source: *Aerospace Structural Metals Handbook*, Vol 4, Code 3715, Battelle Columbus Laboratories, 1975

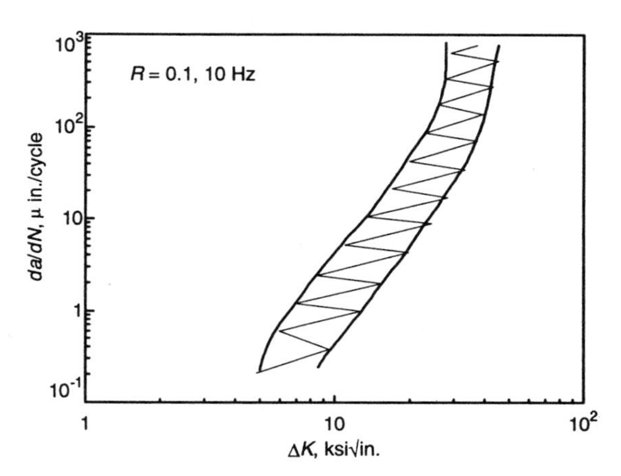

Ti-662: Crack growth range at several temperatures. 13 mm (0.5 in.) mill annealed plate tested at –62 to 82 °C (–80 to 180 °F). Source: *Aerospace Structural Metals Handbook*, Vol 4, Code 3715, Battelle Columbus Laboratories, 1975

Fracture Properties

Import Toughness

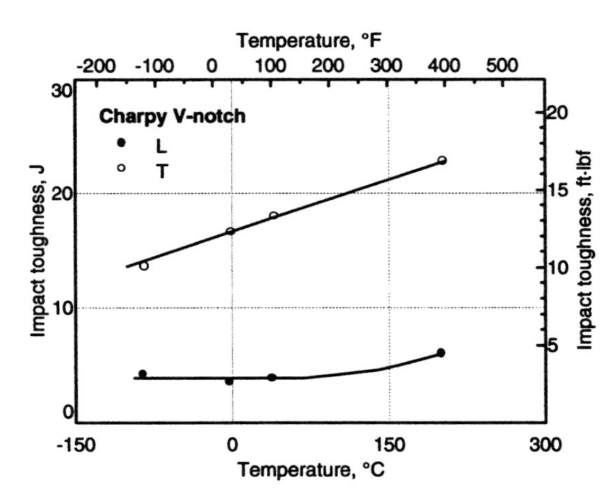

Ti-662: Impact toughness of annealed extrusions. Extrusions were annealed at 705 °C (1300 °F), 40 to 60 min, AC. RT yield strength ~930 MPa (135 ksi). Source: *Aerospace Structural Metals Handbook*, Vol 4, Code 3715, Battelle Columbus Laboratories, 1975

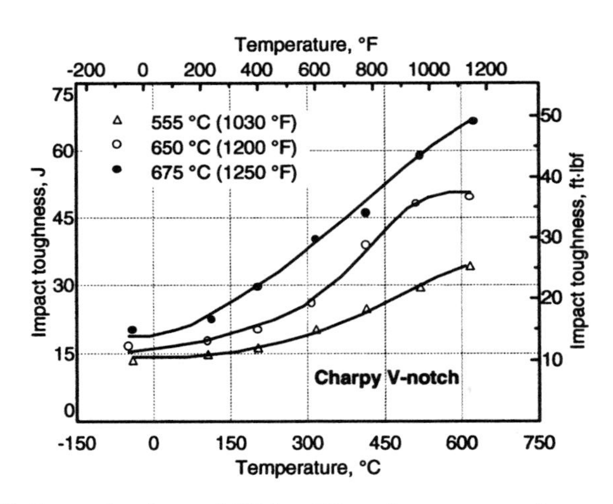

Ti-662: Impact toughness of STA bar. Effect of temperature on impact toughness of bar heat treated at 885 °C (1630 °F), 90 min, WQ and aged 4 h, at temperature indicated. Source: *Aerospace Structural Metals Handbook*, Vol 4, Code 3715, Battelle Columbus Laboratories, 1975

Fracture Toughness

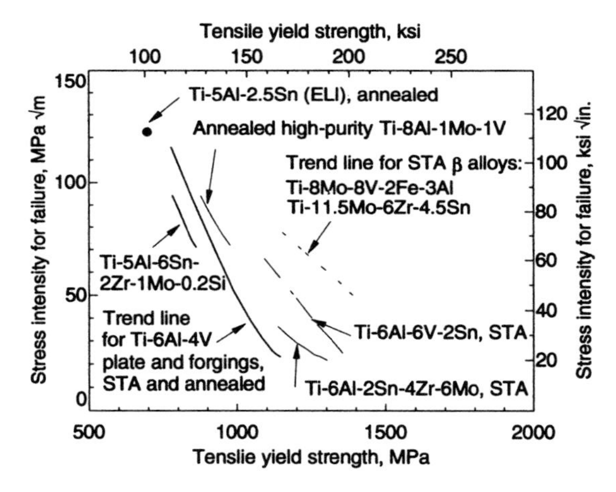

Ti-662: Fracture toughness/yield strength. Source: *Titanium and Titanium Alloys*, MIL-HDBK 697A, 1974

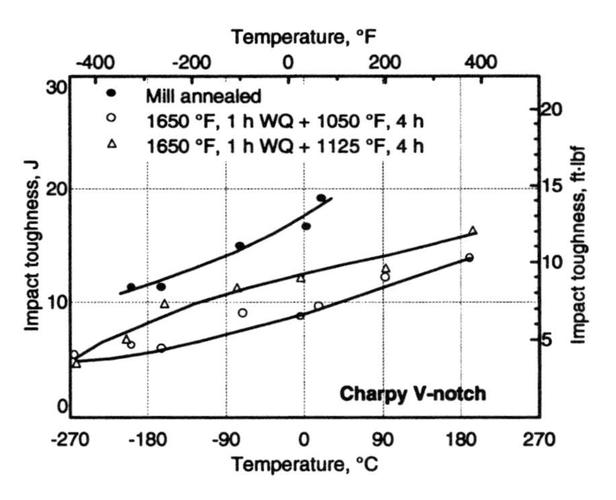

Ti-662: Impact toughness of plate. 25 mm (1 in.) plate composition: 0.081 O_2, 0.018 N_2, 0.006 H_2, 0.015 C, 0.59 Fe. Approximate RT yield strengths: A, 999 MPa (145 ksi); B, 1241 MPa (180 ksi); C, 1172 MPa (170 ksi).

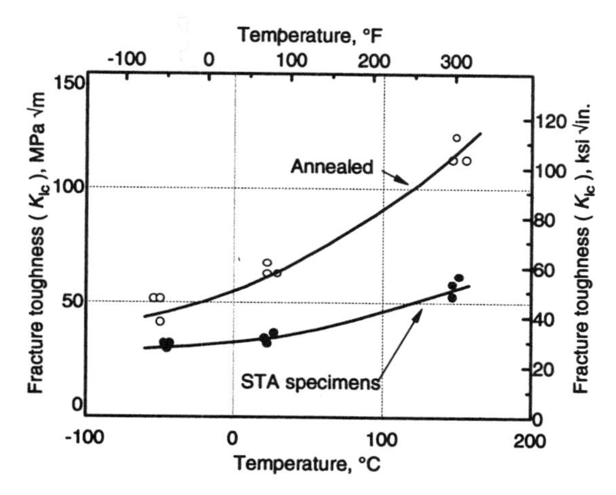

Ti-662: Fracture toughness vs temperature. 96 mm (3.8 in.) square forged bar. Plane-strain fracture toughness determined as per ASTM E399-72. Compact tension tests, LT direction with standard specimen B = 25 mm (1 in.), a = 33 mm (1.3 in). Solution treated and aged specimens were heat treated at 870 °C (1600 °F), 30 min, WQ + 540 °C (1000 °C), 6 h, AC. Source: *Aerospace Structural Metals Handbook*, Vol 4, Code 3715, Battelle Columbus Laboratories, 1975

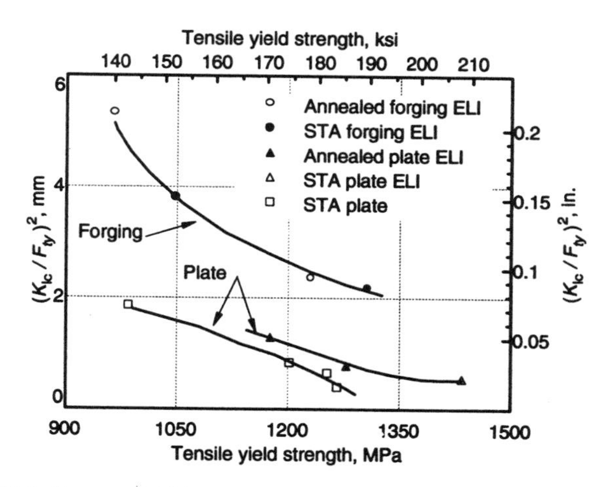

Ti-662: Influence of yield strength on fracture toughness. Source: *Aerospace Structural Metals Handbook*, Vol 4, Code 3715, Battelle Columbus Laboratories, 1975

Ti-662: RT fracture toughness of plate

Condition(a)	Direction	Fracture toughness (b) (K_{Ic}) MPa√m	ksi√in.
870 °C (1600 °F), 1 h, water quench, 565 °C (1050 °F), 4 h, air cool	L-T	32.7 ± 0.55	29.8 ± 0.5
900 °C (1650 °F), 1 h, water quench	L-T	37.3 ± 3.8	34.0 ± 3.5

(a) 25 mm (1 in.) plate. (b) Mean ± standard deviation. Source: J. Gallagher, *Damage Tolerant Design Handbook*, Vol 1, Battelle Columbus Laboratories, 1983

Ti-662: RT fracture toughness of plate, forging, and billet

Thickness mm	in.	Condition	Fracture toughness(b) (K_{Ic}) MPa√m	ksi√in.	Direction
Plate					
9.6	0.38	STOA at 925 °C (1700 °F), 1 h, water quench, 760 °C (1400 °F), 1 h, air cool	47.1 ± 1.3	42.9 ± 1.2	L-T
			50.6 ± 3.4	46.1 ± 3.1	T-L
13	0.50	β anneal at 985 °C (1810 °F), 1 h, argon cool	59.6 ± 2.2	54.3 ± 2.0	T-L
13	0.50	Duplex anneal	71.5 ± 2.2	65.1 ± 2.0	T-L
13	0.50	Mill anneal	38.4 ± 5.7	35.0 ± 5.2	T-L
15.7	0.62	β anneal + STOA at 980 °C (1800 °F), 30 min, air cool, 855 °C (1575 °F), 30 min, water quench, 565 °C (1050 °F), 8 h, air cool	55.0 ± 1.9	50.1 ± 1.8	L-T
32	1.25	STA at 915 °C (1675 °F), 15 min, water quench, 595 °C (1100 °F), 4 h	37.4 ± 4.1	34.1 ± 3.8	T-L
Forging					
96.5	3.80	...	64.4 ± 2.9	58.6 ± 2.7	L-T
96.5	3.80	STA at 870 °C (1600 °F), 30 min, water quench, 540 °C (1000 °F), 6 h, air cool	33.8 ± 0.7	30.8 ± 0.7	L-T
Billet					
55.8	2.20	...	57.4 ± 7.0	52.3 ± 6.4	L-T
55.8	2.20	Mill anneal at 540 °C (1000 °F), 2 h, air cool	62.7 ± 2.4	57.1 ± 2.2	L-T
305	12.00	...	69.0 ± 7.6	62.8 ± 6.9	L-T
		...	62.6 ± 4.0	57.0 ± 3.7	T-L

Source: J. Gallagher, *Damage Tolerant Design Handbook*, Vol 1, Battelle Columbus Laboratories, 1983

Ti-662: RT fracture toughness of plate and forgings

Product form and specimen	Condition	Yield stress MPa	Yield stress ksi	Apparent K_{Ic} MPa$\sqrt{m}$	Apparent K_{Ic} ksi$\sqrt{in.}$
25 mm (1 in.) plate; (three-point bend B = 1 in., a = 1 in., W = 2 in.)(a)	1550 °F, 30 min, WQ + 900 °F, 4 h, AC	1289	187	21	19
	1550 °F 30 min, WQ + 1000 °F, 4 h, AC	1261	183	27	25
	1550 °F, 30 min, WQ + 1100 °F, 4 h, AC	1193	173	34	31
	1550 °F, 30 min, WQ + 1300 °F, 4 h, AC	999	145	43	39
114 × 114 mm (4.5 × 4.5 in.) forging (double edge crack B = 0.5 in., 2a = 1 in., W = 3 in.)(b)	1650 °F, 1 h, WQ + 1050 °F, 4 h, AC	1027	149	66	60
25 mm (1 in.) ELI plate; (three-point bend B = 0.25 in., a = 0.2 in., W = 0.5 in.)(c)	1600 °F, 1 h, WQ + 1050 °F, 4 h, AC	1234	179	33	30
	1650 °F, 1 h, WQ + 1125 °F, 4 h, AC	1179	171	37	34
75 × 228 mm (3 × 9 in.) forging (center crack B = 1 in., 2a = 3 in., W = 9 in.)(d)	Annealed at 1300 °F, 2 h, AC	979	142	61	56
	1575 °F, 1 h, WQ + 1200 °F, 4 h, AC	1186(e)	172(e)	30(e)	28(e)
		1310(e)	190(e)	30(e)	28(e)

(a) J. Strawley, M. Jones, and W. Brown, Jr., "Determination of Plane Strain Fracture Toughness," *Mater. Res. Stand.*, Vol 7, 1967, p 262. (b) R. Bubsey, NASA Lewis Research Center. (c) T. DeSisto and C. Hickey, Jr., "Low Temperature Mechanical Properties and Fracture Toughness of Ti-6Al-6V-2Sn," *Proc. ASTM*, Vol 65, 1965, p 641. (d) J. Shannon, Jr., and W. Brown, Jr., "Thick Section Fracture Toughness," AFML-TDR-64-236, 1964, reported in "A Review of Factors Influencing the Crack Tolerance of Titanium Alloys," in *Applications Related Phenomena in Titanium Alloys*, ASTM STP 432, ASTM, 1968, p 33. (e) At –80 °C (–110 °F)

Ti-6-22-22S
Ti-6Al-2Sn-2Zr-2Mo-2Cr-0.25Si

UNS Number: Unassigned

Compiled by P. Russo, RMI Titanium Co., and R. Boyer, Boeing

Ti-6-22-22S was developed by RMI Titanium Co., with additional development funded through an Air Force Contract in the early 1970s. The alloy was conceived to provide high strength in heavy sections with good fracture toughness and to retain that strength up to moderate temperatures through the addition of silicon. The lack of any production applications precluded further development at that time. Interest in this alloy for fighter aircraft applications, because of its strength advantage over Ti-6Al-4V and good damage tolerance properties, has been revived. A strong effort is underway to develop thermomechanical processing procedures to optimize the strength, toughness, and crack growth rate properties of Ti-6-22-22S in sheet, plate, and forged forms.

Some data report a relatively high elastic modulus, which could be important for certain applications. Sheet can be formed at room temperature and has excellent superplastic forming characteristics.

Effects of Impurities and Alloying. Exceeding impurity limits may result in decreasing the ductility and fracture toughness below required minimums due to the associated increase in strength. As with other α-β titanium alloys, excessive aluminum, oxygen, and nitrogen can reduce ductility and fracture toughness. High amounts of the β stabilizers, chromium and molybdenum, may result in higher strength than desired.

Product Forms. Ti-6-22-22S has been produced in standard wrought product forms such as sheet, plate, bar, and forgings. Sheet exhibits excellent superplastic forming characteristics.

Product Condition. Ti-6-22-22S can be used in the annealed and heat treated conditions; solution treatment and aging can provide significant strengthening. The main emphasis at present, except for sheet, is on a triplex heat treatment involving a β solution treatment with a controlled cooling rate followed by an α-β solution treatment followed by aging to maximize damage-tolerant properties. Sheet should be used in the α-β processed condition.

Applications. There are no production applications for Ti-6-22-22S at this time, but it is bill-of-material for the aft fuselage of the F-22 ATF fighter. The primary interest in this alloy lies in its improved damage tolerance properties with respect to strength in relation to Ti-6Al-4V.

Specifications and Compositions. The only specifications for Ti-6-22-22S to date are those written by Lockheed/Boeing/General Dynamics for the ATF fighter. The composition limits are established as follows (except Si content may be reduced):

	Composition, wt%										
	Al	Sn	Zr	Mo	Cr	Si	Fe	O	C	N	H
Minimum	5.25	1.75	1.75	1.75	1.75	0.20	...	...	...	...	...
Maximum	6.25	2.25	2.25	2.25	2.25	0.27	0.15	0.13	0.04	0.03	125 ppm

Selected References

- H.R. Phelps and J.R. Wood, "Correlation of Mechanical Properties and Microstructures of Ti-6Al-2Sn-2Zr-2Mo-2Cr-0.25Si Titanium Alloy," Proc. 7th Int. Titanium Conf., San Diego, TMS/AIME, June 1992, to be published
- R.R. Boyer and A.E. Caddey, "The Properties of Ti-6Al-2Sn-2Zr-2Mo-2Cr Sheet," Proc. Int. Titanium Conf., San Diego, TMS/AIME, June 1992, to be published

- R.C. Bliss, "Evaluation of Ti-6Al-2Sn-2Zr-2Cr-2Mo-0.23 Si Sheet," Proc. 7th Int. Titanium Conf., San Diego, TMS/AIME, June 1992, to be published
- G.W. Kuhlman *et al.*, "Characterization of Ti-6-22-22S: A High-Strength Alpha-Beta Titanium Alloy for Fracture Critical Applications," Proc. 7th Int. Titanium Conf., San Diego, TMS/AIME, June 1992, to be published
- "Mechanical-Property Data: Ti-6Al-2Zr-2Sn-2Mo-2Cr Alloy Solution Treated and Aged Plate," F33615-72-C-1280, Battelle Columbus Laboratories, Apr 1973
- A.K. Chakrabarti *et al.*, "TMP Conditions-Microstructure-Mechanical Property Relationship in Ti-6-22-22S Alloy," Proc. 7th Int. Titanium Conf., San Diego, TMS/AIME, June 1992, to be published
- G.W. Kuhlman, Beta Processed Ti-6-22-22S Aging Studies, Alcoa Report, Mar 1992
- L.J. Bartlo, H.B. Bomberger, and S.R. Seagle, Deep Hardenable Titanium Alloy, AFML-TR-73-122, Battelle Columbus Laboratories, May 1973
- O.L. Deel, P.E. Ruff, and H. Mindlin, "Engineering Data on New Aerospace Structural Materials," AFML-TR-75-97, Battelle Columbus Laboratories, June 1975

Physical Properties

Phases and Structures

Triplex heat treatment results in a coarse lamellar α structure in a transformed β matrix. Cooling rates from the solution treatments must be controlled within a given window to provide desired strengths. The retained β contains a fine acicular α precipitate due to aging. Very fine, submicron-size silicides have been observed in this alloy.

Sheet can be used in the annealed or solution treated and aged condition. An air cool from the solution treatment temperature provides adequate heat treatment response. The annealed condition consists basically of equiaxed α with intergranular β (see figure). Material in the solution treated and aged condition is very similar to that of Ti-6Al-4V, with equiaxed α in a β matrix. With solution treatments below about 850 °C (1560 °F), β phase at temperature will be retained upon cooling (for sheet gages with an air cool), which provides a strength minimum in the solution treated condition. Solution treating at temperatures above this results in increased amounts of martensite formation as the temperature is increased and higher strengths.

Ti-6-22-22S: Summary of typical physical properties

Beta transus	960 ± 15 °C (1760 ± 25 °F)
Melting (liquidus) point	Not available
Density(a)	4.65 g/cm^3 (0.164 lb/in.^{3})
Electrical resistivity	Not available
Magnetic permeability(a)	Nonmagnetic
Specific heat capacity	Not available
Thermal conductivity	Not available
Thermal coefficient of linear expansion(b)	9.2×10^{-6}/°C (5.1×10^{-6}/°F)

(a) Typical values at room temperature of about 20 to 25 °C (68 to 78 °F). (b) Mean coefficient from room temperature to 425 °C (800 °F)

Elastic Properties

Young's Modulus. The high modulus reported by Battelle (see table) has never been explained and has not been duplicated. The F-22 program is now working with a modulus of 113.8 GPa (16.5 × 10^6 psi).

Poisson's ratio. 0.33

Ti-6-22-22S: Elastic properties of forgings

Young's modulus	
Conventionally processed	122×10^3 MPa (17.7×10^3 ksi)
Beta processed (a)	115×10^3 MPa (16.8×10^3 ksi)
Bulk modulus	117×10^3 MPa (17.0×10^3 ksi)
Shear modulus	46×10^3 MPa (6.7×10^3 ksi)
Poisson's ratio	0.33

(a) Conventional subtransus forging with a triplex heat treatment consisting of β treat/fan cool/ α + β treat/fan or air cool/and aging

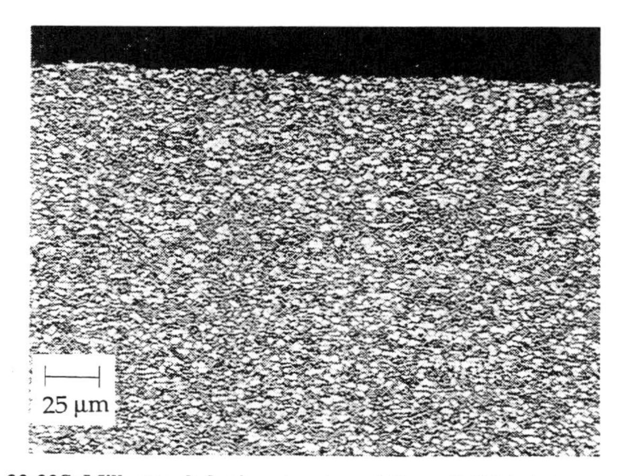

Ti-6-22-22S: Mill annealed microstructure. 1.2 mm (0.050 in.) sheet annealed at 730 °C (1350 °F), 15 min, AC

Ti-6-22-22S: Variation in Young's modulus

Test temperature		Longitudinal		Transverse	
°C	°F	GPa	10^6 psi	GPa	10^6 psi
Room temperature(a)		110	15.9	112	16.2
Room temperature(b)		123	17.9	122	17.8
200	400(b)	110	15.9	112	16.2
315	600(b)	107	15.6	110	16.0
425	800(b)	99	14.4	100	14.6

(a) Beta processed. From G.W. Kuhlman *et al.*, "Characterization of Ti-6-22-22S: A High-Strength Alpha-Beta Titanium Alloy for Fracture Critical Applications," Proc. 7th Int. Titanium Conf., San Diego, TMS/AIME, June 1992, to be published. (b) From "Mechanical-Property Data: Ti-6Al-2Zr-2Sn-2Mo-2Cr Alloy," AFML, Battelle Columbus Laboratories, Apr 1973

Corrosion

Stress-Corrosion Cracking. Boeing has reported the stress-corrosion threshold for this alloy to be about 55 MPa$\sqrt{m}$ (50 ksi$\sqrt{in.}$) in an aqueous 3.5% NaCl solution. Previous work (Battelle, Apr 1973) reported a value of 80% of the tensile yield strength.

Tensile Properties

Because Ti-6-22-22S is an age-hardenable alloy, a range of tensile properties is attainable. The alloy can be heat treated in sections up to 75 to 100 mm (3 to 4 in.) thick. Tensile properties will depend on processing history and on the solution treatment and aging temperature (although strength is not very sensitive to aging temperature over a fairly wide temperature range).

Ti-6-22-22S: Typical mechanical properties for α–β processed STA products

Product	Ultimate tensile strength		Tensile yield strength		Elongation, %	Reduction of area, %
	MPa	ksi	MPa	ksi		
Sheet(a)	1331	193	1193	173	7.5	...
Plate(b)	1204	175	1131	164	12.0	35
Billet(b)	1200	174	1089	158	15.0	41

Source: G.A. Bella, RMI Titanium Co., 8 May 1991. (a) Subtransus ($\beta_T - 50$ °F) solution treatment with 540 °C (1000 °F) age, 8 h. (b) Supratransus and subtransus treat plus aging

Ti-6-22-22S: Typical mechanical properties for β-processed STA products

Product	Ultimate tensile strength		Tensile yield strength		Elongation, %	Reduction of area, %
	MPa	ksi	MPa	ksi		
50 mm (2 in.) plate	1138	165	1020	148	10	17
100 mm (4 in.) plate	1103	160	979	142	10	15
150 mm (6 in.) plate	1076	156	958	139	10	15

Source: J.R. Wood, RMI Titanium Co., 15 Aug 1991. $\beta_T - 28$ °C (50 °F) for 1 h, AC/540 °C (1000 °F) age, 8 h, AC

Ti-6-22-22S: Typical mechanical properties for α–β processed mill annealed products

Product	Ultimate tensile strength		Tensile yield strength		Elongation, %	Reduction of area, %
	MPa	ksi	MPa	ksi		
Sheet	1103	160	1034	150	10	...
Plate	1076	156	1014	147	13	28

Source: G.A. Bella, RMI Titanium Co., 8 May 1991. Mill anneal 730 °C (1350 °F), 2 h, AC

Plate and Forgings

The F-22 program has established a minimum tensile strength of 1035 MPa (150 ksi), and it is felt that the strength should be controlled within the range of 1070-1137 MPa (155 to 165 ksi) to meet the minimum fracture toughness requirement. This has resulted in a cooling rate window (see figure). The effect of slower cooling rates is a coarser lamellar α and lower strength. Oxygen content has the expected influence on strength and toughness.

Sheet

The solution treat temperature has a very strong effect on the mechanical properties of Ti-6-22-22S (see figures). The yield strength minimum is associated with the retention of the maximum amount of β. The effect of superplastic forming temperature is similar to that of solution treat temperature, and superplastic formed parts can be aged to higher strengths (see section on superplastic forming).

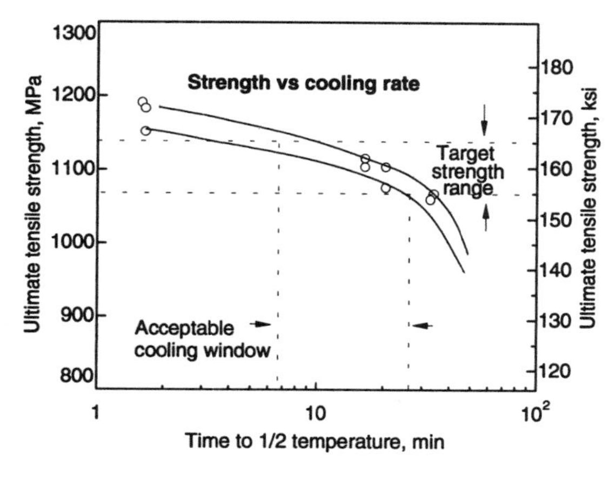

Ti-6-22-22S: Determination of acceptable processing window. Beta processing improves fracture and crack growth resistance, and this study identified through-transus beta forging followed by α-β STA as the optimum TMP route. Beta-processed plate α-β solution treated and aged at 540 °C (1000 °F), 8 h. Time to 1/2 temperature refers to time from ST to 1/2 ST temperature. Source: H.R. Phelps and J.R. Wood, "Correlation of Mechanical Properties and Microstructures of Ti-6Al-2Sn-2Zr-2Mo-2Cr-0.25Si Titanium Alloy," Proc. 7th Int. Titanium Conf., San Diego, TMS/AIME, June 1992, to be published

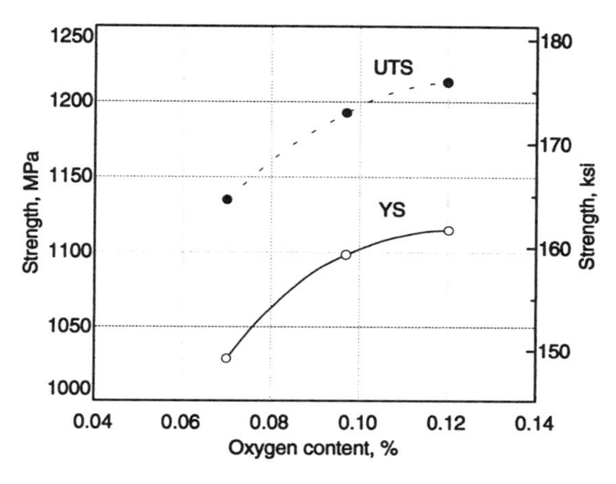

Ti-6-22-22S: Effect of oxygen content on tensile strengths. Forged pancakes α-β forged + α-β solution annealed + 540 °C (1000 °F), 8 h, aged, AC. Source: A.K. Chakrabarti *et al.*, "TMP Conditions-Microstructure-Mechanical Property Relationship in Ti-6-22-22S Alloy," Proc. 7th Int. Titanium Conf., San Diego, TMS/AIME, June 1992, to be published

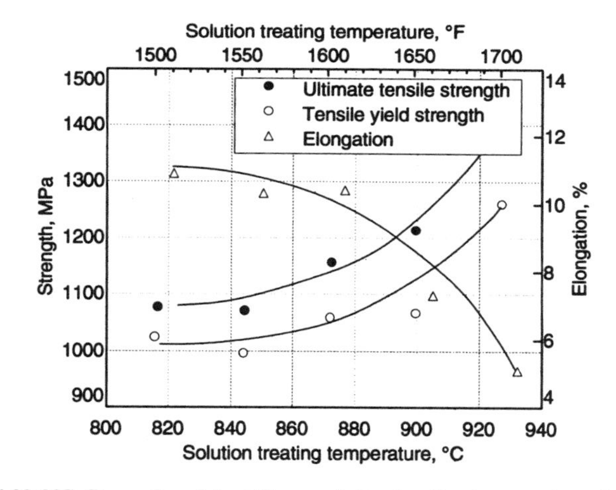

Ti-6-22-22S: Strength and ductility vs solution treating temperature. 1.2 mm (0.050 in.) sheet solution treated 30 min, AC, no aging. Source: R.R. Boyer and A.E. Caddey, "The Properties of Ti-6Al-2Sn-2Zr-2Mo-2Cr Sheet," Proc. 7th Int. Titanium Conf., San Diego, TMS/AIME, June 1992, to be published

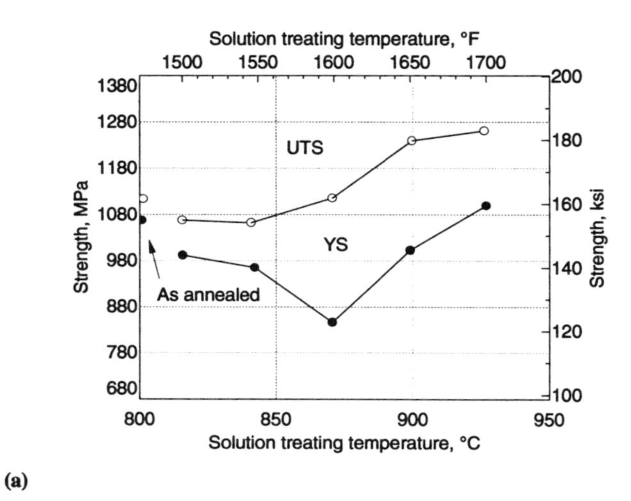

(a)

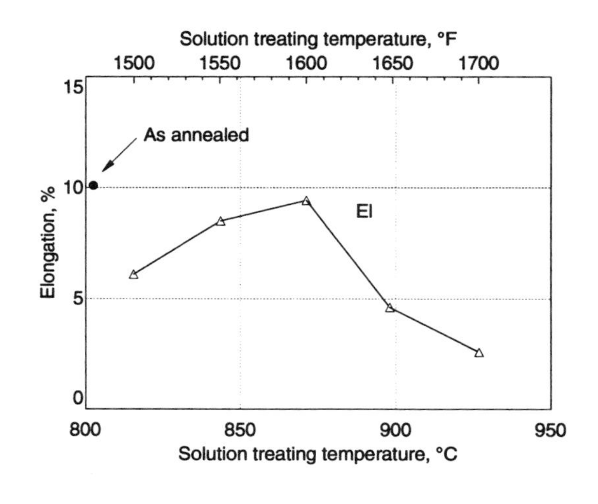

(b)

Ti-6-22-22S: Effect of solution temperature on tensile properties. 1.6 mm (0.063 in.) thick sheet heat treated as indicated. Source: R.C. Bliss, "Evaluation of Ti-6Al-2Sn-2Zr-2Cr-2Mo-0.23Si Sheet," Proc. 7th Int. Titanium Conf., San Diego, TMS/AIME, June 1992, to be published

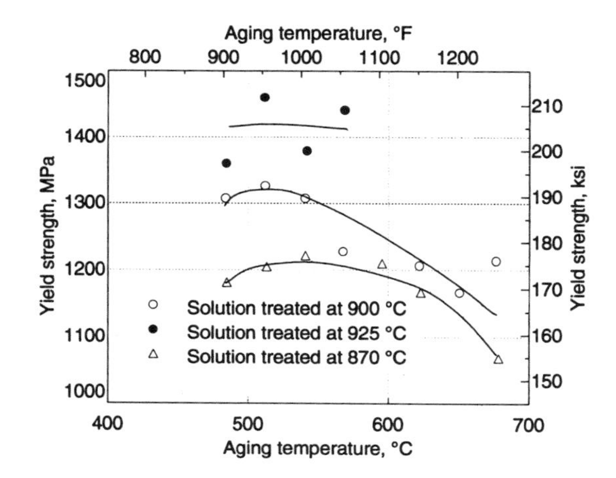

Ti-6-22-22S: Yield strength vs aging temperature. 1.2 mm (0.050 in.) sheet heat treated as indicated. Source: R.R. Boyer and A.E. Caddey, "The Properties of Ti-6Al-2Sn-2Zr-2Mo-2Cr Sheet," Proc. 7th Int. Titanium Conf., San Diego, TMS/AIME, June 1992, to be published

High-Temperature Strength

Ti-6-22-22S: Effect of temperature on tensile, compressive, and shear properties

Room-temperature ratio of tensile strength and compressive and shear strength should be similar for different heat treatments and product forms.

Properties	RT	Temperature 205 °C (400 °F)	315 °C (600 °F)	425 °C (800 °F)
Tension				
Ultimate tensile strength				
Longitudinal, MPa (ksi)	1160 (168.3)	1002 (145.3)	958 (139.0)	910 (132.0)
Transverse, MPa (ksi)	1163 (168.7)	1006 (146.0)	963 (139.7)	910 (132.0)
Tensile yield strength				
Longitudinal, MPa (ksi)	1073 (155.6)	799 (116.0)	737 (107.0)	697 (101.2)
Transverse, MPa (ksi)	1079 (156.6)	825 (119.7)	749 (108.7)	717 (104.0)
Elongation in 25 mm (1 in.)				
Longitudinal, %	18.0	19.5	18.5	21.3
Transverse, %	17.7	19.7	18.2	21.0

(continued)

Ti-6-22-22S: Effect of temperature on tensile, compressive, and shear properties (continued)
Room-temperature ratio of tensile strength and compressive and shear strength should be similar for different heat treatments and product forms.

Properties	RT	Temperature 205 °C (400 °F)	315 °C (600 °F)	425 °C (800 °F)
Tension (continued)				
Reduction of area				
Longitudinal, %	24.8	33.2	34.9	42.1
Transverse, %	26.2	33.7	33.3	41.4
Young's modulus				
Longitudinal, GPa (10^6 psi)	123 (17.9)	109 (15.9)	107 (15.6)	99 (14.4)
Transverse, GPa (10^6 psi)	122 (17.8)	111 (16.2)	110 (16.0)	100 (14.6)
Compression				
Compressive yield strength				
Longitudinal, MPa (ksi)	1170 (169.7)	884 (128.3)	772 (112.0)	728 (105.7)
Transverse, MPa (ksi)	1195 (173.3)	891 (129.3)	793 (115.0)	733 (106.3)
Compressive modulus				
Longitudinal, GPa (10^6 psi)	125 (18.1)	115 (16.7)	109 (15.8)	100 (14.6)
Transverse, GPa (10^6 psi)	127 (18.5)	112 (16.3)	109 (15.8)	100 (14.6)
Ultimate shear strength				
Longitudinal, MPa (ksi)	746 (108.3)	...	...	...
Transverse, MPa (ksi)	744 (108.0)	...	...	...

Note: Specimens were 38 mm (1.5 in.) thick plate in the STA condition heat treated at 950 °C (1740 °F), 1 h, AC + 540 °C (1000 °F), 8 h. Source: "Mechanical-Property Data: Ti-6Al-2Zr-2Sn-2Mo-2Cr Alloy Solution Treated and Aged Plate," F33615-72-C-1280, Battelle Columbus Laboratories, Apr 1973

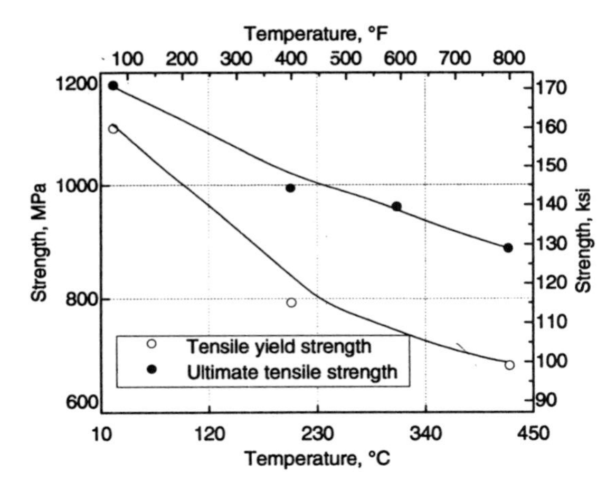

Ti-6-22-22S: High-temperature tensile strength of STA billet. Source: G.A. Bella, RMI Titanium Co., 8 May 1991

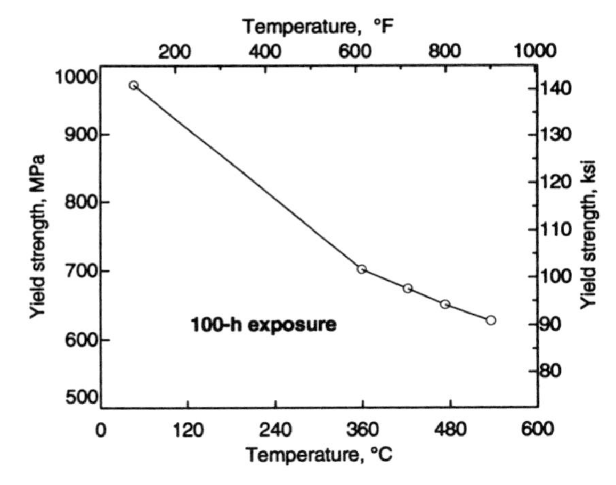

Ti-6-22-22S: High-temperature tensile strength. Forgings heat treated at β_t + 30 °C (50 °F), 30 min, FC + β_t − 40 °C (70 °F), 1 h, AC + 540 °C (1000 °F), 8 h. Source: G.W. Kuhlman *et al.*, "Characterization of Ti-6-22-22S: A High-Strength Alpha-Beta Titanium Alloy for Fracture Critical Applications," Proc. 7th Int. Titanium Conf., San Diego, TMS/AIME, June 1992, to be published

High-Temperature Strength

Creep Strength/Creep Rupture

The limited creep data available on Ti-6-22-22S are presented below. In general, the creep properties are similar to those of Ti-6Al-2Sn-4Zr-6Mo and are superior to Ti-6Al-4V.

Ti-6-22-22S: Stress-rupture and creep properties for STA billet

Property	Temperature 205 °C (400 °F)	315 °C (600 °F)	425 °C (800 °F)
Stress to rupture			
Stress, MPa (ksi)	979 (142)	910 (132)	841 (122)
Time, h	100	100	100
Creep			
Stress, MPa (ksi)	841 (122)	827 (120)	572 (83)
Time, h	100	100	100
Creep, %	0.2	0.2	0.2

Source: O.L. Deel, P.E. Ruff, and H. Mindlin, "Engineering Data on New Aerospace Structural Materials," AFML-TR-75-97, Battelle Columbus Laboratories, June 1975

Ti-6-22-22S: Creep properties of β solution treated and aged forgings

Temperature		Stress		Time, h		
°C	°F	MPa	ksi	0.1%	0.2%	Rupture
370	700	415	60	225	687	687
425	800	345	50	190	606	606
480	900	240	35	69	251	251

Note: Forgings were processed at 30 °C (50 °F) above the β transus temperature, 30 min, fan cooled + 40 °C (70 °F) below the β transus, 1 h, AC + 540 °C (1000 °F), 8 h. Source: G.W. Kuhlman *et al.*, "Characterization of Ti-6-22-22S: High-Strength Alpha-Beta Titanium Alloy for Fracture Critical Applications," Proc. 7th Int. Titanium Conf., San Diego, TMS/AIME, June 1992, to be published

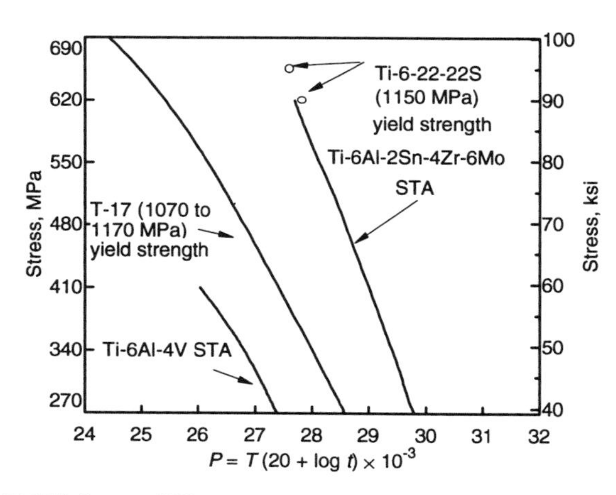

Ti-6-22-22S: Larson-Miller creep curves. Larson-Miller creep curves at 0.2% deformation for specimens heat treated at 950 °C (1740 °F, 1 h, AC + 540 °C (1000 °F), 8 h. Source: L.J. Bartlo, H.B. Bomberger, and S.R. Seagle, "Deep Hardenable Titanium Alloy," AFML-TR-73-122, Battelle Columbus Laboratories, May 1973

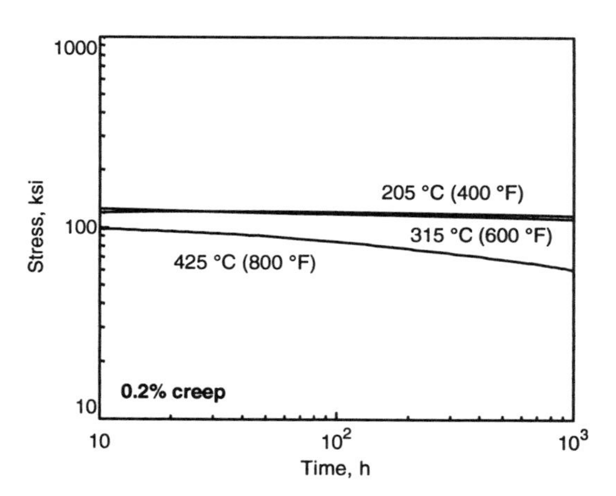

Ti-6-22-22S: Creep of STA plate. 38 mm (1.5 in.)(thick plate; 950 °C (1740 °C), 1 h, AC + 540 °C (1000 °F), 8 h, AC. Source: "Mechanical Property Data: Ti-6Al-2Zr-2Sn-2Mo-2Cr Alloy Solution Treated and Aged Plate," F33615-72-C-1280, Battelle Columbus Laboratories, Apr 1973

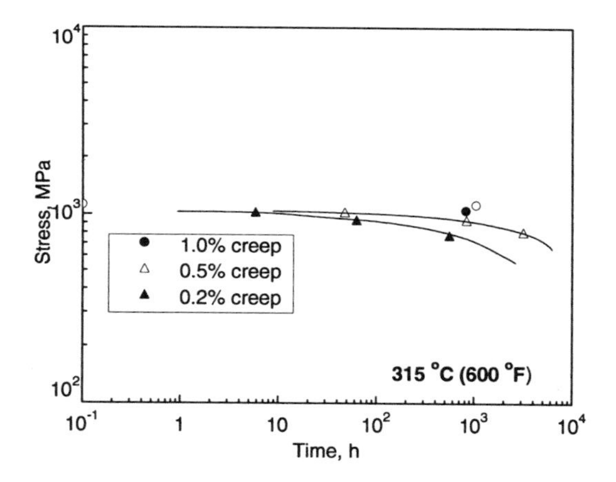

(a)

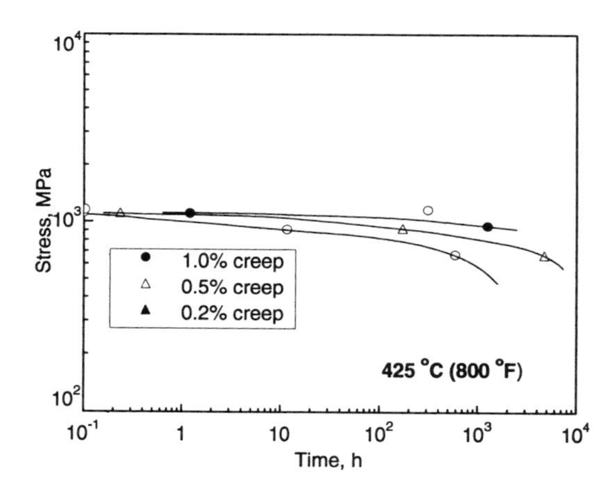

(b)

Ti-6-22-22S: Creep and stress rupture of forged billet. Duplex annealed; test direction, long transverse. Source: O.L. Deel, P.E. Ruff, and H. Mindlin, "Engineering Data on New Aerospace Structural Materials," AFML-TR-75-97, Battelle Columbus Laboratories, June 1975

High- and Low-Cycle Fatigue

The axial fatigue data on duplex annealed (DA) forged billet given below represent data generated in the early 1970s during initial alloy development/characterization and should be representative of α-β processed material. There is some doubt about the 38 mm (1.5 in.) plate data. The last two curves are representative of the β-processed material being studied today.

Ti-6-22-22S: Transverse axial fatigue of STA plate

Fatigue	Temperature RT	205 °C (400 °F)	315 °C (600 °F)
Unnotched, R = 0.1			
10^3 cycles, MPa (ksi)	1158 (168)	1034 (150)	924 (134)
10^5 cycles, MPa (ksi)	930 (135)	848 (123)	799 (116)
10^7 cycles, MPa (ksi)	517 (75)	517 (75)	517 (75)
Notched, K_t = 3.0, R = 0.1			
10^3 cycles, MPa (ksi)	868 (126)	703 (102)	620 (90)
10^5 cycles, MPa (ksi)	413 (60)	379 (55)	344 (50)
10^7 cycles, MPa (ksi)	289 (42)	255 (37)	255 (37)

Note: 38 mm (1.5 in.) plate, 950 °C (1740 °F), 1 h, AC + 540 °C (1000 °F) for 8 h. Source: "Mechanical-Property Data of Ti-6Al-2Sn-2Zr-2Mo-2Cr," AFML, Battelle Columbus Laboratories, Apr 1973

DA Forged Billet

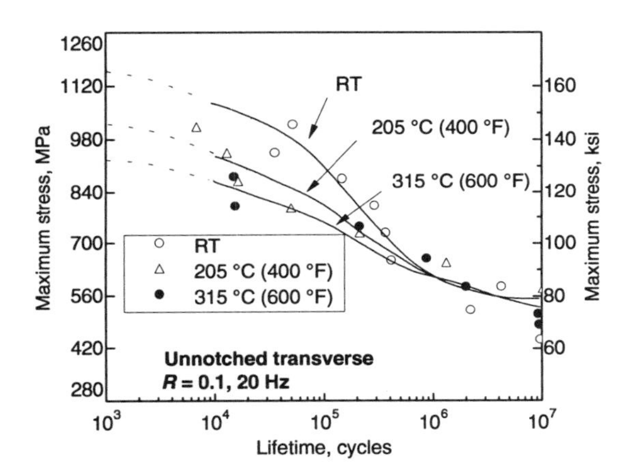

Ti-6-22-22S: Unnotched axial fatigue of DA forged billet. See also accompanying tables on next page.

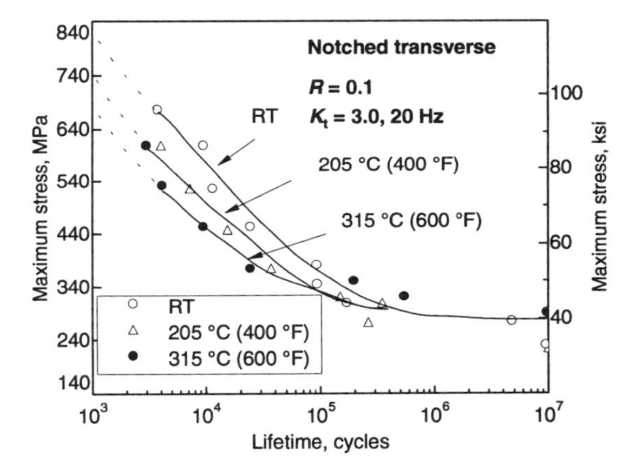

Ti-6-22-22S: Notched axial fatigue of DA forged billet.

Ti-6-22-22S: Unnotched axial fatigue of DA forged billet (R = 0.1)

RT ksi	RT cycles	400 °F ksi	400 °F cycles	600 °F ksi	600 °F cycles
145	52,730	145	6,400	135	(b)
135	37,730	135	12,900	125	15,400
125	159,300	125	15,800	115	14,700
115	303,270	115	47,900	105	218,300
105	392,790	105	212,400	95	836,600
95	429,580	95	1,277,700	95	1,912,100
85	4,527,700	85	10,130,900(a)	75	9,789,300
75	2,268,600			70	13,808,600(a)
65	10,003,500(a)				

(a) Did not fail. (b) Failed on loading. Source: O.L. Deel, P.E. Ruff, and H. Mindlin, "Engineering Data on New Aerospace Structural Materials," AFML-TR-75-97, Battelle Columbus Laboratories, June 1975

Ti-6-22-22S: Notched axial fatigue of DA forged billet ($R = 0.1$, $K_t = 3.0$)

RT		400 °F		600 °F	
ksi	cycles	ksi	cycles	ksi	cycles
95	3,600	85	3,700	85	2,900
85	8,600	75	6,850	75	4,000
75	11,400	65	14,700	65	8,600
65	23,400	55	33,300	55	22,500
55	89,100	47.5	141,200	50	194,600
50	89,900	45	417,400	47.5	527,800
45	153,200	40	237,000	45	10,084,900(a)
40	5,069,900	35	17,270,800(a)		
35	11,645,200(a)				

(a) Did not fail. Source: O.L. Deel, P.E. Ruff, and H. Mindlin, "Engineering Data on New Aerospace Structural Materials," AFML-TR-75-97, Battelle Columbus Laboratories, June 1975

STA Plate

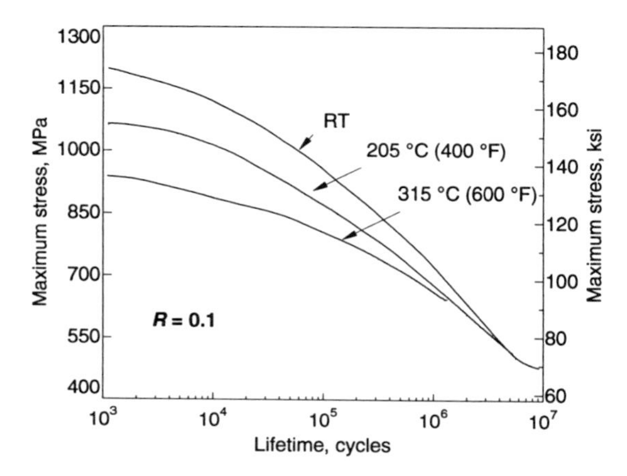

Ti-6-22-22S: Fatigue behavior of unnotched STA plate. 38 mm (1.5 in.) thick plate heat treated at 950 °C (1740 °F), 1 h, AC + 540 °C (1000 °F), 8 h, AC; test direction, transverse; $R = 0.1$. Source: "Mechanical Property Data: Ti-6Al-2Zr-2Sn-2Mo-2Cr Alloy Solution Treated and Aged Plate," F33615-72-C-1280, Battelle Columbus Laboratories, Apr 1973

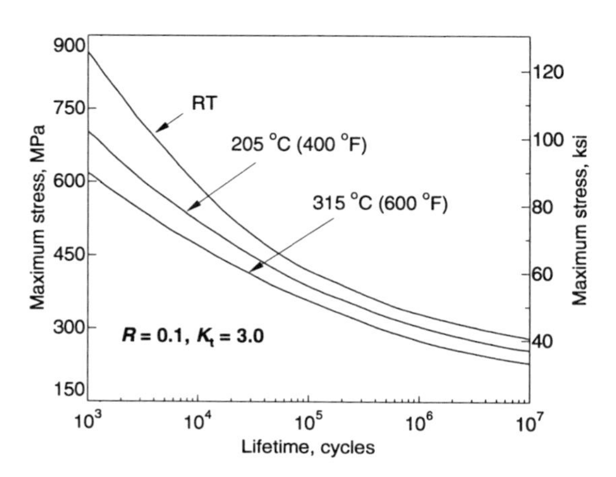

Ti-6-22-22S: Fatigue of notched STA plate. 38 mm (1.5 in.) thick plate heat treated at 950 °C (1740 °F), 1 h, AC + 540 °C (1000 °F), 8 h, AC; test direction, transverse; $R = 0.1$; $K_t = 3.0$. Source: "Mechanical Property Data: Ti-6Al-2Zr-2Sn-2Mo-2Cr Alloy Solution Treated and Aged Plate," F33615-72-C1280, Battelle Columbus Laboratories, Apr 1973

Beta-Processed Material

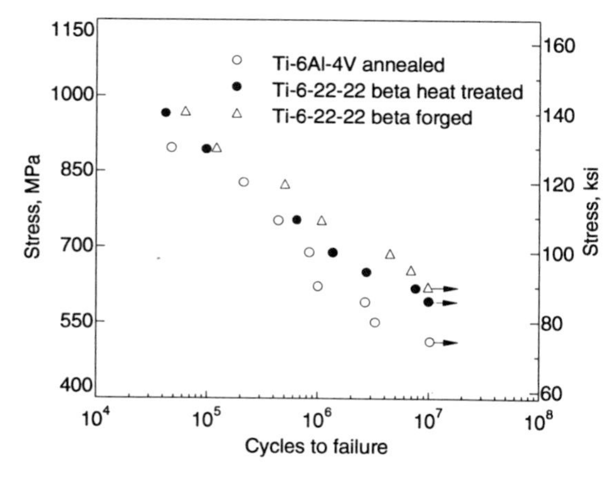

Ti-6-22-22S: Smooth high-cycle fatigue. Close-die forgings; processed as noted. $R = 0.1$; $K_t = 1.0$, 30 Hz. Source: G.W. Kuhlman *et al.*, "Characterization of Ti-6-22-22S: A High-Strength Alpha-Beta Titanium Alloy for Fracture Critical Applications," Proc. 7th Int. Titanium Conf., San Diego, TMS/AIME, June 1992, to be published

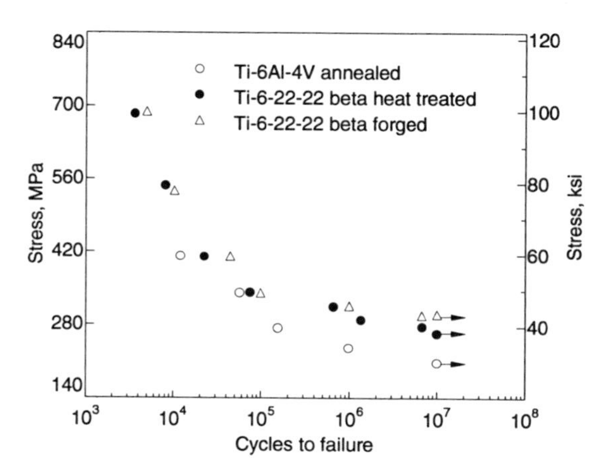

Ti-6-22-22S: Notched high-cycle fatigue. Close-die forgings; processed as noted. $R = 0.1$, $K_t = 3.0$, 30 Hz. Source: "Characterization of Ti-6-22-22S: A High-Strength Alpha-Beta Titanium Alloy for Fracture Critical Applications," Proc. 7th Int. Titanium Conf., San Diego, TMS/AIME, June 1992, to be published

Fatigue-Crack Propagation

The first figures represent early data on α-β processed STA plate, whereas the remaining figures are for β-processed material. The crack propagation rates of the latter are similar to that of β-annealed ELI Ti-6Al-4V. It is readily apparent that the rapid cooling rates, which refine the transformed β structure, detract from fatigue-crack growth resistance. The effect of thermomechanical processing is also illustrated. Basically, the data indicate that the lamellar α-β structure, i.e., β-processed material, provides the slowest crack growth rates.

Billet

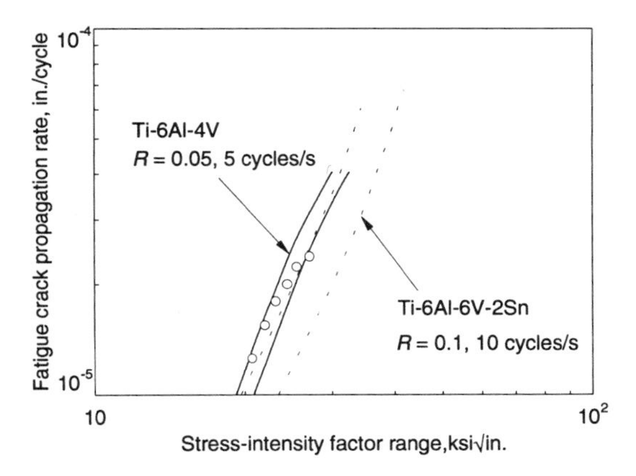

Ti-6-22-22S: Fatigue crack growth rate in forged . 150 mm (6 in.) diam billet; 950 °C (1740 °F), 1 h, water quenched + 540 °C (1000 °F), 8 h, AC; test direction, L-S; yield strength, 1083 MPa (157 ksi); R = 0.044; 25 Hz. Source: L.J. Bartlo, H.B. Bomberger, and S.R. Seagle, "Deep Hardenable Titanium Alloy," AFML-TR-73-122, Battelle Columbus Laboratories, May 1973

STA Plate

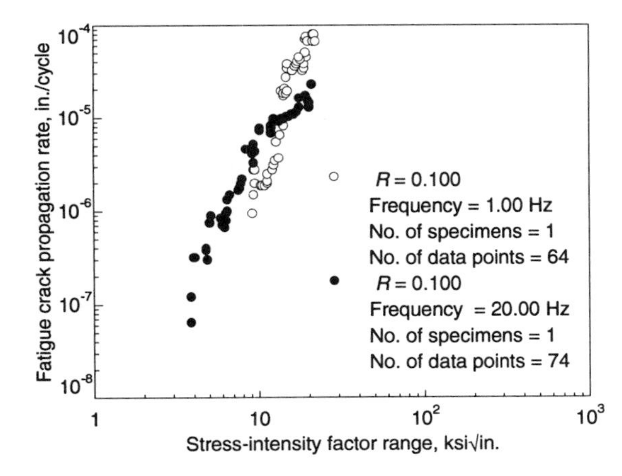

Ti-6-22-22S: Fatigue cracking in 3.5% NaCl of STA plate. 16 mm (5/8 in.) thick plate; test direction, longitudinal transverse; environment, 20 °C (70 °F), 3.5% NaCl; yield strength, 1083 MPa (157 ksi); specimen, 3.8 mm (0.15 in.) thick. Source: *Damage Tolerant Design Handbook*, Part 2, Metals and Ceramic Information Center, Battelle Columbus Laboratories, Jan 1975

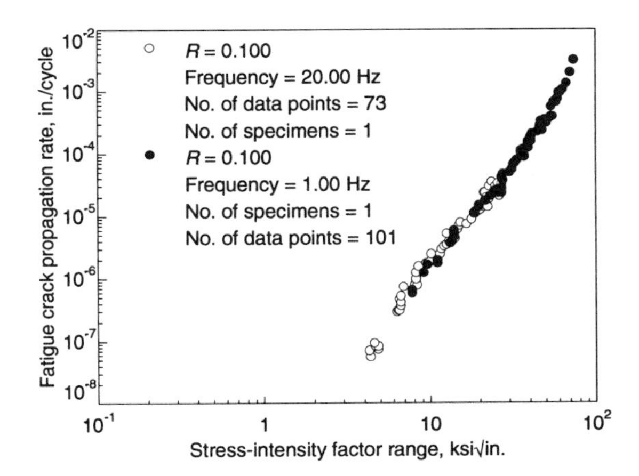

Ti-6-22-22S: Fatigue cracking in air of STA plate. 16 mm (5/8 in.) thick plate; test direction, longitudinal transverse; environment, 20 °C (70 °F), 95% relative humidity; yield strength, 1083 MPa (157 ksi); specimen, 3.8 mm (0.15 in.) thick. Source: *Damage Tolerant Design Handbook*, Part 2, Metals and Ceramic Information Center, Battelle Columbus Laboratories, Jan 1975

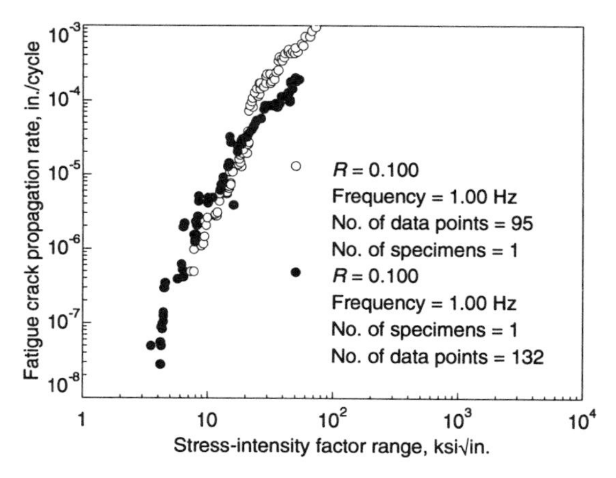

Ti-6-22-22S: Fatigue cracking in 3.5% NaCl of STA plate. 16 mm (5/8 in.) thick plate; test direction, longitudinal transverse; environment, 20 °C (70 °F), 3.5% NaCl; yield strength, 1083 MPa (157 ksi); specimen, 3.8 mm (0.15 in.) thick. Source: *Damage Tolerant Design Handbook,* Part 2, Metals and Ceramic Information Center, Battelle Columbus Laboratories, Jan 1975

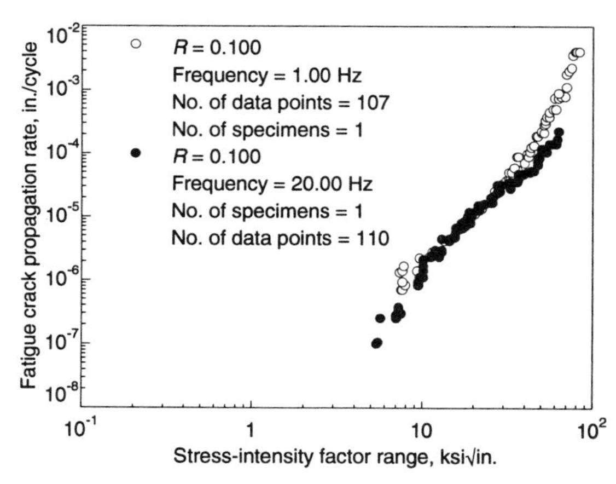

Ti-6-22-22S: Fatigue cracking in 3.5% NaCl of STA plate. 16 mm (5/8 in.) thick plate; test direction, longitudinal transverse; environment, 20 °C (70 °F), 3.5% NaCl; yield strength, 1083 MPa (157 ksi); specimen, 3.8 mm (0.15 in.) thick. Source: *Damage Tolerant Design Handbook,* Part 2, Metals and Ceramic Information Center, Battelle Columbus Laboratories, Jan 1975

Beta-Processed Condition

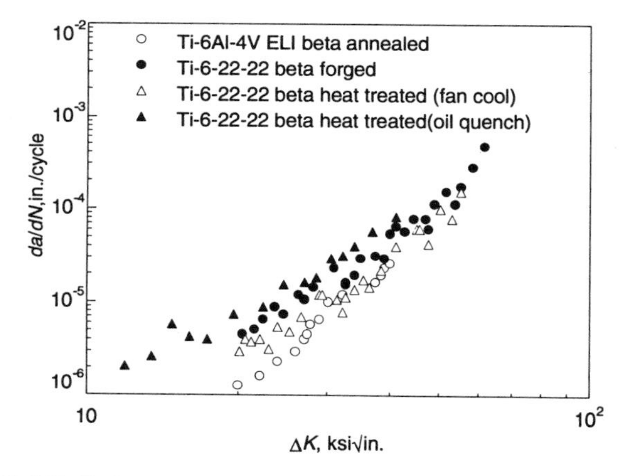

Ti-6-22-22S: Fatigue crack growth rate of forgings. Forged pancakes processed as indicated. $R = 0.1$, 20 Hz, lab air; Ti-6Al-4V ELI, $R = 0.01$. Source: G.W. Kuhlman *et al.*, "Characterization of Ti-6-22-22S: A High-Strength Alpha-Beta Titanium Alloy for Fracture Critical Applications," Proc. 7th Int. Titanium Conf., San Diego, TMS/AIME, June 1992, to be published

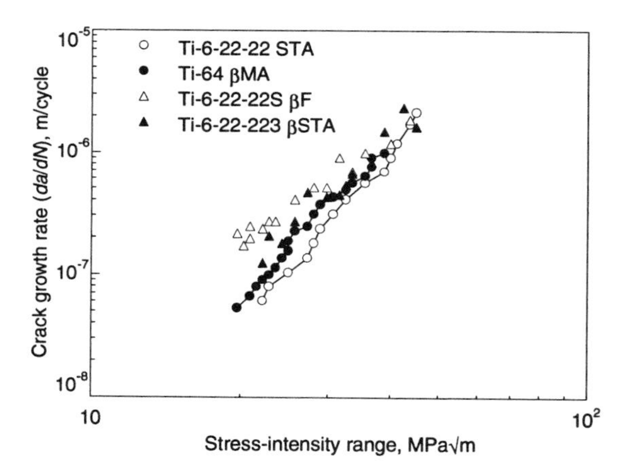

Ti-6-22-22S: Fatigue crack growth rate comparison. R = 0.01, 10 Hz, lab air. Source: A.K. Chakrabarti, R. Pishko, V.M. Sample, and G.W. Kuhlman, "TMP Conditions-Microstructure-Mechanical Property Relation in Ti-6-22-22S Alloy," Proc. 7th Int. Titanium Conf., San Diego, TMS/AIME, June 1992, to be published

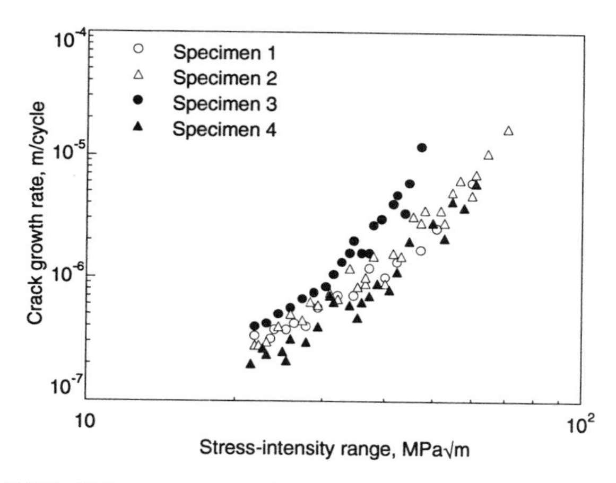

Ti-6-22-22S: Fatigue crack growth rate vs applied stress intensity of forgings. Forged pancakes, triplex beta heat treated with varying cooling rates. Source: A.K. Chakrabarti, R. Pishko, V.M. Sample, and G.W. Kuhlman, "TMP Conditions-Microstructure-Mechanical Property Relation in Ti-6-22-22S Alloy," Proc. 7th Int. Titanium Conf., San Diego, TMS/AIME, June 1992, to be published

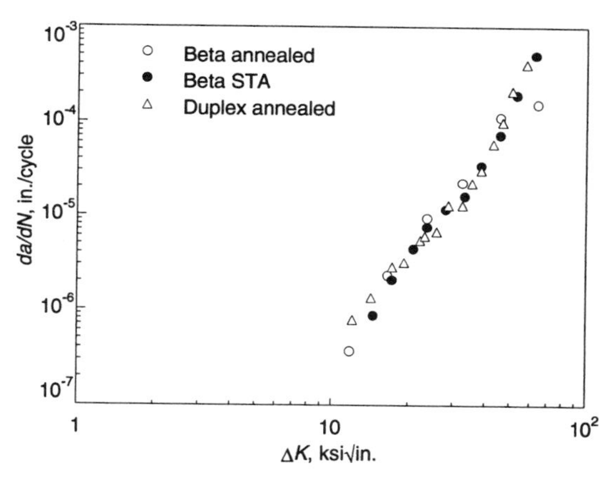

Ti-6-22-22S: Fatigue crack growth rate of plate. Specimens were α-β rolled 14 mm (0.58 in.) plate in three conditions: (1) β annealed at 35 °C (65 °F) above the β transus, 1 h, AC; (2) β solution treated and aged at 35 °C (65 °F) above the β transus, 1 h, AC + 540 °C (1000 °F), 8 h; or (3) duplex heat treated by β annealing + 30 °C (50 °F) below the β transus, 1 h, AC + 540 °C (1000 °F), 8 h. Beta annealed: tensile strength, 930 MPa (135 ksi); K_q, 100 MPa$\sqrt{m}$ (91 ksi$\sqrt{in.}$). Beta solution treated and aged: tensile strength, 1027 MPa (149 ksi); K_q, 81 MPa$\sqrt{m}$ (74 ksi$\sqrt{in.}$). Duplex annealed: tensile strength, 1034 MPa (150 ksi); K_q, 71 MPa$\sqrt{m}$ (65 ksi$\sqrt{in.}$). $R = 0.1$, lab air. Source: H.R. Phelps and J.R. Wood, "Correlation of Mechanical Properties and Microstructures of Ti-6Al-2Sn-2Zr-2Mo-2Cr-0.25Si Titanium Alloy," Proc. 7th Int. Titanium Conf., San Diego, TMS/AIME, June 1992, to be published

3.5% NaCl

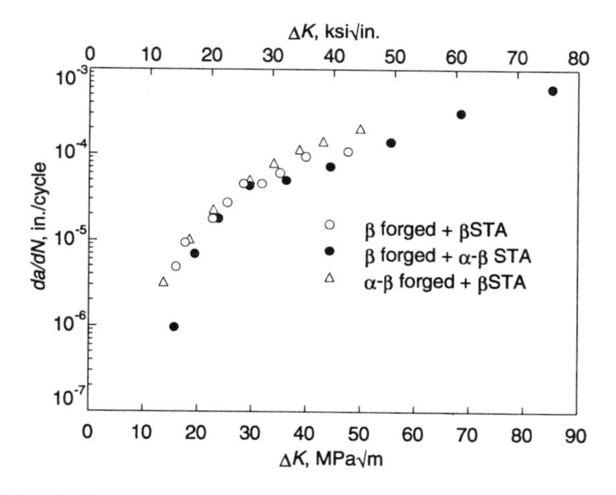

Ti-6-22-22S: Fatigue crack growth rate comparison in 3.5% NaCl. Specimens were α-β rolled 14 mm (0.58 in.) plate in three conditions: (1) β annealed at 35 °C (65 °F) above the β transus, 1 h, AC; (2) β solution treated and aged at 35 °C (65 °F) above the β transus, 1 h, AC + 540 °C (1000 °F), 8 h; or (3) duplex heat treated by β annealing + 30 °C (50 °F) below the β transus, 1 h, AC + 540 °C (1000 °F), 8 h. Specimens were tested in the L-T direction, $R = 0.1$, 2 Hz. Source: H.R. Phelps and J.R. Wood, "Correlation of Mechanical Properties and Microstructures of Ti-6Al-2Sn-2Zr-2Mo-2Cr-0.25Si Titanium Alloy," Proc. 7th Int. Titanium Conf., San Diego, TMS/AIME, June 1992, to be published

Fracture Properties

The fracture properties of Ti-6-22-22S , as for other α + β titanium alloys, are quite dependent on strength and microstructure. In general, the β-processed/α-β annealed and α-β-processed/β annealed conditions produce better toughness than conditions that result in fine α + β microstructures. The effects of thermomechanical processing and oxygen content on fracture toughness are shown below. Similar to the results for fatigue crack propagation rates, the data show that the transformed β structure provides the maximum fracture toughness. It also illustrates an unexplained drop in fracture toughness as the aging temperature is increased. Similar behavior has been observed for plain-stress or mixed mode fracture toughness in sheet, as shown below. It is speculated that this drop in toughness is related to an ordering reaction in the alpha and/or silicide precipitation.

Ti-6-22-22S: Fracture toughness of sheet

Aging temperature		Ultimate tensile strength		Tensile yield strength			Toughness (K_{app})	
°C	°F	MPa	ksi	MPa	ksi	Elongation, %	MPa$\sqrt{m}$	ksi$\sqrt{in.}$
480	900	1275	185	1160	168	11.7	165	150
565	1050	1240	180	1170	170	11.2	109	99
675	1250	1140	165	1078	155	10.4	112	102

Note: Tensile properties are the average of six values each, and toughness values are the averages of two tests each. 1.2 mm (0.05 in.) sheet was solution treated at 900 °C for 30 min, aged 8 h. Source: R.R. Boyer and A.E. Caddey, "The Properties of Ti-6Al-2Sn-2Zr-2Mo-2Cr Sheet," Proc. 7th Int. Titanium Conf., San Diego, TMS/AIME, June 1992, to be published

Ti-6-22-22S: Fracture toughness and impact toughness

Direction	Ultimate tensile strength		Elongation in 25 mm (1 in.),	Charpy V-notch impact toughness		Fracture toughness (K_{Ic})	
	MPa	ksi	%	J	ft · lbf	MPa$\sqrt{m}$	ksi$\sqrt{in.}$
Longitudinal	1160	168.3	18.0	18.8	13.9	...	...
Transverse	1163	168.7	17.7	22.1	16.3	...	...
L-T	...	...	...	...	...	96	88
T-L	...	...	...	...	...	102	93

38 mm (1.5 in.) thick plate, STA condition. Source: AFML, Apr 1973

Ti-6-22-22S: Typical fracture toughness of β-processed STA products

Product	Ultimate tensile strength		Yield strength			Reduction	Fracture toughness (K_{Ic})	
	MPa	ksi	MPa	ksi	Elongation, %	of area, %	MPa$\sqrt{m}$	ksi$\sqrt{in.}$
50 mm (2 in.) plate	1138	165	1020	148	10	17	85	77
100 mm (4 in.) plate	1103	160	979	147	10	15	89	81
150 mm (6 in.) plate	1076	156	958	139	10	15	98	89

Source: J.R. Wood , RMI Titanium Co., 15 Aug 1991

Ti-6-22-22S: Fracture toughness of α + β processed STA products

Product	Ultimate tensile strength		Tensile yield strength			Reduction	Fracture toughness (K_{Ic})	
	MPa	ksi	MPa	ksi	Elongation, %	of area, %	MPa$\sqrt{m}$	ksi$\sqrt{in.}$
50 mm (2 in.) plate	1207	175	1131	164	12	35	67	61
150 mm (6 in.) billet	1200	174	1089	158	15	41	65	72

Source: J.R. Wood, RMI Titanium Co., 15 Aug 1991

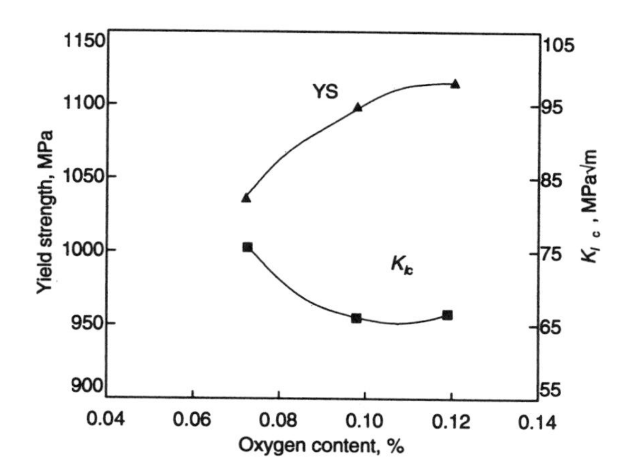

Ti-6-22-22S: Effect of oxygen content on K_{Ic}. Specimens were forged pancakes α-β forged + α-β solution treated + 540 °C (1000 °F), 8 h, aged, AC. Source: A.K. Chakrabarti *et al.*, "TMP Conditions-Microstructure-Mechanical Property Relationship in Ti-6-22-22S Alloy," Proc. 7th Int. Titanium Conf., San Diego, TMS/AIME, June 1992, to be published

Plastic Deformation

Strain Hardening

The *m*-values, indicators of the superplasticity of material, from 790 to 925 °C (1450 to 1700 °F) at two strain rates are illustrated.

Flow Stress

The flow stress over the temperature range of 790 to 925 °C (1450 to 1700 °F) over the range of strain rates from 8×10^{-5} to 2×10^{-4} is illustrated.

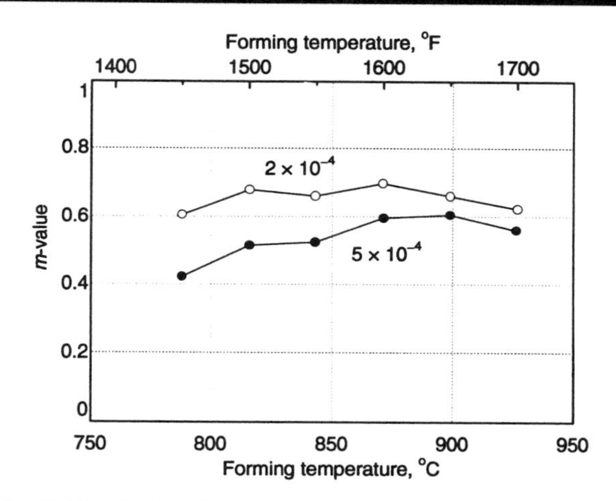

Ti-6-22-22S: Typical *m*-values. Data are shown for two different strain rates. 2.5 mm (0.10 in.) sheet, as annealed, 75% total strain using a step-strain-rate tensile test. Source: R.C. Bliss, "Evaluation of Ti-6-22-22S Sheet," Proc. 7th Int. Titanium Conf., San Diego, TMS/AIME, June 1992, to be published

Ti-3Al-8V-6Cr-4Mo-4Zr (Beta C)

Common Name: Beta C™, 38-6-44
UNS Number: R58640

Ti-3Al-8V-6Cr-4Mo-4Zr (Beta C) is a commercial alloy developed by RMI in the mid-to-late 1960s. It has similar characteristics to Ti-13V-11Cr-3Al (13-11-3), but is easier to melt and shows less segregation. Beta C was developed as an improvement of 13-11-3 which had some melting problems due to a high chromium content. Due to its significant molybdenum content, Beta C exhibits superior resistance to reducing acids and chloride crevice corrosion compared to other high- strength titanium alloys. Currently, Beta C holds a small amount (much less than 1%) of the production market.

Chemistry and Density

Beta C is formulated by depressing the beta transus with the beta isomorphous elements, molybdenum and vanadium, and the sluggish beta eutectoid element, chromium. It is slightly more beta-stabilized than Ti-11.5Mo-6Zr-4.5Sn (Beta III) and less beta-stabilized than Ti-13V-11Cr-3Al.

Density. 4.82 g/cm^3 (0.174 lb/in.^{3})

Product Condition/Microstructure

Beta C is cold rollable and drawable, and is used mainly as bar and wire material for aircraft springs; it has also been explored as a spring material for automotive applications. It constitutes less than 1% of titanium products. Beta C can be heated to high levels above 1380 MPa, 200 ksi—by aging between 480 and 595 °C (900 and 1100 °F). Large variations in tensile strength can be obtained by varying the aging temperature and time. A portion of the beta phase transforms to a finely dispersed alpha during aging. Also, it does not grain-coarsen as rapidly as other beta alloys when heat treated or worked at temperatures above the beta transus.

Applications

Beta C is used in fasteners, springs, torsion bars, and in foil form for making cores for sandwich structures. It is also used for tubulars and casings in oil, gas, and geothermal wells.

Use Limitations. Beta C, like other beta titanium alloys, is highly susceptible to hydrogen pickup and rapid hydrogen diffusion during heating, pickling, and chemical milling. However, because of the much higher solubility of hydrogen in the beta phase than in the alpha phase of titanium, this alloy has a higher tolerance for hydrogen than the alpha or alpha-beta alloys.

Beta C can be welded in the solution-treated condition; however, welding is not recommended after solution treating and aging. Care is necessary in pickling to minimize hydrogen absorption.

Ti-3Al-8V-6Cr-4Mo-4Zr (Beta C): Specifications

			Composition, wt%								
Specification	Designation	Description	Al	Cr	Fe	Mo	N	O	V	Zr	Other
	UNS R58640		3	6		4			8	4	bal Ti
USA											
AMS 4957		Bar Wir CD	3-4	5.5-6.5	0.3	3.5-4.5	0.03	0.12	7.5-8.5	3.5-4.5	H 0.03; C 0.05; OT 0.4; Y 0.005; bal Ti
AMS 4958		Bar Rod STA	3-4	5.5-6.5	0.3	3.5-4.5	0.03	0.12	7.5-8.5	3.5-4.5	H 0.03; C 0.05; OT 0.4; Y 0.005; bal Ti
MIL T-9046J	Code B-3	Sh Strp Plt SHT	3-4	5.5-6.5	0.3	3.5-4.5	0.03	0.12	7.5-8.5	3.5-4.5	H 0.02; C 0.05; OT 0.4; bal Ti
MIL T-9046J	Code B-3	Sh Strp Plt STA	3-4	5.5-6.5	0.3	3.5-4.5	0.03	0.12	7.5-8.5	3.5-4.5	H 0.02; C 0.05; OT 0.4; bal Ti
MIL T-9047G	Ti-3Al-8V-6Cr-4Mo-4Zr	Bar Bil STA	3-4	5.5-6.5	0.3	3.5-4.5	0.03	0.12	7.5-8.5	3.5-4.5	H 0.02; C 0.05; OT 0.4; Y 0.005; bal Ti

Ti-3Al-8V-6Cr-4Mo-4Zr (Beta C): Commercial compositions

			Composition, %								
Specification	Designation	Description	Al	Cr	Fe	Mo	N	O	V	Zr	Other
USA											
Astro	Ti-3Al-8V-6Cr-4Zr-4Mo	Bar Sprg Pip	3-4	5.5-6.5		3.5-4.5	0.03 max	0.14 max	7.5-8.5	3.5-4.5	C 0.05 max; bal Ti
Oremet	Ti-38-6-44										
RMI	3Al-8V-6Cr-4Zr-4Mo	Sh Plt Bar Bil Wir Ex	3-4	5.5-6.5	0.3	3.5-4.5	0.03	0.14	7.5-8.5	3.5-4.5	C 0.05; bal Ti
Teledyne	Tel-Ti-3Al-8V-6Cr-4Mo-4Zr		3-4	5.5-6.5	0.3	3.5-4.5	0.03	0.14	7.5-8.5	3.5-4.5	
Timet	TIMETAL 3-8-6-4-4	Ing Bil STA	3-4	5.5-6.5	0.3	3.5-4.5	0.03	0.14	7.5-8.5	3.5-4.5	H 0.02; C 0.05; bal Ti

Phases and Structures

As a solute-rich β alloy, precipitation of α within the solute-lean β regions (β') of Beta C is slow. Prior cold work accelerates the formation of intragranular α and also reduces the extent of grain boundary α. Peak aging occurs at around 480 °C (900 °F), and smaller quantities of α (in the form of coarse precipitates) are found at higher temperatures. Type 2 α occurs during certain aging treatments. Recrystallization occurs after short times above the β transus, although β grain growth is not a problem. The possibility of a second phase responsible for inhibiting grain growth above the β transus has been suggested (R.A. Wood and R.J. Favor, *Titanium Alloys Handbook*, MCIC-HB-02, Battelle Columbus Laboratories, 1972, Section 1-12, p 72-1).

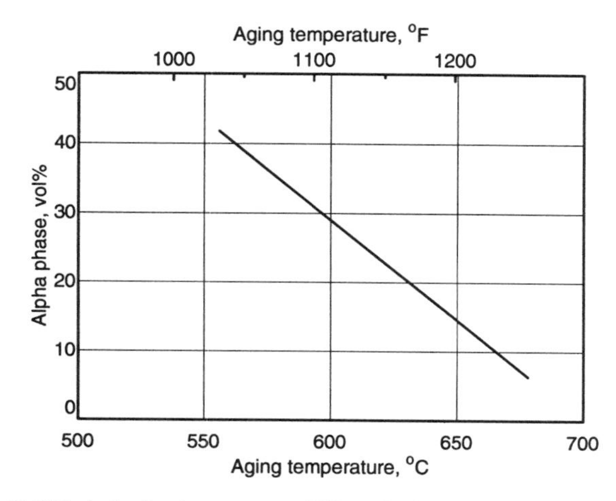

Beta C: Effect of aging temperature. Effect of aging temperature on amount of α phase precipitation in 5 mm (1.25 in.) thick plate solution heat treated at 925 °C (1700 °F) and aged. Source: R.A. Wood and R.J. Favor, *Titanium Alloys Handbook*, MCIC-HB-02, Battelle Columbus Laboratories, 1972, Section 1-12, p 72-1

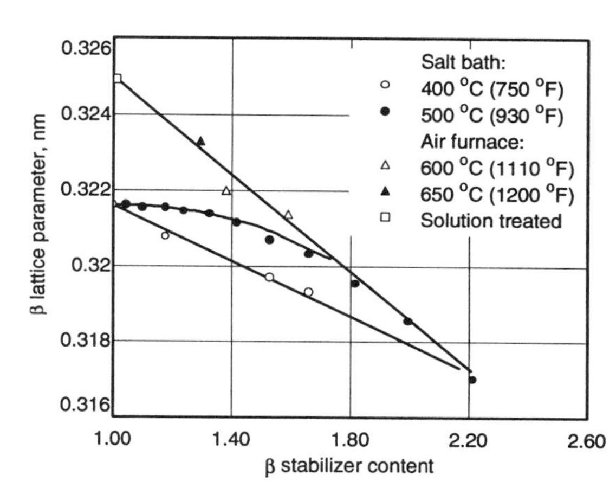

Beta C: Variation of β lattice parameter. Variation of β lattice parameter with β stabilizing alloying element content (normalized to unity at zero volume fraction of α). Source: G.H. Isaac and C. Hammond, The Formation of Type 2 α Phase in Ti-3Al-8V-6Cr-4Zr-4Mo, in *Titanium, Science and Technology*, G. Lütjering, U. Zwicker, and W. Bunk, Ed., Deutsche Gesellschaft für Metallkunde eV, Germany, 1985, p 1608

Beta Transus. 730 °C (1350 °F). The previously published transus temperature of 795 °C (1460 °F) is too high.

Beta C: Lattice parameters of the α and β phases in solution heat treated and aged plate

Aging temperature		Lattice spacing, Å (±0.004 Å)		
		Alpha phase		Beta phase
°C	°F	*a*	*c*	a_o
565	1050	2.948	4.680	3.218
620	1150	2.936	4.682	3.226
675	1250	2.920	4.684	3.229

Note: 32 mm (1.25 in.) plate was solution treated at 925 °C (1700 °F) and aged 8 h. Source: R.A. Wood and R.J. Favor, *Titanium Alloys Handbook*, MCIC-HB-02, Battelle Columbus Laboratories, 1972, Section 1-12, p 72-1

Fatigue Properties

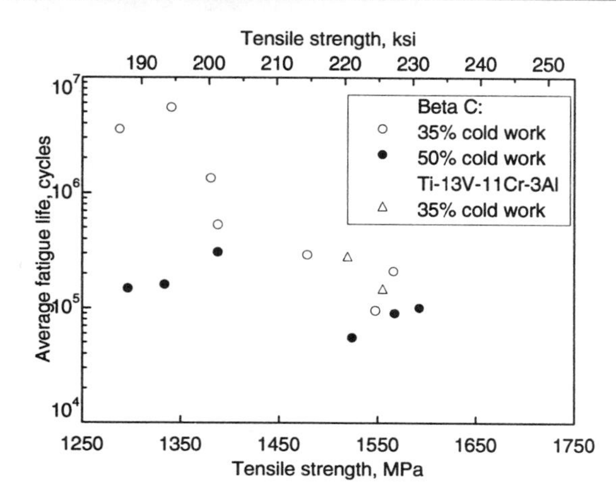

Beta C: Fatigue life of shot peened wire. Shot peening is a critical parameter and shot peen intensities of at least 0.016 to 0.018 A should be used. Higher intensities would provide additional improvement in fatigue life, but a higher intensity call-out could limit the number of shot peening sources available due to equipment limitations. The effects of cold work and tensile strength on the average fatigue life of 10 mm (0.4 in.) diam shot peened to 0.016 to 0.018 A. Each data point represents the log average of six tests. Ti-13V-1Cr-3Al data points included for comparison. 1034 MPa (150 ksi) maximum stress, $R = 0.1$, 30 Hz. Source: *Beta Titanium Alloys in the 1980's*, R.R. Boyer and H.W. Rosenberg, Ed., TMS/AIME, 1984

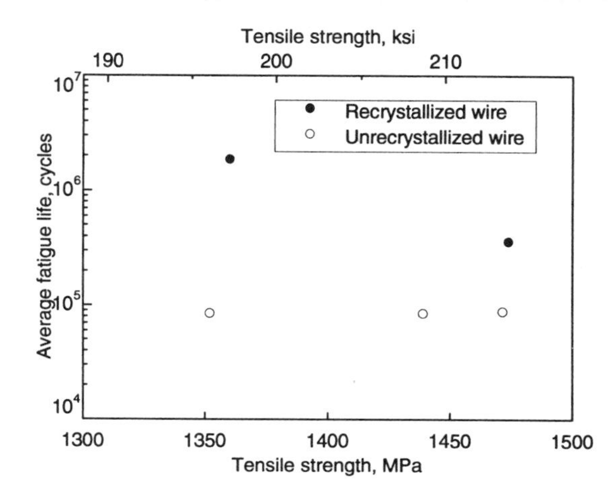

Beta C: Fatigue life of recrystallized wire. Control of grain size is desired, and the wire should be recrystallized during solution treatment. Specimens were 9 mm (0.35 in.) diam wire cold worked 35%. Shot peen intensity of 0.016 to 0.018 A, 1034 MPa (150 ksi) maximum stress, $R = 0.1$, 30 Hz. Source: *Beta Titanium Alloys in the 1980's*, R.R. Boyer and H.W. Rosenberg, Ed., TMS/AIME, 1984

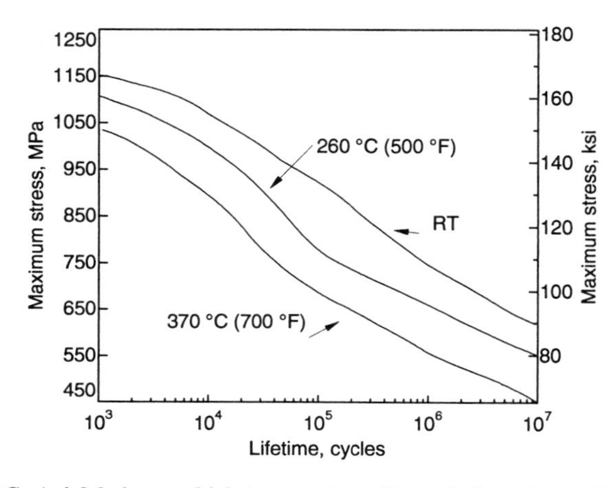

Beta C: Axial fatigue at high temperature. Unnotched specimens (R = 0.1) from 150 mm (6 in.) diam forging, 815 °C (1500 °F) for 15 min, AC, plus aged 12 h at 565 °C (1050 °F), AC. Source: *Beta Titanium Alloys in the 1980's*, R.R. Boyer and H.W. Rosenberg, Ed., TMS/AIME, 1984

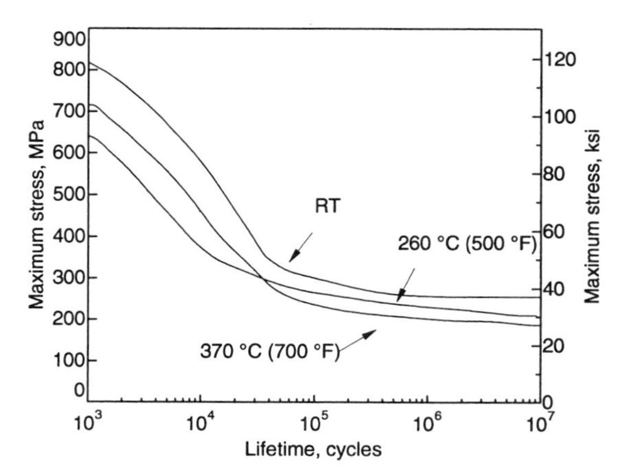

Beta C: Notched fatigue strength at high temperature. Axial fatigue of notched specimens (R = 0.1, K_t = 3.0) from 150 mm (6 in.) diam STA forgings. Source: *Beta Titanium Alloys in the 1980's*, R.R. Boyer and H.W. Rosenberg, Ed., TMS/AIME, 1984

Beta C: Fatigue life at high temperatures

Test condition(a)	Fatigue life, MPa (ksi), at: RT	205 °C (400 °F)	370 °C (700 °F)
Unnotched			
10^3 cycles	1144 (166.0)	1089 (158.0)	1020 (148.0)
10^5 cycles	855 (124.0)	731 (106.0)	634 (92.0)
10^7 cycles	600 (87.0)	551 (80.0)	372 (54.0)
Notched(b)			
10^3 cycles	827 (120.0)	717 (104.0)	634 (92.0)
10^5 cycles	303 (44.0)	248 (36.0)	275 (40.0)
10^7 cycles	275 (40.0)	207 (30.0)	234 (34.0)

(a) Axial fatigue of transverse specimens from 150 mm (6 in.) diam STA forging treated 15 min at 815 °C (1500 °F), AC, plus aging at 565 °C (1050 °F) for 12 h, AC. R = 0.1. (b) K_t = 3.0. Source: *Beta Titanium Alloys in the 1980's*, R.R. Boyer and H.W. Rosenberg, Ed., TMS/AIME, 1984

Fatigue Crack Growth

Crack growth rates in the accompanying figures were determined for Beta C in various conditions (see table). Because Beta C is an attractive alloy for highly corrosive environments such as in sour wells, the effect of aggressive environments on mechanical behavior also is of interest. For the test results presented here (see figures), no noticeable acceleration in crack growth rates was found when going from air to a saltwater environment, or when the frequency was reduced. Because the differences in da/dN–ΔK behavior are insignificant, data are presented as single scatterbands.

A slight tendency toward faster growth rates was observed for LO-Simplex and LO-Duplex (see figure below) as opposed to HI-Simplex and HI-Duplex, presumably as a consequence of the lower ductilities. No effect of the duplex versus the simplex aging treatment was detected. A significant difference, however, was found between aged and unaged material. The value of ΔK_{th} is roughly 3 MPa$\sqrt{m}$ (2.7 ksi$\sqrt{in.}$) for aged material under all testing conditions, as opposed to values of 4 to 5 MPa$\sqrt{m}$ (3.6 to 4.5 ksi$\sqrt{in.}$) for as-SHT material. Correcting for crack closure (ΔK_{eff}) brings the data into accord and reduces $\Delta K_{th,eff}$ to ≈2 MPa$\sqrt{m}$ (1.8 ksi$\sqrt{in.}$), suggesting that the difference between as-SHT and aged material may not be present at high R ratios.

Beta C: Material condition in crack growth tests

Heat treatment(a) (designation)	Tensile yield strength (0.2% offset) MPa	ksi	Ultimate tensile strength MPa	ksi	Elongation, %	Reduction of area, %
800 °C, 30 min, AC (as-SHT 800)	895	130	895	130	22	48
925 °C, 30 min, AC (as-SHT 927)	850	125	850	125	25	62
as-SHT 800 + 535 °C, (8 h) (LO-simplex)	1225	177	1320	190	8	15
as-SHT 800 + 425 °C, (4 h) (LO-duplex)	1220	176	1300	188	10	13
as-SHT 927 + 530 °C (16 h) (HI-simplex)	1140	165	1220	176	12	21
as-SHT 927 + 455 °C (4 h) + 555 (16 h) (HI-duplex)	1075	156	1180	171	14	23

(a) The grain sizes after solution heat treating at 800 and 925 °C were 45 and 160 μm, respectively. The 800 °C SHT did not fully recrystallize the as-hot worked structure, and left approximately 20 vol. % unrecrystallized. Almost no unrecrystallized grains were present after SHT at 925 °C. For both SHT, the 4 hr cycle promotes a somewhat more homogeneous α distribution. Source: H.E. Krugmann and J.K. Gregory, Microstructure and Crack Propagation in Ti-3Al-8V-6Cr-4Mo-4Zr, in Microstructure and Property *Relationships in Titanium Aluminides and Alloys*, Y-W. Kim and R.R. Boyer, Ed., TMS/AIME, 1991, p 551

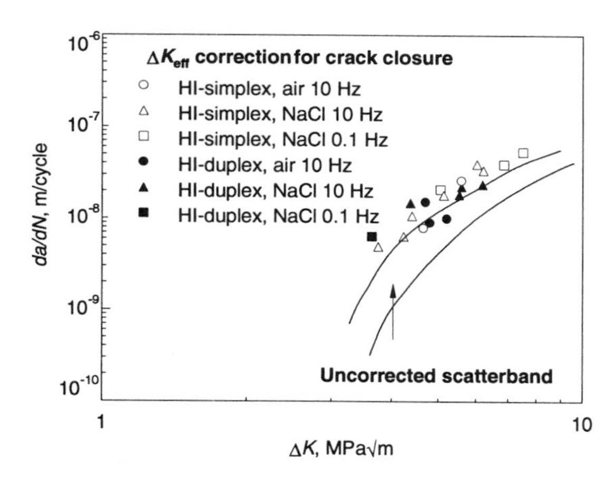

Beta C: Crack growth with high-temperature ST. Source: H.E. Krugmann and J.K. Gregory, Microstructure and Crack Propagation in Ti-3Al-8V-6Cr-4Mo-4Zr, in *Microstructure and Property Relationships in Titanium Aluminides and Alloys*, Y-W. Kim and R.R. Boyer, Ed., TMS/AIME, 1991, p 549-560

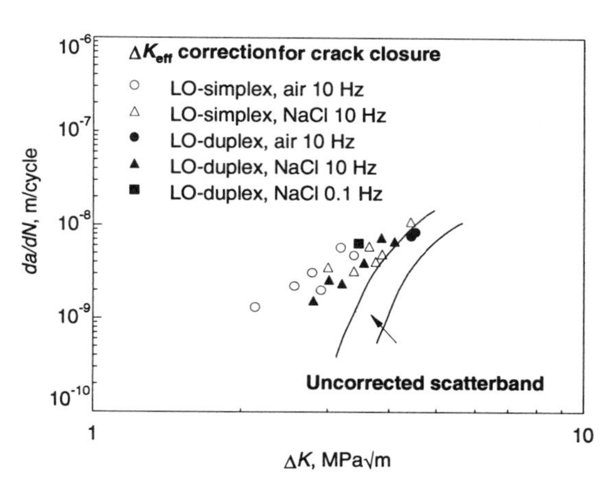

Beta C: Crack growth with low-temperature ST. Source: H.E. Krugmann and J.K. Gregory, Microstructure and Crack Propagation in Ti-3Al-8V-6Cr-4Mo-4Zr, in *Microstructure and Property Relationships in Titanium Aluminides and Alloys*, Y-W. Kim and R.R. Boyer, Ed., TMS/AIME, 1991, p 549-560

Fracture Properties

Beta C: Fracture toughness of bar

Condition	Test direction	Fracture toughness (K_{Ic}) MPa√m	ksi√in.
785 °C (1450 °F), 1 h, AC	C-R	53.7	48.9
+ 550 °C (1025 °F), 24 h, AC	C-R	55.7	50.7
	R-L	53.3	48.5
	R-L	56.7	51.6
840 °C (1550 °F), 1 h, AC	C-R	55.1	50.1
+ 480 °C (900 °F), 24 h, AC	C-R	69.2	63.0
	R-L	57.6	52.4
	R-L	55.2	50.2

Note: Specimens were from 75 mm (3 in.) bar. Source: G. Bella *et al.*, Effects of Processing on Microstructure and Properties of Ti-3Al-8V-6Cr-4Mo-4Zr (Beta C™), in *Microstructure and Property Relationships in Titanium Aluminides and Alloys*, Y-M. Kim and R.R. Boyer, Ed., TMS/AIME, 1991, p 493-510

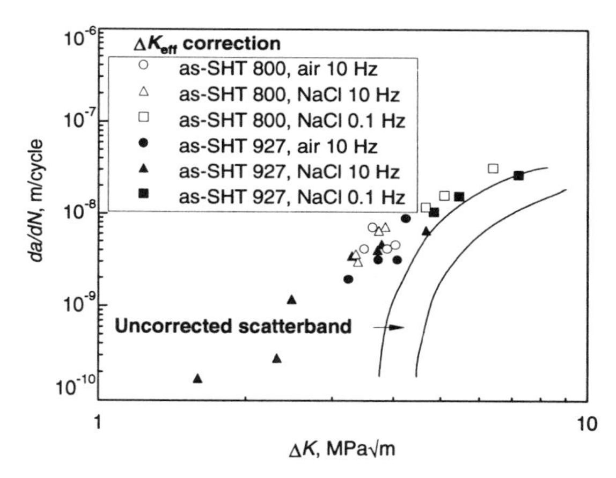

Beta C: Crack growth in solution treated condition. Source: H.E. Krugmann and J.K. Gregory, Microstructure and Crack Propagation in Ti-3Al-8V-6Cr-4Mo-4Zr, in *Microstructure and Property Relationships in Titanium Aluminides and Alloys*, Y-W. Kim and R.R. Boyer, Ed., 1991, p 549-560

Beta C: Fracture toughness of STA billet

Treatment	Test direction	Fracture toughness (K_{Ic}) MPa√m	ksi√in.	Ultimate tensile strength MPa	ksi	Tensile yield strength (2% offset) MPa	ksi	Elongation, %	Reduction of area, %
Water quench	L	96.7	88.0	1189	172.5	1125	163.2	9.5	19.6
	T	62	56.4	1188	172.4	1145	166.0	3.0	5.6
Air cool	L	89.9	81.8	1208	175.2	1150	166.8	9.2	17.4
	T	63.3	57.6	1242	180.2	1184	171.7	3.5	6.6

Note: Specimens were 150 mm (6 in.) billet solution treated 815 °C (1500 °F), 15 min, cooled (WQ,AC), then aged 12 h at 565 °C (1050 °F), AC. Source: RMI Co., reported in *Industrial Applications of Titanium and Zirconium: Fourth Volume*, C.S. Young and J.C. Durham, Ed., ASTM STP 917, 1986, ASTM, p 155

Beta C: Fracture toughness of billet, forging, and plate

Product form/ specimen location	Heat treatment	Direction	Tensile yield strength		Fracture toughness			
					K_Q		K_{Ic}	
			MPa	ksi	MPa$\sqrt{m}$	ksi$\sqrt{in.}$	MPa$\sqrt{m}$	ksi$\sqrt{in.}$
Billet								
150 mm (6 in.) diam	15 min, 815 °C (1500 °F), AC	L	1151	167	...	...	90(a)	82(a)
(midradius specimens)	+ 12 h, 565 °C (1050 °F), AC	T	1186	172	...	...	64	58
	15 min, 815 °C (1500 °F), WQ	L	1124	163	...	...	97	88
	+ 12 h, 565 °C (1050 °F), AC	T	1144	166	...	...	61	56
150 mm (6 in.) (location unknown)	Annealed(b) + aged:							
	8 h, 510 °C (950 °F), AC	T	1330	193	...	...	55(c)	50(c)
	plus exposed(d)	T	...	...	...	...	54	49
	8 h, 565 °C (1050 °F), AC	T	...	...	...	...	69	63
	plus exposed(d)	T	...	...	...	...	70	64
	8 h, 620 °C (1150 °F), AC	T	...	...	...	...	81	74
	plus exposed(d)	T	...	...	...	...	80	73
150 × 150 mm (6 × 6 in.)								
Surface specimens	15 min, 815 °C (1500 °F), AC	(c)	1151	167	...	...	60-66(e)	55-60(e)
Center specimens	+ 12 h, 565 °C (1050 °F), AC	(c)	1151	167	...	...	57-71	52-65
Navaho spar forging								
100 × 150 mm (4 × 6 in.)	15 min, 815 °C (1500 °F), AC							
(forging center speci mens)	+ 12 h, 565 °C (1050 °F), AC	L	1158	168	...	...	58-59(e)	53-54(e)
Plate								
32 mm (1.25 in.)	925 °C (1700 °F), annealed, AC							
(center specimens)	+ 8 h, 565 °C (1050 °F), AC	RW	1137	165	53(f)	48(f)	...	...
	925 °C (1700 °F), annealed, AC							
	+ 8 h, 675 °C (1250 °F), AC	RW	862	125	56(f)	51(f)	...	...
19 mm (0.75 in.)	1 h, 815 °C (1500 °F), AC							
(center specimens)	+ 4 h, 525 °C (975 °F), AC	(c)	1206	175	...	...	90(c)	82(c)

(a) Four-point loaded, slow-bend test. (b) Heat treatment details not given. (c) Test details not given. (d) 1000-h exposure at 285 °C (550 °F) under 172 MPa (25 ksi) load, cooled to room temperature and tested. (e) Slow-bend tests. (f) Compact-tension tests. Source: *Beta Titanium Alloys*, R. Wood, MCIC-72-11, Battelle Columbus Laboratories, 1972

Ti-10V-2Fe-3Al

Common Name: Ti-10-2-3
UNS Number: Unassigned

Ti-10-2-3 is a high-strength titanium-base alloy. Metallurgically it is a near-beta alloy, and it is capable of attaining a wide variety of strength levels depending on selection of heat treatment. Major advantages of this alloy are its excellent forgeability; its high toughness in air and salt-water environments; and its high hardenability, which provides good properties in sections up to 125 mm (5 in.) thick. It is used in the aerospace industry for applications up to 315 °C (600 °F).

A major advantage of Ti-10-2-3 over commercially available alpha-beta compositions of similar strength levels is its toughness in air and salt water environments. This near-beta alloy was developed primarily for high-strength and toughness applications at temperatures up to 315 °C (600 °F) and tensile strengths of 1240 MPa (180 ksi) in order to provide weight savings over steels in airframe forging applications. Of special interest for high-strength forgings for aircraft, it is being used for components by much of the aerospace industry.

Chemistry and Density

Ti-10-2-3 has a near-beta composition and is slightly more beta stabilized than Ti-11.5Mo-6Zr-4.5Sn (Beta III).

Density. 4.65 g/cm^3 (0.168 lb/in.^{3})

Product Forms

Ti-10-2-3 has the best hot-die forgeability of any commercial titanium alloy and is often used for near-net-shape forging applications. Mill products are billet, bar, and plate.

Product Condition/Microstructure

Developed for use in the aerospace industry, Ti-10-2-3 combines many of the advantages of the metastable beta titanium alloys without sacrificing certain inherent alpha-beta characteristics. It shows excellent hardenability in section sizes up to 125 mm (5 in.), but also demonstrates good short-transverse ductility. In the solution-treated and aged condition, this alloy maintains greater than 80% of its room-temperature strength at 315 °C (600 °F) and has creep-stability characteristics similar to those of the alpha-beta alloys at this temperature.

Applications

Ti-10-2-3 is used at temperatures up to 315 °C (600 °F) where medium-to-high strength and high toughness are required in bar, plate, or forged sections up to 125 mm (5 in.) thick. It can be heat treated over a wide strength-toughness range, allowing the tailoring of properties. It is employed for applications requiring uniformity of tensile properties at surface and center locations. Specific applications include aerospace air-frames hot-die and conventional forgings, and other forged parts in a wide variety of components. The major user, Boeing, uses the alloy up to 260 °C (500 °F).

Ti-10V-2Fe-3Al: Specifications and compositions

Specification	Designation	Description	Composition, wt%								Other
			Al	C	Fe	H	N	O	V	Y	
USA											
AMS 4986		Frg STOA	2.6-3.4	0.05	1.6-2.2	0.015	0.05	0.13	9-11	0.005	OT 0.3; bal Ti
AMS 4983A		Frg STA	2.6-3.4	0.05	1.6-2.2	0.015	0.05	0.13	9-11	0.005	OT 0.3; bal Ti
AMS 4984		Frg STA	2.6-3.4	0.05	1.6-2.2	0.015	0.05	0.13	9-11	0.005	OT 0.3; bal Ti
AMS 4987		Frg STOA	2.6-3.4	0.05	1.6-2.2	0.015	0.05	0.13	9-11	0.005	OT 0.3; bal Ti

Ti-10V-2Fe-3Al: Commercial compositions

Specification	Designation	Description	Composition, %								Other
			Al	C	Fe	H	N	O	V	Y	
Japan											
Kobe	KS10-2-3	Bar Frg STA	2.6-3.4		1.6-2.2	0.015	0.05	0.13	9-11		bal Ti
USA											
Timet	TIMETAL 10-2-3	Frg	2.6-3.4		1.6-2.2	0.015	0.05	0.13	9-11		bal Ti

Fatigue (Smooth)

Care must be taken in analyzing this fatigue data and comparing to a given set of conditions as there are so many variables, including specimen geometry, surface finish, R-ratio, and loading condition factors such as load controlled or strain controlled frequency, and wave form.

The high-cycle fatigue strength is a function of the tensile strength as one might expect. Generally the S-N curves are quite flat. Direct aging (with no solution treatment, which can be used over a limited thickness range) has a pronounced advantage over the solution treated and aged condition. This is attributed to two factors, the minimization of grain boundary α and the precipitation of a finer, more uniform dispersion of aged α when using a direct age. A primary α grain size effect has been recently reported (see bottom right figure). The effect of test temperature on fatigue properties is also illustrated. Again, one might expect the higher strength condition to have a lower fatigue debit as a function temperature than the lower strength conditions.

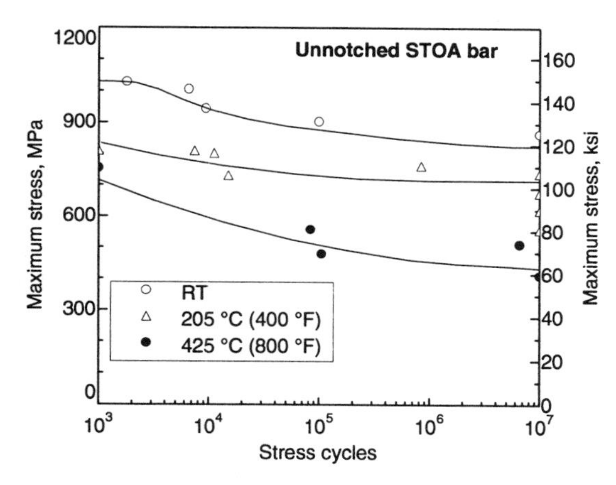

Ti-10V-2Fe-3Al: Effect of temperature on axial fatigue. Axial fatigue of Ti-10V-2Fe-3Al bar stock in the STOA (solution treated and overaged) condition. Specimens were taken from round bars 75 mm (3 in.) in diameter that had been solution treated 1 h at 760 °C (1400 °F), furnace cooled, overaged 8 h at 565 °C (1050 °F), and air cooled. Tests were conducted at a stress ratio of $R = 0.1$ and a frequency of 20 Hz. Source: O. Deel, "Engineering Data on New Aerospace Structural Materials," Air Force Materials Laboratory, AFML-TR-77-198, Wright Patterson AFB, 1977

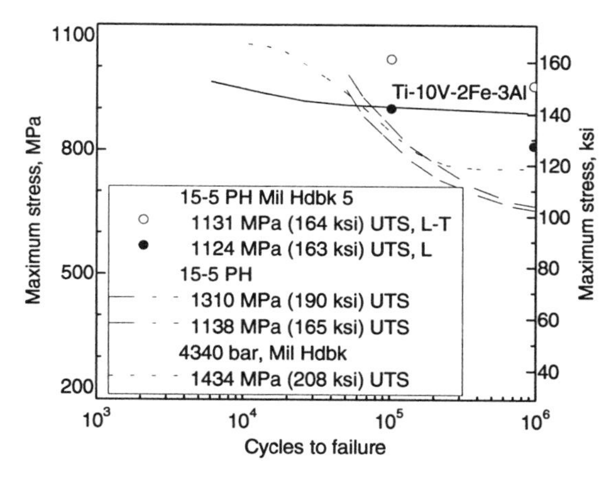

Ti-10V-2Fe-3Al: Comparison of smooth fatigue strengths. Source: *J. of Metals*, March, 1980

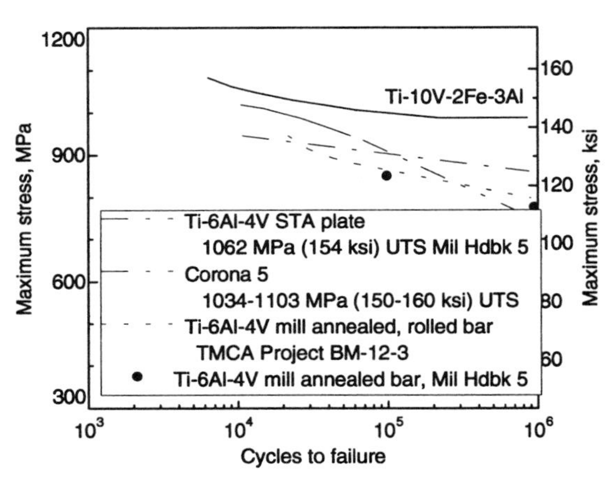

Ti-10V-2Fe-3Al: Comparison of smooth fatigue strengths. Source: *J. of Metals*, March, 1980

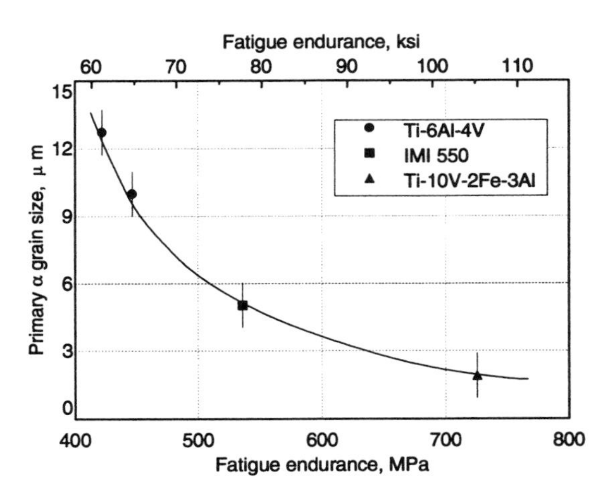

Ti-10V-2Fe-3Al: Fatigue endurance and grain size. Source: D.P. Davies, 7th World Conf on Titanium

Low-Cycle Fatigue

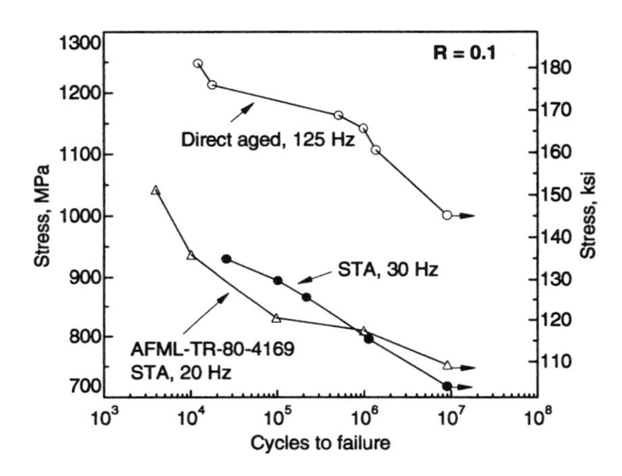

Ti-10V-2Fe-3Al: Fatigue of smooth specimens (1190 MPa UTS). Ti-10V-2Fe-3Al: Solution treated and aged (STA) specimens were taken from β hot die forgings, solution treated at 30 °C (54 °F) below β transus temperature, water quenched, and aged to a strength level of 1190 MPa (175 ksi). Ti-10V-2Fe-3Al direct aged specimens were β hot die forged, post-forge cooled at a rate of 5 °C/s (9 °F/s), and aged to 1190 MPa (172 ksi). Fatigue tests for STA specimens were performed on specimens 3 mm (0.125 in.) in diameter with $K_t = 1$, $R = 0.1$, and frequency of 30 Hz, low stress ground. Fatigue test for direct aged specimens were performed on 3 mm (0.125 in.) diam samples with $K_t = 1$, $R = 0.1$, and frequency of 125 Hz; surfaces were low stress ground and electropolished. Source: G. Kuhlman, A. Chakrabarti, T. Yu, R. Pishko, and G. Terlinde, LCF, Fracture Toughness, and Fatigue/Fatigue Crack Propagation Resistance Optimization in Ti-10V-2Fe-3Al Alloy Though Microstructural Modification, in *Microstructure, Fracture Toughness, and Fatigue Crack Growth Rate in Titanium Alloys,* A. Chakrabarti and J.C. Chesnutt, Ed., TMS/AIME, 1987, p 171

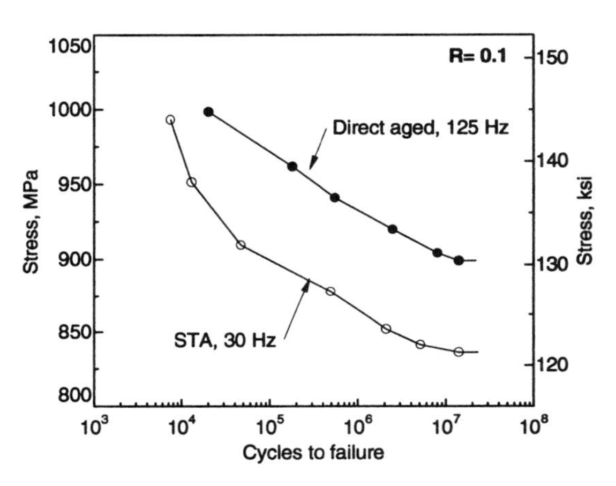

Ti-10V-2Fe-3Al: Fatigue of smooth specimens (965 MPa UTS). Ti-10V-2Fe-3Al solution treated and aged (STA) specimens were taken from β hot die forgings, solution treated at 30 °C (54 °F) below the β transus temperature, water quenched, and aged to a strength level of 965 MPa (140 ksi). Ti-10V-2Fe-3Al direct aged specimens were β hot die forged, post-forge cooled at a ratio of 5 °C/s (9 °F/s), and aged to the desired strength level. Source: G. Kuhlman, A. Chakrabarti, T. Yu, R. Pishko, and G. Terlinde, LCF, Fracture Toughness, and Fatigue/Fatigue Crack Propagation Resistance Optimization in Ti-10V-2Fe-3Al Alloy Through Microstructural Modification, in *Microstructure, Fracture Toughness, and Fatigue Crack Growth Rate in Titanium Alloys,* A. Chakrabarti and J.C. Chesnutt, Ed., TMS/AIME, 1987, p 171

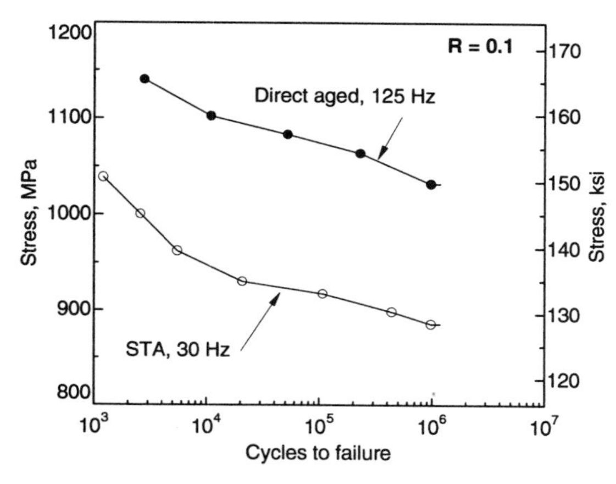

Ti-10V-2Fe-3Al: Fatigue of smooth specimens (1100 MPa UTS). Ti-10V-2Fe-3Al solution treated and aged (STA) specimens were taken from β hot die forgings, solution treated at 30 °C (54 °F) below the β transus temperature, water quenched, and aged to a strength level of 1100 MPa (160 ksi). Ti-10V-2Fe-3Al direct aged specimens were β hot die forged, post-forge cooled at a rate of 5 °C/s (9 °F/s), and aged to the desired strength level. Source: G. Kuhlman, A. Chakrabarti, T. Yu, R. Pishko, and G. Terlinde, LCF, Fracture Toughness, and Fatigue/Fatigue Crack Propagation Resistance Optimization in Ti-10V-2Fe-3Al Alloy Through Microstructural Modification, in *Microstructure, Fracture Toughness, and Fatigue Crack Growth Rate in Titanium Alloys*, A. Chakrabarti and J.C. Chesnutt, Ed., TMS/AIME, 1987, p 171

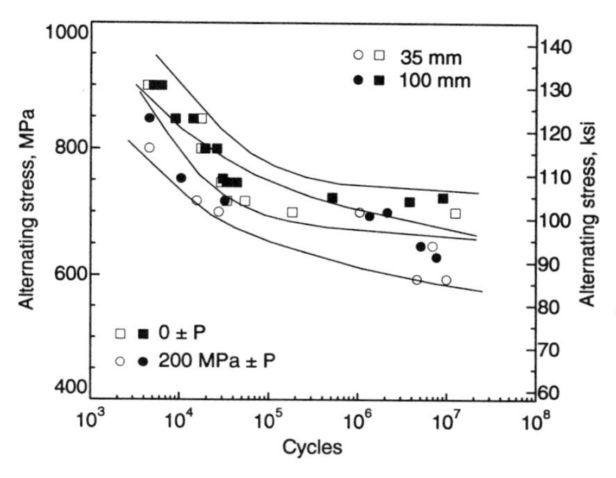

Ti-10V-2Fe-3Al: S/N data at two mean stress levels. The fatigue endurance limit is influenced by the position in the billet, i.e. superior fatigue endurance values were obtained from the outer portion more heavily worked area of the billet ring, although the effect was considered negligible. Source: D.P. Davies, Effect of Heat Treatment on the Mechanical Properties of Ti-10V-2Fe-3Al for Dynamically Critical Helicopter Components, 7th World Conf on Titanium

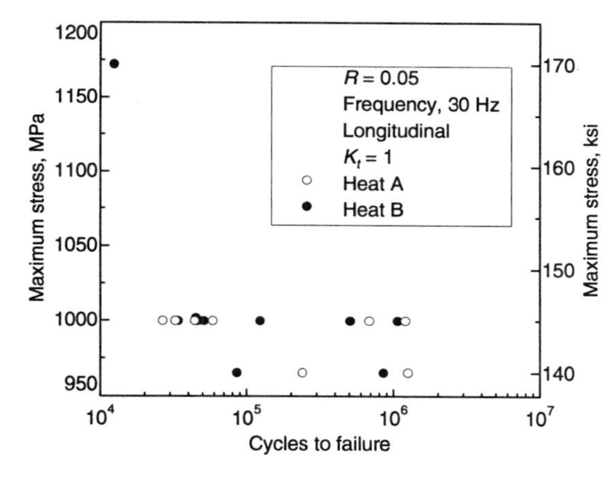

Ti-10V-2Fe-3Al: RT axial fatigue of STA forgings. Boeing 747 lower link fitting, β forged with α-β (≤20%) finish; 775 °C (1435 °F), 2 h, AC + 770 °C (1425 °F), 2 h, WQ + 510 °C (950 °F), 8 h, AC. Source: *Aerospace Structural Metals Handbook*, Vol 4, Code 3726, Battelle, 1972

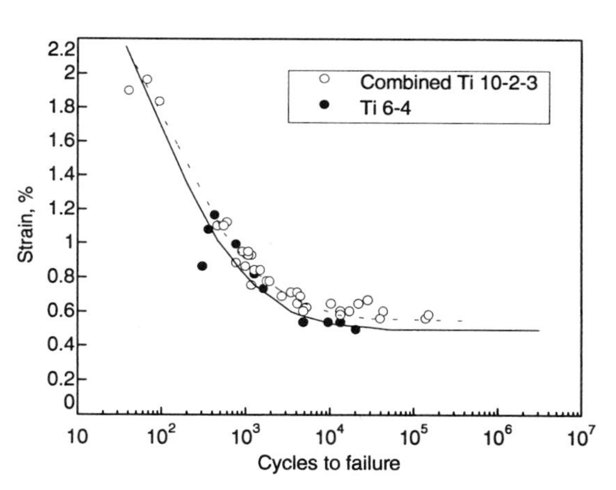

Ti-10V-2Fe-3Al: LCF under strain control. All of the forging heat treat combinations tested in this study cyclically softened. Most of the stress reduction occurred early in the test. For relatively short lives in low-cycle fatigue, the load never completely stabilizes. Ti-10V-2Fe-3Al forgings were processed under four different conditions to an average yield strength of 1103 ± 12 MPa. Low-cycle fatigue testing was performed on a closed loop hydraulic MTS Systems machine according to ASTM E606, "Standard Recommended Practice for Constant Amplitude Low Cycle Fatigue Testing." $R = -1$, and constant strain rate was 0.01/s. Source: R. Carey, R. Boyer, and H. Rosenberg, Fatigue Properties of Ti-10V-2Fe-3Al, in *Titanium, Science and Technology*, Vol 2, G. Lütjering, U. Zwicker, and W. Bunk, Ed., Deutsche Gesellschaft für Metallkunde, e.V., Germany, 1985, p 1261

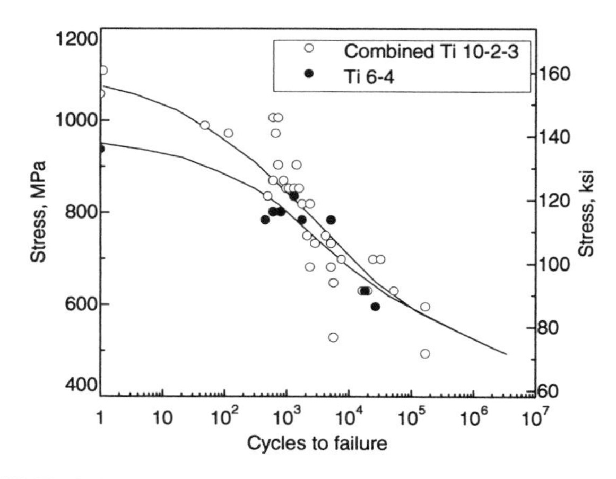

Ti-10V-2Fe-3Al: LCF under load control. Ti-10V-2Fe-3Al forgings were processed under four different conditions to an average yield strength of 1103 ± 12 MPa. Low-cycle fatigue testing was performed on a closed loop hydraulic MTS Systems machine according to ASTM E606, "Standard Recommended Practice for Constant Amplitude Low Cycle Fatigue Testing." $R = -1$. Source: R. Carey, R. Boyer, and H. Rosenberg, Fatigue Properties of Ti-10V-2Fe-3Al, in *Titanium, Science and Technology*, Vol 2, G. Lütjering, U. Zwicker, and W. Bunk, Ed., Deutsche Gesellschaft für Metallkunde, e.V., Germany, 1985, p 1261

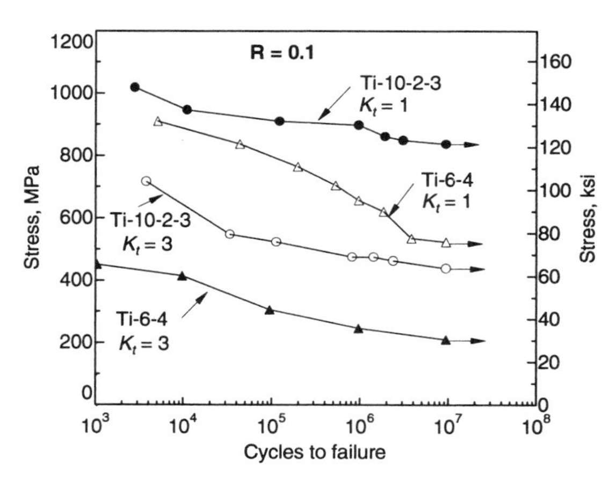

Ti-10V-2Fe-3Al: Notched and smooth fatigue vs Ti-6Al-4V. *S-N* curves for Ti-10V-2Fe-3Al (strength level, 965 MPa, or 140 ksi) and Ti-6Al-4V (strength level, 896 MPa, or 130 ksi). Ti-10V-2Fe-3Al specimens were taken from β hot die forgings, solution treated at 30 °C (54 °F) below the β transus temperature, water quenched, and aged to a strength level of 965 MPa (140 ksi). Ti-6Al-4V isothermal forgings were annealed to a minimum strength level of 896 MPa (130 ksi) with an actual ultimate tensile strength of 1000 MPa (145 ksi). Fatigue tests were performed on specimens 3 mm (0.125 in.) in diameter with $K_t = 1$ or 3, $R = 0.1$, and a frequency of 30 Hz, low stress ground. Source: G. Kuhlman, A. Chakrabarti, T. Yu, R. Pishko, and G. Terlinde, LCF, Fracture Toughness, and Fatigue/Fatigue Crack Propagation Resistance Optimization in Ti-10V-2Fe-3Al Alloy Through Microstructural Modification, in *Microstructure, Fracture Toughness, and Fatigue Crack Growth Rate in Titanium Alloys*, A. Chakrabarti and J.C. Chestnutt, Ed., TMS/AIME, 1987, p 171

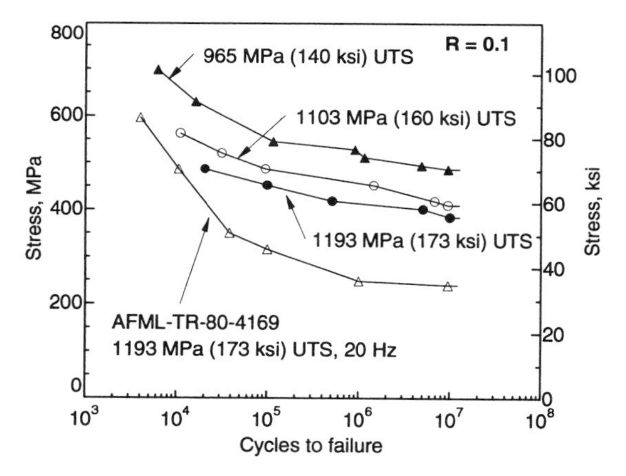

Ti-10V-2Fe-3Al: Fatigue of STA notched ($K_t = 3$) specimens. Ti-10V-2Fe-3Al solution treated and aged (STA) specimens were taken from β hot die forgings, solution treated at 30 °C (54 °F) below the β transus temperature, water quenched, and aged to a strength level of 965 MPa (140 ksi). Fatigue tests for STA specimens were performed on specimens 3 mm (0.125 in.) in diameter with $K_t = 3$, $R = 0.1$, and a frequency of 30 Hz, low stress ground. Source: G. Kuhlman, A. Chakrabarti, T. Yu, R. Pishko, and G. Terlinde, LCF, Fracture Toughness, and Fatigue/Fatigue Crack Propagation Resistance Optimization in Ti-10V-2Fe-3Al Alloy Through Microstructural Modification, in *Microstructure, Fracture Toughness and Fatigue Crack Growth Rate in Titanium Alloys*, A. Chakrabarti and J.C. Chesnutt, Ed., TMS/AIME, 1987, p 171

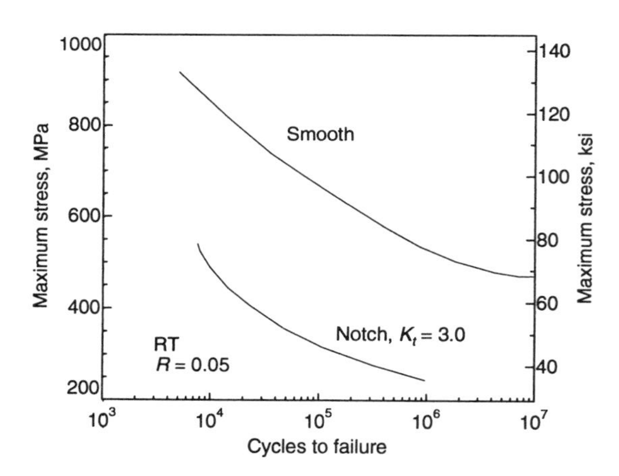

Ti-10V-2Fe-3Al: Smooth and notched fatigue of STA forgings. Heat treatment: 815 °C (1500 °F), 1 h, AC + 620 °C (1150 °F), 8 h, AC. Source: *Aerospace Structural Metals Handbook*, Vol 14, Code 3726, Battelle, 1972

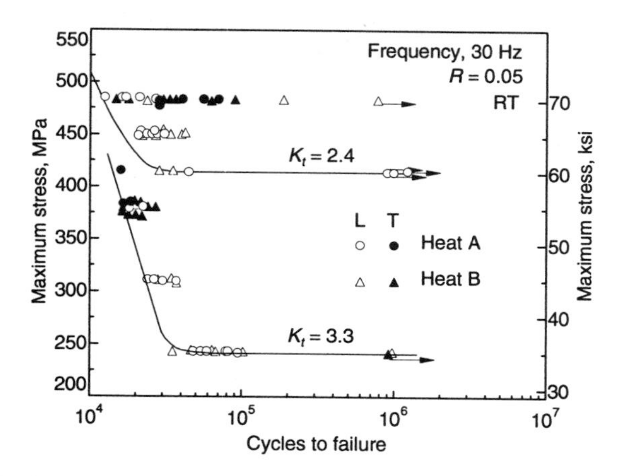

Ti-10V-2Fe-3Al: Notched fatigue of STA forging. Boeing 747 lower link fitting; β forged with α-β (≤20%) finish; 775 °C (1435 °F), 2 h, AC + 770 °C (1425 °F), 2 h, WQ + 510 °C (950 °F), 8 h, AC. Source: *Aerospace Structural Metals Handbook*, Vol 14, Code 3726, Battelle, 1972

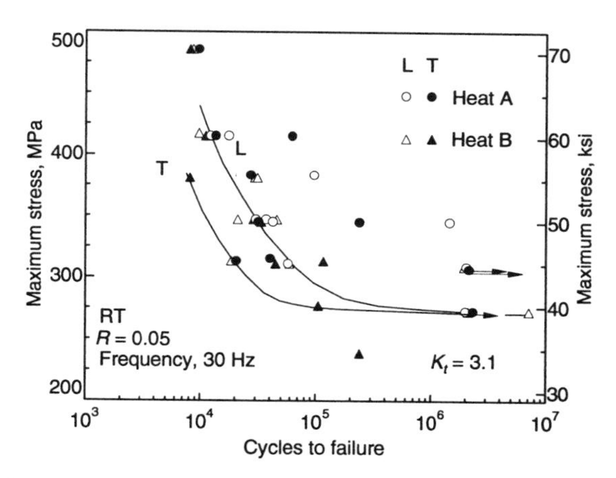

Ti-10V-2Fe-3Al: Fatigue with single-hole notch. Boeing 747 lower link fitting; β forged with α-β (≤20%) finish; 775 °C (1435 °F), 2 h, AC + 770 °C (1425 °F), 2 h, WQ + 510 °C (950 °F), 8 h, AC. Source: *Aerospace Structural Metals Handbook*, Vol 14, Code 3726, Battelle, 1972

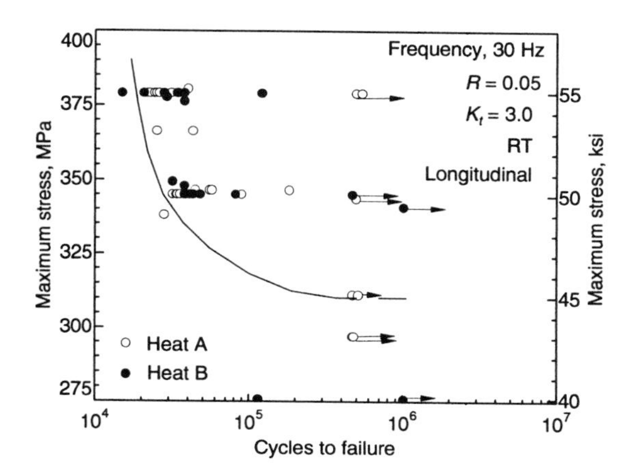

Ti-10V-2Fe-3Al: Fatigue with double-hole notch. Boeing 747 lower link fitting; β forged with α-β (≤20%) finish; 775 °C (1435 °F), 2 h, AC + 1425 °F), 2 h, WQ + 510 °C (950 °F), 8 h, AC. Source: *Aerospace Structural Metals Handbook*, Vol 14, Code 3726, Battelle, 1972

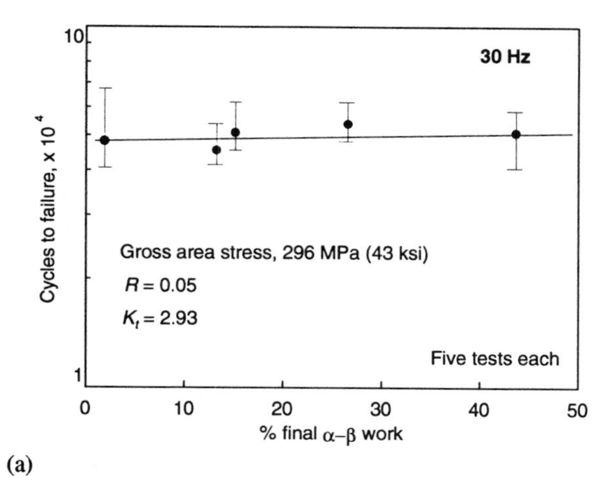

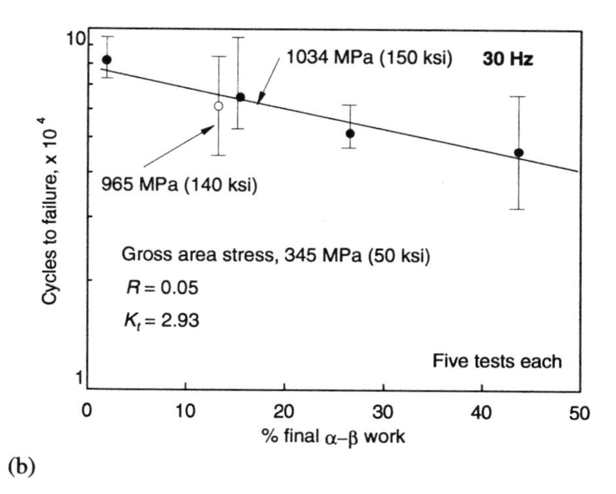

Ti-10V-2Fe-3Al: Notched fatigue performance of forgings. Notched fatigue (cycles to failure) of pancake forgings vs amount of work at (a) strength level of 1310 MPa (190 ksi) and (b) strength level of 865 MPa (140 ksi) and 1034 MPa (150 ksi). Log average lives and scatterband indicated. β transus temperature was 810 °C (1490 °F). Pancake forgings were produced by β forging at temperatures 10 to 25 °C (18 to 45 °F) above the β transus to produce 50 to 70% thickness reduction. Additional reduction of 2 to 58% was accomplished by forging in the α-β range (10 to 25 °C, 18 to 45 °F, below the β transus temperature). Source: R. Boyer and G. Kuhlman, Processing Properties Relationships of Ti-10V-2Fe-3Al, *Metall. Trans. A.*, Vol 18, 1987, p 2095

High-Cycle Notched Fatigue

Room Temperature

The notched fatigue strength at a $K_t = 3$ decreases as the strength level increases. (At all strength levels it is superior to that of Ti-6Al-4V). The drop in fatigue strength as the strength is increased is attributed to increased notch sensitivity at the higher strength levels, leading to earlier crack initiation. Data from several notch geometries are presented. The effect of notch geometry is shown. A round and a flat specimen, with K_t's of 2.4 and 2.5 respectively, show a much larger difference in properties than can be ascribed to the difference in K_t. Microstructure has virtually no effect on the fatigue strength at the high strength level (190 ksi), but does have an influence at lower tensile strength levels for a K_t of 2.93. Higher amounts of α/β work, which result in a more equiaxed primary α, appears to have a negative effect on fatigue strength for the lower strength conditions (140 and 150 ksi).

Elevated Temperature

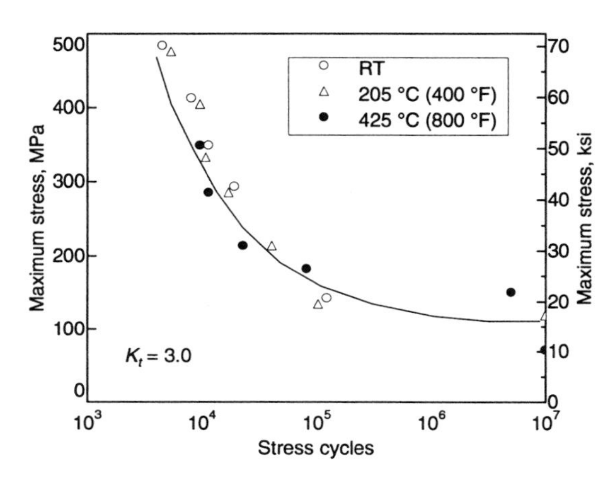

Ti-10V-2Fe-3Al: Fatigue of notched STOA bar. Specimens were taken from round bar 75 mm (3 in.) in diameter, solution treated at 760 °C (1400 °F) for 1 h, furnace cooled, overaged at 565 °C (1050 °F) for 8 h, and air cooled. Fatigue testing performed at $R = 0.1$ and a frequency of 20 Hz. Source: O. Deel, "Engineering Data on New Aerospace Structural Materials," Air Force Materials Laboratory, AFML-TR-77-198, Wright Patterson AFB, 1977

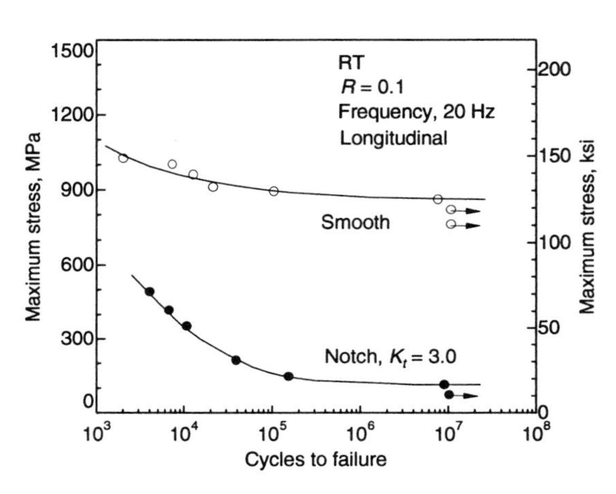

Ti-10V-2Fe-3Al: Smooth and notched fatigue at RT. Solution treated and aged 75 mm (3 in.) diam bar treated at 760 °C (1400 °F), 1 h, FC + 565 °C (1050 °F), 8 h, AC. Source: O. Deel, "Engineering Data on New Aerospace Structural Materials," Air Force Materials Laboratory, AFML-TR-77-198, Wright Patterson AFB, 1977

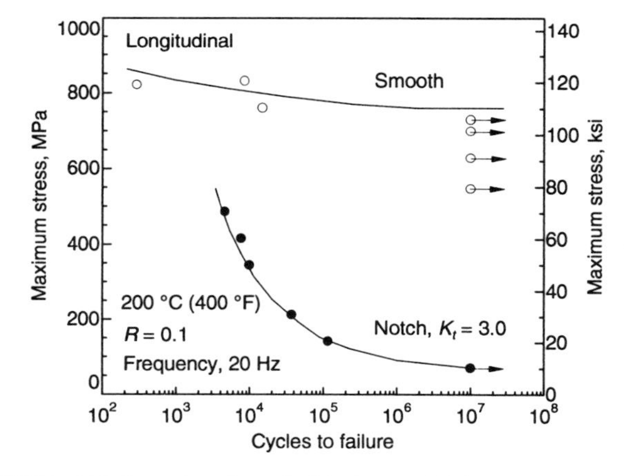

Ti-10V-2Fe-3Al: Smooth and notched fatigue at 200 °C. Solution treated and aged 75 mm (3 in.) diam bar treated at 760 °C (1400 °F), 1 h, FC + 565 °C (1050 °F), 8 h, AC. Source: O. Deel, "Engineering Data on New Aerospace Structural Materials," Air Force Materials Laboratory, AFML-TR-77-198, Wright Patterson AFB, 1977

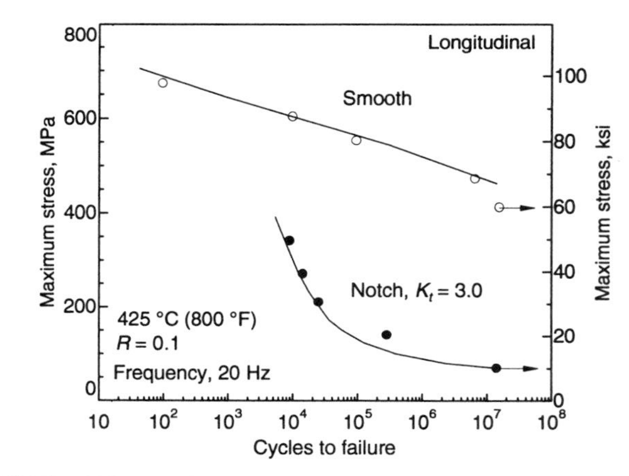

Ti-10V-2Fe-3Al: Smooth and notched fatigue at 425 °C. Solution treated and aged 75 mm (3 in.) diam bar treated at 760 °C (1400 °F), 1 h, FC + 565 °C (1050 °F), 8 h, AC. Source: O. Deel, "Engineering Data on New Aerospace Structural Materials," Air Force Materials Laboratory, AFML-TR-77-198, Wright Patterson AFB, 1977

High-Cycle Fatigue: P/M and Cast

Early work with powder metallurgy compacts indicated a debit in comparison to wrought forgings. Thermomechanical processing can be seen to have an influence. Of the two compaction techniques studied for pre-alloyed powder, the compaction technique does not appear to be important. The blended elemental compact fatigue performance could also be improved by thermomechanical processing and/or the use of low Cl powder.

Ti-10V-2Fe-3Al: Fatigue in notched specimens for several product forms in high-strength and low-strength conditions

Product form	Ultimate tensile strength, MPa	Ultimate tensile strength, ksi	Log average fatigue life
Cast and wrought			
Isothermal forgings	1300-1380	188-200	20 200
	1060-1100	154-159	25 700
Conventional forgings	1230-1350	178-196	50 000
Pancake forgings	1260-1345	183-195	50 000
43% α/β work	1050-1070	152-155	50 000
2% α/β work	1050	152	83 300
Extrusions	1105-1175	160-170	32 500
P/M			
Prealloyed HIP	1345-1415	195-205	16 900
Isothermally forged	1125-1145	163-166	53 300

Note: Test frequency was 30 Hz and $R = 0.05$, with $K_t = 2.93$ and tests performed at stress level of 345 MPa (50 ksi). Source: R. Boyer, D. Eylon, and F. Froes, Comparative Evaluation of Ti-10V-2Fe-3Al Cast, P/M and Wrought Product Forms, Titanium, *Science and Technology*, Vol 2, G. Lütjering, U. Zwicker, and W. Bunk, Ed., Deutsche Gesellschaft für Metallkunde, e.V. Germany, 1985, p 1307

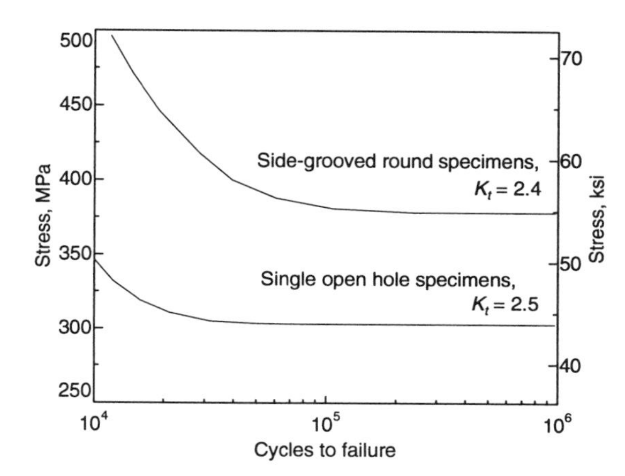

Ti-10V-2Fe-3Al: Effect of notch geometry on fatigue strength. Forgings were heat treated to a strength level of 1241 MPa (180 ksi). Flat sheet-type specimens with holes drilled to a notch factor $K_t = 2.5$ and round side-grooved specimens with $K_t = 2.4$ were used. Heat treating and machining sequences were the same in both cases. Source: R. Carey, R. Boyer, and H. Rosenberg, Fatigue Properties of Ti-10V-2Fe-3Al, in *Titanium, Science and Technology*, Vol 2, G. Lütjering, U. Zwicker, and W. Bunk, Ed., Deutsche Gesellschaft für Metallkunde, e.V., Germany, 1985, p 1261

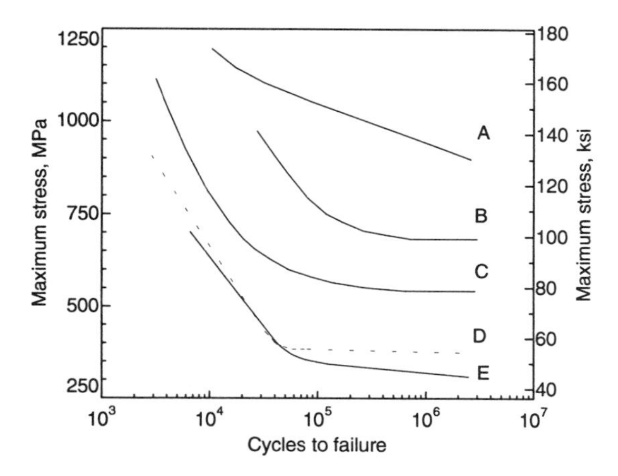

Ti-10V-2Fe-3Al: Fatigue of cast and wrought specimens. Smooth specimen fatigue for (A) cast and wrought plus isothermal forge, ultimate tensile strength = 1300 to 1380 MPa (188 to 200 ksi); (B) prealloyed P/M HIP plus isothermal forge, ultimate tensile strength = 1345 to 1400 MPa (195 to 203 ksi); (C) prealloyed HIP, ultimate tensile strength = 1310 MPa (190 ksi); (D) P/S + HIP, ultimate tensile strength 1228 to 1275 MPa (178 to 185 ksi); (E) P/S, ultimate tensile strength = 1195 MPa (173 ksi). Curves represent data lower limits. Test frequency was 5 Hz and $R = 0.1$. Source: R. Boyer, D. Eylon, and F. Froes, Comparative Evaluation of Ti-10V-2Fe-3Al Cast, P/M and Wrought Product Forms, in *Titanium, Science and Technology*, Vol 2, G. Lütjering, U. Zwicker, and W. Bunk, Ed., Deutsche Gesellschaft für Metallkunde, e.V., Germany, 1985, p 307

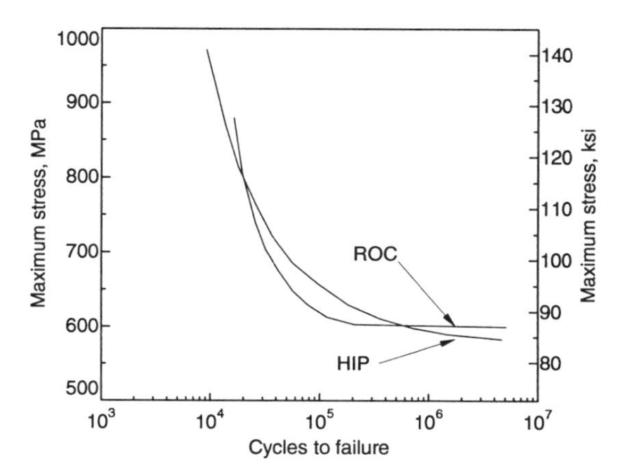

Ti-10V-2Fe-3Al: Fatigue in powder compacts. *S-N* curve for powder compacts consolidated by rapid omnidirectional compaction (ROC) by hot isostatic pressing. Both specimens were heat treated. Chemical composition of the alloy was 3.0 wt% Al, 0.065 wt% C, 2.1 wt% Fe, 0.0063 wt% H, 0.0093 wt% N, 0.1485 wt% O, 9,2 wt% V, and 0.006 wt% W. Processing parameters for consolidation (ROC) were 775 °C (1425 °F) at 830 MPa (120 ksi), 1/2-s dwell, air cool. Processing parameters for HIP were 790 °C (1450 °F) at 103 MPa (15 ksi) for 20 h. Heat treatment was carried out at 745 °C (1365 °F) for 1 h, water quench, and 490 °C (915 °F) for 8 h, air cool. Fatigue tests were performed at room temperature on a servo-hydraulic MTS machine. Constant load triangular waveform cycling was done at $R = 0.1$ and a frequency of 5 Hz. Source: Y. Mahajan, D. Eylon, C. Kelto, and F. Froes, Evaluation of Ti-10V-2Fe-3Al Powder Compacts Produced by the ROC Method, *Metal Powder Rep.*, Oct 1986, p 749

Fatigue Crack Growth

Using conventional processing and heat treatments, the crack growth rate (da/dN) of this alloy is essentially independent of microstructure, strength level and test environment, and, in air, is similar to that of mill annealed Ti-6Al-4V. Aging to produce the omega phase significantly reduced da/dN, but, as mentioned previously, this is not a practical microstructure to use.

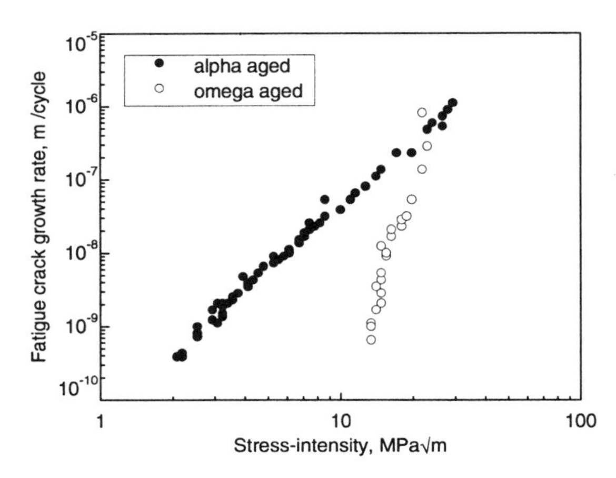

Ti-10V-2Fe-3Al: Crack growth in two aged conditions. Source: T.W. Duerig and J.C. Williams, Overview: Microstructure and Properties of Beta Titanium Alloys, in *Beta Titanium Alloys in the 1980's*, R. Boyer and H. Rosenberg, Ed., TMS/AIME, 1984, p 44

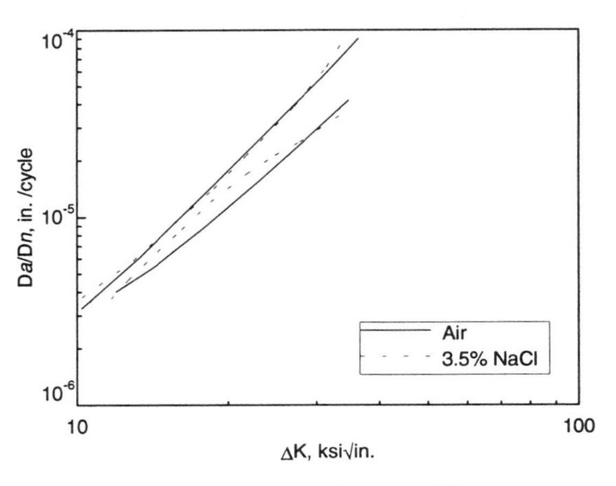

Ti-10V-2Fe-3Al: Crack growth in air and 3.5% NaCl. Superimposed air and 3.5% NaCl fatigue crack propagation rate scatterbands for Ti-10V-2Fe-3Al, R = 0.05, frequency 1-30 Hz, various orientations. Source: R. Boyer, WestTech, 1981

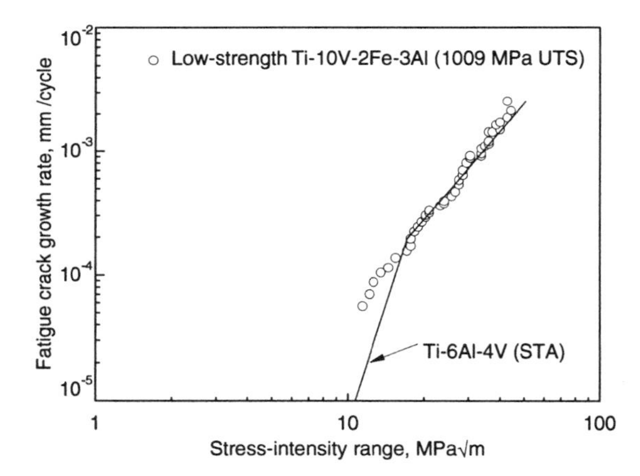

Ti-10V-2Fe-3Al: FCG with low aspect ratio of primary α. Chemical composition of the alloy was 3.2 wt% Al, 0.03 wt% C, 1.8 wt% Fe, 0.005 wt% H, 0.01 wt% N, 0.087 wt% O, and 9.7 wt% V. Material was β forged then α-β worked to effect an additional 55% reduction, followed by heat treatment at 750 °C (1380 °F) for 2 h, and water quench, aged at 550 °C (1020 °F) for 8 h. Ultimate tensile strength was 1009 MPa (146 ksi). Tests were performed in air at room temperature, with Haversine waveform and R = 0.10. Source: G. Yoder, L. Cooley, and R. Boyer, Microstructure/Crack Tolerance Aspects of Notched Fatigue Life in Ti-10V-2Fe-3Al Alloy, in *Microstructure, Fracture Toughness and Fatigue Crack Growth Rate in Titanium Alloys*, A. Chakrabarti and J.C. Chesnutt, Ed., TMS/AIME, 1987, p 209

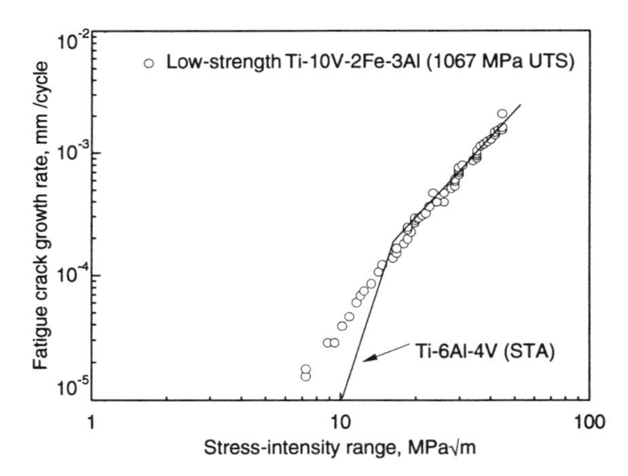

Ti-10V-2Fe-3Al: FCG with high aspect ratio of primary α. Χηεμιχαλ χομποσιτιον οφ τηε αλλοψ ωασ 3.2 ωτ% Αλ, 0.03 ωτ% Χ, 1.8 ωτ% Φε, 0.05 ωτ% Η, 0.01 ωτ% Ν, 0.087 ωτ% Ο, ανδ 9.7 ωτ% ς. Ματεριαλ ωασ β forged then α-β worked to effect an additional 2% reduction, followed by heat treatment at 750 °C (1380 °F) for 2 h, and water quench, aged at 550 °C (1020 °F) for 8 h. Ultimate tensile strength was 1067 MPa (154 ksi). Tests were performed in air at room temperature, with Haversine waveform and R = 0.10. Source: G. Yoder, L. Cooley, and R. Boyer, Microstructure/Crack Tolerance Aspects of Notched Fatigue Life in Ti-10V-2Fe-3Al Alloy, in *Microstructure ,Fracture Toughness and Fatigue Crack Growth Rate in Titanium Alloys*, A. Chakrabarti and J.C. Chesnutt, Ed., TMS/AIME, 1987, p 209

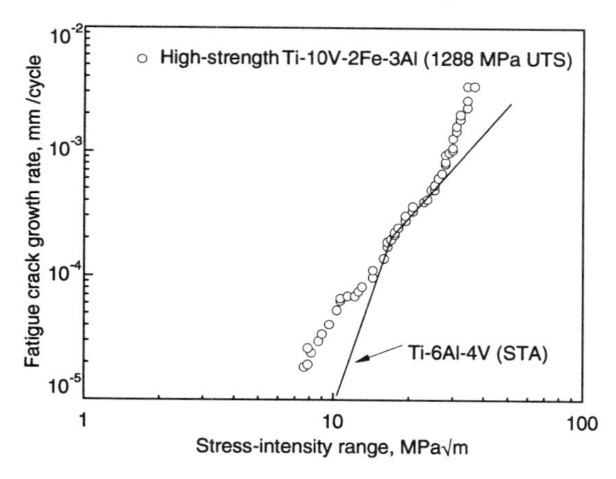

Ti-10V-2Fe-3Al: FCG with high aspect ratio of primary α. Chemical composition of the alloy was 3.2 wt% Al, 0.03 wt% C, 1.8 wt% Fe, 0.05 wt% H, 0.01 wt% N, 0.087 wt% O, and 9.7 wt% V. Material was β forged then α-β worked to effect an additional 2% reduction, followed by heat treatment at 750 °C (1380 °F) for 2 h, and water quench, aged at 495 °C (255 °F) for 8 h. Ultimate tensile strength was 1288 MPa (187 ksi). Tests were performed in air at room temperature, with Haversine waveform and $R = 0.10$. Source: G. Yoder, L. Cooley, and R. Boyer, Microstructure/Crack Tolerance Aspects of Notched Fatigue Life in Ti-10V-2Fe-3Al Alloy, in *Microstructure, Fracture Toughness and Fatigue Crack Growth Rate in Titanium Alloys*, A. Chakrabarti and J.C. Chesnutt, Ed., TMS/AIME, 1987, p 209

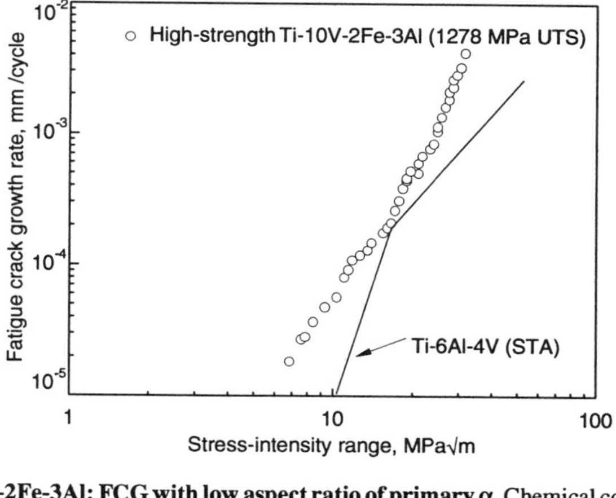

Ti-10V-2Fe-3Al: FCG with low aspect ratio of primary α. Chemical composition was 3.2 wt% Al, 0.03 wt% C, 1.8 wt% Fe, 0.005 wt% H, 0.01 wt% N, 0.087 wt% O, and 9.7 wt% V. Material was β forged then α-β worked to effect an additional 55% reduction, followed by heat treatment at 750 °C (1380 °F) for 2 h, and water quench, aged at 495 °C (255 °F) for 8 h. Ultimate tensile strength was 1278 MPa (185 ksi). Tests were performed in air at room temperature, with Haversine waveform and $R = 0.10$. Source: G. Yoder, L. Cooley, and R. Boyer, Microstructure/Crack Tolerance Aspects of Notched Fatigue Life in Ti-10V-2Fe-3Al Alloy, in *Microstructure, Fracture Toughness and Fatigue Crack Growth Rate in Titanium Alloys*, A. Chakrabarti and J.C. Chesnutt, Ed., TMS/AIME, 1987, p 209

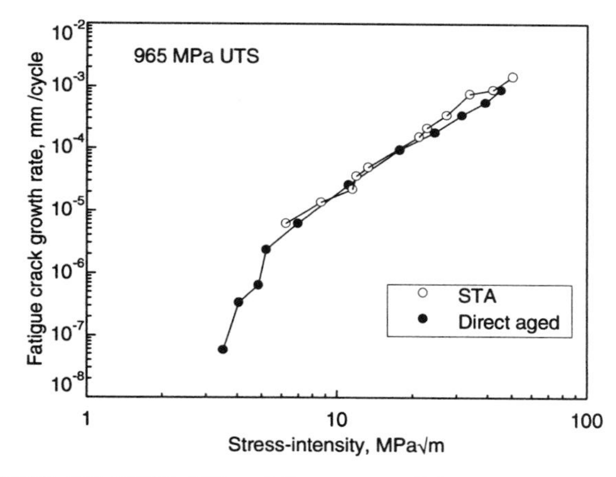

Ti-10V-2Fe-3Al: FCG in STA and direct age conditions. Ti-10V-2Fe-3Al STA specimens were taken from β hot die forgings solution treated at 30 °C (54 °F) below β transus temperature, water quenched, and aged to a strength level of 965 MPa (140 ksi). Ti-10V-2Fe-3Al direct aged specimens were β hot die forged, post-forge cooled at a rate of 5 °C/s (9 °F/s), and aged to the desired strength level. Fatigue crack propagation tests for STA specimens were performed on specimens 6 mm (0.25 in) thick and 37 mm (1.25 in.) in length and width with $R = 0.1$ and a frequency of 30 Hz, compact tension. Fatigue crack propagation tests for direct aged specimens were performed according to ASTM E606 on 6 mm (0.25 in.) diameter specimens, low stress ground, triangular waveform, 20 cycles/min, $R = 0$, and $A = 1.0$, with a frequency of 50 Hz, constant strain. Source: G. Kuhlman, A. Chakrabarti, T. Yu, R. Pishko, and G. Terlinde, LCF, Fracture Toughness and Fatigue/Fatigue Crack Propagation Resistance Optimization in Ti-10V-2Fe-3Al Alloy Through Microstructural Modification, in *Microstructure, Fracture Toughness, and Fatigue Crack Growth Rate in Titanium Alloys*, A. Chakrabarti and J.C. Chesnutt, Ed., TMS/AIME, 1987, p 171

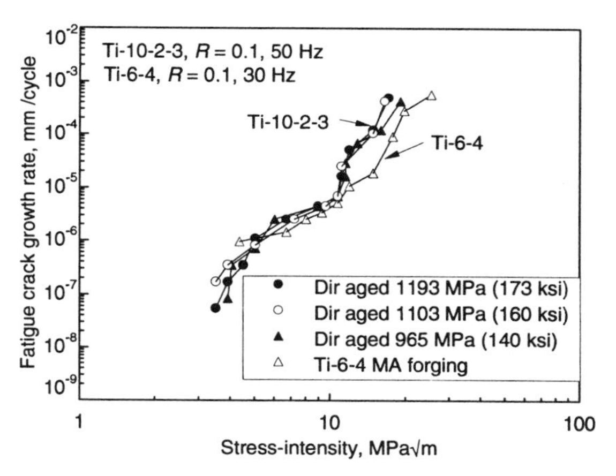

Ti-10V-2Fe-3Al: FCG in direct age condition. Ti-10V-2Fe-3Al direct aged specimens were β hot die forged, post-forge cooled at a rate of 5 °C/s (9 °F/s), and aged to the desired strength level. Fatigue crack propagation tests for direct aged specimens were performed according to ASTM E606 on 6 mm (0.25 in.) diam specimens, low stress ground, triangular waveform, 20 cycles/min, $R = 0$, $A = 1.0$, constant strain. Source: G. Kuhlman, A. Chakrabarti, T. Yu, R. Pishko, and G. Terlinde, LCF, Fracture Toughness, and Fatigue/Fatigue Crack Propagation Resistance Optimization in Ti-10V-2Fe-3Al Alloy Through Microstructural Modification, in *Microstructure Fracture Toughness, and Fatigue Crack Growth Rate in Titanium Alloys*, A. Chakrabarti and J.C. Chesnutt, Ed., TMS/AIME, 1987, p 171

Fracture Toughness

The fracture toughness is strongly dependent on the tensile strength and the microstructure as reported by several authors. The processing, in terms of the amount of α/β work affects the toughness by modification of the morphology of the primary α. Higher amounts of α/β work, following primary working in the β-phase field, changes the primary α to a more globular morphology, which improves ductility at the expense of toughness. There would also appear to be an optimum amount of primary α to achieve maximum toughness (a 10% volume fraction of elongated primary α had significantly higher fracture toughness than 30 vol.%). There seems to be a lot of variation in the toughness reported for powder compacts. There is some evidence that the fracture toughness is related to the volume fraction of defects in P/M products.

Stress Corrosion Resistance. The stress corrosion threshold has been reported to be at least 80% of K_{Ic} except when it is stressed in the short transverse direction, where it is 70% of K_{Ic}.

Ti-10V-2Fe-3Al: Room temperature Charpy impact toughness of STOA bar

Direction	Impact toughness, J	Impact toughness, ft · lb
Longitudinal	35.9	26.5
	40.7	30.0
	40.7	30.0
Average	39.1	28.9
Transverse	27.8	20.5
	26.5	19.5
	23.1	17.0
Average	25.8	19.0

Source: AFML-TR-78-114

Ti-10V-2Fe-3Al: Fracture toughness for several product forms

Product form	Ultimate tensile strength, MPa	Ultimate tensile strength, ksi	Tensile yield strength, MPa	Tensile yield strength, ksi	Elongation, %	Plane-strain fracture toughness, MPa√m	Plane-strain fracture toughness, ksi√in.
High strength condition							
Isothermal forgings	1300-1380	188-200	1200-1255	174-182	3-6	29	26
Conventional forgings	1230-1350	178-196	1145-1280	166-186	4-10	44-60	40-54
Pancake forgings	1275-1310	185-190	1150-1160	167-168	5-8	47	43
Extrusions	1240	179	1170	169	4	...	...
P/M high strength							
Prealloyed, HIP	1310	190	1205	175	9	...	...
Prealloyed, HIP + isothermal forge	1345-1400	195-203	1240-1305	179-189	6-8	28	25
P/S	1195	173	1110	161	3.5	...	...
P/S + HIP	1228-1275	178-185	1185-1245	172-180	7-9	28-29	25-26
Reduced strength condition							
Isothermal forgings	1060-1100	153-159	985-1060	143-153	8-12	70	64
Pancake forgings	965	140	930	135	16	100	91
Extrusions	1110-1170	161-169	1000-1105	145-160	6-7	45-48	41-44
P/M Prealloyed HIP + isothermal forge	1125-1145	163-166	1050-1090	152-158	13-15	55	50
P/M, P/S + HIP	1120-1160	162-168	1070-1105	155-160	9-10	32	29
Castings	1105-1130	160-164	1010-1030	146-149	6-10	...	...
AMS specification (forgings)							
AMS 4984	1190	173	1100	160	4 (in 4D)	44	40
AMS 4986	1100	160	1000	145	6 (in 4D)	60	55
AMS 4987	965	140	895	130	8 (in 4D)	88	80

Source: R. Boyer, D. Eylon, and F. Froes, Comparative Evaluation of Ti-10V-2Fe-3Al Cast, P/M, and Wrought Product Forms, *Titanium Science and Technology*, Vol 2, G. Lütjering, U. Zwicker, and W. Bunk, Ed., Deutsche Gesellschaft für Metallkunde e.V., Germany, 1985 , p 1307

Ti-10V-2Fe-3Al: Typical α / β forged room-temperature tensile properties and fracture toughness of forgings

Forging thickness, mm	Forging thickness, in.	Orientation/ location	Ultimate yield strength (0.2% offset), MPa	Ultimate yield strength (0.2% offset), ksi	Ultimate tensile strength, MPa	Ultimate tensile strength, ksi	Elongation, %	Reduction of area, %	Plane-strain fracture toughness, MPa√m	Plane-strain fracture toughness, ksi√in.
75	3	L/S, MC, C	1256-1263	182-183	1318-1325	191-192	9-11	32-34	40.2	36.58
		LT/S, MS, C	1270-1283	184-186	1332-1339	193-194	8-9	20-30	39.9	36.26
		ST/S, MS, C	1214-1311	176-190	1283-1380	186-200	5-9	12-34	43.2	39.26
		Range	1256-1311	182-190	1283-1380	186-200	5-11	12-34	39~43	36~39
50	2	L/S, MS, C	1249-1256	181-182	1325-1311	190-192	8-11	27-35	35.4	32.22
		LT/S, MS, C	1270-1325	184-192	1346-1394	195-202	5-8	12-27	35.1	31.88
		ST/S, MS, C	1173-1194	170-173	1194-1242	173-180	13-14	46-59	...	...
		Range	1173-1325	170-192	1228-1394	178-202	5-14	12-59	~35	~32
25	1	L/S, MS, C	1256-1283	182-186	1339-1342	194-196	5-9	10-25	30.1	27.32
		LT/S, MS, C,	1221-1241	177-176	1270-1270	184-184	10-13	36-48	30.9	28.05
		Range	1214-1256	176-186	1270-1352	184-196	5-13	10-48	30~31	27-28

Note: L, longitudinal; LT, long transverse; ST, short transverse; S, surface; MS, midsurface; C, center. α + β forging was conducted at 760 °C (1400 °F) with about 60% deformation, followed by hand forging. The alloy was double solution treated and aged. The first solution treatment was performed close to, but below, the beta transus (788 to 802 °C, or 1450 to 1480 °F), followed by a slow cool. The second solution treatment took place at a temperature lower than the first, followed by water quench. Source: C. Chen and R. Boyer, Practical Considerations for Manufacturing High-Strength Ti-10V-2Fe-3Al Alloy Forgings, *J. Metals*, July 1979, p 33

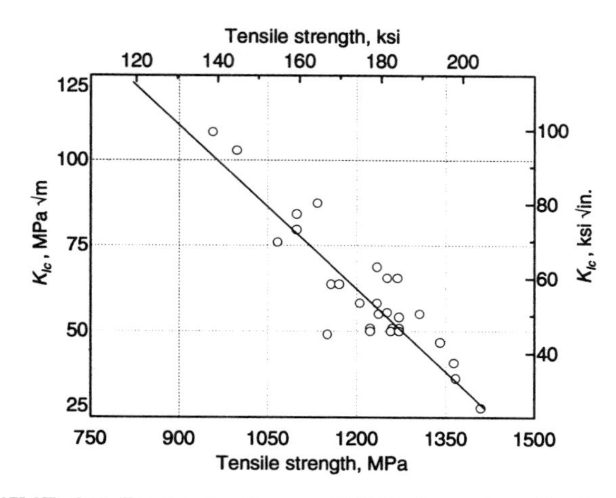

Ti-10V-2Fe-3Al: Fracture toughness vs UTS. Testing performed in air on compact tension specimens according to ASTM E399. Source: G. Kuhlman, "Alcoa Titanium Alloy Ti-10V-2Fe-3Al Forgings, Alcoa Green Letter No. 224," Aluminum Company of American, Forging Division, Cleveland, Aug 1987. With permission

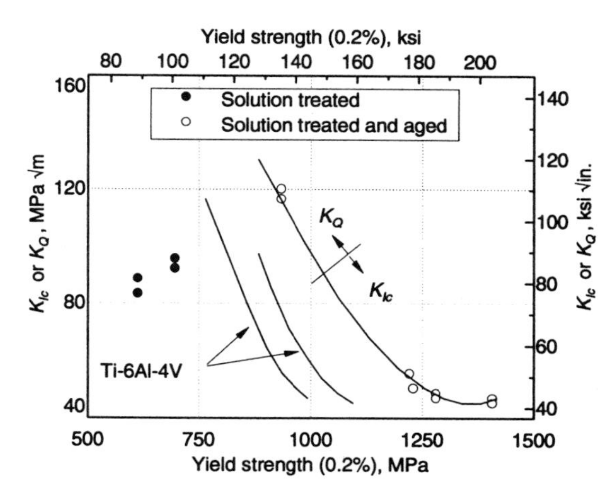

Ti-10V-2Fe-3Al: Fracture toughness vs yield strength. Source: *The Sumitomo Search*, No. 35, Nov 1987, p 21-28

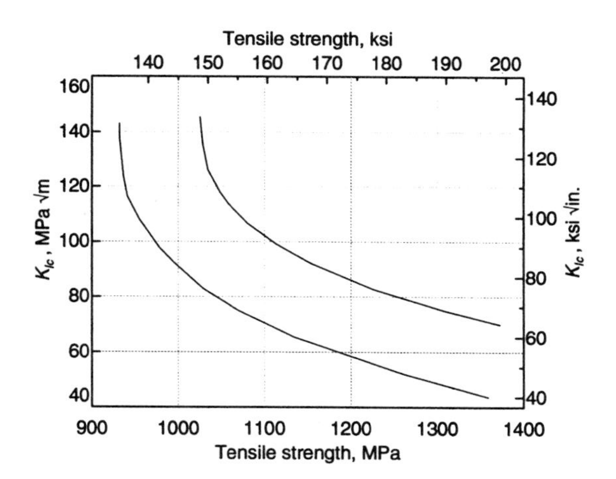

Ti-10V-2Fe-3Al: Plane-strain fracture toughness vs UTS. Data represent a composite of fracture toughness values for β forged die forgings, β forged block forgings, and β forged plus α-β forged die forgings of Ti-10V-2Fe-3Al. Source: *Metals Handbook, Properties and Selection: Stainless Steels, Tool Materials, and Special-Purpose Materials*, Vol 3, 9th ed., American Society for Metals, 1980

Effect of Microstructure and Processing

Ti-10V-2Fe-3Al: Fracture toughness of forgings with different aspect ratios of primary α

Aspect ratio(a)	Tensile yield strength (0.2% offset) MPa	ksi	Ultimate tensile strength MPa	ksi	Elongation, %	Reduction of area, %	Plane-strain fracture toughness MPa√m	ksi√in.
HAR	1149	166.6	1288	186.8	3.0	3.3	46.8	42.6
LAR	1190	172.6	1278	185.4	7.0	12.8	37.0	33.7
HAR	1002	145.4	1067	154.8	9.0	24.3	86.1	78.4
LAR	990	143.6	1009	146.4	15.0	50.7	67.0	61.0

Note: Alloy was forged above the β transus for thickness reduction of 50 to 70%, followed by α-β range forging and heat treatment to indicated strengths. (a) HAR, high aspect ratio; LAR, low aspect ratio of primary alpha. Source: G. Yoder, L. Cooley, and R. Boyer, Microstructure/Crack Tolerance Aspects of Notched Fatigue Life in Ti-10V-2Fe-3Al Alloy, Microstructure, Fracture Toughness, *and Fatigue Crack Growth Rate in Titanium Alloys*, A. Chakrabarti and J.C. Chesnutt, Ed., TMS/AIME, 1987, p 209

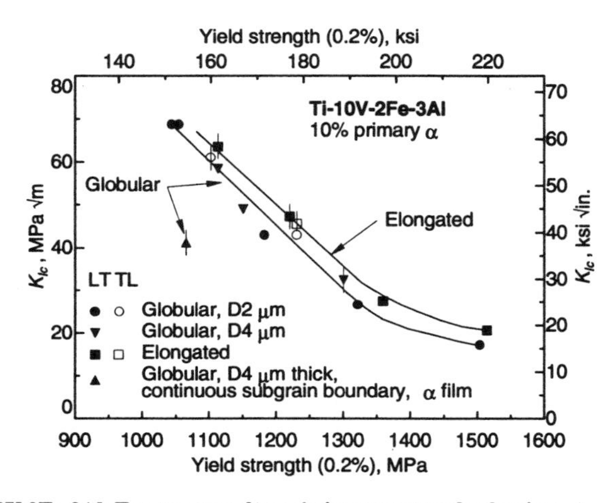

Ti-10V-2Fe-3Al: Fracture toughness/microstructure for forgings. Source: G. Terlinde, H.-J. Rathjen, and K.H. Schwalbe, Microstructure and Fracture Toughness of the Aged β-Ti Alloy Ti-10V-2Fe-3Al, *Metall. Trans. A*, Vol 19, 1988, p 1037

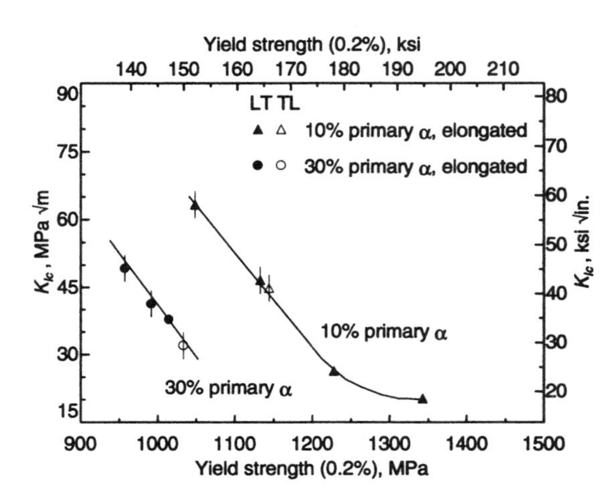

Ti-10V-2Fe-3Al: Effect of elongated α on toughness. Source: *Metall. Trans. A*, Vol 19, 1988, p 1041

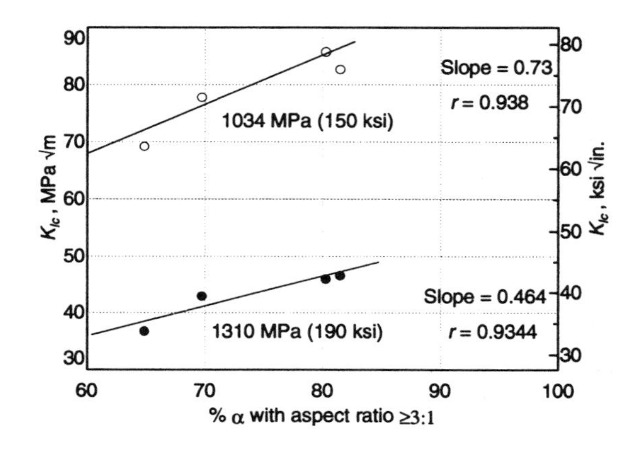

(a)

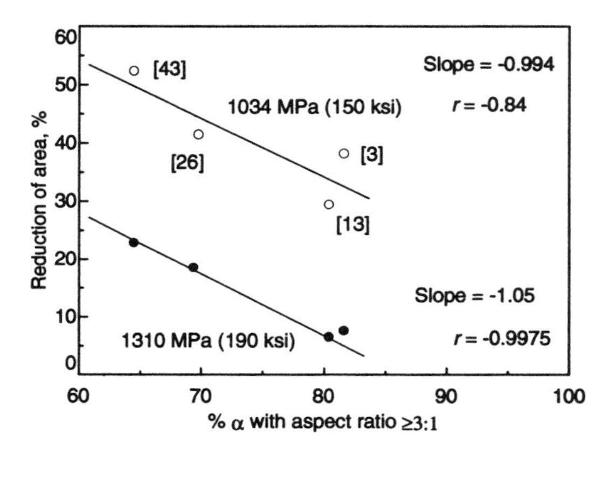

(b)

Ti-10V-2Fe-3Al: Effect of α morphology on toughness/ductility. (a) Toughness. (b) Reduction of area. Numbers in brackets indicate amount of final α-β work. Source: R. Boyer and G. Kuhlman, Processing Properties Relationships of Ti-10V-2Fe-3Al, *Metall. Trans. A*, Vol 18, 1987, p 2095-2103

Effect of Processing

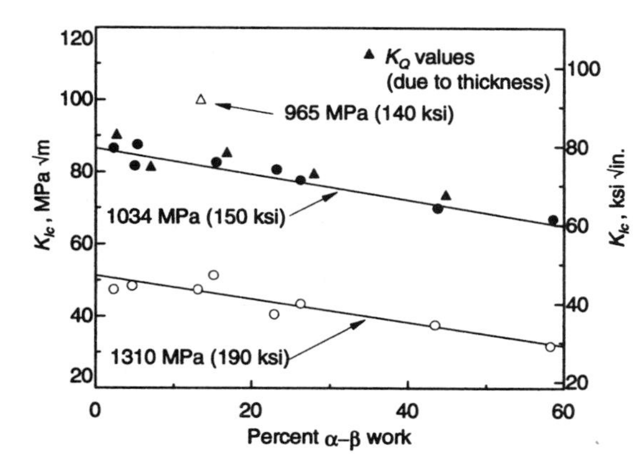

Ti-10V-2Fe-3Al: Fracture toughness of forgings vs final working. Alloy from two heats. No difference in behavior was observed for the two heats. Material was β forged at the β transus temperature plus 10 to 25 °C (18 to 45 °F) for a 50-70% reduction, followed by α-β range forging at the β transus temperature minus 10 to 25 °C (18 to 45 °F). Specimens were solution treated at 750 °C (1380 °F) for 2 h, followed by aging for 8 h at 495 °C (920 °F) for the high-strength condition, and as indicated for the low-strength condition. Source: R. Boyer and G. Kuhlman, Processing Properties Relationships of Ti-10V-2Fe-3Al, *Metall. Trans. A*, Vol 18, 1987, p 2095

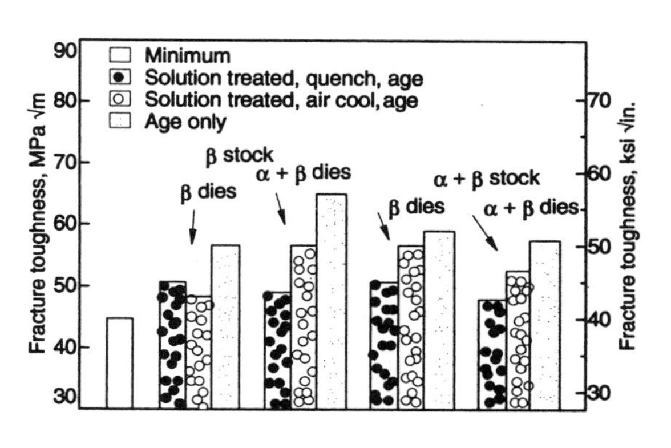

Ti-10V-2Fe-3Al: Fracture toughness vs forging/heat treatment. Source: R. Pishko, T. Yu, and G. Kuhlman, Precision Forging of Titanium Alloy, in *Titanium 1986 Products and Applications*, Titanium Development Association, 1987, p 376

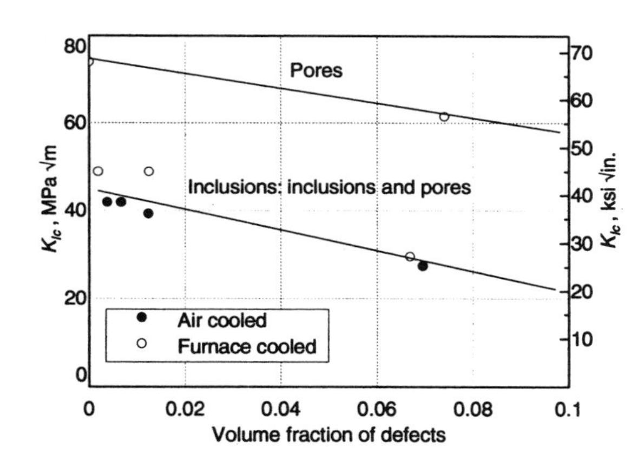

Ti-10V-2Fe-3Al: Toughness vs defect content. Source: N. Moody *et al.*, The Role of Inclusion and Pore Content on the Fracture Toughness of Powder Processed Blended Elemental Ti-10V-2Fe-3Al, in *Microstructure, Fracture Toughness, and Fatigue Crack Growth Rate in Titanium Alloys*, A. Chakrabarti and J.C. Chesnutt, Ed., TMS/AIME, 1987, p 83

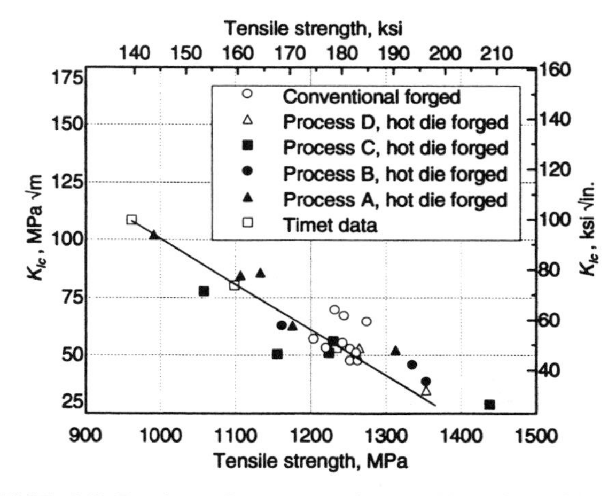

Ti-10V-2Fe-3Al: Toughness from conventional and hot die forging. Source: R. Pishko, T. Yu, and G. Kuhlman, Precision Forging of Titanium Alloy, in *Titanium 1986 Products and Applications*, Titanium Development Association, 1987, p 401

Ti-10V-2Fe-3Al: Comparison of fracture toughness of powder compacts vs wrought alloys

	Powder compact				Wrought alloy			
	Plane-strain fracture toughness		Tensile yield strength		Plane-strain fracture toughness		Tensile yield strength	
Alloy	MPa$\sqrt{m}$	ksi$\sqrt{in.}$	MPa	ksi	MPa$\sqrt{m}$	ksi$\sqrt{in.}$	MPa	ksi
Ti-10V-2Fe-3Al	51	46	854	124	60	55	1100	160
Ti-6Al-6V-2Sn	44	40	910	132	65	59	965	140
Ti-6Al-2Sn-4Zr-6Mo	31	28	1068	155	33	30	1100	160
Ti-6Al-4V	35	32	900	130	66	60	860	125

Note: Compositions of cold isopressed and vacuum sintered billets were as follows. Ti-10-2-3: 2.98 wt% Al, 0.010 wt% C, 2.03 wt% Fe, 0.0008 wt% H, 0.011 wt% N, 0.13 wt% O, 10.25 wt% V. Ti-6-6-2: 6.00 wt% Al, 0.019 wt% C, 0.55 wt% Cu, 0.48 wt% Fe, 0.0016 wt% H, 0.0045 wt% N, 0.27 wt% O, 1.65 wt% Sn, 5.31 wt% V. Ti-6-2-4-6: 6.03 wt% Al, 0.024 wt% C, 0.060 wt% Fe, 0.0009 wt% H, 5.60 wt% Mo, 0.0055 wt% N, 0.32 wt% O, 2.20 wt% Sn, 3.60 wt% Zr. Source: Data for Ti-6-4 powder compact as taken from P. Anderson *et al.*, *Modern Developments in Powder Metallurgy*, Vol 13, 1981. Data for wrought alloys taken from AMS specifications, except Ti-10-2-3 (TIMET data). J. Smugeresky and N. Moody, Properties of High Strength Blended Elemental Powder Metallurgy Titanium Alloys, *Titanium Net Shape Technologies*, F. Froes and D. Eylon, Ed., TMS/AIME, 1984, p 131

Ti-10V-2Fe-3Al: Fracture toughness of powder compacts

		Volume fraction	Tensile yield strength		Ultimate tensile strength		Reduction of area,	Strain to fracture,	Plane-strain fracture toughness	
Condition(a)	Defects	of defects	MPa	ksi	MPa	ksi	%	%	MPa$\sqrt{m}$	ksi$\sqrt{in.}$
~0.15 wt% Cl, 0.38 wt% O										
As sintered, FC	Inclusions, pores	0.067	883	128	966	140	2.1	2.1	29.7	32.6
β annealed, AC	Inclusions, pores	0.069	945	137	961	139	0.9	0.9	27.8	30.5
~0.15 wt% Cl, 0.13 wt% O										
As sintered, FC	Inclusions, pores	0.014	852	123	928	134	12.0	13.0	47.7	52.4
β annealed, AC	Inclusions, pores	0.012	1033	150	1083	157	10.7	11.3	39.5	43.4
HIP (60 MPa, 1000 °C), 1 h, AC	Inclusions	0.0055	977	142	1067	154	21.4	24.0	41.7	45.8
HIP (207 MPa, 750 °C) 850 °C, 1 h, FC	Inclusions	0.0027	928	134	1027	149	21.3	23.9	48.3	53.1(b)
<0.001 wt% Cl(c)										
As sintered, FC	Pores	0.075	786	114	888	128	5.6	5.8	62.9	69.1
HIP (207 MPa, 750 °C), 850 °C, 1 h, FC	None	...	996	144	1102	160	29.0	34.3	75.5(b)	82.9

Note: Fracture toughness values were determined with 15.24 mm (0.6 in.) thick compact tension specimens and with 10.2 mm (0.4 in.) square cross section three-point bend specimens. Both types of specimens were precracked and tested in accordance with ASTM E399 at a rate of 1.5 MPa$\sqrt{m}$/s (1.3 ksi$\sqrt{in.}$/s). (a) Chlorine content of starting titanium powder. (b) One fracture toughness test. Source: N. Moody, W. Garrison, Jr., J. Smugeresky, and J. Costa, The Role of Inclusion and Pore Content on the Fracture Toughness of Powder Processed Blended Elemental Ti-10V-2Fe-3Al, Microstructure, Fracture Toughness, *and Fatigue Crack Growth Rate in Titanium Alloys*, A. Chakrabarti and J.C. Chesnutt, Ed., TMS/AIME, 1987, p 83

Ti-15V-3Cr-3Al-3Sn

Common Name: Ti-15-3
UNS Number: Unassigned

Ti-15-3 was developed during the 1970's on an Air Force contract and was later scaled up to produce titanium strip. It is a solute-rich beta titanium alloy developed primarily to lower the cost of titanium sheet metal parts by reducing processing cost through the capability of being strip producible and its excellent room-temperature formability characteristics. It can also be aged to a wide range of strength levels to meet a variety of applications. Although originally developed as a sheet alloy, it has expanded into other areas such as fasteners, foil, plate, tubing, castings and forgings.

Chemistry and Density

Ti-15-3 is formulated by depressing the beta transus with vanadium and chromium additions. It is less beta-stabilized than Ti-13V-11Cr-3Al.

Density. 4.76 g/cm^3 (0.172 lb/in.^{3})

Product Forms

Ingot, billet, plate, sheet, strip, seamless tube, castings, and welded tube.

Product Conditions/Microstructure

The alloy can be directly aged after forming. However, strength will vary depending upon the amount of cold work in the part. Heating times prior to hot forming should be minimized in order to prevent appreciable aging prior to forming.

Applications

Ti-15-3 is used primarily in sheet metal applications since it is strip-producible, age-hardenable, and highly cold-formable. It is used in a variety of airframe applications, in many cases replacing hot-formed Ti-6Al-4V. Ti-15-3 can also be produced as foil, is an excellent casting alloy, and has also been evaluated for aerospace tankage applications, high-strength hydraulic tubing and fasteners.

Airframe Structures. Ti-15-3 possesses good potential for lowering the manufacturing costs of titanium airframe structures. Studies on its formability led to use as the lower half of the A-10 fuselage frame. Production costs are lower than those for Ti-6-4. Ti-15-3 welded tubing is used for pneumatic ducting, and Ti-15-3 sheet is formed into hemispheres and welded to fabricate fire extinguisher bottles on the Boeing 777. Other potential applications for this material are as seamless tubing, wire, rivets, and foil for honeycomb structures. High-strength castings are in use.

Use Limitations. Ti-15-3, like other beta titanium alloys, is highly susceptible to hydrogen pickup and rapid hydrogen diffusion during heating, pickling, and chemical milling. However, because of the much higher solubility of hydrogen in the beta phase than in the alpha phase of titanium, this alloy has a higher tolerance to hydrogen embrittlement than the alpha or alpha-beta alloys.

Ti-15-3 can be welded in the solution-treated condition; however, welding is not recommended after solution treating and aging. Care is necessary in pickling to minimize hydrogen absorption.

Ti-15V-3Cr-3Al-3Sn: Specifications and Compositions

Specification	Designation	Description	Composition, wt%								
			Al	Cr	Fe	H	N	O	Sn	V	Other
USA											
AMS 4914		Sh Strp SHT	2.5-3.5	2.5-3.5	0.25	0.015	0.05	0.13	2.5-3.5	14-16	C 0.05; OT 0.4; bal Ti
AMS 4914		Sh Strp STA	2.5-3.5	2.5-3.5	0.25	0.015	0.05	0.13	2.5-3.5	14-16	C 0.05; OT 0.4; bal Ti

Fatigue Properties

Ti-15-3: Smooth and notched fatigue

Temperature		Runout stress(a)			
		Smooth		Notched(b)	
°C	°F	MPa	ksi	MPa	ksi
–51	–60	724	105	207	30
24	75	655-758	95-110	207-241	30-35
205	400	655-690	95-100	221-241	32-35

(a) Runout >10^7 cycles, $R = 0.1$, maximum stress shown. (b) $K_t = 3$. Source: *Beta Titanium Alloys in the 1980's*, R.R. Boyer and H.W. Rosenberg, Ed., TMS/AIME, 1984, p 419

Fatigue Crack Growth

Ti-15-3 exhibits crack growth characteristics much like mill annealed Ti-6Al-4V, although Ti-15-3 is not as sensitive to environments such as salt water.

Ti-15-3: Crack growth at $\Delta K = 22$ MPa$\sqrt{m}$ (20 ksi$\sqrt{in.}$)

Evaluating the data at ΔK = 22 MPa$\sqrt{m}$ (20 ksi$\sqrt{in.}$) shows *da/dN* increases slightly as sheet gage increases. The combined salt water plus frequency effect is just at the "detection limit" statistically.

		da/dN at DK = 22 MPa$\sqrt{m}$ (20 ksi$\sqrt{in.}$):	
		10^{-6} in.	10^{-6} mm
Environmental effect	Air at 20 Hz	9.12	232
	Salt at 5 Hz	9.80	249
Gage effect	1.3 mm (0.050 in.)	8.58	218
	2.5 mm (0.100 in.)	10.33	262

Note: Test error for these data was estimated to be 12×10^{-6} mm/cycle (0.49×10^{-6} in./cycle). Source: *Beta Titanium Alloys in the 1980's*, R.R. Boyer and H.W. Rosenberg, Ed., TMS/AIME, 1984, p 419

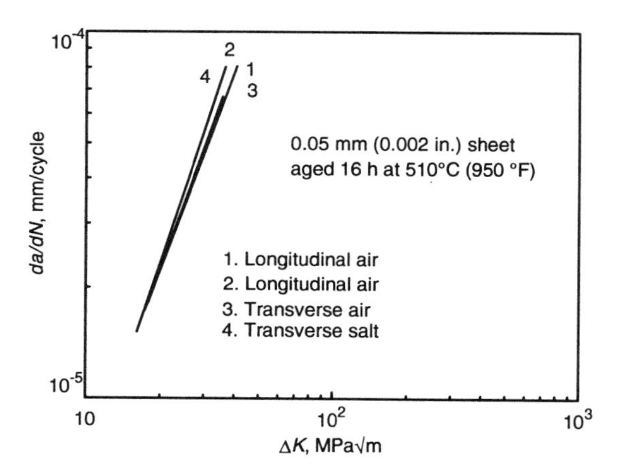

Ti-15-3: Crack growth in air and salt solution. Source: *Beta Titanium Alloys in the 1980's*, R.R. Boyer and H.W. Rosenberg, Ed., TMS/AIME, 1984, p 420

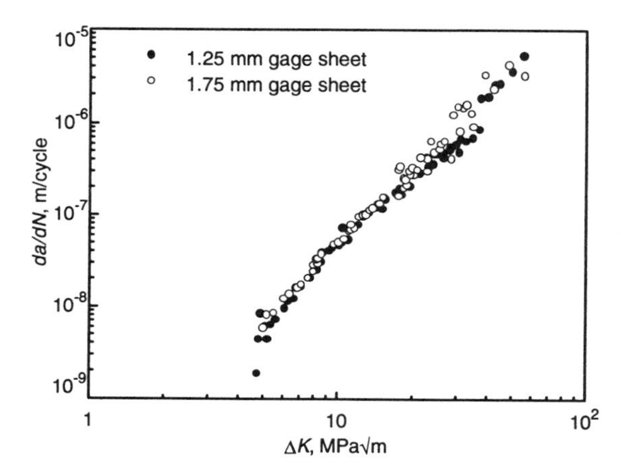

Ti-15-3: Crack growth data for sheet. Specimens were tested in the T-L orientation. Sheet was aged at 540 °C (1000 °F), 8 h. $R = 0.1$; frequency, 30 Hz, at 22 °C (72 °F). Source: *Beta Titanium Alloys in the 1980's*, R.R. Boyer and H.W. Rosenberg, Ed., TMS/AIME, 1984, p 223

Fracture Properties

Ti-15-3: RT fracture toughness of sheet

Gage, mm	Gage, in.	Specimen orientation	Fracture toughness (K_c), MPa√m	Fracture toughness (K_c), ksi√in.
1.27	0.050	L-T	100	91
		T-L	100	91
1.78	0.070	L-T	113	103
		T-L	107	97

Note: Yield strength of 1035 MPa (150 ksi) at RT. Directionality is low, 3 to 4 MPa√m (3 to 4 ksi√in.). Lot-to-lot variations can be up to 11 MPa√m (10 ksi√in.). Source: *Beta Titanium Alloys in the 1980's*, R.R. Boyer and H.W. Rosenberg, Ed., TMS/AIME, 1984, p 416

Ti-15-3: Fracture toughness of STA plate

Heat treatment	Orientation	Tensile yield strength, MPa	Tensile yield strength, ksi	Ultimate tensile strength, MPa	Ultimate tensile strength, ksi	Elongation, %	Fracture toughness (K_{Ic}), MPa√m	Fracture toughness (K_{Ic}), ksi√in.
800 °C (1470 °F), 20 min, AC, 480 °C (895 °F), 14 h, AC	L-T	1253	182	1376	199	6.2	44.3	40.3
	T-L	1304	189	1421	206	6.6	46.8	42.6
800 °C (1470 °F), 20 min, AC, 510 °C (950 °F), 14 h, AC	L-T	1213	176	1337	194	7.8	42.1	38.3
	T-L	1263	183	1382	200	6.9	43.4	39.5

Note: Hot rolled plate had a chemical composition (wt%) of 3.37 Al, 0.004 C, 3.36 Cr, 0.17 Fe, 0.0061 H, 0.0080 N, 0.14 O, 3.04 Sn, and 15.10 V. It was solution treated at 800 °C (1470 °F) for 20 min, air cooled, then aged at 510 °C (950 °F) for 8 or 14 h. Source: C. Ouchi, H. Suenaga, H. Sakuyama, and H. Takatori, Effects of Thermomechanical Processing Variables on Mechanical Properties of Ti-15V-3Cr-3Sn-3Al Alloy Plate, in *Designing With Titanium*, 1986, p 130

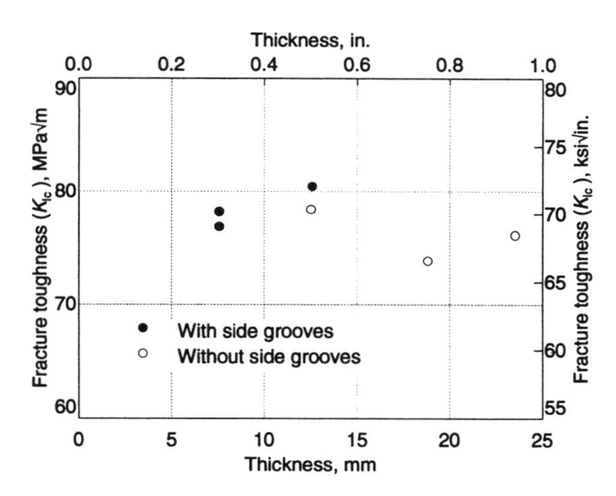

Ti-15-3: Fracture toughness vs sheet thickness. Alloy was heat treated to achieve strength levels of 1055 MPa (153 ksi) yield strength and 1117 MPa (162 ksi) tensile strength, with 7.5% ductility. Compact tension specimens with and without V-shaped side grooves (depth <10% original thickness) were tested on a servohydraulic machine in load control. Source: P. Poulose, Determination of Fracture Toughness from Thin Side-Grooved Specimens, *Eng. Fract. Mech.*, Vol 26, 1987, p 203

TIMETAL® 21S

Ti-15Mo-3Al-2.7Nb-0.25Si
Common Name: Beta-21S
UNS Number: R58210

Tom O'Connell, TIMET

Beta-21S is a very recently developed metastable β alloy that offers the high specific strength and good cold formability of a metastable β alloy, but has been specifically designed for improved oxidation resistance, elevated temperature strength, creep resistance, and thermal stability. Developing commercial applications in forgings include aerospace components and prosthetic devices. For the latter application, with appropriate thermomechanical processing, Beta-21S modulus is comparable to bone. For the former, Beta-21S may be processed to very high strengths with excellent oxidation and corrosion resistance.

Strip is the main product form. Beta-21S is also well suited for metal matrix composites because it can be economically rolled to foil and is compatible with most fibers. Strip is available in gages from 0.3 to 2.5 mm (0.012 to 0.100 in.).

Chemistry. The composition of Beta-21S is based on the objective of obtaining a cold rollable, strip-producible alloy for economical processing into foil form. The key to processing an alloy to foil form is cold rolling of strip product. If an alloy cannot be cold rolled as strip, a hot process on a handmill using cover sheets to form packs for heat retention is the only other viable option. Although the pack process offers the opportunity to cross-roll to minimize texture, it is nonetheless labor intensive and inherently a lower yield process.

In light of the fact that a cold rollable, strip-producible alloy was of primary importance, it was decided that a metastable β alloy was the best approach. This meant that the ordinary obstacles to overcome were the poor oxidation resistance and elevated-temperature mechanical properties of this class of alloy. The initial approach was to concentrate on the Ti-Mo and Ti-Cr systems. Although the Ti-V system is most commonly used for metastable β alloys (e.g., Ti-15V-3Cr-3Sn-3Al and Ti-3Al-8V-6Cr-4Zr-4Mo), vanadium is well known for its detrimental effects on oxidation resistance. Conclusions of chemistry screening on oxidation resistance were as follows:

- Silicon, niobium, hafnium, and tantalum were beneficial additions to the Ti-Mo system, as well as palladium, aluminum, and iron.
- Tin, zirconium, cobalt, yttrium, and iron were not beneficial additions to a Ti-Mo base.
- 20% Mo provides no advantage in oxidation resistance over 15% Mo.
- No additions were found that improve the corrosion resistance of the Ti-Cr series.

Effect of Oxygen. In a study on the effect of oxygen, oxygen levels up to 0.25% were found to have no significant effect on the strength/ductility relationship of aged Beta-21S. Higher oxygen levels degrade ductility. Increasing oxygen decreases the work-hardening capability of annealed sheet material, which could adversely affect some aspects of formability.

Oxygen absorption at the surface during exposure in air at elevated temperature degrades tensile ductility. The magnitude of the effect in sheet is dependent on the exposure time and temperature and on sheet thickness. After a suitable heat treatment, Beta-21S is metallurgically stable for at least 1000 h up to 615 °C (1140 °F).

Product Forms and Conditions. Beta-21S is available as cut sheet, strip, plate, bar, billet, and bloom. It is typically provided in the beta solution treated condition, which precipitates α to provide strengthening on aging. The morphology and distribution of the α depend on the heat-treatment temperature and the oxygen content. Lower heat-treatment temperatures and higher oxygen contents result in homogeneous spheroidal α; higher aging temperatures and lower oxygen result in lath-type α.

Applications. Beta-21S is most useful for applications above 290 °C (550 °F), with thermal stability up to 625 °C (1160 °F) and creep resistance comparable to Ti-6Al-4V. Developing commercial applications include forged prosthetic devices and cold rolled foil for metal matrix composites. Special properties include a modulus that is comparable to bone, improved oxidation resistance up to 650 °C (1200 °F), and resistance to aerospace hydraulic fluids (e.g., Skydrol).The latter properties have led to a number of aircraft engine applications. Excellent corrosion and hydrogen embrittlement resistance have led to chemical and offshore oil use.

Selected References

1. W.M. Parris and P.J. Bania, "Beta-21S: A High-Temperature Metastable Beta Titanium Alloy," Proc. 1990 TDA Int. Conf., Orlando, 1990
2. W.M. Parris and P.J. Bania, "Oxygen Effects on the Mechanical Properties of TIMETAL 21S," Proc. 7th Int. Titanium Conf., San Diego, 1992
3. J.S. Grauman, "A High-Strength Corrosion-Resistant Titanium Alloy," Proc. 1990 TDA Int. Conf., Orlando, 1990
4. J.S. Grauman, "Corrosion Behavior of TIMETAL-21S for Nonaerospace Applications," Proc. 7th Int. Titanium Conf., San Diego, 1992

Beta-21S: Typical composition range

	Composition, wt%									
	Al	Nb	Mo	Si	Fe	C	O_2	N_2	H_2	Ti
Minimum	2.5	2.4	14.0	0.15	0.2	...	0.11	...	...	...
Maximum	3.5	3.0	16.0	0.25	0.4	0.05	0.15	0.05	0.015	...
Aim	3.0	2.8	15.0	0.20	0.3	...	0.13	...	...	bal

Physical Properties

Phases and Structures. The microstructure of Beta-21S consists of recrystallized β grains with occasional unrecrystallized β grains. In addition, titanium silicides are present. The principal aging product is α (close-packed hexagonal α). Omega (ω) also has been observed, though it would not be a problem with proper heat treatment.

Beta-21S: Summary of typical physical properties

Beta transus	~793 to 810 °C (1460-1490 °F)
Melting range(a)	1672 to 1747 °C (3041 to 3177 °F)
Density(b)	4.94 g/cm^3 (0.178 lb/in.^{3})
Elastic modulus,	
Beta annealed	74 to 85 GPA (10.7 to 12.3 × 10^6 psi)
Beta annealed + aged	96.5 to 103.5 GPa (14 to 15 × 10^6 psi)
Electrical resistivity(b)	1.35 μΩ · m
Magnetic permeability	Nonmagnetic
Specific heat capacity(b)	710 J/kg · K (0.17 Btu/lb · °F)
Thermal conductivity(b)	7.5 W/m · K (4.3 Btu/ft · h · °F)
Thermal coefficient of linear expansion(c)	8.5 × 10^{-6}/ °C (4.7 × 10^{-6}/ °F)

(a) Calculated solidus and liquidus temperatures, respectively. (b) Typical values at room temperature of about 20 to 25 °C (68 to 78 °F). (c) Mean coefficient from room temperature to 200 °C (390 °F)

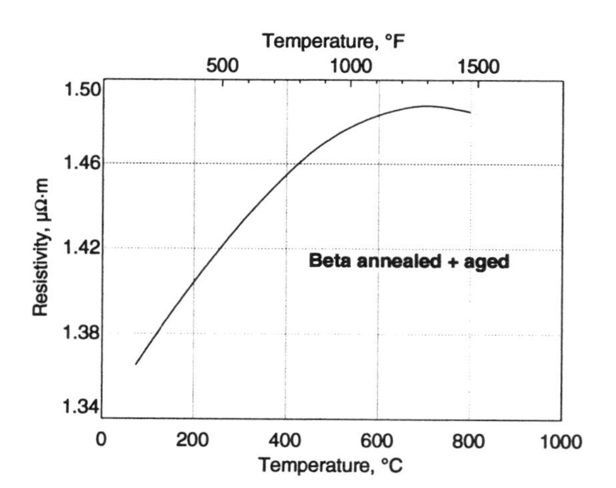

Beta-21S: Electrical resistivity vs temperature. Electrical resistivity (R) between 25 and 750 °C (77 and 1380 °F) fits the expression: $R\,(10^{-8}\,\Omega \cdot m) = 134 + 0.035\,T - 1.11 \times 10^{-5}\,T^2 - 1.27 \times 10^{-8}\,T^3$ Beta annealed plus 600 °C (1110 °F) for 8 h

Corrosion Properties

Molybdenum improves corrosion resistance in reducing media, and this well-known effect is apparent when the corrosion rate of Beta 21S and grade 2 Ti are compared in HCl solution (see figure). However, the increased resistance from molybdenum in reducing media generally comes at the expense of resistance in oxidizing media. In this regard, the possible additive or synergistic effect of alloying on oxidation resistance was considered during the development of Beta 21S. The best overall oxidation resistance occurred with aluminum-silicon additions (see table). This alloying results in a slightly higher repassivation potential compared to other molybdenum-containing titanium alloys (see table on next page).

Crevice corrosion resistance improves with molybdenum additions, and a chloride crevice corrosion test (5% NaCl at 90 °C, pH adjusted to 0.5 and 1.0) indicated a chloride crevice corrosion threshold between pH 0.5 and 1.0.

Hydrogen Damage. Beta-21S retains ductility up to hydrogen levels of 2000 ppm. The percent of retained ductility versus hydrogen content is shown (see figure on next page). High hydrogen levels (2000 ppm) will slow down aging kinetics.

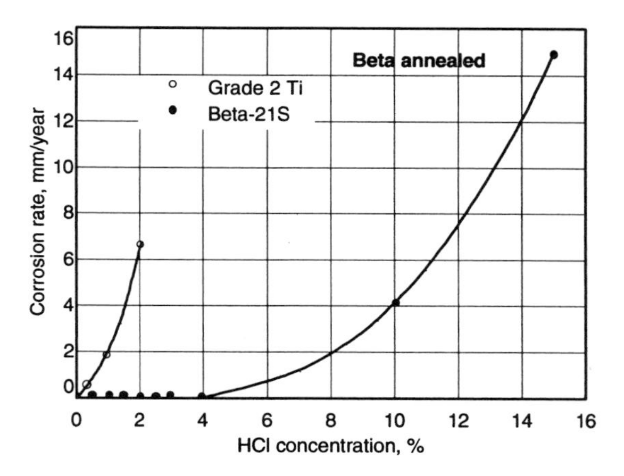

Beta-21S: Corrosion rate as a function of HCl concentration. Boiling HCl, 72-h test. Source: J.S. Grauman, "A New High Strength, Corrosion Resistant Titanium Alloy," TDA Int. Conf., Orlando, 1990

Beta-21S: Oxidation results from alloy development

	Alloy	Weight gain, %
250-g heat		
	Ti-15Mo-5Fe-2Hf	2.40
	Ti-15Mo-5Fe-0.2Si	1.52
	Ti-15Mo-5Fe-2Nb	1.17
	Ti-15Mo-5Fe-2Nb-0.2Si	0.94
	Ti-15Mo-3Nb-1.5Ta-3Al	0.83
	Ti-15Mo-5Nb-0.5Si	0.71
	Ti-15Mo-5Nb-3Al-0.5Si	0.60
8.2-kg heat		
	Ti-15Mo-5Nb-3Al-0.50Si	0.90
	Ti-15Mo-5Nb-0.5Si	0.73
	Ti-15Mo-3Nb-1.5Ta-3Al-0.2Si	0.67
	Ti-15Mo-2Nb-3Al-0.2Si	0.62
250-g buttons		
	Ti-15V-3Cr-3Sn-3Al	>65
	Commercially Pure Ti	7.70
	Ti-15Mo-5Zr	7.70
	Ti-15Mo-3Sn	5.37
	Ti-15Mo-5Co	2.89
	Ti-15Mo-0.1Y	2.73
	Ti-15Mo-5Re	2.68
	Ti-15Mo	2.63
	Ti-15Mo-5Fe	2.10
	Ti-15Mo-3Al	2.00
	Ti-15Mo-0.2Pd	1.79
	Ti-15Mo-0.1Si	1.45
	Ti-15Mo-5Hf	1.41
	Ti-15Mo-0.2Si	1.27
	Ti-15Mo-0.5Si	1.17
	Ti-15Mo-3Ta	1.04
	Ti-20Mo-2Nb	0.99
	Ti-15Mo-2Nb	0.98
	Ti-15Mo-5Nb	0.95
	Ti-15Cr-2Pd	9.76
	Ti-15Cr-3Ta	9.44
	Ti-15Cr-5Nb	7.62
	Ti-15Cr-0.5Si	7.00
	Ti-15Cr-3Sn	4.11
	Ti-15Cr-3Al	3.68
	Ti-15Cr-5Mo	2.90
	Ti-15Cr	2.27

Note: Initial oxidation results from a 48-h exposure at 815 °C (1500 °F) on 1.5 mm (0.02 in.) strip cold rolled from 250-g heat, 8.2-kg heat, and 250-g buttons. Source: W.M. Parris and P.J. Bania, "Beta-21S: A High Temperature Metastable Beta Titanium Alloy," TDA Int. Conf., Orlando, 1990

Beta-21S: Repassivation potential comparison

Alloy	Repassivation potential vs Ag/AgCl
Beta-21S	2.8
Ti grade 2	6.2
Ti-6Al-4V	1.8
Ti-6Al-2Sn-4Zr-6Mo	2.5
Ti-3Al-8V-6Cr-4Zr-4Mo	2.7

Note: The galvanostatic method, boiling 5% NaCl solution pH adjusted to 3.5, was used to measure repassivation. After approximately 1 h of exposure, the test specimen was subjected to a constant current density of 200 mA/cm^2. Source: J.S. Grauman, "A New High Strength, Corrosion Resistant Titanium Alloy," TDA Int. Conf., Orlando, 1990

Beta-21S: General corrosion behavior

Medium	Corrosion rate, mm/year
3% boiling H_2SO_4	0.16
10% $FeCl_3$, boiling	0.01
0.5% HCl, boiling	0.00254
1% HCl, boiling	0.00508
1.5% HCl, boiling	0.01016
2% HCl, boiling	0.01778
2.5% HCl, boiling	0.02794
3% HCl, boiling	0.04064
4% HCl, boiling	0.127
10% HCl, boiling	4.0
15% HCl, boiling	15.0
28% HCl, boiling, deaerated	55.0
10% formic acid, 10% acetic acid, boiling, deaerated	0.0

Note: Beta annealed material. Source: J.S. Grauman, "A New High Strength, Corrosion Resistant Titanium Alloy," TDA Int. Conf., Orlando, 1990

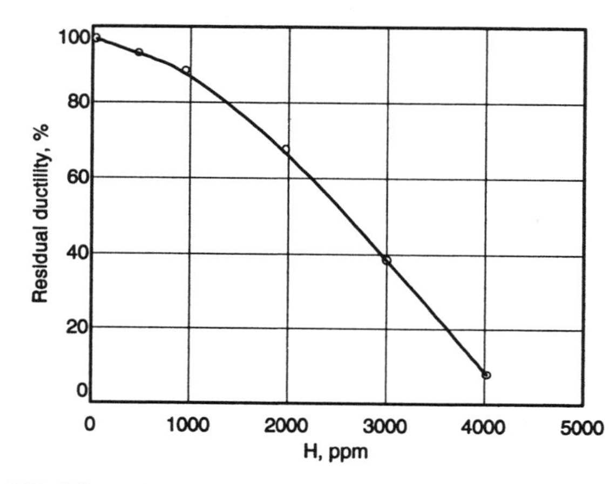

Beta-21S: Effect of hydrogen on residual ductility. Cathodically charged 1.5 mm (0.06 in.) sheet in the beta-annealed condition. Based on bend radius tests on sheet material cathodically charged with H. Residual ductility determined from initial ductility without H charge, which was equivalent to 15% ±3%.

Thermal Properties

Heat Capacity

The specific heat (C_p) for beta-annealed plus aged Beta-21S between 25 and 750 °C (77 and 1380 °F) (see figure) fits the expression:

$$C_p\,(\text{cal/g}\cdot{}^\circ\text{C}) = 0.116 + 4.83 \times 10^{-5}\,(T)$$

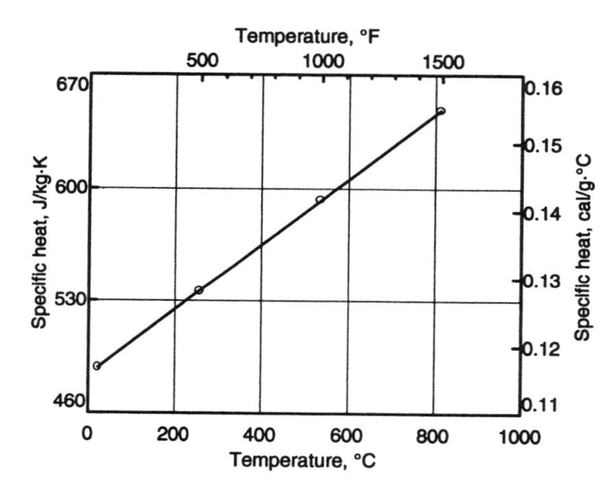

Beta-21S: Specific heat vs temperature. Beta annealed plus 600 °C (1110 °F) for 8 h

Thermal Expansion

The thermal coefficient of linear expansion (α) for beta-annealed plus aged Beta-21S between 25 and 750 °C (77 and 1380 °F) (see figure) follows the expression:

$$\alpha\,(\text{ppm/}{}^\circ\text{C}) = 6.75 + 1.28 \times 10^{-2}\,T - 2.27 \times 10^{-5}\,T^2 + 1.52 \times 10^{-8}\,T^3$$

Thermal Conductivity

The thermal conductivity between 25 and 750 °C (77 and 1380 °F) for beta-annealed plus aged Beta-21S (see figure) fits the equation:

$$Q\,(\text{W/m}\cdot{}^\circ\text{C}) = 7.33 + 1.66 \times 10^{-2}\,T$$

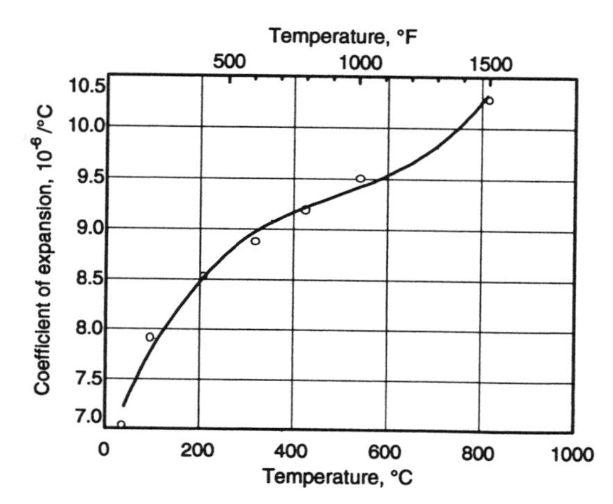

Beta-21S: Thermal coefficient of linear expansion. Beta annealed plus 600 °C (1110 °F) for 8 h

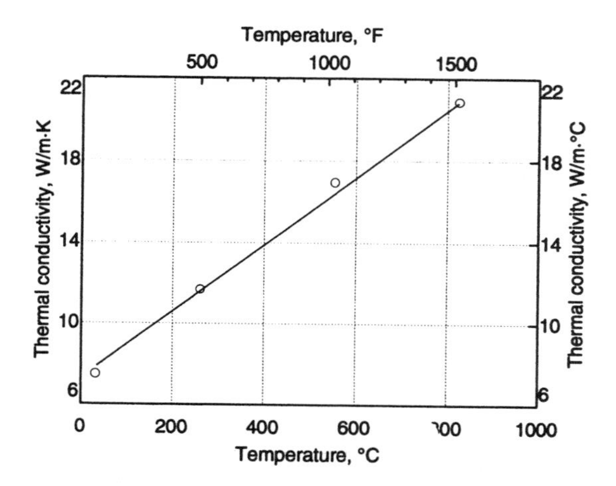

Beta-21S: Thermal conductivity. Beta annealed plus 600 °C (1110 °F) for 8 h

Tensile Properties

See also "Processing" for tensile data.

Although oxygen levels below 0.33 wt% do not appear to significantly affect the strength/ductility relationship, results of tests (see table) on sheet from two heats containing 0.14 and 0.25 wt% oxygen showed a deleterious effect on ductility for the higher oxygen content in the series aged at 595 °C (1100 °F). In the annealed condition, there is another effect of oxygen, which could be important in certain types of forming operations. In the annealed condition, the difference between yield, and ultimate tensile strengths decreased as the oxygen level increased from 42 MPa (6.1 ksi) at 0.09% oxygen to 12 MPa (1.7 ksi) at 0.33% oxygen. This behavior implies a decrease in work-hardening capability with increasing oxygen and, concomitantly, an increase in the tendency to neck locally and fail during stretching or drawing operations.

Beta-21S: RT tensile properties of sheet vs oxygen content

Aging temperature, °C	Aging temperature, °F	Aging time, h	Oxygen content, %	Ultimate tensile strength, MPa	Ultimate tensile strength, ksi	Tensile yield strength, MPa	Tensile yield strength, ksi	Elongation, %
None		...	0.14	880.5	127.7	860.5	124.8	12.0
			0.25	931.5	135.1	914.3	132.6	15.0
480	895	4	0.14	1093.8	158.6	983.9	142.7	11.5
			0.25	1011.5	146.7	975.0	141.4	14.0
		8	0.14	1257.0	182.3	1145.3	166.1	4.0
			0.25	1143.9	165.9	1114.9	161.7	5.0
		16	0.14	1383.2	200.6	1276.9	185.2	5.0
			0.25	1473.5	213.7	1373.5	199.2	4.5
		24	0.14	1428.6	207.2	1319.7	191.4	3.5
			0.25	1529.3	221.8	1454.8	210.9	3.0
540	1000	4	0.14	1297.0	188.1	1199.7	174.0	8.0
			0.25	1381.8	200.4	1297.6	188.2	5.5
		8	0.14	1269.4	184.1	1185.3	171.9	5.0
			0.25	1388.7	201.4	1303.2	189.0	3.5
		16	0.14	1289.4	187.0	1205.3	174.8	6.0
			0.25	1409.3	204.4	1341.1	194.5	3.5
		24	0.14	1268.0	183.9	1192.2	172.9	6.0
			0.25	1388.7	201.4	1336.9	193.9	3.5
595	1100	4	0.14	1103.8	160.1	1024.6	148.6	11.0
			0.25	1180.4	171.2	1108.7	160.8	6.0
		8	0.14	1063.9	154.3	996.3	144.5	11.0
			0.25	1172.2	170.0	1103.9	160.1	7.5
595	1100	16	0.14	1074.2	155.8	999.8	145.0	10.0
			0.25	1194.2	173.2	1128.7	163.7	5.0
		24	0.14	1116.3	161.9	1059.1	153.6	8.0
			0.25	1199.7	174.0	1128.7	163.7	6.5

Note: Cold rolled 50% prior to annealing. 0.14% oxygen annealed at 815 °C (1500 °F), 5 min, AC; 0.25% oxygen annealed at 857 °C (1610 °F), 5 min, AC. Prior to tensile testing, all sheet specimens were descaled and pickled to remove 0.05 mm (0.002 in.) from each surface to remove any material contaminated by oxygen and/or nitrogen during heat treatment. Tensile testing was carried out according to ASTM E8. Gage section for the sheet specimens was 6 mm (0.25 in.) × 25 mm (1 in.) and for the bar specimens 6 mm (0.25 in.) diameter × 25 mm. Source: W.M. Parris and P.J. Bania, "Oxygen Effects on the Mechanical Properties of TIMETAL® 21S," 7th Int. Titanium Conf., July 1992

Beta-21S: RT tensile properties of sheet and bar vs oxygen content

Heat treatment(a)	Oxygen, %	Simulated strip(b): Ultimate tensile strength, MPa	Ultimate tensile strength, ksi	Tensile yield strength, MPa	Tensile yield strength, ksi	Elongation, %	Hot rolled bar: Ultimate tensile strength, MPa	Ultimate tensile strength, ksi	Tensile yield strength, MPa	Tensile yield strength, ksi	Reduction of area, %	Elongation, %
845 °C (1550 °F), 10 min, AC	0.090	813.6	118.0	771.6	111.9	19.8	837.1	121.4	795.7	115.4	66.8	22.5
	0.120	859.8	124.7	819.8	118.9	20.1	847.4	122.9	815.7	118.3	66.2	24.0
	0.130	874.3	126.8	847.4	122.9	17.3	874.3	126.8	843.9	122.4	61.8	23.0
	0.183(c)	900.5	130.6	888.8	128.9	18.4	882.6	128.1	853.6	123.8	63.6	23.3
	0.229(d)	930.8	135.0	912.9	132.4	17.4	899.1	130.4	890.1	129.1	66.1	27.0
	0.334(e)	970.8	140.8	958.4	139.0	21.5	917.0	132.9	913.6	132.5	61.4	26.5
845 °C (1550 °F), 10 min, AC + 480 °C (900 °F), 14 h, AC	0.090	1336.9	193.9	1258.3	182.5	6.8	...	...	...	...	...	...
	0.120	1443.1	209.3	1341.8	194.6	4.1	1431.4	207.6	1352.1	196.1	15.8	7.0
	0.130	1391.4	201.8	1306.6	189.5	3.0	1434.2	208.0	1346.6	195.3	15.9	6.5
	0.183(c)	1447.3	209.9	1375.6	199.5	2.8	1494.8	216.8	1415.5	205.3	12.5	4.5
	0.229(d)	1541.7	223.6	1470.7	213.3	2.3	1583.1	229.6	1501.7	217.8	10.4	4.5
	0.334(e)	1579.6	229.1	1462.4	212.1	3.0	1540.3	223.4	1443.1	209.3	5.0	2.0

(continued)

Beta-21S: RT tensile properties of sheet and bar vs oxygen content (continued)

Heat treatment(a)	Oxygen, %	Simulated strip(b) Ultimate tensile strength, MPa	Simulated strip(b) Ultimate tensile strength, ksi	Simulated strip(b) Tensile yield strength, MPa	Simulated strip(b) Tensile yield strength, ksi	Simulated strip(b) Elongation, %	Hot rolled bar Ultimate tensile strength, MPa	Hot rolled bar Ultimate tensile strength, ksi	Hot rolled bar Tensile yield strength, MPa	Hot rolled bar Tensile yield strength, ksi	Hot rolled bar Reduction of area, %	Hot rolled bar Elongation, %
845 °C (1550 °F), 10 min, AC + 540 °C (1000 °F), 8 h, AC	0.090	1157.0	167.8	1024.6	148.6	9.6	1202.5	174.4	1037.0	150.4	47.6	10.9
	0.120	1314.2	190.6	1232.8	178.8	5.8	1325.2	192.2	1253.5	181.8	24.3	8.3
	0.130	1320.4	191.5	1243.2	180.3	5.8	1326.6	192.4	1254.9	182.0	24.4	8.5
	0.183(c)	1421.7	206.2	1319.7	191.4	1.4	1395.5	202.4	1329.5	192.8	19.0	7.0
	0.229(d)	1434.8	208.1	1377.6	199.8	4.3	1467.3	212.8	1388.0	201.3	19.2	7.8
	0.334(e)	1461.1	211.9	1359.7	197.2	3.4	1425.2	206.7	1332.8	193.3	8.0	4.0
845 °C (1550 °F), 10 min, AC + 595 (1100 °F), 8 h, AC	0.090	937.0	135.9	822.6	119.3	16.8	1045.3	151.6	947.4	137.4	44.2	11.5
	0.120	1068.0	154.9	986.0	143.0	12.5	1103.2	160.0	1010.1	146.5	35.5	14.0
	0.130	1060.5	153.8	987.4	143.2	9.0	1099.8	159.5	1011.5	146.7	35.2	13.5
	0.183(c)	1152.8	167.2	1081.8	156.9	7.9	1166.6	169.2	1084.6	169.2	26.9	12.0
	0.229(d)	1223.2	177.4	1148.0	166.5	8.0	1232.1	178.7	1146.6	166.3	22.5	10.3
	0.334(d)	1289.4	187.0	1194.2	173.2	6.5	1259.0	171.2	1180.4	171.2	16.0	8.3

(a) Annealing time for sheet was 10 min, for bar 1 h. (b) Cold rolled 50% prior to annealing. (c) Annealed 857 °C. (d) Annealed 870 °C. (e) Annealed 885 °C. Source: W.M. Parris and P.J. Bania, "Oxygen Effects on the Mechanical Properties of TIMETAL® 21S," 7th Int. Titanium Conf., July 1992

Beta-21S: Typical room-temperature aged tensile properties

Aging temperature(a), °C	Aging temperature(a), °F	Test direction	Tensile yield strength, MPa	Tensile yield strength, ksi	Ultimate tensile strength, MPa	Ultimate tensile strength, ksi	Elongation, %
540	1000	L	1288	189	1353	196	9.0
		L	1326	192	1394	202	7.5
		T	1346	195	1422	206	6.5
		T	1379	200	1438	208	7.0
		L	1100	159	1179	171	11.0
		T	1185	172	1243	180	11.0
		T	1165	169	1240	179	10.0
Duplex(b)		...	856	124	920	133	18.0
		...	840	122	914	132	20.0

(a) Aged 8 h after beta anneal. (b) 8 h at 690 °C (1275 °F), AC, + 650 °C (1200 °F) for 8 h, AC

Beta-21S: Typical RT beta-annealed tensile properties

Tensile yield strength, MPa	Tensile yield strength, ksi	Ultimate tensile strength, MPa	Ultimate tensile strength, ksi	Elongation, %	Tensile yield strength, MPa	Tensile yield strength, ksi	Ultimate tensile strength, MPa	Ultimate tensile strength, ksi	Elongation, %
Longitudinal					Transverse				
869	126	924	134	11.0	910	132	952	138	10.0
869	126	896	130	9.0	903	131	931	135	9.0
834	121	862	125	12.0	876	127	910	132	10.0
855	124	876	127	12.0	889	129	910	132	10.0
869	126	896	130	12.0	896	130	931	135	10.0
869	126	896	130	14.0	903	131	931	135	11.0
869	126	876	127	15.0	896	130	903	131	12.0
876	127	883	128	12.0	903	131	910	132	12.0
903	131	952	138	11.0	945	137	1007	146	11.0
862	125	896	130	13.0	903	131	938	136	11.0
855	124	896	130	11.0	896	130	938	136	11.0
862	125	896	130	12.0	896	130	924	134	11.0
924	134	952	138	10.0	965	140	993	144	10.0
938	136	986	143	14.0	972	141	1014	147	7.0
938	136	986	143	11.0	979	142	1027	149	8.0
896	130	924	134	10.0	952	138	979	142	9.0

High-Temperature Strength

Beta-21S: High-temperature tensile properties (aged at 540 °C)

Test temperature		Test direction	Tensile yield strength		Ultimate tensile strength		Elongation,
°C	°F		MPa	ksi	MPa	ksi	%
24	75	L	1288	187	1353	196	9.0
		L	1326	192	1394	202	7.5
		T	1346	195	1422	206	6.5
		T	1379	200	1438	208	7.0
205	400	L	1105	160	1200	174	8.5
		L	1096	159	1204	175	9.5
		T	1127	163	1233	179	8.0
		T	1154	167	1249	181	6.0
315	600	L	1041	151	1149	166	8.0
		L	1019	147	1156	167	8.0
		T	1089	158	1197	173	6.0
		T	1050	152	1158	168	7.0
425	800	L	976	141	1090	158	8.0
		L	969	140	1077	156	9.0
		T	1016	147	1132	164	7.0
		T	1005	145	1122	162	6.0
540	1000	L	576	83	838	121	22.0
		L	674	97	849	123	22.0
		T	616	89	867	125	25.0
		T	648	94	886	128	24.5

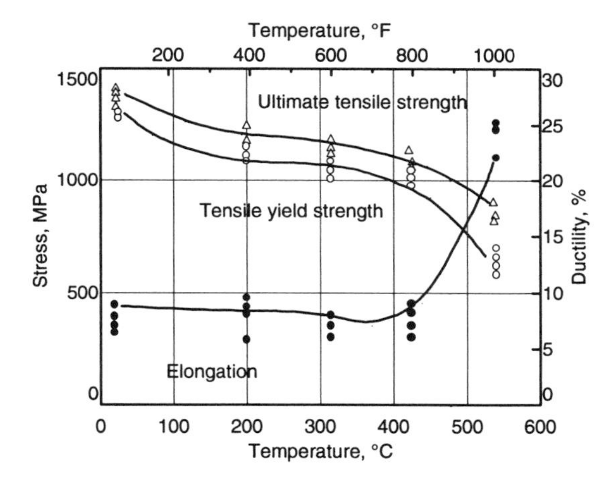

Beta-21S: High-temperature tensile properties. Beta annealed at 845 °C (1550 °F) for 10 min + 540 °C (1000 °F), 8 h

Beta-21S: High-temperature tensile properties (aged at 540 °C)

Test temperature		Test direction	Tensile yield strength		Ultimate tensile strength		Elongation,
°C	°F		MPa	ksi	MPa	ksi	%
24	75	L	1100	159	1179	171	11.0
		T	1185	172	1243	180	11.0
		T	1165	169	1240	179	10.0
205	400	L	893	129	1011	146	12.0
		L	903	130	1020	148	10.0
		T	907	131	1036	150	10.0
		T	944	137	1069	155	10.0
315	600	L	832	121	955	138	10.0
		L	830	120	969	140	10.0
		T	861	125	1001	145	9.0
		T	875	127	994	144	9.0
425	800	L	776	112	909	132	10.0
		L	807	117	925	134	10.0
		T	818	118	946	137	9.0
		T	856	124	967	140	7.0
540	1000	L	598	86	741	107	26.9
		L	587	85	751	109	24.0
		T	613	89	773	112	28.5
		T	633	92	822	119	12.0
		L	598	86	741	107	26.9
		L	587	85	751	109	24.0
		T	613	89	773	112	28.5
		T	633	92	822	119	12.0

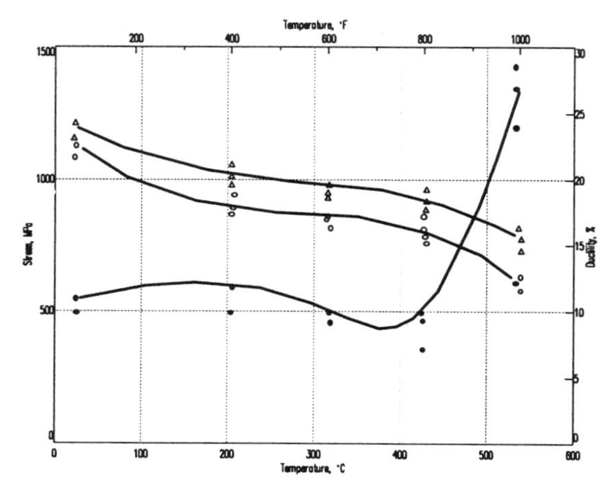

Beta-21S: High-temperature tensile properties. Beta annealed at 845 °C (1550 °F) for 10 min + 595 °C (1100 °F), 8 h

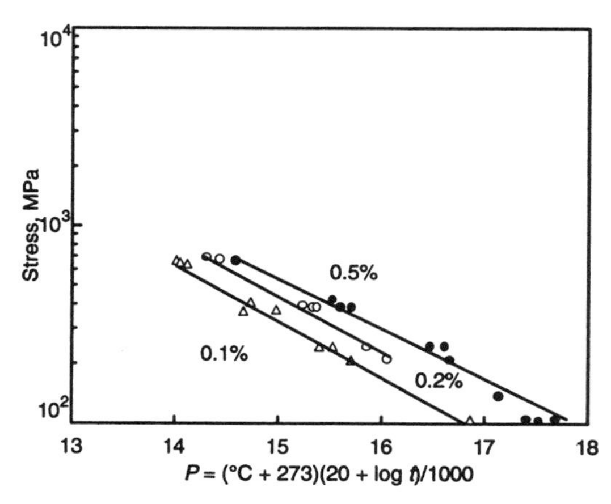

Beta-21S: Creep results in STA material. 1.5 mm (0.06 in.) sheet; beta annealed plus aged 8 h at 540 °C (1000 °F)

Crack Resistance

Beta-21S: Fracture toughness

Heat treatment	Tensile yield strength MPa	Tensile yield strength ksi	Ultimate tensile strength MPa	Ultimate tensile strength ksi	Elongation, %	Test medium	Fracture toughness (K_c) MPa√m	Fracture toughness (K_c) ksi√in.
Beta annealed	865	124	879	127	15	Air	107.5	97.8
							107.5	97.8
						Salt water	107.5	97.8
							106.7	97.1
Beta annealed + 540 °C (1000 °F), 8 h	1220	177	1320	191	6	Air	75.4	68.6
							72.6	66.0
						Salt water	68.5	62.3
							67.8	61.7
Beta annealed + 595 °C (1100 °F), 8 h	1040	151	1130	164	7.5	Air	100.8	91.7
							100.8	91.7

Note: Center notch sheet specimens 1.4 mm (0.05 in.) thick

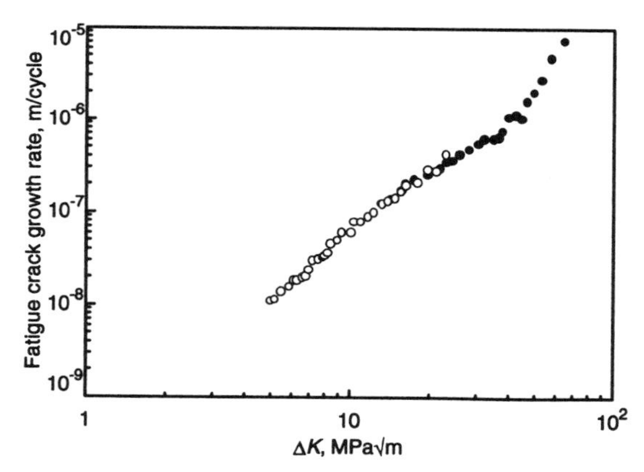

Beta-21S: Fatigue crack growth. As in other beta alloys, microstructure and heat treatment seem to have virtually no effect on crack growth rates of Beta-21S. Solution treated and aged for 8 h at 540 °C (1000 °F). Room-temperature tests at lab air; 29 Hz; $R = 0.1$

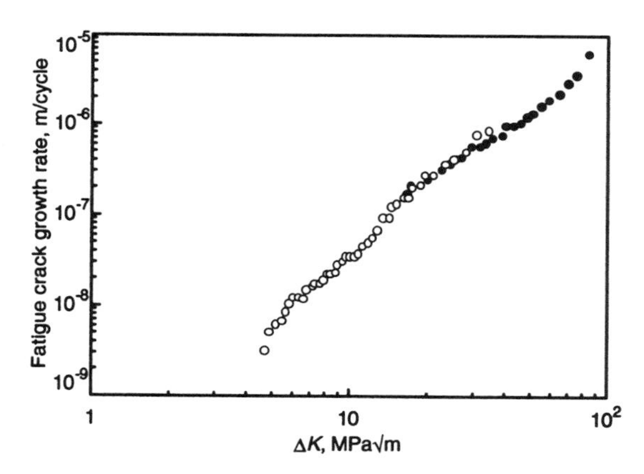

Beta-21S: Fatigue crack growth. Beta annealed (790 °C, 10 min) sheet, 1.4 mm (0.05 in.) center notch sample, tested at RT; $R = 0.1$; 29 Hz in lab air

Processing

Formability. Limited forming data indicate a similarity to Ti-15V-3Cr-3Al-3Sn (Ti-15-3). In addition to tensile tests, some indication of sheet form-ability in annealed material was obtained by bend testing 25 mm (1 in.) wide strip. These specimens were bent 105° around successively smaller radii either until cracking visible at 20× magnification occurred or until the minimum radius of 0.75 mm (0.030 in.) was reached. Sheet in the annealed condition from all heats sustained a 105° bend around a 1.27 mm (0.050 in.) radius without cracking. This translates to a bend ductility of 1 *T* or less for sheet at all oxygen levels. Thus, oxygen contents up to 0.33% had no significant effect on this criterion for formability. However, as shown in the previous section on tensile properties, tensile data indicated a possible oxygen effect on other aspects of sheet formability.

Machining, welding, and brazing of Beta-21S is typical of beta alloys and is considered similar to that of Ti-15-3.

Heat Treatment. In cases where high-temperature exposure is anticipated, a duplex overage is used to retain ductility. The high temperature age "weakens" the grains relative to the grain boundaries and the second age stabilizes the grains against embrittlement. (See table and figures.)

Beta-21S: Selected heat treatments

Treatment	Temperature °C	Temperature °F	Duration	Cooling method
Solution treatment (beta anneal)	800-815	1470-1500	4 min, minimum	...
Aging(a)	480	900	24 h	AC
	540	1000	8 h	AC
	595	1100	8 h	AC
Duplex age				
First stage	690	1275	8 h	AC
Second stage	650	1200	8 h	AC

(a) Three selected aging treatments that cover strength levels likely to be used in commercial applications

Beta-21S: Tensile yield strength vs aging time. Beta annealed 1.5 mm (0.06 in.) sheet aged for indicated times and temperatures

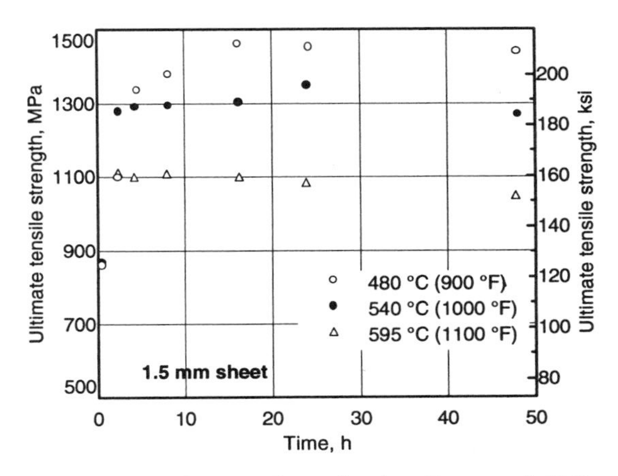

Beta-21S: Ultimate tensile strength vs aging time. Beta annealed 1.5 mm (0.06 in.) sheet aged for indicated times and temperatures

Forging

Beta-21S can be fabricated into all forging product types, although current closed die forging predominates. Beta-21S is a reasonably forgeable alloy when forged above its beta transus, with higher unit pressures (flow stresses), improved forge-ability, and less crack sensitivity in forging than the α–β alloy Ti-6Al-4V. Due to the high alloying content of Beta-21S, flow stresses are higher than those of the near-beta alloy Ti-10V-2Fe-3Al. The desired final microstructure from thermomechanical processing of Beta-21S during forging manufacture is a fine transformed β, with limited grain boundary films and a fine, recrystallized prior β grain size in preparation for final thermal treatments.

Thermomechanical Processing. The very fine microstructural features of Beta-21S achieved in forgings are responsible for its excellent mechanical properties and fatigue resistance. Reheating for subsequent forging operations recrystallizes the alloy from prior hot work, refining the grain size. Beta-21S is generally not subtransus forged because there is no microstructural advantage, and there is a significant increase in unit pressures.

Final thermal treatments for Beta-21S include a simple anneal (or solution anneal) for low-modulus applications or solution anneal and aging for higher strength levels. Forgings may be supplied annealed, solution annealed, and/or fully aged (STA). Annealing or solution annealing generally is conducted at 815 to 870 °C (1500 to 1600 °F). Aging is conducted at 535 to 595 °C (1000 to 1100 °F).

Beta forging working histories for Beta-21S require imparting enough hot work to reach final macrostructure and microstructure objectives. Generally, reductions in any given forging process are 30 to 50% to achieve desired dynamic and static recrystallization. Very low levels of beta reduction are not recommended.

Hydrogen. Beta-21S, as with all beta alloys, has a high affinity for hydrogen. Although Beta-21S forms less α case from heating operations than other alloy classes, therefore requiring less metal removal in chemical pickling (milling) processes, control of chemical removal processes is essential to preclude excessive hydrogen pickup.

Recommended forging metal temperatures range from 790 to 850 °C (1450 to 1560 °F). Recommended die temperatures are summarized in "Technical Note 4: Forging."

Ti-5Al-2Sn-4Zr-4Mo-2Cr-1Fe Beta-CEZ®

Compiled by Y. Combres, CEZUS Centre de Recherches, Ugine, France

Beta-CEZ® is a multifunctional near-β titanium alloy exhibiting high strength, high toughness, and intermediate-temperature creep resistance. Its processing flexibility makes it suitable for a wide range of applications.

Product Forms and Conditions. Typical product forms consist of forged billets in diameters ranging from 150 to 300 mm (6 to 12 in.) and forged or rolled bar in diameters ranging from 10 up to 110 mm (0.4 to 4.3 in.). Rolled plate and sheet are also available in thicknesses ranging from 25 to 3 mm (1 to 0.1 in.) and 500 mm (20 in.) wide. Products are supplied in the forged or solution treated conditions. The microstructure is fine and equiaxed.

Applications. Typical applications include heavy section forgings used for medium-temperature compressor disks in which an optimum combination of strength, ductility, toughness, and creep resistance is required. Beta-CEZ® is a structural alloy with very high strength and a good combination of strength, ductility, and toughness. Near-net shape forgings are possible due to the excellent formability of the alloy. Component applications are as forged parts, springs, and fasteners.

Beta-CEZ®: Chemical composition

Element	Composition, wt%
Aluminum	4.5-5.5
Tin	1.5-2.5
Zirconium	3.5-4.5
Molybdenum	3.5-4.5
Chromium	1.5-2.5
Iron	0.5-1.5
Oxygen	800-1300 ppm
Hydrogen	<150 ppm

Physical Properties

Crystal Structure. In the solution treated and aged condition, the microstructure consists of $\alpha + \beta$ phases. The lattice parameters of the close-packed hexagonal α phase are $a = 2.9287$ Å and $c = 4.6606$ Å, whereas the lattice parameter of the body-centered cubic β phase is $a = 3.2040$ Å.

Grain Structure. The microstructure is typical of β metastable alloys and may be β or $\alpha + \beta$, either equiaxed or lamellar. Highest strength and ductility are achieved with an equiaxed primary α phase and a finely precipitated secondary α phase microstructure. Optimum toughness is obtained with lamellar primary α microstructures.

Transformation Products. The continuous cooling (CCT) diagram is similar to that of Ti-17. Alpha precipitation occurs first at β grain boundaries and secondly inside the grains. For instance, the time difference between grain boundary and intragranular precipitation is about 1 h when cooled at 1 °C/min (1.8 °F/min) from the β field. The transformation of samples cooled from the β field exhibits a coarse α precipitation above 700 °C (1290 °F) and fine acicular precipitation between 700 and 400 °C (1290 and 750 °F). A temperature of > 750 °C (1380 °F) is recommended for solution treatments below the transus, whereas aging treatments are performed below 700 °C (1290 °F)

Chemical Corrosion Resistance. Corrosion resistance in acid or seawater, as well as hydrogen uptake and embrittlement are currently being studied. Data are not available yet.

Beta-CEZ®: Summary of typical physical properties

Beta transus	890 °C (1634 °F)
Solidus	~1550 °C (2820 °F)
Melting (liquidus point)	1602 °C (2916 °F)
Density(a)	4.69 g/cm^3
Specific heat capacity(a)	580 J/kg · K (0.14 Btu/lb · °F)
Thermal conductivity(a)	6.7 W/m · K (3.8 Btu/ft · h · °F)
Thermal coefficient of linear expansion(b)	10×10^{-6}/°C (5.5×10^{-6}/°F)

(a) Typical values at room temperature of about 20 to 25 °C (68 to 78 °F). (b) Mean coefficient from room temperature to 600 °C (1110 °F). See figure.

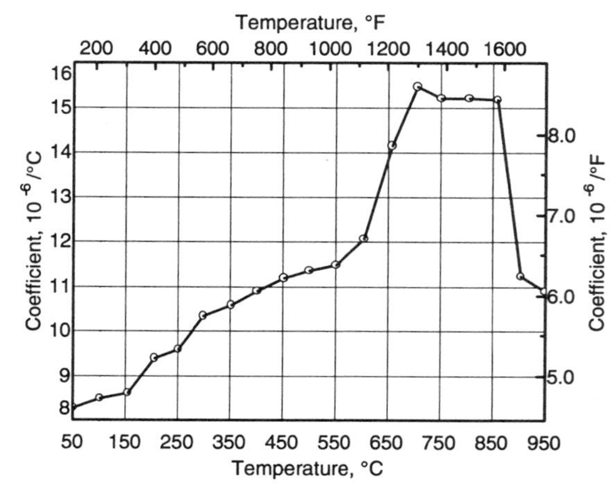

Beta-CEZ®: Thermal coefficient of linear expansion vs temperature

Mechanical Properties

Tensile Properties

Tensile properties depend strongly on microstructure (see table). Forged or rolled bars exhibit an equiaxed microstructure, whereas β processed and "through the β transus" processed pancakes exhibit lamellar and necklaced microstructures, respectively. Beta-CEZ® can maintain a high strength level at high temperatures for both the equiaxed or lamellar microstructures (see figure).

Beta-CEZ®: Young's modulus vs temperature

Temperature		Young's modulus	
°C	°F	GPa	10^6 psi
20	68	122	17
300	570	106	15
400	750	100	14

Beta-CEZ®: Typical tensile properties

Product form	Heat treatment	Ultimate tensile strength, MPa	Ultimate tensile strength, ksi	0.2% yield strength, MPa	0.2% yield strength, ksi	Elongation, %
150 mm (6 in.) diam forged bar	As forged	1040	150	960	139	18
	830 °C (1525 °C), 1 h, WQ + 550 °C (1020 °F), 8 h, AC	1601	232	1518	220	2
	830 °C (1525 °F), 1 h, WQ + 600 °C (1110 °F), 8 h, AC	1283	186	1208	175	11
	860 °C (1580 °F), 1 h, WQ + 550 °C (1020 °F), 8 h, AC	1557	226	1478	214	2
	860 °C (1580 °F), 1 h, WQ + 600 °C (1110 °F), 8 h, AC	1370	198	1304	189	5
12.7 mm (½ in.) diam rolled bar	As rolled	1490	216	1345	195	11
	830 °C (1525 °F), 1 h, WQ + 550 °C (1020 °F), 8 h, AC	1506	218	1460	211	13
	830 °C (1525 °F), 1 h, WQ + 600 °C (1110 °F), 8 h, AC	1373	199	1349	195	15
	860 °C (1580 °F), 1 h, WQ + 550 °C (1020 °F), 8 h, AC	1723	250	1683	244	7
	860 °C (1580 °F), 1 h, WQ + 600 °C (1110 °F), 8 h, AC	1540	223	1485	215	9
25 mm (1 in.) thick rolled plate	As rolled, L	1222	177	1124	163	15
	As rolled, T	1260	182	1163	168	11
	830 °C (1525 °F), 1 h, WQ + 600 °C (1110 °F), 8 h, AC, L	1334	193	1287	186	13
	830 °C (1525 °F), 1 h, WQ + 600 °C (1110 °F), 8 h, AC, T	1351	196	1300	188	12
	860 °C (1580 °F), 1 h, WQ + 600 °C (1110 °F), 8 h, AC, L	1405	203	1338	194	10
	860 °C (1580 °F), 1 h, WQ + 600 °C (1110 °F), 8 h, AC, T	1418	205	1340	194	6
300 mm (12 in.) diam β-processed pancake	600 °C (1110 °F), 8 h, AC	1608	233	1472	213	2
	830 °C (1525 °F), 1 h, WQ + 570 °C (1060 °F), 8 h, AC	1357	197	1171	170	5
	830 °C (1525 °F), 1 h, WQ + 600 °C (1110 °F), 8 h, AC	1326	192	1188	172	6
300 mm (12 in.) diam "through the transus" processed	600 °C (1110 °F), 8 h, AC	1227	178	1138	165	10
	830 °C (1525 °F), 1 h, WQ + 570 °C (1060 °F), 8 h, AC	1314	190	1200	174	10
	830 °C(1525 °F), 1 h, WQ + 600 °C (1110 °F), 8 h, AC	1263	183	1170	169	11

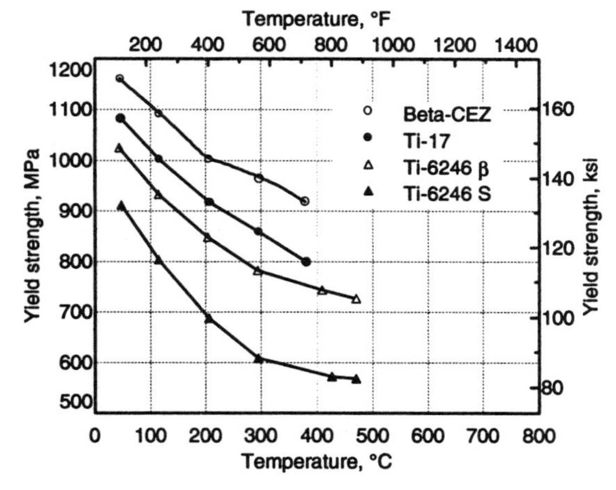

Beta-CEZ®: Yield strength comparison. Specimens were 300 mm (12 in.) diam β processed pancakes with a lamellar microstructure. Source: Ti-17 and Ti-6246 data courtesy of SNECMA.

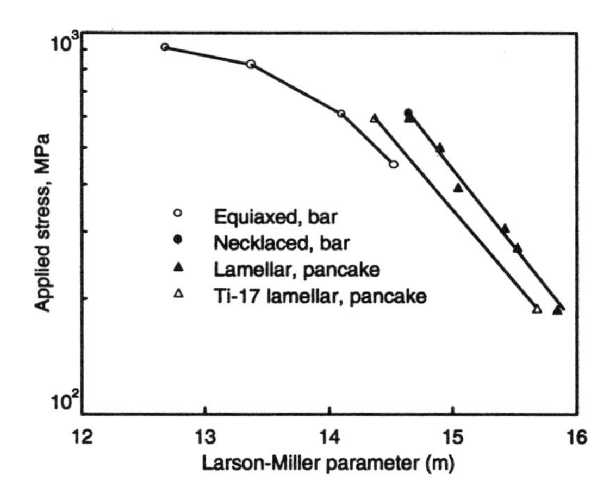

Beta-CEZ®: Creep property comparison. Beta-CEZ® had lamellar, necklaced, and equiaxed microstructures; Ti-17 specimens had lamellar β processed structures. All specimens were 300 mm (12 in.) diameter. $m = T(20 + \log t)10^{-3}$ (T in K; t in h). Source: Ti-17 data courtesy of SNECMA.

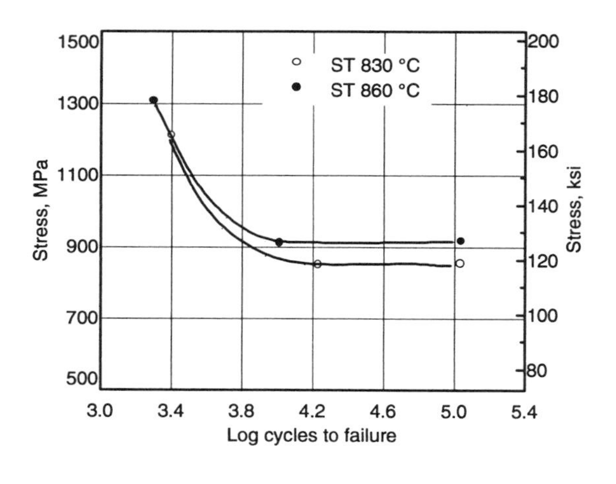

(a)

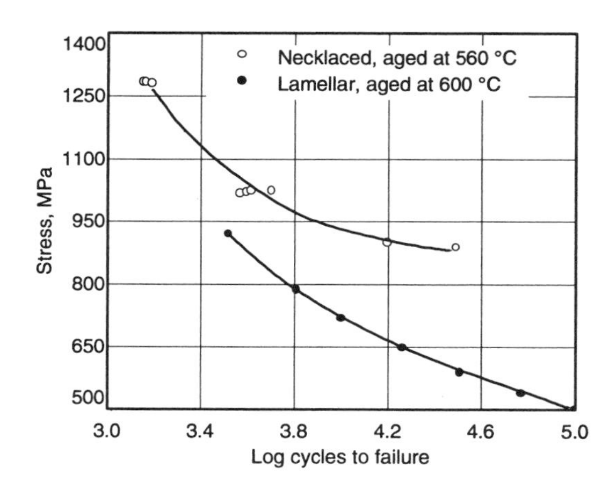

(b)

Beta-CEZ©: Low-cycle fatigue for equiaxed microstructures aged at 600 °C. (a) Specimens were from 150 mm (6 in.) diam forged bar. b) Beta-CEZ®: Low-cycle fatigue for lamellar and necklaced microstructures after solution treating at 830 °C. (b) Necklaced microstructures were 80 mm (3.1 in.) diam "through the β transus" forged bar; lamellar microstructures were 300 mm (12 in.) diam β-processed pancake.

Fatigue

The alloy behaves very well in low-cycle fatigue conditions between 20 °C (68 °F) (700 MPa, or 101 ksi) for 10^4 cycles) and 400 °C (750 °F) (600 MPa, or 87 ksi for 10^4 cycles). For high-cycle fatigue, equiaxed microstructures have a fatigue limit of 900 MPa (130 ksi) for 10^5 cycles at 20 °C (68 °F).

Crack Propagation Resistance

Products with the best toughness exhibit good fatigue crack propagation resistance at 20 °C (68 °F). Typical *da/dN* characteristics are shown here.

Fracture Toughness

Equiaxed microstructures are characterized by toughness ranging from 45 to 55 MPa$\sqrt{m}$ (40 to 50 ksi$\sqrt{in.}$). Toughness of the lamellar structure ranges from 60 to 90 MPa$\sqrt{m}$ (54 to 82 ksi$\sqrt{in.}$), whereas the necklaced microstructure has a toughness ranging from 65 to 95 MPa$\sqrt{m}$ (59 to 86 ksi$\sqrt{in.}$)(see figure).

Low-temperature toughness usually ranges from 30 to 45 MPa$\sqrt{m}$ (27 to 41 ksi$\sqrt{in.}$) at –253 °C (–423 °F).

Beta-CEZ®: Fatigue crack propagation for lamellar or necklaced microstructures

ΔK, MPa$\sqrt{m}$	ΔK, ksi$\sqrt{in.}$	*da/dN*, mm/cycle
6	5.4	10^{-8}
10	9.1	2×10^{-5}
60	54	10^{-3}

Note: Specimens were from 300 mm (12 in.) diameter pancakes in the β processed or "through the β transus" processed condition.

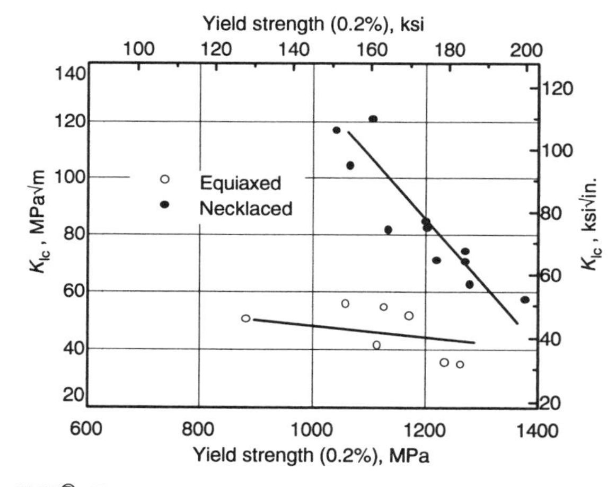

Beta-CEZ®: Fracture toughness vs yield strength comparison. Specimens were 70 mm (2.7 in.) diam α + β rolled bar (equiaxed structure) and 80 mm (3.1 in.) diam "through the β transus" forged bar (necklaced structure).

Fabrication

Forming

Because the strain-rate sensitivity exponent of Beta-CEZ® is rather high compared to conventional alloys (0.3 for Beta-CEZ® versus 0.2 for Ti-6Al-4V), plastic flow is more stable and enhances formability. The metastable nature of the alloy lowers its sensitivity to temperature.

Hot working in the α + β range is recommended at 800 to 860 °C (1470 to 1580 °F) to maintain a fine equiaxed microstructure. In the β range, a temperature around 920 °C (1690 °F) is suggested to obtain a lamellar structure by β processing.

"Through the Transus" Processing is a patented technique that results in a "necklaced" microstructure. It is applied to a 100% β metastable structure below 890 °C (1635 °F). Lamellae in the core of the grains and fine equiaxed grains at the boundaries are thus obtained, which leads to an excellent combination of strength, ductility, and toughness.

Superplastic Forming. The alloy displays superplastic properties between 725 and 775 °C (1340 and 1430 °F); 1000% ductility can be reached at strain rates as high as 8×10^{-4} s^{-1}. Diffusion bonding is being studied.

Superplastic properties of Beta CEZ alloy are obtained at temperatures as low as 725 °C (*Scripta Met*, Vol 29, No. 4, 1993, p 503-508). The as-forged material exhibits a complex microstructure, including the usual β and the globular primary alpha phases, but also a significant amount of acicular alpha. The superplastic behaviour of this unusual microstructure is associated with the breaking up of the acicular alpha in the first steps of deformation, which leads to a very fine mean grain size. The origin of superplasticity at these low temperatures is not yet clearly understood. More detailed investigations are needed, particularly to determine the effective diffusion coefficients in the β phase, since slow and fast diffusing elements (in comparison to Ti atoms) are present in this alloy.

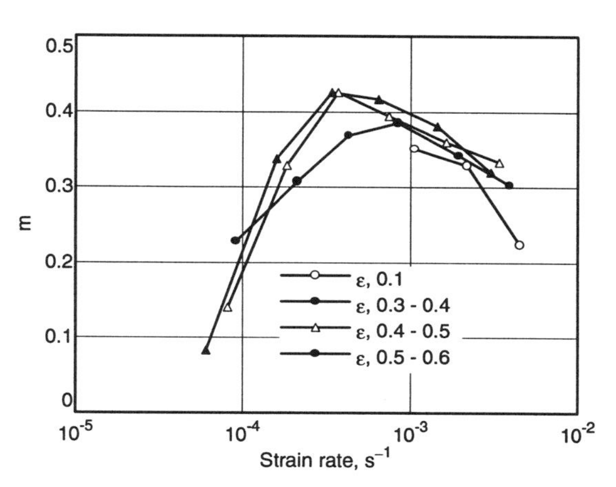

Beta CEZ: Strain-rate sensitivity at 760 °C. Deformation at 760 °C. Source: *Scripta Met,* Vol 29, p 503-508

Heat Treatment

Solution treatment is recommended between 750 and 860 °C (1380 to 1580 °F) from 1 to 4 h.

Aging is recommended between 525 and 650°C (980 to 1200 °F) from 30 min to 8 h. As a function of aging time, the hardness evolves rapidly (see table). Maximum hardness is about 560 HV.

Beta-CEZ®: Hardness kinetics for equiaxed microstructures

Product form	Heat treatment	Aging time, (*t*), min	Hardness (30 kg), HV
150 mm (6 in.) diam forged bar	As forged	0	345
	860 °C (1580 °F), 2 h, WQ + 550 °C (1020 °F), *t*, AC	1	380
		3	440
		10	470
		30	485
		100	480
		300	465
		1000	460
		3000	460

List of Tables and Figures

Aluminum Alloy Fatigue Data

Aluminum Alloy S-N fatigue - Tables

Aluminum Alloy S-N Fatigue - Figures

Aluminum Alloy S-N Data - Tables

Aluminum Alloy S-N Data - Tables

Aluminum Alloy S-N Data - Figures

Magnesium Alloy Fatigue Data

Magesium Alloy Fatigue and Fracture - Tables

Magesium Alloy Fatigue and Fracture - Figures

Magnesium Alloy Fatigue Data - Tables

Mg-Al Casting Alloys

Mg-Al Wrought Alloys

Mg-Zn Alloys

Mg-Th Alloys

Miscellaneous Mg Alloys

Magnesium Alloy Fatigue Data - Figures

Mg-Al Casting Alloys

Mg-Al Wrought Alloys

Mg-Zn Alloys

Mg-Th Alloys

Miscellaneous Alloys

Titanium Alloy Fatigue Data

Titanium Alloys Fatigue and Fracture - Tables

Titanium Alloys Fatigue and Fracture - Figures

Titanium Fatigue Data

Commercially Pure and Modified Titanium - Tables

Commercially Pure and Modified Titanium - Figures

Ti3Al2.5V - Tables

Ti3Al2.5V - Figures

Ti-5Al-2.5Sn - Tables

Ti-5Al-2.5Sn - Figures

Ti-6Al-2Sn-4Zr-2Mo-0.08Si - Tables

Ti-6Al-2Sn-4Zr-2Mo-0.08Si - Figures

Ti-8Al-1Mo-1V - Tables

Ti-8Al-1Mo-1V - Figures

TIMETAL® 1100 - Tables

TIMETAL® 1100 - Figures

IMI 834 Ti-5.8Al-4Sn-3.5Zr-0.7Nb-0.5Mo-0.35Si - Tables

IMI 834 Ti-5.8Al-4Sn-3.5Zr-0.7Nb-0.5Mo-0.35Si - Figures

Ti-5Al-2Sn-2Zr-4Mo-4Cr - Tables

Ti-5Al-2Sn-2Zr-4Mo-4Cr - Figures

Ti-6Al-2S-4Zr-6Mo - Tables

Ti-6Al-2S-4Zr-6Mo - Figures

Ti-6Al-4V - Tables

Ti-6Al-4V - Figures

Ti-6Al-6V-2Sn - Tables

Ti-6Al-6V-2Sn - Figures

Ti-6-22-22S
Ti-6Al-2Sn-2Zr-2Mo-2Cr-0.25Si - Tables

Ti-6-22-22S
Ti-6Al-2Sn-2Zr-2Mo-2Cr-0.25Si - Figures

Ti-3Al-8V-6Cr-4Mo-4Zr (Beta C) - Tables

Ti-3Al-8V-6Cr-4Mo-4Zr (Beta C) - Figures

Ti-10V-2Fe-3Al - Tables

Ti-10V-2Fe-3Al - Figures

Ti-15V-3Cr-3Al-3Sn - Tables

Ti-15V-3Cr-3Al-3Sn - Figures

TIMETAL® 21S - Tables

TIMETAL® 21S - Figures

Ti-5Al-2Sn-4Zr-4Mo-2Cr-1Fe Beta-CEZ® - Tables

Ti-5Al-2Sn-4Zr-4Mo-2Cr-1Fe Beta-CEZ® - Figures

F	future value, future worth
f	inflation rate per year
FW	future worth
g	growth rate for geometric gradient
i	actual interest rate
I	interest amount
i'	real interest rate
I_c	compound interest amount
i_e	effective interest rate
i_s	interest rate per sub-period
I_s	simple interest amount
i°	growth adjusted interest rate
IRR	internal rate of return
IRR_A	actual dollar IRR
IRR_R	real dollar IRR
i^*	internal rate of return
i^*_e	external rate of return
i^*_{ea}	approximate external rate of return
$I_{0,N}$	the value of a global price index at year N, relative to year 0
m	number of sub-periods in a period
MARR	minimum acceptable rate of return
MARR_A	actual dollar MARR
MARR_R	real dollar MARR
MAUT	multi-attribute utility theory
MCDM	multi-criterion decision making
N	number of periods, useful life of an asset
P	present value, present worth, purchase price, principal amount
PCM	pairwise comparison matrix
PW	present worth
$p(x)$	probability distribution
$\Pr\{X = x_i\}$	alternative expression of probability distribution
r	nominal interest rate, rating for a decision matrix
$R_{0,N}$	real dollar equivalent to A_N relative to year 0, the base year
RI	random index
S	salvage value
t	tax rate
UCC	undepreciated capital cost
X	random variable
$\mathbf{w}$	an eigenvector
λ_{max}	the maximun eigenvalue
λ	an eigenvalue
π_{01}	Laspeyres price index

Engineering Economics in Canada

Third Edition

Niall M. Fraser
Open Options Corporation

Elizabeth M. Jewkes
University of Waterloo

Irwin Bernhardt
University of Waterloo – retired

May Tajima
University of Western Ontario

Library and Archives Canada Cataloguing in Publication

Engineering economics in Canada / Niall M. Fraser ... [et al.]. — 3rd ed.

Includes index.
First ed. by Niall M. Fraser, Irwin Bernhardt, Elizabeth M. Jewkes.
ISBN 0-13-126957-7

1. Engineering economy—Canada. I. Fraser, Niall M. (Niall Morris), 1952- II. Fraser, Niall M. (Niall Morris), 1952- . Engineering economics in Canada.

TA177.4.F725 2006 658.15 C2004-905813-4

ISBN 0-13-126957-7

Vice President, Editorial Director: Michael J. Young
Executive Editor: Dave Ward
Marketing Manager: Janet Piper
Developmental Editor: Angela Kurmey
Production Editor: Judith Scott
Copy Editor: Rosemary Tanner
Production Coordinator: Wendy Moran
Manufacturing Coordinator: Susan Johnson
Page Layout: Gerry Dunn and Christine Velakis
Art Director: Julia Hall
Interior Design: Anthony Leung
Cover Design: Anthony Leung
Cover Image: Eyewire

2 3 4 5 10 09 08 07 06

Printed and bound in the United States

Brief Contents

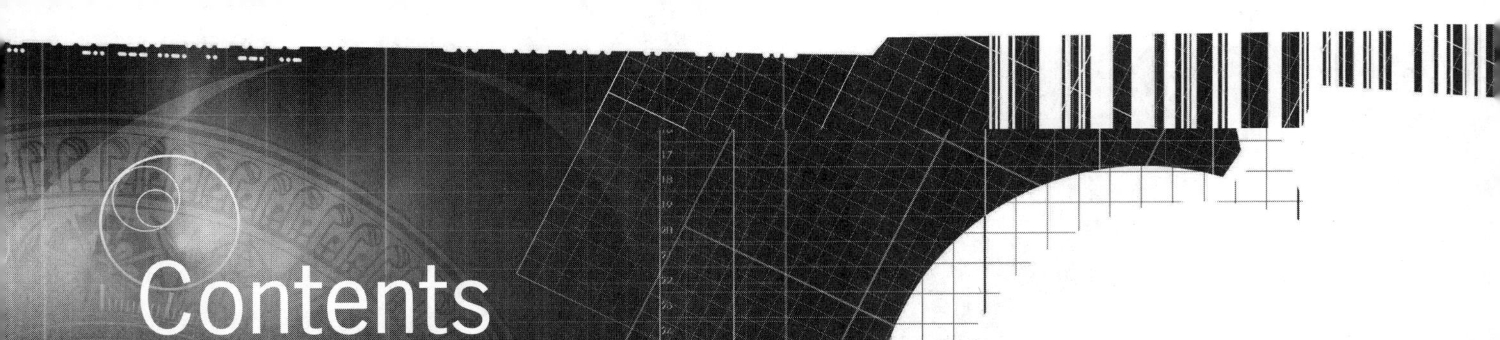
Contents

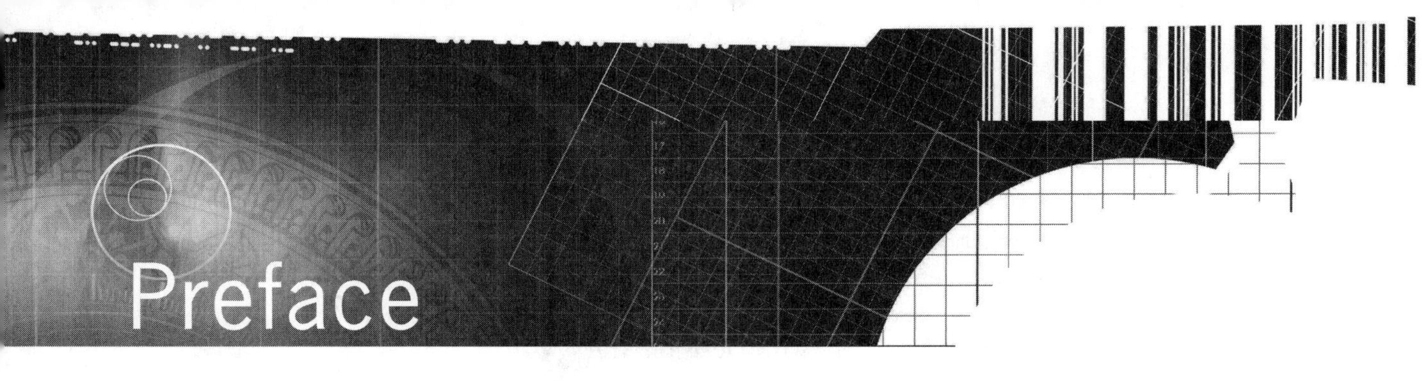

Preface

Courses on engineering economics are traditionally found in engineering curricula in Canada and throughout the world. The courses generally deal with deciding among alternatives with respect to expected costs and benefits. In Canada, the Canadian Engineering Accreditation Board requires that all accredited professional engineering programs provide at least one course in engineering economics. Many engineers have found that a course in engineering economics can be as useful in their practice as any of their more technical courses.

There are several stages to making a good decision. One stage is being able to determine whether a solution to a problem is technically feasible. This is one of the roles of the engineer, who has specialized training to make such technical judgments. Another stage is deciding which of several technically feasible alternatives is best. Deciding among alternatives often does not require the technical competence needed to determine which alternatives are feasible, but it is equally important in making the final choice. Some engineers have found that choosing among alternatives can be more difficult than deciding what alternatives exist.

The role of engineers in Canadian society is changing. In the past, engineers tended to have a fairly narrow focus, concentrating on the technical aspects of a problem and on strictly computational aspects of engineering economics. As a result, many engineering economics texts focused on the mathematics of the subject. Today, engineers are more likely to be the decision makers, and they need to be able to take into account strategic and policy issues.

This book is designed for teaching a course on engineering economics to match engineering practice in Canada today. It recognizes the role of the engineer as a decision maker who has to make and defend sensible decisions. Such decisions must not only take into account a correct assessment of costs and benefits; they must also reflect an understanding of the environment in which the decisions are made.

Canadian engineers have a unique set of circumstances that warrant a text with a specific Canadian focus. Canadian firms make decisions according to norms and standards that reflect Canadian views on social responsibility, environmental concerns, and cultural diversity. This perspective is reflected in the content and tone of much of the material in this book. Furthermore, Canadian tax regulations are complicated and directly affect engineering economic analysis. These regulations and their effect on decision making are covered in detail in Chapter 8.

This book also relates to students' everyday lives. In addition to examples and problems with an engineering focus, there are a number that involve decisions that many students might face, such as renting an apartment, getting a job, or buying a car. Other examples in the text are adapted from familiar sources such as Canadian newspapers and well-known Canadian companies.

Content and Organization

Because the mathematics of finance has not changed dramatically over the past number of years, there is a natural order to the course material. Nevertheless, a modern view of the role of the engineer flavours this entire book and provides a new, balanced exposure to the subject.

Chapter 1 frames the problem of engineering decision making as one involving many issues. Manipulating the cash flows associated with an engineering project is an important process for which useful mathematical tools exist. These tools form the bulk of the remaining chapters. However, throughout the text, students are kept aware of the fact that the eventual decision depends not only on the cash flows, but also on less easily quantifiable considerations of business policy, social responsibility, and ethics.

Chapters 2 and 3 present tools for manipulating monetary values over time. Chapters 4 and 5 show how the students can use their knowledge of manipulating cash flows to make comparisons among alternative engineering projects. Chapter 6 provides an understanding of the environment in which the decisions are made by examining depreciation and the role it plays in the financial functioning of a company and in financial accounting.

Chapter 7 deals with the analysis of replacement decisions. Chapters 8 and 9 are concerned with taxes and inflation, which affect decisions based on cash flows. Chapter 10 provides an introduction to public sector decision making.

Most engineering projects involve estimating future cash flows as well as other project characteristics. Since estimates can be in error and the future unknown, it is important for engineers to take uncertainty and risk into account as completely as possible. Chapter 11 deals with uncertainty, with a focus on sensitivity analysis. Chapter 12 deals with risk, using some of the tools of probability analysis.

Chapter 13 picks up an important thread running throughout the book: a good engineering decision cannot be based only on selecting the least-cost alternative. The increasing influence on decision making of health and safety issues, environmental responsibility, and human relations, among others, makes it necessary for the engineer to understand some of the basic principles of multi-criterion decision making.

New to This Edition

In addition to clarifying explanations, improving readability, updating material, and correcting errors, we have made the following important changes for the second edition:

- We have introduced a "Net Value" feature to each chapter. These boxed sections provide chapter-specific examples of how the Internet can be used as a source of information and guidance for decision making. We have relied heavily on authoritative Canadian sources and have provided relevant URLs.
- Chapter 6 now features a Close-Up box dealing with different types of business ownership and the advantages and disadvantages of each. Commonly used financial measures such as EBIT have been introduced.
- Chapter 7 has been reorganized to allow for a more natural flow describing the replacement decision-making process. The rationale and use of the "One Year Principle" has been clarified.
- Chapter 8 now includes material on personal income taxes and a comparison of the personal tax system in Canada and the corporate tax system. Other, minor changes were made to streamline the presentation of the capital cost allowance system.
- Chapter 10 has been restructured to include separate sections on identifying and eval-

uating social costs and benefits of engineering projects. Market failure is now more precisely defined, and the new Net Value box provides useful links to Canadian benefit-cost analysis (BCA) guides that contain clear and detailed instructions on how to properly perform a BCA.

- Chapter 12 has also been reorganized and expanded to provide additional coverage of probability concepts. As well, a broader coverage of decision-making criteria has been introduced to include dominance screening in addition to the existing expected value criterion. Students in the first or second year of an engineering degree program who have had little or no exposure to probability theory will be able follow the discussion. At the same time, the material on probability theory is concise enough to serve as a helpful review for students in the third or fourth year.
- A companion Web site has been created for the text. It provides supporting materials such as the solutions manual, sample PowerPoint lecture notes, a set of exercises for students, and useful links.
- Minor changes to all other chapters have been made to update and improve the overall flow and presentation of material.

Special Features

We have created special features for this book in order to facilitate the learning of the material and an understanding of its applications:

- **Engineering Economics in Action boxes** near the beginning and end of each chapter recount the fictional experiences of a young engineer at a Canadian company. These vignettes reflect and support the chapter material. The first box in each chapter usually portrays one of the characters trying to deal with a practical problem. The second box demonstrates how the character has solved the problem by applying material discussed in the chapter above. All these vignettes are linked to form a narrative that runs throughout the book. The main character is Naomi, a recent engineering graduate. In the first chapter, she starts her job in the engineering department at Canadian Widget Industries and is given a decision problem by her supervisor. Over the course of the book, Naomi learns about engineering economics on the job as the students learn from the book. There are several characters, who relate to one another in various ways, exposing the students to practical, ethical, and social issues as well as mathematical problems.

Engineering Economics in Action, Part 13A: Don't Box Them In

Naomi and Bill Astad were seated in Naomi's office. "OK," said Naomi. "Now that we know we can handle the demand, it's time to work on the design, right? What is the best design?" She was referring to the self-adjusting vacuum cleaner project for Powerluxe that she and Bill had been working on for several months.

"Probably the best way to find out," Bill answered, "will be to get the information from interviews with small groups of consumers."

"All right," said Naomi. "We have to know what to ask them. I guess the most important step for us is to define the relevant characteristics of vacuums."

"I agree," Bill responded. "We couldn't get meaningful responses about choices if we left out some important aspect of vacuums like suction power. One way to get the relevant characteristics will be to talk to people who have designed vacuums before, and probably to vacuum cleaner sales people, too." They both smiled at the humorous prospect of seeking out vacuum cleaner salespeople, instead of trying to avoid them.

"We're going to need some technical people on the team," Naomi said. "We will have to develop a set of technically feasible possibilities."

- **Close-Up boxes** in the chapters present additional material about concepts that are important but not essential to the chapter.

CLOSE-UP 12.1 Views on Probability

Classical or symmetric probability: This was the first view of probability and relies on games of chance such as dice, where the outcomes are equally likely. For example, if there are m possible outcomes for an uncertain event, since only one can and must occur, and each are assumed to be equally likely, then the chance of each occurring is $1/m$. For example, the probability that a coin toss will result in a "heads" is 1/2 because there are two sides and each is equally likely.

Relative frequency: The outcome of an random event E is observed over a large number of experiments, N. If the number of times the outcome e_i occurs is n_i, then we can estimate the probability of event $p(e_i)$ by n_i/N. More formally, the relative frequency view on probability says that $p(e_i) = \lim_{N\to\infty} n_i / N$ An example of this is flipping a coin 1000 times and discovering that it lands on its edge five times in 1000. An estimate of the probability of the coin landing on its edge is then 5/1000 = 0.005.

Subjective probability: Subjective or personal probability is an attempt to deal with unique

- In each chapter, a **Net Value box** provides a chapter-specific example of how the Internet can be used as a source of information and guidance for decision making.

NET VALUE 13.1

AHP Software

The analytic hierarchy process (AHP) is effective for structuring and analyzing complex, multi-attribute decision-making problems. It has a wide range of applications including vendor selection, risk assessment, strategic planning, resource allocation, and human resources management.

AHP has such broad industrial application areas that software has been developed to support its use. Expert Choice™ software has gained acceptance as a useful tool for companies to help them deal with decision-making situations that might otherwise be too complex to structure and solve.

The Expert Choice Web site (www.expertchoice.com) provides several case studies describing the software's use for a number of large international companies, reference materials for AHP, and access to a trial version of the product.

- At the end of each chapter, a **Canadian Mini-Case**, complete with discussion questions, relates interesting stories about how familiar Canadian companies have used engineering economic principles in practice.

Energy Management for a Quebec Office Building

CANADIAN 12.1 MINI-CASE

In 2003, Public Works and Government Services Canada (PWGSC) moved into their new 35 000-square-metre office space in Gatineau, Quebec. Several years earlier, PWGSC invited developers to tender irrevocable bids to include designing, building, and then leasing the building to PWGSC for a minimum of 15 years. The evaluation of the proposal was based on discounted first cost as well as the annual operating costs of the building which include items such as energy, maintenance, and cleaning costs.

PWGSC had set out extensive requirements on the building, including the quality of materials, the type of elevators, the level of indoor air quality, and a special request that the building operate at least 25% below the energy consumption prescribed in the Model National Energy Code for Building (MNECB).

The developer that won the bid relied on a project team to do extensive analysis of various energy management alternatives for the building design. The team identified alternatives that would reduce the building's energy consumption at least 25% below the MNECB. They then used numerous simulations to evaluate each alternative and decide

- Two **Extended Cases** are provided, one directly following Chapter 6 and the other directly following Chapter 11. They concern complex situations that incorporate much of the material in the preceding chapters. Unlike chapter examples, which are usually directed at a particular concept being taught, the Extended Cases require the students to integrate what they have learned over a number of chapters. They can be used for assignments, class discussions, or independent study.

Extended Case: Part 1

A.1 Introduction

Clem looked up from his computer as Naomi walked into his office. "Hi, Naomi. Sit down. Just let me save this stuff."

After a few seconds Clem turned around showing a grin. "I'm working on our report for the last quarter's operations. Things went pretty well. We exceeded our targets on defect reductions and on reducing overtime. And we shipped everything required—over 90% on time."

Naomi caught a bit of Clem's exuberance. "Sounds like a report you don't mind writing."

"Yeah, well, it was a team job. Everyone did good work. Talking about doing good work, I should have told you this before, but I didn't think about it at the right time. Ed Burns and Anna

Dave Sullivan came in with long strides and dropped into a chair. "Good morning, everybody. It is still morning, barely. Sorry to be late. What's up?"

Clem looked at Dave and started talking. "What's up is this. I want you and Naomi to look into our policy about buying or making small aluminum parts. We now use about 200 000 pieces a month. Most of these, like bolts and sleeves, are cold-formed.

"Prabha Vaidyanathan has just done a market projection for us. If she's right, our demand for these parts will continue to grow. Unfortunately, she wasn't very precise about the *rate* of growth. Her estimate was for anything between 5% and 15% a year. We now contract this work out. But even if growth is only 5%, we may be at the level

Additional Pedagogical Features

- Each chapter begins with a **list of the major sections** to provide an overview of the material that follows.
- **Key terms** are boldfaced where they are defined in the body of the text. For easy reference, all these terms are defined in a Glossary near the back of the book.
- Additional material is presented in **chapter appendixes** at the ends of Chapters 3, 4, 5, 8, 9, and 13.
- Numerous worked-out **Examples** are given throughout the chapters. Although the decisions have often been simplified for clarity, most of them are based on real situations encountered in the authors' consulting experiences.
- Worked-out **Review Problems** near the end of each chapter provide more complex examples that integrate the chapter material.
- A concise prose **Summary** is given for each chapter.
- Each chapter has 30 to 50 **Problems** of various levels of difficulty covering all of the material presented. Like the worked-out Examples, many of the problems have been adapted from real situations.

- A **spreadsheet icon** like the one shown here indicates where Examples or Problems involve spreadsheets, which are available on the Instructor's Resource CD-ROM. The use of computers by engineers is now as commonplace as the use of slide rules was 30 years ago. Students using this book will likely be very familiar with spreadsheet software. Consequently, such knowledge is assumed rather than taught in this book. The spreadsheet Examples and Problems are presented in such a manner that they can be done using any popular spreadsheet program, such as Excel, Lotus 1-2-3, or Quattro Pro.

- **Tables of interest factors** are provided in Appendix A, Appendix B, and Appendix C.
- **Answers to Selected Problems** are provided in Appendix D.
- For convenience, a **List of Symbols** used in the book is given on the inside of the front cover, and a **List of Formulas** is given on the inside of the back cover.

Course Designs

This book is ideal for a one-term course, but with supplemental material it can also be used for a two-term course. It is intended to meet the needs of students in all engineering programs, including, but not limited to, Aeronautical, Chemical, Computer, Electrical, Industrial, Mechanical, Mining, and Systems Engineering. Certain programs emphasizing public projects may wish to supplement Chapter 10, "Public Sector Decision Making," with additional material.

A course based on this book can be taught in the first, second, third, or fourth year of an engineering program. The book is also suitable for college technology programs. No more than high school mathematics is required for a course based on this text. The probability theory required to understand and apply the tools of risk analysis are provided in Chapter 12. Prior knowledge of calculus or linear algebra is not needed, except for working through the appendix to Chapter 13.

This book is also suitable for self-study by a practitioner or anybody interested in the economic aspects of decision making. It is easy to read and self-contained, with many clear examples. It can serve as a permanent resource for practising engineers or anyone involved in decision making.

Supplements

An Instructor's Resource CD-ROM includes sample lecture notes and complete solutions for all problems in the text as well as for the Extended Cases following Chapters 6 and 11. The solutions are also provided in spreadsheet form. They were prepared in Excel format, but they can be read and used by most other popular spreadsheet programs. Spreadsheets from the body of the text and those that can be used to solve some of the end-of-chapter problems are also provided. Each of these spreadsheets is identified in the textbook by a spreadsheet icon. The lecture notes are in PowerPoint format and include all the figures and many of the tables that occur in the text.

The companion Web site for the text provides chapter-by-chapter sample problems for students to work on to reinforce their learning process. Students will also find other resources such as weblinks and pdf files of interest rate tables on the Web site.

Acknowledgments

The authors wish to acknowledge the contributions of a number of individuals who assisted in the development of this text. First and foremost are the hundreds of engineering students at the University of Waterloo who have provided us with feedback on passages they found hard to understand, typographical errors, and examples that they thought could be improved. There are too many individuals to name in person, but we are very thankful to each of them for their patience and diligence.

We are very grateful for the constructive suggestions provided by David Fuller and John Moore as they taught engineering economics to engineering students in the first, second, and third years at the University of Waterloo. We would also like to thank Tim Nye (McMaster University) for his contributions to the second Extended Case, to Yuri Yevdokimov (University of New Brunswick) for his constructive suggestions on the revision of Chapter 10, and to Victor Waese for his review of 8.

During the development process for the new edition, Pearson Education Canada arranged for the anonymous review of parts of the manuscript by instructors of engineering economics courses at other Canadian universities. These reviews were extremely beneficial to us, and many of the best ideas incorporated in the final text originated with these reviewers. We can now thank them by name:

Walid Abdul-Kader, University of Windsor
Lindsay Ashworth, University of New Brunswick
Brent Bertrand, University of Toronto
Michel L. Bilodeau, McGill University
A. Elsawy, University of New Brunswick
Isobel W. Heathcote, University of Guelph
John Jones, Simon Fraser University
Leo Michelis, Ryerson University
John Morrall, University of Calgary
Taraneh Sowlati, University of British Columbia
Claude Théoret, University of Ottawa
Henry Tommy, St. Francis Xavier University
Yuri Yevdokimov, University of New Brunswick

Finally, we want to express our appreciation to the various editors at Pearson Education Canada for their professionalism and support during the writing of this book. Angela Kurmey, our Developmental Editor for this edition, was a tremendous support for the author team. We remain grateful to Maurice Esses, who played a particularly strong role in bringing the first and second editions to completion.

To all of the above, thank you again for your help. To those we may have forgotten to thank, our appreciation is just as great, even if our memories fail us. And to the reader, any errors that remain cannot be blamed on those who helped us. The remaining errors are perhaps the only things for which we can claim sole credit.

Niall M. Fraser
Elizabeth M. Jewkes
Irwin Bernhardt
May Tajima

The Pearson Education Canada **COMPANION WEBSITE**

A Great Way to Learn and Instruct Online

The Pearson Education Canada Companion Website is easy to navigate and is organized to correspond to the chapters in this textbook. Whether you are a student in the classroom or a distance learner you will discover helpful resources for in-depth study and research that empower you in your quest for greater knowledge and maximize your potential for success in the course.

[www.pearsoned.ca/fraser]

PH Companion Website

Engineering Economics in Canada, Third Edition, by Fraser, Jewkes, Bernhardt, and Tajma

Student Resources

The modules in this section provide students with tools for learning course material. These modules include:

- Multiple Choice Quizzes
- Spreadsheets
- Glossary Flashcards
- Weblinks
- Interest Tables

In the quiz modules students can send answers to the grader and receive instant feedback on their progress through the Results Reporter. Coaching comments and references to the textbook may be available to ensure that students take advantage of all available resources to enhance their learning experience.

CHAPTER 1

Engineering Decision Making

Engineering Economics in Action, Part 1A: Naomi Arrives

Naomi's first day on the job wasn't really her first day on the job. Ever since she had received the acceptance letter three weeks earlier, she had been reading and rereading all her notes about the company. Somehow she had arranged to walk past the plant entrance going on errands that never would have taken her that exact route in the past. So today wasn't the first time she had walked through that tidy brick entrance to the main offices of Canadian Widget Industries—she had done it the same way in her imagination a hundred times before.

Clement Sheng, the Engineering Manager who had interviewed Naomi for the job, was waiting for her at the reception desk. His warm smile and easy manner did a lot to break the ice. He suggested that they could go through the plant on the way to her desk. She agreed enthusiastically. "I hope you remember the engineering economics you learned in school," he said.

Naomi did, but rather than sound like a know-it-all, she replied, "I think so, and I still have my old textbook. I suppose you're telling me I'm going to use it."

"Yes. That's where we'll start you out, anyhow. It's a good way for you to learn how things work around here. We've got some projects lined up for you already, and they involve some pretty big decisions for Canadian Widgets. We'll keep you busy."

1.1 Engineering Decision Making

Engineering is a noble profession with a long history. The first engineers supported the military, using practical know-how to build bridges, fortifications, and assault equipment. In fact, the term "civil" engineer was coined to make the distinction between engineers who worked on civilian projects and engineers who worked on military problems.

In the beginning, all engineers had to know was the technical aspects of their jobs. Military commanders, for example, would have wanted a strong bridge built quickly. The engineer would be challenged to find a solution to the technical problem, and would not have been particularly concerned about the costs, safety, or environmental impacts of the project. As years went by, however, the engineer's job became far more complicated.

All engineering projects use resources, such as raw materials, money, labour, and time. Any particular project can be undertaken in a variety of ways, each one calling for a different mix of resources. For example, a standard light bulb requires inexpensive raw materials and little labour, but it is inefficient in its use of electricity and does not last very long. On the other hand, a high-efficiency light bulb uses more expensive raw materials and is more expensive to manufacture, but consumes less electricity and lasts longer. Both products provide light, but choosing which is better in a particular situation depends on how the costs and benefits are compared.

Historically, as the kinds of projects engineers worked on evolved and technology provided more than one way of solving technical problems, engineers were faced more often with having to choose among alternative solutions to a problem. If two solutions both dealt with a problem effectively, clearly the cheaper one was preferred. The practical science of engineering economics was originally developed specifically to deal with determining which of several alternatives was, in fact, the cheapest.

Choosing the cheapest alternative, though, is not the entire story. Though a project might be technically feasible and the cheapest solution to a problem, if the money isn't available to do it, it can't be done. The engineer has to become aware of the financial constraints on the problem, particularly if resources are very limited. In addition, an engineering project can meet all other criteria, but may cause detrimental environmental effects. Finally, any project can be affected by social and political constraints. For example, a large irrigation project called the Garrison Diversion Unit in North Dakota was effectively cancelled because of political action by Canadians and environmental groups, even though over $2 000 000 000 had been spent.

Engineers today must make decisions in an extremely complex environment. The heart of an engineer's skill set is still technical competence in a particular field. This permits the determination of possible solutions to a problem. However, necessary to all engineering is the ability to choose among several technically feasible solutions and to defend that choice credibly. The skills permitting the selection of a good choice are common to all engineers and, for the most part, are independent of which engineering field is involved. These skills form the discipline of engineering economics.

1.2 What Is Engineering Economics?

Just as the role of the engineer in society has changed over the years, so has the nature of engineering economics. Originally, engineering economics was the body of knowledge that allowed the engineer to determine which of several alternatives was economically best—the cheapest, or perhaps the most profitable. In order to make this determination properly, the engineer needed to understand the mathematics governing the relationship between time and money. Most of this book deals with teaching and using this knowledge. Also, for many kinds of decisions the costs and benefits are the most important factors affecting the decision, so concentrating on determining the economically "best" alternative is appropriate.

In earlier times, an engineer would be responsible for making a recommendation on the basis of technical and analytic knowledge, including the knowledge of engineering economics, and then a manager would decide what should be done. A manager's decision would often be different from the engineer's recommendation, because the manager would take into account issues outside the engineer's range of expertise. Recently, however, the trend has been for managers to become more reliant on the technical skills of the engineers, or for the engineers themselves to be the managers. Products are often very complex; manufacturing processes are fine-tuned to optimize productivity; and even understanding the market sometimes requires the analytic skills of an engineer. As a result, it is often only the engineer who has sufficient depth of knowledge to make a competent decision.

Consequently, understanding how to compare costs, although still of vital importance, is not the only skill needed to make suitable engineering decisions. One must also be able to take into account all the other considerations that affect a decision, and to do so in a reasonable and defensible manner.

Engineering economics, then, can be defined as that science which deals with techniques of quantitative analysis useful for selecting a preferable alternative from several technically viable ones.

The evaluation of costs and benefits is very important, and it has formed the primary content of engineering economics in the past. The mathematics for doing this evaluation, which is well developed, still makes up the bulk of studies of engineering economics. However, the modern engineer must be able to recognize the limits and applicability of these economic calculations, and must be able to take into account the inherent complexity of the real world.

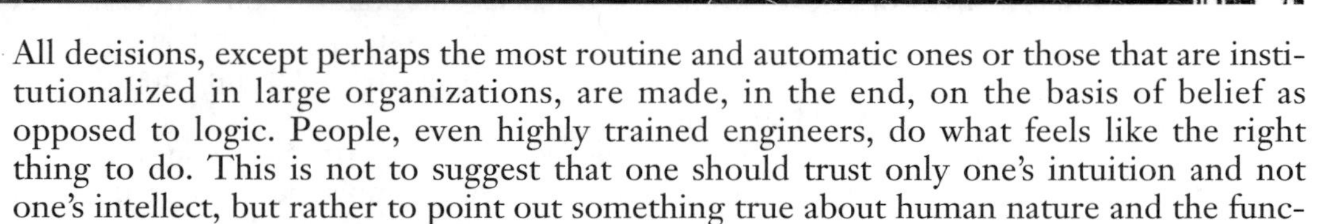

1.3 Making Decisions

All decisions, except perhaps the most routine and automatic ones or those that are institutionalized in large organizations, are made, in the end, on the basis of belief as opposed to logic. People, even highly trained engineers, do what feels like the right thing to do. This is not to suggest that one should trust only one's intuition and not one's intellect, but rather to point out something true about human nature and the function of engineering economics studies.

Figure 1.1 is a useful illustration of how decisions are made. At the top of the pyramid are preferences, which directly control the choices made. Preferences are the beliefs about what is best, and are often hard to explain coherently. They sometimes have an emotional basis and include criteria and issues that are difficult to verbalize.

The next tier is composed of people and politics. Politics in this context means the use of power (intentional or not) in organizations. For example, if the owner of a factory has a strong opinion that automation is important, this has a great effect on engineering decisions on the plant floor. Similarly, an influential personality can affect decision making. It's difficult to make a decision without the support, either real or imagined, of other people. This support can be manipulated, for example, by a persuasive salesperson or a persistent lobbyist. Support might just be a general understanding communicated through subtle messages.

The next tier is a collection of "facts." The facts, which may or may not be valid or verifiable, contribute to the politics and the people, and indirectly to the preferences. At the bottom of the pyramid are the activities that contribute to the facts. These include the history of previous similar decisions, statistics of various sorts, and, among other things, a determination of costs.

In this view of decisions, engineering economics is not very important. It deals essentially with facts and, in particular, with determining costs. Many other facts affect the final

Figure 1.1 Decision Pyramid

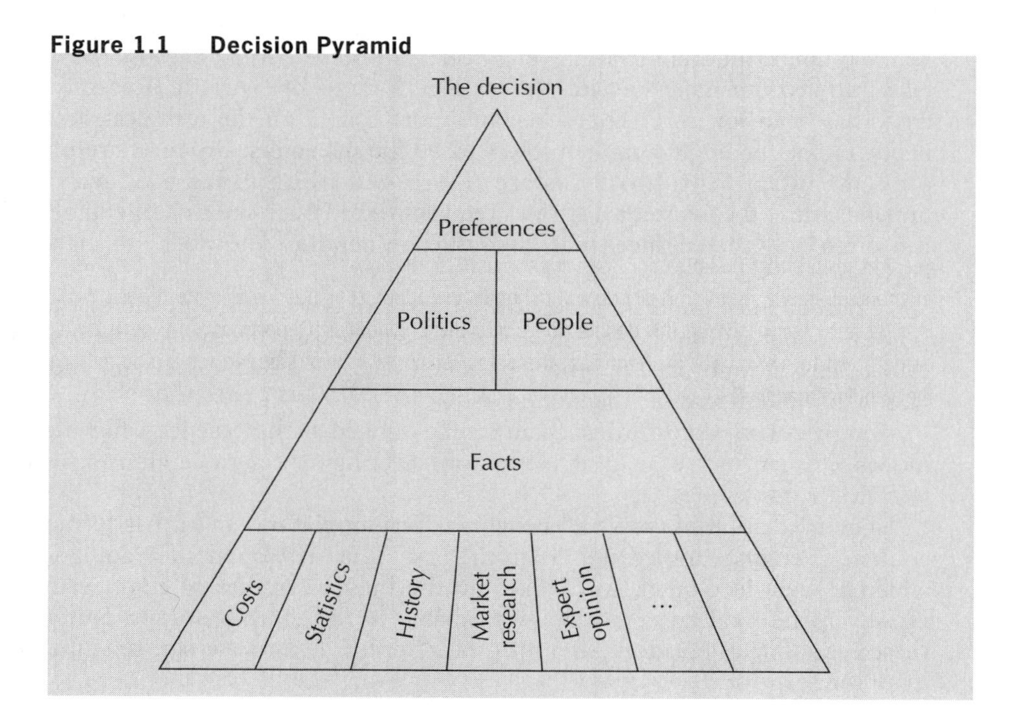

decision, and even then the decision may be made on the basis of politics, personality, or unstated preferences. However, this is an extreme view.

Although preferences, politics, and people can outweigh facts, usually the relationship is the other way around. The facts tend to control the politics, the people, and the preferences. It is facts that allow an individual to develop a strong opinion, which then may be used to influence others. Facts accumulated over time create intuition and experience that control our "gut feeling" about a decision. Facts, and particularly the activities that develop the facts, form the foundation for the pyramid in Figure 1.1. Without the foundation, the pyramid would collapse.

Engineering economics is important because it facilitates the establishment of verifiable facts about a decision. The facts are important and necessary for the decision to be made. However, the decision eventually made may be contrary to that suggested by analysis. For example, a study of several methods of treating effluent might determine that Method A is most efficient and cheapest, but Method B might in fact be chosen because it requires a visible change to the plant which, it is felt, will contribute to the company's image in environmental issues. Such a final decision is appropriate because it takes into account facts beyond those dealt with in the economic analysis.

Engineering Economics in Action, Part 1B: Naomi Settles In

As Naomi and Clement were walking, they passed the loading docks. A honk from behind told them to move over so that a forklift could get through. The operator waved in passing and continued on with the task of moving coils of sheet metal into the warehouse. Naomi noticed shelves and shelves of packaging material, dies, spare parts, and other items that she didn't recognize. She would find out more soon enough. They continued to walk. As they passed a welding area, Clem pointed out the newest recycling project in Canadian Widgets: the water used to degrease the metal was now being cleaned and recycled rather than being used only once.

Naomi became aware of a pervasive, pulsating noise emanating from the distance. Suddenly the corridor opened up to the main part of the plant, and the noise became a bedlam of clanging metal and thumping machinery. Her senses were assaulted. The ceiling was very high, and there were rows of humpbacked metal monsters unlike any presses she had seen before. The tang of mill oil overwhelmed her sense of smell, and she felt the throbbing from the floor knocking her bones together. Clem handed her hearing and eye protectors.

"These are our main press lines." Clem was yelling right into Naomi's ear, but she had to strain to hear. "We go right from sheet metal to finished widgets in 12 operations." A passing forklift blew propane exhaust at her, momentarily replacing the mill-oil odour with hot-engine odour. "Engineering is off to the left there."

As they went through the double doors into the Engineering Department, the din subsided and the ceiling came down to normal height. Removing the safety equipment, they stopped for a moment to get some juice at the vending machines. As Naomi looked around, she saw computers on desks more or less sectioned off by acoustic room dividers. As Clem led her farther, they stopped long enough for him to introduce Naomi to Carole Brown, the receptionist and secretary. Just past Carole's desk and around the corner was Naomi's desk. It was a nondescript metal desk with a long row of empty shelving above. Clem said that her computer would arrive within the week. Naomi noticed that the desk next to hers was empty, too.

"Am I sharing with someone?" she asked.

"Well, you will be. That's for your co-op student."

"My co-op student?"

"Yep. Don't worry, we have enough to do to keep you both busy. Why don't you take a few minutes to settle in, while I take care of a couple of things. I'll be back in, say, fifteen minutes. I'll take you over to Human Resources. You'll need a security pass, and I'm sure they have lots of paperwork for you to fill out."

Clem left. Naomi sat down and opened the briefcase she had carefully packed that morning. Alongside the brownbag lunch was an engineering economics textbook. She took it out and placed it on the empty shelf above the desk. "I thought I might need you," she said to herself. "Now, let's get this place organized!"

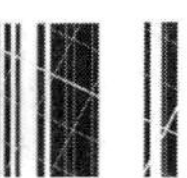

1.4 Dealing with Abstractions

The world is far more complicated than can ever be described in words, or even thought about. Whenever one deals with reality, it is done through models or abstractions. For example, consider the following description:

> Naomi watched the roll of sheet metal pass through the first press. The die descended and punched six oval shapes from the sheet. These "blanks" dropped through a chute into a large metal bin. The strip of sheet metal jerked forward into the die and the press came down again. Pounding like a massive heart 30 times a minute, the machine kept the operator busy full time just providing the giant coils of metal, removing the waste skeleton scrap, and stacking blanks in racks for transport to the next operation.

This gives a description of a manufacturing process that is reasonably complete, in that it permits one to visualize the process. But it is not absolutely complete. For example, how large and thick were the blanks? How big was the metal bin? How heavy was the press? How long did it take to change a die? These questions might be answered, but no matter how many questions are asked, it is impossible to express all of the complexity of the real world. It is also undesirable to do so.

When one describes something, one does so for a purpose. In the description, one selects those aspects of the real world that are relevant to that purpose. This is appropriate, since it would be very confusing if a great deal of unnecessary information were given every time something was talked or written about. For example, if the purpose of the above description were to explain the exact nature of the blanks, there would be considerably less emphasis on the process, and many more details about the blanks themselves.

This process of simplifying the complexities of the real world is necessary for any engineering analysis. For example, in designing a truss for a building, it is usually assumed that the members exhibit uniform characteristics. However, in the real world these members would be lumber with individual variations: some would be stronger than average and some would be weaker. Since it is impractical to measure the characteristics of each piece of wood, a simplification is made. As another example, the various components of an electric circuit, such as resistors and capacitors, have values that differ from their nominal specifications because of manufacturing tolerances, but such differences are often ignored and the nominal values are the ones used in calculations.

Figure 1.2 illustrates the basic process of modelling that applies in so much of what humans do, and applies especially to engineering. The world is too complicated to express completely, as represented by the amorphous shape at the top of the figure. People extract from the real world a simplification (in other words, a model) that captures information useful and appropriate for a given purpose. Once the model is developed, it is used to analyze a situation, and perhaps make some predictions about the real world. The analysis and the predictions are then related back to the real world to make sure the model is valid. As a result, the model might need some modification, so that it more accurately reflects the relevant features of the real world.

Figure 1.2 The Modelling Process

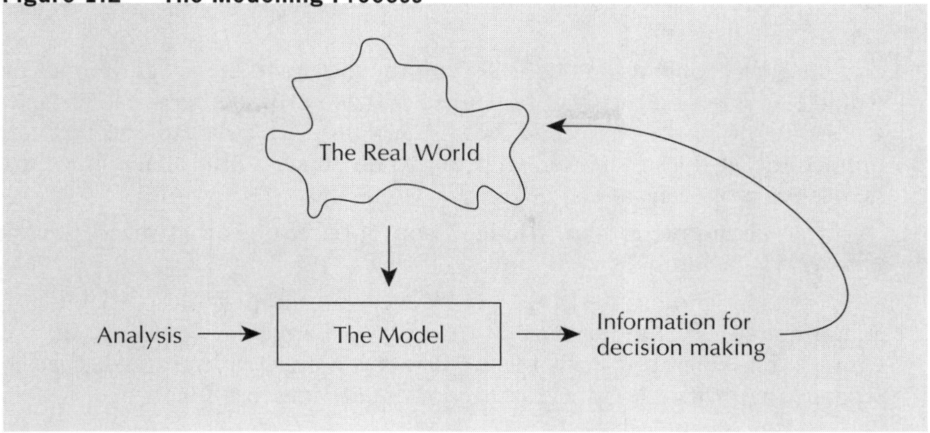

The process illustrated in Figure 1.2 is exactly what is done in engineering economics. The model is often a mathematical one that simplifies a more complicated situation, but does so in a reasonable way. The analysis of the model provides some information, such as which solution to a problem is cheapest. This information must always be related back to the real problem, however, to take into account the aspects of the real world that may have been ignored in the original modelling effort. For example, the economic model might not have included taxes or inflation, and an examination of the result might suggest that taxes and inflation should not be ignored. Or, as already pointed out, environmental, political, or other considerations might modify any conclusions drawn from the mathematical model.

Example 1.1

Naomi's brother Ben has been given a one-year assignment in the Yukon, and he wants to buy a car just for the time he is there. He has three choices, as illustrated in Table 1.1. For each alternative, there is a purchase price, an operating cost (including gas, insurance, and repairs), and an estimated resale value at the end of the year. Which should Ben buy?

Table 1.1 Buying a Car

	1957 Chevy	2002 Ford Focus	1997 BMW 5-Series
Purchase	$12 000	$7 000	$20 000
Operation	$200/month	$100/month	$150/month
Resale	$13 000	$5 000	$20 000

The next few chapters of this book will show how to take the information from Table 1.1 and determine which alternative is economically best. As it turns out, under most circumstances, the Chevy is best. However, in constructing a model of the decision, we must make a number of important assumptions.

For example, how can one be sure of the resale value of something until one actually tries to sell it? Along the same lines, who can tell what the actual maintenance costs will be? There is a lot of uncertainty about future events that is generally ignored in these kinds of calculations. Despite this uncertainty, estimates can provide insights into the appropriate decision.

Another problem for Ben is getting the money to buy a car. Ben is fairly young, and would find it very difficult to raise even $7000, perhaps impossible to raise $20 000. The Chevy might be the best value, but if the money isn't available to take advantage of the opportunity it doesn't matter. In order to do an economic analysis, we may assume that he has the money available.

If an economic model is judged appropriate, does that mean Ben should buy the Chevy? Maybe not.

A person who has to drive to work every morning would probably not want to drive an antique car. It is too important that the car be reliable (especially in the Yukon in the winter). The operating costs for the Chevy are high, reflecting the need for more maintenance than with the other cars, and there are indirect effects of low reliability that are hard to capture in dollars.

If Ben were very tall, he would be extremely uncomfortable in the compact Ford Focus, so that, even if it were economically best, he would hesitate to resign himself to driving with his knees on either side of the steering wheel.

Ben might have strong feelings about the environmental record of one of the car manufacturers, and might want to avoid driving that car as a way of making a statement.

Clearly, there are so many intangibles involved in a decision like this that it is impossible for anyone but Ben himself to make such a personal choice. An outsider can point out to Ben the results of a quantitative analysis, given certain assumptions, but cannot authoritatively determine the best choice for Ben.■

1.5 The Moral Question: Three True Stories

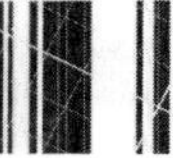

Complex decisions often have an ethical component. Recognizing this component is important for engineers, since society relies on them for so many things. The following three anecdotes concern real Canadian companies—although names and details have been altered for anonymity—and illustrate some extreme examples of the forces acting on engineering decision making.

Example 1.2

The process of making sandpaper is similar to that of making a photocopy. A two-metre-wide roll of paper is coated with glue and given a negative electric charge. It is then passed over sand (of a particular type) that has a positive charge. The sand is attracted to the paper and sticks on the glue. The fact that all of the bits of sand have the same type of charge makes sure that the grains are evenly spaced. The paper then passes through a long, heated chamber to cure the glue. Although the process sounds fairly simple, the machine that does this, called a maker, is very complicated and expensive. One such machine, costing several million dollars, can support a factory employing hundreds of workers.

Preston Sandpapers was the Canadian subsidiary of a large United States firm, and was located in a small town. Its maker was almost 30 years old and desperately needed replacement. However, rather than replace it, the parent company might have chosen to close down the Canadian plant and transfer production to one of the American sister plants.

The chief engineer had a problem. The costs for installing a new maker were extremely high, and it was difficult to justify a new maker economically. However, if he could not do so, the plant would close and hundreds of workers would be out of a job, including perhaps he himself. What he chose to do was lie. He fabricated figures, ignored important costs, and exaggerated benefits to justify the expenditures. The investment was made, and the Canadian plant is still operating.■

Example 1.3

Hespeler Meats is a medium-sized meat processor specializing in deli-style cold cuts and European process meats. Hoping to expand their product offerings, they decided to add a line of canned pâtés. They were eligible for a government grant to cover some of the purchase price of the necessary production equipment.

Government support for manufacturing is generally fairly sensible. Support is usually not given for projects that are clearly very profitable, since the company should be able to justify such an expense itself. On the other hand, support is also usually not given for projects that are clearly not very profitable, because taxpayers' money should not be wasted. Support is directed at projects that the company would not otherwise undertake, but that have good potential to create jobs and expand the economy.

Hespeler Meats had to provide a detailed justification for their canned pâté project in order to qualify for the government grant. Their problem was that they had to predict both the expenditures and the receipts for the following five years. This was a product line with which they had no experience, and which, in fact, had not been offered in North America by any meat processor. They had absolutely no idea what their sales would be. Any numbers they picked would be guesses, but to get the grant they had to give numbers.

What they did was select an estimate of sales that, given the equipment expenditures expected, fell exactly within that range of profitability that made the project suitable for government support. They got the money. As it turned out, the product line was a flop, and the canning equipment was sold as scrap five years later.■

Example 1.4

When a large metal casting is made, as for the engine block of a car, it has only a rough exterior, and often has flash—ragged edges of metal formed where molten metal seeped between the two halves of the mold. The first step in finishing the casting is to grind off the flash, and to grind flat surfaces so that the casting can be held properly for subsequent machining.

Galt Casting Grinders (GCG) made the complex specialized equipment for this operation. It had once commanded the world market for this product, but lost market share to Japanese competitors. The competitors did not have a better product than GCG, but they were able to increase market share by adding fancy display panels with coloured lights, dials, and switches that looked very sophisticated.

GCG's problem was that their idea of sensible design was to omit the features the competitors included (or the customers wanted). GCG reasoned that these features added nothing to the capability of the equipment, but did add a lot to the manufacturing cost and to the maintenance costs that would be borne by the purchaser. They had no doubt that it was unwise, and poor engineering design, to make such unnecessarily complicated displays, so they made no changes.

GCG went bankrupt several years later.■

In each of these three examples, the technical issues are overwhelmed by the non-technical ones. For Preston Sandpapers, the chief engineer was pressured by his social responsibility and self-interest to lie and to recommend a decision not justified by the facts. In the Hespeler Meats case, the engineer had to choose between stating the truth—that future sales were unknown—which would deny the company a very useful grant, and selecting a convenient number that would encourage government support. For Galt Casting Grinders, the issue was marketing. They did not recognize that a product must be more than technically good; it must also be saleable.

Beyond these principles, however, there is a moral component to each of these anecdotes. As guardians of knowledge, engineers have a vital responsibility to society to behave ethically and responsibly in all ways. When so many different issues must be taken into account in engineering decision making, it is often difficult to determine what course of action is ethical.

For Preston Sandpapers, most people would probably say that what the chief engineer did was unethical. However, he did not exploit his position simply for personal gain. He was, to his mind, saving a town. Is the principle of honesty more important than several hundred jobs? Perhaps it is, but when the job holders are friends and family it is understandable that unethical choices might be made.

For Hespeler Meats, the issue is subtler. Is it ethical to choose figures that match the ideal ones to gain a government grant? It is, strictly speaking, a lie, or at least misleading, since there is no estimate of sales. On the other hand, the bureaucracy demands that some numbers be given, so why not pick ones that suit your case?

In the Galt Casting Grinders case, the engineers apparently did no wrong. The ethical question concerns the competitors' actions. Is it ethical to put features on equipment that do no good, add cost, and decrease reliability? In this case and for many other products, this is often done, ethical or not. If it is unethical, the ethical suppliers will sometimes go out of business.

There are no general answers to difficult moral questions. Practising engineers often have to make choices with an ethical component, and can sometimes rely on no stronger foundation than their own sense of right and wrong. More information about ethical issues for engineers can be obtained from provincial professional engineering associations.

NET VALUE 1.1

Professional Engineering Associations and Ethical Decisions

Each of the provincial engineering associations has a Web site that can be a good source of information about engineering practice. At time of publication, the engineering association sites are:

Newfoundland **www.apegn.nf.ca**

Nova Scotia **www.apens.ns.ca**

Prince Edward Island **www.apepei.com**

New Brunswick **www.apenb.nb.ca**

Quebec **www.oiq.qc.ca**

Ontario **www.peo.on.ca**

Manitoba **www.apegm.mb.ca**

Saskatchewan **www.apegs.sk.ca**

Alberta **www.apegga.com**

British Columbia **www.apeg.bc.ca**

Northwest Territories and Nunavut **www.napegg.nt.ca**

For example, check out the Code of Ethics in the Professional Engineers Ontario (PEO) site, the Engineering Practice section of the Association of Professional Engineers of Nova Scotia (APENS) site, and the Disciplinary Decisions section in the OIQ site. Understanding ethics as enforced by the engineering associations is an excellent basis for making your own ethical decisions.

1.6 Uncertainty and Sensitivity Analysis

Whenever people predict the future, errors occur. Sometimes predictions are correct, whether the predictions are about the weather, a ball game, or company cash flow. On the other hand, it would be unrealistic to expect anyone to be always right about something that hasn't happened yet.

Although one cannot expect an engineer to precisely predict the future, approximations are very useful. A Canadian weather forecaster can dependably say that it will not snow in July, for example, even though it may be more difficult to forecast the exact temperature. Similarly, an engineer may not be able to precisely predict the scrap rate of a testing process, but may be able to determine a range of likely rates to help in a decision-making process.

Engineering economics analyses are quantitative in nature, and most of the time the quantities used in economic evaluations are estimates. The fact that we don't have precise values for some quantities may be very important, since decisions may have expensive consequences and significant health and environmental effects. How can the impact of this uncertainty be minimized?

One way to control this uncertainty is to make sure that the information being used is valid and as accurate as possible. The GIGO rule—"garbage in, garbage out"—applies here. Nothing is as useless or potentially dangerous as a precise calculation made from inaccurate data. However, even accurate data from the past is of only limited value when predicting the future. Even with sure knowledge of past events, the future is still uncertain.

Sensitivity analysis involves assessing the effect of uncertainty on a decision. It is very useful in engineering economics. The idea is that, although a particular value for a parameter can be known with only a limited degree of certainty, a range of values can be assessed with reasonable certainty. In sensitivity analysis, the calculations are done several times, varying each important parameter over its range of possible values. Usually only one parameter at a time is changed, so that the effect of each change on the conclusion can be assessed independently of the effect of other changes.

In Example 1.1, Naomi's brother Ben had to choose a car. He made an estimate of the resale value of each of the alternative cars, but the *actual* resale amount is unknown until the car is sold. Similarly, the operating costs are not known with certainty until the cars are driven for a while. Before concluding that the Chevy is the right car to buy (on economic grounds at least), Ben should assess the sensitivity of this decision by varying the resale values and operating costs within a range from the minimum likely amount to the maximum likely amount. Since these calculations are often done on spreadsheets, this assessment is not hard to do, even with many different parameters to vary.

Sensitivity analysis is an integral part of all engineering economics decisions because data regarding future activities are always uncertain. In this text, emphasis is usually given to the structure and formulation of problems rather than to verifying whether the result is robust. In this context, *robust* means that the same decision will be made over a wide range of parameter values. It should be remembered that no decision is properly made unless the sensitivity of that decision to variation in the underlying data is assessed.

A related issue is the number of significant digits in a calculation. Modern calculators and computers can carry out calculations to a large number of decimal places of precision. For most purposes, such precision is meaningless. For example, a cost calculated as $1.0014613076 is of no more use than $1.00 in most applications. It is useful, though, to carry as many decimal places as convenient to reduce the magnitude of accumulated rounding-off errors.

In this book, all calculations have been done to as many significant digits as could conveniently be carried, even though the intermediate values are shown with three to six digits. As a rule, only three significant digits are assumed in the final value. For decision-making purposes, this is plenty.

1.7 How This Book Is Organized

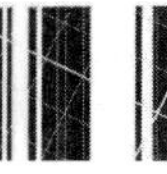

There are twelve chapters remaining in this book. The first block, consisting of Chapters 2 to 5, forms the core material of the book. Chapters 2 and 3 of that block provide the mathematics needed to manipulate monetary values over time. Chapters 4 and 5 deal with comparing alternative projects. Chapter 4 illustrates present worth, annual worth, and payback period comparisons, and Chapter 5 covers the internal rate of return (IRR) method of comparison.

The second block, Chapters 6 to 8, broadens the core material. It covers depreciation and analysis of a company's financial statements, when to replace equipment (replacement analysis), and taxation.

The third block, Chapters 9 to 13, provides supporting material for the previous chapters. Chapter 9 concerns the effect of inflation on engineering decisions, and Chapter 10 explores how decision making is done for projects owned by or affecting the public, rather than an individual or firm. Chapter 11 deals with handling uncertainty about important information through sensitivity analysis, while Chapter 12 deals with situations where exact parameter values are not known, but probability distributions for them are known. Finally, Chapter 13 provides some formal methods for taking into account the intangible components of an engineering decision.

Each chapter begins with a story about Naomi and her experiences at Canadian Widgets. There are several purposes to these stories. They provide an understanding of engineering practice that is impossible to convey with short examples. In each chapter, the story has been chosen to make clear why the ideas being discussed are important. It is also hoped that the stories make the material taught a little more interesting.

There is a two-part Extended Case in the text. Part 1, located between Chapters 6 and 7, presents a problem that is too complicated to include in any particular chapter, but that reflects a realistic situation likely to be encountered in engineering practice. Part 2, located between Chapters 11 and 12, builds on the first case to use some of the more sophisticated ideas presented in the later chapters.

Throughout the text are boxes which contain information that is associated with, and complements, the text material. One set of boxes contains Close-Ups, which focus on topics of relevance to the chapter material. These appear in each chapter in the appropriate section. There are also Net Value boxes, which tie the material presented to internet resources. The boxes are in the middle sections of each chapter. Another set of boxes presents Canadian Mini-Cases, which appear at the end of each chapter, following the problem set. These cases report how engineering economics is used in familiar Canadian companies, and include questions designed for classroom discussion or individual reflection.

End-of-chapter appendices contain relevant but more advanced material. Appendices at the back of the book provide tables of important and useful values and answers to selected chapter-end problems.

Engineering Economics in Action, Part 1C: A Taste of What Is to Come

Naomi was just putting on her newly laminated security pass when Clem came rushing in. "Sorry to be late," he puffed. "I got caught up in a discussion with someone in Marketing. Are you ready for lunch?" She certainly was. She had spent the better part of the morning going through the benefits package offered by Canadian Widgets and was a bit overwhelmed by the paperwork. Dental plan options, pension plan beneficiaries, and tax forms swam in front of her eyes. The thought of food sounded awfully good.

As they walked to the lunchroom, Clem continued to talk. "Maybe you will be able to help out once you get settled in, Naomi."

"What's the problem?" asked Naomi. Obviously Clem was still thinking about his discussion with this person from Marketing.

"Well," said Clem, "currently we buy small aluminum parts from a subcontractor. The cost is quite reasonable, but we should consider making the parts ourselves, because our volumes are increasing and the fabrication process would not be difficult for us to bring in-house. We might be able to make the parts at a lower cost. Of course, we'd have to buy some new equipment. That's why I was up in the Marketing Department talking to Prabha."

"What do you mean?" asked Naomi, still a little unsure. "What does this have to do with marketing?"

Clem realized that he was making a lot of assumptions about Naomi's knowledge of Canadian Widgets. "Sorry," he said, "I need to explain. I was up in Marketing to ask for some demand forecasts so that we would have a better handle on the volumes of these aluminum parts we might need in the next few years. That, combined with some digging on possible equipment costs, would allow us to do an analysis of whether we should make the parts in-house or continue to buy them."

Things made much more sense to Naomi now. Her engineering economics text was certainly going to come in handy.

PROBLEMS

1.1 In which of the following situations would engineering economics analysis play a strong role, and why?

(a) Buying new equipment

(b) Changing design specifications for a product

(c) Deciding the paint colour for the factory floor

(d) Hiring a new engineer

(e) Deciding when to replace old equipment with new equipment of the same type

(f) Extending the cafeteria business hours

(g) Deciding which invoice forms to use

(h) Changing the 8-hour work shift to a 12-hour one

(i) Deciding how much to budget for research and development programs

(j) Deciding how much to donate for the town's new library

(k) Building a new factory

(l) Downsizing the company

1.2 Starting a new business requires many decisions. List five examples of decisions that might be assisted by engineering economics analysis.

1.3 For each of the following items, describe how the design might differ if the costs of manufacturing, use, and maintenance were not important. On the basis of these descriptions, is it important to consider costs in engineering design?

(a) A car

(b) A television set

(c) A light bulb

(d) A book

1.4 Leslie and Sandy, recently married students, are going to rent their first apartment. Leslie has carefully researched the market and has decided that, all things considered, there is only one reasonable choice. The two-bedroom apartment in the building at the corner of University and Erb Streets is the best value for the money, and is also close to school. Sandy, on the other hand, has just fallen in love with the top half of a duplex on Dunbar Road. Which apartment should they move into? Why? Which do you think they will move into? Why?

1.5 Describe the process of using the telephone as you might describe it to a six-year-old using it for the first time to call a friend from school. Describe using the telephone to an electrical engineer who just happens never to have seen one before. What is the correct way to describe a telephone?

1.6 **(a)** Karen has to decide which of several computers to buy for school use. Should she buy the cheapest one? Can she make the best choice on price alone?

(b) Several computers offer essentially the same features, reliability, service, etc. Among these, can she decide the best choice on price alone?

1.7 For each of the following situations, describe what you think you *should* do. In each case *would* you do this?

(a) A fellow student, who is a friend, is copying assignments and submitting them as his own work.

(b) A fellow student, who is *not* a friend, is copying assignments and submitting them as her own work.

(c) A fellow student, who is your only competitor for an important academic award, is copying assignments and submitting them as his own work.

(d) A friend wants to hire you to write an essay for school for her. You are dead broke and the pay is excellent.

(e) A friend wants to hire you to write an essay for school for him. You have lots of money, but the pay is excellent.

(f) A friend wants to hire you to write an essay for school for her. You have lots of money, and the pay is poor.

(g) Your car was in an accident. The insurance adjuster says that the car was totalled and they will give you only the "blue book" value for it as scrap. They will pick up the car in a week. A friend points out that in the meantime you could sell the almost-new tires and replace them with bald ones from the scrap yard, and perhaps sell some other parts, too.

(h) The CD player from your car has been stolen. The insurance adjuster asks you how much it was worth. It was a very cheap one, of poor quality.

(i) The engineer you work for has told you that the meter measuring effluent discharged from a production process exaggerates, and the measured value must be halved for recordkeeping.

(j) The engineer you work for has told you that part of your job is to make up realistic-looking figures reporting effluent discharged from a production process.

(k) You observe unmetered and apparently unreported effluent discharged from a production process.

(l) An engineer where you work is copying directly from a manufacturer's brochure machine-tool specifications to be included in a purchase request. These specifications limit the possible purchase to the particular one specified.

(m) An engineer where you work is copying directly from a manufacturer's brochure machine-tool specifications to be included in a purchase request. These specifications limit the possible purchase to the particular one specified. You know that the engineer's best friend is the salesman for that manufacturer.

1.8 Ciel is trying to decide whether now is a good time to expand her manufacturing plant. The viability of expansion depends on the Canadian economy (an expanding economy means more sales), the relative value of the Canadian dollar (a lower dollar means more exports), and changes in international trade agreements (lower tariffs also mean more exports). These factors, however, may be highly unpredictable. What two things can she do to help make sure she makes a good decision?

1.9 Trevor started a high-tech business two years ago, and now wants to sell out to one of his larger competitors. Two different buyers have made firm offers. They are similar in all but two respects. They differ in price: the Investco offer would result in Trevor's walking away with $2 000 000, while the Venture Corporation offer would give him $3 000 000. The other way they differ is that Investco say they will recapitalize Trevor's company to increase growth, while Trevor thinks that Venture Corporation will close down the business so that it doesn't compete with several of Venture Corporation's other divisions. What would you do if you were Trevor, and why?

1.10 Telekom Company is considering the development of a new type of cell phone based on a brand-new, emerging technology. If successful, Telekom will be able to offer a cell phone that works over long distances and even in mountainous areas. Before proceeding with the project, however, what uncertainties associated with the new technology should they be aware of? Can sensitivity analysis help address these uncertainties?

Imperial Oil v. Quebec

CANADIAN 1.1 MINI-CASE

In 1979, Imperial Oil sold a former petroleum depot in Levis, Quebec, which had been operating since the early 1920s. The purchaser demolished the facilities, and sold the land to a real estate developer. The developer conducted a cleanup that was approved by the Quebec Ministry of the Environment, which issued a certificate of authorization in 1987. Following this, the site was developed and a number of houses were built. However, years later, residents of the subdivision sued the environment ministry claiming there was remaining pollution.

The Ministry, under threat of expensive lawsuits, then ordered Imperial Oil to identify the pollution, recommend corrective action, and potentially pay for the cost of cleanup. In response, Imperial Oil initiated judicial proceedings against the Ministry, claiming violation of principles of natural justice and conflict of interest on its part.

In February 2003, the Supreme Court of Canada ruled that the Ministry had the right to compel Imperial Oil to do the cleanup and save public money, because Imperial was the originator of the pollution and the Minister did not personally benefit.

Source: *Imperial Oil v. Quebec (Minister of the Environment)*, [2003] 2 S.C.R. 624, 2003 SCC 58, Canadian Legal Information Institute (CanLII) site, www.canlii.org/ca/cas/scc/2003/2003scc58.html, accessed September 20, 2004.

Discussion

There is often strong motivation for companies to commit environmental offences. Companies primarily focus on profits, and preventing environmental damage is always a cost. It benefits society to have a clean environment, but almost never benefits the company directly. Sometimes, in spite of the efforts of upper management to be good citizens, the search for profit may result in environmental damage. In a large company it can be difficult for one person to know what is happening everywhere in the firm.

Older companies have an additional problem. An older company may have been producing goods in a certain way for years, and have established ways to dispose of waste. Changes in society make those traditional methods unacceptable. Even when a source of pollution has been identified, it may not be easy to fix. There may be decades of accumulated damage to correct. A production process may not be easily changed in a way that would allow the company to stay in business. Loss of jobs and the effect on the local economy may create strong political pressure to keep the company running.

The government has an important role to offset the profit motive for large companies, for example, by taking action through the courts. Unfortunately, that alone will not be enough to prevent some firms from continuing to cause environmental damage. Economics and politics will occasionally win out.

Questions

1. There are probably several companies in your city or province that are known to pollute. Name some of these. For each:
   - **(a)** What sort of damage do they do?
   - **(b)** How long have they been doing it?
   - **(c)** Why is this company still permitted to pollute?
   - **(d)** What would happen if this company was forced to shut down? Is it ethically correct to allow the company to continue to pollute?

2. Does it make more sense to fine a company for environmental damage or to fine management personally for environmental damage caused by a company? Why?
3. Should the fines for environmental damage be raised enough so that no company is tempted to pollute? Why or why not?
4. Governments can impose fines, give tax breaks, and take other actions that use economics to control the behaviour of companies. Is it necessary to do this whenever a company that pursues profits might do some harm to society as a whole? Why might a company do the socially correct thing even if profits are lost?

2
Time Value of Money

Engineering Economics in Action, Part 2A: A Steal For Steel

"Naomi, can you check this for me?" Terry's request broke the relative silence as Naomi and Terry worked together one Tuesday afternoon. "I was just reviewing our J-class line for Clem, and it seems to me that we could save a lot of money there."

"OK, tell me about it." Since Naomi and Terry had met two weeks earlier, just after Naomi started her job, things had being going very well. Terry, an engineering student at the local university, was on a four-month co-op work term at Canadian Widgets.

"Well, mostly we use the heavy rolled stock on that line. According to the pricing memo we have for that kind of steel, there is a big price break at a volume that could supply our needs for six months. We've been buying this stuff on a week-by-week basis. It just makes sense to me to take advantage of that price break."

"Interesting idea, Terry. Have you got data about how we have ordered before?"

"Yep, right here."

"Let's take a closer look."

"Well," Terry said, as he and Naomi looked over his figures, "the way we have been paying doesn't make too much sense. We order about a week's supply. The cost of this is added to our account. Every six months we pay off our account. Meanwhile, the supplier is charging us 2% of our outstanding amount at the end of each month!"

"Well, at least it looks as if it might make more sense for us to pay off our bills more often," Naomi replied.

"Now look at this. In the six months ending last December, we ordered steel for a total cost of $1 600 000. If we had bought this steel at the beginning of July, it would have only cost $1 400 000. That's a saving of $200 000!"

"Good observation, Terry, but I don't think buying in advance is the right thing to do. If you think about it . . . "

2.1 Introduction

Engineering decisions frequently involve evaluating tradeoffs among costs and benefits that occur at different times. A typical situation is when we invest in a project today in order to obtain benefits from the project in the future. This chapter discusses the economic methods used to compare benefits and costs that occur at different times. The key to making these comparisons is the use of an interest rate. In Sections 2.2 to 2.5, we illustrate the comparison process with examples and introduce some interest and interest rate terminology. Section 2.6 deals with cash flow diagrams, which are graphical representations of the magnitude and timing of cash flows over time. Section 2.7 explains the equivalence of benefits and costs that occur at different times.

2.2 Interest and Interest Rates

Everyone is familiar with the idea of interest from their everyday activities:

From the furniture store ad: *Pay no interest until next year!*

From the bank: *Now 2.6% daily interest on passbook accounts!*

Why are there interest rates? If people are given the choice between having money today and the same amount of money one year from now, most would prefer the money today. If they had the money today, they could do something productive with it in hopes of benefit in the future. For example, they could buy an asset like a machine today, and could use it to make money from their initial investment. Or they may want to buy a consumer good like a new stereo system and start enjoying it immediately. What this means is that one dollar today is worth more than one dollar in the future. This is because a dollar today can be invested for productive use, while that opportunity is lost or diminished if the dollar is not available until some time in the future.

NET VALUE 2.1

Prime Interest Rates

When you hear about the interest rate going up or down in the news, it usually refers to the *prime rate*: the interest rate charged by banks to their most creditworthy customers (e.g., prominent businesspeople). It attracts people's attention because it is the reference point for interest rates charged on many mortgage, personal, and business loans. The prime rate is almost always the same among major banks—the Internet makes it easy to collect and compare interest rate information from banks and other financial institutions.

Contrary to what many people may think, the prime rate is *not* set by the Bank of Canada, the country's central bank. Its role, as defined in the original *Bank of Canada Act* of 1934, is to promote the economic and financial well-being of Canada, but setting the prime rate on behalf of all Canadian banks is not one of its jobs. This doesn't mean, however, that the Bank of Canada is not interested in having control over what the prime rate should be. It does try to influence the rate by setting the target for the short-term interest rates called *overnight rates*, which represent the average interest rate that the Bank of Canada would like to see in the marketplace, that is, the representative of what the prime rate should be. Find more on the Bank of Canada and its role in setting or influencing various interest rates at its Web site at www.bankofcanada.ca.

The observation that a dollar today is worth more than a dollar in the future means that people must be compensated for lending money. They are giving up the opportunity to invest their money for productive purposes now on the promise of getting more money in the future. The compensation for loaning money is in the form of an interest payment, say I. More formally, **interest** is the difference between the amount of money lent and the amount of money later repaid. It is the compensation for giving up the use of the money for the duration of the loan.

An amount of money today, P (also called the *principal amount*), can be related to a *future amount* F by the interest amount I or interest rate i. This relationship is illustrated graphically in Figure 2.1 and can be expressed as $F = P + I$. The interest I can also be expressed as an interest rate i with respect to the principal amount so that $I = Pi$. Thus

$$\begin{aligned} F &= P + Pi \\ &= P(1 + i) \end{aligned}$$

Example 2.1

Samuel bought a one-year guaranteed investment certificate for $5000 from a bank on May 15 last year. The bank was paying 10% on one-year guaranteed investment certificates at the time. One year later, Samuel cashed in his certificate for $5500.

We may think of the interest payment that Samuel got from the bank as compensation

Figure 2.1 Present and Future Worth

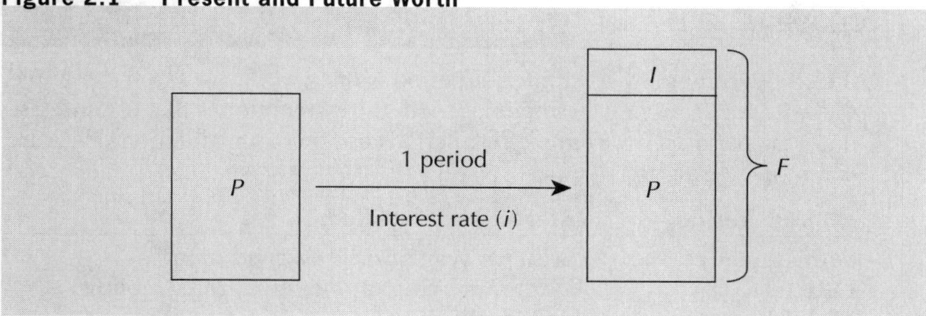

for giving up the use of money. When Samuel bought the guaranteed investment certificate for $5000, he gave up the opportunity to use the money in some other way during the following year. On the other hand, the bank got use of the money for the year. In effect, Samuel lent $5000 to the bank for a year. The $500 interest was payment by the bank to Samuel for the loan. The bank wanted the loan so that it could use the money for the year. (It may have lent the money to someone else at a higher interest rate.) ■

This leads to a formal definition of interest rates. Divide time into periods like days, months, or years. If the right to P at the beginning of a time period exchanges for the right to F at the end of the period, where $F = P(1 + i)$, i is the **interest rate** per time period. In this definition, P is called the **present worth** of F, and F is called the **future worth** of P.

Example 2.1 Restated

Samuel bought a one-year guaranteed investment certificate for $5000 from a bank on May 15 last year. The bank was paying 10% on one-year guaranteed investment certificates at the time. The certificate gave Samuel the right to claim $5500 from the bank one year later.

Notice in this example that there was a transaction between Samuel and the bank on May 15 last year. There was an exchange of $5000 on May 15 a year ago for the right to collect $5500 on May 15 this year. The bank got the $5000 last year and Samuel got the right to collect $5500 one year later. Evidently, having a dollar on May 15 last year was worth more than the right to collect a dollar a year later. Each dollar on May 15 last year was worth the right to collect $5500/5000 = 1.1$ dollars a year later. This 1.1 may be written as $1 + 0.1$ where 0.1 is the interest rate. The interest rate, then, gives the rate of exchange between money at the beginning of a period (one year in this example) and the right to money at the end of the period. ■

The dimension of an interest rate is (dollars/dollars)/time period. For example, a 9% interest rate means that for every dollar lent, 0.09 dollars (or other unit of money) is paid in interest for each time period. The value of the interest rate depends on the length of the time period. Usually, interest rates are expressed on a yearly basis, although they may be given for periods other than a year, such as a month or a quarter. This base unit of time over which an interest rate is calculated is called the **interest period**. Interest periods are described in more detail in Close-Up 2.1. The longer the interest period, the higher the interest rate must be to provide the same return.

Interest concerns the lending and borrowing of money. It is a parameter that allows an exchange of a larger amount of money in the future for a smaller amount of money in the

CLOSE-UP 2.1 Interest Periods

The most commonly used interest period is one year. If we say, for example, "6% interest" without specifying an interest period, the assumption is that 6% interest is paid for a one-year period. However, interest periods can be of any duration. Here are some other common interest periods:

Interest Period	Interest Is Calculated:
Semiannually	Twice per year, or once every six months
Quarterly	Four times a year, or once every three months
Monthly	12 times per year
Weekly	52 times per year
Daily	365 times per year
Continuous	For infinitesimally small periods

present, and vice versa. As we will see in Chapter 3, it also allows us to evaluate very complicated exchanges of money over time.

Interest also has a physical basis. Money can be invested in financial instruments that pay interest, such as a bond or a savings account, and money can also be invested directly in industrial processes or services that generate wealth. In fact, the money invested in financial instruments is also, indirectly, invested in productive activities by the organization providing the instrument. Consequently, the root source of interest is the productive use of money, as this is what makes the money actually increase in value. The actual return generated by a specific productive investment varies enormously, as will be seen in Chapter 4.

2.3 Compound and Simple Interest

We have seen that if an amount, P, is lent for one interest period at the interest rate, i, the amount that must be repaid at the end of the period is $F = P(1 + i)$. But loans may be for several periods. How is the quantity of money that must be repaid computed when the loan is for N interest periods? The usual way is "one period at a time." Suppose that the amount P is borrowed for N periods at the interest rate i. The amount that must be repaid at the end of the N periods is $P(1 + i)^N$, that is

$$F = P(1 + i)^N \tag{2.1}$$

This is derived as shown in Table 2.1.

This method of computing interest is called *compounding*. Compounding assumes that there are N sequential one-period loans. At the end of the first interest period, the borrower owes $P(1 + i)$. This is the amount borrowed for the second period. Interest is required on this larger amount. At the end of the second period $[P(1 + i)](1 + i)$ is owed. This is the amount borrowed for the third period. This continues so that at the end of the $(N - 1)$th period, $P(1 + i)^{N-1}$ is owed. The interest on this over the Nth period is $[P(1 + i)^{N-1}]i$. The total interest on the loan over the N periods is

$$I_c = P(1 + i)^N - P \tag{2.2}$$

I_c is called **compound interest.** It is the standard method of computing interest where interest accumulated in one interest period is added to the principal amount used to

Table 2.1 Compound Interest Computations

Beginning of Period	Amount Lent		Interest Amount	Amount Owed at Period End
1	P	+	Pi	$= P + Pi = P(1 + i)$
2	$P(1 + i)$	+	$P(1 + i)i$	$= P(1 + i) + P(1 + i)i = P(1 + i)^2$
3	$P(1 + i)^2$	+	$P(1 + i)^2 i$	$= P(1 + i)^2 + P(1 + i)^2 i = P(1 + i)^3$
⋮	⋮			
N	$P(1 + i)^{N-1}$	+	$[P(1 + i)^{N-1}]i$	$= P(1 + i)^N$

calculate interest in the next period. The interest period when compounding is used to compute interest is called the **compounding period.**

Example 2.2

If you were to lend $100 for three years at 10% per year compound interest, how much interest would you get at the end of the three years?

If you lend $100 for three years at 10% compound interest per year, you will earn $10 in interest in the first year. That $10 will be lent, along with the original $100, for the second year. Thus, in the second year, the interest earned will be $11 = $110(0.10). The $11 is lent for the third year. This makes the loan for the third year $121, and $12.10 = $121(0.10) in interest will be earned in the third year. At the end of the three years, the amount you are owed will be $133.10. The interest received is then $33.10. This can also be calculated from Equation (2.2):

$$I_c = \$100(1 + 0.1)^3 - \$100 = \$33.10$$

Table 2.2 summarizes the compounding process. ■

If the interest payment for an N-period loan at the interest rate i per period is computed without compounding, the interest amount, I_s, is called *simple interest*. It is computed as

$$I_s = PiN$$

Simple interest is a method of computing interest where interest earned during an interest period is not added to the principal amount used to calculate interest in the next period. Simple interest is rarely used in practice, except as a method of calculating approximate interest.

Table 2.2 Compound Interest Computations for Example 2.2

Beginning of Year	Amount Lent		Interest Amount		Amount Owed at Year End
1	100	+	$100 × 0.1	=	$110
2	110	+	$110 × 0.1	=	$121
3	121	+	$121 × 0.1	=	$133.10

Example 2.3

If you were to lend \$100 for three years at 10% per year simple interest, how much interest would you get at the end of the three years?

The total amount of interest earned on the \$100 over the three years would be \$30. This can be calculated by using $I_s = PiN$:

$$I_s = PiN = \$100(0.10)(3) = \$30 \blacksquare$$

Interest amounts computed with simple interest and compound interest will yield the same results only when the number of interest periods is one. As the number of periods increases, the difference between the accumulated interest amounts for the two methods increases exponentially.

When the number of interest periods is significantly greater than one, the difference between simple interest and compound interest can be very great. In April 1993, a couple in Nevada presented the state government with a \$1000 bond issued by the state in 1865. The bond carried an annual interest rate of 24%. The couple claimed the bond was now worth several trillion dollars (*Newsweek*, August 9, 1993, p. 8). If one takes the length of time from 1865 to the time the couple presented the bond to the state as 127 years, the value of the bond could have been \$732 trillion = $\$1000(1 + 0.24)^{127}$.

If, instead of compound interest, a simple interest rate given by $iN = (24\%)(127) = 3048\%$ were used, the bond would be worth only \$31 480 = \$1000(1 + 30.48). Thus, the difference between compound and simple interest can be dramatic, especially when the interest rate is high and the number of periods is large. The graph in Figure 2.2 shows the difference between compound interest and simple interest for the first 20 years of the bond example. As for the couple in Nevada, the \$1000 bond was worthless after all—a state judge ruled that the bond had to have been cashed by 1872.

The conventional approach for computing interest is the compound interest method rather than simple interest. Simple interest is rarely used, except perhaps as an intuitive (yet incorrect!) way of thinking of compound interest. We mention simple interest primarily to contrast it with compound interest and to indicate that the difference between the two methods can be large.

Figure 2.2 Compound and Simple Interest at 24% per Year for 20 Years

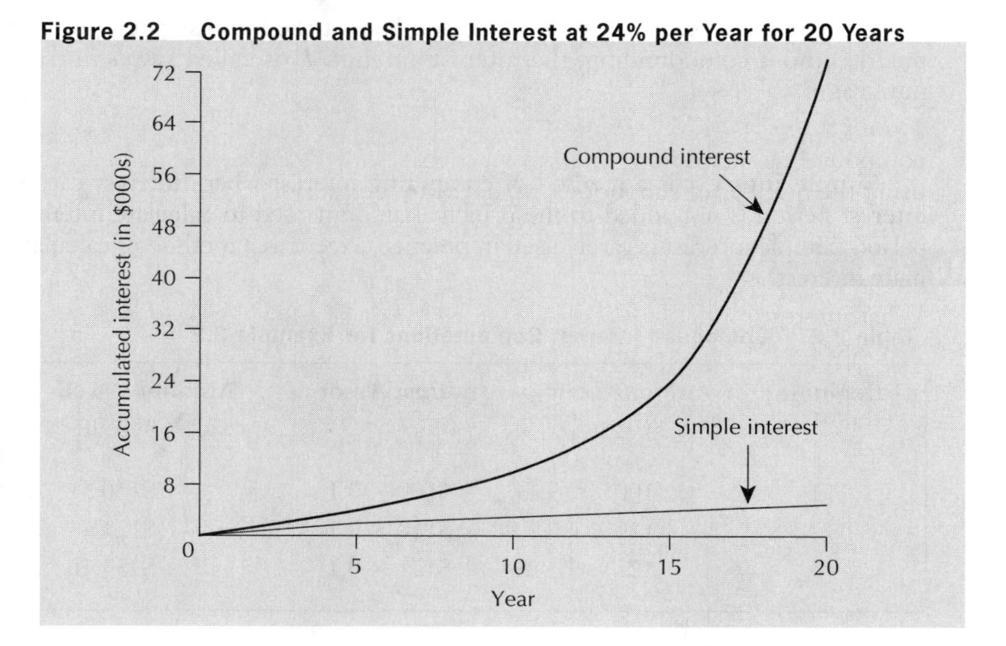

2.4 Effective and Nominal Interest Rates

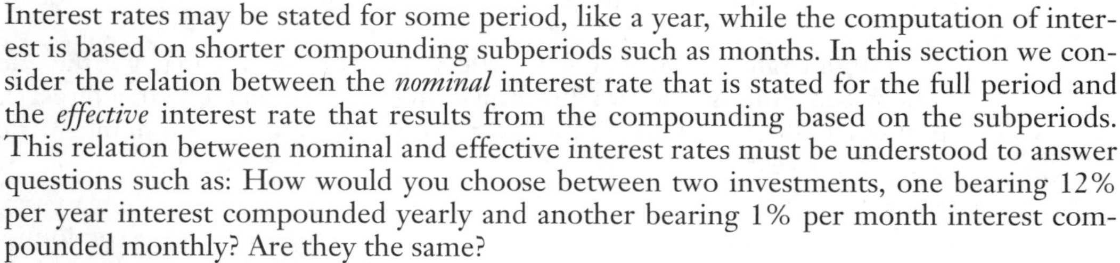

Interest rates may be stated for some period, like a year, while the computation of interest is based on shorter compounding subperiods such as months. In this section we consider the relation between the *nominal* interest rate that is stated for the full period and the *effective* interest rate that results from the compounding based on the subperiods. This relation between nominal and effective interest rates must be understood to answer questions such as: How would you choose between two investments, one bearing 12% per year interest compounded yearly and another bearing 1% per month interest compounded monthly? Are they the same?

Nominal interest rate is the conventional method of stating the annual interest rate. It is calculated by multiplying the interest rate per compounding period by the number of compounding periods per year. Suppose that a time period is divided into m equal subperiods. Let there be stated a nominal interest rate, r, for the full period. By convention, for nominal interest, the interest rate for each subperiod is calculated as $i_s = r/m$. For example, a nominal interest rate of 18% per year, compounded monthly, is the same as

$$0.18/12 = 0.015 \text{ or } 1.5\% \text{ per month}$$

Effective interest rate is the actual but not usually stated interest rate, found by converting a given interest rate with an arbitrary compounding period (normally less than a year) to an equivalent interest rate with a one-year compounding period. What is the effective interest rate, i_e, for the full period that will yield the same amount as compounding at the end of each subperiod, i_s? If we compound interest every subperiod, we have

$$F = P(1 + i_s)^m$$

We want to find the effective interest rate, i_e, that yields the same future amount F at the end of the full period from the present amount P. Set

$$P(1 + i_s)^m = P(1 + i_e)$$

Then

$$(1 + i_s)^m = 1 + i_e$$

$$i_e = (1 + i_s)^m - 1 \tag{2.3}$$

Note that Equation (2.3) allows the conversion between the interest rate over a compounding subperiod, i_s, and the effective interest rate over a longer period, i_e, by using the number of subperiods, m, in the longer period.

Example 2.4

What interest rate per year, compounded yearly, is equivalent to 1% interest per month, compounded monthly?

Since the month is the shorter compounding period, we let $i_s = 0.01$ and $m = 12$. Then i_e refers to the effective interest rate per year. Substitution into Equation 2.3 then gives

$$\begin{aligned} i_e &= (1 + i_s)^m - 1 \\ &= (1 + 0.01)^{12} - 1 \\ &= 0.126825 \\ &\approx 0.127 \text{ or } 12.7\% \end{aligned}$$

An interest rate of 1% per month, compounded monthly, is equivalent to an effective rate of approximately 12.7% per year, compounded yearly. The answer to our previously posed question is that an investment bearing 12% per year interest, compounded yearly, pays less than an investment bearing 1% per month interest, compounded monthly.■

Interest rates are normally given as nominal rates. We may get the effective (yearly) rate by substituting $i_s = r/m$ into Equation (2.3). We then obtain a direct means of computing an effective interest rate, given a nominal rate and the number of compounding periods per year:

$$i_e = \left(1 + \frac{r}{m}\right)^m - 1 \tag{2.4}$$

This formula is suitable only for converting from a nominal rate r to an annual effective rate. If the effective rate desired is for a period longer than a year, then Equation (2.3) must be used.

Example 2.5

Leona the loan shark lends money to clients at the rate of 5% interest per week! What is the nominal interest rate for these loans? What is the effective annual interest rate?

The nominal interest rate is 5% × 52 = 260%. Recall that nominal interest rates are usually expressed on a yearly basis. The effective yearly interest rate can be found by substitution into Equation (2.3):

$$i_e = (1 + 0.05)^{52} - 1 = 11.6$$

Leona charges an effective annual interest rate of about 1160% on her loans.■

Example 2.6

The Cardex Credit Card Company charges a nominal 24% interest on overdue accounts, compounded daily. What is the effective interest rate?

Assuming that there are 365 days per year, we can calculate the interest rate per day using either Equation (2.3) with $i_s = r/m = 0.24/365 = 0.0006575$ or by the use of Equation (2.4) directly. The effective interest rate (per year) is

$$\begin{aligned} i_e &= (1 + 0.0006575)^{365} - 1 \\ &= 0.271 \text{ or } 27.1\% \end{aligned}$$

With a nominal rate of 24% compounded daily, the Cardex Credit Card Company is actually charging an effective rate of about 27.1% per year.■

Although there are laws which may require that the effective interest rate be disclosed for loans and investments, it is still very common for nominal interest rates to be quoted. Since the nominal rate will be less than the effective rate whenever the number of compounding periods per year exceeds one, there is an advantage to quoting loans using the nominal rates, since it makes the loan look more attractive. This is particularly true when interest rates are high and compounding occurs frequently.

2.5 Continuous Compounding

As has been seen, compounding can be done yearly, quarterly, monthly, or daily. The periods can be made even smaller, as small as desired; the main disadvantage in having very small periods is having to do more calculations. If the period is made infinitesimally small, we say that interest is compounded *continuously*. There are situations in which very frequent compounding makes sense. For instance, an improvement in materials handling may reduce downtime on machinery. There will be benefits in the form of increased output that may be used immediately. If there are several additional runs a day, there will be benefits several times a day. Another example is trading on the stock market. Personal and corporate investments are often in the form of mutual funds. Mutual funds represent a changing set of stocks and bonds, in which transactions occur very frequently, often many times a day.

A formula for **continuous compounding** can be developed from Equation (2.3) by allowing the number of compounding periods per year to become infinitely large:

$$i_e = \lim_{m \to \infty}\left(1 + \frac{r}{m}\right)^m - 1$$

By noting from a definition of the natural exponential function, e, that

$$\lim_{m \to \infty}\left(1 + \frac{r}{m}\right)^m = e^r$$

we get

$$i_e = e^r - 1 \qquad (2.5)$$

Example 2.7

Cash flow at the Arctic Oil Company is continuously reinvested. An investment in a new data logging system is expected to return a nominal interest of 40%, compounded continuously. What is the effective interest rate earned by this investment?

The nominal interest rate is given as $r = 0.40$. From Equation (2.5),

$$i_e = e^{0.4} - 1$$
$$= 1.492 - 1 \cong 0.492 \text{ or } 49.2\%$$

The effective interest rate earned on this investment is about 49.2%.■

Although continuous compounding makes sense in some circumstances, it is rarely used. As with effective interest and nominal interest, in the days before calculators and computers, calculations involving continuous compounding were difficult to do. Consequently, discrete compounding is, by convention, the norm. As illustrated in Figure 2.3, the difference between continuous compounding and discrete compounding is relatively insignificant, even at a fairly high interest rate.

Figure 2.3 Growth in Value of $1 at 30% Interest for Various Compounding Periods

2.6 Cash Flow Diagrams

Sometimes a set of cash flows can be sufficiently complicated that it is useful to have a graphical representation. A **cash flow diagram** is a graph that summarizes the timing and magnitude of cash flows as they occur over time.

On a cash flow diagram, the graph's vertical axis is not shown explicitly. The horizontal (X) axis represents time, measured in periods, and the vertical (Y) axis represents the size and direction of the cash flows. Individual cash flows are indicated by arrows pointing up or down from the horizontal axis, as indicated in Figure 2.4. The arrows that point up represent positive cash flows, or receipts. The downward-pointing arrows represent negative cash flows, or disbursements. See Close-Up 2.2 for some conventions pertaining to the beginning and ending of periods.

Figure 2.4 Cash Flow Diagram

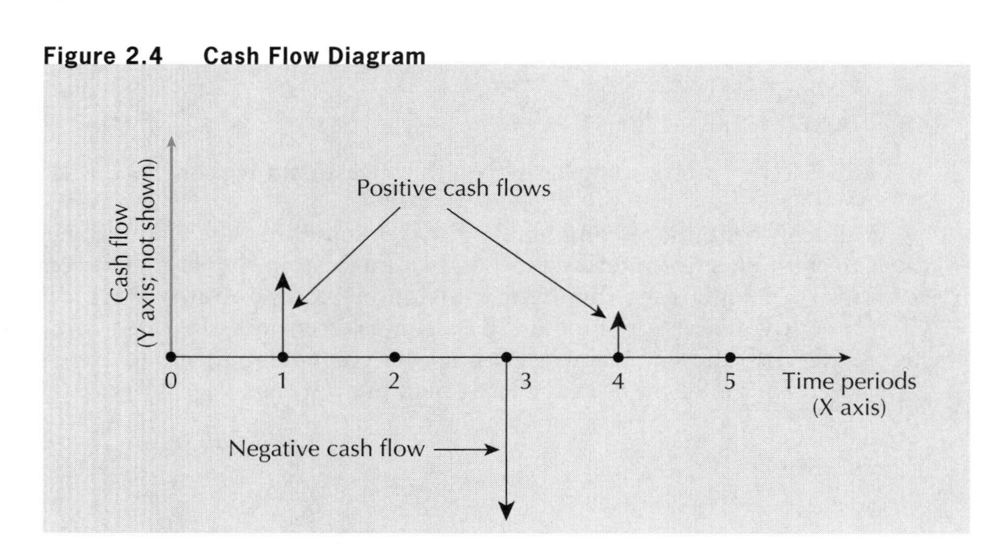

CLOSE-UP 2.2 Beginning and Ending of Periods

As illustrated in a cash flow diagram (see Figure 2.5), the end of one period is exactly the same point in time as the beginning of the next period. Now is time 0, which is the end of period –1 and also the beginning of period 1. The end of period 1 is the same as the beginning of period 2, and so on. *N* years from now is the end of period *N* and the beginning of period ($N + 1$).

Figure 2.5 Beginning and Ending of Periods

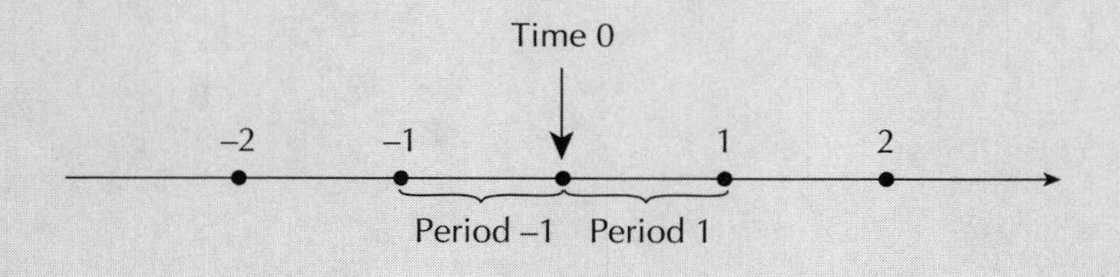

Example 2.8

Consider Ashok, a recent university graduate who is trying to summarize typical cash flows for each month. His monthly income is $2200, received at the end of each month. Out of this he pays for rent, food, entertainment, telephone charges, and a credit card bill for all other purchases. Rent is $700 per month (including utilities), due at the end of each month. Weekly food and entertainment expenses total roughly $120, a typical telephone bill is $40 (due at the end of the first week in the month), and his credit card purchases average $300. Credit card payments are due at the end of the second week of each month.

Figure 2.6 shows the timing and amount of the disbursements and the single receipt over a typical month. It assumes that there are exactly four weeks in a month, and it is now just past the end of the month. Each arrow, which represents a cash flow, is labelled with the amount of the receipt or disbursement.

When two or more cash flows occur in the same time period, the amounts may be shown individually, as in Figure 2.6, or in summary form, as in Figure 2.7. The level of detail used depends on personal choice and the amount of information the diagram is intended to convey.

We suggest that the reader make a practice of using cash flow diagrams when working on a problem with cash flows that occur at different times. Just going through the steps in setting up a cash flow diagram can make the problem structure clearer. Seeing the pattern of cash flows in the completed diagram gives a "feel" for the problem.■

Figure 2.6 Cash Flow Diagram for Example 2.8

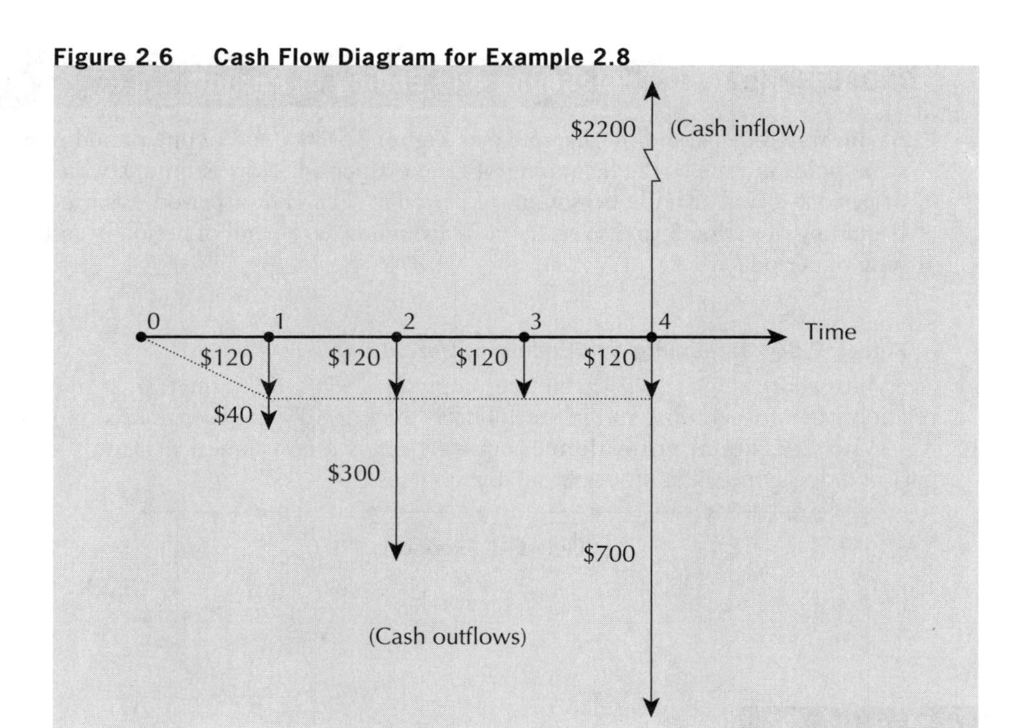

Figure 2.7 Cash Flow Diagram for Example 2.8 in Summary Form

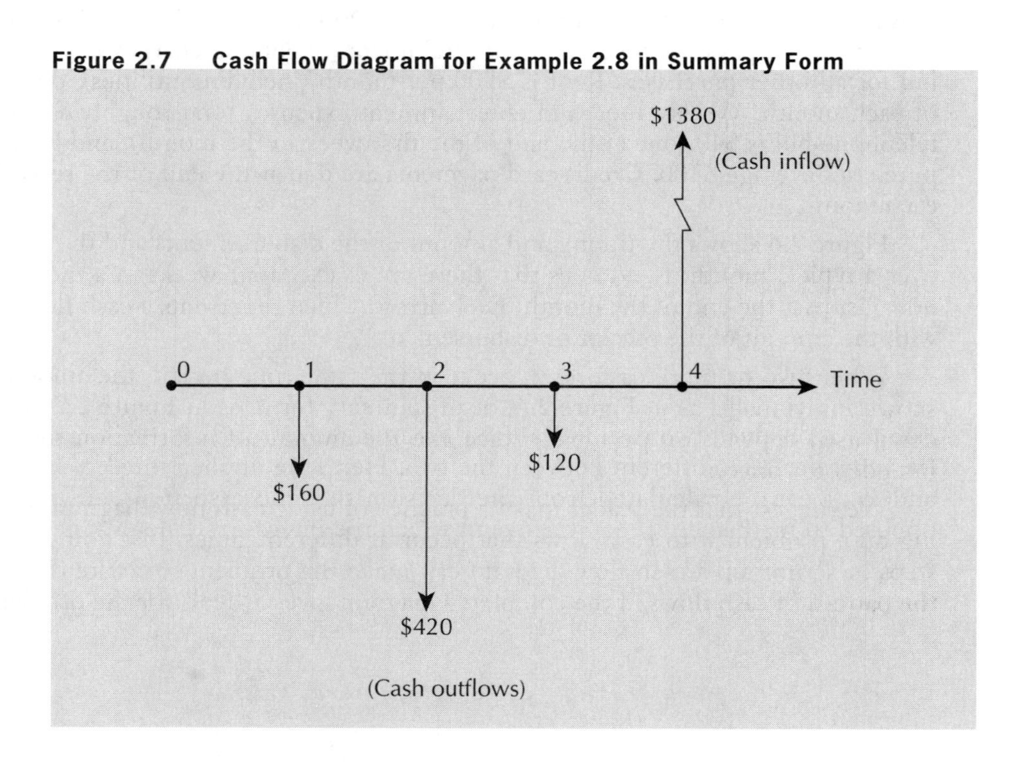

2.7 Equivalence

We started this chapter by pointing out that many engineering decisions involve costs and benefits that occur at different times. Making these decisions requires that the costs and benefits at different times be compared. To make these comparisons, we must be able to say that certain values at different times are *equivalent*. **Equivalence** is a condition that exists when the value of a cost at one time is equivalent to the value of the related benefit received at a different time. In this section we distinguish three concepts of equivalence that may underlie comparisons of costs and benefits at different times.

With **mathematical equivalence**, equivalence is a consequence of the mathematical relationship between time and money. This is the form of equivalence used in $F = P(1 + i)^N$.

With **decisional equivalence**, equivalence is a consequence of indifference on the part of a decision maker among available choices.

With **market equivalence**, equivalence is a consequence of the ability to exchange one cash flow for another at zero cost.

Although the mathematics governing money is the same regardless of which form of equivalence is most appropriate for a given situation, it can be important to be aware of what assumptions must be made for the mathematical operations to be meaningful.

2.7.1 Mathematical Equivalence

Mathematical equivalence is simply a mathematical relationship. It says that two cash flows, P_t at time t and F_{t+N} at time $t + N$, are equivalent with respect to the interest rate, i, if $F_{t+N} = P_t(1 + i)^N$. Notice that if F_{t+N+M} (where M is a second number of periods) is equivalent to P_t, then

$$\begin{aligned} F_{t+N+M} &= P_t(1 + i)^{N+M} \\ &= F_{t+N}(1 + i)^M \end{aligned}$$

so that F_{t+N} and F_{t+N+M} are equivalent to each other. The fact that mathematical equivalence has this property permits complex comparisons to be made among many cash flows that occur over time.

2.7.2 Decisional Equivalence

For any individual, two cash flows, P_t at time t and F_{t+N} at time $t + N$, are equivalent if the individual is indifferent between the two. Here, the implied interest rate relating P_t and F_{t+N} can be calculated from the decision that the cash flows are equivalent, as opposed to mathematical equivalence in which the interest rate determines whether the cash flows are equivalent. This can be illustrated best through an example.

Example 2.9

Bildmet is an extruder of aluminum shapes used in construction. The company buys aluminum from Alpure, an outfit that recovers aluminum from scrap. When Bildmet's purchasing manager, Greta Kehl, called in an order for 1000 kilograms of metal on August 15, she was told that Alpure was having production difficulties and was running behind schedule. Alpure's manager, Masaaki Sawada, said that he could ship the order immediately if Bildmet required it. However, if Alpure shipped Bildmet's order, they would not be able to fill an order for another user whom Mr. Sawada was anxious to impress with Alpure's reliability. Mr. Sawada suggested that, if Ms. Kehl would wait a week until August 22, he would show his appreciation by shipping 1100 kilograms then at the same cost to Bildmet as 1000 kilograms now. In either case, payment would be due at the end of the month. Should Ms. Kehl accept Alpure's offer?

The rate of exchange, 1100 to 1000 kilograms, may be written as $(1 + 0.1)$ to 1, where the $0.1 = 10\%$ is an interest rate for the one-week period. (This is equivalent to an effective interest rate of more than 14 000% per year!) Whether or not Ms. Kehl accepts the offer from Alpure depends on her situation. There is some chance of Bildmet's running out of metal if they don't get supplied for a week. This would require Ms. Kehl to do some scrambling to find other sources of metal in order to ship to her own customers on time. Ms. Kehl would prefer the 1000 kilograms on the 15th to 1000 kilograms on the 22nd. But there is some minimum amount, larger than 1000 kilograms, that she would accept on the 22nd in exchange for 1000 kilograms on the 15th. This amount would take into account both measurable costs and immeasurable costs such as inconvenience and anxiety.

Let the minimum rate at which Ms. Kehl would be willing to make the exchange be 1 kilogram on the 15th for $(1 + x)$ kilograms on the 22nd. In this case, if $x < 10\%$, Ms. Kehl should accept Alpure's offer of 1100 kilograms on the 22nd. ■

In Example 2.9, the aluminum is a capital good that can be used productively by Bildmet. There is value in that use, and that value can be measured by Greta's willingness to postpone receiving the aluminum. It can be seen that interest is not necessarily a function of exchanges of money at different points in time. However, money is a convenient measure of the worth of a variety of goods, and so interest is usually expressed in terms of money.

2.7.3 Market Equivalence

Market equivalence is based on the idea that there is a market for money that permits cash flows in the future to be exchanged for cash flows in the present, and vice versa. Converting a future cash flow, F, to a present cash flow, P, is called borrowing money, while converting P to F is called lending or investing money. The market equivalence of two cash flows P and F means that they can be exchanged, one for the other, at zero cost.

The interest rate associated with an individual's borrowing money is usually a lot higher than the interest rate applied when that individual lends money. For example, the interest rate a bank pays on deposits is lower than what it charges to lend money to clients. The difference between these interest rates provides the bank with income. This means that, for an individual, market equivalence does not exist. An individual can exchange a present worth for a future worth by investing money, but if he or she were to try to borrow against that future worth to obtain money now, the resulting present worth would be less than the original amount invested. Moreover, every time either borrowing or lending occurred, transaction costs (the fees charged or cost incurred) would further diminish the capital.

Example 2.10

This morning, Averill bought a \$5000 one-year guaranteed investment certificate (GIC) at his local bank. It has an effective interest rate of 7% per year. At the end of a year, the GIC will be worth \$5350. On the way home from the bank, Averill unexpectedly discovered a valuable piece of art he had been seeking for some time. He wanted to buy it, but all his spare capital was tied up in the GIC. So he went back to the bank, this time to negotiate a one-year loan for $5000, the cost of the piece of art. He figured that, if the loan came due at the same time as the GIC, he would simply pay off the loan with the proceeds of the GIC.

Unfortunately, Averill found out that the bank charges 10% effective interest per year on loans. Considering the proceeds from the GIC of $5350 one year from now, the amount the bank would give him today is only $5350/1.1 = $4864 (roughly), less any fees applicable to the loan. He discovered that, for him, market equivalence does not hold. He cannot exchange $5000 today for $5350 one year from now, and vice versa, at zero cost.■

Large companies with good records have opportunities that differ from those of individuals. Large companies borrow and invest money in so many ways, both internally and externally, that the interest rates for borrowing and for lending are very close to being the same, and also the transaction costs are negligible. They can shift funds from the future to the present by raising new money or by avoiding investment in a marginal project that would earn only the rate that they pay on new money. They can shift funds from the present to the future by undertaking an additional project or investing externally.

But how large is a "large company"? Established businesses of almost any size, and even individuals with some wealth and good credit, can acquire cash and invest at about the same interest rate, provided that the amounts are small relative to their total assets. For these companies and individuals, market equivalence is a reasonable model assuming that market equivalence makes calculations easier and still generally results in good decisions.

For most of the remainder of this book, we will be making two broad assumptions with respect to equivalence: first, that market equivalence holds, and second, that decisional equivalence can be expressed entirely in monetary terms. If these two assumptions are reasonably valid, mathematical equivalence can be used as an accurate model of how costs and benefits relate to one another over time. In several sections of the book, when we cover how firms raise capital and how to incorporate non-monetary aspects of a situation into a decision, we will discuss the validity of these two assumptions. In the meantime, mathematical equivalence is used to relate cash flows that occur at different points in time.

REVIEW PROBLEMS

REVIEW PROBLEM 2.1

Atsushi has had $800 stashed under his mattress for 30 years. How much money has he lost by not putting it in a bank account at 8% annual compound interest all these years?

ANSWER

Since Atsushi has kept the $800 under his mattress, he has not earned any interest over the 30 years. Had he put the money into an interest-bearing account, he would have far more today. We can think of the $800 as a present amount and the amount in 30 years as the future amount.

Given: $P = \$800$

$i = 0.08$ per year

$N = 30$ years

Formula: $F = P(1 + i)^N$

$= \$800(1 + 0.08)^{30}$

$= \$8050.13$

Atsushi would have $8050.13 in the bank account today had he deposited his $800 at 8% annual compound interest. Instead, he has only $800. He has suffered an opportunity cost of $8050.13 – $800 = $7250.13 by not investing the money.■

REVIEW PROBLEM 2.2

You want to buy a new computer, but you are $1000 short of the amount you need. Your aunt has agreed to lend you the $1000 you need now, provided you pay her $1200 two years from now. She compounds interest monthly. Another place from which you can borrow $1000 is the bank. There is, however, a loan processing fee of $20, which will be included in the loan amount. The bank is expecting to receive $1220 two years from now based on monthly compounding of interest.

(a) What monthly rate is your aunt charging you for the loan? What is the bank charging?

(b) What effective annual rate is your aunt charging? What is the bank charging?

(c) Would you prefer to borrow from your aunt or from the bank?

ANSWER

(a) *Your aunt*

Given: $P = \$1000$

$F = \$1200$

$N = 24$ months (since compounding is done monthly)

Formula: $F = P(1 + i)^N$

The formula $F = P(1 + i)^N$ must be solved in terms of i to answer the question.

$i = \sqrt[N]{F/P} - 1$

$= \sqrt[24]{1200/1000} - 1$

$= 0.007626$

Your aunt is charging interest at a rate of approximately 0.76% per month.

The bank

Given: $P = \$1020$ (since the fee is included in the loan amount)

$F = \$1220$

$N = 24$ months (since compounding is done monthly)

$i = \sqrt[N]{F/P} - 1$

$= \sqrt[24]{1220/1020} - 1$

$= 0.007488$

The bank is charging interest at a rate of approximately 0.75% per month.

(b) The effective annual rate can be found with the formula $i_e = (1 + r/m)^m - 1$, where r is the nominal rate per year and m is the number of compounding periods per year. Since the number of compounding periods per year is 12, notice that r/m is simply the interest rate charged per month.

Your aunt

$$i = 0.007626 \text{ per month}$$

Then

$$\begin{aligned} i_e &= (1 + r/m)^m - 1 \\ &= (1 + 0.007626)^{12} - 1 \\ &= 0.095445 \end{aligned}$$

The effective annual rate your aunt is charging is approximately 9.54%.

The bank

$$i = 0.007488 \text{ per month}$$

Then

$$\begin{aligned} i_e &= (1 + r/m)^m - 1 \\ &= (1 + 0.007488)^{12} - 1 \\ &= 0.09365 \end{aligned}$$

The effective annual rate for the bank is approximately 9.37%.

(c) The bank appears to be charging a lower interest rate than does your aunt. This can be concluded by comparing the two monthly rates or the effective annual rates the two charge. If you were to base your decision only upon who charged the lower interest rate, you would pick the bank, despite the fact they have a fee. However, although you are borrowing \$1020 from the bank, you are getting only \$1000 since the bank immediately gets its \$20 fee. The cost of money for you from the bank is better calculated as

Given: $P = \$1000$

$F = \$1220$

$N = 24$ months (since compounding is done monthly)

$$\begin{aligned} i &= \sqrt[N]{F/P} - 1 \\ &= \sqrt[24]{1220/1000} - 1 \\ &= 0.00832 \end{aligned}$$

From this point of view, the bank is charging interest at a rate of approximately 0.83% per month, and you would be better off borrowing from your aunt.■

REVIEW PROBLEM 2.3

At the end of four years, you would like to have \$5000 in a bank account to purchase a used car. What you need to know is how much to deposit in the bank account now. The account pays daily interest. Create a spreadsheet and plot the necessary deposit today as a function of interest rate. Consider nominal interest rates ranging from 5% to 15% per year, and assume that there are 365 days per year.

ANSWER

From the formula $F = P(1 + i)^N$, we have $\$5000 = P(1 + i)^{365 \times 4}$. This gives

$$P = \$5000 \times \frac{1}{(1+i)^{365\times 4}}$$

Table 2.3 is an excerpt from a sample spreadsheet. It shows the necessary deposit to accumulate $5000 over four years at a variety of interest rates. The following is the calculation for cell B2 (i.e., the second row, second column):

$$\$5000 \times \frac{1}{\left[1 + \left(\frac{A2}{365}\right)\right]^{365\times 4}}$$

Table 2.3 Necessary Deposits for a Range of Interest Rates

Interest Rate (%)	Necessary Deposit ($)
0.05	4114
0.06	3957
0.07	3805
0.08	3660
0.09	3520
0.10	3385
0.11	3256
0.12	3131
0.13	3011
0.14	2896
0.15	2785

Figure 2.8 Graph for Review Problem 2.3

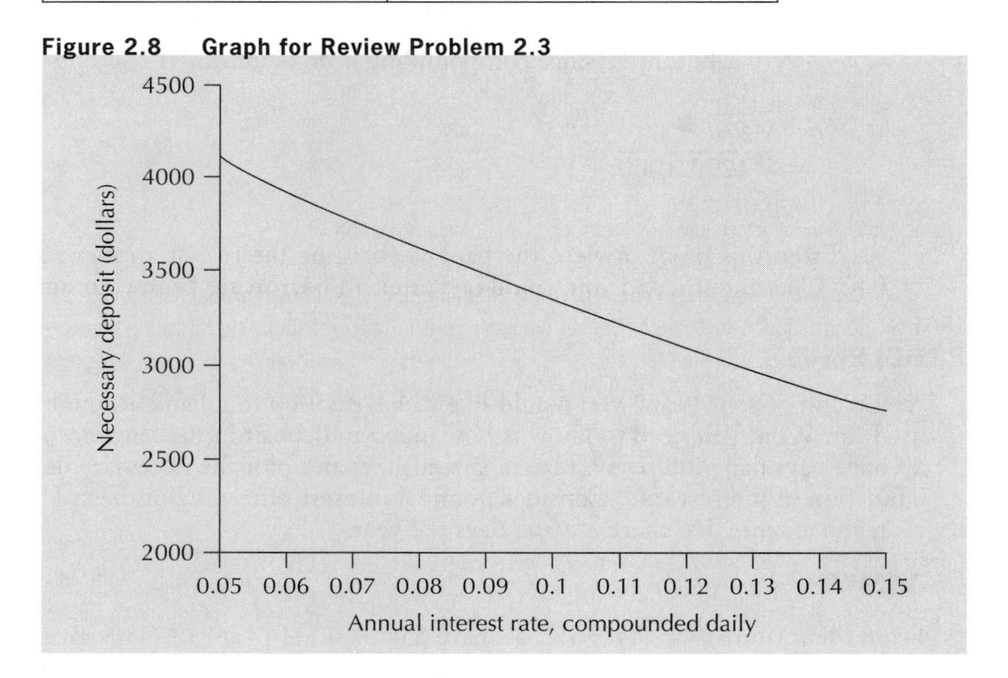

The specific implementation of this formula will vary, depending on the particular spreadsheet program used. Figure 2.8 is a diagram of the necessary deposits plotted against interest rates.■

SUMMARY

This chapter has provided an introduction to interest, interest rate terminology, and interest rate conventions. Through a series of examples, the mechanics of working with simple and compound interest, nominal and effective interest rates, and continuous compounding were illustrated. Cash flow diagrams were introduced in order to represent graphically monetary transactions at various points in time. The final part of the chapter contained a discussion of various forms of cash flow equivalence: mathematical, decisional, and market equivalence. With the assumption that mathematical equivalence can be used as an accurate model of how costs and benefits relate to one another over time, we now move on to Chapter 3, in which equivalence formulas for a variety of cash flow patterns are presented.

Engineering Economics in Action, Part 2B: You Just Have to Know When

Naomi and Terry were looking at the steel orders for the J-class line. Terry thought money could be saved by ordering in advance. "Now look at this," Terry said. "In the six months ending last December, we ordered steel for a total cost of $1 600 000. If we had bought this steel at the beginning of July, it would have cost only $1 400 000. That's a savings of $200 000!"

"Good observation, Terry, but I don't think buying in advance is the right thing to do. If you think about it, the rate of return on our $1 400 000 would be 200 000/1 400 000 or about 14.3% over six months."

"Yes, but that's over 30% effective interest, isn't it? I'll bet we only make 8% or 10% for money we keep in the bank."

"That's true, but the money we would use to buy the steel in advance we don't have sitting in the bank collecting interest. It would have to come from somewhere else, either money we borrow from the bank, at about 14% plus administrative costs, or from our shareholders."

"But it's still a good idea, right?"

"Well, you are right and you are wrong. Mathematically, you could probably show the advantage of buying a six-month supply in advance. But we wouldn't do it for two reasons. The first one has to do with where the money comes from. If we had to pay for six months of steel in advance, we would have a capital requirement beyond what we could cover through normal cash flows. I'm not sure the bank would even lend us that much money, so we would probably have to raise it through equity, that is, selling more shares in the company. This would cost a lot, and throw all your calculations way off."

"Just because it's such a large amount of money?"

"That's right. Our regular calculations are based on the assumption that the capital requirements don't take an extraordinary effort."

"You said there were two reasons. What's the other one?"

"The other reason is that we just wouldn't do it."

"Huh?"

"We just wouldn't do it. Right now the steel company's taking the risk—if we can't pay, they are in trouble. If we buy in advance, it's the other way around—if our widget orders dropped, we would be stuck with some pretty expensive raw materials. We would also have the problem of where to store the steel, and other practical difficulties. It makes sense mathematically, but I'm pretty sure we just wouldn't do it."

Terry looked a little dejected. Naomi continued, "But your figures make sense. The first thing to do is find out why we are carrying that account so long before we pay it off. The second thing to do is see if we can't get that price break, retroactively. We are good customers, and I'll bet we can convince them to give us the price break anyhow, without changing our ordering pattern. Let's talk to Clem about it."

"But, Naomi, why use the mathematical calculations at all, if they don't work?"

"But they do work, Terry. You just have to know when."

PROBLEMS

2.1 Using 12% simple interest per year, how much interest will be owed on a loan of $500 at the end of two years?

2.2 If a sum of $3000 is borrowed for six months at 9% simple interest per year, what is the total amount due (principal and interest) at the end of six months?

2.3 What principal amount will yield $150 in interest at the end of three months when the interest rate is 1% simple interest per month?

2.4 If $2400 interest is paid on a two-year simple-interest loan of $12 000, what is the interest rate per year?

2.5 Simple interest of $190.67 is owed on a loan of $550 after four years and four months. What is the annual interest rate?

2.6 How much will be in a bank account at the end of five years if $2000 is invested today at 12% interest per annum, compounded yearly?

2.7 How much is accumulated in each of these savings plans over two years?

(a) Deposit $1000 today at 10% compounded annually.

(b) Deposit $900 today at 12% compounded monthly.

2.8 Greg wants to have $50 000 in five years. The bank is offering five-year investment certificates that pay 8% nominal interest, compounded quarterly. How much money should he invest in the certificates to reach his goal?

2.9 Greg wants to have $50 000 in five years. He has $20 000 today to invest. The bank is offering five-year investment certificates that pay interest compounded quarterly. What is the minimum nominal interest rate he would have to receive to reach his goal?

2.10 Greg wants to have $50 000. He will invest $20 000 today in investment certificates that pay 8% nominal interest, compounded quarterly. How long will it take him to reach his goal?

2.11 Greg will invest $20 000 today in five-year investment certificates that pay 8% nominal interest, compounded quarterly. How much money will this be in five years?

2.12 You bought an antique car three years ago for $50 000. Today it is worth $65 000.

(a) What annual interest rate did you earn if interest is compounded yearly?

(b) What monthly interest rate did you earn if interest is compounded monthly?

2.13 You have a bank deposit now worth $5000. How long will it take for your deposit to be worth more than $8000 if

(a) the account pays 5% actual interest every half-year, and is compounded?

(b) the account pays 5% nominal interest, compounded semiannually?

2.14 Some time ago, you put $500 into a bank account for a "rainy day." Since then, the bank has been paying you 1% per month, compounded monthly. Today, you checked the balance, and found it to be $708.31. How long ago did you deposit the $500?

2.15 **(a)** If you put $1000 in a bank account today that pays 10% interest per year, how much money could be withdrawn 20 years from now?

(b) If you put $1000 in a bank account today that pays 10% *simple* interest per year, how much money could be withdrawn 20 years from now?

2.16 How long will it take any sum to double itself,

(a) with an 11% simple interest rate?

(b) with an 11% interest rate, compounded annually?

(c) with an 11% interest rate, compounded continuously?

2.17 Compute the effective annual interest rate on each of these investments:

(a) 25% nominal interest, compounded semiannually

(b) 25% nominal interest, compounded quarterly

(c) 25% nominal interest, compounded continuously

2.18 For a 15% effective annual interest rate, what is the nominal interest rate if

(a) interest is compounded monthly?

(b) interest is compounded daily (assume 365 days per year)?

(c) interest is compounded continuously?

2.19 A Studebaker automobile that cost $665 in 1934 was sold as an antique car at $14 800 in 1998. What was the rate of return on this "investment"?

2.20 Clifford has $$X$ right now. In 5 years, X will be $3500 if it is invested at 7.5%, compounded annually. Determine the present value of X. If Clifford invested $$X$ at 7.5%, compounded daily, how much would the value of X be in 10 years?

2.21 You have just won a lottery prize of $1 000 000 collectable in ten yearly installments of $100 000, starting today. Why is this prize not really $1 000 000? What is it really worth today if money can be invested at 10% annual interest, compounded monthly? Use a spreadsheet to construct a table showing the present worth of each installment, and the total present worth of the prize.

2.22 Suppose in Problem 2.21 that you have a large mortgage you want to pay off now. You propose an alternative, but equivalent, payment scheme. You would like $300 000 today, and the balance of the prize in five years when you intend to purchase a large piece of waterfront property. How much will the payment be in five years? Assume that annual interest is 10%, compounded monthly.

2.23 You are looking at purchasing a new computer for your four-year undergraduate program. Brand 1 costs $4000 now, and you expect it will last throughout your program without any upgrades. Brand 2 costs $2500 now and will need an upgrade at the end of two years, which you expect to be $1700. With 8% annual interest, compounded monthly, which is the less expensive alternative, if they provide the same level of service and will both be worthless at the end of the four years?

2.24 The Bank of Edmonton advertises savings account interest as 6% compounded daily. What is the effective interest rate?

2.25 The Bank of Kitchener is offering a new savings account that pays a nominal 7.99% interest, compounded continuously. Will your money earn more in this account than in a daily interest account that pays 8%?

2.26 You are comparing two investments. The first pays 1% interest per month, compounded monthly, and the second pays 6% interest per six months, compounded every six months.

(a) What is the effective semiannual interest rate for each investment?

(b) What is the effective annual interest rate for each investment?

(c) On the basis of interest rate, which investment do you prefer? Does your decision depend on whether you make the comparison based on an effective six-month rate or an effective one-year rate?

2.27 The Bank of Calgary advertises savings account interest as 5.5% compounded weekly and chequing account interest at 7% compounded monthly. What are the effective interest rates for the two types of account?

2.28 Victory Visa, Magnificent Master Card, and Amazing Express are credit card companies that charge different interest on overdue accounts. Victory Visa charges 26% compounded daily, Magnificent Master Card charges 28% compounded weekly, and Amazing Express charges 30% compounded monthly. On the basis of interest rate, which credit card has the best deal?

2.29 April has a bank deposit now worth $796.25. A year ago, it was $750. What was the nominal monthly interest rate on her account?

2.30 You have $50 000 to invest in the stock market and have sought the advice of Adam, an experienced colleague who is willing to advise you, for a fee. Adam has told you that he has found a one-year investment for you that provides 15% interest, compounded monthly.

(a) What is the effective annual interest rate, based on a 15% nominal annual rate and monthly compounding?

(b) Adam says that he will make the investment for you for a modest fee of 2% of the investment's value one year from now. If you invest the $50 000 today, how much will you have at the end of one year (before Adam's fee)?

(c) What is the effective annual interest rate of this investment including Adam's fee?

2.31 May has $2000 in her bank account right now. She had wanted to know how much it would be in one year, so she had calculated and come up with $2140.73. Then she had realized she had made a mistake. She had wanted to use the formula for monthly compounding, but instead, she had used the continuous compounding formula. Redo the calculation for May and find out how much will actually be in her account a year from now.

2.32 Hans now has $6000. In three months, he will receive a cheque for $2000. He must pay $900 at the end of each month (starting exactly one month from now). Draw a single cash flow diagram illustrating all of these payments for a total of six monthly periods. Include his cash on hand as a payment at time 0.

2.33 Margaret is considering an investment that will cost her $500 today. It will pay her $100 at the end of each of the next 12 months, and cost her another $300 one year from today. Illustrate these cash flows in two cash flow diagrams. The first should show each cash flow element separately, and the second should show only the net cash flows in each period.

2.34 Heddy is considering working on a project that will cost her $20 000 today. It will pay her $10 000 at the end of each of the next 12 months, and cost her another $15 000 at the end of each quarter. An extra $10 000 will be received at the end of the project, one year from now. Illustrate these cash flows in two cash flow diagrams. The first should show each cash flow element separately, and the second should show only the net cash flow in each period.

2.35 Illustrate the following cash flows over 12 months in a cash flow diagram. Show only the net cash flow in each period.

Cash Payments	$20 every three months, starting now
Cash Receipts	Receive $30 at the end of the first month, and from that point on, receive 10% more than the previous month at the end of each month

2.36 There are two possible investments, A and B. Their cash flows are shown in the table below. Illustrate these cash flows over 12 months in two cash flow diagrams. Show only the net cash flow in each period. Just looking at the diagrams, would you prefer one investment to the other? Comment on this.

	Investment A	Investment B
Payments	$2400 now and a closing fee of $200 at the end of month 12	$500 every two months, starting two months from now
Receipts	$250 monthly payment at the end of each month	Receive $50 at the end of the first month, and from that point on, receive $50 more than the previous month at the end of each month

2.37 You are indifferent between receiving $100 today and $110 one year from now. The bank pays you 6% interest on deposits and charges you 8% for loans. Name the three types of equivalence and comment (one sentence for each) on whether each exists for this situation and why.

2.38 June has a small house on a small street in a small town. If she sells the house now, she will likely get \$110 000 for it. If she waits for one year, she will likely get more, say \$120 000. If she sells the house now, she can invest the money in one-year guaranteed investment certificate (GIC) that pays 8% interest, compounded monthly. If she keeps the house, then the interest on the mortgage payments is 8% compounded daily. June is indifferent between the two options: selling the house now and keeping the house for another year. Discuss whether each of the three types of equivalence exists in this case.

2.39 Using a spreadsheet, construct graphs for the loan described in (a) below.

(a) Plot the amount owed (principal plus interest) on a simple interest loan of \$100 for N years for $N = 1, 2, \ldots 10$. On the same graph, plot the amount owed on a compound interest loan of \$100 for N years for $N = 1, 2, \ldots 10$. The interest rate is 6% per year for each loan.

(b) Repeat part (a), but use an interest rate of 18%. Observe the dramatic effect compounding has on the amount owed at the higher interest rate.

2.40 **(a)** At 12% interest per annum, how long will it take for a penny to become a million dollars? How long will it take at 18%?

(b) Show the growth in values on a spreadsheet using 10-year time intervals.

2.41 Use a spreadsheet to determine how long it will take for a \$100 deposit to double in value for each of the following interest rates and compounding periods. For each, plot the size of the deposit over time, for as many periods as necessary for the original sum to double.

(a) 8% per year, compounded monthly

(b) 11% per year, compounded semiannually

(c) 12% per year, compounded continuously

2.42 Construct a graph showing how the effective interest rate for the following nominal rates increases as the compounding period becomes shorter and shorter. Consider a range of compounding periods of your choice from daily compounding to annual compounding.

(a) 6% per year

(b) 10% per year

(c) 20% per year

2.43 Today, an investment you made three years ago has matured and is now worth \$3000. It was a three-year deposit that bore an interest rate of 10% per year, compounded monthly. You knew at the time that you were taking a risk in making such an investment because interest rates vary over time and you "locked in" at 10% for three years.

(a) How much was your initial deposit? Plot the value of your investment over the three-year period.

(b) Looking back over the past three years, interest rates for similar one-year investments did indeed vary. The interest rates were 8% the first year, 10% the second, and 14% the third. Plot the value of your initial deposit over time as if you had invested at this set of rates, rather than for a constant 10% rate. Did you lose out by having locked into the 10% investment? If so, by how much?

Student Credit Cards

CANADIAN 2.1 MINI-CASE

Most major banks in Canada offer a credit card service for students. Common features of the student credit cards include no annual fee, a $500 credit limit, and an annual interest rate of 18.5% (as of 2004). Also, the student cards often come with many of the perks available for the general public: purchase security or travel-related insurance, extended warranty protection, access to cash advance, etc. The approval process for getting a card is relatively simple for university and college students so that they can start building a credit history and enjoy the convenience of having a credit card while still in school.

The printed information does not use the term *nominal* or *effective*, nor does it define the compounding period. However, it is common in the credit card business for the annual interest rate to be divided into daily rates for billing purposes. Hence, the quoted annual rate of 18.5% is a nominal rate and the compounding period is daily. The actual effective interest rate is then $(1 + 0.185/365)^{365} - 1 = 0.2032$ or 20.32%.

Discussion

Interest information must be disclosed by law, but lenders and borrowers have some latitude as to how and where they disclose it. Moreover, there is a natural desire to make the interest rate look lower than it really is for borrowers, and higher than it really is for lenders.

In the example of student credit cards, the effective interest rate is 20.32%, roughly 2% higher than the stated interest rate. The actual effective interest rate could even end up being higher if fees such as late fees, over-the-limit fees, and transaction fees are charged.

Questions

1. Go to your local bank branch and find out the interest rate paid for various kinds of savings accounts, chequing accounts, and loans. For each interest rate quoted, determine if it is a nominal or effective rate. If it is nominal, determine the compounding period and calculate the effective interest rate.
2. Have a contest with your classmates to see who can find the organization that will lend money to a student like you at the cheapest effective interest rate, or that will take investments which provide a guaranteed return at the highest effective interest rate. The valid rates must be generally available, not tied to particular behaviour by the client, and not secured to an asset (like a mortgage).
3. If you borrowed $1000 at the best rate you could find and invested it at the best rate you could find, how much money would you make or lose in a year? Explain why the result of your calculation could not have the opposite sign.

CHAPTER 3

Cash Flow Analysis

Engineering Economics in Action, Part 3A: Apples and Oranges

The flyer was slick, all right. The information was laid out so anybody could see that leasing palletizing equipment through the Provincial Finance Company (PFC) made much more sense than buying it. It was something Naomi could copy right into her report to Clem.

Naomi had been asked to check out options for automating part of the Shipping Department. Parts were to be stacked and bound on plastic pallets, then loaded onto trucks and sent to one of Canadian Widget's sister companies. The saleswoman for the company whose equipment seemed most suitable for Canadian Widget's needs included the leasing flyer with her quote.

Naomi looked at the figures again. They seemed to make sense, but there was something that didn't seem right to her. For one thing, if it was cheaper to lease, why didn't everybody lease everything? She knew that some things, like automobiles and airplanes, are often leased instead of bought, but generally companies buy assets. Second, where was the money coming from to give the finance company a profit? If the seller was getting the same amount and the buyer was paying less, how could PFC make money?

"Got a recommendation on that palletizer yet, Naomi?" Clem's voice was cheery as he suddenly appeared at her doorway. Naomi knew that the Shipping Department was the focus of Clem's attention right now and he wanted to get improvements in place as soon as possible.

"Yes, I do. There's really only one that will do the job, and it does it well at a good price. There is something I'm trying to figure out, though. Christine sent me some information about leasing it instead of buying it, and I'm trying to figure out where the catch is. There has got to be one, but I can't see it right now."

"OK, let me give you a hint: apples and oranges. You can't add them. Now, let's get the paperwork started for that palletizer. The shipping department is just too much of a bottleneck." Clem disappeared from her door as quickly as he had arrived, leaving Naomi musing to herself.

"Apples and *oranges*? *Apples* and oranges? Ahh . . . apples and oranges, of course!"

3.1 Introduction

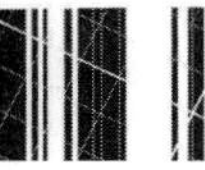

Chapter 2 showed that interest is the basis for determining whether different patterns of cash flows are equivalent. Rather than comparing patterns of cash flows from first principles, it is usually easier to use functions that define *mathematical* equivalence among certain common cash flow patterns. These functions are called *compound interest factors*. We discuss a number of these common cash flow patterns along with their associated compound interest factors in this chapter. These compound interest factors are used throughout the remainder of the book. It is, therefore, particularly important to understand their use before proceeding to subsequent chapters.

This chapter opens with an explanation of how cash flow patterns that engineers commonly use are simplified approximations of complex reality. Next, we discuss four simple, discrete cash flow patterns and the compound interest factors that relate them to each other. There is then a brief discussion of the case in which the number of time periods considered is so large that it is treated as though the relevant cash flows continue indefinitely. Appendix 3A discusses modelling cash flow patterns when the interval between disbursements or receipts is short enough that we may view the flows as being continuous. Appendix 3B presents mathematical derivations of the compound interest factors.

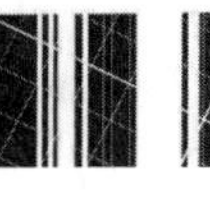

3.2 Timing of Cash Flows and Modelling

The actual timing of cash flows can be very complicated and irregular. Unless some simple approximation is used, comparisons of different cash flow sequences will be very difficult and impractical. Consider, for example, the cash flows generated by a relatively simple operation like a service station that sells gasoline and supplies, and also services cars. Some cash flows, like sales of gasoline and minor supplies, will be almost continuous during the time the station is open. Other flows, like receipts for the servicing of cars, will be on a daily basis. Disbursements for wages may be on a weekly basis. Some disbursements, like those for a manager's salary and for purchases of gasoline and supplies, may be monthly. Disbursements for insurance and taxes may be quarterly or semi-annual. Other receipts and disbursements, like receipts for major repairs or disbursements for used parts, may be irregular.

An analyst trying to make a comparison of two projects with different, irregular timings of cash flows might have to record each of the flows of the projects, then, on a one-by-one basis, find summary equivalent values like present worth that would be used in the comparison. This activity would be very time-consuming and tedious if it could be done, but it probably could not be done because the necessary data would not exist. If the projects were potential rather than actual, the cash flows would have to be predicted. This could not be done with great precision for either size or timing of the flows. Even if the analysis were of the past performances of ongoing operations, it is unlikely that it would be worthwhile to maintain a data bank that contained the exact timing of all cash flows.

Because of the difficulties of making precise calculations of complex and irregular cash flows, engineers usually work with fairly simple models of cash flow patterns. The most common type of model assumes that all cash flows and all compounding of cash flows occur at the ends of conventionally defined periods like months or years. Models that make this assumption are called **discrete models**. In some cases, analysts use models that assume cash flows and their compounding occur continuously over time; such models are called **continuous models**. Whether the analyst uses discrete modelling or continuous modelling, the model is usually an approximation. Cash flows do not occur only at the ends of conventionally defined periods, nor are they actually continuous. We shall emphasize discrete models throughout the book because they are more common and more readily understood by persons of varied backgrounds. Discrete cash flow models are discussed in the main body of this chapter, and continuous models are presented in Appendix 3A, at the end of this chapter.

3.3 Compound Interest Factors for Discrete Compounding

Compound interest factors are formulas that define mathematical equivalence for specific common cash flow patterns. The compound interest factors permit cash flow analysis to be done more conveniently because tables or spreadsheet functions can be used instead of complicated formulas. This section presents compound interest factors for four discrete cash flow patterns that are commonly used to model the timing of receipts and disbursements in engineering economic analysis. The four patterns are:

1. A single disbursement or receipt

2. A set of equal disbursements or receipts over a sequence of periods, referred to as an **annuity**
3. A set of disbursements or receipts that change by a constant *amount* from one period to the next in a sequence of periods, referred to as an **arithmetic gradient series**
4. A set of disbursements or receipts that change by a constant *proportion* from one period to the next in a sequence of periods, referred to as a **geometric gradient series**

The principle of discrete compounding requires several assumptions:

1. Compounding periods are of equal length.
2. Each disbursement and receipt occurs at the end of a period. A payment at time 0 can be considered to occur at the end of period −1.
3. Annuities and gradients coincide with the ends of sequential periods. (Section 3.8 suggests several methods for dealing with annuities and gradients that do not coincide with the ends of sequential periods.)

Mathematical derivations of six of the compound interest factors are given in Appendix 3B at the end of this chapter.

3.4 Compound Interest Factors for Single Disbursements or Receipts

In many situations, a single disbursement or receipt is an appropriate model of cash flows. For example, the salvage value of production equipment with a limited service life will be a single receipt at some future date. An investment today to be redeemed at some future date is another example.

Figure 3.1 illustrates the general form of a single disbursement or receipt. Two commonly used factors relate a single cash flow in one period to another single cash flow in a different period. They are the *compound amount factor* and the *present worth factor*.

Figure 3.1 Single Receipt at End of Period N

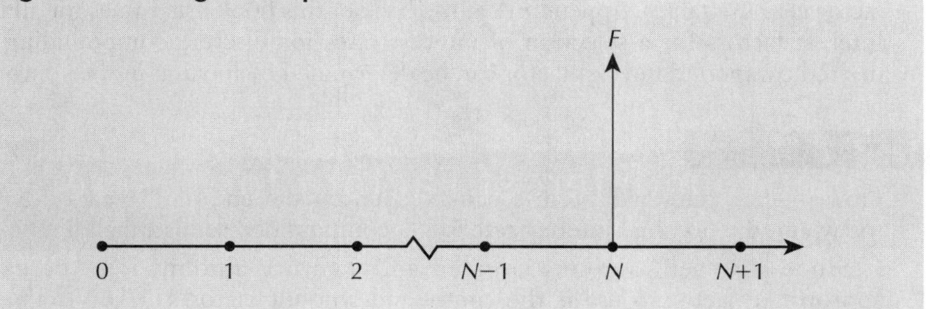

The **compound amount factor**, denoted by $(F/P,i,N)$, gives the future amount, F, that is equivalent to a present amount, P, when the interest rate is i and the number of periods is N. The value of the compound amount factor is easily seen as coming from Equation (2.1), the compound interest equation, which relates present and future values,

$$F = P(1 + i)^N$$

In the symbolic convention used for compound interest factors, this is written

$$F = P(1 + i)^N = P(F/P,i,N)$$

so that the compound amount factor is

$$(F/P,i,N) = (1 + i)^N$$

A handy way of thinking of the notation is (reading from left to right): "What is F, given P, i, and N?"

The compound amount factor is useful in determining the future value of an investment made today if the number of periods and the interest rate are known.

The **present worth factor**, denoted by $(P/F,i,N)$, gives the present amount, P, that is equivalent to a future amount, F, when the interest rate is i and the number of periods is N. The present worth factor is the inverse of the compound amount factor, $(F/P,i,N)$. That is, while the compound amount factor gives the future amount, F, that is equivalent to a present amount, P, the present worth factor goes in the other direction. It gives the present worth, P, of a future amount, F. Since $(F/P,i,N) = (1 + i)^N$,

$$(P/F,i,N) = \frac{1}{(1 + i)^N}.$$

The compound amount factor and the present worth factor are fundamental to engineering economic analysis. Their most basic use is to convert a single cash flow that occurs at one point in time to an equivalent cash flow at another point in time. When comparing several individual cash flows which occur at different points in time, an analyst would apply the compound amount factor or the present worth factor, as necessary, to determine the equivalent cash flows at a common reference point in time. In this way, each of the cash flows is stated as an amount at one particular time. Example 3.1 illustrates this process.

Although the compound amount factor and the present worth factor are relatively easy to calculate, some of the other factors discussed in this chapter are more complicated, and it is therefore desirable to have an easier way to determine their values. The compound interest factors are sometimes available as functions in calculators and spreadsheets, but often these functions are provided in an awkward format that makes them relatively difficult to use. They can, however, be fairly easily programmed in a calculator or spreadsheet.

A traditional and still useful method for determining the value of a compound interest factor is to use tables. Appendix A at the back of this book lists values for all the compound interest factors for a selection of interest rates for discrete compounding periods. The desired compound interest factor can be determined by looking in the appropriate table.

Example 3.1

How much money will be in a bank account at the end of 15 years if \$100 is invested today and the nominal interest rate is 8% compounded semiannually?

Since a present amount is given and a future amount is to be calculated, the appropriate factor to use is the compound amount factor, $(F/P,i,N)$. There are several ways of choosing i and N to solve this problem. The first method is to observe that, since interest is compounded semiannually, the number of compounding periods, N, is 30. The interest rate per six-month period is 4%. Then

$$\begin{aligned} F &= \$100(F/P, 4\%, 30) \\ &= \$100(1 + 0.04)^{30} \\ &= \$324.34 \end{aligned}$$

The bank account will hold \$324.34 at the end of 15 years.

Alternatively, we can obtain the same results by using the interest factor tables.

$$F = \$100(3.2434) \quad \text{(from Appendix A)}$$
$$= \$324.34$$

A second solution to the problem is to calculate the *effective* yearly interest rate and then compound over 15 years at this rate. Recall from Equation (2.4) that the effective interest rate per year is

$$i_e = \left(1 + \frac{r}{m}\right)^m - 1$$

where i_e = the effective annual interest rate
r = the nominal rate per year
m = the number of periods in a year

$$i_e = (1 + 0.08/2)^2 - 1 = 0.0816$$

where $r = 0.08$
$m = 2$

When the effective yearly rate for each of 15 years is applied to the future worth computation, the future worth is

$$F = P(F/P,i,N)$$
$$= P(1 + i)^N$$
$$= \$100(1 + 0.0816)^{15}$$
$$= \$324.34$$

Once again, we conclude that the balance will be \$324.34.■

3.5 Compound Interest Factors for Annuities

The next four factors involve a series of uniform receipts or disbursements that start at the end of the first period and continue over N periods, as illustrated in Figure 3.2. This pattern of cash flows is called an annuity. Mortgage or lease payments and maintenance contract fees are examples of the annuity cash flow pattern. Annuities may also be used to model series of cash flows that fluctuate over time around some average value. Here the average value would be the constant uniform cash flow. This would be done if the fluctuations were unknown or deemed to be unimportant for the problem.

The **sinking fund factor**, denoted by $(A/F,i,N)$, gives the size, A, of a repeated receipt or disbursement that is equivalent to a future amount, F, if the interest rate is i and the number of periods is N. The name of the factor comes from the term **sinking fund**. A sinking fund is an interest-bearing account into which regular deposits are made in order to accumulate some amount.

The equation for the sinking fund factor can be found by decomposing the series of disbursements or receipts made at times 1, 2, . . . , N, and summing to produce a total future value. The formula for the sinking fund factor is

Figure 3.2 Annuity over N Periods

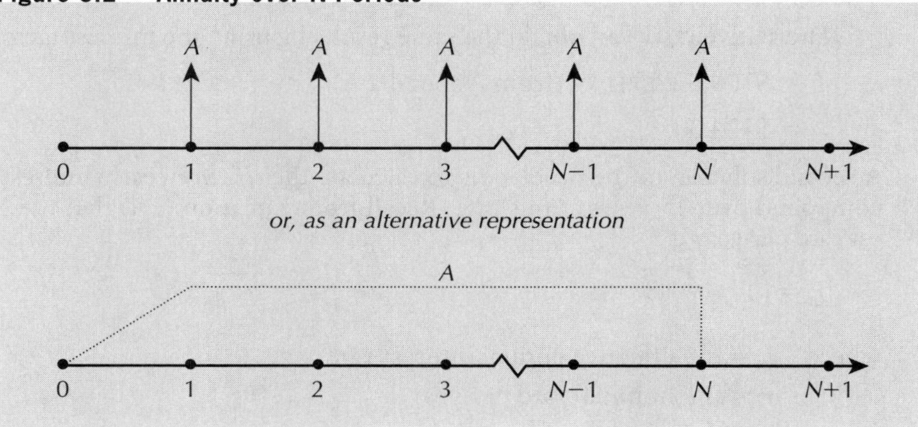

$$(A/F,i,N) = \frac{i}{(1+i)^N - 1}$$

The sinking fund factor is commonly used to determine how much has to be set aside or saved per period to accumulate an amount F at the end of N periods at an interest rate i. The amount F might be used, for example, to purchase new or replacement equipment, to pay for renovations, or to cover capacity expansion costs. In more general terms, the sinking fund factor allows us to convert a single future amount into a series of equal-sized payments, made over N equally spaced intervals, with the use of a given interest rate i.

The **uniform series compound amount factor**, denoted by $(F/A,i,N)$, gives the future value, F, that is equivalent to a series of equal-sized receipts or disbursements, A, when the interest rate is i and the number of periods is N. Since the uniform series compound amount factor is the inverse of the sinking fund factor,

$$(F/A,i,N) = \frac{(1+i)^N - 1}{i}$$

The **capital recovery factor**, denoted by $(A/P,i,N)$, gives the value, A, of the equal periodic payments or receipts that is equivalent to a present amount, P, when the interest rate is i and the number of periods is N. The capital recovery factor is easily derived from the sinking fund factor and the compound amount factor:

$$\begin{aligned}(A/P,i,N) &= (A/F,i,N)(F/P,i,N)\\ &= \frac{i}{(1+i)^N - 1}(1+i)^N\\ &= \frac{i(1+i)^N}{(1+i)^N - 1}\end{aligned}$$

The capital recovery factor can be used to find out, for example, how much money must be saved over N future periods to "recover" a capital investment of P today. The capital recovery factor for the purchase cost of something is sometimes combined with the sinking fund factor for its salvage value after N years to compose the **capital recovery formula**. See Close-Up 3.1.

The **series present worth factor**, denoted by $(P/A,i,N)$, gives the present amount, P, that is equivalent to an annuity with disbursements or receipts in the amount, A,

CLOSE-UP 3.1 Capital Recovery Formula

Industrial equipment and other assets are often purchased at a cost of P on the basis that they will incur savings of A per period for the firm. At the end of their useful life, they will be sold for some salvage value S. The expression to determine A for a given P and S combines the capital recovery factor (for P) with the sinking fund factor (for S):

$$A = P(A/P,i,N) - S(A/F,i,N)$$

Since

$$(A/F,i,N) = \frac{i}{(1+i)^N - 1} = \frac{i}{(1+i)^N - 1} + i - i$$

$$= \frac{i}{(1+i)^N - 1} + \frac{i[(1+i)^N - 1]}{(1+i)^N - 1} - i$$

$$= \frac{i + i(1+i)^N - i}{(1+i)^N - 1} - i = \frac{i(1+i)^N}{(1+i)^N - 1} - i$$

$$= (A/P,i,N) - i$$

then

$$A = P(A/P,i,N) - S[(A/P,i,N) - i]$$

$$= (P - S)(A/P,i,N) + Si$$

This is the capital recovery formula, which can be used to calculate the savings necessary to justify a capital purchase of cost P and salvage value S after N periods at interest rate i.

The capital recovery formula is also used to determine an annual amount which captures the loss in value of an asset over the time it is owned. Chapter 7 treats this use of the capital recovery formula more fully.

where the interest rate is i and the number of periods is N. It is the reciprocal of the capital recovery factor:

$$(P/A,i,N) = \frac{(1+i)^N - 1}{i(1+i)^N}$$

Example 3.2

The Hanover Go-Kart Club has decided to build a clubhouse and track five years from now. It must accumulate \$50 000 by the end of five years by setting aside a uniform amount from its dues at the end of each year. If the interest rate is 10%, how much must be set aside each year?

Since the problem requires that we calculate an annuity amount given a future value, the solution can be obtained using the sinking fund factor where $i = 10\%$, $F = \$50\ 000$, $N = 5$, and A is unknown.

$$A = \$50\ 000(A/F, 10\%, 5)$$

$$= \$50\ 000(0.1638)$$

$$= \$8190.00$$

The Go-Kart Club must set aside \$8190 at the end of each year to accumulate \$50 000 in five years.■

Example 3.3

A car loan requires 30 monthly payments of \$199.00, starting *today*. At an annual rate of 12% compounded monthly, how much money is being lent?

This cash flow pattern is referred to as an **annuity due**. It differs from a standard annuity in that the first of the N payments occurs at time 0 (now) rather than at the end of the first time period. Annuity dues are uncommon—not often will one make the first payment on a loan on the date the loan is received! Unless otherwise stated, it is reasonable to assume that any annuity starts at the end of the first period.

Two simple methods of analyzing an annuity due will be used for this example.

Method 1. Count the first payment as a present worth and the next 29 payments as an annuity:

$$P = \$199 + A(P/A,i,N)$$

where $A = \$199$, $i = 12\%/12 = 1\%$, and $N = 29$.

$$\begin{aligned} P &= \$199 + \$199(P/A, 1\%, 29) \\ &= \$199 + \$199(25.066) \\ &= \$199 + \$4988.13 \\ &= \$5187.13 \end{aligned}$$

The present worth of the loan is the current payment, \$199, plus the present worth of the subsequent 29 payments, \$4988.13, a total of about \$5187.

Method 2. Determine the present worth of a standard annuity at time -1, and then find its worth at time 0 (now). The worth at time -1 is

$$\begin{aligned} P_{-1} &= A(P/A,i,N) \\ &= \$199(P/A, 1\%, 30) \\ &= \$199(25.807) \\ &= \$5135.79 \end{aligned}$$

Then the present worth now (time 0) is

$$\begin{aligned} P_0 &= P_{-1}(F/P,i,N) \\ &= \$5135.79(F/P,1\%,1) \\ &= \$5135.79(1.01) \\ &= \$5187.15 \end{aligned}$$

The second method gives the same result as the first, allowing a small margin for the effects of rounding.■

It is worth noting here that although it is natural to think about the symbol P as meaning a cash flow at time 0, the present, and F as meaning a cash flow in the future, in fact these symbols can be more general in meaning. As illustrated in the last example, we can consider any point in time to be the "present" for calculation purposes, and similarly any point in time to be the "future," provided P is some point in time earlier than F. This observation gives us substantial flexibility in analyzing cash flows.

Example 3.4

Clarence bought a house for \$94 000 in 1982. He made a \$14 000 down payment and negotiated a mortgage from the previous owner for the balance. Clarence agreed to pay the previous owner \$2000 per month at 12% nominal interest, compounded monthly. How long did it take him to pay back the mortgage?

Clarence borrowed only \$80 000, since he made a \$14 000 down payment. The \$2000 payments form an annuity over N months where N is unknown. The interest rate per month is 1%. We must find the value of N such that

$$P = A(P/A,i,N) = A\left(\frac{(1 + i)^N - 1}{i(1 + i)^N}\right)$$

or, alternatively, the value of N such that

$$A = P(A/P,i,N) = P\left(\frac{i(1 + i)^N}{(1 + i)^N - 1}\right)$$

where $P = \$80\ 000$, $A = \$2000$, and $i = 0.01$.

By substituting the known quantities into either expression, some manipulation is required to find N. For illustration, the capital recovery factor has been used.

$$A = P\left(\frac{i(1 + i)^N}{(1 + i)^N - 1}\right)$$

$$\$2000 = \$80\ 000\left(\frac{0.01(1.01)^N}{1.01^N - 1}\right)$$

$$2.5 = \frac{(1.01)^N}{(1.01)^N - 1}$$

$$2.5/1.5 = (1.01)^N$$

$$N[ln(1.01)] = ln(2.5/1.5)$$

$$N = 51.34 \text{ months}$$

It will take Clarence four years and four months to pay off the mortgage. He will make 51 full payments of \$2000 and will be left with only a fraction of a full payment for his 52nd and last monthly installment. Problem 3.34 asks what his final payment will be. Note also that mortgages can be confusing because of the different terms used. See Close-Up 3.2.■

In Example 3.4, it was possible to use the formula for the compound interest factor to solve for the unknown quantity directly. It is not always possible to do this when the number of periods or the interest rate is unknown. We can proceed in several ways. One possibility is to determine the unknown value by trial and error with a spreadsheet. Another approach is to find the nearest values using tables, and then to interpolate linearly to determine an approximate value. Some calculators will perform the interpolation automatically. See Close-Up 3.3 and Figure 3.3 for a reminder of how linear interpolation works.

CLOSE-UP 3.2 Canadian Mortgages

Canadian mortgages can be a little confusing because of the terms used. The interest rate is a nominal rate, usually compounded monthly. The **amortization period** is the duration over which the original loan is calculated to be repaid. The **term** is the duration over which the loan agreement is valid.

For example, Salim has just bought a house for \$135 000. He paid \$25 000 down, and the rest of the cost has been obtained from a mortgage. The mortgage has a nominal interest rate of 9.5% compounded monthly with a 20-year amortization period. The term of the mortgage is three years. What are Salim's monthly payments? How much does he owe after three years?

Salim's monthly payments can be calculated as

$$\begin{aligned} A &= (\$135\,000 - \$25\,000)(A/P, 9.5/12\%, [20 \times 12]) \\ &= \$110\,000(A/P, 0.7917\%, 240) \\ &= \$110\,000(0.00932) \\ &= \$1025.20 \end{aligned}$$

Salim's monthly payments would be about \$1025.20. After three years he would have to renegotiate his mortgage at whatever was the current interest rate at that time. The amount owed would be

$$\begin{aligned} F &= \$110\,000(F/P, 9.5/12\%, 36) - \$1025.20(F/A, 9.5/12\%, 36) \\ &= \$110\,000(1.3283) - \$1025.20(41.47) \\ &\approx \$103\,598 \end{aligned}$$

After three years, Salim still owes \$103 598.

Example 3.5

Clarence paid off an \$80 000 mortgage completely in 48 months. He paid \$2000 per month, and at the end of the first year made an extra payment of \$7000. What interest rate was he charged on the mortgage?

Using the series present worth factor and the present worth factor, this can be formulated for an unknown interest rate:

$$\$80\,000 = \$2000(P/A, i, 48) + \$7000(P/F, i, 12)$$

$$2(P/A, i, 48) + 7(P/F, i, 12) = 80$$

$$2\left[\frac{(1+i)^{48}-1}{i(1+i)^{48}}\right] + 7\left[\frac{1}{(1+i)^{12}}\right] = 80 \tag{3.1}$$

Solving such an equation for i directly is generally not possible. However, using a spreadsheet as illustrated in Table 3.1 can establish some close values for the left-hand side of Equation (3.1), and a similar process can be done using either tables or a calculator. Using a spreadsheet program or calculator, trials can establish a value for the unknown interest rate to the desired number of significant digits.

CLOSE-UP 3.3 Linear Interpolation

Linear interpolation is the process of approximating a complicated function by a straight line in order to estimate a value for the independent variable based on two sample pairs of independent and dependent variables and an instance of the dependent variable. For example, the function f in Figure 3.3 relates the dependent variable y to the independent variable x. Two sample points, (x_1, y_1) and (x_2, y_2), and an instance of y, y^*, are known, but the actual shape of f is not. An estimate of the value for x^* can be made by drawing a straight line between (x_1, y_1) and (x_2, y_2).

Because the line between (x_1, y_1) and (x_2, y_2) is assumed to be straight, the following ratios must be equal:

$$\frac{x^* - x_1}{x_2 - x_1} = \frac{y^* - y_1}{y_2 - y_1}$$

Isolating the x^* gives the linear interpolation formula:

$$x^* = x_1 + (x_2 - x_1)\left[\frac{y^* - y_1}{y_2 - y_1}\right]$$

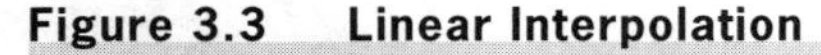

Figure 3.3 Linear Interpolation

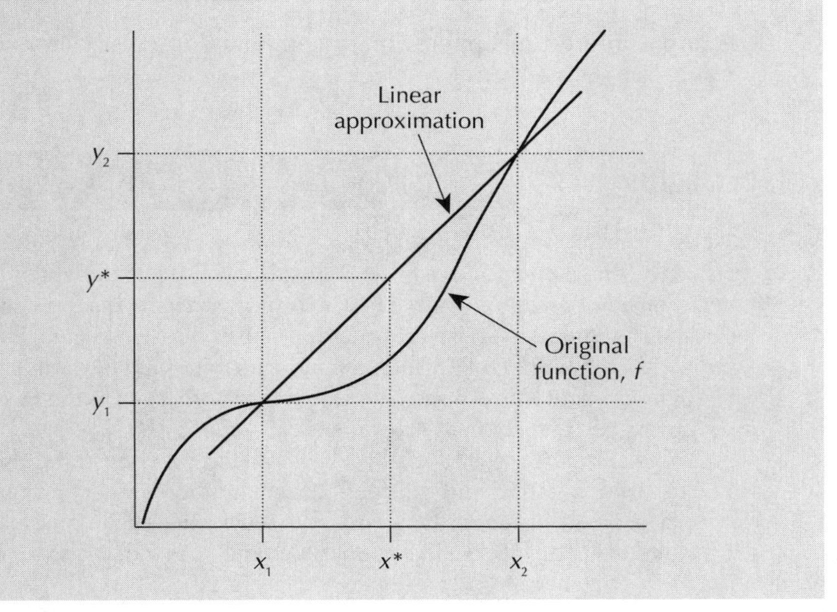

Once the approximate values for the interest rate are found, linear interpolation can be used to find a more precise answer. For instance, working from the values of the interest rate which give the LHS (left-hand side) value closest to the RHS (right-hand side) value of 80, which are 1.1% and 1.2%,

$$i = 1.1 + (1.2 - 1.1)\left[\frac{80 - 80.4141}{78.7209 - 80.4141}\right]$$

$$= 1.1 + 0.02 = 1.12\% \text{ per month}$$

The nominal interest rate was $1.12 \times 12 = 13.44\%$.
The effective interest rate was $(1.0112)^{12} - 1 = 14.30\%$.■

Table 3.1 Trials to Determine an Unknown Interest Rate

Interest Rate i	$2(P/A, i, 48) + 7(P/F, i, 12)$
0.5%	91.7540
0.6%	89.7128
0.7%	87.7350
0.8%	85.8185
0.9%	83.9608
1.0%	82.1601
1.1%	80.4141
1.2%	78.7209
1.3%	77.0787
1.4%	75.4855
1.5%	73.9398

Another interesting application of compound interest factors is calculating the value of a bond. See Close-Up 3.4.

CLOSE-UP 3.4 Bonds

Bonds are investments that provide an annuity and a future value in return for a cost today. They have a *par* or *face* value, which is the amount for which they can be redeemed after a certain period of time. They also have a *coupon rate*, meaning that they pay the bearer an annuity, usually semiannually, calculated as a percentage of the face value. For example, a coupon rate of 10% on a bond with an \$8000 face value would pay an annuity of \$400 each six months. Bonds can sell at more or less than the face value, depending on how buyers perceive them as investments.

To calculate the worth of a bond today, sum together the present worth of the face value (a future amount) and the coupons (an annuity) at an appropriate interest rate. For example, if money can earn 12% compounded semiannually, a bond maturing in 15 years with a face value of \$5000 and a coupon rate of 7% is today worth

$$\begin{aligned} P &= \$5000(P/F, 6\%, 30) + (\$5000 \times 0.07/2)(P/A, 6\%, 30) \\ &= \$5000(0.17411) + \$175(13.765) \\ &= \$3279.43 \end{aligned}$$

The bond is worth about \$3279 today.

3.6 Conversion Factor for Arithmetic Gradient Series

An **arithmetic gradient series** is a series of receipts or disbursements that starts at zero at the end of the first period and then increases by a constant *amount* from period to period. Figure 3.4 illustrates an arithmetic gradient series of receipts. Figure 3.5 shows an arithmetic gradient series of disbursements. As an example, we may model a pattern of increasing operating costs for an aging machine as an arithmetic gradient series if the costs are increasing by (approximately) the same amount each period. Note carefully that the first non-zero cash flow of a gradient occurs at the end of the *second* compounding period, not the first.

Figure 3.4 Arithmetic Gradient Series of Receipts

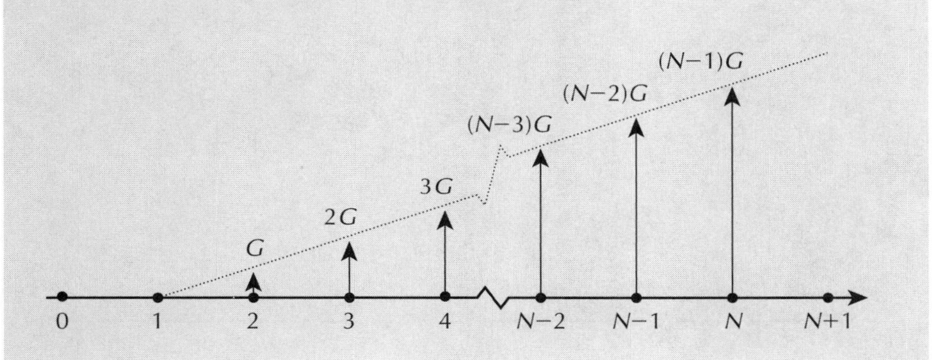

Figure 3.5 Arithmetic Gradient Series of Disbursements

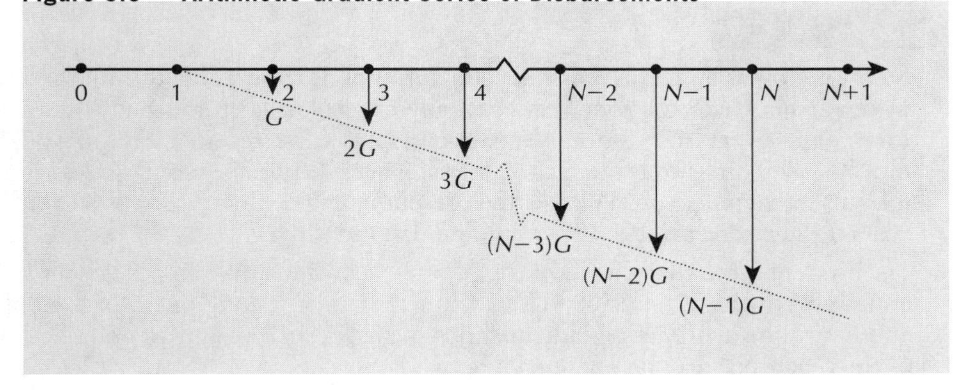

The sum of an annuity plus an arithmetic gradient series is a common pattern. The annuity is a base to which the arithmetic gradient series is added. This is shown in Figure 3.6. A constant-amount increase to a base level of receipts may occur where the increase in receipts is due to adding capacity and where the ability to add capacity is limited. For example, a company that specializes in outfitting warehouses for grocery chains can expand by adding work crews. But the crews must be trained by managers who have time to train only one crew member every six months. Hence, we would have a base amount and a constant amount of growth in cash flows each period.

The **arithmetic gradient to annuity conversion factor**, denoted by $(A/G,i,N)$, gives the value of an annuity, A, that is equivalent to an arithmetic gradient series where the constant increase in receipts or disbursements is G per period, the interest rate is i, and the number of periods is N. That is, the arithmetic gradient series, $0G, 1G, 2G, \ldots, (N-1)G$ is given and the uniform cash flow, A, over N periods is found. Problem 3.29 asks the reader to show that the equation for the arithmetic gradient to annuity factor is

$$(A/G,i,N) = \frac{1}{i} - \frac{N}{(1+i)^N - 1}$$

There is often a base annuity A' associated with a gradient, as illustrated in Figure 3.6. To determine the uniform series equivalent to the *total* cash flow, the base annuity A' must be included to give the overall annuity:

$$A_{\text{tot}} = A' + G(A/G,i,N)$$

Figure 3.6 Arithmetic Gradient Series with Base Annuity

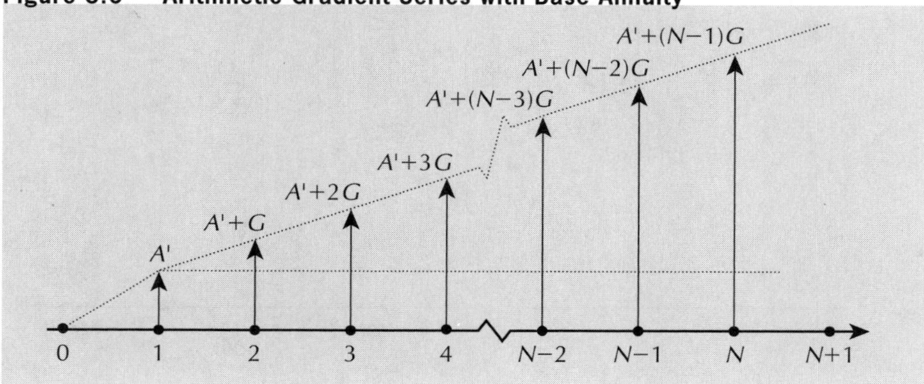

Example 3.6

Susan Ng owns an eight-year-old Jetta automobile. She wants to find the present worth of repair bills over the four years that she expects to keep the car. Susan has the car in for repairs every six months. Repair costs are expected to increase by $50 every six months over the next four years, starting with $500 six months from now, $550 six months later, and so on. What is the present worth of the repair costs over the next four years if the interest rate is 12% compounded monthly?

First, observe that there will be $N = 8$ repair bills over four years and that the base annuity payment, A', is $500. The arithmetic gradient component of the bills, G, is $50, and hence the arithmetic gradient series is $0, $50, $100, and so on. The present worth of the repair bills can be obtained in a two-step process:

Step 1. Find the total uniform annuity, A_{tot}, equivalent to the sum of the base annuity, A' = $500, and the arithmetic gradient series with G = $50 over $N = 8$ periods.

Step 2. Find the present worth of A_{tot}, using the series present worth factor.

The 12% nominal interest rate, compounded monthly, is 1% per month. The effective interest rate per six-month period is

$$i_{\text{6month}} = (1 + 0.12/12)^6 - 1 = 0.06152 \text{ or } 6.152\%$$

Step 1

$$A_{\text{tot}} = A' + G(A/G,i,N)$$
$$= \$500 + \$50\left(\frac{1}{i} - \frac{N}{(1+i)^N - 1}\right)$$
$$= \$500 + \$50\left(\frac{1}{0.06152} - \frac{8}{(1.06152)^8 - 1}\right)$$
$$= \$659.39$$

Step 2

$$P = A_{\text{tot}}(P/A,i,N) = A_{\text{tot}}\left(\frac{(1+i)^N - 1}{i(1+i)^N}\right)$$
$$= \$659.39\left(\frac{(1.06152)^8 - 1}{0.06152(1.06152)^8}\right)$$
$$= \$4070.09$$

The present worth of the repair costs is about $4070.■

3.7 Conversion Factor for Geometric Gradient Series

A **geometric gradient series** is a series of cash flows that increase or decrease by a constant *percentage* each period. The geometric gradient series may be used to model inflation or deflation, productivity improvement or degradation, and growth or shrinkage of market size, as well as many other phenomena.

In a geometric series, the base value of the series is A and the "growth" rate in the series (the rate of increase or decrease) is referred to as g. The terms in such a series are given by $A, A(1 + g), A(1 + g)^2, \ldots, A(1 + g)^{N-1}$ at the ends of periods 1, 2, 3, . . . , N, respectively. If the rate of growth, g, is positive, the terms are increasing in value. If the rate of growth, g, is negative, the terms are decreasing. Figure 3.7 shows a series of receipts where g is positive. Figure 3.8 shows a series of receipts where g is negative.

NET VALUE 3.1

Estimating Growth Rates

Geometric gradient series can be used to model the effect of inflation, deflation, production rate change, and market size change on a future cash flow. When using a geometric gradient series, the relevant growth rate must be estimated. The Internet can be a useful research tool for collecting information such as expert opinions and statistics on trends for national and international activities by product type and industry. For example, a sales growth rate may be estimated by considering a number of factors: economic condition indicators (e.g., gross domestic product, employment, consumer spending), population growth, raw material cost, and even online shopping and business-to-business trading. Information on all of these items can be researched using the Internet. A reliable place to start is the Statistics Canada site (www.statcan.ca). There, you can find a wealth of information about inflation rates, trends in employment and income, international trade statistics, and other economic indicators.

Figure 3.7 Geometric Gradient Series for Receipts with Positive Growth

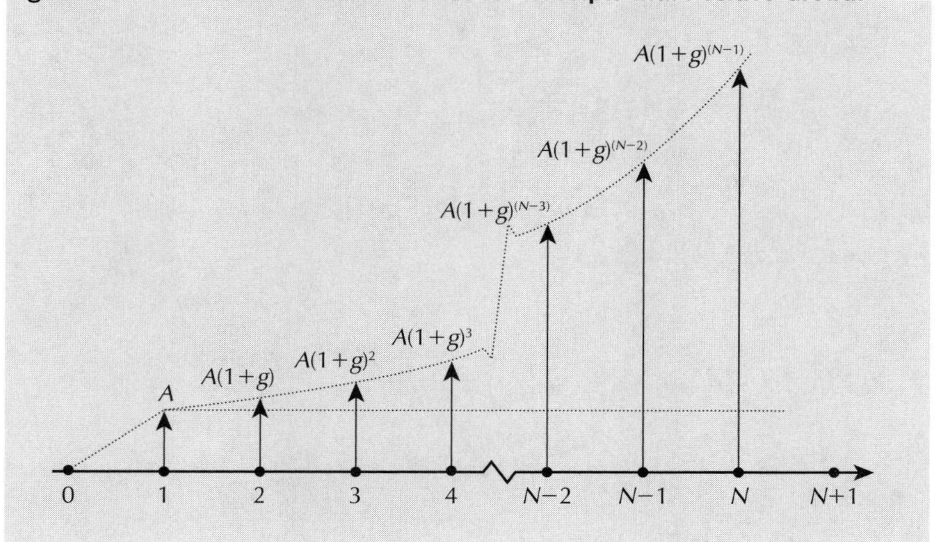

The **geometric gradient to present worth conversion factor**, denoted by $(P/A,g,i,N)$, gives the present worth, P, that is equivalent to a geometric gradient series where the base receipt or disbursement is A, and where the rate of growth is g, the interest rate is i, and the number of periods is N.

The present worth of a geometric series is

$$P = \frac{A}{1+i} + \frac{A(1+g)}{(1+i)^2} + \ldots + \frac{A(1+g)^{N-1}}{(1+i)^N}$$

where A = the base amount
g = the rate of growth
i = the interest rate
N = the number of periods
P = the present worth

We can define a **growth adjusted interest rate**, i°, as

$$i^\circ = \frac{1+i}{1+g} - 1$$

so that

$$\frac{1}{1+i^\circ} = \frac{1+g}{1+i}$$

Then the geometric gradient series to present worth conversion factor is given by

$$(P/A,g,i,N) = \frac{(P/A,i^\circ,N)}{1+g} \text{ or}$$

$$(P/A,g,i,N) = \left(\frac{(1+i^\circ)^N - 1}{i^\circ(1+i^\circ)^N}\right)\frac{1}{1+g}$$

Care must be taken in using the geometric gradient to present worth conversion factor. Four cases may be distinguished:

Figure 3.8 Geometric Gradient Series for Receipts with Negative Growth

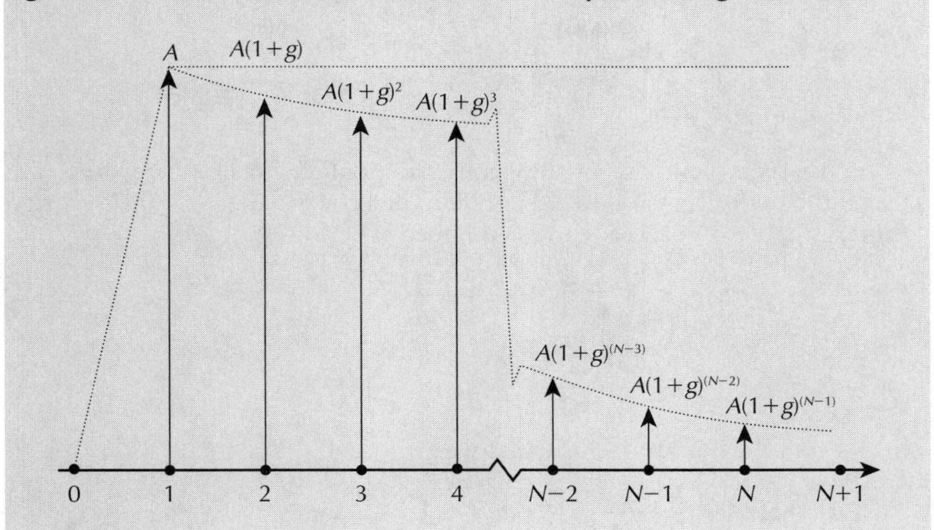

1. $i > g > 0$. *Growth is positive, but less than the rate of interest.* The growth adjusted interest rate, $i°$, is positive. Tables or functions built into software may be used to find the conversion factor.
2. $g > i > 0$. *Growth is positive and greater than the interest rate.* The growth adjusted interest rate, $i°$, is negative. It is necessary to compute the conversion factor directly from the formula.
3. $g = i > 0$. *Growth is positive and exactly equal to the interest rate.* The growth adjusted interest rate $i° = 0$. As with any case where the interest rate is zero, the present worth of the series with constant terms, $A/(1 + g)$, is simply the sum of all the N terms

$$P = N\left(\frac{A}{1 + g}\right)$$

4. $g < 0$. *Growth is negative.* In other words, the series is decreasing. The growth adjusted interest rate, $i°$, is positive. Tables or functions built into software may be used to find the conversion factor.

Example 3.7

Tru-Test is in the business of assembling and packaging automotive and marine testing equipment to be sold through retailers to "do-it-yourselfers" and small repair shops. One of their products is tire pressure gauges. This operation has some excess capacity. Tru-Test is considering using this excess capacity to add engine compression gauges to their line. They can sell engine pressure gauges to retailers for \$8 per gauge. They expect to be able to produce about 1000 gauges in the first month of production. They also expect that, as the workers learn how to do the work more efficiently, productivity will rise by 0.25% per month for the first two years. In other words, each month's output of gauges will be 0.25% more than the previous month's. The interest rate is 1.5% per month. All gauges are sold in the month in which they are produced, and receipts from sales are at the end of each month. What is the present worth of the sales of the engine pressure gauges in the first two years?

We first compute the growth-adjusted interest rate, $i°$:

$$i° = \frac{1+i}{1+g} - 1 = \frac{1.015}{1.0025} - 1 = 0.01247$$

$$i° \approx 1.25\%$$

We then make use of the geometric gradient to present worth conversion factor with the uniform cash flow $A = \$8000$, the growth rate $g = 0.0025$, the growth adjusted interest rate $i° = 0.0125$, and the number of periods $N = 24$.

$$P = A(P/A,g,i,N) = A\left(\frac{(P/A,i°,N)}{1+g}\right)$$

$$P = \$8000\left(\frac{(P/A,\ 1.25\%,\ 24)}{1.0025}\right)$$

From the interest factor tables we get

$$P = \$8000\left(\frac{20.624}{1.0025}\right)$$

$$P = \$164\ 580$$

The present worth of sales of engine compression gauges over the two-year period would be about \$165 000. Recall that we worked with an *approximate* growth-adjusted interest rate of 1.25% when the correct rate was a bit less than 1.25%. This means that \$164 580 is a slight understatement of the present worth.■

Example 3.8

Emery's company, Dry-All, produces control systems for drying grain. Proprietary technology has allowed Dry-All to maintain steady growth in the U.S. market in spite of numerous competitors. Company dividends, all paid to Emery, are expected to rise at a rate of 10% per year over the next 10 years. Dividends at the end of this year are expected to total \$110 000. If all dividends are invested at 10% interest, how much will Emery accumulate in 10 years?

If we calculate the growth adjusted interest rate, we get

$$i° = \frac{1.1}{1.1} - 1 = 0$$

and it is natural to think that the present worth is simply the first year's dividends multiplied by 10. However, recall that in the case where $g = i$ the present worth is given by

$$P = N\left(\frac{A}{1+g}\right) = 10\left(\frac{\$110\ 000}{1.1}\right) = \$1\ 000\ 000$$

Intuitively, dividing by $(1 + g)$ compensates for the fact that growth is considered to start after the end of the first period, but the interest rate applies to all periods. We want the future worth of this amount after 10 years:

$$F = \$1\ 000\ 000\ (F/P,\ 10\%,\ 10) = \$1\ 000\ 000\ (2.5937) = \$2\ 593\ 700$$

Emery will accumulate \$2 593 700 in dividends and interest.■

3.8 Non-standard Annuities and Gradients

As discussed in Section 3.3, the standard assumption for annuities and gradients is that the payment period and compounding period are the same. If they are not, the formulas given in this chapter cannot be applied directly. There are three methods for dealing with this situation:

1. Treat each cash flow in the annuity or gradient individually. This is most useful when the annuity or gradient series is not large.
2. Convert the non-standard annuity or gradient to standard form by changing the compounding period.
3. Convert the non-standard annuity to standard form by finding an equivalent standard annuity for the compounding period. This method cannot be used for gradients.

Example 3.9

How much is accumulated over 20 years in a fund that pays 4% interest, compounded yearly, if $1000 is deposited at the end of every fourth year?

The cash flow diagram for this set of payments is shown in Figure 3.9.

Figure 3.9 Non-standard Annuity for Example 3.9

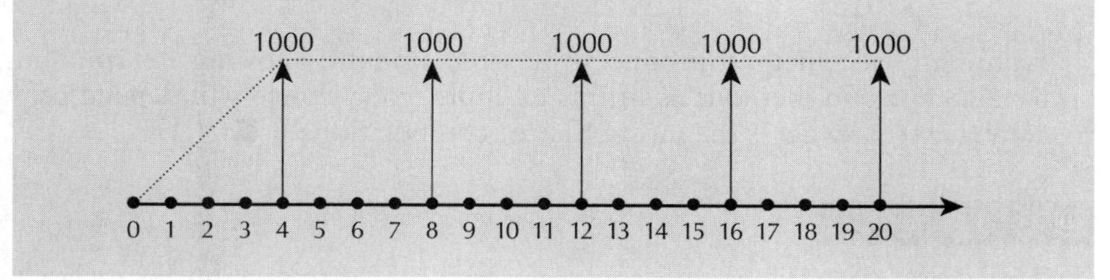

Method 1: Consider the annuities as separate future payments.

Formula: $F = P(F/P,i,N)$

Known values: $P = \$1000$, $i = 0.04$, $N = 16, 12, 8, 4$, and 0

Year	Future Value				
4	$1000($F/P$, 4%, 16)	=	$1000(1.8729)	=	$1873
8	$1000($F/P$, 4%, 12)	=	$1000(1.6010)	=	$1601
12	$1000($F/P$, 4%, 8)	=	$1000(1.3685)	=	$1369
16	$1000($F/P$, 4%, 4)	=	$1000(1.1698)	=	$1170
20	$1000			=	$1000
	Total future value			=	$7013

About $7013 is accumulated over the 20 years.

Method 2: Convert the compounding period from yearly to every four years. This can be done with the effective interest rate formula.

$$i_e = (1 + 0.04)^4 - 1$$
$$= 16.99\%$$

The future value is then

$$F = \$1000(F/A, 16.99\%, 5) = \$1000(7.013)$$
$$= \$7013$$

Method 3: Convert the annuity to an equivalent yearly annuity. This can be done by considering the first payment as a future value over the first four-year period, and finding the equivalent annuity over that period, using the sinking fund factor:

$$A = \$1000(A/F, 4\%, 4)$$
$$= \$1000(0.23549)$$
$$= \$235.49$$

In other words, a \$1000 deposit at the end of the four years is equivalent to four equal deposits of \$235.49 at the end of each of the four years. This yearly annuity is accumulated over the 20 years.

$$F = \$235.49(F/A, 4\%, 20)$$
$$= \$235.49(29.777)$$
$$= \$7012$$

Note that each method produces the same amount, allowing for rounding. When you have a choice in methods as in this example, your choice will depend on what you find convenient, or what is the most efficient computationally.■

Example 3.10

This year's electrical engineering class has decided to save up for a class party. Each of the 90 people in the class is to contribute \$0.25 per day which will be placed in a daily interest (7 days a week, 365 days a year) savings account that pays a nominal 8% interest. Contributions will be made *five* days a week, Monday through Friday, beginning on Monday. The money is put into the account at the beginning of each day, and thus earns interest for the day. The class party is in 14 weeks (a full 14 weeks of payments will be made), and the money will be withdrawn on the Monday morning of the 15th week. How much will be saved, assuming everybody makes payments on time?

There are several good ways to solve this problem. One way is to convert each days' contribution to a weekly amount on Sunday evening/Monday morning, and then accumulate the weekly amounts over the 14 weeks:

Total contribution per day is $\$0.25 \times 90 = \$22.50$

The interest rate per day is $\frac{0.08}{365} = 0.000219$

The effective interest rate for a 1-week period is

$$i = (1 + 0.08/365)^7 - 1 = 0.00154$$

Value of one week's contribution on Friday evening (*annuity due* formula):

$$\$22.50 \times (F/P, 0.08/365, 1) \times (F/A, 0.08/365, 5)$$

On Sunday evening this is worth

$$[\$22.50(F/P, 0.08/365, 1)(F/A, 0.08/365, 5)] \times (F/P, 0.08/365, 2)$$
$$= \$22.50(F/P, 0.08/365, 3)(F/A, 0.08/365, 5)$$

Then the total amount accumulated by Monday morning of the 15th week is given by:

$$[\$22.50(F/P, 0.08/365, 3)(F/A, 0.08/365, 5)](F/A, (1 + 0.08/365)^7 - 1, 14)$$
$$= [\$22.50(1.000\ 658)(5.002\ 19)](14.1406)$$
$$= \$1592.56$$

The total amount saved would be $1592.56.■

3.9 Present Worth Computations When $N \rightarrow \infty$

We have until now assumed that the cash flows of a project occur over some fixed, finite number of periods. For long-lived projects, it may be reasonable to model the cash flows as though they continued indefinitely. The present worth of an infinitely long uniform series of cash flows is called the **capitalized value** of the series. We can get the capitalized value of a series by allowing the number of periods, N, in the series present worth factor to go to infinity:

$$P = \lim_{N \rightarrow \infty} A(P/A,i,N)$$

$$= A \lim_{N \rightarrow \infty} \left[\frac{(1+i)^N - 1}{i(1+i)^N} \right]$$

$$= A \lim_{N \rightarrow \infty} \left[\frac{1 - \frac{1}{(1+i)^N}}{i} \right]$$

$$= \frac{A}{i}$$

Example 3.11

The town of South Battleford is considering building a by-pass for truck traffic around the downtown commercial area. The by-pass will provide merchants and shoppers with benefits that have an estimated value of $500 000 per year. Maintenance costs will be $125 000 per year. If the by-pass is properly maintained, it will provide benefits for a very long time. The actual life of the by-pass will depend on factors like future economic conditions that cannot be forecast at the time the by-pass is being considered. It is, therefore, reasonable to model the flow of benefits as though they continued indefinitely. If the interest rate is 10%, what is the present worth of benefits minus maintenance costs?

$$P = \frac{A}{i} = \frac{\$500\ 000 - \$125\ 000}{0.1} = \$3\ 750\ 000$$

The present worth of benefits net of maintenance costs is $3 750 000.■

REVIEW PROBLEMS

REVIEW PROBLEM 3.1

The benefits of a revised production schedule for a seasonal manufacturer will not be realized until the peak summer months. Net savings will be \$1100, \$1200, \$1300, \$1400, and \$1500 at the ends of months 5, 6, 7, 8, and 9, respectively. It is now the beginning of month 1. Assume 365 days per year, 30 days per month. What is the present worth (PW) of the savings if nominal interest is

(a) 12% per year, compounded monthly?

(b) 12% per year, compounded daily?

ANSWER

(a) $A = \$1100$

$G = \$100$

$i = 0.12/12 = 0.01$ per month $= 1\%$

$$\begin{aligned}
\text{PW(end of period 4)} &= (P/A, 1\%, 5)[\$1100 + \$100(A/G, 1\%, 5)] \\
&= 4.8528[\$1100 + \$100(1.9801)] \\
&= \$6298.98 \\
\text{PW(at time 0)} &= \text{PW(end of period 4)}(P/F, 1\%, 4) \\
&= \$6298.98/(1.01)^4 = \$6053.20
\end{aligned}$$

The present worth is about \$6053.

(b)

$$\begin{aligned}
\text{Effective interest rate } i &= (1 + 0.12/365)^{30} - 1 = 0.0099102 \\
\text{PW(at time 0)} &= \text{PW(end of period 4)}(P/F, i, 4) \\
&= (P/A, i, 5)[\$1100 + \$100(A/G, i, 5)](P/F, i, 4) \\
&= \$4.8547[\$1100 + \$100(1.98023)](0.9613) \\
&= \$6057.80
\end{aligned}$$

The present worth is about \$6058.■

REVIEW PROBLEM 3.2

It is January 1 of this year. You are starting your new job tomorrow, having just finished your engineering degree at the end of last term. Your take-home pay for this year will be \$36 000. It will be paid to you in equal amounts at the end of each month, starting at the end of January. There is a cost-of-living clause in your contract that says that each subsequent January you will get an increase of 3% in your yearly salary (i.e., your take-home pay for next year will be 1.03 × \$36 000). In addition to your salary, a wealthy relative regularly sends you a \$2000 birthday present at the end of each June.

Recognizing that you are not likely to have any government pension, you have decided to start saving 10% of your monthly salary and 50% of your birthday present for your retirement. Interest is 1% per month, compounded monthly. How much will you have saved at the end of five years?

ANSWER

Yearly pay is a geometric gradient; convert your monthly salary into a yearly amount by the use of an effective yearly rate. The birthday present can be dealt with separately.

Salary:

The future worth (FW) of the salary at the end of the first year is

$$\text{FW(salary, year 1)} = \$3000(F/A, 1\%, 12) = \$38\,040.00$$

This forms the base of the geometric gradient; all subsequent years increase by 3% per year. Savings are 10% of salary, which implies that $A = \$3804.00$.

$$A = \$3804.00 \qquad g = 0.03$$

Effective yearly interest rate $i_e = (1 + 0.01)^{12} - 1 = 0.1268$ per year

$$i^\circ = \frac{1 + i_e}{1 + g} - 1 = \frac{1 + 0.1268}{1 + 0.03} - 1 = 0.093981$$

$$\text{PW(gradient)} = A\,(P/A, i^\circ, 5)/(1 + g) = \$3804(3.8498)/1.03$$

$$= \$14\,218$$

$$\text{FW(gradient, end of five years)} = \text{PW(gradient)}(F/P, i_e, 5)$$

$$= \$14\,218(1.1268)^5 = \$25\,827$$

Birthday Present:

The present arrives in the middle of each year. To get the total value of the five gifts, we can find the present worth of an annuity of five payments of \$2000(0.5) as of six months prior to employment:

$$\text{PW}(-6\text{ months}) = \$2000(0.5)(P/A, i_e, 5) = \$3544.90$$

The future worth at $5 \times 12 + 6 = 66$ months later is

$$\text{FW(end of five years)} = \$3544.9(1.01)^{66} = \$6836$$

$$\text{Total amount saved} = \$6836 + \$25\,827 = \$32\,663 \blacksquare$$

REVIEW PROBLEM 3.3

The Easy Loan Company advertises a "10%" loan. You need to borrow \$1000, and the deal you are offered is the following: You pay \$1100 (\$1000 plus \$100 interest) in 11 equal \$100 amounts, starting one month from today. In addition, there is a \$25 administration fee for the loan, payable immediately, and a processing fee of \$10 per payment. Furthermore, there is a \$20 non-optional closing fee to be included in the last payment. Recognizing fees as a form of interest payment, what is the actual effective interest rate?

ANSWER

Since the \$25 administration fee is paid immediately, you are only getting \$975. The remaining payments amount to an annuity of \$110 per month, plus a \$20 future payment 11 months from now.

Formulas: $P = A(P/A,i,N)$, $P = F(P/F,i,N)$

Known values: $P = \$975$, $A = \$110$, $F = \$20$, $N = 11$

$$\$975 = \$110(P/A, i, 11) + \$20(P/F, i, 11)$$

At $i = 4\%$

$$\$110(P/A, 4\%, 11) + \$20(P/F, 4\%, 11)$$
$$= \$110(8.7603) + \$20(0.64958)$$
$$= \$976.62$$

At $i = 5\%$

$$\$110(P/A, 5\%, 11) + \$20(P/F, 5\%, 11)$$
$$= \$110(8.3062) + \$20(0.58469)$$
$$= \$925.37$$

Linearly interpolating gives

$$i = 4 + (5 - 4)(975 - 976.62)/(925.37 - 976.62)$$
$$= 4.03$$

The effective interest rate is then

$$i = (1 + 0.0403)^{12} - 1$$
$$= 60.69\% \text{ per annum (!)}$$

Although the loan is advertised as a "10%" loan, the actual effective rate is over 60%.■

REVIEW PROBLEM 3.4

Ming wants to retire as soon as she has enough money invested in a special bank account (paying 14% interest, compounded annually) to provide her with an annual income of $25 000. She is able to save $10 000 per year, and the account now holds $5000. If she just turned 20, and expects to die in 50 years, how old will she be when she retires? There should be no money left when she turns 70.

ANSWER

Let Ming's retirement age be $20 + x$ so that

$$\$5000(F/P, 14\%, x) + \$10\,000(F/A, 14\%, x) = \$25\,000(P/A, 14\%, 50 - x)$$

Dividing both sides by $5000,

$$(F/P, 14\%, x) + 2(F/A, 14\%, x) - 5(P/A, 14\%, 50 - x) = 0$$

At $x = 5$

$$(F/P, 14\%, 5) + 2(F/A, 14\%, 5) - 5(P/A, 14\%, 45)$$
$$= 1.9254 + 2(6.6101) - 5(7.1232) = -20.4704$$

At $x = 10$

$$(F/P, 14\%, 10) + 2(F/A, 14\%, 10) - 5(P/A, 14\%, 40)$$
$$= 3.7072 + 2(19.337) - 5(7.1050) = 6.8562$$

Linearly interpolating,

$$x = 5 + 5 \times (20.4704)/(6.8562 + 20.4704)$$
$$= 8.7$$

Ming can retire at age 20 + 8.7 = 28.7 years old.■

SUMMARY

In Chapter 3 we considered ways of modelling patterns of cash flows that enable easy comparisons of the worths of projects. The emphasis was on discrete models. Four basic patterns of discrete cash flows were considered:

1. Flows at a single point
2. Flows that are constant over time
3. Flows that grow or decrease at a constant arithmetic rate
4. Flows that grow or decrease at a constant geometric rate

Compound interest factors were presented that defined mathematical equivalence among the basic patterns of cash flows. A list of these factors with their names, symbols, and formulas appears in Table 3.2. The chapter also addressed the issue of how to analyze non-standard annuities and gradients as well as the ideas of capital recovery and capitalized value.

For those who are interested in continuous compounding and continuous cash flows, Appendix 3A contains a summary of relevant notation and interest factors.

Table 3.2 Summary of Useful Formulas for Discrete Models

Name	Symbol and Formula
Compound amount factor	$(F/P,i,N) = (1+i)^N$
Present worth factor	$(P/F,i,N) = \frac{1}{(1+i)^N}$
Sinking fund factor	$(A/F,i,N) = \frac{i}{(1+i)^N - 1}$
Uniform series compound amount factor	$(F/A,i,N) = \frac{(1+i)^N - 1}{i}$
Capital recovery factor	$(A/P,i,N) = \frac{i(1+i)^N}{(1+i)^N - 1}$
Series present worth factor	$(P/A,i,N) = \frac{(1+i)^N - 1}{i(1+i)^N}$
Arithmetic gradient to annuity conversion factor	$(A/G,i,N) = \frac{1}{i} - \frac{N}{(1+i)^N - 1}$
Geometric gradient to present worth conversion factor	$(P/A,g,i,N) = \frac{(P/A,i^\circ,N)}{1+g}$
	$(P/A,g,i,N) = \left(\frac{(1+i^\circ)^N - 1}{i^\circ(1+i^\circ)^N}\right)\frac{1}{1+g}$
	$i^\circ = \frac{1+i}{1+g} - 1$
Capitalized value formula	$P = \frac{A}{i}$
Capital recovery formula	$A = (P - S)(A/P,i,N) + Si$

Engineering Economics in Action, Part 3B: No Free Lunch

This time it was Naomi who stuck her head in Clem's doorway. "Here's the recommendation on the shipping palletizer. Oh, and thanks for the hint on the leasing figures. It cleared up my confusion right away."

"No problem. What did you figure out?" Clem had his "mentor" expression on his face, so Naomi knew he was expecting a clear explanation of the trick used by the leasing company.

"Well, as you hinted, they were adding apples and oranges. They listed the various costs for each choice over time, including interest charges, taxes, and so on. But then, for the final comparison, they added up these costs. When they added, leasing was cheaper."

"So what's wrong with that?" Clem prompted.

"They're adding apples and oranges. We're used to thinking of money as being just money, without remembering that money always has a 'when' associated with it. If you add money at different points in time, you might as well be adding apples and oranges; you have a number but it doesn't mean anything. In order to compare leasing with buying, you first have to change the cash flows into the same money, that is, at the same point in time. That's a little harder to do, especially when there's a complicated set of cash flows."

"So were you able to do it?"

"Yes. I identified various components of the cash flows as annuities, gradients, and present and future worths. Then I converted all of these to a present worth for each alternative and summed them. This is the correct way to compare them. If you do that, buying is cheaper, even when borrowing money to do so. And of course it has to be—that leasing company has to pay for those slick brochures somehow. There's no free lunch."

Clem nodded. "I think you've covered it. Mind you, there are some circumstances where leasing is worthwhile. For example, we lease our company cars to save us the time and trouble of reselling them when we're finished with them. Leasing can be good when it's hard to raise the capital for very large purchases, too. But almost always, buying is better. And you know, it amazes me how easy it is to fall for simplistic cash flow calculations that fail to take into account the time value of money. I've even seen articles in the newspaper quoting accountants who make the same mistake, and you'd think they would know better."

"Engineers can make that mistake, too, Clem. I almost did."

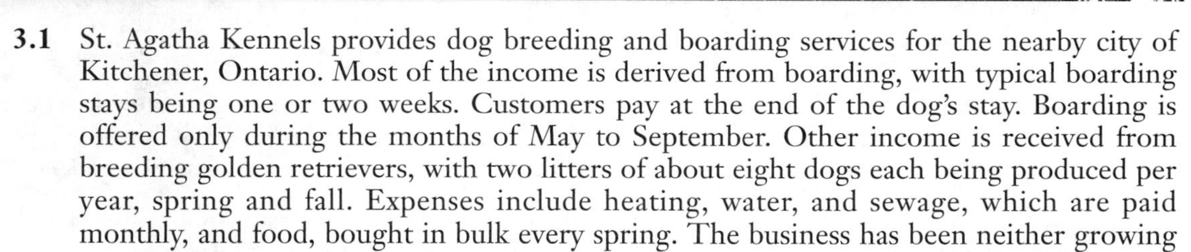

PROBLEMS

3.1 St. Agatha Kennels provides dog breeding and boarding services for the nearby city of Kitchener, Ontario. Most of the income is derived from boarding, with typical boarding stays being one or two weeks. Customers pay at the end of the dog's stay. Boarding is offered only during the months of May to September. Other income is received from breeding golden retrievers, with two litters of about eight dogs each being produced per year, spring and fall. Expenses include heating, water, and sewage, which are paid monthly, and food, bought in bulk every spring. The business has been neither growing nor shrinking over the past few years.

Joan, the owner of St. Agatha Kennels, wants to model the cash flows for the business over the next 10 years. What cash flow elements (e.g., single payments, annuities, gradients) would she likely consider, and how would she estimate their value? Consider the present to be the first of May. For example, one cash flow element is food. It would be modelled as an *annuity due* over 10 years, and estimated by the amount paid for food over the last few years.

3.2 It is September, and Marco has to watch his expenses while he is going to school. Over the next eight months, he wants to estimate his cash flows. He pays rent once a month. He takes the bus to and from school. A couple of times a week he goes to the grocery store for food, and eats lunch in the cafeteria at school every school day. At the end of every four-month term, he will have printing and copying expenses because of reports that will be due. After the first term, over the Christmas holidays, he will have extra expenses for buying presents, but will also get some extra cash from his parents. What cash flow elements (e.g., single payments, annuities, gradients) would Marco likely consider in his estimates? How would he estimate them?

3.3 How much money will be in a bank account at the end of 15 years if \$100 is deposited today and the interest rate is 8% compounded annually?

3.4 How much should you invest today at 12% interest to accumulate \$1 000 000 in 30 years?

3.5 Martin and Marcy McCormack have just become proud parents of Canada's first septuplets. They have savings of $5000. They want to invest their savings so that they can partially support the children's university education. Martin and Marcy hope to provide $20 000 for each child by the time the children turn 18. What does the annual rate of return have to be on the investment for Martin and Marcy to meet their goal?

3.6 You have $1725 to invest. You know that a particular investment will double your money in 5 years. How much will you have in 10 years if you invest in this investment, assuming that the annual rate of return is guaranteed for the time period?

3.7 Morris paid $500 a month for 20 years to pay off the mortgage on his house. If his down payment was $5000 and the interest rate was 6% compounded monthly, how much did the house cost?

3.8 An investment pays $10 000 every five years, starting in seven years, for a total of four payments. If interest is 9%, how much is this investment worth today?

3.9 An industrial juicer costs $45 000. It will be used for five years and then sold to a remarketer for $25 000. If interest is 15%, what net yearly savings are needed to justify its purchase?

3.10 Fred wants to save up for a car. How much must he put in his bank account each month to save $10 000 in two years if the bank pays 6% interest compounded monthly?

3.11 It is May 1. You have just bought $2000 worth of furniture. You will pay for it in 24 equal monthly payments, starting at the end of May next year. Interest is 6% nominal per year, compounded monthly. How much will your payments be?

3.12 What is the present worth of the total of 20 payments, occurring at the end of every four months (the first payment is in four months), which are $400, $500, $600, increasing arithmetically? Interest is 12% nominal per year, compounded continuously.

3.13 What is the total value of the sum of the present worths of all the payments and receipts mentioned in Problem 2.32, at an interest rate of 0.5% per month?

3.14 How much is accumulated in each of the following savings plans over two years?

(a) $40 at the end of each month for 24 months at 12% compounded monthly

(b) $30 at the end of the first month, $31 at the end of the second month, and so forth, increasing by $1 per month, at 12% compounded monthly

3.15 What interest rate will result in $5000 seven years from now, starting with $2300 today?

3.16 Refer back to the Hanover Go-Kart problem of Example 3.2. The club members determined that it is possible to set aside only $7000 each year, and that they will have to put off building the clubhouse until they have saved the $50 000 necessary. How long will it take to save a total of $50 000, assuming that the interest rate is 10%? (*Hint:* Use logarithms to simplify the sinking fund factor.)

3.17 Gwen just bought a satellite dish, which provides her with exactly the same service as cable TV. The dish cost $2000, and the cable service she has now cancelled cost her $40 per month. How long will it take her to recoup her investment in the dish, if she can earn 12% interest, compounded monthly, on her money?

3.18 Yoko has just bought a new computer ($2000), a printer ($350), and a scanner ($210). She wants to take the monthly payment option. There is a monthly interest of 3% on her purchase.

(a) If Yoko pays $100 per month, how long does it take to complete her payments?

(b) If Yoko wants to finish paying in 24 months, how much will her monthly payment be?

3.19 Rinku has just finished her first year of university. She wants to go to Europe when she graduates in three years. By having a part-time job through the school year and a summer job during the summer, she plans to make regular weekly deposits into a savings account, which bears 18% interest, compounded monthly.

(a) If Rinku deposits $15 per week, how much will she save in three years? How about $20 per week?

(b) Find out exactly how much Rinku needs to deposit every week if she wants to save $5000 in three years.

3.20 Seema is looking at an investment in upgrading an inspection line at her plant. The initial cost would be $140 000 with a salvage value of $37 000 after five years. Use the capital recovery formula to determine how much money must be saved every year to justify the investment, at an interest rate of 14%.

3.21 Trenny has asked her assistant to prepare estimates of cost of two different sizes of power plants. The assistant reports that the cost of the 100 MW plant is $20 000 000, while the cost of the 200 MW plant is $36 000 000. If Trenny has a budget of only $30 000 000, estimate how large a power plant she could afford using linear interpolation.

3.22 Enrique has determined that investing $500 per month will enable him to accumulate $11 350 in 12 years, and that investing $800 per month will enable him to accumulate $18 950 over the same period. Estimate, using linear interpolation, how much he would have to invest each month to accumulate exactly $15 000.

3.23 A lottery prize pays $1000 at the end of the first year, $2000 the second, $3000 the third, and so on for 20 years. If there is only one prize in the lottery, 10 000 tickets are sold, and you could invest your money elsewhere at 15% interest, how much is each ticket worth, on average?

3.24 Joseph and three other friends bought a $110 000 house close to the university at the end of August last year. At that time they put down a deposit of $10 000 and took out a mortgage for the balance. Their mortgage payments are due at the end of each month (September 30, last year, was the date of the first payment) and are based on the assumption that Joseph and friends will take 20 years to pay off the debt. Annual nominal interest is 12%, compounded monthly. It is now February. Joseph and friends have made all their fall-term payments and have just made the January 31 payment for this year. How much do they still owe?

3.25 A new software package is expected to improve productivity at Saskatoon Insurance. However, because of training and implementation costs, savings are not expected to occur until the third year of operation. At that time, savings of $10 000 are expected, increasing by $1000 per year for the following five years. After this time (eight years from implementation), the software will be abandoned with no scrap value. How much is the software worth today, at 15% interest?

3.26 Clem is saving for a car in a bank account that pays 12% interest, compounded monthly. The balance is now $2400. Clem will be saving $120 per month from his salary, and once every four months (starting in four months) he adds $200 in dividends from an investment. Bank fees, currently \$10 per month, are expected to increase by \$1 per month henceforth. How much will Clem have saved in two years?

3.27 Yogajothi is thinking of investing in a rental house. The total cost to purchase the house, including legal fees and taxes, is \$115 000. All but \$15 000 of this amount will be mortgaged. He will pay \$800 per month in mortgage payments. At the end of two years, he will sell the house, and at that time expects to clear \$20 000 after paying off the remaining mortgage principal (in other words, he will pay off all his debts for the house, and still have \$20 000 left). Rents will earn him \$1000 per month for the first year, and \$1200 per month for the second year. The house is in fairly good condition now so that he doesn't expect to have any maintenance costs for the first six months. For the seventh month, Yogajothi has budgeted \$200. This figure will be increased by \$20 per month thereafter (e.g., the expected month 7 expense will be \$200, month 8, \$220, month 9, \$240, etc.). If interest is 6% compounded monthly, what is the present worth of this investment? Given that Yogajothi's estimates of revenue and expenses are correct, should Yogajothi buy the house?

3.28 You have been paying off a mortgage in quarterly payments at a 24% nominal annual rate, compounded quarterly. Your bank is now offering an alternative payment plan, so you have a choice of two methods—continuing to pay as before or switching to the new plan. Under the new plan, you would make monthly payments, 30% of the size of your current payments. The interest rate would be 24% nominal, compounded monthly. The time until the end of the mortgage would not change, regardless of the method chosen.

(a) Which plan would you choose, given that you naturally wish to minimize the level of your payment costs? (*Hint:* Look at the costs over a three-month period.)

(b) Under which plan would you be paying a higher effective yearly interest rate?

3.29 Derive the arithmetic gradient conversion to a uniform series formula. (*Hint:* Convert each period's gradient amount to its future value, and then look for a substitution from the other compound amount factors.)

3.30 Derive the geometric gradient to present worth conversion factor. (*Hint:* Divide and multiply the present worth of a geometric series by $[1 + g]$ and then substitute in the growth-adjusted interest rate.)

3.31 Reginald is expecting steady growth of 10% per year in profits from his new company. All profits are going to be invested at 20% interest. If profits for this year (at the end of the year) total \$100 000, how much will be saved at the end of 10 years?

3.32 Reginald is expecting steady growth in profits from his new company of 20% per year. All profits are going to be invested at 10% interest. If profits for this year (at the end of the year) total \$100 000, how much will be saved at the end of 10 years?

3.33 Ruby's business has been growing quickly over the past few years, with sales increasing at about 50% per year. She has been approached by a buyer for the business. She has decided she will sell it for 1/2 of the value of the estimated sales for the next five years. This year she will sell products worth \$1 456 988. Use the geometric gradient factor to calculate her selling price for an interest rate of 5%.

3.34 In Example 3.4, Clarence bought a \$94 000 house with a \$14 000 down payment and took out a mortgage for the remaining \$80 000 at 12% nominal interest, compounded monthly. We determined that he would make 51 \$2000 payments and then a final payment. What is his final payment?

3.35 A new wave-soldering machine is expected to save Yukon Circuit Boards $15 000 per year through reduced labour costs and increased quality. The device will have a life of eight years, and have no salvage value after this time. If the company can generally expect to get 12% return on its capital, how much could it afford to pay for the wave-soldering machine?

3.36 Gail has won a lottery that pays her $100 000 at the end of this year, $110 000 at the end of next year, $120 000 the following year, and so on, for 30 years. Leon has offered Gail $2 500 000 today in exchange for all the money she will receive. If Gail can get 8% interest on her savings, is this a good deal?

3.37 Gail has won a lottery that pays her $100 000 at the end of this year, and increases by 10% per year thereafter for 30 years. Leon has offered Gail $2 500 000 today in exchange for all the money she will receive. If Gail can get 8% interest on her savings, is this a good deal?

3.38 Tina has saved up $20 000 from her summer jobs. Rather than work for a living, she plans to buy an annuity from a trust company and become a beachcomber in Fiji. An annuity will pay her a certain amount each month for the rest of her life, and is calculated at 7% interest, compounded monthly, over Tina's 55 remaining years. Tina calculates that she needs at least $5 per day to live in Fiji, and she needs $1200 for air fare. Can she retire now? How much would she have available to spend each day?

3.39 The Regional Municipality of Kitchener is studying a water supply plan for the tri-city and surrounding area to the end of year 2050. To satisfy the water demand, one suggestion is to construct a pipeline from one of the Great Lakes. Construction would start in 2010 and take five years at a cost of $20 million per year. The cost of maintenance and repairs starts after completion of construction and for the first year is $2 million, increasing by 1% per year thereafter. At an interest rate of 6%, what is the present worth of this project?

Assume that all cash flows take place at year-end. Consider the present to be the end of 2005 / beginning of 2006. Assume that there is no salvage value at the end of year 2050.

3.40 Clem has a $50 000 loan. The interest rate offered is 8% compounded annually, and the repayment period is 15 years. Payments are to be received in equal installments at the end of each year. Construct a spreadsheet (you must use a spreadsheet program) similar to the following table that shows the amount received each year, the portion that is interest, the portion that is unrecovered capital, and the amount that is outstanding (i.e., unrecovered). Also, compute the total recovered capital which must equal the original capital amount; this can serve as a check on your solution. Design the spreadsheet so that the capital amount and the interest rate can be changed by updating only one cell for each. Construct:

(a) the completed spreadsheet for the amount, interest rate, and repayment period indicated

(b) the same spreadsheet, but for $75 000 at 10% interest (same repayment period)

(c) a listing showing the formulas used

Sample Capital Recovery Calculations				
Capital amount			$50 000.00	
Annual interest rate			8.00%	
Number of years to repay			15	
Payment Periods	**Annual Payment**	**Interest Received**	**Recovered Capital**	**Unrecovered Capital**
0				$ 50 000.00
1	$ 5 841.48	$ 4 000.00	$ 1 841.48	$ 48 158.52
2				
.				
.				
.				
15				0.00
Total			50 000.00	

3.41 A software genius has been offered $10 000 per year for the next five years and then $20 000 per year for the following 10 years for the rights to his new video game. At 9% interest, how much is this worth today?

3.42 A bank offers a personal loan called "The Eight Per Cent Plan." The bank adds 8% to the amount borrowed; the borrower pays back 1/12 of this total at the end of each month for a year. On a loan of $500, the monthly payment is 540/12 = $45. There is also an administrative fee of $45, payable now. What is the actual effective interest rate on a $500 loan?

3.43 Coastal Shipping is setting aside capital to fund an expansion project. Funds earmarked for the project will accumulate at the rate of $50 000 per month until the project is completed. The project will take two years to complete. Once the project starts, costs will be incurred monthly at the rate of $150 000 per month over 24 months. Coastal currently has $250 000 saved. What is the minimum number of months they will have to wait before they can start if money is worth 18% nominal, compounded monthly? Assume that

1. Cash flows are all at the ends of months
2. The first $50 000 savings occurs one month from today
3. The first $150 000 payment occurs one month after the start of the project
4. The project must start at the beginning of a month

3.44 A company is about to invest in a joint venture research and development project with another company. The project is expected to last eight years, but yearly payments the company makes will begin immediately (i.e., a payment is made today, and the last payment is eight years from today). Salaries will account for $40 000 of each payment. The remainder of each payment will cover equipment costs and facility overhead. The initial (immediate) equipment and facility cost is $26 000. Each subsequent year the figure will drop by $3000 until a cost of $14 000 is reached, after which the costs will remain constant until the end of the project.

(a) Draw a cash flow diagram to illustrate the cash flows for this situation.

(b) At an interest rate of 7%, what is the total future worth of all project payments at the end of the eight years?

3.45 Shamsir's business has been growing slowly. He has noticed that his monthly profit increases by 1% every two months. Suppose that the profit at the end of this month is $1000. What is the present value of all his profit over the next two years? Annual nominal interest is 18%, compounded monthly.

3.46 Xiaohang is conducting a biochemical experiment for the next 12 months. In the first month, the expenses are estimated to be $1500. As the experiment progresses, the expenses are expected to increase by 5% each month. Xiaohang plans to pay for the experiment by a government grant, which is received in six monthly installments, starting a month after the experiment completion date. Determine the amount of the monthly installment so that the total of the six installments pays for all expenses incurred during the experiment. Annual nominal interest is 12%, compounded monthly.

3.47 City engineers are considering several plans for building municipal aqueduct tunnels. They use an interest rate of 8%. One plan calls for a full-capacity tunnel that will meet the needs of the city forever. The cost is $3 000 000 now, and $100 000 every 10 years thereafter for repairs. What is the total present worth of the costs of building and maintaining the aqueduct?

3.48 The City of Surrey is installing a new swimming pool in the downtown recreation centre. One design being considered is a reinforced concrete pool which will cost $1 500 000 to install. Thereafter, the inner surface of the pool will need to be refinished and painted every 10 years at a cost of $200 000 per refinishing. Assuming that the pool will have essentially an infinite life, what is the present worth of the costs associated with this pool design? The city uses a 5% interest rate.

3.49 Goderich Automotive (GA) wants to donate a vacant lot next door to their plant to the city for use as a public park and ball field. The city will accept only if GA will also donate enough cash to maintain the park forever. The estimated maintenance costs are $18 000 per year and interest is 7%. How much cash must GA donate?

3.50 A 7%, 20-year municipal bond has a $10 000 face value. I want to receive at least 10% compounded semiannually on this investment. How much should I pay for the bond?

3.51 A Paradorian bond pays $500 (Paradorian dollars) twice each year and $5000 five years from now. I want to earn at least 300% *annual* (effective) interest on this investment (to compensate for the very high inflation in Parador). How much should I pay for this bond now?

3.52 If money is worth 8% compounded semiannually, how much is a bond maturing in nine years, with a face value of $10 000 and a coupon rate of 9%, worth today?

3.53 A bond with a face value of $5000 pays quarterly interest of 1.5% each period. Twenty-six interest payments remain before the bond matures. How much would you be willing to pay for this bond today if the next interest payment is due now and you want to earn 8% compounded quarterly on your money?

A New Distribution Station for Kitchener-Wilmot Hydro

CANADIAN 3.1 MINI-CASE

Kitchener-Wilmot Hydro provides electricity to the City of Kitchener, Ontario, and nearby Wilmot Township as a regulated monopoly. It purchases power from the provincial utility Ontario Hydro and distributes it to homes and industries in these areas.

Recently, the utility undertook an investigation of the locations and capacities of the distribution stations in Wilmot Township. These distribution stations are essentially transformers that reduce voltage from 27.6 kilovolts, which is efficient for long-distance distribution, to 8.3 kilovolts, which is better for local distribution. Near each user, it is further transformed to 120 or 600 volts.

A particular problem was distribution station (DS) #5, located northwest of the town of St. Agatha. This station had old technology that required upgrading to match the other distribution stations. Three alternatives were identified. In short, they were:

1. Upgrade DS#5.
2. Build a new distribution station at the town of Philipsburg, using the salvaged transformer from former DS#5.
3. Build a new distribution station at the town of New Prussia, using the salvaged transformer from former DS#5.

Data were gathered concerning current and projected usage and costs for the three alternatives. A present worth analysis was done, using a discount factor (interest rate) of 10%. The least expensive choice turned out to be alternative 1, which was recommended for implementation.

Discussion

In order to compare alternatives, it is necessary to determine the future cash flows associated with each. Of course nobody knows for sure what the future holds, so future cash flows are always estimates. In some cases the estimates will turn out to be significantly different from the actual cash flows, and therefore result in an incorrect decision.

The electricity demands in the distribution area of DS#5 were fast-changing at the time Kitchener-Wilmot Hydro had to make this decision. Former villages were expanding to become suburban communities for the booming twin cities of Kitchener-Waterloo, and affluent country estates were replacing farms. The changing population in turn results in predictable changes in the location and amount of electricity demand. On the other hand, predicting electricity demand itself is tricky—for example, an economic downturn would likely change the growth rate of the Kitchener-Waterloo area.

There are sophisticated ways to deal with uncertainty about future cash flows, some of which are discussed in Chapters 11 and 12. In many cases it makes sense to assume that the future cash flows are treated as if they were certain because there is no particular reason to think they are not. On the other hand, in some cases even estimating future cash flows can be very difficult or impossible.

Questions

1. For each of the following, comment on how sensible it is to estimate the precise value of future cash flows:
   - **(a)** Your rent for the next six months
   - **(b)** Your food bill for the next six months
   - **(c)** Your medical bills for the next six months

(d) A company's payroll for the next six months
(e) A company's raw material costs for the next six months
(f) A company's legal costs for liability lawsuits for the next six months
(g) Canada's costs for funding university research for the next six months
(h) Canada's costs for Employment Insurance for the next six months
(i) Canada's costs for Emergency Management for the next six months

2. Your company is looking at the possibility of buying a new widget grinder for the widget line. The future cash flows associated with the purchase of the grinder are fairly predictable, except for one factor. A significant benefit is achieved with the higher production volume of widgets, which depends on a contract to be signed with a particular important customer. This won't happen for several months, but you must make the decision about the widget grinder now. Discuss some sensible ways of dealing with this issue.

Appendix 3A Continuous Compounding and Continuous Cash Flows

We now consider compound interest factors for continuous models. Two forms of continuous modelling are of interest. The first form has discrete cash flows with continuous compounding. In some cases, projects generate discrete, end-of-period cash flows within an organization in which other cash flows and compounding occur many times per period. In this case, a reasonable model is to assume that the project's cash flows are discrete, but compounding is continuous. The second form has continuous cash flows with continuous compounding. Where a project generates many cash flows per period, we could model this as continuous cash flows with continuous compounding.

We first consider models with discrete cash flows with continuous compounding. We can obtain formulas for compound interest factors for these projects from formulas for discrete compounding simply by substituting the effective continuous interest rate with continuous compounding for the effective rate with discrete compounding.

Recall that for a given nominal interest rate, r, when the number of compounding periods per year becomes infinitely large, the effective interest rate per period is $i_e = e^r - 1$. This implies $1 + i_e = e^r$ and $(1 + i_e)^N = e^{rN}$. The various compound interest factors for continuous compounding can be obtained by substituting $e^r - 1$ for i in the formulas for the factors.

For example, the series present worth factor with discrete compounding is

$$(P/A,i,N) = \frac{(1 + i)^N - 1}{i(1 + i)^N}$$

If we substitute $e^r - 1$ for i and e^{rN} for $(1 + i)^N$, we get the series present worth factor for continuous compounding

$$(P/A,r,N) = \frac{e^{rN} - 1}{(e^r - 1)e^{rN}}$$

Similar substitutions can be made in each of the other compound interest factor formulas to get the compound interest factor for continuous rather than discrete compounding. The formulas are shown in Table 3A.1. Tables of values for these formulas are available in Appendix B at the end of the book.

Table 3A.1 Compound Interest Formulas for Discrete Cash Flow with Continuous Compounding

Name	Symbol and Formula
Compound amount factor	$(F/P,r,N) = e^{rN}$
Present worth factor	$(P/F,r,N) = \frac{1}{e^{rN}}$
Sinking fund factor	$(A/F,r,N) = \frac{e^r - 1}{e^{rN} - 1}$
Uniform series compound amount factor	$(F/A,r,N) = \frac{e^{rN} - 1}{e^r - 1}$
Capital recovery factor	$(A/P,r,N) = \frac{(e^r - 1)e^{rN}}{e^{rN} - 1}$
Series present worth factor	$(P/A,r,N) = \frac{e^{rN} - 1}{(e^r - 1)e^{rN}}$
Arithmetic gradient to annuity conversion factor	$(A/G,r,N) = \frac{1}{e^r - 1} - \frac{N}{e^{rN} - 1}$

Example 3A.1

Yoram Gershon is saving to buy a new sound system. He plans to deposit $100 each month for the next 24 months in the Bank of Montrose. The nominal interest rate at the Bank of Montrose is 0.5% per month, compounded continuously. How much will Yoram have at the end of the 24 months?

We start by computing the uniform series compound amount factor for continuous compounding. Recall that the factor for discrete compounding is

$$(F/A,i,N) = \frac{(1 + i)^N - 1}{i}$$

Substituting $e^r - 1$ for i and e^{rN} for $(1 + i)^N$ gives the series compound amount, when compounding is continuous, as

$$(F/A,r,N) = \frac{e^{rN} - 1}{e^r - 1}$$

The amount Yoram will have at the end of 24 months, F, is given by

$$F = \$1001\left(\frac{e^{(0.005)24} - 1}{e^{0.005} - 1}\right)$$

$$= \$1001\left(\frac{1.127497 - 1}{1.00501 - 1}\right)$$

$$= \$2544.85$$

Yoram will have about $2545 saved at the end of the 24 months.■

The formulas for *continuous cash flows with continuous compounding* are derived using integral calculus. The continuous *series present worth* factor, denoted by $(P/\bar{A},r,T)$, for a continuous flow, $\bar{A}$, over a period length, T, where the nominal interest rate is r, is given by

$$P = \bar{A}\left(\frac{e^{rT} - 1}{re^{rT}}\right)$$

so that

$$(P/\bar{A},r,T) = \frac{e^{rT} - 1}{re^{rT}}$$

It is then easy to derive the formula for the continuous *uniform series compound amount factor*, denoted by $(F/\bar{A},r,T)$, by multiplying the series present worth factor by e^{rT} to get the future worth of a present value, P.

$$(F/\bar{A},r,T) = \frac{e^{rT} - 1}{r}$$

We can get the continuous *capital recovery factor*, denoted by $(\bar{A}/P,r,T)$, as the inverse of the continuous series present worth factor. The *continuous sinking fund factor* $(F/\bar{A},r,T)$ is the inverse of the continuous uniform series compound amount factor. A summary of the formulas for continuous cash flow and continuous compounding is shown in Table 3A.2. Tables of values for these formulas are available in Appendix C at the back of the book.

Table 3A.2 Compound Interest Formulas for Continuous Cash Flow with Continuous Compounding

Name	Symbol and Formula
Sinking fund factor	$(\bar{A},F,r,T) = \frac{r}{e^{rT} - 1}$
Uniform series compound amount factor	$(F/\bar{A},r,T) = \frac{e^{rT} - 1}{r}$
Capital recovery factor	$(\bar{A},/P,r,T) = \frac{re^{rT}}{e^{rT} - 1}$
Series present worth factor	$(P/\bar{A},r,T) = \frac{e^{rT} - 1}{re^{rT}}$

Example 3A.2

Savings from a new widget grinder are estimated to be $10 000 per year. The grinder will last 20 years and will have no scrap value at the end of that time. Assume that the savings are generated as a continuous flow. The *effective* interest rate is 15% compounded continuously. What is the present worth of the savings?

From the problem statement, we know that $\bar{A}$ = $10 000, $i = 0.15$, and $T = 20$. From the relation $i = e^r - 1$, for $i = 0.15$ the interest rate to apply for continuously compounding is $r = 0.13976$. The present worth computations are

$$P = \bar{A}\,(P/\bar{A},r,T)$$

$$= \$10\ 000\left(\frac{e^{(0.13976)20} - 1}{(0.13976)e^{(0.13976)20}}\right)$$

$$= \$67\ 180$$

The present worth of the savings is $67 180. Note that if we had used discrete compounding factors for the present worth computations we would have obtained a lower value.

$$P = A(P/A,i,N)$$
$$= \$10\ 000(6.2593)$$
$$= \$62\ 593$$ ■

REVIEW PROBLEM 3A.1 FOR APPENDIX 3A

Mr. Big is thinking of buying the MQM Grand Hotel in Las Vegas. The hotel has continuous net receipts totalling $120 000 000 per year (Vegas hotels run 24 hours per day). This money could be immediately reinvested in Mr. Big's many other ventures, all of which earn a nominal 10% interest. The hotel will likely be out of style in about eight years, and could then be sold for about $200 000 000. What is the maximum Mr. Big should pay for the hotel today?

ANSWER

$$P = \$120\ 000\ 000(P/A,\ 10\%,\ 8) + \$200\ 000\ 000e^{-(0.1)(8)}$$

$$= \$120\ 000\ 000\ \frac{e^{(0.1)(8)} - 1}{(0.1)e^{(0.1)(8)}} + \$200\ 000\ 000e^{-(0.1)(8)}$$

$$= \$701\ 184\ 547$$

Mr. Big should not pay more than about $700 000 000.■

PROBLEMS

3A.1 An investment in new data logging technology is expected to generate extra revenue continuously for Calgary Petroleum Services. The initial cost is $300 000, but extra revenues total $75 000 per year. If the effective interest rate is 10% compounded continuously, does the present worth of the savings over five years exceed the original purchase cost? By how much does one exceed the other?

3A.2 Desmond earns $25 000 continuously over a year from an investment that pays 8% nominal interest, compounded continuously. How much money does he have at the end of the year?

3A.3 Gina intently plays the stock market, so that any capital she has can be considered to be compounding continuously. At the end of 1999, Gina had $10 000. How much did she have at the beginning of 1999, if she earned a nominal interest rate of 18%?

3A.4 Gina (from Problem 3A.3) had earned a nominal interest rate of 18% on the stock market every year since she started with an initial investment of $100. What year did she start investing?

Appendix 3B Derivation of Discrete Compound Interest Factors

This appendix derives six discrete compound interest factors presented in this chapter. All of them can be derived from the compound interest equation

$$F = P(1 + i)^N$$

3B.1 Compound Amount Factor

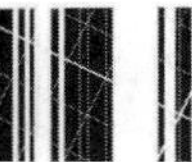

In the symbolic convention used for compound interest factors, the compound interest equation can be written

$$F = P(1 + i)^N = P(F/P,i,N)$$

so that the compound amount factor is

$$(F/P,i,N) = (1 + i)^N \quad (3B.1)$$

3B.2 Present Worth Factor

The present worth factor, $(P/F,i,N)$, converts a future amount F to a present amount P:

$$P = F(P/F,i,N)$$

$$\Rightarrow F = P\left(\frac{1}{(P/F,i,N)}\right)$$

Thus the present worth factor is the reciprocal of the compound amount factor. From Equation (3B.1),

$$(P/F,i,N) = \frac{1}{(1 + i)^N}$$

3B.3 Sinking Fund Factor

If a series of payments A follows the pattern of a standard annuity of N payments in length, then the future value of the payment in the jth period, from Equation (3B.1), is:

$$F = A(1 + i)^{N-j}$$

The future value of all of the annuity payments is then

$$F = A(1 + i)^{N-1} + A(1 + i)^{N-2} + \ldots + A(1 + i)^1 + A$$

Factoring out the annuity amount gives

$$F = A[(1 + i)^{N-1} + (1 + i)^{N-2} + \ldots + (1 + i)^1 + 1] \quad (3B.2)$$

Multiplying Equation (3B.2) by $(1 + i)$ gives

$$F(1 + i) = A[(1 + i)^{N-1} + (1 + i)^{N-2} + \ldots + (1 + i)^1 + 1](1 + i)$$

$$F(1 + i) = A[(1 + i)^N + (1 + i)^{N-1} + \ldots + (1 + i)^2 + (1 + i)] \quad (3B.3)$$

Subtracting Equation (3B.2) from Equation (3B.3) gives

$$F(1+i) - F = A[(1+i)^N - 1]$$

$$Fi = A[(1+i)^N - 1]$$

$$A = F\left[\frac{i}{(1+i)^N - 1}\right]$$

Thus the sinking fund factor is given by

$$(A/F,i,N) = \frac{i}{(1+i)^N - 1} \qquad (3B.4)$$

3B.4 Uniform Series Compound Amount Factor

The uniform series compound amount factor, $(F/A,i,N)$, converts an annuity A into a future amount F:

$$F = A(F/A,i,N)$$

$$\Rightarrow A = F\left(\frac{1}{(F/A,i,N)}\right)$$

Thus the uniform series compound amount factor is the reciprocal of the sinking fund factor. From Equation (3B.4),

$$(F/A,i,N) = \frac{(1+i)^N - 1}{i}$$

3B.5 Capital Recovery Factor

If a series of payments A follows the pattern of a standard annuity of N payments in length, then the present value of the payment in the jth period is

$$P = A\frac{1}{(1+i)^j}$$

The present value of the total of all the annuity payments is

$$P = A\left(\frac{1}{(1+i)}\right) + A\left(\frac{1}{(1+i)^2}\right) + \ldots + A\left(\frac{1}{(1+i)^{N-1}}\right) + A\left(\frac{1}{(1+i)^N}\right)$$

Factoring out the annuity amount gives

$$P = A\left[\left(\frac{1}{(1+i)}\right) + \left(\frac{1}{(1+i)^2}\right) + \ldots + \left(\frac{1}{(1+i)^{N-1}}\right) + \left(\frac{1}{(1+i)^N}\right)\right] \qquad (3B.5)$$

Multiplying both sides of Equation (3B.5) by $(1 + i)$ gives

$$P(1+i) = A\left[1 + \left(\frac{1}{(1+i)}\right) + \ldots + \left(\frac{1}{(1+i)^{N-2}}\right) + \left(\frac{1}{(1+i)^{N-1}}\right)\right] \qquad (3B.6)$$

Subtracting Equation (3B.5) from Equation (3B.6) gives

$$Pi = A\left[1 - \left(\frac{1}{(1+i)^N}\right)\right]$$

$$P = A\left[\frac{(1+i)^N - 1}{i(1+i)^N}\right]$$

$$A = P\left[\frac{i(1+i)^N}{(1+i)^N - 1}\right]$$

Thus the capital recovery factor is given by

$$(A/P,i,N) = \frac{i(1+i)^N}{(1+i)^N - 1} \qquad (3B.7)$$

3B.6 Series Present Worth Factor

The series present worth factor, $(P/A,i,N)$, converts an annuity A into a present amount P:

$$P = A(P/A,i,N)$$

$$\Rightarrow A = P\left(\frac{1}{(P/A,i,N)}\right)$$

Thus the uniform series compound amount factor is the reciprocal of the sinking fund factor. From Equation (3B.7),

$$(P/A,i,N) = \frac{(1+i)^N - 1}{i(1+i)^N}$$

3B.7 Arithmetic and Geometric Gradients

The derivation of the arithmetic gradient to annuity conversion factor and the geometric gradient to present worth conversion factor are left as problems for the student. See Problems 3.29 and 3.30.

CHAPTER 4

Comparison Methods Part 1

Engineering Economics in Action, Part 4A: What's Best?

Naomi waved hello as she breezed by Carole Brown, the receptionist, on her way in from the parking lot one Monday morning. She stopped as Carole caught her eye. "Clem wants to see you right away. Good morning."

After a moment of socializing, Clem got right to the point. "I have a job for you. Put aside the vehicle-life project for a couple of days."

"OK, but you wanted a report by Friday."

"This is more important. You know that drop forging hammer in the South Shop? The beast is about 50 years old. I don't remember the exact age. We got it used four years ago. We were having quality control problems with the parts we were buying on contract and decided to bring production in-house. Stinson Brothers sold it to us cheap when they upgraded their forging operation. Fundamentally the machine is still sound, but the guides are worn out. The production people are spending too much time fiddling with it instead of turning out parts. Something has to be done. I have to make a recommendation to Ed Burns and Anna Kulkowski, who are going to be making decisions on investments for the next quarter. I'd like you to handle it." Ed Burns was the manager of manufacturing, and Anna Kulkowski was, among other things, the president of Canadian Widgets.

"What's the time frame?" Naomi asked. She was shifting job priorities in her mind and deciding what she would need to postpone.

"I want a report by tomorrow morning. I'd like to have a chance to review what you've done and submit a recommendation to Burns and Kulkowski for their Wednesday meeting." Clem sat back and gave Naomi his best big smile.

Naomi's return smile was a bit weak, as she was preoccupied with trying to sort out where to begin.

Clem laughed and continued with "It's really not so bad. Dave Sullivan has done most of the work. But he's away and can't finish. His father-in-law had a heart attack on Friday, and he and Helena have gone to Florida to see him."

"What's involved?" asked Naomi.

"Not much, really. Dave has estimated all the cash flows. He's put everything on a spreadsheet. Essentially, there are three major possibilities. We can refurbish and upgrade the existing machine. We can get a manually operated mechanical press that will use less energy and be a lot quieter. Or we can go for an automated mechanical press.

"Since there is going to be down time while we are changing the unit, we might also want to replace the materials-handling equipment at the same time. If we get the automated press, there is the possibility of going whole hog and integrating materials handling with the press. But even if we automate, we could stay with a separate materials-handling setup.

"Basically, you're looking at a fairly small first cost to upgrade the current beast, versus a large first cost for the automated equipment. But, if you take the high-first-cost route, you will get big savings down the road. All you have to do is decide what's best."

4.1 Introduction

The essential idea of investing is to give up something valuable now for the expectation of receiving something of greater value later. An investment may be thought of as an exchange of resources now for an expected flow of benefits in the future. Business firms, other organizations, and individuals all have opportunities to make such exchanges. A company may be able to use funds to install equipment that will reduce labour costs in the future. These funds might otherwise have been used on another project or returned to the shareholders or owners. An individual may be able to study to become an engineer. Studying requires that time be given up that could have been used to earn money or to travel. The benefit of study, though, is the expectation of a good income from an interesting job in the future.

Not all investment opportunities *should* be taken. The company considering a labour-saving investment may find that the value of the savings is less than the cost of installing the equipment. Not all investment opportunities *can* be taken. The person spending the next four years studying engineering cannot also spend that time getting degrees in law and science.

Engineers play a major role in making decisions about investment opportunities. In many cases, they are the ones who estimate the expected costs of and returns from an investment. They then must decide whether the expected returns outweigh the costs to see if the opportunity is potentially acceptable. They may also have to examine competing investment opportunities to see which is best. Engineers frequently refer to investment opportunities as **projects**. Throughout the rest of this text, the term "project" will be used to mean "investment opportunity."

In this chapter and in Chapter 5, we deal with methods of evaluating and comparing projects, sometimes called **comparison methods**. We start in this chapter with a scheme for classifying groups of projects. This classification system permits the appropriate use of any of the comparison methods. We then turn to a consideration of several widely used methods for evaluating opportunities. The **present worth method** compares projects by looking at the present worth of all cash flows associated with the projects. The **annual worth method** is similar, but converts all cash flows to a uniform series, that is, an annuity. The **payback period method** estimates how long it takes to "pay back" investments. The study of comparison methods is continued in Chapter 5, which deals with the internal rate of return.

We have made six assumptions about all the situations presented in this chapter and in Chapter 5:

1. We have assumed that costs and benefits are always measurable in terms of money. In reality, costs and benefits need not be measurable in terms of money. For example, providing safe working conditions has many benefits, including improvement of worker morale. However, it would be difficult to express the value of improved worker morale objectively in dollars and cents. Such other benefits as the pleasure gained from appreciating beautiful design may not be measurable quantitatively. We shall consider qualitative criteria and multiple objectives in Chapter 13.
2. We have assumed that future cash flows are known with certainty. In reality, future cash flows can only be estimated. Usually the further into the future we try to forecast, the less certain our estimates become. We look at methods of assessing the impact of uncertainty and risks in Chapters 11 and 12.
3. We have assumed that cash flows are unaffected by inflation or deflation. In reality, the purchasing power of money typically declines over time. We shall consider how inflation affects decision making in Chapter 9.
4. Unless otherwise stated, we have assumed that sufficient funds are available to implement all projects. In reality, cash constraints on investments may be very important, especially for new enterprises with limited ability to raise capital. We look at methods of raising capital in Appendix 4A.
5. We have assumed that taxes are not applicable. In reality, taxes are pervasive. We shall show how to include taxes in the decision-making process in Chapter 8.

6. Unless otherwise stated, we shall assume that all investments have a cash outflow at the start. These outflows are called *first costs*. We also assume that projects with first costs have cash inflows after the first costs that are at least as great in total as the first costs. In reality, some projects have cash inflows at the start, but involve a commitment of cash outflows at a later period. For example, a consulting engineer may receive an advance payment from a client, a cash inflow, to cover some of the costs of a project, but to complete the project the engineer will have to make disbursements over the project's life. We shall consider evaluation of such projects in Chapter 5.

4.2 Relations Among Projects

Companies and individuals are often faced with a large number of investment opportunities at the same time. Relations among these opportunities can range from the simple to the complex. We can distinguish three types of connections among projects that cover all the possibilities. Projects may be

1. independent,
2. mutually exclusive, or
3. related but not mutually exclusive.

The simplest relation between projects occurs when they are **independent**. Two projects are independent if the expected costs and the expected benefits of each project do not depend on whether the other one is chosen. A student considering the purchase of a vacuum cleaner and the purchase of a personal computer would probably find that the expected costs and benefits of the computer did not depend on whether he or she bought the vacuum cleaner. Similarly, the benefits and costs of the vacuum cleaner would be the same, whether or not the computer was purchased. If there are more than two projects under consideration, they are said to be independent if all possible pairs of projects in the set are independent. When two or more projects are independent, evaluation is simple. Consider each opportunity one at a time, and accept or reject it on its own merits.

Projects are **mutually exclusive** if, in the process of choosing one, all other alternatives are excluded. In other words, two projects are mutually exclusive if it is impossible to do both or it clearly would not make sense to do both. For example, suppose Bismuth Realty Company wants to develop downtown office space on a specific piece of land. They are considering two potential projects. The first is a low-rise poured-concrete building. The second is a high-rise steel-frame structure with the same capacity as the low-rise building, but it has a small park at the entrance. It is impossible for Bismuth to have both buildings on the same site.

As another example, consider a student about to invest in a computer printer. She can get an inkjet printer or a laser printer, but it would not make sense to get both. She would consider the options to be mutually exclusive.

The third class of projects consists of those that are **related but not mutually exclusive.** For pairs of projects in this category, the expected costs and benefits of one project depend on whether the other one is chosen. For example, Klamath Petroleum may be considering a service station at Fourth Avenue and Main Street as well as one at Twelfth and Main. The costs and benefits from either station will clearly depend on whether the other is built, but it may be possible, and may make sense, to have both stations.

Evaluation of related but not mutually exclusive projects can be simplified by combining them into exhaustive, mutually exclusive sets. For example, the two projects being considered by Klamath can be put into four mutually exclusive sets:

1. Neither station—the "do nothing" option
2. Just the station at Fourth and Main
3. Just the station at Twelfth and Main
4. Both stations

In general, n related projects can be put into 2^n sets including the "do nothing" option. Once the related projects are put into mutually exclusive sets, the analyst treats these sets as the alternatives. We can make 2^n mutually exclusive sets with n related projects by noting that for any single set there are exactly two possibilities for each project. The project may be *in* or *out* of that set. To get the total number of sets, we multiply the n twos to get 2^n. In the Klamath example, there were two possibilities for the station at Fourth and Main—accept or reject. These are combined with the two possibilities for the station at Twelfth and Main, to give the four sets that we listed.

A special case of related projects is where one project is *contingent* on another. Consider the case where Project A could be done alone or A and B could be done together, but B could not be done by itself. Project B is then contingent on Project A because it cannot be taken unless A is taken first. For example, the Athens and Manchester Development Company is considering building a shopping mall on the outskirts of Moncton. They are also considering building a parking garage to avoid long outdoor walks by patrons. Clearly, they would not build the parking garage unless they were also building the mall.

Another special case of related projects is due to resource constraints. Usually the constraints are financial. For example, Bismuth may be considering two office buildings at different sites, where the expected costs and benefits of the two are unrelated, but Bismuth may be able to finance only one building. The two office-building projects would then be mutually exclusive because of financial constraints. If there are more than two projects, then all of the sets of projects that meet the budget form a mutually exclusive set of alternatives.

When there are several related projects, the number of logically possible combinations becomes quite large. If there are four related projects, there are $2^4 = 16$ mutually exclusive sets, including the "do nothing" alternative. If there are five related projects, the number of alternatives doubles to 32. A good way to keep track of these alternatives is to construct a table with all possible combinations of projects. Example 4.1 demonstrates the use of a table.

Example 4.1

The Small Street residential association wants to improve their district. Four ideas for renovation projects have been proposed: (1) converting part of the roadway to gardens, (2) adding old-fashfioned light standards, (3) replacing the pavement with cobblestones, and (4) making the street one-way. However, there are a number of restrictions. The residential association can afford to do only two of the first three projects together. Also, gardens are possible only if the street is one-way. Finally, old-fashioned light standards would look out of place unless the pavement was replaced with cobblestones. The residential association feels it must do something. They do not want simply to leave things the way they are. What mutually exclusive alternatives are possible?

Since the association does not want to "do nothing," only $15 = 2^4 - 1$ alternatives will be considered. These are shown in Table 4.1. The potential projects are listed in rows. The alternatives, which are sets of projects, are in columns. An "x" in a cell indicates that a project is in the alternative represented by that column. Not all logical combinations of projects represent feasible alternatives, as seen in the special cases of contingent alternatives or budget constraints. A last row, below the potential-project rows, indicates whether the sets are feasible alternatives.

Table 4.1 Potential Alternatives for the Small Street Renovation

Potential Alternative	1	2	3	4	5	6	7	8	9	10	11	12	13	14	15
Gardens	x				x	x	x				x	x	x		x
Lights		x			x			x	x		x	x		x	x
Cobblestones			x			x		x		x	x		x	x	x
One-way				x			x		x	x		x	x	x	x
Feasible?	No	No	Yes	Yes	No	No	Yes	Yes	No	Yes	No	No	Yes	Yes	No

The result is that there are seven feasible mutually exclusive alternatives:

1. Cobblestones (alternative 3)
2. One-way street (alternative 4)
3. One-way street with gardens (alternative 7)
4. Cobblestones with lights (alternative 8)
5. One-way street with cobblestones (alternative 10)
6. One-way street with cobblestones and gardens (alternative 13)
7. One-way street with cobblestones and lights (alternative 14) ■

To summarize our investigation of possible relations among projects, we have a threefold classification system: (1) independent projects, (2) mutually exclusive projects, and (3) related but not mutually exclusive projects. We can, however, arrange related projects into mutually exclusive sets and treat the sets as mutually exclusive alternatives. This reduces the system to two categories, independent and mutually exclusive. (See Figure 4.1.) Therefore, in the remainder of this chapter we consider only independent and mutually exclusive projects.

Figure 4.1 Possible Relations Among Projects and How to Treat Them

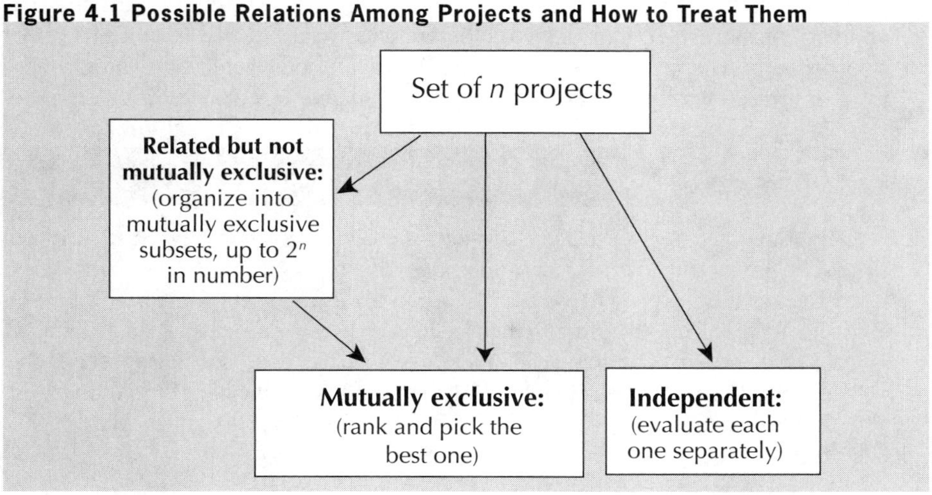

Set of n projects

Related but not mutually exclusive: (organize into mutually exclusive subsets, up to 2^n in number)

Mutually exclusive: (rank and pick the best one)

Independent: (evaluate each one separately)

4.3 Minimum Acceptable Rate of Return (MARR)

A company evaluating projects will set for itself a lower limit for investment acceptability known as the **minimum acceptable rate of return (MARR)**. The MARR is an interest rate that must be earned for any project to be accepted. Projects that earn at least the MARR are desirable, since this means that the money is earning at least as much as can be earned elsewhere. Projects that earn less than the MARR are not desirable, since investing money in these projects denies the opportunity to use the money more profitably elsewhere.

The MARR can also be viewed as the rate of return required to get investors to invest in a business. If a company accepts projects that earn less than the MARR, investors will not be willing to put money into the company. This minimum return required to induce investors to invest in the company is the company's **cost of capital**. Methods for determining the cost of capital are presented in Appendix 4A.

The MARR is thus an opportunity cost in two senses. First, investors have investment opportunities outside any given company. Investing in a given company implies forgoing the opportunity of investing elsewhere. Second, once a company sets a MARR, investing in a given project implies giving up the opportunity of using company funds to invest in other projects that pay at least the MARR.

We shall show in this chapter and in Chapter 5 how the MARR is used in calculations involving the present worth, annual worth, or internal rate of return to evaluate projects. Henceforth, it is assumed that a value for the MARR has been supplied.

4.4 Present Worth (PW) and Annual Worth (AW) Comparisons

The present worth (PW) comparison method and the annual worth (AW) comparison method are based on finding a comparable basis to evaluate projects in monetary units. With the present worth method, the analyst compares project A and project B by computing the present worths of the two projects at the MARR. The preferred project is the one with the greater present worth. The value of any company can be considered to be the present worth of all of its projects. Therefore, choosing projects with the greatest present worth maximizes the value of the company. With the annual worth method, the analyst compares projects A and B by transforming all disbursements and receipts of the two projects to a uniform series at the MARR. The preferred project is the one with the greater annual worth. One can also speak of *present cost* and *annual cost*. See Close-Up 4.1.

CLOSE-UP 4.1 Present Cost and Annual Cost

Sometimes mutually exclusive projects are compared in terms of present cost or annual cost. That is, the best project is the one with minimum present worth of cost as opposed to the maximum present worth. Two conditions should hold for this to be valid: (1) all projects have the same major benefit, and (2) the estimated value of the major benefit clearly outweighs the projects' costs, even if that estimate is imprecise. Therefore, the "do nothing" option is rejected. The value of the major benefit is ignored in further calculations since it is the same for all projects. We choose the project with the lowest cost, considering secondary benefits as offsets to costs.

4.4.1 Present Worth Comparisons for Independent Projects

The alternative to investing money in an independent project is to "do nothing." Doing nothing doesn't mean that the money is not used productively. In fact, it would be used for some other project, earning interest at a rate at least equal to the MARR. However, the present worth of any money invested at the MARR is zero, since the present worth of future receipts would exactly offset the current disbursement. Consequently, if an independent project has a present worth greater than zero, it is acceptable. If an independent project has a present worth less than zero, it is unacceptable. If an independent project has a present worth of exactly zero, it is considered *marginally* acceptable.

Example 4.2

Steve Chen, a third-year electrical engineering student at Seaforth University, has noticed that the networked personal computers provided by the university for its students are frequently fully utilized, so that students often have to wait to get on a machine. The university has a building plan that will create more space for network computers, but the new facilities won't be available for five years. In the meantime, Steve sees the opportunity to create an alternative network in a mall near the campus. The first cost for equipment, furniture, and software is expected to be \$70 000. Students would be able to rent time on computers by the hour and to use the printers at a charge per page. Annual net cash flow from computer rentals and other charges, after paying for labour, supplies, and other costs, is expected to be \$30 000 a year for five years. When the University opens new facilities at the end of five years, business at Steve's network would fall off and net cash flow would turn negative. Therefore, the plan is to dismantle the network after five years. The five-year-old equipment and furniture are expected to have zero value. If investors in this type of service enterprise demand a return of 20% per year, is this a good investment?

The present worth of the project is

$$\begin{aligned} PW &= -\$70\,000 + \$30\,000(P/A, 20\%, 5) \\ &= -\$70\,000 + \$30\,000(2.9906) \\ &= \$19\,718 \\ &\cong \$20\,000 \end{aligned}$$

The project is acceptable since the present worth is greater than zero.

Another way to look at the project is to suppose that, once Steve has set up the network off campus, he tries to sell it. If he can convince potential investors, who demand a return of 20% a year, that the expectation of a \$30 000 per year cash flow for five years is accurate, how much would they be willing to pay for the network? Investors would calculate the present worth of a 20% annuity paying \$30 000 for five years. This is given by

$$\begin{aligned} PW &= \$30\,000(P/A, 20\%, 5) \\ &= \$30\,000(2.9906) \\ &= \$89\,718 \\ &\cong \$90\,000 \end{aligned}$$

Investors would be willing to pay approximately $90 000. Steve will have taken $70 000, the first cost, and used it to create an asset worth almost $90 000. As illustrated in Figure 4.2, the $20 000 difference may be viewed as profit.■

Figure 4.2 Present Worth as a Measure of Profit

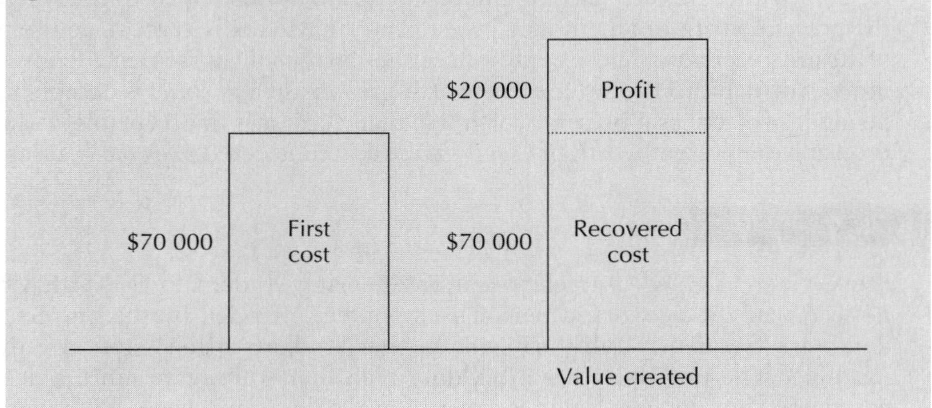

Let us now consider an example in which the benefit of an investment is a reduction in cost.

Example 4.3

A mechanical engineer is considering building automated materials-handling equipment for a production line. On the one hand, the equipment would substantially reduce the manual labour currently required to move items from one part of the production process to the next. On the other hand, the equipment would consume energy, require insurance, and need periodic maintenance.

Alternative 1: Continue to use the current method. Yearly labour costs are $9200.

Alternative 2: Build automated materials-handling equipment with an expected service life of 10 years.

First cost	$15 000	
Labour	$3 300	per year
Power	$400	per year
Maintenance	$2 400	per year
Taxes and insurance	$300	per year

If the MARR is 9%, which alternative is better? Use a present worth comparison.

The investment of $15 000 can be viewed as yielding a positive cash flow of $2800 = $9200 – ($3300 + $400 + $2400 + $300) per year in the form of a reduction in cost.

$$\begin{aligned} \text{PW} &= -\$15\,000 + [\$9200 - (\$3300 + \$400 + \$2400 + \$300)](P/A, 9\%, 10) \\ &= -\$15\,000 + \$2800(P/A, 9\%, 10) \\ &= -\$15\,000 + \$2800(6.4176) \\ &= \$2969.44 \end{aligned}$$

The present worth of the cost savings is approximately \$3000 greater than the \$15 000 first cost. Therefore, the project is worth implementing.■

4.4.2 Present Worth Comparisons for Mutually Exclusive Projects

It is very easy to use the present worth method to choose the best project among a set of mutually exclusive projects *when the service lives are the same*. One just computes the present worth of each project using the MARR. The project with the greatest present worth is the preferred project because it is the one with the greatest profit.

Example 4.4

Fly-by-Night Aircraft must purchase a new lathe. It is considering four lathes, each of which has a life of 10 years with no scrap value.

Lathe	1	2	3	4
First cost	\$100 000	\$150 000	\$200 000	\$255 000
Annual savings	25 000	34 000	46 000	55 000

Given a MARR of 15%, which alternative should be taken?

The present worths are:

Lathe 1: $PW = -\$100\,000 + \$25\,000(P/A, 15\%, 10)$
$= -\$100\,000 + \$25\,000(5.0187) \cong \$25\,468$

Lathe 2: $PW = -\$150\,000 + \$34\,000(P/A, 15\%, 10)$
$= -\$150\,000 + \$34\,000(5.0187) \cong \$20\,636$

Lathe 3: $PW = -\$200\,000 + \$46\,000(P/A, 15\%, 10)$
$= -\$200\,000 + \$46\,000(5.0187) \cong \$30\,860$

Lathe 4: $PW = -\$255\,000 + \$55\,000(P/A, 15\%, 10)$
$= -\$155\,000 + \$55\,000(5.0187) \cong \$21\,029$

Lathe 3 has the greatest present worth, and is therefore the preferred alternative.■

4.4.3 Annual Worth Comparisons

Annual worth comparisons are essentially the same as present worth comparisons, except that all disbursements and receipts are transformed to a uniform series at the MARR, rather than to the present worth. Any present worth P can be converted to an annuity A by the capital recovery factor $(A/P,i,N)$. Therefore, a comparison of two projects *that have the same life* by the present worth and annual worth methods will always indicate the same preferred alternative. Note that, although the method is called annual worth, the uniform series is not necessarily on a yearly basis.

Present worth comparisons make sense because they compare the worth today of each alternative, but annual worth comparisons can sometimes be more easily grasped mentally. For example, to say that operating an automobile over five years has a present cost of $20 000 is less meaningful than saying that it will cost about $5300 per year for each of the following five years.

Sometimes there is no clear justification for preferring either the present worth method or the annual worth method. Then it is reasonable to use the one that requires less conversion. For example, if most receipts or costs are given as annuities or gradients, one can more easily perform an annual worth comparison. Sometimes it can be useful to compare projects on the basis of future worths. See Close-Up 4.2.

NET VALUE 4.1

Car Payment Calculators

The Internet offers useful Web sites when you are thinking of buying a car—to learn more about different makes and models, optional features, prices, what's available (used or new) at which dealer, and financing information if a car is to be purchased and not leased. Major car manufacturers, financial services companies, and car information Web sites make it easy for customers to figure out their financing plans by offering Web-based car payment calculators.

A typical "monthly payment" calculator determines how much a customer pays every month on the basis of the purchase price, down payment, interest rate, and loan term. This is essentially an annuity calculation. An "affordability" calculator, on the other hand, gives a present worth (what price of a car you could afford to buy now) or a future worth (total amount of money that would be spent on a car after all payments are made including interest) based on down payment, monthly payment, interest rate, and loan term. The calculators are useful in making instant comparisons among different cars or car companies, studying what-if scenarios with various payment amounts or lengths of the loan, and determining budget limitations. Similar calculators are available for house mortgage payments.

Example 4.5

Sweat University is considering two alternative types of bleachers for a new athletic stadium:

Alternative 1: Concrete bleachers. The first cost is $350 000. The expected life of the concrete bleachers is 90 years and the annual upkeep costs are $2500.

Alternative 2: Wooden bleachers on earth fill. The first cost of $200 000 consists of $100 000 for earth fill and $100 000 for the wooden bleachers. The annual painting costs are $5000. The wooden bleachers must be replaced every 30 years at a cost of $100 000. The earth fill will last the entire 90 years.

One of the two alternatives will be chosen. It is assumed that the receipts and other benefits of the stadium are the same for both construction methods. Therefore, the greatest net benefit is obtained by choosing the alternative with the lower cost. The university uses a MARR of 7%. Which of the two alternatives is better?

For this example, let us base the analysis on annual worth. Since both alternatives have a life of 90 years, we shall get the equivalent annual costs over 90 years for both at an interest rate of 7%.

CLOSE-UP 4.2 Future Worth

Sometimes it may be desirable to compare projects with the **future worth method**, on the basis of the future worth of each project. This is most likely to be true for cases where money is being saved for some future expense.

For example, two investment plans are being compared to see which accumulates more money for retirement. Plan A consists of a payment of $10 000 today and then $2000 per year over 20 years. Plan B is $3000 per year over 20 years. Interest for both plans is 10%. Rather than convert these cash flows to either present worth or annual worth, it is sensible to compare the future worths of the plans, since the actual dollar value in 20 years has particular meaning.

$$\begin{aligned}\text{FW}_A &= \$10\,000(F/P, 10\%, 20) + \$2000(F/A, 10\%, 20)\\ &= \$10\,000(6.7275) + \$2000(57.275)\\ &= \$181\,825\end{aligned}$$

$$\begin{aligned}\text{FW}_B &= \$3000(F/A, 10\%, 20)\\ &= \$3000(57.275)\\ &= \$171\,825\end{aligned}$$

Plan A is the better choice. It will accumulate to $181 825 over the next 20 years.

Alternative 1: Concrete bleachers

The equivalent annual cost over the 90-year life span of the concrete bleachers is

$$\begin{aligned}\text{AW} &= \$350\,000(A/P, 7\%, 90) + \$2500\\ &= \$350\,000(0.07016) + \$2500\\ &= \$27\,056 \text{ per year}\end{aligned}$$

Alternative 2: Wooden bleachers on earth fill

The total annual costs can be broken into three components: AW_1 (for the earth fill), AW_2 (for the bleachers), and AW_3 (for the painting). The equivalent annual cost of the earth fill is

$$\text{AW}_1 = \$100\,000(A/P, 7\%, 90)$$

The equivalent annual cost of the bleachers is easy to determine. The first set of bleachers is put in at the start of the project, the second set at the end of 30 years, and the third set at the end of 60 years, but the cost of the bleachers is the same at each installation. Therefore, we need to get only the cost of the first installation.

$$\text{AW}_2 = \$100\,000(A/P, 7\%, 30)$$

The last expense is for annual painting:

$$\text{AW}_3 = \$5000$$

The total equivalent annual cost for alternative 2, wooden bleachers on earth fill, is the sum of AW_1, AW_2, and AW_3:

$$\begin{aligned}AW &= AW_1 + AW_2 + AW_3\\ &= \$100\,000[(A/P, 7\%, 90) + (A/P, 7\%, 30)] + \$5000\\ &= \$100\,000(0.07016 + 0.08059) + \$5000\\ &= \$20\,075\end{aligned}$$

The concrete bleachers have an equivalent annual cost of about $7000 more than the wooden ones. Therefore, the wooden bleachers are the better choice.■

4.4.4 Comparison of Alternatives with Unequal Lives

When making present worth comparisons, we must always use the same time period in order to take into account the full benefits and costs of each alternative. If the lives of the alternatives are not the same, we can transform them to equal lives with one of the following two methods:

1. Repeat the *service life* of each alternative to arrive at a common time period for all alternatives. Here we assume that each alternative can be repeated with the same costs and benefits in the future—an assumption known as **repeated lives**. Usually we use the *least common multiple* of the lives of the various alternatives. Sometimes it is convenient to assume that the lives of the various alternatives are repeated indefinitely. Note that the assumption of repeated lives may not be valid where it is reasonable to expect technological improvements.
2. Adopt a specified **study period**—a time period that is given for the analysis. To set an appropriate study period, a company will usually take into account the time of required service, or the length of time they can be relatively certain of their forecasts. The study period method necessitates an additional assumption about *salvage value* whenever the life of one of the alternatives exceeds that of the given study period. Arriving at a reliable estimate of salvage value may be difficult sometimes.

Because they rest on different assumptions, the repeated lives and the study period methods can lead to different conclusions when applied to a particular project choice.

Example 4.6 (Modification of Example 4.3)

A mechanical engineer has decided to introduce automated materials-handling equipment for a production line. She must choose between two alternatives, building the equipment or buying the equipment off the shelf. Each alternative has a different service life and a different set of costs.

Alternative 1: Build custom automated materials-handling equipment.

First cost	$15 000	
Labour	$3 300	per year
Power	$400	per year
Maintenance	$2 400	per year
Taxes and insurance	$300	per year
Service life	10	years

Alternative 2: Buy off-the-shelf standard automated materials-handling equipment.

First cost	$25 000	
Labour	$1 450	per year
Power	$600	per year
Maintenance	$3 075	per year
Taxes and insurance	$500	per year
Service life	15	years

If the MARR is 9%, which alternative is better?

The present worth of the custom system over its 10-year life is

$$\begin{aligned} PW(1) &= -\$15\ 000 - (\$3300 + \$400 + \$2400 + \$300)(P/A, 9\%, 10) \\ &= -\$15\ 000 - \$6400(6.4176) \\ &\cong -\$56\ 073 \end{aligned}$$

The present worth of the off-the-shelf system over its 15-year life is:

$$\begin{aligned} PW(2) &= -\$25\ 000 - (\$1450 + \$600 + \$3075 + \$500)(P/A, 9\%, 15) \\ &= -\$25\ 000 - \$5625(8.0606) \\ &\cong -\$70\ 341 \end{aligned}$$

The custom system has a lower cost for its 10-year life than the off-the-shelf system for its 15-year life, but it would be *wrong* to conclude from these calculations that the custom system should be preferred. The custom system yields benefits for only 10 years, whereas the off-the-shelf system lasts 15 years. It would be surprising if the cost of 15 years of benefits were not higher than the cost of 10 years of benefits. A fair comparison of the costs can be made only if equal lives are compared.

Let us apply the repeated lives method. If each alternative is repeated enough times, there will be a point in time where their service lives are simultaneously completed. This will happen first at the time equal to the least common multiple of the service lives. The least common multiple of 10 years and 15 years is 30 years. Alternative 1 will be repeated twice (after 10 years and after 20 years), while alternative 2 will be repeated once (after 15 years) during the 30-year period. At the end of 30 years, both alternatives will be completed simultaneously. See Figure 4.3.

Figure 4.3 Least Common Multiple of the Service Lives

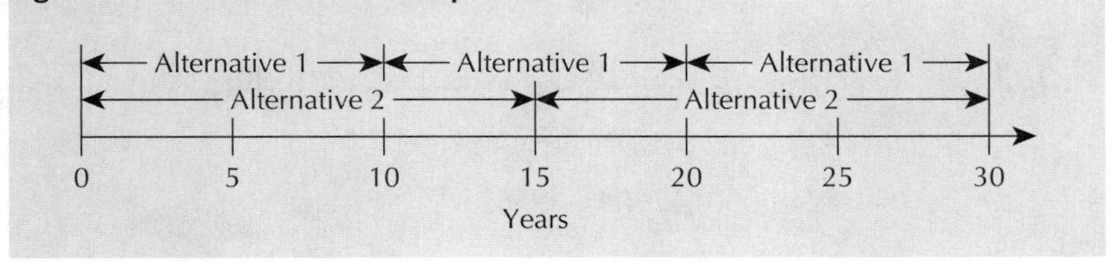

With the same time period of 30 years for both alternatives, we can now compare present worths.

Alternative 1: Build custom automated materials-handling equipment and repeat twice

$$\begin{aligned} PW(1) = &-\$15\ 000 - \$15\ 000(P/F, 9\%, 10) - \$15\ 000(P/F, 9\%, 20) \\ &- (\$3300 + \$400 + \$2400 + \$300)(P/A, 9\%, 30) \end{aligned}$$

$$= -\$15\,000 - \$15\,000(0.42241) - \$15\,000(0.17843) - \$6400\,(10.273)$$

$$\cong -\$89\,760$$

Alternative 2: Buy off-the-shelf standard automated materials-handling equipment and repeat once

$$\text{PW}(2) = -\$25\,000 - \$25\,000(P/F, 9\%, 15)$$

$$- (\$1450 + \$600 + \$3075 + \$500)(P/A, 9\%, 30)$$

$$= -\$25\,000 - \$25\,000(0.27454) - \$5625(10.273)$$

$$\cong -\$89\,649$$

Using the repeated lives method, we find little difference between the alternatives. An annual worth comparison can also be done over a period of time equal to the least common multiple of the service lives by multiplying each of these present worths by the capital recovery factor for 30 years.

$$\text{AW}(1) = -\$89\,760(A/P, 9\%, 30)$$

$$= -\$89\,760(0.09734)$$

$$\cong -\$8737$$

$$\text{AW}(2) = -\$89\,649(A/P, 9\%, 30)$$

$$= -\$89\,649(0.09734)$$

$$\cong -\$8726$$

As we would expect, there is again little difference in the annual cost between the alternatives. However, there is a more convenient approach for an annual worth comparison if it can be assumed that the alternatives are repeated indefinitely. Since the annual costs of an alternative remain the same no matter how many times it is repeated, it is not necessary to determine the least common multiple of the service lives. The annual worth of each alternative can be assessed for whatever time period is most convenient for each alternative.

Alternative 1: Build custom automated materials-handling equipment

$$\text{AW}(1) = -\$15\,000(A/P, 9\%, 10) - \$6400$$

$$= -\$15\,000(0.15582) - \$6400$$

$$\cong -\$8737$$

Alternative 2: Buy off-the-shelf standard automated materials-handling equipment

$$\text{AW}(2) = -\$25\,000(A/P, 9\%, 15) - \$5625$$

$$= -\$25\,000(0.12406) - \$5625$$

$$\cong -\$8726$$

If it cannot be assumed that the alternatives can be repeated to permit a calculation over the least common multiple of their service lives, then it is necessary to use the study period method.

Suppose that the given study period is 10 years, because the engineer is uncertain about costs past that time. The service life of the off-the-shelf system (15 years) is greater than the study period (10 years). Therefore, we have to make an assumption about the salvage value of the off-the-shelf system after 10 years. Suppose the engineer judges that its salvage value will be \$5000. We can now proceed with the comparison.

Alternative 1: Build custom automated materials-handling equipment (10-year study period)

$$\begin{aligned}PW(1) &= -\$15\,000 - (\$3300 + \$400 + \$2400 + \$300)(P/A, 9\%, 10)\\ &= -\$15\,000 - \$6400(6.4176)\\ &\cong -\$56\,073\end{aligned}$$

Alternative 2: Buy off-the-shelf standard automated materials-handling equipment (10-year study period)

$$\begin{aligned}PW(2) &= -\$25\,000 - (\$1450 + \$600 + \$3075 + \$500)(P/A, 9\%, 10)\\ &\quad + \$5000(P/F, 9\%, 10)\\ &= -\$25\,000 - \$5625(6.4176) + \$5000(0.42241)\\ &\cong -\$58\,987\end{aligned}$$

Using the study period method of comparison, alternative 1 has the smaller present worth of costs and is, therefore, preferred.

Note that here the study period method gives a different answer than the repeated lives method gives. The study period method is often sensitive to the chosen salvage value. A larger salvage value tends to make an alternative with a life longer than the study period more attractive, and a smaller value tends to make it less attractive.

In some instances, it may be difficult to arrive at a reliable estimate of salvage value. Given the sensitivity of the study period method to the salvage value estimate, the analyst may be uncertain about the validity of the results. One way of circumventing this problem is to avoid estimating the salvage value at the outset. Instead we calculate what salvage value would make the alternatives equal in value. Then we decide whether the actual salvage value will be above or below the break-even value found. Applying this approach to our example, we set PW(1) = PW(2) so that

$$\begin{aligned}PW(1) &= PW(2)\\ \$56.073 &= -\$25\,000 - \$5625(6.4176) + S(0.42241)\end{aligned}$$

where S is the salvage value.

Solving for S, we find S = \$11 834. Is a reasonable estimate of the salvage value above or below \$11 834? If it is above \$11 834, then we conclude that the off-the-shelf system is the preferred choice. If it is below \$11 834, then we conclude that the custom system is preferable.■

The study period can also be used for the annual worth method if the assumption of being able to indefinitely repeat the choice of alternatives is not justified.

Example 4.7

Joan is renting an apartment while on a one-year assignment in a distant city. The apartment does not have a refrigerator. She can rent one for a \$100 deposit (returned in a year) and \$15 per month (paid at the end of each month). Alternatively, she can buy a used refrigerator for \$300, which she would sell in a year when she leaves. For how much would Joan have to be able to sell the used refrigerator in one year when she leaves, in order to be better off buying the used refrigerator than renting one? Interest is at 6% nominal, compounded monthly.

Let S stand for the unknown salvage value (i.e., the amount Joan will be able to sell the refrigerator for in a year). We then equate the present worth of the rental alternative with the present worth of the purchase alternative for the one-year study period:

$$\text{PW(rental)} = \text{PW(purchase)}$$
$$-\$100 - \$15(P/A, 0.5\%, 12) + \$100(P/F, 0.5\%, 12) = -\$300 + S(P/F, 0.5\%, 12)$$
$$-\$100 - \$15(11.616) + \$100(0.94192) = -\$300 + S(0.94192)$$
$$S = \$127.35$$

If Joan can sell the used refrigerator for more than about $127 after one year's use, she is better off buying it rather than renting one.■

4.5 Payback Period

The simplest method for judging the economic viability of projects is the payback period method. It is a rough measure of the time it takes for an investment to pay for itself. More precisely, the **payback period** is the number of years it takes for an investment to be recouped when the interest rate is assumed to be zero. When annual savings are constant, the payback period is usually calculated as follows:

$$\text{Payback period} = \frac{\text{First cost}}{\text{Annual savings}}$$

For example, if a first cost of $20 000 yielded a return of $8000 per year, then the payback period would be $20 000/$8000 = 2.5 years.

If the annual savings are not constant, we can calculate the payback period by deducting each year of savings from the first cost until the first cost is recovered. The number of years required to pay back the initial investment is the payback period. For example, suppose the saving from a $20 000 first cost is $5000 the first year, increasing by $1000 each year thereafter. By adding the annual savings one year at a time, we see that it would take a just over three years to pay back the first cost ($5000 + $6000 + $7000 + $8000 = $26 000). The payback period would then be stated as either four years (if we assume that the $8000 is received at the end of the fourth year) or 3.25 years (if we assume that the $8000 is received uniformly over the fourth year).

According to the payback period method of comparison, the project with the shorter payback period is the preferred investment. A company may have a policy of rejecting projects for which the payback period exceeds some preset number of years. The length of the maximum payback period depends on the type of project and the company's financial situation. If the company expects a cash constraint in the near future, or if a project's returns are highly uncertain after more than a few periods, the company will set a maximum payback period that is relatively short. As a common rule, a payback period of two years is often considered acceptable, while one of more than four years is unacceptable. Accordingly, government grant programs often target projects with payback periods of between two and four years on the rationale that in this range the grant can justify economically feasible projects that a company with limited cash flow would otherwise be unwilling to undertake.

The payback period need not, and perhaps should not, be used as the sole criterion for evaluating projects. It is a rough method of comparison and possesses some glaring weaknesses (as we shall discuss after Examples 4.8 and 4.9). Nevertheless, the payback period method can be used effectively as a preliminary filter. All projects with paybacks within the minimum would then be evaluated, using either rate of return methods (Chapter 5) or present/annual worth methods.

Example 4.8

Elyse runs a second-hand book business out of her home where she advertises and sells the books over the Internet. Her small business is becoming quite successful and she is considering purchasing an upgrade to her computer system which will give her more reliable uptime. The cost is $5000. She expects that the investment will bring about an annual savings of $2000, due to the fact that her system will no longer suffer long failures and thus she will be able to sell more books. What is the payback period on her investment, assuming that the savings accrue over the whole year?

$$\text{Payback period} = \frac{\text{First Cost}}{\text{Annual Savings}} = \frac{\$5000}{\$2000} = 2.5 \text{ years.}■$$

Example 4.9

Pizza-in-a-Hurry operates a pizza delivery service to its customers with two eight-year-old vehicles, both of which are large, consume a great deal of gas and are starting to cost a lot to repair. The owner, Ray, is thinking of replacing one of the cars with a smaller, three-year-old car that his sister-in-law is selling for $8000. Ray figures he can save $3000, $2000, and $1500 per year for the next three years and $1000 per year for the following two years by purchasing the smaller car. What is the payback period for this decision?

The payback period is the number of years of savings required to pay back the initial cost. After three years, $3000 + $2000 + $1500 = $6500 has been paid back, and this amount is $7500 after four years and $8500 after five years. The payback period would be stated as five years if the savings are assumed to occur at the end of each year, or 4.5 years if the savings accrue continuously throughout the year.■

The payback period method has four main advantages:

1. It is very easy to understand. One of the goals of engineering decision making is to communicate the reasons for a decision to managers or clients with a variety of backgrounds. The reasons behind the payback period and its conclusions are very easy to explain.
2. The payback period is very easy to calculate. It can usually be done without even using a calculator, so projects can be very quickly assessed.
3. It accounts for the need to recover capital quickly. Cash flow is almost always a problem for small-to-medium-sized companies. Even large companies sometimes can't tie up their money in long-term projects.
4. The future is unknown. The future benefits from an investment may be estimated imprecisely. It may not make much sense to use precise methods like present worth on numbers that are imprecise to start with. A simple method like the payback period may be good enough for most purposes.

But the payback period method has three important disadvantages:

1. It discriminates against long-term projects. No houses or highways would ever be built if they had to pay themselves off in two years.
2. It ignores the effect of the timing of cash flows within the payback period. It disregards interest rates and takes no account of the time value of money. (Occasionally, a discounted payback period is used to overcome this disadvantage. See Close-Up 4.3.)
3. It ignores the expected service life. It disregards the benefits that accrue after the end of the payback period.

CLOSE-UP 4.3 **Discounted Payback Period**

In a discounted payback period calculation, the present worth of each year's savings is subtracted from the first cost until the first cost is diminished to zero. The number of years of savings required to do this is the discounted payback period. The main disadvantages of using a discounted payback period include the more complicated calculations and the need for an interest rate.

For instance, in Example 4.8, Elyse had an investment of \$5000 recouped by annual savings of \$2000. If interest were at 10%, the present worth of savings would be:

Year	Present Worth	Cumulative
Year 1	\$2000(P/F, 10%, 1) = \$2000(0.90909) = \$1818	\$1818
Year 2	\$2000(P/F, 10%, 2) = \$2000(0.82645) = \$1653	\$3471
Year 3	\$2000(P/F, 10%, 3) = \$2000(0.75131) = \$1503	\$4974
Year 4	\$2000(P/F, 10%, 4) = \$2000(0.68301) = \$1366	\$6340

Thus the discounted payback period is over 3 years, compared with 2.5 years calculated for the standard payback period.

Example 4.10 illustrates how the payback period method can ignore future cash flows.

Example 4.10

Self Defence Systems is going to upgrade their paper-shredding facility. They have a choice between two models. Model 007, with a first cost of \$50 000 and a service life of seven years, would save them \$10 000 per year. Model MX, with a first cost of \$10 000 and an expected service life of 20 years, would save them \$1500 per year. If the company's MARR is 8%, which model is the better buy?

Using payback period as the sole criterion:

Model 007: Payback period = \$50 000/\$10 000 = 5 years

Model MX: Payback period = \$10 000/\$1500 = 6.6 years

It appears that the 007 model is better.

Using annual worth:

Model 007: AW = –\$50 000(A/P, 8%, 7) + \$10 000 = \$396.50

Model MX: AW = –\$10 000(A/P, 8%, 20) + \$1500 = \$481.50

Here, Model MX is substantially better.

The difference in the results from the two comparison methods is that the payback period method has ignored the benefits of the models that occur after the models have paid themselves off. This is illustrated in Figure 4.4. For Model MX, about 14 years of benefits have been omitted, whereas for model 007, only two years of benefits have been left out.■

Figure 4.4 Flows Ignored by the Payback Period

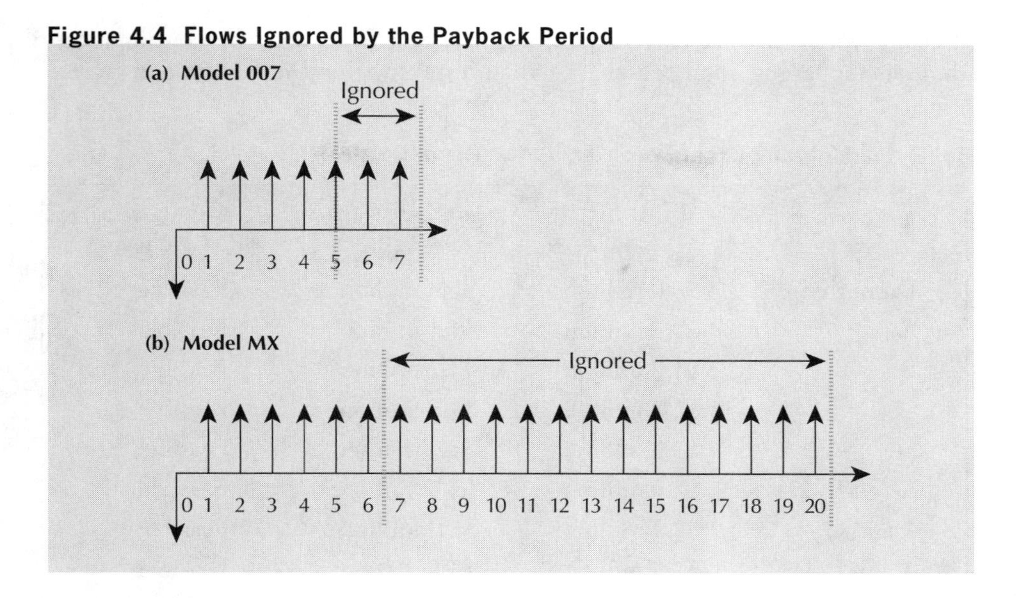

REVIEW PROBLEMS

REVIEW PROBLEM 4.1

Tilson Dairies operates several cheese plants. The plants are all old and in need of renovation. Tilson's engineers have developed plans to renovate all the plants. Each project would have a positive present worth at the company's MARR. Tilson has $3.5 million available to invest in these projects. The following facts about the potential renovation projects are available:

Project	First Cost	Present Worth
A: Renovate plant 1	$0.8 million	$1.1 million
B: Renovate plant 2	$1.2 million	$1.7 million
C: Renovate plant 3	$1.4 million	$1.8 million
D: Renovate plant 4	$2.0 million	$2.7 million

Which projects should Tilson accept?

ANSWER

Table 4.2 shows the possible mutually exclusive projects that Tilson can consider.

Tilson should accept projects A, B, and C. They have a combined present worth of $4.6 million. Other feasible combinations that come close to using all available funds are B and D with a total present worth of $4.4 million, and C and D with a total present worth of $4.5 million.

Note that it is not necessary to consider explicitly the "leftovers" of the $3.5 million budget when comparing the present worths. The assumption is that any leftover part of the budget will be invested and provide interest at the MARR, resulting in a zero present worth for that part. Therefore, it is best to choose the combination of projects that has the largest total present worth and stays within the budget constraint.

Table 4.2 Mutually Exclusive Projects for Tilson Dairies

Project	Total First Cost	Total Present Worth	Feasibility
Do nothing	\$0.0 million	\$0.0 million	Feasible
A	\$0.8 million	\$1.1 million	Feasible
B	\$1.2 million	\$1.7 million	Feasible
C	\$1.4 million	\$1.8 million	Feasible
D	\$2.0 million	\$2.7 million	Feasible
A and B	\$2.0 million	\$2.8 million	Feasible
A and C	\$2.2 million	\$2.9 million	Feasible
A and D	\$2.8 million	\$3.8 million	Feasible
B and C	\$2.6 million	\$3.5 million	Feasible
B and D	\$3.2 million	\$4.4 million	Feasible
C and D	\$3.4 million	\$4.5 million	Feasible
A, B, and C	\$3.4 million	\$4.6 million	Feasible
A, B, and D	\$4.0 million	\$5.5 million	Not feasible
A, C, and D	\$4.2 million	\$5.6 million	Not feasible
B, C, and D	\$4.6 million	\$6.2 million	Not feasible
A, B, C and D	\$5.4 million	\$7.3 million	Not feasible

■

REVIEW PROBLEM 4.2

City engineers are considering two plans for municipal aqueduct tunnels. They are to decide between the two, using an interest rate of 8%.

Plan A is a full-capacity tunnel that will meet the needs of the city forever. Its cost is \$3 000 000 now, and \$100 000 every 10 years for lining repairs.

Plan B involves building a half-capacity tunnel now and a second half-capacity tunnel in 20 years, when the extra capacity will be needed. Each of the half-capacity tunnels costs \$2 000 000. Maintenance costs for each tunnel are \$80 000 every 10 years. There is also an additional \$15 000 per tunnel per year required to pay for extra pumping costs caused by greater friction in the smaller tunnels.

(a) Which alternative is preferred? Use a present worth comparison.

(b) Which alternative is preferred? Use an annual worth comparison.

ANSWER

(a) *Plan A: Full-Capacity Tunnel*

First, the \$100 000 paid at the end of 10 years can be thought of as a future amount which has an equivalent annuity.

$$AW = \$100\,000(A/F, 8\%, 10) = \$100\,000(0.06903) = \$6903$$

Thus, at 8% interest, \$100 000 every 10 years is equivalent to \$6903 every year.

Since the tunnel will have (approximately) an infinite life, the present cost of the lining repairs can be found using the capitalized cost formula, giving a total cost of

$$\text{PW(Plan A)} = \$3\ 000\ 000 + \$6903/0.08 = \$3\ 086\ 288$$

Plan B: Half-Capacity Tunnels

For the first tunnel, the equivalent annuity for the maintenance and pumping costs is

$$\text{AW} = \$15\ 000 + \$80\ 000(0.06903) = \$20\ 522$$

The present cost is then found with the capitalized cost formula, giving a total cost of

$$\text{PW}_1 = \$2\ 000\ 000 + \$20\ 522/0.08 = \$2\ 256\ 525$$

Now, for the second tunnel, basically the same calculation is used, except that the present worth calculated must be discounted by 20 years at 8%, since the second tunnel will be built 20 years in the future.

$$\text{PW}_2 = \{\$2\ 000\ 000 + [\$15\ 000 + \$80\ 000(0.06903)]/0.08\}(P/F, 8\%, 20)$$
$$= \$2\ 256\ 525(0.21455) \cong \$484\ 137$$

$$\text{PW(Plan B)} = \text{PW}_1 + \text{PW}_2 = \$2\ 740\ 662$$

Consequently, the two half-capacity aqueducts are economically preferable.

(b) *Plan A: Full-Capacity Tunnel*

First, the \$100 000 paid at the end of 10 years can be thought of as a future amount that has an equivalent annuity of

$$\text{AW} = \$100\ 000(A/F, 8\%, 10) = \$100\ 000(0.06903) = \$6903$$

Thus, at 8% interest, \$100 000 every 10 years is equivalent to \$6903 every year.

Since the tunnel will have (approximately) an infinite life, an annuity equivalent to the initial cost can be found using the capitalized cost formula, giving a total annual cost of

$$\text{AW(Plan A)} = \$3\ 000\ 000(0.08) + \$6903 = \$246\ 903$$

Plan B: Half-Capacity Tunnels

For the first tunnel, the equivalent annuity for the maintenance and pumping costs is

$$\text{AW} = \$15\ 000 + \$80\ 000(0.06903) \cong \$20\ 522$$

The annual equivalent of the initial cost is then found with the capitalized cost formula, giving a total cost of

$$\text{AW}_1 = \$2\ 000\ 000(0.08) + \$20\ 522 = \$180\ 522$$

Now, for the second tunnel, basically the same calculation is used, except that the annuity must be discounted by 20 years at 8%, since the second tunnel will be built 20 years in the future.

$$\text{AW}_2 = \text{AW}_1(P/F, 8\%, 20)$$
$$= \$180\ 522(0.21455) \cong \$38\ 731$$

$$\text{AW(Plan B)} = \text{AW}_1 + \text{AW}_2$$
$$= \$180\,522 + \$38\,731 = \$219\,253$$

Consequently, the two half-capacity aqueducts are economically preferable.■

REVIEW PROBLEM 4.3

Constantine Fernando, an engineer at Corner Brook Manufacturing, has a $100 000 budget for plant improvements. He has identified four mutually exclusive investments, all of five years' duration, which have the cash flows shown in Table 4.3. For each alternative, he wants to determine the payback period and the present worth. For his recommendation report, he will order the alternatives from most preferred to least preferred in each case. Corner Brook Manufacturing uses an 8% MARR for such decisions.

Table 4.3 Cash Flows for Review Problem 4.3

	Cash Flow at the End of Each Year					
Alternative	0	1	2	3	4	5
A	$−100 000	$25 000	$25 000	$25 000	$25 000	$25 000
B	−100 000	5 000	10 000	20 000	40 000	80 000
C	−100 000	50 000	50 000	10 000	0	0
D	−100 000	0	0	0	0	1 000 000

ANSWER

The payback period can be found by decrementing yearly. The payback periods for the alternatives are then

A: 4 years
B: 4.3125 or 5 years
C: 2 years
D: 4.1 or 5 years

The order of the alternatives from most preferred to least preferred using the payback period method with yearly decrementing is: C, A, D, B. The present worth computations for each alternative are:

A: $\text{PW} = -\$100\,000 + \$25\,000(P/A, 8\%, 5)$
$= -\$100\,000 + \$25\,000(3.9926)$
$= -\$185$

B: $\text{PW} = -\$100\,000 + \$5000(P/F, 8\%, 1) + \$10\,000(P/F, 8\%, 2)$
$+ \$20\,000(P/F, 8\%, 3) + \$40\,000(P/F, 8\%, 4) + \$80\,000(P/F, 8\%, 5)$
$= -\$100\,000 + \$5000(0.92593) + \$10\,000(0.85734)$
$+ \$20\,000(0.79383) + \$40\,000(0.73503) + \$80\,000(0.68059)$
$= \$12\,982$

$$\begin{aligned} \text{C:}\quad \text{PW} &= -\$100\ 000 + \$50\ 000(P/F, 8\%, 1) + \$50\ 000(P/F, 8\%, 2) \\ &\quad + \$10\ 000(P/F, 8\%, 3) \\ &= -\$100\ 000 + \$50\ 000(0.92593) + \$50\ 000(0.85734) \\ &\quad + \$10\ 000(0.79283) \\ &= -\$2908 \end{aligned}$$

$$\begin{aligned} \text{D:}\quad \text{PW} &= -\$100\ 000 + \$1\ 000\ 000(P/F, 8\%, 5) \\ &= -\$100\ 000 + \$1\ 000\ 000(0.68059) \\ &= \$580\ 590 \end{aligned}$$

The order of the alternatives from most preferred to least preferred using the present worth method is: D, B, A, C.■

SUMMARY

This chapter discussed relations among projects, and the present worth, annual worth, and payback period methods for evaluating projects. There are three classes of relations among projects, (1) independent, (2) mutually exclusive, and (3) related but not mutually exclusive. We then showed how the third class of projects, those that are related but not mutually exclusive, could be combined into sets of mutually exclusive projects. This enabled us to limit the discussion to the first two classes, independent and mutually exclusive. Independent projects are considered one at a time and are either accepted or rejected. Only the best of a set of mutually exclusive projects is chosen.

The present worth method compares projects on the basis of converting all cash flows for the project to a present worth. An independent project is acceptable if its present worth is greater than zero. The mutually exclusive project with the highest present worth should be taken. Projects with unequal lives must be compared by assuming that the projects are repeated or by specifying a study period. Annual worth is similar to present worth, except that the cash flows are converted to a uniform series. The annual worth method may be more meaningful, and also does not require more complicated calculations when the projects have different service lives.

The payback period is a simple method that calculates the length of time it takes to pay back an initial investment. It is inaccurate but very easy to calculate.

Engineering Economics in Action, Part 4B: Doing It Right

Naomi stopped for coffee on her way back from Clem's office. She needed time to think about how to decide which potential forge shop investments were best. She wasn't sure that she knew what "best" meant. She got down her engineering economics text and looked at the table of contents. There were a couple of chapters on comparison methods that seemed to be what she wanted. She sat down with the coffee in her right hand and the text on her lap, and hoped for an uninterrupted hour.

One read through the chapters was enough to remind Naomi of the main relevant ideas that she had learned in school. The first thing she had to do was decide whether the investments were independent or not. They clearly were not independent. It would not make sense to refurbish the current forging hammer and replace it with a mechanical press. Where potential investments were not independent, it was easiest to form mutually exclusive combinations as investment options. Naomi came up with seven options. She ranked the options by first cost, starting with the one with the lowest cost:

1. Refurbish the current machine.
2. Refurbish the current machine plus replace the materials-handling equipment.
3. Buy a manually operated mechanical press.
4. Buy a manual mechanical press plus replace the materials-handling equipment.
5. Buy an automated mechanical press.
6. Buy an automated mechanical press plus replace the materials-handling equipment.
7. Buy an automated mechanical press plus integrate it with the materials-handling equipment.

At this point, Naomi wasn't sure what to do next. There were different ways of comparing the options.

Naomi wanted a break from thinking about theory. She decided to take a look at Dave Sullivan's work. She started up her computer and opened up Dave's email. In it Dave apologized for dumping the work on her and invited Naomi to call him in Florida if she needed help. Naomi decided to call him. The phone was answered by Dave's wife, Helena. After telling Naomi that her father was out of intensive care and was in good spirits, Helena turned the phone over to Dave.

"Hi, Naomi. How's it going?"

"Well, I'm trying to finish off the forge project that you started. And I'm taking you up on your offer to consult."

"You have my attention. What's the problem?"

"Well, I've gotten started. I have formed seven mutually exclusive combinations of potential investments." Naomi went on to explain her selection of alternatives.

"That sounds right, Naomi. I like the way you've organized that. Now, how are you going to make the choice?"

"I've just reread the present worth, annual worth, and payback period stuff, and of those three, present worth makes the most sense to me. I can just compare the present worths of the cash flows for each alternative, and the one whose present worth is highest is the best one. Annual worth is the same, but I don't see any good reason in this case to look at the costs on an annual basis."

"What about internal rate of return?"

"Well, actually, Dave, I haven't reviewed IRR yet. I'll need it, will I?"

"You will. Have a look at it, and also remember that your recommendation is for Burns and Kulkowski. Think about how they will be looking at your information."

"Thanks, Dave. I appreciate your help."

"No problem. This first one is important for you; let's make sure we do it right."

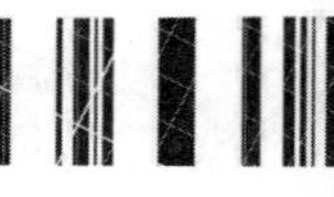

PROBLEMS

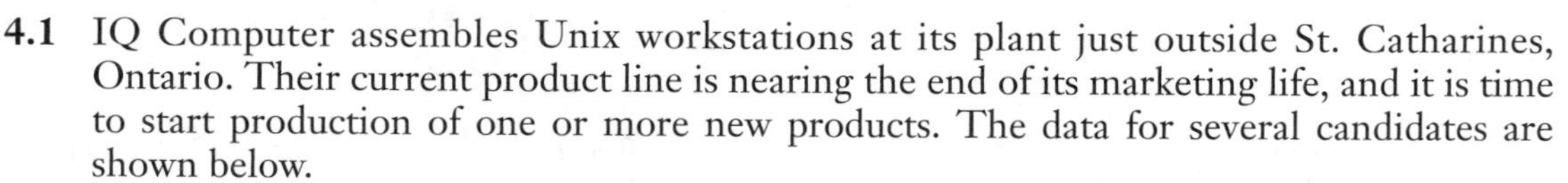

4.1 IQ Computer assembles Unix workstations at its plant just outside St. Catharines, Ontario. Their current product line is nearing the end of its marketing life, and it is time to start production of one or more new products. The data for several candidates are shown below.

The maximum budget for research and development is \$300 000. A minimum of \$200 000 should be spent on these projects. It is desirable to spread out the introduction of new products, so if two products are to be developed together, they should have different lead times. Resource draw refers to the labour and space that are available to the new products; it cannot exceed 100%.

	Potential Product			
	A	B	C	D
Research and development costs	$120 000	$60 000	$150 000	$75 000
Lead time	1 year	2 years	1 year	2 years
Resource draw	60%	50%	40%	30%

On the basis of the above information, determine the set of feasible mutually exclusive alternative projects that IQ Computers should consider.

4.2 The Alabama Alabaster Marble Company (AAM) is considering opening three new quarries. One, designated T, is in Tusksarelooser County; a second, L, is in Lefant County; the third, M, is in Marxbro County. Marble is shipped mainly within a 500-kilometre range of its quarry because of its weight. The market within this range is limited. The returns that AAM can expect from any of the quarries depends on how many quarries AAM opens. Therefore, these potential projects are related.

(a) Construct a set of mutually exclusive alternatives from these three potential projects.

(b) The Lefant County quarry has very rich deposits of marble. This makes the purchase of mechanized cutter-loaders a reasonable investment at this quarry. Such loaders would not be considered at the other quarries. Construct a set of mutually exclusive alternatives from the set of quarry projects augmented by the potential mechanized cutter-loader project.

(c) AAM has decided to invest no more than $2.5 million in the potential quarries. The first costs are as follows:

Project	First Cost
T quarry	$0.9 million
L quarry	$1.4 million
M quarry	$1.0 million
Cutter-loader	$0.4 million

Construct a set of mutually exclusive alternatives that are feasible, given the investment limitation.

4.3 Chatham Automotive has $100 000 to invest in internal projects. The choices are:

Project	Cost
1. Line improvements	$20 000
2. New manual tester	30 000
3. New automatic tester	60 000
4. Overhauling press	50 000

Only one tester may be bought and the press will not need overhauling if the line improvements are not made. What mutually exclusive project combinations are available if Chatham will invest in at least one project?

4.4 The intersection of King and Main Streets needs widening and improvement. The possibilities include

1. Widen King
2. Widen Main
3. Add a left-turn lane on King
4. Add a left-turn lane on Main
5. Add traffic lights at the intersection
6. Add traffic lights at the intersection with advanced green for Main
7. Add traffic lights at the intersection with advanced green for King

A left-turn lane can be installed only if the street in question is widened. A left-turn lane is necessary if the street has traffic lights with an advanced green. The city cannot afford to widen both streets. How many mutually exclusive projects are there?

4.5 Yun is deciding among a number of business opportunities. She can

(a) Sell the X division of her company, Yunco

(b) Buy Barzoo's company, Barco

(c) Get new financing

(d) Expand into the United States

There is no sense in getting new financing unless she is either buying Barco or expanding into the States. She can only buy Barco if she gets financing or sells the X division. She can only expand into the States if she has purchased Barco. The X division is necessary to compete in the United States. What are the feasible projects she should consider?

4.6 Nottawasaga Printing has four printing lines, each of which consists of three printing stations, A, B, and C. They have allocated $20 000 for upgrading the printing stations. Station A costs $7000 and takes 10 days to upgrade. Station B costs $5000 and takes 5 days, and station C costs $3000 and takes 3 days. Due to the limited number of technicians, Nottawasaga can only upgrade one printing station at a time. That is, if they decide to upgrade two Bs, the total downtime will be 10 days. During the upgrading period, the downtime should not exceed 14 days in total. Also, at least two printing lines must be available at all times to satisfy the current customer demand. The entire line will not be available if any of the printing stations is turned off for upgrading. Nottawasaga Printing wants to know which line and which printing station to upgrade. Determine the feasible mutually exclusive combinations of lines and stations for Nottawasaga Printing.

4.7 Margaret has a project with a $28 000 first cost that returns $5000 per year over its 10-year life. It has a salvage value of $3000 at the end of 10 years. If the MARR is 15%,

(a) What is the present worth of this project?

(b) What is the annual worth of this project?

(c) What is the future worth of this project after 10 years?

(d) What is the payback period of this project?

(e) What is the discounted payback period for this project?

4.8 Appledale Dairy is considering upgrading an old ice-cream maker. Upgrading is available at two levels: moderate and extensive. Moderate upgrading costs $6500 now and yields annual savings of $3300 in the first year, $3000 in the second year, $2700 in the third year, and so on. Extensive upgrading costs $10 550 and saves $7600 in the first year. The savings then decrease by 20% each year thereafter. If the upgraded ice-cream maker will last for seven years, which upgrading option is better? Use a present worth comparison. Appledale's MARR is 8%.

4.9 Kiwidale Dairy is considering purchasing a new ice-cream maker. Two models, Smoothie and Creamy, are available and their information is given below.

(a) What is Kiwidale's MARR that makes the two alternatives equivalent? Use a present worth comparison.

	Smoothie	Creamy
First cost	$15 000	$36 000
Service life	12 years	12 years
Annual profit	$4200	$10 800
Annual operating cost	$1200	$3520
Salvage value	$2250	$5000

(b) It turned out that the service life of Smoothie was 14 years. Which alternative is better on the basis of the MARR computed in (a)? Assume that each alternative can be repeated indefinitely.

4.10 Nabil is considering buying a house while he is at university. The house costs $100 000 today. Renting out part of the house and living in the rest over his five years at school will net, after expenses, $1000 per month. He estimates that he will sell the house after five years for $105 000. If Nabil's MARR is 18%, compounded monthly, should he buy the house?

4.11 A software genius is selling the rights to a new video game he has developed. Two companies have offered him contracts. The first contract offers $10 000 at the end of each year for the next five years, and then $20 000 per year for the following 10 years. The second offers ten payments, starting with $10 000 at the end of the first year, $13 000 at the end of the second, and so forth, increasing by $3000 each year (i.e., the tenth payment will be $10 000 + 9 × $3000). Assume the genius uses a MARR of 9%. Which contract should the genius choose? Use a present worth comparison.

4.12 Sam is considering buying a new lawnmower. He has a choice between a "Lawn Guy" mower and a Bargain Joe's "Clip Job" mower. Sam has a MARR of 5%. The salvage value of each mower at the end of its service life is zero.

(a) Determine which alternative is preferable. Use a present worth comparison and the least common multiple of the service lives.

(b) For a four-year study period, what salvage value for the Lawn Guy mower would result in its being the preferred choice? What salvage value for the Lawn Guy would result in the Clip Job being the preferred choice?

	Lawn Guy	Clip Job
First cost	$350	$120
Life	10 years	4 years
Annual gas	$60	$40
Annual maintenance	$30	$60

4.13 Water supply for an irrigation system can be obtained from a stream in some nearby mountains. Two alternatives are being considered, both of which have essentially infinite lives, provided proper maintenance is performed. The first is a concrete reservoir with a steel pipe system and the second is an earthen dam with a wooden aqueduct. Below are the costs associated with each.

Compare the present worths of the two alternatives, using an interest rate of 8%. Which alternative should be chosen?

	Concrete Reservoir	Earthen Dam
First cost	$500 000	$200 000
Annual maintenance costs	2 000	12 000
Replacing the wood portion of the aqueduct each 15 years	N/A	100 000

4.14 CB Electronix needs to expand its capacity. It has two feasible alternatives under consideration. Both alternatives will have essentially infinite lives.

Alternative 1: Construct a new building of 20 000 square metres now. The first cost will be $2 000 000. Annual maintenance costs will be $10 000. In addition, the building will need to be painted every 15 years (starting in 15 years) at a cost of $15 000.

Alternative 2: Construct a new building of 12 500 square metres now and an additional 7500 square metres in 10 years. The first cost of the 12 500-square-metre building will be $1 250 000. The annual maintenance costs will be $5000 for the first 10 years (i.e., until the addition is built). The 7500-square-metre addition will have a first cost of $1 000 000. Annual maintenance costs of the renovated building (the original building and the addition) will be $11 000. The renovated building will cost $15 000 to repaint every 15 years (starting 15 years after the addition is done).

Carry out an annual worth comparison of the two alternatives. Which is preferred if the MARR is 15%?

4.15 Katie's project has a five-year term, a first cost, no salvage value, and annual savings of $20 000 per year. After doing present worth and annual worth calculations with a 15% interest rate, Katie notices that the calculated annual worth for the project is exactly three times the present worth. What is the project's present worth and annual worth? Should Katie undertake the project?

4.16 Newmarket Supermarket wants to replace its cash registers and is currently evaluating two models that seem reasonable. The information on the two alternatives, CR1000 and CRX, is shown in the table.

	CR1000	CRX
First cost	$680	$1100
Annual savings	$245	$440
Annual maintenance cost	$35 in year 1, increasing by $10 each year thereafter	$60
Service life	4 years	6 years
Scrap value	$100	$250

(a) If Newmarket Supermarket's MARR is 10%, which type of cash register should they choose? Use the present worth method.

(b) For the less preferred type of cash register found in (a), what scrap value would make it the preferred choice?

4.17 Midland Metalworking, Inc., is examining a 750-tonne hydraulic press and a 600-tonne molding press for purchase. Midland has only enough budget for one of them. If Midland's MARR is 12% and the relevant information is as given below, which press should they purchase? Use an annual worth comparison.

	Hydraulic Press	Molding Press
Initial cost	$275 000	$185 000
Annual savings	$33 000	$24 500
Annual maintenance cost	$2000, increasing by 15% each year thereafter	$1000, increasing by $350 each year thereafter
Life	15 years	10 years
Salvage value	$19 250	$14 800

4.18 Westmount Waxworks is considering buying a new wax melter for their line of replicas of statues of government leaders. They have two choices of supplier, Finedetail and Simplicity. Their proposals are as follows:

	Finedetail	Simplicity
Expected life	7 years	10 years
First cost	$200 000	$350 000
Maintenance	$10 000/year + $0.05/unit	$20 000/year + $0.01/unit
Labour	$1.25/unit	$0.50/unit
Other costs	$6500/year + $0.95/unit	$15 500/year + $0.55/unit
Salvage value	$5000	$20 000

Management thinks they will sell about 30 000 replicas per year if there is stability in world governments. If the world becomes very unsettled so that there are frequent overturns of governments, sales may be as high as 200 000 units a year. Westmount Waxworks uses a MARR of 15% for equipment projects.

(a) Who is the preferred supplier if sales are 30 000 units per year? Use an annual worth comparison.

(b) Who is the preferred supplier if sales are 200 000 units per year? Use an annual worth comparison.

(c) How sensitive is the choice of supplier to sales level? Experiment with sales levels between 30 000 and 200 000 units per year. At what sales level will the costs of the two melters be equal?

4.19 The City of Hanover is installing a new swimming pool in the downtown recreation centre. Two designs are under consideration, both of which are to be permanent (i.e., lasting forever). The first design is for a reinforced concrete pool which has a first cost of $1 500 000. Every 10 years the inner surface of the pool would have to be refinished and painted at a cost of $200 000.

The second design consists of a metal frame and a plastic liner, which would have an initial cost of $500 000. For this alternative, the plastic liner must be replaced every 5 years at a cost of $100 000, and every 15 years the metal frame would need replacement at a cost of $150 000. Extra insurance of $5000 per year is required for the plastic liner (to cover repair costs if the liner leaks). The city's cost of long-term funds is 5%.

Determine which swimming pool design has the lower present cost.

4.20 Sam is buying a refrigerator. He has two choices. A used one, at $475, should last him about three years. A new one, at $1250, would likely last eight years. Both have a scrap value of $0. The interest rate is 8%.

(a) Which refrigerator has a lower cost? (Use a present worth analysis with repeated lives. Assume operating costs are the same.)

(b) If Sam knew that he could resell the new refrigerator after three years for $1000, would this change the answer in part (a)? (Use a present worth analysis with a three-year study period. Assume operating costs are the same.)

4.21 Val is considering purchasing a new video plasma display panel to use with her notebook computer. One model, the XJ3, costs $4500 new, while another, the Y19, sells for $3200. Val figures that the XJ3 will last about three years, at which point it could be sold for $1000, while the Y19 will last for only two years and will also sell for $1000. Both panels give similar service, except that the Y19 is not suitable for client presentations. If she buys the Y19, about four times a year she will have to rent one similar to the XJ3, at a total year-end cost of about $300. Using present worth and the least common multiple of the service lives, determine which display panel Val should buy. Val's MARR is 10%.

4.22 For Problem 4.21, Val has determined that the salvage value of the XJ3 after two years of service is $1900. Which display panel is the better choice, on the basis of present worth with a two-year study period?

4.23 Tom is considering buying a $24 000 car. After five years, he will be able to sell the car for $8000. Gas costs will be $2000 per year, insurance $600 per year, and parking $600 per year. Maintenance costs for the first year are $1000, rising by $400 per year thereafter.

The alternative is for Tom to take taxis everywhere. This will cost an estimated $6000 per year. Tom will rent a car each year at a total cost (to year-end) of $600 for the family vacation, if he has no car. If Tom values money at 11% annual interest, should he buy the car? Use an annual worth comparison method.

4.24 A new gizmo costs $10 000. Maintenance costs $2000 per year, and labour savings are $6567 per year. What is its payback period?

4.25 Building a bridge will cost $65 million. A round-trip toll of $12 will be charged to all vehicles. Traffic projections are estimated to be 5000 per day. The operating and maintenance costs will be 20% of the toll revenue. Find the payback period (in years) for this project.

4.26 A new packaging machine will save the Greene Cheese Company $3000 per year in reduced spoilage, $2500 per year in labour, and $1000 per year in packaging material. The new machine will have additional expenses of $700 per year in maintenance and $200 per year in energy. If it costs $20 000 to purchase, what is its payback period? Assume that the savings are earned throughout the year, not just at year-end.

4.27 Diana usually uses a three-year payback period to determine if a project is acceptable. A recent project with uniform yearly savings over a five-year life had a payback period of almost exactly three years, so Diana decided to find the project's present worth to help determine if the project was truly justifiable. However, that calculation didn't help either since the present worth was exactly 0. What interest rate was Diana using to calculate the present worth? The project has no salvage value at the end of its five-year life.

4.28 The Biltmore Garage has lights in places that are difficult to reach. Management estimates that it costs about $2 to change a bulb. Standard 100-watt bulbs with an expected life of 1000 hours are now used. Standard bulbs cost $1. A long-life bulb that requires 90 watts for the same effective level of light is available. Long-life bulbs cost $3. The bulbs that are difficult to reach are in use for about 500 hours a month. Electricity costs $0.08/kilowatt-hour payable at the end of each month. Biltmore uses a 12% MARR (1% per month) for projects involving supplies.

(a) What minimum life for the long-life bulb would make its cost lower?

(b) If the cost of changing bulbs is ignored, what is the minimum life for the long-life bulb for them to have a lower cost?

(c) If the solutions are obtained by linear interpolation of the capital recovery factor, will the approximations understate or overstate the required life?

4.29 A chemical recovery system costs $30 000 and saves $5280 each year of its seven-year life. The salvage value is estimated at $7500. The MARR is 9%. What is the net annual benefit or cost of purchasing the chemical recovery system? Use the capital recovery formula.

4.30 Savings of $5600 per year can be achieved through either a $14 000 machine (A) with a seven-year service life and a $2000 salvage value, or a $25 000 machine (B) with a ten-year service life and a $10 000 salvage value. If the MARR is 9%, which machine is a better choice, and for what annual benefit or cost? Use annual worth and the capital recovery formula.

4.31 Ridgley Custom Metal Products (RCMP) must purchase a new tube bender. RCMP's MARR is 11%. They are considering two models:

Model	First Cost	Economic Life	Yearly Net Savings	Salvage Value
T	$100 000	5 years	$50 000	$20 000
A	150 000	5 years	60 000	30 000

(a) Using the *present worth* method, which tube bender should they buy?

(b) RCMP has discovered a third alternative, which has been added to the table below. Now which tube bender should they buy?

Model	First Cost	Economic Life	Yearly Net Savings	Salvage Value
T	$100 000	5 years	$50 000	$ 20 000
A	150 000	5 years	60 000	30 000
X	200 000	3 years	75 000	100 000

4.32 RCMP (see Problem 4.31 (b)) can forecast demand for its products for only three years in advance. The salvage value after three years is $40 000 for model T and $80 000 for model A. Using the study period method, which of the three alternatives is best?

4.33 Using the annual worth method, which of the three tube benders should RCMP buy? The MARR is 11%. Use the data from Problem 4.31 (b).

4.34 What is the payback period for each of the three alternatives from the RCMP problem? Use the data from Problem 4.31 (b).

4.35 Data for two independent investment opportunities are shown below.

	Machine A	Machine B
Initial cost	$15 000	$20 000
Revenues (annual)	$ 9 000	$11 000
Costs (annual)	$ 6 000	$ 8 000
Scrap value	$ 1 000	$ 2 000
Service life	5 years	10 years

(a) For a MARR of 8%, should either, both, or neither machine be purchased? Use the annual worth method.

(b) For a MARR of 8%, should either, both, or neither machine be purchased? Use the present worth method.

(c) What are the payback periods for these machines? Should either, both, or neither machine be purchased, based on the payback periods? The required payback period for investments of this type is three years.

4.36 Xaviera is comparing two mutually exclusive projects, A and B, that have the same initial investment and the same present worth over their service lives. Wolfgang points out that, using the annual worth method, A is clearly better than B. What can be said about the service lives for the two projects?

4.37 Xaviera noticed that two mutually exclusive projects, A and B, have the same payback period and the same economic life, but A has a larger present worth than B does. What can be said about the size of the annual savings for the two projects?

4.38 Two plans have been proposed for accumulating money for capital projects at Thunder Bay Lighting. One idea is to put aside \$10 000 per year, independent of growth. The second is to start with a smaller amount, \$8000 per year, but to increase this in proportion to the expected company growth. The money will accumulate interest at 10%, and the company is expected to grow about 5% per year. Which plan will accumulate more money in 10 years?

4.39 Crystal City Environmental Services is evaluating two alternative methods of disposing of municipal waste. The first involves developing a landfill site near the city. Costs of the site include \$1 000 000 startup costs, \$100 000 closedown costs 30 years from now, and operating costs of \$20 000 per year. Starting in 10 years, it is expected that there will be revenues from user fees of \$30 000 per year. The alternative is to ship the waste out of the province. A United States firm will agree to a long-term contract to dispose of the waste for \$130 000 per year. Using the *annual worth* method, which alternative is economically preferred for a MARR of 11%? Would this likely be the actual preferred choice?

4.40 Peterborough Auto Parts is considering investing in a new forming line for their grille assemblies. For a five-year study period, the cash flows for two separate designs are shown below. Create a spreadsheet which will calculate the present worths for each project for a variable MARR. Through trial and error, establish the MARR at which the present worths of the two projects are exactly the same.

Cash Flows for Grille Assembly Project						
	Automated Line			Manual Line		
Year	Disbursements	Receipts	Net Cash Flow	Disbursements	Receipts	Net Cash Flow
0	\$1 500 000	\$ 0	–\$1 500 000	\$1 000 000	\$ 0	–\$1 000 000
1	50 000	300 000	250 000	20 000	200 000	180 000
2	60 000	300 000	240 000	25 000	200 000	175 000
3	70 000	300 000	230 000	30 000	200 000	170 000
4	80 000	300 000	220 000	35 000	200 000	165 000
5	90 000	800 000	710 000	40 000	200 000	160 000

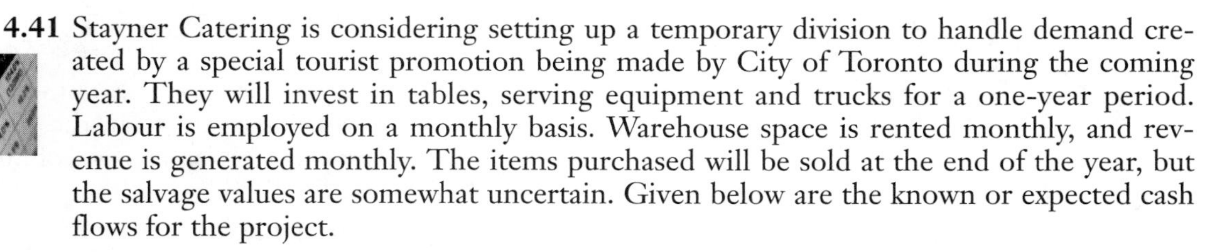

4.41 Stayner Catering is considering setting up a temporary division to handle demand created by a special tourist promotion being made by City of Toronto during the coming year. They will invest in tables, serving equipment and trucks for a one-year period. Labour is employed on a monthly basis. Warehouse space is rented monthly, and revenue is generated monthly. The items purchased will be sold at the end of the year, but the salvage values are somewhat uncertain. Given below are the known or expected cash flows for the project.

Month	Purchase	Labour Expenses	Warehouse Expenses	Revenue
January (beginning)	$200 000			
January (end)		$ 2 000	$3 000	$ 2 000
February		2 000	3 000	2 000
March		2 000	3 000	2 000
April		2 000	3 000	2 000
May		4 000	3 000	10 000
June		10 000	6 000	40 000
July		10 000	6 000	110 000
August		10 000	6 000	60 000
September		4 000	3 000	30 000
October		2 000	3 000	10 000
November		2 000	3 000	5 000
December	Salvage?	2 000	3 000	2 000

For an interest rate of 12% compounded monthly, create a spreadsheet that calculates and graphs the present worth of the project for a range of salvage values of the purchased items from 0% to 100% of the purchase price. Should Stayner Catering go ahead with this project?

4.42 Peterborough Auto Parts has two options for increasing efficiency. They can expand the current building or keep the same building but remodel the inside layout. For a five-year study period, the cash flows for the two options are shown below. Construct a spreadsheet which will calculate the present worth for each option for a variable MARR. By trial and error, determine the MARR at which the present worths of the two options are equivalent.

	Expansion Option			Remodelling Option		
Year	Disbursements	Receipts	Net Cash Flow	Disbursements	Receipts	Net Cash Flow
0	$850 000	$ 0	– $850 000	$230 000	$ 0	– $230 000
1	25 000	200 000	175 000	9000	80 000	71 000
2	30 000	225 000	195 000	11 700	80 000	68 300
3	35 000	250 000	215 000	15 210	80 000	64 790
4	40 000	275 000	235 000	19 773	80 000	60 227
5	45 000	300 000	255 000	25 705	80 000	54 295

4.43 Derek has two choices for a heat-loss prevention system for the shipping doors at Kirkland Manufacturing. He can isolate the shipping department from the rest of the plant, or he can curtain off each shipping door separately. Isolation consists of building a permanent wall around the shipping area. It will cost $60 000 and will save $10 000 in heating costs per year. Plastic curtains around each shipping door will have a total cost of about \$5000, but will have to be replaced about once every two years. Savings in heating costs for installing the curtains will be about \$3000 per year. Use the payback period method to determine which alternative is better. Comment on the use of the payback period for making this decision.

4.44 Assuming that the wall built to isolate the shipping department in Problem 4.43 will last forever, and that the curtains have zero salvage value, compare the annual worths of the two alternatives. The MARR for Kirkland Manufacturing is 11%. Which alternative is better?

4.45 Crystal City Environmental Services is considering investing in a new water treatment system. On the basis of the information given below for two alternatives, a fully automated and a partially automated system, construct a spreadsheet for computing the annual worths for each alternative with a variable MARR. Through trial and error, determine the MARR at which the annual worths of the two alternatives are equivalent.

	Fully Automated System			Partially Automated System		
Year	Disburse-ments	Receipts	Net Cash Flow	Disburse-ments	Receipts	Net Cash Flow
0	\$1 000 000	\$ 0	– \$1 000 000	\$650 000	\$ 0	– \$650 000
1	30 000	300 000	270 000	30 000	220 000	190 000
2	30 000	300 000	270 000	30 000	220 000	190 000
3	80 000	300 000	220 000	35 000	220 000	185 000
4	30 000	300 000	270 000	35 000	220 000	185 000
5	30 000	300 000	270 000	40 000	220 000	180 000
6	80 000	300 000	220 000	40 000	220 000	180 000
7	30 000	300 000	270 000	45 000	220 000	175 000
8	30 000	300 000	270 000	45 000	220 000	175 000
9	80 000	300 000	220 000	50 000	220 000	170 000
10	30 000	300 000	270 000	50 000	220 000	170 000

Rockwell International of Canada

The Light Vehicle Division of Rockwell International of Canada makes seat-slide assemblies for the automotive industry. They have two major classifications for investment opportunities: developing new products to be manufactured and sold and developing new machines to improve production. The overall approach to assessing whether an investment should be made depends on the nature of the project.

In evaluating a new product, they consider the following:

1. *Marketing strategy:* Does it fit the business plan for the company?
2. *Work force:* How will it affect human resources?
3. *Margins:* The product should generate appropriate profits.
4. *Cash flow:* Positive cash flow is expected within two years.

continued

In evaluating a new machine, they consider the following:

1. *Cash flow:* Positive cash flow is expected within a limited time period.
2. *Quality issues:* For issues of quality, justification is based on cost avoidance rather than positive cash flow.
3. *Cost avoidance:* Savings should pay back an investment within one year.

Discussion

All companies consider more than just the economics of a decision. Most take into account the other issues—often called *intangibles*—by using managerial judgement in an informal process. Others, like Rockwell International of Canada, explicitly consider a selection of intangible issues.

The trend today is to carefully consider several intangible issues, either implicitly or explicitly. Human resource issues are particularly important since employee enthusiasm and commitment have significant repercussions. Environmental impacts of a decision can affect the image of the company. Health and safety is another intangible with significant effects.

However, the economics of the decision is usually (but not always) the single most important factor in a decision. Also, economics is the factor that is usually the easiest to measure.

Questions

1. Why do you think Rockwell International of Canada has different issues to consider depending on whether an investment was a new product or a new machine?
2. For each of the issues mentioned, describe how it would be measured. How would you determine if it was worth investing in a new product or new machine with respect to that issue?
3. There are two kinds of errors that can be made. The first is that an investment is made when it should not be, and the second is that an investment is not made when it should be. Describe examples of both kinds of errors for both products and machines (four examples in total) if the issues listed for Rockwell International of Canada are strictly followed. What are some sensible ways to prevent such errors?

Appendix 4A The MARR and the Cost of Capital

For a business to survive, it must be able to earn a high enough return to induce investors to put money into the company. The minimum rate of return required to get investors to invest in a business is that business's **cost of capital**. A company's cost of capital is also its minimum acceptable rate of return for projects, its MARR. This appendix reviews how the cost of capital is determined. We first look at the relation between risk and the cost of capital. Then, we discuss sources of capital for large businesses and small businesses.

4A.1 Risk and the Cost of Capital

There are two main forms of investment in a company, *debt* and *equity*. Investors in a company's debt are lending money to the company. The loans are contracts that give lenders rights to repayment of their loans, and to interest at predetermined interest rates. Investors in a company's equity are the owners of the company. They hold rights to the residual after all contractual payments, including those to lenders, are made.

Investing in equity is more risky than investing in debt. Equity owners are paid only if the company first meets its contractual obligations to lenders. This higher risk means that equity owners require an expectation of a greater return on average than the interest rate paid to debt holders. Consider a simple case in which a company has three possible performance levels—weak results, normal results, and strong results. Investors do not know which level will actually occur. Each level is equally probable. To keep the example simple, we assume that all after-tax income is paid to equity holders as dividends so that there is no growth. The data are shown in Table 4A.1.

Table 4A.1 Cost of Capital Example

	Possible Performance Levels		
	Weak Results	Normal Results	Strong Results
Net operating income ($/year)[1]	40 000	100 000	160 000
Interest payment ($/year)	10 000	10 000	10 000
Net income before tax ($/year)	30 000	90 000	150 000
Tax at 40% ($/year)	12 000	36 000	60 000
After-tax income = Dividends ($/year)	18 000	54 000	90 000
Debt ($)	100 000	100 000	100 000
Value of shares ($)	327 273	327 273	327 273

[1]Net operating income per year is revenue per year minus cost per year.

We see that, no matter what happens, lenders will get a return of 10%:

$$0.1 = \frac{10\,000}{100\,000}$$

Owners get one of three possible returns:

$$5.5\% \left(0.055 = \frac{18\,000}{327\,273}\right),$$

$$16.5\% \left(0.165 = \frac{54\,000}{327\,273}\right), \text{ or}$$

$$27.5\% \left(0.275 = \frac{90\,000}{327\,273}\right)$$

These three possibilities average out to 16.5%. If things are good, owners do better than lenders. If things are bad, owners do worse. But their average return is greater than returns to lenders.

The lower rate of return to lenders means that companies would like to get their capital with debt. However, reliance on debt is limited for two reasons.

1. If a company increases the share of capital from debt, it increases the chance that it will not be able to meet the contractual obligations to lenders. This means the company will be bankrupt. Bankruptcy may lead to reorganizing the company or possibly closing the company. In either case, bankruptcy costs may be high.
2. Lenders are aware of the dangers of high reliance on debt and will, therefore, limit the amount they lend to the company.

4A.2 Company Size and Sources of Capital

Large, well-known companies, like Bell Canada and Molson Breweries, can secure capital both by borrowing and by selling ownership shares at relative ease because there will be ready markets for their shares as well as any debt instruments, like bonds, they may issue. These companies will seek ratios of debt to equity that balance the marginal advantages and disadvantages of debt financing. Hence, the cost of capital for large, well-known companies is a weighted average of the costs of borrowing and of selling shares, which is referred to as the **weighted average cost of capital**. The weights are the fractions of total capital that come from the different sources. If market conditions do not change, a large company that seeks to raise a moderate amount of additional capital can do so at a stable cost of capital. This cost of capital is the company's MARR.

We can compute the after-tax cost of capital for the example shown in Table 4A.1 as follows.

Weighted average cost of capital

$$= 0.1\left(\frac{100\ 000}{427\ 273}\right) + 0.165\left(\frac{327\ 273}{427\ 273}\right) = 0.150$$

This company has a cost of capital of about 15%.

For smaller, less well-known companies, raising capital may be more difficult. Most investors in large companies are not willing to invest in unknown small companies. At startup, a small company may rely entirely on the capital of the owners and their friends and relatives. Here the cost of capital is the opportunity cost for the investors.

If a new company seeks to grow more rapidly than the owners' investment plus cash flow permits, the next source of capital with the lowest cost is usually a bank loan. Bank loans are limited because banks are usually not willing to lend more than some fraction of the amount an owner(s) puts into a business.

If bank loans do not add up to sufficient funds, then the company usually has two options. One option is the sale of financial securities such as stocks and bonds through stock exchanges that specialize in small, speculative companies, like the TSX Venture Exchange. Another option is venture capitalists. Venture capitalists are investors who specialize in investing in new companies. The cost of evaluating new companies is usually high and so is the risk of investing in them. Together, these factors usually lead venture capitalists to want to put enough money into a small company so that they will have enough control over the company.

In general, new equity investment is very expensive for small companies. Studies have shown that venture capitalists typically require the expectation of at least a 35% rate of return after tax. Raising funds on a stock exchange is usually even more expensive than getting funding from a venture capitalist.

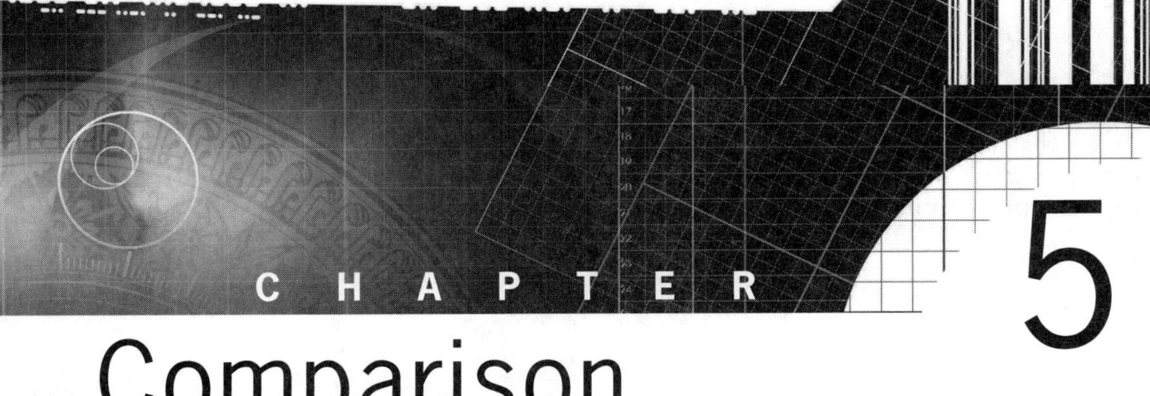

5 Comparison Methods Part 2

Engineering Economics in Action, Part 5A: What's Best? Revisited

Clem had said, "I have to make a recommendation to Ed Burns and Anna Kulkowski for their Wednesday meeting on this forging hammer in the South Shop. I'd like you to handle it." Dave Sullivan, who had started the project, had gone to Florida to see his sick father-in-law. Naomi welcomed the opportunity, but she still had to figure out exactly what to recommend.

Naomi looked carefully at the list of seven mutually exclusive alternatives for replacing or refurbishing the machine. Present worth could tell her which of the seven was "best," but present worth was just one of several comparison methods. Which method should she use?

Dave would help more, if she asked him. In fact, he could no doubt tell her exactly what to do, if she wanted. But this one she knew she could handle, and it was a matter of pride to do it herself. Opening her engineering economics textbook, she read on.

5.1 Introduction

In Chapter 4, we showed how to structure projects so that they were either independent or mutually exclusive. The present worth, annual worth, and payback period methods for evaluating projects were also introduced. This chapter continues on the theme of comparing projects by presenting a commonly used but somewhat more complicated comparison method called the *internal rate of return*, or IRR.

Although the IRR method is widely used, all of the comparison methods have value in particular circumstances. Selecting which method to use is also covered in this chapter. It is also shown that the present worth, annual worth, and IRR methods all result in the same recommendations for the same problem. We close this chapter with a chart summarizing the strengths and weaknesses of the four comparison methods presented in Chapters 4 and 5.

5.2 The Internal Rate of Return

Investments are undertaken with the expectation of a return in the form of future earnings. One way to measure the return from an investment is as a rate of return per dollar invested, or in other words as an interest rate. The rate of return usually calculated for a project is known as the *internal rate of return (IRR)*. The adjective "internal" refers to the fact that the internal rate of return depends only on the cash flows due to the investment. The internal rate of return is that interest rate at which a project just breaks even. The meaning of the IRR is most easily seen with a simple example.

Example 5.1

Suppose $100 is invested today in a project that returns $110 in one year. We can calculate the IRR by finding the interest rate at which $100 now is equivalent to $110 at the end of one year:

$$P = F(P/F, i^*, 1)$$

$$\$100 = \$110(P/F, i^*, 1)$$

$$\$100 = \frac{\$110}{1 + i^*}$$

where i^* is the internal rate of return.

Solving this equation gives a rate of return of 10%. In a simple example like this, the process of finding an internal rate of return is finding the interest rate that makes the present worth of benefits equal to the first cost. This interest rate is the IRR.■

Of course, cash flows associated with a project will usually be more complicated than in the example above. A more formal definition of the IRR is stated as follows. The **internal rate of return (IRR)** on an investment is the interest rate, i^*, such that, when all cash flows associated with the project are discounted at i^*, the present worth of the cash inflows equals the present worth of the cash outflows. That is, the project just breaks even. An equation that expresses this is

$$\sum_{t=0}^{T} \frac{(R_t - D_t)}{(1 + i^*)^t} = 0 \tag{5.1}$$

where

R_t = the cash inflow (receipts) in period t
D_t = the cash outflow (disbursements) in period t
T = the number of time periods
i^* = the internal rate of return

Since Equation (5.1) can also be expressed as

$$\sum_{t=0}^{T} R_t(1 + i^*)^{-t} = \sum_{t=0}^{T} D_t(1 + i^*)^{-t}$$

it can be seen that, in order to calculate the IRR, one sets the disbursements equal to the receipts and solves for the unknown interest rate. For this to be done, the disbursements and receipts must be comparable, as a present worth, a uniform series, or a future worth. That is, use

PW(disbursements) = PW(receipts) and solve for the unknown i^*,
AW(disbursements) = AW(receipts) and solve for the unknown i^*, or
FW(disbursements) = FW(receipts) and solve for the unknown i^*.

The IRR is usually positive, but can be negative as well. A negative IRR means that the project is losing money rather than earning it.

We usually solve the equations for the IRR by trial and error, as there is no explicit means of solving Equation (5.1) for projects where the number of periods is large. A spreadsheet provides a quick way to perform trial-and-error calculations; most spreadsheet programs also include a built-in IRR function.

Example 5.2

Clem is considering buying a tuxedo. It would cost $500, but would save him $160 per year in rental charges over its five-year life. What is the IRR for this investment?

As illustrated in Figure 5.1, Clem's initial cash outflow for the purchase would be \$500. This is an up-front outlay relative to continuing to rent tuxedos. The investment would create a saving of \$160 per year over the five-year life of the tuxedo. These savings can be viewed as a series of receipts relative to rentals. The IRR of Clem's investment can be found by determining what interest rate makes the present worth of the disbursements equal to the present worth of the cash inflows.

Present worth of disbursements = \$500

Present worth of receipts = $\$160(P/A, i^*, 5)$

Setting the two equal,

$$\$500 = \$160(P/A, i^*, 5)$$

$$(P/A, i^*, 5) = \$500/\$160$$

$$= 3.125\text{i}$$

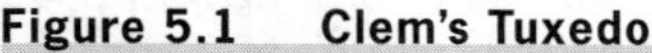

Figure 5.1 Clem's Tuxedo

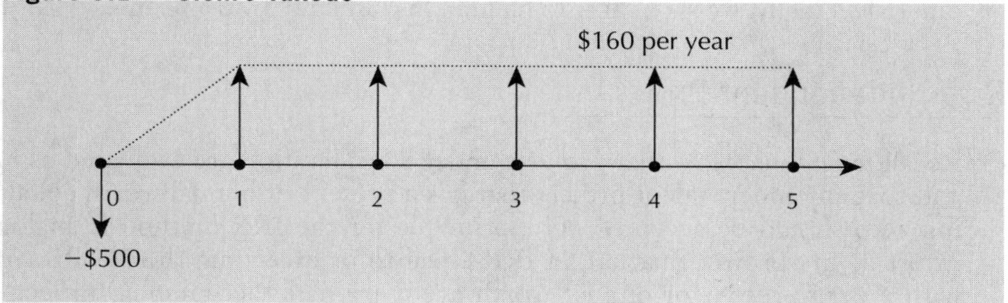

From the interest factor tables, we find that

$$(P/A, 15\%, 5) = 3.3521$$

$$(P/A, 20\%, 5) = 2.9906$$

Interpolating between $(P/A, 15\%, 5)$ and $(P/A, 20\%, 5)$ gives

$$i^* = 15\% + (5\%)[(3.125 - 3.3521)/(2.9906 - 3.3521)]$$

$$= 18.14\%$$

An alternative way to get the IRR for this problem is to convert all cash outflows and inflows to equivalent annuities over the five-year period. This will yield the same result as when present worth was used.

Annuity equivalent to the disbursements = $\$500(A/P, i^*, 5)$

Annuity equivalent to the receipts = \$160

Again, setting the two equal,

$$\$500(A/P, i^*, 5) = \$160$$

$$(A/P, i^*, 5) = \$160/\$500$$

$$= 0.32$$

From the interest factor tables,

$$(A/P, 15\%, 5) = 0.29832$$

$$(A/P, 20\%, 5) = 0.33438$$

An interpolation gives

$$i^* = 15\% + 5\%[(0.32 - 0.29832)/(0.33438 - 0.29832)]$$
$$\cong 18.0\%$$

Note that there is a slight difference in the answers, depending on whether the disbursements and receipts were compared as present worths or as annuities. This difference is due to the small error induced by the linear interpolation.■

5.3 Internal Rate of Return Comparisons

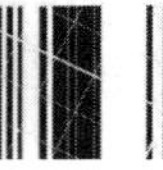

In this section, we show how the internal rate of return can be used to decide whether a project should be accepted. We first show how to use the IRR to evaluate independent projects. Then we show how to use the IRR to decide which of a group of mutually exclusive alternatives to accept. We then show that it is possible for a project to have more than one IRR. Finally, we show how to handle this difficulty by using an *external rate of return*.

5.3.1 IRR for Independent Projects

Recall from Chapter 4 that projects under consideration are evaluated using the MARR, and that any independent project that has a present or annual worth equal to or exceeding zero should be accepted. The principle for the IRR method is analogous. We will invest in any project that has an IRR equal to or exceeding the MARR. Just as projects with a zero present or annual worth are marginally acceptable, projects with IRR = MARR have a marginally acceptable rate of return (by definition of the MARR).

Also analogous to Chapter 4, when we perform a rate of return comparison on several independent projects, the projects must have equal lives. If this is not the case, then the approaches covered in Section 4.4.4 (Comparison of Alternatives with Unequal Lives) must be used.

Example 5.3

The High Society Baked Bean Co. is considering a new canner. The canner costs \$120 000, and will have a scrap value of \$5000 after its 10-year life. Given the expected increases in sales, the total savings due to the new canner compared, with continuing with the current operation, will be \$15 000 the first year, increasing by \$5000 each year thereafter. Total extra costs due to the more complex equipment will be \$10 000 per year. The MARR for High Society is 12%. Should they invest in the new canner?

The cash inflows and outflows for this problem are summarized in Figure 5.2. We need to compute the internal rate of return in order to decide if High Society should buy the canner. There are several ways we can do this. In this problem, equating annual outflows and receipts appears to be the easiest approach, because most of the cash flows are already stated on a yearly basis.

$$\$5000(A/F, i^*, 10) + \$15\,000 + \$5000(A/G, i^*, 10)$$
$$- \$120\,000(A/P, i^*, 10) - \$10\,000 = 0$$

Dividing by 5000,

$$(A/F, i^*, 10) + 1 + (A/G, i^*, 10) - 24(A/P, i^*, 10) = 0$$

Figure 5.2 High Society Baked Bean Canner

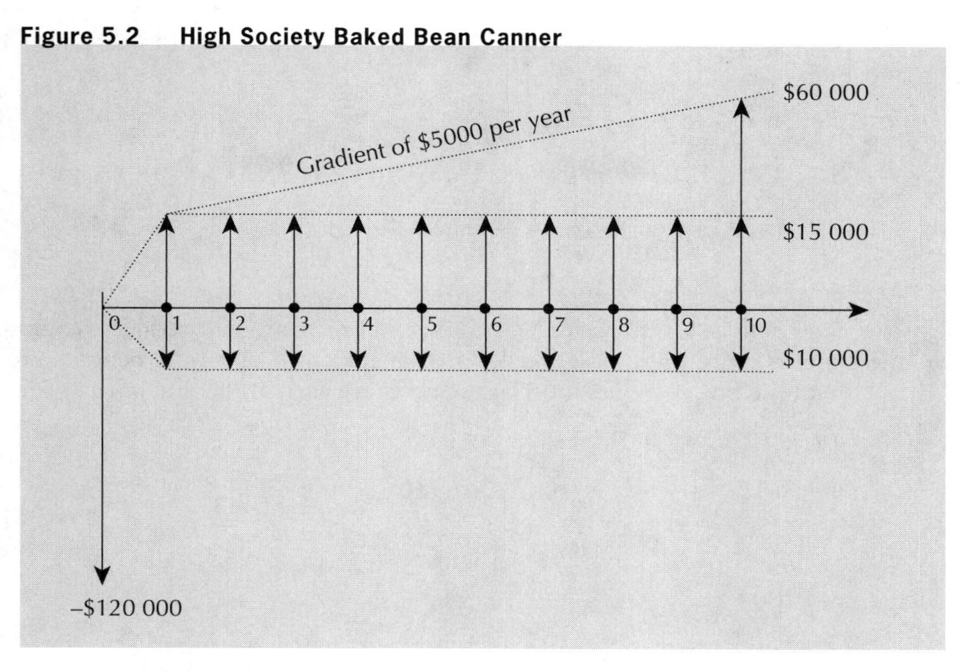

The IRR can be found by trial and error alone, by trial and error and linear interpolation, or by a spreadsheet IRR function. A trial-and-error process is particularly easy using a spreadsheet, so this is often the best approach. A good starting point for the process is at zero interest. A graph (Figure 5.3) derived from the spreadsheet indicates that the IRR is between 13% and 14%. This may be good enough for a decision, since it exceeds the MARR of 12%.

If finer precision is required, there are two ways to proceed. One way is to use a finer grid on the graph, for example, one that covers 13% to 14%. The other way is to interpolate between 13% and 14%. We shall first use the interest factor tables to show

Figure 5.3 Estimating the IRR for Example 5.3

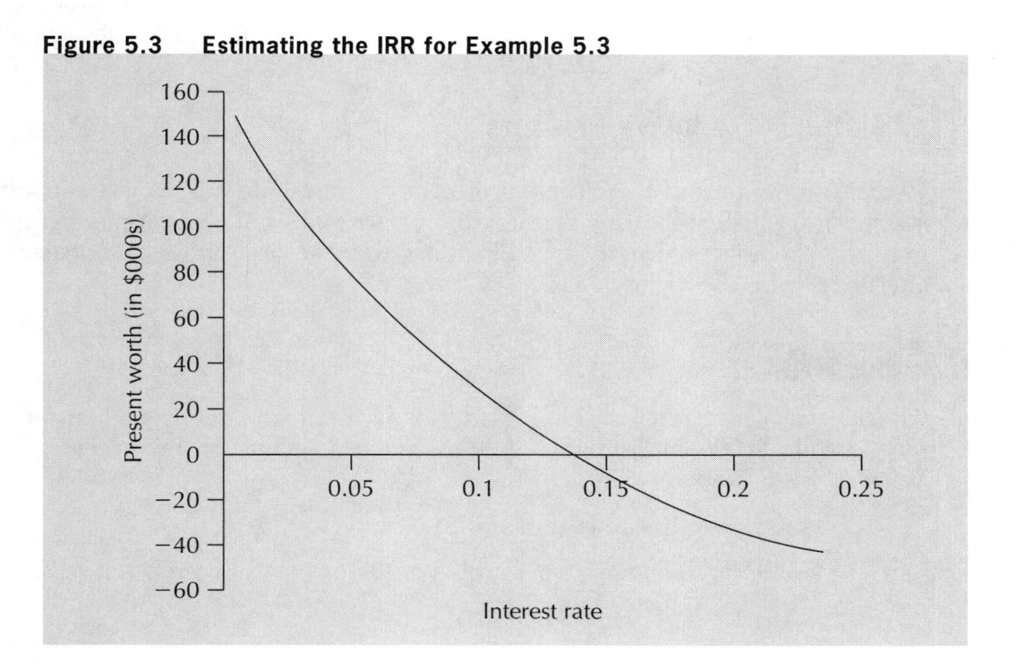

that the IRR is indeed between 13% and 14%. Next we shall interpolate between 13% and 14%.

First, at $i = 13\%$, we have

$$(A/F, 13\%, 10) + 1 + (A/G, 13\%, 10) - 24(A/P, 13\%, 10)$$

$$= 0.05429 + 1 + 3.5161 - 24(0.18429)$$

$$\cong 0.1474$$

The result is a bit too high. A higher interest rate will reduce the annual worth of the benefits more than the annual worth of the costs, since the benefits are spread over the life of the project while most of the costs are early in the life of the project.

At $i = 14\%$, we have

$$(A/F, 14\%, 10) + 1 + (A/G, 14\%, 10) - 24(A/P, 14\%, 10)$$

$$= 0.05171 + 1 + 3.4489 - 24(0.19171)$$

$$\cong -0.1004$$

This confirms that the IRR of the investment is between 13% and 14%. A good approximation to the IRR can be found by linearly interpolating:

$$i^* = 13\% + (0 - 0.1474)/(0.1004 - 0.1474)$$

$$\cong 13.6\%$$

The IRR for the investment is approximately 13.6%. Since this is greater than the MARR of 12%, the company should buy the new canner. Note again that it was not actually necessary to determine where in the range of 13% to 14% the IRR fell. It was enough to demonstrate that it was 12% or more.■

In summary, if there are several independent projects, the IRR for each is calculated separately, and those having an IRR equal to or exceeding the MARR should be chosen.

5.3.2 IRR for Mutually Exclusive Projects

Choice among mutually exclusive projects using the IRR is a bit more involved. Some insight into the complicating factors can be obtained from an example that involves two mutually exclusive alternatives. It illustrates that the best choice is not necessarily the alternative with the highest IRR.

Example 5.4

Consider two investments. The first costs $1 today and returns $2 in one year. The second costs $1000 and returns $1900 in one year. Which is the preferred investment? Your MARR is 70%.

The first project has an IRR of 100%:

$$-\$1 + \$2(P/F, i^*, 1) = 0$$

$$(P/F, i^*, 1) = \$1/\$2 = 0.5$$

$$i^* = 100\%$$

The second has an IRR of 90%:

$$-\$1000 + \$1900(P/F, i^*, 1) = 0$$

$$(P/F, i^*, 1) = \$1000/\$1900 = 0.52631$$

$$i^* = 90\%$$

If these were independent projects, both would be acceptable since their IRRs exceed the MARR. If one of the two projects must be chosen, it might be tempting to choose the first project, the alternative with the larger rate of return. However, this approach is incorrect because it can overlook projects that have a rate of return equal to or greater than the MARR, but don't have the maximum IRR. In the example, the correct approach is to first observe that the least cost investment provides a rate of return that exceeds the MARR. The next step is to find the rate of return on the more expensive investment to see if the *incremental* investment has a rate of return equal to or exceeding the MARR. The incremental investment is the additional $999 that would be invested if the second investment was taken instead of the first:

$$-(\$1000 - \$1) + (\$1900 - \$2)(P/F, i^*, 1) = 0$$

$$(P/F, i^*, 1) = \$999/\$1898 = 0.52634$$

$$i^* = 89.98\%$$

Indeed, the incremental investment has an IRR exceeding 70% and thus the second investment should be chosen.■

The next example illustrates the process again, showing this time how an incremental investment is not justified.

Example 5.5

Monster Meats can buy a new meat slicer system for $50 000. They estimate it will save them $11 000 per year in labour and operating costs. The same system with an automatic loader is $68 000, and will save approximately $14 000 per year. The life of either system is thought to be eight years. Monster Meats has three feasible alternatives:

tAlternative	First Cost	Annual Savings
"Do nothing"	$ 0	$ 0
Meat slicer alone	50 000	11 000
Meat slicer with automatic loader	68 000	14 000

Monster Meats uses a MARR of 12% for this type of project. Which alternative is better?

We first consider the system without the loader. Its IRR is 14.5%, which exceeds the MARR of 12%. This can be seen by solving for i^* in

$$-\$50\ 000 + \$11\ 000(P/A, i^*, 8) = 0$$

$$(P/A, i^*, 8) = \$50\ 000/\$11\ 000$$

$$(P/A, i^*, 8) = 4.545$$

From the interest factor tables, or by trial and error with a spreadsheet,

$$(P/A, 14\%, 8) = 4.6388$$

$$(P/A, 15\%, 8) = 4.4873$$

By interpolation or further trial and error,

$$i^* \cong 14.5\%$$

The slicer alone is thus economically justified and is better than the "do nothing" alternative.

We now consider the system with the slicer and loader. Its IRR is 12.5%, which may be seen by solving for i^* in

$$-\$68\,000 + \$14\,000(P/A, i^*, 8) = 0$$

$$(P/A, i^*, 8) = \$68\,000/\$14\,000$$

$$(P/A, i^*, 8) = 4.857$$

$$(P/A, 12\%, 8) = 4.9676$$

$$(P/A, 13\%, 8) = 4.7987$$

$$i^* \cong 12.5\%$$

The IRR of the meat slicer and automatic loader is about 12.5%, which on the surface appears to meet the 12% MARR requirement. But, on the incremental investment, Monster Meats would be earning only 7%. This may be seen by looking at the IRR on the *extra*, or *incremental*, $18 000 spent on the loader.

$$-(\$68\,000 - \$50\,000) + (\$14\,000 - \$11\,000)(P/A, i^*, 8) = 0$$

$$-\$18\,000 + \$3000(P/A, i^*, 8) = 0$$

$$(P/A, i^*, 8) = \$18\,000/\$3000$$

$$(P/A, i^*, 8) = 6$$

$$i^* \cong 7\%$$

This is less than the MARR; therefore, Monster Meats should not buy the automated loader.

When the IRR was calculated for the system including the loader, the surplus return on investment earned by the slicer alone essentially subsidized the loader. The slicer investment made enough of a return so that, even when it was coupled with the money-losing loader, the whole machine still seemed to be a good buy. In fact, the extra $18 000 would be better spent on some other project at the MARR or higher. The relation between the potential projects is shown in Figure 5.4.■

Figure 5.4 Monster Meats

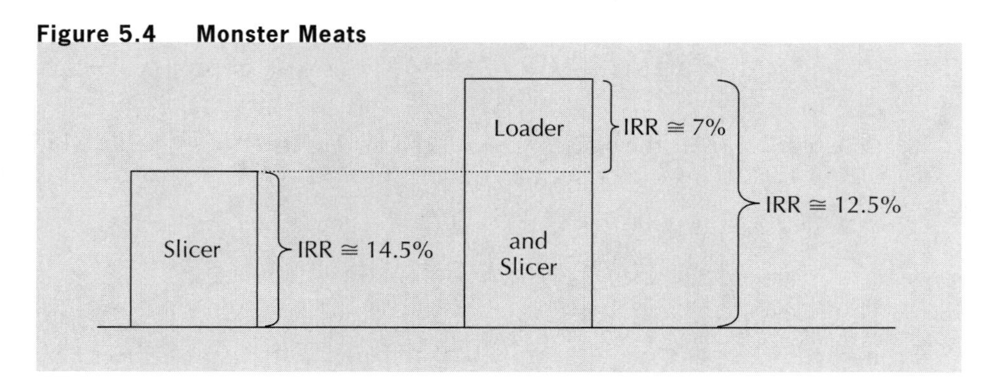

The fundamental principle illustrated by the two examples is that, to use the IRR to compare two or more mutually exclusive alternatives properly, we cannot make the decision on the basis of the IRRs of individual alternatives alone; we must take the IRRs of the *incremental* investments into consideration. In order to properly assess the worthiness of the incremental investments, it is necessary to have a systematic way to conduct pair-wise comparisons of projects. Note that before undertaking a systematic analysis of mutually exclusive alternatives with the IRR method, you should ensure that the alternatives have equal lives. If they do not have equal lives, then the methods of Section 4.4.4 (study period or repeated lives methods) must be applied first to set up comparable cash flows.

The first step in the process of comparing several mutually exclusive alternatives using the IRR is to order the alternatives from the smallest first cost to the largest first cost. Since one alternative must be chosen, accept the alternative with the smallest first cost (which may be the "do nothing" alternative with $0 first cost) as the *current best alternative* regardless of its IRR exceeding the MARR. This means that the current best alternative may have an IRR *less* than the MARR. Even if that's the case, a proper analysis of the IRRs of the incremental investments will lead us to the *correct* best overall alternative. For this reason, we don't have to check the IRR of any of the individual alternatives.

The second step of the analysis consists of looking at the incremental investments of alternatives that have a higher first cost than the current best alternative. Assume that there are n projects and they are ranked from 1 (the current best) to n, in increasing order of first costs. The current best is "challenged" by the project ranked second. One of two things occurs:

1. The incremental investment to implement the challenger does not have an IRR at least equal to the MARR. In this case, the challenger is excluded from further consideration and the current best is challenged by the project ranked third.
2. The incremental investment to implement the challenger has an IRR at least as high as the MARR. In this case, the challenger replaces the current best. It then is challenged by the alternative ranked third.

The process then continues with the next alternative challenging the current best until all alternatives have been compared. The current best alternative remaining at the end of the process is then selected as the best overall alternative. Figure 5.5 on page 137 summarizes the incremental investment analysis for the mutually exclusive projects.

Example 5.6 (Reprise of Example 4.4)

Fly-by-Night Aircraft must purchase a new lathe. It is considering one of four new lathes, each of which has a life of 10 years with no scrap value. Given a MARR of 15%, which alternative should be chosen?

Lathe	1	2	3	4
First cost	$100 000	$150 000	$200 000	$255 000
Annual savings	25 000	34 000	46 000	55 000

The alternatives have already been ordered from lathe 1, which has the smallest first cost, to lathe 4, which has the greatest first cost. Since one lathe must be purchased, accept lathe 1 as the current best alternative. Calculating the IRR for lathe 1, although not necessary, is shown as follows:

$$\$100\,000 = \$25\,000(P/A, i^*, 10)$$

$$(P/A, i^*, 10) = 4$$

An approximate IRR is obtained by trial and error with a spreadsheet.

$$i^* \cong 21.4\%$$

The current best alternative is then challenged by the first challenger, lathe 2, which has the next-highest first cost. The IRR of the incremental investment from lathe 1 to lathe 2 is calculated as follows:

$$(\$150\,000 - \$100\,000) - (\$34\,000 - \$25\,000)(P/A, i^*, 10) = 0$$

or

$$[\$150\,000 - \$34\,000(P/A, i^*, 10)] - [\$100\,000 - \$25\,000(P/A, i^*, 10)] = 0$$

$$(P/A, i^*, 10) = \$50\,000/\$9000 = 5.556$$

An approximate IRR is obtained by trial and error.

$$i^* \cong 12.4\%$$

Since the IRR of the incremental investment falls below the MARR, lathe 2 fails the challenge to become the current best alternative. The reader can verify that lathe 2 alone has an IRR of approximately 18.7%. Even so, lathe 2 is not considered a viable alternative. In other words, the incremental investment of $50 000 could be put to better use elsewhere. Lathe 1 remains the current best and the next challenger is lathe 3.

As before, the incremental IRR is the interest rate at which the present worth of lathe 3 less the present worth of lathe 1 is 0:

$$[\$200\,000 - \$46\,000(P/A, i^*, 10)] - [\$100\,000 - \$25\,000(P/A, i^*, 10)] = 0$$

$$(P/A, i^*, 10) = \$100\,000/\$21\,000 = 4.762$$

An approximate IRR is obtained by trial and error.

$$i^* \cong 16.4\%$$

The IRR on the incremental investment exceeds the MARR, and therefore lathe 3 is preferred to lathe 1. Lathe 3 now becomes the current best. The new challenger is lathe 4. The IRR on the incremental investment is

$$[\$255\,000 - \$55\,000(P/A, i^*, 10)] - [\$200\,000 - \$46\,000(P/A, i^*, 10)] = 0$$

$$(P/A, i^*, 10) = \$55\,000/\$9000 = 6.11$$

$$i^* \cong 10.1\%$$

The additional investment from lathe 3 to lathe 4 is not justified. The reader can verify that the IRR of lathe 4 alone is about 17%. Once again, we have a challenger with an IRR greater than the MARR, but it fails as a challenger because the incremental investment from the current best does not have an IRR at least equal to the MARR. The current best remains lathe 3. There are no more challengers, and so the best overall alternative is lathe 3.■

In the next section, the issue of multiple IRRs is discussed, and methods for identifying and eliminating them are given. Note that the process described in Figure 5.5 requires that a single IRR (or ERR, as discussed later) be determined for each incremental investment. If there are multiple IRRs, they must be dealt with for *each* increment of investment.

Figure 5.5 Flowchart for Comparing Mutually Exclusive Alternatives

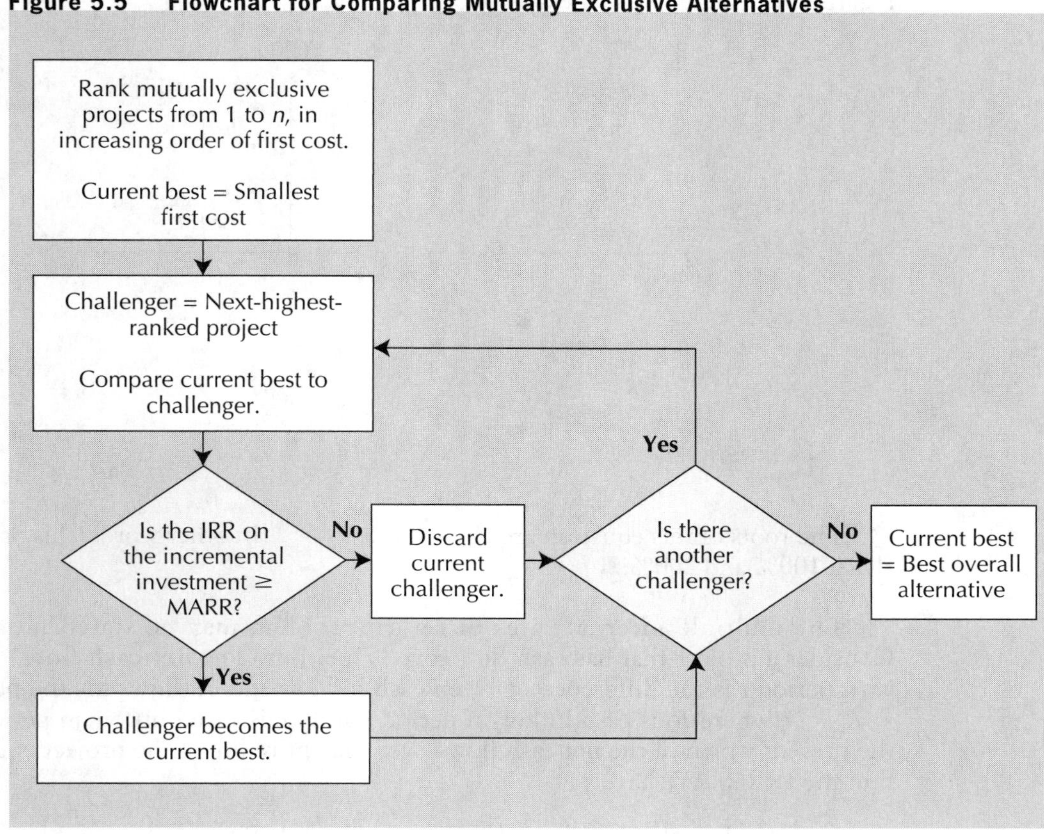

5.3.3 Multiple IRRs

A problem with implementing the internal rate of return method is that there may be more than one internal rate of return. Consider the following example.

Example 5.7

A project pays \$1000 today, costs \$5000 a year from now, and pays \$6000 in two years. (See Figure 5.6.) What is its IRR?

Equating the present worths of disbursements and receipts and solving for the IRR gives the following:

$$\$1000 - \$5000(P/F, i^*, 1) + \$6000(P/F, i^*, 2) = 0$$

Recalling that $(P/F,i^*,N)$ stands for $1/(1 + i^*)^N$, we have

$$1 - \frac{5}{1 + i^*} + \frac{6}{(1 + i^*)^2} = 0$$

$$(1 + i^*)^2 - 5(1 + i^*) + 6 = 0$$

$$(1 + 2i^* + i^{*2}) - 5i^* + 1 = 0$$

$$i^{*2} - 3i^* + 2 = 0$$

$$(i^* - 1)(i^* - 2) = 0$$

Figure 5.6 Multibple IRR Example

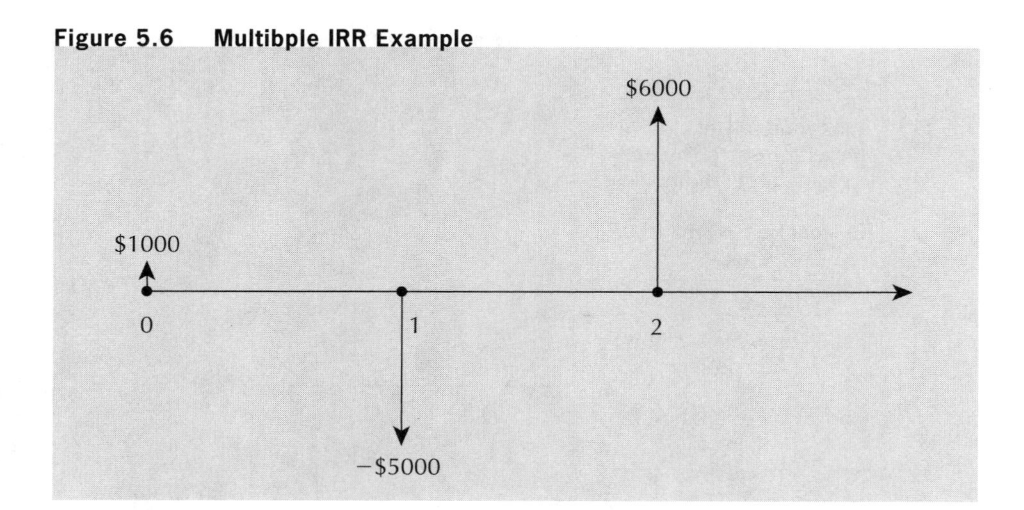

The roots of this equation are $i^*_1 = 1$ and $i^*_2 = 2$. In other words, this project has two IRRs: 100% and 200%!■

The multiple internal rates of return problem may be stated more generally. Consider a project that has cash flows over T periods. The **net cash flow**, A_t, associated with period t is the difference between cash inflows and outflows for the period (i.e., $A_t = R_t - D_t$ where R_t is cash inflow in period t and D_t is cash outflow in period t). We set the present worth of the net cash flows over the entire life of the project equal to zero to find the IRR(s). We have

$$A_0 + A_1(1 + i)^{-1} + A_2(1 + i)^{-2} + \dots + A_T(1 + i)^{-T} = 0 \quad (5.2)$$

Any value of i that solves Equation (5.2) is an internal rate of return for that project. That there may be multiple solutions to Equation (5.2) can be seen if we rewrite the equation as

$$A_0 + A_1x + A_2x^2 + \dots + A_Tx^T = 0 \quad (5.3)$$

where $x = (1 + i)^{-1}$.

Solving the Tth degree polynomial of Equation (5.3) is the same as solving for the internal rates of return in Equation (5.2). In general, when finding the roots of Equation (5.3), there may be as many positive real solutions for x as there are sign changes in the coefficients, the A's. Thus, there may be as many IRRs as there are sign changes in the A's.

We can see the meaning of multiple roots most easily with the concept of **project balance**. If a project has a sequence of net cash flows $A_0, A_1, A_2, \dots, A_T$, and the interest rate is i', there are $T + 1$ project balances, $B_0, B_1, B_2, \dots, B_T$, one at the end of each period t, $t = 0, 1, \dots, T$. A project balance, B_t, is the accumulated future value of all cash flows, up to the end of period t, compounded at the rate, i'. That is,

$$B_0 = A_0$$

$$B_1 = A_0(1 + i') + A_1$$

$$B_2 = A_0(1 + i')^2 + A_1(1 + i') + A_2$$

$$B_T = A_0(1 + i')^T + A_1(1 + i')^{T-1} + \dots + A_T$$

Table 5.1 shows the project balances at the end of each year for both 100% and 200% interest rates for the project in Example 5.7. The project starts with a cash inflow of \$1000. At a 100% interest rate, the \$1000 increases to \$2000 over the first year. At the end of the first year, there is a \$5000 disbursement, leaving a negative project balance of \$3000. At 100% interest, this negative balance increases to \$6000 over the second year. This negative \$6000 is offset exactly by the \$6000 inflow. This makes the project balance zero at the end of the second year. The project balance at the end of the project is the future worth of all the cash flows in the project. When the future worth at the end of the project life is zero, the present worth is also zero. This verifies that the 100% is an IRR.

Table 5.1 Project Balances for Example 5.7

End of Year	At $i' = 100\%$	At $i' = 200\%$
0	\$1000	\$1000
1	\$1000(1 + 1) − \$5000 = −\$3000	\$1000(1 + 2) − \$5000 = −\$2000
2	−\$3000(1 + 1) + \$6000 = 0	−\$2000(1 + 2) + \$6000 = 0

Now consider the 200% interest rate. Over the first year, the \$1000 inflow increases to \$3000. At the end of the first year, \$5000 is paid out, leaving a negative project balance of \$2000. This negative balance grows at 200% to \$6000 over the second year. This is offset exactly by the \$6000 inflow so that the project balance is zero at the end of the second year. This verifies that the 200% is also an IRR!

Looking at Table 5.1, it's actually fairly obvious that an important assumption is being made about the initial \$1000 received. The IRR computation implicitly assumes that the \$1000 is *invested* during the first period at either 100% or 200%, one of the two IRRs.

NET VALUE 5.1

Capital Budgeting and Financial Management Resources

The Internet can be an excellent source of information about the project comparison methods presented in Chapters 4 and 5. A broad search for materials on the PW, AW, and IRR comparison methods might use keywords such as "financial management" or "capital budgeting." Such a search can yield very useful supports for practising engineers or financial managers responsible for making investment decisions. For example, one can find short courses on the process of project evaluation using present worth or IRR methods and how to determine the cost of capital in the Education Centre at http://www.globeadvisor.com. Investopedia (www.investopedia.com) provides online investment resources, including definitions for a wide number of commonly used financial terms.

With more focused keywords such as "IRR," one might find examples of how the internal rate of return is used in practice in a wide range of engineering and other applications. Some consulting companies publish promotional white papers on their Web sites that provide examples of project evaluation methods. For example, cautionary words on the use of the IRR method and advocacy for the use of ERR (also known as the *modified internal rate of return* or *MIRR*) can be found at McKinsey's online newsletter, *McKinsey on Finance* (www.corporatefinance.mckinsey.com).

Source: J. Kelleher and J. MacCormack, "Internal Rate of Return: A Cautionary Tale," *McKinsey on Finance*, Summer 2004, www.corporatefinance.mckinsey.com/_downloads/knowledge/MOF/2004_no12/IRR%20caution.pdf, accessed October 10, 2004.

However, during the first period, the project is not an investment. The project balance is positive. The project is *providing* money, not using it. This money cannot be reinvested immediately in the project. It is simply cash on hand. The $1000 must be invested elsewhere for one year if it is to earn any return. It is unlikely, however, that the $1000 provided by the project in this example would be invested in something else at 100% or 200%. More likely, it would be invested at a rate at or near the company's MARR.

5.3.4 External Rate of Return Methods

To resolve the multiple IRR difficulty, we need to consider what return is earned by money associated with a project that is not invested in the project. The usual assumption is that the funds are invested elsewhere and earn an *explicit rate of return* equal to the MARR. This makes sense, because when there is cash on hand that is not invested in the project under study, it will be used elsewhere. These funds would, by definition, gain interest at a rate at least equal to the MARR. The **external rate of return (ERR)**, denoted by i^*_e, is the rate of return on a project where any cash flows that are not invested in the project are assumed to earn interest at a predetermined explicit rate (usually the MARR). For a given explicit rate of return, a project can have only one value for its ERR.

It is possible to calculate a precise ERR that is comparable to the IRRs of other projects using an explicit interest rate when necessary. Because the cash flows of Example 5.7 are fairly simple, let us use them to illustrate how to calculate the ERR precisely.

Example 5.8 (Example 5.7 Revisited: ERR)

A project pays $1000 today, costs $5000 a year from now, and pays $6000 in two years. What is its rate of return? Assume that the MARR is 25%.

The first $1000 is not invested immediately in the project. Therefore, we assume that it is invested outside the project for one year at the MARR. Thus, the cumulative cash flow for year 1 is

$$\$1000(F/P, 25\%, 1) - \$5000 = \$1250 - \$5000 = -\$3750$$

With this calculation, we transform the cash flow diagram representing this problem from that in Figure 5.7(a) to that in Figure 5.7(b). These cash flows provide a single (precise) ERR, as follows:

$$-\$3750 + \$6000(P/F, i^*_e, 1) = 0$$

$$(P/F, i^*_e, 1) = \$3750/\$6000 = 0.625$$

$$\frac{1}{1+i^*_e} = 0.625$$

$$1+i^*_e = \frac{1}{0.625} = 1.6$$

$$i^*_e = 0.6$$

$$ERR = 60\% \blacksquare$$

In general, computing a precise ERR can be a complex procedure because of the difficulty in determining exactly when the explicit interest rate should be applied. In order to do such a calculation, project balances have to be computed for trial ERRs. In periods in which project balances are positive for the trial ERR, the project is a source of funds.

Figure 5.7 Multiple IRR Solved

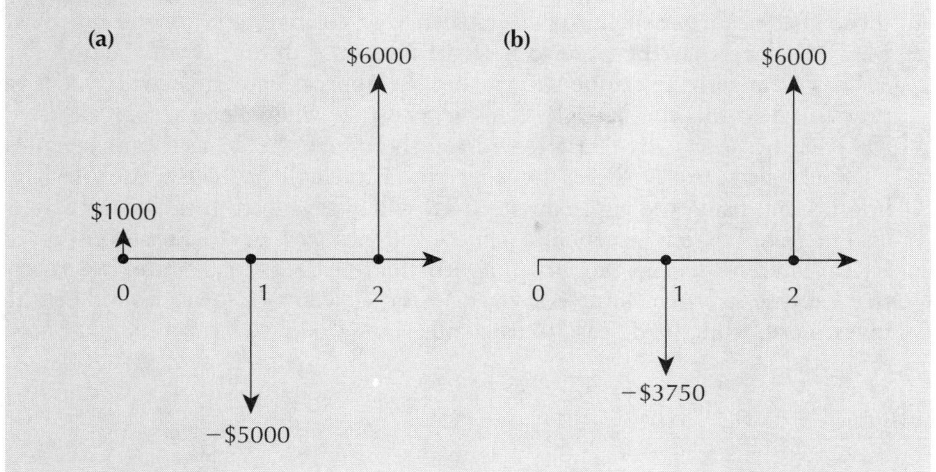

These funds would have to be invested outside the project at the MARR. During periods when the project balance is negative for the trial ERR, any receipts would be invested in the project and will typically yield more than the MARR. Whether the project balances are negative or positive will depend on the trial ERRs. This implies that the calculation process requires much experimenting with trial ERRs before an ERR is found that makes the future worth zero. A more convenient, but approximate, method is to use the following procedure:

1. Take all *net* receipts forward at the MARR to the time of the last cash flow.
2. Take all *net* disbursements forward at an unknown interest rate, i^*_{ea}, also to the time of the last cash flow.
3. Equate the future value of the receipts from Step 1 to the future value of the disbursements from Step 2 and solve for i^*_{ea}.
4. The value for i^*_{ea} is the *approximate ERR* for the project.

Example 5.9 (Example 5.7 Revisited Again: An Approximate ERR)

To approximate the ERR, we compute the interest rate that gives a zero future worth at the end of the project when all receipts are brought forward at the MARR. In Example 5.7, the \$1000 is thus assumed to be reinvested at the MARR for two years, the life of the project. The disbursements are taken forward to the end of the two years at an unknown interest rate, i^*_{ea}. With a MARR of 25%, the revised calculation is

$$\$1000(F/P, 25\%, 2) + \$6000 = \$5000(F/P, i^*_{ea}, 1)$$

$$(F/P, i^*_{ea}, 1) = [\$1000(1.5625) + \$6000]/\$5000$$

$$(F/P, i^*_{ea}, 1) = 1.5125$$

$$1 + i^*_{ea} = 1.5125$$

$$i^*_{ea} = 0.5125 \text{ or } 51.25\%$$

$$ERR \cong 51\% \blacksquare$$

The ERR calculated using this method is an approximation, since all receipts, not just those that occur when the project balance is positive, are assumed to be invested at the MARR. Note that the precise ERR of 60% is different from the approximate ERR of 51%. Fortunately, it can be shown that the approximate ERR will always be between the precise ERR and the MARR. This means that whenever the precise ERR is above the MARR, the approximate ERR will also be above the MARR and whenever the precise ERR is below the MARR, the approximation will be below the MARR as well. This implies that using the approximate ERR will always lead to the correct decision. It should also be noted that an acceptable project will earn *at least* the rate given by the approximate ERR. Therefore, even though an approximate ERR is inaccurate, it is often used in practice because it provides the correct decision as well as a lower bound on the return on an investment, while being easy to calculate.

5.3.5 When to Use the ERR

The ERR (approximate or precise) must be used whenever there are multiple IRRs possible. Unfortunately, it can be difficult to know in advance whether there will be multiple IRRs. On the other hand, it is fortunate that most ordinary projects have a structure that precludes multiple IRRs.

Most projects consist of one or more periods of outflows at the start, followed only by one or more periods of inflows. Such projects are called **simple investments**. The cash flow diagram for a simple investment takes the general form shown in Figure 5.8. In terms of Equations (5.2) and (5.3), there is only one change of sign, from negative to positive in the A's, the sequence of coefficients. Hence, a simple investment always has a unique IRR.

Figure 5.8 The General Form of Simple Investments

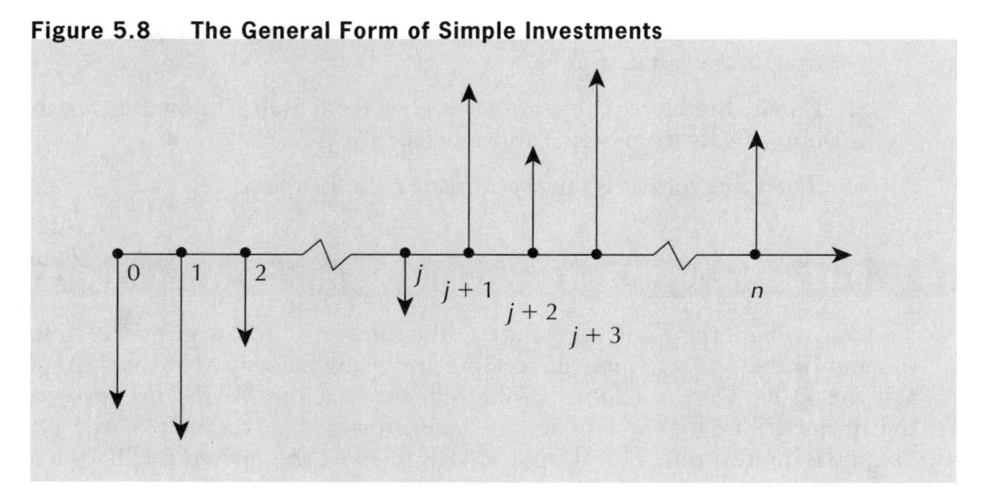

If a project is not a simple investment, there may or may not be multiple IRRs—there is no way of knowing for sure without further analysis. In practice, it may be reasonable to use an approximate ERR whenever the project is not a simple investment. Recall from Section 5.3.4 that the approximate ERR will always provide a correct decision, whether its use is required or not, since it will understate the true rate of return.

However, it is generally desirable to compute an IRR whenever it is possible to do so, and to use an approximate ERR only when there may be multiple IRRs. In this way, the computations will be as accurate as possible. When it is desirable to know for sure whether there will be only one IRR, there are several steps of analysis that can be undertaken. These are covered in detail in Appendix 5A.

To reiterate, the approximate ERR can be used to evaluate any project, whether it is a simple investment or not. However, the approximate ERR will tend to be a less accurate rate than the IRR. The inaccuracy will tend to be similar for projects with cash flows of a similar structure, and either method will result in the same decision in the end.

5.4 Rate of Return and Present/Annual Worth Methods Compared

A comparison of rate of return and present/annual worth methods leads to two important conclusions:

1. The two sets of methods, when properly used, give the same decisions.
2. Each set of methods has its own advantages and disadvantages.

Let us consider each of these conclusions in more detail.

5.4.1 Equivalence of Rate of Return and Present/Annual Worth Methods

If an independent project has a unique IRR, the IRR method and the present worth method give the same decision. Consider Figure 5.9. It shows the present worth as a function of the interest rate for a project with a unique IRR. The maximum of the curve lies at the vertical axis (where the interest rate = 0) at the point given by the sum of all undiscounted net cash flows. (We assume that the sum of all the undiscounted net cash flows is positive.) As the interest rate increases, the present worth of all cash flows after the first cost decreases. Therefore, the present worth curve slopes down to the right. To determine what happens as the interest rate increases indefinitely, let us recall the general equation for present worth

$$PW = \sum_{t=0}^{T} A_t(1 + i)^{-t} \quad (5.4)$$

where

i = the interest rate
A_t = the net cash flow in period t
T = the number of periods

Letting $i \rightarrow \infty$, we have

$$\lim_{i \rightarrow \infty} \frac{1}{(1 + i)^t} = 0 \text{ for } t = 1, 2, \ldots, T$$

Therefore, as the interest rate becomes indefinitely large, all terms in Equation (5.4) approach zero except the first term (where $t = 0$), which remains at A_0. In Figure 5.9, this is shown by the asymptotic approach of the curve to the first cost, which, being negative, is below the horizontal axis.

The interest rate at which the curve crosses the horizontal axis (i^* in Figure 5.9), where the present worth is zero, is, by definition, the IRR.

Figure 5.9 Present Worth (PW) as a Function of Interest Rate *(i)* for a Simple Investment

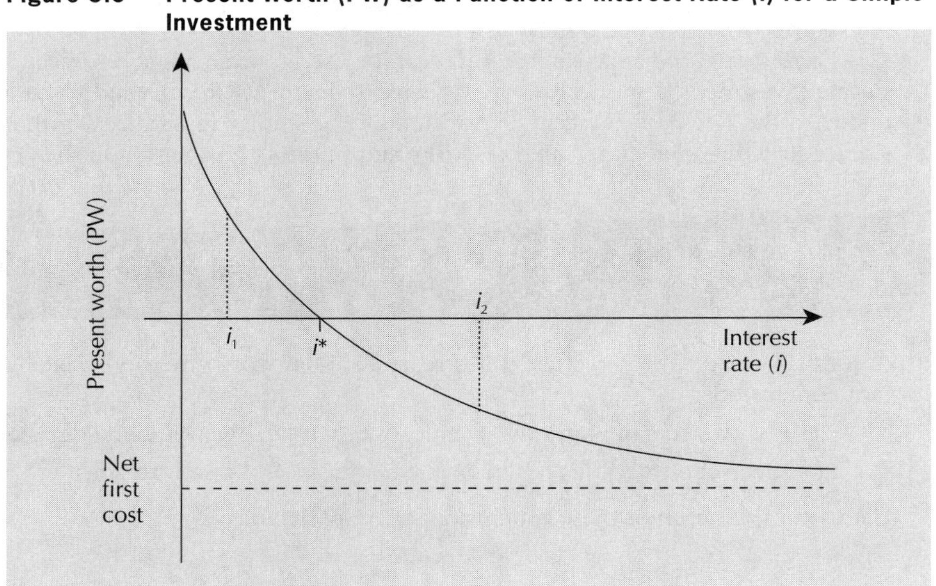

To demonstrate the equivalence of the rate of return and the present/annual worth methods for decision making, let us consider possible values for the MARR. First, suppose the MARR = i_1, where $i_1 < i^*$. In Figure 5.9, this MARR would lie to the left of the IRR. From the graph we see that the present worth is positive at i_1. In other words, we have

IRR > MARR

and

PW > 0

Thus, in this case, both the IRR and PW methods lead to the same conclusion: Accept the project.

Second, suppose the MARR = i_2, where $i_2 > i^*$. In Figure 5.9, this MARR would lie to the right of the IRR. From the graph we see that, at i_2, the present worth is negative. Thus we have

IRR < MARR

and

PW < 0

Here, too, the IRR and the PW method lead to the same conclusion: Reject the project.

Now consider two simple, mutually exclusive projects, A and B, where the first cost of B is greater than the first cost of A. If the increment from A to B has a unique IRR, then we can readily demonstrate that the IRR and PW methods lead to the same decision. See Figure 5.10(a), which shows the present worths of projects A and B as a function of the interest rate. Since the first cost of B is greater than that of A, the curve for project B asymptotically approaches a lower present worth than does the curve for project A as the interest rate becomes indefinitely large, and thus the two curves must cross at some point.

To apply the IRR method, we must consider the increment (denoted by B – A). The present worth of the increment (B – A) will be zero where the two curves cross. This point of intersection is marked by the interest rate, i^*. We have plotted the curve for the increment (B – A) in Figure 5.10(b) to clarify the relationships.

Figure 5.10 Present Worth as a Function of Interest Rate (i) for Two Simple, Mutually Exclusive Projects

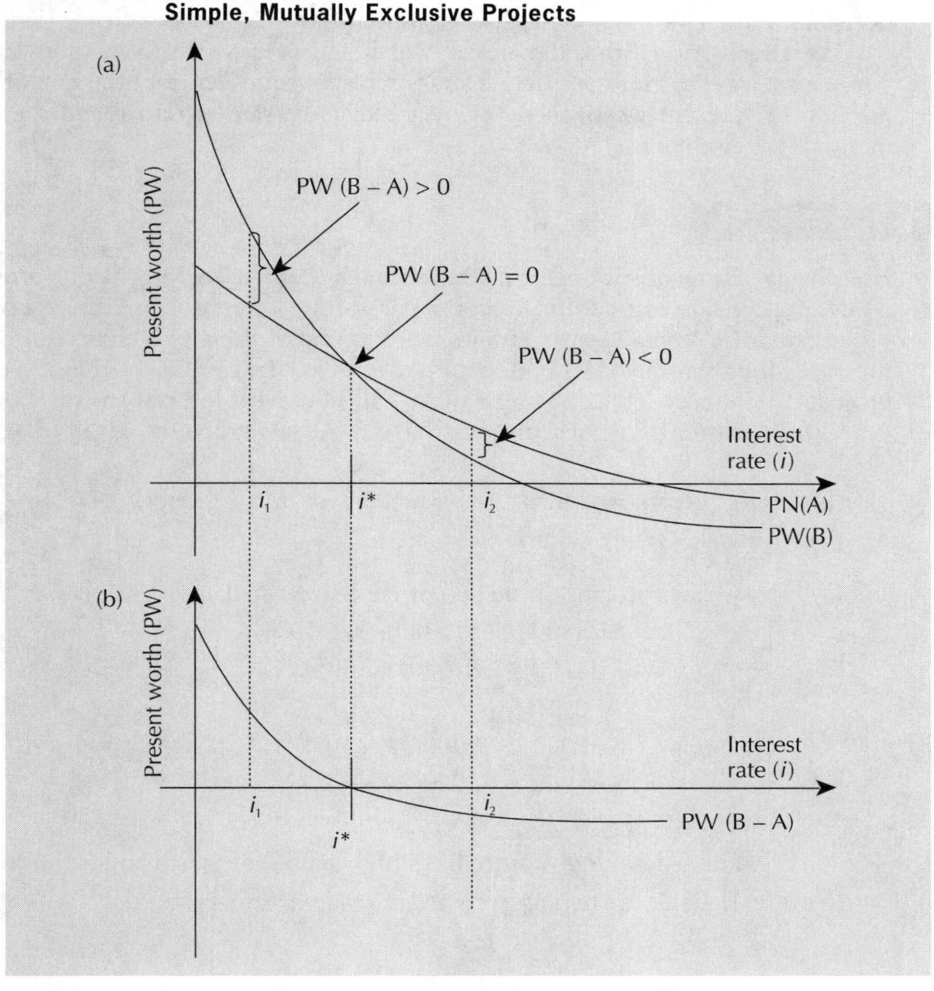

Let us again deal with possible values of the MARR. First, suppose the MARR (i_1) is less than i^*. Then, as we see in Figure 5.10(b), the present worth of (B – A) is positive at i_1. That is, the following conditions hold:

IRR(B – A) > MARR

and

PW(B – A) > 0

Thus, according to both the IRR method and the PW method, project B is better than project A.

Second, suppose the MARR = i_2, where $i_2 > i^*$. Then we see from Figure 5.10(b) that the present worth of the increment (B – A) is negative at i_2. In other words, the following conditions hold:

IRR(B – A) < MARR

and

PW(B – A) < 0

Thus, according to both methods, project A is better than project B.

In a similar fashion, we could show that the approximate ERR method gives the same decisions as the PW method in those cases where there may be multiple IRRs.

We already noted that the annual worth and present worth methods are equivalent. Therefore, by extension, our demonstration of the equivalence of the rate of return methods and the present worth methods means that the rate of return and the annual worth methods are also equivalent.

Example 5.10

Bracebridge Enterprises operates a resort in the tourist area north of Toronto, Ontario. They are considering adding either a parasailing operation or canoe rentals to their other activities. Available space limits them to one of these two choices. The initial costs for parasailing will be \$100 000, with net returns of \$15 000 for the 15-year life of the project. Initial costs for canoeing will be \$10 000, with net returns of \$2000 for its 15-year life. Assume that both projects have a \$0 salvage value after 15 years, and the MARR is 10%.

(a) Using present worth analysis, which alternative is better?

(b) Using IRR, which alternative is better?

(a) The present worths of the two projects are calculated as follows:

$$\begin{aligned}PW_{para} &= -\$100\,000 + \$15\,000(P/A, 10\%, 15)\\ &= -\$100\,000 + \$15\,000(7.6061)\\ &= \$14\,091.50\end{aligned}$$

$$\begin{aligned}PW_{can} &= -\$10\,000 + \$2000(P/A, 10\%, 15)\\ &= -\$10\,000 + \$2000(7.6061)\\ &= \$5212.20\end{aligned}$$

The parasailing venture has a higher present worth and is thus preferred.

(b) The IRRs of the two projects are calculated as follows:

Parasailing

$$\$100\,000 = \$15\,000(P/A, i^*, 15)$$

$$(P/A, i^*, 15) = \$100\,000/\$15\,000 = 6.67 \rightarrow i^*_{para} = 12.4\%$$

Canoeing

$$\$10\,000 = \$2000(P/A, i^*, 15)$$

$$(P/A, i^*, 15) = 5 \rightarrow i^*_{can} = 18.4\%$$

One might conclude that, because IRR_{can} is larger, Bracebridge Enterprises should invest in the canoeing project, but this is *wrong*. When done correctly, a present worth analysis and an IRR analysis will always agree. The error here is that the parasailing project was assessed without consideration of the increment from the canoeing project. Checking the IRR of the increment (denoted by the subscript "can-para"):

$$(\$100\,000 - \$10\,000) = (\$15\,000 - \$2000)(P/A, i^*, 15)$$

$$(P/A, i^*, 15) = \$90\,000/\$13\,000 = 6.923 \rightarrow i^*_{can-para} = 11.7\%$$

Since the increment from the canoeing project also exceeds the MARR, the larger parasailing project should be taken.■

5.4.2 Why Choose One Method over the Other?

Although rate of return methods and present worth/annual worth methods give the same decisions, each set of methods has its own advantages and disadvantages. The choice of method may depend on the way the results are to be used and the sort of data the decision makers prefer to consider. In fact, many companies, by policy, require that several methods be applied so that a more complete picture of the situation is presented. A summary of the advantages and disadvantages of each method is given in Table 5.2.

Rate of return methods state results in terms of *rates*, while present/annual worth methods state results in absolute figures. Many managers prefer rates to absolute figures because rates facilitate direct comparisons of projects whose sizes are quite different. For example, a petroleum company comparing performances of a refining division and a distribution division would not look at the typical values of present or annual worth for projects in the two divisions. A refining project may have first costs in the range of hundreds of *millions* of dollars, while distribution projects may have first costs in the range of *thousands* of dollars. It would not be meaningful to compare the absolute profits between a refining project and a distribution project. The absolute profits of refining projects will almost certainly be larger than those of distribution projects. Expressing project performance in terms of rates of return permits understandable comparisons. A disadvantage of rate of return methods, however, is the possible complication that there may be more than one rate of return. Under such circumstances, it is necessary to calculate an ERR.

In contrast to a rate of return, a present or annual worth computation gives a direct measure of the profit provided by a project. A company's main goal is likely to earn profits for its owners. The present and annual worth methods state the contribution of a project toward that goal. Another reason that managers prefer these methods is that present worth and annual worth methods are typically easier to apply than rate of return methods.

For completeness of coverage, we note that the payback period method may not give results consistent with rate of return or present/annual worth methods as it ignores the time value of money and the service life of projects. It is, however, a method commonly used in practice due to its ease of use and intuitive appeal.

Table 5.2 Advantages and Disadvantages of Comparison Methods

Method	Advantages	Disadvantages
IRR	Facilitates comparisons of projects of different sizes Commonly used	Relatively difficult to calculate Multiple IRRs may exist
Present worth	Gives explicit measure of profit contribution	Difficult to compare projects of different sizes
Annual worth	Annual cash flows may have familiar meanings to decision makers	Difficult to compare projects of different sizes
Payback period	Very easy to calculate Commonly used Takes into account the need to have capital recovered quickly	Discriminates against long-term projects Ignores time value of money Ignores the expected service life

Example 5.11

Each of the following scenarios suggests a best choice of comparison method:

1. Edward has his own small firm that will lease injection-molding equipment to make polyethylene containers. He must decide on the specific model to lease. He has estimates of future monthly sales.

   The annual worth method makes sense here, because Edward's cash flows, including sales receipts and leasing expenses, will probably all be on a monthly basis. As a sole proprietor, Edward need not report his conclusions to others.

2. Ramesh works for a large power company and must assess the viability of locating a transformer station at various sites in the city. He is looking at the cost of the building lot, power lines, and power losses for the various locations. He has fairly accurate data about costs and future demand for electricity.

   As part of a large firm, Ramesh will probably be obliged to use a specific comparison method. This would probably be IRR. A power company makes many large and small investments, and the IRR method allows them to be compared fairly. Ramesh has the data necessary for the IRR calculations.

3. Sehdev must buy a relatively inexpensive log splitter for his agricultural firm. There are several different types that require a higher or lower degree of manual assistance. He has only rough estimates of how this machine will affect future cash flows.

   This relatively inexpensive purchase is a good candidate for the payback period method. The fact that it is inexpensive means that extensive data gathering and analysis are probably not warranted. Also, since future cash flows are relatively uncertain, there is no justification for using a particularly precise comparison method.

4. Ziva will be living in Inuvik for six months, testing her company's equipment under hostile weather conditions. She needs a field office and must determine which of the following choices is economically best: (1) renting space in an industrial building, (2) buying and outfitting a trailer, (3) renting a hotel room for the purpose.

   For this decision, a present worth analysis would be appropriate. The cash flows for each of the alternatives are of different types, and bringing them to present worth would be a fair way to compare them. It would also provide an accurate estimate to Ziva's firm of the expected cost of the remote office for planning purposes.■

REVIEW PROBLEMS

REVIEW PROBLEM 5.1

Wei-Ping's consulting firm needs new quarters. A downtown office building is ideal. The company can either buy or lease it. To buy the office building will cost \$6 000 000. If the building is leased, the lease fee is \$400 000 payable at the beginning of each year. In either case, the company must pay city taxes, maintenance, and utilities.

Wei-Ping figures that the company needs the office space for only 15 years. Therefore, they will either sign a 15-year lease or buy the building. If they buy the building, they will then sell it after 15 years. The value of the building at that time is estimated to be $15 000 000.

What rate of return will Wei-Ping's firm receive by buying the office building instead of leasing it?

ANSWER

The rate of return can be calculated as the IRR on the incremental investment necessary to buy the building rather than lease it.

The IRR on the incremental investment is found by solving for i^* in

$$(\$6\,000\,000 - \$400\,000) - \$15\,000\,000(P/F, i^*, 15) = \$400\,000(P/A, i^*, 14)$$

$$4(P/A, i^*, 14) + 150(P/F, i^*, 15) = 56$$

For $i^* = 11\%$, the result is

$$4(P/A, 11\%, 14) + 150(P/F, 11\%, 15)$$
$$= 4(6.9819) + 150(0.20900)$$
$$= 59.2781$$

For $i^* = 12\%$,

$$4(P/A, 12\%, 14) + 150(P/F, 12\%, 15)$$
$$= 4(6.6282) + 150(0.1827)$$
$$= 53.9171$$

A linear interpolation between 11% and 12% gives the IRR

$$i^* = 11\% + (59.2781 - 56)/(59.2781 - 53.9171) = 11.6115\%$$

By investing their money in buying the building rather than leasing, Wei-Ping's firm is earning an IRR of about 11.6%.■

REVIEW PROBLEM 5.2

The Real S. Tate Company is considering investing in one of four rental properties. Real S. Tate will rent out whatever property they buy for four years and then sell it at the end of that period. The data concerning the properties is shown below.

Rental Property	Purchase Price	Net Annual Rental Income	Sale Price at the End of Four Years
1	$100 000	$ 7 200	$100 000
2	120 000	9 600	130 000
3	150 000	10 800	160 000
4	200 000	12 000	230 000

On the basis of the purchase prices, rental incomes, and sale prices at the end of the four years, answer the following questions:

(a) Which property, if any, should Tate invest in? Real S. Tate uses a MARR of 8% for projects of this type.

(b) Construct a graph that depicts the present worth of each alternative as a function of interest rates ranging from 0% to 20%. (A spreadsheet would be helpful in answering this part of the problem.)

(c) From your graph, determine the range of interest rates for which your choice in part (a) is the best investment. If the MARR were 9%, which rental property would be the best investment? Comment on the sensitivity of your choice to the MARR used by the Real S. Tate Company.

ANSWER

(a) Since the "do nothing" alternative is feasible and it has the least first cost, it becomes the current best alternative. The IRR on the incremental investment for property 1 is given by:

$$-\$100\,000 + \$100\,000(P/F, i^*, 4) + \$7200(P/A, i^*, 4) = 0$$

The IRR on the incremental investment is 7.2%. Because this is less than the MARR of 8%, property 1 is discarded from further consideration.

Next, the IRR for the incremental investment for property 2, the alternative with the next-highest first cost, is found by solving for i^* in

$$-\$120\,000 + \$130\,000(P/F, i^*, 4) + \$9600(P/A, i^*, 4) = 0$$

The interest rate that solves the above equation is 9.8%. Since an IRR of 9.8% exceeds the MARR, property 2 becomes the current best alternative. Now the incremental investments over and above the first cost of property 2 are analyzed.

Next, property 3 challenges the current best. The IRR in the incremental investment to property 3 is

$$(-\$150\,000 + \$120\,000) + (\$160\,000 - \$130\,000)(P/F, i^*, 4)$$
$$+ (\$10\,800 - \$9600)(P/A, i^*, 4) = 0$$
$$-\$30\,000 + \$30\,000(P/F, i^*, 4) + \$1200(P/A, i^*, 4) = 0$$

This gives an IRR of only 4%, which is below the MARR. Property 2 remains the current best alternative and property 3 is discarded.

Finally, property 4 challenges the current best. The IRR on the incremental investment from property 2 to property 4 is

$$(-\$200\,000 + \$120\,000) + (\$230\,000 - \$130\,000)(P/F, i^*, 4)$$
$$+ (\$12\,000 - \$9600)(P/A, i^*, 4) = 0$$
$$-\$80\,000 + \$100\,000(P/F, i^*, 4) + \$2400(P/A, i^*, 4) = 0$$

The IRR on the incremental investment is 8.5%, which is above the MARR. Property 4 becomes the current best choice. Since there are no further challengers, the choice based on IRR is the current best, property 4.

(b) The graph for part (b) is shown in Figure 5.11.

(c) From the graph, one can see that property 4 is the best alternative provided that the MARR is between 0% and 8.5%. This is the range of interest rates over which property 4 has the largest present worth.

If the MARR is 9%, the best alternative is property 2. This can be seen by going back to the original IRR computations and observing that the results of the analysis are essentially the same, except that the incremental investment from property 2 to property 4 no longer has a return exceeding the MARR. This can be confirmed from the diagram (Figure 5.11) as well, since the property with the largest present worth at 9% is property 2.

With respect to sensitivity analysis, the graph shows that, for a MARR between 0% and 8.5%, property 4 is the best choice and, for a MARR between 8.5% and 9.8%, property 2 is the best choice. If the MARR is above 9.8%, no property has an acceptable return on investment, and the "do nothing" alternative would be chosen.■

Figure 5.11 Present Worths for Review Problem 5.2

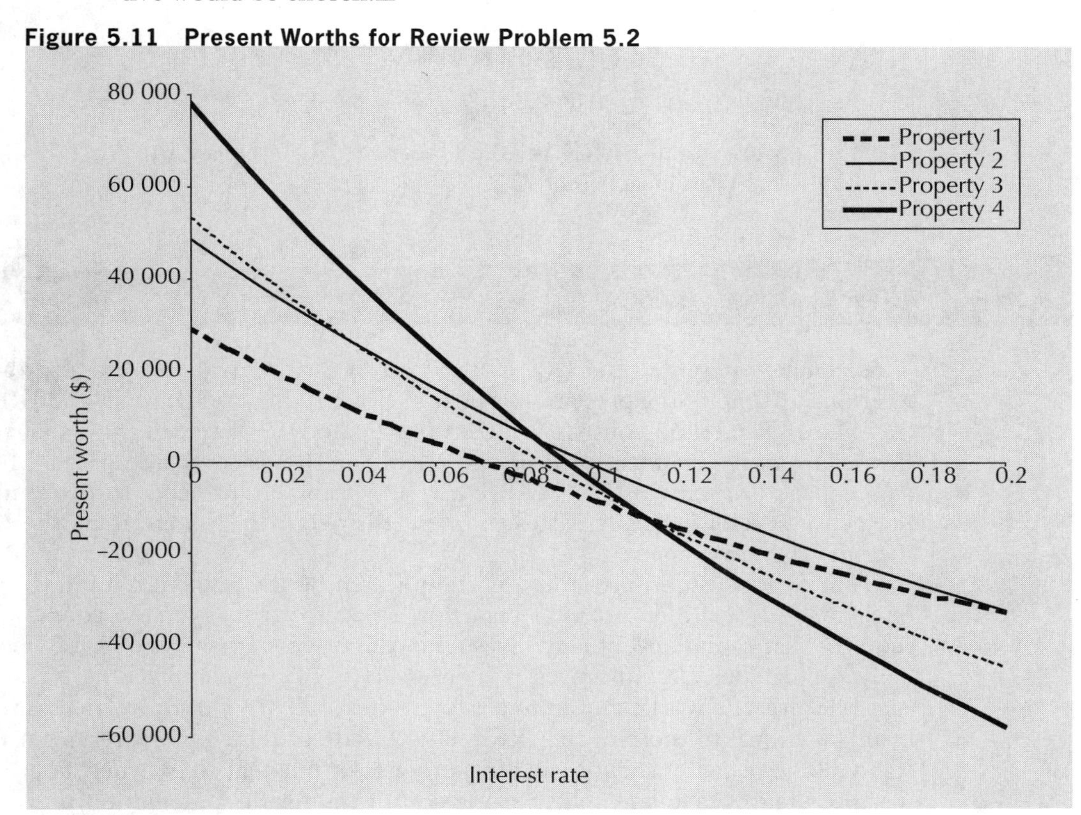

REVIEW PROBLEM 5.3

You are in the process of arranging a marketing contract for a new Java applet you are writing. It still needs more development, so your contract will pay you \$5000 today to finish the prototype. You will then get royalties of \$10 000 at the end of each of the second and third years. At the end of each of the first and fourth years, you will be required to spend \$20 000 and \$10 000 in upgrades, respectively. What is the (approximate) ERR on this project, assuming a MARR of 20%? Should you accept the contract?

ANSWER

To calculate the approximate ERR, set

FW(receipts @ MARR) = FW(disbursements @ ERR)

$$\$5000(F/P, 20\%, 4) + \$10\,000(F/P, 20\%, 2) + \$10\,000(F/P, 20\%, 1) = \$20\,000(F/P, i^*_{ea}, 3) + \$10\,000$$

$$\$5000(2.0736) + \$10\,000(1.44) + \$10\,000(1.2) = \$20\,000(F/P, i^*_{ea}, 3) + \$10\,000$$

$$(F/P, i^*_{ea}, 3) = 1.3384$$

$$\text{At ERR} = 10\%, (F/P, i, 3) = 1.3310$$

$$\text{At ERR} = 11\%, (F/P, i, 3) = 1.3676$$

Interpolating:

$$i^*_{ea} = 10\% + (1.3384 - 1.3310)(11 - 10)/(1.3676 - 1.3310)$$

$$= 10\% + 0.0074/0.0366 \cong 10.2\%$$

The (approximate) ERR is 10.2%. Since this is below the MARR of 20%, the contract should not be accepted.■

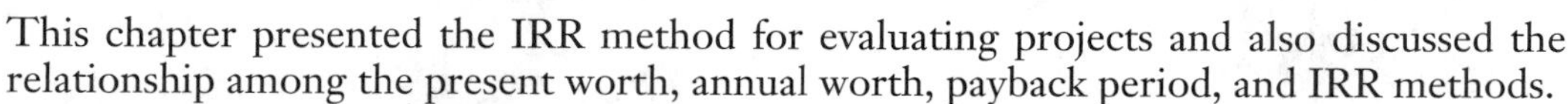

SUMMARY

This chapter presented the IRR method for evaluating projects and also discussed the relationship among the present worth, annual worth, payback period, and IRR methods.

The IRR method consists of determining the rate of return for a sequence of cash flows. For an independent project, the calculated IRR is compared with a MARR, and if it is equal to or exceeds the MARR it is an acceptable project. To determine the best project of several mutually exclusive ones, it is necessary to determine the IRR of each increment of investment.

The IRR selection procedure is complicated by the possibility of having more than one rate of return because of a cash flow structure that, over the course of a project, requires that capital, eventually invested in the project at some point, be invested externally. Under such circumstances, it is necessary to calculate an ERR.

The present worth and annual worth methods are closely related, and both give results identical to those of the IRR method. Rate of return measures are readily understandable, especially when comparing projects of unequal sizes, whereas present/annual worth measures give an explicit expression of the profit contribution of a project. The main advantage of the payback period method is that it is easy to implement and understand, and takes into account the need to have capital recovered quickly.

Engineering Economics in Action, Part 5B: The Invisible Hand

"Hello." Dave's voice was clear enough over the phone that he could have been in his office down the hall, but Naomi could tell from his relaxed tone that the office was not on his mind.

"Hi, Dave, it's Naomi. Can I bend your ear about that drop forge project again?"

"Oh, hi, Naomi. Sure, what have you got?"

"Well, as I see it, IRR has got to be the way to go. Of course, present worth or annual worth will give the same answer, but I'm sure Ed Burns and Anna Kulkowski would prefer IRR. They have to compare potential investments across different parts of the organization. It's kind of hard to compare net present worths of investments in information systems, where you rarely get above a first cost of $100 000, with forge investments where you can easily get up to a few hundred thousand first cost. And, as I said before, the drop forge operation isn't one in which the annual cost has any particular significance."

There was a short pause. Naomi suddenly regretted speaking as if she was so sure of herself—but, darn it, she was sure on this one.

"Exactly right," Dave replied. Naomi could feel an invisible hand pat her on the back. "So how exactly would you proceed?"

"Well, I have the options ranked by first cost. The first one is just refurbishing the existing machine. There is no test on that one unless we are willing to stop making our own parts, and Clem told me that was out . . . "

Dave interjected with "You don't mean that you're automatically going to refurbish the existing machine, do you?"

"No, no. The simple refurbishing option is the base. I then go to the next option which is to refurbish the drop forging hammer and replace the materials-handling system. I compare this with the just-refurbish option by looking at the incremental first cost. I will check to see if the additional first cost has an IRR of at least 15% after tax, which, Clem tells me, is the minimum acceptable rate of return. If the incremental first cost has an IRR of at least 15%, the combination of refurbishing and replacing the materials-handling system is better than just refurbishing. I then consider the next option, which is to buy the manually operated mechanical press with no change in materials handling. I look at the incremental investment here and see if its IRR is at least 15%. To go back a step, if the IRR of replacing materials handling plus refurbishing the old machine did not pay off at 15%, I would have rejected that and compared the manually operated mechanical press with the first option, just refurbishing the old machine. I then work my way, option by option, up to the seventh. How does that sound?"

"Well, that sounds great, as far as it goes. Have you checked for problems with multiple IRRs?"

"Well, so far each set of cash flows has been a simple investment, but I will be careful."

"I would also compute payback periods for them in case we are having cash flow problems. If the payback is too long, they may not necessarily take an option even with the incremental IRR being above their 15% MARR."

Naomi considered this for a second. "One other question, Dave. What should I do about intangibles?"

"You mean the noise from the forging hammer?"

"Yes. It's important, but you can't evaluate it in dollars and cents."

"Just remind them of it in your report. If they want a more formal analysis, they'll come back to you."

"Thanks, Dave. You've been a big help."

As Naomi hung up the phone, she couldn't help smiling ruefully to herself. She had ignored the payback period altogether—after all, it didn't take either interest or service life into account. "I guess that's what they call practical experience," she said to herself as she got out her laptop.

PROBLEMS

5.1 What is the IRR for a \$1000 investment that returns \$200 at the end of each of the next

(a) 7 years?

(b) 6 years?

(c) 100 years?

(d) 2 years?

5.2 New windows are expected to save \$400 per year in energy costs over their 30-year life for Nottawasaga Fabricating. At an initial cost of \$8000 and zero salvage value, are they a good investment? NF's MARR is 8%.

5.3 An advertising campaign will cost $2 000 000 for planning and $400 000 in each of the next six years. It is expected to increase revenues permanently by $400 000 per year. Additional revenues will be gained in the pattern of an arithmetic gradient with $200 000 in the first year, declining by $50 000 per year to zero in the fifth year. What is the IRR of this investment? If the company's MARR is 12%, is this a good investment?

5.4 Aline has three contracts from which to choose. The first contract will require an outlay of $100 000 but will return $150 000 one year from now. The second contract requires an outlay of $200 000 and will return $300 000 one year from now. The third contract requires an outlay of $250 000 and will return $355 000 one year from now. Only one contract can be accepted. If her MARR is 20%, which one should she choose?

5.5 Refer to Review Problem 4.3. Assuming the four investments are independent, use the IRR method to select which, if any, should be chosen. Use a MARR of 8%.

5.6 Antigonish Footwear can invest in one of two different automated clicker cutters. The first, A, has a $10 000 first cost. A similar one with many extra features, B, has a $40 000 first cost. A will save $5000 per year over the cutter now in use. B will save $15 000 per year. Each clicker cutter will last five years. If the MARR is 10%, which alternative is better? Use an IRR comparison.

5.7 CB Electronix must buy a piece of equipment to place electronic components on the printed circuit boards it assembles. The proposed equipment has a 10-year life with no scrap value.

The supplier has given CB several purchase alternatives. The first is to purchase the equipment for $850 000. The second is to pay for the equipment in 10 equal instalments of $135 000 each, starting one year from now. The third is to pay $200 000 now and $95 000 at the end of each year for the next 10 years.

(a) Which alternative should CB choose if their MARR is 11% per year? Use an IRR comparison approach.

(b) Below what MARR does it make sense for CB to buy the equipment now for $850 000?

5.8 The following table summarizes information for four projects:

Project	First Cost	IRR on Overall Investment	IRR on Increments of Investment Compared with Project 1	2	3
1	$100 000	19%			
2	175 000	15%	9%		
3	200 000	18%	17%	23%	
4	250 000	16%	12%	17%	13%

The data can be interpreted in the following way: The IRR on the incremental investment between project 4 and project 3 is 13%.

(a) If the projects are independent, which projects should be undertaken if the MARR is 16%?

(b) If the projects are mutually exclusive, which project should be undertaken if the MARR is 15%? Indicate what logic you have used.

(c) If the projects are mutually exclusive, which project should be undertaken if the MARR is 17%? Indicate what logic you have used.

5.9 There are several mutually exclusive ways Rimouski Dairy can meet a requirement for a filling machine for their creamer line. One choice is to buy a machine. This would cost \$65 000 and last for six years with a salvage value of \$10 000. Alternatively, they could contract with a packaging supplier to get a machine free. In this case, the extra costs for packaging supplies would amount to \$15 000 per year over the six-year life (after which the supplier gets the machine back with no salvage value for Rimouski). The third alternative is to buy a used machine for \$30 000 with zero salvage value after six years. The used machine has extra maintenance costs of \$3000 in the first year, increasing by \$2500 per year. In all cases, there are installation costs of \$6000 and revenues of \$20 000 per year. Using the IRR method, determine which is the best alternative. The MARR is 10%.

5.10 Project X has an IRR of 16% and a first cost of \$20 000. Project Y has an IRR of 17% and a first cost of \$18 000. The MARR is 15%. What can be said about which (if either) of the two projects should be undertaken if (a) the projects are independent and (b) the projects are mutually exclusive?

5.11 Charlie has a project for which he had determined a present worth of \$56 740. He now has to calculate the IRR for the project, but unfortunately he has lost complete information about the cash flows. He knows only that the project has a five-year service life and a first cost of \$180 000, that a set of equal cash flows occurred at the end of each year, and that the MARR used was 10%. What is the IRR for this project?

5.12 Lucy's project has a first cost P, annual savings A, and a salvage value of \$1000 at the end of the 10-year service life. She has calculated the present worth as \$20 000, the annual worth as \$4000, and the payback period as three years. What is the IRR for this project?

5.13 Patti's project has an IRR of 15%, first cost P, and annual savings A. She observed that the salvage value *S* at the end of the five-year life of the project was exactly half of the purchase price, and that the present worth of the project was exactly double the annual savings. What was Patti's MARR?

5.14 Jerry has an opportunity to buy a bond with a face value of \$10 000 and a coupon rate of 14%, payable semiannually.

(a) If the bond matures in five years and Jerry can buy one now for \$3500, what is his IRR for this investment?

(b) If his MARR for this type of investment is 20%, should he buy the bond?

5.15 The following cash flows result from a potential construction contract for Estevan Engineering:

1. Receipts of \$500 000 at the start of the contract and \$1 200 000 at the end of the fourth year
2. Expenditures at the end of the first year of \$400 000 and at the end of the second year of \$900 000
3. A net cash flow of \$0 at the end of the third year

Using an appropriate rate of return method, for a MARR of 25%, should Estevan Engineering accept this project?

5.16 Samiran has entered into an agreement to develop and maintain a computer program for symbolic mathematics. Under the terms of the agreement, he will pay $90 000 in royalties to the investor at the end of the fifth, tenth, and fifteenth years, with the investor paying Samiran $45 000 now, and then $65 000 at the end of the twelfth year.

Samiran's MARR for this type of investment is 20%. Calculate the ERR of this project. Should he accept this agreement, on the basis of these disbursements and receipts alone? Are you sure that the ERR you calculated is the only ERR? Why? Are you sure that your recommendation to Samiran is correct? Justify your answer.

5.17 Refer to Problem 4.12. Find which alternative is preferable using the IRR method and a MARR of 5%. Assume that one of the alternatives must be chosen. Answer the following questions by using present worth computations to find the IRRs. Use the least common multiple of service lives.

(a) What are the cash flows for each year of the comparison period (i.e., the least common multiple of service lives)?

(b) Are you able to conclude that there is a single IRR on the incremental investment? Why or why not?

(c) Which of the two alternatives should be chosen? Use the ERR method if necessary.

5.18 Refer to Example 4.6 in which a mechanical engineer has decided to introduce automated materials-handling equipment to a production line. Use a present worth approach with an IRR analysis to determine which of the two alternatives is best. The MARR is 9%. Use the repeated lives method to deal with the fact that the service lives of the two alternatives are not equal.

5.19 Refer to Problem 4.20. Use an IRR analysis to determine which of the two alternatives is best. The MARR is 8%. Use the repeated lives method to deal with the unequal service lives of the two alternatives.

5.20 Refer to Problem 4.21. Val has determined that the salvage value of the XJ3 after two years of service is $1900. Using the IRR method, which display panel is the better choice? Use a two-year study period. She must choose one of the alternatives.

5.21 Yee Swian has received an advance of $2000 on a software program she is writing. She will spend $12 000 this year writing it (consider the money to have been spent at the end of year 1), and then receive $10 000 at the end of the second year. The MARR is 12%.

(a) What is the IRR for this project? Does the result make sense?

(b) What is the precise ERR?

(c) What is the approximate ERR?

5.22 Zhe develops truss analysis software for civil engineers. He has the opportunity to contract with at most one of two clients who have approached him with development proposals. One contract pays him $15 000 immediately, and then $22 000 at the end of the project three years from now. The other possibility pays $20 000 now and $5000 at the end of each of the three years. In either case, his expenses will be $10 000 per year. For a MARR of 10%, which project should Zhe accept? Use an appropriate rate of return method.

5.23 The following table summarizes cash flows for a project:

Year	Cash Flow at End of Year
0	−\$5000
1	3000
2	4000
3	−1000

(a) Write out the expression you need to solve to find the IRR(s) for this set of cash flows. Do not solve.

(b) What is the maximum number of solutions for the IRR that could be found in part (a)? Explain your answer in one sentence.

(c) You have found that an IRR of 14.58% solves the expression in part (a). Compute the project balances for each year.

(d) Can you tell (without further computations) if there is a unique IRR from this set of cash flows? Explain in one sentence.

5.24 Orillia Properties screens various projects using the payback period method. For renovation decisions, the minimum acceptable payback period is five years. Renovation projects are characterized by an immediate investment of P dollars which is recouped as an annuity of A dollars per year over 20 years. They are considering changing to the IRR method for such decisions. If they changed to the IRR method, what MARR would result in exactly the same decisions as their current policy using payback period?

5.25 Six mutually exclusive projects, A, B, C, D, E, and F, are being considered. They have been ordered by first costs so that project A has the smallest first cost, F the largest. The data in the table below applies to these projects. The data can be interpreted as follows: the IRR on the incremental investment between project D and project C is 6%. Which project should be chosen using a MARR of 15%?

	IRR on	IRR on Increments of Investment				
	Overall	Compared with Project				
Project	Investment	A	B	C	D	E
A	20%					
B	15%	12%				
C	24%	30%	35%			
D	16%	18%	22%	6%		
E	17%	16%	19%	15%	16%	
F	21%	20%	21%	19%	18%	11%

5.26 Three mutually exclusive designs for a by-pass are under consideration. The by-pass has a 10-year life. The first design incurs a cost of \$1.2 million for a net savings of \$300 000 per annum. The second design would cost \$1.5 million for a net savings of \$400 000 per annum. The third has a cost of \$2.1 million for a net savings of \$500 000 per annum.

For each of the alternatives, what range of values for the MARR results in its being chosen? It is not necessary that any be chosen.

5.27 Linus's project has cash flows at times 0, 1, and 2. He notices that for a MARR of 12%, the ERR falls exactly halfway between the MARR and the IRR, while for a MARR of 18%, the ERR falls exactly 1/4-way between the MARR and the IRR. If the cash flow is $2000 at time 2 and negative at time 0, what are the possible values of the cash flow at time 1?

5.28 Three construction jobs are being considered by Crystal City Construction (see the following table). Each is characterized by an initial deposit paid by the client to CCC, a yearly cost incurred by CCC at the end of each of three years, and a final payment to CCC by the client at the end of three years. CCC has the capacity to do only one of these contracts. Use an appropriate rate of return method to determine which they should do. Their MARR is 10%.

Job	Deposit ($)	Cost per Year ($)	Final Payment ($)
1	100 000	75 000	200 000
2	150 000	100 000	230 000
3	175 000	150 000	300 000

5.29 Kenora Karavans is considering three investment proposals. Each of them is characterized by an initial cost, annual savings over four years, and no salvage value, as illustrated in the following table. They can only invest in two of these proposals. If their MARR is 12%, which two should they choose?

Proposal	First Cost ($)	Annual Savings ($)
A	40 000	20 000
B	110 000	30 000
C	130 000	45 000

5.30 Development projects done by Produits Trois Rivières are subsidized by a government grant program. The program pays 30% of the total cost of the project (costs summed without discounting, i.e., the interest rate is zero), half at the beginning of the project and half at the end, up to a maximum of $100 000. There are two projects being considered. One is a customized checkweigher for cheese products, and the other is an automated production scheduling system. Each project has a service life of five years. Costs and benefits for both projects, not including grant income, are shown below. Only one can be done, and the grant money is certain. PTR has a MARR of 15% for projects of this type. Using an appropriate rate of return method, which project should be chosen?

	Checkweigher	Scheduler
First cost	$30 000	$10 000
Annual costs	$5 000	12 000
Annual benefits	$14 000	17 000
Salvage value	$8 000	0

5.31 Jacob is considering the replacement of the heating system for his building. There are three alternatives. All are natural-gas-fired furnaces, but they vary in energy efficiency. Model A is leased at a cost of $500 per year over a 10-year study period. There are installation charges of $500 and no salvage value. It is expected to provide energy savings of $200 per year. Model B is purchased for a total cost of $3600, including installation. It has a salvage value of $1000 after 10 years of service, and is expected to provide energy savings of $500 per year. Model C is also purchased, for a total cost of $8000, including installation. However, half of this cost is paid now, and the other half is paid at the end of two years. It has a salvage value of $1000 after 10 years, and is expected to provide energy savings of $1000 per year. For a MARR of 12% and using a rate of return method, which heating system should be installed? One model must be chosen.

5.32 Calgary Cartage leases trucks to service its shipping contracts. Larger trucks have cheaper operating costs if there is sufficient business, but are more expensive if they are not full. CC has estimates of monthly shipping demand. What comparison method(s) would be appropriate for choosing which trucks to lease?

5.33 The bottom flaps of shipping cartons for Yonge Auto Parts are fastened with industrial staples. Yonge needs to buy a new stapler. What comparison method(s) would be appropriate for choosing which stapler to buy?

5.34 Joan runs a dog kennel. She is considering installing a heating system for the interior runs which will allow her to operate all year. What comparison method(s) would be appropriate for choosing which heating system to buy?

5.35 A large Canadian food company is considering replacing a scale on its packaging line with a more accurate one. What comparison method(s) would be appropriate for choosing which scale to buy?

5.36 Mona runs a one-person company producing custom paints for hobbyists. She is considering buying printing equipment to produce her own labels. What comparison method(s) would be appropriate for choosing which equipment to buy?

5.37 Peter is the president of a rapidly growing company. There are dozens of important things to do, and cash flow is tight. What comparison method(s) would be appropriate for Peter to make acquisition decisions?

5.38 Lemuel is an engineer working for Hydro One. He must compare several routes for transmission lines from the Darlington nuclear plant to new industrial parks north of Toronto. What comparison method(s) is he likely to use?

5.39 Vicky runs a music store that has been suffering from thefts. She is considering installing a magnetic tag system. What comparison method(s) would be best for her to use to choose among competing leased systems?

5.40 Thanh's company is growing very fast and has a hard time meeting its orders. An opportunity to purchase additional production equipment has arisen. What comparison method(s) would Thanh use to justify to her manager that the equipment purchase was prudent?

The Galore Creek Project

NovaGold Resources is a former gold exploration company that has recently been transforming itself into a gold producer. Its first independent development is the Galore Creek Project, located 150 kilometres northeast of Stewart, British Columbia. It is also involved as a partner with Placer Dome in another project, and with Rio Tinto in a third. Galore Creek is expected to produce an average of 7650 kilograms of gold, 51 030 kilograms of silver, and 5 670 000 kilograms of copper over its first five years.

In a news release, NovaGold reported that an independent engineering services company calculated that the project would pay back the US$500 million mine capital costs in 3.4 years of a 23-year life. They also calculated a pre-tax rate of return of 12.6% and an undiscounted after-tax NPV of US$329 million. All of these calculations were done at long-term average metal prices. At then-current metal prices the pre-tax rate of return almost doubles to 24.3% and the NPV (Net Present Value = Present Worth) increases to US$1.065 billion.

Source: "Higher Grades and Expanded Tonnage Indicated by Drilling at Galore Creek Gold-Silver-Copper Project," news release, August 18, 2004, NovaGold Resources Inc. site, www.novagold.net, accessed October 12, 2004.

Discussion

Companies have a choice of how to calculate the benefits of a project in order to determine if it is worth doing. They also have a choice of how to report the benefits of a project to others.

NovaGold is a publicly traded company. Because of this, when a large and very important project is being planned, not only does NovaGold want to make good business decisions, but it also must maintain strong investor confidence and interest.

In this news release, payback period, IRR and NPV were used to communicate the value of the Galore Creek project. However, you need to look carefully at the wording to ensure that you can correctly interpret the claims about the economic viability of the project.

Questions

1. "[A]n independent engineering services company calculated that the project would pay back the US$500 million mine capital costs in 3.4 years of a 23-year life." There are a variety of costs associated with any project. The payback period here is calculated with respect to "mine capital costs." This suggests that there might be "non-mine" capital costs—for example, administrative infrastructure, transportation system, etc. It also means that operating costs are not included in this calculation. What do you think is the effect of calculating the payback period on "mine capital costs" alone?
2. "They also calculated a pre-tax rate of return of 12.6%. . . ." Taxes reduce the profit from an enterprise, and correspondingly reduce the rate of return. As will be seen in Chapter 8, Canadian companies generally pay about 1/2 of their profit to taxes. Thus if the pre-tax rate is 12.6%, the after-tax rate will be about 6.3%. Does 6.3% seem to you a sufficient return for a capital-intensive, risky project of this nature, given other investment opportunities available?
3. "[A]nd an 'undiscounted' after-tax NPV of US$329 million." The term "undiscounted" means that the present worth of the project was calculated with an interest rate of 0%. Using a spreadsheet, construct a graph showing the present worth of the project for a range of interest rates from 0% to 20%, assuming the annual returns for

the project are evenly distributed over the 23-year life of the project. Does the reported value of $US329 million fairly represent a meaningful NPV for the project?

4. The returns for the Galore Creek Project are much more attractive at then-current metal prices, which were significantly higher than long-term average metal prices. Which metal prices are more sensible to use when evaluating the worth of the project?
5. Did NovaGold report its economic evaluation of the Galore Creek Project in an ethical manner?

Appendix 5A Tests for Multiple IRRs

When the IRR method is used to evaluate projects, we have to test for multiple IRRs. If there are undetected multiple IRRs, an IRR might be calculated that seems correct, but is in error. We consider three tests for multiple IRRs, forming essentially a three-step procedure. In the first test, the signs of the cash flows are examined to see if the project is a simple investment. In the second test, the present worth of the project is plotted against the interest rate to search for interest rates at which the present value is zero. In the third test, project balances are calculated. Each of these tests has three possible outcomes:

1. There is definitely a unique IRR and there is no possibility for multiple IRRs.
2. There are definitely multiple IRRs because two or more IRRs have been found.
3. The test outcome is inconclusive; a unique IRR or multiple IRRs are both possible.

The tests are applied sequentially. The second test is applied only if the outcome of the first test is not clear. The third test is applied only if the outcomes of the first two are not clear. Keep in mind that, even after all three tests have been applied, the test outcomes may remain inconclusive.

The first test examines whether the project is simple. Recall that most projects consist of one or more periods of outflows at the start, followed only by one or more periods of inflows; these are called simple investments. Although simple investments guarantee a single IRR, a project that is not simple may have a single IRR or multiple IRRs. Some investment projects have large cash outflows during their lives or at the ends of their lives that cause net cash flows to be negative after years of positive net cash flows. For example, a project that involves the construction of a manufacturing plant may involve a planned expansion of the plant requiring a large expenditure some years after its initial operation. As another example, a nuclear electricity plant may have planned large cash outflows for disposal of spent fuel at the end of its life. Such a project may have a unique IRR, but it may also have multiple IRRs. Consequently we must examine such projects further.

Where a project is not simple, we go to the second test. The second test consists of making a graph plotting present worth against interest rate. Points at which the present worth crosses or just touches the interest-rate axis (i.e., where present worth = 0) are IRRs. (We assume that there is at least one IRR.) If there are more than one such point, we know that there are more than one IRR. A convenient way to produce such a graph is using a spreadsheet. See Example 5A.1.

Example 5A.1 (Example 5.7 Restated)

A project pays $1000 today, costs $5000 a year from now, and pays $6000 in two years. Are there multiple IRRs?

Table 5A.1 was obtained by computing the present worth of the cash flows in Example 5.7 for a variety of interest rates. Figure 5A.1 shows the graph of the values in Table 5A.1.

Table 5A.1 Spreadsheet Cells Used to Construct Figure 5A.1

Interest Rate, *i*	Present Worth ($)
0.6	218.8
0.8	74.1
1.0	0.0
1.2	−33.1
1.4	−41.7
1.6	−35.5
1.8	−20.4
2.0	0.0
2.2	23.4
2.4	48.4

While finding multiple IRRs in a plot ensures that the project does indeed have multiple IRRs, failure to find multiple IRRs does not necessarily mean that multiple IRRs do not exist. *Any* plot will cover only a finite set of points. There may be values of the interest rate for which the present worth of the project is zero that are not in the range of interest rates used.

Where the project is not simple and a plot does not show multiple IRRs, we apply the third test. The third test entails calculating the project balances. As we mentioned

Figure 5A.1 Illustration of Two IRRs for Example 5A.1

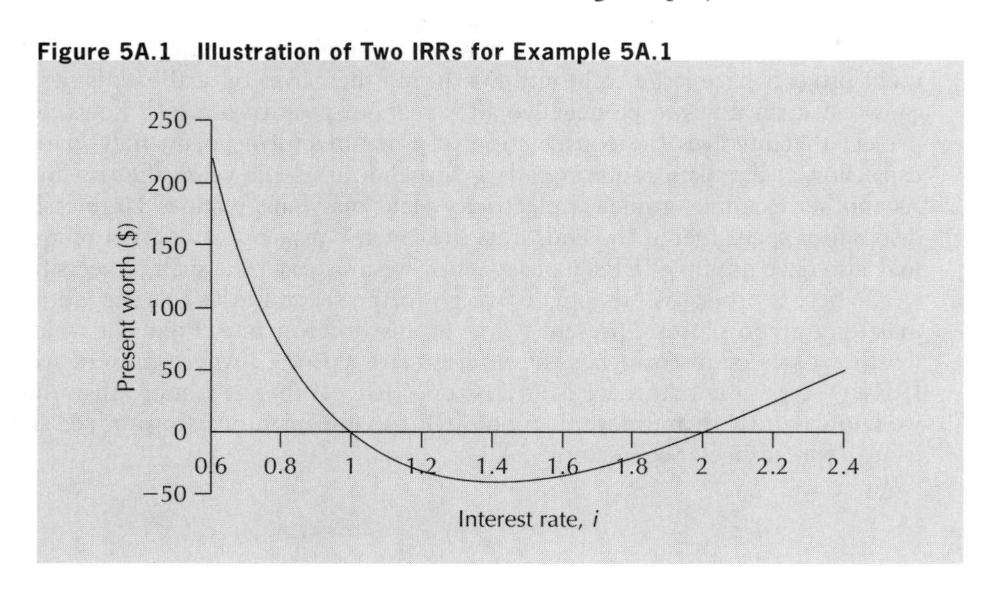

earlier, project balances refer to the cumulative net cash flows at the end of each time period. For an IRR to be unique, there should be no time when the project balances, computed using that IRR, are positive. This means that there is no extra cash not reinvested in the project. This is a sufficient condition for there to be a unique IRR. (Recall that it is the cash generated by a project, but not reinvested in the project, that creates the possibility of multiple IRRs.)

We now present three examples. All three examples involve projects that are not simple investments. In the first, a plot shows multiple IRRs. In the second, the plot shows only a single IRR. This is inconclusive, so project balances are computed. None of the project balances is positive, so we know that there is a single IRR. In the third example, the plot shows only one IRR, so the project balances are computed. One of these is positive, so the results of all tests are inconclusive.■

Example 5A.2

Wellington Woods is considering buying land that they will log for three years. In the second year, they expect to develop the area that they clear as a residential subdivision that will entail considerable costs. Thus, in the second year, the net cash flow will be negative. In the third year, they expect to sell the developed land at a profit. The net cash flows that are expected for the project are:

End of Year	Cash Flow
0	−$100 000
1	440 000
2	− 639 000
3	306 000

The negative net cash flow in the second period implies that this is not a simple project. Therefore, we apply the second test. We plot the present worth against interest rates to search for IRRs. (See Figure 5A.2.) At 0% interest, the present worth is a small positive amount, $7000. The present worth is then 0 at 20%, 50%, and 70%. Each of these values is an IRR. The spreadsheet cells that were used for the plot are shown in Table 5A.2.

In this example, a moderately fine grid of two percentage points was used. Depending on the problem, the analyst may wish to use a finer or coarser grid.

The correct decision in this case can be obtained by the approximate ERR method. Suppose the MARR is 15%; then the approximate ERR is the interest rate that makes the future worth of outlays equal to the future worth of receipts when the receipts earn 15%. In other words, the approximate ERR is the value of i that solves

$$\$100\ 000(1 + i)^3 + \$639\ 000(1 + i) = \$440\ 000(1.15)^2 + \$306\ 000$$

$$\$100\ 000(1 + i)^3 + \$639\ 000(1 + i) = \$887\ 900$$

Try 15% for i^*_{ea}. Using the tables for the left-hand side of the above equation, we have

$$\$100\ 000(F/P, 15\%, 3) + \$639\ 000(F/P, 15\%, 1)$$

$$= \$100\ 000(1.5208) + \$639\ 000(1.15)$$

$$= \$887\ 000 < \$887\ 900$$

Figure 5A.2 Wellington Woods Present Worth

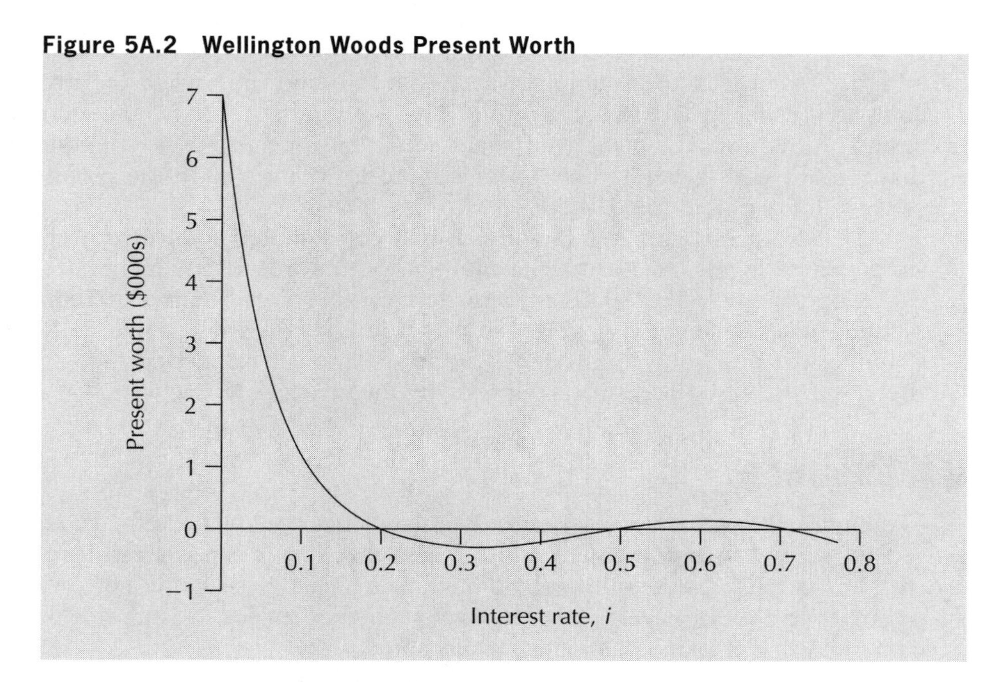

Thus, the approximate ERR is slightly above 15%. The project is (marginally) acceptable by this calculation because the approximate ERR, which is a conservative estimate of the correct ERR, is above the MARR.■

Table 5A.2 Spreadsheet Cells Used to Generate Figure 5A.2

Interest Rate	Present Worth	Interest Rate	Present Worth	Interest Rate	Present Worth
0%	$7000.00	28%	− 352.48	56%	$ 79.65
2%	5536.33	30%	− 364.13	58%	92.49
4%	4318.39	32%	− 356.87	60%	97.66
6%	3310.12	34%	− 335.15	62%	94.84
8%	2480.57	36%	− 302.77	64%	83.79
10%	1803.16	38%	− 263.01	66%	64.36
12%	1255.01	40%	− 218.66	68%	36.44
14%	816.45	42%	− 172.11	70%	0.00
16%	470.50	44%	− 125.39	72%	− 44.96
18%	202.55	46%	− 80.20	74%	− 98.41
20%	0.00	48%	− 38.00	76%	− 160.24
22%	− 148.03	50%	0.00	78%	− 230.36
24%	− 250.91	52%	32.80		
26%	− 316.74	54%	59.58		

Example 5A.3

Investment in a new office coffeemaker has the following effects:

1. There is a three-month rental fee of $40 for the equipment, payable immediately and in three months.
2. A rebate of $30 from the supplier is given immediately for an exclusive six-month contract.
3. Supplies will cost $20 per month, payable at the beginning of each month.
4. Income from sales will be $30 per month, received at the end of each month.

Will there be more than one IRR for this problem?

We apply the first test by calculating net cash flows for each time period. The net cash flows for this project are as follows:

End of Month	Receipts	Disbursements	Net Cash Flow
0	+ $30	− $40 + (− $20)	− $30
1	+ 30	− 20	+ 10
2	+ 30	− 20	+ 10
3	+ 30	− 40 + (− 20)	− 30
4	+ 30	− 20	+ 10
5	+ 30	− 20	+ 10
6	+ 30	0	+ 30

As illustrated in Figure 5A.3, the net cash flows for this problem do not follow the pattern of a simple investment. Therefore, there may be more than one IRR for this problem. Accordingly, we apply the second test.

A plot of present worth against the interest rate is shown in Figure 5A.4. The plot starts with a zero interest rate where the present worth is just the arithmetic sum of all the net cash flows over the project's life. This is a positive $10. The present worth as a

Figure 5A.3 Net Cash Flows for the Coffeemaker

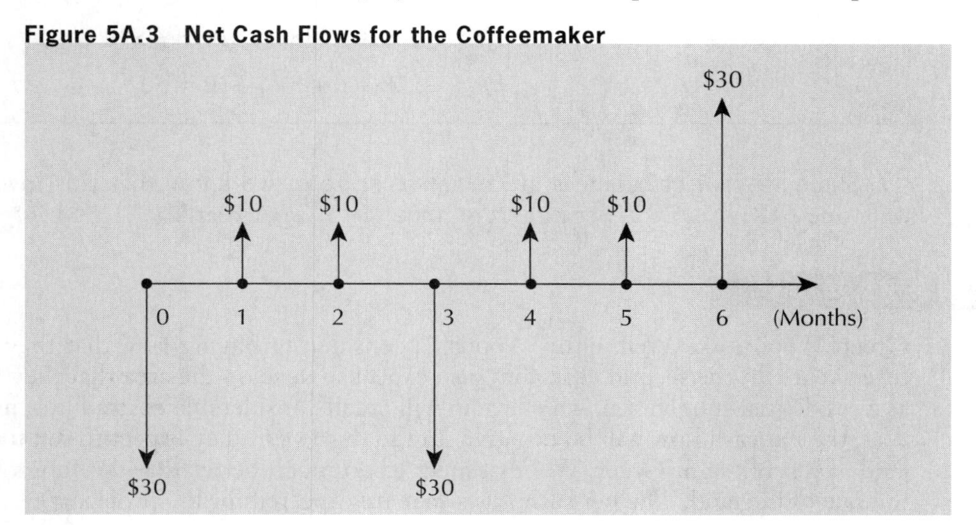

function of the interest rate decreases as the interest rate increases from zero. The plot continues down, and passes through the interest-rate axis at about 5.8%. There is only one IRR in the range plotted. We need to apply the third test by computing project balances at the 5.8% interest rate.

Figure 5A.4 IRR for New Coffeemaker

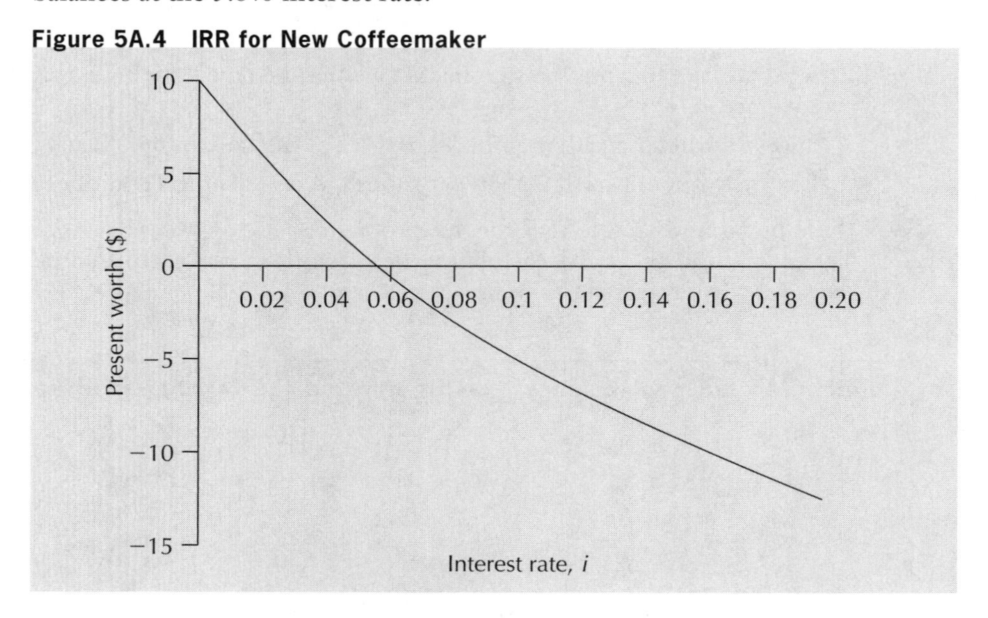

The project balances at the 5.8% interest rate are as follows:

Month	Project Balance
0	$B_0 = -\$30$
1	$B_1 = -\$30.0(1.058) + \$10 = -\$21.7$
2	$B_2 = -\$21.7(1.058) + \$10 = -\$13.0$
3	$B_3 = -\$13.0(1.058) - \$30 = -\$43.7$
4	$B_4 = -\$43.7(1.058) + \$10 = -\$36.3$
5	$B_5 = -\$36.3(1.058) + \$10 = -\$28.4$
6	$B_6 = -\$28.4(1.058) + \$30 = \$0$

Since all project balances are negative or zero, we know that this investment has only one IRR. It is 5.8% per month or about 69.3% per year.■

Example 5A.4

Green Woods, like Wellington Woods, is considering buying land that they will log for three years. In the second year, they also expect to develop the area that they have logged as a residential subdivision, which again will entail considerable costs. Thus, in the second year, the net cash flow will be negative. In the third year, they expect to sell the developed land at a profit. But Green Woods expect to do much better than Wellington Woods in the sale of the land. The net cash flows that are expected for the project are:

Year	Cash Flow
0	−$100 000
1	455 000
2	− 667 500
3	650 000

The negative net cash flow in the second period implies that this is not a simple project. We now plot the present worth against interest rate. See Figure 5A.5. At zero interest rate, the present worth is a positive $337 500.

Figure 5A.5 Present Worths for Example 5A.4

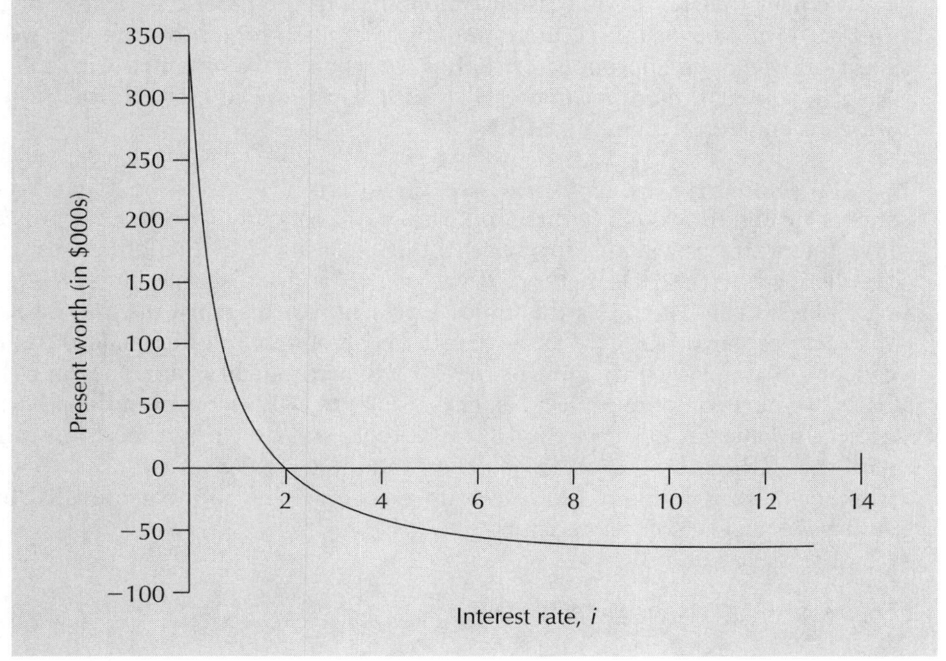

The present worth falls as the interest rate rises. It crosses the interest-rate axis at about 206.4%. There are no further crossings of the interest-rate axis in the range plotted, but since this is not conclusive we compute project balances.

Year	Project Balance
0	$B_0 = -\ \$100\,000$
1	$B_1 = -\ \$100\,000(3.064) + \$455\,000 = \$148\,600$
2	$B_2 = -\ \$148\,600(3.064) - \$667\,500 = -\ \$2\,121\,900$
3	$B_3 = -\ \$2\,121\,900(3.064) + \$650\,000 = -\ \$1500$

We note that the project balance is positive at the end of the first period. This means that a unique IRR is *not* guaranteed. We have gone as far as the three tests can take us. On the basis of the three tests, there may be only the single IRR that we have found, 206.4%, or there may be multiple IRRs.

In this case, we use the approximate ERR to get a decision. Suppose the MARR is 30%. The approximate ERR, then, is the interest rate that makes the future worth of outlays equal to the future worth of receipts when the receipts earn 30%. That is, the approximate ERR is the value of i that solves the following:

$$\$100\,000(1 + i)^3 + \$667\,500(1 + i) = \$455\,000(1.3)^2 + \$650\,000$$

Trial and error with a spreadsheet gives the approximate ERR as $i^*_{ea} \cong 57\%$. This is above the MARR of 30%. Therefore, the investment is acceptable.

It is possible, using the precise ERR, to determine that the IRR that we got with the plot of present worth against the interest rate, 206.4%, is, in fact, unique. The precise ERR equals the IRR in this case. Computation of the precise ERR may be cumbersome, and we do not cover this computation in this book. Note, however, that we got the same decision using the approximate ERR as we would have obtained with the precise ERR. Also note that the approximate ERR is a conservative estimate of the precise ERR, which is equal to the unique IRR.■

To summarize, we have discussed three tests that are to be applied sequentially, as shown in Figure 5A.6. The first and easiest test to apply is to see if a project is a simple investment. If it is a simple investment, there is a single IRR, and the correct decision can be obtained by the IRR method. If the project is not a simple investment, we apply the second test, which is to plot the project's present worth against the interest rate. If the plot shows at least two IRRs, we know that there is not a unique IRR, and the correct decision can be obtained with the approximate ERR method. If a plot does not show multiple IRRs, we next compute project balances using the IRR found from the plot. If none of the project balances is positive, the IRR is unique, and the correct decision can be obtained with the IRR method. If one or more of the project balances are positive, we don't know whether there is a unique IRR. Accordingly, we use the approximate ERR method which always will yield a correct decision.

Figure 5A.6 Tests for Multiple IRRs

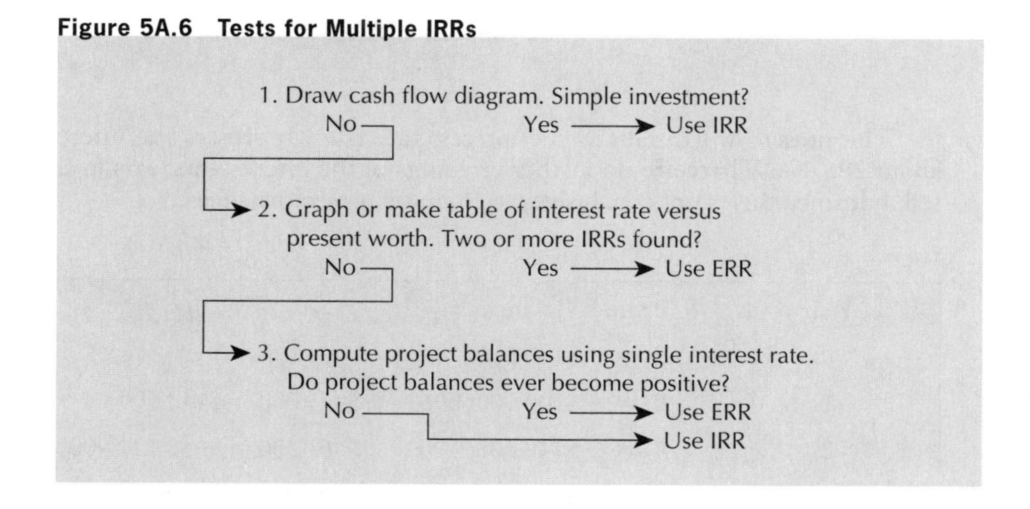

PROBLEMS FOR APPENDIX 5A

5A.1 A five-year construction project for Wawa Engineering receives staged payments in years 2 and 5. The resulting net cash flows are as follows:

Year	Cash Flow
0	−$300 000
1	− 500 000
2	700 000
3	− 400 000
4	− 100 000
5	900 000

The MARR for Wawa Engineering is 15%.

(a) Is this a simple project?

(b) Plot the present worth of the project against interest rates from 0% to 100%. How many times is the interest-rate axis crossed? How many IRRs are there?

(c) Calculate project balances over the five-year life of the project. Can we conclude that the IRR(s) observed in (b) is (are) the only IRR(s)? If so, should the project be accepted?

(d) Calculate the approximate ERR for this project. Should the project be accepted?

5A.2 For the cash flows associated with the projects below, determine whether there is a unique IRR, using the project balances method.

	Project		
End of Period	1	2	3
0	−$3000	−$1500	$ 600
1	900	7000	− 2000
2	900	− 9000	500
3	900	2900	500
4	900	500	1000

5A.3 A mining opportunity in a third-world country has the following cash flows:

1. $10 000 000 is received at time 0 as an advance against expenses.
2. Costs in the first year are $20 000 000, and in the second year $10 000 000.
3. Over years 3 to 10, annual revenues of $5 000 000 are expected.

After 10 years, the site reverts to government ownership. MARR is 30%.

(a) Is this a simple project?

(b) Plot the present worth of the project against interest rates from 0% to 100%. How many times is the interest-rate axis crossed? How many IRRs are there?

(c) Calculate project balances over the 10-year life of the project. Can we conclude that the IRR(s) observed in (b) is (are) the only IRR(s)? If so, should the project be accepted?

(d) Calculate the approximate ERR for this project. Should the project be accepted?

(e) Calculate the exact ERR for this project. Should the project be accepted?

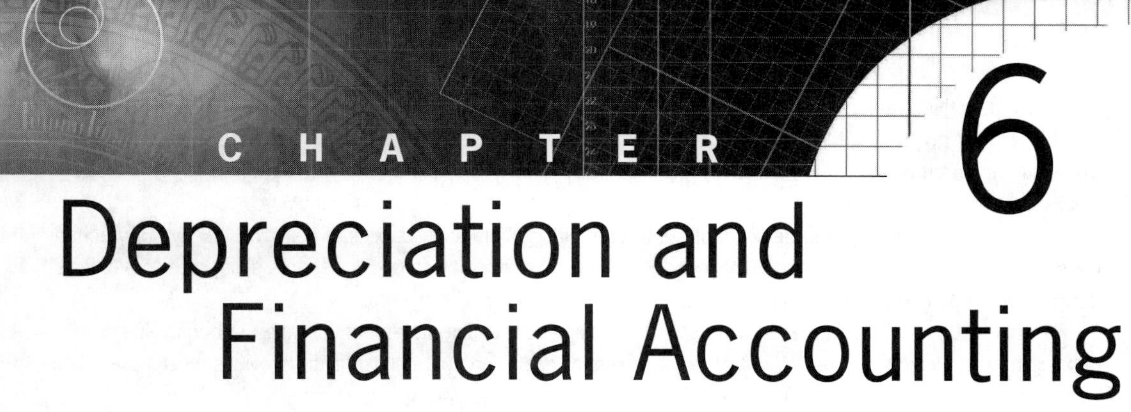

6 Depreciation and Financial Accounting

Engineering Economics in Action, Part 6A: The Pit Bull

Naomi liked to think of Terry as a pit bull. Terry had this endearing habit of finding some detail that irked him, and not letting go of it until he was satisfied that things were done properly. Naomi had seen this several times in the months they had worked together. Terry would sink his teeth into some quirk of Canadian Widgets' operating procedures and, just like a fighting dog, not let go until the fight was over.

This time, it was about the disposal of some computing equipment. Papers in hand, he quietly approached Naomi and earnestly started to explain his concern. "Naomi, I don't know what Bill Fisher is doing, but something's definitely not right here. Look at this."

Terry displayed two documents to Naomi. One was an accounting statement showing the book value of various equipment, including some CAD/CAM computers that had been sold for scrap the previous week. The other was a copy of a sales receipt from a local salvage firm for that same equipment.

"I don't like criticizing my fellow workers, but I really am afraid that Bill might be doing something wrong." Bill Fisher was the buyer responsible for capital equipment at Canadian Widgets, and he also disposed of surplus assets. "You know the CAD/CAM workstations they had in Engineering Design? Well, they were replaced recently and sold. Here is the problem. They were only three years old, and our own accounting department estimated their value as about $5000 each." Terry's finger pointed to the evidence on the accounting statement. "But here," his finger moving to the guilty figure on the sales receipt, "they were actually sold for $300 each!" Terry sat back in his chair. "How about that!"

Naomi smiled. Unfortunately, she would have to pry his teeth out of this one. "Interesting observation, Terry. But you know, I think it's probably OK. Let me explain."

6.1 Introduction

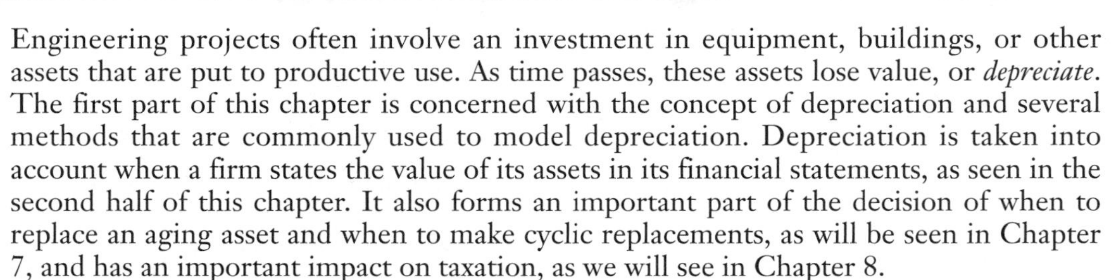

Engineering projects often involve an investment in equipment, buildings, or other assets that are put to productive use. As time passes, these assets lose value, or *depreciate*. The first part of this chapter is concerned with the concept of depreciation and several methods that are commonly used to model depreciation. Depreciation is taken into account when a firm states the value of its assets in its financial statements, as seen in the second half of this chapter. It also forms an important part of the decision of when to replace an aging asset and when to make cyclic replacements, as will be seen in Chapter 7, and has an important impact on taxation, as we will see in Chapter 8.

With the growth in importance of small technology-based enterprises, many engineers have taken on broad managerial responsibilities that include financial accounting. Financial accounting is concerned with recording and organizing the financial data of businesses. The data cover both *flows over time*, like revenues and expenses, and *levels*, like an enterprise's resources and the claims on those resources at a given date. Even engineers who do not have broad managerial responsibilities need to know the elements of financial accounting to understand the enterprises with which they work.

In the second part of this chapter, we explain two basic financial statements used to summarize the financial dimensions of a business. We then explain how these statements can be used to make inferences about the financial health of the firm.

6.2 Depreciation and Depreciation Accounting

6.2.1 Reasons for Depreciation

An asset starts to lose value as soon as it is purchased. For example, a car bought for $20 000 today may be worth $18 000 next week, $15 000 next year, and $1000 in 10 years. This loss in value, called **depreciation**, occurs for several reasons.

Use-related physical loss: As something is used, parts wear out. An automobile engine has a limited life span because the metal parts within it wear out. This is one reason why a car diminishes in value over time. Often, use-related physical loss is measured with respect to *units of production*, such as thousands of kilometres for a car, hours of use for a light bulb, or thousands of cycles for a punch press.

Time-related physical loss: Even if something is not used, there can be a physical loss over time. This can be due to environmental factors affecting the asset or to endogenous physical factors. For example, an unused car can rust and thus lose value over time. Time-related physical loss is expressed in units of time, such as a 10-year-old car or a 40-year-old sewage treatment plant.

Functional loss: Losses can occur without any physical changes. For example, a car can lose value over time because styles change so that it is no longer fashionable. Other examples of causes of loss of value include legislative changes, such as for pollution control or safety devices, and technical changes. Functional loss is usually expressed simply in terms of the particular unsatisfied function.

6.2.2 Value of an Asset

Models of depreciation can be used to estimate the loss in value of an asset over time, and also to determine the remaining value of the asset at any point in time. This remaining value has several names, depending on the circumstances.

Market value is usually taken as the actual value an asset can be sold for in an open market. Of course, the only way to determine the actual market value for an asset is to sell it. Consequently, the term *market value* usually means an *estimate* of the market value. One way to make such an estimation is by using a depreciation model that reasonably captures the true loss in value of an asset.

Book value is the depreciated value of an asset for accounting purposes, as calculated with a depreciation model. The book value may be more or less than market value. The depreciation model used to arrive at a book value might be controlled by regulation for some purposes, such as taxation, or simply by the desirability of an easy calculation scheme. There might be several different book values for the same asset, depending on the purpose and depreciation model applied. We shall see how book values are reported in financial statements later in this chapter.

Scrap value can be either the actual value of an asset at the end of its physical life (when it is broken up for the material value of its parts) or an estimate of the scrap value calculated using a depreciation model.

Salvage value can be either the actual value of an asset at the end of its useful life (when it is sold) or an estimate of the salvage value calculated using a depreciation model.

It is desirable to be able to construct a good model of depreciation in order to state a book value of an asset for a variety of reasons:

1. In order to make many managerial decisions, it is necessary to know the value of owned assets. For example, money may be borrowed using the firm's assets as collateral. In order to demonstrate to the lender that there is security for the loan, a credible estimate of the assets' value must be made. A depreciation model permits this to be done. The use of depreciation for this purpose is explored more thoroughly in the second part of this chapter.
2. One needs an estimate of the value of owned assets for planning purposes. In order to decide whether to keep an asset or replace it, you have to be able to judge how much it is worth. More than that, you have to be able to assess how much it will be worth at some time in the future. The impact of depreciation in replacement studies is covered in Chapter 7.
3. Government tax legislation requires that taxes be paid on company profits. Because there can be many ways of calculating profits, strict rules are made concerning how to calculate income and expenses. These rules include a particular scheme for determining depreciation expenses. This use of depreciation is discussed more thoroughly in Chapter 8.

To match the way in which certain assets depreciate and to meet regulatory or accuracy requirements, many different depreciation models have been developed over time. Of the large number of depreciation schemes available (see Close-Up 6.1), straight-line and declining-balance are certainly the most commonly used in Canada. Straight-line depreciation is popular primarily because it is particularly easy to calculate. The declining-balance method is required by Canadian tax law for determining corporate taxes, as is discussed in Chapter 8. In the United States, tax laws prior to 1954 required the use of straight-line depreciation, and between 1954 and 1981, several other methods were permitted. Under current United States law, things are more complicated, but the main depreciation methods used are straight-line and declining-balance. Consequently, these are the only depreciation methods presented in detail in this book.

CLOSE-UP 6.1 Depreciation Methods

Method	Description
Straight-line	The book value of an asset diminishes by an equal *amount* each year.
Declining-balance	The book value of an asset diminishes by an equal *proportion* each year.
Sum-of-the-years'-digits	An accelerated method, like declining-balance, in which the depreciation rate is calculated as the ratio of the remaining years of life to the sum of the digits corresponding to the years of life.
Double-declining-balance	A declining-balance method in which the depreciation rate is calculated as $2/N$ for an asset with a service life of N years.
150%-declining-balance	A declining-balance method in which the depreciation rate is calculated as $1.5/N$ for an asset with a service life of N years.
Units-of-production	Depreciation rate is calculated per unit of production as the ratio of the units produced in a particular year to the total estimated units produced over the asset's lifetime.

6.2.3 Straight-Line Depreciation

The **straight-line method of depreciation** assumes that the rate of loss in value of an asset is constant over its useful life. This is illustrated in Figure 6.1 for an asset worth \$1000 at the time of purchase and \$200 eight years later. Graphically, the book value of the asset is determined by drawing a straight line between its first cost and its salvage or scrap value.

Figure 6.1 Book Value under Straight-Line Depreciation (\$1000 Purchase and \$200 Salvage Value After Eight Years)

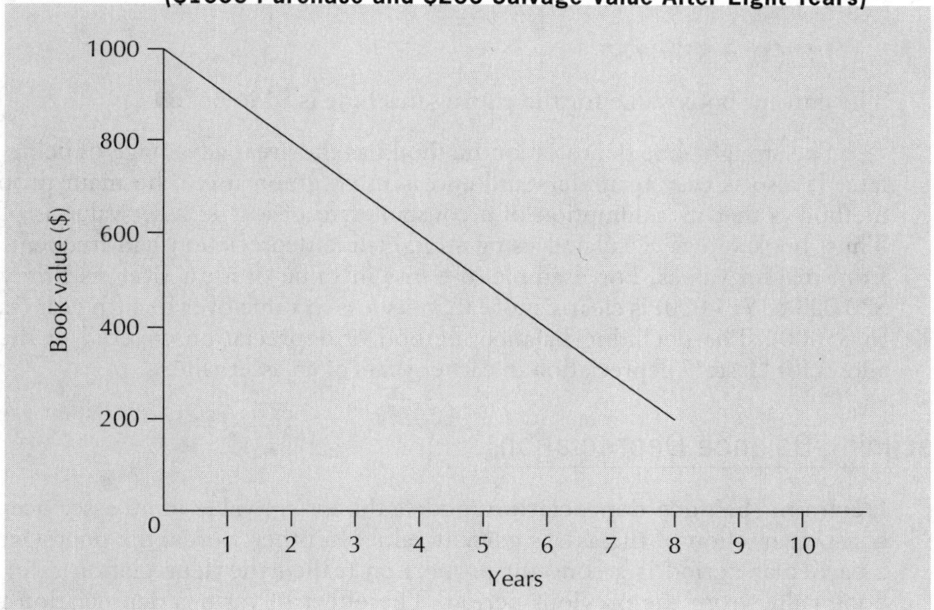

Algebraically, the assumption is that the rate of loss in asset value is constant and is based on its original cost and salvage value. This gives rise to a simple expression for the depreciation charge per period. We determine the depreciation per period from the asset's current value and its estimated salvage value at the end of its useful life, N periods from now, by

$$D_{sl}(n) = \frac{P - S}{N} \tag{6.1}$$

where

$D_{sl}(n)$ = the depreciation charge for period n using the straight-line method

P = the purchase price or current market value

S = the salvage value after N periods

N = the useful life of the asset, in periods

Similarly, the book value at the end of any particular period is easy to calculate:

$$BV_{sl}(n) = P - n\left[\frac{P - S}{N}\right] \tag{6.2}$$

where

$BV_{sl}(n)$ = the book value at the end of period n using straight-line depreciation

Example 6.1

A laser cutting machine was purchased four years ago for \$380 000. It will have a salvage value of \$30 000 two years from now. If we believe a constant rate of depreciation is a reasonable means of determining book value, what is its current book value?

From Equation (6.2), with P = \$380 000, S = \$30 000, N = 6, and n = 4,

$$BV_{sl}(4) = \$380\,000 - 4\left[\frac{\$380\,000 - \$30\,000}{6}\right]$$

$$BV_{sl}(4) = \$146\,667$$

The current book value for the cutting machine is \$146 667.■

The straight-line depreciation method has the great advantage of being easy to calculate. It also is easy to understand and is in common use. The main problem with the method is that its assumption of a constant rate of loss in asset value is often not valid. Thus, book values calculated using straight-line depreciation will frequently be different from market values. For example, the loss in value of a car over its first year (say from \$20 000 to \$15 000) is clearly more than its loss in value over its fifth year (say from \$6000 to \$5000). The declining-balance method of depreciation covered in the next section allows for "faster" depreciation in earlier years of an asset's life.

6.2.4 Declining-Balance Depreciation

Declining-balance depreciation models the loss in value of an asset over a period as a constant fraction of the asset's current value. In other words, the depreciation charge in a particular period is a constant proportion (called the depreciation rate) of its closing book value from the previous period. The effect of various depreciation rates on book values is illustrated in Figure 6.2.

Figure 6.2 Book Value Under Declining-Balance Depreciation ($1000 Purchase with Various Depreciation Rates)

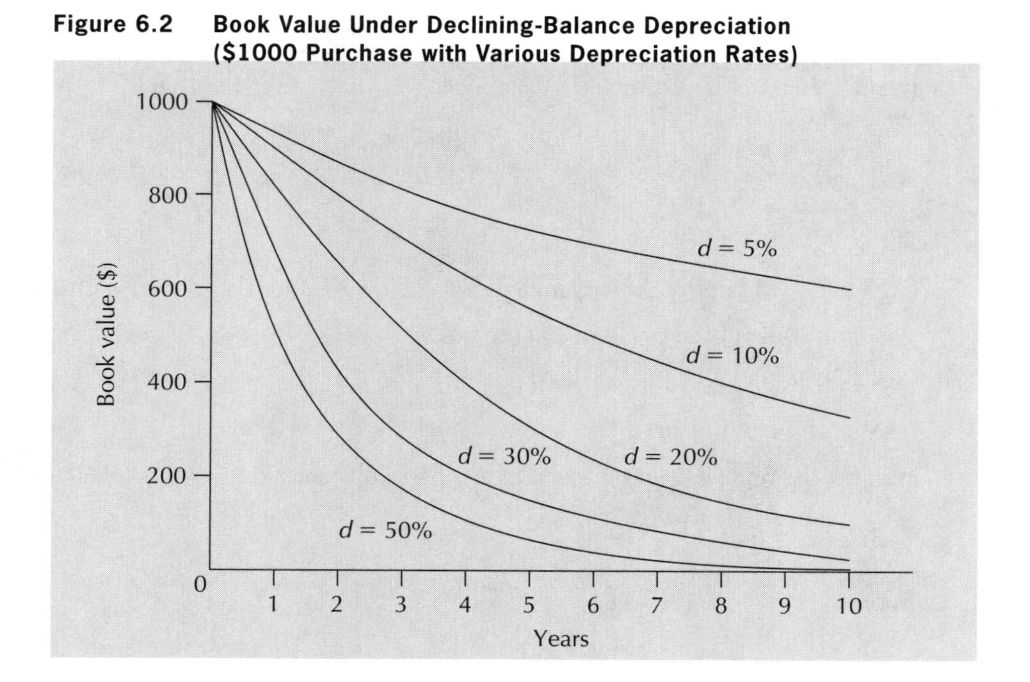

Algebraically, the depreciation charge for period n is simply the depreciation rate multiplied by the book value from the end of period $(n - 1)$. Noting that $BV_{db}(0) = P$,

$$D_{db}(n) = BV_{db}(n - 1) \times d \tag{6.3}$$

where

$D_{db}(n)$ = the depreciation charge in period n using the declining-balance method

$BV_{db}(n)$ = the book value at the end of period n using the declining-balance method

P = the purchase price or current market value

d = the depreciation rate

Similarly, the book value at the end of any particular period is easy to calculate, by noting that the remaining value after each period is $(1 - d)$ times the value at the end of the previous period.

$$BV_{db}(n) = P(1 - d)^n \tag{6.4}$$

In order to use the declining-balance method of depreciation, we must determine a reasonable depreciation rate. By using an asset's current value, P, and a salvage value, S, n periods from now, we can use Equation (6.4) to find the declining balance rate that relates P and S.

$$BV_{db}(n) = S = P(1 - d)^n$$

$$(1 - d) = \sqrt[n]{\frac{S}{P}}$$

$$d = 1 - \sqrt[n]{\frac{S}{P}} \tag{6.5}$$

Example 6.2

Paquita wants to estimate the scrap value of a smokehouse 20 years after purchase. She feels that the depreciation is best represented using the declining-balance method, but she doesn't know what depreciation rate to use. She observes that the purchase price of the smokehouse was \$245 000 three years ago, and an estimate of its current salvage value is \$180 000. What is a good estimate of the value of the smokehouse after 20 years?

From Equation (6.5),

$$d = 1 - \sqrt[n]{\frac{S}{P}}$$

$$= 1 - \sqrt[3]{\frac{\$180\ 000}{\$245\ 000}}$$

$$= 0.097663$$

Then, by using Equation (6.4), we have

$$BV_{db}(20) = \$245\ 000(1 - 0.097663)^{20} = \$31\ 372$$

An estimate of the salvage value of the smokehouse after 20 years using the declining-balance method of depreciation is \$31 372.■

The declining-balance method has a number of useful features. For one thing, it matches the observable loss in value that many assets have over time. The rate of loss is expressed in one parameter, the depreciation rate. It is relatively easy to calculate, although perhaps not quite as easy as the straight-line method. In particular, it is required to be used in Canada for taxation purposes, as discussed in detail in Chapter 8.

Example 6.3

Sherbrooke Data Services has purchased a new mass storage system for \$250 000. It is expected to last six years, with a \$10 000 salvage value. Using both the straight-line and declining-balance methods, determine the following:

(a) The depreciation charge in year 1

(b) The depreciation charge in year 6

(c) The book value at the end of year 4

(d) The accumulated depreciation at the end of year 4

This is an ideal application for a spreadsheet solution. Table 6.1 illustrates a spreadsheet that calculates the book value, depreciation charge, and accumulated depreciation for both depreciation methods over the six-year life of the system.

Table 6.1 Spreadsheet for Example 6.3

	Straight-Line Depreciation		
Year	Depreciation Charge	Accumulated Depreciation	Book Value
0			\$250 000
1	\$40 000	\$ 40 000	210 000
2	40 000	80 000	170 000
3	40 000	120 000	130 000
4	40 000	160 000	90 000
5	40 000	200 000	50 000
6	40 000	240 000	10 000

	Declining-Balance Depreciation		
Year	Depreciation Charge	Accumulated Depreciation	Book Value
0			\$250 000
1	\$103 799	\$103 799	146 201
2	60 702	164 501	85 499
3	35 499	200 000	50 000
4	20 760	220 760	29 240
5	12 140	232 900	17 100
6	7 100	240 000	10 000

The depreciation charge for each year with the *straight-line* method is $40 000:

$D_{sl}(n) = (\$250\ 000 - \$10\ 000)/6 = \$40\ 000$

The depreciation rate for the *declining-balance* method is

$$d = 1 - \sqrt[n]{\frac{S}{P}} = 1 - \sqrt[6]{\frac{\$10\ 000}{\$250\ 000}} = 0.4152$$

The detailed calculation for each of the questions is as follows:

(a) Depreciation charge in year 1

$D_{sl}(1) = (\$250\ 000 - \$10\ 000)/6 = \$40\ 000$

$D_{db}(1) = BV_{db}(0)d = \$250\ 000(0.4152) = \$103\ 799.11$

(b) Depreciation charge in year 6

$D_{sl}(6) = D_{sl}(1) = \$40\ 000$

$D_{db}(6) = BV_{db}(5)d = \$250\ 000(0.5848)^5(0.4152) = \$7099.82$

(c) Book value at the end of year 4

$BV_{sl}(4) = \$250\ 000 - 4(\$250\ 000 - \$10\ 000)/6 = \$90\ 000$

$BV_{db}(4) = \$250\ 000(1 - 0.4152)^4 = \$29\ 240.17$

(d) Accumulated depreciation at the end of year 4

Using the straight-line method: $P - BV_{sl}(4) = \$160\ 000$

Using the declining-balance method: $P - BV_{db}(4) = \$220\ 759.83$ ■

In summary, depreciation affects economic analyses in several ways. First, it allows us to estimate the value of an owned asset, as illustrated in the above examples. We shall see in the next part of this chapter how these values are reported in a firm's financial statements. Next, the capability of estimating the value of an asset is particularly useful in replacement studies, which is the topic of Chapter 7. Finally, in Chapter 8, we cover aspects of the Canadian tax system that affect decision making; in particular we look at the effect of depreciation expenses.

6.3 Elements of Financial Accounting

How well is a business doing? Can it survive an unforeseen temporary drop in cash flows? How does a business compare with others of its size in the industry? Answering these questions and others like them is part of the accounting function. The accounting function has two parts, financial accounting and management accounting. **Financial accounting** is concerned with recording and organizing the financial data of a business, which include revenues and expenses, and an enterprise's resources and the claims on those resources. **Management accounting** is concerned with the costs and benefits of the various activities of an enterprise. The goal of management accounting is to provide managers with information to help in decision making.

Engineers have always played a major role in management accounting, especially in a part of management accounting called *cost* accounting. They have not, for the most part, had significant responsibility for financial accounting until recently. With the growth in importance of small technology-based enterprises, many engineers have taken on broad managerial responsibilities that include financial accounting. Even engineers who do not have broad managerial responsibilities need to know the elements of financial accounting to understand the enterprises in which they work. Management accounting is not covered in this text because it is difficult to provide useful information without taking more than a single chapter. Instead, we focus on financial accounting.

The object of financial accounting is to provide information to internal management and interested external parties. Internally, management uses financial accounting information for processes such as budgeting, cash management, and management of long-term debt. External users include actual and potential investors and creditors who wish to make rational decisions about an enterprise. External users also include government agencies concerned with taxes and regulation.

Areas of interest to all these groups include an enterprise's revenues and expenses, and assets (resources held by the enterprise) and liabilities (claims on those resources).

In the next few sections, we discuss two basic summary financial statements that give information about these matters: the *balance sheet* and the *income statement*. These statements form the basis of a financial report, which is usually produced on a monthly, quarterly, or yearly basis. Following the discussion of financial statements, we shall consider the use of information in these statements when making inferences about an enterprise's performance compared with industry standards and with its own performance over time.

6.3.1 Measuring the Performance of a Firm

The flow of money in a company is much like the flow of water in a network of pipes or the flow of electricity in an electrical circuit, as illustrated in Figure 6.3. In order to measure the performance of a water system, we need to determine the flow through the system and the pressure in the system. For an electrical circuit, the analogous parameters are current and voltage. Flow and current are referred to as *through variables*, and are measured with respect to time (flow is litres per second and current is amperes, which are coulombs per second). Pressure and voltage are referred to as *across variables*, and are measured at a point in time.

The flow of money in an organization is measured in a similar way with the income statement and balance sheet. The income statement represents a *through variable* because it summarizes revenues and expenses over a period of time. It is prepared by listing the revenues earned during a period and the expenses incurred during the same period, and by subtracting total expenses from total revenues, arriving at a net income. An income statement is always associated with a particular period of time, be it a month, quarter, or year.

The balance sheet, in contrast to the income statement, is a snapshot of the financial position of a firm at a particular point in time, and so represents an *across variable*. The financial position is summarized by listing the assets of the firm, its liabilities (debts), and the equity of the owner or owners.

6.3.2 The Balance Sheet

A **balance sheet** (also called a *position statement*) is a snapshot of an enterprise's financial position at a particular point in time, normally the last day of an accounting period. A firm's financial position is summarized in a balance sheet by listing its assets, liabilities, and owners' equity. The heading of the balance sheet gives the name of the enterprise and the date.

Figure 6.3 Through and Across Variables

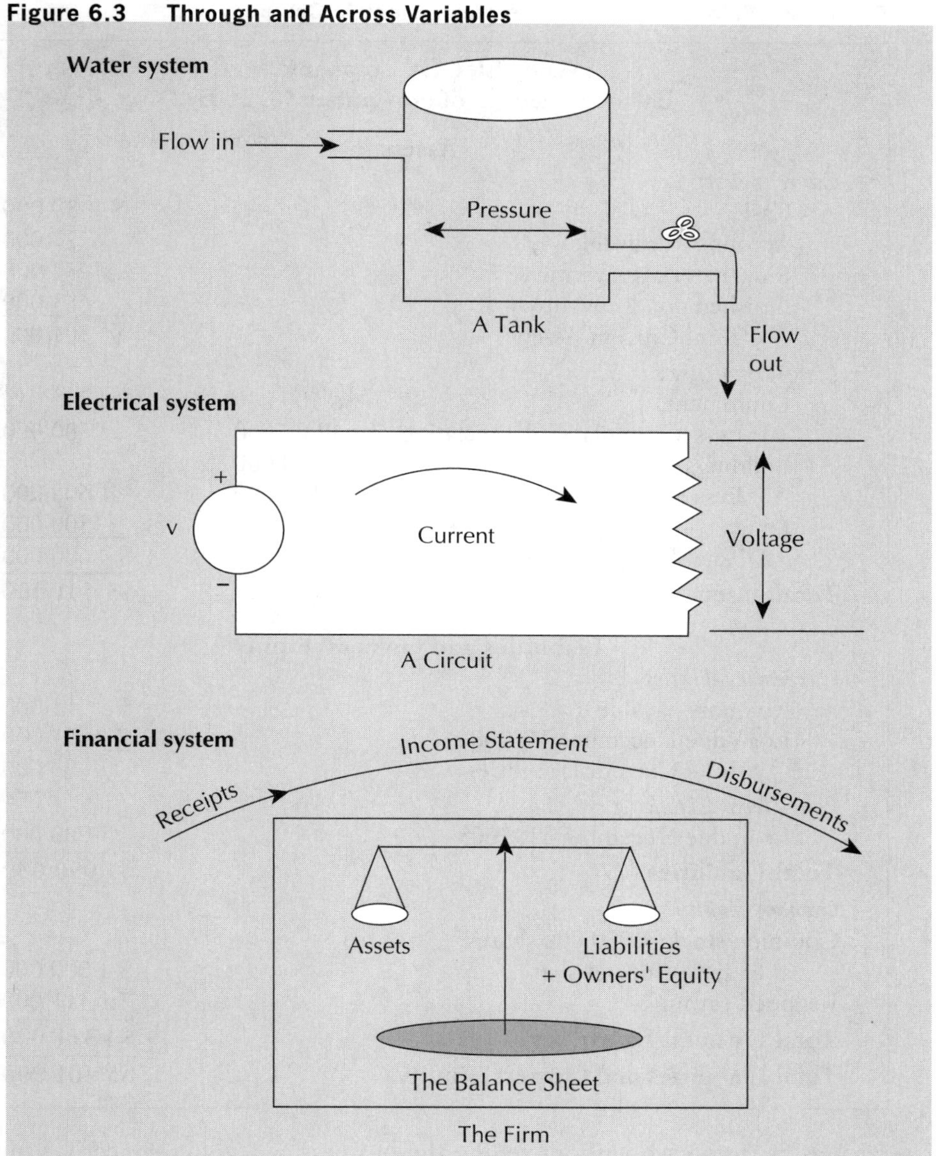

Example 6.4

Table 6.2 shows a balance sheet for the Major Electric Company, a manufacturer of small electrical appliances.

The first category of financial information in a balance sheet reports the **assets** of the enterprise. These are the economic resources owned by the enterprise, or more simply, everything that the enterprise owns. Assets are classified on a balance sheet as current assets and long-term assets. **Current assets** are cash and other assets that could be converted to cash within a relatively short period of time, usually a year or less. Inventory and accounts receivable are examples of non-cash current assets. **Long-term assets** (also called fixed assets) are assets that are not expected to be converted to cash in the short term, usually taken to be one year. Indeed, it may be difficult to convert long-

Table 6.2 Balance Sheet for the Major Electric Company

Major Electric Company Balance Sheet as of November 30, 2005		
Assets		
Current Assets		
Cash		$ 39 000
Accounts receivable		27 000
Raw materials inventory		52 000
Finished goods inventory		683 000
Total Current Assets		$ 801 000
Long-Term Assets		
Equipment	$6 500 000	
Less accumulated depreciation	4 000 000	2 500 000
Buildings	1 750 000	
Less accumulated depreciation	150 000	1 600 000
Land		500 000
Total Long-Term Assets		$ 4 600 000
Total Assets		**$5 401 000**
Liabilities and Owners' Equity		
Current Liabilities		
Accounts payable		$ 15 000
Loan due December 31, 2005		75 000
Total Current Liabilities		90 000
Long-Term Liabilities		
Loan due December 31, 2008		1 000 000
Total Liabilities		**$1 090 000**
Owners' Equity		
Common stock: 1 000 000 shares at $3 par value per share		$ 3 000 000
Retained earnings		1 311 000
Total Owners' Equity		**$4 311 000**
Total Liabilities and Owners' Equity		**$5 401 000**

term assets into cash without selling the business as a going concern. Equipment, land, and buildings are examples of long-term assets.

An enterprise's **liabilities** are claims, other than those of the owners, on a business's assets, or simply put, everything that the enterprise owes. Debts are usually the most important liabilities on a balance sheet. There may be other forms of liabilities as well. A commitment to the employees' pension plan is an example of a non-debt liability. As with assets, liabilities may be classified as current or long-term. **Current liabilities** are liabilities that are due within some short period of time, usually a year or less. Examples of current liabilities are debts that are close to maturity, accounts payable to suppliers, and taxes due. **Long-term liabilities** are liabilities that are not expected to draw on the business's current assets. Long-term loans and bonds issued by the business are examples of long-term liabilities.

The difference between a business's assets and its liabilities is the amount due to the owners—their equity in the business. That is, owners' equity is what is left over from assets after claims of others are deducted. We have, therefore,

CLOSE-UP 6.2 Types of Business Ownerships

There are three basic ways to structure a business organization.

Sole proprietorship is a business owned by one person. It is the simplest and least regulated form of business to start (in essence, all you need is a business name registered under the *Business Name Act*). This form of business accounts for the largest number of businesses operated in Canada. Under a sole proprietorship, the owner keeps all the profits, but at the same time, has *unlimited liability*; that is, the owner is personally responsible for all business debts and the creditors can come after even his or her personal assets in order to recoup debts.

Partnership is a business owned by two or more owners (partners). In a *general partnership*, the partners run the business together and share all profits and losses (unlimited liability) according to a partnership agreement. In a *limited partnership*, some partners are involved only as investors (limited partners) and they let one or more general partners take charge of day-to-day operation. The limited partners have *limited liability*; that is, they are only liable for up to the amount of their investment, and their personal assets are protected from the creditors of business debts.

Corporation is owned by shareholders. The shareholders elect the board of directors, and the board is responsible for selecting the managers to run the business in the interest of shareholders. A corporation is set up as a business entity, with its own rights and responsibilities, separate from the owners. This means that the corporation is responsible for its own debts, and the owners have limited liability (up to the amount of their investment). Canadian companies can be incorporated under the *Canada Business Corporations Act* or provincial law.

Generally speaking, the main advantage of incorporating a business is its relative ease in raising capital for a growth opportunity. Sole proprietorships and partnerships, although easier to start up than corporations, are both disadvantaged by personal liability issues, limited availability of equity, and difficulty of ownership transfer, which all contribute to potential difficulties in raising sufficient funds for growth.

$$\text{Owners' equity} = \text{Assets} - \text{Liabilities}$$

or

$$\text{Assets} = \text{Liabilities} + \text{Owners' equity}$$

Owners' equity is the interest of the owner or owners of a firm in its assets. For the basic types of ownership structure in business organizations, see Close-Up 6.2. Owners' equity usually appears as two components on a balance sheet of a corporation. The first is the par value of the owners' shares. When an enterprise is first organized, it is authorized to issue a certain number of shares. **Par value** is the price per share set by the corporation at the time the shares are originally issued. At any time after the first sale, the shares may be traded at prices that are greater than or less than the par value, depending on investors' expectations of the return that will be earned by the business in the future. There is no reason to expect the market price to equal the par value for very long after the shares are first sold. Nonetheless, the amount recorded in the balance sheet is the original par value. New shares sold anytime after the first issue may have a par value of their own, distinct from those of the original issue.

The second part of owners' equity usually shown on the balance sheet is retained earnings. **Retained earnings** includes the cumulative sum of earnings from normal operations, in addition to gains (or losses) from transactions like the sale of plant assets or investments the proceeds of which have been reinvested in the business (i.e., not paid out as dividends). Firms retain earnings mainly to expand operations through the purchase of additional assets. Contrary to what one may think, retained earnings do not represent cash. They may be invested in assets such as equipment and inventory.

The balance sheet gets its name from the idea that the total assets are equal in value to or *balanced by* the sum of the total liabilities and the owners' equity. A simple way of thinking about it is that the capital used to buy each asset has to come from debt (Liabilities) and/or equity (Owners' Equity). At a company's startup, the original shareholders may provide capital in the form of equity, and there will also likely be debt capital. The general term used to describe the markets in which short or long-term debt and equity are exchanged is the **financial** market. The capital provided through a financial market finances the assets and working capital for production. As the company undertakes its business activities, it will make or lose money. As it does so, assets will rise or fall, and equity and/or debt will rise or fall correspondingly. For example, if the company makes a profit, the profits will either pay off debts, be invested in new assets, or be paid as dividends to shareholders. Figure 6.4 provides an overview of the sources and uses of capital in an organization.■

Figure 6.4 Cash Flow Relationship Between the Company's Assets and the Financial Markets

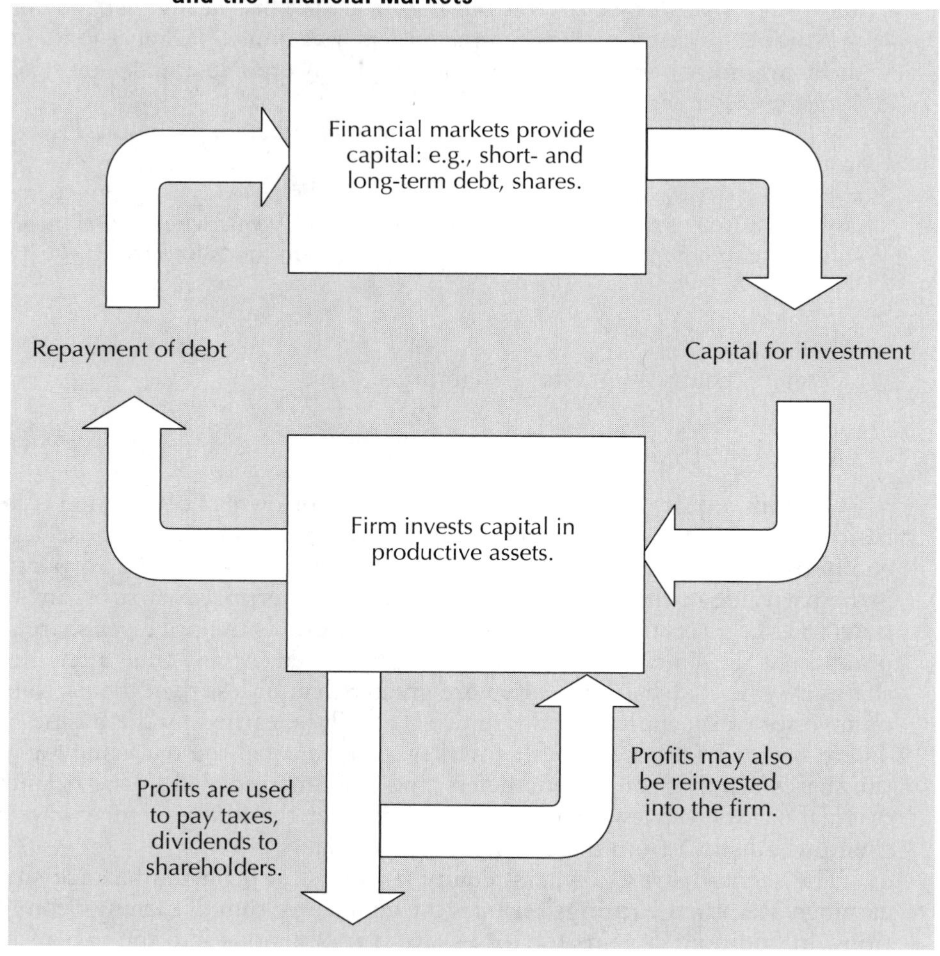

Example 6.5

Ian Claymore is the accountant at Major Electric. He has just realized that he forgot to include in the balance sheet for November 30, 2005, a government loan of $10 000 to help in the purchase of a $25 000 test stand (which he also forgot to include). The loan is due to be repaid in two years. When he revises the statement, what changes should he make?

The government loan is a long-term liability because it is due in more than one year. Consequently, an extra $10 000 would appear in long-term liabilities. This extra $10 000 must also appear as an asset for the balance to be maintained. The $25 000 value of the test stand would appear as equipment, increasing the equipment amount to $6 525 000. The $15 000 extra must come from a decrease in cash from $39 000 to $24 000.

Depending on the timing of the purchase, depreciation for the test stand might also be recognized in the balance sheet. Depreciation would reduce the net value of the equipment by the depreciation charge. The same amount would be balanced in the liabilities section by a reduction in retained earnings.■

6.3.3 The Income Statement

An **income statement** summarizes an enterprise's revenues and expenses over a specified accounting period. Months, quarters, and years are commonly used as reporting periods. As with the balance sheet, the heading of an income statement gives the name of the enterprise and the reporting period. The income statement first lists revenues, by type, followed by expenses. Expenses are then subtracted from revenues to give income (or profit) before taxes. Income taxes are then deducted to obtain net income. See Close-Up 6.3 for a measure used for operating profit.

The income statement summarizes the revenues and expenses of a business over a period of time. However, it does not directly give information about the generation of cash. For this reason, it may be useful to augment the income statement with a **statement of changes in financial position** (also called a cash flow statement, a statement of sources and uses of funds, or a funds statement), which shows the amounts of cash generated by a company's operation and by other sources, and the amounts of cash used for investments and other non-operating disbursements.

CLOSE-UP 6.3 Earnings Before Interest and Income Tax

Income before taxes *and* before interest payments, that is, total revenue minus operating expenses (all expenses except for income tax and interest), is commonly referred to as the **earnings before interest and income tax** (EBIT). EBIT measures the company's operating profit, which results from making sales and controlling operating expenses. Due to its focus on operating profit, EBIT is often used to judge whether there is enough profit to recoup the cost of capital (see Appendix 4A).

Example 6.6

The income statement for the Major Electric Company for the year ended November 30, 2005, is shown in Table 6.3.

We see that Major Electric's largest source of revenue was the sale of goods. The firm also earned revenue from the sale of management services to other companies. The

Table 6.3 Income Statement for the Major Electric Company

Major Electric Company **Income Statement** for the Year Ended November 30, 2005		
Revenues		
Sales	$7 536 000	
Management fees earned	106 000	
Total Revenues		**$7 642 000**
Expenses		
Cost of goods sold	$6 007 000	
Selling costs	285 000	
Administrative expenses	757 000	
Interest paid	86 000	
Total Expenses		**$7 135 000**
Income before taxes		$507 000
Taxes (at 40%)		202 800
Net Income		**$304 200**

largest expense was the cost of the goods sold. This includes the cost of raw materials, production costs, and other costs incurred to produce the items sold. Sometimes firms choose to list cost of goods sold as a *negative revenue*. The cost of goods sold will be subtracted from the sales to give a net sales figure. The net sales amount is the listed revenue, and the cost of goods sold does not appear as an expense.

The particular entries listed as either revenues or expenses will vary, depending on the nature of the business and the activities carried out. All revenues and expenses appear in one of the listed categories. For Major Electric, for example, the next item on the list of expenses, selling costs, includes delivery cost and other expenses such as salespersons' salaries. Administrative expenses include those costs incurred in running the company that are not directly associated with manufacturing. Payroll administration is an example of an administrative expense.

Subtracting the expenses from the revenues gives a measure of profit for the company over the accounting period, one year in the example. However, this profit is taxed at a rate that depends on the company's particular characteristics. For Major Electric, the tax rate is 40%, so the company's profit is reduced by that amount.■

Example 6.7

Refer back to Example 6.5. Ian Claymore also forgot to include the effects of the loan and test stand purchase in the income statement shown in Example 6.6. When he revises the statement, what changes should he make?

Neither the loan nor the purchase of the asset appears directly in the income statement. The loan itself is neither income nor expense; if interest is paid on it, this is an expense. The test stand is a depreciable asset, which means that only the depreciation for the test stand appears as an expense.■

6.3.4 Estimated Values in Financial Statements

The values in financial statements appear to be authoritative. However, many of the values in financial statements are estimates based on the **cost principle of accounting**. The cost principle of accounting states that assets are to be valued on the basis of their cost as opposed to market or other values. For example, the $500 000 given as the value of the land held by Major Electric is what Major Electric paid for the land. The market value of the land may now be greater or less than $500 000.

The value of plant and equipment is also based on cost. The value reported in the balance sheet is given by the initial cost minus accumulated depreciation. If Major Electric tried to sell the equipment, they might get more or less than this because depreciation models only approximate market value. For example, if there were a significant improvement in new equipment offered now by equipment suppliers compared with when Major Electric bought their equipment, Major Electric might get less than the $2 500 000 shown on the balance sheet.

Consider the finished goods inventory as another example of the cost principle of accounting. The value reported is Major Electric's manufacturing cost for producing the items. The implicit assumption being made is that Major Electric will be able to sell these goods for at least the cost of producing them. Their value may be reduced in later balance sheets if it appears that Major Electric cannot sell the goods easily.

The value shown for accounts receivable is clearly an estimate. Some fraction of accounts receivable may never be collected by Major. The value in the balance sheet reflects what the accountant believes to be a conservative estimate based on experience.

In summary, when examining financial statement data, it is important to remember that many reported values are estimates. Most firms include their accounting methods and assumptions within their periodic reports to assist in the interpretation of the statements.

6.3.5 Financial Ratio Analysis

Performance measures are calculated values that allow conclusions to be drawn from data. Performance measures drawn from financial statements can be used to answer such questions as:

1. Is the firm able to meet its short-term financial obligations?
2. Are sufficient profits being generated from the firm's assets?
3. How dependent is the firm on its creditors?

Financial ratios are one kind of performance measure that can answer these questions. They give an analyst a framework for asking questions about the firm's liquidity, asset management, leverage, and profitability. Financial ratios are ratios between key amounts taken from the financial statements of the firm. While financial ratios are simple to compute, they do require some skill to interpret, and they may be used for different purposes. For example, internal management may be concerned with the firm's ability to pay its current liabilities or the effect of long-term borrowing for a plant expansion. An external investor may be interested in past and current earnings to judge the wisdom of investing in the firm's stock. A bank will assess the riskiness of lending money to a firm before extending credit.

To properly interpret financial ratios, analysts commonly make comparisons with ratios computed for the same company from previous financial statements (a **trend analysis**) and with industry standard ratios. This is referred to as **financial ratio analysis**.

Industry standards can be obtained from catalogues routinely published by Dun & Bradstreet Canada and by Statistics Canada. *Financial and Taxation Statistics for Enterprises*, a Statistics Canada publication, lists financial data from the balance sheets and income statements as well as selected financial ratios for numerous industries. Examples of a past industry-total balance sheet and income statement (in millions of dollars) for the electronic products manufacturing industry are shown in Tables 6.4 and 6.5. These statistics allow an analyst to compare an individual firm's financial statements with those of the appropriate industry.

We shall see in the next section how the financial ratios derived from industry-total financial data can be used to assess the health of a firm.

NET VALUE 6.1

The Canadian Securities Administrators (CSA)

Securities regulators from each province and territory in Canada have teamed up to form the Canadian Securities Administrators (CSA) for the purposes of coordinating securities regulation across the country.

The CSA (www.csa-acvm.ca) also helps educate Canadians about the securities industry, the stock markets, and how to protect investors from scams by providing a variety of educational materials on securities and investing.

The CSA Web site also provides links to the provincial regulatory bodies' sites, information about Canada's various stock exchanges, and access to several electronic databases, including SEDAR.

SEDAR (the System for Electronic Document Analysis and Retrieval) is used by public companies to electronically file most securities-related information with the Canadian securities regulatory authorities. Electronic filing with SEDAR has been required for public companies since 1997. The SEDAR system allows users to gain access to public company information such as company profiles, news releases, annual reports, and financial statements. SEDAR can be easily searched, and thus is useful for researching a company's financial performance.

6.3.6 Financial Ratios

Numerous financial ratios are used in the financial analysis of a firm. Here we shall introduce six commonly used ratios to illustrate their use in a comparison with ratios from the industry-total data and in trend analysis. To facilitate the discussion, we shall use Tables 6.6 and 6.7, which show the balance sheets and income statements for Electco Electronics, a small electronics equipment manufacturer.

The first two financial ratios we address are referred to as **liquidity ratios**. Liquidity ratios evaluate the ability of a business to meet its current liability obligations. In other words, they help us evaluate its ability to weather unforeseen fluctuations in cash flows. A company that does not have a reserve of cash, or other assets that can be converted into cash easily, may not be able to fulfil its short-term obligations.

A company's net reserve of cash and assets easily converted to cash is called its working capital. **Working capital** is simply the difference between total current assets and total current liabilities:

$$\text{Working capital} = \text{Current assets} - \text{Current liabilities}$$

Table 6.4 Example of Industry-Total Balance Sheet (in Millions of $)

Electronic Product Manufacturing Balance Sheet	
Assets	
Assests	
Cash and deposits	$ 2 547
Accounts receivable and accrued revenue	9 169
Inventories	4 344
Investments and accounts with affiliates	10 066
Portfolio investments	1 299
Loans	839
Capital assets, net	4 966
Other assets	2 898
Total assets	$36 128
Liabilities and Owners' Equity	
Liabilities	
Accounts payable and accrued liabilities	$ 8 487
Loans and accounts with affiliates	3 645
Borrowings:	
Loans and overdrafts from banks	1 408
Loans and overdrafts from others	1 345
Bankers' acceptances and paper	232
Bonds and debentures	2 487
Mortgages	137
Deferred income tax	(247)
Other liabilities	1 931
Total liabilities	$19 426
Owner's Equity	
Share capital	9 177
Contributed surplus and other	415
Retained earnings	7 110
Total owners' equity	$16 702
Current assets	$17 663
Current liabilities	11 692

The adequacy of working capital is commonly measured with two ratios. The first, the **current ratio**, is the ratio of all current assets relative to all current liabilities. The current ratio may also be referred to as the **working capital ratio**.

$$\text{Current ratio} = \frac{\text{Current assets}}{\text{Current liabilities}}$$

Electco Electronics had a current ratio of $4314/$2489 = 1.73 in 2003 (Table 6.6). Ordinarily, a current ratio of 2 is considered adequate, although this determination may depend a great deal on the composition of current assets. It also may depend on industry standards. In the case of Electco, the industry standard is 1.51, which is listed in Table 6.8. It would appear that Electco had a reasonable amount of liquidity in 2003 from the industry's perspective.

Table 6.5 Example of Industry-Total Income Statement (Millions of $)

Electronic Product Manufacturing Income Statement	
Operating revenues	**$50 681**
Sales of goods and services	49 357
Other operating revenues	1 324
Operating expenses	**48 091**
Depreciation, depletion, and amortization	1 501
Other operating expenses	46 590
Operating profit	**2 590**
Other revenue	**309**
Interest and dividends	309
Other expenses	**646**
Interest on short-term debt	168
Interest on long-term debt	478
Gains (Losses)	**(122)**
On sale of assets	25
Others	(148)
Profit before income tax	**2 130**
Income tax	715
Equity in affiliates' earnings	123
Profit before extraordinary gains	**1 538**
Extraordinary gains/losses	—
Net profit	**1 538**

A second ratio, the acid-test ratio, is more conservative than the current ratio. The **acid-test ratio** (also known as the **quick ratio**) is the ratio of quick assets to current liabilities:

$$\text{Acid-test ratio} = \frac{\text{Quick assets}}{\text{Current liabilities}}$$

The acid-test ratio recognizes that some current assets, for example, inventory and prepaid expenses, may be more difficult to turn into cash than others. *Quick assets* are cash, accounts receivable, notes receivable, and temporary investments in marketable securities—those current assets considered to be highly *liquid*.

The acid-test ratio for Electco for 2003 was ($431 + $2489)/$2489 = 1.17. Normally, an acid-test ratio of 1 is considered adequate, as this indicates that a firm could meet all its current liabilities with the use of its *quick* current assets if it were necessary. Electco appears to meet this requirement.

The current ratio and the acid-test ratio provide important information about how liquid a firm is, or how well it is able to meet its current financial obligations. The extent to which a firm relies on debt for its operations can be captured by what are called **leverage** or **debt-management ratios**. An example of such a ratio is the **equity ratio**. It is the ratio of total owners' equity to total assets. The smaller this ratio, the more dependent the firm is on debt for its operations and the higher are the risks the company faces.

Table 6.6 Balance Sheets for Electco Electronics for Years Ended 2003, 2004, and 2005

Electco Electronics **Year-End Balance Sheets** *(in thousands of dollars)*			
	2003	**2004**	**2005**
Assets			
Current assets			
Cash	431	340	320
Accounts receivable	2489	2723	2756
Inventories	1244	2034	2965
Prepaid services	150	145	149
Total current assets	4314	5242	6190
Long-term assets			
Buildings and equipment (net of depreciation)	3461	2907	2464
Land	521	521	521
Total long-term assets	3982	3428	2985
Total Assets	8296	8670	9175
Liabilities			
Current liabilities			
Accounts payable	1493	1780	2245
Bank overdraft	971	984	992
Accrued taxes	25	27	27
Total current liabilities	2489	2791	3264
Long-term liabilities			
Mortgage	2489	2455	2417
Total Liabilities	4978	5246	5681
Owners' Equity			
Share capital	1825	1825	1825
Retained earnings	1493	1599	1669
Total Owners' Equity	3318	3424	3494
Total Liabilities and Owners' Equity	8296	8670	9175

$$\text{Equity ratio} = \frac{\text{Total owners' equity}}{\text{Total liabilities} + \text{Total equity}}$$

$$= \frac{\text{Total owners' equity}}{\text{Total assets}}$$

The equity ratio for Electco in 2003 was $3318/$8296 = 0.40 and for the industry was 0.46 as shown in Table 6.8. Electco has paid for roughly 60% of its assets with debt; the remaining 40% represents equity. This is close to the industry practice as a whole and would appear acceptable.

Table 6.7 Income Statements for Electco Electronics for Years Ended 2003, 2004, 2005

Electco Electronics Income Statements *(in thousands of dollars)*			
	2003	**2004**	**2005**
Revenues			
Sales	12 440	11 934	12 100
Total Revenues	12 440	11 934	12 100
Expenses			
Cost of goods sold (excluding depreciation)	10 100	10 879	11 200
Depreciation	692	554	443
Interest paid	346	344	341
Total Expenses	11 138	11 777	11 984
Profit before taxes	1 302	157	116
Taxes (at 40%)	521	63	46
Profit before extraordinary items	781	94	70
Extraordinary gains/losses	70		
Profit after taxes	**851**	**94**	**70**

Table 6.8 Industry-Standard Ratios and Financial Ratios for Electco Electronics

Financial Ratio	**Industry Standard**	**Electco Electronics**		
		2003	**2004**	**2005**
Current ratio	1.51	1.73	1.88	1.90
Quick ratio	—	1.17	1.10	0.94
Equity ratio	0.46	0.40	0.39	0.38
Inventory-turnover ratio	11.36	10.00	5.87	4.08
Return-on-total-assets ratio	4.26%	9.41%	1.09%	0.76%
Return-on-equity ratio	9.21%	23.53%	2.75%	2.00%

Another group of ratios is called the **asset-management** or **efficiency ratios**. They assess how efficiently a firm is using its assets. Inventory-turnover ratio is an example. **Inventory-turnover ratio** specifically looks at how efficiently a firm is using its resources to manage its inventories. This is reflected in the number of times that its inventories are replaced (or turned over) per year. The inventory-turnover ratio provides a measure of whether the firm has more or less inventory than normal.

$$\text{Inventory-turnover ratio} = \frac{\text{Sales}}{\text{Inventories}}$$

Electco's turnover ratio for 2003 was \$12 440/\$1244 = 10 turns per year. This is reasonably close to the industry standard of 11.36 turns per year as shown in Table 6.8. In

2003, Electco invested roughly the same amount in inventory per dollar of sales as the industry did, on average.

Two points should be observed about the inventory-turnover ratio. First, the sales amount in the numerator has been generated over a period of time, while the inventory amount in the denominator is for one point in time. A more accurate measure of inventory turns would be to approximate the average inventory over the period in which sales were generated.

A second point is that sales refer to market prices, while inventories are listed at cost. The result is that the inventory-turnover ratio as computed above will be an overstatement of the true turnover. It may be more reasonable to compute inventory turnover based on the ratio of cost of goods sold to inventories. Despite this observation, traditional financial analysis uses the sales-to-inventories ratio.

The next two ratios give evidence of how productively assets have been employed in producing a profit. The **return-on-assets (ROA)** or **net-profit ratio** is the first example of a **profitability ratio**:

$$\text{Return-on-assets ratio} = \frac{\text{Net income (before extraordinary items)}}{\text{Total assets}}$$

Electco had a return on assets of \$781/\$8296 = 0.0941 or 9.41% in 2003. Table 6.8 shows that the industry-total ROA for 2003 was 4.26%. Note the comments on extraordinary items in Close-Up 6.4.

The second example of a profitability ratio is the **return-on-equity** (ROE) **ratio**:

$$\text{Return-on-equity ratio} = \frac{\text{Net income (before extraordinary items)}}{\text{Total equity}}$$

The return-on-equity ratio looks at how much profit a company has earned in comparison to the amount of capital that the owners have tied up in the company. It is often compared to how much the owners could have earned from an alternative investment and used as a measure of investment performance. Electco had a ROE of \$781/\$3318 = 0.2354 or 23.54% in 2003, whereas the industry-standard ROE was 9.21% as shown in Table 6.8. The year 2003 was an excellent one for the owners at Electco from their investment point of view.

Overall, Electco's performance in 2003 was similar to that of the electronic product manufacturing industry as a whole. One exception is that Electco generated higher profits than the norm; it may have been extremely efficient in its operations that year.

The rosy profit picture painted for Electco in 2003 does not appear to extend into 2004 and 2005, as a trend analysis shows. Table 6.8 shows the financial ratios computed for 2004 and 2005 with those of 2003 and the industry standard.

For more convenient reference, we have summarized the six financial ratios we have dealt with and their definitions in Table 6.9.

CLOSE-UP 6.4 Extraordinary Items

Extraordinary items are gains and losses that do not typically result from a company's normal business activities, are not expected to occur regularly, and are not recurring factors in any evaluations of the ordinary operations of the business. For example, cost or loss of income caused by natural disasters (floods, tornados, ice storms, etc.) would be extraordinary loss. Revenue created by the sale of a division of a firm is an example of extraordinary gain. Extraordinary items are reported separately from regular items and are listed net of applicable taxes.

Table 6.9 A Summary of Financial Ratios and Definitions

Ratio	Definition	Comments
Current ratio (Working capital ratio)	$\frac{\text{Current assets}}{\text{Current liabilities}}$	A liquidity ratio
Acid-test ratio (Quick ratio)	$\frac{\text{Quick assets}}{\text{Current liabilities}}$	A liquidity ratio (Quick assets = Current assets − Inventories − Prepaid items)
Equity ratio	$\frac{\text{Total equity}}{\text{Total assets}}$	A leverage or debt-management ratio
Inventory-turnover ratio	$\frac{\text{Sales}}{\text{Inventories}}$	An asset management or efficiency ratio
Return-on-assets ratio (Net-profit ratio)	$\frac{\text{Net income}}{\text{Total assets}}$	A profitability ratio (excludes extraordinary items)
Return-on-equity ratio	$\frac{\text{Net income}}{\text{Total equity}}$	A profitability ratio (measure of investment performance; excludes extraordinary items)

Electco's return on assets has dropped significantly over the three-year period. Though the current and quick ratios indicate that Electco should be able to meet its short-term liabilities, there has been a significant buildup of inventories over the period. Electco is not selling what it is manufacturing. This would explain the drop in Electco's inventory turns.

Coupled with rising inventory levels is a slight increase in the cost of goods sold over the three years. From the building and equipment entries in the balance sheet, we know that no major capital expenditures on new equipment have occurred during the period. Electco's equipment may be aging and in need of replacement, though further analysis of what is happening is necessary before any conclusions on this issue can be drawn.

A final observation is that Electco's accounts receivable seems to be growing over the three-year period. Since there may be risks associated with the possibility of bad debt, Electco should probably investigate the matter.

In summary, Electco's main problem appears to be a mismatch between production levels and sales levels. Other areas deserving attention are the increasing cost of goods sold and possible problems with accounts receivable collection. These matters need investigation if Electco is to recover from its current slump in profitability.

We close the section on financial ratios with some cautionary notes on their use. First, since financial statement values are often approximations, we need to interpret the financial ratios accordingly. In addition, accounting practices vary from firm to firm and may lead to differences in ratio values. Wherever possible, look for explanations of how values are derived in financial reports.

A second problem encountered in comparing a firm's financial ratios with the industry-standard ratios is that it may be difficult to determine what industry the firm best fits into. Furthermore, within every industry, large variations exist. In some cases, an analyst may construct a relevant "average" by searching out a small number of similar firms (in size and business type) that may be used to form a customized industry average.

Finally, it is important to recognize the effect of seasonality on the financial ratios calculated. Many firms have highly seasonal operations with natural high and low periods of activity. An analyst needs to judge these fluctuations in context. One solution to this problem is to adjust the data seasonally through the use of averages. Another is to collect financial data from several seasons so that any deviations from the normal pattern of activity can be picked up.

Despite our cautionary words on the use of financial ratios, they do provide a useful framework for analyzing the financial health of a firm and for answering many questions about its operations.

REVIEW PROBLEMS

REVIEW PROBLEM 6.1

Joan is the sole proprietor of a small lawn-care service. Last year, she purchased an eight-horsepower chipper-shredder to make mulch out of small tree branches and leaves. At the time it cost $760. She expects that the value of the chipper-shredder will decline by a constant amount each year over the next six years. At the end of six years, she thinks that it will have a salvage value of $100.

Construct a table that gives the book value of the chipper-shredder at the end of each year, for six years. Also indicate the accumulated depreciation at the end of each year. A spreadsheet may be helpful.

ANSWER

The depreciation charge for each year is

$$D_{sl}(n) = \frac{(P - S)}{N} = \frac{\$760 - \$100}{6} = \$110 \qquad n = 1, \ldots, 6$$

This is the requested table:

Year	Depreciation Charge	Book Value	Accumulated Depreciation
0		$760	
1	$110	650	$110
2	110	540	220
3	110	430	330
4	110	320	440
5	110	210	550
6	110	100	660

REVIEW PROBLEM 6.2

A three-year-old extruder used in making plastic yogurt cups has a current book value of \$168 750. The declining-balance method of depreciation with a rate $d = 0.25$ is used to determine depreciation charges. What was its original price? What will its book value be two years from now?

ANSWER

Let the original price of the extruder be P. The book value three years after purchase is \$168 750. This means that the original price was

$$BV_{db}(3) = P(1 - d)^3$$

$$\$168\ 750 = P(1 - 0.25)^3$$

$$P = \$400\ 000$$

The original price was \$400 000.

The book value two years from now can be determined either from the original purchase price (depreciated for five years) or the current value (depreciated for two years):

$$BV_{db}(5) = \$400\ 000(1 - 0.25)^5 = \$94\ 921.88$$

or

$$BV_{db}(2) = \$168\ 750(1 - 0.25)^2 = \$94\ 921.88$$

The book value two years from now will be \$94 921.88.

REVIEW PROBLEM 6.3

You have been given the following data from the Fine Fishing Factory for the year ending December 31, 2005. Construct an income statement and a balance sheet from the data.

Accounts payable	\$ 27 500
Accounts receivable	32 000
Advertising expense	2 500
Bad debts expense	1 100
Buildings, net	14 000
Cash	45 250
Common stock	125 000
Cost of goods sold	311 250
Depreciation expense, buildings	900
Government bonds	25 000
Income taxes	9 350
Insurance expense	600
Interest expense	500
Inventory, December 31, 2005	42 000
Land	25 000
Machinery, net	3 400
Mortgage due May 30, 2007	5 000
Office equipment, net	5 250
Office supplies expense	2 025
Other expenses	7 000

Prepaid expenses	3 000
Retained earnings	?
Salaries expense	69 025
Sales	421 400
Taxes payable	2 500
Wages payable	600

ANSWER

Solving this problem consists of sorting through the listed items and identifying which are balance sheet entries and which are income statement entries. Then, assets can be separated from liabilities and owners' equity, and revenues from expenses.

Fine Fishing Factory **Balance Sheet** **as of December 31, 2005**	
Assets	
Current assets	
Cash	$ 45 250
Accounts receivable	32 000
Inventory, December 31, 2005	42 000
Prepaid expenses	3 000
Total current assets	$122 250
Long-term assets	
Land	25 000
Government bonds	25 000
Machinery, net	3 400
Office equipment, net	5 250
Buildings, net	14 000
Total long-term assets	$ 72 650
Total assets	**$194 900**
Liabilities	
Current liabilities	
Accounts payable	$ 27 500
Taxes payable	2 500
Wages payable	600
Total current liabilities	$ 30 600
Long-term liabilities	
Mortgage due May 30, 2007	5 000
Total long-term liabilities	$ 5 000
Total liabilities	**$ 35 600**
Owners' Equity	
Common stock	$125 000
Retained earnings	34 300
Total owners' equity	**$159 300**
Total liabilities and owners' equity	**$194 900**

Fine Fishing Factory Income Statement for the Year Ending December 31, 2005	
Revenues	
Sales	$421 400
Cost of goods sold	311 250
Net revenue from sales	$110 150
Expenses	
Salaries expense	69 025
Bad debts expense	1 100
Advertising expense	2 500
Interest expense	500
Insurance expense	600
Office supplies expense	2 025
Other expenses	7 000
Depreciation expense, buildings	900
Depreciation expense, office equipment	850
Total Expenses	$ 84 500
Income before taxes	$ 25 650
Income taxes	9 350
Income after taxes	**$ 16 300**

REVIEW PROBLEM 6.4

Perform a financial ratio analysis for the Major Electric Company using the balance sheet and income statement from Sections 6.3.2 and 6.3.3. Industry standards for the ratios are as follows:

Ratio	Industry Standard
Current ratio	1.80
Acid-test ratio	0.92
Equity ratio	0.71
Inventory-turnover ratio	14.21
Return-on-assets ratio	7.91%
Return-on-equity ratio	11.14%

ANSWER

The ratio computations for Major Electric are:

$$\text{Current ratio} = \frac{\text{Current assets}}{\text{Current liabilities}} = \frac{\$801\,000}{\$90\,000} = 8.9$$

$$\text{Acid-test ratio} = \frac{\text{Quick assets}}{\text{Current liabilities}} = \frac{\$66\,000}{\$90\,000} = 0.73$$

$$\text{Equity ratio} = \frac{\text{Total equity}}{\text{Total assets}} = \frac{\$4\,311\,000}{\$5\,401\,000} = 0.7982 \cong 0.80$$

$$\text{Inventory-turnover ratio} = \frac{\text{Sales}}{\text{Inventories}} = \frac{\$7\,536\,000}{\$683\,000}$$
$$= 11.03 \text{ turns per year}$$

$$\text{Return-on-total-assets ratio} = \frac{\text{Profits after taxes}}{\text{Total assets}} = \frac{\$304\,200}{\$5\,401\,000}$$
$$= 0.0563 \text{ or } 5.63\% \text{ per year}$$

$$\text{Return-on-equity ratio} = \frac{\text{Net income}}{\text{Total equity}} = \frac{\$304\,200}{\$4\,311\,000}$$
$$= 0.0706 \text{ or } 7.06\% \text{ per year}$$

A summary of the ratio analysis results follows:

Ratio	Industry Standard	Major Electric
Current ratio	1.80	8.90
Acid-test ratio	0.92	0.73
Equity ratio	0.71	0.80
Inventory-turnover ratio	14.21	11.03
Return-on-assets ratio	7.91%	5.63%
Return-on-equity ratio	11.14%	7.06%

Major Electric's current ratio is well above the industry standard and well above the general guideline of 2. The firm appears to be quite liquid. However, the acid-test ratio, with a value of 0.73, gives a slightly different view of Major Electric's liquidity. Most of Major Electric's current assets are inventory; thus, the current ratio is somewhat misleading. If we look at the acid-test ratio, Major Electric's quick assets are only 73% of their current liabilities. The firm may have a difficult time meeting their current debt obligations if they have unforeseen difficulties with their cash flow.

Major Electric's equity ratio of 0.80 indicates that it is not heavily reliant on debt and therefore is not at high risk of going bankrupt. Major Electric's inventory turns are lower than the industry norm, as are its ROA and ROE.

Taken together, Major Electric appears to be in reasonable financial shape. One matter they should probably investigate is why their inventories are so high. With lower inventories, they could improve their inventory turns and liquidity, as well as their return on assets.

SUMMARY

This chapter opened with a discussion of the concept of depreciation and various reasons why assets lose value. Two popular depreciation models, straight-line and declining-balance, were then presented as methods commonly used to determine book value of capital assets and depreciation charges.

The second part of the chapter dealt with the elements of financial accounting. We first presented the two main financial statements: the balance sheet and the income statement. Next, we showed how these statements can be used to assess the financial health of a firm through the use of ratios. Comparisons with industry norms and trend analysis are normally part of financial analysis. We closed with cautionary notes on the interpretation of the ratios.

The significance of the material in this chapter is twofold. First, it sets the groundwork for material in Chapters 7 and 8, replacement analysis and taxation. Second, and perhaps more importantly, it is increasingly necessary for all engineers to have an understanding of depreciation and financial accounting as they become more and more involved in decisions that affect the financial positions of the organizations in which they work.

Engineering Economics in Action, Part 6B: Usually the Truth

Terry had shown Naomi what he thought was evidence of wrongdoing by a fellow employee. Naomi said, "Interesting observation, Terry. But you know, I think it's probably OK. Let me explain. The main problem is that you are looking at two kinds of evaluation here, book value and market value. The book value is what an asset is worth from an accounting point of view, while the market value is what it sells for."

Terry nodded. "Yes, I know that. That's true about anything you sell. But this is different. We've got a $5000 estimate against a $300 sale. You can't tell me that our guess about the sales price can be that far out!"

"Yes, it can, and I'll tell you why. That book value is an estimate of the market value that has been calculated according to very particular rules. The Canadian tax rules require us to use a declining-balance depreciation method for almost all of our assets. They also specify which declining-balance rate to use. The depreciation charge is called the CCA, which stands for 'capital cost allowance.' For most equipment it's 20%. Now, the reality is that things decline in value at different rates, and computers lose value really quickly. We could, for our own purposes, determine a book value for any asset that is a better estimate of its market value, but sometimes it's too much trouble to keep one set of figures for tax reasons and another for other purposes. So often everything is given a book value according to the tax rules, and consequently sometimes the difference between the book value and the market value can be a lot."

"But surely the government can see that computers in particular don't match that 20% rate. Or are they just ripping us off?"

"Well, they can see that. Until a few years ago, the CCA rate for computers was 20%. But because it was painfully obvious that this rate was too low, it was revised to 30%. This is still too low, as you can see from the sale of our own computers, but it is better than it was."

Terry smiled ruefully, "So our accounting statements don't really show the truth?"

Naomi smiled back, "I guess not, if by 'truth' you mean market value. But usually they're close. Usually."

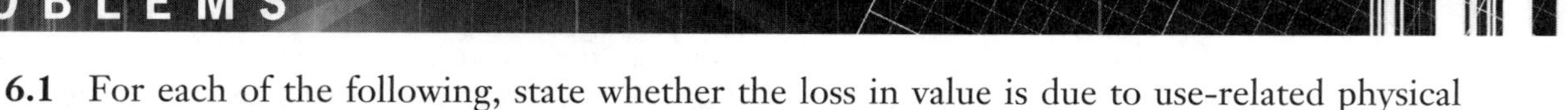

PROBLEMS

6.1 For each of the following, state whether the loss in value is due to use-related physical loss, time-related physical loss, or functional loss:

(a) Albert sold his two-year-old computer for $500, but he paid $4000 for it new. It wasn't fast enough for the new software he wanted.

(b) Beatrice threw out her old tennis shoes because the soles had worn thin.

(c) Claudia threw out her old tennis shoes because she is jogging now instead.

(d) Day-old bread is sold at half-price at the neighbourhood bakery.

(e) Egbert sold his old lawnmower to his neighbour for $20.

(f) Fred picked up a used overcoat at the thrift store for less than 10% of the new price.

(g) Greg notices that newspapers cost $0.50 on the day of purchase, but are worth less than $0.01 each as recyclable newsprint.

(h) Harold couldn't get the price he wanted for his house because the exterior paint was faded and flaking.

6.2 For each of the following, state whether the value is a market value, book value, scrap value, or salvage value:

(a) Inta can buy a new stove for $800 at Joe's Appliances.

(b) Jack can sell his used stove to Inta for $200.

(c) Kitty can sell her used stove to the recyclers for $20.

(d) Liam can buy Jack's used stove for $200.

(e) Mick is adding up the value of the things he owns. He figures his stove is worth at least $200.

6.3 A new industrial sewing machine costs in the range of $5000 to $10 000. Technological change in sewing machines does not occur very quickly, nor is there much change in the functional requirements of a sewing machine. A machine can operate well for many years with proper care and maintenance. Discuss the different reasons for depreciation and which you think would be most appropriate for a sewing machine.

6.4 Communications network switches are changing dramatically in price and functionality as changes in technology occur in the communications industry. Prices drop frequently as more functionality and capacity are achieved. A switch only six months old will have depreciated since it was installed. Discuss the different reasons for depreciation and which you think would be most appropriate for this switch.

6.5 Ryan owns a five-hectare plot of land in the countryside. He has been planning to build a cottage on the site for some time, but has not been able to afford it yet. However, five years ago, he dug a pond to collect rainwater as a water supply for the cottage. It has never been used, and is beginning to fill in with plant life and garbage that has been dumped there. Ryan realizes that his investment in the pond has depreciated in value since he dug it. Discuss the different reasons for depreciation and which you think would be most appropriate for the pond.

6.6 A company that sells a particular type of Web-indexing software has had two larger firms approach it for a possible buyout. The current value of the company, based on recent financial statements, is $4.5 million. The two bids were for $4 million and $7 million, respectively. Both bids were bona fide, meaning they were real offers. What is the market value of the company? Its book value?

6.7 An asset costs $14 000 and has a scrap value of $3000 after seven years. Calculate its book value using straight-line depreciation

(a) After one year

(b) After four years

(c) After seven years

6.8 An asset costs $14 000. At a depreciation rate of 20%, calculate its book value using the declining-balance method

(a) After one year

(b) After four years

(c) After seven years

6.9 (a) An asset costs $14 000. What declining-balance depreciation rate would result in the scrap value of $3000 after seven years?

(b) Using the depreciation rate from part (a), what is the book value of the asset after four years?

6.10 Using a spreadsheet program, chart the book value of a $14 000 asset over a seven-year life using declining-balance depreciation (d = 0.2). On the same chart, show the book value of the $14 000 asset using straight-line depreciation with a scrap value of $3000 after seven years.

6.11 Using a spreadsheet program, chart the book value of a $150 000 asset for the first 10 years of its life at declining-balance depreciation rates of 5%, 20%, and 30%.

6.12 A machine has a life of 30 years, costs $245 000, and has a salvage value of $10 000 using straight-line depreciation. What depreciation rate will result in the same book value for both the declining-balance and straight-line methods at the end of year 20?

6.13 A new press brake costs Medicine Hat Steel $780 000. It is expected to last 20 years, with a $60 000 salvage value. What rate of depreciation for the declining-balance method will produce a book value after 20 years that equals the salvage value of the press?

6.14 (a) Using straight-line depreciation, what is the book value after four years for an asset costing $150 000 that has a salvage value of $25 000 after 10 years? What is the depreciation charge in the fifth year?

(b) Using declining-balance depreciation with d = 20%, what is the book value after four years for an asset costing $150 000? What is the depreciation charge in the fifth year?

(c) What is the depreciation rate using declining-balance for an asset costing $150 000 that has a salvage value of $25 000 after 10 years?

6.15 Julia must choose between two different designs for a safety enclosure, which will be in use indefinitely. Model A has a life of three years, a first cost of $8000, and maintenance of $1000 per year. Model B will last four years, has a first cost of $10 000, and has maintenance of $800 per year. A salvage value can be estimated for Model A using a depreciation rate of 40% and declining-balance depreciation, while a salvage value for Model B can be estimated using straight-line depreciation and the knowledge that after one year its salvage value will be $7500. Interest is at 14%. Using a present worth analysis, which design is better?

6.16 Adventure Airline's new baggage handling conveyor costs $250 000 and has a service life of 10 years. For the first six years, depreciation of the conveyor is calculated using the declining-balance method at the rate of 30%. During the last four years, the straight-line method is used for accounting purposes in order to have a book value of 0 at the end of the service life. What is the book value of the conveyor after 7, 8, 9, and 10 years?

6.17 Molly inherited $5000 and decided to start a lawn-mowing service. With her inheritance and a bank loan of $5000, she bought a used ride-on lawnmower and a used truck. For five years, Molly had a gross income of $30 000, which covered the annual operating costs and the loan payment, both of which totalled $4500. She spent the rest of her income personally. At the end of five years, Molly found that her loan was paid off but the equipment was wearing out.

(a) If the equipment (lawn mower and truck) was depreciating at the rate of 50% according to the declining-balance method, what is its book value after five years?

(b) If Molly wanted to avoid being left with a worthless lawnmower and truck and with no money for renewing them at the end of five years, how much should she have saved annually toward the second set of used lawnmower and used truck of the same initial value as the first set? Assume that Molly's first lawnmower and truck could be sold at their book value. Use an interest rate of 7%.

6.18 Enrique is planning a trip around the world in three years. He will sell all of his possessions at that time to fund the trip. Two years ago, he bought a used car for $12 500. He observes that the market value for the car now is about $8300. He needs to know how much money his car will add to his stash for his trip when he sells it three years from now. Use the declining-balance depreciation method to tell him.

6.19 Ben is choosing between two different industrial fryers using an annual worth calculation. Fryer 1 has a five-year service life and a first cost of $400 000. It will generate net year-end savings of $128 000 per year. Fryer 2 has an eight-year service life and a first cost of $600 000. It will generate net year-end savings of $135 000 per year. If the salvage value is estimated using declining-balance depreciation with a 20% depreciation rate, and the MARR is 12%, which fryer should Ben buy?

6.20 Dick noticed that the book value of an asset he owned was exactly $500 higher if the value was calculated by straight-line depreciation over declining-balance depreciation, exactly halfway through the asset's service life, and the scrap value at the end of the service life was the same by either method. If the scrap value was $100, what was the purchase price of the asset?

6.21 A company had net sales of $20 000 last month. From the balance sheet for the end of last month, and an income statement for the month, you have determined that the current ratio was 2.0, the acid-test ratio was 1.2, and the inventory turnover was two per month. What was the value of the company's current assets?

6.22 In the last quarter, the financial-analysis report for XYZ Company revealed that the current, quick, and equity ratios were 1.9, 0.8, and 0.37, respectively. In order to improve the firm's financial health based on these financial ratios, the following strategies are considered by XYZ for the current quarter:

(i) Reduce inventory

(ii) Pay back short-term loans

(iii) Increase retained earnings

(a) Which strategy (or strategies) is effective for improving each of the three financial ratios?

(b) If only one strategy is considered by XYZ, which one seems to be most effective? Assume no other information is available for analysis.

6.23 The end-of-quarter balance sheet for XYZ Company indicated that the current ratio was 1.8 and the equity ratio was 0.45. It also indicated that the long-term assets were twice as much as the current assets, and half of the current assets were highly liquid. The total equity was $68 000. Since the current ratio was close to 2, XYZ feels that the company had a reasonable amount of liquidity in the last quarter. However, if XYZ wants more assurance, which financial ratio would provide further information? Using the information provided, compute the appropriate ratio and comment on XYZ's concern.

Chicoutimi Metals
Consolidated Balance Sheets
December 31, 2003, 2004, and 2005
(in thousands of dollars)

Assets

	2005	**2004**	**2003**
Current Assets			
Cash	19	19	24
Accounts receivable	779	884	1176
Inventories	3563	3155	2722
	4361	4058	3922
Fixed Assets			
Land	1136	1064	243
Buildings and equipment	2386	4682	2801
	3552	5746	3044
Other Assets	3413	....	3
Total Assets	8296	9804	6969

Liabilities and Owners' Equity

	2005	**2004**	**2003**
Current Liabilities			
Due to bank	1431	1 929	2040
Accounts payable	1644	1 349	455
Wages payable	341	312	333
Income tax payable	562	362	147
Long-Term Debt	2338	4743	2528
Total Liabilities	6316	8695	5503
Owners' Equity			
Capital stock	1194	1191	1091
Retained earnings	786	(82)	375
Total Owner's Equity	1980	1109	1466
Total Liability and Owners' Equity	8296	9804	6969

Chicoutimi Metals
Income Statement
for the Years Ending December 31, 2003, 2004, and 2005
(in thousands of dollars)

	2005	2004	2003
Total revenue	9355	9961	8470
Less: Costs	8281	9632	7654
Net revenue	1074	329	816
Less:			
Depreciation	447	431	398
Interest	412	334	426
Income taxes	117	21	156
Net income from operations	98	(457)	(164)
Add: Extraordinary item	770		(1832)
Net income	868	(457)	(1996)

6.24 A potentially very large customer for Chicoutimi Metals wants to fully assess the financial health of the company in order to decide whether to commit to a long-term, high-volume relationship. You have been asked by the company president, Roch, to review the company's financial performance over the last three years and make a complete report to him. He will then select from your report information to present to the customer. Consequently, your report should be as thorough and honest as possible.

Research has revealed that in your industry (sheet metal products), the average value of the current ratio is 2.79, the equity ratio is 0.54, the inventory turnover is 4.9, and the net-profit ratio is 3.87. Beyond that information, you have access to only the balance sheet and the income statement shown here, and should make no further assumptions. Your report should be limited to the equivalent of about 300 words.

6.25 The Chicoutimi Metals income statement and balance sheets shown for Problem 6.24 were in error. A piece of production equipment was sold for $100 000 cash in 2005 and was not accounted for. Which items on these statements must be changed, and (if known) by how much?

6.26 The Chicoutimi Metals income statement and balance sheet shown for Problem 6.24 were in error. An extra $100 000 in sales was made in 2005 and not accounted for. Only half of the payments for these sales have been received. Which items on these statements must be changed, and (if known) by how much?

6.27 At the end of last month, Estevan Manufacturing had $45 954 in the bank. It owed the bank, because of the mortgage, $224 000. It also had a working capital loan of $30 000. Its customers owed them $22 943, and it owed its suppliers $12 992. The company owned property worth $250 000. It had $123 000 in finished goods, $102 000 in raw materials, and $40 000 in work in progress. Its production equipment was worth $450 000 when new (partially paid for by a large government loan due to be paid back in three years) but had accumulated a total of $240 000 in depreciation, $34 000 worth last month.

The company has investors who put up $100 000 for their ownership. It has been reasonably profitable; this month the gross income from sales was $220 000, and the cost of the sales was only $40 000. Expenses were also relatively low; salaries were $45 000 last month, while the other expenses were depreciation, maintenance at $1500, advertising at $3400, and insurance at $300. In spite of $32 909 in accrued taxes (they pay taxes at 55%), the company had retained earnings of $135 000.

Construct a balance sheet (at the end of this month) and income statement (for this month) for Estevan Manufacturing. Should the company release some of its retained earnings through dividends at this time?

6.28 Brandon Industries bought land and built its plant 20 years ago. The depreciation on the building is calculated using the straight-line method, with a life of 30 years and a salvage value of $50 000. Land is not depreciated. The depreciation for the equipment, all of which was purchased at the same time the plant was constructed, is calculated using declining-balance at 20%. Brandon currently has two outstanding loans, one for $50 000 due December 31, 2005, and another one for which the next payment is due in four years.

Brandon Industries Balance Sheet as of June 30, 2005		
Assets		
[ ]		
Cash		$ 350 000
Accounts receivable		2 820 000
Inventories		2 003 000
Prepaid services		160 000
Total Current Assets		[ ]
Long-Term Assets		
Building	$200 000	
Less accumulated depreciation	[ ]	[ ]
Equipment	$480 000	
Less accumulated depreciation	[ ]	[ ]
Land		540 000
Total Long-Term Assets		[ ]
Total Assets		[ ]
Liabilities and Owners' Equity		
Current Liabilities		
Accounts payable		$ 921 534
[ ]		[ ]
Accrued taxes		29 000
Total [ ]		[ ]
Long-Term Liabilities		
Mortgage		$1 200 000
[ ]		318 000
Total Long-Term Liabilities		[ ]
Total [ ]		[ ]
Owners' Equity		
Capital stock		$1 920 000
[ ]		[ ]
Total Owners' Equity		[ ]
Total Liabilities and Owners' Equity		[ ]

During April 2005, there was a flood in the building because a nearby river overflowed its banks after unusually heavy rain. Pumping out the water and cleaning up the basement and the first floor of the building took a week. Manufacturing was suspended during this period, and some inventory was damaged. Because of lack of adequate insurance, this unusual and unexpected event cost the company $100 000 net.

(a) Fill in the blanks and complete a copy of the balance sheet and income statement here, using any of the above information you feel necessary.

Brandon Industries **Income Statement for the Year Ended June 30, 2005**		
Income		
Gross income from sales	$8 635 000	
Less [ ]	7 490 000	[ ]
Total income		[ ]
[ ]		
Depreciation		70 000
Interest paid		240 000
Other expenses		100 000
Total expenses		[ ]
Income before taxes		[ ]
Taxes at 40%		[ ]
[ ]		[ ]
[ ]		[ ]
Net income		[ ]

(b) Show how information from financial ratios can indicate whether Brandon Industries can manage an unusual and unexpected event such as the flood without threatening its existence as a viable business.

6.29 Movit Manufacturing has the following alphabetized income statement and balance sheet entries from year 2005. Construct an income statement and a balance sheet from the information given.

Accounts payable	$ 7 500
Accounts receivable	15 000
Accrued wages	2 850
Cash	2 100
Common shares	150
Contributed capital	3 000
Cost of goods sold	57 000
Current assets	
Current liabilities	
Deferred income taxes	2 250
Depreciation expense	750

General expense	8 100
GIC's	450
Income taxes	1 800
Interest expense	1 500
Inventories	18 000
Land	3 000
Less: Accumulated depreciation	10 950
Long-term assets	
Long-term bonds	4 350
Long-term liabilities	
Mortgage	9 450
Net income after taxes	2 700
Net income before taxes	4 500
Net plant and equipment	7 500
Net sales	76 500
Operating expenses	
Owners' equity	
Prepaid expenses	450
Selling expenses	4 650
Total assets	46 500
Total current assets	36 000
Total current liabilities	15 000
Total expenses	15 000
Total liabilities and owners' equity	46 500
Total long-term assets	10 500
Total long-term liabilities	16 050
Total owners' equity	15 450
Working capital loan	4 650

6.30 Calculate for Movit Manufacturing in Problem 6.29 the financial ratios listed in the table below. Using these ratios and those provided for 2003 and 2004, conduct a short analysis of Movit's financial health.

Movit Manufacturing
Financial Ratios

Ratio	2005	2004	2003
Current ratio		1.90	1.60
Acid-test ratio		0.90	0.75
Equity ratio		0.40	0.55
Inventory-turnover ratio		7.00	12.00
Return-on-assets ratio		0.08	0.10
Return-on-equity ratio		0.20	0.18

6.31 Fraser Phraser operates a small publishing company. He is interested in getting a loan for expanding his computer systems. The bank has asked Phraser to supply them with his financial statements from the past two years. His statements appear below. Comment on Phraser's financial position with regard to the loan based on the results of financial ratio analysis.

Fraser Phraser Company
Comparative Balance Sheets
for the Years Ending in 2004 and 2005
(in thousands of dollars)

	2004	**2005**
Assets		
Current Assets		
Cash	22 500	1 250
Accounts receivable	31 250	40 000
Inventories	72 500	113 750
Total Current Assets	126 250	155 000
Long-Term Assets		
Land	50 000	65 000
Plant and equipment	175 000	250 000
Less: Accumulated depreciation	70 000	95 000
Net Plant and equipment	105 000	155 000
Total Long-Term Assets	155 000	220 000
Total Assets	281 250	375 000
Liabilities and Owners' Equity		
Current Liabilities		
Accounts payable	26 250	55 000
Working capital loan	42 500	117 500
Total Current Liabilities	68 750	172 500
Long-Term Liabilities		
Mortgage	71 875	57 375
Total Long-Term Liabilities	71 875	57 375
Owners' Equity		
Common shares	78 750	78 750
Retained earnings	61 875	66 375
Total Owners' Equity	140 625	145 125
Total Liabilities and Owners' Equity	281 250	375 000

Fraser Phraser
Income Statements
for Years Ending in 2004 and 2005
(in thousands of dollars)

	2004	2005
Revenues		
Sales	156 250	200 000
Cost of goods sold	93 750	120 000
Net revenue from sales	62 500	80 000
Expenses		
Operating expenses	41 875	46 250
Depreciation expense	5 625	12 500
Interest expense	3 750	7 625
Total expenses	51 250	66 375
Income before Taxes	11 250	13 625
Income taxes	5 625	6 813
Net income	5 625	6 813

6.32 A friend of yours is thinking of purchasing shares in Petit Ourson Ltée in the near future, and decided that it would be prudent to examine its financial statements for the past two years before making a phone call to his stockbroker. The statements are shown below.

Your friend has asked you to help him conduct a financial ratio analysis. Fill out the financial ratio information on a copy of the third table below. After comparison with industry standards, what advice would you give your friend?

Petit Ourson Ltée
Comparative Balance Sheets
for the Years Ending 2004 and 2005
(in thousands of dollars)

	2004	2005
Assets		
Current Assets		
Cash	500	375
Accounts receivable	1125	1063
Inventories	1375	1563
Total Current Assets	3000	3000
Long-Term Assets		
Plant and equipment	5500	6500
Less: Accumulated depreciation	2500	3000
Net Plant and Equipment	3000	3500
Total Long-Term Assets	3000	3500
Total Assets	6000	6500

Liabilities and Owners' Equity		
Current Liabilities		
Accounts payable	500	375
Working capital loan	000	375
Total Current Liabilities	500	750
Long-Term Liabilities		
Bonds	1500	1500
Total Long-Term Liabilities	1500	1500
Owners' Equity		
Common shares	750	750
Contributed capital	1500	1500
Retained earnings	1750	2000
Total Owners' Equity	4000	4250
Total Liabilities and Owners' Equity	6000	6500

Petit Ourson Ltée
Income Statements
for the Years Ending in 2004 and 2005
(in thousands of dollars)

	2004	**2005**
Revenues		
Sales	3000	3625
Cost of goods sold	1750	2125
Net revenue from sales	1250	1500
Expenses		
Operating expenses	75	100
Depreciation expense	550	500
Interest expense	125	150
Total expenses	750	750
Income Before Taxes	500	750
Income taxes	200	300
Net Income	300	450

6.33 Construct an income statement and a balance sheet from the scrambled entries for Paradise Pond Company from the years 2004 and 2005 shown in the table below.

Petit Ourson Ltée
Financial Ratios

	Industry Norm	**2004**	**2005**
Current ratio	4.50		
Acid-test ratio	2.75		
Equity ratio	0.60		
Inventory-turnover ratio	2.20		
Return-on-assets ratio	0.09		
Return-on-equity ratio	0.15		

Paradise Pond Company	**2004**	**2005**
Accounts receivable	$ 675	$ 638
Less: Accumulated depreciation	1500	1800
Accounts payable	300	225
Bonds	900	900
Cash	300	225
Common shares	450	450
Contributed capital	900	900
Cost of goods sold	1750	2125
Depreciation expense	550	500
Income taxes	200	300
Interest expense	125	150
Inventories	825	938
Net plant and equipment	1800	2100
Net revenue from sales	1250	1500
Operating expenses	075	100
Plant and equipment	3300	3900
Profit after taxes	300	450
Profit before taxes	500	750
Retained earnings	1050	1200
Sales	3000	3625
Total assets	3600	3900
Total current assets	1800	1800
Total current liabilities	300	450
Total expenses	750	750
Total liabilities and owners' equity	3600	3900
Total long-term assets	1800	2100
Total long-term liabilities	900	900
Total owners' equity	2400	2550
Working capital loan	000	225

Capital Expenditure or Business Expense?

CANADIAN 6.1 MINI-CASE

When a Calgary shopping centre began to experience considerable congestion on the roadways that provided access to the centre, it approached the local municipality to see if the roadways close to the mall could be improved. An agreement was struck with the municipality, and Oxford Shopping Centres Ltd. paid the municipality $450 050 to make the improvements in lieu of any increase in local tax rates which might have otherwise been charged by the municipality.

Was the $450 050 a business expense or a capital expenditure? The issue at hand was whether Oxford Shopping Centres Ltd. could claim the $450 050 as a current expense in the year in which it was paid, or whether the amount had to be amortized over a period of years, with only a fraction of the total amount claimed as an expense each year. The owners of the shopping centre argued that the aim was to increase the popularity of the mall and that the outlay related to the business as a whole rather than an improvement to the physical premises. With this logic, Oxford Shopping Centres Ltd. was allowed to claim the entire amount as an expense in the year in which it was paid.

Discussion

Calculating depreciation is made difficult by many factors. First, the value of an asset can change over time in many complicated ways. Age, wear and tear, and functional changes all have their effects, and often unpredictably. A 30-year old VW Beetle, for example, can suddenly increase in value because a new Beetle is introduced by the manufacturer.

A second complication is created by tax laws that make it desirable for companies to depreciate things quickly, while at the same time restricting the way they calculate depreciation. As will be seen in Chapter 8, tax laws require that companies, at least for tax purposes, use declining-balance depreciation with a specified depreciation rate.

A third complication is that, in real life, it is sometimes hard to determine what is an asset and what is not. In the case of the Calgary shopping centre, it was unclear whether the road improvements were part of the asset—the physical shopping centre—or simply the equivalent of maintenance expenses.

In the end, depreciation calculations are simply estimates that are useful only with a clear understanding of the assumptions underlying them.

Questions

1. For each of the following, indicate whether the straight-line method or the declining-balance method would likely give the most accurate estimate of how the asset's value changes over time, or explain why neither method is suitable.

(a) A $20 bill

(b) A $2 bill

(c) A pair of shoes

(d) A haircut

(e) An engineering degree

(f) A Van Gogh painting

2. What differences would have occurred to the balance sheet and income statement for Oxford Shopping Centres Ltd. if the $450 050 expense had been considered a capital expenditure (depreciating at, for example, 5% per year) as opposed to an expense? Would the company's profit be greater or less in the first year? How does the total profit over the life of the asset compare?

Welcome to the Real World

A.1 Introduction

Clem looked up from his computer as Naomi walked into his office. "Hi, Naomi. Sit down. Just let me save this stuff."

After a few seconds Clem turned around showing a grin. "I'm working on our report for the last quarter's operations. Things went pretty well. We exceeded our targets on defect reductions and on reducing overtime. And we shipped everything required—over 90% on time."

Naomi caught a bit of Clem's exuberance. "Sounds like a report you don't mind writing."

"Yeah, well, it was a team job. Everyone did good work. Talking about doing good work, I should have told you this before, but I didn't think about it at the right time. Ed Burns and Anna Kulkowski were really impressed with the report you did on the forge project."

Naomi leaned forward. "But they didn't follow my recommendation to get a new manual forging press. I assumed there was something wrong with what I did."

"I read your report carefully before I sent it over to them. If there had been something wrong with it, you would have heard right away. Trust me. I'm not shy. It's just that we were a little short of cash at the time. We could stay in business with just fixing up the guides on the old forging hammer. And Burns and Kulkowski decided there were more important things to do with our money.

"If I didn't have confidence in you, you wouldn't be here this morning. I'm going to ask you and Dave Sullivan to look into an important strategic issue concerning whether we continue to buy small aluminum parts or whether we make them ourselves. We're just waiting for Dave to show up."

"OK. Thanks, Clem. But please tell me next time if what I do is all right. I'm still finding my way around here."

"You're right. I guess that I'm still more of an engineer than a manager."

Voices carried into Clem's office from the corridor. "That sounds like Dave in the hall saying hello to Carole," Naomi observed. "It looks like we can get started."

Dave Sullivan came in with long strides and dropped into a chair. "Good morning, everybody. It is still morning, barely. Sorry to be late. What's up?"

Clem looked at Dave and started talking. "What's up is this. I want you and Naomi to look into our policy about buying or making small aluminum parts. We now use about 200 000 pieces a month. Most of these, like bolts and sleeves, are cold-formed.

"Prabha Vaidyanathan has just done a market projection for us. If she's right, our demand for these parts will continue to grow. Unfortunately, she wasn't very precise about the *rate* of growth. Her estimate was for anything between 5% and 15% a year. We now contract this work out. But even if growth is only 5%, we may be at the level where it pays for us to start doing this work ourselves.

"You remember we had a couple of engineers from Hamilton Tools looking over our processes last week? Well, they've come back to us with an offer to sell us a cold-former. They have two possibilities. One is a high-volume job that is a version of their Model E2. The other is a low-volume machine based on their Model E1.

"The E2 will do about 2000 pieces an hour, depending on the sizes of the parts and the number of changeovers we make. The E1 will do about 1000 pieces an hour."

Naomi asked, "About how many hours per year will these formers run?"

"Well, with our two shifts, I think we're talking about 3600 hours a year for either model."

Dave came in with "Hold it. If my third-grade arithmetic still works, that sounds like either 3.6 million or 7.2 million pieces a year. You say that we are using only 2.4 million pieces a year."

Clem answered, "That's right. Ms. Vaidyanathan has an answer to that one. She says we can sell excess capacity until we need it ourselves. Again, unfortunately, she isn't very precise about what this means. We now pay about five cents a piece. Metal cost is in addition to that. We pay for that by weight. She says that we won't get as much as five cents because we don't have the market connections. But she says we should be able to find a broker so that we net somewhere between three cents and four cents a piece, again plus metal."

Naomi spoke. "That's a pretty wide range, Clem."

"I know. Prabha says that she couldn't do any better with the budget Burns and Kulkowski gave her. For another $5000, she says that she can narrow the range on *either* the growth rate or the potential prices for selling pieces from any excess capacity. Or, for about $7500, she could do both. I spoke to Anna Kulkowski about this. Anna says that they won't approve anything over $5000. One of the things I want you two to look at is whether or not it's necessary to get more information. If you do recommend spending on market research, it has to be for just one of either the selling price range or the growth rate.

"I have the proposal from Hamilton Tools here. It has information on the two formers. This is Wednesday. I'd like a report from you by Friday afternoon so that I can look at it over the weekend.

"Did I leave anything out?"

Naomi asked, "Are we still working with a 15% after-tax MARR?"

Clem hesitated. "This is just a first cut. Don't worry about details on taxes. We can do a more precise calculation before we actually make a decision. Just bump up the MARR to 25% before tax. That will about cover our 40% marginal tax rate."[1]

Dave asked, "What time frame should we use in our calculations?"

"Right. Use 10 years. Either of these models should last at least that long. But I wouldn't want to stretch Prabha's market projections beyond 10 years."

Dave stood up and announced, "It's about a quarter to one." He turned to Naomi. "Do you want to start on this over at the Grand China Restaurant? It's past the lunch rush, and we'll be able to talk while we eat. I think Clem will buy us lunch out of his budget."

Clem interjected, "All right, Dave. Just don't order the most expensive thing on the menu."

Naomi laughed. "I'm glad we have one big spender around here."

A.2 Problem Definition

About 40 minutes later, Dave and Naomi were most of the way through their main courses. Dave suggested that they get started. He took a pad from his briefcase and said that he would take notes. Naomi agreed to let him do that.

Dave started with "OK. What are our options?"

"Well, I did a bit of arithmetic on my calculator while you were on the phone before lunch. It looks as though, even if the demand growth rate is only 5%, a single small former will not have enough capacity to see us through 10 years. This means that there are four options. The first is a sequence of two low-capacity formers. The second is just a high-capacity former. The third, which would kick in only if the growth rate is high, would be a sequence of three low-capacity formers. The fourth is a low-capacity and a high-capacity. I'm not sure of the sequence for that."

Dave thought for a bit. "I don't think so. I assume that, even if we put in our own former, we could contract out requirements that our own shop couldn't handle. That might be the way to go. That is, you wouldn't want to add to capacity if there was only one year left in the 10-year horizon. There probably would not be enough unsatisfied output requirement to amortize the first cost of a second small former."

"That sounds as though there are a whole bunch of options. It looks like we're in for a couple of long nights." Naomi sounded dejected.

"Well, maybe not."

"Maybe not what?"

"Maybe we won't have to spend those long nights. I think we can look at just three options at the start. We could look at a sequence of two small formers. We could look at a small former followed by outsourcing when capacity is exhausted. And we could look at a large former, possibly followed by outsourcing if capacity is exhausted. If we can rule out the two-small-formers option, we can certainly rule out three small formers or a big former combined with a small one."

"Smart, Dave. How should we proceed?"

"Well, there are a couple of ways of doing this. But, given that we have only 10 years to look at,

1. Businesses in Canada and the United States are required to pay tax on their earnings. Determining the effect of taxes on before-tax earnings may involve extensive computation. Managers frequently approximate the effect of taxes on decisions by increasing the MARR. However, it is good practice to do precise calculations before actually making a decision.

it's probably easiest to use a spreadsheet to develop cash flow sequences for the three options. The two options in which only a single machine is bought at time zero are pretty straightforward. The one with a sequence of two small formers is a bit more complicated. One of us can do the two easy ones. The other can do the sequence. For each option we need to look at, say, nine outcomes: three possible demand growth rates—5%, 10%, and 15%—times three possible prices—3¢, 3.5¢, and 4¢—for selling excess capacity. That should be enough to show us what's happening. What part do you want to do?"

"I'll take the hard one, the sequence of two. I'd like the practice. I'll let you check my analysis when I'm finished."

"OK. That's fine. But notice that the two-low-capacity-formers sequence is not a simple investment, so let's stick with present worths at this stage. Also, we are going to have to make some decision about how we record the timing of the purchase of the second former if we run out of capacity during a year. I suggest that we assume the second former is bought at the end of the year before we run out of capacity."

"OK."

Dave continued with "We need to put together a simple table on the specifications for these two machines. I'll do that, since you are doing the hard job with cash flows. Why don't we go back to the plant. I'll make up the table. I'll get you a copy later this afternoon."

A.3 Crunch Time

Naomi was sitting in her office thinking about structuring the cash flows for the two-small-machine sequence. Dave knocked and came in.

He handed Naomi a sheet of paper.

"Here's the table. Shall we meet tomorrow morning to compare results?" (See Table A.1.)

Naomi glanced at the sheet of paper, and motioned Dave to stay. "This is a bit new to me. Would it be okay if we get all of our assumptions down before we go too far?"

"Sure, we can do that," replied Dave as he grabbed a seat beside Naomi's desk. Naomi already had a pad and pencil out.

Dave went on, "OK, first, what's our goal?"

"Well, right now we are faced with a 'make or buy' decision," suggested Naomi. "We need to find out if it is cheaper to make these parts or to continue to buy them, and if making them is cheaper we need to choose which machine or machines we should buy."

"Pretty close, Naomi, but I think that may be more than we need at this point. Remember, Clem is primarily interested in whether this project is worth pursuing as a full-blown proposal to top management. So we can use a greatly simplified model now to answer that question, and get into the details if Clem decides it's worth pursuing."

"Right. If we look at the present worths of the three options we came up with at lunch, we should see quickly enough if investing in our own machines is feasible. Positive present worths indi-

Table A.1 Specifications for the Two Cold Formers

Characteristics	Model E1 (Small)	Model E2 (Large)
First cost ($)	125 000	225 000
Hourly running cost ($)	35.00	61.25
Average number of pieces/hour	1000	2000
Hours/year	3600	3600
Depreciation	20%/year declining balance, for both machines	
Market facts	Buying price: $0.05/piece plus cost of metal Selling price: $0.03 to $0.04 per piece plus metal Demand growth: 5% to 15%	

cate buying machines is the best course; negative values would suggest it's probably not worth pursuing further."

Dave nodded. "OK," continued Naomi, "so we want to do a 'first approximation' calculation of present worths of each of our three options. Now, what assumptions should we use?"

"Let's see. We want to consider only a 10-year study period for the moment. I think we can assume the machines will have no salvage value at the end of the study period, so we don't have to worry about estimating depreciation and salvage values. This will be a conservative estimate in any case."

"Next," Dave continued, "we will ignore tax implications, and use a before-tax MARR of 25%. The 'sequence of machines' option has a complex cash flow, but by only considering present worths we can avoid dealing with possible multiple IRRs."

"You know," interjected Naomi, "it just occurred to me that the small machine makes 1000 pieces per hour at an operating cost of $35."

"And this is news to you?" said a bemused Dave.

"No, Dave. What I mean is that the operating cost is $0.035 per part on this machine, while one scenario calls for us selling excess capacity at only $0.03. We'll be losing half a cent on every part made for sale."

"Hmmm. You know, you're right. And look at this. Even the bigger machine's operating cost is $0.030625 per piece. We'd be losing money on that one too."

"When we run the different cases, it looks like we'll have to consider whether excess capacity will be sold or whether the machine will just be shut down. Just when I thought this was going to be easy . . . ," trailed Naomi.

"Maybe, but maybe not. Look at this another way. If the small machine sells its excess capacity at a loss of half a cent per piece, that comes to five bucks per hour. On the other hand, if the machine is shut down, we'll have an operator standing around at a cost of a lot more than that."

"But can't an idle operator be given other work to do?"

"Again, maybe, but maybe not. With the job security clause the union has now, if we lay the operator off for any period of time, the company has to pay most of the wages anyway, so there isn't much savings that way. On the other hand, the operator would probably be idled at unpredictable times, so other departments would have a difficult job of scheduling work for him or her to do. I'm not saying it's impossible, but it's a job in itself to figure out if we can place the operator in productive work when trying to idle the machine."

"Boy, they didn't talk about these problems in engineering school!"

"Welcome to the real world, Naomi, where nothing is simple."

"Then how about if we do this?" Naomi continued. "We assume that since an idle operator is probably more costly to the company than $5 per hour, the machines will run a full 3600 hours per year each regardless of whether the excess capacity is being sold at a price above operating costs or not."

"Sounds good, Naomi. In fact, we could claim an indirect benefit of this otherwise unprofitable operation since our operators will gain more experience with doing quick setups and Statistical Process Control. That should make the brass happy," Dave said with a quick wink.

"I guess we can also assume that a machine is purchased and paid for at the beginning of a year while savings and operating costs accrue at the end of the year," continued Naomi. "And that the demand for parts is constant for each month within a year but grows by a fixed proportion from one year to the next."

Naomi finished scribbling down the assumptions, including the ones they had discussed with Clem earlier that day. Turning to Dave, she said "so how does this look?"

"Looks fine to me. Should we get together tomorrow and compare results?"

"OK. What time?"

"Why don't we exchange results first thing in the morning and then meet about nine-thirty in my office?"

"That's fine. See you then."

As Dave left, Naomi looked more closely at the table he had given her (Table A.1). "Time to crunch those numbers," she said to herself.

QUESTIONS

1. Construct spreadsheets for calculating present worths of the three proposals. For each proposal, you need to calculate PWs for each of

5%, 10%, and 15% demand growth and $0.03, $0.035, and $0.04 selling price (nine combinations in all). Present the results in tabular and/or graphical format to support your analysis. A portion of a sample spreadsheet layout is given in Table A.2.

2. Write a memo to Clem presenting your findings. The goal of the analysis is to determine if bringing production in-house appears feasible, and if so, which machine purchase sequence(s) should be studied in further detail. The memo should contain a tentative recommendation about which option looks best and what additional research, if any, should be done. Keep the memo as concise as possible.

Table A.2 Portion of a Present Worth Spreadsheet

Sequence: E1 Machine purchased at time 0					
Projected Demand Growth	5%	/year			
Projected Selling Price	$0.03	/piece			
Year	0	1	2	3	4
Projected Demand (units)		2 400 000	2 520 000	2 646 000	2 778 300
Capacity (units)		3 600 000	3 600 000	3 600 000	3 600 000
Purchases (units)		0	0	0	0
Sales (units)		1 200 000	1 080 000	954 000	821 700
Operating Cost	$0	–$126 000	–$126 000	–$126 000	–$126 000
Purchasing Savings	$0	$120 000	$126 000	$132 300	$138 915
Sales Revenue	$0	$36 000	$32 400	$28 620	$24 651
Capital Expenditure	–$125 000	$0	$0	$0	$0
Net Cash Flow	–$125 000	$30 000	$32 400	$34 920	$37 566
Present Worth	$8 490				

CHAPTER 7

Replacement Decisions

Engineering Economics in Action, Part 7A: You Need the Facts

"You know the 5-stage progressive die that we use for the Admiral Motors rocker arm contract?" Naomi was speaking to Terry, her co-op student, one Tuesday afternoon. "Clem asked me to look into replacing it with a 10-stage progressive die that would reduce the hand finishing substantially. It's mostly a matter of labour cost saving, but there is likely to be some quality improvement with the 10-stage die as well. I would like you to come up with a ballpark estimate of the cost of switching to the 10-stage progressive die."

Terry asked, "Don't you have the cost from the supplier?"

"Yes, but not really," said Naomi. "The supplier is Hamilton Tools. They've given us a price for the machine, but there are a lot of other costs involved in replacing one production process with another."

"You mean things like putting the machine in place?" Terry asked.

"Well, there's that," responded Naomi. "But there is also a lot more. For example, we will lose production during the changeover. That's going to cost us something."

"Is that part of the cost of the 10-stage die?"

"It's part of the first cost of switching to the 10-stage die," Naomi said. "If we decide to go ahead with the 10-stage die and incur these costs, we'll never recover them—they are sunk. We have already incurred those costs for the 5-stage die and it's only two years old. It still has lots of life in it. If the first costs of the 10-stage die are large, it's going to be hard to make a cost justification for switching to the 10-stage die at this time."

"OK. How do I go about this?" Terry asked.

Naomi sat back and chewed on her yellow pencil for about 15 seconds. She leaned forward and began. "Let's start with order-of-magnitude estimates of what it's going to cost to get the 10-stage die in place. If it looks as if there is no way that the 10-stage die is going to be cost-effective now, we can just stop there."

"It sounds like a lot of fuzzy work," said Terry.

"Terry, I know you like to be working with mathematical models. I'm also sure that you can read the appropriate sections on replacement models in an engineering economics book. But none of those models is worth anything unless you have data to put in it. You need the models to know what information to look for. And once you have the information, you will make better decisions using the models. But you do need the facts."

7.1 Introduction

Survival of businesses in a competitive environment requires regular evaluation of the plant and equipment used in production. As these assets age, they may not provide adequate quality, or their costs may become excessive. When a plant or equipment is evaluated, one of four mutually exclusive choices will be made:

1. An existing asset may be kept in its current use without major change.
2. An existing asset may be overhauled so as to improve its performance.
3. An existing asset may be removed from use without replacement by another asset.
4. An existing asset may be replaced with another asset.

This chapter is concerned with methods of making choices about possible replacement of long-lived assets. While the comparison methods developed in Chapters 4 and 5 for choosing between alternatives are often used for making these choices, the issues of replacement deserve a separate chapter for several reasons. First, the relevant costs for making replacement decisions are not always obvious, since there are costs associated with taking the replaced assets out of service that should be considered. This was ignored in the studies in Chapters 4 and 5. Second, the service lives of assets were provided to the reader in Chapters 4 and 5. As seen in this chapter, the principles of replacement allow the calculation of these service lives. Third, assumptions about how an asset might be replaced in the future can affect a decision now. Some of these assumptions are implicit in the methods analysts use to make the choices. It is therefore important to be aware of these assumptions when making replacement decisions.

The chapter starts with an example to introduce some of the basic concepts involved in replacement decisions. Following this is a discussion of the reasons why a long-lived asset may be replaced. We then develop the idea of the *economic life* of an asset—the service life that minimizes the average cost of using the asset. This is followed by a discussion of replacement with an asset that differs from the current asset, in which the built-in cost advantage of existing assets is revealed. Finally, we look at the case of replacement where there may be a series of replacements for a current asset, each of which might be different.

We shall not consider the implications of taxes for replacement decisions in this chapter. This is postponed until Chapter 8. We shall assume in this chapter that no future price changes due to inflation are expected. The effect of expected price changes on replacement decisions will be considered in Chapter 9.

7.2 A Replacement Example

We introduce some of the basic concepts involved in replacement decisions through an example.

Example 7.1

Sergio likes hiring engineering students to work in his landscaping business during the summer because they are such hard workers and have a lot of common sense. The students are always complaining about maintenance problems with the lawnmowers, which are subject to a lot of use and wear out fairly quickly. His routine has been to replace the machines every five years. Clarissa, one of the engineering student workers, has suggested that replacing them more often might make sense, since so much time is lost when there is a breakdown, in addition to the actual repair cost.

"I've checked the records, and have made some estimates that might help us figure out the best time to replace the machines," Clarissa reports. "Every time there is a breakdown, the average cost to fix the machine is $60. In addition to that, there is an average loss of two hours of time at $20 per hour. As far as the number of repairs required goes, the pattern seems to be zero repairs the first season we use a new lawnmower. However, in the second season, you can expect about one repair, two repairs in the third season, four repairs in the fourth, and eight in the fifth season. I can see why you only keep each mower five years!"

"Given that the cost of a new lawnmower has averaged about $600 and that they decline in value at 40% per year, and assuming an interest rate of 5%, I think we have enough information to determine how often the machines should be replaced," Clarissa concludes authoritatively.

How often should Sergio replace his lawnmowers? How much money will he save?

To keep things simple for now, let's assume that Sergio has to have lawns mowed every year, for an indefinite number of years into the future. If he keeps each lawnmower for, say, three years rather than five years, he will end up buying lawnmowers more frequently, and his overall capital costs for owning the lawnmowers will increase.

However, his repair costs should decrease at the same time since newer lawnmowers require fewer repairs. We want to find out which replacement period results in the lowest overall costs—this is the replacement period Sergio should adopt and is referred to as the *economic life* of the lawnmowers.

We could take any period of time as a study period to compare the various possible replacement patterns using a *present worth* approach. The least common multiple of the service lives method (see Chapter 3), suggests that we would need to use a $3 \times 4 \times 5 =$ 60-year period if we were considering service lives between one and five years. This is an awkward approach in this case, and even worse in situations where there are more possibilities to consider. It makes much more sense to use an *annual worth* approach. Furthermore, since we typically analyze the costs associated with owning and operating an asset, annual worth calculated in the context of a replacement study is commonly referred to as **equivalent annual cost (EAC)**. In the balance of the chapter, we will therefore use EAC rather than annual worth. However, we should not lose sight of the fact that EAC computations are nothing more than the annual worth approach with a different name adopted for use in replacement studies.

Returning to Sergio's replacement problem, if we calculate the EAC for the possibility of a one-year, two-year, three-year (and so on) replacement period, the pattern that has the lowest EAC would indicate which is best for Sergio. It would be the best because he would be spending the least, keeping both the cost of purchase and the cost of repairs in mind.

This can be done in a spreadsheet, as illustrated in Table 7.1. The first column, "Replacement Period," lists the possible replacement periods being considered. In this case, Sergio is looking at periods ranging from one to five years. The second column lists the salvage value of a lawnmower at the end of the replacement period, in this case estimated using the declining-balance method with a rate of 40%. This is used to compute the entries of the third column. "EAC Capital Costs" is the annualized cost of purchasing and salvaging each lawnmower, assuming replacement periods ranging from one to five years. Using the capital recovery formula (refer back to Close-Up 3.1), this annualized cost is calculated as:

$$\text{EAC(capital costs)} = (P - S)(A/P, i, N) + Si$$

where

EAC(capital costs) = annualized capital costs for an N-year replacement period

P = purchase price

S = salvage value of the asset at the end of N years

Table 7.1 Total Equivalent Annual Cost Calculations for Lawnmower Replacement Example

Replacement Period (Years)	Salvage Value	EAC Capital Costs	Annual Repair Costs	EAC Repair Costs	EAC Total
1	$360.00	$270.00	$0.00	$0.00	$270.00
2	216.00	217.32	100.00	48.78	266.10
3	129.60	179.21	200.00	96.75	275.96
4	77.76	151.17	400.00	167.11	318.27
5	46.66	130.14	800.00	281.64	411.79

S can be in turn calculated as

$$BV_{db}(N) = P(1 - d)^N$$

For example, in the case of a three-year replacement period, the calculation is

$$\text{EAC(capital costs)} = [\$600 - \$600(1 - 0.40)^3](A/P, 5\%, 3) + [\$600(1 - 0.40)^3](0.05)$$

$$= (\$600 - \$129.60)(0.36721) + (\$129.60)(0.05)$$

$$\cong \$179.21$$

The "average" annual cost of repairs (under the heading "EAC Repair Costs"), assuming for simplicity that the cash flows occur at the end of the year in each case, can be calculated as:

$$\text{EAC(repairs)} = [(\$60 + \$40)(P/F, 5\%, 2) + (\$60 + \$40)(2)(P/F, 5\%, 3)](A/P, 5\%, 3)$$

$$= [\$100(0.90703) + \$200(0.86584)](0.36721)$$

$$\cong \$96.75$$

The total EAC is then:

$$\text{EAC(total)} = \text{EAC(capital costs)} + \text{EAC(repairs)} = \$179.22 + \$96.75 \cong \$275.96$$

Examining the EAC calculations in Table 7.1, it can be seen that the EAC is minimized when the replacement period is two years (and this is only slightly cheaper than replacing the lawnmowers every year). So this is saying that if Sergio keeps his lawnmowers for two years and then replaces them, his average annual costs over time will be minimized. We can also see how much money Clarissa's observation can save him, by subtracting the expected yearly costs from the estimate of what he is currently paying. This shows that on average Sergio saves \$411.79 – \$266.10 = \$145.69 per lawnmower per year by replacing his lawnmowers on a two-year cycle over replacing them every five years.■

The situation illustrated in Example 7.1 is the simplest case possible in replacement studies. We have assumed that the physical asset to be replaced is identical to the one replacing it, and such a sequence continues indefinitely into the future. By **asset**, we mean any machine or resource of value to an enterprise. An existing physical asset is called the **defender**, because it is currently performing the value-generating activity. The potential replacement is called the **challenger**. However, it is not always the case that the challenger and the defender are the same. It is generally more common that the defender is outmoded or less adequate than the challenger, and also that new challengers can be expected in the future.

This gives rise to several cases to consider. Situations like Sergio's lawnmower problem are relatively uncommon—we live in a technological age where it is unlikely that a replacement for an asset will be identical to the asset it is replacing. Even lawnmowers improve in price, capability, or quality in the space of a few years. A second case, then, is that of a defender that is different from the challenger, with the assumption that the replacement will continue in a sequence of identical replacements indefinitely into the future. Finally, there is the case of a defender different from the challenger, which is itself different from another replacing it further in the future. All three of these cases will be addressed in this chapter.

Before we look at these three cases in detail, let's look at why assets have to be replaced at all and how to incorporate various costs into replacement decision.

7.3 Reasons for Replacement or Retirement

If there is an ongoing need for the service an asset provides, at some point it will need replacement. **Replacement** becomes necessary if there is a cheaper way to get the service the asset provides, or if the service provided by the existing asset is no longer adequate.

An existing asset is **retired** if it is removed from use without being replaced. This can happen if the service that the asset provides is no longer needed. Changes in customer demand, changes in production methods, or changes in technology may result in an asset no longer being necessary. For example, the growth in the use of compact discs for audio recordings has led manufacturers of cassette tapes to retire production equipment since the service it provided is no longer needed.

There may be a cheaper way to get the service provided by the existing asset for several reasons. First, productive assets often deteriorate over time because of wearing out in use, or simply because of the effect of time. As a familiar example, an automobile becomes less valuable with age (older cars, unless they are collectors' cars, are worth less than newer cars with the same mileage) or if it has high mileage (the kilometres driven reflect the wear on the vehicle). Similarly, production equipment may become less productive or more costly to operate over time. It is usually more expensive to maintain older assets. They need fixing more often, and parts may be harder to find and cost more.

Technological or organizational change can also bring about cheaper methods of providing service than the method used by an existing asset. For example, the technological changes associated with the use of computers have improved productivity. Organizational changes, both within a company and in markets outside the company, can lead to lower-cost methods of production. A form of organizational change that has become very popular is the specialist company. These companies take on parts of the production activities of other companies. Their specialization may enable them to have lower costs than the companies can attain themselves. See Close-Up 7.1.

The second major reason why a current asset may be replaced is inadequacy. An asset used in production can become inadequate because it has insufficient capacity to meet growing demand or because it no longer produces items of high-enough quality. A company may have a choice between adding new capacity parallel to the existing capacity, or replacing the existing asset with a higher-capacity asset, perhaps one with more

CLOSE-UP 7.1 Specialist Companies

Specialist companies concentrate on a limited range of very specialized products. They develop the expertise to manufacture these products at minimal cost. Larger firms often find it more economical to contract out production of low-volume components instead of manufacturing the components themselves.

In some industries, the use of specialist companies is so pervasive that the companies apparently manufacturing a product are simply assembling it; the actual manufacturing takes place at dozens or sometimes hundreds of supplier firms.

A good example of this is the automotive industry. In North America, auto makers support an extremely large network of specialist companies, linked by computer. A single specialist company might supply brake pads, for example, to all three major auto manufacturers. Producing brake pads in huge quantities, the specialist firm can refine its production process to extremes of efficiency and profitability.

advanced technology. If higher quality is required, there may be a choice between upgrading an existing piece of equipment or replacing it with equipment that will yield the higher quality. In either case, contracting out the work to a specialist is one possibility.

In summary, there are two main reasons for replacing an existing asset. First, an existing asset will be replaced if there is a cheaper way to get the service that the asset provides. This can occur because the asset ages or because of technological or organizational changes. Second, an existing asset will be replaced if the service it provides is inadequate in either quantity or quality.

7.4 Capital Costs and Other Costs

When a decision is made to acquire a new asset, it is essentially a decision to purchase the capacity to perform tasks or produce output. **Capacity** is the ability to produce, often measured in units of production per time period. Although production requires capacity, it is also important to understand that just acquiring the capacity entails costs that are incurred whether or not there is actual production. Furthermore, a large portion of the capacity cost is incurred early in the life of the capacity. There are two main reasons for this:

1. Part of the cost of acquiring capacity is the expense incurred over time because the assets required for that capacity gradually lose their value. This expense is often called the **capital cost** of the asset. It is incurred by the difference between what is paid for the assets required for a particular capacity and what the assets could be resold for some time after purchase. The largest portion of the capital costs typically occurs early in the lives of the assets.
2. Installing a new piece of equipment or new plant sometimes involves substantial up-front costs, called **installation costs**. These are the costs of acquiring capacity, excluding the purchase cost, which may include disruption of production, training of workers, and perhaps a reorganization of other processes. Installation costs are not reversible once the capacity has been put in place.

For example, if Sergio bought a new lawnmower to accommodate new landscaping clients, rather than for replacement, he would be increasing the capacity of his lawnmowing service. The capital cost of the lawnmower in the first year would be its associated loss in market value over that year. The installation cost would probably be negligible.

It is worth noting that, in general, the total cost of a new asset includes both the installation costs and the cost of purchasing the asset. When we compute the capital costs of an asset over a period of time, the first cost (usually denoted by P) includes the installation costs. However, when we compute a salvage value for the asset as it ages, we do *not* include the installation costs as part of the depreciable value of the asset, since these costs are expended upon installation and cannot be recovered at a later time.

The large influence of capital costs associated with acquiring new capacity means that, once the capacity has been installed, the *incremental* cost of continuing to use that capacity during its planned life is relatively low. This gives a defender a cost advantage during its planned life over a challenger.

In addition to the capital and installation costs, the purchase of an asset carries with it future **operating and maintenance costs**. Operating costs might include electricity, gasoline, or other consumables, and maintenance might include periodic servicing and repairs. Also, it is worth noting that a challenger may also give rise to changes in revenues as well as changes in costs that should not be neglected.

The different kinds of costs discussed in this section can be related to the more general ideas of fixed and variable costs. **Fixed costs** are those that remain the same, regardless of actual units of production. For example, capital costs are usually fixed—once an asset is purchased, the cost is incurred even if there is zero production. **Variable costs** are costs that change depending on the number of units produced. The costs of the raw materials used in production are certainly variable costs, and to a certain degree operating and maintenance costs are as well.

With this background, we now look at the three different replacement cases:

1. Defender and challenger are identical, and repeat indefinitely
2. Challenger repeats indefinitely, but is different from defender
3. Challenger is different from defender, but does not repeat

NET VALUE 7.1

Estimating Salvage Values and Scrap Values

An operating asset can have considerable value as long as it continues to perform the function for which it is intended. However, if it is replaced, its salvage value depends much on what is done with it. It is likely that the salvage value is different from whatever depreciation value is calculated for accounting or taxation purposes. However, to make a good replacement decision, it is desirable to have an accurate estimate of the actual salvage value of the asset.

The Internet can be very helpful in estimating salvage values. A search for the asset by type, year, and model may reveal a similar asset in a used market. Or a more general search may reveal a broker for used assets of this nature who might be contacted to provide an estimate of salvage value based on the broker's experience with similar assets. In both of these cases, not only can a salvage value be determined, but also a channel for disposing of the asset is found. Even if the asset is scrapped, its value can be estimated from posted values for metal scrap available on the Web.

7.5 Defender and Challenger Are Identical

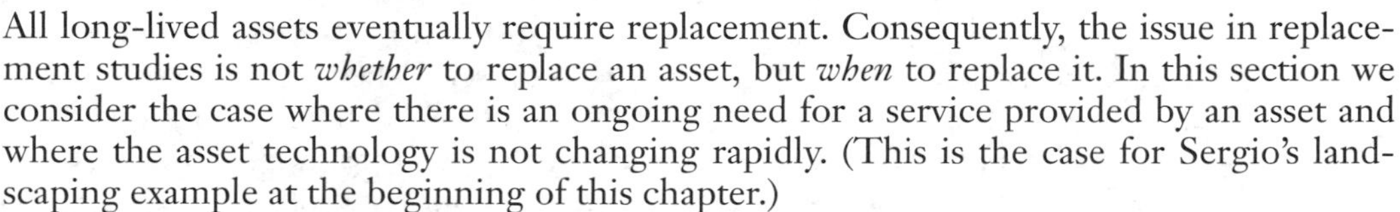

All long-lived assets eventually require replacement. Consequently, the issue in replacement studies is not *whether* to replace an asset, but *when* to replace it. In this section we consider the case where there is an ongoing need for a service provided by an asset and where the asset technology is not changing rapidly. (This is the case for Sergio's landscaping example at the beginning of this chapter.)

There are several assumptions made:

1. The defender and challenger are assumed to be technologically identical. It is also assumed that this remains true for the company's entire planning horizon.
2. The lives of these identical assets are assumed to be short relative to the time horizon over which the assets are required.
3. Relative prices and interest rates are assumed to be constant over the company's time horizon.

These assumptions are quite restrictive. In fact, there are only a few cases where the assumptions strictly hold (cable used for electric power delivery is an example). Nonetheless, the idea of economic life of an asset is still useful to our understanding of replacement analysis.

Assumptions 1 and 2 imply that we may model the replacement decision as being repeated an indefinitely large number of times. The objective is then to determine a minimum-cost lifetime for the assets, a lifetime that will be the same for all the assets in the sequence of replacements over the company's time horizon.

We have seen that the relevant costs associated with acquiring a new asset are the capital costs, installation costs (which are often pooled with the capital costs), and operating and maintenance costs. It is usually true that operating and maintenance costs of assets—plant or equipment—rise with the age of the asset. Offsetting increases in operating and maintenance costs is the fall in capital costs per year that usually occurs as the asset life is extended and the capital costs are spread over a greater number of years. The rise in operating and maintenance costs per year, and the fall in capital costs per year as the life of an asset is extended, work in opposite directions. In the early years of an asset's life, the capital costs per year (although decreasing) usually, but not always, dominate total yearly costs. As the asset ages, increasing operating and maintenance costs usually overtake the declining annual capital costs. This means that there is a lifetime that will minimize the *average* cost (adjusting for the time value of money) per year of owning and using long-lived assets. This is referred to as the **economic life** of the asset.

These ideas are illustrated in Figure 7.1. Here we see the capital costs per period decrease as the number of periods the asset is kept increases because assets tend to depreciate in value by a smaller amount each period of use. On the other hand, the operating and maintenance costs per period increase because older assets tend to need more repairs and have other increasing costs with age. The economic life of an asset is found at the point where the rate of increase in operating and maintenance costs per period equals the rate of decrease in capital costs per period, or equivalently where the total cost per period is minimized.

Figure 7.1 Cost Components for Replacement Studies

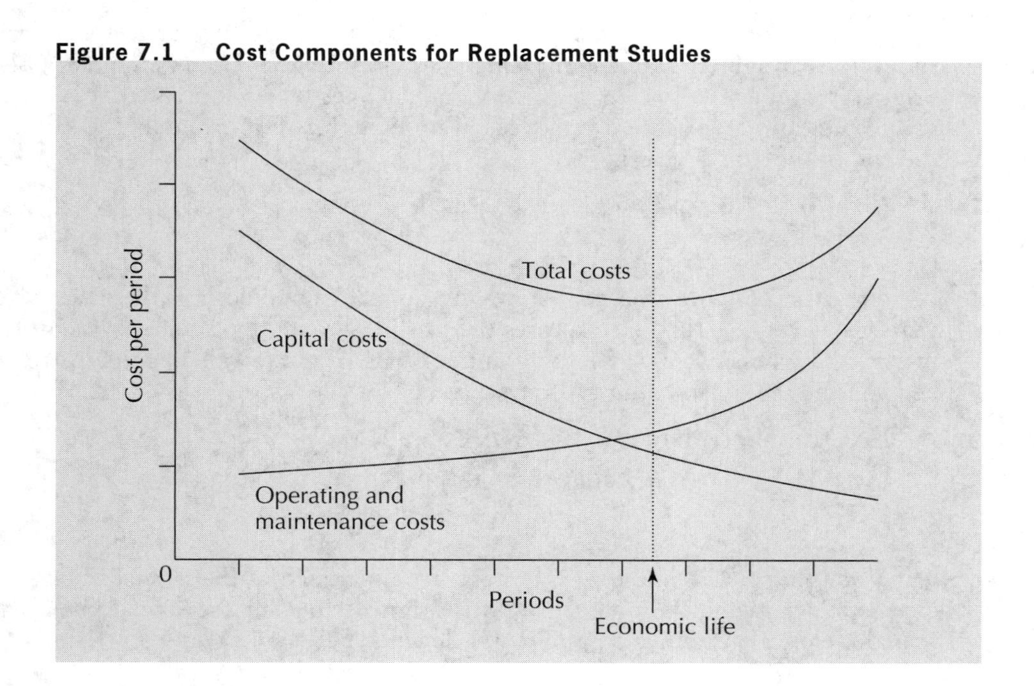

Example 7.2

The Jiffy Printer Company produces printers for home use. Jiffy is considering installing an automated plastic molding system to produce parts for the printers. The molder itself costs \$20 000 and the installation costs are estimated to be \$5000. Operating and maintenance costs are expected to be \$30 000 in the first year and to rise at the rate of 5% per year. Jiffy estimates depreciation with a declining-balance model using a rate of 40%, and uses a MARR of 15% for capital investments. Assuming that there will be an ongoing need for the molder, and assuming that the technology does not change (i.e., no cheaper or better method will arise), how long should Jiffy keep a molder before replacing it with a new model? In other words, what is the economic life of the automated molding system?

Determining the economic life of an asset is most easily done with a spreadsheet. Table 7.2 shows the development of the equivalent annual costs for the automated plastic molding system of Example 7.2.

In general, the EAC for an asset has two components:

$$\text{EAC} = \text{EAC(capital costs)} + \text{EAC(operating and maintenance)}$$

If

P = the current value of an asset = (for a new asset) purchase price + installation costs

S = the salvage value of an asset N years in the future

then

$$\text{EAC(capital costs)} = [P - (P/F, i, N)S]\,(A/P, i, N)$$

which in Close-Up 3.1 was shown to be equivalent to

$$\text{EAC(capital costs)} = (P - S)(A/P, i, N) + Si$$

In the first column of Table 7.2 is the life of the asset, in years. The second column shows the salvage value of the molding system as it ages. The equipment costs \$20 000 originally, and as the system ages the value declines by 40% of current value each year, giving the estimated salvage values listed in Table 7.2. For example, the salvage value at the end of the fourth year is:

$$\begin{aligned} BV_{db}(4) &= \$20\,000(1 - 0.4)^4 \\ &\cong \$2592 \end{aligned}$$

The next column gives the equivalent annual capital costs if the asset is kept for N years, $N = 1, \ldots, 10$. This captures the loss in value of the asset over the time it is kept in service. As an example of the computations, the equivalent annual capital cost of keeping the molding system for four years is

$$\begin{aligned} \text{EAC(capital costs)} &= (P - S)(A/P, 15\%, 4) + Si \\ &= (\$20\,000 + \$5000 - \$2592)(0.35027) + \$2592(0.15) \\ &\cong \$8238 \end{aligned}$$

Note that the installation costs have been included in the capital costs, as these are expenses incurred at the time the asset is originally put into service. Table 7.2 illustrates that the equivalent annual capital costs decline as the asset's life is extended. Next, the

Table 7.2 Computation of Total Equivalent Annual Costs of the Molding System with a MARR = 15%

Life in Years	Salvage Value	EAC Capital Costs	EAC Operating and Maintenance Costs	EAC Total
0	$20 000.00			
1	12 000.00	$16 750.00	$30 000.00	$46 750.00
2	7 200.00	12 029.07	30 697.67	42 726.74
3	4 320.00	9 705.36	31 382.29	41 087.65
4	2 592.00	8 237.55	32 052.47	40 290.02
5	1 555.20	7 227.23	32 706.94	39 934.17
6	933.12	6 499.33	33 344.56	39 843.88
7	559.87	5 958.42	33 964.28	39 922.70
8	335.92	5 546.78	34 565.20	40 111.98
9	201.55	5 227.34	35 146.55	40 373.89
10	120.93	4 975.35	35 707.69	40 683.04

equivalent annual operating and maintenance costs are found by converting the stream of operating and maintenance costs (which are increasing by 5% per year) into a stream of equal-sized annual amounts. Continuing with our sample calculations, the EAC of operating and maintenance costs when the molding system is kept for four years is

$$\begin{aligned}&\text{EAC(operating and maintenance costs)}\\&= \$30\,000\,[(P/F, 15\%, 1) + (1.05)(P/F, 15\%, 2) + (1.05)^2(P/F, 15\%, 3)\\&\quad + (1.05)^3(P/F, 15\%, 4)]\,(A/P, 15\%, 4)\\&\cong \$32\,052\end{aligned}$$

Notice that the equivalent annual operating and maintenance costs increase as the life of the molding system increases.

Finally, we obtain the total equivalent annual cost of the molding system by adding the equivalent annual capital costs and the equivalent annual operating and maintenance costs. This is shown in the last column of Table 7.2. We see that at a six-year life the declining equivalent annual installation and capital costs offset the increasing operating and maintenance costs. In other words, the economic life of the molder is six years, with a total EAC of $39 844.■

While it is *usually* true that capital cost per year falls with increasing life, it is not always true. Capital costs per year can rise at some point in the life of an asset if the decline in value of the asset is not smooth or if the asset requires a major overhaul.

If there is a large drop in the value of the asset in some year during the asset's life, the cost of holding the asset over that year will be high. Consider the following example.

Example 7.3

An asset costs \$50 000 to buy and install. The asset has a resale value of \$40 000 after installation. It then declines in value by 20% per year until the fourth year when its value drops from over \$20 000 to \$5000 because of a predictable wearing-out of a major component. Determine the equivalent annual capital cost of this asset for lives ranging from one to four years. The MARR is 15%.

The computations are summarized in Table 7.3. The first column gives the life of the asset in years. The second gives the salvage value of the asset as it ages. The asset loses 20% of its previous year's value each year except in the fourth, when its value drops to \$5000. The last column summarizes the equivalent annual capital cost of the asset. Sample computations for the third and fourth years are:

EAC(capital costs, three-year life)

$$= (P - S)(A/P, 15\%, 3) + Si$$
$$= (\$40\,000 + \$10\,000 - \$20\,480)(0.43798) + \$20\,480(0.15)$$
$$\cong \$16\,001$$

EAC(capital costs, four-year life)

$$= (P - S)(A/P, 15\%, 4) + Si$$
$$= (\$40\,000 + \$10\,000 - \$5000)(0.35027) + \$5000(0.15)$$
$$\cong \$16\,512$$

Table 7.3 EAC of Capital Costs for Example 7.3

Life in Years	Salvage Value	EAC Capital Costs
0	\$40 000	
1	32 000	\$25 500
2	25 600	18 849
3	20 480	16 001
4	5 000	16 512

The large drop in value in the fourth year means that there is a high cost of holding the asset in the fourth year. This is enough to raise the average capital cost per year.■

In summary, when we replace one asset with another with an identical technology, it makes sense to speak of its economic life. This is the lifetime of an individual asset that will minimize the average cost per year of owning and using it. In the next section, we deal with the case where the challenger is different from the defender.

7.6 Challenger Is Different from Defender; Challenger Repeats Indefinitely

The decision rule that minimizes cost in the case where a defender is faced by a challenger that is expected to be followed by a sequence of identical challengers is as follows:

1. Determine the economic life of the challenger and its associated EAC.
2. Determine the remaining economic life of the defender and its associated EAC.
3. If the EAC of the defender is greater than the EAC of the challenger, replace the defender now. Otherwise, do not replace now.

In many cases, the computations in step 2 can be reduced somewhat. For assets that have been in use for several years, the yearly capital costs will typically be low compared to the yearly operating costs—the asset's salvage value won't change much from year to year but the operating costs will continue to increase. If this is true, as it often is for assets being replaced, the **One Year Principle** can be used. This principle states that if the capital costs for the defender are small compared to the operating costs, and the yearly operating costs are monotonically increasing, the economic life of the defender is one year and its total EAC is the cost of using the defender for one more year.

The Principle thus says that if the EAC of keeping the defender one more year exceeds the EAC of the challenger at its economic life, the defender should be replaced immediately. The advantage of the One Year Principle is that there is no need to find the service life for the defender that minimizes the EAC—it is known in advance that the EAC is minimized in a one-year period. The Principle is particularly useful because for most assets the operating costs increase smoothly and monotonically as the asset is kept for longer and longer periods of time, while the capital costs decrease smoothly and monotonically. For a defender that has been in use for some time, the EAC(operating costs) will typically dominate the EAC(capital costs), and thus the total EAC will increase over any additional years that the asset is kept. This is illustrated in Figure 7.2, which can be compared to Figure 7.1 earlier in this chapter.

Figure 7.2 Cost Components for Certain Older Assets

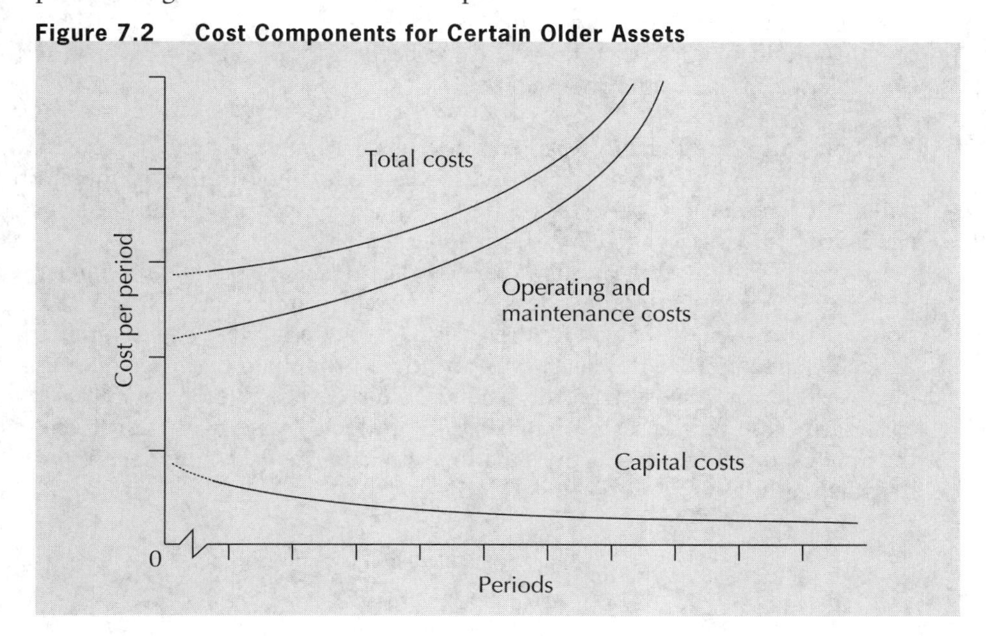

For older assets that conform to this pattern of costs, it is only necessary to check whether the defender needs replacing due to costs over the next year, because in subsequent years the case for the defender will only get worse, not better. If there is an uneven yearly pattern of operating costs, the One Year Principle cannot be used, because the current year might give an unrealistic value due to the particular expenses in that year.

Example 7.4

An asset is three years old. Yearly operating and maintenance costs are currently $5000 per year, increasing by 10% per year. The salvage value for the asset is currently $60 000 and is expected to be $50 000 one year from now. Can the One Year Principle be used?

The capital costs are not low (and thus the EAC(capital costs) for any given service life are not low) compared to the operating and maintenance costs. Even though costs have a regular pattern, the One Year Principle cannot be used.■

Example 7.5

An asset is 10 years old. Yearly operating and maintenance costs are currently $5000 per year, increasing by 10% per year. The salvage value of the asset is currently $8000 and is expected to be $7000 one year from now. Can the One Year Principle be used?

The capital costs are low compared to the operating and maintenance costs, and costs have a regular pattern. The One Year Principle can be used.■

Example 7.6

An asset is 10 years old. Operating and maintenance costs average $5000 per year, increasing by 10% per year. However, most of the operating and maintenance costs consist of a periodic overhaul costing $15 000 that occurs every three years. The salvage value of the asset is currently $8000 and is expected to be $7000 one year from now. Can the One Year Principle be used?

The capital costs are low compared to the operating and maintenance costs but the operating and maintenance costs do not have an even pattern from year to year. The One Year Principle cannot be used.■

The One Year Principle can be used when it is clear that the key conditions—low capital costs and a regular year-to-year pattern of monotonically increasing operating and maintenance costs—are satisfied. Where a situation is ambiguous, it is always prudent to fully assess the EAC of the defender.

To fully explore the case of a defender being replaced by a challenger that repeats, as well as to explore some other ideas useful in replacement analysis, the next three subsections cover examples to illustrate the analysis steps. In the first, we examine the situation of replacing subcontracted capacity with in-house production. This is an example of replacing a service with an asset. In the second example, the issue of sunk costs is examined by considering the replacement of an existing productive asset with a different challenger. In the final example, we look at the situation of making replacement decisions when there are irregular cash flows.

7.6.1 Converting from Subcontracted to In-House Production

Example 7.7

Currently, the Jiffy Printer Company of Example 7.2 pays a custom molder \$0.25 per piece (excluding material costs) to produce parts for their printers. Demand is forecast to be 200 000 parts per year. Jiffy is considering installing the automated plastic molding system described in Example 7.2 to produce the parts themselves. Should they do so now?

In Jiffy's situation, the *defender* is the current technology: a subcontractor. The *challenger* is the automated plastic molding system. In order to decide whether Jiffy is better off with the defender or the challenger, we need to compute the unit cost (cost per piece) of production with the automated molder and compare it to the unit cost for the subcontracted parts. If the automated system is better, the challenger wins and should replace the defender.

From Example 7.2:

EAC(molder) = \$39 844

Dividing the EAC by the expected number of parts needed per year allows us to calculate the unit cost as \$39 838/200 000 = \$0.1992.

When the unit cost of in-house production is compared with the \$0.25 unit cost of subcontracting the work, in-house production is cheaper, so Jiffy should replace the subcontracting with an in-house automated plastic molding system.■

This example has illustrated the basic idea behind a replacement analysis when we are considering the purchase of a new asset as a replacement to current technology. The cost of the replacement must take into account the capital costs (including installation) and the operating and maintenance costs over the life of the new asset.

In the next subsection, we see how some costs are no longer relevant in the decision to replace an *existing* asset.

7.6.2 The Irrelevance of Sunk Costs

Once an asset has been installed and has been operating for some time, the costs of installation and all other costs incurred up to that time are no longer relevant to any decision to replace the current asset. These costs are called **sunk costs**. Only those costs that will be incurred in keeping and operating the asset from this time on are relevant. This is best illustrated with an example.

Example 7.8

Two years have passed since the Jiffy Printer Company from Example 7.7 installed an automated molding system to produce parts for their printers. At the time of installation, they expected to be producing about 200 000 pieces per year, which justified the investment. However, their actual production has turned out to be only about 150 000 pieces per year. Their unit cost is \$39 844/150 000 = \$0.2656 rather than the \$0.1992 they had expected. They estimate the current market value of the molder at \$7200. In this case, maintenance costs do not depend on the actual production rate. Should Jiffy sell the molding system and go back to buying from the custom molder at \$0.25 per piece?

In the context of a replacement problem, Jiffy is looking at replacing the existing system (the defender) with a different technology (the challenger). Since Jiffy already has the molder and has already expended considerable capital on putting it into place, it may be better for Jiffy to keep the current molder for some time longer. Let us calculate the cost to Jiffy of keeping the molding system for one more year. This may not be the optimal length of time to continue to use the system, but if the cost is less than \$0.25 per piece it is cheaper than retiring or replacing it now.

The reason that the cost of keeping the molder an additional year may be low is that the capital costs for the two-year-old system are now low compared with the costs of putting the capacity in place. The capital cost for the third year is simply the loss in value of the molder over the third year. This is the difference between what Jiffy can get for the system now, at the end of the second year, and what they can get a year from now when the system will be three years old. Jiffy can get \$7200 for the system now. Using the declining-balance depreciation rate of 40% to calculate a salvage value, we can determine the capital cost associated with keeping the molder over the third year.

Applying the capital recovery formula from Chapter 3, the EAC for capital costs is

$$\begin{aligned}\text{EAC(capital costs, third year)} &= (P - S)(A/P, 15\%, 1) + Si \\ &= [\$7200 - 0.6(\$7200)](1.15) + 0.6(\$7200)(0.15) \\ &\cong \$3960\end{aligned}$$

Recall that the operating and maintenance costs started at \$30 000 and rose at 5% each year. The operating and maintenance costs for the third year are

$$\begin{aligned}\text{EAC(operating and maintenance, third year)} &= \$30\,000(1.05)^2 \\ &\cong \$33\,075\end{aligned}$$

The total cost of keeping the molder for the third year is the sum of the capital costs and the operating and maintenance costs:

$$\begin{aligned}\text{EAC(third year)} &= \text{EAC(capital costs, third year)} + \\ &\quad\ \text{EAC (operating and maintenance, third year)} \\ &= \$3960 + \$33\,075 \\ &\cong \$37\,035\end{aligned}$$

Dividing the annual costs for the third year by 150 000 gives us a unit cost of \$0.247 for molding in-house during the third year. Not only is this lower than the average cost over a six-year life of the system, it is also lower than what the subcontracted custom molder charges. Similar computations would show that Jiffy could go two more years after the third year with in-house molding. Only then would the increase in operating and maintenance costs cause the total unit cost to rise above \$0.25.

Given the lower demand, we see that installing the automated molding system was a mistake for Jiffy. The average lifetime costs for in-house molding were greater than the cost of subcontracting, but once the system was installed, it was not optimal to go back to subcontracting immediately. This is because the capital cost associated with the purchase and installation of an asset (which becomes sunk after its installation) is disproportionately large as compared with the cost of using the asset once it is in place.■

That a defender has a cost advantage over a challenger, or over contracting out during its planned life, is important. It means that if a defender is to be removed from service during its life for cost reasons, the average lifetime costs for the challenger or the costs of contracting out must be considerably lower than the average lifetime costs of the defender.

Just because well-functioning defenders are not often retired for cost reasons does not mean that they will not be retired at all. Changes in markets or technology may make the quantity or quality of output produced by a defender inadequate. This can lead to their retirement or replacement even when they are functioning well.

7.6.3 When Capital or Operating Costs Are Non-monotonic

Sometimes operating costs do not increase smoothly and monotonically over time. The same can happen to capital costs. When the operating or capital costs are not smooth and monotonic, the One Year Principle does not apply. The reason that the Principle does not apply is that there may be periodic or one-time costs that occur over the course of the next year (as in the case where periodic overhauls are required). These costs may make the cost of keeping the defender for one more year greater than the cost of installing and using the challenger over its economic life. However, there may be a life longer than one year over which the cost of using the defender is less that the cost of installing and using a challenger. Consider this example concerning the potential replacement of a generator.

Example 7.9

The Colossal Construction Company uses a generator to produce power at remote sites. The existing generator is now three years old. It cost $11 000 when purchased. Its current salvage value is $2400, expected to fall to $1400 next year and $980 the year after, and to continue declining at 30% of current value per year. Its ordinary operating and maintenance costs are now $1000 per year and are expected to rise by $500 per year. There is also a requirement to do an overhaul costing $1000 this year and every third year thereafter.

New fuel-efficient generators have been developed, and Colossal is thinking of replacing its existing generator. It is expected that the new generator technology will be the best available for the foreseeable future. The new generator sells for $9500. Installation costs are negligible. Other data for the new generator are summarized in Table 7.4.

Table 7.4 Salvage Values and Operating Costs for New Generator

End of Year	Salvage Value	Operating Cost
0	$9500	
1	8000	$1000
2	7000	1000
3	6000	1200
4	5000	1500
5	4000	2000
6	3000	2000
7	2000	2000
8	1000	3000

Should Colossal replace the existing generator with the new type? The MARR is 12%.

We first determine the economic life for the challenger. The calculations are shown in Table 7.5.

Table 7.5 Economic Life of the New Generator

End of Year	Salvage Value	Operating Costs	EAC
1	$8000	$1000	$3640.00
2	7000	1000	3319.25
3	6000	1200	3236.50
4	5000	1500	**3233.07***
5	4000	2000	3290.81
6	3000	2000	3314.16
7	2000	2000	3318.68
8	1000	3000	3393.52

*Lowest equivalent annual cost.

Sample calculations for the EAC of keeping the challenger for one, two, and three years are as follows:

$$\begin{aligned}\text{EAC(1 year)} &= (P - S)(A/P, 12\%, 1) + Si + \$1000\\ &= (\$9500 - \$8000)(1.12) + \$8000(0.12) + \$1000\\ &\cong 3640\end{aligned}$$

$$\begin{aligned}\text{EAC(2 years)} &= (P - S)(A/P, 12\%, 2) + Si + \$1000\\ &= (\$9500 - \$7000)(0.5917) + \$7000(0.12) + \$1000\\ &\cong \$3319\end{aligned}$$

$$\begin{aligned}\text{EAC(3 years)} &= (P - S)(A/P, 12\%, 3) + Si + \$1000 + \$200(A/F, 12\%, 3)\\ &= (\$9500 - \$6000)(0.41635) + \$6000(0.12) + \$1000\\ &\quad + \$200(0.29635)\\ &\cong \$3237\end{aligned}$$

As the number of years increases, this approach for calculating the EAC becomes more difficult, especially since in this case the operating costs are neither a standard annuity nor an arithmetic gradient. An alternative is to calculate the present worths of the operating costs for each year. The EAC of the operating costs can be found by applying the capital recovery factor to the sum of the present worths for the particular service period considered. This approach is particularly handy when using spreadsheets.

By either calculation, we see in Table 7.5 that the economic life of the generator is four years.

Next, to see if and when the defender should be replaced, we calculate the costs of keeping the defender for one more year. Using the capital recovery formula:

EAC (keep defender 1 more year)

$$= \text{EAC(capital costs)} + \text{EAC(operating costs)}$$
$$= (\$2400 - \$1400)(A/P, 12\%, 1) + \$1400(0.12) + \$2000$$
$$\cong \$3288$$

The equivalent annual cost of using the defender one more year is $3288. This is more than the yearly cost of installing and using the challenger over its economic life. Since the operating costs are not smoothly increasing, we need to see if there is a longer life for the defender for which its costs are lower than for the challenger. This can be done with a spreadsheet, as shown in Table 7.6.

We see that, for an additional life of three years, the defender has a lower cost per year than the challenger, when the challenger is kept over its economic life. Therefore, the defender should not be replaced at this time. Next year a new evaluation should be performed.

Table 7.6 Equivalent Annual Cost of Additional Life for the Defender

Additional Life in Years	Salvage Value	Operating Costs	EAC
0	$2400		
1	1400	$2000	$3288
2	980	1500	**2722**
3	686	2000	2630
4	480	3500	2872
5	336	3000	2924
6	235	3500	3013

7.7 Challenger Is Different from Defender; Challenger Does Not Repeat

In this section, we no longer assume that challengers are alike. We recognize that future challengers will be available and we expect them to be better than the current challenger. We must then decide if the defender should be replaced by the current challenger. Furthermore, if it is to be replaced by the current challenger, *when* should the replacement occur? This problem is quite complex. The reason for the complexity is that, if we believe that challengers will be improving, we may be better off skipping the current challenger and waiting until the next improved challenger arrives. The difficulties are outlined in Example 7.10.

Example 7.10

Rita is examining the possibility of replacing the kiln controllers at the Burnaby Insulators plant. She has information about the existing controllers and the best replacement on the market. She also has information about a new controller design that will be available in three years. Rita has a five-year time horizon for the problem. What replacement alternatives should Rita consider?

One way to determine the minimum cost over the five-year horizon is to determine the costs of *all* possible combinations of the defender and the two challengers. This is impossible, since the defender and challengers could replace one another at any time. However, it is reasonable to consider only the combinations of the period length of one year. Any period length could be used, but a year is a natural choice because investment decisions tend, in practice, to follow a yearly cycle. These combinations form a mutually exclusive set of investment opportunities (see Section 4.2). If no time horizon were given in the problem, we would have had to assume one, to limit the number of possible alternatives.

The possible decisions that need to be evaluated in this case are shown in Table 7.7.

Table 7.7 Possible Decisions for Burnaby Insulators

Decision Alternative	Defender Life in Years	First Challenger Life in Years	Second Challenger Life in Years
1	5	0	0
2	4	1	0
3	4	0	1
4	3	2	0
5	3	1	1
6	3	0	2
7	2	3	0
8	2	2	1
9	2	1	2
10	1	4	0
11	1	3	1
12	1	2	2
13	0	5	0
14	0	4	1
15	0	3	2

For example, the first row in Table 7.7 (Alternative 1) means to keep the defender for the whole five-year period. Alternative 2 is to keep the defender for four years, and then purchase the challenger four years from now, and keep it for one year. Alternative 15 is to replace the defender now with the first challenger, keep it three years, then replace it with the second challenger, and keep the second challenger for the remaining two years.

To choose among these possible alternatives, we need information about the following for the defender and both challengers:

1. Costs of installing the challengers
2. Salvage values for different possible lives for all three kiln controllers
3. Operating and maintenance costs for all possible ages for all three

With this information, the minimum-cost solution is obtained by computing the costs for all possible decision alternatives. Since these are mutually exclusive projects, any of the comparison methods of Chapters 4 and 5 are appropriate, including Present Worth, Annual Worth, or IRR. The effects of sunk costs are already included in the enumeration of the various replacement possibilities, so looking at the benefits of keeping the defender is already automatically taken into account.■

The difficulty with this approach is that the computational burden becomes great if the number of years in the time horizon is large. On the other hand, it is unlikely that information about a future challenger will be available under normal circumstances. In Example 7.10, Rita knew about a controller that wouldn't be available for three years. In real life, even if somehow Rita had inside information on the supplier research and marketing plans, it is unlikely that she would be confident enough of events three years away to use the information with complete assurance. Normally, if the information were available, the challenger itself would be available, too. Consequently, in many cases it is reasonable to assume that challengers in the planning future will be identical to the current challenger, and the decision procedure to use is the simpler one presented in the previous section.

REVIEW PROBLEMS

REVIEW PROBLEM 7.1

Kenwood Limousines runs a fleet of vans which ferry people from several outlying cities to a major international airport. New vans cost $45 000 each and depreciate at a declining-balance rate of 30% per year. Maintenance for each van is quite expensive, because they are in use 24 hours a day, seven days a week. Maintenance costs, which are about $3000 the first year, double each year the vehicle is in use. Given a MARR of 8%, what is the economic life of a van?

ANSWER

Table 7.8 shows the various components of this problem for replacement periods from one to five years. It can be seen that the replacement period with the minimum equivalent annual cost is two years. Therefore, the economic life is two years.

Table 7.8 Summary Computations for Review Problem 7.1

Year	Salvage Value	Maintenance Costs	Equivalent Annual Costs		
			Capital	Maintenance	Total
0	$45 000				
1	31 500	$ 3 000	$17 100	$ 3 000	$20 100
2	22 050	6 000	14 634	4 442	**19 076**
3	15 435	12 000	12 707	6 770	19 477
4	10 805	24 000	11 189	10 594	21 783
5	7 563	48 000	9 981	16 970	26 951

As an example, the calculation for a three-year period is:

EAC(capital costs)

$= (\$45\ 000 - \$15\ 435)(A/P, 8\%, 3) + \$15\ 435(0.08)$

$= \$29\ 565(0.38803) + \$15\ 435(0.08)$

$\cong \$12\ 707$

EAC(maintenance costs)

$= [\$3000(F/P, 8\%, 2) + \$6000(F/P, 8\%, 1) + \$12\ 000]\ (A/F, 8\%, 3)$

$= [\$3000(1.1664) + \$6000(1.08) + \$12\ 000)](0.30804)$

$\cong \$6770$

EAC(total) $=$ EAC(capital costs) + EAC(maintenance costs)

$= \$12\ 707 + \$6770 = \$19\ 477$ ■

REVIEW PROBLEM 7.2

Canadian Widgets makes rocker arms for car engines. The manufacturing process consists of punching blanks from raw stock, forming the rocker arm in a 5-stage progressive die, and finishing in a sequence of operations using hand tools. A recently developed 10-stage die can eliminate many of the finishing operations for high-volume production.

The existing 5-stage die could be used for a different product, and in this case would have a salvage value of \$20 000. Maintenance costs of the 5-stage die will total \$3500 this year, and are expected to increase by \$3500 per year. The 10-stage die will cost \$89 000, and will incur maintenance costs of \$4000 this year, increasing by \$2700 per year thereafter. Both dies depreciate at a declining-balance rate of 20% per year. The net yearly benefit of the automation of the finishing operations is expected to be \$16 000 per year. The MARR is 10%. Should the 5-stage die be replaced?

ANSWER

Since there is no information about subsequent challengers, it is reasonable to assume that the 10-stage die would be repeated. The EAC of using the 10-stage die for various periods is shown in Table 7.9.

A sample EAC computation for keeping the 10-stage die for two years is as follows:

EAC(capital costs, two-year life)

$= (P - S)(A/P, 10\%, 2) + Si$

$= (\$89\ 000 - \$56\ 960)(0.57619) + \$56\ 960(0.10)$

$\cong \$24\ 157$

EAC(maintenance costs, two-year life)

$= [\$4000(F/P, 10\%, 1) + \$6700](A/F, 10\%, 2)$

$= [\$4000(1.1) + \$6700](0.47619)$

$\cong \$5286$

Table 7.9 EAC Computations for the Challenger in Review Problem 7.2

Life in Years	Salvage Value	Maintenance Costs	Equivalent Annual Costs		
			Capital	Maintenance	Total
0	$89 000				
1	71 200	$ 4 000	$26 700	$ 4 000	$30 700
2	56 960	6 700	24 157	5 286	29 443
3	45 568	9 400	22 021	6 529	28 550
4	36 454	12 100	20 222	7 729	27 951
5	29 164	14 800	18 701	8 887	27 589
6	23 331	17 500	17 411	10 004	27 415
7	18 665	20 200	16 314	11 079	**27 393**
8	14 932	22 900	15 377	12 113	27 490

EAC(total, two-year life)

= $24 157 + $5286

= $29 443

Completing similar computations for other lifetimes shows that the economic life of the 10-stage die is seven years and the associated equivalent annual costs are $27 393.

The next step in the replacement analysis is to consider the annual cost of the 5-stage die (the defender) over the next year. This cost is to be compared with the economic life EAC of the 10-stage die, that is, $27 393. Note that the cost analysis of the defender should include the benefits generated by the 10-stage die as an operating cost for the 5-stage die as this $16 000 is a cost of *not* changing to the 10-stage die. Since the capital costs are low, and operating costs are monotonically increasing, the One Year Principle applies. The EAC of the capital and operating costs of keeping the defender one additional year are found as follows:

Salvage value of 5-stage die after one year = $20 000(1 – 0.2) = $16 000

EAC(capital costs, one additional year)

$= (P - S)(A/P, 10\%, 1) + Si$

= ($20 000 – $16 000)(1.10) + $16 000(0.10)

≅ $6000

EAC(maintenance and operating costs, one additional year)

= $3500 + $16 000

≅ $19 500

EAC(total, one additional year)

= $19 500 + $6000

≅ $25 500

The 5-stage die should not be replaced this year because the EAC of keeping it one additional year ($25 500) is less than the optimal EAC of the 10-stage die ($27 393). The knowledge that the 5-stage die should not be replaced this year is usually sufficient for the immediate replacement decision. However, if a different challenger appears in the future, we would want to reassess the replacement decision.

It may also be desirable to estimate when in the future the defender might be replaced, even if it is not being replaced now. This can be done by calculating the equivalent annual cost of keeping the defender additional years until the time we can determine when it should be replaced. Table 7.10 summarizes those calculations for additional years of operating the 5-stage die.

Table 7.10 EAC Computations for Keeping the Defender Additional Years

Additional Life in Years	Salvage Value	Maintenance and Operating Costs	Equivalent Annual Costs		
			Capital	Operating	Total
0	$20 000				
1	16 000	$19 500	$6 000	$19 500	$25 500
2	12 800	23 000	5 429	21 167	26 595
3	10 240	26 500	4 949	22 778	27 727
4	8 192	30 000	4 544	24 334	28 878
5	6 554	33 500	4 202	25 836	30 038
6	5 243	37 000	3 913	27 283	31 196
7	4 194	40 500	3 666	28 677	32 343
8	3 355	44 000	3 455	30 018	33 473

As an example of the computations, the EAC of keeping the defender for two additional years is calculated as

Salvage value of 5-stage die after two years = $16 000(1 – 0.2) = $12 800

EAC(capital costs, two additional years)

$$= (P - S)(A/P, 10\%, 2) + Si$$

$$= (\$20\,000 - \$12\,800)(0.57619) + \$12\,800(0.10)$$

$$\cong \$5429$$

EAC(maintenance and operating costs, two additional years)

$$= [\$19\,500(F/P, 10\%, 1) + (\$16\,000 + \$7000)](A/F, 10\%, 2)$$

$$= [\$19\,500(1.1) + \$23\,000](0.47619)$$

$$\cong \$21\,167$$

EAC(total, two additional years)

$$= \$5429 + \$21\,167 = \$26\,595$$

Further calculations in this manner will predict that the defender should be replaced at the end of the second year, given that the challenger remains the same during this time. This is because the EAC of keeping the defender for two years is less than the optimal EAC of the 10-stage die, but keeping the defender three years or more is more costly.■

REVIEW PROBLEM 7.3

Avril bought a computer three years ago for $3000, which she can now sell on the open market for $300. The local Mr. Computer store will sell her a new HAL computer for $4000, including the new accounting package she wants. Her own computer will probably last another two years, and then would be worthless. The new computer would have a salvage value of $300 at the end of its economic life of five years. The net benefits to Avril of the new accounting package and other features of the new computer amount to $800 per year. An additional feature is that Mr. Computer will give Avril a $500 trade-in on her current computer. Interest is 15%. What should Avril do?

ANSWER

There are a couple of things to note about this problem. First, the cost of the new computer should be taken as $3800 rather than $4000. This is because, although the price was quoted as $4000, the dealer was willing to give Avril a $500 trade-in on a used computer that had a market value of only $300. This amounts to discounting the price of the new computer to $3800. Similarly, the used computer should be taken to be worth $300, and not $500. The $500 figure does not represent the market value of the used computer, but rather the value of the used computer combined with the discount on the new computer. One must sometimes be careful to extract from the available information the best estimates of the values and costs for the various components of a replacement study.

First, we need to determine the EAC of the challenger over its economic life. We are told that the economic life is five years and hence the EAC computations are as follows:

$$\begin{aligned} \text{EAC(capital costs)} &= (\$3800 - \$300)(A/P, 15\%, 5) + \$300(0.15) \\ &= \$3500(0.29832) + \$45 \\ &\cong \$1089 \\ \text{EAC(operating costs)} &= \$0 \\ \text{EAC(challenger, total)} &= \$1089 \end{aligned}$$

Now we need to check the equivalent annual cost of keeping the existing computer one additional year. A salvage value for the computer for one year was not given. However, we can check to see if the EAC for the defender over two years is less than for the challenger. If it is, this is sufficient to retain the old computer.

$$\begin{aligned} \text{EAC(capital costs)} &= (\$300 - 0)(A/P, 15\%, 2) + \$0(0.15) \\ &= \$300(0.61512) + \$0 \\ &\cong \$185 \\ \text{EAC(operating costs)} &= \$800 \\ \text{EAC(defender, total over 2 years)} &= \$985 \end{aligned}$$

Avril should hang on to her current computer because its EAC over two years is less than the EAC of the challenger over its five-year economic life.■

SUMMARY

This chapter is concerned with replacement and retirement decisions. Replacement can be required because there may be a cheaper way to provide the same service, or the nature of the service may have changed. Retirement can be required if there is no longer a need for the asset.

If an asset is replaced by a stream of identical assets, it is useful to determine the economic life of the asset, which is the replacement interval that provides the minimum annual cost. The asset should then be replaced at the end of its economic life.

If there is a challenger which is different from the defender, and future changes in technology are not known, one can determine the minimum EAC of the challenger and compare this with the cost of keeping the defender. If keeping the defender for any period of time is cheaper than the minimum EAC of the challenger, the defender should be kept. Often it is sufficient to assess the cost of keeping the defender for one more year.

Defenders that are still functioning well have a significant cost advantage over challengers or over obtaining the service performed by the defender from another source. This is because there are installation costs and because the capital cost per year of an asset diminishes over time.

Where future changes in technology are expected, decisions about when and whether to replace defenders are more complex. In this case, possible replacement decisions must be enumerated as a set of mutually exclusive alternatives and subjected to any of the standard comparison methods.

Figure 7.3 provides a summary of the overall procedure for assessing a replacement decision.

Figure 7.3 The Replacement Decision Making Process

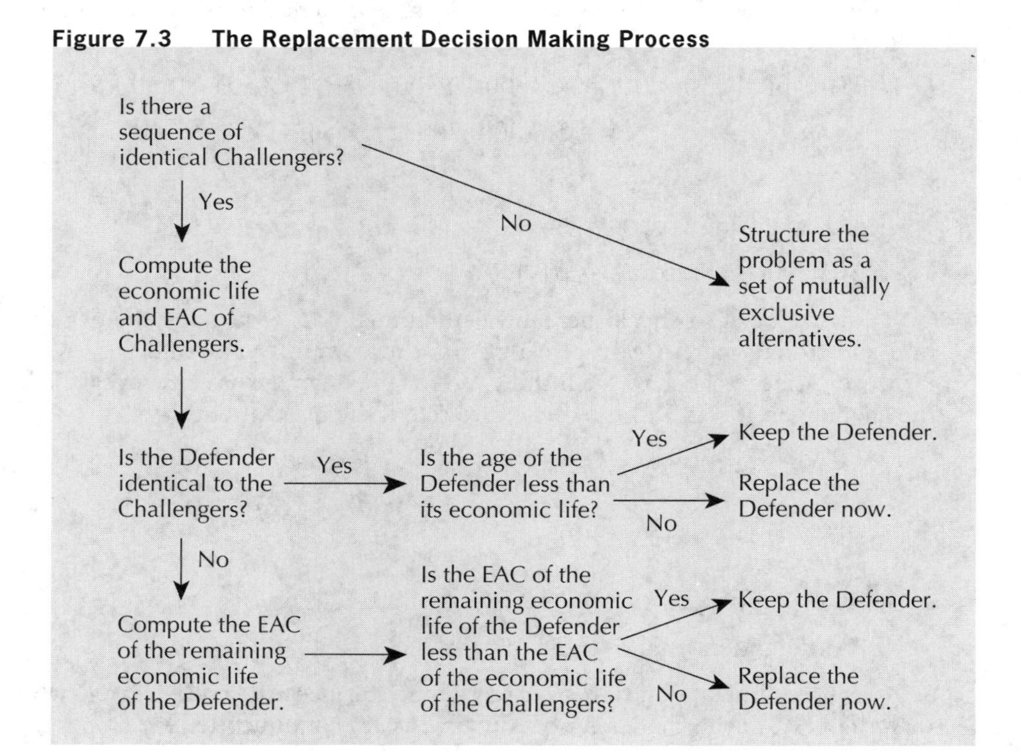

Engineering Economics in Action, Part 7B: Decision Time

Naomi, Dave, and Clem were meeting in Clem's office. They had just finished a discussion of their steel-ordering policy. Clem turned to Naomi and said, "OK. Let's look at the 10-stage progressive die. Where does that stand?"

Naomi said, "It looks possible. Did you get a chance to read Terry's report?"

"Yes, I did," Clem answered. "Was it his idea to use the 5-stage die for small runs so that we don't have to take a big hit from scrapping it?"

"Actually, it was," Naomi said.

"The kid may be a little intense," Clem said, "but he does good work. So where does that leave us?"

"Well, as I said, it looks possible that the 10-stage die will pay off," Naomi responded.

"We have to decide what the correct time horizon is for making the analysis. Then we need more precise estimates of the costs and salvage value for the 10-stage die."

Clem turned in his chair and asked, "What do you think, Dave?"

Dave straightened himself in his chair and said, "I really don't know. How much experience has Hamilton Tools had at making dies this complicated?"

Naomi answered "Not much. If we took them up on their proposal, we'd be their second or third customer."

"What do you have in mind, Dave?" Clem asked.

Dave said, "Well, if it's only the second or third time they've done something like this, I think we can expect some improvements over the next couple of years. So maybe we ought to wait."

"That makes sense," Clem responded. "I'd like you two to work on this. Give Tan Wang at Hamilton Tools a call. He'll know if anything is in the works. Get him to give you an estimate of what to expect. Then I want you to consider some possibilities. You know: 'Replace now.' 'Wait one year.' 'Wait two years.' And so on. Don't make it too complicated. Then, evaluate the different possibilities. I want a recommendation for next week's meeting. It's getting to be decision time."

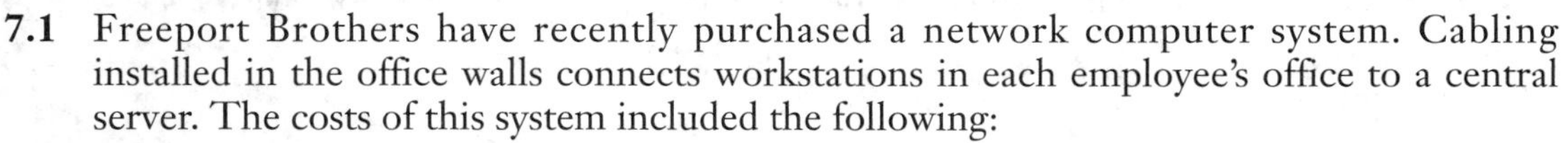

PROBLEMS

7.1 Freeport Brothers have recently purchased a network computer system. Cabling installed in the office walls connects workstations in each employee's office to a central server. The costs of this system included the following:

5 workstations	\$4500 each
New server	\$6000
60-metre cable	\$11.40 per metre
Cabling hardware	\$188
Workstation software	\$1190 per workstation
Server software	\$1950
10 hours of hardware installer time	\$20 per hour
25 hours of software installer time	\$60 per hour

If Freeport Brothers wanted to calculate a replacement interval for such a computer system, what would be the capital cost for the first year at a depreciation rate of 30%?

7.2 Last year, Clairbrook Canning Co. bought a fancy colour printer which cost \$20 000, for special printing jobs. Fast changes in colour printing technology have resulted in almost identical printers being available today for about 1/4 of the cost. Should CCC consider selling their printer and buying one of the new ones?

7.3 Maryhill Mines has a pelletizer that they are considering for replacement. Every three years it is overhauled at considerable cost. It is due for an overhaul this year. Evelyn, the company's mining engineer, has calculated that the sum of the operating and capital costs for this year for the pelletizer are significantly more than the EAC for a new pelletizer over its service life. Should the existing pelletizer be replaced?

7.4 Determine the economic life for each of the items listed below. Salvage values can be estimated by the declining-balance method using an annual rate of 20%. The MARR is 8%.

	Purchase	Installation	Operating
Item 1	\$10 000	\$2000	\$300 first year, increasing by \$300 per year
Item 2	\$20 000	\$2000	\$200 first year, increasing by \$200 per year
Item 3	\$30 000	\$3000	\$2000 first year, increasing by \$2000 per year

7.5 A new bottle-capping machine costs \$45 000, including \$5000 for installation. The machine is expected to have a useful life of eight years with no salvage value at that time (assume straight-line depreciation). Operating and maintenance costs are expected to be \$3000 for the first year, increasing by \$1000 each year thereafter. Interest is 12%.

(a) Construct a spreadsheet that has the following headings: Year, Salvage Value, Maintenance Costs, EAC(Capital Costs), EAC(Operating Costs), and EAC(Total Costs). Compute the EAC(Total Costs) if the bottle capper is kept for n years, $n = 1, \ldots, 8$.

(b) Construct a chart showing the EAC(Capital Costs), EAC(Operating Costs), and EAC(Total Costs) if the bottle capper were to be kept for n years, $n = 1, \ldots, 8$.

(c) What is the economic life of the bottle capper?

7.6 Gerry likes driving small cars and buys nearly identical ones whenever the old one needs replacing. Typically, he trades in his old car for a new one costing about \$15 000. A new car warranty covers all repair costs above standard maintenance (standard maintenance costs are constant over the life of the car) for the first two years. After that, his records show an average repair expense (over standard maintenance) of \$2500 in the third year (at the end of the year), increasing by 50% per year thereafter. If a 30% declining-balance depreciation rate is used to estimate salvage values, and interest is 8%, how often should Gerry get a new car?

7.7 Gerry (see Problem 7.6) has observed that the cars he buys are somewhat more reliable now than in the past. A better estimate of the repair costs is \$1500 in the third year, increasing by 50% per year thereafter, with all other information in Problem 7.6 being the same. Now how often should Gerry get a new car?

7.8 For each of the following cases, determine whether the One Year Principle would apply.

(a) A defender has been in use for seven years and has negligible salvage value. Operating cost are $400 per year for electricity. Once every five years it is overhauled at a cost of $1000.

(b) A defender has been in use for seven years and has negligible salvage value. Operating cost are $400 per year for electricity. Once a year it is overhauled at a cost of $1000.

(c) A defender has been in use for two years and has negligible salvage value. Operating cost are $400 per year for electricity. Once a year it is overhauled at a cost of $1000.

(d) A defender has been in use for seven years and has current salvage value of $4000. Its value one year from now is estimated to be $4000. Operating cost are $400 per year for electricity. Once a year it is overhauled at a cost of $1000.

(e) A defender has been in use for seven years and has current salvage value of $4000. Its value one year from now is estimated to be $2000. Operating cost are $400 per year for electricity. Once a year it is overhauled at a cost of $1000.

7.9 If the operating costs for an asset are $\$500 \times 2^n$ and the capital costs are $\$10\ 000 \times (0.8)^n$, where n is the life in years, what is the economic life of the asset?

7.10 A roller conveyor system used to transport cardboard boxes along an order-filling line costs $100 000 plus $20 000 to install. It is estimated to depreciate at a declining-balance rate of 25% per year over its 15-year useful life. Annual maintenance costs are estimated to be $6000 for the first year, increasing by 20% every year thereafter. In addition, every third year, the rollers must be replaced at a cost of $7000. Interest is at 10%.

(a) Construct a spreadsheet that has the following headings: Year, Salvage Value, Maintenance Costs, EAC(Capital Costs), EAC(Maintenance Costs), and EAC(Total Costs). Compute the EAC(Total Costs) if the conveyor were to be kept for n years, $n = 1, \ldots, 15$.

(b) Construct a chart showing the EAC(Capital Costs), EAC(Maintenance Costs), and EAC(Total Costs) if the conveyor were to be kept for n years, $n = 1, \ldots, 15$.

(c) What is the economic life of the roller conveyor system?

7.11 Brockville Brackets (BB) has a three-year-old robot that welds small brackets onto car-frame assemblies. At the time the robot was purchased, it cost $300 000 and an additional $50 000 was spent on installation. BB acquired the robot as part of an eight-year contract to produce the car-frame assemblies. The useful life of the robot is 12 years, and its value is estimated to decline by 20% of current value per year, as shown in the table below. Operating and maintenance costs estimated when the robot was purchased are also shown in the table on the next page.

BB has found that the operating and maintenance costs for the robot have been higher than anticipated. At the end of the third year, new estimates of the operating and maintenance costs are shown in the second table on the next page.

BB has determined that the reason their operating and maintenance costs were in error was that they positioned the robot too close to existing equipment so that the mechanics could not easily and quickly repair it. BB is considering moving the robot farther away from some adjacent equipment so that mechanics can get easier access for repairs. To move the robot will cause BB to lose valuable production time, which they estimate to have a cost of $25 000. However, once complete, the move will lower maintenance costs to what they originally had expected for the remainder of the contract

Defender, When New		
Life (Years)	**Salvage Value**	**Operating and Maintenance Costs**
0	$300 000	
1	240 000	$40 000
2	192 000	40 000
3	153 600	40 000
4	122 880	40 000
5	98 304	44 000
6	78 643	48 400
7	62 915	53 240
8	50 332	58 564
9	40 265	64 420
10	32 212	70 862
11	25 770	77 949
12	20 616	85 744

Costs for 3-Year-Old Defender		
Additional Life (Years)	**Salvage Value**	**Operating and Maintenance Costs**
0	$153 600	
1	122 880	$50 000
2	98 304	55 000
3	78 643	60 500
4	62 915	66 550
5	50 332	73 205

(e.g., $40 000 for the fourth year, increasing by 10% per year thereafter). Moving the robot will not affect its salvage value.

If BB uses a MARR of 15%, should they move the robot? If so, when? Remember that the contract exists only for a further five years.

7.12 Consider Brockville Brackets from Problem 7.11 but assume that they have a contract to produce the car assemblies for an indefinite period. If they do not move the robot, their operating and maintenance costs will be higher than expected. If they do move the robot (at a cost of $25 000), their operating and maintenance costs are expected to be what they originally expected for the robot. Furthermore, BB expects to be able to obtain new versions of the existing robot for an indefinite period in the future; each is expected to have an installation cost of $50 000.

(a) Construct a spreadsheet table showing the EAC(total costs) if BB keeps the current robot in its current position for n more years, $n = 1, \ldots, 9$.

(b) Construct a spreadsheet table showing the EAC(total costs) if BB moves the current robot and then keeps it for n more years, $n = 1, \ldots, 9$.

(c) Construct a spreadsheet table showing the EAC(total costs) if BB is to buy a new robot and keep it for n years, $n = 1, \ldots, 9$.

(d) On the basis of your answers for parts (a) through (c), what do you advise BB to do?

7.13 Nico has a 20-year-old oil-fired hot air furnace in his house. He is considering replacing it with a new high-efficiency natural gas furnace. The oil-fired furnace has a scrap value of $500, which it will retain indefinitely. A maintenance contract costs $300 per year, plus parts. Nico estimates that parts will cost $200 this year, increasing by $100 per year in subsequent years. The new gas furnace will cost $4500 to buy and $500 to install. It will save $500 per year in energy costs. The maintenance costs for the gas furnace are covered under guarantee for the first five years. The market value of the gas furnace can be estimated from straight-line depreciation with a salvage value of $500 after 10 years. Using a MARR of 10%, should the oil furnace be replaced?

7.14 A certain machine costs $25 000 to purchase and install. It has salvage values and operating costs as shown in the table below. The salvage value of $20 000 listed at time 0 reflects the loss of the installation costs at the time of installation. The MARR is 12%.

(a) What is the economic life of the machine?

Costs and Salvage Values for Various Lives		
Life in Years	**Salvage Value**	**Operating Cost**
0	$20 000.00	
1	16 000.00	$ 3 000.00
2	12 800.00	3 225.00
3	10 240.00	3 466.88
4	8 192.00	3 726.89
5	6 553.60	4 006.41
6	5 242.88	4 306.89
7	4 194.30	4 629.90
8	3 355.44	4 977.15
9	2 684.35	5 350.43
10	2 147.48	5 751.72
11	1 717.99	6 183.09
12	1 374.39	6 646.83
13	1 099.51	7 145.34
14	879.61	7 681.24
15	703.69	8 257.33
16	562.95	8 876.63
17	450.36	9 542.38
18	360.29	10 258.06

(b) What is the equivalent annual cost over that life?

Now assume that the MARR is 5%.

(c) What is the economic life of the machine?

(d) What is the equivalent annual cost over that life?

(e) Explain the effect of decreasing the MARR.

7.15 Jack and Jill live on a hill in Deep Cove. Jack is a self-employed house painter who works out of their house. Jill works in Burnaby, to which she regularly commutes by car. The car is a four-year-old Japanese import. Jill could commute by bus. They are considering selling the car and getting by with the van Jack uses for work.

The car cost $12 000 new. It dropped about 20% in value in the first year. After that it fell by about 15% per year. The car is now worth about $5900. They expect it to continue to decline in value by about 15% of current value every year. Operating and other costs are about $2670 per year. They expect this to rise by about 7.5% per year. A commuter pass costs $112 per month, and is not expected to increase in cost.

Jack and Jill have a MARR of 10%, which is what Jack earns on his business investments. Their time horizon is two years because Jill expects to quit work at that time.

(a) Will commuting by bus save money?

(b) Can you advise Jack and Jill about retiring the car?

7.16 Ener-G purchases new turbines at a cost of $100 000. Each has a 15-year useful life and must be overhauled periodically at a cost of $10 000. The salvage value of a turbine

Defender, When New, Overhaul Every Five Years		
Life (Years)	**Salvage Value**	**Operating and Maintenance Costs**
0	$100 000	
1	85 000	$15 000
2	72 250	20 000
3	61 413	25 000
4	52 201	30 000
5	44 371	45 000
6	37 715	20 000
7	32 058	25 000
8	27 249	30 000
9	23 162	35 000
10	19 687	50 000
11	16 734	25 000
12	14 224	30 000
13	12 091	35 000
14	10 277	40 000
15	8 735	45 000

declines 15% of current value each year, and operating and maintenance costs (including the cost of the overhauls) of a typical turbine are as shown in the table on the previous page (the costs for the fifth and tenth years include a $10 000 overhaul, but an overhaul is not done in the fifteenth year since this is the end of the turbine's useful life).

(a) Construct a spreadsheet that gives, for each year, the EAC(operating and maintenance costs), EAC(capital costs), and EAC(total costs) for the turbines. Interest is 15%. How long should Ener-G keep each turbine before replacing it, given a five-year overhaul schedule? What are the associated equivalent annual costs?

(b) If Ener-G were to overhaul its turbines every six years (at the same cost), the salvage value and operating and maintenance costs would be as shown in the table below. Should Ener-G switch to a six-year overhaul cycle?

Defender, When New, Overhaul Every Six Years		
Life (Years)	**Salvage Value**	**Operating and Maintenance Costs**
0	$100 000	
1	85 000	$15 000
2	72 250	20 000
3	61 413	25 000
4	52 201	30 000
5	44 371	35 000
6	37 715	50 000
7	32 058	20 000
8	27 249	25 000
9	23 162	30 000
10	19 687	35 000
11	16 734	40 000
12	14 224	55 000
13	12 091	25 000
14	10 277	30 000
15	8 735	35 000

7.17 The Burnaby Machine Company makes small parts under contract for manufacturers in the Vancouver area. The company makes a group of metal parts on a turret lathe for a local ski manufacturer. The current lathe is now six years old. It has a planned further life of three years. The contract with the ski manufacturer has three more years to run as well. A new, improved lathe has become available. The challenger will have lower operating costs than the defender.

The defender can now be sold for $1200 in the used-equipment market. The challenger will cost $25 000 including installation. Its salvage value after installation, but before use, will be $20 000. Further data for the defender and the challenger are shown in the tables on the next page.

Defender		
Additional Life in Years	Salvage Value	Operating Cost
0	$1 200	
1	600	$20 000
2	300	20 500
3	150	21 012.50

Challenger		
Life in Years	Salvage Value	Operating Cost
0	20 000	
1	14 000	13 875
2	9 800	14 360.63
3	6 860	14 863.25

Burnaby Machine is not sure if the contract it has with the ski company will be renewed. Therefore, Burnaby wants to make the decision about replacing the defender with the challenger using a three-year study period. Burnaby Machine uses a 12% MARR for this type of investment.

(a) What is the present worth of costs over the next three years for the defender?

(b) What is the present worth of costs over the next three years for the challenger?

(c) Now suppose that Burnaby did not have a good estimate of the salvage value of the challenger at the end of three years. What minimum salvage value for the challenger at the end of three years would make the present worth of costs for the challenger equal to that of the defender?

7.18 Suppose, in the situation described in Problem 7.17, Burnaby Machine believed that the contract with the ski company would be renewed. Burnaby also believed that all challengers after the current challenger would be identical to the current challenger. Further data concerning these challengers are given on the next page. Recall that a new challenger costs $25 000 installed.

Burnaby were also advised that machines identical to the defender would be available indefinitely. New units of the defender would cost $17 500, including installation. Further data concerning new defenders are shown in the table on the next page. The MARR is 12%.

(a) Find the economic life of the challenger. What is the equivalent annual cost over that life?

(b) Should the defender be replaced with the challenger or with a new defender?

(c) When should this be done?

Challenger		
Life in Years	Salvage Value	Operating Cost
0	$20 000.00	
1	14 000.00	$13 875.00
2	9 800.00	14 369.63
3	6 860.00	14 863.25
4	4 802.00	15 383.46
5	3 361.40	15 921.88
6	2 352.98	16 479.15
7	1 647.09	17 055.92
8	1 152.96	17 652.87
9	807.07	18 270.73
10	564.95	18 910.20
11	395.47	19 572.06
12	276.83	20 257.08

Defender When New		
Life in Years	Salvage Value	Operating Cost
0	$15 000.00	
1	9 846.45	$17 250.00
2	6 463.51	17 681.25
3	4 242.84	18 123.28
4	2 785.13	18 576.36
5	1 828.24	19 040.77
6	1 200.11	19 516.79
7	600.00	20 004.71
8	300.00	20 504.83
9	150.00	21 017.45
10	150.00	21 542.89
11	150.00	22 081.46
12	150.00	22 633.49
13	150.00	23 199.33

7.19 You own several copiers that are currently valued at $10 000 all together. Annual operating and maintenance costs for all copiers are estimated at $9000 next year, increasing by 10% each year thereafter. Salvage values decrease at a rate of 20% per year.

You are considering replacing your existing copiers with new ones that have a suggested retail price of $25 000. Operating and maintenance costs for the new equipment will be $6000 over the first year, increasing by 10% each year thereafter. The salvage value of the new equipment is well approximated by a 20% drop from the suggested retail price per year. Furthermore, you can get a trade-in allowance of $12 000 for your equipment if you purchase the new equipment at its suggested retail price. Your MARR is 8%. Should you replace your existing equipment now?

7.20 An existing piece of equipment has the following pattern of salvage values and operating and maintenance costs:

Defender					
Additional Life (Years)	Salvage Value	Maintenance Costs	EAC Capital Costs	EAC Operating and Maintenance Costs	EAC Total
0	$10 000				
1	8 000	$2 000	$3 500	$2 000	$5 500
2	6 400	2 500	3 174	2 233	5 407
3	5 120	3 000	2 905	2 454	5 359
4	4 096	3 500	2 682	2 663	5 345
5	3 277	4 000	2 497	2 861	5 359
6	2 621	4 500	2 343	3 049	5 391
7	2 097	5 000	2 214	3 225	5 439
8	1 678	5 500	2 106	3 391	5 497
9	1 342	6 000	2 016	3 546	5 562

A replacement asset is being considered. Its relevant costs over the next nine years are:

Challenger					
Additional Life (Years)	Salvage Value	Maintenance Costs	EAC Capital Costs	EAC Operating and Maintenance Costs	EAC Total
0	$12 000				
1	9 600	$1 500	$4 200	$1 500	$5 700
2	7 680	1 900	3 809	1 686	5 495
3	6 144	2 300	3 486	1 863	5 349
4	4 915	2 700	3 219	2 031	5 249
5	3 932	3 100	2 997	2 189	5 186
6	3 146	3 500	2 811	2 339	5 150
7	2 517	3 900	2 657	2 480	5 137
8	2 013	4 300	2 528	2 613	5 140
9	1 611	4 700	2 419	2 737	5 156

There is a need for the asset (either the defender or the challenger) for the next nine years.

(a) What replacement alternatives are there?

(b) What replacement timing do you recommend?

7.21 The Brunswick Table Top Company makes tops for tables and desks. The company now owns a seven-year-old planer that is experiencing increasing operating costs. The defender has a maximum additional life of five years. The company is considering replacing the defender with a new planer.

The new planer would cost $30 000 installed. Its value after installation, but before use, would be about $25 000. The company has been told that there will be a new-model planer available in two years. The new model is expected to have the same first costs as the current challenger. However, it is expected to have lower operating costs. Data concerning the defender and the two challengers are shown in the tables below. Brunswick Table has a 10-year planning period and uses a MARR of 10%.

Defender		
Additional Life in Years	**Salvage Value**	**Operating Cost**
0	$4 000	
1	3 000	$20 000
2	2 000	25 000
3	1 000	30 000
4	500	35 000
5	500	40 000

(a) What are the combinations of planers that Brunswick can use to cover requirements

First Challenger		
Life in Years	**Salvage Value**	**Operating Cost**
0	$25 000	
1	20 000	$16 800
2	16 000	17 640
3	12 800	18 522
4	10 240	19 448
5	8 192	20 421
6	6 554	21 442
7	5 243	22 514
8	4 194	23 639
9	3 355	24 821
10	2 684	26 062

Second Challenger		
Life in Years	Salvage Value	Operating Cost
0	$25 000	
1	20 000	$12 000
2	16 000	12 600
3	12 800	13 230
4	10 240	13 892
5	8 192	14 586
6	6 554	15 315
7	5 243	16 081
8	4 194	16 885
9	3 355	17 729
10	2 684	18 616

for the next 10 years? For example, Brunswick may keep the defender one more year, then install the first challenger and keep it for nine years. Notice that the first challenger will not be installed after the second year when the second challenger becomes available. You may ignore combinations that involve installing the first challenger after the second becomes available. Recall also that the maximum additional life for the defender is five years.

(b) What is the best combination?

7.22 You estimate that your two-year-old car is now worth $12 000 and that it will decline in value by 25% of its current value each year of its eight-year remaining useful life. You also estimate that its operating and maintenance costs will be $2100, increasing by 20% per year thereafter. Your MARR is 12%.

(a) Construct a spreadsheet showing (1) additional life in years, (2) salvage value, (3) operating and maintenance costs, (4) EAC(operating and maintenance costs), (5) EAC(capital costs), and (6) EAC(total costs). What additional life minimizes the EAC(total costs)?

(b) Now you are considering the possibility of painting the car in three years' time for $2000. Painting the car will increase its salvage value. By how much will the salvage value have to increase before painting the car is economically justified? Modify the spreadsheet you developed for part (a) to show this salvage value and the EAC(total costs) for each additional year of life. Will painting the car extend its economic life?

7.23 A long-standing principle of computer innovations is that computers double in power for the same price, or, equivalently, halve in cost for the same power, every 18 months. Barrie Data Services (BDS) owns a single computer that is at the end of its third year of service. BDS will continue to buy computers of the same power as its current one. Its current computer would cost $80 000 to buy today, excluding installation. Given that a new model is released every 18 months, what replacement policy should BDS adopt for computers over the next three years? Other facts to be considered are:

1. Installation cost is 15% of purchase price.
2. Salvage values are computed at a declining-balance depreciation rate of 50%.
3. Annual maintenance cost is estimated as 10% of accumulated depreciation or as 15% of accumulated depreciation per 18-month period.
4. BDS uses a MARR of 12%.

7.24 A water pump to be used by the city's maintenance department costs $10 000 new. A running-in period, costing $1000 immediately, is required for a new pump. Operating and maintenance costs average $500 the first year, increasing by $300 per year thereafter. The salvage value of the pump at any time can be estimated by the declining-balance rate of 20%. Interest is at 10%. Using a spreadsheet, calculate the EAC for replacing the pump after one year, two years, etc. How often should the pump be replaced?

7.25 The water pump from Problem 7.24 is being considered to replace an existing one. The current one has a salvage value of $1000 and will retain this salvage value indefinitely.

(a) Operating costs are currently $2500 per year and rise by $400 per year. Should the current pump be replaced? When?

(b) Operating costs are currently $3500 per year and rise by $200 per year. Should the current pump be replaced? When?

7.26 Chatham Automotive purchased new electric forklifts to move steel automobile parts two years ago. These cost $75 000 each, including the charging stand. In practice, it was found that they did not hold a charge as long as claimed by the manufacturer, so operating costs are very high. This also results in their currently having a salvage value of about $10 000.

Chatham is considering replacing them with propane models. The new ones cost $58 000. After one year, they have a salvage value of $40 000, and thereafter decline in value at a declining-balance depreciation rate of 20%, as does the electric model from this time on. The MARR is 8%. Operating costs for the electric model are $20 000 over the first year, rising by 12% per year. Operating costs for the propane model initially will be $10 000 over the first year, rising by 12% per year. Should Chatham Automotive replace the forklifts now?

7.27 Suppose that Chatham Automotive (Problem 7.26) can get a $14 000 trade-in value for their current electric model when they purchase a new propane model. Should they replace the electric forklifts now?

7.28 A joint former cost $60 000 to purchase and $10 000 to install seven years ago. The market value now is $33 000, and this will decline by 12% of current value each year for the next three years. Operating and maintenance costs are estimated to be $3400 this year, and are expected to increase by $500 per year.

(a) How much should the EAC of a new joint former be over its economic life to justify replacing the old one sometime in the next three years? The MARR is 10%.

(b) The EAC for a new joint former turns out to be $10 300 for a 10-year life. Should the old joint former be replaced within the next three years? If so, when?

(c) Is it necessary to consider replacing the old joint former more than three years from now, given that a new one has an EAC of $10 300?

7.29 Northwest Aerocomposite manufactures fibreglass and carbon fibre fairings. Their largest water-jet cutter will have to be replaced some time before the end of four years. The old cutter is currently worth \$49 000. Other cost data for the current and replacement cutters can be found in the tables below. The MARR is 15%. What is the economic life of the new cutter, and what is the equivalent annual cost for that life? When should the new cutter replace the old?

Challenger		
Life in Years	**Salvage Value**	**Operating and Maintenance Costs**
0	\$90 000	
1	72 000	\$12 000
2	57 600	14 400
3	46 080	17 280
4	36 864	20 736
5	29 491	24 883
6	23 593	29 860
7	18 874	35 832
8	15 099	42 998
9	12 080	51 598

Defender		
Life in Years	**Salvage Value**	**Operating and Maintenance Costs**
0	\$49 000	
1	36 500	\$17 000
2	19 875	21 320
3	15 656	26 806
4	6 742	33 774

7.30 The water pump from Problem 7.24 has an option to be overhauled once. It costs \$1000 to overhaul a three-year-old pump and \$2000 to overhaul a five-year-old pump. The major advantage of an overhaul is that it reduces the operating and maintenance costs to \$500, which will increase again by \$300 per year thereafter. Should the pump be overhauled? If so, should it be overhauled in three years or five years?

7.31 Northwest Aerocomposite in Problem 7.29 found out that their old water-jet cutter may be overhauled at a cost of \$14 000 now. The cost information for the old cutter after an overhaul is as shown in the table on the next page.

Defender with an Overhaul		
Life (Years)	**Salvage Value**	**Operating and Maintenance Costs**
0	$55 000	
1	40 970	$16 500
2	22 310	20 690
3	17 574	26 013
4	7 568	32 775

Should Northwest overhaul the old cutter? If an overhaul takes place, when should the new cutter replace the old? Assume that the cost information for the replacement cutter is as given in Problem 7.29.

7.32 Tiny Bay Freight Company (TBFC) wants to begin business with one delivery truck. After two years of operation, the company plans to increase the number of trucks to two, and after four years, plans to increase the number to three. TBFC currently has no trucks. The company is considering purchasing one type of truck that costs $30 000. The operating and maintenance costs are estimated to be $7200 per year. The resale value of the truck will decline each year by 40% of the current value. The company will consider replacing a truck every two years. That is, the company may keep a truck for two years, four years, six years, and so on. TBFC's MARR is 12%.

(a) What are the possible combinations for purchasing and replacing trucks over the next five years so that TBFC will meet their expansion and will have three trucks in hand at the end of five years?

(b) Which purchase/replacement combination is the best?

Problems 7.33 through 7.36 are concerned with the economic life of assets where there is a sequence of identical assets. The problems explore the sensitivity of the economic life to four parameters: the MARR, the level of operating cost, the rate of increase in operating cost, and the level of first cost. In each problem there is a pair of assets. The assets differ in only a single parameter. The problem asks you to determine the effect of this difference on the economic life and to explain the result. All assets decline in value by 20% of current value each year. Installation costs are zero for all assets. Further data concerning the four pairs of assets are given in the table that follows.

Asset Number	First Cost	Initial Operating Cost	Rate of Operating Cost Increase	MARR
A1	$125 000	$30 000	12.5%/year	5%
B1	125 000	30 000	12.5%/year	25%
A2	100 000	30 000	$2000/year	15%
B2	100 000	40 000	$2000/year	15%
A3	100 000	30 000	5%/year	15%
B3	100 000	30 000	12.5%/year	15%
A4	75 000	30 000	5%/year	15%
B4	150 000	30 000	5%/year	15%

7.33 Consider Assets A1 and B1. They differ only in the MARR.

(a) Determine the economic lives for Assets A1 and B1.

(b) Create a diagram showing the EAC(capital), the EAC(operating), and the EAC(total) for Assets A1 and B1.

(c) Explain the difference in economic life between A1 and B1.

7.34 Consider Assets A2 and B2. They differ only in the level of initial operating cost.

(a) Determine the economic lives for Assets A2 and B2.

(b) Create a diagram showing the EAC(capital), the EAC(operating), and the EAC(total) for Assets A2 and B2.

(c) Explain the difference in economic life between A2 and B2.

7.35 Consider Assets A3 and B3. They differ only in the rate of increase of operating cost.

(a) Determine the economic lives for Assets A3 and B3.

(b) Create a diagram showing the EAC(capital), the EAC(operating), and the EAC(total) for Assets A3 and B3.

(c) Explain the difference in economic life between A3 and B3.

7.36 Consider Assets A4 and B4. They differ only in the level of first cost.

(a) Determine the economic lives for Assets A4 and B4.

(b) Create a diagram showing the EAC(capital), the EAC(operating), and the EAC(total) for Assets A4 and B4.

(c) Explain the difference in economic life between A4 and B4.

7.37 This problem concerns the economic life of assets where there is a sequence of identical assets. In this case there is an opportunity to overhaul equipment. Two issues are explored. The first concerns the optimal life of equipment. The second concerns the decision of whether to replace equipment that is past its economic life. Consider a piece of equipment that costs $40 000 to buy and install. The equipment has a maximum life of 15 years. Overhaul is required in the fourth, eighth, and twelfth years. The company uses a MARR of 20%. Further information is given in the table on the next page.

(a) Show that the economic life for this equipment is seven years.

(b) Suppose that the equipment is overhauled in the eighth year rather than replaced. Show that keeping the equipment for three more years (after the eighth year), until it next comes up for overhaul, has lower cost than replacing the equipment immediately.

Hint for (b): The comparison must be done fairly and carefully. Assume that under either plan the replacement is kept for its optimal life of seven years. It is easier to compare the plans if they cover the same number of years. One way to do this is to consider an 11-year period as shown on the next page.

First, show that the present worth of costs over the 11 years is lower under Plan A than under Plan B. Second, point out that the equipment that is in place at the end of the eleventh year is newer under Plan A than under Plan B.

Year	Salvage Value	Operating Cost	Overhaul Cost
0	$15 000		
1	12 000	$2 000	
2	9 600	2 200	
3	7 680	2 420	
4	7 500	2 662	$2 500
5	6 000	2 000	
6	4 800	2 200	
7	3 840	2 420	
8	4 500	2 662	32 500
9	3 600	2 000	
10	2 880	2 800	
11	2 304	3 920	
12	2 000	5 488	17 500
13	1 200	4 000	
14	720	8 000	
15	432	16 000	

Year	Plan A	Plan B
0		
1	Defender	Replacement #1
2	Defender	Replacement #1
3	Defender	Replacement #1
4	Replacement #1	Replacement #1
5	Replacement #1	Replacement #1
6	Replacement #1	Replacement #1
7	Replacement #1	Replacement #1
8	Replacement #1	Replacement #2
9	Replacement #1	Replacement #2
10	Replacement #1	Replacement #2
11	Replacement #2	Replacement #2

(c) Why is it necessary to take into account the age of the equipment at the end of the 11-year period?

7.38 Northfield Metal Works is a Manitoba household appliance parts manufacturer that has just won a contract with a major appliance company to supply replacement parts to service shops. The contract is for five years. Northfield is considering using three existing manual punch presses or a new automatic press for part of the work. The new press would cost $225 000 installed. Northfield is using a five-year time horizon for the project. The MARR is 25% for projects of this type. Further data concerning the two options are shown below.

Automatic Punch Press		
Life in Years	**Salvage Value**	**Operating Cost**
0	$125 000	
1	100 000	$25 000
2	80 000	23 750
3	64 000	22 563
4	51 200	21 434
5	40 960	20 363

Hand-Fed Press		
Additional Life in Years	**Salvage Value**	**Operating Cost**
0	$10 000	
1	9 000	$25 000
2	8 000	25 000
3	7000	25 000
4	6 000	25 000
5	5 000	25 000

Note that the hand-fed press values are for each of the three presses. Costs must be multiplied by three to get the costs for three presses. Northfield is not sure of the salvage values for the new press. What salvage value at the end of five years would make the two options equal?

Used Cars Versus New Cars

CANADIAN 7.1 MINI-CASE

A typical Canadian motorist now drives a car at least five years old with close to 100 000 kilometres on it. On the basis of a survey done by the Canadian Automobile Association (CAA), people are keeping their cars longer because they are built better and fewer people can afford the cost of driving a new car.

One reason people are driving older cars is that the total cost of repairs for old cars is not that much more than for newer cars. A 2002 study by the CAA showed that average maintenance costs for cars more than about five years old remained fairly constant, at about $1000 per year.

Another reason people are driving older cars is that the total cost of car ownership goes down with the length of time you keep the vehicle. A 2003 Runzheimer International analysis of the cost of operating new and used vehicles revealed that a typical four-year-old car was more than $13 000 cheaper over a four-year period than a new one (www.runzheimer.com/coprc/news/scripts/100903.asp). "New car payments are the decisive factor," according to Greg Harper, vice-president of Business Vehicle Services at Runzheimer. "Even though the new model vehicle has a much greater trade-in value after four years and you save on repairs and maintenance, the monthly car payments—which total slightly over $17 500 in principal and interest over four years—more than counterbalance these other factors."

Discussion

People buy and sell things for many reasons. A car might be desirable because it exhibits a sporty image, is known to be reliable, or is big enough for a large family. Purchase price, resale value, and operating costs are also very important, but people often make purchase decisions that are not economically wise.

In particular, the cost of a new car is hard to justify in many cases. A new car depreciates very quickly in terms of resale value over the first year of use. However, the functional difference between a new car and a year-old car is often very small. On the other hand, if nobody bought new cars, there wouldn't be any old cars available to buy!

Questions

1. Draw a graph of the purchase or resale value of a particular make and model of car over time. Estimate the values you use from personal experience, newspaper want ads, Internet sites, etc.
2. For the same make and model of car, estimate the operating cost per year, including insurance, repairs, fuel, etc., over the same number of years as in Question 1.
3. If you were going to buy this make and model, brand-new or used, and keep it for four years, what would be the most economical four-year period to do so?
4. Are there reasons why the answer in Question 3 would not be correct in practice?

CHAPTER 8

Taxes

Engineering Economics in Action, Part 8A: It's in the Details

"Details, Terry. Sometimes it's all in the details." Naomi pursed her lips and nodded sagely. Terry and Naomi were sitting in the coffee room together. The main break periods for the line workers had passed, so they were alone except for a maintenance person on the other side of the room who was enjoying either a late breakfast or an early lunch.

"Uh, OK, Naomi. What is?"

"Well," Naomi replied, "you know that rocker arm die deal? The one where we're upgrading to a 10-stage die? The rough replacement study you did seems to have worked out OK. We're going to do something, sometime, but now we have to be a little more precise."

"OK. What do we do?" Terry was interested now. Terry liked things precise and detailed.

"The main thing is to make sure we are working with the best numbers we can get. I'm getting good cost figures from Tan Wang at Hamilton Tools for this die and future possibilities, and we'll also work out our own costs for the changeover. Your cost calculations are going to have to be more accurate, too."

"You mean to more significant digits?" Naomi couldn't tell whether Terry was making a joke or the idea of more significant digits really did thrill him. She decided it was the former.

"Ha, ha. No, I mean that we had better explicitly look at the tax implications of the purchase. Our rough calculations ignored taxes, and taxes can have a significant effect on the choice of best alternative. And when lots of money is at stake, the details matter."

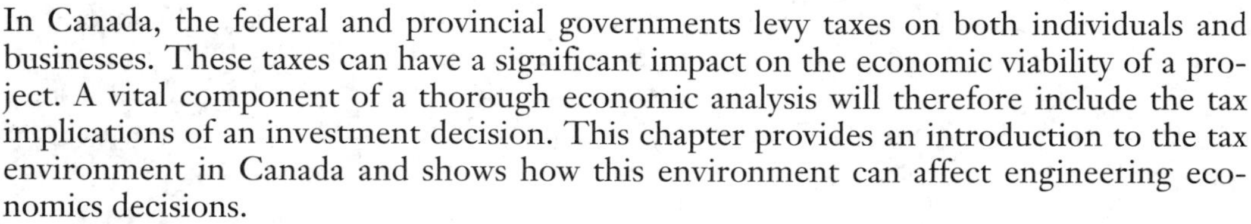

8.1 Introduction

In Canada, the federal and provincial governments levy taxes on both individuals and businesses. These taxes can have a significant impact on the economic viability of a project. A vital component of a thorough economic analysis will therefore include the tax implications of an investment decision. This chapter provides an introduction to the tax environment in Canada and shows how this environment can affect engineering economics decisions.

Personal income taxes and corporate taxes are significant sources of government revenue. They provide to government a portion of the income received from individuals and profits earned by companies. As a result, the federal and provincial governments can pay for social services, health services, infrastructure such as highways and dams, the military and other government services, or what are called public goods and services.

When a firm makes an investment, the income from the project will affect the company's cash flows. If the investment yields a profit, the profits will be taxed. Since the taxes result as a direct consequence of the investment, they reduce the net profits associated with that investment. In this sense, taxes associated with a project are a disbursement. If the investment yields a loss, the company may be able to offset the loss from this project against the profits from another, and end up paying less tax overall. As a result, when evaluating a loss-generating project, the net savings in tax can be viewed as a negative disbursement. Income taxes thus reduce the benefits of a successful project, while at the same time reducing the costs of an unsuccessful project.

Calculating the effects of income taxes on the viability of a project can be complicated. Prior to exploring the details of such calculations, it is helpful to look at the context of corporate income taxes. In the next two sections we will compare corporate income taxes to the more familiar personal income taxes, and also describe how corporate tax rates are determined.

8.2 Personal Income Taxes and Corporate Income Taxes Compared

There are substantial differences between personal income taxes and corporate income taxes. Most adults are familiar with the routine of filing income tax forms, due on April 30 of each year. The procedure is generally very simple; an employer provides a statement reporting income, income tax already paid, and other amounts for the previous year. These amounts are assembled on the tax form to determine the total tax owed or the amount of a refund, and submitted to the Canada Revenue Agency for processing.

Corporate taxes are similar in some ways, but there are several substantial differences. One main difference has to do with the tax rate. Personal income taxes are **progressive**, meaning that the rate of taxation increases at higher income levels. For example, if taxable income is very low, an individual will pay no income taxes. At about $40 000, an individual might be expected to pay about 25% of taxable income as income tax. A very-high-income individual might pay around 50% of taxable income as income tax. Although the exact rate of taxation changes from year to year, and individual provinces have different rates across Canada, individual income taxes are always progressive.

In contrast, corporate taxes are levied, with some exceptions, at a constant rate independent of income level. Again, rates can change over time and across provinces, but the rate will be the same whether a company makes a small profit or a very large profit.

Another difference between personal income tax and corporate income tax has to do with how the tax is calculated. An individual's taxes are reduced by **tax credits**, which are real or nominal costs that are not taxed or are taxed at a reduced rate. For example, excess medical expenses, tuition fees, CPP (Canada Pension Plan), and Employment Insurance premiums all contribute to tax credits, along with a substantial "basic personal amount." A corporation's taxes are calculated in quite a different manner. Net income for tax purposes is calculated by subtracting **expenses**, which are either real costs associated with performing the corporation's business or a portion of the capital expense for an asset, from the gross income. Consequently, a company that makes no profit (income less expenses) may pay no income taxes.

NET VALUE 8.1

Canada Revenue Agency Web site

The Canada Revenue Agency Web site (www.cra-arc.gc.ca) is a very rich and complete source of information about tax rules. However, because there are so many different types of users and special cases it is not always easy to find the information you are looking for.

There is a built-in search engine that can be very helpful, but may give you more hits than are useful, and the order of responses may not be convenient. For example, a recent key phrase of "CCA rates" gave CCA rates of interest to farmers and fishers, along with several other more specific rates, before presenting a link to a description of CCA rates generally applicable in business.

Another approach may be to seek out publications directed at your area of interest. For example, the publication TA4002(E) *Business and Professional Income* provides very good information about taxes for businesses that would be of interest to anyone who would like to start their own business.

Finally, an individual's tax situation is quite different in the nature of its complexity than the situation for a company. In particular, a company's taxes will usually be complicated by issues concerning **capital expenses**, which are purchases of assets of significant value. Such assets have a strong effect on the income taxes paid by a company, but how they are treated for tax purposes is complex. This is usually not an issue of concern to individuals.

8.3 Corporate Tax Rates

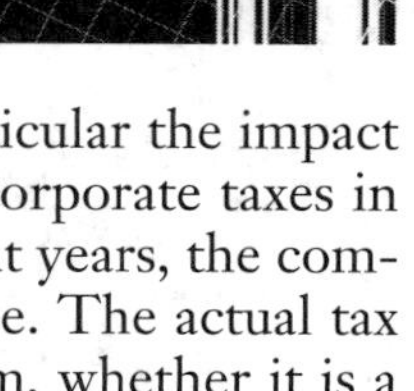

In this chapter, we are concerned only with corporate taxes, and in particular the impact of corporate income taxes on the viability of an engineering project. Corporate taxes in Canada have a federal component and a provincial component. In recent years, the combined tax rate has fallen from about 50% to close to 40% of net income. The actual tax rate applied is fairly complicated and can depend on the size of the firm, whether it is a manufacturer, its location, and a variety of other factors, but generally ranges from 17% to 52%. Our concern here is with the basic approach used in determining the impact of taxes on a project. For special tax rules, it is best to check with the Canada Revenue Agency or a tax specialist.

There are special tax advantages given to smaller companies such as a technically oriented startup. This may be of particular interest to engineers interested in starting their own firm rather than working for a larger company. The **Small Business Deduction** applies to small Canadian-controlled private corporations, and provides at the federal level a tax credit of 16% on the first $300 000 in income every year (lesser amounts prior to 2005). Under current corporate tax rates this reduces the effective tax rate for a small Canadian company to less than 20%. The purpose of the deduction is to provide more after-tax income for reinvestment and expansion. Also, since an entrepreneur is likely to have a personal tax rate higher than 20%, this gives a strong motivation to incorporate a new company.

It is also worth noting that tax rules can change suddenly. As seen below, the current corporate tax rule structure has applied only since 1981, and there have been significant changes almost every year. Also, the Canadian tax rules are unique to Canada, so that the methods presented in this chapter may not apply elsewhere. For comparison, Close-Up 8.1 reviews the tax procedures used in the United States.

CLOSE-UP 8.1 United States Tax Rules

In the U.S. tax system, all depreciable assets are designated as belonging to a "Modified Accelerated Cost Recovery System (MACRS) Class." The MACRS Class determines the declining-balance rate (usually double-declining-balance or 150% declining-balance) and the **recovery period**, which is the designated service life for depreciation calculation purposes.

The declining-balance method is not used for the entire recovery period. All assets are required to attain a book value of $0 at the end of the recovery period. Since, under a declining-balance depreciation method, a book value of $0 is never reached, at an appropriate point the depreciation method switches from declining-balance to straight-line.

8.4 Before- and After-Tax MARR

Taxes have a significant effect on engineering decision making, so much that they cannot be ignored. In this text so far, it seems that they have been ignored, since no specific tax calculations have been done. In fact, they have been implicitly incorporated into the computations through the use of a before-tax MARR, though we have not called it such.

The basic logic is as follows: since taxes have the effect of reducing profits associated with a project, we need to make sure that we set the MARR for project acceptability accordingly. If we do not explicitly account for the impact of taxes in the project cash flows, then we need to set a MARR high enough to recognize that taxes will need to be paid. This is the *before-tax* MARR. If, on the other hand, the impact of taxes are explicitly accounted for in the cash flows of a project (i.e., reduce the cash flows by the tax rate), then the MARR used for the project should be lower, since the cash flows already take into account the payment of taxes. This is the *after-tax* MARR.

In fact, we can express an approximate relationship

$$\text{MARR}_{\text{after-tax}} \cong \text{MARR}_{\text{before-tax}} \times (1 - t) \tag{8.1}$$

where t is the corporate tax rate. The *before-tax* MARR means that the MARR has been chosen high enough to provide an acceptable rate of return without explicitly considering taxes. In other words, since all profits are taxed at the rate t, the *before-tax* MARR has to include enough returns to meet the *after-tax* MARR and, in addition, provide the amount to be paid in taxes. As we can see from Equation (8.1), the after-tax MARR will generally be lower than the before-tax MARR. We will see later in this chapter how the relationship given in Equation (8.1) is a simplification but a reasonable approximation of the effect of taxes. In practice, the before- and after-tax MARRs are often chosen independently and are not directly related by Equation (8.1). Generally speaking, if a MARR is given without specifying whether it is on a before- or after-tax basis, it can be assumed to be a before-tax MARR.

Example 8.1

Prince Rupert Gold Mines (PRGM) has been selecting projects for investments on the basis of a before-tax MARR of 12%. Sherri feels that some good projects have been missed because the effects of taxation on the projects have not been examined in enough detail, so she proposed reviewing the projects on an after-tax basis. What would be a good choice of after-tax MARR for her review? PRGM pays 45% corporate taxes.

Although the issue of selecting an after-tax MARR is likely to be more complicated, a reasonable choice for Sherri would be to use Equation (8.1) as a way of calculating an after-tax MARR for her review. This gives

$$\text{MARR}_{\text{after-tax}} \cong 0.12 \times (1 - 0.45) = 0.066 = 6.6\%$$

A reasonable choice for after-tax MARR would be 6.6%.■

8.5 The Capital Cost Allowance (CCA) System

When a firm buys a depreciable asset for use in its business, a **capital expense** is incurred. It is the expenditure associated with the purchase of a long-term depreciable asset. (Almost all tangible assets are depreciable. The primary exception to this is land.) Since the asset deteriorates through the passage of time, the firm must deduct the capital expense over a period of years. This is done by claiming a depreciation expense each year of the asset's useful life, as its value declines. The depreciation is recorded by accountants in the balance sheet as a reduction in the book value of the asset. It is also recorded as an expense on the income statement. Depreciation thus reduces the before-tax income even though, in reality, there has been no cash expense.

In general, a firm will want to "write off" (i.e., depreciate) an investment as fast as possible. This is because depreciation is considered an expense and offsets revenue so as to reduce net income. Since net income is taxed, taxes can be deferred or reduced by depreciating assets quickly. The effect of deferring the taxes can be considerable. To counter this effect, the Canadian tax system defines a specific amount of depreciation that companies may claim in any year for any one depreciable asset. This amount is called the *capital cost allowance (CCA)*. In this section, we demonstrate how to apply CCA rules to investment decisions and compare the CCA to depreciation claimed for accounting records.

Example 8.2

In the imaginary country of Monovia, companies can depreciate their capital asset purchases as fast as they want. Clive Cutler, owner of Monovia Manufacturing, has just bought $200 000 worth of equipment. Two spreadsheets (Tables 8.1 and 8.2) illustrate the effect of different depreciation strategies over the next five years assuming the following:

1. Income is $300 000 per year.
2. Expenses excluding depreciation are $100 000 per year.
3. The tax rate is 50%.
4. Available cash is invested at 10% interest.
5. The salvage value of the equipment after five years is $0.

Table 8.1 Full Depreciation in One Year

Year	1	2	3	4	5
Income	$300 000	$300 000	$300 000	$300 000	$300 000
Expenses excluding depreciation	100 000	100 000	100 000	100 000	100 000
Depreciation expense	200 000	0	0	0	0
Net income	0	200 000	200 000	200 000	200 000
Taxes	0	100 000	100 000	100 000	100 000
Profit	0	100 000	100 000	100 000	100 000
Cash	200 000	100 000	100 000	100 000	100 000
Accumulated cash	200 000	320 000	452 000	597 200	756 920

Table 8.2 Straight-Line Depreciation over Five Years

Year	1	2	3	4	5
Income	$300 000	$300 000	$300 000	$300 000	$300 000
Expenses excluding depreciation	100 000	100 000	100 000	100 000	100 000
Depreciation expense	40 000	40 000	40 000	40 000	40 000
Net income	160 000	160 000	160 000	160 000	160 000
Taxes	80 000	80 000	80 000	80 000	80 000
Profit	80 000	80 000	80 000	80 000	80 000
Cash	120 000	120 000	120 000	120 000	120 000
Accumulated cash	120 000	252 000	397 200	556 920	732 612

Table 8.1 illustrates the case in which the equipment is fully depreciated in the first year, although it generates revenue over its five-year life. In Table 8.2, straight-line depreciation is used over the five-year life.

When we look at the effects of depreciation on economic analyses, it is important to distinguish between expenses that represent a cash outflow and expenses that do not. Purchasing an asset, like a piece of equipment, will produce a cash outflow at the time the purchase is made. In particular, the balance sheet will reflect a transfer out of current assets (cash) and a transfer into fixed assets (equipment) and perhaps to current liabilities (bank loan).

Depreciation, on the other hand, does not actually represent a cash outflow, although it is recorded as an expense in the income statement. For example, in Table 8.1, writing off (depreciating) the entire cost of the equipment in its first year produced a depreciation expense of $200 000 in that year. There was no actual cash outflow due to the depreciation (although there was for the actual purchase of the asset), but depreciation caused the net income to be reduced to zero for that year, even though $200 000 in cash was actually available. Investing the $200 000 for the second year at 10% interest produces an accumulated cash amount of $220 000. Adding this to the profit of $100 000 for the second year gives accumulated cash of $320 000 at the end of the second year. Continuing in this fashion produces accumulated cash of $756 920 at the end of the five-year period.

In contrast, if the equipment is depreciated on a straight-line basis over five years, only $732 612 in cash is accumulated. This can be seen by working through the expenses, net income, taxes, and profit for each year. For example, in year 1, a (straight-line) depreciation expense of $200 000/5 = $40 000 is claimed. This reduces net income by $40 000, to $160 000, and leaves after-tax profits of $80 000. Now, as before, the depreciation expense of $40 000 is not a cash outflow, so the cash actually available to invest at the end of the first year is the $80 000 profit *plus* $40 000. Since the depreciation expense is constant with the straight-line method, a cash amount of $120 000 will be available for investment at the end of each of the five years. Thus, at the end of five years, the accumulated cash will be $732 612. This is $24 308 less than when the equipment was fully depreciated in the first year, because taxes were delayed by depreciating more of the asset's value earlier. The extra income that was available earlier for investment allowed more interest to accumulate over the five-year period. The $24 308 is significant, and illustrates why faster depreciation is preferred to slower depreciation.■

As seen in Chapter 6, there are several generally accepted depreciation methods. The most prevalent methods in Canada are straight-line and declining-balance. For the purposes of preparing financial statements for investors, a firm may use any one (or all) of the generally accepted methods for calculating depreciation expenses, provided that the method used is the same from period to period. However, if companies had the freedom to depreciate as they wanted to for tax purposes, they would depreciate their assets immediately, since in that way they would get the largest benefit because of tax savings.

Governments have a different perspective. They would prefer to receive taxes as quickly as possible, and they would want companies to depreciate assets as slowly as possible to keep taxable income as high as possible and produce the most taxes. In order to limit the depreciation amount that companies use for tax purposes, the Canadian government has established a maximum level of capital cost expense (i.e., depreciation) that a company can claim each year. This maximum amount is referred to as the firm's **capital cost allowance (CCA)**. The system set up to allow firms to compute their capital cost allowance is called the **capital cost allowance (CCA) system**. This system, established by the Canadian government, specifies the amount and timing of depreciation expenses on capital assets. According to the CCA system, the declining-balance method of depreciation must be used for claiming capital costs associated with most tangible assets. Straight-line depreciation is only used for certain intangible assets. We are mainly concerned with tangible assets, so our discussion will focus on declining-balance depreciation.

The CCA system specifies the maximum rate a company can use to depreciate its assets for tax calculations; this is referred to as the **capital cost allowance (CCA) rate**. To implement the CCA system, a firm's assets are grouped by **capital cost allowance (CCA) asset class** for which a specified CCA rate is used to compute the CCA. For example, all assets classified as office equipment (desks, chairs, filing cabinets, copiers, and the like) are grouped together, and depreciation expenses are based on the total remaining undepreciated cost of all assets in that class. Some examples of CCA rates and CCA asset classes are given in Table 8.3.

In addition to these standard rates, sometimes, as part of government policy, special rates are set to encourage certain kinds of investments. This is a form of *incentive*. Close-up 8.2 discusses the general idea of government incentives.

Figure 8.1 illustrates how the remaining value of an asset subject to taxes diminishes within the standard range of CCA rates.

While capital cost allowance and depreciation are conceptually similar, it is important to distinguish between the two terms. Recall from Chapter 6 that, in determining net income, we deduct depreciation expenses from the revenues to arrive at net income. This

Table 8.3 Sample CCA Rates and Classes

CCA Rate (%)	Class	Description
4 to 10	1, 3, 6	Buildings and additions
20	8	Machinery, office furniture, and equipment
25	9	Aircraft, aircraft furniture, and equipment
30	10	Passenger vehicles, vans, trucks, computers, and systems software
40	16	Taxis, rental cars, and freight trucks; coin-operated video games
100	12	Dies, tools, and instruments that cost less than $200; computer software other than systems software

CLOSE-UP 8.2 Incentives

Federal and provincial governments sometimes try to influence corporate behaviour through the use of *incentives*. These incentives include grants to certain types of projects, to projects undertaken in particular geographic areas, or to projects providing employment to certain categories of people.

Other incentives take the form of tax relief. For example, currently the CCA rate for pollution control equipment is 50%, and in the recent past it has been as high as 100%. The ability to depreciate pollution equipment quickly makes it a more desirable investment for a company, and a beneficial investment for society as a whole.

The exact form of incentives changes from year to year as governments change and as the political interests of society change. In most companies there is an individual or department that keeps track of possible programs affecting company projects.

Incentives must be considered when assessing the viability of a project. Grant incentives provide an additional cash flow to the project that can be taken into account like any other cash flow element. Tax incentives may be more difficult to assess since sometimes, for example, they use other forms of depreciation, or may result in different tax rates for different parts of the project.

is the net income for accounting purposes. For tax purposes, we need to determine taxable net income. To establish taxable net income, we start with the accounting net income, add back the depreciation expense for accounting purposes, and then deduct the CCA. Such accounting adjustments are common in determining the amount of tax that a company needs to pay. Given the complexities of the tax system, it is possible that net income for accounting purposes differs from net income for tax purposes by a large amount. For our purposes, we need only to distinguish between the depreciation for accounting purposes and the capital cost allowance, which is depreciation for tax purposes.

Figure 8.1 Effect of Different CCA Rates

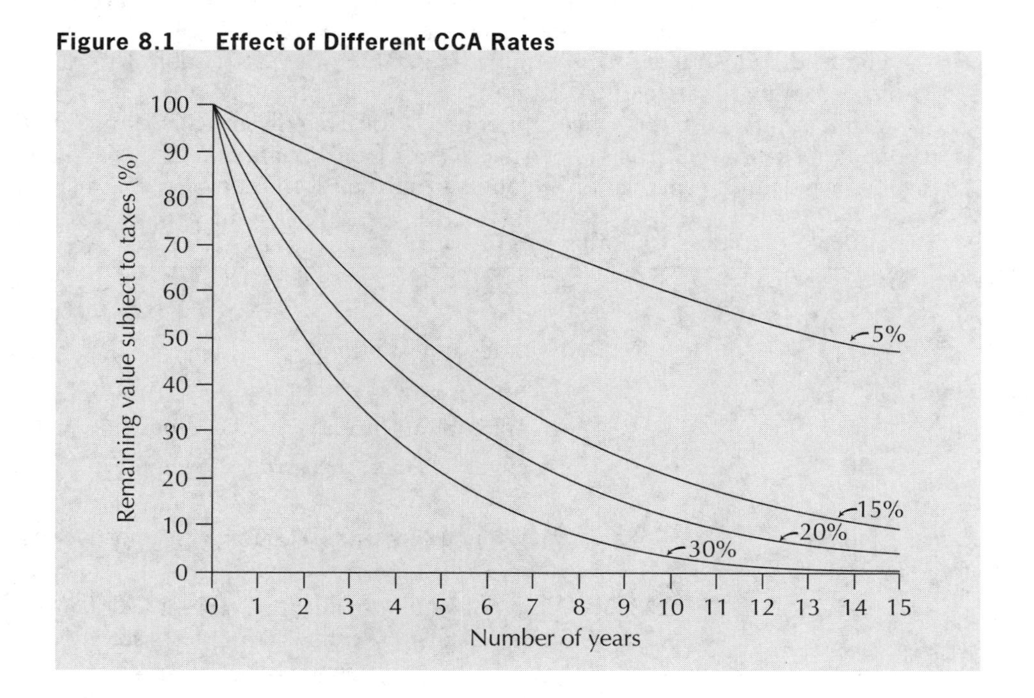

8.6 Undepreciated Capital Cost and the Half-Year Rule

The basis for calculating the capital cost allowance for assets in a particular asset class is the total *undepreciated capital cost (UCC)* of the assets included in that class. The capital cost of an asset when it is purchased is the total cost of acquiring the asset. This includes the purchase price, installation cost, transportation cost, legal fees, accounting costs, and possibly other costs over and above the purchase price. As the asset is depreciated, companies keep track of the undepreciated portion of the original capital cost through an **undepreciated capital cost (UCC)** account. The UCC is the remaining book value for the assets subject to depreciation for taxation purposes, which may or may not differ from the market or salvage value.

The undepreciated capital cost for each asset is not usually recorded individually within a class; instead, assets in each class are pooled and only one account is maintained for each asset class. The capital cost allowance for a particular asset class is then calculated from the CCA rate for that class and its UCC.

Prior to November 13, 1981, a company was allowed to include in its base for the calculation of capital cost allowance the full purchase price of an asset purchased within the year, regardless of when the asset was purchased during the taxation year. Consequently, there was considerable motivation for companies to purchase assets at the end of their fiscal year. Recognizing the substantial tax losses brought about in this manner, the Canadian government changed the rules, effective Friday, November 13, 1981. Since that date, only half of the capital cost of acquiring an asset is considered in the CCA in the year of purchase of that asset, while the other half is included in the following year. This is commonly referred to as "the half-year rule" in the CCA system.

To see the effect of this change and to illustrate the UCC account, consider a company that has just purchased a $1 000 000 piece of equipment. For simplicity, we will assume that this equipment is the only equipment in its class. The CCA rate for the equipment is 20%, and the company's tax rate is 50%. Table 8.4 shows the company's UCC amounts for the first four years of the asset's life, assuming that the purchase occurred before the 1981 tax rule change. Table 8.5 shows equivalent figures, assuming that the purchase occurred after the 1981 tax change.

To explain some of the amounts, we will start with Table 8.4, showing the pre-1981 rules. The UCC at the end of the year in which the asset was bought, prior to claiming the CCA, was the purchase cost of the asset. The CCA rate for the equipment was 20%.

Table 8.4 UCC Amounts Using Pre-1981 CCA Rules

Year	Adjustments to UCC from Purchases and Dispositions	Base UCC Amount for Capital Cost Allowance	Capital Cost Allowance	Remaining UCC	Tax Savings Due to the CCA
1	$1 000 000	$1 000 000	$200 000	$800 000	$100 000
2	0	800 000	160 000	640 000	80 000
3	0	640 000	128 000	512 000	64 000
4	0	512 000	102 400	409 600	51 200

Thus the company could claim a capital cost allowance of 20% of the $1 000 000 for the first year, leaving a UCC of $800 000 at the end of the first year. In the second year, the CCA amount was 20% of the UCC from the end of the previous year: 20% of $800 000 = $160 000. The UCC of the asset thus declined by 20% of the *current* book value each year as the CCA rate was applied to the undepreciated capital cost from the previous year.

Table 8.5 shows what happens to the UCC account if it is assumed that the purchase occurred *after* the 1981 CCA tax regulation change. In the year of purchase, the full first cost of $1 000 000 can be added to the UCC account, but only half of that amount is subject to a CCA claim. Thus the CCA amount in the first year is 20% of $500 000, leaving a balance of $900 000 of undepreciated capital cost. The CCA amount for the second year is then 20% of $900 000, or $180 000. The remainder of the CCA calculations are computed as usual.

Table 8.5 UCC Amounts Using Post-1981 CCA Rules

Year	Adjustments to UCC from Purchases and Dispositions	Base UCC Amount for Capital Cost Allowance	Capital Cost Allowance	Remaining UCC	Tax Savings Due to the CCA
1	$1 000 000	$500 000	$100 000	$900 000	$50 000
2	0	900 000	180 000	720 000	90 000
3	0	720 000	144 000	576 000	72 000
4	0	576 000	115 200	460 800	57 600

Notice that the CCA expenses generate tax savings by reducing taxable income. At a 10% interest rate, the present worth of the tax savings using the pre-1981 rules is $240 079, while that for the post-1981 rules is only $213 271.

Since the change in the tax law pertaining to "the half-year rule" in 1981, there still remains an incentive to purchase equipment at the end of the (fiscal) year. However, the incentive has been reduced as the tax effects have been diminished.

The previous example illustrated a simple case in which only one asset was purchased. In fact, a company typically purchases assets over time and disposes of them when they are no longer required. It is important to note that, if an asset is disposed of in the same year as another one in the same CCA class is purchased, the disposal amount (for the class) is subtracted from the purchase amount (for the class) *before* applying the half-year rule. For any given year, the UCC balance can be calculated as follows:

$$UCC_{opening} + \text{additions} - \text{disposals} - CCA = UCC_{ending}$$

Table 8.6 summarizes the effect of the half-year rule under various circumstances.

To illustrate the use of UCC accounts when several assets of the same class are acquired and then disposed of, consider Example 8.3.

Table 8.6 Half-Year Rule Summary

Component	Treatment
Purchase	Add only 1/2 of the purchase cost of an asset to the Base UCC Amount for its CCA class in the year of purchase. After the CCA calculation, add the other 1/2 to the remaining UCC (note that the second 1/2 is not considered an acquisition in the following year).
Disposition	Subtract the full amount received for a disposition of an asset from the Base UCC Amount for its CCA class.
Purchases and Dispositions in Same Year	Subtract total dispositions from total purchases for a CCA class. If the remainder is positive, treat it as a Purchase. If the remainder is negative, treat it as a Disposition.

Example 8.3

Egonomical Corporation, an injection-molding firm, is planning to set up business. It will purchase two used injection molders for $5000 each in 2006, a new, full-featured molder for $20 000 in 2007, and a computer controller for the new molder for $5000 in 2011. One used molder will be salvaged for $2000 in 2011, and the other for the same amount in 2012. If the CCA rate for all these assets is 20%, determine the balance in the UCC account in years 2006 to 2013.

Table 8.7 illustrates the calculations for the UCC balance for Example 8.3. It can be assumed that the original balance is zero, since the company is just starting up. In 2006, purchases totalling $10 000 were made. However, only half of that amount, $5000, is used for the CCA calculations because of the half-year rule. At 20%, the CCA claim is then $1000. The UCC account for that class is increased by the full amount of the purchase, so subtracting the $1000 from the UCC results in a balance of $9000. In 2007, a purchase of $20 000 increases the amount subject to the UCC to $19 000 (since only half of the cost of the new purchase can be included in 2007), resulting in a CCA claim for that year of $3800. The UCC balance is calculated as $9000 + $20 000 – $3800 = $25 200. For years

Table 8.7 UCC Computations with Several Changes in Asset Holdings

Year	Adjustments to UCC from Purchases and Dispositions	Base UCC Amount for Capital Cost Allowance	Capital Cost Allowance	Remaining UCC
2006	$10 000	$ 5 000	$1 000	$ 9 000
2007	20 000	19 000	3 800	25 200
2008	0	25 200	5 040	20 160
2009	0	20 160	4 032	16 128
2010	0	16 128	3 226	12 902
2011	3 000	14 402	2 880	13 022
2012	(2 000)	11 022	2 204	8 817
2013	0	8 817	1 763	7 054

2008 to 2010, the CCA claim is simply 20% of the UCC balance for the previous year, since no acquisitions or disposals are made.

In 2011, two transactions occur. A computer controller is purchased for \$5000, and a molder is salvaged for \$2000. This results in a net positive adjustment to the UCC account of \$3000. It is this \$3000 which is subject to the half-year rule. Thus the UCC amount for CCA calculations is half of this amount plus the UCC balance from the previous year: \$1500 + \$12 902 = \$14 402. The UCC balance at the end of 2011 includes the whole \$3000 amount: \$12 902 + \$3000 – \$2880 = \$13 022. The negative adjustment in 2012 is not subject to the half-year rule, since the rule only applies to net purchases over the year, so the UCC amount for CCA calculations is \$2000 less than the previous year's balance. In 2013, there are no adjustments to the UCC account other than the CCA claim, leaving a final balance of \$7054.■

8.7 The Capital Cost Tax Factor

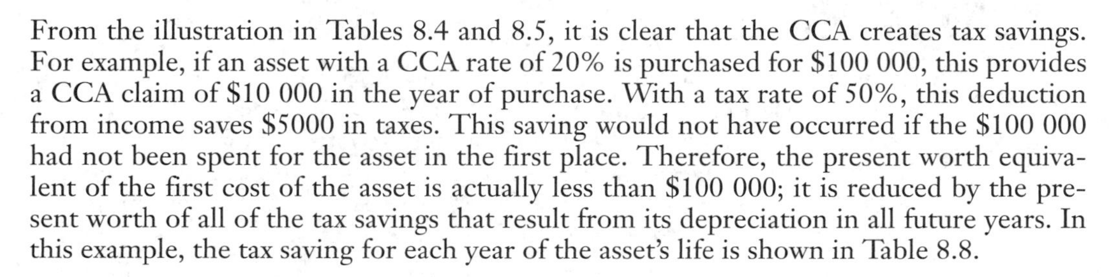

From the illustration in Tables 8.4 and 8.5, it is clear that the CCA creates tax savings. For example, if an asset with a CCA rate of 20% is purchased for \$100 000, this provides a CCA claim of \$10 000 in the year of purchase. With a tax rate of 50%, this deduction from income saves \$5000 in taxes. This saving would not have occurred if the \$100 000 had not been spent for the asset in the first place. Therefore, the present worth equivalent of the first cost of the asset is actually less than \$100 000; it is reduced by the present worth of all of the tax savings that result from its depreciation in all future years. In this example, the tax saving for each year of the asset's life is shown in Table 8.8.

Table 8.8 Tax Savings Due to the Capital Cost Allowance (50% Tax Rate)

Year	Base UCC Amount for Capital Cost Allowance	Capital Cost Allowance	Remaining UCC	Tax Savings Due to the CCA
1	\$50 000	\$10 000	\$90 000	\$5 000
2	90 000	18 000	72 000	9 000
3	72 000	14 400	57 600	7 200
4	57 600	11 520	46 080	5 760
5	46 080	9 216	36 864	4 608

The present worth of the savings is

$$\text{Present worth of tax savings} = \$5000(P/F, i, 1) + \$9000(P/F, i, 2) + \$7200(P/F, i, 3) + \$5760(P/F, i, 4) + \$4608(P/F, i, 5) + \cdots$$

The present worth of the tax savings essentially reduces the first cost of the investment because making the investment and depreciating it over time brings about tax benefits. The **capital cost tax factor (CCTF)** is a value that summarizes the effect of the benefit of future tax savings due to CCA and allows analysts to take these benefits into account when calculating the present worth equivalent of an asset. The CCTF is constant for a given CCA rate, interest rate, and tax rate, and allows the determination of the present worth independently of the actual first cost of the asset. This makes it a very useful element.

Because of the change in tax laws on November 13, 1981, there are two CCTFs: the old CCTF ($CCTF_{old}$) and the new CCTF ($CCTF_{new}$). The $CCTF_{old}$ is valid for purchases made before November 13, 1981, and, as seen later in this chapter, is also used for the salvage of assets regardless of the time at which the salvage is made (assuming there are no acquisitions in the year of disposal). The $CCTF_{new}$ is used for capital purchases on or after November 13, 1981.

As shown in detail in Appendix 8A, the $CCTF_{old}$ is given by

$$CCTF_{old} = 1 - \frac{td}{(i + d)}$$

where

t = taxation rate
d = CCA rate
i = after-tax interest rate

The $CCTF_{new}$ is also derived in Appendix 8A as

$$CCTF_{new} = 1 - \frac{td\left(1 + \frac{i}{2}\right)}{(i + d)(1 + i)}$$

For an asset purchased after 1981, the present worth of the first cost (at the time of purchase) is found by multiplying the first cost by the $CCTF_{new}$. This takes into account the tax benefits forever. When an asset is salvaged or scrapped, we need to terminate the remaining stream of tax savings. This is done by applying the $CCTF_{old}$. Examples in Section 8.8 will clarify the process.

Example 8.4

An automobile purchased this year by Lestev Corporation for $25 000 has a CCA rate of 30%. Lestev is subject to 43% corporate taxes and the corporate (after-tax) MARR is 12%. What is the present worth of the first cost of this automobile, taking into account the future tax savings resulting from depreciation?

The car is purchased this year, so the $CCTF_{new}$ applies. The $CCTF_{new}$ is calculated as

$$\begin{aligned} CCTF_{new} &= 1 - (0.43)(0.3)(1 + 0.06)/[(0.12 + 0.3)(1 + 0.12)] \\ &= 0.709311 \end{aligned}$$

The present worth of the first cost of the car is then calculated as

$$\begin{aligned} PW &= 0.709311(\$25\ 000) \\ &= \$17\ 733 \end{aligned}$$

The present worth of the first cost of the car, taking into account all future tax savings due to depreciation, is about $17 733. The tax benefit due to claiming CCA has effectively reduced the cost of the car from \$25 000 to \$17 733 in terms of present worth.■

It may seem strange that the effective cost of purchasing an asset is less than its first cost. However, bear in mind that the first cost is not the only effect that the purchase of an asset has on cash flows. The purchase will also likely generate savings. These savings are income, which is also taxed. Taking taxes into account when determining the present worth or annual worth of an asset will affect the present or annual worth *positively* because of the tax benefits resulting from future CCA, but also *negatively* because of the taxation of future savings.

8.8 Components of a Complete Tax Calculation

As discussed in Section 8.4, in evaluating projects with the explicit consideration of taxes, it is important to recognize that there is a difference between a before-tax MARR and an after-tax MARR. A before-tax MARR is chosen to reflect the fact that taxes are not explicitly taken into account in the economic calculations, and conversely the after-tax MARR is used when taxes are explicitly taken into account.

Evaluating the economic impact of purchasing a depreciable asset goes beyond the impact of taxes on the first cost. There are two other components in a complete economic analysis. First, we need to assess the tax implications of the savings or additional expenses brought about by the asset over its useful life. Second, when the asset is disposed of, we no longer can take advantage of its capital cost allowance and thus must terminate the stream of tax savings resulting from depreciating the asset.

Each of these components has a tax effect that has to be taken into account when doing a cash flow analysis such as determining present worth or annual worth. A summary of the procedure for a present worth computation is shown in Table 8.9.

Table 8.9 Components of a Complete Present-Worth Tax Calculation

Component	Treatment
First cost	Multiply by the $CCTF_{new}$.
Savings or expenses	Multiply by $(1 - t)$. Convert to present worth.
Salvage value	Multiply by the $CCTF_{old}$. Convert to present worth.

First cost: As presented in Section 8.7, the first cost of an asset purchased after 1981 is reduced by the tax savings due to CCA. Multiply the first cost by the $CCTF_{new}$ to find the after-tax first cost.

Savings or expenses: Reduce savings or expenses by the tax rate by multiplying by $(1 - t)$. There is an assumption that the company is making a profit, so that taxes are paid on all the savings at the rate t, and expenses will reduce taxes at the rate t.

Salvage value: Apply $CCTF_{old}$. When an asset is disposed of, the salvage value reduces the UCC for the *full* amount in the year of disposal (at least in the absence of a corresponding purchase in the same year). The effect of reducing the UCC is the same in magnitude but opposite in sign as increasing the UCC. The $CCTF_{new}$ has built into it a delay in depreciating half of the value of the asset, whereas the full effect occurs immediately when disposing of an asset. Consequently, the $CCTF_{old}$ is the one to use at the time an asset is sold.

Note that technically, when assets are disposed of in a given year, they are netted against any additions for the year before the half-year rule is applied. However, in project analysis, we generally want to evaluate the project independently, at least in preliminary evaluation. Thus, when we determine the salvage value we do not consider the effects of other additions or disposals that the company may also be planning for the same year. Nevertheless, it is worth noting that significant tax advantages can be made by properly planning the timing of investment additions and disposals. Our goal is to decide on the merits of the project on a more basic level at this time.

Example 8.5

The owner of a spring water bottling company in Erbsville has just purchased an automated bottle capper. What is the after-tax present worth of the new automated bottle capper if it costs \$10 000 and saves \$4000 per year over its five-year life? Assume a \$2000 salvage value and a 50% tax rate. The after-tax MARR is 12%.

A CCA rate is not given in this question. As production equipment, the new bottle capper can be assumed to be in CCA Class 8, with a rate of 20%.

The present worth of the first cost (assuming that the purchase took place after November 13, 1981) must take into account the tax benefits of CCA. The after-tax first cost is

$$\text{PW(first cost)} = -\$10\,000(\text{CCTF}_{\text{new}})$$

where the CCTF_{new} is calculated as

$$\text{CCTF}_{\text{new}} = 1 - \frac{(0.5)(0.2)(1 + 0.06)}{(0.12 + 0.2)(1 + 0.12)}$$

$$= 0.70424$$

Therefore, the present worth of the first cost is

$$\text{PW(first cost)} = -\$10\,000(0.70424)$$

$$\cong -\$7042$$

The annual savings are taxed at 50%, so the present worth of the savings is

$$\text{PW(annual savings)} = \$4000(P/A, 12\%, 5)(1 - t)$$

$$= \$4000(3.6047)(0.5)$$

$$\cong \$7209$$

The salvage value is not simply \$2000 five years from now. It reduces the UCC and thus diminishes the tax benefits resulting from the CCA on the original purchase. The after-tax benefits can be determined by applying the pre-November 13, 1981, CCTF:

$$\text{PW(salvage value)} = \$2000(P/F, 12\%, 5)\text{CCTF}_{\text{old}}$$

$$\text{CCTF}_{\text{old}} = 1 - \frac{(0.5)(0.2)}{(0.12 + 0.2)}$$

$$= 0.6875$$

$$\text{PW(salvage value)} = \$2000(0.56743)(0.6875)$$

$$\cong \$780$$

Summing the present worths,

$$PW = PW(\text{first cost}) + PW(\text{annual savings}) + PW(\text{salvage value})$$
$$= -\$7042 + \$7209 + \$780$$
$$= \$947$$

The present worth, after taxes, for the new bottle capper is \$947.■

Example 8.5 illustrated the complete effect of taxes for a present worth analysis. Similar adjustments are made to an annual worth computation, as illustrated in Table 8.10.

Table 8.10 Components of a Complete Annual-Worth Tax Calculation

Component	Treatment
First cost	Multiply by the $CCTF_{new}$. Convert to annual worth.
Savings or expenses	Multiply by $(1 - t)$.
Salvage value	Multiply by the $CCTF_{old}$. Convert to annual worth.

Example 8.6

A small device used to test printed circuit boards has a first cost of \$45 000. The tester is expected to reduce labour costs and improve the defect detection rate so as to bring about a saving of \$23 000 per year. Additional operating costs are expected to be \$7300 per year. The salvage value of the tester will be \$5000 in five years. With an after-tax MARR of 12%, a CCA rate of 20%, and a tax rate of 42%, what is the annual worth of the tester, taking into account the effect of taxes?

The basic process for adjusting for tax effects in an annual worth comparison is similar to a present worth analysis. First, we apply the $CCTF_{new}$ to the first cost, and convert it into an annual amount over five years. Next, the annual savings and expenses are multiplied by $(1 - t)$. Finally, the salvage value at the end of five years is multiplied by $CCTF_{old}$ and then converted into an annual amount:

$$AW(\text{tester}) = -\$45\,000(A/P, 12\%, 5)CCTF_{new} + (\$23\,000 - \$7300)(1 - t) + \$5000(A/F, 12\%, 5)CCTF_{old}$$

Using $d = 0.20$, $t = 0.42$, and $i = 0.12$, we have

$$CCTF_{new} = 0.7516$$
$$CCTF_{old} = 0.7375$$

Therefore

$$AW(\text{tester}) = -\$45\,000(0.27741)(0.7516) + (\$23\,000 - \$7300)(0.58) + \$5000(0.15741)(0.7375)$$
$$\cong \$304$$

The annual worth, taking into account taxes, is \$304.■

As the previous two examples show, taking taxes into account for present worth and annual worth analyses is relatively straightforward. IRR computations are a bit more involved, however, as the next example illustrates.

Example 8.7

Find the after-tax IRR for the testing equipment described in Example 8.6.

First, observe that, if we solve for i in AW(receipts) – AW(disbursements) = 0, or PW(receipts) – PW(disbursements) = 0, the resulting rate will be an after-tax IRR as the amounts will have been adjusted for taxes. Since Example 8.6 was expressed in terms of annual amounts, we will find the after-tax IRR by solving for i in AW(receipts) – AW(disbursements) = 0, using the operations listed in Table 8.10:

$$(\$23\,000 - \$7300)(1 - t) + \$5000(A/F, i, 5)\text{CCTF}_{\text{old}} - \$45\,000(A/P, i, 5)\text{CCTF}_{\text{new}} = 0$$

In order to solve the above equation, a trial-and-error approach is necessary because the interest rate i appears in the capital cost tax factors as well as in the compound interest factors. Table 8.11 shows the result of this process obtained through the use of a spreadsheet.

Table 8.11 Trial and Error Process for Finding the After-Tax IRR

i	AW(receipts) – AW(disbursements)
0.10	$997.5485
0.11	651.3784
0.12	304.3646
0.13	– 43.6201
0.14	– 392.682

From the spreadsheet computations, we can see that the after-tax IRR on the testing equipment is between 12% and 13%. Additional trial-and-error iterations with the spreadsheet program give an after-tax IRR of 12.87%.■

8.9 Approximate After-Tax Rate-of-Return Calculations

The IRR is probably one of the most popular ways to assess the desirability of an investment. Unfortunately, as we saw in Section 8.8, a detailed analysis can be somewhat involved. However, an approximate IRR analysis when taxes are explicitly considered can be very easy. The formula to use is:

$$\text{IRR}_{\text{after-tax}} \cong \text{IRR}_{\text{before-tax}} \times (1 - t)$$

The reasons for this are exactly the same as described in Section 8.4 for the before- and after-tax MARR. It is an approximation that works because the IRR represents the percentage of the total investment that is net income. Since the tax rate is applied to net income, it correspondingly reduces the IRR by the same proportion. It is not exactly correct because it assumes that expenses offset receipts in the year that they occur, rather than being spread out over time (as CCA deductions) as required by the Canadian tax laws.

It should therefore be noted that this after-tax IRR will tend to be somewhat *higher* than it would be if calculated in a perfectly accurate manner. Thus, if the after-tax IRR is close to the after-tax MARR, a more precise calculation is required.

Example 8.8

What is the approximate IRR on the testing equipment described in Example 8.6?

First, we find the $IRR_{before\text{-}tax}$ by solving for i in

$$AW(receipts) - AW(disbursements) = 0$$

$$(\$23\,000 - \$7300) + \$5000(A/F, i, 5) - \$45\,000(A/P, i, 5) = 0$$

Through trial and error, we find that the $IRR_{before\text{-}tax}$ is 23.8%. The $IRR_{after\text{-}tax}$ is then approximately 13.8%:

$$\begin{aligned} IRR_{after\text{-}tax} &\cong IRR_{before\text{-}tax} \times (1 - t) \\ &= 0.238(1 - 0.42) \\ &= 0.13804 \end{aligned}$$

Notice that it is a little higher than the precise IRR of 12.87%.■

When doing an after-tax IRR computation in practice, the approximate after-tax IRR can be used as a first pass on the IRR computation. If the approximate after-tax IRR turns out to be close to the after-tax MARR, a precise after-tax IRR computation may be required to make a fully informed decision about the project.

Example 8.9

Waterloo Industries pays 40% corporate income taxes. Their after-tax MARR is 18%. A project has a before-tax IRR of 24%. Should the project be approved? What would your decision be if the after-tax MARR were 14%?

$$\begin{aligned} IRR_{after\text{-}tax} &\cong IRR_{before\text{-}tax} \times (1 - t) \\ &= 0.24(1 - 0.40) \\ &= 0.144 \end{aligned}$$

The after-tax IRR is approximately 14.4%. For an after-tax MARR of 18%, the project should not be approved. However, for an after-tax MARR of 14%, since the after-tax IRR is an approximation, a more detailed examination would be advisable.■

In summary, we can simplify after-tax IRR computations by using an easy approximation. The approximate after-tax IRR may be adequate for decision making in many cases, but in others a detailed after-tax analysis may be necessary.

REVIEW PROBLEMS

REVIEW PROBLEM 8.1

Angus and his sister Oona operate a small charter flight service that takes tourists on sightseeing tours over the beautiful Margaree River on Cape Breton Island. At the end of 2001, they had one four-seater plane in the aircraft asset class with a UCC of \$30 000. In 2002, they purchased a second plane for \$50 000. Business was going well in 2003, so they sold the old plane they had in 2001 for \$15 000 and bought a newer version for \$64 000. What was the UCC balance in the aircraft asset class at the end of 2004? The CCA rate for aircraft is 25%.

ANSWER

Table 8.12 shows the fluctuation in the UCC balance over time. At the end of 2001, the UCC for the aircraft asset class was \$30 000. In 2002, half of the capital cost of the airplane purchased in 2002 (= \$25 000) contributed to the CCA calculation. The CCA rate of 25% gave a CCA amount of \$13 750 and resulted in a UCC balance of \$66 250 at the end of 2002. In 2003, the net positive adjustment to the UCC due to the capital cost of \$64 000 for the new plane and \$15 000 benefit from the sale of the old plane was \$49 000. Half of this amount, \$24 500, contributed to the CCA calculation. After subtracting the CCA amount of \$22 688 for 2002, the remaining UCC was \$92 563 (= 66 250 + \$49 000 – \$22 688). Finally, in 2004, there were no further adjustments to the UCC, and after the CCA was deducted the closing UCC account balance was \$69 422.■

Table 8.12 Summary of UCC Computations for Review Problem 8.1

Year	Adjustments to UCC from Purchases or Dispositions	Base UCC Amount for Capital Cost Allowance	CCA Allowance	Remaining UCC
2001	\$30 000			
2002	50 000	\$55 000	\$13 750	\$66 250
2003	49 000	90 750	22 688	92 563
2004	0	92 563	23 141	69 422

REVIEW PROBLEM 8.2

David Cosgrove has just started a management consulting firm that he operates out of his home at Paradise Lake. As part of his new business, David is considering buying a new \$30 000 van, which will be used 100% of the time for earning business income. He estimates that the expenses associated with operating the van will be \$3000 per year in gas, \$1200 per year for insurance, \$600 annually for parking, and maintenance costs of \$1000 for the first year, rising \$400 per year thereafter. He expects to keep the van for five years. At the end of this time, he estimates a salvage value of \$6000. The CCA rate for vans is 30%.

The alternative for David is to lease the van. With a lease arrangement, he will have to pay for parking, gas, and insurance, but the leasing company will pay for the repairs. The lease costs are \$10 500 per year.

David estimates his after-tax cost of capital to be 12% per year and his tax rate is 40%. On the basis of an annual worth analysis over the five years, should David buy the van or lease it?

ANSWER

The approach will be to find the after-tax annual worth of each alternative. Since the parking, insurance, and gas costs are the same for both alternatives, we can exclude them from the analysis.

The after-tax annual costs of purchasing the van are

$$\begin{aligned}\text{AW(van)} &= \$30\,000(A/P, 12\%, 5)\text{CCTF}_{\text{new}} - \$6000(A/F, 12\%, 5)\text{CCTF}_{\text{old}} \\ &\quad + [\$1000 + \$400(A/G, 12\%, 5)](1 - t)\end{aligned}$$

We can calculate that

$$\text{CCTF}_{\text{new}} = 1 - \frac{td\left(1 + \frac{i}{2}\right)}{(i + d)(1 + i)}$$

$$= 1 - \frac{0.4(0.3)(1 + 0.12/2)}{(0.12 + 0.30)(1 + 0.12)}$$

$$= 0.7296$$

$$\text{CCTF}_{\text{old}} = 1 - \frac{td}{(i + d)}$$

$$= 1 - \frac{0.4(0.3)}{(0.12 + 0.30)}$$

$$= 0.71429$$

Thus

$$\text{AW(van)} = \$30\,000(0.27741)(0.7296) - \$6000(0.15741)(0.71429) + [\$1000 + \$400(1.7745)](1 - 0.4)$$

$$\cong \$6423$$

The annual cost of purchasing and operating the van over a five-year period is a little over $6400.

There is a large difference between buying and leasing. When we lease, we do not have a depreciable asset on which to claim depreciation expenses. We only have lease payment expenses. Therefore, the impact of taxes on the lease expense is simply to multiply the leasing costs by $(1 - t)$:

$$\text{AW(lease)} = \$10\,500(1 - 0.4)$$

$$= \$10\,500(0.6)$$

$$= \$6300$$

The after-tax annual cost of leasing is $6300. It is less expensive to lease the van than it is to buy it. Therefore, David should lease the van. It is, however, worth noting that this example is a simplified version of real life, since numerous tax rules relating to the eligibility of expenses for automobiles have been ignored for illustration purposes.■

REVIEW PROBLEM 8.3

Putco does subcontracting for an electronics firm that assembles printed circuit boards. Business has been good lately, and Putco is thinking of purchasing a new IC chip-placement machine. It has a first cost of $450 000 and is expected to save them $125 000 per year in labour and operating costs compared with the manual system they have now. A similar system that also automates the circuit board loading and unloading process costs $550 000 and will save about $155 000 per year. The life of either system is expected to be four years. The salvage value of the $450 000 machine will be $180 000, and that of the $550 000 machine will be $200 000. Putco uses an after-tax MARR of 9% to make decisions about such projects. On the basis of an IRR comparison, which alternative (if either) should they choose? Putco pays taxes at a rate of 40%, and the CCA rate for the equipment is 20%.

ANSWER

Putco has three mutually exclusive alternatives:

1. Do nothing
2. Buy the chip-placement machine
3. Buy a similar chip-placement machine with an automated loading and unloading process

Following the procedure from Chapter 5, the projects are already ordered, on the basis of first cost. We therefore begin with the first alternative: the before-tax (and thus the after-tax) IRR of the "do nothing" alternative is 0%. Next, the before-tax IRR on the incremental investment to the second alternative can be found by solving for i in

$$-\$450\,000 + \$125\,000(P/A, i, 4) + \$180\,000(P/F, i, 4) = 0$$

By trial and error, we obtain an $\text{IRR}_{\text{before-tax}}$ of 15.92%. This gives an approximate $\text{IRR}_{\text{after-tax}}$ of $0.1592(1 - 0.40) = 0.0944$ or 9.44%. With an after-tax MARR of 9%, it would appear that this alternative is acceptable, though a detailed after-tax computation may be in order. We need to solve for i in

$$(-\$450\,000)\text{CCTF}_{\text{new}} + \$125\,000(P/A, i, 4)(1 - t) + \$180\,000(P/F, i, 4)\text{CCTF}_{\text{old}} = 0$$

Doing so gives an $\text{IRR}_{\text{after-tax}}$ of 9.5%. Since this exceeds the required after-tax MARR of 9%, this alternative becomes the current best. We next find the $\text{IRR}_{\text{after-tax}}$ on the incremental investment required for the third alternative. The $\text{IRR}_{\text{before-tax}}$ is first found by solving for i in

$$-(\$550\,000 - \$450\,000) + (\$155\,000 - \$125\,000)(P/A, i, 4)$$
$$+ (\$200\,000 - \$180\,000)(P/F, i, 4) = 0$$

This gives an $\text{IRR}_{\text{before-tax}}$ of 7.13%, or an approximate $\text{IRR}_{\text{after-tax}}$ of 4.28%. This is sufficiently below the required after-tax MARR of 9% to warrant rejection of the third alternative without a detailed incremental $\text{IRR}_{\text{after-tax}}$ computation. Putco should therefore select the second alternative.■

REVIEW PROBLEM 8.4

David Cosgrove (from Review Problem 8.2) is still thinking over whether to buy a van. Assuming he remains in business for the foreseeable future, he will need a vehicle for transportation indefinitely, whether he owns or leases it. In his original analysis, he assumed that the van would be replaced at the end of five years. Because appearances are important to David, he would not consider keeping a vehicle for longer than five years, but he now recognizes that the economic life of the van may be *shorter* than five years. Assuming that the van depreciates in value by a constant proportion each year, determine how frequently David should replace it. The CCA rate is 30% and his tax rate is 40%.

ANSWER

The first step in the solution is to recognize that David is facing a cyclic replacement problem, since it is reasonable to assume that he will replace each van with one similar to the previous, indefinitely. We now need to assess the annual cost of replacing a van each year, every two years, and so on, up to replacement every five years. Before proceeding, however, we need to determine the depreciation rate to use so that we can determine the approximate value of the van when it is n years old for $n = 1, 2, 3$ and 4.

Referring back to Chapter 6, we have for the declining-balance method of depreciation

$$d = 1 - \sqrt[n]{\frac{S}{P}} = 1 - \sqrt[5]{\frac{\$6000}{\$30\ 000}} = 0.27522$$

Using the formula $BV_{db}(n) = P(1 - d)^n$, we find that the book value of the van at the end of each year is:

End of Year	Book Value
0	$30 000
1	21 743
2	15 759
3	11 422
4	8 278
5	6 000

Note that these are book values, not the UCC balances. The book values are estimates of the market value, which is needed to judge when the asset should be replaced. A UCC balance is similar to a book value, but is used for calculating the CCA only.

Now the annual worth computations can be done using the CCTF values calculated in Review Problem 8.2:

AW(replace every year)

$$= \text{AW(capital recovery)} + \text{AW(operating)}$$

$$= \$30\ 000(A/P, 12\%, 1)\text{CCTF}_{\text{new}} - \$21\ 743(A/F, 12\%, 1)\text{CCTF}_{\text{old}} + (\$5800)(1 - t)$$

$$= \$12\ 463$$

AW(replace every two years)

$$= \$30\ 000(A/P, 12\%, 2)\text{CCTF}_{\text{new}} - \$15\ 759(A/F, 12\%, 2)\text{CCTF}_{\text{old}} + \$5800(1 - t) + \$400(A/F, 12\%, 2)(1 - t)$$

$$= \$11\ 235$$

AW(replace every three years)

$$= \$30\ 000(A/P, 12\%, 3)\text{CCTF}_{\text{new}} - \$11\ 422(A/F, 12\%, 3)\text{CCTF}_{\text{old}} + \$5800(1 - t) + [\$400(F/P, 12\%, 1) + \$800](A/F, 12\%, 3)(1 - t)$$

$$= \$10\ 397$$

AW(replace every four years)

$$= \$30\ 000(A/P, 12\%, 4)\text{CCTF}_{\text{new}} - \$8278(A/F, 12\%, 4)\text{CCTF}_{\text{old}} + \$5800(1 - t) + [\$400(F/P, 12\%, 2) + \$800(F/P, 12\%, 1) + \$1200](A/F, 12\%, 4)(1 - t)$$

$$= \$9775$$

AW(replace every five years)

$$= \$30\,000(A/P, 12\%, 5)\text{CCTF}_{\text{new}} - \$6000(A/F, 12\%, 5)\text{CCTF}_{\text{old}} + \$5800(1 - t) + [\$400(F/P, 12\%, 3) + \$800(F/P, 12\%, 2) + \$1200(F/P, 12\%, 1) + \$1600](A/F, 12\%, 5)(1 - t)$$

$$= \$9303$$

On the basis of these calculations, it is best for David to replace the van at the end of every five years. Its economic life may be longer than five years, but as far as David is concerned, a five-year-old van has reached the end of its useful life and must be replaced. ■

SUMMARY

Income taxes can have a significant effect on engineering economics decisions. In particular, taxes reduce the effective cost of an asset, the savings generated, and the value of the sale of an asset.

In this chapter, we provided a basic introduction to the Canadian capital cost allowance (CCA) system and the use of undepreciated capital cost (UCC) accounts. The CCA rate is a declining-balance rate that is mandated for use in calculating the depreciation expenses for capital assets. These depreciation expenses are then used in determining the amount of taxes owing for the year. Assets are designated as belonging to a particular CCA class. The book values for taxation purposes calculated for all the assets in each class are accumulated into a UCC account.

The future CCA claims that arise from the purchase of an asset are benefits that reduce the after-tax first cost. The CCTF_{new} permits the quick calculation of the net effect of these benefits, while similarly the CCTF_{old} permits the calculation of the net effect of future loss of CCA claims for assets that are sold or scrapped.

It was noted that, for after-tax calculations, an after-tax MARR must be used. After-tax calculations were illustrated for present worth, annual worth, and IRR evaluations. An approximate IRR comparison method was also given.

The review problems at the end of the chapter illustrated how taxes affect present worth and annual worth comparisons, replacement analysis, and internal rate of return computations.

Engineering Economics in Action, Part 8B: The Work Report

"So what is this, anyhow?" Clem was looking at the report that Naomi had handed him. "A consulting report?"

"Sorry, chief, it is a bit thick." Naomi looked a little embarrassed. "You see, Terry has to do a work report for his university. It's part of the co-op program that they have. He got interested in the 10-stage die problem and asked me if he could make that study his work report. I said OK, subject to its perhaps being confidential. I didn't expect it to be so thick, either. But he's got a good executive summary at the front."

"Hmm . . . " The room was quiet for a few minutes while Clem read the summary. He then leafed through the remaining parts of the report. "Have you read this through? It looks really quite good."

"I have. He has done a very professional job—er, at least what seems to me to be very professional." Naomi suddenly remembered that she hadn't yet gained her professional engineer's designation. She also hadn't been working at Canadian Widgets much longer than Terry. "I gathered most of the data for him, but he did an excellent job of analyzing it. As you can see from the summary, he set up the replacement problem as a set of mutually exclusive alternatives, involving upgrading the die now or later and even more than once. He did a nice job on the taxes, too."

"How did he handle the taxes?"

"Quite well. I had to hold his hand a bit to make sure he understood how the UCC accounts work, but once he had that everything else seemed to fall into place. He reduced the purchase price by the benefits of future CCA claims. The installation cost and future savings were reduced by the taxation rate, and the salvage values were reduced for loss of future CCA claims."

"Did he understand about the old and new CCTFs?"

"Yes, he did."

"Not bad. I think we've got a winner there, Naomi. Let's make sure we get him back for his next work term."

Naomi nodded. "What about his work report, Clem? Should we ask him to keep it confidential?"

Clem laughed, "Well, I think we should, and not just because there are trade secrets in the report. I don't want anyone else knowing what a gem we have in Terry!"

PROBLEMS

8.1 Go to the Canada Revenue Agency Web site and search for the *T2 Corporation—Income Tax Guide*. Find the section in the *Guide* dealing with CCA rates. Identify each of the following according to their CCA class(es) and CCA rate(s).

(a) New software for inventory control purposes costing $70 000

(b) Communications network switching equipment which costs $56 000

(c) A heated, rigid-frame greenhouse worth $57 000

(d) A small tractor with attachments for earth-moving and snow removal worth $24 000

(e) Cash register and barcode scanning device used for point-of-sales data collection. Cost is $12 000 for the cash register and $200 for the scanning device.

8.2 The MARR generally used by Collingwood Caskets is a before-tax MARR of 14%. Vincent wants to do a detailed calculation of the cash flows associated with a new planer for the assembly line. What would be an appropriate after-tax MARR for him to use if Collingwood Caskets pays

(a) 40% corporate taxes?

(b) 50% corporate taxes?

(c) 60% corporate taxes?

8.3 Enrique has just completed a detailed analysis of the IRR of a waste-water treatment plant for Gimli Meat Products. The 8.7% after-tax IRR he calculated compared favourably with a 5.2% after-tax MARR. For reporting to upper management, he wants to present this information as a before-tax IRR. If Gimli Meat Products pays 53% corporate taxes, what figures will Enrique report to upper management?

8.4 A company's first year's operations (in 1976) can be summarized as follows:

Revenues: $110 000

Expenses (except CCA): $65 000

Their capital asset purchases in the first year totalled $100 000. With a CCA rate of 20% and a tax rate of 55%, how much income tax did they pay?

8.5 A company's first year's operations (in 2004) can be summarized as follows:

Revenues: $110 000

Expenses (except CCA): $65 000

Their capital asset purchases in the first year totalled $100 000. With a CCA rate of 20% and a tax rate of 55%, how much income tax did they pay?

8.6 What is the after-tax present worth of a chip placer if it costs $55 000 and saves $17 000 per year? After-tax interest is at 10%. Assume the device will be sold for a $1000 salvage value at the end of its six-year life. The CCA rate is 20%, and the corporate income tax rate is 54%.

8.7 Quebec Widgets is looking at a $400 000 digital midget rigid widget gadget (CCA Class 8). It is expected to save $85 000 per year over its 10-year life, with no scrap value. Their tax rate is 45%, and their after-tax MARR is 15%. On the basis of an approximate IRR, should they invest in this gadget?

8.8 Canada Widgets is looking at a $400 000 digital midget rigid widget gadget (CCA Class 8). It is expected to save $85 000 per year over its 10-year life, with no scrap value. Their tax rate is 45%, and their after-tax MARR is 15%. On the basis of an exact IRR, should they invest in this gadget?

8.9 The UCC balance for a firm's automobile fleet at the end of 2004 was $10 000. There was one truck in service at this time. At the beginning of 2005, they purchased two trucks for a total of $50 000. At the beginning of 2007, they purchased another truck for $20 000. At the beginning of 2008, the truck owned in 2004 was sold for $3000. The CCA rate for automobiles is 30%. What was the firm's UCC balance at the end of 2008?

8.10 Go to the Canada Revenue Agency Web site and find the form T2S(8), a worksheet for calculating UCC balances. Use the sheet to make the following calculations (separately). In all cases, there are no adjustments (e.g., for GST rebates or for investment tax credits).

(a) The UCC balance at the end of the previous year is $10 000. Assets purchased in Class 10 for the current year amount to $30 000. Find the UCC at the end of the year.

(b) The UCC balance at the end of the previous year is $10 000. Assets purchased in Class 10 for the current year amount to $20 000. Dispositions were $5 000. Find the UCC at the end of the year.

(c) The UCC balance at the end of the previous year for Class 8 was $20 000. This year, an asset worth $20 000 was added to Class 8 and another worth $15 000 was disposed of. What is the year-end UCC for this year?

8.11 Churchill Metal Products opened for business in 1984. Over the following years, their transactions for CCA Class 8 assets consisted of the following:

Date	Item	Activity	Amount
March 11, 1984	Machine 1	Purchase	$ 50 000
April 24, 1984	Machine 2	Purchase	150 000
November 3, 1987	Machine 3	Purchase	250 000
November 22, 1987	Machine 1	Sale	10 000
May 20, 1991	Machine 4	Purchase	60 000
August 3, 1996	Machine 5	Purchase	345 000
September 12, 1997	Machine 3	Sale	45 000

What CCA amount did Churchill Metal Products claim for the 20% UCC account in 1998?

8.12 Calculate the $CCTF_{old}$ and $CCTF_{new}$ for each of the following:

(a) Tax rate of 50%, CCA of 20%, and an after-tax MARR of 9%

(b) Tax rate of 35%, CCA of 30%, and an after-tax MARR of 12%

(c) Tax rate of 55%, CCA of 5%, and an after-tax MARR of 6%

8.13 Use a spreadsheet program to create a chart showing how the values of the $CCTF_{old}$ and the $CCTF_{new}$ change for after-tax MARRs of 0% to 30%. Assume a fixed tax rate of 50% and a CCA rate of 20%.

8.14 What is the approximate after-tax IRR on a project for which the first cost is $12 000, savings are $5000 in the first year and $10 000 in the second year, and taxes are at 40%?

8.15 What is the exact after-tax IRR on a project for which the first cost is $12 000, savings are $5000 in the first year and $10 000 in the second year, taxes are at 40%, and the CCA rate is 30%? What would the exact after-tax IRR have been if this analysis had been done in 1975?

8.16 What is the total after-tax annual cost of a machine with a first cost of $45 000 and operating and maintenance costs of $0.22 per unit produced? It will be sold for $4500 at the end of five years. Production is 750 units per day, 250 days per year. The CCA rate is 30%, the after-tax MARR is 20%, and the corporate income tax rate is 40%.

8.17 Refer to Problem 5.9. Rimouski Dairy has a corporate tax rate of 40% and the filling machine for the dairy line has a CCA rate of 30%. The firm has an after-tax MARR of 10%. On the basis of the exact IRR method, determine which alternative Rimouski Dairy should choose.

8.18 In 1965, the Sackville Furniture Company bought a new band saw for $360 000. Aside from depreciation expenses, their yearly expenses totalled $1 300 000 versus $1 600 000 in income. How much tax (at 50%) would they have paid in 1965 if they had been permitted to use each of the following depreciation schemes?

(a) Straight-line, with a life of 10 years and a 0 salvage value

(b) Straight-line, with a life of five years and a 0 salvage value

(c) Declining-balance, at 20%

(d) Declining-balance, at 40%

(e) Fully expensed that year

8.19 Mulroney Brothers Salvage had several equipment purchases in the '70s. Their first asset was a tow truck bought in 1972 for $25 000. In 1974, a van was purchased for $14 000. A second tow truck was bought in 1977 for $28 000, and the first one was sold the following year for $5000. Using a 30% CCA rate (automobiles, trucks, and vans), what was the balance of their UCC account at the end of 1979?

8.20 Chrétien Brothers Salvage had several equipment purchases in the '80s. Their first asset was a tow truck bought in 1982 for $25 000. In 1984, a van was purchased for $14 000. A second tow truck was bought in 1987 for $28 000, and the first one was sold the following year for $5000. Using a 30% CCA rate (automobiles, trucks, and vans), what was the balance of their UCC account at the end of 1989?

8.21 Whitehorse Construction has just bought a crane for $380 000. At a CCA rate of 20%, what is the present worth of the crane, taking into account the future benefits resulting from the CCA? Whitehorse has a tax rate of 35% and an after-tax MARR of 6%.

8.22 Hull Hulls is considering the purchase of a 30-tonne hoist. The first cost is expected to be $230 000. Net savings will be $35 000 per year over a 12-year life. It will be salvaged for $30 000. If their after-tax MARR is 8% and they are taxed at 45%, what is the present worth of this project?

8.23 Kanata Konstruction is considering the purchase of a truck. Over its five-year life, it will provide net revenues of $15 000 per year, at an initial cost of $65 000 and a salvage value of $20 000. KK pays 35% in taxes, the CCA rate for trucks is 30%, and their after-tax MARR is 12%. What is the annual cost or worth of this purchase?

8.24 A new binder will cost Revelstoke Printing $17 000, generate net savings of $3000 per year over a seven-year life, and be salvaged for $1000. Revelstoke's before-tax MARR is 10%, they are taxed at 40%, and the binder has a 20% CCA rate. What is their approximate after-tax IRR on this investment? Should the investment be made?

8.25 A new binder will cost Revelstoke Printing $17 000, generate net savings of $3000 per year over a seven-year life, and be salvaged for $1000. Revelstoke's before-tax MARR is 10%, they are taxed at 40%, and the binder has a 20% CCA rate. What is their exact after-tax IRR on this investment? Should the investment be made?

8.26 A slitter for sheet sandpaper owned by Abbotsford Abrasives (AA) requires regular maintenance costing $7500 per year. Every five years it is overhauled at a cost of $25 000. The original capital cost was $200 000, with an additional $25 000 in non-capital expenses that occurred at the time of installation. The machine has an expected life of 20 years and a $15 000 salvage value. The machine will not be overhauled at the end of its life. AA pays taxes at a rate of 45% and expects an after-tax rate of return of 10% on investments. Recalling that the CCA rate for production equipment is 20%, what is the after-tax annual cost of the slitter?

8.27 Rodney has discovered that, for the last three years, his company has been classifying as Class 8 items costing between $100 and $200 that should be in CCA Class 12. If an estimated $10 000 of assets per year were misclassified, what is the present worth today of the cost of this mistake? Assume that the mistake can only be corrected for assets bought in the future. Rodney's company pays taxes at 50% and their after-tax MARR is 9%.

8.28 Identify each of the following according to their CCA class(es) and CCA rate(s):

(a) A soldering gun costing $75

(b) A garage used to store spare parts

(c) A new computer

(d) A 100-tonne punch press

(e) A crop dusting attachment for a small airplane

(f) An oscilloscope worth exactly $200

8.29 Roch bought a $100 000 machine (Machine A) on November 12, 1981. As a CCA Class 8 asset, what was its book value, measured as its contribution to the UCC for that class, at the end of 1991? Roch purchased an identical $100 000 machine (Machine B) on November 14, 1981. What was its book value at the end of 1991?

8.30 A chemical recovery system costs $30 000 and saves $5280 each year of its seven-year life. The salvage value is estimated at $7500. The after-tax MARR is 9% and taxes are at 45%. What is the net after-tax annual benefit or cost of purchasing the chemical recovery system?

8.31 CB Electronix needs to expand its capacity. It has two feasible alternatives under consideration. Both alternatives will have essentially infinite lives.

Alternative 1: Construct a new building of 20 000 square metres now. The first cost will be $2 000 000. Annual maintenance costs will be $10 000. In addition, the building will need to be painted every 15 years (starting in 15 years) at a cost of $15 000.

Alternative 2: Construct a new building of 12 500 square metres now and an addition of 7 500 square metres in 10 years. The first cost of the 12 500-square-metre building will be $1 250 000. The annual maintenance costs will be $5000 for the first 10 years (i.e., until the addition is built). The 7500-square-metre addition will have a first cost of $1 000 000. Annual maintenance costs of the renovated building (the original building and the addition) will be $11 000. The renovated building will cost $15 000 to repaint every 15 years (starting 15 years after the addition is done).

Given a CCA rate of 5% for the buildings, a corporate tax rate of 45%, and an after-tax MARR of 15%, carry out an annual-worth comparison of the two alternatives. Which is preferred?

The following Ridgely Custom Metal Products (RCMP) case is used for Problems 8.32 to 8.37. RCMP must purchase a new tube bender. There are three models:

Model	First Cost	Economic Life	Yearly Net Savings	Salvage Value
T	$100 000	5 years	$50 000	$20 000
A	150 000	5 years	60 000	30 000
X	200 000	3 years	75 000	100 000

RCMP's after-tax MARR is 11% and the corporate tax rate is 52%. A tube bender is a CCA Class 8 asset.

8.32 Using the present worth method and the least-cost multiple of the service lives, which tube bender should they buy?

8.33 RCMP realizes that it can forecast demand for its products for only three years in advance. The salvage value for model T after three years is $40 000 and for model A, $80 000. Using the present worth method and a three-year study period, which of the three alternatives is now best?

8.34 Using the annual worth method, which tube bender should Ridgely buy?

8.35 What is the approximate after-tax IRR for *each* of the tube benders?

8.36 What is the exact after-tax IRR for *each* of the tube benders?

8.37 Using the approximate after-tax IRR comparison method, which of the tube benders should Ridgely buy? (*Reminder:* You must look at the incremental investments.)

8.38 Refer to Problem 5.9. Rimouski Dairy has a corporate tax rate of 40% and the filling machine for the dairy line has a CCA rate of 30%. They have an after-tax MARR of 10%. Using an approximate IRR approach, determine which alternative Rimouski Dairy should choose.

8.39 Salim is considering the purchase of a backhoe for his pipeline contracting firm. The machine will cost $110 000, last six years with a salvage value of $20 000, and reduce annual

Lease or Loan: A Small Business Perspective

CANADIAN 8.1 MINI-CASE

Small business owners face many decisions about how to finance the growth of their business. When these decisions involve capital assets, one fundamental decision is whether the assets should be leased or purchased. The purpose of this Mini-Case is to outline some of the advantages and disadvantages of leasing and purchasing (via a loan) and to illustrate the tax implications of each alternative.

A lease is a contract that grants, for payment, use of capital assets such as equipment or real estate for a specified time. The user of the assets is referred to the lessee, the provider of the assets, the lessor. One of the main advantages of a lease to a small business is that it generally involves lower monthly payments than a loan. This is primarily for two reasons. First, the entire value of the asset is not paid for, and second, sales taxes are only paid for on the lease payments, not the full value of the asset as they would be for a purchased asset financed through a loan. Another advantage of a lease is that lease financing is done at a fixed interest rate, allowing a small business to predict its cash flows better than with a loan with fluctuating interest rates (e.g., a line of credit).

A disadvantage of a lease is that the total amount of cash paid out tends to be higher than if the lessee had purchased the asset in the first place. This is because leases are often computed using a higher interest rate than for a loan. The higher rate, in part, compensates the lessor for taking such risks as estimating the salvage value of the asset at the end of the lease.

There are two primary types of leases—capital and operating leases. With a capital lease the lessee is considered to have both borrowed funds and purchased a depreciable asset, while with an operating lease the lessee is considered to have borrowed funds but the lessor maintains ownership of the assets. In a capital lease, then, there is an additional tax benefit—the lessor can claim both CCA deductions and interest-rate deductions, while with a loan only the interest expense on the loan is tax-deductible.

The choice between a lease or a loan to acquire the use of fixed assets will depend on the circumstances of the individual small business owner, and on both the qualitative considerations of the flexibility a lease provides and a quantitative consideration of its financial implications.

The Canada Small Business Financing (CSBF) Program is intended to help small businesses obtain loans and capital leases for fixed assets of value up to $250 000. For further information, see www.cbsc.org/ontario/english/index.cfm under "Financing."

Discussion

To more fully understand the financial implications to a small business of a lease or a loan to acquire a capital asset, consider the case of a small business operator who wants to purchase furniture and equipment for a new office space. The total cost of the furniture and equipment is $40 000, and they fall into CCA Class 8 with a CCA rate of 20%. The operator is considering a three-year, 12% capital or operating lease versus using the company line of credit (at 10%) for a loan of the same term.

Questions

1. Assuming that the assets have a salvage value of $15 000 at the end of three years, use the capital recovery formula to show that the monthly payments for a 36-month lease are $1029.44. For each month of the lease, compute the principal and interest paid and the balance still owed. See a sample computation below.

Month Ending	Monthly Payment	Interest Paid	Principal Paid	Outstanding Amount Owed	Annual Interest Paid
0				$40 000.00	
1	1 029.44	466.67	562.77	39 437.23	
2	1 029.44	460.10	569.34	38 867.89	
12	1 029.44	390.08	639.36	32 796.07	5 149.36
24	1 029.44	294.60	734.84	24 516.30	4 073.51
36	1 029.44	184.85	844.59	5 000.00	2 836.99

2. What would be the monthly payments if the operator were to purchase the assets on her line of credit (10% interest)? Assume, for simplicity, that the loan is repaid over three years (the term of the lease). Produce a table like the one above to show the total interest payments for each month of the three-year loan.
3. What is the total of the lease payments (the payout) each year? What are the total loan payments each year? Why is the total payout of the loan higher than for the lease?
4. With a capital lease, the small business owner can claim both the interest expense and the capital cost allowance as tax deductions each year. Construct a table summarizing the following for each year of the lease: (a) the total annual lease payments, (b) the total annual interest payments, (c) the UCC balance, (d) the CCA allowance, (e) the tax savings due to lease interest payments, (f) the tax savings due to the CCA

allowance, and finally (g) the after tax cash flows = (a) – (e) – (f). What is the present worth of the annual after-tax cash flows (at 12%)?

5. With an operating lease, the small business owner can deduct the lease payments as expenses. Construct a table showing the after-tax cash flows associated with an operating lease. What is the present worth of the annual after-tax cash flows (at 12%)?
6. With a loan, the small business owner can claim interest expenses and the capital cost allowance as tax deductions every year. On the basis of your approach in Question 4, construct a table that will allow you to compute the present worth of the annual after-tax cash flows (at 10%).
7. At the end of the three years of a loan, the small business owner owns the fixed assets. What is the present worth of the estimated salvage value of the furniture and equipment? What is the overall present worth of the costs associated with the loan?
8. What choice (lease or loan) would you recommend? Why? What factors other than present worth might you consider in your recommendation?

maintenance, insurance, and labour costs by $30 000 per year. The after-tax MARR is 9%, and Salim's corporate tax rate is 55%. What is the exact after-tax IRR for this investment? What is the approximate after-tax IRR for this investment? Should Salim buy the backhoe?

Appendix 8A Deriving the Capital Cost Tax Factors

The change in tax laws of November 13, 1981, has made the formula for the CCTF a little complicated. To derive the CCTF formula, it is easiest to look at the situation before the laws were changed.

Before November 13, 1981, the tax benefit that could be obtained for a depreciable asset with a CCA rate d and a first cost P, when the company was paying tax at rate t is

Ptd for the first year

$Ptd(1 - d)$ for the second year

$Ptd(1 - d)^{N-1}$ for the Nth year

Taking the present worth of each of these benefits and summing gives

$$\text{PW(benefits)} = Ptd\left(\frac{1}{(1+i)} + \frac{(1-d)}{(1+i)^2} + \cdots \frac{(1-d)^{N-1}}{(1+i)^N} + \cdots\right)$$

$$= \frac{Ptd}{(1+i)}\left(1 + \frac{(1-d)}{(1+i)} + \frac{(1-d)^2}{(1+i)^2} + \cdots + \frac{(1-d)^N}{(1+i)^N} + \cdots\right)$$

Noting that for $q < 1$

$$\lim_{n \to \infty} (1 + q + q^2 + \cdots + q^n) = \frac{1}{1-q}$$

and

$$\frac{1-d}{1+i} < 1$$

then

$$\text{PW(benefits)} = \frac{Ptd}{1+i}\left[\frac{1}{1-\dfrac{(1-d)}{(1+i)}}\right]$$

$$= \frac{Ptd}{(1+i)}\left[\frac{1}{\dfrac{(1+i)}{(1+i)}-\dfrac{(1-d)}{(1+i)}}\right]$$

$$= \frac{Ptd}{(1+i)}\left(\frac{(1+i)}{(i+d)}\right)$$

$$= \frac{Ptd}{(i+d)}$$

If we subtract the present worth of the tax benefits from the first cost, it will give us the present worth of the asset, taking into account all tax benefits from depreciation forever.

$$\text{PW(asset)} = P - \frac{Ptd}{i+d}$$

$$= P\left(1 - \frac{td}{i+d}\right)$$

The factor $1 - \dfrac{td}{(i+d)}$ is called the *old* capital cost tax factor (CCTF_{old}), and was the formula in use before November 13, 1981.

The new tax rules imply that since November 13, 1981, only half of the first cost of an asset can be used in the CCA calculations for the first year. By recognizing that the net effect of the new law is to delay the tax benefits of half of the first cost by one year, the present worth of the benefits is then

$$\text{PW(benefits)} = 0.5\,\frac{Ptd}{i+d} + 0.5\left(\frac{Ptd}{1+d}\right)\left(\frac{1}{1+i}\right)$$

$$= 0.5\,\frac{Ptd}{i+d}\left(1 + \frac{1}{1+i}\right)$$

$$= 0.5\,\frac{Ptd}{i+d}\left(\frac{1+i}{1+i} + \frac{1}{1+i}\right)$$

$$= 0.5\,\frac{Ptd}{i+d}\left(\frac{2+i}{1+i}\right)$$

$$= P\,\frac{td\left(1+\dfrac{i}{2}\right)}{(i+d)(1+i)}$$

And the present worth of the asset itself is

$$\text{PW}(asset) = P - P\frac{td\left(\frac{1+i}{2}\right)}{(i+d)(1+i)}$$

$$= P\left[1 - \frac{td\left(1+\frac{i}{2}\right)}{(i+d)(1+i)}\right]$$

Thus the CCTF_{new} is

$$\text{CCTF}_{new} = 1 - \frac{td\left(1+\frac{i}{2}\right)}{(i+d)(1+i)}$$

CHAPTER 9

Inflation

Engineering Economics in Action, Part 9A: The Inflated Expert

Terry left Canadian Widgets to go back for his final term of school. Naomi and Terry had worked through several projects over Terry's last few work terms, and Naomi had been increasingly taking part in projects involving sister companies of Canadian Widgets; all were owned by Canadian Conglomerate Inc., often referred to as "head office."

"There's a guy from head office to see you, Naomi." It was Carole announcing the expected visitor, Bill Astad. Bill was one of the company troubleshooters. His current interest concerned a sister company, Mexifab, a maquiladora on the Mexican border with Texas. (A maquiladora is an assembly plant that manufactures finished goods in northern Mexico under special tariff and tax rules.) After a few minutes of socializing, Bill explained the concern.

"It's the variability in the Mexican inflation rate that causes the problems. Mexico gets a new president every six years, and usually, about the time the president changes, the economy goes out of whack. And we can't price everything to U.S. or Canadian dollars. We do some of that, but we are located in Mexico and so we have to use Mexican money for a lot of our transactions.

"I understand from Anna Kulkowski that you know something about how to treat problems like that," Bill continued.

Naomi smiled to herself. She had written a memo a few weeks earlier pointing out how Canadian Widgets had been missing some good projects by failing to take advantage of the current very low inflation rates, and suddenly she was the expert!

"Well," she said, "I might be able to help. What you can do is this."

9.1 Introduction

Prices of goods and services bought and sold by individuals and firms change over time. Some prices, like those of agricultural commodities, may change several times a day. Other prices, like those for sugar and paper, change infrequently. While prices for some consumer goods and services occasionally decrease (as with high-tech products), on average it is more typical for prices to increase over time. In fact, on a yearly basis, average prices of consumer goods and services in Canada have risen in every year but one since 1940.

Inflation is the increase, over time, in average prices of goods and services. It can also be described as a decrease in the purchasing power of money over time. While Canada has experienced inflation in most years since World War II, there have been short periods when average prices in Canada have fallen. A decrease, over time, in average prices is called **deflation**. It can also be viewed as an increase in the purchasing power of money over time.

Because of inflation or deflation, prices are likely to change over the lives of most engineering projects. These changes will affect cash flows associated with the projects. Engineers may have to take predicted price changes into account during project evaluation to prevent the changes from distorting decisions.

In this chapter, we shall discuss how to incorporate an expectation of inflation into project evaluation. We focus on inflation because it has been the dominant pattern of price changes since the beginning of the twentieth century. The chapter begins with a discussion of how inflation is measured. We then show how to convert cash flows that occur at different points in time into dollars with the same purchasing power. We then consider how inflation affects the MARR, the internal rate of return, and the present worth of a project.

9.2 Measuring the Inflation Rate

The **inflation rate** is the rate of increase in average prices of goods and services over a specified time period, usually a year. If prices of all goods and services moved up and down together, determining the inflation rate would be trivial. If all prices increased by 2% over a year, it would be clear that the average inflation rate would also be 2%. But prices do not move in perfect synchronization. In any period, some prices will increase, others will fall, and some will remain about the same. For example, candy bars are about ten times as costly now as they were in the 1960s, but standard television sets are about the same price or cheaper.

Because prices do not move in perfect synchronization, a variety of methods have been developed to measure the inflation rate. Statistics Canada tracks movement of average prices for a number of different collections of goods and services and calculates inflation rates from the changes in prices in these collections over time.

NET VALUE 9.1

Statistics Canada

A prime source of Canadian statistics, including CPI and inflation rates, is Statistics Canada. The latest statistics are readily available from Statistics Canada (www.statcan.ca). Four main categories for Canadian statistics are the economy, land, people, and state. Each of these categories includes information that can be useful in engineering economics. For example, the "economy" category includes the latest economic indicators such as gross domestic product (GDP) and personal spending on consumer goods and services, and manufacturing and construction figures such as new housing price index, energy supply and demand, and the number of manufacturing shipments. The "land" category includes information on environment (air pollution, expenditures on environmental protection by industry, etc.), which may be useful in estimating the impact of environmental factors in engineering projects. In the "people" category, statistics on population, labour, and employment can be found. Lastly, in the "state" category, statistics specific to the public sector (e.g., employment, assets and liabilities, and revenue and expenditures) can be found.

One set of prices tracked by Statistics Canada consists of goods and services bought by Canadian consumers. This set forms the basis of the **consumer price index (CPI)**. The CPI for a given period relates the average price of a fixed "basket" of these goods in the given period to the average price of the same basket in a *base period*. The current CPI has 1992 as the base year. The base year index is set at 100. The index for any other year indicates the number of dollars needed in that year to buy the fixed basket of goods that cost $100 in 1992. Figure 9.1 shows the CPI for the period 1961 to 2003. It shows that the basket of goods, which cost $100 in 1992 (the base year), would have cost about $20 in 1961 and over $120 in 2003. More information about the CPI can be found at the Statistics Canada site (www.statcan.ca).

A national inflation rate can be estimated directly from the CPI by expressing the changes in the CPI as a year-by-year percentage change. This is probably the most commonly used estimate of a national inflation rate. Figure 9.2 shows the national inflation rate for the period from 1961 to 2003 as derived from the CPI quantities in Figure 9.1.

Figure 9.1 Canadian CPI 1961–2003

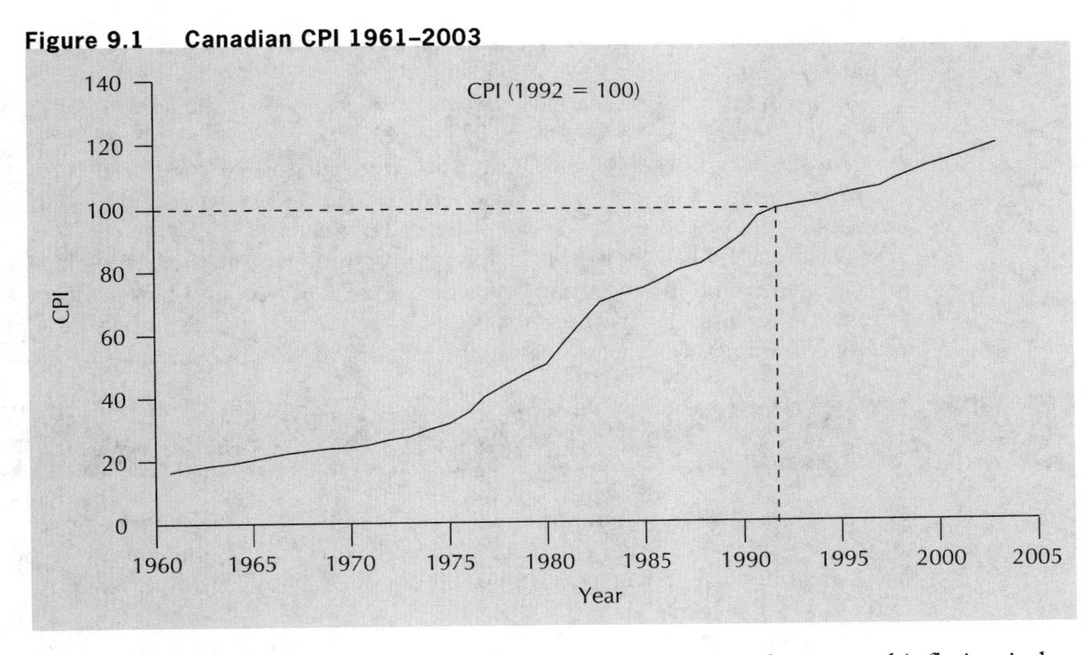

It is important to note that, although the CPI is a commonly accepted inflation index, many different indexes are used to measure inflation. The value of an index depends on the method used to compute the index and the set of goods and services for which the index measures price changes. To judge whether an index is appropriate for a particular purpose, the analyst should know how the goods and services for which he or she is estimating inflation compare with the set of goods and services used to compute the index.

Figure 9.2 Canadian Inflation Rate 1961–2003

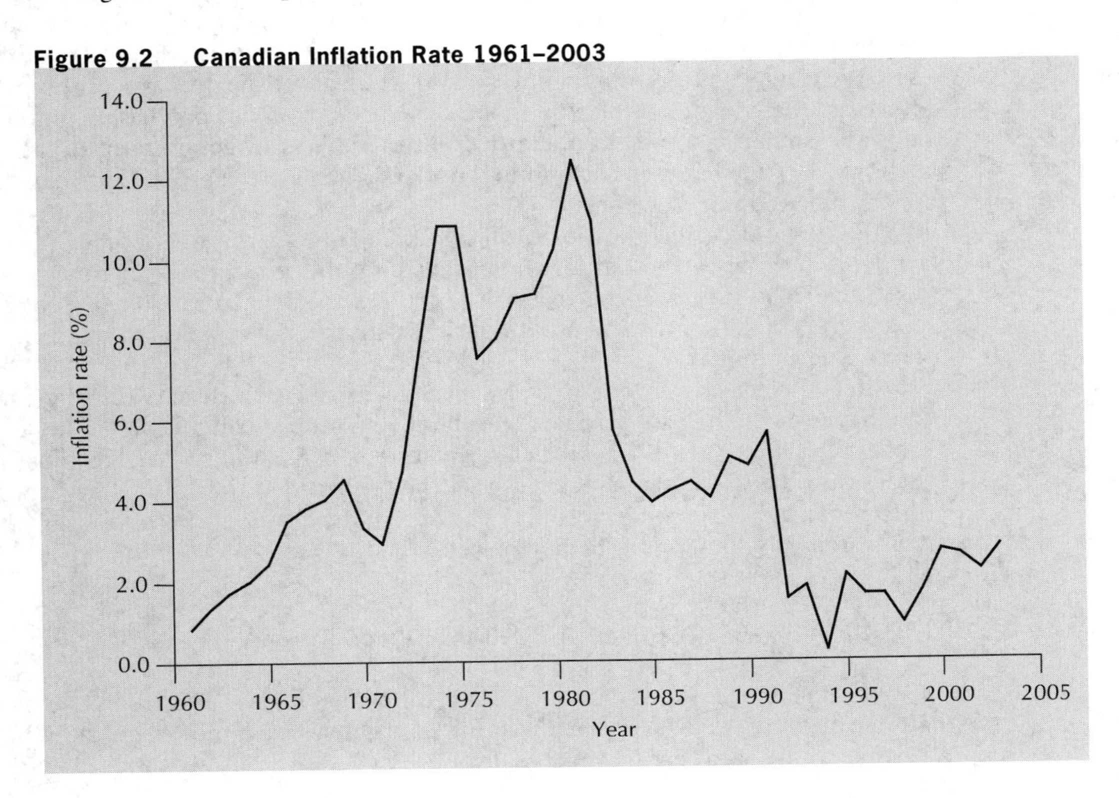

For this reason, we provide Appendix 9A, in which we illustrate the computation of one popularly used index.

Inflation rates in Canada have varied considerably over the last 40 years from a high of 12.4% in 1981 to a low of 0.2% in 1994 (see Figure 9.2). Low expected rates of inflation may be safely ignored, given the typical imprecision of predicted future cash flows. However, when expected inflation is high, it is necessary to include inflation in detailed economic calculations to avoid rejecting good projects.

Throughout the rest of this chapter, we assume that an analyst is able to obtain estimates for expected inflation rates over the life of a project and that project cash flows will change at the same rate as average prices. Consequently, the cash flows for a project can be assumed to increase at approximately the rate of inflation per year.

9.3 Economic Evaluation with Inflation

When prices change, the amount of goods a dollar will buy changes too. If prices fall, more goods may be bought for a given number of dollars, and the value of a dollar has risen. If prices rise, fewer goods may be bought for a given number of dollars, and the value of a dollar has fallen.

In project evaluation, we cannot make comparisons of dollar values across time without taking the price changes into account. We want dollars, not for themselves, but for what we can get for them. Workers are not directly interested in the money wages they will earn in a job. They are interested in how many hours of work it will take to cover expenses for their families, or how long it will take them to accumulate enough to make down payments on houses. Similarly, investors want to know if making an investment now will enable them to buy more real goods in the future, and by how much the amount they can buy in the future will increase. To know if an investment will lead to an increase in the amount they can buy in the future, they must take into account expected price changes.

We can take price changes into account in an approximate way by measuring the cash flows associated with a project in monetary units of constant purchasing power called **real dollars** (sometimes called **constant dollars**). This is in contrast to the **actual dollars** (sometimes called **current** or **nominal dollars**), which are expressed in the monetary units at the time the cash flows occur.

For example, if a photocopier will cost $2200 one year from now, the $2200 represents *actual* dollars since that is the amount that would be paid at that time. If inflation is expected to be 10% over the year, the $2200 is equivalent to $2000 real dollars. Although the term "dollar" is used by convention when speaking of inflation, the principles apply to any monetary unit.

Real dollars always need to be associated with a particular date (usually a year), called the **base year**. The base year need not be the present; it could be any time. People speak of "2000 dollars" or "1985 dollars" to indicate that real dollars are being used as well as indicating the base year associated with them. Provided that cash flows which occur at different times are converted into real dollars with the same base year, they can be compared fairly in terms of buying power.

9.3.1 Converting Between Real and Actual Dollars

Converting actual dollars in year N into real dollars in year N relative to a base year 0 is straightforward, provided that the value of a global price index like the CPI at year N relative to the base year is available. Let

A_N = actual dollars in year N

$R_{0,N}$ = real dollars equivalent to A_N relative to year 0, the base year

$I_{0,N}$ = the value of a global price index (like the CPI) at year N, relative to year 0

Then the conversion from actual to real dollars is

$$R_{0,N} = \frac{A_N}{I_{0,N}/100} \tag{9.1}$$

Note that in Equation (9.1), $I_{0,N}$ is divided by 100 to convert it into a fraction because of the convention that a price index is set at 100 for the base year.

Transforming actual dollar values into real dollars gives only an approximate offset to the effect of inflation. The reason is that there may be no readily available price index that accurately matches the "basket" of goods and services being evaluated. Despite the fact that available price indexes are approximate, they do provide a reasonable means of converting actual cash flows to real cash flows relative to a base year.

An alternative means of converting actual dollars to real dollars is available if we have an estimate for the average yearly inflation between now (year 0) and year N. Let

A_N = actual dollars in year N

$R_{0,N}$ = real dollars equivalent to A_N relative to year 0, the base year

f = the inflation rate per year, assumed to be constant from year 0 to year N

Then the conversion from actual dollars in year N to real dollars in year N relative to the base year 0 is

$$R_{0,N} = \frac{A_N}{(1+f)^N}$$

When the base year is omitted from the notation for real dollars, it is understood that the current year (year 0) is the base year, as in

$$R_N = \frac{A_N}{(1+f)^N} \tag{9.2}$$

Equation (9.2) can also conveniently be written in terms of the present worth compound interest factor

$$R_N = A_N(P/F,f,N) \tag{9.3}$$

Note that here A_N is the actual dollar amount in year N, that is, a future value. It should not be confused with an annuity A.

Example 9.1

Elliot Weisgerber's income rose from $40 000 per year in 2000 to $42 000 in 2003. At the same time the CPI rose from 113.5 in 2000 to 122.3 in 2003. Was Elliot worse off or better off in 2003 compared with 2000?

We can convert Elliot's actual dollar income in 2000 and 2003 into real dollars. This will tell us if his total purchasing power increased or decreased over the period from 2000 to 2003. Since the current base year for the CPI is 1992, we shall compare his 2000 and 2003 incomes in terms of real 1992 dollars.

His real incomes in 2000 and 2003 in terms of 1992 dollars, using Equation (9.1), were

$$R_{92,00} = \$40\,000/1.135 = \$35\,242$$

$$R_{92,03} = \$42\,000/1.223 = \$34\,342$$

Even though Elliot's actual dollar income rose between 2000 and 2003, his purchasing power fell, since the real dollar value of his income, according to the CPI, fell about 3%.■

Example 9.2

The cost of replacing a storage tank one year from now is expected to be $2 000 000. If inflation is assumed to be 5% per year, what is the cost of replacing the storage tank in real (today) dollars?

First, note that the $2 000 000 is expressed in actual dollars one year from today. The cost of replacing the tank in real (today) dollars can be found by letting

$A_1 = \$2\,000\,000 =$ the actual cost one year from the base year (today)

$R_1 =$ the real dollar cost of the storage tank in one year

$f =$ the inflation rate per year

Then, with Equation (9.2)

$$R_1 = \frac{A_1}{1+f} = \frac{\$2\,000\,000}{1.05} = \$1\,904\,762$$

Alternatively, Equation (9.3) gives

$$R_1 = A_1(P/F, 5\%, 1) = \$2\,000\,000\,(0.9524) = \$1\,904\,762$$

The $2 000 000 actual cost is equivalent to $1 904 762 real (today) dollars at the end of one year.■

Example 9.3

The cost of replacing a storage tank in 15 years is expected to be $2 000 000. If inflation is assumed to be 5% per year, what is the cost of replacing the storage tank 15 years from now in real (today) dollars?

The cost of the tank 15 years from now in real dollars can be found by letting

$A_{15} = \$2\,000\,000 =$ the actual cost 15 years from the base year (today)

$R_{15} =$ the real dollar cost of the storage tank in 15 years

$f =$ the inflation rate per year

Then, with the use of Equation (9.2), we have

$$R_{15} = \frac{A_{15}}{(1+f)^{15}} = \frac{\$2\,000\,000}{(1.05)^{15}} = \$962\,040$$

Alternatively, Equation (9.3) gives

$$R_{15} = A_{15}(P/F, 5\%, 15) = \$2\,000\,000\,(0.48102) = \$962\,040$$

In 15 years, the storage tank will cost \$962 040 in real (today) dollars. Note that this \$962 040 is money to be paid 15 years from now. What this means is that the new storage tank can be replaced at a cost that would have the same purchasing power as about \$962 040 today.■

Now that we have the ability to convert from actual to real dollars using an index or an inflation rate, we turn to the question of how inflation affects project evaluation.

9.4 The Effect of Correctly Anticipated Inflation

The main observation made in this section is that engineers must be aware of potential changes in price levels over the life of a project. We shall see that when future inflation is expected over the life of a project, the MARR needs to be increased. Engineers need to recognize this effect of inflation on the MARR to avoid rejecting good projects.

9.4.1 The Effect of Inflation on the MARR

If we expect inflation, the number of actual dollars that will be returned in the future does not tell us the value, in terms of purchasing power, of the future cash flow. The purchasing power of the earnings from an investment depends on the *real* dollar value of those earnings.

The **actual interest rate** is the stated or observed interest rate based on actual dollars. If we wish to earn interest at the actual interest rate, i, on a one-year investment, and we invest \$$M$, the investment will yield \$$M(1 + i)$ at the end of the year. If the inflation rate over the next year is f, the real value of our cash flow is \$$M(1 + i)/(1 + f)$. We can use this to define the *real* interest rate. The **real interest rate**, i', is the interest rate that would yield the same number of real dollars in the absence of inflation as the actual interest rate yields in the presence of inflation.

$$M(1 + i') = M\left(\frac{1 + i}{1 + f}\right)$$

$$i' = \frac{1 + i}{1 + f} - 1 \qquad (9.4)$$

We may see terms like "real rate of return" or "real discount rate." These are just special cases of the real interest rate.

The definition of the real interest rate can be turned around by asking the following question: If an investor wants a real rate of return, i', over the next year, and the inflation rate is expected to be f, what actual interest rate, i, must be realized to get a real rate of return of i'?

The answer can be obtained with some manipulation of the definition of the real interest rate in Equation (9.4):

$$i = (1 + i')(1 + f) - 1 \text{ or, equivalently, } i = i' + f + i'f \qquad (9.5)$$

Therefore, an investor who desires a real rate of return i' and who expects inflation at a rate of f will require an actual interest rate $i = i' + f + i'f$. This has implications for the actual MARR used in economic analyses. The **actual MARR** is the minimum acceptable rate of return when cash flows are expressed in actual dollars. If investors expect inflation, they require higher actual rates of return on their investments than if inflation were not expected. The actual MARR then will be the real MARR plus an upward adjustment that reflects the effect of inflation. The **real MARR** is the minimum acceptable rate of return when cash flows are expressed in real, or constant, dollars.

If we denote the actual MARR by $MARR_A$ and the real MARR by $MARR_R$, we have from Equation (9.5)

$$MARR_A = MARR_R + f + MARR_R \times f \qquad (9.6)$$

Note that if $MARR_R$ and f are small, the term $MARR_R \times f$ may be ignored and $MARR_A = MARR_R + f$ can be used as a "back of the envelope" approximation.

The real MARR can also be expressed as a function of the actual MARR and the expected inflation rate:

$$MARR_R = \frac{1 + MARR_A}{1 + f} - 1 \qquad (9.7)$$

Figure 9.3 shows the Canadian experience with inflation, the actual prime interest rate, and the real interest rate for the 1961–2003 period. From 1961 to 1971, when inflation was moderate and stable, the real interest rate was also stable, except for one blip in 1967. In the 1970s, conditions were very different when inflation exploded. This was due partly to large jumps in energy prices. Real interest rates were negative for the period 1972 to 1975 and 1977 to 1978. In the 1980s and early 1990s, real interest rates were quite high. The rest of the 1990s and early 2000s experienced lower inflation rates and real interest rates.

Figure 9.3 Canadian Inflation Rate and Actual and Real Interest Rates 1961–2003

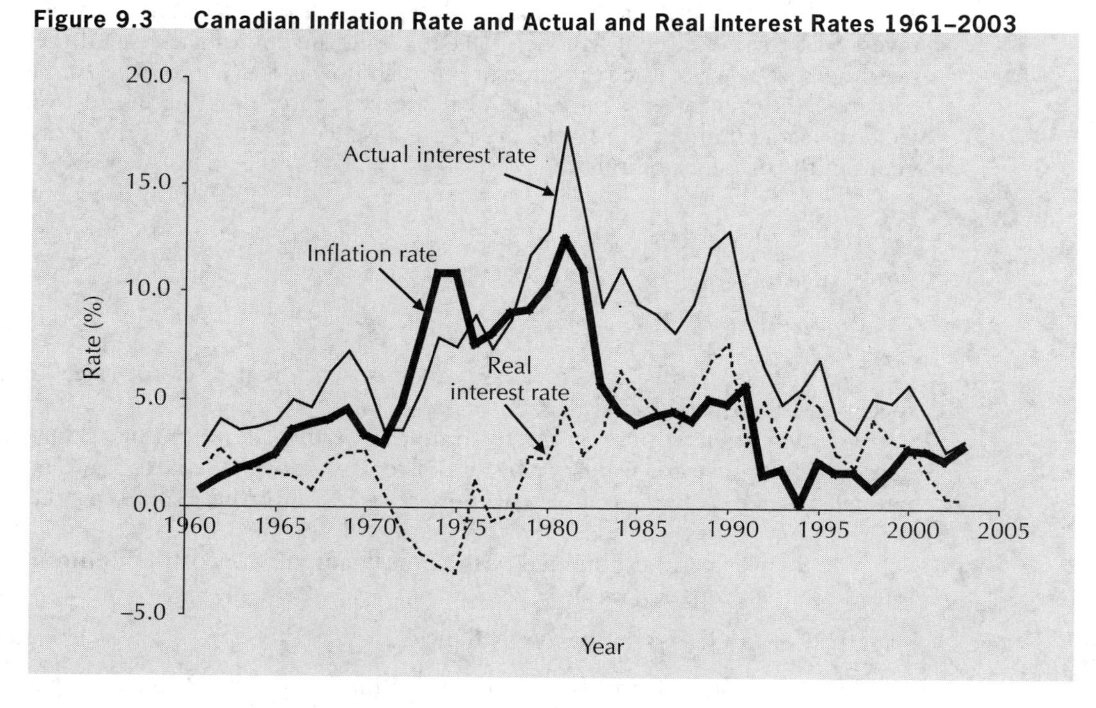

The high inflation rates of the 1970s were very unusual. Inflation in the range of 2% to 4% per year is more typical of Canadian experience. Averages of real interest rates and actual inflation rates over subperiods are shown in Table 9.1.

Table 9.1 Average Canadian Real and Actual Interest Rates

Period	Average Real Interest Rate (%)	Average Actual Interest Rate (%)
1961–1971	1.82	2.75
1972–1981	0.08	8.47
1982–1991	4.84	5.29
1992–2003	2.95	1.81

Example 9.4

Security Trust is paying 12% on one-year guaranteed investment certificates (GICs). The inflation rate is expected to be 5% over the next year. What is the real rate of interest? For a $5000 GIC, what will be the real dollar value of the amount received at the end of the year?

The real interest rate is

$$i' = \frac{1+i}{1+f} - 1 = \frac{1.12}{1.05} - 1 = 0.067, \text{ or } 6.7\%$$

A $5000 GIC will return $5600 at the end of the year. The real value of the $5600 in *today's* dollars is $5600/1.05 = $5333. This is the same as if there were no inflation and the investment earned 6.7% interest.■

Example 9.5

Susan got a $1000 present from her aunt on her 16th birthday. She has noticed that Security Trust offers 6.5% on one-year guaranteed investment certificates (GICs). Her mother's *National Post* indicates that analysts are predicting an inflation rate of about 3.5% for the coming year. Susan's real MARR for such investments is 4%. If the analysts are correct, what is Susan's actual MARR? Should she invest?

If the analysts are correct, Susan's actual MARR is

$$\begin{aligned} \text{MARR}_\text{A} &= \text{MARR}_\text{R} + f + \text{MARR}_\text{R} \times f \\ &= 0.04 + 0.035 + (0.04)(0.035) \\ &= 0.0764 \end{aligned}$$

Susan's actual MARR is about 7.64%. Since the actual interest rate on the GIC is only 6.5%, she should not invest in the GIC.■

9.4.2 The Effect of Inflation on the IRR

The effect of expected inflation on the **actual internal rate of return** of a project is similar to the effect of inflation on the actual MARR. Suppose that we are considering a T-year investment. The **actual internal rate of return** on the project, IRR_A, is the rate of return of the project on the basis of actual dollar cash flows. It can be found by solving for i^* in

$$\sum_{t=0}^{T} \frac{A_t}{(1+i^*)^t} = 0$$

where

A_t = the actual cash flow in period t (receipts – disbursements)
T = the number of time periods
i^* = the actual internal rate of return

Suppose further that a yearly inflation rate of f is expected over the T-year life of the project. In terms of real dollars (with a base year of the time of the first cost), the actual cash flow in period t can be written as $A_t = R_t(1 + f)^t$ where R_t refers to the *real* dollar amount equivalent to the cash flow A_t. The expression which gives the actual internal rate of return can be rewritten as

$$\sum_{t=0}^{T} \frac{R_t(1 + f)^t}{(1 + i^*)^t} = 0 \tag{9.8}$$

In contrast, the **real internal rate of return** for the project, IRR_R, is the rate of return obtained on the real dollar cash flows associated with the project. It is the solution for i' in

$$\sum_{t=0}^{T} \frac{R_t}{(1 + i')^t} = 0 \tag{9.9}$$

What is the relationship between IRR_R and IRR_A? We have from Equations (9.4) and (9.5)

$$\frac{1}{1 + i'} = \frac{1 + f}{1 + i} \quad \text{or} \quad i = i' + f + i'f$$

and thus, analogous to Equation (9.5),

$$IRR_A = IRR_R + f + IRR_R \times f \tag{9.10}$$

Or, analogous to Equation (9.4), the real IRR can be expressed in terms of the actual IRR and the inflation rate:

$$IRR_R = \frac{1 + IRR_A}{1 + f} - 1 \tag{9.11}$$

In summary, the effect of inflation on the IRR is that the actual IRR will be the real IRR plus an upward adjustment that reflects the effect of inflation.

Example 9.6

Consider a two-year project that has a \$10 000 first cost and that is expected to bring about a saving of \$15 000 at the end of the two years. If inflation is expected to be 5% per year and the real MARR is 13%, should the project be undertaken? Base your answer on an IRR analysis.

From the information given, $A_0 = -\$10\ 000$, $A_2 = \$15\ 000$, and $f = 0.05$. The actual IRR can be found by solving for i in

$$A_0 + \frac{A_2}{(1 + i)^2} = 0$$

$$\$10\ 000 = \$15\ 000/(1 + i)^2$$

which leads to an actual IRR of 22.475%.

The real IRR is then

$$\text{IRR}_\text{R} = \frac{1 + \text{IRR}_\text{A}}{1 + f} - 1$$

$$= \frac{1 + 0.22475}{1 + 0.05} - 1$$

$$= 0.1664 \text{ or } 16.64\%$$

Since the real IRR exceeds the real MARR, the project should be undertaken.■

In conclusion, the impact of inflation on the actual MARR and the actual IRR is that both have an adjustment for expected inflation implicitly included in them. The main implication of this observation is that, since both the actual MARR and the actual IRR increase in the same fashion, any project that was acceptable without inflation remains acceptable when inflation is expected. Any project that was unacceptable remains unacceptable.

9.5 Project Evaluation Methods with Inflation

The engineer typically starts a project evaluation with an observed (actual) MARR and projections of cash flows. As we have seen, the actual MARR has two parts, the real rate of return on investment that investors require to put money into the company, plus an adjustment for the expected rate of inflation. The engineer usually observes only the sum and not the individual parts.

As for the projected cash flows, these are typically based on current prices. Because the projected cash flows are based on prices of the period in which evaluations are being carried out, they are in *real* dollars. They do not incorporate the effect of inflation. In this case, the challenge for the engineer is to correctly analyze the project when he or she has an *actual* MARR (which incorporates inflation implicitly) and *real* cash flows (which do not take inflation into account).

Though it is common to do so, the engineer or analyst does not always start out with an *actual* MARR and *real* cash flows. The cash flows may already have inflation implicitly factored in (in which case the cash flows are said to be actual amounts). To carry out a project evaluation properly, the analyst must know whether inflation has been accounted for already in the MARR and the cash flows or whether it needs to be dealt with explicitly.

As a brief example, consider a one-year project that requires an investment of \$1000 today and that yields \$1200 in one year. The actual MARR is 25%. Whether the project is considered acceptable will depend on whether the \$1200 in one year is understood to be the actual cash flow in one year or if it is the real value of the cash flow received in one year.

If the \$1200 is taken to be the actual cash flow, the actual internal rate of return, IRR_A, is found by solving for i^* in

$$-\$1000 + \frac{\$1200}{(1 + i^*)} = 0$$

$$i^* = \text{IRR}_\text{A} = 20\%$$

Hence, the project would not be considered economical. However, if the \$1200 is taken to be the real value of the cash flow in one year, and inflation is expected to be 5% over the year, then the actual internal rate of return is found by solving for i^* in:

$$-\$1000 + \frac{\$1200\,(1 + 0.05)}{(1 + i^*)} = 0$$

$$i^* = \text{IRR}_\text{A} = 26\%$$

Hence, the project *would* be considered acceptable. As seen in this example, the economic viability of the project may depend on whether the $1200 in one year has inflation implicitly factored in (i.e., is taken to be the actual amount). This is why it is important to know what type of cash flows you are dealing with.

If the engineer has an estimate of inflation, there are two equivalent ways to carry out a project evaluation properly. The first is to work with actual values for cash flows and actual interest rates. The second is to work with real values for cash flows and real interest rates. *The two methods should not be mixed.*

These two methods of dealing with expected inflation, as well as two incorrect methods, are shown in Table 9.2.

Table 9.2 Methods of Incorporating Inflation into Project Evaluation

1. Real MARR and Real Cash Flows
The real MARR does not include the effect of expected inflation. Cash flows are determined by today's prices. **Correct**
2. Actual MARR and Actual Cash Flows
The actual MARR includes the effect of anticipated inflation. Cash flows include increases due to inflation. **Correct**
3. Actual MARR and Real Cash Flows
The actual MARR includes the effect of anticipated inflation. Cash flows are determined by today's prices. **Incorrect:** Biased against investments
4. Real MARR and Actual Cash Flows
The real MARR does not include the effect of expected inflation. Cash flows include increases due to inflation. **Incorrect:** Biased in favour of investments

The engineer must have a forecast of the inflation rate over the life of the project in order to adjust the MARR or cash flows for inflation. The best source of such forecasts may be the estimates of experts. Financial publications like the *Report on Business* of the *Globe and Mail* regularly report such predictions for relatively short periods of up to one year. Because there is evidence that even the short-term estimates are not totally reliable, and estimates for longer periods will necessarily be imprecise, it is good practice to determine a range of possible inflation values for both long- and short-term projects. The engineer should test for sensitivity of the decision to values in the range. The subject of sensitivity analysis is addressed more fully in Chapter 11. Close-Up 9.1 discusses atypical patterns of price changes that may be specific to certain industries.

CLOSE-UP 9.1 **Relative Price Changes**

Engineers usually expect prices associated with a project to move with the general inflation rate. However, there are situations in which it makes sense to expect prices associated with a project to move differently from the average. This can happen when there are atypical forces affecting either the supply or the demand for the goods.

For example, reductions in the availability of logs in North America caused a decrease in the supply of wood for construction, furniture, and pulp and paper in the 1980s. Average wood prices in Canada more than doubled between 1986 and 1995. This was about twice the increase in the CPI over that period. Since then, average wood prices have been decreasing while the CPI continues to increase. Another example is the price of computers. Product development and increases in productivity have led to increases in the supply of computers. This, in turn, has led to reductions in the relative price of computing power.

Changes in the relative prices of the goods sold by a specific industry will generally not have a noticeable effect on a MARR because investors are concerned with the overall purchasing power of the dollars they receive from an investment. Changes in the relative prices of the goods of one industry will not have much effect on investors' abilities to buy what they want.

Because the relative price changes will not affect the MARR, the analyst must incorporate expected relative price changes directly into the expected cash flows associated with a project. If the rate of relative price change is expected to be constant over the life of the project, this can be done using a geometric gradient to present worth conversion factor.

Example 9.7

Jagdeep can put his money into an investment that will pay him \$1000 a year for the next four years and \$10 000 at the end of the fifth year. Inflation is expected to be 5% over the next five years. Jagdeep's real MARR is 8%. What is the present worth of this investment?

The present worth may be obtained with real dollar cash flows and a real MARR or with actual dollar cash flows and an actual MARR.

The first solution approach will be to use real dollars and MARR_R. The real dollar cash flows in terms of *today's* dollars are

$$R_1, R_2, R_3, R_4, R_5 = \frac{A_1}{(1+f)}, \frac{A_2}{(1+f)^2}, \frac{A_3}{(1+f)^3}, \frac{A_4}{(1+f)^4}, \frac{A_5}{(1+f)^5}$$

$$= \frac{\$1000}{(1.05)}, \frac{\$1000}{(1.05)^2}, \frac{\$1000}{(1.05)^3}, \frac{\$1000}{(1.05)^4}, \frac{\$10\ 000}{(1.05)^5}$$

The present worth of the real cash flows, discounted by $\text{MARR}_R = 8\%$, is

$$\text{PW} = \frac{\$1000}{(1.05)(1.08)} + \frac{\$1000}{(1.05)^2(1.08)^2} + \frac{\$1000}{(1.05)^3(1.08)^3}$$

$$+ \frac{\$1000}{(1.05)^4(1.08)^4} + \frac{\$10\ 000}{(1.05)^5(1.08)^5}$$

$$\cong \$8282$$

The present worth of Jagdeep's investment is about \$8282.

Alternatively, the present worth can be found in terms of actual dollars and MARR_A:

$$\text{PW} = \frac{\$1000}{(1 + \text{MARR}_A)} + \frac{\$1000}{(1 + \text{MARR}_A)^2} + \frac{\$1000}{(1 + \text{MARR}_A)^3}$$

$$+ \frac{\$1000}{(1 + \text{MARR}_A)^4} + \frac{\$10\ 000}{(1 + \text{MARR}_A)^5}$$

where

$$\text{MARR}_A = \text{MARR}_R + f + \text{MARR}_R \times f$$

Note that this is the sum of a four-period annuity with equal payments of \$1000 for four years and a single payment of \$10 000 in period 5. With this observation, the present worth computation can be simplified by the use of compound interest formulas:

$$\text{PW} = \$1000(P/A, \text{MARR}_A, 4) + \$10\ 000(P/F, \text{MARR}_A, 5)$$

With a real MARR of 8% and an inflation rate of 5%, the actual MARR is then

$$\begin{aligned} \text{MARR}_A &= \text{MARR}_R + f + \text{MARR}_R \times f \\ &= 0.08 + 0.05 + (0.08)(0.05) \\ &= 0.134 \end{aligned}$$

and the present worth of Jagdeep's investment is

$$\begin{aligned} \text{PW} &= \$1000(P/A, 13.4\%, 4) + \$10\ 000(P/F, 13.4\%, 5) \\ &\cong \$8282 \end{aligned}$$

The present worth of Jagdeep's investment is about \$8282, as was obtained through the use of the real MARR and a conversion from actual to real dollars.■

Though there are two distinct means of correctly adjusting for inflation in project analysis, the norm for engineering analysis in Canada is to make comparisons with the actual MARR. One reason this is done has to do with how a MARR is chosen. As discussed in Chapter 3, the MARR is based on, among other things, the cost of capital. Since lenders and investors recognize the need to have a return on their investments higher than the expected inflation rate, they will lend to or invest in companies only at a rate that exceeds the inflation rate. In other words, the cost of capital already has inflation included. A MARR based on this cost of capital already includes, to some extent, inflation.

Consequently, if inflation is fairly static (even if it is high), an *actual* dollar MARR is sensible and will arise naturally. On the other hand, if changes in inflation are foreseen, or if sensitivity analysis specifically for inflation is desired, it may be wise to set a *real* dollar MARR and recognize an inflation rate explicitly in the analysis.

Example 9.8

Lethbridge Communications is considering an investment in plastic molding equipment for its product casings. The project involves \$150 000 in first costs and is expected to generate net savings (in actual dollars) of \$65 000 per year over its three-year life. They forecast an inflation rate of 15% over the next year, and then inflation of 10% thereafter. Their real dollar MARR is 5%. Should this project be accepted on the basis of an IRR analysis?

In this problem, the inflation rate is not constant over the life of the project, so it is easiest to consider the cash flows for each year separately and to work in real dollars. First, as shown in Table 9.3, the actual cash flows are converted into real cash flows.

Table 9.3 Converting from Actual to Real Dollars for Lethbridge Communications

Year	Actual Dollars	Real Dollars	
0	−\$150 000	−\$150 000	
1	65 000	56 522	= \$65 000(P/F, 15%, 1) = \$65 000(0.86957)
2	65 000	51 384	= \$65 000(P/F, 15%, 1)(P/F, 10%, 1) = \$65 000(0.86957)(0.90909)
3	65 000	46 713	= \$65 000(P/F, 15%, 1)(P/F, 10%, 2) = \$65 000(0.86957)(0.82645)

Then, the real IRR can be found by solving for i' in

$$\$56\,522(P/F, i', 1) + \$51\,384(P/F, i', 2) + \$46\,713(P/F, i', 3) = \$150\,000$$

At $i' = 1\%$, LHS (left-hand side) = \$151 673

At $i' = 2\%$, LHS = \$148 821

The real IRR is between 1% and 2%. This is less than the real dollar MARR of 5%, so the project should not be undertaken.■

Example 9.9

New Glasgow Resources (NGR) has been offered a contract to sell land to the government at the end of 20 years. The contract states that NGR will get \$500 000 after 20 years from today, with no costs or benefits in the intervening years. A financial analyst for the firm believes that the inflation rate will be 4% for the next two years, rise to 15% for the succeeding 10 years, and then go down to 10%, where it will stay forever. NGR's real dollar MARR is 10%. What is the present worth of the contract?

In this case, it is easiest to proceed by calculating the actual dollar MARR for each of the different inflation periods:

$$\text{MARR}_A\text{, years 13 to 20} = 0.10 + 0.10 + (0.10)(0.10) = 0.21 \text{ or } 21\%$$

$$\text{MARR}_A\text{, years 3 to 12} = 0.10 + 0.15 + (0.10)(0.15) = 0.265 \text{ or } 26.5\%$$

$$\text{MARR}_A\text{, years 0 to 2} = 0.10 + 0.04 + (0.10)(0.04) = 0.144 \text{ or } 14.4\%$$

With the individual MARRs, the present worth of the \$500 000 for each of years 12, 2, and 0 can be found:

$$\text{PW(year 12)} = \$500\,000(P/F, 21\%, 8) = \$500\,000 \times 1/(1.21)^8 = \$108\,815$$

$$\text{PW(year 2)} = \$108\,815\ (P/F, 26.5\%, 10) = \$108\,815 \times 1/(1.265)^{10} = \$10\,370$$

$$\text{PW(year 0)} = \$10\,370\ (P/F, 14.4\%, 2) = \$10\,370 \times 1/(1.144)^2 = \$7924$$

The present worth of the contract is approximately $7924.■

Example 9.10

Bildmet is an extruder of aluminum shapes used in construction. They are experiencing a high scrap rate of 5%. The manager, Greta Kehl, estimates that reprocessing scrap costs about $0.30 per kilogram. The high scrap rate is due partly to operator error. Ms. Kehl believes that a short training course for the operator would reduce the scrap rate to about 4%. The course would cost about $1100. Bildmet is now working with a before-tax actual MARR of 22%. Past experience suggests that operators quit their jobs after about five years; the correct time horizon for the retraining project is therefore five years. The data pertaining to the training course are summarized in Table 9.4. Should Bildmet retrain its operator?

Table 9.4 Training Course Data

Output (kilograms/year)	125 000
Scrap (kilograms/year)	6 250
Reprocessing cost ($/kilogram)	0.30
Scrap cost ($/year)	1875
Savings due to training ($/year)	375
First cost of training ($)	1 100
Inflation rate (%/year)	5
Actual MARR (%/year)	22

First, note that the actual MARR i = 22% incorporates an estimate by investors of inflation of f = 5% per year over the next five years. If this estimate of future inflation is correct, Ms. Kehl needs to make an adjustment to take inflation into account. Either the projected annual saving from reduced scrap needs to be increased by the 5% rate of inflation, or she needs to reduce the MARR to its real value. We shall illustrate the first approach with actual cash flows and the actual MARR.

Increasing savings to take inflation into account leads to projected (actual) savings as shown in the Table 9.5.

For example, using Equation (9.2), the expected saving in year 3 is $\$375(1 + f)^3 = \$434.11$. The present worth of the savings in year 3 is $\$434.11/(1 + i)^3 = \$239.07$.

The present worth of the savings over the five-year time frame, when discounted at the actual MARR of 22%, is $1222. This makes the project viable since its cost is $1100.

Table 9.5 Savings Due to the Training Course

Year	Actual Savings	Present Worth
1	\$393.75	\$322.75
2	413.44	277.77
3	434.11	239.07
4	455.81	205.75
5	478.61	177.08

We note that the same result could have been reached by working with the real MARR and the constant cost savings of \$375 per year. MARR_R is given by

$$\text{MARR}_R = 1 + \frac{\text{MARR}_A}{1+f} - 1 = \frac{1.22}{1.05} - 1 = 0.1619$$

The present worth of the real stream of returns, when these are discounted by the real MARR, is given by

$$\text{PW} = \$375(P/A, 0.1619, 5) \cong \$1222$$

which is the same result as the one obtained with the actual MARR and actual cash flows.■

REVIEW PROBLEMS

REVIEW PROBLEM 9.1

Athabaska Engineering was paid \$100 000 to manage a construction project in 1985. How much would the same job have cost in 2005 if the average annual inflation rate between 1985 and 2005 were 5%?

ANSWER

The compound amount factor can be used to calculate the value of 100 000 1985 dollars in 2005 dollars, using the inflation rate as an interest rate:

$$\begin{aligned} \text{2005 dollars} &= \$100\,000(F/P, 5\%, 20) \\ &= \$100\,000(2.6533) \\ &= \$265\,330 \end{aligned}$$

The same job would have cost about \$265 330 in 2005 dollars.■

REVIEW PROBLEM 9.2

A computerized course drop-and-add program is being developed for a local community college. It will cost \$300 000 to develop and is expected to save \$50 000 per year in administrative costs over its 10-year life. If inflation is expected to be 4% per year for the next 10 years and a real MARR of 5% is required, should the project be adopted?

ANSWER

First, assuming that \$50 000 in administrative costs are actual dollars, we can calculate the actual IRR for the project. The actual IRR is the solution for i in

$$\$300\,000 = \$50\,000(P/A, i, 10)$$

$$(P/A, i, 10) = 6$$

$$\text{For } i = 11\%, (P/A, i, 10) = 5.8892$$

$$\text{For } i = 10\%, (P/A, i, 10) = 6.1445$$

The actual IRR of 10.55% is found by interpolating between these two points. We then convert the actual IRR into a real IRR to determine if the project is viable:

$$IRR_R = \frac{1 + IRR_A}{1 + f} - 1 = \frac{1.1055}{1.04} - 1 = 0.06298 \text{ or } 6.3\%$$

Since the real IRR of 6.3% exceeds the MARR of 5%, the project should be undertaken.■

REVIEW PROBLEM 9.3

Robert is considering purchasing a bond with a face value of \$5000 and a coupon rate of 8%, due in 10 years. Inflation is expected to be 5% over the next 10 years. Robert's real MARR is 10%, compounded semiannually. What is the present worth of this bond to Robert?

ANSWER

This problem can be done with either real interest and real cash flows or actual interest and actual cash flows. It is somewhat easier to work with actual cash flows, so we must first convert the real interest rate given to an actual interest rate.

Robert's annual real MARR is $(1 + 0.10/2)^2 - 1 = 0.1025$. (Recall that the 10% is a nominal rate, compounded semiannually.)

If annual inflation is 5%, Robert's actual *annual* MARR is

$$\begin{aligned} MARR_A &= MARR_R + f + MARR_R \times f \\ &= 0.1025 + 0.05 + (0.1025)(0.05) \\ &= 0.15763 \text{ or } 15.763\% \end{aligned}$$

The present worth of the \$5000 Robert will get in 10 years is then

$$\begin{aligned} PW &= \$5000(P/F, MARR_A, 10) \\ &= \$5000(0.23138) \cong \$1157 \end{aligned}$$

Next, the bond pays an annuity of $\$5000 \times 0.08/2 = \$200$ every six months. To convert the annuity payments to their present worth, we need an actual six-month MARR. This can be obtained with a six-month inflation rate and Robert's six-month real MARR of 10%/2 = 5%. With f = 5% per annum, the inflation rate per six-month period can be calculated with

$$\begin{aligned} f_{12} &= (1 + f_6)^2 - 1 \\ f_6 &= (1 + f_{12})^{1/2} - 1 \\ &= (1 + 0.05)^{1/2} - 1 = 0.0247 \text{ or } 2.47\% \end{aligned}$$

The actual MARR per six-month period is then given by

$$\text{MARR}_A = \text{MARR}_R + f + \text{MARR}_R \times f = 0.05 + 0.0247 + (0.05)(0.0247)$$
$$= 0.07593 \text{ or } 7.593\%$$

The present worth of the dividend payments is

$$\text{PW(dividends)} = \$200(P/A, 7.59\%, 20)$$
$$= \$200(10.125)$$
$$= \$2025$$

Finally,

$$\text{PW(bond)} = \$1157 + \$2025$$
$$= \$3182$$

The present worth of the bond is $3182.■

REVIEW PROBLEM 9.4

Trimfit, a Southern Ontario manufacturer of automobile interior trim, is considering the addition of a new product to their line. Data concerning the project are given below. Should Trimfit accept the project?

New Product Line Information	
First cost ($)	11 500 000
Planned output (units/year)	275 000
Actual MARR	20%
Range of possible inflation rates	0% to 4%
Study period	10 years

Current-Year Prices ($/unit)	
Raw materials	16.00
Labour	6.25
Product sales price	32.00

ANSWER

First, we note that the expected net revenue per unit (not counting amortization of first costs) is $9.75 = $32 – $16 – $6.25. The project is potentially viable.

In doing the project evaluation, we can proceed with either actual dollars or real dollars. Since we do not know what the inflation rate will be, the easiest way to account for inflation is to keep all prices in real dollars and adjust the actual MARR to a real MARR by using values for the inflation rate within the potential range given. The project can then be evaluated with one of the standard methods. Since many of the figures are given in terms of annual amounts, an annual worth analysis will be carried out. Inflation rates of 0%, 1%, and 4% will be used. The results are shown below in Table 9.6.

Table 9.6 Annual Worth Computations for Trimfit

Annual Worth Comparisons for Various Inflation Rates					
Inflation Rate per Year	Real MARR	Fixed Cost per Year ($)	Variable Cost per Year ($)	Revenue per Year ($)	Annual Worth (Profit) per Year ($)
0%	20.00%	2 743 012	6 118 750	8 800 000	– 61 762
1%	18.81%	2 633 122	6 118 750	8 800 000	48 128
4%	15.38%	2 325 083	6 118 750	8 800 000	356 167

In Table 9.6, the annual worth of the project depends on the inflation rate assumed. Since the actual MARR of 20% implicitly includes anticipated inflation, different trial inflation rates imply different values for the real MARR. For example, at 1% inflation, the real MARR implied is

$$\text{MARR}_R = \frac{1 + \text{MARR}_A}{1 + f} - 1 = \frac{1.20}{1.01} - 1 = 0.1881 \text{ or } 18.81\%$$

The fixed cost per year is obtained by finding the annual amount over 10 years equivalent to the first cost when the appropriate real MARR is used. For example, with 1% inflation, the fixed cost per year is

$$A = P(A/P, \text{MARR}_R, 10) = \$11\ 500\ 000 \left(\frac{0.1881\ (1.1881)^{10}}{(1.1881)^{10}} - 1\right)$$

$$= \$2\ 633\ 122$$

Next, the variable cost per year is the sum of the raw material cost and the labour cost per unit multiplied by the total expected output per year, that is, $22.25 × 275 000 = $6 118 750. Revenue per year is the sales price multiplied by the expected output: $32 × 275 000 = $8 800 000. Notice that the variable cost and the revenue are the same for all three values of the inflation rate. This is because they are given in real dollars.

Finally, the annual worth of the project is determined by the revenue per year less the fixed and variable costs per year. The annual worth is negative for zero inflation, but is positive for both 1% and 4% inflation rates. Since periods of at least 10 years in which there has been zero inflation have been rare in the twentieth century, it is probably safe to assume that there will be some inflation over the life of the project. Therefore, the project appears to be acceptable, since its annual worth will be positive if inflation is at least 1%.■

SUMMARY

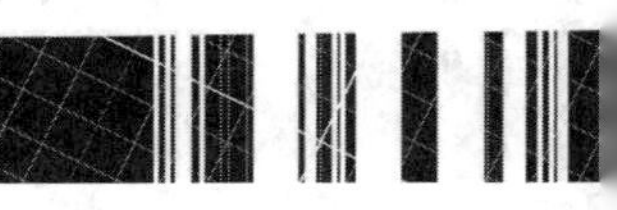

In this chapter, the concept of inflation was introduced, and we considered the impact that inflation has on project evaluation. We began by discussing methods of measuring inflation. The main result here was that there are many possible measures, all of which are only approximate.

The concept of actual cash flows and interest rates and real cash flows and interest rates was introduced. Actual dollars are in currency at the time of payment or receipt, while real dollars are constant over time and are expressed with respect to a base year. Compound amount factors can be used to convert single payments between real and actual dollars.

Most of the chapter was concerned with the effect of correctly anticipated inflation on project evaluation and on how to incorporate inflation into project evaluation correctly. We showed that, where engineers have no reason to believe project prices will behave differently from average prices, project decisions are the same with or without correctly anticipated inflation. Finally, we pointed out that predicting inflation is very difficult. This implies that engineers should work with ranges of values for possible future inflation rates. The engineer should test for sensitivity of decisions to possible inflation rates.

Engineering Economics in Action, Part 9B: Exploiting Volatility

Bill Astad of head office had been asking Naomi about how to deal with the variable inflation rates experienced by a sister company in Mexico. "OK, Naomi, let's see if I have this straight. For long-term projects of, say, six years or more, it makes sense to use a single inflation figure—the average rate. I can just add that to the real MARR to get an actual MARR. Boy, it's easy to get confused between the real and the actual. But I do understand the principle."

"And the short-term projects?" Naomi prompted.

"For the short-term ones, it makes more sense to break them up into time periods. For each period, select a 'best guess' inflation rate, and do a stepwise calculation from period to period. So the inflation rate in the middle of the presidential cycle would be relatively low, while near the changeover time it would be a higher estimate. Of course, the actual values used would depend on the political and economic situation at the time the decision is made. I understand that one, too, but it is complicated."

"I agree," said Naomi. "I guess we're lucky things are more predictable here."

"We are," Bill replied. "On the other hand, if we can make good decisions in spite of a volatile economy in Mexico, Mexifab may have an advantage over its competitors. Thanks for your help, Naomi."

PROBLEMS

9.1 Which of the following are real dollars and which are actual dollars?

(a) Allyson has been promised a \$10 000 inheritance when her Uncle Bill dies.

(b) Bette's auto insurance will pay the cost of a new windshield if her current one breaks.

(c) Cory's meal allowance while he is in university is \$2000 per term.

(d) Dieter's company promises that its prices will always be the same as they were in 1975.

(e) Engworth will construct a house for Zolda, and Zolda will pay Engworth $150 000 when the house is finished.

(f) Fran's current salary is $3000 per month.

(g) Greta's retirement plan will pay her $1500 per month, adjusted for the cost of living.

9.2 Find the real dollars (with today as the base year) corresponding to the actual dollars shown below, for a 4% inflation rate.

(a) $400 three years from now

(b) $400 three years ago

(c) $10 next year

(d) $350 983 ten years from now

(e) $1 one thousand years ago

(f) $1 000 000 000 three hundred years from now

9.3 Find the present worth today in real dollars corresponding to the actual dollars shown below, for a 4% inflation rate and a 4% interest rate.

(a) $400 three years from now

(b) $400 three years ago

(c) $10 next year

(d) $350 983 ten years from now

(e) $1 one thousand years ago

(f) $1 000 000 000 three hundred years from now

9.4 An investment pays $10 000 in five years.

(a) If inflation is 10% per year, what is the real value of the $10 000 in today's dollars?

(b) If inflation is 10% and the real MARR is 10%, what is the present worth?

(c) What actual dollar MARR is equivalent to a 10% real MARR when inflation is 10%?

(d) Compute the present worth using the actual dollar MARR from part (c).

9.5 An annuity pays $1000 per year for 10 years. Inflation is 6% per year.

(a) If the real MARR is 8%, what is the actual dollar MARR?

(b) Using the actual dollar MARR from part (a), calculate the present worth of the annuity.

9.6 An annuity pays $1000 per year for 12 years. Inflation is 6% per year. The annuity costs $7500 now.

(a) What is the actual dollar internal rate of return?

(b) What is the real internal rate of return?

9.7 A bond pays $10 000 per year for the next ten years. The bond costs $90 000 now. Inflation is expected to be 5% over the next 10 years.

(a) What is the actual dollar internal rate of return?

(b) What is the real internal rate of return?

9.8 Inflation is expected to average about 4% over the next 50 years. How much would we expect to pay 50 years from now for each of the following?

(a) $1.59 hamburger

(b) $15 000 automobile

(c) $180 000 house

9.9 The average Canadian now has assets totalling $38 000. If the average real wealth per Canadian remains the same, and if inflation averages 5% in the future, when will the average Canadian become a millionaire?

9.10 How much is the present worth of $10 000 ten years from now under each of the following patterns of inflation, if interest is at 5%? On the basis of your answers, is it generally reasonable to use an average inflation rate in economic calculations?

(a) Inflation is 4%.

(b) Inflation is 0% for five years, and then 8% for five years.

(c) Inflation is 8% for five years, and then 0% for five years.

(d) Inflation is 6% for five years and then 2% for five years.

(e) Inflation is 0% for nine years and then 40% for one year.

9.11 The actual dollar MARR for Jungle Products Ltd. of Parador is 300%. The inflation rate in Parador is 250%. What is the company's real MARR?

9.12 Krystyna has a long-term consulting contract with an insurance company that guarantees her $25 000 per year for five years. She believes inflation will be 3% this year and 5% next year, and then will stay at 10% indefinitely. Krystyna's real dollar MARR is 12%. What is the present worth of this contract?

9.13 I have a bond that will pay me $2000 every year for the next 30 years. My first payment will be a year from today. I expect inflation to average 3% over the next 30 years. My real MARR is 10%. What is the present worth of this bond?

9.14 Ken will receive a $15 000 annual payment from a family trust. This will continue until Ken is 30; he is now 20. Inflation averages 4%, and Ken's real MARR is 8%. If the first payment is a year from now and a total of 10 payments are to be made, what is the present worth of his remaining income from the trust?

9.15 Inflation in Russistan currently averages 40% per month. It is expected to diminish to 20% per month following the presidential elections 12 months from now. The Russistan Oil Company (ROC) has just signed an agreement with the Canadian Petroleum Group for the sale of future shipments. The ROC will receive 500 million rubles per month over the next two years, and also 500 million rubles per month indexed to inflation (i.e., real rubles). If the ROC has a real MARR of 1.5% per month, what is the total present worth of this contract?

9.16 The widget industry maintains a price index for a standard collection of widgets. The base year was 1992 until 2002, when the index was recomputed with 2002 as the base year. The following data concerning prices for the years 2000 to 2003 are available:

Year	Price Index 1992 Base	Price Index 2002 Base
2000	125	N/A
2001	127	N/A
2002	130	100
2003	N/A	110

What was the percentage increase in prices of widgets between 2000 and 2003?

9.17 A group of farmers in Inverness is considering building an irrigation system from a water supply in some nearby mountains. They want to build a concrete reservoir with a steel pipe system. The first cost would be \$200 000 with (actual) annual maintenance costs of \$2000. They expect the irrigation system will bring them \$22 000 per year in additional (actual) revenues due to better crop production. Their real dollar MARR is 4% and they anticipate inflation to be 3% per year. Assume that the reservoir will have a 20-year life.

(a) Using the actual cash flows, find the actual IRR on this project.

(b) What is the actual MARR?

(c) Should they invest?

9.18 Refer to Problem 9.17.

(a) Convert the actual cash flows into real cash flows.

(b) Find the present worth of the project using the real MARR.

(c) Should they invest?

9.19 Go to the Statistics Canada site (www.statcan.ca) and find the document "Your Guide to the Consumer Price Index."

(a) Summarize, in several paragraphs, what the CPI is, misconceptions about the CPI, and four of the price indexes used other than the CPI.

(b) Summarize, in several paragraphs, what the commodities in the CPI basket are, the relative importance of the commodities in the CPI basket, how the CPI basket is updated, and how prices are collected for the CPI.

9.20 Go to the Statistics Canada site (www.statcan.ca) and find *CPI Indexes for Canada and the Provinces.*

(a) For the most recent time period reported, find the province in Canada with the highest "all items" CPI.

(b) For the province identified in (a), which of the subcategories (e.g., food, transportation, etc.) appears to be contributing the most to the overall CPI index?

(c) Discuss why you think this province has the highest CPI.

9.21 Bosco Consulting of Calgary is considering a potential contract with the Upper Sobonian government to advise them on exploration for oil in Upper Sobonia. Bosco would make an investment of 1 500 000 Sobonian zerts to set up a Sobonian office in 2005. The Upper Sobonian government would pay Bosco 300 000 zerts in 2006. In the years 2007 to 2012, the actual zerts value of the payments would increase at the rate of inflation in Upper Sobonia. The following data are available concerning the project:

Investment in Upper Sobonia	
Expected Sobonian inflation rate (2005–2012)	15%/year
Expected Canadian inflation rate (2005–2012)	3%/year
Value of Sobonian zerts in 2005	$0.25
Expected decline in value of zerts (2005–2012)	10%/year
First cost in 2005 (zerts)	1 500 000
Cash flows in 2006–2012 (real 2006 zerts)	300 000
Bosco's actual dollar MARR	22%

(a) Construct a table with the following items:
Real (2005) zert cash flows
Actual zert cash flows
Actual dollar cash flows
Real dollar cash flows

(b) What is the present worth in 2005 dollars of this project?

9.22 Bildkit, an Alberta building products company, is considering an agreement with a distributor in Maloria to supply kits for constructing houses in Maloria. Sales would start next year. The expected receipts from the sale of the kits next year is 30 000 000 Malorian yen. The number of units sold is expected to grow by 10% per year over the life of the contract. The actual yen price is expected to grow at the rate of Malorian inflation.

There will be a first cost for Bildkit. As well, there will be operating costs over the life of the contract. Operating cost per unit will be constant in real dollars over the life of the contract. Since the number of units sold will rise by 10% per year, real operating costs will rise by 10% per year. Actual operating costs per unit will rise at the rate of inflation in Canada.

The value of the Malorian yen is expected to increase over the life of the contract. Data concerning the proposed contract are shown in the table below.

Bildkit in Maloria			
Receipts in first year (actual yen)	30 000 000	First cost now (actual $)	200 000
Growth of receipts (real yen)	10%/year	Operating cost in first year of operation (actual $)	350 000
Malorian inflation rate	1%/year	Canadian inflation rate	3%
Value of yen year 0 ($)	0.015	Actual dollar MARR	22%
Rate of increase in value of yen	2%	Study period	8 years

(a) What is the present worth of receipts in dollars?

(b) What is the present worth of the cash outflows in dollars?

9.23 Leftway Information Systems of Saint John, New Brunswick, is considering a contract with the Ibernian government to supply consulting services over a five-year period. The following real Ibernian pounds cash flows are expected:

Cash Flows in Year 2005 Ibernian Pounds	
First cost	1 800 000
Net revenue 2006 to 2010	550 000

Further information is in the table below:

Expected Ibernian inflation rate	10%
Value of Ibernian pound in 2005 (Canadian $)	1.25
Expected annual rate of decline in the value of the Ibernian pound	5%
Expected Canadian inflation rate	2.50%
Leftway's real MARR	15%

(a) What is the real Ibernian pound internal rate of return on this project? (*Hint:* Canada can be ignored in answering this question.)

(b) What is the actual pound internal rate of return? (*Hint:* Canada can be ignored in answering this question.)

(c) Use the internal rate of return in Canadian dollars to decide if Leftway should accept the proposed contract.

9.24 Sonar warning devices are being purchased by the St. James Bay department store chain to help trucks back up at store loading docks. The total cost of purchase and installation is $220 000. There are two types of saving from the system. Faster turnaround time at the congested loading docks will save $50 000 per year in today's dollars. Reduced damage to the loading docks will save $30 000 per year in today's dollars. St. James Bay has an observed actual dollar MARR of 18%. The sonar system has a life of four years. Its scrap value in today's dollars is $20 000. The inflation rate is expected to be 6% per year over the next four years.

(a) What is St. James Bay's real MARR?

(b) What is the real internal rate of return? (This is most easily done with a spreadsheet.)

(c) Compute the actual internal rate of return using Equation (9.10).

(d) Compute the actual internal rate of return from the actual dollar cash flows. (This is most easily done with a spreadsheet.)

(e) What is the present worth of the system?

9.25 Johnson Products, a manufacturer in Wolfville, Nova Scotia, now buys a certain part for its chain saws. They are considering the production of the part in-house. They can install a production system that would have a life of five years with no salvage value. They believe that over the next five years the real price of purchased parts will remain fixed. They expect the real price of labour and other inputs to production in Wolfville to rise over the next five years. Further information about the situation is in the table below.

Annual cost of purchase ($/year)	750 000
Expected real change in cost of purchase	0%
Expected real change in labour cost	4%
Expected real change in other operating costs	2%
Labour cost/unit (first year of operation) ($)	10.5
Other operating cost/unit (first year of operation) ($)	9
In-house first cost ($)	200 000
Use rate (units/year)	25 000
Actual dollar MARR	20%
Study period (years)	5

(a) Assume inflation is 2% per year in the first year of operation. What will be the actual dollar cost of labour for in-house production in the second year?

(b) Assume inflation is 2% per year in the first two years of operation. What will be the actual dollar cost of other operating inputs for in-house production in the third year?

(c) Assume that inflation averages 2% per year over the five-year life of the project. What is the present worth of costs for purchase and for in-house production?

9.26 Lifewear, a Winnipeg manufacturer of women's sports clothes, is considering adding a line of skirts and jackets. The production would take place in a part of their factory that is now not being used. The first output would be available in time for the 2006 fall season. The following information is available:

New Product Line Information	
First cost in 2005 ($)	15 500 000
Planned output (units/year)	325 000
Observed, actual dollar MARR before tax	0.25
Study period	6 years
Year 2005 Prices ($/unit)	
Materials	12
Labour	7.75
Output	35

(a) What is the real internal rate of return?

(b) What inflation rate will make the real MARR equal to the real internal rate of return?

(c) Calculate the present worths of the project under three possible future inflation rates. Assume that the inflation rate will be 1%, 2%, or 3% per year.

(d) Decide if Lifewear should add this new line of skirts and jackets. Explain your answer.

9.27 Century Foods, a Saskatoon producer of frozen meat products, is considering a new plant near Calgary for its sausage rolls and frozen meat pies. The company has estimates of production cost and selling prices in the first year. It expects the real value of operating costs per unit to fall because of improved operating methods. It also expects competitive pressures to cause the real value of product prices to fall. The following data are available:

Century Food Plant Data	
Output price in 2006 ($/box)	22
Operating cost in 2006 ($/box)	15.5
Planned output rate (boxes/year)	275 000
Fall in real output price	1.5% per year
Fall in real operating cost per box	1.0% per year
First cost in 2005 ($)	7 500 000
Study period	10 years
Actual dollar MARR before tax	20%

(a) Assume that there is zero inflation. What is the present worth, in 2005, of the project?

(b) Assume that there is zero inflation. What is the internal rate of return? (This is most easily done with a spreadsheet.)

(c) At what inflation rate would the actual dollar internal rate of return equal 20%?

(d) Should Century Foods build the new plant? Explain your answer.

9.28 Metcan Ltd.'s Newfoundland smelter produces its own electric power. The plant's power capacity exceeds its current requirements. Metcan has been offered a contract to sell excess power to a nearby utility company. Metcan would supply the utility company with 17 500 megawatt-hours per year (MWh/a) for 10 years. The contract would specify a price of $22.75 per megawatt-hour for the first year of supply. The price would rise by 1% per year after this. This is independent of the actual rate of inflation over the 10 years.

Metcan would incur a first cost to connect its plant to the utility system. There would also be operating costs attributable to the contract. Metcan believes these costs would track the actual inflation rate. The terms of the contract and Metcan's costs are shown in the tables on the next page.

(a) Find the present worth of the contract under the assumption that there is no inflation over the life of the contract.

(b) Find the present worths of the contract under four assumptions: inflation is (i) 1% per year, (ii) 2% per year, (iii) 3% per year, and (iv) 4% per year.

Metcan Sale of Power	
Output price in 2006 ($/MWh)	22.75
Price adjustment (2007–2015)	1% per year
Power to be supplied (MWh/a)	17 500
Contract length	10 years
Metcan's Costs	
First cost in 2005 ($)	175 000
Operating cost in 2000($)	332 500
Actual dollar MARR before tax	20%

(c) Should Metcan accept the contract?

9.29 Clarkwood is a British Columbia wood products manufacturer. They are considering a modification to their production line that would enable an increase in their output. One of Clarkwood's concerns is that the price of wood is rising more rapidly than inflation. They expect that because of this their operating cost per unit will rise at a rate 4% higher than the rate of inflation. That is, if the rate of inflation is f, Clarkwood's operating cost will rise at the rate $f_c = 1.04(1 + f) - 1$. However, competitive pressures from plastics will prevent the prices of their products from rising more than 1% above the inflation rate. The particulars of the project are shown in the table below.

Clarkwood's Project	
Output price in 2006 ($/unit)	30
Price increase	2% above inflation
Operating cost in 2006 ($/unit)	24
Operating cost increase	4% above inflation
Expected output due to project (units/year)	50 000
First cost in 2005 ($)	900 000
Observed actual dollar MARR	0.25
Time horizon (years)	10

(a) Find the present worth of the project under the assumption of zero inflation.

(b) Find the present worth of the project under these assumptions: the expected inflation is (i) 1% per year, and (ii) 2% per year.

(c) Should Clarkwood accept the project?

9.30 Smooth-Top is a British Columbia manufacturer of desktops. They are considering an increase of their capacity. Consulting engineers have submitted two routes to accomplish this: (1) install a new production line that would produce wood desktops finished with hardwood veneer and (2) install a new production line that would produce wood desktops finished with simulated wood made from hard plastic.

Smooth-Top is concerned about the price of hardwood veneer. They believe the price of veneer will rise over the next 10 years. However, they believe the price of veneer-finished desktops will rise by less than the rate at which the price of veneer rises. Information about the two potential projects is in the following table.

Smooth-Top Desktop Project	
Plastic-finish real price and real cost change	0%
Veneer-finish expected real price change	1%
Veneer-finish expected real cost change	5%
Wood cost/unit ($)	12.5
Plastic cost/unit ($)	9
Wood price/unit ($)	32
Plastic price/unit ($)	26
Wood first cost ($)	2 050 000
Plastic first cost ($)	2 700 000
Wood output rate (units/year)	30 000
Plastic output rate (units/year)	45 000
Study period	10 years
Actual dollar MARR	25%

(a) Compute the present worth of each option under the assumption that the real price of hardwood-finished desktops and real cost of hardwood veneer do not change (rather than as stated in the table). Assume zero inflation.

(b) Compute the present worth of each option under the assumption that the real price of hardwood-finished desktops and the real value of hardwood veneer desktop operating costs increase as indicated in the table. Assume that inflation is expected to be 2% over the study period.

9.31 Belmont Grocers has a distribution centre in Fredericton. The manual materials-handling system at the centre has deteriorated to the point at which it must be either replaced or substantially refurbished. Replacement with an automated system would cost about $240 000. Refurbishing the manual system would cost about $50 000. In either case, capital expenditures would take place this year. Operating either the new system or the refurbished system would begin next year. It is expected that either the new system or the refurbished system will operate for 10 years with no further capital expenditures. Belmont is concerned that labour costs in Fredericton may rise in real terms over the next 10 years. The range of increases in real terms that appears possible is from 4% to 7% per year. Inflation rates between 2% and 4% are expected over the next 10 years. Complete data on the two alternatives are given in the table below.

Materials Handling Data	
Automated expected real operating cost change	0%
Manual expected real operating cost change	4% to 7%
Manual operating cost/unit (first year of operation) ($)	10.5
Automated operating cost/unit (first year of operation) ($)	9
Manual first cost ($)	50 000
Automated first cost ($)	240 000
Output rate (units/year)	15 000
Actual dollar MARR	20%
Study period	10 years
Possible inflation rates	2% to 4%

(a) Find the total costs per unit for each of the two alternatives under the assumption of zero inflation and no increase in costs for the manual system.

(b) Make a recommendation about which alternative to adopt. Base the recommendation on the present worth of costs for the two systems under various assumptions concerning inflation and the rate of change in the real operating cost of the manual system. Explain your recommendation.

9.32 The United Gum Workers have a cost-of-living clause in their contract with Mont-Gum-Ery Foods. The contract is for two years. The contract states that, if the inflation rate in the first year exceeds 1%, wages in the second year will increase by the inflation rate of the first year. Does this clause increase or decrease risk? Explain.

9.33 Free Wheels Manitoba has a plant that assembles bicycles in Louisbourg, Manitoba. The plant now has a small cafeteria for the workers. The kitchen equipment is in need of substantial overhaul. Free Wheels has been offered a contract by Besteats to supply food to their workers. The particulars of the situation are shown in the table. Should Free Wheels continue with the in-house food service or contract the service to Besteats?

Food Service: In-House Versus Contract	
Food service labour (hours/year)	6 000
Wage rate (real, time 1, $/hour)	7.5
Overhead cost (real, time 1, $/year)	18 000
Kitchen equipment first cost (actual, time 0, $)	25 000
Contract cost, years 1 to 3 (actual $)	55 000
Contract cost, years 4 to 6 (actual $)	63 700
Actual dollar MARR	22%
Expected annual inflation rate	5%
Study period (years)	6

Economic Comparison of High Pressure and Conventional Pipelines: Associated Engineering

CANADIAN 9.1 MINI-CASE

Associated Engineering of Toronto conducted an evaluation of sources of water supply for an Ontario municipality. One of the considerations was the choice of high-pressure or conventional pipelines for transmitting treated water from one of the Great Lakes to the municipality.

Conventional pipelines, most often made of concrete, have a limited maximum tensile strength, which for analysis purposes was taken to be 200 pounds per square inch (psi). High-pressure pipe, made of steel, can withstand up to 60 000 psi, although the pipe examined by Associated Engineering had a strength of 42 000 psi.

The major advantage of the steel pipe is that fewer pumping stations are needed than with the concrete pipe. The distance to be pumped is 85 kilometres; this requires either one pumping station for high-pressure pipe or six pumping stations for concrete pipe.

Each pipeline type was analyzed over a range of pipeline diameters ranging from 24" to 72". Construction costs included the pipe, pumping stations, and a reception reservoir, with the time of the cost taken to be the commissioning date of 2025. Operating and maintenance costs starting in 2026 were included, and administration, engineering fees, contingencies, and taxes were also accounted for.

The best alternative was chosen on the basis of a present worth comparison with a 4% discount rate. In the analysis, real 1993 dollars were used and an inflation rate of 2% was assumed for the period of study. The result was that a 360-diameter high-pressure pipeline was economically best, at a present cost $7.5 million lower than for the best conventional pipeline.

Discussion

Estimating future inflation is difficult. The average inflation in Canada over the last 50 years has been about 4%, but over the last five years it has been 2% to 2.5%. For some other five-year periods, inflation has averaged over 10%. How can we estimate future inflation?

One way is to simply assume that inflation will remain at the current value. This is probably wrong; as has been seen, inflation typically changes over time. However, there are factors that are controlling the inflation rate. Lacking knowledge of any reason why these controlling factors might change, the current rate seems to be a reasonable choice.

A second approach is to use the long-term average. Knowing that inflation will change over time suggests that the long-term average is a good choice even if inflation is lower or higher than the average right now. After all, those controlling factors have changed in the past and are likely to change again.

A third way is to take into account the controlling factors for inflation. These include government policy: a government committed to social welfare is likely to induce more inflation than one committed to fiscal responsibility. Trends in business and consumer behaviour affect inflation: large labour contract increases presage inflation, as does high consumer borrowing. Social trends like the aging of the baby boomers also has an effect on inflation.

Understanding the effect of the controlling factors for inflation in detail is very difficult. So usually decision makers make a broad judgment based on both the current inflation rate and the historical average, and perhaps informed by a general understanding of the contributing factors.

Questions

1. How significant would the difference have been to the savings of the high-pressure pipeline if an inflation rate of 4% had been used instead? Assume the only difference

between the concrete- and steel-pipe systems was the capital cost, expended in 2025. Would the decision be any different? Could it be different for any assumed inflation rate?

2. Design two cash flow structures for projects that start in 2025, such that the present worth in the current year at a discount rate of 4% is higher for one project than the other at an inflation rate of 2%, but lower at an inflation rate of 4%. Is there a significant opportunity to control the best choice in a decision situation by selecting the appropriate inflation rate?
3. Why would the analysts have chosen to separate the inflation rate from the discount rate for this problem, rather than combining them into an actual dollar discount rate? Do you think the analysts estimated the actual dollar cost of the alternatives in 2025, or would they have used the real costs?

Appendix 9A Computing a Price Index

We can represent changes in average prices over time with a **price index**. A price index relates the average price of a given set of goods in some time period to the average price of the same set of goods in another period. Commonly used price indexes work with weighted averages because simple averages do not reflect the differences in importance of the various goods and services in which we are interested.

Many different ways of weighting changes in prices may be used, and each method leads to a different price index. We shall discuss only the most commonly used index, the **Laspeyres price index**. It can be explained as follows.

Suppose there are n goods in which we are interested. We want to represent their prices at a time, t_1, relative to a **base period**, t_0, the period from which the expenditure shares are calculated.

The prices of the n goods at times t_0 and t_1 are denoted by $p_{01}, p_{02}, \ldots, p_{0n}$ and $p_{11}, p_{12}, \ldots, p_{1n}$. The quantities of the n goods purchased at t_0 are denoted by $q_{01}, q_{02}, \ldots, q_{0n}$. The share, s_{0j}, of good j in the total expenditure for the period, t_0, is defined as

$$s_{0j} = \frac{p_{0j}q_{0j}}{p_{01}q_{01} + p_{02}q_{02} + \ldots + p_{0n}q_{0n}}$$

Note that

$$\sum_{j=1}^{n} s_{0j} = 1$$

A Laspeyres price index, π_{01}, is defined as a weighted average of relative prices.

$$\pi_{01} = \left(\frac{p_{11}}{p_{01}}s_{01} + \frac{p_{12}}{p_{02}}s_{02} + \ldots + \frac{p_{1n}}{p_{0n}}s_{0n}\right) \times 100$$

The term in the brackets is a weighted average because the weights (the expenditure shares in the base period) sum to one. The relative prices are the prices of the individual goods in period t_1 relative to the base period, t_0. The weighted average is multiplied by 100 to put the index in percentage terms.

Example 9A.1

A student uses four foods for hamburgers: (1) ground beef, (2) hamburger buns, (3) onions, and (4) breath mints. Suppose that, in one year, the price of ground beef fell by 10%, the price of buns fell by 1%, the price of onions rose by 5%, and the price of breath mints rose by 50%.

The price and quantity data for the student's hamburger are shown in Table 9A.1.

Table 9A.1 Price and Quantity Data for Hamburger

	Quantity at t_0	Price at t_0 ($)	Price at t_1 ($)
Ground beef (kg)	0.25	3.5/kg	3.15/kg
Buns	1	0.40	0.396
Onions	1	0.20	0.21
Breath mints	1	0.10	0.15

The Laspeyres price index is calculated in four steps:

1. Compute the base period expenditure for each ingredient.
2. Compute the share of each ingredient in the total base period expenditure.
3. Compute the relative price for each ingredient.
4. Use the shares to form a weighted average of the relative prices.

These computations are shown in Table 9A.2.

Table 9A.2 The Laspeyres Price Index Calculation

	Price at t_0 ($)	Share at t_0	Relative Price	Weighted Relative Price
Ground beef (kg)	0.875	0.556	0.900	0.500
Buns	0.400	0.254	0.990	0.251
Onions	0.200	0.127	1.050	0.133
Breath mints	0.100	0.063	1.500	0.095
Sums	1.575	1.000		0.980

As an example of the computations, the price of the ground beef per hamburger at t_0 is found by multiplying the price per kilogram by the weight of the hamburger used: $3.50/kg $\times$ 0.25 kg = $0.875. Similar computations for each of the other ingredients lead to a total cost of $1.575 per hamburger. The ground beef then represents a share of 0.875/1.575 = 0.556 of the total cost. The relative price for the hamburger is 3.15/3.5 = 0.9 and thus the weighted relative price is $0.556 \times 0.9 = 0.50$. Similar computations for the other ingredients lead to a total weighted average of 0.98. After multiplying by 100, the Laspeyres price index is 98 (it is understood that this is a percentage). Therefore, the cost of the hamburger ingredients at t_1 was 2% lower than in the base period.■

Statistics Canada compiles many Laspeyres price indexes. The consumer price index (CPI) is a Laspeyres price index in which the weights are the shares of urban consumers' budgets in the base year, currently 1992. Another well-known Laspeyres price index is the gross national product (GNP) deflator. For the GNP deflator, the weights are the shares of total output in the base year (also 1992).

The CPI and the GNP deflator are global indexes in that they represent an economy-wide set of prices. As well, Statistics Canada produces Laspeyres price indexes by sector. For example, there are price indexes for durable consumer goods, for exports, and for investment by businesses. It is up to the analyst to know the composition of the different indexes and to decide which is best for his or her purposes.

Example 9A.2

We can classify consumer goods and services into four classes: durable goods, semi-durable goods, non-durable goods, and services. Assume the classes had the following prices in 1996 and 2003:

	Price ($)	
Category	**1996**	**2003**
Durable	2.421	2.818
Semi-durable	2.849	3.715
Non-durable	4.926	6.404
Services	4.608	6.263

Quantities in 1996 were

Quantity in 1996 (Units)	
Durable	21.304
Semi-durable	11.315
Non-durable	19.159
Services	31.422

Find the Laspeyres price index for 2003 with 1996 as a base. We first calculate the relative prices:

	Price ($)		
Category	**1996**	**2003**	**Relative Price**
Durable	2.421	2.818	1.164
Semi-durable	2.849	3.715	1.304
Non-durable	4.926	6.404	1.300
Services	4.608	6.263	1.359

We next determine expenditure shares in 1996:

Category	Expenditure	Share
Durable	51.583	0.1597
Semi-durable	32.235	0.0998
Non-durable	94.381	0.2922
Services	144.801	0.4483
Total	323.000	

We then multiply the relative prices by the shares and sum. We get the index by multiplying the sum by 100. For example, the term for durable goods is given by 1.164(0.1597) = 0.186.

	Index
Durable	0.186
Semi-durable	0.130
Non-durable	0.380
Services	0.609
Total	1.305

This gives a Laspeyres price index of 130.5.■

CHAPTER 10

Public Sector Decision Making

Engineering Economics in Action, Part 10A: New Challenges in Lotus Land

"Hi, Naomi. How's it going in Lotus Land?" Naomi could easily imagine Bill's feet up on his desk, leaning back in his chair, telephone wedged against his ear. Naomi was in Vancouver, checking into alternative plans for how Edgemont Pulp Mill, a sister company of Canadian Widgets, was dealing with the proposed Domestic Emission Trading (DET) System for controlling greenhouse gases (GHG) under the Kyoto agreement. The Kyoto agreement committed Canada to reduce GHGs, and the forest industry was expected to reduce emissions by 15% by 2008–2010. The DET system was a mechanism proposed by the federal government to make the reductions as economically efficient at possible.

Naomi answered, "Lotus Land is great. I can look out my window and see green grass and rhododendrons with new buds. Not bad for late February. Things look pretty good at Edgemont Pulp Mill, too, but we're up against some tough decisions. Edgemont has spent over $40 million in capital improvement in the past few years. Some of this was to reduce the levels of AOX in response to government regulation."

Toxic forms of AOX (absorbable organic halides) are created when elemental chlorine is used in the pulp bleaching process. In 1992, British Columbia passed a law requiring pulp mills to reduce AOX levels in their wastewater to zero by the end of 2002. In the ensuing decade, most mills, including Edgemont, moved to the use of elemental chlorine-free chemicals in the bleaching process. The result was a dramatic drop in AOX levels, but not complete elimination. "Edgemont was pretty relieved when the B.C. government softened up on its requirements back in 2002 because it was going to prove expensive to comply with the zero-AOX law."

"So, what's the problem, then?" Bill interjected.

"Well," Naomi began, "Edgemont also invested in capital improvements to reduce their CO_2 emissions significantly. The problem is, the 15% additional improvement the government is seeking ignores improvements we made before 1999, which is when Edgemont did their big changes. Also, its capital equipment is relatively new, so replacing it in the near future to meet the new requirements really doesn't make good economic sense. It isn't right that Edgemont should be penalized for being an environmental leader. Maybe if we did a broadly focused benefit-cost analysis, we could demonstrate that what the federal government is proposing is flawed. Can we do that?" asked Naomi.

"I'm not sure," Bill responded. "Let's explore the idea. Talk to the people at the mill today. I'll try to get some ideas from the people here. See if you can get a flight back this afternoon. I'll try to set up a meeting for us with Anna Kulkowski in the next couple of days. I suspect she may have some helpful ideas."

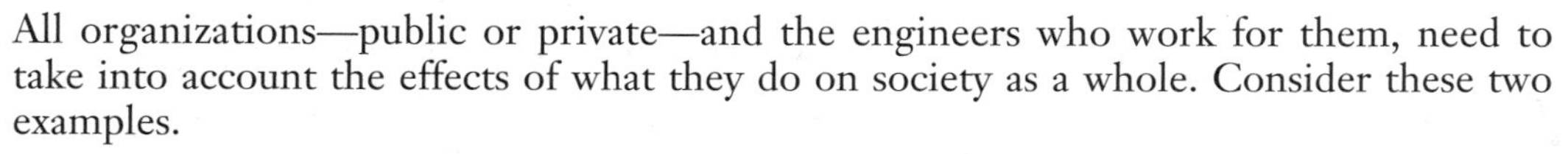

10.1 Introduction

All organizations—public or private—and the engineers who work for them, need to take into account the effects of what they do on society as a whole. Consider these two examples.

1. MacMillan Bloedel (MB), a British Columbia–based forest products company, was faced with a campaign in the spring of 1995 that included boycotts of their customers, demonstrations, and even a bomb threat against MB's headquarters. The campaign was a protest against MB's timber cutting practices in the Clayoquot area of Vancouver Island. The protesters were concerned about the impact of MB's clear-cutting on the ecological balance in Clayoquot. By the spring of 1995, the boycott had already cost MB a $5 million contract to supply a

British subsidiary of the Scott Paper Company, and other customers were being pressured to stop buying from MB. In 1999, following long years of protest, MB signed agreements to phase out clear-cutting of old-growth forests. When Weyerhaeuser, a Washington-based forest products company, took over MB shortly thereafter, they agreed to the more stringent forest management practices. However, there continues to be substantial controversy over Weyerhaeuser's logging due to sharply polarized viewpoints on the social costs and benefits associated with forest management practices.

2. The extraction of coal and the production of steel once was the mainstay of the Nova Scotian economy. The centre of the steel prosperity, Sydney, is now the site of one of North America's largest environmental disasters—the Sydney Tar Ponds. The contaminants in the tar ponds are mainly the result of steel-making operations—the core of which is the coke oven. As coal is heated in a coke oven, tar and gases are separated off from the desired coke. Over more than 80 years, these toxic wastes were discharged into an estuary of Sydney Harbour in an area known as Muggah Creek. The water in the area has been seriously contaminated with arsenic, lead, benzene, and other toxins such as PCBs (polychlorinated biphenyls), many of which are believed to cause cancer. Though concerns about air and ground pollution had been evident for decades, little money was spent to reduce the large volumes of water and air pollution. The steel mills, owned privately until the 1960s, were finally closed in 2000 after 40 years of operation by Sydney Steel (SYSCO), a provincial Crown corporation.

   Residents of the area say that when it rains, puddles in the area turn fluorescent green. Others who live near the ponds report orange ooze in their basements, massive headaches, nosebleeds, and serious breathing problems. Sydney has one of the highest rates of cancer, birth defects, and miscarriages in Canada. In March 2004, residents filed a $1 billion lawsuit, the largest class-action suit to date in Canada, for damages allegedly caused by years of pollution from the SYSCO operations. In May 2004, the federal and provincial governments agreed to commit $400 million to assist with the cleanup. It remains to be seen what the outcome of the lawsuit will be and how the cleanup operation will remediate the long-polluted site.

The MB and Sydney Tar Ponds examples illustrate a recent phenomenon that has important implications for engineers. It is not enough to produce goods and services at a cost that customers are willing to pay. Engineers must also pay attention to broader social values. This is because the market prices that guide most production decisions may not reflect all the social benefits and costs of engineering decisions adequately. Where markets fail to reflect all social benefits and costs, society uses other means of attaining social values.

Up to this point in our coverage of engineering economics, we have been concerned with project evaluation methods for private sector organizations. The private sector's frame of reference is profitability of the firm itself—a well-understood objective. To be profitable, firms sell goods and services to customers with the goal of realizing a return on investment. The focus in project evaluation is usually on the impact it has on the financial well-being of the firm, even if the impact of the project may have effects on individuals or groups external to the firm. For example, building a new pulp mill will provide additional production capacity, but it also may have an impact on local water quality. The firm may limit its economic analysis to the direct impact of the plant on its own profitability. This would include, for example, the construction and operating costs of the mill and the additional revenues associated with expanded production capacity. It may not consider the effect their emissions may have on the local fisheries.

When we consider the broader social context, the concept of profitability is extremely difficult to define because the frame of reference is society at large. This is particularly true for projects where there may not be an open market in which customers have free choice over what products they buy. Costs and benefits for these projects can be difficult to identify and value in a quantifiable manner. For example, if Transport Canada were to consider construction of a new highway to improve traffic flow in a congested area, the analysis should take into account, in addition to the costs of construction, the benefits and costs to society at large. This includes assigning a value to social benefits such as increased safety and reduced travel time for business and recreational travellers. It also includes assigning values to social costs such as increased noise to neighbouring houses or alternately, the costs of noise abatement.

In this chapter, we shall look at the social aspects of engineering decision making. First, we shall consider the reasons markets may fail in such areas as the environment and health. We shall also consider different methods that society may use to correct these failures. Second, we shall consider decision making in the public sector. Here we shall be concerned mainly with government projects or government-supported projects.

10.2 Market Failure

A **market** is a group of buyers and sellers linked by trade in a particular product or service. The prices in a market that guide most production decisions usually reflect all the social benefits and costs of engineering decisions adequately. This is not always true, however. When prices do not reflect all social benefits and costs of a decision, we say that there has been *market failure*. When this occurs, society will seek a means of correcting the failure. In this section, we define market failure and give examples of its effects. Then we discuss a number of ways in which society seeks remedies for market failure.

10.2.1 Market Failure Defined

Most decisions in the private sector lead to market behaviour that have desirable outcomes. This is because these decisions affect mainly those people who are party to those decisions. Since people can generally freely choose to participate in markets, it is reasonable to assume that the individuals who participate must somehow benefit by their actions. In most cases, this is the end of the story. The market is **efficient** in that decisions are made so that it is impossible to find a way for at least one person to be better off and no person to be worse off.

In some cases, however, decisions have important effects on people who are not party to the decisions. In these cases, it is possible that the gains to the decision makers, and any others who might benefit from the decisions, are less than the losses imposed on those who are affected by the decisions. Such situations are clearly undesirable. These decisions are instances of market failure. **Market failure** occurs when a market, left on its own, fails to make decisions in which resources are allocated efficiently. When decisions are made in which aggregate benefits to all persons who benefit from the decision are less than aggregate costs imposed on persons who bear the costs that result from the decision, the decision is inefficient and hence market failure has occurred. Market failure can occur if the decision maker does not correctly take into account the gains and losses imposed on others by the consequences of a decision.

There are several reasons for market failure. First, there may be no market through which those affected by the decision can induce the decision maker to take their situations into account. However, losses may exceed gains even when there is a market, if the market has insufficient information about the gains and losses resulting from decisions. Market failure can also occur whenever a single buyer or seller (a monopolist) can influence prices or output. Market failure can even occur when someone decides *not* to do something that would create benefits to others. The market would fail if the cost of creating the benefits was less than the value of the benefits.

Acid rain is an example of the effects of market failure. The burning of high-sulphur coal by thermal-electric power plants is believed to be one of the causes of acid rain. These plants could burn low-sulphur fuels, but they do not, partly because low-sulphur fuels are more expensive than high-sulphur fuels. If a market existed through which those affected by acid rain could buy a reduction in power plant sulphur emissions, they could try to make a deal with the power plants. They would be able to offer the power plants enough to offset the higher costs of low-sulphur fuel and still come out ahead. But there is no such market. The reason for this is that there is no single private individual or group whose loss from acid rain is large enough to make it worthwhile to offer the power plants payment to reduce sulphur emissions. It would require a large number of those affected by acid rain to form a coalition to make the offer. There are markets for electric power and for coal. However, they do not lead to socially desirable decisions about the use of power or coal. This is because the market prices for power and coal do not reflect the costs related to acid rain. If the prices for power and high-sulphur coal reflected these costs, less power would be used, and less of it would be made with high-sulphur coal. Both would reduce acid rain.

The damage to the residents of Sydney due to air pollution and toxic waste is another example of the effects of market failure. The market failed to take into account the health and environmental costs in the price of steel. In the early years of steel production, there was little information about the deleterious health effects of air pollution and of the toxic coke oven by-products. However, as early as the 1960s the government started to exert pressure on the Dominion Steel and Coal Company (DOSCO), then operating the steel mill, to reduce its emissions. DOSCO balked, however, indicating that the estimated cost would put them out of business. Not long after, DOSCO announced its intention to close the steel mills due to its inability to compete in international markets. This led the local residents to lobby DOSCO and the provincial government to keep the mills open. They claimed that closing the mill would have a devastating effect on the local economy. Sydney Steel (SYSCO), a Crown corporation, took over the operation of the steel mill. SYSCO continued to contribute to the pollution problem, and it was only in 1982, when contamination in the local lobster catch forced the fisheries to close, that the effects of market failure really started to become evident. The loss to the fishermen certainly had not been factored into the costs of running the steel mills. Over the 1980s other costs associated with the steel mills, unaccounted for in the price of steel, began to become more evident. Health problems such as birth defects, cancer, severe headaches, and lung disease in local residents were the subject of numerous scientific studies. The government spent $250 million between 1980 and 2000 in several unsuccessful attempts to clean up the toxic wastes. Finally, a decision was made in 2000 to close the steel mill as it became evident that the true costs of the mill far exceeded the benefits to the steel mill and to the local economy. Estimates range from $400 million to almost a billion dollars. The Sydney Tar Ponds are now acknowledged to be one of North America's worst-contaminated industrial sites.

We can see how market failure has caused socially undesirable outcomes such as acid rain and contaminated industrial sites. When markets fail, as in cases such as these, society will seek to remedy these problems through a variety of mechanisms. These remedies are the subject of the next section.

10.2.2 Remedies for Market Failure

There are three main formal methods of eliminating or reducing the impact of market failure:

1. Policy instruments used by the government
2. The ability of persons or companies adversely affected by the actions of others to seek compensation in the courts
3. Government provision of goods and services

We shall discuss the first two methods in this section. Government provision of goods and services is discussed in the next section under decision making in the public sector.

The first and most common means of trying to deter or reduce the effects of market failure is the use of what are commonly called **policy instruments**—government-imposed rules intended to modify behaviour.

One class of policy instruments are *regulations* such as standards, bans, permits, or quotas. The regulations are backed by penalties for non-compliance. Canada has regulations concerning such widely differing areas as product labelling, automobile emissions, and the use of bodies of water to dispose of waste. It has a permit system, with associated fees, for various environmentally harmful emissions.

A challenge associated with developing regulations is that they may be inefficient. For example, suppose that we wish to improve the quality of a lake by reducing the amount of effluents dumped into it. These effluents may contain material with excessive biological oxygen demand (BOD). A typical regulation to control dumping would require all those who dump to meet the same BOD standards. But the costs of meeting these standards are likely to differ among the producers of the effluent. To attain a regulated reduction in BOD in their effluent, some producers will have to make expensive changes to their procedures, while others can respond with a lower cost. The most efficient way to obtain the reduction in BOD in the lake is to have those with low effluent-cleaning costs make the greatest reduction in BOD. Uniform regulation is not likely to do this.

A second class of policy instruments meant to overcome market failure is to offer *monetary incentives* or *deterrents* to induce desired behaviour. Monetary incentives or deterrents, often more efficient than regulations, may be subsidies or special tax treatments. For example, referring back to effluent dumping, there could be subsidies for the installation of equipment that would reduce the amount of effluent produced. In this way, producers for whom the cost of reducing BOD is low will do so since this would be cheaper than paying fees. By setting an appropriate subsidy, the desired reduction can be attained.

Policy instruments are constantly being evaluated and modified. The reason is that market failure is a complex issue and policy makers may not fully understand its causes, or the implications of implementing a given policy. Nonetheless, their use is widespread and they can be highly effective in mitigating the effects of market failure.

A second formal means of reducing the effects of market failure is *litigation*. The use of courts as a means of reducing the health and safety effects of market failure has grown in both Canada and the United States. Since the 1970s in Canada (the exact year depends on the province), courts have held that regular sellers of a product implicitly guarantee that the product is fit for reasonable use. Where the cost of reducing a risk in the use of their product is less than the objectively estimated expected loss, sellers are supposed to reduce the risk. Moreover, these sellers are held legally responsible for having expertise in the production and use of the products. It is not enough for sellers to say they did not know that use of the product was risky. Sellers are supposed to make reasonable efforts to determine potential risks in the use of the products they sell.

The third formal method of reducing the effects of market failure is *government provision of goods and services*. This provision may be direct, as in the case of police services, or indirect, as in the case of health care given by physicians. Market failure is remedied when public sector analysts take into account all parties affected by a decision through a comprehensive assessment of total costs and benefits of a decision. Health care provision, transportation, municipal services, and electric and gas utilities are some examples of goods and services provided by the public sector. Each service requires numerous economic decisions to be made in the best interests of the public. This is of sufficient importance for us to devote a separate section of this chapter to the topic.

In addition to formal methods for dealing with market failure, *informal methods* can be used. For example, the boycotts against MB mentioned in the introduction of this chapter may have been an effective means of getting MB to alter its clear-cutting behaviour. Another informal method is information dissemination to industry and to the general public, encouraging or discouraging certain behaviours.

10.3 Decision Making in the Public Sector

This section is devoted to the decision-making process for public provision of goods and services. Public (government) production in Canada has occurred mainly in two classes of goods and services. The first class includes those services for which there is no market because it is not practical to require people to pay for the service. Police and fire protection, defence, and the maintenance of city streets are examples of government services that it is not practical for users to pay for.

The second class includes those services for which scale economies make it inefficient to have more than one provider. Where there is only a single provider of a service, there is no market competition to enforce efficiency and low prices. There is a danger that the single provider, called a *monopolist*, will charge excessive prices and/or be inefficient. To ameliorate this potential problem, governments are often the provider.

For example, local deliveries of natural gas and electric power are situations where economies of scale are important enough that having more than one provider is not efficient. These services may be provided by publicly owned monopolies. For example, Saskatchewan Power, a Crown corporation, is the principal supplier of electricity in Saskatchewan.

An alternative to government provision of services where there is one provider is for the government to monitor and regulate the performance of a private monopolist. For example, Ontario Power Generation (OPG) is a private Ontario-based company whose primary business is the generation and sale of electricity to customers in Ontario and in interconnected markets.

Privatization of traditionally government-run functions has been a growing trend in Canada in the past decade. Proponents of privatization argue that the private sector is more efficient than the government in its operations. Those against privatization argue that private organizations will not take into account all the social costs and benefits of their actions and thus contribute to the potential market failure. Close-Up 10.1 describes **public-private partnership** (P3) projects, in which private sector companies invest in public sector projects and recoup their investments through user-based tolls, fees, or tariffs.

In this section, we shall consider **benefit-cost analysis** (BCA), a method of project evaluation widely used in the public sector. BCA provides a general framework for assessing the gains and losses associated with alternative projects when a broad societal view is necessary. This method is commonly used for public projects because the government is

CLOSE-UP 10.1 Public-Private Partnerships

Traditionally, large-scale public infrastructure projects, such as roads, bridges, power plants, and public utilities, have been financed, built, and maintained by government. In recent years, Canada has moved increasingly to what is referred to as "public-private partnerships," or "P3s."

A public-private partnership (P3) is a cooperative arrangement between the private and public sectors for the provision of infrastructure or services. In a P3, the public sector maintains an oversight and quality-assessment role while the private sector is responsible for provision of the service or project.

Numerous categories of P3s exist. Each is categorized on the basis of the extent of public and private sector involvement and the allocation of financial responsibility between the two.

The use of P3s in Canada is growing. Some high-profile examples are the Confederation Bridge construction between Nova Scotia and Prince Edward Island and Ontario's 407 ETR toll highway. Numerous airports, including Canada's two largest (Toronto and Vancouver), are now run by independent operating agencies outside the federal government. P3s are also used for water services in cities such as Goderich, Edmonton, Halifax, Moncton, London, and Winnipeg.

Source: "The Canadian Experience," Strategis site, November 17, 2003 (**strategis.ic.gc.ca/epic/internet/inpupr-bdpr.nsf/en/h_qz01552e.html**), accessed October 1, 2004.

NET VALUE 10.1

Canadian Benefit-Cost Analysis Guides

Various branches of the Canadian government make benefit-cost analysis guides available to assist analysts and managers with the task of evaluating public projects. Several are listed below.

Benefit Cost Analysis Guide (Treasury Board of Canada): This guide provides a framework for submissions to the Treasury Board of Canada for projects that have major social, economic, or environmental impacts. The guide indicates the steps to be taken in a thorough benefit-cost analysis and sets out best practices at each stage. It also provides insights into identifying and measuring costs and benefits associated with public projects. One section provides examples of estimating values when no market prices exist. Among the examples given are how to estimate the value of travel-time savings, job creation, and health and safety. The guide can be found at www.tbs-sct.gc.ca.

Guide to Benefit-Cost Analysis in Transport Canada (Transport Canada): This guide provides a framework for evaluating alternative projects when they have a transportation focus. It also gives several illustrative examples of benefit-cost analysis applied to airport refurbishment, the acquisition of depth-sounding catamarans in the St. Lawrence River, and consolidation of maintenance work centres for air navigation electronic systems. The guide can be found at www.tc.gc.ca.

Benefit Cost Analysis Guide for Regulatory Programs (Privy Council Office): This guide provides a framework for analysts who wish to evaluate proposals to introduce or change Government of Canada regulations. Regulations are a form of government intervention that have the force of law. They typically have penalties for non-compliance. Examples are income tax regulations, dangerous chemicals and noxious liquid substances regulations, and motor vehicle safety regulations. The guide can be found at www.pco-bcp.gc.ca.

responsible for applying public resources in a manner that balances the costs and benefits of a project in such a way to produce the greatest overall social benefit. BCA can also be used for project evaluation in the private sector where the broader impact on society must be taken into account.

The starting point for a BCA is a clear understanding of what alternative courses of action, or projects, are being considered. Once these are established, there are two main sets of issues to resolve. The first is to identify and measure the benefits associated with each project, and to clearly understand to whom these benefits accrue. The second is to identify and measure the costs associated with each project, and to establish who pays the costs. Once these issues are resolved, then the projects can be fairly evaluated by a variety of comparison methods. Benefit-cost ratios are a commonly used comparison method for public sector projects, though the methods of Chapters 4 and 5 are also applicable.

10.3.1 The Point of View Used for Project Evaluation

In carrying out a benefit-cost analysis, it is important to clearly establish what point of view is to be taken, and to use this point of view consistently. The basic questions to address are: "Who will benefit from the project?" and "Who will pay for the project?" By identifying these two points of view clearly at the outset of the evaluation, confusion about what to include and what not to include can be reduced. The point of view defines who is "in" and who is "out."

Generally, it is members of society who are the users and beneficiaries of project services, and the government, the sponsor, who pays for the project. The evaluation will thus take into account the impact of the project on both the users and the sponsors. For example, if a provincial government is considering a highway improvement project to reduce traffic congestion and to improve safety, the point of view taken would focus on the impact of the project on the provincial finances, and on the benefits to the users most affected by the highway improvement. Therefore the analysis would include government costs such as construction and maintenance expenses. Savings (reduction in costs) to the government could include new property tax income if land values were to go up due to increased economic activity in the area. Social benefits could include factors such as reduced travel time, reduced vehicle operating costs, and increased safety for users of the goods and services provided by the project. Social costs (also called disbenefits) include the costs of travel delays and disruption during the construction period, increased traffic noise in nearby neighbourhoods, and possibly negative environmental impacts, all of which can affect users. Careful definition of the point of view taken ensures that all the effects of the project are taken into account.

One of the challenges associated with the analysis of public projects is that there may be several reasonable points of view, each with a different set of users or sponsors to be included in the analysis. Consider the decision to fund remediation of the Sydney Tar Ponds. An analyst providing advice to the Minister of the Environment might need to look at the project from the point of view of Canadian society, from the perspective of the province of Nova Scotia, and from the point of view of local residents and environmental groups. The point of view chosen delineates whose costs and benefits are to be taken into account in a benefit-cost analysis. The fact that several points of view may be taken can lead to ambiguity and controversy over the results. Nonetheless, BCA is a common framework for analysis.

Another important factor to consider in project evaluation is the frame of reference for measuring the impacts of the project. The analysis should consider costs and benefits associated with the project based on the difference between what would occur *with* the project versus what would occur *without* the project. In other words, we concern ourselves with the marginal benefits and costs associated with the project so that its impact is fairly measured.

10.3.2 Identifying and Measuring the Costs of Public Projects

Once we have a clear view of who will benefit from a project (the users) and who will pay for a project (the sponsor), the next step in a BCA is to identify and measure the various costs and benefits associated with the project. The fundamental questions are "In what ways will users benefit from the project, and by how much will they benefit?" and "What costs will be incurred by the sponsor, and how much will these costs be?"

This section deals with identifying and measuring the costs of public projects to the project sponsor. For the most part, these costs are straightforward to identify and measure. This is not the case with benefits, which will be covered in the next section.

The sponsor costs broadly include all of the resources, goods, and services required to develop, implement, and maintain a project or program. These can be generally classified into the initial capital costs and ongoing operating and administration costs. If the project creates savings, these are deducted from costs to produce a net cost to the sponsor.

For the highway-widening project mentioned in previous sections, the following is a classification of some of the sponsor's costs and savings:

Costs to the Sponsor
- Construction costs
- Operating and maintenance costs
- Administrative costs

Savings to the Sponsor
- Increased tax revenues due to higher land values

Some costs can be *directly* attributed to a project, while others may be stimulated indirectly by the project. These are referred to as *direct* and *indirect* costs, respectively. For example, in the above project, construction and operating costs are direct costs, and increased tax revenues are indirect savings.

While identifying and valuing the costs of a public project can be relatively straightforward, the same cannot generally be said about identifying and valuing social benefits.

10.3.3 Identifying and Measuring the Benefits of Public Projects

This section deals with identifying and measuring the benefits associated with public projects. We begin providing an overview of the process, and then indicate some of the challenges associated with this portion of a benefit-cost analysis.

The benefits of a project to society include the value of all goods and services that result from the project or program. Generally, the benefits of a public project will be positive. However, some of the effects from a project may be negative. These are referred to as social costs, and are subtracted from the benefits to obtain a net measure of social benefits.

For the example of a highway improvement project, the following is a classification of some of the social benefits and social costs:

Social Benefits
- Reduced travel time for business and recreational users
- Increased safety
- Reduced vehicle operating costs for business and recreational users

Social Costs
- Increased noise and air pollution
- Disruption to the local environment
- Disruption of traffic flows or business transactions during construction
- Loss of business elsewhere due to traffic rerouting onto the new highway

As with sponsor costs, some of the social benefits or social costs can be *directly* attributed to a project, while others may be stimulated indirectly by the project. These are referred to as *direct* and *indirect* benefits, respectively. For example, direct benefits of an improved highway may be decreased travel time and increased safety. Indirect benefits could include lost business to other areas if traffic is rerouted from these areas onto the new highway. A thorough benefit-cost analysis should always include the direct benefits, and will include indirect benefits if they have an important effect on the overall project.

The task of identifying and measuring the social benefits and costs of public projects can be challenging. The challenge arises because the benefits may not be reflected in the monetary flows of the project. We are concerned with the real effects of a project, but the cash flows may or may not reflect all these real effects. For example, in the highway improvement project, there are cash flows for the wages of the workers who construct the road. This is a real cost to the government, and the wages reflect these costs well. In contrast, consider the intangible social cost of disruption of traffic during road construction. These costs are not reflected in the cash flows of a project, but are nonetheless an important cost of putting the road in place.

Beyond the challenge of identifying certain costs and benefits is that of assigning a value or measure to each. For goods and services that are distributed through markets, we have prices to measure values. Valuation is relatively straightforward for these items. However, many public projects create or use goods and services for which there is no market in which prices reflect values. These intangible, non-market goods and services, are challenging to value.

Consider the highway improvement project mentioned earlier. Obvious sponsor costs are the labour, materials, and equipment used for the project. These costs can be relatively easy to estimate, as there are markets for these goods and services through which appropriate prices have been established. However, there are several intangible social costs and benefits that are somewhat more difficult to measure. First, there are the social costs associated with temporary travel-time disruptions due to construction. Measuring the cost of traffic disruption during the work requires valuing the time delays incurred by car passengers and trucks. There are approximations to these delay costs based on earnings per hour of passengers and the hourly cost of running trucks. The approximation for the value of travellers' time is based on the idea that a person who can earn, for example, $35 an hour working should be willing to pay $35 an hour for time saved travelling to work. The disruption costs may be large enough that they make it more efficient to have the work done at night, despite the fact that this will raise the explicit construction cost. The intangible social benefits of reduced travel time once the project is completed can also be factored into the benefit-cost analysis in a similar way.

An intangible social benefit associated with many public projects is improved health and safety. A common method for assessing health and safety benefits is to estimate the reduction in the number and type of injuries expected with a project or program and then to put a value on these injuries. A variety of methods exist, but one approach is to use lost wages, treatment costs or insurance claims as the basis for estimating the value of certain injuries. Estimates of the value of a human life using this method range from several hundred thousand to several million dollars.

Other examples of intangible items commonly valued in a benefit cost-analysis are the value of noise abatement and the value of the environment. Transport Canada's *Guide to Benefit-Cost Analysis* and the Treasury Board's *Benefit-Cost Analysis Guide* (see Net Value 10.1) both contain useful information on these and other non-market costs and benefits associated with typical public projects, and helpful examples of how to value impacts where there are no market prices.

Two basic methods have been developed to value intangible social benefits and costs. The first is *contingent valuation*. This method uses surveys to ask members of the public what they would be willing to pay for the good or service in question. For example, members of the public may be asked how much they would be willing to pay for each minute of reduced travel time due to a highway improvement. Contingent valuation has the benefit of being a direct approach to valuation, but has the drawback that individuals may not provide accurate or truthful responses.

The second commonly used method for assessing the value of non-market goods and services is called the *hedonic price* method. This method involves deducing indirectly individuals' valuations based on their behaviour in other markets for goods and services. For example, the value of noise pollution due to a busy road may be estimated by examining the selling price of houses near other busy roads and comparing these prices to those of similar houses away from busy roads. The value of the noise pollution cannot be higher than the difference in prices; otherwise, the owners would move. The advantage of this method is that members of the public reveal their true preferences through their behaviour. The disadvantage of the hedonic price method is that it requires reasonably sophisticated knowledge.

The following example provides further illustration of how the measurement of costs and benefits without market prices can be challenging, and also how these challenges can be dealt with.

Example 10.1

Consider the construction of a bridge across a narrow part of a lake that gives access to a provincial park. The major benefit of the bridge will be reduced travel time to get to the park from a nearby urban centre. This will lower the cost of camping trips at the park. As well, more people are expected to use the park because of the lower cost per visit. How can these benefits be measured?

Data concerning the number of week-long visits and their costs are shown in Table 10.1.

First, the reduction in cost for the 8000 visits that would have been made even without the bridge creates a straightforward benefit:

Table 10.1 Average Cost per Visit and Number of Visits per Year

	Without Bridge	With Bridge
Travel cost per visit ($)	140.00	87.50
Use of equipment ($)	50.00	50.00
Food cost per visit ($)	100.00	100.00
Total cost per visit ($)	290.00	237.50
Number of visits/year	8000	11 000

Travel cost saving on 8000 visits = ($140 − $87.50) × 8000 = $420 000

There is a benefit of $420 000 per year from reduced travel cost on the 8000 visits that would have been taken even without the bridge.

Next, we see that the number of visits to the park is expected to rise from 8000 per year to 11 000 per year. But how much is this worth? We do not have prices for park visits, but we do have data that enable estimates of actual costs to campers. These costs may be used to infer the value of visits to campers.

We see that before the bridge, the cost of a week-long park visit, including travel and other costs, averaged $290.

It is reasonable to assume that a week spent camping was worth at least $290 to anyone who incurred that cost. The average cost of a weeklong visit to a park would fall from $290 per visit to $237.50 per visit if the bridge were built. We are concerned with the value of the incremental 3000 visits per year. Clearly, none of these visits would be made if the cost were $290 per trip. And each of them is worth at least $237.50 or else the trip would not have been taken. The standard approximation in cases like this is halfway between the highest and lowest possible values. This gives an aggregate benefit of the increased use of the park of

$$\frac{(\$290.00 + \$237.50)}{2} \times 3000 = \$791\ 250$$

Therefore, the value of the incremental 3000 visits per year is estimated as approximately $791 000 per year. However, there is also a cost of $237.50 per visit. The net benefit of the incremental 3000 visits is therefore

$$\$791\ 250 - \$237.50(3000) = \$78\ 750$$

The total value of benefits of the bridge is the sum of the reduced travel cost plus increased use:

$$\$420\ 000 + \$78\ 750 = \$498\ 750$$

The total value of the benefits yielded by the bridge is almost $500 000 per year. These benefits are the shaded area shown in Figure 10.1. These benefits must then be weighed against the costs of the bridge.■

Figure 10.1 Benefits from Bridge

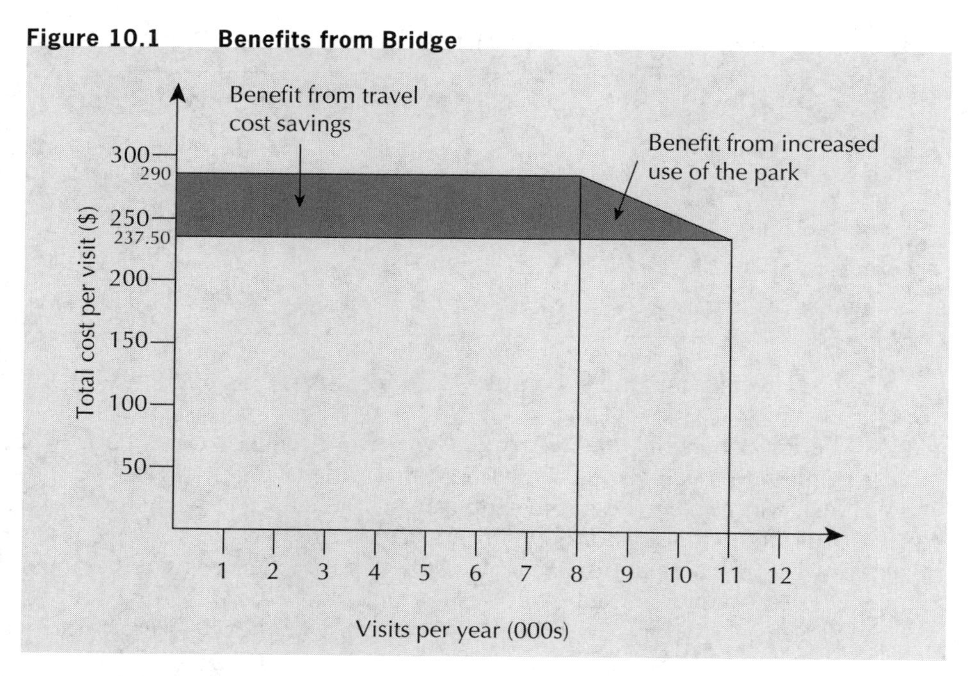

10.3.4 Benefit-Cost Ratios

Once we have identified and obtained measures for the costs and benefits associated with a public project, we can evaluate its economic viability. The same comparison methods that are used for private sector projects are appropriate for government sector projects. That is, we can use the present worth, annual worth, and internal-rate-of-return performance measures in the same ways in both the private and government sectors. It is important to emphasize this, because other methods based on ratios of benefits to costs have been used frequently in government project evaluations, almost to the exclusion of present worth, annual worth, and internal-rate-of-return methods. Because of the prevalent use of these ratios, this section is devoted to a discussion of several benefit-cost ratios that are commonly used in public sector decision making. We then point out several problems associated with the use of benefit-cost ratios so that the reader is aware of them and understands the correct way of using them.

Benefit-cost ratios can be based on either the present worths or the annual worths of benefits and costs of projects. The resulting ratios are equivalent in that they lead to the same decisions. We shall discuss the ratios in terms of present worths, but the reader should be aware that everything we say about ratios based on present worths applies to ratios based on annual worths.

The conventional **benefit-cost ratio**, denoted BCR, is given by the ratio of the present worth of net users' benefits (social benefits less social costs) to the present worth of the net sponsor's costs for a project. That is,

$$\text{BCR} = \frac{\text{PW(users' benefits)}}{\text{PW(sponsors' costs)}}$$

A **modified benefit-cost ratio**, also in common use, denoted BCRM, is given by the ratio of the present worth of net benefits minus the present worth of operating costs to the present worth of capital costs, that is,

$$\text{BCRM} = \frac{\text{PW(users' benefits)} - \text{PW(sponsors' operating costs)}}{\text{PW(sponsors' capital costs)}}$$

Though the modified benefit-cost ratio is less used than the conventional benefit-cost ratio, it has the advantage that it provides a measure of the net gain per dollar invested by the project sponsor.

In general, a project is considered desirable if its benefit-cost ratio exceeds one, which is to say, its benefits exceed its costs.

Example 10.2

The town of Helen Lake, Manitoba has limited parking for cars near its main shopping street. The town plans to pave a lot for parking near Main Street. The main beneficiaries will be the merchants and their customers. The present worth of expected benefits is \$3 000 000. The cost to the town of buying the lot, clearing, paving, and painting is expected to be \$500 000. The present worth of expected maintenance costs over the lifetime of the project is \$50 000. In the period in which the lot is being cleared and paved, there will be some disbenefits to the local merchants and customers due to disruption of traffic in the Main Street area. The present worth of the disruption is expected to be about \$75 000. What is the benefit-cost ratio?

$$\text{BCR} = \frac{\text{PW(users' benefits)}}{\text{PW(sponsors' costs)}} = \frac{\$3\ 000\ 000 - \$75\ 000}{\$500\ 000 + \$50\ 000} = 5.3$$

We can see that the benefit-cost ratio exceeds 1, and thus the proposal is economically justified.■

Example 10.3

A fire department in a medium-sized city is considering a new dispatch system for its firefighting equipment responding to calls. The system would select routes, taking into account recently updated traffic conditions. This would reduce response times. An indirect effect would be a reduction in required fire-fighting equipment. The present worth of benefits and costs are:

Users' Benefits

PW(benefits) = \$37 500 000

Sponsors' Costs

PW(operating costs) = \$3 750 000

PW(capital costs for the city) = \$13 500 000

PW(reduced equipment requirements) = \$2 250 000

What is the benefit-cost ratio for the dispatch system? What is the modified benefit-cost ratio for the dispatch system? Is the project economically justifiable?

$$\text{BCR} = \frac{\$37\ 500\ 000}{(\$13\ 500\ 000 + \$3\ 750\ 000 - \$2\ 250\ 000)} = 2.50$$

$$\text{BCRM} = \frac{(\$37\ 500\ 000 - \$3\ 750\ 000)}{(\$13\ 500\ 000 - \$\ 2\ 250\ 000)} = 3.00$$

We see for this example that both the conventional and modified cost-benefit ratio are greater than 1, indicating that the project, under either criterion, is economically justified.■

Examples 10.3 and 10.4 demonstrate the basic use of benefit-cost ratios. It is worth noting that if a conventional benefit-cost ratio is greater (less) than one, it follows that the modified benefit-cost ratio will be greater (less) than one. This means that the two types of benefit-cost ratios will lead to the same decision. We present both because the reader may encounter either ratio.

For *independent* projects then, the following decision rule may be used: *Accept all projects with a benefit-cost ratio greater than one*. In other words, accept a project if

$$\text{BCR} = \frac{\text{PW(users' benefits)}}{\text{PW(sponsors' costs)}} > 1, \text{ or}$$

$$\text{BCRM} = \frac{\text{PW(users' benefits)} - \text{PW(sponsors' operating costs)}}{\text{PW(sponsors' capital costs)}} > 1$$

This rule, using either the benefit-cost ratio or the modified benefit-cost ratio, is equivalent to the rule that all projects with a present worth of benefits greater than the present worth of costs should be accepted. This is, then, the same rule that was presented in Chapter 4, which accepts a project if its present worth is positive. This is easily shown. If

$$\frac{\text{PW(users' benefits)}}{\text{PW(sponsors' costs)}} > 1$$

it follows that

$$\text{PW(users' benefits)} > \text{PW(sponsors' costs)}$$

which is equivalent to

$$\text{PW(users' benefits)} - \text{PW(sponsors' costs)} > 0$$

Also recall that, as shown in Chapter 5, the present/annual worth method and the internal rate of return method give the same conclusion if an independent project has a unique IRR.

To use benefit-cost ratios to choose among *mutually exclusive* projects, we must evaluate the increment between projects, just as we did in Chapter 5 with the internal rate of return method. Suppose we have two mutually exclusive projects, X and Y, with present worths of benefits, B_X and B_Y, and present worths of costs, C_X and C_Y. We first check to see if the individual benefit-cost ratios are greater than one. We discard a project with a benefit-cost ratio of less than one. If both projects have benefit-cost ratios greater than one, we rank the projects in ascending order by the present worths of costs. Suppose $C_X \geq C_Y$. We then form the ratio of the differences in benefits and costs,

$$\text{BCR(X} - \text{Y)} = \frac{B_X - B_Y}{C_X - C_Y}$$

If this ratio is greater than one, we choose project X. If it is less than one, we choose project Y. If $C_X = C_Y$ (in which case the ratio is undefined), we choose the project with the greater present worth of benefits. If $B_X = B_Y$, we choose the project with the lower present worth of costs.

This rule is the same as comparing two mutually exclusive projects using the internal rate of return method, which was presented in Chapter 5. In order to choose a project, we saw in Chapter 5 that not only the IRR of the individual project, but also the IRR of the incremental investment, must exceed the MARR. This rule is also the same as choosing the project with the largest present worth, as presented in Chapter 4, provided the present worth is positive.

The following example, concerning two mutually exclusive projects, summarizes our discussion of benefit-cost ratios and illustrates their use.

Example 10.4

A medium-sized city is considering increasing its airport capacity. At the current airport, flights are frequently delayed and congestion at the terminal has limited the number of flights. Two mutually exclusive alternatives are being considered. Alternative A is to construct a new airport 65 kilometres from the city. Alternative B is to enlarge and otherwise upgrade the current airport that is only 15 kilometres from the city. The advantage of a new airport is that there are essentially no limits on its size. The disadvantage is that it will require travellers to spend additional travel time to and from the airport.

There are two disadvantages for upgrading and enlarging the current airport. One disadvantage is that the increase in size is limited by existing development. The second disadvantage is that the noise of airplanes in a new flight path to the current airport will reduce the value of homes near that flight path. A thousand homes will be affected, with the average loss in value about $25 000. Such losses are large enough so that it is not reasonable to expect that the owners' losses will be offset by gains elsewhere. If the city wishes to ensure that their losses are offset, it must compensate these owners. Note that such compensation would not be an additional social cost. It would merely be a transfer from taxpayers (through the government) to the affected owners. Benefit and cost data are shown in Table 10.2.

Table 10.2 Airport Benefits and Costs

Effect	Alternative A New Airport (millions of $)	Alternative B Current Airport (millions of $)
Improved service/year	55	28.5
Increased travel cost/year	15	0
Cost of highway improvements	50	10
Construction costs	150	115
Reduced value of houses	0	25

The city will use a MARR of 10% and a 10-year time horizon for this project. What are the benefit-cost ratios for the two alternatives? Which alternative should be accepted?

Before we start the computation of benefit-cost ratios, we need to convert all values to present worths. Two effects are shown on an annual basis: improved service, which occurs under both alternatives, and increased travel cost, which appears only with the new airport. We get the present worths (in millions of $) by multiplying the relevant terms by the series present worth factor:

$$\text{PW(improved service of A)} = \$55(P/A, 10\%, 10)$$
$$= \$55(6.1446) = \$337.95$$

$$\text{PW(improved service of B)} = \$28.5(6.1446) = \$175.10$$

$$\text{PW(increased travel cost of A)} = \$15(6.1446) = \$92.17$$

The remainder of the costs are already in terms of present worth.
The benefit-cost ratio for alternative A is

$$\text{BCR(alternative A)} = \frac{\$337.95}{(\$150 + \$50 + \$92.17)} \cong 1.16$$

and for alternative B is

$$\text{BCR(alternative B)} = \frac{\$175.10}{(\$10 + \$115 + \$25)} = 1.1673$$

First, we note that both benefit-cost ratios exceed one. Both alternatives are viable. Since they are mutually exclusive, we must choose one of the two. To decide which alternative is better, we must compute the benefit-cost ratio of the incremental investment between the alternatives. We use the alternative with the smallest present worth of costs as the starting point (alternative B), and compute the benefit-cost ratio associated with the marginal investment:

$$\begin{aligned}\text{BCR(A} - \text{B)} &= (B_A - B_B)/(C_A - C_B) \\ &= \frac{(\$337.95 - \$92.17 - \$175.10)}{(\$50 + \$150) - (\$10 + \$115 + \$25)} = 1.4136\end{aligned}$$

The ratio of the benefits of the new airport minus the benefits of the current airport modification over the difference in their costs is greater than one. We interpret this to mean that the benefit-cost ratio ranks the new airport ahead of the current airport.

We also note that the same ranking would result if we were to compare the present worths of the two alternatives. The present worth of the new airport (in millions of $) is

$$\text{PW(alternative A)} = \$337.95 - \$92.17 - \$150 - \$50 = \$45.78$$

The present worth of modifying the current airport (in millions of $) is

$$\text{PW(alternative B)} = \$175.10 - \$10 - \$115 - \$25 = \$25.10$$

In addition to the quantifiable aspects of each of the two airports, recall that the new airport is preferred in terms of the unmeasured value of space for future growth. As well, the new airport does not entail the difficulties associated with compensating home owners for loss in value of their homes. Together with the benefit-cost ratios (or equivalently, the present worths), this means that the city should build the new airport.■

A word of caution before we conclude this section. There may be some ambiguity in how to correctly calculate benefit-cost ratios, either conventional or modified. The reason is that for some projects, it may not be clear whether certain positive effects of projects are benefits to the public or reductions in cost to the government, or whether certain negative effects are disbenefits to the public or increases in costs to the government. As a result, benefit-cost ratios for a given project may not be unique. This lack of uniqueness means that a *comparison of the benefit-cost ratios of two projects is meaningless*. It is for this reason that the Treasury Board of Canada's *Benefit-Cost Analysis Guide* recommends the use of present worth or annual worth for comparisons rather than benefit-cost ratios. Despite this recommendation, benefit-cost ratios are still used extensively for project comparisons, and since they are still in use we present readers with enough material to properly construct and interpret them.

To illustrate the ambiguity in constructing benefit-cost ratios, consider Example 10.4. The reduction in the value of residential properties could have been treated as a disbenefit to the public instead of a cost to the government for compensating homeowners. As a result, there are two reasonable benefit-cost ratios for alternative B. The second is

$$\text{BCR}_2(\text{alternative B}) = \frac{\$175.10 - \$25}{(\$10 + \$115)} = 1.2008$$

The lack of uniqueness of benefit-cost ratios does not mean that the ratios cannot be used, but it does mean that some care needs to be given to their correct application. Comparison methods based on benefit-cost ratios remain valid because the comparison methods depend only on whether the benefit-cost ratio is less than or greater than one. Whether it is greater or less than one does not depend upon how positive and negative effects are classified. This is clearly illustrated in the following example.

Example 10.5

A certain project has present worth of benefits, B, and present worth of costs, C. As well, there is a positive effect with a present worth of d; the analyst is unsure of whether this positive effect is a benefit or a reduction in cost. There are two possible benefit-cost ratios,

$$\text{BCR}_1 = \frac{B + d}{C}$$

and

$$\text{BCR}_2 = \frac{B}{C - d}$$

For a ratio to exceed one, the numerator must be greater than the denominator. This means that, for BCR_1,

$$\text{BCR}_1 = \frac{B + d}{C} > 1 \quad \Leftrightarrow \quad B + d > C \quad \Leftrightarrow \quad B > C - d$$

But this is the same as

$$\frac{B}{C - d} = \text{BCR}_2 > 1$$

Consequently, any BCR that is greater than one or less than one will be so regardless of whether any positive effects are treated as positive benefits or as negative costs. A similar analysis would show that the choice of classification of a negative effect as a cost or as a reduction in benefits does not affect whether the benefit-cost ratio is greater or less than one.■

To conclude this section, we can note several things about the evaluation of public sector projects. First, the comparison methods developed in Chapters 4 and 5 are fully applicable to public sector projects. Second, the reader may encounter the use of benefit-cost ratios for public projects, despite the fact that benefit-cost ratios can be ambiguous. Next, the reader should be wary of decisions based on the absolute magnitude of benefit-cost ratios, as these ratios may be changed by reclassification of some of the positive effects of projects as benefits or as reductions in cost, or by reclassification of some of the negative effects as reductions in benefits or increases in costs. However, since the classifications do not affect whether a ratio is greater or less than one, it is possible to use benefit-cost ratios to reach the same conclusions as those reached using the methods of Chapters 4 and 5.

10.3.5 The MARR in the Public Sector

There are significant differences between the private sector and public (government) sector of society with respect to investment. As observed in the previous sections, public sector organizations provide a mechanism for resources to be allocated to projects believed beneficial to society in general. These include projects for which scale economies make it inefficient to have more than one supplier, and projects in markets that would otherwise suffer market failure. The government also regulates markets and collects and redistributes taxes toward the goal of maximizing social benefits. Typical projects include health, safety, education programs, cultural development, and infrastructure development. Profits generated by public projects are not taxed.

Private institutions and individuals, in contrast to public institutions, are more concerned with generating wealth (profits) and are taxed by the government on this wealth. We would therefore expect the MARR for a public institution to be lower than that of a private institution, because the latter has a substantial extra expense acting to reduce its profits.

In evaluating public sector projects, the MARR is used in the same way as in evaluating private sector projects—it captures the time value of money. In the private sector, the MARR expresses the minimum rate of return required on projects, taking into account that those profits will ultimately be taxed. The MARR used for public projects, often called the "social discount rate," reflects the more general investment goal of maximizing social benefits. Of course, as with private projects, public projects will sometimes be chosen on the basis of lowest present cost (or the equivalent) rather than highest present worth.

There is substantial debate as to what an appropriate social discount rate should be. Several viewpoints can shed some light on this issue.

The first is simply that the MARR should be the interest rate on capital borrowed to fund a project. In the public sector, funds are typically raised by issuing government bonds. Hence, the current bond rate might be an appropriate MARR to use. The second is that the MARR used to evaluate public projects should take into account that government spending on public projects consumes capital that might otherwise be used by taxpayers for private purposes. This viewpoint says that funds taken away from individuals and private organizations in the form of taxes might otherwise be used to fund private projects. This line of thinking might lead to a social discount rate that is the same as the MARR for the taxpayers. The two viewpoints are lower and upper bounds, respectively, to what can be seen as reasonable MARRs to apply when evaluating public projects.

In practice, the two viewpoints are captured by the *Benefit-Cost Analysis Guide*'s recommendation that public projects be evaluated at a 10% (real) social discount rate and that the evaluation should include a sensitivity analysis at lower and upper bounds of 8% and 12% percent, respectively. (See Close-Up 10.2.)

CLOSE-UP 10.2 MARR Used in the Canadian Public Sector

Public projects in Canada are evaluated using various MARRs. Some examples of the MARR applied in the public sector are shown in the table below. Due to uncertainty about what the *correct* MARR is, sensitivity analysis is recommended for public projects. For example, in the *Benefit-Cost Analysis Guide*, the Treasury Board suggests that one should use 10% as the base case and vary it in the range of 8% to 12%. The table provides guidelines suggested by other government agencies.

Project Type	Government Level	MARR	Suggested Range of Values
Benefit-Cost Analysis Guide	Treasury Board of Canada	10%	8%–12%
Pharmaceutical	Canadian Coordinating Office for Health Technology Assessment (CCOHTA)	5%	0%–3%
Land and resource management planning	Government of British Columbia	8%	6%–10%
Agricultural	Government of Alberta	13%	—
Assessment of damages in personal injury and fatal accident litigation	Provincial governments (B.C., Sask., Man., Ont., N.B., N.S., P.E.I.)	2.5–3.5%	—

REVIEW PROBLEMS

REVIEW PROBLEM 10.1

This review problem is adapted from an example in the Treasury Board's *Benefit-Cost Analysis Guide* (1976).

There are periodic floods in the spring and drought conditions in the summer that cause losses in a 15 000-square-kilometre Prairie river basin that has a population of 50 000 people. The area is mostly farmland, but there are several towns. Several flood control and irrigation alternatives are being considered:

1. Dam the river to provide flood control, irrigation, and recreation.
2. Dam the river to provide flood control and irrigation without recreation.
3. Control flooding with a joint Canada–United States water control project on the river.
4. Develop alternative land uses that would not be affected by flooding.

The constraints faced by the government are the following:

1. The project must not reduce arable land.
2. Joint Canada–United States projects are subject to delays caused by legal and political obstacles.
3. Damming of the river in the United States will cause damage to wildlife refuges.
4. The target date for completion is three years.

Taking into account the constraints, alternatives 3 and 4 above can be eliminated, leaving two:

1. Construct a dam for flood control, irrigation, and recreation.
2. Construct a dam for flood control and irrigation only.

A number of assumptions were made with respect to the dam and the recreational facilities:

1. An earthen dam will have a 50-year useful life.
2. Population and demand for recreational facilities will grow by 3.25% per year.
3. A three-year planning and construction period is reasonable for the dam.
4. Operating and maintenance costs for the dam will be constant in real dollars.
5. Recreational facilities will be constructed in year 2.
6. It will be necessary to replace the recreational facilities every 10 years. This will occur in years 12, 22, 32, and 42. Replacement costs will be constant in real dollars.
7. Operating and maintenance costs for the recreational facilities will be constant in real dollars.
8. The real dollar opportunity cost of funds used for this project is estimated to be in the range of 5% to 15%.

The benefits and costs of the two projects are shown in Tables 10.3 and 10.4.

Notice that the benefits and costs are estimated averages. For example, the value of reduced flood damages will vary from year to year, depending on such factors as rainfall and snowmelt. It is not possible to predict actual values for a 50-year period.

(a) What is the present worth of building the dam only? What is the benefit-cost ratio? What is the modified benefit-cost ratio? Use 10% as the MARR.

(b) What is the present worth of building the dam plus the recreational facilities? Use 10% as the MARR.

(c) What is the benefit-cost ratio for building the dam and recreation facilities together? What is the modified benefit-cost ratio?

(d) Which project, 1 or 2, is preferred, on the basis of your benefit-cost analysis? Use 10% as the MARR.

Table 10.3 Estimated Average Benefits of the Two Projects

Year	Flood Damage Reduction	Irrigation Benefits	Recreation Benefits
0	$ 0	$ 0	$ 0
1	0	0	0
2	0	0	0
3	182 510	200 000	27 600
4	182 510	200 000	28 497
⋮	⋮	⋮	⋮
52	182 510	200 000	132 288

Table 10.4 Estimated Average Costs of the Two Projects

Year	Dam Construction	Operating and Maintenance Dam	Recreation Construction	Operating and Maintenance Recreation
0	$ 300 000	$ 0	$ 0	$ 0
1	750 000	0	0	0
2	1 500 000	0	50 000	0
3	0	30 000	0	15 000
4	0	30 000	0	15 000
⋮	⋮	⋮	⋮	⋮
11	0	30 000	0	15 000
12	0	30 000	20 000	15 000
13	0	30 000	0	15 000
⋮	⋮	⋮	⋮	⋮
21	0	30 000	0	15 000
22	0	30 000	20 000	15 000
23	0	30 000	0	15 000
⋮	⋮	⋮	⋮	⋮
31	0	30 000	0	15 000
32	0	30 000	20 000	15 000
33	0	30 000	0	15 000
⋮	⋮	⋮	⋮	⋮
41	0	30 000	0	15 000
42	0	30 000	20 000	15 000
43	0	30 000	0	15 000
⋮	⋮	⋮	⋮	⋮
52	0	30 000	0	15 000

ANSWER

(a) We need to determine the present worth of benefits and costs of the dam alone. There are two benefits from the dam alone. They are those resulting from reduced flood damage and those associated with the benefits of irrigation. Both are approximated as annuities that start in year 3. We get the present worths of these benefits by multiplying the annual benefits by the series present worth factor and the present worth factor. The present worth of benefits resulting from reduced flood damage is

$$\text{PW(flood control)} = \$182\,510(P/A, 10\%, 50)(P/F, 10\%, 2)$$

$$= \frac{\$182\,510(9.99148)}{(1.1)^2}$$

$$= \$1\,495\,498$$

Similar computations give the present worth of irrigation as

$$\text{PW(irrigation)} = \$1\ 638\ 812$$

There are two costs for the dam: capital costs that are incurred at time 0 and over years 1 and 2, and operating and maintenance costs that are approximated as an annuity that begins in year 3. Capital costs are given by

$$\begin{aligned}\text{PW(dam, capital cost)} &= \$300\ 000 + \$750\ 000(P/F, 10\%, 1) \\ &\quad + \$1\ 500\ 000(P/F, 10\%, 2) \\ &= \$2\ 221\ 487\end{aligned}$$

The present worth of operating and maintenance costs is obtained in the same way as the present worths of flood control and irrigation benefits. The result is

$$\text{PW(operating and maintenance)} = \$245\ 822$$

The present worth of the dam alone is

$$\begin{aligned}\text{PW(dam)} &= \$1\ 495\ 498 + \$1\ 638\ 812 - (\$2\ 221\ 488 + \$245\ 822) \\ &= \$667\ 001\end{aligned}$$

The benefit-cost ratio for the dam is

$$\text{BCR(dam)} = \frac{\$1\ 495\ 498 + \$1\ 638\ 812}{\$2\ 221\ 488 + \$245\ 822} = 1.27$$

The modified benefit-cost ratio is given by

$$\text{BCRM(dam)} = \frac{\$1\ 495\ 498 + \$1\ 638\ 812 - \$245\ 822}{\$2\ 221\ 488} = 1.30$$

The present worth of the dam alone is positive, and both benefit-cost ratios are greater than 1. The dam alone appears to be economically viable.

(b) We already have the present w orths of benefits and costs for the dam alone. Therefore, we need only compute the present worths of benefits and costs for the recreation facilities. The capital costs for the recreation facilities consist of five outlays, in years 2, 12, 22, 32, and 42. The present worth of capital costs for the recreational facilities is given by

$$\begin{aligned}&\text{PW(recreation facilities, capital cost)} \\ &\quad = \$50\ 000(P/F, 10\%, 2) \\ &\quad\quad + \$20\ 000[(P/F, 10\%, 12) + (P/F, 10\%, 22) \\ &\quad\quad + (P/F, 10\%, 32) + (P/F, 10\%, 42)] \\ &\quad = \$51\ 464\end{aligned}$$

Operating and maintenance costs are estimated as an annuity that starts in year 3. The computation is the same as that of similar annuities that were used for the benefits and operating and maintenance costs of the dam. Thus, the present worth of recreation operating and maintenance costs is

$$\text{PW(recreation facilities, operating and maintenance)} = \$122\ 911$$

To obtain the present worth of the benefits from recreation, we need to define a growth-adjusted interest rate with $i = 10\%$ and $g = 3.25\%$ per year.

$$i° = \frac{1 + i}{1 + g} - 1 = \frac{1 + 0.10}{1 + 0.0325} - 1 = 0.0653$$

We then use this to get the present worth geometric gradient series factor,

$$(P/A,g,i,N) = \left(\frac{(1 + i°)^N - 1}{i°(1 + i°)^N}\right)\left(\frac{1}{1 + g}\right)$$

$$= \left(\frac{(1.0653)^{50} - 1}{0.0653(1.0653)^{50}}\right)\left(\frac{1}{1.0325}\right) = 14.19$$

To bring this to the end of year 0, we multiply by (*P*/*F*, 10%, 2). We then multiply by the initial value to get the present worth of recreation benefits.

$$\begin{aligned}\text{PW(recreation benefits)} &= [(P/A, 3.25\%, 10\%, N)(\$27\ 600)\ (P/F, 10\%, 2)] \\ &= 14.19(\$27\ 600)(1.1)^{-2} \\ &= \$323\ 679\end{aligned}$$

Another way to get this result is to use a spreadsheet. First, a column that contains the benefits in each year is created. The benefits start at $27 600 in the third year and grow at 3.25% each year. Each benefit is then multiplied by the appropriate present worth factor in another column to obtain its present worth. The individual present worths are then summed to obtain the overall total of $323 679. Some of the spreadsheet computations are shown in Table 10.5.

Table 10.5 Spreadsheet Computations

Year	Recreation Benefits	PW of Recreation Benefits
0	$ 0.00	$ 0.00
1	0.00	0.00
2	0.00	0.00
3	28 600.00	20 736.29
4	28 497.00	19 463.83
5	29 423.15	18 269.46
6	30 379.40	17 148.38
⋮	⋮	⋮
52	132 288.48	931.33
Total PW		323 678 .84

The total present worth of the recreation facilities is

$$\begin{aligned}\text{PW(total recreation facilities)} &= \$323\ 679 - \$51\ 464 - \$122\ 911 \\ &= \$149\ 304\end{aligned}$$

The present worth of the dam plus recreation facilities is obtained by adding the present worth of the dam alone to the present worth of the recreation facility. The final result is given by

$$\begin{aligned}\text{PW(dam and recreation facility)} &= \$667\,001 + \$149\,304 \\ &= \$816\,305\end{aligned}$$

In present worth terms, the present worth of the dam *and* recreation facility exceeds that of the dam alone, so the dam and the recreation facility should be chosen.

(c) The benefit-cost ratio for the dam and recreation facilities together is

BCR(dam and recreation)

$$= \frac{(\$1\,495\,498 + \$1\,638\,812 + \$323\,679)}{(\$2\,221\,488 + \$245\,822 + \$51\,464 + \$122\,911)} = 1.31$$

The modified benefit-cost ratio is

BCRM(dam and recreation)

$$= \frac{(\$1\,495\,498 + \$1\,638\,812 + \$323\,679 - \$245\,822 - \$122\,911)}{(\$2\,221\,488 + \$51\,464)}$$

$$= 1.36$$

The dam and recreation facilities project appears to be viable, since the benefit-cost ratios are greater than one.

(d) On the basis of benefit-cost ratios, which of the two projects, project 1 or project 2, should be chosen? The dam and recreation facility is more costly, so the correct benefit-cost ratio for comparing the two is

$$\text{BCR} = \frac{\text{Benefits(dam and recreation facility)} - \text{Benefits(dam)}}{\text{Costs(dam and recreation facility)} - \text{Costs(dam)}}$$

$$= \frac{\$323\,679}{(\$51\,464 + \$122\,911)} = 1.86$$

The ratio exceeds one, and hence the dam and recreation facilities project should be chosen. This is consistent with the original present worth computations.■

SUMMARY

Chapter 10 concerns decision making in the public sector. We started by considering why markets may fail to lead to efficient decisions. We presented three formal methods by which society seeks to remedy market failure, one of which is to have production by government. Next, we laid out three issues in decision making about government production. First, we saw that the identification and measurement of costs and benefits in the public sector are more difficult than in the private sector. Identification may be difficult because there may not be cash flows that reflect the costs or benefits. Measurement may be difficult because there are no prices to indicate values. Second, we discussed the use of benefit-cost ratios in public sector project evaluation. While it is possible to use benefit-cost ratios so as to give the same conclusions as those obtained using the comparison methods discussed in Chapters 4 and 5, care must be taken because there may be ambiguity in the way in which costs and benefits are classified. Third and last, we considered the MARR for government sector investments. The result of this discussion is that it is wise to use a range of MARRs corresponding to the opportunity cost of funds used in the public sector.

Engineering Economics in Action, Part 10B: Look at It from Their Side

"How was your trip, Naomi?" Anna Kulkowski asked as she walked into the conference room. Bill and Naomi were already there waiting. "Well, I'm here," responded Naomi, "but my jets are still lagging."

"You'll get used to it." Anna looked at Bill. "So, what's it going to cost to meet the CO_2 targets?"

"Naomi got some back-of-the-envelope figures from the people at Edgemont," Bill answered. "They're talking over $50 million between now and 2010, primarily to put in place a full cogeneration power plant to use biomass waste. If we amortize that over 20 years it's going to add about 20% to total costs per tonne of pulp. The DET system isn't going to help much, either. If we buy CO_2 credits from elsewhere, it will cost us just about the same amount, because all of the mills are pretty much in the same boat. Worse, we have an image problem to deal with. Buying credits labels us as a polluter. In B.C., the environmental groups will make sure everybody knows that, and it won't help us sell our products at all. We have to be seen to be clean."

"Well," Naomi continued, "I did talk to the people at the mill yesterday before coming home. Things may not be quite as bad as I first told Bill. First, even if costs do go up by 20%, we may still be able to compete by finding niches that demand 'environmentally friendly' pulp. Our investment in very low AOX takes us partially there, and the state-of-the-art biomass power plant can be an effective selling point. However, there will still be offshore mills that will be able to underprice us, even in the niche markets."

"That doesn't sound like a big help," Anna said. "What's the other reason for hope?"

"The other reason is that we may be able to make a reasonable argument for modifying the proposed regulatory system," Naomi said. "The form of the regulation doesn't make sense when a wider viewpoint is taken."

"Why not?" Anna asked.

"Well," said Naomi, "There has been an intense focus on meeting the terms of the Kyoto agreement, and the DET trading system. Also, there is a lot of attention paid to general preservation of the environment, particularly in B.C. But when you look at the overall effect that the proposed regulations have on the people affected, there are ways to increase the benefits and reduce the costs, while still meeting the needs of both the politicians and the environment."

"What do you mean in particular, Naomi?" Anna asked.

"Two things in particular," Naomi said. "In the first case, environmental leaders like Edgemont shouldn't be penalized. This is a terrible message to send. I don't know if we can make an economic argument about that one, but it's real."

"And the other?" Anna prompted.

"The other is the timing of capital improvements. I'm sure a benefit-cost analysis would show that it just doesn't make sense, economically or environmentally, for Edgemont to scrap its relatively new recovery boilers just to meet the 2008–2010 Kyoto reporting deadline. I propose that we lobby the government for some sensible changes, backed up by proper analysis."

"That sounds interesting, Naomi," Anna said. "Why don't you write this up. I'm going to Ottawa in a few weeks. I'll see if I can get a meeting with some people in the environment ministry. I'll try the idea out on them. It's worth a shot."

"By the way, Naomi," Anna continued, "good work!"

PROBLEMS

10.1 The following data are available for a project:

Present worth of benefits	\$17 000 000
Present worth of operating and maintenance costs	\$5 000 000
Present worth of capital costs	\$6 000 000

(a) Find the benefit-cost ratio.

(b) Find the modified benefit-cost ratio.

10.2 The following data are available for two mutually exclusive projects:

	Project A	Project B
PW(benefits)	\$19 000 000	\$15 000 000
PW(operating and maintenance costs)	5 000 000	8 000 000
PW(capital cost)	5 000 000	1 000 000

(a) Compute the benefit-cost ratios for both projects.

(b) Compute the modified benefit-cost ratios for both projects.

(c) Compute the benefit-cost ratio for the increment between the projects.

(d) Compute the present worths of the two projects.

(e) Which is the preferred project? Explain.

10.3 The following data are available for two mutually exclusive projects:

	Project A	Project B
PW(benefits)	\$17 000 000	\$17 000 000
PW(operating and maintenance costs)	5 000 000	11 000 000
PW(capital cost)	6 000 000	1 000 000

(a) Compute the benefit-cost ratios for both projects.

(b) Compute the modified benefit-cost ratios for both projects.

(c) Compute the benefit-cost ratio for the increment between the projects.

(d) Compute the present worths of the two projects.

(e) Which is the preferred project? Explain.

10.4 The following data are available for two mutually exclusive projects:

	Project A	Project B
PW(benefits)	\$17 000 000	\$15 000 000
PW(operating and maintenance costs)	5 000 000	8 000 000
PW(capital cost)	6 000 000	3 000 000

(a) Compute the benefit-cost ratios for both projects.

(b) Compute the modified benefit-cost ratios for both projects.

(c) Compute the benefit-cost ratio for the increment between the projects.

(d) Compute the present worths of the two projects.

(e) Which is the preferred project? Explain.

10.5 There are two beef packing plants, A and B, in the town of Reybourne, Saskatchewan. Both plants dump partially treated liquid waste into Lake Jeannette. The two plants together dump over 33 000 kilograms of BOD5 per day. (BOD5 is the amount of oxygen used by microorganisms over five days to decompose the waste.) This is more than half the total BOD5 dumped into Lake Jeannette. Reybourne town council wants to reduce the BOD5 dumped by the two plants by 10 000 kilograms per day.

The following data are available concerning the two plants:

	Outputs of the Two Plants	
	Steers/Day	BOD5/Steer (kg)
Plant A	20 000	1.0
Plant B	9 000	1.5

The costs of making reductions in BOD5 per steer are shown on the next page:

Reduction (kg/Steer)	Incremental Cost of Reducing BOD ($/kg/Steer) 0.1	0.2	0.3	0.4	0.5	0.6	0.7
Plant A	0.05	0.08	0.12	0.25	0.45	0.65	0.95
Plant B	0.15	0.15	0.15	0.15	0.15	0.35	0.45

For example, to reduce the BOD5 of Plant A by 0.25 kilograms per steer, the cost is calculated as

$$(0.1 \times 0.05) + (0.1 \times 0.08) + (0.05 \times 0.12) = 0.019/\text{steer}$$

The council is considering three methods of inducing the plants to reduce their BOD5 dumping: (1) a regulation that limits BOD5 dumping to 0.81 kilograms/steer, (2) a tax of $0.16/kilogram of BOD5 dumped, and (3) a subsidy paid by the town to the plants of $0.16/kilogram reduction from their current levels in BOD5 dumped.

(a) Verify that, if both plants reduce their BOD5 dumping to 0.81 kilograms/steer, there will be a 10 000 kilograms/day reduction in BOD5 dumped. What will this cost?

(b) Under a tax of $0.16/kilogram, how much BOD5 will Plant A dump? How much will Plant B dump? (Assume that outputs of steers would not be affected by the tax.) Verify that this will lead to more than a 10 000 kilograms/day reduction in BOD5. What will this cost?

(c) Under a subsidy of $0.16/kilogram reduction in BOD5, how much will Plant A dump? How much will Plant B dump? Verify that this will lead to more than a 10 000 kilograms/day reduction in BOD5. What will this cost?

(d) Explain why the tax and subsidy schemes lead to the same behaviour by the meat packing plants.

(e) Explain why the tax and subsidy schemes have lower costs for the company than the regulation.

10.6 There are three petrochemical plants, A, B, and C, in Port Jayne, Ontario. The three plants produce Good Stuff. Unfortunately, they also dump Bad Stuff into the air. Data concerning their outputs of Stuff are shown below:

	Outputs Good (kg/Day)	Bad/Good (cL/kg)
Plant A	17 000	10
Plant B	11 000	15
Plant C	8 000	18

The town council wants to reduce the dumping of Bad Stuff by 150 000 centilitres per day. Costs for reducing the concentration of Bad Stuff in output are shown on the next page.

The council is considering two methods: (1) Require all plants to meet the performance level of the best-practice plant, Plant A, which is 10 centilitres of Bad Stuff per kilogram of Good Stuff. (2) Impose a tax of $0.20/centilitre of Bad Stuff dumped.

Reduction (cL/kg)	Incremental Cost of Reducing Bad Stuff/Good Stuff ($/cL/kg) 1	2	3	4	5	6	7	8
Plant A	0.02	0.032	0.048	0.1	0.18	0.26	0.38	0.57
Plant B	0.06	0.06	0.063	0.068	0.075	0.193	0.27	0.405
Plant C	0.25	0.25	0.25	0.25	0.25	0.25	0.25	0.375

(a) What will be the reduction in dumping of Bad Stuff under the best-practice regulation? What will be the cost of this reduction?

(b) Under the tax, how much Bad Stuff will be dumped from the three plants combined? What will be each plant's reduction in dumping per kilogram of Good Stuff?

(c) How much will the reduction of dumping cost for each company under the tax?

10.7 In the summer of 2004, the Kitchener-Waterloo area often experienced the worst air pollution in Canada, affecting the health of hundreds of thousands of people. Explain how air pollution is an example of market failure. Give an example of how each of the four remedies for market failure listed in Subsection 10.2.2 might be applied to the case of air pollution.

10.8 The Canadian fishing industry has been devastated in recent years because of overfishing. On the east coast, cod fishing has almost disappeared, while on the west coast, salmon fishing has been considerably reduced. How is overfishing an example of market failure? Give an example of how each of the four remedies for market failure listed in Subsection 10.2.2 might be applied to the case of overfishing.

10.9 Consider these situations in which cutting of trees is relevant:

1. The Brown family owns a view house in West Vancouver. Their neighbours across the street, the Smith family, have trees on their lot that are obstructing the Browns' view of the Lions Gate Bridge. The Browns are the only ones affected by the Smiths' trees. The Browns have asked the Smiths to top their trees, but the Smiths refuse to do so. The Browns have also offered to pay the Smiths for the topping and an additional $500 to cover any loss they might feel because their tall trees were topped. The Smiths still refuse.
2. The Brown family, the Green family, the White family, and the Blue family own view houses in West Vancouver. Their neighbours across the street, the Smith family, have trees on their lot that are obstructing everyone's view of the Lions Gate Bridge. The Browns, Greens, Whites, and Blues have asked the Smiths to top their trees, but the Smiths refuse to do so.
3. Timber companies on Vancouver Island sometimes use clear-cutting on old-growth forests. Environmentalists have asked the companies to change this practice because it leads to reduced biodiversity.

Why is there no market failure in the first situation involving the Smiths and the Browns? Why may there be market failure in the situation with several families? Why is there market failure in the third situation involving the timber companies?

10.10 Technical changes in electricity supply and information transmission have made it efficient for consumers of both services to be served by suppliers using different technologies and operating in different locations. Does this increase or decrease the need for government regulation in these industries?

10.11 An electric utility company is considering a reengineering of a major hydroelectric facility. The project would yield greater capacity and lower cost per kilowatt-hour of power. As a result of the project, the price of power would be reduced. This is expected to increase the quantity of power demanded. The following data are available:

Effect of Reduced Price of Power	
Current price ($/kWh)	0.07
Current consumption (kWh/year)	9 000 000
New price ($/kWh)	0.05
Expected consumption (kWh/year)	12 250 000

What is the annual benefit to consumers of power from this project?

10.12 Brisbane and Johnsonburg are two Prairie towns separated by the Wind River. Traffic between them crosses the river by a ferry run by the Johnsonburg Ferry Company, who charge a toll. The province is considering building a bridge somewhat upstream from the ferry crossing; there would be no toll on the bridge. Travel time between the towns would be about the same with the bridge as with the ferry because of the bridge's upstream location. The following information is available concerning the crossing:

Ferry/Bridge Information	
Ferry crossings (number/year)	60 000
Average cost of ferry trip ($/crossing)	1
Ferry fare ($/crossing)	1.5
Bridge toll ($/crossing)	0
Expected bridge crossings (number/year)	90 000
EAC of bridge ($/year)	85 000

Note that all data are on an annual basis. The cost of the bridge is given as the equivalent annual cost of capital and operating costs. We assume that all bridge costs are independent of use, that is, there are no costs that are due to use of the bridge. The average cost per crossing of the ferry includes capital cost and operating cost.

(a) If the bridge were built, what would be the annual benefits to travellers?

(b) How much would the owners of the Johnsonburg Ferry Company lose if the bridge were built?

(c) What would be the effect on taxpayers if the bridge were built? (Assume that Johnsonburg Ferry pays no taxes.)

(d) What would be the net social gains or losses if the bridge were built? Take into account the effects on travellers, Johnsonburg Ferry owners, and taxpayers.

(e) Would the net social gains or losses be improved if there were a toll for crossing the bridge?

10.13 It is common for municipalities to provide snowplow service for public roads. The major benefit of such services is to allow the convenient movement of vehicles over public roads following snow accumulation, at a cost of the snowplow (capital and operating) and driver. Are there other costs and benefits? List all you can think of, along with how they could be measured?

10.14 Most provinces and states provide travel information kiosks alongside major highways just across the border from neighbouring provinces and states. These kiosks provide maps and brochures on attractions, and some will make hotel and campground reservations. There are obvious costs, such as staffing costs and building capital and maintenance costs. What other costs and benefits are associated with this government service? How can these costs and benefits be measured?

10.15 The Ontario government is considering putting a carpool parking lot at a new interchange near Cambridge, Ontario, on the main east-west highway through the province, the 401. The parking lot allows commuters to meet in separate cars, park all but one, and proceed in one car to a joint destination. Studies estimate that an average of 200 cars will be parked at the lot on weekdays, with 1/4 of that number on weekends. The average commuting distance from that intersection is 75 kilometres, and the marginal cost of driving an average car is $0.28 per kilometre.

(a) If it is assumed that all the cars that are parked in the lot would otherwise have been driven to work, how much will be saved by this commuter parking lot per year? Assume an interest rate of 0.

(b) How could you find out how many would have been parked somewhere else?

(c) How could you calculate the benefit to all drivers of having fewer cars on the road?

10.16 A medium-sized city in Alberta (population 250 000, 45 000 families) is considering introducing a recycling program. The program would require them to separate newspaper, cardboard, and cans from their regular waste so that it could be collected weekly, sorted, and sold, rather than put into the local landfill site. The program would also require households to separate "wet," or compostable, waste, which they would then be responsible for composting. The city would provide free composting units to the households. What kinds of potential costs and benefits can you identify for this project? How might this information be gathered?

10.17 Consider Problem 10.5. We saw that an effluent tax enabled the same reduction in BOD5 as a regulation, but with a lower cost. We did not consider the distribution of this cost.

(a) How much tax does Plant A pay under the tax of $t=$ $0.6667 per kilogram dumped? How much does Plant B pay?

(b) What are the two total effluent costs (tax plus cleaning cost) for Plant A? For Plant B?

(c) Compare these costs with the costs under regulation.

(d) Suppose that the province used the proceeds of the tax to provide benefits that had equal value to each plant. How great would these benefits have to be to ensure that both plants would be better off with the combination of tax and benefits?

10.18 A four-day school week has been advocated as a means of reducing education costs. School days would be longer so as to maintain the same number of hours per week as under the current five-day-a-week system. The main cost savings that are expected are in school cleaning and maintenance and in school bus operation. The main effects that are anticipated are

(a) Reduced school cleaning and maintenance.
(b) Reduced use of school buses on the off-day.
(c) Reduced driving to school by parents, students, and staff on the off-day.
(d) Reduction in public transportation use on the off-day.
(e) Some high school students and school staff will seek part-time work for the off-day.
(f) Reduced absences by students and staff. This is mainly because some required personal activities could be scheduled for the off-day.
(g) Reduced subsidized school lunch requirements.
(h) Greater need for day care on the off-day for working parents. About a third of elementary schools could be opened for day care. The costs would be covered by fees.
(i) Learning by elementary students may be reduced because of their limited attention spans.
(j) Lower school taxes.

Which of these effects are benefits? Who receives the benefits? Which are costs? Who bears the costs? Which are neither costs nor benefits?

10.19 A new suburban development project is being planned near Petroville, Alberta. There is now a two-lane road from the site of the development, along the river, to Petroville. The new development will require additional road capacity. Two alternatives are being considered. The first is to upgrade the existing road to four wide modern lanes. The second is to build a new four-lane highway through Beaver Hill tunnel. The following data are available concerning the two routes:

Route	Distance	First Cost	Operating and Maintenance Costs per Year
River	20 km	$21 million	$90 000
Tunnel	10 km	$45 million	$130 000

The planning period is 40 years for both routes. The MARR is 10%. Cars will travel at 100 kilometres per hour along either route. Operating cost for the cars is expected to be about $0.25 per kilometre along either route. About 400 000 trips per year are expected on either route.

(a) Which route should be built? Use only the data given. Use annual worth to make your decision.

(b) What important benefit of the Beaver Hill tunnel route has been left out of the analysis?

(c) Do you need more information about travellers to determine if the benefit that has been left out of the analysis from part (b) would change the recommendation? Explain.

(d) How would the possibility of collecting a toll on the tunnel route affect your recommendation?

10.20 The Principality of Upper Pigovia has just one export, pig crackling. The crackling is produced in two plants, Old Gloria and New Gloria. Both plants give off a delightful, mouth-watering odour while in operation. This odour has created a health problem. The citizens' appetites are huge, and, consequently, so are the citizens. Princess Piglet has decreed that the daily emission of odour from the two plants must be reduced by 7000 Odour Units (OU) per day. The Princess prides herself on her even-handedness. The decree specifies that each plant is to reduce its emission by 3500 OU per day. The following table shows the incremental cost per 1000 OU of attaining various levels of odour reduction in the two plants. To help in interpreting the table, note that the cost for Old Gloria to remove 3000 OU per day would be 2(U$25) + U$30 = U$80.

Quantity of Odour Removed	Incremental Cost of Removing Odour in $U/1000 OU			
(1000 OU/Day)	0 to 2	Over 2	Over 4	Over 5
Old	25	30	40	50
New	20	20	25	30

(a) What is the cost per day of implementing Princess Piglet's decree?

(b) Can you suggest a tax scheme that will yield the same reduction in odour emission as the decree at a lower cost?

(c) Can you suggest a subsidy scheme that will yield the same reduction in odour emission as the decree, but at a lower cost?

10.21 A provincial ministry of transportation is considering the construction of a bridge over a narrow point in a lake. Traffic now goes around the lake. The bridge will save 30 kilometres in travel distance. Three alternatives are being considered: (1) do nothing, (2) build the bridge, and (3) build the bridge and charge a toll. If the bridge is built, the present road will be maintained, but its use will decline. One effect of the reduced use of the present road is a loss of revenue by businesses along that road. Available data is given in the following table.

Costs and Benefits	Do Nothing ($)	Bridge with Toll ($)	Bridge Without Toll ($)
First cost	0	46 000 000	46 000 000
Annual road and/or bridge operating and maintenance costs	160 000	80 000	10 000
Annual vehicle operating cost	3 300 000	100 000	100 000
Annual driver and passenger time cost	2 500 000	500 000	500 000
Annual accident cost	500 000	10 000	10 000
Annual revenue lost by roadside businesses	0	1 000 000	1 000 000
Annual toll revenues	N/A	1 200 000	N/A

(a) Identify social benefits and costs of the bridge.

(b) Which of the costs and revenues in the table are neither social benefits nor social costs?

(c) The table makes an implicit assumption about the effect of the toll on bridge traffic. This assumption is probably incorrect. What is the assumption?

(d) How would you expect the toll to affect benefits and costs? Explain.

10.22 An example concerning the effect of a flood control project is found in the benefit-cost analysis chapter of an imaginary engineering economics text. The benefits of the project are stated as:

Benefit	Cost
Prevented losses due to floods in the Conestogo River Basin	\$480 000/year
Annual worth of increased land value in the Conestogo River Basin	\$48 000

Comment on these two items.

10.23 A province is considering the construction of a bridge. The bridge would cross a narrow part of a lake near a provincial park. The major benefit of the bridge would be reduced travel time to travel to a campsite from a nearby urban centre. This lowers the cost of camping trips at the park. As well, an increase in the number of visits resulting from the lower cost per visit is expected.

Data concerning the number of weeklong visits and their costs are shown below:

	Number of Visits and Average Cost per Visit to Park	
Inputs	**Without Bridge**	**With Bridge**
Travel cost (\$)	140	87.5
Use of equipment (\$)	50	50
Food (\$)	100	100
Total (\$)	290	237.5
Number of visits/year	8000	11 000

The following data are available as well:

1. The bridge will take one year to build.
2. The bridge will have a 25-year life once it is completed. This means that the time horizon for computations is 26 years.
3. Construction cost for the bridge is \$3 750 000. Assume that this cost is incurred at the beginning of year 1.
4. Annual operating and maintenance costs for the bridge are given by

   $\$7500 + 0.25q$

   where 7500 is the fixed operating and maintenance cost per year and q is the number of crossings.
5. Operating and maintenance costs are incurred at the end of each year over which the bridge is in operation. This is at the ends of years 2, 3, . . . , 26.
6. The MARR is 10%.

(Notice that the annual benefits for this project were computed as part of the discussion of Example 10.2.)

(a) Compute the present worth of the project.

(b) Compute the benefit-cost ratio.

(c) Compute the modified benefit-cost ratio.

10.24 Consider the bridge project of Problem 10.23. There, we assumed there would be no toll for crossing. Now suppose the province is considering a toll of $7 per round trip over the bridge. They estimate that, if the toll is charged, the number of park visits will rise to only 10 600 per year instead of 11 000.

(a) Compute the present worth of the project if the toll is charged.

(b) Why is the present worth of the project reduced by the toll?

10.25 The town of Migli Lake, Manitoba, has a new subdivision, Paradise Mountain, at its outskirts. The town wants to encourage the growth of Paradise Mountain by improving transportation between Paradise Mountain and the centre of Migli Lake. Two alternatives are being considered: (1) new buses on the route between Paradise Mountain and Migli Lake centre and (2) improvement of the road between Paradise Mountain and Migli Lake centre.

Both projects will have as their main benefit improved transportation between Paradise Mountain and Migli Lake centre. Rather than measure the value of this benefit directly to the city, engineers have estimated the benefit in terms of an increase in the value of land in Paradise Mountain. That is, potential residents are expected to show their evaluations of the present worth of improved access to the town centre by their willingness to pay more for homes in Paradise Mountain.

The road improvement will entail construction cost and increased operating and maintenance costs. As well, the improved road will require construction of a parking garage in the centre of Migli Lake. The new buses will have a first cost as well as operating and maintenance costs. Information about the two alternatives is shown below.

(a) Compute the benefit-cost ratio of both alternatives. Is each individually viable?

	Road Improvement	New Buses
First cost	$15 000 000	$ 4 500 000
PW (operating and maintenance cost)	5 000 000	12 000 000
Parking garage cost	4 000 000	
Estimated increased land value	26 000 000	18 000 000

(b) Using an incremental benefit-cost ratio approach, which of the two alternatives should be chosen?

(c) Compute the present worths of the two alternatives. Compare the decision based on present worths with the decisions based on benefit-cost ratios.

10.26 A provincial government is considering a new two-lane road through a mountainous area. The new road would improve access to a city from farms on the other side of the mountains. The improved access would permit farmers to switch from grains to perishable soft fruits that would be either frozen at an existing plant near the city or sold in the city. Two routes are being considered. Route A is more roundabout. Even though the speed on route B would be less than that on Route A, the trip on Route B would take less time. Almost all vehicles using either road would go over the full length of the road. A Department of Transport engineer has produced information shown in the table below. The province uses a 10% MARR for road projects. The road will take one year to build. The province is using a 21-year time horizon for this project, since it is not known what the market for perishable crops will be in the distant future. Comment on the engineer's list of benefits; there may be a couple of errors.

Costs and Benefits of the New Road

	Route A	Route B
Properties		
Distance (km)	24	16
Construction cost ($)	53 400	75 000
Operating and maintenance cost per year ($)	60	45
Resurfacing after 10 years of use ($)	3 100	2 350
Road Use		
Number of vehicles per year	1 000 000	1 200 000
Vehicle cost per km ($)	0.3	0.3
Speed (kph)	100	80
Value of time per vehicle hour ($)	15	15
Benefits		
Increased crop value per year ($)	13 500	18 000
Increased land value ($)	104 484.6	139 312.8
Increased tax collections per year ($)	811.21	1 081.61

10.27 Consider the road project in Problem 10.26. (*Note:* Correct for the errors mentioned.)

(a) Compute a benefit-cost ratio for Route A with road use costs counted as a cost.

(b) Compute a benefit-cost ratio for Route A with road use costs counted as a reduction in benefits.

(c) In what way are the two benefit-cost ratios consistent, even though the numerical values differ?

(d) Make a recommendation as to what the province should do regarding these two roads. Explain your answer briefly.

10.28 Find the net present worths for the dam and the dam plus recreation facilities considered in Review Problem 10.1. Use a MARR of 15%. Make a recommendation as to which option should be adopted.

10.29 The recreation department for Port Elgin is trying to decide how to develop a piece of land. They have narrowed the choices down to tennis courts or a swimming pool. The swimming pool will cost $2.5 million to construct, and will cost $300 000 per year to operate, but will bring benefits of $475 000 per year over its 25-year expected life. Tennis courts would cost $200 000 to build, cost $20 000 per year to operate, and bring $60 000 per year in benefits over their 8-year life. Both projects are assumed to have a salvage value of zero. The appropriate MARR is 5%.

(a) Which project is preferable? Use a BCR and an annual worth approach.

(b) Which project is preferable? Use a BCRM and an annual worth approach.

10.30 The environmental protection agency of a county would like to preserve a piece of land as a wilderness area. The owner of the land will lease the land to the county for a 20-year period for the sum of $1 750 000, payable immediately. The protection agency estimates that the land will generate benefits of $150 000 per year, but they will forgo $20 000 per year in taxes. Assume that the MARR for the county is 5%.

(a) Calculate a BCR using annual worth and classify the forgone taxes as a cost to the government.

(b) Repeat part (a), but consider the forgone taxes a reduction in benefits.

(c) Using your results from part (b), determine the most the county would be willing to pay for the land (within $10 000) if they accept projects with a BCM of 1 or more.

10.31 The data processing centre at a local government tax centre has been plagued recently by the increasing incidence of repetitive strain injuries in the workplace. Health and Safety consultants have recommended to management that they invest in upgrading computer desks and chairs at a cost of $500 000. They advise that this would reduce the number and severity of medical costs by $70 000 per year and that productivity losses and sick leaves could be reduced by a further $80 000/year. The furniture has a life of eight years with zero scrap value. The city uses a MARR of 9%. Should the centre purchase the furniture? Use a benefit-cost analysis.

10.32 Several new big-box stores have created additional congestion at an intersection in north Winnipeg. City engineers have recommended the addition of a turn lane, a computer-controlled signal, and sidewalks, at an estimated cost of $1.5 million. The annual maintenance costs at the new intersection will be $8000, but users will save $50 000 per year due to reduced waiting time. In addition, accidents are expected to decline, representing a property and medical savings of $175 000 per year. The renovation is expected to handle traffic adequately over a 10-year period. The city uses a MARR of 5%.

(a) What is the BCR of this project?

(b) What is the BCRM of the project?

(c) Comment on whether the project should be done.

10.33 What determines the MARR used on government-funded projects?

10.34 How will a decrease in the tax rate on investment income affect the MARR used for evaluating government-funded projects?

10.35 How does an expectation of inflation affect the MARR for public sector projects?

10.36 A provincial department of transportation has $16 500 000 that it can commit to highway safety projects. The goal is to maximize the total life-years saved per dollar. The potential projects are: (1) flashing lights at 10 railroad crossings; (2) flashing lights and gates at the same 10 railroad crossings; (3) widening the roadway on an existing rural bridge from 3 to 3.5 metres; (4) widening the roadway on a second and third rural bridge from 3 to 3.5 metres; (5) reducing the density of utility poles on rural roads from 30 to 15 poles per kilometre; and (6) building runaway lanes for trucks on steep downhills.

	Highway Safety Projects	Total Cost ($)	Life-Years Saved per Year
1	Flashing lights	450 000	14
2	Flashing lights and gates	750 000	20
3	Widening bridge #1	1 200 000	14
4	Widening bridge #2	700 000	10
5	Widening bridge #3	1 100 000	18
6	Pole density reduction	3 000 000	96
7	Runaway lane #1	6 000 000	206
8	Runaway lane #2	6 000 000	156

The data for the flashing-lights and flashing-lights-with-gates projects reflect the costs and benefits for the entire set of 10 crossings. Portions of the projects may also be completed for individual crossings at proportional reductions in costs and savings. At any single site, the lights and lights-with-gates projects are mutually exclusive. Any fraction of the reduction of utility pole density project can be carried out. Data concerning costs and safety effects of the projects are shown above. (A life-year saved is one year of additional life for one person.)

Advise the department of transportation how best to commit the $16 500 000. Assume that the money must be used to increase highway safety.

10.37 A provincial department of transportation is considering widening lanes on major highways from 6 to 7.5 metres. The objective is to reduce the accident rate. Accidents have both material and human costs. The following data are available for highway section XYZ:

Lane Widening on Section XYZ	
Accidents per 100 000 000 vehicle-km in 6 m lanes	150
Accidents per 100 000 000 vehicle-km in 7.5 m lanes	90
Serious personal injuries per accident	10%
Average non-human cost per accident ($)	2500
Annual road use (vehicles)	7 500 000
First cost per kilometre ($)	175 000
Operating and maintenance costs per km/year ($)	7500
Project life (years)	25
MARR	10%

(a) Compute the present worth of costs of lane-widening.

(b) Compute the present worth of savings of non-human accident costs.

(c) What minimum value for a serious personal injury would justify the project?

10.38 The federal government is considering three flood control projects in Manitoba. Projects A and B consist of permanent dikes along the Rat River near Winnipeg. Project C is a dam. The dam will have recreation and irrigation benefits as well as the flood control benefits. Facts about the three projects are shown in the following table. Each project has a life of 25 years. The MARR is 10%.

	Project A	Project B	Project C
First cost (millions of $)	25	32	52
Annual benefits (millions of $)	3.3	4.2	7.1
Annual operating and maintenance costs (millions of $)	0.3	0.3	0.5

(a) Use present worth to choose the best project.

(b) Compute the benefit-cost ratios for the three projects.

(c) Use incremental benefit-cost to choose the best project.

(d) It is possible that the dam (Project C) will cause some wildlife damage. This damage might be prevented by an additional expenditure of $3 000 000 when the dam is constructed. Does this change the choice of best project? Compute the benefit-cost ratio of Project C assuming that the additional $3 000 000 is an addition to the first cost.

10.39 A provincial highway department needs to upgrade a 20 kilometre rural road to accommodate increased traffic. There will be 5000 cars per day in the first year after upgrading. The number of cars will increase each year by 200 cars per day for the next 20 years. A modern two-lane road will be adequate for current traffic levels, but it will be inadequate after about 10 years. At that time, a four-lane road will be required. A four-lane road will permit greater speed even at the current traffic level.

Two alternatives are being considered. One proposal is for a modern two-lane road now, followed by addition of lanes to make a four-lane road after 10 years. The second alternative is for a four-lane road now. The planning horizon for both alternatives is 20 years. Costs for the alternatives are shown in the table below. All values are in real dollars.

	Two Lanes	Four Lanes
First cost (millions of $)	20	32
Adding lanes after 10 years (millions of $)	18	0
Annual operating and maintenance costs (years 1–10) (in $000s)	30	40
Annual operating and maintenance costs (years 11–20) (in $000s)	40	40

The average value of travel time for each car is estimated at $40 an hour. Speeds will be as shown on the next page.

(a) Which alternative has a greater present worth if the MARR is 5%?

(b) Which alternative has a greater present worth if the MARR is 20%?

(c) Suppose the real after-tax rate of return on government savings bonds is 5% and the real before-tax rate of return on investment in the private sector is 20%. Use your answers to parts (a) and (b) to decide which alternative is better. Explain your answer.

Years	Two Lanes	Four Lanes
1 to 5	80 kmh	100 kmh
6 to 10	70 kmh	100 kmh
11 to 20	N/A	100 kmh

10.40 A small municipality needs to upgrade the town dump to meet provincial environment standards. Two alternatives are being considered. Alternative A has a first cost of \$420 000 and annual operating and maintenance costs of \$52 500. Alternative B has a first cost of \$315 000 and annual operating and maintenance costs of \$74 000. Both alternatives have 15-year lives with no salvage. An increase in dumping fees for households and business in the town is expected to yield \$50 000 per year.

(a) Which alternative has the lower cost if the MARR is 5%? Use annual worth.

(b) Which alternative has the lower cost if the MARR is 20%? Use annual worth.

(c) The after-tax return on government savings bonds is 5%. The average rate of return before taxes in private sector investment is 20%. Which alternative should be chosen?

Ontario's Approach to Reducing Acid Rain

CANADIAN 10.1 MINI-CASE

Fossil-fuel-burning electric power plants in the United States and Canada produce air pollution as a byproduct of their operation. The pollutants produced by Midwestern U.S. and Ontario plants are linked to high levels of acid rain in the northeastern United States, Ontario, and Quebec as well as a growing number of "smog" days in Ontario. The primary pollutants associated with acid rain are nitrogen oxide and sulphur dioxide (NO_X, SO_2). Acid rain is linked to increased fish morbidity rates, destruction of forests, and building decay. Overall increases in smog levels have led to a host of respiratory illnesses such as asthma and lung cancer in the northeastern United States and Ontario.

Air quality is a public good. Though there are markets for fossil fuels and electricity, they fail to take into account the public costs of smog, acid rain, and climate change. Government regulation has been used as a mechanism to produce more socially desirable outcomes.

The U.S. Environmental Protection Agency (EPA) uses a market-based "Cap and Trade" policy instrument to permit plants to adopt the most cost-efficient approach to reduce their SO_2 emissions. The EPA sets caps, or limits, on allowable total emissions levels, and then issues a limited number of permits that are auctioned off annually. These permits can be traded or saved. In 2003, a permit for the atmospheric release of one tonne of SO_2 cost approximately US\$170. Between 1990 and 2002, this policy reduced the level of SO_2 emissions by approximately 32%.

In 2002, Ontario introduced a "Cap, Credit, and Trade" system for controlling SO_2 emissions. Ontario Power Generation, one of the largest SO_2 emitters, is currently the only organization with an emissions cap on SO_2. OPG is permitted to allocate its emission allowances to its plants in whatever manner is most cost-efficient. Plants in Ontario and the northeastern United States that are not capped can create emission reduction credits by voluntarily reducing their emissions. Both emission allowances and emission reduction credits can be traded or saved should an emitter be unable to meet its targets.

Sources: "EPA's Clean Air Markets—Acid Rain Program," Environmental Protection Agency site, www.epa.gov/airmarkets/arp/index.html; "Air," Ontario Ministry of the Environment site, www.ene.gov.on.ca/air.htm. Both accessed October 2, 2004.

Discussion

Despite the early successes of these systems, some issues remain. First, these regulations may be successful in reducing overall emission levels, but they may not address local "hot spots." Second, the global targets themselves may not be sufficient to truly reflect the social costs of acid rain. Finally, there may be more efficient ways of distributing the allowances so that the overall cost of operating the system is lowered.

Government regulatory programs such as the "Cap, Credit, and Trade" system are established to remediate the impact of market failure. These programs are intended to act as an incentive to allow plants to adopt the most cost-efficient means of emissions reductions. But despite the best intentions of the government, individuals or companies can exploit poorly designed programs for their own gain. At the best of times, such as with SO_2 emission controls in Ontario, the inherent process of establishing the system incurs inefficiencies. On the other hand, the benefits to society of correcting situations of market failure can nevertheless be enormous.

Questions

List some of the services provided by your municipal, provincial, or federal government. For each of these services:

1. Describe the result if the service were provided instead by private individuals or firms. Does the government service correct a potential market failure?
2. Estimate the direct cost per served individual for the government service. Would this cost go up or down if the service were provided privately?
3. Are there other costs that would be incurred by society as a whole if the service were provided privately? Estimate these costs if you can.
4. Is society generally better off with or without government provision of this service? Can you estimate the quantitative value of the benefit?

CHAPTER 11

Dealing with Uncertainty: Sensitivity Analysis

Engineering Economics in Action, Part 11A: Filling a Vacuum

"I have something new for you, Naomi. It's going to require some imagination and disciplined thinking at the same time." Anna's tone indicated to Naomi that something interesting was coming.

"As you may know, Canadian Widgets has been working toward getting into consumer products for some time." Anna continued, "An opportunity has come up to cooperate on a new vacuum cleaner with Powerluxe. They have some potential designs for a vacuum with electronic sensing capabilities. It will sense carpet height and density. It will then adjust the power and the angle of the head to optimize cleaning on a continuous basis. Our role in this would be to design the manufacturing system and to do the actual manufacturing for North America. Sound interesting so far?"

"Yes, Ms. Kulkowski," Naomi answered. Naomi couldn't help being respectful in front of Anna, who was Canadian Widget's president, among other things. "What would my role be?"

"For one thing, you will be working with Bill Astad from head office. I think you know him." Naomi did; she had given Bill some advice on handling variable rates of inflation a few weeks earlier. "First, we want to establish some idea of the demand for the product. We have to see how this might affect our manufacturing capacity and capital costs to determine if the whole idea is even feasible. Later on, we will have to make some design decisions, but the general feasibility comes first."

"Do we have any market studies from Powerluxe?" Naomi asked.

"Yes, but they seem to be guesswork and magic, not hard figures. After all, no one has sold a product like this before."

Naomi looked pensive. "Sounds as though we'll need to do some sensitivity analysis on this one." She muttered more to herself than Anna. Then to Anna, "Right. We'll do what we can. Thanks for the nice opportunity, Anna. This is interesting!"

11.1 Introduction

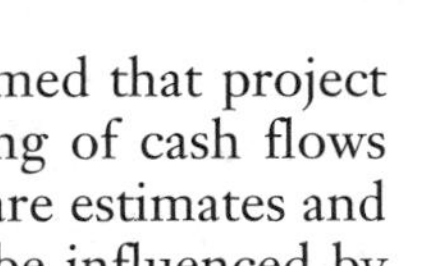

To this point in our coverage of engineering economics, we have assumed that project parameters such as prices, interest rates, and the magnitude and timing of cash flows have values that are known with certainty. In fact, many of these values are estimates and are subject to some uncertainty. Since the results of an evaluation can be influenced by variations in uncertain parameters, it is important to know how *sensitive* the outcome is to variations in these parameters.

There are several reasons why there may be uncertainty in estimating project parameter values. Technological change can unexpectedly shorten the life of a product or piece of equipment. A change in the number of competing firms may affect sales volume or market share or the life of a product. In addition, the general economic environment may affect inflation and interest rates and overall activity levels within an industry. All of these factors may result in cash flows different from what was expected in both timing and size, or in other changes to the parameters of an evaluation.

Making decisions under uncertainty is challenging because the overall impact of uncertainty on project evaluation may initially not be well understood. **Sensitivity analysis** is an approach to project evaluation that can be used to gain a better understanding of how uncertainty affects the outcome of the evaluation by examining how sensitive the outcome is to changes in the uncertain parameters. It can help a manager decide whether it is worthwhile to get more accurate data as part of a more detailed evaluation, or whether it may be necessary to control or limit project uncertainties. The problem of decision making under *risk*, where there is information about the probability distribution of parameter values, is discussed in the next chapter.

Economic analyses are not complete unless we try to assess the potential effects of project uncertainties on the outcomes of the evaluations. Because parameter estimates can be so hard to determine, analysts usually consider a range of possible values for uncertain components of a project. There is then naturally a range of values for present worth, annual worth, or whatever the relevant performance measure is. In this way, the analyst can get a better understanding for the range of possible outcomes and can make better decisions.

In this chapter, we will consider three basic sensitivity analysis methods commonly used by analysts. These methods are used to better understand the effect that uncertainties or errors in parameter values have on economic decisions. The first method is the use of *sensitivity graphs*. Sensitivity graphs illustrate the sensitivity of a particular measure (e.g., present worth or annual worth) to one-at-a-time changes in the uncertain parameters of a project. Sensitivity graphs can reveal key parameters that have a significant impact on the performance measures of interest and hence we should be particularly careful to get good estimates for these key parameters.

The second method for sensitivity analysis is the use of *break-even analysis*. Break-even analysis can answer such questions as "What production level is necessary in order for the present worth of the project to be greater than zero?" or "Below what interest rate will the project have a positive annual worth?" Break-even analysis can also give insights into comparisons between projects. With break-even analysis, we can answer questions like "What scrap value for the proposed forklift will cause us to be indifferent between replacing the old forklift and not replacing it?"

The third method we will introduce is called *scenario analysis*. Both sensitivity graphs and break-even analyses have the drawback that we can look at parameter changes only one at a time. Scenario analysis allows us to look at the overall impact of different sets of parameter values, referred to as "scenarios," on project evaluation. In other words, they allow us to look at the impact of varying several parameters at a time. In this way, the analyst comes to understand the range of possible economic outcomes.

Each of these three sensitivity analysis methods—sensitivity graphs, break-even analysis, and scenario analysis—tries to assess the sensitivity of an economic measure to uncertainties in estimates in the various parameters of the problem. A thorough economic evaluation should include aspects of all three types of analysis.

11.2 Sensitivity Graphs

The first sensitivity analysis tool we will look at is the sensitivity graph. Sensitivity graphs are used to assess the effect of one-at-a-time changes in key parameter values of a project on an economic performance measure. We usually begin with a "base case" where all the estimated parameter values are used to evaluate the present worth, annual worth, or IRR of a project, whatever the appropriate measure is. We then vary parameters above and below the base case one at a time, *holding all other parameters fixed*. A graph of the changes in a performance measure brought about by these one-at-a-time parameter changes is called a **sensitivity graph**. From the graph, the analyst can see which parameters have a significant impact on the performance measure and which do not.

Example 11.1

Cogenesis Corporation is replacing their current steam plant with a 6-megawatt cogeneration plant that will produce both steam and electric power for their operations. The new plant will use wood as a source of fuel, which will eliminate the need for Cogenesis to purchase a large amount of electric power from a public utility. To move to the new system, Cogenesis will have to integrate a new turbogenerator and cooling tower with their current system. The estimated first cost of the equipment and installation is \$3 000 000, though there is some uncertainty surrounding this estimate. The plant is expected to have a 20-year life and no scrap value at the end of this life. In addition to the first cost, the turbogenerator will require an overhaul with an estimated cost of \$35 000 at the end of years 4, 8, 12, and 16. The cooling tower will need an overhaul at the end of 10 years. This is expected to cost \$17 000.

The cogeneration system is expected to have higher annual operating and maintenance costs than the current system, and will require the use of chemicals to treat the water used in the new plant. These incremental costs are estimated to be \$65 000 per year. The incremental annual costs of wood fuel are estimated to be \$375 000. The cogeneration plant will save Cogenesis from having to purchase 40 000 000 kilowatt-hours of electricity per year at \$0.025 per kilowatt-hour, an annual savings of \$1 000 000. Cogenesis uses a MARR of 12%. What is the present worth of the incremental investment in the cogeneration plant? What is the impact of a 5% and 10% increase and decrease in each of the parameters of the problem?

$$\begin{aligned}
&\text{PW(cogeneration plant)} \\
&\quad = -\$3\,000\,000 - (\$65\,000 + \$375\,000 - \$1\,000\,000)(P/A, 12\%, 20) \\
&\qquad - \$17\,000(P/F, 12\%, 10) \\
&\qquad - \$35\,000[(P/F, 12\%, 4) + (P/F, 12\%, 8) + (P/F, 12\%, 12) \\
&\qquad + (P/F, 12\%, 16)] \\
&\quad = \$1\,126\,343
\end{aligned}$$

The present worth of the incremental investment is \$1 126 343. On the basis of this assessment, the project appears to be economically viable.

In order to better understand the situation, analysts for Cogenesis have also completed some sensitivity graphs that indicate how sensitive the present worth is to changes in some of the parameters. In particular, they feel that some of the cash flows may turn out to be different from their estimates, and they would like to get a feel for what impact these errors may have on the evaluation of the cogeneration plant. To investigate, they have labelled their current estimates the "base case" and have generated other cash flow estimates that are 5% and 10% above and below the base case for each major cash flow category. These are summarized in Table 11.1.

For example, the initial investment may be more than the estimate of \$3 000 000 if they run into unforeseen difficulties in the installation. Or the savings in electricity costs may be overestimated if the cost per kilowatt-hour drops in the future. The analysts would like to get a better understanding of which of these changes would have the greatest impact on the evaluation of the plant.

To keep the illustration simple, we will consider changes to the initial investment; annual chemical, operations, and maintenance costs; the MARR; and the savings in electrical costs. Each of these is varied one at a time, leaving all other cash flow estimates at the base case values. For example, if the initial investment is 10% below the initial estimate of \$3 000 000, and all other estimates are as in the base case, the present worth of

Table 11.1 Summary Data for Example 11.1

Cost Category	−10%	−5%	Base Case	+5%	+10%
Initial investment	$2 700 000	$2 850 000	$3 000 000	$3 150 000	$3 300 000
Annual chemical, operations, and maintenance costs	58 500	61 750	65 000	68 250	71 500
Cooling tower overhaul (after 10 years)	15 300	16 150	17 000	17 850	18 700
Turbogenerator overhauls (after 4, 8, 12, and 16 years)	31 500	33 250	35 000	36 750	38 500
Annual wood costs	337 500	356 250	375 000	393 750	412 500
Annual savings in electricity costs	900 000	950 000	1 000 000	1 050 000	1 100 000
MARR	0.108	0.114	0.12	0.126	0.132

the project will be $1 426 343 (see the first row of Table 11.2, under –10%). Similarly, if the first cost is 10% more than the original estimate, the present worth drops to $826 343.

Interest rate uncertainty will almost always be present in an economic analysis. If Cogenesis' MARR increases by 10% (with all other parameters at their base case values), the present worth of the project drops to $835 115, about the same impact as if the first cost ended up being 10% more than expected. Other variations are shown in Table 11.2. A sensitivity graph, shown in Figure 11.1, illustrates the impact of one-at-a-time parameter variations on the present worth.

Small changes in the annual chemical, operations, and maintenance costs do not have much of an impact on the present worth of the project, as can be seen from Table 11.2 and Figure 11.1. What appears to have the greatest impact on the viability of the project is the savings in electricity costs. A 10% drop in the savings causes the present worth of the project to drop to about one-third of the base case estimate. This change could occur because of a drop in electricity rates or a drop in demand. Alternatively, the present worth of the project increases to almost $1 900 000 if the savings are higher than anticipated. This could, once again, occur because of a change in either rates or demand for power. Clearly, if Cogenesis is to expend effort in getting better forecasts, it should be for energy consumption and power rates.

One final point about this example should be noted. If management feels that, individually, the cash flow estimates will fall within the ±10% range, the investment looks economically viable (i.e., yields a positive present worth) and they should go ahead with it.■

As we can see from Example 11.1, the benefit of a sensitivity graph is that it can be used to select key parameters in an economic analysis. It is easy to understand and communicates a lot of information in a single diagram. There are, however, several shortcomings of sensitivity graphs. First, they are valid only over the range of parameter values in the graph. The impact of parameter variations outside the range considered may not be simply a linear extrapolation of the lines in the graph. If you need to assess

Table 11.2 Present Worth of Variations from Base Case in Example 11.1

Cost Category	−10%	−5%	Base Case	+5%	+10%
Initial investment	$1 426 343	$1 276 343	$1 126 343	$ 976 343	$ 826 343
Annual chemical, operations, and maintenance costs	1 174 894	1 150 619	1 126 343	1 102 067	1 077 792
Cooling tower overhaul (after 10 years)	1 126 890	1 126 617	1 126 343	1 126 069	1 125 796
Turbogenerator overhauls (after 4, 8, 12, and 16 years)	1 131 450	1 128 897	1 126 343	1 123 789	1 121 236
Annual wood costs	1 406 447	1 266 395	1 126 343	986 291	846 239
Savings in electricity costs	379 399	752 871	1 126 343	1 499 815	1 873 287
MARR	1 456 693	1 286 224	1 126 343	976 224	835 115

Figure 11.1 Sensitivity Graph for Example 11.1

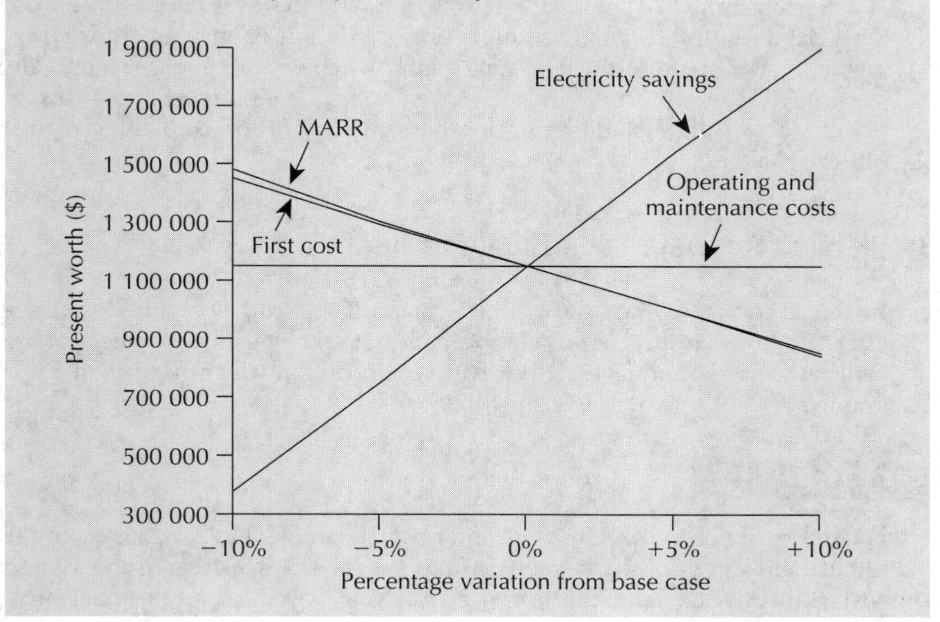

the impact of greater variations, the computations should be redone. Second, and probably the greatest drawback of sensitivity graphs, is that they do not consider the possible interaction between two or more parameters. You cannot simply "add up" the impact of individual changes when several parameters are varied, producing an interaction effect. We will come back to this issue in the section on scenario analysis, where we do consider entire "packages" of changes from the base case.

11.3 Break-Even Analysis

In this section, we cover a second type of sensitivity analysis called break-even analysis. Once again, we are trying to answer the question of what impact changes (or errors) in parameter estimates will have on the economic performance measures we use in our analyses, or on a decision made on the basis of an economic performance measure. In general, **break-even analysis** is the process of varying a parameter of a problem and determining what parameter value causes the performance measure to reach some threshold or "break-even" value. In Example 11.1, we saw that an increase in the MARR caused the present worth of the cogeneration plant to decrease. If the MARR were to increase sufficiently, the project might have a zero present worth. A break-even analysis could answer the question "What MARR will result in a zero present worth?" This analysis would be particularly useful if Cogenesis were uncertain about the MARR and wanted to find a threshold MARR above which the project would not be viable. Other such break-even questions could be posed for the cogeneration problem, to try to get a better understanding of the impact of changes in parameter values on the economic analysis.

Break-even analysis can also be used in the comparison of two or more projects. We have already seen in Chapter 4 that the best choice among mutually exclusive alternatives may depend on the interest rate, production level, or a variety of other problem parameters. Break-even analysis applied to multiple projects can answer questions like "Over what range of interest rates is project A the best choice?" or "For what output level are we indifferent between two projects?" Notice that we are varying a single parameter in two or more projects and asking when the performance measure for the projects meets some threshold or break-even point. The point of doing this analysis is to try to get a better understanding of how sensitive a decision is to changes in the parameters of the problem.

11.3.1 Break-Even Analysis for a Single Project

In this section, we show how break-even analysis can be applied to a single project to illustrate how sensitive a project evaluation is to changes in project parameters. We will continue with Example 11.1 to expand upon the information provided by the sensitivity graphs.

Example 11.2

Having completed the sensitivity graph in Example 11.1, management recognizes that the present worth of the cogeneration plant is quite sensitive to the savings in electricity costs, the MARR, and the initial costs. Since there is some uncertainty about these estimates, they want to explore further the impact of changes in these parameters on the viability of the project. You are to carry out a break-even analysis for each of these parameters to find out what range of values results in a viable project (i.e., PW > 0) and to determine the "break-even" parameter values which make the present worth of the project zero. You are also to construct a graph to illustrate the present worth of the project as a function of each parameter.

First, Figure 11.2 shows the present worth of the project as a function of the MARR. It shows that the break-even MARR is 17.73%. In other words, the project has a positive present worth for any MARR less than 17.73% (all other parameters fixed) and a negative present worth for a MARR more than 17.73%. Notice that the break-even interest rate is, in fact, the IRR for the project.

Figure 11.2 Break-Even Chart for the MARR

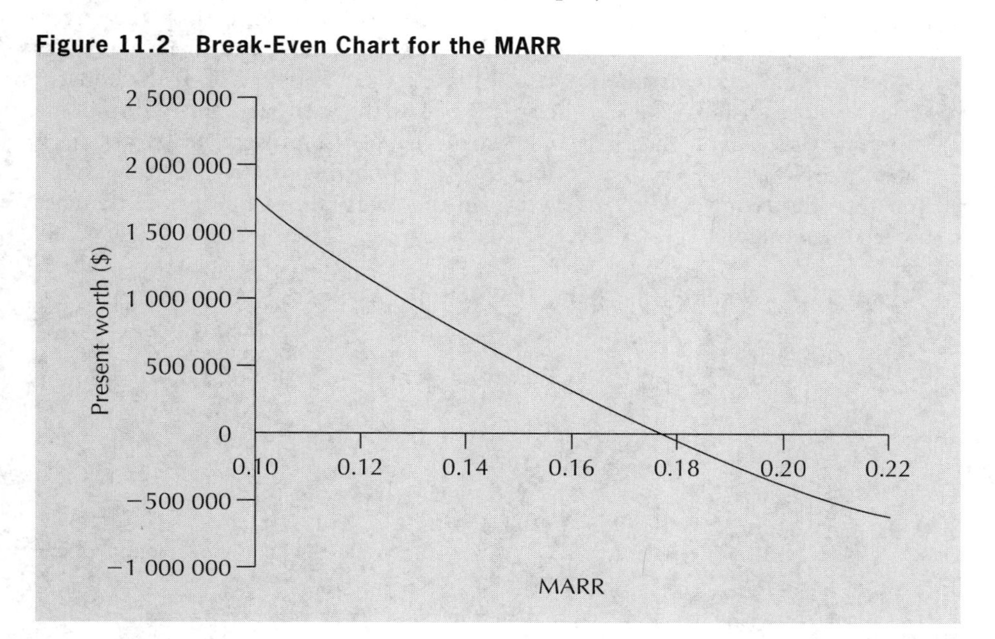

A similar break-even chart for the first cost, Figure 11.3, shows that the first cost can be as high as $4 126 350 before the present worth declines to zero. Assuming that all other cost estimates are accurate, the project will be viable as long as the first cost is below this break-even amount. One issue management should assess is the likelihood that the first cost will exceed $4 126 350.

Figure 11.3 Break-Even Chart for First Cost

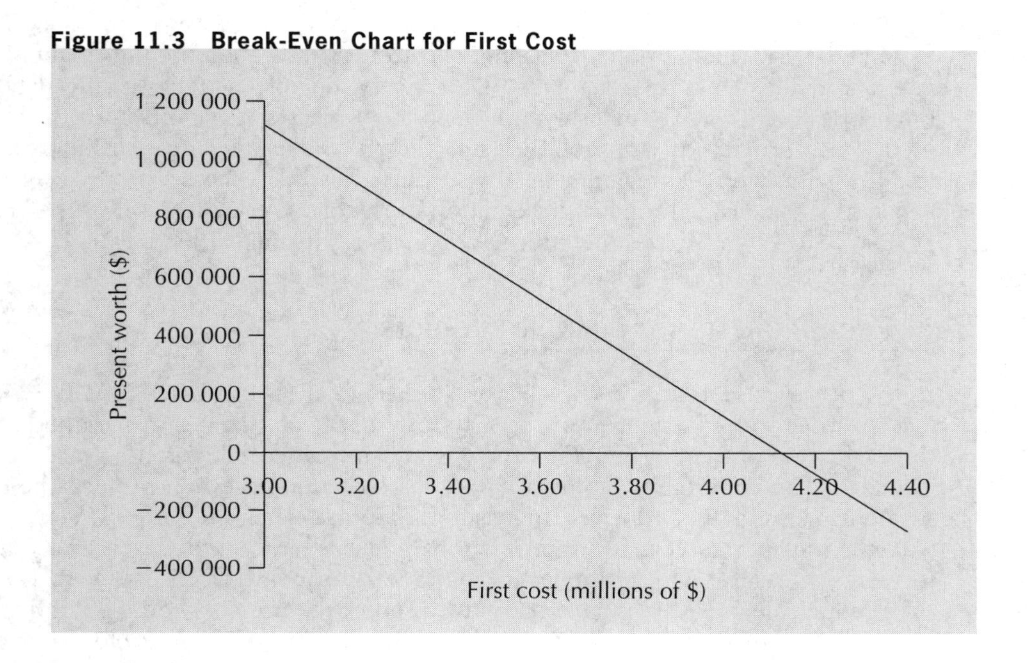

Finally, a break-even chart for the savings in electrical power costs is shown in Figure 11.4. We have already seen from the sensitivity graph that the viability of the project is very sensitive to the savings in electricity produced by the cogeneration plant. Provided that the annual savings are above $849 207, the project is viable. Below this break-even level, the present worth of the project is negative. If the actual saving in electrical power costs is likely to be much below the estimate, this will put the project's viability at risk. Given the particular sensitivity of the present worth to the savings, it may be worthwhile to spend additional time looking into the two factors that make up these savings: the cost per kilowatt-hour and the total kilowatt-hours of demand provided for by the new plant.

Figure 11.4 Break-Even Chart for Electricity Savings

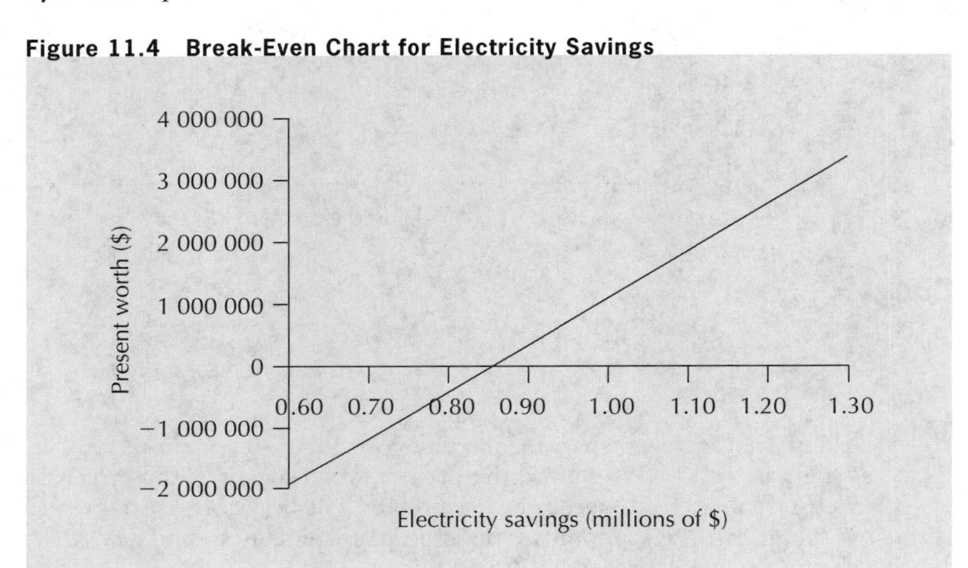

■

Break-even analysis done for a single project expands upon the information sensitivity graphs provide. It has the advantage that it is easy to apply and allows us to determine the range of values for a parameter within which the project is viable or some other criteria are met. It can provide us with break-even parameter values that give an indication of how much a parameter can change from its original estimate before the project's viability becomes a concern. Graphical presentation of the break-even analysis, as in Figures 11.2, 11.3, and 11.4, summarizes the information in an easily understood way

11.3.2 Break-Even Analysis for Multiple Projects

In the previous section, we saw how break-even analysis can be applied to a single project in order to understand more clearly the impact of changes in parameter values on the evaluation of the project. This analysis may influence a decision on whether the project should be undertaken. When there is a choice among several projects, be they independent or mutually exclusive, the basic question remains the same. We are concerned with the impact that changes in problem parameters have on the relevant economic performance measure, and, ultimately, on the decision made with respect to the projects. With one project, we are concerned with whether the project should be undertaken and how changes in parameter values affect this decision. With multiple projects, we are concerned about how changes in parameter values affect which project or projects are chosen.

For multiple independent projects, assuming that there are sufficient funds to finance all projects, break-even analysis can be carried out on each project independently, as was done for a single project in the previous section. This will lead to insights into how robust a decision is under changes in the parameters.

For mutually exclusive projects, the best choice will seldom stand out as clearly superior from all points of view. Even if we have narrowed down the choices, it is still likely that the best choice may depend on a particular interest rate, level of output, or first cost. A break-even comparison can reveal the range over which each alternative is preferred and can show the break-even points where we are indifferent between two projects. Break-even analysis will provide a decision maker with further information about each of the projects and how they relate to one another when parameters change.

Example 11.3

Westmount Waxworks (see Problem 4.18) is considering buying a new wax melter for their line of replicas of statues of government leaders. They have two choices of suppliers, Finedetail and Simplicity. The proposals are as follows:

	Finedetail Wax Melter	Simplicity Wax Melter
Expected life	7 years	10 years
First cost	$200 000	$350 000
Maintenance	$10 000/year + $0.05/unit	$20 000/year + $0.01/unit
Labour	$1.25/unit	$0.50/unit
Other costs	$6 500/year + $0.95/unit	$15 500/year + $0.55/unit
Salvage value	$5 000	$20 000

The marketing manager has indicated that sales have averaged 50 000 units per year over the past five years. In addition to this information, management thinks that they will sell about 30 000 replicas per year if there is stability in world governments. If the world becomes very unsettled so that there are frequent overturns of governments, sales may be as high as 200 000 units per year. There is also some uncertainty about the "other costs" of the Simplicity wax melter. These include energy costs and an allowance for scrap. Though the costs are estimated to be $0.55 per unit, the Simplicity model is a new technology, and the costs may be as low as $0.45 per unit or as high as $0.75 per unit. Westmount Waxworks would like to carry out a break-even analysis on the sales volume and on the "other costs" of the Simplicity wax melter. They want to know which the preferred supplier would be as sales vary from 30 000 per year to 200 000 per year. They also wish to know which is the preferred supplier if the "other costs" per unit for the Simplicity model are as low as $0.45 per unit or as high as $0.75 per unit. Westmount Waxworks uses an after-tax MARR of 15% for equipment projects. Their tax rate is 40% and the CCA rate for such equipment is 30%.

Assuming that the "other costs" of the Simplicity wax melter are $0.55 per unit, a break-even chart that shows the present worth of the projects as a function of sales levels can give much insight into the supplier selection. Table 11.3 gives the annual cost of each of the two alternatives, and Figure 11.5 shows the break-even chart for sales level. A sample computation for the Finedetail wax melter at the 60 000 sales level is

$$\begin{aligned}\text{AW(Finedetail)} &= \text{CCTF}_{new}(\$200\,000)(A/P, 15\%, 7) \\ &\quad - \text{CCTF}_{old}(\$5000)(A/F, 15\%, 7) + (1 - t)[\$10\,000 \\ &\quad + \$6500 + (\$0.05 + \$1.25 + \$0.95)(\text{sales level})] \\ &= 0.75073(\$200\,000)(0.24036) - 0.73333(\$5000)(0.09036) \\ &\quad + (1 - 0.4)[\$16\,500 + \$2.25(60\,000)] \\ &\cong \$126\,658\end{aligned}$$

where

$$\text{CCTF}_{old} = 1 - \frac{td}{i+d} = 1 - \frac{0.4(0.3)}{0.15 + 0.30} \cong 0.73333$$

$$\text{CCTF}_{new} = 1 - \frac{td\left(1 + \frac{i}{2}\right)}{(i+d)(1+i)} = 1 - \frac{(0.4)(0.3)(1+0.075)}{(0.15+0.30)(1+0.15)} \cong 0.75073$$

Table 11.3 Annual Cost as a Function of Sales

Sales (Units)	Annual Costs ($)	
	Finedetail	Simplicity
20 000	72 658	85 651
60 000	126 658	111 091
100 000	180 658	136 531
140 000	234 658	161 971
180 000	288 658	187 411
220 000	342 658	212 851

Figure 11.5 Break-Even Chart for Sales Level

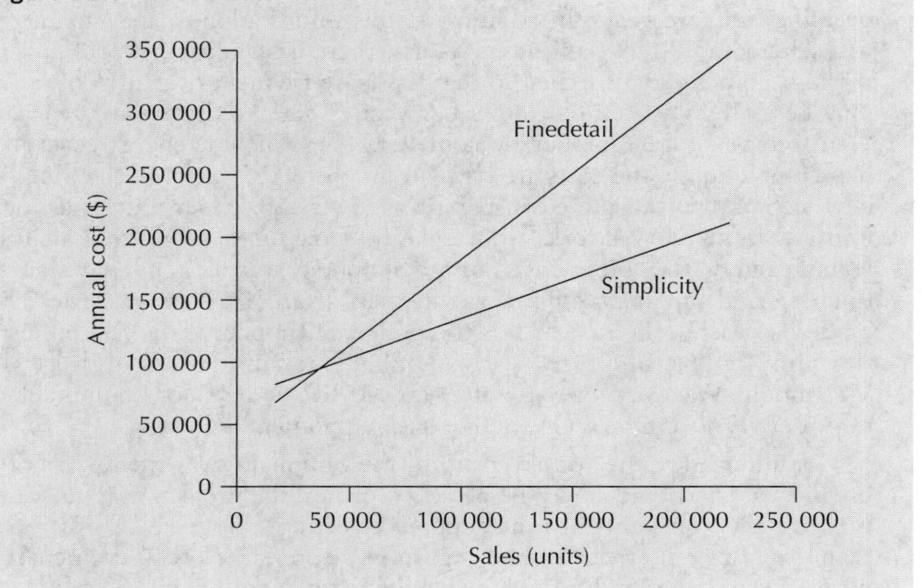

If sales are 30 000 units per year, the Finedetail wax melter is slightly preferred to the Simplicity melter. At a sales level of 200 000 units per year, the preference is for the Simplicity wax melter. Interpolation of the amounts in Table 11.3 indicates that the break-even sales level is 38 199 units. That is to say, for sales below 38 199 per year, Finedetail is preferred, and Simplicity is preferred for sales levels of 38 199 units and above.

Since 30 000 units per year is the lowest sales will likely be, and sales have averaged 50 000 units per year over the past five years, it appears that the Simplicity wax melter would be the preferred choice, assuming that its "other costs" per unit is $0.55. The robustness of this decision may be affected by the other types of costs, such as maintenance and labour, of the Simplicity melter.

To assess the sensitivity of the choice of wax melter to the variable other costs of Simplicity, a break-even analysis similar to that for sales level can be carried out. We can vary the "other costs" from the estimate of $0.45 per unit to $0.75 per unit and observe the effect on the preferred wax melter. Table 11.4 gives the annual costs for the two wax melters as a function of the "other costs" of the Simplicity model for sales levels of 30 000, 50 000, and 200 000 units per year. In each case, we see that the best choice is not sensitive to the "other costs" of the Simplicity wax melter. In fact, for a sales level of 30 000 units per year, the break-even "other cost" is less than $0.25, as shown in Figure 11.6. This means that the other cost per unit would have to be lower than $0.25 for the best choice to change from Finedetail to Simplicity. For a sales level of 200 000 per year, the break-even "other cost" is much higher, at $1.51 per unit, and for a sales level of 50 000 units per year the break-even cost per unit is $0.83. For both of the latter sales levels, the Simplicity model is preferred.

Table 11.4 Annual Cost as a Function of Simplicity's Other Costs per Unit

Other Costs per Unit ($)	Sales = 30 000 Units/Year Annual Costs ($)		Sales = 50 000 Units/Year Annual Costs ($)		Sales = 200 000 Units/Year Annual Costs ($)	
	Finedetail	Simplicity	Finedetail	Simplicity	Finedetail	Simplicity
0.45	86 158	90 211	113 158	101 731	315 658	188 131
0.55	86 158	92 011	113 158	104 731	315 658	200 131
0.65	86 158	93 811	113 158	107 731	315 658	212 131
0.75	86 158	95 611	113 158	110 731	315 658	224 131

Having done the break-even analysis for both sales level and "other costs" per unit for the Simplicity wax melter, it would appear that the Simplicity model is the better choice if sales are at all likely to exceed the break-even sales level of 38 199. Historically, sales have exceeded this amount. Even if sales in a particular year fall below the break-even level, the Simplicity wax melter does not have annual costs far in excess of those of the Finedetail model, so the decision would appear to be robust with respect to possible sales levels. Similarly, the decision is not sensitive to the other cost per unit of the Simplicity wax melter.■

We have seen in this section that break-even analysis for either a single project or multiple projects is a simple tool and that it can be used to extract insights from a modest amount of data. It communicates threshold (break-even) parameter values where preference changes from one alternative to another or where a project changes from being economically justified to not justified. Break-even analysis is a popular means of assessing the impact of errors or changes in parameter values on an economic performance measure or a decision.

Figure 11.6 Break-Even Chart for Simplicity's Other Costs Per Unit (Sales = 30 000 Units)

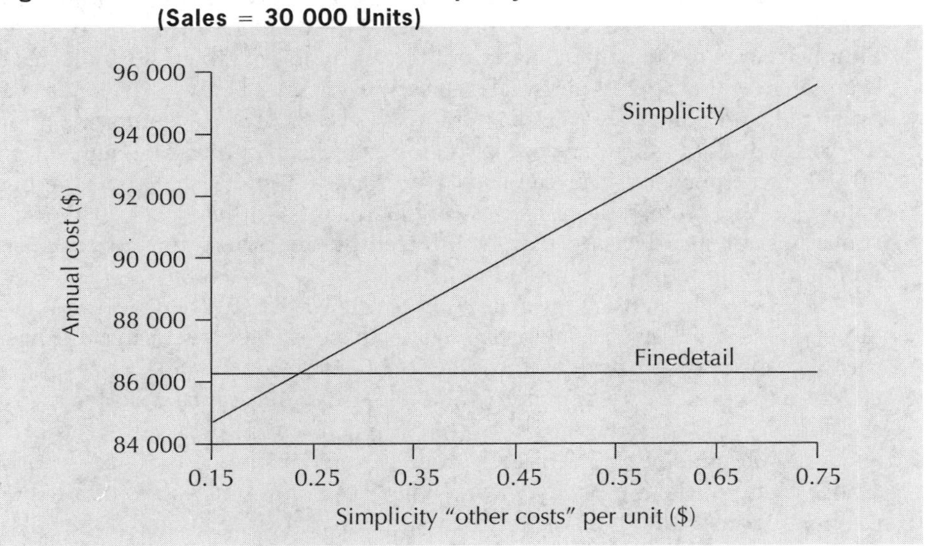

The main disadvantage of break-even analysis is that it cannot easily capture interdependencies among variables. Although we can vary one or two parameters at a time and graph the results, more complicated analyses are not often feasible. This disadvantage can be overcome to some degree by what is referred to as scenario analysis, the subject of the next section.

11.4 Scenario Analysis

The third type of sensitivity analysis tool that we will look at is **scenario analysis**, which is the process of examining the consequences of several possible sets of variables associated with a project. Scenario analysis recognizes that many estimates of cash flows or other project parameters may vary from what is projected. It is useful to look at several "what if" scenarios in order to understand the effect of changes in values of whole sets of parameters. Commonly used scenarios are the "optimistic" (or "best case") outcome, the "pessimistic" (or "worst case") outcome, and the "expected" (or "most likely") outcome. The best-case and worst-case outcomes can, in some sense, capture the entire range of possible outcomes for a project or a comparison among projects and provide an enriched view of the decision.

Example 11.4

Cogenesis (refer to Example 11.1) wishes to do a scenario analysis of their cogeneration problem in order to decide whether the project should be undertaken. They have come up with optimistic, pessimistic, and expected estimates of each of the parameters for their decision problem in order to get a better understanding of the possible range of present worth outcomes for the cogeneration plant. The three scenarios and the associated estimates are summarized in Table 11.5.

The scenarios capture combinations of parameter estimates which reflect the worst, best, and expected outcomes for the project. In contrast with sensitivity graphs and break-even analysis, scenario analysis allows entire groups of parameters to be changed at one time.

Table 11.5 Present Worth of Cogeneration Plant Scenarios

Cost Category	Pessimistic Scenario	Expected Scenario	Optimistic Scenario
Initial investment ($)	3 300 000	3 000 000	2 700 000
Annual chemical, operations, and maintenance costs ($)	75 000	65 000	60 000
Cooling tower overhaul (after 10 years) ($)	21 000	17 000	13 000
Turbogenerator overhauls (after 4, 8, 12, and 16 years) ($)	40 000	35 000	30 000
Additional annual wood costs ($)	400 000	375 000	350 000
Savings in annual electricity costs ($)	920 000	1 000 000	1 080 000
MARR	0.13	0.12	0.11
Present worth of cogeneration plant ($)	–234 639	1 126 343	2 583 848

Evaluation of each scenario reveals that the present worth of the cogeneration plant will be negative if all parameters take on their worst-case values, and hence, the project is not advisable. The major problem is that the savings in electricity costs are insufficient to make up for the high first cost of the project. In contrast, both the expected-case and best-case scenarios lead to positive present worths, and hence, the project would be viable. (To put the present worths into context, the expected-case and best-case scenarios have IRRs of 17.73% and 24.29%, respectively.) From an overall point of view, there is some risk that the cogeneration project will have a negative present worth, but this will occur only if the worst-case scenario does occur. Even if the worst-case outcome does occur, the loss is not huge compared with the potential gain in the other two cases. What Cogenesis needs to do if they wish to look further into the project's viability is assess the risk (or likelihood) that the worst outcome will occur. Decision making under risk is discussed further in Chapter 12.■

As we can see from Example 11.4, scenario analysis allows us to look at the effect of multiple changes in parameter values on an individual project's viability. It can also be used to evaluate the effect of scenarios in a case where there are several alternatives.

Example 11.5

Westmount Waxworks has carried out a scenario analysis for three possible outcomes they feel represent pessimistic, optimistic, and expected outcomes for sales levels and the Simplicity wax melter's other costs per unit. The scenarios and the annual costs of the two wax melters are summarized in Table 11.6. From the scenario analysis, we see that the Simplicity wax melter is the preferred choice for the expected and optimistic scenarios. The Finedetail wax melter is preferred only if the pessimistic scenario occurs. In terms of the opportunity cost of making the wrong choice, it is far larger if the optimistic outcome occurs ($315 658 – $188 301 = $127 357) than if the pessimistic outcome occurs ($95 611 – $86 158 = $9453).■

As was seen in Examples 11.4 and 11.5, scenario analysis allows us to take into account the interrelationships among parameters when making a choice by examining likely groupings of parameter values in scenarios. The most commonly used scenarios are the pessimistic, optimistic, and expected outcomes. The use of scenarios allows an analyst to capture the range of possible outcomes for a project or group of projects. Done in

Table 11.6 Scenario Analysis for Westmount Waxworks

	Pessimistic Scenario	Expected Scenario	Optimistic Scenario
Sales level (units)	30 000	50 000	200 000
Other costs per unit (Simplicity)	$0.75	$0.55	$0.45
Annual cost: Finedetail	$86 158	$113 158	$315 658
Annual cost: Simplicity	$95 611	$104 731	$188 131

combination with sensitivity graphs and break-even analysis, a great deal of information can be obtained regarding the economic viability of a project.

The one drawback common to each of the three sensitivity analysis methods covered in this chapter is that they do not capture the likelihood that a parameter will take on a certain value or the likelihood that a certain scenario will occur. This information can further guide a decision maker and is often crucial to assessing the risk of the worst case outcome. Chapter 12 will describe how these concerns are addressed.

REVIEW PROBLEMS

The following case is the basis of Review Problems 11.1 through 11.3.

Burnaby Insurance Inc. is considering two independent energy efficiency improvement projects. Each has a lifetime of 10 years and will have a scrap value of zero at the end of this time. Burnaby can afford to do both if both are economically justified. The first project involves installing high-efficiency motors in their air conditioning system. High-efficiency units use about 7% less electricity than the current motors, which represents annual savings of 70 000 kilowatt-hours. They cost $28 000 to purchase and install and will require maintenance costs of $700 annually.

The second project involves installing a heat exchange unit in the current ventilation system. During the winter, the heat exchange unit transfers heat from warm room air to the cold ventilation air before the air is sent back into the building. This will save about 2 250 000 cubic feet of natural gas per year. In the summer, the heat exchange unit removes heat from the hot ventilation air before it is added to the cooler room air for recirculation. This saves about 29 000 kilowatt-hours of electricity annually. Each heat exchange unit costs $40 000 to purchase and install and annual maintenance costs are $3200.

Burnaby Insurance would like to evaluate the two projects, but there is some uncertainty surrounding what the electricity and natural gas prices will be over the life of the project. Current prices are $0.07 per kilowatt-hour for electricity and $3.50 per thousand cubic feet of natural gas, but some changes are anticipated. They use a MARR of 10%.

REVIEW PROBLEM 11.1

Construct a sensitivity graph to determine the effect that a 5% and 10% drop or increase in the cost of electricity and the cost of natural gas would have upon the present worth of each project.

ANSWER

Table 11.7 gives the costs of electricity and natural gas with 5% and 10% increases and decreases from the base case of $0.07 per kilowatt-hour for electricity and $3.50 per

1000 cubic feet of natural gas. The table also shows the present worths of the two energy efficiency projects as the costs vary. A sample calculation for the heat exchange unit with base case costs is

$$\begin{aligned}\text{PW(Heat exchanger)} &= -\$40\,000 + (P/A, 10\%, 10) \\ &\quad \times [\$29\,000(0.07) + \$2250(3.50) - \$3200] \\ &\cong \$1199\end{aligned}$$

Table 11.7 Costs Used as the Basis of the Sensitivity Graph for Review Problem 11.1

	–10%	–5%	0%	+5%	+10%
Cost of electricity ($/kWh)	0.063	0.0665	0.07	0.0735	0.077
Cost of natural gas ($/1000 cubic feet)	3.15	3.325	3.5	3.675	3.85
PW of high-efficiency motor ($)					
With changes to electricity costs	–5204	–3698	–2193	–687	818
With changes to natural gas cost	–2193	–2193	–2193	–2193	–2193
PW of heat exchanger ($)					
With changes to electricity costs	–48	576	1199	1823	2447
With changes to natural gas cost	–3640	–1220	1199	3619	6038

Figure 11.7 is a sensitivity graph for the high-efficiency motor. It graphically illustrates the effect of changes in the costs of electricity and natural gas on the present worth of a motor. The high-efficiency motor is not economically viable at the current prices for electricity and gas. Only if there is an increase of almost 10% in electricity costs for the life of the project will the motor produce sufficient savings for the project to have a positive present worth.

Figure 11.7 Sensitivity Graph for the High-Efficiency Motor

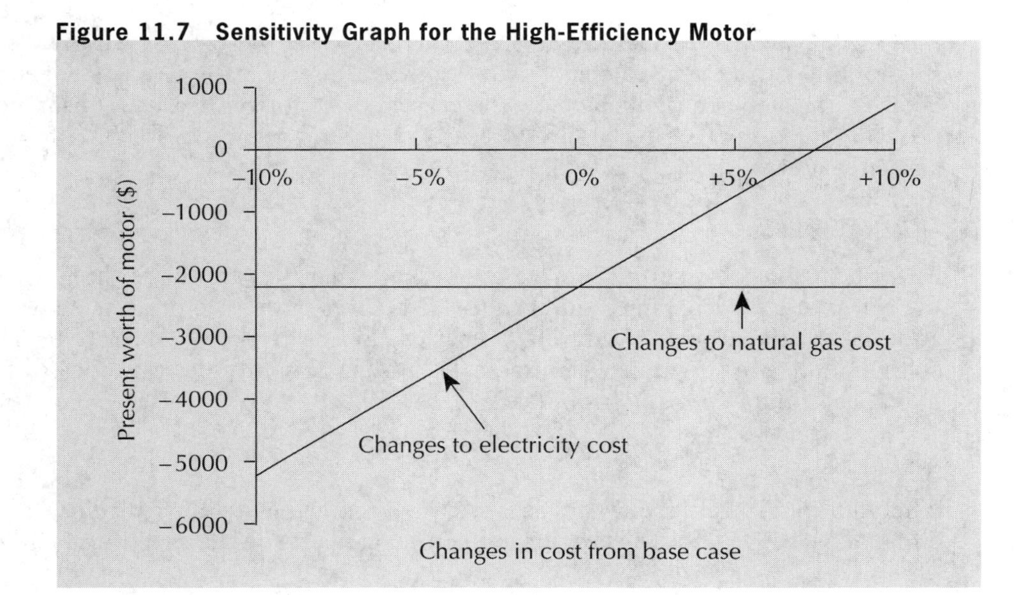

Figure 11.8 is the sensitivity graph for the heat exchange unit. The heat exchange unit has a positive present worth for the current prices, but the present worth is quite sensitive to the price of natural gas. A drop in the price of natural gas in the range of only 2% to 3% (reading from the graph) will cause the project to have a negative present worth.■

Figure 11.8 Sensitivity Graph for the Heat Exchanger

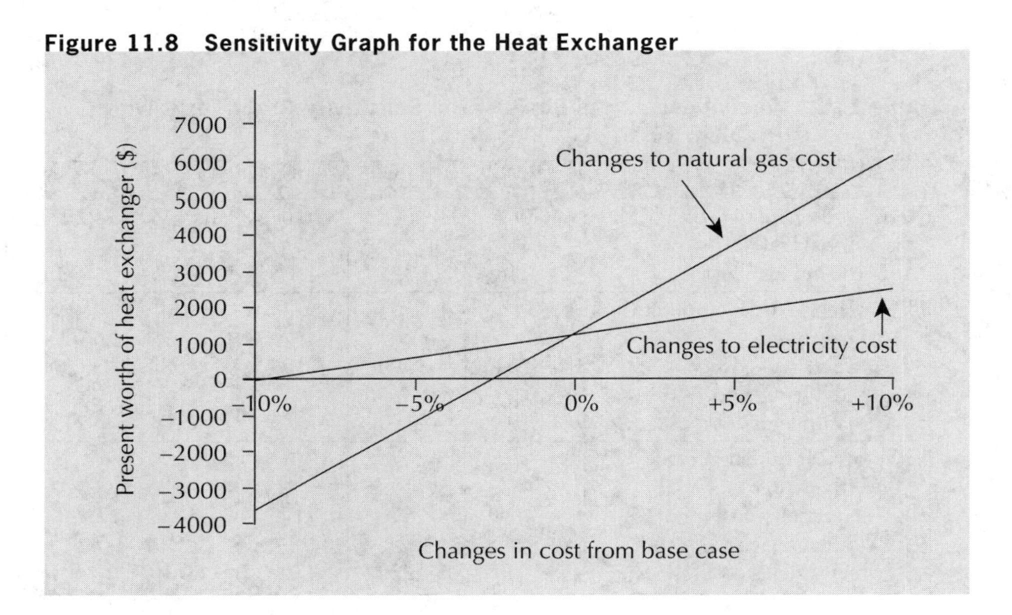

REVIEW PROBLEM 11.2

Refer to Review Problem 11.1. How much of a drop in the cost of natural gas will result in the heat exchange unit's having a present worth of zero? Construct a break-even graph to illustrate this break-even cost.

ANSWER

By varying the cost of natural gas from the base case, the break-even graph shown in Figure 11.9 can be constructed. The break-even cost of natural gas is $3.41 per 1000 cubic feet, which is not much below the current price for gas. Burnaby Insurance should probably look more seriously into forecasts of natural gas prices for the life of the heat exchange unit.■

REVIEW PROBLEM 11.3

Analysts at Burnaby Insurance have established what they think are three scenarios for the prices of electricity and natural gas over the lives of the two projects under consideration in Review Problem 11.1. The scenarios along with the appropriate present worth computations are summarized in Table 11.8. What insight does this add to the investment decision for Burnaby Insurance?

ANSWER

The additional insight that the scenario analysis brings to Burnaby Insurance is the effect of changes in both electricity and natural gas costs on the two proposed projects.

Figure 11.9 Break-Even Chart for the Cost of Natural Gas

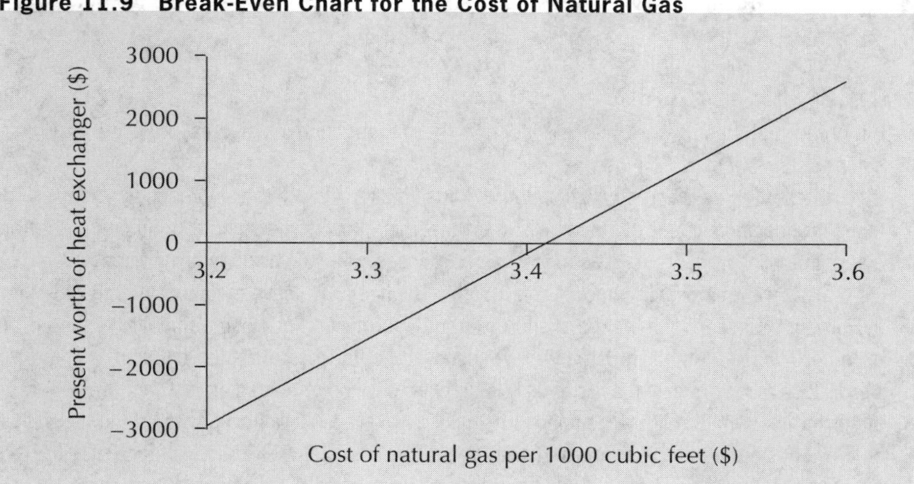

Table 11.8 Scenario Analysis for Burnaby Insurance

	Pessimistic Scenario	Expected Scenario	Optimistic Scenario
Cost of electricity ($/kWh)	0.063	0.070	0.077
Cost of natural gas ($/1000 cubic feet)	3.35	3.50	3.65
PW of high efficiency motor ($)	–5204	–2193	818
PW of heat exchange unit ($)	–2122	1199	4520

Sensitivity graphs and the break-even analysis can look only at the effect of one-at-a-time parameter changes on the present worth computations. It appears that the high-efficiency motor is a bad investment, as its present worth is not much above zero even if the optimistic scenario occurs. The heat exchange unit appears to be a better investment, but even that has a chance of having a negative present worth if the pessimistic outcome occurs. What Burnaby really needs to know is the likelihood of each of these scenarios occurring, or some other means of assessing the likelihood of what energy prices will be in the future.■

SUMMARY

In this chapter, we considered three basic methods used by analysts in order to better understand the effect that uncertainties in estimated cash flows have on economic decisions. The first was the use of sensitivity graphs. Sensitivity graphs illustrate the sensitivity of a particular measure (e.g., present worth or annual worth) to changes in one or more of the parameters of a project. The second method was the use of break-even analysis for evaluating both individual projects and comparisons among projects. Finally, scenario analysis allowed us to look at the overall impact of a variety of outcomes, usually optimistic, expected, and pessimistic.

Engineering Economics in Action, Part 11B: Where the Risks Lie

Bill Astad and Naomi were working through the market demand figures provided by Powerluxe for the new self-adjusting vacuum.

"These figures are pretty ambiguous," Bill said. "We have three approaches: a set of opinions taken from focus groups and surveys of customers, the same thing from dealers and distributors, and an analysis of trends in a set of parallel products such as fuzzy-logic appliances. Like Anna said, nothing hard."

"What we really want to know," said Naomi, "is whether we have the capacity to handle the manufacturing for the product. Based on the surveys and the trend information, let's come up with three scenarios: low demand, expected demand, and high demand. If we behave according to expected demand, and the true demand is low, we will lose money because our capital investments won't be recouped as fast, and we may have passed up other opportunities. Similarly, if the demand is high, we will lose by having to pay overtime, paying for contracting out, or losing customers. But if we make money in all three cases, there really isn't much of a problem."

"And if it turns out we don't make money in all three cases?" Bill asked. "What then?"

"I know it's a lot of work, Bill, but let's do it and find out," Naomi replied. "At minimum, we will know where our risks lie."

PROBLEMS

11.1 Identify possible parameters that are involved in economic analysis for the following situations:

(a) Buying new equipment

(b) Supplying products to a foreign country with a high inflation rate

11.2 For the following examples of parameters, how would you assign a reasonable base case and a range of variation so that you can carry out sensitivity analysis? Assign specific numerical figures wherever you can.

(a) Canadian inflation rate

(b) Canadian-American exchange rate

(c) Expected annual savings from a new piece of equipment similar to the one you already have

(d) Expected annual revenue from an Internet-based business

(e) Salvage value of a personal computer

11.3 Which sensitivity analysis method may be appropriate for analyzing the following uncertain situations?

(a) Calgary Cartage leases trucks to service its shipping contracts. Larger trucks have cheaper operating costs if there is sufficient business, but are more expensive if they are not full. Calgary Cartage is not certain about their future demand.

(b) Joan runs a dog kennel. She is considering installing a heating system for the interior runs that will allow her to operate all year. Joan is not sure how much the annual heating expenses will be.

(c) Pushpa runs a one-person company producing custom paints for hobbyists. She is considering buying printing equipment to produce her own labels. However, she is not sure if she will have enough orders in the future to justify the purchase of the new equipment.

(d) Lemuel is an engineer working for Ontario Hydro. He is estimating the total cost for building transmission lines from the Darlington Nuclear Plant to new industrial parks north of Toronto. Lemuel is uncertain about the construction cost (per kilometre) of transmission lines.

(e) Thanh's company is growing very fast and has a hard time meeting its orders. An opportunity to purchase additional production equipment has arisen. She is not certain if the company will continue to grow at the same rate in the future, and she is not even certain how long the growth may last.

11.4 The Hanover Go-Kart Club has decided to build a clubhouse and track several years from now. The club needs to accumulate $50 000 by setting aside a uniform amount at the end of every year. They believe it possible to set aside $7000 every year at 10% interest. They wish to know how many years it will take to save $50 000 and how sensitive this result is to a 5% and a 10% increase or decrease in the amount saved per year and in the interest rate. Construct a sensitivity graph to illustrate the situation.

11.5 A new software package is expected to improve productivity at Saskatoon Insurance. However, because of training and implementation costs, savings are not expected to occur until the third year of operation. Annual savings of approximately $10 000 are expected, increasing by about $1000 per year for the following five years. After this time (eight years from implementation), the software will be abandoned with no scrap value. Construct a sensitivity graph showing what would happen to the present worth of the software with 7.5% and 15% increases and decreases in the interest rate, the $10 000 base savings, and the $1000 savings gradient. MARR is 15%.

11.6 The Regional Municipality of Kitchener is studying a water supply plan for the area to the end of the year 2050. To satisfy the water demand, one suggestion is to construct a pipeline from one of the Great Lakes. It is now the end of 2005. Construction would start in the year 2010 (five years from now) and take five years to complete at a cost of $20 million per year. Annual maintenance and repair costs are expected to be $2 million and will start the year following project completion (all costs are based on current estimates). From a predicted inflation rate of 3% per year, and the real MARR, city engineers have determined that a MARR of 7% per year is appropriate. Assume that all cash flows take place at the end of the year and that there is no salvage value at the end of 2050.

(a) Find the present worth of the project.

(b) Construct a sensitivity graph showing the effects of 5% and 10% increases and decreases in the construction costs, maintenance costs, and inflation rate. To which is the present worth most sensitive?

11.7 The city of Surrey is installing a new swimming pool in the downtown recreation centre. One design being considered is a reinforced concrete pool that will cost $6 000 000 to install. Thereafter, the inner surface of the pool will need to be refinished and painted

every 10 years at a cost of $40 000 per refinishing. Assuming that the pool will have essentially an infinite life, what is the present worth of the costs associated with the pool design? The city uses a MARR of 5%. If the installation costs, refinishing costs, and MARR are subject to 5% or 10% increases or decreases, how is the present worth affected? To which parameter is the present worth most sensitive?

11.8 You and two friends are thinking about setting up a grocery delivery service for local residents to finance your last two years at university. In order to start up the business, you will need to purchase a car. You have found a used car that costs $6000 and you expect to be able to sell it for $3000 at the end of two years. Insurance costs are $600 for each six months of operation, starting now. Advertising costs (e.g., flyers, newspaper advertisements) are estimated to be $100 per month, but these might vary as much as 20% above or below the $100, depending on the intensity of your advertising. The big questions you have now are how many customers you will have and how much of a service fee to charge per delivery. You estimate that you will have 300 deliveries every month, and are thinking of setting a $2-per-delivery fee, payable at the end of each month. The interest rate over the two-year period is expected to be 8% per year, compounded monthly, but may be 20% above or below this figure.

Using equivalent monthly worth, construct a sensitivity graph showing how sensitive the monthly worth of this project will be to the interest rate, advertising costs, and the number of deliveries you make each month. To which parameter is the equivalent monthly worth most sensitive?

11.9 Timmons Testing (TT) does subcontracting work for printed circuit board manufacturers. They perform a variety of specialized functional tests on the assembled circuit boards. TT is considering buying a new probing device that will assist the technicians in diagnosing functional defects in the printed circuit boards. Two vendors have given them quotes on first costs and expected operating costs over the life of their equipment.

	Vendor A	Vendor B
Expected life	7 years	10 years
First cost	$200 000	$350 000
Maintenance costs	$10 000/year + $0.05/unit	$20 000/year + $0.01/unit
Labour costs	$1.25/unit	$0.50/unit
Other costs	$6 500/year + $0.95/unit	$15 000/year + $0.55/unit
Salvage value	$5 000	$20 000

Production levels vary for TT. They may be as low as 20 000 boards per year or as high as 200 000 boards per year if a contract currently under negotiation comes through. They expect, however, that production quantities will be about 50 000 boards. Timmons Testing uses a MARR of 15% for equipment projects, and will be using an annual worth comparison for the two devices.

Timmons Testing is aware that the equipment vendors have given them estimates only for costs. In particular, TT would like to know how sensitive the annual worth of each device is to the first cost, annual fixed costs (maintenance + other), variable costs (maintenance + labour + other), and the salvage value.

(a) Construct a sensitivity graph for Vendor A's device, showing the effects of 5% and 10% decreases and increases in the first cost, annual fixed costs, variable costs, and the salvage value. Assume an annual production level of 50 000 units.

(b) Construct a sensitivity graph for Vendor B's device, showing the effects of 5% and 10% decreases and increases in the first cost, annual fixed costs, variable costs, and the salvage value. Assume an annual production level of 50 000 units.

11.10 Manitoba Metalworks would like to implement a local area network (LAN) for file transfer, email, and database access throughout its facility. Two feasible network topologies have been identified, which they have labelled Alternative A and Alternative B. The three main components of costs for the network are (1) initial hardware and installation costs, (2) initial software development costs, and (3) software and hardware maintenance costs. The installation and hardware costs for both systems are somewhat uncertain, as prices for the components are changing and Manitoba Metalworks are not sure of the installation costs for the LAN hardware. The costs for each alternative are summarized below.

	Alternative A	Alternative B
Initial hardware and installation costs ($):		
Optimistic estimate	70 000	86 000
Average estimate	92 500	105 500
Pessimistic estimate	115 000	125 000
Initial software cost ($)	138 750	158 250
Annual maintenance costs ($)	9 250	10 550
Annual benefits ($):		
Optimistic estimate	80 000	94 000
Average estimate	65 000	74 000
Pessimistic estimate	50 000	54 000

Benefits from the LAN are increased productivity because of faster file transfer times, reduced data redundancy, and improved data accuracy because of the database access. The benefits were difficult to quantify and are stated below as only a range of possible values and an average.

Manitoba Metalworks uses a 15% MARR and has established a 10-year study period for this decision. They wish to compare the projects on the basis of annual worth.

(a) Construct a sensitivity graph for Alternative A. For the base case, use the average values for the initial hardware cost and the annual benefits. Each graph should indicate the effect of a 5% and a 10% drop or increase in the initial hardware cost and the annual benefits. Which of the two factors most affects the annual worth of Alternative A?

(b) Construct a sensitivity graph for Alternative B. For the base case, use the average values for the initial hardware cost and the annual benefits. Each graph should indicate the effect of a 5% and a 10% drop or increase in the initial hardware cost and the annual benefits. Which of the two factors most affects the annual worth of Alternative B?

11.11 Refer back to Problem 11.8.

(a) Assuming base case figures for advertising costs and interest rates, what is the break-even number of deliveries per month? Construct a graph showing the break-even number.

(b) Assuming base case figures for advertising costs and number of deliveries per month, what is the break-even interest rate? Construct a graph illustrating the break-even interest rate.

11.12 Refer back to Problem 11.4. Members of the Go-Kart Club do not wish to wait for more than five years to build their clubhouse. They have decided to start a fundraising campaign to increase their ability to save each year between $7000 and whatever is necessary to have $50 000 saved in five years. Construct a table and a graph that illustrate how the number of years they must wait depends on the amount they save each year. What additional funds per year will allow them to save $50 000 in five years? Use a 10% interest rate.

11.13 Refer back to Problem 11.10 in which Manitoba Metalworks is considering two LAN alternatives.

(a) For Alternative A, by how much will the installation cost have to rise before the annual worth becomes zero? In other words, what is the break-even installation cost? Is the break-even level within or above the range of likely values Manitoba Metalworks has specified?

(b) What is the break-even annual benefit for Alternative A? Use the average installation costs. Is the break-even level within or above the range of likely values Manitoba Metalworks has specified?

11.14 Repeat Problem 11.13 for Alternative B.

11.15 Refer back to Timmons Testing, Problem 11.9.

(a) TT charges $3.25 per board tested. Assuming that costs are as in Vendor A's estimates, what production level per year would allow TT to break even if they select Vendor A's equipment? That is, for what production level would annual revenues equal annual costs? Construct a graph showing total revenues and total costs for various production levels, and indicate on it the break-even production level.

(b) Repeat (a) for Vendor B's equipment.

11.16 The Bountiful Bread Company produces home bread-making machines. Currently, they pay a custom molder $0.19 per piece (not including material costs) for the clear plastic face on the control panel. Demand for the bread-makers is forecast to be 200 000 machines per year, but there is some uncertainty surrounding this estimate. Bountiful is considering installing a plastic molding system to produce the parts themselves. The molder costs $20 000 plus $7000 to install, and has an expected life of six years. Operating and maintenance costs are expected to be $30 000 in the first year and to rise at the rate of 5% per year. Bountiful estimates its capital costs using a declining-balance depreciation model with a rate of 40%, and uses a MARR of 15% for such investments.

Determine the total equivalent annual cost of the new molder. What is the cost per unit, assuming that production is 200 000 units per year? Also, determine the break-even production quantity. That is, what is the production quantity below which it is better to continue to purchase parts and above which it is better to purchase the molder and make the parts in-house?

11.17 Trenton Trucking (TT) is considering the purchase of a new $65 000 truck. The truck is expected to generate revenues between $12 000 and $22 000 each year, and will have a salvage value of $20 000 at the end of its five-year life. TT pays taxes at the rate of 35%. The CCA rate for trucks is 30%, and TT's after-tax MARR is 12%. Find the annual worth of the truck if the annual revenues are $12 000, and for each $1000 revenue increment up to $22 000. What is the break-even annual revenue? Provide a graph to illustrate the break-even annual revenue.

11.18 A new bottle-capping machine costs $45 000, including $5000 for installation. Operating and maintenance costs are expected to be $3000 for the first year, increasing by $1000 each year thereafter. The salvage value is calculated by straight-line depreciation where a value of 0 is assumed at the end of the service life.

(a) Construct a spreadsheet that computes the equivalent annual cost (EAC) for the bottle capper. What is the economic life if the expected service life is 6, 7, 8, 9, or 10 years? Interest is 12%.

(b) How sensitive is the economic life to the different length of service life? Construct a sensitivity graph to illustrate this point.

11.19 A chemical plant is considering installing a new water purification system which costs $21 500. The expected service life of the system is 10 years and the salvage value is computed using the declining-balance method with a depreciation rate of 20%. The operating and maintenance costs are estimated to be $5 per hour of operation. The expected savings is $10 per operating hour.

(a) Find the annual worth of the new water purification system if the current operating hours are 1500 per year on average. MARR is 10%.

(b) What is the break-even level of operating hours? Construct a graph showing the annual worth for various levels of operating hours.

11.20 Antigonish Footwear can invest in one of two different automated clicker cutters. The first, A, has a $10 000 first cost. A similar one, B, with many extra features, has a first cost of $40 000. A will save $5000 per year over the cutter now in use. B will save between $12 000 and $15 000 per year. Each clicker cutter will last five years and have a zero scrap value.

(a) If the MARR is 10%, and B will save $15 000 per year, which alternative is better?

(b) B will save between $12 000 and $15 000 per year. Determine the IRR for the incremental investment from A to B for this range, in increments of $500. Plot savings of B versus the IRR of the incremental investment. Over what range of savings per year is your answer from part (a) valid? What is the break-even savings for alternative B such that below this amount, A is preferred and above this amount, B is preferred?

11.21 Sam is considering buying a new lawnmower. He has a choice between a "Lawn Guy" model or a Bargain Joe's "Clip Job" model. Sam has a MARR of 5%. The mowers' salvage values at the end of their respective service lives is zero. Sam has collected the following information about the two mowers.

Although Sam has estimated the maintenance costs of the Clip Job at $60, he has

	Lawn Guy	Clip Job
First cost	$350	$120
Life	10 years	4 years
Annual gas	$60	$40
Annual maintenance	$30	$60

heard that the machines have had highly variable maintenance costs. One friend claimed that her Clip Job had maintenance costs comparable to those of the Lawn Guy, but another said the maintenance costs could be as high as $80 per year. Construct a table that shows the annual worth of the Clip Job for annual maintenance costs varying from $30 per year to $80 per year. What Clip Job maintenance costs would make Sam indifferent between the two mowers, on the basis of annual worth? Construct a graph showing the break-even maintenance costs. Which mower would you recommend to Sam?

11.22 Ganesh is considering buying a $24 000 car. After five years, he thinks he will be able to sell the car for $8000, but this is just an estimate that he is not certain about. He is confident that gas will cost $2000 per year, insurance $800 per year, and parking $600 per year, and that maintenance costs for the first year will be $1000, rising by $400 per year thereafter.

The alternative is for Ganesh to take taxis everywhere. This will cost an estimated $7000 per year. If he has no car, Ganesh will rent a car for the family vacation every year at a total (year-end) cost of $1000. Ganesh values money at 11% annual interest. If the salvage value of the car is $8000, should he buy the car? Base your answer on annual worth. Determine the annual worth of the car for a variety of salvage values so that you can help Ganesh decide whether this uncertainty will affect his decision. For what break-even salvage value will he be indifferent between taking taxis and buying a car? Construct a break-even graph showing the annual worth of both alternatives as a function of the salvage value of the car. What advice would you give Ganesh?

11.23 Ridgely Custom Metal Products (RCMP) must purchase a new tube bender. They are considering two alternatives that have the following characteristics:

	Model T	Model A
First cost	$100 000	$150 000
Economic life	5 years	5 years
Yearly savings	$50 000	$62 000
Salvage value	$20 000	$30 000

Construct a break-even graph showing the present worth of each alternative as a function of interest rates between 6% and 20%. Which is the preferred choice at 8% interest? Which is the preferred choice at 16% interest? What is the break-even interest rate?

11.24 Julia must choose between two different designs for a safety enclosure. Model A has a life of three years, has a first cost of $8000, and requires maintenance of $1000 per year. She believes that a salvage value can be estimated for Model A using a depreciation rate of between 30% and 40% and declining-balance depreciation. Model B will last four years, has a first cost of $10 000, and has maintenance costs of $700 per year. A salvage value for Model B can be estimated using straight-line depreciation and the knowledge that after one year the salvage value will be $7500. Interest is at 11%. Which of the two models would you suggest Julia choose? What break-even depreciation rate for Model A will make her indifferent between the two models? Construct a sensitivity graph showing the break-even depreciation rate.

11.25 Your neighbour, Kelly Strome, is trying to make a decision about his growing home-based copying business. He needs to acquire colour copiers able to handle maps and other large documents. He is looking at one set of copiers that will cost $15 000 to purchase. If he purchases the equipment, he will need to buy a maintenance contract that will cost $1000 for the first year, rising by $400 per year afterward. He intends to keep the copiers for five years, and expects to salvage them for $2500. The CCA rate for office equipment is 20%.

Rather than buy the copiers, Kelly could lease them for $5500 per year with no maintenance fee. His business volume has varied over the past few years, and his tax rate has varied from a low of 20% to a high of 40%. Kelly's current cost of capital is 8%. Kelly has asked you for some help in deciding what to do. He wants to know whether he should lease or buy the copiers, and, moreover, he wants to know the impact of his tax rate on the decision. Evaluate both alternatives for him for a variety of tax rates between 20% and 40% so that you can advise him confidently. What do you advise?

11.26 Alberta Insurance wants to introduce a new accounting software package for their human resources department. A small-scale version is sufficient and economical if the number of employees is less than 50. A large-scale version is effective for managing 80 employees or more. All relevant information on the two packages is shown in the following table. Alberta Insurance's business is growing, and the number of employees has increased from 10 to 40 in the past three years. Construct a graph showing the annual worth of the two software packages as a function of the number of employees ranging from 40 to 100. On the basis of break-even analysis, which accounting package is a better choice for Alberta Insurance? MARR is 12%.

Parameter		Small-Scale	Large-Scale
First cost ($)		6000	10 000
Training cost at the time of installation ($)		1500	3500
Service life (years)		5	5
Salvage value		0	0
Expected annual savings ($ per employee) if the average number of employees over the next 5 years is:	Less than 50	200	250
	Between 50 and 80	170	300
	Greater than 80	120	400

11.27 Refer back to Problem 11.10 in which Manitoba Metalworks is looking at several LAN alternatives. Conduct a scenario analysis for each alternative, using the pessimistic, expected (average), and optimistic outcomes for both installation costs and annual benefits. Which of the two alternatives would you choose? Why?

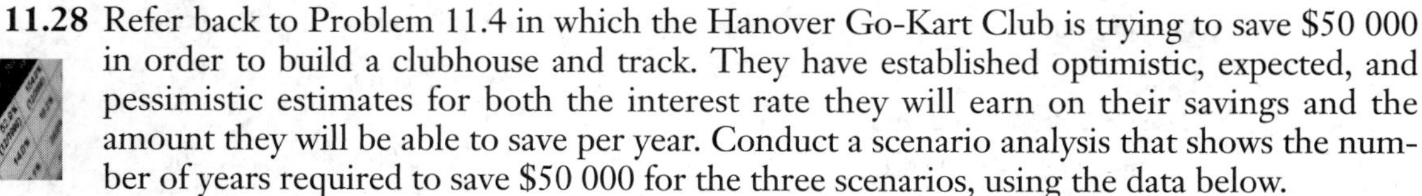

11.28 Refer back to Problem 11.4 in which the Hanover Go-Kart Club is trying to save $50 000 in order to build a clubhouse and track. They have established optimistic, expected, and pessimistic estimates for both the interest rate they will earn on their savings and the amount they will be able to save per year. Conduct a scenario analysis that shows the number of years required to save $50 000 for the three scenarios, using the data below.

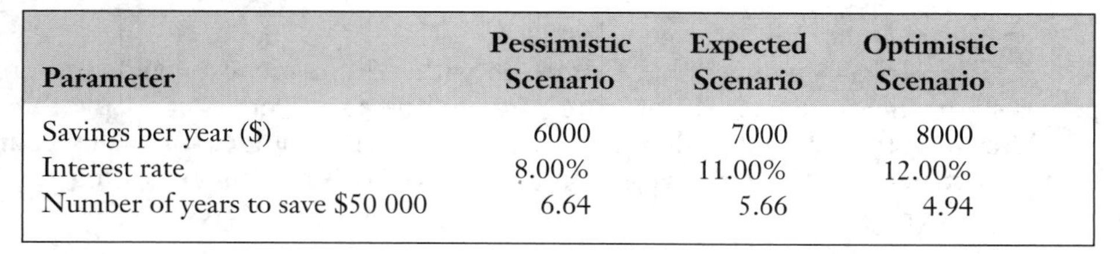

Parameter	Pessimistic Scenario	Expected Scenario	Optimistic Scenario
Savings per year ($)	6000	7000	8000
Interest rate	8.00%	11.00%	12.00%
Number of years to save $50 000	6.64	5.66	4.94

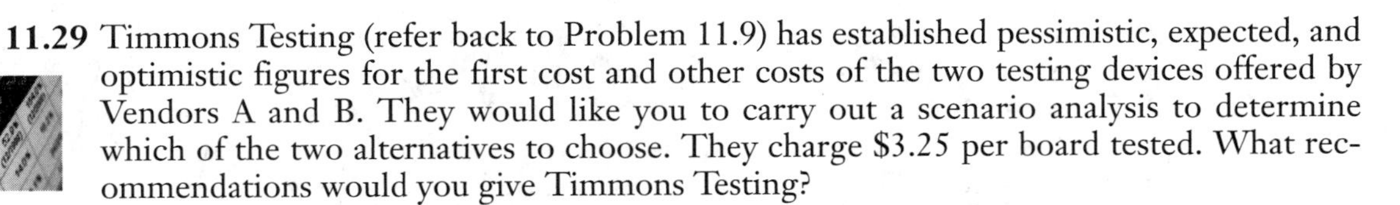

11.29 Timmons Testing (refer back to Problem 11.9) has established pessimistic, expected, and optimistic figures for the first cost and other costs of the two testing devices offered by Vendors A and B. They would like you to carry out a scenario analysis to determine which of the two alternatives to choose. They charge $3.25 per board tested. What recommendations would you give Timmons Testing?

	Alternative A		
Parameter	**Pessimistic Scenario**	**Expected Scenario**	**Optimistic Scenario**
First cost ($)	220 000	200 000	190 000
Annual fixed costs ($)	18 000	16 500	13 000
Annual variable costs ($ per board)	2.35	2.25	2.20
Salvage value ($)	2 000	5 000	7 000
Annual production volume (boards)	40 000	50 000	80 000

	Alternative B		
Parameter	**Pessimistic Scenario**	**Expected Scenario**	**Optimistic Scenario**
First cost ($)	365 000	350 000	320 000
Annual fixed costs ($)	45 000	35 500	25 000
Annual variable costs ($ per board)	1.100	1.060	1.010
Salvage value ($)	17 000	20 000	23 000
Annual production volume (boards)	40 000	50 000	80 000

11.30 The Bountiful Bread Company (Problem 11.16) currently pays a custom molder $0.19 per piece (not including material costs) for the clear plastic face on the control panel of bread-maker machines they manufacture. Demand for the bread-makers is estimated at 200 000 machines per year, but there is some uncertainty surrounding this estimate. Bountiful is considering installing a plastic molding system to produce the parts themselves. Installation costs are $7000, and the molder has an expected life of six years. Operating and maintenance costs are somewhat uncertain, but are expected to be $30 000 in the first year and to rise at the rate of 5% per year. Bountiful estimates its capital costs with a declining-balance depreciation model with a rate of 40%, and uses a MARR of 15% for such investments.

Parameter	Pessimistic Scenario	Expected Scenario	Optimistic Scenario
First cost ($)	25 000	20 000	18 000
Base annual operating and maintenance costs ($)	35 000	30 000	27 000
Production volume (units)	170 000	200 000	240 000

The project engineers have come up with pessimistic, expected, and optimistic figures for the first cost, operating and maintenance costs, and production levels. Determine the total equivalent annual cost of the new molder for each scenario and then the cost per unit. What advice would you give to Bountiful regarding the purchase of the molder?

BC Hydro Electricity Planning

BC Hydro is a commercial Crown corporation owned by the Province of British Columbia and regulated by the B.C. Utilities Commission. It is one of the largest electric utilities in Canada and serves more than 1.6 million customers. In 2004, BC Hydro completed a 20-year Integrated Electricity Plan (IEP) that provides strategic direction for meeting future electricity needs for the province. The process of developing the IEP included

- Reviewing the economic and regulatory context in which BC Hydro conducts its planning
- Considering the outlook for electricity supply and demand
- Compiling an inventory of resource options
- Building and comparing alternative resource portfolios to meet electricity requirements

The action plan in the IEP identified initiatives that provide reliable, least-cost electricity supply in an environmentally responsible manner.

The portfolio modelling process involved constructing various alternatives for the quantity and timing of resource additions. They had a wide variety of options available including small and large hydroelectric projects, private sector electricity generation, coal, natural gas, wind, and geothermal and solar generation options, among others. Each alternative provided a different level of cost, dependability and reliability, and environmental and social impact, and a diverse set of risks and uncertainties.

Once alternative schedules for new resource additions were determined, a hydrological system simulation model (HYSIM) was used to evaluate the expected annual electricity generation from each alternative. From this, the annual variable cost for each portfolio was determined. The annual variable costs were then consolidated with capital and fixed costs in BC Hydro's Multi-Attribute Portfolio Analysis (MAPA) model.

The MAPA model calculates the net present worth of each portfolio over the 20-year study period in real dollars. Sensitivity analysis was then used to determine the effect on the present worth of a portfolio due to changes in financial factors such as the discount rate and fuel and capital costs. MAPA was also used to run a variety of scenarios so that BC Hydro could examine the impact of using different planning assumptions on the long-term performance of a portfolio. Some of the scenarios examined were high and low electricity demand, alternative sets of gas and electricity prices, and two scenarios for the costs associated with greenhouse gas emissions.

Among the key conclusions were that Power Smart (BC Hydro's initiatives to encourage energy efficiency by its customers) and Resource Smart (BC Hydro's improvement program for existing facilities) were both economical and should be continued, and that a modest acquisition program from the private sector should be pursued. They also determined that assumptions surrounding the Burrard Thermal Generating Station (whose future was uncertain) had a significant impact on their resource portfolio. Other insights had to do with keeping several resource expansion options open for the future.

Source: "2004 Integrated Electricity Plan," BC Hydro site, May 13, 2004, www.bchydro.com/info/epi/epi8970.html, accessed October 10, 2004.

Discussion

Engineering design often assumes that the world is much simpler than it really is. When an engineer designs a roof truss, for example, he or she often assumes that the lumber making up the truss behaves in a standard, predictable manner. Similarly, the engineer who designs a circuit will assume that the electrical components will behave according to their nominal values. But lumber is a natural product, and individual pieces will be weaker or stronger than expected. Electrical components, similarly, will have actual values and behaviour different, in general, from their nominal values and mathematical models.

Good engineers understand this and design accordingly. The truss-builder specifies a certain grade of lumber, or makes sure that redundant support is built into the design. The circuit designer similarly specifies the tolerances of significant components, or designs the circuit in a robust way.

The role of sensitivity analysis in economic studies is exactly the same. We don't know exactly the cash flow of a project just as we don't know exactly the behaviour of a piece of wood or a circuit component. We want to design the project to control the uncertainty of the economic elements as well as the physical ones. In the BC Hydro Electricity planning process, sensitivity analysis was able to identify robust alternatives for long-term provision of electricity for the province of British Columbia.

Questions

1. For your next job—summer, co-op, or after graduation—what salary do you expect to make? What is the lowest you would likely get? The highest? What steps could you take now to reduce the uncertainty of that salary amount?
2. Milo is considering buying a car to drive to school and back every day, and for recreational purposes. His expenses will include the car loan payment, gas, insurance, maintenance, and repairs. He is very concerned about the cost of the car, especially since his income is very limited. What considerations should he bear in mind to reduce the uncertainty of his future costs?
3. Derek has been assigned the task of designing a parking facility for an insurance company. He must keep in mind a number of different issues, including land acquisition costs, building costs (if a parking building is required), expected usage, fee method (monthly fees, hourly fees, or in-and-out fees), whether the company will subsidize the facility in part or completely, etc. His boss is particularly concerned about reducing the uncertainty of the future cash flows associated with the project. How would you advise Derek?

Back to the Real World

1. A Morning Meeting

Carole smiled and said a cheerful "Good morning" as Naomi and Dave came up the hallway. "Clem's not here yet, but you might as well have a seat in his office."

"Just like a manager to be late for his own meeting," observed Naomi with a smile, as she sank into her favourite green paisley chair. "Looks like we know who'll be buying the coffee this morning."

No sooner had they sat down when Clem came bounding in with a breathless "Sorry I'm late."

"No problem, Clem. Oh, don't forget the sugar in my coffee this time."

"I deserve that, Naomi," replied Clem unexpectedly. "Tell you what. You guys did such a good job on that cold-former evaluation that I'll even spring for donuts today."

Dave raised his eyebrows in mock surprise. "What is this, Clem, an early Christmas?"

"Don't be so sure, Dave," Naomi interjected. "I have a feeling we're being buttered up for something." Looking at Clem, she continued, "Do I feel some onerous task coming on?"

"'Opportunity,' Naomi. We only have 'opportunities' around here," corrected Clem, smiling. "And speaking of opportunities, it looks like we have a good one in this project." He was pointing to Naomi and Dave's cold-former report. "I have a feeling that Anna Kulkowski and Ed Burns will be very happy to get this kind of cost savings."

"So we now do a full engineering economics evaluation for buying a cold-former?" asked Naomi.

"Right. In fact, I'd like you to do evaluations of both the single E1 and single E2 options. No matter what you said in your report, I think that looking at these more carefully is worthwhile."

Dave thought for a minute. "Why both? If we get a better estimate of selling price from Prabha, we'll know whether the E1 or E2 is the better option."

"Yes," replied Clem, "but I am going to recommend performing a market study on selling prices to Ed Burns right after this meeting. With the data you have here, I'm sure he'll approve the $5000 expense right away. The trouble is, it will be at least two weeks before Prabha can give us an answer. I'd like to be able to make a decision as soon as we get it."

"So . . . you'd like us to prepare an evaluation for each machine . . . leaving the selling price as a variable . . . so that when we get the better estimate of what the price will be, we can make an immediate decision about which machine to buy. That sounds like a 'break-even' analysis, Clem," finished Naomi.

"Can't you just see those mental wheels spinning?" said Clem to Dave with a grin. "That's exactly what I'd like. Not only would a break-even analysis show us which machine would be best at whatever selling price we end up getting, but it would also give us an idea of how sensitive the benefits of this project are to the price at which we sell excess production."

Naomi and Dave were quiet for a few seconds while they thought this over.

Dave glanced up. "Would this mean that if we show enough extra benefit from selling our excess capacity, Marketing would be asked to put some effort into getting better prices for us?"

"That's right, Dave. A break-even analysis would give us the data to demonstrate to management how much the company would benefit and what the payoff would be as a function of the price Marketing gets for those parts."

Naomi broke in. "But for which measure should we do a break-even analysis: present worth, IRR, or payback period?"

"Anna and Ed base their decisions mostly on PW and IRR values. I know they will ask how each behaves as a function of selling price, so we should do a break-even analysis on both."

"Both?"

"Both. We'll still have to calculate payback periods as per the company capital justification procedure, but we can calculate it over a few values of selling price and demand growth if that's more convenient than a break-even analysis."

"'A few values of demand growth'?" echoed Naomi.

"That's right. We still only have a poor estimate of the demand growth rate. So, just like you've already done, we will need to do all the analyses over three possible growth rates: 5, 10, and 15%. It didn't look like the optimal decision was affected by demand growth in your analysis, but we'd better check anyway."

Naomi and Dave glanced at each other, both recognizing a few long work days coming. "Anything else?" asked Dave, with only the slightest hint of sarcasm.

"Glad you asked, Dave. Of course, these will all be after-tax calculations, and you'll have to take salvage values into account. Oh, and write up your results in a full engineering report; the President and everyone between her and us are going to read it. Am I forgetting anything?"

"Inflation, too?" Dave couldn't keep the sour tone out of his voice completely.

"No," Clem laughed. "You're off the hook on that one. Inflation looks like it will continue to stay low for the foreseeable future, so we will ignore it."

"Are we limited to just two options, or can we look for better alternatives?" Naomi asked.

"Good question, Naomi. These two are currently the most likely options to produce the best solution, depending on the selling price we get. We definitely need to do a full study of them. If you want, you can look for a better solution; if you find one it will certainly be a feather in your cap. On the other hand, remember that you still have all your other work to do."

"Don't worry, Clem, we certainly remember that we have other work to do, too," Dave said, dryly but with a smile.

"Well, I've got one last question, Clem," Dave said, a few seconds later. "Who did you say was buying the coffee today?"

"And donuts, too!" piped in Naomi.

2. Down to Details

Later that morning, Naomi met Dave in his office to divide up the work. His office was enclosed, and thankfully so, because the walls protected the rest of the offices from Dave's unique organizing style.

"Looks like we're going to be crunching numbers for some time," Dave said, while clearing a chair for Naomi to sit in.

"Maybe not, Dave. We've got most of the spreadsheets for the calculations already set up. We'll have to include salvage values and tax effects. It turns out that the spreadsheet already has PW and IRR financial functions built in."

Dave continued to clear space on the table in front of Naomi as they talked. "That will certainly save a lot of trial-and-error to come up with IRRs, but what about all the cases we have to do to make break-even graphs?"

Naomi gestured at Dave's computer, almost the only thing in the office not hidden by paper. "I've found a feature in Excel called 'Data Tables'—I suppose most spreadsheet programs have something like it. It allows us to vary the contents of several spreadsheets based on data in a table. It looks like that will allow us to calculate all the values we need in no time."

"Sounds like I know who'll be doing the spreadsheets this time," said Dave. "How about if I concentrate on writing the report?" Dave had a clump of papers in his hand that he had not yet repositioned, and was pacing about the room looking for a place to put them.

"OK. Oh, before I forget, this equipment falls into the CCA Asset Class 8? CCA rate is 20%?"

"That's right. We can also get a good estimate of salvage value with a declining-balance depreciation rate of 20%. By the way, what was that you said to Clem about better options?"

"Oh, I've got some ideas. I'll let you know if they pan out."

Finally Dave sat down with a clear workspace in front of both of them. "Fair enough. According to my stomach, it must be lunchtime. Want to see if Clem wants to go over to the Grand China?" He jumped out of his chair again, after spending no more than three seconds sitting down.

"Sure," said Naomi. "Do we have to spread these papers out again before we leave?"

QUESTIONS

1. Using data provided in both parts of the Extended Case, update your spreadsheets to include salvage value, tax effects, PW, IRR, and payback calculations. PW, IRR, and payback analyses are required for each project (single E1 machine or single E2 machine purchased at time 0). Summarize PW, IRR, and payback values in tables for each of 5, 10, and 15% demand growth and $0.03, 0.035, and 0.04 selling prices per piece.

2. Perform a break-even analysis for each of PW and IRR for each project over the range of selling prices from \$0.03 to \$0.04. Repeat for demand growth rates of 5, 10, and 15% (i.e., 2 comparison methods × 2 projects × 3 growth rates = 12 break-even calculations). Present the results graphically.
3. On the basis of your current analysis results, can you make a clear recommendation to the company's management which of E1 and E2 should be purchased? If so, why? If not, why not?
4. Write a full engineering economics report to the President, Anna Kulkowski, about this project. The report should follow the format of an engineering report. See the guidelines given below. In the report, you should present the results of your analyses so that company's management can make a defensible decision about which machine they should purchase.

OPTIONAL

5. Currently we only consider purchasing one machine, either E1 or E2, at time 0. Find out if a better solution exists by varying the number of machines purchased at a time and/or the timing of the purchase (i.e., time 1, 2, etc.). Include your findings in the report.

Guidelines for an Engineering Report

A Typical Engineering Report May Include:

Cover Letter ("Here is the report you ordered" etc.)

Title Page

Table of Contents

Summary (of the contents of the report)

Introduction (e.g., description of the problem, background, and purpose)

Body (e.g., procedure, calculations, results, possible errors, and unanswered questions)

Conclusions

Recommendations

References

Appendixes (important but too lengthy or disorderly to be included in the body)

Notes for Excel Users (Similar Issues Exist for Other Spreadsheet Programs)

Circular references: When calculating after-tax IRRs, you might run across a problem in Excel with "Circular Reference" warnings. This occurs because you use the IRR rate to find the CCTFs, which you then use to calculate the IRR. The way to get around this problem is to go to the Excel "Tools" menu, then select "Options." Select the "Calculation" tab and click on the word "Iteration," then "OK." This will set the worksheet to perform iterative calculations that will converge for the value of the IRR. You might try pressing the F9 (recalculate) key a few times to make sure the values have converged.

IRR and NPV (present worth) built-in functions: Excel has built-in functions for calculating IRR and PW (Excel uses the term "NPV: Net Present Value" for present worth). Check Excel's Help file for details. Be careful with the NPV function: it calculates a PW for one period in the future. You have to adjust the value for time 0.

Data tables: You can run analyses for many different cases very quickly by using the Data Table feature, found under the "Data/Table . . . " menu.

CHAPTER 12

Dealing with Risk: Probability Analysis

Engineering Economics in Action, Part 12A: Trees from Another Planet

The coffee cups had been cleaned up, but the numerous brown rings that remained reported the hours of work Bill and Naomi had put in calculating the economic effects of manufacturing the new self-adjusting vacuum cleaner in partnership with the Powerluxe company.

"So, the bottom line is: if the demand is low, we lose money, but otherwise we gain. And more than that, if demand is high, we gain big-time." Bill was a bit plaintive because he knew that this did not solve their problems—he had hoped that they would make money no matter what. "So what it really comes down to is this: What is the chance that demand is low?"

Naomi looked into space for a second, and then thoughtfully started, "No, Bill, it's a little more complicated than that. It's . . . "

Just then, she was interrupted by Clem, who stepped in the lunchroom door. Glancing at the papers spread over the table, he remarked, "Not much of a lunch! Anyhow, I've been talking to Ms. Kulkowski, and she said that you should take into account in your Powerluxe project the potential for a competitive product."

"You mean another vacuum cleaner, just like the—what are they calling it, the 'Adaptamatic'—with the sensing capability?" asked Naomi.

"Yeah," replied Clem. "I guess the big shots are worried that those guys over at Erie Gadgets have some sort of similar thing under study. Well, that's all I had to tell you. Work hard." Clem winked and disappeared as quickly as he arrived.

Bill leaned back in his chair with his arms in the air. "For heaven's sake! We don't have any chance of coming to a clear recommendation now!"

"No, I think we're OK, Bill. In fact, since you mention it, 'chance' is pretty important here. So are trees."

Bill looked at Naomi as if she were from another planet. "Trees! Trees?"

12.1 Introduction to Uncertainty and Risk

In our day-to-day life, we often encounter situations where we don't know for sure what events will happen in the future. We talk about the "chance" that the weather will be rainy tomorrow, the "likelihood" that our favourite hockey team will win the Stanley Cup, or the "probability" of a railcar spill. Of course, once the event occurs, we know the outcome: either it did or did not rain, our team won or didn't, or the railcar spill happened or didn't happen.

In carrying out an economic analysis of a project, it is often the case that the engineer must estimate various parameters of a situation—the life of an asset, the interest rate, the magnitude and timing of cash flows, or factors such as the likely success of a new product. Chapter 11 dealt with sensitivity analysis tools appropriate for decision making under uncertainty—situations in which we can characterize a range of possible outcomes, but do not have a probability assessment associated with each outcome. Sensitivity analysis allowed us to look at the impact of this lack of probability information in a general way. Chapter 12, on the other hand, deals with decision making under *risk*. Decisions made under risk are those where the analyst can characterize a possible range of future outcomes and has available an estimate of the probability of each outcome. The term "risk" is also often used to refer to the probability distribution of outcomes associated with a project, or the probability of an undesirable outcome (this last definition of risk is not utilized in this text, though it is commonly used in the financial literature). Knowledge of the probability distribution of outcomes often permits us to draw more authoritative conclusions than we can draw using sensitivity analysis alone.

This chapter deals with a variety of approaches used by engineers to structure and make decisions when the alternatives involve risk. The first approach, decision trees, is commonly used to decompose a problem clearly into its decision alternatives and uncertain events. After presenting the use of decision trees, the chapter focuses on some frequently used decision criteria as a basis upon which alternatives can be evaluated. Of these criteria, expected value is the most commonly used, but dominance concepts such as mean-variance, outcome, and stochastic dominance are also useful for either screening alternatives or selecting the most preferable alternative. Finally, the chapter closes with a brief coverage of Monte Carlo simulation, a powerful approach for analyzing complex problems that have a large number of decision alternatives or many uncertain components. This chapter is intended only to provide introductory coverage of commonly used methods. More advanced approaches are beyond the scope of this book.

12.2 Basic Concepts of Probability

In this section, we will define more precisely what we mean by "chance" and "likelihood" so that we are able to make useful predictions in the context of engineering economics. The branch of mathematics that formalizes this common notion of "chance" is called probability theory.

Suppose you are concerned with the market success of a new fuel-cell technology, but you are uncertain about its outcome. It could be that the fuel cell fails to gain market acceptance, it may produce adequate sales, or it could gain market dominance. In the terminology of decision analysis, "market success" is referred to as a *random variable*. A **random variable** is a parameter or variable that can take on a number of possible outcomes. Only one of these outcomes will eventually occur, but which one will occur is unknown at the time a decision is being made. For example, if market success is considered to be a random variable, then the set of possible outcomes are failure to gain market acceptance, adequate sales, or market dominance. To construct a model of this uncertainty, you will need to know the probability associated with each possible outcome. This is accomplished through a probability distribution function.

Consider a random variable X (e.g., the market success of a fuel cell) that can take on m discrete outcomes $x_1, x_2, \ldots, x_m$. If these events are mutually exclusive (if one occurs, another cannot) and collectively exhaustive (one of them must occur), a **probability distribution function** $p(x)$ is a set of numerical measures $p(x_i)$ such that:

$$0 \leq p(x_i) \leq 1 \quad \text{for } i = 1, \ldots, m$$

and

$$\sum_{i=1}^{m} p(x_i) = 1$$

with the intuitive interpretation that the higher $p(x_i)$, the more likely it is that x_i will occur.

The first statement above says that the **probability** associated with each outcome must be positive, and must be between 0 and 1 (inclusive). Intuitively, this means that any outcome cannot have a chance of occurring less than 0%, nor more than 100%. The second statement above says that since the outcomes are mutually exclusive and collectively exhaustive, that only one of the outcomes will occur, and that one *must* occur.

Over the years, various views on **probability** have emerged. Each is appropriate in different circumstances. Close-Up 12.1 summarizes different views on probability and when they are useful.

CLOSE-UP 12.1 Views on Probability

Classical or symmetric probability: This was the first view of probability and relies on games of chance such as dice, where the outcomes are equally likely. For example, if there are m possible outcomes for an uncertain event, since only one can and must occur, and each are assumed to be equally likely, then the chance of each occurring is $1/m$. For example, the probability that a coin toss will result in a "heads" is 1/2 because there are two sides and each is equally likely.

Relative frequency: The outcome of a random event E is observed over a large number of experiments, N. If the number of times the outcome e_i occurs is n_i, then we can estimate the probability of event $p(e_i)$ by n_i/N. More formally, the relative frequency view on probability says that $p(e_i) = \lim_{N \to \infty} n_i / N$. An example of this is flipping a coin 1000 times and discovering that it lands on its edge five times in 1000. An estimate of the probability of the coin landing on its edge is then 5/1000 = 0.005.

Subjective probability: Subjective or personal probability is an attempt to deal with unique events which cannot be repeated and hence can't be given a frequency interpretation. In rough terms, subjective probability can be interpreted as the odds one would personally give in betting on an event, or it may be a matter of human judgment and intuition as formed by physical relationships and experimental results. An example of this is a person who judges that the chance of winning a coin toss with one of the authors of this text is very low, say 1/1000, because, in their experience, the authors usually cheat.

Axiomatic probability: One of the problems associated with defining probabilities is that the definition of probability requires using probability itself. To get around this circular logic, axiomatic probability makes no attempt to define probability, but simply states the rules or axioms it follows. Other properties can then be derived from the basic axioms.

Each of the above methods may be correct, given the circumstances. When the physics of a process suggest a clear judgment of probability, the classical approach makes sense. Where formal experimentation is possible, the relative frequency method may be justified. In many real-world cases, subjective probability supported by historical information and other data is frequently used.

12.3 Random Variables and Probability Distributions

When the number of outcomes for a random variable X is discrete, $p(x)$ is referred to as a **discrete probability distribution function** (PDF). Examples of discrete random variables are the number of good items in a batch of 100 tested products, the number of car accidents at an intersection each year, the number of days since the last plant shutdown or the number of bugs found in software testing. Whether $p(x)$ is estimated using the classical, the relative frequency, or the subjective approach, the same terminology for the probability distribution function, $p(x)$, is used.

Various symbols may be used to define a random variable. The normal convention is to capitalize the symbol used for the random variable, and to use subscript lowercase letters to denote its various outcomes. For example, for a discrete random variable X, outcomes are denoted by $x_1, x_2, x_3, \ldots$ and its probability distribution function is $p(x)$. The probability that X takes on the value x_1 is written $\Pr(X = x_1) = p(x_1)$.

Example 12.1

Suppose that you are testing solder joints on a printed circuit board and that you are interested in determining the probability distribution function for the random variable X, the number of open joints in three tested joints (to keep it simple). Prior to testing, you don't know how many open joints there will be. Since there are three joints and each will be either open or closed, X is a discrete random variable which can take on four possible values: $x_1 = 0, x_2 = 1, x_3 = 2$, and $x_4 = 3$.

Note that there are eight distinct test result sequences that can occur. Denoting the result of a single test by O for open and C for closed, the set of possible test results are: (O,O,O), (O,O,C), (O,C,O), (C,O,O), (O,C,C), (C,C,O), (C,O,C), and (C,C,C). We must look through the set of individual test results to see which corresponds to $x_1 = 0$, $x_2 = 1, x_3 = 2$, and $x_4 = 3$ open solder joints.

You know from previous data collection efforts that the result of a single test is uncertain. The probability that a single tested joint will be open is 20%. In other words, the outcome of the test is a random variable, say Y, where the result of a single test can have two outcomes: $y_1 = \text{O} = \text{open}$ and $y_2 = \text{C} = \text{closed}$, and $\Pr(Y = \text{O}) = p(\text{O}) = 0.2$ and $\Pr(Y = C) = p(C) = 0.8$. Further, suppose it is reasonable to assume that the quality of a solder joint does not change from joint to joint (i.e., the test results are independent of one another). Then the probability that a test sequence results in three open joints is calculated by

$$\Pr(X = 3) = p(x_4) = p(\text{O}) \times p(O) \times p(\text{O}) = (0.2) \times (0.2) \times (0.2) = 0.008$$

Similar calculations, shown in Table 12.1, yield the probabilities for each of the eight possible test sequences.

Table 12.1 Probability Corresponding to the Outcomes of the Solder-Joint Testing with a 20% Chance of an Open Joint

Test Sequence	Number of "Opens"	Probability
(O,O,O)	3	$0.008 = 0.2 \times 0.2 \times 0.2$
(O,O,C)	2	$0.032 = 0.2 \times 0.2 \times 0.8$
(O,C,O)	2	$0.032 = 0.2 \times 0.8 \times 0.2$
(C,O,O)	2	$0.032 = 0.8 \times 0.2 \times 0.2$
(O,C,C)	1	$0.128 = 0.2 \times 0.8 \times 0.8$
(C,C,O)	1	$0.128 = 0.8 \times 0.8 \times 0.2$
(C,O,C)	1	$0.128 = 0.8 \times 0.2 \times 0.8$
(C,C,C)	0	$0.512 = 0.8 \times 0.8 \times 0.8$

Finally, the probability distribution function of X, the number of "open" joints in the three tests, is

$$\Pr(X = 0) = p(x_1) = 0.512$$
$$\Pr(X = 1) = p(x_2) = 0.384$$
$$\Pr(X = 2) = p(x_3) = 0.096$$
$$\Pr(X = 3) = p(x_4) = 0.008$$

Note that the two important properties of probabilities hold: $\geq p(x_i) \geq 0$ for all i, and $\sum p(x_i) = 1$. ■

It is often useful to display a probability distribution function in graphical format. Such a graph is referred to as a histogram. Figure 12.1 shows the probability distribution function associated with the solder joint testing results.

Figure 12.1 Probability Distribution for the Solder Joint Example

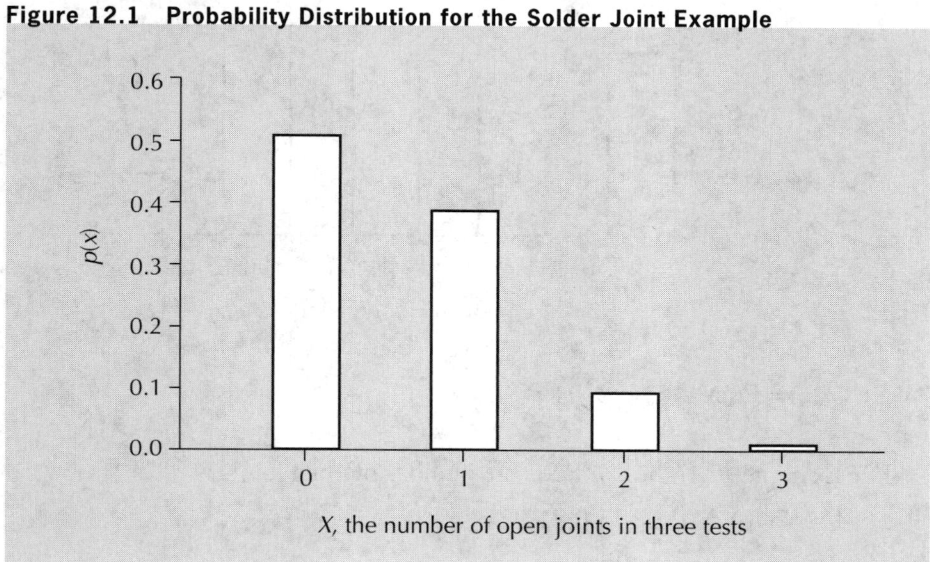

In contrast to a discrete random variable, a continuous random variable can take on any real value over a defined interval. For example, daily demand for drinking water in a municipality might be anywhere between 10 and 200 million litres. The actual amount consumed—the outcome—is a continuous random variable with a minimum value of 10 million litres and a maximum value of 200 million litres.

In this chapter, we focus on applications of discrete random variables in engineering economics analysis. We do not use continuous random variables because proper treatment requires more advanced mathematical concepts such as differential and integral calculus. Also, continuous random variables can be well approximated as discrete random variables by grouping the possible output values into a number of categories or ranges. For example, rather than treating demand for drinking water as a continuous random variable, demand could be characterized as high, medium, or low. Figure 12.2 shows an example of a probability distribution associated with future demand for water, approximated by a discrete random variable denoted by D.

Another way to characterize a random variable is through its **cumulative distribution function (CDF)**. If X is the random variable of interest, and x is a specific value, then the cumulative distribution function (for a discrete random variable) is defined as follows:

Figure 12.2 Probability Distribution Function of the Demand for Drinking Water

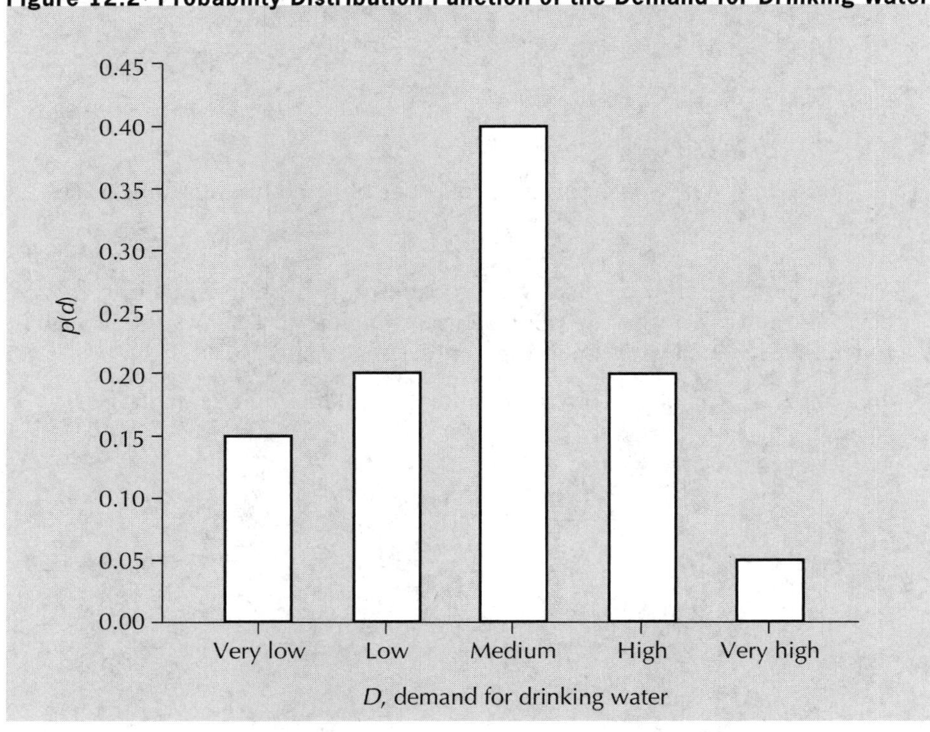

$$P(x) = \Pr(X \le x) = \sum_{x_i \le x} p(x_i)$$

For Example 12.1, the CDF for the number of open joints is:

$$\Pr(X \le 0) = P(x_1) = 0.512$$

$$\Pr(X \le 1) = P(x_2) = 0.896$$

$$\Pr(X \le 2) = P(x_3) = 0.992$$

$$\Pr(X \le 3) = P(x_4) = 1.000$$

Figure 12.3 shows the **cumulative distribution function** associated with the solder joint testing example.

The probability distribution of a random variable contains a great deal of information that can be useful for decision-making purposes. However, certain summary statistics are often used to capture an overall picture rather than working with the entire distribution. One particularly useful summary statistic is the **expected value**, or **mean**, of a random variable. The expected value of a discrete random variable X, $E(X)$, that can take on values $x_1, x_2, \ldots, x_m$ is defined as follows:

$$E(X) = \sum_{i=1}^{m} x_i p(x_i)$$

You will no doubt observe that computing the expected value of a random variable is much like computing the centre of mass for an object. The expected value is simply the centre of the probability "mass."

Figure 12.3 Cumulative Distribution Function for the Solder Joint Example

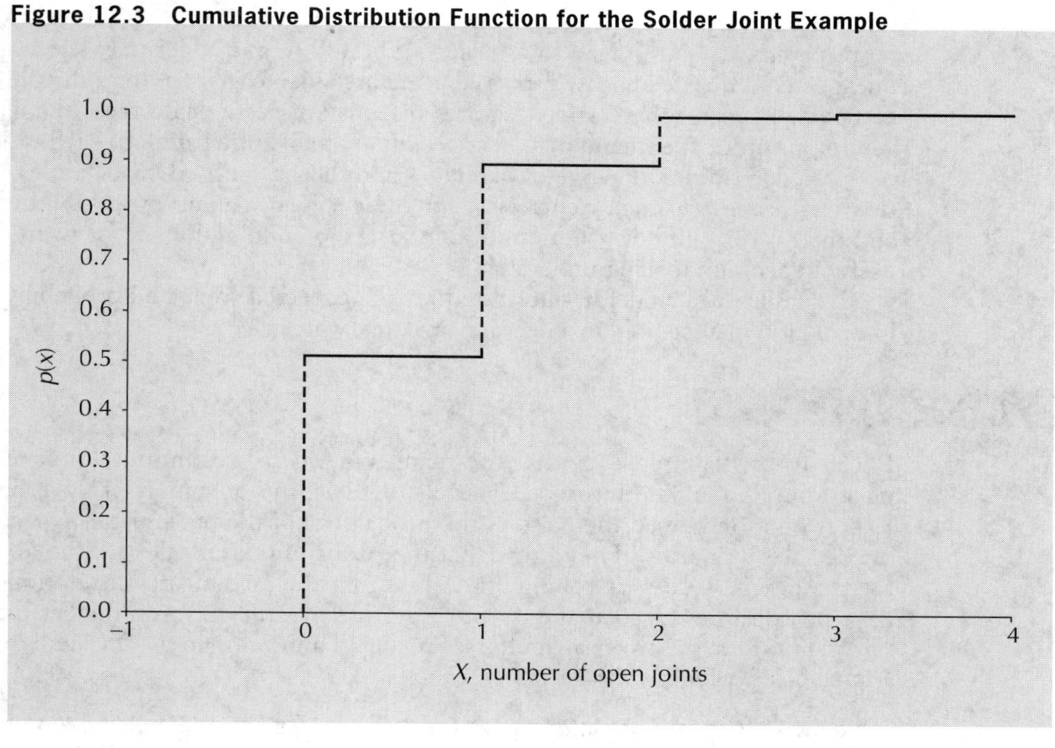

Other useful summary statistics are the variance and standard deviation of a random variable. Both capture the degree of spread or dispersion in a random variable around the mean. The **variance** of a discrete random variable X is

$$Var(X) = \sum_{i=1}^{m} p(x_i)(x_i - E(X))^2$$

and the square root of the variance is the **standard deviation.** Distributions that have outcomes far above or below the mean will have higher variances than those with outcomes clustered near the mean.

The mean and variance of the number of open solder joints from Example 12.1 can be easily calculated:

$$E(X) = \sum_{i=1}^{m} x_i p(x_i)$$

$$= 0 \times 0.512 + 1 \times 0.384 + 2 \times 0.096 + 3 \times 0.008 = 0.60$$

The mean number of open solder joints is 0.6.

$$Var(X) = \sum_{i=1}^{m} p(x_i)(x_i - E(X))^2$$

$$= 0.512 \times (0 - 0.6)^2 + 0.384 \times (1 - 0.6)^2 + 0.096 \times (2 - 0.6)^2 + 0.008 \times (3 - 0.6)^2 = 0.48$$

The variance of the number of open solder joints is 0.48.

How are random variables, probability distributions, and their summary measures relevant to engineering economics? In conducting engineering economic studies, costs and benefits can be influenced by several uncertain factors. Often, an engineer does not know the outcome of a project or may not know with certainty the actual value of one or more parameters important to a project. The life of a product may shorten or lengthen unexpectedly due to market forces, the maintenance costs may be difficult to state with certainty, a car may have safety features that may increase or decrease the liability for the vendor, or product demand may be more or less than anticipated. In each case, if the engineer can determine the range of outcomes and their associated probabilities, this information can present a much richer view for the decision-making process. Random variables and their probability distributions are, in fact, the building blocks for many tools that are useful in decision making under risk.

Examples 12.2 and 12.3 illustrate how the expected value and probability distribution information may be used in a decision-making context.

Example 12.2

Recall from Example 11.5 that the management of Westmount Waxworks had some uncertainty about the future sales levels of their line of statues of government leaders. Expert opinion helped them assess the probability of the pessimistic, expected, and optimistic sales scenarios. They think that the probability that sales will be 50 000 per year for the next few years is roughly 50% and that the pessimistic and optimistic scenarios have probabilities of 20% and 30%, respectively. Table 12.2 reproduces the annual cost information for the two wax melters, Finedetail and Simplicity. On the basis of expected annual costs, which is the best choice?

Table 12.2 Annual Cost Information for the Finedetail and Simplicity Wax Melters

Scenario	Annual Cost for Finedetail	Annual Cost for Simplicity	Probability
Pessimistic	\$ 85 314	\$ 94 381	0.2
Expected	112 314	103 501	0.5
Optimistic	314 814	186 901	0.3

The sales level can be represented by a discrete random variable, X. The possible values for X are: x_1 = pessimistic, x_2 = expected and x_3 = optimistic.

The expected annual cost of the Finedetail wax melter is:

E(Finedetail, annual cost)

$$= (\$85\ 314)p(x_1) + (\$112\ 314)p(x_2) + (\$314\ 814)p(x_3)$$

$$= (\$85\ 314)(0.2) + (\$112\ 314)(0.5) + (\$314\ 814)(0.3)$$

$$= \$167\ 663$$

The expected annual cost of the Simplicity wax melter is:

E(Simplicity, annual cost)

$$= (\$94\ 381)p(x_1) + (\$103\ 501)p(x_2) + (\$186\ 901)p(x_3)$$

$$= (\$94\ 381)(0.2) + (\$103\ 501)(0.3) + (\$186\ 901)(0.4)$$

$$= \$126\ 697$$

The expected annual cost of the Simplicity wax melter is lower than that of the Finedetail. Hence, the Simplicity melter is preferred.■

Example 12.3

Regional Express is a small courier service company operating in Southern Ontario. At the Toronto office, all parcels from the surrounding regions are collected, sorted, and distributed to the appropriate destinations. Regional Express is considering the purchase of a new computerized sorting device for their Toronto office. The device is so new—in fact, it is still under continuous improvement—that its maximum capacity is somewhat uncertain at the present time. They are told that the possible capacity can be 40 000, 60 000, or 80 000 parcels per month, regardless of the size of the parcels. They have estimated the probabilities corresponding to the three capacity levels. Table 12.3 shows this information. What is the expected capacity level for the new sorting device? Regional Express is growing steadily, so such a computerized sorting device will be a necessity in the future. However, if Regional Express currently deals with an average of 50 000 parcels per month, should they seriously consider purchasing the device now or should they wait?

Table 12.3 Probability Distribution Function for Capacity Levels of the New Sorting Device

i	Capacity Level (Parcels/Month)	$p(x_i)$
1	40 000	0.3
2	60 000	0.6
3	80 000	0.1

If the discrete random variable X denotes the capacity of the device, then the expected capacity level $E(X)$ is

$$\begin{aligned} E(X) &= (40\,000)p(x_1) + (60\,000)p(x_2) + (80\,000)p(x_3) \\ &= (40\,000)(0.3) + (60\,000)(0.6) + (80\,000)(0.1) \\ &= 56\,000 \text{ parcels per month} \end{aligned}$$

The expected capacity level exceeds the average monthly demand of 50 000 parcels per month, so according to the expected value analysis alone, Regional Express should consider buying the sorting device now. However, by studying the probability distribution, we see that there is a 30% probability that the capacity level may fall below 50 000 parcels per month. Perhaps Regional Express should include this information in their decision making, and ask themselves whether a 30% chance of not meeting their demand is too risky or costly for them if they decide to purchase the sorting device.■

In summary of this section, engineers can be faced with a variety of uncertain events in project evaluation. When the outcomes of each event can be characterized by a probability distribution, this greatly enhances the analyst's ability to develop a deeper understanding of the risks associated with various decisions. This section provided an introduction to random variables and probability distributions as a starting point for an analysis of project risk.

12.4 Structuring Decisions with Decision Trees

Many different types of uncertainties exist in decision making. When an economic analysis becomes complex due to these uncertainties, formal analysis methods can help in several ways. First, formal methods can help by providing a means of decomposing a problem and structuring it clearly. Second, formal methods can help by suggesting a variety of decision criteria to help with the process of selecting a preferred course of action. This section provides an introduction to decision trees, which are a graphical means of structuring a decision-making situation where the uncertainties can be characterized by probability distributions. Other means of structuring decisions such as influence diagrams are available, but are beyond the scope of our coverage.

Decision trees help decompose and structure problems characterized by a sequence of one or more decisions and event outcomes. For example, a judgment about the chance of a thunderstorm tomorrow will affect your decision to plan for a picnic tomorrow afternoon. Similarly, the success or failure of a new product may largely depend on future demand for the product. As another example, a decision on the replacement interval for an asset relies on an assessment of its economic life, which can be highly uncertain if the equipment employs an emerging technology.

When a decision is influenced by outcomes of one or more random events, the decision maker must anticipate what those outcomes might be as part of the process of analysis. This section presents a useful tool for structuring such problems, called a decision tree. It is particularly suited to decisions and events that have a natural sequence in time or space.

A **decision tree** is a graphical representation of the logical structure of a decision problem in terms of the sequence of decisions to be made and outcomes of chance events. It provides a mechanism to decompose a large and complex problem into a sequence of small and essential components. In this way, a decision tree clarifies the options a decision maker has and provides a framework with which to deal with the risk involved.

Example 12.4 introduces the overall approach to constructing a decision tree. A detailed explanation of the components and structure of the decision tree is included in the example.

Example 12.4

Edwin Electronics (EE) has a factory for assembling TVs in Midland, Ontario. One of the key components is the TV screen. EE does not currently produce TV screens onsite; they are outsourced to a supplier in Barrie. Recently, EE's industrial engineering team asked if they should continue outsourcing the TV screens or produce them in-house. They realized that it was important to consider the uncertainty in demand for the company's TVs. If the future demand is low, outsourcing seems to be the reasonable option in order to save production costs. On the other hand, if the demand is high, then it may be worthwhile to produce the screens onsite, thus getting economies of scale. EE's engineers analyzed the effect of the demand uncertainty in their decision making. They represented their decision problem in a graphical manner with a decision tree. Figure 12.4 represents EE's decision tree.

There are four main components in a decision tree: decision nodes, chance nodes, branches, and leaves. A decision node represents a decision to be made by the decision maker and is denoted by a square in the tree diagram. In Figure 12.4, the single square node represents the decision to produce or outsource TV screens (Node 1). A chance node represents an event whose outcome is uncertain, but which has to be considered during decision making. The outcome of a chance node is a discrete random variable, as

Figure 12.4 Decision Tree for Edwin Electronics

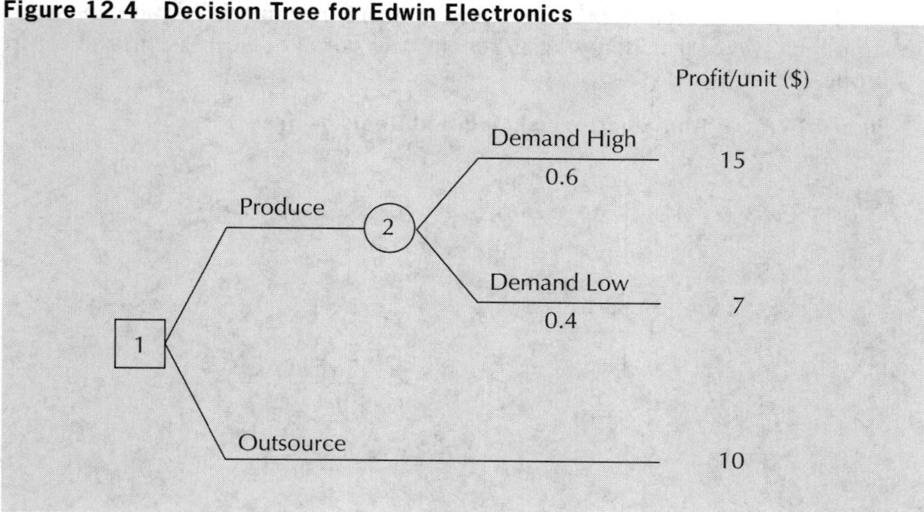

it has a number of distinct outcomes and each outcome has an associated probability. The circle in the diagram denotes a chance node. The chance node in Figure 12.4 represents the uncertain demand for TV screens (Node 2). The branches of a tree are the lines connecting nodes depicting the sequence of possible decisions and chance events. Finally, the leaves indicate the values, or payoffs, associated with each branch of the decision tree.

A decision tree grows from left to right and usually begins with a decision node. The leftmost decision node represents an immediate decision faced by the decision maker. The branches extending from a decision node represent the decision options available for the decision maker at that node, whereas the branches extending from a chance node represent the possible outcomes of the chance event. Each branch extending from a chance node has an associated probability. In Edwin Electronics' case, the two decision options, to produce or outsource, are represented by the two branches extending from the decision node. The two branches from the chance node indicate that the future demand may be high or low. It is important in decision making that all branches out of a node, whether a decision node or chance node, constitute a set of mutually exclusive and collectively exhaustive consequences. In other words, when a decision is made, exactly one option is taken, or when uncertainty is resolved, exactly one outcome occurs as a result.

Whenever a chance node follows a decision node, as in Figure 12.4, it implies that the decision maker must anticipate the outcome of future uncertain events in decision making. On the other hand, when a decision node follows a chance node, it implies that a decision must be made assuming that a particular outcome of a chance event has occurred. Finally, the rightmost branches lead to the leaves of the decision tree, indicating all possible outcomes of the overall decision situation represented by the tree. Each leaf has an associated valuation, referred to as a "payoff"; quite typically, the payoff is a monetary value. Edwin Electronics uses profit per TV unit as their performance measure.■

The decision tree for Edwin Electronics from Figure 12.4 can be modified to show more complex decision situations.

Example 12.5

The EE engineering team from Example 12.4 has realized that the cost per TV screen may vary in the future, especially since EE is subject to purchasing conditions set by the Barrie supplier. How does this affect their decision tree?

Figure 12.5 includes the additional uncertainty in the TV screen cost charged by the supplier. The cost may increase, remain the same, or decrease in the future, as shown at Node 3.■

Figure 12.5 Edwin Electronics' Modified Decision Tree 1

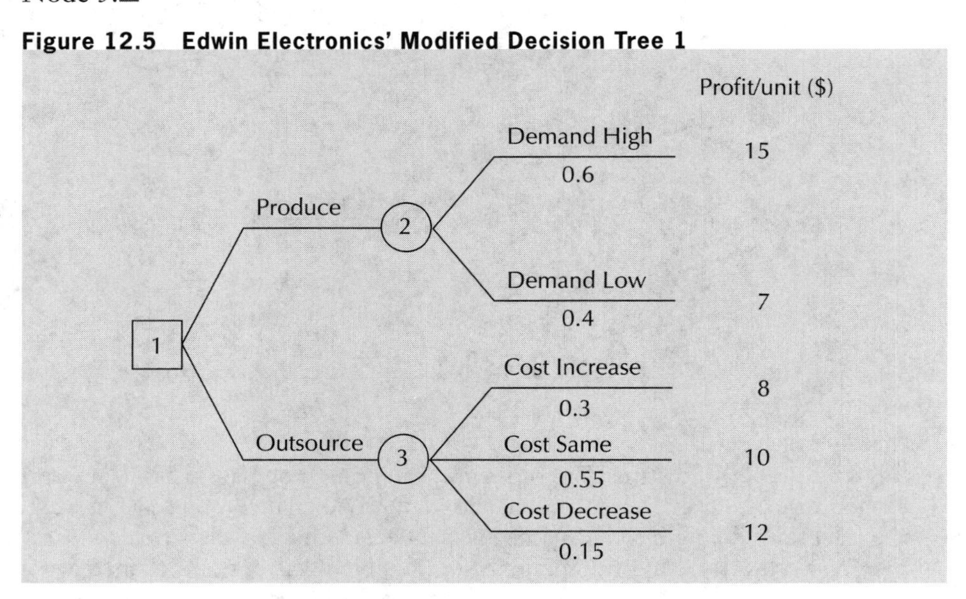

Example 12.6

The EE engineering team then considered increasing in-house production of the TV screens if the demand is high. This raises an additional uncertainty about the ability of the market to absorb the increased production. How does this change their decision tree?

Figure 12.6 Edwin Electronics' Modified Decision Tree 2

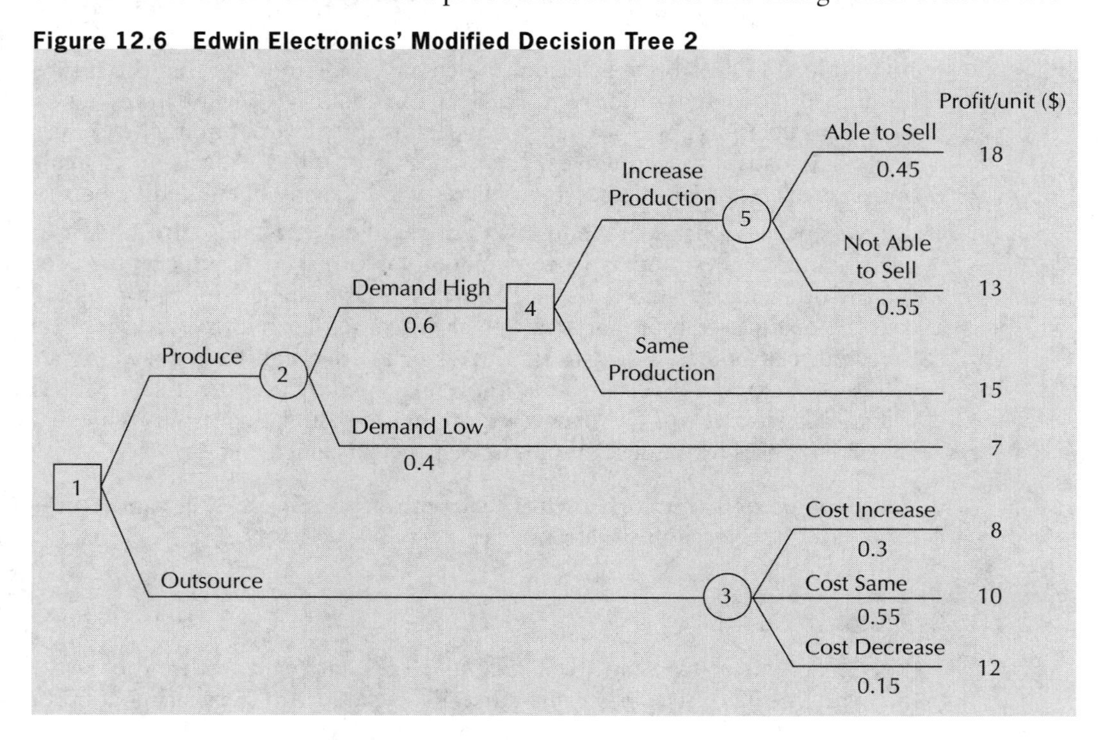

Figure 12.6 further modifies Figure 12.5 to include a new decision component and a chance event (see Nodes 4 and 5). The decision component indicates the choice of whether to increase production. If production is increased, a new chance node captures the uncertainty about their ability to sell the excess production.■

In summary, a decision tree is a graphical representation of the logical structure of a decision problem, showing the sequence of decisions and outcomes of chance events. It provides decision makers with a mechanism to structure and communicate a decision-making situation when risk is present. Decision tree analysis provides a framework with which a decision maker can deal with the uncertainty involved in a decision.

12.5 Decision Criteria

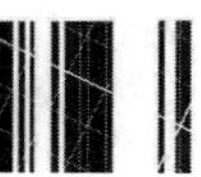

Once a complete decision tree is structured, an analyst is in a better position to select a preferred alternative from a set of possible choices influenced by uncertain events. But how do we understand the risks associated with these uncertain events? And even if we can quantify the risks, how are choices then made? This section deals with several commonly used decision criteria for situations that involve uncertainty. Each has strengths and weaknesses that will be pointed out through a series of examples.

12.5.1 Expected Value

One criterion for selecting among alternatives when there are risky outcomes is to pick the alternative that has the highest expected value, EV, as previously presented in Examples 12.3. If the units of value associated with the rightmost branches of a decision tree are measured in dollars, then this criterion may be referred to as *expected monetary value*, or EMV.

When a decision problem has been structured with a decision tree, finding the expected value of a particular decision is obtained by a procedure referred to as "rolling back" the decision tree. The procedure is as follows:

1. *Structure the problem:* Develop a decision tree representing the decision situation in question.
2. *Rollback:* Execute the **rollback procedure** (also known as **backward induction**) on the decision tree from right to left as follows:

   (a) At each chance node, compute the expected value of the possible outcomes. The resulting expected value becomes the value associated with the chance node and the branch on the left of that node (if there is one).

   (b) At each decision node, select the option with the best expected value (best may be highest value or lowest cost depending on the context). The best expected value becomes the value associated with the decision node and the branch on the left of that node (if there is one). For the option(s) not selected at this time, indicate their termination by a double-slash (//) on the corresponding branch.

   (c) Continue rolling back until the leftmost node is reached.

3. *Conclusion:* The expected value associated with the final node is the expected value of the overall decision. Tracing forward (left to right), the non-terminated decision options indicate the set of recommended decisions at each subsequent node.

Example 12.7

Carry out a decision tree analysis on Edwin Electronics' modified tree in Figure 12.5 using the expected value criterion.

Since the decision tree is already provided, step 1 is complete. The rollback procedure described in step 2 has two phases in this case. First, the tree is rolled back to each of the chance nodes as in step 2(a) (phase 1). The expected values at Nodes 2 and 3 are computed as follows:

$$EV(2) = 0.6(\$15) + 0.4(\$7) = \$11.80$$

$$EV(3) = 0.3(\$8) + 0.55(\$10) + 0.15(\$12) = \$9.70$$

Figure 12.7 shows the rollback so far.

Figure 12.7 Phase 1 of Step 2: Rolling Back to the Chance Nodes

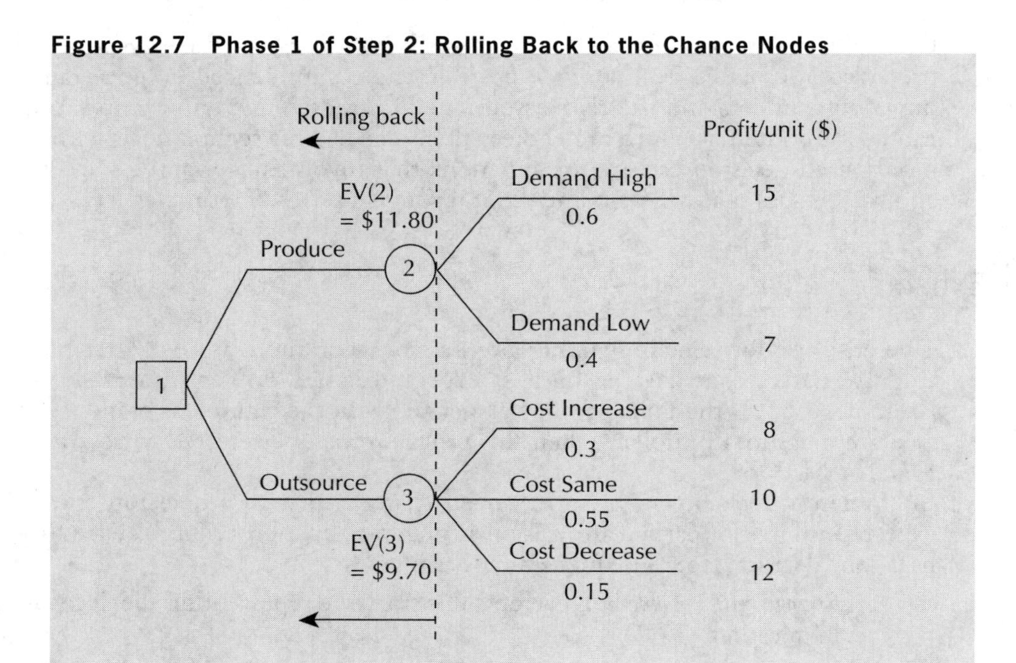

Next, the tree is further rolled back to the decision node as in step 2(b) (phase 2). The expected value at Node 1 is then EV(1) = $11.80, which is equal to EV(2) since EV(2) is higher than EV(3). Figure 12.8 shows this result. As for step 3, the following conclusion is made: the expected value of the overall decision is $11.80 per unit and the recommended decision is to produce TV screens in-house.■

Example 12.8

Perform decision tree analysis on Edwin Electronics' second modified tree in Figure 12.6.

The result of this analysis is shown in Figure 12.9. The overall expected profit for this tree is $11.95 per unit. The recommended decision is to produce TV screens in-house, and if the demand is high, the production level should be increased.■

Figure 12.8 Phase 2 of Step 2: Rolling Back to the Decision Nodes

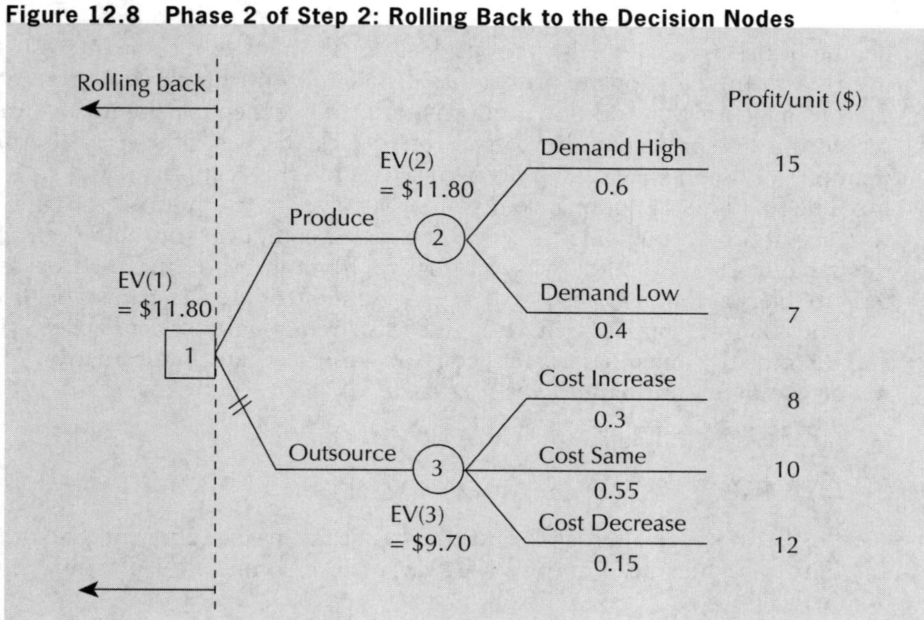
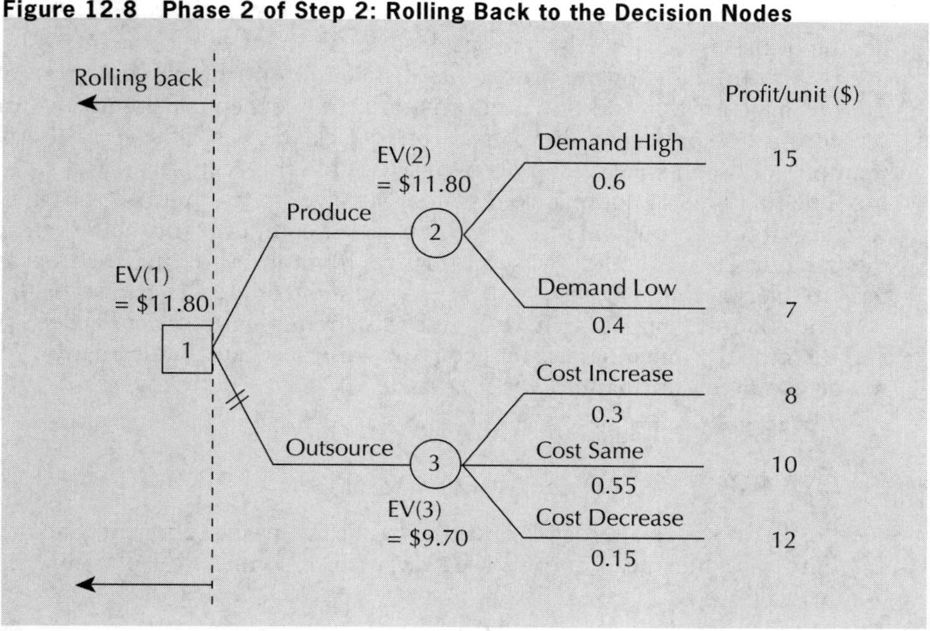

Figure 12.9 Completed Analysis for EE's Modified Decision Tree 2

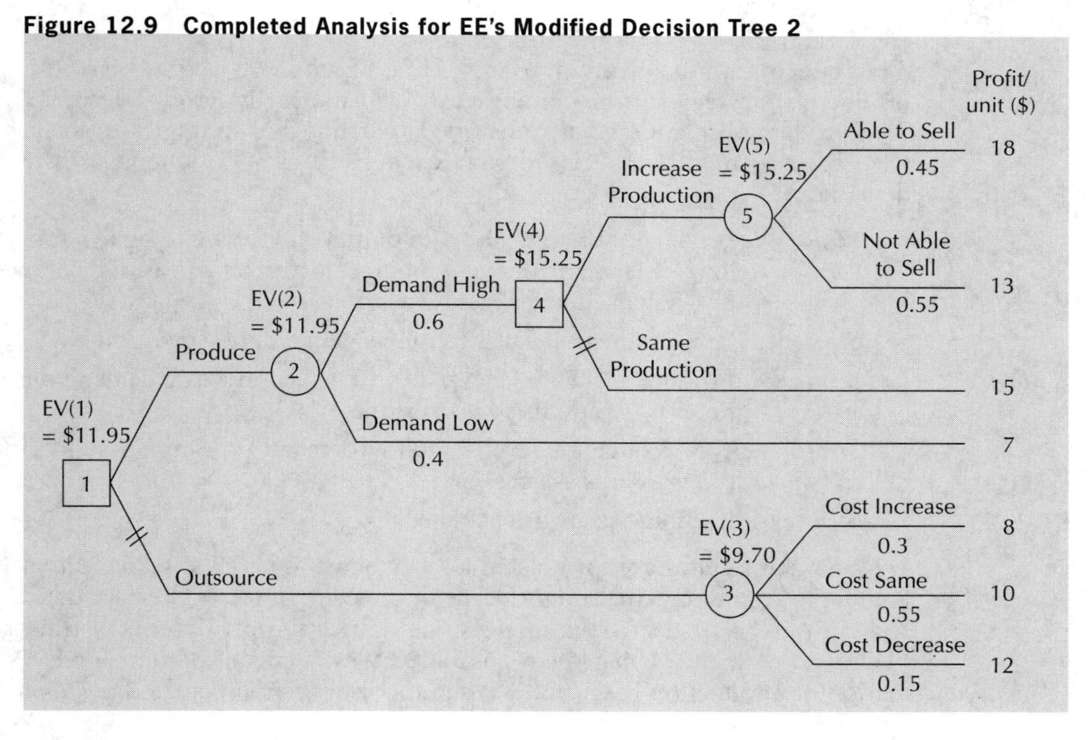

12.5.2 Dominance

Comparing several risky alternatives on the basis of the expected value criterion is straightforward, because it is easily computed and intuitive. However, expected value is only a summary measure and it does not consider the dispersion of the outcomes associated with a decision. Two decision alternatives with the same expected value may have very different ranges of possible outcome values. A decision maker sensitive to the degree of risk associated with various decision alternatives may be interested in quantifying risk on the basis of the probability distribution of the outcomes rather than simply the mean. With this additional information, the concept of dominance can be used to screen out less preferred alternatives, or to pick the best of several alternatives. Three examples of dominance reasoning are discussed in this section.

The first type of dominance reasoning involves measuring risk through the mean and variance of project outcomes. If a decision maker computes both the mean and variance of the outcomes for several decision alternatives, both measures may be used as decision criteria to select among projects, or to screen out some of the alternatives being considered.

For example, suppose that an engineer is attempting to select between two projects X and Y where the outcomes are monetary (i.e., more is better). Alternative X is said to have **mean-variance dominance** over alternative Y

(a) If $\text{EV}(X) \geq \text{EV}(Y)$ and $\text{Var}(X) < \text{Var}(Y)$, or

(b) If $\text{EV}(X) > \text{EV}(Y)$ and $\text{Var}(X) \leq \text{Var}(Y)$

Furthermore, an alternative is said to be mean-variance efficient if no other alternative has both a higher mean and a lower variance. Example 12.9 illustrates mean-variance dominance.

Example 12.9

The Tireco Tire Company produces a line of automobile tires. Tireco is planning its production strategy for the coming year. Demand for its products varies over the year. The production planners have identified four possible strategies to meet demand, but the expected profits associated with each strategy will depend on the outcome of demand:

Strategy 1: Produce at the same level of output all year, building inventory in times of low demand and drawing on inventory during peak demand periods. Do not vary production staff levels.

Strategy 2: Vary production levels to follow demand. Use overtime as necessary.

Strategy 3: Produce at the same level all year, subcontracting demand during peak times to avoid excessive overtime.

Strategy 4: Vary production levels to follow demand by using a second shift of regular time workers as necessary.

Strategy 5: Combine strategies 3 and 4.

Marketing staff have formulated a set of demand patterns taking into account possible market conditions that might occur in the coming year. They are not sure what pattern might occur, but have made some subjective probability estimates as to the likelihood of each. Management combined these demand pattern forecasts with information on production costs, and has summarized their anticipated profits for each strategy-demand pattern as shown in Table 12.4.

The marketing department is not sure what demand pattern might occur, but they have estimated subjectively that the probability that demand takes on pattern 1 is 20%; pattern 2, 50%; and pattern 3, 30%. From these estimates, the mean and the variance associated with each strategy were calculated and are also shown in Table 12.4.

It is natural that Tireco would want to maximize its expected profits. However, they might also like to minimize the variance of the strategy chosen at the same time so that the strategy chosen does not have a large degree of risk associated with possible outcome values. In view of these two objectives, several strategies can be removed from the above list with dominance reasoning:

1. Strategies 2 and 3 have the same mean profit, but strategy 2 has a higher variance. Strategy 2 is mean-variance dominated by strategy 3, and thus can be removed from consideration.
2. Strategy 5 has a lower mean and a higher variance when compared to strategy 1. Strategy 5 can be removed from consideration because it is mean-variance dominated by strategy 1.
3. Strategy 4 has a lower mean and higher variance when compared to strategy 1. Strategy 4 is thus mean-variance dominated by strategy 1.

Table 12.4 Tireco Tire Company Forecast Profit Contributions ($000s)

Production Strategy		**Demand Pattern**			
	1		2		3
Mean Variance					
1	420	310	600	401	12 169
2	280	340	630	380	16300
3	500	290	425	380	8775
4	600	275	390	395.5	19 812.25

Of the five strategies, only strategies 1 and 3 remain. While strategy 1 has the highest mean, it also has the highest variance, so a choice between the two cannot be made on the basis of the mean and variance alone. This set of two alternatives is thus the efficient set, because it can be reduced no further by mean-variance dominance reasoning. A choice between the two will require management to assess its willingness to trade off mean profits with the variability in profits.■

Mean-variance analysis is useful when the mean and variance capture well the distribution of possible outcomes for a decision. It is commonly used for screening investment opportunities. If more information about the distribution of outcomes is needed by a decision maker, it may be preferable to compare alternatives using dominance concepts that take into account the full distribution of outcomes.

The second type of dominance reasoning commonly used to choose between risky decision alternatives is to compare their full sets of outcomes directly. When outcome dominance exists, one or more of the alternatives can be removed from consideration. **Outcome dominance** of alternative X over alternative Y can occur in one of two ways. The first is when the worst outcome for alternative X is at least as good as the best outcome for alternative Y. Consider a slight variation of Example 12.5, in which Edwin Electronics must decide between in-house production and outsourcing. Suppose that the data for the problem is the same as earlier presented, but that the profit per unit if they produce in-house and demand is low is $13 rather than $7 (refer to Figure 12.5). With this small change, inspection of the range of possible outcomes now shows that the profit per unit for the "produce" decision is better than for the "outsource" decision, for all possible outcomes. The "produce" decision is said to dominate the "outsource" decision due to outcome dominance.

The second way in which outcome dominance can occur is when one alternative is at least as preferred to another for each outcome, and is better for at least one outcome. In Example 12.9, strategy 1 has outcome dominance over strategy 5 because it has a higher profit contribution for each demand pattern.

Outcome dominance can be useful in screening out alternatives that are clearly worse than others among a set of choices. Though outcome dominance is straightforward to apply, it may not screen out or remove many alternatives.

The third common type of dominance reasoning that can be used to screen or order risky decision alternatives is stochastic dominance, as illustrated in Example 12.10.

Example 12.10

Suppose in Example 12.5 that the profit per unit for EE when the decision to produce is taken and demand is low is $10. Figure 12.10 shows the probability distribution functions, also referred to as **risk profiles**, of the outcomes for the two decision alternatives that EE is considering.

A look at the cumulative distribution functions, also known as **cumulative risk profiles**, for the two alternatives provides further insights. This is seen in Figure 12.11.

The cumulative risk profile for the "outsource" decision either overlaps with or lies to the left of and above the cumulative risk profile of the "produce" decision. This means that for all outcomes, the probability that the "outsource" decision gives a lower profit per unit is equal to or greater than the corresponding probability for the "produce" decision. The produce strategy is said to dominate the outsource strategy according to (first-order) stochastic dominance. A more precise definition is as follows: If two decision alternatives *a* and *b* have outcome cumulative distribution functions $F(x)$ and $G(x)$, respectively, then alternative *a* is said to have first order **stochastic dominance** over alternative *b* if $F(x) \geqslant G(x)$ for all x. In other words, alternaive *a* is more likely to give a higher (better) outcome than alternative *b* for all possible outcomes.

Figure 12.10 Illustration of Risk Profiles for Example 12.10

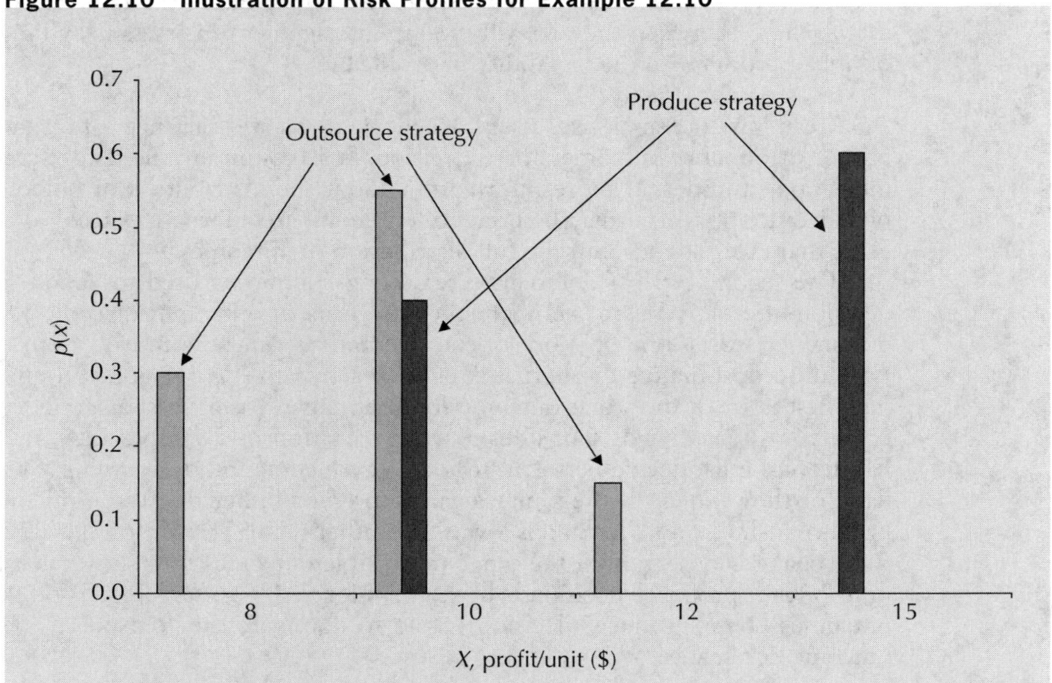

Figure 12.11 Illustration of First-Order Stochastic Dominance for Example 12.10

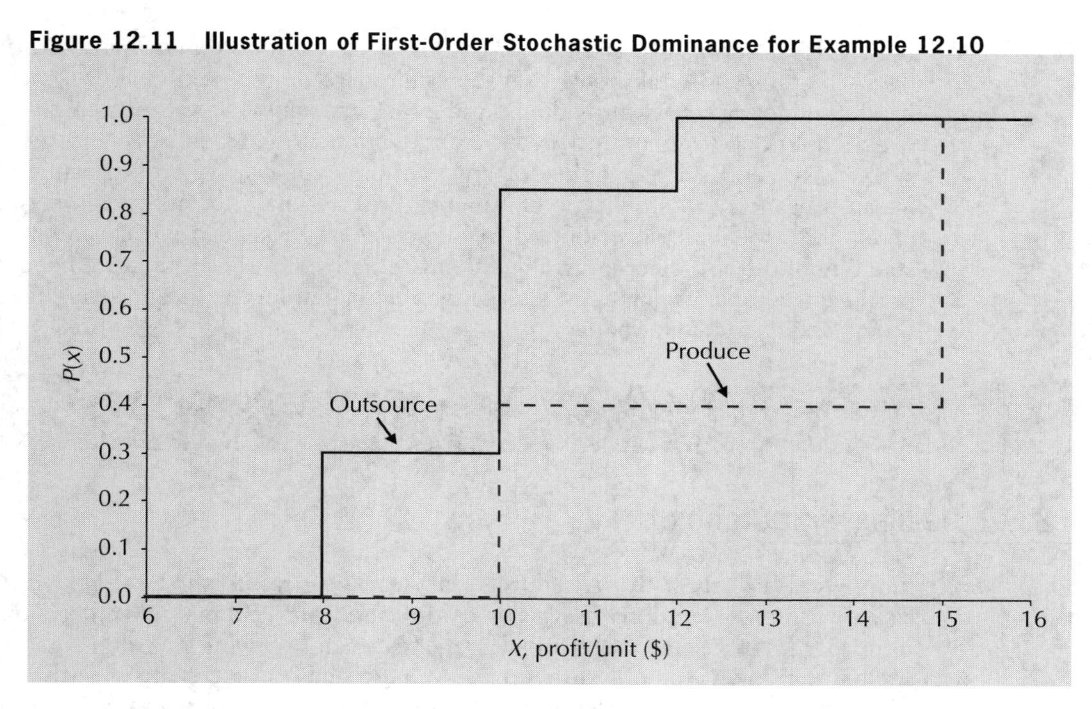

While first-order stochastic dominance and outcome dominance can be used to screen alternatives, it is often the case that they are not able to provide a definitive "best" alternative. For example, the cumulative risk profiles for the original EE problem from Example 12.6 show that the cumulative risk profiles intersect (see Figure 12.12). The produce strategy is preferred using the EV criterion, but no definitive preference can be stated using ideas of deterministic or stochastic dominance. Despite this limitation with the use of cumulative risk profiles for ordering alternatives, they can be very useful in making statements such as "Alternative A is more likely to produce a profit in excess of $1 000 000 than alternative B" or "Project C is more likely to suffer a loss than project D."■

Figure 12.12 Cumulative Risk Profiles for the Decision Alternatives in Example12.6

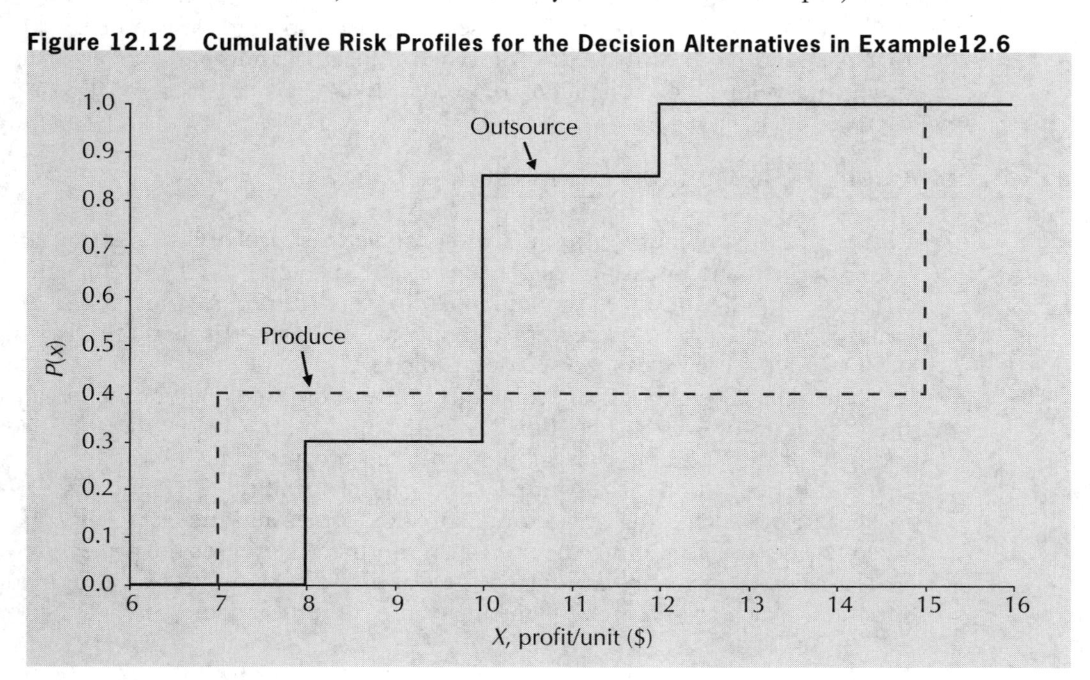

In this section, several decision criteria were introduced and illustrated. Each has its own strengths and weaknesses, and the criteria are best viewed as approaches to decision making under risk, each providing a different perspective. The expected value criterion is widely used due to its intuitive appeal and computational simplicity. It is a powerful tool and a very good starting point for any economic analysis where uncertainty is present. Mean-Variance and stochastic dominance criteria take into account more information about the probability distribution of outcomes and thus can enrich an analyst's understanding of the risk inherent in a decision alternative. However, they are not always able to produce a complete ranking of alternatives. They tend to be more useful in screening out dominated decision alternatives.

12.6 Monte Carlo Simulation

12.6.1 Dealing with Complexity

This section deals with procedures that have been found to be effective in analyzing risk in complex decisions. Large or complicated projects may have decision trees with hundreds or thousands of terminal branches, each involving multiple sources of risk. In such situations, it may be difficult to evaluate decision alternatives with decision trees alone. For instance, the present worth computation for a project may require a number of inputs, such as the initial cost, a series of revenues and savings, operating and maintenance costs, and salvage value. When one or more of these inputs are random variables, the present worth inherits this randomness and may be very difficult to characterize.

When dealing with large and complex decision trees, a method called **Monte Carlo simulation** can be very useful. The basic idea behind the approach is to evaluate alternative decision strategies by randomly sampling branches of the decision tree, and to thereby assemble probability distributions (risk profiles) for relevant performance measures. As with any sampling procedure, the results will not be perfect, but the accuracy of the results will increase as the sample size increases. The main strength of Monte Carlo simulation is that it allows us to analyze the combined impact of multiple sources of uncertainty in order to develop an overall picture of overall risk.

12.6.2 Probability Distribution Estimation

Monte Carlo simulation attempts to construct the probability distribution of an outcome performance measure of a project (e.g., present worth) by repeatedly sampling from the input random variable probability distributions. This is a useful technique when it would otherwise be very difficult to obtain the probability distribution of the performance measure of interest for a project.

Since one basic notion of probability comes from the long-run frequency of an event, the sample frequency distribution generated by Monte Carlo simulation can be a good estimate of the probability distribution of the outcome. This is true, of course, provided that a sufficient sample size is used. Once the probability distribution is estimated, summary statistics such as the range of possible outcomes and the expected value can be analyzed to provide insight into the possible performance level for the project.

In Monte Carlo simulation, the probability distributions of the individual random variables, which are the input elements of the overall outcome, are assumed to be known in advance. By randomly sampling values for the random variables from the specified probability distributions, Monte Carlo simulation produces a sample of the overall performance measure for the project. This process can be seen as imitating the randomness in the performance measure as a result of the randomness of the project inputs. Incidentally, the name "Monte Carlo simulation" refers to the casino games at Monte Carlo, which symbolize the random behaviour. The use and practicality of Monte Carlo simulation have greatly increased with the widespread availability of application software such as spreadsheets (e.g., Excel and Lotus), spreadsheet add-ons including @Risk and Crystal Ball, and special-purpose simulation languages.

12.6.3 The Monte Carlo Simulation Approach

The following five steps are taken in developing a Monte Carlo simulation model:

1. *Analytical model:* Identify all input random variables that affect the outcome performance measure of the project in question. Develop the equation(s) for computing the outcome (denoted by Y) from a particular realization of the input random variables. This is sometimes referred to as "the deterministic version" of the model.
2. *Probability distributions:* Establish an appropriate probability distribution for each input random variable.
3. *Random sampling:* Sample a value for each input random variable from its associated probability distribution. The following is a random sampling procedure for discrete random variables:
   (a) For each discrete random variable, create a table similar to Table 12.5 containing the possible outcomes, their associated probabilities, and the corresponding random-number assignment ranges.
   (b) For each input random variable, generate a random number Z (see Close-Up 12.2). Find the range to which Z belongs from Table 12.6 and assign the appropriate outcome.
   (c) Substitute the sample values of the random variables into the expression for the outcome measure, Y, and compute the value of Y. This forms one sample point in the procedure. Table 12.6 can be the basis for a spreadsheet application of random sampling.

Table 12.5 Random Number Assignment Ranges

i	Outcome x_i	Probability $p(x_i)$	Random Number (Z) Assignment Range
1	x_1	$p(x_1)$	$0 \le Z \le p(x_1)$
2	x_2	$p(x_2)$	$p(x_1) \le Z < p(x_1) + p(x_2)$
⋮	⋮	⋮	⋮
$m - 1$	x_{m-1}	$p(x_{m-1})$	$p(x_1) + \ldots + p(x_{m-2}) \le Z < p(x_1) + \ldots + p(x_{m-1})$
m	x_m	$p(x_m)$	$p(x_1) + \ldots + p(x_{m-1}) \le Z < 1$

Table 12.6 Random Sampling Table

Step 3(b) (do for all random variables)		Step 3(c)
Random Number	Value of X	Value of Y
Generate Z	Assign value to X (one of x_i's) using Table 12.5 from step 3(a)	Compute Y using the analytical model from step 1

4. *Repeat sampling:* Continue sampling until a sufficient sample size is obtained for the value of Y.
5. *Summary:* Summarize the frequency distribution of the sample outcomes using a histogram. Summary statistics, like the range of possible outcomes and expected value, can also be calculated from the sample outcomes.

NET VALUE 12.1

Monte Carlo Simulation Analysis Software

Monte Carlo simulation analysis often involves a large amount of data and requires a sophisticated, multi-step analytical process with solution presentation aids such as graphics and charts. There are a number of commercial products available for Monte Carlo simulation, and the Internet is a good place to find out more about these products and even download a trial version of the software.

Two examples of commercially available Monte Carlo simulation software packages are Crystal Ball (www.decisioneering.com/crystal_ball) and @Risk (www.palisade.com). They are both designed as add-ins to spreadsheet software in order to ensure ease of use and compatibility with packages most engineers use routinely. They contain features designed to provide speedy and versatile calculation power to help answer questions such as "What is the probability that a project will be profitable?" or "How sensitive is our decision to interest rates or inflation?" There are other features to facilitate the selection of model input variables (such as a library of useful probability distributions) and to assist with the preparation and presentation of simulation results.

One point remains to be clarified. In step 4 of Monte Carlo simulation, random sampling continues until a sufficient sample size is obtained. What is sufficient? The law of large numbers tells us that as the number of samples approaches infinity, the frequencies will converge to the true underlying probabilities. As a practical guideline, one should aim for a sample size of at least 100 in order to obtain reasonable results. One way to ensure the validity of the end result is to monitor the expected value of the frequency distribution and continue sampling until some stability appears in it. More rigorous guidelines exist for selecting the appropriate number of samples (e.g., confidence interval methods) but these are beyond the scope of this text.

CLOSE-UP 12.2 Generating Random Numbers

In order to generate random numbers, a source of independent and uniformly distributed random numbers between 0 and 1 is required. These uniformly distributed random numbers can then be converted to random numbers from any probability distribution via a variety of methods, one of which is a table look-up as illustrated in Table 12.6. Today, random-number generation is easily achieved by using calculator functions or application software such as Excel or Lotus, which have built-in random-number generators for a limited number of probability distributions. Specialized commercial software such as @Risk or Crystal Ball provide a wider range of built-in random-number generators.

Example 12.11

Pharma-Excel, a pharmaceutical company based in Halifax, is considering the worth of a R&D project that involves improvement of vitamin pills. Since this is a new research domain for Pharma, they are not certain about the related costs and benefits. As a part of the initial feasibility study, Pharma-Excel estimated probability distributions for the first cost and annual revenue as seen in Table 12.7, which are assumed to be independent quantities. Simulate the present worth of this project on the basis of a 10-year study period using the Monte Carlo method. Does the project seem viable? Pharma-Excel's MARR is 15% for this type of project.

As the first step, the analytical expression for the present worth of the project is developed. We use the following expression for computing the present worth of the project:

$$\begin{aligned} \text{PW} &= -(\text{First Cost}) + (\text{Annual Revenue})(P/A, 15\%, 10) \\ &= -(\text{First Cost}) + (\text{Annual Revenue})(5.0188) \end{aligned}$$

Table 12.7 First Cost and Annual Revenue for Pharma-Excel's Research Project

First Cost	Probability	Annual Revenue	Probability
$1 000 000	0.2	$ 100 000	0.125
1 250 000	0.2	350 000	0.125
1 500 000	0.2	600 000	0.125
1 750 000	0.2	850 000	0.125
2 000 000	0.2	1 100 000	0.125
		1 350 000	0.125
		1 600 000	0.125
		1 850 000	0.125

The probability distributions for the first cost and annual revenue are provided in Table 12.7 (step 2). On the basis of these distributions, random numbers are assigned to particular intervals on the 0 to 1 probability scale (step 3(a)). Tables 12.8 and 12.9 summarize the random-number assignment for each random variable.

Table 12.8 Random Number Assignment Ranges for the First Cost

First Cost	Probability	Random Number (Z_1) Assignment Range
$1 000 000	0.2	$0 \leq Z_1 < 0.2$
1 250 000	0.2	$0.2 \leq Z_1 < 0.4$
1 500 000	0.2	$0.4 \leq Z_1 < 0.6$
1 750 000	0.2	$0.6 \leq Z_1 < 0.8$
2 000 000	0.2	$0.8 \leq Z_1 < 1$

Table 12.9 Random Number Assignment Ranges for the Annual Revenue

Annual Revenue	Probability	Random Number (Z_2) Assignment Range
$100 000	0.125	$0 \leq Z_2 < 0.125$
350 000	0.125	$0.125 \leq Z_2 < 0.25$
600 000	0.125	$0.25 \leq Z_2 < 0.375$
850 000	0.125	$0.375 \leq Z_2 < 0.5$
1 100 000	0.125	$0.5 \leq Z_2 < 0.625$
1 350 000	0.125	$0.625 \leq Z_2 < 0.75$
1 600 000	0.125	$0.75 \leq Z_2 < 0.875$
1 850 000	0.125	$0.875 \leq Z_2 < 1$

Following steps 3(b) and 3(c), the simulation results are obtained for a sample size of 200. Table 12.10 presents partial results. The frequency distribution shown in Figure 12.13 is also generated on the basis of the simulation results. Note: Each bar in the histogram includes all values up to that value that were not included in the previous bar.

Table 12.10 Partial Results for the Monte Carlo Simulation

Sample Number	Random Number (Z_1)	First Cost	Random Number (Z_2)	Annual Revenue	Present Worth
1	0.076162	$1 000 000	0.605155	$1 100 000	$4 520 680
2	0.728782	1 750 000	0.293282	600 000	1 261 280
3	0.29656	1 250 000	0.747692	1 350 000	5 525 380
4	0.940748	2 000 000	0.327516	600 000	1 011 280
5	0.384964	1 250 000	0.788017	1 600 000	6 780 080
⋮	⋮	⋮	⋮	⋮	⋮

From the simulation results, the average present worth of the research project over the 10-year study period is \$3 152 437 with a maximum value of \$8 284 780 and a minimum value of –\$1 498 120. The project exhibited a negative present worth roughly 18% of the time. From these figures, the project seems to be viable, because it has a good chance of having a positive present worth that might be as high as \$8 million.■

Figure 12.13 Frequency Distribution for Pharma-Excel's Research Project

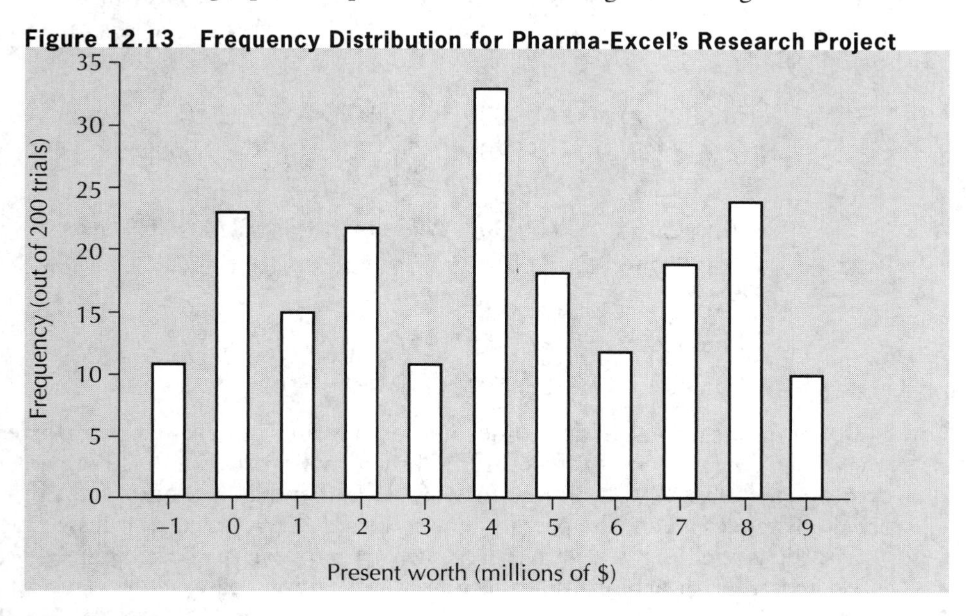

Example 12.12

While considering the cogeneration plant project outlined in Example 11.1, Cogenesis Corporation wishes to determine the probability distribution of the project's present worth to better assess the probability of a negative present worth. The three random variables they wish to investigate are the initial investment, the savings in electricity costs, and the extra wood costs. Previously, management determined that the range of possible values for the initial costs was between \$2 800 000 and \$3 300 000, the range for savings in electricity costs was \$920 000 to \$1 080 000 per year, and the additional wood costs were \$350 000 to \$400 000 per year. To the best of their knowledge, management thinks that any outcome between the lower and upper bounds for the first cost, electricity savings, and additional wood costs is equally likely. Table 12.11 shows the dis-

Table 12.11 Discrete Probability Distributions for Cogenesis' Plant Project

Initial Cost	Probability	Electricity Savings	Probability	Additional Wood Costs	Probability
\$2 800 000	1/6	\$ 920 000	0.2	\$350 000	1/6
2 900 000	1/6	960 000	0.2	360 000	1/6
3 000 000	1/6	1 000 000	0.2	370 000	1/6
3 100 000	1/6	1 040 000	0.2	380 000	1/6
3 200 000	1/6	1 080 000	0.2	390 000	1/6
3 300 000	1/6			400 000	1/6

crete approximation of the probability distribution functions created for the three random variables.

By randomly sampling repeatedly from each of these distributions, we can construct a probability distribution of the present worth of the project. Assuming that all other parameters are fixed at their expected scenario values, the following expression is used for computing the present worth. A portion of the Monte Carlo simulation results is shown in Table 12.12.

$$\begin{aligned}\text{PW} &= -(\text{First Cost}) + (\$65\ 000 + \text{Wood Costs} - \text{Electricity Savings})\\ &\quad \times (P/A, 12\%, 20) - \$17\ 000(P/F, 12\%, 10) - \$35\ 000[(P/F, 12\%, 4)\\ &\quad + (P/F, 12\%, 8) + (P/F, 12\%, 12) + (P/F, 12\%, 16)]\\ &= -(\text{First Cost}) + (\$65\ 000 + \text{Wood Costs} - \text{Electricity Savings})\\ &\quad \times (7.4694) - \$17\ 000(0.32197) - \$35\ 000(0.63552 + 0.40388\\ &\quad + 0.25668 + 0.16312)\\ &= -(\text{First Cost}) + 7.4694(\$65\ 000 + \text{Wood Costs}\\ &\quad - \text{Electricity Savings}) - \$56\ 545.49\end{aligned}$$

By sampling a total of 200 times and computing a present worth for each sample, management arrived at the histogram shown in Figure 12.14. The expected present worth was \$1 007 816 with a minimum of \$42 032 and a maximum of \$2 110 606. There were no instances where the present worth turned out to be zero or less. The probability of this project yielding a negative present worth appears to be negligible—assuming, of course, that the probability distributions for the input parameters have been specified correctly!■

Table 12.12 Partial Results for the Cogenesis Monte Carlo Simulation

Initial Cost	Electricity Savings	Additional Wood Costs	Present Worth
\$3 200 000	\$1 080 000	\$360 000	\$1 635 912
2 800 000	920 000	350 000	915 502
3 000 000	960 000	380 000	790 196
2 900 000	920 000	370 000	666 114
⋮	⋮	⋮	⋮

Figure 12.14 Frequency Distribution for Cogenesis' Plant Project

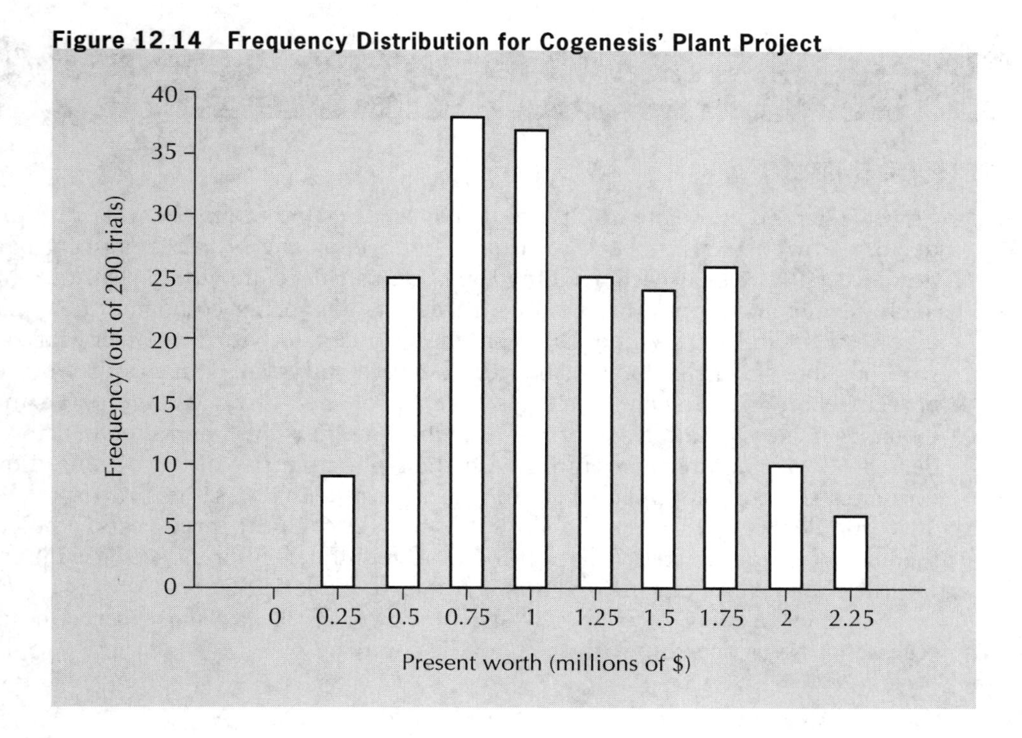

12.7 Application Issues

Decision trees are a powerful tool for structuring decisions when one or more sources of risk are present and there is a sequence of decisions to be made. When decision trees get very large, or when there are numerous sources of uncertainty, Monte Carlo simulation is useful for approximating the probability distribution function of performance measures associated with a given decision strategy. Both methods provide a way to structure decision making under uncertainty using probability theory. By explicitly modelling the components of the decision-making process, engineers and analysts can structure a decision clearly and better understand the implications of the risk in decision making.

The major drawback of probability-based methods is that the probability distributions of each of the random variables must be specified. This can be a real challenge when the probabilities are highly subjective or there is a lack of historical or experimental data upon which to base probability assessments. Therefore, we must be aware of this grey area when we interpret the output and try to put the results of the analysis in context.

Despite this drawback, as engineers we recognize that every decision has elements of uncertainty and risk. The tools discussed in this chapter provide a framework for analyzing risk and gaining insights that would otherwise be ignored. In conjunction with the sensitivity analysis methods covered in Chapter 11, probability theory is a powerful tool for assessing project viability and the implications of the risk associated with every engineering project.

REVIEW PROBLEMS

REVIEW PROBLEM 12.1

Power Tech is a company in Ottawa that specializes in building power-surge-protection devices. Power Tech has been focusing its efforts on the North American market until now. Recently, a deal with a Chinese manufacturing company has surfaced. If Power Tech decides to become partners with this manufacturing company, their market will expand to include Asia. They are, however, concerned with the uncertainty associated with possible change in North American demand and Asian demand. From studying the current economy, Power Tech feels that the chance of no change or an increase in demand in North America over the next three years is 60% and the chance of demand decrease is 40%. After discussions with their potential partners in China, Power Tech estimates that Asian demand may increase (or remain the same) with a probability of 30% and decrease with a probability of 70% over the next three years. They have estimated the revenue increase that can be expected under different scenarios if they establish the partnership; this information is shown in Table 12.13.

Conduct a decision tree analysis for Power Tech and make a recommendation regarding the partnership with the Chinese company.

Table 12.13 Expected Revenue Increase for Power Tech

	North American Demand	Asian Demand	Revenue Increase (millions of $)
Partnership with Chinese company	increase	increase	2
	increase	decrease	0.75
	decrease	increase	0.5
	decrease	decrease	−1
No partnership with Chinese company	increase	increase	0.75
	increase	decrease	0.5
	decrease	increase	0.1
	decrease	decrease	0.3

ANSWER

The result of the analysis is shown in Figure 12.15. The expected value calculations at each chance node are shown below.

The first phase of rollback (in millions of dollars):

$$EV(4) = 0.3(2) + 0.7(0.75) = 1.125$$
$$EV(5) = 0.3(0.5) + 0.7(-1) = -0.55$$
$$EV(6) = 0.3(0.75) + 0.7(0.5) = 0.575$$
$$EV(7) = 0.3(0.1) + 0.7(0.3) = 0.24$$

The second phase of rollback (in millions of dollars):

$$EV(2) = 0.6(1.125) + 0.4(-0.55) = 0.455$$
$$EV(3) = 0.6(0.575) + 0.4(0.24) = 0.441$$

Figure 12.15 Decision Tree Analysis for Power Tech (Review Problem 12.1)

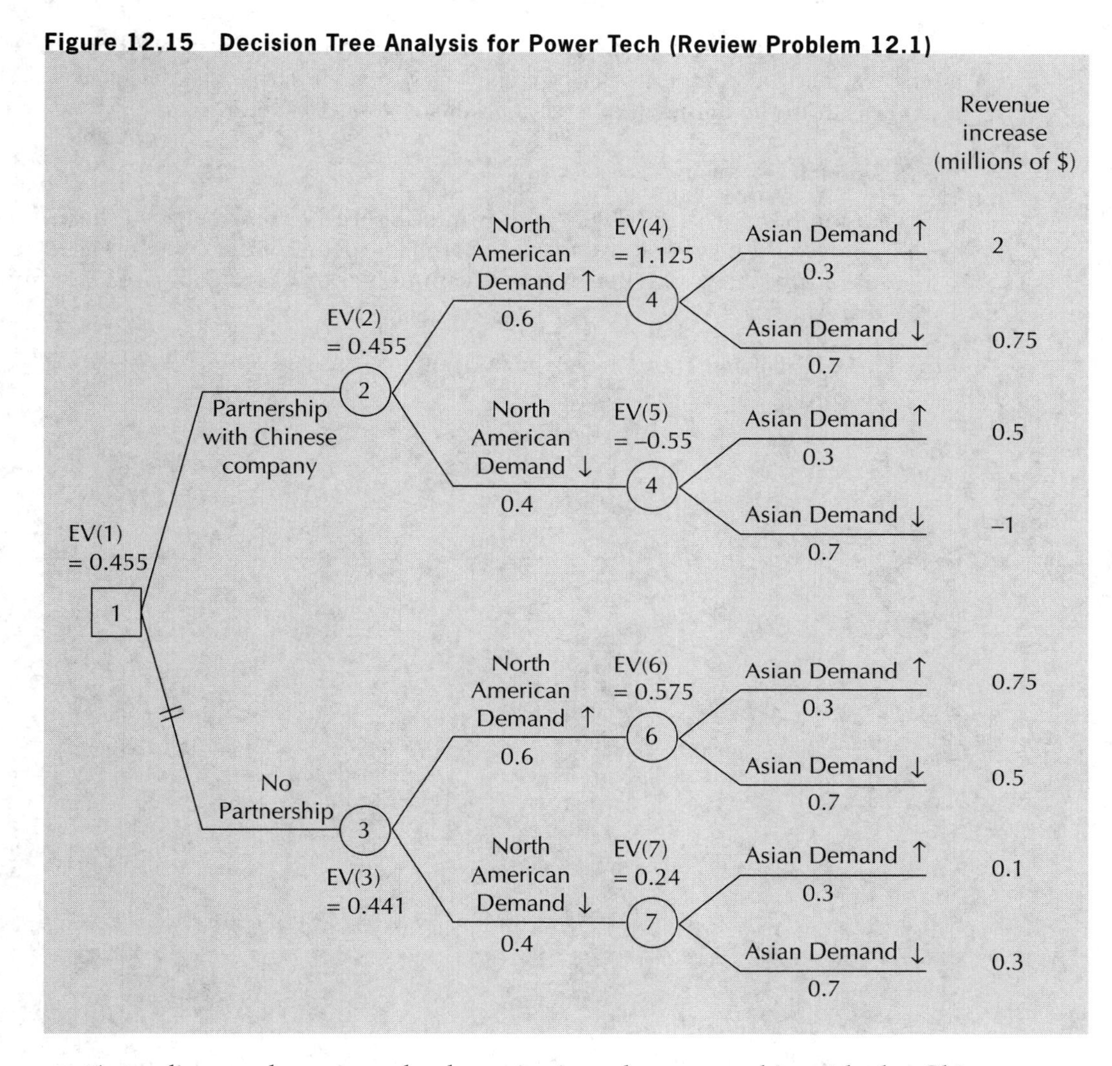

According to the expected value criterion, the partnership with the Chinese company is recommended, since the expected value for forming the partnership is higher than not forming it. However, Power Tech should also note that these expected values have only a marginal difference. It is perhaps wise to collect more information regarding other aspects of this proposed partnership.■

REVIEW PROBLEM 12.2

A telephone company in London, Ontario called LOTell thinks the introduction of an Internet service package for residential customers would give it a competitive advantage over its competitors. However, a survey of the potential growth of Internet home users would take at least three months. LOTell has two options at present: first, to introduce the Internet package without the survey result in order to make sure that no competitors are present at the time of market entry, and second, to wait for the survey result in order to minimize the risk of failing to attract enough customers. If LOTell decides to wait for the survey result, there are three possible outcomes: the market growth is rapid (30% probability), steady (40%), or slow (30%). Depending on the survey result, LOTell may decide to introduce or not introduce the new Internet service. If it decides to launch the new service after the survey, which is three months from now, then there is a 70% chance that the competitors will come up with a similar service package. What decision will result in the highest expected market share for LOTell?

ANSWER

A decision tree for the problem is shown in Figure 12.16, which also shows the results of the analysis. The recommended decision is to introduce the Internet service now because it produces a higher expected gain in market share compared to waiting for the survey results.■

Figure 12.16 Decision Tree Analysis for LOTell

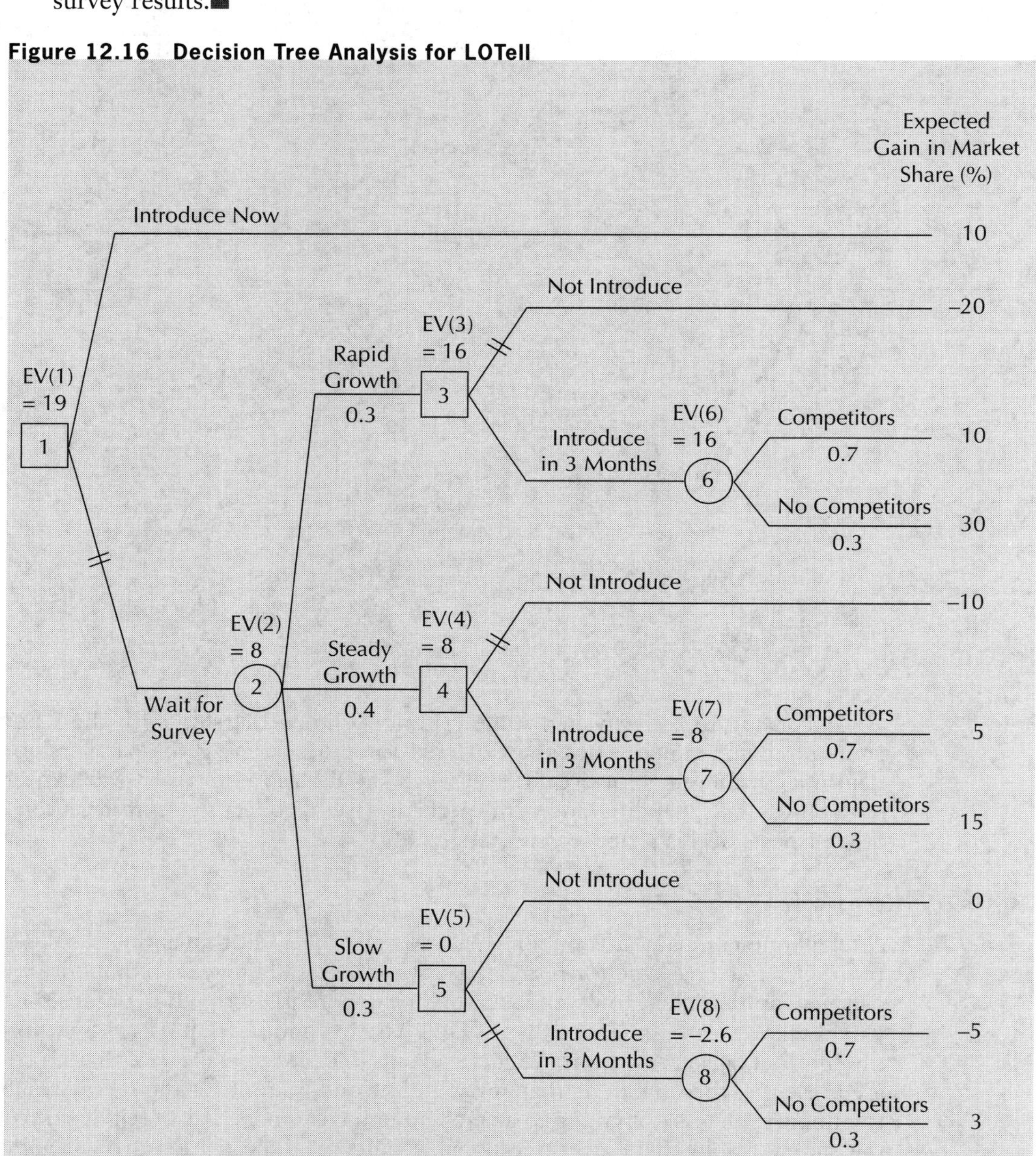

REVIEW PROBLEM 12.3

A land developer in Saskatoon is building a new set of condominiums, which when rented will have all utilities included in the monthly rental fees paid by tenants. The developer is looking at using tankless gas water heaters as an alternative to standard gas water heaters. Tankless water heaters provide hot water without using a storage tank. Cold water passes through a pipe to the heater, and a gas burner heats the water on demand. Tankless water heaters can be cost-effective because they do not incur the standby heat loss associated with regular tank-based gas hot water heaters. Though tankless heaters cost more to install, the manufacturers claim that they can save a substantial amount on water heating bills. The first cost of a standard gas water heater is $600 and it is expected to have a service life of 10 years. A comparable tankless unit costs $1500 installed and is expected to have a service life of 20 years. The annual hot water heating costs for a small family (low demand) is approximately $480, and for a large family (high demand) $960. From past experience, the developer knows that 75% of tenants will fall into the high-demand category. The developer does not have any experience with tankless heaters, but the manufacturers indicate that energy savings, compared to a regular tank gas water heater, can be 20% to 30%. The developer has decided to model this uncertainty by a discrete distribution with a 50% probability that the savings will be 20% and a 50% probability that the savings will be 30%. Their MARR is 15%.

(a) Draw a decision tree for this situation. Which of the two alternatives minimizes the expected annual worth of costs?

(b) If the developer is concerned about risk, can he find a preferred alternative using mean-variance dominance reasoning?

(c) Is one alternative preferred to the other on the basis of outcome dominance?

(d) Compare the cumulative risk profiles for the two alternatives. Does first-order stochastic dominance between the two alternatives exist?

ANSWER

(a) Figure 12.17 shows the completed decision tree for the developer. To carry out the decision analysis, the annual worth of costs for each branch of the tree must be calculated. For example, the annual worth of costs for the regular heater when demand is low is

$$\begin{aligned}\text{AW(standard heater, low demand)} &= \$600\ (A/P,\ 15\%,\ 10) + \$480 \\ &= \$599.55\end{aligned}$$

and the annual worth of costs for the tankless heater, when demand is high and savings are 20% is

$$\begin{aligned}\text{AW(tankless heater, high demand, savings)} &= \$1500(A/P,\ 15\%,\ 20) + (1-0.20)(960) \\ &= \$1007.64\end{aligned}$$

On the basis of these annual worths, the decision tree can be "rolled back" to demonstrate that the tankless heater has expected annual costs of $869.64 and the standard heater has expected annual costs of $959.55. The tankless heater is preferred according to the EV criterion.

(b) The mean annual costs of each alternative can be taken from part (a). The variance of the annual worth of costs for the standard heater is

$$\begin{aligned}\text{Var(standard)} &= 0.75 \times (1079.55 - 959.55)^2 + 0.25 \times (599.55 - 959.55)^2 \\ &= 43\ 200\end{aligned}$$

Figure 12.17 Decision Tree for Heater

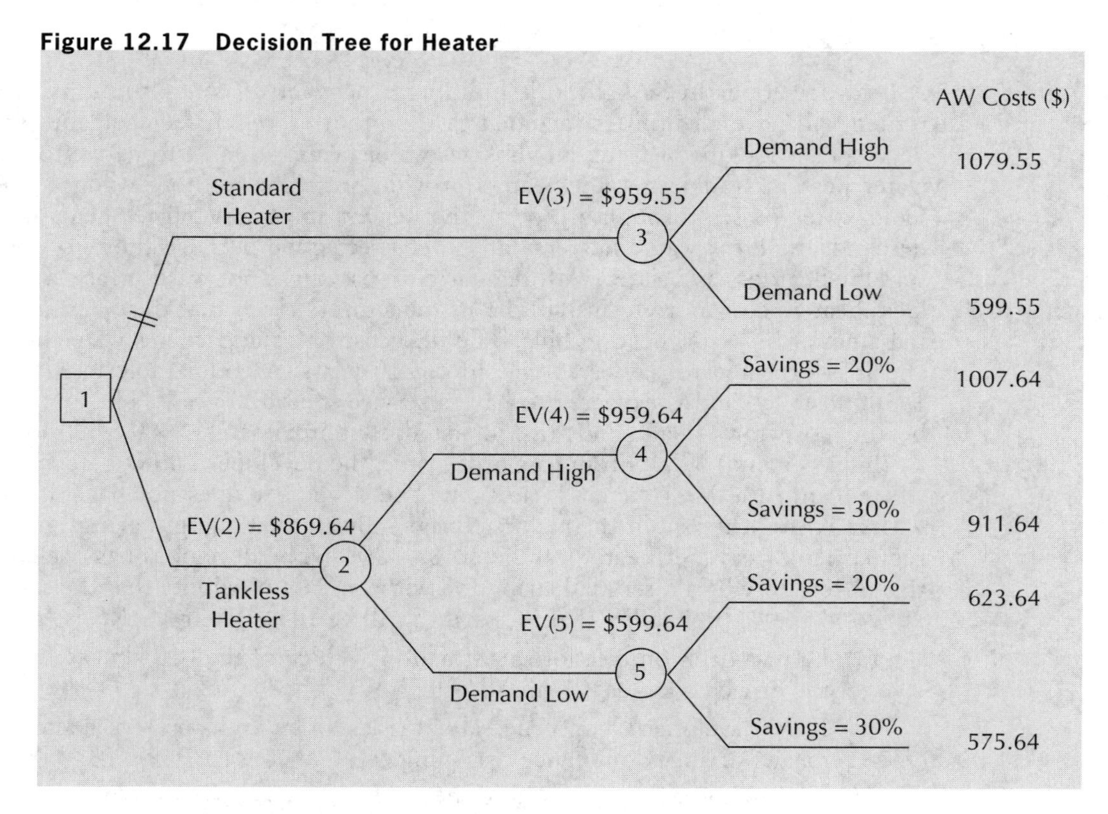

This is calculated on the basis of the fact that for the decision to purchase a standard gas hot water heater, the only uncertainty is the demand level. It will be high with probability 75%, and low with probability 25%.

The variance of the annual worth of costs for the tankless heater is

$$\begin{aligned}\text{Var(tankless)} &= 0.75 \times 0.50 \times (1007.64 - 869.64)^2 + 0.75 \times 0.50 \times \\ &\quad (911.64 - 869.64)^2 + 0.25 \times 0.5 \times (623.64 - 869.64)^2 \\ &\quad + 0.25 \times 0.50 \times (575.64 - 869.64)^2 \\ &= 26\,172\end{aligned}$$

This is calculated on the basis of the fact that for the decision to purchase a tankless gas water heater, there are four possible outcomes. They are {demand is high, savings = 20%}, {demand is high, savings = 30%}, {demand is low, savings = 20%}, and finally {demand is low, savings = 30%}. The respective probabilities are $0.75 \times 0.50 = 0.375$, $0.75 \times 0.50 = 0.375$, $0.25 \times 0.5 = 0.125$, and $0.25 \times 0.5 = 0.125$.

The tankless heater is preferred to the standard gas water heater by mean-variance reasoning, because it has both lower expected annual worth of costs and lower variance of annual worth of costs.

(c) Neither type of outcome dominance exists. For example, the worst outcome for the tankless heater is $1007.64. This is not better than the best outcome for the standard heater ($599.55). The converse also holds in that the worst outcome for the standard heater ($1079.55) is not better than the best outcome for the tankless heater ($575.64). For the second type of outcome dominance, the possible outcomes associated with the two uncertain events must be examined (see part (b)). By directly examining the outcomes for each, it can be seen that the tankless heater is better for each outcome except when demand is low and savings is 20%. Therefore the second type of outcome dominance does not hold.

(d) Figure 12.18 provides the cumulative risk profiles for the standard and tankless heaters. The figures demonstrate that first-degree stochastic dominance cannot screen out either alternative because they overlap one another.■

Figure 12.18 Cumulative Risk Profiles for Standard and Tankless Heaters

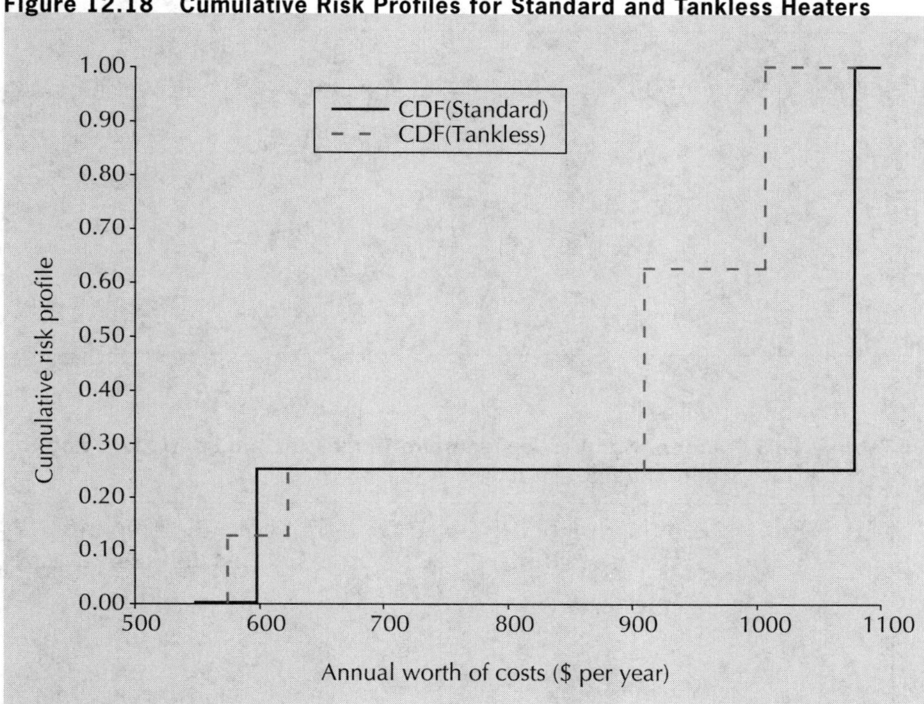

REVIEW PROBLEM 12.4

Orla is in the car rental business. The cars she uses have a useful life of two to four years. The cars can be traded in for a new car at the dealer. The trade-in values vary from $1000 to $5000 depending on the condition of the car. Orla has estimated the probability distributions for the length of useful life and the trade-in value for a typical car. Her estimates are shown in Table 12.14. Orla wants to find out the typical annual cost of owning a car for doing this business using the Monte Carlo simulation method. Assume that the cost of a new car is $15 000 and the annual maintenance cost is $800. Her MARR is 12%.

ANSWER

Orla first comes up with the following annual cost expression for her problem:

$$\begin{aligned} AC &= (\text{Cost of New Car})(A/P, 12\%, \text{Useful Life}) + (\text{Maintenance Cost}) \\ &\quad - (\text{Trade-In Value})(A/F, 12\%, \text{Useful Life}) \\ &= \$15\,000(A/P, 12\%, \text{Useful Life}) + 800 \\ &\quad - (\text{Trade-In Value})(A/F, 12\%, \text{Useful Life}) \end{aligned}$$

Then, she figures out the random-number assignment for the useful life and the trade-in value of a car (see Table 12.15), and performs a Monte Carlo simulation and collects 200 trials. The partial result of the simulation is shown in Table 12.16 and the histogram of the frequency distribution is presented in Figure 12.19.

Table 12.14 Probability Distributions for the Useful Life and the Trade-in Value of a Rental Car

Useful Life (Years)	Probability	Trade-in Value	Probability
2	0.4	\$1000	0.2
3	0.3	2000	0.2
4	0.3	3000	0.2
		4000	0.2
		5000	0.2

Table 12.15 Random Number Assignment Ranges for the Useful Life and the Trade-in Value

Useful Life	Random Number (Z_1) Assignment Range	Trade-in Value	Random Number (Z_2) Assignment Range
2	$0 \leq Z_1 < 0.4$	\$1000	$0 \leq Z_2 < 0.2$
3	$0.4 \leq Z_1 < 0.7$	2000	$0.2 \leq Z_2 < 0.4$
4	$0.7 \leq Z_1 < 1$	3000	$0.4 \leq Z_2 < 0.6$
		4000	$0.6 \leq Z_2 < 0.8$
		5000	$0.8 \leq Z_2 < 1$

Table 12.16 Partial Results for the Monte Carlo Simulation

Sample Number	Random Number (Z_1)	Useful Life	Random Number (Z_2)	Trade-in Value	Annual Cost
1	0.522750	3	0.325129	2000	5652.54
2	0.809623	4	0.423085	3000	4310.81
3	0.124285	2	0.799329	4000	6988.68
4	0.104359	2	0.207269	2000	7932.08
5	0.961704	4	0.108746	1000	4729.28
⋮	⋮	⋮	⋮	⋮	⋮

Figure 12.19 Frequency Distribution for Orla's Annual Cost

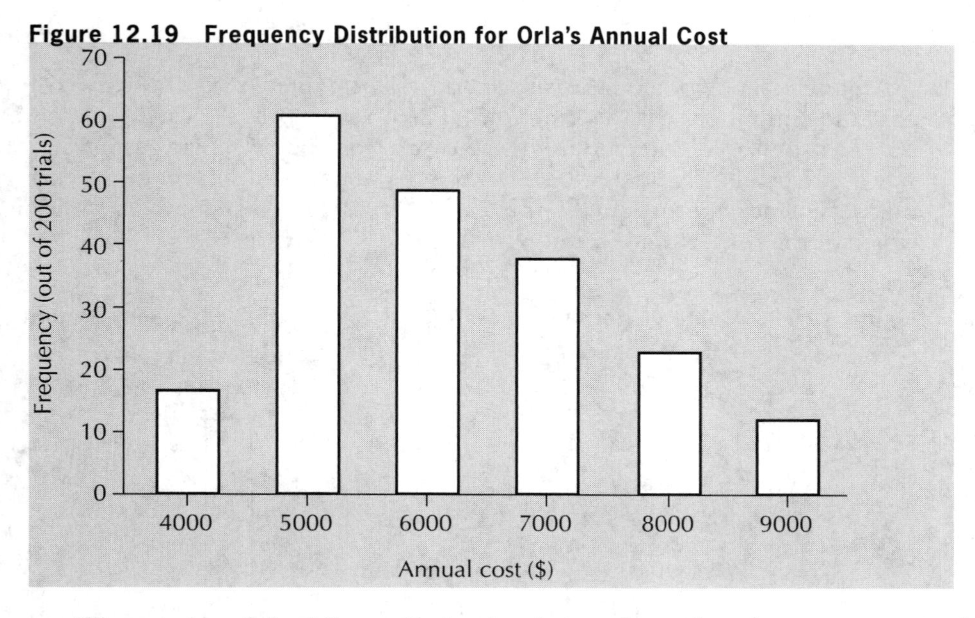

The results of the Monte Carlo simulation show that the average annual cost for each car is $5706. In the 200 samples, the annual cost ranged from $3892 to $8404.■

REVIEW PROBLEM 12.5

Burnaby Insurance (see Review Problems 11.1 to 11.3) has consulted several energy experts in order to further understand the implications of electricity and natural gas price changes on their two energy efficiency projects. They have estimated that the cost of electricity can range from $0.063 per kilowatt-hour to $0.077 per kilowatt-hour, and the price of natural gas can range from $3.35 to $3.65 per 1000 cubic feet. Both probability distributions (discrete approximations) are shown in Table 12.17. Carry out a Monte Carlo simulation to determine the probability distribution of the present worth of the two energy efficiency projects.

Table 12.17 Probability Distributions for the Cost of Electricity and the Price of Natural Gas

Electricity Cost (per kWh)	Probability	Natural Gas Price (per 1000 cubic feet)	Probability
$0.063	0.125	$3.35	1/7
0.065	0.125	3.40	1/7
0.067	0.125	3.45	1/7
0.069	0.125	3.50	1/7
0.071	0.125	3.55	1/7
0.073	0.125	3.60	1/7
0.075	0.125	3.65	1/7
0.077	0.125		

ANSWER

To conduct the Monte Carlo simulation, 300 samples were drawn from the electricity cost and natural gas cost distributions. The present worth of each project was computed for each simulated outcome using the expressions below, and a histogram of the results was constructed. Figure 12.20 shows the frequency distribution of present worth of the high-efficiency motor and Figure 12.21 shows the frequency distribution of present worth of the heat exchange unit.

Figure 12.20 Monte Carlo Simulation Results for the High-Efficiency Motor

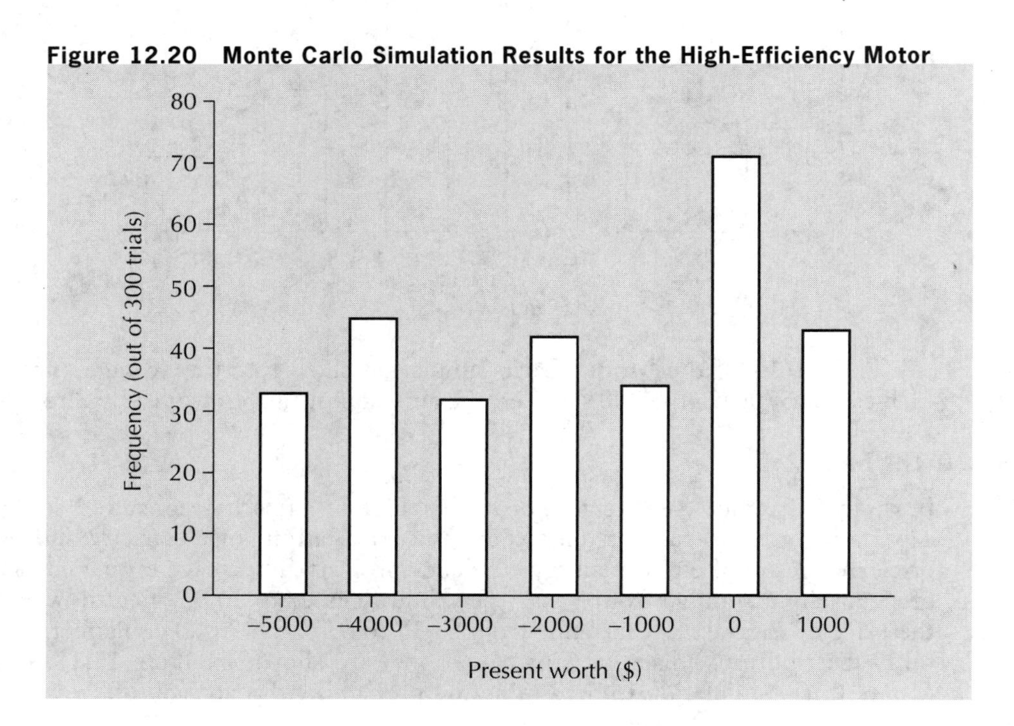

Figure 12.21 Monte Carlo Simulation Results for the Heat Exchange Unit

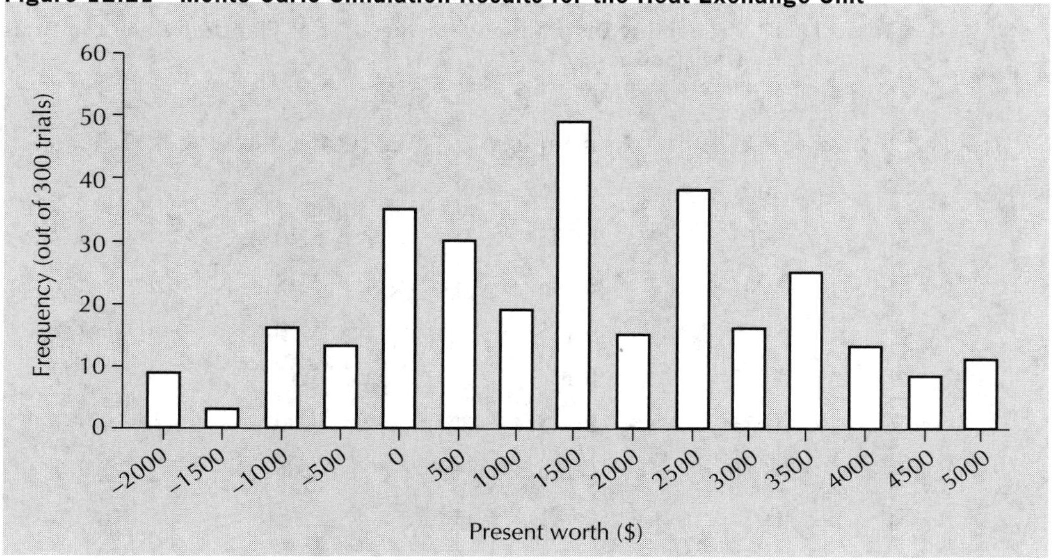

PW (High Efficiency motor)

$$= -\$28\,000 + (P/A, 10\%, 10)[\$70\,000(\text{Electricity}) - \$700]$$
$$= -\$28\,000 + (6.1446)[\$70\,000(\text{Electricity}) - \$700]$$

PW(Heat exchanger)

$$= -\$40\,000 + (P/A, 10\%, 10)[\$29\,000(\text{Electricity}) + \$2250(\text{Natural Gas}) - \$3200]$$
$$= -\$40\,000 + (6.1446)[\$29\,000(\text{Electricity}) + \$2250(\text{Natural Gas}) - \$3200]$$

It is clear from the simulations that the high-efficiency motor has roughly a 40% chance of having a positive present worth and hence is not a good investment. The heat exchange unit, on the other hand, has only a small chance of having a negative present worth, and hence looks like a better choice for Burnaby Insurance.■

SUMMARY

In this chapter, some basic probability theory was used to assist in dealing with decision making under uncertainty. Basic concepts such as random variables, probability distributions, and expected value were introduced to help understand decision tree analysis and Monte Carlo simulation analysis. A decision tree is a graphical representation of the logical structure of a decision problem in terms of the sequence of decisions and outcomes of chance events. An analysis based on a decision tree can therefore account for the sequential nature of decisions.

A variety of criteria can be used to select a preferred decision from a set of alternatives. Expected monetary value is a commonly used criterion that can form the starting point of any economic analysis. If a decision maker wants to more fully characterize the risk associated with a decision, mean-variance or stochastic dominance decision criteria can be applied.

Monte Carlo simulation is useful for estimating the probability distribution of the economic outcome of a project when there are multiple sources of uncertainty involved in a project. The probability distribution of the outcome performance measure can then be used to assess the economic viability of the project.

Engineering Economics in Action, Part 12B: Chances Are Good

"So let me get this straight," Clem started, a few days later. "You structured this as a tree. We—Canadian Widgets—have the first decision whether to proceed with development of the Adaptamatic at all. Then the next node represents the probability that the Adaptamatic will have a competitor. Then there is our choice of going into production or not. Then there are the customers, which we represent as chance nodes, who can have low, medium, and high demand. Finally, there are the outcomes, each of which has a dollar value associated with it. Is that right?"

"Perfect," Naomi said.

"And the dollar values came from . . . ?"

"Well, some reasonable assumptions about the marketplace," Naomi replied. "We had enough information to quantify three levels of demand. Also, we figured we had as much consumer acceptance as any competitor because of our association with Powerluxe, so when a competitive product was in the marketplace, we assumed that we would have the same market share as anyone else."

"What about the probabilities?" Clem continued.

"That was harder." Bill answered this one. Naomi couldn't help but smile. A few days ago, she didn't know anything about marketing, and Bill didn't know about decision trees, but now she was answering the questions about marketing, and he was answering the ones about decision trees! Go figure.

"We had no hard information on probabilities," Bill continued. "We set it all up as a spreadsheet so that we could adjust the probabilities freely. First we put in our best guesses—subjective probabilities. We talked to Prabha up in Marketing and to a bunch of people at Powerluxe, and refined that down. Finally, we tested it for a whole range of possibilities. Bottom line: chances are good that we will make a killing."

Clem looked at the decision tree and at the table Bill gave him that reported the expected value of proceeding with the Adaptamatic development under various assumptions. "Are you sure this is right? How come we make more money if we have a competitor than if we don't?"

"Good observation, Clem." Bill was beaming. This was his area of expertise. "That is interesting, isn't it? That's what's called 'building the category' in marketing. The principle is that when you have a new product, you have to spend a lot of resources educating the consumer. If you have to do that yourself, it's really costly. But a competitor can work for you by taking a share of those education costs and actually making it cheaper for you. That's what happens here."

"It was the decision tree approach that revealed this dynamic in this case," Bill continued. "By ourselves, we could lose money if demand was low, even though the expected value was fairly high. However, with a competitor on the market, we can almost guarantee making money because the efforts of our competitor reduce our marketing costs and expand the market at the same time. Combine them both in the decision tree, and we have a high expected value with almost no risk. Cool, eh?"

"Really cool." Clem glanced at Naomi with an appraising look. "Speaking of chances—chances are good that Anna's going to like this. Nice job."

PROBLEMS

12.1 An investment has a possible rate of return of 7%, 10%, and 15% over five years. The probabilities of attaining these rates, estimated on the basis of the current economy, are 0.65, 0.25, and 0.1, respectively. If you have $10 000 to invest,

(a) What is the expected rate of return from this investment?

(b) What is the variance of the rate of return on this investment?

12.2 Rockies Adventure Wear, Inc. sells athletic and outdoor clothing through catalogue sales. They want to upgrade their order-processing centre so that they have less chance of losing customers by putting them on hold. The upgrade may result in a processing capacity of 30, 40, 50, or 60 calls per hour with the probabilities of 0.2, 0.4, 0.3, and 0.1, respectively. Market research indicates that the average number of calls that Rockies may receive is 50 per hour.

(a) How many customers are expected to be lost per hour due to the lack of processing capacity?

(b) What is the variance of the loss due to the lack of processing capacity?

12.3 Power Tech builds power-surge-protection devices. One of the components, a plastic molded cover, can be produced from two automated machines, A1 and X1000. Each machine produces a number of defects with probabilities shown in the following table.

A1		X1000	
No. of Defects (out of 100)	Probability	No. of Defects (out of 100)	Probability
0	0.3	0	0.25
1	0.28	1	0.33
2	0.15	2	0.26
3	0.15	3	0.1
4	0.1	4	0.05
5	0.02	5	0.01

(a) Which machine is better with regard to the expected number of defective products?

(b) If Power Tech wants to take the variance of the number of defects into account, can mean-variance dominance be used to select a preferable machine?

12.4 Lightning City is famous for having many thunderstorms during the summer months (from June to August). One of the CB Electronix factories is located in Lightning City. They have collected information, shown in the table below, regarding the number of blackouts caused by lightning.

Number of Blackouts (Per Month)	Probability (Summer Months)	Probability (Non-summer Months)
0	0	0.45
1	0.4	0.4
2	0.25	0.15
3	0.2	0
4	0.1	0
5	0.05	0

For the first three blackouts in a month, the cost due to suspended manufacturing is $800 per blackout. For the fourth and fifth blackout, the cost increases to $1500 per blackout. A local insurance company offers protection against lightning-related expenses. The monthly payment is $500 for complete annual coverage. Assume that the number of blackouts in any month is independent of those in any other month.

(a) What is the expected cost related to blackouts over the summer months? Over the non-summer months? Should CB consider purchasing the insurance policy ?

(b) What is the variance of costs related to blackouts over the summer months? Over the non-summer months? Can mean-variance dominance reasoning be used to decide whether to purchase the insurance policy?

12.5 A new wave-soldering machine is expected to generate monthly savings of either $8000, $10 000, $12 000, or $14 000 over the next two years. The manager is not sure about the likelihood of the four savings scenarios, so she assumes that they are equally likely. What is the present worth of the expected monthly savings? Use MARR of 12%, compounded monthly, for this problem.

12.6 Regional Express is a small courier service operating in Southern Ontario. By introducing a new computerized tracking device, they anticipate some increase in revenue, currently estimated at $2.75 per parcel. The possible new revenue ranges from $2.95 to $5.00 per parcel with probabilities shown in the table below. Assuming that Regional's monthly capacity is 60 000 parcels and the monthly operating and maintenance costs are $8000, what is the present worth of the expected revenue over 12 months? Regional's MARR is 12%, compounded monthly.

Revenue per parcel	$2.95	$3.25	$3.50	$4.00	$5.00
Probability	0.1	0.35	0.3	0.15	0.1

12.7 Katrina is thinking about buying a car. She figures her monthly payments will be $90 for insurance, $30 for gas, and $20 for general maintenance. The car she would like to buy may last for 4, 5, or 6 years before a major repair, with probabilities of 0.4, 0.4, and 0.2, respectively. Calculate the present worth of the monthly expenses over the expected life of the car (before a major repair). Katrina's MARR is 10%, compounded monthly.

12.8 Pharma-Excel is a pharmaceutical company based in Halifax. They are currently studying the feasibility of a research project that involves improvement of vitamin C pills. To examine the optimistic, expected, and pessimistic scenarios for this project, they gathered the data shown below. What is the expected annual cost of the vitamin C project? Assume Pharma-Excel's MARR is 15%. Note that the lead time is different for each scenario.

	Optimistic	**Expected**	**Pessimistic**
Research and development costs (at the end of research)	$75 000	$240 000	$500 000
Lead time to production (years)	1	2	3
Probability	0.15	0.5	0.35

12.9 Mega City Hospital is selling lottery tickets. All proceeds go to their cancer research program. Each ticket costs $100, but the campaign catchphrase promises a 1-in-1000 chance of winning the first prize. The first prize is a "dream" house, which is worth $250 000. On the basis of decision tree analysis, is buying a ticket worthwhile?

12.10 See Problem 12.9. Determine the price of a ticket so that not buying a ticket is the preferred option and determine the chance of winning so that not buying a ticket is the preferred option.

12.11 Randall at Churchill Circuits (CC) has just received an emergency order for one of CC's special-purpose circuit boards. Five are in stock at the moment. However, when they were tested last week, two were defective but were mixed up with the three good ones. There is not enough time to retest the boards before shipment to the customer. Randall can either choose one of the five boards at random to ship to the customer or he can obtain a proven non-defective one from another plant. If the customer gets a bad board, the total incremental cost to CC is $10 000. The incremental cost to CC of getting the board from another plant is $5000.

(a) What is the chance that the customer gets a bad board if Randall sends them one of the five in stock?

(b) What is the expected value of the decision to send the customer one of the five boards in stock?

(c) Draw a decision tree for Randall's decision. On the basis of EV, what should he do?

12.12 St. Jacobs Cheese Factory (SJCF) is getting ready for a busy tourist season. SJCF wants to either increase production or produce the same amount as last year, depending on the demand level for the coming season. SJCF estimates the probabilities for high, medium, and low demands to be 0.4, 0.35, and 0.25, respectively, on the basis of the number of tourists forecasted by the local recreational bureau. If SJCF increases production, the expected profits corresponding to high, medium, and low demands are $750 000, $350 000, and $100 000, respectively. If SJCF does not increase production, the expected profits are $500 000, $400 000, and $200 000, respectively.

(a) Construct a decision tree for SJCF. On the basis of EV, what should they do?

(b) Construct a cumulative risk profile for both decision alternatives. Does either outcome or first-order stochastic dominance exist?

12.13 LOTell, a telephone company in London, has two options for their new Internet service package: they can introduce a combined rate for the residential phone line and the Internet access or they can offer various add-on Internet service rates in addition to the regular phone rate. LOTell can only afford to introduce one of the packages at this point. The expected gain in market share by introducing the Internet service would likely differ for different market growth rates. LOTell has estimated that if they introduce the combined rate, they would gain 30%, 15%, and 3% of the market share with rapid, steady, or slow market growth. If they introduce the add-on rates, they gain 15%, 10%, and 5% of the market share with rapid, steady, or slow market growth.

(a) Construct a decision tree for LOTell. If they wish to maximize expected market share growth, which package should LOTell introduce to the market now?

(b) Can either the combined rate or the add-on rate alternative be eliminated from consideration due to dominance reasoning?

12.14 Brockville Brackets (BB) uses a robot for welding small brackets onto car-frame assemblies. BB's R&D team is proposing a new design for the welding robot. The new design should provide substantial savings to BB by increasing efficiency in the robot's mobility. However, the new design is based on the latest technology, and there is some uncertainty associated with the performance level of the robot. The R&D team estimates that the new robot may exhibit high, medium, and low performance levels with the probabilities of 0.35, 0.55, and 0.05 respectively. The annual savings corresponding to high-, medium-, and low-performance levels are $500 000, $250 000, and $150 000 respectively. The development cost of the new robot is $550 000.

(a) On the basis of a five-year study period, what is the present worth of the new robot for each performance scenario? Assume BB's MARR is 12%.

(b) Construct a decision tree. On the basis of EV, should BB approve the development of a new robot?

12.15 Refer to Review Problem 12.1. Power Tech is still considering the partnership with the Chinese manufacturing company. Their analysis in Review Problem 12.1 has shown that the partnership is recommended (by a marginal difference in the expected revenue increase between the two options). Power Tech now wants to further examine the possible shipping delay and quality control problems associated with the partnership. Power Tech estimates that shipping may be delayed 40% of the time due to the distance. Independently of the shipping problem, there may be a quality problem 25% of the time due to communication difficulties and lack of close supervision by Power Tech. The payoff information is estimated as shown below. Develop a decision tree for Power Tech's shipping and quality control problems and analyze it. On the basis of EV, what is the recommendation regarding the possible partnership?

Shipping Problem	Quality Problem	Gain in Annual Profit
No shipping delay	Acceptable quality	$ 200 000
No shipping delay	Poor quality	25 000
Shipping delay	Acceptable quality	100 000
Shipping delay	Poor quality	−100 000

12.16 Refer to Problem 12.2. Rockies Adventure Wear, Inc. has upgraded its order-processing centre in order to improve the processing speed and customer access rate. Before completely switching to the upgraded system, Rockies has an option of testing it. The test will cost Rockies $50 000, which includes the testing cost and loss of business due to shutting down their business for a half-day. If Rockies does not test the system, there is a 55% chance of severe failure ($150 000 repair and loss of business costs), a 35% chance of minor failure ($35 000 repair and loss of business costs), and a 10% chance of no failure. If Rockies tests the system, the result can be favourable with the probability of 0.34, which requires no modification, and not favourable with the probability of 0.66. If the test result is not favourable, Rockies has two options: minor modification and major modification. The minor modification costs $5000 and the major modification costs $30 000. After the minor modification, there is still a 15% chance of severe failure ($150 000 costs), a 45% chance of minor failure ($35 000 costs), and a 40% chance of no failure. Finally, after the major modification, there is still a 5% chance of severe failure, a 30% chance of minor failure, and a 65% chance of no failure. What is the recommended action for Rockies, using a decision tree analysis?

12.17 Refer to Problem 12.8. As a part of Pharma-Excel's feasibility study, they want to include information on the acceptance attitude of the public toward the new vitamin C product. Regardless of the optimistic, expected, and pessimistic scenarios on research and development, there is a chance the general public may not feel comfortable with the new product because it is based on a new technology. They estimate that the likelihood of the public accepting the product (and purchasing it) is 33.3% and not accepting it is 66.7%. The expected annual profit after the research is $1 000 000 if the public accepts the new product and $200 000 if the public does not accept it.

(a) Calculate the annual worth for all possible combinations of three R&D scenarios (optimistic, expected, and pessimistic) and two scenarios on public reaction (accept or not accept). Pharma-Excel's MARR is 15%.

(b) Using the annual worth information as the payoff information, build a decision tree for Pharma's problem. Should they proceed with the development of this new vitamin C product?

12.18 Baby Bear Beads (BBB) found themselves confronting a decision problem when a packaging line suffered a major breakdown. Ross, the manager of Maintenance, Rita, plant manager, and Ravi, the company president met to discuss the problem.

Ross reported that the current line could be repaired, but the cost and result were uncertain. He estimated that for $40 000, there was a 75% chance the line would be as good as new. Otherwise, an extra $100 000 would have to be spent to achieve the same result.

Rita's studies suggested that for $90 000, the whole line might be replaced by a new piece of equipment. However, there was a 40% chance an extra $20 000 might be required to modify downstream operations to accept a slightly different package size.

Ravi, who had reviewed his sales projections, revealed that there was a 30% chance the production line would no longer be required anyway, but that this wouldn't be known until after a replacement decision was made. Rita then pointed out that there was a 80% chance the new equipment she proposed could easily be adapted to other purposes, so that the investment, including the modifications to downstream operations, could be completely recovered even if the line was no longer needed. On the other hand, the repaired packing line would have to be scrapped with essentially no recovery of the costs.

The present worth of the benefit of having the line running is $150 000. Use decision tree analysis to determine what BBB should do about the packaging line.

12.19 Refer to Review Problem 12.1. Power Tech feels comfortable about their probability estimate regarding the change in North American demand. However, they would like to examine the probability estimate for Asian demand more carefully. Perform sensitivity analysis on the probability that Asian demand increases. Try the following values, {0.1, 0.2, 0.3, 0.4, 0.5}, in which 0.3 is the base case value. Analyze the result and give a revised recommendation as to Power Tech's possible partnership.

12.20 Refer to Review Problem 12.2. LOTell is happy with the decision recommendation suggested by the previous decision tree analysis considering information on market growth. However, LOTell feels that the uncertainty in market growth is the most important factor in their overall decision regarding the introduction of the Internet service package. Hence, they wish to examine the sensitivity of the probability estimates for the market growth. Answer the following questions on the basis of the decision tree developed for Review Problem 12.2.

(a) Let p_1 be the probability of rapid market growth, p_2 be the probability of steady growth, and p_3 be the probability of slow market growth. Express the expected value at Node 2, EV(2), in terms of p_1, p_2, and p_3.

(b) If $EV(2) < 10$, then the option to introduce the package now is preferred. Using the expression of EV(2) that was developed in (a), graph all possible values of p_1 and p_2 that lead to the decision to introduce the package now. (You will see that p_3 from part (a) is not involved.) What can you observe from the graph regarding the values of p_1 and p_2?

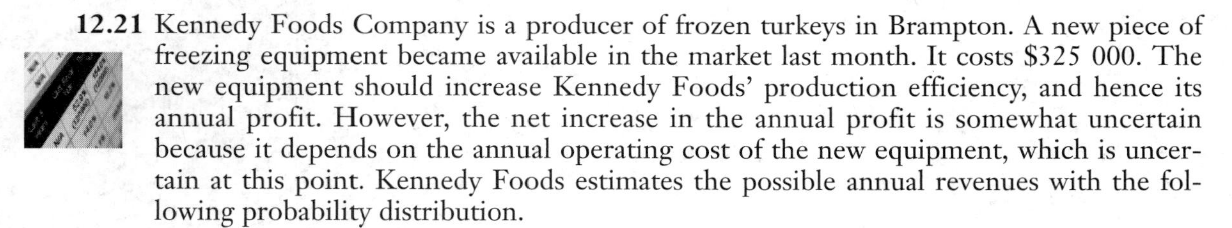

12.21 Kennedy Foods Company is a producer of frozen turkeys in Brampton. A new piece of freezing equipment became available in the market last month. It costs $325 000. The new equipment should increase Kennedy Foods' production efficiency, and hence its annual profit. However, the net increase in the annual profit is somewhat uncertain because it depends on the annual operating cost of the new equipment, which is uncertain at this point. Kennedy Foods estimates the possible annual revenues with the following probability distribution.

Net Increase in Annual Revenue	Probability
$25 000	0.1
30 000	0.35
35 000	0.4
40 000	0.15

(a) Express the present worth of this investment in analytical terms. Use a 10-year study period and a MARR of 15%.

(b) Show the random-number assignment ranges that can be used in Monte Carlo simulation.

(c) Carry out a Monte Carlo simulation of 100 trials. What is the expected present worth? What are the maximum and minimum PW in the sample frequency distribution? Construct a histogram of the present worth. Is it worthwhile for Kennedy Foods to purchase the new freezing equipment?

12.22 Refer back to Problem 12.3. Power Tech has decided to use the X1000 model exclusively for producing plastic molded covers. The revenue for each non-defective unit is $0.10. For the defective units, rework costs are $0.15 per unit.

(a) Show the analytical expression for the total revenue.

(b) Using the same probability distribution in Problem 12.3, create a table showing the random number assignment ranges for the number of defective units.

(c) Assume that the X1000 model produces molded covers in batches of 100. Carry out a Monte Carlo simulation of 100 production runs. What is the average total revenue? What are the maximum and minimum revenue in the 100 trials? Construct a histogram of the total revenue and comment on your results.

12.23 Ron-Jing is starting her undergraduate studies in September and needs a place to live over the next four years. Her friend, Nabil, told her about his plan to buy a house and rent out part of it. She thinks it may be a good idea to buy a house too, as long as she can get a reasonable rental income and can sell the house for a good price in four years. A fair-sized house, located a 15-minute walk from the university, costs $120 000. She estimates that the net rental income, after expenses, will be $1050 per month, which seems reasonable. She is, however, concerned about the resale value. She figures that the resale value will depend on the housing market in four years. She estimates the possible resale values and their likelihoods as shown in the table below. Using the Monte Carlo method, simulate 100 trials of the present worth of investing in the house. Use a MARR of 12%. Is this a viable investment for her?

Resale Value ($)	Probability
$100 000	0.2
110 000	0.2
120 000	0.2
130 000	0.2
140 000	0.2

12.24 An oil company owns a tract of land that has good potential for containing oil. The size of the oil deposit is unknown, but from previous experience with land of similar characteristics, the geological engineers predict that an oil well will yield between 0 (a dry well) and 100 million barrels per year over a five-year period. The following probability distribution for well yield is also estimated. The cost of drilling a well is $10 million, and the profit (after deducting production costs) is $0.50 per barrel. Interest is 10% per year. Carry out a Monte Carlo simulation of 100 trials and construct a histogram of the resulting net present worth of investing in drilling a well. Comment on your results. Do you recommend drilling?

Annual Number of Barrels (in Millions)	Probability
0	0.05
10	0.1
20	0.1
30	0.1
40	0.1
50	0.1
60	0.1
70	0.1
80	0.1
90	0.1
100	0.05

12.25 Refer to Problem 9.24. Before they purchase sonar warning devices to help trucks back up at store loading docks, St. James Bay department store is reexamining the original estimates of two types of annual savings generated by installing the devices. St. James feels that the original estimates were somewhat optimistic, and they want to include probability information in their analysis. The table below shows their revised estimates, with probability distributions, of annual savings from faster turnaround time and reduced damage to the loading docks. The sonar system costs $220 000 and has a life of four years and a scrap value of $20 000. Carry out a Monte Carlo simulation and generate 100 random samples of the present worth of investing in the sonar system. Assume that St. James' MARR is 18% and ignore the inflation. Make a recommendation regarding the possible purchase based on the frequency distribution.

Savings from Faster Turnaround	Probability	Savings from Reduced Damage	Probability
$38 000	0.2	$24 000	0.25
41 000	0.2	26 000	0.25
44 000	0.2	28 000	0.25
47 000	0.2	30 000	0.25
50 000	0.2		

12.26 A fabric manufacturer has been asked to extend a line of credit to a new customer, a dress manufacturer. In the past, the mill has extended credit to customers. Although most pay back the debt, some have defaulted on the payments and the fabric manufacturer has lost money. Previous experience with similar new customers indicates that 20% of customers are bad risks, 50% are average in that they will pay most of their bills, and 30% are good customers in that they will regularly pay their bills. The average length of business affiliation with bad, average, and good customers is 2, 5, and 10 years, respectively. Previous experience also indicates that, for each group, annual profits have the following probability distribution.

Bad Risk		Average Risk		Good Risk	
Annual Profit	**Probability**	**Annual Profit**	**Probability**	**Annual Profit**	**Probability**
$−50 000	1/7	$10 000	0.2	$20 000	1/7
−40 000	1/7	15 000	0.2	25 000	1/7
−30 000	1/7	20 000	0.2	30 000	1/7
−20 000	1/7	25 000	0.2	35 000	1/7
−10 000	1/7	30 000	0.2	40 000	1/7
0	1/7			45 000	1/7
10 000	1/7			50 000	1/7

Construct a spreadsheet with the headings as shown in the table below. Generate 100 random trials, and construct a frequency distribution of the present worth of extending a line of credit to a customer. Use a MARR of 10% per year.

Sample Number	Random Number 1	Risk Rating	Years of Business	Random Number 2	Annual Profit	PW
1						
2						
3						
4						
5						
⋮	⋮	⋮	⋮	⋮	⋮	⋮

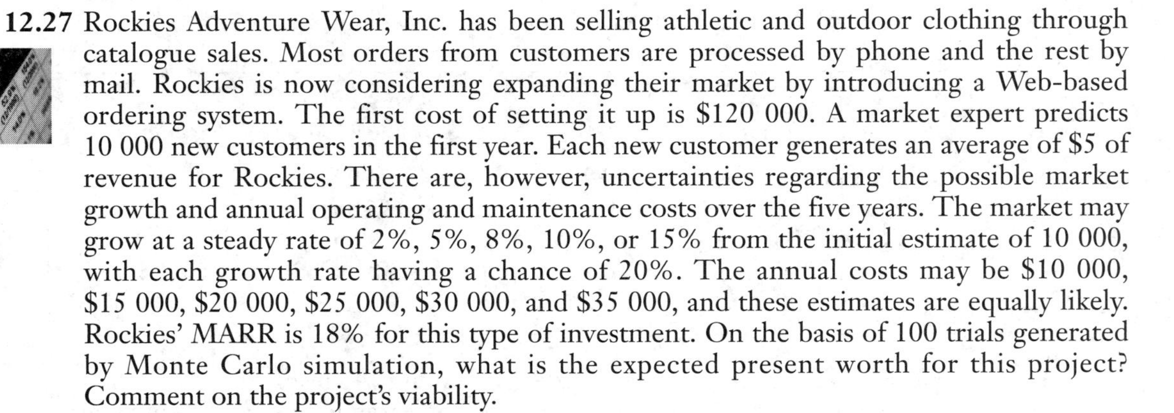

12.27 Rockies Adventure Wear, Inc. has been selling athletic and outdoor clothing through catalogue sales. Most orders from customers are processed by phone and the rest by mail. Rockies is now considering expanding their market by introducing a Web-based ordering system. The first cost of setting it up is $120 000. A market expert predicts 10 000 new customers in the first year. Each new customer generates an average of $5 of revenue for Rockies. There are, however, uncertainties regarding the possible market growth and annual operating and maintenance costs over the five years. The market may grow at a steady rate of 2%, 5%, 8%, 10%, or 15% from the initial estimate of 10 000, with each growth rate having a chance of 20%. The annual costs may be $10 000, $15 000, $20 000, $25 000, $30 000, and $35 000, and these estimates are equally likely. Rockies' MARR is 18% for this type of investment. On the basis of 100 trials generated by Monte Carlo simulation, what is the expected present worth for this project? Comment on the project's viability.

12.28 Hitomi is considering buying a new lawnmower. She has a choice of a Lawn Guy or a Clip Job. Her neighbour, Sam, looked at buying a mower himself a while ago, and gave her the following information on the two types of mowers.

	Lawn Guy	Clip Job
First cost	$350	$120
Life (years)	10	4
Annual gas	$60	$40
Annual maintenance	$30	$60
Salvage value	0	0

Due to the long life of a Lawn Guy mower and the uncertainty of future gas prices, Hitomi is reluctant to use a single estimate for its life or the annual cost of gas. As for the Clip Job, she is not sure about the annual maintenance cost, since that model has a relatively short life and it may break down easily. With help from a friend who works at a hardware store, she comes up with the probabilistic estimates shown below for the Lawn Guy's expected life and the cost of gas, and Clip Job's annual maintenance cost. Hitomi's MARR is 5%. Find the expected annual cost for each mower using the Monte Carlo simulation method. Perform at least 100 trials. Which mower is preferred?

Lawn Guy				Clip Job	
Life	Probability	Gas	Probability	Maintenance	Probability
7	0.25	$50	0.2	$50	0.2
8	0.25	60	0.3	60	0.5
9	0.25	70	0.4	70	0.2
10	0.25	80	0.1	80	0.1

12.29 Mountain Beer Brewery (MBB) in Mill Bay, British Columbia, currently buys 250 000 beer labels every year from a local label-maker. The label-maker charges MBB $0.075 per piece. A demand forecast indicates that MBB's annual demand may grow up to 400 000 in the near future, so MBB is considering making the labels themselves. If they did so, they would purchase a high-quality colour photocopier that costs $6000 and lasts for five years with no salvage value at the end of its life. The operating cost of the photocopier would be $4900 per year, including the cost of colour cartridges and special paper used for labels. MBB would also have to hire a label designer. The cost of labour is estimated to be $0.04 per label produced. On the basis of the following probabilistic estimate on the future demand, simulate 200 Monte Carlo trials of the present worth of costs for (a) continuing to purchase from the local label-maker and (b) making their own labels. MBB's MARR is 12%. Make a recommendation as to whether MBB should consider making their own labels.

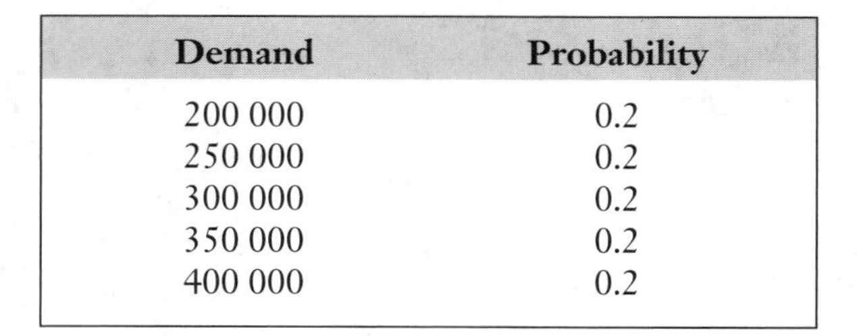

Demand	Probability
200 000	0.2
250 000	0.2
300 000	0.2
350 000	0.2
400 000	0.2

12.30 Refer to Problem 11.29. Timmons Testing (TT) has now established probabilistic information on the annual variable costs and annual production volume for the two testing devices offered by Vendors A and B. TT would like to perform a Monte Carlo simulation analysis before they decide which of the two alternatives they should choose. First cost, annual fixed costs, and salvage value information are presented in the table below in addition to the probabilistic estimates. TT charge $3.25 per board tested. Their MARR is 15%. Compare the expected annual worth and possible cost ranges for the two alternatives. What would you recommend to TT now?

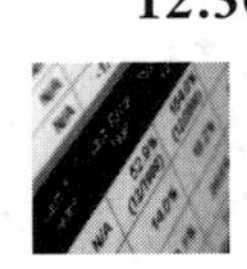

Alternative A				Alternative B			
Annual Variable Costs ($/Board)	Probability	Production Volume (Boards)	Probability	Annual Variable Costs ($/Board)	Probability	Production Volume (Boards)	Probability
2.20	0.2	40 000	0.05	1.01	0.1	40 000	0.05
2.25	0.3	45 000	0.15	1.03	0.2	45 000	0.05
2.30	0.3	50 000	0.25	1.06	0.4	50 000	0.2
2.35	0.2	55 000	0.2	1.10	0.3	55 000	0.2
		60 000	0.15			60 000	0.2
		65 000	0.05			65 000	0.1
		70 000	0.05			70 000	0.1
		75 000	0.05			75 000	0.05
		80 000	0.05			80 000	0.05

	A	B
First cost ($)	200 000	350 000
Annual fixed costs ($)	16 500	35 500
Salvage value ($)	5000	20 000
Life	7	10

Energy Management for a Quebec Office Building

CANADIAN 12.1 MINI-CASE

In 2003, Public Works and Government Services Canada (PWGSC) moved into their new 35 000-square-metre office space in Gatineau, Quebec. Several years earlier, PWGSC invited developers to tender irrevocable bids to include designing, building, and then leasing the building to PWGSC for a minimum of 15 years. The evaluation of the proposal was based on discounted first cost as well as the annual operating costs of the building which include items such as energy, maintenance, and cleaning costs.

PWGSC had set out extensive requirements on the building, including the quality of materials, the type of elevators, the level of indoor air quality, and a special request that the building operate at least 25% below the energy consumption prescribed in the Model National Energy Code for Building (MNECB).

The developer that won the bid relied on a project team to do extensive analysis of various energy management alternatives for the building design. The team identified alternatives that would reduce the building's energy consumption at least 25% below the MNECB. They then used numerous simulations to evaluate each alternative and decide which was the best from the point of view of its return on investment. Several examples of the alternatives considered are provided below.

One of the first energy-saving alternatives the team identified was to recycle heat within the building. They had decided on large central office spaces that usually need to be air-conditioned all year. To improve the energy efficiency of the building, the team discovered that the excess heat gains (e.g., from lighting, equipment, people) from the central areas of the building were more than enough to make up for the heat losses in the periphery of the building. A variety of simulations were run to compare heat recycling to two other alternatives for air conditioning—an electric chiller alone, and an electric chiller with "free cooling." The team was able to show that recycling heat gain from the central zone to the perimeter zones brought the predicted annual energy use down to 4 236 000 kilowatt-hours per year, as against 6 000 000 kilowatt-hours per year with an electric chiller and "free cooling" and 5 990 000 kilowatt-hours with an electric chiller without "free cooling."

Another set of interesting alternatives had to do with the source of energy used for the building. Hydro-Quebec charges for electricity consumption based on peak demand (kilowatt-hours), and total energy use (kilowatt-hours). Several of the alternatives considered were aimed at reducing costs by shifting heating costs into off-peak hours. Simulations were used to compare a number of heating alternatives: natural gas hot-water heating, natural gas hot-water heating during peak hours with electric baseboard heating during off-peak hours, the use of high-efficiency boilers for water heating, and the use of "heat accumulators." Heat accumulators use a mass of bricks that is heated with electricity during off-peak periods, with the stored heat extracted during peak periods.

The results of the simulations showed that the $130 000 heat accumulators could save $13 000 per year, but that it created a load of 25 tonnes in the building. Consequently this alternative was not considered further. The use of high-efficiency natural gas water heating during peak hours and electric baseboards during off-peak hours produced savings of $28 800 per year as against regular-efficiency natural gas hot water heating.

Another energy-saving strategy was to use energy-efficient lighting fixtures throughout the building. This resulted in savings of 14% on the total annual energy consumption.

The predicted energy consumption of the building was 50% more efficient than the MNECB guidelines, a savings of approximately $275 000 per year for a 35 000-square-metre building.

Source: *Canadian Consulting Engineer*, March/April 2003.

Discussion

A computer simulation involves constructing a software model that captures the essential structure of a complex product or process. The model may be built in a spreadsheet package, a general-purpose programming language, or a special-purpose software package suited to the application. For the energy management application described above, the project team constructed a software model of energy consumption for the office building, and then used it to simulate (predict) what energy consumption would result from numerous "what if" alternatives. Contrast this usage of the word "simulation" to that of a Monte Carlo model discussed in this chapter. In the context of Monte Carlo simulation, an analyst makes assumptions about the probability distributions of uncertain outcomes and then repeatedly samples from these distributions to predict the overall uncertainty associated with a decision alternative.

Many large projects are simulated by computer before construction or implementation, as was done with the office building described above. Not only can the design requirements be taken into account, but also potential problems can be discovered before irreversible decisions are made.

In planning the PWGSC office building, it was possible to reduce annual energy consumption from $550 000, the normal energy consumption for a 35 000-square-metre building, to $275 000. The saving of $275 000 per year due to the detailed energy simulations was worth the cost. Not all simulation projects will save this much money, but in any case where costs are large, the relatively small cost of a simulation can be a very good investment. Because simulations never capture all natural processes completely, the analyst must understand the sources of error and use that understanding in the larger context of the analysis. Simulation experts often use sensitivity analysis, as presented in Chapter 11, to test which sources of error have the most influence on the simulation results.

Questions

1. Many projects can be implemented without an initial simulation study. Name a few of these, and explain why decisions about their acceptance can be made without a simulation study.
2. What are the likely characteristics of projects where simulation makes sense? Name a few projects of this nature.
3. In addition to the direct cost of a simulation study, what are other negative consequences of such a study?
4. In addition to a better formulation of design requirements and the ability to find potential problems, can you think of other benefits provided by a simulation?
5. Fredricka faces a $20 000 000 project to build a production plant in Mexico. A consultant offers to provide her with a simulation to model the plant's performance in the North American marketplace. How would she judge how much to spend for such a simulation?

C H A P T E R 13

Qualitative Considerations and Multiple Criteria

Engineering Economics in Action, Part 13A: Don't Box Them In

Naomi and Bill Astad were seated in Naomi's office. "OK," said Naomi. "Now that we know we can handle the demand, it's time to work on the design, right? What is the best design?" She was referring to the self-adjusting vacuum cleaner project for Powerluxe that she and Bill had been working on for several months.

"Probably the best way to find out," Bill answered, "will be to get the information from interviews with small groups of consumers."

"All right," said Naomi. "We have to know what to ask them. I guess the most important step for us is to define the relevant characteristics of vacuums."

"I agree," Bill responded. "We couldn't get meaningful responses about choices if we left out some important aspect of vacuums like suction power. One way to get the relevant characteristics will be to talk to people who have designed vacuums before, and probably to vacuum cleaner salespeople, too." They both smiled at the humorous prospect of seeking out vacuum cleaner salespeople, instead of trying to avoid them.

"We're going to need some technical people on the team," Naomi said. "We will have to develop a set of technically feasible possibilities."

"Exactly," Bill replied. "Moreover, we need to have working models of the feasible types. That is, we can't just ask questions about attributes in the abstract. Most people would have a hard time inferring actual performance from numbers about weight or suction power, for example. Also, consumers are not directly interested in these measurements. They don't care what the vacuum weighs. They care about what it takes to move it around and go up and down stairs. This depends on several aspects of the cleaner. It includes weight, but also the way the cleaner is balanced and the size of the wheels."

"That makes sense," Naomi said. "I assume that we would want to structure the interviews to make use of some form of MCDM approach."

"Huh?" Bill said.

13.1 Introduction

Most of this book has been concerned with making decisions based on a single economic measure such as present worth, annual worth, or internal rate of return. This is natural, since many of the decisions that are made by an individual, and most that are made by businesses have the financial impact of a project as a primary consideration. However, rarely are costs and benefits the only consideration in evaluating a project. Sometimes other considerations are paramount.

For decisions made by and for an individual, cost may be relatively unimportant. One individual may buy vegetables on the basis of their freshness, regardless of the cost. A dress or suit may be purchased because it is fashionable or attractive. A car may be chosen for its comfort and not its cost.

Traditionally, firms were different from individuals in this way. It was felt that all decisions for a firm *should* be made on the basis of the costs and benefits as measured in money (even if they sometimes were not, in practice), since the firm's survival depended solely on being financially competitive.

Society has changed, however. Companies now make decisions that apparently involve factors that are very difficult to measure in monetary terms. Money spent by firms on charities and good causes provides a benefit in image that is very hard to quantify. Resource companies that demonstrate a concern for the environment incur costs with no clear financial benefit. Companies that provide benefits for employees beyond statute or collective agreement norms gain something that is hard to measure.

The fact that firms are making decisions on the basis of criteria other than only money most individuals would hail as a good thing. It seems to be a good thing for the companies, too, since those that do so tend to be successful. However, it can make the process of decision making more difficult, because there is no longer a single measure of value.

Money has the convenient feature that, in general, more is better. For example, of several mutually exclusive projects (of identical service lives), the one with the highest present worth is the best choice. People prefer a higher salary to a lower one. However, if there are reasons to make a choice other than just the cost, things get somewhat more difficult. For example, which is better: the project with the higher present worth but that involves clear-cutting a forest, or the one with lower present worth but that preserves the forest? Does a high salary compensate for working for a company that does business with a totalitarian government?

Although such considerations have had particular influence in recent years, the problem of including both qualitative *and* quantitative criteria in engineering decisions has always been present. This leads to the question of how a decision maker deals with multiple objectives, be they quantitative or qualitative. There are three basic approaches to the problem:

1. Model and analyze the costs alone. Leave the other considerations to be dealt with on the basis of experience and managerial judgment. In other words, consider the problem in two stages. First, treat it as if cost were the only important criterion. Subsequently, make a decision based on the refined cost information—the economic analysis—and all other considerations. The benefit of this approach is its simplicity; the methods for analyzing costs are well established and defensible. The liability is that errors can be made, since humans have only a limited ability to process information. A bad decision can be made, and, moreover, it can be hard to explain why a particular decision was made.

2. Convert other criteria to money, and then treat the problem as a cost-minimization or profit-maximization problem. Before environmental issues were recognized as being so important, the major criterion that was not easily converted to money was human health and safety. Elaborate schemes were developed to measure the cost of a lost life or injury so that good economic decisions were made. For example, one method was to estimate the money that a worker would have made if he or she had not been injured. With an estimate like this, the cost of a project in lives and injuries could be compared with the profits obtained. A benefit of this approach is that it does take non-monetary criteria into account. A drawback is the difficult and politically sensitive task of determining the cost of a human life or the cost of cutting down a 300-year-old tree.

3. Use a **multi-criterion decision making (MCDM)** approach. There are several MCDM methods that explicitly consider multiple criteria and guide a decision maker to superior or optimal choices. The benefit of MCDM is that all important criteria can be explicitly taken into account in an appropriate manner. The main drawback is that MCDM methods take time and effort to use.

In recent years and in many circumstances, looking at only the monetary costs and benefits of projects has become inappropriate. Consequently, considerable attention has been focused on how best to make a choice under competing criteria. The first two approaches listed above still have validity in some circumstances; in particular, when non-monetary criteria are relatively unimportant, it makes sense to look at costs alone.

However, it is necessary to use an MCDM method of some sort in much of engineering decision making today.

In this chapter, we focus on three useful MCDM approaches. The first, *efficiency*, permits the identification of a subset of superior alternatives when there are multiple criteria. The second approach, *decision matrixes*, is a version of multi-attribute utility theory (MAUT) that is widely practiced. The third, the *analytic hierarchy process (AHP)*, is a relatively new but popular MAUT approach. It should be noted that all of these methods make assumptions about the tradeoffs among criteria that may not be suitable in particular cases. They should not be applied blindly, but critically and with a strong dose of common sense.

13.2 Efficiency

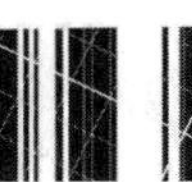

When dealing with a single criterion like cost, it is usually clear which alternatives are better than others. The rule for a present worth analysis of mutually exclusive alternatives (with identical service lives) is, for example, that the highest present worth alternative is best.

All criteria can be measured in some way. The scale might be continuous, such as "weight in kilograms" or "distance from home," or discrete, such as "number of doors" or "operators needed." The measurement might be subjective, such as a rating of "excellent," "very good," "good," "fair," and "poor," or conform to an objective physical property such as voltage or luminescence.

Once measured, the value of the alternative can be established with respect to that criterion. It may be that the smaller or lower measurement is better, as is often the case with cost, or that the higher is better, as with a criterion such as "lives saved." Sometimes a target is desired, for example, a target weight or room temperature. In this case, the criterion could be adjusted to be the distance from the target, with a shorter distance being better.

Consequently, given one criterion, we can recognize which of several alternatives is best. However, once there are more than one criteria, the problem is more difficult. This is because an alternative can be highly valued with respect to one criterion and lowly valued with respect to another.

Example 13.1

Simcoe Meats will be replacing its effluent treatment system. It has evaluated several alternatives, shown in Figure 13.1. Two criteria were considered, present worth and discharge purity. Which alternatives can be eliminated from further consideration?

Consider alternatives A and E in Figure 13.1. Alternative E *dominates* alternative A because it is less costly and it provides purer discharge. If these were the only criteria to consider in making a choice, one would always choose E over A. Similarly, one can eliminate F, B, and H, all of which are dominated by D and other alternatives.

Now consider alternative G, which has the same cost as E but has poor discharge purity. One would still always choose E over G, since, for the purity criterion, E is better at the same price.

Three alternatives now remain, E, D, and C. E is cheapest, but provides the least purity output of the three. C is the most expensive, but provides the greatest purity, while D is in the middle. Certainly none of these dominates the others. There is a natural tendency to focus on D, since it seems to balance the two criteria, but this really depends on the relative importance of the criteria to the decision maker. For example, if cost were very important, E might be the best choice, since the difference in purity between E and C may be considered relatively small.■

Figure 13.1 Selecting an Effluent Treatment

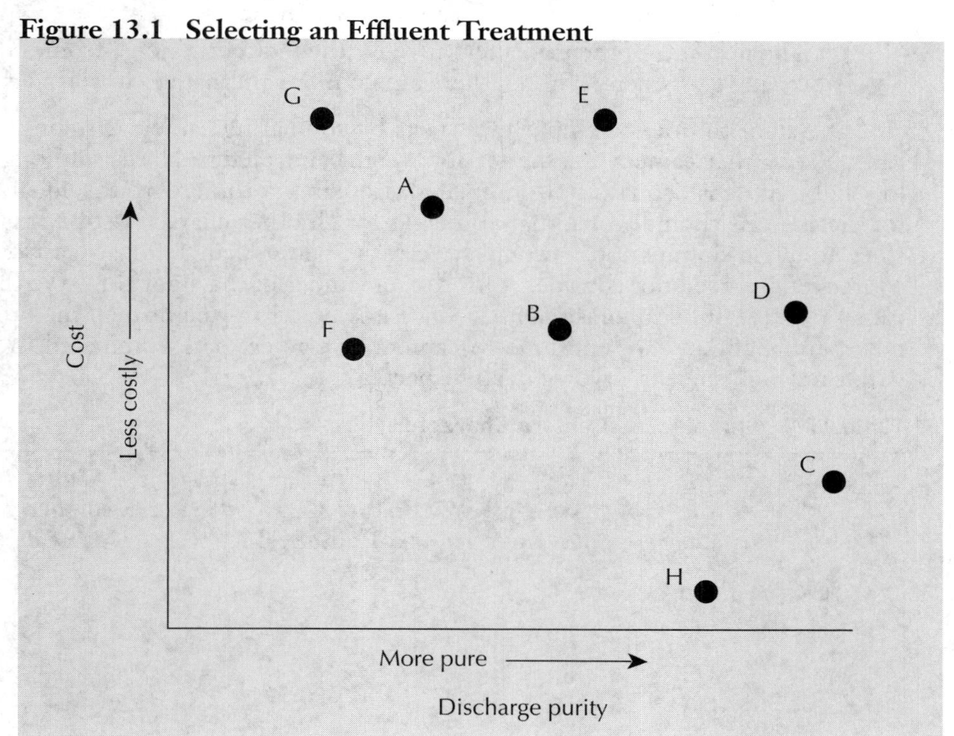

Decisions that involve only two criteria can be simplified graphically as done for Example 13.1, but when there are more than two criteria graphical methods become more difficult.

An alternative is **efficient** if no other alternative is valued as at least equal to it for all criteria and preferred to it for at least one. If an alternative is not efficient it is **inefficient**; this is the same as a **dominated** alternative.

Example 13.2

Skiven is evaluating surveillance cameras for a security system. The criteria he is taking into account are price, weight, picture clarity, and low-light performance. The details for the 10 models are shown in Table 13.1. Skiven wants a camera with low cost, low weight, a high score for picture clarity, and a high score for low-light performance. Which models can be eliminated from further consideration?

To determine the efficient alternatives, the following algorithm can be used:

1. Order the alternatives according to one criterion, the *index criterion*. The cameras for Example 13.2 are already ordered by cost, so cost can be the index criterion.
2. Start with the second most preferred alternative for the index criterion. Call this the *candidate alternative*.
3. Compare the candidate alternative with each of the alternatives that are more preferred for the index criterion. (For the first candidate alternative, there is only one.)

4. If any alternative equals or exceeds the candidate for all criteria, and exceeds it for at least one, the candidate is dominated, and can be eliminated from further consideration. If no alternative equals or exceeds the candidate for all criteria and exceeds it for at least one, the candidate is efficient.
5. The next most preferred alternative for the index criterion becomes the new candidate; go to step 3. Stop if there are no more alternatives to consider.

The algorithm for Example 13.2 starts by comparing camera 2 against camera 1. It can be seen that camera 2 is better for weight and picture clarity, although worse for low-light and cost, so it is not dominated. Looking at camera 3, it is equal to camera 1 for picture clarity, and worse than camera 1 for all other characteristics. It is dominated, since to avoid domination it would have to be better than 1 for at least one criterion. Moreover, we need not consider it for the remainder of the algorithm. We then compare camera 4 with only cameras 1 and 2. Since it is better in weight than the other two, it is not dominated, and we continue to camera 5. Camera 5 is dominated in comparison with camera 4, since it is worse in all respects.

Table 13.1 Surveillance Camera Characteristics

Camera	Price ($)	Weight (Grams)	Picture Clarity (10-Point Scale)	Low-Light Performance (10-Point Scale)
1	230	900	3	6
2	243	640	5	4
3	274	910	3	5
4	313	433	5	7
5	365	450	2	4
6	415	330	6	6
7	418	552	7	5
8	565	440	3	6
9	590	630	7	4
10	765	255	9	5

Carrying through in this manner shows that camera 3 is dominated by camera 1, camera 5 is dominated by camera 4, camera 8 is dominated by camera 6, and camera 9 is dominated by camera 7. The set of efficient alternatives consists of cameras 1, 2, 4, 6, 7, and 10. This set clearly includes the best choice, since there is no reason to choose a dominated alternative.■

Usually there is more than one alternative in the efficient set. Sometimes reducing the number of alternatives to be considered makes the problem easier to solve through intuition or judgment, but usually it is desirable to have some clear method for selecting a single alternative. One popular method is to use decision matrixes.

13.3 Decision Matrixes

Usually not all of the criteria that can be identified for a decision problem are equally important. Often cost is the most important criterion, but in some cases another criterion, safety, for example, might be most important. As suggested with Example 13.1, the choice of the most important criterion will have a direct effect on which alternative is best.

One approach to choosing the best alternative is to put numerical weights on the criteria. For example, if cost were most important, it would have a high weight, while a less important criterion might be given a low weight. If criteria are evaluated according to a scale that can be used directly as a measure of preference, then the weights and preference measures can be combined mathematically to determine a best alternative. This approach is called **multi-attribute utility theory (MAUT)**.

Many different specific techniques for making decisions are based on MAUT. This section deals with decision matrixes, which are commonly used in engineering studies. The subsequent section reviews the analytic hierarchy process, a MAUT method of increasing popularity.

In a **decision matrix**, the rows of the matrix represent the criteria and the columns the alternatives. There is an extra column for the weights of the criteria. The cells of the matrix (other than the criteria weights) contain an evaluation of the alternatives on a scale from 0 to 10, where 0 is worst and 10 is best. The weights are chosen so that they sum to 10.

The following algorithm can be used:

1. Give a weight to each criterion to express its relative importance: the higher the weight, the more important the criterion. Choose the weight values so that they sum to 10.
2. For each alternative, give a rating from 0 to 10 of how well it meets each criterion. A rating of 0 is given to the worst possible fulfilment of the criterion and 10 to the best possible.
3. For each alternative, multiply each rating by the corresponding criterion weight, and sum to give an overall score.
4. The alternative with the highest score is best. The value of the score can be interpreted as the percentage of an ideal solution achieved by the alternative being evaluated.
5. Carry out some sensitivity analysis with respect to weights or rating estimates to verify the indicated decision or to determine under which conditions different choices are made.

Example 13.3

Skiven is evaluating surveillance cameras for a security system. The criteria he is taking into account, in order of importance for him, are low-light performance, picture clarity, weight, and price. The details for the six efficient models are shown in Table 13.2. Which model is best?

In order to follow the steps given above, we need to determine the criteria weights. It is usually fairly easy for a decision maker to determine which criteria are more important than others, but generally more difficult to specify particular weights. There exist many formal methods for establishing such weights in a rigorous way, but in practice,

Table 13.2 Efficient Set of Surveillance Camera Alternatives

Camera	Price ($)	Weight (Grams)	Picture Clarity (10-Point Scale)	Low-Light Performance (10-Point Scale)
1	230	900	3	6
2	243	640	5	4
4	313	433	5	7
6	415	330	6	6
7	418	552	7	5
10	765	255	9	5

estimating weights on the basis of careful consideration or a discussion with the decision maker is sufficient. Recall that a sensitivity analysis forms part of the overall decision process, and this compensates somewhat for the imprecision of the weights.

Skiven suggests that weights of 1, 1.5, 3.5, and 4 for price, weight, picture clarity, and low-light performance, respectively, are appropriate weights for this problem. These weights are listed as the second column of Table 13.3.

Table 13.3 Decision Matrix for Example 13.3

Criterion	Criterion Weight	Alternatives 1	2	4	6	7	10
Price	1.0	10.0	9.8	8.4	6.5	6.5	0.0
Weight	1.5	0.0	4.0	7.2	8.8	5.4	10.0
Clarity	3.5	3.0	5.0	5.0	6.0	7.0	9.0
Low-light performance	4.0	6.0	4.0	7.0	6.0	5.0	5.0
Score	10.0	44.5	49.3	64.8	64.8	59.1	66.5

The ratings for each alternative for picture clarity and low-light performance are already on a scale from 0 to 10, so those ratings can be used directly. To select ratings for the price and weight, two different measures could be used:

1. *Normalization:* The rating r for the least preferred alternative (α) is 0 and the most preferred (β) is 10. For each remaining measure (γ) the rating r can be determined as

$$r = 10 \times \frac{\gamma - \alpha}{\beta - \alpha}$$

For this problem, the rating of alternative 6 for price would be

$$r_{6,price} = 10 \times \frac{415 - 765}{230 - 765} = 6.54$$

The advantage of normalization is that it provides a mathematical basis for the rating evaluations. One disadvantage is that the rating may not reflect the value as perceived by the decision maker. A second disadvantage is that it may overrate the best alternative and underrate the worst, since these are set to the extreme values. A third disadvantage is that the addition or deletion of a single alternative (the one with the highest or lowest evaluation for a criterion) will change the entire set of ratings.

2. *Subjective evaluation:* Ask the decision maker to rate the alternatives on the 0 to 10 scale. For example, asked to rate alternative 6 for cost, Skiven might give it a 7. The advantages of subjective evaluation include that it is relatively immune to changes in the alternative set, and that it may be more accurate since it includes perceptions of worth that cannot be directly calculated from the criteria measures. Its main disadvantage is that people often make mistakes and give inconsistent evaluations.

For the ratings shown in Table 13.3, the normalization process was used. The overall score is then calculated by summing for each alternative the rating for a criterion multiplied by the weighting for that criterion. From Table 13.3, the total score for alternative 1 is calculated as

$$1 \times 10 + 1.5 \times 0 + 3.5 \times 3 + 4 \times 6 = 44.5$$

It can be seen in Table 13.3 that the highest score is for alternative 10. This means essentially that the greatest total benefit is achieved if alternative 10 is taken.

Also note that a "perfect" alternative, that is, one that rated 10 on every criterion, would have a total score of 100. Thus the 66.5 score for alternative 10 means that it is only about 66.5% of the score of a perfect alternative. The practice of making weights sum to 10 and rating the alternatives on a scale from 0 to 10 is done specifically so that the resulting score can be interpreted as a percentage of the ideal; if this is not desired, any relative weights or rating scale can be used.

Alternative 10 is the best choice for the particular weights and ratings given, but there should be some sensitivity analysis done to verify its robustness. There are several ways to do this sensitivity analysis, but the most sensible is to vary the weights of the criteria to see how the results change. This is easy to do when a spreadsheet is being used to calculate the scores.

Table 13.4 shows a range of criteria weights and the corresponding alternative scores. It can be seen that cameras 4 and 6 also can be identified as best in some of the criteria weight possibilities. For the final recommendation, it may be necessary to review these results with Skiven to let him determine which of the weight possibilities are most appropriate for him.■

Table 13.4 Sensitivity Analysis for the Surveillance Camera

Criterion	Criterion Weights						
Price	1	1	1	1	2	2.5	1
Weight	1.5	2	1	2	2	2.5	1
Picture clarity	3.5	3	3	2	2	2.5	4
Low-light performance	4	4	5	5	4	2.5	4
Alternative	**Alternative Scores**						
Camera 1	44.5	43.0	49.0	46.0	50.0	47.5	46.0
Camera 2	49.3	48.8	48.8	47.8	53.6	57.0	49.8
Camera 4	64.8	65.9	**65.7**	**67.9**	**69.4**	67.2	63.7
Camera 6	64.8	66.2	63.4	66.2	66.8	**68.4**	63.4
Camera 7	59.1	58.3	57.9	56.3	57.8	59.7	59.9
Camera 10	**66.5**	**67.7**	62.0	63.6	58.0	60.0	**66.0**

As has been seen in Example 13.3, the decision matrix approach structures information about multiple objectives of the problem. An additive utility model permits the calculation of an overall score for each alternative. A comparison of the scores permits the best one to be selected. Doing a sensitivity analysis may reveal promising alternatives from relatively small changes in the alternative weight assumptions.

13.4 The Analytic Hierarchy Process

The **analytic hierarchy process (AHP)** is also a MAUT approach. It offers two features beyond what is done in decision matrixes. First, it provides a mechanism for structuring the problem that is particularly useful for large, complex decisions. Second, it provides a better method for establishing the criteria weights.

NET VALUE 13.1

AHP Software

The analytic hierarchy process (AHP) is effective for structuring and analyzing complex, multiattribute decision-making problems. It has a wide range of applications including vendor selection, risk assessment, strategic planning, resource allocation, and human resources management.

AHP has such broad industrial application areas that software has been developed to support its use. Expert Choice™ software has gained acceptance as a useful tool for companies to help them deal with decision-making situations that might otherwise be too complex to structure and solve.

The Expert Choice Web site (www.expertchoice.com) provides several case studies describing the software's use for a number of large international companies, reference materials for AHP, and access to a trial version of the product.

AHP is somewhat more complicated to carry out than decision matrixes. In order to describe the procedure, we first list the basic steps. Example 13.4, which follows the list of steps, explains in more detail the operations at each step. The basic steps of AHP are as follows:

1. Identify the decision to be made, called the **goal**. Structure the goal, criteria, and alternatives into a hierarchy, as illustrated in Figure 13.2. The criteria could be more than one level (not illustrated in Figure 13.2) to provide additional structure to very complex problems.
2. Perform pairwise comparisons for alternatives. **Pairwise comparison** is an evaluation of the importance or preference of a pair of alternatives. Comparisions are made for all possible pairs of alternatives *with respect to each criterion*. This is done by giving each pair of alternatives a value according to Table 13.5 for their relationship for each criterion. These values are placed in a **pairwise comparison matrix (PCM)**.
3. **Priority weights** for the alternatives are calculated by normalizing the elements of the PCM and averaging the row entries. (The columns of priority weights together form a *priority matrix*.)

Figure 13.2 AHP Hierarchy

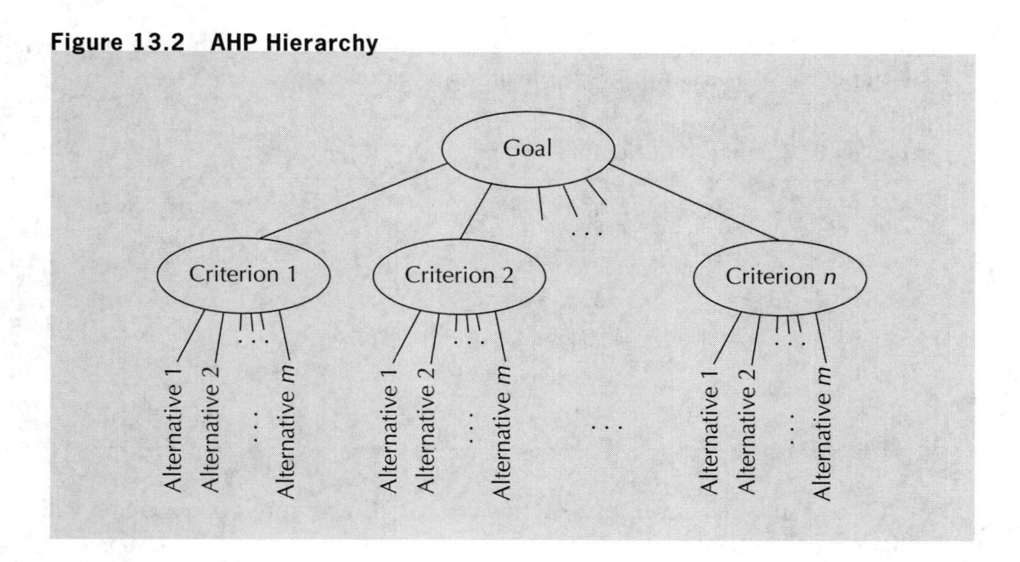

Table 13.5 The AHP Value Scale for Comparison of Two Alternatives or Two Criteria (A and B)

Value	Interpretation
1/9 = 0.111	Extreme preference/importance of B over A
1/7 = 0.1429	Very strong preference/importance of B over A
1/5 = 0.2	Strong preference/importance of B over A
1/3 = 0.333	Moderate preference/importance of B over A
1	Equal preference/importance of A and B
3	Moderate preference/importance of A over B
5	Strong preference/importance of A over B
7	Very strong preference/importance of A over B
9	Extreme preference/importance of A over B
Intermediate values	*For more detail between above values*

4. Perform pairwise comparisons for criteria. As in step 2, all pairs of criteria are compared using the AHP value scale (Table 13.5). A PCM is determined, as in step 3, and priority weights are calculated for the criteria.

5. Alternative priority weights are multiplied by the corresponding criteria priority weights and summed to give an overall alternative ranking. (That is, the priority matrix of alternatives is multiplied by the column of criteria priority weights to give a column of overall evaluations of the alternatives.)

The following example illustrates this process.

Example 13.4

Oksana is examining the cooling of a laboratory at Beaconsfield Pharmaceuticals. She has determined that a 12 000 BTU (British thermal unit) per hour cooling unit is suitable, but there are several models available with different features. The available quantitative data concerning the choices are shown in Table 13.6. The energy efficiency rating is a standard measure of power consumption efficiency.

Table 13.6 Cooling Unit Features

Model	Price	Energy Efficiency Rating
1	$640	9.5
2	$600	9.1
3	$959	10.0
4	$480	9.0
5	$460	9.0

Oksana also has several subjective criteria to take into account. She will use AHP to help her make this decision.

The first step for this problem is to structure the hierarchy. The goal is clear: to choose a cooling unit. The alternatives are also known, and are listed in Table 13.6. After some consideration, Oksana concludes that the following are critical to her consideration:

(1) Cost
(2) Energy consumption
(3) Loudness
(4) Perceived comfort

The resulting hierarchy is illustrated in Figure 13.3.

The second step is to construct a PCM for the alternatives with respect to each criterion. For illustration purposes, we will do this step for the criterion "perceived comfort" only.

Oksana first considers two of the alternatives only, say 1 and 2, with respect to "perceived comfort." She gives the preferred one (which is alternative 1) a rating from the scale shown in Table 13.5. In this case, Oksana judges that alternative 1 is moderately

Figure 13.3 The AHP Hierarchy for Example 13.4

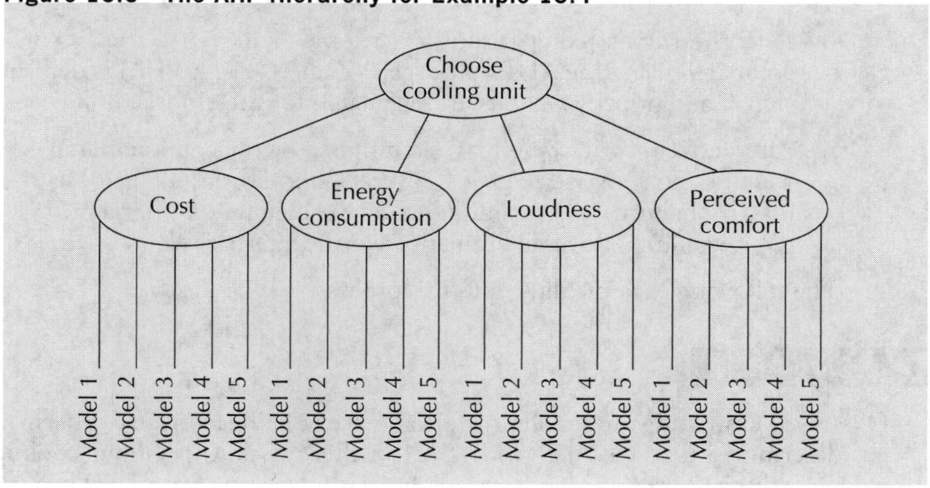

better than 2; the rating is a 3. This appears in the PCM shown in Figure 13.4 in row 1, column 2, corresponding to alternative 1 and alternative 2. Correspondingly, the reciprocal, 1/3, is put in row 2, column 1. This can be interpreted loosely as indicating that alternative 1 is three times as desirable as alternative 2 for perceived comfort, and correspondingly criterion 2 is 1/3 as desirable as alternative 1.

As another example, consider the comparison of alternatives 3 and 4. Alternative 4 is strongly preferred to 3 in "perceived comfort," so a 5 appears in row 4, column 3, and 1/5 appears in row 3, column 4. Similar comparisons of all pairs complete the PCM in Figure 13.4 for "perceived comfort." In summary, row 1 of Figure 13.4 shows the results of comparing alternative 1 with alternatives 1, 2, 3, 4, and 5. Row 2 shows the results of comparing alternative 2 with alternatives 1, 2, 3, 4, and 5, and so on. Note that an alternative is equally preferred to itself, so all main diagonal entries are 1. PCMs for the other three criteria are developed in exactly the same manner.

Figure 13.4 PCM for Perceived Comfort

$$\begin{bmatrix} 1 & 3 & 5 & 1 & 3 \\ \frac{1}{3} & 1 & 3 & \frac{1}{3} & 1 \\ \frac{1}{5} & \frac{1}{3} & 1 & \frac{1}{5} & \frac{1}{3} \\ 1 & 3 & 5 & 1 & 3 \\ \frac{1}{3} & 1 & 3 & \frac{1}{3} & 1 \end{bmatrix}$$

The next step is to determine priority weights for the alternatives. This is done by first normalizing the columns of the PCM and then averaging the rows. To normalize the columns, sum the column entries, and divide each entry by this sum. To average the rows, sum the rows (after normalizing) and divide by the number of entries per row.

For example, the sum of column 1 of the PCM in Figure 13.4 is 2.866. Then the normalized entries for the column will be 1/2.866, 0.333/2.866, etc. The complete normalized PCM for "perceived comfort" is shown in Figure 13.5. The priority weights are then calculated as the average of each row of the normalized PCM, and are also illustrated in Figure 13.5.

Figure 13.5 Normalized PCM for Perceived Comfort

$$\underbrace{\begin{bmatrix} 0.349 & 0.360 & 0.294 & 0.349 & 0.360 \\ 0.116 & 0.120 & 0.176 & 0.116 & 0.120 \\ 0.070 & 0.040 & 0.059 & 0.070 & 0.040 \\ 0.349 & 0.360 & 0.294 & 0.349 & 0.360 \\ 0.116 & 0.120 & 0.176 & 0.116 & 0.120 \end{bmatrix}}_{\text{Normalized PCM}} \quad \underbrace{\begin{bmatrix} 0.342 \\ 0.130 \\ 0.056 \\ 0.342 \\ 0.130 \end{bmatrix}}_{\text{Average}}$$

A similar process can be carried out for the other three criteria. The four columns (one for each criterion) of priority weights form a priority matrix, shown in Figure 13.6. The first column of this matrix consists of the priority weights for cost, the second for energy efficiency, the third for noise, and the fourth for perceived comfort.

Figure 13.6 Priority Matrix for Example 13.4

$$\begin{bmatrix} 0.90 & 0.256 & 0.033 & 0.342 \\ 0.114 & 0.230 & 0.468 & 0.130 \\ 0.031 & 0.338 & 0.282 & 0.056 \\ 0.383 & 0.088 & 0.086 & 0.342 \\ 0.383 & 0.088 & 0.131 & 0.130 \end{bmatrix}$$

The next step is to construct a PCM for the criteria themselves. This is done in the same manner as for the alternatives for each criterion, except that now one rates the criteria in pairwise comparisons with each other. The PCM Oksana creates is illustrated as Figure 13.7, with the rows and columns in the order: cost, energy consumption, loudness, and perceived comfort. For example, energy consumption is moderately more important than energy cost. Thus, there is a 3 in row 2, column 1 and a 1/3 in row 1, column 2. The normalized PCM and row averages are shown in Figure 13.8.

Figure 13.7 PCM for Goal

$$\begin{bmatrix} 1 & \frac{1}{3} & 5 & 1 \\ 3 & 1 & 7 & 3 \\ \frac{1}{5} & \frac{1}{7} & 1 & \frac{1}{5} \\ 1 & \frac{1}{3} & 5 & 1 \end{bmatrix}$$

Figure 13.8 Normalized PCM and Average Values for Goal

$$\begin{bmatrix} 0.192 & 0.184 & 0.278 & 0.192 \\ 0.577 & 0.553 & 0.389 & 0.577 \\ 0.038 & 0.079 & 0.056 & 0.038 \\ 0.192 & 0.184 & 0.278 & 0.192 \end{bmatrix} \qquad \overset{\textbf{Average}}{\begin{bmatrix} 0.212 \\ 0.524 \\ 0.053 \\ 0.212 \end{bmatrix}}$$

The order of the rows and columns of Figures 13.7 and 13.8 are: cost, energy efficiency, noise, and perceived comfort. Thus, the criterion with the highest priority rating is noise, at 0.524, then cost and perceived comfort identical at 0.212, and finally energy efficiency last at 0.053.

The final stage of the process consists of determining an overall score for each alternative. Note that the entire process of AHP has essentially led to the development of a decision matrix: the priority ratings for the criteria are the weights, while the priority ratings for the alternatives are the ratings of the alternatives for the criteria. Consequently, the final score is determined by multiplying each alternative priority rating by the appropriate criterion rating and then summing.

This can also be viewed as matrix multiplication of the priority matrixes for the alternatives by the column of priority weights of the criteria, as shown in Figure 13.9. The interpretation of the column vector on the right in Figure 13.9 is a ranking of the alternatives. The best alternative is number 1, followed by number 3, 4, and 2; number 5 is the worst.

Figure 13.9 Final Alternative Scores for Example 13.4

$$\begin{bmatrix} 0.090 & 0.256 & 0.033 & 0.342 \\ 0.114 & 0.230 & 0.468 & 0.130 \\ 0.031 & 0.338 & 0.282 & 0.056 \\ 0.383 & 0.088 & 0.086 & 0.342 \\ 0.383 & 0.088 & 0.131 & 0.130 \end{bmatrix} \times \begin{bmatrix} 0.212 \\ 0.524 \\ 0.053 \\ 0.212 \end{bmatrix} = \begin{bmatrix} 0.227 \\ 0.197 \\ 0.211 \\ 0.204 \\ 0.162 \end{bmatrix}$$

In conclusion, the best cooling unit is model 1. Oksana should buy this one for the laboratory.■

13.5 The Consistency Ratio for AHP

The subjective evaluation of the PCMs can be inconsistent. For example, Joe can say that alternative 1 is five times as important as alternative 2, and alternative 2 is five times as important as alternative 3, but then claim that alternative 1 is only twice as important as alternative 3. Or he might even say alternative 1 is less important than alternative 3.

The fact that the construction of PCMs includes redundant information is useful because it helps get a good estimate of the best rating for the alternative. However, there has to be a check made that the decision maker is being consistent.

A measure called the **consistency ratio** (to measure the consistency of the reported comparisons) can be calculated for any PCM. The consistency ratio ranges from 0 (perfect consistency) to 1 (no consistency). A consistency ratio of 0.1 or less is considered acceptable in practice. The calculation of the consistency ratio is briefly reviewed in Appendix 13A.

REVIEW PROBLEMS

REVIEW PROBLEM 13.1

Contrex makes thermostat controls for baseboard heaters. As part of the control manufacturing process, a two-centimetre steel diaphragm is fitted to a steel cup. The diaphragm is used to open a safety switch rapidly to avoid arcing across the contacts. Currently the cup is seam-welded, which is both expensive and a source of quality problems. The company wants to explore the use of adhesives to replace the welding process. Table 13.7 lists the ones examined, along with various properties for each.

High-temperature resistance is desirable, as are tensile bond strength and pressure resistance. Fast curing speeds are desirable to reduce work-in-progress inventory storage costs, and, of course, cheaper material costs are important.

Contrex wants to select a single adhesive type for comparison experiments against the current seam-welding method.

Table 13.7 Possible Adhesives for Thermostat Control

Adhesive	Maximum Temperature (°C)	Tensile Bond Strength (kPa)	Pressure Resistance (kPa)	Curing Speed	Cost
Acrylic	106	21 000	3738	Medium	Cheap
Silicone	200	3 150	560	Slow	Medium
Cyanoacrylate	250	3 500	630	Fast	Cheap
Methacrylate A	225	28 000	4984	Slow	Expensive
Methacrylate B	225	7 000	1246	Medium	Expensive

(a) Are any of the listed adhesives in Table 13.7 inefficient?

(b) Discussions with management indicate that the criteria can be weighted as follows:

Temperature resistance	1.5
Bond strength	1.5
Pressure resistance	2.5
Curing speed	3.5
Cost	1.0
Total	10.0

Create a decision matrix for this problem using the above weights. Normalize the data in Table 13.7 to estimate the ratings of each alternative for each criterion. Use only the efficient alternatives. Use the maximum temperature figures to measure temperature resistance. For curing speed, set fast as 8, medium as 5, and slow as 2, while for cost, set expensive as 2, medium as 5, and cheap as 10. Under these conditions, which is the recommended adhesive?

(c) The analytic hierarchy process (AHP) was performed for this problem. The hierarchy is shown in Figure 13.10. The priority matrix in Figure 13.11 represents the results from PCMs calculated for the different criteria. The rows of

Figure 13.11 correspond to the alternatives acrylic, cyanoacrylate, methacrylate A, and methacrylate B, respectively, while the columns correspond to temperature resistance, bond strength, pressure resistance, curing speed, and cost, respectively. A PCM for the goal is shown in Figure 13.12, with the criteria in the same order as for Figure 13.11. With this information, what is the best adhesive to recommend for the experiment?

Figure 13.10 The AHP Hierarchy for Review Problem 13.1(c)

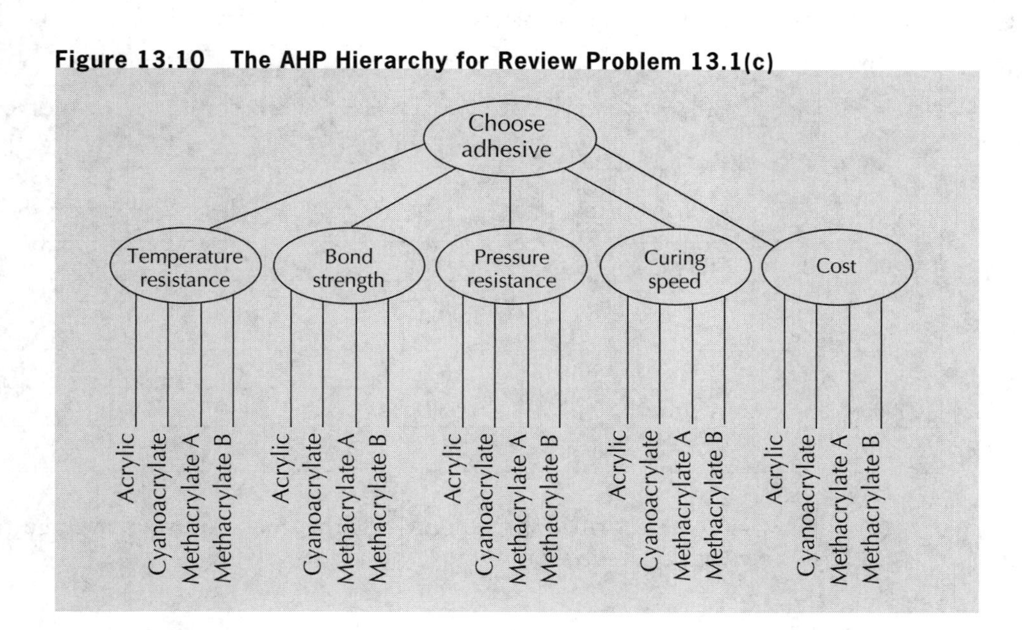

Figure 13.11 Priority Matrix for Review Problem 13.1(c)

$$\begin{bmatrix} 0.095 & 0.368 & 0.392 & 0.213 & 0.341 \\ 0.331 & 0.077 & 0.067 & 0.502 & 0.278 \\ 0.287 & 0.445 & 0.438 & 0.061 & 0.188 \\ 0.287 & 0.110 & 0.103 & 0.224 & 0.193 \end{bmatrix}$$

Figure 13.12 PCM for Goal for Review Problem 13.1(c)

$$\begin{bmatrix} 1 & 1 & \frac{1}{2} & \frac{1}{3} & 2 \\ 1 & 1 & \frac{1}{2} & \frac{1}{3} & 2 \\ 2 & 2 & 1 & \frac{1}{2} & 3 \\ 3 & 3 & 2 & 1 & 4 \\ \frac{1}{2} & \frac{1}{2} & \frac{1}{3} & \frac{1}{4} & 1 \end{bmatrix}$$

ANSWER

(a) It can be observed that silicone is dominated by cyanoacrylate in all criteria, and therefore is inefficient.

(b) As shown in Table 13.8, under the weighting and rating conditions specified, the methacrylate A adhesive is clearly best.

Table 13.8 Decision Matrix of Adhesives

Criterion	Criterion Weight	Alternatives			
		Acrylic	Cyanoacrylate	Methacrylate A	Methacrylate B
Temperature resistance	1.5	0.0	10.0	8.3	8.3
Bond strength	1.5	7.1	0.0	10.0	1.4
Pressure resistance	2.5	7.1	0.0	10.0	1.4
Curing speed	3.5	5.0	8.0	2.0	5.0
Cost	1.0	10.0	10.0	2.0	2.0
Score	10.0	56.1	53.0	61.4	37.6

(c) First we have to normalize the PCM for the goal, and then average to get the criteria weights, as shown in Figure 13.13.

Figure 13.13 Normalized PCM for Goal for Review Problem 13.1(c)

$$\text{Normalized PCM} \quad \begin{bmatrix} 0.133 & 0.133 & 0.115 & 0.138 & 0.167 \\ 0.133 & 0.133 & 0.115 & 0.138 & 0.167 \\ 0.267 & 0.267 & 0.231 & 0.207 & 0.250 \\ 0.4 & 0.4 & 0.462 & 0.414 & 0.333 \\ 0.067 & 0.067 & 0.077 & 0.130 & 0.083 \end{bmatrix} \quad \text{Average} \quad \begin{bmatrix} 0.137 \\ 0.137 \\ 0.244 \\ 0.402 \\ 0.079 \end{bmatrix}$$

Then we multiply the priority matrix by the average criteria weights, as illustrated in Figure 13.14.

Since the second alternative in Figure 13.14 has the highest net weight, it is preferred. The recommended adhesive using AHP is the cyanoacrylate. This disagrees with the result obtained using decision matrixes.

Figure 13.14 Calculating Alternative Weights for Review Problem 13.1(c)

$$\begin{bmatrix} 0.095 & 0.368 & 0.392 & 0.213 & 0.341 \\ 0.331 & 0.077 & 0.067 & 0.502 & 0.278 \\ 0.287 & 0.445 & 0.438 & 0.061 & 0.188 \\ 0.287 & 0.110 & 0.103 & 0.224 & 0.193 \end{bmatrix} \times \begin{bmatrix} 0.137 \\ 0.137 \\ 0.244 \\ 0.402 \\ 0.079 \end{bmatrix} = \begin{bmatrix} 0.272 \\ 0.296 \\ 0.247 \\ 0.185 \end{bmatrix}$$

SUMMARY

In this chapter, three approaches for dealing explicitly with multiple criteria were presented. The first, *efficiency*, allows the identification of alternatives that are not dominated by others. An alternative is dominated if there is another alternative at least as good with respect to all criteria, and better in at least one.

The second approach, *decision matrixes*, is in wide usage. Decision matrixes are a multi-attribute utility theory (MAUT) method in which criteria are subjectively weighted and then multiplied by subjectively evaluated criteria ratings to give an overall score. The weights sum to 10 and the criteria ratings are on a scale from 0 to 10, resulting in an overall score that could range from 0 to 100. The alternative with the highest score is best, and the value of the score can be considered a percentage of an "ideal" alternative.

The third approach presented in this chapter is the *analytic hierarchy process*. AHP is also a MAUT method. Pairwise comparisons are used to extract criterion weights in a more rigorous manner than for decision matrixes. Pairwise comparisons similarly are used to rate alternatives. Multiplying criterion weights and alternative ratings gives an overall evaluation for each alternative.

Engineering Economics in Action, Part 13B: Moving On

Three months later Naomi was seated in Anna Kulkowski's office. The Powerluxe project report had been submitted two days before.

"Naomi, the work you and Bill did and your report are first-rate," Anna said. "We're going to start negotiations with Powerluxe to bring the Adaptamatic line of vacuum cleaners to market. I had no idea anybody could make such clear recommendations on such a complex problem. Congratulations on a good job."

Naomi thought back on all the people who had helped her in her almost two years at Canadian Widgets: how Clem had taught her practical problem solving; how Dave had shown her the ropes; how Terry had helped her realize the benefits of attention to detail. Bill had shown her how the real world mixed engineering with marketing, business, and government. Anna, too, had shown her how to manage people. "Thank you very much, Ms. Kulkowski," Naomi responded, a small break in her voice betraying her emotion.

"Do you enjoy this kind of work?" Anna asked.

"Yes, I do," Naomi replied. "It's exciting to see how the engineering relates to everything else."

"I think we have a new long-term assignment for you," Anna said. "This is just the first step for Canadian Widgets in developing new products. We have a first-rate team of engineers. We want to make better use of them. Ed Burns is going to head up a product development group. He read your report on the Adaptamatic line of vacuum cleaners and was quite impressed. He and I would like you in that development group. We need someone who understands the engineering and can relate it to markets. What do you say?"

"I'm in!" Naomi had a big grin on her face.

A few days later, Naomi answered the phone.

"Hey, Naomi, it's Terry. I hear you got promoted!"

"Hi, Terry. Nice to hear from you. Well, the money's about the same, but it sure will be interesting. How are things with you?" Naomi had fond memories of working with Terry.

"Well, I graduate next month, and I have a job. Guess where?"

Naomi knew exactly where. Clem had told her about the interviews. It wasn't really fair to the other candidates—Clem had decided to hire Terry as soon as he had applied. "Here? Really?"

PROBLEMS

13.1 The table below shows information about alternative choices. Criteria C and E are to be minimized, while all the rest are to be maximized. Which of the alternatives can be eliminated from further consideration?

	Criteria				
Alternative	**A**	**B**	**C**	**D**	**E**
1	340	5	11	1.2	1
2	570	8	22	3.3	1
3	410	9	22	3.2	2
4	120	4	36	0.9	3
5	122	1	46	1.3	2
6	345	8	47	0.6	3
7	119	4	57	1.1	2
8	554	2	89	2.1	3
9	317	9	117	0.9	1
10	129	5	165	1.5	3

13.2 The following is a partially completed pairwise comparison matrix. Complete it. What are the corresponding priority weights?

$$\begin{bmatrix} - & \frac{1}{2} & - & \frac{1}{9} \\ - & - & - & - \\ 4 & 2 & - & \frac{1}{2} \\ - & 4 & - & - \end{bmatrix}$$

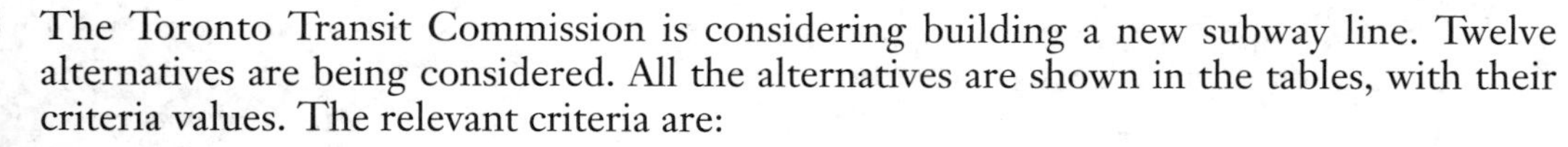

13.3 The Toronto Transit Commission is considering building a new subway line. Twelve alternatives are being considered. All the alternatives are shown in the tables, with their criteria values. The relevant criteria are:

C1 Population and jobs served per kilometre
C2 Projected daily traffic per kilometre
C3 Capital cost per kilometre (in millions of dollars)
C4 IRR
C5 Structural effect on urbanization

It is desirable to have high population served, high traffic, low capital cost per kilometre, and a high IRR. Criterion C5 concerns the benefits for urban growth caused by the subway location, and is measured on a scale from 0 to 10, with the higher values being preferred.

	Alternative Subway Lines					
Criterion	**L1**	**L2**	**L3**	**L4**	**L5**	**L6**
C1	81 900	31 800	11 500	31 100	23 000	16 100
C2	25 500	11 600	7 100	10 500	10 200	3 500
C3	65	45	29	35	40	10
C4	8.6	6.3	4.5	14.1	13.	11.8
C5	3	7	4	6	5	4

	Alternative Subway Lines					
Criterion	**L7**	**L8**	**L9**	**L10**	**L11**	**L12**
C1	13200	28 200	36 500	24 400	18 400	13 900
C2	3 700	7 400	10 300	7 100	4 700	3 100
C3	32	30	13.2	40	43	25
C4	3.9	6	3	3.7	3.7	5.8
C5	9	10	6	9	5	4

(a) Which of these alternatives are efficient?

(b) Establish ratings for each of the efficient alternatives from part (a) for a decision matrix through normalization. The weights of the five criteria are C1: 1.5, C2: 2, C3: 2.5, C4: 3, C5: 1. Construct a decision matrix and determine the best subway route.

(c) If the criteria weights were C1: 2, C2: 2, C3: 2, C4: 2, C5: 2, would the recommended alternative be different?

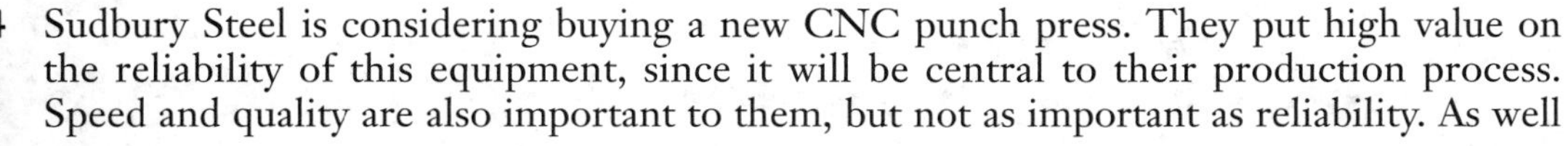

13.4 Sudbury Steel is considering buying a new CNC punch press. They put high value on the reliability of this equipment, since it will be central to their production process. Speed and quality are also important to them, but not as important as reliability. As well as examining these factors, the company will make sure that the equipment will be easily adaptable to changes in production.

Construct a decision matrix for the alternatives listed and appropriate criteria. Select appropriate weightings for the criteria (state any assumptions), and determine the preferred alternative. Do a reasonable sensitivity analysis to determine the conditions under which the choice of alternative may change. A spreadsheet program should be used.

The alternatives are as follows:

1. Name: Accumate Plasmapress
   Cost: $428 600
   Service: Average
   Reliability: Average
   Speed: High
   Quality: Good
   Flexibility: Excellent
   Size: Average
   Other: None
2. Name: Weissman Model 4560
   Cost: $383 765
   Service: Below average
   Reliability: Average
   Speed: High
   Quality: Fair
   Flexibility: Not good
   Size: Average
   Other: Many tool stations (desirable), but small tools (not desirable)
3. Name: A. D. Hockley Model 661-84
   Cost: $533 725
   Service: Untested
   Reliability: Average
   Speed: Slow
   Quality: Very good
   Flexibility: Very good
   Size: Very compact
   Other: Small turret (not desirable)
4. Name: Frammit Manu-Centre 1500/45
   Cost: $393 000
   Service: Average
   Reliability: Below average
   Speed: Average
   Quality: Average
   Flexibility: Average
   Size: Average

Other:	Has perforating and coining feature (very desirable) Poor torch design causes too rapid wear (undesirable)

5.

Name:	Frammit Manu-Centre Lasertool 1250/30/1500
Cost:	$340 056
Service:	Average
Reliability:	Average
Speed:	Exceptionally fast
Quality:	Exceptionally good
Flexibility:	Average
Size:	Undesirably small
Other:	Cannot handle heavy-gauge metal (not desirable)

6.

Name:	Boxcab 3025/12P CNC
Cost:	$405 232
Service:	Untested
Reliability:	Excellent
Speed:	Average
Quality:	Very good
Flexibility:	Average
Size:	Average
Other:	None

13.5 Complete each of the following pairwise comparison matrixes:

(a)
$$\begin{bmatrix} - & 1 & - & - \\ - & - & - & 2 \\ \frac{1}{5} & 3 & - & - \\ 3 & - & \frac{1}{2} & - \end{bmatrix}$$

(b)
$$\begin{bmatrix} - & 9 & - & 1 & 3 \\ - & - & 3 & - & - \\ \frac{1}{7} & - & - & \frac{1}{4} & 2 \\ - & 6 & - & - & - \\ - & 1 & - & 2 & - \end{bmatrix}$$

13.6 For each of the PCMs in Problem 13.5, compute priority weights for the alternatives.

13.7 (a) Francis has several job opportunities for his co-op work term. He would like a job with good pay that is close to home, contributes to his engineering studies, and is with a smaller company. Which of the opportunities listed below should be removed from further consideration?

	Job	Pay	Home	Studies	Size
		Criteria			
1.	Spinoff Consulting	1700	2	3	5
2.	Nub Automotive	1600	5	3	500
3.	Soutel	2200	80	4	150
4.	Provincial Hydro	1800	100	3	3000
5.	Fitzsimon Associates	1700	100	1	20
6.	General Auto	2000	150	2	2500
7.	Ring Canada	2200	250	5	300
8.	Jones Mines	2700	500	3	20
9.	Resources, Inc.	2700	2000	2	40

Pay: Monthly salary in dollars
Home: Distance from home in kilometres
Studies: Contribution to engineering studies, 0 = none, 5 = a lot
Size: Number of employees at that location

(b) Francis feels that the following weights can represent the importance of his four criteria:

Pay:	4.0
Home:	2.5
Studies:	2.0
Size:	1.5

Using normalization to establish the ratings, which job is best?

Problems 13.8 to 13.23 are based on the following situation:

John is considering the selection of a consultant to provide ongoing support for a computer system. John wishes to contract with one consultant only, based on the criteria of cost, reliability, familiarity with equipment, location, and quality. Cost is measured by the quoted daily rate. Reliability and quality are measured on a qualitative scale based on discussions with references and interviews with the consultants. Familiarity with equipment has three possibilities: none, some, or much. Location is measured by distance from the consultant's office to the plant site in kilometres.

The specific data for the consultants are as follows:

Criterion	A	B	C	D	E
			Consultants		
Cost	$500	$500	$450	$600	$400
Reliability	Good	Good	Excellent	Good	Fair
Familiarity	None	Some	Some	Much	Some
Location	3	1	5	2	1
Quality	Excellent	Good	Fair	Excellent	Good

13.8 Is any choice of consultant inefficient?

13.9 John can assign weights to the criteria as follows:

Criterion	Weight
Cost	2
Reliability	3
Familiarity	1
Location	1
Quality	3

Using the following tables to convert qualitative evaluations to numbers and using the formula for determining ratings by normalization, construct a decision matrix for choosing a consultant. Which consultant is best?

For Reliability and Quantity	
Description	**Value**
Excellent	5
Very Good	4
Good	3
Fair	2
Poor	1

For Familiarity	
Description	**Value**
None	1
Some	2
Much	3

13.10 Draw an AHP hierarchy for this problem.

13.11 John considers a cost difference of \$100 or more to be of strong importance, and a difference of \$50 to be of moderate importance. Construct a PCM for the criterion *cost*.

13.12 John considers the difference between good reliability and fair reliability to be of moderate importance, and the difference between good and excellent to be of strong importance. He considers the difference between fair and excellent to be of very strong importance. Construct a PCM for the criterion *reliability*.

13.13 John considers the difference between no familiarity with the computing equipment and some familiarity to be of strong importance, and the difference between none and much familiarity to be of extreme importance. He considers the difference between some and much to be of strong importance. Construct a PCM for the criterion *familiarity*.

13.14 John considers each kilometre of distance worth two units on the AHP value scale. For example, the value of the location of consultant D over consultant C is $(5 - 2) \times 2 = 6$, and a 6 would be placed in the fourth row, third column, of the PCM for the criterion *location*. Construct the complete PCM for the criterion *location*.

13.15 John considers the difference between fair quality and good quality to be of strong importance, and the difference between good and excellent to be of very strong importance. He considers the difference between fair and excellent to be of extreme importance. Construct a PCM for the criterion *quality*.

13.16 Calculate the priority weights for the criterion *cost* from the PCM constructed in Problem 13.11.

13.17 Calculate the priority weights for the criterion *reliability* from the PCM constructed in Problem 13.12.

13.18 Calculate the priority weights for the criterion *familiarity* from the PCM constructed in Problem 13.13.

13.19 Calculate the priority weights for the criterion *location* from the PCM constructed in Problem 13.14.

13.20 Calculate the priority weights for the criterion *quality* from the PCM constructed in Problem 13.15.

13.21 Construct the priority matrix from the answers to Problems 13.16 to 13.20.

13.22 A partially completed PCM for the criteria is shown below. Complete the PCM and calculate the priority weights for the criteria.

$$\begin{bmatrix} - & \frac{1}{3} & - & 3 & - \\ - & - & - & - & 1 \\ \frac{1}{3} & \frac{1}{7} & - & 1 & - \\ - & \frac{1}{6} & - & - & - \\ 1 & - & 7 & 7 & - \end{bmatrix}$$

13.23 Determine the overall score for each alternative. Which consultant is best?

Problems 13.24 to 13.31 are based on the following case:

Fabian has several job opportunities for his co-op work term. The pay is in dollars per month. The distance from home is in kilometres. Relevance to studies is on a five-point scale, 5 meaning very relevant to studies and 0 meaning not relevant at all. The company size refers to the number of employees at the job location.

In general, Fabian wants a job with good pay, that is close to home, contributes to his engineering studies, and is with a smaller company.

		Criteria			
	Job	**Pay**	**Distance from Home**	**Relevance to Studies**	**Company Size**
1.	Spinoff Consulting	1700	2	3	5
2.	Soutel	2200	80	4	150
3.	Ring Canada	2200	250	5	300
4.	Jones Mines	2700	500	3	20

13.24 Draw an AHP hierarchy for this problem.

13.25 **(a)** Complete the PCM below for *Pay*.

$$\begin{bmatrix} - & \frac{1}{5} & \frac{1}{5} & \frac{1}{7} \\ - & - & 1 & \frac{1}{5} \\ - & - & - & \frac{1}{5} \\ - & - & - & - \end{bmatrix}$$

(b) What are the priority weights for *Pay*?

13.26 (a) Complete the PCM below for *Distance from Home*.

$$\begin{bmatrix} - & - & - & - \\ \frac{1}{5} & - & - & - \\ \frac{1}{7} & \frac{1}{5} & - & - \\ \frac{1}{9} & \frac{1}{7} & \frac{1}{2} & - \end{bmatrix}$$

(b) What are the priority weights for *Distance from Home*?

13.27 (a) Complete the PCM below for *Studies*.

$$\begin{bmatrix} - & - & \frac{1}{3} & - \\ 2 & - & - & 2 \\ - & 2 & - & - \\ 1 & - & \frac{1}{3} & - \end{bmatrix}$$

(b) What are the priority weights for *Studies*?

13.28 (a) Complete the PCM below for *Size*.

$$\begin{bmatrix} - & - & - & 1 \\ \frac{1}{7} & - & - & \frac{1}{7} \\ \frac{1}{9} & \frac{1}{2} & - & \frac{1}{9} \\ - & - & - & - \end{bmatrix}$$

(b) What are the priority weights for *Size*?

13.29 Form a priority matrix for the PCMs in Problems 13.25 to 13.28.

13.30 Given the following PCM for the criteria, calculate the priority weights for the criteria.

$$\begin{bmatrix} 1 & 3 & 5 & 5 \\ \frac{1}{3} & 1 & 3 & 3 \\ \frac{1}{5} & \frac{1}{3} & 1 & 2 \\ \frac{1}{5} & \frac{1}{3} & \frac{1}{2} & 1 \end{bmatrix}$$

13.31 Using the results of Problems 13.29 and 13.30, calculate the priority weights for the goal. Which job is Fabian's best choice? His second best choice?

Northwind Stoneware

CANADIAN 13.1 MINI-CASE

Northwind Stoneware of Kitchener, Ontario makes consumer stoneware products. Stoneware is fired in a kiln, which is an enclosure made of a porous brick having heating elements designed to raise the internal temperature to over 1200°C. Clay items such as stoneware can be hardened by firing them, following a particular temperature pattern called a firing curve.

Quality and cost problems led Northwind to examine better ways to control the firing curve. Alternatives available to them included

1. Direct human control (the temperature sensitivity of the human eye cannot be matched by any automatic control)
2. Use of a KilnSitter, a mechanical switch that shuts off the kiln at a preset temperature
3. Use of a pyrometer, an electrical instrument for measuring heat, with a programmable controller
4. Use of a pyrometer and a computer

The criteria used to determine the best choice included installation cost, effectiveness, reliability, energy savings, maintenance costs, and other applications.

A decision matrix evaluation method was used. Under a variety of criteria weights, the use of a pyrometer and a computer was the recommended choice. The exception was when installation cost was given overwhelming weight; in this case the use of a KilnSitter was recommended.

Discussion

Real life is always more complicated than any model used for analysis. In reading about a case in which decisions are made, the complexity of the real decision process is always hidden, because the very process of describing the situation is itself a model. In describing something, choices are made about what is important and worth describing and what is unimportant and not worth describing. In real life, one has to go through a process of separating from a great mass of information exactly what is important and what is not important.

The process of solving the quality and cost problems for Northwind Stoneware first involved collecting a great deal of information about the manufacturing process. Many people were interviewed, production was observed, and technical details of stoneware chemistry were researched. Market analysis was required to identify possible solutions to the process. Several person-weeks of work were required to gather the information to create the decision matrix mentioned above.

We often concentrate on the mechanics of using a decision tool instead of the mechanics of getting good data to use as input to the tool. In many cases, getting good data is 90% of the solution.

Questions

Think of an everyday decision situation that you have been involved in recently: it might be deciding which novel to buy, which apartment to rent, or which movie to go to.

1. Write down a one-page description of the decision. Include where you were, what choices you had, what you chose to do, and why.

2. Think about how you decided what to write for Question 1. Was it easy to know what to include in your description, and what not to include?
3. Can you immediately identify alternatives and criteria, as formally defined in this chapter, from your one-page description? If not, why not?
4. Construct a formal decision matrix model for your problem, augmenting the alternatives and criteria if necessary. Fill in the weights and ratings. Comment on any difficulties you have coming up with the weights and ratings.
5. Calculate the overall score for your decision matrix. Did you actually choose the one with the highest score? If not, why not?

Appendix 13A Calculating the Consistency Ratio for AHP

The **consistency ratio** (*CR*) for AHP provides a measure of the ability of a decision maker to report preferences over alternatives or criteria. Calculating the *CR* can be done without understanding the concepts underlying it, but some background can be helpful. In this appendix, a brief overview of the basis for the *CR* is given, but the main purpose is to present an algorithm for calculating the *CR*. For the background information, some understanding of linear algebra is desirable, but the algorithm for calculating the *CR* can be easily followed without this background information.

Recall that in AHP a pairwise comparison matrix (PCM) is developed by comparing, for example, the alternatives with respect to one criterion. If alternative x is compared with y and found to be twice as preferred, and y is compared with z and is three times as preferred, then it is easy to deduce that x should be six times as preferred as z. However, in AHP every pairwise comparison is made giving several judgments about the same relationship. Humans will not necessarily be perfectly consistent in reporting preferences; one of the strengths of AHP is that, by getting redundant preference information, the quality of the information is improved.

Observe that if a decision maker were perfectly consistent, the columns of a PCM would be multiples of each other. They would differ in scale because for each column n, the nth row is fixed as 1, but the relative values would be constant.

Recall that an eigenvalue λ is one of n solutions to the equation

$$\mathbf{Aw} = \lambda \mathbf{w}$$

where $\mathbf{A}$ is a square $n \times n$ matrix and $\mathbf{w}$ is an $n \times 1$ eigenvector.

There is one eigenvector corresponding to each eigenvalue λ. A PCM is unusual in that, in addition to being a square matrix, all of the entries are positive, the corresponding entries across the main diagonal are reciprocals, and there are 1s along the main diagonal. It can be shown that, in this case, if the decision maker is perfectly consistent, there will be a single non-zero eigenvalue, and $n - 1$ eigenvalues of value 0.

In practice, a PCM is not perfectly consistent. However, assuming that the error is relatively small, there should be one large eigenvalue λ_{max} and $n - 1$ small ones. Further, it can be shown that, with small inconsistencies, $\lambda_{max} > n$.

The *CR* is developed from the difference between λ_{max} and n. First, this difference is divided by $n - 1$, effectively distributing it over the other, supposedly zero, eigenvalues. This gives the **consistency index (CI)**.

$$CI = \frac{\lambda_{max} - n}{n - 1}$$

Second, the *CI* is divided by the *CI* of a random matrix of the same size, called a *random index* (*RI*), to form the consistency ratio (*CR*). The *RI* for matrixes of up to 10 rows and columns is shown in Table 13A.1; these were developed by averaging the *CI*s for hundreds of randomly generated matrixes.

$$CR = \frac{CI}{RI}$$

The idea is that, if the *PCM* were completely random, we would expect the *CI* to be equal to the *RI*. Thus, a random *PCM* would have a *CR* of 1. However, a consistent *PCM*, having a *CI* of 0, would also have a *CR* of 0. *CR*s of between 0 and 1 indicate how consistent or random the PCM is, in a manner that is independent of the size of the matrix.

Table 13A.1 Random Indexes for Various Sizes of Matrixes

Size ($n \times n$)	Random Index
2	0
3	0.58
4	0.90
5	1.12
6	1.24
7	1.32
8	1.41
9	1.45
10	1.49

The *CR* is actually a well-designed statistical measure of the deviation of a particular PCM from perfect consistency. As a common rule, an upper limit of 0.1 is usually used. Thus, if $CR \leq 0.1$, the PCM is *acceptably* consistent. If the $CR > 0.1$, the PCM should be reevaluated by the decision maker.

This background can be used to construct an algorithm for calculating the *CR* for any PCM. Essentially, one determines λ_{max} and its associated eigenvector, called the **principal eigenvector**. The general algorithm for a PCM **A** of size $n \times n$ is as follows:

1. Find **w**, the eigenvector associated with λ_{max}, from normalizing the following:

$$\mathbf{w} = \lim_{k \to \infty} \frac{\mathbf{A}^k \mathbf{e}}{n}$$

where **e** is a column vector $[111 \ \ldots \ 111]^T$.

In other words, form a sequence of powers of the matrix **A**. Normalize each of the resulting matrixes by dividing each element of **A** by the sum of the elements from the corresponding column. Form **w** by summing each row of normalized **A** and dividing this by n. Eventually, **w** will not noticeably change from one power of **A** to the next-higher power. This value of **w** is the desired principal eigenvector.

In the main part of this chapter, we calculated the priority weights for a PCM. That procedure was an approximate method of determining **w**; the priority weights and the eigenvector associated with λ_{max} are the same thing. The procedure mentioned here is more accurate, but more difficult to compute and time-consuming.

2. Since **w** is a vector, λ_{max} can be found by solving $\mathbf{Aw} = \lambda_{max}\mathbf{w}$ for any element of **w**, or equivalently, any row of **A**. Thus, for any i, compute

$$\lambda_{max} = \frac{\sum_j (a_{ij} \times w_j)}{w_i}$$

3. Calculate the *CI* from

$$CI = \frac{\lambda_{max} - n}{n - 1}$$

4. With reference to Table 13A.1, calculate the CR from

$$CR = \frac{CI}{RI}$$

If $CR \leq 0.1$, the *PCM* is acceptably consistent.

Example 13A.1

Does the PCM of Figure 13.4 meet the requirement for a consistency ratio of less than or equal to 0.1?

The PCM of Figure 13.4 is reproduced as Figure 13A.1, with successive powers of the original matrix, normalized versions of the powers, and the calculated **w**. The calculations were done using a spreadsheet program; most popular spreadsheet programs can automatically find powers of matrixes.

It can be seen that **w** converges very quickly to a stable set of values. Normally, the less consistent a PCM is, the longer it will take to converge. Also, note that the elements of **w** are very close to the priority weights that were calculated for this PCM in Section 13.4.

To determine λ_{max}, we must now solve $\mathbf{Aw} = \lambda_{max}\mathbf{w}$ for one row of **A**. Selecting the first row of **A** results in the following expression:

$$[1\ 3\ 5\ 1\ 3] \begin{bmatrix} 0.343 \\ 0.129 \\ 0.055 \\ 0.343 \\ 0.129 \end{bmatrix} = \lambda_{max} \times 0.343$$

Or, equivalently,

$$\lambda_{max} = \frac{(1 \times 0.343) + (3 \times 0.129) + (5 \times 0.055) + (1 \times 0.343) + (3 \times 0.129)}{0.343}$$

$$= 5.0583$$

The consistency index can then be calculated from

$$CI = \frac{\lambda_{\max} - n}{n - 1}$$

$$= \frac{5.0583 - 5}{5 - 1}$$

$$= 0.0146$$

Figure 13A.1 Calculating the Principal Eigenvector w

	A					**Normalized**					**w**
	1	3	5	1	3	0.349	0.36	0.29	0.349	0.36	0.342
	0.333	1	3	0.333	1	0.116	0.12	0.18	0.116	0.12	0.130
$\mathbf{A}^1$	0.2	0.333	1	0.2	0.333	0.07	0.04	0.06	0.07	0.04	0.056
	1	3	5	1	3	0.349	0.36	0.29	0.349	0.36	0.342
	0.333	1	3	0.33	1	0.116	0.12	0.18	0.116	0.12	0.130
	5	13.67	33	5	13.667	0.34	0.346	0.35	0.34	0.346	0.343
	1.933	5	13.3	1.93	5	0.132	0.126	0.13	0.132	0.126	0.129
$\mathbf{A}^2$	0.822	2.2	5	0.82	2.2	0.056	0.056	0.05	0.056	0.056	0.055
	5	13.67	33	5	13.667	0.34	0.346	0.35	0.34	0.346	0.343
	1.933	5	13.3	1.93	5	0.132	0.126	0.13	0.132	0.126	0.129
	25.711	68.33	165	25.7	68.333	0.343	0.343	0.34	0.343	0.343	0.343
	9.667	25.71	61.7	9.67	25.711	0.129	0.129	0.13	0.129	0.129	01.29
$\mathbf{A}^3$	4.111	1	26.4	4.11	11	0.055	0.055	0.06	0.055	0.055	0.055
	25.711	68.33	165	25.7	68.333	0.343	0.343	0.34	0.343	0.343	0.343
	9.667	25.71	61.7	9.67	25.711	0.129	0.129	0.13	0.129	0.129	01.29
	129.98	345.9	832	130	345.93	0.343	0.343	0.34	0.343	0.343	0.349
	48.807	130	313	48.8	129.98	0.129	0.129	0.13	0.129	0.129	0.129
$\mathbf{A}^4$	20.84	55.47	134	20.8	55.474	0.055	0.055	0.06	0.055	0.055	0.055
	129.98	345.9	832	130	345.93	0.343	0.343	0.34	0.343	0.343	0.349
	48.807	130	313	48.8	129.98	0.129	0.129	0.13	0.129	0.129	0.129
	657	1749	4207	657	1749.1	0.343	0.343	0.34	0.343	0.343	0.343
	246.79	657	1581	247	657	0.129	0.129	0.13	0.129	0.129	0.129
$\mathbf{A}^5$	105.37	280.5	675	105	280.5	0.055	0.055	0.06	0.055	0.055	0.055
	657	1749	4207	657	1749.1	0.343	0.343	0.34	0.343	0.343	0.343
	246.79	657	1581	247	657	0.129	0.129	0.13	0.129	0.129	0.129

As seen in Table 13A.1, the RI for a matrix of five rows and columns is 1.12. We can then calculate the consistency ratio as

$$CR = \frac{CI}{RI} = \frac{0.0146}{1.12} = 0.013$$

Clearly, the CR is very much less than 0.1. It can thus be concluded that the original PCM is acceptably consistent.■

PROBLEMS FOR APPENDIX 13A

13A.1 Does the following PCM meet the requirement for a consistency ratio of less than or equal to 0.1? What are the values of the consistency index and the consistency ratio?

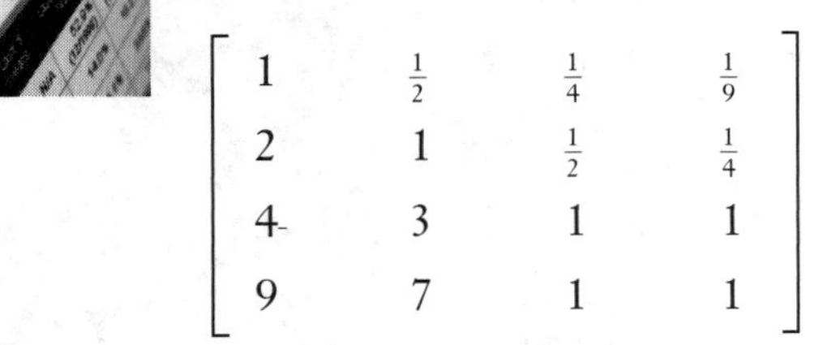

$$\begin{bmatrix} 1 & \frac{1}{2} & \frac{1}{4} & \frac{1}{9} \\ 2 & 1 & \frac{1}{2} & \frac{1}{4} \\ 4 & 3 & 1 & 1 \\ 9 & 7 & 1 & 1 \end{bmatrix}$$

13A.2 Does the following PCM meet the requirement for a consistency ratio of less than or equal to 0.1? What are the values of the consistency index and the consistency ratio?

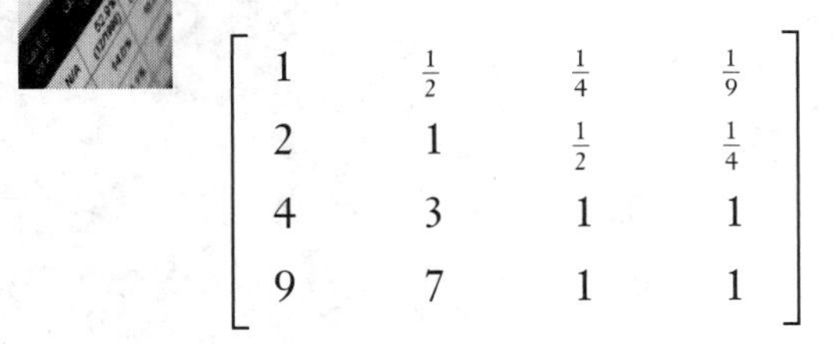

$$\begin{bmatrix} 1 & \frac{1}{2} & \frac{1}{4} & \frac{1}{9} \\ 2 & 1 & \frac{1}{2} & \frac{1}{4} \\ 4 & 3 & 1 & 1 \\ 9 & 7 & 1 & 1 \end{bmatrix}$$

13A.3 Does the following PCM meet the requirement for a consistency ratio of less than or equal to 0.1? What are the values of the consistency index and the consistency ratio? How does an accurate evaluation of the principal eigenvector for this PCM compare with the priority weights calculated in Problem 13.30?

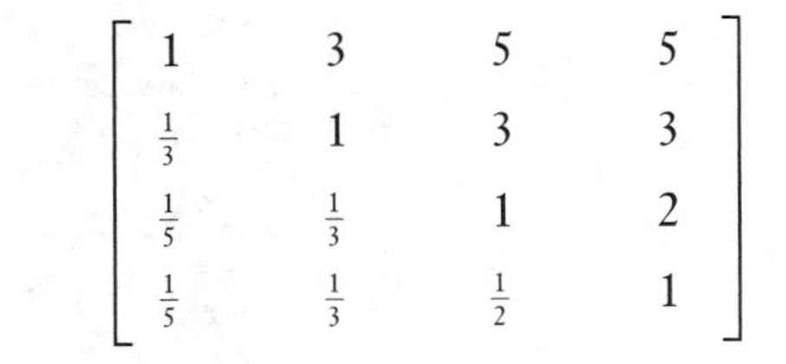

$$\begin{bmatrix} 1 & 3 & 5 & 5 \\ \frac{1}{3} & 1 & 3 & 3 \\ \frac{1}{5} & \frac{1}{3} & 1 & 2 \\ \frac{1}{5} & \frac{1}{3} & \frac{1}{2} & 1 \end{bmatrix}$$

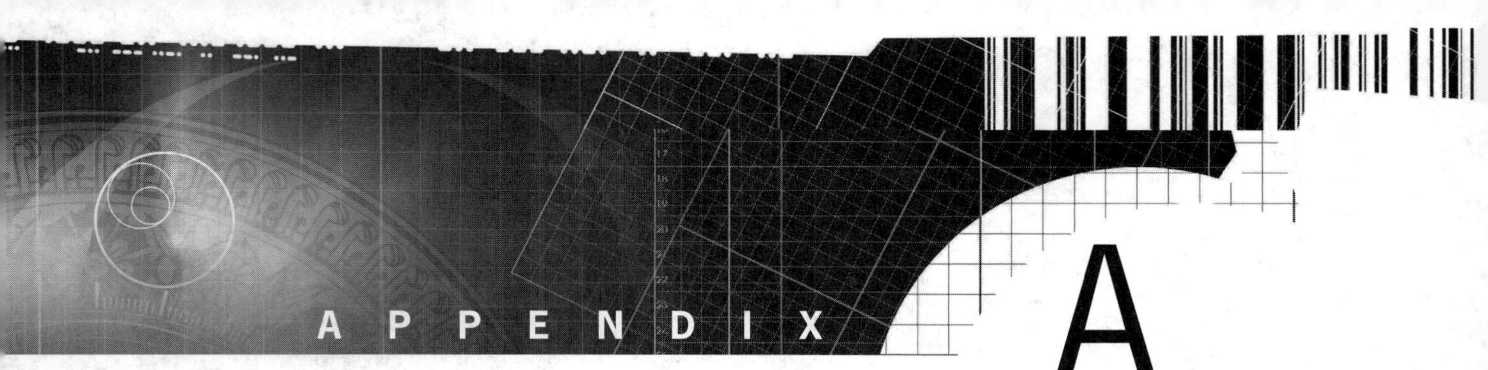

A

Compound Interest Factors for Discrete Compounding, Discrete Cash Flows

$i = 0.5\%$ **Discrete Compounding, Discrete Cash Flows**

	SINGLE PAYMENT		UNIFORM SERIES				
	Compound Amount Factor	Present Worth Factor	Sinking Fund Factor	Uniform Series Factor	Capital Recovery Factor	Series Present Worth Factor	Arithmetic Gradient Series Factor
N	*(F/P,i,N)*	*(P/F,i,N)*	*(A/F,i,N)*	*(F/A,i,N)*	*(A/P,i,N)*	*(P/A,i,N)*	*(A/G,i,N)*
1	1.0050	0.99502	1.0000	1.0000	1.0050	0.99502	0.00000
2	1.0100	0.99007	0.49875	2.0050	0.50375	1.9851	0.49875
3	1.0151	0.98515	0.33167	3.0150	0.33667	2.9702	0.99667
4	1.0202	0.98025	0.24813	4.0301	0.25313	3.9505	1.4938
5	1.0253	0.97537	0.19801	5.0503	0.20301	4.9259	1.9900
6	1.0304	0.97052	0.16460	6.0755	0.16960	5.8964	2.4855
7	1.0355	0.96569	0.14073	7.1059	0.14573	6.8621	2.9801
8	1.0407	0.96089	0.12283	8.1414	0.12783	7.8230	3.4738
9	1.0459	0.95610	0.10891	9.1821	0.11391	8.7791	3.9668
10	1.0511	0.95135	0.09777	10.228	0.10277	9.7304	4.4589
11	1.0564	0.94661	0.08866	11.279	0.09366	10.677	4.9501
12	1.0617	0.94191	0.08107	12.336	0.08607	11.619	5.4406
13	1.0670	0.93722	0.07464	13.397	0.07964	12.556	5.9302
14	1.0723	0.93256	0.06914	14.464	0.07414	13.489	6.4190
15	1.0777	0.92792	0.06436	15.537	0.06936	14.417	6.9069
16	1.0831	0.92330	0.06019	16.614	0.06519	15.340	7.3940
17	1.0885	0.91871	0.05651	17.697	0.06151	16.259	7.8803
18	1.0939	0.91414	0.05323	18.786	0.05823	17.173	8.3658
19	1.0994	0.90959	0.05030	19.880	0.05530	18.082	8.8504
20	1.1049	0.90506	0.04767	20.979	0.05267	18.987	9.3342
21	1.1104	0.90056	0.04528	22.084	0.05028	19.888	9.8172
22	1.1160	0.89608	0.04311	23.194	0.04811	20.784	10.299
23	1.1216	0.89162	0.04113	24.310	0.04613	21.676	10.781
24	1.1272	0.88719	0.03932	25.432	0.04432	22.563	11.261
25	1.1328	0.88277	0.03765	26.559	0.04265	23.446	11.741
26	1.1385	0.87838	0.03611	27.692	0.04111	24.324	12.220
27	1.1442	0.87401	0.03469	28.830	0.03969	25.198	12.698
28	1.1499	0.86966	0.03336	29.975	0.03836	26.068	13.175
29	1.1556	0.86533	0.03213	31.124	0.03713	26.933	13.651
30	1.1614	0.86103	0.03098	32.280	0.03598	27.794	14.126
31	1.1672	0.85675	0.02990	33.441	0.03490	28.651	14.601
32	1.1730	0.85248	0.02889	34.609	0.03389	29.503	15.075
33	1.1789	0.84824	0.02795	35.782	0.03295	30.352	15.548
34	1.1848	0.84402	0.02706	36.961	0.03206	31.196	16.020
35	1.1907	0.83982	0.02622	38.145	0.03122	32.035	16.492
40	1.2208	0.81914	0.02265	44.159	0.02765	36.172	18.836
45	1.2516	0.79896	0.01987	50.324	0.02487	40.207	21.159
50	1.2832	0.77929	0.01765	56.645	0.02265	44.143	23.462
55	1.3156	0.76009	0.01584	63.126	0.02084	47.981	25.745
60	1.3489	0.74137	0.01433	69.770	0.01933	51.726	28.006
65	1.3829	0.72311	0.01306	76.582	0.01806	55.377	30.247
70	1.4178	0.70530	0.01197	83.566	0.01697	58.939	32.468
75	1.4536	0.68793	0.01102	90.727	0.01602	62.414	34.668
80	1.4903	0.67099	0.01020	98.068	0.01520	65.802	36.847
85	1.5280	0.65446	0.00947	105.59	0.01447	69.108	39.006
90	1.5666	0.63834	0.00883	113.31	0.01383	72.331	41.145
95	1.6061	0.62262	0.00825	121.22	0.01325	75.476	43.263
100	1.6467	0.60729	0.00773	129.33	0.01273	78.543	45.361

$i = 1\%$ **Discrete Compounding, Discrete Cash Flows**

	SINGLE PAYMENT		UNIFORM SERIES				
	Compound Amount Factor	Present Worth Factor	Sinking Fund Factor	Uniform Series Factor	Capital Recovery Factor	Series Present Worth Factor	Arithmetic Gradient Series Factor
N	*(F/P,i,N)*	*(P/F,i,N)*	*(A/F,i,N)*	*(F/A,i,N)*	*(A/P,i,N)*	*(P/A,i,N)*	*(A/G,i,N)*
1	1.0100	0.99010	1.0000	1.0000	1.0100	0.99010	0.00000
2	1.0201	0.98030	0.49751	2.0100	0.50751	1.9704	0.49751
3	1.0303	0.97059	0.33002	3.0301	0.34002	2.9410	0.99337
4	1.0406	0.96098	0.24628	4.0604	0.25628	3.9020	1.4876
5	1.0510	0.95147	0.19604	5.1010	0.20604	4.8534	1.9801
6	1.0615	0.94205	0.16255	6.1520	0.17255	5.7955	2.4710
7	1.0721	0.93272	0.13863	7.2135	0.14863	6.7282	2.9602
8	1.0829	0.92348	0.12069	8.2857	0.13069	7.6517	3.4478
9	1.0937	0.91434	0.10674	9.3685	0.11674	8.5660	3.9337
10	1.1046	0.90529	0.09558	10.462	0.10558	9.4713	4.4179
11	1.1157	0.89632	0.08645	11.567	0.09645	10.368	4.9005
12	1.1268	0.88745	0.07885	12.683	0.08885	11.255	5.3815
13	1.1381	0.87866	0.07241	13.809	0.08241	12.134	5.8607
14	1.1495	0.86996	0.06690	14.947	0.07690	13.004	6.3384
15	1.1610	0.86135	0.06212	16.097	0.07212	13.865	6.8143
16	1.1726	0.85282	0.05794	17.258	0.06794	14.718	7.2886
17	1.1843	0.84438	0.05426	18.430	0.06426	15.562	7.7613
18	1.1961	0.83602	0.05098	19.615	0.06098	16.398	8.2323
19	1.2081	0.82774	0.04805	20.811	0.05805	17.226	8.7017
20	1.2202	0.81954	0.04542	22.019	0.05542	18.046	9.1694
21	1.2324	0.81143	0.04303	23.239	0.05303	18.857	9.6354
22	1.2447	0.80340	0.04086	24.472	0.05086	19.660	10.100
23	1.2572	0.79544	0.03889	25.716	0.04889	20.456	10.563
24	1.2697	0.78757	0.03707	26.973	0.04707	21.243	11.024
25	1.2824	0.77977	0.03541	28.243	0.04541	22.023	11.483
26	1.2953	0.77205	0.03387	29.526	0.04387	22.795	11.941
27	1.3082	0.76440	0.03245	30.821	0.04245	23.560	12.397
28	1.3213	0.75684	0.03112	32.129	0.04112	24.316	12.852
29	1.3345	0.74934	0.02990	33.450	0.03990	25.066	13.304
30	1.3478	0.74192	0.02875	34.785	0.03875	25.808	13.756
31	1.3613	0.73458	0.02768	36.133	0.03768	26.542	14.205
32	1.3749	0.72730	0.02667	37.494	0.03667	27.270	14.653
33	1.3887	0.72010	0.02573	38.869	0.03573	27.990	15.099
34	1.4026	0.71297	0.02484	40.258	0.03484	28.703	15.544
35	1.4166	0.70591	0.02400	41.660	0.03400	29.409	15.987
40	1.4889	0.67165	0.02046	48.886	0.03046	32.835	18.178
45	1.5648	0.63905	0.01771	56.481	0.02771	36.095	20.327
50	1.6446	0.60804	0.01551	64.463	0.02551	39.196	22.436
55	1.7285	0.57853	0.01373	72.852	0.02373	42.147	24.505
60	1.8167	0.55045	0.01224	81.670	0.02224	44.955	26.533
65	1.9094	0.52373	0.01100	90.937	0.02100	47.627	28.522
70	2.0068	0.49831	0.00993	100.68	0.01993	50.169	30.470
75	2.1091	0.47413	0.00902	110.91	0.01902	52.587	32.379
80	2.2167	0.45112	0.00822	121.67	0.01822	54.888	34.249
85	2.3298	0.42922	0.00752	132.98	0.01752	57.078	36.080
90	2.4486	0.40839	0.00690	144.86	0.01690	59.161	37.872
95	2.5735	0.38857	0.00636	157.35	0.01636	61.143	39.626
100	2.7048	0.36971	0.00587	170.48	0.01587	63.029	41.343

i = 1.5% **Discrete Compounding, Discrete Cash Flows**

	SINGLE PAYMENT		UNIFORM SERIES				Arithmetic Gradient Series Factor
	Compound Amount Factor	Present Worth Factor	Sinking Fund Factor	Uniform Series Factor	Capital Recovery Factor	Series Present Worth Factor	
N	(F/P,i,N)	(P/F,i,N)	(A/F,i,N)	(F/A,i,N)	(A/P,i,N)	(P/A,i,N)	(A/G,i,N)
1	1.0150	0.98522	1.0000	1.0000	1.0150	0.98522	0.00000
2	1.0302	0.97066	0.49628	2.0150	0.51128	1.9559	0.49628
3	1.0457	0.95632	0.32838	3.0452	0.34338	2.9122	0.99007
4	1.0614	0.94218	0.24444	4.0909	0.25944	3.8544	1.4814
5	1.0773	0.92826	0.19409	5.1523	0.20909	4.7826	1.9702
6	1.0934	0.91454	0.16053	6.2296	0.17553	5.6972	2.4566
7	1.1098	0.90103	0.13656	7.3230	0.15156	6.5982	2.9405
8	1.1265	0.88771	0.11858	8.4328	0.13358	7.4859	3.4219
9	1.1434	0.87459	0.10461	9.5593	0.11961	8.3605	3.9008
10	1.1605	0.86167	0.09343	10.703	0.10843	9.2222	4.3772
11	1.1779	0.84893	0.08429	11.863	0.09929	10.071	4.8512
12	1.1956	0.83639	0.07668	13.041	0.09168	10.908	5.3227
13	1.2136	0.82403	0.07024	14.237	0.08524	11.732	5.7917
14	1.2318	0.81185	0.06472	15.450	0.07972	12.543	6.2582
15	1.2502	0.79985	0.05994	16.682	0.07494	13.343	6.7223
16	1.2690	0.78803	0.05577	17.932	0.07077	14.131	7.1839
17	1.2880	0.77639	0.05208	19.201	0.06708	14.908	7.6431
18	1.3073	0.76491	0.04881	20.489	0.06381	15.673	8.0997
19	1.3270	0.75361	0.04588	21.797	0.06088	16.426	8.5539
20	1.3469	0.74247	0.04325	23.124	0.05825	17.169	9.0057
21	1.3671	0.73150	0.04087	24.471	0.05587	17.900	9.4550
22	1.3876	0.72069	0.03870	25.838	0.05370	18.621	9.9018
23	1.4084	0.71004	0.03673	27.225	0.05173	19.331	10.346
24	1.4295	0.69954	0.03492	28.634	0.04992	20.030	10.788
25	1.4509	0.68921	0.03326	30.063	0.04826	20.720	11.228
26	1.4727	0.67902	0.03173	31.514	0.04673	21.399	11.665
27	1.4948	0.66899	0.03032	32.987	0.04532	22.068	12.099
28	1.5172	0.65910	0.02900	34.481	0.04400	22.727	12.531
29	1.5400	0.64936	0.02778	35.999	0.04278	23.376	12.961
30	1.5631	0.63976	0.02664	37.539	0.04164	24.016	13.388
31	1.5865	0.63031	0.02557	39.102	0.04057	24.646	13.813
32	1.6103	0.62099	0.02458	40.688	0.03958	25.267	14.236
33	1.6345	0.61182	0.02364	42.299	0.03864	25.879	14.656
34	1.6590	0.60277	0.02276	43.933	0.03776	26.482	15.073
35	1.6839	0.59387	0.02193	45.592	0.03693	27.076	15.488
40	1.8140	0.55126	0.01843	54.268	0.03343	29.916	17.528
45	1.9542	0.51171	0.01572	63.614	0.03072	32.552	19.507
50	2.1052	0.47500	0.01357	73.683	0.02857	35.000	21.428
55	2.2679	0.44093	0.01183	84.530	0.02683	37.271	23.289
60	2.4432	0.40930	0.01039	96.215	0.02539	39.380	25.093
65	2.6320	0.37993	0.00919	108.80	0.02419	41.338	26.839
70	2.8355	0.35268	0.00817	122.36	0.02317	43.155	28.529
75	3.0546	0.32738	0.00730	136.97	0.02230	44.842	30.163
80	3.2907	0.30389	0.00655	152.71	0.02155	46.407	31.742
85	3.5450	0.28209	0.00589	169.67	0.02089	47.861	33.268
90	3.8189	0.26185	0.00532	187.93	0.02032	49.210	34.740
95	4.1141	0.24307	0.00482	207.61	0.01982	50.462	36.160
100	4.4320	0.22563	0.00437	228.80	0.01937	51.625	37.530

i = 2%

Discrete Compounding, Discrete Cash Flows

	SINGLE PAYMENT		UNIFORM SERIES				
	Compound Amount Factor	Present Worth Factor	Sinking Fund Factor	Uniform Series Factor	Capital Recovery Factor	Series Present Worth Factor	Arithmetic Gradient Series Factor
N	(*F*/*P*,*i*,*N*)	(*P*/*F*,*i*,*N*)	(*A*/*F*,*i*,*N*)	(*F*/*A*,*i*,*N*)	(*A*/*P*,*i*,*N*)	(*P*/*A*,*i*,*N*)	(*A*/*G*,*i*,*N*)
1	1.0200	0.98039	1.0000	1.0000	1.0200	0.98039	0.00000
2	1.0404	0.96117	0.49505	2.0200	0.51505	1.9416	0.49505
3	1.0612	0.94232	0.32675	3.0604	0.34675	2.8839	0.98680
4	1.0824	0.92385	0.24262	4.1216	0.26262	3.8077	1.4752
5	1.1041	0.90573	0.19216	5.2040	0.21216	4.7135	1.9604
6	1.1262	0.88797	0.15853	6.3081	0.17853	5.6014	2.4423
7	1.1487	0.87056	0.13451	7.4343	0.15451	6.4720	2.9208
8	1.1717	0.85349	0.11651	8.5830	0.13651	7.3255	3.3961
9	1.1951	0.83676	0.10252	9.7546	0.12252	8.1622	3.8681
10	1.2190	0.82035	0.09133	10.950	0.11133	8.9826	4.3367
11	1.2434	0.80426	0.08218	12.169	0.10218	9.787	4.8021
12	1.2682	0.78849	0.07456	13.412	0.09456	10.575	5.2642
13	1.2936	0.77303	0.06812	14.680	0.08812	11.348	5.7231
14	1.3195	0.75788	0.06260	15.974	0.08260	12.106	6.1786
15	1.3459	0.74301	0.05783	17.293	0.07783	12.849	6.6309
16	1.3728	0.72845	0.05365	18.639	0.07365	13.578	7.0799
17	1.4002	0.71416	0.04997	20.012	0.06997	14.292	7.5256
18	1.4282	0.70016	0.04670	21.412	0.06670	14.992	7.9681
19	1.4568	0.68643	0.04378	22.841	0.06378	15.678	8.4073
20	1.4859	0.67297	0.04116	24.297	0.06116	16.351	8.8433
21	1.5157	0.65978	0.03878	25.783	0.05878	17.011	9.2760
22	1.5460	0.64684	0.03663	27.299	0.05663	17.658	9.7050
23	1.5769	0.63416	0.03467	28.845	0.05467	18.292	10.1320
24	1.6084	0.62172	0.03287	30.422	0.05287	18.914	10.5550
25	1.6406	0.60953	0.03122	32.030	0.05122	19.523	10.9740
26	1.6734	0.59758	0.02970	33.671	0.04970	20.121	11.391
27	1.7069	0.58586	0.02829	35.344	0.04829	20.707	11.804
28	1.7410	0.57437	0.02699	37.051	0.04699	21.281	12.214
29	1.7758	0.56311	0.02578	38.792	0.04578	21.844	12.621
30	1.8114	0.55207	0.02465	40.568	0.04465	22.396	13.025
31	1.8476	0.54125	0.02360	42.379	0.04360	22.938	13.426
32	1.8845	0.53063	0.02261	44.227	0.04261	23.468	13.823
33	1.9222	0.52023	0.02169	46.112	0.04169	23.989	14.217
34	1.9607	0.51003	0.02082	48.034	0.04082	24.499	14.608
35	1.9999	0.50003	0.02000	49.994	0.04000	24.999	14.996
40	2.2080	0.45289	0.01656	60.402	0.03656	27.355	16.889
45	2.4379	0.41020	0.01391	71.893	0.03391	29.490	18.703
50	2.6916	0.37153	0.01182	84.579	0.03182	31.424	20.442
55	2.9717	0.33650	0.01014	98.587	0.03014	33.175	22.106
60	3.2810	0.30478	0.00877	114.05	0.02877	34.761	23.696
65	3.6225	0.27605	0.00763	131.13	0.02763	36.197	25.215
70	3.9996	0.25003	0.00667	149.98	0.02667	37.499	26.663
75	4.4158	0.22646	0.00586	170.79	0.02586	38.677	28.043
80	4.8754	0.20511	0.00516	193.77	0.02516	39.745	29.357
85	5.3829	0.18577	0.00456	219.14	0.02456	40.711	30.606
90	5.9431	0.16826	0.00405	247.16	0.02405	41.587	31.793
95	6.5617	0.15240	0.00360	278.08	0.02360	42.380	32.919
100	7.2446	0.13803	0.00320	312.23	0.02320	43.098	33.986

i = 3% **Discrete Compounding, Discrete Cash Flows**

	SINGLE PAYMENT		UNIFORM SERIES				Arithmetic Gradient Series Factor
	Compound Amount Factor	Present Worth Factor	Sinking Fund Factor	Uniform Series Factor	Capital Recovery Factor	Series Present Worth Factor	
N	(*F/P,i,N*)	(*P/F,i,N*)	(*A/F,i,N*)	(*F/A,i,N*)	(*A/P,i,N*)	(*P/A,i,N*)	(*A/G,i,N*)
1	1.0300	0.97087	1.0000	1.0000	1.0300	0.97087	0.00000
2	1.0609	0.94260	0.49261	2.0300	0.52261	1.9135	0.49261
3	1.0927	0.91514	0.32353	3.0909	0.35353	2.8286	0.98030
4	1.1255	0.88849	0.23903	4.1836	0.26903	3.7171	1.4631
5	1.1593	0.86261	0.18835	5.3091	0.21835	4.5797	1.9409
6	1.1941	0.83748	0.15460	6.4684	0.18460	5.4172	2.4138
7	1.2299	0.81309	0.13051	7.6625	0.16051	6.2303	2.8819
8	1.2668	0.78941	0.11246	8.8923	0.14246	7.0197	3.3450
9	1.3048	0.76642	0.09843	10.159	0.12843	7.7861	3.8032
10	1.3439	0.74409	0.08723	11.464	0.11723	8.5302	4.2565
11	1.3842	0.72242	0.07808	12.808	0.10808	9.2526	4.7049
12	1.4258	0.70138	0.07046	14.192	0.10046	9.9540	5.1485
13	1.4685	0.68095	0.06403	15.618	0.09403	10.635	5.5872
14	1.5126	0.66112	0.05853	17.086	0.08853	11.296	6.0210
15	1.5580	0.64186	0.05377	18.599	0.08377	11.938	6.4500
16	1.6047	0.62317	0.04961	20.157	0.07961	12.561	6.8742
17	1.6528	0.60502	0.04595	21.762	0.07595	13.166	7.2936
18	1.7024	0.58739	0.04271	23.414	0.07271	13.754	7.7081
19	1.7535	0.57029	0.03981	25.117	0.06981	14.324	8.1179
20	1.8061	0.55368	0.03722	26.870	0.06722	14.877	8.5229
21	1.8603	0.53755	0.03487	28.676	0.06487	15.415	8.9231
22	1.9161	0.52189	0.03275	30.537	0.06275	15.937	9.3186
23	1.9736	0.50669	0.03081	32.453	0.06081	16.444	9.7093
24	2.0328	0.49193	0.02905	34.426	0.05905	16.936	10.095
25	2.0938	0.47761	0.02743	36.459	0.05743	17.413	10.477
26	2.1566	0.46369	0.02594	38.553	0.05594	17.877	10.853
27	2.2213	0.45019	0.02456	40.710	0.05456	18.327	11.226
28	2.2879	0.43708	0.02329	42.931	0.05329	18.764	11.593
29	2.3566	0.42435	0.02211	45.219	0.05211	19.188	11.956
30	2.4273	0.41199	0.02102	47.575	0.05102	19.600	12.314
31	2.5001	0.39999	0.02000	50.003	0.05000	20.000	12.668
32	2.5751	0.38834	0.01905	52.503	0.04905	20.389	13.017
33	2.6523	0.37703	0.01816	55.078	0.04816	20.766	13.362
34	2.7319	0.36604	0.01732	57.730	0.04732	21.132	13.702
35	2.8139	0.35538	0.01654	60.462	0.04654	21.487	14.037
40	3.2620	0.30656	0.01326	75.401	0.04326	23.115	15.650
45	3.7816	0.26444	0.01079	92.720	0.04079	24.519	17.156
50	4.3839	0.22811	0.00887	112.80	0.03887	25.730	18.558
55	5.0821	0.19677	0.00735	136.07	0.03735	26.774	19.860
60	5.8916	0.16973	0.00613	163.05	0.03613	27.676	21.067
65	6.8300	0.14641	0.00515	194.33	0.03515	28.453	22.184
70	7.9178	0.12630	0.00434	230.59	0.03434	29.123	23.215
75	9.1789	0.10895	0.00367	272.63	0.03367	29.702	24.163
80	10.641	0.09398	0.00311	321.36	0.03311	30.201	25.035
85	12.336	0.08107	0.00265	377.86	0.03265	30.631	25.835
90	14.300	0.06993	0.00226	443.35	0.03226	31.002	26.567
95	16.578	0.06032	0.00193	519.27	0.03193	31.323	27.235
100	19.219	0.05203	0.00165	607.29	0.03165	31.599	27.844

i = 4%

Discrete Compounding, Discrete Cash Flows

	SINGLE PAYMENT		UNIFORM SERIES				Arithmetic Gradient Series Factor
	Compound Amount Factor	Present Worth Factor	Sinking Fund Factor	Uniform Series Factor	Capital Recovery Factor	Series Present Worth Factor	
N	(*F*/*P*,*i*,*N*)	(*P*/*F*,*i*,*N*)	(*A*/*F*,*i*,*N*)	(*F*/*A*,*i*,*N*)	(*A*/*P*,*i*,*N*)	(*P*/*A*,*i*,*N*)	(*A*/*G*,*i*,*N*)
1	1.0400	0.96154	1.0000	1.0000	1.0400	0.96154	0.00000
2	1.0816	0.92456	0.49020	2.0400	0.53020	1.8861	0.49020
3	1.1249	0.88900	0.32035	3.1216	0.36035	2.7751	0.97386
4	1.1699	0.85480	0.23549	4.2465	0.27549	3.6299	1.4510
5	1.2167	0.82193	0.18463	5.4163	0.22463	4.4518	1.9216
6	1.2653	0.79031	0.15076	6.6330	0.19076	5.2421	2.3857
7	1.3159	0.75992	0.12661	7.8983	0.16661	6.0021	2.8433
8	1.3686	0.73069	0.10853	9.2142	0.14853	6.7327	3.2944
9	1.4233	0.70259	0.09449	10.583	0.13449	7.4353	3.7391
10	1.4802	0.67556	0.08329	12.006	0.12329	8.1109	4.1773
11	1.5395	0.64958	0.07415	13.486	0.11415	8.7605	4.6090
12	1.6010	0.62460	0.06655	15.026	0.10655	9.3851	5.0343
13	1.6651	0.60057	0.06014	16.627	0.10014	9.9856	5.4533
14	1.7317	0.57748	0.05467	18.292	0.09467	10.563	5.8659
15	1.8009	0.55526	0.04994	20.024	0.08994	11.118	6.2721
16	1.8730	0.53391	0.04582	21.825	0.08582	11.652	6.6720
17	1.9479	0.51337	0.04220	23.698	0.08220	12.166	7.0656
18	2.0258	0.49363	0.03899	25.645	0.07899	12.659	7.4530
19	2.1068	0.47464	0.03614	27.671	0.07614	13.134	7.8342
20	2.1911	0.45639	0.03358	29.778	0.07358	13.590	8.2091
21	2.2788	0.43883	0.03128	31.969	0.07128	14.029	8.5779
22	2.3699	0.42196	0.02920	34.248	0.06920	14.451	8.9407
23	2.4647	0.40573	0.02731	36.618	0.06731	14.857	9.2973
24	2.5633	0.39012	0.02559	39.083	0.06559	15.247	9.6479
25	2.6658	0.37512	0.02401	41.646	0.06401	15.622	9.9925
26	2.7725	0.36069	0.02257	44.312	0.06257	15.983	10.331
27	2.8834	0.34682	0.02124	47.084	0.06124	16.330	10.664
28	2.9987	0.33348	0.02001	49.968	0.06001	16.663	10.991
29	3.1187	0.32065	0.01888	52.966	0.05888	16.984	11.312
30	3.2434	0.30832	0.01783	56.085	0.05783	17.292	11.627
31	3.3731	0.29646	0.01686	59.328	0.05686	17.588	11.937
32	3.5081	0.28506	0.01595	62.701	0.05595	17.874	12.241
33	3.6484	0.27409	0.01510	66.210	0.05510	18.148	12.540
34	3.7943	0.26355	0.01431	69.858	0.05431	18.411	12.832
35	3.9461	0.25342	0.01358	73.652	0.05358	18.665	13.120
40	4.8010	0.20829	0.01052	95.026	0.05052	19.793	14.477
45	5.8412	0.17120	0.00826	121.03	0.04826	20.720	15.705
50	7.1067	0.14071	0.00655	152.67	0.04655	21.482	16.812
55	8.6464	0.11566	0.00523	191.16	0.04523	22.109	17.807
60	10.520	0.09506	0.00420	237.99	0.04420	22.623	18.697
65	12.799	0.07813	0.00339	294.97	0.04339	23.047	19.491
70	15.572	0.06422	0.00275	364.29	0.04275	23.395	20.196
75	18.945	0.05278	0.00223	448.63	0.04223	23.680	20.821
80	23.050	0.04338	0.00181	551.24	0.04181	23.915	21.372
85	28.044	0.03566	0.00148	676.09	0.04148	24.109	21.857
90	34.119	0.02931	0.00121	827.98	0.04121	24.267	22.283
95	41.511	0.02409	0.00099	1012.8	0.04099	24.398	22.655
100	50.505	0.01980	0.00081	1237.6	0.04081	24.505	22.980

i = 5% **Discrete Compounding, Discrete Cash Flows**

	SINGLE PAYMENT		UNIFORM SERIES				
	Compound Amount Factor	Present Worth Factor	Sinking Fund Factor	Uniform Series Factor	Capital Recovery Factor	Series Present Worth Factor	Arithmetic Gradient Series Factor
N	(*F/P,i,N*)	(*P/F,i,N*)	(*A/F,i,N*)	(*F/A,i,N*)	(*A/P,i,N*)	(*P/A,i,N*)	(*A/G,i,N*)
1	1.0500	0.95238	1.0000	1.0000	1.0500	0.95238	0.00000
2	1.1025	0.90703	0.48780	2.0500	0.53780	1.8594	0.48780
3	1.1576	0.86384	0.31721	3.1525	0.36721	2.7232	0.96749
4	1.2155	0.82270	0.23201	4.3101	0.28201	3.5460	1.4391
5	1.2763	0.78353	0.18097	5.5256	0.23097	4.3295	1.9025
6	1.3401	0.74622	0.14702	6.8019	0.19702	5.0757	2.3579
7	1.4071	0.71068	0.12282	8.1420	0.17282	5.7864	2.8052
8	1.4775	0.67684	0.10472	9.5491	0.15472	6.4632	3.2445
9	1.5513	0.64461	0.09069	11.027	0.14069	7.1078	3.6758
10	1.6289	0.61391	0.07950	12.578	0.12950	7.7217	4.0991
11	1.7103	0.58468	0.07039	14.207	0.12039	8.3064	4.5144
12	1.7959	0.55684	0.06283	15.917	0.11283	8.8633	4.9219
13	1.8856	0.53032	0.05646	17.713	0.10646	9.3936	5.3215
14	1.9799	0.50507	0.05102	19.599	0.10102	9.8986	5.7133
15	2.0789	0.48102	0.04634	21.579	0.09634	10.380	6.0973
16	2.1829	0.45811	0.04227	23.657	0.09227	10.838	6.4736
17	2.2920	0.43630	0.03870	25.840	0.08870	11.274	6.8423
18	2.4066	0.41552	0.03555	28.132	0.08555	11.690	7.2034
19	2.5270	0.39573	0.03275	30.539	0.08275	12.085	7.5569
20	2.6533	0.37689	0.03024	33.066	0.08024	12.462	7.9030
21	2.7860	0.35894	0.02800	35.719	0.07800	12.821	8.2416
22	2.9253	0.34185	0.02597	38.505	0.07597	13.163	8.5730
23	3.0715	0.32557	0.02414	41.430	0.07414	13.489	8.8971
24	3.2251	0.31007	0.02247	44.502	0.07247	13.799	9.2140
25	3.3864	0.29530	0.02095	47.727	0.07095	14.094	9.5238
26	3.5557	0.28124	0.01956	51.113	0.06956	14.375	9.8266
27	3.7335	0.26785	0.01829	54.669	0.06829	14.643	10.122
28	3.9201	0.25509	0.01712	58.403	0.06712	14.898	10.411
29	4.1161	0.24295	0.01605	62.323	0.06605	15.141	10.694
30	4.3219	0.23138	0.01505	66.439	0.06505	15.372	10.969
31	4.5380	0.22036	0.01413	70.761	0.06413	15.593	11.238
32	4.7649	0.20987	0.01328	75.299	0.06328	15.803	11.501
33	5.0032	0.19987	0.01249	80.064	0.06249	16.003	11.757
34	5.2533	0.19035	0.01176	85.067	0.06176	16.193	12.006
35	5.5160	0.18129	0.01107	90.320	0.06107	16.374	12.250
40	7.0400	0.14205	0.00828	120.80	0.05828	17.159	13.377
45	8.9850	0.11130	0.00626	159.70	0.05626	17.774	14.364
50	11.467	0.08720	0.00478	209.35	0.05478	18.256	15.223
55	14.636	0.06833	0.00367	272.71	0.05367	18.633	15.966
60	18.679	0.05354	0.00283	353.58	0.05283	18.929	16.606
65	23.840	0.04195	0.00219	456.80	0.05219	19.161	17.154
70	30.426	0.03287	0.00170	588.53	0.05170	19.343	17.621
75	38.833	0.02575	0.00132	756.65	0.05132	19.485	18.018
80	49.561	0.02018	0.00103	971.23	0.05103	19.596	18.353
85	63.254	0.01581	0.00080	1245.1	0.05080	19.684	18.635
90	80.730	0.01239	0.00063	1594.6	0.05063	19.752	18.871
95	103.03	0.00971	0.00049	2040.7	0.05049	19.806	19.069
100	131.50	0.00760	0.00038	2610.0	0.05038	19.848	19.234

i = 6%

Discrete Compounding, Discrete Cash Flows

	SINGLE PAYMENT		UNIFORM SERIES				
	Compound Amount Factor	Present Worth Factor	Sinking Fund Factor	Uniform Series Factor	Capital Recovery Factor	Series Present Worth Factor	Arithmetic Gradient Series Factor
N	(*F*/*P*,*i*,*N*)	(*P*/*F*,*i*,*N*)	(*A*/*F*,*i*,*N*)	(*F*/*A*,*i*,*N*)	(*A*/*P*,*i*,*N*)	(*P*/*A*,*i*,*N*)	(*A*/*G*,*i*,*N*)
1	1.0600	0.94340	1.0000	1.0000	1.0600	0.94340	0.00000
2	1.1236	0.89000	0.48544	2.0600	0.54544	1.8334	0.48544
3	1.1910	0.83962	0.31411	3.1836	0.37411	2.6730	0.96118
4	1.2625	0.79209	0.22859	4.3746	0.28859	3.4651	1.4272
5	1.3382	0.74726	0.17740	5.6371	0.23740	4.2124	1.8836
6	1.4185	0.70496	0.14336	6.9753	0.20336	4.9173	2.3304
7	1.5036	0.66506	0.11914	8.3938	0.17914	5.5824	2.7676
8	1.5938	0.62741	0.10104	9.8975	0.16104	6.2098	3.1952
9	1.6895	0.59190	0.08702	11.491	0.14702	6.8017	3.6133
10	1.7908	0.55839	0.07587	13.181	0.13587	7.3601	4.0220
11	1.8983	0.52679	0.06679	14.972	0.12679	7.8869	4.4213
12	2.0122	0.49697	0.05928	16.870	0.11928	8.3838	4.8113
13	2.1329	0.46884	0.05296	18.882	0.11296	8.8527	5.1920
14	2.2609	0.44230	0.04758	21.015	0.10758	9.2950	5.5635
15	2.3966	0.41727	0.04296	23.276	0.10296	9.7122	5.9260
16	2.5404	0.39365	0.03895	25.673	0.09895	10.106	6.2794
17	2.6928	0.37136	0.03544	28.213	0.09544	10.477	6.6240
18	2.8543	0.35034	0.03236	30.906	0.09236	10.828	6.9597
19	3.0256	0.33051	0.02962	33.760	0.08962	11.158	7.2867
20	3.2071	0.31180	0.02718	36.786	0.08718	11.470	7.6051
21	3.3996	0.29416	0.02500	39.993	0.08500	11.764	7.9151
22	3.6035	0.27751	0.02305	43.392	0.08305	12.042	8.2166
23	3.8197	0.26180	0.02128	46.996	0.08128	12.303	8.5099
24	4.0489	0.24698	0.01968	50.816	0.07968	12.550	8.7951
25	4.2919	0.23300	0.01823	54.865	0.07823	12.783	9.0722
26	4.5494	0.21981	0.01690	59.156	0.07690	13.003	9.3414
27	4.8223	0.20737	0.01570	63.706	0.07570	13.211	9.6029
28	5.1117	0.19563	0.01459	68.528	0.07459	13.406	9.8568
29	5.4184	0.18456	0.01358	73.640	0.07358	13.591	10.103
30	5.7435	0.17411	0.01265	79.058	0.07265	13.765	10.342
31	6.0881	0.16425	0.01179	84.802	0.07179	13.929	10.574
32	6.4534	0.15496	0.01100	90.890	0.07100	14.084	10.799
33	6.8406	0.14619	0.01027	97.343	0.07027	14.230	11.017
34	7.2510	0.13791	0.00960	104.18	0.06960	14.368	11.228
35	7.6861	0.13011	0.00897	111.43	0.06897	14.498	11.432
40	10.286	0.09722	0.00646	154.76	0.06646	15.046	12.359
45	13.765	0.07265	0.00470	212.74	0.06470	15.456	13.141
50	18.420	0.05429	0.00344	290.34	0.06344	15.762	13.796
55	24.650	0.04057	0.00254	394.17	0.06254	15.991	14.341
60	32.988	0.03031	0.00188	533.13	0.06188	16.161	14.791
65	44.145	0.02265	0.00139	719.08	0.06139	16.289	15.160
70	59.076	0.01693	0.00103	967.93	0.06103	16.385	15.461
75	79.057	0.01265	0.00077	1300.9	0.06077	16.456	15.706
80	105.80	0.00945	0.00057	1746.6	0.06057	16.509	15.903
85	141.58	0.00706	0.00043	2343.0	0.06043	16.549	16.062
90	189.46	0.00528	0.00032	3141.1	0.06032	16.579	16.189
95	253.55	0.00394	0.00024	4209.1	0.06024	16.601	16.290
100	339.30	0.00295	0.00018	5638.4	0.06018	16.618	16.371

i = 7% **Discrete Compounding, Discrete Cash Flows**

	SINGLE PAYMENT		UNIFORM SERIES				
	Compound Amount Factor	Present Worth Factor	Sinking Fund Factor	Uniform Series Factor	Capital Recovery Factor	Series Present Worth Factor	Arithmetic Gradient Series Factor
N	**(F/P,i,N)**	**(P/F,i,N)**	**(A/F,i,N)**	**(F/A,i,N)**	**(A/P,i,N)**	**(P/A,i,N)**	**(A/G,i,N)**
1	1.0700	0.93458	1.0000	1.0000	1.0700	0.93458	0.00000
2	1.1449	0.87344	0.48309	2.0700	0.55309	1.8080	0.48309
3	1.2250	0.81630	0.31105	3.2149	0.38105	2.6243	0.95493
4	1.3108	0.76290	0.22523	4.4399	0.29523	3.3872	1.4155
5	1.4026	0.71299	0.17389	5.7507	0.24389	4.1002	1.8650
6	1.5007	0.66634	0.13980	7.1533	0.20980	4.7665	2.3032
7	1.6058	0.62275	0.11555	8.6540	0.18555	5.3893	2.7304
8	1.7182	0.58201	0.09747	10.260	0.16747	5.9713	3.1465
9	1.8385	0.54393	0.08349	11.978	0.15349	6.5152	3.5517
10	1.9672	0.50835	0.07238	13.816	0.14238	7.0236	3.9461
11	2.1049	0.47509	0.06336	15.784	0.13336	7.4987	4.3296
12	2.2522	0.44401	0.05590	17.888	0.12590	7.9427	4.7025
13	2.4098	0.41496	0.04965	20.141	0.11965	8.3577	5.0648
14	2.5785	0.38782	0.04434	22.550	0.11434	8.7455	5.4167
15	2.7590	0.36245	0.03979	25.129	0.10979	9.1079	5.7583
16	2.9522	0.33873	0.03586	27.888	0.10586	9.4466	6.0897
17	3.1588	0.31657	0.03243	30.840	0.10243	9.7632	6.4110
18	3.3799	0.29586	0.02941	33.999	0.09941	10.059	6.7225
19	3.6165	0.27651	0.02675	37.379	0.09675	10.336	7.0242
20	3.8697	0.25842	0.02439	40.995	0.09439	10.594	7.3163
21	4.1406	0.24151	0.02229	44.865	0.09229	10.836	7.5990
22	4.4304	0.22571	0.02041	49.006	0.09041	11.061	7.8725
23	4.7405	0.21095	0.01871	53.436	0.08871	11.272	8.1369
24	5.0724	0.19715	0.01719	58.177	0.08719	11.469	8.3923
25	5.4274	0.18425	0.01581	63.249	0.08581	11.654	8.6391
26	5.8074	0.17220	0.01456	68.676	0.08456	11.826	8.8773
27	6.2139	0.16093	0.01343	74.484	0.08343	11.987	9.1072
28	6.6488	0.15040	0.01239	80.698	0.08239	12.137	9.3289
29	7.1143	0.14056	0.01145	87.347	0.08145	12.278	9.5427
30	7.6123	0.13137	0.01059	94.461	0.08059	12.409	9.7487
31	8.1451	0.12277	0.00980	102.07	0.07980	12.532	9.9471
32	8.7153	0.11474	0.00907	110.22	0.07907	12.647	10.138
33	9.3253	0.10723	0.00841	118.93	0.07841	12.754	10.322
34	9.9781	0.10022	0.00780	128.26	0.07780	12.854	10.499
35	10.677	0.09366	0.00723	138.24	0.07723	12.948	10.669
40	14.974	0.06678	0.00501	199.64	0.07501	13.332	11.423
45	21.002	0.04761	0.00350	285.75	0.07350	13.606	12.036
50	29.457	0.03395	0.00246	406.53	0.07246	13.801	12.529
55	41.315	0.02420	0.00174	575.93	0.07174	13.940	12.921
60	57.946	0.01726	0.00123	813.52	0.07123	14.039	13.232
65	81.273	0.01230	0.00087	1146.8	0.07087	14.110	13.476
70	113.99	0.00877	0.00062	1614.1	0.07062	14.160	13.666
75	159.88	0.00625	0.00044	2269.7	0.07044	14.196	13.814
80	224.23	0.00446	0.00031	3189.1	0.07031	14.222	13.927
85	314.50	0.00318	0.00022	4478.6	0.07022	14.240	14.015
90	441.10	0.00227	0.00016	6287.2	0.07016	14.253	14.081
95	618.67	0.00162	0.00011	8823.9	0.07011	14.263	14.132
100	867.72	0.00115	0.00008	12 382.0	0.07008	14.269	14.170

$i = 8\%$

Discrete Compounding, Discrete Cash Flows

	SINGLE PAYMENT		UNIFORM SERIES				
	Compound Amount Factor	Present Worth Factor	Sinking Fund Factor	Uniform Series Factor	Capital Recovery Factor	Series Present Worth Factor	Arithmetic Gradient Series Factor
N	*(F/P,i,N)*	*(P/F,i,N)*	*(A/F,i,N)*	*(F/A,i,N)*	*(A/P,i,N)*	*(P/A,i,N)*	*(A/G,i,N)*
1	1.0800	0.92593	1.0000	1.0000	1.0800	0.92593	0.00000
2	1.1664	0.85734	0.48077	2.0800	0.56077	1.7833	0.48077
3	1.2597	0.79383	0.30803	3.2464	0.38803	2.5771	0.94874
4	1.3605	0.73503	0.22192	4.5061	0.30192	3.3121	1.4040
5	1.4693	0.68058	0.17046	5.8666	0.25046	3.9927	1.8465
6	1.5869	0.63017	0.13632	7.3359	0.21632	4.6229	2.2763
7	1.7138	0.58349	0.11207	8.9228	0.19207	5.2064	2.6937
8	1.8509	0.54027	0.09401	10.637	0.17401	5.7466	3.0985
9	1.9990	0.50025	0.08008	12.488	0.16008	6.2469	3.4910
10	2.1589	0.46319	0.06903	14.487	0.14903	6.7101	3.8713
11	2.3316	0.42888	0.06008	16.645	0.14008	7.1390	4.2395
12	2.5182	0.39711	0.05270	18.977	0.13270	7.5361	4.5957
13	2.7196	0.36770	0.04652	21.495	0.12652	7.9038	4.9402
14	2.9372	0.34046	0.04130	24.215	0.12130	8.2442	5.2731
15	3.1722	0.31524	0.03683	27.152	0.11683	8.5595	5.5945
16	3.4259	0.29189	0.03298	30.324	0.11298	8.8514	5.9046
17	3.7000	0.27027	0.02963	33.750	0.10963	9.1216	6.2037
18	3.9960	0.25025	0.02670	37.450	0.10670	9.3719	6.4920
19	4.3157	0.23171	0.02413	41.446	0.10413	9.6036	6.7697
20	4.6610	0.21455	0.02185	45.762	0.10185	9.8181	7.0369
21	5.0338	0.19866	0.01983	50.423	0.09983	10.017	7.2940
22	5.4365	0.18394	0.01803	55.457	0.09803	10.201	7.5412
23	5.8715	0.17032	0.01642	60.893	0.09642	10.371	7.7786
24	6.3412	0.15770	0.01498	66.765	0.09498	10.529	8.0066
25	6.8485	0.14602	0.01368	73.106	0.09368	10.675	8.2254
26	7.3964	0.13520	0.01251	79.954	0.09251	10.810	8.4352
27	7.9881	0.12519	0.01145	87.351	0.09145	10.935	8.6363
28	8.6271	0.11591	0.01049	95.339	0.09049	11.051	8.8289
29	9.3173	0.10733	0.00962	103.97	0.08962	11.158	9.0133
30	10.063	0.09938	0.00883	113.28	0.08883	11.258	9.1897
31	10.868	0.09202	0.00811	123.35	0.08811	11.350	9.3584
32	11.737	0.08520	0.00745	134.21	0.08745	11.435	9.5197
33	12.676	0.07889	0.00685	145.95	0.08685	11.514	9.6737
34	13.690	0.07305	0.00630	158.63	0.08630	11.587	9.8208
35	14.785	0.06763	0.00580	172.32	0.08580	11.655	9.9611
40	21.725	0.04603	0.00386	259.06	0.08386	11.925	10.570
45	31.920	0.03133	0.00259	386.51	0.08259	12.108	11.045
50	46.902	0.02132	0.00174	573.77	0.08174	12.233	11.411
55	68.914	0.01451	0.00118	848.92	0.08118	12.319	11.690
60	101.26	0.00988	0.00080	1253.2	0.08080	12.377	11.902
65	148.78	0.00672	0.00054	1847.2	0.08054	12.416	12.060
70	218.61	0.00457	0.00037	2720.1	0.08037	12.443	12.178
75	321.20	0.00311	0.00025	4002.6	0.08025	12.461	12.266
80	471.95	0.00212	0.00017	5886.9	0.08017	12.474	12.330
85	693.46	0.00144	0.00012	8655.7	0.08012	12.482	12.377
90	1018.9	0.00098	0.00008	12724.0	0.08008	12.488	12.412
95	1497.1	0.00067	0.00005	18702.0	0.08005	12.492	12.437
100	2199.8	0.00045	0.00004	27485.0	0.08004	12.494	12.455

$i = 9\%$ **Discrete Compounding, Discrete Cash Flows**

	SINGLE PAYMENT		UNIFORM SERIES				Arithmetic Gradient Series Factor
	Compound Amount Factor	Present Worth Factor	Sinking Fund Factor	Uniform Series Factor	Capital Recovery Factor	Series Present Worth Factor	
N	(F/P,i,N)	(P/F,i,N)	(A/F,i,N)	(F/A,i,N)	(A/P,i,N)	(P/A,i,N)	(A/G,i,N)
1	1.0900	0.91743	1.0000	1.0000	1.0900	0.91743	0.00000
2	1.1881	0.84168	0.47847	2.0900	0.56847	1.7591	0.47847
3	1.2950	0.77218	0.30505	3.2781	0.39505	2.5313	0.94262
4	1.4116	0.70843	0.21867	4.5731	0.30867	3.2397	1.3925
5	1.5386	0.64993	0.16709	5.9847	0.25709	3.8897	1.8282
6	1.6771	0.59627	0.13292	7.5233	0.22292	4.4859	2.2498
7	1.8280	0.54703	0.10869	9.2004	0.19869	5.0330	2.6574
8	1.9926	0.50187	0.09067	11.028	0.18067	5.5348	3.0512
9	2.1719	0.46043	0.07680	13.021	0.16680	5.9952	3.4312
10	2.3674	0.42241	0.06582	15.193	0.15582	6.4177	3.7978
11	2.5804	0.38753	0.05695	17.560	0.14695	6.8052	4.1510
12	2.8127	0.35553	0.04965	20.141	0.13965	7.1607	4.4910
13	3.0658	0.32618	0.04357	22.953	0.13357	7.4869	4.8182
14	3.3417	0.29925	0.03843	26.019	0.12843	7.7862	5.1326
15	3.6425	0.27454	0.03406	29.361	0.12406	8.0607	5.4346
16	3.9703	0.25187	0.03030	33.003	0.12030	8.3126	5.7245
17	4.3276	0.23107	0.02705	36.974	0.11705	8.5436	6.0024
18	4.7171	0.21199	0.02421	41.301	0.11421	8.7556	6.2687
19	5.1417	0.19449	0.02173	46.018	0.11173	8.9501	6.5236
20	5.6044	0.17843	0.01955	51.160	0.10955	9.1285	6.7674
21	6.1088	0.16370	0.01762	56.765	0.10762	9.2922	7.0006
22	6.6586	0.15018	0.01590	62.873	0.10590	9.4424	7.2232
23	7.2579	0.13778	0.01438	69.532	0.10438	9.5802	7.4357
24	7.9111	0.12640	0.01302	76.790	0.10302	9.7066	7.6384
25	8.6231	0.11597	0.01181	84.701	0.10181	9.8226	7.8316
26	9.3992	0.10639	0.01072	93.324	0.10072	9.9290	8.0156
27	10.245	0.09761	0.00973	102.72	0.09973	10.027	8.1906
28	11.167	0.08955	0.00885	112.97	0.09885	10.116	8.3571
29	12.172	0.08215	0.00806	124.14	0.09806	10.198	8.5154
30	13.268	0.07537	0.00734	136.31	0.09734	10.274	8.6657
31	14.462	0.06915	0.00669	149.58	0.09669	10.343	8.8083
32	15.763	0.06344	0.00610	164.04	0.09610	10.406	8.9436
33	17.182	0.05820	0.00556	179.80	0.09556	10.464	9.0718
34	18.728	0.05339	0.00508	196.98	0.09508	10.518	9.1933
35	20.414	0.04899	0.00464	215.71	0.09464	10.567	9.3083
40	31.409	0.03184	0.00296	337.88	0.09296	10.757	9.7957
45	48.327	0.02069	0.00190	525.86	0.09190	10.881	10.160
50	74.358	0.01345	0.00123	815.08	0.09123	10.962	10.430
55	114.41	0.00874	0.00079	1260.1	0.09079	11.014	10.626
60	176.03	0.00568	0.00051	1944.8	0.09051	11.048	10.768
65	270.85	0.00369	0.00033	2998.3	0.09033	11.070	10.870
70	416.73	0.00240	0.00022	4619.2	0.09022	11.084	10.943
75	641.19	0.00156	0.00014	7113.2	0.09014	11.094	10.994
80	986.55	0.00101	0.00009	10951.0	0.09009	11.100	11.030
85	1517.9	0.00066	0.00006	16855.0	0.09006	11.104	11.055
90	2335.5	0.00043	0.00004	25939.0	0.09004	11.106	11.073
95	3593.5	0.00028	0.00003	39917.0	0.09003	11.108	11.085
100	5529.0	0.00018	0.00002	61423.0	0.09002	11.109	11.093

i = 10% **Discrete Compounding, Discrete Cash Flows**

	SINGLE PAYMENT		UNIFORM SERIES				Arithmetic Gradient Series Factor
	Compound Amount Factor	Present Worth Factor	Sinking Fund Factor	Uniform Series Factor	Capital Recovery Factor	Series Present Worth Factor	
N	(*F*/*P*,*i*,*N*)	(*P*/*F*,*i*,*N*)	(*A*/*F*,*i*,*N*)	(*F*/*A*,*i*,*N*)	(*A*/*P*,*i*,*N*)	(*P*/*A*,*i*,*N*)	(*A*/*G*,*i*,*N*)
1	1.1000	0.90909	1.0000	1.0000	1.1000	0.90909	0.00000
2	1.2100	0.82645	0.47619	2.1000	0.57619	1.7355	0.47619
3	1.3310	0.75131	0.30211	3.3100	0.40211	2.4869	0.93656
4	1.4641	0.68301	0.21547	4.6410	0.31547	3.1699	1.3812
5	1.6105	0.62092	0.16380	6.1051	0.26380	3.7908	1.8101
6	1.7716	0.56447	0.12961	7.7156	0.22961	4.3553	2.2236
7	1.9487	0.51316	0.10541	9.4872	0.20541	4.8684	2.6216
8	2.1436	0.46651	0.08744	11.436	0.18744	5.3349	3.0045
9	2.3579	0.42410	0.07364	13.579	0.17364	5.7590	3.3724
10	2.5937	0.38554	0.06275	15.937	0.16275	6.1446	3.7255
11	2.8531	0.35049	0.05396	18.531	0.15396	6.4951	4.0641
12	3.1384	0.31863	0.04676	21.384	0.14676	6.8137	4.3884
13	3.4523	0.28966	0.04078	24.523	0.14078	7.1034	4.6988
14	3.7975	0.26333	0.03575	27.975	0.13575	7.3667	4.9955
15	4.1772	0.23939	0.03147	31.772	0.13147	7.6061	5.2789
16	4.5950	0.21763	0.02782	35.950	0.12782	7.8237	5.5493
17	5.0545	0.19784	0.02466	40.545	0.12466	8.0216	5.8071
18	5.5599	0.17986	0.02193	45.599	0.12193	8.2014	6.0526
19	6.1159	0.16351	0.01955	51.159	0.11955	8.3649	6.2861
20	6.7275	0.14864	0.01746	57.275	0.11746	8.5136	6.5081
21	7.4002	0.13513	0.01562	64.002	0.11562	8.6487	6.7189
22	8.1403	0.12285	0.01401	71.403	0.11401	8.7715	6.9189
23	8.9543	0.11168	0.01257	79.543	0.11257	8.8832	7.1085
24	9.8497	0.10153	0.01130	88.497	0.11130	8.9847	7.2881
25	10.835	0.09230	0.01017	98.347	0.11017	9.0770	7.4580
26	11.918	0.08391	0.00916	109.18	0.10916	9.1609	7.6186
27	13.110	0.07628	0.00826	121.10	0.10826	9.2372	7.7704
28	14.421	0.06934	0.00745	134.21	0.10745	9.3066	7.9137
29	15.863	0.06304	0.00673	148.63	0.10673	9.3696	8.0489
30	17.449	0.05731	0.00608	164.49	0.10608	9.4269	8.1762
31	19.194	0.05210	0.00550	181.94	0.10550	9.4790	8.2962
32	21.114	0.04736	0.00497	201.14	0.10497	9.5264	8.4091
33	23.225	0.04306	0.00450	222.25	0.10450	9.5694	8.5152
34	25.548	0.03914	0.00407	245.48	0.10407	9.6086	8.6149
35	28.102	0.03558	0.00369	271.02	0.10369	9.6442	8.7086
40	45.259	0.02209	0.00226	442.59	0.10226	9.7791	9.0962
45	72.890	0.01372	0.00139	718.90	0.10139	9.8628	9.3740
50	117.39	0.00852	0.00086	1163.9	0.10086	9.9148	9.5704
55	189.06	0.00529	0.00053	1880.6	0.10053	9.9471	9.7075
60	304.48	0.00328	0.00033	3034.8	0.10033	9.9672	9.8023
65	490.37	0.00204	0.00020	4893.7	0.10020	9.9796	9.8672
70	789.75	0.00127	0.00013	7887.5	0.10013	9.9873	9.9113
75	1271.9	0.00079	0.00008	12 709.0	0.10008	9.9921	9.9410

i = 11% **Discrete Compounding, Discrete Cash Flows**

	SINGLE PAYMENT		UNIFORM SERIES				
	Compound Amount Factor	Present Worth Factor	Sinking Fund Factor	Uniform Series Factor	Capital Recovery Factor	Series Present Worth Factor	Arithmetic Gradient Series Factor
N	(*F*/*P*,*i*,*N*)	(*P*/*F*,*i*,*N*)	(*A*/*F*,*i*,*N*)	(*F*/*A*,*i*,*N*)	(*A*/*P*,*i*,*N*)	(*P*/*A*,*i*,*N*)	(*A*/*G*,*i*,*N*)
1	1.1100	0.90090	1.0000	1.0000	1.1100	0.90090	0.00000
2	1.2321	0.81162	0.47393	2.1100	0.58393	1.7125	0.47393
3	1.3676	0.73119	0.29921	3.3421	0.40921	2.4437	0.93055
4	1.5181	0.65873	0.21233	4.7097	0.32233	3.1024	1.3700
5	1.6851	0.59345	0.16057	6.2278	0.27057	3.6959	1.7923
6	1.8704	0.53464	0.12638	7.9129	0.23638	4.2305	2.1976
7	2.0762	0.48166	0.10222	9.783	0.21222	4.7122	2.5863
8	2.3045	0.43393	0.08432	11.859	0.19432	5.1461	2.9585
9	2.5580	0.39092	0.07060	14.164	0.18060	5.5370	3.3144
10	2.8394	0.35218	0.05980	16.722	0.16980	5.8892	3.6544
11	3.1518	0.31728	0.05112	19.561	0.16112	6.2065	3.9788
12	3.4985	0.28584	0.04403	22.713	0.15403	6.4924	4.2879
13	3.8833	0.25751	0.03815	26.212	0.14815	6.7499	4.5822
14	4.3104	0.23199	0.03323	30.095	0.14323	6.9819	4.8619
15	4.7846	0.20900	0.02907	34.405	0.13907	7.1909	5.1275
16	5.3109	0.18829	0.02552	39.190	0.13552	7.3792	5.3794
17	5.8951	0.16963	0.02247	44.501	0.13247	7.5488	5.6180
18	6.5436	0.15282	0.01984	50.396	0.12984	7.7016	5.8439
19	7.2633	0.13768	0.01756	56.939	0.12756	7.8393	6.0574
20	8.0623	0.12403	0.01558	64.203	0.12558	7.9633	6.2590
21	8.949	0.11174	0.01384	72.265	0.12384	8.0751	6.4491
22	9.934	0.10067	0.01231	81.214	0.12231	8.1757	6.6283
23	11.026	0.09069	0.01097	91.15	0.12097	8.2664	6.7969
24	12.239	0.08170	0.00979	102.17	0.11979	8.3481	6.9555
25	13.585	0.07361	0.00874	114.41	0.11874	8.4217	7.1045
26	15.080	0.06631	0.00781	128.00	0.11781	8.4881	7.2443
27	16.739	0.05974	0.00699	143.08	0.11699	8.5478	7.3754
28	18.580	0.05382	0.00626	159.82	0.11626	8.6016	7.4982
29	20.624	0.04849	0.00561	178.40	0.11561	8.6501	7.6131
30	22.892	0.04368	0.00502	199.02	0.11502	8.6938	7.7206
31	25.410	0.03935	0.00451	221.91	0.11451	8.7331	7.8210
32	28.206	0.03545	0.00404	247.32	0.11404	8.7686	7.9147
33	31.308	0.03194	0.00363	275.53	0.11363	8.8005	8.0021
34	34.752	0.02878	0.00326	306.84	0.11326	8.8293	8.0836
35	38.575	0.02592	0.00293	341.59	0.11293	8.8552	8.1594
40	65.001	0.01538	0.00172	581.83	0.11172	8.9511	8.4659
45	109.53	0.00913	0.00101	986.6	0.11101	9.0079	8.6763
50	184.56	0.00542	0.00060	1668.8	0.11060	9.0417	8.8185
55	311.00	0.00322	0.00035	2818.2	0.11035	9.0617	8.9135

i = 12%

Discrete Compounding, Discrete Cash Flows

	SINGLE PAYMENT		UNIFORM SERIES				
	Compound Amount Factor	Present Worth Factor	Sinking Fund Factor	Uniform Series Factor	Capital Recovery Factor	Series Present Worth Factor	Arithmetic Gradient Series Factor
N	(*F*/*P*,*i*,*N*)	(*P*/*F*,*i*,*N*)	(*A*/*F*,*i*,*N*)	(*F*/*A*,*i*,*N*)	(*A*/*P*,*i*,*N*)	(*P*/*A*,*i*,*N*)	(*A*/*G*,*i*,*N*)
1	1.1200	0.89286	1.0000	1.0000	1.1200	0.89286	0.00000
2	1.2544	0.79719	0.47170	2.1200	0.59170	1.6901	0.47170
3	1.4049	0.71178	0.29635	3.3744	0.41635	2.4018	0.92461
4	1.5735	0.63552	0.20923	4.7793	0.32923	3.0373	1.3589
5	1.7623	0.56743	0.15741	6.3528	0.27741	3.6048	1.7746
6	1.9738	0.50663	0.12323	8.1152	0.24323	4.1114	2.1720
7	2.2107	0.45235	0.09912	10.089	0.21912	4.5638	2.5515
8	2.4760	0.40388	0.08130	12.300	0.20130	4.9676	2.9131
9	2.7731	0.36061	0.06768	14.776	0.18768	5.3282	3.2574
10	3.1058	0.32197	0.05698	17.549	0.17698	5.6502	3.5847
11	3.4785	0.28748	0.04842	20.655	0.16842	5.9377	3.8953
12	3.8960	0.25668	0.04144	24.133	0.16144	6.1944	4.1897
13	4.3635	0.22917	0.03568	28.029	0.15568	6.4235	4.4683
14	4.8871	0.20462	0.03087	32.393	0.15087	6.6282	4.7317
15	5.4736	0.18270	0.02682	37.280	0.14682	6.8109	4.9803
16	6.1304	0.16312	0.02339	42.753	0.14339	6.9740	5.2147
17	6.8660	0.14564	0.02046	48.884	0.14046	7.1196	5.4353
18	7.6900	0.13004	0.01794	55.750	0.13794	7.2497	5.6427
19	8.6128	0.11611	0.01576	63.440	0.13576	7.3658	5.8375
20	9.6463	0.10367	0.01388	72.052	0.13388	7.4694	6.0202
21	10.804	0.09256	0.01224	81.699	0.13224	7.5620	6.1913
22	12.100	0.08264	0.01081	92.503	0.13081	7.6446	6.3514
23	13.552	0.07379	0.00956	104.60	0.12956	7.7184	6.5010
24	15.179	0.06588	0.00846	118.16	0.12846	7.7843	6.6406
25	17.000	0.05882	0.00750	133.33	0.12750	7.8431	6.7708
26	19.040	0.05252	0.00665	150.33	0.12665	7.8957	6.8921
27	21.325	0.04689	0.00590	169.37	0.12590	7.9426	7.0049
28	23.884	0.04187	0.00524	190.70	0.12524	7.9844	7.1098
29	26.750	0.03738	0.00466	214.58	0.12466	8.0218	7.2071
30	29.960	0.03338	0.00414	241.33	0.12414	8.0552	7.2974
31	33.555	0.02980	0.00369	271.29	0.12369	8.0850	7.3811
32	37.582	0.02661	0.00328	304.85	0.12328	8.1116	7.4586
33	42.092	0.02376	0.00292	342.43	0.12292	8.1354	7.5302
34	47.143	0.02121	0.00260	384.52	0.12260	8.1566	7.5965
35	52.800	0.01894	0.00232	431.66	0.12232	8.1755	7.6577
40	93.051	0.01075	0.00130	767.09	0.12130	8.2438	7.8988
45	163.99	0.00610	0.00074	1358.2	0.12074	8.2825	8.0572
50	289.00	0.00346	0.00042	2400.0	0.12042	8.3045	8.1597
55	509.32	0.00196	0.00024	4236.0	0.12024	8.3170	8.2251

i = 13% **Discrete Compounding, Discrete Cash Flows**

	SINGLE PAYMENT		UNIFORM SERIES				
	Compound Amount Factor	Present Worth Factor	Sinking Fund Factor	Uniform Series Factor	Capital Recovery Factor	Series Present Worth Factor	Arithmetic Gradient Series Factor
N	(*F/P,i,N*)	(*P/F,i,N*)	(*A/F,i,N*)	(*F/A,i,N*)	(*A/P,i,N*)	(*P/A,i,N*)	(*A/G,i,N*)
1	1.1300	0.88496	1.0000	1.0000	1.1300	0.88496	0.00000
2	1.2769	0.78315	0.46948	2.1300	0.59948	1.6681	0.46948
3	1.4429	0.69305	0.29352	3.4069	0.42352	2.3612	0.91872
4	1.6305	0.61332	0.20619	4.8498	0.33619	2.9745	1.3479
5	1.8424	0.54276	0.15431	6.4803	0.28431	3.5172	1.7571
6	2.0820	0.48032	0.12015	8.3227	0.25015	3.9975	2.1468
7	2.3526	0.42506	0.09611	10.405	0.22611	4.4226	2.5171
8	2.6584	0.37616	0.07839	12.757	0.20839	4.7988	2.8685
9	3.0040	0.33288	0.06487	15.416	0.19487	5.1317	3.2014
10	3.3946	0.29459	0.05429	18.420	0.18429	5.4262	3.5162
11	3.8359	0.26070	0.04584	21.814	0.17584	5.6869	3.8134
12	4.3345	0.23071	0.03899	25.650	0.16899	5.9176	4.0936
13	4.8980	0.20416	0.03335	29.985	0.16335	6.1218	4.3573
14	5.5348	0.18068	0.02867	34.883	0.15867	6.3025	4.6050
15	6.2543	0.15989	0.02474	40.417	0.15474	6.4624	4.8375
16	7.0673	0.14150	0.02143	46.672	0.15143	6.6039	5.0552
17	7.9861	0.12522	0.01861	53.739	0.14861	6.7291	5.2589
18	9.0243	0.11081	0.01620	61.725	0.14620	6.8399	5.4491
19	10.197	0.09806	0.01413	70.749	0.14413	6.9380	5.6265
20	11.523	0.08678	0.01235	80.947	0.14235	7.0248	5.7917
21	13.021	0.07680	0.01081	92.470	0.14081	7.1016	5.9454
22	14.714	0.06796	0.00948	105.49	0.13948	7.1695	6.0881
23	16.627	0.06014	0.00832	120.20	0.13832	7.2297	6.2205
24	18.788	0.05323	0.00731	136.83	0.13731	7.2829	6.3431
25	21.231	0.04710	0.00643	155.62	0.13643	7.3300	6.4566
26	23.991	0.04168	0.00565	176.85	0.13565	7.3717	6.5614
27	27.109	0.03689	0.00498	200.84	0.13498	7.4086	6.6582
28	30.633	0.03264	0.00439	227.95	0.13439	7.4412	6.7474
29	34.616	0.02889	0.00387	258.58	0.13387	7.4701	6.8296
30	39.116	0.02557	0.00341	293.20	0.13341	7.4957	6.9052
31	44.201	0.02262	0.00301	332.32	0.13301	7.5183	6.9747
32	49.947	0.02002	0.00266	376.52	0.13266	7.5383	7.0385
33	56.440	0.01772	0.00234	426.46	0.13234	7.5560	7.0971
34	63.777	0.01568	0.00207	482.90	0.13207	7.5717	7.1507
35	72.069	0.01388	0.00183	546.68	0.13183	7.5856	7.1998
40	132.78	0.00753	0.00099	1013.7	0.13099	7.6344	7.3888
45	244.64	0.00409	0.00053	1874.2	0.13053	7.6609	7.5076
50	450.74	0.00222	0.00029	3459.5	0.13029	7.6752	7.5811
55	830.45	0.00120	0.00016	6380.4	0.13016	7.6830	7.6260

i = 14%

Discrete Compounding, Discrete Cash Flows

	SINGLE PAYMENT		UNIFORM SERIES				Arithmetic Gradient Series Factor
	Compound Amount Factor	Present Worth Factor	Sinking Fund Factor	Uniform Series Factor	Capital Recovery Factor	Series Present Worth Factor	
N	(*F*/*P*,*i*,*N*)	(*P*/*F*,*i*,*N*)	(*A*/*F*,*i*,*N*)	(*F*/*A*,*i*,*N*)	(*A*/*P*,*i*,*N*)	(*P*/*A*,*i*,*N*)	(*A*/*G*,*i*,*N*)
1	1.1400	0.87719	1.0000	1.0000	1.1400	0.87719	0.00000
2	1.2996	0.76947	0.46729	2.1400	0.60729	1.6467	0.46729
3	1.4815	0.67497	0.29073	3.4396	0.43073	2.3216	0.91290
4	1.6890	0.59208	0.20320	4.9211	0.34320	2.9137	1.3370
5	1.9254	0.51937	0.15128	6.6101	0.29128	3.4331	1.7399
6	2.1950	0.45559	0.11716	8.5355	0.25716	3.8887	2.1218
7	2.5023	0.39964	0.09319	10.730	0.23319	4.2883	2.4832
8	2.8526	0.35056	0.07557	13.233	0.21557	4.6389	2.8246
9	3.2519	0.30751	0.06217	16.085	0.20217	4.9464	3.1463
10	3.7072	0.26974	0.05171	19.337	0.19171	5.2161	3.4490
11	4.2262	0.23662	0.04339	23.045	0.18339	5.4527	3.7333
12	4.8179	0.20756	0.03667	27.271	0.17667	5.6603	3.9998
13	5.4924	0.18207	0.03116	32.089	0.17116	5.8424	4.2491
14	6.2613	0.15971	0.02661	37.581	0.16661	6.0021	4.4819
15	7.1379	0.14010	0.02281	43.842	0.16281	6.1422	4.6990
16	8.1372	0.12289	0.01962	50.980	0.15962	6.2651	4.9011
17	9.2765	0.10780	0.01692	59.118	0.15692	6.3729	5.0888
18	10.575	0.09456	0.01462	68.394	0.15462	6.4674	5.2630
19	12.056	0.08295	0.01266	78.969	0.15266	6.5504	5.4243
20	13.743	0.07276	0.01099	91.025	0.15099	6.6231	5.5734
21	15.668	0.06383	0.00954	104.77	0.14954	6.6870	5.7111
22	17.861	0.05599	0.00830	120.44	0.14830	6.7429	5.8381
23	20.362	0.04911	0.00723	138.30	0.14723	6.7921	5.9549
24	23.212	0.04308	0.00630	158.66	0.14630	6.8351	6.0624
25	26.462	0.03779	0.00550	181.87	0.14550	6.8729	6.1610
26	30.167	0.03315	0.00480	208.33	0.14480	6.9061	6.2514
27	34.390	0.02908	0.00419	238.50	0.14419	6.9352	6.3342
28	39.204	0.02551	0.00366	272.89	0.14366	6.9607	6.4100
29	44.693	0.02237	0.00320	312.09	0.14320	6.9830	6.4791
30	50.950	0.01963	0.00280	356.79	0.14280	7.0027	6.5423
31	58.083	0.01722	0.00245	407.74	0.14245	7.0199	6.5998
32	66.215	0.01510	0.00215	465.82	0.14215	7.0350	6.6522
33	75.485	0.01325	0.00188	532.04	0.14188	7.0482	6.6998
34	86.053	0.01162	0.00165	607.52	0.14165	7.0599	6.7431
35	98.100	0.01019	0.00144	693.57	0.14144	7.0700	6.7824
40	188.88	0.00529	0.00075	1342.0	0.14075	7.1050	6.9300
45	363.68	0.00275	0.00039	2590.6	0.14039	7.1232	7.0188
50	700.23	0.00143	0.00020	4994.5	0.14020	7.1327	7.0714
55	1348.2	0.00074	0.00010	9623.1	0.14010	7.1376	7.1020

i = 15% **Discrete Compounding, Discrete Cash Flows**

	SINGLE PAYMENT		UNIFORM SERIES				
	Compound Amount Factor	Present Worth Factor	Sinking Fund Factor	Uniform Series Factor	Capital Recovery Factor	Series Present Worth Factor	Arithmetic Gradient Series Factor
N	(*F/P,i,N*)	(*P/F,i,N*)	(*A/F,i,N*)	(*F/A,i,N*)	(*A/P,i,N*)	(*P/A,i,N*)	(*A/G,i,N*)
1	1.1500	0.86957	1.0000	1.0000	1.1500	0.86957	0.00000
2	1.3225	0.75614	0.46512	2.1500	0.61512	1.6257	0.46512
3	1.5209	0.65752	0.28798	3.4725	0.43798	2.2832	0.90713
4	1.7490	0.57175	0.20027	4.9934	0.35027	2.8550	1.3263
5	2.0114	0.49718	0.14832	6.7424	0.29832	3.3522	1.7228
6	2.3131	0.43233	0.11424	8.7537	0.26424	3.7845	2.0972
7	2.6600	0.37594	0.09036	11.067	0.24036	4.1604	2.4498
8	3.0590	0.32690	0.07285	13.727	0.22285	4.4873	2.7813
9	3.5179	0.28426	0.05957	16.786	0.20957	4.7716	3.0922
10	4.0456	0.24718	0.04925	20.304	0.19925	5.0188	3.3832
11	4.6524	0.21494	0.04107	24.349	0.19107	5.2337	3.6549
12	5.3503	0.18691	0.03448	29.002	0.18448	5.4206	3.9082
13	6.1528	0.16253	0.02911	34.352	0.17911	5.5831	4.1438
14	7.0757	0.14133	0.02469	40.505	0.17469	5.7245	4.3624
15	8.1371	0.12289	0.02102	47.580	0.17102	5.8474	4.5650
16	9.3576	0.10686	0.01795	55.717	0.16795	5.9542	4.7522
17	10.761	0.09293	0.01537	65.075	0.16537	6.0472	4.9251
18	12.375	0.08081	0.01319	75.836	0.16319	6.1280	5.0843
19	14.232	0.07027	0.01134	88.212	0.16134	6.1982	5.2307
20	16.367	0.06110	0.00976	102.44	0.15976	6.2593	5.3651
21	18.822	0.05313	0.00842	118.81	0.15842	6.3125	5.4883
22	21.645	0.04620	0.00727	137.63	0.15727	6.3587	5.6010
23	24.891	0.04017	0.00628	159.28	0.15628	6.3988	5.7040
24	28.625	0.03493	0.00543	184.17	0.15543	6.4338	5.7979
25	32.919	0.03038	0.00470	212.79	0.15470	6.4641	5.8834
26	37.857	0.02642	0.00407	245.71	0.15407	6.4906	5.9612
27	43.535	0.02297	0.00353	283.57	0.15353	6.5135	6.0319
28	50.066	0.01997	0.00306	327.10	0.15306	6.5335	6.0960
29	57.575	0.01737	0.00265	377.17	0.15265	6.5509	6.1541
30	66.212	0.01510	0.00230	434.75	0.15230	6.5660	6.2066
31	76.144	0.01313	0.00200	500.96	0.15200	6.5791	6.2541
32	87.565	0.01142	0.00173	577.10	0.15173	6.5905	6.2970
33	100.70	0.00993	0.00150	664.67	0.15150	6.6005	6.3357
34	115.80	0.00864	0.00131	765.37	0.15131	6.6091	6.3705
35	133.18	0.00751	0.00113	881.17	0.15113	6.6166	6.4019
40	267.86	0.00373	0.00056	1779.1	0.15056	6.6418	6.5168
45	538.77	0.00186	0.00028	3585.1	0.15028	6.6543	6.5830
50	1083.7	0.00092	0.00014	7217.7	0.15014	6.6605	6.6205
55	2179.6	0.00046	0.00007	14524.0	0.15007	6.6636	6.6414

$i = 20\%$ **Discrete Compounding, Discrete Cash Flows**

	SINGLE PAYMENT		UNIFORM SERIES				
	Compound Amount Factor	Present Worth Factor	Sinking Fund Factor	Uniform Series Factor	Capital Recovery Factor	Series Present Worth Factor	Arithmetic Gradient Series Factor
N	(*F*/*P*,*i*,*N*)	(*P*/*F*,*i*,*N*)	(*A*/*F*,*i*,*N*)	(*F*/*A*,*i*,*N*)	(*A*/*P*,*i*,*N*)	(*P*/*A*,*i*,*N*)	(*A*/*G*,*i*,*N*)
1	1.2000	0.83333	1.0000	1.0000	1.2000	0.83333	0.00000
2	1.4400	0.69444	0.45455	2.2000	0.65455	1.5278	0.45455
3	1.7280	0.57870	0.27473	3.6400	0.47473	2.1065	0.87912
4	2.0736	0.48225	0.18629	5.3680	0.38629	2.5887	1.2742
5	2.4883	0.40188	0.13438	7.4416	0.33438	2.9906	1.6405
6	2.9860	0.33490	0.10071	9.9299	0.30071	3.3255	1.9788
7	3.5832	0.27908	0.07742	12.916	0.27742	3.6046	2.2902
8	4.2998	0.23257	0.06061	16.499	0.26061	3.8372	2.5756
9	5.1598	0.19381	0.04808	20.799	0.24808	4.0310	2.8364
10	6.1917	0.16151	0.03852	25.959	0.23852	4.1925	3.0739
11	7.4301	0.13459	0.03110	32.150	0.23110	4.3271	3.2893
12	8.9161	0.11216	0.02526	39.581	0.22526	4.4392	3.4841
13	10.699	0.09346	0.02062	48.497	0.22062	4.5327	3.6597
14	12.839	0.07789	0.01689	59.196	0.21689	4.6106	3.8175
15	15.407	0.06491	0.01388	72.035	0.21388	4.6755	3.9588
16	18.488	0.05409	0.01144	87.442	0.21144	4.7296	4.0851
17	22.186	0.04507	0.00944	105.93	0.20944	4.7746	4.1976
18	26.623	0.03756	0.00781	128.12	0.20781	4.8122	4.2975
19	31.948	0.03130	0.00646	154.74	0.20646	4.8435	4.3861
20	38.338	0.02608	0.00536	186.69	0.20536	4.8696	4.4643
21	46.005	0.02174	0.00444	225.03	0.20444	4.8913	4.5334
22	55.206	0.01811	0.00369	271.03	0.20369	4.9094	4.5941
23	66.247	0.01509	0.00307	326.24	0.20307	4.9245	4.6475
24	79.497	0.01258	0.00255	392.48	0.20255	4.9371	4.6943
25	95.396	0.01048	0.00212	471.98	0.20212	4.9476	4.7352
26	114.48	0.00874	0.00176	567.38	0.20176	4.9563	4.7709
27	137.37	0.00728	0.00147	681.85	0.20147	4.9636	4.8020
28	164.84	0.00607	0.00122	819.22	0.20122	4.9697	4.8291
29	197.81	0.00506	0.00102	984.07	0.20102	4.9747	4.8527
30	237.38	0.00421	0.00085	1181.9	0.20085	4.9789	4.8731
31	284.85	0.00351	0.00070	1419.3	0.20070	4.9824	4.8908
32	341.82	0.00293	0.00059	1704.1	0.20059	4.9854	4.9061
33	410.19	0.00244	0.00049	2045.9	0.20049	4.9878	4.9194
34	492.22	0.00203	0.00041	2456.1	0.20041	4.9898	4.9308
35	590.67	0.00169	0.00034	2948.3	0.20034	4.9915	4.9406

i = 25% **Discrete Compounding, Discrete Cash Flows**

	SINGLE PAYMENT		UNIFORM SERIES				Arithmetic Gradient Series Factor
	Compound Amount Factor	Present Worth Factor	Sinking Fund Factor	Uniform Series Factor	Capital Recovery Factor	Series Present Worth Factor	
N	(*F*/*P*,*i*,*N*)	(*P*/*F*,*i*,*N*)	(*A*/*F*,*i*,*N*)	(*F*/*A*,*i*,*N*)	(*A*/*P*,*i*,*N*)	(*P*/*A*,*i*,*N*)	(*A*/*G*,*i*,*N*)
1	1.2500	0.80000	1.0000	1.0000	1.2500	0.80000	0.00000
2	1.5625	0.64000	0.44444	2.2500	0.69444	1.4400	0.44444
3	1.9531	0.51200	0.26230	3.8125	0.51230	1.9520	0.85246
4	2.4414	0.40960	0.17344	5.7656	0.42344	2.3616	1.2249
5	3.0518	0.32768	0.12185	8.2070	0.37185	2.6893	1.5631
6	3.8147	0.26214	0.08882	11.259	0.33882	2.9514	1.8683
7	4.7684	0.20972	0.06634	15.073	0.31634	3.1611	2.1424
8	5.9605	0.16777	0.05040	19.842	0.30040	3.3289	2.3872
9	7.4506	0.13422	0.03876	25.802	0.28876	3.4631	2.6048
10	9.3132	0.10737	0.03007	33.253	0.28007	3.5705	2.7971
11	11.642	0.08590	0.02349	42.566	0.27349	3.6564	2.9663
12	14.552	0.06872	0.01845	54.208	0.26845	3.7251	3.1145
13	18.190	0.05498	0.01454	68.760	0.26454	3.7801	3.2437
14	22.737	0.04398	0.01150	86.949	0.26150	3.8241	3.3559
15	28.422	0.03518	0.00912	109.69	0.25912	3.8593	3.4530
16	35.527	0.02815	0.00724	138.11	0.25724	3.8874	3.5366
17	44.409	0.02252	0.00576	173.64	0.25576	3.9099	3.6084
18	55.511	0.01801	0.00459	218.04	0.25459	3.9279	3.6698
19	69.389	0.01441	0.00366	273.56	0.25366	3.9424	3.7222
20	86.736	0.01153	0.00292	342.94	0.25292	3.9539	3.7667
21	108.42	0.00922	0.00233	429.68	0.25233	3.9631	3.8045
22	135.53	0.00738	0.00186	538.10	0.25186	3.9705	3.8365
23	169.41	0.00590	0.00148	673.63	0.25148	3.9764	3.8634
24	211.76	0.00472	0.00119	843.03	0.25119	3.9811	3.8861
25	264.70	0.00378	0.00095	1054.8	0.25095	3.9849	3.9052
26	330.87	0.00302	0.00076	1319.5	0.25076	3.9879	3.9212
27	413.59	0.00242	0.00061	1650.4	0.25061	3.9903	3.9346
28	516.99	0.00193	0.00048	2064.0	0.25048	3.9923	3.9457
29	646.23	0.00155	0.00039	2580.9	0.25039	3.9938	3.9551
30	807.79	0.00124	0.00031	3227.2	0.25031	3.9950	3.9628
31	1009.7	0.00099	0.00025	4035.0	0.25025	3.9960	3.9693
32	1262.2	0.00079	0.00020	5044.7	0.25020	3.9968	3.9746
33	1577.7	0.00063	0.00016	6306.9	0.25016	3.9975	3.9791
34	1972.2	0.00051	0.00013	7884.6	0.25013	3.9980	3.9828
35	2465.2	0.00041	0.00010	9856.8	0.25010	3.9984	3.9858

$i = 30\%$ **Discrete Compounding, Discrete Cash Flows**

	SINGLE PAYMENT		UNIFORM SERIES				Arithmetic Gradient Series Factor
	Compound Amount Factor	Present Worth Factor	Sinking Fund Factor	Uniform Series Factor	Capital Recovery Factor	Series Present Worth Factor	
N	*(F/P,i,N)*	*(P/F,i,N)*	*(A/F,i,N)*	*(F/A,i,N)*	*(A/P,i,N)*	*(P/A,i,N)*	*(A/G,i,N)*
1	1.3000	0.76923	1.0000	1.0000	1.3000	0.76923	0.00000
2	1.6900	0.59172	0.43478	2.3000	0.73478	1.3609	0.43478
3	2.1970	0.45517	0.25063	3.9900	0.55063	1.8161	0.82707
4	2.8561	0.35013	0.16163	6.1870	0.46163	2.1662	1.1783
5	3.7129	0.26933	0.11058	9.0431	0.41058	2.4356	1.4903
6	4.8268	0.20718	0.07839	12.756	0.37839	2.6427	1.7654
7	6.2749	0.15937	0.05687	17.583	0.35687	2.8021	2.0063
8	8.1573	0.12259	0.04192	23.858	0.34192	2.9247	2.2156
9	10.604	0.09430	0.03124	32.015	0.33124	3.0190	2.3963
10	13.786	0.07254	0.02346	42.619	0.32346	3.0915	2.5512
11	17.922	0.05580	0.01773	56.405	0.31773	3.1473	2.6833
12	23.298	0.04292	0.01345	74.327	0.31345	3.1903	2.7952
13	30.288	0.03302	0.01024	97.625	0.31024	3.2233	2.8895
14	39.374	0.02540	0.00782	127.91	0.30782	3.2487	2.9685
15	51.186	0.01954	0.00598	167.29	0.30598	3.2682	3.0344
16	66.542	0.01503	0.00458	218.47	0.30458	3.2832	3.0892
17	86.504	0.01156	0.00351	285.01	0.30351	3.2948	3.1345
18	112.46	0.00889	0.00269	371.52	0.30269	3.3037	3.1718
19	146.19	0.00684	0.00207	483.97	0.30207	3.3105	3.2025
20	190.05	0.00526	0.00159	630.17	0.30159	3.3158	3.2275
21	247.06	0.00405	0.00122	820.22	0.30122	3.3198	3.2480
22	321.18	0.00311	0.00094	1067.3	0.30094	3.3230	3.2646
23	417.54	0.00239	0.00072	1388.5	0.30072	3.3254	3.2781
24	542.80	0.00184	0.00055	1806.0	0.30055	3.3272	3.2890
25	705.64	0.00142	0.00043	2348.8	0.30043	3.3286	3.2979
26	917.33	0.00109	0.00033	3054.4	0.30033	3.3297	3.3050
27	1192.5	0.00084	0.00025	3971.8	0.30025	3.3305	3.3107
28	1550.3	0.00065	0.00019	5164.3	0.30019	3.3312	3.3153
29	2015.4	0.00050	0.00015	6714.6	0.30015	3.3317	3.3189
30	2620.0	0.00038	0.00011	8730.0	0.30011	3.3321	3.3219
31	3406.0	0.00029	0.00009	11 350.0	0.30009	3.3324	3.3242
32	4427.8	0.00023	0.00007	14 756.0	0.30007	3.3326	3.3261
33	5756.1	0.00017	0.00005	19 184.0	0.30005	3.3328	3.3276
34	7483.0	0.00013	0.00004	24 940.0	0.30004	3.3329	3.3288
35	9727.9	0.00010	0.00003	32 423.0	0.30003	3.3330	3.3297

i = 40% **Discrete Compounding, Discrete Cash Flows**

	SINGLE PAYMENT		UNIFORM SERIES				
	Compound Amount Factor	Present Worth Factor	Sinking Fund Factor	Uniform Series Factor	Capital Recovery Factor	Series Present Worth Factor	Arithmetic Gradient Series Factor
N	(*F*/*P*,*i*,*N*)	(*P*/*F*,*i*,*N*)	(*A*/*F*,*i*,*N*)	(*F*/*A*,*i*,*N*)	(*A*/*P*,*i*,*N*)	(*P*/*A*,*i*,*N*)	(*A*/*G*,*i*,*N*)
1	1.4000	0.71429	1.0000	1.0000	1.4000	0.71429	0.00000
2	1.9600	0.51020	0.41667	2.4000	0.81667	1.2245	0.41667
3	2.7440	0.36443	0.22936	4.3600	0.62936	1.5889	0.77982
4	3.8416	0.26031	0.14077	7.1040	0.54077	1.8492	1.0923
5	5.3782	0.18593	0.09136	10.946	0.49136	2.0352	1.3580
6	7.5295	0.13281	0.06126	16.324	0.46126	2.1680	1.5811
7	10.541	0.09486	0.04192	23.853	0.44192	2.2628	1.7664
8	14.758	0.06776	0.02907	34.395	0.42907	2.3306	1.9185
9	20.661	0.04840	0.02034	49.153	0.42034	2.3790	2.0422
10	28.925	0.03457	0.01432	69.814	0.41432	2.4136	2.1419
11	40.496	0.02469	0.01013	98.739	0.41013	2.4383	2.2215
12	56.694	0.01764	0.00718	139.23	0.40718	2.4559	2.2845
13	79.371	0.01260	0.00510	195.93	0.40510	2.4685	2.3341
14	111.12	0.00900	0.00363	275.30	0.40363	2.4775	2.3729
15	155.57	0.00643	0.00259	386.42	0.40259	2.4839	2.4030
16	217.80	0.00459	0.00185	541.99	0.40185	2.4885	2.4262
17	304.91	0.00328	0.00132	759.78	0.40132	2.4918	2.4441
18	426.88	0.00234	0.00094	1064.70	0.40094	2.4941	2.4577
19	597.63	0.00167	0.00067	1491.58	0.40067	2.4958	2.4682
20	836.68	0.00120	0.00048	2089.21	0.40048	2.4970	2.4761
21	1171.36	0.00085	0.00034	2925.89	0.40034	2.4979	2.4821
22	1639.90	0.00061	0.00024	4097.24	0.40024	2.4985	2.4866
23	2295.86	0.00044	0.00017	5737.14	0.40017	2.4989	2.4900
24	3214.20	0.00031	0.00012	8033.00	0.40012	2.4992	2.4925
25	4499.88	0.00022	0.00009	11 247.0	0.40009	2.4994	2.4944
26	6299.83	0.00016	0.00006	15 747.0	0.40006	2.4996	2.4959
27	8819.76	0.00011	0.00005	22 047.0	0.40005	2.4997	2.4969
28	12 348.0	0.00008	0.00003	30 867.0	0.40003	2.4998	2.4977
29	17 287.0	0.00006	0.00002	43 214.0	0.40002	2.4999	2.4983
30	24 201.0	0.00004	0.00002	60 501.0	0.40002	2.4999	2.4988
31	33 882.0	0.00003	0.00001	84 703.0	0.40001	2.4999	2.4991
32	47 435.0	0.00002	0.00001	118 585.0	0.40001	2.4999	2.4993
33	66 409.0	0.00002	0.00001	166 019.0	0.40001	2.5000	2.4995
34	92 972.0	0.00001	0.00000	232 428.0	0.40000	2.5000	2.4996
35	130 161.0	0.00001	0.00000	325 400.0	0.40000	2.5000	2.4997

i = 50%

Discrete Compounding, Discrete Cash Flows

	SINGLE PAYMENT		UNIFORM SERIES				
	Compound Amount Factor	Present Worth Factor	Sinking Fund Factor	Uniform Series Factor	Capital Recovery Factor	Series Present Worth Factor	Arithmetic Gradient Series Factor
N	(*F*/*P*,*i*,*N*)	(*P*/*F*,*i*,*N*)	(*A*/*F*,*i*,*N*)	(*F*/*A*,*i*,*N*)	(*A*/*P*,*i*,*N*)	(*P*/*A*,*i*,*N*)	(*A*/*G*,*i*,*N*)
1	1.5000	0.66667	1.0000	1.0000	1.5000	0.66667	0.00000
2	2.2500	0.44444	0.40000	2.5000	0.90000	1.1111	0.40000
3	3.3750	0.29630	0.21053	4.7500	0.71053	1.4074	0.73684
4	5.0625	0.19753	0.12308	8.1250	0.62308	1.6049	1.0154
5	7.5938	0.13169	0.07583	13.1875	0.57583	1.7366	1.2417
6	11.3906	0.08779	0.04812	20.781	0.54812	1.8244	1.4226
7	17.0859	0.05853	0.03108	32.172	0.53108	1.8829	1.5648
8	25.6289	0.03902	0.02030	49.258	0.52030	1.9220	1.6752
9	38.443	0.02601	0.01335	74.887	0.51335	1.9480	1.7596
10	57.665	0.01734	0.00882	113.330	0.50882	1.9653	1.8235
11	86.498	0.01156	0.00585	170.995	0.50585	1.9769	1.8713
12	129.746	0.00771	0.00388	257.493	0.50388	1.9846	1.9068
13	194.620	0.00514	0.00258	387.239	0.50258	1.9897	1.9329
14	291.929	0.00343	0.00172	581.86	0.50172	1.9931	1.9519
15	437.894	0.00228	0.00114	873.79	0.50114	1.9954	1.9657
16	656.841	0.00152	0.00076	1311.68	0.50076	1.9970	1.9756
17	985.261	0.00101	0.00051	1968.52	0.50051	1.9980	1.9827
18	1477.89	0.00068	0.00034	2953.78	0.50034	1.9986	1.9878
19	2216.84	0.00045	0.00023	4431.68	0.50023	1.9991	1.9914
20	3325.26	0.00030	0.00015	6648.51	0.50015	1.9994	1.9940
21	4987.89	0.00020	0.00010	9973.77	0.50010	1.9996	1.9958
22	7481.83	0.00013	0.00007	14 962.0	0.50007	1.9997	1.9971
23	11 223.0	0.00009	0.00004	22 443.0	0.50004	1.9998	1.9980
24	16 834.0	0.00006	0.00003	33 666.0	0.50003	1.9999	1.9986
25	25 251.0	0.00004	0.00002	50 500.0	0.50002	1.9999	1.9990
26	37 877.0	0.00003	0.00001	75 752.0	0.50001	1.9999	1.9993
27	56 815.0	0.00002	0.00001	113 628.0	0.50001	2.0000	1.9995
28	85 223.0	0.00001	0.00001	170 443.0	0.50001	2.0000	1.9997
29	127 834.0	0.00001	0.00000	255 666.0	0.50000	2.0000	1.9998
30	191 751.0	0.00001	0.00000	383 500.0	0.50000	2.0000	1.9998
31	287 627.0	0.00000	0.00000	575 251.0	0.50000	2.0000	1.9999
32	431 440.0	0.00000	0.00000	862 878.0	0.50000	2.0000	1.9999
33	647 160.0	0.00000	0.00000	1 294 318.0	0.50000	2.0000	1.9999
34	970 740.0	0.00000	0.00000	1 941 477.0	0.50000	2.0000	2.0000
35	1 456 110.0	0.00000	0.00000	2 912 217.0	0.50000	2.0000	2.0000

B

Compound Interest Factors for Continuous Compounding, Discrete Cash Flows

$r = 1\%$ **Continuous Compounding, Discrete Cash Flows**

	SINGLE PAYMENT		UNIFORM SERIES				Arithmetic Gradient Series Factor
	Compound Amount Factor	Present Worth Factor	Sinking Fund Factor	Uniform Series Factor	Capital Recovery Factor	Series Present Worth Factor	
N	(*F/P,r,N*)	(*P/F,r,N*)	(*A/F,r,N*)	(*F/A,r,N*)	(*A/P,r,N*)	(*P/A,r,N*)	(*A/G,r,N*)
1	1.0101	0.99005	1.0000	1.0000	1.0101	0.99005	0.00000
2	1.0202	0.98020	0.49750	2.0101	0.50755	1.97025	0.49750
3	1.0305	0.97045	0.33001	3.0303	0.34006	2.94069	0.99333
4	1.0408	0.96079	0.24626	4.0607	0.25631	3.90148	1.48750
5	1.0513	0.95123	0.19602	5.1015	0.20607	4.85271	1.98000
6	1.0618	0.94176	0.16253	6.1528	0.17258	5.79448	2.47084
7	1.0725	0.93239	0.13861	7.2146	0.14866	6.72687	2.96000
8	1.0833	0.92312	0.12067	8.2871	0.13072	7.64999	3.44751
9	1.0942	0.91393	0.10672	9.3704	0.11677	8.56392	3.93334
10	1.1052	0.90484	0.09556	10.4646	0.10561	9.46876	4.41751
11	1.1163	0.89583	0.08643	11.5698	0.09648	10.36459	4.90002
12	1.1275	0.88692	0.07883	12.6860	0.08888	11.25151	5.38086
13	1.1388	0.87810	0.07239	13.8135	0.08244	12.12961	5.86004
14	1.1503	0.86936	0.06688	14.9524	0.07693	12.99896	6.33755
15	1.1618	0.86071	0.06210	16.1026	0.07215	13.85967	6.81340
16	1.1735	0.85214	0.05792	17.2645	0.06797	14.71182	7.28759
17	1.1853	0.84366	0.05424	18.4380	0.06429	15.55548	7.76012
18	1.1972	0.83527	0.05096	19.6233	0.06101	16.39075	8.23098
19	1.2092	0.82696	0.04803	20.8205	0.05808	17.21771	8.70018
20	1.2214	0.81873	0.04539	22.0298	0.05544	18.03644	9.16772
21	1.2337	0.81058	0.04301	23.2512	0.05306	18.84703	9.63360
22	1.2461	0.80252	0.04084	24.4848	0.05089	19.64954	10.09782
23	1.2586	0.79453	0.03886	25.7309	0.04891	20.44408	10.56039
24	1.2712	0.78663	0.03705	26.9895	0.04710	21.23071	11.02129
25	1.2840	0.77880	0.03538	28.2608	0.04543	22.00951	11.48054
26	1.2969	0.77105	0.03385	29.5448	0.04390	22.78056	11.93813
27	1.3100	0.76338	0.03242	30.8417	0.04247	23.54394	12.39407
28	1.3231	0.75578	0.03110	32.1517	0.04115	24.29972	12.84835
29	1.3364	0.74826	0.02987	33.4748	0.03992	25.04798	13.30098
30	1.3499	0.74082	0.02873	34.8112	0.03878	25.78880	13.75196
31	1.3634	0.73345	0.02765	36.1611	0.03770	26.52225	14.20128
32	1.3771	0.72615	0.02665	37.5245	0.03670	27.24840	14.64895
33	1.3910	0.71892	0.02571	38.9017	0.03576	27.96732	15.09498
34	1.4049	0.71177	0.02482	40.2926	0.03487	28.67909	15.53935
35	1.4191	0.70469	0.02398	41.6976	0.03403	29.38378	15.98208
40	1.4918	0.67032	0.02043	48.9370	0.03048	32.80343	18.17104
45	1.5683	0.63763	0.01768	56.5475	0.02773	36.05630	20.31900
50	1.6487	0.60653	0.01549	64.5483	0.02554	39.15053	22.42613
55	1.7333	0.57695	0.01371	72.9593	0.02376	42.09385	24.49262
60	1.8221	0.54881	0.01222	81.8015	0.02227	44.89362	26.51868
65	1.9155	0.52205	0.01098	91.0971	0.02103	47.55684	28.50455
70	2.0138	0.49659	0.00991	100.869	0.01996	50.09018	30.45046
75	2.1170	0.47237	0.00900	111.142	0.01905	52.49997	32.35670
80	2.2255	0.44933	0.00820	121.942	0.01825	54.79223	34.22354
85	2.3396	0.42741	0.00750	133.296	0.01755	56.97269	36.05128
90	2.4596	0.40657	0.00689	145.232	0.01694	59.04681	37.84024
95	2.5857	0.38674	0.00634	157.779	0.01639	61.01978	39.59075
100	2.7183	0.36788	0.00585	170.970	0.01582	63.21206	41.30316

$r = 2\%$

Continuous Compounding, Discrete Cash Flows

	SINGLE PAYMENT		UNIFORM SERIES				Arithmetic Gradient Series Factor
	Compound Amount Factor	Present Worth Factor	Sinking Fund Factor	Uniform Series Factor	Capital Recovery Factor	Series Present Worth Factor	
N	(*F/P,r,N*)	(*P/F,r,N*)	(*A/F,r,N*)	(*F/A,r,N*)	(*A/P,r,N*)	(*P/A,r,N*)	(*A/G,r,N*)
1	1.0202	0.98020	1.0000	1.0000	1.0202	0.98020	0.00000
2	1.0408	0.96079	0.49500	2.0202	0.51520	1.94099	0.49500
3	1.0618	0.94176	0.32669	3.0610	0.34689	2.88275	0.98667
4	1.0833	0.92312	0.24255	4.1228	0.26275	3.80587	1.47500
5	1.1052	0.90484	0.19208	5.2061	0.21228	4.71071	1.96001
6	1.1275	0.88692	0.15845	6.3113	0.17865	5.59763	2.44168
7	1.1503	0.86936	0.13443	7.4388	0.15463	6.46699	2.92003
8	1.1735	0.85214	0.11643	8.5891	0.13663	7.31913	3.39505
9	1.1972	0.83527	0.10243	9.7626	0.12263	8.15440	3.86674
10	1.2214	0.81873	0.09124	10.9598	0.11144	8.97313	4.33511
11	1.2461	0.80252	0.08209	12.1812	0.10230	9.77565	4.80016
12	1.2712	0.78663	0.07448	13.4273	0.09468	10.56228	5.26190
13	1.2969	0.77105	0.06803	14.6985	0.08824	11.33333	5.72032
14	1.3231	0.75578	0.06252	15.9955	0.08272	12.08911	6.17543
15	1.3499	0.74082	0.05774	17.3186	0.07794	12.82993	6.62723
16	1.3771	0.72615	0.05357	18.6685	0.07377	13.55608	7.07573
17	1.4049	0.71177	0.04989	20.0456	0.07009	14.26785	7.52093
18	1.4333	0.69768	0.04662	21.4505	0.06682	14.96553	7.96283
19	1.4623	0.68386	0.04370	22.8839	0.06390	15.64939	8.40144
20	1.4918	0.67032	0.04107	24.3461	0.06128	16.31971	8.83677
21	1.5220	0.65705	0.03870	25.8380	0.05890	16.97675	9.26882
22	1.5527	0.64404	0.03655	27.3599	0.05675	17.62079	9.69759
23	1.5841	0.63128	0.03459	28.9126	0.05479	18.25207	10.12309
24	1.6161	0.61878	0.03279	30.4967	0.05299	18.87086	10.54533
25	1.6487	0.60653	0.03114	32.1128	0.05134	19.47739	10.96431
26	1.6820	0.59452	0.02962	33.7615	0.04982	20.07191	11.38005
27	1.7160	0.58275	0.02821	35.4435	0.04842	20.65466	11.79253
28	1.7507	0.57121	0.02691	37.1595	0.04711	21.22587	12.20178
29	1.7860	0.55990	0.02570	38.9102	0.04590	21.78576	12.60780
30	1.8221	0.54881	0.02457	40.6963	0.04477	22.33458	13.01059
31	1.8589	0.53794	0.02352	42.5184	0.04372	22.87252	13.41017
32	1.8965	0.52729	0.02253	44.3773	0.04274	23.39981	13.80654
33	1.9348	0.51685	0.02161	46.2738	0.04181	23.91666	14.19971
34	1.9739	0.50662	0.02074	48.2086	0.04094	24.42328	14.58969
35	2.0138	0.49659	0.01993	50.1824	0.04013	24.91987	14.97648
40	2.2255	0.44933	0.01648	60.6663	0.03668	27.25913	16.86302
45	2.4596	0.40657	0.01384	72.2528	0.03404	29.37579	18.67137
50	2.7183	0.36788	0.01176	85.0578	0.03196	31.29102	20.40283
55	3.0042	0.33287	0.01008	99.2096	0.03028	33.02399	22.05883
60	3.3201	0.30119	0.00871	114.850	0.02891	34.59205	23.64090
65	3.6693	0.27253	0.00757	132.135	0.02777	36.01089	25.15068
70	4.0552	0.24660	0.00661	151.237	0.02681	37.29471	26.58991
75	4.4817	0.22313	0.00580	172.349	0.02600	38.45635	27.96040
80	4.9530	0.20190	0.00511	195.682	0.02531	39.50745	29.26404
85	5.4739	0.18268	0.00452	221.468	0.02472	40.45853	30.50278
90	6.0496	0.16530	0.00400	249.966	0.02420	41.31910	31.67864
95	6.6859	0.14957	0.00355	281.461	0.02375	42.09777	32.79365
100	7.3891	0.13534	0.00316	316.269	0.02336	42.80234	33.84990

$r = 3\%$ **Continuous Compounding, Discrete Cash Flows**

	SINGLE PAYMENT		UNIFORM SERIES				Arithmetic Gradient Series Factor
	Compound Amount Factor	Present Worth Factor	Sinking Fund Factor	Uniform Series Factor	Capital Recovery Factor	Series Present Worth Factor	
N	(*F*/*P*,*r*,*N*)	(*P*/*F*,*r*,*N*)	(*A*/*F*,*r*,*N*)	(*F*/*A*,*r*,*N*)	(*A*/*P*,*r*,*N*)	(*P*/*A*,*r*,*N*)	(*A*/*G*,*r*,*N*)
1	1.0305	0.97045	1.0000	1.0000	1.0305	0.97045	0.00000
2	1.0618	0.94176	0.49250	2.0305	0.52296	1.91221	0.49250
3	1.0942	0.91393	0.32338	3.0923	0.35384	2.82614	0.98000
4	1.1275	0.88692	0.23886	4.1865	0.26932	3.71306	1.46251
5	1.1618	0.86071	0.18818	5.3140	0.21864	4.57377	1.94002
6	1.1972	0.83527	0.15442	6.4758	0.18488	5.40904	2.41255
7	1.2337	0.81058	0.13033	7.6730	0.16078	6.21962	2.88009
8	1.2712	0.78663	0.11228	8.9067	0.14273	7.00625	3.34265
9	1.3100	0.76338	0.09825	10.1779	0.12871	7.76963	3.80025
10	1.3499	0.74082	0.08705	11.4879	0.11750	8.51045	4.25287
11	1.3910	0.71892	0.07790	12.8378	0.10835	9.22937	4.70055
12	1.4333	0.69768	0.07028	14.2287	0.10073	9.92705	5.14328
13	1.4770	0.67706	0.06385	15.6621	0.09430	10.60411	5.58107
14	1.5220	0.65705	0.05835	17.1390	0.08880	11.26115	6.01393
15	1.5683	0.63763	0.05359	18.6610	0.08404	11.89878	6.44189
16	1.6161	0.61878	0.04943	20.2293	0.07989	12.51756	6.86494
17	1.6653	0.60050	0.04578	21.8454	0.07623	13.11806	7.28311
18	1.7160	0.58275	0.04253	23.5107	0.07299	13.70081	7.69641
19	1.7683	0.56553	0.03964	25.2267	0.07010	14.26633	8.10485
20	1.8221	0.54881	0.03704	26.9950	0.06750	14.81515	8.50845
21	1.8776	0.53259	0.03470	28.8171	0.06516	15.34774	8.90722
22	1.9348	0.51685	0.03258	30.6947	0.06303	15.86459	9.30119
23	1.9937	0.50158	0.03065	32.6295	0.06110	16.36617	9.69038
24	2.0544	0.48675	0.02888	34.6232	0.05934	16.85292	10.07479
25	2.1170	0.47237	0.02726	36.6776	0.05772	17.32528	10.45445
26	2.1815	0.45841	0.02578	38.7946	0.05623	17.78369	10.82939
27	2.2479	0.44486	0.02440	40.9761	0.05486	18.22855	11.19962
28	2.3164	0.43171	0.02314	43.2240	0.05359	18.66026	11.56517
29	2.3869	0.41895	0.02196	45.5404	0.05241	19.07921	11.92605
30	2.4596	0.40657	0.02086	47.9273	0.05132	19.48578	12.28230
31	2.5345	0.39455	0.01985	50.3869	0.05030	19.88033	12.63393
32	2.6117	0.38289	0.01890	52.9214	0.04935	20.26323	12.98098
33	2.6912	0.37158	0.01801	55.5331	0.04846	20.63480	13.32346
34	2.7732	0.36059	0.01717	58.2243	0.04763	20.99540	13.66140
35	2.8577	0.34994	0.01639	60.9975	0.04685	21.34534	13.99484
40	3.3201	0.30119	0.01313	76.1830	0.04358	22.94587	15.59532
45	3.8574	0.25924	0.01066	93.8259	0.04111	24.32346	17.08739
50	4.4817	0.22313	0.00875	114.324	0.03920	25.50917	18.47499
55	5.2070	0.19205	0.00724	138.140	0.03769	26.52971	19.76232
60	6.0496	0.16530	0.00603	165.809	0.03649	27.40811	20.95382
65	7.0287	0.14227	0.00505	197.957	0.03551	28.16415	22.05405
70	8.1662	0.12246	0.00425	235.307	0.03470	28.81487	23.06771
75	9.4877	0.10540	0.00359	278.702	0.03404	29.37496	23.99955
80	11.0232	0.09072	0.00304	329.119	0.03349	29.85703	24.85433
85	12.8071	0.07808	0.00258	387.696	0.03303	30.27196	25.63678
90	14.8797	0.06721	0.00219	455.753	0.03265	30.62908	26.35156
95	17.2878	0.05784	0.00187	534.823	0.03232	30.93647	27.00324
100	20.0855	0.04979	0.00160	626.690	0.03205	31.20103	27.59626

$r = 4\%$

Continuous Compounding, Discrete Cash Flows

	SINGLE PAYMENT		UNIFORM SERIES				
	Compound Amount Factor	Present Worth Factor	Sinking Fund Factor	Uniform Series Factor	Capital Recovery Factor	Series Present Worth Factor	Arithmetic Gradient Series Factor
N	(F/P,r,N)	(P/F,r,N)	(A/F,r,N)	(F/A,r,N)	(A/P,r,N)	(P/A,r,N)	(A/G,r,N)
1	1.0408	0.96079	1.0000	1.0000	1.0408	0.96079	0.00000
2	1.0833	0.92312	0.49000	2.0408	0.53081	1.88391	0.49000
3	1.1275	0.88692	0.32009	3.1241	0.36090	2.77083	0.97334
4	1.1735	0.85214	0.23521	4.2516	0.27602	3.62297	1.45002
5	1.2214	0.81873	0.18433	5.4251	0.22514	4.44170	1.92006
6	1.2712	0.78663	0.15045	6.6465	0.19127	5.22833	2.38345
7	1.3231	0.75578	0.12630	7.9178	0.16711	5.98411	2.84021
8	1.3771	0.72615	0.10821	9.2409	0.14903	6.71026	3.29036
9	1.4333	0.69768	0.09418	10.6180	0.13499	7.40794	3.73391
10	1.4918	0.67032	0.08298	12.0513	0.12379	8.07826	4.17089
11	1.5527	0.64404	0.07384	13.5432	0.11465	8.72229	4.60130
12	1.6161	0.61878	0.06624	15.0959	0.10705	9.34108	5.02517
13	1.6820	0.59452	0.05984	16.7120	0.10065	9.93560	5.44252
14	1.7507	0.57121	0.05437	18.3940	0.09518	10.50681	5.85339
15	1.8221	0.54881	0.04964	20.1447	0.09045	11.05562	6.25780
16	1.8965	0.52729	0.04552	21.9668	0.08633	11.58291	6.65577
17	1.9739	0.50662	0.04191	23.8633	0.08272	12.08953	7.04734
18	2.0544	0.48675	0.03870	25.8371	0.07951	12.57628	7.43255
19	2.1383	0.46767	0.03585	27.8916	0.07666	13.04395	7.81143
20	2.2255	0.44933	0.03330	30.0298	0.07411	13.49328	8.18401
21	2.3164	0.43171	0.03100	32.2554	0.07181	13.92499	8.55034
22	2.4109	0.41478	0.02893	34.5717	0.06974	14.33977	8.91045
23	2.5093	0.39852	0.02704	36.9826	0.06785	14.73829	9.26438
24	2.6117	0.38289	0.02532	39.4919	0.06613	15.12118	9.61219
25	2.7183	0.36788	0.02375	42.1036	0.06456	15.48906	9.95392
26	2.8292	0.35345	0.02231	44.8219	0.06312	15.84252	10.28960
27	2.9447	0.33960	0.02099	47.6511	0.06180	16.18211	10.61930
28	3.0649	0.32628	0.01976	50.5958	0.06058	16.50839	10.94305
29	3.1899	0.31349	0.01864	53.6607	0.05945	16.82188	11.26092
30	3.3201	0.30119	0.01759	56.8506	0.05840	17.12307	11.57295
31	3.4556	0.28938	0.01662	60.1707	0.05743	17.41246	11.87920
32	3.5966	0.27804	0.01572	63.6263	0.05653	17.69049	12.17971
33	3.7434	0.26714	0.01488	67.2230	0.05569	17.95763	12.47456
34	3.8962	0.25666	0.01409	70.9664	0.05490	18.21429	12.76379
35	4.0552	0.24660	0.01336	74.8626	0.05417	18.46089	13.04745
40	4.9530	0.20190	0.01032	96.8625	0.05113	19.55620	14.38452
45	6.0496	0.16530	0.00808	123.733	0.04889	20.45296	15.59182
50	7.3891	0.13534	0.00639	156.553	0.04720	21.18717	16.67745
55	9.0250	0.11080	0.00509	196.640	0.04590	21.78829	17.64976
60	11.0232	0.09072	0.00407	245.601	0.04488	22.28044	18.51721
65	13.4637	0.07427	0.00327	305.403	0.04409	22.68338	19.28820
70	16.4446	0.06081	0.00264	378.445	0.04345	23.01328	19.97102
75	20.0855	0.04979	0.00214	467.659	0.04295	23.28338	20.57366
80	24.5325	0.04076	0.00173	576.625	0.04255	23.50452	21.10378
85	29.9641	0.03337	0.00141	709.717	0.04222	23.68558	21.56867
90	36.5982	0.02732	0.00115	872.275	0.04196	23.83381	21.97512
95	44.7012	0.02237	0.00093	1070.82	0.04174	23.95517	22.32948
100	54.5982	0.01832	0.00076	1313.33	0.04157	24.05454	22.63760

r = 5% **Continuous Compounding, Discrete Cash Flows**

	SINGLE PAYMENT		UNIFORM SERIES				Arithmetic Gradient Series Factor
	Compound Amount Factor	Present Worth Factor	Sinking Fund Factor	Uniform Series Factor	Capital Recovery Factor	Series Present Worth Factor	
N	*(F/P,r,N)*	*(P/F,r,N)*	*(A/F,r,N)*	*(F/A,r,N)*	*(A/P,r,N)*	*(P/A,r,N)*	*(A/G,r,N)*
1	1.0513	0.95123	1.0000	1.0000	1.0513	0.95123	0.00000
2	1.1052	0.90484	0.48750	2.0513	0.53877	1.85607	0.48750
3	1.1618	0.86071	0.31681	3.1564	0.36808	2.71677	0.96668
4	1.2214	0.81873	0.23157	4.3183	0.28284	3.53551	1.43754
5	1.2840	0.77880	0.18052	5.5397	0.23179	4.31431	1.90011
6	1.3499	0.74082	0.14655	6.8237	0.19782	5.05512	2.35439
7	1.4191	0.70469	0.12235	8.1736	0.17362	5.75981	2.80042
8	1.4918	0.67032	0.10425	9.5926	0.15552	6.43013	3.23821
9	1.5683	0.63763	0.09022	11.0845	0.14149	7.06776	3.66780
10	1.6487	0.60653	0.07903	12.6528	0.13031	7.67429	4.08923
11	1.7333	0.57695	0.06992	14.3015	0.12119	8.25124	4.50252
12	1.8221	0.54881	0.06236	16.0347	0.11364	8.80005	4.90774
13	1.9155	0.52205	0.05600	17.8569	0.10727	9.32210	5.30491
14	2.0138	0.49659	0.05058	19.7724	0.10185	9.81868	5.69409
15	2.1170	0.47237	0.04590	21.7862	0.09717	10.29105	6.07534
16	2.2255	0.44933	0.04184	23.9032	0.09311	10.74038	6.44871
17	2.3396	0.42741	0.03827	26.1287	0.08954	11.16779	6.81425
18	2.4596	0.40657	0.03513	28.4683	0.08640	11.57436	7.17205
19	2.5857	0.38674	0.03233	30.9279	0.08360	11.96111	7.52215
20	2.7183	0.36788	0.02984	33.5137	0.08111	12.32898	7.86463
21	2.8577	0.34994	0.02760	36.2319	0.07887	12.67892	8.19957
22	3.0042	0.33287	0.02558	39.0896	0.07685	13.01179	8.52703
23	3.1582	0.31664	0.02376	42.0938	0.07503	13.32843	8.84710
24	3.3201	0.30119	0.02210	45.2519	0.07337	13.62962	9.15986
25	3.4903	0.28650	0.02059	48.5721	0.07186	13.91613	9.46539
26	3.6693	0.27253	0.01921	52.0624	0.07048	14.18866	9.76377
27	3.8574	0.25924	0.01794	55.7317	0.06921	14.44790	10.05510
28	4.0552	0.24660	0.01678	59.5891	0.06805	14.69450	10.33946
29	4.2631	0.23457	0.01571	63.6443	0.06698	14.92907	10.61695
30	4.4817	0.22313	0.01473	67.9074	0.06600	15.15220	10.88766
31	4.7115	0.21225	0.01381	72.3891	0.06509	15.36445	11.15168
32	4.9530	0.20190	0.01297	77.1006	0.06424	15.56634	11.40912
33	5.2070	0.19205	0.01219	82.0536	0.06346	15.75839	11.66006
34	5.4739	0.18268	0.01146	87.2606	0.06273	15.94108	11.90461
35	5.7546	0.17377	0.01078	92.7346	0.06205	16.11485	12.14288
40	7.3891	0.13534	0.00802	124.613	0.05930	16.86456	13.24346
45	9.4877	0.10540	0.00604	165.546	0.05731	17.44844	14.20240
50	12.1825	0.08208	0.00458	218.105	0.05586	17.90317	15.03289
55	15.6426	0.06393	0.00350	285.592	0.05477	18.25731	15.74801
60	20.0855	0.04979	0.00269	372.247	0.05396	18.53311	16.36042
65	25.7903	0.03877	0.00207	483.515	0.05334	18.74791	16.88218
70	33.1155	0.03020	0.00160	626.385	0.05287	18.91519	17.32453
75	42.5211	0.02352	0.00123	809.834	0.05251	19.04547	17.69786
80	54.5982	0.01832	0.00096	1045.39	0.05223	19.14694	18.01158
85	70.1054	0.01426	0.00074	1347.84	0.05201	19.22595	18.27416
90	90.0171	0.01111	0.00058	1736.20	0.05185	19.28749	18.49313
95	115.584	0.00865	0.00045	2234.87	0.05172	19.33542	18.67508
100	148.413	0.00674	0.00035	2875.17	0.05162	19.37275	18.82580

r = 6% **Continuous Compounding, Discrete Cash Flows**

	SINGLE PAYMENT		UNIFORM SERIES				Arithmetic Gradient Series Factor
	Compound Amount Factor	Present Worth Factor	Sinking Fund Factor	Uniform Series Factor	Capital Recovery Factor	Series Present Worth Factor	
N	(*F*/*P*,*r*,*N*)	(*P*/*F*,*r*,*N*)	(*A*/*F*,*r*,*N*)	(*F*/*A*,*r*,*N*)	(*A*/*P*,*r*,*N*)	(*P*/*A*,*r*,*N*)	(*A*/*G*,*r*,*N*)
1	1.0618	0.94176	1.0000	1.0000	1.0618	0.94176	0.00000
2	1.1275	0.88692	0.48500	2.0618	0.54684	1.82868	0.48500
3	1.1972	0.83527	0.31355	3.1893	0.37538	2.66396	0.96002
4	1.2712	0.78663	0.22797	4.3866	0.28981	3.45058	1.42508
5	1.3499	0.74082	0.17675	5.6578	0.23858	4.19140	1.88019
6	1.4333	0.69768	0.14270	7.0077	0.20454	4.88908	2.32539
7	1.5220	0.65705	0.11847	8.4410	0.18031	5.54612	2.76072
8	1.6161	0.61878	0.10037	9.9629	0.16221	6.16491	3.18622
9	1.7160	0.58275	0.08636	11.5790	0.14820	6.74766	3.60195
10	1.8221	0.54881	0.07522	13.2950	0.13705	7.29647	4.00797
11	1.9348	0.51685	0.06615	15.1171	0.12799	7.81332	4.40435
12	2.0544	0.48675	0.05864	17.0519	0.12048	8.30007	4.79114
13	2.1815	0.45841	0.05234	19.1064	0.11418	8.75848	5.16845
14	2.3164	0.43171	0.04698	21.2878	0.10881	9.19019	5.53633
15	2.4596	0.40657	0.04237	23.6042	0.10420	9.59676	5.89490
16	2.6117	0.38289	0.03837	26.0638	0.10020	9.97965	6.24424
17	2.7732	0.36059	0.03487	28.6755	0.09671	10.34025	6.58445
18	2.9447	0.33960	0.03180	31.4487	0.09363	10.67984	6.91564
19	3.1268	0.31982	0.02908	34.3934	0.09091	10.99966	7.23792
20	3.3201	0.30119	0.02665	37.5202	0.08849	11.30085	7.55141
21	3.5254	0.28365	0.02449	40.8403	0.08632	11.58451	7.85622
22	3.7434	0.26714	0.02254	44.3657	0.08438	11.85164	8.15248
23	3.9749	0.25158	0.02079	48.1091	0.08262	12.10322	8.44032
24	4.2207	0.23693	0.01920	52.0840	0.08104	12.34015	8.71986
25	4.4817	0.22313	0.01776	56.3047	0.07960	12.56328	8.99124
26	4.7588	0.21014	0.01645	60.7864	0.07829	12.77342	9.25460
27	5.0531	0.19790	0.01526	65.5452	0.07709	12.97131	9.51008
28	5.3656	0.18637	0.01416	70.5983	0.07600	13.15769	9.75782
29	5.6973	0.17552	0.01316	75.9639	0.07500	13.33321	9.99796
30	6.0496	0.16530	0.01225	81.6612	0.07408	13.49851	10.23066
31	6.4237	0.15567	0.01140	87.7109	0.07324	13.65418	10.45605
32	6.8210	0.14661	0.01062	94.1346	0.07246	13.80079	10.67429
33	7.2427	0.13807	0.00991	100.956	0.07174	13.93886	10.88553
34	7.6906	0.13003	0.00924	108.198	0.07108	14.06889	11.08992
35	8.1662	0.12246	0.00863	115.889	0.07047	14.19134	11.28761
40	11.0232	0.09072	0.00617	162.091	0.06801	14.70461	12.18092
45	14.8797	0.06721	0.00446	224.458	0.06629	15.08484	12.92953
50	20.0855	0.04979	0.00324	308.645	0.06508	15.36653	13.55188
55	27.1126	0.03688	0.00237	422.285	0.06420	15.57520	14.06541
60	36.5982	0.02732	0.00174	575.683	0.06357	15.72980	14.48619
65	49.4024	0.02024	0.00128	782.748	0.06311	15.84432	14.82876
70	66.6863	0.01500	0.00094	1062.26	0.06278	15.92916	15.10600
75	90.0171	0.01111	0.00069	1439.56	0.06253	15.99202	15.32913
80	121.510	0.00823	0.00051	1948.85	0.06235	16.03858	15.50782
85	164.022	0.00610	0.00038	2636.34	0.06222	16.07307	15.65026
90	221.406	0.00452	0.00028	3564.34	0.06212	16.09863	15.76333
95	298.867	0.00335	0.00021	4817.01	0.06204	16.11756	15.85273
100	403.429	0.00248	0.00015	6507.94	0.06199	16.13158	15.92318

r = 7% **Continuous Compounding, Discrete Cash Flows**

	SINGLE PAYMENT		UNIFORM SERIES				Arithmetic Gradient Series Factor
	Compound Amount Factor	Present Worth Factor	Sinking Fund Factor	Uniform Series Factor	Capital Recovery Factor	Series Present Worth Factor	
N	(*F/P,r,N*)	(*P/F,r,N*)	(*A/F,r,N*)	(*F/A,r,N*)	(*A/P,r,N*)	(*P/A,r,N*)	(*A/G,r,N*)
1	1.0725	0.93239	1.0000	1.0000	1.0725	0.93239	0.00000
2	1.1503	0.86936	0.48251	2.0725	0.55502	1.80175	0.48251
3	1.2337	0.81058	0.31029	3.2228	0.38280	2.61234	0.95337
4	1.3231	0.75578	0.22439	4.4565	0.29690	3.36812	1.41262
5	1.4191	0.70469	0.17302	5.7796	0.24553	4.07281	1.86030
6	1.5220	0.65705	0.13891	7.1987	0.21142	4.72985	2.29645
7	1.6323	0.61263	0.11467	8.7206	0.18718	5.34248	2.72114
8	1.7507	0.57121	0.09659	10.3529	0.16910	5.91369	3.13444
9	1.8776	0.53259	0.08262	12.1036	0.15513	6.44628	3.53643
10	2.0138	0.49659	0.07152	13.9812	0.14403	6.94287	3.92721
11	2.1598	0.46301	0.06252	15.9950	0.13503	7.40588	4.30688
12	2.3164	0.43171	0.05508	18.1547	0.12759	7.83759	4.67555
13	2.4843	0.40252	0.04885	20.4711	0.12136	8.24012	5.03334
14	2.6645	0.37531	0.04356	22.9554	0.11607	8.61543	5.38039
15	2.8577	0.34994	0.03903	25.6199	0.11154	8.96536	5.71683
16	3.0649	0.32628	0.03512	28.4775	0.10762	9.29164	6.04282
17	3.2871	0.30422	0.03170	31.5424	0.10421	9.59587	6.35849
18	3.5254	0.28365	0.02871	34.8295	0.10122	9.87952	6.66402
19	3.7810	0.26448	0.02607	38.3549	0.09858	10.14400	6.95958
20	4.0552	0.24660	0.02373	42.1359	0.09624	10.39059	7.24533
21	4.3492	0.22993	0.02165	46.1911	0.09416	10.62052	7.52146
22	4.6646	0.21438	0.01979	50.5404	0.09229	10.83490	7.78815
23	5.0028	0.19989	0.01811	55.2050	0.09062	11.03479	8.04559
24	5.3656	0.18637	0.01661	60.2078	0.08912	11.22116	8.29397
25	5.7546	0.17377	0.01525	65.5733	0.08776	11.39494	8.53348
26	6.1719	0.16203	0.01402	71.3279	0.08653	11.55696	8.76434
27	6.6194	0.15107	0.01290	77.4998	0.08541	11.70803	8.98674
28	7.0993	0.14086	0.01189	84.1192	0.08440	11.84889	9.20088
29	7.6141	0.13134	0.01096	91.2185	0.08347	11.98023	9.40697
30	8.1662	0.12246	0.01012	98.8326	0.08263	12.10268	9.60521
31	8.7583	0.11418	0.00935	106.999	0.08185	12.21686	9.79582
32	9.3933	0.10646	0.00864	115.757	0.08115	12.32332	9.97900
33	10.0744	0.09926	0.00799	125.150	0.08050	12.42258	10.15495
34	10.8049	0.09255	0.00740	135.225	0.07990	12.51513	10.32389
35	11.5883	0.08629	0.00685	146.030	0.07936	12.60143	10.48603
40	16.4446	0.06081	0.00469	213.006	0.07720	12.95288	11.20165
45	23.3361	0.04285	0.00325	308.049	0.07575	13.20055	11.77687
50	33.1155	0.03020	0.00226	442.922	0.07477	13.37508	12.23466
55	46.9931	0.02128	0.00158	634.315	0.07408	13.49807	12.59571
60	66.6863	0.01500	0.00110	905.916	0.07361	13.58473	12.87812
65	94.6324	0.01057	0.00077	1291.34	0.07328	13.64581	13.09734
70	134.290	0.00745	0.00054	1838.27	0.07305	13.68885	13.26638
75	190.566	0.00525	0.00038	2614.41	0.07289	13.71918	13.39591
80	270.426	0.00370	0.00027	3715.81	0.07278	13.74055	13.49462
85	383.753	0.00261	0.00019	5278.76	0.07270	13.75561	13.56947
90	544.572	0.00184	0.00013	7496.70	0.07264	13.76622	13.62598
95	772.784	0.00129	0.00009	10 644.0	0.07260	13.77370	13.66846
100	1096.63	0.00091	0.00007	15 110.0	0.07257	13.77897	13.70028

r = 8%

Continuous Compounding, Discrete Cash Flows

	SINGLE PAYMENT		UNIFORM SERIES				Arithmetic Gradient Series Factor
	Compound Amount Factor	Present Worth Factor	Sinking Fund Factor	Uniform Series Factor	Capital Recovery Factor	Series Present Worth Factor	
N	(*F*/*P*,*r*,*N*)	(*P*/*F*,*r*,*N*)	(*A*/*F*,*r*,*N*)	(*F*/*A*,*r*,*N*)	(*A*/*P*,*r*,*N*)	(*P*/*A*,*r*,*N*)	(*A*/*G*,*r*,*N*)
1	1.0833	0.92312	1.0000	1.0000	1.0833	0.92312	0.00000
2	1.1735	0.85214	0.48001	2.0833	0.56330	1.77526	0.48001
3	1.2712	0.78663	0.30705	3.2568	0.39034	2.56189	0.94672
4	1.3771	0.72615	0.22085	4.5280	0.30413	3.28804	1.40018
5	1.4918	0.67032	0.16934	5.9052	0.25263	3.95836	1.84044
6	1.6161	0.61878	0.13519	7.3970	0.21848	4.57714	2.26758
7	1.7507	0.57121	0.11095	9.0131	0.19424	5.14835	2.68169
8	1.8965	0.52729	0.09290	10.7637	0.17619	5.67564	3.08288
9	2.0544	0.48675	0.07899	12.6602	0.16227	6.16239	3.47127
10	2.2255	0.44933	0.06796	14.7147	0.15125	6.61172	3.84700
11	2.4109	0.41478	0.05903	16.9402	0.14232	7.02651	4.21022
12	2.6117	0.38289	0.05168	19.3511	0.13496	7.40940	4.56110
13	2.8292	0.35345	0.04553	21.9628	0.12882	7.76285	4.89980
14	3.0649	0.32628	0.04034	24.7920	0.12362	8.08913	5.22653
15	3.3201	0.30119	0.03590	27.8569	0.11918	8.39033	5.54147
16	3.5966	0.27804	0.03207	31.1770	0.11536	8.66836	5.84486
17	3.8962	0.25666	0.02876	34.7736	0.11204	8.92503	6.13689
18	4.2207	0.23693	0.02586	38.6698	0.10915	9.16195	6.41781
19	4.5722	0.21871	0.02332	42.8905	0.10660	9.38067	6.68785
20	4.9530	0.20190	0.02107	47.4627	0.10436	9.58256	6.94726
21	5.3656	0.18637	0.01908	52.4158	0.10237	9.76894	7.19628
22	5.8124	0.17204	0.01731	57.7813	0.10059	9.94098	7.43518
23	6.2965	0.15882	0.01572	63.5938	0.09901	10.09980	7.66421
24	6.8210	0.14661	0.01431	69.8903	0.09760	10.24641	7.88363
25	7.3891	0.13534	0.01304	76.7113	0.09632	10.38174	8.09372
26	8.0045	0.12493	0.01189	84.1003	0.09518	10.50667	8.29475
27	8.6711	0.11533	0.01086	92.1048	0.09414	10.62200	8.48698
28	9.3933	0.10646	0.00992	100.776	0.09321	10.72845	8.67068
29	10.1757	0.09827	0.00908	110.169	0.09236	10.82673	8.84614
30	11.0232	0.09072	0.00831	120.345	0.09160	10.91745	9.01360
31	11.9413	0.08374	0.00761	131.368	0.09090	11.00119	9.17336
32	12.9358	0.07730	0.00698	143.309	0.09026	11.07849	9.32566
33	14.0132	0.07136	0.00640	156.245	0.08969	11.14986	9.47078
34	15.1803	0.06587	0.00587	170.258	0.08916	11.21573	9.60898
35	16.4446	0.06081	0.00539	185.439	0.08868	11.27654	9.74051
40	24.5325	0.04076	0.00354	282.547	0.08683	11.51725	10.30689
45	36.5982	0.02732	0.00234	427.416	0.08563	11.67860	10.74256
50	54.5982	0.01832	0.00155	643.535	0.08484	11.78676	11.07380
55	81.4509	0.01228	0.00104	965.947	0.08432	11.85926	11.32302
60	121.510	0.00823	0.00069	1446.93	0.08398	11.90785	11.50878
65	181.272	0.00552	0.00046	2164.47	0.08375	11.94043	11.64610
70	270.426	0.00370	0.00031	3234.91	0.08360	11.96227	11.74685
75	403.429	0.00248	0.00021	4831.83	0.08349	11.97690	11.82030
80	601.845	0.00166	0.00014	7214.15	0.08343	11.98672	11.87352
85	897.847	0.00111	0.00009	10 768.0	0.08338	11.99329	11.91189
90	1339.43	0.00075	0.00006	16 070.0	0.08335	11.99770	11.93942
95	1998.20	0.00050	0.00004	23 980.0	0.08333	12.00066	11.95910
100	2980.96	0.00034	0.00003	35 779.0	0.08332	12.00264	11.97311

r = 9% **Continuous Compounding, Discrete Cash Flows**

	SINGLE PAYMENT		UNIFORM SERIES				Arithmetic Gradient Series Factor
	Compound Amount Factor	Present Worth Factor	Sinking Fund Factor	Uniform Series Factor	Capital Recovery Factor	Series Present Worth Factor	
N	(*F*/*P*,*r*,*N*)	(*P*/*F*,*r*,*N*)	(*A*/*F*,*r*,*N*)	(*F*/*A*,*r*,*N*)	(*A*/*P*,*r*,*N*)	(*P*/*A*,*r*,*N*)	(*A*/*G*,*r*,*N*)
1	1.0942	0.91393	1.0000	1.0000	1.0942	0.91393	0.00000
2	1.1972	0.83527	0.47752	2.0942	0.57169	1.74920	0.47752
3	1.3100	0.76338	0.30382	3.2914	0.39800	2.51258	0.94008
4	1.4333	0.69768	0.21733	4.6014	0.31150	3.21026	1.38776
5	1.5683	0.63763	0.16571	6.0347	0.25988	3.84789	1.82063
6	1.7160	0.58275	0.13153	7.6030	0.22570	4.43063	2.23880
7	1.8776	0.53259	0.10731	9.3190	0.20148	4.96323	2.64241
8	2.0544	0.48675	0.08931	11.1966	0.18349	5.44998	3.03160
9	2.2479	0.44486	0.07547	13.2510	0.16964	5.89484	3.40654
10	2.4596	0.40657	0.06452	15.4990	0.15869	6.30141	3.76743
11	2.6912	0.37158	0.05568	17.9586	0.14986	6.67298	4.11449
12	2.9447	0.33960	0.04843	20.6498	0.14260	7.01258	4.44793
13	3.2220	0.31037	0.04238	23.5945	0.13656	7.32294	4.76801
14	3.5254	0.28365	0.03729	26.8165	0.13146	7.60660	5.07498
15	3.8574	0.25924	0.03296	30.3419	0.12713	7.86584	5.36913
16	4.2207	0.23693	0.02924	34.1993	0.12341	8.10277	5.65074
17	4.6182	0.21654	0.02603	38.4200	0.12020	8.31930	5.92011
18	5.0531	0.19790	0.02324	43.0382	0.11741	8.51720	6.17755
19	5.5290	0.18087	0.02079	48.0913	0.11497	8.69807	6.42339
20	6.0496	0.16530	0.01865	53.6202	0.11282	8.86337	6.65794
21	6.6194	0.15107	0.01676	59.6699	0.11093	9.01444	6.88154
22	7.2427	0.13807	0.01509	66.2893	0.10926	9.15251	7.09452
23	7.9248	0.12619	0.01360	73.5320	0.10777	9.27869	7.29723
24	8.6711	0.11533	0.01228	81.4568	0.10645	9.39402	7.49000
25	9.4877	0.10540	0.01110	90.1280	0.10527	9.49942	7.67318
26	10.3812	0.09633	0.01004	99.6157	0.10421	9.59574	7.84712
27	11.3589	0.08804	0.00909	109.997	0.10327	9.68378	8.01215
28	12.4286	0.08046	0.00824	121.356	0.10241	9.76424	8.16862
29	13.5991	0.07353	0.00747	133.784	0.10165	9.83778	8.31685
30	14.8797	0.06721	0.00679	147.383	0.10096	9.90498	8.45719
31	16.2810	0.06142	0.00616	162.263	0.10034	9.96640	8.58995
32	17.8143	0.05613	0.00560	178.544	0.09978	10.02254	8.71547
33	19.4919	0.05130	0.00509	196.358	0.09927	10.07384	8.83405
34	21.3276	0.04689	0.00463	215.850	0.09881	10.12073	8.94600
35	23.3361	0.04285	0.00422	237.178	0.09839	10.16358	9.05164
40	36.5982	0.02732	0.00265	378.004	0.09682	10.32847	9.49496
45	57.3975	0.01742	0.00167	598.863	0.09584	10.43361	9.82070
50	90.0171	0.01111	0.00106	945.238	0.09523	10.50065	10.05692
55	141.175	0.00708	0.00067	1488.46	0.09485	10.54339	10.22624
60	221.406	0.00452	0.00043	2340.41	0.09460	10.57065	10.34639
65	347.234	0.00288	0.00027	3676.53	0.09445	10.58803	10.43088
70	544.572	0.00184	0.00017	5771.98	0.09435	10.59911	10.48983
75	854.059	0.00117	0.00011	9058.30	0.09428	10.60618	10.53069
80	1339.43	0.00075	0.00007	14 212.0	0.09424	10.61068	10.55884
85	2100.65	0.00048	0.00004	22 295.0	0.09422	10.61356	10.57813
90	3294.47	0.00030	0.00003	34 972.0	0.09420	10.61539	10.59128
95	5166.75	0.00019	0.00002	54 853.0	0.09419	10.61655	10.60022
100	8103.08	0.00012	0.00001	86 033.0	0.09419	10.61730	10.60627

r = 10% **Continuous Compounding, Discrete Cash Flows**

	SINGLE PAYMENT		UNIFORM SERIES				
	Compound Amount Factor	Present Worth Factor	Sinking Fund Factor	Uniform Series Factor	Capital Recovery Factor	Series Present Worth Factor	Arithmetic Gradient Series Factor
N	$(F/P,r,N)$	$(P/F,r,N)$	$(A/F,r,N)$	$(F/A,r,N)$	$(A/P,r,N)$	$(P/A,r,N)$	$(A/G,r,N)$
1	1.1052	0.90484	1.0000	1.0000	1.1052	0.90484	0.00000
2	1.2214	0.81873	0.47502	2.1052	0.58019	1.72357	0.47502
3	1.3499	0.74082	0.30061	3.3266	0.40578	2.46439	0.93344
4	1.4918	0.67032	0.21384	4.6764	0.31901	3.13471	1.37535
5	1.6487	0.60653	0.16212	6.1683	0.26729	3.74124	1.80086
6	1.8221	0.54881	0.12793	7.8170	0.23310	4.29005	2.21012
7	2.0138	0.49659	0.10374	9.6391	0.20892	4.78663	2.60329
8	2.2255	0.44933	0.08582	11.6528	0.19099	5.23596	2.98060
9	2.4596	0.40657	0.07205	13.8784	0.17723	5.64253	3.34227
10	2.7183	0.36788	0.06121	16.3380	0.16638	6.01041	3.68856
11	3.0042	0.33287	0.05248	19.0563	0.15765	6.34328	4.01976
12	3.3201	0.30119	0.04533	22.0604	0.15050	6.64448	4.33618
13	3.6693	0.27253	0.03940	25.3806	0.14457	6.91701	4.63814
14	4.0552	0.24660	0.03442	29.0499	0.13959	7.16361	4.92598
15	4.4817	0.22313	0.03021	33.1051	0.13538	7.38674	5.20008
16	4.9530	0.20190	0.02661	37.5867	0.13178	7.58863	5.46081
17	5.4739	0.18268	0.02351	42.5398	0.12868	7.77132	5.70856
18	6.0496	0.16530	0.02083	48.0137	0.12600	7.93662	5.94373
19	6.6859	0.14957	0.01850	54.0634	0.12367	8.08618	6.16673
20	7.3891	0.13534	0.01646	60.7493	0.12163	8.22152	6.37798
21	8.1662	0.12246	0.01468	68.1383	0.11985	8.34398	6.57790
22	9.0250	0.11080	0.01311	76.3045	0.11828	8.45478	6.76690
23	9.9742	0.10026	0.01172	85.3295	0.11689	8.55504	6.94542
24	11.0232	0.09072	0.01049	95.3037	0.11566	8.64576	7.11388
25	12.1825	0.08208	0.00940	106.327	0.11458	8.72784	7.27269
26	13.4637	0.07427	0.00844	118.509	0.11361	8.80211	7.42228
27	14.8797	0.06721	0.00758	131.973	0.11275	8.86932	7.56305
28	16.4446	0.06081	0.00681	146.853	0.11198	8.93013	7.69541
29	18.1741	0.05502	0.00612	163.297	0.11129	8.98515	7.81975
30	20.0855	0.04979	0.00551	181.472	0.11068	9.03494	7.93646
31	22.1980	0.04505	0.00496	201.557	0.11013	9.07999	8.04593
32	24.5325	0.04076	0.00447	223.755	0.10964	9.12075	8.14851
33	27.1126	0.03688	0.00403	248.288	0.10920	9.15763	8.24458
34	29.9641	0.03337	0.00363	275.400	0.10880	9.19101	8.33446
35	33.1155	0.03020	0.00327	305.364	0.10845	9.22121	8.41851
40	54.5982	0.01832	0.00196	509.629	0.10713	9.33418	8.76204
45	90.0171	0.01111	0.00118	846.404	0.10635	9.40270	9.00281
50	148.413	0.00674	0.00071	1401.65	0.10588	9.44427	9.16915
55	244.692	0.00409	0.00043	2317.10	0.10560	9.46947	9.28264
60	403.429	0.00248	0.00026	3826.43	0.10543	9.48476	9.35924
65	665.142	0.00150	0.00016	6314.88	0.10533	9.49404	9.41046
70	1096.63	0.00091	0.00010	10 418.0	0.10527	9.49966	9.44444
75	1808.04	0.00055	0.00006	17 182.0	0.10523	9.50307	9.46683

r = 11% **Continuous Compounding, Discrete Cash Flows**

	SINGLE PAYMENT		UNIFORM SERIES				Arithmetic Gradient Series Factor
	Compound Amount Factor	Present Worth Factor	Sinking Fund Factor	Uniform Series Factor	Capital Recovery Factor	Series Present Worth Factor	
N	(*F*/*P*,*r*,*N*)	(*P*/*F*,*r*,*N*)	(*A*/*F*,*r*,*N*)	(*F*/*A*,*r*,*N*)	(*A*/*P*,*r*,*N*)	(*P*/*A*,*r*,*N*)	(*A*/*G*,*r*,*N*)
1	1.1163	0.89583	1.0000	1.0000	1.1163	0.89583	0.00000
2	1.2461	0.80252	0.47253	2.1163	0.58881	1.69835	0.47253
3	1.3910	0.71892	0.29741	3.3624	0.41369	2.41728	0.92681
4	1.5527	0.64404	0.21038	4.7533	0.32666	3.06131	1.36297
5	1.7333	0.57695	0.15858	6.3060	0.27486	3.63826	1.78115
6	1.9348	0.51685	0.12439	8.0393	0.24067	4.15511	2.18154
7	2.1598	0.46301	0.10026	9.9741	0.21654	4.61813	2.56437
8	2.4109	0.41478	0.08241	12.1338	0.19869	5.03291	2.92993
9	2.6912	0.37158	0.06875	14.5447	0.18503	5.40449	3.27852
10	3.0042	0.33287	0.05802	17.2360	0.17430	5.73736	3.61047
11	3.3535	0.29820	0.04941	20.2401	0.16568	6.03556	3.92615
12	3.7434	0.26714	0.04238	23.5936	0.15866	6.30269	4.22597
13	4.1787	0.23931	0.03658	27.3370	0.15286	6.54200	4.51035
14	4.6646	0.21438	0.03173	31.5157	0.14801	6.75638	4.77973
15	5.2070	0.19205	0.02764	36.1803	0.14392	6.94843	5.03457
16	5.8124	0.17204	0.02416	41.3873	0.14044	7.12048	5.27536
17	6.4883	0.15412	0.02119	47.1998	0.13746	7.27460	5.50257
18	7.2427	0.13807	0.01863	53.6881	0.13490	7.41267	5.71673
19	8.0849	0.12369	0.01641	60.9308	0.13269	7.53636	5.91832
20	9.0250	0.11080	0.01449	69.0157	0.13077	7.64716	6.10787
21	10.0744	0.09926	0.01281	78.0407	0.12909	7.74642	6.28588
22	11.2459	0.08892	0.01135	88.1151	0.12763	7.83534	6.45287
23	12.5535	0.07966	0.01006	99.3610	0.12634	7.91500	6.60934
24	14.0132	0.07136	0.00894	111.915	0.12521	7.98636	6.75579
25	15.6426	0.06393	0.00794	125.928	0.12422	8.05029	6.89273
26	17.4615	0.05727	0.00706	141.570	0.12334	8.10756	7.02063
27	19.4919	0.05130	0.00629	159.032	0.12257	8.15886	7.13998
28	21.7584	0.04596	0.00560	178.524	0.12188	8.20482	7.25122
29	24.2884	0.04117	0.00499	200.282	0.12127	8.24599	7.35482
30	27.1126	0.03688	0.00445	224.571	0.12073	8.28288	7.45120
31	30.2652	0.03304	0.00397	251.683	0.12025	8.31592	7.54080
32	33.7844	0.02960	0.00355	281.949	0.11982	8.34552	7.62400
33	37.7128	0.02652	0.00317	315.733	0.11945	8.37203	7.70121
34	42.0980	0.02375	0.00283	353.446	0.11911	8.39579	7.77278
35	46.9931	0.02128	0.00253	395.544	0.11881	8.41707	7.83909
40	81.4509	0.01228	0.00145	691.883	0.11772	8.49449	8.10288
45	141.175	0.00708	0.00083	1205.52	0.11711	8.53916	8.27905
50	244.692	0.00409	0.00048	2095.77	0.11676	8.56493	8.39490
55	424.113	0.00236	0.00027	3638.80	0.11655	8.57980	8.47009

$r = 12\%$ **Continuous Compounding, Discrete Cash Flows**

	SINGLE PAYMENT		UNIFORM SERIES				Arithmetic Gradient Series Factor
	Compound Amount Factor	Present Worth Factor	Sinking Fund Factor	Uniform Series Factor	Capital Recovery Factor	Series Present Worth Factor	
N	*(F/P,r,N)*	*(P/F,r,N)*	*(A/F,r,N)*	*(F/A,r,N)*	*(A/P,r,N)*	*(P/A,r,N)*	*(A/G,r,N)*
1	1.1275	0.88692	1.0000	1.0000	1.1275	0.88692	0.00000
2	1.2712	0.78663	0.47004	2.1275	0.59753	1.67355	0.47004
3	1.4333	0.69768	0.29423	3.3987	0.42172	2.37122	0.92019
4	1.6161	0.61878	0.20695	4.8321	0.33445	2.99001	1.35061
5	1.8221	0.54881	0.15508	6.4481	0.28258	3.53882	1.76148
6	2.0544	0.48675	0.12092	8.2703	0.24841	4.02557	2.15307
7	2.3164	0.43171	0.09686	10.3247	0.22435	4.45728	2.52566
8	2.6117	0.38289	0.07911	12.6411	0.20660	4.84018	2.87962
9	2.9447	0.33960	0.06556	15.2528	0.19306	5.17977	3.21532
10	3.3201	0.30119	0.05495	18.1974	0.18245	5.48097	3.53320
11	3.7434	0.26714	0.04647	21.5176	0.17397	5.74810	3.83374
12	4.2207	0.23693	0.03959	25.2610	0.16708	5.98503	4.11743
13	4.7588	0.21014	0.03392	29.4817	0.16142	6.19516	4.38480
14	5.3656	0.18637	0.02921	34.2405	0.15670	6.38154	4.63641
15	6.0496	0.16530	0.02525	39.6061	0.15275	6.54684	4.87283
16	6.8210	0.14661	0.02190	45.6557	0.14940	6.69344	5.09464
17	7.6906	0.13003	0.01906	52.4767	0.14655	6.82347	5.30246
18	8.6711	0.11533	0.01662	60.1673	0.14412	6.93880	5.49687
19	9.7767	0.10228	0.01453	68.8384	0.14202	7.04108	5.67850
20	11.0232	0.09072	0.01272	78.6151	0.14022	7.13180	5.84796
21	12.4286	0.08046	0.01116	89.6383	0.13865	7.21226	6.00583
22	14.0132	0.07136	0.00980	102.067	0.13729	7.28362	6.15274
23	15.7998	0.06329	0.00861	116.080	0.13611	7.34691	6.28926
24	17.8143	0.05613	0.00758	131.880	0.13508	7.40305	6.41597
25	20.0855	0.04979	0.00668	149.694	0.13418	7.45283	6.53344
26	22.6464	0.04416	0.00589	169.780	0.13339	7.49699	6.64221
27	25.5337	0.03916	0.00520	192.426	0.13269	7.53616	6.74280
28	28.7892	0.03474	0.00459	217.960	0.13208	7.57089	6.83574
29	32.4597	0.03081	0.00405	246.749	0.13155	7.60170	6.92152
30	36.5982	0.02732	0.00358	279.209	0.13108	7.62902	7.00059
31	41.2644	0.02423	0.00317	315.807	0.13066	7.65326	7.07342
32	46.5255	0.02149	0.00280	357.071	0.13030	7.67475	7.14043
33	52.4573	0.01906	0.00248	403.597	0.12997	7.69381	7.20202
34	59.1455	0.01691	0.00219	456.054	0.12969	7.71072	7.25859
35	66.6863	0.01500	0.00194	515.200	0.12944	7.72572	7.31050
40	121.510	0.00823	0.00106	945.203	0.12855	7.77878	7.51141
45	221.406	0.00452	0.00058	1728.72	0.12808	7.80791	7.63916
50	403.429	0.00248	0.00032	3156.38	0.12781	7.82389	7.71909
55	735.095	0.00136	0.00017	5757.75	0.12767	7.83266	7.76841

r = 13% **Continuous Compounding, Discrete Cash Flows**

	SINGLE PAYMENT		UNIFORM SERIES				Arithmetic Gradient Series Factor
	Compound Amount Factor	Present Worth Factor	Sinking Fund Factor	Uniform Series Factor	Capital Recovery Factor	Series Present Worth Factor	
N	(*F*/*P*,*r*,*N*)	(*P*/*F*,*r*,*N*)	(*A*/*F*,*r*,*N*)	(*F*/*A*,*r*,*N*)	(*A*/*P*,*r*,*N*)	(*P*/*A*,*r*,*N*)	(*A*/*G*,*r*,*N*)
1	1.1388	0.87810	1.0000	1.0000	1.1388	0.87810	0.00000
2	1.2969	0.77105	0.46755	2.1388	0.60637	1.64915	0.46755
3	1.4770	0.67706	0.29106	3.4358	0.42988	2.32620	0.91358
4	1.6820	0.59452	0.20355	4.9127	0.34238	2.92072	1.33827
5	1.9155	0.52205	0.15164	6.5948	0.29046	3.44277	1.74189
6	2.1815	0.45841	0.11750	8.5103	0.25633	3.90118	2.12473
7	2.4843	0.40252	0.09353	10.6918	0.23236	4.30370	2.48718
8	2.8292	0.35345	0.07589	13.1761	0.21472	4.65716	2.82968
9	3.2220	0.31037	0.06248	16.0053	0.20131	4.96752	3.15272
10	3.6693	0.27253	0.05201	19.2273	0.19084	5.24005	3.45683
11	4.1787	0.23931	0.04367	22.8966	0.18250	5.47936	3.74260
12	4.7588	0.21014	0.03693	27.0753	0.17576	5.68950	4.01065
13	5.4195	0.18452	0.03141	31.8341	0.17024	5.87402	4.26162
14	6.1719	0.16203	0.02684	37.2536	0.16567	6.03604	4.49618
15	7.0287	0.14227	0.02303	43.4255	0.16186	6.17832	4.71503
16	8.0045	0.12493	0.01982	50.4542	0.15865	6.30325	4.91888
17	9.1157	0.10970	0.01711	58.4586	0.15593	6.41295	5.10844
18	10.3812	0.09633	0.01480	67.5743	0.15363	6.50928	5.28441
19	11.8224	0.08458	0.01283	77.9556	0.15166	6.59386	5.44753
20	13.4637	0.07427	0.01114	89.7780	0.14997	6.66814	5.59848
21	15.3329	0.06522	0.00969	103.242	0.14851	6.73335	5.73798
22	17.4615	0.05727	0.00843	118.575	0.14726	6.79062	5.86669
23	19.8857	0.05029	0.00735	136.036	0.14618	6.84091	5.98528
24	22.6464	0.04416	0.00641	155.922	0.14524	6.88507	6.09441
25	25.7903	0.03877	0.00560	178.568	0.14443	6.92384	6.19468
26	29.3708	0.03405	0.00489	204.359	0.14372	6.95789	6.28670
27	33.4483	0.02990	0.00428	233.729	0.14311	6.98779	6.37104
28	38.0918	0.02625	0.00374	267.178	0.14257	7.01404	6.44825
29	43.3801	0.02305	0.00328	305.269	0.14210	7.03709	6.51885
30	49.4024	0.02024	0.00287	348.650	0.14170	7.05733	6.58333
31	56.2609	0.01777	0.00251	398.052	0.14134	7.07511	6.64216
32	64.0715	0.01561	0.00220	454.313	0.14103	7.09071	6.69578
33	72.9665	0.01370	0.00193	518.384	0.14076	7.10442	6.74459
34	83.0963	0.01203	0.00169	591.351	0.14052	7.11645	6.78899
35	94.6324	0.01057	0.00148	674.447	0.14031	7.12702	6.82934
40	181.272	0.00552	0.00077	1298.53	0.13960	7.16340	6.98125
45	347.234	0.00288	0.00040	2493.97	0.13923	7.18239	7.07317
50	665.142	0.00150	0.00021	4783.90	0.13904	7.19231	7.12785
55	1274.11	0.00078	0.00011	9170.36	0.13894	7.19748	7.15994

r = 14%

Continuous Compounding, Discrete Cash Flows

	SINGLE PAYMENT		UNIFORM SERIES				Arithmetic Gradient Series Factor
	Compound Amount Factor	Present Worth Factor	Sinking Fund Factor	Uniform Series Factor	Capital Recovery Factor	Series Present Worth Factor	
N	(*F*/*P*,*r*,*N*)	(*P*/*F*,*r*,*N*)	(*A*/*F*,*r*,*N*)	(*F*/*A*,*r*,*N*)	(*A*/*P*,*r*,*N*)	(*P*/*A*,*r*,*N*)	(*A*/*G*,*r*,*N*)
1	1.1503	0.86936	1.0000	1.0000	1.1503	0.86936	0.00000
2	1.3231	0.75578	0.46506	2.1503	0.61533	1.62514	0.46506
3	1.5220	0.65705	0.28790	3.4734	0.43818	2.28219	0.90697
4	1.7507	0.57121	0.20019	4.9954	0.35046	2.85340	1.32596
5	2.0138	0.49659	0.14824	6.7460	0.29851	3.34998	1.72235
6	2.3164	0.43171	0.11416	8.7598	0.26443	3.78169	2.09652
7	2.6645	0.37531	0.09028	11.0762	0.24056	4.15700	2.44894
8	3.0649	0.32628	0.07278	13.7406	0.22305	4.48328	2.78015
9	3.5254	0.28365	0.05950	16.8055	0.20978	4.76694	3.09076
10	4.0552	0.24660	0.04919	20.3309	0.19946	5.01354	3.38141
11	4.6646	0.21438	0.04101	24.3861	0.19128	5.22792	3.65282
12	5.3656	0.18637	0.03442	29.0507	0.18470	5.41429	3.90573
13	6.1719	0.16203	0.02906	34.4162	0.17933	5.57632	4.14092
14	7.0993	0.14086	0.02464	40.5881	0.17491	5.71717	4.35918
15	8.1662	0.12246	0.02097	47.6874	0.17124	5.83963	4.56135
16	9.3933	0.10646	0.01790	55.8536	0.16818	5.94609	4.74824
17	10.8049	0.09255	0.01533	65.2469	0.16560	6.03864	4.92069
18	12.4286	0.08046	0.01315	76.0518	0.16342	6.11910	5.07952
19	14.2963	0.06995	0.01130	88.4804	0.16158	6.18905	5.22555
20	16.4446	0.06081	0.00973	102.777	0.16000	6.24986	5.35957
21	18.9158	0.05287	0.00839	119.221	0.15866	6.30272	5.48237
22	21.7584	0.04596	0.00724	138.137	0.15751	6.34868	5.59471
23	25.0281	0.03996	0.00625	159.896	0.15653	6.38864	5.69731
24	28.7892	0.03474	0.00541	184.924	0.15568	6.42337	5.79087
25	33.1155	0.03020	0.00468	213.713	0.15495	6.45357	5.87608
26	38.0918	0.02625	0.00405	246.828	0.15433	6.47982	5.95356
27	43.8160	0.02282	0.00351	284.920	0.15378	6.50265	6.02392
28	50.4004	0.01984	0.00304	328.736	0.15332	6.52249	6.08772
29	57.9743	0.01725	0.00264	379.137	0.15291	6.53974	6.14552
30	66.6863	0.01500	0.00229	437.111	0.15256	6.55473	6.19780
31	76.7075	0.01304	0.00198	503.797	0.15226	6.56777	6.24505
32	88.2347	0.01133	0.00172	580.505	0.15200	6.57910	6.28769
33	101.494	0.00985	0.00150	668.740	0.15177	6.58895	6.32614
34	116.746	0.00857	0.00130	770.234	0.15157	6.59752	6.36077
35	134.290	0.00745	0.00113	886.980	0.15140	6.60497	6.39193
40	270.426	0.00370	0.00056	1792.90	0.15083	6.62991	6.50606
45	544.572	0.00184	0.00028	3617.21	0.15055	6.64230	6.57173
50	1096.63	0.00091	0.00014	7290.91	0.15041	6.64845	6.60888
55	2208.35	0.00045	0.00007	14689.0	0.15034	6.65151	6.62960

$r = 15\%$ **Continuous Compounding, Discrete Cash Flows**

	SINGLE PAYMENT		UNIFORM SERIES				Arithmetic Gradient Series Factor
	Compound Amount Factor	Present Worth Factor	Sinking Fund Factor	Uniform Series Factor	Capital Recovery Factor	Series Present Worth Factor	
N	(*F/P,r,N*)	(*P/F,r,N*)	(*A/F,r,N*)	(*F/A,r,N*)	(*A/P,r,N*)	(*P/A,r,N*)	(*A/G,r,N*)
1	1.1618	0.86071	1.0000	1.0000	1.1618	0.86071	0.00000
2	1.3499	0.74082	0.46257	2.1618	0.62440	1.60153	0.46257
3	1.5683	0.63763	0.28476	3.5117	0.44660	2.23915	0.90037
4	1.8221	0.54881	0.19685	5.0800	0.35868	2.78797	1.31369
5	2.1170	0.47237	0.14488	6.9021	0.30672	3.26033	1.70289
6	2.4596	0.40657	0.11088	9.0191	0.27271	3.66690	2.06846
7	2.8577	0.34994	0.08712	11.4787	0.24895	4.01684	2.41096
8	3.3201	0.30119	0.06975	14.3364	0.23159	4.31803	2.73106
9	3.8574	0.25924	0.05664	17.6565	0.21847	4.57727	3.02947
10	4.4817	0.22313	0.04648	21.5139	0.20832	4.80040	3.30699
11	5.2070	0.19205	0.03847	25.9956	0.20030	4.99245	3.56446
12	6.0496	0.16530	0.03205	31.2026	0.19388	5.15775	3.80276
13	7.0287	0.14227	0.02684	37.2522	0.18868	5.30003	4.02281
14	8.1662	0.12246	0.02258	44.2809	0.18442	5.42248	4.22554
15	9.4877	0.10540	0.01907	52.4471	0.18090	5.52788	4.41191
16	11.0232	0.09072	0.01615	61.9348	0.17798	5.61860	4.58286
17	12.8071	0.07808	0.01371	72.9580	0.17554	5.69668	4.73935
18	14.8797	0.06721	0.01166	85.7651	0.17349	5.76389	4.88231
19	17.2878	0.05784	0.00994	100.645	0.17177	5.82173	5.01264
20	20.0855	0.04979	0.00848	117.933	0.17031	5.87152	5.13125
21	23.3361	0.04285	0.00725	138.018	0.16908	5.91437	5.23898
22	27.1126	0.03688	0.00620	161.354	0.16803	5.95125	5.33666
23	31.5004	0.03175	0.00531	188.467	0.16714	5.98300	5.42507
24	36.5982	0.02732	0.00455	219.967	0.16638	6.01032	5.50497
25	42.5211	0.02352	0.00390	256.565	0.16573	6.03384	5.57706
26	49.4024	0.02024	0.00334	299.087	0.16518	6.05408	5.64200
27	57.3975	0.01742	0.00287	348.489	0.16470	6.07151	5.70042
28	66.6863	0.01500	0.00246	405.886	0.16430	6.08650	5.75289
29	77.4785	0.01291	0.00212	472.573	0.16395	6.09941	5.79997
30	90.0171	0.01111	0.00182	550.051	0.16365	6.11052	5.84215
31	104.585	0.00956	0.00156	640.068	0.16340	6.12008	5.87989
32	121.510	0.00823	0.00134	744.653	0.16318	6.12831	5.91362
33	141.175	0.00708	0.00115	866.164	0.16299	6.13539	5.94374
34	164.022	0.00610	0.00099	1007.34	0.16283	6.14149	5.97060
35	190.566	0.00525	0.00085	1171.36	0.16269	6.14674	5.99453

r = 20%

Continuous Compounding, Discrete Cash Flows

	SINGLE PAYMENT		UNIFORM SERIES				Arithmetic Gradient Series Factor
	Compound Amount Factor	Present Worth Factor	Sinking Fund Factor	Uniform Series Factor	Capital Recovery Factor	Series Present Worth Factor	
N	*(F/P,r,N)*	*(P/F,r,N)*	*(A/F,r,N)*	*(F/A,r,N)*	*(A/P,r,N)*	*(P/A,r,N)*	*(A/G,r,N)*
1	1.2214	0.81873	1.0000	1.0000	1.2214	0.81873	0.00000
2	1.4918	0.67032	0.45017	2.2214	0.67157	1.48905	0.45017
3	1.8221	0.54881	0.26931	3.7132	0.49071	2.03786	0.86755
4	2.2255	0.44933	0.18066	5.5353	0.40206	2.48719	1.25279
5	2.7183	0.36788	0.12885	7.7609	0.35025	2.85507	1.60677
6	3.3201	0.30119	0.09543	10.4792	0.31683	3.15627	1.93058
7	4.0552	0.24660	0.07247	13.7993	0.29387	3.40286	2.22548
8	4.9530	0.20190	0.05601	17.8545	0.27741	3.60476	2.49289
9	6.0496	0.16530	0.04385	22.8075	0.26525	3.77006	2.73435
10	7.3891	0.13534	0.03465	28.8572	0.25606	3.90539	2.95148
11	9.0250	0.11080	0.02759	36.2462	0.24899	4.01620	3.14594
12	11.0232	0.09072	0.02209	45.2712	0.24349	4.10691	3.31943
13	13.4637	0.07427	0.01776	56.2944	0.23917	4.18119	3.47363
14	16.4446	0.06081	0.01434	69.7581	0.23574	4.24200	3.61019
15	20.0855	0.04979	0.01160	86.2028	0.23300	4.29178	3.73072
16	24.5325	0.04076	0.00941	106.288	0.23081	4.33255	3.83675
17	29.9641	0.03337	0.00764	130.821	0.22905	4.36592	3.92972
18	36.5982	0.02732	0.00622	160.785	0.22762	4.39324	4.01101
19	44.7012	0.02237	0.00507	197.383	0.22647	4.41561	4.08188
20	54.5982	0.01832	0.00413	242.084	0.22553	4.43393	4.14351
21	66.6863	0.01500	0.00337	296.683	0.22477	4.44893	4.19695
22	81.4509	0.01228	0.00275	363.369	0.22415	4.46120	4.24320
23	99.4843	0.01005	0.00225	444.820	0.22365	4.47125	4.28312
24	121.510	0.00823	0.00184	544.304	0.22324	4.47948	4.31750
25	148.413	0.00674	0.00150	665.814	0.22290	4.48622	4.34706
26	181.272	0.00552	0.00123	814.228	0.22263	4.49174	4.37243
27	221.406	0.00452	0.00100	995.500	0.22241	4.49626	4.39415
28	270.426	0.00370	0.00082	1216.91	0.22222	4.49995	4.41273
29	330.300	0.00303	0.00067	1487.33	0.22208	4.50298	4.42859
30	403.429	0.00248	0.00055	1817.63	0.22195	4.50546	4.44211
31	492.749	0.00203	0.00045	2221.06	0.22185	4.50749	4.45362
32	601.845	0.00166	0.00037	2713.81	0.22177	4.50915	4.46340
33	735.095	0.00136	0.00030	3315.66	0.22170	4.51051	4.47170
34	897.847	0.00111	0.00025	4050.75	0.22165	4.51163	4.47874
35	1096.63	0.00091	0.00020	4948.60	0.22160	4.51254	4.48471

r = 25% **Continuous Compounding, Discrete Cash Flows**

	SINGLE PAYMENT		UNIFORM SERIES				
	Compound Amount Factor	Present Worth Factor	Sinking Fund Factor	Uniform Series Factor	Capital Recovery Factor	Series Present Worth Factor	Arithmetic Gradient Series Factor
N	(*F*/*P*,*r*,*N*)	(*P*/*F*,*r*,*N*)	(*A*/*F*,*r*,*N*)	(*F*/*A*,*r*,*N*)	(*A*/*P*,*r*,*N*)	(*P*/*A*,*r*,*N*)	(*A*/*G*,*r*,*N*)
1	1.2840	0.77880	1.0000	1.0000	1.2840	0.77880	0.00000
2	1.6487	0.60653	0.43782	2.2840	0.72185	1.38533	0.43782
3	2.1170	0.47237	0.25428	3.9327	0.53830	1.85770	0.83505
4	2.7183	0.36788	0.16530	6.0497	0.44932	2.22558	1.19290
5	3.4903	0.28650	0.11405	8.7680	0.39808	2.51208	1.51306
6	4.4817	0.22313	0.08158	12.2584	0.36560	2.73521	1.79751
7	5.7546	0.17377	0.05974	16.7401	0.34376	2.90899	2.04855
8	7.3891	0.13534	0.04445	22.4947	0.32848	3.04432	2.26867
9	9.4877	0.10540	0.03346	29.8837	0.31749	3.14972	2.46046
10	12.1825	0.08208	0.02540	39.3715	0.30942	3.23181	2.62656
11	15.6426	0.06393	0.01940	51.5539	0.30342	3.29573	2.76958
12	20.0855	0.04979	0.01488	67.1966	0.29891	3.34552	2.89206
13	25.7903	0.03877	0.01146	87.2821	0.29548	3.38429	2.99641
14	33.1155	0.03020	0.00884	113.072	0.29287	3.41449	3.08488
15	42.5211	0.02352	0.00684	146.188	0.29087	3.43801	3.15955
16	54.5982	0.01832	0.00530	188.709	0.28932	3.45633	3.22229
17	70.1054	0.01426	0.00411	243.307	0.28814	3.47059	3.27481
18	90.0171	0.01111	0.00319	313.413	0.28722	3.48170	3.31860
19	115.584	0.00865	0.00248	403.430	0.28650	3.49035	3.35499
20	148.413	0.00674	0.00193	519.014	0.28595	3.49709	3.38514
21	190.566	0.00525	0.00150	667.427	0.28552	3.50234	3.41003
22	244.692	0.00409	0.00117	857.993	0.28519	3.50642	3.43053
23	314.191	0.00318	0.00091	1102.69	0.28493	3.50961	3.44737
24	403.429	0.00248	0.00071	1416.88	0.28473	3.51208	3.46117
25	518.013	0.00193	0.00055	1820.30	0.28457	3.51401	3.47246
26	665.142	0.00150	0.00043	2338.32	0.28445	3.51552	3.48166
27	854.059	0.00117	0.00033	3003.46	0.28436	3.51669	3.48916
28	1096.63	0.00091	0.00026	3857.52	0.28428	3.51760	3.49526
29	1408.10	0.00071	0.00020	4954.15	0.28423	3.51831	3.50020
30	1808.04	0.00055	0.00016	6362.26	0.28418	3.51886	3.50421
31	2321.57	0.00043	0.00012	8170.30	0.28415	3.51930	3.50745
32	2980.96	0.00034	0.00010	10 492.0	0.28412	3.51963	3.51007
33	3827.63	0.00026	0.00007	13 473.0	0.28410	3.51989	3.51219
34	4914.77	0.00020	0.00006	17 300.0	0.28408	3.52010	3.51389
35	6310.69	0.00016	0.00005	22 215.0	0.28407	3.52025	3.51526

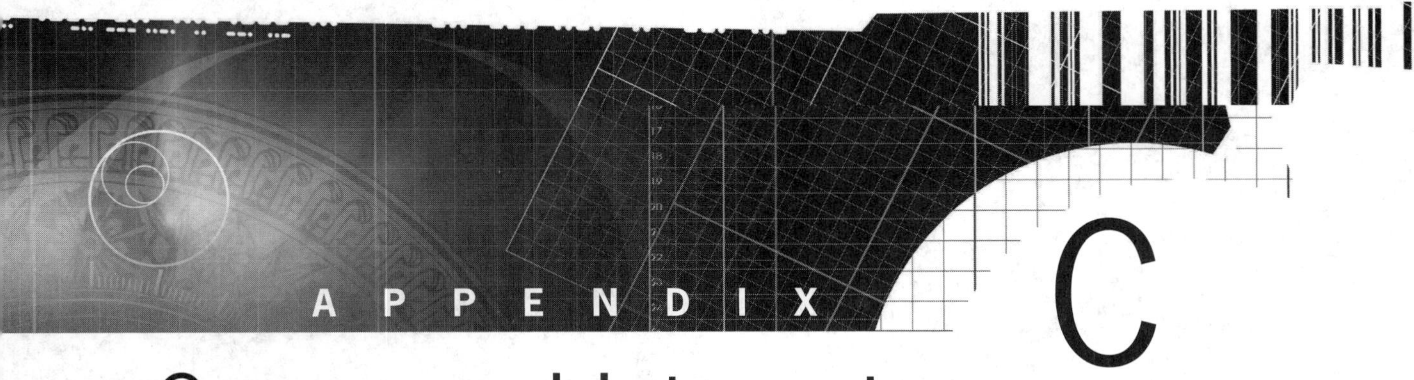

C

Compound Interest Factors for Continuous Compounding, Continuous Compounding Periods

$r = 1\%$ **Continuous Compounding, Continuous Compounding Periods**

	Sinking Fund Factor	Uniform Series Factor	Capital Recovery Factor	Series Present Worth Factor
T	$(A/F,r,T)$	$(F/A,r,T)$	$(A/P,r,T)$	$(P/A,r,T)$
1	0.99501	1.0050	1.0050	0.99502
2	0.49502	2.0201	0.50502	1.9801
3	0.32836	3.0455	0.33836	2.9554
4	0.24503	4.0811	0.25503	3.9211
5	0.19504	5.1271	0.20504	4.8771
6	0.16172	6.1837	0.17172	5.8235
7	0.13792	7.2508	0.14792	6.7606
8	0.12007	8.3287	0.13007	7.6884
9	0.10619	9.4174	0.11619	8.6069
10	0.09508	10.5171	0.10508	9.5163
11	0.08600	11.6278	0.09600	10.4166
12	0.07843	12.7497	0.08843	11.3080
13	0.07203	13.8828	0.08203	12.1905
14	0.06655	15.0274	0.07655	13.0642
15	0.06179	16.1834	0.07179	13.9292
16	0.05763	17.3511	0.06763	14.7856
17	0.05397	18.5305	0.06397	15.6335
18	0.05071	19.7217	0.06071	16.4730
19	0.04779	20.9250	0.05779	17.3041
20	0.04517	22.1403	0.05517	18.1269
21	0.04279	23.3678	0.05279	18.9416
22	0.04064	24.6077	0.05064	19.7481
23	0.03867	25.8600	0.04867	20.5466
24	0.03687	27.1249	0.04687	21.3372
25	0.03521	28.4025	0.04521	22.1199
26	0.03368	29.6930	0.04368	22.8948
27	0.03226	30.9964	0.04226	23.6621
28	0.03095	32.3130	0.04095	24.4216
29	0.02972	33.6427	0.03972	25.1736
30	0.02858	34.9859	0.03858	25.9182
31	0.02752	36.3425	0.03752	26.6553
32	0.02652	37.7128	0.03652	27.3851
33	0.02558	39.0968	0.03558	28.1076
34	0.02469	40.4948	0.03469	28.8230
35	0.02386	41.9068	0.03386	29.5312
40	0.02033	49.1825	0.03033	32.9680
45	0.01760	56.8312	0.02760	36.2372
50	0.01541	64.8721	0.02541	39.3469
55	0.01364	73.3253	0.02364	42.3050
60	0.01216	82.2119	0.02216	45.1188
65	0.01092	91.5541	0.02092	47.7954
70	0.00986	101.375	0.01986	50.3415
75	0.00895	111.700	0.01895	52.7633
80	0.00816	122.554	0.01816	55.0671
85	0.00746	133.965	0.01746	57.2585
90	0.00685	145.960	0.01685	59.3430
95	0.00631	158.571	0.01631	61.3259
100	0.00582	171.828	0.01582	63.2121

$r = 2\%$ **Continuous Compounding, Continuous Compounding Periods**

	Sinking Fund Factor	Uniform Series Factor	Capital Recovery Factor	Series Present Worth Factor
T	$(A/F,r,T)$	$(F/A,r,T)$	$(A/P,r,T)$	$(P/A,r,T)$
1	0.99003	1.0101	1.0100	0.99007
2	0.49007	2.0405	0.51007	1.9605
3	0.32343	3.0918	0.34343	2.9118
4	0.24013	4.1644	0.26013	3.8442
5	0.19017	5.2585	0.21017	4.7581
6	0.15687	6.3748	0.17687	5.6540
7	0.13309	7.5137	0.15309	6.5321
8	0.11527	8.6755	0.13527	7.3928
9	0.10141	9.8609	0.12141	8.2365
10	0.09033	11.0701	0.11033	9.0635
11	0.08128	12.3038	0.10128	9.8741
12	0.07373	13.5625	0.09373	10.6686
13	0.06736	14.8465	0.08736	11.4474
14	0.06189	16.1565	0.08189	12.2108
15	0.05717	17.4929	0.07717	12.9591
16	0.05303	18.8564	0.07303	13.6925
17	0.04939	20.2474	0.06939	14.4115
18	0.04615	21.6665	0.06615	15.1162
19	0.04326	23.1142	0.06326	15.8069
20	0.04066	24.5912	0.06066	16.4840
21	0.03832	26.0981	0.05832	17.1477
22	0.03619	27.6354	0.05619	17.7982
23	0.03424	29.2037	0.05424	18.4358
24	0.03246	30.8037	0.05246	19.0608
25	0.03083	32.4361	0.05083	19.6735
26	0.02932	34.1014	0.04932	20.2740
27	0.02793	35.8003	0.04793	20.8626
28	0.02664	37.5336	0.04664	21.4395
29	0.02544	39.3019	0.04544	22.0051
30	0.02433	41.1059	0.04433	22.5594
31	0.02328	42.9464	0.04328	23.1028
32	0.02231	44.8240	0.04231	23.6354
33	0.02140	46.7396	0.04140	24.1574
34	0.02054	48.6939	0.04054	24.6692
35	0.01973	50.6876	0.03973	25.1707
40	0.01632	61.2770	0.03632	27.5336
45	0.01370	72.9802	0.03370	29.6715
50	0.01164	85.9141	0.03164	31.6060
55	0.00998	100.208	0.02998	33.3564
60	0.00862	116.006	0.02862	34.9403
65	0.00749	133.465	0.02749	36.3734
70	0.00655	152.760	0.02655	37.6702
75	0.00574	174.084	0.02574	38.8435
80	0.00506	197.652	0.02506	39.9052
85	0.00447	223.697	0.02447	40.8658
90	0.00396	252.482	0.02396	41.7351
95	0.00352	284.295	0.02352	42.5216
100	0.00313	319.453	0.02313	43.2332

$r = 3\%$ **Continuous Compounding, Continuous Compounding Periods**

	Sinking Fund Factor	Uniform Series Factor	Capital Recovery Factor	Series Present Worth Factor
T	*(A/F,r,T)*	*(F/A,r,T)*	*(A/P,r,T)*	*(P/A,r,T)*
1	0.98507	1.0152	1.0151	0.98515
2	0.48515	2.0612	0.51515	1.9412
3	0.31856	3.1391	0.34856	2.8690
4	0.23530	4.2499	0.26530	3.7693
5	0.18537	5.3945	0.21537	4.6431
6	0.15212	6.5739	0.18212	5.4910
7	0.12838	7.7893	0.15838	6.3139
8	0.11060	9.0416	0.14060	7.1124
9	0.09679	10.3321	0.12679	7.8874
10	0.08575	11.6620	0.11575	8.6394
11	0.07673	13.0323	0.10673	9.3692
12	0.06923	14.4443	0.09923	10.0775
13	0.06290	15.8994	0.09290	10.7648
14	0.05748	17.3987	0.08748	11.4318
15	0.05279	18.9437	0.08279	12.0791
16	0.04870	20.5358	0.07870	12.7072
17	0.04509	22.1764	0.07509	13.3168
18	0.04190	23.8669	0.07190	13.9084
19	0.03905	25.6089	0.06905	14.4825
20	0.03649	27.4040	0.06649	15.0396
21	0.03418	29.2537	0.06418	15.5803
22	0.03209	31.1597	0.06209	16.1050
23	0.03019	33.1239	0.06019	16.6141
24	0.02845	35.1478	0.05845	17.1083
25	0.02686	37.2333	0.05686	17.5878
26	0.02539	39.3824	0.05539	18.0531
27	0.02404	41.5969	0.05404	18.5047
28	0.02279	43.8789	0.05279	18.9430
29	0.02163	46.2304	0.05163	19.3683
30	0.02055	48.6534	0.05055	19.7810
31	0.01955	51.1503	0.04955	20.1815
32	0.01861	53.7232	0.04861	20.5702
33	0.01774	56.3745	0.04774	20.9474
34	0.01692	59.1065	0.04692	21.3135
35	0.01615	61.9217	0.04615	21.6687
40	0.01293	77.3372	0.04293	23.2935
45	0.01050	95.2475	0.04050	24.6920
50	0.00862	116.056	0.03862	25.8957
55	0.00713	140.233	0.03713	26.9317
60	0.00594	168.322	0.03594	27.8234
65	0.00498	200.956	0.03498	28.5909
70	0.00419	238.872	0.03419	29.2515
75	0.00353	282.925	0.03353	29.8200
80	0.00299	334.106	0.03299	30.3094
85	0.00254	393.570	0.03254	30.7306
90	0.00216	462.658	0.03216	31.0931
95	0.00184	542.926	0.03184	31.4052
100	0.00157	636.185	0.03157	31.6738

$r = 4\%$ **Continuous Compounding, Continuous Compounding Periods**

	Sinking Fund Factor	Uniform Series Factor	Capital Recovery Factor	Series Present Worth Factor
T	*(A/F,r,T)*	*(F/A,r,T)*	*(A/P,r,T)*	*(P/A,r,T)*
1	0.98013	1.0203	1.0201	0.98026
2	0.48027	2.0822	0.52027	1.9221
3	0.31373	3.1874	0.35373	2.8270
4	0.23053	4.3378	0.27053	3.6964
5	0.18067	5.5351	0.22067	4.5317
6	0.14747	6.7812	0.18747	5.3343
7	0.12379	8.0782	0.16379	6.1054
8	0.10606	9.4282	0.14606	6.8463
9	0.09231	10.8332	0.13231	7.5581
10	0.08133	12.2956	0.12133	8.2420
11	0.07237	13.8177	0.11237	8.8991
12	0.06493	15.4019	0.10493	9.5304
13	0.05865	17.0507	0.09865	10.1370
14	0.05329	18.7668	0.09329	10.7198
15	0.04865	20.5530	0.08865	11.2797
16	0.04462	22.4120	0.08462	11.8177
17	0.04107	24.3469	0.08107	12.3346
18	0.03794	26.3608	0.07794	12.8312
19	0.03514	28.4569	0.07514	13.3083
20	0.03264	30.6385	0.07264	13.7668
21	0.03039	32.9092	0.07039	14.2072
22	0.02835	35.2725	0.06835	14.6304
23	0.02650	37.7323	0.06650	15.0370
24	0.02482	40.2924	0.06482	15.4277
25	0.02328	42.9570	0.06328	15.8030
26	0.02187	45.7304	0.06187	16.1636
27	0.02057	48.6170	0.06057	16.5101
28	0.01937	51.6214	0.05937	16.8430
29	0.01827	54.7483	0.05827	17.1628
30	0.01724	58.0029	0.05724	17.4701
31	0.01629	61.3903	0.05629	17.7654
32	0.01540	64.9160	0.05540	18.0491
33	0.01458	68.5855	0.05458	18.3216
34	0.01381	72.4048	0.05381	18.5835
35	0.01309	76.3800	0.05309	18.8351
40	0.01012	98.8258	0.05012	19.9526
45	0.00792	126.241	0.04792	20.8675
50	0.00626	159.726	0.04626	21.6166
55	0.00498	200.625	0.04498	22.2299
60	0.00399	250.579	0.04399	22.7321
65	0.00321	311.593	0.04321	23.1432
70	0.00259	386.116	0.04259	23.4797
75	0.00210	477.138	0.04210	23.7553
80	0.00170	588.313	0.04170	23.9809
85	0.00138	724.103	0.04138	24.1657
90	0.00112	889.956	0.04112	24.3169
95	0.00092	1092.53	0.04092	24.4407
100	0.00075	1339.95	0.04075	24.5421

$r = 5\%$ **Continuous Compounding, Continuous Compounding Periods**

	Sinking Fund Factor	Uniform Series Factor	Capital Recovery Factor	Series Present Worth Factor
T	$(A/F,r,T)$	$(F/A,r,T)$	$(A/P,r,T)$	$(P/A,r,T)$
1	0.97521	1.0254	1.0252	0.97541
2	0.47542	2.1034	0.52542	1.9033
3	0.30896	3.2367	0.35896	2.7858
4	0.22583	4.4281	0.27583	3.6254
5	0.17604	5.6805	0.22604	4.4240
6	0.14291	6.9972	0.19291	5.1836
7	0.11931	8.3814	0.16931	5.9062
8	0.10166	9.8365	0.15166	6.5936
9	0.08798	11.3662	0.13798	7.2474
10	0.07707	12.9744	0.12707	7.8694
11	0.06819	14.6651	0.11819	8.4610
12	0.06082	16.4424	0.11082	9.0238
13	0.05461	18.3108	0.10461	9.5591
14	0.04932	20.2751	0.09932	10.0683
15	0.04476	22.3400	0.09476	10.5527
16	0.04080	24.5108	0.09080	11.0134
17	0.03732	26.7929	0.08732	11.4517
18	0.03426	29.1921	0.08426	11.8686
19	0.03153	31.7142	0.08153	12.2652
20	0.02910	34.3656	0.07910	12.6424
21	0.02692	37.1530	0.07692	13.0012
22	0.02495	40.0833	0.07495	13.3426
23	0.02317	43.1639	0.07317	13.6673
24	0.02155	46.4023	0.07155	13.9761
25	0.02008	49.8069	0.07008	14.2699
26	0.01873	53.3859	0.06873	14.5494
27	0.01750	57.1485	0.06750	14.8152
28	0.01637	61.1040	0.06637	15.0681
29	0.01532	65.2623	0.06532	15.3086
30	0.01436	69.6338	0.06436	15.5374
31	0.01347	74.2294	0.06347	15.7550
32	0.01265	79.0606	0.06265	15.9621
33	0.01189	84.1396	0.06189	16.1590
34	0.01118	89.4789	0.06118	16.3463
35	0.01052	95.0921	0.06052	16.5245
40	0.00783	127.781	0.05783	17.2933
45	0.00589	169.755	0.05589	17.8920
50	0.00447	223.650	0.05447	18.3583
55	0.00341	292.853	0.05341	18.7214
60	0.00262	381.711	0.05262	19.0043
65	0.00202	495.807	0.05202	19.2245
70	0.00156	642.309	0.05156	19.3961
75	0.00120	830.422	0.05120	19.5296
80	0.00093	1071.96	0.05093	19.6337
85	0.00072	1382.11	0.05072	19.7147
90	0.00056	1780.34	0.05056	19.7778
95	0.00044	2291.69	0.05044	19.8270
100	0.00034	2948.26	0.05034	19.8652

r = 6% **Continuous Compounding, Continuous Compounding Periods**

	Sinking Fund Factor	Uniform Series Factor	Capital Recovery Factor	Series Present Worth Factor
T	(*A*/*F*,*r*,*T*)	(*F*/*A*,*r*,*T*)	(*A*/*P*,*r*,*T*)	(*P*/*A*,*r*,*T*)
1	0.97030	1.0306	1.0303	0.97059
2	0.47060	2.1249	0.53060	1.8847
3	0.30423	3.2870	0.36423	2.7455
4	0.22120	4.5208	0.28120	3.5562
5	0.17150	5.8310	0.23150	4.3197
6	0.13846	7.2222	0.19846	5.0387
7	0.11495	8.6994	0.17495	5.7159
8	0.09739	10.2679	0.15739	6.3536
9	0.08380	11.9334	0.14380	6.9542
10	0.07298	13.7020	0.13298	7.5198
11	0.06419	15.5799	0.12419	8.0525
12	0.05690	17.5739	0.11690	8.5541
13	0.05078	19.6912	0.11078	9.0266
14	0.04558	21.9394	0.10558	9.4715
15	0.04111	24.3267	0.10111	9.8905
16	0.03723	26.8616	0.09723	10.2851
17	0.03384	29.5532	0.09384	10.6568
18	0.03085	32.4113	0.09085	11.0067
19	0.02821	35.4461	0.08821	11.3363
20	0.02586	38.6686	0.08586	11.6468
21	0.02376	42.0904	0.08376	11.9391
22	0.02187	45.7237	0.08187	12.2144
23	0.02017	49.5817	0.08017	12.4737
24	0.01863	53.6783	0.07863	12.7179
25	0.01723	58.0282	0.07723	12.9478
26	0.01596	62.6470	0.07596	13.1644
27	0.01480	67.5515	0.07480	13.3684
28	0.01374	72.7593	0.07374	13.5604
29	0.01277	78.2891	0.07277	13.7413
30	0.01188	84.1608	0.07188	13.9117
31	0.01106	90.3956	0.07106	14.0721
32	0.01031	97.0160	0.07031	14.2232
33	0.00961	104.046	0.06961	14.3655
34	0.00897	111.510	0.06897	14.4995
35	0.00837	119.436	0.06837	14.6257
40	0.00599	167.053	0.06599	15.1547
45	0.00432	231.329	0.06432	15.5466
50	0.00314	318.092	0.06314	15.8369
55	0.00230	435.211	0.06230	16.0519
60	0.00169	593.304	0.06169	16.2113
65	0.00124	806.707	0.06124	16.3293
70	0.00091	1094.77	0.06091	16.4167
75	0.00067	1483.62	0.06067	16.4815
80	0.00050	2008.51	0.06050	16.5295
85	0.00037	2717.03	0.06037	16.5651
90	0.00027	3673.44	0.06027	16.5914
95	0.00020	4964.46	0.06020	16.6109
100	0.00015	6707.15	0.06015	16.6254

r = 7% **Continuous Compounding, Continuous Compounding Periods**

	Sinking Fund Factor	Uniform Series Factor	Capital Recovery Factor	Series Present Worth Factor
T	(*A/F,r,T*)	(*F/A,r,T*)	(*A/P,r,T*)	(*P/A,r,T*)
1	0.96541	1.0358	1.0354	0.96580
2	0.46582	2.1468	0.53582	1.8663
3	0.29956	3.3383	0.36956	2.7059
4	0.21663	4.6161	0.28663	3.4888
5	0.16704	5.9867	0.23704	4.2187
6	0.13411	7.4566	0.20411	4.8993
7	0.11070	9.0331	0.18070	5.5339
8	0.09325	10.7239	0.16325	6.1256
9	0.07976	12.5373	0.14976	6.6773
10	0.06905	14.4822	0.13905	7.1916
11	0.06036	16.5681	0.13036	7.6712
12	0.05318	18.8052	0.12318	8.1184
13	0.04716	21.2046	0.11716	8.5354
14	0.04206	23.7779	0.11206	8.9241
15	0.03768	26.5379	0.10768	9.2866
16	0.03390	29.4979	0.10390	9.6246
17	0.03061	32.6726	0.10061	9.9397
18	0.02772	36.0774	0.09772	10.2335
19	0.02517	39.7292	0.09517	10.5075
20	0.02291	43.6457	0.09291	10.7629
21	0.02090	47.8462	0.09090	11.0011
22	0.01910	52.3513	0.08910	11.2231
23	0.01749	57.1830	0.08749	11.4302
24	0.01603	62.3651	0.08603	11.6232
25	0.01472	67.9229	0.08472	11.8032
26	0.01353	73.8837	0.08353	11.9711
27	0.01246	80.2767	0.08246	12.1275
28	0.01148	87.1332	0.08148	12.2735
29	0.01058	94.4869	0.08058	12.4095
30	0.00977	102.374	0.07977	12.5363
31	0.00902	110.833	0.07902	12.6546
32	0.00834	119.905	0.07834	12.7649
33	0.00771	129.635	0.07771	12.8677
34	0.00714	140.070	0.07714	12.9636
35	0.00661	151.262	0.07661	13.0529
40	0.00453	220.638	0.07453	13.4170
45	0.00313	319.087	0.07313	13.6735
50	0.00218	458.792	0.07218	13.8543
55	0.00152	657.044	0.07152	13.9817
60	0.00107	938.376	0.07107	14.0715
65	0.00075	1337.61	0.07075	14.1348
70	0.00053	1904.14	0.07053	14.1793
75	0.00037	2708.09	0.07037	14.2107
80	0.00026	3848.95	0.07026	14.2329
85	0.00018	5467.90	0.07018	14.2485
90	0.00013	7765.31	0.07013	14.2595
95	0.00009	11 025.0	0.07009	14.2672
100	0.00006	15 652.0	0.07006	14.2727

$r = 8\%$ **Continuous Compounding, Continuous Compounding Periods**

	Sinking Fund Factor	Uniform Series Factor	Capital Recovery Factor	Series Present Worth Factor
T	$(A/F,r,T)$	$(F/A,r,T)$	$(A/P,r,T)$	$(P/A,r,T)$
1	0.96053	1.0411	1.0405	0.96105
2	0.46107	2.1689	0.54107	1.8482
3	0.29493	3.3906	0.37493	2.6672
4	0.21213	4.7141	0.29213	3.4231
5	0.16266	6.1478	0.24266	4.1210
6	0.12985	7.7009	0.20985	4.7652
7	0.10657	9.3834	0.18657	5.3599
8	0.08924	11.2060	0.16924	5.9088
9	0.07587	13.1804	0.15587	6.4156
10	0.06528	15.3193	0.14528	6.8834
11	0.05670	17.6362	0.13670	7.3152
12	0.04964	20.1462	0.12964	7.7138
13	0.04373	22.8652	0.12373	8.0818
14	0.03874	25.8107	0.11874	8.4215
15	0.03448	29.0015	0.11448	8.7351
16	0.03081	32.4580	0.11081	9.0245
17	0.02762	36.2024	0.10762	9.2917
18	0.02484	40.2587	0.10484	9.5384
19	0.02240	44.6528	0.10240	9.7661
20	0.02024	49.4129	0.10024	9.9763
21	0.01833	54.5694	0.09833	10.1703
22	0.01662	60.1555	0.09662	10.3494
23	0.01510	66.2067	0.09510	10.5148
24	0.01374	72.7620	0.09374	10.6674
25	0.01252	79.8632	0.09252	10.8083
26	0.01142	87.5559	0.09142	10.9384
27	0.01043	95.8892	0.09043	11.0584
28	0.00953	104.917	0.08953	11.1693
29	0.00872	114.696	0.08872	11.2716
30	0.00798	125.290	0.08798	11.3660
31	0.00731	136.766	0.08731	11.4532
32	0.00670	149.198	0.08670	11.5337
33	0.00615	162.665	0.08615	11.6080
34	0.00564	177.254	0.08564	11.6766
35	0.00518	193.058	0.08518	11.7399
40	0.00340	294.157	0.08340	11.9905
45	0.00225	444.978	0.08225	12.1585
50	0.00149	669.977	0.08149	12.2711
55	0.00099	1005.64	0.08099	12.3465
60	0.00066	1506.38	0.08066	12.3971
65	0.00044	2253.40	0.08044	12.4310
70	0.00030	3367.83	0.08030	12.4538
75	0.00020	5030.36	0.08020	12.4690
80	0.00013	7510.56	0.08013	12.4792
85	0.00009	11 211.0	0.08009	12.4861
90	0.00006	16 730.0	0.08006	12.4907
95	0.00004	24 965.0	0.08004	12.4937
100	0.00003	37 249.0	0.08003	12.4958

$r = 9\%$ **Continuous Compounding, Continuous Compounding Periods**

	Sinking Fund Factor	Uniform Series Factor	Capital Recovery Factor	Series Present Worth Factor
T	$(A/F,r,T)$	$(F/A,r,T)$	$(A/P,r,T)$	$(P/A,r,T)$
1	0.95567	1.0464	1.0457	0.95632
2	0.45635	2.1913	0.54635	1.8303
3	0.29036	3.4440	0.38036	2.6291
4	0.20769	4.8148	0.29769	3.3592
5	0.15836	6.3146	0.24836	4.0264
6	0.12570	7.9556	0.21570	4.6361
7	0.10255	9.7512	0.19255	5.1934
8	0.08535	11.7159	0.17535	5.7028
9	0.07212	13.8656	0.16212	6.1682
10	0.06166	16.2178	0.15166	6.5937
11	0.05322	18.7915	0.14322	6.9825
12	0.04628	21.6076	0.13628	7.3378
13	0.04050	24.6888	0.13050	7.6626
14	0.03564	28.0602	0.12564	7.9594
15	0.03150	31.7492	0.12150	8.2307
16	0.02794	35.7855	0.11794	8.4786
17	0.02487	40.2020	0.11487	8.7052
18	0.02221	45.0343	0.11221	8.9122
19	0.01987	50.3218	0.10987	9.1015
20	0.01782	56.1072	0.10782	9.2745
21	0.01602	62.4374	0.10602	9.4325
22	0.01442	69.3638	0.10442	9.5770
23	0.01300	76.9425	0.10300	9.7090
24	0.01173	85.2349	0.10173	9.8297
25	0.01060	94.3082	0.10060	9.9400
26	0.00959	104.236	0.09959	10.0408
27	0.00869	115.099	0.09869	10.1329
28	0.00787	126.984	0.09787	10.2171
29	0.00714	139.989	0.09714	10.2941
30	0.00648	154.219	0.09648	10.3644
31	0.00589	169.789	0.09589	10.4287
32	0.00535	186.825	0.09535	10.4874
33	0.00487	205.466	0.09487	10.5411
34	0.00443	225.862	0.09443	10.5901
35	0.00403	248.178	0.09403	10.6350
40	0.00253	395.536	0.09253	10.8075
45	0.00160	626.638	0.09160	10.9175
50	0.00101	989.079	0.09101	10.9877
55	0.00064	1557.50	0.09064	11.0324
60	0.00041	2448.96	0.09041	11.0609
65	0.00026	3847.05	0.09026	11.0791
70	0.00017	6039.69	0.09017	11.0907
75	0.00011	9478.43	0.09011	11.0981
80	0.00007	14 871.0	0.09007	11.1028
85	0.00004	23 329.0	0.09004	11.1058
90	0.00003	36 594.0	0.09003	11.1077
95	0.00002	57 397.0	0.09002	11.1090
100	0.00001	90 023.0	0.09001	11.1097

$r = 10\%$ **Continuous Compounding, Continuous Compounding Periods**

	Sinking Fund Factor	Uniform Series Factor	Capital Recovery Factor	Series Present Worth Factor
T	*(A/F,r,T)*	*(F/A,r,T)*	*(A/P,r,T)*	*(P/A,r,T)*
1	0.95083	1.0517	1.0508	0.95163
2	0.45167	2.2140	0.55167	1.8127
3	0.28583	3.4986	0.38583	2.5918
4	0.20332	4.9182	0.30332	3.2968
5	0.15415	6.4872	0.25415	3.9347
6	0.12164	8.2212	0.22164	4.5119
7	0.09864	10.1375	0.19864	5.0341
8	0.08160	12.2554	0.18160	5.5067
9	0.06851	14.5960	0.16851	5.9343
10	0.05820	17.1828	0.15820	6.3212
11	0.04990	20.0417	0.14990	6.6713
12	0.04310	23.2012	0.14310	6.9881
13	0.03746	26.6930	0.13746	7.2747
14	0.03273	30.5520	0.13273	7.5340
15	0.02872	34.8169	0.12872	7.7687
16	0.02530	39.5303	0.12530	7.9810
17	0.02235	44.7395	0.12235	8.1732
18	0.01980	50.4965	0.11980	8.3470
19	0.01759	56.8589	0.11759	8.5043
20	0.01565	63.8906	0.11565	8.6466
21	0.01395	71.6617	0.11395	8.7754
22	0.01246	80.2501	0.11246	8.8920
23	0.01114	89.7418	0.11114	8.9974
24	0.00998	100.232	0.10998	9.0928
25	0.00894	111.825	0.10894	9.1792
26	0.00802	124.637	0.10802	9.2573
27	0.00720	138.797	0.10720	9.3279
28	0.00647	154.446	0.10647	9.3919
29	0.00582	171.741	0.10582	9.4498
30	0.00524	190.855	0.10524	9.5021
31	0.00472	211.980	0.10472	9.5495
32	0.00425	235.325	0.10425	9.5924
33	0.00383	261.126	0.10383	9.6312
34	0.00345	289.641	0.10345	9.6663
35	0.00311	321.155	0.10311	9.6980
40	0.00187	535.982	0.10187	9.8168
45	0.00112	890.171	0.10112	9.8889
50	0.00068	1474.13	0.10068	9.9326
55	0.00041	2436.92	0.10041	9.9591
60	0.00025	4024.29	0.10025	9.9752
65	0.00015	6641.42	0.10015	9.9850
70	0.00009	10 956.0	0.10009	9.9909
75	0.00006	18 070.0	0.10006	9.9945
80	0.00003	29 800.0	0.10003	9.9966
85	0.00002	49 138.0	0.10002	9.9980
90	0.00001	81 021.0	0.10001	9.9988
95	0.00001	133 587.0	0.10001	9.9993
100	0.00000	220 255.0	0.10000	9.9995

$r = 11\%$ **Continuous Compounding, Continuous Compounding Periods**

	Sinking Fund Factor	Uniform Series Factor	Capital Recovery Factor	Series Present Worth Factor
T	$(A/F,r,T)$	$(F/A,r,T)$	$(A/P,r,T)$	$(P/A,r,T)$
1	0.94601	1.0571	1.0560	0.94696
2	0.44702	2.2371	0.55702	1.7953
3	0.28135	3.5543	0.39135	2.5552
4	0.19902	5.0246	0.30902	3.2360
5	0.15002	6.6659	0.26002	3.8459
6	0.11767	8.4981	0.22767	4.3923
7	0.09485	10.5433	0.20485	4.8817
8	0.07796	12.8264	0.18796	5.3202
9	0.06504	15.3749	0.17504	5.7129
10	0.05489	18.2197	0.16489	6.0648
11	0.04674	21.3953	0.15674	6.3800
12	0.04010	24.9402	0.15010	6.6624
13	0.03461	28.8973	0.14461	6.9154
14	0.03002	33.3145	0.14002	7.1420
15	0.02615	38.2453	0.13615	7.3450
16	0.02286	43.7494	0.13286	7.5269
17	0.02004	49.8936	0.13004	7.6898
18	0.01762	56.7522	0.12762	7.8357
19	0.01553	64.4083	0.12553	7.9665
20	0.01371	72.9547	0.12371	8.0836
21	0.01212	82.4948	0.12212	8.1885
22	0.01074	93.1442	0.12074	8.2825
23	0.00952	105.032	0.11952	8.3667
24	0.00845	118.302	0.11845	8.4422
25	0.00751	133.115	0.11751	8.5097
26	0.00668	149.650	0.11668	8.5703
27	0.00595	168.108	0.11595	8.6245
28	0.00530	188.713	0.11530	8.6731
29	0.00472	211.713	0.11472	8.7166
30	0.00421	237.388	0.11421	8.7556
31	0.00376	266.048	0.11376	8.7905
32	0.00336	298.040	0.11336	8.8218
33	0.00300	333.753	0.11300	8.8499
34	0.00268	373.618	0.11268	8.8750
35	0.00239	418.119	0.11239	8.8975
40	0.00137	731.372	0.11137	8.9793
45	0.00078	1274.32	0.11078	9.0265
50	0.00045	2215.38	0.11045	9.0538
55	0.00026	3846.48	0.11026	9.0695
60	0.00015	6 674.0	0.11015	9.0785
65	0.00009	11 574.0	0.11009	9.0838
70	0.00005	20 067.0	0.11005	9.0868
75	0.00003	34 788.0	0.11003	9.0885
80	0.00002	60302.218	0.11002	9.0895
85	0.00001	104525.668	0.11001	9.0901
90	0.00001	181176.095	0.11001	9.0905
95	0.00000	314030.679	0.11000	9.0906
100	0.00000	544301.288	0.11000	9.0908

r = 12% **Continuous Compounding, Continuous Compounding Periods**

	Sinking Fund Factor	Uniform Series Factor	Capital Recovery Factor	Series Present Worth Factor
T	*(A/F,r,T)*	*(F/A,r,T)*	*(A/P,r,T)*	*(P/A,r,T)*
1	0.94120	1.0625	1.0612	0.94233
2	0.44240	2.2604	0.56240	1.7781
3	0.27693	3.6111	0.39693	2.5194
4	0.19478	5.1340	0.31478	3.1768
5	0.14596	6.8510	0.26596	3.7599
6	0.11381	8.7869	0.23381	4.2771
7	0.09116	10.9697	0.21116	4.7357
8	0.07446	13.4308	0.19446	5.1426
9	0.06171	16.2057	0.18171	5.5034
10	0.05172	19.3343	0.17172	5.8234
11	0.04374	22.8618	0.16374	6.1072
12	0.03726	26.8391	0.15726	6.3589
13	0.03192	31.3235	0.15192	6.5822
14	0.02749	36.3796	0.14749	6.7802
15	0.02376	42.0804	0.14376	6.9558
16	0.02062	48.5080	0.14062	7.1116
17	0.01794	55.7551	0.13794	7.2498
18	0.01564	63.9261	0.13564	7.3723
19	0.01367	73.1390	0.13367	7.4810
20	0.01197	83.5265	0.13197	7.5774
21	0.01050	95.2383	0.13050	7.6628
22	0.00922	108.443	0.12922	7.7387
23	0.00811	123.332	0.12811	7.8059
24	0.00714	140.119	0.12714	7.8655
25	0.00629	159.046	0.12629	7.9184
26	0.00554	180.386	0.12554	7.9654
27	0.00489	204.448	0.12489	8.0070
28	0.00432	231.577	0.12432	8.0439
29	0.00381	262.164	0.12381	8.0766
30	0.00337	296.652	0.12337	8.1056
31	0.00298	335.537	0.12298	8.1314
32	0.00264	379.379	0.12264	8.1542
33	0.00233	428.811	0.12233	8.1745
34	0.00206	484.546	0.12206	8.1924
35	0.00183	547.386	0.12183	8.2084
40	0.00100	1004.25	0.12100	8.2648
45	0.00054	1836.72	0.12054	8.2957
50	0.00030	3353.57	0.12030	8.3127
55	0.00016	6117.46	0.12016	8.3220
60	0.00009	11 154.0	0.12009	8.3271
65	0.00005	20 330.0	0.12005	8.3299
70	0.00003	37 051.0	0.12003	8.3315
75	0.00001	67 517.0	0.12001	8.3323
80	0.00001	123 032.0	0.12001	8.3328
85	0.00000	224 185.0	0.12000	8.3330
90	0.00000	408 498.0	0.12000	8.3332
95	0.00000	744 339.0	0.12000	8.3332
100	0.00000	1 356 282.0	0.12000	8.3333

r = 13% **Continuous Compounding, Continuous Compounding Periods**

	Sinking Fund Factor	Uniform Series Factor	Capital Recovery Factor	Series Present Worth Factor
T	(*A*/*F*,*r*,*T*)	(*F*/*A*,*r*,*T*)	(*A*/*P*,*r*,*T*)	(*P*/*A*,*r*,*T*)
1	0.93641	1.0679	1.0664	0.93773
2	0.43781	2.2841	0.56781	1.7611
3	0.27255	3.6691	0.40255	2.4842
4	0.19061	5.2464	0.32061	3.1191
5	0.14199	7.0426	0.27199	3.6766
6	0.11003	9.0882	0.24003	4.1661
7	0.08758	11.4179	0.21758	4.5960
8	0.07107	14.0709	0.20107	4.9734
9	0.05851	17.0923	0.18851	5.3049
10	0.04870	20.5331	0.17870	5.5959
11	0.04090	24.4515	0.17090	5.8515
12	0.03459	28.9140	0.16459	6.0759
13	0.02942	33.9960	0.15942	6.2729
14	0.02514	39.7835	0.15514	6.4460
15	0.02156	46.3745	0.15156	6.5979
16	0.01856	53.8805	0.14856	6.7313
17	0.01602	62.4286	0.14602	6.8485
18	0.01386	72.1634	0.14386	6.9513
19	0.01201	83.2496	0.14201	7.0417
20	0.01043	95.8749	0.14043	7.1210
21	0.00907	110.253	0.13907	7.1906
22	0.00790	126.627	0.13790	7.2518
23	0.00688	145.274	0.13688	7.3055
24	0.00601	166.511	0.13601	7.3526
25	0.00524	190.695	0.13524	7.3940
26	0.00458	218.237	0.13458	7.4304
27	0.00401	249.602	0.13401	7.4623
28	0.00350	285.322	0.13350	7.4904
29	0.00307	326.000	0.13307	7.5150
30	0.00269	372.327	0.13269	7.5366
31	0.00235	425.084	0.13235	7.5556
32	0.00206	485.166	0.13206	7.5722
33	0.00181	553.588	0.13181	7.5869
34	0.00158	631.510	0.13158	7.5997
35	0.00139	720.249	0.13139	7.6110
40	0.00072	1386.71	0.13072	7.6499
45	0.00038	2663.34	0.13038	7.6702
50	0.00020	5108.78	0.13020	7.6807
55	0.00010	9793.12	0.13010	7.6863
60	0.00005	18 766.0	0.13005	7.6892
65	0.00003	35 954.0	0.13003	7.6907
70	0.00001	68 879.0	0.13001	7.6914
75	0.00001	131 948.0	0.13001	7.6919
80	0.00000	252 759.0	0.13000	7.6921
85	0.00000	484 177.0	0.13000	7.6922
90	0.00000	927 467.0	0.13000	7.6922
95	0.00000	1 776 608.0	0.13000	7.6923
100	0.00000	3 403 172.0	0.13000	7.6923

$r = 14\%$ **Continuous Compounding, Continuous Compounding Periods**

	Sinking Fund Factor	Uniform Series Factor	Capital Recovery Factor	Series Present Worth Factor
T	(*A*/*F*,*r*,*T*)	(*F*/*A*,*r*,*T*)	(*A*/*P*,*r*,*T*)	(*P*/*A*,*r*,*T*)
1	0.93163	1.0734	1.0716	0.93316
2	0.43326	2.3081	0.57326	1.7444
3	0.26822	3.7283	0.40822	2.4497
4	0.18650	5.3619	0.32650	3.0628
5	0.13810	7.2411	0.27810	3.5958
6	0.10635	9.4026	0.24635	4.0592
7	0.08411	11.8890	0.22411	4.4621
8	0.06780	14.7490	0.20780	4.8123
9	0.05544	18.0387	0.19544	5.1168
10	0.04582	21.8229	0.18582	5.3815
11	0.03820	26.1756	0.17820	5.6116
12	0.03207	31.1825	0.17207	5.8116
13	0.02707	36.9418	0.16707	5.9855
14	0.02295	43.5666	0.16295	6.1367
15	0.01954	51.1869	0.15954	6.2682
16	0.01668	59.9524	0.15668	6.3824
17	0.01428	70.0350	0.15428	6.4818
18	0.01225	81.6328	0.15225	6.5681
19	0.01053	94.9735	0.15053	6.6432
20	0.00906	110.319	0.14906	6.7085
21	0.00781	127.970	0.14781	6.7652
22	0.00674	148.274	0.14674	6.8146
23	0.00583	171.629	0.14583	6.8575
24	0.00504	198.494	0.14504	6.8947
25	0.00436	229.396	0.14436	6.9272
26	0.00377	264.942	0.14377	6.9553
27	0.00327	305.829	0.14327	6.9798
28	0.00283	352.860	0.14283	7.0011
29	0.00246	406.959	0.14246	7.0196
30	0.00213	469.188	0.14213	7.0357
31	0.00185	540.768	0.14185	7.0497
32	0.00160	623.105	0.14160	7.0619
33	0.00139	717.815	0.14139	7.0725
34	0.00121	826.757	0.14121	7.0817
35	0.00105	952.070	0.14105	7.0897
40	0.00052	1924.47	0.14052	7.1164
45	0.00026	3882.66	0.14026	7.1297
50	0.00013	7825.95	0.14013	7.1363
55	0.00006	15 767.0	0.14006	7.1396
60	0.00003	31 758.0	0.14003	7.1413
65	0.00002	63 959.0	0.14002	7.1421
70	0.00001	128 805.0	0.14001	7.1425
75	0.00000	259 389.0	0.14000	7.1427
80	0.00000	522 353.0	0.14000	7.1428
85	0.00000	1 051 897.0	0.14000	7.1428
90	0.00000	2 118 268.0	0.14000	7.1428
95	0.00000	4 265 676.0	0.14000	7.1428
100	0.00000	8 590 023.0	0.14000	7.1429

$r = 15\%$ **Continuous Compounding, Continuous Compounding Periods**

	Sinking Fund Factor	Uniform Series Factor	Capital Recovery Factor	Series Present Worth Factor
T	(*A*/*F*,*r*,*T*)	(*F*/*A*,*r*,*T*)	(*A*/*P*,*r*,*T*)	(*P*/*A*,*r*,*T*)
1	0.92687	1.0789	1.0769	0.92861
2	0.42874	2.3324	0.57874	1.7279
3	0.26394	3.7887	0.41394	2.4158
4	0.18246	5.4808	0.33246	3.0079
5	0.13429	7.4467	0.28429	3.5176
6	0.10277	9.7307	0.25277	3.9562
7	0.08075	12.3843	0.23075	4.3337
8	0.06465	15.4674	0.21465	4.6587
9	0.05249	19.0495	0.20249	4.9384
10	0.04308	23.2113	0.19308	5.1791
11	0.03566	28.0465	0.18566	5.3863
12	0.02971	33.6643	0.17971	5.5647
13	0.02488	40.1913	0.17488	5.7182
14	0.02093	47.7745	0.17093	5.8503
15	0.01767	56.5849	0.16767	5.9640
16	0.01497	66.8212	0.16497	6.0619
17	0.01270	78.7140	0.16270	6.1461
18	0.01081	92.5315	0.16081	6.2186
19	0.00921	108.585	0.15921	6.2810
20	0.00786	127.237	0.15786	6.3348
21	0.00672	148.907	0.15672	6.3810
22	0.00574	174.084	0.15574	6.4208
23	0.00492	203.336	0.15492	6.4550
24	0.00421	237.322	0.15421	6.4845
25	0.00361	276.807	0.15361	6.5099
26	0.00310	322.683	0.15310	6.5317
27	0.00266	375.983	0.15266	6.5505
28	0.00228	437.909	0.15228	6.5667
29	0.00196	509.856	0.15196	6.5806
30	0.00169	593.448	0.15169	6.5926
31	0.00145	690.567	0.15145	6.6029
32	0.00124	803.403	0.15124	6.6118
33	0.00107	934.500	0.15107	6.6194
34	0.00092	1086.81	0.15092	6.6260
35	0.00079	1263.78	0.15079	6.6317
40	0.00037	2682.86	0.15037	6.6501
45	0.00018	5687.06	0.15018	6.6589
50	0.00008	12 047.0	0.15008	6.6630
55	0.00004	25 511.0	0.15004	6.6649
60	0.00002	54 014.0	0.15002	6.6658
65	0.00001	114 355.0	0.15001	6.6663
70	0.00000	242 097.0	0.15000	6.6665
75	0.00000	512 526.0	0.15000	6.6666

r = 20% **Continuous Compounding, Continuous Compounding Periods**

	Sinking Fund Factor	Uniform Series Factor	Capital Recovery Factor	Series Present Worth Factor
T	*(A/F,r,T)*	*(F/A,r,T)*	*(A/P,r,T)*	*(P/A,r,T)*
1	0.90333	1.1070	1.1033	0.90635
2	0.40665	2.4591	0.60665	1.6484
3	0.24327	4.1106	0.44327	2.2559
4	0.16319	6.1277	0.36319	2.7534
5	0.11640	8.5914	0.31640	3.1606
6	0.08620	11.6006	0.28620	3.4940
7	0.06546	15.2760	0.26546	3.7670
8	0.05059	19.7652	0.25059	3.9905
9	0.03961	25.2482	0.23961	4.1735
10	0.03130	31.9453	0.23130	4.3233
11	0.02492	40.1251	0.22492	4.4460
12	0.01995	50.1159	0.21995	4.5464
13	0.01605	62.3187	0.21605	4.6286
14	0.01295	77.2232	0.21295	4.6959
15	0.01048	95.4277	0.21048	4.7511
16	0.00850	117.663	0.20850	4.7962
17	0.00691	144.821	0.20691	4.8331
18	0.00562	177.991	0.20562	4.8634
19	0.00458	218.506	0.20458	4.8881
20	0.00373	267.991	0.20373	4.9084
21	0.00304	328.432	0.20304	4.9250
22	0.00249	402.254	0.20249	4.9386
23	0.00203	492.422	0.20203	4.9497
24	0.00166	602.552	0.20166	4.9589
25	0.00136	737.066	0.20136	4.9663
26	0.00111	901.361	0.20111	4.9724
27	0.00091	1102.03	0.20091	4.9774
28	0.00074	1347.13	0.20074	4.9815
29	0.00061	1646.50	0.20061	4.9849
30	0.00050	2012.14	0.20050	4.9876
31	0.00041	2458.75	0.20041	4.9899
32	0.00033	3004.23	0.20033	4.9917
33	0.00027	3670.48	0.20027	4.9932
34	0.00022	4484.24	0.20022	4.9944
35	0.00018	5478.17	0.20018	4.9954
40	0.00007	14 900.0	0.20007	4.9983
45	0.00002	40 510.0	0.20002	4.9994
50	0.00001	110 127.0	0.20001	4.9998
55	0.00000	299 366.0	0.20000	4.9999
60	0.00000	813 769.0	0.20000	5.0000
65	0.00000	2 212 062.0	0.20000	5.0000
70	0.00000	6 013 016.0	0.20000	5.0000
75	0.00000	16 345 082.0	0.20000	5.0000

r = 25% **Continuous Compounding, Continuous Compounding Periods**

	Sinking Fund Factor	Uniform Series Factor	Capital Recovery Factor	Series Present Worth Factor
T	(*A*/*F*,*r*,*T*)	(*F*/*A*,*r*,*T*)	(*A*/*P*,*r*,*T*)	(*P*/*A*,*r*,*T*)
1	0.88020	1.1361	1.1302	0.88480
2	0.38537	2.5949	0.63537	1.5739
3	0.22381	4.4680	0.47381	2.1105
4	0.14549	6.8731	0.39549	2.5285
5	0.10039	9.9614	0.35039	2.8540
6	0.07180	13.9268	0.32180	3.1075
7	0.05258	19.0184	0.30258	3.3049
8	0.03913	25.5562	0.28913	3.4587
9	0.02945	33.9509	0.27945	3.5784
10	0.02236	44.7300	0.27236	3.6717
11	0.01707	58.5705	0.26707	3.7443
12	0.01310	76.3421	0.26310	3.8009
13	0.01008	99.1614	0.26008	3.8449
14	0.00778	128.462	0.25778	3.8792
15	0.00602	166.084	0.25602	3.9059
16	0.00466	214.393	0.25466	3.9267
17	0.00362	276.422	0.25362	3.9429
18	0.00281	356.069	0.25281	3.9556
19	0.00218	458.337	0.25218	3.9654
20	0.00170	589.653	0.25170	3.9730
21	0.00132	758.265	0.25132	3.9790
22	0.00103	974.768	0.25103	3.9837
23	0.00080	1252.76	0.25080	3.9873
24	0.00062	1609.72	0.25062	3.9901
25	0.00048	2068.05	0.25048	3.9923
26	0.00038	2656.57	0.25038	3.9940
27	0.00029	3412.24	0.25029	3.9953
28	0.00023	4382.53	0.25023	3.9964
29	0.00018	5628.42	0.25018	3.9972
30	0.00014	7228.17	0.25014	3.9978
31	0.00011	9282.29	0.25011	3.9983
32	0.00008	11 920.0	0.25008	3.9987
33	0.00007	15 307.0	0.25007	3.9990
34	0.00005	19 655.0	0.25005	3.9992
35	0.00004	25 239.0	0.25004	3.9994
40	0.00001	88 102.0	0.25001	3.9998
45	0.00000	307 516.0	0.25000	3.9999
50	0.00000	1 073 345.0	0.25000	4.0000
55	0.00000	3 746 353.0	0.25000	4.0000

D

Answers to Selected Problems

CHAPTER 1

CHAPTER 2

2.1	$120
2.3	$5000
2.5	8%
2.7(a)	$1210
2.9	18.75%
2.11	$29 719
2.13 (a)	5 years
2.15 (a)	$6728
2.17 (a)	26.6%
2.19	5%
2.21	$665 270
2.23	Brand 2 about $51 less
2.25	$i_{e\ \text{continuous}} = 8.318\%$
2.27	$i_{e\ \text{weekly}} = 5.65\%$
2.29	0.5%
2.31	$2140
2.37	Decisional equivalence holds
2.41(a)	105 months
2.43 (b)	Lost $60 by locking in

CHAPTER 3

3.3	$317.22
3.5	20.3%
3.7	$74 790
3.9	$9252 per year
3.11	$94.13
3.13	$2664
3.15	11.7%
3.17	5.8 years
3.19 (b)	$26.44 per week
3.21	162.5 MW
3.23	$3.98
3.25	$34 616
3.27	$3339
3.31	$3 598 109
3.33	$8 013 275
3.35	$74 514
3.37	No, P = $3 670 261
3.39	$85.9 million
3.41	$122 316
3.43	27 months
3.45	$21 098
3.47	$3 086 287
3.49	$257 143
3.51	No more than $504
3.53	Up to $4587
3A.1	Purchase cost > savings by $1701
3A.3	$8353

CHAPTER 4

4.1	BC, CD
4.5	A, AB, BC, ABC, BCD
4.7 (a)	–$2164
4.9 (a)	18.1%
4.11	Second offer, PW = $137 000
4.13	Earthen dam, PW = $396 038
4.15	No, PW = –$16
4.17	Hydraulic press, AW = $24 716
4.19	Plastic liner, PW = $1 100 000
4.21	XJ3, PW = –$6565
4.23	No, AC(car) = $10 126
4.25	3.71 years
4.27	About 20%
4.29	$134
4.31 (a)	T, PW = $96 664
4.33	T, AW = $26 154
4.35 (a)	Only B, AW = $157.46

4.39	Landfill site, AW = $125 351
4.43	Curtains, payback period = 1.67 years

CHAPTER 5

5.1 (a)	9.2%
5.3	12.4%
5.5	B and D
5.7 (b)	7.58%
5.9	New machine
5.11	21.7%
5.13	3.19%
5.15	Approximate ERR > 30%
5.17 (c)	Clip Job
5.19	Used refrigerator
5.21 (b)	2.46%
5.23 (b)	2
5.25	E
5.27	$119
5.29	A and C
5.31	B
5.33	Payback period
5.35	IRR or present worth
5.37	Payback period
5.39	Annual worth
5A.1 (d)	No, 12.6%
5A.3 (e)	No, 21.3%

CHAPTER 6

6.1 (a)	Functional loss
6.3	Use-related loss
6.5	Time-related loss
6.7 (b)	$7714
6.9 (b)	$5806
6.13	12%
6.15	A, PW = –$22 321
6.17 (a)	$312.50
6.19	Fryer 1, AW = $37 668
6.21	$25 000
6.23	acid test = 0.9
6.25	Current assets, fixed assets, depreciation, income taxes, net income (operations), and net income
6.27	Equity ratio = 0.296
6.29	Total assets = $46 500, total equity = $15 450, net income = $2700
6.31	e.g., current = 0.9, equity = 0.39, ROA = 2%
6.33	For 2005, total assets = $3900, total equity = $1350, net income = $450

CHAPTER 7

7.1	$10 920
7.3	No
7.5 (c)	6 years
7.7	Every 3 years
7.9	4.7 years
7.11	Move robot immediately
7.13	Yes
7.15 (a)	Yes
7.17 (c)	$12 879
7.19	Replace immediately
7.21 (b)	Defender two years, followed by Challenger 2
7.23	Replace the computer now, and keep the replacement for three years
7.25 (a)	Replace old pump after four years
7.27	No
7.29	Challenger EAC is $39 452 with an economic life of five years
7.31	Overhaul old cutter and replace after four years
7.33 (a)	A1 = 4 years, B1 = 7 years
7.35 (a)	A3 = 12 years, B3 = 4 years

CHAPTER 8

8.1 (a) Class 10, 30%

8.3 $IRR_{before\text{-}tax}$ = 18.5%

8.5 $19 250

8.7 $IRR_{after\text{-}tax}$ = 9.3%

8.9 $26 779

8.11 $51 929

8.15 6.23%, 6.72%

8.17 Leasing

8.19 $10 242

8.21 $280 588

8.23 Annual cost of $1647

8.25 $IRR_{after\text{-}tax}$ = 3.7%, $MARR_{after\text{-}tax}$ = 6.0%

8.27 $4199

8.29 A = $10 737, B = $12 080

8.31 Alternative 2, AC = $204 562

8.33 T, PW = $9974

8.35 20.7%, 15.1%, 11.7%

8.37 T

8.39 $IRR_{after\text{-}tax}$ 5 8.5%

CHAPTER 9

9.1 (a) Actual

9.3 (a) $292

9.5 (a) 14.5%

9.7 (b) 22.9%

9.9 67 years

9.11 14.3%

9.13 $14 683

9.15 11 216 million rubles

9.17 (a) 7.75%

9.21 (a) –$125 532

9.23 (a) About 16%

9.25 (b) $243 547

9.27 (c) 2.13%

9.29 (b) $998 for f = 2%

CHAPTER 10

10.1 (a) 1.55

10.1 (b) 2

10.3 (c) 0

10.5 (a) $1599 per day

10.11 $212 500

10.15 (a) $1 201 200 per year

10.17 (d) $1333 per day

10.19 (a) River route

10.23 (b) 1.07

10.25 (a) 1.08, 1.1

10.27 (a) 0.75

10.29 (a) Tennis courts, BCR = 1.18

10.31 Yes, BCR = 1.66

10.37 (c) $34 510

10.39 (a) 2 lanes

CHAPTER 11

11.3 (a) Break-even analysis for multiple projects

11.5 PW(base) = $34 617

11.7 PW(base) = $6 636 000, most sensitive to the first cost

11.9 (a) AW(base) = $176 620, most sensitive to the first cost

11.11 (a) 179 deliveries per month

11.13 (a) $141 045, outside of the likely values

11.15 (b) 47 375 boards per year

11.17 $17 535

11.19 (a) $4146

11.19 (b) Break-even operating hours = 671

11.21 Break-even maintenance cost $61.50

11.23 Break-even interest rate 11.2%

11.25 Lease if tax rate is below 27%

CHAPTER 12

12.1 (a) \$855

12.3 (b) X1000, Var(defects) = 1.36/100 units^2

12.5 \$233 673

12.7 \$6384

12.9 E(buy ticket) = 150

12.11 (c) E(send stock) = 4000

12.13 (a) E(combined rate) = 15.9%

12.15 E(partnership) = \$113 750

12.17 (a) AW(optimistic, public accept) = \$925 000

12.17 (b) E(new product) = \$107 975

12.21 (c) Average PW ≃ –\$160 600

12.23 Average PW ≃ –\$5600

12.25 Average PW ≃ –\$20 300

12.27 Average PW ≃ –\$15 700

CHAPTER 13

13.1 4, 5, 6, 7, 8, and 10

13.3 (b) L4, 60.2

13.5 (a)

$$\begin{bmatrix} 1 & 1 & 5 & \frac{1}{3} \\ 1 & 1 & \frac{1}{3} & 2 \\ \frac{1}{5} & 3 & 1 & 2 \\ 3 & \frac{1}{2} & \frac{1}{2} & 1 \end{bmatrix}$$

13.7 (a) 1, 3, 7, 8

13.9 D, 57.5

13.11

$$\begin{bmatrix} 1 & 1 & \frac{1}{3} & 5 & \frac{1}{5} \\ 1 & 1 & \frac{1}{3} & 5 & \frac{1}{5} \\ 3 & 3 & 1 & 5 & \frac{1}{3} \\ \frac{1}{5} & \frac{1}{5} & \frac{1}{5} & 1 & \frac{1}{5} \\ 5 & 5 & 3 & 5 & 1 \end{bmatrix}$$

13.17 $[0.13\ \ 0.13\ \ 0.56\ \ 0.13\ \ 0.05]^T$

13.23 C, 0.277

13.25 (b) $[0.05\ \ 0.17\ \ 0.17\ \ 0.61]^T$

13.27 (b) $[0.141\ \ 0.263\ \ 0.455\ \ 0.141]^T$

13.29

$$\begin{bmatrix} 0.05 & 0.61 & 0.14 & 0.44 \\ 0.17 & 0.26 & 0.26 & 0.07 \\ 0.17 & 0.08 & 0.46 & 0.05 \\ 0.61 & 0.05 & 0.14 & 0.44 \end{bmatrix}$$

13.31 $[0.24\ \ 0.20\ \ 0.17\ \ 0.40]^T$

13.A1 $CR = 0.1343$

13.A3 $CR = 0.0393$

Glossary

acid-test ratio: The ratio of quick assets to current liabilities. Quick assets are cash, accounts receivable, and marketable securities — those current assets considered to be highly *liquid*. The acid-test ratio is also known as the quick ratio.

actual dollars: Monetary units at the time of payment.

actual interest rate: The stated, or observed, interest rate based on actual dollars. If the real interest rate is i' and the inflation rate is f, the actual interest rate i is found by: $i = i' + f + i'f$

actual internal rate of return, IRR_A: The internal rate of return on a project based on actual dollar cash flows associated with the project; also the real internal rate of return which has been adjusted upwards to include the effect of inflation.

actual MARR: The minimum acceptable rate of return for *actual dollar* cash flows. It is the real MARR adjusted upwards for inflation.

amortization period: The duration over which a loan is repaid. It is used to compute periodic loan payment amounts.

analytic hierarchy process (AHP): A multi-attribute utility theory (MAUT) approach used for large, complex decisions, which provides a method for establishing the criteria weights.

annual worth method: Comparing alternatives by converting all cash flows to a uniform series, i.e., an annuity.

annuity: A series of uniform-sized receipts or disbursements that start at the end of the first period and continue over a number, N, of regularly spaced time intervals.

annuity due: An annuity whose first of N receipts or disbursements is immediate, at time 0, rather than at the end of the first period.

arithmetic gradient series: A series of receipts or disbursements that start at the end of the first period and then increase by a constant amount from period to period.

arithmetic gradient to annuity conversion factor: Denoted by $(A/G,i,N)$, gives the value of an annuity, A, that is equivalent to an arithmetic gradient series where the constant increase in receipts or disbursements is G per period, the interest rate is i, and the number of periods is N.

asset-management ratios: Financial ratios that assess how efficiently a firm is using its assets. Asset management ratios are also known as efficiency ratios. Inventory turnover is an example.

assets: The economic resources owned by an enterprise.

backward induction: See **rollback procedure.**

balance sheet: A financial statement which gives a snapshot of an enterprise's financial position at a particular point in time, normally the last day of an accounting period.

base period: A particular date associated with *real dollars* that is used as a reference point for price changes; also the period from which the expenditure shares are calculated in a Laspeyres price index.

base year: The year on which real dollars are based.

benefit-cost analysis (BCA): A method of project evaluation widely used in the public sector that provides a general framework for assessing the gains and losses associated with alternative projects when a broad societal view is necessary.

benefit-cost ratio (BCR): The ratio of the present worth of net user's benefits (social benefits less social costs) to the present worth of the net sponsor's costs for a project. That is,

$$\text{BCR} = \frac{\text{PW(user's benefits)}}{\text{PW(sponsor's costs)}}$$

bond: An investment that provides an annuity and a future value in return for a cost today. It has a "par" or "face" value, which is the amount for which it can be redeemed after a certain period of time. It also has a "coupon rate," meaning that the bearer is paid an annuity, usually semi-annually, calculated as a percentage of the face value.

book value: The depreciated value of an asset for accounting purposes, as calculated with a depreciation model.

break-even analysis: The process of varying a parameter of a problem and determining what parameter value causes the performance measure to reach some threshold or "break-even" value.

capacity: The ability to produce, often measured in units of production per time period.

capital cost: The depreciation expense incurred by the difference between what is paid for the assets required for a particular capacity and what the assets could be resold for some time after purchase.

capital cost allowance (CCA): The maximum depreciation expense allowed for tax purposes on all assets belonging to an asset class.

capital cost allowance (CCA) asset class: A categorization of assets for which a specified CCA rate is used to compute CCA. Numerous CCA asset classes exist in the CCA system.

capital cost allowance (CCA) rate: The maximum depreciation rate allowed for assets in a designated asset class within the CCA system.

capital cost allowance (CCA) system: The system established by the Canadian government whereby the amount and timing of depreciation expenses on capital assets is controlled.

capital cost tax factor (CCTF): A value that summarizes the effect of the future benefit of tax savings due to the CCA. It allows analysts to take these benefits into account when calculating the value of an asset. The *new* CCTF takes into account the "half-year rule" where only half of the capital cost of an asset can be used to calculate the CCA in the first year. This rule came into effect in November 1981. The *old* CCTF takes into account the entire amount of a capital expense in one year.

capital expense: The expenditure associated with the purchase of a long-term depreciable asset.

capital recovery factor: Denoted by $(A/P,i,N)$, gives the value, A, of the periodic payments or receipts that is equivalent to a present amount, P, when the interest rate is i and the number of periods is N.

capital recovery formula: A formula that can be used to calculate the savings necessary to justify a capital purchase based on first cost and salvage value.

capitalized value: The present worth of an infinitely long series of uniform cash flows.

cash flow diagram: A chart that summarizes the timing and magnitude of cash flows as they occur over time. The X axis represents time, measured in periods, and the Y axis represents the size and direction of the cash flows. Individual cash flows are indicated by arrows pointing up (positive cash flows, or receipts) or down (negative cash flows, or disbursements).

challenger: A potential replacement for an existing asset. See **defender**.

comparison methods: Methods of evaluating and comparing projects, such as present worth, annual worth, payback, and IRR.

compound amount factor: Denoted by $(F/P,i,N)$, gives the future amount, F, that is equivalent to a present amount, P, when the interest rate is i and the number of periods is N.

compound interest: The standard method of computing interest where interest accumulated in one interest period is added to the principal amount used to calculate interest in the next period.

compound interest factors: Functions that define the mathematical equivalence of certain common cash flow patterns.

compounding period: The interest period used with the compound interest method of computing interest.

consistency ratio: A measure of the consistency of the reported comparisons in a PCM. The consistency ratio ranges from 0 (perfect consistency) to 1 (no consistency). A consistency ratio of 0.1 or less is considered acceptable.

constant dollars: See **real dollars**.

consumer price index (CPI): The CPI relates the average price of a standard set of goods and services in some base period to the average price of the same set of goods and services in another period. Currently, Statistics Canada uses a base year of 1992 for the CPI.

continuous compounding: Compounding of interest which occurs continuously over time, i.e., as the length of the compounding period tends toward zero.

continuous models: Models that assume all cash flows and all compounding of cash flows occur continuously over time.

cumulative distribution function (CDF): A function that gives the probability that the random variable X takes on a value less than or equal to x, for all x. The CDF is often denoted by $F(x)$ and is defined by $F(x) = Pr(X \leq x)$ for all values of x. Also know as **cumulative risk profile**.

cumulative risk profile: See **cumulative distribution function**.

corporation: A business owned by shareholders.

cost of capital: The minimum rate of return required to induce investors to invest in a business.

cost principle of accounting: A principle of accounting which states that assets are to be valued on the basis of their cost as opposed to market or other values.

current assets: Cash and other assets that could be converted to cash within a relatively short period of time, usually a year or less.

current dollars: See **actual dollars**.

current liabilities: Liabilities that are due within some short period of time, usually a year or less.

current ratio: The ratio of all current assets to all current liabilities. It is also known as the working capital ratio.

decision matrix: A multi-attribute utility theory (MAUT) method in which the rows of a matrix represent criteria, and the columns alternatives. There is an extra column for the weights of the criteria. The cells of the matrix (other than the criteria weights) contain an evaluation of the alternatives.

decisional equivalence: Decisional equivalence is a consequence of indifference on the part of a decision maker among available choices.

decision tree: A graphical representation of the logical structure of a decision problem in terms of a sequence of decisions and chance events.

declining-balance method of depreciation: A method of modelling depreciation where the loss in value of an asset in a period is assumed to be a constant proportion of the asset's current value.

defender: An existing asset being assessed for possible replacement. See **challenger**.

deflation: The decrease, over time, in average prices. It can also be described as the increase in the purchasing power of money over time.

debt-management ratios: See **leverage ratios**.

depreciation: The loss in value of a capital asset.

discrete models: Models that assume all cash flows and all compounding of cash flows occur at the ends of conventionally defined periods like months or years.

discrete probability distribution function: A probability distribution function in which the random variable is discrete.

dominated: An alternative that is not efficient. See **inefficient**.

economic life: The service life of an asset that minimizes its total cost of use.

earnings before interest and taxes (EBIT): A measure of a company's operating profit (revenues less operating expenses), which results from making sales and controlling operating expenses.

effective interest rate: The actual but not usually stated interest rate, found by converting a given interest rate (with an arbitrary compounding period, normally less than a year) to an equivalent interest rate, with a one-year compounding period.

efficiency ratios: See **asset-management ratios**.

efficient: An alternative is efficient if no other alternative is valued as at least equal to it for all criteria and preferred to it for at least one.

efficient market: A market is efficient when decisions are made so that it is impossible to find a way for at least one person to be better off and no person would be worse off.

engineering economics: Science that deals with techniques of quantitative analysis useful for selecting a preferable alternative from several technically viable ones.

equity ratio: A financial ratio which is the ratio of total owners' equity to total assets. The smaller this ratio is, the more dependent the firm is on debt for its operations and the higher are the risks the company faces.

equivalence: A condition that exists when the value of a cost at one time is equivalent to the value of the related benefit at a different time.

equivalent annual cost (EAC): An annuity that is mathematically equivalent to a more complex set of cash flows.

expected value: A summary statistic of a random variable which gives its mean or average value. Also known as the **mean**.

expensed: Term applied to an asset with a CCA rate of 100%. For all intents and purposes, this is the same as treating the cost of the asset as an operating cost rather than a capital cost.

expenses: Either real costs associated with performing a corporation's business or a portion of the capital expense for an asset.

external rate of return (ERR): The rate of return on a project where any cash flows that are not invested in the project are assumed to earn interest at a predetermined rate (such as the MARR).

extraordinary item: A gain or loss which does not typically result from a company's normal business activities and is therefore not a recurring item.

financial accounting: The process recording and organizing the financial data of a business. The data cover both flows over time, like revenues and expenses, and levels, like an enterprise's resources and the claims on those resources, at a given date.

financial ratio analysis: Comparison of a firm's financial ratios with ratios computed for the same firm from previous financial statements and with industry standard ratios.

financial market: Market for the exchange of capital and credit in the economy.

financial ratios: Ratios between key amounts taken from the financial statements of a firm. They give an analyst a framework for answering questions about the firm's liquidity, asset management, leverage, and profitability.

fixed costs: costs that remain the same, regardless of actual units of production.

future worth: See the definition of **interest rate**.

future worth method: Comparing alternatives by taking all cash flows to future worth.

geometric gradient series: A set of disbursements or receipts that change by a constant *proportion* from one period to the next in a sequence of periods.

geometric gradient to present worth conversion factor: Denoted by $(P/A,g,i,N)$, gives the present worth, P, that is equivalent to a geometric gradient series where the base receipt or disbursement is A, and where the rate of growth is g, the interest rate is i, and the number of periods is N.

goal: The decision to be made in AHP.

growth adjusted interest rate, $i°$: $i° = (1+i)/(1+g)$ so that $1/(1+i°) = (1+i)/(1+g)$ where i is the interest rate and g is the growth rate. The growth adjusted interest rate is used in computing the geometric gradient to present worth conversion factor.

income statement: A financial statement which summarizes an enterprise's revenues and expenses over a specified accounting period.

independent projects: Two projects are independent if the expected costs and the expected benefits of each of the projects do not depend on whether or not the other one is chosen.

inefficient: An alternative that is not efficient.

inflation: The increase, over time, in average prices of goods and services. It can also be described as the decrease in the purchasing power of money over time.

inflation rate: The rate of increase in average prices of goods and services over a specified time period, usually a year; also, the rate of decrease in purchasing power of money over a specified time period, usually a year.

installation costs: Costs of acquiring capacity (excluding the purchase cost) which may include disruption of production, training of workers, and perhaps a reorganization of other production.

interest: The compensation for giving up the use of money.

interest period: The base unit of time over which an interest rate is quoted. The interest period is referred to as the compounding period when compound interest is used.

interest rate: If the right to P at the beginning of a time period exchanges for the right to F at the end of the period, where $F = P(1+i)$, i is the interest rate per time period. In this definition, P is called the *present worth* of F, and F is called the *future worth* of P.

internal rate of return (IRR): That interest rate, i^*, such that, when all cash flows associated with a project are discounted at i^*, the present worth of the cash inflows equals the present worth of the cash outflows.

inventory-turnover ratio: A financial ratio that captures the number of times that a firm's inventories are replaced (or turned over) per year. It provides a measure of whether the firm has more or less inventory than normal.

Laspeyres price index: A commonly used price index which measures weighted average changes in prices of a set of goods and services over time as compared with the prices in a base period. The weights are the expenditure shares in the base period. The weights are then converted to percentages by multiplying by 100.

leverage ratios: Financial ratios that provide information about how liquid a firm is, or how well it is able to meet its current obligations. Also know as **debt-management ratios**.

liabilities: Claims, other than those of the owners, on a business's assets.

liquidity ratio: A financial ratio that evaluates the ability of a business to meet its current liability obligations. The current ratio and quick ratio are two examples of liquidity ratios.

long-term assets: Assets that are not expected to be converted to cash in the short term, usually taken to be one year.

long-term liabilities: Liabilities that are not expected to draw on the business's current assets.

management accounting: The process of analyzing and recording the costs and benefits of the various activities of an enterprise. The goal of management accounting is to provide managers with information to help in decision making.

market: A group of buyers and sellers linked by trade in a particular product or service.

market equivalence: The ability to exchange one cash flow for another at zero cost.

market failure: Condition in which output or consumption decisions are made in which aggregate benefits to all persons who benefit from the decision are less than aggregate costs imposed on persons who bear costs that result from the decision.

market value: Usually taken as the actual value an asset can be sold for in an open market.

mathematical equivalence: An equivalence of cash flows due to the mathematical relationship between time and money.

mean: See **expected value**.

mean-variance dominance: Alternative X dominates alternative Y in a mean-variance sense if: $EV(X) \geq EV(Y)$ and Var(X) Y) and $Var(X) \leq Var(Y)$.

minimum acceptable rate of return (MARR): An interest rate that must be earned for any project to be accepted.

modified benefit-cost ratio: The ratio of the present worth (or annual worth) of benefits minus the present worth (or annual worth) of operating costs to the present worth (or annual worth) of capital costs, that is, $BCRM = \frac{PW(\text{user's benefits}) - PW(\text{sponsor's operating costs})}{PW(\text{sponsor's capital costs})}$

Monte Carlo simulation: A procedure that constructs the probability distribution of an outcome performance measure of a project by repeatedly sampling from the input random variable probability distributions.

multi-attribute utility theory (MAUT): An MCDM approach in which criteria weights and preference measures are combined mathematically to determine a best alternative.

multi-criterion decision making (MCDM): Methods that explicitly take into account multiple criteria and guide a decision maker to superior or optimal choices.

mutually exclusive projects: Projects are mutually exclusive if, in the process of choosing one, all the other alternatives are excluded.

net cash flow: The difference between cash inflows and outflows for the period. The net cash flow, A_t, is given by $A_t = R_t - D_t$, where R_t is cash inflow in period t, and D_t is cash disbursed in period t.

net-profit ratio: See **return-on-assets (ROA) ratio.**

nominal dollars: See **actual dollars.**

nominal interest rate: The conventional method of stating the annual interest rate. It is calculated by multiplying the interest rate per compounding period by the number of compounding periods per year.

one year principle: This principle states that if the EAC (capital costs) for the defender are small compared to the EAC (operating costs), and the yearly operating costs are monotonically increasing, the economic life of a defender is one year and its total EAC is the cost of using the defender for one more year.

operating and maintenance costs: Ongoing costs to operate and maintain an asset over its useful life. Includes costs such as electricity, gasoline, parts, repair and insurance.

outcome dominance: Alternative X is said to have outcome dominance over alternative Y when a) the worst outcome for alternative X is at least as good as the best outcome for alternative Y, or b) X is at least as preferred to Y for each outcome, and is better for at least one outcome.

owners' equity: The interest of the owner or owners of a firm in its assets.

pairwise comparison: An evaluation of the importance or preference of a pair of criteria (or alternatives), based on an AHP value scale.

pairwise comparison matrix (PCM): A device for storing pairwise comparison evaluations.

partnership: A business owned by two or more owners (partners).

par value: The price per share set by a firm at the time the shares are originally issued.

payback period: The period of time it takes for an investment to be recouped when the interest rate is assumed to be zero.

payback period method: A method used for comparing alternatives by comparing the periods of time required for the investments to pay for themselves.

performance measures: Calculated values that allow conclusions to be drawn from data.

policy instrument: A remedy for market failure in which the government imposes rules intended to modify behaviour of business organizations.

present worth: See **interest rate.**

present worth factor: Denoted by $(P/F,i,N)$, gives the present amount, P, that is equivalent to a future amount, F, when the interest rate is i and the number of periods is N.

present worth method: Comparing alternatives by taking all cash flows to present worth.

price index: A number, usually a percentage, that relates prices of a given set of goods and services in some period, t_1, to the prices of the same set of goods and service in another period, t_0.

principal eigenvector: The eigenvector associated with the largest eigenvalue, λ_{max}, of a PCM.

priority weights: Weights calculated for alternatives by normalizing the elements of a PCM and averaging the row entries.

probability: The limit of a long-run proportion or relative frequency; see Close-Up 12.1 for other views of probability.

probability distribution function: A set of numerical measures (probabilities) associated with outcomes of a random variable. Also known as a risk profile.

profitability ratios: Financial ratios that give evidence of how productively assets have been employed in producing a profit. Return on total assets (or net-profit ratio) is an example of a profitability ratio.

progressive tax rate: A taxation rate that increases as the income level increases.

project: A term used throughout this text to mean "investment opportunity."

project balance: If a project has a sequence of net cash flows A_0, A_1, A_2,..., A_T, and the interest rate is i', there are $T + 1$ project balances, B_0, B_1,..., B_T, one at the end of each period t, $t = 0,1,..., T$. A project balance, B_t, is the cumulative future value of all cash flows, up to the end of period t, compounded at the rate, i'.

public-private partnership (P3): A cooperative arrangement between the private and public sectors for the provision of infrastructure or services.

quick ratio: See **acid-test ratio.**

random variable: A random variable is a parameter or variable that can take on a number of possible outcomes. The probability of each outcome is given by its probability distribution.

real dollars: Monetary units of constant purchasing power.

real interest rate: The interest rate, i', is the interest rate that would yield the same number of real dollars in the absence of inflation as the actual interest rate yields in the presence of inflation at the rate *f*. It is given by $i' = (1+i)/(1+f) - 1$

real internal rate of return: The internal rate of return on a project based on real dollar cash flows associated with the project.

real MARR: The minimum acceptable rate of return when cash flows are expressed in real, or constant, dollars.

recovery period: The designated service life for depreciation calculation purposes in U.S. tax law.

related but not mutually exclusive projects: For pairs of projects in this category, the expected costs and benefits of one project depend on whether the other one is chosen.

repeated lives: Used for comparing alternatives with different service lives, based on the assumption that alternatives can be repeated in the future, with the same costs and benefits, as often as necessary. The life of each alternative is repeated until a common total time period is reached for all alternatives.

replacement: An asset may be replaced if there is cheaper way to get the service the asset provides or of the service provided by the asset is no longer adequate.

retained earnings: The cumulative sum of earnings from normal operations, in addition to gains (or losses) from transactions such as the sale of plant assets or investments that have been reinvested in the business, i.e., not paid out as dividends.

retire: To remove an asset from use without replacement.

return-on-assets (ROA) ratio: A financial ratio that captures how productively assets have been employed in producing a profit. It is also known as the net-profit ratio.

return-on-equity (ROE) ratio: A financial ratio that captures how much profit a company has earned in comparison to the amount of capital that the owners have tied up in the company.

risk profile: See **probability distribution function**.

rollback procedure: A procedure in decision tree analysis that computes an expected value at each chance node and selects a preferred alternative at each decision node; also known as backward induction.

salvage value: Either the actual value of an asset at the end of its useful life (when it is sold), or an estimate of the salvage value calculated using a depreciation model.

scenario analysis: The process of examining the consequences of several possible sets of variable values associated with a project.

scrap value: Either the actual value of an asset at the end of its physical life (when it is broken up for the material value of its parts), or an estimate of the scrap value calculated using a depreciation model.

sensitivity analysis: Methods that assess the sensitivity of an economic measure to uncertainties in estimates in the various parameters of a problem.

sensitivity graph: A graph of the changes in a performance measure, holding all other variables fixed.

series present worth factor: Denoted by $(P/A,i,N)$, gives the present amount, P, that is equivalent to an annuity, A, when the interest rate is i and the number of periods is N.

simple interest: A method of computing interest where interest earned during an interest period is not added to the principal amount used to calculate interest in the next period. Simple interest is rarely used, except as a method of calculating approximate interest.

simple investment: A project that consists of one or more cash outflows at the beginning, followed only by one or more cash inflows.

sinking fund: Interest-bearing account into which regular deposits are made in order to accumulate some amount.

sinking fund factor: Denoted by $(A/F,i,N)$, gives the size, A, of a repeated receipt or disbursement that is equivalent to a future amount, F, when the interest rate is i and the number of periods is N.

small business deduction: A tax deduction which reduces the effective tax rate for a small Canadian company to less than 20%.

sole proprietorship: A business owned by one person.

specialist company: A firm that concentrates on manufacturing a limited range of very specialized products.

standard deviation: The square root of the variance. A measure of dispersion for a random variable.

statement of changes in financial position: A financial statement that shows the amounts of cash generated by a company's operations and by other sources, and the amounts of cash used for investments and other non-operating disbursements.

stochastic dominance: If two decision alternatives a and b have outcome cumulative distribution functions $F(x)$ and $G(x)$, respectively, then alternative a is said to have first order stochastic dominance over alternative b if $F(x) \geq G(x)$ for all x.

straight-line depreciation: A method of modelling depreciation which assumes that the rate of loss in value of an asset is constant over its useful life.

study period: A period of time over which alternative projects are compared.

sunk costs: Costs that were incurred in the past and are no longer relevant in replacement decisions.

tax credits: Real or nominal costs that are not taxed or are taxed at a reduced rate.

term: The duration over which a loan agreement is valid.

trend analysis: A form of financial analysis which traces the financial ratios of a firm over several accounting periods.

undepreciated capital cost (UCC): The remaining book value of assets subject to depreciation for taxation purposes. For any given year, the UCC balance can be calculated as follows: $UCC_{opening}$ + additions - disposals - CCA = UCC_{ending}

uniform series compound amount factor: Denoted by $(F/A,i,N)$, gives the future value, F, that is equivalent to a series of equal-sized receipts or disbursements, A, when the interest rate is i and the number of periods is N.

variable costs: Costs that change depending on the number of units produced.

variance: Variance measures the dispersion of a random variable about its mean.

weighted average cost of capital: A weighted average of the costs of borrowing and of selling shares. The weights are the fractions of total capital that come from the different sources.

working capital: The difference between total current assets and total current liabilities.

working capital ratio: See **current ratio**.

Index

D

E

V

W

List of Formulas

After-tax IRR:

$$\text{IRR}_{\text{after-tax}} \cong \text{IRR}_{\text{before-tax}} \times (1 - t)$$

After-tax MARR:

$$\text{MARR}_{\text{after-tax}} \cong \text{MARR}_{\text{before-tax}} \times (1 - t)$$

Benefit-cost ratio:

$$\text{BCR} = \frac{\text{PW (benefits)}}{\text{PW(costs)}}$$

Book Value, Declining Balance:

$$\text{BV}_{db}(n) = P(1 - d)^n$$

Book Value, Straight Line:

$$BV_{sl}(n) = P - n\left(\frac{P - S}{N}\right)$$

Capital Cost Tax Factor (new):

$$\text{CCTF}_{\text{new}} = 1 - \frac{td\left(1 + \frac{i}{2}\right)}{(i + d)(1 + i)}$$

Capital Cost Tax Factor (old):

$$\text{CCTF}_{\text{old}} = 1 - \frac{td}{(i + d)}$$

Capitalized Value:

$$P = \frac{\text{A}}{i}$$

Capital Recovery Formula:

$$A = (P - S)(A/P, i, N) + Si$$

Compound Interest:

$$F = P(1 + i)^N$$

Compound Interest Factors:

- **Compound Amount Factor**

$$(F/P, i, N) = (1 + i)^N$$

- **Present Worth Factor**

$$(P/F, i, N) = \frac{1}{(1 + i)^N}$$

- **Sinking Fund Factor**

$$(A/F, i, N) = \frac{i}{(1 + i)^N - 1}$$

- **Uniform Series Compound Amount Factor**

$$(F/A, i, N) = \frac{(1 + i)^N - 1}{i}$$

- **Capital Recovery Factor**

$$(A/P, i, N) = \frac{i(1 + i)^N}{(1 + i)^N - 1}$$

- **Series Present Worth Factor**

$$(P/A, i, N) = \frac{(1 + i)^N - 1}{i(1 + i)^N}$$

- **Arithmetic Gradient to Annuity Conversion Factor:**

$$(A/G, i, N) = \frac{1}{i} - \frac{N}{(1 + i)^N - 1}$$

- **Geometric Gradient Series to Present Worth Factor**

$$(P/A, g, i, N) = \frac{(P/A, i^\circ, N)}{1 + g}$$

$$(P/A, g, i, N) = \left[\frac{(1 + i^\circ)^N - 1}{i^\circ(1 + i^\circ)^N}\right]\frac{1}{1 + g}$$

Depreciation Amount, Straight Line:

$$D_{sl}(n) = \frac{P - S}{N}$$

Depreciation Amount, Declining Balance:

$$D_{db}(n) = BV_{db}(n - 1) \times d$$